HANDBOOK *of* METAL ETCHANTS

Editors

Perrin Walker
William H. Tarn

CRC Press
Taylor & Francis Group
Boca Raton London New York

CRC Press is an imprint of the
Taylor & Francis Group, an **informa** business

CRC Press
Taylor & Francis Group
6000 Broken Sound Parkway NW, Suite 300
Boca Raton, FL 33487-2742

First issued in paperback 2019

ISBN-13: 978-0-8493-3623-2 (hbk)
ISBN-13: 978-0-367-40308-9 (pbk)
Library of Congress Card Number 90-15046

Library of Congress Cataloging-in-Publication Data

CRC handbook of metal etchants/editors, Perrin Walker, William H. Tarn
p. cm.
Includes bibliographical references and index.
ISBN 0-8493-3623-6
1. Etching reagents—Handbooks, manuals, etc. I. Walker, Perrin. II. Tarn, William H.
TN690.7.C73 1991
617.7—dc20 90-15046
CIP

Visit the Taylor & Francis Web site at
http://www.taylorandfrancis.com

and the CRC Press Web site at
http://www.crcpress.com

PREFACE

This book is a collection of cleaning and etching solutions extracted from the open technical literature from about 1940 to the present. It began when the authors and co-workers became involved with the development of germanium and silicon semiconductor devices as solutions for cleaning, removal, polishing and defect development were needed and, over the years, has expanded to include compound semiconductors, other device materials, and all metals or metallic compounds used in the processing and assembly of devices in their various shapes and physical forms . . . the entire field of inorganic materials and their chemical processing . . . with, here in this single volume, all major metal elements, and the majority of metallic compounds currently involved in our High Technology fields.

The collection includes several solutions as originally formulated by the authors and co-workers as new compounds were being developed, many of which have been published or presented by ourselves or other workers in the chemical processing field.

With the continual influx of new scientists, engineers and technicians into the Metals, Solid State, Electronic or Silicate and Ceramic fields, there is always a ''need-to-know'' availability of chemical solutions already developed. Even at that, old solutions appear in the literature as ''new'' development, approximately every 10 to 15 years, mainly due to a lack of time or the nonavailability of a technical library for extensive literature search. This is not to castigate such ''new'' development, as it often includes a greater understanding of chemical and material reactions utilizing advanced equipment capability that was not originally available.

In any event, many books have been published over the years with regard to chemical processing of materials, often on a particular metal or metallic group . . . but not all-inclusive . . . and more often involved with the material processing with but a short, selected list of chemical solutions and limited discussion of their application. There are exceptions, such as the ASTM E407-70 pamphlet — no longer in print — that has an extensive numbered list of formulations developed for the ''metal'' industries . . . mainly for irons, steels and copper alloys. The A/B Metals Digest published by Buehler Ltd. is another example, but this too is concerned with the so-called ''heavy metal'' industries, some natural minerals, and is largely involved with the preparation of metallographic specimens for study . . . not Solid State and Electronic material processing. The ASM Metals Handbook Series of some 12 volumes probably presents one of the most extensive collections of data with regard to metal processing — again — the metal industries with etchant solutions for general material processing and plating.

Many of the worldwide technical journals today are specifically published for Solid State and Electronics, but there is such a vast amount of literature available that it is all but impossible for any one individual to stay abreast of the new data and developments. The format of presentation varies by Journal and again may be more concerned with development other than chemical processing — a newly developed compound often extensively studied for crystallographic structure with little or no mention of chemical solutions.

An understanding of crystallography is essential, yes, but any new material requires eventual development of basically four types of chemical solutions: (i) cleaning; (ii) general removal; (iii) polishing, and (iv) defect development etchants. Fortunately, the vast majority of metals and metallic compounds form in the Isometric (Cubic) System, and a brief introduction to Crystallography is presented in Appendix F.

There are many excellent books on the market that discuss the crystal defect state in detail with particular regard to Solid State materials, but the field of mineralogy (geology ''hard-rock'') established much of the basic data on natural minerals — Dana & Ford *A Textbook of Mineralogy* — not only includes an extensive section on Crystallography, but describes all known minerals — over 4000 — their crystal structure, refractive index, specific gravity, formula, pyrolitic data (including etchants), natural occurrence, and major industrial use. Many artificially grown compounds are found as natural minerals, several are included in this book, where they have been studied and/or grown. Sphalerite, ZnS, as one example — zinc sulfide — as its structure is basic to many compound semiconductors . . . Isometric System, Tetrahedral Class.

Though this is a book of cleaning and etching solutions, crystal structure cannot be overemphasized, as much of chemical processing is directly related to the crystal structure of the materials in both general study and device fabrication. A knowledge of bulk crystal plane directions, as an example, is essential in the structuring of many specific microelectronic devices.

The objective of this book is to present at least one solution for all major metals and metallic compounds in current use and/or under development for immediate, hands-on application, and can be used as a "starting place" for further specific solution development.

Hopefully, those who have an immediate need-to-know will find it of use.

Perrin Walker
William H. Tarn
Los Angeles, California
1990

THE EDITORS

Perrin Walker is currently retired after 30 years in the Solid State industry as a scientist and engineer, and has over 20 years military service.

He holds a B.A. degree with honors in Geology from the University of Virginia, Charlottesville, Virginia, 1953 with work toward an M.S. degree at UCLA in 1954. He has been a member of Sigma Gamma Epsilon, the Electrochemical Society, and the N.R.A.

With his crystallography and chemistry background, much of his work has been in R & D on semiconductor and quartz crystal surface chemistry and chemical processing of all inorganic Solid State materials. He also has been involved in processing parts and equipment design, and all phases of device design, testing and fabrication, to include sales. He has been the line engineer in charge of half a dozen government device development contracts, supervisor of operations, to include an MIC Laboratory, and Manager of Quartz Crystal R & D.

Mr. Walker holds patents in semiconductor processing, and has presented papers in both the semiconductor and geology fields. He is still involved with studies in both Solid State and Geology, actively pursuing a career as a freelance writer (both technical and non-technical), is a published cartoonist, also with published sketches and drawings of macro- and micro-fossils in geology.

William H. Tarn is recently retired after being employed for 45 years in the research and development areas of a panoply of disciplines. The disciplines each required, at some point in time, the chemical reaction of a surface etchant against an often obstinate and uncooperative metallic surface; hence his interest in chemical etches.

Mr. Tarn worked during World War II under research and development O.S.R.D. contracts and holds a patent for a 'Resilient Mechanical Wheel' that required neither rubber nor etching. The Synthetic Liquid Fuel Program of the Department of the Interior and the infant semiconductor industry were later sites of surface chemistry activities that led to etch use in preparation and sensitization of metallic surfaces for assault by intermediate or terminal chemical agents.

Mr. Tarn holds several patents and has been the co-author of several papers in these fields and on these subjects; he also states emphatically that although he has accidentally skin tested approximately 20% of the etches herein, there are no commercial cosmetic applications for the mixtures and advises great care be used in their application.

PUBLISHING STATEMENT

Cleaning and etching solutions and general data presented are for information, and direct application in chemical processing by those operators with a basic understanding of chemistry and physics of materials. The solution discussion sections are as applied by the referenced author(s) published articles, and reflects those applications only. With the continual advance in purity of chemicals, materials, and solvents it should be understood that results discussed are relative to the period of time, locale, chemicals and chemicals involved, and that results may vary within the known limits of chemical reactions. Information is accurate and reliable to the best of our knowledge according to known applications and practices, but the authors make no guarantee as to completeness or accuracy and assume no liability in that regard.

NOTE ON SAFETY

Many of the etching solutions discussed in this book represent a safety hazard, not only from the type chemicals involved, but their temperature, type reactions, methods of applications, and so forth.

Cryogenic liquids can be dangerous due to reduced temperature, being inert, or explosive; all cyanides are extremely toxic; fluorine compounds are extremely dangerous as they attack nerve tissue, yet do not show an immediate ''burn'' reaction like most acids; chlorine gas and silanes are irritants on exposed skin, unbreathable, with attack of mucous membranes. Alkalies are most dangerous when being used hot; and the strong acids give immediate skin burns.

In mixing acids and salts as solutions, the end products of reaction should be known, to include reaction results from the material involved.

Safety procedures are not specifically discussed in this book, but individuals should be sufficiently knowledgeable and instructed prior to handling metals, chemicals, gases and solutions in this regard.

DISCLAIMER

Specific etchant solutions are patented or a part of device patents, which may include chemical processing as part of the disclosure, and other chemical processes or solutions can be classified as Company Proprietary. There are a few such patented solutions — with patent number — included in this etchant assembly but to the best of our knowledge . . . no company proprietary information. What may be classified as company proprietary, per se, for a given operation, is often already in the open literature — there are only certain chemicals and mixtures applicable to a particular metal or metallic compound. This is particularly true in the area of clean/etch sequences and etch sequences, the latter involved with multilayer or composite structures — several such sequences are shown in this text as they have appeared in the open literature. There are cases of ''new'' etchants or chemical processes developed, which are already in the open literature — sometimes — for several years prior to the ''new'' development and, in many instances, the same or closely similar solutions are developed simultaneously by different experimenters in generate laboratories and areas of the world, such that it is sometimes difficult to determine priority of development. This disclaimer is to state that there has been no attempt to release any information not already available in the metals and metallic compounds processing areas.

All information herein is as extracted from open literature, and is as originally written by the authors, from the source references cited.

The Editors

TABLE OF CONTENTS

INTRODUCTION

In the development or fabrication of any metal or metallic compound, chemical processing is involved at one or more steps — from the study of a newly developed material — to a final market product. This includes wet chemical etching (WCE) — liquid solutions; dry chemical etching (DCE) — ionized gases; a solid metal, such as copper as a reactive agent; molten fluxes at elevated temperature; molecular gases, such as in hydrogen firing; and electrolytic solutions that include the use of electric current. Plating of metals or compounds is a separate subject, not covered to any extent in this book, though chemical preparation of surfaces and a few plating solutions are presented where they have particular application in Solid State processing.

A major factor in chemical processing involves the physical structure of the material on which it is applied — in many instances, specific solutions have been developed for processing of . . . colloidal, amorphous, crystalline, and single crystal structure. In addition, any one solution can have a number of different applications: general removal, polishing, preferential, structuring, selective . . . development of pinholes in an oxide thin film . . . and temperature alone can alter a solution from a slow cleaning system to polishing or preferential. Water is the primary diluant for solution reactivity control, but both alcohols and solvents are similarly used as well as other acids as buffering agents. Other special additives, such as surfactants and chelating agents for foaming and de-foaming, glycerin or ethylene glycol for viscosity control, and soaps for improved solution surface wetting, may be included in a solution formulation. Gases may be bubbled through a solution: chlorine as an active agent or inert nitrogen for a stirring action only. Solvents alone, such as trichloroethylene, TCE, are used in liquid form or as hot vapor degreasers; acids, such as HF or HCl, also are used as liquids or in hot vapor form.

Surface oxidation can be of major concern in chemical processing as all metals and many metallic compounds always have a passivating native oxide on their surfaces. Such oxides are considered surface contamination, as are oils, greases, dirt, or other organic residues, for they can mask a surface and produce erratic etching results. Further, if such contamination remains on a surface that is subjected to some form of heat treatment during processing, it can diffuse into the material and alter both physical and electrical parameters. HF or other fluorine-containing compounds are the primary oxide removers, but surfaces may be intentionally oxidized as a surface cleaning step (boiling water, HNO_3, H_2O_2); and many etching solutions can leave a trace oxide after use, and are followed by a final HF strip. Needless to say, oxides, nitrides, borides, and carbides are controllably grown on surfaces as final passivating element or as masks against etching, diffusion, metallization, epitaxy growth, and so forth, and may be an active element in a device construction.

The physical properties of a metal or compound are assembled to present factors that can be important to chemical processing, such as melting points, chemical reactivity, crystal structure and so forth. This includes their use as bulk material; thin or thick films; as a mixture, such as alloys; and in multilayer construction, where each individual metal/compound layer may require a specially designed solution that will attack that material but not the others in the assembly. There are several collective sections in Chapter 2 as solutions containing chlorine, fluorine, bromine, chromium, etc., with selected etchants formulas, listed by material from A to Z, brief application and format reference number. These can be used for a quick summary of solution effect where a user is working with multilayer or composite systems.

Chapter 1 discusses in some detail the etchant presentation used in Chapter 2 — all elements and gases, and major compounds start with: (1) Physical Properties; (2) General: associated natural minerals and general industrial use; (3) Technical Application: development and studies, usage and devices, with emphasis on Solid State, and (4) Etching: a brief mention of applicable chemicals. Three formats are used in presenting each numbered solution in order to assemble all pertinent information in an easily read manner for consistency of descriptions. The bulk of Chapter 1 is a discussion of the items listed in the Formats: Etch Name, Time, Temperature, Type, Composition, Discussion, Reference, and so forth. All etchants have an assigned number using the chemical formula or an acronym: AL-xxxx, for Aluminum; GAS-xxxx, for Gallium Arsenide, etc. Some natural minerals are included by name, particularly where they are of importance with regard to their artificial counterparts or, like Alpha-Quartz, as a device.

Chapter 2 — Etchant Solution — is the alphabetical list of solutions, and includes some techniques that are not strictly etchants, per se, but special forming or fabricating methods, such as sphere forming or the cutting and use of cylinders, bars, and oriented single crystal cubes, rhomboids, octahedrons, and so forth. Many such forms are used in the study of etch rates on concave and convex surfaces . . . the finite crystal form of a sphere . . . establishing bulk plane directions for controlled etching of channels, via holes, etc. This includes similar oxidation rates on crystallographic planes using such forms. Gases are discussed in some detail — as gas or as cryogenic liquids — with their application in chemical processing, and includes their growth and study as single crystals. There is a section on Mounting Materials, including some organics, such as waxes, resins, and lacquers (photo resists), etc., as they are involved in material process handling, mainly cleaning solutions.

Over 80 of the major elements and gases — over 600 metallic compounds, for a total of approximately 700 items with at least one cleaning or etching solution are presented. Where there is a large number of solutions involved, such as under silicon (400+); gallium arsenide (300+) etc., a "Selection Guide" is included: solution mixture, such as $HF:HNO_3$ or $H_2SO_4:H_2O_2:H_2O$, with application — polish, preferential, etc. — and the assigned etchant numbers are shown.

Discussion sections of each Format include the material structure and formula: a-Si:H, GaAs:Cr, (100) (SI), clean/etch sequences used, and other materials to which the solution was applied. The lower case letter following an Etchant Number (AL-0027a) signifies another solution was used on the material in the referenced article. The discussion is "as applied" by the reference article; and it may vary with the period of development and use due to the improvement of material or higher purity chemicals now available.

The objective of this book is to present hands-on applicable solutions from the open literature for all major metals and compounds in a single reference volume to reduce time for an extensive literature search and it includes most of the current high technology materials.

A number of technical journals, both domestic and foreign; books; magazines; industrial technical brochures and pamphlets; and technical conferences have been used in the assembly of this book for the specific data. Much of the referencing done throughout uses the assigned solution number as developed in Chapter 2, as the Formats include the specific, say, journal/volume number/page/year. In most cases, the article title is not included, because only the type material and chemistry of concern is presented. Several of the tables, such as for material hardness, or the Physical Properties lists, are compiled from up to half a dozen sources, to include data developed by the authors, and do not reflect a pure abstract of any one source.

The end-of-book Appendix H is a complete list of materials, their formulas, and Etchant Format letter acronyms used in Chapter 2.

JOURNAL REFERENCES*

American Journals (or U.S. publication):

Acta Crystallographica (Acta Crystallogr) — 1958: 1959: 1962
Acta Metallurgia (Acta Metall) — 1955—1966
Abstr J Metall — 1955: 1958
J Am Ceram Soc — 1959—1961: 1968: 1977
J Am Chem Soc — 1941
Am Mineral — 1950
J Appl Phys — 1954—1985
J Appl Phys Lett — 1966: 1981—1984
Aluminum — 1942
Aluminum (Budapest) — 1955
J Appl Chem — 1967
Proc Am Electropl Soc — 1956
Proc Am Soc Test Mater — 1924
Bell Systems Tech Pub — 1956—1959
Bull Am Phys Soc — 1960
Chem Ind — 1963
J Chem Phys — 1940—1962: 1977: 1980
J Chem Soc — 1929
J Cryst Growth — 1979—1985
J Crystallogr — 1984
Corrosion — 1970
J Electrochem Abstr — 1961
Electrochem Acta — 1961
J Electrochem Soc (JES) — 1955—1985
J Electrochem Tech — 1964—1966
J Electron — 1956—1958
J Electron Control — 1957
J Electron Mater — 1983—1985
IBM J — 1961
IBM J Res Dev — 1956—1957; 1966
IEEE Trans Electron Div — 1978
Proc IEEE — 1982
Ind Eng Chem (Ann) — 1948
Inst Phys Conf Ser — 1979
Int J Appl Radiat Iso — 1961; 1969
IRE Trans Nucl Sci — 1960—1961; 1966
J Inst Met — 1913; 1947/48; 1953; 1958/59
Met Abstr — 1958
Met Prog — 1948; 1949; 1956; 1958
Met Rev — 1905; 1951
J Met — 1953; 1956
Met Alloys — 1930
J Mater Sci — 1976
Trans Metall Soc AIME — 1961; 1962; 1967
Met Finish — 1951
NASA Tech Briefs — 1985

* Singular or inclusive year dates. Standard abbreviations used.

Phillips Res Rep — 1959
Phillips Tech Rev — 1960; 1961
Philosophical Magazine (Phil Mag) — 1960—1066
Phys Rev — 1955—1969; 1984
Phys Rev Lett — 1958
J Phys Chem Solids — 1940; 1956—1965; 1983—1985
Plating — 1950; 1953
Proc Phys Soc — 1958; 1960
Prog Semicon — 1965
Proc Inst Electr Eng — 1959
Proc Am Electropl Soc — 1949
RCA Rev — 1969—1984
Rev Sci Instrum — 1951—1969; 1980
Semicond Prod Solid State Technol — 1964
S.E.R.L. Tech J — 1960
Solid State Electron — 1960—1972; 1982—1985
Solid State Technol — 1973—1977
Sylvania Technol — 1958
Thin Solid Films — 1976; 1982—1985
Trans AIME — 1944; 1956—1957
Trans Electrochem Soc — 1942
J Vac Sci Res — 1984
J Vac Sci Technol — 1976—1985

Foreign Journals

Acta Phys Pol — 1958
Alta Freq — 1958
Br J Appl Phys — 1958; 1965/66
Cesk Cas Fys — 1958
CR Acad Sci (Paris) — 1959
Czech J Phys — 1958
Compt Vend — 1935
Delk Akad Nauk SSSR — 1947; 195/59
Fiz Met Metalloved — 1957
Fizverdogo Tela — 1959
Izv Akad Nauk SSSR Ser Fiz — 1957
J Phys Soc Jpn — 1956—1985
Jpn J Appl Phys — 1957; 1962/63; 1980—1985
Jerkontorets Ann 126 — 1942
Kristallografiya — 1957—1959
Metalen — 1954
Met Corros — 1943
Metallurgica — 1957
Nature (London) — 1958
Nippon Kinzoku Gakkaishi — 1955
Phys Status Solidi — 1969—1972
Proc R Soc — 1948
Sov Microelectron — 1984
Sov Phys (Solid State) — 1959—1960
Sov Phys Tech Phys — 1958
Sov Phys Crystallogr — 1960

J Phys Soc Jpn — 1976—1985
Jpn J Appl Phys — 1957; 1962/63; 1980—1985
Z Angew Phys — 1957
Z Analikhum — 1948
Zavod Lab — 1947
Z Anorg Allg Chem — 1959
Z Phys — 1938; 1954; 1957/58
Z Phys Chem — 1942
Z Metallkd — 1954—1961
Z Naturforsch — 1957—1959
Zh Tekh Fiz — 1956/57

Magazines

Research & Development, October 1966
Electric World, February 1962
Solid State Technology, March 1983
Kodak Data Book, 1966
Chemical Engineering, March 1962
Science News, 1987—1988
Solid State Technology, May 1984
Scientific American, 1976
Semiconductor International, April 1981
Defense Electronics, 1985
Science, September 1985
Semiconductor Electronics, June 1967
Aviation Week, May 1957
Power, December 1961
Microwave & RF, August 1984
Iron Age, 1941

Pamphlets/Brochures

Bull: 116B/Rev-HPD/VB/TM/0883/1K — Material Research Corp (MRC)
ASTM E407-70
Metals Catalog 1035-AMD-CHG/ID/09997/30M SSS — (MRC)
Bull: ADV 211-10M-11-59 — Crucible Steel Co. of Am., Pittsburgh, PA
Bull: Photo Resists and Materials — Scientific Gas Products Co. CA, 1972
Pam: SM-102R (Silox) — Applied Materials Tech., Inc., 1970
Bull: OVB-4 11/i/64 — Corning Glass Works
Bull: A-1-8 ''Micro Etch M-01-006'' — R & G Enterprises, CA, 1983
Bull: 1092 AMD/ID/JBA/5K/0482/GUILD/1982 — (MRC)
Pam: ''Platinum'' — Baker & Sons, Inc. (no date)
Pam: ''Bright Gold'' — Hanvia Chem & Mfg Co., Newark, NJ (no date)
Pam: ''Silvaloy'' — Engelhard Industries, Inc. (no date)
Bull: H681-10M — ''Metal Finishing Products'' — Heatbath Corp, 1984
Data Sheets: 557, 585; 593 — Chas. Pfizer & Co., Inc. (1958/59)
Sales Catalog — VWR
Bull: 13332-1 — Ashland Chemical Co
Brochure: ''AZ Series Positive Photo Resists for Semiconductors and Microelectronics'', Am Hoerchst Corp., 1983
Bull: CD-101-5M-SC-282 — Stackpole Corp., 1982

Bull: B-11D — The American Brass Co. (no date)
Pam: "Molybdenum" 406.6.71 — Schwartzkopf Development Corp. (no date)
Bull: 5 M 1/1/83C — Hycomp Inc.
Tin R & D Council Bulletins — 1949
A/B Metals Digest — Buehler Ltd, 1983
Pam: "Tungsten", 395/1./71 — Schwartzkopf Development Corp
Brochure: "Magnesium and Alloys" — Magnesium Electron Inc., Farmington, NJ (no date)
Pam: "A Dictionary of Carbide Terms" — Adams Carbide Corp., Kenilworth, NJ
Wise, W M — "The Platinum Metals", The Int Nickel Co., Inc., 1954

Meetings & Reports

Final Rep (15 Apr 66), Control 65-0388f (Si)
Lincoln Labs Tech Rep
Electrochem Soc Meet — WDC, 12—16 May 1957
Electrochem Soc Meet — Pittsburgh, PA, Oct 13, 1955
Electrochem Soc Meet — Cleveland, OH, 30 September—4 October 1956
Electrochem Soc Met — Dallas, TX, Spring 1967
Am Phys Soc Meet — Pittsburgh, PA, March 15—17, 1956
Int Water Conf, Pittsburgh, PA, October 1961
Am Soc Mech Eng Meet — New York, NY, November—December 1958

Books

1. Dana, E S & Ford, W E — *A Textbook of Mineralogy*, 4th ed, John Wiley & Sons, New York, 1950
2. Ruben, S — *Handbook of the Elements*, 2nd ed, Howard W Sams, Indianapolis, IN, 1967
3. Hodgman, C D — *Handbook of Chemistry and Physics,* 27th ed, Chemical Rubber Co., Cleveland, OH, 1943
4. Weast, R C — *Handbook of Chemistry and Physics,* 65th ed, CRC Press, Boca Raton, FL, 1985
5. *Research Chemicals and Materials,* Alfa Catalog, 1983/84
6. *Refractory Ceramics & Materials Handbook,* Coltronics Corp., Brooklyn, NY, 1985
7. Moss, T S — *Photoconductivity of the Elements,* Academic Press, New York, 1952
8. Fisher, J C et al — *Dislocations and Mechanical Properties of Crystals,* John Wiley & Sons, New York, 1957
9. Gatos, H C et al — *The Surface Chemistry of Metals and Semiconductors,* John Wiley & Sons, New York, 1960
10. Lancaster, O — *Electron Spin Response in Semiconductors,* Plasma Press, New York, 1967
11. Gray, T J et al — *The Defect Solid-State,* Interscience, New York, 1957
12. Bennet, C E, *Physics Without Mathematics,* College Outline Series, Barns and Noble, New York, 1951
13. Gatos, H C — *Properties of Elemental and Compound Semiconductors,* Vol 5, Interscience, New York, 1960
14. Roe, J H — *Properties of Chemistry,* 7th ed, C V Mosby, St Louis, MO, 1950
15. Harper, C A — *Handbook of Materials and Processes for Electronics,* McGraw-Hill, New York, 1970
16. Tegart W C McG — *The Electrolytic & Chemical Etching of Metals in Research and Industry,* Pergamon Press, London, 1956

17. Scholler, W R & Powell, A R — *Analysis of Minerals and Ores of the Rare Earth Elements,* Chas Griffin, London, 1940
18. Berglund, T & Deardon, W H — *Metallographer's Handbook of Etching,* Pitman & Sons, London, 1931
19. Massel, L I & Glang, R — *Handbook of Thin Film Technology,* McGraw-Hill, New York, 1970
20. Oberg, E & Jones, F D — *Machinery's Handbook,* 4th ed, The Industrial Press, New York, 1951
21. Holland, L — *The Properties of Glass Surfaces,* John Wiley & Sons, New York, 1964
22. Koll, W H — *Handbook of Materials and Techniques for Vacuum Devices,* Reinhold, New York, 1967
23. Skeist, I — *Epoxy Resins,* Reinhold, New York, 1959
24. Simonds, H R et al — *Handbook of Plastics,* 2nd ed, Reinhold, New York, 1979
25. Schwentner, N et al — *Rare Gas Solids,* Vol 3, Academic Press, New York, 1970
26. Ghandi, S K — *VLSI Fabrication Principles,* John Wiley & Sons, New York, 1983
27. Willardson, R K & Beer, A C — *Semiconductors & Metals,* Vol 7, Academic Press, New York, 1971
28. Bondi, F J — *Transistor Technology,* Vol III, D Van Nostrand, New York, 1958
29. Bridgers, H E et al — *Transistor Technology,* Vol I, D Van Nostrand, New York, 1958
30. Faust, J W — *The Surface Chemistry of Metals and Semiconductors,* John Wiley & Sons, New York, 1950
31. Coblenz, A & Owens, H L — *Transistors — Theory and Applications,* McGraw-Hill, New York, 1955
32. Brechenridge, R C — *Photoconductivity Conference,* John Wiley & Sons, New York, 1956
33. Nye, J F — *Physical Properties of Crystal,* Oxford University Press, London, 1967
34. Read, W T Jr — *Dislocations in Crystals,* McGraw-Hill, New York, 1953
35. Zhdanov, G S & Brown, A P — *Crystal Physics,* Academic Press, New York, 1965
36. Moss, R S — *Optical Properties of Semiconductors,* Academic Press, New York, 1959
37. Lowenhein, F A — *Modern Electroplating,* 3rd ed, John Wiley & Sons, New York, 1974
38. Gordon, J — *Tool Engineer's Handbook,* McGraw-Hill, New York, 1949
39. Thomas, O — *Transmission Electron-Microscopy of Metals,* John Wiley & Sons, New York, 1962
40. ASM — *Metals Handbook Series,* Vol 2, 8th ed, 1974
41. Grubel, R C — *Metallurgy of Elemental and Compound Semiconductors,* Vol 12, Interscience, New York, 1961
42. Holmes, P J — *The Electrochemistry of Semiconductors,* Academic Press, New York, 1962
43. Gilman, J J & Johnston, W S — *Dislocations in Crystals,* John Wiley & Sons, New York, 1956
44. Willis, R — *Rare Metals Handbook,* Reinhold, New York, 1954
45. Eisenberg, D & Kadsmann, W — *The Structure and Properties of Water,* Oxford University Press, 1969
46. Hausmann, E et al — *Physics,* 2nd ed, D Van Nostrand, 1944
47. *1982 Almanac and Yearbook,* Reader's Digest Association, Pleasantville, NY, 1982
48. Jeans, J H — *Electricity and Magnetism,* 5th ed, Cambridge University Press, England, 1933
49. Winchell, A N — *Elements of Optical Mineralogy* — Part III: Determinative Tables, 2nd ed, John Wiley & Sons, NY, 1951
50. Bunn, C W — *Chemical Crystallography,* Oxford University Press, London, 1946

51. Bijvoet, J M et al — *X-ray Analysis of Crystals,* Interscience, New York, 1951
52. Vossen, J L & Kern, W — *Thin Film Processing,* Academic Press, New York, 1978
53. Johnston, W C & Gilman, J J — *Dislocations and Mechanical Properties of Crystals,* John Wiley & Sons, New York, 1957
54. Phillips, A B — *Transistor Engineering,* McGraw-Hill, New York, 1962
55. Ducommun Catalog, 1968
56. Kehl, G L — *Principles of Metallographic Laboratory Practices,* McGraw-Hill, New York, 1950
57. A/B Metal Digests 1976—1978, Buehler Ltd
58. Foster, W & Alyea, H N — *An Introduction to General Chemistry,* 3rd ed, Van Nostrand, New York, 1947
59. Collie, M J — *Etching Compositions and Processes,* Noyes Data Corp., 1982 (patents)

Chapter 1

MATERIAL FORMAT SECTION

INTRODUCTION

The cleaning and etching solutions are those that have been developed for the processing and study of artificially grown single crystals. Some natural single crystal minerals are included where they have been studied in conjunction with their artificial counterparts or are representative of the crystallographic structure of materials as a class, such as sphalerite, ZnS and wurtzite, ZnS for binary compounds; chalcopyrite, $CuFeS_2$, for ternary and sylvanite, $3Cu_2S{\cdot}V_2S_5$ for quaternary compounds in addition to spinel, $MgAl_2O_4$; rutile, TiO_2; corundum, Al_2O_3; wolframite, $(Fe,Mn)WO_4$ and others for binary and ternary oxides. There are a hundred or more such compounds that have been presented in the literature with regard to their methods of growth or in study of their crystallographic classification, but do not include cleaning or etching solutions and are, thus, not included.

It should be realized that there are over 4000 natural single crystal minerals that have been studied and classified in the field of mineralogy, to include evaluation for pyrolytic and solubility reactions. This information is in book reference (1) *A Textbook of Mineralogy* and in the section on "Physical Constants of Inorganic Compounds" — book references (3) and (4) *The Handbook of Chemistry and Physics* published by CRC Press, which lists elements and compounds with brief notations as to solubility in water, acids, alkalies, alcohols, or some mixed acids, but no specific solution formulas and application data, and do not include many etch mixtures that have been developed in the metals and solid state technologies. The latter text also includes the name of the natural mineral where it is referenced to a chemical compound, and both texts are excellent references for general solubility information.

In addition to single crystals, many elements and compounds are now being grown and used in other crystallographically classified forms, such as colloidal, amorphous and crystalline. The latter are sometimes referred to as "polycrystalline" and may be as sub-micron, micro-, or macro-crystallites or as a bi-crystal. All of these "forms" may be used separately or in combination as multiple layers of metallic compounds/oxides/nitrides/metals, etc. Such variations are included in this list of cleaning and etching solutions under their primary single crystal element or compound and, in many cases, they require special formulation mixtures for controlled etching and processing.

Another classification is the effect of solutions on the material. This includes some of the following: (i) general removal; (ii) polishing; (iii) preferential; (iv) stain; (v) decoration as metal diffusion to develop defect structures; (vi) plating; (vii) selective, and others.

Where the solution is used for a physical action: (i) step-etch; (ii) thinning; (iii) etch-stop; (iv) pinhole; (v) via hole; (vi) channel, etc., the material itself may be in a specific form, such as a wafer, cylinder, bar, wire or as oriented cubes, octahedrons, tetrahedrons or rhomboids and spheroids. It may be as a single crystal sphere, a bi-crystal or polycrystalline sphere used for etching to finite crystal form, oxidation reactions of spheres are studied or they are used to fabricate a device. Some metal alloys spheres are included where they are of specific importance in fabrication or have been grown as a single crystal, such as brass.

Special sections include borides, carbides, nitrides, and silicides and even a brief section on plastics — lucite, epoxies, and polyimides used in device assembly, mainly for cleaning or removal of such materials.

In some instances, even though the basic etch format is used, there is no etchant shown as it is a growth or forming technique thought worthy of inclusion, such as the forming of spheres. Ice, as an example, is grown and studied using a cryostat microscope stage and "etching" is done by varying cryotemperature, and ice spheres are made by spraying water into liquid nitrogen, argon, or oxygen.

Many metals and compounds are used as substrates, which may or may not be involved with a final operating device. Ceramics and dielectrics, such as alumina, beryllia, quartz, and sapphire, are metallized and photo resist patterned in the fabrication of electronic circuits — pressed powder, fused or single crystal substrates — or parts may be used as stand-offs or heat sinks in addition to their electronic applications. Semiconductors, such as gallium arsenide (GaAs) and indium phosphide (InP), are used as substrates for layered thin film construction of heterostructure or heterojunction devices. Many single crystal compounds, such as sodium chloride (NaCl), potassium chlorides (KCl), and magnesium oxide (MgO), are used for thin film metallization studies, with the films deposited, then removed by the float-off technique for study under a transmission electron microscope (TEM). The natural mineral muscovite, one of the micas, is used in a similar manner.

Where the thin film is an absorber and nonreflective for TEM study, graphite coatings with gold thin films are used to fabricate replicas of the surface and can require special chemical removal techniques, such as for float-off.

Where a single crystal wafer is being prepared for processing, it may be cut from an oriented ingot, mechanically lapped and polished, and then cleaned and etched, or it may be cleaved, used as-cleaved or mechanically lapped, etc. In any case, cleaning and etching sequences have been developed for processing. Many such sequences being assembled for specific device line processes are considered company proprietary, but several also have been published in the open literature and are described in the discussion portion of etch formats. Any basic clean/etch sequence may consist of any of the following, alone, in combination, and not necessarily in the order shown: (i) chem/mech lap polish to remove previous cutting/mechanical lap damage; (ii) degreasing; (iii) etching steps; (iv) solvent cleaning steps and a final (v) solvent or acid dip clean.

There also is a degree of cross-referencing involved in the discussion section of etch formats where the solution has been applied to more than one metal or compound in the article referenced. That same etchant and reference will occur under each individual metal or compound with its own, specific etchant number. For example: aqua regia ($3HCl:1HNO_3$) will appear under gold (letter prefix: AU-xxxx); it also will appear under platinum (letter prefix: PT-xxxx); where both metals were under study in the article. This leads to a certain amount of repetition, but each etchant has a distinctive number assigned which is associated with the element or compound. There also is a degree of repitition involved with particular solutions, such as the bromine-methanol, BRM, mixtures. The concentration of bromine varies from about 0.5 to 20% and, regardless of the general acronym BRM, each concentration is a specific solution as applied, and so listed.

Where the same concentration, or etch mixture, appears in more than one author/article, they too have an etch number assigned and immediately follow the primary etch format as an additional reference, and may show another use of the solution.

Repetition also occurs with certain etchants where they are applied for a different purpose. Hydrofluoric acid (HF) is an example, as it may be used for (i) removal; (ii) pinholing; (iii) thinning, etc., such that it will appear as a separately numbered etch solution in each application case.

A few compounds only have been deposited as a thin film, some amorphous, others as a single crystal, still others as polycrystalline films. Several are included whether an etch solution is included or not, because of the tremendous development, application and interest in thin film technology.

There are many aspects of cleaning and etching not mentioned in the above discussion, but we believe the etchant formats are self-explanatory. The following sections describe the methods used to present the materials of Chapter 2: Etchant Section. The introduction sections to each element, metal or compound vary in length according to their light or heavy usage, and the three basic formats have been organized to present a consistent approach for presenting data. The format items are described in some detail.

A. INTRODUCTION TO MATERIALS

All of the major elements and compounds are introduced with the following: (i) physical properties, (ii) general description, (iii) technical application, and (iv) etching, general.

(i) Physical properties — The physical properties of the element or compound, and some electronic properties where applicable. This may be extensive or brief depending upon the availability of information found in the literature. Some new compounds show no physical properties, and trinary or quaternary compounds may follow their binary compounds with no individual list of physical properties. Crystal structure is shown by crystal system-class according to standard crystallographic notation. Cleavage (fracture) is shown as many single crystal compounds are being used as-cleaved, or cleaved sections are used in both study and processing.

(ii) General — Occurrence as a natural single crystal mineral or element and associated minerals. If it is not a natural crystal, then artificial compounds; general industrial use to include other applications such as in medicine, food preservation and so forth. The basic gases are discussed and, where they are major constituents of etchants, they also appear under the applicable material in the etch formulation. Some natural single crystals are under their mineral name, such as magnetite, Fe_3O_4 and/or included under their chemical name, such as zinc sulfide, ZnS and sphalerite, ZnS, where the two compounds have been involved in comparative studies.

(iii) Technical application — Primarily as solid state material. If not a semiconductor, the compound application in semiconductor processing, or other use as a material or device in high technology. This includes use as other than a single crystal structure, such as a colloid, amorphous or crystalline form. It should be realized that both semiconductors and metals, as well as other metallic compounds are in use, alone, or in combination in all of their possible physical forms. Other materials have use in semiconductor processing, but also in other applications, such as oxides or ceramics or as dielectric material as capacitors or resistors, or their use as light frequency filters, lenses, lasers, and fiber optics.

(iv) Etching — General solubility in acids, alkalies, alcohols, halogens. As single or mixed acids; ionic gases; molten fluxes, and so forth.

Immediately following the Physical Properties list is the General Section. The ''General'' section is headed by the material name and is introduced by the name and formula of the material: example: CALCIUM CARBONATE, $CaCO_3$ or CALCITE, $CaCO_3$

Following the (iv) etching section, the formatted etchant section is introduced with the following heading: example: CALCIUM CARBONATE ETCHANTS

B. ETCHANT FORMATS

The three formats used have been designed for consistency of presentation in order to include all pertinent data in a concise, readable and repetitive form. As will be seen, all information is not present for many of the etchants as it may not have been included in the reference article. In many cases, it is easy to assume, say, that a solution was used at room temperature — the normal application of that particular mixture — but unless it was so stated in the referenced article, the temperature (TEMP) is left blank.

I. Wet chemical etching (WCE) format
(1) Etch Number
(2) ETCH NAME: (5) TIME:
(3) TYPE: (6) TEMP:
(4) COMPOSITION: (7) RATE:
(formula)
(8) DISCUSSION:
(9) REF:

II. Electrolytic etching (EE) format
(1) Etch Number (5) TIME:
(2) ETCH NAME: (6) TEMP:
(3) TYPE: (10) ANODE:
(4) COMPOSITION: (11) CATHODE:
(formula) (12) POWER:
(8) DISCUSSION: (7) RATE:
(9) REF:

III. Dry chemical etching (DCE) format
(1) Etch Number (5) TIME:
(2) ETCH NAME: (6) TEMP:
(3) TYPE: (13) GAS FLOW:
(4) COMPOSITION: (14) PRESSURE:
(formula) (12) POWER:
(8) DISCUSSION: (7) RATE:
(9) REF:

The WDC format is used as the general format, such as for use of molecular gas as against DCE ionized gases. Also includes molten flux etches, special forming and fabrication techniques that may not include a specific etchant.

C. FORMAT ITEM DESCRIPTION

(1) ETCH NUMBER

Every referenced etchant has an assigned number under each element, metal or compound, and consists of three items: (1) capital letter designation of the material; (2) an individual number, and (3) lower case letter where more than one etchant appears for the material in the referenced article.

Up to four capital letters are used and are the chemical symbols for the material, element, compound, etc., where it is a singular element or a binary compound. In some instances, these are "common usage" acronyms, such as GAS for gallium arsenide, (GaAs) as GAS-xxxx. For yttrium iron garnet, $Y_3Fe_2O_{12}$, the acronym YIG-xxxx. For trinary or quaternary compounds, where a chemical formula is in conflict, a special acronym is used, such as for molybdenum selenium sulfide, $MoSeS_2$ (acronym: MOSS-). And there is the special case of carbon, C-; diamond, D-; and a third for natural or artificial graphite, GR-, all three being forms of carbon.

All element or compound numbers start: -0000, e.g., AL-0001; then BE-0009; SI-0274; and so forth. This leaves sufficient numbering space for any currently foreseeable new etchant solutions as they are added to the handbook.

The use of the lower case letter following the number is limited to those references where more than one solution has been used in the article on that particular element, compound, etc. The full reference will appear under the "a" designated solution only; whereas the "b", "c", etc. solutions will be shown as: REF: Ibid.

Example: CD-0004a .... REF: Doe, J — *J Appl Phys*, 57,112(1985)
CD-0004b .... REF: Ibid.

A complete example of a full format and additional reference is shown following the description of REF.

(2) ETCH NAME

If it is a singular acid, alkali, salt, the full name is used, e.g., hydrochloric acid. It also is shown where the singular acid, alkali, etc. is water diluted, e.g., hydrochloric acid, dilute. It is not shown if dilution is with other than water — alcohols, solvents, etc.

The vast majority of solutions show no name as they are mixtures. On the other hand, where a name is in common usage it is shown and may be a modification of an original formulation. The following is an example of one such solution:

ETCH NAME: Camp #4	The original solution development
ETCH NAME: CP4	Common usage acronym
ETCH NAME: CP4A	CP4 without bromine
ETCH NAME: CP4, modified	CP4 with bromine replaced by iodine, etc.
ETCH NAME: CP4, modified	CP4 with a different bromine content

A similar case is the bromine-methanol (BRM) classification. The acronym is used in the literature without regard to bromine concentration and no attempt has been made to classify these solutions as modifications. As in many instances the referenced article only says: " . . . a BRM solution was used . . . ", which leaves a bit to be desired.

There are some poorly named solutions, such as the "1:1:1", but it is still shown as an ETCH NAME: per article reference. The use of the name "White Etch" is another example, as it has been used to describe HF:HNO_3, and HF-HNO_3:HAc solutions — none of which are white, but transparent, clear and colorless solutions.

The buffered hydrofluoric acid solutions (BHF) are in the same variation category as are the BRM solutions — even HF, dilute has been called BHF, or KF has been used in place of NH_4F, etc. A general standard in use is 1HF:1NH_4F(40%), though the reference may show only " . . . BHF . . . " used. Unless it has been shown in the article, the BHF formulation is not shown under COMPOSITION.

(3) TYPE

Divided into two parts: (1) a "chemical" designation and (2) an "application" designation. The objective of this two-part approach is to show how an acid or solution mixture is used for more than one application as described in the various referenced articles.

Terms used are as brief and concise as possible and the "chemical" descriptive term may not necessarily agree with pure chemistry nomenclature, but all terms are chosen for general repetitive consistency.

Some of the more commonly applied terms are shown below in the two categories:

Chemical	**Application**
Acid	Cleaning, polish
Alkali	Damage removal, polish, jet
Alcohol	Dislocation, preferential
Gas	Isotropic, oxide etch
Halogen	Anisotropic, oxide removal
Ionic gas	Junction, selective
Metal	Macro-etch, sphere

Chemical	Application
Molten flux	Micro-etch, structure
Salt	Mesa, stain
Solvent	Orientation, step
Electrolytic	Forming, thinning

There are others in both categories.

All electrolytic solutions are chemically designated "electrolytic" and followed by an application term even though common practice would say: "electropolish", as example; whereas many of the electrolytic solutions described have far more than just the polish application.

This is not a plating handbook, but there are some plating solutions shown, primarily where they have been used in developing p-n junctions, as metal contact plating, or are of particular use in Solid State or metal processing.

Three examples of TYPE: designation are shown below:

TYPE: Acid, polish
TYPE: Alkali, cleaning
TYPE: Electrolytic, thinning

(4) COMPOSITION

Divided into two parts: (1) volume/quantity used and (2) chemical formula of constituents.

In many instances, the volume/quantity was not shown in the referenced article and is therefore designated as an "x". Where it is obvious that the volume/quantity such as a single acid, alkali, etc. is not critical, it too is shown as "x":

x .... HF, conc.	or	x .... HNO_3	or	x .... 2% Br_2
		x .... H_2O		x .... MeOH

The third example above is very common for the BRM and BHF solutions where the total volume is not critical, provided the percent concentration of bromine, fluorides, etc. is maintained.

Many mixed solutions are shown by volume percentage. This is common to the metals industries and some early solid state articles:

6% .... $HClO_4$
35% .... butyl cellusolve
59% .... MeOH

Where dry chemical compounds are used, such as ferric chloride, $FeCl_3$ or potassium permanganate, $KMNO_4$, they are usually formulated as a water diluted solution, specifically as gram-weight per liter of water: g/l. Both liquid acids or dry chemicals may be mixed by viscosity, using the Baumé scale, °Bé, which should not be confused with the temperature scales "degree" designation, e.g., °F or °C, although it is called "degrees Baumé". Using ferric chloride as an example:

x .... $FeCl_3$ (35°Bé) or x .... 20% $FeCl_3$ or x .... x% $FeCl_3$

It also may be shown as

112 g .... $FeCl_3$	*or	50 g .... $FeCl_3$
10 ml .... HCl		1000 ml .... H_2O
50 ml .... H_2O		

* as a g/l solution

Liquid acids, solvents, etc. are usually measured and used by milliliter, ml, volume as commercially supplied with water diluted concentrations shown as 70% HNO_3; 30% H_2O_2; 49% HF etc.

The most common method of designating volume/quantity of a solution mixture, where the total volume is neither critical nor required, is to use a numerical ratio of constituents:

1 .... HF	or	3 .... HCl	or	1 .... Br_2
5 .... HNO_3		1 .... HNO_3		10 .... MeOH
12 .... CH_3COOH (HAc)				

Where a mixed acid is used as a "constituent" in a particular formulation, it is shown as:

1 .... *CP4
3 .... HNO_3
6 .... H_2O

*CP4: 30HF:50HNO_3:30HAc:0.6Br_2

Saturated solutions may be shown in two contexts:

50 ml .... HF	or	100 ml .... HCl
20 ml .... NH_4F, sat. sol.		x .... *Br_2

*saturate with bromine

To differentiate between using an acid in concentrated form as against its use in vapor form: ("conc." shown only when referenced.)

x .... HF, conc. or x .... HF, vapor

When a specific mixture is described, it is usually in milliliters, ml (liquid) or grams, g (gm still seen) for dry chemicals:

250 ml .... 30% KOH	or	20 g .... KOH	or	10 ml .... HF
		100 ml .... H_2O		30 ml .... HNO_3
				100 ml .... HAc

In some of the older literature, milliliters appear as centiliters, cc, but the two terms are equivalent. Some volumes are shown in cubic centimeters, cm^3, such as 150 cm^3 HCl, which is equivalent to 150 ml HCl. There are some large volume solutions measured in gallons and ounces, etc. The latter more often are in the metals and plating area, rather than Solid State.

Solutions also may be shown as a Normal or Molar solution:

1 .... 12 *N* HCl or x .... *EDTA 2 *M*
1 .... 17 *N* HAc x .... NH_4OH for pH 3.5
1 .... 1 *N* $K_2Cr_2O_7$

*Ethylene diamine tetraacetic disodium salt

In the second example, the EDTA solution is available as a standard molar mixture and, in this case, a specific acid/base pH value is established using ammonium hydroxide.

Cleaning solvents are used for liquid immersion or (hot) vapor:

x .... TCE or x .... TCE, vapor or x .... TCE, spray

Where metals, or metal salts are used for either developing p-n junctions by plating, or as a defect decoration agent with in-diffusion:

50 g .... $Cu(NO_3)_2$ or x .... Au
100 ml .... H_2O
\+ .... intense light

There are a number of clean/etch sequences that have been developed, many of them are company proprietary to a specific line operation, but several have been published in the open literature and follow a general pattern of any one or more of the following, and not necessarily in the order shown below:

(1) Chem/mech lap to remove damage, such as BRM on a polishing cloth
(2) Degrease in TCE, Freon, etc., as vapor, spray, or liquid (RT/hot)
(3) Alcohol rinse with acetone, methyl or ethyl alcohol
(4) Acid etch — a slow polish and/or cleaning solution
(5) Alkali etch — as an etch or surface conditioner (optional)
(6) Water rinse — DI, Hi-Q, etc., [also after (1) and (4)]
(7) Alcohol rinse [as in (3)]
(8) HCl + water rinse (often 50% HCl)
(9) HF + water rinse (50% HF common)
(10) Nitrogen blow dry — may be He, Ar, air, etc. (RT/warm/hot) or IR lamps, other heaters, or air dry
(11) As an additional process, final cleaning or drying in vacuum, in a furnace with hot gases, etc.

These sequences are shown in the discussion section of formats where they may be a major operation step. Some references may not include any clean/etch steps, but only the primary solutions associated with the material study even though one would normally prepare the surfaces by using one or more of the clean/etch steps shown.

A gas or gas mixture may be used as an etch in either its molecular or ionic form, alone. A molecular gas also may be bubbled through a liquid and act only as a carrying vehicle for the liquid; for stirring action of the solution only; or be an active element in the solution reaction. The molecular gases are shown by their normal diatomic representation; whereas the ionic gases as in the example for argon, below. In some cases, the ionization (+) symbol is not present, but it will be apparent that ionic gases are applied, because the dry chemical etching (DCE) format it used. The last example shown below is of this type:

x .... N_2 or x .... Cl_2 or x .... Ar^+ ion or x .... CF_4
x .... 5% O_2

Heat in vacuum, alone, is used for both preferential etching as well as for annealing, e.g., thermal heat treatment. A vacuum system may be back-filled to a partial pressure with gas, but when gases are used, it is usually as an oven or furnace system. Oxygen or O_2/N_2 or CO_2/N_2 for oxidation; N_2 or N_2/Ar for nitridization; H_2 or forming gas (FG) 85% N_2/15% H_2 as a reducing atmosphere, all along with heat. Where heat is the primary operator, it is represented as an etch under composition; as shown below; but if molecular gases are involved in addition to heat, the gas is shown:

x .... heat or x .... N_2, hot

There are special cases where pressure, with or without heat, have been used to initiate defects under various conditions of temperature, or under vacuum using vapor pressure change as the agent:

x ... pressure or x .... vacuum

Another special case is the use of light. It may be white light or referenced to a color and/or specific frequency spectrum, used alone for drying; as a work light, such as etching silver under red light; as a damage induction/reaction vehicle; in conjunction with plating; or to enhance etching:

20 g .... $Cu(NO_2)_3$ or x .... light or 5 .... HF
100 ml .... H_2O 2 .... HNO_3
\+ .... *intense light + ... *light

*Mercury lamp *Sodium (5890 Å) yellow

Where acid-slurries are used for lapping and etching, the acid, solvent carrier, and abrasive are shown. Occasionally, only an abrasive where it is the only agent used for polishing the material:

50 g .... 0.5 μm Al_2O_3 or x .... diamond, powder
5 ml .... HNO_3
50 ml .... ethylene glycol (EG)

Acid-saw systems are under the metal or other material used as the cutting agent (iron wire, cotton thread, tantalum rod, etc.) and may include abrasives to improve slicing and cutting action.

Where two or more solutions are shown under COMPOSITION there are three methods of designation using the lettering or numbering systems as shown in the following three examples:

(1) If two or more solutions are to be mixed together:
(i) 4 .... HF (ii) 10 g .... $AgNO_3$
3 .... HNO_3 100 ml .... H_2O

mix: 1 part (i):3 parts (ii), when ready to use.

(2) Where two or more solutions are to be used in order:
(A) 1 .... HF (then) (B) x .... MeOH
6 .... H_2O

(3) When two or more solutions are to be used separately:
(1) 1 .... HF (and/or) (2) 1 .... HF
1 .... H_2O 50 .... H_2O

Where water is used as the diluent or rinse it is shown as H_2O, without regard to purity designation, but should be considered as distilled water (DI H_2O) unless otherwise specified:

x .... HF or x .... H_2O, salt or x .... H_2O, tap
x .... H_2O

There are many types of water, but only two are normally used in all industrial processing, as either natural water with only solid matter removed, or as purified water, e.g., DI water. The latter most commonly distilled on-site, and may be double-, even triple-distilled with names such as "Hi-Q". Occasionally, it is known as demineralized water (dm-H_2O). Referenced use is as fresh, potable (drinking) water, not saline, e.g., salt water, except as noted . . . when used as a quenching medium or direct etchant, itself.

Although water is a neutral solvent, it is applied as an etching solution on water-soluble minerals and artificial compounds. See the section on Water for a more extensive discussion, to include "cold etchants" as mixtures that may contain ice, or snow, or be cryogenic liquids.

(5) TIME

Appears only when given in the referenced article. Usually a single time and temperature; occasionally repetitive time steps or two times at two different temperatures, as separate etching levels; sequence etching times and temperatures and so forth. The time can vary from seconds to minutes to hours depending upon the material and application. Some examples are

TIME: 12 sec or TIME: 10 sec, 3 times or TIME: 4 h or
TIME: 2 min or 15 sec
RT 80°C

In metallographic sectioning, etching for micro- or macro-structure, terms such as "dip" or "swab" may be shown under TIME. These are a method of applying a solution, but were used as a time referent in the article. In defect and structure etching, final surface cleaning or tuning device electrical parameters, an initial etch time may be shown and followed by one or more dip periods to obtain optimum results. This is particularly applicable to photo resist metallization patterning on circuit substrates and is shown in the discussion section of the format.

A time may be shown, and no temperature. In such cases, the solution was probably used at room temperature but again, information is included only if it appears in the referenced article.

(6) TEMP

Appears only when shown in the referenced article. In most cases, when a solution is used at other than room temperature the reference includes temperature. Fortunately, most etching is done at room temperature even though there may be an increase in temperature during the etching period due to the exothermic nature of the solution or the heating effects of an RF plasma in dry chemical etching.

Temperature can have a major affect on etching results with any given solution. The hydroxides, KOH and NaOH, for example, when used at room temperature as 2—10% concentration liquid solutions, are used for surface cleaning; as 20—50% liquid solutions they are heavy scale removers on many metals; and when used hot to boiling are progressively preferential. When used as a molten flux at elevated temperature (360°C), they are also-preferential. All liquid etch solutions become progressively more surface preferential and slow below room temperature, even though they may be a polishing solution at room temperature, and the reverse is true. Much above 40°C, an exothermic solution becomes too rapid for good polishing control. For this reason, in batch etching semiconductor wafers, a large volume of solution is used — allowing it to cool between etch periods — or etching is done in a cycle-controlled etching bath using water, glycols or Freons chilling piping to dissipate heat.

There are three primary temperature scales in common use:

(1) Centigrade, °C	Also called Celsius. Used in scientific works and as the atmospheric scale in Europe . . . also now in the Americas.
(2) Fahrenheit, °F	Widely used in the metals and plating industries and as the atmospheric scale in the U.S., in addition to Centigrade.
(3) Kelvin, K	Scientific scale based on absolute zero. Absolute, °A, is equivalent. Note that "K" does not use the degree (°) notation, although it still does appear with the degree sign in some publications.

Correspondence between these scales is as follows:

Comparative Temperatures Table

Reaction	°F	°C	°A or K		Correspondence
Boiling	212	100		373	°F = 9 × °C/5 + 32
		50		323	
			20	293	
Room temp		72	22.2	298	°C = (°F − 32) × 5/9;
Freezing	32	0	273		
	−40	−40			
		−200	73		
Zero		−273	0		

Under the temperature heading of the etchant formats, room temperature is shown as RT whenever possible, even though "room temperature" has been shown in articles from 20 to 25°C. It should be noted that many of the electrical measurements in the Physical Properties sections are made at 20 or 25°C.

Verbal terms also are used in describing a temperature, such as cool, warm, hot, or boiling.

The following arbitrary temperature levels have been established for use in this text for consistency:

Cryogenic: −270 to −30°C — Liquified gases, such as LN_2, LHe, LOX, etc. Normally used to chill other liquids, though they can be a direct etchant under special conditions.
Freezing: −30 to 0°C — Certain acids or acid/ice-snow mixtures. Freezing level can produce a slush etch solution with cryogenic liquids for chilling.

Cold: 0 to +10°C — Some acid mixtures, alone; others housed in a bath of ice water, acetone/ice, methanol/ice, etc.
Cool: 10 to 20°C — Acid mixtures chilled with cold water or other cold solvents, cryogenic liquids, etc.
RT: 22.2°C or 72°F — Standard atmosphere and pressure (760 mmHg), much etching done at room temperature.
Warm: RT to 40°C — Heat with light, gases or electricity (*Note:* "Chrome Etch" is widely used at a controlled 30°C).
Hot: 40 to 80°C — Varies with liquid and mixture, but below boiling point.
Boiling:
  Low boil: 80 to 100°C — Slow solution movement; fine, sporadic bubbles.
  Medium boil: 100 to 130°C — Convection currents in solution; continual surface movement.
  High boil: above 130°C — Severe convection currents in solution; massive bubbles explosive to surface; and violent, erratic surface. Some solutions, even above 130°C, show a roiling or rolling boil without severe, massive bubbling.

Note: Most hot etching is done at low or medium boil.

Vapor: above 100°C — Specific temperature dependent on solvent or gas used.
Vacuum +/− gas: to 1000°C or more; vacuum as low as 10^{-10} Torr.

(7) RATE

Not listed in formats unless it is referenced in the article as the majority of solutions are used without regard to removal rates.

In processing single crystal materials, such as the semiconductors, removal rates relative to the wafer surface as well as bulk crystallographic planes can be a critical factor of the device design and structuring. Because of this, whenever a new single crystal compound is developed, much of the initial studies are involved in establishing the etching of planar, concave, and convex surfaces. Convex surfaces are studied with spheres and hemispheres — the fast etch planes; concave surfaces — slow etch planes — are developed by controlled pit damaging and etching with pit shape and bulk planes controlled by the planar surface orientation, such that a (111) surface develops triangle type pits and a (100), square pits. In both concave and convex etching, the type etchant solution used is an additional variable — hexagonal pits can be etched on (111) surfaces (superimposed negative and positive tetrahedrons); one solution will etch a sphere to finite crystal form as a dodecahedron, (110); another as a cube, (100) and yet, another as an octahedron, (111) or tetrahexahedron, (hk0). All of the above are in addition to defect study — dislocations, slip, stacking faults, twinning — and other structures.

Removal rates on single crystals and materials in general, where only small amounts of material are removed, are commonly measured in angstroms per unit time: 50 Å/sec; 1200 Å/min. In the metals, ceramic, and glass industries — corrosion and similar studies — rate measurement may include a total volume gram-weight loss with time. The etch rate and time for some more chemically inert metals, such as tungsten, titanium, or molybdenum, which have extremely slow attack rates, may be only in mils-per-year (mpy).

Where material removal rates are more substantial, measurements are in millimicrons, mμ; in microns, μm; or in mils (1 mil = .001″), with the time in seconds (sec), minutes (min) or hours (h). Rate is always shown relative to time. Some examples are

10 Å/sec or 20 μm/min or 1.5 mils/h or 4 μ/10 min

The following table shows the correspondence of thickness measurements of some of the more widely used units:

Thickness Measurement Table

Angstroms (Å)	Micro-inch (μ'')	Micron (μ)	Mil (mil)	Inch (")
254	1			
10,000	1			
25,400	100	2.5	0.1	.0001
254,000	1000	25	1	.001

The micro-inch measure is common for metal plating thickness, though not limited to plating, with 100 μ" a relative standard, say, for copper, nickel, or chromium; whereas gold is 25—50 μ" in most commercial applications, but supplied as 150 μ" or 300 μ" as gold plated alumina substrates in Solid State circuit processing. These thickness measurements become important in all material processing, as the rate of removal of a specific metal/compound/multilayer on a substrate is critical. Rate vs. thickness is experimentally established, initially, for controlled patterning of thin films in device circuit fabrication. In etching for line-width and depth, an etch solution will attack vertically and laterally at an equal rate, such that initial rubilith pattern widths calculations must take the lateral etch factor into consideration to prevent subsequent undercutting, lifting, peeling of the thin film, or loss of electrical characteristics. In such fabrication, rates of removal must be known, and are an extremely critical factor in microwave device fabrication.

As already said, removal rates are only shown in the etch formats when listed in the referenced articles as a major item, such as the initial study of a new compound. As a secondary function, it may appear in the discussion section of the format along with other processing data.

(8) DISCUSSION

Varies from a few words to a fairly extensive discussion depending upon the application. Many of the earlier solutions that were, say, used for defect etching, plane attack data, etc. are brief; whereas with more current development involving metallization, multiple layers, selective etching, and so forth, to include clean/etch sequences, the discussion is more extensive. In most instances, the discussion starts with the material formula; single crystal orientation where applicable; or other structure, such as amorphous; the physical specimen used, such as a wafer, blank, etc. and the study or fabrication involved where appropriate — with results, say, a new refractive index, dielectric constant, or the test involved, and so forth. Clean/etch sequences or special processing applications are included. Where the referenced article shows the study of other materials, these are shown by chemical formula for cross-indexing purposes.

Some of the general sections, such as Cleaning (solutions/methods) may be a specific solution application or a general discussion that can include other fields than metals/metallic compounds, such as the use of borax . . . a cleaner, a solder flux, a preservative . . . in industry, in home use, in medicine, and so forth.

The introductory sections — ''Etching'', as an example, describe the basic etching methods . . . wet chemical etching (WCE), or dry chemical etching (DCE), and molten flux etching (MFE), or electrolytic etching, (EE) . . . without inclusion of a specific component clean or etch solution.

In other cases, the WCE etchant format is used for convenience, but involves a special fabrication procedure, such as the fabrication of spheres by lathe cutting, melt forming, or ''race track'' grinding. The latter starts by using pre-cut single crystal cubes.

(9) REFERENCES

Standard practice for presentation with accepted abbreviations are used. Author(s) with last name, first, then initials — then the journal abbreviation (journal, volume number, page, and year). For books, the full title is given, followed by publisher, year, and page; and magazines with month/year/page.

Where there are more than two authors for the article, we have taken the liberty of showing the primary author followed by "et al.". One horrendous example: a full page of 52 authors, and only a page and a half of text! Several foreign journals often list four to eight authors — again, first author shown only, for sake of brevity, otherwise a drastic increase in total book volume which we feel is unwarranted, but no reflection against the input or status of those other authors of the article!

Some sections have an open reference . . . less than 2% as unknown or unavailable references, while others may be as "N/A" for common usage solutions, such as aqua regia; or in the section "Mounting Materials", and similar sections, where only a general discussion is involved.

A consolidation of reference materials used are shown here in Chapter 1 following this introduction, such as the *Journal of Applied Physics* with only year-inclusive dates. Specific articles are referenced following each etchant format as part of the format: REF: xxxx and, where other authors/articles have used the solution, they immediately follow the primary format as an additional references.

As referenced, the article title is not included in most cases, again, for brevity, and because the chemical solutions are the primary concern in this book; whereas the article may have been on another aspect of material developed and study with only minor description of chemical processing.

One example of a primary reference, followed by an additional reference is

*REF: Belyustin, A V — *Sov Phys-Cryst.* 5,143(1969)
SI-0247: Richards, T L & Crocker, M — *J Electrochem Soc*, 121,14(1956)

*This immediately follows the discussion section of the etchant format.

An example of a complete Chapter 2 text etchant format and additional reference is shown below:

AL-0001
ETCH NAME: Hydrochloric acid TIME: 1 min
TYPE: Acid, removal/clean TEMP: RT
COMPOSITION:
1 .... HCl
2 .... H_2O
DISCUSSION:
Al, as thin films . . . etc.
REF: Doe, J — *J Apply Phys*, 27,142(1976)
AL-0007: John, J K — *J Electrochem Soc*, 112,219(1977)
Al, (100) wafers . . . etc.

ELECTROLYTIC FORMAT

As a method of etching, electrolytic etching is more widely used in the metal industries than in solid state and electronics device fabrication. But if the iron industry says ''electro-polish'' here, we say ''electrolytic, polish'', using the word polish as an application term because there are a dozen or more such applications: cleaning, forming, thinning, oxidizing, preferential, to mention only a few.

(10) ANODE

Where anodic etching is used, the anode is understood to be the material involved — gallium arsenide, silicon, manganese, etc., but it is not listed unless actually stated in the referenced article.

In special cases there may be more than one anode, or cathode and where a semiconductor p-n junction is being etched the specimen p- or n-type zone may be the anode.

The ''etch'' solution, proper, may be either a true etchant or a plating solution — again, primarily in developing p-n junctions — such as using copper or silver to selectively plate-out the p- or n-type side of a junction. It may include intense white light (copper) or be operated in the dark (silver). It should be understood that this book is not a Plating Handbook, the only presentation of plating is with regard to junction delineation, or material plating for contact metallization of electronic devices. There are several electrolytic solutions as anodic etching action formulas used, alone, or in conjunction with cathodic ''etching''.

According to standard electrolytic operation the anode is the positive (+) terminal.

(11) CATHODE

Although there is occasional reference to ''cathodic etching'', it is not etching in the sense of removal of material, but de-plating or conversion of a surface.

The cathode may be the same material as the anode in special cases, but it is more often a base metal such as lead, steel, stainless steel, iron, copper, nickel, or platinum that is, effectively, inert to the etch solution involved and does not directly affect the anode material being etched. Both anodes and cathodes may be in several forms, such as a bar, a single wire, a wire mesh, etc.

There is much developmental work in progress in the anodic/cathodic oxidation of surfaces, in particular, silicon in potassium hydroxide, KOH, solutions. It is a combination of oxidizing and de-oxidizing the surface by switching from anode-to-cathode. This operation is included due to its importance in the study of metal and metallic compound surfaces.

The cathode material is shown only if it is given in the referenced article and includes the form: rod, wire, screen or special shape, when described in the article.

According to standard electrolytic operation the cathode is the negative (−) terminal.

(12) POWER

Not shown unless it is given in the referenced article; may be voltage or amperage only, preferably both.

Most semiconductor materials are etched in the milliampere (mA) range; whereas many metals, ceramics and dielectrics such as glasses are etched in the ampere (A) range. In both cases, voltage (V) can range from tenths, say 0.2 V, to several volts — 5 V, 20 V, etc. Standard practice is to show amperage per square centimeter, A/cm^2 of surface area: 0.5 mA/cm^2 and so on. Most power is DC, occasionally AC, or a combination of both.

Unfortunately, many references simply say ''anodic etching'' was used with no details given, as the article is not concerned with chemical processing of a material as a major factor in the study presented.

DRY CHEMICAL ETCH FORMAT

Note: Gases used in their molecular form, such as H_2 as a furnace reduction gas or N_2 and Ar as inert atmospheres, are shown under the wet chemical etch format, WFCE. Dry chemical etching (DCE), using the dry format (DF), applies to the use of gases in their ionic form, as obtained using RF, RF magnetron, or DC power. Special applications may include a combination of both RF and DC operation.

(13) GAS FLOW

Usually shown as cc/min, occasionally as SCFM:

10 cc/min or 0.5 SCFM

The gas or gas mixture is shown by chemical/application under TYPE as TYPE: Ionized gas, selective (and others) with the formula(s) under COMPOSITION:

x .... Ar^+ ion or x .... CF_4
x .... O_2 (5%)

In the second example, that ionization is not shown as the dry chemical etching format is self-evident. In addition, the Ar^+ (ionic symbol) is used, rather than Ar only, which is representative of a molecular gas. The additional descriptor of PRESSURE also is distinctively DCE processing, although RF or DC power application is not; both RF and DC power also are used in sputter evaporation of metals, epitaxy growth of materials and plating.

(14) PRESSURE

Usually shown as milliTorr of over-pressure of the gas(es) when referenced to DCE processing, but can be shown in atmosphere, (atm):

0.5 mTorr (mT) or 0.2 Pa (Pascal) or 0.4 atm

Ion gas etching includes ion milling as an operating system, or RF plasma cleaning or etching systems. RF plasma cleaners using N_2 or O_2 as ionic gases are being widely used in Solid State device processing, and specific vacuum systems are designed to handle chloride gas etching of aluminum oxide thin films with $AlCl_3$ or BCl_3 as ionic gases. In all cases, the systems operate with an input gas over- or back-pressure.

Pressure as an individual function/operation without regard to dry chemical etching (DCE) appears in the wet chemical etching (WCE) — classification, with pressure shown either in atmospheres or as pounds per square inch (psi).

Many materials are pressurized in the study of their high pressure crystallographic forms and to establish their transition temperatures. Standard hexagonal ice, for example, under pressure, becomes cubic (isometric system, normal (cubic) class), and there are six high pressure forms of silicon.

In chemical vapor deposition, CVD, material growth, specific pressure designated systems are used: low pressure, as LPCVD; high pressure, as HPCVD, and several others.

The growth of artificial alpha-quartz is done at a minimum of 30,000 psi, and in geology studies of rock magma formation, better than 300,000 atm.

Pressure alone is used as an ''etchant'' for preferentially developing defects in such materials as ice, alkaline halides (ionic crystals) or single crystal rare gases (vary vapor pressure in the vacuum system).

In dry chemical etching (DCE) a number of methods and terms have been developed with their own acronyms. RF or DC plasma etching, PE, is the most general; reactive ion etching, RIE, and several others. These may be shown in the format as an ETCH NAME, as a TYPE: chemical designator; or referred in the DISCUSSION section. Additional anodes and cathodes may be used in conjunction with the RF plasma or additional electric bias is applied to specimens. These are shown in an expanded format — a combination of the electrolytic and dry chemical etching formats.

With reference to the terms isotropic and anisotropic as used in wet chemical etching, WCE, they are polishing or preferential etchants, respectively. Isotropic/polish are solutions that attack crystallographic planes at an equal rate in all crystal plane directions producing planar, flat surfaces; whereas anisotropic/preferential solutions attack crystallographic planes at different rates producing structure.

In dry chemical etching, the terms refer to structuring development without regard to their preferential or nonpreferential crystal plane attack . . . ionic gases can be either isotropic or anisotropic in character. From a pure chemical reaction point of view this variation of terms can be confusing, because of this difference in interpretation between wet or dry etching. In physics, geology, and optics the two terms were originally defined with regard to light propagation through solids, and are basic to crystallography, e.g., the refractive index of materials, etc.

METHODS OF ETCHING

There have been over 200 terms used in the literature to describe a method of etching, and generally fall into one of the three format categories described in the previous section: wet or dry chemical and electrolytic etching. The use of a molten flux, here, is considered a form of wet chemical etching as the chemical flux is in liquid form, and also is used for growing single crystals, such as garnets and ferrites.

In addition to the acid and base definition using the Sorensen pH scale with water neutral at pH 7; acids with pH below 7; and bases (also called alkalies) above pH 7, there are many other categories. Water itself can be considered an acid-like etching solution; alcohols and solvents, though usually considered as cleaning solvents, can act as an etchant on certain materials; molecular gases can be used in conjunction with acids or used alone, such as in hydrogen firing; vacuum as a "vapor" etchant on certain materials; molecular gases can be used in conjunction with acids or used alone, such as in hydrogen firing; vacuum and vapor pressure in vacuum have been used as etching vehicles; thermal heat treatment, alone, or with gases is used both as a part of material fabrication as well as to etch. The halogens are an etchant category of their own, with the bromine-methyl alcohol (BRM) solutions in wide use.

Some highly specialized methods include the use of pressure, ice, cryogenic fluids, and even abrasives. The latter, for example, are used for surface cleaning. Wire-saw cutting is another area used in conjunction with an acid, with an abrasive slurry or acid-slurry. Many of the specialized methods or techniques are categorized under wet chemical etching (WCE) using the wet format (WF).

The use of etching solutions is still referred to as a magic art rather than a science by many people, particularly when it involves developing a new solution as, though known chemicals are used, results are often obtained by trial and error: varying solution constituents; varying constituent concentrations; temperature variations; different methods of application. Other factors include the material being etched and its structure; chemical reactivity or contamination involved.

With time and usage many solutions have become industrial standards, such as aqua regia ($3HCl:1HNO_3$). When it was first formulated in the Middle Ages in Europe, the alchemists of the period thought it to be the universal etch for all metals and metallic compounds which, as we know today, is far from the case. It was the first solution developed that would etch gold, and also will dissolve metals of the platinum group — platinum, iridium, osmium. All of these metals etch slowly in comparison to gold in aqua regia, even when used hot to boiling. It does etch a few other metals and compounds, can be either a removal/polish or preferential solution and, on still other metals, is a surface cleaning solution only. It is particularly useful for removal of heavy metals contamination on surfaces, e.g., copper, iron, etc. The aqua regia solution varies in constituent concentrations, and may be diluted with water or alcohols.

Other acid mixtures have been developed for one material and a specific result, such as CP4 — initially Camp #4 — developed as a light figure orientation etch for (100) oriented germanium wafers, yet, even on germanium, it can be both a preferential or polishing solution. The original solution was 15 ml HF:30 ml HNO_3:15 ml HAc:0.6 ml Br_2. Since its development it has been used on several semiconductor materials and other metals and compounds. Without bromine it is called CP4A, and is considered a better polishing etch for either germanium or silicon than CP4. Further, as CP4-modified, the bromine has been replaced with I_2, $Cu(NO_2)_3$, $Ag(NO_2)_3$, etc. to improve preferential action for a particular type of defect. And depending upon the material involved, CP4 may be applied as a surface cleaning solution.

Nital and Picral — HNO_3:EOH and $(NO_2)3C_6H_2OH$:EOH, respectively, have long been major solutions on iron, iron alloys and steels, both as polish and preferential etchants. The acid concentrations are varied, the ethanol replaced by methanol, and both solutions are now used on other metals, such as the wide range of brass and bronze . . . both major copper alloys since antiquity: bronze (the Bronze Age) around 5000 B.C.; brass in use by the Egyptians, possibly, circa 3000 B.C.

The Sirtl etch (1HF:2CrO_3), developed in the early 1960s, has been a major preferential etchant for dislocation and defect development, initially on silicon, now on many other metals and metallic compounds. Since its inception, a dozen other HF:CrO_3-type solutions have been developed for improved defect enhancement in general or for specific type defects. It is interesting to note that like Sirtl etch, many of these solutions have been named after their developers: Schimmel, Sopori, Wright, etc., to include variations and modifications of their original formulations. These solutions are strictly preferential, as are many of the "iodine" etches — there are several iodine etches that have been developed over the years, all called "iodine etchant" although they are different formulas.

The HF:HNO_3, HF:HNO_3:H_2O or HF:HNO_3:HAc solutions are the original oxidation-reduction etchants developed in the Solid State semiconductor industry for general removal, polishing, and preferential attack on wafers, and are used on most metal and metallic compounds. Heavily diluted with either water or acetic acid (HAc), they act as surface cleaning, staining or passivating solutions. The mixture of HF:HNO_3, high in either acid, act as surface staining solutions.

There are a wide range of mixtures with H_2O_2 replacing HNO_3 as the oxidizing agent . . . the peroxide containing etches . . . with similar uses to those containing HNO_3. Another oxidizer type preferential solution is Kalling's etch (HF:$KMnO_4$) developed in the iron and steel industry, and used on semiconductor wafers and other metals.

Caro's etch (1H_2SO_4:3H_2O_2) is primarily a surface cleaning solution on a wide range of metals and metallic compounds, and there are many modifications of that base formula, to include water dilution. These solutions should be handled with care as they become self-heating upon being activated and, once reaction is initiated, will reach a constant boiling level of about 175°C.

Sulfuric acid, H_2SO_4, alone, is a glass cleaner and conditioner, and has been long used in preparing soda-lime glass plates for metallization as chrome glass masks for photolithography. It also is a general removal and polishing solution on many metals, particularly, as an electrolytic etch in the metal industries. As a mixture of H_2SO_4:$K_2Cr_2O_7$ it has long been known as the "glass cleaner" solution in many laboratories, and it is an extremely powerful oxidizer.

The two major alkalies, KOH and NaOH, have several applications depending upon solution concentration, and/or temperature. A 2—10% mixture, at RT, is used for surface cleaning. At 30—40% (RT to about 40°C, warm), good for general removal. As 20—50% solutions, hot to boiling, they become increasingly preferential. As electrolytic solutions, they are good for general removal, polishing, shaping, surface oxidation or selective structuring. As a molten flux, liquified pellets at 360°C, they are highly preferential and used for dislocation development in silicon and other single crystals. Other molten fluxes, such as carbonate and nitrate mixtures are applied as cleaning or etch solutions in general metals processing. Still other fluxes are used in the growth of single crystals, such as ferrites and garnets, and borax has already been mentioned as a flux for alloying.

The primary oxide or nitride remover is hydrofluoric acid, HF, applied as a concentrated solution; or water, even alcohol or solvent diluted; used from RT to boiling; or as hot vapor. Because of HF's rapid attack of oxides, buffered hydrofluoric acid (BHF) mixtures have been developed to better control removal rates. A fairly standard solution is 1HF:1NH_4F(40%), though there are other fluorine compounds used, such as KF and NaF, or NH_4F.HF.

Concentrated, hot phosphoric acid, H_3PO_4, also is both an oxide and nitride removal solution and, as it attacks silicon dioxide at a slower rate than it does silicon nitride, it can be used as a stop-etch where the nitride is on top, and the reduced attack rate of the oxide effectively "stops" the reaction. With the oxide over the nitride, the oxide will act an etch mask for photolithographic patterning of the nitride.

Hydrochloric acid (HCl) concentrated, water diluted, or diluted with alcohols, is used as a general surface cleaner. As a hot vapor in epitaxy systems, it is the primary glass tube and graphite susceptor cleaning method. In final rinse cleaning of semiconductor wafers, say, prior to metallization, a two-step process is often used — dip in 50% HCl and water rinse — then dip in 50% HF, and final water rinse, then N_2 blow dry.

The bromine-methanol (BRM) solutions are being widely used on many different metals and metallic compounds as a chemical/mechanical (chem/mech) removal and polishing solution at RT; for general polishing by immersion; or high concentrations of bromine as preferential solutions. When a wafer is patterned with Apiezon-W (black wax), BRM solutions have been used for selective etching structure. Most polishing is done with solutions of less than 5% Br_2, and preferential etching between 5—20% concentration. In conjunction with black wax patterning, solutions have been applied for thinning of specimens for transmission electron microscopy (TEM) study.

It should be noted that any chemical polish solution can be used for thinning, but a reasonably slow solution that will produce a flat, planar surface is preferred. For thick specimens, a rapid etch may be used for initial thickness reduction, followed by final thinning with a slower more controllable etchant such as BRM. Iodine can replace the bromine in BRM solutions with less evaporation loss, and similar results.

Ammonia, NH_3 and ammonium hydroxide, NH_4OH at standard liquid concentration are the most widely applied neutralizers for acid waste sumps before disposal into sewer systems. They may be the sole constituent of an etch on alkaline halides or as an additive constituent in an etchant mixture. Both liquids can be used to neutralize acid burns on the skin, and solid crystalline ammonia is the medical "smelling salts".

Water, alone, is a neutral solution (pH 7 on the Sorenson Scale), and the major quenching medium following etching. It is considered the universal solvent in chemistry and geology as, with time, it will eventually dissolve all compounds. In material processing, it can act as an etching solution on water soluble compounds, such as the alkaline halides (NaCl, KCl, etc.) and, when so applied, is referred to as an "acid" under TYPE in the formats. Sea water, containing dissolved salts, either natural or artificially compounded, has been used as an etchant for single crystal ice; as a quenching medium in metal processing; and even used to etch single crystal sodium chloride (salt), itself . . . the primary saline compound in ocean water.

The "Chrome Etch" solutions contain cerric ammonium nitrate or sulfate, with or without small amounts of HNO_3, HCl, or H_3PO_4. They are commercially available as premixed solutions, and were developed specifically for the controlled etching of chromium thin films on glass masks in the fabrication of chrome photo masks for photolithographic processing. They are commonly used at 30°C. The masks are made with iron oxide in place of chromium or nickel, today, for improved handling. The solutions used on chrome are quite rapid with about 2000 Å metallization, plus a 400—600 Å anti-reflective (AR) coating and the nitrate is preferred, as it has been shown that the sulfate solutions may leave an unwanted film on the glass plate after etching. Note that these solutions also can be used on nickel.

Though alcohols and solvents are mainly used for surface cleaning, often as a final rinse following water quenching of an acid etch for their water absorbing qualities, they can act as etchants on particular materials — in particular alkaline halide crystals.

The following methods of etching are in alphabetical order without regard to chemicals and/or processes involved, and represent the wide variety of terms used throughout the industrial cleaning and etching of materials.

METHODS OF ETCHING INDEX

-A-

1. Abrasive clean etch
2. Acid etch
3. Acid-slurry etch
4. Adhesion etch
5. Aged etch
6. Agitation etch
7. Alcohol etch
8. Alkali etch
9. Alloy etch
10. Air etch
11. Angle etch
12. Anisotropic etch
13. Anodic etch
14. Argon etch
15. Atmospheric etch
16. Atomizer etch
17. Autoclave etch

-B-

18. Barrel etch
19. Base etch
20. Basic etch
21. Basket etch
22. Batch etch
23. Beaker etch
24. Blank etch
25. Boiling bead etch
26. Bomb etch
27. Bottle etch
28. Brine etch
29. Brush etch
30. Bulk etch

-C-

31. Cap etch
32. Cascade etch
33. Cathodic etch
34. Caustic etch
35. Centrifuge etch
36. Chem/mech etch
37. Channel etch
38. Cleave etch
39. Cleaning etch
40. Cold etch
41. Conditioning etch
42. Contamination etch
43. Control etch
44. Cool etch
45. Corrosion etch
46. Coupon etch
47. Cover etch
48. Crucible etch
49. Cryogenic etch
50. Crystallographic etch
51. Cup etch
52. Cutting etch
53. Cyclic etch

-D-

54. Damage removal etch
55. Decoration etch
56. Defect etch
57. Definition etch
58. Degreasing etch
59. Descaling etch
60. Dicing etch
61. Dip etch
62. Dislocation etch
63. Dissolutionment etch
64. Drip etch
65. Drop etch
66. Dry chemical etch (DCE)
67. Dry etch
68. Dry ice etch

-E-

69. Electrolytic etch
70. Electron beam etch
71. Etch-back etch
72. Etch/clean sequences
73. Etch-stop etch

-F-

74. Failure etch
75. Fatigue etch
76. Figure etch
77. Finish etch
78. Finite form etch
79. Flame etch
80. Flash etch
81. Flat etch
82. Float-off etch
83. Flood etch

84. Flush etch
85. Flux etch
86. Form etch
87. Forming etch
88. Free etch
89. Freeze etch
90. Fume etch

-G-

91. Gas etch
92. Gate etch
93. General etch
94. Graphic etch
95. Gravity etch
96. Groove etch

-H-

97. Halogen etch
98. Heat etch
99. Heat-tint etch
100. Heavy etch
101. Hot etch

-I-

102. Ice etch
103. Immersion etch
104. Induced damage etch
105. Initial etch
106. Ion (gas) etch
107. Irradiation etch
108. Isotropic etch

-J-

109. Jar etch
110. Jet etch
111. Junction etch

-K-

112. Key etch
113. Krypton etch

-L-

114. Laser etch
115. Layer etch
116. Lift-off etch
117. Light etch
118. Light figure etch
119. Light figure orientation etch
120. Lineation etch
121. Liquid etch
122. Long etch
123. Loose etch
124. Low etch

-M-

125. Macro etch
126. Magnetic etch
127. Matte etch
128. Medium etch
129. Melt-away etch
130. Melt-back etch
131. Mesa etch
132. Metal etch
133. Metallographic etch
134. Metallurgical etch
135. Micro etch
136. Milling etch
137. Minimum etch
138. Mist etch
139. Molar etch
140. Molten flux etch

-N-

141. Named etch
142. Native oxide etch
143. Neutral etch
144. Nitride etch
145. Normal etch

-O-

146. Oil etch
147. Orientation etch
148. Oxide cleaning etch
149. Oxide etch
150. Oxide removal etch
151. Ozone etch

-P-

152. Particulate etch
153. Passivation etch
154. Pattern etch
155. Phase etch
156. Photo resist etch
157. Pickling etch

158. Pinhole etch
159. Pit etch
160. Planar etch
161. Plane etch
162. Plasma etch
163. Polar etch
164. Polarity etch
165. Pole etch
166. Polish etch
167. Preferential etch
168. Pre-plate etch
169. Pressure etch
170. Priming etch
171. Profile etch
172. Pylon etch

-Q-

173. Quality etch
174. Quantity etch

-R-

175. Rapid etch
176. Raster etch
177. Rate etch
178. Reactive (ion) etch
179. REDOX etch
180. Relief etch
181. Removal etch
182. RF plasma etch
183. Rolling etch
184. Rough etch

-S-

185. Sand etch
186. Satin etch
187. Saturation etch
188. Saw (acid) etch
189. Scale etch
190. Seeded etch
191. Segregate etch
192. Selective etch
193. Sequence etch
194. Series etch
195. Shaping etch
196. Shim etch
197. Short etch
198. Sizing etch
199. Slosh etch
200. Slow etch
201. Slush etch
202. Snow etch
203. Soak etch
204. Solid etch
205. Solution etch
206. Solvent etch
207. Spark etch
208. Sphere etch
209. Spin etch
210. Sputter etch
211. Squirt etch
212. Stacking fault etch
213. Stagnant etch
214. Stain etch
215. Standard etch
216. Steam etch
217. Step-etch
218. Still etch
219. Stir etch
220. Stop-etch (etch stop)
221. Strain etch
222. Stress etch
223. Strong etch
224. Structure etch
225. Subtractive etch
226. Swab etch
227. Swirl etch

-T-

228. Temperature etch
229. Thermal etch
230. Thimble etch
231. Thinning etch
232. Trim etch
233. Tumble etch

-U-

234. Ultrasonic etch
235. Used etch

-V-

236. Vacuum etch
237. Vapor etch
238. Vapor degrease
239. Via-hole etch
240. Vibration etch

METHOD DEFINITIONS

1. ABRASIVE CLEAN

Dry or wet abrasive powders used as a method of cleaning material surfaces. Applied as a dry powder under gas pressure; in a water slurry; acid-slurry; as sandpaper; or in other medium, such as grease, or a resinoid disc cutting blade. Silicon carbide, SiC, is a major compound for general processing, original manufacturing name "carborundum"; garnet powder, as the W-0 to W-16 series, graded by abrasive size/grit, with W-5 (600 grit or 16 μm) a general standard surface finish on materials. Surface machining shown in micro-inches, μ″, (250 Å = 1 μ″) with 60 μ″ a standard finish. Abrasive hardness (Mohs scale of 1 to 10); particle physical size and shape; mesh size in millimeters, microns, or inches — shape as blocky, round, slivers, etc., influence surface finish and cleaning.

---

2. ACID ETCH

The designation pH is the abbreviation used to describe the hydrogen-ion concentration of a solution. On the Sorensen Scale, pH values run from 0 to 14 and can be defined as pH 7 is a neutral solution (water). Below pH 7, solution is acid, and the lower the value the greater the hydrogen-ion concentration. Above pH 7 the solution is basic, and the higher the value, the greater the hydroxyl-ion concentration.

Strong acids that are widely used can be represented by sulfuric acid and nitric acid. Weak acids, and this can be confusing, may contain the OH^- hydroxyl radical, such as glacial acetic acid, CH_3COOH (HAc). Even water, the neutral solution at pH 7, is sometimes called an "acid" where it is used as an etching solution on material like the alkaline halides, such as sodium chloride, NaCl or potassium bromide, KBr and similar compounds and salts that are soluble in water.

Almost all known inorganic solids, whether they are an element, natural mineral or artificially grown material can be etched in either a single acid or acid mixture, and the material solubility is controlled by concentration with dilution — usually by water — and by solution temperature.

Semiconduetors and many metal compounds require mixed acids to produce an oxidation-reduction reaction — $HF:HNO_3$ solutions as an example — where the nitric acid oxidizes the material and the hydrofluoric acid dissolves the oxide formed. Other acids, such as hydrochloric acid, HCl or sulfuric acid, H_2SO_4, convert the material to a water soluble chloride or sulfide, respectively. And, as is the case of many iodine, I_2, solutions, the material removed forms water insoluble compounds that can be removed later as a sludge for material recovery, e.g., gold.

Most acids are obtained and used in liquid form as a specific concentration in water: sulfuric acid, 95%; hydrogen peroxide, 30%, and so forth. Solution mixing and use for etching is mostly done using the commercial concentration supplied as a physical volume quantity, regardless of absolute acid concentration. An exception is in titration of solutions for determination of constituents, certain percentage evaluations of mixed materials, and similar chemical tests and studies.

As wet chemical etching, acids and acid mixtures can be used in several physical ways: by immersion, jet, spray, hot vapor, etc., as well as for specific results: cleaning, polishing, surface structuring, defect development, or thinning, etc. Such methods and/or techniques form the bulk of this Methods of Etching section.

Acids also are used as a liquid electrolyte (electrolytic etching [EE]), including electric current, and the specimen etched by immersion, jet spraying, with similar specific results as in wet chemical etching (WCE).

---

3. ACID SLURRY ETCH

A form of mechanical polish lapping where an acid is mixed with an abrasive. This reduces the amount of work damage introduced by the abrasive into the subsurface of material being lapped or polished and can improve surface finish. The etch solution can be an acid or a base as a chemical mixture, such as SiC:H_2O:Glycerin:10%HF, or Al_2O_3:H_2O:10%KOH.

Acid slurries are of particular interest in polishing materials with a Mohs hardness greater than H = 7, such as glass or ceramic compounds for these materials tend to produce an amorphous, smeared layer during normal abrasive lapping, and such layers show different etching characteristics than that of the material bulk. This can affect subsequent processing as a masking surface film (referred to as orange-peel), or initiate the introduction of pin-holes in an oxide, etc. The acid action in an acid-slurry will reduce such effects.

Soft, malleable and ductile metals, such as pure gold, lead, copper and silver, are not recommended for acid-slurry etching as they also tend to smear.

---

4. ADHESION ETCH

Etching of thin films to relieve stress and evaluate the film failure, as lifting, peeling, crazing, etc. Etching may include heat treatment to initiate or enhance failure.

---

5. AGED ETCH

Has two meanings: (1) any solution mixed and allowed to stand for a specified period of time, or (2) any solution used more than one time may be called an aged or used solution. Aqua regia is an example of an etch requiring an aging time before use. See Stagnant Etch; Used Etch.

---

6. AGITATION ETCH

Any solution used with some form of physical movement. It can be movement of the solution, of the part, or a combination of both, and there are various forms of movement:

1. Solution
   (i) Rotational stirring — a rod in solution, hand or electrically operated
   (ii) Magnetic stirring — magnet in solution with an exterior magnetic flux
   (iii) Ultrasonic motion using transducer
   (iv) Rocking, as on a shaker table
   (v) Convection, as with boiling solutions
   (vi) Hand swirling, a form of rotational stirring
   (vii) As spray on a rotating part
   (viii) Gas bubbled through the solution as an agitation mechanism
2. Part
   (i) Vertical motion
   (ii) Horizontal motion
   (iii) Spinning motion
   (iv) Solution rotation, as in angle etching
   (v) Rocking motion, or various combinations of any of the above

Solutions that produce heavy evolution of gas bubbles are commonly agitated in some manner to prevent bubbles from adhering to surfaces being etched to prevent the appearance of "footprints".

---

7. ALCOHOL ETCH

Alcohols are normally considered cleaning or rinsing solutions rather than etchants, but they will attack certain alcohol soluble materials.

Three widely used alcohols are ethyl alcohol (ethanol) with the acronym of EOH; methyl alcohol (methanol or wood alcohol) with the acronym of MeOH; and isopropyl alcohol (rubbing alcohol) with acronyms of either ISO or IPA. In addition, acetone (Ace acronym) is included here because of its wide use as a cleaning solution in photolithography, even though it is a ketone, not an alcohol.

All alcohols are poisonous with the exception of ethanol (drinking alcohol) which is supplied commercially as denatured alcohol with 5% MeOH or ethylene glycol added to make it nonpotable. Absolute alcohol is 100% EOH (200 proof), but most ethanol is 95% EOH. 5% H_2O (190 proof). Note that most drinking alcohol is 80 proof (whiskeys, rum, brandy, gin, vodka) with proof-gallon as an initial 50% water content.

In addition to being used as a pure alcohol etch, many etching solutions contain alcohol as a constituent — Nital, HNO_3:EOH, as an example.

---

8. ALKALI ETCH

Alkaline solutions are greater than pH 7, are the base or basic chemical solutions according to the Sorenson scale of pH as they contain the hydroxyl-radical, (OH^-). They may be in liquid form, such as ammonium hydroxide, NH_4OH or in solid form as potassium hydroxide, KOH, or sodium hydroxide, NaOH. Where ammonium hydroxide is used already in liquid form, the solid hydroxides must be dissolved in water for a specific percent concentration liquid. As a solution, they can be mixed as a Molar solution concentration, e.g., moles per liter of water and shown as, say, 1 *M* KOH or 0.5 *M* NaOH (one mole of any substance is the total of the molecular weights of the elements in the compound that is equivalent to one gram-molecular weight in one liter of water — (1/l — or 1000 ml). A computation example is

Potassium hydroxide, KOH

K =	39.102 (at wt)
O =	15.999 (at wt)
H =	1.008 (at wt)
Total	56.109 g

This says that approximately 56 g of KOH in 1 liter of water is a 1 Molar (1 *M*) solution. It is the gram-mole of the substance. There are tables available that show concentration as a percent solution, such as 10% KOH (109.2 g KOH), which is roughly equivalent to a 2 *M* solution.

It also may be shown as a Baumé, °Bé, solution by specific gravity, such that 12.2 °Bé is equivalent to a 10% solution with a specific gravity of 1.0918, regardless of the chemical(s) involved. Note the degree (°) sign is used with the Baumé scale, but it is not a temperature scale like Centigrade, °C, or Fahrenheit, °F.

The term "hydroxide" is often used in place of alkali or base, and all are equivalent, having the same meaning. In the etchant section of Chapter 2 as the chemical descriptor word, it is shown as TYPE: Alkali, polish/removal/preferential, etc.

In general, hydroxides are weak etches at room temperature and primarily used for surface cleaning in concentrations of 2 to about 15%; when used hot to boiling, usually in concentrations between 20 to 50%, they are removal etchants and can be surface preferential — developing crystallographic structure.

Alkalies also are used as molten fluxes for preferential etching, such as on silicon; or as growth fluxes for high temperature type single crystals, such as garnets and ferrites.

---

9. ALLOY ETCH

Any solution used to clean an alloyed joint or structure, such as water for flux removal. The use of an alloy material acting as the etching agent, such as an AlSi(5%) alloy bead forming a pit in a single crystal wafer, then etch removed for study of the convex pit side-slopes and their crystallographic planes.

---

10. AIR ETCH

Alternate term for Atmospheric Etch. Actual etching is due to oxygen, salt (chlorine), water moisture, or other contaminants in the air such sulfur, acids . . . smog. See Atmospheric Etch.

---

11. ANGLE ETCH

There are two meanings: (1) specimen: the part is lapped at a specified angle and then etched. Used to measure diffusion depths; to develop p-n junctions, or epitaxy layer structure in semiconductor devices; or defects and structure in metal alloys, or (2) containers: the etch container is placed on a rotating spindle, say, at a 45° angle. Specimens are mounted on discs of Teflon or SST with Apiezon-W (black wax), placed face-up on the container bottom, with etching done at rotation speeds of between 130 to 300 rpm. The method used to produce extremely planar wafer surface. Also used to etch-thin material for transmission electron microscopy (TEM) study, and to polish spheres, cylinders or other forms by slow rotation during etching.

---

12. ANISOTROPIC ETCH

Another term for Preferential Etch. Any etch that attacks crystallographic planes at different rates. Anisotropy and isotropy are terms related to the physical state of matter and in mineralogy are defined with regard to the propagation of light through solids. Recently, the term has been applied in dry chemical etching (DCE) of selective structures whether the solution used is a preferential/anisotropic or polish/isotropic solution. See Dry Chemical Etch; Selective Etch; Preferential Etch.

---

13. ANODIC ETCH

Electrolytic etching with the specimen as the anode for cleaning, removal, polishing, structuring, and may include switching from anode-to-cathode. See Electrolytic Etch; Cathode Etch.

---

14. ARGON ETCH

Primary: (1) Argon ion, Ar^+, cleaning or etching of surfaces. Secondary: (2) The etching of single crystal argon.

(1) Argon ion (Ar^+) etching and cleaning of surfaces in vacuum has become a widely used technique in processing immediately prior to metallization or epitaxy growth. RF power levels vary from 50 to 500 KeV and the electron energy can be sufficient to cause subsurface damage.
(2) Argon and other gases have been grown as single crystals in vacuum under pressure and cryogenic conditions. Once grown, preferential etching of the solid ingot has been accomplished by varying vapor pressure in the system. Note that molecular argon is used for parts drying or as an inert processing atmosphere, and may preferentially etch a specimen surface at high temperature.

---

15. ATMOSPHERE ETCH

Air, nominally 24% oxygen and 75% nitrogen with water vapor, contaminating gases or other compounds will attack all known inorganic materials with time — hours, days, years — millions of years, geologically.

In Solid State materials processing, atmospheric etching is not widely used although

Clean Rooms have controlled atmosphere, temperature, and humidity for optimum device quality control.

The metals industries, on the other hand, are vitally concerned with the atmospheric corrosion of metals, alloys and materials with on-going studies in progress for development of improved stability, reliability and corrosion resistance of all materials to weathering — the most general term applied to effects of atmospheric etching reactions in geology.

There are specific component tests, such as salt spray, gas absorption, and corrosion, that are standard evaluation techniques in both Solid State and many other industries that use air as an active medium or carrying medium for other vapors. See Air Etch; Gas Etch; Dry Etch.

---

16. ATOMIZER ETCH

Use of an atomizer to apply a fine mist/spray to etch or clean a surface. Also used to fine-tune electrical parameters on a semiconductor device with exposed junctions. A perfume atomizer has been used with a weak, polish acid mixture to etch, or pressure spray cans containing acetone or other solvents for cleaning pressure with nitrogen gas.

---

17. AUTOCLAVE ETCH

A closed metal container capable of handling high pressure. Called autoclaves, they have long been used for steam cleaning of medical tools. The type autoclave application can be by process, material, or study involved, to include temperature and pressure levels:

Low pressure: General cleaning of parts. 2—10 psi.
Medium pressure: Artificial alpha-quartz crystal growth. 30,000—50,000 psi.
High pressure: Compression of single crystal metals and compounds in study of high density structures. 300,000—500,000 psi.
Cryogenic: Artificial growth of single crystal ice at liquid nitrogen (−196°C), or liquid helium (−2 K) temperature. Etch by varying vapor pressure, say, of a vacuum system. 2—5 psi (ice); to 30,000 psi or more for high pressure crystallographic forms of single crystals (cryogenic to elevated temperatures).

Steam oxidation of silicon is being done in low pressure autoclaves for both an oxide thin film or, with HF removal of the steam generated oxide, as a method of cleaning surfaces.

---

18. BARREL ETCH

A barrel-like container or bottle for cleaning and etching, used closed and rotated horizontally on roller bars; open-topped and mounted on a rotating spindle similar to an angle etch beaker; or close-topped on a shaker table.

The roller bar method, using ceramic jars and abrasives, has been a long standing way of polishing gem stones, and called tumbling. The same system is used for general cleaning of small parts with dry abrasives with the tumbler unit capable of handling from 2—5 lb of material. Machine shops and plating shops may use cement mixer-sized systems for parts cleaning. See Bottle Etch; Angle Etch.

---

19. BASE ETCH

Term used in three ways: (1) Any chemical solution with pH greater than pH 7 is a chemical base, or basic solution as determined by the hydroxyl-radical concentration, OH^- radical. (2) The first etch solution used in a clean/etch sequence or the primary solution in such a sequence. (3) A specific etch mixture with established characteristics against which other solutions are evaluated. This has included modification of the original mixture and comparison of reactivity changes. See Alkali Etch; Basic Etch.

---

20. BASIC ETCH

Any solution with pH >7 as established by the hydroxyl-ion concentration. The chemical compound is a base as against an acid. Occasionally refers to a primary solution in an etching sequence or process. See Base Etch; Alkali Etch.

---

21. BASKET ETCH

The etching of wafers, dice or other parts in some form of a basket holder that is submerged in an etch solution; passed through an etch solution; held in hot vapor; or held for spraying. The latter two largely for cleaning, as in vapor degreasers. Basket design and material depend upon acid mixtures and/or temperature. Ceramic and platinum for high temperature; Teflon, polyethylene, and other plastics, or glass for nonreactivity (glass not with fluorine-containing solutions). There are many Teflon and polyethylene basket designs for processing Solid State materials; steel basket trays for vapor degreasers; wire mesh baskets, and so forth. Teflon will handle temperatures to 200°C, but some plastics will melt in vapor degreasers where temperature levels can range from 80°C (TCE) to about 120°C (PCE/Perk).

CAUTION: Where heavy etching is done, a basket can mask the material surface causing erratic results. Stir the solution and/or rotate the basket against the solution flow direction. See Agitation Etch: Rotation Etch; Stir Etch; Angle Etch; Batch Etch.

---

22. BATCH ETCH

The etching or cleaning of any two or more wafers, specimens or parts in a solution at one time. A large glass or polyethylene beaker or similar container with wafers, or parts held in a basket immersed in the solution is the simplest form of batch etching. High volume production systems may have two containers — one etch, one wash — a trough through which parts are passed, or a centrifuge with immersion, spray etching, water quenching, and air or nitrogen drying sequences. Such systems can be manual or automatic cycling, with wafers or parts in a basket holder or fed from a cassette through a belt-spinner-belt to re-cassette loading. Cassette systems are designed for processing wafers and dielectric substrate circuits though photo resist applications; PCB boards though solder alloy furnaces. Plating is done with multiple parts or hangers, and there is "free etching", where parts are dropped loose in the etch solution.

---

23. BEAKER ETCH

Any open-topped container used to hold an etch or cleaning solution. Quartz, Pyrex, polyethylene, or Teflon beakers from 200 to 1000 ml are common. Thick walled beakers (Pyrex) are recommended for high temperature. Beakers are widely used for individual wafer/part or small batch processing of materials, to include plating. Teflon and polyethylene beakers are used if solutions contain fluorine compounds. Metal beakers of steel and platinum or ceramic are used in high temperature etching, and as thimbles, cups or crucibles for molten flux etching or as single crystal growth fluxes.

---

24. BLANK ETCH

Any etch solution used on a relatively small flat part, such as a 1 × 1 × .010″ ceramic or metal substrate. Also called a flat or coupon. Heavily used in the metals industries for a wide variety of evaluation tests (coupons), or in alpha-quartz frequency crystal processing (blanks). See Coupon Etch.

---

25. BOILING BEAD ETCH

Ceramic, glass, or metal beads of various sizes placed in the bottom of an etch beaker to introduce and control a bubble/agitation action of a hot to boiling solution. Bead material should be inert against etch attack, and not contribute contamination.

---

26. BOMB ETCH

A metal container capable of holding high pressure, and used to hold an etching or cleaning solution/gas. See Autoclave Etch.

---

27. BOTTLE ETCH

Use as a closed bottle for etching. Ink patterned and taped silicon wafers have been etched in this manner with solutions of HF:HNO_3:HAc in 500 ml polyethylene bottles. With the etch solution and wafers loaded, the bottle cap attached, the bottle is then placed horizontally on a shaker table for the etch period.

CAUTION: With repeated use the polyethylene will harden and crack. If the cap is not firmly tightened, it can come loose during etching. With insufficient acid or too many wafers, excessive fumes and pressure are generated and the bottle can rupture explosively. Even with good shaker table agitation wafers tend to overlap, stick together, and produce erratic etching results. See Batch Etch; Barrel Etch.

---

28. BRINE ETCH

Natural or artificial salt water used as an etch solution. Not common; some test evaluation on metals and compounds in corrosion study. Quenching in brine solutions is used in fabricating steel alloys and other metal mixtures to obtain specific structure and other physical characteristics. The salt spray test might be considered as a type of brine etch. In the metals industries coupons are exposed to seacoast salt environments in on-going corrosion tests. See Water Etch; Atmosphere Etch.

---

29. BRUSH ETCH

The use of a brush to apply an etch solution to a material surface. It may be for light cleaning/etching, to developed etched structure, to fine-tune electrical parameters of a device, or to produce a particular surface finish. In the latter case, a wire brush may be used with or without electric current applied, for a ''brush'' surface finish. See Swab Etch; Matte Etch; Satin Etch.

---

30. BULK ETCH

After mechanical lap and polish of a material surface, the surface is etched to remove the residual subsurface damage down to the undamaged bulk metal. Or use of a heavy etch for general bulk removal, which may be measured by total gram-weight loss of a specimen. Occasionally, with regard to a large volume of etching or cleaning solution. See Rough Etch; Removal Etch.

---

31. CAP ETCH

Special terms applied to Solid State processing and etching single crystals with thin films.

(1) After ion implantation of a semiconductor wafer when a CVD silicon dioxide or nitride coating is then deposited on the implanted surface, it is called a ''cap''. After an annealing cycle to stabilize the implanted element the cap is etch removed with solutions of BHF, HF, or RF plasma (ionized gas).

(2) In molecular beam epitaxy (MBE), the top thin film layer also is referred to as a cap, as an epitaxy or oxide/nitride material layer. If the latter, a suitable fluorine containing etch is used; whereas, if it is an epitaxy material layer, it may be a selective acid or ionized gas etch.

(3) Metallized thin films are sometimes referred to as capping layers, such as when being pattern etched by photolithographic processing.
(4) The use of a capped crucible etching container may be referred to as a capped method of etching, as in molten flux growth of single crystals with subsequent etch removal from the flux. See Cover Etch.

---

32. CASCADE ETCH

Though often called etching, more commonly a method of final water quenching and washing following an etch period, with reference to the container structure used: a rectangular trough divided into three or four progressively lower sections, usually fabricated from polyethylene sheet or similar high purity plastic. Commercially available systems are about 6″ wide, each trough section about 8″ long, and progressively more shallow. High purity water (18 MΩ) enters the top section, flowing downward through lower sections. To use a cascade, the part to be quenched and rinsed is placed in the lowest trough first, then moved upward to the final top trough. Repeat rinse steps until the megohm meter remains at 18 MΩ in the upper rinse trough.

---

33. CATHODIC ETCH

Electrolytic etching is normally anodic, but switching from anode to cathode acts as a deplating and removal system, not an actual method of etching. The term has been used relative to studying the growth effects of hydrated oxides on metals, and in sputter etching with ionized gases. See Sputter Etch; Dry Chemical Etch; Anodic Etch; Ion Etch.

---

34. CAUSTIC ETCH

An alkali or hydroxide solution with a pH >7. See Alkali Etch; Hydroxide Etch; Base Etch.

---

35. CENTRIFUGE ETCH

Rapid spinning motion of a part during etching or cleaning. A photo resist spinner has been used for acetone cleaning during photo resist application and etching of semiconductor wafers with the holder platen rotating at about 3500 rpm. There are automatic systems designed to clean, etch and dry wafers through automatic cycles with wafers handled from cassettes.

A specialized system, including directional magnetic flux to produce a negative (−) or positive (+) gravity effect during etching has been used to form pits and via-holes in brass and silicon wafers, and referred to as a centrifuge etching method.

Centrifuge systems are used to mixed glass-frits as a liquified slurry, which are then spun onto surfaces and melt-fired (nominal 600—800°C).

---

36. CHEM/MECH ETCH

Chemical/Mechanical etching is being more and more widely used in single crystal wafer processing to remove residual subsurface damage introduced by previous cutting, lapping, or polishing steps. The bromine:methyl alcohol (BRM) solutions are used to remove damage with a slow polish action.

---

37. CHANNEL ETCH

The etching of a groove into a surface. Used in selective structuring of single crystal devices.

(1) Wafer dicing: Semiconductor wafers are channel etched into individual dice shape and

then, without demounting from the mounting substrate channels are oxidized or metallized as a wrap-around edge coating of the dice.

(2) Wafer channeling: Surfaces coated with oxide/photo resist with channel pattern etch opened in ⟨110⟩ directions on (100) oriented wafers. Dry chemical etching (DCE) or wet chemical etching (WCE) used to channel etch. Method used to fabricate single channels, grids, saw-tooth gratings and so forth. Also for diffusion depth profiling; to observe change in dislocation density, with depth, or dislocation movement in bulk. The channels can have a curved radius, be "V" shaped, or with vertical or sloped side walls and flat bottom.

(3) Snap-Strate: Etch, sandblast, or saw cut a channel into, but not through, the material. The wafer/substrate can then be handled in processing as a unit and, after final structuring, individual units are "snapped" apart. Substrate commonly Alumina, Al_2O_3.

---

38. CLEAVE ETCH

The etch polishing of a cleaved single crystal wafer surface to remove residual cleavage steps.

---

39. CLEANING ETCH

Any solution or gas used to clean a surface with minimum etch removal of the material. See Solvent Etch.

---

40. COLD ETCH

Any solution used between about 2—10°C. Selective preferential etching to observe structure, or fine-tuning electrical parameters. There are "cold etches" using snow in their formulas and chemical mixtures, some as eutectics, that form a specific cold temperature. These are used directly or to chill other etch solutions. See Temperature Etch.

---

41. CONDITIONING ETCH

Any solution used to prepare a surface for subsequent processing. Very common in the plating of metals, such as zincating aluminum before metal deposition.

---

42. CONTAMINATION REMOVAL ETCH

Any solution used to clean a surface of unwanted solid material, such as dirt, oil, or grease; or the removal of ionic radicals, such as F^+, Cl^-, Cr^{++}, etc., remaining on a surface after etching. See Alcohol Etch; Solvent Etch.

---

43. CONTROL ETCH

Any etching for a specified time period or at a particular temperature. RF plasma cleaning done without regard to time, but to a specified temperature stop-point. A solution may be established as a standard control mixture, against which other solutions are evaluated.

---

44. COOL ETCH

Any solution used between about 10—20°C. Most often used as a slow cleaning and polishing etch. See Temperature Etch.

---

45. CORROSION ETCH

The attack or alteration of any material surface from the action of a solid, liquid, or gas. All metals and alloys are subjected to atmospheric corrosion evaluation; high temperature gas reaction; or molten metals and chemicals used as an etching medium, in determining the stability of materials under varying conditions.

---

46. COUPON ETCH

Etching of a specimen cut as a flat form sheet material. Term common to metal flats used in atmospheric corrosion studies, and other metal tests in the metal industries, but not widely used in Solid State processing. See Blank Etch; Wafer Etch.

---

47. COVER ETCH

As a specialized term where methyl alcohol is over a BRM mixture: MeOH/Br_2:MeOH. The part to be etched is slowly lowered into the BRM solution, etched, then slowly withdrawn through the pure methanol ''cover'' to quench etching action without exposure to air.

---

48. CRUCIBLE ETCH

The use of graphite, ceramic, or high temperature metals, such as platinum, as a cup to hold an etching solution. Most commonly used in molten flux etching, such as defect development in silicon with KOH pellets at about 360°C; or crystal growing from such fluxes, followed by their etch removal. See Cup Etch.

---

49. CRYOGENIC ETCH

The use of cryogenic liquids as an etching medium, such as liquid chlorine, LCl_2 at −102°C. Also refers to the use of cryogenic liquid as a chilling agent for other acid solutions, such as liquid nitrogen, LN_2. See Temperature Etch.

---

50. CRYSTALLOGRAPHIC ETCH

Any solution that will develop single crystal plane structure by preferential attack. Specific solutions have been developed on most metals and metallic compounds during evaluation development, and include etching spheres (convex), and pits (concave) surfaces. See Finite Form Etch; Dissolutionment Etch; Anisotropic Etch.

---

51. CUP ETCH

Any open container when used to hold an etching solution may be referred to as a cup. Usually a small ceramic, graphite, or high temperature metal (Pt, Mo, Ti, Ta) used as a crucible for either an etch solution or solid molten metal, and may include a cap or cover. See Beaker Etch; Crucible Etch.

---

52. CUTTING ETCH

The use of an etch solution, alone, or a wire soaked in the etch to cut material. See Acid-saw Etch; Acid-slurry Etch.

---

53. CYCLIC ETCH

Etching with sequential periods of time. May include water or alcohol quenching between etch cycles, or the removal to air and return to etch without quenching. See Sequence Etch; Step Etch.

---

54. DAMAGE REMOVAL ETCH

Any etch solution used to remove either the surface or subsurface damage present or induced by previous abrasive lapping, etc. It is usually a slow polish type etch, such as bromine:methanol that both remove the damaged zone and polish simultaneously. See Chem/Mech Etch.

---

55. DECORATION ETCH

The use of a metal thin film diffused or alloyed into a single crystal surface to enhance defect structure for observation. Other materials used are fluorescent dyes that may or may not be biased electrically to develop color patterns (liquid crystal, etc.); and, though not an etch, carbon or iron powders have been brushed on preferentially etched surfaces to enhance defect observation. See Stain Etch; Heat-tint Etch.

---

56. DEFECT ETCH

A general term denoting the etch development of any bulk or surface anomaly in a material whether or not it is single crystal, colloidal, amorphous or crystalline in structure. Much of the etching study on materials is concerned with the recognition and elimination of a wide range of defects, several of which have developed specific etching names as shown below.

See:	Dislocation Etch	Stacking Fault Etch	Slip Etch
	Preferential Etch	Polarity Etch	Crystallographic Etch
	Anisotropic Etch	Thermal Etch	
	Lineation Etch	Vacuum Etch	

---

57. DEFINITION ETCH

Term used in two ways: (1) an etch used to develop a particular structure, such as fine-line definition, and (2) a solution developed for a particular purpose — a definitive etch.

---

58. DEGREASING ETCH

Usually a solvent for removal of oils or greases, rather than etch solution, although an etch may be used. Laboratory glass is etch cleaned in a solution of H_2SO_4:$K_2Cr_2O_7$, and soda-lime glass plates used for chrome photo resist masks are scrub cleaned with soap.

---

59. DESCALING ETCH

An etch used to remove heavy contamination from metal surfaces. Term widely used in metal processing, but not in Solid State development where most materials are supplied as nominally clean parts.

---

60. DICING ETCH

Any etch solution used to cut and separate discrete devices or units from a semiconductor or other type material in wafer or thin sheet form. See Channel Etch.

---

61. DIP ETCH

The etching or cleaning of a specimen for a very short period of time — the "time" is difficult to define but can be arbitrarily said to be between 1—3 sec. This form of etching has been used for (1) a final surface cleaning step; (2) to develop and optimize a defect or structure; (3) to fine-tune electrical or frequency parameters of a device, etc. It may be a single "dip" or a series of such dips, depending on requirements.

---

62. DISLOCATION ETCH

The preferential etch development of structure in a single crystal material that can be related to crystallographically oriented defects associated with bulk structure or surface defects. Dislocations can be introduced during ingot growth; by heat treatment, alone, or in conjunction with alloying, diffusion, epitaxy, cutting, and lapping; by controlled bending or striking of a surface in defect studies; or inadvertent damaging from process handling, etc. The study of dislocations in single crystals is probably one of the most important areas of developmental work in the Solid State field, because of the major effects defects produce on device parameters. See Defect Etch.

---

63. DISSOLUTIONMENT ETCH

In the broadest context the term dissolution or dissolutionment etch refers to any solution that will dissolve a material. In the study of single crystals, it is used with specific reference to the etching of spheres to finite crystal form (FCF) with development of crystal facets (planes) on convex surfaces. Also in determining the general etching for characteristics of all metals and metallic compounds.

---

64. DRIP ETCH

To apply a single, or series of droplets on a surface. Usually applied to a limited area. Iron specimens have been etched in this manner with DI water. See Pinhole Etch; Structure Etch; Selective Etch.

---

65. DROP ETCH

Term has been used in three ways: (1) the free etching or cleaning of a part by physically dropping free into the solution; (2) placing of a drop of solution on a surface to etch and/or plate a semiconductor p-n junction; develop pinholes in oxide/nitride thin films, etc., and (3) dropping of a part down through a column of solution with distance, time, and temperature used to control fabrication, e.g., the shot tower technique used in sphere forming.

---

66. DRY CHEMICAL ETCH (DCE)

The use of an ionized gas for cleaning or etching surfaces. Ar^+ ion cleaning of single crystal wafer surfaces has become a standard technique in processing. Also used for selective etching of structure in device design. Gases used may be inert, like argon; or they may be a reactive species, such as BCl_3. The latter type called reactive ion etching (RIE). DCE is one of the three major etch formats used to describe etching.

---

67. DRY ETCH

Not considered a true form of etching, yet drying can affect the surface or bulk of a specimen. Water removal from a surface can cause crazing, cracking or leave stains. Bulk removal can alter both chemical formula and crystal structure. Heat treatment or annealing and hydrogen firing are two methods. The latter a method of surface cleaning in a reducing atmosphere.

---

68. DRY ICE ETCH

Solid CO_2 used as a direct etching medium. Also used as a mixture with alcohols or acetone for chilling another etch solution, or for removal of water vapor from process gases . . . drying the gas by chilling and freeze-out of the water vapor as ice.

---

69. ELECTROLYTIC ETCH (EE)

The use of electric current applied to any etching or cleaning solution. Many such solutions are shown in the following section for different metals and metallic compounds using the electrolytic etchant (EE) format for presentation of data. The specimen being etched in the anode and a metal (such as Cu, Pb, Fe) as the cathode. Additional anodes and cathodes may be included, there can be anode-cathode switching, or the specimen separately electrically biased. Both liquids and molecular gases are used.

---

70. ELECTRON BEAM ETCH

E-beam or EB vacuum systems are used for metal evaporation from a rotatable copper hearth containing from one to four crucibles. The beam is magnetically bent up and around into the crucible, a 270° beam being a standard, today. The E-beam can be used to etch remove and vaporize metals on a specimen surface at controlled rates by varying power input. Also used to anneal a specimen (heat treatment), in addition to being a method of metal evaporation.

---

72. ETCH/CLEAN SEQUENCES
    See Clean/Etch Sequences under Composition.

---

73. ETCH-STOP
    An etch solution that will attack one material but not another in a multilayer thin film structure. Hot H_3PO_4 will etch remove Si_3N_4 much more rapidly than SiO_2, such that the oxide can work as an etch-stop mechanism. The method is used in structuring devices, and in the removal and thinning of layers for TEM study.

---

74. FAILURE ETCH
    Any solution use in the study of device failure, and used to etch develop, surface stain or otherwise expose the causes of failure.

---

75. FATIGUE ETCH
    Specific solutions used on metals and their alloys in the study of material failure due to fatigue, such as cracking from bending or crazing from atmospheric corrosion. This is a major test evaluation method of study applied to metals and alloys with specimens prepared by metallographic techniques for observation after the test period. See Metallographic Etch.

---

76. FIGURE ETCH
    Any form of defect or structure developed in a surface by etching, regardless of the type etchant, e.g., gas, liquid, solid. Figures also can be formed by temperature, pressure, or direct flame. As a pressure formed figure, called a ''percussion figure'' on mica, (0001) surfaces developing as a six-rayed star. Star line pattern representative of bulk prism plane directions and used for orienting the micas.

---

77. FINISH ETCH
    The final etch used on a material surface. Term primarily used in the metals industries, and can apply generally or to a specific etch or technique developed to produce a particular surface finish. See Matte Etch; Satin Etch.

---

78. FINITE FORM ETCH
    Preferential etching of single crystal spheres of any metal or metallic compound to produce a solid with crystallographically oriented exterior planes that are the fast etching planes of a convex surface. Planes developed vary with solution mixture, as cube, (100); octahedron, (111); dodecahedron, (110); or tetrahexahedron, (hk0) as most common forms, e.g., also called finite crystal form etching, and primarily applies to materials that form in the isometric (cubic) system. See Dissolutionment Etch; Sphere Etch.

---

79. FLAME ETCH
    Use of a propane torch or similar gas torch to produce etching action on a surface. Has been used to develop surface etch figures on high melting point temperature metals and alloys. Also used for surface cleaning of metals and their alloys during brazing operations.

---

80. FLASH ETCH
    Any very rapid etch applied for a short period of time. It can be a liquid solution or an electrical spark. The latter used in spectrographic analysis. See Dip Etch; Electric Spark Etch.

---

81. FLAT ETCH

Either to etch a surface to be planar and flat, or etching of a material sheet in the form of a ''flat''. See Blank Etch; Coupon Etch.

---

82. FLOAT-OFF ETCH

The removal of thin film layers from a surface for microscope study by TEM, SEM, etc. Etch solution attacks the substrate, but not the film. Sodium chloride, NaCl, (100) substrates use water, H_2O; silicon Si, wafers when used as a substrate in this manner, use $HF:HNO_3$ to dissolve the silicon substrate for film removal. Mica, (0001), MgO, (111) and other materials used as substrates have their own etch mixtures, such as HF for quartz and glass blanks. This is a separated and distinct method . . . the float-off technique . . . as against the lift-off technique used for photo resist and excess metallization removal in device fabrication (with acetone).

---

83. FLOOD ETCH

To rapidly cover a specimen surface with an etching or cleaning solution, usually with reference to washing or quenching acid reaction, rather than etching. An etching container is flooded with water to stop etch action. See Flush Etch.

---

84. FLUSH ETCH

To cover a surface with a moving liquid etch solution. Used as a light surface cleaner, or to produce an etch-washed pattern for a decorative effect. Also refers to water quenching an etch solution by flushing to stop etch action. See Flood Etch; Matte Etch; Satin Etch.

---

85. FLUX ETCH

Use of a molten metal or solid chemical compound for etching. May also refer to the use of a flux, such as borax, in metal alloying and brazing, and solutions used to clean and remove residual borax after joint fabrication. See Molten Flux Etch.

---

86. FORM ETCH

Either etching a material to a specific shape, or etching of a particular shape. Spheres are etched to finite crystal form; rectangular bars are control-thinned as electronic reeds; silicon is via hole pattern etched as an inking mask; many metal shim stocks are electroformed or pattern etched as evaporation masks; tantalum is etch formed as an antenna, etc. See Finite Form Etch; Sphere Etch.

---

87. FORMING ETCH

Preferential, selective, or electrolytic etching of a specimen to a bulk shape, or structuring a surface by etching pits, via holes, channels. Diamond is etched with a saw-tooth structure as a filter element.

---

88. FREE ETCH

The etching of specimens by dropping them loose into a solution without being held in any manner. See Bottle Etch.

---

89. FREEZE ETCH

The use of a solution below 0°C to effect etching action, or, the quenching of a material from an elevated temperature into a liquid bath solution.

---

90. FUME ETCH

Use of hot acid or solvent vapors for etching or cleaning action. Hot HCl/H_2 vapors are used in epitaxy systems for general cleaning of quartz tubes, graphite susceptors and specimens; hot HF vapors for oxide removal or hot water vapor (steam) for hydrated oxide growth conversion of a surface followed by HF Strip as a surface cleaning step. See Vapor Etch; Vapor Degreasing.

---

91. GAS ETCH

The use of a gas to produce etching action in its molecular form, such as argon, Ar; hydrogen, H_2; nitrogen, N_2, etc. It usually includes heat and/or pressure. Argon, as an example, under pressure and heat in a vacuum system has been shown to preferential etch semiconductor surfaces producing hillocks and pits similar to those observed with wet chemical etching. See Ion Etch; Dry Chemical Etch (both as ionic gases, not molecular).

---

92. GATE ETCH

Specialized term denoting channel etching for the active area of a Schottky Barrier device, such as a field effect transistor (FET). The channels are either wet chemically etched or electron photolithographically etched on the submicron width scale. An active gate metal, either aluminum or gold is then pattern evaporated in the channel, and may be followed with RF plasma final etch forming with ionized gases (aluminum with BCl_3).

---

93. GENERAL ETCH

Any etch solution can be so called when used without other specific definition, such as a general removal etch; and there is an etch solution with the name General Etch.

---

94. GRAPHIC ETCH

A repetitive pattern of figures etched onto a surface. Such patterns can be controllably produced for decorative effects or developed in nature — meteorites have a distinctive etched structure called Widmanstatten Figures that appear much like a grossly twinned single crystal ingot; also the Neumann Line structure associated with twin lamellae — several of the quartz-feldspar granites show a fine graphic structure of a clear quartz network enclosing the softer pink or white feldspar. See Matte Etch; Satin Etch.

---

95. GRAVITY ETCH

A highly specialized method of etching where the specimen is subjected to either a positive (+) or negative (−) gravitational force by centrifugal action during preferential etching to form controlled structures. Brass and gallium arsenide have been via hole and pit patterned in this manner. A specimen may be dropped down through an etch solution column for a controlled distance for forming or cleaning.

---

96. GROOVE ETCH

In single crystal wafer structuring of devices, the etching of a channel in surfaces. It may be as a "V" groove, have a bottom curvature of known radius; as a saw-tooth series of ridges; have vertical side walls with a flat bottom, and side walls may be specific crystal planes.

---

97. HALOGEN ETCH

Refers to a chemical class of elements as etching agents: fluorine, chlorine, bromine, and iodine. In Solid State processing, the term is most often applied to the bromine:methyl alcohol (BRM) solutions. Chlorine, as a gas, is bubbled through other etch solutions for added etching action; hot HCl or HF vapors are used as etchants; and iodine can replace bromine in the BRM solutions, and is used in several designated/named "iodine" etches. Low concentration bromine BRM solutions are polish etchants, hot vapors are primarily cleaning agents, and the iodine solutions are mainly preferential.

---

98. HEAT ETCH

The use of heat only, to effect an etching or cleaning action. Used under vacuum conditions to flash clean specimen surfaces. Used in air to oxidize a specimen surface. Occasionally referenced as heating of a liquid or gas for etching action. See Thermal Etch; Heat-tint Etch; Hot Etch.

---

99. HEAT-TINT ETCH

Heating of a specimen on a hot plate in air to produce oxide colors. Used in the study of metallographic specimens. A crystalline material varies in color by rate of oxidation of individual crystallite grains according to their crystallographic plane orientation. A single crystal material or alloy can be differentiated by phase structure or internal crystallographic plane directions by similar color variation. In an alloy mixture or similar specimen, different metal or compounds can be determined by their characteristic oxide color when heat tinted for a specific time. Single crystal spheres of metals and compounds have been widely studied by controlled furnace oxidation in the study of oxidation kinetics and growth rates related to internal crystal plane location and orientation. See Heat Etch; Thermal Etch.

---

100. HEAVY ETCH

Any etch solution used to remove a large volume of material. Measurement is done by mils of surface removal, or by total specimen gram-weight loss. See Rough Etch; Removal Etch.

---

101. HOT ETCH

Any solution used for etching or cleaning above room temperature as either a liquid or gas, and may be shown in °F, °C, or K. Also shown as warm, hot, or boiling. See Temperature Etch.

---

102. ICE ETCH

Solid ice has been used as an etching medium in the sense of the freeze-out of a hot, liquid metal poured on the ice surface. Mixed with water or alcohol it is used to cool other etch solutions and, along with snow, has been used as a constituent in formulating a series of "cold etches" for specific temperature levels. Also can refer to the etching of single crystal ice specimens grown under cryogenic conditions in cold cryostats. See Water section, Chapter 2 for Cold Etchants.

---

103. IMMERSION ETCH

Complete submersion of a specimen in a liquid etch solution, or in a molten flux solid chemical solute. It is the most common form of etching as wet chemical etching (WCE) or electrolytic etching (EE), which are two of the Etchant Formats used in the next section. (Dry chemical etching (DCE) is the third format.)

---

104. INDUCED DAMAGE ETCH

Preferential etching to develop defects or structure in a surface that has been subjected to some form of damage. Such damage may be a controlled scratch or a point of damage introduce by a diamond-type stylus to develop specific defects, or to initiate the etch forming of a pit, via hole, channel, etc. Also has been used with reference to residual subsurface damage remaining after cutting or mechanical lap and polish. Such damage being removed by chem/mech polishing or straight chemical etching in Solid State device fabrication. See Defect Etch.

---

105. INITIAL ETCH

Any solution that is the first in a series, or the first etch applied in a specific material process.

---

106. ION ETCH

A gas in its ionized rather than molecular state used to clean or etch a surface. RF plasma N^+ or O^+ cleaning systems are widely used in Solid State material processing. Argon, as ionic Ar^+, also is widely used in Solid State processing to final clean surfaces under vacuum immediately prior to metallization and compound growth or deposition, and can introduce subsurface damage. Other ionic gases, such as helium (He^+), xenon (Xe^+), etc., are used in material irradiation damage and thin film adhesion studies. Electron irradiation from a TEM microscope or an electron-beam in vacuum system; lasers, for annealing or alternating materials; and nuclear particles have all been used as forms of ion etching. See Argon Etch; Dry Chemical Etch; Irradiation Etch; Particle Etch.

---

107. IRRADIATION ETCH

The use of ionized gases or radiation particles to affect etching action. May also be used to induce damage into surfaces in the broader study of irradiation effects. Note that a single crystal, when irradiated, will revert toward the noncrystalline, amorphous state, and the reverse — an amorphous material will tend toward single crystal or the crystalline state. Such alternations of the immediate subsurface of Solid State materials is of major consideration in device structuring, particularly, when an active diffused junction is in the <1 μm region, as such altered zones can severely affect the electronic functions of devices.

---

108. ISOTROPIC ETCH

Another name for a polish etch solution. The etch attacks all crystal planes at an equal rate, producing a flat, planar surface. The opposite is anisotropic (preferential) etching. The terms anisotropic and isotropic originally applied to the propagation of light through a solid mineral. They also are now used in dry chemical etching (DCE) with regard to shaping a pit, channel or via hole with ionized gases without reference to the polish (isotropic) or preferential (anisotropic) nature of a liquid etch solution.

---

109. JAR ETCH

(Jug). The use of any closed vessel for etching in which parts or specimens are immersed. See Beaker Etch; Bottle Etch.

---

110. JET ETCH

A fine stream of liquid under pressure applied as an etchant, and most often for shaping the exterior of a solid specimen, structuring a material surface, etch cutting a hole through a wafer, etc. May be a single or multiple jet system with or without electric current. There are jet systems combining etch/cut action for slicing or dicing material with high velocity gas, steam, water, or an acid. There also are hot-melt electrical plasma composites deposited on surfaces . . . not etch action, but used in constructing Solid State substrates.

---

111. JUNCTION ETCH

As a semiconductor term, the development of a p-n junction by staining, etching or selective plating. In the heavy metals industry, weld joints can be developed or cleaned by etching or light sandblasting. In the plating and cladding industries, etching to develop interface joints between layers can be done with a jet. See Layer Etch; Stain Etch; Selective Etch; Jet Etch.

---

112. KEY ETCH

A primary etch solution within a clean/etch sequence, or any solution applied in a processing step that is considered a fundamental solution. A solution around which other solutions or processing revolve.

---

113. KRYPTON ETCH

The use of ionized krypton, Kr^+, as an etch or material damaging agent by irradiation. The use of vapor pressure change to etch single crystal krypton grown under pressure and cryogenic vacuum conditions.

---

114. LASER ETCH

Use of electrons propagated by light at a controlled frequency and power level to effect an etching action. Several types of lasers are used to etch channels and other structures for operational Solid State devices; for material cutting, such as preparing circuit substrates; as annealing to increase crystallite size on dendritic crystals, or in conversion of carbon to a diamond-like-compound (DLC). See Electron-Beam Etch; Flash Etch; Irradiation Etch; Ion Etch.

---

115. LAYER ETCH

As applied in Solid State processing, the etch development of epitaxy multilayer structures in device fabrication or in material studies. As a more general term, the etch removal of a specific layer of material from a dissimilar material. It may be total removal as in oxide or nitride stripping from a semiconductor surface, or selective removal through a photo resist or similar surface coating to develop patterns for subsequent device fabrication. See Junction Etch.

---

116. LIFT-OFF ETCH

A specific technique developed to remove thin film metallization from a photolithographically prepared wafer surface. Wafers are soaked, sprayed and/or lightly scrubbed with a plastic foam Q-tip in acetone. The acetone dissolves the photo resist layer used for patterning which loosens excess metal by lift-off, and exposes the metallized pattern for further device processing. This is a separate and distinct operation, though similar to the float-off technique. See Float-off Etch.

---

117. LIGHT ETCH

The term has been used in two contexts: (1) as a physical term to differentiate between light-medium-heavy etching; or in the sense of a slow, minimum removal etch, and (2) the use of light alone, or in conjunction with solutions to enhance etching reaction or for selective plating action. White light most widely used, such as a strobe light for semiconductor p-n junction plating with copper. See Electron Beam Etch; Laser Etch.

---

118. LIGHT FIGURE ETCH

The use of preferential etchants to develop surface etch pit and dislocations on the surface of a cut single crystal ingot face or wafer and used to crystallographically orient the surface by reflected light. This is standard practice in the processing of single crystal ingots of all materials. It should be noted that many single crystal ingots can be cleaved into wafers on preferred fracture planes, which does not always require etching to obtained surface figures for light figure orientation (LFO). After etching face-cut silicon ingots in boiling KOH, 5 min, as an example, the ingot is mounted on a ceramic block, then on an x-y-z positioner and a pinhole light is reflected off of the surface back into a black box. The pattern that appears at the back of the box is then used to orient the ingot for cutting wafers. Orientation accuracy can be $<^{1}/_{2}^{\circ}$. See Preferential Etch; Light Figure Orientation Etch.

---

119. LIGHT FIGURE ORIENTATION ETCH

The same as item 118. As initially developed in Solid State semiconductor ingot processing, the etching of silicon in hot KOH solutions; germanium in $KOH:I_2$, or acid solutions, such as CP4. The reflected surface defect pattern are used to orient the ingot for wafer slicing. See Light Figure Etch.

---

120. LINEATION ETCH

In semiconductor development the term was originally applied to distinctly aligned dislocation patterns associated with stress induced during the rotation/pull growth of Czochralski (CZ) single crystal ingots. The patterns are very distinctive on (111) and (100) oriented wafer surfaces: a series of over-lapping dislocations reducing in size from the wafer periphery toward wafer center, and disappearing before reaching the wafer center. On (111) surfaces, as three lines at 120° and associated with the bulk ⟨211⟩ directions; whereas on (100) surfaces, four such lines at 90° and associated with ⟨110⟩ bulk directions. With improved growth control, this form of defect rarely observed today.

Any crystallographically oriented line segment — representative of slip in a single crystal surface — may be termed lineation or slip. Where a series of such line segments are closely parallel — called stacking faults (SFs). This type of defect can be introduced by processing — oxidation, diffusion, epitaxy, ion implantation, and is often associated with oxide and nitride thin films. For these latter, $HF:CrO_3$ preferential etchants have been tailored for stacking fault study. See Defect Etch.

---

121. LIQUID ETCH

Any wet chemical etch (WCE) solution used to effect etch action. Also the use of a solid chemical molten flux etch, such as KOH pellets at 360°C. See Acid Etch; Alkali Etch; Solution Etch; Molten Etch.

---

122. LONG ETCH

A time period for etching, as against a short or dip time period. Solution is either a slow etch on the material, or a cleaning solution with an extended soak time. This latter is common in metal processing, such as in soak etching or conditioning a surface prior to plating.

---

123. LOOSE ETCH

Any solution used where parts are dropped free and not held in any way in the solution during the etch period.

---

124. LOW ETCH

Applied with specific reference to the level of a solution above a part being etched, where insufficient covering acid will not block the in-diffusion of atmosphere over the open container. If the solution level is too shallow, it also can cause erratic etching results.

---

125. MACRO ETCH

Defects and structure observable on a surface with the unaided eye after etching. The term is widely used in the metallographic preparation of metals and alloys in the metals industries, and in geologic study of thin sections. It has occasionally been used with reference to preferential etching of semiconductor and other single crystal surfaces. See Preferential Etch; Structure Etch; Anisotropic Etch; Heat-Tint Etch.

---

126. MAGNETIC ETCH

Not a real etching method, yet has been used in two ways: (1) magnetic powder has been brushed on wafer surfaces of ferromagnetic materials to develop domain structure, such as barium titanate and ferrites; (2) iron or carbon powder brushed on a preferentially etched wafer surface to accentuate defects and etched patterns. There are magnet-stir hot plates where a Teflon or plastic-coated magnet is dropped into the solution, then solution rotation and heat controlled by the hot plate dials. Barium titanate, being fabricated as an ultrasonic transducer, is electrically (magnetic flux) poled in water to orient domains. See Gravity Etch.

---

127. MATTE ETCH

A surface finish etch used for decorative purposes on metal surfaces, such as copper or nickel. The surface has a low-profile grain-like structure with a dull sheen, and can be as a semi-matte finish. See Satin Etch; Silk Etch.

---

128. MEDIUM ETCH

Used in two ways: (1) with regard to time, as a slow, medium, or fast etch, and (2) with regard to solution strength, as weak, medium, or strong.

---

129. MELT-AWAY ETCH

The separation of a thin film from a substrate for microscope study by heat liquefying and removal of the substrate. Etch removal of a metal thin film or alloyed pre-form, wire, etc., to observe the pit formed in the material surface by the metal. Semiconductor wafers with alloyed p-n junctions have been etched from the back to observe the buried junction-front in the bulk wafer. Sodium chloride, (100) substrates have been heated to the liquid state in order to remove a thin film for TEM study, rather than the usual use of water to dissolve the substrate/film interface.

---

130. MELT-BACK ETCH

A specialized Solid State term relative to wafer surface etching during epitaxy growth, such as using indium on indium phosphide, InP, or gallium on gallium arsenide, GaAs during liquid phase epitaxy (LPE). The method is used to clean the surfaces to reduce the growth of defects in the epitaxy film.

---

131. MESA ETCH

The etching of a roughly cylindrical column or pylon on a single crystal surface as a p-n junction device structure. In Solid State, as a single mesa, it is the mesa diode. As an array of mesas in the fabrication of SCRs, they are elements for power distribution and control in the electrical operation of the device. Mesas are commonly formed with a slightly preferential etch such that the mesa side slopes have a degree of crystallographic orientation. Using dry chemical etching techniques, the mesa sides can be more cylindrical without crystal facets. See Pylon Etch; Structure Etch.

---

132. METAL ETCH

In the broadest sense, any solution used to etch any metal. More commonly, the use of a liquified metal as an etch medium on high temperature, chemically inert metals, such as molybdenum, tantalum, or titanium. Occasionally refers to the metallizing of a material surface with in-diffusion to decorate defects. See Decoration Etch; Molten Flux Etch.

---

133. METALLOGRAPHIC ETCH

Term primarily used in the preparation of metals and alloys as specimens for defect and structure analysis using preferential etches. Common in the metals industries and in geology, and most Solid State companies maintain a metallographic laboratory for material inspection and evaluation of processing. The etch can be macro- or micro-etch as a size definition. See Macro Etch; Micro Etch; Preferential Etch.

---

134. METALLURGICAL ETCH

Not common, but any etch applied to a metal or alloy specimen with reference to the field of metallurgy or metallurgical engineering as a science. Also with reference to the use of a metallurgical microscope for material observation.

---

135. MICRO ETCH

Any defect, figure, or structure etched on a surface that cannot be easily observed by the unaided eye, and requires a microscope for proper viewing. Both macro- and micro-etch are terms widely used in the study of metallographic specimens in general metal processing, not as often used in Solid State processing. See Metallurgical Etch; Macro Etch.

---

136. MILLING ETCH

In Solid State it refers to the use of an ion milling vacuum system where ionized argon (Ar^+) or nitrogen (N^+) is used to etch remove and pattern thin film metallization on circuit substrates or active devices. In general metal processing, it occasionally means the use of a lathe or cutting mill for a combination of cutting, etching, or shaping of a part, and may be as an electrolytic etch with the lathe head as the cathode.

---

137. MINIMUM ETCH

An etch used for a short period of time, or one that removes little or no material during the etching period. See Slow Etch.

---

138. MIST ETCH

The use of an etch as a spray of finely divided particles as from an atomizer. The method has been used for final etch cleaning of a surface; to fine-tune and optimize electrical characteristics of an exposed p-n junction semiconductor device; or to develop defects, structure, or figures in a surface for optimum clarity observation. Used in metallographic specimen etching to develop fine structure of crystallites, phases etc. See Vapor Etch.

---

139. MOLAR ETCH

Any solution mixed by its molecular weight. The total of the atomic weights of the compound elements is one molecular gram-weight of the substance as dissolved in one liter of water (1 liter = 1000 ml):

HCl:	H = 1.008 (at wt)	KOH:	K =	39.102 (at wt)
	Cl = 35.453		O =	15.999
			H =	1.000
Total	36.461 g	Total		56.109 g

Using potassium hydroxide, KOH, as an example, approximate solutions would be: (1) 1 *M* KOH = 56 g/l; (2) 2 *M* KOH = 112 g/l; or (3) a 0.5 *M* KOH solution about 28 g/l.

A Molar solution may be used as the etching solution by itself, or be only one constituent of an etch mixture. See Normal Etch.

---

140. MOLTEN FLUX ETCH

Any metal or compound liquidized at or slightly above its melting point without the inclusion of water or other liquid solvent. In Solid State processing, as an example, KOH, NaOH, or a 1KOH:1NaOH (eutectic) mixture are used as dislocation etches in their molten state at about 365°C. Other molten fluxes of nitrates and carbonates ($NaNO_3$ or KCO_3, examples) are used in the metal industry for etching, cleaning or structuring of materials. These fluxes also are used for growth of certain compounds, such as single crystal garnets. See Flux Etch; Melt-back Etch.

---

141. NAMED ETCH

Many etch solutions have a number, letter, chemical, individual's name, or a combination of such. Where names have been given to a specific etch solution, it is shown after ETCH NAME in the etchant formats used in Chapter 2. The following is a list of some of the variations in nomenclature that have appeared in the literature:

Name	Example
Chemical name	Single acid/chemical: HCl; H_2SO_4; KOH, etc.
Chemical name + letter	*Iodine A; Tri-iodide etch
Chemical acronym	Mixed solutions: bromine-methanol (BRM), or **buffered hydrofluoric acid (BHF)
Volume ratio	*** 1:1:1
Letter(s)/volume ratio	BK-112; BK-213
Letter(s)/number	CP4; SR2
Individual's name	Billig's etch; Sirtl etch; Wright etch; Dash etch; Makuri's reagent, etc.
Individual's name + number	Ellis #1; Camp #2
Color	White etch****
Individual's name(s), acronym	A/B etch*****; BJ etch
Two component mixture	AB etch*****
Company abbreviation	RCA etch*****
Company abbreviation + number	RCA #1; GE-3
Company + chemical symbol	WAg
Number only	See ASTM 407-70 pamphlet

*There are several "iodine etch" mixtures of varying composition, several developed many years ago for a specific metal or compound, such that the general term "iodine etch" can be any one of several solutions.

**Both the BHF and BRM designations should be considered general terms, not a specific etch solution as there are, literally, several hundred possible compositions, and many different mixtures are shown by the acronym only.

***The use of a numerical volume ratio, alone, is not recommended. From the example show — 1:1:1 — there are many acid mixtures with such a ratio.

A letter designation along with the ratio — BK-112 — is an improvement as it becomes a specific etchant nomenclature, but the user still needs to know the acids involved.

****The use of the name "White etch" is misleading. When $HF:HNO_3$ solutions were first being applied to silicon etching several of them were termed "White etch", even though the solutions are clear and transparent, not white in color.

*****The use of the terms AB or A/B can be confusing. In the former case, it has been used to describe a two-part solution with an "A" portion and a "B" portion used separately; but also has been shown as A-B — with the total quantity of the two solutions mixed together when ready for use, or, a specified ratio of the solutions mixed. And another conflict with the use of an "A" and "B" nomenclature is the A/B etch named after Abrahams and Bioucchi.

The use of the term "RCA etch" is also questionable, as several hundred etch solutions have been developed at RCA over the years. In this case, the "RCA etch" usually refers to the "A" and "B" solutions developed for cleaning silicon by Kern and Puotinenr. Further, some developers have used only the "A" or "B" portion of those two solutions without letter designation, yet still call out "RCA etch".

---

142. NATIVE OXIDE ETCH

Almost all inorganic metals and compounds become surface passivated by an oxide when exposed to air. Such oxides are called native oxides as they are a normal attribute of the surface and not artificially produced. The removal of this type of oxide can be critical to metal processing, such as preparing aluminum surfaces for plating as well as copper, nickel, iron, and steels. It is of major importance in preparing semiconductor wafers for etching, metallization, diffusion, etc., as such residual oxides can affect device characteristics.

Any etch solution used to remove such oxides is called a native oxide removal etch to differentiate it from a solution used to etch an oxide thin film or material, such as titanium dioxide, TiO_2 or quartz, SiO_2, and called an "oxide etch". It should be realized that a "native" oxide only occurs under natural atmospheric conditions, even though it is called such in Solid State processing where residual surface oxide remains after chemical treatment. See Oxide Etch

---

143. NEUTRAL ETCH

A very slow, nonreactive etch, when used as a cleaning solution, or the use of water as a pH 7 neutral wash or quenching solvent.

---

144. NITRIDE ETCH

There are several artificially grown nitride compounds, such as silicon nitride, Si_3N_4 and aluminum nitride, AlN. The metal industries use a nitridization process to condition metal surfaces and the Solid State industry is developing and applying both oxide and nitride surface thin films in processes. In either case, any etch solution used to remove or pattern a nitride is called a nitride etch, and many also can be used on oxides.

---

145. NORMAL ETCH

An etch solution mixed on the basis of the total valence of the metallic radical ions in solution. To obtain the number of grams of a compound for a Normal solution: divide one gram-molecular-weight (1 mole) by the total valence number of the element and radical: (See #139 for computation of the gram mole.). Examples of Normal solution computations are

(1) Sodium hydroxide, NaOH:
Na^{+1} = metallic valence of "1"
(OH^{-1} = hydroxyl valence of "1" [ref. only]) and 40 (Na at wt) = 40 g or 1 gram-mole = 40 g

1 Normal (*N*) solution. In this instance, it also happens to be a 1 Molar (*M*) solution.

(2) Sulfuric acid, H_2SO_4
$H_2{}^{+1}$ = metallic valence of "2"
($SO_4{}^{-1/2}$ = sulfate valence of "2" [ref. only]) and 98 (total at wt elements) = 49 g or 1 gram-mole = 49 g

In this case, 49 g/l of water (1000 ml H_2O) of sulfuric acid is a 1 *N* solution. It also happens to be a 0.5 *M* solution.

Note: If a 50 ml *N* solution of a chemical base, such as sodium hydroxide, is mixed with an equal quantity of an acid *N* solution, such as sulfuric acid, the resulting solution will be neutral (pH 7, like water).

Both ammonia, NH_3, and ammonium hydroxide, NH_4OH are the primary chemical base solutions used to neutralize acid sumps before effluent discharge into waste disposal sewer lines.

Many etchants are mixed with both *N* and *M* constituents, even combining acids and bases, but equal quantities of *N* and *M* solutions are not mixed or else a neutral solution will result without etch action.

---

146. OIL ETCH

Petroleum base oils can act as etchants on some metals, even though they are normally only thought of as coolants in metal cutting, and similar processing. For critical materials, such as semiconductor wafers and assembly or test parts, such oil coolants with various additives for rust prevent, foaming, etc. can be severely degraded because of chemical attack, residual films, and other anomalies, such that silicones are used as replacement liquids for petroleum oils.

---

147. ORIENTATION ETCH

A preferential solution used to determine the single crystal orientation of a specimen surface by development of surface etch pits. See Light Figure Etch; Optical Orientation Etch; Figure Etch.

---

148. OXIDE CLEANING ETCH

A solution developed to clean an oxide surface with minimum or no removal. It can be acid, alkali, or solvent.

---

149. OXIDE ETCH

Any solution used to etch a metallic oxide material. See the immediately following terms for special case applications.

---

150. OXIDE REMOVAL ETCH

In Solid State processing, often refers to removal of a native oxide on a material surface prior to further processing. It also is used where a deposited oxide thin film (SiO_2, Al_2O_3) is being pattern etched or removed. See Native Oxide Etch.

---

151. OZONE ETCH

Ozone, O_3, is an extremely strong oxidizing agent. In Solid State and some metal processing it is used, by itself, as a surface cleaner. Caution should be exercised as concentrations greater than 1% in air can be hazardous to health. There are ozone producing commercial cleaning systems used in metal processing for material surface cleaning. Similar units are used in movie theaters or other commercial offices and buildings as an air freshener, and for cigarette smoke removal. (Ozone ash trays are on the market for home or office use.)

---

152. PARTICULATE ETCH

Any solution used to remove a material matrix and expose embedded particles without affecting the particles. This is common in some ore processing operations where gangue material is separated, and in some material studies. The etch may expose the particulate, only, or be used to remove the particles for separate microscope study.

---

153. PASSIVATION ETCH

The term is used in two ways: (1) a solution developed to remove a passivating thin film from a surface, such as a native oxide, or (2) a solution that will introduce a surface film passivation. An iodine solution has been used on diamonds in the latter case as passivation against etching with H_3PO_4, and an I_2:MeOH rinse applied as an ionic surface contamination removal system on silicon wafers prior to diffusion.

---

154. PATTERN ETCH

Any etch solution that will develop structure in or on a surface and there are several different applications and methods: (i) to develop defects in surfaces; (ii) to differentiate between elements, structure, or minerals in a mixture material; (iii) to etch through a masking layer, such as photo resist patterned thin film oxides to remove the oxide down to the substrate in the desired pattern; (iv) etching via holes, or (v) circuit pattern etching of substrates. See Photo Resist Etch; Thermal Etch; Defect Etch; Figure Etch; Vacuum Etch; Structure Etch.

---

155. PHASE ETCH

Common to metal and metal alloy etching of steel in the etch development of alpha-, beta-, or delta-phase structure, and the recognition of martensite, carbide, and similar crystal structures.

---

156. PHOTO RESIST ETCH

Photo resist lacquers, such as the AZ- series, COP-, or PMMA types used in device and circuit fabrication of semiconductor devices, have their own solutions called developers, and used after UV exposure of the resists in fabricating patterns. The developers are designed for each type of commercial photo resist formulation, many contain hydroxide, such that caution should be observed if the material being processed in particularly vulnerable to attack by alkaline solutions.

Note that in removing photo resists before or after metallization of the specimens or wafers, the most widely used solvent is acetone (a ketone) by soaking, spraying, or light scrubbing of the specimen surfaces.

---

157. PICKLING ETCH

Term common in metal processing and plating. Metal surfaces are soak-cleaned for conditioning, such as for removal of scale, or other type contamination. See Scale Etch; Conditioning Etch.

---

158. PINHOLE ETCH

Small, roughly circular defects in a deposited thin film are referred to as "pinholes" and may or may not go completely through the film. Any solution or method used to locate and observe such pinholes is called a pinhole etch. Such pinholes can be created by contamination on a substrate surface, be due to insufficient cohesion within the growing thin film, or from entrapment of particles in the film. Oxide and nitride thin films are particularly prone to pinholing in Solid State processing, and are the subject of much study. The term also is applied to the etching of a controlled pinhole, such as for thickness measurement of the film, diffusion depth profiling, or observation and study of epitaxy layer structures. See Structure Etch; Profile Etch.

---

159. PIT ETCH

A preferential etch used to develop dislocations or surface damage pits in single crystal wafers or similar specimens as wet chemical etching (WCE). This includes controlled damage pit development as a device structure. Dry chemical etching (DCE) through a photo resist, metal, or oxide/nitride thin film mask, as well as WCE, also for device fabrication. Note that dislocation pits conform to crystallographic structure and bulk plane directions, such as the sharp triangular pit on a (111) wafer surface, and do not increase in size to any extent with extended etching; whereas a surface damage pit does not conform to crystal planes and directions, expand in size with extended etching with a flat pit bottom that may be heavily terraced, and will disappear when the bulk, undamaged material surface is reached. Undamaged bulk surfaces can be recognized by their usual high reflectivity, and common near-mammillary and near-hexagon structure, particularly recognizable on (111) oriented surfaces. The (100) wafer bulk surface is more block-like in structure, coincident with the square outline dislocation and surface etch pits. See Preferential Etch; Via Hole Etch; Selective Etch.

---

160. PLANAR ETCH

A polish etch that produces a very flat, highly reflective surface. Now a major surface finish for three-dimensional layered electronic device and circuit substrate structuring.

---

161. PLANE ETCH

Preferential solution used to develop a crystal facet, or plane in a single crystal material. Occasionally with reference to etching a planar surface. See Planar Etch; Crystallographic Etch; Structure Etch.

---

162. PLASMA ETCH

The use of ionized gas particles to effect cleaning or etching. Either RF or DC plasmas are used, RF more common in microwave, high frequency electronic device fabrication. See Dry Chemical Etching; RF Plasma Etch.

---

163. POLAR ETCH

Any etch used on a polar material, such as the compound semiconductors and, in particular, the (111) oriented surfaces. Gallium arsenide, as an example: the positive (111)Ga [(111)A] surface vs. the negative $(\overline{111})$As[$(\overline{111})$B] surface show different etching phenomena. One surface will etch preferentially with defects; whereas the opposed surface, usually the negative $(\overline{111})$B, will be erratic, and it is difficult to develop an etch solution that will equally polish both surfaces. Note that in compound semiconductors there also may be different electronic characteristics. See Polarity Etch.

---

164. POLARITY ETCH

Any solution used to develop the preferential etching characteristics of polar compounds, such as GaAs, InP, AlSb, or their associated trinary and quaternary forms. See Polar Etch.

---

165. POLE ETCH

A specialized term applied to the etching of magnetizable structure in materials such as barium titanate, Ba_2TiO_3, where the magnetic domains are aligned by polarization with electrolytic solutions and a magnetic flux. Garnet memory devices as a computer chip are similarly poled. The term also is applied in the etching of single crystal spheres for finite crystal form, when the solution produces only crystallographic ''pole'' figures at axial points, rather than developing exterior facets (planes) as a finite crystal solid form.

---

166. POLISH ETCH

Any etch solution that attacks a material surface at an equal rate in all crystal plane directions without regard to their orientation. This is the opposite of preferential etching. In all Solid State single crystal etching, and much metal etching, a polished surface is required in the fabrication of devices and parts, such that polish etching is of great importance, and a major criterion in processes. It also is now called isotropic etching. See Isotropic Etch.

---

167. PREFERENTIAL ETCH

Any etch solution that will attack crystallographic planes at different rates, and produce structure as controlled by those planes. It is the opposite of polish etching. Much of the structuring and selective etching of semiconductors wafers and similar materials as devices is done with preferential etching, as well as in crystallographic study of single crystals. Such device processing includes formation of pits, channels, "V" grooves, via holes, saw-tooth structures. This was the only method of selective structuring semiconductor devices (wet chemical etching or electrolytic etching) until the fairly recent advent of dry chemical etching. Preferential etching also is now called anisotropic etching. See Anisotropic Etch.

---

168. PRE-PLATE ETCH

Any acid, alkali, salt, or alcohol solution used to clean or condition a material surface prior to plating. See Pickling Etch; Descaling Etch.

---

169. PRESSURE ETCH

The forming of material, or alteration of a surface by direct pressure, alone, and may include heat or a gas atmosphere when a furnace or vacuum is used. Natural pressure produces etch figures on meteor surfaces during atmospheric entry; six-rayed star percussion figures are formed on (0001) oriented mica sheets by striking; controlled point pressure damaging of single crystal wafer surface is used in forming etched pits or grooves in device structuring. See Autoclave Etch.

---

170. PRIMING ETCH

The term is applied with two meanings: (1) the type preparation of a surface prior to plating, such as a priming etch for conditioning, and (2) the adding of a small piece of the material to be etched to the etching solution prior to use. The latter method is used on highly reactive solutions, such as $1HF:3HNO_3$, to obtain an even etch rate from the beginning. See Seeded Etch; Conditioning Etch.

---

171. PROFILE ETCH

The term is used in two ways: (1) selective etching of any form of structure in or on material, and (2) profile etching. The latter includes etching through thin film or metal masks; etching to observe and measure diffusion depths; to study epitaxy layer structures. Many wafers are processed with (100) surface orientation and, for profile study of structures, are cross-sectioned by cleaving in a ⟨110⟩ bulk plane direction. See Angle Etch.

---

172. PYLON ETCH

Any etch used to form vertical, roughly cylindrical structures on a surface. A slightly preferential etch will produce facetted side-wall slopes with WCE; whereas smooth, unfacetted walls can be fabricated by DCE etching. See Mesa Etch.

---

173. QUALITY ETCH

Applied with regard to the purity of acids and chemicals used, such as electronic grade vs. commercial grade liquids or gases.

---

174. QUANTITY ETCH

The etching or cleaning of a number of parts at a single time, or the use of a large volume of solution. See Batch Etch; Volume Etch.

---

175. RAPID ETCH

Any fast etching solution as against a slow etch. As an example, the 1HF:$3HNO_3$ mixture is the most rapid etch solution of this two-component etching system.

---

176. RASTER ETCH

Electron beam (E-beam) etching using high intensity electrons; electron lithography, where a computer is used for beam positioning and exposure of photo resist patterns; and laser etching may be referred to as raster etching or annealing, as has been irradiation with ionic gases or radiation with particles in the sense of using a controlled beam of energized particles, electrons, etc., to effect an etching action. The term relates to the raster tracking element of a computer screen.

---

177. RATE ETCH

Used in two ways: (1) the time period for any solution required to obtain desired results, or (2) the physical reactivity time/rate of specific mixtures. The latter most important when using exothermic solutions. Determining etch rates is a major factor in processing.

---

178. REACTIVE ION ETCH

A form of dry chemical etching (DCE) where one or more of the ionized gases is a reactive gas, such as BCl_3 used in etching aluminum oxide surface films. Common acronym: RIE. This form of etching is increasingly used for selective etch structuring of electronic devices that contain layered epitaxy and metallized structure. Note that there are a number of specifically designed dry ionized gas etching systems with their own special acronyms, though all operate in a similar manner. See Dry Etch; Gas Etch; Dry Chemical Etching (DCE).

---

179. REDOX ETCH

Originally referred to as oxidation-reduction reaction, where one acid is a reducing agent, and the other an oxidizer. HF:HNO_3 solutions are an atypical example, where HF acts as the reducer and HNO_3 the oxidizer. As a REDOX etching system, the term is applied to selective etching with pH control of the solution.

---

180. RELIEF ETCH

Etching of any form of structure on a surface. It can be as a raised mesa, pit, channel, or via-hole completely through the material. See Figure Etch; Structure Etch; Selective Etch.

---

181. REMOVAL ETCH

Any solution that will dissolve and reduce the thickness or weight of a material. All etches are removal type and many are classified by their etch-rate of removal. See Rough Etch; Heavy Etch.

---

182. RF PLASMA ETCH

Another term for dry chemical etching (DCE), and used as a general term for any form of ionized gas etching or cleaning.

---

183. ROLLING ETCH

Any solution used with a rolling motion. Single crystal spheres are etch polished in a beaker held at about 45° and hand swirled to produce a slow rolling action of the spheres.

Occasionally used with reference to a boiling solution that has a rolling or roiling motion. It also has been used with regard to allowing a specimen to roll and tumble down an incline during an etching period, or the etching of an extruded, rolled metal sheet . . . to include etching in or against the rolled direction.

---

184. ROUGH ETCH

Has been applied with two meanings: (1) general reduction in thickness or size without regard to surface finish and measured in mils of depth removed from a surface or by total gram-weight loss of a specimen and (2) a controlled etch to roughen a surface. Glass microscope slides have had one side roughed with HF vapor etching to improve thin film gold adhesion for TEM study of the film growth and structure characteristics, and some metal parts are surface finished with a named roughness, such as a matte or satin finish. See Removal Etch; Matte Etch; Satin Etch.

---

185. SAND ETCH

The term has been applied with reference to the use of a dry abrasive to effect removal action with the abrasive under gas pressure (nitrogen) applied by spray or jet. The method also is used to clean, roughen, or condition a surface.

Abrasives are usually considered as lapping and polishing compounds, although fine jets of sand have been used to fabricate screws, to drill holes through material, or form cavities (pits) in a surface. The use of S.S. White Dental Unit was one of the original methods used to dice silicon and germanium wafers, and the units are still used for sand cleaning of surfaces, e.g., bead blasting technique for cleaning metal parts. See Abrasive Etch.

---

186. SATIN ETCH

Specialized term used in the metal industry where an etch solution produces a surface finish that has the appearance of satin cloth. The surface structure contains variable length and width lines, roughly parallel with some cross-hatching, and with the reflective sheen of satin cloth. The method is called satin finishing, and is used as a decorative finish on copper, nickel, brass and other metals. See Matte Etch.

---

187. SATURATION ETCH

A solution containing the maximum amount of a dissolved chemical in water, alcohol or solvent at room temperature and standard pressure. If such a solution is mixed above room temperature, and under pressure, it is called a super-saturated solution. The latter solution is not used to any extent in material processing, but saturated solutions can be used in six contexts:

1. A chemical is mixed as a saturated solution. In the Etchant Section, Chapter 2, such solutions are shown under COMPOSITION as:

   Example: 50 ml .... CrO_3, sat. sol.

2. One constituent of a solution may be "saturated" with a second. This is shown as:

   Example: 100 ml .... MeOH
   x .... *I_2

   * saturate MeOH with I_2

3. When steam is used for cleaning it may be under pressure, and referred to as "saturated steam".
4. Where a cloth pad is used for chem/mech polishing or damage removal from surfaces, the pad may be "saturated" with the etch solution.
5. Single crystals that are water or solvent soluble are grown from saturated solutions. High purity sodium chloride, NaCl, is artificially grown by evaporation from such solutions, cut as (100) oriented substrates, and used for thin film metal evaporation or epitaxy growth of compounds in material studies.
6. After repeated use of an etchant, removal rate control can be lost due to the solution becoming "saturated" with by-products that no longer allow etch reaction to continue. One mixture constituent can be completely used, leaving the solution "saturated" with a nonreactive constituent. This is common to solutions containing hydrogen peroxide, H_2O_2, where the peroxide rapidly depletes.

---

188. SAW (ACID) ETCH

When wafers are cut from an ingot by a wire that is wetted with acid, it is called the acid-saw technique (AST). The wire can be of iron, SST, rayon, plastic thread, etc.

In cutting, lapping, or polishing a surface mechanically there is always a subsurface damaged zone remaining with damage depth determined by abrasive grit size and other factors. Solid State wafers, such as silicon or gallium arsenide, are often chem/mech polished with bromine-methanol (BRM) solutions to remove this damaged zone. This has been referred to as saw damage removal etching.

---

189. SCALE ETCH

Term used in metal processing where slow etch solutions are used to remove surface contamination — oxidation, oils, dirt — that are called "scale". See Descaling Etch; Soak Etch.

---

190. SEEDED ETCH

Any solution to which is added a small piece of the material to be etched. This technique is used with solutions that initially show a rapid rate of attack in the first few seconds, then a more controllable linear rate. A mixture of $1HF:3HNO_3$ is of this type. By initially seeding such a solution, allowing the piece to completely be dissolved, the etchant becomes more controllable for linear controlled removal.

---

191. SEGREGATE ETCH

Any etch used to remove a matrix material to expose a particulate or segregate embedded. This may be as a contaminant included during ingot growth or a new compound due to regrowth, such as is observed in fabricating silicides as blocking layers in Solid State devices. See Particulate Etch.

---

192. SELECTIVE ETCH

Either wet chemical etching (WCE) or dry chemical etching (DCE), where the solution or gas is used to structure a surface, or to remove specific material layers in a heterojunction/heterostructure device. In the latter case, the etch will attack one material layer and not another, and by suitable masking of surfaces with a thin film oxide, photo resist, or metal, then exposing a pattern, pits, channels, via-holes, or other structure can be etched selectively.

---

193. SEQUENCE ETCH

A single solution, or a series of different solutions used in a consecutive order of etching steps. Clean/etch sequences are of this type, and can include vapor degreasing, etch removal of subsurface damage, acid etching, alkali etching, with water and/or alcohol rinses following etching, or the rinses as individual steps.

The term also is applied to the etching of different layers of heterostructure devices where different etch mixtures are used again consecutively. See Series Etch; Step Etch; Stop Etch.

---

194. SERIES ETCH

Term has been used in two ways: (1) a single etchant used two or more times in sequence, such as in dip etching or (2) different acids, alkalies, and alcohols used in a sequence. See Sequence Etch; Chem/Mech Etch.

---

195. SHAPING ETCH

Any solution used to etch form a solid material. May be electrolytic etching, and term has been used with reference to electroforming.

---

196. SHIM ETCH

Either the etching of shim-stock material, such as thin nickel sheet pattern etching for use as an evaporation mask; or the etch thinning of a material to shim thickness, e.g., at or under about 0.010″.

---

197. SHORT ETCH

Any etch solution used for a brief period of time, as against a long period, such as a soak etch. See Dip Etch.

---

198. SIZING ETCH

In Solid State material processing the term applies to the etch reduction of physical size to a particular dimension, such as in thinning of a specimen for TEM microscope study. In general material processing, there is a gelatinous substance called ''size'', much like a weak glue which is used as a surface coating with applications similar to those in photolithographic processing with photo resist lacquers.

In paper manufacture, size is added to the pulp to prevent ink from running, and is the more accurate meaning, where tree rosins and alums are added as the sizing compounds.

---

199. SLOSH ETCH

Any etch solution used with either movement of the solution or the part. See Agitation Etch.

---

200. SLOW ETCH

The removal rate of a solution as against a rapid or fast rate. Many polish etch solutions are designed to be slow for maximum planarity of surfaces, and for the prevention of erratic surface anomalies that can occur from a too rapid etch. See Soak Etch.

---

201. SLUSH ETCH

Etches containing ice or snow as cold solutions.

---

202. SNOW ETCH

Some cold etch solution use snow as one constituent to establish a specific temperature. As a group, they are sometimes referred to as the ''snow etches''. See section on Water, Chapter 2, for ''Cold Etchants''.

---

203. SOAK ETCH

A slow cleaning or etching solution where the part remains immersed for an extended period of time. Common to the metal and plating industries for surface preparation. See Conditioning Etch.

---

204. SOLID ETCH

The term has been used in two ways: (1) the etching of any solid material, or (2) molten flux etching with a liquified solid chemical compound, such as KOH pellets. See Molten Flux Etch.

---

205. SOLUTION ETCH

A term used for wet chemical etching (WCE) where the etchant is a liquid; whereas dry chemical etching (DCE) uses ionized gases.

---

206. SOLVENT ETCH

The use of a chemical solvent as a cleaning or etching solution as against an acid, alkali, or alcohol. A few metallic compounds can only be etched in a solvent. Trichloroethylene, TCE; trichloroethane, TCA; and Freons are the primary vapor degreasing solvents used for parts cleaning.

---

207. SPARK ETCH

The use of an electric spark to generate an etching action. As a cutting method, called spark erosion; as a etching method an electrically activated wire loop in alcohol is used to observe pinholes in oxide and nitride thin films by the appearance of bubbles come from the substrate surface as the wire passes over in the alcohol solution. Metal spheres, both single crystal and polycrystalline, have been formed from chips of material by electrical sparking from a copper pot under argon, and spark vaporization of material from a surface is a standard method for spectrographic analysis.

---

208. SPHERE ETCH

The term has been used in two ways: (1) the slow polish etching of a material sphere, or (2) the preferential etching of a single crystal sphere to finite crystal form (FCF). Single crystal spheres have been widely used to establish etch and oxidation rates on convex surfaces in metal and metallic compound development for device structuring applications. See Finite Form Etch.

---

209. SPIN ETCH

Etching of a specimen with the solution and/or part rotating. See Agitation Etch; Photo Resist Etch.

---

210. SPUTTER ETCH

A term for RF or DC plasma etching. See RF Plasma Etch.

---

211. SQUIRT ETCH

Any solution used in the form of a liquid jet or spray. Term often applied when a polyethylene bottle know as a "squirt bottle" is used, and there are many designed jet etch systems. See Jet Etch; Mist Etch; Spray Etch.

---

212. STACKING FAULT ETCH

Specialized dislocation term used in single crystal processing. Specific preferential etches have been developed to accentuate this form of defect, which can be common to oxide and nitride thin films. The defect appears as a series of short, parallel slip lines in a translucent to transparent thin section, and are a three-dimensional defect relative to x,y,z crystallographic axes. They occur due to inherent stress factors from the difference between coefficients of expansion of oxide/nitride thin films and the substrate materials on which they are deposited. The stacking faults can be in both the oxide/nitride or in the immediate substrate surface near the interface between the two compounds (SiO_2/Si interface, as an example). See Defect Etch; Preferential Etch.

---

213. STAGNANT ETCH

The term is applied to an etch mixture that has been allowed to stand before use. The H_2SO_4:H_2O_2 mixtures have been so used, with effective depletion of the hydrogen peroxide. CP4 also has been allowed to sit 24 h before use with the effective vaporization loss of the bromine fraction. Aqua regia (1HCl:3HNO_3) is an example of a solution that requires aging before use, but it is not a stagnant solution. See Aged Etch.

---

214. STAIN ETCH

Any liquid solution or gas, such as air or oxygen, producing a coloration action on a surface with minimum etch removal. HF:HNO_3 solutions, either high in HF on high in HNO_3, will stain a surface rather than etch. Other solutions, such as those containing salts of copper, silver or gold, also will produce stains. Both HF and HNO_3 develop red/blue/yellow color; whereas metal stains are grey to black.

Staining has been used for depth profiling of diffused p-n junctions; in delineating exposed planar junctions; or to observe epitaxy layer structures in cleaved ⟨110⟩ cross-sections, which are stained after cleaving. Spheres have been oxidized in oxide growth rate studies relative to crystal planes, and metallographic samples are heat-tinted by oxidation in air on a hot plate. See Junction Etch; Metallographic Etch.

---

215. STANDARD ETCH

The term has been used in three ways: (1) a solution developed for a particular process becomes a standard for that process; (2) a solution used as a reference in material or chemical analysis, and (3) a solution that has become standard for a particular use, such as CP4 as a polish etch, or Sirtl etch for defects.

---

216. STEAM ETCH

The use of water at its boiling point as a cleaning or etching solution. Steam cleaning under pressure for cleaning buildings, clothing, and small parts in autoclaves. In material processing the steam actively getters and uses a portion of the material surface in forming the oxide, such as a silicon wafer surface atoms being used to form silicon dioxide. See Autoclave Etch.

---

217. STEP ETCH

Any solution used to form an etched step in a material surface. The method is used in device structuring; to measure a layer thickness; to profile diffused junction depths; to trace defects in a material bulk; and may be single or multiple steps. A specimen surface is successively coasted with stripes of Apiezon-W (black wax) or photo resist coated in a similar manner, being etched to a controlled depth between each coating to form the steps. See Junction Etch; Structure Etch; Selective Etch.

---

218. STILL ETCH

A solution used without movement of either the solution or the part being etched. Some very slow polishing etches have been used in this manner.

---

219. STIR ETCH

Any solution used with a rotational motion. Magnetic stirring hot plates; hand swirling;

a hand or electrically operated stirring rod in solution, etc. In etching wafer surfaces for planarity, rotation speed is controlled so as to prevent flow patterns being developed.

---

220. STOP ETCH

Also called: "Etch-Stop." Any etch solution that will attack one material and not another in a layer-type structure. A fairly recent term as applied to the selective etching of semiconductor heterostructure devices, although the method has been in existence for many, many years in both the metals and semiconductor industries without having a specific name applied to the process, such as the etch removal or pattern etching of an oxide with HF that does not attack the underlying substrate. The solution may not be a complete "stop", as with an Si_3N_4 thin film on SiO_2 etched with H_3PO_4 where the oxide etches at a much slower rate, effectively working as an etch-stop. See Oxide Etch; Selective Etch; Layer Etch.

---

221. STRAIN ETCH

Similar to Stress Etch. Terms often used in combination as stress and strain studies. The bending of a thin material, such as a semiconductor wafer, will develop dislocations due to strain. Such defects have been studied as both a positive (+) or negative (−) strain direction relative to (100), (111)A or $(\overline{111})$B faces, and $\langle 110\rangle$ directions. See Stress Etch.

---

222. STRESS ETCH

A preferential solution used to develop stress figures in a material. Stress may be from normal wear-and-tear; induced by heat treatment; physically induced by tension, compression or torque with or without heat. The latter method used in material studies. See Strain Etch.

---

223. STRONG ETCH

A general term applied to the use of concentrated acids or alkalies, as against a diluted solution, or a solution using weak acids.

---

224. STRUCTURE ETCH

Any solution used to develop physical structure whether it is a defect in the material, a selectively etched pit, mesa, channel or the shaping of a solid. See Defect Etch; Mesa Etch; Dislocation Etch; Sphere Shaping Etch; Channel Etch; Stacking Fault Etch; Selective Etch; Groove Etch.

---

225. SUBTRACTIVE ETCH

Term applies to the selective etching of multiple thin film layers. The removal of one layer without affecting others. See Stop Etch; Selective Etch; RIE Etch.

---

226. SWAB ETCH

Use of cotton or plastic foam type material on a stick. Medical Q-tips are used and, in Solid State processing, the cotton has been replaced with plastic foam to prevent reaction from the glue used to attach the cotton. Used with acetone for cleaning photo resist lacquers from surfaces. In processing metallographic specimens, swab etching is a common term and method of etching surface structure. Some metallographic solutions used on irons, steels, and copper in the Etchant Section, Chapter 2 show TIME as "swab", when shown for "time" in the referenced article.

---

227. SWIRL ETCH

Any solution used with a rotating motion. See Stir Etch; Agitation Etch; Angle Etch.

---

228. TEMPERATURE ETCH

Temperature of cleaning or etching solutions vary with solution mixture, application requirements, and may be as a solid, liquid, or gas. May also refer to the use of temperature, alone, as a (heat/thermal) etch agent, under vacuum or furnace conditions. Much WCE etching is done at room temperature, but many solutions are exothermic, thus heating the solution during the etching period. To control such reactions, water or solvent cooling coils surround a water bath holding the etch solution vessel have been designed to limit such temperature rise. And temperature can be a major control factor for particular etch mixtures . . . at room temperature (RT) a 1HF:8HNO_3 solution is a good polish etch . . . but at 8°C becomes preferential . . . and at 50°C too rapid and erratic for control. See the previous section discussion on TEMP in Etchant Formats for a description of the various levels of temperature as used in this book.

---

229. THERMAL ETCH

Term has been applied as (1) the use of heat in vacuum; (2) the use of heat in furnaces with a gas atmosphere. In the latter case, also called heat treatment (metals), and annealing (Solid State), though both terms have been used in all material processing areas. Pure heat alone can develop etch figures or structures on surfaces. See Heat Etch; Heat-Tint Etch.

---

230. THIMBLE ETCH

Any cup shaped vessel used to hold an etch solution. Specifically referenced to using small, ceramic, high temperature metal (platinum), or graphite crucibles to contain highly reactive, hot chemicals or liquid metals, such as KOH pellets at 360°C for dislocation etching of silicon wafers. See Cup Etch; Beaker Etch.

---

231. THINNING ETCH

Any solution used to reduce thickness of a specimen. Widely used with reference to preparing specimens for transmission electron microscopy (TEM) studies.

---

232. TRIM ETCH

Any etch used to reduce thickness, width, or length of a part, specimen, or device by small increments, only. This can include laser trimming, such as in fine-tuning paired diode devices for electrical matching parameters. See Sizing Etch; Thinning Etch.

---

233. TUMBLE ETCH

A closed-end cylinder — called a ''tumbler'' — has long been used to polish rock and mineral specimens, or for surface cleaning of parts. Dry abrasive, a wet slurry or acid-slurry abrasive is added to specimens, then the tumbler rotated horizontally on a set of parallel bars. Rotation can be with a hand-operated crank but today tumblers are electrically operated. For parts clean/etch cycles, time is in minutes; whereas gem stone polishing is in days. Specimens and parts can be tumble etched in a beaker with the solution being stirred and parts floating free. See Spin Etch; Bottle Etch; Agitation Etch.

---

234. ULTRASONIC ETCH

Any solution used with an ultrasonic generator to develop agitation during an etching or cleaning period. There are small cup to very large basin-type systems. The latter include vapor degreasing systems. The cup types are used for cleaning and etching small parts with an etch beaker seated in a water bath that transmits the vibration frequency to the solution. Most systems are fixed frequency, but variable frequency units are available. A barium titanate, Ba_2TiO_3, transducer is used to translate physical motion of the titanate into electrical frequency, much like piezoelectric quartz crystal radio frequency blanks. See Vapor Degreaser Etch; Vibration Etch.

---

235. USED ETCH

Any solution applied for more than one period of time. In high volume processing where several parts are batch etched at one time, large volumes of etch solutions are used to sustain solution action without constituent depletion, and ensure good repetitive, temperature independent etching without having to replenish with fresh solution after each etch period. In metal plating, the plating baths are automatically monitored, and additional solution added when plating rate reduces to a pre-set level. See Cyclic Etch; Aged Etch; Stagnant Etch.

---

236. VACUUM ETCH

The use of pure vacuum to effect etching action by varying the vapor pressure. This method has been used to preferentially etch single crystal gases grown under cryogenic and pressure conditions in vacuum.

---

237. VAPOR DEGREASING ETCH

Vapor degreasing has wide use in cleaning parts with a combination of hot liquid, hot spray, and a hot vapor head in which parts are held. In the large systems, the hot liquid tank may include an ultrasonic transducer. Size ranges from a beaker on a hot plate to large tanks with an overhead crane hoist. Parts are lowered and held in the hot vapor head until they are deemed clean; they are then slowly removed and are found clean and dry. Although trichloroethylene (TCE), dichloroethylene (DCE, Perk), or trichloroethane (TCA) have been used in the past, Freon solvents are being used as replacements due to the carcinogenic nature of chlorinated solvents. See Vapor Etch.

---

238. VAPOR ETCH

Any solution used in vapor form to effect an etching or cleaning action. HNO_3, H_2O, and H_2O_2 are all used to oxidize a surface, then the oxide stripped with HF as one method of surface cleaning. In epitaxy system, hot HCl vapor is used to clean the quartz tube walls, the susceptor carrier and, in some cases, the material surface about to be deposited upon. See Atmosphere Etch; Corrosion Etch; Vapor Degreasing Etch.

---

239. VIA HOLE ETCH

Specialized term in semiconductor processing where a crystallographically orient hole is selectively etched through a wafer. The side walls of the hole are then metallized through from top to bottom surface, often referred to as wrap-around plating for an electrical ground plane. See Selective Etch.

---

240. VIBRATION ETCH

Any form of vibration used in conjunction with an etching or cleaning solution. The two most common systems are an ultrasonic transducer or a shaker table. Vibration alone is a standard assembly test vehicle, and can cause part or assembly failure. Such vibration tests include acceleration/deceleration, and g-force generators with an x,y,z spin component.

---

241. WASH ETCH

Used with reference to the pouring of a solution across a surface. Not widely used for etching, as it tends to cause surface channelling. Standard water quenching of parts after etching is sometimes referred to as "wash" clean, but is not an actual etching method.

---

242. WATER ETCH

Though water is considered neutral, with pH 7, the major liquid used for quenching an etch solution, and for general washing and rinsing, it can act as an acid etch on water soluble compounds. Such compounds can be polished, preferential etched, or selective etched with water only. Sodium chloride, NaCl, is one such compound that is widely used as a (100) oriented substrate for thin film metal evaporation and epitaxy growth in morphological studies with thin films removed by the float-off technique for TEM observation.

High purity water, as well as brine and atmospheric moisture, will slowly corrode many metals and alloys. The metal industry uses coupons of new materials in corrosion tests by exposure to salt atmosphere along seacoasts, and study irons and steels by dripping high purity water onto surfaces over extended periods of time.

In the study of ice, it is grown as single crystals and water is used to polish or preferential etch surfaces. See Brine Etch; Atmosphere Etch.

---

243. WEAK ETCH

The use of a highly diluted etch mixture or a singular liquid, such as ammonium hydroxide, NH_4OH and ammonia, NH_3, which are both weak bases.

---

244. WET CHEMICAL ETCHING (WCE)

One of the three major divisions in chemical etching. It is the use of any liquid to effect cleaning or etching action and still the most widely used method of etching. See Liquid Etch.

---

245. WHITE ETCH

The term has been applied to mixtures of $HF:HNO_3$ with or without HAc/H_2O. It is not recommended, as these solutions are clear and transparent, not an opaque white. See Named Etch.

---

246. WORK DAMAGE ETCH

Any polish etch used to remove residual sub-surface damage remaining after cutting and mechanical lap and polish in preparing a wafer or similar specimen. The bromine:methanol (Br_2:MeOH) or BRM solutions are currently used on many metals and metallic compounds for this purpose, as they not only remove such damage, but act as slow polishing solutions.

---

247. X-RAY ETCH

Though X-rays are a photographic technique used in both medicine and material studies, X-rays, as well as other particles, have been used in material surface studies. Fluorite, CaF_2, will show color changes when subjected to X-rays, such that here it is referred to as a method of etching. Radiation or irradiation using ionized gases or atomic particles can produce similar effects in material, as well as cause internal, bulk damage. Again, not a true etching phenomenon, but of major importance in material processing. All space hardware is subjected to radiation evaluation as one high reliability test, and X-ray may be included.

---

248. YELLOW ROOM ETCH

Any solution used in a Yellow Room for the processing of material by photolithographic techniques. Photo resist lacquers are affected by exposure to white light, such that all processing is done under yellow light with humidity controlled at 40% ± 5 RH, and temperature between 70—72°F for optimum results. Too much humidity and photo resist will not harden properly; whereas too low a humidity, the lacquer will harden too rapidly and crack. If the temperature approaches 80°F, the resist also will not cure or harden properly even under controlled oven bake out conditions.

NAMED ETCHANTS

In the past it used to be fairly common practice to assign a name or number to an etching solution when it was developed, often the developer's name, or a number associated, for example, with the mixture components, a company, and so forth. Some approaches used are discussed in the Methods of Etching section under "Named Etches".

The ASTM E407-70 pamphlet, as originally assembled and published, is a list of solutions developed for the commercial metals industries — copper, iron, steel, aluminum, and their alloys — many with both a name and assigned number, such as Nital (HNO_3:EOH), ASTM #74 — Fry's Reagent, ASTM #79, etc. The pamphlet does not assign solutions for specific metals as does this book, but where the solution has been used in Solid State development and known to be from the pamphlet, it is referenced by ASTM number in addition to our assigned metal-associated number.

The *Buehler Ltd A/B Metals Digest* series is another excellent source. Although each booklet is individualistic, they cover equipment and metal processing mainly with regard to metallographic study of specimens, and include geologic mineral cross-sections to some degree. Several of the booklets include excellent photographs, to include data on the solution used to develop micro- and macro-structure of the specimen shown. Many of their solutions used are from the ASTM pamphlet, but also include solutions developed by the Buehler organization.

The ASM Metals Handbook Series — some ten books — cover all aspects of industrial metal processing, fairly comprehensive from aluminum to zirconium. Volume 2 is, in large part, the etching and cleaning book of the series as applied to large volume commercial processing. Much of general metals processing includes surface preparation solutions, electrolytic etching, thermal heat treatment, and plating solutions. For plating, which is not covered to any extent in this book: Lowenheim, F A — *Modern Electroplating*, sponsored by the Electrochemical Society (John Wiley & Sons, publisher) is an excellent reference.

Many cleaning and etching solutions have been developed in the Solid State area of which semiconductors was an initiating technology in the late 1940s. In some instances solutions have been carried over from the general metals industries, but many others have been formulated with particular regard to single crystal technology and today all of the associated physical structures (colloidal, amorphous, crystalline), as well as associated metals, ceramics, glasses, and other compounds. Again, several solutions have been named by their developer, such as the Camp #1, #2, #3, #4, etc. series. As a common usage acronym, the Camp #4, as an example, is best known as CP4 or CP4A (without bromine). Preferential defect solutions containing chromium salts have largely been named for their developers: Sirtl Etch among the first, then Sopori, Schimmel, Wright, and so forth. And there is the well known WAg etch (Westinghouse Silver etch). Two other solution types bear mention: the bromine:methyl alcohol (BRM), and buffered hydrofluoric acid (BHF) series. Both are in wide current use referred to by their acronyms, but have a wide range of component mixtures. BRM solutions below 5% BR_2 are polish types; above, can be preferential. A more or less standard BHF mixture is 1HF:1NH_4F (40%), but even dilute HF has been referred to as a BHF solution, or a mixture may contain NaF, KF, yet all are still called BHF as a general acronym.

The following list includes most of the named and/or numbered solutions as discussed and applied in the Etchant Section, Chapter 2 accompanied by the assigned Etchant Numbers for referencing purposes.

Etchant name	Material references
"A" etch	LIF-0010a; NACL-0007e; TA-0003a
A-B or A/B etch	GASB-0014c; GAS-0059; INAS-0005d; INP-0051
A/B, modified	INP-0021d
AB etch (RCA etch)	SIO-0043
Allen's etch	INSB-0001d
Aqua regia	ALAU-0002; WB-0001b; CDSE-0005b; CDTE-0018b; CAW-000b; CUIS-0003; GA-0002; GAS-0056a; GAP-0011; AU-0001; IR-0001; MO-0008a; PMA-0001a; OS-0001b; PD-0001; PT-0001; QTZ-0003; RU-0001; SIO-0052; SIN-0016; TAN-0001; TE-0002; TH-0002; SN-0005b; TIO-0004; TIW-0006; W-0001d; V-0002a; ZNTE-0005b; ZR-0007; ZHPD-0001; ZRRH-0001
Aqua regia, dilute	GAS-0044c; AU-0005; NICR-0003; PD-0002b; PTSB-0001; TE-0008
Aqua regia, modified	GAS-0056b; GAP-0022e; NI-0005
Barber etch	NACL-0007b
BCK-111	INP-0031a
BHF	SI-0020; SI-0021a; SI-0022; SI-0023; SI-0028; SIO-0009; SIO-0011a; SIO-0013; SIO-0071; SIN-0008a; SIN-0009; TIN-0003; SIC-0010e
BHF, modified	SIN-0008b; SIN-0014; SIN-0023
Bichromate finish dip	BRA-0008b; CU-0008c
Billig's etch	GE-0139b
BJ etch	GE-0027
BOE (BHF)	SIO-0017; SIN-0006
BPK-221	INP-0031b
Brilliant etch	CU-0011a
Bright dip	BRA-0008d
Brite dip	AL-0013; CU-0028; CUO-0005
BRM	ALAS-0001; CDTE-0008; CDTE-0017b; CUIS-0001b; GAS-0001a; GAS-0006; GAS-0173b; GAP-0006; GAP-0009; GE-0038; INSB-0028; INSB-0022j; INAS-0002; INP-0001; INP-0006; INP-0009; INP-0062c; FE-0102; FE-0113; FE-0100; MN-0002, SIC-0005b; SI-0001
Bromine etch (for arsenic)	AS-0002b
BSG	SIO-0048
Camp #2 (Superoxol, CP2)	GE-0065a; SI-0165
Camp #3 (CP3)	GE-0065b
Camp #4 (See CP4)	
Camp #8 (See CP8)	
Caro's etch	GAS-0125; SI-0103; SI-0169b
Chrome etch	CR-0006; CRO-0002; CRO-0004; CRO-0005

Chrome etch, modified	CR-0015; CR-0016
Chrome dislocation etch	SI-0038a
Chrome regia	AU-0017; SI-0019; SIN-0018
Cook's etch	NACL-0007c
Copper etch	SI-0153c; SI-0153d
Copper dislocation etch	SI-0038b
CP4 (Camp #4)	GE-0065c; AU-0016; INSB-0001f; INAS-0009; SI-0173; SIGE-0001; TE-0005; ZNO-0004
CP4, dilute	GE-0201b; INSB-0021
CP4, modified	SB-0001g; SB-0001h; GE-0219; INSB-0006f; LIF-0001a; TE-0003; INAS-0003; INAS-0006d
CP4, variety	INSB-0021; SI-0055
CP4A	GE-0063; INSB-0029; INSB-0022c; FEGE-0002; SI-0102d
CP8	SI-0062
Dash etch	SB-0001a; SI-0046
Dash etch, modified	SB-0001b; SB-0001d; SB-0001e; SB-0001f; SI-0153b; SI-0046a; SI-0154a
DE-100	GE-0047; SIO-0023
Dash copper decoration etch	SI-0154a
E Etch	CDTE-0006a; ZN-0008d
EAg1	CDTE-0006b
EAg2	CDTE-0006c
EDTA	CA-0003b
Erhard's etch	SI-0158
Ellis #1	GE-0150q; GE-0060c
Ellis #5	GE-0150o
Ellis #7	GE-0150r
Etch #1	INSB-0014a
Etch #2	INSB-0014b
Ferric cyanide etch	GE-0150s; GE-0060d
Freeze etch	BE-0004b
Fry's reagent	FE-0112; FESI-0008
Flick's etch	AL-0017
General etch	CU-0011b
Glass cleaner etch	SIN-0015
Gold etch (on silicon)	SI-0064
Glyceregia	SI-0018
H etch	INP-0021c; INP-0051
Healy's junction etch	SI-0069
Hypo	AGCL-0007a
H 100	INSB-0022f
Iodate etch	PBTE-0002a
Iodine etch	AL-0031; GE-0218; GE-0137c; GE-0205; FE-0109; PBTE-0002b; SI-0240; SI-0242a
Iodine A	GE-0185
Jacquet's etch	AL-0048; BRA-0014
Jewitt-Wise etch	NI-0004a
Kalling's etch	ST-0001b; YXM-0001
Keller's etch	AL-0023

Chapter 2

ETCHANT SECTION

Refer to Appendix C (Metals and Metallic Compounds with Reference Acronyms) for the alphabetic list of all materials in this chapter. We have made no attempt to categorize formats or etchants by chemicals or type format — three types: (1) Wet Chemical Etching, (2) Dry Chemical Etching, and (3) Electrolytic Etching as in most instances there is only one to a half-dozen referenced and formated etchants on any particular material.

There are exceptions, such as for aluminum, gallium arsenide, germanium, indium antimonide, and silicon, where the number of specific etchants is extensive (silicon has over 450 referenced items). In this case, a Selection Index has been included following the general introduction which lists chemical formulas, solution applications, and the material format numbers.

Where certain elements such as chlorine, fluorine, sulfur, or phosphorus are widely used in etchant formulations a section titled: ''Etchants Containing xxxx'' precedes the formats . . . an A—Z selection of etch formulas — material — application — format reference number. This is a quick reference for those interested in associative reactions with regard to multilayer thin film or composite constructions. This includes some special groups, such as aqua regia, the buffered hydrofluoric acid (BHF) solutions, bromine:methanol (BRM) category, or silver and chromium mixtures that are specifically preferential in character.

Some binary and trinary compounds follow their parent element without a general introduction, and a few natural minerals are shown where they have been specifically involved in Solid State development.

The list of solutions is far from complete, but it will give the user at least a starting place for immediate application or further development to included additional data obtainable from the referenced article accompanying each etchant format.

ADHESION

General: Many natural minerals contain thin films of alteration products on their surfaces or grow as incrustations on dissimilar minerals as a surface coating. Magmatic intrusion and percolation of mineralized waters into cracks of solidified rock formations form solid veins and, if introduced into a vug or void, can form fine crystals as an encrustation within the void: ''dinosaur eggs'' are silica, SiO_2 are concretionary segregates found in compressed sedimentary rocks — they may be a solid multicolored agate — or contain fine single crystals of clear quartz, occasionally tinted as purple amethyst, yellow citron, etc. as an incrustation growth in the hollow concretion.

All encrustations and vein-type deposits are of geologic interest in the study of mineral and rock formation, but are of little concern with regard to adhesion, other than in separation of gangue materials, such as in metal ore processing. Layer adhesion is of importance in making jewelry, as layer separation during fabrication can destroy the item. Moonstones, as one example, are similar in many respects to the ''dinosaur eggs'' in formation, but are solid colloidal to amorphous in internal structure with a hard white outer coating of magnesium and calcium sulfate with extremely good adhesion. With the coating removed by etching, the moonstone cut and polished, it shows fine light halos due to its noncrystalline structure.

Technical Application: Thin film adhesion is of major concern in all material processing, regardless of the method of deposition: vacuum evaporation, epitaxy growth, electrolytic plating, to include all phases of subsequent processing, temperature effects, in particular.

In Solid State processing most of the metal elements have been deposited, singly, as multilayers of metallization; as metallic compounds, with or without subsequent recrystallization (silicides); or as oxides, nitrides, carbides, and borides. Many have been studied for adhesion as single crystals and similar structure of thin films. In both adhesion and physical properties studies or interface reactions, the substrate may be removed for thin film microscope study by SEM or TEM.

Only a few direct studies of adhesion are shown here, but many are mentioned in the format discussion sections throughout this chapter under the specific metals and metallic compounds involved.

Etching: Varies with material and type adhesion test involved.

ADHESION ETCHANTS

AD-0001
ETCH NAME: van der Graaf
TYPE: Particle, conditioning
COMPOSITION:
x .... electron

TIME:
TEMP:
POWER: 1013—1019 cm^{-3} or 1—5 MeV cm^{-2}

DISCUSSION:

Au, Pd, and Ag thin films deposited on different substrates in a study of adhesion and the effects of van der Graaf power and ionic species impingement on the metal surfaces. Film thicknesses were 500 Å, and electrons applied were He^+, Cl^+, H^+, and F^+, and used on different metal/substrate combinations. Substrates were Teflon, SiO_2, CaF_2, Al_2O_3. Si, InP, GaAs, W, and ferrite. The silver/silicon samples were processed after van der Graaf treatment as follows: (1) remove silver with HNO_3, (2) rough thin silicon with HF:HNO_3, (3) final thin with CP4A. Adhesion was generally improved by about 30%, with center areas of films being thinner relative to the flux dose level. After electron treament all films passed the standard tape test.

Etch thinning was done in order to observe the metal/substrate interface by microscope examination. (*Note:* CP4A is 3HF:5HNO_3:3HAc.)

REF: Werner, B T et al — *Thin Solid Films,* 104,163(1983)

PHYSICAL PROPERTIES OF AIR, N_2/O_2 + TRACE GASES

Classification	Gas
Atomic numbers	7 & 16
Atomic weight	58 (nominal)
Melting point (°C — solid) >200 atms	−200 (variable)
Boiling point (°C — liquid)	−180 (variable)
Density (g/cm³) 22°C (dry)	137
Hardness (Mohs — scratch) solid	1—2
Crystal structure (isometric — normal)	(100) cube
Color (gas)	Colorless
(liquid/solid)	Bluish
Cleavage (cubic — solid)	(001)

AIR, N_2/O_2

General: The term atmosphere refers to the gaseous envelop of aeriform fluid that surrounds the earth and commonly called air. It is a mixture of gases not a compound, containing 21% oxygen, 78% nitrogen with trace rare gases in addition to dust particles and water vapor. Dust and water content vary seasonally and with locale, as do other compounds. Volcanic areas release chlorine and sulfur which combine as hydrochloric acid, HCl and sulfuric acid, H_2SO_4 or hydrogen sulfide, H_2S; industrial areas release a variety of contaminants, now called "smog", though "smog" has been associated with the cities of man since ancient times, as far back as 4000 B.C. Decaying vegetation releases ammonia, NH_3; swamps produce methane gas, CH_4 — it is "foxfire" seen burning at night — man and the animals release carbon dioxide, CO_2, as they breath and utilize the oxygen in the air; whereas plants use the CO_2 and release oxygen, O_2 . . . a symbiotic ecology. The jungle areas of the Congo River in Africa, and the Amazon River in South America are major sources for oxygen and, with the deforestation of the Amazon basin for farming, there may be a major shift in the O_2/CO_2 cycle in the atmosphere.

Air is clear and does not appear to have weight. The standard measurement is 760 mmHg at 0°C, as measured in a tube about 30″ in height and 8 cm in diameter. This weight is approximately 15 pounds per square inch (psi) for a total pressure on the human body of some 18 tons. This is called 1 atmosphere (atm) of pressure, and the amount of pressure reduces with increased height above sea level as the air envelope thins out to the vacuum of space. At standard pressure water boils at 100°C; at 720 mmHg it boils at 98.72°C. At about 17,000 feet on Mount Blanc water boils at 85°C.

At normal temperatures it is a clear gas, but as a liquid (−180°C) or solid (−200°C), it has a pale bluish color tint. The pressure liquefaction and fractionation of air is a major industry for the fabrication of liquid air, L_{air} (−180°C) — liquid oxygen, LOX (−183°C), and liquid nitrogen, LN_2 (−195°C). All of the gases have been grown in solid form, to include single crystal structure for morphological study, e.g., helium excepted.

Water vapor in the air is of critical importance, not only for the seasonal planting and harvesting of crops, but today, the weather patterns of the atmosphere are of vital concern to aircraft traffic control and space vehicle launching. Wind-shear has been responsible for several airplane crashes; the explosion of the Challenger shuttle has been attributed, in part, to freezing temperatures at launch time; and a satellite launch into a rainstorm was destroyed by lightning strikes. Relative humidity (RH) is the measure of water vapor content in the air (100% = maximum saturation). The higher the temperature and humidity the more uncomfortable for human beings, as it reduces the perspiration ability to cool the body. The most comfortable zone is 70—72°F temperature with humidity between 45—60%. Very low humidity, for example, 10% in desert areas, can cause dehydration; in Arctic areas such low humidity, even though the land is covered with ice, can cause severe frostbite and even death from breathing . . . dry, freezing, dehydrated air. Temperature, humidity, and particulate content of air can be critical in industrial processes, such as electronic assemblies and semiconductor device fabrication. The construction of "Clean Rooms" has become a major industry itself, using air conditioning and filtration control for a nominal 70°F, 45% RH, and less than 10% particulate content in the air . . . the particulates measured in micron size. A single piece of dust landing at a critical point of a device structure can destroy the functional capability.

With the beginning development of man's metal culture in the Bronze Age — somewhere around 5000 B.C. — we started using air as an industrial application: smelting of ores and fabrication of bronze. The Iron Age began about 1350 B.C. when the Hittites of Asia Minor developed iron smelting — requiring higher temperatures than bronze, leading to the use of bellows. The Egyptians had already developed kiln firing and soda-lime glass (around 3000 B.C.) initially used as a glaze on pottery, and the same formula is still in use for ordinary glass. Iron tools may have been in use by the Egyptians about this same time.

Sailing ships have been major power bases in the Mediterranean Sea for centuries as trading and ruling civilizations rose and fell (Crete, Egypt, Greece, Phoenicia, Roma). Wooden sailing ship power culminated in the mid-1800s when the British Empire ruled the oceans of the world after Spain, Portugal, Holland, and France. Great Britain was the major sea power after the defeat of the Spanish Armada in 1588 until after World War I, controlling much of the world for over 250 years. Even today, large diesel-powered transport ships have had metal sails added to increase speed; fishing fleets still use both sail as well as engines; sport sailing is a worldwide industry. Sailing . . . the world still uses wind power for this purpose.

There is a wide range of air sports: the Chinese developed the flying of kites in ancient times; the French developed hot-air balloons and were racing across Paris in the mid-1800s; Count Zeppelin in Germany invented the first lighter-than-air dirigibles — they were used against England in World War I and culminated as a transport airship with the unfortunate burning of the Hindenburg at Lakehurst, NJ in 1936. A few are still in use today, using helium rather than hydrogen for lift — the Goodyear blimps as an example — and are under consideration again for heavy transport of materials.

Man has always wanted to fly. It goes back to the legend of Icarius with his wax wings during the Cretian dynastic period around 1300 B.C. The Wright brothers are credited with designing and flying the first fixed-wing air machine — the airplane — in the early 1900s, though similar development was being made in Europe in the late 1800s. Small, open-cockpit biplanes and monoplanes were used in World War I — the Americans as the U.S. Army Air Force (did not become a separate branch of the military services until shortly before World War II). These small planes were used for barnstorming around the U.S. in the 1920s and 1930s, the first airmail delivery in the 1930s, and the "clipper ship" flying-boats were crossing the Pacific to the Philippines as early as 1935.

Balloons were used in the American Civil War, and World Wars I and II, mainly for reconnaissance observation, and today weather balloons are an element in weather fore-casting. Ballooning as a sport is still in existence, both for height — over 13 miles in 1935 — 15 miles in 1988 — and flown for cross-country distance flying, to include attempts to cross the Atlantic Ocean. The Japanese sent incendiary balloons across the Pacific during World War II into the American northwest — the 555 PIR were used as smoke jumpers — the only black parachute unit at that time. The use of parachutes was first envisioned in Italy in the 1400s, were used for free-fall jumping during the airplane barnstorming days, but the Germans used the first massed airborne parachute troops in the invasion of Holland — 1st German Airborne Division, under the command of General Kurt Student. The largest airborne drop was by the Allies when they jumped into Holland in September of 1945. Elements of the 82nd Airborne Division jumped in Korea, and 101st Airborne Division elements were in Vietnam.

Skydiving has become a civilian sport, as well as being used by the military for small unit drops, and new parachute designs are under continual development.

The development of large aircraft in the 1930s culminated with their use in World War II. The Douglas DC-6 — the military C-47 — is probably one of the best known transport aircraft, and many are still in service today. The B-24 and British Halifax bombers; Germany Messerschmitt . . . British Spitfire . . . American P-51, P-38, P-47, or the Navy Corsairs, as fighter aircraft, and, not the least, the Japanese Zero, to mention only a few. The Germans developed the first jet aircraft toward the end of World War II, and this has led to the present-day military and commercial jet aircraft that have tied much transportation of the world together in hours/days rather than months.

The helicopter bears mention as an aircraft in its own right. This did not become a viable aircraft until the late 1940s, and the first real service (which was military) was in Korea for medical evacuation. By the time of the Vietnam war there were wings of transport and attack helicopters. Have you ever heard of "Puff . . . the Magic Dragon"? It was a helicopter built as a flying gun platform, capable of decimating some 100 square yard of area at a time with a single burst of concentrated fire-power. Military branches of all countries today have both transport and attack helicopters, and there are civilian general transport and construction helicopters, as well as medical-evacuation choppers.

As a source of mechanical power, windmills have been used in Holland for centuries, and several designs were brought to the Americas in the 17th and 18th centuries. Single and multiple sets of wind vanes are now in use for production of electricity using wind power as an alternate energy source.

Wind tunnels are used in evaluating air turbulence in aircraft wing and body design, astronauts have used large fans to simulate freefall in space, and air-lift vehicles are under development and evaluation.

Air compressors supply air in many industrial operations, such as for air cylinders or for switches and valves. They supply the air in gas stations for filling tires, and pressurized air for general cleaning in metal workshops, etc. There are now back-pack units and vehicle mounted systems for street cleaning or yard maintenance. Pressurized air is used for materials and devices under pressure testing. **CAUTION:** Air compressors should be well maintained to prevent oil from the compressor pumps leaking into air lines and contaminating parts being cleaned or under test.

Technical Application: Air, both as a gas and as liquid, has many applications in metal and Solid State processing. Liquid air is mainly used for conversion to gaseous, pure air for use as an oxidizing atmosphere or for surface cleaning and drying. Air drying can be simply allowing a part to dry in the room atmosphere at room temperature; on a hot plate; in an air-circulating oven at elevated temperature; or under IR lamps. It can be supplied under pressure in chemical sinks for "blow-off" drying of parts. Where dangerous chemical gases are in use, such as in epitaxy equipment areas, a combination of air-conditioning and exhaust fans are designed to completely evacuate and replace the room volume of air in 1 min in the event of a gas leak. Dry air is supplied in pressurized cylinders alone, or as a gas mixture for special applications (A-1 cylinders at 3600 psi — 9″ diameter, 5 ft high with a two-stage regulator). It also is supplied as wet-air for high purity steam oxidation. In hydrogen belt furnace operation a nitrogen curtain (a series of glass cloth hangers at front and end entrance with nitrogen gas blown down into the curtain zone) are used to prevent air entering the hot furnace, mixing with hydrogen and exploding.

Metallographic samples after polishing are heated in air on a hot plate for "color tinting". The resulting oxidation colors vary with the metal mixtures as the crystallites have different orientations that oxidize at different rates with varying distinctive colors. Where it is a single crystal metal with different phases, the phases can be differentiated by color in a similar manner. When a metal thin film has a preferred deposit orientation there will be a difference in color: Titanium is an example, (1010) surfaces are brilliant reds/blues; whereas (0001) surfaces are tan/yellow-brown. The basal (0001) surface gives good adhesion; the (1010) prism plane tends to peel. Air is supplied by compressors for operation of valves and equipment, or to hoses for heavy duty air cleaning or pressure testing of parts.

As a gas mixture air is not grown as a single crystal, although oxygen, nitrogen, and the rare gases have been grown under vacuum cryogenic and pressure conditions for morphological study.

Etching: N/A

AIR ETCHANTS

AIR-0001

ETCH NAME: Pressure | TIME:

TYPE: Pressure, preferential | TEMP: -196°C

COMPOSITION:

x ... pressure, vapor

DICUSSION:

N_2/O_2 grown as a single crystal in vacuum under pressure and cryogenic conditions is difficult to grow because of the difference in vapor pressure between oxygen and nitrogen.

Varying vapor pressure under vacuum can develop etch figures and defects on surfaces of the solid.

REF: Schwentner, N et al — *Rare Gas Solids,* Vol 3, Academic Press, New York, 1971

PHYSICAL PROPERTIES OF ALUMINUM, Al

Classification	Metal
Atomic number	13
Atomic weight	30
Melting point (°C)	660
Boiling point (°C)	2467
Density (g/cm^3)	2.699
Thermal conductance (cal/sec)(cm^2)(°C/cm) 20°C	0.50
Specific heat (cal/g) 25°C	0.215
Latent heat of fusion (cal/g)	94.5
Heat of fusion (k-cal/g-atom)	2.55
Heat of vaporization (k-cal/g-atom)	69.9
Atomic volume (W/D)	10.0
1st ionization energy (K-cal/g-mole)	138
1st ionization potential (eV)	5.98
Electronegativity (Paulings)	1.5
Covalent radius (angstroms)	1.18
Ionic radius (angstroms)	0.51 (Al^{+3})
Coefficient of linear thermal expansion ($\times 10^{-6}$ cm/cm/°C) 20°C	22.4 (23.9)
Electrical resistivity (micro-ohms-cm)	2.655
Cross section (barns)	0.215 (0.23)
Magnetic susceptibility (cgs $\times 10^{-6}$)	0.6
Tensile strength (psi)	30,000
Vapor pressure (°C)	1749
Hardness (Mohs — scratch)	2—2.9
(Knoop — kgf/mm^{-2})	100—140
Crystal structure (isometric — normal)	(100) cube, fcc
Color (solid)	Silver white
Cleavage (cubic)	(001)

ALUMINUM, Al

General: Aluminum does not occur as a native element, but is found as a constituent in over 200 single crystal minerals. In order, oxygen, silicon, and aluminum are the three most abundant elements in the world, and the oxides SiO_2 and Al_2O_3 are the most inert and stable under normal atmospheric conditions. The mineral bauxite, $Al_2O_3.2H_2O$ — a hydrated aluminum oxide — is the major ore. It is most often found in clay-like deposits called laterites as a mixture with clays and iron oxides, and may contain large amounts of a colloidal form of aluminum oxide. Bauxite's mode of origin is not completely understood, but it is known to form under tropical conditions from the prolonged weathering of aluminum bearing rocks, and as a sedimentary colloidal precipitate which can be associated with volcanic activity. As the mineral corundum it is mostly a constituent in rock forming minerals of the Chlorate Group, but can be an original constituent of igneous rocks and pegmatites. It is not used as an ore of aluminum, but its three varieties are of commercial importance: (1) gem stones: sapphire (blue); ruby (red), and as "Oriental" — topaz (yellow), emerald (green), amethyst (purple); (2) corundum, as a powder for abrasives, and (3) emery, also as an admixture with magnetite, Fe_3O_4. When corundum is cut normal to the z-axis and polished encabachon, it may produce a six-rayed star — the star sapphire, with top quality a fine blue-grey.

Pure aluminum metal is second only to gold in malleability, sixth in ductility, and one of the softest of metals (H = 2—3); whereas corundum is second only to diamond in hardness (H = 9 and H = 10, respectively, on the Mohs Scale of hardness). The metal is silver-white in color and, as is highly electropositive, when in contact with other metals, will corrode rapidly. It has about 60% of the electrical carrying capacity of copper, and has been used for high power transmission lines as it is much lighter than copper. It also has been used in finer wire form for house current power and lighting, but the wire tend to oxidize with time, say, over 20 years to the oxide — a nonconductor and industrially considered a high temperature ceramic. The conversion of aluminum to aluminum oxide has a high heat of formation which is strongly exothermic: Al + Fe_2O_3, when heated to form Al_2O_3 + Fe produces 198,000 calories of heat (thermite), and used both for welding and as an explosive. Aluminum foil, when ignited in photoflash bulbs, produces heat equivalent to that of the sun surface (some 8000 K) with intense white light.

By far the greatest use of aluminum is as a construction metal. Physical strength is increased with the addition of small amounts of other metals, such that the majority of "aluminum" materials are in some form of an alloy. As an example, duralumin — containing Cu, Mg, and Mn — after heat treatment has the strength of mild steel and, when duralumin sheets are clad with pure aluminum, becomes even more weather resistant to corrosion. Applications include airplane skins; car and truck bodies; engine blocks; building girders and house siding; window and door frames, including the wire mesh screening; cooking pots and pans; as well as aluminum foil, and hundreds of small products, such as screws, bolts, rods, tweezers, and so forth. Evaporated on glass, it is a low-cost replacement for silver mirrors, and evaporated or sprayed on glass or plastic as an anti-reflective (AR) coating. In sports: baseball bats, water and snow skis, the frame for back-packs, and tennis racket frames. Aluminum paint is either oil-base or plastic-base, e.g., acrylic spray paint. In some plastics it is added as flakes for "glitter" with wide use in jewelry items. Military use includes dropping of thin aluminum foil to confuse radar and heat-seeking missiles; as weapon frames, and as a constituent in both flares and incendiary bombs. In the clothing industry for some everyday wearing apparel but, more importantly, in the construction of fire retardant and chemical protective suiting and space suits. There are medical applications of compounds, such as the hydroxide, $Al(OH)_3$ as a salve. The hydroxide also is used as a mordant for setting dyes in the clothing industry, and as a clarifier in water treatment. Alums — double sulfates of aluminum salts and sodium, potassium or ammonium salts — also have important medical and chemical applications.

In the pottery industry — clays have been in use since earliest times in the Middle East, circa 12,000 B.C. — and, with the Egyptian development of soda-lime glass around 3500 B.C., the pottery industry has become known as the ceramic pottery industry, e.g., the inclusion of glass as a surface glaze. Archaeologists date the rise and changes of ancient civilizations by the designs, and pigments used on glazed pottery. Note that many clays contain alumina, such that the pottery industry includes aluminum compounds, but aluminum metal has only been available in quantity since the discovery of the electrolytic method for processing the ore within the past century.

Technical Application: Pure aluminum metal is a p-type dopant in Solid State processing of silicon, and was used as wire or cut pre-forms, then as an evaporated thin film to fabricate the first silicon diodes and transistors that led to development of the semiconductor industry in the early 1950s. The metal is still used on silicon and other compound semiconductors, although aluminum chloride, $AlCl_3$, is used as a gaseous diffusant source. And today, an AlSi(2—5%) alloy is more widely used as a pre-form than the pure metal as an aluminum alloy device. Aluminum is evaporated as electrical contact pads on many devices, with aluminum wire ultrasonically bonded to the pads. Ultrasonic vibration is used to break through the thin native oxide that always grows and stabilizes on aluminum metal surfaces exposed to air.

Aluminum should not be used in direct contact with gold as with time and/or heat the two metals form distinctive purple compounds that are extremely brittle and destroy the electrical integrity of an operating device. The reaction is called "purple plague". Walker (author) first recognized purple plague as a problem during a failure analysis of a high power silicon diode, and L. Bernstein carried on the study with several excellent papers and reports, circa 1960.

Where aluminum and gold are used in a device structure, they are separated by evaporated layers of other metals, such as Pd/Cr for a complete layer structure as Au/Pd/Cr/Al. Such layers are not only a vertical separation, but are horizontally off-step separated, as both gold and aluminum can show slow "creep" characteristics on a device surface under operating load, particularly, at elevated temperature.

The chemical processing of thin film aluminum is a basic problem as, when pure aluminum is exposed to air, it forms a passivating, stable oxide layer on the surface, and the oxide is impervious to most liquid chemical processing solutions for controlled removal and patterning. The aluminum deposits as a microcrystalline layer, crystallite size controlled to some degree by rate and temperature of evaporation, such that the ensuing oxide also is crystalline in structure. Acid attack initiates along these oxide crystallite boundaries, undercutting the "oxide" with a more rapidly attack of the underlying pure aluminum. This can produce the easily recognizable "Swiss cheese" effect, and remaining oxide on a surface will create a high resistance zone if other metals are subsequently over-evaporated, reduce electrical efficiency, and, in many cases, poor metal adhesion with eventual peeling and loss of metal contact. RIE — reactive ion etching — using an RF plasma with gas mixtures, such as BCl_3 or $AlCl_3$, are now in use for improved etch removal of the residual oxide, and most aluminum structures are processed with these gases in current device fabrication.

In quartz crystal radio frequency blank processing the evaporated metal electrodes are usually gold, silver or aluminum. Aluminum is given preference due to its lesser mass loading effect on crystal frequency.

SELECTION GUIDE: Al

(1) Acetone: *Note:* Applies to lift-off of all thin film metallization deposited on photo resist lacquers.
  (i) Lift-off: AL-0088
(2) H_2O: *Note:* Applies to all thin films deposited on soluble substrates.
  (i) Float-off: AL-0095
(3) Br_2:MeOH: (BRM)
  (i) Removal/Polish: AL-0082a
(4) HCl, conc: (Electrolytic)
  (i) Preferential: AL-0060; -0082b ;-0098
(5) HCL:H_2O
  (i) Removal: AL-0003b; -0004
  (ii) Polish: AL-0034
  (iii) Preferential: AL-0005
(6) HCl:$FeCl_3$:H_2O:
  (i) Cutting: AL-0083
  (ii) Preferential: AL-0003
(7) H_2SO_4, conc: (electrolytic)
  (i) Anodizing: AL-0020b
(8) Air: O_2/N_2 w/H_2O (electrolytic)
  (i) Anodizing: AL-0040; -0049; -0050; -0052

(9) H_2SO_4:H_3PO_4:
  (i) Polish: AL-0041a
(10) H_2SO_4:H_3PO_4:HNO_3:
  (i) Polish: AL-0035; -0097; -0045
(11) H_3PO_4:HAc: (electrolytic)
  (i) Polish: AL-0024; -0087; -0090
(12) H_3PO_4:HNO_3:H_2O:
  (i) Cleaning: AL-0033
  (ii) Removal/Polish: AL-0009; -0033b
(13) H_3PO_4:HNO_3:HAc:H_2O:
  (i) Removal/Polish: AL-0010; -0014; -0084
  (ii) Selective: AL-0031
(14) H_3PO_4:H_2O_2:H_2O:
  (i) Removal/Polish: AL-0011; -0067
(15) H_3PO_4:H_2O:Glycerin:
  (i) Removal/Polish: AL-0007
(16) H_3PO_4:HNO_3:Al-Brite:
  (i) Polish: AL-0013
(17) H_3PO_4:Cr_2O_3:
  (i) Oxidation removal: AL-0002
(18) H_2SO_4:$Na_2Cr_2O_7$:H_2O:
  (i) Cleaning: AL-0032
(19) HF:H_2O:
  (i) Preferential: AL-0023b;
  (ii) Removal: AL-0091
(20) HF:NH_4F:HF: (BHF)
  (i) Cleaning: AL-0030a
(21) HF:HCl:H_2O:
  (i) Preferential: AL-0017; -0020a
(22) HF:HCl:MeOH/EOH: (electrolytic)
  (i) Polish: AL-0025b
(23) HF:HNO_3:HCl:H_2O:
  (i) Cleaning: AL-0081
  (ii) Preferential: AL-0030; -0020a; -0058; -0059a
  (iii) Polish: AL-0030
(24) HF:HNO_3:HCl:Glycerin: (electrolytic)
  (i) Polish: AL-0027
(25) HNO_3:MeOH: (electrolytic)
  (i) Polish: AL-0027
(26) $FeCl_3$:H_2O:
  (i) Removal: AL-0004
(27) $CuCl_3$:H_2O:
  (i) Polish: AL-0022a
(28) $HClO_4$:MeOH: (electrolytic)
  (i) Polish: AL-0037; -0038; -0039a; -0042; -0033; -0035
(29) $HClO_4$:EOH:H_2O:Butyl Cellusolve: (electrolytic)
  (i) Polish: AL-0044
(30) $HClO_4$:HAc: (electrolytic)
  (i) Polish/Removal: AL-0046; -0049; -0048
(31) $HClO_4$:EOH:Glycerin: (electrolytic)
  (i) Thinning: AL-0047; -0091

(32) $NaOH:H_2O$: (electrolytic)
  (i) Removal/Polish: AL-0064
(33) $NaOH:H_2O$:(Glycerin)
  (i) Removal/Polish: AL-0033a; -0093; -0062; -0077; -0078
(34) $NaOH:HNO_3:H_2O$:
  (i) Cleaning: AL-0030
(35) $NaOH:HNO_3$:MeOH:
  (i) Preferential: AL-0041b; -0045; -0055a
(36) $KOH/NaOH:H_2O$:
  (i) Cleaning: AL-0001; -0061; -0062; -0066
  (ii) Oxidation: AL-0002
  (iii) Removal: AL-0003a; -0062; -0063; -0065
(37) $KOH:H_2O$: (electrolytic)
  (i) Cleaning/Removal: AL-0064
(38) $KOH:K_3Fe(CN)_6:K_2B_4O_7.4H_2O$:
  (i) Removal: Al-0019a; -0084
(39) KI:MeOH:
  (i) Selective: AL-0094
(40) CCl4:
  (i) Removal: AL-0091; -0092
(41) Br_2/HBr:
  (i) Removal: AL-0092

GAS: (molecular)

(1) Air (oxidation): AL-0061; -0049; -0030; -0052
(2) Chlorine, Cl_2 (removal): AL-0092

GAS: (ionic) (DCE)

(1) $BCl_3/AlCl_3$ (selective): AL-0051; -0089($+O_2$)
(2) BBr_3 (removal/polish): AL-0056; -0059b; -0054
(3) $SiCl_4$ (removal): AL-0089 ($+O_2$)
(4) Neon, Ne (preferential): AL-0040

ABRASIVES:

(1) SiC, paper (polish/forming): AL-0076
(2) Al_2O_3, powder (polish): AL-0100

ANODIZING/OXIDATION:

(1) H_2SO_4 (electrolytic): AL-0020b
(2) Air, N_2/O_2 + H_2O: AL-0040; -0049; -0050; -0052; -0030; -0061
(3) Water, H_2O (cleaning): AL-0101
(4) $KOH/NaOH:H_2O$ (cleaning): AL-0002

NAMED/NUMBERED ETCHANTS:

(1) BHF:($HF:NH_4HF.HF$) (cleaning): AL-0030a
(2) BRM: (Br_2:MeOH) (removal/polish): AL-0082a
(3) Flick's Etch ($HF:HCl:H_2O$) (preferential): Al-0027; -0023c
(4) General Etch ($HF:H_2O$) (preferential): Al-0023d
(5) Iodine Etch (I_2:MeOH) (structuring): Al-0031
(6) Jacquet's Etch ($HClO_4$:HAc — electrolytic) (removal): AL-0048
(7) Keller's Etch ($HF:HNO_3:HCl:H_2O$) (preferential): AL-0023a; -0043
(8) Nital (HNO_3:EOH/MeOH) (polish): AL-0022b; -0065a; -0061

THERMAL

(1) $+H_2$ or N_2/H_2 (reducing atm) (cleaning/annealing): AL-0102
(2) $+O_2$ (oxidizing atm) (cleaning/anodizing): See Air or Anodizing
(3) + Vacuum (preferential): AL-0039a-b

ELECTRICITY:

(1) Thin Film (creep): AL-0041

FATIGUE:

(1) See Thermal-vacuum
(2) Pressure: AL-0096

Both pure beryllium and an aluminum/beryllium alloy, AlBe (30%) have been evaporated to further reduce the mass-loading effect.

Aluminum has been evaporated on telescope lenses, then overcoated with a thin layer of silica, SiO_2 for wear resistance. The aluminum used as an absorber and reflector. And such lenses are periodically stripped for recoating.

Aluminum parts are used in package assemblies and, the combination of aluminum and alumina are important as radiation resistant packages. Aluminum foil is used as a grounded wrapping for static sensitive devices, and as a temporary shield in vacuum evaporator bell jars.

Single crystals of pure aluminum, as well as several alloys, have been grown for general morphological study, electrical evaluation and a wide range of defect, corrosion and fatigue failure studies. These studies include the other physical structures of aluminum: colloidal, amorphous and crystalline.

Etching: Soluble in HCl, HNO_3, H_2SO_4, alkalies and acid mixtures. Gas mixtures as an RF Plasma.

ALUMINUM ETCHANTS

AL-0033a
ETCH NAME: Sodium hydroxide — TIME:
TYPE: Alkali, removal — TEMP:
COMPOSITION:
x .... 1 *M* NaOH
DISCUSSION:

Al thin film removal from vacuum evaporation systems. Used for removal of metal build-up from stainless steel, glass, copper, and ceramic surfaces. Can damage titanium and silver surfaces. Another notes that hydroxide cleaning can leave residual contamination on parts unless thoroughly washed. (*Note:* An excellent article on cleaning procedures for vacuum deposition equipment.)

REF: Nichols, D R — *Solid State Technol,* December 1979

AL-0033b
ETCH NAME: — TIME:
TYPE: Acid, removal — TEMP:
COMPOSITION:
20 .... H_3PO_4
2 .... HNO_3
5 .... H_2O

DISCUSSION:

Al thin film removal from vacuum evaporation systems. Use for removal from steels, glass and ceramics. Do not use on copper parts.

REF: Ibid.

AL-0003

ETCH NAME: Hydrochloric acid — TIME: 2—10 min

TYPE: Acid, removal — TEMP: RT

COMPOSITION:

(1) x .... HCl, conc.

(2) 1 .... HCl
1—10 .... H_2O

DISCUSSION:

Al thin film removal from vacuum evaporation systems. Used to remove multiple layer films of aluminum/nickel/chromium/gold and gold germanium from steel, glass and ceramic — dilute solutions more commonly used, particularly on copper parts. Used to soak and scrub parts with a stainless steel brush. Used with lint-free cloth dampened with solution to wipe base plates and other in-place fixturing after light scrapping to remove heavy build-up. Such surfaces should be well wiped down with a water soaked cloth after acid cleaning and follow with methyl alcohol. Vacuum pump-down after cleaning to at least 10^{-6} Torr before system use.

REF: Walker, P — personal application, 1960—1980

AL-0005: Hurden, M J & Averbach, B L — *Acta Metall,* 9,237(1961)

Al (111) and (100) wafers and ingots. A 50% solution was used to "heavy" etch specimens in a study of dislocation density in deformed aluminum.

AL-0033c

ETCH NAME: Potassium hydroxide — TIME: Variable

TYPE: Alkali, removal — TEMP: RT to 180°F

COMPOSITION:

x .... 8—10% KOH (NaOH)

DISCUSSION:

Al and Al alloys. At room temperature solution is slow and used as a surface cleaning etch. Used hot for heavy removal. RT solutions used for removal of aluminum evaporation onto bell jars and fixtures of vacuum systems — glass, SST, and ceramic material.

REF: Ibid.

Al-0060: Walker, P & Menth, M — personal application, 1981

Al, evaporated film deposits in vacuum systems. Solution used for cleaning systems. Follow with heavy water washing and IR lamp drying. Vacuum bake system before use. Also used sodium hydroxide.

Al-0002: Stirland, D J & Bicknell, R W — *J Electrochem Soc,* 106,481(1959)

Al, specimens used in a study of surface oxidation. A 3% NaOH solution, RT, 3 min used to clean surfaces. After anodizing, a solution of 35 ml H_3PO_4:20 g Cr_2O_3/l used to remove the oxide.

Al-0001: DeSorbo, W — *Phys Rev,* 111,810(1958)

Al, single crystal specimens used in a study of imperfection at low temperature. Used a solution of 15% NaOH, warm as a pickling solution to clean surfaces.

AL-0061: Walker, P — personal application, 1964

Al and AlSi (5%) alloy as spherical pellets used in fabricating Silicon Sphere Alloy Zener diodes. Used a 1% KOH solution to clean wafers with multiple alloyed junctions after fabrication prior to DC leakage evaluation. Contour plots of leakage of devices on wafers used to determine segregation zones in ingots with wafers cut and numbered consecutively for traceability.

AL-0062: Tarn, W H & Walker, P — personal application, 1976

Al, thin film evaporation on soda-lime glass used for photo resist masks and special lenses. A 30% NaOH solution, warm (40°C) used to remove aluminum and clean the glass.

AL-0063: Menth, M — personal communication, 1982

Al, thin films evaporated on telescope lenses for reflectivity control. A 15—20% NaOH solution at RT or warm used to remove aluminum prior to re-work of old lenses.

Al-0064: Tegart, W J McG — *The Electrolytic and Chemical Polishing of Metals in Research and Industry,* Pergamon Press, London, 1956

Al and Al alloys. Alkali solutions used for etching and cleaning. Includes a discussion of chemical and electrolytic etching as preferred over mechanical lapping of metals in order to preserve bulk characteristics.

AL-0065: ASTE Committee — *Tool Engineers Handbook,* McGraw-Hill, New York, 1949

Al, alloy specimens, and other soft metals such as tin and zinc. Buffer alkali solutions with silicates to reduce etch attack. A 5% NaOH solution plus carbonates, silicates, phosphates, or borates as buffers recommended for general etching.

AL-0066: Campos, R & Walker, P — personal application, 1961—1963

Al alloy piping and parts on 2000 gallon capacity cryogenic trailers. 5—10% NaOH solutions used for corrosion removal cleaning of surfaces with plastic brush scrubbing, and heavy water wash after etching. Final drying with warm nitrogen gas.

AL-0019a

ETCH NAME: TIME:

TYPE: Acid, removal TEMP:

COMPOSITION:

x .... KOH

x .... $K_3Fe(CN)_6$

x .... $K_2B_4O_7.4H_2O$

DISCUSSION:

Al, thin films on semiconductor wafers. Use COP type photo resists to coat surfaces and develop patterns using standard photolithographic techniques. Use solution shown to remove exposed aluminum. COP resists etch more slowly than the aluminum.

REF: Turner, J K — *J Vac Sci Technol,* 15,962(1978)

AL-0010

ETCH NAME: TIME:

TYPE: Acid, removal TEMP:

COMPOSITION:

1520 ml .... H_3PO_4

120 ml .... HAc

50 ml .... HNO_3

300 ml .... H_2O

DISCUSSION:

Al, thin films deposited on GaAs and silicon (100) wafers. Use AZ-type photo resists to coat and open patterns with standard photolithographic techniques. The solution was used to remove aluminum and will attack both wafer and aluminum. After etching, rinse in DI water; MeOH and nitrogen blow dry.

REF: Shipley Sales Brochures, 1975—1980

AL-0014
ETCH NAME: TIME:
TYPE: Acid, removal TEMP:
COMPOSITION:
80 ml .... H_3PO_4
5 ml .... HAc
5 ml .... HNO_3
10 ml .... H_2O
DISCUSSION:
Al thin films evaporated as part of a layer metallization structure on Si(100) substrates, as Ni/TiN/Al/Si(100). After photo resist patterning, etch remove Ni and TiN, then use solution shown for removal of aluminum layer.
REF: Ballard, N L et al — *J Electron Mater,* 13,327(1984)

AL-0009
ETCH NAME: TIME:
TYPE: Acid, removal TEMP: 40°C
COMPOSITION: RATE: 1500 Å/min
80 ml .... H_3PO_4
5 ml .... HNO_3
0—20 ml .... H_2O
DISCUSSION:
Al specimens. Used as a general removal etch with agitation. (*Note:* Rate shown is a guideline, only as water content varies.)
REF:

AL-0011
ETCH NAME: Krumm etch TIME:
TYPE: Acid, removal TIME: 35°C
COMPOSITION: RATE: 100 Å/sec
8 .... H_3PO_4
1 .... H_2O_2
1 .... H_2O
DISCUSSION:
Al thin films evaporated on GaAs, (100) wafer substrates. Solution used to remove aluminum and in photo resist pattern etching. Solution will also attack gallium arsenide at a more rapid rate than aluminum.
REF: Siracusa, M — personal communication, 1979

AL-0067
ETCH NAME: TIME:
TYPE: Aid, removal TEMP: RT
COMPOSITION:
2 .... H_3PO_4
1 .... H_2O_3
3 .... H_2O
DISCUSSION:
Al thin films evaporated on silicon and gallium arsenide, (111) wafers. Solution used as a general removal etch in studies of metal/substrate interfaces.
REF: Topas, B — personal communication, 1970

AL-0007
ETCH NAME:
TYPE: Acid, removal
COMPOSITION:
30 ml .... H_3PO_4
10 ml .... H_2O
50 ml .... glycerin
TIME:
TEMP: 80°C
RATE: 60 Å/sec
DISCUSSION:

Al thin films evaporated on GaAs, (100) substrates used in fabricating GaAs FETs. Solution used to remove the aluminum gates of devices in studying the Al/GaAs interface.

REF: Walker, P — personal development, 1979

AL-0013
ETCH NAME: Brite dip
TYPE: Acid, polish
COMPOSITION:
x .... H_3PO_4
x .... HNO_3
x .... Al-brite
TIME:
TEMP: RT
DISCUSSION:

Al specimens and alloys. Mixtures of these acids are polish type solutions for aluminum. The addition of Al-Brite to the solutions improves the surface polish and gives better etch control.

REF: Bulletin #H681-10M — *Metal Finishing Products,* Heatbath Corp., 1984

AL-0032
ETCH NAME:
TYPE: Acid, cleaning
COMPOSITION:
27 .... H_2SO_4
3 .... $Na_2Cr_2O_7$
70 .... H_2O
TIME: 10 min
TEMP: 150°F
DISCUSSION:

Al alloys as sheet material and as an evaporated thin film on other materials. Solution is the preferred surface cleaning etch in preparing surfaces for plastic film adhesion, giving the best adhesion results. After etching, rinse in cold water, then hot water, and air dry. (*Note:* Solution is a form of the standard glass cleaner mixture.)

REF: Skeist, I — *Epoxy Resins,* Reinhold, New York, 1959, pp 190—192

AL-0024
ETCH NAME:
TYPE: Electrolytic, polish
COMPOSITION:
2 .... H_3PO_4
7 .... CH_3CHOOH (HAc)
TIME: 3—5 min
TEMP: RT
ANODE: Al
CATHODE:
POWER: 10, 20, 30 & 40 V
DISCUSSION:

Al specimens used in a study of interfacial structures. Solution used to polish material prior to etch developing structure.

REF: Randall, S T & Bernard, A — *J Appl Phys,* 34,1210(1964)

AL-0004
ETCH NAME: Ferric chloride — TIME:
TYPE: Salt, removal — TEMP: 110°F
COMPOSITION:
x .... $FeCl_3$
Note: A standard solution is 35° Baumé
DISCUSSION:

Al and Al alloys. A general etch for aluminum and several heavy metals, such as irons and steels. It has been used as a general removal or polishing solution as well as a preferential etch depending upon concentration and method of application. Use at RT, hot to boiling. After etching, wash heavily in water until all traces of the brown colored solution is removed.

REF: Meyer, W M & Brown, S H — *Proc Am Electropl Soc,* 36,163(1949)

AL-0029
ETCH NAME: Boron trichloride — TIME:
TYPE: Ionized gas, removal — TEMP:
COMPOSITION: — GAS FLOW:
(1) x .... BCl_3 — PRESSURE:
(2) x .... $CHCl_3$ — POWER:
(3) x .... Cl_2/Ar
DISCUSSION:

Al thin film evaporations used in semiconductor device fabrication. Gas used for reactive ion etching (RIE) for controlled removal of aluminum after photo resist patterning. Better removal control of Al_2O_3/Al than by wet chemical etching.

REF: Bruce, R H & Malafsky, G P — *J Electrochem Soc,* 130,1369(1983)

AL-0022a
ETCH NAME: Cupric chloride — TIME:
TYPE: Salt, polish — TEMP:
COMPOSITION:
x .... $CuCl_2$
x .... H_2O
DISCUSSION:

Al single crystal specimens. Wafers were polished in this solution prior to X-ray studies of surface and bulk structure.

REF: Walker, C B — *Phys Rev,* 103,547(1956)

AL-0023a
ETCH NAME: Keller's etch — TIME: Dip
TYPE: Acid, microetch — TEMP: RT
COMPOSITION: — ASTM: #3
2 ml .... HF
5 ml .... HNO_3
3 ml .... HCl
190 ml .... H_2O
DISCUSSION:

Al, alloy and cast aluminum specimens. Solution used as a general micro-etch in the study of surface structure.

REF: *A/B Metal Digest,* 21(2),23(1983) — Buehler Ltd
AL-0043: ASTM E407-70

Reference for Keller's Reagent as #3.

AL-0023b
ETCH NAME: Hydrofluoric acid TIME: Swab/dip
TYPE: Acid, macro-etch TEMP: RT
COMPOSITION:
1 ml .... HF
199 ml .... H_2O
DISCUSSION:
Al, low alloy content specimens. Solution used as a macro-etch by swabbing the surface or immersion dipping to develop surface structure.
REF: Ibid.
AL-0021: Tokumaru, Y & Okada, Y — *Jpn J Appl Phys Lett,* 23,123(1984)
Al thin film aluminum deposited on both Si, (100) and GaAs, (100): Cr semi-insulating (SI) wafers. Solution used to remove aluminum after photo resist patterning.

AL-0030
ETCH NAME: TIME: (1) 5 sec or (2) 15 sec
TYPE: Acid, polish, preferential TEMP: (1) 0°C (2) 8°C
COMPOSITION:
(1) 4 .... HF
35 .... HNO_3
61 .... HCl

(2) 1 .... solution (1)
100 ml .... H_2O
DISCUSSION:
Al single crystal specimens. Solution (1) used as a polish etch; solution (2) is slightly preferential. Ice/acetone = 8°C, and dry ice/acetone = 0°C used for the two cold temperature applications.
REF: Tucker, G E G & Murphy, P C — *J Inst Met,* 8,235(1953)

AL-0017a-b
ETCH NAME: Flick's etch TIME: 10—20 sec
TYPE: Acid, macro-etch TEMP: RT
COMPOSITION:
(1) 10 ml .... HF
5 ml .... HCl
90 ml .... H_2O

(2) x .... HCl, conc.
DISCUSSION:
Al, cast alloys. Etch used for preparation of metallographic specimens as a macro-structure solution. After etching with solution (1) remove black smut that remains by dipping in concentrated hydrochloric acid (2) and rinse in DI water.
REF: *A/B Met Dig,* 21(2),23(1983), Buehler Ltd

AL-0020a
ETCH NAME: TIME:
TYPE: Acid, preferential TEMP:
COMPOSITION:
3% .... HF
47% .... HNO_3
50% .... HCl
DISCUSSION:
Al specimens. Specimens were anodized and etched in solution shown in a study of dislocation loops and patterns developed in aluminum by oxidation.
REF: Bassett, G A & Edeleanu, C — *Phil Mag,* 5,709(1960)

AL-0020b

ETCH NAME: Sulfuric acid
TYPE: Electrolytic, anodizing
COMPOSITION:
10% .... H_2SO_4

TIME:
TEMP:
ANODE: Al
CATHODE: Pt
POWER: 15 V

DISCUSSION:

Al specimens. Specimens were anodized and etched in a study of dislocation loops and patterns. Also used 3% NaCl at 1.5 V. A third anodizing solution was 4 ml H_2SO_4:12 g $Na_2H_3PO_4$:100 ml H_2O at 30 V.

REF: Ibid.

AL-0037

ETCH NAME:
TYPE: Electrolytic, polish
COMPOSITION:
1 .... $HClO_4$
5 .... MeOH

TIME:
TEMP: 0 to -10°C
ANODE: Al
CATHODE:
POWER: 20 V

DISCUSSION:

Al, (100) wafers used in an oxidation study. Both single crystals and tri-crystals used. Polish surfaces in solution shown prior to oxidation.

REF: Doherty, D E & Davis, K S — *J Appl Phys,* 34,619(1963)

AL-0038: Alden, J H & Backofen, W A — *Acta Metall,* 9,352(1961)

Al, (100) wafers used in a study of fatigue crack formation. Solution used prior to stressing specimens as an electropolish using 60 V power.

AL-0039a: Noggles, T S — *J Appl Phys,* 28,913(1957)

Al, (100) wafers used in a study of thermal etch patterning surfaces. Solution used to electropolish surfaces prior to thermal annealing. Polishing done at less than -10°C with power at about 16 A/cm^2.

AL-0042: Alden, J H — *Rev Sci Instr,* 31,897(1960)

Al, single crystal specimens used in a material failure study. Solution used to electropolish specimens at -50°C with solution continually stirred.

AL-0015: Mehdizadeh, P & Block, R J — *J Electrochem Soc,* 119,1090(1972)

Al, single crystal ingots grown by boat method (Bridgman) using a graphite crucible. After cutting for surface orientation, specimens were electropolished in a perchloric acid/alcohol solution prior to studying slip caused by oxidizing surfaces.

AL-0107: McGrath, J T & Waldron, G W J — *Phil Mag,* 9,249(1964)

Al:Mg(1%) ingots and wafers used in a dislocation study. Solution used as an electropolish with EOH replacing MeOH in the formula shown.

AL-0018

ETCH NAME: Iodine etch
TYPE: Halogen, structuring
COMPOSITION:
(1) 10 g .... I_2
100 ml .... MeOH
(2) x .... MeOH

TIME: 3—8 min
TEMP: 70°C
RATE: 1 mil/min

DISCUSSION:

Al thin films and crystalline aluminum sheet. Specimens were anodized, then a channel pattern was cut through the oxide with a diamond scribe. Solution shown was used to etch channels in the aluminum. Rinse in MeOH to remove residual iodine. KMER photo resist

was also used for pattern definition and oxide removed with H_3PO_4 to expose the aluminum.
REF: Walker, P & Schwartz, B — personal application, 1957

AL-0027
ETCH NAME: Nital, modified — TIME:
TYPE: Electrolytic, polish — TEMP:
COMPOSITION: — ANODE: Al
1 .... HNO_3 — CATHODE:
2 .... EOH — POWER:
DISCUSSION:
Al, single crystal specimens used in a study of twin boundaries. Solution used to polish specimens prior to preferential etching.
REF: Aust, K T — *Trans Met Soc AIME,* 221,758(1961)

AL-0022b
ETCH NAME: Nital, modified — TIME:
TYPE: Electrolytic, polish — TEMP:
COMPOSITION: — ANODE: Al
1 .... HNO_3 — CATHODE:
2 .... EOH — POWER:
ASTM: #74
DISCUSSION:
Al, single crystal specimens. Solution used to polish surfaces for a lattice vibration X-ray study.
REF: Ibid.
AL-0064a: ASTM E407-70
Reference for ASTM #74. Solution shown as: 1—5 ml HNO_3:100 ml EOH/MeOH.
AL-0108: Grooskreutz, J C & Shaw, G G — *J Appl Phys,* 35,2194(1984)
Al, (100) specimens used in a study of anodic structure on oriented surfaces. Various mixtures of Nital as HNO_3:MeOH were used.

AL-0025a
ETCH NAME: — TIME: 7—10 min
TYPE: Electrolytic, polish — TEMP: 0—8°C
COMPOSITION: — ANODE: Al
2% .... HF — CATHODE:
15% .... HNO_3 — POWER:
42% .... HCl
15% .... glycerin
DISCUSSION:
Al, single crystal specimens. Solution used to electropolish surfaces.
REF: Hone, A & Pearson, E C — *Met Prog,* 53,363(1948)

AL-0006
ETCH NAME: — TIME:
TYPE: Acid, cleaning — TEMP:
COMPOSITION:
x .... 5% NaOH
x .... 50% HNO_3
DISCUSSION:
Al, polycrystalline sheet. Coupons cut from sheet and used in a study of anodization. Clean surfaces with solution shown before oxidizing.
REF: Ibid.

AL-0031
ETCH NAME: TIME:
TYPE: Acid, removal TEMP:
COMPOSITION:
80 ml .... H_3PO_4
5 ml .... HNO_3
5 ml .... HAc
10 ml .... H_2O
DISCUSSION:

Al, thin film evaporation layer used in a diode metallization structure as Ni/TiN/Al on a Si, (100) substrate. After photolithographic processing for diode pattern structure and metal layer evaporation, each metal was selectively removed — aluminum with the solution shown.
REF: Martin, T L et al — *J Electron Mater,* 3,309(1984)

AL-0039b
ETCH NAME: Heat TIME:
TYPE: Thermal, preferential TEMP:
COMPOSITION:
x .... heat
DISCUSSION:

Al, (100) wafers used in a study of surface etch pits developed from concentration of vacancy clusters during vacuum thermal annealing. Surfaces were etch polished prior to thermal anneal. See AL-0039a.
REF: Ibid.
AL-0012: Pierce, J M & Thomas, T L — *Appl Phys Lett,* 39,165(1981)

Al, 0.2 μm thick thin films evaporated on Si, (100) heated substrates (360°C) pre-coated with thin film SiO_2. Evaporated aluminum film was photo resist patterned as conductors and overcoated with p-SiO_2 (BSG). After aluminum wire bonding, current was applied at different temperatures on discrete conductors and electromigration was observed as chains of single crystal aluminum crystallites.

AL-0040
ETCH NAME: Neon TIME:
TYPE: Ionized gas, preferential TEMP:
COMPOSITION: GAS FLOW:
x .... Ne^+ ions PRESSURE:
POWER:
DISCUSSION:

Al, single crystal specimens ion bombarded to develop and observe etch figures on surfaces. Other materials studied were Bi, Cd, Mg, Cu, Sn, and Zn.
REF: Yurasova, V E — *Kristallografiya,* 2,770(1957)

AL-0023c
ETCH NAME: Flick's Etch TIME: 10—20 sec
TYPE: Acid, micro-etch TEMP: RT
COMPOSITION:
(1) 10 ml .... HF (2) x .... HNO_3
5 ml .... HCl
90 ml .... H_2O
DISCUSSION:

Al, specimens and cast alloys. Solution used as a structure etch. Used HNO_3 to remove smudge remaining after etching.
REF: Ibid.

AL-0023d
ETCH NAME: General
TIME: Dip/swab
TYPE: Acid, micro-etch
TEMP: RT
COMPOSITION:
1 ml .... HF
199 ml .... H_2O
DISCUSSION:
Al, specimens and alloys with low aluminum content. Solution used as a macro-etch by swabbing or dipping specimens.
REF: Ibid.

AL-0025b
ETCH NAME:
TIME: 7—10 min
TYPE: Electrolytic, polish
TEMP: 0—8°C
COMPOSITION:
ANODE: Al
4% .... HF
CATHODE:
70% .... HCl
POWER:
25% .... MeOH/EOH
DISCUSSION:
Al, single crystal specimens. Solution used to electropolish surfaces in solution shown.
REF: Ibid.

AL-0026
ETCH NAME:
TIME:
TYPE: Acid, cleaning
TEMP:
COMPOSITION:
x .... HNO_3
x .... $NH_4F.HF$
DISCUSSION:
Al, polycrystalline sheet. Coupons cut and used in a study of aluminum anodization. Clean surfaces with solution shown before oxidizing.
REF: Sharp, D J et al — *Thin Solid Films,* 111,227(1984)

AL-0034
ETCH NAME: Hydrochloric acid
TIME:
TYPE: Electrolytic, polish
TEMP:
COMPOSITION:
ANODE:
1 .... 30% HCl
CATHODE:
3 .... H_2O
POWER:
DISCUSSION:
Al, specimens. A general electropolish for aluminum.
REF: Berglund, T & Dearden, W H — *Metallographer's Handbook of Etching,* Pitman & Sons, London, 1931

AL-0035
ETCH NAME:
TIME: $^1/_2$—2 min
TYPE: Acid, polish
TEMP: 85°C
COMPOSITION:
25 ml .... H_2SO_4
70 ml .... H_3PO_4
5 ml .... HNO_3

DISCUSSION:

Al, specimens used in developing polished surfaces. The solution will also polish Al:Cu and Al:Si alloys.

REF: Herengue, J & Segond R — *Rev Met,* 48,262(1951) (in French)

AL-0097: Herengue, J — *Rev Alum,* 30,261(1953)

Al specimens and alloys. A discussion of chemical polishing.

AL-0036

ETCH NAME: TIME:

TYPE: Acid, cleaning TEMP:

COMPOSITION:

95% .... H_3PO_4

5% .... HNO_3

DISCUSSION:

Al, high purity slugs cleaned in this solution before being used to grow AlSb. After etch cleaning, rinse with in DI water.

REF: Allred, W P et al — *J Electrochem Soc,* 107,117(1960)

AL-0103

ETCH NAME: Hydrochloric acid TIME:

TYPE: Electrolytic, preferential TEMP:

COMPOSITION: ANODE: Al

x .... HCl, conc. CATHODE:

POWER:

DISCUSSION:

Al, specimens. Solution will develop etch figures.

REF: Mahl, M & Stranski, L — *Z Phys Chem,* 257,53(1942)

AL-8104

ETCH NAME: Sodium hydroxide, dilute TIME:

TYPE: Alkali, polish TEMP: RT

COMPOSITION: RATE: 1 μm/h

2 g .... NaOH

1000 ml .... H_2O

DISCUSSION:

Al, cold-rolled specimens used in a study of the effect of thickness on dislocations. Solution used with agitation to produce a highly polished surface without any pitting prior to dislocation etching.

REF: Ham, R K & Wright, M G — *Phil Mag,* 9,937(1964)

AL-0105

ETCH NAME: TIME:

TYPE: Acid, preferential TEMP:

COMPOSITION:

1 .... HCl

1 .... H_2O

x .... *$FeCl_3$

*Saturate solution with $FeCl_3$

DISCUSSION:

Al, (100) wafer surfaces preferentially etched in this solution.

REF: Jacquet, P — *Comp Rend,* 201,1473(1935)

AL-0050

ETCH NAME: Sodium hydroxide
TYPE: Electrolytic, removal
COMPOSITION:
x .... 2.5 *M* NaOH

TIME:
TEMP:
ANODE: Al
CATHODE:
POWER:
RATE: 0.005 $mg/cm^2/h$

DISCUSSION:

Al:Ni specimen surface coatings produced by electrodeposition or diffusion of nickel. Solution shown used as a cleaning etch.

REF: Couch, D E & Connor, J H — *J Electrochem Soc,* 107,272(1960)

AL-0051

ETCH NAME: Carbon tetrachloride
TYPE: Solvent, removal
COMPOSITION:
x .... CCl_4

TIME:
TEMP: Boiling

DISCUSSION:

Al, specimens. A study of the reaction of carbon tetrachloride on aluminum. Aluminum reacts rapidly in concentrated CCl_4. A number of additives were used to inhibit the rate.

REF: Minford, J D et al — *J Electrochem Soc,* 106,185(1959)

AL-0058

ETCH NAME:
TYPE: Acid, preferential
COMPOSITION:
15% .... HF
15% .... HCl
15% .... HNO_3, fuming
25% .... H_2O

TIME:
TEMP:

DISCUSSION:

Al, specimens. Solution used to develop etch figures. (*Note:* "Fuming" concentration and color not specified.)

REF: Tucker, C M — *Met Alloys,* 1,655(1930)

AL-0059a

ETCH NAME:
TYPE: Acid, preferential
COMPOSITION:
10 .... HF
46 .... HCl
15 .... HNO_3
26 .... H_2O

TIME:
TEMP:

DISCUSSION:

Al, specimens. Solution used as a preferential etch on aluminum. (*Note:* Does not specify fuming nitric acid; otherwise similar to AL-0058.)

REF: Barrett, C & Levenson, L H — *Trans Am Inst Min Met,* 137,112(1940) (England)

AL-0041a
ETCH NAME: TIME: 2 min
TYPE: Acid, polish TEMP: 70°C
COMPOSITION:
10 .... H_2SO_4
90 .... H_3PO_4
DISCUSSION:
Al, (001) wafers used in a study of lithium precipitation along dislocations. After polishing in this solution — coat surfaces with 0.1% Li (usually in oil), heat and bubbles will appear along dislocation lines.
REF: Murray, G T — *J Appl Phys,* 32,1014(1961)

AL-0041b
ETCH NAME: TIME: 10 sec
TYPE: Acid, dislocation TEMP: 10°C
COMPOSITION:
2 ml .... HF
50 ml .... HNO_3
32 ml .... HCl
50 ml .... MeOH
DISCUSSION:
Al, (001) wafers. Following treatment show in Al-0041a, develop dislocation density with this solution — lithium created "bubbles" vary but are about equal in number to dislocation pits — all pits were on one side of wafer and showed movement after annealing at 575°C for 9 h.
REF: Ibid.

AL-0045
ETCH NAME: TIME: Seconds
TYPE: Acid, preferential TEMP: RT
COMPOSITION:
3 .... HF
47 .... HNO_3, fuming
80 .... HCl
DISCUSSION:
Al, (100) and other orientations used in a study of thermal etching. After thermal annealing at 460 to 610°C use solution shown as a dislocation etch. (*Note:* Red or yellow fuming, not shown.)
REF: Foss, D & Herbjornsen, O H — *Phil Mag,* 13,945(1966)
AL-0055a; Lacombe, P & Beaujard, L — *Inst Met J,* 132,1(1947—48)
Al, high purity specimens used in a study of etch figures and structure. Solution used at 10°C, $^1/_2$—1 min. If iron or other metal ions are in the HCl, it will affect controlled etching. With addition of water in the solution shown etch figures increase in number but decrease in size with water increase. The HCl produces primarily localized attack and HF uniformally dissolves aluminum.

AL-0044
ETCH NAME: TIME: 2 min
TYPE: Electrolytic, polish TEMP:
COMPOSITION: ANODE: Al
CATHODE:
POWER: 25—35 V/cm^2

78 ml .... $HClO_4$
700 ml .... EOH
120 ml .... H_2O
100 ml .... butyl cellusolve

DISCUSSION:

Al, (1000) specimens used in a study of surface etching effects of 8 KeV Ar^+ ion bombardment. Polish surfaces with solution shown before bombardment. Specimens of zinc and gold also were studied but etched in other solutions.

REF: Cunningham, R L et al — *J Appl Phys,* 31,839(1960)

AL-0016
ETCH NAME: TIME: 30—120 sec
TYPE: Acid, polish TEMP: 65°C
COMPOSITION:

25 ml .... H_2SO_4
70 ml .... H_3PO_4
5 ml .... HNO_3

DISCUSSION:

Al, and aluminum alloys. Used as a general cleaning and polishing etch. See AL-0035.

REF: Tegart, W J McG — *The Electrolytic & Chemical Etching of Metals,* Pergamon Press, London, 1956

AL-0046
ETCH NAME: TIME:
TYPE: Electrolytic, polish TEMP:
COMPOSITION: ANODE:
CATHODE:
POWER:

12 .... $HClO_4$
88 .... HAc

DISCUSSION:

Al, (001) wafers and other orientations. Solution used to polish specimens in a study of stress-induced desorption. Also studied molybdenum and magnesium.

REF: Fouerstein, S & John, I W — *J Appl Phys,* 40,3334(1969)

AL-0047
ETCH NAME: TIME:
TYPE: Electrolytic, thinning TEMP: 0°C (ice)
COMPOSITION: ANODE:
CATHODE:
POWER:

1 .... $HClO_4$
1 .... glycerol
7 .... EOH

DISCUSSION:

Al, single crystal wafers. Specimens etch thinned in this solution for a study of electrical resistivity of dislocations.

REF: Rider, J G & Foxon, T B — *Phil Mag,* 13,289(1966)

AL-0049
ETCH NAME:
TYPE: Electrolytic, polish
COMPOSITION:
20 .... $HClO_4$
80 .... HAc

TIME: 1—2 h
TEMP: 20°C
ANODE:
CATHODE:
POWER: 35 V & 500 mA/cm^2

DISCUSSION:
Al, single crystal specimens. Polish surfaces in this solution. Specimens were subjected to various environments in a study of surface etch pit development.
REF: Kasen, M K et al — *Phil Mag,* 13,453(1966)

AL-0048
ETCH NAME: Jacquet's etch
TYPE: Electrolytic, removal
COMPOSITION:
345 ml .... $HClO_4$
655 ml .... HAc

TIME: 30 min
TEMP:
ANODE:
CATHODE:
POWER:

DISCUSSION:
Al, single crystal specimens cut from an ingot that was slow cooled. Specimens were polished in this solution with a removal of about 30 μm per side. Used in a dislocation density study as related to slow cooling.
REF: Nes, E & Nost, B — *Phil Mag,* 13,855(1966)

AL-0026
ETCH NAME: Air
TYPE: Gas, oxidation
COMPOSITION:
x .... air

TIME: 15—48 h
TEMP: 550°C

DISCUSSION:
Al, single crystal sphere. Sphere electropolished prior to oxidation. No change in 15 h sitting in air; faint patterns after 48 h of oxidation. Developed foggy pole figures at axial points of (100); (110) and least on (111).
REF: Gwathmey, A T et al — *Mater Advisory Comm Aeronaut Tech Notes* #1460(1948)
AL-0050: Bedair, S M et al — *J Appl Phys,* 39,4026(1968)
Al, (100) wafers used in a study of oxidation. Forms an amorphous oxide film below 600°C. Above 600°C film becomes crystalline Al_2O_3.
AL-0052: Bond, H E & Harvey, K B — *J Appl Phys,* 34,440(1963)
Al, specimens. Work was done with synthetic sapphire that was polished on a maple lap with diamond dust in oil. With 1% by weight ZrO_2 in the material, dislocations are seen on the (0001) basal plane. Oxide growth on aluminum can cause subsurface damage and pitting where O_2 moves to the Al_2O_3/Al interface pitting the metallic aluminum.

AL-0053
ETCH NAME: Boron trichloride
TYPE: Gas, selective (RIE)
COMPOSITION:
x .... BCl_3

TIME:
TEMP:
GAS FLOW:
PRESSURE:
POWER:

DISCUSSION:
Al, thin films deposited on GaAs, (100) wafer substrates. Layered structures were: (1) Al/Ni/SiO_2 or (2) Al/Ni/SiN_x. Surfaces were photo resist processed for patterning to include

via-holes through the gallium arsenide in fabricating monomicrowave ICs. Both BCl_3 and CF_4:O_2 were used for removal of aluminum. Removal of nickel referred to as a nickel lift-off technique.
REF: Gelsgberger, A E & Claytor, P R — *J Vac Sci Technol,* A(3),863(1985)

AL-0054
ETCH NAME: Boron tribromide
TYPE: Gas, removal (RIE)
COMPOSITION:
x .... BBr_3

TIME:
TEMP:
GAS FLOW: 40 cc
PRESSURE:
POWER: 125 W

DISCUSSION:

Al, and Al_2O_3/AlN thin films. The oxide was formed by anodization of aluminum deposited on SST plates or aluminum evaporated on silicon (100) wafers containing an SiO_2 thin film: Al/SiO_2/Si(100) structure, which was given a 1 min treatment in nitrogen prior to and after BBr_3 etching. RIE ionized gas etching used in studying etch reactions of aluminum oxide, aluminum nitride and aluminum.
REF: Landaner, A & Hess, D W — *J Vac Sci Technol,* A(3),962(1985)
AL-0056: Jacquet, P — *Met Corros,* 18,198(1943)

Al, specimens. A polish etch for aluminum. Use at 22—25 V and 0.8—2.5 A/cm^2 or 3—4 A/cm^2. Agitate during etching to prevent irregularities developing as the degree of polish affects development of etch figures. Should be preferentially etched immediately after cleaning surfaces to prevent growth of aluminum oxide and its masking effects.
AL-0059b: Ibid.

Al, specimens. By raising the current of this solution toward the end of the etch period both aluminum and copper can be etched.

AL-0076
ETCH NAME: Abrasive
TYPE: Abrasive, polish
COMPOSITION:
x .... abrasive, paper

TIME: 30—60 min
TEMP: RT

DISCUSSION:

Al, specimens and alloys. Mechanical polish of specimens for metallographic study. Cement an abrasive cloth in a bowl and use an electromagnet for vibration. (*Note:* This is similar to sphere grinding methods using a cylinder "race track" with nitrogen gas as the movement agent against abrasive paper.)
REF: Krill, F M — *Met Abstr,* 8,641(1958)
AL-0080: Krill, F M — *Met Prog,* 700,81(1956)

Al, specimens and alloys, similar work as in AL-0076.

AL-0077
ETCH NAME:
TYPE: Acid, machining
COMPOSITION:
4 mg .... NaOH
1000 ml .... H_2O
200 ml .... glycerin

TIME:
TEMP:

DISCUSSION:

Al, specimens. A method of chem/mech etching and forming parts. System has an abrasive wheel above an etch tank with wheel working-edge vertical. Specimen is held against the wheel as it is rotated through the etch solution in the tank below. (*Note:* This is

similar to using a grinding wheel with a liquid coolant.)
REF: Fegredo, D M & Greenough, O B — *J Inst Met,* 87,1(1958—59)
AL-0078: Yamamoto, M & Watanabe, J — *Sci Rep Res Inst Tohoku Univ Ser. A,* 8,230(1956)
Al, specimens. A report on acid saw cutting of materials.

AL-0081
ETCH NAME: TIME: 15 sec
TYPE: Acid, cleaning TEMP: RT
COMPOSITION:
2 .... HF
3 .... HNO_3
9 .... HCl
5 .... H_2O
DISCUSSION:
Al material used as the metal source in growing AlGaAsP single crystals. Aluminum was cleaned in the solution shown before melting with GaAs, GaP, and GaAsP material to produce aluminum doped single crystal ingots.
REF: Fujimoto, A — *Jpn J Appl Phys,* 22,109(1984)

AL-0082a
ETCH NAME: BRM TIME:
TYPE: Halogen, removal TEMP: Warm
COMPOSITION:
x .... 10% Br_2
x .... MeOH
DISCUSSION:
Al foil with an Al_2O_3 thin film grown in a solution of 40% $NH_4B_5O_8.4H_2O$:60% ethylene glycol at RT. The BRM solution was used to dissolve the foil leaving oxide flakes. The oxide also was stripped from foil with 5% H_3PO_4:3%CrO_3 (ALO-0020). Aluminum foil was used in a study of AC etching.
REF: Dyer, C K & Alwitt, R S — *J Electrochem Soc,* 128,300(1981)

AL-0082b
ETCH NAME: Hydrochloric acid TIME:
TYPE: Electrolytic, preferential TEMP: 60°C
COMPOSITION: ANODE: Al
x .... 1 *M* HCl CATHODE: Carbon w/25 mA/cm^2 + 5 Hz sinusoidal current applied
POWER: rms 700 mA/cm^2
DISCUSSION:
Al, foil used in a study of change with AC etching. The solution produces a high density of cubic pits. After etching, rinse in DI water, then MeOH. Pre-treat foil before electrolytic etching with x%NaOH, 10 min and follow with heavy water rinse.
REF: Ibid.

AL-0083a-b
ETCH NAME: TIME: 40 min
TYPE: Acid, cutting TEMP: RT
COMPOSITION:

(1) 200 ml .... HCl
50 g .... $FeCl_3$
250 ml .... H_2O

(2) 1 .... HNO_3
1 .... H_2O

DISCUSSION:

Al, specimens $^1/_8''$ thick. Terylene thread soaked in solution (1) used to cut aluminum, brass, and copper. Solution (2) used to cut tin and zinc. (*Note:* Excellent review of metal growth methods.)

REF: Honeycombe, R W K — *Met Rev,* 4,1(1959)

AL-0019b
ETCH NAME: TIME:
TYPE: Acid, removal TEMP:
COMPOSITION:

(1) x .... H_3PO_4
x .... HNO_3
x .... HAc

(2) x .... x% KOH
x .... $K_3Fe(CN)_6$
x .... $K_2B_4O_7.4H_2O$

DISCUSSION:

Al thin film on (100) silicon wafers. Solutions used for pattern etching aluminum with photolithography. Solution (1) will cause COP type photo resists to lift; solution (2) will not affect the resist.

REF: Reynolds, R — *J Vac Sci Technol,* 15,962(1978)

AL-0089
ETCH NAME: RIE
TYPE: Gas, removal
COMPOSITION:

(1) x .... $SiCl_4$
(2) x .... $SiCl_4/O_2$
(3) x .... $BaCl_3/O_2$
(4) x .... O_2

TIME:
TEMP:
GAS FLOW: 24 SCCM
PRESSURE: 75 mTorr (10 Pa)
ANODE:
CATHODE: −180 V
POWER: 0.26 W/cm^2

DISCUSSION:

Al thin films deposited on silicon substrates using tungsten as an evaporation film mask. A study of sputter etching of aluminum thin films. Gas mixtures (2) and (3) with O_2 plasma develop SiO_2 films as does pure O_2 (4). Etch rate of the SiO_2 and W films are approximately equal. $SiCl_4$ etch rate was 700 Å/min for aluminum.

REF: Degenkolb, E O — *J Electrochem Soc,* 129,1150(1982)

AL-0088
ETCH NAME: Acetone TIME: Variable
TYPE: Ketone, lift-off TEMP: RT to warm
COMPOSITION:

x .... CH_3COCH_3

DISCUSSION:

Al thin films evaporated on glass substrates for four different thicknesses as step-evaporation through photo resist mask patterns. A study of aluminum structure and reactions.

Acetone used to lift-off excess aluminum and photo resist after evaporation.
REF: Mizuno, K — *J Phys Soc Jpn,* 4,1434(1984)

AL-0092
ETCH NAME: TIME:
TYPE: Gas, removal TEMP:
COMPOSITION:
(1) x .... Br_2 (2) x .... Cl_2 (3) x .... CCl_4 (4) x .... CBr_4
DISCUSSION:
Al thin films evaporated on quartz substrates in a study of etching aluminum thin films by E-beam sputter in ultra high vacuum (UHV).
REF: Park, S et al — *J Vac Sci Technol,* A(2),790(1985)

AL-0087
ETCH NAME: TIME:
TYPE: Electrolytic, polish TEMP:
COMPOSITION: ANODE:
CATHODE:
POWER:
DISCUSSION:
Al as high purity, dislocation free specimens used in a general study of such material. Solution used not shown.
REF: Howe, S & Elbaum, C — *J Appl Phys,* 32,742(1961)
AL-0090: Elbaum, C — *J Appl Phys,* 31,1413(1960)
Reference for electropolish solution used in AL-0087.

AL-0091
ETCH NAME: TIME:
TYPE: Acid, thinning TEMP:
COMPOSITION:
DISCUSSION:
Al specimens chemically etch thinned for study of dislocations by transmission electron microscopy (TEM). Thinning solution not shown.
REF: Robinson, D L — *J Appl Phys,* 29,1635(1958)

AL-0093
ETCH NAME: Sodium hydroxide TIME:
TYPE: Alkali, removal TEMP:
COMPOSITION:
x .... 0.2 *M* NaOH
\+ .... light
DISCUSSION:
Al thin films evaporated on SiO_2, Al_2O_3 and ZrO_2 substrates in a study of surface protection against chemical attack. Also studied silver thin films, TiN, Si:H and PbS. Silver thin films removed with 1 *M* HNO_3 + light.
REF: Martin, P J et al — *J Vac Sci Technol,* A(2),341(1984)

AL-0094
ETCH NAME:
TYPE: Electrolytic, selective
COMPOSITION:
6.4 g .... KI
0.32/l .... MeOH

TIME: 30 sec
TEMP:
ANODE: Al
CATHODE: Al
POWER: 0.3 A/mm^2 & 30 V

DISCUSSION:
Al alloys as sheet, plate and rod used in a study of precipitates in the material (alloys #3003, #5182, #2124, #7475). Prepare surfaces by lap polish using 1 μm diamond paste, then selectively etch to expose precipitates for SEM or TEM observation. In alloy #5182, dispersoids were Mn and Mg_2Si; in #2124, Al_2CoMg; in #7475, Cr.
REF: Hower, J — *Metallography (MEIJAP),* 15,247(1983)

AL-0095
ETCH NAME: Water
TYPE: Acid, float-off
COMPOSITION:
x .... H_2O

TIME:
TEMP: RT

DISCUSSION:
Al, evaporated on KCl, (100) and (111) cleaved substrates as oriented thin films. Some films were converted to AlN by N^+ ion implantation. Al on (111) KCl was hexagonal wurtzite (fcc) structure; on (100) was cubic (100). Water was used to liquefy the KCl/Al interface for film float-off and study by TEM. (*Note:* AlN is hexagonal system.)
REF: Kimura, K et al — *Jpn J Appl Phys,* 23,1145(1984)

AL-0096
ETCH NAME:
TYPE: Mechanical, deformation
COMPOSITION:
x .... pressure

TIME:
TEMP:

DISCUSSION:
Al specimens used in a study of surface deformation in aluminum due to fatigue.
REF: Grooskreutz, J C & Gosselin, C M — *J Appl Phys,* 31,1127(1960)

AL-0098
ETCH NAME: Hydrochloric acid
TYPE: Electrolytic, forming
COMPOSITION:
x .... HCl

TIME:
TEMP: Elevated
ANODE:
CATHODE: Pb
POWER:

DISCUSSION:
Al specimens as Al:Si(20%) alloyed with the following metals: Na, Mg, Zn, Cr, Mn, Cu, Cd, Sn, Pb, Sb, Bi, Fe, Ni, and Co. Specimens were electrolyzed in HCl to form single crystal silicon in the anodic slime. Three silicon crystal structures were obtained: (1) with Na, as (111) and (100) granules; (2) with Mg or Mn, as prismatic crystals (111) and (110): and (3) as plate-like (111) with negative (111) predominate with all other metal additives. A study of the crystal structure of silicon in aluminum-silicon alloys. (*Note:* Pure aluminum alloyed into silicon to form p-n junctions produces an Al:Si mixture containing needle to blade-like silicon elements in a fine grained aluminum matrix.)
REF: Obinata, I & Komatso, N — *Metall Abstr,* 8,94(1958)

AL-0100a
ETCH NAME: Abrasives TIME:
TYPE: Mineral, polish TEMP: RT
COMPOSITION:
x .... Al_2O_3, powder
x .... H_2O
DISCUSSION:

Al, and alloys. Solution for general cleaning and polishing. Alumina, as fine ground powder in a water slurry with or without glycerin: (1) as sub-micron grits, Linde A and B; or (2) as alumina in a colloidal suspension of water. With addition of a 1—5% KOH solution these become an acid slurry, and slurry particles of alumina slowly reduce to a fine paste during use. Linde abrasives used with and without KOH for aluminum parts surface cleaning and polishing. Both colloidal alumina and silica for fine surface polishing. Used on aluminum and aluminum alloy specimens in material studies and in cleaning vacuum parts.
REF: Walker, P & Menth, M — Studies/Vacuum Parts (1981—1985)

AL-0101b
ETCH NAME: Water TIME:
TYPE: Acid, oxidation TEMP: Hot to boiling
COMPOSITION:
x .... H_2O
DISCUSSION:

Al specimens. DI water used to form a hydrated oxide on aluminum, then remove with HF or BHF solution for surface cleaning. Used in preparing aluminum slugs prior to use in metal evaporation, and for general parts cleaning of aluminum and aluminum alloys.
REF: Ibid.

ALUMINUM ALLOYS, AlM

General: Aluminum does not form metal alloys in nature and does not occur as a native element, although there are well over 200 single crystal minerals containing aluminum as oxides, silicates, phosphates, and sulfates. There also are mineral hydrates.

As artificial aluminum alloys, in the general formula shown above, M = Au, Ag, Be, Zn, and other metal elements. Pure aluminum is soft, like pure copper and gold, such that it is not used as the pure metal for construction parts. Several alloys have specific names, such as aluminum-silicon #43 (5% Si); magalum, with 10% Mg; and there are "zeppelin" braces, angles and channels containing small percentages of Si, Fe, Sn, Mn, and Zn. Beryllium-aluminum, containing 38% Be is one of the lightest known alloys with a density (lb/in^3) of 0.075 g/cm^3 as compared to pure beryllium, 0.067 g/cm^3, and pure aluminum, 0.98 g/cm^3. Aluminum-bronze is another type of alloy. There are both high temperature aluminum brazing alloys and low temperature solders. It is worth noting that aluminum oxide is very difficult to braze or alloy itself, due to the presence of the inert oxide layer that is always present on surfaces after being exposed to air.

Technical Application: Both pure aluminum and aluminum-silicon(5%) are used as a p-type dopant sources on silicon, and evaporated contact pads are still in use for alloyed silicon diodes and transistor devices, such as the silicon Sphere Alloy Zener process that uses Al:Si(5%) pellets. Silicon SCRs, when fabricated with an aluminum pre-form, use aluminum contact wires. Al:Be(38%) has been used as an evaporated electrode on quartz crystal frequency blanks to reduce mass loading effects. Al_x:Au_y alloys produce "purple plague", as a brittle compound that will eventually break and destroy electrical continuity in devices, although it is used as a purple gold alloy for jewelry.

Aluminum alloy materials are basic construction elements in much Solid State processing equipment — electrical transfer plates; cooling plates; lapping plates; high power feed-thrus in vacuum systems; RF shielding; test blocks and many other similar applications. In combination with other metals and ceramics, as package material for discrete devices as well as circuit assemblies for both minimum weight and radiation protection. There are several aluminum alloy semiconductor compounds, such as aluminum arsenide, AlAs; aluminum antimonide, AlSb; and aluminum phosphide, AlP. In addition to compound semiconductors, several other aluminum alloys have been grown as single crystals for general morphological study, such as Al:Au, Al:Ag, Al:Cu, Al:Ni, Al:Zn etc.

Etching: Soluble in HCl, H_2SO_4, alkalies, and mixed acids.

ALUMINUM ALLOYS

ALUMINUM:BERYLLIUM ETCHANTS

ALBE-0001

ETCH NAME: Potassium hydroxide — TIME: 2—3 min

TYPE: Alkali, cleaning — TEMP: 30—40°C (warm)

COMPOSITION:

x .... 30% KOH (NaOH)

DISCUSSION:

Al:Be(38%), as 0.060 diameter polycrystalline wire. Wire was etch cleaned in this solution followed by heavy water washing and drying under an IR heat lamp prior to use as an evaporation source for metal contact electrodes on quartz radio frequency crystals. Evaporation was done with "U" shaped clips of wire hung on tungsten coils and, after evaporation, the system was pumped to the 10^{-7} Torr range for 20—30 min before opening the bell jar to eliminate beryllium vapors. Material was used in evaluating improved frequency stability and levels due to reduced mass loading effects.

REF: Walker, P — personal development, 1969

ALUMINUM:CERIUM ETCHANTS

ALCE-0001

ETCH NAME: Potassium hydroxide — TIME:

TYPE: Alkali, removal — TEMP: Hot

COMPOSITION:

x .... 30% KOH

DISCUSSION:

$CeAl_3$ single crystals grown by arc melting and aging at temperature. Used in a material growth study. Also developed $CeCu_6$ and $LaCu_6$. Growth was as Czochralski (CZ) ingots. Included studies on CeB_6 and $CeCu_2Si_2$. No etch shown. (*Note:* Solution shown is for general aluminum alloy etching.)

REF: Onuki, Y et al — *J Phys Soc Jpn,* 4,1210(1984)

ALUMINUM:COPPER ETCHANTS

ALCU-0001a

ETCH NAME: Sodium hydroxide — TIME: 30 min

TYPE: Alkali, removal — TEMP:

COMPOSITION:

x .... 10% NaOH

DISCUSSION:

Al:Cu, single crystal specimens used in a study of particle dispersion due to hardening. After lap/polish of surfaces, specimens were etched in this solution to remove work damage.

Keller's reagent also was used as the etch.
REF: Dew-Hughes, D & Robertson, W D — *Acta Metall,* 8,147(1960)

ALCU-0002
ETCH NAME: TIME:
TYPE: Acid, macro-polish TEMP: Boiling
COMPOSITION:
5 .... HNO_3
15 .... CH_3COOH (HAc)
80 .... H_3PO_4
DISCUSSION:
Al:Cu, single crystal specimens used in a study of the mechanisms of hardening and aging. The solution will polish surfaces and develop slip lines.
REF: Dew-Hughes, D & Robertson, W D — *Acta Metall,* 8,156(1960)

ALCU-0003
ETCH NAME: TIME: 30 sec to 2 min
TYPE: Acid, polish TEMP: 85°C
COMPOSITION:
25 ml .... H_2SO_4
70 ml .... H_3PO_4
5 ml .... HNO_3
DISCUSSION:
Al:Cu alloy specimens. Used as a polish etch on Al:Cu; Al:Si, and Al.
REF: Heregue, J & Segond, R — *Rev Metall,* 48,262(1951) (in French)

ALCU-0001b
ETCH NAME: Keller's reagent TIME:
TYPE: Acid, removal/polish TEMP:
COMPOSITION: ASTM: #3
2 ml .... HF
3 ml .... HCl
5 ml .... HNO_3
190 ml .... H_2O
DISCUSSION:
Al:Cu, single crystal specimens used in a study of particulate dispersion in hardening. Solution used to remove work damage after mechanical lap and polish.
REF: Ibid.
ALCU-0004: ASTM E407-70
Reference for ASTM number for Keller's reagent.

ALUMINUM:GOLD ETCHANTS

ALAU-0001
ETCH NAME: TIME:
TYPE: Electrolytic, structure TEMP:
COMPOSITION: ANODE:
x .... $HClO_4$ CATHODE:
x .... EOH POWER:
DISCUSSION:
Al:Au(2%) specimens. Solution used to develop precipitates in this alloy. Before electrolytic etching, remove native oxide with etch solution shown under ALO-0014 under

Aluminum Oxide Etchants. Also refers to "Lenore's Solution" referencing Thomas, O — *Transmission Electron-Microscopy of Metals,* John Wiley & Sons, New York, 1962, 161 (*Note:* This alloy is a form of purple gold used in the jewelry trade.)
REF: Von Heimendahl, M — *Acta Metall,* 15,1441(1967)

ALAU-0002
ETCH NAME: Aqua regia, dilute
TYPE: Acid, removal
COMPOSITION:
3 .... HCl
1 .... HNO_3
10 .... H_2O

TIME:
TEMP: RT

DISCUSSION:

Al:Au, alloy reaction product observed on the side-walls of high power silicon diodes that had been aluminum evaporated and gold evaporated on opposed wafer surfaces prior to dicing. Reaction appeared after power burn-in as a failure mode, e.g., "purple plague". Solution was used in a series of short-period etch-steps in the initial studies of these brittle compounds. This was the initial observation of this failure mode.
REF: Walker, P — personal application, 1958
ALAU-0003: Bernstein, L — personal communication, 1958

Al:Au thin films with various single crystal structures or as amorphous reactive films. Additional study of purple plague and effects on device failure. Several articles written on this subject (1959—1962).

ALUMINUM:NICKEL ETCHANTS

ALNI-0001a
ETCH NAME: Sodium hydroxide
TYPE: Alkali, removal
COMPOSITION:
x .... 2.5 *M* NaOH

TIME:
TEMP:
RATE: 0.005 $mg/cm^2/h$

DISCUSSION:

Al:Ni, alloy thin films deposited by electrodeposition or by nickel diffusion into aluminum. Solution used for general removal.
REF: Cough, D E & Connor, J H — *J Electrochem Soc,* 107,272(1960)

ALNI-0001b
ETCH NAME:
TYPE: Fused salt, removal
COMPOSITION:
400 g .... NaCl
560 g .... KCl
150 g .... Na_3AlF_6 (Cryolite)

TIME:
TEMP: 700—800°C
ANODE: Graphite
CATHODE:
POWER: 2—10 A/cm^2

DISCUSSION:

Al:Ni, alloy thin films. A form of molten flux etching and cleaning of high temperature melting point metals and alloys. Cryolite, Na_3AlF_6, is a natural mineral used in processing aluminum ore as a primary reducing flux.
REF: Ibid.

ALNI-0001c
ETCH NAME:
TYPE: Fused salt, removal
COMPOSITION:
99 g .... $AlCl_3$
200 g .... NaCl
TIME:
TEMP: 160—180°C
ANODE: Carbon or tungsten
CATHODE:
POWER: 1—4 A/cm^2
DISCUSSION:

Al:Ni, alloy thin film coatings. Crucible and cover were Pyrex, rather than graphite as in ALNI-0001a. Method used to etch the alloy.

REF: Ibid.

ALNI-0002
ETCH NAME:
TYPE: Acid, defect
COMPOSITION:
x .... 10% $HClO_4$
x .... butyl cellusolve
TIME:
TEMP: 5°C
DISCUSSION:

$AlNi_2$, single crystal specimens used in a study of phase change with recrystallization. Solution used to develop structure.

REF: Calvagrac, Y & Fayard, M — *Acta Metall,* 14,783(1966) (in French)

ALUMINUM:SILICON ETCHANTS

ALSI-0001a
ETCH NAME: Hydrochloric acid
TYPE: Acid, cleaning
COMPOSITION:
(1) 1 .... HCl
1 .... H_2O
(2) 1 .... HCl
10 .... H_2O
TIME: 10—30 sec
TEMP: RT
DISCUSSION:

Al:Si(5%), foil pre-form and spherical pellets. Both solutions have been used to clean parts before furnace alloying of silicon (111) wafers in a hydrogen atmosphere or in forming gas (85% N_2/15% H_2). The alloy produces less stress in the silicon than pure aluminum and reduces the formation of dislocations along the alloy front. Solution (2) preferred for light cleaning after alloying.

REF: Walker, P — personal application, 1966

ALSI-0002
ETCH NAME:
TYPE: Acid, polish
COMPOSITION:
25 ml .... H_3PO_4
70 ml .... H_2SO_4
5 ml .... HNO_3
TIME: 30 sec to 2 min
TEMP: 85°C
DISCUSSION:

Al:Si specimens. Solution used as a polish etch on Al:Si, Al:Cu, and pure aluminum.

REF: Heregue, J & Segond, R — *Rev Met,* 48,262(1951) (in French)

ALSI-0003
ETCH NAME: TIME:
TYPE: Acid, preferential TEMP:
COMPOSITION:
5 .... HCl
10 .... HNO_3
84 .... H_2O
DISCUSSION:

Al:Si, as an alloy solid solution. Solution used to study the orientation re-growth of silicon in the aluminum-rich solid. Silicon showed a preferred precipitate (111) and (100) orientation.

REF: Rosenbaum, H S et al — *Acta Metall,* 7,678(1959)

ALSI-0004
ETCH NAME: TIME:
TYPE: Acid, selective TEMP: RT
COMPOSITION:
1 .... HF
3 .... HNO_3
50 .... HAc
DISCUSSION:

Al:Si, alloy re-growth p-n junctions in n-type silicon wafers, 3—10 Ω cm resistivity. After aluminum evaporation and alloying into emitter and collector etched pits, excess aluminum was hand-lap removed to define circular pit areas. The silicon wafer or individual die were etched with the solution shown to remove aluminum and expose the internal alloy front in a study of spiking defects in the bulk silicon caused by aluminum alloying.

REF: Walker, P & Waters, W P — device development, 1957

ALSI-0005
ETCH NAME: Hydrofluoric acid TIME:
TYPE: Acid, micro-etch TEMP: RT
COMPOSITION:
x .... 0.5% HF
DISCUSSION:

Al:Si(12%) alloy subjected to different processing modifications. Solution was used to develop the various structure alterations observed.

REF: A/B *Met Dig,* 21(2),19(1983) — Buehler Ltd
ALSI-0007: Hume-Rothery, W — *Inst Met Proc,* 46,239(1931)

Al:Si material. A general study of macro-etching of alloys.

ALSI-0006
ETCH NAME: RF plasma TIME:
TYPE: Ionized gas, selective removal TEMP:
COMPOSITION: GAS FLOW:
x .... CF_4 PRESSURE: 1.7 Pa (Argon)
x .... O_2 POWER: 0.5 W/cm^2
RATE: 100 A/min @ 200°C
DISCUSSION:

AlSi, thin film layers as interconnects on silicon devices along with phosphorus doped silicon dioxide [(PSG) phosphorus silica glass]; p-SiN and AlSiCu as multilayer structures.

RF plasma etching was used in multilevel structuring of these electrical interconnects.
REF: Kotani, H et al — *J Electrochem Soc,* 130,645(1983)

ALSI-0001b
ETCH NAME: Hydrogen — TIME:
TYPE: Gas, cleaning — TEMP: 500°C
COMPOSITION:
x .... H_2
DISCUSSION:

AlSi(5%) spheres .010 diameter used to fabricate Sphere Alloy Zener diodes. The silicon wafers were held in a graphite assembly for furnace alloying, and the spheres were hydrogen fired as a cleaning step immediately before alloying. After alloying, individual diodes in a hexagonal array were measured for leakage current and plotted to determine segregation and defect areas in the silicon wafer. Results used to improve Float Zone (FZ) ingot growth of zener material.

REF: Ibid.

ALSI-0001c
ETCH NAME: — TIME: 2—3 sec
TYPE: Acid, selective — TEMP: RT
COMPOSITION:
1 .... HF
10 .... HNO_3
5 .... H_2O
DISCUSSION:

Si, (111) wafers used to alloy Al:Si(5%), .010 diameter spheres as Sphere Alloy Zener diodes. After alloying, prior to testing and dicing, wafers were lightly etched in the solution shown to optimize electrical parameters.

REF: Ibid.

ALSI-0001d
ETCH NAME: Forming gas — TIME:
TYPE: Gas, alloying — TEMP: 800°C
COMPOSITION:
x .... 85% N_2
x .... 15% H_2
DISCUSSION:

Al:Si(5%) and ''quad-alloy'' (Al:Si:Ag:In) flat preforms about 10 mil thick used in assembly of silicon SCRs. Belt furnace used to alloy pre-forms to moly discs prior to alloying SCR wafers. Both forming gas (FG) and pure hydrogen were used with a graphite heater strip to strip/remove remaining metal alloy from moly discs for re-use of discs.

REF: Ibid (1974)

ALUMINUM:SILVER ETCHANTS

ALAG-0001
ETCH NAME: — TIME:
TYPE: Alkali/acid, removal — TEMP:
COMPOSITION:
(1) x .... 3 *N* NaOH — (2) x .... HNO_3

DISCUSSION:

Al:Ag, as both polycrystalline and single crystal ingots. Wafers were cut and lapped, then etched in solution (1), followed by washing in solution (2). The material was used in a study of precipitates in cold-worked specimens. Both Al:Ag and Al:Zn alloys were studied.
REF: Jan, J P — *J Appl Phys,* 26,1291(1955)

ALUMINUM:ZINC ETCHANTS

ALZN-0001
ETCH NAME: TIME:
TYPE: Alkali/acid, removal TEMP:
COMPOSITION:
(1) x .... 3 *N* NaOH (2) x .... HNO_3
DISCUSSION:

Al:Zn, as both polycrystalline and single crystal ingots. Wafers were cut and lapped, then etched in solution (1), followed by washing in solution (2). The material was used in a study of precipitates in cold-worked specimens. Both Al:Ag and Al:Zn alloys were studied.
REF: Jan, J P — *J Appl Phys,* 26,1291(1955)

ALZN-0002
ETCH NAME: TIME:
TYPE: Electrolytic, polish TEMP: 30°C
COMPOSITION: ANODE: Al:Zn
3% .... $HClO_4$ CATHODE:
30% .... butoxyethanol POWER: 25 V
5% .... MeOH
DISCUSSION:

Al:Zn(10%), alloy sheet. Blanks were punched from sheet and then annealed. A twin-jet system was used to electropolish specimens.
REF: Nicholls, A W & Jones, L P — *J Phys Chem Solids,* 44,696(1983)

ALUMINUM:ZINC:COPPER ETCHANTS

AZC-0001
ETCH NAME: Heat TIME:
TYPE: Thermal, preferential TEMP:
COMPOSITION:
x .... heat
DISCUSSION:

Al:Zn,Cu, alloy specimens were homogenized by annealing and aging. Cubic shaped spiral etch patterns observed parallel to the cubic axis after thermal annealing. (*Note:* The z-axis direction as ⟨001⟩.)
REF: Bulnov, W N & Shehegolrva, T V — *Fiz Metalloved,* 5,566(1957)

ALUMINUM IRON ETCHANTS

ALFE-0001
ETCH NAME: TIME:
TYPE: Solvent, cleaning TEMP:
COMPOSITION:
80 .... acetone
20 .... toluene

DISCUSSION:

AlFe alloy specimens used for selective oxidation. Degrease specimens in solution shown before oxidizing.

REF: Grace, R E & Seybolt, A U — *J Electrochem Soc,* 106,582(1958)

PHYSICAL PROPERTIES OF ALUMINUM ANTIMONIDE, AlSb

Classification	Antimonide
Atomic numbers	13 & 51
Atomic weight	148.73
Melting point (°C)	600—700
Boiling point (°C)	
Density (g/cm³)	4.67
Hardness (Mohs — scratch)	3—4
Crystal structure (isometric — tetrahedral)	(111) tetrahedron
Color (solid)	Silver grey
Cleavage (dodecahedral)	(110)

ALUMINUM ANTIMONIDE, AlSb

General: Does not occur as a natural compound, although there are other metal antimonides of copper, nickel, and silver as mineral species, as well as mixed metal antimonides. There is no application of the compound in the metal industries, at the present time, but some use in the fabrication of glass and silicates.

Technical Application: The material is a III—V compound semiconductor that has been grown by the Horizontal Bridgman (HB), and Czochralski (CZ) methods. Like the arsenide, the antimonide oxidizes readily when exposed to moisture in air. Both materials are processed and held under liquid solvents or in inert gas atmospheres to prevent oxidation. It also has been deposited as a thin film layer by CVD. As a thin film layer element of a device, it has been fabricated as solar cells.

Etching: Mixed acids of $HF:HNO_3$, $HF:H_2O_2$, and $HCl:HNO_3$.

ALUMINUM ANTIMONIDE ETCHANTS

ALSB-0001a

ETCH NAME: TIME:

TYPE: Acid, polish TEMP:

COMPOSITION:

x .... H_2O_2

DISCUSSION:

AlSb as cut wafers. All processing and handling of aluminum antimonide should be under an inert, dry atmosphere such as argon or nitrogen, as this compound is attacked by atmospheric moisture. Mechanical lap and polish with abrasive slurry under oil. Etch polish with the solution shown.

REF: Herczog, A et al — *J Electrochem Soc,* 105,533(1958)

ALSB-0001b
ETCH NAME: TIME:
TYPE: Acid, polish TEMP:
COMPOSITION:
x .... HF
x .... HNO_3
x .... CH_3COOH (HAc)
DISCUSSION:
AlSb as cut wafers. Solution used to polish wafers. See additional discussion under ALSB-0001a.
REF: Ibid.

ALSB-0002a
ETCH NAME: TIME: 1 min
TYPE: Acid, polish TEMP: RT
COMPOSITION:
4 .... HCl
4 .... HNO_3
1 .... H-tartaric acid
DISCUSSION:
AlSb wafers. Used as a polish etch in studying copper diffusion in aluminum antimonide. Follow etching with a 2—3 sec dip in 1HCl:1HNO_3 to remove smut residue left on surfaces after etching.
REF: Wieber, R H et al — *J Appl Phys,* 31,600(1960)

ALSB-0003
ETCH NAME: TIME: 5 min
TYPE: Acid, polish TEMP:
COMPOSITION:
1 .... HNO_3
1 .... HCl
DISCUSSION:
AlSb wafers. Wafers were mechanically lapped with $303^1/_2$ abrasive grit on an iron lap with kerosene as the liquid carrier, before etch polishing with the solution shown. Material used in a study of elastic constants.
REF: Bulef, D I & Menes, M — *J Appl Phys,* 31,46(1960)
ALSB-0002b: Wieber, E H — *J Appl Phys,* 31,608(1960)
AlSb, wafers used in a study of copper diffusion in the material. Solution used at RT, 2—3 sec as a final cleaning/polish etch.

ALSB-0001c
ETCH NAME: TIME:
TYPE: Acid, polish TEMP:
COMPOSITION:
x .... HF
x .... HNO_3
DISCUSSION:
AlSb wafers. Various concentrations of HF and HNO_3 are polish solutions for this material. See additional discussion under ALSB-0001a.
REF: Ibid.

ALSB-0006
ETCH NAME: TIME:
TYPE: Acid, macro-etch TEMP:
COMPOSITION:
1 .... HCl
1 .... HNO_3
DISCUSSION:

AlSb wafers. The referenced author describes this as a macro-etch of aluminum antimonide (See Bulef & Menes ALSB-0003). Fibers of Al_3Ta, 1—5 µm in diameter, were observed in the AlSb after etching as Al_3Ta is insoluble in this etch solution. Under high magnification the surface etch pits are similar to those observed in other compound semiconductors.

REF: Gorton, H C — *J Electrochem Soc,* 107,248(1960)

ALSB-0007
ETCH NAME: TIME:
TYPE: Acid, stain TEMP:
COMPOSITION:
10 g .... $FeCl_3$
100 ml .... HCl
1000 ml .... H_2O
DISCUSSION:

AlSb/GaSb wafers from an ingot grown as a solid solution alloy single crystal. Wafers were lapped mechanically under anhydrous benzene. The acid solution shown was used to etch stain surfaces in a structure study.

REF: Miller, J F et al — *J Electrochem Soc,* 107,527(1960)

ALSB-0008a-b
ETCH NAME: TIME: (1) 1 min (2) 2 sec
TYPE: Acid, preferential TEMP: (1) 25°C (RT) (2) 25°C (RT)
COMPOSITION:

(1) 1 .... HF
1 .... H_2O_2
1 .... H_2O

(2) 1 .... HNO_3
1 .... HCl

DISCUSSION:

AlSb, (111) wafers. Solutions used for preferential etching to develop etch pits and other structures in a general study of aluminum antimonide growth structures. Rotation of tetrahedral etch pits show polarity between (111)A and $(\overline{111})$B surfaces of wafers.

REF: Gatos, H C & Lavine, M C — *J Electrochem Soc,* 107,427(1960)

ALSO-0009
ETCH NAME: TIME:
TYPE: Acid, polish/removal TEMP:
COMPOSITION:
1 .... HNO_3
1 .... HCl
DISCUSSION:

AlSb wafers, as grown p-n junction diodes. Etch was used to develop dendritic filaments of Al_3Te_2 (?). Filaments were 1—5 µm long and parallel to the growth direction of the ingot. (See ALSB-0003).

REF: Gorton, H C — *J Electrochem Soc,* 107,248(1960)

ALSB-0010a-b

ETCH NAME: TIME:

TYPE: Acid, cleaning TEMP:

COMPOSITION:

(1) 95% .... H_3PO_4
5% .... HNO_3

(2) 30 .... HF
50 .... HNO_3
30 .... HAc
0.5 .... Br_2

DISCUSSION:

AlSb, ingot material preparation. Both solutions used to remove oxides and for general cleaning. Solution (1) used on aluminum: solution (2) (CP4) on antimony. Both materials rinsed in DI water after etching prior to mixing for ingot growth.

REF: Allred, W P et al — *J Electrochem Soc,* 107,117(1960)

PHYSICAL PROPERTIES OF ALUMINUM ARSENIDE, AlAs

Classification	Arsenide
Atomic numbers	13 & 33
Atomic weight	100.98
Melting point (°C)	800—900
Boiling point (°C)	
Density (g/cm)	
Hardness (Mohs — scratch)	6—7
Crystal structure (isometric — tetrahedral)	(111) tetrahedron
Color (solid)	Grey silver
Cleavage (dodecahedral)	(110)

ALUMINUM ARSENIDE, AlAs

General: Does not occur as a natural compound, although there are four minerals as arsenates of aluminum with iron, manganese, or copper, all of which are hydrated. There are natural mineral arsenides of bismuth, cobalt, copper, iron, nickel, and platinum. There is no use of the compound in the metal industries though as an artificial compound or as the individual elements they are used as additives in the glass industry.

Technical Application: The material is a III—V compound semiconductor, and has been grown by the Horizontal Bridgman (HB), Czochralski (CZ) methods, or from a Molten Flux (MF), to include a mixed GaAs:AlAs crystal. The material is hygroscopic to the degree that it will react with moisture when exposed to air, oxidizing rapidly. The oxide then forms a stable passivating surface film. To prevent oxidation, wafers are processed and held under inert atmospheres, such as nitrogen, argon, or helium.

As a Solid State device, the compound has been fabricated as electroluminescent diodes and solar cells.

Etching: Mixed acids, HF:HNO_3, and halogens.

ALUMINUM ARSENIDE ETCHANTS

ALAS-0001

ETCH NAME: BRM TIME:

TYPE: Halogen, polish TEMP:

COMPOSITION:

x .... 0.5% Br_2
x .... MeOH

DISCUSSION:

AlAs, thin films deposited on (100) gallium arsenide substrates. The substrates were polished in the solution shown prior to growth of AlAs. The AlAs surface will oxidize on exposure to air and stabilize within 24 h. AlAs also can be etched in the solution shown. Process and hold under N_2 to prevent oxidation.

REF: Gordon, N L et al — *J Electrochem Soc,* 119,992(1972)

ALAS-0002

ETCH NAME: Air — TIME:

TYPE: Gas, oxidation — TEMP: RT

COMPOSITION:

x .... Air

DISCUSSION:

AlAs,(110) wafers cleaved from an oriented ingot, and as single crystal spheres. Surfaces allowed to stabilize in air. A major study of microcleavage, bonding character, and surface structure in materials of tetrahedral form as aluminum, gallium, indium antimonides, arsenides, and phosphides.

REF: Wolff, G A & Broder, J D — *Acta Crystallogr,* 12,313(1959)

PHYSICAL PROPERTIES OF ALUMINUM NITRIDE, AlN

Classification	Nitride
Atomic numbers	13 & 7
Atomic weight	40.98
Melting point (°C)	2400
Boiling point (°C)	
Density (g/cm^3)	3.05
Refractive index (n=)	1.9—2.2
Band gap (eV)	6.3
Coefficient of thermal expansion ($\times 10^{-6}$ cm/cm/°C)	4.2/5.3// or ⊥ c-axis
Hardness (Mohs — scratch)	7+
(Vicker's — kfg/mm^2)	3500
Crystal structure (hexagonal — hemimorphic)	(11$\bar{2}$0) prism
Color (solid)	Yellowish
Cleavage (prismatic — distinct)	(11$\bar{2}$0)

ALUMINUM NITRIDE, AlN

General: Does not occur as a natural compound. There are no metallic nitrides in nature, although there are many nitrates and nitrites of importance. The metal industries have used the nitridization process for years: firing metals above 1000°C in a nitrogen atmosphere to form a surface nitrided skin that acts as a surface hardening agent, and is more stable to corrosion.

Technical Application: The material has been deposited as a thin film by evaporation of aluminum in a nitrogen atmosphere. In Solid State device development, it is under evaluation as a protective surface coating similar to that of silicon dioxide and nitride. It has been grown epitaxially as both a crystalline structure and single crystal by RF sputtering of an aluminum plate in a nitrogen atmosphere. It is both a dielectric and piezoelectric material similar to alpha-quartz, SiO_2 and, as an epitaxy thin film, has been fabricated as a surface acoustic wave device.

Etching: HF; hot H_3PO_4, and hot alkalies.

ALUMINUM NITRIDE ETCHANTS

ALN-0001
ETCH NAME: Hydrofluoric acid TIME: Variable
TYPE: Acid, removal/cleaning TEMP: RT
COMPOSITION:
(1) x .... HF, conc. (2) 1 .... HF
1 .... H_2O
(3) 1 .... HF
10 .... H_2O
DISCUSSION:
AlN, thin films deposited on (111) silicon wafers by RF plasma sputter deposition. Used in a development study of the material. Films were 1000—2000 Å thick as deposited. Films were then photolithographically processed and etch patterned with the solutions shown in a general study of etching and cleaning of this material. This was some of the original work done on AlN thin films deposited by DC sputtering of a silicon target in nitrogen.
REF: Hersch, N & Walker, P — original development, 1966—1967

ALN-0002
ETCH NAME: Hydrofluoric acid TIME: Variable
TYPE: Acid, removal TEMP: Boiling
COMPOSITION: RATE: 170 mils/year
1 .... HF
1 .... H_2O
DISCUSSION:
AlN, specimens. The removal rate shown was computed after etching aluminum nitride in this solution. AlN is unaffected by mineral acids.
REF: Taylor, K M & Lenie, C — *J Electrochem Soc,* 107,308(1960)
ALN-0003: Kohn, J A et al — *Am Mineral,* 41,355(1950)
AlN, specimens. Referenced in ALN-0002, above.
ALN-0004: Renner, Von Th. — *Z Anorg Algem Chem,* 298,22(January 1959)
AlN, specimens. Referenced in AlN-0002, above.
ALN-0005: Long, G & Fuster, L M — *J Am Ceram Soc,* 42,53(1959)
AlN, specimens. Referenced in ALN-0002, above.

AlN-0006
ETCH NAME: Phosphoric acid TIME:
TYPE: Acid, removal TEMP: Hot
COMPOSITION:
x .... H_3PO_4
DISCUSSION:
AlN, thin films deposited as surface passivation on (100) gallium arsenide substrates as: AlN/SiO_2/GaAs(100). Solution used after photo resist patterning to remove the AlN layer.
REF: Naki, K & Ozeki, M — *J Cryst Growth,* 68,200(1984)
ALN-0008a: Chu, T L & Kelm, R W Jr — *J Electrochem Soc,* 122,995(1975)
AlN, thin film deposits. Solution used as a general etch.
ALN-0013: Pauleau, Y et al — *J Electrochem Soc,* 129,1045(1982)
AlN, thin films grown by CVD in a study of composition, kinetics and growth mechanisms. Growth at 45°C gave a rough surface; growth at 850°C a smooth surface. Etch was used at 65°C and removal rate varies with growth temperature of the deposit.

ALN-0007
ETCH NAME: Heat — TIME:
TYPE: Thermal, preferential — TEMP:
COMPOSITION:
x .... heat (electron)
DISCUSSION:

AlN, thin films formed after aluminum evaporation on NaCl (100) substrates by implantation of N^+ ions. The thin films were removed from the substrate by water float-off for morphological study with a high power electron microscope (HVEM) which was used as an annealing vehicle in observing physical changes in the thin film. AlN is hexagonal wurtzite in structure; has a Vicker's hardness of 3500 kgf/mm^2; a band gap of 6.3 eV, and a melting point of 2400°C.

REF: Rauschbach, B et al — *Thin Solid Films,* 109,37(1983)

ALN-0008b
ETCH NAME: Sodium hydroxide — TIME:
TYPE: Alkali, removal — TEMP: RT to hot (80°C)
COMPOSITION:
x .... 10—30% NaOH (KOH)
DISCUSSION:

AlN, amorphous thin films deposited on Si and GaAs substrates. Solution used as a general etch. (*Note:* 2—5% concentrations can be used for surface cleaners at RT and as pin-hole development solutions with AlN.)

REF: Ibid.

ALN-0009
ETCH NAME: Water — TIME:
TYPE: Acid, float-off — TEMP:
COMPOSITION:
x .... H_2O
DISCUSSION:

AlN, thin film (100) and (111) oriented were fabricated for TEM study as follows: potassium chloride, KCl, single crystal substrates of (100) and (111) orientation were mechanically polished; polish etched in water, rinsed in acetone, and nitrogen dried. Single crystal aluminum was evaporated: on (100) KCl was (100) single and sphalerite structure (isometric); on (111) KCl was (111) wurtzite structure (hexagonal). The aluminum was ion implanted with nitrogen to form aluminum nitride films. Films were removed for TEM study by milling the substrates with water.

REF: Kimura, K et al — *Jpn J Appl Phys,* 223,1145(1984)
ALN-0015: Ritajima, M et al — *J Electrochem Soc,* 128,1588(1981)

AlN, thin films deposited on molybdenum substrates by RF reactive ion plating (RIP). AlN forms at 100—1200°C; at 750°C MoO_3 is formed. Above 1200°C N_2O_3 develops. Authors say AlN is unstable in water vapor.

ALN-0010
ETCH NAME: Hydrofluoric acid — TIME:
TYPE: Acid, removal — TEMP: RT
COMPOSITION:
x .... HF, conc.
DISCUSSION:

AlN, thin films deposited on GaAs:Zn doped, (100) wafers used in a study of Zn^+ ion implantation (I^2) with annealing. After implant, wafers were surface coated with either AlN

or Si_3N_4 as a cap prior to annealing. Both thin films were removed after anneal with the solution shown.
REF: Moser, K et al — *J Appl Phys,* 57,5470(1985)
ALN-0011: Delavignette, P et al — *J Appl Phys,* 32,1098(1961)
AlN, specimens. Electron microscope used for direct observation of dislocations and stacking faults.
ALN-0012: Barrett, N J et al — *J Appl Phys,* 57,5470(1985)
AlN, deposited as a thin film on GaAs ion implanted with zinc in a study of annealing. Both AlN and Si_3N_4 were used as "caps" after ion implant. After annealing cycles, AlN removed with HF.

ALN-0013
ETCH NAME: TIME:
TYPE: Acid, removal TIME:
COMPOSITION:
x .... acids/alkalies/halogens
DISCUSSION:
AlN, single crystals used in an evaluation of cohesive energy features of tetrahedral semiconductors. Etchant solutions vary with the compounds listed below:

AlN	CdTe	BN	Ge
AlAs	BeO	HgSe	Si
AlSb	*CuCl	HgTe	ZnS
AlP	*CuBr	GaP	ZnSe
*Ag1	*CuI	GaAs	ZnTe
*$AgGaSe_2$	*$CuGaSe_2$	GaSb	*$ZnSiF_2$
*$AgInSe_2$	$CuInSe_2$	InP	*$ZnSnAs_2$
*$AgGaTe_2$	*$CuGaTe_2$	InAs	*MgS
*$AgInTe_2$	$CuInTe_2$	InSb	*MgSe
CdS	*$CaSiF_2$	Sn, grey	*MgTe

*This reference also listed under these individual compounds to include etchant solutions. Use solutions as shown under the individual compounds.

REF: Aresti, A et al — *J Phys Chem Solids,* 45,361(1984)

PHYSICAL PROPERTIES OF ALUMINUM OXIDE, Al_2O_3

Classification	Oxide
Atomic numbers	13 & 8
Atomic weight	102
Melting point (°C)	1725
Boiling point (°C)	2250 (2050)
Softening point (°C)	2050
Density (g/cm^3)	3.9 (*3.955—4.10)
Thermal conductivity (cal/sec/cm/°C) 25°C	0.077
Coefficient of thermal expansion ($\times 10^{-6}$ cm/cm/°C)	6.6
Specific heat (cal/g/°C) 25°C	0.25
Compressive strength (kg/cm^2)	24,000
Compressive strength (dynes/cm^2 $\times 10^{-10}$)	1.2

Compressive strength (dynes/cm^2 × 10^9 @ 750°C N_2 atm)	1.0
Modulus of rupture (kg/cm^2)	4800
Dielectric constant (1 kHz) 25—500°C	9.87—10.93
Electrical resistance (ohms mm^2/cm) 500°C/1000°C	10^{11} × 10^6
Young's modulus (× 10^6)	55
Flexural strength (psi × 10^3)	50
Impact resistance (Charpy — in/lb)	7
Dielectric strength (volts/mil)	230
Tensile strength (psi × 10^3)	30
Dissociation factor (1 kHz) 25—500°C	0.0007
Loss factor (1 kHz) 25—500°C	0.0076
Refractive index (n=)	1.77 (1.65)
Hardness (Mohs — scratch)	9
(Knoop — kgf/mm^2)	2000 (1370)
(Vicker's — kgf/mm^2)	1650
Crystal structure (hexagonal — rhombohedral) alpha	$(10\bar{1}1)$ rhomb
(isometric — normal) beta	(100) cube
Color (solid — powder/sxtl)	White/black/grey/brown/blue
Cleavage (basal-parting only, or cubic)	(0001) or (001)

Note: Artificial alumina, 99.98%, except where noted (*).

ALUMINUM OXIDE, Al_2O_3

General: Occurs in nature as the mineral corundum. Though it is hexagonal system, rhombohedral division, normal class, it often occurs in elongated pseudohexagonal prisms with a laminated structure; also massive, granular with near rectangular parting (pseudo-cleavage) on (0001) and $(10\bar{1}1)$. Fracture is uneven to conchoidal (glassy materials). It is next to diamond (H = 10) in hardness, vs. H = 9 for corundum with a high specific gravity (density) G = 3.95—4.10. Color range is blue, red, yellow, brown, and grey. It is opaque, translucent to transparent as sapphire. Refractive index, n = about 1.76, varying with axial measurement direction. There are three major divisions, with subspecies, all of which are important in industry. As single crystals: (1) sapphire (blue); (2) ruby (red); (3) Oriental topaz (yellow); (4) Oriental emerald (green); (5) Oriental amethyst (purple) and (6) star sapphire, with an asterated structure on the (0001) due to cavities parallel to bulk prism planes. Sapphire and ruby are the two precious gemstones. The others are semi-precious gem stones, though star sapphire with fine color (bluish) may be classed as precious.

Industrially, as artificial materials both ruby and sapphire were initially grown as ingots (boules) by the Verneiul process in the late 1800s for gem stones and, the ruby laser, developed in the mid-1950s, was the first commercial laser unit. It is worth nothing that natural and artificial rubies can be differentiated by internal structure observable with a 20× lens. Internal growth lines in natural material are angular; in artificial stones, curved. This is true for any other mineral.

Corundum: (common) the opaque varieties with dull colors of blue, grey, brown, and black. It has been produced artificially with a melting point of 2050°C, as single crystals from several liquid fluxes, such as lead oxide or sodium sulfide. The artificial abrasive alundum is made by heating bauxite (a hydrated alumina ore) in an electric furnace at 5000 to 6000°C. It occurs in four forms, but the alpha-corundum type is the only one stable at high temperatures.

Alumina, Al_2O_3 — corundum — is one of the most inert compounds known, second only to quartz, SiO_2 in nature. It is not acted upon by single acids, though some acid mixtures have been developed. When finely powdered and treated with cobalt solutions it gives a beautiful blue color.

In industry it is classified as a high temperature ceramic and is fabricated as pressed powder sheet, rod, tube, and other forms from a mixture of powdered alumina with silica as a binder. For very high purity alumina, 99.6%, the silica is removed by heat treatment. There are a wide variety of alumina-based ceramics containing other metal oxides that have been developed for special applications. Highest purity alumina is white in color, followed by grey to light tan, and there is a black variety containing graphite. It has wide industrial applications as an insulator: telephone lines, furnace liners and tubes, electrical stand-offs, thermocouple beads, and rods. Spheres are used as boiling beads in hot acid solutions. A major application is as a lapping and polishing abrasive.

Emery is a natural mixture of corundum and magnetite or hematite, both iron ores. Usually dark black with gradation of grain size from very fine to coarse and occasionally containing embedded single crystals of corundum. The primary industrial use is an abrasive (Turkish emery, best quality), or common emery (one form as Arkansas stone).

Technical Application: There are several major uses of alumina and alumina-based ceramics in Solid State processing and device fabrication:

1. General parts: As either or both electrical and heat insulators in vacuum systems and electronic test equipment; furnace tubes or bead for insulation on thermocouples in diffusion and epitaxy systems; inert work surfaces or holders for chemical processing or electrical assembly; in etch solutions as boiling beads; and other applications.
2. Lapping compound: From general lapping grits (180 to 350 mesh) to fine powder polishing compounds (0.2 to 0.5 μm) such as Linde A and B used for slurry paste polishing. Used for mechanical lap/polish of all metals and metallic compounds.
3. Cutting/sawing: Discs and blades with embedded alumina grit or melt-fired onto the cutting edge of a blade to improve wear resistance. These have been used to wafer dice semiconductor and dielectric elements.
4. Solid state packages: A major element in package construction for hermetically sealed devices. Combined with aluminum, a radiation hard package. Alumina is usually the body of the package brazed onto a metal base that acts as a an electrical insulator.
5. Thin film: Evaporated or sputtered (CVD or RF plasma) as an amorphous coating on material surfaces as a dielectric insulator or passivation layer — it resists ionizing radiation. Also used, like silicon dioxide, as a chemical etching or diffusion mask in conjunction with photo resist. Also as an active memory element in semiconductor device assemblies. Applied to aluminum surfaces — anodizing — to produce an inert, hard surface for wear protection. Anodization can be in different colors: red, yellow, blue, brown, and black with a variety of applications, such as color coded buttons.
6. Substrates: High purity alumina, >99.6%, as pressed powder blanks (0.005 to 0.050 thick) are metallized, photo resist patterned, and etched as electrical circuits for a wide range of Solid State devices. It should be noted that high purity alumina with a 1—2 μ″ polish finish can be difficult to metallize without peeling of the metallization during subsequent cutting or temperature processing like glass or other highly polished surfaces. As-fired 4 μ″ substrates do not have this problem, but high frequency microwave devices require highly polished surfaces for optimum electrical parameters.
7. Resistors: A wide range of discrete resistors and resistor networks are fabricated as discrete alumina elements or on substrates. As size and thicknesses increase the resistance values increase: 0.005/.014 thick, to about 0.060/0.080 as squares are nominal sizes.
8. Capacitors: Fabrication is similar to that of resistors and both can be special mixture formulations with other elements and oxides to obtain special electrical characteristics.

9. Amorphous ruby: Ruby parts are fabricated from a liquid melt poured into molybdenum forms to produce flats, rods, wire bonding tips, and other parts. Bond tips are used in wire bonding (gold, silver, and aluminum) of discrete devices and device assemblies. Flats, rods, tubes are used for a variety of applications including filter elements. Ruby spheres are excellent hard, inert, and high-temperature resistant ballbearings.

Regardless of the many and varied forms and applications of alumina, it is still one of the most difficult materials to etch. As will be seen in the following list of etchants where most are cleaning solutions.

Etching: Difficult. Hot HCl and H_3PO_4 and some hot gases. RIE with RF plasma gases.

ALUMINUM OXIDE ETCHANTS

ALO-0001a
ETCH NAME: TIME:
TYPE: Acid, removal TEMP: RT & 80°C
COMPOSITION:
35 ml .... H_3PO_4
20 g .... CrO_3/l
DISCUSSION:

Al_2O_3, thin films anodized on aluminum surfaces. Anodize 5 min at 20°C and 500 V in 3% boric acid: 0.05% Borax. A study of film structure and Al_2O_3/Al interfaces.
REF: Stirland, D J & Bicknell, R W — *J Electrochem Soc,* 106,481(1959)
ALO-0021: Dyer, C K & Alwitt, R S — *J Electrochem Soc,* 128,300(1981)

Al_2O_3, thin films grown on aluminum foil and used in a study of AC etching of foil. Oxide grown in 40% $NH_4B_5O_8.4H_2O$: ethylene glycol, RT. Solution shown used to strip oxide. BRM solution (ALO-0020b) used to dissolve foil leaving oxide flakes. Solution was: 5% H_3PO_4:3% CrO_3.
ALO-0001b: Ibid.

Al_2O_3, thin films as native oxide on single crystal aluminum ingots and wafers. Solution used to remove oxide.

ALO-0006
ETCH NAME: Water TIME:
TYPE: Solvent, cleaning TEMP:
COMPOSITION:
x .... H_2O
DISCUSSION:

Al_2O_3 and $Al_2P_xO_y$ thin films deposited on n-type InP, (100) wafer substrates to form MIS diodes in an electrical parameter study. Alumina not etched — used as an insulator between InP and additional surface metallization. Refractive index shown as n = 1.48—1.50. (*Note:* Natural sapphire n = 1.765.)
REF:

ALO-0022
ETCH NAME: Hydrogen TIME: To 1 h
TYPE: Gas, cleaning TEMP: 800—1000°C
COMPOSITION:
x .... H_2
DISCUSSION:

Al_2O_3, pressed powder blanks with 2 μ" (polished) and 4 μ" (as fired) surface finish. Blanks used in fabricating microwave electronic circuits with Au/TiW (10% Ti) metallization

by evaporation sputter and electrolytic gold up-plating. Up-plated parts subjected to 400°C, 1—5 min on a hot plate in air, show blistering of metallization. Hydrogen firing used to clean alumina surfaces prior to initial metallizing. Showed major reduction of blistering, but did not eliminate. The 2 μ″ finish appears to contained entrapped oils from diamond paste polishing. See AU-0024 for additional discussion of this study.
REF: Walker, P & Valardi, N — personal development, 1985

ALO-0004
ETCH NAME: RCA etch (A-B etch) TIME: 15—20 min
TYPE: Acid, cleaning TEMP: 80°C
COMPOSITION:

(1)		(2)	
1—2	NH_4OH	1—2	HCl
1—2	H_2O_2	1—2	H_2O_2
4—6	H_2O	4—6	H_2O

DISCUSSION:
Al_2O_3 pressed powder blanks with 2 μ″ (polished) and 4 μ″ (as-fired) surface finish metallized and photo resist patterned/etched in the fabrication of microelectronic circuits. The solutions, as shown, were originally developed for cleaning of silicon surfaces (Kern & Puotinen). (Walker) used the system for cleaning alumina, quartz, sapphire, and silicon: place parts in a Teflon holder introduced into solution (1); transferred still wet to solution (2); then rinse 2 min in running DI water and N_2 blow dry. Method used in a study of metal adhesion on alumina. Immediately after cleaning in the two solutions, parts were held under MeOH until introduced into vacuum for metallization and were still wet with alcohol, or parts were DI water rinse, 1 min; HF dip, 30 sec; DI water rinse, 30 sec and N_2 blow dry immediately before introduction into vacuum. Both methods gave good adhesion of sputtered metallization.
REF: Kern, W & Puotinen, D A — *RCA Rev*, 69,187(1979)
ALO-0015: Walker, P — personal application, 1980
Al_2O_3 pressed powder blanks used for microelectronic circuit fabrication. Solutions used to clean surface prior to multilayer metallization.

ALO-0005
ETCH NAME: Hydrochloric acid TIME: 2—5 min
TYPE: Acid, cleaning TEMP: RT or 70°C
COMPOSITION:

(1)		(2)	
x	HCl, conc.	1	HCl
		1	H_2O

DISCUSSION:
Al_2O_3 pressed powder substrates, 2 and 4 μ″ surface finish, used in as study of metallization. Etch soak substrates in either solution, rinse in running DI water, 2 min, and N_2 blow dry. This cleaning gave good sputtered metal adhesion. Also used on fused quartz and sapphire, and single crystal sapphire substrates.
REF: Walker, P — personal application, 1980—1985

ALO-0021
ETCH NAME: Hydrochloric acid TIME:
TYPE: Acid, preferential TEMP: Boiling
COMPOSITION:
x .... 6 *N* HCl
DISCUSSION:
Alpha-Al_2O_3 (0001) wafers used in a study of impurity penetration along dislocation lines. It was shown that traces of both NaCl and $Mg(NO_2)_2$ will contaminate the surface,

penetrate along dislocation lines, and defect structure can be developed with the solution shown.
REF: Tucker, R N & Gibbs, P — *J Appl Phys,* 29,1375(1958)

ALO-0007
ETCH NAME: Steam TIME: 6+ min
TYPE: Acid, cleaning TEMP: 100°C
COMPOSITION:
x .... H_2O, steam
DISCUSSION:
Al_2O_3 substrate blanks used for fabrication of metallized circuits in semiconductor device applications. Substrates are panel mounted facing two vertical jets. The top jet contains water plus surfactants; the lower jet, pure water. Jets are moved up and down continuously spraying the substrates for about 6 min. Rack mounted parts are then placed in vacuum ovens at greater than 100°C for at least 28 min. Cleaning is done prior to Au/Cu/Cr metallization. Authors say that this will remove both fingerprints and carbon.
REF: Rogelstad, T & Matarese, G — *J Vac Sci Technol,* A3,516(1985)

ALO-0009
ETCH NAME: Argon TIME:
TYPE: Ionized gas, cleaning TEMP:
COMPOSITION: GAS FLOW:
x .... Ar^+, ions PRESSURE:
POWER:
DISCUSSION:
Al_2O_3 substrate blanks used for plasma assisted physical vapor deposition (PAPVD). Ceramics, oxides, nitrides, carbides, and sulfides were cleaned by reactive sputtering (RS) prior to metal deposition by PAPVD or by activated reactive evaporation (ARE). Over 25 substrates and metallizations described.
REF: Bunshah, R F & Deshpandey, C C — *J Vac Sci Technol,* A3,553(1985)

ALO-0010
ETCH NAME: TIME:
TYPE: Acid, anodizing TEMP:
COMPOSITION: ANODE:
x .... 5% NaOH CATHODE:
x .... 50% HNO_3 POWER:
DISCUSSION:
Al_2O_3 anodized coating on aluminum sheet. The aluminum sheet was cleaned in the above solution prior to anodization. An HNO_3:NH_4HF solution also used. After etching, rinse in DI H_2O, then EOH.
REF: Sharp, D J — *Thin Solid Films,* 111,337(1985)

ALO-0011
ETCH NAME: Sulfamic acid TIME:
TYPE: Acid, cleaning TEMP:
COMPOSITION:
x .... NH_2SO_3H
DISCUSSION:
Al_2O_3, specimens and other ceramics. Solution is a cleaning acid for ceramics. Solutions

are strongly acidic but with pH values lower than formic, citric, oxalic, and phosphoric acid. Also used for descaling and pickling of metals.
REF: Bulletin 13332-1, Ashland Chemical Co.

ALO-0012
ETCH NAME: Air TIME: 30 min, minimum
TYPE: Gas, cleaning TEMP: 200°C
COMPOSITION:
x .... air
DISCUSSION:

Al_2O_3, substrates and parts both metallized and unmetallized material. If materials are stored for prolonged periods of time, their surfaces can become hydrated with water vapors. Where photolithography is to be used, at least an air oven bake-out is recommended before applying AZ-type photo resist lacquers to obtain better adhesion of resist.

REF: Brochure: *AZ 1300 Series Positive Photo Resists for Semiconductors and Microelectronics,* American Hoerchst Corp, 1983
ALO-0013: Walker, P — personal application, 1983

Al_2O_3, pressed powder substrates "as received" and after metallization with RF sputtered TiW:Au, Ti:Au or Cr:Au. Substrates were air oven baked at 150°C for 20 min; 4, 8, and 24 h prior to gold up-plating in a study of blisters and cracking of metallization. TiW:Au showed worst case blistering which was reduced but not eliminated with extended baking time. Alumina substrates shipped and stored in individual polyethylene envelopes have surface oils(?) contamination. A 2 μ″ finish surface contains entrapped oil from diamond paste polishing and shows different blister characteristics than the as-fired 4 μ″ surface. Evaluation done at 400°C on a hot plate in air. Best results were obtained when as-received blanks were hydrogen fired at 800 to 1000°C prior to metallization.

ALO-0002b
ETCH NAME: Air TIME: 9 min
TYPE: Gas, cleaning TEMP: 1470°C
COMPOSITION:
x .... air
DISCUSSION:

Beta-Al_2O_3:Na doped substrates were high temperature fired to clean surfaces.

REF: Ibid.

ALO-0003
ETCH NAME: Argon + air TIME:
TYPE: Gas, preferential TEMP: 1500°C
COMPOSITION:
x .... Ar + air
DISCUSSION:

Al_2O_3:Zr(1%) doped single crystal ingots grown by the Verneuil Process. Wafers were cut basal, (0001) and mechanically polished with diamond paste on a maple wood lap with olive oil as the liquid carrier. No dislocations were observed in the ingot material as-grown. After firing in the atmosphere and at temperature shown — surfaces show asterated sixfold symmetry patterns and hexagonal loops. (*Note:* Cylindrical cavities parallel to prism planes produce asterism in natural corundum and polished encabachon become star sapphires.)

REF: Bond, H E et al — *J Appl Phys,* 34,440(1964)

ALO-0002c
ETCH NAME: Phosphoric acid — TIME: 1 h
TYPE: Acid, removal/cleaning — TEMP: 150°C
COMPOSITION:
x .... H_3PO_4
DISCUSSION:
Beta-Al_2O_3:Na doped substrates were surface cleaned with this solution in addition to air firing.
REF: Ibid.
ALO-0017:
Al_2O_3:MgO(1/2%) and Al_2O_3:TiO(11%) ceramic substrates and parts were cleaned in the solution shown at 200°C.

ALO-0014
ETCH NAME: — TIME: 30 sec
TYPE: Acid, oxide removal — TEMP: 60°C
COMPOSITION:
"A" 35 .... H_3PO_4 — "B" x .... H_2O
10 .... HNO_3
Mix: 1 part "A" to 3 parts "B".
DISCUSSION:
Al_2O_3 native oxide films on Al:Au alloys. This solution used to remove native oxide from specimens before preferential etching to observe precipitates in the alloys. See ALAU-0001 under Aluminum Alloys.
REF: Von Heimendahl, M — *Acta Metall,* 15,1441(1967)

ALO-0016
ETCH NAME: Laser — TIME:
TYPE: Photochemical, forming — TEMP:
COMPOSITION:
x .... Excimer laser
DISCUSSION:
Al_2O_3, thin films deposited on silicon, (100) wafers as a $(11\bar{2}0)$ oriented thin film. Solution was CCl_4:($Al_2CH_3O_6$) and an Excimer laser was used to photochemically deposit aluminum oxide as single crystal sapphire from the liquid solution.
REF: Ehrlich, D J & Tsao, J Y — *J Vac Sci Technol,* A(3),904(1985)

ALO-0018
ETCH NAME: — TIME:
TYPE: Oxide, passivation — TEMP:
COMPOSITION:
x .... Al_2O_3
DISCUSSION:
Al_2O_3 thin film deposition on InGaAsP/InP (100) LED devices as an anti-reflective (AR) coating as an improvement over Si_3N_4. Thickness was 1805 Å onto TO-18 type package headers with LEDs previously bonded.
REF: Chin, A K et al — *J Vac Sci Technol,* B(1),72(1983)

ALO-0019
ETCH NAME: TIME:
TYPE: Electrolytic, preferential TEMP: 0°C
COMPOSITION: ANODE:
x .... H_3PO_4 CATHODE:
x .... EOH POWER: 20 V
DISCUSSION:

Al_2O_3, thin films anodized on (100) oriented single crystal aluminum blanks used in a study of the alumina layer structure. Solution produced ridging on (110), hillocks on (111) and (100). (*Note:* The blanks must have been beta-Al_2O_3.)

REF: Grooskreutz, J C & Shaw, G G — *J Appl Phys,* 35,2194(1964)

ALO-0023
ETCH NAME: TIME:
TYPE: Acid, cleaning TEMP:
COMPOSITION:
x .... H_3PO_4
x .... H_2O
DISCUSSION:

Al_2O_3 thin films DC reactively sputtered on (111) silicon wafers with a poly-Si epitaxy layer. Two sputter conditions used: "A" — developed 10 Ω cm resistivity; "B" — 40 Ω cm (down vs. up sputtering). After Al metallization wafers were photolithographic processed as MOS structures. Prior to metallization, wafers were heat treated with different schedules in He atmosphere (bubble He through LN_2 to remove O_2 contamination). No etch shown. (*Note:* Etch shown is a general solution for removal and cleaning.)

REF: Chen, M-C — *J Electrochem Soc,* 118,591(1971)

ALO-0024
ETCH NAME: TIME:
TYPE: Molten, flux, removal TEMP: 1200 K
COMPOSITION:
x .... $NaSO_4$
x .... Na_2O_2
DISCUSSION:

Alpha-Al_2O_3 powdered material used in a reaction study with development of a phase diagram for Na-Al-S-O with material in flux shown. At 1 atm, Na_2O is a base; SO_3 is the acid component. Other solubility studies included: Cr_2O_3 beta-Al_2O_3, and NiO.

REF: Jose, P D et al — *J Electrochem Soc,* 132,735(1985)

ALO-0025: Liang, W W & Elliot, J F — *J Electrochem Soc,* 125,572(1978)

Al_2O_3 material. Similar work as in ALO-0024.

ALO-0026
ETCH NAME: TIME:
TYPE: Vacuum, cleaning TEMP:
COMPOSITION:
x .... vacuum
DISCUSSION:

Al_2O_3 as single crystal blanks used in a study of ceramic microstructure and metal adhesion to surfaces by smear/rubbing. (0001), $(11\bar{2}0)$ and $(10\bar{1}0)$ surfaces studied. Friction coefficient and adhesion was anisotropic vs. load on (0001) and $(11\bar{2}0)$. Blanks were degassed in vacuum due to inherent surface oxygen present. The (0001) is a glide plane; $(11\bar{2}0)$ and other prism planes are slip planes, and can show plastic deformation. Smeared on metals

evaluated included Cu, Ni, Rh, Co, and Be. Surface attachment to some degree was due to presence of oxygen bonding. Also studied ferrites (FER-0002) and SiC (SIC-0021).
REF: Buckley, D H — *J Vac Sci Technol,* A3(3),762(1985)

PHYSICAL PROPERTIES OF ALUMINUM PHOSPHATE, $AlPO_4$

Classification	Phosphate
Atomic numbers	13, 15, & 8
Atomic weight	121.29
Melting point (°C)	1540
Boiling point (°C)	
Density (g/cm^3)	2.57
Refractive index (n=)	1.54—1.57
Hardness (Mohs — scratch)	4—5
Crystal structure (hexagonal — rhombohedral)	$(10\bar{1}1)$ rhomb
Color (solid)	Greyish
Cleavage (rhombohedral)	(1101)

ALUMINUM PHOSPHATE, $AlPO_4$

General: Does not occur as a pure natural compound, but there are some 40 minerals, mostly as hydrous phosphates of aluminum, potassium, or calcium. The mineral eggonite, $AlPO_4.2H_2O$ is representative of the most pure variety, and there are half a dozen minerals of the same hydrated formula. All are of minor occurrence; whereas bauxite, $Al(OH)_3$, is the chief ore of aluminum as a complex admixture with clays occurring in large tropical area laterite deposits.

As a phosphate there is little use in the metal industry, though the sulfates and oxides have applications. Aluminum metal and its alloys is one of the major industrial metals, equal in importance to copper and iron. The phosphide, AlP, is used as a phosphor-aluminum additive in alloys, and has been grown as a single crystal. All natural phosphates are important fertilizers.

Technical Application: No use in Solid State processing at present, though it has been grown as a single crystal, as has AlP, both for general material studies.

Etching: Soluble in acids and alkalies.

ALUMINUM PHOSPHATE ETCHANTS

ALPH-0001
ETCH NAME: TIME:
TYPE: TEMP:
COMPOSITION:
x .... HCl, conc.
DISCUSSION:

$AlPO_4$, single crystals grown by a hydrothermal temperature gradient process from H_3PO_4 in sealed glass ampoules, and used in a material study. No etch shown. (*Note:* Solution shown is a general etch for phosphates.)
REF: Requardt, A & Lehmann, G — *J Phys Chem Solids,* 46,107(1985)

ALUMINUM PHOSPHIDE ETCHANTS

ALP-0001
ETCH NAME: Water — TIME:
TYPE: Acid, polish — TEMP:
COMPOSITION:
x .... H_2O
DISCUSSION:
AlP, single crystal wafers. Solution used to remove and etch polish the material.
REF: Reynolds, A J et al — *Acta Metall,* 14,119(1966)

PHYSICAL PROPERTIES OF AMALGAM, HgAg

Classification	Metal
Atomic numbers	47 & 80
Atomic weight	388.5 (variable)
Melting point (°C)	960.5 (Ag)
Boiling point (°C)	357 (Hg)
Density (g/cm^3)	13.75—14.1
Hardness (Mohs — scratch)	3—3.5
Crystal structure (isometric — normal)	(110) dodecahedron
Color (solid)	Silver-white
Cleavage (dodecahedral — partial)	(110)

AMALGAM, HgAg

General: Occurs as a native metal compound, and found in both mercury and silver deposits, usually as small grains or fine crystals. Rare. Quantity of constituents varies widely from AgHg, Ag_5Hg_3, $Ag_{36}Hg$. There are several named minerals by percentage of silver and locale: ordinary amalgam, AgHg (26.4% Ag) or Ag_2Hg (35% Ag); arquerite, $Ag_{12}Hg$ (86.6% Ag); kongsbergite, $Ag_{32}Hg$ or $Ag_{16}Hg$ with over 90% silver. Heating on charcoal will volatilize the mercury, leaving a bead of silver. In mineralogy the term "amalgam" is reserved for AgHg compounds, though native gold occurs as an admixture of AuAg — electrum — also called white or argentiferous gold.

Native amalgam has minor use in industry due to scarcity as a distinct compound as it is a minor fraction of the ore bodies mined for their mercury or silver content. Artificial amalgams are formed as mercury combines with most metals: (1) combining with easy with tin, zinc, gold, sodium and (2) using sodium amalgam with metal salts of iron, nickel, cobalt, manganese or platinum, forms the so-called heavy metal amalgams.

There are a number of mercury containing alloys used as low temperature solders, but as amalgams . . . called amalgam even though it is electrum . . . the best known application is in dentistry, where gold and silver are compounded in small quantities in a mortar and pestle as tooth fillings . . . no mercury . . . yet still called an amalgam.

Technical Application: Natural or artificial amalgams have little use in the Solid State industry per se, though there are a number of binary and trinary compound semiconductors as selenides, tellurides and antimonides. Mercury itself is a dopant species and can be pyrolitically diffused from an HgCl source. Pure mercury has been rubbed on metal surfaces to form a temporary amalgams for mounting or as an electrical contact, and there are high intensity mercury light bulbs.

Etching: HNO_3 and aqua regia (AuAg).

AMALGAM ETCHANTS

HGAG-0001
ETCH NAME: Nitric acid — TIME:
TYPE: Acid, removal — TEMP: RT
COMPOSITION:
x .... HNO_3, conc.
DISCUSSION:

HgAg, natural single crystal. Concentrated acid used to dissolve the mineral for quantitative analysis. Water diluted acid solutions used to clean surfaces. Where pure mercury was applied as a film contact for electrical testing, concentrated acid was used to removal and clean contacts after tests. Mercury films applied to natural metal oxides — hematite, magnetite and siderite (iron oxides) — removed with dilute acid to observe structural effects as developed by the amalgam reaction.

REF: Walker, P — mineralogy studies, 1951—1953

HGZN-0001
ETCH NAME: — TIME:
TYPE: Acid, cut — TEMP:
COMPOSITION:
1 .... HNO_3
1 .... H_2O_2
1 .... alcohol
DISCUSSION:

ZnHg specimens acid saw cut with this solution, and used in a study of embrittlement cracking of this compound. After cutting, wash in water. Dry in alcohol, then with ether.

REF: Westwood, A R C — *Phil Mag*, 9,199(1964)

PHYSICAL PROPERTIES OF ANTIMONY, Sb

Classification	Semi-metal
Atomic number	51
Atomic weight	121.75
Melting point (°C)	630.5
Boiling point (°C)	1380
Density (g/cm³)	6.618 (6.691)
Thermal conductance (cal/sec)(cm²)(°C/cm)	0.045
Specific heat (cal/g) 25°C	0.0494
Latent heat of fusion (cal/g)	38.3
Coefficient of linear thermal expansion:	
(micro-inch °C) 20°C	8^{-11}
($\times 10^{-6}$ cm/cm/°C) 20°C	8.5
Electrical resistivity (micro-ohms-cm)	39.1
1st ionization energy (K-cal/g-mole)	199
1st ionization potential (eV)	8.64
Electronegativity (Pauling's)	1.9
Covalent radius (angstroms)	1.40
Ionic radius (angstroms)	0.62 (Sb^{+5})
Cross section (barns)	5.7
Magnetic susceptibility (cgs $\times 10^{-6}$) 18°C	99
Vapor pressure (°C)	1223

Hardness (Mohs — scratch)	3—3.5
Crystal structure (hexagonal — rhombohedral)	$(10\bar{1}1)$ rhomb
Color (solid)	Tin white
Cleavage (basal — perfect)	(0001)

ANTIMONY, Sb

General: Occurs as a native element in veins associated with silver and arsenic ores although the mineral stibnite, Sb_2S_3, which occurs as a primary mineral in quartz and granitic rock is the major ore of antimony. The native antimony has a specific gravity (density) of G = 6.7, is higher than the purified material (G = 6.6) due to trace presence of silver and iron. It is sectile, brittle, and is slightly malleable when heated. It has a metallic luster with a tin-white color and has a perfect (0001) cleavage, which is not common to pure metallic type elements. Surfaces tarnish readily, particularly in the presence of sulfur.

It is not used as a pure metal in heavy industry, although there are several applications as an additive to irons, steels, aluminum, brass, and bronze. The pure metal is used in special low melting point glasses, and as a black coloring pigment in both glass and clay pottery.

It is one of the elements known to ancient man, and both antimony and arsenic compounds have been used in black facial make-up, particularly as eye shadow, but are no longer in use today due to their toxicity though there are several medical applications.

Technical Application: Although antimony it is not as widely used as phosphorus, it is an n-type dopant in silicon.

It also is used to grow several compound semiconductors such as gallium antimonide, GaSb; indium antimonide, InSb; and aluminum antimonide, AlSb.

Antimony is used as a constituent in several medium temperature solders and some high temperature brazing alloys. Both used in Solid State package assemblies.

Antimony has been grown as a single crystal for morphology and defect studies.

Etching: Aqua regia; hot concentrated H_2SO_4, and mixed acids of $HF:HNO_3$.

ANTIMONY ETCHANTS

SB-0001a

ETCH NAME: Dash etch — TIME: 5—10 min

TYPE: Acid, polish — TEMP: RT

COMPOSITION: — RATE: 1 mil/min

1 .... HF

3 .... HNO_3

12 .... CH_3COOH (HAc)

DISCUSSION:

Sb, (0001) specimens cleaved after freezing in liquid nitrogen, LN_2, or cleaved under LN_2. Specimens were used in a polish etch development and dislocation etch study of antimony. Solution shown produces a good polish with no gas evolution.

REF: Wernick, J H et al — *J Appl Phys,* 29,1013(1958)

SB-0001b

ETCH NAME: Dash etch, modified — TIME: 5—10 min

TYPE: Acid, polish — TEMP: RT

COMPOSITION:

1 .... HF

3 .... HNO_3

12 .... HAc

1 .... Br_2

DISCUSSION:

Sb,(0001) specimen cleaved in LN_2. Solution shown produces a fair polish with no gas evolution. See SB-0001a for additional discussion.

REF: Ibid.

SB-0001c

ETCH NAME: Dash, etch, modified — TIME: 5—10 min

TYPE: Acid, polish — TEMP: RT

COMPOSITION:

1 .... HF
3 .... HNO_3
6 .... HAc

DISCUSSION:

Sb, (0001) specimen cleaved in LN_2. Solution shown produces a fair polish with no gas evolution. See SB-0001a for further discussion. (*Note:* This solution is a modification of the original Dash, etc.)

REF: Ibid.

SB-0001d

ETCH NAME: Dash etch, modified — TIME: 5—10 min

TYPE: Acid, polish — TEMP: RT

COMPOSITION:

1 .... HF
5 .... HNO_3
12 .... HAc

DISCUSSION:

Sb, (0001) specimen cleaved in LN_2. Produces a fair polish with no gas evolution. See SB-0001a for further discussion.

REF: Ibid.

SB-0001e

ETCH NAME: Dash etch, modified — TIME: 1 min

TYPE: Acid, polish — TEMP: RT

COMPOSITION:

1 .... HF
5 .... HNO_3
6 .... HAc

DISCUSSION:

Sb, (0001) specimens cleaved in LN_2. Produces a very good polish with no gas evolution. See SB-0001a for further discussion.

REF: Ibid.

SB-0001f

ETCH NAME: Dash etch, modified — TIME:

TYPE: Acid, removal — TEMP: RT

COMPOSITION:

3 .... HF
5 .... HNO_3
12 .... HAc
1 .... Br_2

DISCUSSION:

Sb, (0001) specimen cleaved in LN_2. Produces a rough and uneven surface and etches

very rapidly with gas evolution. See SB-0001a for further discussion. (*Note:* Compare results with SB-0001b.)
REF: Ibid.

SB-0004
ETCH NAME: TIME:
TYPE: Acid, removal/clean TEMP:
COMPOSITION:
2 .... HF
3 .... HNO_3
5 .... H_2O
DISCUSSION:
Sb, (0001) wafers. Cut wafers under glycol thallate with a high speed cutting wheel to prevent chipping. Mechanically lap with 600 grit, then 000 and 0000 aluminum abrasive paper. Use solution shown as a light etch to remove damage and debris. Specimens used in a study of self-diffusion.
REF: Rosolowski, J H et al — *J Appl Phys,* 31,3027(1964)

SB-0005
ETCH NAME: Sulfuric acid TIME:
TYPE: Acid, polish TEMP: Hot
COMPOSITION:
x .... H_2SO_4
DISCUSSION:
Sb, single crystal wafers of different orientations used in a study of etch pits. Solution can be used to polish surfaces prior to preferential etching.
REF: Hiramatsu, H & Shigeta, J — *J Phys Soc Jpn,* 13,1404(1958)

SB-0001g
ETCH NAME: CP4, modified TIME: 1—2 sec
TYPE: Acid, polish TEMP: RT
COMPOSITION:
3 .... HF
5 .... HNO_3
3 .... HAc
1 .... Br_2
DISCUSSION:
Sb, (0001) wafers cleaved under LN_2. Solution produces a very good polish with some pitting and gas evolution. See SB-0001a for added discussion. (*Note:* Original CP4 solution contained 0.6 Br_2 and was developed for germanium.)
REF: Ibid.
SB-0002: Allred, W P et al — *J Electrochem Soc,* 107,117(1960)
Sb, high purity slugs etch cleaned in CP4 solution before using to grown single crystal aluminum antimonide, AlSb. After etching rinse in DI water.

SB-0001h
ETCH NAME: CP4, modified TIME: 1—2 sec
TYPE: Acid, polish TEMP: RT
COMPOSITION:
2 .... HF
5 .... HNO_3
4 .... HAc
1 .... Br_2
DISCUSSION:
Sb,(0001) wafers cleaved under LN_2. Solution produces an uneven polish with both pits and gas evolution. See SB-0001a for added discussion.
REF: Ibid.

SB-0003
ETCH NAME: TIME:
TYPE: Cleave, defect TEMP: −195°C
COMPOSITION:
x .... cleave
DISCUSSION:
Sb, (0001) wafers cleaved under LN_2, and used as cleaved. Surface will show direction of trigonal axes and may show 60° twinning. Specimens used in a galvanometric properties study at liquid helium temperatures.
REF: Steele, M C — *Phys Rev,* 99,1751(1955)

PHYSICAL PROPERTIES OF ARGON, Ar

Classification	Inert gas
Atomic number	18
Atomic weight	39.948
Melting point (°C)	−189.2
Boiling point (°C)	−185.7
Density (g/cm³) −233°C	1.65
(g/cm³) 0°C $\times 10^{-3}$	1.784
Specific heat (cal/g/°C) 20°C	0.125
Latent heat of fusion (cal/g)	6.7
Thermal conductivity ($\times 10^{-4}$ cal/cm²/cm/°C/sec) 20°C	0.389
1st ionization potential (eV)	15.577
Cross section (barns)	0.63
Vapor pressure (°C)	200.5
Chemical reactivity	None
Hardness (Mohs — scratch)	2—2.5
Crystal structure (isometric — normal)	(100) cube, fcc
Color (gas)	Colorless
(liquid)	Clear
(solid)	Clear/bluish
Cleavage (cubic — solid)	(001)

ARGON, Ar

General: A natural element in air which contains about 0.94%. It is separated by fractionation of liquid air and is about 2½ times more soluble in water than nitrogen though it does not combine with any other element as a compound; whereas nitrogen forms nitrides, nitrates, and nitrites. It is an inert, colorless gas at standard temperature and pressure (22.2°C and 760 mmHg) and is best recognized by its characteristic spectra lines in the red end of the spectrum. Symbol may be shown as A only, though Ar is standard (Gr: argon = inactive).

In industry it is used for its inert properties in electric light bulbs and fluorescent tubes. Also used as an inert atmosphere for storage and chemical processing of metals and compounds that react to oxygen or water vapor in air, and as an inert furnace atmosphere in metal processing.

Technical Application: Argon is not directly used in fabrication of Solid State device structural components, but it is used as an inert atmosphere in vacuum or furnace systems, such as RF sputter and epitaxy. It also is used as various gas mixture: Ar/H_2 for a reducing atmosphere in these systems; Ar/N_2 for the sputtering of metal nitrides, such as Si_3N_4 or AlN; Ar/O_2 for evaporation or sputter of metal oxides, such as SiO_2 or BeO, and as $Ar/N_2/O_2$ for oxynitrides and, as an RF ionized gas, Ar^+, for surface cleaning prior to metallization of a specimen. It is widely used in this latter process, but can induce subsurface damage depending upon RF power levels.

Chemical processing of some of the compound semiconductors, such as aluminum arsenide and aluminum antimonide, AlAs and AlSb, respectively, which oxidize in air, are handled under argon or nitrogen atmospheres.

Argon has been grown as a single crystal in a vacuum system under cryogenic conditions and pressure for morphological study. It is fabricated as the argon laser.

Etching: As single crystal, vary vacuum vapor pressure.

ARGON ETCHANTS

AR-0001

ETCH NAME: Pressure | TIME:

TYPE: Pressure, dislocation | TEMP: 2 K (L_{He})

COMPOSITION:

x .... pressure, vapor

DISCUSSION:

Ar, single crystal ingots grown in vacuum under cryogen conditions and pressure. Develop structure and defects by varying vacuum pressure. (*Note:* The reference describes methods and techniques for growing single crystal rare gases.)

REF: Schwentner, N — *Rare Gas Solids,* Vol 3, Academic Press, New York, 1960

PHYSICAL PROPERTIES OF ARSENIC, As

Classification	Semi-metal
Atomic number	33
Atomic weight	74.92
Melting point (°C)	817 (28 atm)
Boiling point (°C)	613 (sublimes)
Density (g/cm³)	5.72
Specific heat (cal/g) 20°C	0.082
Heat of fusion (k-cal/g-atom)	88.5
Heat of sublimation (cal/g)	102
Atomic volume (W/D)	13.1
1st ionization energy (K-cal/g-mole)	231
1st ionization potential (eV)	9.8
Electronegativity (Pauling's)	2.0
Covalent radius (angstroms)	1.20

Ionic radius (angstroms)	0.46 (As^{+5})
Linear coefficient of expansion ($\times 10^{-6}$) 20°C	4.7
Electrical resistivity (micro ohms-cm) 20°C	33.3 (35)
Magnetic susceptibility (cgs $\times 10^{-6}$) 20°C	5.5
Cross section (barns)	4.5
Vapor pressure (°C)	518
Hardness (Mohs — scratch)	3.5
Crystal structure (hexagonal — rhombohedral)	$(10\bar{1}1)$ rhomb (near-cube)
Color (solid)	Tin white
Cleavage (basal — perfect)	(0001)

ARSENIC, As

General: Occurs in nature as the pure element, but not in quantity. Usually found in association with metallic veins of silver, cobalt, and nickel ores and with cinnabar, realgar, and orpiment. The latter two minerals are sulfides, As_2O_3, noted for their brilliant red or yellow color, respectively. It is believed that amorphous arsenic (black) was first recognized by Albertus Magnus in 1250 A.D. It is silver-white like tin when first cleaved, but oxidizes to dull grey and, when sublimed by heating, re-solidifies as both a crystalline and amorphous structure. The pure metal is not considered dangerous, but several compounds are extremely hazardous.

In industry the metallic element is used as an additive for many alloys of iron and steel, and as a hardening agent in bronze. Many compounds are used as insecticides, such as Paris Green and both Adamsite and Lewisite were used as gases in World War I. Black amorphous arsenic was used as a cosmetic eye shadow in ancient times, but is no longer used due to its health hazard. When the metal is ignited in air it burns to the trioxide, which is noted for an intense garlic odor, and is extremely toxic. It also is used in glass fabrication as a coloring agent, and for low melting point glasses.

Technical Application: Arsenic has several important applications in Solid State device fabrication, both as an n-type dopant in silicon and in the growth of several compound semiconductors.

The most important of the compound semiconductors is gallium arsenide, GaAs; common acronym: GAS-. Aluminum arsenide, AlAs, is not widely used, due to its hygroscopic nature; indium arsenide, InAs has similar problems.

In growth of arsenic compounds, such as epitaxy GaAs thin films or, when used as a diffusion dopant, the element is obtained using arseine, AsH_3, to include growth by OMCVD and MBE. It has been used as a low melting point glass in fabrication of silicon radiation detectors as a paint-on coating over exposed planar junctions.

Arsenic has been grown as a single crystal for general morphological and defect studies.

Etching: HNO_3, halogens, alkalies and mixed acids.

ARSENIC ETCHANTS

AS-0001a
ETCH NAME: Sodium hydroxide — TIME:
TYPE: Alkali, removal — TEMP: Hot
COMPOSITION:
x .... 5—20% NaOH
DISCUSSION:

As,(0001) wafers cleaved from a bulk mass. Solution used as a general removal etch. Specimens used in a study of the DeHass-van Alphan effect.

REF: Berlincourt, T G — *Phys Rev*, 99,1716(1955)

AS-0001b
ETCH NAME: Nitric acid TIME:
TYPE: Acid, removal/polish TEMP: RT
COMPOSITION:
x .... HNO_3, conc.
DISCUSSION:
As, (0001) cleaved specimens. Solution used as a general removal and polishing etch.
REF: Ibid.
AS-0003: Adachi, H & Hartnagel, H L — *J Vac Sci Technol,* 19,427(1961)
As, as a constituent in single crystal GaAs, p-type wafers fabricated as light emitting diodes with Al or Au dot contacts. Arsenic is removed with HNO_3; whereas gallium is removed with alkalies producing a surface enriched with Ga or As, respectively. Light emission was good or improved with basic (alkali) solutions or by annealing at 350°C in N_2 for 5 min. Strong acid treatment reduced or eliminated emission.

AS-0002a
ETCH NAME: TIME: 10 sec
TYPE: Halogen, dislocation TEMP: RT
COMPOSITION:
x .... 10% I_2
x .... MeOH
DISCUSSION:
As, (111)(?) specimens cleaved in air at RT. Wafers were polished with 1HCl:2HNO_3:12HAc. Solution shown was used to develop dislocations. Rinse in MeOH to remove residual iodine with a final DI water wash. (*Note:* Arsenic is hexagonal-rhombohedral with perfect basal (0001) cleavage. It is occasionally observed as a near-cubic rhombohedron, such that the (111) cleavage shown is questionable. The probability is that the specimen was near-cubic in appearance.)
REF: Sheity, M N & Tailor, J B — *J Appl Phys,* 39,3717(1968)

AS-0002b
ETCH NAME: Bromine etch TIME: 45 sec
TYPE: Acid, dislocation TEMP: RT
COMPOSITION:
1 .... HF
2 .... HNO_3
1 .... HCl
24 .... HAc (CH_3COOH)
1 .... Br_2
DISCUSSION:
As, (111)(?) specimens cleaved in air at RT. Solution used as a dislocation etch. See note under AS-0002a for additional discussion.
REF: Ibid.

AS-0002c
ETCH NAME: TIME: 1—2 min
TYPE: Acid, polish TEMP: RT
COMPOSITION:
1 .... HCl
2 .... HNO_3
12 .... HAc

DISCUSSION:

As, (111)(?) specimens cleaved in air at RT. Solution used as a polish etch prior to dislocation etching. See note under AS-0002a for additional discussion.

REF: Ibid.

PHYSICAL PROPERTIES OF BARIUM, Ba

Classification	Alkaline metal
Atomic number	56
Atomic weight	137.36
Melting point (°C)	850 (725)
Boiling point (°C)	1640 (1140)
Density (g/cm^3)	3.5
Specific heat (cal/g) 25°C	0.068
Heat of fusion (k-cal/g-atom)	1.83
Heat of vaporization (k-cal/g-atom)	35.7
Atomic volume (W/D)	39
1st ionization energy (K-cal/g-mole)	120
1st ionization potential (eV)	5.21
Electronegativity (Pauling's)	0.9
Covalent radius (angstroms)	1.98
Ionic radius (angstroms)	1.34 (Ba^{+2})
Electrical resistivity (micro-ohms-cm)	50
Work function (eV)	2.48
Cross section (barns)	1.2
Vapor pressure (°C)	1301
Hardness (Mohs — scratch)	8.5—9
Crystal structure (isometric — normal)	(100) cube, bcc
Color (solid)	Silver white
Cleavage (cubic)	(001)

BARIUM, Ba

General: Does not occur as a native element. Barite, $BaSO_4$ (heavy spar), is the major ore and, when transparent, resembles calcite, $CaCO_3$ (Iceland spar). Barite can form in a white ''barite rose'' structure, equated to the term ''rosette'' used in material processing. The pure metal has had little use in metal processing, but several compounds are of importance: the sulfate is a white pigment called ''blanc fixe'' used to weight paper; the carbonate is a rat poison; and nitrates are green coloring in pyrotechnics. Other barium salts have medical applications, such as $Ba(OH)_3$ solutions used to evaluate the intestinal tract as it is opaque to X-ray. The hydroxide also is used in the refining of sugar.

Technical Application: The metal has not been used in Solid State processing, although barium fluoride, BaF_2, is used as a substrate, and is under evaluation as a thin film. In electronic processing of materials, barium titanate, Ba_2TiO_3, is the primary transducer element for ultrasonic generators.

Etching: Soluble in water, single acids, and alcohols.

BARIUM ETCHANTS

BA-0001a

ETCH NAME: Ethyl alcohol — TIME:

TYPE: Alcohol, polish — TEMP: RT

COMPOSITION:

x .... EOH

DISCUSSION:

Ba, specimens. A general removal and polish etch for barium.

REF: Hodgman, C D et al — *Handbook of Chemistry and Physics,* 27th ed, Chemical Rubber Co., Cleveland, OH, 1943, 348

BA-0001b
ETCH NAME: Hydrochloric acid — TIME:
TYPE: Acid, polish — TEMP: RT
COMPOSITION:
x .... HCl
DISCUSSION:
Ba, specimens. This and other single acids can be used as general etchants for barium.
REF: Ibid.

PHYSICAL PROPERTIES OF BARIUM FLUORIDE, BaF_2

Classification	Fluoride
Atomic numbers	56 & 9
Atomic weight	175.36
Melting point (°C)	1280
Boiling point (°C)	2137
Density (g/cm^3)	4.83
Ionic radius (angstroms)	1.43
Surface free-energy (ergs cm^{-2}) 273°C	393 [on (111)]
−195°C	280 [on (111)]
Hardness (Mohs — scratch)	2—3
(Knoop — kgf mm^{-2})	82
Crystal structure (isometric — normal)	(100) cube
Color (solid)	Colorless
Cleavage (cubic — distinct)	(111)

BARIUM FLUORITE, BaF_2

General: Does not occur in nature as a compound, although there are several metallic fluorides, such as fluorite, CaF_2, and sellaite, MgF_2. As an artificially compounded material there is major industrial use in the manufacture of white enamels, and it is used in medicine as both an antiseptic and as embalming fluid.

Technical Application: Barium fluoride is one of the few single crystals that show (111) cleavage like diamond and fluorite, as most others in the isometric system — normal class show cubic, (001) cleavage. It has been used as a (111) cleaved substrate in the study of metal thin films, with the deposited films removed by float-off technique with HCl for TEM study.

As a deposited thin film, it has been converted to the oxide, BaO, as a more stable surface coating. Bulk single crystals are used for their optical properties as prisms, lenses, or filter elements. As a (100) or (110) deposited thin film on semiconductor wafers, it has been used as an optical coating in device structuring.

Etching: Soluble in water, acids, and NH_4Cl.

BARIUM FLUORIDE ETCHANTS

BAF-0005b
ETCH NAME: Sulfuric acid — TIME:
TYPE: Acid, removal — TEMP:

COMPOSITION:

x .... H_2SO_4

DISCUSSION:

BaF_2 specimens. This acid will etch barium fluoride very slowly and incompletely.

REF: Ibid.

BAF-0005c

ETCH NAME: Oxygen — TIME:

TYPE: Ionized gas, conversion — TEMP:

COMPOSITION: — GAS FLOW:

x .... O^+ ions — PRESSURE:

POWER:

DISCUSSION:

BaF_2 single crystal specimens. Etching in an RF plasma of oxygen will convert the BaF_2 surface, in part, to BaO which is an even more inert and stable surface thin film than the fluoride.

REF: Ibid.

BAF-0005a

ETCH NAME: Hydrofluoric acid — TIME:

TYPE: Acid, removal — TEMP:

COMPOSITION:

(1) x .... HF, conc. (2) x .... x% NH_4F

DISCUSSION:

BaF_2 specimens. The material is soluble in both of the fluorine solutions shown. (*Note:* BHF solutions also apply.)

REF: *Encyclopedia of Chemical Technology,* Vol 10, John Wiley & Sons, New York, 1980

BAF-0006: Tu, C W — *J Vac Sci Technol,* B2,24(1984)

BaF_2 thin films. Material was deposited by MBE on indium phosphide (100) and (110) substrates. Films were deposited as single crystal. Also were co-deposited with SrF_2 as $Ba_xSr_{1-x}F_2$ thin films. Other materials studied were SrF_2 and CaF_2.

BAF-0007

ETCH NAME: Water — TIME:

TYPE: Acid, polish — TEMP: RT or warm

COMPOSITION:

x .... H_2O

DISCUSSION:

BaF_2, (111) wafers used as substrates for metal deposition of aluminum, chromium, and silver in a study of the mechanical properties of optical films. Water can be used to lightly clean and polish barium fluoride surfaces.

REF: Pulker, H K — *Thin Solid Films,* 89,191(1982)

BAF-0003

ETCH NAME: — TIME: 4 h

TYPE: Acid, polish — TEMP: RT

COMPOSITION:

"A"		"B"		"C"	
30	H_2SO_4	40	HCl	80	Aerosol OT
70	H_2O	60	H_2O	20	HAc (CH_3COOH)

DISCUSSION:

BaF_2, (111) wafers were cleaved in air at RT. Polish time is related to time required to reduce and remove cleavage steps. The polishing process was as follows:

1. Glue wafer to a polishing pad.
2. Mount pad on a brass holder with paraffin — heat up slowly to prevent wafer cracking.
3. Saturate the polish pad with water and add 4 drops of solution "A" and 6 drops of solution "B", using an eye dropper.
4. Lap at 132 rpm — adding H_2O:HCl as needed.
5. Demount and clean in solution "C" (Aerosol OT supplier is Fischer Scientific.)

REF: Bis, R F et al — *J Appl Phys,* 47,736(1976)

BAF-0004
ETCH NAME: Hydrochloric acid — TIME:
TYPE: Acid, float-off — TEMP:
COMPOSITION:
x .... HCl (30%)
DISCUSSION:

BaF_2, (111) wafers cleaved in air and used as a substrate for Hot Wall Epitaxy (HWE) growth of PbSnTe. The substrate was held at 250°C during epi growth. The lead tin telluride film was removed by hydrochloric etch undercutting using the float-off technique. Film thicknesses were from 70 to 2500 Å and used in a morphological study of the compound.

REF: Suyman, H C et al — *J Cryst Growth,* 70,393(1984)

BAF-0009
ETCH NAME: Hydrochloric acid, dilute — TIME:
TYPE: Acid, float-off — TEMP:
COMPOSITION:
x .... HCl
x .... H_2O
DISCUSSION:

BaF_2, (111) wafers cleaved in air and used as a substrate for epitaxy growth of InGaSb. Substrate was vapor degreased in TCE, then rinsed in liquid MeOH, and then acetone. After epitaxy the thin film was float-off removed with dilute HCl.

REF: Eltoukhy, A H & Greene, J E — *J Appl Phys,* 50,505(1979)

BAF-0008
ETCH NAME: Nitrogen — TIME:
TYPE: Gas, cleaning — TEMP: RT
COMPOSITION:
x .... N_2
DISCUSSION:

BaF_2, (111) specimens cut and polished as circular lenses. Nitrogen blow off was used to remove any particulate matter from surfaces before evaporation of thin film chromium, aluminum, or gold with no liquid treatment.

REF: Walker, P — personal application, 1976

PHYSICAL PROPERTIES OF BARIUM TITANATE, Ba_2TiO_3

Classification	Oxide
Atomic numbers	56, 22 & 8
Atomic weight	372.57
Melting point (°C)	
Boiling point (°C)	
Density (g/cm³)	

Ferroelectric activity (KeV/cm)	2—450
Hardness (Mohs — scratch)	6—7
Crystal structure (orthorhombic) 286 K (RT)	(100) a-pinacoid
(hexagonal — rhombohedral) 193 K	$(10\bar{1}1)$ rhomb
(isometric — normal) 292 K	(100) cube
Color (solid)	Grey-white
Cleavage (cubic)	(001)

BARIUM TITANATE, Ba_2TiO_3

General: Does not occur as a natural compound. There are minerals as mixed titano-silicates containing barium, such as benitoite, $BaTiSi_2O_9$. The primary use of the artificially grown titanate is as the transducer element for ultrasonic generators. Ultrasonics, in conjunction with vapor degreaser systems, are a major method of cleaning for parts, materials, and equipment from a small beaker-sized ultrasonic unit — to large, walk-in vapor degreasing modules.

Technical Application: The Solid State industry is a major user of ultrasonic equipment in chemical processing of materials. Relatively large floor model vapor degreasers are used with ultrasonic generators for material and parts cleaning. Smaller tabletop units are used for vapor or liquid cleaning, and for active bubble stirring etching solutions in processing semiconductor wafers, circuit substrates, and allied materials.

Barium titanate is grown and used as a single crystal, and has been studied for general morphology with specific attention to ferroelectric properties. It also has been studied in its two other pressure/temperature related crystal forms, and it has been fabricated as a hollow polycrystalline sphere for evaluation of the pressure effects on the Curie temperature.

Etching: HCl, and hot H_3PO_4.

BARIUM TITANATE ETCHANTS

BAT-0001
ETCH NAME: Hydrochloric acid — TIME: 4 min or 4 h
TYPE: Acid, preferential — TEMP: RT to 0°C
COMPOSITION:
x .... HCl, conc.
DISCUSSION:

Ba_2TiO_3 (111) and (100) wafers. Solution used at the two times/temperatures shown to develop magnetic domain structures. (*Note:* Photographs are excellent.)

REF: Camerson, D P — *IBM Res Dev,* 1,2(1957)

BAT-0008a: Schlosser, H & Drougard, M E — *J Appl Phys,* 32,1227(1961)

Ba_2TiO_3 specimens used in a study of surface layer structure above the Curie Point. Solution used as a dip etch for final surface cleaning.

BAT-0012: Hooten, J A & Merz, W J — *Phys Rev,* 988,409,(1955)

Ba_2TiO_3 single crystals. Solution used at RT with several minutes soak. Develops domain related pole figures, and shows that the nose end of an ingot etches more rapidly than the tail end. (*Note:* Nose and tail more often called seed end and tail end.)

BAT-0002
ETCH NAME: Phosphoric acid — TIME:
TYPE: Acid, polish — TEMP: 155°C
COMPOSITION:
x .... H_3PO_4
DISCUSSION:

Ba_2TiO_3 (100) wafers used in a study of ferroelectric domains activated from 2 to 450

KeV/cm. After etching in the solution shown, pole in water. Use 4 *N* NaCl solution for pulse poling — add 0.1% HF to solution to develop 180° domain walls: pulse 5 sec — wait 15 sec — repulse, and repeat as needed. Domain walls will move with each pulse sequence.
REF: Stadler, H L & Zachmanidis, P J — *J Appl Phys,* 34,3255(1963)
BAT-0003: Stadler, H L — *J Appl Phys,* 34,571(1961)

Ba_2TiO_3 (100) wafers used in a study of etch hillock formation. Etch slightly in H_3PO_4 at 150°C — AC pole specimens in water — wash in acetone — wash in aqua regia, RT, 5 sec, and final etch in 1% HF, RT, 30 sec to develop grain boundaries. Then, to develop hillocks: (1) DC pole in water and (2) etch in 1% HF, RT, 30 sec. Hillocks, ad hoc, are negatively charged areas near domain walls. Mounds and pyramids were observed on (100) surfaces, that were aligned in ⟨110⟩ directions. Changes in resistivity were proportional to strain.
BAT-0004: Pearson, G L & Feldman, W J Jr — *Phys Chem Solids,* 9,28(1958)

Reference for method of developing grain boundaries shown in BAT-0003.
BAT-0005: Myerhofer, D — *Phys Rev,* 112,413(1958)

Ba_2TiO_3 (100) wafers prepared as very thin specimens and used in domain studies. Solution was used at 120°C as a thinning and polishing etch. Rinse in DI water, then alcohol.
BAT-0006: Miller, R C — *Phys Rev,* 111,736(1958)

Ba_2TiO_3, (100) wafers. Solution used at 155°C as a thinning and polishing etch, then DC pole in water with a Pt electrode.
BAT-0007: Miller, R C & Savage, A — *Phys Rev,* 112,755(1958)

Ba_2TiO_{33}, (100) wafers. Solution used at 155°C, 10—20 min, rinse in DI water, then in alcohol before DC poling.
BAT-0008b: Ibid.

Ba_2TiO_3 specimens. Solution used at 140°C, 15 min to 1 h as a polishing and thinning etch. Follow with HCl final cleaning dip: BAT-0008a.
BAT-0009: Last, J T — *Phys Rev,* 105,1740(1957)

Ba_2TiO_3 specimens. Solution used above 130°C (Curie Point of material) to prevent selective domain etching.
BAT-0010: Last, J T — *Rev Sci Instr,* 28,720(1957)

Ba_2TiO_3 specimens. Wafers were mounted on Teflon or silicone tape, mechanically lapped and then etch thinned with solution at 130°C. Removal rate shown as 1 µm/min.

BAT-0011
ETCH NAME: TIME:
TYPE: Cut, forming TEMP:
COMPOSITION:
x .... pressure
DISCUSSION:

Ba_2TiO_3 polycrystalline hemispheres cemented together to form a sphere. The specimen was used in studying the effects of pressure on the Curie temperature.
REF: Jaffe, H et al — *Phys Rev,* 105,57(1957)

BAT-0013
ETCH NAME: TIME:
TYPE: Thermal, transition TEMP:
COMPOSITION:
x .... heat
DISCUSSION:

Ba_2TiO_3 single crystals used in an electrostatic study of ferroelectric phases. Crystal structure change was rhombohedral at 193 K to orthorhombic at 268 K (RT), to cubic at 293 K.

REF: Smolander, J & Ahtee, M — *J Phys Chem Solids,* 44,1(1983)

PHYSICAL PROPERTIES OF BARIUM TUNGSTATE, $BaWO_4$

Classification	Oxide
Atomic numbers	56, 74 & 8
Atomic weight	385.28
Melting point (°C)	1200 approx.
Boiling point (°C)	
Density (g/cm³)	5.04
Hardness (Mohs — scratch)	4—5
Crystal structure (tetragonal — pyramidal)	(111) pyramid
Color (solid)	Clear to whitish
Cleavage (octahedral)	p(111)

BARIUM TUNGSTATE, $BaWO_4$

General: Does not occur as a natural compound, although there are other minerals as tungstates of copper or lead alone or with molybdenum. Scheelite, $CaWO_4$ is an important tungsten ore, as is wolframite, $(Fe,Mn)WO_4$. There is no use of barium tungstate in industry, and the natural minerals are primarily used as a source of tungsten. They do have some application as coloring agents in glass.

Technical Application: Single crystal barium tungstate as a Solid State material has been evaluated for general morphology and compound characteristics, but has found no use as an electronic device, to date.

Etching: H_3PO_4.

BARIUM TUNGSTATE ETCHANTS

BWO-0001

ETCH NAME: Phosphoric acid TIME:

TYPE: Acid, polish TEMP:

COMPOSITION:

x .... H_3PO_4, conc.

DISCUSSION:

$BaWO_4$ single crystal specimens used in a study of Raman frequency shifts and temperature dependence. Material has the scheelite structure: tetragonal system — pyramidal class. Other tungstates studied were $CaWO_4$ and $SrWO_4$.

REF: Degreniere, S et al — *J Phys Chem Solids,* 45,1105(1984)

PHYSICAL PROPERTIES OF BERYLLIUM, Be

Classification	Alkaline metal
Atomic number	4
Atomic weight	9.0122
Melting point (°C)	1283 (1278)
Boiling point (°C)	2970
Density (g/cm³)	1.85
Thermal conductance (cal/sec)(cm²)(°C/cm) 0°C	0.440
Specific heat (cal/g) 0°C	0.41 (0.45 @ 20°C)
Latent heat of fusion (cal/g)	260

Heat of fusion (cal/g)	260—275
Heat of vaporization (k-cal/g-atom)	73.9
Atomic volume (W/D)	5.0
1st ionization energy (K-cal/g-moles)	215
1st ionization potential (eV)	9.32
Electronegativity (Pauling's)	1.5
Covalent radius (angstroms)	0.90
Ionic radius (angstroms)	0.35 (Be^{+2})
Coefficient of linear thermal expansion ($\times 10^{-6}$ cm/cm/°C) 25°C	11.5
Electrical conductance (micro-ohms^{-1})	0.25
Electrical resistivity ($\times 10^{-6}$ ohms cm) 20°C	4
Electron work function (eV)	3.92
Cross section (barns)	0.009
Hardness (Mohs — scratch)	6—7
Crystal structure (hexagonal — normal)	$(10\bar{1}0)$ prism
Color (solid)	Silver-grey
Cleavage (prismatic, poor)	$(10\bar{1}0)$

BERYLLIUM, Be

General: Does not occur as a free element in nature. There are about 20 beryllium-containing minerals with none as major occurrences, though several are quite widely distributed. The beryllium aluminum silicates are the main sources of ore: beryl, $Be_3Al_2(SiO_3)_6$ is representative as both an ore and as the precious gem stone sapphire, deep blue; aquamarine, pale blue; emerald, bright green, as well as a yellow variety called yellow beryl. The pure metal resembles magnesium in appearance and chemical properties and is one of the few metals that will scratch glass, H = 6.5—7.

It is the lightest of known metals and is finding increasing use in industry, particularly as an alloy with aluminum for lightweight construction, and other metals where it increases fatigue endurance and corrosion resistance. Beryllium compounds have a sweet odor and many are considered hazardous to health, the oxide in particular due to dusting during cutting.

Technical Application: Beryllium metal is not widely used in Solid State processing although it has use as a dopant species in some compound semiconductors. It has been used as the pure metal and as an aluminum alloy for metal electrode deposition on quartz radio frequency crystals to reduce mass loading effects on frequency, although special handling is required due to the toxicity of beryllium and its compounds.

Etching: Soluble in dilute acids and alkalies.

Note: Caution should be observed in handling beryllium and its compounds due to toxicity and health hazard. There are OSHA regulation requirements that must be met for those processing the material.

BERYLLIUM ETCHANTS

BE-0001

ETCH NAME: Potassium hydroxide — TIME: 2—3 min

TYPE: Alkali, cleaning/removal — TEMP: Warm (30—40°C)

COMPOSITION:

x .... *30% KOH or NaOH

*450 g/l.

DISCUSSION:

Be, as 0.060 diameter wire. Solution was used to etch clean wire prior to material evaporation as electrodes on AT-cut quartz crystal blanks. After etching, heavy water wash, and IR lamp dry. "U" shaped clips of Be wire were hung on tungsten coils and evaporated in a small oil-pump vacuum system with a 6" diameter/high Pyrex bell jar. After metallization, pump system to at least 10^{-6} Torr and hold 20—30 min to remove Be vapors, before opening.

REF: Walker, P — personal development, 1968

BE-0002

ETCH NAME: Hydrochloric acid — TIME:

TYPE: Acid, removal — TEMP:

COMPOSITION:

x .... HCl

DISCUSSION:

Be, as an evaporated thin film used is a study of reactions with silicon and oxygen at different temperatures. Be was deposited on a (111) silicon substrate containing a layer of SiO_2. The silicon substrate was degreased in TCE: then acetone, methanol, and water rinsed. Then dipped in 10% HF and water rinsed immediately prior to CVD deposition of 4000 Å of SiO_2 followed by beryllium metal evaporation. Wafers were annealed at different temperatures in a study of reactions. Above 400°C Be reacts with SiO_2 forming an insoluble residue, possibly BeO, in part.

REF: Moore, J B & McCaldin, J C — *J Electrochem Soc,* 124,625(1977)

BE-0003

ETCH NAME: Phosphoric acid — TIME: 60 sec

TYPE: Acid, preferential — TEMP: 175°C

COMPOSITION:

x .... H_3PO_4, conc.

DISCUSSION:

Be specimens. Solution used as a defect and structure development etch.

REF:

BE-0004a

ETCH NAME: — TIME:

TYPE: Electrolytic, thinning — TEMP:

COMPOSITION: — ANODE:

1% .... HCl — CATHODE:

2% .... HNO_3 — POWER:

97% .... ethylene glycol (EG)

DISCUSSION:

Be, thin film deposits thinned with this solution for electron microscope study of dislocations.

REF: Baird, J D et al — *Nature (London),* 182,1660(1958)

BE-0004b

ETCH NAME: Freeze etch — TIME:

TYPE: Ice, preferential — TEMP: −196°C

COMPOSITION:

x .... ice

DISCUSSION:

Be, thin films deposited on ice under vacuum at -196°C (LN_2) temperature. Produces a ''freeze-out'' structure of the beryllium.

REF: Ibid.

BE-0005

ETCH NAME: Heat — TIME:

TYPE: Thermal, forming — TEMP:

COMPOSITION:

x .... heat

DISCUSSION:

Be, polycrystalline spheres from 0.1 to 1.5 mm diameter formed by arc melting under an He atmosphere. Fifty percent of smaller sizes were single crystal. Use chips and flakes in a copper pot that was water cooled, striking the material with sporadic bursts of power to splatter form spheres.

REF: Ray, A E & Smith, J F — *Acta Metall,* 11,310(1958)

BE-0006

ETCH NAME: — TIME:

TYPE: Acid, removal — TEMP:

COMPOSITION:

x .... 6 *N* HCl

x .... NH_4OH (1 mg equivalent)

DISCUSSION:

Be, specimens used in a study of neutron-transfer reactions associated with nitrogen bombardment.

REF: Halbert, M L et al — *Phys Rev,* 106,251(1957)

BE-0007

ETCH NAME: — TIME:

TYPE: Electrolytic, polish — TEMP:

COMPOSITION: — ANODE: Be

4 ml .... HCl — CATHODE:

20 ml .... HNO_3 — POWER: 15 V

4 ml .... H_2SO_4

200 ml .... ethylene glycol (EG)

DISCUSSION:

Be, single crystal specimens used in a study of micro-strain. Solution shown used to polish ingots before straining.

REF: Lawley, A et al — *Acta Metall,* 14,1339(1966)

BE-0008

ETCH NAME: — TIME:

TYPE: Electrolytic, removal — TEMP:

COMPOSITION: — ANODE: Be

100 ml .... H_3PO_4 — CATHODE:

30 ml .... EOH — POWER:

30 ml .... H_2SO_4

30 ml .... glycol

DISCUSSION:

Be, (001), (100), and (110) cut wafers used as substrates for growth of BeO. After cutting, wafers were lapped down with 3-μm diamond grit using a wheel. Solution shown

was used to polish and remove residual damage prior to oxidation to convert surfaces to BeO. (*Note:* Glycol was probably ethylene glycol.)
REF: Scott, V D — *Acta Crystallogr,* 12,136(1959)
BE-0009: Mott, B W & Haines, H R — *J Inst Met,* 80,628(1951)
Be, specimens. Solution used as a general etchant for beryllium.

BERYLLIUM ALLOYS, BeM_x

General: Beryllium metallic alloys do not occur in nature. There are only nine minerals listed as beryllium minerals, though it appears as a trace element in several others. The most important ore is beryl, an aluminum silicate, and it occurs as an oxide, sulfide, and phosphate.

Industrially the element is used as an alloy with copper, aluminum, iron, etc., to improve wear, corrosion resistance, and reduce alloy weight.

Technical Application: Beryllium copper shim stock is used in Solid State processing for its spring qualities. In vacuum evaporation it is cut into small ''fingers'' to hold semiconductor wafers and dielectric substrates in position on a flat plate during metallization and, because of its wear characteristics, does not lose it spring capabilities under such heat application. Similar fingers or plate holder designs are used for holding parts during electrical testing, metal plating, and in chemical processing where the material is not attacked by solutions being used.

An aluminum:beryllium mixture (38% Be) has been used as an evaporation alloy on quartz crystal radio frequency blanks to reduce metal mass loading effects for improved frequency stability. Caution should be observed in such operations due to the toxicity of beryllium compounds — there is a distinctive sweet odor.

Etching: Single acids of HCl, H_3PO_4, and alkalies; HCl:HNO_3 mixtures.

BERYLLIUM ALLOYS

BERYLLIUM:COPPER ETCHANTS

BECU-0001
ETCH NAME: Hydrochloric acid, dilute — TIME: 1—2 min
TYPE: Acid, cleaning — TEMP: RT
COMPOSITION:
1 .... HCl
20 .... H_2O
DISCUSSION:

Be:Cu, strips 0.005 to 0.030 thick. Material cut as finger clips to hold wafers in position by spring action on evaporation holder plates during metallization of parts. Also used as spring contacts to hold specimens in position in test fixtures. After cutting fingers to shape lightly etch clean in the solution shown. Water rinse, methyl alcohol rinse, and air dry or nitrogen blow dry. After use, with excess metal build-up on fingers from evaporation, peel away metal with tweezers and etch clean as needed.
REF: Fahr, F — personal communication, 1978

BECU-0002
ETCH NAME: Acetone — TIME: 1—2 min
TYPE: Ketone, cleaning — TEMP: RT
COMPOSITION:
x .... acetone
DISCUSSION:

Be:Cu, strips 0.003 to 0.040 thick. Material cut and used as either finger clips to hold

wafers/substrates in place during vacuum metallization or as holding clips on wafer test assembly units. In evaporation, excess metal is peeled away and/or wire brushed, degreased with TCE, then rinsed in acetone and nitrogen blown dry. Test holder fingers degreased and rinsed as required. "U" clips on tungsten coils similarly cleaned prior to use as an evaporation source.
REF: Walker, P — personal application, 1968/1978

ALUMINUM BERYLLIUM ETCHANTS

BEAL-0001
ETCH NAME: Potassium hydroxide — TIME: 2—3 min
TYPE: Alkali, cleaning — TEMP: 30—40°C
COMPOSITION:
x .... 30% KOH (NaOH)
DISCUSSION:
Al:Be(38%), as 0.060 diameter polycrystalline wire. Solution used to clean wire prior to use as an evaporation alloy on quartz crystal radio frequency blanks to reduce mass loading. After etching, wash heavily in DI water and dry under a heat lamp. "U" clips were hung on tungsten coils for evaporation and, after evaporation, system was pumped to 10^{-6} Torr for at least 30 min to cool the system and eliminate beryllium vapors due to toxicity.
REF: Walker, P — personal development, 1969

PHYSICAL PROPERTIES OF BERYLLIUM OXIDE, BeO

Classification	Oxide
Atomic numbers	4 & 8
Atomic weight	25.02
Melting point (°C)	2550
Boiling point (°C)	
Density (g/cm^3)	2.017 (3.025)
Refractive index (n=)	1.733—1.719
Limit of application (°C)	2400
Specific heat (mean) (J/kg °C)	2180
Coefficient of thermal linear expansion ($\times 10^{-6}$ cm/cm/°C) 25—800°C	7.5
Thermal conductivity (W/m °C) 1000 °C	29
Electrical resistivity ($\times 10^{8}$) 600°C	4
($\times 10^{12}$) 2100°C	8
Hardness (Mohs — scratch)	9
Crystal structure (hexagonal — hemimorphic)	$(21\bar{3}3)$ 3rd order pyramid
Color (solid)	White
Cleavage (basal or pedion)	(0001)

BERYLLIUM OXIDE, BeO

General: Occurs as the mineral bromellite, BeO, and noted from only one location in Sweden. Industrially, the oxide is classified as a high temperature refractory ceramic. It is used as pressed powder alone or mixed with alumina as an insulator, for crucibles, furnace linings, and tubing.

Technical Application: Beryllium oxide is a II—VI compound semiconductor, but in Solid State processing its major use has been as a ceramic substrate for circuit fabrication, particularly for high frequency microwave device application. It is fabricated as a pressed powder blank (like alumina), and there is ongoing development to reduce grain size for improved frequency circuit assemblies, to include use as a packaging material. It has been grown and studied as a single crystal, and deposited as a thin film, directly, or converted from a nitride to an oxide by thermal oxidation in evaluation as a surface coating similar to silicon dioxide. **CAUTION:** Beryllium and its compounds are toxic and a health hazard.

Etching: H_2SO_4, H_3PO_4, HF, and alkalies.

BERYLLIUM OXIDE ETCHANTS

BEO-0001a
ETCH NAME: Phosphoric acid TIME: 60 sec
TYPE: Acid, preferential TEMP: 175°C
COMPOSITION:
x .... H_3PO_4
DISCUSSION:

BeO, (0001) wafer. Wafers cut from a single crystal and lap/polished with diamond paste. The material shows polarity like compound semiconductors. The (0001)A-beryllium surface is very slowly attacked in this solution. The $(000\bar{1})$B-oxide surface is rapidly etched. The solution develops etch pits and possibly dislocation pits on the (0001)A, but the authors says that further work is needed to develop dislocations. A damaged point on the $(000\bar{1})$B surface developed a four-sided, square prism etch pit. Crystals tend to grow as pyramids with a preferred (0001) basal plane.

REF: Austerman, S B — *J Appl Phys,* 34,339(1963)

BEO-0001b
ETCH NAME: Sulfuric acid TIME:
TYPE: Acid, preferential TEMP:
COMPOSITION:
x .... H_2SO_4, conc.
DISCUSSION:

BeO, (0001) wafers. Solution used as a preferential etch in a similar manner as shown under BEO-0001a for phosphoric acid.

REF: Ibid.

BEO-0003
ETCH NAME: Sulfuric acid, dilute TIME: 5—10 min
TYPE: Acid, cleaning TEMP: RT
COMPOSITION:
1 .... H_2SO_4
1 .... H_2O
DISCUSSION:

BeO, pressed powder blanks. After cutting to size for use as substrates for device mounting of power diodes, parts were soaked clean in this solution. Follow with heavy DI water washing, MeOH rinse, and dry under IR lamp. Also used for cleaning of package parts in device assembly.

REF: Walker, P — personal application, 1963—1964

BEO-0007: Scott, V D — *Acta Crystallogr,* 12,136(1959)

BeO, thin films grown on single crystal beryllium wafers cut (001); (100) and (110). Material used in a growth and structure study of BeO. A 15% sulfuric acid solution developed fine cracks parallel to ⟨001⟩ directions. Material cleaved along these line directions.

BEO-0004
ETCH NAME: Potassium hydroxide
TYPE: Alkali, removal
COMPOSITION:
x .... 10—30% KOH
TIME:
TEMP:
DISCUSSION:
BeO, (0001) wafers and pressed powder substrates. Solution is a general etch for the material.
REF: Grossmann, J & Herman, D S — *J Electrochem Soc,* 116,674(1969)
BEO-0005a: Shehata, M T & Kelly, R — *J Electrochem Soc,* 122,1359(1975)
BeO, (0001) wafers. Solution used as a general etch.

BEO-0006a
ETCH NAME: Hydrochloric acid
TYPE: Acid, removal
COMPOSITION:
x .... HCl, conc.
TIME:
TEMP: 120°C
DISCUSSION:
BeO, (0001) wafers and pressed powder substrates. Referred to as a general etch for the material.
REF: Ehman, M F — *J Electrochem Soc,* 121,1240(1974)

BEO-0006b
ETCH NAME:
TYPE: Acid, removal
COMPOSITION:
x .... H_3PO_4
x .... H_2SO_4
TIME:
TEMP: Boiling
DISCUSSION:
BeO, (0001) wafers and pressed powder substrates. Solution referred to as a general etch for the material.
REF: Ibid.
BEO-0005b: Ibid.
BeO, (0001) wafers. Solution used as a general etch.

BEO-0002
ETCH NAME:
TYPE: Acid, dislocation
COMPOSITION:
1 .... HF
6 .... H_2O
TIME: 10 min or 30—60 min
TEMP: Boiling
DISCUSSION:
BeO, (0001) single crystal wafers used in a study of dislocation development. Solution develops very sharply defined hexagonal pits after 10 min. With extended etching 1st and 2nd order prisms and pyramidal planes are developed.
REF: Vandervoort, A R & Barmors, W L — *J Appl Phys,* 37,4483(1966)

BEO-0008
ETCH NAME:
TYPE: Mechanical, defect
COMPOSITION:
x .... stress
TIME:
TEMP:

DISCUSSION:

BeO, (0001) single crystals used in a study of dislocation, slip and fracture. An intron hardness tester was used to point damage surfaces at various temperature levels to induce defects. (*Note:* This is a Knoop hardness tester unit using a specially designed diamond wedge tip.)

REF: Bentlo, G G & Miller, K T — *J Appl Phys,* 38,4248(1967)

PHYSICAL PROPERTIES OF BISMUTH, Bi

Classification	Semi-metal
Atomic number	83
Atomic weight	208.98
Melting point (°C)	271.3
Boiling point (°C)	1560
Density (g/cm^3)	9.8
Thermal conductance (cal/sec)(cm^2)(°C/cm) 20°C	0.020
Specific heat (cal/g) 20°C	0.0294
Latent heat of fusion (cal/g)	12.5
Heat of fusion (k-cal/g-atom)	12.5
Heat of vaporization (cal/g)	204.3
Atomic volume (W/D)	21.3
1st ionization energy (K-cal/g-mole)	185
1st ionization potential (eV)	9.0
Electronegativity (Pauling's)	1.9
Covalent radius (angstroms)	1.46
Ionic radius (angstroms)	0.96 (Bi^{+3})
Coefficient of linear thermal expansion ($\times 10^{-6}$ cm/cm/°C)	13.3
Electrical resistivity (micro ohms-cm)	106.8
Vapor pressure (°C)	1271
(mmHg) 1200°C	100
Modulus of elasticity (psi $\times 10^4$)	4.6
Cross section (barns)	0.034
Poisson ratio	0.33
Shear modulus (psi $\times 10^4$)	1.8
Hardness (Mohs — scratch)	2.56
Crystal structure (hexagonal — rhombohedral)	$(10\bar{1}1)$ rhomb
Color (solid)	Grey/reddish-silver
Cleavage (basal — perfect)	(0001)

BISMUTH, Bi

General: Occurs as a native element as a vein mineral commonly associated with lead, silver, nickel, and cobalt ores. There are a number of bismuth-containing minerals, mostly as silver and lead sulfides and arsenides. The most important ore mineral is bismuthinite, Bi_2S_3. Bismuth is a poor conductor of electricity, is diamagnetic, and expands upon solidification in crystalline form. Bismuth will burn in air with a brilliant blue flame with production of heavy, yellow oxide fumes. Although it is hexagonal — rhombohedral it is usually found in reticulated, arborescent masses and as grains. It has perfect (0001) cleavage, is sectile, and very brittle, but when heated is slightly malleable. For many years it was confused with lead and tin as all are silver-white in color, but bismuth has a reddish hue.

Rarely used in industry as the pure metal, but is a constituent in many alloys —

particularly low melting solders such as those used in sprinkler heads for fire protection. It has special use as a casting form in low temperature fabrication of parts.

Technical Application: Not used in Solid State processing and fabrication of silicon, germanium, and gallium arsenide devices though it is an n-type dopant in some compound semiconductors.

Bismuth is used in growing II—V compound semiconductors such as bismuth selenide, Bi_2Se_3 and bismuth telluride, Bi_2Te_3. It has been grown as a single crystal for structural data and general morphological studies.

Etching: HNO_3, aqua regia; hot H_2SO_4; slowly in HCl, HF, and halogens.

BISMUTH ETCHANTS

BI-0001
ETCH NAME: Nitric acid — TIME:
TYPE: Acid, cleaning — TEMP:
COMPOSITION:
x .... HNO_3, conc.
DISCUSSION:

Bi, single crystal specimens from ingots grown by the Bridgman technique. Bars cut with a jeweler's saw. (0001) bar ends were mechanically polished. The solution was used as a ''sizing'' etch to reduce bar dimensions and to etch clean the bars.

REF: Gallo, C F — *J Appl Phys,* 34,144(1963)
BI-0005: Mullins, W W — *Acta Metall,* 4,421(1956)

Bi, single crystal specimens used in a study of induced grain boundary motion by application of magnetic flux. Solution used as a general removal etch.

BI-0002
ETCH NAME: Hydrofluoric acid, dilute — TIME: 15—20 min
TYPE: Acid, damage removal — TEMP: RT
COMPOSITION:
2 .... HF
98 .... H_2O
DISCUSSION:

Bi, (0001) wafers used in a study of slip and failure from applied tension and its orientation dependence. After mechanical lap and polish, solution used to remove residual damage. Up to 0.2 mm damage removed.

REF: Garber, R I et al — *Sov Phys (Solid State),* 3,832(1960)

BI-0003
ETCH NAME: Tri-iodide, modified — TIME:
TYPE: Halogen, polish — TEMP: RT to hot
COMPOSITION:
x .... KI, sat. sol.
x .... I_2
x .... H_2O
DISCUSSION:

Bi, (0001) wafers cleaved under liquid nitrogen, LN_2, to minimize strain. Etchant used as a KI saturated solution to polish wafers.

REF: Abeles, B & Meiboom, S — *Phys Rev,* 101,544(1956)

BI-0004a
ETCH NAME:
TYPE: Electrolytic, polish
COMPOSITION:
35 g .... KI
1 g .... I_2
10 ml .... HCl
200 ml .... H_2O
TIME: 60 sec, 10 cycles
TEMP:
ANODE: Bi
CATHODE:
POWER: 1 A/cm^2
RATE: 0.015″/60 sec

DISCUSSION:

Bi, (0001) wafers cut and sandblasted as $^5/_8$ diameter discs used in a study of anomalous skin effects. Solution used to etch polish to within 1 μm of surface plane planar orientation for strain-free surface. After etching, rinse in 1HCl:1EOH to remove surface film left on material from electrolytic etching. (*Note:* This is an acidified tri-iodide solution; film remaining is residual iodine. See: Gold.)

REF: Smith, G F — *Phys Rev,* 115,1561(1959)

BI-0015a
ETCH NAME:
TYPE: Acid, polish
COMPOSITION:
3 .... HAc
1 .... H_2O_2
TIME:
TEMP:

DISCUSSION:

Bi, single crystal specimens. Solution used as a general cleaning and polishing etch for bismuth.

REF: Rutherford, R J — *Proc Am Soc Test Mater,* 24,739(1924)

BI-0006a
ETCH NAME:
TYPE: Acid, polish
COMPOSITION:
6 .... HNO_3
6 .... HAc
1 .... H_2O
TIME: 1—5 min
TEMP: RT

DISCUSSION:

Bi, (0001) wafers used in a study of dislocations. Solution used to polish surfaces prior to preferential etching.

REF: Lovell, L C & Wernick, J H — *J Appl Phys,* 30,234(1959)

BI-0013: Yim, W M & Stofke, E J — *J Appl Phys,* 38,5210(1967)

Bi:BiMn (4%), single crystals. The BiMn segments were as ordered filaments in the hexagonal bismuth matrix. Ingot was a eutectic crystal. Solution used to polish surfaces in a study of crystal morphology.

BI-0006b
ETCH NAME:
TYPE: Halogen, dislocation
COMPOSITION:
x .... 1% I_2
x .... MeOH
TIME: 15 sec
TEMP: RT

DISCUSSION:

Bi, (0001) wafers used in a dislocation study. After polishing in solution shown under BI-0006a, this solution was used to develop dislocations. (*Note:* This mixture is used on

silicon, germanium, compound semiconductors, and other metals as a polishing, preferential, selective or surface conditioning solution. Reactions are similar to the bromine-methanol (BRM) mixtures.)
REF: Ibid.

BI-0015b
ETCH NAME: Hydrochloric acid
TYPE: Acid, cleaning
COMPOSITION:
(1) x .... HCl
(2) 1 .... HCl
1 .... EOH
TIME:
TEMP:
DISCUSSION:

Bi, single crystal specimens. Both solutions used as general cleaning etches after etching specimens in iodine solutions to remove residual iodine films from surfaces and as a general etch. (*Note:* Residual surface iodine can be removed by washing in methanol.)
REF: Ibid.
RI-0004b: Ibid.

Bi, (0001) wafers used in a study of anomalous skin effects. The 1HCl:1EOH solution was used to clean residual iodine from surfaces after electrolytic etch polishing in a tri-iodide type solution.

BI-0007
ETCH NAME: Nitric acid, dilute
TYPE: Acid, preferential
COMPOSITION:
x .... HNO_3
x .... H_2O
TIME:
TEMP:
DISCUSSION:

Bi, specimens used in a study of galvanomagnetic properties in longitudinal magnetic fields. Pits developed by this solution were used to orient the specimens.
REF: Babiskin, J — *Phys Rev,* 107,981(1957)
BI-0008: Connell, R A & Marcus, J A — *Phys Rev,* 107,940(1957)

Bi, single crystal specimens used in studying the low temperature galvanomagnetic effects. Etch clean in the solution shown.
BI-0009: Hurle, D'T J — *Br J Appl Phys,* 11,336(1960)

Bi, (111) wafers. A $1HNO_3$:$1H_2O$ solution used at RT, 30 sec as a light figure orientation etch. Quench with DI water slowly at end of etch period to reduce formation of hydrated oxide.

BI-0010
ETCH NAME: Neon
TYPE: Ionized gas, preferential
COMPOSITION:
x .... Ne^+ ion
TIME:
TEMP:
GAS FLOW:
PRESSURE:
POWER:
DISCUSSION:

Bi, single crystal specimens. Neon ion bombardment used to develop structure and orientation figures. Other metals studied were Al, Cd, Co, Mg, Cu, Sn, and Zn.
REF: Yurasova, V E — *Kristallografiya,* 2,770(1957)

BI-0012
ETCH NAME: TIME:
TYPE: Cleave, defect TEMP: -195°C
COMPOSITION:
x .... cleave
DISCUSSION:
Bi, (0001) wafers cleaved in LN_2 and used "as cleaved" in a study of thermodynamic properties at liquid helium temperatures. Surface defects on cleaved wafers can be used for bulk orientation of the specimens.
REF: Steele, M C & Babiskin, J — *Phys Rev,* 98,359(1955)

BI-0014a
ETCH NAME: Nitric acid TIME:
TYPE: Acid, cutting TEMP: RT
COMPOSITION:
x .... HNO_3
DISCUSSION:
Bi and bismuth alloy specimens used in a cyclotron absorption study. Wafers were cut from ingots by acid saw cutting using a linen string saturated in nitric acid. Specimen surfaces mechanically lap polished on 2/0 emery paper with a solution of 2 kerosene:1 MeOH after being cut.
REF: Kalt, J K et al — *Phys Rev,* 114,1396(1959)

BI-0014b
ETCH NAME: TIME:
TYPE: Acid, polish TEMP: 70—100°C
COMPOSITION:
1 .... HCl
2 .... HNO_3
3 .... H_2O
DISCUSSION:
Bi and bismuth alloy specimens etched polished in this solution after acid saw cutting (BI-0014a). Residual film remaining after etching in solution shown removed by a 5 sec dip in concentrated HNO_3 at RT.
REF: Ibid.

BI-0016
ETCH NAME: Nitric acid TIME:
TYPE: Acid, removal TEMP:
COMPOSITION:
x .... 20% HNO_3
DISCUSSION:
Bi, specimens. Material used in a study of ultrasonic attenuation of bismuth at low temperatures. Etching showed a polygonized subsurface layer about 10^1 thick is formed during lapping as a fine crystallite zone.
REF: Reneker, D H — *Phys Rev,* 115,303(1959)

BISMUTH ALLOYS, BiM_x

General: Do not occur in nature as bismuth metallic compounds, though there are a number of bismuth minerals, mainly as sulfides and oxides, with or without other metal elements.

Bismuth metal has a low melting point (27°C) and, as an industrial alloy with other metals, has wide use as a low temperature solder with specific melting temperatures controlled by the amount of additive metal: Wood's Metal (50Bi:25Pb:12.5Sn:12Cd) or Rose Metal (50Bi:40Pb:22.9Sn) as examples.

Technical Application: No direct application in the fabrication of Solid State devices, though used as a low temperature alloy solder. Available from suppliers in rod, wire, or cut pre-forms. Wire may be used as a fuse in package assembly design as part of a protection circuit.

Several bismuth alloys have been grown as single crystals for general morphological and/or defect studies.

Etching: Varies with compound — aqua regia types, cyanides as acid mixtures.

BISMUTH ALLOYS

BISMUTH:ANTIMONY ETCHANTS

BISB-0001
ETCH NAME: TIME:
TYPE: Acid, dislocation TEMP:
COMPOSITION:
7 .... HNO_3
4 .... tartaric acid, sat. sol.
1 .... H_2O
DISCUSSION:

BiSb, single crystals Te doped. Wafers were cleaved (111) under LN_2 from CZ grown ingots with no further treatment prior to dislocation etching. Increasing the Te concentration increases the dislocation content. Sn doped crystals also studied.
REF: Zemskov, V S et al — *J Cryst Growth,* 71,243(1985)

BISMUTH:CADMIUM ETCHANTS

BICD-0001
ETCH NAME: TIME:
TYPE: Acid, preferential TEMP:
COMPOSITION:
x .... $K_3Fe(CN)_6$
x .... EOH
DISCUSSION:

Bi:Cd, alloy and single crystal specimens. After mechanical polishing, specimens were preferentially etched in the solution shown to develop defect structure.
REF: Savas, M A & Smith, R W — *J Cryst Growth,* 71,66(1985)

BISMUTH:TIN ETCHANTS

BISN-0001
ETCH NAME: TIME:
TYPE: Acid, preferential TEMP:
COMPOSITION:
x .... $K_3Fe(CN)_6$
x .... EOH
DISCUSSION:

BiSn, alloy and single crystal specimens. After mechanical polishing, specimens were preferentially etched in the solution shown to develop defect structure.
REF: Savas, M A & Smith, R W — *J Cryst Growth,* 71,66(1985)

BIS-0002
ETCH NAME:
TYPE: Acid, polish
COMPOSITION:
95% .... HCl
5% .... HNO_3
TIME:
TEMP:
DISCUSSION:

BiSb alloys used in a study of the temperature dependence of electrical properties. After polishing with solution shown specimens were annealed.

REF: Jain, A L — *Phys Rev,* 114,1518(1959)

BISMUTH GERMANATE, $Bi_{14}Ge_3O_{12}$

General: Does not occur as a natural compound and has no industrial applications at present.

Technical Application: Both this germanate and the silicate compound are under development for applications similar to those of artificial garnets, such as YAG — yttrium aluminum garnet, $Y_3Al_5O_{12}$, with isometric system, structure as a cube or dodecahedron. With the formula $Bi_{12}GeO_{20}$ the material is called BGO, the silicon counterpart, BSO, and both also are cubic structure, Space Group 123.

Etching: Mixed acids and halogens.

BISMUTH SILICATE ETCHANTS

BSO-0001
ETCH NAME:
TYPE: Acid, removal
COMPOSITION:
x .... HF
x .... HNO_3
TIME:
TEMP:
DISCUSSION:

$Bi_{12}SiO_{20}$ (BSO) single crystal material under development as a compound with similar applications to those of artificial garnets (YIG, YAG etc.). See discussion under Bismuth Germanate (BIGE-0001) for additional information.

REF: Sterudner, R & Zmija, J — *J Phys Chem Solids,* 7,803(1985)

BISMUTH GERMANATE ETCHANTS

BIGE-0001
ETCH NAME:
TYPE: Acid, removal
COMPOSITION:
x .... HF
x .... HNO_3
TIME:
TEMP:
DISCUSSION:

$Bi_{12}GeO_{20}$ (BGO) single crystal material grown by flux method has cubic structure with Space Group 123. With silicon replacing germanium (BSO) the material is of similar structure. This was a material development study for compounds similar to artificial garnets (YIG and YAG). (*Note:* Solution shown is for reference only and can include glacial acetic acid,

HAc, or water, H_2O. BRM solutions (x% Br_2:MeOH) and iodine solutions (tri-iodide used on gold), etc. also apply.)
REF: Sterudner, R & Zmija, J — *J Phys Chem Solids,* 7,803(1985)

PHYSICAL PROPERTIES OF BISMUTH SELENIDE, Bi_2Se_3

Classification	Selenide
Atomic numbers	83 & 34
Atomic weight	654.9
Melting point (°C)	710
Boiling point (°C)	
Density (g/cm³)	6.25—6.98
Hardness (Mohs — scratch)	2.5—3.5
Crystal structure (orthorhombic — normal)	(010) b-pinacoid
Color (solid)	Bluish grey
Cleavage (pinacoidal — distinct)	b(010)

BISMUTH SELENIDE, Bi_2Se_3

General: Occurs as the mineral guanajuatite (frenzelite, selenobismitite), and selenium may be replaced in part by sulfur. As a mineral specie it is of rare occurrence and has had no application in the metal industries.

Technical Application: Bismuth selenide is a V—VI compound semiconductor, and has been grown as a single crystal in developing etching characteristics and semiconducting properties. It is shown as being used as a basal (0001) cleaved wafer, which would be hexagonal system, though the natural mineral is listed as being orthorhombic system with b-pinacoid b(010) cleavage.

Etching: HCl and aqua regia. Halogens.

BISMUTH SELENIDE ETCHANTS

BISE-0001a
ETCH NAME: Hydrochloric acid, dilute
TIME:
TYPE: Acid, removal/polish
TEMP: RT
COMPOSITION:
1 .... HCl
1 .... H_2O
DISCUSSION:
Bi_2Se_3, (0001) wafers. Solution used to remove lapping damage and polish surfaces.
REF: Faust, J W — *J Electrochem Soc,* 105,252C(1958)

BISE-0001b
ETCH NAME:
TIME:
TYPE: Acid, polish
TEMP: RT
COMPOSITION:
1 .... HCl
2 .... HNO_3
DISCUSSION:
Bi_2Se_3, (0001) wafers. Solution used for general cleaning and polishing of surfaces. (*Note:* Aqua regia, 3HCl:1HNO_3. The solution shown is a modification, and aqua regia-type solutions reactivity can be reduced by diluting with water.)
REF: Ibid.

BISE-0002
ETCH NAME: Water
TYPE: Electrolytic, oxidizing
COMPOSITION:
x .... H_2O

TIME:
TEMP: Hot
ANODE:
CATHODE:
POWER:

DISCUSSION:
Bi_2Se_3, (0001) cleaved wafers. Materials were anodized in a study of oxidation reactions. Also studied p-type Bi_2Te_3; p-type GaSe, and GaTe.
REF: Moritani, A et al — *J Appl Phys*, 126,1191(1979)

BISE-0003
ETCH TIME: BRM
TYPE: Halogen, polish
COMPOSITION:
x .... x% Br_2
x .... MeOH

TIME:
TEMP: RT

DISCUSSION:
Bi_2Se_3 single crystal ingot grown in a study of new semiconducting compounds. Other compounds were Ag_2Se, Li_3Bi, TlSe, Tl_2Se_3, SnSe, $SnSe_2$, In_2Tl_3, $AgInTe_2$, and In_2Te_3. (*Note:* The term BRM for this solution was not in use prior to about 1965.)
REF: Mooser, E & Pearson, W B — *Phys Rev*, 101,492(1956)

PHYSICAL PROPERTIES OF BISMUTH TELLURIDE, Bi_2Te_3

Classification	Telluride
Atomic numbers	83 & 52
Atomic weight	800.83
Melting point (°C)	573
Boiling point (°C)	
Density (g/cm^3)	7.642
Hardness (Mohs — scratch)	1.5—2
Crystal structure (hexagonal — rhombohedral)	$(10\bar{1}1)$ rhomb (bladed)
Color (solid)	Steel grey
Cleavage (basal — perfect)	(0001)

BISMUTH TELLURIDE, Bi_2Se_3

General: In nature the pure mineral is known as tetradymite, free of sulfur, as the normal formula shows trace sulfur with a density of G = 7.2—7.6, as $Bi_2(Te,S)_3$. There are three other similar minerals, all of which occur in association with gold-bearing quartz veins. There is no use in the metal industries, although there is major use of metallic tellurium.

Technical Application: Bismuth telluride is a V—VI compound semiconductor, and has been evaluated for etching characteristics and semiconducting properties, along with both lead and zinc tellurides. It has been fabricated for its optoelectronic characteristics as a laser and electroluminescent diode, both in its single crystal form and as a deposited thin film element.

Etching: Single acids and mixed acids of HF:HNO_3, HFl:HNO_3.

BITE-0001a
ETCH NAME:
TYPE: Acid, preferential

TIME: 1—2 min
TEMP: RT

COMPOSITION:

1 .... HCl
2 .... HNO_3
6 .... H_2O

DISCUSSION:

Bi_2Te_3 (0001) wafers. Solution used to develop etch pits on single crystal wafers. Pits were truncated on three alternate sides. Following etching, rinse in water and dry on filter paper. (*Note:* See SI-0092a-b for discussion of truncated hexagonal pit development.)

REF: Toramoto, I & Takayangi, S — *J Appl Phys,* 32,118(1961)

BITE-0001b

ETCH NAME: — TIME: 1—2 min

TYPE: Acid, preferential — TEMP: RT

COMPOSITION:

10 ml .... HNO_3
10 ml .... HCl
40 ml .... H_2O
1 g .... I_2

Note: Mix iodine in HCl and heat to 60°C, then cool to RT before adding other constituents.

DISCUSSION:

Bi_2Te_3, (0001) wafers. This etch will develop triangular etch pits. After etching, rinse specimens in ethyl alcohol. (*Note:* Methyl alcohol could be used as a final rinse as it is a good solvent for residual iodine.)

REF: Ibid.

BITE-0002

ETCH NAME: — TIME:

TYPE: Acid, removal — TEMP:

COMPOSITION:

1 .... HNO_3
1 .... HCl
2 .... H_2O

DISCUSSION:

n-Bi_2Te_3, (0001) wafers. Described as a "removal" etch for n-type bismuth telluride. (*Note:* BT-0001a.)

REF: Drabble, T R et al — *Proc Phys Soc,* 71,568(1958)

BITE-0003

ETCH NAME: Water — TIME:

TYPE: Acid, oxidation — TEMP: Hot

COMPOSITION: — ANODE:

x .... H_2O — CATHODE:

POWER:

DISCUSSION:

p-Bi_2Te_3, (0001) cleaved wafers. Also p-GaSe, p-GaTe and Bi_2Se_3. Materials were anodized in a study of oxidation reactions.

REF: Moritani, A et al — *J Appl Phys,* 126,1191(1979)

BITE-0001c

ETCH NAME: Nitric acid — TIME: 1—2 min

TYPE: Acid, preferential — TEMP: RT

COMPOSITION:

x .... 30% HNO_3

DISCUSSION:

Bi_2Te_3 (0001) wafers. Used to develop hexagonal etch pits in bismuth telluride specimens. Pits were truncated similar to those obtained in etch BITE-0001a.

REF: Ibid.

BITE-0004

ETCH NAME: Nitric acid, dilute — TIME:

TYPE: Acid, removal — TEMP:

COMPOSITION:

1 .... HNO_3

1 .... H_2O

DISCUSSION:

Bi_2Te_3, (0001) wafers used in developing structural-cell data and the coefficients of expansion of this material. After mechanical lap and polish, solution shown was used to remove residual damage, e.g., strained surfaces.

REF: Francombe, M H — *Br J Appl Phys,* 11,415(1960)

BT-0005: Francombe, M H — *Br J Appl Phys,* 9,415(1958)

Bi_2Te_3 (0001) wafers studied and etched as shown under BITE-0004.

PHYSICAL PROPERTIES OF BISMUTH TRIOXIDE, Bi_2O_3 (BISMITE)

Classification	Oxide
Atomic numbers	83 & 8
Atomic weight	466
Melting point (°C)	860
Boiling point (°C)	
Density (g/cm³)	8.5
Refractive index (n =)	1.81—2.00
Hardness (Mohs — scratch)	2—3
Crystal structure (hexagonal — rhombohedral)	$(10\bar{1}1)$ rhomb
(orthorhombic) artificial	(hk0) prism
Color (solid)	Straw yellow
Cleavage (basal)	(0001)

BISMUTH TRIOXIDE, Bi_2O_3

General: The natural mineral bismite, Bi_2O_3, is probably an hydroxide, and only occurs in nature as an earthy coating and small scales. In chemistry it is listed as the hydroxide: $Bi_2O_3.3H_2O$. It has little use in the metal industries, but in medicine is known as pearl white, pearl powder, etc. Bismuth metal has wide use as a low melting point solder constituent.

Technical Application: The compound has had little use in Solid State processing to date, though it is under evaluation as a thin film for its superconducting characteristics. As a mixed compound: $BaPb_{1-x}Bi_xO_3$ (BPB), it is a superconducting oxide used in the development of Boundary Josephson Junction (BJJ) devices.

Etching: Soluble in single acids.

BISMUTH TRIOXIDE ETCHANTS

BIO-0001

ETCH NAME: Hydrochloric acid — TIME:

TYPE: Acid, removal — TEMP:

COMPOSITION:

x .... HCl, conc.

DISCUSSION:

Bi_2O_3, deposited as a thin film in study of superconducting oxides. Solution used as a removal and patterning etchant.

REF: Moriwaki, K et al — *Jpn J Appl Phys*, 23,L115(1984)

BLISTERS

General: Blisters occur in natural rock formations as cavities or vugs, often produced by gas expansion (water vapor or others) during solidification. Such structure is common to AA lava . . . jagged type lava, such as in northern California ... as against Poi-Poi lava — rope-like — as represented by the Mauna Loa volcano in the Hawaiian Islands. Pumice is a solidified froth of silica, SiO_2, as a volcanic extrusion from an acidic magma with a high void density that produces a very lightweight rock used in construction and as a lapping stone. The so-called ''dinosaur eggs'' are concentinary nodules that can range from pea size to greater than a foot in diameter, fist-size being most common, and are a segregation solidification of silica most often found in sedimentary beds. When completely filled, they produce varicolor agates and, if only partly filled, may contain fine clear or colored single crystals of quartz. Cavities in other rocks may be similar as partly or completely filled by subsequent intrusion of siliceous waters producing vein deposits. In obsidian (natural glass) there may be segregated blebs of calcite, $CaCO_3$ . . . ''snowflake obsidian''. Clear, single crystal quartz sometimes contains bubbles or voids with some crystal plane structure containing either entrapped gas or liquids, and the gas/liquid is used in determining the age and atmospheric conditions of formation. Incrustations as drusy coatings on rocks and minerals may show blisters or contain voids with similar entrapped gases and liquids, though less common. Basalt, a common volcanic rock, often contains many small cavities filled with silica blebs or other segregated minerals.

Blisters on the surface of metals and metallic compounds are an unwanted phenomenon in metal processing — the slow weathering of iron to iron oxide (rust) — very often forms a blistered surface, such that most construction irons and steels are coated with red-lead to reduce or prevent oxidation. Most metals and compounds are measured for their porosity — the number of voids in the material sometimes intentionally produced for a specific product.

These are all natural forms as blisters or associated blebs, voids that can be involved with blister formation.

Technical Application: There is ongoing study of the formation of blisters and development of voids in solid metals or compounds and thin films as they occur under different processing conditions.

Any irradiation of a surface can produce both surface blisters and internal voids. Ion implantation (I^2); radioactive particle or ionic gas irradiation, etc., such as Ar^+ ion cleaning — all can introduce physical damage within a material that can lead to blisters. If the material has a crystalline grain structure, such as in thin film metal or epitaxy compound layers, there can be grain-boundary in-or-out-diffusion of both liquids and gases creating voids internally or blisters on surfaces.

Much concern involves adhesion of thin films, where they are evaporated, sputtered, epitaxially grown, or plated, as all such process steps may introduce entrapped gases or liquids, such that subsequent heat treatment will develop surface blisters with lifting, peeling, and loss of adhesion. And it often is not apparent until material is heat treated above 400 to 500°C . . . blisters can be observed to start as low as 150°C in a thin film metallization heated on a hot plate in air. It should be noted that if the part is heat treated in an air oven, in vacuum, or a furnace — N_2, H_2 or other atmosphere — blisters may not appear, yet will appear when the same part is placed on the hot plate in air. Plating solutions can develop blisters spontaneously when the plating metal concentration becomes low or if gas and liquids are entrapped at the part/plating interface or within the plating metal film . . . again, sometimes not apparent until heat treatment. And the preparation of surfaces prior to any thin film deposition can be extremely critical: 2 μm″ finish alumina substrates are an example . . . they are polished with diamond paste, and paste oil can be entrapped in the fine surface cracks . . . hydrogen firing at 800 to 900°C was found to be the best method of cleaning. Any organic or other contamination on a surface can be a factor causing subsequent blistering.

The interaction of metal elements in processing can introduce stress and strain with the formation of new compounds due to solid-solid diffusion. Silicides are an example of controlled formation of a new compound but, if they fail to operate as a buffer layer and react with other elements, such as with aluminum in a multilayer structure, another crystallite form can develop with expansion causing swelling, stress, and strain within the bulk and eventual blister or void formation. Where a silicon dioxide layer is used as a diffusion mask, as against boron, it can develop a "rosette" crystalline structure in the otherwise amorphous film as a borosilicate glass, and such structures have been referred to as blisters.

With time under operating electrical load or during temperature evaluation of multilayer devices, there also can be solid-solid diffusion. In one example, where cobalt is a metal constituent, it was shown to migrate to wire bond device contacts causing embrittlement and contact failure. Silver, in particular, but also gold and aluminum thin films have been known to "creep" with time and temperature or under electric bias with subsequent device failure, to include swelling, blistering, and contact void formation. The occurrence of "purple plague" as crystal forms of Al_xAu_y has long been known as a device failure mechanism, again, swelling, blistering, embrittlement, and operational failure.

Note: Purple Gold (70Au:21Al) is used in the jewelry trade and as a decorative finish.

BLISTER ETCHANTS

BLIS-0001

ETCH NAME: Water — TIME:

TYPE: Acid, diffusion — TEMP: RT to hot

COMPOSITION:

x .... H_2O, liquid/vapor

DISCUSSION:

GaAs, (100) wafers doped p-type (Ge) and n-type (Si) and used as substrates for sequential sputter of Ti and Pt, then annealed at 450°C, 2 min in an open tube with FG will produce blisters 5—6 μm diameter with pinholes in blister tops. To clean substrates (1) oxidize and strip with HF; (2) boil in chloroform; (3) boil in acetone; (4) boil in MeOH; (5) N_2 dry; (6) swab with 1HCl:1H_2O; and (7) NH_4OH rinse to remove residual oxide.

Surface treatment had no effect on blister occurrence; no blisters with Ti, only. TiPt film is polycrystalline with grain boundaries and microvoids, and Pt will catalyze H_2 and O_2 in water vapor at elevated temperature . . . nucleating blister-type points in film from water/gas expansion . . . plastic deformation and rupture of blisters can result. As Au/Ti/Pt the film will blister but not rupture due to gold malleability. A solder bonding flux

of $ZnCl_2$:NH_4Cl:H_2O produces HCl with both corrosion and blisters. Heating above 375—450°C develops blisters and gas-induced delamination of films.
REF: Henein, G — *Thin Solid Films,* 109,155(1983)

BLIS-0002
ETCH NAME: Oxygen | TIME:
TYPE: Gas, diffusion | TEMP: Elevated
COMPOSITION:
x .... Air, H_2, O_2
DISCUSSION:
Au thin films. Gas diffusion into films forms bubbles along grain boundaries with 1.8 eV activation energy with gas entrapped in voids during film growth. Voids unstable below a few hundred angstroms in size, but stabilize by gas filling as bubbloids or poroid bubbles.
REF: Andrew, R & Lloyd, J R — *Thin Solid Films,* 88,125(1982)

BLIS-0003
ETCH NAME: Irradiation | TIME:
TYPE: Proton, damage | TEMP:
COMPOSITION:
x .... H^+, H_2, H_3
DISCUSSION:
Al alloy #6061-Tg sheet. Proton irradiation at 100 and 200 KeV at −196, −100, +100, and +200°C. Large and small blisters, round and elongated appear aligned in the sheet roll direction after irradiation. At 200°C blisters appear only at grain boundaries. By heating 10 min at 300°C blisters appear at grain boundaries and on general surfaces. After scratching a surface, then 100 KeV rad and 20 min at 250°C anneal, blisters appear along cracks — increase in size as circular blisters with added 10 min at 350°C. Proton beam was 48H:$32H_2$$2OH_3$. Entrapped hydrogen gas with heat expansion was cause of blister occurrence in the solid material surface.
REF: Milack, T et al — *Thin Solid Films,* 88,2805(1982)
BLIS-0004: Daniels, R A & Cooley, F — *Thin Solid Films,* 88,2815(1982)
Similar work to that of Milack, BLIS-0003.

BLIS-0005
ETCH NAME: Air | TIME: 6 min
TYPE: Gas, blister forming | TEMP: 300°C
COMPOSITION:
x .... H_2, O_2, air
DISCUSSION:
Au/TiW thin films on Al film deposited on (111) silicon wafers. At temperature and time shown Au alters from fcc to bcc cubic structure and, when heated in air, becomes silvery in color due to aluminum diffusion into gold. Gold on Al/Si (111) developed round, doughnut spots (collapsed ring blisters?). As Al/TiW/Si with final Au coating on Al at 300°C becomes a silvery colored surface with blisters due to gas grain boundary diffusion (GBD). Nitrided TiW films convert from bcc to fcc. Addition of N_2 improves barrier effect against Al solid-solid diffusion. Some films were TiO_3WO_7.
REF: Nowicki, R S et al — *Thin Solid Films,* 53,195(1978)

BLIS-0006
ETCH NAME: Helium
TYPE: Ionized gas ion, implantation
COMPOSITION:
x .... He^+, ion

TIME:
TEMP: 900°C
GAS FLOW:
PRESSURE:
POWER:

DISCUSSION:
Nb, (111) blanks He^+ ion implanted at 0.5—1.5 MeV. Blister shapes strongly dependent on temperature. At 900°C developed "crow-foot" blisters with three ⟨112⟩ directional prongs and preferred (211), (112), (121) orientation. At lower temperatures only round dome blisters were observed with size increase as temperature reduced toward RT. Blister density varied with the direction of implant beam. Blisters caused by expansion of implanted He^+.
REF: Das, S K & Kaminsky, M — *J Appl Phys,* 44,2520(1973)

BLIS-0007
ETCH NAME: Oxygen
TYPE: Gas, crystallization
COMPOSITION:
x .... O_2 (air)

TIME:
TEMP: 500—550°C

DISCUSSION:
Al/TiW/PtSi/c-Si substrate metallized structure. With heating all of the following crystallites can occur: Al_2Pt, $Al_{12}W$, WSi_2 and TiWSi. With O_2 present: Al/TiWO. Surfaces developed pinhole blisters due to expansion and stress of Al_2Pt regrowth, formation, etc., as well as from gas grain boundary diffusion action. Al deposited at 250°C + O_2 developed an Al/O/TiW structure.
REF: Canali, C et al — *Thin Solid Films,* 88,9(1982)

BLIS-0008
ETCH NAME: Boron
TYPE: Metal, blister forming
COMPOSITION:
x .... 0.05—0.2% B

TIME:
TEMP: To 800°C

DISCUSSION:
Ni:B thin films deposited as columnar growth structure. Anneal 4 h at 800°C to form Ni_2B, developing angular voids along grain boundaries with brown Ni_2B crystallites growing at grain boundaries. Annealed 4 h and evaluated at 100°C steps from 400 to 800°C in vacuum and showed little change up to 600°C. The bubbles and dislocations appeared at grain boundaries with deep cracks from the surface with 0.2% boron. With CVD growth of Ni:B films, bubbles were very fine.
REF: Skibo, M & Greulich, F A — *Thin Solid Films,* 113,224(1984)

BLIS-0009
ETCH NAME: Kovar
TYPE: Metal, conversion
COMPOSITION:
x .... Fe, Ni, Co

TIME:
TEMP: 200°C

DISCUSSION:
Au/Fe, Au/Ni and Au/Co thin film metallization. Annealed at temperature shown, metals will diffuse into overlay gold. Under oxidizing conditions Fe + SiO_2 will develop Fe_2O_3. The study was for observation of metal diffusion reactions in Kovar used in device package construction. It was seen that solid-solid diffusion into gold can develop surface blisters.
REF: Schwartz, W E et al — *Thin Solid Films,* 114,349(1984)

BLIS-0010
ETCH NAME: Metal
TYPE: Metal, blister forming
COMPOSITION:
x .... Ga or Sm
TIME:
TEMP: 200°C and above
DISCUSSION:

Au:Ga and Au:Sm thin films deposited onto NaCl, (100) substrates. The Ib group of elements will diffuse rapidly into group IIIa and IVa elements. Au/Ga and Au/Sm act as diffusion couples with diffusion stress differentials causing large blisters — 1.5 mm size as dome blisters with a "T"-shaped crack rupture on top surface of blisters. Blisters tend to align along cleavage steps on the NaCl surfaces. Segregated white spots appear in films as Au_2Ga as grains, other areas were orange in color as pure Au grains. Blister cracks were due to Kirkendall voids and volume change was caused by interdiffusion.

REF: Nakahara, S & Kinsbron, E — *Thin Solid Films,* 113,15(1984)
BLIS-0011; Nakahara, S & McCoy, R J — *Thin Solid Films,* 88,285(1982)

Pd/Sn and $PdSn_2$ using tin and Sn substrates. Similar results observed as described in BLIS-0010.

BLIS-0012
ETCH NAME: Air
TYPE: Gas, blister forming
COMPOSITION:
x .... Air
TIME: 2—5 min
TEMP: To 400°C
DISCUSSION:

Au/TiW thin films sputter deposited on 2 and 4 μm″ finish alumina blanks. As sputter deposited with about 2000 Å Au/500 Å TiW(Ti 10%) no blisters were present at RT but, after 400°C on a hot plate in air, both circular and elongated blisters appeared. Blisters also occurred with size and density variable under the following process conditions: (1) Up-plating gold from a standard gold cyanide bath . . . (2) HCl cleaning rinse, H_2O rinse, then direct to gold up-plate . . . (3) HCl rinse and H_2O rinse, then electroless nickel 500 Å coating, followed directly by gold up-plate . . . (4) no cleaning of sputtered gold surface prior to Au or Au/Ni plating . . . (5) air oven cleaning evaporated gold surfaces at 150°C for 2, 4, 6 and 24 h prior to Au or Au/Ni plate . . . (6) hydrogen firing of alumina at 800°C prior to initial Au/TiW sputter evaporation. Fused quartz blanks similarly processed as a control against the 2 and 4 μm″ alumina blanks. The 2 μm″ alumina showed worst-case blistering with size, shape (round, elongated or combination), and density-variable with surface treatment before gold up-plating. Presence of HCl developed both fine blisters and black liquid extrusion from ruptured blisters, or appeared at grain boundaries, with limited, erratic distribution vs. total blister density. Air oven bake reduced the number of large blisters, but increased the density of small blisters. A thick nickel layer (about 2000 Å) developed large, circular blisters with complete pop-out and silvery discoloration of Au surface. Hydrogen firing of alumina blanks prior to any metal evaporation reduced but did not eliminate all blisters with up-plated Au. Au or Ti, only, did not produce blisters. TiW, only, developed cracks and peeling associated with crack locations in the alumina blank surfaces. With sputtered films, only, and soaked time at 400°C in air on a hot plate, a glassy WO_x dark grey film with some spectrum colors appeared on surface of gold without blistering. The 4 μm″ alumina finish blanks showed a much lesser blister density with larger size than with 2 μm″ finish. The presence of entrapped gas, oils, and liquids, and the solid-solid diffusion of TiW were considered as the causes of blistering. Blisters developed on quartz substrates only when TiW was present, not with other metal combinations. Since these initial

experiments were performed, it has been shown that slow, step-heating of Au/TiW thin films will prevent solid-solid diffusion of the TiW fractions that cause blistering.
REF: Walker, P & Velardi, N — personal development, 1984—1985

BLIS-0013
ETCH NAME: Hydrofluoric acid TIME:
TYPE: Acid, removal TEMP: RT
COMPOSITION:
x .... HF, conc.
DISCUSSION:
a-Si:H thin films deposited on SST substrates, then aluminum evaporated as p-i-n diodes. Blisters, pinholes, and lifting observed due to deposit conditions and/or substrate surface preparation. The film surface roughness affects solar cell efficiency.
REF: Yacobi, B G et al — *J Electron Mater,* 13,843(1984)

BONDING

General: Many natural minerals themselves, as individual components or as mixture components, form solid bonds. Sandstone is an example of such a two-component bonding system where, under pressure, heat and with time, sand grains are compacted and cemented by the intrusion of colloidal iron oxide. The reason that most sandstone is tan to brown in color is due to the presence of the brown iron oxide. The hydrothermal action of hot waters percolating through subsurface rocks can alter a mineral completely to another compound, cause only surface alteration, or deposit a different mineral compound as a firmly attached surface coating. Limonite pseudomorphs after pyrite, where the sulfur in the pyrite is leached away and replaced by oxygen — still retaining the pyrite crystal form as a face striated cube — is an example of complete alternation, called a limonite pseudomorph after pyrite. Moonstones — colloidal to amorphous silica nodules found along seashores — are often surface coated with a very hard coating of calcium/magnesium sulfate or mixed oxide, white in color, and silica is a common drusy surface coating on many rocks and minerals which may be fine crystals, in part. Both air and water can oxidize the surface of a mineral as a metal oxide surface coating. Petrified wood or wood opal are special cases, where siliceous waters have entered the cellular structure of the buried wood forming a hard silica rock — cat's eye or tiger eye semiprecious gem stones are similarly silicified asbestos minerals. Many types of rock are compacted mixtures of different minerals, some acting as the bonding medium, such as mica or hornblende granite or graphic granite. The latter as silica in an open cellular structure which is filled with the softer feldspars, which are white, pink, and grey-blue colors of the feldspars. When polished, they make fine ornamental stones.

In industry, much of metal fabrication involves some form of bonding within the metal or metal mixture. Irons and steels are heat treated to develop specific structural bonding, such as the alpha, beta, and delta phases; formation of marcasite; iron carbide as Fe_3C, etc.

Other metals are bonded together by pressure and heat — called cladding — such as a copper/gold strip. Metal evaporation or plating form thin film coatings intimately bonded to a surface. Coldweld sealing uses direct pressure, only — 5 to 20 tons — to form copper-copper or nickel-nickel bonds by metal cross-diffusion. Like the natural siliceous waters penetrating wood, plastic resins and lacquers are used to impregnate linen cloth to form a hard, insulating board with the plastic/lacquer as the bonding medium. Shatterproof glass is fabricated by bonding two glass plates with a thin sheet of plastic between as a sandwich, or a metal wire grid is embedded in the glass for structural rigidity, sometimes as an electrical heater system for deicing glass panes.

There are many laminated products — pressed wood, as plywood, as one example — again, using resins, etc. as the binding medium. Paper is "loaded" with barium sulfate powder as a binder to increase weight; colloidal silica is used as a filler and binder in many applications; in the fabrication of fire brick, both silica and graphite are used as binders. Needless to say, there are many types of glue used to bond similar or different materials together, and Teflon coatings are applied to cooking ware for a nonstick surface.

Technical Application: In Solid State processing probably the most important bonding applications are wire bonding of devices, substrate circuits, and package assemblies. Several of the etchant formats in this section cover such bonding as methods, rather than etching solutions.

Many of the bonding areas described in the General section above apply to Solid State with regard to products used in both equipment operation and device assemblies. The following Etchant Section here includes some bonding techniques as well as cleaning solutions — see the Mounting Materials section for additional information.

Etching: Acids, alkalies, alcohols, solvents, variable with material.

BONDING ETCHANTS

BB-0001

ETCH NAME: Ball bonding — TIME: Seconds

TYPE: Wire, bonding — TEMP: RT to 150°C

COMPOSITION:

x .... Au, Al, Ag, Cu wire

DISCUSSION:

Ball bonding technique. Wire is fed down through a vertical capillary tube with a ball formed by a hydrogen flame cut-off at the tube tip. Lower tube so that the preformed wire ball is in contact with the contact pad to be bonded. With a combination of time and pressure, with or without heat and ultrasonic vibration, a hemispherical bond contact about 3—4 × the wire diameter is formed on the contact pad. Wire diameters range from .0007 to about .010. Dead soft .0007 gold wire is common for microelectronic circuitry, and Au:Be(2—5%) used with automatic bonders where a stiffer wire is required. Wire is supplied on spools of 100 to 500 ft. If cleaning is necessary, vapor degrease with TCE or rinse in alcohols with air, oven bake at 125°C. If bond wire fails to stick: (1) RF plasma clean device or substrate with O^+ or N^+ ion plasmas, 5—10 min, and (2) check hardness of the metal pad surface. In the first case, the surface may be contaminated with oils; whereas in the second case, prior processing may have hardened the pad, such that a soft wire bond will not stick. Aluminum pads can have an oxide layer (use ultrasonic in bonding). Gold pads as an Au/AuGe/Ni multilayer metallization, a common contact pad in high frequency microelectronic devices, can be excessively hardened by the annealing step following metallization. If this occurs, the only correction is to add an additional soft gold evaporated layer before the wafer is diced into individual units as, once diced, the entire dice lot will be rejects. Ball bonding is still widely used, but is limited to a pad area larger than the formed ball bond, such that it is of limited use in high frequency microelectronic circuitry where bond pads are under 5 mils square in area. After bonding, spray clean contacts with acetone; rinse in MeOH, EOH, Freons, or other solvents. The capillary tubes used for wire feed have a rounded tip end that should be free of chips and kept clean, e.g., solvent clean as needed, or replace. Bonding tips are of glass, ruby, or titanium and tungsten.

REF: Harman, G G — *Solid State Technol*, 1984, 186

B8-0002

ETCH NAME: Wedge bonding — TIME: Seconds

TYPE: Wire, bonding — TEMP: RT to 150°C

COMPOSITION:

x .... Al, wire

DISCUSSION:

Wedge bonding technique. A solid wedge-tipped rod of tungsten, titanium, steel, etc. Tip face is flat with slightly curved edges to prevent cutting action during pressure bonding. Place wire end in position on aluminum bonding pad of device, lower wedge in contact, and apply controlled pressure and ultrasonic vibration with or without heating of part. Special application: bonding .010 aluminum wire to aluminum alloy pads of SCRs. Ultrasonic vibration needed to insure breakthrough of native aluminum oxide films for a good metal/metal bond. Prior to using new aluminum wire, heat treat about 30 min at about 450°C in forming gas (FG) for dead-soft condition, as wire-work hardens rapidly during bonding. Rinse in alcohols and air dry as needed after bonding.

REF: Topas, B et al — personal application, 1970—1976

BB-0003

ETCH NAME: Foot bonding — TIME: Seconds

TYPE: Wire, bonding — TEMP: RT to 150°C

COMPOSITION:

x .... Au, wire

DISCUSSION:

Foot bonding technique (sometimes called wedge bonding). Bonding tip is a capillary tube with an upper round shaft for position holding, being cut down to a rectangular tip face — flat face or with convex or concave center — back-end sharp ledge; front-end curved upward. Lower back of tip has extended portion with angled hole for wire to be drawn down and across tip face toward the front, facing the operator. Bond flat about .005 width (.002 special) and .0005 to .0007 length. Bonding is similar to ball bonding without H_2 ball-forming cut-off. Tip material is of tungsten, titanium, ruby, etc. Gold wire, .0007 diameter common; AuBe for automatic bonders, supplied on 100 or 500 ft spools. With time, wire-work hardens on spool and shows excessive breakage.

Bonding is done like ball bonding with the wire brought down in contact with the gold bonding pad of the device, then pressure, with or without ultrasonic vibration (minimum) is applied. The bond formed is elongated — a .0007 wire widening to about .003 (with an initial .002 tip width) with much the appearance of a human foot, hence the method name. Forward bonding most common, where the initial bond is on the device, second bond on a substrate or package pad, but back-bonding can be done . . . the reverse of forward bonding. Some of these food bonders have the capability of "stitch bonding", where a multiple line of bonds is made before final bond-end break off. With the relatively flat foot bond, it is possible to make bonds-on-top-of-bonds . . . "stack bonding". As many as four wires have been so stack bonded with device pad-pad, pad-substrate, pad-package bonds coming from a single contact pad of the device.

REF: Marich, L A et al — personal communication/application, 1980—1986

BB-0004

ETCH NAME: Weld bonding — TIME:

TYPE: Part, bonding — TEMP: Elevated

COMPOSITION:

x .... Cu

DISCUSSION:

Weld bonding technique. Two opposed carbon electrodes or metal tips similar to a soldering iron, and may be a ring-shaped metal electrode. The wire to be bonded is held on bond area, and welding is accomplished from an electric spark from the carbon electrodes, or an electric pulse from a metal tip. The weld is usually as a cup-shaped joint with some weld splatter and metal erosion points of the surface being bonded. The standard welding techniques use a gas torch with a welding rod material fed against the part or parts being

joined, such as in bead-welding a seam, or spot welding. An electrode spark is also used to vaporize a metal/compound surface for spectrographic analysis of constituent elements. Glass welding uses torches in a similar manner to metal welding, and requires controlled annealing after welding. Where fluxes such as borox are used in welding, clean with water or commercial solvents designed for the particular flux. Alloy brazing is a form of high temperature welding and used with alloys above 800°C. Below 800°C it is called soldering.
REF: Walker, P & Tarn, W H — personal application (all), 1950/1985

BB-0005
ETCH NAME: Alloy bonding — TIME: Variable
TYPE: Wire, bonding — TEMP: RT and 800°C
COMPOSITION:
x .... Au, Al, etc.
DISCUSSION:

Alloy bonding technique. There are hundreds of alloys available with temperatures ranging from about 50 to 800°C (800 to 2500°C for brazing rather than alloying). Two of the more common solders are #62 60Pb/40Sn, (m.p. 180°C), and #63 (m.p. 220°C) Pb/Sn/Ag(5%) . . . the former also called 60/40 solder and the most common general solder. Bonding is done with a solder iron using solid or rosin-core wire, or a solder pre-form and heat. Parts may be pre-tinned for assembly, such as for a PCB, and passed through a furnace or a gas torch, such as propane; acetylene or oxy-hydrogen may be used depending upon temperature required. Occasionally used to solder wire directly to a discrete device, more often used for final attachment of wires internally or externally in package assemblies. Where a resin core solder or flux is applied in alloying, bond area should be solvent or alcohol cleaned after bonding to remove residual flux. *Note:* Remaining flux and some commercial solvents may contain chemicals that will attack assembled components, but good water washing is often sufficient for cleaning.
REF: Tarn, W H & Walker, P — personal application, 1950/1985

BB-0006
ETCH NAME: Epoxy bonding — TIME: 30—60 min
TYPE: Parts, bonding — TEMP: 120—150°C
COMPOSITION:
x .... Ag-epoxy or polyimides
DISCUSSION:

Epoxy paste technique. As an electrical contact both epoxies and polyimides are loaded with fine particles of Ag, Au, Al, Cu, Ni, Pd, Pt, etc. as a paste. Used unloaded for general bonding of parts where electrical continuity is not required. Pastes are applied to the wire/bond pad, the device or wire positioned, then the assembly is oven-cured with time and temperature depending upon the particular compound mixture. Widely used in microelectronic circuit assemblies for testing, but not in final high reliability packaging. Cure schedules range from about 10 to 45 min at 120 to 150°C, and there is ongoing development of higher temperature pastes (>250°C). Epoxy type pastes do not require cleaning after cure. Alloy pastes with fine beads of the alloy in a resin base are similarly used with heat melting, and do require final cleaning. See BB-0005.
REF: Marich, L A & Rodriguez, S — personal communication, 1979

BB-0007
ETCH NAME: — TIME: 10—20 sec
TYPE: Acid, cleaning — TEMP: RT

COMPOSITION:

1 .... H_2O_2

1 .... HCl

DISCUSSION:

Pb/Sn type solder joints attaching wire leads to discrete silicon diodes. Copper, dumet, aluminum, and silver wires attached vertically by soldering to the device surface. Operation done by hand or with a carbon/metal holding fixture and parts passed through a furnace in an N_2 or forming gas atmosphere. Solution used for light cleaning of formed bonds, following with water rinse, then alcohol rinsing and air dry.

REF: Walker, P — personal application/development, 1950—1970

BB-0008

ETCH NAME: Alcohols/solvents TIME: 10—30 sec

TYPE: Alcohol, cleaning TEMP: RT to hot

COMPOSITION:

(1) x .... MeOH, EOH (2) x .... TCE, TCA, Freons

DISCUSSION:

Pb/Sn solder joints or other metal alloys. Both alcohols and solvents used as general cleaners for wire alloying to parts. Water only is often sufficient for removal of residual solder fluxes — borax a major flux — used at RT or hot. Some solder fluxes contain fluorine compounds, such that particular care should be exercised during cleaning (CaF_2 standard metal flux).

REF: Fahr, F — personal communication, 1957

BB-0009

ETCH NAME: Cladding TIME:

TYPE: Pressure, bonding TEMP: RT to molten

COMPOSITION:

x .... pressure

DISCUSSION:

Metals, plastics, cloth, and wood. Dissimilar metal sheets or strips are formed with pressure/heat by passing thin sheets/strips through rollers with sufficient temperature to cause metal-metal migration at the interface to form a bond by pressure extrusion. A similar process is used with plastics or woods using glues or other binders, and called veneer. Plasticizing may include additional adhesives, or applied as impregnation. Linen cloth sheets are plastic impregnated under pressure to form a nonconductive board (PCBs); sawdust, plant and hair fibers, asbestos, glass wool, wood pulp, mica flakes, are all used with various chemical binders, and press-formed as sheet, blocks, or other shapes. Mica and glass wool products are replacing asbestos for insulation. Plasticized veneer, for relatively chemically inert table tops in laboratories, is replacing natural soapstone, or marbles. Glass-to-metal tube as a graded seal is made for fabricating metal-to-glass assemblies; metal wires are glass-beaded for insulated feed-thrus in package fabrication with melt/press insertion. Mylar (red, blue, white) sheets are pressure clad to clear acetate sheet, as rubylith for photolithographic circuit fabrication or, mylar with a sticky back is used to hold a wafer in place for scribe & break wafer dicing.

REF: Siracusa, M et al — personal communication, 1980

BB-0010

ETCH NAME: Coldweld TIME:

TYPE: Pressure, bonding TEMP: RT

COMPOSITION: PRESSURE: 2—20 tons

x .... pressure

DISCUSSION:

Copper or nickel clad/plated shim steel or pure nickel Solid State packages with a seal lip or rim used for package sealing of both SCR Hockey Puk, and quartz radio frequency packages. A coldweld sealing system is rated for tons of direct pressure, with 5—7 tons as nominal. There is an upper and lower die operated by a hydraulic press and the two dies may be surrounded by individual cylinders that mesh for a vacuum seal pulled to about 10^{-3} Torr, and capable of gas back-fill after pulling vacuum. Bottom and top die are individually designed for each specific package configuration. The package body to be sealed is seated in the lower die, the cover positioned on the package, then the upper die activated to lower and pressure seal. This produces a copper/copper or nickel/nickel metal seal, and eliminates any weld splatter as observed in resistance welded packages. Small quartz crystal and semiconductor TO-5 type packages have been sealed at 2—5 tons pressure; larger silicon SCR Hockey Puk packages at 6—10 tons pressure. Quartz crystals packages have been sealed in a vacuum system at the 10^{-10} Torr level or, sealed after vacuum pull and backfill with He or N_2/He (15%), to reduce mass loading effects using helium as a lightweight atmosphere within the package. Package surfaces to be sealed should be degreased and dry before assembly.

REF: Walker, P — personal application, 1968—1969

BB-0011

ETCH NAME: Resistance weld
TYPE: Electric, weld
COMPOSITION:
x .... power

TIME: Seconds
TEMP: RT
PRESSURE: 10—200 psi
POWER: 1000—1500 V

DISCUSSION:

A standard Raytheon type resistance welder uses a capacitor bank discharge to effect the weld on steel package parts, which may or may not be gold plated. The system contains a copper bottom package holder and top head for pressure and power discharge, and there is the possibility of weld splatter inside the package. These systems are still a major method of sealing Solid State packages. Another system uses a powered roller head somewhat like a glass cutter tool that, as it is rolled across the sealing surface, produces a bead of microspot welds for sealing, and the head and/or package may be heated during weld passes. It has been used on circular packages, but it is now more widely used on steel square and flat paks in Solid State assembly. This operation requires four individual weld passes with 90° rotation after each two parallel weld passes. The weld seal produces a very fine weld-bead with minimum internal package weld splatter.

REF: Walker, P — personal application, 1962—1980

BB-0012

ETCH NAME: Alloy pre-form
TYPE: Metal, sealing
COMPOSITION:
x .... AuGe, AuSn, etc.

TIME:
TEMP: Variable

DISCUSSION:

Pure metals or metal mixtures cut as pre-forms and used for device/part mounting or package sealing. Operation similar to that discussed under BB-0005 (alloy bonding) as a heated bonding mechanism. Shaped In, Cu, Au pre-form flats or wires used as vacuum system crush-seals, with indium also used for sealing cryopumps. Crush-seals do not use heat, and the pre-form is replaced whenever the system seal is opened and requires re-sealing.

REF: Walker, P et al — personal application, 1958—1985

BB-0013
ETCH NAME: Metal evaporation
TYPE: Metal, cladding
TIME: Variable
TEMP: Variable
COMPOSITION:
x .... metal
DISCUSSION:

Metal evaporation when applied as a surface cladding. Aluminum, copper, carbon, nickel, iron, etc. have all been used as a thin film metallization on a variety of materials: on metals, plastics, mylar, glass, and so forth. Carbon is evaporated on plastic or mylar sheeting and, with a sticky-back, used as a sunscreen on windows; aluminum is evaporated on plastic or rubber toy balloons, as well as weather balloons for its heat dissipation and reflective capability. Aluminum and nickel are replacing silver as mirrors, including telescope lenses, and are also used on clothing. Copper and aluminum flakes or powder are added to plastic acrylic-type spray paints as a "glitter" coating and surface sealer. Glassy metals and nitrides are evaporated for specific colors of the oxides or nitrides: red, blue, green, gold, etc. Electrolytic plating has been used in a similar manner for color, and may include color additives in solution, such as in the anodizing of aluminum: black, red, green, blue, and so forth. Spray coating with metal powders are applied to surfaces as a thin film bonded material for both decorative effects and in fabricating Solid State substrates or active devices, such as a carbon film resistor. Both brass and bronze are used for their yellow color as replacements for gold, as direct spray coatings or mixed with liquid plastics like aluminum and copper.

REF: Walker, P — personal application, 1955—1985 (mirrors)

PHYSICAL PROPERTIES OF BORON, B

Classification	Metal
Atomic number	5
Atomic weight	10.811
Melting point (°C)	2100 (2300)
Boiling point (°C)	2800 (2550) sub.
Density (g/cm^3)	2.34
Specific heat (cal/g) 25°C	0.309
Heat of fusion (k-cal/g-atom)	5.3
Heat of vaporization (k-cal/g-atom)	128
Atomic volume (W/D)	4.6
1st ionization energy (K-cal/g-mole)	191
1st ionization potential (eV)	8.298
Electronegativity (Pauling's)	2.0
Covalent radius (angstroms)	0.82
Ionic radius (angstroms)	0.23 (B^{+3})
Electrical resistivity ($\times 10^{-6}$ ohms cm) 20°C	4
0°C	1.8
Electron work function (eV)	4.5
Hardness (Mohs — scratch)	9.3
Crystal structure (hexagonal — normal)	$(10\bar{1}0)$ prism, hcp
Color (solid)	Grey-white
Cleavage (basal — poor)	(0001)

BORON, B

General: Does not occur as a native element, but is widely found as a constituent in several minerals, such as beryl and chrysoberyl, both beryllium aluminates and major precious gem stones, both as "Cat's Eye", or yellow and emerald green mock emerald. The borates form a distinctive mineral group as oxygen salts, some 50 individual minerals, borax, $Na_2B_4O_7.10H_2O$, being one of the best known. Boric acid, H_3BO_3 is of common occurrence in hot springs and associated with volcanic areas.

The pure metal has little or no use in industry, though nitrides and borides have application as dielectrics and high temperature ceramics with similar characteristics to those of aluminum oxide. Boric acid has long been used as a mild antiseptic, and borax is a general cleaner and metal alloy flux, as well as a water softener.

Technical Application: In Solid State processing boron has been the major p-type dopant for silicon wafers since their original development for semiconductor devices. Boric acid was first used, then boracine, and then diborane, B_2H_6, as a gas diffusion source. With ion implantation, I^2, B^+ as an ionic species is now used. The diborane also is used for doping silicon dioxide, SiO_2, for use as a p-type doping drive-in source, or as a thin film passivating layer with the acronym of BSG. Boron nitride (BN) also is used as a drive-in source with silicon wafers sandwiched between BN discs. As BCl_3 as an RF plasma, it is a dry chemical etching (DCE) method for etch removal or patterning of aluminum thin films. It is the best method to date for such processing due to the inherent difficulty of controlled etching of aluminum oxide in device fabrication.

Etching: HNO_3, H_2SO_4, molten salts, and metals.

BORON ETCHANTS

B-0001

ETCH NAME: Hydrofluoric acid — TIME:

TYPE: Acid, removal — TEMP: RT

COMPOSITION:

x .... HF, conc.

DISCUSSION:

B, as an amorphous surface layer remaining after ion implantation of boron in silicon (100) and (111) wafers. HF used to remove the glassy layer remaining on surfaces after implant. Also used on phosphorus and arsenic implantation glassy layers. Acronyms: PSG and ASG, respectively.

REF: Prussin, S et al — *J Appl Phys,* 57,181(1985)

B-0002a

ETCH NAME: Heat — TIME:

TYPE: Thermal, preferential — TEMP: 2000°C

COMPOSITION:

x .... heat

DISCUSSION:

B, single crystal ingot. During growth and cooling from 2000°C, heat will produce macroscopic surface structure similar to that observed by chemical etching: hexagonal, spiral, pyramidal pits, and oval plateaus.

REF: Talley, C P — *Nature (London),* 182,1593(1958)

B-0002b

ETCH NAME: Nitric acid — TIME:

TYPE: Acid, removal — TEMP: RT

COMPOSITION:

x .... HNO_3, conc.

DISCUSSION:

B, single crystal ingots. Solution used as a general etch for boron. Sulfuric acid also can be used.

REF: Ibid.

B-0002c

ETCH NAME: Beryllium	TIME:
TYPE: Metal, removal/preferential	TEMP: Molten
COMPOSITION:	
x .... Be	

DISCUSSION:

B, as boron nitride test blanks. Metallic beryllium, uranium, nickel, and platinum will attack boron. See Boron Nitride.

REF: Ibid.

B-0003

ETCH NAME: RF plasma	TIME:
TYPE: Ionized gas, growth	TEMP:
COMPOSITION:	GAS FLOW:
x .... BCl_3	PRESSURE:
	POWER:

DISCUSSION:

B, grown as thin films. Describes a method of evaporating the films and use of ion bombardment to vaporize the metal and melt the boron as a thin film surface layer.

REF: Barnes, D et al — *Atomic Energy Res Establ* (Harwell Mem R/M), 125,4(1951)

PHYSICAL PROPERTIES OF BORON CARBIDE, B_4C

Classification	Ceramic
Atomic numbers	5 & 6
Atomic weight	55
Melting point (°C)	2350
Boiling point (°C)	3500
Density (g/cm^3)	2.50
Application limits (°C in air)	540
(°C in inert atm)	2260
Mean specific heat (J/kg °C) 25—1000°C	2090
Coefficient of linear thermal expansion ($\times 10^{-6}$/cm/cm°C) 25—800°C	5.7
Thermal conductivity (w/m °C) 800°C)	17.3
Hardness (Mohs — scratch)	9.3
Crystal structure (isometric — normal)	(100) cube
Color (solid)	Black
Cleavage (cubic)	(001)

BORON CARBIDE, B_4C

General: Carbides do not normally occur as natural compounds, but are artificially fabricated using many different metals. There is a single occurrence of a mineral carbide called moissanite, CSi, which has been found as small green hexagonal platelets in meteoric iron at Cañon Diablo, Arizona. Note that artificial silicon carbide is shown as SiC.

Industrially, boron carbide is one of the high temperature refractory ceramics, though not as widely used as silicon carbide, but with similar applications as fire brick, furnace tubes, etc. In powder form, it has been used as a lapping abrasive.

Technical Application: Not used in Solid State processing to any extent, to date, other than as a lapping and polishing abrasive. There are possible applications as a pressed powder substrate for circuit fabrication or in package construction.

Etching: Insoluble in acids. Soluble in fused salts.

BORON CARBIDE ETCHANTS

BC-0001

ETCH NAME: Potassium hydroxide | TIME:

TYPE: Molten flux, removal | TEMP: 365°C

COMPOSITION:

x .... KOH, pellets

DISCUSSION:

B_4C, as pressed powder blanks. Material can be etched in fused salts: alkaline hydroxides, and carbonates.

REF: *A Dictionary of Carbide Terms* — Adams Carbide Corp., Kenilworth, NJ

BORIDES, MBx

General: Do not occur in nature as metallic boron compounds, though boron is a constituent element in many minerals as hydrated silicates, phosphates, sulfides, and carbonates. Borax, $Na_2B_4O_7.10H_2O$, is representative of the borate mineral group.

The borides are artificially grown in industry as high temperature ceramics with a Mohs hardness generally greater than H = 8. In the general formula shown: M = Cr, Mo, Nb, U, V, Ta, W, and Zr; and Bx = B, B_2, B_{12}, or as a pentaborate, B_2O_5. The materials are used as crucibles, furnace tubes, and liners, etc. for their refractive capabilities and general chemical inertness. Some are used in powder form as lapping and polishing abrasives; others for their dielectric properties in sheet, rod, or flats that are metallized as capacitors or resistors.

Technical Application: No major use in Solid State device fabrication, although there is possible application as circuit substrates and packaging assemblies. Some borides are under evaluation as thin film surface coatings with applications similar to those of SiO_2 and Al_2O_3, or as discrete Solid State dielectric elements, sputter evaporated as thin films and planar surface capacitors or resistors in circuit fabrication.

Most of the borides shown in the following section have been grown as single crystal specimens for general morphological studies. See Boron Nitride and Boron Carbide.

Etching: Varies with compound — soluble in HNO_3, H_2SO_4, H_2O_2; alkalies or mixed acids; and aqua regia.

Note: The following are the borides listed:

Cerium boride, CeB_6
Chromium boride, Cr_3B_2
Molybdenum boride, Mo_2B_5
Niobium boride, Nb_3B_3
Rare earth borides, $R_3Ni_7B_2$
Silicon boride, SiB_6
Tantalum boride, TaB_2
Titanium boride, TiB_2
Tungsten boride, WB
Uranium boride, UB_2
Vanadium boride, VB_2
Zirconium boride, ZrB_2

BORIDE ETCHANTS

CERIUM BORIDE ETCHANTS

CEB-0001
ETCH NAME: Heat
TIME:
TYPE: Thermal, annealing
TEMP: Elevated
COMPOSITION:
x .... heat
DISCUSSION:
CeB_6, ingots grown by Czochralski (CZ) method. Also grown were $CeCu_6$, and $LaCu_6$. Both of these compounds are orthorhombic in structure. $CeAl_3$ was grown by arc melting, and annealed 1 week at 900 to 1000°C. Also grew $CeCu_2Si_2$. All materials were used in a general structural study.
REF: Onuki, Y et al — *J Phys Soc Jpn,* 4,1210(1984)

CHROMIUM BORIDE ETCHANTS

CRB-0001a
ETCH NAME: Perchloric acid
TIME:
TYPE: Acid, removal
TEMP: Hot
COMPOSITION:
x .... $HClO_4$
DISCUSSION:
Cr_3B_2 specimens. Solution is a general etch for this boride. Material used in a composition study. Other chromium borides were Cr_2B, Cr_3B_4, and CrB_2O. Compounds will not etch in HF, HCl, HNO_3, or alkalies.
REF: *Ceramics and High Temperature Materials Handbook,* Vol 85, #1, Coltronics Corp., 1982

CRB-0001b
ETCH NAME: Hydrogen peroxide
TIME:
TYPE: Acid, removal
TEMP:
COMPOSITION:
x .... H_2O_2
DISCUSSION:
Cr_3B_2 specimens. This acid is a general etch for all compounds of this material, as well as other peroxide chemical solutions.
REF: Ibid.

CRB-0001c
ETCH NAME: Sulfuric acid
TIME:
TYPE: Acid, removal
TEMP: Hot to boiling
COMPOSITION:
x .... H_2SO_4
DISCUSSION:
Cr_3B_2 specimens. Very slow etch attack with this solution, but the higher the metal content, the greater the attack.
REF: Ibid.

MOLYBDENUM BORIDE ETCHANTS

MOB-0001a
ETCH NAME: Nitric acid — TIME:
TYPE: Acid, removal — TEMP:
COMPOSITION:
x .... HNO_3
DISCUSSION:
Mo_2B_5 specimens. Solution is a general etch. Also used on Mo_2B; and MoB. Will not etch in HCl. (*Note:* Mixed HF:HNO_3 will etch.)
REF: Ibid. CRB-0001a

MOB-0001b
ETCH NAME: Sulfuric acid — TIME:
TYPE: Acid, removal — TEMP: Hot
COMPOSITION:
x .... H_2SO_4
DISCUSSION:
Mo_2B_5 specimens. Solution is a general etch.
REF: Ibid. CRB-0001a

MOB-0001c
ETCH NAME: Sodium hydroxide — TIME:
TYPE: Molten flux, removal — TEMP: 365°C
COMPOSITION:
x .... NaOH (KOH)
DISCUSSION:
Mo_2B_5 specimens. Solution is a general etch. It may be preferential on single crystal material.
REF: Ibid. CRB-0001a

MOB-0002
ETCH NAME: Acetone — TIME:
TYPE: Ketone, cleaning — TEMP:
COMPOSITION:
x .... CH_3COCH_3
DISCUSSION:
MoB surface penetration film developed in a study of boriding metals at less than 670°C. Other metals evaluated were Ni, Fe, Co, Ti, Nb, and Hastalloy B. See Nickel Boride for details of cleaning procedure of substrates and borate baths used in process evaluated. See NIB-0002.
REF: Koyama, K et al — *J Electrochem Soc,* 126,147(1978)

NIOBIUM BORIDE ETCHANTS

NBB-0001a
ETCH NAME: Sulfuric acid — TIME:
TYPE: Acid, removal — TEMP: Hot to boiling
COMPOSITION:
x .... H_2SO_4

DISCUSSION:

Nb_3B_3 specimens. Solution is a slow etch on both Nb_3B_3 and NbB. These borides will not etch in: HCl; HNO_3 or aqua regia (1HCl:1HNO_3).

REF: Ibid. CRB-0001a

NBB-0001b

ETCH NAME: Potassium hydroxide — TIME:

TYPE: Molten flux, removal — TEMP: 365°C

COMPOSITION:

x .... KOH (NaOH)

DISCUSSION:

Nb_3B_3 specimens. A general etch for this material. (*Note:* Eutectic mixtures as 1NaOH:1KOH will also etch, and may be slightly preferential on single crystals.)

REF: Ibid. CRB-0001a

NBB-0001c

ETCH NAME: Hydrofluoric acid — TIME:

TYPE: Acid, removal — TEMP:

COMPOSITION:

x .... HF

DISCUSSION:

Nb_3B_3 specimens. This is a slow etch on this material.

REF: Ibid. CRB-0001a

NBB-0001d

ETCH NAME: Potassium sulfate — TIME:

TYPE: Acid, removal — TEMP:

COMPOSITION:

x .... K_2SO_4

x .... H_2O

DISCUSSION:

NB_3B_3 specimens. Several sulfates will etch this material. Some carbonates will also act as etchants.

REF: Ibid. CRB-0001a

RARE EARTH BORIDE ETCHANTS

RNIB-0001

ETCH NAME: Nitric acid — TIME:

TYPE: Acid, removal — TEMP:

COMPOSITION:

x .... HNO_3

DISCUSSION:

RNi_6B_2 buttons were grown by arc melting under argon, then each specimen was annealed 2 weeks at 900°C. In the general formula shown, "R" stands for rare earth, though other workers use "RE" as the acronym. Here, R = Nd, Ce, Gd, Eu, and Yb. Material used in a magnetics and structural study.

REF: Felner, I — *J Phys Chem Solids,* 44,43(1983)

SILICON BORIDE ETCHANTS

SIB-0001
ETCH NAME: Water — TIME:
TYPE: Acid, cleaning — TEMP:
COMPOSITION:
x .... H_2O
DISCUSSION:
SiB_6 specimens fabricated as 0.5 mm diameter spheres. Water used to clean spheres prior to study of the unit cell space group. The material is orthorhombic system. Weissenberg photographs were made on the b- and c-axes.
REF: Adamsky, R F — *Acta Crystallogr*, 11,144(1958)

TANTALUM BORIDE ETCHANTS

TAB-0001a
ETCH NAME: Potassium hydroxide — TIME:
TYPE: Alkali, removal — TEMP:
COMPOSITION:
x .... x% KOH (NaOH)
x .... H_2O
DISCUSSION:
TaB_2, TaB, and Ta_3B_4 specimens. This material can be etched in various concentrations of these alkalies. These borides do not etch in HCl, HNO_3, or aqua regia ($3HCl:1HNO_3$).
REF: Ibid. CRB-0001a

TAB-0001b
ETCH NAME: — TIME:
TYPE: Acid, removal — TEMP:
COMPOSITION:
x .... HF
x .... H_2SO_4
DISCUSSION:
TaB_2 and other specimens. Solutions of these acids are slow etchants on this material.
REF: Ibid. CRB-0001a

TAB-0001c
ETCH NAME: Sodium peroxide — TIME:
TYPE: Acid, removal — TEMP:
COMPOSITION:
x .... x% Na_2O_2
x .... H_2O
DISCUSSION:
TaB_2 and other specimens. The solution is a general etch for this material. Bisulfates and carbonates will also act as etchants.
REF: Ibid. CRB-0001a

TITANTIUM BORIDE

TIB-0001
ETCH NAME: Water TIME:
TYPE: Acid, cleaning TEMP:
COMPOSITION:
x .... H_2O
DISCUSSION:
TiB_2 and TiC single crystal wafers used in a study of elastic constants. Both materials were either cleaved (001) or diamond saw cut. Lap polish surfaces with diamond paste. Degrease with TCE, rinse in alcohol and water. Authors say that both materials appear to be harder than diamond. No etch shown.
REF: Gilman, J J & Roberts, B W — *J Appl Phys,* 32,1406(1961)

TUNGSTEN BORIDE ETCHANTS

WB-0001a
ETCH NAME: Hydrofluoric acid TIME:
TYPE: Acid, removal TEMP:
COMPOSITION:
x .... HF
DISCUSSION:
WB_2, W_2B_2, W_2B_5, and Beta-WB specimens. A general etch for these materials. WB will not etch in HCl; hot H_2SO_4 or HNO_3.
REF: Ibid. CRB-0001a

WB-0001b
ETCH NAME: Aqua regia TIME:
TYPE: Acid, removal TEMP:
COMPOSITION:
3 .... HCl
1 .... HNO_3
DISCUSSION:
WB_2 and other tungsten borides can be etched in aqua regia.
REF: Ibid. CRB-0001a

WB-0001c
ETCH NAME: Sodium hydroxide TIME:
TYPE: Alkali, removal TEMP:
COMPOSITION:
x .... x% NaOH (KOH)
x .... H_2O
DISCUSSION:
WB_2 and other tungsten borides can be etched in various solution concentrations of these alkalies.
REF: Ibid. CRB-0001a

URANIUM BORIDE ETCHANTS

UB-0001a
ETCH NAME: Sodium peroxide
TYPE: Acid, removal
COMPOSITION:
x .... Na_2O_2
x .... H_2O
TIME:
TEMP:
DISCUSSION:
UB_4, UB_2, and UB_{12} specimens. Difficult to etch controllably as the materials decompose in peroxides.
REF: Ibid. CRB-0001a

UB-0001b
ETCH NAME: Nitric acid
TYPE: Acid, removal
COMPOSITION:
x .... HNO_3
TIME:
TEMP:
DISCUSSION:
UB_4, UB_2, and UB_{12} specimens. Difficult to etch controllably as the materials decompose in acids.
REF: Ibid. CRB-0001a

UB-0002
ETCH NAME:
TYPE: Cut, forming
COMPOSITION:
x .... electricity
TIME:
TEMP:
DISCUSSION:
UB, (110) cut single crystal specimens spark cut for fabrication as spheres. Compound has the ThB_4 structure and grows with a preferred (110) orientation. The material is brittle and fractures easily on (001) with cubic cleavage. Density: 3.98 g/cm^3.
REF: Menovsky, A et al — *J Crystal*, 9,70(1984)

VANADIUM BORIDE ETCHANTS

VB-0001a
ETCH NAME: Nitric acid
TYPE: Acid, removal
COMPOSITION:
x .... HNO_3
TIME:
TEMP:
DISCUSSION:
VB_2 and VB specimens. A general etch for these materials. Does not etch in HCl, HF, H_2SO_4.
REF: Ibid. CRB-0001a

VB-0001b
ETCH NAME: Potassium hydroxide
TYPE: Alkali, removal
COMPOSITION:
x .... x% KOH (NaOH)
x .... H_2O
TIME:
TEMP:

DISCUSSION:

VB_2 and VB specimens. This material can be etched in various liquid concentrations of alkalies.

REF: Ibid. CRB-0001a

VB-0001c

ETCH NAME: Potassium nitrate — TIME:

TYPE: Acid, removal — TEMP:

COMPOSITION:

x .... x% KNO_3

x .... H_2O

DISCUSSION:

VB_2 and VB specimens. Solutions of nitrates will etch this material. Also carbonates, bisulfates, and peroxides.

REF: Ibid. CRB-0001a

ZIRCONIUM BORIDE ETCHANTS

ZRB-0001a

ETCH NAME: Potassium hydroxide — TIME:

TYPE: Alkali, removal — TEMP:

COMPOSITION:

x .... KOH (NaOH)

x .... H_2O

DISCUSSION:

ZrB, ZrB_2, Zr_3B_4, and ZrB_{12} specimens can be etched in various concentrations of the alkalies. Will not etch in HCl or HNO_3.

REF: Ibid. CRB-0001a

ZRB-0001b

ETCH NAME: Sodium peroxide — TIME:

TYPE: Acid, removal — TEMP:

COMPOSITION:

x .... Na_2O_2

DISCUSSION:

ZrB and other zirconium borides. Etching is violent with peroxides.

REF: Ibid. CRB-0001a

PHYSICAL PROPERTIES OF BORON NITRIDE, BN

Classification	Nitride
Atomic numbers	5 & 7
Atomic weight	24.8
Melting point (°C)	2730
Boiling point (°C)	2.25
Density (g/cm^3)	
Application limit (°C)	650
Specific heat (mean)(J/kg °C) 25—1000°C	1570
Coefficient of linear thermal expansion ($\times 10^{-6}$ cm/cm/°C) 25—800°C	7.5 // c-axis
Thermal conductivity (W/m °C) 900°C	26

Electrical resistivity (ohms-cm $\times 10^{13}$) 25°C	1.7 // c-axis
(ohms-cm $\times 10^{10}$) 480°C	2.3 // c-axis
Refractive index (n=)	1.54—1.68
Dielectric constant (e=)	4—8
Energy band gap (eV)	3—10
Lattice constant (angstroms)	3.615
Hardness (Mohs — scratch)	8+
Crystal structure (hexagonal — normal) alpha	(1010) prism
(isometric — normal) beta	(100) cube
Color (solid)	White
Cleavage (dodecahedral)	(110)

BORON NITRIDE, BN

General: Does not occur as a natural compound, although there are several boron-containing minerals: sassolite, $B(OH)_3$, borax, $Na_2B_4O_7.10H_2O$, as examples. Boric acid, H_3BO_3, occurs in hot springs associated with volcanic action.

As a nitride it is an industrial ceramic with a high melting point, used as a crucible under reducing atmosphere conditions (H_2 or FG) for the processing of liquid metals, such as steels, aluminum and aluminum alloys, tin, bismuth, and others. In an oxidizing atmosphere BN converts to B_2O_3 above 650°C.

Technical Application: In Solid State processing pressed powder BN discs have been used as a diffusion source of boron as a p-type dopant in silicon, with silicon wafers sandwiched between the discs (boric acid was the first such dopant, then diborane, B_2H_6 as a gas.)

The nitride has been CVD grown as amorphous thin films, also for boron doping drive-in, or for use as a dielectric, e.g., capacitor circuit element. Under pressure, with elevated temperature, amorphous films have been converted to hexagonal or cubic single crystal structure. The latter was fabricated as a high temperature functioning diode. BN has been converted to B_2O_3 thin films and applied as passivation surface coatings. Also conversion to boron carbide, B_4C by treatment with CCl_4 vapor at elevated temperature.

Etching: Limited solubility in HCl, HF, H_2SO_4, H_2O_2, H_3PO_4.

BORON NITRIDE ETCHANTS

BN-0002a

ETCH NAME: Phosphoric acid | TIME:

TYPE: Acid, removal/cleaning | TEMP: RT to hot

COMPOSITION:

x .... H_3PO_4, conc.

DISCUSSION:

BN, as pressed powder blanks and parts. Solution used as a general cleaning and material removal etch.

REF: Bulletin C-941 eff 6/81(716/731/3221) — The Carborundum Company

BN-0009: Rand, M J & Roberts, J F — *J Electrochem Soc,* 115,423(1968)

BN, as a thin film amorphous coating on silicon, gallium arsenide, and other semiconductors. Solution used hot for removal of BN through photolithographic patterned gold masks; also used SiO_2 and Si_3N_4 as pattern masks.

BN-0010: Kim, C et al — *J Electrochem Soc,* 131,1384(1984)

BN, as thin film amorphous coating deposition by CVD on (100), n-type silicon wafers, then overcoat with Si_3N_4 to prevent out-diffusion of boron. Wafer heated to drive-in boron as p-type dopant. Remove silicon nitride with HF and remaining boron nitride with hot phosphoric acid.

BN-0002b
ETCH NAME: Sulfuric acid
TYPE: Acid, removal/cleaning
COMPOSITION:
x .... 20% H_2SO_4
TIME: To 1 h
TEMP: RT
RATE: 10.7 mg/cm³
DISCUSSION:

BN, as pressed powder test blanks. Dilute solutions of H_2SO_4 will attack boron nitride; whereas concentrated solution will not.

REF: Ibid.

BN-0002c
ETCH NAME: Nitric acid
TYPE: Acid, removal
COMPOSITION:
x .... HNO_3, conc. (fuming)
TIME: To 1 h
TEMP: RT
RATE: 8.9 mg/cm³
DISCUSSION:

BN, as pressed powder test blanks. "Fuming" nitric acid will attack boron nitride. (*Note:* Standard concentration is "white" fuming 70%; then "yellow" 72%; and "red" 74% fuming. When fuming nitric is shown, it is usually red fuming.)

REF: Ibid.

BN-0002d
ETCH NAME: Hydrofluoric acid
TYPE: Acid, removal
COMPOSITION:
x .... HF, conc.
TIME: To 1 h
TEMP: RT
RATE: 17 mg/cm³
DISCUSSION:

BN, as pressed powder test blanks. Solution will attack boron nitride.

REF: Ibid.

BN-0001b: Ibid.

BN, as pressed powder discs for boron drive-in of sandwiched silicon wafers. Soak in HF, rinse heavily in running DI water, and dry under IR lamp. Use HF 5 min, RT, for cleaning BN.

BN-0001a
ETCH NAME: Paper
TYPE: Abrasion, cleaning
COMPOSITION:
x .... filter paper
TIME:
TEMP: RT
DISCUSSION:

BN, as pressed powder discs. Light hand lapping on Whattman #5 filter paper used to clean surfaces, followed with N_2 blow off. Discs used in a metal plating study. Results were poor due to metal failure to adhere to the powdered surface.

REF: Walker, P & Dokko, P — personal application, 1985

BN-0002e
ETCH NAME: Sodium hydroxide
TYPE: Alkali, removal/cleaning
COMPOSITION:
x .... 20% NaOH
TIME: To 1 h
TEMP: RT
RATE: 8.9 mg/cm³

DISCUSSION:

BN, as pressed powder test blanks. Lower concentrations used for cleaning; higher concentrations for more rapid removal.

REF: Ibid.

BN-0002f

ETCH NAME: Lead oxide — TIME:

TYPE: Molten flux, removal — TEMP: Molten

COMPOSITION:

x .... PbO

DISCUSSION:

BN, as pressed powder test blanks. Molten flux metal oxides attack BN with conversion to B_2O_3.

REF: Ibid.

BN-0003

ETCH NAME: Hydrogen peroxide — TIME:

TYPE: Acid, removal — TEMP: RT to hot

COMPOSITION:

x .... H_2O_2

x .... H_2O

DISCUSSION:

BN, as amorphous thin films on silicon. Various concentrations used in etch patterning film through metal masks.

REF: Hirayama, M & Shomo, J — *J Electrochem Soc,* 122,1671(1975)

BN-0004

ETCH NAME: Argon — TIME: 5 h

TYPE: Gas, crystallization — TEMP: 1000°C

COMPOSITION:

x .... Ar

DISCUSSION:

BN, single crystal films. Initially deposited on copper substrates as clear, amorphous coatings. Heat treatment as shown above showed conversion to single crystal hexagonal BN. BN deposited much below 600°C were unstable in moist atmospheres and devitrified; thick films showed spontaneous delamination from the substrate during cooling after deposition. Films deposited at low temperature also swelled and partly dissolved in water, and thick films tended toward white, opaque, rather than clear.

REF: Motojima, S et al — *Thin Solid Films,* 88,269(1982)

BN-0005: Szmidt, J et al — *Thin Solid Films,* 110,7(1983)

BN, single crystal thin films grown by reactive pulse plasma on silicon, (111) and (110) orientated wafers, p- and n-type. Films were called Borazone and consisted mainly of beta-BN.

BN-0006

ETCH NAME: — TIME:

TYPE: Metal, removal — TEMP: Molten

COMPOSITION:

x .... Be, U, Ni, Pt

DISCUSSION:

BN, (100) as cubic boron nitride. Single crystals with grown junctions fabricated under 55 K atm pressure at 1700°C. For n-type, doped with Be; for p-type, doped with Si. Fabricated

as a diode the BN appears capable of operating up to 530°C and possibly 1300°C. Diamond also shows semiconducting properties, but is more difficult to fabricate. (*Note:* Silicon carbide, Si_3C, has been fabricated as a high temperature diode.)
REF: Mishima, C et al — *Science,* October 1987
BN-0007: Greenberg, J (Ed) — *Sci News,* 132,241(1987)
BN, (100) new development as a diode reported as shown in BN-0006.
BN-0008: Merbarki, M et al — *J Cryst Growth,* 61,636(1983)
BN, (0001) specimens grown as hexagonal single crystal platelets were yellow to colorless. BN has a layer structure like graphite parallel to the c-axis. Growth was with boron in a silicon flux under nitrogen in an induction furnace at 1850°C.

PHYSICAL PROPERTIES OF BORON PHOSPHIDE, BP

Classification	Phosphide
Atomic numbers	5 & 15
Atomic weight	41.79
Melting point (°C)	1300
Boiling point (°C)	
Density (g/cm^3)	
Electrical resistivity (ohms-cm)	0.41 (100) 5.64 (111)
Carrier concentration (n) ($\times 10^{17}$ cm^{-3})	1.0 (100) 3.07 (111)
Electron mobility (u) ($cm^2/V/sec$)	145 (100) 36.5 (111)
Lattice constant (angstroms $\times 10^{-6}$)	4.539 (100) 4.536 (111)
Hardness (Mohs — scratch)	9+
(Vicker's — kg/mm^2)	3000-4000
Crystal structure (isometric — normal)	(100) cube
Color (solid)	Red-orange
Cleavage (cubic)	(001)

BORON PHOSPHIDE, BP

General: Does not occur as a natural compound, though there are many boron minerals as phosphates, arsenates, carbonates, and sulfates. The borate group, as hydrous salts, such as borax, has major industrial use. BP is not normally used in metal processing, although it can be a phosphorus additive in irons and steels, brass, bronze, etc.

Technical Application: Boron phosphide is a III—V compound semiconductor with the common (111) polarity of such compounds. The individual elements are silicon dopants: boron, p-type; phosphorus, n-type. They have been occasionally used for dual-doping, but this is not recommended as there can be drastic alteration of the material resistivity with subsequent heat processing.

The material has been grown by the Czochralski (CZ) method; and in a molten flux, both methods as single crystals. It also has been deposited as an oriented thin film. It does not oxidize at elevated temperature, and is under evaluation as a high temperature diode, similar to silicon carbide and boron nitride. As a thin film it has been fabricated as a photovoltaic device.

Etching: Alkali electrolytic solutions. Molten flux salts.

BORON PHOSPHIDE ETCHANTS

BP-0001
ETCH NAME:
TIME:
TYPE: Molten flux, preferential
TEMP: Molten
COMPOSITION:
3 .... NaOH
1 .... Na_2O_2
DISCUSSION:

BP, (100) and (111) grown as single crystal platelets by a flux method using Cu_3P or $Ni_{12}P_5O$ as the flux at high temperature and under high pressure. Remove crystals from flux by etching in mixtures of $HF:HNO_3$. The (111) crystals are deep red in color; (100) crystals, orange/red. The molten flux etch shown was used to develop dislocations and other structure. The (111) surface is wavy, representative of slight misorientation. Growth temperature was 1300°C with pressure of 18 atm under argon.

REF: Kumashiro, Y et al — *J Cryst Growth,* 70,515(1984)

BP-0002: Kumashiro, Y et al — *J Cryst Growth,* 70,507(1984)

BP, (100) and (111) single crystal thin films grown by CVD on silicon substrates of like orientation. The BP crystal plates were removed from the silicon by etching away the silicon in $HF:HNO_3$. A form of float-off for film study under TEM.

BP-0003
ETCH NAME: Sodium hydroxide
TIME:
TYPE: Electrolytic, removal
TEMP:
COMPOSITION:
ANODE: BP
x .... 10% NaOH
CATHODE:
POWER: 0.1—10 A/cm^2

DISCUSSION:

BP, single crystal wafers. Solution used in a dark room under red light for general removal and polishing — higher concentrations may show some preferential attack. P-type etches more than 100× faster than n-type.

REF: Chu, T L & Chu, S S — *J Electrochem Soc,* 123,259(1976)

PHYSICAL PROPERTIES OF BORON TELLURIDE, B_2Te_3

Classification	Telluride
Atomic numbers	5 & 52
Atomic weight	404.47
Melting point (°C)	~1800
Boiling point (°C)	
Density (g/cm^3)	
Hardness (Mohs — scratch)	7—8
Crystal structure (isometric — normal)	(100) cube
Color (solid)	Grey/black
Cleavage (cubic)	(001)

BORON TELLURIDE, B_2Te_3

General: Does not occur as a natural metallic telluride, although there are several other natural telluride and selenide minerals. Most boron minerals are silicates or phosphides, and there is the natural occurrence of boric acid, $B(OH)_3$, as the single crystal sassolite or extracted from the water of hot springs. The telluride has little use in the metals industry other than as a possible compound semiconductor.

Technical Application: Boron telluride is a III—IV compound semiconductor and has been quite widely studied, alone, as an isomorphous series with selenides, and as telluride-selenide-sulfide mixed crystals. There is no major application as Solid State devices at the present time, although it may be used as a thin film material in heterostructural devices.

Etching: Mixed acids of HF:HNO_3:HAc/H_2O; dichromates alone, or as acid mixtures.

BORON TELLURIDE ETCHANTS

BTF-0001
ETCH NAME: Potassium chromate — TIME: 10 min
TYPE: Acid, preferential — TEMP: 95°C
COMPOSITION:

x .... 0.5 *M* $K_2Cr_2O_7$

DISCUSSION:

B_2Te_3, (111) wafers used in a dislocation study. Solution will develop dislocations on the (111)A — boron surface only.

REF: Harper, C A — *Handbook of Materials and Processes for Electronics,* McGraw-Hill, New York, 1970, 7

BTE-0002: Gatos, H C & Lavine, M C — *Lincoln Lab Tech Rep,* 293, January 1963

B_2Te_3, single crystal material used in a general development study.

PHYSICAL PROPERTIES OF BORON TRIFLUORIDE, BF_3

Classification	Fluoride
Atomic numbers	5 & 9
Atomic weight	67.82
Melting point (°C)	−127
Boiling point (°C)	−101
Density (g/cm^3) liquid	2.99
Hardness (Mohs — scratch)	1.5—2
Crystal structure (isometric — normal)	(100) cube
Color (solid)	Colorless
(gas)	Colorless
Cleavage (cubic)	(001)

BORON TRIFLUORIDE, BF_3

General: Does not occur as a natural compound, although there are other metallic fluorides, such as fluorite, CaF_2, and cryolite, Na_2AlF_6, both of which have important optical properties and use as fluxes in metal processing. Note that cryolite has a refractive index (n = 1.33×) that is the closest of any mineral to that of water (n = 1.00), such that it "disappears" when introduced into water. The trifluoride is not used in the metal industries but, as a fluoride compound, could be used as a flux in soldering similar to fluorite applications.

Technical Application: Boron trifluoride is under study and development in Solid State processing as a thin film with single crystal orientation. It has possible optical characteristics and applications, and is under evaluation for defect structure.

Etching: H_2SO_4: mixed acids of HF:HNO_3.

BORON TRIFLUORIDE ETCHANTS

BTF-0001
ETCH NAME: TIME:
TYPE: Acid, preferential TEMP:
COMPOSITION:
1 .... HF
3 .... HNO_3
DISCUSSION:

BF_3, (100) oriented thin films grown by CVD. Solution used to develop defects in a structure study.

REF: Caine, E J — *J Electron Mater,* 13,341(1984)

BRASS: 70Cu:30Zn

General: Does not occur as a natural alloy, although there are some carbonates and arsenates. The minerals rosasite, $(Cu,Zn)CO_3.(Cu,Zn)OH_3$, and barthite, $3ZnO.CuO.3As_2O_5.H_2O$ are representative of such hydrates.

Brass is of ancient origin, circa 4500 B.C., as some possible artifacts of brass have been found in Egyptian tombs of the Pharaoh dynastic period, but they are rare and were pre-dated by the discovery of bronze, circa 5000 B.C. There are still individual bronzes and brasses today, but far more trinary mixtures of Cu:Zn:Sn (Bronze, Cu:Sn) which are more commonly called brass. There are other brass mixtures containing Pb, Al, Ni, Mn, P, and Fe, in addition to tin, Sn. Brass is more widely used as it is not as brittle as bronze, and is more easily fabricated. A few of the better known brasses are shown below:

Formula	Name
85Cu:15Zn	Red brass
67-72Cu:28-32Zn:xPb:xFe	Cartridge or spring brass
67Cu:33Zn	Yellow brass (ordinary)
61Cu:39Zn	Wire brass

In addition to the above, there are some well-known named brasses, such as muntz metal, pewter, and German silver. All brasses, as well as several other metal alloys, are heat treated for a specific hardness, such as $^1/_4$- or $^1/_2$-hard, as against the dead-soft heat treatment of aluminum or gold wire used in Solid State device bonding. Brass is a fair electrical conductor and has excellent thermal transfer characteristics, such that it is used as a heat sink for mounting devices in electrical testing. It also is quite resistant to atmospheric corrosion, and can be used outside under variable weather conditions. Brasses are available in a wide range of forms — rod, tube, sheet, wire, thin shim stock, bulk material for shaping or cutting — and finely powdered brass is added to plastics and paint as a coloring pigment. It can be drawn or extruded into shaping, such as for cartridge casings, which is not the case of bronze.

Technical Application: Brass is not used in the direct fabrication of Solid State devices, but it is used in package designs, as general heat sinks, or device test blocks. Where critical devices, such as high frequency microelectronic gallium arsenide FETs are being tested under elevated temperature conditions, there can be a problem of out-diffusion of zinc from the brass into the device under test. This can alter or destroy device characteristics. To prevent this from occurring, copper plate the brass surfaces to a sufficient thickness, such that the copper will act as a blocking layer against zinc diffusion, vs. the temperature level being used.

There are many uses of brass parts in material processing. Shim stock has been used as a pattern mask for applying ink to a semiconductor wafer that is then etch-patterned, or the patterned shim used as a metal evaporation mask. Rods are used as either holders for probes in testing, or as the probes themselves. Brass platens have been in use for many years for lapping materials — rocks and minerals — as well as many other compounds. An abrasive slurry is added directly to the platen surface, or a felt-type pad is glued to the surface, then the pad saturated with an etch solution; or an abrasive slurry is drip-fed onto the pad with the platen rotated. Examples are the Buehler, Speed-Fam, or Strausbaugh lapping/polishing machines.

Alpha-brass has been grown as single crystals for morphological studies.

Etching: HCl, H_2SO_4, H_3PO_4, HNO_3, aqua regia, and other mixed acids.

BRASS ETCHANTS

BRA-0001a

ETCH NAME: TIME:

TYPE: Acid, removal TEMP: RT

COMPOSITION:

(1) 16 .... HNO_3
160 .... H_2O

(2) (6) .... KCl
100 .... H_2O

DISCUSSION:

Brass, cartridge, or eyelet. A general two-step etch for cleaning brass. Etch with the dilute nitric acid solution, water rinse; then follow with the salt solution, and water rinse.

REF: Oberg, E & Jones, F D — *Machinery Handbook,* 4th ed, The Industrial Press, New York, 1968

BR-0002a: Walker, P & Campos, R — personal application, 1961

Brass/bronze piping on cryogenic trailers. Used the (2) solution as 6 g KCl:100 ml H_2O for light cleaning of parts.

BRA-0008a

ETCH NAME: Scale dip or fire-off dip TIME:

TYPE: Acid, removal TEMP: RT

COMPOSITION:

1 .... HNO_3
1 .... H_2O

DISCUSSION:

Brass specimens and parts. As a ''scale'' dip, the solution is used to remove oxide prior to a brite dip. As a ''fire-off'' dip, solution is used as a general removal etch for brass, bronze, copper, and copper alloys.

REF: Bulletin #B-11D — The American Brass Co.

BRA-0004: Maddin, R & Asher, W R — *Rev Sci Instr,* 21,881(1957)

Brass specimens. A Saran plastic thread wetted with the solution mixture was used as an acid-saw to cut specimens.

BRA-0005: McGuire, T R — Yale University, Ph.D. dissertation (1949)

Brass, specimen. Acid-saw cutting method reference from BR-0004.

BRA-0001b

ETCH NAME: TIME:

TYPE: Acid, removal TEMP: RT

COMPOSITION:

4 .... HCl
1 .... HNO_3

DISCUSSION:
Brass specimens, as cartridge or eyelet. A general etch for both brass and bronze. (*Note:* Solution is very close to aqua regia.)
REF: Ibid.
BRA-0002: Ibid.
Brass and bronze cryogenic trailer piping and parts were etch cleaned in this solution.

BRA-0002c
ETCH NAME: Sulfuric acid, dilute TIME:
TYPE: Acid, removal TEMP:
COMPOSITION:
1 .... H_2SO_4
10 .... H_2O
DISCUSSION:
Brass and bronze cryogenic trailer piping and parts. Solution used for corrosion removal.
REF: Ibid.

BRA-0008b
ETCH NAME: Bichromate finish dip TIME:
TYPE: Acid, finish TEMP:
COMPOSITION:
12 oz .... H_2SO_4
4 oz .... $NaCr_2O_3$
1 gal .... H_2O
DISCUSSION:
Brass, bronze, and copper parts. Solution used to remove scale (oxide). Produces a semi-matte finish.
REF: Ibid.

BRA-0008c
ETCH NAME: Matte dip TIME:
TYPE: Acid, finish TEMP:
COMPOSITION:
2 .... H_2SO_4
1 .... HNO_3
x .... *ZnO or ZnS

*x = saturate the solution

DISCUSSION:
Brass, bronze, and copper alloys. Solution will produce a matte finish on parts.
REF: Ibid.

BRA-0006
ETCH NAME: TIME:
TYPE: Acid, removal TEMP:
COMPOSITION:
x .... NH_4OH
x .... H_2O_2
DISCUSSION:
Brass specimens and parts. Used as a general etch for brass. Solution reaction reduced with the addition of water as a cleaning etch.

REF: Bergland, T & Deardon, W H — *Metallographer's Handbook of Etching,* Pitman & Sons, London, 1931

BRA-0013a
ETCH NAME: Phosphoric acid
TYPE: Acid, preferential
COMPOSITION:
x .... H_3PO_4
x .... H_2O
TIME:
TEMP: Hot
DISCUSSION:

CuZn(30%) specimens used in a study of striations and slip. Use very dilute solutions for defect development. Solution cannot be used electrolytically as it will polish specimens. A $(NH_4)_2SO_4$:NH_3 solution did not work. Mechanical hand polishing used to reduce damage depth, and slip did not show below a damage depth of about $^1/_4$ mm.
REF: McLeon, D — *J Inst Met,* 95(1947—1948)
BRA-0019: Boswell, F W C & Weinberg, F — *J Appl Phys,* 31,1835(1960)

Alpha-brass specimens. Specimen blanks were lapped down and etch thinned for direction observation of dislocation movement. A 16 mm movie camera with a 2″ focal length, 1.6F lens was used to record the results. (*Note:* A similar movie was made while etching silicon wafers in 1957.)

BRA-0013b
ETCH NAME:
TYPE: Acid, preferential
COMPOSITION:
x .... $2NH_4Cl.CuCl_2.H_2O$
x .... KOH (NaOH)
TIME:
TEMP:
DISCUSSION:

CuZn(30%) specimens. Solution will develop striations and slip. See BRA-0013a for further discussion.
REF: Ibid.

BRA-0014
ETCH NAME: Jacquet's etch
TYPE: Electrolytic, polish
COMPOSITION:
345 ml .... $HClO_4$
655 ml .... HAc
TIME:
TEMP:
ANODE: Brass
CATHODE:
POWER:
DISCUSSION:

Brass specimens. Solution used as a general electrolytic etch.
REF: Barrett, C & Levenson, L H — *Trans Am Inst Min Met (England),* 137,112(1940)

BRA-0015
ETCH NAME: Sodium chloride
TYPE: Electrolytic, preferential
COMPOSITION:
x .... 5% NaCl
TIME:
TEMP:
ANODE: Brass
CATHODE:
POWER:
DISCUSSION:

Brass specimens. Solution used to develop etch figures.
REF: Fesch, C H & Whyte, S — *J Inst Met,* 904,10(1913)

BRA-0007
ETCH NAME: Ferric chloride TIME:
TYPE: Acid, removal TEMP: RT to hot
COMPOSITION:
x .... 30—35°Bé $FeCl_3$
x .... H_2O
DISCUSSION:

Brass, bronze, and copper alloy parts. Solution used as a general etch for metals to include irons, steel, and nickel, and may be for surface cleaning only, polishing, or preferential. (*Note:* Ferric chloride solutions are mixed as Normal; percent as shown; or by Baume, °Bé. The latter is a measure of solution viscosity. A common mixture is as 35°Bé.)
REF: N/A

BR-0008d
ETCH NAME: Bright dip TIME:
TYPE: Acid, removal TEMP: RT
COMPOSITION:
2 gal .... H_2SO_4
1 gal .... HNO_3
1/2 fl oz .... HCl
1—2 qt .... H_2O
DISCUSSION:

Brass, bronze, and copper alloys. Surfaces are usually oxide stripped with a scale dip or pickling solution, such as the solution shown, prior to a brite dip.
REF: Ibid.

BRA-0008d
ETCH NAME: Pickling solution TIME:
TYPE: Acid, clean TEMP: RT to hot
COMPOSITION:
x .... 12—15% H_2SO_4
DISCUSSION:

Brass, bronze, and copper alloys. Solution used to soak and remove oxide and scale prior to further treatment of material.
REF: Ibid.

BRA-0009
ETCH NAME: Phosphoric acid TIME:
TYPE: Electrolytic, polish TEMP:
COMPOSITION: ANODE: Brass
x .... 35% H_3PO_4 CATHODE:
POWER:
DISCUSSION:

Brass, as single crystal alpha-brass. Wafers were cut and polished in the solution shown. Specimens were used in a study of slip modes.
REF: Wilsdorf, H & Fourier, J J — *Acta Metall,* 4,271(1956)
BR-0010: Wilsdorf, H — *Z Metallkd,* 45,14(1954)
BRA-0012: Fourie, J T — *Acta Metall,* 8,88(1960)

Alpha-brass as single crystal specimens used in a study of slip during plastic deformation. Power used was 2.67 V.

BRA-0016
ETCH NAME: TIME:
TYPE: Pressure, preferential TEMP:
COMPOSITION:
x .... deformation
DISCUSSION:
Alpha-brass single crystals deformed in a study of secondary slip. Also studied copper crystals. Secondary slip nucleates at the end of primary slip lines when internal stress fields are high.
REF: Michell, T E & Thornton, P R — *Phil Mag,* 10,314(1964)

BRA-0017
ETCH NAME: TIME: 120 min
TYPE: Acid, cutting TEMP: RT
COMPOSITION:
(1) 200 ml .... HCl
50 g .... $FeCl_3$
250 ml .... H_2O
(2) 1 .... HNO_3
1 .... H_2O
DISCUSSION:
Brass specimens $^3/_{16}''$ thick are soaked in solution (1) used to cut brass, aluminum, and copper. Solution (2) used to cut tin and zinc. (*Note:* An excellent review of metal growth methods.)
REF: Honeycombe, R W K — *Met Rev,* 4,1(1959)

BRA-0018
ETCH NAME: Nitric acid TIME:
TYPE: Acid, cleaning TEMP: RT
COMPOSITION:
1 .... 45% HNO_3
1 .... H_2O
DISCUSSION:
Alpha-brass [(Cu:Zn) with 29.38% Cu)]. Cut as 12 mm square blanks and plated with copper containing radioactive tracer $^{29}Cu^{64}$ used in a study of the mechanisms of Cu_2O formation. Brass was cleaned: (1) degrease with TCE; (2) clean in solution shown; and (3) air dry. After plating, blanks were water soaked for 15 h to oxide tarnish surfaces. See CUO-0016.
REF: Birley, S S & Tromans, D — *J Electrochem Soc,* 118,636(1971)

PHYSICAL PROPERTIES OF BROMINE, Br_2

Classification	Halogen
Atomic number	35
Atomic weight	159.83
Melting point (°C)	−7.3 (−7.2)
Boiling point (°C)	58.78
Density (g/cm^3) 20°C	3.12
(g/cm^3) 59°C	2.93
Thermal conductivity ($\times 10^{-4}$ $cal/cm^2/cm/°C/sec$) 0°C	0.091
Latent heat of fusion (cal/g)	16.2
Specific heat (cal/g/°C)	0.107
Solubility (100 ml H_2O) 50°C	52 ml

1st ionization potential (eV)	11.84
Ionic radius (angstroms)	1.96 (Br^{+1})
Cross section (barns)	6.7
Vapor pressure (°C)	9.3
Crystal structure (orthorhombic—normal)	(100) a-pinacoid
Color (liquid/solid)	Dark red-brown
Cleavage (basal — solid)	(001)

BROMINE, Br

General: Bromine and mercury are the only two elements that are liquid at standard temperature and pressure (22.2°C and 76 mmHg). Mercury is found as a solid, native element in nature, but bromine is not, being primarily extracted from sea water and from the soil of alluvial plains in the form of an alkaline halide. It is a dark red-brown liquid with an offensive odor and a member of the halogen group of elements: fluorine and chlorine are gases; bromine is a liquid; and iodine, a solid. It is more reactive than iodine but less reactive than either chlorine or fluorine, with fluorine being the most reactive element known. All of the halogens form diatomic molecules as F_2, Cl_2, Br_2, and I_2, even though they are occasionally shown in the literature as monatomic, e.g., Br, Cl, etc.

The most important use of bromine is as ethylene dibromide, $C_2H_4Br_2$, which is added to leaded ethyl gasoline to remove by-products of tetraethyl lead. It also is used for medicinal purposes, KBr or NaBr as sedatives; in silver emulsions (AgBr) for photography; and in the manufacture of "tear gas" as bromine-acetone. It is poisonous if ingested and can produce sores on the skin which are difficult to heal. The vapors can be a severe irritant to eyes, nose, and throat.

As a low percentage mixture in methyl alcohol (up to 3%), it is a major polish and removal etch for many metals and metallic compounds, and can be a preferential etchant above 3%, used to about 20% in MeOH. These Br_2:MeOH mixtures have acquired the acronym BRM. The exact bromine concentration that alters from polish to preferential can vary to some degree by method of application, such that it is not accurately defined. BRM etchants are often used on semiconductor wafer surfaces after cutting and prior to final cleaning as a chemical/mechanical (chem/mech) lapping solution to remove residual subsurface damage remaining after cutting and mechanical lap/polish with abrasives. The solutions also are used for general removal, as a jet for thinning, etc. In all cases, when used as an etch solution, bromine can be replaced by iodine with similar results and, as bromine, is so volatile, iodine is often preferred when solutions are to be used above room temperature. Bromine has been mixed with other alcohols or solvents, and shows similar results as with BRM solutions.

It is used as an additive to other etch solutions, such as Camp #4, now known as CP4. The original CP4 contained 0.6 ml Br_2 and, when used without bromine, is called CP4A. There are several CP4 modifications where the Br_2 has been replaced by I_2, $Cu(NO_2)_3$, $AgNO_3$, etc.

Technical Application: Bromine is used in etching solutions as described in the preceding General section. It has been solidified and studied by freezing.

Etching: Soluble in MeOH.

The following is a selected list of BRM solutions as applied to several semiconductor materials, metals and other compounds. Additional solutions are shown under the individual elements/compounds in this Etchant Section:

Formula	Material	Use	Ref.
0.5% Br_2:MeOH	AlAs, GaAs	Polish	AS-0001
10 ml P Etch:10 ml Br_2 "PBr Etch" (P: 10 ml HCl:10 ml HNO_3:5 ml H_2O)	CdTe	Preferential	CDTE-0006a
0.05% Br_2:MeOH	GaAs	Polish	GAS-0006
1% Br_2:MeOH	GaAs	Polish	GAS-0003a
1.5% Br_2:MeOH	GaAs	Polish/thinning	GAS-0004b
5% Br_2:MeOH	GaAs	Selective	GAS-0009
1% Br_2:EOH	GaAs	Preferential	GAS-0013
2% Br_2:MeOH	GaAs	Oxide removal	GAS-0132
20% Br_2:MeOH	GaP	Polish	GAP-0007
Br_2, conc.	GaP	Preferential	GAP-0010
HBr, conc.	GGG	Preferential	GGG-0001b
250 ml 30% KOH:25 ml Br_2	Gs, Si	Preferential	GE-0012b
30HF:50HNO_3:30HAc:0.6Br_2 "CP4"	Ge, Si, InAs, InSb	Polish/preferential	GE-0065c
1% Br_2:MeOH	CdInTe	Polish/clean	CIT-0001
1% Br_2:MeOH	CdP_2	Selective	CDP-0001
4% Br_2:MeOH	CdSe	Polish (0001)	CDSE-0002
1% Br_2:MeOH:x1MHCl	CdSe	Clean	CDSE-0004
10% Br_2:MeOH	CdTe	Polish	CDTE-0010
2% Br_2:MeOH	CdTe	Polish/removal	CDTE-0009
2% Br_2:Glycol	CdTe	Smut removal (Te)	CDTE-0011
1 1 *M* $FeCl_3$:1 12 *M* HCl:1 1.25 *M* HBr	Cu	Dislocation	CU-0015
6K_2SO_4:1H_2O_2:1H_2O then 1% Br_2:MeOH	$CuInSe_2$	Polish/clean	CUIS-0003
2% Br_2:MeOH	GaSb	Polish (100)	GASB-0004c
1Br_2:10MeOH	GaSb	Preferential	GASB-0008a
5—10% Br_2:MeOH	InAs	Dislocation	INAS-0005
0.5% Br_2:MeOH	InP	Step-etch (100)	INP-0007
9 *N* HBr (48%)	InP	Preferential	INP-0026
5HF:1Br_2	InP	Preferential	INP-0017b
2$H_3PO_4$1Br_2	InP	Preferential	INP-0038
2H_2SO_4:1HBr (47%)	InP	Dislocation	INP-0051
xBr_2:MeOH	Fe	Oxide removal	FE-0100
1HBr:1H_2O + Cr, pcs	SnO_2	Removal (thin film)	SNO-0002
$CBrF_2$, ionized	W	RF plasma	W-0011
xBr_2:MeOH	ZnSe	Polish/removal	ZNSE-0001
xBr_2:CS_2	ZnSe	Polish/removal	ZNSE-0001

BROMINE ETCHANTS

BR-0001

ETCH NAME: Methyl alcohol | TIME:

TYPE: Alcohol, preferential | TEMP: 0°C to RT

COMPOSITION:

x .... CH_3OH (MeOH)

DISCUSSION:

Br_2, specimens solidified by freezing the liquid to 0°C with liquid nitrogen, LN_2. A

drop of MeOH on the surface was used to develop etch figures and crystalline structure of the solidified mass. Used in a general etching study of bromine vs. iodine as applied on (111) silicon wafers. A piece of solid bromine was placed on wafer surfaces and allowed to liquefy. A similar piece of iodine was used with just sufficient heating to liquefy, and surface artifacts were compared.
REF: Walker, P & Waters, W P — personal application, 1957

BRONZE, Cu:Sn

General: Bronze does not occur in nature as a specific single crystal, although native copper often contains small amounts of iron, silver, bismuth, lead, antimony, and tin. It was probably the natural occurrence of a copper-tin alloy that led to the discovery and development of bronze as pure copper was extensively mined throughout southern Europe and the Near East in ancient times, e.g., 5000 B.C. or earlier. Pure copper is soft and malleable and the two copper carbonate minerals, which are usually associated with copper vein deposits, are easily recognized by their brilliant colors: azurite (deep blue) and malachite (bright green); both are easily reduced to pure copper at temperatures between 200 to 250°C. Those two minerals also have been in use as semi-precious gem stones since ancient times and are still used in jewelry.

Although the use of gold and copper pre-dates the use of bronze, it was the development of bronze that led to the "Bronze Age" and the beginning of man's metal culture. Pure copper is too soft, but bronze can be cast, shaped, and honed to a fine cutting edge far superior to that of flint or bone, as the latter fracture, chip, and sliver easily. Although bronze is brittle in comparison to irons and steels, it is highly resistant to weathering and corrosion. Bronze artifacts have been discovered in archaeological digs, pre-dating the Christian era, and are still usable as weapons or tools with little wear or alteration.

Today there are many types of bronze with specific names, such as aluminum-bronze; manganin; sea water bronze; phosphor-bronze; bearing bronze, and others. With the development of brass, Cu:Zn, which is not brittle like bronze, and the addition of other metal elements in the two mixtures, the terms brass and bronze have become synonymous in many instances as they are mixtures of Cu:Zn:Sn plus other metal constituents.

Bronze still has many industrial uses, such as the sea water bronze for its resistance to salt water and phosphor-bronze which, when used as thin sheet, has excellent spring characteristics. Because of the high copper content, both bronze and brass are excellent electrical and heat conductors — copper telephone wires are an alloy rather than pure copper, and bronze/brass tools are used in working with magnetic materials as they are nonmagnetic and unaffected by the magnetic flux. Cryogenic liquid piping and fittings are made of bronze as the material structure is tough, and unaffected by continual severe changes in temperature, such as from RT down to -196°C (LN_2), and back up to 100°C on a hot day.

Technical Application: Bronze is not used in direct Solid State material fabrication, although phosphor-bronze spring clips are widely used to hold semiconductor wafers in position during metallization in vacuum systems, or as holders in device test fixtures. Bronze/brass parts also are used as heating or cooling blanks; as abrasive lapping platens; as electrical probes; as an electronic assembly packaging material; as electrical pin contacts; inking and evaporation masks; and electrical test blocks.

There is reference to brass having been grown and studied as a single crystal, but we have no similar reference for bronze, although it is probable that it has been so grown and studied.

Etching: HNO_3, H_2SO_4, and acid mixtures.

BRONZE ETCHANTS

BRO-0001
ETCH NAME:
TIME:
TYPE: Acid, removal
TEMP:
COMPOSITION:
5 .... HCl
100 .... HNO_3
DISCUSSION:
Cu:Sn(30%), standard bronze specimens. Solution is a general etch for bronze. (*Note:* This is a dilute form of aqua regia and can be used on both bronzes and brasses.)
REF: Oberg, E & Jones, F D — *Machinery Handbook,* 4th ed, The Industrial Press, New York, 1951

BRO-0002a
ETCH NAME:
TIME: Variable
TYPE: Acid, cleaning
TEMP: RT
COMPOSITION:
5 .... HCl
100 .... HNO_3
150 .... H_2O
DISCUSSION:
Bronze and brass cryogenic trailer piping. Solution used as a general surface cleaner for removal of dirt and grease from these parts. After etching, heavy water wash and pat dry with toweling. Where severe corrosion was present, parts were lightly scrubbed in solution with a stainless steel brush.
REF: Campos, R & Walker, P — AF Contract to Clean Cryogenic Trailers, 1961

BRO-0002b
ETCH NAME: Sulfuric acid
TIME: 1—10 min
TYPE: Acid, cleaning
TEMP: RT
COMPOSITION:
1 .... H_2SO_4
10 .... H_2O
DISCUSSION:
Bronze and brass cryogenic trailer piping and parts. Solution used as a corrosion removal etch to include light scrubbing with a stainless steel brush. Water wash after etching and pat dry with toweling.
REF: Ibid.

BRO-0003
ETCH NAME: Ferric chloride
TIME:
TYPE: Salt, macroetch
TEMP:
COMPOSITION:
x .... 5% $FeCl_3$
x .... H_2O
DISCUSSION:
Bronze artifact. Solution used as a macro-etch on bronze arrowheads found at the biblical Migdol Fort, dated in the 6th century A.D. The etch developed a dendritic cast structure. (*Note:* Various concentrations of ferric chloride are used in structure development etching of both bronzes and brasses.)
REF: Peiles, J et al — *Sci Am,* 244,61(August 1976)

BRO-0004a
ETCH NAME:
TYPE: Electrolytic, macroetch
COMPOSITION:
1 g .... CrO_3
90 ml .... H_2O
TIME:
TEMP:
ANODE:
CATHODE:
POWER: 1.5—2 VDC
DISCUSSION:
88—96Cu:2.3—10.5Al:xFe:xSn, as aluminum bronze. Solution used to develop structure in this material.
REF: *A/B Met Dig,* 21(22),23(1983) — Buehler Ltd

BRO-0005
ETCH NAME:
TYPE: Salt, anisotropic
COMPOSITION:
x .... *$FeCl_3$
TIME: 15, 30, or 90 min
TEMP: RT

*1.4 g/cm^3

DISCUSSION:
Bronze:Mg, as magnesium bronze. Solution was used in a centrifuge to produce a positive (+) and negative (−) gravity during the etching period. Used to form pits and via-holes in both bronze and gallium arsenide specimen sheets or wafers.
REF: Kuiken, N K & Tilburg, R P — *J Electrochem Soc,* 130,1722(1983)

BRO-0004b
ETCH NAME:
TYPE: Acid, microstructure
COMPOSITION:
x .... NH_4OH
x .... H_2O_2
TIME:
TEMP:
DISCUSSION:
Bronze specimens. Solution used to develop microstructure. Also used on copper.
REF: Ibid.

BRO-0004c
ETCH NAME: Chromic acid
TYPE: Electrolytic, preferential
COMPOSITION:
x .... 1% CrO_3
TIME:
TEMP:
ANODE: Bronze
CATHODE:
POWER:

DISCUSSION:
Red brass specimens. Solution used to develop structure in a cast alloy that was furnace cooled.
REF: Ibid.

BRO-0004d
ETCH NAME: Chromic acid
TYPE: Acid, preferential
COMPOSITION:
1 g .... CrO_3
99 ml .... H_2O
TIME: Swab
TEMP: RT

DISCUSSION:

Bronze, as aluminum bronze (See BRO-0004a). Solution used as a wet chemical etch (WCE) rather than electrolytic. Apply with a swab wetted in solution to specimen surfaces to develop structure.

REF: Ibid.

BRO-0004e

ETCH NAME: ALN3-1 — TIME: Dip

TYPE: Acid, micro-, macroetch — TEMP: RT

COMPOSITION:

5 ml .... HNO_3
5 ml .... H_2SO_4
4 g .... CrO_3
1 g .... NH_4Cl
90 ml .... H_2O

DISCUSSION:

Red brass specimens. Solution used to develop microstructure.

REF: Ibid.

PHYSICAL PROPERTIES OF CADMIUM, Cd

Classification	Transition metal
Atomic number	48
Atomic weight	112.4
Melting point (°C)	320.9
Boiling point (°C)	765
Density (g/cm^3)	8.65
Thermal conductance (cal/sec)(cm^2)(°C/cm) 20°C	0.22
Specific heat (cal/g) 20°C	0.055
Latent heat of vaporization (cal/g)	286.4
Atomic volume (W/D)	13.1
1st ionization energy (K-cal/g-mole)	207
1st ionization potential (eV)	8.99
Electronegativity (Pauling's)	1.7
Covalent radius (angstroms)	1.48
Ionic radius (angstroms)	0.97 (Cd^{+2})
Linear coefficient of thermal expansion ($\times 10^{-6}$ cm/cm/°C) 20°C	29.8
Electrical resistivity (micro-ohms-cm) 20°C	6.83
Electron work function (eV)	4.07
Vapor pressure (mmHg in °C)	611
Cross section (barns)	2450
Hardness (Mohs — scratch)	2
Crystal structure (hexagonal — normal) alpha	$(10\bar{1}0)$ prism, hcp
(isometric — normal) beta	(100) cube
Color (solid)	White-silver
Cleavage (basal/cubic)	(0001)/(001)

CADMIUM, Cd

General: Does not occur as a native element. There are three cadmium minerals of which only greenockite, CdS, is of importance as an ore, though cadmium occurs as a trace constituent in several minerals. Greenockite does occur as single crystals, but is most common as a drusy coating associated with zinc deposits, e.g., sphalerite, ZnS. It is a by-product of zinc ore refining, volatilizing first and collected as the brown oxide. As the pure metal it is bluish-silver white, very soft, with a relatively low melting point and, like silver, tarnishes in air (presence of sulfur). When heated in air, it burns to the oxide.

It has industrial uses similar to those of zinc and, as a surface plating, is more corrosion resistant than zinc to salt (NaCl) spray, though it is slowly attacked by salts in a natural sea water environment, and several cadmium compounds are toxic. It is widely used in the plating of bolts and screws as, due in part to its softness, it has excellent low friction binding ability. It is a constituent in many low melting point alloys, and used as a bearing alloy for its low coefficient of friction and resistance to fatigue. It is in the standard cells used for EMF measurement, in vehicle batteries with zinc, and as the "cadmium battery" for electronics, such as transistor radios and hearing aids. As the sulfide, CdS, it is a bright yellow pigment for glass, enamel, and paint.

Technical Application: In Solid State processing, it is a p-type dopant in several compound semiconductors. Like indium, it is used alone or as a constituent in low melting solders for device package assemblies, though it is not recommended for space hardware. As the pure metal pre-form, rings or flats, it is used as a metal crush-seal in vacuum equipment, such as in cryo-pumps.

It is a metal element in a number of binary and trinary compound semiconductors such as cadmium sulfide, CdS, cadmium selenide, CdSe, and cadmium telluride, CdTe, as well as ZnCdS, CdHgTe, and trinary materials. It also has been grown and studied as a single crystal for general morphology, defects, and superconductivity.

Etching: HCl, HNO_3, mixed acids, and halogens.

CADMIUM ETCHANTS

CD-0001a
ETCH NAME: TIME:
TYPE: Acid, polish TEMP:
COMPOSITION:
1 .... HNO_3
2 .... CH_3COOH (HAc)
2 .... H_2O_2
DISCUSSION:

Cd, (111) and (100) single crystal wafers used in a morphological study. Wafers were etch polished in the solution shown.

REF: Gilman, J J & Johnston, W S — *Dislocations in Crystals,* John Wiley & Sons, New York, 1956

CD-0015a: DeCarlo, V J — *Met Abstr* 8,94(1958)

Cd specimens. Development work on polishing of both pure cadmium and zinc.

CD-0015b: DeCarlo, V J & Gilman, J J — *Trans Am Inst Min Met Eng,* 206,511(1956)

Cd specimens. Similar work to that of CD-0015a.

CD-0002
ETCH NAME: Neon TIME:
TYPE: Ionized gas, preferential TEMP:
COMPOSITION: GAS FLOW:
x .... Ne^+ ions PRESSURE:
POWER:
DISCUSSION:

Cd, single crystal specimens. Neon ion bombardment used to develop structure and orientation figures. Other metals studied were Al, Bi, Co, Mg, Cu, Sn, and Zn.

REF: Yurasova, V E — *Kristallografiya,* 2,1770(1957)

CD-0003a
ETCH NAME: TIME: 5—10 sec, then 30—60 sec
TYPE: Acid, polish TEMP: 20°C RT
COMPOSITION:
(1) 75 ml .... HNO_3, fuming
25 ml .... H_2O
(2) 10 ml .... HNO_3, fuming
70 ml .... HAc
DISCUSSION:

Cd, specimens. Use etch (1) to polish and follow with (2) as a wash; rinse in DI water and blow dry with nitrogen.

REF: Tegart, W J McG — *The Electrolytic & Chemical Etching of Metals,* Pergamon Press, London, 1956

CD-0007: McAfee, J — *Aust Eng Yearbook,* (1944)

Cd specimens. Solution used to polish cadmium. Author refers to using a dip etching technique.

CD-0004
ETCH NAME: Silicone — TIME:
TYPE: Plastic, forming — TEMP: 120°C
COMPOSITION:
x .... silicone
DISCUSSION:

Cd spheres from 44 to 1200 μm diameter formed by whipping the molten metal in a warm silicone oil bath. A study of size effects of material on superconductivity.

REF: Hein, R A & Steel, M V — *Phys Rev,* 105,877(1957)

CD-0005
ETCH NAME: Tri-iodide etch — TIME: Variable
TYPE: Halogen, removal/polish — TEMP: 40—80°C
COMPOSITION:
4 g .... I_2
20 g .... KI
80 ml .... H_2O
20 ml .... MeOH
DISCUSSION:

Cd specimens and alloys. Solution used to clean cadmium pellets and slugs prior to use in evaporation, follow with MeOH rinse, DI water rinse, and N_2 blow-dry. Also used to remove cadmium alloys or to clean surfaces after alloying.

REF: Fahr, F — personal communication, 1978

CD-0006
ETCH NAME: Ammonium hydroxide — TIME: 10—15 min
TYPE: Alkali, cleaning — TEMP: RT
COMPOSITION:
x .... NH_4OH
DISCUSSION:

Cd powder used for epitaxy growth of CdSe and $(Cd,Se)1_{-x}Zn_x$. Material was cleaned as follows: (1) soak in solution shown, (2) rinse in DI water, and (3) vacuum dry at 50°C. Once cleaned, the material can be stored for up to 2 weeks without requiring re-cleaning surfaces.

REF: Kim, S U & Parks, M J — *Jpn J Appl Phys,* 223,1070(1984)

CD-0008
ETCH NAME: — TIME:
TYPE: Mechanical, defect — TEMP:
COMPOSITION:
x .... stress
DISCUSSION:

Cd dislocation free single crystals used in a study of non-basal glide of (1011) oriented wafers in the $\langle 12\bar{1}0\rangle$ direction.

REF: Price, P B — *J Appl Phys,* 32,746(1961)

CD-0009: Price, P B — *J Appl Phys,* 32,1750(1961)

Cd, dislocation free single crystals used in a study of non-basal glide $(10\bar{2}2)$ of $(10\bar{2}2)$ oriented wafers in the $\langle 11\bar{2}3\rangle$ direction.

CD-0010: Coleman, R V & Sears, O W — *Acta Metall,* 5,131(1957)

Cd, single crystals. Describes a method of growing crystals.

CD-0011
ETCH NAME: Hydrochloric acid
TYPE: Acid, thinning
COMPOSITION:
x .... HCl
x .... H_2O
TIME:
TEMP:
DISCUSSION:

Cd specimens used in a study of the de-Hass-Van-Allen effect in cadmium at high magnetic fields. Solution used to reduce thickness of specimens.

REF: Grasse, A D C — *Phil Mag,* 17,847(1964)

CD-0012a: Kratechuil, P & Hemela, J — *Acta Metall,* 14,1757(1966)

Cd single crystal specimens used in a study of dislocations developed in deformed crystals. After annealing to remove oxide, crystals were etch cleaned/polished in HCl at RT, 1 min before preferential etching.

CD-0012b
ETCH NAME:
TYPE: Acid, dislocation
COMPOSITION:
160 g .... CrO_3
20 g .... Na_2SO_4
500 ml .... H_2O
TIME: 5 min
TEMP: RT
DISCUSSION:

Cd single crystals subjected to deformation used in a study of dislocations developed from strain. After etching in solution shown for 5 min, rinse in DI water. Solution used to develop dislocations.

REF: Ibid.

CD-0003b
ETCH NAME: Tri-iodide
TYPE: Halogen, polish
COMPOSITION:
1 g .... I_2
3 g .... KI
10 ml .... H_2O
TIME:
TEMP: RT to hot
DISCUSSION:

Cd, as single crystals and alloy specimens. Solution used for general removal and polishing. After etching, rinse in MeOH. See CD-0005.

REF: Ibid.

CD-0013
ETCH NAME:
TYPE: Mechanical, orientation
COMPOSITION:
x .... indentation
TIME:
TEMP: RT
DISCUSSION:

Cd single crystal wafer used in a study of lattice and grain boundary diffusion. Indent surfaces with a diamond point, and direction of cracks developed can be used to crystallographically orient the material.

REF: Wajda, E S — *Acta Metall,* 3,39(1955)

CD-0014: Lang, L G & Hien, N C — *Phys Rev,* 110,1002(1958)

Cd single crystal wafers cut from ingots by an acid saw technique.

CD-0015
ETCH NAME: Trichloroethylene
TYPE: Solvent, cleaning
TIME: 1—2 min
TEMP: Hot
COMPOSITION:
x .... TCE
DISCUSSION:

Cd cut pre-forms and alloy wire. Parts were vapor degreased in TCE before use in alloying microelectronic devices and package assemblies. After degreasing, parts were stored in a nitrogen dry box until ready for use.

REF: Marich, L A et al — personal communication, 1983

PHYSICAL PROPERTIES OF CADMIUM ANTIMONIDE, CdSb

Classification	Antimonide
Atomic numbers	48 & 51
Atomic weight	234.17
Melting point (°C)	1000
Boiling point (°C)	
Density (g/cm³)	
Hardness (Mohs — scratch)	6—7
Crystal structure (isometric — normal)	(100) cube
Color (solid)	Metallic grey
Cleavage (cubic)	(001)

CADMIUM ANTIMONIDE, CdSb

General: Does not occur as a natural metallic compound, although there is native antimony, and half a dozen minerals as arsenides, sulfides, not including several antimonates, etc. The most important mineral is stibnite, Sb_2S_3, as an ore which may contain both gold and silver.

There is no industrial application at present other than being evaluated as a compound semiconductor.

Technical Application: Cadmium antimonide is a II—V compound semiconductor and has been evaluated for characterization of its physical and semiconducting properties.

Etching: HCl, other single acids, mixed acids, or halogens.

CADMIUM ANTIMONIDE ETCHANTS

CDSB-0001
ETCH NAME: Hydrochloric acid
TYPE: Acid, polish
TIME:
TEMP:
COMPOSITION:
x .... HCl, conc.
DISCUSSION:

CdSb, (100) wafers used in a study of indium doping acting as a donor impurity. With increase of indium, diamagnetism becomes temperature dependent. Develops two types of holes (In and Sb) as local donor sites with 3 Sb sp-hybrid bonds and mixed conduction.

REF: Pilat, I M — *Phys Rev*, 110,354(1959)

PHYSICAL PROPERTIES OF CADMIUM ARSENIDE, Cd_3As_2

Classification	Arsenide
Atomic numbers	48 & 33
Atomic weight	486.05
Melting point (°C)	>1000
Boiling point (°C)	
Density (g/cm³)	~6
Hardness (Mohs — scratch)	6—7
Crystal structure (isometric — normal)	(100) cube
(hexagonal — normal)	$(10\bar{1}0)$ prism
Color (solid)	Greyish
Cleavage (basal)	(0001)

CADMIUM ARSENIDE, Cd_3As_2

General: Does not occur in nature as a metallic compound and, at present has no industrial application other than as a possible semiconductor compound.

Technical Application: Cadmium arsenide is a II—V compound semiconductor and has been evaluated for its physical and semiconducting characteristics alone and as a mixed single crystal of Cd_3As_2:Zn_3As_2.

Etching: HCl, other acids, and mixed acids.

CADMIUM ARSENIDE ETCHANTS

CDAS-0001

ETCH NAME: Hydrochloric acid — TIME:

TYPE: Acid, polish — TEMP:

COMPOSITION:

x .... HCl, conc.

DISCUSSION:

Cd_3As_2 single crystal specimens and Cd_3As_2:Zn_3As_2 as mixed single crystals used in a study of the anomalous thermal conductivity of these materials. Solution shown will polish both materials.

REF: Spitzer, D P et al — *J Appl Phys,* 37,3795(1966)

PHYSICAL PROPERTIES OF CADMIUM FLUORIDE, CdF_2

Classification	Fluoride
Atomic numbers	48 & 9
Atomic weight	150.4
Melting point (°C)	1100
Boiling point (°C)	1758
Density (g/cm³)	6.64
Hardness (Mohs — scratch)	2—3
Crystal structure (isometric — normal) alpha	(100) cube, bcc
(isometric — tetrahedral) beta	(111) tetrahedron
Color (solid)	White
Cleavage (cubic/octahedral)	(001)/(111)

CADMIUM FLUORIDE, CdF_2

General: Does not occur in nature as a compound due to its solubility in water. The alpha form is body-centered cubic (bcc), and the normal room temperature structure; whereas the beta form is the high temperature structure. Atomic structure is similar to fluorite, CaF_2, with a unit cell of eight fluorine atoms surrounding a central Cd or Ca atom (bcc). The main source of cadmium as the pure metal comes from sphalerite, ZnS, as a constituent trace element in the ore (about 0.2%), and there are three minor cadmium minerals: as a sulfide, oxide, and carbonate.

There has been little use as the fluoride in industry to date, but as a sulfide (CdS) it is one of the best and most stable yellow pigments used in paints, and for the coloring of glass. Cadmium metals as an alloy with bismuth are low temperature alloys.

Technical Application: No major use in Solid State processing to date, although it has been grown and studied as a single crystal both for general morphology and, when doped, for its semiconducting properties. As a CVD thin film it has been evaluated as a dielectric, as it has optical properties similar to fluorite, but is less used due to water solubility. Thin films also have been used for cadmium drive-in doping of semiconductor materials.

Etching: Soluble in water and acids.

CADMIUM FLUORIDE ETCHANTS

CDF-0001a
ETCH NAME: Water — TIME:
TYPE: Acid, polish/removal — TEMP: RT
COMPOSITION:
x .... H_2O
DISCUSSION:

CdF_2 specimens. As the material is slightly soluble in cold water, water can be used to polish surfaces. (*Note:* Mix with alcohol to reduce etch rate.)

REF: Hodgman, C D et al — *Handbook of Chemistry and Physics,* 27th ed, Chemical Rubber Co., Cleveland, OH, 1943, 356

CDF-0004: Sullivan, P W — *J Vac Sci Technol,* B2,202(1984)

CdF_2 as CVD deposited thin films on GaAs wafers used in a study of dielectric coatings. After photo resist patterning, film was E-beam annealed (EBA). Fluorine desorbs during annealing, and the remaining CdF_x can be removed by water washing. Also studied SrF_2 (SRF-0001).

CDF-0001b
ETCH NAME: Hydrofluoric acid — TIME:
TYPE: Acid, polish/removal — TEMP: RT
COMPOSITION:

(1) x .... HF, conc.

(2) x .... HF
x .... H_2O

(3) x .... HF
x .... MeOH/EOH

DISCUSSION:

CdF_2 specimens. HF solutions can be used to polish surfaces, or for general removal. Solution (3) is best for rate control, as CdF_2 is insoluble in alcohols.

REF: Ibid.

CDF-0003
ETCH NAME: Heat — TIME:
TYPE: Thermal, annealing — TEMP: 600—800°C
COMPOSITION:
x .... heat
DISCUSSION:

CdF_2 single crystal ingots, as both doped and undoped materials. Used in a study of

expansion coefficients and semiconducting properties. During annealing, if oxygen is present a brownish CdO_x film will form on the ingot surface.
REF: Acuna, L A & Fortiz, M — *J Phys Chem Solids,* 46,401(1985)

CADMIUM INDIUM SELENIDE, $CdIn_2Se_4$

General: Does not occur as a natural compound. See cadmium selenide for additional general discussion.

Technical Application: Cadmium indium selenide is a ternary semiconductor compound with polar characteristics similar to binary compounds. It is a ternary chalcogenide under development and application for its semiconductor capabilities both as a bulk single crystal element as well as an amorphous glassy material. This compound, and other chalcogenides, have been used to fabricate circuits as an alterable amorphous-semiconductor memory element for computer applications.

Etching: Aqua regia and additive mixtures of aqua regia solutions.

CADMIUM INDIUM SELINIDE ETCHANTS

CDISE-0001a
ETCH NAME: Aqua regia, variety TIME:
TYPE: Acid, polish TEMP:
COMPOSITION:
4 .... HCl
1 .... HNO_3
DISCUSSION:

$CdIn_2Se_4$ single crystal specimens used to fabricate photoelectrochemical cells from ternary chalcogenides. First, mechanically polish with 0.5 μm alumina. Etch in solution shown; follow with "poly sulfide". (See CDISE-0001b) at RT, seconds, then rinse in 10% KCN.
REF: Tennes, R et al — *J Electrochem Soc,* 129,1506(1982)

CDISE-0001b
ETCH NAME: TIME:
TYPE: Acid, polish TEMP:
COMPOSITION:
1 .... *aqua regia
10 .... 2 *M* KOH + S/Na_2S

*3HCl:1HNO_3

DISCUSSION:

$CdIn_2Se_4$ single crystal specimens. Solutions of aqua regia diluted with KOH, S, or Na_2S used as alternate polishing solutions. (See CDISE-0001a)
REF: Ibid.

CDISE-0002
ETCH NAME: TIME:
TYPE: TEMP:
COMPOSITION:

DISCUSSION:

Chalcogenides used as amorphous-semiconductor elements for alterable memory elements in computer applications. (*Note:* Excellent discussion of this subject.)

REF: Adler, D — Amorphous Semiconductor Devices, *Sci Am,* December 1979

CADMIUM INDIUM TELLURIDE, $CdIn_2Te_4$

General: Does not occur in nature as a native compound, although there are a number of selenide and telluride metallic compounds, but not with cadmium. There are no industrial applications, other than as a compound semiconductor.

Technical Application: The material is a II—III—VI trinary compound semiconductor and is under development as a possible laser diode or for photovoltaic devices.

Etching: Soluble in mixed acids of $HCl:HNO_3$ or $HF:HNO_3$ types, and halogens.

CADMIUM INDIUM TELLURIDE ETCHANTS

CDITE-0001

ETCH NAME: BRM — TIME: 10—30 sec

TYPE: Halogen, polish — TEMP: RT

COMPOSITION:

x .... 1% Br_2

x .... MeOH

DISCUSSION:

$CdIn_2Te_4$ as cut wafers from both p- and n-type doped polycrystalline ingots. Specimens were used in a study of general morphology. Surfaces were chem/mech polished in solution shown to remove native oxide, lap damage, and contamination. Band gap: 0.88 eV. Dielectric constant: 200. Space group crystal structure: S_4.

REF: Ou, S S et al — *J Appl Phys,* 57,354(1985)

PHYSICAL PROPERTIES OF CADMIUM IODIDE, CdI_2

Classification	Iodide
Atomic numbers	48 & 53
Atomic weight	366.25
Melting point (°C)	388
Boiling point (°C)	713
Density (g/cm^3)	5.67
Hardness (Mohs — scratch)	2—3
Crystal structure (hexagonal — normal)	$(10\bar{1}0)$ prism
Color (solid)	Brownish
Cleavage (basal — perfect)	(0001)

CADMIUM IODIDE, CdI_2

General: Does not occur as a natural compound as it is soluble in water. The mineral greenockite, CdS, is the most stable compound and is found in association with sphalerite, ZnS. Cadmium is a minor constituent in the zinc sulfide, and is collected during refining of the ore. Heated in air pure cadmium metal will burn to a brown powder of cadmium oxide, CdO.

There are no industrial applications for the iodide, though the sulfide as ''cadmium yellow'' is one of the most stable pigments in paints and as a glass coloring agent.

Technical Application: Cadmium iodide is a II—VIII compound semiconductor and under evaluation. It is a layered structure similar to CdS and ZrS_2 with distinctive basal (0001) cleavage.

It has been grown as a single crystal by zone refining under argon for general study and semiconductor characterization. Processing is done under argon as wafers react to moisture in the air.

Etching: Easily soluble in hot and cold water and most acids.

CADMIUM IODIDE ETCHANTS

CDI-0001
ETCH NAME: Water — TIME:
TYPE: Acid, polish — TEMP:
COMPOSITION:
x .... H_2O
DISCUSSION:

CdI_2, (0001) wafers cleaved from an ingot grown by zone refining under argon. Water can be used to polish surfaces. Material is under general study.

REF: Unnilrishnan, N V et al — *J Phys Chem Solids,* 45,1205(1984)

PHYSICAL PROPERTIES OF CADMIUM OXIDE, CdO

Classification	Oxide
Atomic numbers	48 & 8
Atomic weight	124.41
Melting point (°C)	900 del
Boiling point (°C)	
Density (g/cm^3)	8.192
Hardness (Mohs — scratch)	4—5
Crystal structure (isometric — normal)	(100) cube
Color (solid)	Brown/black
Cleavage (cubic)	(001)

CADMIUM OXIDE, CdO

General: Occurs as a natural compound and called cadmium oxide, CdO. It has been observed as minute octahedrons and as a black, brilliant coating on smithsonite, $ZnCO_3$ (calamine), which contains trace cadmium and indium. Cadmium metal is easily converted to a brown oxide by heating in air to about 350°C. There is no industrial application for the oxide, although cadmium metal is an alloy in steels, and has wide use in low temperature solders.

Technical Application: The oxide has had no use in Solid State processing to date, although there are a number of compound semiconductors containing cadmium as selenides and tellurides. The oxide has been studied for general morphology as a single crystal, but is considered an unwanted compound in processing cadmium-containing compound semiconductors.

Etching: HCl and NH_4OH easily soluble.

CADMIUM OXIDE ETCHANTS

CDO-0001
ETCH NAME: Ammonium hydroxide — TIME: 5—15 sec
TYPE: Acid, cleaning — TEMP: RT
COMPOSITION:
x .... NH_4OH
DISCUSSION:
CdO as a surface oxide on cadmium pellets used for metal evaporation and epitaxy, or as cadmium alloys used for soldering. Solution used to clean surfaces.
REF: Fahr, F — personal communication, 1978

CD0-0002
ETCH NAME: Hydrochloric acid — TIME:
TYPE: Acid, removal — TEMP:
COMPOSITION:
x .... HCl, conc.
DISCUSSION:
CdO native oxide on cadmium used for epitaxy growth of HgCdTe films in a study of chemical etching and oxidation. Remove cadmium oxides with HCl.
REF: Aspenes, D E & Arwin, H — *J Vac Sci Technol,* A(2),1309(1984)

PHYSICAL PROPERTIES OF CADMIUM PHOSPHIDE, CdP_2

Classification	Phosphide
Atomic numbers	48 & 15
Atomic weight	192.37
Melting point (°C)	>1000
Boiling point (°C)	
Density (g/cm³)	4.5
Hardness (Mohs — scratch)	6—7
Crystal structure (isometric — tetrahedral)	(111) tetrahedron
Color (solid)	Grey-black
Cleavage (octahedral)	(111)

CADMIUM PHOSPHIDE, CdP_2

General: Does not occur as a natural mineral, although there are many phosphates with a PO_4^- radical. The artificial compound has been used industrially as an additive in making steels, but has no other major applications.

Technical Application: The compound is under evaluation in Solid State as a surface thin film deposited on single crystal indium phosphide, InP, for possible photo diode applications.

CADMIUM PHOSPHIDE ETCHANTS

CDP-0001
ETCH NAME: BRM — TIME:
TYPE: Halogen, removal — TEMP: RT
COMPOSITION:
x .... 1% Br_2
x .... MeOH

DISCUSSION:

CdP_2 deposited as a thin film on InP, (100) wafers as a photo diode, and studied for noise generation. Solution shown used to etch-form mesa structures in the thin film.

REF: Susa, N — *Appl Phys Lett,* 39,168(1981)

PHYSICAL PROPERTIES OF CADMIUM SELENIDE, CdSe

Classification	Selenide
Atomic numbers	48 & 34
Atomic weight	191.3
Melting point (°C)	~1000
Boiling point (°C)	
Density (g/cm^3)	~6
Band gap (eV)	1.8
Hardness (Mohs — scratch)	5—6
Crystal structure (hexagonal — normal)	$(10\bar{1}0)$ prism
Color (solid)	Yellowish
Cleavage (basal — distinct)	(0001)

CADMIUM SELENIDE, CdSe

General: Does not occur as a natural mineral although there are several selenium metallic compounds, such as aguilarite, $Ag_2(S,Se)$; berzelianite, Cu_2Se; tiemannite, HgSe. There are no industrial applications other than as a compound semiconductor.

Technical Application: Cadmium selenide is a II—V compound semiconductor with hexagonal structure. It is a polar compound with regard to prism, $(10\bar{1}0)$ faces, and has distinctive basal (0001) cleavage. Most wafer processing is done with the (0001) surface, though second order prism faces, $(11\bar{2}0)$ have been used.

It has been grown as a single crystal ingot by the Horizontal Bridgman (HB), Czochralski (CZ), and Float Zone (FZ) methods. Epitaxy growth on substrates is by Vapor Transport (VT), and Vapor Phase Epitaxy (VPE). Also developed as the trinary compound: $CdS_{1-x}Se_x$.

A variety of devices have been fabricated from both compounds, such as photoconductors, solar cells, piezoelectric devices, photoconductors, and laser diodes. See CdS for other piezoelectric applications.

Etching: Aqua regia and its varieties and in mixed acids with HNO_3. Halogens.

CADMIUM SELENIDE ETCHANTS

CDSE-0001a

ETCH NAME: Aqua regia, variety — TIME:

TYPE: Acid, preferential — TEMP:

COMPOSITION:

1 .... HCl

3 .... HNO_3

DISCUSSION:

CdSe, (0001) wafers used in a study of etch pit and structure development of the material. Authors say that both (0001)A and $(000\bar{1})$B surfaces develop similar etch figures and a heavy selenium film remains after etching. The Se film can be removed with H_2SO_4. (*Note:* This is an extensive article on etching of II—IV compound semiconductors.)

REF: Warekois, E P et al — *J Appl Phys,* 33,690(1962)

CDSE-0002a
ETCH NAME: BRM — TIME: 5 sec
TYPE: Halogen, cleaning — TEMP: RT
COMPOSITION:
x .... 4% Br_2
x .... MeOH
DISCUSSION:

CdSe, (0001), $(10\bar{1}0)$, and $(11\bar{2}0)$ used in a study of surface etching and morphology on the stability of CdSe:S_x photoelectrochemical cells. Dip in BRM solution; rinse in polysulfide solution, then DI H_2O, and repeat BRM dip. This was one of several etch treatments evaluated. This was treatment #1.

REF: Hodes, O & Manassen, J et al — *J Electrochem Soc*, 128,2325(1981)

CDSE-0002b
ETCH NAME: Aqua regia — TIME: 15 sec
TYPE: Acid, cleaning — TEMP: RT
COMPOSITION:
3 .... HCl
1 .... HNO_3
DISCUSSION:

CdSe, (0001) and other orientations (CDSE-0002a). This is treatment #3: dip in fresh aqua regia; rinse in $S_x^=$, then DI H_2O, and repeat aqua regia.

REF: Ibid.

CDSE-0005b: Gatos, H C & Lavine, M C — *Lincoln Lab Tech Rep*, 293, January 1963.

CdSe, (0001) used in a defect study. Aqua regia developed sharply bevelled pits on the (0001)A surface, but not on $(000\bar{1})$B, cadmium (A) and selenium (B) surfaces, respectively.

CDSE-0002c
ETCH NAME: Chrome regia — TIME: 5 sec
TYPE: Acid, cleaning — TEMP: RT
COMPOSITION:
6 .... CrO_3
10 .... HCl
4 .... H_2O
DISCUSSION:

CdSe, (0001) and other orientations (CDSE-0002a). This is treatment #2. Dip in solution shown; rinse in $S_x^=$ solution, then rinse in DI H_2O and repeat chrome regia dip. (*Note:* This is one variety of the chrome regia etchants.)

REF: Ibid.

CDSE-0002d
ETCH NAME: — TIME:
TYPE: Acid, cleaning — TEMP: RT
COMPOSITION:
(1) 3 .... HCl — (2) 6 .... CrO_3
1 .... HNO_3 — 10 .... HCl
DISCUSSION:

CdSe, (0001) and other orientations (CDSE-0002a). This is treatment #2, as a combination of treatments #3, then #2. Dip in aqua regia; $S_x^=$ solution rinse: water rinse; repeat aqua regia — follow with chrome regia dip; $S_x^=$ rinse; water rinse, and repeat chrome regia.

REF: Ibid.

CDSE-0002e
ETCH NAME: Photo etch
TYPE: Electrolytic, cleaning
COMPOSITION:
9.7 .... HCl
0.3 .... HNO_3
90 .... H_2O

TIME: 4—5 sec
TEMP: RT
ANODE:
CATHODE:
POWER:

DISCUSSION:

CdSe, (0001) and other orientations (CDSE-0002a). This is treatment #5. Solution shown was used as a photo etch under AMI illumination and was short circuited with a counter carbon electrode. Use aqua regia #3 treatment, then the electrolytic solution shown, rinse in $S_x^=$ solution, and water rinse.

REF: Ibid.

CDSE-0002f
ETCH NAME: Aqua regia, variety
TYPE: Acid, cleaning
COMPOSITION:
97% .... HCl
3% .... HNO_3

TIME: 4—5 sec
TEMP: RT

DISCUSSION:

CdSe as a deposited polycrystalline thin film on a titanium substrate. This is treatment #6 using a reduced activity aqua regia solution to reduce etch rate on the thin polyfilm. Follow treatment #3 procedure (CDSE-0002b) using solution shown above; then the photo etch step shown under CDSE-0002e.

REF: Ibid.

CDSE-0002g
ETCH NAME:
TYPE: Electrolytic, reactive
COMPOSITION:
x .... 1 *M* KOH
x .... Na_2S
x .... S
x .... H_2O

TIME: To 20 h
TEMP: 35°C
ANODE:
CATHODE:
POWER:

DISCUSSION:

CdSe, (0001), $(10\bar{1}0)$ and $(11\bar{2}0)$ wafers used to study the overgrowth of CdS from a polysulfide electrolyte and decay of photoelectric current following different surface treatments to include polycrystalline thin film CdSe on titanium substrates. The electrolyte shown was used as the evaluation vehicle, e.g., S will replace Se in CdSe until CdS film stabilizes in thickness and stabilizes the CdSe/CdS interface.

REF: Ibid.

CDSE-0001b
ETCH NAME:
TYPE: Acid, preferential
COMPOSITION:
1 .... HCl
30 .... HNO_3
10 .... CH_3COOH (HAc)

TIME: 8 sec
TEMP: 40°C

DISCUSSION:

CdSe, (0001) wafers used in a study of etch pit and structure development of II—IV

compound semiconductors. The (0001) surface developed aligned hexagonal pits in the solution shown; both (0001)Cd and $(000\bar{1})$Se surfaces were coated with selenium after etching, which can be removed by dipping in H_2SO_4.
REF: Ibid.

CDSE-0003
ETCH NAME: Aqua regia, dilute — TIME:
TYPE: Acid, polish — TEMP:
COMPOSITION:
3 .... HCl
1 .... HNO_3
12 .... H_2O
DISCUSSION:
CdSe polycrystalline thin films deposited on titanium electrodes. The dilute aqua regia solution shown was used to clean and etch polish surfaces. (See CDSE-0002f).
REF: Tomkiewicz, M et al — *J Electrochem Soc,* 129,2016(1982)

CDSE-0004
ETCH NAME: BRM — TIME:
TYPE: Halogen, cleaning — TEMP: RT
COMPOSITION:
x .... 1% Br_2
x .... MeOH
DISCUSSION:
CdSe,$(11\bar{2}0)$, n-type wafers used in a study of photocurrent decay in polysulfide electrolytes for corrosion and stability. Back surfaces of wafers were metallized for ohmic contact with front face exposed in an electrolytic solution. A cotton swab soaked in the solution shown was used to wipe-clean surfaces prior to each measurement in the electrolyte. See CDSE-0002g.
REF: Frese, K & Canfield, D — *J Electrochem Soc,* 132,1649(1985)

CDSE-0005a
ETCH NAME: — TIME: 8 sec
TYPE: Acid, preferential — TEMP: 40°C
COMPOSITION:
0.1 .... HCl
30 .... HNO_3
20 .... 18 *N* H_2SO_4
DISCUSSION:
CdSe, (0001) wafers used in a study of etch pit development. The (0001)A developed hexagonal etch pits; the $(000\bar{1})$B developed no pits. Both surfaces were coated with selenium films after etching. Rinse in H_2SO_4 to remove residual selenium.
REF: Gatos, H C & Lavine, M C — *Lincoln Lab Tech Rep,* 293, January 1963

CDSE-0001c
ETCH NAME: — TIME: 8 sec
TYPE: Acid, preferential — TEMP: 40°C
COMPOSITION:
1 .... HCl
10 .... CH_3COOH(HAc)
30 .... HNO_3
1 .... 8 *N* H_2SO_4

DISCUSSION:

CdSe, (0001) wafers used in an etch pit development study on II—IV compound semiconductors. Solution develops hexagonal etch pits on (0001)A, only, not on $(000\bar{1})$B. Remove residual selenium films with dip in H_2SO_4.

REF: Ibid.

CDSE-0006

ETCH NAME: TIME: 20 sec

TYPE: Acid, removal TEMP: RT

COMPOSITION:

5 .... HCl

1 .... HNO_3

DISCUSSION:

CdSe, (0001) or $(11\bar{2}0)$ wafers cleaved from Czochralski (CZ) grown ingots. Solution used as a general removal/polish etch. First, mechanically lap with Linde A (0.3 μm) alumina, then etch in solution shown. Also worked with CdS.

REF: Heller, A et al — *J Electrochem Soc,* 124,697(1977)

CDSE-0007

ETCH NAME: Ammonium hydroxide TIME: (1) 10—15 min (2) Heat

TYPE: Acid, clean TEMP: RT 300°C

COMPOSITION:

(1) x .... NH_4OH (2) x .... heat

DISCUSSION:

CdSe and $(Cd,Se)_xZn_{1-x}$ single crystal thin films grown in a material study. Prepare metals by (1) cadmium powder: soak clean in solution shown; DI H_2O rinse; dry at 50°C, (2) selenium powder: bake at 300°C, then regrind. Authors caution that improper cleaning can cause detonation.

REF: Kim, S U & Park, M J — *Jpn J Appl Phys,* 23,1070(1984)

CDSE-0008

ETCH NAME: Nitric acid TIME: (1) Wash (2) 15 min

TYPE: Acid, cleaning TEMP: RT 400°C

COMPOSITION:

(1) x .... 25% HNO_3 (2) x .... heat

DISCUSSION:

CdSe thin films co-evaporated on Ti (0001) substrates as polycrystalline films. Heat treat in air at 400°C, 15 min, to homogenize films. Clean CdSe surfaces with HNO_3: then in 1 *M* S:5 *M* Na_2S:1 *M* KOH to remove residual selenium.

REF: Haak, R et al — *J Electrochem Soc,* 131,2709(1984)

CDSE-0009a

ETCH NAME: Aqua regia TIME:

TYPE: Acid, cleaning TEMP:

COMPOSITION:

(1) 3 .... HCl (2) x .... 10% KCN

1 .... HNO_3

DISCUSSION:

CdSe, thin films electrodeposited on titanium substrates. Solution used was 0.1 *M* $CdSO_4$:0.01 *M* SeO_2:0.5 *M* H_2SO_4 at 6 mA/cm^3 power. CdSe films were etch cleaned in

aqua regia; cyanide solution (2) used to remove residual Se with DI water rinse. Fabricated as photoelectrochemical solar cells (PECs) with n-type CdSe films.
REF: Silberstein, R P et al — *J Vac Sci Technol,* 19,406(1981)

CDSE-0009b
ETCH NAME: Xenon TIME:
TYPE: Light, reactive TEMP:
COMPOSITION:
x .... Xe
DISCUSSION:
CdSe $(10\bar{1}0)$ cleaved wafers used in a study of insulated gate FETs and IGFETs with evaporated aluminum. Studied contact potential difference (CPD) and time resolved charge injection (TRCI) at the aluminum interface. Wafers were cleaved in an ultra high vacuum (UHV) at 2×10^{-10} Torr. CdS wafers similarly studied. Flooding surfaces with high intensity xenon light shows potential and injection effects in Al, CdSe and CdS.
REF: Ibid.

CADMIUM SILICON ARSENIDE, $CdSiAs_2$

General: Does not occur as a natural compound; the primary cadmium minerals are mainly sulfides, although there are several arsenates, but none contain cadmium. There is no industrial use, other than as a compound semiconductor.

Technical Application: The material is a II—IV—V compound semiconductor, and has been evaluated for its characteristics as a possible laser, photoconductor, and photo-diode. The material has been grown as a single crystal, and as a thin film layer in heterojunction structures.

Etching: HCl, and mixed acids: $HCl:HNO_3$ or $HF:HNO_3$.

CADMIUM SILICON ARSENIDE ETCHANTS

CSA-0001
ETCH NAME: TIME:
TYPE: Acid, polish TEMP:
COMPOSITION:
1 .... HF
3 .... HNO_3
1 .... H_2O
DISCUSSION:
$CdSiAs_2$, (001) and (111) wafers cut from an n-type, indium doped ingot grown by Vapor Transport (VT) in a tin flux. Wafers were etch polished in the solution shown before silver and gold contacts were evaporated.
REF: Avirovic, N et al — *J Cryst Growth,* 67,185(1984)

PHYSICAL PROPERTIES OF CADMIUM SULFIDE, CdS

Classification	Sulfide
Atomic numbers	48 & 16
Atomic weight	144.5
Melting point (°C) 100 atms	1750
Boiling point (°C)	
Density (g/cm³)	4.82 (4.9—5.0)
Refractive index (n=)	2.506—2.529

Energy band gap (eV)	2.45
Hardness (Mohs — scratch)	3—3.5
Crystal structure (hexagonal — hemimorphic) alpha	m(10$\bar{1}$0) prism
(isometric — tetrahedral) beta	(111) tetrahedron
Color (solid)	Yellow
Cleavage (prismatic — distinct)	a(11$\bar{2}$0)

CADMIUM SULFIDE, CdS

General: There are three natural compounds with greenockite, CdS, the more important, as it is fairly abundant. It is rarely found as crystals, but common as coatings or incrustations on other minerals, primarily zinc compounds, e.g., sphalerite, ZnS (zinc blende).

It is an industrial ore of cadmium, and the natural compound is also used as a yellow pigment in paints. In glass optics there are applications for its piezoelectric properties for filter type devices alone or in conjunction with quartz, SiO_2. (See Cadmium Selenide, CdSe, for similar applications.)

Technical Application: Cadmium sulfide is a II—V compound semiconductor with hexagonal structure. It is a polar compound with regard to prism faces, (10$\bar{1}$0), and has distinctive (0001) cleavage, such that wafer processing is usually done with (0001) basal surfaces. The second order prism face, (11$\bar{2}$0) has been used.

As a piezoelectric material it has been dual-evaporated in alternate orientation layer structure: CdS/SCd/CdS . . . n. As many as 15 thin film layers have been so deposited on alpha-quartz radio frequency blanks for both frequency crystals and electronic filter-type structures.

CdS is grown as a single crystal ingot in both the alpha and beta-CdS form, or epitaxially deposited as an amorphous thin film, a-CdS. It also occurs naturally as an amorphous film. There is a trinary compound as $Zn_{x-1}Cd_xS$.

Although it is relatively soft (H = 3—3.5) it has been fabricated as Solid State devices, such as photoconductors, solar cells, thin film transistors, and laser diodes.

Etching: Easily etched in most acids and NH_4OH.

CADMIUM SULFIDE ETCHANTS

CDS-0001a

ETCH NAME: Hydrochloric acid — TIME:

TYPE: Acid, damage removal — TEMP:

COMPOSITION:

x .... HCl, conc.

DISCUSSION:

CdS, (0001) and (10$\bar{1}$3) wafers. Solution used to remove cutting and lapping damage from surfaces. Material used in a defect study.

REF: Scranton, R A — *J Appl Phys,* 50,842(1979)

CDS-00020: Heller, A et al — *J Electrochem Soc,* 123,697(1977)

CdS, (11$\bar{2}$0) cleaved wafers. Solution used 20 sec at RT to lightly polish clean surfaces as a general etch. First lap polish with Linde A (0.3 μm alumina), then apply solution.

CDS-0017: Richard D et al — *J Vac Sci Technol,* A2(2),132(1984)

CdS thin films doped with O_2 and deposited by spray pyrolysis on glass and sapphire substrates, or substrates pre-coated with SnO_2 or ITO. CdS surfaces were cleaned with HCl after deposition.

CDS-0019a: Chikawa, J & Namayama, T — *J Appl Phys,* 35,2492(1964)

CdS single crystal specimens used in a study of dislocations, structure and growth mechanisms. Specimens were first polished in HCl before thermal etch treatment to develop defect and structure.

CDS-0005b
ETCH NAME: Hydrochloric acid TIME: 10 sec
TYPE: Acid, polish TEMP: RT
COMPOSITION:
x .... 9 *N* HCl
DISCUSSION:
CdS, (100) wafers cut perpendicular and parallel to the single crystal ingot c-axis used in a study of copper diffusion as insulating layers on CdS. Solution was used to polish surface prior to copper evaporation.
REF: Ibid.

CDS-0001b
ETCH NAME: Potassium dichromate TIME: 10 min
TYPE: Acid, polish/preferential TEMP: 95°C
COMPOSITION:
x .... 0.5 *M* $K_2Cr_2O_7$
x .... 16 *N* H_2SO_4
DISCUSSION:
CdS, (111) wafers. Solution used to etch polish both CdS and ZnS wafers with similar results. Produces a highly polished surface containing dislocation etch pits on $(\overline{1}\overline{1}\overline{1})$B (sulfur) surface; shallow disc-like structure on the (111)A (cadmium) surface or $(\overline{1}\overline{1}\overline{1})$B (zinc) surface.
REF: Ibid.
CDS-0005a: Simheny, M et al — *J Appl Phys,* 39,152(1968)
CdS, (100) wafers cut perpendicular and parallel to the single crystal c-axis used in a study of copper diffusion as insulating layers in CdS. Solution used as a preferential etch at 90°C for 5—10 min to observe defects induced by copper.

CDS-0011a
ETCH NAME: TIME: 2 min
TYPE: Acid, preferential TEMP: RT (25°C)
COMPOSITION:
6 .... HNO_3, fuming
6 .... CH_3COOH (HAc)
1 .... H_2O
DISCUSSION:
CdS, (0001) wafers. The $(000\overline{1})$S surface develops hexagonal etch pits, but only a sulfur film on (0001)Cd. (*Note:* HNO_3, white fuming = 70%; yellow fuming = 72%; red fuming = 74%. This is a major article on etching II—VI compounds.)
REF: Warekois, E P et al — *J Appl Phys,* 33,690(1962)
CDS-0004a: Gatos, H C & Lavine, M C — *Lincoln Lab Tech Rep,* 293, January 1963
CdS, (0001) wafers used in a similar defect etching study. (*Note:* A major article on etching of this compound.)

CDS-0002
ETCH NAME: TIME: 1—2 min
TYPE: Acid, removal TIME: RT
COMPOSITION:
1 .... HCl
1 .... CH_3COOH (HAc)
1 .... H_2O
DISCUSSION:
CdS, (0001) wafers used with particular interest on the (0001)A (cadmium) surface.

Work also done with thin film deposits on alpha-quartz substrates as Ag/CdS/quartz. Mechanically polish CdS wafers, then remove lapping damage with solution shown. (*Note:* The Ag/CdS/quartz assembly is a radio frequency device structure.)
REF: Lubbert, C — *J Appl Phys,* 47,366(1976)

CDS-0011b
ETCH NAME: TIME:
TYPE: Acid, preferential TEMP:
COMPOSITION:
1 .... HCl
1 .... HNO_3
DISCUSSION:
CdS, (0001) wafers. This solution develops conical etch pits on (0001)S surfaces that approach hexagonal form. The (0001)Cd surface is rough and granular.
REF: Ibid.

CDS-0003
ETCH NAME: Hydrochloric acid TIME: 1 min
TYPE: Acid, preferential TEMP: Hot
COMPOSITION:
x .... HCl vapor
DISCUSSION:
CdS, (0001) wafers used in an etch pit study. Wafers were wetted in water prior to being held in HCl vapor to develop etch pits.
REF: Eland, A J — *Phillips Tech Rev,* 22,266(1960—1961)
CDS-0012: Pisarerko, Zh D & Sheinkman, M K — *Sov Phys (Solid State),* 3838(1960)
CdS, (0001) cleaved wafers used in a study of dislocations. Hold wafers 4—5 cm above 20—30% HCl liquid in the vapors at 100°C for 1—2 min, then wash in DI H_2O. Other etchants evaluated, but this solution gave the best results.
CDS-0013: Reynolds, D C & Czyzak, S J — *J Appl Phys,* 31,95(1960)
CdS, (0001) wafers used in a study of dislocations. Used a boiling solution of HCl with wafers held in hot vapors, and followed with DI H_2O rinse. The (0001)A surface does not develop pits; the $(000\bar{1})$B surface develops pits. Two dislocation density materials evaluated: 10^1 and $10^7/cm^2$.

CDS-0022
ETCH NAME: Perchloroethylene TIME:
TYPE: Solvent, cleaning TEMP:
COMPOSITION:
x .... CCl_2:CCl_3 (PCE, Perk)
DISCUSSION:
CdS, (100) wafers. Mechanical lap and polish and then clean with the following sequence: (1) decrease in PCE, (2) wash in HCl, (3) DI water rinse, and (4) rinse in acetone.
REF: Sherohman, J W — *J Electrochem Soc,* 128,1817(1981)

CDS-0006a
ETCH NAME: TIME: 2—3 min
TYPE: Acid, preferential TEMP: RT
COMPOSITION:
x .... H_2SO_4
x .... 0.3 *M* $KMnO_4$

DISCUSSION:

CdS, (100) wafers. Referred to as a new optical quality etch for II—V compounds. Mixture is a dark green color. The permanganate should be added slowly as there is violent reaction with the sulfuric acid. Produces smaller etch pits (?) and a cleaner surface than the "Chromate Etch". (See CDS-0002.) This etch was compared to HCl; H_3PO_4 and dilute HNO_3 which also are etches for II—V compounds. (*Note:* See Silicon Dioxide section for comparative standard glass cleaner solution.)

REF: Rowe, J E & Forma R A — *J Appl Phys,* 39,1917(1968)

CDS-0007a

ETCH NAME: Hydrochloric acid — TIME:

TYPE: Acid, preferential, vapor — TEMP: Hot

COMPOSITION:

x .... HCl, vapors

DISCUSSION:

CdS, (0001) wafers used in a study of defects and dislocations. Wafers were suspended in HCl vapors. The (0001)A surface develops dislocation etch pits; none on $(000\bar{1})$B.

REF: Woods, J — *Br J Appl Phys,* 11,296(1960)

CDS-0008: Reynolds, R C & Green, L C — *J Appl Phys,* 29,559(1958)

This reference cited for use of HCl, vapor.

CDS-0007b

ETCH NAME: — TIME: 10 min

TYPE: Acid, preferential — TEMP: 80°C

COMPOSITION:

5 ml .... H_2SO_4

1250 ml .... H_2O

1 g .... Cr_2O_5

DISCUSSION:

CdS, (0001) and $(10\bar{1}0)$ wafers used in a study of defects and dislocations. Place wafers in solution at RT and bring up to 80°C; rinse in DI water, then in MeOH. Etch pits results were as follows:

(0001)A: Hexagonal pits with variable side lengths. Etch for 30 sec only and pit sides are of equal length.

$(000\bar{1})$B: Conical pits only.

$(10\bar{1}0)$: Triangle pits — one side perpendicular to the c-axis; apex in the $(000\bar{1})$B direction.

$(11\bar{2}0)$: No pits developed.

REF: Ibid.

CDS-0009: Votava, E — *Seitz Naturforsch,* 13a,542(1958)

CdS specimens. Used HCl by immersion, rather than as hot vapor. Results were similar to those obtained in CDS-0007a.

CDS-0010

ETCH NAME: Hydrochloric acid, dilute — TIME:

TYPE: Acid, cleaning — TEMP: RT

COMPOSITION:

x .... HCl

x .... H_2O

DISCUSSION:

CdS, (0001) wafers used to fabricate photojunctions by diffusing copper into insulating

CdS. Wafer surfaces were rinsed in the solution shown and then in DI water prior to copper diffusion.
REF: Rockemuehl, R R et al — *J Appl Phys,* 32,1324(1961)
CDS-0014: Sullivan, M V & Bracht, W R — *J Electrochem Soc,* 114,295(1967)
CdS, (111) wafers. Solution used as 30HCl:70H_2O as a chem/mech polish for both the (111)Cd and $(\overline{111})$S surfaces using rotation on a pellon type cloth.

CDS-0015a
ETCH NAME: TIME:
TYPE: Acid, polish TEMP: RT
COMPOSITION:
13.7 g .... KCl
0.5 ml .... HCl
100 ml .... H_2O
DISCUSSION:
CdS, (111) wafers. Solution used as a chem/mech polish at 58 rpm rotation on a pellon-type pad. Used to polish the (111)Cd surface.
REF: Pritchard, A A & Wagener, S — *J Electrochem Soc,* 124,961(1977)

CDS-0015b
ETCH NAME: TIME:
TYPE: Acid, polish TEMP: RT
COMPOSITION:
13.3 g .... KCl
16 ml .... HCl
100 ml .... H_2O
DISCUSSION:
CdS, (111) wafers. Solution used as in CDS-0015a to polish the $(\overline{111})$S wafer surface.
REF: Ibid.

CDS-0016
ETCH NAME: TIME:
TYPE: Acid-slurry, polish TEMP: RT
COMPOSITION:
90 ml .... HNO_3
10 g .... $AgCl_3$
1000 ml .... H_2O
300 ml .... silica
DISCUSSION:
CdS, (0001) wafer. Solution used to chem/mech polish wafers at 240 rpm with pellon-type pad.
REF: Pickhardt, V Y & Smith, D L — *J Electrochem Soc,* 124,961(1977)

CDS-0011c
ETCH NAME: TIME: 5—15 min
TYPE: Acid, polish TEMP: 95°C
COMPOSITION:
x .... 16 *N* H_2SO_4
x .... 0.5 *N* $K_2Cr_2O_7$

DISCUSSION:

CdS, $(10\bar{1}0)$ wafers. Solution is a general polish for (111)Cd surfaces, but develops etch pits on the $(\overline{111})$B surfaces.

REF: Ibid.

CDS-0007c

ETCH NAME: TIME: 10 min

TYPE: Acid, preferential TEMP: 80°C

COMPOSITION:

1 ml .... H_2SO_4
0.08 g .... Cr_2O_3
100 ml .... H_2O

DISCUSSION:

CdS, $(10\bar{1}0)$ wafers. Solution used as a preferential etch.

REF: Ibid.

CDS-0018

ETCH NAME: TIME:

TYPE: Acid, preferential TEMP:

COMPOSITION:

2 .... HCl
5 .... EOH

DISCUSSION:

CdS, (0001) wafers used in a study of electrochemical coupling. Solution used to develop pits and defects.

REF: Wilson, R B — *J Appl Phys*, 37,1932(1966)

CDS-0019b

ETCH NAME: Heat TIME: (1) 2—5 sec

TYPE: Thermal, dislocation TEMP: (1) RT

COMPOSITION:

(1) x .... HCl, conc. (2) x .... heat

DISCUSSION:

CdS, single crystal specimens used in a study of dislocations, structure and growth mechanisms. Specimens polished in HCl before thermal etching to develop dislocations and structures.

REF: Chikawa, J & Nakayama, T — *J Appl Phys*, 35,2492(1964)

CDS-0021

ETCH NAME: TIME: $^1/_2$ h

TYPE: Acid, removal/cleaning TEMP: Boiling

COMPOSITION:

1 .... 5% NaOH
1 .... 10% NaCN

DISCUSSION:

CdS wafers copper plated and diffused used in a study of the dependence of hole ionization energy relative to imperfections and impurity concentration. Solution used to wash and remove excess copper remaining after diffusion; follow with water rinsing until pH paper is neutral.

REF: Bube, R H & Dreeben, P B — *Phys Rev*, 115,1578(1959)

CDS-0023
ETCH NAME: Xenon
TYPE: Light, reactive
COMPOSITION:
x .... Xe

TIME:
TEMP:

DISCUSSION:

CdS, (0001) wafers and CdSe (10$\bar{1}$0) cleaved wafer in UHV. Wafers fabricated as insulated gate devices with aluminum gates for IGFETs. High intensity xenon light will affect the activity of aluminum in both materials. See: CDSE-0009a-b.

REF: Sinerstein, R P et al — *J Vac Sci Technol,* 19(3),406(1981)

PHYSICAL PROPERTIES OF CADMIUM TELLURIDE, CdTe

Classification	Telluride
Atomic numbers	48 & 52
Atomic weight	240
Melting point (°C)	1041
Boiling point (°C)	
Density (g/cm^3)	6.20
Energy band gap (eV)	1.45
Hardness (Mohs — scratch)	5—6
Crystal structure (isometric — tetrahedral)	(111) tetrahedron
Color (solid)	Black
Cleavage (cubic — distinct)	(001)

CADMIUM TELLURIDE, CdTe

General: Does not occur as a natural compound. There are several tellurides and tellurates with iron and other metals. Cadmium telluride has had little use in industrial metal processing, other than as a compound semiconductor.

Technical Application: Cadmium telluride is a II—V compound semiconductor with the sphalerite, ZnS, structure. It is similar to other polar type compounds with positive (111) water surfaces as (111)Cd, and negative ($\overline{111}$)Te opposed wafer surfaces which show different etching characteristics, such that the material is more often used in the (100) orientation for device processing, though the polarity effects have been widely studied. It has been grown as a single crystal by the Horizontal Bridgman (HB), Czochralski (CZ), and Float Zone (FZ) methods. Also grown as an epitaxy thin film by Vapor Transport (VT), Chemical Vapor Deposition (CVD), and by Vapor Phase Epitaxy (VPE). It is isomorphous with InSb, and further grown as a trinary semiconductor: $Cd_xHg_{1-x}Te$. Devices made from both the binary and trinary compounds include photoconductors, photodiodes, and laser diodes.

Etching: Mixed acids of HF:HNO_3, HCl:HNO_3, and halogens.

CADMIUM TELLURIDE ETCHANTS

CDTE-0006b
ETCH NAME: EAg1 etch
TYPE: Acid, preferential
COMPOSITION:
10 ml .... *E etch
5 mg .... $AgNO_3$

TIME:
TEMP:

*See CDTE-0015b

DISCUSSION:

CdTe, (100), (111) and (110) wafers. The (110) wafers were cleaved; the (100) and (111) were saw cut. All orientations were mechanically polished with "chrome green" (CrO) as 1—3 μm grit. Several etches developed as polish and preferential solutions for studying CdTe polarity and defects. (*Note:* Cr_2O_3 is the normal green form of chromium oxide.)

REF: Inoue, M et al — *J Appl Phys,* 33,2578(1962)

CDTE-0031: Iwanaga, H J — *J Cryst Growth,* 62,690(1979)

CdTe, (111) and (110) wafers used in a study of pit structures. Solution used at RT, 3 min. The (111)A surface develops triangle pits; $(\overline{111})$B also developed triangle pits, but deeper than on (111)A. Isosceles triangles with pointed bottoms developed on the (110)B, and similar pits with flat bottoms on the (110)A. Etch time 3 min at RT.

CDTE-0006a

ETCH NAME: E etch — TIME:

TYPE: Acid, polish — TEMP: RT

COMPOSITION:

10 ml .... HNO_3
20 ml .... H_2O
4 g .... $K_2Cr_2O_7$

DISCUSSION:

CdTe, (100), (111), and (110) wafers used in development of defect etches and a study of polarity effects on CdTe. This solution is a polishing etch and was used before preferential etchants were applied. See CDTE-0006b-f.

REF: Ibid.

CDTE-0017d: Ibid.

CdTe, (111) wafers. Solution used at RT with agitation and produced a mirror polish. Rate: 4.2 μm/min. See CDTE-0017a.

CDTE-0006c

ETCH NAME: EAg2 etch — TIME:

TYPE: Acid, preferential — TEMP: RT

COMPOSITION:

10 ml .... *E Etch
10 mg .... $AgNO_3$

*See CDTE-0006a

DISCUSSION:

CdTe, (100), (111), and (110) wafers. See CDTE-0006a,b for discussion.

REF: Ibid.

CDTE-0006d

ETCH NAME: P etch — TIME:

TYPE: Acid, polish — TEMP: RT

COMPOSITION:

10 ml .... HCl
10 ml .... HNO_3
5 ml .... H_2O

DISCUSSION:

CdTe, (111), (100) and (110) wafers. Solution used as a polishing etch prior to pref-

erential etching for defects and polarity effects. See other CDTE-0006 references for additional discussion.
REF: Ibid.
CDTE-0017e: Ibid.
CdTe, (111) wafers. Solution produces pits as used at RT with agitation. Rate: 12.5 μm/min. See CDTE-0017a.

CDTE-0006e
ETCH NAME: PBr etch TIME:
TYPE: Halogen, preferential TEMP: RT
COMPOSITION:
(1) 10 ml .... P etch (2) 10 ml .... P etch
10 mg .... Br_2 5 mg .... Br_2
DISCUSSION:
CdTe, (111), (100), and (110) wafers. Both solutions develop etch pits. They also can be used as p-n junction delineation solutions. See other CDTE-0006 references for further discussion.
REF: Ibid.

CDTE-0006f
ETCH NAME: Heat TIME:
TYPE: Thermal, preferential TEMP: 600°C
COMPOSITION:
x .... heat
DISCUSSION:
CdTe, (111), (100), and (110) wafers used in developing polish and preferential etch solutions and other methods of developing structure. Wafers were placed in vacuum at 5^{-10} mmHg. Thermal etching develops etch structures that can be correlated with chemical liquid etching. See other CDTE-0006 references.
REF: Ibid.

CDTE-0012a
ETCH NAME: TIME:
TYPE: Acid, float-off TEMP:
COMPOSITION:
x .... HNO_3
x .... H_2O
DISCUSSION:
CdTe, thin films grown on muscovite mica (0001) cleaved substrates. After growth the films were removed from the mica by the float-off technique with the solution shown for TEM study of structure.
REF: Davenere, A et al — *J Cryst Growth,* 70,452(1984)

CDTE-0001a
ETCH NAME: TIME: 2 min
TYPE: Acid, preferential TEMP: RT
COMPOSITION:
3 .... HF
2 .... H_2O_2
1 .... H_2O
DISCUSSION:
CdTe, (111) wafers. The (111)Cd surface shows a near-polish with cup-like figures that

disappear upon further etching. The $(\overline{111})$Te surface develops triangular etch pits. The Te film that remains after etching can be removed by dipping in HCl.
REF: Warekois, E P et al — *J Appl Phys,* 33,690(1962)
CDTE-0017f: Ibid.

CdTe, (111) wafers. Solution used at RT with agitation. Polished the (111)Cd surface. Rate: 3.4 μm/min. See CDTE-0017a.

CDTE-0001b
ETCH NAME:
TIME: 8—10 min
TYPE: Acid, polish
TEMP: RT
COMPOSITION:
1 .... HF
1 .... HNO_3
DISCUSSION:

CdTe, (111) wafers. Described as a good polish etch for both wafer surfaces, (111)A and $(\overline{111})$B, and does not develop etch pits. The grey tellurium film remaining after etching can be removed by dipping in HCl.
REF: Ibid.

CDTE-0001c
ETCH NAME:
TIME:
TYPE: Acid, preferential
TEMP:
COMPOSITION:
2 .... HF
3 .... HNO_3
1 .... H_2O
DISCUSSION:

CdTe, (111) wafers. Solution will develop triangle etch pits on the $(\overline{111})$B face but a semi-polish on (111)A.
REF: Ibid.

CDTE-0002
ETCH NAME:
TIME:
TYPE: Acid, polish
TEMP:
COMPOSITION:
(1) x .... H_2SO_4
x .... $K_2Cr_2O_7$
(2) x .... Br_2
x .... MeOH
DISCUSSION:

CdTe, (111) n-type wafers doped with either aluminum or chlorine. Wafers were first mechanically polished, then etch polished in the number (1) solution shown. After etching, rinse in the BRM number (2) solution.
REF: Vazquez-Lopez, C et al — *J Appl Phys,* 50,5390(1979)
CDTE-0017g: Ibid.

CdTe, (111) wafers. Solution mixture was $3gH_2SO_4$:$7gK_2Cr_2O_7$, and was used at RT with agitation. Produced polished surfaces. Rate: 2.3 μm/min. See CDTE-0017a.
CDTE-0024b: Ibid.

CdTe, (111) wafers and other orientations used in a study of anodic oxidation. Etch clean wafers in solution (1) before oxidation.

CDTE-0004
ETCH NAME: TIME: (1) 1 min (2) 30 sec
TYPE: Acid, preferential TEMP: RT RT
COMPOSITION:

(1) 3 .... HF
2 .... H_2O_2
1 .... H_2O

(2) x .... 0.1% Br_2
x .... MeOH

DISCUSSION:

CdTe, (111) wafers and ingots. Solution used to develop defect in both wafers and ingots in a study of twinning. Etch in solution (1) and DI water rinse; follow with solution (2) and DI water rinse. The BRM solution will polish surfaces accentuating the defects developed by the acid solution. See CDTE-0001a.

REF: Vere, A W et al — *J Electron Mater,* 12,551(1983)

CDTE-0005a
ETCH NAME: TIME:
TYPE: Acid, isotropic & defect TEMP:
COMPOSITION:

(1) x .... 0.5% Br_2
x .... MeOH

(2) 5 ml .... HNO_3
10 ml .... H_2O
2 g .... $K_2Cr_2O_7$

(3) 42 .... HF
20 .... H_2O_2
29 .... H_2O

DISCUSSION:

CdTe, (111) wafers. Mechanical lap with abrasives on a glass plate in the following order: (1) 240-grit SiC; (2) 600-grit SiC and (3) 0.3 μm alumina. Then use solution (1) to chem/mech polish by dripping solution onto a polishing pad with hand lap or mechanical rotation. Wax coat surfaces with a pattern opened area with the black wax (Apiezon-W) acting as a dike. Use etch solution (2) to chemically etch-mill pits into the bulk CdTe surface. Then use solution (3) as a defect etch to develop dislocations in the previously etched pits. This will develop defects in the material bulk. Remove the dark Te film remaining in pits after defect etching with alcohol. (*Note:* See Silicon, SI0092a-k for similar bulk defect study.)

REF: Weirauch, D F — *J Electrochem Soc,* 132,250(1985)

CDTE-0005b
ETCH NAME: TIME: See discussion
TYPE: Acid, isotropic defect TEMP: RT
COMPOSITION:

(1) "A" 5 ml .... HNO_3
10 ml .... H_2O
2 g .... $K_2Cr_2O_7$

"B" x .... H_2O

(2) 42 .... HF
29 .... H_2O_2
29 .... H_2O

DISCUSSION:

CdTe, (111) wafers. Mechanically polish: (1) 5 μm, then (2) 0.3 μm alumina. Plate surface with gold using $AuCl_3$, 15 sec. (*Note:* Standard electrolytic gold plating bath or electroless?). Mix solution (1) as 4 "A":1"B", and etch polish surfaces after first scratching through gold film to induce microcracks in the CdTe. Then polish etch, RT, 2 sec into scratched areas. Use solution (2) as a defect etch, RT, 20 sec. Remove grey Te film in etched areas with solution (1). Cleave wafers on ⟨110⟩ for observation and study of scratched damaged zones in cross-section.

REF: Ibid.

CDTE-0007
ETCH NAME: TIME:
TYPE: Acid, polish TEMP:
COMPOSITION:
3 .... HF
2 .... H_2O_2
1 .... H_2O
DISCUSSION:

CdTe, (111) wafers. Authors say that this solution leaves a high Te content surface film after etching and suggest that high electropositive elements tend to pile-up at the surface for reasons unknown. See CDTE-0001a, where solution was developed and used as a preferential etch.

REF: Reiyi, V T et al — *Thin Solid Films,* 113,157(1984)

CDTE-0008
ETCH NAME: BRM TIME:
TYPE: Halogen, polish TEMP: RT
COMPOSITION:
x .... Br_2
x .... MeOH
DISCUSSION:

CdTe, (111) wafers grown by the Horizontal Bridgman (HB) method, and used for epitaxy growth of CdTe on both (111)A and $(\overline{111})$B oriented surfaces. Wafers were mechanically lapped and then polished in a BRM solution prior to epitaxy. Epitaxy growth was by Vapor Transport (VT) using a vertical, three-zone furnace under argon with wafers at the bottom. Epitaxy would grow only on the (111)A surface and showed triangular steps with (110) slopes. *Note:* Argon was used to thermally etch both surfaces while still in the growth system: the (111)A showed terracing; whereas $(\overline{111})$B showed random etching only.

REF: Kuwamoto, H — *J Cryst Growth,* 69,204(1984)

CDTE-0009
ETCH NAME: BRM TIME: 5 min
TYPE: Halogen, removal/polish TEMP: RT
COMPOSITION:
x .... 2% Br_2
x .... MeOH
DISCUSSION:

CdTe, (111) wafers, n-type, high resistivity. Mechanically polish with 0.3 μm alumina, then etch with the solution shown to remove about 20 μm. A Te rich surface film is left. Rinse in MeOH to remove residual bromine, DI water rinse, and N_2 blow dry. Place wafers in vacuum, and hold 8 h at 900°C with 0.5 atm Cd back-pressure to suppress Cd vacancies. Then evaporate indium pads for ohmic contact. See CDTE-0005a and CDTE-0002.

REF: Nozaki, S et al — *J Electron Mater,* 14,137(1985)

CDTE-0021: Sagar, A et al — *J Appl Phys,* 39,5336(1968)

CdTe, (111) wafers. Used a 0.5% Br_2:MeOH solution as a general polishing etch, RT, up to 45 min. Clean Te surface deposits that remain with CS_2.

CDTE-0010
ETCH NAME: BRM — TIME:
TYPE: Halogen, polish — TEMP: RT
COMPOSITION:
x .... 10% Br_2
x .... MeOH
DISCUSSION:

CdTe, (111) wafers, p-type. After mechanical lap and polish, degrease in TCE, and rinse in DI water. Use solution shown to etch polish surfaces.
REF: Selim, F A & Kroger, E A — *J Electrochem Soc,* 124,401(1977)

CDTE-0011a
ETCH NAME: — TIME:
TYPE: Acid, dislocation — TEMP:
COMPOSITION:

(1)	(2)	(3)
x .... x% Br_2	3 .... HF	x .... 2% Br_2
x .... MeOH	2 .... H_2O_2	x .... glycol
	3 .... H_2O	

DISCUSSION:

CdTe, (111) wafers used as substrates for epitaxy growth of CdTe on both (111)Cd and $(\overline{111})$Te surfaces by LPE (liquid phase epitaxy) — deposited substrate surfaces were rich in Te. In preparing substrates, use solution (1) as a polish etch. After epitaxy, use solution (2) to develop dislocations in the epitaxy layers. Follow with solution (3) to remove black ''smut'' remaining after dislocation etching by dip cleaning, RT, 5 sec, then MeOH rinse, DI water rise, and N_2 blow dry. Dislocation etch (2) develops star-shaped pits and low angle grain boundaries in the epitaxy layers that show correspondence between the two polar wafer faces.
REF: Astles, M et al — *J Electron Mater,* 13,167(1984)

CDTE-0011b
ETCH NAME: — TIME:
TYPE: Acid, preferential — TEMP: RT
COMPOSITION:
1 .... HCl
1 .... HNO_3
DISCUSSION:

CdTe, (111) wafers used in an etch reaction study. The (111)Cd surface develops triangular etch pits with a surface background of shallow, triangular etch figures. The $(\overline{111})$Te surface is covered with a Te film in which triangular figures appear to have been etched.
REF: Ibid.

CDTE-0012b
ETCH NAME: BRM — TIME:
TYPE: Halogen, polish — TEMP: RT
COMPOSITION:
x .... 1% Br_2
x .... MeOH
DISCUSSION:

CdTe, (111) wafers used for epitaxy growth of CdTe/Bi/CdTe (111). Muscovite mica, cleaved (0001), also used as a substrate — blow mica clean with N_2, and apply no liquids.

Etch polish CdTe substrates in solution shown before epitaxy. Material used in a study of superlattice capability.
REF: Davenere, A et al — *J Cryst Growth,* 70,452(1984)
CDTE-0020: Pessa, M et al — *J Vac Sci Technol,* A2(2),418(1984)

CdTe, (111), p-type wafers used as substrates for epitaxy growth of CdTe thin films. Wafers were chem/mech polished in a BRM solution. Final clean in vacuum with Ar^+ ion bombardment for 10 min at 2.5 KeV power.
CDTE-0021: Bhat, I & Ghandhi, S K — *J Electrochem Soc,* 131,1923(1984)

CdTe, (100) wafers cut 3°-off toward (110), and used for epitaxy growth of HgTe thin films. Clean substrates as follows: (1) degrease with solvents, and (2) etch in 2% Br_2:MeOH to remove 20 μm of surface depth and polish simultaneously.

CDTE-0013
ETCH NAME:
TYPE: Acid, polish
COMPOSITION:
3 .... HF
2 .... H_2O_2
2 .... H_2O

TIME:
TEMP: RT

DISCUSSION:

CdTe (111) wafers used as substrates for epitaxy growth of HgCdTe on (111)Cd and $(\overline{111})$Te surfaces in a study of epitaxy morphology and lattice mismatch. Solution used to etch polish wafers prior to epitaxy. The negative Te surface produced the better lattice match.
REF: Edwall, A D et al — *J Appl Phys,* 55,145(1984)

CDTE-0016a
ETCH NAME: Nitric acid, dilute
TYPE: Acid, removal/polish
COMPOSITION:
x .... HNO_3
x .... H_2O

TIME: 2 min
TEMP: 30°C
RATE: 100 μm/min

DISCUSSION:

CdTe, (111) wafers used in a phase equilibria and properties study. Solution used as a general removal and polishing etch. It leaves a layer of Te and Te-oxides on surfaces that can be removed by the solution shown under CDTE-0016b.
REF: Phillips Res Rep, #4,361(1969)

CDTE-0016b
ETCH NAME:
TYPE: Alkali, removal
COMPOSITION:
x .... 10% NaOH
x .... $Na_2S_2O_4$
x .... H_2O

TIME:
TEMP: 60°C

DISCUSSION:

CdTe, (111) wafers used in a properties study. Solution used to remove Te and Te-oxides remaining as surface films on wafers after etch polishing in nitric acid. See CDTE-0016a. Also see Tellurium section for additional etchants.
REF: Ibid.
CDTE-0017h: Ibid.

CdTe, (111) wafers. Solution mixture was: 10% NaOH:x Na_2SO_4, used at RT with agitation. Produced a bright surface. Rate: 1—2 μm/min. See CDTE-0017a.

CDTE-0017a
ETCH NAME: Sodium hydroxide TIME:
TYPE: Alkali, polish TEMP: RT
COMPOSITION: RATE: 1.6 μm/min
x .... 50% NaOH
DISCUSSION:

CdTe, (111) wafers used in an etching reaction and rate experiment. Solution produced a bright surface. In this work, HCl and H_2SO_4 at RT showed no removal. Agitate solution during etching. Rate determined by weight loss and should be used as a guide.
REF: Gaugash, P & Milnes, A G — *J Electrochem Soc,* 128,924(1981)

CDTE-0017b
ETCH NAME: BRM TIME:
TYPE: Halogen, preferential TEMP: RT
COMPOSITION: RATE: 2.2 μm/min
x .... 0.5% Br_2
10 mg .... $AgNO_3$
x .... MeOH
DISCUSSION:

CdTe, (111) wafers used in an etch reaction and rate experiment. Solution was medium red in color and produced pits on (111) surfaces. Was also used as a p-n junction etch. See CDTE-0017a for further discussion.
REF: Ibid.

CDTE-0017c
ETCH NAME: Nitric acid TIME:
TYPE: Acid, burn TEMP:
COMPOSITION: RATE: 3.5 μm/min
x .... HNO_3 conc.
DISCUSSION:

CdTe, (111) wafers. Solution left a dark precipitate on surface that could be removed easily by light mechanical rubbing. See CDTE-0017a for additional discussion.
REF: Ibid.

CDTE-0017i
ETCH NAME: TIME:
TYPE: Acid, polish TEMP: RT
COMPOSITION: RATE: 8 μm/min
1 .... HCl
50 .... HNO_3
18 .... H_2SO_4
10 .... HAc
DISCUSSION:

CdTe, (111) wafers used in an etch and rate experiment. Solution produced polished surfaces. See CDTE-0017a.
REF: Ibid.

CDTE-0017j
ETCH NAME:
TYPE: Acid, preferential
COMPOSITION:
2 .... 45% HF
1 .... HNO_3
1 .... HAc
DISCUSSION:

TIME:
TEMP: RT
RATE: 60 μm/min

CdTe, (111) wafers. Solution used with agitation and produced pits. See CDTE-0017a for additional discussion.

REF: Ibid.

CDTE-0018a
ETCH NAME: BRM
TYPE: Halogen, polish
COMPOSITION:
1 .... Br_2
20 .... MeOH
DISCUSSION:

TIME:
TEMP:

CdTe, (100) wafers used for fabrication of CdTe:HgTe heterojunctions. Solution used to polish CdTe surfaces. Produces a clean, highly reflective surface.

REF: Almasi, G S & Smith, A C — *J Appl Phys,* 39,233(1968)

CDTE-0018b
ETCH NAME: Aqua regia, dilute
TYPE: Acid, polish
COMPOSITION:
3 .... HCl
1 .... HNO_3
1 .... HAc
DISCUSSION:

TIME:
TEMP:

CdTe, (111) wafers used to growth CdTe:HgTe heterojunctions. Solution may leave a black smudge on surface, but it does not affect device performance.

REF: Ibid.

CDTE-0019
ETCH NAME:
TYPE: Acid, structure
COMPOSITION:
1 .... HF
2 .... H_2O_2
2 .... H_2O
DISCUSSION:

TIME:
TEMP:

CdTe, (111) wafers cut from an ingot grown by the Vertical Bridgman (VB) technique. Specimens were cleaned in the following sequence: 13% Br_2:MeOH, RT, 3 min to etch remove work damage, and DI water rinse. EAg-1 etch (CDTE-0006b) used to develop etch pits on (111)Cd surfaces. Substructure was developed by the solution shown. This structure was cellular due to subgrains from supercooling (?), or from high temperature stress (?) during growth. Slow ingot growth will reduce such structure.

REF: Oda, O et al — *J Cryst Growth,* 71,273(1985)

CDTE-0022
ETCH NAME: Isopropyl alcohol
TYPE: Alcohol, cleaning
COMPOSITION:
x .... C_2H_7OH(ISO, IPA)
TIME:
TEMP: RT
DISCUSSION:

CdTe, growth as single crystal ingot by Czochralski (CZ) method is difficult due to twinning and low angle grain boundary development. At less than 50°C, ingot is poly-CdTe. Between 50—150°C good single crystal, but some ⟨122⟩ directional defects. CdTe is isomorphous with InSb. CdTe thin films grown on InSb by MBE showed high perfection of films. Clean surfaces with alcohol as shown. (*Note:* MBE = molecular beam epitaxy.)

REF: Farrow, R F C et al — *Appl Phys Lett,* 39,954(1981)

CDTE-0023: Yellin, N et al — *J Electron Mater,* 14,85(1985)

CdTe ingot growth. Initial poly-CdTe ingot grown by Vertical Unseeded Vapor Growth (VUVG). Place poly-ingot in closed quartz ampoule under excess Te/Cd vapor for Vapor Transport (VT) conversion to single crystal.

CDTE-0024a
ETCH NAME:
TYPE: Electrolytic, oxidation
COMPOSITION:
x .... 1 *N* KOH
x .... MeOH
\+ .... light
TIME:
TEMP:
ANODE: CdTe
CATHODE:
POWER: 1 mA/cm^2
DISCUSSION:

CdTe, (111) and other orientations. Wafers used in a study of anodic oxidation. After etch cleaning wafers in H_2SO_4:$K_2Cr_2O_7$, evaporate gold on wafer back prior to anodic oxidation, shown above. Then evaporate nickel as for gated FET type devices. Voltage anneal with a potential drop across wafer.

REF: Talasek, R T & Syooaios, A J — *J Electrochem Soc,* 132,887(1985)

CDTE-0024a:

CdTe, (111) wafers, and other orientations, used in a study of anodic oxidation.

CADMIUM MERCURIC TELLURIDE ETCHANTS

CDHT-0001
ETCH NAME:
TYPE: Acid, polish
COMPOSITION:
1 .... HCl
6 .... HNO_3
TIME:
TEMP:
DISCUSSION:

CdTe:HgTe specimens. Solution used as a polish etch on this material. After etching, rinse in 1HCl:1MeOH.

REF: Gatos, H C & Lavine, M C — *Prog Semicond,* 9,1(1965)

CDHT-0002: Agajanian, A H — *Solid State Technol,* 16,73(1973)

$Cd_{1-x}Hg_xTe$ specimens. Reference has additional etch solutions for this compound.

CDHT-0003: Agajanian, A H — *Solid State Technol,* 18,61(1975)

$Cd_{1-x}Hg_xTe$ specimens. Reference for additional etch solutions for this material.

CDHT-0004: Agajanian, A H — *Solid State Technol,* 20,36(1977)

$Cd_{1-x}Hg_xTe$ specimens. Reference for additional etch solutions for this material.

PHYSICAL PROPERTIES OF CALCIUM, Ca

Classification	Alkaline metal
Atomic number	20
Atomic weight	40.08
Melting point (°C)	845
Boiling point (°C)	1420 (1487)
Density (g/cm^3)	1.55
Thermal conductance (cal/sec)(cm^2)(°C/cm)	0.3
Specific heat (cal/g) 25°C	0.149
Latent heat of fusion (cal/g)	52
Heat of vaporization (cal/g)	1000
Atomic volume (W/D)	29.9
1st ionization energy (K-cal/g-mole)	141
1st ionization potential (eV)	6.11
Electron work function (eV)	2.706
Electronegativity (Pauling's)	1.0
Covalent radius (angstroms)	1.74
Ionic radius (angstroms)	0.99 (Ca^{+2})
Linear coefficient of thermal expansion ($\times 10^{-6}$ cm/cm/°C) 20°C	22.3
Electrical conductance (micro-ohms^{-1})	0.218
Electrical resistivity ($\times 10^{-6}$ ohms cm) 0°C	3.91
Modulus of elasticity (psi $\times 10^6$)	3.2—3.8
Tensile strength (psi)	6900
Cross section (barns)	0.43
Vapor pressure (°C)	1207
Hardness (Mohs — scratch)	1.5—2
(Brinnel — kfg/mm^2)	16—18
Crystal structure (isometric — normal)	(100) cube, fcc
Color (solid)	Silver-white
Cleavage (cubic)	(001)

CALCIUM, Ca

General: Does not occur as a free element in nature but is widely distributed in carbonate minerals — calcite, $CaCO_3$ and aragonite, $CaCO_3$ as the primary single crystals — in massive form as chalk, limestone, dolomite (containing magnesium and calcium), marl. sea shells. As a calcium sulfate, the minerals gypsum, $CaSO_4.2H_2O$ and the dehydrated anhydrite, $CaSO_4$. The sulfates are known as Plaster of Paris. The name — calcium — is from Latin calx = lime, hence the name for limestone. Both calcium carbonate and calcium hydroxide occur in natural ground waters and produce "hard" water. Calcium oxide, CaO, is "quick-lime" and, as slack lime, $Ca(OH)_2$. Lime as mortar was in use by the Romans in the 1st century A.D. and a wall fresco has been discovered in Asia Minor dated to about 1600 B.C. Fluorite, CaF_2, and the phosphate apatite are of importance.

Calcium is the fifth element in abundance and, in addition to its occurrence in inorganic minerals, is found in leaves, teeth, and bones, such that it is an essential ingredient for mammillary growth. Both calcium and phosphorus are supplied by milk.

Metallic calcium is silver-white in color although it turns yellowish when exposed to air as it reacts with nitrogen. It will burn if heated in air, is electropositive like alkali metals, and is used as a deoxidizer in metal alloys and steel. In its oxide, sulfate, carbonate, and fluoride forms it is used as mortar, cement, plaster, as a flux in glass and metals, in paper making, sugar refining, water purification, and in agriculture to neutralize acid soil. When limestone is pressurized and heated, the natural metamorphic process, it becomes marble. Both materials have been used since ancient times for construction, objets d'art, or utensils, such as plates and drinking cups. Alabaster is a pure white form of marble, and there are both marbles and limestones that are color banded. The limestone called onyx is black, and there are many other limestone forms and names.

Both natural calcium, derived from the carbonates, and organic calcium, derived from seashells, are used medicinally in pill form for human consumption. As a dry powder both calcium and magnesium are available in the chemical laboratory and, when mixed as a paste, can be applied to the skin to neutralize acid burns.

Technical Application: Metallic calcium is not used in Solid State processing to any extent, though it may be an element in tungstates, molybdates, and garnets that are artificially fabricated. It has been grown as the nitride, Ca_2N_3.

Etching: Deliquesces in water to the hydroxide. Soluble in acids; slowly soluble in alcohols.

CALCIUM ETCHANTS

CA-0001a
ETCH NAME: Hydrochloric acid — TIME:
TYPE: Acid, removal — TEMP: RT
COMPOSITION:
x .... HCl, conc.
DISCUSSION:

Ca specimens. Solution used in the general study of the metal. Process and hold under argon to prevent reaction with air (nitrogen).

REF: Foster, W & Alyea, H N — *An Introduction to General Chemistry,* 3rd ed, D Van Nostrand, New York, 1947, 395

CA-0001b
ETCH NAME: Air — TIME:
TYPE: Gas, reactive — TEMP: RT to hot
COMPOSITION:
x .... air
DISCUSSION:

Ca, as pure metal specimens. Reacted in air to develop a yellowish, stable coating of the oxynitride, $Ca_2N_3O_x$. See CAN-0001 for the pure nitride formation.

REF: Ibid.

CA-0001c
ETCH NAME: Water — TIME:
TYPE: Acid, reactive — TEMP: RT to hot
COMPOSITION:
x .... H_2O
DISCUSSION:

Ca metal. Reacted in water to produce the hydroxide, $Ca(OH)_2$, called "slacklime". When "lime" (CaO) is reacted with water, it is called "quicklime". Both limes, mixed with gypsum and clay, are cement.

REF: Ibid.

CA-0001d

ETCH NAME: Sulfur	TIME:
TYPE: Metal, reactive	TEMP: Hot

COMPOSITION:

x .... S

DISCUSSION:

Ca metal. Powdered and mixed with sulfur to form anhydrite, then water added to form gypsum as Plaster of Paris, $CaSO_4.nH_2O$. It is still used for plaster body casts; major use is in construction as general plaster, plaster wallboard; for holding small stones in mineral cutting; shaped as objets d'art.

REF: Ibid.

PHYSICAL PROPERTIES OF CALCIUM CARBONATE, $CaCO_3$

Classification	Carbonate
Atomic numbers	20, 6 & 8
Atomic weight	100
Melting point (°C)	1339
Boiling point (°C)	898.6
Density (g/cm³)	1.71
Refractive index (n=)	w = 1.65849; e = 1.48625
Hardness (Mohs — scratch)	3
Crystal structure (hexagonal — rhombohedral)	$(10\bar{1}1)$ rhomb
Color (solid)	Clear/white
Cleavage (rhomb — perfect)	$(10\bar{1}1)$

CALCIUM CARBONATE, $CaCO_3$

General: There are two primary natural minerals — calcite and aragonite — both representative of a group of minerals with the general formula MCO_3. The Calcite Group is hexagonal-rhombohedral in structure with M = Ca, Mg, Fe, Mn, Zn, and Co. The Aragonite Group is orthorhombic in structure with M = Ca, Ba, Sr, and Pb. Within each group the minerals are isomorphous, giving rise to a wide variety of subspecies.

Calcite is the more stable of these two calcium carbonates and is discussed here. Common name: Iceland Spar. It has perfect (1011) rhombic cleavage, though twinning is common and forms a bi-crystal with c(0001) as a common vertical axial plane. When pure, it is transparent and clear, although it may have pale shades of red, green, blue, yellow, etc. As a transparent, flat-faced rhomb, the single crystal acts as a lens, but is more common in compact masses (chalk) or limestone, or as stalactites, stalagmites, and nodular forms, such as are found in caves or around geysers. It is one of the most widely distributed minerals even as a small fraction in volcanic rocks (''snowflake'' obsidian — natural glass containing ''flowers'' of calcite). Common and widespread in sedimentary rocks of varied character — as vein cement — to massive formations of chalk and limestone. Under pressure and heat, metamorphic compaction, it is marble. It is deposited from lime-bearing waters, the carbonates, vs. siliceous-bearing waters that form the silicates. Also found as calc-sinter or travertine around hot springs and as calcareous rock from sea life shells (organic derivation). Formation temperature varies from about 90°C (springs/caves) to 800—900°C (volcanic).

Calcium carbonates have wide industrial use in the manufacture of mortars and cements; as a building and ornamental material; as a flux in metallurgy; as blackboard ''chalk'' in glass (soda-lime); in fertilizers, animal feed; and in paint as whitewash. For human consumption it is a de-acidifier as aspirin, Maalox and, in paste form, used to neutralize acids burns (also magnesia — the mineral magnesite, $MgCO_3$ — a member of the Calcite Group).

As a single crystal, due to its high birefringence, it is used as a polarizing prism in microscopes and should be handled carefully to prevent scratch damage.

Technical Application: No current use in Solid State processing and fabrication though it is under evaluation as a cleaved substrate for thin film metal evaporation and semiconductor epitaxy growth morphology studies, similar to the present uses of sodium chlorides, NaCl.

As a constituent in glass — soda-lime — this glass is the primary type used to fabricate photo resist glass masks. As a constituent in alloy fluxes, it is used in some solder applications for device assemblies. Both calcium and magnesium are kept in chemical laboratories for neutralization of acid burns as already mentioned.

Natural single crystals have been extensively studied for their optical characteristics.

Etching: Soluble in HCl with effervescence, other acids, and slightly soluble in water.

CALCIUM CARBONATE ETCHANTS

CAC-0001
ETCH NAME: Hydrochloric acid — TIME: 5—10 sec
TYPE: Acid, polish — TEMP: RT
COMPOSITION:
x .... HCl, conc.
DISCUSSION:

$CaCO_3$ $(10\bar{1}1)$ cleaved substrates from natural "Iceland Spar". Specimens were used in a study of surface energies of calcite. After cleaving, surfaces were etch polished and rinsed in alcohol. Solution shown used as a rapid polish with effervescence of the material.
REF: Gilman, J J — *J Appl Phys,* 31,2208(1960)

CAC-0002
ETCH NAME: Hydrochloric acid, dilute — TIME: Variable
TYPE: Acid, polish — TEMP: RT
COMPOSITION:
1 .... HCl
10 .... H_2O
DISCUSSION:

$CaCO_3$, as polarizing prisms. Solution used to lap/polish remove scratches from microscope prisms using a felt pad. Also to clean marble surfaces.
REF: Walker, P — personal application, 1952
CAC-0005: Gross, K A — *Phil Mag,* 12,801(1965)

$CaCO_3$, $(10\bar{1}1)$ cleaved wafers used in an energy study of deformed and annealed natural calcite using X-ray techniques. A 10% HCl solution was used to polish surfaces.

CAC-0003a
ETCH NAME: Nitric acid, dilute — TIME: 90 sec
TYPE: Acid, dislocation — TEMP: RT
COMPOSITION:
1 .... HNO_3
x .... H_2O
DISCUSSION:

$CaCO_3$, r$(10\bar{1}1)$ cleaved wafers and other orientations. A study of dislocations and plastic deformation evaluating a number of acids. Alkalies are too slow at room temperature and mineral acids show an excessive rate of attack with CO_2 evolution. Weak acids with an inert carrying liquid are best. If a dislocation moves away from the original etch pit due to stress,

the pit becomes flat-bottomed and, with further etching, a sharp-bottomed pit will appear at the new dislocation point. (*Note:* A very fine article with excellent pictures.)
REF: Keith, R E & Gilman, J J — *Acta Metall,* 9,1(1961)

CAC-0003b
ETCH NAME: Formic acid — TIME: 15 sec
TYPE: Acid, dislocation — TEMP: RT
COMPOSITION:
x .... HCOOH (90%)
DISCUSSION:
$CaCO_3$, r($10\bar{1}1$) cleaved wafers. See CAC-0003a.
REF: Ibid.
CAC-0004: Keith, R E & Gliman, J J — *Acta Metall,* 8,1(1960)
$CaCO_3$, r($10\bar{1}1$) cleaved wafers used in a dislocation pit etch development study. Formic acid gave best results with only one type pit.

CAC-0003b.1
ETCH NAME: Formic acid, dilute — TIME: 5 sec
TYPE: Acid, dislocation — TEMP: RT
COMPOSITION:
1 .... HCOOH (90%)
1 .... H_2O
DISCUSSION:
$CaCO_3$, r($10\bar{1}1$) cleaved wafers. See CAC-0003a.
REF: Ibid.

CAC-0003b.2
ETCH NAME: Formic acid, dilute — TIME: $1^1/_2$ sec
TYPE: Acid, dislocation — TEMP: RT
COMPOSITION:
1 .... HCOOH (90%)
10 .... H_2O
DISCUSSION:
$CaCO_3$, r($10\bar{1}1$) cleaved wafers. See CAC-0333a.
REF: Ibid.

CAC-0003b.3
ETCH NAME: Formic acid, dilute — TIME: 45 sec
TYPE: Acid, dislocation — TEMP: RT
COMPOSITION:
1 .... HCOOH (90%)
1 .... EOH
DISCUSSION:
$CaCO_3$, r($10\bar{1}1$) cleaved wafers. See CAC-0003a.
REF: Ibid.

CAC-0003b.4
ETCH NAME: Formic acid, dilute — TIME: 90 sec
TYPE: Acid, dislocation — TEMP: RT
COMPOSITION:
1 .... HCOOH (90%)
1 .... glycerin

DISCUSSION:

$CaCO_3$, r(10$\bar{1}$1) cleaved wafers. See CAC-0003a.

REF: Ibid.

CAC-0003c.1

ETCH NAME: Glacial acetic acid — TIME: 360 sec

TYPE: Acid, dislocation — TEMP: RT

COMPOSITION:

x .... CH_3COOH (HAc)

DISCUSSION:

$CaCO_3$, r(10$\bar{1}$1) cleaved wafers. See CAC-0003a.

REF: Ibid.

CAC-0003c.2

ETCH NAME: Acetic acid, dilute — TIME: 75 sec

TYPE: Acid, dislocation — TEMP: RT

COMPOSITION:

1 .... CH_3COOH (HAc)
1 .... H_2O

DISCUSSION:

$CaCO_3$, r(10$\bar{1}$1) cleaved wafers. See CAC-0003a.

REF: Ibid.

CAC-0003c.3

ETCH NAME: Acetic acid, dilute — TIME: 30 sec

TYPE: Acid, dislocation — TEMP: RT

COMPOSITION:

1 .... CH_3COOH (HAc)
10 .... H_2O

DISCUSSION:

$CaCO_3$, r(10$\bar{1}$1) cleaved wafers. See CAC-0003a.

REF: Ibid.

CAC-0003d

ETCH NAME: Propionic acid — TIME: 6 min

TYPE: Acid, cleaning — TEMP: RT

COMPOSITION:

x .... CH_3CH_2COOH

DISCUSSION:

$CaCO_3$, r(10$\bar{1}$1) cleaved wafers. Solution showed no attack in 6 min, but did give some cleaning action. See CAC-0003a.

REF: Ibid.

CAC-0003d.1

ETCH NAME: Propionic acid, dilute — TIME: 120 sec

TYPE: Acid, dislocation — TEMP: RT

COMPOSITION:

1 .... CH_3CH_2COOH
1 .... H_2O

DISCUSSION:

$CaCO_3$, r(10$\bar{1}$1) cleaved wafers. See CAC-0003a.

REF: Ibid.

CAC-0003e
ETCH NAME: Lactic acid — TIME: 15 sec
TYPE: Acid, dislocation — TEMP: RT
COMPOSITION:
x .... $CH_3CHOHCOOH$ (85%)
DISCUSSION:
$CaCO_3$, r(10$\bar{1}$1) cleaved wafers. CAC-0003a.
REF: Ibid.

CAC-0003f
ETCH NAME: Maleic acid — TIME: 10 sec
TYPE: Acid, dislocation — TEMP: RT
COMPOSITION:
x .... HOOCCH:CHCOOH, sat. sol.
x .... H_2O
DISCUSSION:
$CaCO_3$, r(10$\bar{1}$1) cleaved wafers. See CAC-0003a.
REF: Ibid.

CAC-0003g
ETCH NAME: Tartaric acid — TIME: 10 sec
TYPE: Acid, dislocation — TEMP: RT
COMPOSITION:
x .... $HOOC(CHOH)_2COOH$, sat. sol.
x .... H_2O
DISCUSSION:
$CaCO_3$, r(10$\bar{1}$1) cleaved wafers. Developed dislocations with a fresh solution. Used after 2 months, there was no reaction. See CAC-0003a.
REF: Ibid.

CAC-0003h
ETCH NAME: EDTA — TIME:
TYPE: Acid, cleaning — TEMP: RT
COMPOSITION:
x .... $(HOOC\text{-}CH_2)2NCH_2CH_2$
DISCUSSION:
$CaCO_3$, (10$\bar{1}$1) cleaved wafers. There was no reaction with this solution. Also no pit forming reaction with oxalic acid. See CAC-0003a.
REF: Ibid.
CAC-0014: Keith, R E & Gilman, J J — *Acta Metall,* 8,1(1960)
$CaCO_3$, (10$\bar{1}$1) cleaved wafers used in a dislocation pit etch development study. Formic acid gave best results producing only one type pit.

PHYSICAL PROPERTIES OF CALCIUM FLUORIDE, CaF_2

Classification	Fluoride
Atomic numbers	20 & 9
Atomic weight	76.08
Melting point (°C)	1360
Boiling point (°C)	
Density (g/cm³)	3.18 (3.01—3.25)

Refractive index (n =)	1.4339
Hardness (Mohs — scratch)	4
(Knoop — kgf mm^{-2})	163
Ionic radius (angstroms)	1.06
Surface free energy (ergs cm^{-2}) −273°C	543 (111)
−195°C	450 (111)
Crystal structure (isometric — normal)	(100) cube
Color (solid)	Yellow-green/variable
Cleavage (octahedral-distinct)	o(111)

CALCIUM FLUORIDE, CaF_2

General: Occurs as the mineral fluorite, is fairly common, and found as a vein deposit, alone, or with silver, zinc, etc. Like diamond, it is one of the few minerals with distinctive (111) cleavage, even though it is in the normal class of the isometric system, where most materials shown cubic (001) cleavage. It is noted for its wide range of colors that can vary within a crystal, or be banded. Colors can be altered by heat, X-rays, UV light, pressure, etc., and may show blue fluorescence.

It was in use as a solder flux before the discovery of fluorine, and is the major ore for fluorine, used in the making of hydrogen fluoride (HF), e.g., hydrofluoric acid. Fluorite as a powder has major use in making opalescent glass and enamel, such as for cookware. It is used in jewelry making for its colors, even though the mineral is relatively soft.

Technical Application: There is no direct use in Solid State device fabrication, although it is under evaluation as a surface coating for its optical properties. It is used as a microscope prism for its high birefringence, and has applications as a filter element or as lenses. There is still use of the powdered material as a solder flux in package fabrication.

Fluorite has been widely studied as the natural single crystal due to its transparency for observing the physical, electrical, and electronic effects of various treatments, such as X-ray, pressure, etc., already mentioned in the proceeding General section.

Etching: Slow in acids, water — easily in ammonium salts.

CALCIUM FLUORIDE ETCHANTS

CAF-0001

ETCH NAME: Water — TIME:

TYPE: Acid, removal — TEMP:

COMPOSITION:

x .... H_2O

DISCUSSION:

CaF_2, (100) thin films deposited on GaAs, (100) substrates. After deposition, wafers were photolithographically processed for an open square pattern, then E-beam annealed. After annealing, water was used to wash away the CaF_2 thin films. Also processed SrF_2 (SRF-0001) with dilute HCl used to wash-away the strontium compound.

REF: Sullivan, P W — *J Vac Sci Technol,* B2,202(1984)

CAF-0004: Tu, C W — *J Vac Sci Technol,* B2,24(1984)

CaF_2, (100) and (110) thin films deposited on InP substrates of similar orientation. Thin films were grown by MBE, including SrF_2, BF_2, and $Ba_xSr_{1-x}F_2$ co-deposited from the two compounds. Water used to wash away CaF_2; dilute HCl for SrF_2; HF for BaF_2 and the mixed compound.

CAF-0002
ETCH NAME: Sulfamic acid TIME:
TYPE: Acid, dislocation TEMP:
COMPOSITION:
6 g .... H_2NSO_3H
100 ml .... H_2O
DISCUSSION:

CaF_2, (111) cleaved wafers used in a dislocation study. Specimens were studied as-cleaved without further surface conditioning prior to preferential etching in the solution shown.

REF: Steijn, R P — *J Appl Phys,* 34,419(1964)

CAF-0003
ETCH NAME: Ethyl alcohol TIME: 5 min
TYPE: Alcohol, cleaning TEMP: RT
COMPOSITION:
x .... EOH
DISCUSSION:

CaF_2, (100) cleaved wafers. Both natural fluorite and artificially grown material were used in an X-ray reflection study. Specimens prepared by immersion in alcohol with light scrubbing on surfaces using a cotton swab during the cleaning period.

REF: Barton, V P — *Acta Crystallogr,* 11,848(1958)
CAF-0008: Walker, P et al — personal application, 1968

CaF_2, natural octahedral, (111) crystals prepared for light irradiation study. Specimens polished on a Buehler metallographic wheel at high rpm with a pellon-type pad soaked with alcohol. Intense light of different frequencies colors used to develop internal defect. Similar light frequency effects evaluated on alpha-quartz radio frequency blanks. Artificial material fabricated as lenses was wash-cleaned in alcohols prior to metallization with Au:Cr thin films.

CAF-0005
ETCH NAME: Nitric acid TIME:
TYPE: Acid, preferential TEMP:
COMPOSITION:
x .... 0.2 *N* HNO_3
DISCUSSION:

CaF_2, (111) cleaved wafers used in a study of cleavage and dislocation pits as compared to diamonds. Solution shown developed trigon pits like those on natural diamond faces, but no etch would artificially develop similar pits on diamond. The CaF_2 was difficult to cleave on (111).

REF: Patel, A R et al — *Phil Mag,* 9,951(1964)
CAF-0007: Amelinck, S et al — *Physica,* 23,270(1975)

CaF_2, (111) oriented wafers used in a study of etch pits. Pits can occurs on natural fluorite faces under normal growth conditions, and can appear on both natural and artificially grown material cleaved surfaces without etching.

CAF-0006a
ETCH NAME: Ammonia TIME: Seconds
TYPE: Base, polish TEMP: RT
COMPOSITION:
x .... NH_3

DISCUSSION:

CaF_2 as natural fluorite crystals as octahedron, (111) solid forms. Solution used to clean and polish (111) faces of bulk crystals. As (111) cleaved sections, solution used to polish thin specimens for microscope study.

REF: Walker, P — mineralogy study, 1952

CAF-0006b

ETCH NAME: Ammonium hydroxide — TIME: Seconds

TYPE: Base, cleaning — TEMP: RT

COMPOSITION:

x .... NH_4OH

DISCUSSION:

CaF_2 specimens as artificially grown and cut lenses. Specimen surfaces were lightly polish cleaned by hand lap on a hard felt surface saturated with the solution prior to vacuum metallization with Au/Cr thin films. Water wash and nitrogen blow dry after cleaning.

REF: Walker, P — personal application, 1975

CAF-0009

ETCH NAME: Sulfuric acid — TIME:

TYPE: Acid, preferential — TEMP:

COMPOSITION:

x .... H_2SO_4, conc.

DISCUSSION:

CaF_2, (111) wafers cleaved from natural fluorite crystals. Solution developed triangular etch pits with 1:1 correspondence between the opposed positive and negative (111) faces of wafers.

REF: Pandya, N S & Pandya, J K — *Curr Sci,* 27,437(1958)

CALCIUM SILICON FLUORIDE ETCHANTS

CASF-0001

ETCH NAME: — TIME:

TYPE: Acid, removal — TEMP:

COMPOSITION:

x .... HF

x .... HNO_3

DISCUSSION:

$CaSiF_2$ single crystals were used in evaluating cohesive energy features of tetrahedral semiconductors. Forty materials were studied.

REF: Aresti, A et al — *J Phys Chem Solids,* 45,361(1984)

CALCIUM TIN FLUORIDE ETCHANTS

CATF-0001

ETCH NAME: Ammonium hydroxide — TIME:

TYPE: Base, removal/cleaning — TEMP:

COMPOSITION:

x .... NH_4OH

DISCUSSION:

$CaSnF_2$ deposited as a thin film in forming a semiconductor-dielectric-semiconductor

(SIS) structure. Compound referred to as a spinel, deposited as a Ge/$CaSnF_2$ multifilm structure on silicon (111) substrates.
REF: Tu, G W et al — *J Vac Sci Technol,* B2,212(1984)

PHYSICAL PROPERTIES OF CALCIUM MOLYBDATE, $CaMoO_4$

Classification	Oxide
Atomic numbers	20, 42, & 8
Atomic weight	200
Melting point (°C)	1200
Boiling point (°C)	
Density (g/cm³)	4.35
Refractive index (n=)	1.974—1.984
Hardness (Mohs — scratch)	3—3.5
Crystal structure (tetragonal — normal)	(111) pyramid
Color (solid)	Yellowish
Cleavage (basal)	(001)

CALCIUM MOLYBDATE, $CaMoO_4$

General: Does not occur as a pure compound, but the mineral powellite, Ca(W,Mo)O_4 is found as a mixed element compound with about 10% $CaWO_4$. The pure molybdate has been referred to as powellite in the literature.

The pure compound can be made by roasting molybdenite, MoS_2 with coke and lime, the material then used in making other molybdenum compounds. The pure metal is used in fabricating high speed steels, and other compounds in the dyeing of silk or other fabrics.

Technical Application: Both natural and artificial single crystals have been studied for general morphology; electronic and electrical properties; and crystallographic structure changes with pressure and temperature.

Etching: Soluble in acids.

CALCIUM MOLYBDATE ETCHANTS

CAMO-0001
ETCH NAME: Pressure — TIME:
TYPE: Pressure, alternation — TEMP:
COMPOSITION:
x .... pressure
DISCUSSION:

$CaMoO_4$ as single crystals were studied in a high pressure chemistry test of molybdnates and tungstates at about 56 Pa pressure. Other materials evaluated were $CaWO_4$, $PbWO_4$, and $CdMoO_4$. No etch shown, but H_3PO_4 is a general removal solution on tungstates.
REF: Hazen, R M et al — *J Phys Chem Solids,* 46,253(1985)

PHYSICAL PROPERTIES OF CALCIUM NITRIDE, Ca_2N_3

Classification	Nitride
Atomic numbers	20 & 7
Atomic weight	148.26
Melting point (°C)	900
Boiling point (°C)	

Density (g/cm^3)	2.63
Hardness (Mohs — scratch)	5—6
Crystal structure (isometric — normal) alpha	(100) cube
(tetragonal — normal) beta	(110) prism
Color (solid)	Brown/black
Cleavage (cubic/basal)	(001)

CALCIUM NITRIDE, Ca_2N_3

General: Does not occur as a natural compound even though there are over 250 calcium minerals as fluorides, sulfides, carbonates, silicates, etc. (Nitrogen alone does not form pure nitride compounds.)

The nitride has no application in industry at present, though several calcium compounds are of importance, such as calcite, $CaCO_3$.

Technical Application: Not used in Solid State processing at present, though it has been grown as a single crystal thin film on calcium substrates in material evaluation. There are possible applications as a dielectric for its optical properties, or as a surface protective coating similar to those of other nitride compounds.

Etching: Soluble in dilute acids.

CALCIUM NITRIDE ETCHANTS

CAN-0001

ETCH NAME: Nitric acid — TIME:

TYPE; Acid, removal — TEMP: RT

COMPOSITION:

x .... HNO_3

x .... H_2O

DISCUSSION:

Ca_2N_3 thin films deposited on calcium substrates in a study of the use of high purity nitrogen to reduce contamination. Deposited at 600°C, nitride is tetragonal — at 675°C is cubic structure. Also worked with magnesium nitride (MGN-0001), and references work done with lithium nitride (LIN-0001). If O_2 or air is present during deposition in the N_2 atmosphere, white spots of CaO or MgO appear in the otherwise black nitride films.

REF: Aubry, J & Streiff, R — *J Electrochem Soc,* 118,650(1971)

PHYSICAL PROPERTIES OF CALCIUM TUNGSTATE, $CaWO_4$

Classification	Tungstate
Atomic numbers	20, 74, & 8
Atomic weight	289.92
Melting point (°C)	>1200
Boiling point (°C)	
Density (g/cm^3)	6.06 (5.9—6.1) R
Refractive index (n=)	1.918—1.934
Crystal structure (tetragonal — pyramidal)	(111) pyramid
Color (solid)	Yellow/brownish
Cleavage (pyramidal — distinct)	p(111)

CALCIUM TUNGSTATE, $CaWO_4$

General: Occurs in nature as the mineral scheelite, $CaWO_4$, which is a major ore of tungsten. The mineral forms under pneumatolytic or hydrothermal conditions in pegamatites, or as ore veins in association with granites. It is commonly yellow to brown in color, and fluoresces a bright yellow-white under a black light. Fused with salt of phosphorus it melts to a glass at about 1200°C with a fine blue color.

Industrially it is a major ore of tungsten, and the black light is used during grinding to insure complete segregation of the gangue material. Also used as a fine blue pigment in glass and enamels.

Technical Application: No application in Solid State processing at this time, though it has been grown and studied as a single crystal for defects and general morphology, to include possibly frequency applications.

Etching: Hot H_3PO_4, and mixed acids.

CALCIUM TUNGSTATE ETCHANTS

CAW-0001a

ETCH NAME: Phosphoric acid — TIME:

TYPE: Acid, removal — TEMP: 250°C

COMPOSITION:

x .... H_3PO_4

DISCUSSION:

$CaWO_4$, (001) wafers. Ingots were grown with (100) orientation and wafers were cut as (001) basal orientation. Solution used to remove work damage after mechanical lap and polish.

REF: Lockayne, B — *Phil Mag,* 10,911(1964)

CAW-0001b

ETCH NAME: Aqua regia — TIME:

TYPE: Acid, preferential — TEMP: 55°C

COMPOSITION:

3 .... HCl

1 .... HNO_3

DISCUSSION:

$CaWO_4$, (001) wafers used in a study of deformation and slip. Solution used to develop slip patterns and dislocation etch pits.

REF: Ibid.

CAW-0002: Lockayne, B — *Br J Appl Phys,* 16,423(1965)

$CaWO_4$, (001) wafers. Ingots were grown with (100) orientation, and and cut (001) basal as free of low-angle grain boundaries. After mechanical lap, solution used to etch polish and develop dislocation distributions.

CAW-0003a

ETCH NAME: — TIME:

TYPE: Acid, polish — TEMP: 200°C

COMPOSITION: — RATE: 0.009—0.06 μm/sec

3 .... H_3PO_4

1 .... CrO_3, sat. sol.

DISCUSSION:

$CaWO_4$, (100) wafers. Solution used to etch polish specimens in a study of dislocations.

REF: Ibid.

CAW-0003b
ETCH NAME: TIME: 2—25 min
TYPE: Acid, dislocation TEMP: RT
COMPOSITION: RATE: .003 µm/sec
1 .... HF
2 .... CrO_3, sat. sol.
DISCUSSION:
$CaWO_4$, (001) wafers. Solution used to develop dislocation pits and pits showed a 1:1 correspondence from positive to negative wafer surfaces.
REF: Ibid.

CAW-0004
ETCH NAME: Phosphoric acid TIME:
TYPE: Acid, polish TEMP: Hot
COMPOSITION:
x .... H_3PO_4, conc.
DISCUSSION:
$CaWO_4$, single crystal specimens used in a study of Raman frequency shifts and temperature dependence. Other tungstates studied: $SrWO_4$ and $BaWO_4$.
REF: Degreniers, S et al — *J Phys Chem Solids,* 45,1105(1984)

CAW-0005a
ETCH NAME: Phosphoric acid TIME: 15 min
TYPE: Acid, polish TEMP: 300°C
COMPOSITION:
x .... H_3PO_4
DISCUSSION:
$CaWO_4$, single crystal specimens used in a study of plastic deformation. Solution used to remove residual lap damage and polish, though it can be rough if etching is over 15 min. Use a Pt basket to hold specimens, place in acid at RT, then bring up to 300°C boiling point — reduce to 150°C before removal to boiling water quench. If quenched above 150°C, specimen surface cracks.
REF: Arbel, A & Stokes, R J — *J Appl Phys,* 36,1460(1965)

CAW-0005b
ETCH NAME: TIME: 15—20 min
TYPE: Acid, dislocation TEMP: 100°C
COMPOSITION:
1 .... H_3PO_4
1 .... NH_4Cl, sat. sol.
2 .... H_2O
DISCUSSION:
$CaWO_4$, single crystal specimens used in a study of plastic deformation. Solution used to develop defects after material stressing. Place specimen in solution at RT, raise to 100°C for etching period, then quench directly into water at RT.
REF: Ibid.

CAW-0006
ETCH NAME: TIME:
TYPE: Pressure, alteration TEMP:
COMPOSITION:
x .... pressure

DISCUSSION:

$CaWO_4$ single crystals used in a high pressure chemistry test of materials above 56 Pa pressure. Both tungstates and molybdates were studied: $PbWO_4$, $CaWO_4$, $PbMO_4$, and $CdMo_4$. $CaWO_4$ noted as the mineral scheelite.

REF: Hazen, R M et al — *J Phys Chem Solids*, 46,253(1985)

PHYSICAL PROPERTIES OF CALIFORNIUM, Cf

Classification	Actinide
Atomic number	98
Atomic weight	251
Melting point (°C)	
Boiling point (°C)	
Density (g/cm^3)	
Ionic radius (angstroms)	0.98 (Cf^{+3})
Half-life (years)	800
Cross section (barns)	3000
Electrochemical equivalents (g/amp-h)	3.1
Valence electron potential (eV)	44.1
Hardness (Mohs — scratch)	3—4
Crystal structure (isometric — normal)	(100) cube
Color (solid)	Grey
(gas)	Green
Cleavage (cubic)	(001)

CALIFORNIUM, Cf

General: A naturally occurring gas as one of the radioactive decay series. No use in industry at the present time.

Technical Application: No use in Solid State processing to date. It has been studied in solid form.

Etching: Soluble in HCl.

CALIFORNIUM ETCHANTS

CF-0001

ETCH NAME: Hydrochloric acid — TIME:

TYPE: Acid, removal — TEMP:

COMPOSITION:

x .... HCl, conc.

DISCUSSION:

Cf specimens. Solution used to dissolve the material in a study of the properties of the isotope $^2Fm_{52}$ (fermium).

REF: Friedman, A M et al — *Phys Rev*, 101,1472(1956)

CARBIDES, M_xC

General: There is only one known occurrence of a carbide mineral in nature — the mineral moissanite, CSi, found as small green hexagonal platelet in the meteoric iron from Cañon Diablo, Arizona — though the material was originally produced as artificial carborundum, SiC.

Both carbides and borides are part of the ceramic industry of today, though the ancient pottery industry is still referred to as ''ceramic pottery'' where silicate glaze (SiO_2) with or without mineral additives for color as enamels have been in use since about 3500 B.C. when the Egyptians first developed soda-lime glass. There are many industrial applications for the high temperature ceramics, such as for furnace lining fire-bricks, crucibles, and, when mixed with carbon, for radiation resistant materials used in atomic nuclear equipment construction. Ceramics also have been called "Cermets". See General discussion under Ceramics, this chapter, for further discussion.

Technical Application: Several carbides are used as a ceramic in Solid State processing as circuit substrates, in package construction, or as electrical stand-off insulators. As an active discrete device, they are designed as resistors and capacitors, or may be co-deposited as a planar type dielectric in a substrate circuit.

Silicon carbide, SiC, when boron doped, is a semiconductor capable of operating as a diode above 500°C. Boron nitride and diamond also show semiconducting characteristics at elevated temperatures.

The individual carbides are not listed, here, as several are of sufficient importance as both ceramics and dielectrics that they are under their own listed sections following their parent metals. See the Material Index for numbers. Borides are as a collective listed heading. Also see aluminum oxide, Al_2O_3; silicon dioxide, SiO_2; and boron nitride, BN. Many of these compounds are used as ceramics or as dielectrics.

Etching: Difficult. See separate compounds for solutions.

PHYSICAL PROPERTIES OF CARBON, C

Classification	Ceramic
Atomic number	6
Atomic weight	12.01
Melting point (°C)	3727 (3550)
Boiling point (°C)	4830 (4827)
Density (g/cm^3)	2.26 (2.09—2.23)
Thermal conductance (cal/sec)(cm^2)(°C/cm)	0.057
Specific heat (cal/g) 25°C	0.165
Heat of vaporization (k-cal/g-atom)	171.7
Atomic volume (W/D)	5.3
1st ionization energy (K-cal/g-mole)	260
1st ionization potential (eV)	11.264
Electronegativity (Pauling's)	2.5
Covalent radius (angstroms)	0.77
Ionic radius (angstroms)	0.16 (C^{+4})
Coefficient of thermal linear expansion	0.6—4.3
($\times 10^{-6}$ cm/cm/°C) 20°C ($K^{-1} \times 10^{-6}$)	3.0
Electron work function (eV)	4.82
Cross section (barns)	0.0034
Vapor pressure (°C)	4373
Electricity resistivity ($\times 10^{-6}$ ohms cm) 0°C	1375
Bond dissociation energy (kJ mol-1)	606
Cohesive energy (kJ mol-1)	713
Tensile strength (kgf cm-2 $\times 10^{-4}$)	7.7
Thermal endurance coefficient [F(cmK min$^{-1/2}$ $\times 10^{-4}$)]	2.04
Relative F match	52

Hardness (Mohs — scratch) carbon (hard)	6—7
graphite, natural	1—2
(Vicker's — kgf mm_2) — carbon (vitreous)	2200—2859
(Scleroscope — pressure strike)	10—100
(Shore-ball bounce) — carbon (vitreous)	120—150
Crystal structure (hexagonal — rhombohedral)	(0001) plates 6-sided
Color (solid/powder)	Coal black
Cleavage (basal — perfect)	(0001)

CARBON, C

General: Occurs in nature as the mineral graphite, C, (plumbago, black lead). Industrially called carbon, the high pressure form is diamond. Carbon can be recognized by its extreme softness, greasy feeling, black color and metallic luster. The two terms are interchangeable — graphite or carbon. It is the ''lead'' in pencils as it was first thought to be a form of lead (Pb). It is in the hexagonal system crystallographically like micas with the prominent basal (0001) cleavage known as micaceous cleavage, which is more apparent in graphite than in artificially grown carbons. The monolayer hexagonal planes contain the atomic ''benzene ring'' structure with strong binding energy within the monolayer, but weak binding energy between monolayers. In natural graphite these benzene-type carbon rings are vertically aligned directly above each other from layer to layer; whereas in carbon the rings are off-set between alternate monolayers. Because of this structure variation many types of carbon materials can be artificially fabricated, often constructed to improve the binding energy between monolayers. In fabrication at 500°C, grain spacing is 10—20 Å, whereas at 2500—3000°C, 70—1000 Å spacing. Further, with controlled granular structure, the material can be cut parallel or perpendicular to grain orientations, thus producing different characteristics.

In addition to being mined as natural graphite, carbon is obtained from petroleum, natural asphalt and tar; from coal — anthracite (hard), and bituminous (soft); from natural gases, such as methane; and from the burning of vegetation. When partially fired to remove water vapor and gaseous impurities, but below the ignition point, it is called ''coke'', hardwoods similarly processed as charcoal. The charcoal burners of ancient times were a race among themselves, processing hardwood trees as a coke-type fuel for the metals and glass industries, as well as for cooking fires throughout Europe, the Middle East, and Asia. Cooking was being done prior to 1350 B.C. (Iron Age) for the smelting of bronze and today coke is a primary fuel in the fabrication and processing of steels. As lampblack, a near-liquid by-product of oil, or as finely divided carbon powder down to colloidal structure, it is a black pigment for glass, enamels, cement, etc. It also is still the black ''lead'' in pencils, the term ''lead'' being a misnomer, as it is graphite or carbon, not lead.

In addition to its use as a fuel or additive to metal products, carbon parts have many industrial uses for its high temperature, and both electrical and heat transfer capabilities. It is extremely inert to chemical attack and, coupled with its high temperature capability, is used as a crucible for processing liquid metals, such as in semiconductor ingot growth pots, as a metal evaporation crucible, or as crucibles in heavy metal processing. For its heat transfer capabilities and as a bulk item, it is used as graphite susceptor plates, alone, or SiC coated in epitaxy systems; as heater strips for alloying parts, or disassembly and recovering of rejected parts; as furnace linings and heater rods or furnace tubes, and as boats, flats, etc., as holding vehicles for parts in alloy furnaces.

As a wire it was one of the first light bulb filaments — is still so used in special applications — or wires can be embedded in glass or plastics as a heater grid, and are used as automotive ignition wires for spark plugs. As a pair of rods, it is used for spark gap welding or vaporizing a metal surface for spectrographic analysis of elements. Also as rods, due to its low neutron cross-section, carbons are the damping rods in atomic piles in nuclear energy plants. As powder, in addition to being a pigment, it is used as a dry lubricant like molybdenum sulfides and, sprayed on glass or plastic, acts as a sunscreen. As a powdered mixture in alcohol (Aqua Dag is one commercial product) it is used as a spray or paint-on vehicle on metal surfaces as the replication technique: after drying, the carbon thin film can be removed for microscope study or, coated with thin film gold for reflectivity, for a scanning electron microscope (SEM) study. The dry powder has been brushed on preferentially etched surfaces to enhance defects for microscope examination, as has magnetite and barium titanate. The latter include magnetization of the particles in surface studies of barium titanate, or magnetic nickels, irons, etc.

Although this book is concerned with inorganic materials, carbon is a major element in organic chemistry — C, H, O, N — the four elements of our carbon-base life forms . . . humans, animals, and vegetation. There are thousands of chemicals, solvents, and gases containing these elements involved in both organic chemistry and medicine.

Technical Application: Although carbon alone is not used to any extent in Solid State device fabrication at present, it is under development as a thin film colloid (c-C) or amorphous (a-C) deposited structure, as grown by CVD methods using a variety of sources for the carbon, such as carbon tetrachloride, CCl_4, benzene, C_6H_6, and other such compounds. Deposition is between 400 to 600°C and, with subsequent temperature annealing, the carbon films are being converted to a crystalline diamond-like carbon (DLC) film; also referred to by the acronym i-C. Additional laser anneal has been used to increase the size of crystallites and, eventually, it is hoped to achieve a complete single crystal diamond structure. As diamond it is the hardest of known substances and is chemically inert, it has tremendous applications as a thin film surface coating, in addition to the diamond semiconducting characteristics.

Carbon thin films also are deposited on silicon, Si, wafers for the fabrication of thin film silicon carbide, SiC. See Silicon Carbide section for additional discussion. There are carbon composite materials in use or under development in both the industrial and Solid State areas as parts and device structures. See sections on carbides, graphite, and diamond.

Etching: Acid resistant to most chemicals under normal conditions. Dissolves in boiling HNO_3, KOH, H_2SO_4, H_3PO_4, and in molten fluxes of alkalies and metals.

CARBON ETCHANTS

C-0001a

ETCH NAME: Potassium hydroxide — TIME:

TYPE: Molten flux, removal — TEMP: 350°C

COMPOSITION:

x .... KOH (NaOH)

DISCUSSION:

C, as bulk graphite or thin film deposit. Molten flux of alkalies will attack carbon and graphite, and may be slightly preferential.

REF: Kohl, W H — *Handbook of Materials and Techniques for Vacuum Devices*, Reinhold, New York, 1967

C-0001b

ETCH NAME: Potassium hydroxide — TIME:

TYPE: Alkali, removal — TEMP: Boiling

COMPOSITION:

x .... 50% KOH (NaOH)

DISCUSSION:

C, as bulk graphite or thin film deposit. Hot, concentrated solutions of alkalies will dissolve graphite and carbon.

REF: Ibid.

C-0001c
ETCH NAME: Nitric acid TIME:
TYPE: Acid, removal TEMP:
COMPOSITION:
x .... HNO_3, conc.
DISCUSSION:
C, as bulk graphite or thin film deposit. Nitric acid will dissolve graphite or carbon forming mellitic acid, hydrocyanic acid or CO_2 + N_2O, depending upon etching conditions.
REF: Ibid.
C-0002a: Bulletin CD-101-5M-SC-282, Stackpole Corp., 1982
C, as graphite or carbon specimens. Nitric acid will dissolve material and produce mellitic acid, hydrocyanic acid with both CO_2 and NO.

C-0001d
ETCH NAME: TIME:
TYPE: Acid, reduction TEMP:
COMPOSITION:
20 ml .... HNO_3
40 ml .... H_2SO_4
20 g .... $KClO_4$
DISCUSSION:
C, as bulk graphite or thin film deposit. This solution volume will dissolve one gram of carbon forming graphitic acid.
REF: Ibid.

C-0002b
ETCH NAME: Sulfuric acid TIME:
TYPE: Acid, removal TEMP: Boiling
COMPOSITION:
x ... H_2SO_4, conc.
DISCUSSION:
C, as both carbon and graphite parts. Solution will form mellitic acid or benzene pentacarboxylic acid with CO_2 and SO_2.
REF: Ibid.
C-0006a: Halbert, M L et al — *Phys Rev,* 106,251(1957)
C, specimens used in a study of neutron-transfer reactions by nitrogen ion bombardment. Carbon was etched in "fuming" sulfuric acid. (*Note:* Only reference to sulfuric acid as "fuming".)

C-0003
ETCH NAME: Phosphoric acid TIME:
TYPE: Acid, polish TEMP: 240°C
composition:
x .... H_3PO_4, conc.
DISCUSSION:
C, as natural graphite, cleaved as (0001) basal specimens and studied for magnetic domains. Mechanical polish with carborundum paste (SiC), then with 8 μm diamond paste. Use silica crucible to hold specimens during etching to reduce thickness. Solution also used to etch thin and to pinhole specimens for electron microscope (EM) study. (*Note:* Material was called "plumbite", which is a common name for graphite, e.g., plumbago.)
REF: Grundy, P J — *Br J Appl Phys,* 16,409(1965)

C-0004
ETCH NAME: Argon
TYPE: Ionized gas, preferential
COMPOSITION:
x .... Ar^+ ions

TIME: 15 min
TEMP:
GAS: Argon
PRESSURE:
POWER:

DISCUSSION:

C, (0001) cleaved pyrolytic graphite specimens. Argon ionic bombardment produces conical etch pits.

REF: Tarpinian, A & Gaza, G E — *J Appl Phys,* 31,1657(1960)

C-0005
ETCH NAME: Water
TYPE: Acid, cleaning
COMPOSITION:
x .... H_2O

TIME:
TEMP: RT

DISCUSSION:

C, as thin films deposits on Si, SiO_2, Al, Al_2O_3, KBr, NaCl, ZnMn, GaAs and InP. Deposition was by RF Plasma using butane, C_4H_{10}, with the glow discharge from ultra-pure carbon electrodes. Growth rate: 8—360 Å/min. Film was transparent and amorphous in structure. Water used to wash and clean surfaces to observe hydrophobic/hydrophilic reaction. Film referred to as a diamond-like carbon (DLC) film.

REF: Zelez, J — *RCA Rev,* 43,665(1982)

C-0006b
ETCH NAME:
TYPE: Acid, removal
COMPOSITION:
x .... H_3PO_4
x .... CrO_3
x .... NaCN

TIME:
TEMP:

DISCUSSION:

C specimens. Solution described as a general etch for carbon. Other chromates can be used, and NaCN added for an equivalent 1 ml of NH_4.

REF: Ibid.

C-0007a
ETCH NAME:
TYPE: Acid, cleaning
COMPOSITION:
50 ml .... HF
150 ml .... HNO_3
100 ml .... CH_3COOH (HAc)

TIME: 1—3 min
TEMP: RT

DISCUSSION:

C, as epitaxy reactor plates and susceptors, or for Silox system deposition of SiO_2 thin films. Graphite susceptors used in metallic compound growth, such as silicon on silicon, gallium arsenide, indium antimonide, etc. They may be plain graphite or coated with silicon carbide. The solution shown is used to etch clean plate surfaces to remove silicon and/or silica growth. Heavily water wash after acid cleaning and wipe dry with toweling; then vacuum oven bake at 130°C and 10^{-3} Torr pressure overnight. **CAUTION:** Where phos-

phorus gases have been used as a dopant during silicon growth, it can ignite under liquid during cleaning.
REF: Walker, P et al — personal application, 1970—1975

C-0007b
ETCH NAME: Hydrogen/chlorine TIME: 1—3 min
TYPE: Gas, cleaning TEMP: 1100°C
COMPOSITION:
(1) x .... H_2 (2) x .... HCl, vapor
DISCUSSION:

C, as epitaxy reactor susceptor plates for Si/Si(111) epitaxy growth. Susceptors with or without SiC coating. The hot hydrogen in reactor system used to clean graphite surfaces prior to introduction of HCl vapors for additional cleaning.
REF: Ibid.
C-0007c: Ibid.

C, as furnace alloy boats and disc parts. Lightly lap parts on Whattman filter paper to clean, then fire through belt furnace in forming gas (85% N_2:15% H_2) at 450°C prior to using newly fabricated parts, or after cleaning old boats and discs. Firing can be done in a hydrogen humpback furnace.

C-0008
ETCH NAME: Carbonization TIME:
TYPE: Thermal, forming TEMP: Elevated
COMPOSITION:
x .... heat
DISCUSSION:

C, as a wood product by charring. Hard woods are partly burned to remove volatiles, the resultant material called charcoal. When natural coal is so processed, it is called "coke". Charcoal has been a fuel for glass and metal fabrication since ancient times, and also is used as a cooking fuel. Coke is widely used in metal processing industries today, as a primary fuel such as for steel manufacture.

In the fractionation of petroleum, the pure carbon fraction is called lampblack and used as a black pigment. Metal surfaces are carbonized with hot CO_2 treatment for hardening and when carbon is added to iron it becomes steel. The carbonization process has been a fundamental step in the manufacture of irons and steels since early times. Two notable steels are Damascus and Toledo steel, both were widely used during the Middle Ages in Europe.

In Solid State processing of epoxies and polyimides used for electrical contact, one method of removal is to heat to the char-point with subsequent scrape-off. Carbon thin films are deposited and, with either heat treatment or laser annealing, converted to diamond-like carbon (DLC). Carbon in an alcohol solution is painted onto material surfaces, hardened by evaporation removal of the alcohol carrier, the resultant film used for surface replication. Molasses has been used as an etching solution constituent on quartz frequency crystal blanks, but can leave carbon flakes on surfaces after vacuum metallization. Carbon powder has been brushed onto material surfaces to enhance surface defects and structure for microscope study.

There are several major products uses for carbon, such as electrical ignition wires; damping rods in atomic reactors; as an additive to high temperature ceramics for both electrical and high temperature capabilities; as powder sprayed on plastic sheeting for sun-screen applications; as a heater strip or rod; as a black pigment in paint, glass, cement.
REF: N/A

PHYSICAL PROPERTIES OF CARBON DIOXIDE, CO_2

Classification	Gas
Atomic numbers	6 & 8
Atomic weight	44.01
Melting point (°C) 5.2 atms	−56.6
Boiling point (°C) solid	−78.5
Density (g/cm³) solid −79°C	1.56
(g/l) liquid, −37°C	1.101
(g/l) liquid, 0°C	1.97
Refractive index (n=)	1.38—1.46
Solubility (100 ml H_2O) RT	1.45 g
(100 ml H_2O) 0°C	1.79 g
Hardness (Mohs — scratch) solid	1.5—2
Crystal structure (isometric — normal) solid	(100) cube
Color (gas)	Colorless
(crystalline solid)	White
Cleavage (cubic)	(001)

CARBON DIOXIDE, CO_2

General: Occurs in nature as a fraction of air, in waters, and in the soil. In certain areas it can be associated as a gas from subsurface magmas and volcanoes, or be released into the atmosphere from earth fissures. It is a by-product of combustion; comes from decaying matter; and is the gas resulting from fermentation, such as beers and wines. Animal life breathes air, extracts the oxygen, and release carbon dioxide; whereas vegetation uses carbon dioxide, releasing oxygen back into the atmosphere . . . a symbiotic life cycle.

The decomposition of certain carbonate minerals or reduction of their metallic ores can release the gas. Like nitrogen, it will not support combustion — it is a by-product of combustion — and can be lethal when oxygen is not present. Skin contact should not be made with dry ice as it can cause burn-like blisters.

As a commercially prepared food product, it is dissolved in water as soda water, seltzer water, or carbonated water. In addition to the natural fermentation of beers and wines . . . the British "bubbly" is champagne . . . it is added to soft drinks for effervescence as a carbonated beverage.

Industrially, as solid dry ice (−79°C) it is used as a temporary refrigerant, such as for transportation of epoxy and polyamide pastes that are shipped frozen for extended shelf-life. Where used in large volumes, it usually is transported as a cryogenic liquid at around −37°C. In chemical processing, dry ice has been used as a freezing mixture in acetone (−100 to −80°C) for the removal of water vapor from other gases, or as a chilling medium in etching, as a hardening agent, or as an oxidizer.

Technical Application: Not used directly in Solid State device design, though it is occasionally used as a drying gas or cleaning agent at room temperature like nitrogen, argon, and helium, but at elevated temperatures can be reactive as an oxidizer or carbonizer. CO_2 has been bubbled through etching solutions for stirring action only or as an active element for oxidation and etch action. The gas also has been used as a hot furnace gas as an oxidizing agent, usually in combination with air and/or oxygen, or as a mixed gas with the carbon fraction forming carbides, specifically silicon carbide on silicon surfaces.

As dry ice in laboratories, it is used as a temporary refrigerant for the storage of such materials as gallium, bromine, epoxy, and polyamide formulations.

Dry ice is one of the few compounds that sublimes directly from the solid to the gaseous state without passing through a liquid state, though it can be liquefied under pressure as a cryogenic gas. It has been studied as a crystalline mass of dry ice and grown as a single crystal for morphological studies.

Etching: Solid sublimes without melting.

CARBON DIOXIDE ETCHANTS

COD-0001

ETCH NAME: Liquid nitrogen — TIME:

TYPE: Cryogenic gas, forming — TEMP: Freezing (−196°C)

COMPOSITION:

x .... LN_2

DISCUSSION:

CO_2, grown as a single crystal using a cold temperature cryostat in order to measure the refractive index. Index varies with wavelength, and is between n = 1.38 and 1.46.

REF: Tempelmeyer, K E et al — *J Appl Phys,* 39,2968(1968)

COD-0002

ETCH NAME: Methyl alcohol — TIME:

TYPE: Alcohol, chilling — TEMP:

COMPOSITION:

(1) x .... MeOH (2) x .... acetone

DISCUSSION:

CO_2, as solid "dry ice". Used as a mixture of chunks in methanol or acetone it is a "cold solution" used for removal of water vapor in gases. It has been used for drying of argon from pressurized A-1 cylinders where argon was used to backfill sealed quartz tubes containing silicon wafers for closed tube gallium diffusion in fabrication SCRs. Both liquids shown above can act as structure etchants on solid CO_2.

REF: Walker, P & Orwin, H — personal application, 1974

COD-0003

ETCH NAME: Carbon dioxide — TIME:

TYPE: Gas, oxidation — TEMP:

COMPOSITION:

x .... O_2

x .... CO_2

DISCUSSION:

CO_2 is used as a furnace gas mixture to oxidize chromium metal to chromium trioxide in a study of Cr_2O_3.

REF: Torkington, R S & Vaughn, J G — *J Vac Sci Technol,* A(3),795(1985)

CERAMICS

General: The term is still applied to the pottery industry which has been in existence for some 10,000 years, and it includes many natural minerals, such as corundum, Al_2O_3, and quartz, SiO_2, as two crystal varieties; and as a group, both the feldspar and clay minerals. Aluminum silicates as orthoclase or microcline, $KAlSi_3O_8$, and albite, $NaAlSi_3O_8$ are found in some 60% of all rocks and, with weathering, break down forming clay minerals: kaolinite, $Al_2H_2(SiO_4)_2$, as an admixture with sand, iron, and other metal oxides is common clay — from the feldspars. Kaolin is a high purity form of kaolinite which, when heated, burns to a white enamel and is used for fine "bone" china, pottery, and porcelain.

Fuller's Earth is a form of clay that absorbs water, as does montmorillonite. Bentonite is a colloidal clay that will swell in water, and is used as a packing around piping in oil fields, and as a filler for rubber products or binder in manufacturing of lead pencils, e.g., carbon powder.

In making pottery, finely powdered clays will form a wet plastic mass that can be shaped, then hardened by heating to remove the water content. The wet mass is called slip. For general clay products, such as pottery, porcelain, brick or tile, and tile piping, it is kiln fired at 1400°C. First firing is a porous product called bisque; prior to second firing the surfaces are coated with a feldspar silicate paste to make the clay water tight (a glaze); prior to third firing, as for enamels, coat or paint with metallic oxides for color — hematite, Fe_2O_3, as red ocher; cobalt for blue; chromium trioxide for green, etc.

Hard porcelain uses a high silicate content clay, is fired at above 1600°C without further treatment, and it forms porous pottery ware, such as terra cotta, bricks, crockery, etc. This type of clay, when fired at a lower temperature along with red ocher (hematite) is standard brick fabrication and, if NaCl (halite, common salt) is added, produces a glaze. Refractory ceramics, such as fire-brick, contain both magnesium and chromium oxides for high temperature stability. It was the application of silicates and glass as a surface glaze on clay pottery (Egyptians with development of soda-lime glass, circa 3500 B.C.) that produced the name: ceramic pottery.

The production of high quality refractory ceramics has become separated from the ancient and original pottery field in modern times as the ceramics industry, which includes both general products such as fire-brick and tile, as well as high purity alumina or beryllia substrate used in Solid State circuit fabrication.

Many composite ceramics are now manufactured with variety of metal oxides, some of which include carbon or graphite. As an example, most alumina substrates are pure Al_2O_3, as a white ceramic, but there is a black alumina containing carbon which can be electrically active rather than an insulator.

Alumina is supplied as pressed powder blanks; as amorphous, fused alumina, called fused sapphire; or as single crystal sapphire. The sapphires are clear and transparent, and can only be differentiated from similar quartz blanks by material hardness . . . H = 7 (quartz) . . . H = 9 (sapphire). Both sapphire and quartz single crystal blanks used as substrates are usually oriented as basal (0001) surfaces.

Transparent, slightly yellow/brown tinted garnet (YIG, YAG, etc.) blanks are available in (001) or (110) orientation, cubic, or dodecahedral, respectively. Red fused alumina — ruby — is manufactured by pouring a liquid melt into high temperature resistant metal forms such as molybdenum. A variety of products are so fabricated, such as tubes, rods, flats (substrates) wire bonding tips, or further processed as high temperature ball-bearings (ruby spheres).

Although the glass industry is a separate entity . . . the clay, pottery, silicate, glass, and ceramic industries are inter-related . . . and glass is not normally considered a ceramic, but the Solid State industry used both silica and quartz blanks as substrate. Clear, fused amorphous silica, or oriented single crystal quartz. The latter, as alpha-quartz; processed as radio frequency crystals from both natural quartz and artificial boules. In glass applications as such substrates, the material is often grouped with and called a ceramic or dielectric.

The ferrites are another group of materials that are distinct, but can be classified as ceramic type materials with regard to some applications in Solid State processing. They are supplied as flats or blanks like other substrates as mixtures of iron, carbon, and other metal additives. They are coal black in color with highly polished surfaces, and specifically formulated for fabrication of microelectronic circuit resistors at different frequencies. All of these substrates, including the ferrites, are metallized, then photolithographically patterned as circuits or discrete devices.

Technical Application: All of the ceramic material mentioned above have wide use and application in Solid State circuit fabrication, in addition to their normal applications as furnace fire-brick; furnace tubes and liners; insulated stand-offs in both vacuum systems and test fixtures, or in device packages; as tubes or beads for high temperature thermocouples, or crucibles; as chemical solution boiling beads; the pressed powder alumina as a tool sharpener; as the cutting tool, itself; or as lapping abrasives, to mention only a few such uses.

In device fabrication the ceramics are primarily insulators — for mounting of discrete devices, after metallizing and circuit patterning of the substrate. Several also are used as discrete devices, such as the ferrites as resistors, or glass and ceramics as dielectrics, and for filter elements.

Only those ceramics referred to as cermets are shown here. The others are listed under their individually named sections.

Etching: See individual ceramics, borides, glass, etc. Etch solutions vary with type material, and all are generally difficult to etch.

CERMET ETCHANTS

CMT-0001a
ETCH NAME: TIME:
TYPE: Acid, removal TEMP: RT
COMPOSITION:
1 .... HF
5 .... HNO_3
60 .... H_3PO_4
DISCUSSION:

Cr:SiO (30%) as a deposited thin film surface coating. Solution used for both surface cleaning and material removal. Over 30% SiO limits the etch rate. (*Note:* There is a wide variety of cermet materials.)

REF: Miassel, L I & Glang, R — *Handbook of Thin Film Technology*, McGraw-Hill, New York, 1970.

CMT-0001b
ETCH NAME: TIME:
TYPE: Acid, removal TEMP: 50—60°C
COMPOSITION:
20 g .... $K_2Fe(CN)_6$
10 g .... NaOH
100 ml .... H_2O
DISCUSSION:

Cr:SiO (30%) as a deposited thin film coating. Solution used for both surface cleaning and removal of the material.

REF: Ibid.

CERAMIC ETCHANTS

CERM-0001
ETCH NAME: TIME:
TYPE: Metal, contacts TEMP:
COMPOSITION:
x .... $PdCl_2$
x .... H_2O
x .... $SnCl_2$
DISCUSSION:

Ceramics. Dipping parts in a stannous palladium chloride solution prior to electroless

nickel plating produces a light Pd coating that aids in the gripping power of nickel. Nickel plating evaluated at 95°C with a 0.001 mm thick deposit.
REF: Turner, D R & Sauer, H A — *J Electrochem Soc,* 107,250(1960)
CERM-0002: Pearlstein, F — *Met Finish,* 53,59(1955)
Reference cited in CERM-0001 for palladium solution.

PHYSICAL PROPERTIES OF CERIUM, Ce

Classification	Lanthanide
Atomic number	58
Atomic weight	140.13
Melting point (°C)	795
Boiling point (°C)	3468
Density (g/cm^3)	6.67 (6.90)
Thermal conductance (cal/sec)(cm^2)(°C/cm)	0.026
Specific heat (cal/g) 25°C	0.042
Heat of fusion (k-cal/g-atom)	1.2
Latent heat of fusion (cal/g)	8.5
Heat of vaporization (k-cal/g-mole)	95
Atomic volume (W/D)	21.0
1st ionization energy (K-cal/g-mole)	159
1st ionization potential (eV)	6.54
Electronegativity (Pauling's)	1.1
Covalent radius (angstroms)	1.65
Ionic radius (angstroms)	1.07 (Ce^{+3})
Electrical conductance (micro-ohms^{-1})	0.013
Electrical resistivity (10^{-6} ohms cm) 20°C	75
Poisson's ratio	0.248
Young's modulus ($\times 10^6$)	6.3
Tensile strength (psi)	15.000
Yield strength (psi)	13.200
Magnetic susceptibility (10^{-6} emu/mole)	2.430
Electron work function (eV)	2.84
Cross section (barns)	0.7
Hardness (Mohs — scratch)	1—2
(Vickers — kgf/mm^2)	24
Crystal structure (isometric — normal)	(100) cube, fcc
Color (solid)	Steel-grey met.
Cleavage (cubic)	(001)

CERIUM, Ce

General: Does not occur as a native element though there are over 50 minerals containing cerium. These include fluorides, oxides, carbonates, phosphates, niobates, and tantalates. As a group, there is the Cerium Group of metals, which also are members of the Rare Earths.

One major use of cerium metal in industry is as an alloy with iron producing "flints" . . . the cerium producing sparks for cigarette and gas lighters, gas ignition sparkers, etc. As cerium oxide it is used to make incandescent gas mantles, producing an intense white light with only slow disintegration of the material.

Technical Application: Though not used to any extent in Solid State processing to date, it has possible application as a doping element in compound semiconductors or as an element in manmade garnets.

Etching: Soluble in dilute acids.

CERIUM ETCHANTS

CE-0001

ETCH NAME: Nitric acid — TIME:

TYPE: Acid, removal — TEMP: RT

COMPOSITION:

x .... HNO_3

x .... H_2O

DISCUSSION:

Ce specimens. Material can be etched in dilute acids.

REF: Hodgman, C D et al — *Handbook of Chemistry and Physics,* 27th ed, Chemical Rubber Co., Cleveland, OH, 1943, 364

CE-0002: Walker, P — Mineral study, 1954

Ce as an element fraction in the mineral monazite, (Ce,La,Di)PO_4. H_2SO_4, HCl and HNO_3 used in a separation evaluation of constituents of the single crystal elements of the material.

PHYSICAL PROPERTIES OF CERIUM DIOXIDE, CeO_2

Classification	Oxide
Atomic numbers	58 & 8
Atomic weight	172.13
Melting point (°C)	1950
Boiling point (°C)	
Density (g/cm^3)	7.3
Hardness (Mohs — scratch)	5—6
Crystal structure (isometric — normal)	(100) cube
Color (solid)	White/yellow
Cleavage (cubic)	(001)

CERIUM DIOXIDE, CeO_2

General: Does not occur in nature as an oxide though there are some 50 cerium containing minerals of the Rare Earth group, and there is the allied Cerium Group of metal elements.

The oxide can be fabricated as a thread, then woven into a cloth-like material used as a gas mantle that produces an intense white light at incandescent heat, and is still used for lanterns, such as the Coleman lantern using white gasoline as the gas source.

Technical Application: Not used in Solid State processing at present, but may have application as a thin film protective coating on semiconductors similar to that of other oxides and nitrides, or as a high temperature dielectric/ceramic material.

Etching: H_2SO_4, HNO_3.

CERIUM OXIDE ETCHANTS

CEO-0001

ETCH NAME: Sulfuric acid — TIME:

TYPE: Acid, removal — TEMP:

COMPOSITION:

x .... H_2SO_4

DISCUSSION:

CeO_2 or Ce_2O_3 specimens. Both materials can be dissolved in this solution.

REF: Hodgman, C D et al — *Handbook of Chemistry and Physics,* 27th ed, Chemical Rubber Co., Cleveland, OH, 1943, 368

PHYSICAL PROPERTIES OF CESIUM, Cs

Classification	Alkali metal
Atomic number	55
Atomic weight	132.9
Melting point (°C)	28.61
Boiling point (°C)	690
Density (g/cm^3) 18°C/20°C	1.892/1.873
Thermal conductance (cal/sec)(cm^2)(°C/cm) liquid	0.0440
Specific heat (cal/g) 20°C	0.052 (0.048)
Heat of fusion (cal/g)	3.913
Latent heat of fusion (cal/g)	3.766
Heat of vaporization (cal/g)	146
Atomic volume (W/D)	70
1st ionization energy (K-cal/g-mole)	90
1st ionization potential (eV)	3.89
Electronegativity (Pauling's)	0.7
Covalent radius (angstroms)	2.35
Ionic radius (angstroms)	1.67 (Ce^{+1})
Electron work function (eV)	1.89
Magnetic susceptibility (cgs $\times 10^{-6}$)	0.10
Coefficient of thermal linear expansion ($\times 10^{-6}$ cm/cm/°C) 20°C	97
Electrical resistivity (ohms cm $\times 10^{-6}$) 0°C	20
Cross section (barns)	30
Vapor pressure (°C)	509
Hardness (Mohs — scratch)	4—5
Crystal structure (isometric — normal)	(100) cube, bcc
Color (solid)	Silver-white
(gas)	Blue (spectra)
Cleavage (cubic — perfect)	(001)

CESIUM, Cs

General: Does not occur as a native element. A primary ore is the mica lepidolite, and the rare pollucite mineral. It also is extracted from certain mineral springs as a halogen. It is an alkaline metal with characteristics similar to potassium, and has a high affinity for oxygen. Due to the latter, it is used as an oxygen "getter" in radio tubes and photo-electric cells, and could be used in ion-pump vacuum systems as a replacement for titanium as a getter. It is used as a catalyst in hydrogenation of certain organic compounds in chemical processing.

Technical Application: There has been no use of the pure element in Solid State processing, to date, though it has been used as an additive to special garnets, ferrites, etc. It is under evaluation as a nitride, CsN, for both a surface coating and for optical characteristics.

Etching: Water and acids.

CESIUM ETCHANTS

CS-0001
ETCH NAME: Water
TYPE: Acid, removal
TIME:
TEMP: RT to hot
COMPOSITION:
x .... H_2O
DISCUSSION:

Cs metal specimens. Like many of the rare earth elements, the material reacts with water with a recognizable hissing sound, converting to the hydroxide with release of hydrogen, such that it is used to hydrogenate other compounds and materials, particularly in organic chemistry.

REF: Foster, W & Alyea, H N — *An Introduction to General Chemistry,* 3rd ed, D Van Nostrand, New York, 1948, 422

PHYSICAL PROPERTIES OF CESIUM BROMIDE, CsBr

Classification	Bromide
Atomic number	55 & 35
Atomic weight	213
Melting point (°C)	636
Boiling point (°C)	1300
Density (g/cm^3)	4.44 (4.5)
Refractive index (n =)	1.6984
Wavelength limits (micron)	0.2—40
Young's modulus (psi $\times 10^6$)	2,3
Coefficient of linear thermal expansion ($\times 10^{-6}$/cm/cm/°C)	48
Hardness (Mohs — scratch)	2—3
Crystal structure (isometric — normal)	(100) cube
Color (solid)	Clear
Cleavage (cubic — perfect)	(001)

CESIUM BROMIDE, CsBr

General: Does not occur as a solid natural compound, though cesium metal is extracted from hot springs as a bromide or iodide in solution.

Not used in industry as a compound in metal processing, but does have some medical applications.

Technical Application: Not used in Solid State processing though it can be used as a bromine-type etchant solution on semiconductors and other metallic compounds.

It has been evaluated as a single crystal for general morphology, and for defect studies. The material has been used for its optical and filter element properties of infrared transmission even though it is hygroscopic.

Etching: Soluble in cold water and acids.

CESIUM BROMIDE ETCHANTS

CEBR-0001
ETCH NAME: Ethyl alcohol
TYPE: Alcohol, polish
TIME:
TEMP: RT
COMPOSITION:
x .... EOH

DISCUSSION:

CeBr, (100) specimens. Material is hygroscopic and soft. Polish surfaces with alcohols. Used for infra-red transmission.

REF:

CEBR-0002

ETCH NAME: TIME:

TYPE: Metal, decoration TEMP: Hot

COMPOSITION:

(1) x .... $HAuBr_4$ (2) x .... Au (3) x .... Ag

DISCUSSION:

CeBr, (001) specimens. Used in a study of dislocation development by metal decoration. Coat surfaces with solution (1) or evaporate metals shown in (2) and (3). Heat treat and anneal for metal decoration drive-in. ''Colloidal'' gold will diffuse into CeBr and deposit at dislocation sites.

REF: Amelinck, S — *Phil Mag,* 3,307(1958)

CEBR-0003

ETCH NAME: TIME:

TYPE: Alcohol, polish TEMP: RT

COMPOSITION:

25 .... MeOH

1 .... H_2O

DISCUSSION:

CsBr, (001) wafers used to measure the elastic constants using ultrasonic resonance. Mechanically rough lap with A/O 302 grit; polish lap with 303$^1/_2$ grit; then Linde A; and then sapphire dust on cloth. Use kerosene as the carrier liquid for mechanical lap/polish. Etch polish in solution shown and rinse in isopropyl alcohol (IPA), then rinse in carbon tetrachloride, CCl_4. See Cesium Iodide.

REF: Bulef, D I & Menes, M — *J Appl Phys,* 31,1010(1960)

PHYSICAL PROPERTIES OF CESIUM CHLORIDE, CeCl

Classification	Chloride
Atomic numbers	58 & 17
Atomic weight	168.37
Melting point (°C)	645
Boiling point (°C)	1290
Density (g/cm^3)	3.97
Refractive index (n=)	1.6418
Hardness (Mohs — scratch)	1.5—2.5
Crystal structure (isometric — normal)	(100) cube
Color (solid)	Colorless
Cleavage (cubic)	(001)

CESIUM CHLORIDE, CsCl

General: Does not occur as a natural solid compound though cesium metal is extracted from hot springs and ocean water. The latter may be as a CsCl saline solution.

There is no use of the compound in industrial metal processing, but it may have some medical applications.

Technical Application: No use in Solid State processing to date, though it can be used as an etching solution on metals and compounds as a chloride.

It has been grown as a single crystal by solution evaporation and by molten flux growth for properties study.

Etching: Soluble in water and alcohols.

CESIUM CHLORIDE ETCHANTS

CSCL-0001
ETCH NAME: Water
TYPE: Acid, removal
COMPOSITION:
x .... H_2O
TIME:
TEMP: RT
DISCUSSION:

CsCl grown as single crystal by solution evaporation, and from a molten flux. Also as CsCl:Na_2CO_3(10%) crystals. Water can be used to remove and polish specimen surfaces. Materials used in a growth methods and properties study.

REF: Avakian, P & Smakula, A — *J Appl Phys,* 31,1720(1960)

PHYSICAL PROPERTIES OF CESIUM DIOXIDE, Cs_2O

Classification	Oxide
Atomic numbers	55 & 8
Atomic weight	281.82
Melting point (°C)	300 del.
Boiling point (°C)	
Density (g/cm³)	4.30
Hardness (Mohs — scratch)	5—6
Crystal structure (isometric — normal)	(100) cube
Color (solid)	Red-orange
Cleavage (cubic)	(001)

CESIUM DIOXIDE, Cs_2O

General: Does not occur as a natural compound as it is soluble in water though it may occur as a surface temporary coating on other cesium minerals under very dry conditions.

As cesium metal has a high affinity for oxygen like titanium, it is used as an oxygen "getter" in vacuum tubes, and could be used in ion pump vacuum systems.

Technical Application: No use in Solid State processing to date, though the hydroxide, CsOH, has been used in etching formulations.

This mono-oxide, Cs_2O, has been grown and studied as a single crystal from molten fluxes. Specimens were (111) cleaved with difficulty, as the ⟨100⟩/⟨110⟩ planes direction or (001) basal plane are the more normal cleavage planes in the isometric system — normal (cubic) class.

Etching: Soluble in water and alcohols.

CESIUM DIOXIDE ETCHANTS

CSO-0001
ETCH NAME: Sodium hydroxide
TYPE: Alkali, cleaning
COMPOSITION:
x .... 1 *N* NaOH
TIME:
TEMP:

DISCUSSION:

Cs_2O, (111) wafers cleaved with difficulty from molten flux grown single crystals (lithium ditungstate flux), and used in a compound study. Solution used to clean surfaces after cleaving.

REF: Finch, C B & Clerk, G W — *J Appl Phys,* 37,3910(1966)

CESIUM PLATINIDE ETCHANTS

CEPT-0001

ETCH NAME: Kalling's etch — TIME: 10—60 sec

TYPE: Acid, polish — TEMP: RT

COMPOSITION:

100 ml .... HCl

5 g .... $CuCl_3$

100 ml .... EOH

DISCUSSION:

CePt specimens arc melted in fabrication, and used in a study of metallic compounds. Solution used as a general removal and polishing etchant. Other compounds evaluated: $CeSi_2$, Ho_2O_{17}, Ho_2Fe_{17}, $Ho_2Co_{14}Fe_3$, UCe_2, UPt_3, UNi_2, UCo_2, V_3Si, V_3Ge, Zr_2Nl_7, $MoSi_2$, $TiSi_2$. Crystals grown by the modified Bridgman method were $Y_2(CoM)_{17}$ where M = Al, Fe, Cu, and Ni.

REF: Slepowronsky, M et al — *J Cryst Growth,* 65,293(1983)

PHYSICAL PROPERTIES OF CESIUM IODIDE, CsI

Classification	Iodide
Atomic numbers	55 & 53
Atomic weight	259.83
Melting point (°C)	621
Boiling point (°C)	1280
Density (g/cm^3)	4.52
Refractive index (n=)	1.7876
Hardness (Mohs — scratch)	2—3
Crystal structure (isometric — normal)	(100) cube
Color (solid)	Clear
Cleavage (cubic)	(001)

CESIUM IODIDE, CsI

General: Does not occur as a natural compound, but may be found as a liquid fraction in hot springs and ground water like bromide and chloride.

Not used in industrial metal processing, but can have medical applications.

Technical Application: Not used in Solid State processing, though it can be used as an etchant on metals and compounds.

It has been grown and studied for physical properties as a single crystal.

Etching: Soluble in water and alcohols.

CESIUM IODIDE ETCHANTS

CSI-0001
ETCH NAME: TIME:
TYPE: Alcohol, polish/removal TEMP: RT to warm
COMPOSITION:
25 .... MeOH
1 .... H_2O
DISCUSSION:

CsI, (100) oriented single crystal wafers used to measure elastic constants with ultrasonic resonance frequencies. Mechanical rough lap with A/O 303 abrasive, then polish lap with A/O 302½, both as slurries on an iron lap platen. Mechanically fine polish with Linde A alumina (iron lap), then with diamond dust on a cloth, all using kerosene as the liquid carrier. For final chemical polish, use solution shown, rinse in isopropyl alcohol (IPA), then carbon tetrachloride, CCl_4. Preparation method can be used on bromides and chlorides.
REF: Bulef, D I & Menes, M — *J Appl Phys*, 31,1010(1960)

PHYSICAL PROPERTIES OF CHLORINE, Cl^2

Classification	Halogen
Atomic number	17
Atomic weight	35.45
Melting point (°C)	−102 (−100.98)
Boiling point (°C)	−33.7 (−34.6)
Density (g/cm³) −33.6°C	1.57
(g/l — liquid) 0°C	3.214
Thermal conductivity ($\times 10^{-4}$ cal/cm²/cm/°C sec) 20°C	0.182
Latent heat of fusion (cal/g)	21.6
Specific heat (cal/g/°C) 20°C	0.116
Solubility (100 ml H_2O) 20°C	215 ml
1st ionization potential (eV)	13.01
Ionic radius (angstroms)	1.81
Cross section (barns)	33
Vapor pressure (°C)	−71.7
Refractive index (n=) gas	−.000768
Critical temperature (C.T. °C)	140
Critical pressure (C.P. atms)	83.9
Crystal structure (tetragonal — normal)	(110) prism
Color (solid)	Yellow-greenish
Cleavage (basal)	(001)

PHYSICAL PROPERTIES OF CHLORINE HYDRATE, $Cl_2.6H_2O$

Classification	Hydrate
Atomic numbers	17, 1, & 8
Atomic weight	170.01
Melting point (°C)	9.6, del.
Boiling point (°C)	
Density (g/cm³)	1.23
Formation temperature (°C in water)	0

Hardness (Mohs — scratch) solid	1.5—2.5
Crystal structure (hexagonal — rhombohedral)	r($10\bar{1}1$) rhomb
Color (solid)	Light yellow
Cleavage (rhombic)	($10\bar{1}1$)

CHLORINE CONTAINING ETCHANTS

Chlorine is a greenish-yellow gas at standard temperature and pressure (22.2°C and 760 mmHg) very similar to that of fluorine. Chlorine has a disagreeable odor and, if breathed, is excessively irritating to the membranes of the eyes, nose, and throat. In contact with moisture in air it produces heavy white fumes of hydrogen chloride, HCl, which also is a severe irritant and can cause suffocation, even death. Fluorine and chlorine are the gaseous elements of the halogen group, which include bromine (liquid) and iodine (solid) with reducing atomic weight and reactivity in the order shown. As chlorine can be liquefied at 0°C under 6 atm of pressure, it was one of the first gases to be liquefied as a cryogenic gas. If chlorine is introduce into water containing ice, and is chilled, it will solidify as the greenish-yellow hydrate, $Cl_2.8H_2O$.

In the metal industries all metals and alloys are subjected to corrosion studies by other elements and compounds, which include the effect of chlorine at various temperatures and, as chlorine will combine with most elements, it is of major concern in developing acid resistant alloys. In semiconductor processing, silicon has been etch polished or thinned by bubbling chlorine through water under intense light. Both germanium and silicon have been etched electrolytically at 300°C with a jet of gaseous chlorine.

Chlorine forms a number of useful acids and compounds in addition to several dangerous gases: hydrogen chlorine, HCl (hydrochloric acid or muriatic acid). As a solution in water, it was discovered by Glauber (circa 1648) from the interaction of sodium chloride, NaCl (common salt) and sulfuric acid, H_2SO_4. Priestley (1772) collected gaseous hydrogen chloride and named it ''marine-acid air'' because it was produced from salt, and it is now called muriatic acid.

Both hydrochloric and sulfuric acids occur in nature associated with volcanic activity — the volcano releases chlorine and sulfur as gas — which combine with hydrogen and salt in the air or, if submerged in the ocean, from the salt water. Hydrochloric acid is used as a general solvent of materials; in the preparation of chlorides and chlorine; in the cloth dyeing industry; in medicine as a disinfectant; and plays a major role in human digestion . . . 0.2 to 0.4% HCl in gastric juices.

The use of gas in warfare during World War I led to the outlawing of the use of gas in war by the Geneva Convention, which has been accepted by most humane nations of the world. Gas has been used in warfare since ancient times and, unfortunately, is still being used to some extent. The Japanese attempted gas warfare against the Chinese, on a small scale, during World War II — fortunately, the winds were blowing in the wrong direction, or changed direction, and the Japanese got the worst of it.

Anthrax, a virulent cattle disease, was evaluated during World War II by the British — it not only kills mammilary life but as a bacteria permeates the ground and remains lethal for years. Recently it was reported the Russians used gas in Afghanistan, and Iran has accused Iraq of gas warfare.

Some of the gases developed and used during World War I are described below, mainly as they can be a by-product of metal and metallic compound processing when chlorine and its compounds are used, and some knowledge of their danger should be considered:

- Cl_2 — Chlorine. A 1 part chlorine to 10,000 parts of air concentration causes severe breathing difficulty which will incapacitate a man within about 5 min. The Germans used chlorine for the first time on April 22, 1915. The gas is wind-borne, about 2.5 times heavier than air, and is carried as a ground hugging cloud across the surface.

One possible reason for the Axis powers not using gas in Europe during World War II could be due to the fact that, though variable by locale and season, the prevailing winds coming off the North Sea and English Channel blow inland — across Germany.

- Chlorine gas is used as a jet etch, as a furnace gas in metal corrosion study, and is bubbled through other solutions as part of an etching system. HCl and $HClO_4$ are acid etches, alone, or in combination with other acids. HCl, from pressurized cylinders, and di-, tri-, and chlorosilanes from liquid sources are used in silicon epitaxy growth, and both can produce heavy chlorine vapors if released to the atmosphere.
- Chlorinated water, Cl_2:H_2O, as well as HCl and $HClO_4$, are used as disinfectants in water. Chlorinated lime, $CaOCl_2$, is a bleaching agent in the wood and textile industries. Chloramine, NH_4Cl, will destroy bacteria, and is used in sewage disposal. Chloral, CCl_3CHO, is a medical hypnotic, Chloroform, $CHCl_3$, is a medical anesthetic alone or with dissolved iodine, I_2, as a solution; and also has been used as a metal surface cleaner. Carbon tetrachloride, CCl_4, has been used in dry cleaning, and is still used in ''pyrene'' type fire extinguishers, or as a metal cleaning solvent. Many of the chlorinated solvents, such as trichloroethylene, TCE, are being replaced due to their carcinogenic nature. Ethylene chloride, C_2H_5Cl, is a medical anesthetic like chloroform, and occasionally has been used as a cleaning and drying solvent for metals.
- $COCl_3$ — Phosgene. Primarily a respiratory irritant, and more poisonous than chlorine. It also was a war gas during World War I. Caution should be observed in mixing etch solutions with hydrochloric acid and other chlorine compounds, as certain mixtures can produce phosgene. Phosphene, PH_3, as a gas, is used as an n-type dopant in silicon device fabrication, and should not be confused with phosgene, though it too is dangerous if breathed.
- $(C_3CCH_4Cl)_2S$ — Mustard gas. Severe skin burns; attacks lungs, larynx, and bronchial tubes. Probably the best known of the World War I gases as so many soldiers were affected. Hitler, for example, was severely gassed near the end of World War I; in fact, he was in hospital at the end of the war, and had gastritis problems for the rest of his life, possibly one of the contributing causes to his final physical degeneration . . . the treatments he was given over the years also contributed.
- CCl:NO_3 — Nitrochloroform, or chloropicrin. Induces coughing, vomiting, and unconsciousness. The compound sodium hypochlorite, NaClO, as a 0.5% solution is used to reduce the effects of gangrene, which was referred to as trench foot in both World War I and World War II, and has been a major cause of death or loss of limbs associated with wars since ancient time. Secondary effects of all of these war gases are asthma, weakened heart, and gastritis.

Mention also should be made of ''tear gas'' — a lachrymator — which is a slight inhalation irritant, and mainly affects the eyes and tear glands. There are several concentration mixtures of bromine-acetone which comprise the family of tear gases. They are not permanently debilitating and are widely used throughout the world for crowd or mob dispersal. Interestingly enough, tear gas has an astringent, slightly pleasing odor and, in light concentrations, many individuals can withstand the ''tearing'' effect for several minutes, although the eyes become irritated and inflamed. Many mixtures of bromine-methyl alcohol (BRMs) are used as etchants for metals and metallic compounds, but do not produce the tear gas effect unless acetone is used with the mixtures. Bromine itself is poisonous and can be a severe irritant. Methyl alcohol is wood alcohol, also poisonous if ingested.

Other than acids and gases, chlorine forms a number of important compounds. Sodium chloride, the mineral halite, NaCl, and potassium chloride, the mineral sylvite, KCl, both occur in sea water and as sublimation products around hot springs and volcanoes. Ammonium chloride, the mineral salt ammoniac (chloramine) NH_4Cl, used in sewage disposal, also is found as a sublimate around volcanic fumaroles. Halite is common salt and essential to the human diet. Sylvite, bitter in taste, which is not retained by the body as is common salt, is commercially called "sea salt", and used as a dietary replacement for true salt. In Solid State materials study, both NaCl and KCl are used as single crystal wafers, and cut as (100) oriented substrates for thin film evaporation of metals and epitaxy growth of compounds. The deposited thin films are removed by water as the float-off technique for study by transmission electron microscopy (TEM). In addition, both of these salt chlorides are used in etch solutions, as molten fluxes, or in solder fluxes.

The following is a selected list of chlorine-containing etchants. Aqua regia solutions are shown in a separate list. See the Etchant Section under individual materials for additional solutions.

Formula	Material	Use	Ref.
HCl, conc.	Al	Removal	AL-0003b
1HCl:1—20H_2O	Al	Heavy removal	AL-0003b
4HF:35HNO_3:61HCl	Al	Polish	AL-0030
4HF:35HNO_3:61HCl:100H_2O	Al	Preferential	AL-0030
1$HClO_4$:5MeOH	Al	Electropolish	AL-0017
×$HClO_4$:×EOH	AlAu(2%)	Electrolytic, defect	ALAU-0001
1HCl:1H_2O	AlSi(5%)	Cleaning	ALSI-0001
4HCl:4HNO_3:1 H-Tartaric acid	AlSb	Polish	ALSB-0002a
1HCl:1HNO_3	AlSb	Macroetch	ALSB-0006
35 ml $HClO_4$:20 mg CrO_3/l	Al_2O_3	Oxide removal	ALO-0001b
×6 *N* HCl	Al_2O_3	Preferential	ALO-0016
1HCl:2HNO_3:12HAc	As	Polish	AS-0003c
HCl, conc.	Ba	Polish/Removal	BA-0001b
HCl, conc.	Ba_2TiO_3	Structuring	BAT-0001
HCl, conc.	Be	Removal	BE-0002
10 ml HCl:1gI_2:200 ml H_2O	Bi	Electropolish	BI-0004
HCl, conc.	Bi_2O_3	Removal	BIO-0001
1HCl:2HNO_3:6H_2O	Bi_2Te_3	Preferential	BITE-0001a
3HCl:1HNO_3 "Aqua Regia"	WB_2	Removal	WB-0001a
4HCl:1HNO_3	Brass Bronze	Removal	BR-0001b
HCl, conc.	Cd_3As_2	Polish	CDAS-0001
1HCl:1HNO_3	CdSe	Preferential	CDSE-0001a
3HCl:1HNO_3:10H_2O	CdSe	Polish	CDSE-0003
1HCl:30HNO_3:30HAc	CdSe	Preferential	CDSE-0001b
HCl, conc.	CdS	Damage removal	CDS-0001a
1HCl:1HNO_3	CdS	Preferential	CDS-0011b
HCl, vapor	CdS	Preferential	CDS-0001
10 ml HCl:10 ml HNO_3:5 ml H_2O "P Etch"	CdTe	Polish	CDTE-0006d
10 ml P Etch:10 mg Br_2 "PBr Etch"	CdTe	Preferential	CDTE-0006e
HCl, conc.	Ca	Removal/polish	CA-0001
HCl, conc.	$CaCO_3$	Polish	CAO-0001
3HCl:1HNO_3 "Aqua Regia"	$CaWO_4$	Preferential	CAW-0001b
HCl, conc.	Cf	Removal	CF-0001
3HCl:1H_2O_2	Cr	Removal	CR-0003
1HCl:2$FeCl_3$(35° Bé)	Cr	Removal/polish	CR-0010
500 ml HCl:500 ml EOH	Co	Electropolish	CO-0001

Formula	Material	Use	Ref.
$1HCl:3FeCl_3$, sat. sol.	Cu	Polish	CU-0025
$1HCl:1H_2O:1FeCl_3$ sat. sol.	Cu	Preferential	CU-0013b
10 ml HCl:1000 ml HNO_3	Cu_2O	Removal	CUO-0014b
HCl, conc.	Ga	Cleaning	GA-0002
HCl, conc.	GaSb	Removal	GASB-0001a
1 ml HCl:1 ml H_2O_2:2 ml H_2O	GaSb	Preferential	GASB-0008b
HCl, vapor	GaAs	Cleaning	GAS-0028
$1HCl:1HNO_3$:8glycerin	GaAs	Polish	GAS-0050
$2HCl:1HNO_3:2H_2O$	GaAs	Preferential	GAS-0057
$1HCl:2HNO_3$	GaP	Polish	GAP-0002
HCl, conc.	GaP	Removal	GAP-0014
$3HCl:1HNO_3$ "Aqua Regia"	GaP	Selective	GAP-0011
$1HCl:1HClO_4:6HNO_3$	GaP	Cleaning	GAP-0017
13 *N* HCl	Ge	Preferential	GE-0132a
× % KCl	Ge	Electrolytic, defect	GE-0133a
× % NaCl	Ge	Electrolytic, shaping	GE-0129e
$7HCl:1HNO_3$: + glycerin	Ge	Removal/polish	GE-0173
$2HCl:1HNO_3$	GeAs	Removal/polish	GEAS-0001b
$3HCl:1HNO_3$ "Aqua Regia"	Au	Removal	AU-0001
HCl, conc.	In	Removal	IN-0003
$1HCl:2NaClO_3:3H_2O$	InSb	Polish	INSB-0001e
$2HCl:1HNO_3$	InSb	Removal	INSB-0002d
× HCl:x0.2 *N* $FeCl_3$ "Etch #1"	InSb	Preferential	INSB-0014a
1 *M* HCl	InAs	Removal	INAS-0001b
HCl, conc.	InP	Removal	INP-0020a
$6HCl:1HNO_3:6H_2O$	InP	Polish	INP-0021a
$40HCl:80HNO_3:1Br_2$	InP	Preferential	INP-0021b
$3HCl:1H_3PO_4$	InP	Selective	INP-0035
$3HCl:1HNO_3$ "Aqua Regia"	Ir	Removal	IR-0001
$45HCl:5HNO_3$:40MeOH	Fe	Preferential	FE-0103b
$HClO_4$:HAc	Fe	Electrolytic, thin	FE-0104
$1HCl:1H_2SO_4:1H_2O$	Fe	Cleaning	FE-0108
$xHClO_4$:xHAc	Fe:Si	Electrolytic, form	FESI-0007
$20HCl:20HNO_3$:60MeOH	Fe:Be	Electrolytic, thin	FEBE-0001
1HCl:3 Thiourea	PbS	Dislocation	PBS-0001
HCl, conc.	$PbZrO_3$	Polish	PBZO-0001a
$xHCl:xH_2O$	Mg	Acid saw cutting	MG-0001
$1HCl:5H_2O$	Fe_3O_4	Preferential	FE-3001
HCl, conc.	MnO_2	Removal	MNO-0002
HCl, conc.	Mo	Cleaning	MO-0013b
× HCl:xHNO_3	Mo	Removal	MO-0001g
$3HCl:1HNO_3$ "Aqua Regia"	Mo	Polish	MO-0008a
$1HCl:1HNO_3$	HgSe	Preferential	HGSE-0001b
$1HCl:50HNO_3$:10HAc: 20 10 *N* H_2SO_4	HgSe	Polish	HGSE-0001a
$1HCl:6HNO_3:1H_2O$	HgTe	Polish	HGTE-0001a
$1HCl:1HNO_3$	HgTe	Preferential	HGTE-0001b
$1HCl:1HNO_3:1H_2O$	HgCdTe	Stain, defects	HGCT-0001
$10HCl:33HNO_3:67H_2O$	Ni	Clean	NI-0004b
$3HCl:1HNO_3$ "Aqua Regia"	Ni:Cr	Removal	NICR-0003

Formula	Material	Use	Ref.
45HCl:5HNO_3:50MeOH	Ni:Cr	Preferential	NICR-0004a
×HCl:×H_2O	Nb	Cleaning	NB-0004a
3HCl:1HNO_3 "Aqua Regia"	Os	Removal	OS-0001b
3HCl:1HNO_3 "Aqua Regia"	Pd	Cleaning	PD-0001
3HCl:1HNO_3 "Aqua Regia"	Pt	Removal	PT-0001
1HCl:1HNO_3	$PtSb_2$	Preferential	PTSB-0001a
1HCl:1HNO_3:1H_2O	$PtSb_2$	Removal	PTSB-0002
3HCl:1HNO_3 "Aqua Regia"	SiO_2(Qtz)	Cleaning	QTZ-0003
3HCl:1HNO_3 "Aqua Regia"	Ru	Removal	RU-0001
HCl, conc.	Sc	Polish	SC-0001a
×HCl:×HNO_3	Sc	Removal	SC-0001c
1HCl:3HNO_3	Si	Defect	SI-0171
×KF:×KCl:2H_2O	Si	Electropolish	SI-0186a
HCl, conc.	Si	Cleaning	SI-0017
40HF:37HCl	SiO_2	Cleaning	SIO-0038b
3HCl:1HNO_3 "Aqua Regia"	SiO_2	Cleaning	SIO-0051
3HCl:1HNO_3 "Aqua Regia"	Si_3N_3	Cleaning	SIN-0016
500 ml HCl:100 ml 10% CrO_3	Si_3N_4	Cleaning/pin-holing	SIN-0018
HCl, conc.	AgCl	Cleaning	AGCL-0003
NaCl, sat. sol.	NaCl	Polish, (100)	NACL-0002b
70HCl:30H_2O	NaCl	Polish, (100)	NACL-0005
1HCl:50HAc:1H_2O: 1$FeCl_3$ sat. sol.	NaCl	Preferential, (100)	NACL-0007a
1HCl:1H_2O	Steel	Preferential	ST-0001a
5 ml HCl:1 g picric acid: 100 ml EOH	Steel	Preferential	ST-0001e
1HCl:2HNO_3	Steel	Cleaning/removal	ST-0007b
4HCl:1HNO_3	Steel	Removal	ST-0007a
1HCl:1OH$_2$O	SrF_2	Removal	SRF-0001
3HCl:1HNO_3 "Aqua Regia"	TaN	Polish	TAN-0001
1HCl:1CrO_3:3H_2O	Te	Preferential	TE-0001c
3HCl:1HNO_3 "Aqua Regia"	Te	Polish	TE-0002
HCl, conc.	Tl	Removal, slow	TL-0004
3HCl:1HNO_3 "Aqua Regia"	Th	Polish	TH-0002
100 ml HCl:10NH_4NO_3: 500H_2O:×$CuSO_4$	Sn	Defect, (100)	SN-0002a
×5%HCl:×IPA	Sn	Removal	SN-0001a
3HCl:1HNO_3 "Aqua Regia"	Sn	Preferential, (100)	SN-0005b
20$HClO_4$:70HAc	Sn	Electropolish	SN-0007
1HCl:5H_2O	SnO_2	Electrolytic, removal	SNO-0001b
x20% HCl	Ti	Cleaning	Ti-0001a
HCl, conc.	Ti	Removal	TI-0004c
3HCl:1HNO_3 "Aqua Regia"	TiO_2	Cleaning	TiO-0004
3HCl:1HNO_3 "Aqua Regia"	W	Removal	W-0001d
×$HClO_4$:xH_3PO_4	W	Polish	W-0018b
3HCl:1HNO_3 "Aqua Regia"	V	Removal	V-0002a
1HCl:1H_2O	Zn	Removal	ZN-0001a
1HCl:1HNO_3	ZnS	Preferential	ZNS-0002c
3HCl:1HNO_3 "Aqua Regia"	ZnTe	Preferential	ZNTE-0005b

Formula	Material	Use	Ref.
$3HCl:4HNO_3$	ZnTe	Polish	ZNTE-0002c
$3HCl:1HNO_3$ "Aqua Regia"	Zr	Removal	ZR-0007
$1HClO_4:9HAc$	ZrO	Electropolish	ZRO-0002

AQUA REGIA

This solution was originally developed in the Middle Ages of Europe as $3HCl:1HNO_3$ and, at the time, was thought to be a universal etchant for all materials. It was the first solution capable of dissolving gold, in addition to the Platinum Group metals: Pt, Os, Rh, Re. Of course, since those time it has been found not to be the ultimate etchant, but is in use today on a number of metals and metallic compounds . . . as the 3:1 mixture . . . with other constituent variations, to include water dilution and other chemical additives — applied for cleaning, general removal, polishing, or preferential depending upon the material.

In addition, there are two other "regia" termed formulations: (1) chrome regia, with CrO_3 replacing the HNO_3, and (2) glyceregia, where aqua regia is diluted with glycerin. *Note:* Any solution containing nitric acid and glycerin should be handled with extreme care . . . it is the precursor of nitroglycerin. **CAUTION:** HNO_3:MeOH, red fuming, was one of the first rocket fuels, and can be explosive. These two regia etchants are shown at the end of the following list of aqua regia and its varieties.

Although aqua regia has been shown in the previous "Chlorine Containing . . . " list, it is felt that it would be useful to present the following list covering both aqua regia and its variations, as applied to a number of metals and metallic compounds. The use of aqua regia and its variations is not limited to the materials as shown. An example: if $AgCdTe_2$ can be etched, so could $AgTe_2$ . . . though we have no reference available for the latter.

Note that a polish etchant can be a general removal etchant; whereas, a removal etchant is not necessarily a polish etchant. Also, both types, can be used for pattern etching of thin films and/or structure etching through a mask, e.g., pits, channels, etc. The preferential etchants can be for general defect development or for a specific type dislocation, such as stacking faults. Where aqua regia is used for surface passivation, it can be either chlorination or oxidation, and these factors apply to other etchant solutions.

Aqua regia and variations

Formula	Material	Use	Ref.
$3HCl:11HNO_3:10H_2O$	Al_xAu_y	Step-etch "purple plague"	AIAU-0002
$5HCl:1OHNO_3:84H_2O$	Al:Si	Preferential	ALSI-0003
$4HCl:3HNO_3$	AlSb	Preferential/polish	ALSB-0002a
$1HCl:1HNO_3$	AlSb	Polish/macro-etch	ALSB-0003/ 0006
$1HCl:2HNO_3:3H_2O$	Bi	Polish	BI-0014b
$20HCl:1NHO_3$	BiSb	Polish	BISB-0002
$1HCl:2HNO_3$	Bi_2Se_3	Polish	BISE-0001a
$1HCl:12HNO_3$	Bi_2Te_3	Removal	BITE-0002
$1HCl:2HNO_3:6H_2O$	Bi_2Te_3	Preferential	BITE-0001a
$5HCl:1000HNO_3$	Bronze	Removal	BRO-0001
$4HCl:1HNO_3$	Brass	Removal	BRA-0001b
$4HCl:1HNO_3$	$CdInSe_2$	Polish	CDISE-0001a
$1HCl:1HNO_3$	CdSe	Preferential	CDSF-0001a

Formula	Material	Use	Ref.
$3HCl:1HNO_3$	CdSe	Cleaning	CDSE-0002b
$3HCl:1HNO_3$	CdSe	Preferential	CDSE-0005b
$3HCl:1HNO_3:12H_2O$	CdSe	Polish	CDSE-0003
$5HCl:1HNO_3$	CdSe	Removal	CDSE-0006
$1HCl:1HNO_3$	CdS	Preferential	CDS-0001b
$2HCl:2HNO_3:1H_2O$	CdTe	Polish	CDTE-0006d
$1HCl:1HNO_3$	CdTe	Preferential	CDTE-0011b
$3HCl:1HNO_3:1HAc$	CdTe	Polish	CDTE-0018b
$1HCl:6HNO_3$	CdTe:HgTe	Polish	CDHT-0001
$3HCl:1HNO_3$	$CaWO_4$	Preferential	CAW-0001b
$3HCl:1HNO_3$	$CaWO_4$	Polish	CAW-0002
$10HCl:1000HNO_3$	Cu_2O	Removal	CUO-0019
$3HCl:1HNO_3$	$CuGaSe_2$	Polish	CGSE-0001
$3HCl:1HNO_3$	$CuInSe_2$	Polish/stain	CISE-0001
$1HCl:1HNO_3:1H_2O$	$CuInSe_2$	Polish	CISE-0002
$1HCl:1HNO_3$	$CuGaTe_2$	Removal	CGTE-0001
$1Hcl:1HNO_3:2H_2O$	$CuInTe_2$	Thinning	CUIT-0001b
$3HCl:1HNO_3$	Apatite	Preferential	FAP-0001a
$3HCl:1HNO_3$	Ga	Preferential	GA-0002
$1HCl:2HNO_3:2H_2O$	GaAs	Polish	GAS-0052
$3HCl:3HNO_3:2H_2O$	GaAs	Polish	GAS-0044d
$3HCl:1HNO_3$	GaAs	Polish	GAS-0056a
$2HCl:1HNO_3:2H_2O$	GaAs	Preferential	GAS-0057
$5HCl:1HNO_3$	GaAs	Polish/junction	GAS-0056c
$2HCl:3HNO_3:2HF$	GaAs	Polish	GAS-0005a
$1HCl:2HNO_3:1H_2O$	GaAs	Removal	GAS-0140d
$1HCl:1HNO_3:2H_2O$	GaAs	Preferential	GAS-0167c
1—$2HCl:1HNO_3$	GaSb	Removal	GASB-0001c
$30HCl:1HNO_3$	GaSb	Surface passivation	GASB-0007b
$1HCl:2HNO_3:1H_2O$	GaP	Polish	GAP-0002
$2HCl:1HNO_3:2H_2O$	GaP	Polish	GAP-0008
$3HCl:1HNO_3$	GaP	Selective	GAP-0011
$3HCl:1HNO_3$	GaP	Cleaning	GAP-0012
$1HCl:6HNO_3:1HClO_4$	GaP	Cleaning	GAP-0017
$3HCl:1HNO_3$	GaP	Preferential	GAP-0022b
$1HCl:1HNO_3$ + Ag/Fe	GaP	Dislocation	GAP-0022e
$7HCl:1HNO_3$: + Gly	Ge	Removal/polish	GE-01773
$3HCl:1HNO_3$	GeSn	Removal	GESN-0001
$2HCl:1HNO_3$	GeAs	Removal, slow	GEAS-0001b
$2HCl:1HNO_3$	Glass:PdNiP	Clean/removal	PDNP-0001
$3HCl:1HNO_3$	Au	Removal/pattern	AU-0001a
$3HCl:1HNO_3$:20—$50H_2O$	Au	Cleaning	AU-0003b
$3HCl:1HNO_3:xH_2O$	Au:Cs	Removal	AUCS-0001
1—$5HCl:1HNO_3$:4—$6H_2O$	InSb	Polish/preferential	INSB-0001b,d
$2HCl:5HNO_3:4HAc$	InSb	Polish	INSB-0001c
$2HCl:2HNO_3:3H_2O/25H_2O$	InSb	Polish	INSB-0004
1—$2HCl:1HNO_3$	InP	Removal	INP-0020b
$6HCl:1HNO_3:6H_2O$	InP	Polish	INP-0021a
$40HCl:80HNO_3:1Br_2$	InP	Preferential	INP-0021b
1—2 HCl:1—$2HNO_3$	InP	Selective	INP-0062g

Formula	Material	Use	Ref.
1HCl:3/1.5/2/2.5 HNO_3	InP	Selective/step-etch	INP-0064a,k
1HCl:1HNO_3	InP	Polish	INP-0070
3HCl:1HNO_3	Ir/IrV	Removal/patterning	IR-0001
45HCl:5HNO_3:50H_2O	Fe	Preferential	FE-0003b
10HCl:1HF:5HNO_3	Fe	Dislocation	FE-0011
1HCl:1HNO_3:2H_2O	Fe	Electropolish	FE-0030
1HCl:1HNO_3:3MeOH	Fe:Be	Electro-clean	FFBE-0001
1HCl:1HNO_3	MgTe	Removal	MGTE-0001
1HCl:50HNO_3:10HAc:20 10 *N* H_2SO_4	HgSe	Polish	HGSE-0001a
11HCl:1HNO_3	HgSe	Preferential	HGSE-0001b
6HCl:2HNO_3:3H_2O	HgSe	Preferential	HGSE-0001c
1HCl:1HNO_3:1H_2O	HgTe	Polish	HGTE-0001a
1HCl:1HNO_3	HgTe	Preferential	HGTE-0001b
1HCl:1HNO_3:1H_2O	HgCdTe	Defect/strain	HGCT-0001
2HCl:1HNO_3:3H_2O	HgCdTe	Preferential	HGCT-0002a
xHCl:xHNO_3	Mo	Removal	MO-0001g
3HCl:1HNO_3	Mo	Polish	MO-0008a
1HCl:1HNO_3	Mo	Cleaning	MO-0010
xHCl:xHNO_3:xH_2O/MeOH/EOH	Mo	Removal	MO-0022b
1HCl:2HNO_3	$Fe_2Mo_3O_8$	Removal	FEMO-0001
3HCl:1HNO_3:5H_2O	MoS_2	Removal	MO-0004
4HCl:1HNO_3	Ni	Removal	NI-0005
10HCl:33HNO_3:67H_2O	Ni	Cleaning	NI-0004b
3HCl:1HNO_3	Permalloy	Polish	PMA-0001a
1HCl:1HNO_3:3H_2O	Nichrome	Removal/cleaning	NICR-0006
3HCl:1HNO_3	Nichrome	Removal	NICR-0003
45HCl:5HNO_3:50MeOH	Nichrome	Preferential	NICR-0004a
4HCl:1HNO_3	$RENi_2$	Removal	REN-0001
1HCl:3HNO_3	Os	Removal/cleaning	OS-0001b
3HCl:1$HNO_3$3H_2O	Pd	Removal/cleaning	PD-0001
3HCl:1HNO_3	PdAg	Removal/cleaning	PDAG-0001
3HCl:1HNO_3	PdAu	Removal/cleaning	PDAU-0001
3HCl:1HNO_3:xH_2O	Pd:H	Cleaning	PDH-0001
3HCl:1HNO_3	Pt	Removal/cleaning	PT-0001
3HCl:1HNO_3:xH_2O	Pt:Pd	Preferential/removal	PTPD-0001
3HCl:1HNO_3:xH_2O	Pt:Au	Preferential/removal	PTAU-0001
3HCl:1HNO_3	Pt:Rh	Cleaning	PTRH-0001
3HCl:1HNO_3:xH_2O	PtH	Cleaning	PTH-0001
1HCl:1HNO_3:xH_2O	$PtSb_2$	Preferential	PTSB-0001a-b
3HCl:1HNO_3	SiO_2 (Qtz)	Cleaning	QTZ-0003
3HCl:1HNO_3	Rh	Cleaning	RH-0001
3HCl:1HNO_3	Ru	Removal	RU-0001
xHCl:xHNO_3	Sc	Removal	SC-0001c
3HCl:1HNO_3	$TiWSi_2$	Removal	TWSI-0001
1HCl:1HNO_3	Si	Cleaning	SI-0108a
3HCl:1HNO_3	Si	Cleaning	SI-0108c
3HCl:1HNO_3	SiO_2	Cleaning	SIO-0052
3HCl:1HNO_3	Si_3N_4	Cleaning	SIN-0007
1HCl:1HNO_3:1H_2O	Ag	Removal	AG-0007

Formula	Material	Use	Ref.
1HCl:3HNO_3:6H_2O	Pb:Sn:Ag	Removal/cleaning	AGPS-0001b
1HCl:1HNO_3:	$AgSbTe_2$	Polish	SAT-0001
2$K_2S_2O_7$, sat. sol.	$AgFeTe_2$	Polish	SIT-0001
15 ml HCl:10 ml HNO_3: 10 ml HAc:5 ml glycerin	Steel, 300	Electrolytic	ST-0001i
4HCl:1HNO_3	Steel	Removal (hi-speed)	ST-0007a
1HCl:2HNO_3	Steel, hard	Removal/cleaning	ST-0007b
3HCl:1HNO_3	TaN	Polish	TAN-0001
3HCl:1HNO_3	Te	Removal/polish	TE-0002
3HCl:1HNO_3:1H_2O	Te	Polish	TE-0008
3HCl:1HNO_3	Th	Polish	TH-0002
3HCl:1HNO_3	Sn	Preferential	SN-0005b
11HCl:1HNO_3: 20$K_2Cr_2O_7$, sat. sol.	SnTe	Polish	SNTE-0001
3HCl:1HNO_3	TiO_2	Cleaning	TIO-0004a
3HCl:1HNO_3	TiW	Removal	TIW-0006
3HCl:1HNO_3	W	Removal	W-0001d
3HCl:1HNO_3	V	Removal	V-0002a
1HCl:1HNO_3	ZnS	Preferential	ZNS-0002b
3HCl:1HNO_3	ZnTe	Preferential	ZNTE-0005b
3HCl:4HNO_3	ZnTe	Polish	ZNTE-0002c
3HCl:1HNO_3	Zr	Removal	ZR-0007
3HCl:1HNO_3	Zr_2Pd	Polish/removal	ZRPD-0001
3HCl:1HNO_3	Zr_3Rh	Polish/removal	ZPRH-0001
	Chrome regia		
3HCl:1CrO_3 (10—20%)	Au	Preferential	AU-0017
5HCl:1CrO_3(10%)	Si	Preferential	SI-0019
5HCl:1CrO_3(10%)	Si_3N_3	Preferential	SIN-0002e
0.5HCl:5 gCrO_3:50 ml H_2O	Ag	Polish	AG-0012b
0.01HCl:5 gCrO_3:25 ml H_2O	Ag	Preferential	AG-0012c
1HCl:1CrO_3:3H_2O	Ta	Preferential	TA-0001c
1HCl:1CrO_3:1H_2O	Ta	Polish/acid saw cut	TA-0002a-b
1HCl:1CrO_3:3H_2O	Te	Acid saw cut	TE-0012a
1HCl:1CrO_3:3H_2O	Te	Polish	TE-0001c
3000 ml 6 *N* HCl:20 g Cr, metal	SnO_2	Removal	SNO-0005
	Glyceregia		
20—50 ml HCl:10 ml HNO_3: 30 ml glycerin	Steel	Macro-etch/cleaning	ST-0018
7HCl:1HNO_3:x glycerin	Ge	Polish/removal	GE-0173
1HCl:1HNO_3:8 glycerin	GaAs	Preferential	GAS-0050
1HCl:1HNO_3:4 glycerin	PbAg PbSn:PbTi	Preferential	PBAG-0001

CHLORINE, Cl_2

General: Chlorine gas is too active to remain in the free state although it can be a temporary product of volcanic eruptions which immediately converts to hydrochloric acid,

HCl. Sulfur also is a volcanic by-product, converting to sulfuric acid, H_2SO_4. These two gases are the primary corrosive acid producers in the atmosphere whether from volcanic action or as industrial pollutants. Acid rain is one result, another is smog. Note that smog has been in existence since ancient times.

Chlorine combines to form many metallic compounds as natural minerals, such as cerargyrite, AgCl (horn silver). Probably the best known and one of the most widely used compounds is halite, NaCl (sodium chloride) as common salt, and "sea salt", the mineral sylvite, KCl. The latter used as a dietary replacement for common salt. Man has extracted salt from oceans since ancient times by evaporating sea water in shallow basins — the method is still used — though both salt and sulfur are mined in quantity, today, from salt domes or similar sulfur domes, and from tremendous salt beds buried in the earth from the evaporation of ancient shallow seas that go back in age for millions of years.

Sodium chloride, as halite, is one of the most important — and abundant — minerals in the world. It is a major material in the chemical industry, used as a concentrated brine in metal processing, as well as being separated for the sodium metal content and gaseous chlorine. Chlorine is a greenish-yellow gas with a disagreeable smell, and is extremely irritating to membranes of the nose, eyes, and throat. Dispersed in the air as a cloud, usually as HCl, as a heavy white vapor, it is easily recognized by its astringent action. It can liquefy at 0°C under only 6 atm of pressure, and was one of the first gases to be liquefied. With added water and cooled with ice, it will form a highly unstable greenish-yellow crystalline hydrate.

The gas will combine to form compounds with most elements, other than oxygen, nitrogen, or carbon, and there are many industrial uses for both metallic sodium and gaseous chlorine; the latter in the production of hydrochloric acid.

The gas, usually in combination with water, as a mixture of hydrochloric and hypochlorous acids yields oxygen in a highly active state. Hypochlorous acid, HClO is bleach. When used as a disinfectant, the released oxygen kills bacteria. Perchloric acid, $HClO_4$, is a powerful oxidizing agent, used as a replacement for nitric acid, HNO_3, in etch solutions. It is not an oxidizer in dilute solution and its perchlorate salts are the mostly stable salts of the oxy-acids of chlorine.

Technical Application: Pure chlorine gas has been used under pressure as an electrolytic jet etch, or bubbled through an etch solution as part of the reaction process in Solid State device fabrication.

The acids of chlorine, as mentioned in the General section above, and shown in the preceding list of A—Z materials with formulas of chlorine containing etchants have wide usage in metal and metallic compound processing as liquid acids.

In addition, using the chlorine as a cleaning gas, hydrochloric acid is used in epitaxy systems, and several other chloride compounds have been in use as doping element carriers in semiconductor diffusion for several years. Phosphorus oxy-chloride, $POCl_3$ (pockle) for n-type phosphorus; boron trichloride, BCl_3, as a gas, for p-type boron, or as dry chemical etching (DCE) in the selective etching of thin film aluminum for patterning structure.

The following list of chlorine etchants includes growth as a single crystal for morphological studies.

Etching: N/A

CHLORINE ETCHANTS

CL-0001
ETCH NAME: Pressure | TIME:
TYPE: Pressure, preferential | TEMP:
COMPOSITION:
x .... pressure

DISCUSSION:

Cl_2 grown as a single crystal in vacuum and under cryogenic temperature and pressure conditions. Preferentially etch crystal surfaces by varying vapor pressure of the vacuum.

REF: Schwentner, N et al — *Rare Gas Solids,* Vol 3, Academic Press, New York, 1970

CL-0002b

ETCH NAME: Chlorine — TIME:

TYPE: Gas, etching — TEMP: RT to hot

COMPOSITION:

x .... Cl_2

x .... acid solutions

DISCUSSION:

Cl_2, as gas bubbled through an acid solution has been used to create both a stirring and/ or etching action.

REF: Ibid.

CL-0002a

ETCH NAME: Hydrochloric acid — TIME: 1—5 min

TYPE: Gas, cleaning — TEMP: 800—1200°C

COMPOSITION:

x .... HCl, vapor

DISCUSSION:

HCl, as hot vapor with dissociation of H_2 and Cl_2 with the acid chlorine vapor used for general cleaning action in epitaxy systems. This includes cleaning of the quartz tube, graphite susceptor plate, and materials to be epitaxially deposited, such as silicon (111) wafer substrates for Si, Ge, GaAs thin film growth in device fabrication.

REF: Walker, P et al — personal application, 1970—1985

CL-0003

ETCH NAME: Chlorine — TIME:

TYPE: Electrolytic, gas jet — TEMP:

COMPOSITION: — ANODE:

x .... Cl_2 — CATHODE:

x .... H_2O — POWER:

DISCUSSION:

Cl_2 gas used as a jet under pressure for electrolytic forming of etched holes in germanium wafers. See Germanium section for reference.

REF:

CL-0004

ETCH NAME: Chlorine water — TIME:

TYPE: Acid, disinfectant — TEMP: RT

COMPOSITION:

x .... Cl_2 (HCl)

x .... H_2O

DISCUSSION:

Cl_2, as gas dissolved in water . . . chlorinated water, chlorine-water. Used as a general disinfectant in water, for cleaning metal surfaces, and in drinking water to kill bacteria. Pure gas can be used, but more often HCl or a dry chemical chlorine compound, such as a hypochlorate, is used. As a very dilute solution of HCl it is used in medicine for human consumption as an acidifier.

REF: Roe, J H — *Principles of Chemistry,* 7th ed, C V Mosby, St Louis, MO, 1950

PHYSICAL PROPERTIES OF CHROMIUM, Cr

Classification	Transition metal
Atomic number	24
Atomic weight	52
Melting point (°C)	1875 (1890)
Boiling point (°C)	2582
Density (g/cm^3)	7.2
Thermal conductance (cal/sec)(cm^2)(°C/cm)	0.16
Specific heat (cal/g) 25°C	0.11
Latent heat of fusion (cal/g)	21.6
Heat of fusion (k-cal/g-atom)	3.3
Heat of vaporization (k-cal/g-atom)	76.635
Atomic volume (W/D)	7.23
1st ionization energy (K-cal/g-mole)	156
1st ionization potential (eV)	6.76
Electronegativity (Pauling's)	1.6
Covalent radius (angstroms)	1.18
Ionic radius (angstroms)	0.63 (Cr^{+3})
Coefficient of linear thermal expansion ($\times 10^{-6}$ cm/cm/°C) 20°C	6.2
Electrical resistivity (micro-ohms-cm)	12.9
Magnetic susceptibility (emu $\times 10^{-6}$)	3.6
Electron work function (eV)	4.37
Cross section (barns)	3.1
Vapor pressure (°C)	2139
Hardness (Mohs — scratch)	9
(Knoop — 100 g load)	700—1200
(Knoop) — kgf mm^{-2})	940
Crystal structure (isometric — normal)	(100) cube bcc
Color (solid)	Steel grey/silvery
Cleavage (cubic — poor)	(001)

CHROMIUM CONTAINING ETCHANTS

Several chromium compounds are used as the oxidizing agent in a number of etch solutions. Two widely used are chromium trioxide, CrO_3 and potassium dichromate, $K_2Cr_2O_7$, both used alone or as an additive to other acids. In another instance, metallic chromium is added to solutions. Such solutions have been used on a wide range of metals and metal compounds in addition to oxides and nitrides as surface cleaners to remove contamination, for polishing, thinning, and as preferential or dislocation etchants.

It should be noted that the chromic and chromous ions — Cr^{++} and Cr^{+++} are extremely tenacious and difficult to remove from surfaces even with prolonged and heavy water washing. This is particularly true in cleaning quartz and glassware with the standard "glass cleaner" solution of H_2SO_4:$K_2Cr_2O_7$. It has been shown that quartz diffusion tubes etched in the glass cleaner still contained traces of chromium ions even after 3 days of heavy water washing. This is due to chromium's affinity for oxygen forming Cr_2O_3 and other Cr-complexed surfaces. A final rinse of glassware in a "chrome etch" solution containing either ceric ammonium nitrate or sulfate at 30—40°C is recommended where subsequent diffusion can contaminate parts with chromium which, in turn, can degrade the eventual electrical characteristics of a device.

There are nearly 100 etch solutions shown in this Etchant Section that contain chromium, about half being preferential/dislocation types and the other half polish/removal, all used for thinning or structuring. As has been mentioned previously, many etchants are named after their developer and this is particularly true for the dislocation etchants containing chromium that have been initially formulated for silicon. Sirtl's etch, circa 1962, was one of the first and, over the years, several modifications of the Sirtl etch have been introduced for specific enhancement of particular types of dislocations and defects. Such named chromium containing, preferential etchants are Secco, Schimmel, Wright, Erhard's, and others.

The "chrome etch", available commercially mixed, was formulated for controlled removal of chromium and chromium oxide, initially, for the processing of chrome glass masks (CGMs) used in photolithography, and is not included in the following "Chromium Containing Etchants". Chrome regia has already been listed under the "Chlorine Containing Etchants".

The following is a selected list of some of the more prominent etchants containing chromium as applied to metals and compounds. Additional solutions are shown under their individual materials.

Formula	Material	Use	Ref.
$27H_2SO_4$:$70H_2O$;$3Na_2Cr_2O_7$, sat. sol.	Al	Cleaning	AL-0032
35 ml H_3PO_4:20 g CrO_3/l	Al_2O_3	Oxide removal	AL-0001a
0.5 *M* $K_2Cr_2O_7$	B_2Te_3	Preferential	BTE-0001
12 oz H_2SO_4:4 oz$Na_2Cr_2O_7$:1 gal H_2O	Brass	Matte finish/clean	BRA-0008b
5 ml H_2SO_4:1 gCr_2O_5:1250 ml H_2O	CdS	Preferential	CDS-0007b
10 ml HNO_3:10 ml H_2O:4 g $K_2Cr_2O_7$	E Etch CdTe	Polish	CDTE-0006a
10 ml E etch:5 mg $AgNO_3$	CdTe	Dislocation	CDTE-0006b
$3H_3PO_4$:$1CrO_3$, sat. sol.	$CaWO_4$	Polish	CAW-0003a
xH_3PO_4:xCrO_3:xNaCN	Cu	Removal (dissolves)	C-0006b
$6H_2SO_4$:$12CrO_2$:$82H_2O$	Cu	Cleaning	CU-0017
12 oz H_2SO_4:4 oz $Na_2Cr_2O_7$:1 gal H_2O	Cu	Oxide removal/matte	CU-0027b
1 12 *N* HCl:1 17 *N* HAc:1 1 *N* $K_2Cr_2O_7$	GaAs	Polish	GAS-0041
1 ml HF:2 ml H_2O:1 g CrO_3:8 mg $AgNO_3$	A/B GaAs	Dislocation	GAS-0059
0.5 *M* $Cr_2(SO_4)_3$:2 mg Ag	Ge	Electrolytic	GE-0040
$1HNO_3$:$20K_2Cr_2O_7$, sat. sol.	GeTe	Polish	GETE-0001
1000 ml H_2SO_4:100 g $K_2Cr_2O_7$, sat. sol.	Au	Electropolish	AU-0009
1HCl:$1CrO_3$:$3H_2O$	Te	Preferential	TE-0001c
xHAc:xCrO_3:xH_2O	Fe/Ni	Electropolish	FE-0012b
1HF:$2CrO_3$(33%) Sirtl etch	Si	Preferential	SI-0039
2HF:1 0.75 *M* CrO_3:$1.5H_2O$ Schimmel	Si	Preferential	SI-0017
2HF:1 1.15 *M* $K_2Cr_2O_7$ Secco	Si	Preferential	SI-0036
1HF:$1CrO_3$(%?) Erhard's	Si	Preferential	SI-0158
60HF:$30CrO_3$(1 g/l):$30HNO_3$:60HAc:$60H_2O$:0.2 g $Cu(NO_3)_2$ Wright	Si	Preferential	SI-0045
1HF;$1CrO_3$(38%) Syer	Si	Preferential	SI-0132
1% $Na_2Cr_2O_7$	Si	Cleaning	SI-0066b
xH_2SO_4:x$K_2Cr_2O_7$, sat. sol.	Ta	Cleaning	TA-0003a

Formula	Material	Use	Ref.
$1HNO_3$:$11HCl$:$20K_2Cr_2O_7$, sat. sol.	SnTe	Polish	SNTE-0001
1 g Cr:20 ml HBr:20 ml H_2O	SnO_2	Removal	SNO-0001a
300 ml HAc:30 ml H_2O:25 g CrO_3	U	Electropolish	U-0002a
$xHNO_3$:$xCrO_3$:xAl_2O_3 powder	U	Lap and polish	U-0002e
67 ml H_2SO_4:310 ml H_3PO_4:120 ml H_2O: 78 g CeO_3	UC	Polish	UC-0001
16 g CrO_3:5 g $NaSO_4$:15 g H_2O	Zn	Polish	ZN-0008c
160 g CrO_3:500 ml H_2O	Zn	Preferential	ZN-0009
$40H_2SO_4$:$60K_2Cr_2O_7$, sat. sol.	ZnSe	Cleaning	ZNSE-0001b
16 *N* H_2SO_4:5 *M* $K_2Cr_2O_7$, sat. sol.	ZnS	Preferential	ZNS-0001a
0.5 *M* $K_2Cr_2O_7$	ZnS	Dislocation	ZNS-0002
1000 ml H_2SO_4:300 ml $K_2Cr_2O_7$, sat. sol.	SiO_2, Qtz	Cleaning	SI-
Glass cleaner	Si_3N_4		

CHROMIUM, Cr

General: Does not occur as a native element. There are some eight minerals containing chromium, several lead chromates, but the primary ore is the mineral chromite, $FeCr_2O_4$. The formula is variable as iron can be replaced with magnesium, and chromium with aluminum. As the pure chromium, it is one of the hardest known metals (H = 9), and is extremely stable under normal atmospheric conditions.

In the metals industry chromium ore is reduced as an iron alloy called ferrochromium (like ferromanganese and ferrosilicon) and, this alloy added to iron, produces stainless steel (SST), and other special steels. It is a hardening agent in both iron and other metal alloys.

The metal can be easily plated electrolytically and is best known as chrome plate. It forms a hard thin film coating that can be highly polished, and has long been used on vehicles — bumpers, trim, and engine parts. Large, vertical RF plasma sputter systems have been used for chrome plating of razor blades. Such a system has a vertical, central rod of chromium and, for a single sputter cycle, thousands of blades are racked around the inner barrel facing the rod which can be 4 ft long and an inch or more in diameter. Many small parts and tools are chrome plated electrolytically or by RF sputter for general surface protection or wear resistance.

Chromium salts, such as the potassium and sodium chromates are highly colored and extremely active oxidizers. Used as reducing agents to produce chromium hydroxide, $Cr(OH)_3$, they are "chrome tannage" as used in the treatment of leather goods, chiefly gloves of calf skin as glazed kid. Because of their brilliant yellow color, some salts are called "chrome yellow" and used as a paint pigment, and in some special glasses, ceramics, and enamels.

Potassium dichromate mixed with sulfuric acid produces a deep yellow/red solution that turns coal black with usage. It has long been referred to as the "glass cleaner" solution, and used to clean laboratory glassware. Caution should be observed in handling as it is an extremely active oxidizing solution, to include human skin. Sulfuric acid can react violently with water, particularly when hot — always slowly add sulfuric to water; not water to sulfuric.

Technical Application: Metallic chromium has several important uses in Solid State and semiconductor processing. It is a dopant element in some compound semiconductors, but a primary use is in thin film, multilayer metallization as the initial evaporation metal on a highly polished surface where gold, platinum, and palladium are used as the top metal. The precious metals do not adhere well to polished surfaces as they are extremely inert to oxidation; on the other hand, chromium has a high affinity for oxygen, also alloys readily with most other metals and compounds, such that it is an excellent tie-down metal as a thin film. Chromium evaporates from the solid phase, not becoming liquid, but going directly to the gas phase — it sublimes — and redeposits as a crystalline structure. Because of this rough crystalline surface, like a lapped rough surface, other metals adhere well. As a multilayer, Au/Cr has long been a general standard for metallization pads and there are others, such as Pd/Cr, Pt/Cr or Au/Pd/Cr, Au/Pt/Cr, etc.

In chrome evaporation, either a tungsten rod plated with chrome is used as a one-shot, throw-away source or chunk chrome pieces/powder in a tungsten boat is the source. In either case, after use, with subsequent exposure to air (oxygen), the surface will oxidize to green chromium trioxide, Cr_2O_3, such that, if the source is to be re-used, it requires heavy out-gassing to remove the oxide before further metallization. If not sufficiently out-gassed, the initial evaporated layer will be a thin film of resistive, glassy trioxide.

For photolithographic patterning, soda-lime glass blanks are chromium evaporated (approximately 2000 Å thick) — the chrome photo resist mask. They commonly have an anti-reflective (AR) coating — oxygen introduced into the evaporation bell jar to form an oxide layer about 400 Å thick (tan color) after chromium has been evaporated, and before opening the bell jar. Such AR coatings have been deposited up to brilliant first order blue or carmine red. These chrome glass masks with or without AR coating have been used as mirrors in a variety of processing applications. One interesting application for a chrome glass plate without AR coating is for a permanent fingerprint identification system. Skin oils act as a mask and activator on chrome, and very clear fingerprints can be developed by light etching in a ceramic ammonium nitrate/sulfate solution. This was developed by these authors and, though not used at present, is a viable approach for better definition than inking.

High purity ingots of chromium using a Vapor Transport (VT) growth method with an iodine atmosphere gas carrier are supplied commercially as a mass of solid chromium crystallites, not a single crystal. Some of the larger, individual crystals, which are single crystal units have been chipped out for morphological study of the metal.

Etching: Easily in HCl or dilute HNO_3, or H_2SO_4. "Chrome Etch" is ceric ammonium nitrate or sulfate.

CHROMIUM ETCHANTS

CR-0001a
ETCH NAME: Hydrochloric acid TIME:
TYPE: Acid, removal TEMP: RT
COMPOSITION:
x .... HCl, conc.
DISCUSSION:

Cr, thin film deposits as a plated surface coating. Solution used to develop cracks in a study of brittle chromium surfaces. CrH will crack subsurface bulk Cr, and cracks propagate by corrosion and stress.

REF: Smith, W H — *Acta Metall,* 103,51(1956)

CR-0002a Walker, P — personal application, 1960—1980

Cr, evaporation film build-up on vacuum glass and metal bell jars or internal fixtures. Remove parts, soak in solution, and scrub with a SST brush. Black smut remaining can be removed with water and wiping with toweling. Can be used on steels, glass, ceramic, and copper parts.

CR-0017a: Nichols, D R — *Solid State Technol,* December 1979

Cr, evaporation deposits in vacuum systems. Solution used for cleaning of bell jars and fixtures. (*Note:* This is an excellent article on the maintenance and cleaning of metal vacuum evaporator systems.)

CR-0002b
ETCH NAME: Hydrochloric acid, dilute TIME:
TYPE: Acid, removal TEMP: RT
COMPOSITION:
1 .... HCl
1—20 .... H_2O
DISCUSSION:

Cr, evaporation deposits in vacuum systems and crystalline chrome pieces. Dilute solutions 1:1 to 1:5 used to clean vacuum base plates and bell jars without removal from the system. Soak heavy toweling with solution and soak/scrub surfaces. In chrome glass mask fabrication "dusting" from chromium particles can be a major cause of film pin-holing and, with eight to ten evaporation runs per day, fixturing may require cleaning on a daily basis.

After breaking down large chunks of high purity chromium for use as an evaporation source, pieces were soaked in the more dilute solutions, DI water washed, and dried under an IR lamp.

REF: Ibid.

CR-0003
ETCH NAME: TIME:
TYPE: Acid, removal TEMP: RT
COMPOSITION:
3 .... HCl
1 .... H_2O_2
DISCUSSION:

Cr specimens. Referred to as a general etch for chromium.

REFS: *Metals Catalogue, MRC* (1984)

CR-0017b
ETCH NAME: TIME:
TYPE: Acid, cleaning TEMP: RT
COMPOSITION:
1 .... HCl
1 .... glycerin
DISCUSSION:

Cr, as evaporated deposits in vacuum systems. Solution used to remove chrome deposits from stainless steel, glass, ceramic, and copper parts. Do not use on iron.

REF: Ibid.

CR-0017c
ETCH NAME: TIME:
TYPE: Acid, cleaning TEMP: RT
COMPOSITION:
5 g .... $KMnO_4$
7.5 g .... NaOH
30 ml .... H_2O
DISCUSSION:

Cr as evaporation deposits in vacuum systems. Solution used to remove chrome deposits from steel, glass and ceramic. Do not use on aluminum.

REF: Ibid.

CR-0005
ETCH NAME: TIME:
TYPE: Acid, patterning TEMP: Warm
COMPOSITION:

453.6 g .... $2NH_4NO_3.Ce(NO_3)_3.4H_2O$
125 ml .... H_3PO_4
2500 ml .... H_2O

DISCUSSION:

Cr, thin film deposits on glass substrates. Solution used for patterning chrome glass masks as photo resist masks. (*Note:* This is one form of the "Chrome Etch" using ceric ammonium nitrate. Nitric acid can replace the phosphoric.)

REF: Angle, D L et al — *Semiconductor Int,* April 1981, 179—196

CR-0006
ETCH NAME: Chrome etch TIME: Variable
TYPE: Acid, patterning TEMP: RT or 30°C
COMPOSITION: RATE: 22 Å/sec

1 g .... $2NH_4NO_3.Ce(NO_3)_3.4H_2O$
10 ml .... HNO_3
100 ml .... H_2O

DISCUSSION:

Cr, thin film evaporation on soda-lime glass used in fabricating photo resist chrome glass masks. Chrome thickness 2000—2500 Å with or without a Cr_2O_3 anti-reflective (AR) coating 400 Å thick (light tan color). Without nitric acid, rate was 40 Å/second. Solution as shown was used as a patterning etch with intense white light (halogen) flooding the surface. Pattern edges etch sharply vertical. Small, half-moon structures along etched edges, termed "mouse-nips", are defect structures that appear to be related to the chrome/glass interface — possibly glass surface defects or due to initial growth nucleation of Cr metal/ Cr_2O_3 anomalies.

Solution diluted to 500 ml and used at room temperature without intense light showed a rate of 1—3 Å/sec. Solution used as a slow step-etch in studying the internal structure of the chromium deposits and the chrome/glass interface. Random black spots were observed on the glass surface after chrome removal which could be removed with nitric acid, e.g., probably either CrO or CrO_2, both of which form black oxides. Low spots in the polished glass surfaces can contain an amorphous silica smear from surface polishing? Work compared to commercial "Chrome Etch" (KTI).

REF: Walker, P & Tarn, W — personal development, Optifilm Co., 1974—1975

CR-0007
ETCH NAME: Chrome etch TIME: Variable
TYPE: Acid, patterning TEMP: 28°C
COMPOSITION:

1 g .... $2NH_4SO_4.Ce(SO_4)_2.2H_2O$
5 ml .... HNO_3
25 ml .... H_2O

DISCUSSION:

Cr, thin film evaporations on glass substrates used for chrome glass mask fabrication for photo resist applications.

REF: Bulletin 1116B, Rev-HPD/VB/TM/83/1K, MRC Corp., 1975
CR-0004: Trost, E — personal communication, 1974

Cr, thin film evaporation on soda-lime glass used for chrome mask fabrication. Prefers ceric ammonium nitrate rather than the sulfate, as the sulfate can leave a scum on glass surfaces.

CR-0015
ETCH NAME: Chrome etch, variation
TYPE: Acid, removal
COMPOSITION:
 12 g .... 2 *N* $H_4NO_3.Cd(NO_3)_3.4H_2O$
 60 ml .... H_2O
TIME: 1—2 min
TEMP: RT
DISCUSSION:
 Au:Cr, thin films deposited by E-beam evaporation on alumina substrates used in a pattern etching study of fine line microelectronic circuit definition. Standard photo resist techniques used with 1350J lacquer. After patterning, gold was removed with tri-iodide etch solution and the chromium with commercial "chrome etch" solution (KTI) to develop the circuit. Chromium also removed with the solution shown for comparative purposes. Both solutions were satisfactory for one mil line definition with the solution shown etching more rapidly. This solution also used for general stripping of chromium from substrates, follow with nitric acid, RT, 1—2 min, DI water rinse 1—2 min, MeOH rinse, and nitrogen blow dry.
REF: Walker, P — personal application, 1983

CR-0010
ETCH NAME:
TYPE: Acid, removal
COMPOSITION:
 2 .... *$FeCl_3$
 1 .... HCl
*30—35°Bé
TIME:
TEMP:
DISCUSSION:
 Cr specimens. Described as a general etch for chromium.
REF:

CR-0001c
ETCH NAME:
TYPE: Electrolytic, polish
COMPOSITION:
 64% .... H_3PO_4
 15% .... H_2SO_4
 21% .... H_2O
TIME:
TEMP:
ANODE:
CATHODE:
POWER:
DISCUSSION:
 Cr, thin film deposits as a plated surface coating. Solution used to polish specimens before etching with HCl in a study of cracks in brittle chromium surfaces.
REF: Ibid.

CR-0018
ETCH NAME: Oxygen
TYPE: Gas, oxidation
COMPOSITION:
 (1) x .... O_2 (2) x .... CO_2
TIME:
TEMP:

DISCUSSION:

Cr, thin films evaporated on glass microscope slides used in a study of chromium adhesion and oxidation. Scratch test used a steel stylus with a 0.125 radius to produce parallel scratches cutting vertically into the underlying glass with different gram-weight loads. Cuts observed under transmitted light. Oxidation was done in vacuum under varied back pressure and included treatment with H_2, Ar and methane in addition to O_2 and CO_2 to form Cr_2O_3. It was noted that adhesion tenacity increased with time. See SIO-0040 for preparation of glass substrates. (*Note:* Improved adhesion with time has been observed on both soda-lime glass and alumina substrates.)

REF: Torkington, R S & Vaughn, J G — *J Vac Sci Technol,* A(3)795(1985)

CR-0019

ETCH NAME: RF plasma

TYPE: Ionized gas, removal

COMPOSITION:

68 .... He

30 .... Cl_2

20 .... O_2

TIME: 15 min

TEMP:

GAS FLOW:

PRESSURE: 0.5 Torr

POWER: 100 W

RATES: See discussion

DISCUSSION:

Cr, thin films used as a mask along with PMMA photo resist for photolithographic patterning. Pre-bake PMMA at 150°C, 30 min, and develop in 1 MIBK:3 IPA at 22—23°C, 150 sec for a 3000 Å PMMA layer. Etch rates were: (1) 35 Å/min for PMMA; (2) 110 Å/min for Cr.

REF: Sewell, H — *J Vac Sci Technol,* 15,920(1978)

CR-0008

ETCH NAME: Kodak EB-5

TYPE: Acid, patterning

COMPOSITION:

x .... NaOH

x .... $K_3Fe(CN)_6$

TIME: 1—3 min

TEMP: RT

DISCUSSION:

Cr, thin films evaporated on glass as photo masks or on silicon wafers coated with SiO_2. Used in an evaluation of different epoxy polymers as E-beam (EB) photo resists. Solution shown used to etch pattern deposited chromium — polybutadiene as photo resist coating showed 300× higher sensitivity to EB than KTFR. Develop coating with cyclohexanon at 180°C.

REF: Hirai, T et al — *J Electrochem Soc,* 118,669(1971)

CR-0016

ETCH NAME: Chrome etch, variety

TYPE: Acid, patterning

COMPOSITION:

164 g .... $Ce(NH_4)_2(NO_3)_6$

42 ml .... $HClO_4$

1000 ml .... H_2O

TIME:

TEMP:

DISCUSSION:

Cr, evaporated thin films. Solution used to remove chromium films evaporated on glass masks. Metallization was Cr/As_2S_3/Glass, with the arsenic trisulfide used to improve photo resist pattern definition. After pattern etching chromium with the solution shown, pattern etch remove As_2S_3 with 5% NaOH. (*Note:* See discussion under CR-0006).

REF: Mednikarov, B — *Solid State Technol,* 1984, 177

CHROMIUM ALLOYS, Cr_xM_y

General: Does not occur in nature as metallic alloy compound. There are about a dozen chromium minerals of which chromite, $FeO.Cr_2O_3$ and crocoite, $PbCrO_4$, are the chief chromium ores.

Industrially, chromium with nickel forms high temperature brazing alloys (soldering below, and brazing above 800°C). There are a wide range of such mixtures as nichromes for specific temperature applications, and are discussed in more detail under Nickel Alloys. Chromel A is 80% Ni:20% Cr, used in steel brazing, and the same mixture has other names. As paired wires — chromel-alumel — it is a widely used thermocouple in the medium temperature range with power/temperature data tables available. There are several iron/nickel/chromium alloys as trinary mixtures in addition to other chromium:metal binary systems.

Technical Application: Other than as general brazing alloys in Solid State processing — mainly for package construction — nichrome is sputter evaporated as a contact pad and, as a thin film with photolithographic patterning, used as an etching mask, e.g., silicon via hole and pit etching with KOH solutions.

Only one chromium alloy is shown here, others are under Nickel and Iron Alloy sections. All alloys have been the subject of study as both mixtures and/or in single crystal form.

Etching: Varies with alloy mixture, but mainly as mixed acids of the oxidation/reduction types ($HF:HNO_3$, etc.).

CHROMIUM ALLOYS

CHROMIUM TITANIUM ETCHANTS

CRTI-0001
ETCH NAME: TIME:
TYPE: Acid, removal TEMP:
COMPOSITION:
x .... HF
x .... HNO_3
x .... glycerin
DISCUSSION:

Cr:Ti, alloy specimen. Solution is a general etch for chromium, titanium and their alloys. REF: Berglund, T & Deardon, W H — *Metallographer's Handbook of Etching,* Pitman & Sons, London, 1931

PHYSICAL PROPERTIES OF CHROMIUM TRIOXIDE, Cr_2O_3

Classification	Metal oxide
Atomic numbers	24 & 16
Atomic weight	152.03
Melting point (°C)	1990
Boiling point (°C)	
Density (g/cm^3)	5.21
Hardness (Mohs — scratch)	8—9
(Knoop — kgf mm^{-2})	~2000
Crystal structure (hexagonal — normal)	$(10\bar{1}0)$ prism
Color (solid)	Green
Cleavage (basal — perfect)	(0001)

CHROMIUM TRIOXIDE, Cr_2O_3

General: Does not occur as a natural compound. The most important ore of chromium is the mineral chromite, $FeO.Cr_2O_3$.

Chromium trioxide is a glass-like material and used industrially for its green color as a paint or glass coloring pigment. (*Note:* Iron imparts a light greenish tinge to glass.) Potassium chromate, $K_2Cr_2O_7$, crystals added to sulfuric acid, H_2SO_4, is the standard ''glass cleaner'' solution.

Technical Application: Not used in the fabrication of Solid State devices at the present time. It has possible application as a thin film surface protectant to include optical properties for anti-reflection. It is used on chrome glass masks as an anti-reflective (AR) coating.

In powder form it is a lapping and polishing abrasive called ''chrome green'' and, as an additive to etch solutions, is mainly preferential. There are other chromium oxides: CrO (black); CrO_2 (brown/black); CrO_3 (red); and Cr_2O_3 (bright green). The latter most often recognized as a reaction coating on chromium metal used as an evaporation source in metallization of devices.

Note that CrO_2 thin films have been used as magneto-optics memory devices.

Single crystals of chromium trioxide have been grown and studied for general morphology and defect structure.

Etching: Insoluble in water, acids, alkalies and alcohols soluble in the ''chrome etch'' formulations containing ceric ammonium nitrate or sulfate. Soluble in fused alkalies. CrO_2 soluble in HNO_3.

CHROMIUM TRIOXIDE ETCHANTS

CRO-0001
ETCH NAME: Chrome etch — TIME:
TYPE: Acid, removal — TEMP: RT to 30°C
COMPOSITION:
1 g .... 2 *N* $H_4SO_4.Ce(SO_4)_2.2H_2O$
5 ml .... HNO_3
25 ml .... H_2O
DISCUSSION:

Cr_2O_3 as amorphous thin films grown on chromium metallized glass masks. The solution will etch both the trioxide and the chromium.
REF: Bulletin 1116B, Rev-HPD/VB/TM/883/1K, MRC Corp 1975

CRO-0002
ETCH NAME: Chrome etch — TIME:
TYPE: Acid, removal — TEMP: 30°C
COMPOSITION:
1 g .... 2 *N* $H_4NO_3.Ce(NO_3)_3.4H_2O$
10 ml .... HNO_3
100 ml .... H_2O
DISCUSSION:

Cr_2O_3 as amorphous thin films grown on chromium metallized glass masks. The solution will etch both the trioxide and chromium.
REF: Walker, P & Tarn, W H — personal development and application, 1976
CRO-0003: Troost, E — personal communication (1976)

Cr_2O_3, amorphous thin films grown on chromium metallized glass masks. Recommends the use of the nitrate rather than the sulfate as the sulfate compound can leave a residual film on the glass.

CRO-0004
ETCH NAME: Chrome etch TIME:
TYPE: Acid, removal TEMP: Warm
COMPOSITION:
453.6 g .... 2 *N* $H_4NO_3.Ce(NO_3)_3.4H_2O$
125 ml .... H_3PO_4
2500 ml .... H_2O
DISCUSSION:

Cr_2O_3 amorphous thin films and thin film chromium deposited on glass substrates. Solution is a general etch for both trioxide and metal.

REF: Smith, R K — *J Appl Phys,* 34,1442(1964)

CRO-0005
ETCH NAME: Chrome etch TIME:
TYPE: Acid, removal TEMP:
COMPOSITION:
164 g .... $Ce(NH_4)_2(NO_3)_6$
42 ml .... $HClO_4$
1000 ml .... H_2O
DISCUSSION:

Cr_2O_3 as a thin film contamination on evaporated chromium. Solution used as a cleaning and removal etch.

REF: Mednikarov, B — *Solid State Technol,* 1984, 177

CRO-0006a
ETCH NAME: Sodium hydroxide TIME: 5—7 min
TYPE: Molten flux, dislocation TEMP: 550°C
COMPOSITION:
x .... NaOH, pellets
DISCUSSION:

Cr_2O_3, (0001) wafers both cut and cleaved from single crystal boules. Cut wafers were mechanically polished on a wooden lap with diamond paste. Wafers with $(10\bar{1}1)$ orientation were prismatically $[(10\bar{1}0)]$ cleaved in air at RT, and used as-cleaved. After etching in the fused alkali, leach away hydroxide with water to recover specimens. Then wash specimens in hot HCl to clean surfaces, then wash in DI water and dry in air. On basal (0001) surfaces, etch pits were triangular with edges parallel to $(11\bar{2}0)$. On $(10\bar{1}1)$ pits were five sided structures. Dislocations were normal to (0001), random and along grain boundaries. Density measured to be $5 \times 10^5/cm^2$. Potassium hydroxide, KOH, gave less distinct structures.

REF: Brower, W S & Farabaugh, E N — *J Appl Phys,* 36,1489(1965)

CRO-0006b
ETCH NAME: Hydrochloric acid TIME:
TYPE: Acid, cleaning TEMP: Hot
COMPOSITION:
x .... HCl, conc.
DISCUSSION:

Cr_2O_3, (0001) and $(10\bar{1}1)$ wafers. Solution used to clean surfaces after preferential etching. After cleaning, wash in DI water and air dry. See CRO-0006a for additional discussion.

REF: Ibid.

CRO-0007
ETCH NAME: Chrome etch
TYPE: Acid, removal
COMPOSITION:
x .... Chrome etch
TIME:
TEMP:
DISCUSSION:

P_2Cr_5 as a thin film deposited by E-beam evaporation (EBE) and evaluated for photovoltaic effect. (*Note:* See CRO-0001 for chrome etch solution referenced.)

REF: Toda, K & Crita, M — *J Appl Phys,* 57,5325(1985)

CRO-0008
ETCH NAME: Chromium trioxide
TYPE: Material, growth
COMPOSITION:
x .... Cr_2O_x
TIME:
TEMP: 1770 K
PRESSURE: 4 × 10^{13} atm
DISCUSSION:

Cr_2O_3 (111), grown as a single crystal from pressed chromium powder pellets. Furnace was an aluminum tube with a boat pre-coated with Cr_2O_x, then processed at temperature and pressure shown in a $CO/O_2(2\%)$ atmosphere. The O_2 partial pressure affected material density. At RT large grains were n-type; at high temperature grains converted to p-type. (*Note:* Cr_2O_3 is hexagonal, not isometric (cubic). (111) should be $(10\bar{1}0)$ prism?)

REF: Young, R W A et al — *J Electrochem Soc,* 132,884(1985)

CR-0009: Torkington, R S & Vaughn, J G — *J Vac Sci Technol,* A(3),795(1985)

Cr_2O_3 thin films grown by oxidizing chromium metal specimens in an O_2 or CO_2 atmosphere. (*Note:* When the metal is used as an evaporation metallization source, it commonly oxidizes after evaporation with a green trioxide coating.)

CLEANING, GENERAL

General: There are many aspects to the cleaning of surfaces as related to material processing or part and equipment maintenance. Selection of the particular solvent, acid, or alcohol to use is dependent upon the metals, metallic compounds, and other materials involved, and may be a combination of solutions and steps in what are called clean/etch sequences . . . using liquids, hot vapors, ionic gases, molecular gases, and molten fluxes. Cleaning may include pressure, such as in steam cleaning, or as steam drying.

Water alone is considered the universal solvent and, as it is neutral with a pH 7 on the Sorenson Scale, it is the general washing and final rinse solution used after any acid or alkali cleaning step, often followed by an alcohol rinse to remove residual surface water, then with drying in air, nitrogen, argon, or helium. Where heavy dirt and oil contamination are present on parts, they are water washed, to include scrubbing with a soap or detergent solution with or without chelating and sequestering agents. In the processing of critical parts, such as missile and space hardware, both Ivory and Castille soaps are considered as two of the highest purity compounds for detergent cleaning solutions, and Joy liquid soap has been used in cleaning of semiconductor wafers and electronic hardware for the same reason. Ionic and nonionic agents, or a wetting agent such as Glyptal may be added to such detergent solutions, and wetting agents have been added to acid etch solutions, as well, for a cleaning assist action. Laboratory glassware and quartzware are cleaned in either soap solutions or acid solutions, followed by heavy water washing and drying. And in the processing of soda-lime glass blanks as photo resist masks, one step in the cleaning process includes soap solution scrubbing. All soaps and detergents are cleaners or degreasers for removal of dirt, grease, or oil, and used on all materials, not just metals and metallic compounds.

Most small parts as used in the electronics and Solid State industries are free of heavy contamination, such as dirt, and only require light degreasing prior to any chemical treatment.

Such parts are delivered in clean containers, which may be as plastic bags sealed under nitrogen, plastic envelopes, and boxes, or may be coated with a removable plastic coating, such as lathe cutting or drilling tools; diamond saw blades, or lucite sheeting with a sticky-back paper covering. In some cases, there may be oils, greases and organics from these wrappings and carriers introduced on such surfaces, even though they are relatively clean.

The first step in cleaning such parts, and this applies in particular to semiconductor wafers and their associated metals and assembly parts, is to degrease, liquid or vapor, with solvents. Vapor degreasers have been developed for this purpose from a large glass beaker of boiling TCE to cabinet sized units with boiling liquid, hot vapor, spray heads, with or without ultrasonic vibration, and may include overhead chain hoists to handle heavy single or multiple parts in a basket. Parts are placed in a stainless steel basket — lowered into the boiling solvent for heavy cleaning, or just into the hot vapor head above the hot solvent for general and fine cleaning. The basket is then slowly removed from the vapor head with automatic drying of the hot part. If streaking is observed on surfaces, return parts to the hot vapor as often as needed until clean. The solvent is in a closed, recirculating system, and the working, hot chamber is separated from the return spill-over chamber by a wall with continual spill-over feed of cleaned solvent to the working chamber section. The hot working chamber is surrounded with water cooling coils near the top, hot rising vapors cool and collect on the stainless steel side walls as droplets and return to the chamber bottom — with minimum drag-loss of the solvent to the atmosphere when parts are removed slowly. Both trichloroethylene (TCE) and perchloroethylene (PCE — Perk) are being replaced with either trichloroethane (TCA) or Freons as the solvents due to the carcinogenic nature of chlorinated solvents. Freons are being more widely used as they are inert and leave no residue on surfaces, such as possible Cl^+ ions from TCE and Perk.

Acid and alkali solutions may be as wet chemical etching (WCE) or as electrolytic etching (EE) for liquid solutions, or vapors; or as dry chemical etching (DCE) for ionic gases as RF plasmas. These may be as surface cleaning only, or include minimum polish removal. BRM solutions (Br_2:MeOH) are coming into wide use as bromine (also iodine) clean etch, and polish almost all metals and metallic compounds. Such BRM solutions may be used prior to degreasing or after degreasing as either a chem/mech lap and polish to remove subsurface damage from previous cutting or lapping steps, or as a free-etch for general surface cleaning and polishing. Both acids and alkalies may be used to oxidize a surface as a cleaning step, such as boiling water, nitric acid or hydrogen peroxide, then followed by stripping the oxide with HF. Electrolytic oxidation with KOH or H_2SO_4 and other oxidizing solutions are used in a similar manner, or used to permanently anodize a surface.

Clean/etch sequences follow the general order: etch damage removal; degrease; solution etch clean; alcohol rinse; final acid dip etch clean. There are a number of such sequences described in the Discussion section of Etchant Formats, and can involve any one or all of the steps just mentioned.

Much processing equipment and facility equipment require on-going maintenance cleaning — water lines in particular — and such items as boilers and DI water stills. Hard water and the electrolytic reaction between dissimilar metals cause compounds, such as calcium and magnesium sulfates, or aluminum hydroxide gel to deposit out in piping, and in boilers as scale. High purity water, such as DI water, is an excellent medium for algae growth, such that the water can become contaminated with both flakes of scale from piping as well as skeletons of algae. Improper pipe assembly from weld joints can contribute both particulate matter and unwanted oil-type contamination from fluxes used, and the piping, itself, can slowly disintegrate, introduce particles. Stainless steel piping used in nuclear reactors is internally etch polished to remove proturberances to prevent such particulates occurring.

If not properly maintained, air compressors can introduce oil into air lines, and all equipment is prone to collection of dust and oils from the atmosphere; insulation on high power electric lines are spray washed, particularly, in arid areas to remove dust and dirt build-up and prevent arcing: RF generators and their associated lines and insulation, such as used in epitaxy systems, require cleaning even in a clean room environment, also to prevent arcing. Pressurized cans of Freon are used for light cleaning in addition to other solvent or water washing. Water piping is flushed with hydrogen peroxide or potassium permanganate solutions, followed by heavy water flushing, and boilers and stills are etch cleaned with acid solutions . . . HCl, citric acid, etc.

This Cleaning section includes general references for parts and equipment cleaning, and plant maintenance or water systems. Specific applications on metals and metallic compounds will be found under the individually numbered etchants as part of the Discussion section of their formats.

In addition to metal processing technology there are several general methods for cleaning of materials, or approaches used in cleaning parts and surfaces in the home, in small businesses, and similar areas not classified, strictly, as an industrial operation. Several of these are not actual etchants, but a material or method applied in a particular area of work, on specific materials, and so forth. It should be realized that the modern home uses many chemicals on a daily basis, such that some discussion of these items is felt worth mentioning.

Note that several of the items have no reference listed due to the commonality of usage. For those desiring further information, consult any chemical text book, see the information brochure supplied with a product, or contact the manufacturer as they often have pamphlets and brochures available on their specific products.

CLE-0001
ETCH NAME: Citric acid — TIME:
TYPE: Acid, cleaning — TEMP:
COMPOSITION:
x .... $(COOH)CH_2C(OH)(COOH)CH_2COOH$
DISCUSSION:

Organic acids used for cleaning of power plant equipment. A comparison of formic and citric acids for cleaning water systems.

REF: Loucks, C M et al — *Am Soc Mech Eng Meet,* New York, November—December 1958

CLE-0002: Data Sheet #585 — Chas. Pfizer & Co., Inc. (1958)

Analytical methods of Citrosolv process for determination of dissolved Fe^{++} and Fe^{+++} or copper assay after equipment cleaning.

CLE-0003: Alfano, S — *Electric World,* 5 February 1962

The use of ammoniated citric acid for removal of boiler oxides.

CLE-0004: Data Sheet #557 — Chas. Pfizer & Co., Inc. (1958)

Chemical cleaning of equipment with citric and ammoniated citric acid solutions.

CLE-0005
ETCH NAME: — TIME:
TYPE: Sequestrant, cleaning — TEMP:
COMPOSITION:
x .... acids/alkalies
\+ .... sequestrants
DISCUSSION:

The use of sequestering and chelating agents added to solutions used for the cleaning of water systems.

REF: Loucks, C M — *Power,* December 1961

CLE-0006
ETCH NAME: TIME:
TYPE: Acid, cleaning TEMP:
COMPOSITION:
x .... acid(s)
DISCUSSION:

A general presentation of the chemistry and cleaning of boilers and heat exchangers in plant maintenance.

REF: Loucks, C M — *Chem Eng*, 5 March 1962

CLE-0007: Data Sheet #593 — Chas. Pfizer & Co., Inc. (1958)

Use of alkaline solutions by immersion or electrolytic etch removal of scale, and for paint stripping.

CLE-0008: Alfano, S & Bell, W E — Int Water Conf, October 1961, Pittsburgh, PA.

Removal of magnetite and copper from boiler surfaces by chemical etching.

CLE-0009
ETCH NAME: Water TIME: Variable
TYPE: Acid, cleaning TEMP: RT to boiling
COMPOSITION:
x .... H_2O
DISCUSSION:

H_2O. Water is the universal solvent, neither an acid nor a base, being neutral with pH 7 on the Sorenson Scale. It is used to quench and stop etching action by soaking; by a dip rinsing; by holding a part in running water in a beaker, or running water as a cascade system; or as a spray. Commercial on industrial water is so called, and used for general cleaning; deionized (DI) water is deionized by boiling and collecting the resulting steam for high purity cleaning requirements. Water can be used chilled; at room temperature (RT); or warm, hot to boiling. The latter is used as boiling water immersion or steam or oxidize a surface. The hydrated oxide may then strip with HF as a cleaning step. It also is used as a cleaning, removal, or polishing solution on soluble compounds, such as the halides.

Water vapor — As a fine mist (fog) for light surface cleaning applied by spraying. Several metallic compounds, such as AlAs as hygroscopic, absorbing water moisture in the atmosphere, and alkaline chemical compounds (KOH, NaOH) deliquesce and liquefy with exposure to air. Pure iron and several iron alloys will oxidize to iron oxides (brown "rust") under normal atmosphere conditions, the brown discoloration observed on "tin" cans, cast iron cooking utensils. Construction girders are coated with red lead to minimize such oxidation.

Steam — Hot water vapor under pressure is used as a cleaning vehicle, as well as a power source for electricity generation or mechanical operation, even as a cutting medium on metals and alloys. Dry cleaning uses steam for pressing, there are steam irons, steam tables in restaurants, building walls, vehicle engines are spray steam cleaned, medical instruments are steam cleaned using an autoclave, these are only a few examples of the use of steam as a cleaning agent.

Water treatment — Natural ground waters contain minerals and organic contamination, as well as industrial wastes, such that water treatment is a major industry, itself. Water can be hard, (arid areas); soft (mountain areas); acidic, with HCl in volcanic areas; acidic with tannic acid in swamps; there are hot sulfur spring and ground waters; and saline water along seacoasts.

Industrial or commercial water . . . filtered to remove solid matter . . . so called "water with the rocks removed" is standard tap water. It still contains dissolved minerals, and is aerated to oxygenize for drinking water, potable water, as against saline nonpotable water. It is the general cooling or wash water in industry, business, and the home.

Such water may be additionally treated to remove calcium and magnesium sulfates (hard water), with further chemical additives for soft water. Most soaps alone, or detergents, are water softeners to some degree, with or without specific additives in their formulas.

High purity water is distilled by boiling, the resulting steam collected in a separate container (still distillation). In Solid State processing this is referred to as DI water, may be double-, even triple-distilled, and include ion exchange beds containing resins to remove both positive and negative ionic contamination. Although distilled water is potable, it lacks the dissolved mineral content of natural water essential to animal life, and tastes "flat". This DI water is used in car batteries and steam irons, as well as for critical cleaning of electronic parts. And it may be just demineralized, not distilled, sometimes shown as dmH_2O, as against DIH_2O.

The greatest water usage cycle starts with (1) agriculture (greatest volume of use); (2) industrial, which may be prior to, or following agricultural use; and (3) human and animal consumption.

The treatment of waste water is of major concern throughout the industrialize world. All major cities have water waste disposal and treatment plants where both aerobic and anaerobic bacteria are used to clarify the water — it is actually sufficiently clean for use as drinking water — although, if it is not discarded as run-off water, it is usually limited for use in agriculture . . . beginning the usage cycle all over again.

REF: Eisenberg, D & Kadsmann, W — *The Structure and Properties of Water,* Oxford University Press, New York, 1969

CLE-0010a

ETCH NAME: Kerosene TIME: Variable

TYPE: Hydrocarbon, cleaning TEMP: RT

COMPOSITION:

x .... C_{10}—C_{16}

DISCUSSION:

Kerosene and gasoline are hydrocarbon derivatives obtained by the fractionation of crude oil, and both are used for cleaning of metals, or as fuels. Petroleum ether, C_4—C_7, also is a solvent with a lower boiling range, e.g., 35—80°C vs. 175—300°C and 40—225°C for kerosene and gasoline, respectively. They are still used as general degreasing solvents at room temperature to warm but, due to flammability, are being replaced by other nonflammable and nonexplosive solvents, such as the Freons.

One clean/etch sequence used for iron is (1) etch with $NaCO_3$; (2) rinse in kerosene/benzene, and (3) wash in alcohol.

REF: Foster, W & Alyea, H N — *An Introduction to General Chemistry,* 3rd ed, D Van Nostrand, New York, 1947, 652

CLE-0035: Measor, J C & Afzulpurkar, K K — *Phil Mag,* 10,817(1964)

Fe, specimens used in a study of oxidation rates, for clean/etch sequence shown for iron under CLE-0010.

CLE-0023

ETCH NAME: Air TIME:

TYPE: Gas, cleaning TEMP: RT to hot

COMPOSITION:

x .... N_2/O_2(24%)

DISCUSSION:

Natural atmospheric air or air in pressurized cylinders, as dry air, is used for both drying and surface cleaning to remove light contamination, such as dust, and there are the household and heavy duty industrial vacuum cleaners for floors, rugs, and furniture. Industrial clean rooms filter in-coming air, and may include electrostatic generators to remove particulate

dust content where critical parts are being fabricated. Such rooms are closely monitored for a particulate count, in some cases, to less than 10 ppm. Air-conditioning units can include water spray to control humidity, as well as electrostatic elements. Air compressors supply pressurized air for cleaning, in addition to their use for mechanical or electrical operations.
REF: N/A

CLE-00024
ETCH NAME: Cryogenics TIME:
TYPE: Liquid gas, cleaning TEMP: − 100°C or below
COMPOSITION:
x .... LN_2, and others
DISCUSSION:

LN_2 or other subzero gases or chemical liquids have been used to soak or spray clean surfaces of materials. Pressurized cans of Freon are used to spray clean electronic parts, as both a gas or cold liquid, and such cold sprays are used on specimens surfaces during microscopic examination for both their chilling action and cleaning function. See the section on Water for additional information as "Cold Etchants".
REF: N/A

CLE-0010b
ETCH NAME: Benzene TIME:
TYPE: Aromatic, cleaning TEMP: RT to warm
COMPOSITION:
x .... C_6H_6
DISCUSSION:

C_6H_6. Benzene is a coal derivative, as are its homologs, toluene (methyl benzene) and xylene (dimethyl benzene) with general formula C_nH_{n-6}. All are used as general cleaning solvents, alone, or in a clean/etch sequence. In some instances, they are the specific solvent applied to a metallic compound due to the material reactivity with water or other quenching and cleaning solvents. Some benzene compounds are used as hardening agents for photo resist lacquers in photolithographic processing, such as for gallium arsenide devices. These are called the "aromatic compounds" due to their sweet smell. They are relatively heavy liquids and, after use, parts are usually rinsed in alcohol to remove the residual solvents. See CLE-0010.
REF: Ibid.

CLE-0010c
ETCH NAME: Alcohol TIME: Variable
TYPE: Alcohol, cleaning TEMP: RT to hot vapor
COMPOSITION:
x .... MeOH, EOH, ISO, etc.
DISCUSSION:

Alcohols are used alone, as a general cleaning solvent; as one or more steps in clean/etch sequences; or as the final rinse after etching and water quenching. Some compounds that are reactive to water are directly quenched in alcohol or the alcohol used as the etchant. Under normal cleaning conditions after acid etching, first water quench, then follow with alcohol rinse, with or without nitrogen blow drying, or simple air drying. Note that solutions containing nitric acid should not be directly quenched with methyl alcohol . . . the combination of red fuming nitric and methanol was one of the first rocket fuels.

Alcohols are derived from different sources and three of the more widely used in industry are methyl, ethyl, and isopropyl. Methyl alcohol (methanol) with the acronym of MeOH, is wood alcohol, as it was initially distilled from wood pulp. Ethyl alcohol (ethanol) with acronym EOH, is grain alcohol, though it can be extracted from other vegetation by reduction of starch. It is the nonpoisonous, drinking alcohol; whereas the other alcohols discussed here are poisonous. Isopropyl alcohol with acronyms ISO or IPA is rubbing alcohol.

Industry ethanol contains 5% methanol or ethylene glycol to make it poisonous and nonpotable, although 95% (5% water) pure ethanol is sometimes used. Absolute alcohol is 100%, classified as 200 proof-gallon as drinking alcohol, and used in medicine.

All alcohols will absorb water, such that they are used to remove residual water remaining on a surface following water quench/rinse after etching, with MeOH being more widely used. They also evaporate rapidly, producing a chilling effect. When alcohol is used as a spray, this effect has been used to cool a specimen surface during microscope study. Hot vapors can be used for degreasing or final cleaning like Freons, TCE, TCA, etc., but alcohols are more commonly used as liquids with nitrogen blow-off for drying. Both MeOH and EOH are used as part of etch formulations for either solution viscosity control or cleaning action. Nital (HNO_3:EOH) is a widely used cleaning/polishing/preferential etch on irons and steels, as one example. Many cleaning solutions, as well as shaving lotions and perfumes contain alcohol.

REF: Ibid.

CLE-0015

ETCH NAME: Acids — TIME:

TYPE: Acid, cleaning — TEMP: RT to hot vapor

COMPOSITION:

x .... acid or acid mixtures

DISCUSSION:

Acids and acid mixtures are used for both etching and cleaning, with specific mixtures depending on the material being processed. Weak acids, such as acetic acid (3—5% in vinegar) or an iodine solution are both cleaners and medical antiseptics. Strong acids, such as H_2SO_4 or HCl, even concentrated, are cleaners for the more chemically inert metals, or they are used in dilute form on other metals and metallic compounds.

REF: N/A

CLE-0016

ETCH NAME: Halogens — TIME:

TYPE: Halogen, cleaning — TEMP: RT to boiling or gas

COMPOSITION:

x .... I_2, Br_2, Cl_2, F_2

DISCUSSION:

The halogen group of four elements includes fluorine, the most reactive element known, and the only halogen not used on human tissue as a cleaner or disinfectant. Fluorine, as hydrofluoric acid, HF, is the primary reducing acid for removal of metal oxides . . . glass (SiO_2) is etch polished in liquid HF, and frosted in the vapors. Bromine, Br_2, is poisonous, can cause goiter, so it is not used on humans, but is a widely used cleaning and etching solution when mixed with methanol — Br_2:MeOH — the BRM solutions, and bromine mixed with acetone is "tear gas". Pure choline gas is occasionally used as a cleaning and etching agent, and several of the chlorine compounds are discussed under the Chlorine section. Iodine, as a mixture with potassium iodide, KI, and ethanol is tincture of iodine, and a major medical disinfectant. The I_2:KI:H_2O solutions in metal processing are the tri-iodide etchants for gold, and other metallic compounds. Similar solutions are general disinfectants for plastics, glass in the laboratory, or in the medical facility.

REF: N/A

CLE-0010d
ETCH NAME: Acetone TIME:
TYPE: Ketone, cleaning TEMP: RT to hot
COMPOSITION:
x .... $(CH_3)_2CO$
DISCUSSION:

Acetone has a high affinity for water absorption, such that it is used as a final rinse after etching and water quenching to remove residual water from surfaces. It is used, alone, as the only solvent rinse or in conjunction with alcohols. As a solvent, acetone is the primary agent for dissolving photo resist lacquers, used from a pressurized can as a spray to clean specimen surfaces prior to photo resist application, to clean equipment (spinners, etc.) used in application, and as a soak/spray after photo resist patterning and thin film metallization . . . the lift-off technique.

As a mixture with ice it is used as both a chilling solution for other etch mixtures, and as a drying agent for removal of water vapor from other gases, such as pressurized argon. As a direct spray, it also is used for its chilling action, and has been used on specimen study under a microscope to reduce heat distortion, or reduce electron impingement damage effects using a transmission electron microscope (TEM).

Where acetone is used in a Yellow Room in photolithographic processing, temperature and humidity are controlled, nominally 72°F and 40% RH. This control is not only critical for photo resist lacquers, but in the use of acetone. If humidity is much above 50% RH, an acetone spray will absorb water vapor from the air and deposit droplets on the specimen surface being cleaned . . . contaminating the surface . . . not cleaning it. See the section on Water for "Cold Etchants".

REF: Ibid.

CLE-0014
ETCH NAME: Trichloroethylene TIME: Variable
TYPE: Solvent, cleaning TEMP: RT to hot vapor
COMPOSITION:
x .... TCE, Perk, TCA
DISCUSSION:

TCE is one of the chlorinated solvents, all of which are carcinogenic to some degree and, though they are still used for general cleaning and degreasing, are being replaced with Freons and other more inert solvents. Vapor degreasers are used with hot liquid, vapor or spray in cleaning all metals, metallic compounds, and general parts. See the Discussion sections of individual materials for the use of these solvents in specific processing.

REF: N/A

CLE-0017
ETCH NAME: Argon TIME: 1—10 min
TYPE: Ionized gas, cleaning TEMP: RT
COMPOSITION: ANODE: Material
x .... Ar^+ ions CATHODE: Pb, Pt, C etc.
GAS PRESSURE: cc/minute as milliTorr (mT)
POWER: KeV range
DISCUSSION:

Argon as ionized gas. Ar^+ ions, are developed by an RF plasma under vacuum, which may be DC power or as RF magnetron (ring magnet on material target being used as the depositing metal(s) to control plasma). These are RF sputter vacuum systems. A basic production type system consists of (1) four target holders above a rotatable specimen holding platen; (2) standard vacuum hardware; (3) RF and/or DC controls; (4) a horizontal feeding input/exit section. This latter capable of being brought up-to-air; whereas the working vacuum chamber is held under at least a 10^{-6} Torr vacuum. Three target locations can be used for multilayer deposition of different metals/alloys, the fourth location for argon ion cleaning prior to metal deposition. Power levels vary with materials used, but a nominal range is from 100—150 KeV with a 0.5 mT argon over-pressure (back pressure).

Irradiation: Although argon has been the main ionized gas used for cleaning, all of the inert gases, some reactive gases, and nuclear particles have all been used in bulk and thin film material studies. The latter, thin films, studied both for adhesion and structure alteration; bulk materials and thin films for induced damage, compound regrowth, gas entrapment, bubble or blister formation, and so forth. All such ionic gases and particles can introduce sub-surface damage into a specimen, even Ar^+ ions when being used for surface cleaning, such that power levels vs. material being cleaned can be critical.

Other ionized gases and irradiation studies are shown under the individual metals and metallic compounds.

REF: Farnsworth, H E et al — *J Appl Phys,* 29,1150(1958)

CLE-0018

ETCH NAME: Hydrogen/oxygen — TIME:

TYPE: Gas, cleaning — TEMP: 600—100°C, nominal

COMPOSITION

x .... H_2, O_2, etc.

DISCUSSION:

H_2 in its molecular state, as against ionized H^+, is the reducing atmosphere for removal of oxides from metal surfaces, referred to as hydrogen firing. It also is used for surface cleaning of high temperature metals and alloys, or ceramics and dielectrics, such as alumina and beryllia substrates prior to metallization as microelectronic circuits. Both hydrogen and forming gas (FG) (85% N_2:15% H_2), as a nonflammable mixture, are used in Solid State and metal processing of solder assemblies to prevent oxidation of parts and subsequent adhesion failure. Heat treatment of metals and alloys is usually done in an inert atmosphere, such as nitrogen or argon, but it can be done with hydrogen for specific hydrogen embrittlement of the surface. Note that platinum and palladium as powders — platinum black or palladium black — will absorb large volumes of hydrogen, and are used for such purpose as hydrogen storage systems.

Oxygen: Pure O_2; O_2/Ar, O_2/N_2, or O_2/CO_2 are used to grow metal oxides under hot gas furnace conditions, and may include water vapor as in growing SiO_2 thin films on silicon. Such oxides are "hard" oxides, as compared to the hydrated forms grown from boiling water, HNO_3, or H_2O_2, and from electrolytic KOH solutions. Where the hard oxides are used as etch masks and for surface passivation, the hydrated oxides are often used as a surface cleaning step . . . deposited, then stripped with HF or BHF.

As an oxy-hydrogen torch, the gas mixture is one of the most widely used methods of brazing metals or working glass, other than acetylene or propane torch. As a hot flame, such torches can be used for brazing, annealing, or surface cleaning in the assembly of parts, although the cleaning function is of only minor concern.

REF: N/A

CLE-0019
ETCH NAME: Dry chemical etching
TYPE: RF plasma, etch/clean
COMPOSITION:
x .... CF_4
x .... O_2 (5%)

TIME: Variable
TEMP: RT
ANODE: Material
CATHODE: Variable by material
GAS PRESSURE: cc/min in milliTorr (mT)
POWER: KeV range

DISCUSSION:

The mixture shown is a generally application gas mixture for many metals and metallic compounds in both the etching and cleaning of surfaces. There are a number of other singular gases or mixtures designed for specific materials, and all gases are used in their ionic species form as against molecular (CLE-0018), as an RF plasma. Where the gas is reactive, such as BCl_3, the system is called Reactive Ion Etching, RIE, and there are several other specialized systems with their own acronyms.

Dry chemical etching (DCE) is one of the major divisions of etching as used in this text, and is more often applied for selective structuring of Solid State devices than as a cleaning method.

REF: Bollinger, D et al — *Solid State Technol,* March 1984, 117

CLE-0020
ETCH NAME: Indium
TYPE: Metal, cleaning
COMPOSITION:
x .... In

TIME: Variable
TEMP: Molten

DISCUSSION:

In, as a molten metal is used as a surface cleaner and light material remover in liquid phase epitaxy (LPE). It is specifically used with indium phosphide, InP, wafers during the growth of additional thin film InP. In LPE systems the wafer substrates are attached to a graphite plate that is wiped horizontally across the surface of the molten metal to effect epitaxy thin film growth. In order to reduce defects being introduced into the growing film from the substrate, the initial operation uses what is called the melt-back technique, where the substrate surface is cleaned with the molten metal before controlled growth is initiated. Further melt-back may be done on the growing thin film during growth, also to improve the quality of the film. Similar LPE melt-back is done with other materials, such as gallium melt-back is used in growing gallium arsenide thin films on gallium arsenide substrates.

Molten Flux: High melting temperature and chemically inert metals, such as titanium, molybdenum, tantalum, and tungsten are surface cleaned or etched in molten metals. The flux can be a metal or mixture of chemical compounds, applied to any material, often as a preferential etch in addition to cleaning. Such fluxes also are used in the growth of single crystals, such as garnets and ferrites. See CLE-0029 — Hone: CLE-0022 — Abrasion.

REF: N/A

CLE-0021
ETCH NAME:
TYPE: Methods, cleaning
COMPOSITION:
x .... solvents/acids/alkalies

TIME:
TEMP:

DISCUSSION:

Methods of surface cleaning described for (1) solvent liquids/vapors; (2) acids — HCl; H_2O_2; HNO_3; HAc, alone or as mixtures; (3) alkalies — both hydroxides and/or carbonates;

(4) heat treatment — vacuum, wet or dry atmospheres of N_2, H_2, O_2 or air; (5) aqueous oxidation — H_2O_2 and H_2O_2:NH_4OH, and (6) ultrasonic — detergents, solvents, acids, or alkalies. Also electrolytic solutions, and the use of abrasives included.
REF: *Bell Tel Tech Pub,* 3143(1958)

CLE-0022
ETCH NAME: Abrasion — TIME:
TYPE: Abrasive, cleaning — TEMP: RT
COMPOSITION:
x .... abrasive grit/paper etc.
DISCUSSION:

Abrasive lapping of a surface for general cleaning. Sand for scouring; sandpapers, such as SiC or emery; smooth paper or rough filter paper only; colloidal liquid suspensions of SiO_2 or Al_2O_3. Applicable to use on several materials, other than those usually thought of as abrasives, such as metals, glass, woods, and even some plastics.

Pads of steel, copper, and plastic with or without detergents are used to remove heavy contamination on industrial equipment and parts, as well as in general cleaners of pots and pans in a household. Hand brushes of steel, brass, plastic, or copper are used in a similar manner. Grinding wheels and arbor brushes; bead-blast cleaning with sand under pressure; steel, glass and plastic beads rotating to clean surfaces as in removing old paint from small parts.

Abrasives are most often used in slurry form. In Solid State material processing a mixture of W-5 garnet (15 μm grit) in water and glycerin is widely used for lapping wafers with final water washing. Finely powdered white alumina (Linde A and Linde B) and diamond grits of less than 1 μm size are used for final surface polishing and cleaning.
REF: N/A

CLE-0025
ETCH NAME: Borax — TEMP:
TYPE: Borate, cleaning — TEMP: RT to hot
COMPOSITION:
x .... $Na_2B_4O_7.10H_2O$
x .... H_2O
DISCUSSION:

Borax, as the natural mineral, has been in use for over 3000 years with several uses: for washing and cleaning of cloths and dishes; as an antiseptic and preservative; as a solvent for metallic oxides in soldering and brazing; as a general flux in alloying metals. In washing it is a water softener for hard water. The Arabs and Egyptians were using both borax and niter (sodium carbonate) solutions in mummification, and as a flux for gold soldering, as far back as 4500 B.C. during the Egyptian Dynastic Period. "20 Mule Team Borax" was mined from the California Death Valley area beginning in the late 1800s.

In metal processing, to include Solid State materials, borax has been used for surface cleaning, and as a constituent in some cleaning and etching solutions with other acid components.
REF: Dana, E & Ford, W E — *A Textbook of Mineralogy,* John Wiley & Sons, New York, 1950, 743

CLE-0027
ETCH NAME: Broom — TIME:
TYPE: Broom, cleaning — TEMP: RT
COMPOSITION:
x .... straw, wood, plastic, etc.

DISCUSSION:

Brooms and mops have been in use since most ancient times for general removal of litter on floors, tables, walls, etc. The first brooms were wood twigs, rushes, or leather strips tied together on a wooden handle with plaited grass or leather as a tying rope. Large palm leaves are still in use in the Pacific area, as are tree leaves in Africa and southern Asia.

The use of straw and feathers also is of ancient origin, both still in use today as the standard household short- or long-handle brooms for floors and tables or a feather duster. Horse hair brooms have a long history of use and today plastic brooms of many sizes and shapes are standard in both industry and household use, to include large push brooms. The latter may be of cloth, with or without oil for additional dust collection or for polishing action, such as heavy-duty floor polishers to include waxing, or vehicle driven street sweepers with their circular, rotating brushes.

The standard mop is similar to a broom, but is an assembly of fairly thick ($^1/_8$″ diameter) cloth strings or strips that can be used as a dry oil, or wet mop. The latter is used with a mop bucket to hold water and a roller squeezer, and the bucket may be on wheels. There are squeegee type mops with foam plastic inserts for throw-away and replacement when the plastic becomes too stiff and loses absorbency.

Common terms are dusting, mopping, scrubbing, and brushing. The soft camel's hair brushes have special application in painting and cosmetics, and there is a special static removal brush used on engineering vellum, or in static sensitive areas. The brush body contains a replaceable strip of radioactive polonium, Po, behind the camel's hair brush section (about 2″ wide) with a handle. Continual rubbing across a paper or plastic/wood surface or cloth, even a metal surface can build up a static charge of electricity . . . it can make paper lift and curl when the hand approaches . . . and an electric spark discharge can occur. This is a common occurrence in households during cold or dry weather, when an individual walks across a rug — building up a static charge — then reaches for a door handle . . . zap . . . a static discharge. The static brush, when wiped across a surface, will remove the static charge. There are other, larger static eliminators available that blow cold or hot air across the polonium strip to remove the charges and there are smaller "spike charge" units electrically activated to eliminate static. Most houses have static arresters (against lightning), metal wire and rods buried in the soil, e.g., grounding. Static electricity is used to collect dust in some homes and many businesses as part of the air-conditioning or heating duct systems. See CLE-0033 for use of ozone, O_3, for elimination of smoke, CLE-0028 for use of a brush.

The legendary Witch's Broom is most often depicted with wooden twigs as the broom section, and there is a plant called Witch's Broom, the tassels sometimes used as a dusting broom. The tail of a fox is called a brush, and has been used as such, as have been other animal furs.

N/A

CLE-0028

ETCH NAME: Brush | TIME:

TYPE: Brush, cleaning | TEMP: RT

COMPOSITION:

x .... bristles (pig), plastic, etc.

DISCUSSION:

Brush is another name for broom (CLE-0027) but is discussed here separately due to some more specific applications than just general cleaning as associated with brooms and mops. As a hand brush, notably for personal grooming, pig's bristles have been in use since ancient times, today, plastic fibers and other fiber-like materials are more prevalent.

Most well known as a brush and comb set for personal use, the comb is made of —

originally — wood or tortoise shell — now, also of plastic, metal, etc. With metal bristles, one type of brush is known as a Curry Comb used for grooming horses; another, with a wooden handle and steel bristles, as a heavy duty scrubbing brush on metal parts — removal of scale (oxidation and other contamination products), old paint, thin film metal plating, extraneous thin film metal build-up on parts due to metal evaporation. Smaller steel or brass bristle brushes are used for general scouring of pots and pans in cooking, or to "brush pattern" a softer surface of metal, plastic, leather, wood for decorative purposes. Plastic fiber brushes in a wood or plastic housing are similarly used for scrubbing. Scouring pads — fine metal or plastic mesh — with or without soap detergent in the pad, are general scrubbing brushes.

A fox's tail is called a brush, used as an item of apparel, or used as a dusting brush like a feather duster. Other forms of special brushes are part of manicure sets, such as the finger-like chamois skin brush used for polishing nails, and small eye brushes for make-up shadowing. The whisk broom is a hand-sized "broom" of straw for general surface cleaning, notably used for removing lint from clothing by "whisking". There is a special dusting brush — a wooden handle/holder about $^1/_2$″ wide — the brush portion about 2″ in depth — total length about 1 ft . . . originally designed for use on a drafting table to remove dust and detritus from erasers, pencils, etc. used in drawing, also now used as a general table duster. There are long metal bristled brushes with the bristles set in a slightly mobile rubber/plastic base, used for grooming and carding the fur of cats and dogs, or as a carding comb used for separating flax fibers. Also the horse curry comb already mentioned under CLE-0027.

REF: N/A

CLE-0029

ETCH NAME: Hone — TIME:

TYPE: Stone, cleaning — TEMP: RT

COMPOSITION:

x .... rock, metal, etc.

DISCUSSION:

Hones are used for cleaning and sharpening, and may be used dry, water wetted, or impregnated with oil. Arkansas stone is a hard, fine grained and dense aluminum oxide rock long used for sharpening metal knives . . . by "honing". Natural emery is a mixture of aluminum oxide and magnetite, magnetic iron ore — Al_2O_3 and Fe_3O_4, respectively, and similar to Arkansas stone in use. Turkish emery is considered of highest quality, and both types of stones are also referred to as whetstones. And today, artificial silicon carbide, SiC, known commercially as carborundum, also is used as a hone. All of these rock type materials have a Mohs hardness rating of H = 9/9+, second only to diamond and capable of cleaning, cutting, or polishing all other materials. There also are artificially hardened or fabricated metals, metal alloys, carbides, and borides. The carbides may be as shaped parts for use with a drill press or hand drill, and used for cutting, shaping, or cleaning of other materials.

In some countries the itinerant knife grinder still comes around with his foot-powered grinding wheel; straight razors are sharpened with a combination of a leather strop and stone hone; and a meat carving knife set includes a serrated metal tong for sharpening. Hand and electric hones use a series of hard steel discs across which the knife edge is drawn for sharpening.

The term "stoning" is sometimes applied to the use of rocks as a cleaning or surface conditioning agent, such as wood being stoned to develop a grain finish structure.

REF: N/A

CLE-0010e
ETCH NAME: Ether TIME:
TYPE: Ether, cleaning TEMP: RT to warm
COMPOSITION:
x .... $C_2H_5OC_2H_5$
DISCUSSION:

Ether (ethyl ether) is made by treating ethyl alcohol with sulfuric acid to produce a colorless, volatile liquid with a pleasing odor. It has long been used as a general anesthetic, but has been largely replaced by other less inflammable and explosive chemical compounds as ether will ignite and burn in air. It is an excellent solvent for fats, gums, and resins, and may be a constituent in some household cleaning solvents, though ammonia, NH_3, is more widely used (Windex).

Ether is used in many chemical processes, and is a fuel additive in high performance racing cars. It is occasionally used in Solid State chemical processing as a cleaning and dry solvent similar to alcohols or nitrogen gas.

REF: Ibid.

CLE-00031
ETCH NAME: Rubber TIME:
TYPE: Organic, cleaner TEMP: RT
COMPOSITION:
x .... rubbers
DISCUSSION:

Rubber from the South American rubber tree was imported to England in the early 1800s as Indian rubber as it was found to erase pencil marks from paper by rubbing. It is still so used, to include "gum rubber" as 1″ squares, 2″ long yellowish bars of a soft rubber widely used in engineering drafting. Other than as a paper pencil mark remover, there are many rubber products, today, from vulcanized vehicle tires to plastic rubber products. In industry two artificial rubbers are Buta-A and Viton used as gasket seals in vacuum equipment, and there are others such as Buna-N, isoprene, neoprene, etc.

British gum is dextrin, $C_6H_{10}O_5$, as a by-product in the reduction of starch for its sugar content, when mixed with water is called mucilage (glue). Animal glues are from hooves and horns; acrylics from plastics; white paste glue from flour and water . . . none of which are cleaners . . . but can be dissolved with some of the alcohols and other fluid cleaners described here. Methyl ethyl ketone, MEK, is one such solvent.

REF: N/A

CLE-0032
ETCH NAME: Sand TIME:
TYPE: Rock, cleaner TEMP: RT
COMPOSITION:
x .... SiO_2, grains
DISCUSSION:

Sand as water-washed, roughly rounded grains of silicon dioxide, SiO_2. When mixed as a water slurry, sand has been used since ancient times to scour and clean cooking utensils. Under pressure, using dry air or nitrogen as the carrier gases, it is called sandblasting or bead blasting, as it may be artificially fabricated "beads" of glass. In either case, pressurized sand particles are used in many areas and industries to clean material surfaces from removing grime from rock building faces to paint removal from parts; to general cleaning or roughening a metal surface; or to develop grain structure in woods with both cleaning and polishing action. S.S. White Dental units were initially developed with a rubber tube and small exit nozzle using finely powdered sand under pressure to clean teeth. They have been adapted for general industrial use as a method of cleaning, cutting, or shaping of parts. One example of such use is in the fabrication of screws, as these mobile nozzles are one of the few methods capable of cutting a re-entrant angle, or to form a circular hollow within a solid material, etc. They also are used in the fabrication of jewelry and in the pattern frosting of glass, rather than using hydrofluoric acid, HF, which is particularly dangerous. These S.S. White units were initially used in the early days of the semiconductor field for both dicing wafers and as general surface cleaners and, more recently, for angle shaping exposed p-n junctions around the wafer edge.

There are many other natural rocks that, when powdered, are used as cleaning, removal, or polishing abrasives. Selection depends on the Mohs hardness, abrasive size, and particle shape, e.g., sand has a rounded shape — quartz forms sharp slivers with faster cutting action — as colloidal silica, the submicron-sized spherical particles produce a fine polish. All silicas have a Mohs hardness of H = 7, but it is the particle shape that controls the cutting, polishing, or cleaning action.

Corundum, Al_2O_3 (H = 9) has sharp, flat grains; garnets (H = 5) have blunt grains that break down in size during use, giving finer and finer cleaning and polishing action; limestone, $CaCO_3$, (H = 4—5) has soft grains; diamond, (H = 10), the hardest material, has angular grains and gives fine polishing action, and is most commonly used in an oil paste. Artificial borides and carbides or cubic boron nitride, B_4N, and silicon carbide, SiC, with a hardness of H = 9+, can have a blocky grain structure with rapid cutting action.

Lava soap contains a high percentage of silica grains that were originally obtained from pumice, a froth-like, physically light rock extrusion from volcanic magmas. Pumice is used as a whetstone for cleaning other materials, as well as a lightweight building stone.
REF: N/A

CLE-0010f
ETCH NAME: Ozone — TIME:
TYPE: Gas, cleaning — TEMP: RT to warm
COMPOSITION:
x .... O_3
DISCUSSION:

O_3, as ozone, occurs naturally in the atmosphere in small quantity. Fluctuation of the ozone-oxygen level in the high altitudes over the Antarctic are of concern, as ozone acts as a screen against ultraviolet light rays from the sun. It can be stable for years at very cold temperatures, months at room temperature, but decomposes rapidly above 100°C; almost instantaneously and explosively at 300°C. Ozone can be formed from normal O_2 by passing air through an electric discharge, e.g., the pungent odor associated with lightning storms. The rapid quenching of flames from a burning wood fire produces some ozone, and it is part of some chemical processes; 1 ppm in air can be injurious to health.

Ozone is an extremely powerful oxidizer, and electrically powered ozone generators are used in buildings such as theaters, restaurants, etc. as air fresheners. There are small cigarette ash tray units available that will dissipate the smoke from a burning cigarette. Larger units, with air being blown through an electric discharge, are used to clean metal surfaces in industry: the part is held in the ozonized air for a few seconds to clean.
REF: Ibid.

CLE-0034
ETCH NAME: Soap — TIME:
TYPE: Organic, cleaning — TEMP: RT to hot
COMPOSITION:
x .... $3C_{17}H_{35}COO$

DISCUSSION:

Soap is made by treating fats and oils with sodium or potassium hydroxide. Sodium soaps are hard, while potassium soaps are soft. Other oils may be added, such as lanolin (from sheep) in Dove soap. Other oils are used, such as palm oil (Palmolive soap) — or scented with chemicals and natural herbals — sandlewood, lemon, Attar of Roses, are all used. Two of the highest purity soaps are Ivory and Castille soap, both recommended for the cleaning of critical parts, such as missiles and space hardware. Lava soap contains a high percentage of powder pumice (volcanic silica froth rock), specifically designed for removal of heavy grease, oil, and grime; anyone who has been in the U.S. military is familiar with the semisoft, yellow blocks of lye soap, with a high sodium hydroxide content for improved chemical cutting action. Both KOH and NaOH may be referred to as "lye", and sodium hydroxide as "caustic soda". In water solution, alone, they have a soapy feeling and, in high concentration, can cause blisters and burns on the skin, though they are not quite as dangerous as acids. The hydroxides are chemical bases, and when mixed as Normal solutions will neutralize a Normal acid solution.

There are many liquid soaps that grade into skin creams with added palm, lanolin, or coconut oils. Both liquid and dry powdered soaps used for dishes and clothes washing are called detergent soaps, and contain chelating or sequestering agents that improve wetting action, control sudsing action for foaming or de-foaming. Pure oils alone have long been used to clense skin. The ancient Greeks rubbed themselves with olive oil, then scraped it off with a stigil. Today there are many oils and creams as sunscreen lotions, and Chapstick is a thickened petroleum jelly similar to Vasoline. The latter may be mentholated. Egyptian records dating back to 4000 B.C. list hundreds of formulas as their priesthood were the original chemists and medical practitioners in the Mediterranean area.

REF: Ibid.

CLE-0036

ETCH NAME: Vacuum TIME:

TYPE: Vacuum, cleaning TEMP: RT

COMPOSITION:

x .... vacuum

DISCUSSION:

Vacuum is created by the removal of air and other gases from a container. A mechanical vacuum pump (oil pump — roughing pump) is constructed with internal moving vanes and chambers that physically pull air from a closed glass or metal chamber through an attached hose to create a vacuum in the chamber. They are nominally rated for a 10^{-3} Torr vacuum level (1 μm), and are used as the initial pump in industrial vacuum systems. These mechanical pumps are called roughing pumps, and there are similar cryo-pumps, chilled with LN_2, that operate the LHe without oil.

Vacuum systems as used in Solid State and metal processing range from small R & D type units to large production systems, but all operate in a similar manner. Once the initial pumpdown is completed with a roughing pump, the system is switched over to a high vacuum pump — hot oil; cryo-pump with helium and charcoal getter; turbo-pump; ion pump — with the cryo-pumps currently most prominent as oil-free systems. There are large, walk-in vacuum chambers, such as those used in testing satellites; and belt systems, where a continuous roll of material, such as mylar, can be fed through and metallized. Vacuum work is most often done on the 10^{-5} to 10^{-6} Torr, but modern systems are capable of levels to 10^{-10} and 10^{-11} Torr. Note that space vacuum is considered to be on the order of 10^{-18} Torr. There are RF/DC sputter systems; ion milling systems; and RF dry chemical etching systems, using ionized gases. The two latter systems are used for structuring material surfaces and multilayer thin films.

Any vacuum system can be used with vacuum, alone, as a drying and cleaning agent, or in can include heat in inorganic material processing. And organic material, such as foodstuffs, are now vacuum packed, the operation combining a drying action with vacuum sealing the product in a plastic bag. This may be as vacuum freezing, but quick-freezing vegetables is a different process as moisture is retained.

Other vacuum systems include the household vacuum cleaner, and there are units capable of either vacuum or air blowing by including a reversing switch. Large industrial systems may be for cleaning floors, vacuum removal of water and, where oil spills in the ocean occur, for vacuum removal of oil. Vehicle mounted systems are used in street cleaning with a combination of vacuum and air blowing, and rotating brushes.

REF: N/A

CLE-0037

ETCH NAME: Oil | TIME:

TYPE: Oil, cleaning | TEMP: RT to warm

COMPOSITION:

x .... oil

DISCUSSION:

Oil used as a cleaning solution. Mineral oil from petroleum or coal; animal oil, such as lanolin from sheep; or vegetable oils, such as olive, peanut, linseed, safflower, or cottonseed oil. Olive oil has been used in the Middle East and Mediterranean are as a body cleansing oil since at least 3000 B.C. — rub oil into the body, then scrape excess away with a wooden or bone stigil. Oil also is worked into leather to make it pliant, as well as for cleaning action: saddle soap is a soft, high oil content soap used alone for cleaning, or following by waxing. Linseed oil is rubbed into wood to bring out grain structure, as well as for cleaning and polishing. Cottonseed oil is used in large quantities for the making of soaps. Oil mops are made of cloth pads saturated with oil, used for cleaning and polishing floors, and oiled hand clothes on furniture, table tops, etc. The oil aids in dust and dirt removal, and leaves a thin protective surface coating to prevent drying out. Light oils, such as 3-in-1 oil, are used to both remove dust and dirt from parts as well as a lubricant; there are various grades and thickness of car engine oils, some with chemical detergents added to improve cleaning action. There are plastic compounds that grade from a thin oil, to grease consistency, to solids, all of which have been used as cleaners in addition to their properties such as lubricants.

Oils can be homogenized or thickened to cream consistency, such as cold cream used as a facial cleanser . . . lanolin and coconut oil widely used as the base ingredient in such facial creams . . . suntan lotions are similar from a light oil to cream consistency with additional sunscreen additives. There is a new product containing small spheres of artificial sponge in a suntan lotion cream, the sponges helping to retain body oils and perspiration. Natural sponges have long been used, soaked in water or oil for cleaning . . . called "sponging", and, today there are plastic sponges used in a similar manner.

Thickened oils also become greases: grease-paints in various colors for facial decoration . . . used in religious ceremonies since earliest times, the American Indian "war paint", today the circus clown . . . removal is often with cold cream. Cosmoline is a heavy grease used to coat weapons and other metal parts for long-term storage to prevent rusting, and soft iron lapping platens are coated with a light oil or glycerin also as a rust preventative.

Spermiceti — whale oil — is actually a wax-like substance obtained from the head of the sperm whale. Its primary use, again, since ancient times, has been for illumination . . . an oil lamp. The spermiceti flame was used to establish the International Candle-Power unit as a light intensity measurement.

Some nut oils are used in cosmetics. Coconut oil has been mentioned, but almond oil is widely used; peanut oil most often used in cooking. Oil from citrus fruits — orange, lemon, lime, tangerine — for food flavoring, in air fresheners, facial creams, soaps . . . the colored rinds called "zests" as food additives. Many flowers produce oils, such as violets and roses. The latter as Attar of Roses, both used as air fresheners, in soaps, etc. Aromatic wood oils: Oil of Cedar was in use in biblical days in Asia Minor from the Cedars of Lebanon, and Sandlewood pre-dates even the use of cedar, circa 3000 B.C. Both used as air fresheners . . . sandlewood as incense . . . cedar for both fine odor and as a preservative against insect pests for storage of clothing . . . the cedar chest.

Waxes are another category derived from oils. Paraffin is from petroleum; beeswax from the honeybee honeycomb; carnauba wax from the conifer tree of the same name; castor oil from the castor bean; beechnut oil from the beechnut tree; myrtl oil from the bush myrtl; neat's foot oil from cow hooves. All of these natural oils can be used in wax formulations. And there are many artificially derived chemicals as oils or waxes, such as banana oil. Such oils are mixed with waxes, such as beeswax:paraffin for polishing and cleaning, as wax produces a more long-lasting, often water repellant surface finish than oil alone.

REF: N/A

CLE-0038

ETCH NAME: — TIME:

TYPE: Alkali, cleaning — TEMP: RT to boiling

COMPOSITION:

x .... 2—10% KOH/NaOH

DISCUSSION:

Hydroxide solutions of low concentration are general metal cleaners and, in electroplating, may be used as a pickling bath or for surface conditioning. Other base chemicals (>pH 7, Sorenson Scale) may be used with the hydroxides or by themselves, such as carbonates and borates. After alkali cleaning, wash parts thoroughly until the soapy feeling of surfaces is removed. Higher solution concentrations, particularly when used hot to boiling, can be preferential etchants on metals, and will produce severe skin burns.

Some household cleaners, such as Drano, contain hydroxide and should not be handled with bare hands.

REF: N/A

CLE-0039

ETCH NAME: — TIME:

TYPE: Electrolytic, cleaning — TEMP:

COMPOSITION: — ANODE:

x .... x% KOH — CATHODE:

or x .... acids — POWER:

DISCUSSION:

Hydroxides are used as surface oxidizing solutions with deplating by anode/cathode reversal, or the oxide is stripped with HF as a method of surface cleaning. Acids such as HCl and H_2SO_4 also may be used as electrolytic surface cleaners, or used as wet chemical etching (WCE) cleaners without electric current. HNO_3 and H_2O_2 will be oxidizing like the alkalies.

REF: N/A

CLE-0040

ETCH NAME: Ammonia — TIME:

TYPE: Base, cleaning — TEMP: RT

COMPOSITION:

x .... NH_3/NH_4OH

DISCUSSION:

Both ammonia and ammonium hydroxide are general surface cleaners, and also used to neutralize acids. Chemical acid sinks are drained into an acid sump, which is then neutralized with ammonia before effluent release into a city sewer system. Note that alcohols are disposed of separately, not down an acid drain, as alcohol/acid mixtures can be toxic or explosive, and alcohols will kill bacterial action used in water waste disposal operations.

Crystalline ammonia is the medicinal ''smelling salts''. The tarnish observed on silver, often due in part to the presence of sulfur compounds in the air, is easily removed with ammoniated compounds supplied as a paste or saturated cloth. There are similar compounds for brass, bronze, and copper, such as ''Brasso''.

REF: N/A

CLE-0041

ETCH NAME: Zeolite TIME:

TYPE: Mineral, cleaning TEMP: RT

COMPOSITION:

x .... $(Ca,Na_2)O.Al_2O_3.9SiO_2:6H_2O$

DISCUSSION:

There are some 30 zeolite minerals as natural hydrated silicates, the formula shown is for the mineral mordenite which is typical as a calcium, sodium, or aluminum silicate. They are used in granular form as drying compounds. Commercial Dri-Rite is dyed light blue in color and, as it absorbs moisture from the air, it turns light pink. Oven baked above 125°C to remove collected water, granules return to the blue color, and material can be re-used. Place a small amount in the bottom of a glass desiccator jar, cover and seal with a vacuum grease, and material will remain dry without oxidizing. Hygroscopic compounds, such as AlAs, are held in desiccators. In food processing, desiccator drying, to include heat and a light vacuum to draw off liquids is one preparation method, and zeolites are used in shipping packages or cartons as a drying compound against possible breaking or spills.

REF: N/A

CLE-0042

ETCH NAME: Hydrofluoric acid TIME: Variable

TYPE: Acid, cleaning TEMP: RT to hot

COMPOSITION:

x .... HF

DISCUSSION:

Hf is the primary reducing and oxide removal acid used in Solid State material processing. It is used alone, as a concentrated etchant, water or alcohol diluted, or mixed with other fluorine compounds, as buffered hydrofluoric acid (BHF) solutions. It will remove the ''native oxide'' that automatically forms on metal surfaces as a passivating layer when exposed to atmosphere, and has long been used in the etch patterning of glass — immerse to polish — vapors for frosting. It is considered a cleaning solution for stripping of thin film oxides only, otherwise it is the reducing acid in oxidation/reduction solution mixture, with HNO_3, H_2O_2 as the oxidizers. Caution should be used in handling any fluorine containing acid or compound due to health hazard.

REF: N/A

CLE-0043

ETCH NAME: Heat TIME:

TYPE: Thermal, cleaning TEMP: >22.2°C

COMPOSITION:

x .... heat

DISCUSSION:

Thermal heat treatment of metals and metallic compounds used as a cleaning or drying agent, alone, under vacuum conditions, or with inert gas atmospheres under furnace conditions. Pure heat also is used in fabricating metal alloys (irons, steels, coppers, aluminums, etc.) for specific structure, and characteristics such as material strength, ductility, hardness, and so forth. With temperature as the main factor, may act as a preferential structure and defect etchant. With reactive atmospheres, in addition to cleaning action, to passivate or condition surfaces by nitridization, oxidation, sulfurization, halogenation, etc. Drying with heat includes the use of white light or IR lamps, such as for drying automotive paints or general parts after water washing or acid treatment.

Clothing is dried and cleaned by hanging in sunlight; salt is collected by evaporation of sea water; foodstuffs are dried either in the sun or by hanging in a cool, dry storage area. These are all methods of cleaning and drying with the use of some form of heat.

REF: N/A

PHYSICAL PROPERTIES OF COBALT, Co

Classification	Transition metal
Atomic number	27
Atomic weight	58.93
Melting point (°C)	1495
Boiling point (°C)	2900
Density (g/cm^3)	8.92 (8.71)
Thermal conductance (cal/sec)(cm^2)(°C/cm)	0.1652
Specific heat (cal/g) 25°C	0.1056
Latent heat of fusion (cal/g)	58.14 (62)
Heat of vaporization (cal/g)	1500
Atomic volume (W/D)	6.7
1st ionization energy (k-cal/g-mole)	181
1st ionization potential (eV)	7.86
Electronegativity (Pauling's)	1.8
Covalent radius (angstroms)	1.16
Ionic radius (angstroms)	0.72 (Co^{+2})
Coefficient of linear thermal expansion ($\times 10^{-6}$ cm/cm/°C)	12.5
Electrical resistivity (ohms-cm)	6.24
Curie temperature (°C)	1121
Poisson ratio	0.32
Young's modulus (psi $\times$ 10^6)	30.6
Tensile strength (psi $\times$ 10^3) — annealed	37.0
Compressibility (cm^2/kg $\times$ 10^{-6})	0.50
Electron work function (eV)	4.18
Cross section (barns)	37
Hardness (Mohs — scratch)	8+
(Brinell — kgf/mm)	125
Crystal structure (hexagonal — normal)	$(10\bar{1}0)$ prism, hcp
Color (solid)	Grey-reddish
(gas/pigment)	Cobalt blue
Cleavage (prismatic)	$(10\bar{1}0)$

COBALT, Co

General: Does not occur as a native element. There are some 20 cobalt minerals, the most important being sulfides and arsenides. Cobaltite, CoAsS and smaltite, $CoAs_3.NiAs_3$, are ores, though no cobalt minerals occur in large quantity. They are associated with nickel, iron and copper deposits commonly as metasomatic contact deposits. Both metallic nickel and cobalt can show slight, natural magnetism, but magnetite, Fe_3O_4, is black magnetic iron as primary magnetic natural mineral. Irons and steels can be artificially magnetized, but are not magnetic under normal processing conditions.

Primary industrial use of cobalt is as an alloy with iron in the production of steels, including stainless steel, SST. Stellite, an alloy of cobalt, chromium, and tungsten is used for lathe tools and, as it will hold a sharp edge at red heat, is used for medical knives, scalpels, etc. It is a constituent in many of the iron magnets. Cobalt produces a deep blue color, "cobalt blue", used as a pigment in paints and glass. The salts of cobalt are highly colored, particularly pink or blue. Solution saturated paper will turn from pink to blue with the approach of wet weather and, as a writing fluid, is "invisible ink" — pale pink, changing to blue when heated.

As the isotope cobalt-60, ^{60}Co, it has a radioactive half-life of about 5.75 years, and is used in nuclear studies and testing. **CAUTION**: Some alloys containing cobalt, when subjected to radiation, can become radioactive with conversion of the cobalt fraction to ^{60}Co.

Cobalt, nickel, and chromium can all be easily electroplated as hard thin film surface coatings, though nickel and chromium are more widely used due to quantity availability.

Technical Application: The metal is used as an n-type dopant in some compound semiconductors, and is under evaluation as a silicide: CoSi, Co_2Si and $CoSi_2$, for use as a blocking layer in heterostructures. As already mentioned, cobalt-60, ^{60}Co, is used in radiation testing of Solid State devices, usually by converting lead, Pb, pellets — for alpha particle radiation — with a 24-h half-life.

It has been grown as a single crystal for general morphological and defect studies, and also has been deposited as an epitaxy thin film, e.g., primarily in the development of silicides.

One interesting study: brush iron powder on a cobalt surface, and the iron will align itself according to the magnetic domains in the cobalt specimen.

Etching: Easily etched in acids, most commonly as electrolytic solutions: HCl, H_3PO_4, and mixed acids.

COBALT ETCHANTS

CO-0001

ETCH NAME:	TIME: 2—3 min
TYPE: Electrolytic, polish	TEMP: RT
COMPOSITION:	ANODE: Co
500 ml .... HCl	CATHODE: SS
500 ml .... EOH	POWER: 2 V

DISCUSSION:

Co, (0001) wafers and with other orientations used in a study of growth habits. After etching, wash in boiling DI water, then rinse in alcohol and dry with hot air.

REF: Cliffe, D R & Farr, J P G — *J Electrochem Soc,* 111,299(1964)

CO-0002

ETCH NAME: Iron	TIME:
TYPE: Metal, preferential	TEMP: RT

COMPOSITION:

x .... Fe, powder

DISCUSSION:

Co, (0001) wafers and other orientations used in a structure study. As cobalt is slightly magnetic, by brushing iron powder across specimen surfaces, the iron will align itself developing magnetic zones/domains.

REF: Hall, E O — *Acta Metall,* 6,110(1958)

CO-0003

ETCH NAME:

TYPE: Electrolytic, polish

COMPOSITION:

23 .... $HClO_4$

77 .... HAc

TIME:

TEMP: 5°C

ANODE:

CATHODE:

POWER:

DISCUSSION:

Co, (0001) wafers used to study stacking faults developed by nuclear magnetic resonance. Study was done with an electron microscope (EM). Solution used to polish and etch-thin specimens for microscope study.

REF: Morgan, J T — *Phil Mag,* 9,607(1964)

CO-0004

ETCH NAME: Neon

TYPE: Ionized gas, preferential

COMPOSITION:

x .... Ne^+ ion

TIME:

TEMP:

GAS FLOW:

PRESSURE:

POWER:

DISCUSSION:

Co specimens. Ne^+ ion bombardment used in a study of etch figures obtained on metals by irradiation treatment. Other metals evaluated were Al, Bi, Cd, Mg, Cu, Sn, and Zn.

REF: Yurasova, V E — *Kristallografiya,* 2,770(1957)

CO-0005

ETCH NAME: Phosphoric acid

TYPE: Electrolytic, polish

COMPOSITION:

x .... H_3PO_4

TIME:

TEMP:

ANODE:

CATHODE:

POWER:

DISCUSSION:

Co spheres 5/16″ in diameter polished in the solution shown prior to oxidation experiments. Oxidation was done in air at 400 and 450°C. As cobalt has a structure transition point at 420°C, as expected, the oxidation patterns differentiated between the two structures: below 420°C, hexagonal, hcp, and above, cubic, fcc, oxidation figures were observed.

REF: Kehrer, V J & Leidheiser, H — *J Chem Phys,* 21,570(1953)

PHYSICAL PROPERTIES OF COBALT OXIDE, CoO

Classification	Oxide
Atomic numbers	27 & 8
Atomic weight	74.9
Melting point (°C)	1935
Boiling point (°C)	
Density (g/cm³)	6.47
Hardness (Mohs — scratch)	6—7

Crystal structure (isometric — normal)	(100) cube
Color (solid)	Green/brown
(pigment)	Cobalt blue
Cleavage (cubic)	(001)

COBALT OXIDE, CoO

General: Does not occur as a natural compound, although there are over 15 cobalt minerals. The most important ores are as sulfides and arsenates. The oxide has no major use as a compound in metal processing, but cobalt is an important additive in the making of steels. The oxide does have application as a coloring agent (cobalt blue) in the glass and ceramics industries, and is used as a hydrogenation catalyst. There are four cobalt oxides, the monoxide being used as the referent here.

Technical Application: The oxides are not used in Solid State processing, but the monoxide has been grown and studied as a single crystal.

Etching: HF and acid mixtures.

COBALT OXIDE ETCHANTS

COO-0001
ETCH NAME: Hydrofluoric acid TIME: 5 min
TYPE: Acid, preferential TEMP: RT
COMPOSITION:
x .... HF
DISCUSSION:

CoO, (100) wafers. Cut and mechanically lap polish with diamond paste. Solution will develop dislocations and sub-boundaries.

REF: Nehring, V W et al — *J Am Ceram Soc,* 14,328(1952)

COO-0002
ETCH NAME: Heat TIME:
TYPE: Thermal, preferential TEMP: 1000 K
COMPOSITION:
x .... Heat
DISCUSSION:

CoO, (100) wafers cleaved and strain annealed for use in a study of magnetic susceptibility for both CoO and NiO. Note that temperature is shown in degrees Kelvin. Pure heat used to develop defects and domain structures.

REF: Singer, J R — *Phys Rev,* 104,929(1956)

COO-0003
ETCH NAME: Oxygen TIME:
TYPE: Gas oxidation TEMP:
COMPOSITION:
x .... O_2
DISCUSSION:

CoO, (100) wafers used in a study of oxidation to Co_3O_4. Prepare specimens by (1) abrading surface with 800-grit SiC; (2) polishing with 0.5 diamond grit in kerosene, and (3) washing in acetone with ultrasonic agitation prior to oxidizing. Process used in a study of oxidation mechanisms.

REF: Przybylski, K & Smeltzer, W W — *J Electrochem Soc,* 128,896(1981)

PHYSICAL PROPERTIES OF COBALT SULFIDE, Co_3S_4

Classification	Sulfide
Atomic numbers	27 & 16
Atomic weight	305.06
Melting point (°C)	>1200
Boiling point (°C)	
Density (g/cm³)	4.85 (4.8—5)
Hardness: (Mohs — scratch)	5.5
Crystal structure (isometric — normal)	(100) cube, fcc
Color (solid)	Steel grey/reddish
Cleavage (octahedral)	(111)

COBALT SULFIDE, Co_3S_4

General: Occurs as the mineral linnaerite, Co_3S_4, often associated with iron and copper ore deposits, and may be an admixture with these as sulfides. Fairly widely distributed as an ore of cobalt with structure analogous to the spinel group. No direct industrial use other than as an ore of cobalt.

Technical Application: Not used in Solid State processing at present, though it is a spinel-type mineral and may have similar applications as an operating device. The reference shown is for Co_9S_8, and there are other cobalt sulfides, such as jaipurite, CoS as a natural mineral mono-sulfide.

Etching: Soluble in acids.

COBALT SULFIDE ETCHANTS

COS-0001

ETCH NAME: Sulfurous acid — TIME:

TYPE: Acid, preferential — TEMP:

COMPOSITION:

x .... H_2SO_3, conc.

DISCUSSION:

Co_9S_8, fabricated as a single crystal sphere 0.28 mm in diameter, used in a study of crystal structure. The solution shown was used to develop defects and structure.

REF: Geller, S — *Acta Metall*, 15,1195(1962)

PHYSICAL PROPERTIES OF COLEMANITE, $Ca_2B_6O_{11}.5H_2O$

Classification	Borate
Atomic numbers	25, 5, 8, 2
Atomic weight	396.4
Melting point (°C)	~1200
Boiling point (°C)	
Density (g/cm³)	2.42
Hardness (Mohs — scratch)	4—4.5
Crystal structure (monoclinic — normal)	(110) prism
Color (solid)	Clear/whitish
Cleavage (b-pinacoidal)	(010)

COLEMANITE, $Ca_2B_6O_{11}.5H_2O$

General: Occurs as a natural mineral associated with borax, and was first discovered in Death Valley, California. It is mined along with borax and has similar medical and industrial applications as a soldering and welding flux; as a washing and cleaning detergent; and as an antiseptic or preservative.

Technical Application: There has been no use in Solid State processing to date, although it has been evaluated for ferroelectric and pyroelectric properties, as both the natural and synthetic crystal.

Etching: Hot HCl.

COLEMANITE ETCHANTS

COL-0001

ETCH NAME: Hydrochloric acid — TIME:

TYPE: Acid, removal — TEMP: Hot

COMPOSITION:

x .... HCl, conc.

DISCUSSION:

$Ca_2B_6O_{11}.5H_2O$, (010) cleaved wafers. Both the natural and artificial compound were studied for ferroelectric and pyroelectric properties. The material cleaves easily on (010), and is soluble in HCl.

REF: Wieler, H H et al — *J Appl Phys,* 33,1720(1962)

PHYSICAL PROPERTIES OF COLUMBIUM, Cb (See NIOBIUM, Nb)

COLUMBIUM, Cb

General: Does not occur as a native metal element. It is considered to be a rare element which commonly is found in pegmatites as the mineral series columbite-tantalite. It is quite widely distributed and varies from nearly pure columbite to pure tantalite depending upon locale. When first discovered it was called columbium but, today, is called niobium.

In industry it has been used as the pure metal and as an alloy constituent in steels for its high temperature, hardness, and chemically inert characteristics. Items shown here are for references published as columbium. See section on Niobium for physical properties and additional solutions and applications.

Technical Application: The metal, as niobium, has some specific device development applications in Solid State device fabrication as the Josephson Junction device, Nb_2Sn_3.

Etching: Hot H_2SO_4, HF and mixed acids.

COLUMBIUM ETCHANTS

CB-0001

ETCH NAME: — TIME:

TYPE: Acid, polish — TEMP:

COMPOSITION:

2 .... HF

4 .... H_2SO_4

2 .... HNO_3

2 .... HF

DISCUSSION:

Cb, single crystal specimens used in a study of anodic oxidation. Solution used to clean and polish surfaces prior to oxidation. Anodic solution used for oxidation was: 0.1% H_3PO_4 at 5.5 mA/cm^2.

REF: Bakish, R — *J Electrochem Soc,* 107,653(1960)

CB-0002a
ETCH NAME: TIME:
TYPE: Electrolytic, polish TEMP: RT
COMPOSITION: ANODE:
90 ml .... H_2SO_4 CATHODE:
10 ml .... HF POWER: 0.1 A/cm^2
DISCUSSION:

Cb, single crystal specimens used in a study of grain boundaries and etch pits. Solution used to clean and polish surfaces prior to preferential etching. Use mild agitation during etching. Solution also used on tantalum.

REF: Micheal, A B & Huegel, F J — *Acta Metall,* 5,339(1957)

CB-0002b
ETCH NAME: TIME:
TYPE: Acid, preferential TEMP: RT
COMPOSITION:
10 ml .... HF
10 ml .... H_2SO_4
10 ml .... H_2O
x drops .... H_2O_2
DISCUSSION:

Cb, single crystal specimens used in a study of grain boundaries and etch pits. Solution used by immersion with agitation to develop sub-boundaries and etch pits. Etch pit density increased with time and pits reduced in size.

REF: Ibid.

CB-0003
ETCH NAME: TIME:
TYPE: Electrolytic, polish TEMP:
COMPOSITION: ANODE:
10 .... HF CATHODE:
90 .... H_2O_4 POWER:
DISCUSSION:

Ta:Cb. single crystal alloys. Five crystals grown with increasing amounts of columbium: 20, 30, 40, 60, and 80%. Material used in a study of alloy strength. Solution used to clean and polish specimens.

REF: Peters, B C & Hendrickson, A A — *Acta Metall,* 14,1121(1966)

PHYSICAL PROPERTIES OF COPPER, Cu

Classification	Transition metal
Atomic number	29
Atomic weight	63.57
Melting point (°C)	1083
Boiling point (°C)	2595
Density (g/cm^3)	8.94 (8.8—8.9)
Thermal conductance (cal/sec)(cm^2)(°C/cm)	0.94
Specific heat (cal/g) 20°C	0.0918
Latent heat of fusion (cal/g)	48.9 (50.6)
Heat of vaporization (k-cal/g-atom)	72.8
Atomic volume (W/D)	7.1

1st ionization energy (k-cal/g-atom)	178.
1st ionization potential (eV)	7.723
Electronegativity (Pauling's)	1.9
Covalent radius (angstroms)	1.17
Ionic radius (angstroms)	0.96 (Cu^{+1})
Coefficient of linear thermal expansion ($\times 10^{-6}$ cm/cm/°C) 20°C	16.42
Electrical conductance (micro ohm^{-1})	0.593
Electrical resistivity ($\times 10^{-6}$ ohms cm) 20°C	1.673
Electron work function (eV)	4.47
Cross section (barns)	3.8
Vapor pressure (°C)	2207
Tensile strength (psi) annealed	30,000
Hardness (Mohs — scratch)	2.5—3
Crystal structure (isometric — normal)	(100) cube, fcc
Color (solid)	Copper-red
(flame)	Emerald green (oxide)
Cleavage (parting, only)	None

COPPER, Cu

General: Occurs as a native element, usually of secondary origin and found in many localities but seldom in commercial quantities. It is found in beds and veins associated with copper compounds, such as malachite, green copper ore, and azurite, blue copper ore, both being hydrated carbonates of copper. The iron pyrites containing copper and sulfur, and the mineral cuprite, Cu_2O, all occur in copper veins. The symbol, Cu, is from the Latin cuprum, which comes from the Greek cyprium (Cyprian bronze) as, in ancient times, the island of Cyprus was noted for its copper mines. It is thought that copper was the first metal employed by man, after gold, as both occur in fair quantity in the free state of southern Europe and the Mediterranean area. The two sulfides chalcopyrite, $CuFeS_2$ and chalcocite, Cu_2S represent about 75% of the copper ore mined in the world.

Pure copper is ductile and malleable, like gold, and an excellent conductor of heat and electricity as are both gold and silver, all first used as body ornaments, plates, and other utensils.

Bronze, Cu:Sn, was the first metal alloy used by man, leading him out of the Stone Age into the Bronze Age about 5000 B.C. Brass, Cu:Zn, is not as brittle as bronze, and artifacts are said to have been discovered in Egyptian tombs, circa 4000 B.C. but, with the discovery of tin in Great Britain — the Tin Isles of ancient Rome — bronze was the major metal alloy until the discovery of iron smelting by the Hittites of Asia Minor around 1350 B.C. Today there are a large number of alloy mixtures of both bronze and brass.

About one third of the copper mined in the world is used for electrical wiring as an alloy to improve hardness. It is fabricated as rods, tubing, sheet, and as bulk material for a wide number of products. As sheeting, it is used in building construction, such as a roof covering or as rain gutters which, with time and exposure to the atmosphere, can turn green — verdigris — a copper sulfate. It also is available as high purity copper (OFHC copper) used in electronics and Solid State processing. Copper is easily electroplated, and many products are so plated. Gold, silver, and copper are known as the "coinage" metals, although nickel is now a lower cost replacement for silver.

Technical Application: In Solid State processing copper is used as an n-type dopant in some compound semiconductors. As a thin film metallization along with other metals, such as gold, used in device construction as a contact pad on semiconductor devices as well as for circuits on substrate dielectrics. It is as multilayer mixture such as Au/Cu; Au/Cu/Ni; Au/Cu/Cr; or as Cu/Au may be a series of alternating layers of the two thin films. Used in a similar manner on metals, such as aluminum, iron, steel, nickel, and molybdenum fabricated as test parts and package elements.

Thin film deposits can be by standard vacuum evaporation, E-beam evaporation, RF sputter from a solid target or by electrolytic plating. The latter method is used in plating PC boards (PCBs) and some of the metal parts used in test and package assembly.

Solid copper parts, as copper alloys or OFHC copper, to include wire or strap contacts, or copper pins are used for both their electrical and heat dissipation characteristics. As a clad material, such as copper/nickel coldweld or resistance weld packages, the copper is the seal metal.

Single crystal copper has been grown, as have several copper alloys, all studies for their general physical characteristics, and trinary compound semiconductors containing copper are under study: $CuInSe_2$; $CuInS_2$; and $CuInTe_2$, as examples.

Etching: Soluble in HNO_3, hot H_2SO_4, slowly in HCl and NH_4OH. $FeCl_3$, mixed acids, and ionized gases.

COPPER ETCHANTS

CU-0001

ETCH NAME: Sodium hydroxide — TIME:

TYPE: Alkali, cleaning — TEMP:

COMPOSITION:

(1) x .... x% NaOH

(2) 2 .... H_2SO_4
1 .... HNO_3

DISCUSSION:

Cu, wire and coupons. Solutions used to clean copper wire prior to plating. First, clean in solution (1), and DI water wash; follow with (2) acid mixture, and DI water rinse. In plating, initially deposit copper to 0.001″ before nickel plating.

REF: Bailey, G C & Ehitich, A C — *J Appl Phys,* 50,453(1979)

CU-0002

ETCH NAME: Nitric acid — TIME: 1 min

TYPE: Acid, removal — TEMP: RT

COMPOSITION:

x .... HNO_3, conc.

DISCUSSION:

Cu, single crystal ingots cut as cylinder shaped specimens. Nitric acid was used to remove cutting work damage from surfaces.

REF: Reshon, D D Jr — *J Appl Phys,* 35,1262(1964)

CU-0003a

ETCH NAME: Nitric acid, dilute — TIME: 1—10 min

TYPE: Acid, cleaning — TEMP: RT

COMPOSITION:

1 .... HNO_3
20 .... H_2O

DISCUSSION:

Cu piping. Solution used for cleaning piping on 2000 gallon LN_2 and LOX transport trailers. Included steel, brass, and bronze parts. Parts were soaked to remove general contamination, heavily water washed and blown dry with nitrogen. Where parts were stored or prepared for shipment they were sealed with warm nitrogen filled plastic bags.

REF: Walker, P & Campos, R — personal application, 1962

CU-0003b
ETCH NAME: Hydrochloric acid, dilute TIME: 1—10 min
TYPE: Acid, cleaning TEMP: RT
COMPOSITION:
1 .... HCl
10 .... H_2O
DISCUSSION:
Cu piping on cryogenic transport tailers. Solution used for general cleaning of steel, bronze, and brass parts. See discussion under CU-0003a.
REF: Ibid.

CU-0008a
ETCH NAME: Nitric acid, dilute TIME:
TYPE: Acid, cleaning TEMP:
COMPOSITION:
1 .... HNO_3
1 .... H_2O
DISCUSSION:
Cu and Cu alloys. Solution used to remove copper oxide from parts prior to etch polishing with a Copper Brite Dip solution (CU-0028).
REF: Bulletin B-11D, The American Brass Co.

CU-0004a
ETCH NAME: Hydrochloric acid, dilute TIME:
TYPE: Acid, cleaning TEMP:
COMPOSITION:
1 .... HCl
1 .... H_2O
DISCUSSION:
Cu and Cu alloys. Pure copper was annealed and cold drawn. Specimens were used in a study of low temperature transport properties. Solution used to clean specimens.
REF: Powell, R L et al — *Phys Rev,* 115,314 (1959)

CU-0010
ETCH NAME: Nitric acid, dilute TIME:
TYPE: Acid, removal TEMP:
COMPOSITION:
x .... HNO_3
x .... H_2O
DISCUSSION:
Cu single crystal specimens used in irradiation effect studies. Solution was used as a general removal etch.
REF: Vook, R & Wert, C — *Phys Rev,* 109,1529(1958)
CU-0009a: Hurdon, M J & Averbach, B L — *Acta Metall,* 9,237(1961)
Cu single crystal specimens. Solution referred to as a ''heavy etch'' for copper and used to clean surface prior to studying dislocation density in deformed specimens. Ferric chloride was used to develop dislocations and other structures after deformation.
CU-0089: Magnuson, G D & Carlston, C E — *J Appl Phys,* 34,3267(1963)
Cu single crystal wafers used in an argon bombardment damage study. A weak solution was used to remove lap damage prior 10 KeV Ar^+ ion sputter.

CU-0011a
ETCH NAME: Brilliant etch TIME: Dip
TYPE: Acid, macro-etch TEMP: RT
COMPOSITION:
50 ml .... HNO_3
0.5 ml .... $AgNO_3$
50 ml .... H_2O
DISCUSSION:
Cu and Cu alloys. Solution used to develop macro-structures on metallographic specimens.
REF: A/B Metals Digest, 21 #2,23(1983) — Buehler Ltd

CU-0012
ETCH NAME: Silica TIME: Variable
TYPE: Mineral, cleaning TEMP: RT
COMPOSITION:
x .... SiO_2, sand
x .... H_2O
DISCUSSION:
Cu parts as sheet, pans and piping. For general removal of heavy contamination, such as oxidation and dirt from copper and copper alloy surfaces. Method shown has been used for hand scouring of parts. As a dry abrasive using nitrogen for a pressure jet, used for surface cleaning and removal of other metal plating on copper parts, such as from power electrodes in vacuum systems. Also used on other metal parts (steels, stainless steel, nickel) of vacuum systems to produce a rough surface for better adhesion of evaporating metals and reduction of chromium dusting during evaporation on glass photo resist masks.
REF: Tarn, W H — personal communication, 1975

CU-0013a
ETCH NAME: Ferric chloride TIME:
TYPE: Salt, removal/preferential TEMP:
COMPOSITION:
x .... $FeCl_3$, sat. sol.
DISCUSSION:
Cu single crystal spheres used in a study of surface plane development. Parts were first cleaned in dilute nitric acid before etching in this solution.
REF: Economou, N A & Trivich, D — *Electrochem Acta,* 3,292(1961)
CU-0009b: Ibid.
Cu single crystal specimens. A ferric chloride solution was used to develop dislocations and other structure after specimens were subjected to deformation.

CU-0014
ETCH NAME: Ferric chloride, dilute TIME:
TYPE: Salt, removal/preferential TEMP: Hot
COMPOSITION: ASTM #100
(1) 300 g .... $FeCl_3$ (2) 90 g .... $FeCl_3$
1000 ml .... H_2O 200 ml .... H_2O
DISCUSSION:
Cu and Cu alloy parts. Both solutions used as general removal etchants on copper parts.
REF: Mackliet, C A — *Phys Rev,* 109,1964(1958)
CU-0009c: Ibid.

Cu single crystal specimens used in a study of dislocation density after specimen deformation. A dilute ferric chloride solution was used to develop dislocations and other structures. Also studied aluminum single crystals to include X-ray study for bulk deformation observations.
CU-0080a: ASTM E407-70
Cu, specimens. Standard ASTM #100 solution is 10 g $FeCl_3$:90 ml H_2O.

CU-0015
ETCH NAME: TIME:
TYPE: Acid, dislocation TEMP:
COMPOSITION:
1 .... 1 *M* $FeCl_3.H_2O$
1 .... 1 *M* HCl
1 .... 0.25 *M* HBr
DISCUSSION:
Cu single crystal wafers of various orientations. Solution used to etch dislocations and structures in single crystal copper surfaces prior to irradiation studies. No change observed in dislocations after irradiation, but initial background terrace structure changed to a finely pitted surface. Required several annealing periods, with etching after each anneal, to remove surface pits.
REF: Young, F W Jr — *J Appl Phys,* 33,749(1962)

CU-0003c
ETCH NAME: TIME:
TYPE: Acid, cleaning TEMP:
COMPOSITION:
6 .... H_2SO_4
12 .... CrO_3
82 .... H_2O
DISCUSSION:
Cu cryogenic trailer piping, including brass, bronze and stainless steels. Solution used as a ''Brite Copper'' dip. See Cu-0028 for other similar solutions.
REF: Ibid.

CU-0043
ETCH NAME: TIME: 5—10 min
TYPE: Acid, clean/condition TEMP: RT
COMPOSITION:
5 .... $FeCl_3$ (40%)
10 .... HNO_3
82 .... H_2O
DISCUSSION:
Cu and Cu alloys as sheet or foil were cleaned in this solution prior to being plastic laminated. Metal surface cleanliness is essential to obtain a good plastic/metal adhesion seal. This solution gave the best adhesion results. Several types of plastic sheet were evaluated and also were applied to aluminum.
REF: Skeist, I — *Epoxy Resins,* Reinhold, New York, 1950, 190

CU-0018
ETCH NAME: TIME:
TYPE: Acid, preferential TEMP:
COMPOSITION:
x .... H_2SO_4
x .... $Cu(NO_2)_3$
DISCUSSION:
Cu single crystal wafers of different orientations and spheres. Various solution concentrations were evaluated as the pH of solutions affect characteristics of different crystal planes.
REF: Gwathmey, A T et al — *Electrochem Soc Meet, WDC*, 12—16 May, 1957

CU-0019a
ETCH NAME: Copper sulfate, dilute TIME:
TYPE: Salt, preferential TEMP:
COMPOSITION:
x .... $CuSO_4$
x .... H_2O
DISCUSSION:
Cu OFHC copper specimens. Solution used to develop structure caused by fatigue cracking.
REF: Bendler, H A — *Acta Metall*, 8,402(1960)

CU-0020
ETCH NAME: TIME:
TYPE: Acid, preferential TEMP:
COMPOSITION:
x .... $CuSO_4$
x .... H_2SO_3
DISCUSSION:
Cu single crystal wafers of different orientations and spheres were etched in various concentrations of this solution in a study of crystallographic face orientation.
REF: Gwathmey, A T & Benton, A F — *J Chem Phys*, 8,431(1940)

CU-0011b
ETCH NAME: General etch TIME: Swab
TYPE: Acid, micro-etch TEMP: RT
COMPOSITION:
50 ml .... NH_4OH
50 ml .... H_2O_2
DISCUSSION:
Cu and Cu alloy specimens. Metallographic specimen surfaces are lightly scrubbed with a swab soaked in the solution shown to develop microstructure in material studies.
REF: A/B Metals Digest, 21 12,23(1983) — Buehler Ltd

CU-0022a
ETCH NAME: Nitric acid, dilute TIME:
TYPE: Acid, polish TEMP:
COMPOSITION:
x .... HNO_3
x .... H_2O
DISCUSSION:
Cu single crystal specimens as spark-cut rods 6.5 cm long. Degrease, thermally anneal,

then electropolish in phosphoric acid. After subjecting samples to strain, chemically polish in the solution shown. This solution produces a high polish on (111) surfaces and, although it does attack other crystal planes, the (111) is well defined. After polishing specimens they were preferentially etched to develop dislocations and structures caused by strain.
REF: Basinski, Z S & Basinski, S J — *Phil Mag,* 9,51(1964)
CU-0023: Magnuson, G D et al — *J Appl Phys,* 32,369(1961)

Cu single crystal spheres with a (110) orientation cut flats. Solution used to polish spheres prior to Hg^+ ion bombardment studies. During bombardment approximately 20 μm of material was removed and redeposited on the surfaces in hillock form.

CU-0024a
ETCH NAME: TIME: 30 sec
TYPE: Acid, dislocation TEMP: RT
COMPOSITION:

(A) 1 .... Br_2 (B) 45 .... HCl
30 .... HAc 250 .... H_2O

DISCUSSION:

Cu, (111), within 2—3° orientation. Mix the two solutions together when ready to use. Solution will develop triangular surface etch pits that increase in size linearly with etch time. (*Note:* Surface damage etch pits increase in size; but dislocation etch pits normally do not.)
REF: Livingston, J D — *J Appl Phys,* 3,1071(1960)

CU-0025
ETCH NAME: TIME: 5—10 min
TYPE: Acid, dislocation TEMP: RT
COMPOSITION:

20 ml .... HCl
1 ml .... HAc
4 g .... $FeCl_3.6H_2O$
8 drops .... Br_2
150 ml .... H_2O

DISCUSSION:

Cu single crystal wafers of various orientations. Specimens were point damaged using a sapphire stylus and then deformed by high temperature annealing prior to etching in the solution shown to develop dislocations and stress structures.
REF: Bailey, J M & Gwathmey, A T — *J Appl Phys,* 31,215(1960)

CU-0026
ETCH NAME: TIME:
TYPE: Acid, polish TEMP: RT
COMPOSITION:

75 ml .... $FeCl_3$, sat. sol.
25 ml .... HCl

DISCUSSION:

Cu, (100) wafers. Solution has been used as both a polish etch, as well as a preferential type etch to develop internal (010) plane directions in a (100) surface.
REF: Burns, J — *Phys Rev,* 119,102(1960)
CU-0010b: Ibid.

Cu single crystal wafers. Solution used as a polish etch.

CU-0008b
ETCH NAME: Sulfuric acid, dilute TIME:
TYPE: Acid, oxide removal TEMP:
COMPOSITION:
x .... H_2SO_4 (12—15%)
DISCUSSION:
Cu and Cu alloys. Solution used as a pickling solution for the removal of scale and oxides from copper surfaces.
REF: Ibid.

CU-0008c
ETCH NAME: Dichromate finish dip TIME:
TYPE: Acid, polish TEMP:
COMPOSITION:
12 oz .... H_2SO_4
4 oz .... $NaCrO_4$
1 gal .... H_2O
DISCUSSION:
Cu and Cu alloys. Solution will remove heavy oxides and leave a matte type surface finish.
REF: Ibid.

CU-0027
ETCH NAME: TIME:
TYPE: Acid, removal TEMP:
COMPOSITION:
(1) 1 .... HNO_3
20 .... H_2O_2
(2) 40 .... NH_3
10 .... H_2O_2
DISCUSSION:
Cu and Cu alloys. Both solutions shown as general removal/polish etchants for copper.
REF: Berglund, T & Dearden, WH — *Metallographer's Handbook of Etching,* Pitman & Sons, London, 1931

CU-0005a
ETCH NAME: Phosphoric acid TIME:
TYPE: Electrolytic, polish TEMP: 70—178°C
COMPOSITION: ANODE:
x .... H_3PO_4 CATHODE:
POWER:
DISCUSSION:
Cu single crystal spheres used in a study of oxidation reactions. Electropolish in solution shown; chemical clean in $xHNO_3$:EOH; and H^+ ion clean in vacuum prior to oxidation.
REF: Young, F W Jr et al — *Acta Metall,* 4,145(1956)
CU-0006: Young, F W Jr — *J Appl Phys,* 32,192(1961)
Cu single crystal waters and spheres used in a wide range study of the effect of different etch solutions on development of dislocation etch pits.
CU-0007: Young, F W Jr — *Bull Am Phys Soc,* 5,190(1960)
Cu single crystal wafers and spheres used in a study of dislocation variation in structure as related to (111), (110) and (100) surface planes. Spheres contained oriented flats.

CU-0028

ETCH NAME: Copper Brite dip — TIME: Seconds

TYPE: Acid, polish — TEMP: RT

COMPOSITION:

(1)		(2)	(3)	(4)	(5)	(6)	(7)
3	H_2SO_4	3	7	5	600 ml	2 gal	250 ml
1	HNO_3	1	7	3	250 ml	1 gal	125 ml
0	HCl	0	0	5	20 ml	$^1/_2$ fl oz	4 ml
1	H_2O	10	1	2	130 ml	1—2 qt	250 ml

DISCUSSION:

Cu specimens and parts. The following cleaning procedure is recommended: (1) Degrease: (i) hot TCE — DI H_2O rinse — MeOH rinse — N_2 blow dry, then (2) NH_4OH dip at RT to remove native oxide — DI H_2O rinse — MeOH rinse, and N_2 blow dry. All of the solutions are extremely rapid — dip in solution — immediate water quench and rinse. Produces a highly reflective, polished surface that may not be planar due to rapidity of attack. Not recommended where critical dimensional tolerances are required. Solution (1) used to clean copper vacuum system parts and copper electrical test holders for microelectronic devices. Also to clean copper plated PC boards (PCBs).

REF: (1) Fahr, F — personal communication, 1979

CU-0029a: (2) Walker, P & Fahr, F — personal application, 1979—1980

Cu, OFHC copper parts used as microelectronic device test carriers. Solution used to remove residual "burn" marks from machine cutting. Solution was too rapid and erratic to maintain critical dimensions. Hydrogen firing at 300 to 400°C was more satisfactory. Parts cleaned prior to Au/Ni plating.

CU-0030: (3) Westinghouse Data Sheet (date, number unknown) Cu specimens and parts. Solution applied for general polishing.

CU-0031: (4) *ASTM Metals Handbook #2,* 8th ed, 1974 Cu specimens and parts. Solution applied as a general polishing etch.

CU-0032: (5) Kohl, W H — *Handbook of Materials and Techniques for Vacuum Devices,* Reinhold, New York, 1967 Cu specimens parts. Applied as a general polishing solution.

CU-0008e: (6) Bulletin B-11D, the American Brass Company (no date) Cu specimens and parts. Described as a general polishing solution for copper and copper alloys.

CU-0087: (7) Chen, H Y — *J Electrochem Soc,* 118,681(1971)

Cu, OFHC copper discs used as cathodes for electrolytic plating of gold from gel electrolytes. Degrease copper parts with TCE, then acetone and Brite Dip 10 sec, RT. Two cellulose compounds used for gelling in both acidic and cyanide gold solutions. 0.5 μm thick deposits evaluated. Deposit rates of gold were (1) at 0.15 mA/in^2, 2 μm/h, and (2) at 0.15 mA/in^2, 0.6 μm/h. Electrographic porocity meter and pore printing used to evaluate deposits — pore print at 8 V and 800 lb/in^2 pressure.

CU-0029b: Ibid.

Cu, OFHC copper parts, after Au/Ni plating (Au 2000 Å:Ni 500 Å). Solution (1) used at 8°C (ice/acetone) to evaluate completeness of metallization and/or pinholing, grain structure, and grain boundary diffusion of liquids.

CU-0008d
ETCH NAME: Matte dip
TYPE: Acid, finishing
COMPOSITION:
2 .... H_2SO_4
1 .... HNO_3
x .... *ZnO or ZnS

TIME:
TEMP:

*Saturate the solution

DISCUSSION:
Cu and Cu alloy parts. For decorative purposes the solution will produce a matte type surface finish with dull reflectivity and beaded structure.
REF: Ibid.

CU-0032
ETCH NAME: Ammonium persulfate
TYPE: Salt, cleaning
COMPOSITION:
x .... 2—5% $(NH_4)_2S_2O_8$

TIME: 15 min
TEMP: RT

DISCUSSION:
Cu, (111) single crystal blanks and other orientations. Solution used to clean surfaces prior to oxidation experiments (Cu_2O).
REF: Baur, J P et al — *Acta Metall,* 103,273(1956)
CU-0019c: Ibid.
Cu, OFHC copper single crystal specimens used in a study of the formation of fatigue cracks using an electron microscope (EM) for observation. An ammonium sulfate solution was used to develop cracks.

CU-0024b
ETCH NAME:
TYPE: Electrolytic, polish
COMPOSITION:
60 .... H_3PO_4
40 .... H_2O

TIME:
TEMP:
ANODE:
CATHODE:
POWER:

DISCUSSION:
Cu, (111) wafers used in a study of etch pits associated with dislocations. Solution shown was used to electropolish specimens before preferential etching.
REF: Livingston, J D — *J Appl Phys,* 31,1071(1960)

CU-0024c
ETCH NAME:
TYPE: Acid, preferential
COMPOSITION:
45 .... HCl
30 .... HAc
250 .... H_2O
1 .... Br_2

TIME: 30 sec
TEMP: RT

DISCUSSION:
Cu, (111) wafers. Solution used to develop etch pits associated with dislocations. Mix the Br_2 + HAc, first; then add to the HCl:H_2O. Pits observed contained flat bottoms and

varied from small to large in size as sharply defined triangles. Solution referred to as a modification of a Lovell & Wernick etch.
REF: Ibid.

CU-0069
ETCH NAME: TIME:
TYPE: Acid, cleaning TEMP:
COMPOSITION:
x .... HNO_3
x .... H_2O
DISCUSSION:
Cu single crystal specimens used in a study of recovery in electron irradiated copper. Dilute nitric acid was used to clean and wash surfaces.
REF: Corbett J W et al — *Phys Rev,* 114,1452(1959)
CU-0019b0: Bendler, H A — *Acta Metall,* 8,402(1960)
Cu single crystals of OFHC copper used in an electron microscope study of fatigue cracks and their formation. A dilute nitric acid solution was used as a crack selective etchant. Also used ammonium persulfate solutions.

CU-0061a
ETCH NAME: Nitric acid TIME:
TYPE: Electrolytic, thin/polish TEMP: $-20°C$ or less
COMPOSITION: ANODE: Cu
x .... HNO_3 (33%) CATHODE:
POWER:
DISCUSSION:
Cu foil. Solution used was a double-jet to polish and thin copper for TEM study. Copper specimens were also electroless nickel plated; palladium and platinum immersion plated; and evaporated with gold in a study of thin films and deposition reactions. Final coating of all metals with electroless nickel also was done as part of a thin film adhesion study. Nickel plates more rapidly on Pd than unplated Cu.
REF: Flis, J & DuOette, J — *J Electrochem Soc,* 131,254(1984)

CU-0022b
ETCH NAME: Phosphoric acid TIME:
TYPE: Electrolytic, polish TEMP:
COMPOSITION: ANODE:
x .... H_3PO_4 CATHODE:
POWER:
DISCUSSION:
Cu single crystal specimens. First, spark-cut as rods, then etch polish in the solution shown prior to thermal annealing to reduce dislocation content.
REF: Ibid.
CU-0074a Inman, M C & Barr, L W — *Acta Metall,* 8,112(1960)
Cu single crystals were cut as cylinders and electropolished in the solution shown in a study of antimony diffusion into copper.
CU0024d: Ibid.
Cu, (111) wafers cut within 2—3° of plane. Wafers used in a dislocation study. First, etch polish in solution shown before dislocation etching (CU-0024a).

CU-0061b
ETCH NAME:
TYPE: Electrolytic, polish
COMPOSITION:
x .... HNO_3
x .... MeOH
TIME:
TEMP: 0°C
ANODE: Cu
CATHODE:
POWER: 6 V
DISCUSSION:

Cu, foil. See discussion under CU-0061a. Solution used to electropolish copper before various plating sequences in a study of nickel adhesion.

REF: Ibid.

CU-0062a
ETCH NAME: Toluene
TYPE: Ester, cleaning
COMPOSITION:
x .... $C_6H_5CH_3$
TIME:
TEMP:
DISCUSSION:

Cu specimens. Solution used in a study of surface cleaning. Toluene is excellent for removal of fatty acids, such as palmitic, stearic and oleic acids from copper and gold surfaces. Authors say that less than one monolayer of contamination remains. Fatty acids were ^{51}C isotope tagged.

REF: Fatzer, G D et al — *Int J Appl Radiat Isot,* 10,167(1961)

CU-0063
ETCH NAME: Ethyl alcohol
TYPE: Alcohol, cleaning
COMPOSITION:
x .... EOH
TIME:
TEMP:
DISCUSSION:

Cu, blanks as plated substrates used for a metallization, stress and failure study. First, ultrasonically clean in EOH, then Ar^+ ion sputter clean in vacuum prior to CVD deposition of Ni, Ti and Al_2O_3 with copper substrates at 200°C, and increased to 350°C. Metallizations were (1) Al_2O_3/Cu; (2) Al_2O_3/Ti/Cu, and (3) Al_2O_3/Ti/Ni/Cu. Tensile strain up to 10% was observed. After metallization, temperature cycle increased to 600°C. Alumina showed parallel cracks after deposition and was amorphous, a-Al_2O_3, below 400—450°C. Without a metal interface between alumina and copper, alumina peels from copper.

REF: Jarvinen R et al — *Thin Solid Films,* 114,311(1984)

CU-0064: Walker, P & Velardi, N — personal application, 1985

Cu, OFHC copper parts. After gold plating of parts and DI water washing, final rinse with acetone, then either methanol or ethanol and nitrogen blow dry. Used for final cleaning and drying of parts.

CU-0065a
ETCH NAME: Trichloroethylene
TYPE: Solvent, cleaning
COMPOSITION:
x .... TCE
TIME: Minutes
TEMP: 75—80°C
DISCUSSION:

Cu, OFHC parts, semiconductor packages, and other metals and compounds. TCE is a general degreasing solvent used as a liquid, hot or cold, with or without ultrasonic agitation, as a spray or hot vapor. The latter as vapor degreasers. Vapor degreaser systems can be from a beaker on a hot plate to large tanks with overhead cranes for handling large parts, and operate with a combination of hot liquid (in an ultrasonic tank), hot vapor head above working tank, and a mobile spray nozzle. Where the part is cleaned in a vapor degreaser, the part will dry as it is slowly pulled from the hot vapor head. Where used as a hot or cold liquid, DI water wash after degreasing, then rinse in acetone, then MeOH or EOH, and nitrogen blow dry. TCE is considered carcinogenic and is being replaced by 1-1-1 trichloroethane, TCA. TCE also has been used for gross leak testing of hermetically sealed packages.

REF: Walker, P — personal application, 1970—1985

CU-0065b

ETCH NAME: Freon TA — TIME: Minutes

TYPE: Solvent, cleaning — TEMP: 75 to 85°C

COMPOSITION:

x .... Freon TA

DISCUSSION:

Cu, OFHC copper parts, semiconductors and other metals or compounds. Solution used in a similar manner as that described for TCE (CU-0065a). There are several Freon mixtures with various alcohol additives to include aziotropes, such as Freon TF, Freon TM, etc. Freon is a Dupont registered trade name, and other manufacturers have similar solutions with their own trade names, such as the Genusolves (Allied Chemical), etc. Freons evaporate completely, leaving no residual or contamination from the solvent on cleaned surfaces, such that they are of major use in the cleaning of semiconductor and assembly parts. Freons are also used as special mixtures for gross leak testing of hermetically sealed packages as, if there is a leak, the solution completely evaporates; whereas TCE and similar solvents so used can leave a residue within the package.

REF: Ibid.

CU-0033

ETCH NAME: — TIME: 3—10 min

TYPE: Acid, polish/clean — TEMP: RT

COMPOSITION:

20 g .... $2NH_4NO_3.Ce(NO_3)_3.4H_2O$

10 ml .... HNO_3

150 ml .... H_2O

DISCUSSION:

Cu, wire and OFHC copper parts. This solution is similar to commercial "chrome etch" formulations and was developed for use on chromium evaporated on soda-lime glass as photo resist masks, but it can be used on both nickel as well as copper thin films and parts. The solution is a very slow polishing and cleaning etch for copper and useful where part dimensions are critical. By varying the nitric acid content, removal rates can be closely controlled on the angstroms/minute level.

REF: Walker, P — personal development and application, 1975

CU-0067

ETCH NAME: — TIME:

TYPE: Acid, preferential — TEMP:

COMPOSITION:

1 .... HNO_3

10 .... H_2O

DISCUSSION:

Cu, bicrystals used in a study of grain boundary diffusion. Specimens wet cut and lapped with 600-grit alumina and then etched in the solution shown to develop grain boundary structure.

REF: Austin, A E & Richard, N A — *J Appl Phys,* 32,1462(1961)

CU-0034

ETCH NAME: Nitric acid — TIME: 1—10 min

TYPE: Acid, cleaning — TEMP: RT

COMPOSITION:

(1) x .... HNO_3

(2) 1 .... HNO_3
25 .... H_2O

DISCUSSION:

Copper polycrystalline parts. Solutions used in cleaning cryogenic piping and parts of 200 gallon LN_2 transport trailers. Also used as a general cleaner for copper and copper alloys.

REF: Campo, F & Walker, P — cryogenic parts cleaning, 1962

CU-0035

ETCH NAME: Potassium cyanide — TIME:

TYPE: Acid, removal — TEMP:

COMPOSITION:

x .... 20% KCN

DISCUSSION:

Cu, copper thin films. Solution used to remove copper from Ge, Si, and GaAs surfaces.

REF: Valardi, N — personal communication, 1984

CU-0036a

ETCH NAME: Nitric acid, dilute — TIME: 5 min

TYPE: Acid, polish — TEMP: RT

COMPOSITION:

1 .... HNO_3
1 .... H_2O

DISCUSSION:

Cu single crystal specimens used in a study of plastic flow at 90 and 170°C. Solution used as a polishing etch.

REF: Conrad, H — *Acta Metall,* 6,338(1958)

CU-0036b

ETCH NAME: — TIME: 45 min

TYPE: Electrolytic, polish — TEMP: – 10 to 0°C

COMPOSITION: — ANODE:

1 .... HNO_3 — CATHODE:
2 .... MeOH — POWER:

DISCUSSION:

Cu single crystal specimens. After etching rinse in (1) tap water, (2) DI water, (3) acetone rinse, and (4) air dry.

REF: Ibid.

CU-0037
ETCH NAME: TIME:
TYPE: Acid, sizing TEMP:
COMPOSITION:
x .... 30% HNO_3
DISCUSSION:
Cu single crystal specimens used in a study of surface effects on plastic properties. Solution used to etch specimens to required dimensions. (*Note:* This use of the term "sizing" reflects a physical action, not the use of chemicals as size in reducing ink running in paper.)
REF: Rosi, F D — *Acta Metall,* 5,349(1957)

CU-0038a
ETCH NAME: TIME:
TYPE: Electrolytic, polish TEMP:
COMPOSITION: ANODE:
x .... H_3PO_4 CATHODE:
POWER:
DISCUSSION:
Cu single crystal spheres used in a study of oxidation on smooth convex surfaces. Spheres were polished after cutting with the solution shown, then etched to remove oxide before controlled oxidation.
REF: Harris, W W et al — *Acta Metall,* 5,574(1957)

CU-0038b
ETCH NAME: TIME: 5—10 min
TYPE: Electrolytic, oxide removal TEMP:
COMPOSITION: ANODE:
x .... KCl, sat. sol. CATHODE:
POWER: 30—50 mA/cm^3 & 5—10 V
DISCUSSION:
Cu single crystal spheres. Solution used to remove native oxide left after electropolishing in phosphoric acid before controlled oxidation.
REF: Ibid.
CU-0039: Evans, U R & Stockdale, J — *J Chem Soc,* 2,651(1929)
Reference for etchants shown in CU-0038a and CU-0038b.
CU-0040: Phelps, R T et al — *Ind Eng Chem Anal Ed,* 181,391(1948)
Cu specimens. Used solution to remove native oxide from parts.
CU-0041a: Greenfield, I G & Wildorf, H G F — *J Appl Phys,* 32,827(1961)
Cu specimens. Used solution to clean copper surfaces of residual oxide.

CU-0041a
ETCH NAME: Phosphoric acid, dilute TIME:
TYPE: Electrolytic, polish TEMP:
COMPOSITION: ANODE:
1 .... H_3PO_4 CATHODE:
1 .... H_2O POWER:
DISCUSSION:
Cu single crystal specimens used in a study of the effects of neutron irradiation on plastic deformation. Solution used to polish specimens before irradiation.
REF: Greenfield, I & Wilsdorf, H G F — *J Appl Phys,* 32,827(1961)

CU-0024e
ETCH NAME: Phosphoric acid, dilute
TYPE: Electrolytic, polish
COMPOSITION:
60 .... H_3PO_4
40 .... H_2O
TIME:
TEMP:
ANODE:
CATHODE:
POWER:
DISCUSSION:

Cu, (111) wafers used in an etch pit and dislocation study. Solution used to polish surfaces prior to preferential etching (See CU-0024a).

REF: Ibid.

CU-0044: Moser, R T & Whitmore, J — *J Phys Chem Solids,* 31,115(1960)

Cu single crystal hemispheres grown or machined from a rod. Specimens were thermally etched at high temperature in vacuum to develop facets, spirals, and etch pits. Solution shown used to polish specimens prior to thermal treatment.

CU-0042
ETCH NAME: Neon
TYPE: Ionized gas, preferential
COMPOSITION:
x .... Ne^+ ions
TIME:
TEMP:
GAS FLOW:
PRESSURE:
POWER:
DISCUSSION:

Cu single crystal specimens. Neon ion bombardment used to develop structure and orientation figures. Other metals studied were Al, Bi, Cd, Co, Mg, Sn, and Zn.

REF: Yurasova, V E — *Kristallografiya,* 2,770(1957)

CU-0024g
ETCH NAME:
TYPE: Acid, dislocation
COMPOSITION:
45 .... HCl
30 .... HAc
250 .... H_2O
1 .... Br_2
TIME: 30 sec
TEMP: RT
DISCUSSION:

Cu, (111) single crystal wafers within 2—3° of orientation. Solution develops triangular etch pits and pits increase in size, linearly, with etch time. (See CU-0043a.)

REF: Ibid.

CU-0045
ETCH NAME:
TYPE: Acid, polish
COMPOSITION:
1 .... H_3PO_4
1 .... HNO_3
1 .... HAc
TIME:
TEMP: 70°C
DISCUSSION:

Cu single crystal specimens. Etch polish in the solution shown. Specimens used in a study of ductile fracture.

REF: Saimoto, S et al — *Phil Mag,* 12,319(1965)

CU-0046: Ebeling, R & Ashby, M F — *Phil Mag,* 13,437(1966)

Cu single crystal specimens. Etch polished as shown above. Specimens used in a study of dispersion hardening.

CU-0047
ETCH NAME: TIME: 1—2 min
TYPE: Acid, removal TEMP: 60—70°C
COMPOSITION:
33 ml .... H_3PO_4
33 ml .... HNO_3
33 ml .... HAc
DISCUSSION:

Cu specimens used in an etching study. The solution will remove copper and copper oxide, and is particularly good for oxide removal.

REF: de Jong, J J — *Metalen,* 9,2,(1954) (in Dutch)

CU-0073: Tegart, WJ McG — *The Electrolytic and Chemical Etching of Metals,* Pergamon Press, England, 1956

Cu specimens. Solution described as a polish etch for copper, and a removal etch for Cu_2O oxide.

CU-0048
ETCH NAME: TIME:
TYPE: Acid, oxidation TEMP:
COMPOSITION:
x .... $CuSO_4$
\+ .... O_2
DISCUSSION:

Cu single crystal sphere ground to shape and electropolished with (111), (100), (110), (311) and (210) flats lapped and polished. Mixture shown was used to oxidize surfaces in a study of growth spirals in Cu_2O.

REF: Miller, G T & Lawless, K R — *J Appl Phys,* 29,863(1958)

CU-0011c
ETCH NAME: ALN3-1 TIME: Dip
TYPE: Acid, macro-, micro-etch TEMP: RT
COMPOSITION:
5 ml .... HNO_3
5 ml .... H_2SO_4
4 g .... CrO_3
1 g .... NH_4Cl
90 ml .... H_2O
DISCUSSION:

Cu specimens and copper alloys. Solution is both a macro- or micro-etch depending upon type alloy and used in a material structure study.

REF: Ibid.

CU-0011d
ETCH NAME: Geard #1 TIME: Swab
TYPE: Acid, preferential TEMP: RT
COMPOSITION:
5 ml .... HCl
20 g .... $FeCl_3$
100 ml .... H_2O

DISCUSSION:

Cu, specimens and copper alloys. A general micro- or macro-etch on most alloys of copper. Apply by swabbing the surface.

REF: Ibid.

CU-0069

ETCH NAME: Smog — TIME:

TYPE: Gas, corrosion — TEMP:

COMPOSITION:

x .... Air (+ SO_2, H_2S, O_3, HCl, Cl_2, vapor)

x .... H_2O, vapor (as RH)

DISCUSSION:

Cu specimens used in a study of atmospheric corrosion using an artificially developed smog with variable water vapor as humidity factor. Also studied silver with same smog atmosphere, but in dry air.

REF: Rice, D W et al — *J Electrochem Soc,* 128,275(1981)

CU-0050b

ETCH NAME: — TIME:

TYPE: Electrolytic, polish — TEMP:

COMPOSITION: — ANODE:

x .... HNO_3 — CATHODE:

x .... EOH — POWER:

DISCUSSION:

Cu single crystal spheres. Solution used as a polish etch. See CU-0050a for discussion.

REF: Ibid.

CU-0051

ETCH NAME: — TIME:

TYPE: Thermal, forming — TEMP:

COMPOSITION:

x .... heat

DISCUSSION:

Cu single crystal spheres formed by asymmetric cooling of copper on a tungsten ribbon in high vacuum. Used in an oxidation study of tungsten carbide growth on a copper sphere.

REF: Menael-Kopp, C — *Z Naturforsch,* 122,1003(1957)

CU-0060

ETCH NAME: Ammonium peroxydisulfate — TIME: Variable

TYPE: Salt, polish — TEMP: 45°C

COMPOSITION:

x .... x% $(NH_4)_2S_2O_8$

+ .... $HgCl_2/PdCl_2$

DISCUSSION:

Cu specimens used in a study of the effects of the solution shown. The addition of metal salts will increase the removal rate. (*Note:* Solution also called ammonium persulfate.)

REF: Schlabach, T D & Diggery, B A — *Electrochem Technol,* 2,118(1964)

CU-0061
ETCH NAME:
TYPE: Electrolytic, polish
COMPOSITION:
600 ml .... H_3PO_4
12.9—19 g .... CuO
400 ml .... H_2O

TIME: 15—40 h
TEMP:
ANODE: Cu specimen
CATHODE: Cu sheet
POWER: 1.0—1.1 V
RATE: 0.01—0.2 µm/min

DISCUSSION:
Cu specimens. First, dip specimens in HNO_3, rinse in water and while still wet introduce into solution shown above. Hold specimen surface in a horizontal position during etching. After etching, with potential still applied, rinse in H_3PO_4, 30 sec. With 12.5 g CuO and high purity copper specimens gave the best results. There were some parallel ripples on surfaces less than 100 Å in height.
REF: Powers, R W — *Electrochem Technol,* 2,274(1964)

CU-0068
ETCH NAME: Chlorine
TYPE: Gas, removal
COMPOSITION:
x .... Cl_2, vapor

TIME:
TEMP: See discussion

DISCUSSION:
Cu, (100) single crystal wafers cut within 3° of orientation, with some twinning observed, were used in a study of the surface reactions with chlorine gas. Three general reactions were observed at different temperature levels: (1) less than 150°C, CuCl growth; (2) 150—580°C, Cu_3Cl_3, as a gas, and (3) above 650°C, CuCl as gas.
REF: Winters, H F — *J Vac Sci Technol,* A(3),786(1985)

CU-0013b
ETCH NAME:
TYPE: Acid, preferential
COMPOSITION:
1 .... HCl
1 .... H_2O
x .... *$FeCl_3$

TIME:
TEMP:

*Saturate the solution.

DISCUSSION:
Cu, single crystal sphere 1 cm diameter. After polishing with nitric acid and then ferric chloride, the solution shown was used to preferentially etch. Developed (001) pole figures.
REF: Ibid.

CU-0052
ETCH NAME: Copper sulfate
TYPE: Acid, oxidation
COMPOSITION:
x .... 50 g/l $CuSO_4.5H_2O$ pH: 3.8

TIME: Variable
TEMP: RT

DISCUSSION:
Cu, single crystal spheres $^5/_8$ and $^3/_4$" diameter with (100), (110), (111), (012), and (311) flats. Solution used to oxidize surfaces in a study of epitaxy relationships of Cu_2O. In 5 min a general orientation pattern will form; 24—60 h oxide will show pitting; 5 sec to 90

h and Cu_2O crystals form. After oxidizing spheres in the solution, wash in DI water and blow dry with nitrogen.
REF: Lawless, K R & Miller, G T Jr — *Acta Crystallogr,* 12,594(1959)

CU-0053
ETCH NAME: TIME:
TYPE: Thermal, forming TEMP:
COMPOSITION:
x .... heat
DISCUSSION:
Cu, single crystal spheres formed by melting the tip of single crystal wire. Wire was 1—2 mm diameter; spheres, 2—3 mm diameter. Other materials used were Ag and Au.
REF: Rose, D K & Gerisder, H — *J Electrochem Soc,* 110,350(1963)
CU-0086: Young, F W & Gwathmey, A T — *J Appl Phys,* 31,225(1960)
Cu, single crystal specimens. Heating in ultra high vacuum (UHV) develops facets and spiral etch pits.

CU-0071b
ETCH NAME: Water TIME:
TYPE; Acid, float-off TEMP:
COMPOSITION:
x .... H_2O
DISCUSSION:
Cu, thin films evaporated on NaCl, (100) substrates in a general study of thin film defects by TEM. After film deposition, copper was removed by dissolving the NaCl surface with water using the float-off technique. Copper thin films deposited on mica, (0001) substrates were removed with 10% HF. A study of copper structure.
REF: Ibid.

CU-0071a
ETCH NAME: TIME:
TYPE: Deformation, defect TEMP:
COMPOSITION:
x .... stress
DISCUSSION:
Cu, single crystals used in a study of secondary slip caused by stress deformation. Secondary slip nucleates at the end of primary slip lines when the internal stress fields are high. Also studied single crystal alpha-brass.
REF: Mitchell, T E & Thornton, P R — *Phil Mag,* 10,314(1964)

CU-0062b
ETCH NAME: TIME:
TYPE: Acid, cleaning TEMP:
COMPOSITION:
1 .... H_2SO_4
1 .... *$K_2Cr_2O_7$

*100 g/l

DISCUSSION:

Cu, single crystals used in a cleaning study. The solution shown was radiotracer tagged with ^{35}S and ^{51}Cr isotopes and used as a surface cleaning etch.

REF: Fatzer, O D et al — *Int J Appl Radiat & Isot,* 10,167(1961)

CU-0062c

ETCH NAME: TIME: 2 min

TYPE: Electrolytic, polish TEMP: 120°C

COMPOSITION: ANODE: Cu

x .... x% CrO_3 CATHODE:

POWER:

DISCUSSION:

Cu specimens. See discussion under CU-0062a and CU-0062b.

REF: Ibid.

CU-0085

ETCH NAME: TIME: 60 min

TYPE: Acid, cutting TEMP: RT

COMPOSITION:

(1) 200 ml .... HCl
50 g .... $FeCl_3$
250 ml .... H_2O

(2) 1 .... HNO_3
1 .... H_2O

DISCUSSION:

Cu specimens $^1/_8$″ thick. Terylene thread soaked in solution (1) used to cut copper, aluminum and brass. Solution (2) used for tin and zinc. (*Note:* Excellent review of metal growth methods.)

REF: Honeycombe, R W K — *Metall Rev,* 4,1(1959)

CU-0087

ETCH NAME: Argon TIME:

TYPE: Ionized gas, cleaning TEMP:

COMPOSITION: GAS FLOW:

x .... Ar^+ ions PRESSURE:

POWER:

DISCUSSION:

Cu, (111) wafers used as substrates for thin film deposition of copper by condensation of metallic vapor which can produce twinning. Substrates were Ar^+ ion bombardment cleaned prior to copper deposition. Epitaxy growth of copper evaporated at an angle as a function showed that the greater the angle, the greater the increase of twinning observed. (*Note:* Growth of silica at an angle will produce a columnar growth structure rather than normal amorphous thin films.)

REF: Lafourcado, L et al — *C R Acad Sci (Paris),* 249,230(1959)

CU-0088: Yurasova, V E — *Zh Tekh Fiz,* 28,1966(1958)

Cu, single crystal specimens. Cathode positive sputtering as ion bombardment used in a study of surface micro-relief and erosion. Etch pits observed in Cu and Zn and slip traces on deformed Zn.

COPPER ALLOYS

General: Copper occurs as a native element as well as in a number of single crystal metallic compounds: horsfordite, Cu_6Sb; mohawkite, Cu_3As; rickardite, Cu_4Te_3; barzelianite, Cu_2Se. The two major copper sulfides are chalcocite, Cu_2S and covellite, CuS. The copper iron sulfides: bornite, $CuFeS_4$ and chalcopyrite, $CuFeS_2$. The latter mineral structure is a base for the artificial ternary chalcogenides. Possible the two most well known copper minerals, due to their brilliant colors, are the copper carbonates: malachite (green) and azurite (blue), both being primary copper ores as well as semi-precious gem stones. Cuprite, Cu_2O, is the major oxide of copper, also an ore as are all the copper alloy compounds mentioned. They occur in vein deposits alone, or are associated with iron.

There are many artificial alloys of copper, the two most well known being bronze, Cu:Sn and brass, Cu:Zn. Mixtures of Cu:Sn:Zn are used for bell metal and gun metal, and there are special mixtures such as phosphor-bronze; silicon bronze; manganese bronze. There are similar alloys of brass: cartridge brass, wire brass, aluminum brass, etc., and mixtures of Cu:W are some of the alloys referred to as heavy metal. As pure copper is soft, ductile and malleable it is usually used in some form of alloy for strengthening purposes, although OFHC copper, the highest purity form of copper, has application for its low electrical resistance and heat dissipation characteristics.

Technical Application: There are many uses for pure copper and copper alloys in Solid State material processing as evaporated contact pads in device fabrication, as package assembly parts and, of major importance today, as PCBs (laminated plastic linen boards, copper plated) used to hold electronic circuit assemblies for computers and similar products. Such boards are capable of being inserted and removed in a variety of equipment, and are widely used in the miniaturization of operating functions.

OFHC copper and brass are both used as test blocks for mounting semiconductor devices. Heavy metal, Cu:W alloys, are used as physical weights for belt furnace alloying of parts; phosphor-bronze, Cu:Sn:P, sheet is cut and used as holding spring clips for both metal evaporation of wafers on a flat plate, or devices on a test plate, as is Cu:Be. There are epoxy and polyimide pastes containing copper powder as one constituent for use as electrical contact materials, e.g., ''loaded'' epoxies.

Several binary and trinary copper alloys have been grown and studied as single crystals for general morphology and structure. There are also trinary and quarternary sulfide and selenide compound semiconductors containing copper.

Etching: Soluble in HNO_3, H_3PO_4, and mixed acids, variable with type alloy.

COPPER ALLOY ETCHANTS

COPPER:ANTIMONY ETCHANTS

CUSB-0001
ETCH NAME: Nital | TIME:
TYPE: Acid, preferential | TEMP:
COMPOSITION:
x .... HNO_3
x .... EOH/MeOH (H_2O)
DISCUSSION:

Cu_2Sb, and $SbAu_2Sb$ specimens. Structure could be observed on material without etching. Solution shown was used to accentuate defects and structure.

REF: Savas, M A & Smith, R W — *J Cryst Growth,* 71,66(1985)

COPPER:BERYLLIUM ETCHANTS

CUBE-0001
ETCH NAME: TIME:
TYPE: Acid, structure TEMP:
COMPOSITION:
100 ml .... H_2O_2
35 ml .... 20% KOH
125 ml .... H_2O
DISCUSSION:
Cu:Be:Ni alloy specimens. Specimens used in a study of the effects of pressure on age hardening. Solution used to develop defect and structure.
REF: Phillips, V A — *Acta Metall,* 9,216(1961)

CUBE-0002
ETCH NAME: Nitric acid, dilute TIME:
TYPE: Acid, cleaning TEMP: RT
COMPOSITION:
100 ml .... HNO_3
100 ml .... H_2O
DISCUSSION:
Cu:Be, spring shim stock, 0.010—0.015″ thick, cut and used as spring clips to hold semiconductor wafers on SST plates during metallization. After cutting and shaping, clips were degreased in hot TCE vapors, rinsed in MeOH, and N_2 blown dry. Solution shown used to lightly etch clean clips before use: (1) etch; (2) DI water rinse; (3) MeOH rinse; and (4) N_2 blow dry. Same method used to clean clips after metal depositing with a combination of physical scrap/removal of loose metallization build-up, and etch cleaning.
REF: Fahr, F & Walker, P — personal application, 1980

COPPER:CERIUM ETCHANTS

CUCE-0001
ETCH NAME: Ferric chloride TIME:
TYPE: Acid, removal TEMP:
COMPOSITION:
x .... 25% $FeCl_3$
DISCUSSION:
$CeCu_6$, grown as a single crystal ingot by the Czochralski (CZ) method. Material is orthorhombic in structure and was used in a material growth study. Also developed $LaCu_6$ CZ grown ingots, and $CeAl_3$ by arc melting and aging at temperature. CeB_6 and $CeCu_2Si_2$ are also grown as single crystals. No etchant shown. (*Note:* Ferric chloride solutions are general etchants for copper and copper alloys or compounds.)
REF: Onuki, Y et al — *J Phys Soc Jpn,* 4,1210(1984)

COPPER:GALLIUM ETCHANTS

CUGA-0001a
ETCH NAME: TIME:
TYPE: Acid, preferential TEMP:
COMPOSITION:
x .... $FeCl_3$
x .... EOH

DISCUSSION:

Cu:Ga alloy specimens used in a study of transformation due to strain. Specimens were subjected to strain at elevated temperatures, quenched directly into an ice/brine solution, or a 10% NaOH solution. Solution shown used to develop defects and structure. Also studied Cu:Zn and Cu:Ga:Ge alloys.

REF: Massaski, T B — *Acta Metall,* 6,243(1958)

CUGA-0001b

ETCH NAME: Ammonium hydroxide — TIME:

TYPE: Base, preferential — TEMP:

COMPOSITION:

x .... NH_4OH

x .... H_2O

DISCUSSION:

Cu:Ga alloy specimens. Solution used to develop defects and structure. See CUGA-0001a for additional discussion.

REF: Ibid.

COPPER:GERMANIUM ETCHANTS

CUGE-0001

ETCH NAME: Phosphoric acid — TIME:

TYPE: Electrolytic, polish — TEMP:

COMPOSITION: — ANODE:

x ... H_3PO_4 — CATHODE;

POWER:

DISCUSSION:

Cu:Ge alloys of varying compositions grown and used in a study of optical absorption at 4.2 K. Solution was erratic for germanium concentration alloys above 7% Ge, as the material will not polish. Other alloys were studied.

REF: Rayne, J A — *Phys Rev,* 121,456(1961)

COPPER:GOLD ETCHANTS

CUAU-0001

ETCH NAME: — TIME:

TYPE: Electrolytic, polish — TEMP:

COMPOSITION: — ANODE:

x .... HAc — CATHODE:

x .... CrO_3 — POWER:

DISCUSSION:

Cu_3Au single crystal specimens used in a study of long range order from strain hardening. Solution used to polish specimens prior to straining.

REF: Davis, R G & Steloff, N S — *Phil Mag,* 12,297(1965)

CUAU-0002

ETCH NAME: Phosphoric acid — TIME:

TYPE: Electrolytic, polish — TEMP:

COMPOSITION: — ANODE:

x .... H_3PO_4 — CATHODE:

POWER:

DISCUSSION:

Cu_3Au single crystal specimens used in a study of ordering strength and dislocations in superlattice materials. Solution used to polish specimens. The material Ni_3Mn also was studied.

REF: Markoikowski, M J & Miller, D S — *Phil Mag,* 6,871(1961)

COPPER:LANTHANUM ETCHANTS

CULA-0001

ETCH NAME: Ferric chloride TIME:

TYPE: Acid, removal TEMP:

COMPOSITION:

x .... 25% $FeCl_3$

DISCUSSION:

$LaCu_6$ grown as single crystal ingots by the Czochralski (CZ) method. See discussion under Copper:Cerium (CECU-0001) for further discussion.

REF: Onuki, Y et al — *J Phys Soc Jpn,* 4,1210(1984)

COPPER:NICKEL ETCHANTS

CUNI-0001

ETCH NAME: Nitric acid TIME:

TYPE: Acid, preferential TEMP:

COMPOSITION:

x .... HNO_3

DISCUSSION:

CuNi single crystal specimens fabricated as spheres. Solution will develop pole figures at (100), (110), and (111) plane locations, and is not actually preferential as a finite crystal form etchant.

REF: Schmunk, R E & Smith, C S — *Acta Metall,* 8,396(1960)

COPPER PALLADIUM ETCHANTS

CUPD-0001

ETCH NAME: Nitric acid TIME:

TYPE: Acid, removal TEMP: RT to hot

COMPOSITION:

1 .... HNO_3

1 .... H_2O

DISCUSSION:

Cu:Pd thin films co-evaporated on Pyrex blanks in a material study. Quartz was evaluated as a substrate material, but it will form SiPd. Solution used to etch remove or etch pattern thin film after annealing cycles.

REF: van Langeveld, A D — *Thin Solid Films,* 109,179(1983)

COPPER:ZINC ETCHANTS

CUZN-0001

ETCH NAME: TIME:

TYPE: Acid, preferential TEMP:

COMPOSITION:

x .... $FeCl_3$

x .... EOH

DISCUSSION:

Cu:Zn alloy specimens used in a study of transformations due to strain. Solution used to develop defects and structure. See CUGA-0001a for additional discussion.

REF: Massaski, T B — *Acta Metall,* 6,243(1958)

PHYSICAL PROPERTIES OF COPPER BROMIDE, $CuBr_2$

Classification	Bromide
Atomic numbers	29 & 35
Atomic weight	223.4
Melting point (°C)	498
Boiling point (°C)	
Density (g/cm³)	3.0×
Hardness (Mohs — scratch)	2—3
Crystal structure (monoclinic-normal)	(100) a-pinacoid
Color (solid)	Black
Cleavage (basal)	(001)

COPPER BROMIDE, $CuBr_2$

General: Does not occur as a natural compound due to high solubility in water, but can occur in solution and be extracted from ocean waters and some hot springs like other bromides, chlorides and some iodides. There is no major industrial use.

Technical Application: Copper bromide has been studied in association with some 40 other compounds for semiconductor characteristics and, even though water soluble, may have useful optical characteristics like some other bromides.

Etching: Water and alcohols.

COPPER BROMIDE ETCHANTS

CUBR-0001

ETCH NAME: Ethyl alcohol — TIME:

TYPE: Alcohol, removal — TEMP: RT

COMPOSITION:

x .... EOH (MeOH)

DISCUSSION:

CuBr single crystals used in an evaluation of cohesive energy features of tetrahedral semiconductor materials. Solution shown was used as a general polish and removal etchant on surfaces. Some 40 materials were studied.

REF: Aresti, A et al — *J Phys Chem Solids* 45,361(1984)

PHYSICAL PROPERTIES OF COPPER CHLORIDE, $CuCl_2$

Classification	Chloride
Atomic numbers	29 & 17
Atomic weight	134.48
Melting point (°C)	498
Boiling point (°C)	993
Density (g/cm³)	3.045
Hardness (Mohs — scratch)	2—3
Crystal structure (isometric — normal)	(100) cube
Color (solid)	Yellow
Cleavage (cubic)	(001)

COPPER CHLORIDE, $CuCl_2$

General: Does not occur in natural as a solid compound due to water solubility, but can be extracted from ocean and ground waters in solution, particularly in areas associated with copper mining. May occur as a mixed component in arid areas with potassium chloride, carbonates, and sulfates. There is no major use in industry, other than as a copper source in electrolytic copper plating or as a component in etch solutions.

Technical Application: In Solid State processing it has been used as a copper doping source in some compound semiconductors and, as in the general metal industries, for electrolytic copper plating. The chloride also has been used as a constituent in some preferential etchants, and for copper plate-out from solution with subsequent heat treatment drive-in for defect decoration with copper. It has been studied along with other halides for semiconducting properties, and can have applications as an optical filter.

Etching: Water and alcohols.

COPPER CHLORIDE ETCHANTS

CUCL-0001
ETCH NAME: Ethyl alcohol — TIME:
TYPE: Alcohol, removal — TEMP: RT
COMPOSITION:
x .... EOH (MeOH)
DISCUSSION:

CuCl as single crystals were used in an evaluation of cohesive energy features of tetrahedral semiconductors. Alcohols were used for general removal and polishing of surfaces. Forty materials were studied.

REF: Aresti, A et al — *J Phys Chem Solids,* 45,361(1984)

PHYSICAL PROPERTIES OF COPPER IODIDE, CuI

Classification	Iodide
Atomic numbers	29 & 53
Atomic weight	190.4
Melting point (°C)	605
Boiling point (°C)	1200
Density (g/cm³)	5.62
Refractive index (n=)	2.346
Hardness (Mohs — scratch)	2.5
Crystal structure (isometric — tetrahedral)	(111) tetrahedron
Color (solid)	Brown
Cleavage (dodecahedral)	(110)

COPPER IODIDE, CuI

General: Occurs in nature as the mineral marshite, CuI, as a minor compound associated with copper deposits. There is little use of the natural mineral in the metal industry due to scarcity, but the artificial compound has both chemical and medical applications.

Technical Application: There is no use in Solid State device fabrication, to date, other than occasional use as a constituent in etching solutions. The compound has been characterized for semiconducting properties, and has possible application in optics as a filter element.

Etching: KI, KCN, and NH_4OH

COPPER IODIDE ETCHANTS

CUI-0001

ETCH NAME: Potassium iodide TIME:

TYPE: Halogen, removal TEMP: RT

COMPOSITION:

x .... KI

x .... H_2O (EOH)

DISCUSSION:

CuI grown as single crystals in an evaluation of cohesive energy features of tetrahedral semiconductors. Solution shown used as a general removal and surface polishing etch. Forty materials were evaluated.

REF: Aresti, A et al — *J Phys Chem Solids* 45,361(1984)

PHYSICAL PROPERTIES OF COPPER OXIDE, Cu_2O

Classification	Oxide
Atomic numbers	29 & 8
Atomic weight	143.14
Melting point (°C)	1235
Boiling point (°C)	1800
Density (g/cm³)	5.85—6.15
Refractive index (n=)	2.849
Hardness (Mohs — scratch)	3.5—4
Crystal structure (isometric — plagiohedral)	(111) octahedron
Color (solid)	Black (CuO)
(solid)	Copper-red (Cu_2O)
(flame)	Emerald green
Cleavage (cubic — interrupted)	(001)

COPPER OXIDE, Cu_2O

General: As the natural mineral cuprite, Cu_2O (red copper ore), it is a primary ore of copper, and also occurs as the monoxide: tenorite, CuO. The latter, as black incrustations whereas cuprite is brilliant conchineal-red. Both colors appear on pure copper parts with exposure to air, but the green "verdigris" is copper sulfate from sulfur in the atmosphere. Other than industrial ores of copper, both forms are important as coloring agents in paints and glass, and the mineral is used for its flame color in pyrotechnics.

Technical Application: Though not used in general Solid State material processing, both the polycrystalline and single crystal forms are used to fabricated discrete rectifiers, similar to selenium rectifiers. Cu_2O has been grown by oxidation of OFHC copper surfaces and, on alpha-brass, by copper plating and water soak-tarnishing as a copper hydrated thin film. Both methods used in material studies. Most of the listed references here are for removal of "native oxide" films in copper material processing.

Etching: HCl (CuO). HNO_3, NH_4OH, and acid mixtures (Cu_2O).

COPPER OXIDE ETCHANTS

CUO-0001
ETCH NAME: Ethylenediamine TIME:
TYPE: Acid, oxide removal TEMP:
COMPOSITION:
x .... 1% ethylenediamine
DISCUSSION:

Cu_2O, as a native oxide thin film on surfaces. This amino acid will remove the oxide but will not etch the pure copper metal.

REF: Stiegler, J O & Noggles, T S — *J Appl Phys,* 31,1827(1960)

CUO-0002
ETCH NAME: Nitric acid, dilute TIME:
TYPE: Acid, removal TEMP:
COMPOSITION:
x .... HNO_3
x .... H_2O
DISCUSSION:

Cu_2O as a native oxide on copper surfaces. Dilute solutions are used to remove native oxides and etch polish copper surfaces. Cleaned specimens were control oxidized at 1030 to 1095°C for 3—4 h and then annealed for 40 to 150 h. Large area single crystals (10 to 40 mils) were formed with size increasing as annealing time was increased.

REF: Toth, R S et al — *J Appl Phys,* 31,1117(1960)
CUO-0003: Bulletin B-11D, The American Brass Co.

Cu, and Cu alloys. Used a 1:1 mixture for removal of native oxide and to etch polish copper surfaces.

CUO-0004: Walker, P & Campos, R — personal application, 1961—1962

Cu alloys as cryogenic trailer piping. Used a 1:20 solution for removal of surface contamination. Solution applied by soaking and scrubbing with an SST brush. Heavy DI water wash after etch cleaning and wipe dry with lint-free toweling.

CUO-0005
ETCH NAME: Copper Brite dip TIME:
TYPE: Acid, polish TEMP:
COMPOSITION:
3—7 .... H_2SO_4
1—7 .... HNO_3
0—5 .... HCl
1—10 .... H_2O
DISCUSSION:

Cu, and Cu alloys. There are several specific mixtures of this "Brite Dip" which is sometimes referred to as a "scale" removal etch. Used to remove oil/grease contamination or copper oxide, and as surface polish etches for copper. Solutions are extremely reactive, producing a bright copper surface. They should not be used on parts requiring close dimensional control as etching proceeds by pinholing and undercutting of the surface contamination, such that surface planarity and tolerances cannot be maintained. See CU-0028 for other specific mixtures.

REF: Walker, P & Fahr, F — personal application, 1978—1983

CUO-0006a
ETCH NAME: Ammonium hydroxide TIME:
TYPE: Hydroxide, oxide removal TEMP: RT to hot
COMPOSITION:
x .... NH_4OH
DISCUSSION:

Cu_2O, native oxide. Solution will remove both Cu_2O and CuO oxides and slowly etch copper.

REF: Hodgman, C D et al — *Handbook of Chemistry and Physics,* Chemical Rubber Co., Cleveland, OH, 1943, 380

CUO-0007: Walker, P — personal application, 1980—1985

OFHC copper parts used in test assemblies of microelectronic devices. Surfaces were cleaned by soaking parts in solution shown, then wash in DI water, and final rinse with MeOH, and N_2 blow dry.

CUO-0017: Horn, F N — *J Appl Phys* 32,900(1961)

CuO single crystals doped with iron and grown by the Czochralski (CZ) method in a study of ferrites. Other ferrites grown and studied were Fe_3O_4, and iron doped ZnO, Ga_2O_3, and MnO.

CUO-0008
ETCH NAME: Phosphoric acid TIME:
TYPE: Acid, polish TEMP: 130°C
COMPOSITION: RATE: 1 μm/min
x .... H_3PO_4
DISCUSSION:

Cu_2O, material use as rectifiers in a study of etching effects on device operating parameters. Etching will reduce the forward current while a lapped surface will increase current.

REF: Makovskii, F A & Usachev, B P — *Zh Tekh Fiz,* 27,2786(1957)

CUO-0009
ETCH NAME: TIME: 1—2 min
TYPE: Acid, removal TEMP: 60—70°C
COMPOSITION:
33 ml .... H_3PO_4
33 ml .... HNO_3
33 ml .... HAc
DISCUSSION:

Cu_2O, as a native oxide on copper specimens. Solution will etch both copper oxide and copper, but is particularly good for removal of oxides.

REF: de Jong, J J — *Metalen,* 9,2(1954) (in Dutch)

CUO-0015: Tegart, W J McG — *The Electrolytic and Chemical Etching of Metals,* Pergamon Press, England, 1956

Cu_2O, native oxide on copper surfaces. Solution can be used to etch both copper and copper oxides and is very good for oxide removal.

CUO-0010
ETCH NAME: Ammonium sulfate TIME: 15 min
TYPE: Acid, cleaning TEMP: RT
COMPOSITION:
x .... 2—5% $(NH_4)_2SO_4$

DISCUSSION:

Cu_2O as thin films grown on OFHC copper by oxidation. Clean copper with solution shown before oxidation. After etching, wash in DI water.

REF: Baur, J P et al — *Acta Metall,* 103,273(1956)

CUO-0011

ETCH NAME: Oxygen

TYPE: Gas, preferential

COMPOSITION:

x .... O_2

TIME: 10 min

TEMP: 250°C

PRESSURE: 10 mmHg

DISCUSSION:

Cu, single crystal sphere. Used in a study of oxidation rates as related to crystallographic faces and plane zones.

REF: Gwathmey, A T et al — *J Chem Phys,* 8,431(1940)

CUO-0012: Gwathmey, A T et al — *J Phys Chem,* 46,969(1942)

Cu, single crystal spheres with cut and oriented flats. Oxidation rate varies with crystallographic faces in the following decreasing order: (100)<(210)<(111)<(110)<(311). Thickness variation can be as much as fivefold. Rate varies with orientation, temperature, and oxygen pressure.

CUO-0013

ETCH NAME: Oxygen

TYPE: Gas, preferential

COMPOSITION:

x .... O_2

TIME: Variable

TEMP: Variable

DISCUSSION:

Cu, single crystal spheres. A comprehensive article of oxidation of materials — thickness, epitaxy orientation, rates. Other materials were Ni, Fe, Nb, Co, Al, Cr, Ge, Ag, Mg, Pd, Zn, Cd, Be, Pb, Sb, Ta, Mn, Fe-Ni, brass, and U. Included cleaved or cut wafer surfaces of some of the materials.

REF: Gwathmey, A T & Lawless, K R — Gatos, H C Ed — *The Surface Chemistry of Metals and Semiconductors,* John Wiley & Sons, New York, 1960, 483

CUO-0014a

ETCH NAME:

TYPE: Acid, removal

COMPOSITION:

300 ml .... H_2SO_4

400 ml .... HNO_3

5 ml .... HCl

295 ml .... H_2O

TIME:

TEMP: RT

DISCUSSION:

Cu specimens. Described as a copper oxide removal etch. (*Note:* See CU-0028, Copper Brite Dip solutions. This is a variation of the CU-0032 solution.)

REF: Kohl, W H — *Handbook of Materials and Techniques for Vacuum Devices,* Reinhold, New York, 1967

CUO-0014b
ETCH NAME: TIME: 2—3 sec
TYPE: Acid, removal TEMP: RT
COMPOSITION:
1000 ml .... HNO_3
10 ml .... HCl
DISCUSSION:
Cu specimens. Described as a copper oxide removal etch.
REF: Ibid.

CUO-0006b
ETCH NAME: Ammonium chloride TIME:
TYPE: Salt, removal TEMP: RT
COMPOSITION:
x .... x% NH_4Cl
DISCUSSION:
Cu_2O specimens, or as a native oxide on copper. Described as a copper oxide removal etchant.
REF: Ibid.

CUO-0016
ETCH NAME: Water TIME: 15 h
TYPE: Acid, oxidation TEMP: RT
COMPOSITION:
x .... H_2O
DISCUSSION:
Cu_2O thin films plated from solution onto alpha-brass, Cu:Zn with 29.38% Cu. Plating solution was: 0.04 *M* $CuSO_4$:1.57 $(NH_4)_2SO_4$ for pH 7 balanced with NH_4OH, and containing tracer $^2O^{64}Cu$. After plating, 12 mm blanks were soaked in water for 15 h to oxide tarnish, then DI water washed, and sealed in polyethylene bags for radiation count. A study of the mechanisms fo Cu_2O formation.
REF: Birley, S S & Tromans, D — *J Electrochem Soc,* 118,636(1971)

PHYSICAL PROPERTIES OF COPPER PHOSPHIDE, Cu_3P

Classification	Phosphide
Atomic numbers	29 & 15
Atomic weight	221.73
Melting point (°C)	~1200
Boiling point (°C)	
Density (g/cm³)	6.4—6.8
Hardness (Mohs — scratch)	6—7
Crystal structure (isometric — normal)	(100) cube
Color (solid)	Grey/black
Cleavage (cubic)	(001)

COPPER PHOSPHIDE, Cu_3P

General: Does not occur in nature as a pure metallic phosphide, although the mineral libethenite, $Cu_2(OH)PO_4$, and tsumebite, $Pb,Cu(PO_4)_x$ are minor phosphate minerals. Artificial copper phosphide is fabricated in industry as phosphor-bronze, and used for its spring qualities.

Technical Application: As phosphor-bronze the material is widely used in Solid State processing as spring clips for holding semiconductor wafers or parts during vacuum metallization, and as general holding clips during device testing.

The material has been grown as a single crystal with germanium doping for evaluation as a compound semiconductor.

Etching: Soluble in HNO_3 and mixed acids ($HF:HNO_3$).

COPPER GERMANIUM PHOSPHIDE ETCHANTS

CUGP-0001
ETCH NAME: TIME:
TYPE: Acid, removal TEMP: RT
COMPOSITION:
x .... HF
x .... HNO_3
x .... H_2O
DISCUSSION:

$CuGe_2P_3$ single crystal ingots grown by the Horizontal Bridgman (HB) method as a p-type semiconductor. Specimens were cut as parallelapipeds (001), (110)/(110), and mechanically polished in a material properties evaluation. (*Note:* Solution shown can be used to etch clean phosphor-bronze.)

REF: McDonald, J E — *J Phys Chem Solids,* 8,951(1985)

PHYSICAL PROPERTIES OF COPPER SELENIDE, Cu_2Se

Classification	Selenide
Atomic numbers	29 & 34
Atomic weight	206.10
Melting point (°C)	>1200
Boiling point (°C)	
Density (g/cm³)	6.71
Hardness (Mohs — scratch)	3—4
Crystal structure (isometric — normal)	(100) cube
Color (solid)	Silver-white
Cleavage (cubic)	(001)

COPPER SELENIDE, Cu_2Se

General: Occurs in nature as the mineral berzelianite, Cu_2Se and klockmannite, CuSe. The variety umangite, $CuSe.Cu_2Se$ is dark cherry-red in color, and there are other trinary or quaternary constituent copper selenides, such as crookesite, containing thallium and silver.

The main industrial use of these minerals is as minor ores of copper and selenium.

Technical Application: Trinary copper selenides are under evaluation as tetrahedral compound semiconductors in Solid State, but not used as devices at the present time. The following two copper selenides are trinary compounds.

Etching: Soluble in mixed acids.

COPPER GALLIUM SELENIDE ETCHANTS

CGSE-0001
ETCH NAME: Aqua regia, variety TIME:
TYPE: Acid, polish TEMP: RT
COMPOSITION:
1 .... HCl
1 .... HNO_3
DISCUSSION:
$CuGaSe_2$ single crystal evaluated for cohesive energy features of tetrahedral semiconductors. Forty materials were studied. (*Note:* See copper indium selenide for addition solutions.)
REF: Aresti, A et al — *J Phys Chem Solids,* 45,361(1984)

COPPER INDIUM SELENIDE ETCHANTS

CISE-0001
ETCH NAME: TIME: 2 min
TYPE: Acid, polish/stain TEMP: RT
COMPOSITION:
1 .... HCl
1 .... HNO_3
DISCUSSION:
$CuInSe_2$ single crystal specimens. Crystalline ingot grown by Horizontal Bridgman (HB) process, then large single crystal grains were cut out and mechanically lapped with 0.3 μm alumina. After indium diffusion, the solution shown was used to acid stain the p-n junction after angle lapping.
REF: Shih, T et al — *J Appl Phys,* 66,420(1984)
CISE-0002b: Ibid.
$CuInSe_2$ singe crystal wafers. Annealed under Se vapor to produce a p-type layer as a p-n junction. Solution used as a junction stain.
CISE-0005: Ciszek, T F — *J Electron Mater,* 14,451(1985)
$CuInSe_2$ ingots grown by LEC or the Bridgman/Stockberger process with slow directional solidification for large grains. Material is p- or n-type depending on the charge. Preferred crystal growth habit is (112) tetragonal.

CISE-0002a
ETCH NAME: Nitric acid TIME:
TYPE: Acid, junction TEMP:
COMPOSITION:
x .... HNO_3
DISCUSSION:
$CuInSe_2$ wafers annealed under Se to produce a p-type layer as a p-n junction. Solution used as a junction stain.
REF: Tell, B et al — *J Appl Phys,* 48,2477(1977)

CISE-0003
ETCH NAME: TIME:
TYPE: Acid/halogen, cleaning TEMP:
COMPOSITION:

(1) 6 .... H_2SO_4
1 .... H_2O_2
1 .... H_2O

(2) x .... 1% Br_2
x .... MeOH

DISCUSSION:

$CuInSe_2$ p-type wafers. Grown from a melt by directional freezing with a preferred (112) tetragonal orientation. Specimens were cleaned and polished sequentially in these two acid solutions to remove carbon and oxygen contamination.

REF: von Bardeleben, H J — *J Appl Phys,* 66,320(1984)
CISE-0004: von Bardeleben, H J — *J Appl Phys,* 66,586(1984)

PHYSICAL PROPERTIES COPPER SULFIDE, CuS

Classification	Sulfide
Atomic numbers	29 & 16
Atomic weight	159.29
Melting point (°C)	103
Boiling point (°C)	220 del.
Density (g/cm³)	4.6
Refractive index (n=)	1.45 (Na)
Hardness (Mohs—scratch)	1.5—2
Crystal structure (hexagonal—normal)	(1010) prism
Color (solid)	Indigo blue
Cleavage (basal)	(0001)

COPPER SULFIDE, CuS

General: Occurs as the mineral covellite, CuS in association with other copper vein ores. It is of secondary in origin and, occasionally, as a volcanic sublimate. Industrial use is as a minor copper ore in association with the other vein ores.

Technical Application: No direct use in Solid State processing, but has been artificially grown as a single crystal. Both the natural and artificial materials have been studied for general morphological data.

Etching: Soluble in HNO_3, HCl, H_2SO_4, and KCN.

COPPER SULFIDE ETCHANTS

CUS-0001
ETCH NAME: Nitric acid TIME:
TYPE: Acid, removal TEMP: RT
COMPOSITION:

1 .... HNO_3
10 .... H_2O

DISCUSSION:

CuS specimens as the natural mineral covellite. Solution used as a slow removal etch of the purplish tarnish on surfaces in a general study of the copper minerals.

REF: Walker, P — mineral study, 1953

COPPER GALLIUM SULFIDE ETCHANTS

CUGS-0001
ETCH NAME: Hydrochloric acid
TYPE: Acid, cleaning
COMPOSITION:
x .... HCl

TIME:
TEMP:

DISCUSSION:

$CuGaS_2$ single crystals grown at less than 630°C in quartz ampoules with tin as a metal solvent. Clean copper with HCl. Clean quartz ampoule with boiling HCl and rinse in EOH. SnS platelets nucleate on the tin during compound grown and have a metallic luster. The $CuGaS_2$ crystals are transparent with red color. A $Ga_4SnS_{7\text{-}8}$ crystal was transparent and colorless. A $CuGaSn_2$ crystal was orange in color. $CuGa_3S_5$ is cubic and $CuGa_5S_8$ is sphalerite structure isometric system, tetrahedral class. (*Note:* Reference to "cubic" is the isometric system, normal class commonly call the cubic system.)

REF: Tsubaki, K & Sugiyama, K — *J Electron Mater*, 12,43(1983)

PHYSICAL PROPERTIES OF COPPER DISULFIDE, Cu_2S

Classification	Sulfide
Atomic numbers	29 & 16
Atomic weight	159.20
Melting point (°C)	1100
Boiling point (°C)	
Density (g/cm³)	5.5—5.8
Hardness (Mohs— scratch)	2.5—3
Crystal structure (orthorhombic — normal)	(110) a-pinacoid
(isometric — normal) >91°C	(100) cube
Color (solid)	Lead grey
Cleavage (pinacoidal — indistinct)	(110)
(octahedral — poor)	(111)

COPPER SULFIDE, Cu_2S

General: Occurs in nature as the mineral chalcocite, Cu_2S and is found as a vein deposit along with other copper ore minerals. It is mined for both the copper and sulfur content.

Other than as copper ores in industry, the chemical $CuSO_4$ "Blue Vitrol" has important use in agriculture as an algicide, and additional use in chemical processing.

Technical Application: Copper sulfide, Cu_2S, doped with indium is a trinary compound semiconductor, and under evaluation for general properties and device applications.

The pure sulfides as both CuS and Cu_2S, natural and artificial minerals, have been studied for general morphology. The $CuInS_2$ trinary material in the following section is in the crystallographic tetragonal system, and under evaluation as a trinary compound semiconductor.

Etching: Soluble in single acids, and mixed acids.

COPPER INDIUM SULFIDE ETCHANTS

CIS-0001a
ETCH NAME: Nitric acid TIME: 3—4 min
TYPE: Acid, dislocation TEMP: RT
COMPOSITION:
x .... HNO_3
DISCUSSION:
$CuInS_2$, (112) wafers. Grown as small single crystals with preferred (112) surface orientation. Nitric acid was used to develop dislocations on the (112)B sulfur surface, as material is a polar compound. A sulfur film is left on the surface after etching. (*Note:* Remove sulfur with sulfur disulfide, CS_2.)
REF: Thiel, F A — *J Electrochem Soc,* 129,1570(1982)

CIS-0001b
ETCH NAME: BRM TIME: 10—15 sec
TYPE: Halogen, dislocation TEMP: RT
COMPOSITION:
x .... 2% Br_2
x .... MeOH
DISCUSSION:
$CuInS_2$, (112) wafers. Solution used to develop dislocations on the (112)A positive CuIn polar surface.
REF: Ibid.

CIS-0001c
ETCH NAME: TIME: 45 min
TYPE: Acid, dislocation TEMP: RT
COMPOSITION:
3 .... H_2SO_4
1 .... H_2O_2
1 .... H_2O
DISCUSSION:
$CuInS_2$ (112) wafers. Solution used to develop dislocations on the (112)A positive CuIn polar surface.
REF: Ibid.

CIS-0002
ETCH NAME: TIME: 10 sec
TYPE: Acid, polish TEMP: RT
COMPOSITION:
1 .... HCl
1 .... HNO_3
1 .... H_2O
DISCUSSION:
$CuInS_2$ wafer. Crystals are tetragonal system. Mechanically polish wafers with diamond grit, then etch polish with the mixture shown, and rinse in water.
REF: Barradas, R T et al — *J Phys & Chem Solids,* 45,1185(1984)

CIS-0003
ETCH NAME: Aqua regia — TIME:
TYPE: Acid, polish — TEMP:
COMPOSITION:
1 .... HNO_3
3 .... HCl
DISCUSSION:
$CuInS_2$ wafers. Ingots grown by travelling heater method (THM) using starter powdered material with an indium flux in a sealed quartz ampoule. Cut wafers were mechanically polished with 0.06 μm alumina. Then degreased in TCE, rinsed in acetone, then MeOH, and etch polish with aqua regia.
REF: Hsu, J et al — *J Cryst Growth* 70,427(1984)

CIS-0004a
ETCH NAME: Nitric acid, dilute — TIME:
TYPE: Acid, polish — TEMP:
COMPOSITION:
1 .... HNO_3
3 .... H_2O
DISCUSSION:
$CuInS_2$, n-type wafers. Etch polish in this solution.
REF: Russak, M & Creter, C — *J Electrochem Soc,* 132,1741(1985)

CIS-0004b
ETCH NAME: — TIME:
TYPE: Acid, photo-etch polish — TEMP:
COMPOSITION:
x .... HCl
x .... HNO_3
x .... H_2O
DISCUSSION:
$CuInS_2$ n-type wafers. Solution used with intense white light to photo-etch polish surfaces.
REF: Ibid.
CIS-0005: Tenne, R P & Hodges, P — *Appl Phys Lett* 37,428(1980)
$CuInS_2$ wafers. Used solution as a polishing etch.
CIS-0006: Takeuchi, S et al — *J Phys & Chem Solids* 8,887(1985)
$CuIn_5S_8$, (111) oriented single crystal ingots grown by a Growth Freeze method. Material was n-type, and growth was (111) preferential. Wafers were cut and mechanically lapped and polished in a study of semiconducting properties.

COPPER PHOSPHO-SULFIDE, Cu_6PS_5:I/Cl/Br

General: Does not occur as a natural compound, although there are several copper sulfide minerals containing iron, bismuth, arsenic, and antimony, but not phosphorus. The structure has been referred to as the silver sulfo-germanate, argyodite, $4Ag_2S.GeS_2$ — isometric with octahedral, (111) habit, and Spinel Law twinning, e.g., 180° rotation on o(111) as bi-crystal or multiple twin lamellae. Sphalerite, ZnS, shows similar twinning lamellae, but is isometric system, tetrahedral class, not normal (cubic) class.

There is no industrial use of the compound to date, although copper sulfide minerals are important ores for both elements. There are many copper phosphate compounds in chemistry.

Technical Application: The compound has been grown by Vapor Transport (VT) from vapor/liquid/solid (VLS) materials with halogen doping: Cl_2, Br_2, and I_2. As undoped Cu_3PS_4, crystals are transparent yellow; doped with iodine, Cu_6PS_5:I, were transparent red crystals. Electrically the material is a conductor, and has been studied for general morphology.

Etching: soluble in HNO_3 and NH_4OH.

COPPER PHOSPHO-SULFIDE ETCHANTS

CPS-0001
ETCH NAME: Nitric acid — TIME:
TYPE: Acid, removal — TEMP:
COMPOSITION:
x .... HNO_3
DISCUSSION:

Cu_6PS_5:I, single crystal platelets grown by Vapor Transport (VT) with a CuI liquid flux. Crystals are transparent, red in color, forming in the isometric system, normal (cubic) class, as high temperature 43m space group. Nitric acid was used as a general etching solution.

REF: Tributsch, H & Betz, G — *J Electrochem Soc,* 131,640(1984)

CPS-0002
ETCH NAME: Ammonium hydroxide — TIME:
TYPE: Base, removal — TEMP:
COMPOSITION:
x .... NH_4OH
DISCUSSION:

Cu_6PS_5:halogens. Halogens: Cl_2, Br_2, I_2. Crystal grown by Vapor Transport (VT) between 870—900°C and above 970°C. A vapor/liquid/solid (VLS) flux was used: $PSCl_3$ (liquid); Co_2S (black powder). Cu_3PS_4 was grown as yellow, transparent crystals. Materials were conductors, and structure was referred to the natural mineral argyodite, $4Ag_2S.GeS_2$. Solution shown can be used as a removal etchant. With the flux shown, crystals of Cu_6PS_5:Cl also were grown.

REF: Fiechter, S et al — *J Cryst Growth,* 61,275(1983)

PHYSICAL PROPERTIES OF COPPER TELLURIDE, Cu_4Te_3

Classification	Telluride
Atomic numbers	29 & 43
Atomic weight	551.28
Melting point (°C)	1000
Boiling point (°C)	
Density (g/cm³)	7.4
Hardness (Mohs — scratch)	3.5
Crystal structure (isometric — tetrahedral)	(112) tetrahedron
Color (solid)	Purple/black
Cleavage (tetrahedral)	(112)

COPPER TELLURIDE, Cu_4Te_3

General: Occurs as the natural mineral rickardite, Cu_4Te_3, and weissite, Cu_5Te_3. Both minerals occur in massive form associated with pyrite and tellurium ores in minor quantity. There is no industrial use other than as minor ores of tellurium and copper.

Technical Application: Copper telluride doped with either gallium or indium has been evaluated as a possible compound semiconductor of the I—VII type. Grown as a single crystal these trinary compounds show a distinctive (112) growth habit.

Etching: Variations of aqua regia.

COPPER TELLURIDE ETCHANTS

CUTE-0001
ETCH NAME: Aqua regia, variety TIME:
TYPE: Acid, removal TEMP: RT
COMPOSITION:
1 .... HCl
1 .... HNO_3
DISCUSSION:

Cu_4Te_3 specimens as the natural mineral rickardite. Solution used as a surface cleaning and light removal etchant in a material study.

REF: Walker, P — mineralogy study, 1953

COPPER GALLIUM TELLURIDE ETCHANTS

CGTE-0001
ETCH NAME: Aqua regia, variety TIME:
TYPE: Acid, removal TEMP:
COMPOSITION:
x .... HCl
x .... HNO_3
DISCUSSION:

$CuGaTe_2$, single crystals used in an evaluation of cohesive energy features of tetrahedral semiconductors. Forty materials were studied.

REF: Aresti, A et al — *J Phys & Chem Solids,* 45,361(1984)

COPPER INDIUM TELLURIDE, $CuInTe_2$

General: Does not occur as a natural metallic compound, though there are other binary and trinary telluride minerals. There is no application of the compound in the metal industries.

Technical Application: Copper indium telluride has been grown as a ternary compound semiconductor, and is under development and study for its semiconducting characteristics.

Etching: Soluble in mixed acids of $HCl:HNO_3$ with or without water.

COPPER INDIUM TELLURIDE ETCHANTS

CUIT-0001a
ETCH NAME: Hydrofluoric acid, dilute TIME:
TYPE: Acid, float-off TEMP:
COMPOSITION:
1 .... HF
1 .... H_2O
DISCUSSION:

$CuInTe_2$, thin films. The thin films were deposited on glass substrates by CVD. Films were removed from the glass by undercut etching of the glass — the float-off technique for removal of films from substrates and study under a transmission electron microscope (TEM).

REF: Kazmnerski, L L — *J Vac Sci Technol,* 14,769(1977)

CUIT-0001b

ETCH NAME: Aqua regia, variety
TYPE: Acid, thinning
COMPOSITION:
1 .... HCl
1 .... HNO_3
2 .... H_2O

TIME:
TEMP: RT
RATE: 1000 Å/min

DISCUSSION:

$CuInTe_2$, thin films. Thin films were deposited on glass by CVD. Before removal from the glass, films were etch thinned in the solution shown. After float-off (CUIT-0001a) specimens were studied morphologically using an electron microscope (EM).

REF: Ibid.

CUIT-0002

ETCH NAME: Aqua regia, variety
TYPE: Acid, removal
COMPOSITION:
x .... HCl
x .... HNO_3

TIME:
TEMP:

DISCUSSION:

$CuInTe_2$, single crystals used in an evaluation of cohesive energy features of tetrahedral semiconductors. Forty materials were studied.

REF: Aresti, A et al — *J Phys & Chem Solids,* 45,361(1984)

PHYSICAL PROPERTIES OF CORUNDUM, Al_2O_3

Classification	Oxide
Atomic numbers	13 & 8
Atomic weight	101.94
Melting point (°C)	2050
Boiling point (°C)	2250
Density (g/cm³)	3.95—4.10
Refractive index (n=)	1.75—1.76
Hardness (Mohs — scratch)	9
Crystal structure (hexagonal — rhombohedral)	$(10\bar{1}1)$ rhomb
Color (solid)	Grey/black
Cleavage (basal)	(0001)

CORUNDUM, Al_2O_3

General: The natural mineral is widely distributed and associated with chlorate type rocks, as well as limestone and dolomite. The two latter in massive amounts, such as in the chalk cliffs of Dover, England. As emery, $Al_2O_3.Fe_3O_4$, it is a major abrasive. Clear, color tinted varieties are gem stones: ruby (red); sapphire (blue), etc. It is still used as a lapping and polishing abrasive. See Aluminum Oxide, Al_2O_3, for additional discussion.

Technical Application: The natural mineral has little application in Solid State processing other than as a polishing abrasive, but the artificially pressed powder or single crystal (sapphire) blanks have major use as circuit substrates, and other high temperature ceramic applications. Also there are amorphous ruby parts artificially fabricated. See Aluminum Oxide for an expanded discussion.

Natural corundum has been studied for many years, and was first grown as an artificial single crystal in the late 1800s by the Verneiul process for artificial rubies and sapphires.

Etching: Single acids, mixed acids, and molten fluxes.

CORUNDUM ETCHANTS

COR-0001
ETCH NAME: Metal — TIME:
TYPE: Metal, removal — TEMP:
COMPOSITION:
x .... Si
DISCUSSION:

Al_2O_3, (0001) natural corundum discs. Metallic silicon vacuum evaporated on corundum will etch the material.

REF: Reynolds, F H & Elliot, A B M — *Phil Mag,* 13,1073(1966)

COR-0002
ETCH NAME: Potassium bisulfate — TIME:
TYPE: Molten flux, removal — TEMP: 590°C
COMPOSITION:
x .... K_2SO_4, pellets
DISCUSSION:

Al_2O_3, as natural corundum. Material will dissolve in potassium bisulfate when used as a molten flux. Has been used in the metallographic study of single crystal compounds.

REF: Dana, E S & Ford, W E — *A Textbook of Mineralogy*, 4th ed, John Wiley & Sons, New York, 1950, 482

COR-0003
ETCH NAME: Hydrochloric acid — TIME:
TYPE: Acid cleaning — TEMP: Hot
COMPOSITION:
x .... HCl, conc.
DISCUSSION:

Al_2O_3, natural single crystals cut and polished en cabachon on the z-axis (0001) basal plane as star sapphires. Wash in HCl after polishing, then in H_2O for general cleaning.

REF: Walker, P — mineralogy development, 1953

PHYSICAL PROPERTIES OF CRONSTEDITE, $4FeO.2Fe_2O_3.3SiO_2.4H_2O$

Classification	Silicate
Atomic numbers	26, 14, 1, 8
Atomic weight	859
Melting point (°C)	>1000
Boiling point (°C)	
Density (g/cm³)	3.34—3.45
Refractive index (n=)	1.80
Hardness (Mohs — scratch)	3.5
Crystal structure (monoclinic — normal)	(hk0) prism
Color (solid)	Black/brownish
Cleavage (basal — perfect)	(001)

CRONSTEDITE, $4FeO.2Fe_2O_3.3SiO_2.4H_2O$

General: Occurs as a minor natural mineral in association with other iron ores in England and Brazil. No use in industry other than as a minor ore of iron.

Technical Application: The natural mineral has been studied in Solid State for general morphology and defects. In thin section, transmitted light is brilliant emerald-green, and the material may have application as light frequency lenses and filters.

Etching: H_2SO_4, HF.

CRONSTEDITE ETCHANTS

CRON-0001
ETCH NAME: Hydrofluoric acid, dilute — TIME:
TYPE: Acid, dislocation — TEMP: RT
COMPOSITION:
x .... HF
x .... H_2O
DISCUSSION:

$4FeO.2Fe_2O_3.2SiO_2.4H_2O$, (001) cleaved wafers used in a general material study. Solution used to develop dislocations on the cleaved surfaces. Authors say their specimens were brittle, although natural material is shown to be elastic in thin section. They also referred to chronstedite as a kaolin-type mineral, e.g., kaolinite, $Al_2O_3.2SiO_2.2H_2O$.

REF: Steadman, R & Pugh, J D — *Phil Mag,* 12,969(1965)

PHYSICAL PROPERTIES OF CRISTOBALITE, SiO_2

Classification	Oxide
Atomic numbers	14 & 8
Atomic weight	~60
Melting point (°C)	>1200
Boiling point (°C)	
Density (g/cm³)	2.27
Refractive index (n=)	1.486
Hardness (Mohs — scratch)	7
Crystal structure (tetragonal — normal) alpha	(110) prism
(isometric — normal) beta	(111) octahedron
Color (solid)	White
Cleavage (cubic — poor)	(001)

CRISTOBALITE, SiO_2

General: A naturally occurring high temperature form of silica and considered as a polymer of quartz. Occurs in small white octahedrons, often twinning according to the Spinel Law. The alpha to beta forms have a transition temperature range between 195 to 275°C with initial formation temperature >1200°C. The alpha-cristobalite is tetragonal; whereas beta-cristobalite is isometric.

There is no industrial use of this form of silica as a natural product due to scarcity. See silicon dioxide, SiO_2 for major discussion.

Technical Application: There is no use of cristobalite in Solid State processing, although silicon dioxide, SiO_2, as a thin film, or its various mixtures and forms as glass (silica) have major applications.

Artificially grown cristobalite has been observed as minute crystals in quartz epitaxy tubes as a by-product of oxidizing other metals and metallic compounds. Both natural and artificial crystals have been studied.

Etching: HF, KF, NH_4F

CRISTOBALITE ETCHANTS

CRIS-0001
ETCH NAME: Heat — TIME:
TYPE: Thermal, conversion — TEMP: 195—275°C
COMPOSITION:
x .... heat
DISCUSSION:

SiO_2, as alpha-cristobalite carried through inversion point to beta-cristobalite on a hot-stage microscope using polarized light for observation. Authors say that transformation time was less than 0.1 sec.

REF: Krisement, O & Tromel, G — *Z Naturforsch,* 14a(7),685(1959)

CRIS-0002
ETCH NAME: Hydrofluoric acid — TIME: Minutes
TYPE: Acid, preferential — TEMP: RT
COMPOSITION:
(1) x .... HF, conc. (2) x .... HF, vapor (hot)
DISCUSSION:

SiO_2, as natural single crystal alpha-cristobalite and beta-cristobalite used in a general study of natural SiO_2 compounds. Solutions used to polish and preferential etch specimens in a structure study.

REF: Walker, P — mineralogy study, 1952—1953

CRIS-0003
ETCH NAME: — TIME:
TYPE: Thermal oxidation — TEMP:
COMPOSITION:
x .... air
DISCUSSION:

Si, (100) wafers cut from Czochralski (CZ) ingots used in a study of re-dissolution of precipitated oxide. Material oxidized at 1000°C for 256 days. Precipitated dislocation complexes (PDCs) appeared as plate-like structures on (100) surfaces parallel to ⟨110⟩ directions. Precipitates were cristobalite. (*Note:* Crystallites of cristobalite have been observed in quartz furnace tubes as a by-product after silicon oxidation at elevated temperatures.)

REF: Shimura, F — *J Appl Phys Lett,* 39,987(1981)

PHYSICAL PROPERTIES OF DEUTERIUM, D

Classification	Heavy hydrogen (^3H)
Atomic number	H_2 isotope
Atomic weight (protium)	1.07825
*(deuterium)	2.00147 (D_2)
(tritium)	2.01605
Melting point (°C)	−252.7 (101.42)*
Boiling point (°C)	−2254.3 (3.82)*
Density (g/cm³)	1.11 (D_2O)*
Hardness (Mohs — scratch) solid	1—2
Crystal structure (isometric — normal)	(100) cube
Color gas/liquid	Colorless
(solid)	White
Cleavage (cubic — solid)	(001)

DEUTERIUM, D

General: Occurs in nature in all substances containing hydrogen in a ratio of about 5000:1 as normal (light) hydrogen H_2 to heavy hydrogen, ^3H or D. The latter is called deuterium and, as a transmutation particle — a deuteron. By repeated electrolysis of water, using nickel-iron electrodes, the light hydrogen off-gases and deuterium is concentrated about sixfold in the residue. This was the original source of "heavy water", D_2O, initially thought to be essential in the design of an atomic bomb. High concentrations of deuterium also can be obtained from metal processing areas where electrolytic cleaning tanks containing sodium or potassium hydroxide have been in operation for years as collected in a sludge concentrate.

Deuteron particles are used in cyclotrons and particle accelerators for bombardment producing protons, alpha particles or neutrons. In medicine, heavy water, D_2O, has been used as a tracer compound in the study of water elimination by the human body. It was shown that deuterium appeared in perspiration within hours of ingestion, but required about 30 days for complete elimination in urine.

Technical Application: Deuterium is not used in general processing and fabrication in Solid State, though it has been used in chemical processing as a tracer element. Many metals and metallic compounds are now being hydrogenated for special purpose applications as thin films or in the development of hydrogen storage batteries — a-Si:H. a-Ge:H, a-SiN:H using normal light hydrogen.

Deuterium and the other isotopes of hydrogen shown under the Physical Properties section have applications in atomic physics, radiation chemistry, and in irradiation studies. Several of their compounds are shown here in this section on deuterium, such as TiD_2, ScD_2, and $ErTi_2$:D_2.

Etching: See parent metal element for applicable chemicals.

DEUTERIDE ETCHANTS

TID-0001
ETCH NAME: TIME:
TYPE: Acid, removal TEMP:
COMPOSITION:
x .... HNO_3
DISCUSSION:

TiD_2 as thin films. Molybdenum blanks were used as substrates for metallization with

titanium and scandium. The materials were furnace treated with heavy hydrogen, ^3H, to form deuterides, $ErTi_2$ was also deuterided.
REF: Adachi, T et al — *J Vac Sci Technol,* 19(1),119(1981)

SCD-0001
ETCH NAME: TIME:
TYPE: Acid, removal TEMP:
COMPOSITION:
x .... HF
x .... HNO_3
DISCUSSION:
ScD_2 as thin films. See TID-0001 for discussion.
REF: Ibid.

DEU-0001
ETCH NAME: Deutron
TYPE: Particle, transmutation TEMP:
COMPOSITION:
x .... deutrons
DISCUSSION:
Deutron particles used in cyclotrons to produce protons, alpha particles, and neutrons. Used in the study of transmutation of elements and materials.
REF: Foster, W & Alyea, H N — *An Introduction to General Chemistry,* 3rd ed, D Van Nostrand, New York, 1947, 77 & 300

TRIT-0001
ETCH NAME: TIME:
TYPE: Acid, cleaning TEMP: RT
COMPOSITION:
1 .... H_2SO_4
1 .... HCl
1 .... H_2O
DISCUSSION:
SST tubing used in atomic reactors for handling tritium — a heavy isotope of hydrogen. Internal bore of tubing etch cleaned in this type of solution to remove burrs and draw marks for a highly polished internal surface to prevent spontaneous ignition of tritium during flow passage. After etching, rinse heavily with water, then with MeOH, and use the heat of a torch flame to heat and dry tube lengths. See FE-0024 for general tube cleaning.
REF: Mills, T — personal communication, 1972

PHYSICAL PROPERTIES OF DIAMOND, C

Classification	Nonmetal (carbon)
Atomic number	6 (carbon/graphite)
Atomic weight	12 (carbon/graphite)
Melting point (°C)	1900 (to carbon)
Boiling point (°C)	2000 (to CO_2 — in air)
Density (g/cm³)	3.516—3.525
Refractive index (n=)	2.4195 (1.3—1.4)
Dielectric constant (e=)	5.68—5.70
Absorption constant (cm^{-1}):	

Wavelength (microns) — 0.405	0.12 (type II xtls)
(microns) — 0.226	14.8 (type II xtls)
Absorption band (microns — 0.4 eV)	4.8
Volume expansion ($\times 10^{-6}$) 28—105°C	4.36
Hole mobility (eV)	45
Activation energy (eV)	5+
Electron mobility (cm sec^{-1}/V cm^{-1})	800
Carrier lifetime ($\times 10^{-9}$ sec)	9
Energy gap (eV)	7 (5.6)
Lattice constant (angstroms)	3.567
Hardness (Mohs — scratch)	10
Crystal structure (isometric — normal)	(111) octahedron
Color (solid)	Colorless (+ pastel hues)
Cleavage (octahedral — perfect)	(111)

DIAMOND, C

General: Occurs as a native element as the high pressure form of carbon or graphite. Diamond crystal structure is the atypical octahedron, o(111), which is called the "diamond" form. Both of the elemental semiconductors, silicon and germanium, are called "diamond" crystallographic structure. The elements sulfur, carbon, and selenium are physically and chemically related to the transition class of metals, such as carbon, manganese, molybdenum, etc. Diamond and fluorite, CaF_2, both have (111) cleavage, rather than cubic (100) as is common to the isometric — normal class. Because of its high refractive index, producing high birefringence, it is the most prized gem stone in the world and is found as transparent stones in various colors, with the clear, blue-white being "top-water"; yellow most common; but also green, orange, brown, and black (high carbon content, and called "bort"), and many diamonds have carbon flecks internally. Diamonds are found in "blue ground" pipes in Africa; one occurrence in the U.S.; in stream gravel in Brazil, Southeast Asia, India, and the northern mid-continental U.S. The DeBeers syndicate controls diamond production and sales throughout the world.

Diamonds have been grown artificially using a high pressure trigon system. At high temperature, in oxygen, diamond will burn to CO_2; in vacuum at 1900°C it will convert to graphite. "Industrial diamonds" are those diamonds that do not meet gem-grade quality, and are primarily used as finely powdered lapping and polishing abrasives either in an oil paste, in a solvent slurry, embedded in metal, rubber, or plastics, and as a "diamond" blade for cutting. It is the hardest known element or compound and is used in determining hardness of minerals and other materials in the Mohs hardness scale of 1—10, with diamond H = 10. Using the Mohs-Wooddell scale: Diamond H = 42.5, which is the more accurate. Some of the artificial carbides and borides approach diamond in hardness (H = 9+), and the Knoop hardness scale (kfg/mm^2) using the pressure of a shaped diamond stylus gives an even more accurate measure of hardness.

Technical Application: Oriented single crystal diamonds are used as mounting substrates in semiconductor device assembly as both heat sinks and active elements in the assembly structure.

Diamond abrasive in an oil paste is still a major lapping and polishing medium for all metals and metallic compounds. As diamond grit impregnated metal or resinoid blades, saw blades have been in use for cutting rocks and minerals since the 1800s, and are the primary method of cutting Solid State ingots as wafers, and the dicing of wafers as discrete devices, as well as the cutting of ceramics as circuit or dielectric substrates. As diamond embedded blocks and other shapes in rubber, metals, plastics, etc., used for general lapping, honing, polishing, and similar applications. As a single crystal material it is fabricated for optoelectric device applications, and is under study for various reactions, such as irradiation-produced color centers, similar to those observed in NaCl; for its brilliant phosphorescence with applied electrical discharge under vacuum; and reactions in a similar manner in ultraviolet light.

Doped with boron, it is an n-type semiconductor, capable of operating as a diode above 500°C like both SiC and BN. And carbon thin films are being deposited, then converted by annealing to a diamond-like-carbon (DLC) structure with possible applications as both an electrically active thin film or surface passivator. Such DLC structure also is referred to as an i-C structure. In this book there are three sections: (1) Carbon, C-; (2) Diamond, D-; and Graphite, GR-, as the basic types of this material.

Etching: Insoluble in acids and alkalies. Soluble by dry chemical etching (DCE); and in molten fluxes.

DIAMOND ETCHANTS

D-0001
ETCH NAME:
TYPE: Ionized gas, removal
COMPOSITION:
x .... Xe^+
x .... NO_2, fumes
TIME:
TEMP:
GAS FLOW:
PRESSURE:
POWER:
DISCUSSION:
D, (111) wafers used to fabricate photoelectric devices. Degrease and clean surfaces with water and alcohol prior to photo resist patterning. Gases shown used to etch a grating structure in the diamond surface by ion beam assisted etching (IBAE), which is a form of reactive ion etching (RIE).
REF: Efemow, N N et al — *J Vac Sci Technol,* B3,416(1985)

D-0002
ETCH NAME: Hydrofluoric acid
TYPE: Acid, cleaning
COMPOSITION:
x .... HF, conc.
TIME: xx h
TEMP: Hot
DISCUSSION:
D, (111) wafers used in irradiation damage studies. HF used to clean surfaces prior to irradiation, follow with through washing in water after etching.
REF: Primak, W et al — *Phys Rev,* 103,1184(1956)

D-0003
ETCH NAME: Boron
TYPE: Metal, implantation
COMPOSITION:
x .... B
TIME:
TEMP:
DISCUSSION:
D, (111) wafers cut 6°-off toward (110) used in a study of diamond semiconducting properties. Boron was ion implanted for a 5×10^{15} concentration at 200 KeV for maximum amorphous implantation (I^2) structure.
REF: Maby, F W et al — *J Appl Phys Lett,* 39,157(1981)

D-0004
ETCH NAME:
TYPE: Electron, crystallization
COMPOSITION:
x .... laser
TIME:
TEMP:

DISCUSSION:

D, crystallized from thin film carbon deposits by laser annealing. The resulting crystallization is called diamond-like-carbon (DLC) structure and shown to be high in hydrogen. Tensile cracking was observed with greatest stress at crystallite edges. Compressive stress causes inward bulging of the crystallites.

REF: Nir, D — *Thin Solid Films,* 112,41(1984)

D-0005

ETCH NAME: Tungsten — TIME:

TYPE: Metal, preferential — TEMP: 1200—1500°C

COMPOSITION:

x .... W

DISCUSSION:

D, (111) platelets used in a study of plasticity. Tungsten metal used to etch diamond surfaces at elevated temperature.

REF: Evans, T & Wild, R K — *Phil Mag,* 12,479(1965)

D-0006

ETCH NAME: Oxygen (air) — TIME:

TYPE: Gas, preferential — TEMP: 1000°C, nominal

COMPOSITION:

x .... O_2, moist air

DISCUSSION:

D, (100), (111), and (110) oriented wafers used in a defect development study. Heating in moist air will develop structure, but will not develop structure in O_2/H_2O or O_2/N_2 pure atmospheres. Below 1000°C trigon pits show a positive (+) orientation; above 1000°C a negative (−) orientation. Above 1000°C pits tend to be more hexagonal to rounded in shape rather than sharply triangular. Pits observed on all orientations below 1000°C, but not on (100) above that temperature.

REF: Evans, P R & Sauter, T — *Contemp Phys,* 2,217(1961)

D-0007

ETCH NAME: Abrasion — TIME:

TYPE: Abrasive, hardness — TEMP:

COMPOSITION:

x .... SiC

DISCUSSION:

D, (111) specimens used in a study of the relative hardness of crystallographic directions in the (111) surface. There are eight (111) octahedral exterior surface planes; four bulk (100) cubic planes at 45°; and two dodecahedral planes at 90° in each (111) surface. Author says all planes are resistant to abrasion. (*Note:* Hardness differences can be observed using the geologic Sclerometer test — draw a diamond stylus across a single crystal oriented surface in a known bulk plane direction under a specified load to determine directional bulk plane hardness. This is a standard geological mineral test.)

REF: Wilks, E M — *Phil Mag,* 6,701(1961)

D-0010

ETCH NAME: Soap — TIME: 5 min, nominal

TYPE: Detergent, cleaning — TEMP: Hot

COMPOSITION:

x .... Joy soap

DISCUSSION:

D, (111) oriented small parts used as stand-offs in device assembly. Units as squares or rectangles, and others with an angled surface section. Prior to metallization by evaporation or plating (Au/Cr or Au/Ni, respectively) as thin films, parts were degreased in hot TCE vapors, then scrubbed and soaked in a soap solution, then followed by heavy DI water washing, and N_2 blow dry. Some parts were photo resist patterned after cleaning, and used as both active and passive device structures.

REF: Fahr, F & Walker, P — personal application, 1980/1984

D-0008

ETCH NAME: Methane TIME:

TYPE: Gas, growth TEMP:

COMPOSITION:

x .... CH_4

DISCUSSION:

i-C, initially deposited as carbon from methane on glass predeposited with thin films of either gold or aluminum. Deposition of carbon was by RF plasma and films were annealed to develop a diamond-like-carbon (DLC) structure, which is also called the i-C structure. The carbon films were then evaporated with additional metals and patterns developed by standard photolithography to produce MIM (metal-insulator-metal) device test structures for evaluation of dielectric properties of the i-C films.

REF: Lamb, J D & Woolam, J A — *J Appl Phys,* 57,5420(1985)

D-0009: Szmidt, J et al — *Thin Solid Films,* 110,7(1983)

D, thin films deposited by reactive pulse plasma of graphite onto silicon, (111) and (110) substrates. Used in a comparative study of diamond and boron nitride as surface coating dielectrics.

D-0011

ETCH NAME: TIME:

TYPE: Thermal, preferential TEMP:

COMPOSITION:

x .... heat

DISCUSSION:

D, (111) single crystal specimens. A study of the etching of trigon pits on diamond surfaces.

REF: Patel, A R & Patel, S M — *Br J Appl Phys,* 1,1445(1968)

D-0012: Mitchell, E W J — *J Phys Chem Solids,* 8,444(1959)

D, specimen. A review of work done on diamonds.

D-0013

ETCH NAME: TIME:

TYPE: Pressure, fabrication TEMP:

COMPOSITION:

x .... pressure

DISCUSSION:

D, fabricated as artificial diamond. A Pressure Trigon system consisting of three large pressure cylinders at 60° aimed at a central point capable of pressures in excess of 100,000 psi. Place a small carbon sample at the center point on a holding base and apply pressure to structurally condense the carbon to form a diamond.

REF: *Sci Am,* (May 1978)

D-0014
ETCH NAME: TIME:
TYPE: Electroless, plating TEMP:
COMPOSITION:
x .... nickel
DISCUSSION:

D, as polycrystalline 4 μm sized particles embedded in nickel. Material fabricated by the Shock Sinter Method (SSM). Plated specimen discs used in a study of possible catalytic face orientation reaction of diamonds during electroless nickel plating with a phosphate type bath.

REF: Feldstein, N & Lancsek, T S — *J Electrochem Soc,* 131,3026(1985)
D-0015: Cowan, et al — U.S. Patent 3,401,019
Reference for Shock Sinter Method of fabricating polycrystalline diamonds.

D-0016
ETCH NAME: Iodine TIME:
TYPE: Halogen, passivation TEMP: Elevated
COMPOSITION:
x .... I_2
x .... H_2O/MeOH
DISCUSSION:

D, as single crystal elements. Iodine washed onto surfaces as a surface passivation with some adsorption. This acts to reduce chemical attack by acids, specifically, H_3PO_4.

REF:

DIELECTRICS, M_xO or M_xN

General: Many of the natural oxide or silicate minerals can be considered as dielectric compounds, and several of these natural minerals or their artificial counterparts are so used. Quartz, SiO_2, and corundum, Al_2O_3 are representative of natural oxide dielectrics; zircon, $ZrSiO_4$, as a silicate; sellaite, MgF_2, as a fluoride dielectric, as well as the natural halides. The mineral rutile, TiO_2, usually as transparent, clear artificial titanium dioxide, is a major dielectric compound used in fabricating microelectronic circuits, as well as being used as the artificial gem stone "titania".

In the general metal industries, the majority of the natural minerals are used as ores for their metal content, rather than as dielectrics, though several are used as surface coatings for passivation, corrosion resistance, etc. All of the nitrides are artificial, and the nitridization of metal surfaces with hot nitrogen gas furnace treatment has long been used for hardening and corrosion resistance of surfaces and metal nitrides show dielectric properties just as do the oxides and silicates.

Technical Application: All of the minerals mentioned, nitrides and artificial compounds specifically designed for dielectric levels are used in Solid State device processing. They may be as discrete resistors or capacitors or as thin film coatings fabricated as planar surface elements on a device or on a circuit substrate.

As the majority of dielectric materials also have electrical and light frequency characteristics, they are used as filter elements, as well as capacitors or resistors, and there are ferrite materials used as resistors in microelectronic fabrication that are specifically designed for microwave frequency levels.

All of the dielectric materials may be fabricated as pressed powder blanks (crystalline), as deposited amorphous thin films, as single crystal films, or as hydrogenated thin films, such as colloidal c-SiN:H, etc. TiO_2, as an example, is supplied as a pressed powder blank or an oriented single crystal blank — parallel or normal to the c-axis, each with different dielectric constant values.

The dielectric materials are not listed here but appear following their parent metals, such as aluminum oxide, after aluminum, etc.

Etching: See individual compounds.

PHYSICAL PROPERTIES OF DYSPROSIUM, Dy

Classification	Lanthanide
Atomic number	66
Atomic weight	162.50
Melting point (°C)	1407
Boiling point (°C)	2600
Density (g/cm^3)	8.54
Thermal conductance (cal/sec)(cm^2)(°C/cm)	0.024
Specific heat (cal/g) 25°C	0.041
Latent heat of fusion (cal/g)	25.2
Heat of vaporization (k-cal/g-atom)	67
Atomic volume (W/D)	19.0
1st ionization energy (k-cal/g-mole)	157
1st ionization potential (eV)	6.8
Covalent radius (angstroms)	1.56
Ionic radius (angstroms)	0.92 (Dy^{+3})
Electrical resistivity (micro ohms-cm) 298 K	92.6
Compressibility (cm^2/kg × 10^{-6})	2.55
Magnetic moment (Bohr magnetons)	10.64
Tensile strength (psi)	35,700
Yield strength (psi)	32,600
Magnetic susceptibility (× 10^{-6} emu/mole)	99,800
Transformation temperature (°C)	1384
Coefficient of linear thermal expansion (× 10^{-6}/cm/cm/°C) 25°C	8.6
Cross section (barns)	940
Hardness (Mohs — scratch)	3—4
Crystal structure (hexagonal — normal)	$(10\bar{1}0)$ prism, hcp
Color (solid)	Grey
Cleavage (basal)	(0001)

DYSPROSIUM: Dy

General: Does not occur as a native element. It is a member of the rare earth yttrium group in the lanthanide series, and is mainly found in pegmatites in minerals such as gadolinite and fergusonite with other rare earths. The metal salts are noted for their brilliant yellow color, but there is little use of the metal in industry due to scarcity.

Technical Application: There has been no use of the metal in Solid State processing, to date, though it has been grown as a single crystal for general morphological study. Spheres have been used to evaluate magnetic properties, and material defects by etching.

Etching: Mixed acids with HNO_3 + HAc/HCl/HF/H_2O_2.

DYSPROSIUM ETCHANTS

DY-0001

ETCH NAME: TIME:

TYPE: Acid, preferential TEMP:

COMPOSITION:

4 .... HNO_3

6 .... HAc

DISCUSSION:

Dy, as single crystal spheres used in a study of magnetic properties. Solution shown used to develop defects and grain structure.

REF: Behrendt, D R et al — *Phys, Rev,* 109,1544(1958)

DY-0002: Bond, W L — *Rev Sci Instr,* 22,344(1951)

Reference for method of making single crystal spheres.

DY-0003: Bond, W L — *Rev Sci Instr,* 25,410(1954)

Dy, specimens. Reference for crystal structure of the material, and other physical data.

ELECTRIDE, $C_xH_yO_zM$

General: A new class of crystalline materials called "Electrides" has been developed from original development of alkalides. It is a cross between a metal and semiconductor fabricated from an organic solvent molecule of cylindrical or rope-like structure enclosing an alkali cation in a cavity. Initial development required fabrication below −20°C, but a new solvent has produced stable compounds up to 45°C. The compounds are highly reactive and can spontaneously disintegrate, but single crystals of several millimeters in size have been grown by special vacuum evaporation and drying techniques.

The compounds have shown optical, opto-electric (including infrared), and magnetic properties similar to those of Solid State semiconductors and, with further stability development, may well lead to a new class of semiconductor compounds. Some work with other organic semiconductors also is included.

Technical Application: None at present.

Etching: N/A at present.

ELECTRIDE ETCHANTS

ETD-0001
ETCH NAME: TIME:
TYPE: TEMP:
COMPOSITION:
DISCUSSION:

Electride, $C_xH_yO_zM$, where M = Li, Na, P, K, Rb, Ce, the alkali metals. An alkali anion, such as potassium is enclosed in the cavity of a cylindrical or rope-like solvent molecule structure, and shows possible semiconducting type properties. These compounds are a cross between metals and semiconductors. Vary in color from black to dark blue.

REF: Dye, J L — *Sci Am,* (Sept 1987), 66

ORG-0002
ETCH NAME: TIME:
TYPE: Pressure, forming TEMP:
COMPOSITION:
x .... pressure
DISCUSSION:

$C_{14}H_{10}$, as single crystal anthracene were grown and evaluated as an organic semiconductor. The material contains three benzene rings, and can be a derivative from both petroleum and coal.

REF: Matte, H & Pick, H — *Z Physik,* 134,566(1953)

ORG-0003
ETCH NAME: TIME:
TYPE: Pressure, forming TEMP:
COMPOSITION:
x .... pressure
DISCUSSION:

$C_{10}H_3$ as single crystal naphthalene were grown and evaluated as an organic semiconductor. The material contains two benzene rings, and can be a derivative of petroleum and coal.

REF: Pick, H & Wissman, W — *Z Physik,* 6,959(1955)

ORG-0001
ETCH NAME: TIME:
TYPE: Pressure, forming TEMP:
COMPOSITION:
x .... pressure
DISCUSSION:
Atomatic substances, such as derivatives from petroleum and coal fabricated as single crystals, are predicted to have semiconducting properties.
REF: Many, A et al — *J Chem Phys,* 23,1733L(1955)

EPOXY & POLYIMIDE

General: All plastics, epoxies, and polyimides are artificial compounds formed by the polymerization of simple hydrocarbons and their derivatives. Epoxies are liquid paste formulations as a singular compound that is self-hardening at room temperature or as an "A" and "B" component, proportionally mixed when ready for use, and both types oven cured. The epoxy compounds, alone, are nonconductive sealing compounds used for mounting parts on substrates or for sealing packages. As a loaded-epoxy — metal powder additives — they are conductive pastes used in the electronic and Solid State industries for the assembly of devices and parts. Silver epoxy, Ag-epoxy, has high electrical conductivity and is widely used, although gold, aluminum and copper are also available. As all epoxies can slowly self-harden at room temperature, they are often refrigerated with only the quantity to be used for a specific assembly mixed at room temperature when ready to be used.

Some mixtures have a working life of 4 to 8 h, once mixed, while others may set in 5—10 sec. Both epoxies and polyimides are air oven cured between 100—150°C for 20 to 50 min. Some polyimides cure up to 200°C, and 250—300°C capability is under development.

Storage-life is from 6 months to 1 or 2 years, even 3 years at room temperature, and some epoxies have still been useful after 3 years. Both epoxy and polyimide can be refrigerated at +40°C, and some are kept frozen from 0 to −40°C. They are often shipped from the supplier in dry ice (solid CO_2 at −78°C). Refrigerating extends working-life and shelf-life, even though they are still listed for 6 months to 1 year usability.

There is no physical contact danger with most epoxies and polyimides, but there can be some processing problems. If overcured, particles may flake. Where silicon diodes were mounted under nitrogen in a glass package, with time, silver epoxy dried and flaked causing device contact failure, such that the use of epoxies and polyimides are not recommended for high reliability type assemblies. Where silver epoxy is used and the mounting substrate then put through an oxygen RF plasma cleaner system, the epoxy will turn coal-black on the surface . . . a surface silver oxide is formed. It is usually insufficient to affect electrical contact integrity on test samples, but such discoloration often is not acceptable for a finished product. The black oxide can be removed by rinsing assemblies in ammonium hydroxide.

Technical Application: Both unloaded and loaded epoxies and polyimides are used in Solid State devices assemblies for electrical testing. Discrete semiconductor devices, resistors, capacitors, etc. are mounted on circuit substrates with a small dot of epoxy, or for package sealing, to include a pressure extruded package.

Where there is an epoxy mounted device failure under test, the device and epoxy can be removed and replaced by heat-charring the epoxy (heat to about 200°C), then scraping away the device/epoxy without affecting other components of the circuit. There is equipment available for this purpose . . . it consists of a hot stage, a jet of hot nitrogen, and a metal tong for scraping. For complete removal and cleaning of a circuit substrate, acid etching can be used.

Etching: Mixed acids or heat.

EPOXY ETCHANTS

GOLD EPOXY

EPAU-0001a
ETCH NAME:
TYPE: Acid, removal
COMPOSITION:
30 ml .... HNO_3
10 ml .... HCl
100 ml .... H_2O
TIME: 1—3 min
TEMP: RT
DISCUSSION:

Au, spherulitic particles in an epoxy matrix as a contact paste, such as Epo-Tech #81. Used as an electrical contact in device assembly. Where parts are unaffected or for general removal, the etchant shown has been used for removal and cleaning. Water content was variable, depending upon desired soak time.

REF: Walker, P — personal development/application, 1981

EPAU-0001b
ETCH NAME: Heat
TYPE: Thermal, removal
COMPOSITION:
x .... heat
TIME:
TEMP: 300—400°C
DISCUSSION:

Au, pellets in an epoxy matrix used as an electrical contact paste, such as Epo-Tech #81. Heat part until epoxy carbonizes (chars) and then scrape surface to remove remaining char. See Carbon section for specific etchants.

REF: Ibid.

SILVER EPOXY

EPAG-0001a
ETCH NAME:
TYPE: Acid, removal
COMPOSITION:
1 .... HF
3 .... HNO_3
TIME: 1—3 min
TEMP: RT
DISCUSSION:

Ag, pellets in an epoxy matrix used as an electrical contact paste, such as Ablestik #519. Widely used in assembly of semiconductor devices for electrical test evaluation and in package assembly. Solution used to etch remove silver epoxy.

REF: Walker, P — personal development/application, 1980

EPAG-0001b
ETCH NAME: Heat
TYPE: Thermal, removal
COMPOSITION:
x .... heat
TIME:
TEMP: 300—400°C
DISCUSSION:

Ag, pellets in an epoxy matrix used as an electrical contact paste in device and parts

assembly for test evaluation or in a final packaging of units. Heat part until epoxy carbonizes, then scrape surface for removal.
REF: Ibid.

ALUMINUM EPOXY

EPAL-0001
ETCH NAME: Sulfuric acid — TIME:
TYPE: Acid, removal — TEMP: Hot
COMPOSITION:
x .... H_2SO_4
DISCUSSION:
Al, pellets in an epoxy matrix or boron doped. Used with n-type silicon solar cells. Can be used as a contact paste on aluminum where #62 solder is used (180°C m.p.). Ti doped epoxy used for p-type silicon. (*Note:* Solution is a general removal solution for plastics.)
REF: Brochure: Electroscience Lab, NJ, 1983

PHYSICAL PROPERTIES OF ERBIUM, Er

Classification	Lanthanide
Atomic number	68
Atomic weight	167.28
Melting point (°C)	1497
Boiling point (°C)	2900
Density (g/cm^3)	9.05
Thermal conductance (cal/sec)(cm^2)(°C/cm)	0.023
Specific heat (cal/g) 25°C	0.040
Heat of fusion (k-cal/g-atom)	4.1
Latent heat of fusion (cal/g/°C)	24.5
Heat of vaporization (k-cal/g-atom)	67
Atomic volume (W/D)	18.1
Covalent radius (angstroms)	1.57
Ionic radius (angstroms)	0.89 (Er^{+3})
Electrical resistivity (micro ohms-cm) 298 K	86.0 (107) (10^7)
Coefficient of thermal linear expansion ($\times 10^{-6}$ cm/cm/°C) 25°C	9.2
Compressibility (cm^{2}kg $\times$ 10^{-6})	2.39
Cross section (barns)	166
Magnetic moment (Bohr magnetons)	9.5
Tensile strength (psi)	42,400
Yield strength (psi)	38,700
Magnetic susceptibility ($\times 10^{-6}$ emu/mole)	44,100
Hardness (Mohs — scratch)	4—5
Crystal structure (hexagonal — normal)	$(10\bar{1}0)$ prism, hcp
Color (solid)	Dark grey
Cleavage (basal)	(0001)

ERBIUM, Er

General: Does not occur as a native element. It is a member of the rare earth yttrium group in the lanthanide series (elements 63—71). Most rare earths are found in acidic pegmatite dikes: the mineral gadolinite, for the yttrium group; monazite, for the cerium group; or allinite, for both groups, e.g., cerium group includes elements 57—63. All of the rare earth elements are considered quite active metals similar to alkalines: they burn in air to their oxides; the oxides dissolve in water to hydroxides with hydrogen evolution; and their salts are brilliantly colored, often yellow. As erbium is one of the rarer rare earths, it has found little use in industry.

Technical Application: There has been only minor use in Solid State processing, to date. It has been grown as a single crystal in a general study of the growth of rare earth elements with thermal etching used to develop defects and structure. It also has been deposited by RF (magnetron) sputter and CVD as a thin film on silicon with conversion to a silicide: ErSi, Er_2Si, and $ErSi_2$ for evaluation as a blocking layer in device construction. As an additive element, it has been used in the growth of artificial garnets with possible device applications.

Etching: Mixed acids $HF:HNO_3$; $HF:H_2O_2$, etc. Thermal heat treatment.

ERBIUM ETCHANTS

ER-0001
ETCH NAME: Heat — TIME:
TYPE: Thermal, preferential — TEMP:
COMPOSITION:
x .... heat
DISCUSSION:

Er, single crystal specimens. A growth method study of rare earth elements. After crystal growth, annealing developed defects, low angle grain boundaries, and other structure. Additional metals studied were dysprosium, gadolinium, holmium, thulium, terbium, and yttrium.

REF: Nigh, H E — *J Appl Phys,* 34,3323(1963)

ER-0002
ETCH NAME: — TIME:
TYPE: Acid, removal — TEMP:
COMPOSITION:
x .... HF
x .... HNO_3
DISCUSSION:

Er, as an evaporated thin film with Si on (110) GaAs wafers Si doped n-type used in a study of ion beam etching. See ERSI-0001 for heat formation of the silicide, under Silicides. Also GAS-0247.

REF: Wu, C S et al — *J Electrochem Soc,* 132,918(1985)

ERBIUM DIHYDRIDE ETCHANTS

ERH-0001
ETCH NAME: — TIME:
TYPE: Acid, Float-off — TEMP:
COMPOSITION:
x .... H_2O
DISCUSSION:

ErH_2 and ErH_3 formed by epitaxy growth after evaporation of erbium on NaCl substrates. In hydrogen at 298 K (RT) material was poly-ErH_2; at 488 K using different NaCl substrate surface orientations: (1) ErH_2 (100) on (100) NaCl; (2) ErH_3 (100) on (100), (110) and (111) NaCl. Heating the dihydride in H_2 will convert from the isometric-normal (cubic) (100) fcc thin film orientation to hexagonal-normal (0001) hcp ErH_3 structure (trihydride). Presence of H_2 in NaCl surfaces influences dihydride growth. The (100) NaCl substrates were cleaved;

whereas the (110) and (111) substrates were cut and polished. Both substrate polishing before Er metallization and float-off of the hydride compounds for TEM study can be done with water.

REF: Rahman Khan, M S — *Thin Solid Films,* 113,207(1984)

ETCHING

General: Etching is the dissolutionment or dissolving of any material by any medium: liquid, gas or solid. In chemistry, water (H_2O) is considered the universal solvent as it will dissolve all natural minerals with time to the state of a water soluble salt.

Salt itself can be an etching solution and, as brine (salt water), a major solvent of water soluble compounds as reflected by the ocean waters of the world. Subsurface ground waters can contain elements, such as sulfur, iron, or silica (siliceous water) and, as such hot waters percolate up through rock fissures and strata, they become mineral formers.

In nature, not only are all minerals formed from solution, but are often altered after forming by the action of gases and liquids. As atmospheric action, this process is called weathering in geology or oxidation in chemistry and material processing. As oxidation, iron to iron oxides (rust); as sulfurization, the green verdigris (copper sulfate) that appears on copper. Corrosion is another term applied to atmospheric weathering, particularly in metal and alloy evaluations, which are subjected to seacoast environments (salt air) as well as being submerged in salt water. They also are furnace evaluated under various hot gas conditions, all for corrosion effects and chemical stability. Smog has been in existence for centuries wherever man has built a city or established an industry, such as smoke from hibachi cooking fires in Japan, peat smoke in the British Isles, or exhaust smoke from the smelting of metal ores . . . most open-hearth operations are no longer in existence due to such pollution.

Needless to say, the controlled cleaning and etching of materials is the subject of this book. Acids, alkalies, alcohols, solvents, solid metals, or compounds are all involved, and certain names have been established as references for different applications. Wet chemical etching (WCE), for the use of liquid chemicals; dry chemical etching (DCE), for the use of ionized gases; electrolytic etching (EE), where electric current is included; or molten flux etching (MFE), using liquified metals or chemical compounds at elevated temperature, are the four main divisions. There are other primary etching methods, and many subcategories beyond these four, such as the difference between a molecular and ionized gas, the use of pressure, only, or the use of vacuum, etc. This section has been assembled where the referenced articles are general in nature, or the reference may be shown as "N/A" due to common usage.

Technical Application: See specific metals and metallic compounds by individual names for additional detailed solutions and their applications.

Etching: Generalized. See individual elements and compounds for specific solutions.

ETCHING ETCHANTS

DIRECT PRESSURE ETCHING (DPE)

PRE-0001
ETCH NAME: Pressure — TIME:
TYPE: Pressure, forming — TEMP: RT to molten
COMPOSITION:

x .... pressure

DISCUSSION:

Direct pressure can be an etching agent, though more often as a forming vehicle. In furnace reduction of metal ores, the hot, liquid metal is extracted at atmospheric pressure, and poured as a raw ingot called pigs in the metal industry. Subsequent alloys and parts are re-melted as mixtures, and can be poured into forms for rough shape with final lathe grinding/ cutting to shape. Such molds often include pressure as part of the shaping technique. Some oxides, such as ruby, are directly hot formed from a melt in pressurized molds as rods, sheet, etc. Pressure extrusion — metal through diamond hole-dies (wire) — through heated rollers (metal/glass sheet) — into a form (plastic packages, lamp shades, etc.). This may include vacuum forming such as heating a plastic sheet, then pulling a light vacuum to draw the material to a desired shape (TV modules, tote boxes, etc.). It may be as direct pressure molding, such as in the compaction of powders (medical pills, metal or plastic parts), and can include heat to melt the material during forming. Extrusion molding is used to form a plastic compound around a device or assembly as a package.

Pressure figures on certain compounds, such as the micas, are obtained by striking the surface. On micas this produces a six-rayed percussion figure on an (0001) basal plane surface. Meteoric iron surfaces have classified structure developed by a combination of heat and pressure during passage through the atmosphere.

Single crystals are subjected to increasingly high pressure in Solid State study of high pressure crystallographic structures — silicon has at least six such crystallographic structures — irons and steels have alpha, beta and delta phase structures — ice is normally hexagonal system, but has a high pressure transition form that is isometric (cubic) system, and many other metals and compounds, both natural and artificial, have similar transition pressures/ temperatures with an alteration of the atomic lattice to that of another crystal system. In geology and physics, pressures on the order of 300,000 atm are used on rocks in the study of magma formation. In Solid State, liquified sodium chloride, NaCl, has been pressurized between two glass plates in a material study, and many ferrite compounds are similarly pressure studied. Carbon has been converted to diamond by direct pressure only.
REF: See individual metals and compounds for specific applications.

DRY CHEMICAL ETCHING (DCE)

DCE-0001
ETCH NAME: RF plasma
TYPE: Ionized gas, cleaning/defect
COMPOSITION:
x .... CF_4
x .... O_2 (5%)

TIME:
TEMP:
GAS FLOW:
PRESSURE:
POWER:

DISCUSSION:

Gas, ionized. As singular or mixed gases applied by an RF or DC plasma to effect etching action for cleaning, general preferential etching or surface structuring. The mixture shown has general usage, and there are several others with specific design applications.

Argon, as Ar^+ ion cleaning, is a fairly standard practice before sputter metallization, alloy, or metallic compound deposition. Most vacuum sputter systems have one to four target locations with one location used for argon ion cleaning. Specimens are placed on a rotating platen below the targets, Ar^+ ion cleaned, then the platen rotated beneath sputter targets for metal or compound deposition. These systems may include a magnet on targets, and are then called RF magnetron sputter systems.

Where a gas is reactive, such as Cl^+ or BCl_3, the method is called reactive ion etching (RIE), and there are other acronyms for similar specialized systems. The mixture shown can be classified in this category.

RF plasma cleaning systems using N^+ or O^+ ions are widely used in Solid State material processing, such as for surface cleaning of semiconductor wafers, or their substrate circuit assemblies. This form of cleaning removes organic contamination from surfaces, and has been used for etch removal of germanium thin films.

Ion milling also uses an RF plasma of argon. In this case, the vacuum system has a tillable and/or rotatable copper plate on which a substrate is waxed down with a low vapor pressure wax, such as Apiezon M, and a controlled jet of ionized argon used to etch pattern the thin film metallization for a circuit substrate, or similar pattern etch a device structure on a wafer. Ion bombardment is a general term for similar action, as is irradiation. The latter referring to both gas ions (H^+, He^+, Xe^+, Ne^+, N^+, O^+, etc.) and ionized particles: protons, neutrons, deutrons, electrons, etc. All such etching is done under vacuum conditions, though electron irradiation can be done with an electron microscope (EM), or transmission electron microscope (TEM).

An allied operation is ion implantation, I^2, where ionized gases are implanted into a semiconductor wafer to form p-n junctions, or p^+ and n^+ layers, such as in forming semi-insulated structures (SI). Both gallium arsenide, as GaAs:Cr, (100) (SI), and indium phosphide, InP:Fe (100) (SI) wafers, as examples.

All ionized gases can produce sub-surface damage in materials or thin films and, as irradiation, many metals and metallic compounds have been studied for such effects. This includes effects on thin film adhesion, amorphatization, and high reliability radiation testing with Van der Graaf or Triga facilities. Dry chemical etching (DCE) is strictly the use of ionized gases, not gases used in their molecular state. The latter are included here in this text as a subsection of wet chemical etching (WCE).

REF: Bierlein, T K & Mastel, B — *Rev Sci Instrum,* 30,832(1959)

REF: See individual metals/compounds for specific applications.

ELECTROLYTIC ETCHING, EE

EE-0001

ETCH NAME: Electrolytic etching — TIME:

TYPE: Electrolytic, polish — TEMP:

COMPOSITION: — ANODE:

(1) x .... acids/salts — CATHODE:

(2) x .... alkalies — POWER:

(3) x .... alcohol

DISCUSSION:

Electrolytic etching. Any liquid solution used with an electric current applied between the positive (+) anode and negative (−) cathode to effect anodic etching action where the part being processed is the anode. Although cathodic etching is referred to in the literature, as such, it is a deplating or film removal type action, not true etching of a bulk part. Switching from anode-to-cathode is done in some etching sequences, as in growing a hydrated oxide on a metal surface such as silicon. The power applied varies with the type material, and metals usually are in the milliampere range; whereas with oxides, nitrides, and silicates, effectively, the insulator materials require amperage levels of current. In electrolytic etching, both an ampere and voltage level are used and should be shown, although very often only the amperes are shown. Many industrial metals and alloys are best electrolytically etched, but most of the Solid State semiconductor type materials use wet chemical etching. Electrolytic etching may include the use of molecular gas bubbled through a solution, often used as a jet etch system.

Metal plating from solution is another method of using electrical current applied to a liquid mixture containing a metal salt, such as gold, nickel, copper or chromium sulfides, chlorides, cyanides, etc. This is not a plating book, per se, such that only a few plating

systems are shown where they are widely used in processing Solid State materials. Electrolytic gold from cyanide baths with electroless nickel under plating, Au/Ni thin films, used in both device and substrate circuit fabrication; or Au, Ag and Cu plating of p-n device junctions, as two examples. (*Note:* All electrolytic solutions in this book appear in the etchant format shown, above. The first "chemical" descriptor term of TYPE is always "Electrolytic", followed by an "Application" term, as shown above. This approach has been used due to the number of electrolytic applications involved, even though standard practice would say: electropolish, electroplating, etc.)
REF: N/A
EE-0002: Niggins, J K — *J Electrochem Soc,* 106,999(1959)
Electrolytic etching in a study of anodic dissolution of metals and electropolishing.
EE-0003: Rowland, P R — *Nature* 171,931(1963)
A study of the mechanisms of electropolishing.
EE-0004: Lowenheim, F A — *Modern Electroplating,* 3rd ed, John Wiley & Sons, New York, 1974, 52
The electroplating of metals. Includes specific cleaning chemicals and procedures for preparation of materials for plating, as well as plating formulas and processes.
REF: See individual metals/compounds where solutions are listed in this Electrolytic Format.

GAS ETCHING (GE)

GE-0001
ETCH NAME: Hydrogen — TIME:
TYPE: Gas, removal/growth — TEMP: 500—1500°C
COMPOSITION:
x .... H_2
DISCUSSION:
Gas or gas mixtures used in their molecular state to effect a cleaning or etching action. Hydrogen firing (reducing atmosphere) — also called furnace firing — is used to remove surface oxides or scale; in the alloy of assembly of parts, to prevent oxidation; and with pressure included, to hydrogenate materials (Si:H, SiC:H, etc.). Forming gas (FG) 85% N_2:15% H_2 is used in a similar manner, but is nonflammable or explosive.

Inert gases (N_2, Ar, He, etc.) are used as furnace gases in general processing and, at elevated temperature, some are used for their reactive qualities, such as nitrogen for surface nitridization, or oxygen to grow an oxide thin film . . . combined, these two gases will form oxynitrides. In the open atmosphere, at RT to hot, inert gases — N_2, most commonly — but also Ar or He are used as pressure jet blow-off drying systems, or parts are dried in air, with and without heating (IR lamps, hot plates, etc.). Flame torches are another use of gases, with acetylene; oxyhydrogen; and propylene among the leaders. Such torches are used for soldering, brazing, drying, or cleaning of parts and assemblies. Other gases are used as hot vapors from acids, such as chlorine from hydrochloric acid, or oxygen from water, nitric acid and hydrogen peroxide. Chlorine to clean surfaces, oxygen to either oxidize or as a surface cleaning step with the oxide removed with hydrofluoric acid.

In any case, this is the use of gases in their molecular state, not as dry chemical etching (DCE) with the gas in its ionized state.
REF: See individual materials, general formats.

MOLTEN FLUX ETCHING (MFE)

MFE-0001
ETCH NAME: Molten flux etching
TYPE: Salt, preferential
COMPOSITION:
x .... KOH, NaCl, Fe etc.
TIME: Variable
TEMP: Elevated
DISCUSSION:

Molten flux etching is the use of a salt (NaCl); a metal (Fe, Cu, etc.); an alkali (KOH, NaOH, etc.) from their solid state, liquified at their melting point temperature to effect removal, polishing or development of defects. They can be mixtures: KOH:NaOH (eutectic), or NaCl:$NaCO_3$, etc. High temperature, chemically inert metals, such as titanium or molybdenum, are so etched for both cleaning and polishing of surfaces. On single crystals, such as silicon, molten flux alkalies are used for defect development. Molten metals may be the primary etching mediums on certain other metals, or be diffused into another to decorate and enhance dislocations, defects, and structure.

Molten fluxes also are used for single crystal growth of materials, such as garnets and ferrites.

The general wet chemical etching (WCE) etchant format is used to present molten fluxes, as well as metals, by name, when used as an etching medium or method of decorating defects.

Note that tin and zinc are plated from molten fluxes, as tin plating or galvanized zinc. The part is dipped into the molten metal, then withdrawn with a thin coated surface.

REF: See individual materials for specific applications.

THERMAL ETCHING (THE)

THE-0001
ETCH NAME: Thermal etching
TYPE: Thermal, removal/defect
COMPOSITION:
x .... heat
TIME:
TEMP: Elevated
DISCUSSION:

Thermal etching of metals and compounds is the use of pure heat as an etching medium. The reference shown below as a study of the influence on surface energy of metals. But thermal etching also is used to develop etch figures on any metal surface or, if a single crystal, to develop dislocations, defects, and structure. Pure thermal etching is done under vacuum conditions; otherwise a gas is included, such as in furnace heat treatment and annealing. The use of heat only can be for surface cleaning and drying and, where a specimen is heated on a hot plate in air, it is called heat-tinting . . . oxide colors used to differentiate sample structure in metallographic study of metals subjected to various test conditions of strain, fatigue, etc.

REF: Moore, A J — *Acta Metall,* 6,293(1958)

REF: See individual metals/compounds shown as heat and thermal etching in the wet chemical etching format.

VACUUM ETCHING (VE)

VE-0001

ETCH NAME:

TYPE: Vacuum, preferential

COMPOSITION:

x .... vacuum

TIME:

TEMP: Cryogenic to elevated

VACUUM: 10^{-4} to 10^{-10} Torr

DISCUSSION:

Vacuum, used as a preferential etching vehicle, and often includes a temperature level. As thermal heat treatment, ingots, wafers, or parts are subjected to elevated temperatures up to some 2000°C under vacuum; gases are grown as single crystals under pressure cryogenic conditions in vacuum, and the changes in vapor pressure of the system used to preferentially etch structure. It has been observed that some metals, under an ultra high vacuum (UHV), have shown surface etching effects as pits or hillocks. Vacuum also is used as a drying medium in metal processing, as well as in food processing.

REF: See individual materials, general formats.

WET CHEMICAL ETCHING (WCE)

WDC-0001

ETCH NAME: Wet chemical etching

TYPE: Chemical, removal/defect

COMPOSITION:

x .... chemicals, solvents, alcohols

TIME: Variable

TEMP: Cold, RT to hot

DISCUSSION:

Wet chemical etching is the use of liquids for etching, cleaning, or development of structure, as acids, bases, alcohols or solvents. This is the primary method of processing all metals and metallic compounds, and may include gases, such as bubble chlorine through an etch; or intense light, as in plating out from an etch solution to accentuate multilayer structure or p-n junctions in semiconductor devices.

The majority of solutions shown in this book are WCE with the format as shown above. The format also is used for all chemical processing not covered by DCE; EE; or the other methods discussed in this section.

REF: See individual solutions under the specific materials.

PHYSICAL PROPERTIES OF EUROPIUM, Eu

Classification	Lanthanide
Atomic number	63
Atomic weight	152
Melting point (°C)	826
Boiling point (°C)	1439
Density (g/cm^3)	5.26
Specific heat (cal/g) 25°C	0.0395
Heat of fusion (k-cal/g-atom)	2.2
Latent heat of fusion (cal/g)	16.5
Heat of vaporization (k-cal/g-atom)	42
Atomic volume (W/D)	28.9
1st ionization energy (K-cal/g-mole)	131

1st ionization potential (eV)	5.67
Covalent radius (angstroms)	1.85
Ionic radius (angstroms)	0.98 (Eu^{+3})
Electrical resistivity (micro-cm) 298 K	91.0
Compressibility ($cm^2/kg \times 10^{-6}$)	8.29
Neutron cross section (barns)	4600
Magnetic moment (Bohr magnetons)	7.12
Magnetic susceptibility ($\times 10^{-6}$ emu/mole)	33,100
Coefficient of thermal linear expansion ($\times 10^{-6}$ cm/cm/°C) 20°C	26
Hardness (Mohs — scratch)	2—3
Crystal structure (isometric — normal)	(100) cube, bcc
Color (solid)	Steel-grey
Cleavage (cubic)	(000)

EUROPIUM, Eu

General: Although it occurs in nature it is extremely rare, as a member of the Rare Earth group of elements, and usually assigned to the yttrium sub-group even when found with cerium sub-group elements. Gadolinite, $Be_2FeY_2Si_2O_{10}$, is variable in formula, and with a high cerium oxide content called cergadolinite. As ''gadolinite earths'' the variety is high in yttrium sub-group elements. Often occurs in pegmatites with other rare earth bearing minerals. Monazite, $(Ce,La,Di)PO_4$ is similarly variable in rare earth content, and there are some commercial deposits as monazite sands. Monazite often contains a high percentage of thorium oxide, ThO_2, mined chiefly for that compound in India.

All rare earths are considered quite active metals, similar to calcium: they burn in air to their oxides: the oxides dissolve in water with a hissing sound, like quicklime; and hydroxides are alkaline. Europium has little use in industry due to scarcity.

Technical Application: No major use in Solid State processing at present, though there is a europium containing artificially grown garnet which may have use as an optical device, filter element, etc.

Europium has been grown as a single crystal and studied for its physical properties, to include paramagnetic behavior.

Etching: Air conversion to oxide.

EUROPIUM ETCHANTS

EU-0001

ETCH NAME: Argon — TIME:

TYPE: Gas, storage — TEMP: RT

COMPOSITION:

x .... Ar

DISCUSSION:

Eu specimens. Samples processed under argon. Fresh surfaces prepared by cutting with a knife, then wrapped in tantalum foil and stored in a silica capsule under argon to maintain a bright surface. Specimens used in a study of paramagnetic behavior.

REF: Colvin, R V et al — *Phys Rev,* 122,14(1961).

PHYSICAL PROPERTIES OF EUROPIUM OXIDE, Eu_2O_3

Classification	Rare earth
Atomic numbers	63 & 8
Atomic weight	352
Melting point (°C)	
Boiling point (°C)	
Density (g/cm³)	7.42
Hardness (Mohs — scratch)	4—5
Crystal structure (isometric — normal)	(100) cube
Color (solid)	Pale rose
Cleavage (cubic)	(001)

EUROPIUM OXIDE, Eu_2O_3

General: Does not occur as a natural compound by itself, though it may be an oxide, in part, as extracted from the mineral gadolinite (see Europium for further discussion).

Technical Application: Not used in Solid State processing at present, though europium metal is one constituent in some artificial garnets. The oxide is obtained by burning the metal in air, and the oxide is hygroscopic. The oxide is pale rose in color and called europia.

Etching: Soluble in water.

EUROPIUM OXIDE ETCHANTS

EUO-0001

ETCH NAME: Water — TIME:

TYPE: Acid, removal — TEMP:

COMPOSITION:

x .... H_2O

DISCUSSION:

Eu_2O_3 specimens. Material will dissolve in water with a hissing sound like quicklime, and resulting hydroxide is alkaline.

REF: Foster, W & Alyea, H N — *An Introduction to General Chemistry*, 3rd ed, D Van Nostrand, New York, 1941, 423

EUROPIUM SULFIDE ETCHANTS

EUS-0001

ETCH NAME: — TIME:

TYPE: — TEMP:

COMPOSITION:

DISCUSSION:

EuS, single crystal spheres fabricated for general morphological study. No etch shown. Europium is a rare earth element, and both the element and its compounds are highly reactive. As this is a sulfur compound, it can probably be etched in CS_2 like sulfur.

REF: Franzblau, M L et al — *J Appl Phys*, 38,4462(1967)

FERRIC OXIDE, Fe_2O_3

General: Occurs in nature as the iron mineral hematite, Fe_2O_3 — common name red iron ore, and is a major ore of iron. It is found in rocks of all ages and in several forms, from single crystals resplendent black in color to massive earthy "red ocher"; in botryoidal shape as "kidney ore"; as "oolitic hematite", originating as a probable colloidal precipitate from iron bearing ocean waters around a central sand-grain core, forming much like a pearl in an oyster.

Although it is not magnetic, it may appear to be so when it is an admixture of hematite and magnetite, Fe_3O_4 — magnetic black iron ore. Both occur with the mineral corundum, Al_2O_3 as emery, Al_2O_3:Fe_2O_3/Fe_3O_4. Hematite can be distinguished by its cherry-red streak, infusibility in a bunsen burner flame, and can be reduced to a grey magnetic powder with soda in an RF flame.

Industrially it is a major ore of iron. As powder, it is a red paint pigment, and a coloring agent in glass, ceramics and enamels. Also as a powder, it is "red rouge", used both in cosmetics and as a lapping and polishing abrasive with a Mohs hardness of about H = 6. It can be fabricated artificially by treating ferric chloride with steam at high temperature; or by the action of air and hydrochloric acid on iron; and formed from several molten fluxes.

Technical Application: As ferric oxide (iron sesquioxide) it has no direct application in Solid State device fabrication. It is used as a lapping abrasive, as a thin film in the fabrication of photo resist masks; and as an optical filter. As MFe_3O_4, it is fabricated as microwave and memory devices; as $M_3Fe_5O_{12}$ (garnet) for bubble memory devices, microwave filters, and surface wave devices and as $PbFe_{12}O_{19}$. As already mentioned, also as thin film, semitransparent photo resist mask.

Etching: Slowly in HCl.

See Hematite, Fe_2O_3

FERRIC OXIDE ETCHANTS

FEO-1000

ETCH NAME: Hydrochloric acid — TIME:

TYPE: Acid, removal — TEMP: Warm

COMPOSITION:

x .... HCl, conc.

DISCUSSION:

Fe_2O, thin film deposited on soda-lime glass as a photo resist mask. Film patterned by photolithography and etched in hydrochloric acid in fabricating photo masks for semiconductor processing. More widely used than chromium as it is harder, less prone to physical handling damage, and more stable.

REF: Lipman, R — personal communication, 1983

FERRITE, FeM_xM_y

General: The word is from Latin: ferrum, hence the symbol for iron, Fe, and the word "iron" is Anglo-Saxon. Native iron does occur, occasionally in large ore bodies, but more often as small grains in other rocks of basic origin. Iron occurs with two valences: Ferric, Fe^{++} and Ferrous, Fe^{+++}, and magnetic iron (extracted from magnetite, Fe_3O_4) contains iron of both valences, such that it is called ferrosoferric iron.

In iron processing the term ferrite refers to soft iron with less than 0.04% carbon, also called wrought iron . . . the original form of iron first smelted by the Hittites in Asia Minor (Turkey) around 1350 B.C. and is still used for decorative purposes. In metallographic specimens it can be recognized as a constituent by its soft dove-grey color.

Ferrite also refers to a group of metallic compounds as a class of materials . . . even metal oxides containing iron have been called ferrites. In effect, any material containing iron may be referred to as a ferritic compound, ferrous or ferric, depending upon iron valence. All ferrite compounds are of interest for their ferromagnetic, ferro-electric, or ferro-optic capabilities with several developed as operational devices in Solid State.

Technical Application: In Solid State development of semiconductor device assemblies, a class of ferrites has been designed as resistor materials with particular metal additives, such as zinc, manganese, etc., for specific power and/or frequency levels associated with the microelectronic circuitry. They are supplied as square blanks — dark black, highly polished surfaces in thickness ranges of 0.0005 to 0.025, and 1 × 1″ square or larger. The blanks are metallized by plating (Au/Ni) or evaporated/sputtered (Au/Cr and other combinations), then cut to the required size as a discrete resistor element. Metallization of ferrite of this type can be difficult for, like other highly polished surfaces, metal adhesion can be a problem. Roughing the face by abrasion (400-grit SiC) can be used when possible (not for high frequency), or lightly oxidizing — 10 min at 200°C on a hot plate in air — will aid in the thin film metal adhesion.

Most ferrites used as resistor elements are crystalline, pressed powder blanks; others have been grown as single crystals, such as iron carbide, Fe_3C. See Iron Alloys, Iron, and Steel sections for additional discussion and etchant solutions.

Etching: Nital, Picral, and mixed acids.

FERRITE ETCHANTS

HCFE-0001
ETCH NAME: Kalling's etch — TIME: 10—60 sec
TYPE: Acid, polish — TEMP: RT
COMPOSITION:
100 ml .... HCl
5 g .... $CuCl_3$
100 ml .... EOH
DISCUSSION:

$Ho_2Co_{14}Fe_3$ specimens arc melt fabricated, and used in a study of metallic compounds. Solution used as a general removal and polishing etchant. Also studied Ho_2Fe_{17}. See Cesium Platinide for other materials studied.

REF: Slepowronsky, M et al — *J Cryst Growth,* 65,293(1983)

MAGT-0004
ETCH NAME: Hydrochloric acid — TIME:
TYPE: Acid, removal — TEMP:
COMPOSITION:
x .... HCl, conc.
DISCUSSION:

Fe_3O_4, specimens from ingots grown by the Czochralski (CZ) method. A study of single crystal iron ferrites grown by this method. Other iron doped materials were ZnO, Ga_2O_3, CuO, and MnO.

REF: Horn, F N — *J Appl Phys* 32,900(1961)
FER-0001a: Walker, P — mineral study, 1950—1964

Fe_3O_4 natural single crystal specimens used in a study of iron oxides. Solution used concentrated, and as $1HCl{:}1H_2O$ for surface cleaning. Also studied other natural iron specimens of hematite, Fe_2O_3, limonite, $2Fe_2O_3.3H_2O$ and goethite, $Fe_2O_3.H_2O$ [FeO(OH)].

FER-0001b
ETCH NAME: Trichloroethylene TIME:
TYPE: Solvent, cleaning TEMP: 85°C
COMPOSITION:
x .... TCE, vapor
DISCUSSION:

Fe:Mn:Zn, pressed powder blanks used as resistor material for microelectronic circuits. Blanks were highly polished 1 × 1 × .015/.025″ size. After vapor degreasing in TCE, bake out in air oven at 150°C for 30 min prior to metallizing with Au/Ni plating or Au/Cr evaporation.

REF: Walker, P et al — personal application, 1980—1985

FER-0002: Buckley, D H — *J Vac Sci Technol,* A3(3),762(1985)

Fe:Mn:Zn and Fe:Ni:Zn ferrites used in a study of ceramic microstructure and adhesion of thin films. The iron absorbs O_2 even in vacuum or under argon. The study applied sliding contacts across the metal thin film surfaces. The metal film remains in place, and bonding was assisted by the presence of O_2. Cu, Ni, Rh, Co, and Be were rubbed across surfaces in this evaluation.

FER-0003
ETCH NAME: TIME:
TYPE: Thermal, defect TEMP: Elevated
COMPOSITION:
x .... heat
DISCUSSION:

Ferrite specimens as crystalline material used in a study to determine directions of magnetization in polycrystalline ferrites. Thermal etching was better than chemical etching to develop well defined grain boundaries with the remainder of the surfaces relatively smooth. To thermal etch: first, mechanically polish surface, then hold in furnace for 10 min at the same temperature and atmosphere used for annealing/sintering.

REF: Callaby, D R — *Phys Rev,* S31.375S(1961)

FER-0004
ETCH NAME: TIME:
TYPE: Acid, cleaning TEMP:
COMPOSITION:
x .... HF
x .... HNO_3
x .... EOH
DISCUSSION:

$MFeM_2$, material grown in evaluating new semiconducting type compounds as ferrites. M = Ag, Cu, and M_2 = S, Se, Te. Solution shown can be used for general cleaning.

REF: Zhuze, U P et al — *Zh Tekh Fiz,* 28,233(1958)

MAGT-0005
ETCH NAME: TIME:
TYPE: Acid, removal TEMP: (1) Boiling
(2) Molten
COMPOSITION:
(1) x .... HCl, conc. (2) x .... $NaSO_4$, fused
DISCUSSION:

Fe_3O_4 single crystal material grown in a "skull melter" with RF generator. This was an isomorphic ingot series from Fe_3O_4 to $FeTiO_3$ (natural mineral is ilmenite). The titano-

magnetites are grown by oxygen control (fugacity, fO_2). During growth, O_2 concentration was buffered with CO_2/CO. With addition of tin or tin foil to either solution shown above, and containing dissolved ferritic material, a brown-violet to fine blue or violet color is observed in solution depending on the amount of titanium present. (*Note:* This color reaction in solutions containing Ti has been a primary method used in determinative pyrolytic mineralogy of ilmenite for many years.)
REF: Aragon, R et al — *J Cryst Growth,* 61,221(1983)

COFE-0001
ETCH NAME: Nital — TIME:
TYPE: Acid, removal — TEMP:
COMPOSITION:
x .... HNO_3
x .... EOH
DISCUSSION:

CoFeO, (100) wafers cut from ingots grown by the Verneuil arc-image furnace method. After cutting and mechanically polishing, re-anneal at 1200°C several days in air for ferrite material $CoO_4(Co_{1-c}Fe_c)O$.
REF: Hoshino, K & Peterson, R L — *J Phys Chem Solids,* 46,229(1985)

RFE-0001
ETCH NAME: Nital — TIME:
TYPE: Acid, removal — TEMP:
COMPOSITION:
x .... HNO_3
x .... EOH
DISCUSSION:

RFe_2, as rare earth ferrites. R = Sm, Gd, Tb, Dy, Ho, and Er. Materials were fabricated in an arc melt furnace under argon. Then ingots were annealed 200 h in sealed quartz ampoules.
REF: Klimker, H et al — *J Phys Chem Solids,* 46,157(1985)

PHYSICAL PROPERTIES OF FLUORAPATITE, $(CaF)Ca_4(PO_4)_3$

Classification	Phosphate
Atomic numbers	20, 9, 15, & 8
Atomic weight	480
Melting point (°C)	1200
Boiling point (°C)	
Density (g/cm^3)	3.17—3.23
Refractive index (n=)	1.630—1.648
Hardness (Mohs — scratch)	5
Crystal structure (hexagonal — tripyramidal)	$(21\bar{3}0)$ prism, 3rd order
Color (solid)	Colorless to colored
Cleavage (basal — imperfect)	(0001)

FLUORAPATITE, $(CaF)Ca_4(PO_4)_3$

General: As the natural mineral "apatite", there are two major subspecies: fluor- and chloroapatite, with the former more common, and there are a number of additional varieties of differing constituent composition among, the two subspecies. It is quite widely distributed, often occurring in iron and tin mines. It has a variety of colors — blue, yellow, green, pink, brown and white. The fine transparent varieties have been used as gem stones, even though the mineral is relatively soft for this purpose. Along with other phosphate rocks and minerals, the apatites have primary use as industrial fertilizers.

Technical Application: Apatite-type single crystals have been grown and doped for use as long-pulse CW laser devices, both as a phosphate, and as a silicate oxy-apatite (no phosphate), but otherwise similar in crystal structure. The fluorapatite, rather than chloroapatite, is the base material, and doping elements are rare earths. The silicate types are being developed as improved Q-switch lasers.

Etching: Aqua regia; fluorine salts, and molten fluxes.

FLUORAPATITE ETCHANTS

FAP-0001a
ETCH NAME: Aqua regia — TIME: 1—3 min
TYPE: Acid, preferential — TEMP: RT
COMPOSITION:
3 .... HCl
1 .... HNO_3
DISCUSSION:

$Ca_5(PO_4)_3F$:Nd as artificially grown single crystals doped with neodymium, and used as long pulse CW lasers. In this study a series of silicate oxy-apatites were grown as possible materials for improved Q-switch lasers. General formula of these compounds: $MeLn_4(SiO_4)_3O$ with Me = Mg, Ca, and Ln = Y, La, Gd. $SrLa_4(SiO_4)_3O$ and $CaLa_4(SiO_4)_3O$ discussed in detail. Crystals were melt grown in an iridium boat under argon, and some IrO flakes were observed on the single crystal wafers. Polish wafers in molten eutectic flux: NaF:KF at 350°C prior to preferential etching with aqua regia. Aqua regia develops pits at dislocation sites. (*Note:* Formula for natural fluorapatite is $(CaF)Ca_4(PO_4)_3$, and as chloroapatite is $(CaCl)Ca_4(PO_4)_3$).

REF: Hopkins, R H et al — *J Electrochem Soc,* 118,637(1971)

FAP-0001b
ETCH NAME: — TIME:
TYPE: Molten flux, polish — TEMP: 350°C
COMPOSITION:
1 .... NaF
1 .... KF
DISCUSSION:

$Ca_5(PO_4)_3F$ and silicate oxy-apatites. Single crystal specimen wafers were polished in the eutectic molten flux shown prior to dislocation etching with aqua regia. See FAP-0001a for further discussion.

REF: Ibid.

PHYSICAL PROPERTIES OF FLUORINE, F_2

Classification	Halogen
Atomic number	9
Atomic weight	38 (F2)
Melting point (°C)	−233
Boiling point (°C)	−187

Density (g/cm^3 — solid) −187°C	1.108
(g/l — liquid) 18°C	1.69
Thermal conductance ($\times 10^{-4}$ $cal/cm^2/cm/°C/sec$) 20°C	0.579
Specific heat (cal/g/°C) 20°C	0.18
Latent heat of fusion (cal/g)	10.1
1st ionization potential (eV)	17.42
Ionic radius (angstroms)	1.33 (F)
Cross section (barns)	0.010
Vapor pressure (°C)	202.7
Refractive index (n=)	1.000195
Hardness (Mohs — scratch) solid	1—2
Crystal structure (isometric — normal) solid	(100) cube
Color (solid)	Yellow-green
(liquid)	Yellow-green
(gas)	Pale yellow
Cleavage (cubic — solid)	(0001)

FLUORINE CONTAINING ETCHANTS

Fluorine is a pale greenish yellow gas at standard temperature and pressure (22.2°C and 760 mmHg) that can be condensed as a pale-yellow liquid with a boiling point of −187°C. It is the most reactive element known and a member of the halogen group of elements which includes chlorine, bromine, and iodine, in order of reactivity. Chlorine, too, is a gas; bromine a liquid; and iodine a solid. Fluorine does not occur as a free element in nature although traces of hydrogen fluoride are occasionally found in association with volcanic activity. The best known fluorine mineral is fluorite, CaF_2, which has been used as a flux material in metallurgy as "fluorspar" even before the discovery of fluorine and its establishment as a separate element in 1886 by Moissan. Prior to its discovery, fluorspar was known to contain calcium and some other element that reacted much like chlorine. Because of its high reactivity with elements such as sulfur, phosphorus, carbon, silicon, and boron — it will catch fire in fluorine, although most metals, with the exception of gold and platinum, will burn in the presence of the gas.

Of all of the compounds of fluorine, the best known is probably hydrogen fluoride, HF — better known as hydrofluoric acid. It is the primary etch for glass, with silicon tetrafluoride, SiF_4, as a gaseous by-product. Glass is etched by either immersion in the liquid (a smooth surface) or by vapor (a rough surface). The latter is widely used for pattern etching of glass objects — drinking glassware, plates, and objets d'art. Microscope slides have been vapor etched to roughen the surface for improved adhesion of thin film metal deposits where, after deposition, the metal film is removed by the float-off technique, e.g., immersion in HF. Gold thin films have been prepared in this manner for morphological and defect study under high power magnification.

In processing semiconductor wafers, silicon dioxide, SiO_2, and silicon nitride, Si_3N_4 are deposited as thin films — 1000 to 3000 Å thick — and used as masks against subsequent pattern etching, diffusion or epitaxy growth. The oxide/nitride surfaces are photolithographically processed with patterns developed by UV exposure of a photo resist coating. The underlying oxide or nitride then removed by etching with hydrofluoric acid or a mixture of $HF:NH_4F$ to expose the semiconductor surface for diffusion, etching, epitaxy etc. The $HF:NH_4F$ solutions are called buffered hydrofluoric acid (BHF) etchants. The ammonium fluoride, NH_4F, is commonly used as a 40% concentration and added to concentrated HF, although it is sometimes added as a saturated solution (sat. sol.). This NH_4F saturated solution in HF is a primary etchant for frequency tuning of quartz crystal blanks used for radiofrequency devices. As the concentration of NH_4F varies by individual preference or application, the acronym BHF should be considered a general classification term, not a specific etch mixture. Such solutions also have been referred to as buffered oxide etch

(BOE), though BHF is more common. The mixture ratio of HF to NH_4F varies from 1HF:1NH_4F(40%) to 1HF:100NH_4F, sat. sol., and may include dilution with water, and the ammonium fluoride may be replaced with ammonium bifluoride, $NH_4F.HF$.

Other fluoride compounds are used as etching and cleaning solutions on both semiconductors and other metals and compounds such as sodium and potassium fluoride, NaF and KF, respectively. These halide-type compounds may be used as solutions for wet chemical or electrolytic etching or as molten flux etching.

Of all the HF acid mixtures, solutions of HF:HNO_3; HF:HNO_3:H_2O or HF:HNO_3:HAc — oxidation-reduction systems — are the most widely used etch solutions for a wide range of material processing where they have no, or limited, solubility in single acids. Understandably, there are hundreds of possible ratio combinations with such mixtures and every laboratory, establishes their own particular mixtures as required for a specific process step. Much development and study has been done and is still being done, on these solutions and many of the mixtures referenced in this Etchant Section are of this type. Generally speaking, all of these solutions are polishing-type etchants within certain limits: when high in HF or HNO_3 they are self-limiting where excessive reduction or oxidation produces a passivation reaction. Such solutions are used surface "stain" etch as they produce recognizable color patterns. The HF type stain is difficult to remove as it is, in effect, a fluorine-burn; whereas the HNO_3 stain is an oxide and easily removed with HF. The most rapid mixture is 1HF:3HNO_3, and either water, H_2O or glacial acetic acid, CH_3COOH (HAc or GLA) is used as an inhibitor or rate control agent. Although glycerin is sometimes added to these solutions as a viscosity control agent, it is not recommended as nitric acid and glycerin are the precursor mix for nitroglycerin. If viscosity control is required, ethylene glycol, CH_2OHCH_2OH, is recommended. The excessive addition of water or acetic acid to HF:HNO_3 mixtures can severely alter polishing action to the point where controlled etch reaction is highly erratic and results are poor. Most of the solutions are used at room temperature, due to their relatively high reactivity and exothermic nature and, for batch etching of semiconductor wafers, etching is often done in a controlled temperature bath.

These solutions, as well as most other etchants, become progressively preferential as temperature is reduced toward 0°C, regardless of their polishing capability at room temperature. Several mixtures are used at 8°C, chilled with a mixture of ice in acetone; others chilled with liquid nitrogen for use at 0°C. Such cold etchants are used for surface cleaning, or as slow preferential etchants for structure, electrical tuning of a p-n junction device, etc.

Other widely used HF mixtures are the HF:H_2O_2:H_2O solutions where the nitric acid has been replaced as the oxidizer by hydrogen peroxide, H_2O_2. Hydrogen peroxide is supplied commercially as a 30% solution and, chemically, is called "superoxol". This should not be confused with the etchant that also is called superoxol (Camp #2): 1HF:1H_2O_2:4H_2O (GE-0065a). In addition, there is an etch solution called "Peroxide Etch": 1 mlHF:1 mlH_2O_2:1HAc (see GE-0060f), which should not be confused with the concentrated or dilute pure H_2O_2.

The hydrofluoric/hydrogen peroxide etchants are polishing solutions on most semiconductor and other materials, and used in a similar manner to those hydrofluoric/nitric acids. As H_2O_2 it is an extremely violent and reactive oxidizer, in comparison to others, it has a tendency to dissociate rapidly once in solution with heavy oxygen bubble evolution, particularly when used above room temperature. Because of this, such mixtures can become rapidly depleted in the peroxide, such that the peroxide may be added only when ready for solution use, and use is stopped when complete dissociation in reached. There are always exceptions . . . some solutions are mixed, allowed to sit and "age", then used as a "stagnant" solution.

Potassium permanganate, $KMnO_4$, and chromium trioxide, CrO_3, or other chromium compounds, such as sodium dichromate, $Na_2Cr_2O_7$, etc., are oxidizers and, when mixed with HF, are used as defect/dislocation etchants on many semiconductor materials. Several

$HF:CrO_3$ solutions have been developed, and are discussed in the preceding section on Chromium Containing Etchants. $HF:I_2$ and $HF:Br_2$ with varying halogen concentrations are used as defect/dislocation etchants and discussed in sections similar to that for chromium.

CP4 (Camp #4) etch: $30HF:50HNO_3:30HAc:0.6Br_2$ was originally developed as a defect and light orientation figure (LOF) etch for (100) germanium ingots and wafers, and used at room temperature. It also is a polishing solution, and has been used on a number of other semiconductor, metals, and compounds as a removal, polish, thinning, cleaning or step-etch at room temperature, below room temperature, hot to boiling. Without bromine, it is called CP4A and, as CP4, modified, the bromine has been replaced with I_2, $Cu(NO_2)_3$, $AgNO_3$, etc.

The following list of HF-type etches is presented to show the wide variety of metals and compounds to which such solutions are applied. Many additional solutions are shown under their appropriate etchant section headings.

Formula	Material	Use	Ref.
HF, conc.	Al	Removal	AL-0021
$xHF:xH_2O$	Al	Metallographic macro-etch	AL-0016
$30HF:47HNO_3:50HCl$	Al	Preferential	AL-0020a
$xHF: H_2O$	AlN	Removal & cleaning	ALN-0002
$xHF:xHNO_3:xHAc$	AlSb	Polish	ALSB-0001b
$xHF:xH_2O_2$	AlSb	Polish	ALSB-0001a
$30HF:50HNO_3:30HAc:0.6Br_2$	AlSb	Oxide removal & cleaning	ALSB-0010
HF, conc.	B	Removal	B-0001
HF, conc.	BaF_2	Removal	BAF-0005
$1HF:3HNO_3$	BF_3	Dislocation	BF-0001
$xHF:xH_2O$	Bi	Damage, removal	BI-0002
15 ml HF:15 ml HNO_3:45 ml HCl: 25 ml H_2O	Au	Preferential	AU-0011
$1HF:20CrO_3$, sat. sol.	$CaWO_4$	Dislocation	CAW-0003b
HF, conc.	CdF_2	Polish	CDF-0001b
$1HF:3HNO_3:3H_2O$	$CdSiAs_2$	Polish	CSA-0001
$1HF:1HNO_3$	CdTe	Polish	CDTE-0001e
$2HF:3HNO_3:1H_2O$	CdTe	Preferential	CDTE-0001b
$3HF:2H_2O_2:1H_2O$	CdTe	Preferential	CDTE-0001f
HF, conc.	CoO	Preferential	COO-0001
HF, conc.	D (diamond)	Cleaning	D-0002
$1HF:4HNO_3:5H_2O$	Fe_2C	Thinning	FEC-0001
HF, conc.	GaAs	Selective/removal	GAS-0024
HF, vapor	GaAs	Cleaning	GAS-0129
$1HF:9H_2O$	GaAs	Cleaning	GAS-0112
$2HF:3HNO_3:2HCl$	GaAs	Polish	GAS-0145a
$1HF:1HNO_3:1HAc$	GaAs	Cleaning	GAS-0037a
$1HF:3H_2O_2$	GaAs	Preferential	GAS-0049
$1HF:1H_2O_2:50H_2O$	GaAs	Selective/step-etch	GAS-0051a
$5HF:1H_2O_2:1H_2O$	GaAs	Polish	GAS-0053b
$1HF:1H_2O_2:4H_2O$ "superoxol"	GaAs	Preferential	GAS-0167a
$1HF:1HNO_3$	GaAs	Cleaning	GAS-0047a
$1HF:3HNO_3:2H_2O$	GaAs	Polish, rapid	GAS-0006d

Formula	Material	Use	Ref.
1 ml HF:2 ml H_2O:6 mg $AgNO_3$: "A/B" 1 g CrO_3	GaAs	Preferential	GAS-0059
1HF:5HNO_3:10HAc	GaSb	Polish	GASB-0002
1HF:1HNO_3	GaSb	Polish	GASB-0001e
1HF:1HNO_3	GaP	Polish	GAP-0001
HF, conc.	Ge	Fracture: HF vs. air medium	GE-0064
1HF:8H_2O	Ge	Preferential	GE-0150e
xHF:xHNO_3:xHCl:xH_2O_2	Ge	Preferential	GE-0170
1HF-3HNO_3:12HAc "Dash Etch"	Ge	Preferential	GE-0065a
1HF:3HNO_3:8HAc	Ge	Polish/thinning	GE-0171
1HF:1H_2O_2:4H_2O "superoxol"	Ge	Preferential	GE-0002
30HF:50HNO_3:30HAc:0.6Br_2 CP4	Ge	Preferential polish	GE-0065c
15 ml HF:30 ml HNO_3:33 ml HAc:80 mg I_2	Ge	Preferential	GE-0185
1HF:1HNO_3	Ge	p-n junction etch	GE-0033
4HF:7HNO_3:2H_2O	Ge	Polish cleaning	GE-0036
HF, conc.	GeN	Removal	GEN-0002
1HF:10H_2O	GeN	Removal	GEN-0002
1HF:1HNO_3	GeAs	Removal	GEAS-0001a
10 ml HF:45 ml HNO_3:45 ml H_2O	Hf	Cleaning	HF-0001
1HF:1H_2O_2:20H_2O	Hf	Removal	HF-0002
1HF:1H_2O_2:4H_2O "superoxol"	InAs	Preferential	INAS-0006a
74HF:75HNO_3:15HAc:0.6Br_2	InAs	Preferential	INAS-0006d
1 ml HF:2 ml H_2O:8 mg $AgNO_3$: 1 gCrO_3 A/B Etch	InAs	Preferential	INAS-0005d
HF, conc.	InSb	Oxide removal	INSB-0027
10HF:25HNO_3:20HAc	InSb	Polish	INSB-0001a
1HF:5H_2O_2:xH_2O	InSb	Polish	INSB-0009
1HF:1H_2O_2:4H_2O "superoxol"	InSb	Preferential	INSB-0024a
30HF:50HNO_3:30HAc:0.6Br_2 CP4	InSb	Polish	INSB-0001f
1HF:1HNO_3	InSb	Removal	INSB-0002e
1HF:1HNO_3:6H_2O	InSb	Preferential	INSB-0005a
1HF:1H_2O_2:2H_2O	InSb	Dislocation	INSB-0025c
8HF:100H_2O	InP	Native oxide removal	INP-0045a
1 ml HF:2 ml H_2O:8 mg $AgNO_3$: 1 g CrO_3	InP	Preferential	INP-0051
1HF:10HBr	InP	Preferential	INP-0027
5HF:1MeOH	InP	Dislocation	INP-0037a
5HF:1Br_2	InP	Dislocation	INP-0037b
100 ml HF:100 ml HNO_3:160 ml HAc: 2 ml Br_2	LiF	Dislocation	LIF-0001e
1HF:2HNO_3	$LiTaO_3$	Preferential	LITA-0002
1HF:2HNO_3	$LiNbO_3$	Preferential	LINB-0002
xHF:xH_2O	MgO	Preferential	MGO-0010
xHF:xH_2O_2	Mo	Removal	MO-0001a
xHF:xHNO_3	Mo	Removal	MO-0001f
xHF:xMnO_4	Mo	Removal	MO-0001c
3.5 ml HF:96 ml H_2SO_4:0.5 ml HNO_3:xCrO_3	Mo	Polish	MO-0008c

Formula	Material	Use	Ref.
HF, conc.	NB	Cleaning	NB-0004b
HF, conc.	Nb	Electropolish	NB-0007
$1HF:4HNO_3$	Nb	Polish	NB-0002
HF, conc.	$NbAl^y$	Removal	NBAL-0001
$1HF:1HNO_3:10HAc$	Nb_3Se	Cleaning	NBSE-0001
HF, conc.	Nb_3B_3	Removal	NBB-0001c
HF, vapor	Ni	Removal	NI-0004b
$1HF:1H_2O_2$	$PtSb_2$	Preferential	PTSB-0001e
$1HF:1HNO_3:1H_2O$	$PtSb_2$	Preferential	PTSB-0001c
$3HF:5HNO_3:3HAc$	$PtSb_2$	Preferential	PTSB-0001f
$8HF\text{-}2HNO_3$	Re	Preferential	RE-0001
$1HF:3HNO_3:12HAc$ "Dash Etch"	Sb	Polish	SB-0001a
$1HF:2HNO_3:24HAc:1Br_2$	Sb	Dislocation	SB-0003b
HF, conc. + intense light	Si	Junction stain	SI-0188a
HF, conc.	Si	Oxide removal	SI-0002a
HF, vapor	Si	Cleaning	SI-0009
$1HF:10H_2O$	Si	Cleaning	SI-0004
$1HF:3HNO_3:12HAc$ "Dash Etch"	Si	Preferential	SI-0044
$1HF:5HNO_3:1HAc$	Si	Thinning	SI-0074
$1HF;5HNO_3:1HAc$	Si	Polish	SI-0076
$1HF:3HNO_3:1HAc$	Si	Dislocation	SI-0075
$10HF:50HNO_3:30HAc:0.6Br_2$ CP4	Si	Polish/preferential	SI-0173
$2HF:15HNO_3:5HAc$	Si	Structure forming	SI-0173a
$1HF:3HNO_3$	Si	Polish	SI-0046b
$1HF:1HNO_3:50H_2O$	Si	Removal/polish	SI-0073
xHF:xMeOH	Si	Cleaning	SI-0016
1HF:1.5 *M* $CrO_3:1H_2O$	Si	Defect	SI-0030
2HF:0.15 *M* $K_2Cr_2O_7$ "Secco"	Si	Preferential/dislocation	SI-0036
2HF:1 0.75 *M* $CrO_3:1.5H_2O$ "Shimmel"	Si	Preferential/dislocation	SI-0017
$1HF:3CrO_3$ (33%) "Sirtl"	Si	Preferential/dislocation	SI-0039
50 ml HF:50 ml HNO_3:100 mg $KMnO_4$	Si	Polish	SI-0125
$4HF:2Cu(NO_3)_2$	Si	Junction stain	SI-0188b
$xHF:xI_2$	Si	Preferential	SI-0159
10 ml HF:2 mg I_2:20 ml MeOH	Si	Preferential	SI-0204
$1HF:9NH_4F(40\%)$ "BHF"	Si	Cleaning/slow polish	SI-0021a
5 mlHF:20 mgNH_4F:200 mlH_2O "BHF"	Si	Oxide/nitride removal	SI-0026
30 mlHF:8.4 gNaF/l	Si	Electropolish, jet	SI-0174b
$xHF:xH_2O$	SiC	Cleaning	SIC-0001
HF, conc.	SiC	Cleaning	SIC-0004
$2HF:3HNO_3$	SiC	Removal	SIC-0008
$2HF:1CrO_3$ (33%) "Sirtl"	SiC	Polish	SIC-0012a
1HF:xKF, sat. sol.	SiC	Electrolytic, microstructure	SIC-0013b
$1HF:1H_2O$	SiO_2(Qtz)	Polish removal	QTZ-0001
$1HF:3HNO_3$	SiO_2(Qtz)	Cleaning, quartzware	QTZ-0010

Formula	Material	Use	Ref.
$1HF:1HNO_3$	SiO_2(Qtz)	Thinning, alpha-quartz	QTZ-0006
$1HF:20HNO_3:20EG$	SiO_2(Qtz)	Tuning, alpha-quartz	QTZ-0007
50 ml HF:50 g $NH_4F.HF$:100 ml H_2O: 50 ml Gly	SiO_2(Qtz)	Polish, alpha-quartz	QTZ-0011
HF, conc.	SiO_2	Patterning/pinholing	SIO-0004
HF, vapor	SiO_2	Removal/roughening	SiO-0006
$1HF:100H_2O$	SiO_2	Cleaning/oxide removal	SIO-0007
$1HF:2HNO_3:60H_2O$ "P Etch"	SiO_2	Removal	SIO-0011a
$1HF:100NH_4F$, sat. sol. "BHF"	SiO_2	Removal	SIO-0009
1HF:6NH4F (40%) "BHF"	SiO_2	Removal	SIO-0011b
$1HF:10NH_4F:15H_2O$ "BHF"	SiO_2	Step-etch	SIO-0014
200 ml HF:9 g CrO_3:100 ml H_2O	SiO_2	Dislocation	SIO-0021a
$xHF:xNH_4F:xH_2O$ "BOE"	SiO_2	Removal	SIO-0006
HF, vapor	Si_3N_4	Removal	SIN-0002a
HF, conc.	Si_3N_4	Removal/pinholing	SIN-0002c
$15HF:10H_3PO_4:60EOH$	Si_3N_4	Preferential	SIN-0007
$1HF:1H_2O$	Ta	Cleaning	TA-0008
$2HF:5HNO_3:3HAc$ "CP4A"	Ta	Acid saw cut/polish	TA-0003
$3HF:5HNO_3:1HAc$	Ta	Electropolish	TA-0002d
$1HF:2HNO_3:1H_2O$	Ta	Removal	TA-0001
$1HF:1NH_4F$ (20%)	Ta	Preferential	TA-0004b
$2HF:2HNO_3:5H_2SO_4$	Ta	Preferential	TA-0004b
$1.5HF:2HNO_3:5H_2SO_4$	Ta	Polish	TA-0009
$1HF:20H_2O$	$TaSi_2$	Cleaning	TASI-0001
$x\%NH_4F$	ThO_2	Preferential	THO-0002
$1HF:4HNO_3:5H_2O$	Ti	Removal	TI-0006
10 ml HF:90 ml H_2O	Ti	Removal	TI-0011a
$3HF:16H_3PO_4:1H_2O$	Ti	Polish	TI-0010
$xHF:xHCl:xH_2O$	Ti	Removal	TI-0003
1HF:1MeOH	Ti	Electropolish	TI-0013
$1HF:1HNO_3:6H_2O$	TiC	Electrolytic, preferential	TIC-0002
$xHF:xHNO_3:xHAc$	TiC	Electropolish	TIC-0003
10 ml HF:45 ml HNO_3:45 ml H_2O_2	TiN	Cleaning	TIN-0004
13HF:452 g NH_4F:625 mH_2O	TiN	Removal	TIN-0003
$1HF:24NH_4F:5H_2O$	TiO_2	Removal	TIO-0005
$1HF:1HNO_3:10H_2O$	V	Electrolytic, preferential	V-0001
HF, conc.	V	Removal	V-0002c
$1HF:1H_2O_2:4H_2O$ "superoxol"	V_3Si	Polish	VSI-0001a
$15HF:4H_2O$	V_3Si	Dislocation	VSI-0001c
15 ml HF:100 ml HNO_3	V_3Si	Cleaning	VSI-0002
HF, conc.	W	Removal	W-0003b
$1HF:1H_2O_2$	W	Removal	W-0003b
$xHF:xHNO_3:xH_2O$	W	Polish	W-0002b
$3HF:2HNO_3:1HAc$	W	Cleaning	W-0002d
$xHF:xHNO_3$	W	Polish	W-0018d
HF, conc.	WB_2	Removal	WB-0001a

Formula	Material	Use	Ref.
$1HF:1HNO_3$	ZnS	Polish	ZNS-0001b
$3HF:2H_2O_2:1H_2O$ "Warekois"	ZnS	Preferential	ZNS-0002a
$1HF:1HNO_3$	$ZnSiP_2$	Polish	ZSP-0001
$3HF:2H_2O_2:1H_2O$ "Warekois"	ZnTe	Preferential	ZNTE-0002
$4HF:30HNO_3$. then xH_2SO_4	Zr	Removal	ZR-0003
$1HF:4HNO_3:2H_2O$	Zr	Electropolish	ZR-0004b
50 ml HF:50 ml HNO_3:50 ml H_2O	Zr	Removal	ZR-0009
8 ml HF:50 ml HNO_3:50 ml H_2O	Zr	Preferential, macroetch	ZR-0008
$xHF:xNH_4F$	Zr	Polish	ZR-0009
10 ml HF:45 ml HNO_3:45 ml H_2O_2	ZrN	Cleaning	ZRN-0001
HF, conc.	ZrO	Removal	ZRO-0001b

ETCH NAME: BUFFERED HYDROFLUORIC ACID (BHF)

General: Glass is both a naturally occurring mineral, as obsidian, as well as a manmade product since the Egyptians first developed soda-lime glass around 3500 B.C., and the same formula is still in use today. Glass has been in use as a pottery glaze (ceramic pottery) since those ancient times, and the present glass industry is now a major industry of its own within the original silicate industries, which include clay and ceramics also, today, as separate industries.

In Solid State, the semiconductor industry, in particular, has been using silicon dioxide, SiO_2 (glass, silica), in device fabrication since about 1950. The common term is "oxide", still used with reference to silicon dioxide, only, although several other metallic oxides are now in use in a similar manner, such as Al_2O_3, TiO_2, etc. This now also includes nitrides, such as silicon nitride, Si_3N_4, and oxynitrides as oxides and nitrides form isomorphous series.

Such oxides and nitrides are deposited as thin films in many types of device fabrication as a final surface passivation, or in processing as a mask against etching, metallization, or irradiation damage; and, when doped with boron, phosphorus, or arsenic, as a drive-in diffusion dopant source. Such doped oxides have acquired their own acronyms such as BSG, PSG, ASG, and PBSG.

Hydrofluoric acid (HF) has long been the primary solution for etching glass. When any glass is immersed in the liquid, it will be etch polished; whereas if it is held in the hot vapors, the surface will be rough or frosted. HF is supplied commercially in a standard 49% solution. Always in polyethylene bottles, as it attacks glass, and most processing laboratories use it from 5-gallon bottles. For general removal of thin film oxides and nitrides, both concentrated and water diluted hydrofluoric acid are used, but are too rapid for patterning or controlled removal, as most thin films are only on the order of 1500 to 3000 Å thick. In pattern etching of any thin film, whether an oxide, nitride, or metal, all acid solutions will etch down vertically through the film, as well as horizontally at an equal rate. This must be compensated for in fabricating fine-line patterns to prevent destructive undercutting of the metal structure.

As HF is too rapid for such fine control, buffered hydrofluoric acid solutions have been developed with the general acronym of BHF which is applied to many HF + fluorine compound mixtures. Such mixtures also have been referred to as a buffered oxide etch (BOE), but BHF is more commonly applied.

Because of mixture variations the "BHF" term should be accepted with caution, unless the specific formulation is shown, as even $HF:H_2O$ solutions have been referred to as BHF solutions. The most widely used fluorine compound used to buffer is ammonium fluoride, NH_4F as $HF:NH_4F$ (40%), but others are used (KF, NaF), only, or in addition to NH_4F. Also, the ammonium bifluoride ($NH_4F.HF$) compound has been used.

The use of BHF solutions is not limited to silicon dioxide, or silicon nitride, as they are applicable to any glassy metal compound. The following list covers several applications on other metals and metallic compounds. See the individual etchant sections for further usage.

Formula	Material	Use	Ref.
xHF:xH_2O	Al	Removal	AL-0021
1 ml HF:199 ml H_2O	Al	Macro-etch	AL-0016
1HF:1H_2O or 1HF:10H_2O	AlN	Cleaning/removal	ALN-0001
HF, conc. or x% NH_4F	BaF_2	Removal	BAF-0005
2HF:98H_2O	Bi	Damage removal	BI-0002
xHF:xH_2O	CdF_2	Polish	CDF-0001b
1HF:9H_2O	GaAs	Cleaning	GAS-0112
10^{-6} 0.1 *N* KF	Ge	Electrolytic polish	GE-0046
1HF:10H_2O	GeN	Removal	GEN-0002
5HF:1MeOH	InP	Dislocation	INP-0037a
2HF:10H_2O	InP	Native oxide removal	INP-0045a
xHF:xH_2O	MgO	Preferential	MGO-0003
1HF:1H_2O	SiO_2 (Qtz)	Removal	QTZ-0001
1HF:10H_2O	Si	Cleaning	SI-0004
1HF:16H_2O	Si	Native oxide removal	SI-0132a
xHF:xMeOH	Si	Cleaning	SI-0016
1HF(40%):1NH_4F(40%) BHF	Si	Native oxide removal	SI-0020
1HF:9NH_4F(40%) BHF	Si	Native oxide removal	SI-0022
	SiO_2	Controlled removal	
	Si_3N_4	Controlled removal	
2HF:13NH_4F BHF	Si	Native oxide removal	SI-0023
5 ml HF:20 g:NH_4F:300 ml H_2O BHF	Si	Native oxide removal	SI-0026
	Si_3N_4	Controlled removal	
xHF:xNH_4F:xH_2O BHF	Si	Native oxide removal	SI-0028
	SiO_2	Controlled removal	
	Si_3N_4	Controlled removal	
30 ml HF:8,4 g NaF/l	Si	Electrolytic jet polish	SI-0174b
xHF:xEOH (MeOH or glycols)	Si	Electrolytic polish	SI-0041c
2 ml HF:4.3 g NaF:96 ml H_2O	Si	Native oxide removal	SI-0211
108 ml HF:350 g NH_4F:1000 ml H_2O	Si	Controlled removal	SI-0211
50 ml HF:50 g NH_4F.HF:100 ml H_2O:	Si	Native oxide removal	SI-0220
50 ml Gly	SiO_2	Controlled removal	
	Si_3N_4	Controlled removal	
	Si_3(ON)x	Controlled removal	
xHF:xH_2O	SiC	Cleaning	SIC-0001
xBHF	SiC	Removal	SIC-0005a
xHF:xKF, sat. sol. BHF	SiC	Electrolytic stain	SIC-0013b
1HF:20H_2O	SiO	Controlled removal	SIO-0003
1HF:100NH_4F, sat. sol. BHF	SiO_2	Controlled removal	SIO-0009
	Si_2N_4	Controlled removal	
1HF:6 NH_4F(40%) BHF	SiO_2	Controlled removal	SIO-00011

Formula	Material	Use	Ref.
	BSG	As above	
	PSG	As above	
	BPSG	As above	
53HF:37HCl	SiO_2	Cleaning, Corning #7720	SIO-0038b
xHF:xNH_4F:xH_2O BOE	Si_3N_4	Removal	SIN-0006
15 ml HF:200 ml NH_4.HF, sat. sol.	Si_3N_4	Controlled removal	SIN-0014
	$Si_3O_xN_y$	As above	
	SiO_2	As above	
50 ml HF:50 g NH_4F:HF:100 ml H_2O:	Si_3N_4	Controlled removal study	SIN-0008b
10 ml Gly	$Si_3O_xN_y$	As above	
1HF:$1NH_4F$(20%)	Ta	Preferential	TA-0004b
1HF:$1H_2O$	Ta	Cleaning	TA-0008
20 gNH_4F:100 ml H_2O	Ta	Polish	TA-0003h
x x%NH_4F	ThO_2	Preferential	THO-0002
1HF:$1H_2O$	Ti	Cleaning	TI-0002
1HF:1MeOH	Ti	Electrolytic polish	TI-0013
2HF:$25NH_4F$:$5H_2O$	TiO_2	Controlled removal	TIO-0006
134 ml HF:452 g NH_4F:625 ml H_2O	TiN	Controlled removal	TIN-0003
	TiO_2	As above	
1HF:$1H_2O$	W	Removal	W-0003b
xHF:xNH_4F	Zr	Polish	ZR-0009

FLUORINE, F_2

General: May occur as a free element in nature associated with volcanic eruptions, but will immediately combine with other elements such as hydrogen or sulfur to form more stable compounds.

In industry it is used as a corrosive gas in the study of metals and alloys for stability against chemical attack, as are chlorine and other mixed gases under various temperature conditions. Both drinking water and toothpastes are fluorinated as a medical specific against tooth decay.

Technical Application: The free gas is rarely used in Solid State processing, although it can be bubbled through a liquid etchant as an additional etching assist element to include stirring action. It has been grown as a single crystal under pressure and cryogenic conditions in vacuum for morphological study.

Etching: Vapor pressure under cryogenic conditions in vacuum.

FLUORINE ETCHANTS

F-0001

ETCH NAME: TIME:

TYPE: Pressure, preferential TEMP:

COMPOSITION:

x .... pressure

DISCUSSION:

F_2, as a gas can be grown as a single crystal under pressure and cryogenic conditions in a vacuum system. Vary vapor pressure to preferentially etch in ingot.

REF: Schwentner, N et al — *Rare Gas Solids,* Vol 3, Academic Press, New York, 1960

FRESONITE, $Ba_2Si_2TiO_8$

General: This mineral was found as a minor occurrence near Fresno, California in the U.S. sometime after 1950. There is no use in industry, other than as a very minor ore of titanium.

Technical Application: The material has been studied in Solid State development as a possible piezoelectric material similar to alpha-quartz.

Etching: Mixed acids of $HF{:}HNO_3$

FRESONITE ETCHANTS

FRE-0001

ETCH NAME: TIME:

TYPE: Acid, removal TEMP:

COMPOSITION:

x .... HF

x .... HNO_3

DISCUSSION:

$Ba_2Si_2TiO_2$ single crystal material. Crystals are space group 4 mm^2. Various concentrations of the acid mixture shown are recommended from work done on $(Ag,Ti)_xSiO_4$. The crystals have piezoelectric properties similar to those of alpha-quartz.

REF: Richards, R L et al — *J Appl Phys,* 49,6025(1978)

PHYSICAL PROPERTIES OF GADOLINIUM, Gd

Classification	Lanthanide
Atomic number	64
Atomic weight	157.25
Melting point (°C)	1312
Boiling point (°C)	3000
Density (g/cm^3)	7.89
Thermal conductance (cal/sec)(cm^2)(°C/cm)	0.021
Specific heat (cal/g) 25°C	0.071
Heat of fusion (k-cal/g-atom)	3.70
Latent heat of fusion (cal/g)	23.5
Heat of vaporization (k-cal/g-atom)	72
Atomic volume (W/D)	19.9
1st ionization energy (K-cal/g-mole)	142
1st ionization potential (eV)	6.16
Electronegativity (Pauling's)	1.1
Covalent radius (angstroms)	1.81
Ionic radius (angstroms)	0.62 (Gd^{+3})
Electrical resistivity (micro ohms-cm) 25°C	131 (134)
Compressibility (cm^3/kg $\times$ 10^{-6})	2.56
Cross section (barns)	46,000
Magnetic moment (Bohr magnetons)	7.95
Tensile strength (psi)	27,800
Yield strength (psi)	25,100
Magnetic susceptibility ($\times 10^{-6}$ emu/mole)	356,000
Transformation temperature (°C)	1260
Coefficient of linear thermal expansion ($\times 10^{-6}$ cm/cm/°C) 20°C	4
Hardness (Mohs — scratch)	4
Crystal structure (hexagonal — normal)	$(10\bar{1}0)$ prism, hcp
Color (solid)	Grey
Cleavage (basal)	(0001)

GADOLINIUM, Gd

General: Does not occur as a native element. It is one of the Rare Earths and classified in the yttrium sub-group. The primary mineral is gadolinite, $BeOFeY_2Si_2O_{10}$, which is variable in formula, and contains small quantities of many of the rare earth elements: Ce, La, Gd, Cs, etc. Also found in cerium earth minerals of which there are over 50 as silicates and phosphates. Monazite, $(Ce,La,Di)PO_4$, being a major phosphate commercially mined from monazite sands.

No major use in industrial metal processing due to limited quantities, occasionally used as a doping element in irons and steels; in solder alloys; and the rare earth oxides are used as coloring agents in glass and enamel.

Technical Application: No major use in Solid State processing to date. It has been used as a constituent in some artificial garnets.

Gadolinium has been grown as a single crystal in a study of growth methods, defects, and structure, along with other rare earth elements.

Etching: Acids. Air conversion to oxide.

GADOLINIUM ETCHANTS

GD-0001

ETCH NAME: Heat | TIME:

TYPE: Thermal, preferential | TEMP: 800—1000°C

COMPOSITION:

x .... heat

DISCUSSION:

Gd, single crystal specimens. Describes a method of growing Rare Earth elements as single crystals. Thermal annealing was done to develop defects, low angle grain boundaries, and structure. Other elements grown were dysprosium, holmium, terbium, erbium, thulium, and yttrium.

REF: Nigh, H E — *J Appl Phys,* 34,3323(1963)

GADOLINIUM NITRIDE ETCHANTS

GDN-0001

ETCH NAME: Nitric acid | TIME:

TYPE: Acid, removal | TEMP:

COMPOSITION:

x .... HNO_3

DISCUSSION:

GdN_{12} specimens formed from alloy arc melted buttons. Wrap in tantalum sheet and anneal 3—9 weeks at 600—900°C and 2 × 10^{-7} Torr vacuum. Forms in the Cubic Laves structure. A study of Rare Earth (RE) nitrides. Other nitrides formed were La, Nd, Sm, Tb, Dy, Ho, Er, Y, Pr, and Ce. The SmN_{12} was formed in a silica capsule under 0.5 atm argon to prevent Sm evaporation. Nitrides were used in a thermal expansion study.

REF: Ibarra, M R et al — *J Phys Chem Solids,* 45,789(1984)

GADOLINIUM TERBIUM IRON ETCHANTS

CDTF-0001

ETCH NAME: Nitric acid | TIME:

TYPE: Acid, removal | TEMP:

COMPOSITION:

1 .... HNO_3

1 .... H_2O

DISCUSSION:

GdTbFe thin films co-sputtered on inch square glass plates with 10—20 rpm holder rotation during metallization, and films were amorphous structure. See Iron for other etchants. (*Note:* A similar rotation method is used for evaporation of chromium for chrome glass masks used in photolithographic processing.)

REF: Taki, J — *J Appl Phys,* 55,2799(1984)

PHYSICAL PROPERTIES OF GALLIUM, Ga

Classification	Metal
Atomic number	31
Atomic weight	69.72
Melting point (°C)	29.78
Boiling point (°C)	2403

Density (g/cm^3)	5.907
Thermal conductance (cal/sec)(cm^2)(°C/cm)	0.08
Specific heat (cal/g) 25°C	0.088
Heat of fusion (cal/g)	19.18
Heat of vaporization (cal/g)	950—1020
Atomic volume (W/D)	11.8
1st ionization energy (K-cal/g-mole)	138
1st ionization potential (eV)	6.00
Electronegativity (Pauling's)	1.6
Covalent radius (angstroms)	1.26
Ionic radius (angstroms)	0.62 (Ga^{+3})
Vapor pressure (760 mmHg °C)	2400 (1784)
Cubic coefficient of expansion ($\times 10^{-5}$) 20°C	5.4
Coefficient of linear thermal expansion ($\times 10^{-6}$ cm/cm/°C)	18
Electrical resistivity (micro-ohms-cm) 0°C	53.4
Electron work function (eV)	3.96
Cross section (barns)	3.0
Hardness (Mohs — scratch)	1.5—2.5
Crystal structure (hexagonal — rhombohedral)	$(10\bar{1}1)$ rhomb
Color (solid)	Grey/black
Cleavage (rhombic — poor)	$(10\bar{1}1)$

GALLIUM, Ga

General: Does not occur as a native element. It is a rare element found as a trace constituent in sphalerite, ZnS, other sulfides, and almost always in bauxite, essentially $Al_2O_3 \cdot 2H_2O$, as an admixture with clays. It is part of the aluminum family of elements — Al, Ga, In, Tl — as the reaction of its salts resembles those of aluminum. Besides mercury, cesium, and rubidium, gallium is the only metal that can be liquefied at or near room temperature (Ga m.p. = 29.78°C vs. RT = 22.2°C). Because of this low melting point, high purity gallium used in the Solid State and electronic fields is usually stored under refrigeration.

A primary industrial use is as a low temperature solder, as pure gallium or alloyed with lead, zinc, silver, etc. As it is liquid from near room temperature to 1600°C, it is used as the liquid filler in high temperature thermometers.

Technical Application: Both gallium and indium were two of the first p-type dopants used in the initial development of germanium as a semiconductor, and gallium is a major constituent in several compound semiconductors, such as gallium antimonide, GaSb; gallium phosphide, GaP and gallium arsenide, GaAs. Silicon has been the major semiconductor material since the mid-1950s, but gallium arsenide devices have become increasingly important since about the mid-1960s when they first appeared as a viable product.

Pure gallium and its alloys are used in Solid State and electronics as solders in fabricating assemblies, as have been indium solders. Both gallium and indium are used as paste smear contacts on molybdenum discs to mount and hold wafers during molecular beam epitaxy (MBE) or as a general holder of specimens for microscope study.

References here do not include single crystal gallium, though it can be so grown under vacuum cryogenic conditions like gases.

Etching: Soluble in acids, alkalies, and mixed acids. Clean with alcohols.

GALLIUM ETCHANTS

GA-0001b
ETCH NAME: Methyl alcohol TIME:
TYPE: Alcohol, cleaning TEMP: 0 to 8°C
COMPOSITION:
*x .... MeOH

* Chill with LN_2

DISCUSSION:

Ga as solid material. Because of the low melting point of gallium, it is usually stored at cold temperature in a refrigerator in sealed ampoules so it will not adsorb contaminating gases. In the liquid state it readily adsorbs oxygen. In order to desorb oxygen from gallium, place the gallium in methyl alcohol chilled with liquid nitrogen, LN_2, at or below the temperatures shown.

REF: Ibid.

GA-0001a
ETCH NAME: Hydrochloric acid TIME:
TYPE: Acid, cleaning TEMP:
COMPOSITION:
x .... HCl, conc.
DISCUSSION:

Ga as solid material. Concentrated hydrochloric acid was used to clean gallium before mixing with zinc. The Ga/Zn mixture was used as a molten flux in LPE growth of zinc selenide, ZnSe.

REF: Fujitz, S et al — *J Appl Phys,* 50,1079(1979)

GA-0002
ETCH NAME: Aqua regia TIME: 3 min
TYPE: Acid, preferential TEMP: 10°C
COMPOSITION:
3 .... HCl
1 .... HNO_3
DISCUSSION:

Ga specimens were used in a study of superconducting transition. Samples were hot wire cut and etched in the solution shown to develop grain boundaries. Zinc hemispheres also studied.

REF: Cochran, J F & Mapother, D E — *Phys Rev,* 121,1688(1961)

GA-0003
ETCH NAME: Potassium hydroxide TIME:
TYPE: Alkali, removal TEMP:
COMPOSITION:
x .... x% KOH (NaOH)
DISCUSSION:

Ga, as a constituent in single crystal GaAs p-type wafers fabricated as light emitting diodes with Al or Au dot contact pads. Where As is removed with HNO_3; Ga is removed with alkalies. Diodes studied for the effects of chemical surface treatment and annealing on

light emission. Strong acids reduce or eliminate emission; basic solutions are good and 350°C in N_2, 5 min increases efficiency. Alkalies also remove $Ga(OH)_3$ and Ga_2O_3.
REF: Adachi, R & Hartnagel, H L — *J Vac Sci Technol,* 19(3),427(1961)

PHYSICAL PROPERTIES OF GALLIUM ANTIMONIDE, GaSb

Classification	Antimonide
Atomic numbers	31 & 51
Atomic weight	191.5
Melting point (°C)	712
Boiling point (°C)	
Density (g/cm³)	5.60
Lattice constant (angstroms)	6.09
Energy gap (eV)	0.74
Refractive index (n=)	3.34
Hardness (Mohs — scratch)	6—7
Crystal structure (isometric — tetrahedral)	(111) tetrahedron
Color (solid)	Silver-grey
Cleavage (dodecahedral)	(110)

GALLIUM ANTIMONIDE, GaSb

General: Does not occur as a natural compound, though antimony does occur as a native element in minor quantities, and forms many minerals such as allemontite, $SbAs_2$, and stibnite, Sb_2S_3. The latter is of greater importance as an ore, and there are several other antimonates. There is no industrial use of the compound.

Technical Application: Gallium antimonide is a III—V compound semiconductor with the sphalerite, ZnS structure. It is a (111) surface polar compound with (111)Ga and $(\overline{111})$Sb surfaces showing different etching characteristics, such that it is usually processed as a (100) oriented wafer like similar polar compounds. Single crystal ingots have been grown by the Horizontal Bridgman (HB); Czochralski (CZ); and the Float Zone (FZ) methods. Thin film epitaxy growth has been by Vapor Transport (VT) and Chemical Vapor Deposition (CVD). It also has been grown as a trinary compound with arsenic, aluminum or indium. Both the binary and trinary materials as wafers or thin films have been fabricated as diodes, transistors, laser diodes, or electroluminescent diodes and photo cathodes.

Etching: Water, HCl, HNO_3, halogens and mixed acids of HF:HNO_3, HCl:HNO_3.

GALLIUM ANTIMONIDE ETCHANTS

GASB-0001a
ETCH NAME: Hydrochloric acid — TIME:
TYPE: Acid, removal — TEMP: Hot (80°C)
COMPOSITION:
x .... HCl, conc.
DISCUSSION:

GaSB, (111) and (100) wafers used in a development etch study. When used hot, solution will attack gallium antimonide with moderate to vigorous reaction.
REF: Robbins, H & Schwartz, B — *J Electrochem Soc,* 107,108(1960)

GASB-0001b
ETCH NAME: Nitric acid TIME:
TYPE: Acid, removal TEMP: RT
COMPOSITION:

(1) x .... HNO_3, conc.

(2) 2 .... HNO_3
1 .... H_2O

DISCUSSION:

GaSb, (111) and (100) wafers used in a development etch study. There are three commercial concentrations of nitric acid: standard, 70%, sometimes called "white" fuming; 72%, "yellow" fuming; and 74%, "red" fuming. All three concentrations were evaluated in addition to the dilute solution shown. In all cases, solutions showed moderate to vigorous reaction.

REF: Ibid.

GASB-0005
ETCH NAME: TIME: 1—5 min
TYPE: Acid, polish TEMP: RT
COMPOSITION:

1 .... HF
9 .... HNO_3

DISCUSSION:

GaSb, (100) undoped wafers used in a study of surface oxidation. Surfaces were etch polished in this solution before oxidation. Oxidation was in air at 300 and 400°C for 1 or 2 h for a Ga_2O_3/GaSb structure. Little or no Sb appears in the oxide but does build up at the Ga_2O_3/GaSb interface.

REF: Harmon, L S et al — *Jpn J Appl Phys,* 23,1534(1985)
GASB-0009a: Fuller, C S & Allison, H W — *J Electrochem Soc,* 109,880(1962)
GaSb, (111) wafers. Solution described as a general polish etch.

GASB-0001c
ETCH NAME: TIME:
TYPE: Acid, removal TEMP: RT
COMPOSITION:

(1) 2 .... HCl

(2) 1 .... HCl
1 .... HNO_3

DISCUSSION:

GaSb, (111), and (100) wafers. Etch reaction is shown to be moderate to vigorous with both formulations. Used in a development study of etchants for GaSb.

REF: Ibid.

GASB-0000d
ETCH NAME: TIME:
TYPE: Acid, removal TEMP: Hot
COMPOSITION:

1 .... H_2O_2 (30%)
1 .... NaOH (20%)

DISCUSSION:

GaSb, (111) and (100) wafers. A moderate etch reaction. Used in a development study of etchants for GaSb.

REF: Ibid.

GASB-0002
ETCH NAME: TIME: 20 sec
TYPE: Acid, polish TEMP: RT
COMPOSITION:

(1) 1 .... HF
5 .... HNO_3
10 .... HAc

(2) Rinse: HCl, hot

(3) HOLD: under HCl

DISCUSSION:

GaSb, (111) and (100) wafers used as substrates for epitaxy growth of AlGaSb thin films. Substrates were mechanically polished with 0.25 μm diamond paste before polish etching. Wafers were placed in the epitaxy vacuum system still wet with HCl, and the acid was allowed to boil off in an H_2 atmosphere with heat at 300°C for several hours of *in situ* cleaning prior to deposition of AlGaSb.

REF: Nenow, D et al — *J Cryst Growth* 3,489(1984)

GASB-0003: Wada T et al — *J Cryst Growth,* 3,439(1984)

GaSb, (111), and (100) wafers. Used this solution formulation as a general polishing etch.

GASB-0004a
ETCH NAME: Water TIME:
TYPE: Acid, removal TEMP:
COMPOSITION:

x .... H_2O

DISCUSSION:

GaSb, (100) undoped wafers used as substrates for MBE growth of GaSb/GaSb. A study of substrate surface preparation with different etching mixtures. Substrates were first lapped with SiC powder on a pad, then with 0.3 μm alumina. Water polishing of surface was used to remove lap/polish damage remaining before polish etching.

REF: Kodama, M et al — *J Electrochem Soc,* 132,659(1985)

GASB-0004b
ETCH NAME: TIME:
TYPE: Acid, polish TEMP:
COMPOSITION:

2 .... HF
18 .... HNO_3
40 .... CH_3COOH (HAc)

DISCUSSION:

GaSb, (100) undoped wafers used as substrates for MBE growth of GaSb/GaSb. See GASB-0004a. The base etch shown above was followed by: (1) hot HCl, 10 min; (2) 2% Br_2:MeOH, RT, 1 min; (3) 1HNO_3:30HCl, at 5°C, 1 min. After etching, the wafers were annealed in an MBE vacuum system using REED at 20 KeV prior to epitaxy growth GaSb/GaSb. Prior to etching, after water polish substrates were degreased in TCE, then acetone, then MeOH. After etching, rinse in MeOH and spin dry.

GASB-0004c
ETCH TIME: BRM TIME: 2 min
TYPE: Acid, polish TEMP: RT
COMPOSITION:

x .... 2% Br_2
x .... MeOH

DISCUSSION:

GaSb, (100) undoped wafers used as substrates. See both GASB-0004a and GASB-0004b, above, for details. Used to etch polish substrate surfaces.

REF: Ibid.

GASB-0009b: Ibid.

GaSb, (111) wafers. Solutions between 1—20% Br_2 used as general chem/mech polish etchants.

GASB-0006a

ETCH NAME: BRM | TIME: 20 sec + 10 sec

TYPE: Halogen, polish | TEMP: RT

COMPOSITION:

x .... 0.05% Br_2

x .... MeOH

DISCUSSION:

p-GaSb, (111) wafers used in a surface cleaning study, as follows: use lens paper saturated with solution as a lapping pad — lap for 20 sec — then dilute with MeOH — then continue lapping for 10 sec. Rinse in the solution shown and rinse in water. Authors state that percentage of bromine is not critical. (*Note:* With the bromine concentration shown, percentage would not be critical.) This surface study also included Si, Ge, GaP, GaAs, InP, InAs and InSb.

REF: Aspenes, D E & Studna, A A — *J Vac Sci Technol,* 20,488(1982)

GASB-0001e

ETCH NAME: | TIME:

TYPE: Acid, removal | TEMP: Hot

COMPOSITION:

1 .... HF

1 .... HNO_3

DISCUSSION:

GaSb, (111) and (100) wafers. Solution gives moderate to vigorous reaction.

REF: Ibid.

GASB-0007a

ETCH NAME: | TIME:

TYPE: Acid, passivating | TEMP: RT

COMPOSITION:

x .... HCl

DISCUSSION:

GaSb, (100), both undoped and Te-doped wafers used in a cleaning study before MBE thin film deposition. Spin-on HCl at 3500 rpm to passivate surfaces after acid etching.

REF: Kodama, M et al — *Jpn J Appl Phys,* 23,1657(1984)

GASB-0007b

ETCH NAME: | TIME: 1 min

TYPE: Acid, removal | TEMP: 5°C

COMPOSITION:

1 .... HNO_3

30 .... HCl

DISCUSSION:

GaSb, (100), undoped and Te-doped wafers used in a cleaning study before MBE thin film deposition. This etch removed O_2 contamination from surfaces. Passivate surfaces with HCl (GASB-0007a).

REF: Ibid.

GASB-0007c

ETCH NAME: TIME: 40 sec

TYPE: Acid, polish TEMP: RT

COMPOSITION:

2 .... HF

18 .... HNO_3

40 .... HAc

DISCUSSION:

GaSb, (100) un-doped and Te-doped wafers used in a cleaning study before MBE thin film deposition. This solution leaves a high O_2 content on surface. Follow with $1HNO_3$:30HCl (GASB-0007b), which reduces O_2 content and leaves a highly reflective surface. Passivate surface with HCl (GASB-0007a).

REF: Ibid.

GASB-0008a

ETCH NAME: BRM TIME: 20 sec

TYPE: Halogen, preferential TEMP:

COMPOSITION:

1 .... Br_2

10 .... MeOH

DISCUSSION:

GaSb, (111) wafers. This solution of bromine-methanol (BRM) developed shallow etch pits on the (111)A surface.

REF: Harper, C A — *Handbook of Materials and Processes for Electronics,* McGraw-Hill, New York, 1970, 7

GASB-0008b

ETCH NAME: TIME: 1 min

TYPE: Acid, preferential TEMP:

COMPOSITION:

1 ml .... HCl

1 ml .... H_2O_2

2 ml .... H_2O

DISCUSSION:

GaSb, (111) wafers. The solution develops etch figures.

REF: Ibid.

GASB-0010

ETCH NAME: TIME: 15 sec

TYPE: Acid, polish/preferential TEMP: RT

COMPOSITION:

1 .... HF

2 .... HNO_3

1 .... HAc

DISCUSSION:

GaSb, (111) wafers and other orientations. Solution used as a rapid polish etch and will develop etch pits.

REF: Gatos, H C & Lavine, M C — *J Electrochem Soc,* 107,427(1960)

GASB-0011

ETCH NAME: TIME:

TYPE: Acid, preferential TEMP: RT

COMPOSITION:

1 .... HF

1 .... HNO_3

1 .... H_2O

DISCUSSION:

GaSb, (111) wafers. Solution used to polish $(\overline{111})$Sb faces.

REF: Faust, J W & Sager, A — *J Appl Phys,* 31,331(1960)

GASB-0012

ETCH NAME: TIME:

TYPE: Acid, selective TEMP: 50°C

COMPOSITION: RATE: 10 Å/sec

1 .... H_2O_2

10 .... citric acid (50%)

[(COOH)CH_2C(OH)(COOH)CH_2COOH]

DISCUSSION:

GaSb, (100), p-type wafers with surfaces patterned by photolithography. Solution used to selectively etch pits in the material surface and also can be used to form mesas and other etched structures.

REF: Otsubo, M et al — *J Electrochem Soc,* 123,676(1976)

GASB-0013

ETCH NAME: TIME:

TYPE: Acid, selective TEMP:

COMPOSITION:

3 .... HF

5 .... HNO_3

3 .... HAc

DISCUSSION:

GaSb, (111) wafers used as substrates for growth of heterojunctions as: GaIn/GaInSb/GaSb(111). After MBE growth wafers were cleaved ⟨110⟩ and the solution used to develop deposited layer structure.

REF: Mroczkowski, K S et al — *J Electrochem Soc,* 117,750(1968)

GASB-0014a

ETCH NAME: TIME: 1 min

TYPE: Acid, polish TEMP: RT

COMPOSITION:

1 .... HF

19 .... HNO_3

30 .... CH_3COOH (HAc)

DISCUSSION:

GaSb, (100) wafers within 0.2° of the crystal plane used as substrates for LPE growth of AlGaSb. Growth habit is sensitive to orientation. Substrates were cleaned as follows: (1)

degrease with organic solvents; (2) etch in solution shown; (3) quench in 18 MΩ DI water and (4) MeOH wash. Under vacuum, prior to epitaxy growth: (1) bake at 700°C up to 24 h; (2) heat clean at 550°C, 1 h; (3) melt-back, 10—15 sec, at 10°C above growth temperature, with a 1—2 h soak prior to melt back with gallium. The etch solution leaves an oxide on the substrate surface and exposure to air will oxidize surfaces. Surfaces were studied by AES profiling in vacuum at 6×10^{-9} Torr. Sputter etching with argon was done at 1 kV under 5×10^{-5} Torr vacuum. In profiling, substrates were: (1) cleaved ⟨110⟩; (2) etched in solution shown and DI water rinsed; (3) then thermal cycled 1 h, at 400°C; (4) 1 h, at 550°C; and (5) Ga melt-back at 400°C. (*Note:* Excellent article on growth temperatures and morphology of AlGaSb surfaces.)
REF: Takeda, Y et al — *J Electron Mater,* 13,855(1984)

GASB-0014b
ETCH NAME: Hydrochloric acid — TIME: x min
TYPE: Acid, oxide removal — TEMP: RT
COMPOSITION:
x .... HCl, conc.
DISCUSSION:

GaSb, (100) wafers used as substrates for LPE growth of AlGaSb. Solution used to remove oxide from previous etching or exposure to air. Soak wafers in HCl, then blow off acid with nitrogen — or — HCl soak, then MeOH rinse and nitrogen blow dry. Both methods are effective for removing oxides. See GASB-0014a.
REF: Ibid.

GASB-0014c
ETCH NAME: A-B etch, dilute — TIME:
TYPE: Acid, stain — TEMP: RT
COMPOSITION:
1 .... A-B Etch
10 .... H_2O
DISCUSSION:

GaSb, (100) wafers Te-doped and used as substrates for LPE growth of AlGaSb. After epitaxy deposition, wafers were cleaved ⟨110⟩ and stained in the solution shown to develop AlGaSb/GaSb interface and defect structures in the thin film layer. (*Note:* The A-B solution shown is the Abrahams and Buiocchi preferential etch normally shown as A/B. The A-B designation is sometimes used for the Kern and Puotinen — RCA etch.)
REF: Ibid.
GASB-0017: Abrahams M S & Buiocchi, C J — *J Appl Phys,* 36,2855(1965)

Reference for A-B Etch shown in GASB-0014c (See GAS-0059). Solution mixture is 1 ml HF:2 ml H_2O:1 g CrO_3:8 g $AgNO_3$.

GASB-0015a
ETCH NAME: — TIME: 1 min
TYPE: Acid, preferential — TEMP: RT
COMPOSITION:
1 .... HF
9 .... HNO_3
20 .... HAc

DISCUSSION:

GaSb, (111) wafers used in a study of growth striations. Mechanically polish surfaces before structure development with solution shown. See GASB-0015b for (211) wafer orientation use.

REF: Tohno, S-I & Katsui, A — *J Electrochem Soc,* 128,1614(1981)

GASB-0015b

ETCH NAME: TIME: 1—2 min

TYPE: Acid, preferential TEMP: RT

COMPOSITION:

1 .... $KMnO_4$

0.05 .... HF

20 .... HAc

DISCUSSION:

GaSb, (211) wafer orientation. Solution used to develop growth striations.

REF: Ibid.

GASB-0016

ETCH NAME: TIME:

TYPE: Acid, preferential TEMP:

COMPOSITION:

(A)		(B)		(C)	
40 ml	HF	4 g	Cu_2O_3	1	A
3 g	$AgNO_3$	40 ml	H_2O	1	B
				40	H_2O

DISCUSSION:

GaSb, (100) wafers cleaved and used as substrates for epitaxy growth of GaAlSb. Mix solution "C" when ready for use as a preferential etchant on the epitaxy layer. Intrinsic GaSb material is p-type; Te dopant for n-type.

REF: Merbarki, P et al — *J Cryst Growth,* 61,636(1983)

PHYSICAL PROPERTIES OF GALLIUM ARSENIDE, GaAs

Classification	Arsenide
Atomic numbers	31 & 33
Atomic weight	145
Melting point (°C)	1260 (1230)
Boiling point (°C)	
Density (g/cm^3)	5.37
Thermal conductance (W/cm/°C) 300°F	0.46
Coefficient of linear thermal expansion [L/L T(°C^{-1}) $\times 10^{-6}$]	5.6
Specific heat (J/g/°C)	0.35
Lattice constant (angstroms)	5.654 (5.63)
Dielectric constant (e =)	10.9
Vapor pressure (°C) 1050°C	1
Refractive index (n =)	3.53
Formation temperature (°C)	600—800
Energy band gap (eV)	1.35
Hardness (Mohs — scratch)	6—7
Crystal structure (isometric — normal)	(100) cube, fcc

Color (solid)	Grey-silver
Cleavage (cubic)	(001)

GALLIUM ARSENIDE, GaAs

General: Does not occur as a natural compound, although there are several other metallic arsenides of iron, copper, cobalt, nickel and platinum whose minerals may contain sulfur as well as arsenic. The only industrial application of gallium arsenide is as a compound semiconductor.

Technical Application: Gallium arsenide is a III—V compound semiconductor with the sphalerite, ZnS structure, even though it is referred to as cubic, fcc. It is a (111) surface polar compound with (111)Ga and $(\overline{111})$As showing different etching characteristics, such that it is usually wafer processed in (100) orientation as are other such polar compounds.

Germanium and silicon — the two elemental semiconductors — were first developed in the late 1940s, and gallium arsenide was one of the first compound semiconductors developed in the mid-1950s. It was initially grown by the Horizontal Bridgman (HB) method. It is still so grown, but much larger wafers (now up to 8" diameter) are available as Czochralski (CZ) ingots.

As an epitaxy thin film it is grown by horizontal or vertical CVD, to include OMCVD, or by MBE.

Development of gallium arsenide has been much slower than that of silicon, which is still the major semiconductor that developed the present electronic industry as it is known today. The compound semiconductors, as a group, are more brittle than either germanium or silicon, more problematical in handling and, though gallium arsenide has better high frequency characteristics than the elemental semiconductors, the control of doping to obtain operational devices was an initial problem factor, as was temperature control . . . both gallium and arsenic can vaporize from the solid compound during processing . . . and, in addition, there was far more emphasis on silicon technology up through the late 1960s.

With development of the Schottky Barrier type devices, gallium arsenide has been fabricated as a field effect transistor (FET) with a gate, source, and drain configuration — the base wafer semi-insulating (SI) — doped with chromium, and the active gold or aluminum gate at the compound/metal interface. The material for such devices are shown here as GaAs:Cr, (100 (SI) wafers.

Devices are now fabricated as low noise or power FETS: the LN-FET or P-FET, respectively. This technology has been extended to other compound semiconductors, such as an InP:Fe, (100) (SI) wafer. And today, with the development of molecular beam epitaxy (MBE), multilayered structures are under device development as the heterojunction or heterostructure devices, which include different epitaxy layered elements or compounds, plus oxides, nitrides, metals grown on substrates of GaAs, etc. Some of the devices fabricated are FETs, Gunn and IMPATT diodes, thyristors, rectifiers, and GaAs:Al trinary compounds are being developed as solar cells with up to 30% efficiency, as compared to silicon with only about 18%. Other trinary compounds with phosphorus and indium are under evaluation.

Gallium arsenide is grown in single crystal ingot form and used, directly, as a device; deposited as a single crystal or polycrystalline thin film; or used as a substrate for other epitaxy growth in device fabrication. This latter use, as a chromium-doped semi-insulating substrate, is probably the most widely used wafer form for current device processing, although there is much on-going development of the trinary AlGaAs material for an improved solar cell.

Etching: Mixed acids $HF:HNO_3$ + HCl or H_2O_2. Alkali + halogen; halogens, molten fluxes. Gases as DCE and thermal etching.

SELECTION GUIDE: GaAs

(1) Br_2:MeOH (BRM)
- (i) Oxide Removal: GAS-0132
- (ii) Polish: GAS-0001a; -0184a; -0005a; -0012a; -0013a; -0004a; -0200; -0006; -0007a; -0011; -0202; -0010
- (iii) Preferential: GAS-0002; -0173b
- (iv) Selective: GAS-0009
- (v) Thinning: GAS-0004a; -0008

(2) Br_2:EOH
- (i) Polish: GAS-0015
- (ii) Preferential: GAS-0013

(3) I_2:MeOH
- (i) Polish: GAS-0015

(4) $NH_4OH:H_2O_2$
- (i) Polish: GAS-0007c; -0006b
- (ii) Selective: GAS-0018; -0023; -0025
- (iii) Staining: GAS-0022

(5) $NH_4OH:H_2O_2:H_2O$
- (i) Oxidation: GAS-0181
- (ii) Polish: GAS-0007c; -0006b
- (iii) Removal: GAS-0001b
- (iv) Step-etch: GAS-0022
- (v) Junction: GAS-0021

(6) HCl, conc.
- (i) Cleaning: GAS-0028; -0207; -0029; -0018b; -0118; -0199
- (ii) Preferential: GAS-0192
- (iii) Oxide Removal: GAS-0114
- (iv) Removal: GAS-0182b; -0140

(7) HCl, vapor
- (i) Oxide Removal: GAS-0074

(8) $HCl:H_2O$
- (i) Cleaning: GAS-0004a; -0117
- (ii) Removal: GAS-0001c; -0030

(9) HCl:EOH
- (i) Oxide removal: GAS-0001d

(10) $HCl:FeCl_3$
- (i) Preferential: GAS-0174

(11) $HCl:HNO_3$
- (i) Dislocation: GAS-0186; -0189
- (ii) Junction: GAS-0056b
- (iii) Polish: GAS-0056a; -0179
- (iv) Removal: GAS-0190; -0140c; -0197

(12) $HCl:HNO_3:H_2O$
- (i) Preferential: GAS-0167c; -0052; 0044c; -0144a
- (ii) Polish: GAS-0057; -0148; -0176; -0178
- (iii) Junction: GAS-0056a

(13) $HCl:HNO_3$:Glycerin
- (i) Polish: GAS-0050; -0006e; -0051

(14) $HCl:H_2O_2:H_2O$
- (i) Polish: GAS-0104; -0141
- (ii) Removal: GAS-0186
- (iii) Structure: GAS-0060

(15) H_2O & $H_2O:O_2$
- (i) Oxidation: GAS-0001e; -0181

(16) NaOCl (Clorox)
- (i) Polish: GAS-0012b; -0113; -0114; -0184; -0195a
- (ii) Preferential: GAS-0173a

(17) $HCl:HAc:K_2Cr_2O_7$
- (i) Polish: GAS-0041

(18) HF, conc.
- (i) Oxide removal: GAS-0191
- (ii) Selective: GAS-0024

(19) HF, vapor
- (i) Cleaning: GAS-0129

(20) $HF:H_2O$
- (i) Cleaning: GAS-0112

(21) $HF:HNO_3$
- (i) Cleaning: GAS-0047b
- (ii) Removal: GAS-0140d

(22) $HF:HNO_3:H_2O$
- (i) Cleaning: GAS-0047a
- (ii) Polish: GAS-0044c; -0006d; -0046; -0045b; -0146b; -0147; -0211a
- (iii) Preferential: GAS-0146d; -0146e; -0177
- (iv) Staining: GAS-0211a

(23) $HF:HCl:HNO_3$
- (i) Polish: GAS-0055; -0145a

(24) $HF:HNO_3:HAc$
- (i) Cleaning: GAS-0038b

(25) $HF:HNO_3:AgNO_3:+H_2O$
- (i) Preferential: GAS-0146c; -0048; -0119b

(26) $HF:H_2O_2$
- (i) Preferential: GAS-0049; -0053a; -0144d

(27) $HF:H_2O_2:H_2O$
- (i) Junction: GAS-0212b
- (ii) Polish: GAS-0053b
- (iii) Preferential: GAS-0045d; -0044f; -0168; -0144e-f; -0157; -0167a; -0146f; -0144b
- (iv) Structure: GAS-0054; -0160

(28) $HF:H_2O_2:H_2SO_4$
- (i) Polish: GAS-0088
- (ii) Thinning: GAS-0088

(29) $HF:CrO_2:H_2O$
- (i) Dislocation: GAS-0192

(30) HNO_3, conc.
- (i) Oxidation: GAS-0139
- (ii) Removal: GAS-0140b
- (iii) Staining: GAS-0042

(31) $HNO_3:H_2O$
- (i) Junction: GAS-0180
- (ii) Polish: GAS-0145b
- (iii) Preferential: GAS-0043; -0044b; -0183b; -0044a; -0045a; -0120; -0146a
- (iv) Removal: GAS-0140b

(32) HNO_3:Tartaric acid
- (i) Preferential: GAS-0167b

(33) HNO_3:H_2O:$AgNO_3$
  (i) Preferential: GAS-0045c
(34) H_3PO_4:HNO_3
  (i) Polish: GAS-0110a
(35) H_3PO_4:H_2O_2:H_2O
  (i) Selective: GAS-0108a
  (ii) Removal: GAS-0105; -0106
  (iii) Thinning: GAS-0108b
  (iv) Via-hole: GAS-0109
(36) H_3PO_4:H_2O_2:MeOH
  (i) Polish: GAS-0110b
  (ii) Selective: GAS-0096b
(37) H_3PO_4:H_2O_2:H_2SO_4
  (i) Dislocation: GAS-0004d; -0110c
(38) Vacuum
  (i) Cleaning: GAS-0156
(39) Argon, Ar
  (i) Cleaning: GAS-0184
  (ii) Thinning: GAS-0186b
(40) H_2SO_4:H_2O: + O_2
  (i) Cleaning: GAS-0149
  (ii) Oxidation: GAS-0181
  (iii) Polish: GAS-0187b
(41) H_2SO_4:Polypropaline glycol
  (i) Dislocation: GAS-0121
(42) H_2SO_4:H_2O_2
  (i) Cleaning: GAS-0150; -0194
  (ii) Removal: GAS-0150
(43) H_2SO_4:H_2O_2:H_2O
  (i) Cleaning: GAS-0066; -0069; -0074; -0131; -0198; -0191a; -0195b; -0200; -0077; -0088; -0081; -0082; -0083; -0084; -0085; -0086; -0087; -0091; -0098; -0099; -0100; -0102; -0103; -0125; -0128; -0151; -0152; -0153; -0154; -0155; -0165; -0170; -0171; -0172; -0182a
  (ii) Oxide removal: GAS-0114
  (iii) Junction: GAS-0099; -0169; -0193b
  (iv) Polish: GAS-0062; -0063; -0065; -0069; -0070; -0073; -0075; -0130; -0056b; -0077; -0007h; -0079; -0089; -0091; -0101; -0182a; -0158; -0159; -0160; -0212a
  (v) Preferential: GAS-0064; -0116; -0164
  (vi) Removal: GAS-0068; -0070; -0126; -0212c
  (vii) Selective: GAS-0063a,b-m; -0006g; -0072; -0076; -0078
  (viii) Step-etch: GAS-0067; -0089; -0090
  (ix) Thinning: GAS-0090
(44) I_2:KI:H_2O
  (i) Selective: GAS-0061e-f
(45) Quinone:Hydroquinone
  (i) Selective: GAS-0060
(46) $Ce(SO_4)_2$:$Ce(NO_3)_3$ + H_2O
  (i) Selective: GAS-0061b
(47) $K_3Fe(CN)_6$:$K_4Fe(CN)_6$ + H_2O
  (i) Selective: GAS-0061c-d

(48) A/B Etch
  (i) Dislocation: GAS-0059; -0130; -0119a; -0181
(49) $KOH:H_2O$
  (i) Junction: GAS-0210; -0212d
(50) NaOH/KOH, molten flux
  (i) Preferential: GAS-0032; -0033; -0034; -0035; -0157; -0163b; -0119b; -0122; -0123; -0124; -0127; -0162; -0163
(51) NaOH:KOH, molten flux
  (i) Preferential: GAS-0036
(52) KOH:NaOH:Gly/EG
  (i) Preferential: GAS-0036
(53) $KOH:H_2O + O_2$
  (i) Oxidation: GAS-0181
(54) $KOH:K_2Fe(CN)_6:H_2O$
  (i) Polish: GAS-0040
  (ii) Preferential: GAS-0019
  (iii) Staining: GAS-0037; -0038b
(55) $NaOH:H_2O_2 + H_2O$
  (i) Junction: GAS-0187
  (ii) Polish: GAS-0175; -0115
  (iii) Preferential: GAS-0031; -0183a; -0185; -0187; -0142; -0167d; -0173c
  (iv) Selective: GAS-0188a
  (v) Removal: GAS-0143
(56) $MeOH:Cl_2$
  (i) Polish: GAS-0187a

GALLIUM ARSENIDE ETCHANTS

GAS-0001a
ETCH NAME: BRM — TIME:
TYPE: Halogen, polish — TEMP: RT
COMPOSITION:
  x .... x% Br_2
  x .... MeOH
DISCUSSION:

GaAs:Cr, (100) (SI) wafers prepared as substrates for epitaxy growth of GaAs. The following cleaning procedure was used: (1) etch polish in solution shown; (2) damage removal/polish in $H_2SO_4:H_2O_2:H_2O$; (3) boil in DI H_2O to develop oxide; (4) remove oxide with $HCl:H_2O$; (5) repeat boiling water and (6) remove oxide with HCl:EOH. (*Note:* Bromine concentration not shown.)

REF: Lewis, B F et al — *J Phys Chem,* 45,419(1984)

GAS-0184a: Bhat, R — *J Electron Mater,* 14,433(1985)

GaAs:Cr, (100) (SI) wafers oriented 6°-off (100) toward (111)A, also (111), and (110) wafers. Wafers used as substrates for OMCVD epitaxy growth of both GaAs and AlGaAs layers. Prepare substrates by (1) chem/mech polish in solution shown to remove lap damage; (2) degrease, and (3) etch clean in $5H_2SO_4:1H_2O_2:1H_2O$.

GAS-0005: Davis, G A et al — *J Cryst Growth,* 69,141(1984)

GaAs, (100) wafers cut 2°-off toward (110), used as substrates for epitaxy growth of GaAs layers. After mixing BRM solution, allow it to sit 10 min before use. Clean substrates as follows: (1) chem/mech polish with solution shown; (2) degrease in boiling TCE, and rinse with MeOH, then (3) etch clean in $10H_2SO_4:1H_2O_2$. Immediately before loading in vacuum, rinse in DI H_2O. (*Note:* Allowing the BRM solution to sit 10 min before use produces a ''stagnant'' solution as some of the bromine will gas-off while diffusing into the methanol. It has been shown by other authors that bromine diffusion in methanol is complete in about 10 min.)

GAS-0012a: Palmateer, S C & Eastman, L F — *J Vac Sci Technol,* B2,188(1984)

GaAs:Cr, (100) (SI) wafers cut from ingots grown by Bridgman method and by HPLEC or LPLEC. After cutting, wafers were chem/mech polished in the solution shown, annealed 24 h in H_2, then re-polished in the solution shown or in Clorox. Wafers were then polished in $5H_2SO_4$:$1H_2O_2$:$1H_2O$ just before loading in an MBE vacuum system for epitaxy growth of GaAs.

GAS-0214: Inarey, S & MacLaurin, B — *J Vac Sci Technol,* A2(2),358(1985)

GaAs, (100), n-type wafers used as substrates for Ge growth and Au metallizing. Clean GaAs: (1) chem/mech polish with BRM solution; (2) MeOH rinse; (3) HCl dip: (4) DI H_2O rinse; and (5) in vacuum MBE system Ar^+ ion sputter clean prior to Ge growth.

GAS-0237: Kolodziejski, L A et al — *J Vac Sci Technol,* B3(2),714(1985)

GaAs, (111) wafers used as substrates for MBE deposition of CdTe and CdMnTe thin films as dilute, magnetic semiconductor superlattices. BRM solutions can be used as removal and polishing etchants for these compounds. There was a 14% mismatch lattice of CdMnTe/GaAs.

GAS-0238: Qadri, S B et al — *J Vac Sci Technol,* B3(2),718,(1985)

GaAs, (110) wafers used as substrates for MBE deposition of Fe thin films for a structural study of films by X-ray. Films were (110) oriented with a 1.34% lattice mismatch. BRM solutions can be used as removal and polishing etchants for these materials.

GAS-0002

ETCH NAME: BRM — TIME:

TYPE: Halogen, preferential — TEMP:

COMPOSITION:

x .... 1% Br_2

x .... MeOH

DISCUSSION:

GaAs, (111)A wafer surfaces etched in the solution shown by immersion. Surface etch pits will have (322) pit side slopes if the bromine concentration is greater than 1%. The plane slope angle was measured as 10° from (111)A surface.

REF: Kozi, L A & Rode, L A — *J Electrochem Soc,* 122,1511(1975)

GAS-0003a

ETCH NAME: BRM — TIME:

TYPE: Halogen, polish — TEMP: RT

COMPOSITION: — RATE: 1 μm/min on (110)

x .... 1% Br_2

x .... MeOH

DISCUSSION:

GaAs, (110), (111), (100) wafers used in a general etch rate study. Etch ratio shown as: 6:5:4.6:1 for (110):$(\overline{111})$As:(100):(111)Ga. Solution used as an immersion etch to establish etch rates. (*Note:* (111)Ga = (111)A and $(\overline{111})$As = $(\overline{111})$B in general crystallographic notation of polar compounds.)

REF: Ghandhi, S K — *VLSI Fabrication Procedures,* John Wiley & Sons, New York, 1983

GAS-0004a: Budutt, R et al — *J Electrochem Soc,* 128,1573(1981)

GaAs, (100), n-type wafers used as substrates for growth of GaAs layers by LPE. Substrates were prepared as follows: (1) Syntron system used to chem/mech polish with solution shown; (2) degrease; (3) etch clean in $1HCl$:$2H_2O$ and DI water rinse. After LPE growth of (GaAl)As, substrate wafer was thinned with 1.5% Br_2:MeOH for microscope

study of the epitaxy layer. Other wafers were cleaved ⟨110⟩ and defect etched in H_3PO_4:H_2SO_4:H_2O_2, RT, 1—2 min. Defects are similar to stacking faults.
GAS-0200: Cochran, L & Gomez, A — personal communication, 1979

GaAs:Cr, (100) (SI) wafers used in the fabrication of GaAs FET devices. Wafer back-weighted with steel discs and chem/mech polished on a shaker table using the solution shown.

GAS-0006a
ETCH NAME: BRM — TIME: 20 sec + 10 sec
TYPE: Halogen, polish — TEMP: RT
COMPOSITION:
x .... 0.05% Br_2
x .... MeOH
DISCUSSION:

GaAs, (100), n-type wafers used in a surface cleaning study. Cleaning was done as follows: (1) chem/mech hand lap on lens paper soaked with solution shown and, after 20 sec dilute with MeOH, and continue lapping an additional 10 sec; (2) rinse in NH_4OH:H_2O (AMH solution); (3) rinse in BRM solution shown, and (4) final DI water rinse. Method also used on: Si, Ge, GaP, GaSb, InP, InAs, and InSb.
REF: Aspenes, D E & Studa, A A — *J Vac Sci Technol*, 20,488(1982)

GAS-0004b
ETCH NAME: BRM — TIME:
TYPE: Halogen, thinning — TEMP: RT
COMPOSITION:
x .... 1.5% Br_2
x .... MeOH
DISCUSSION:

GaAs, (100) wafer substrates used for (GaAl)As epitaxy growth by LPE. After epitaxy the GaAs substrate was etch thinned and removed leaving the (GaAl)As thin film free for TEM study. Stacking faults (SFs) start near the substrate/epitaxy interface. Micro-cracks were observed in the thin sections and attributed to probable tensile stress release.
REF: Ibid.

GAS-0007a
ETCH NAME: BRM — TIME:
TYPE: Halogen, isotropic — TEMP: RT
COMPOSITION: — RATE: See discussion
x .... 1—3% Br_2
x .... MeOH
DISCUSSION:

GaAs, (111), (100), and (110) wafers used in a general etch rate study of various concentrations of bromine-methanol solutions. 1—2% Br_2 concentrations are preferential on both (111)A and (111) faces. For a 1% Br_2 concentration, etch rate increased as: (110)>(111)B>(100)>(111)A with (110) being most rapid. (*Note:* This is an excellent review article of GaAs etching.)
REF: Kern, W — *RCA Rev*, 39,278(1978)
GAS-0011: Christon, A et al — *Semiconductor International*, May 1979, 59

GaAs, (100) and InP, (100) wafers used in the characterization of defects by ASLEEP, SEM, and scanning AES. A 1% Br_2 solution was used at RT, 1—5 min for polishing.

Depending upon application, bromine content can vary from 0.1 to 5% with good polishing results.

GAS-0202: Walker, P — personal application, 1980

GaAs:Cr, (100) (SI) wafers used in fabricating GaAs LN-FET devices. A 2% Br_2 solution was used as a spray cleaning etch after metallization pattern evaporation. After etching, rinse in MeOH and N_2 blow dry.

GAS-0010: Tuck, E — *J Mater Sci,* 10,321(1976)

GaAs, (100) and GaP, (100) wafers used in a general polish etch study of both materials. A $1Br_2$:9MeOH mixture gave an approximate $^1/_4$ μm/min removal rate. Described as only one of several bromine concentrations evaluated.

GAS-0008

ETCH NAME; BRM | TIME:

TYPE: Halogen, thinning | TEMP:

COMPOSITION:

x .... 5% Br_2

x .... MeOH

DISCUSSION:

GaAs, (100), and InP, (100) wafers used for evaporation of gold thin films. After annealing of metallized wafers, they were black wax patterned and the GaAs and InP substrates were etch thinned in the solution shown for TEM study of line defects at the substrate/gold interfaces and in the gold thin film.

REF: Nakahara, S et al — *J Electrochem Soc,* 131,1917(1984)

GAS-0009

ETCH NAME: | TIME:

TYPE: Halogen, selective | TEMP: RT

COMPOSITION:

x .... 5% Br_2

x .... MeOH

DISCUSSION:

GaAs, (100), (111) and (110) wafers. Controlled pits were etched into wafer surfaces selectively through photo resist patterned SiO_2 masks. Etch rates for the different plane surfaces were shown to be similar to rates described by Kern (GAS-0007a).

REF: Tarui, Y et al — *J Electrochem Soc,* 118,118(1971)

GAS-0013

ETCH NAME: BRE | TIME:

TYPE: Halogen, preferential | TEMP: RT

COMPOSITION: | RATE: See discussion

x .... 1% Br_2

x .... EOH

DISCUSSION:

GaAs, (100), (111) and (110) wafers used in a general study of etch rates and surface reactions. The solution is preferential on all orientations and etch rates are similar to those described by Kern (GAS-0007a) for BRM solutions. This says that bromine solutions with either methanol (BRM) or ethanol (BRE) show equivalent rates and reactions.

REF:

GAS-0014
ETCH NAME: BRE — TIME:
TYPE: Halogen, polish — TEMP: RT
COMPOSITION: — RATE: 8 μm/min
x .... 2% Br_2
x .... EOH
DISCUSSION:

GaAs, (111) wafers. Specimen surfaces were polish etched with a jet of the solution shown with specimens rotating, as on a photo resist spinner. Authors say this is nonpreferential on (111) orientations.

REF:
GAS-0260: Bertrand, P A — *J Electrochem Soc,* 132,923(1985)

GaAs:Zn, (100) wafers, p-type, used in a study of photochemical oxidation. Solution shown used 10 sec at RT, with MeOH rinse to remove native oxide; following with EOH rinse and HCl, RT, 15 min and DI H_2O rinse. Hold specimens in a glove box in air at RT for several months to develop a native oxide.

GAS-0003b
ETCH NAME: — TIME:
TYPE: Acid, removal — TEMP:
COMPOSITION:
20 .... NH_4OH
7 .... H_2O_2
973 .... H_2O
DISCUSSION:

GaAs, (100), (111) and (110) wafers used in a general etch rate study. Etch rates are much slower with this solution than those obtained with bromine-methanol. See GAS-0003a.

REF: Ibid.

GAS-0001c
ETCH NAME: Hydrochloric acid — TIME:
TYPE: Acid, oxide removal — TEMP:
COMPOSITION:

(1) 1 .... HCl
1 .... H_2O_2
1 .... H_2O

(2) 1 .... HCl
1 .... H_2O

DISCUSSION:

GaAs:Cr, (100) (SI) wafers used as substrates for epitaxy growth of GaAs layers. As part of the substrate cleaning process, wafers were boiled in DI water and then stripped with the solution shown or with 1HCl:1EOH.

REF: Ibid.
GAS-0030a: Siracusa, M — personal communication, 1978

GaAs:Cr, (100) (SI) wafers used in fabricating FET devices. Solution used as a dip cleaning step with DI water rinse prior to wafer metallization. The HCl is followed with a dip in 1HF:1H_2O and DI water rinse, immediately before placing in the vacuum system. N_2 blow dry, in both cases. (*Note:* Many authors refer to using this 50% mixture for cleaning purposes.)

GAS-0247: Wu, C S et al — *J Electrochem Soc,* 132,918(1985)

GaAs:Si, (100) n-type wafers used as substrates for metallization with Er and Si in a study of ion beam etching. Clean substrates: (1) in acetone; (2) in isopropanol (ISO); (3) wash in solution (1), and (4) rinse in DI H_2O prior to metal evaporation.

GAS-0004c
ETCH NAME: Hydrochloric acid TIME:
TYPE: Acid, cleaning TEMP:
COMPOSITION:
1 .... HCl
2 .... H_2O
DISCUSSION:
GaAs, (100) n-type wafers used as substrates for LPE growth of GaAlAs layers. Solution used as part of substrate cleaning process following degreasing. See GAS-0004a for additional discussion.
REF: Ibid.

GAS-0007c
ETCH NAME: TIME:
TYPE: Acid, isotropic TEMP:
COMPOSITION:
x .... NH_4OH
x .... H_2O_2
DISCUSSION:
GaAs, (100) wafers, and other orientations. Solution used as a general polish etch similar to BRM solutions.
REF: Ibid.

GAS-0015
ETCH NAME: TIME:
TYPE: Halogen, polish TEMP: RT
COMPOSITION:
$^1/_4$ g .... I_2
250 ml .... MeOH
DISCUSSION:
GaAs:Cr, (100) (SI) wafers. The amount of iodine can be varied. The mixture as shown was used as a surface polish and cleaning solution similar to BRM solutions prior to metal evaporation. Wafers were used in the fabrication of Low Noise GAS-FETs. Also used as a final dip-rinse of GaAs and silicon, (111) wafers prior to diffusion, and for cleaning alumina, quartz, and sapphire circuit substrates prior to metallization. In both cases, remaining iodine on the surfaces was used to complex remaining ionic contamination from prior etching and cleaning steps to obtain a molecularly clean surface for improved diffusion depth control, and/or better metal adhesion.
REF: Walker, P — personal development, 1965/1980

GAS-0016
ETCH NAME: TIME:
TYPE: Acid, polish TEMP:
COMPOSITION:
3 .... *CP4
1 .... MeOH
1 .... H_2O
*CP4: 2HF:5HNO_3:2HAc: 1% Br_2
DISCUSSION:
GaAs wafers. A general polish etch for gallium arsenide wafers. After mixing the solution it should be allowed to age before use. (*Note:* This is a form of ''stagnant'' etching.)
REF: TI Rep

GAS-0182a
ETCH NAME: TIME:
TYPE: Acid, junction TEMP:
COMPOSITION:
1 .... HF
5 .... HNO_3
DISCUSSION:
GaAs, (100) wafers doped with germanium under arsenic pressure and used in a study of the effect of arsenic during doping. Solution used as a p-n junction staining and development etch.
REF: McCaldin, J O & Harada, K — *J Appl Phys*, 31,2065(1960)

GAS-0018a
ETCH NAME: TIME:
TYPE: Acid, selective TEMP:
COMPOSITION:
(1) 1 .... NH_4OH (2) x .... HCl, conc. (3) x .... Ar^+ ions
20 .... H_2O_2
DISCUSSION:
GaAs, (100), n-type, wafers used as substrates for LPE growth of Double Heterostructures (DHs) as follows: n-GaAs cap/InGaAsP/InGaP/GaAs substrate. Solution (1) was used to selectively remove the substrate of DH fabricated diodes that had failed life test, and also remove the n-GaAs cap. The InGaAsP layer was selectively removed with concentrated HCl, and the n-InGaP layer was argon ion sputter removed.
REF: Veda, O et al — *J Appl Phys*, 57,1523(1985)

GAS-0019
ETCH NAME: Ammonium hydroxide TIME:
TYPE: Acid, junction stain TEMP:
COMPOSITION:
1 .... NH_4OH
5 .... H_2O
DISCUSSION:
GaAs, (100) Zn-doped, p-type wafers used as substrates for MBE growth of GaAlAs layers which were alternately doped with Ge, then Mg (both p-type dopants). Wafers were cleaved ⟨110⟩ and stained with the solution shown to develop both diffused junctions as well as layer structure.
REF: Schwartz, B et al — *J Electrochem Soc*, 131,1703(1984)
GAS-0020: Schwartz, B et al — *J Electrochem Soc*, 131,1674(1984)
GaAs wafers. Similar work as in GAS-0019.
GAS-0006b: Ibid.
GaAs, (100), n-type wafers used in a cleaning study. After BRM etching, an $NH_4OH:H_2O$ solution was used as a rinse and called the AMH solution.

GAS-0030b
ETCH NAME: 0°C etch TIME:
TYPE: Acid, cleaning TEMP: 0°C
COMPOSITION:
100 ml .... NH_4OH
1000 ml .... H_2O

DISCUSSION:

GaAs:Cr, (100) (SI) wafers used in the fabrication of GaAs FET devices. This solution is widely used both to clean the GaAs surfaces as well as to clean evaporated aluminum patterns on wafers, and the ammonium hydroxide concentration may vary from 10 to 200 ml. The solution can be chilled with either LN_2 or dry ice, CO_2.

REF: Siracusa, M — personal communication, 1978

GAS-0021

ETCH NAME: TIME:

TYPE: Acid, junction TEMP:

COMPOSITION: RATE: 0.16 μm/min

3 .... NH_4OH

1 .... H_2O_2

150 .... H_2O

DISCUSSION:

GaAs, (100) wafers zinc diffused. Solution used to develop the p-n diffused junction. Wafers were cleaved, ⟨110⟩ to expose junction prior to etching.

REF: Jett-Field, R & Ghandhi, S R — *J Electrochem Soc,* 129,1566(1982)

GAS-0022

ETCH NAME: TIME:

TYPE: Acid, step-etch TEMP:

COMPOSITION:

3 .... NH_4OH

1 .... H_2O_2

1 .... H_2O

DISCUSSION:

GaAs:CR, (100) wafers within $^1/_2$° of plane, and un-doped wafers used as substrates for MOCVD epitaxy of GaAs/GaAs(100). Epitaxy GaAs was doped with Sn. The solution shown was used for step-etching the epitaxy layer during C-V profiling.

REF: Parsons, J A & Krajenbrink, F G — *J Cryst Growth,* 68,60(1984)

GAS-0006c

ETCH NAME: TIME:

TYPE: Acid, isotropic TEMP:

COMPOSITION:

1 .... NH_4OH

700 .... H_2O_2

DISCUSSION:

GaAs, (100) wafers and other orientations. Solution has been applied as a general polishing etch using a saturated pellon pad with rotation.

REF: Ibid.

GAS-0023

ETCH NAME: TIME:

TYPE: Acid, selective TEMP:

COMPOSITION:

x .... H_2O_2

x .... NH_4OH for pH 7

DISCUSSION:

GaAs, (100) substrates with epitaxy layer growth of GaAlAs. This solution was used

to selectively remove the GaAs substrate for thin section study of the general morphology and structure of GaAlAs.
REF: Logan, R A & Reinhart, F K — *J Appl Phys,* 44,4172(1973)

GAS-0024
ETCH NAME: Hydrofluoric acid — TIME:
TYPE: Acid, selective — TEMP:
COMPOSITION:
x .... HF
DISCUSSION:
GaAs, $(\overline{111})$B and (100) both n-type and un-doped wafers used as substrates for epitaxy growth of Be doped GaAlAs, p-type. The GaAlAs layers were selectively removed from the GaAs substrates with hydrofluoric acid.
REF: Masu, K et al — *J Electrochem Soc,* 129,1623(1982).

GAS-0025
ETCH NAME: — TIME:
TYPE: Acid, selective — TEMP:
COMPOSITION:
x .... NH_4OH
x .... H_2O_2
DISCUSSION:
GaAs, (100) Si-doped wafers used as substrates for LPE growth of GaAlAs layers. This solution was used to selectively remove the GaAs substrate. The author refers the etch to Ralogan & Reinhart — *J Appl Phys,* 44,4127(1973).
REF: Swaninathan, V et al — *J Electrochem Soc,* 129,1563(1982)

GAS-0026
ETCH NAME: — TIME:
TYPE: Electrolytic, oxidation — TEMP:
COMPOSITION: — ANODE: GaAs
x .... *EDTA, 2 *M* — CATHODE: Pt foil
x .... NH_4OH — POWER: To 1000 VDC

*Ethylene diamine tetraacetic disodium salt

DISCUSSION:
GaAs, (100) and GaAs, (111) wafers. Both materials were anodically oxidized in a study of oxidation defects. Mixtures of Br:MeOH also were used to develop defects.
REF: Elliot, C R & Regnault, J C — *J Electrochem Soc,* 127,1557(1980)

GAS-0077
ETCH NAME: Hydrochloric acid, vapor — TIME:
TYPE: Acid, oxide removal — TEMP: Hot
COMPOSITION:
x .... HCl, vapor
DISCUSSION:
GaAs, (100) wafers used as substrates for LPE growth of epitaxy GaAs layers. After epitaxy, wafers were capped with Si_3N_4 or PSG mask layers, which were photo resist patterned for selective beryllium ion implantation. Prior to oxide/nitride capping, wafers were cleaned by: (1) degreasing in boiling solvents; (2) soaking in hot water to form an oxide and (3) removing the oxide with hot HCl vapors. After oxide/nitride coating and

patterning, HCl vapor used to final clean GaAs exposed surfaces prior to Be implant. Be goes to Ga sites and SiO_2 getters Be.
REF: Campbell, P M & Ballga, D J — *J Electrochem Soc,* 123,186(1985)

GAS-0028a
ETCH NAME: Hydrochloric acid, vapor TIME:
TYPE: Acid, cleaning TEMP: Hot
COMPOSITION:
x .... HCl, vapor
DISCUSSION:

GaAs, (100), Si-doped wafers used in a study of oxide and nitride coatings on GaAs surfaces. Hot HCl vapors used to clean substrates prior to CVD deposition of SiO_2 or RF plasma deposition of SiN_x. Patterned photo resist openings were fabricated and oxides or nitrides were removed with either: (1) BHF ($5NH_4F$:1HF) or (2) RIE, using CF_4 gas. The GaAs acts as an etch-stop against both BHF or CF_4.
REF: Blaauw, C et al — *J Electron Mater,* 13,251(1984)
GAS-0207: Zelinsky, T — personal communication, 1978

GaAs:Cr, (100) (SI) wafers used as substrates for CVD and MOCVD growth of GaAs. With wafers in the vacuum system, hot HCl vapors are used to final clean surfaces prior to epitaxy deposition. (*Note:* This is a standard method for cleaning of CVD and Epitaxy systems. The substrates, quartz tube system, and graphite susceptors are all etch cleaned with the temperature ranges from 600 to 1250°C depending upon materials.

GAS-0029
ETCH NAME: Hydrochloric acid TIME:
TYPE: Acid, cleaning/selective TEMP: Boiling
COMPOSITION:
x .... HCl, conc.
DISCUSSION:

GaAs, (100) wafers used for epitaxy growth of InGaAs. Substrates were cleaned before epitaxy by: (1) degreasing in TCE; (2) rinsing in acetone; (3) rinsing in methanol, and (4) etching clean in HCl with final DI water rinse.
REF: Penna, T C et al — *J Cryst Growth,* 67,27(1984)
GAS-0018b: Ibid.

GaAs, (100) wafers used for epitaxy growth of a Double Heterostructure (DH) device. The InGaAsP layer was selectively removed with boiling HCl.

GAS-0182b
ETCH NAME: Hydrochloric acid TIME: 10 min
TYPE: Acid, oxide removal TEMP: 30°C
COMPOSITION:
x .... HCl, conc.
DISCUSSION:

GaAs, (100) and (111) wafers. Solution used in a cleaning sequence as the final step prior to ZnS epitaxy deposition. See GAS-0182a for additional discussion.
REF: Ibid.

GAS-0140a-b
ETCH NAME: Hydrochloric acid TIME:
TYPE: Acid, removal TEMP: RT
COMPOSITION:
(1) x .... HCl, conc. (2) 2 .... HCl
1 .... H_2O

DISCUSSION:

GaAs, (111) wafers and dice used in an etch development experiment. Both solutions will attack this orientation slowly at RT.

REF: Robbins, H & Schwartz, B — development work, 1960

GAS-0008

ETCH NAME: Hydrochloric acid — TIME: 30—60 sec

TYPE: Acid, preferential — TEMP: RT

COMPOSITION:

x .... HCl, conc.

DISCUSSION:

GaAs, (111) and (100) wafers. Wafers were cleaved ⟨110⟩ and etched in the solution shown to develop triangle etch pits to establish (011) or $(0\bar{1}1)$ bulk plane directions. If point of triangle in down, away from the upper (100) or (111)A surface, the bulk plane is (011); whereas, if the point is 180° rotated bulk plane is $(0\bar{1}1)$. The $(0\bar{1}1)$ is preferred for selective channel and pit structuring.

REF: Pennington, P & Walker, P — development work, 1957

GAS-0118

ETCH NAME: Hydrochloric acid — TIME:

TYPE: Acid, cleaning — TEMP: RT

COMPOSITION:

x .... HCl, conc.

DISCUSSION:

GaAs, (100), Te-doped, n-type wafers used as substrates in a study of leakage caused by cleaning. Cleaning sequence was: (1) degrease in TCE; (2) rinse in acetone; (3) rinse in methanol; (4) wash with HCl and (5) rinse in MeOH. Wafers were then argon ion cleaned 15 min with power varying from 100 to 1000 eV. Both aluminum and gold dots were evaporated for electrical test evaluations.

REF: Kwan, P et al — *Solid State Electron,* 26,125(1983)

GAS-0117

ETCH NAME: Hydrochloric acid, dilute — TIME:

TYPE: Acid, cleaning — TEMP: 70°C

COMPOSITION:

3 .... HCl

7 .... H_2O

DISCUSSION:

GaAs, (100) wafers used as substrates for LPE growth of GaAs, Sn-doped and Si ion implanted thin films. Substrates were cleaned prior to epitaxy by: (1) degrease in boiling TCE; (2) boiling IPA; (3) wash in 15% HCl and DI water rinse; (4) clean in $NH_4OH:H_2O_2$, RT, 2 min with DI water rinse and (5) N_2 blow dry. Wafers were used in a study of Ni/AuGe and Au/Ni/AuGe evaporated metal contacts.

REF: Marlow, G S et al — *Solid State Electron,* 26,259(1983)

GAS-0001d

ETCH NAME: — TIME:

TYPE: Acid, oxide removal — TEMP: RT

COMPOSITION:

1 .... HCl

1 .... EOH

DISCUSSION:

GaAs:Cr, (100) (SI) wafers used as substrates for GaAs epitaxy growth. Solution used as the final step in a cleaning sequence of substrates before epitaxy. Wafers were oxidized in boiling water, the oxide stripped with the solution shown. See GAS-0001a for additional discussion.

REF: Ibid.

GAS-0012b

ETCH NAME: Clorox — TIME:

TYPE: Acid, polish — TEMP: RT

COMPOSITION:

x .... *NaClO

*Standard Clorox contains about $5^1/_2$% NaClO

DISCUSSION:

GaAs:Cr, (100) (SI) wafers used for epitaxy growth of GaAs thin films Solution used as a replacement for BRM polishing solutions. See GAS-0012a for added discussion. (*Note:* Clorox is a highly reactive bleaching agent for paper and textiles.)

REF: Ibid.

GAS-0001e

ETCH NAME: Water — TIME:

TYPE: Acid, oxidation/cleaning — TEMP: Boiling

COMPOSITION:

x .... H_2O

DISCUSSION:

GaAs:Cr, (100) (SI) wafers used for epitaxy growth of GaAs thin films. Boiling water used to oxidize substrate surfaces as part of a cleaning sequence prior to epitaxy. The oxide was removed with either 1HCl:1H_2O or 1HCl:1EOH. See GAS-0001a for additional discussion.

REF: Ibid.

GAS-0031

ETCH NAME: — TIME:

TYPE: Alkali, preferential — TEMP:

COMPOSITION:

50 ml .... 5% NaOH

10 ml .... H_2O_2

DISCUSSION:

GaAs wafers. The authors say that this solution mixture will develop grain structure.

REF: Waslgov, R T et al — *J Electrochem Soc,* 129,547(1982)

GAS-0032a

ETCH NAME: Potassium hydroxide — TIME: 4 min

TYPE: Alkali, dislocation — TEMP: Molten

COMPOSITION:

x .... KOH, molten

DISCUSSION:

GaAs, (100) wafers and other orientations. This molten flux was used to develop dis-

locations and were compared to dislocations obtained by electrolytic etching with H_2SO_4 (the SSA etch).
REF: Nagata, K et al — *J Electrochem Soc,* 128,2347(1981)
GAS-0033: Mo, P et al — *J Cryst Growth,* 65,243(1983)

Used to develop dislocations in GaAs ingots. The "A/B" etch was used to develop stria, applied at RT with light. (See GAS-0059.).
GAS-0034: Katz, L E — *J Electrochem Soc,* 124,425(1977)

GaAs:Cr, (100) (SI), and n-GaAs, (100): Si-doped wafers used as substrates for GaAs, Sn-doped epitaxy thin film growth. GaAs substrates were etch cleaned in $3H_3PO_4$:$1H_2O_2$:$1H_2O$ prior to epitaxy. The molten KOH was used to develop dislocations in the Sn-doped epitaxy layers.
GAS-0157: Kimura, H et al — *J Cryst Growth,* 70,185(1984)

GaAs, (100) wafers. Ingots grown by LEC and HPLEC (20 atm argon) with and without indium doping. Molten KOH used to observe etch pit density (EPD). Pits show a cellular network in un-doped GaAs; whereas pits align in ⟨110⟩ directions with In-doped GaAs. Rosette patterns observed at seed end of ingot, but these figures decrease toward tail.
GAS-0128b: Elliot, A G et al — *J Cryst Growth,* 70,169(1984)

GaAs, (100) wafers with low dislocation density, doped with In, Si, and Te were grown by LEC, and used in a study of dislocation density. Molten flux KOH was used 10 min at 425°C to develop dislocations.
GAS-0163b: Grabmair, J G & Watson, C B — *Phys Status Solidi,* 32,K13(1969)

GaAs:Cr, (100), and (111) wafers used in a study of dislocation density. Wafers were cut from ingots grown by LEC and HB. Molten flux KOH was evaluated for dislocation development between 300 and 450°C. Below 360°C results were poor, possibly due to water in the KOH flux.
GAS-0219b: Mori, H & Takagashi, S — *J Cryst Growth,* 69,23(1984)

GaAs:Cr, (100) (SI) wafers cut $^1/_2$°-off toward (110). Wafers used in a study of dislocations. Chem/mech polish wafers with BRM solution prior to KOH molten flux etching for dislocations.
GAS-0225b: Motsuto, M & Mike, H — *J Electrochem Soc,* 124,441(1977)

GaAs:Cr, (100) (SI), or Si doped wafers used as substrates for epitaxy growth study of tin doped GaAs. Molten flux used to develop dislocations in epitaxy layers.

GAS-0035
ETCH NAME: TIME:
TYPE: Alkali, dislocation TEMP: 160°C
COMPOSITION:

1 .... KOH or x .... KOH
1 .... NaOH

DISCUSSION:

GaAs, (100) wafers and other orientations. The 1KOH:1NaOH mixture is a eutectic molten flux. Authors show that if the wafers are immediately immersed in ethylene glycol for 20 min after molten flux etching the surface oxidation is reduced and the EG slowly dissolves residual KOH/NaOH leaving a less damaged surface for defect/dislocation study.
REF: Lessoff, H & Gorman, R — *J Electron Mater,* 14,203(1985)
GAS-0163b: Grabmair, H J G & Watson, C B — *Phys Status Solidi,* 32,K13(1969)

GaAs:Cr, (100) and (111) (SI) wafers, and n-type wafers used in a dislocation study. The molten flux was in a platinum crucible with wafers held in a platinum wire basket. Etch times were (1) 15—120 sec for (100), and (2) 1—30 sec for (111).
GAS-0119b: Ibid.

GaAs, (100), and (111) wafers cut within $^1/_2$° of planes from ingots grown by CZ and LEC methods. Surfaces were polished with $1Br_2$:MeOH, and then etched in KOH at 360°C, 15 min, in a dislocation study. Compare results with the A/B Etch (GAS-0059), and the RC-1 Etch (GAS-0048).

GAS-0183a
ETCH NAME: TIME: 2—5 min
TYPE: Acid, preferential TEMP: RT
COMPOSITION:
1—5 .... 2% NaOH
1 .... H_2O_2
DISCUSSION:
GaAs, (111) wafers used in a study of precipitates induced by zinc diffusion. Solution produces sharp triangular etch pits.
REF: Black, J F & Jungbluth, E D — *J Electrochem Soc,* 114,181(1967)

GAS-0185
ETCH NAME: TIME:
TYPE: Acid, preferential TEMP:
COMPOSITION:
1 .... 15% NaOH
1 .... 5% H_2O_2
DISCUSSION:
GaAs, (100) and (111) wafers doped with Se, Te, Zn, and Pd used in a study of stress effects and electrical measurements. Solution used as a dislocation etch.
REF: Black, J & Lubin, P — *J Appl Phys,* 35,2462(1964)

GAS-0036
ETCH NAME: TIME: 5—20 min
TYPE: Alkali, preferential TEMP: Boiling
COMPOSITION:
(1) 250 ml .... 30% KOH
1000 ml .... MeOH
(2) 250 ml .... 30% KOH
100 ml .... ethylene glycol/glycerin
DISCUSSION:
GaAs, (111) wafers. Initial, general study and development of etchants for gallium arsenide. These solutions were shown to be preferential. Spheres were etched to finite crystal form in additional wafer studies.
REF: Walker, P — personal development, 1957

GAS-0017
ETCH NAME: TIME:
TYPE: Acid, junction stain TEMP:
COMPOSITION:
x .... KOH
x .... $K_3Fe(CN)_6$
x .... H_2O
DISCUSSION:
GaAs, (100) wafers, zinc diffused. Wafers were cleaved in ⟨110⟩ and stained with the solution shown to develop p-n junctions.
REF: Roedel, R J et al — *J Electrochem Soc,* 131,1726(1984)
GAS-0038a: Radiker R H & Lowen, J — *J Electrochem Soc,* 107,26(1960)
GaAs wafers fabricated as diffused diodes. Solution shown was used as a junction etch.

GAS-0039
ETCH NAME: TIME:
TYPE: Acid, anisotropic TEMP:
COMPOSITION:
12 g .... KOH
8 g .... $K_3Fe(CN)_6$
100 ml .... H_2O
DISCUSSION:

GaAs, (100) wafers used as substrates for epitaxy growth of InGaAsP layers. Material was cleaved ⟨110⟩ and etched in solution shown to develop layer structure.

REF: Bolkhovityanova, E et al — *J Electron Mater,* 12,525(1983)

GAS-0040
ETCH NAME: TIME: 1 min
TYPE: Acid, preferential TEMP: RT
COMPOSITION:
2 g .... KOH
8 g .... $K_3Fe(CN)_6$
100 ml .... H_2O
DISCUSSION:

GaAs, (100) p-type wafers used as substrates for epitaxy growth of p-lnP. After epitaxy, wafers were cleaved ⟨110⟩ and etched to develop layer structure.

REF: Takeda, Y et al — *Jpn J Appl Phys,* 23,84(1983)

GAS-0041
ETCH NAME: TIME:
TYPE: Acid, polish TEMP: RT and 60°C
COMPOSITION:
1 .... 12 *N* HCl
1 .... 17 *N* CH_3COOH (HAc)
*1 .... 1 *N* $K_2Cr_2O_7$

*14.7 g/100 cm H_2O

DISCUSSION:

GaAs wafers used in a chemical etching study. Authors say there are two etch regions, which they classify as A and B. The A region is rate limited, and the solution is light yellow in color. The B region is diffusion limited and dark brown in color.

REF: Adachi, S & Os, K — *J Electrochem Soc,* 131,126(1984)

GAS-0042
ETCH NAME: Nitric acid TIME:
TYPE: Acid, stain TEMP:
COMPOSITION:
x .... HNO_3, conc.
DISCUSSION:

GaAs and Si (100) wafers used as substrates for epitaxy growth of Ge/GaAs and GaP/Si. Wafers were cleaved ⟨110⟩ and stained to observe layer structure.

REF: Rosztoczy, F E & Stein, W W — *J Electrochem Soc,* 119,1119(1972)

GAS-0043
ETCH NAME: Nitric acid, dilute TIME: $2^1/_2$ h
TYPE: Acid, preferential TEMP: RT
COMPOSITION:
1 .... HNO_3
2 .... H_2O
DISCUSSION:

GaAs, grown as a (111) ingot oriented 15° off-(111) and Te doped. After growth ingot was cut lengthwise on (110), and solution used to develop growth patterns. Patterns showed anomalous nonconcentric freeze-out structures and high impurity concentration areas indicative of super-cooling of the ingot during growth similar to that observed with both Ge and InSb ingots. Striations observed were caused by Te segregation. The solution was changed every 15 min during the etch period.

REF: LeMay, C Z — *J Appl Phys,* 34,439(1964)
GAS-0044b: Schell, H A — *Z Metallkd,* 48,158(1957)

GaAs, (111) wafers used in a study of surface polarity. Solution will develop dislocation pits on the (111)Ga surfaces but not on $(\overline{111})$As.

GAS-0183b: Black, J F & Jungbluth, E D — *J Electrochem Soc,* 114,181(1967)

GaAs, (111) wafers used in a study of precipitates induced by zinc diffusion. Solution used to develop dislocation pits.

GAS-0211b: Ibid.

GaAs, (100) wafers cut from ingots doped with germanium, and used in a study of a Travelling Solvent growth method. Solution was used for dislocations and defects.

GAS-0044a
ETCH NAME: Schell etch TIME:
TYPE: Acid, preferential TEMP: RT
COMPOSITION:
1 .... HNO_3
3 .... H_3O
DISCUSSION:

GaAs, (111) wafers. Solution developed as a dislocation etch for polarity study. Develops dislocation pits on (111)Ga surfaces.

REF: Schell, H A — *Z Metallkd,* 48,158(1957)
GAS-0045a: Richards, J L & Crocker, A J — *J Appl Phys,* 31,611(1960)

GaAs, (111) wafers. Solution developed dislocation on the (111)A surfaces, and terraces on the $(\overline{111})$B. A sphere etched in the solution produce the finite crystal form of a tetrahexahedron, (hk0).

GAS-0044c
ETCH NAME: TIME:
TYPE: Acid, polish TEMP:
COMPOSITION:
1 .... HF
1 .... HNO_3
1 .... H_2O
DISCUSSION:

GaAs, (111) wafers. Solution used as a polishing etch prior to dislocation etching of polar surfaces.

REF: Ibid.

GAS-0006d
ETCH NAME: TIME:
TYPE: Acid, isotropic TEMP:
COMPOSITION:
1 .... HF
3 .... HNO_3
2 .... H_2O
DISCUSSION:
GaAs, wafers. Referred to as a rapid polish etch.
REF: Ibid.
GAS-0046: White, J C & Roth, W C — *J Appl Phys*, 30,956(1959)
GaAs, (111) wafers. The solution polishes the (111)A (gallium) surface, and produced etch pits on the $(\overline{111})$B (arsenic) surface.
GAS-0045b: Ibid.
GaAs, (111) wafers. Results as for GAS-0046, above. Sphere etched to finite crystal form was a rhombic dodecahedron, (hh0).

GAS-0047a
ETCH NAME: TIME:
TYPE: Acid, cleaning TEMP:
COMPOSITION:
1 .... HF
9 .... HNO_3
10 .... H_2O
DISCUSSION:
GaAs, (100) and (111) wafers. The latter orientation used as (111)A and $(\overline{111})$B. All wafers used as substrates for VPE growth of ZnTe and ZnS. Surfaces were mechanically lapped, then prepared as follows: (1) degrease in boiling TCE; (2) rinse in acetone, then MeOH; (3) etch in $3H_2SO_4$:$1H_2O$:$1H_2O$ to remove lap damage. Final etch clean in solution shown prior to epitaxy.
REF: Matsumoto, T & Ishida, T — *J Cryst Growth*, 67,135(1984)

GAS-0045c
ETCH NAME: TIME:
TYPE: Acid, preferential TEMP:
COMPOSITION:
1 .... HNO_3
3 .... H_2O
\+ .... 1% $AgNO_3$
DISCUSSION:
GaAs, (111) wafers. Solution produces etch pits on both (111)A and $(\overline{111})$B surfaces. Sphere etched in solution does not produce a finite crystal form.
REF: Ibid.

GAS-0048
ETCH NAME: RC-1 TIME: 3 min
TYPE: Acid, dislocation TEMP: RT
COMPOSITION:
2 .... HF
3 .... HNO_3
\+ .... 22.4×10^{-3} *M* $AgNO_3$

DISCUSSION:

GaAs, (111) wafers. Solution produces triangular etch pits on the $(\overline{111})$B surface similar to the Schell etch. Produces conical etch pits on the (111)A surface. For comparison etched with the Schell etch, 15 min, and the RC-1 etch, for $1^1/_2$ min. They both shown 1-to-1 correspondence of etch pits. Used $1HF:1HNO_3:1H_2O$ solution to polish surfaces and to etch thin specimens.

REF: Abrahams, M S — *J Appl Phys,* 35,3626(1964)

GAS-0119a Ibid.

GaAs, (100) and (111) wafers from ingots grown by LEC and CZ. Solution used in a dislocation study.

GAS-0049a

ETCH NAME: TIME:

TYPE: Acid, preferential TEMP:

COMPOSITION:

1 .... HF

3 .... H_2O_2

DISCUSSION:

GaAs, single crystal spheres. A sphere etched in this solution produced the finite crystal form of a tetrahexahedron, (hk0).

REF: Richards, J L — *J Appl Phys,* 31,604(1960)

GAS-0050

ETCH NAME: TIME:

TYPE: Acid, isotropic TEMP:

COMPOSITION:

1 .... HNO_3

1 .... HCl

8 .... Glycerol

DISCUSSION:

GaAs, wafers. Used as an etch polish solution. High viscosity etches preferred for chemical polishing as attack tends to reduce high points more evenly, and produces a more planar surface.

REF: Packard, R D — *J Electrochem Soc,* 112,112(1965)

GAS-0006s, Ibid.

GAS-0051: Walker, P — personal development, 1979

GaAs:Cr, (100) (SI) wafers. Etch polished wafers in the solution as shown to remove lapping damage prior to ion implantation with Si^+ ions. Used both glycerin and ethylene glycol in varying amounts for viscosity control to optimize surface planarity.

GAS-0052

ETCH NAME: TIME:

TYPE: Acid, polish TEMP:

COMPOSITION:

1 .... HNO_3

2 .... HCl

2 .... H_2O

DISCUSSION:

GaAs, (111) wafers. Described as a polish etch for both GaAs and GaP.

REF: Uragski, T et al — *J Electrochem Soc,* 123,580(1976)

GAS-0167: Gatos, H C & Lavine, M C — Lincoln Lab Tech Rep 293, Jan 1963

GaAs, (111) wafers. Etch time 10 min. Mix solution and use fresh. Pits are developed on the (111)A surfaces.

GAS-0044d
ETCH NAME: Aqua regia — TIME:
TYPE: Acid, polish — TEMP:
COMPOSITION:
1 .... HNO_3
3 .... HCl
2 .... H_2O
DISCUSSION:
GaAs, (111) wafers. Described as a polishing etch for GaAs.
REF: Ibid.

GAS-0053a
ETCH NAME: — TIME:
TYPE: Acid, step-etch — TEMP: 2°C
COMPOSITION:
1 .... HF
1 .... H_2O_2
50 .... H_2O
DISCUSSION:
GaAs, (100) wafers, were Mg^+ ion implanted. This mixture was used to step-etch after ion implantation to study the implantation depth profile.
REF: Yu, P W — *J Appl Phys,* 48,2434(1977)

GAS-0053b
ETCH NAME: — TIME:
TYPE: Acid, polish — TEMP:
COMPOSITION:
5 .... HF
1 .... H_2O_2
1 .... H_2O
DISCUSSION:
GaAs, (100) wafers used as substrates for Mg^+ ion implantation. Solution used to etch clean GaAs prior to implantation.
REF: Ibid.

GAS-0045d
ETCH NAME: — TIME:
TYPE: Acid, preferential — TEMP:
COMPOSITION:
1 .... HF
2 .... H_2O_2
4—6 .... H_2O
DISCUSSION:
GaAs, (111) wafers. Solution will produce etch pits on the (111)A but only terraces on (111)B. A sphere etched in the solution produce the finite crystal form of a tetrahexahedron, (hk0).
REF: Ibid.

GAS-0054
ETCH NAME: TIME:
TYPE: Acid, structure TEMP:
COMPOSITION:
1 .... HF
1 .... H_2O_2
10 .... H_2O
DISCUSSION:
GaAs, (100) wafers, and other single crystal material substrates used for epitaxy growth of GaAs. After epitaxy, substrates were cleaved ⟨110⟩ and etched to develop layer structure. Authors say that this etch will not develop stria.
REF: Willardson, R K & Beer, A C — *Semiconductors and Metals,* Vol 7, Academic Press, New York, 1971, 160

GAS-0044f
ETCH NAME: TIME:
TYPE: Acid, preferential TEMP:
COMPOSITION:
1 .... HF
1 .... H_2O_2
2 .... H_2O
DISCUSSION:
GaAs, (111) wafers. Described as a preferential etch for GaAs.
REF: Ibid.

GAS-0056a
ETCH NAME: Aqua regia TIME:
TYPE: Acid, polish TEMP:
COMPOSITION:
3 .... HCl
1 .... HNO_3
DISCUSSION:
GaAs, (111) wafers. Aqua regia can be used as a polish etch.
REF: *S.E.R.L. Tech J,* 10,86(January 1960)

Gas-0057
ETCH NAME: TIME: 4 min
TYPE: Acid, preferential TEMP: RT
COMPOSITION:
2 .... HCl
1 .... HNO_3
2 .... H_2O
DISCUSSION:
GaAs, (111) wafers used in a defect etching study. The solution will develop macroscopic spiral etch pits and figures on (111)A surfaces, but no figures on $(\overline{111})$B.
REF: Little, C L & McCarty, R T — *J Electrochem Soc,* 120,419(1973)
GAS-0186: White, J G & Roth, W C — *J Appl Phys,* 30,956(1959)
GaAs, (111) wafers used in a study of polar surfaces. Solution used at RT, 10 min. It will develop both dislocations and surface damage etch pits.

GAS-0047b
ETCH NAME: TIME:
TYPE: Acid, cleaning TEMP:
COMPOSITION:
1 .... HF
1 .... HNO_3
DISCUSSION:
GaAs, (100) wafers used as substrates for VPE growth of ZnTe and ZnS. Substrate preparation was: (1) mechanical lap/polish; (2) degrease in boiling TCE; (3) rinse in acetone, then methanol. Immediately prior to epitaxy, etch clean with solution shown.
REF: Ibid.

GAS-0056b
ETCH NAME: TIME: 30 sec
TYPE: Acid, junction TEMP: RT
COMPOSITION:
3 .... *"51" etch
2 .... H_2O

*5HCl:1HNO_3

DISCUSSION:
GaAs, (111) wafers. Used to develop p-n junctions.
REF: Ibid.
GAS-0189: Edmonds, J T — *J Appl Phys,* 3,1428(1960)
GaAs, (111) n-type wafers used in a heat treatment study. Solution used as a structure development etch after heat treatment.

GAS-0056c
ETCH NAME: 51 Etch TIME: 30 sec
TYPE: Acid, polish/junction TEMP: RT/cold
COMPOSITION:
1 .... HNO_3
5 .... HCl
DISCUSSION:
GaAs, (111) wafer. Solution used at room temperature is a general polish etch for gallium arsenide. Used cold, it is a p-n junction etch. Solution should be allowed to "age" before using as it is a form of aqua regia. (*Note:* Many acid mixtures become increasingly preferential below RT, and can be used to develop structures, p-n junctions, etc.)
REF: Ibid.
GAS-0190: Cardona, M — *Phys Rev,* 121,752(1961)
GaAs, (111) wafers used in a material study. Solution used as a general etch; follow etching with DI water rinsing. Also used on InAs.

GAS-0038b
ETCH NAME: 1:1:1 TIME:
TYPE: Acid, cleaning TEMP:
COMPOSITION:
1 .... HF
1 .... HNO_3
1 .... HAc

DISCUSSION:

GaAs, (111) wafers used to fabricate diffused diodes. Etch clean diffused diode surfaces in solution shown. This is a general polishing solution for both silicon and germanium. (*Note:* It is not good practice to "name" an etch solution by the volume of its constituents as there are many 1:1:1 acid mixtures.)

REF: Ibid.

GAS-0059

ETCH NAME: A/B — TIME: 10 min

TYPE: Acid, dislocation — TEMP: 60°C (4× slower at RT)

COMPOSITION:

1 ml .... HF

2 ml .... H_2O

1 g .... CrO_3

8 mg .... $AgNO_3$

DISCUSSION:

GaAs, (111), (100) and (110) wafers. During use the solution requires constant stirring to prevent precipitation of $AgCrO_4$. This was a major study of dislocation etching of GaAs. All wafers were undoped, cut within 0.5° of the surface plane, and ingots were grown by a modified Bridgman method. Alpha- and beta-type dislocation were obtained by bending wafers — positive (+) vs. negative (−) (111) surfaces. It was noted that a cleaved (111) surface etched 1.5× faster than a mechanically lap/polished surface, probably due to the cleavage steps. Eliminate CrO_3 for etching the $(\overline{111})$B surface. Also included is etch study of GaAs/GaAs epitaxy layers. Etch rates shown as a ratio: 1:0.71:0.69:0.5 for (100)-(111)A-(110)-$(\overline{111})$B. (*Note:* This is a widely used dislocation etch commonly referred to as the A/B etch. It should not be confused with the A-B, two-step cleaning etch for silicon developed by Kern & Puotinen, which is variously shown as the A-B Etch, or the RCA Etch.)

REF: Abrahams, M S & Buiocchi, C J — *J Appl Phys,* 36,2855(1965)

GAS-0130a

ETCH NAME: A/B Etch, modified — TIME:

TYPE: Acid, dislocation — TEMP:

COMPOSITION:

1 .... A/B Etch

5 .... H_2O

DISCUSSION:

GaAs, (100) ingot and wafers. Ingot grown by HPLEC and M-LEC (magnetic liquid encapsulated Czochralski). Solution used with a 100 W halogen lamp as a photo etch. Stria were observed on HPLEC grown material, but none on M-LEC.

REF: Namajima, M et al — *Jpn J Appl Phys,* 24,L65(1985)

GAS-0060

ETCH NAME: — TIME:

TYPE: Acid, mesa etch — TEMP:

COMPOSITION:

80 .... HCl

4 .... H_2O_2

1 .... H_2O

DISCUSSION:

GaAs, (100) wafers used as substrates for epitaxy growth of heterostructures, as: p-

GaAs:Zn/p-GaAlAs/n-GaAs/GaAs (100) substrate. Solution was used to form mesa structures in the p-GaAs:Zn doped layer.
REF: Reynolds, R T Jr et al — *J Appl Phys*, 56,1968(1984)
GAS-0226: Kular, S S et al — *Solid State Electron*, 27,83(1984)

GaAs:Cr, (100) (SI) wafers used as substrates for zinc implantation, then coated with Si_3N_4. Nitride patterned with photolithographic techniques and aluminum metallized in fabricating Hall samples. The Hall samples then used in a Hall measurement study. Clean the nitrided surfaces with solution shown prior to aluminum evaporation.

GAS-0061a
ETCH NAME: TIME:
TYPE: Acid, selective TEMP:
COMPOSITION: pH: 1 or 10
x .... $C_6H_4O_2$ (Quinone)
x .... $C_4H_6O_2.H_2O$ (Hydroquinone)
DISCUSSION:

GaAs, (100) wafers used as substrates for epitaxy growth of GaAlAs/GaAs. Lower pH values will etch GaAlAs. Higher pH values will etch GaAs.
REF: Tijburg, R P & Dongen, T — *J Electrochem Soc*, 122,687(1976)

GAS-0061b
ETCH NAME: TIME:
TYPE: Acid, selective TEMP:
COMPOSITION: pH: 2
x .... 0.0025 *M*/l $Ce(SO_4)_2.4H_2O$
x .... 0.0025 *M*/l $Ce(NO_3)_3.6H_2O$
DISCUSSION:

GaAs, (100) wafers used as substrates for epitaxy growth of GaAlAs/GaAs. Solution is selective for p-type GaAlAs.
REF: Ibid.

GAS-0061c
ETCH NAME: TIME:
TYPE: Acid, selective TEMP:
COMPOSITION: pH: 7
x .... 0.1 *M*/l $K_3Fe(CN)_6$
x .... 0.1 *M*/l $K_4Fe(CN)_6$
DISCUSSION:

GaAs, (100) wafers used as substrates for epitaxy growth of GaAlAs/GaAs. Solution is selective for GaAs.
REF: Ibid.

GAS-0061d
ETCH NAME: TIME:
TYPE: Acid, selective TEMP:
COMPOSITION: pH: 9.7
x .... 0.225 *M*/l $K_3Fe(CN)_6$
x .... 0.225 *M*/l $K_4Fe(CN)_6$
DISCUSSION:

GaAs, (100) wafers used as substrates for epitaxy growth of GaAlAs/GaAs. Solution is selective for $Ga_{0.7}Al_{0.5}3As$.
REF: Ibid.

GAS-0061e
ETCH NAME: TIME:
TYPE: Acid, selective TEMP:
COMPOSITION: pH: 9.4
x .... 0.3 *M*/l KI
x .... 0.04 *M*/l I_2
DISCUSSION:

GaAs, (100) wafers used as substrates for epitaxy growth of GaAlAs/GaAs. Solution is selective for GaAs without aluminum present.

REF: Ibid.

GAS-0061f
ETCH NAME: TIME:
TYPE: Acid, selective TEMP:
COMPOSITION: pH: 9
x .... 0.3 *M*/l KI
x .... 0.1 *M*/l I_2
DISCUSSION:

GaAs, (100) wafers used as substrates for epitaxy growth of GaAlAs/GaAs. Solution is selective for $Ga_{1-x}Al_xAs$ (x = 0.15). Also used at pH 11 for etching InGaP/GaP or GaAlAs.

REF: Ibid.

GAS-0007b
ETCH NAME: TIME:
TYPE: Acid, polish TEMP: RT
COMPOSITION:
x .... H_2SO_4
x .... H_2O_2
x .... H_2O
DISCUSSION:

GaAs wafers. A general etch solution for polishing gallium arsenide. Composition variations are used. Used as an immersion etch; as a chem/mech etch on a polishing pad with rotation; angle etch, etc.

REF: Kern, W — *RCA Rev,* 39,278(1978)

GAS-0063b: Shaw, D W — *J Electrochem Soc,* 128,874(1981)

GaAs wafers used in a study of the reaction of hydrogen peroxide containing solutions on GaAs. Etch solution was used in a beaker at a 30° angle, and rotated at 135 rpm. Solution produced two different angle side slopes in the etched pits, and such pit slope angles vary with changes in the etch composition. (*Note:* See SI-0092a-k & SI-0300 for more details on pit side slope angles vs. etch compositions.)

GAS-0064
ETCH NAME: TIME:
TYPE: Acid, preferential TEMP:
COMPOSITION:
1 .... H_2SO_4
1 .... H_2O_2
8 .... H_2O
DISCUSSION:

GaAs, (111) wafers. Solution produced grain structure.

REF: Connell, F et al — *Solid State Electron,* 1, 97(1960)

GAS-0065
ETCH NAME: TIME:
TYPE: Acid, polish TEMP:
COMPOSITION:
x .... H_2SO_4
x .... H_2O_2
x .... H_2O
DISCUSSION:

GaAs wafers. Used a beaker at a tilt angle with rotation to polish gallium arsenide.
REF: Sullivan, P T & Pompliano, R S — *J Electrochem Soc*, 122,764(1976)

GAS-0063a
ETCH NAME: TIME:
TYPE: Acid, selective TEMP: RT
COMPOSITION: RATE: 14.6 μm/min
1 .... H_2SO_4
8 .... H_2O_2
1 .... H_2O
DISCUSSION:

GaAs, (100), n-type wafers polished on both sides were used in a study of acid hydrogen peroxide solutions. Parameters were etch rate, morphology, mask undercutting, and relative anisotropy. Silicon nitride was deposited and photolithographically patterned as the etch masks. Patterns were both as strips in ⟨110⟩ direction orientation, for channels; and as circular openings for pits. Etching was with an angled rotation beaker system: beaker angle 30°, with rotation at 135 rpm. Rotation was reversed half way through etch periods. Nine sulfuric acid-hydrogen peroxide, and four hydrochloric acid-hydrogen peroxide solutions were evaluated. Solutions were mixed fresh and used within 2 h. Wafers were ⟨110⟩ cross-sectioned to profile etch pit side slopes, and show the variation obtained for (011) and $(01\overline{1})$.
REF: Shaw, D W — *J Electrochem Soc*, 128,875(1981)

GAS-0063c
ETCH NAME: TIME:
TYPE: Acid, selective TEMP: RT
COMPOSITION: RATE: 1.2 μm/min
1 .... H_2SO_4
8 .... H_2O_2
40 .... H_2O
DISCUSSION:

GaAs, (100), n-type wafers. Solution developed re-entrant double angle (011) and shallow concave, curve $(01\overline{1})$ pit side slopes. See GAS-0063a for added discussion.
REF: Ibid.

GAS-0063d
ETCH NAME: TIME:
TYPE: Acid, selective TEMP: RT
COMPOSITION: RATE: 0.54 μm/min
1 .... H_2SO_4
8 .... H_2O_2
80 .... H_2O

DISCUSSION:

GaAs, (100), n-type wafers. Solution developed re-entrant double angle (011) and sharp, flat positive slopes $(01\overline{1})$ pit sides. See GAS-0063a for added discussion.

REF: Ibid.

GAS-0063e

ETCH NAME:
TYPE: Acid, selective
COMPOSITION:
1 .... H_2SO_4
8 .... H_2O_2
160 .... H_2O

TIME:
TEMP: RT
RATE: 0.25 μm/min

DISCUSSION:

GaAs, (100), n-type wafers. Solution developed re-entrant double angle (011) and sharp, flat positive slope $(01\overline{1})$ pit sides. See GAS-0063a for added discussion.

REF: Ibid.

GAS-0063f

ETCH NAME:
TYPE: Acid, selective
COMPOSITION:
1 .... H_2SO_4
8 .... H_2O_2
1000 .... H_2O

TIME:
TEMP: RT
RATE: 0.038 μm/min

DISCUSSION:

GaAs, (100), n-type wafers. Solution developed re-entrant double angle (011) and sharp, flat positive slope $(01\overline{1})$ pit sides. The re-entrant leg lengths on these solutions from GAS-00063a to GAS-0063e vary progressively from a long negative upper slope to a long positive lower slope with the 1:8:80 solution showing equal positive/negative lengths. See GAS-0063a for added discussion.

REF: Ibid.

GAS-0063g

ETCH NAME:
TYPE: Acid, selective
COMPOSITION:
1 .... H_2SO_4
1 .... H_2O_2
8 .... H_2O

TIME:
TEMP: RT

DISCUSSION:

GaAs, (100), n-type wafers. Solution developed re-entrant double angle (011) and sharp, flat positive slope $(01\overline{1})$ pit sides. Re-entrant slope angles similar to GAS-0063b with upper leg longer negative slope. See GAS-0063a for added discussion.

REF: Ibid.

GAS-0063h

ETCH NAME:
TYPE: Acid, selective
COMPOSITION:
4 .... H_2SO_4
1 .... H_2O_2
5 .... H_2O

TIME:
TEMP: RT
RATE: 5.0 μm/min

DISCUSSION:

GaAs, (100), n-type wafers. Solution developed re-entrant angle with soft positive curve at bottom ($01\bar{1}$) and sharp, flat positive slope (011) pit sides. See GAS-0063a for added discussion.

REF: Ibid.

GAS-00631

ETCH NAME:
TYPE: Acid, selective
COMPOSITION:

8 .... H_2SO_4
1 .... H_2O_2
1 .... H_2O

TIME:
TEMP: RT
RATE: 1.2 μm/min

DISCUSSION:

GaAs, (100), n-type wafers. Solution developed short, negative upper slope with shallow curved bottom slope ($01\bar{1}$) and sharp, flat positive slope (011) pit sides. See GAS-0063a for added discussion.

REF: Ibid.

GAS-0063j

ETCH NAME:
TYPE: Acid, selective
COMPOSITION:

3 .... H_2SO_4
1 .... H_2O_2
1 .... H_2O

TIME:
TEMP: RT
RATE: 5.9 μm/min

DISCUSSION:

GaAs, (100), n-type wafers. Solution developed both pit slopes (011) and ($01\bar{1}$) similar to those of GAS-0061h. See GAS-0063a for added discussion.

REF: Ibid.

GAS-0063k

ETCH NAME:
TYPE: Acid, selective
COMPOSITION:

1 .... HCl
4 .... H_2O_2
40 .... H_2O

TIME:
TEMP: RT
RATE: 0.22 μm/min

DISCUSSION:

GaAs, (100), n-type wafers. Both pit slopes (011) and ($01\bar{1}$) similar to those of GAS-0063a. See GAS-0063e for slope comparisons and GAS-0063a for added discussion.

REF: Ibid.

GAS-0063i

ETCH NAME:
TYPE: Acid, selective
COMPOSITION:

1 .... HCl
1 .... H_2O_2
9 .... H_2O

TIME:
TEMP: RT
RATE: 0.20 μm/min

DISCUSSION:

GaAs, (100), n-type wafers. Single, sharp negative slope (011) and single, positive slope $(01\bar{1})$. See GAS-0063a for added discussion.

REF: Ibid.

GAS-0063m

ETCH NAME: TIME:

TYPE: Acid, selective TEMP: RT

COMPOSITION: RATE: 5.0 μm/min

40 .... HCl

4 .... H_2O_2

1 .... H_2O

DISCUSSION:

GaAs, (100), n-type wafers. Both (011) and $(01\bar{1})$ show positive curves in pit slopes. See GAS-0063a for added discussion.

REF: Ibid.

GAS-0063n

ETCH NAME: TIME:

TYPE: Acid, selective TEMP: RT

COMPOSITION: RATE: 11 μm/min

80 .... HCl

4 .... H_2O_2

1 .... H_2O

DISCUSSION:

GaAs, (100), n-type wafers. Both (011) and $(01\bar{1})$ show positive curves in pit slopes. In all of these solutions etch undercut was greater on the (011). GAS-0063k solution showed the least undercutting of both (011) and $(01\bar{1})$. See GAS-0063a for added discussion.

REF: Ibid.

GAS-0191

ETCH NAME: Hydrofluoric acid TIME:

TYPE: Acid, mask removal TEMP:

COMPOSITION:

x .... HF

DISCUSSION:

GaAs, (100) wafers ion implanted with zinc, then coated with AlN and Si_3N_4 prior to an annealing study. Both coatings were removed with HF after annealing before diffusion depth study of zinc.

REF: Barrett, N J — *J Appl Phys,* 57,5470(1985)

GAS-0192

ETCH NAME: TIME:

TYPE: Acid, dislocation TEMP: 0°C

COMPOSITION: RATE: 1.25 μm/min

1 .... HF

3 .... CrO_3

2 .... H_2O

DISCUSSION:

GaAs, (100) wafers used as substrates for OMVPE growth of GaInAs and GaInP layers, and used in a defect study of the epitaxy thin films. Solution used to develop dislocations and defects. An He-Ne laser at 0°C was used to illuminate the solution during the etch period.

REF: Kuo, C P — *J Appl Phys,* 57,5428(1985)

GAS-0193: Wayher, J L et al — *J Cryst Growth,* 63,285(1983)

Reference for the etching method shown in GAS-0192.

GAS-0066

ETCH NAME: TIME:

TYPE: Acid, cleaning TEMP:

COMPOSITION:

1 .... H_2SO_4

10 .... H_2O

DISCUSSION:

GaAs:Te, (100) n-type wafer substrates were oxidized with SiO_2, then evaporated with SnO_2 in fabricating MIS solar cells. Substrates were degreased ultrasonically with TCE, then MeOH; and then etch cleaned in the solution shown prior to oxidation/metallization.

REF: Brinker, P J et al — *J Electrochem Soc,* 128,1968(1981)

GAS-0068

ETCH NAME: TIME:

TYPE: Acid, damage removal TEMP:

COMPOSITION:

(1)		(2)	
18	H_2SO_4	1	H_2SO_4
1	H_2O_2	1	H_2O_2
1	H_2O	125	H_2O

DISCUSSION:

GaAs:Cr, (100) (SI) wafers used as substrates for MBE growth of GaAs/GaAs. Substrates were etched in solution (1) to remove lap/polish damage; then etch cleaned in the slower (2) solution immediately before epitaxy.

REF: Hiyamizu, S et al — *J Electrochem Soc,* 127,1562(1980)

GAS-0069

ETCH NAME: TIME: 30 sec

TYPE: Acid, polish/cleaning TEMP: RT

COMPOSITION:

(1)		(2)	
3	H_2SO_4	20	H_2SO_4
1	H_2O_2	1	H_2O_2
1	H_2O	1	H_2O

DISCUSSION:

GaAs:Cr, (100) (SI) wafers. Wafers were etch polished in solution (1). Then after photo resist patterning and acetone removal of the photo resist, surfaces were final cleaned with solution (2).

REF: Tokumitsu, E et al — *J Appl Phys,* 55,3163(1984)

GAS-0070
ETCH NAME: TIME:
TYPE: Acid, polish/removal TEMP:
COMPOSITION:
90 .... H_2SO_4
5 .... H_2O_2
5 .... H_2O
DISCUSSION:

GaAs, (100) n^+ wafers used as substrates for an anodic etching study. After mechanical polishing, remove 9—11 μm of surface material with the etch solution shown.

REF: Nagata, K et al — *J Electrochem Soc,* 128,2247(1981)

GAS-0006g
ETCH NAME: TIME:
TYPE: Acid, selective TEMP:
COMPOSITION:
1 .... H_2SO_4
8 .... H_2O_2
1 .... H_2O
DISCUSSION:

GaAs, (100) wafers used as substrates for Gunn diode fabrication. This solution was used to selectively etch channels in wafer surfaces. Also used to develop layer structure in ⟨110⟩ cleaved sections of MBE layers deposited in a study of superlattice device structures.

REF: Kern, W — *RCA Rev,* 39,278(1978)

GAS-0071
ETCH NAME: TIME:
TYPE: Acid, step-etch TEMP: RT
COMPOSITION:
1 .... H_2SO_4
1 .... H_2O_2
75 .... H_2O
DISCUSSION:

GaAs, (100) wafers Be diffused. Used Apiezon-W, black wax, to coat surfaces, then used the solution shown to step-etch for development of the Be diffusion depth profile.

REF: McLevige, W V et al — *J Appl Phys,* 48,3342(1977)

GAS-0072a
ETCH NAME: TIME:
TYPE: Acid, selective TEMP: 30°C (20—50°C)
COMPOSITION:
1 .... H_2SO_4
2.5 .... H_2O_2
50 .... H_2O
DISCUSSION:

GaAs:Cr, (100) (SI), or n^+ diffused wafers used in a development study of selectively etched pit shapes. Etching was done through photo resist and/or SiO_2 masks. A 5:1:1 mixture produces rough pit bottoms at mask edges. Mixtures between 3:1:1 and 6:1:1 show preferential etching at mask edges. The composition shown produced good, flat bottomed pits. Etching is linear with both time and temperature, and ultrasonic stirring will increase rate with a nominal rate of 0.38 μm/min at 30°C. MBE growth was done into the etched pits.

pits. Pits were prepared by etching through an SiO_2 mask, and the SiO_2 overhang remaining after pit forming was removed with HF prior to epitaxy.
REF: Li, A-Z — *J Electrochem Soc,* 130,2027(1983)

GAS-0073
ETCH NAME: TIME:
TYPE: Acid, polish/clean/selective TEMP:
COMPOSITION:
3 .... H_2SO_4
1 .... H_2O_2
1 .... H_2O
DISCUSSION:

GaAs, (100), n-type wafers. Solution used for general polish and cleaning surfaces.
REF: Nakamura, T & Katoda, T — *J Appl Phys,* 55,3064(1984)
GAS-0047c: Matsumoto, T & Ishida, T — *J Cryst Growth* 67,135(1984)

GaAs, (100), (111)A, and $(\overline{111})$B wafers were used as substrates for VPE growth of ZnTe and ZnS. Prepare wafers by: (1) degrease in boiling TCE; (2) rinse in acetone, then MeOH, and (3) etch in solution shown. After etching, follow with a final etch in $1HF{:}9HNO_3{:}10H_2O$.
GAS-0075: Serl, J — *J Electrochem Soc,* 120,1417(1973)

GaAs, (100) wafers. As a polish etch the rate shown as 5 μ/min.
GAS-0076: Kohn-Kuhnenfeld, F — *J Electrochem Soc,* 119,1063(1972)

GaAs, (100) wafers used in a photo etching study. Zinc-doped wafers were grown by Float Zone (FZ) and Te- or Cr-doped (SI) wafers were grown by the Czochralski (CZ) method. An HBO 200 W Hg lamp was used as the photo light source during 5 min etching at RT. Material resistivity produced different structures: n-type wafers produced etched pits; whereas p-type wafers formed a mesa; and semi-insulating wafers showed an etch "foot-print", but neither a pit nor mesa. The solution will also develop stria and dislocations, and the cone type structures observed can be caused by gas bubbles adhering to surfaces during the etch period.
GAS-0130: Namajima, M et al — *Jpn J Appl Phys,* 24,L65(1985)

GaAs, (100) wafers cut from ingots grown from a BN crucible by HPLEC and M-LEC (magnetic field LEC). Wafers were mechanically polished and etched at 60°C in the 3:1:1 etch solution. The A/B etch as $1A/B{:}5H_2O$ was used as a photo etch with a 100 W halogen lamp (for A/B Etch, see: GAS-0059). Material used in a study of stria and dislocations.
GAS-0131: Chung, Y et al — *Thin Solid Films,* 104,193(1983)

GaAs, (100), n-type wafers used as substrates. AgSe evaporated on wafer backs for contact, and annealed at 450°C. Then Ge_3N_4 grown on front surface. Before depositions, etch clean at RT with 3:1:1 solution, DI water rinse, and N_2 dry. Material used in a study of interface density of Ge_3N_4/GaAs. The nitride is unstable at high temperature; N_2 anneal is better than H_2 anneal. Latter is leaky. N_2 anneal densifies; H_2 anneal increases film porosity with segregation of free Ge.
GAS-0198: Kraulte, H et al — *J Electron Mater,* 12,215(1983)

GaAs, (100) wafers cut within 1° of plane. Wafers were cleaned: (1) degrease; (2) etch in HCl and (3) etch solution shown.
GAS-0056d: Ibid.

GaAs, (111) wafers. Solution used warm as a polish etch with a removal of 5 μm/min.
GAS-0223: Reep, D H & Ghandi, S K — *J Electrochem Soc,* 131,2697(1984)

GaAs:Cr, (100) wafers cut 3°-off toward (110), and used as substrates for GaAs epitaxy thin film growth. Clean wafers: (1) degrease in TCE, and rinse in acetone, then MeOH; (2) etch in solution shown at 60°C, 1 min; (3) DI H_2O rinse, and (4) N_2 blow dry.
GAS-0224: Ploog, K et al — *J Electrochem Soc,* 128,400(1981)

GaAs, (100) wafers used as substrates for MBE growth of GaAs thin films doped with S or Be. Material used in a study of structure obtained from periodic impurity doping. Substrate cleaning: (1) diamond paste lap/polish; (2) polish on lens paper saturated with NaClO; (3) rinse in TCE, then MeOH, then double-distilled H_2O; (4) boil in HCl, twice; (5) free etch in solution shown at 48°C, 1 min; (6) rinse in double-distilled H_2O, and (7) N_2 blow dry. The etch solution was allowed to sit after mixing, then used as a "stagnant" etch.

GAS-0225a: Motsuto, M & Mike, H — *J Electrochem Soc,* 124,441(1977)

GaAs:Cr, (100) wafers Si doped. Used as substrates in a study of tin doped GaAs epitaxy layers. Clean in solution shown before epitaxy. Develop epitaxy defects in molten KOH flux at 360°C, 5 min.

GAS-0229: Lafere, W H et al — *Solid State Electron,* 25,389(1982)

GaAs, (100) n-type wafers used as substrates in the study of MIS Schottky barrier type devices. Clean substrates: (1) in DI H_2O with ultrasonic agitation; (2) degrease in boiling TCE, then rinse in acetone, then MeOH; (3) clean in solution shown at 80°C, 90 sec; (4) DI H_2O rinse, and N_2 blow dry. Metallize with Al, Ag, Au, or Sn, and evaluate electron trapping in the native oxide as related to type metal contact deposited.

GAS-0077

ETCH NAME: TIME:

TYPE: Acid, polish/cleaning TEMP: RT

COMPOSITION:

n-type	p-type
3	2 .... H_2SO_4
1	1 .... H_2O_2
1	1 .... H_2O

DISCUSSION:

GaAs:Cr, (100) (SI) wafers used in a study of surface cleaning. The two etches shown will leave a thin Ga_2O_3 oxide on surfaces which can be removed by vacuum bake at 550°C, 10—30 min, as Ga oxides decompose between 370 to 570°C. By polishing with Br_2:MeOH and DI H_2O rinse, surfaces will be passivated. The following solutions and operations were evaluated:

(1) H_2O_2 or HNO_3 used, alone, etch and clean is erratic.
(2) Br_2:MeOH; H_2SO_4:H_2O_2:H_2O; HF:HNO_3; and HF:H_2O_2 solutions all leave a slight oxide.
(3) In reducing carbon and oxygen surface contamination prior to vacuum deposition of metals, epitaxy growth, or diffusion the following sequence was used:
    (a) mechanical lap and polish, then:
        (i) TCE — bp
        (ii) HAc — bp
        (iii) EOH — 60°C
        (iv) DI H_2O — 60°C
        (v) HCl — dip
        (vi) H_2SO_4 — 70°C — use ultrasonic with heavy stirring + white light and follow with DI H_2O rinse and immediately into vacuum.

REF: Munzo-Yague, A et al — *J Electrochem Soc,* 128,149(1981)

GAS-0007h
ETCH NAME: TIME: 3 min
TYPE: Acid, isotropic TEMP: 50°C
COMPOSITION:
4 .... H_2SO_4
1 .... H_2O_2
1 .... H_2O
DISCUSSION:

GaAs, (100) wafers. This is a rapid polishing solution and leaves a residual oxide film on surface approximately 5 Å thick. Follow with successive dips in HF, then highly $NaOH:H_2O_2:H_2O$ at 30°C to reduce residual oxide to less than 10 Å thick.

REF: Ibid.

GAS-0079: Langmuir, M E — *J Electrochem Soc,* 129,1704(1982)

GaAs, (100), n-type, Si-doped, and Zn diffused wafers. The solution was used as a general etch polish.

GAS-0080: Takagishi, S & Mori, H — *Jpn J Appl Phys Lett,* 23,L100(1984)

GaAs:Cr, (100) wafer substrates etched cleaned in the solution shown. Time was 2 min at 50°C.

GAS-0193a: Lewis, C R et al — *J Electron Mater,* 12,507(1983).

GaAs, (100) wafers cut within 2° of plane. Wafers were Mg doped during MBE layer growth. Solution used to pre-clean wafers before epitaxy. After epitaxy and Mg junction forming, solution also used to delineate p-n junctions.

GAS-0078
ETCH NAME: TIME:
TYPE: Acid, selective TEMP:
COMPOSITION:
4 .... H_2SO_4
1 .... H_2O_2
1 .... H_2O
DISCUSSION:

GaAs, (100) wafers cut within $\pm 1/2°$ of plane, Te-doped. LPE grown wafers were p^+, Ge-doped. CZ grown wafers were p^+, Zn doped and Bridgman (boat) grown wafers were p^+, Cd-doped. 1500 Å of SiO_2 was deposited and photo resist patterned, then "V" channels etched into wafer surfaces with the solution shown. Remove photo resist with acetone, then remove oxide with HF. Follow with epitaxy n-type GaAs growth into "V" channels.

REF: Sankaran, R — *J Appl Phys,* 126,1241(1979)

GAS-0081
ETCH NAME: TIME: $1^1/_2$ min
TYPE: Acid, cleaning TEMP: RT
COMPOSITION:
5 .... H_2SO_4
1 .... H_2O_2
1 .... H_2O
DISCUSSION:

GaAs, (100) wafers used in a study of chemical oxidation state depth profiling. Wafers were etch cleaned in the solution shown. Follow with 1 *M* HCl, 1 min; then running DI H_2O, 5 min: MeOH rinse; and N_2 blow dry. Allow wafers to sit in air for 12 months prior to study.

REF: Kohiki, S et al — *Jpn J Appl Phys Lett,* 23,L15(1984)

GAS-0082: Johns, R C — *Thin Solid Films,* 96,285(1982)

GaAs, (100), n-type wafers. Solution used to etch clean surfaces. Time was 3 min at RT. Follow with Br_2:MeOH, 2 min at RT, and then DI H_2O rinse with N_2 blow dry.
GAS-0083: Yoneda, K et al — *J Cryst Growth,* 67,125(1984)

GaAs:Cr, (100) (SI) and $(\overline{111})$B wafers were used as substrates for MBE ZnS thin film epitaxy. After degreasing use the etch shown to clean surfaces and DI water rinse. Follow with HCl soak, 10 min at 30°C, to remove native oxide. Final clean in vacuum, 15 min at 620°C before epitaxy deposition. Time in the solution shown was 10 min at 60°C.
GAS-0030c: Yu, P W — *J Appl Phys,* 48,2434(1977)

GaAs, (100) wafers used for Mg ion implantation. Wafers cleaned in the solution for 15 sec at RT before ion implant.
GAS-0085: Patterson, A M — *J Electron Mater,* 13,621(1984)

GaAs:Cr, (100) (SI) wafers. Complete cleaning cycle was: (1) Teepol detergent, boiling; (2) DI H_2O, rinse; (3) degrease with hot TCE; (4) etch in solution shown, 30 sec, RT; (5) boiling HCl dip; and (6) DI H_2O rinse. Author says this produces a hydrophobic surface.
GAS-0195b: Ibid.

GaAs, (100) wafers cut within 2—3° of plane. Wafers were cleaned and polished: (1) polish in xNaClO:xH_2O, (2) degrease with solvents; (3) etch in HCl with DI H_2O; and (4) final clean in solution shown with DI H_2O rinse.
GAS-0200: Khiki, S et al — *Jpn J Appl Phys,* 23,L15(1984)

GaAs, (100) wafers used in a study of the chemical states on surfaces. Etch in solution shown $1^1/_2$ min, and DI H_2O rinse. Then clean substrates: (1) 1 *M* HCl, 1 min, RT; (2) wash in running DI H_2O, 5 min, RT; (3) MeOH rinse, and N_2 blow dry. Let wafers sit in air for 12 months before evaluating.
GAS-0165: Bhat, R — *J Electron Mater* 14,433(1985)

GaAs, (100) wafers cut 6°-off plane toward (111)A. Also used (110) and (111) wafers. All wafers were OMCVD deposited with GaAs and AlGaAs epitaxy thin films. Wafers were prepared by: (1) chem/mech polish with Br_2:MeOH to remove subsurface work damage; (2) degrease in solvents; (3) then etch polish in solution shown. Material used in a study of the thin film morphology.

GAS-0086
ETCH NAME: — TIME: 20 sec
TYPE: Acid, clean — TEMP: RT
COMPOSITION:
7 .... H_2SO_4
1 .... H_2O_2
1 .... H_2O
DISCUSSION:

GaAs, (100), undoped wafers. Degrease with solvents and then etch clean in the above solution prior to MBE growth of a 1000 Å GaAs buffer layer followed by an SrF_2 dielectric coating.
REF: Sullivan, P W — *J Vac Sci Technol,* B2,202(1984)
GAS-0231: Laidiy, W D et al — *J Vac Sci Technol,* B(1),155(1983)

GaAs wafers. Chem/mech polish in BRM solution, then final clean with solution shown.

GAS-8087
ETCH NAME: — TIME:
TYPE: Acid, clean — TEMP:
COMPOSITION:
1 .... H_2SO_4
1 .... H_2O_2
1 .... H_2O

DISCUSSION:

GaAs, (100), Te-doped wafer used in a study of oxygenated surfaces. First, clean in MeOH then in HCl to remove native oxide before etching in the solution shown, and DI H_2O rinse. Anneal in UHV 30 min at 800 K before oxidation.

REF: Szuber, J — *Thin Solid Films,* 111,309(1984)

GAS-0262: Leung, S et al — *J Electrochem Soc,* 132,898(1985)

GaAs, (100) n-type wafers cut from ingots grown by LEC under a B_2O_3 atmosphere, or grown by Horizontal Bridgman (HB). Clean substrates: (1) degrease in TCE, rinse in acetone, then MeOH; (2) soak in 1HCl:1H_2O at RT, 5 min, DI H_2O rinse; (3) etch clean in solution shown for 2 min, at RT, and (4) final water rinse and N_2 dry. Co-evaporate Au/Ga for 20 nm thickness. Annealing metallization at 225°C will develop orthorhombic structured Au_2Ga, though material was annealed over a range of 130—400°C. To thin for TEM study: (1) epoxy two metallized wafers face-to-face; (2) section cut in ⟨110⟩ direction; (3) mechanically lap specimen wafer surfaces down, and (3) final thin by Ar^+ ion milling with a 6 KeV power setting for observation of GaAs/metal interfaces.

GAS-0007d

ETCH NAME: TIME:

TYPE: Acid, isotropic/thinning TEMP: RT and 40°C

COMPOSITION:

1 .... H_2SO_4
4 .... H_2O_2
1 .... HF

DISCUSSION:

GaAs, (100) wafers and other low index planes. A very rapid polish etch. Also used to remove mechanical lap damage and as a general removal/thinning solution.

REF: Kern, W — *RCA Rev,* 39,278(1978)

GAS-0090

ETCH NAME: TIME:

TYPE: Acid, thinning/step-etch TEMP: RT

COMPOSITION:

1 .... H_2SO_4
1 .... H_2O_2
100 .... H_2O

DISCUSSION:

GaAs, (100) wafers ion implanted with Si, Zn, and Be in a capless anneal study. Two, 1000 W tungsten-halogen lamps used for annealing wafers with implant side up covered with an Si wafer, or a substrate of quartz or sapphire to prevent arsenic depositing on test surfaces during anneal in an AsH_3/H_2, or AsH_3/Ar atmosphere. Above solution used to thin or step-etch for SEM study of implanted atom distribution.

REF: Lie, S C & Narayan, S Y — *J Electron Mater,* 13,897(1984)

GAS-0091

ETCH NAME: TIME: 30 sec

TYPE: Acid, polish/cleaning TEMP: RT

COMPOSITION:

2 .... H_2SO_4
1 .... H_2O_2
1 .... H_2O

DISCUSSION:

GaAs:CR, (100) (SI), and Si-doped wafers used as substrates for LPE growth of InGaAsP layers. After wafer dicing, individual die were etch cleaned in the solution shown.
REF: Hiramatsu, K et al — *Jpn J Appl Phys,* 23,68(1984)

GAS-0098
ETCH NAME: TIME: 3 min
TYPE: Acid, cleaning/junction TEMP: 50°C
COMPOSITION:

4 .... H_2SO_4
1 .... H_2O_2
1 .... H_2O

DISCUSSION:

GaAs:Cr, (100) (SI), and Si-doped wafers used as substrates for MOVPE growth of InGaP to produce p-n junction diodes. Substrates were degreased and then etched in the above solution prior to epitaxy. Substrates heated in vacuum in H_2 at 500°C for final cleaning prior to epitaxy deposition.
REF: Iwamoto, T et al — *J Cryst Growth,* 68,27(1984)
GAS-0193b: Ibid.

GaAs, (100), wafers cut 2°-off toward (110), and doped with Sn, Zn, or Cr were used as substrates for epitaxy OMVPE growth of both Mg thin films and Mg-doped GaAs layers to form p-n junctions. The solution shown was used to etch clean substrates prior to epitaxy and to delineate junctions after epitaxy.
GAS-0100: Waldrop, J R — *J Phys Chem Solids,* 45,44(1984)

GaAs, (100), n- and p-doped wafers used in a study of native oxides. The n- and p-type wafers were etched together in the above solution for 30 sec at RT to surface clean. Final cleaning to remove native oxide was done in vacuum at 550°C.

GaAs, (100), wafers cut within ± $^1/_2$°, and (100), 2°-off plane toward (110) from both Horizontal Bridgman (HB) and Czochralski (CZ) grown ingots were used in a dislocation study. The above solution was used for 2 min at 50°C to chem/mech polish substrates. Dislocation pits were developed with molten KOH.
GAS-0222: Hoke, W E & Laborrier, W C — *J Vac Sci Technol,* B2,272(1984)

GaAs, (100) wafers used to grow epitaxy MBE GaAs thin film layers. Clean wafers: (1) chem/mech polish in BRM solution, and MeOH rinse; (2) degrease in TCE, then rinse in acetone, then in MeOH; (3) etch solution shown; (4) DI water rinse, and (5) N_2 blow dry. Epitaxy layers were Si doped n-type; Be doped p-type.

GAS-0072b
ETCH NAME: TIME:
TYPE: Acid, cleaning TEMP:
COMPOSITION:

6 .... H_2SO_4
1 .... H_2O_2
1 .... H_2O

DISCUSSION:

GaAs, (100), wafers, Si or Be doped wafers were cleaned in this solution.
REF: Li, A-Z — *J Electron Mater,* 12,71(1983)
GAS-0221: Mitsunaga, K et al — *J Vac Sci Technol,* B2,256(1984)

GaAs:Cr, (100) (SI) wafers used as substrates for MBE growth of heterostructure layers of AlGaAs, and n-type GaAs. Clean wafers: (1) degrease in TCE, then rinse in acetone, then ISO; (2) etch in solution shown. After epitaxy the solution was used to etch-strip epitaxy layers, and called a transverse junction strip (TJS) system. (*Note:* TJS is a form of channel etching to measure epitaxy layer thicknesses and observe p-n junctions.)

GAS-0103
ETCH NAME: TIME:
TYPE: Acid, clean TEMP:
COMPOSITION:
90 .... H_2SO_4
5 .... H_2O_2
5 .... H_2O
DISCUSSION:

GaAs, (100) wafers used as substrates for deposition of AlN by CVD. SiO_2 was deposited as a mask using photo resist to pattern, open, and etch with HF. The exposed GaAs in windows was lightly etch clean in the above solution before AlN deposition.

REF: Nake, K & Ozeski, M — *J Cryst Growth,* 68,200(1984)

GAS-0104
ETCH NAME: TIME:
TYPE: Acid, polish TEMP:
COMPOSITION:
20 .... HCl
2 .... H_2O_2
1 .... H_2O
DISCUSSION:

GaAs, (100) wafers as substrates used for gold deposition in a study of the gold/GaAs interface reaction. Substrates were mechanically lapped and then etched in the above solution. Ar^+ ion milling was used for final cleaning prior to gold sputter deposition. Ion milling also used to thin substrates for TEM study. Both the (100) surface and cleaved ⟨110⟩ cross-sections studied.

REF: Yoshie, T et al — *Thin Solid Films,* 111,149(1984)

GAS-0105
ETCH NAME: Phosphoric acid TIME:
TYPE: Acid, removal TEMP: Hot
COMPOSITION:
40 .... H_3PO_4
1 .... H_2O_2
40 .... H_2O
DISCUSSION:

GaAs, (100) wafers used as substrates for LPE growth of p-type, Ge-doped GaAlAs layers. The etch shown was used to etch remove material to near the junction, and the substrates were Ar^+ ion thinned for TEM study.

REF: Veda, O et al — *J Appl Phys,* 50,764(1979)

GAS-0106
ETCH NAME: TIME:
TYPE: Acid, removal TEMP:
COMPOSITION:
3 .... H_3PO_4
1 .... H_2O_2
50 .... H_2O

DISCUSSION:

GaAs, (100) wafers used as substrates for Si^+ and S^+ ion implantation and capless anneal. Even under an As overpressure at 800°C during anneal, thermal etch pits formed. The higher the ion energy dose, the higher the number of pits formed.

REF: Kaskara, J et al — *J Appl Phys,* 50,541(1979)

GAS-0129b: Heiblum, M et al — *Solid State Electron,* 25,185(1982)

GaAs, (100) wafers, and GaAs epitaxy thin film deposited wafers were used in a study of ohmic contacts with AuGe/Ni metallization. After contact deposition as dot structures, surfaces were cleaned in the solution shown. Reaction produced an etched moat around metal contact dots which showed a resistivity variations vs. the adjoining surface.

GAS-0004d

ETCH NAME: TIME: 1—2 min

TYPE: Acid, dislocation TEMP: RT

COMPOSITION:

100 .... H_3PO_4

100 .... H_2SO_4

1 .... H_2O_2

DISCUSSION:

GaAs, (100) wafers used as substrates for LPE growth of GaAlAs. This etch used to develop Stacking Faults (SFs) at the substrate interface.

REF: Ibid.

GAS-0108a

ETCH NAME: TIME: 2 min

TYPE: Acid, selective TEMP: RT

COMPOSITION:

1 .... H_3PO_4

1 .... H_2O_2

20 .... H_2O

DISCUSSION:

GaAs, (100) wafers used as substrates for epitaxy growth of AlGaAs double heterojunction (DH) structures. The etch shown was used to define mesas and channels.

REF: Inoue, K & Sakaki, R — *Jpn J Appl Phys Lett,* L61(1984)

GAS-0108b

ETCH NAME: TIME:

TYPE: Acid, thinning TEMP:

COMPOSITION:

3 .... H_3PO_4

1 .... H_2O_2

75 .... H_2O

DISCUSSION:

GaAs, (100) wafers used as substrates for epitaxy growth of AlGaAs double heterojunction (DH) structures. The AlGaAs layers were etch thinned in the solution shown to remove about 200 Å in forming gate patterns.

REF: Ibid.

GAS-0109
ETCH NAME: TIME: 35 min
TYPE: Acid, via hole TEMP: RT
COMPOSITION:
75 .... H_3PO_4
100 .... H_2O_2
25 .... H_2O
DISCUSSION:

GaAs, (100) wafers were 200 μm thick and etched in the solution shown to develop via holes. Stirring the acid will produce a faster etch rate.

REF: Yengalla, S P & Ghosh, C L — *J Electrochem Soc,* 130,1377(1983)

GAS-0007e
ETCH NAME: TIME:
TYPE: Acid, polish TEMP: 60°C
COMPOSITION:
49 .... H_3PO_4
11 .... HNO_3
DISCUSSION:

GaAs, (111)A and $(\overline{111})$B wafers. Solution will etch polish both surface orientations equally.

REF: Kern, W — *RCA Rev,* 39,278(1978)

GAS-0007f
ETCH NAME: TIME:
TYPE: Acid, anisotropic TEMP:
COMPOSITION:
1 .... H_3PO_4
1 .... H_2O_2
3 .... MeOH
DISCUSSION:

GaAs, (100) wafers with epitaxy grown heterostructure. Etch will produce selective etch structure.

REF: Ibid.

GAS-0096b: Kukken, N K & Tilburg, R P — *J Electrochem Soc,* 130,1722(1983)

GaAs, (100) wafers selectively etched through a mask to develop pit and via hole structures using positive (+) or negative (−) gravity during the etch period. Brass material also etched in this manner.

GAS-0007i
ETCH NAME: TIME:
TYPE: Acid, preferential TEMP:
COMPOSITION:
5 .... H_3PO_4
5 .... H_2SO_4
2 .... H_2O_2
DISCUSSION:

GaAs, (100) wafers used as substrates for selective etching of pit structure using an artificial gravity during the etch period. This solution develops pits selectively.

REF: Ibid.

GAS-0007j
ETCH NAME: TIME:
TYPE: Acid, preferential TEMP: Variable
COMPOSITION:
10 .... *citric acid
1 .... H_2O_2
1 .... *H_2O

*Concentration variable

DISCUSSION:
GaAs, (100), (111), (110), (211) wafers. Rate varies with: (1) plane orientation; (2) solution composition; (3) temperature and (4) agitation of solution. Low concentrations of citric acid are limited by diffusion reaction. Higher concentrations are rate limited.
REF: Ibid.

GAS-0017
ETCH NAME: Hydrofluoric acid, dilute TIME:
TYPE: Acid, cleaning TEMP:
COMPOSITION:
x .... 10% HF
DISCUSSION:
GaAs, (100) n/n^+, Si-doped wafers used in a study of amorphous metal deposition. Substrates were cleaned by degreasing in TCE, then rinsing in IPA. Immediately before placing wafers in vacuum for metallization, HF rinse in the solution shown, water rinse, and N_2 dry.
REF: Wickenden, D K et al — *Solid State Electron,* 27,515(1984)

GAS-0007k
ETCH NAME: TIME:
TYPE: Acid, polish TEMP:
COMPOSITION:
x .... NaOCl
x .... H_2O
DISCUSSION:
GaAs wafers of various orientations. Solution concentration can be varied. Used for wafer polishing by rotation on an impregnated pellon-type pad.
REF: Ibid.
GAS-0114: Frese, J W Jr & Morrison, S R — *J Appl Phys,* 126,1235(1979)
GaAs, (111), Cd-doped, and (110), Sn-doped wafers used in an electrolytic oxidation study. Native oxides were removed from surfaces prior to controlled oxidation by soak in 1 *M* HCl, or by: (1) etching 3 min, at 80°C in $3H_2SO_4$:$1H_2O_2$:$1H_2O$, then (2) in HCl, conc. for 15 sec at RT.

GAS-0115
ETCH NAME: TIME:
TYPE: Acid, polish TEMP: RT
COMPOSITION:
1 .... 2% NaOH
1 .... 1.25% H_2O_2
x .... H_2O

DISCUSSION:

GaS, (100), n-type wafers used as substrates in a surface treatment study. Degrease in boiling TCE. Rinse in acetone and MeOH. Etch rate for the solution shown is 1000 Å/min. Treatments of study were (1) $NaOH/H_2O_2$; (2) HCl, conc.; (3) Ar^+ ion sputter; and (4) sitting in air. Ar^+ ion cleaning produced subsurface damage to 100 Å depth.

REF: Huber, E & Hartnagel, H L — *Solid State Electron,* 27,589(1984)

GAS-0116

ETCH NAME: TIME: 4 min

TYPE: Acid, preferential TEMP: 20°C

COMPOSITION:

10 .... H_2SO_4	to	1 .... H_2SO_4
1 .... H_2O_2		1 .... H_2O_2
1 .... H_2O		16 .... H_2O

DISCUSSION:

GaAs, wafers used in a general etch study of etch characteristics of solutions containing hydrogen peroxide. Variations in solution concentrations shown above change a pit side-slope/bottom from an edge-groove bottom to a flat edge-bottom with a relative 90° vertical pit slope. A good cleaning/removal solution sequence is: (1) the above solution as an 8:1:1 mixture, 3 min at 60°C; (2) 3 min in HCl, conc.; (3) 3 min in H_2O, and (4) a final wash with hot chloroform. Other etch systems described.

REF: Kohn, E — *J Electrochem Soc,* 127,505(1980)

GAS-0120

ETCH NAME: Schell etch TIME:

TYPE: Acid, preferential TEMP:

COMPOSITION:

3 .... HNO_3

1 .... H_2O

DISCUSSION:

GaAs, (111) wafers Cr, Te, and Zn doped. This etch was used to develop dislocations.

REF: Weiss, B L et al — *J Appl Phys,* 48,3614(1977)

GAS-0032b

ETCH NAME: SSA TIME: 10—15 min

TYPE: Electrolytic, dislocation TEMP:

COMPOSITION: ANODE: Al strip to GaAs

5 .... H_2SO_4 CATHODE: Pt

95 .... propylene glycol POWER: 1.5 mA/cm^3

RATE: 0.3 µm/10 min

DISCUSSION:

GaAs wafers. Etch was used as an anodic dislocation etch on gallium arsenide. Wafers were first mechanically polished; then chemically etch polished in: $90H_2SO_4:5H_2O_2:5H_2O$ to remove lap damage. All electrolytic, (SSA) etching was done in the dark with solution stirring. A molten KOH etch was used for dislocation etch result comparison.

REF: Ibid.

GAS-0186
ETCH NAME: TIME:
TYPE: Acid, damage removal TEMP:
COMPOSITION:
20 .... HCl
2 .... H_2O_2
1 .... H_2O
DISCUSSION:

GaAs, (100) wafers used as substrates for gold evaporation in a study of the Au/GaAs interface. Mechanically lap surfaces and then etch polish in the solution shown. In vacuum system, Ar^+ ion sputter clean prior to gold evaporation. In preparing specimens for TEM study, etch thin by Ar^+ ion sputter removal of the substrate from the back un-alloyed side to near GaAs/Au interface.

REF: Yoshie, T et al — *Thin Solid Films,* 111,201(1984)

GAS-0119a
ETCH NAME: A/B etch TIME: 30 sec to 25 min
TYPE: Acid, dislocation TEMP: RT
COMPOSITION:
1 ml .... HF
1 g .... CrO_3
8 mg .... $AgNO_3$
DISCUSSION:

GaAs, (100) and (111) wafers cut within $^1/_2$° of plane. Cut from ingots grown by Czochralski (CZ) or Liquid Encapsulated Czochralski (LEC) method. Wafers were polished with $1Br_2$:MeOH to remove lap damage, and then studied for dislocations. Three preferential solutions were evaluated against each other. The solution shown [see GAS-0059] against: molten flux KOH at 360°C, 15 min [see GAS-0032], and the RC-1 Etch, 45 sec, RT [see GAS-0048].

REF: Hope, D A O & Cockayne, B — *J Cryst Growth,* 67,153(1984)

GAS-0125
ETCH NAME: Caro's etch TIME: 5 min
TYPE: Acid, cleaning TEMP:
COMPOSITION:
10 .... H_2SO_4
1 .... H_2O_2
1 .... H_2O
DISCUSSION:

GaAs, (100), n-type wafers grown by LEC as ingots. Clean wafers by: (1) degrease in TCE at 70°C; (2) rinse in acetone, then MeOH: (3) etch clean in solution shown, and (4) DI H_2O rinse. (*Note:* Original Caro's Etch mixture is $1H_2SO_4$:$1H_2O_2$. Once these type solutions are activated they become self-heating, and will reach a steady-state boiling point of about 175°C. See: SI-0104.)

REF: Bhat, I et al — *Solid State Electron,* 27,121(1984)

GAS-0128: Iliadis, A & Singer, K E — *Solid State Electron,* 26,7(1983)

GaAs:Cr, (100) (SI) wafers with thin film S-doped GaAs epitaxy used in a study of Ge reactions from AuGe(13%) metallization. Clean wafers by: (1) etch in solution shown 90 sec, RT; (2) DI H_2O rinse; (2) HCl rinse 90 sec, RT; (3) DI H_2O rinse; (4) N_2 blow dry prior to vacuum metallization of AuGe.

GAS-0127: Kular, S S et al — *Solid State Electron,* 27,83(1984)

GaAs:Cr, (100) (SI) wafers with Zn^+ ion implantation used in the preparation of Hall samples. Clean wafers in solution shown prior to Si_3N_4 thin film masking, followed by aluminum dot metallization.

GAS-0126
ETCH NAME: TIME:
TYPE: Acid, thinning TEMP:
COMPOSITION:
x .... H_3O_4
x .... H_2O_2
x .... H_2O
DISCUSSION:

GaAs, (100) wafers fabricated as Schottky barrier diodes for a heat treatment study. Wafers were etch thinned in solution shown, then an n-type GaAs buffer layer was grown by VPE, followed by E-beam metallization with Ti, then Al.

REF: Wada, Y & Chino, K I — *Solid State Electron,* 26,559(1983)

GAS-0129
ETCH NAME: Hydrofluoric acid TIME:
TYPE: Acid, cleaning TEMP: Hot
COMPOSITION:
x .... HF, vapor
DISCUSSION:

GaAs:Cr, (100) (SI) wafers used as substrates for epitaxy growth of un-doped (naturally p-type) epitaxy buffer layers, then a doped GaAs layer. Solution shown used to clean epitaxy surfaces before metallization: (1) HF, vapor, (2) DI water rinse, and (3) N_2 blow dry. SiO_2 evaporated as a thin film mask, and photolithographically processed with patterns opened with $1HF:3H_2O$ before metallizing.

REF: Heiblum, M et al — *Solid State Electron,* 25,185(1982)

GAS-0181
ETCH NAME: A/B etch TIME:
TYPE: Acid, preferential TEMP:
COMPOSITION:
12 ml .... HF
24 ml .... H_2O
12 g .... CrO_3
24 mg .... $AgNO_3$
DISCUSSION:

GaAs, (111) wafers with zinc diffusion. Solution used as a dislocation etch in studying zinc reactions. Also worked with GaP. (*Note:* See GAS-0059 for original A/B etch.)

REF: Cohon, M M & Bedard, F D — *J Appl Phys,* 39,75(1968)

GAS-0132
ETCH NAME: TIME: 15 sec
TYPE: Halogen, oxide removal TEMP: RT
COMPOSITION:
x .... 2% Br_2
x .... MeOH
DISCUSSION:

GaAs, (100), Zn-doped, p-type wafers used as substrates for chemical solution growth of TiO_2 films. Remove native oxide with solution shown, rinse with MeOH and hold under

MeOH. Plate Ti with Ti-isoproxide, or Ti-ethylhexoxide in isopropyl alcohol for 1 min. Then hold over water (95% RH) for about 1 week to obtain a 500—2000 Å TiO_2 film. Etch film in $2HF:2NH_4F:5H_2O$; rinse in IPA, then in MeOH under a nitrogen atmosphere. Ar^+ ion sputter etch cleaning produced craters and contamination. Chemical etch cleaning may produce some pits.
REF: Bertrand, P A & Fleischauer, P D — *Thin Solid Films,* 103,167(1983)

GAS-0186b
ETCH NAME: Argon
TYPE: Ionized gas, thinning
COMPOSITION:
x .... Ar^+ ions

TIME:
TEMP:
GAS FLOW:
PRESSURE:
POWER:

DISCUSSION:
GaAs, (100) wafers used in a study of the Au/GaAs interface. Argon ion etching was used to thin the wafers for TEM study. See GAS-0186a for additional discussion.
REF: Ibid.

GAS-0187
ETCH NAME:
TYPE: Acid, junction/defect
COMPOSITION:
50 ml .... NaOH
10 ml .... H_2O_2

TIME: 15 sec
TEMP: RT

DISCUSSION:
GaAs wafers. Solution used to develop grain boundaries. Time shown is for use as a junction etch on solar cells devices that were fabricated in a study of spectral characteristics.
REF: Nasledov, D N & Tsarenkov, B A — *Sov Phys Solid State,* 1,1346(1959)

GAS-0141
ETCH NAME:
TYPE: Acid, polish jet
COMPOSITION:
40 .... HCl
4 .... H_2O_2
1 .... H_2O

TIME:
TEMP:

DISCUSSION:
GaAs, (111), n-type and undoped material cut as rectangular bars, used in a study of dislocations induced by compression. Compression in ⟨123⟩ crystallographic directions was 2% at 550°C in argon. After compression, the bar was sliced on ⟨111⟩ and the solution shown was used to jet polish thin the specimen for SEM study. Under microscope, the material was stretched in the ⟨123⟩ direction at 120—500°C, and dislocations were observed to glide. Compression also used to introduce dislocations in Horizontal Bridgman (HB) grown silicon.
REF: Meeda, K et al — *J Appl Phys,* 66,554(1984)

GAS-0142
ETCH NAME:
TYPE: Acid, preferential
COMPOSITION:
10 ml .... H_2O_2
50 ml .... 5% NaOH

TIME:
TEMP: Boiling
RATE: 10—15 μm/min

DISCUSSION:

GaAs, (111) wafers used in an etch development study. Etch will develop grain boundaries.

REF: Nasledov, D N et al — *Sov Phys-Tech Phys,* 3,726(1958)

GAS-0140b

ETCH NAME: TIME:

TYPE: Acid, removal TEMP: RT

COMPOSITION:

1 .... 20% NaOH

1 .... H_2O_2

DISCUSSION:

GaAs, (111) wafers used in an etch development study. Solution will attack material at room temperature.

REF: Robbins, H & Schwartz, B — *Tech Rep* 1960

GAS-0049b

ETCH NAME: TIME:

TYPE: Acid, polish TEMP:

COMPOSITION:

2 .... HCl

1 .... HNO_3

2 .... H_2O

DISCUSSION:

GaAs, (111) wafers and spheres used in an etch development study of Horizontal Bridgman (HB) grown material. Solution shown used to etch polish wafer surfaces and spheres before dislocation and defect etching.

REF: Richards, J L — *J Appl Phys,* 31,604(1960)

GAS-0005a

ETCH NAME: TIME:

TYPE: Acid, polish TEMP:

COMPOSITION:

2 .... HF

2 .... HCl

3 .... HNO_3

DISCUSSION:

GaAs, (111), n-type, 5—30 Ω cm resistivity wafers used in an etch development study. Initial reaction of solution is slow but after warming during reaction becomes rapid. Add DI H_2O to slow reaction or use cooled (8—10°C) with etch solution in a bath of acetone/ice.

REF: Stopek, S — personal communication, 1956

GAS-0045c

ETCH NAME: TIME:

TYPE: Acid, preferential TEMP:

COMPOSITION:

1 .... HNO_3

3 .... H_2O

DISCUSSION:

GaAs, (111) wafers and spheres. The (111)A surfaces develop etch pits; and $(\overline{111})$B

surfaces show terraces, only. A sphere etched to finite crystal form is a tetrahexahedron, (hk0). (*Note:* Solution is called Schell etch.)
REF: Richards, J L & Crocker, A J — *J Appl Phys,* 31,611(1960)

GAS-0045c
ETCH NAME: TIME:
TYPE: Acid, polish TEMP:
COMPOSITION:
1 .... HF
3 .... HNO_3
2 .... H_2O
DISCUSSION:
GaAs, (111) wafers and spheres. Solution will polish both the (111)A and $(\overline{111})$B surfaces. A sphere etched to finite crystal form is a polished tetrahexahedron, (hk0).
REF: Ibid.

GAS-0045d
ETCH NAME: TIME:
TYPE: Acid, preferential TEMP:
COMPOSITION:
1 .... HNO_3
1 .... $AgNO_3$
3 .... H_2O
DISCUSSION:
GaAs, (111) wafers and spheres. Both the (111)A and $(\overline{111})$B surfaces develop pits. A sphere etched in this solution shows no finite crystal form.
REF: Ibid.

GAS-0045e
ETCH NAME: TIME:
TYPE: Acid, preferential TEMP:
COMPOSITION:
1 .... HF
7 .... HNO_3
8—12 .... H_2O
DISCUSSION:
GaAs, (111) wafers and spheres. Both the (111)A and $(\overline{111})$B surfaces develop pits. A sphere etched to finite crystal form is a highly terraced rhombic dodecahedron, (hh0).
REF: Ibid.

GAS-0045f
ETCH NAME: TIME:
TYPE: Acid, preferential TEMP:
COMPOSITION:
1 .... HF
3 .... HNO_3
0—4 .... H_2O
DISCUSSION:
GaAs, (111) wafers and spheres. The (111)A surface will polish in this solution; whereas the $(\overline{111})$B surface will develop pits. The finite crystal form is a rhombic dodecahedron (hh0).
REF: Ibid.

GAS-0005b
ETCH NAME: TIME:
TYPE: Acid, polish TEMP: 60°C
COMPOSITION:
1 .... HNO_3
2 .... H_2O
DISCUSSION:
GaAs, (111), n-type, 5—30 Ω cm resistivity wafers used in an etch development study. When used hot, this solution can be very rapid and slightly preferential.
REF: Ibid.

GAS-0147
ETCH NAME: TIME:
TYPE: Acid, polish TEMP:
COMPOSITION:
1 .... HF
1 .... HNO_3
1 .... H_2O
DISCUSSION:
GaAs, (100) wafers used in a study of zinc diffusion. Solution used to etch clean wafers prior to diffusion to remove subsurface damage from lapping and polishing. Zinc was diffused at 650—700°C, 3—72 h (66 h gave an 8 μm diffusion depth). A KOH:$K_2Fe(CN)_6$ solution was used as a junction etch on ⟨110⟩ cleaved cross-sections.
REF: Lowen, J & Rediker, R H — *J Electrochem Soc*, 107,26(1960)

GAS-0140c
ETCH NAME: Nitric acid TIME:
TYPE: Acid, removal TEMP: RT and hot
COMPOSITION:
(1) x .... HNO_3 (2) 2 .... HNO_3 (3) 1 .... HNO_3
1 .... H_2O 2 .... H_2O
DISCUSSION:
GaAs, (111) wafers used in an etch development study. Both white and red fuming nitric acid were evaluated. Red fuming nitric etches more rapidly. Both solutions attack GaAs at RT.
REF: Ibid.

GAS-0140d
ETCH NAME: TIME:
TYPE: Acid, removal TEMP: RT or hot
COMPOSITION:
(1) 1 .... HCl (2) 2 .... HCl
1 .... HNO_3 1 .... HNO_3
DISCUSSION:
GaAs, (111) wafers used in an etch development study. The 1:1 solution is moderate to vigorous at RT. The 2:1 solution shows slow reaction at both RT and hot temperatures.
REF: Ibid.

GAS-0140e
ETCH NAME: TIME:
TYPE: Acid, removal TEMP: RT
COMPOSITION:
1 .... HF
1 .... HNO_3
DISCUSSION:
GaAs, (111) wafers used in an etch development study. Solution will attack GaAs with a moderate to vigorous reaction at RT.
REF: Ibid.

GAS-0149
ETCH NAME: TIME:
TYPE: Acid, cleaning TEMP:
COMPOSITION:
6 .... H_2SO_4
1 .... H_2O
DISCUSSION:
GaAs:Cr, (100) (SI) wafers used as substrates for GaAs growth by MBE. Wafers were etch cleaned in this solution and then edge-cleaved into square blanks in ⟨110⟩ directions to eliminate the etch-curved edges before MBE.
REF: Van Hove, J M et al — *J Vac Sci Technol,* B3,563(1985)

GAS-0150
ETCH NAME: TIME:
TYPE: Acid, cleaning/removal TEMP:
COMPOSITION:
1 .... H_2SO_4
1 .... H_2O_2
DISCUSSION:
GaAs:Cr, (100) (SI) wafers, or un-doped wafers used as substrates for SrF_2 and CaF_2 thin film deposition in a study of the films growth habit. Prepare wafers by: (1) degrease in solvents; (2) rinse in acetone, then MeOH, then DI H_2O; (3) use etch shown to remove residual lap damage and clean surfaces; (4) rinse in DI H_2O, and (5) N_2 blow dry.
REF: Sullivan P W et al — *J Vac Sci Technol,* B3,500(1985)

GAS-0151
ETCH NAME: TIME:
TYPE: Acid, cleaning TEMP:
COMPOSITION:
7 .... H_2SO_4
1 .... H_2O_2
1 .... H_2O
DISCUSSION:
GaAs, (100) wafers used in a study of the atomic surface structure of epitaxy GaAs deposited by MBE. Degrease wafers and etch clean with solution shown before MBE.
REF: Croydon, W F et al — *J Vac Sci Technol,* B3,604(1985)
GAS-0249: Wang, W I — *J Vac Sci Technol,* B1(3),630(1983)
GaAs:Cr, (110) wafers used as substrates for epitaxy growth of GaAlAs thin films. Solution used to etch clean surfaces. An Ar^+ ion laser, 1—100 W power, was used for PL measurements.

GAS-0152
ETCH NAME: TIME:
TYPE: Acid, cleaning TEMP:
COMPOSITION:
5 .... H_2SO_4
1 .... H_2O_2
1 .... H_2O
DISCUSSION:

GaAs, (100) and (111) wafers used as substrates for MBE growth of germanium thin films. Etch clean with the solution shown, rinse in H_2O and N_2 blow dry. Argon ion sputter clean in vacuum before Ge deposition.

REF: Katnani, A A et al — *J Vac Sci Technol,* B3,608(1985)

GAS-0153
ETCH NAME: TIME:
TYPE: Acid, cleaning TEMP:
COMPOSITION:
20 .... H_2SO_4
1 .... H_2O_2
1 .... H_2O
DISCUSSION:

GaAs, (100) wafers used as substrates for MBE deposition of AlGaAs in a morphology study. Degrease substrates with solvents, then etch with solution shown, and N_2 blow dry.

REF: Stall, R A et al — *J Vac Sci Technol,* B3,524(1985)

GAS-0087b
ETCH NAME: TIME:
TYPE: Acid, cleaning TEMP:
COMPOSITION:
1 .... H_2SO_4
1 .... H_2O_2
1 .... H_2O
DISCUSSION:

GaAs, (100) Te-doped, n-type wafers used in a study of the oxygenation of surfaces. Degrease with MeOH. Dip in HCl to remove native oxide and then etch in solution shown. Anneal in UHV, 30 min at 800°C before oxidizing.

REF: Szuber, J — *Thin Solid Films,* 112,309(1984)

GAS-0049c
ETCH NAME: TIME:
TYPE: Acid, preferential TEMP:
COMPOSITION:
1 .... HF
2 .... H_2O_2
5 .... H_2O
DISCUSSION:

GaAs, (111) wafers and spheres used in an etch development study. The (111)A surface produces etch pits, and $(\overline{111})$B surface produces terraces.

REF: Ibid.

GAS-0049d
ETCH NAME: TIME:
TYPE: Acid, preferential TEMP:
COMPOSITION:
1 .... HF
3 .... H_2O_2
DISCUSSION:
GaAs, single crystal sphere. Etched in this solution to finite crystal form as a tetrahexahedron, (hk0).
REF: Ibid.

GAS-0049e
ETCH NAME: TIME:
TYPE: Acid, preferential TEMP:
COMPOSITION:
1 .... HF
2 .... H_2O_2
5 .... H_2O
DISCUSSION:
GaAs, (111) wafers and spheres cut from ingots grown by Horizontal Bridgman (HB). The (111)A surface develops etch pits, and the $(\overline{111})$B develops only terraces.
REF: Ibid.

GAS-0045g
ETCH NAME: TIME:
TYPE: Acid, preferential TEMP:
COMPOSITION:
1 .... HF
2 .... H_2O_2
6 .... H_2O
DISCUSSION:
GaAs, (111) wafers and spheres. The (111)A surface produces pits, and $(\overline{111})$B surface is terraced. The sphere etches to a finite crystal form of a tetrahexahedron, (hk0).
REF: Ibid.

GAS-0155
ETCH NAME: TIME: 2 min
TYPE: Acid, cleaning TEMP: RT
COMPOSITION:
5 .... H_2SO_4
1 .... H_2O_2
1 .... H_2O
DISCUSSION:
GaAs:Cr, (100) (SI) wafers used as substrates for MBE thin film growth of GaAs. Degrease substrates in TCE, then rinse in acetone, then MeOH. Etch clean with solution shown, then rinse in DI H_2O, and N_2 blow dry before MBE.
REF: Salerno, J P et al — *J Vac Sci Technol,* B3,618(1985)
AS-0239: Katnani, A A et al — *J Vac Sci Technol,* B3(2),608(1985)
GaAs, (100) and (111) wafers used as substrates for MBE growth of Ge thin films. Clean substrates: (1) etch in solution shown, with DI H_2O rinse and N_2 blow dry, (2) Ar^+ ion sputter clean in MBE vacuum system immediately prior to germanium deposition.

GAS-0156
ETCH NAME: Vacuum
TYPE: Vacuum, cleaning
COMPOSITION:
x .... vacuum

TIME:
TEMP:

DISCUSSION:
GaAs, (110) wafers were cleaved under UHV to obtain very clean surfaces and, with no further treatment, still in vacuum, were then evaporated with Ag, Al, and Au. (*Note:* Cleaved wafers can have microstructure on surfaces, and are often fine etch polished before metallization.)
REF: Newman, N et al — *J Vac Sci Technol,* A3,996(1985)

GAS-0157
ETCH NAME:
TYPE: Acid, preferential
COMPOSITION:
2% .... HF
1.2% .... H_2O_2
96.8% .... H_2O

TIME:
TEMP: RT
RATE: 1000 Å/min

DISCUSSION:
GaAs, (100), n-type wafers cleaved and used in a study of etch treatments. Clean wafers: (1) degrease in boiling TCE; (2) rinse in hot acetone, and (3) rinse in hot MeOH. Wafers were dip etched cleaned in concentrated HCl immediately before being placed in vacuum. Wafers were Ar^+ ion cleaned, then stored in air. Wafer backs were AuGe sputter metallized, then re-etched to clean front surfaces. It was noted that Ar^+ ion bombardment produced a damage depth of 100 Å. (*Note:* Care should be taken in using hot acetone due to its fairly low flash ignition point.)
REF: Huber, E & Hartnagel, H L — *Solid State Electron,* 27,589(1984)

GAS-0159
ETCH NAME:
TYPE: Acid, polish
COMPOSITION:
2—3 .... H_2SO_4
1 .... H_2O_2
1 .... H_2O

TIME:
TEMP: 70°C

DISCUSSION:
GaAs, (100) wafers used in a study of preparing carbon free GaAs surfaces. Surfaces analyzed by AES and RHEED. Mix solution fresh and use with heavy ultrasonic stirring.
REF: Munoz-Yagus, A — *J Electrochem Soc,* 128,149(1981)

GAS-0160
ETCH NAME:
TYPE: Acid, photo-selective
COMPOSITION:
1 .... HF
1 .... H_2O_2
10 .... H_2O

TIME: 2 min
TEMP: RT

DISCUSSION:
GaAs, (110), (111), and (211) wafers and GaAs/GaAs epitaxy used in a study of ridge/valley growth structure that shows variable resistivity. Wafers were saw cut, mechanically lapped and polished and then cleaved. Illuminate etch solution during etching with a tungsten

lamp. Develops "Valley traces type II" striations showing variable resistivity. (*Note:* Similar resistivity variations have been observed in silicon, germanium, and several compound semiconductors. In some cases, as concentric rings of ridge/valley structure. In other cases, as segregation areas of high doping elements. Concentric ring/zone variation can be a problem in critical etching of large volume, small devices on any wafer, particularly, where wafer sizes are greater than 2" in diameter. Resistivity increases from center toward wafer periphery, and etch rates vary across wafer surfaces.)
REF: Lu, Y C & Bauser, E — *J Cryst Growth,* 71,305(1985)

GAS-0161
ETCH NAME: TIME:
TYPE: Acid, polish TEMP:
COMPOSITION:
8 .... H_2SO_4
1 .... H_2O_2
1 .... H_2O
DISCUSSION:
GaAs; Zn, (100) wafers cut 2—3°-off plane toward (110). Wafers were solvent cleaned prior to etch polishing in the solution shown.
REF: Lewis, C R & Ludowise, M J — *J Electron Mater,* 12,749(1984)

GAS-0166
ETCH NAME: TIME:
TYPE: Acid, clean TEMP:
COMPOSITION:
3 .... NH_4OH
1 .... H_2O_2
70 .... H_2O
DISCUSSION:
GaAs, (100) wafers cut 2°-off plane toward (110). Wafers used in an evaluation of the new design of a quartz envelope heater for MBE growth of GaAs/GaAs (100). GaAs substrates were degreased and then etch cleaned in the solution shown before MBE. Substrates were undoped, and cut from ingots grown by LEC. After MBE, cleave wafers ⟨110⟩, and stain in a 1:1:8 A/B solution to develop dislocations and layer structure. (*Note:* See GAS-0059 for A/B solution.)
REF: Boidish, S I et al — *J Electron Mater,* 14,586(1985)

GAS-0167a
ETCH NAME: Superoxol TIME:
TYPE: Acid, preferential TEMP:
COMPOSITION:
1 .... HF
1 .... H_2O_2
4 .... H_2O
DISCUSSION:
GaAs, (111) wafers used in a polarity study of III—V compound semiconductors. Solution develops triangle pits on the (111)A, but only variable structure on $(\overline{111})$B. Other compounds studied were: InSb, InAs, and GaSb.
REF: Faust, J W & Sayar, A — *J Appl Phys,* 31,331(1960)

GAS-0167b
ETCH NAME: TIME:
TYPE: Acid, preferential TEMP:
COMPOSITION:
3 .... HNO_3
1 .... 40% tartaric acid
DISCUSSION:

GaAs, (111) wafers used in a polarity study. Solution develops etch pits on (111)A surfaces.

REF: Ibid.

GAS-0167c
ETCH NAME: TIME:
TYPE: Acid, preferential TEMP:
COMPOSITION:
1 .... HCl
1 .... HNO_3
2 .... H_2O
DISCUSSION:

GaAs, (111) wafers used in a polarity study. See GAS-0057 and GAS-0052 for use of similar solutions. The first reference is called preferential; the latter a polish etchant.

REF: Ibid.

GAS-0167d
ETCH NAME: Sodium hydroxide TIME:
TYPE: Alkali, preferential TEMP:
COMPOSITION:
1 .... 40% NaOH
3 .... H_2O
DISCUSSION:

GaAs, (111) wafers used in a polarity study. Solution develops etch pits on both positive and negative surfaces.

REF: Ibid.

GAS-0168
ETCH NAME: TIME:
TYPE: Acid, preferential TEMP:
COMPOSITION:
1 .... HF
2 .... H_2O_2
5 .... H_2O
DISCUSSION:

GaAs, specimens cut as cylinders and hemispheres. Specimens were etched preferentially in the solution shown in developing shapes. Modified free energy theorems used to predict equilibrium growth and etching shapes. Also used germanium, etching with superoxol, a 1:1:4 mixture of the solution shown.

REF: Jaccodine, R J — *J Appl Phys,* 33,2663(1962)

GAS-0169
ETCH NAME: TIME:
TYPE: Acid, junction TEMP: 30°C
COMPOSITION:
1 .... H_2SO_4
50 .... H_2O_2
1000 .... H_2O
DISCUSSION:

GaAs, (100) wafer Zn-doped were used as substrates for epitaxy growth of GaAs to form a p-n junction diode. Solution shown was used to delineate the p-n junction.
REF: Szubor, J M & Singer, K E — *J Vac Sci Technol,* B3,794(1985)

GAS-0170
ETCH NAME: TIME:
TYPE: Acid, cleaning TEMP:
COMPOSITION:
15 .... H_2SO_4
1 .... H_2O_2
1 .... H_2O
DISCUSSION:

GaAs:Cr, (100) (SI) wafers used for MBE growth of GaAs/GaAs (100). Solution used to etch clean wafer prior to epitaxy.
REF: Kobayashi, K et al — *J Vac Sci Technol,* B3,753(1985)

GAS-0171
ETCH NAME: TIME:
TYPE: Acid, cleaning TEMP:
COMPOSITION:
4 .... H_2SO_4
1 .... H_2O_2
1 .... H_2O
DISCUSSION:

Ga, (100) wafers used as substrates for selective deposition of tungsten. Cleaning procedure was: (1) chem/mech polish with xBr_2:xMeOH; (2) clean in solution shown; (3) dip in 1HCl:1H_2O, and (4) in vacuum with an As_4 overpressure. Thermally clean at 625°C prior to tungsten deposition for 1000 Å metal thickness. Surface was photo resist processed for pattern openings and the tungsten etched by RF plasma using CF_4:O_2(5%), 3 min at 50 W power, and 1.5 Torr pressure with an additional 10 sec at 0.5 Torr and 100 VDC on the substrate to clean the GaAs surface.
REF: Harbison, J P & Derkits, G E Jr — *J Vac Sci Technol,* B3,743(1985)

GAS-0172
ETCH NAME: TIME:
TYPE: Acid, cleaning TEMP:
COMPOSITION:
x .... H_2SO_4
x .... H_2O_2
x .... H_2O
DISCUSSION:

GaAs:Cr (100) (SI) wafers used as substrates for MBE growth of thin film epitaxy as GaAs/GaAs. Wafer cleaning procedure before MBE was: (1) degrease in solvents; (2) clean

in solution shown; (3) DI water rinse. Final clean in vacuum at 580°C, 5 min with a 1×10^{-6} Torr back-pressure of As_4.
REF: Shimizu, S et al — *J Vac Sci Technol,* B1,554(1985)

GAS-0173a
ETCH NAME: Sodium hypochlorate TIME:
TYPE: Acid, preferential TEMP:
COMPOSITION:
x .... NaClO
DISCUSSION:

GaAs, (111)A wafer surfaces used in a study of facet development through holes opened in an SiO_2 mask. After photolithographic opening of the mask, surfaces were etched in the solution shown to develop controlled side slopes in etched pits. On (100) processed wafer surfaces, either $1Br_2$:1000MeOH or 0.7 *M* H_2O_2:1.0 *M* NaOH solutions were used to etch form the pits.
REF: Shaw, D W — *J Electrochem Soc,* 115,777(1968)

GAS-0173b
ETCH NAME: BRM TIME:
TYPE: Halogen, preferential TEMP:
COMPOSITION:
1 .... Br_2
1000 .... MeOH
DISCUSSION:

GaAs, (100) wafers, and other orientations, except (111)A, were used as substrates in a facetted hole etching study. See GAS-0173a for added discussion.
REF: Ibid.

GAS-0173c
ETCH NAME: TIME:
TYPE: Acid, preferential TEMP:
COMPOSITION:
x .... 0.7 *M* H_2O_2
x .... 1 *M* NaOH
DISCUSSION:

GaAs, (100) wafers and other orientations, except (111)A, used in a study of etched pit facet development through a silica mask to define a hole. See GAS-0173a for additional discussion.
REF: Ibid.

GAS-0174
ETCH NAME: TIME: 10 min
TYPE: Acid, preferential TEMP: 82°C
COMPOSITION:
x .... 0.2 *N* $FeCl_3$
x .... 6 *N* HCl
DISCUSSION:

GaAs, (111) wafers used in a study of surface etching characteristics of (111)A vs. $(\overline{111})$B polarity. Other compound semiconductors studied using other specific etchants were InSb, InAs, GaSb, AlSb, InP, and InSb. The direction of positive (+) and negative (−) tetrahedrons (111) pit directions were used to establish the A vs. B wafer face. (*Note:*

Tetrahedrons show 180°C rotation of their triangle pit orientations on opposed positive and negative wafer sides from (111)A to $(\overline{111})$B. This is true for all (111) oriented materials.)
REF: Gatos, H C & Lavine, M C — *J Electrochem Soc,* 107,427(1960)

GAS-0175
ETCH NAME: TIME:
TYPE: Acid, polish TEMP: Hot
COMPOSITION:
x .... 4% NaOH
x .... 5% H_2O_2
DISCUSSION:
GaAs, (111) wafers fabricated as Esaki diodes for high frequency applications. Solution used for polishing and finishing diodes. Other solutions evaluated included aqua regia and mixtures of HF:HNO_3. The solution shown gave best polishing and cleaning results.
REF: Burrus, C A — *J Appl Phys,* 32,1031(1961)

GAS-0176
ETCH NAME: TIME:
TYPE: Acid, preferential TEMP:
COMPOSITION:
x .... HCl
x .... HNO_3
x .... H_2O
DISCUSSION:
GaAs, (111) wafers used in a study of electrical properties. Author says that if the solution shown is used without dilution it will chemically polish.
REF: Detweiler, D P — *Phys Rev,* 97,1575(1955)

GAS-0177
ETCH NAME: TIME:
TYPE: Acid, preferential TEMP:
COMPOSITION:
1 .... HF
3 .... HNO_3
2 .... H_2O
DISCUSSION:
GaAs, (111) wafers used in a polarity etching study. The solution shown develops etch pits on negative $(\overline{111})$B, arsenic surface, but no pits on the positive (111)A gallium surface.
REF: White, J C & Roth W C — *J Appl Phys,* 31,611(1960)

GAS-0179
ETCH NAME: TIME:
TYPE: Acid, polish TEMP:
COMPOSITION:
5 .... HCl
1 .... HNO_3
DISCUSSION:
GaAs, (111) wafers used in a heat treatment study of the material. Solution was used to polish n-type GaAs. Water wash after etching.
REF: Edmonds, J T — *J Appl Phys,* 31,142b(1960)

GAS-0180
ETCH NAME: Nitric acid TIME:
TYPE: Acid, junction TEMP:
COMPOSITION:
x .... HNO_3
x .... H_2O
DISCUSSION:

GaAs, as thin film epitaxy grown on germanium substrates to form p-n junctions. The solution shown was used to delineate the junction. Also studied GaAs/GaP and GaP/Ge as epitaxy thin films grown on substrates as devices.

REF: Weinstein, M et al — *J Electrochem Soc,* 111,674(1964)

GAS-0181
ETCH NAME: Potassium hydroxide TIME:
TYPE: Alkali, reactivity TEMP:
COMPOSITION:

(1)	(2)	(3)	(4)
x .... 1 *M* KOH	1 .... NH_4OH	x .... H_2O	1 .... H_2SO_4
+ .... O_2	1 .... H_2O	3 .... H_2O	+ .... O_2/N_2
	+ .... O_2/N_2	+ .... O_2/N_2	

DISCUSSION:

GaAs, (100) wafers used in a study of acid vs. base surface treatment. All solutions shown were used with oxygen or nitrogen bubbled through the etchant during the test period. HCl also was evaluated and was similar to (4). Solution (2) was no better in preventing surface degradation than water, (3), and also produced surface micro-roughness requiring re-polish with acid solutions. Rapid O_2 increase on surfaces with (4) remained unless reduced with a bromine-methanol treatment as re-polishing. Surfaces more susceptible to acid attack than with bases.

REF: Aspenes, D E — *J Vac Sci Technol,* A(3),1018(1985)

GAS-0182a
ETCH NAME: TIME: 10 min
TYPE: Acid, polish/clean TEMP: 60°C
COMPOSITION:
5 .... H_2SO_4
1 .... H_2O_2
1 .... H_2O
DISCUSSION:

GaAs, (100) wafers, Cr-O doped semi-insulating (SI) substrates used for ZnS epitaxy growth. GaAs, $(\overline{111})$B wafers, arsenic doped also used. After chem/mech lap with Br_2:MeOH and degreasing in solvents, wafers were cleaned as follows: (1) etch in solution shown; (2) rinse in DI H_2O; (3) etch 10 min in HCl at 30°C to remove oxide, and (4) final water rinse and drying. Gallium phosphide wafers also were used as substrates.

REF: Yoneda, K et al — *J Cryst Growth,* 67,125(1984)

GAS-0184
ETCH NAME: Sodium hypochlorite TIME: 2 h
TYPE: Acid, polish/removal TEMP: RT
COMPOSITION: RATE: 3 mils/2 h
x .... x% NaClO
DISCUSSION:

GaAs:Cr, (100) (SI) wafers used as substrates for epitaxy growth of InAs. Use solution shown on a paper pad to polish wafers. Just before epitaxy: (1) etch clean in

$5H_2SO_4$:$1H_2O_2$:$1H_2O$, RT, 5 min then add $5H_2O$ for an additional 5 min all with beaker rotating. After epitaxy, cleave ⟨110⟩ and stain etch, RT, 1 sec in solution shown. To obtain defect density, etch in 1HF:$3HNO_3$.
REF: Cronin, G R et al — *J Electrochem Soc,* 113,1336(1966)

GAS-0187a
ETCH NAME: TIME:
TYPE: Halogen, polish TEMP:
COMPOSITION:
x .... x% Cl_2
x .... MeOH
DISCUSSION:
GaAs, (111) and (100) wafers used in a study of photoluminescence and material growth. Solution used as a polish etch with chlorine gas bubbled through MeOH. Also used $2H_2SO_4$:H_2O.
REF: Panaish, M B et al — *Solid State Electron,* 9,311(1966)

GAS-0187b
ETCH NAME: Sulfuric acid TIME:
TYPE: Acid, polish TEMP:
COMPOSITION:
x .... H_2SO_4
x .... H_2O
DISCUSSION:
GaAs, (111) and (100) wafers used in a study of photoluminescence and material growth. Solution used as a polish etch.
REF: Ibid.

GAS-0188a
ETCH NAME: TIME:
TYPE: Acid, selective TEMP:
COMPOSITION:
x .... 1 *M* NaOH
x .... 0.7 *M* H_2O_2
DISCUSSION:
GaAs, (111) and (100) wafers used in a study of etching and etch rates through an SiO_2 surface coating mask. All of the following solutions showed etch undercut at the SiO_2 mask edge: (1) $1Br_2$:1000MeOH; (2)$5H_2SO_4$:$1H_2O_2$:$1H_2O$; (3) 3% $HClO_4$; and (4) HCl vapor in vacuum at 800°C. The holes were deeper around and near the SiO_2 mask edges. Solution shown did not etch holes with such undercutting and edge depth variation.
REF: Shaw, D W — *J Electrochem Soc,* 113,958(1966)

GAS-0194
ETCH NAME: TIME:
TYPE: Acid, cleaning sequence TEMP:
COMPOSITION:
(1) degrease (2) 7 .... H_2SO_4
1 .... H_2O_2
DISCUSSION:
GaAs:Cr, (100) (SI) wafers used as substrates for epitaxy growth of single crystal $ZnGeAs_2$. Substrates were degreased successively in TCE, then rinsed in acetone, and MeOH. Then etch in solution shown, water rinse, and final N_2 jet blow dry. Substrates

were mounted on $ZnGeAs_2$ coated Mo heater blocks with indium and heated to 450—520°C in the sputter system before MBE growth.
REF: Shah, S I & Greene, J E — *J Cryst Growth,* 68,537(1984)

GAS-0193c
ETCH NAME: TIME:
TYPE: Acid, junction TEMP:
COMPOSITION:
4 .... H_2SO_4
1 .... H_2O_2
1 .... H_2O
DISCUSSION:
GaAs, (100) wafers within 2° of plane used as substrates for OMVPE growth of Mg-doped GaAs layers. After epitaxy, cleave ⟨110⟩ and use solution shown to stain develop junction. Also used as a cleaning/polishing etch on initial GaAs substrates prior to epitaxy.
REF: Lewis, C R et al — *J Electron Mater,* 12,507(1983)

GAS-0195a
ETCH NAME: Sodium hypochlorate TIME:
TYPE: Acid, polish TEMP:
COMPOSITION:
x .... x% NaClO
DISCUSSION:
GaAs, (100) wafers cut within 2—3° of plane. Use solution to polish etch wafers, then clean as follows: (1) degrease; (2) clean in HCl; (3) etch clean in $5H_2SO_4$:$1H_2O_2$:$1H_2O$; and (4) rinse in DI water.
REF: Abrokwah, J I C — *J Electron Mater,* 12,681(1983)

GAS-0197
ETCH NAME: TIME:
TYPE: Acid, dissolve TEMP:
COMPOSITION:
x .... HCl
x .... HNO_3
DISCUSSION:
GaAs wafers grown by Horizontal Bridgman (HB) technique and then dissolved in an aqua regia type solution for a study of iron contamination.
REF: Udagawa, T V et al — *J Electron Mater,* 12,563(1983)

GAS-0028b
ETCH NAME: TIME:
TYPE: Acid, cleaning sequence TEMP:
COMPOSITION:
(1) x .... acetone (2) x .... MeOH, vapor (3) x .... HCl
DISCUSSION:
GaAs, (100) wafers used as substrates for deposition of CVD SiO_2 and SiN_x, then an SiO_2 cap layer used as masks for zinc diffusion into GaAs in a study of pinhole reduction. Wafers were degreased and etch cleaned with solvents as shown.
REF: Blaauw, C et al — *J Electron Mater,* 13,251(1984)

GAS-0045h
ETCH NAME: TIME:
TYPE: Acid, preferential TEMP:
COMPOSITION:
1 .... HF
1—2 .... H_2O_2
4—6 .... H_2O
DISCUSSION:

GaAs, (111) as single crystal wafers and spheres. The (111)A face developed etch pits; the $(\overline{1}\overline{1}\overline{1})$B, terrace, structure. The sphere etched to a finite crystal form as a tetrahexahedron, (hk0).

REF: Ibid.

GAS-0049e
ETCH NAME: TIME:
TYPE: Acid, preferential TEMP:
COMPOSITION:
1 .... HF
2 .... H_2O_2
5 .... H_2O
DISCUSSION:

GaAs, (111) wafers cut from ingots grown by horizontal zone melting. Solution developed etch pits on (111)A surfaces; terraces on $(\overline{1}\overline{1}\overline{1})$B. (*Note:* Growth method is now called Horizontal Bridgman (HB).)

REF: Ibid.

GAS-0049f
ETCH NAME: TIME:
TYPE: Acid, preferential TEMP:
COMPOSITION:
1 .... HF
3 .... H_2O_2
DISCUSSION:

GaAs single crystal spheres. Solution developed a finite crystal form as a tetrahexahedron, (hk0).

REF: Ibid.

GAS-0210
ETCH NAME: Potassium hydroxide TIME:
TYPE: Electrolytic, junction TEMP:
COMPOSITION: ANODE: GaAs
x .... KOH CATHODE:
x .... H_2O POWER:
DISCUSSION:

GaAs, (100) wafers fabricated as diodes in a study of electroluminescent properties with negative resistance. Solution used to electropolish and develop p-n junctions.

REF: Wieser, K & Levitt, R S — *J Appl Phys,* 35,2431(1964)
GAS-0212d; Ibid.

GaAs, (100) wafers used in an X-ray analysis of diffusion induced defects. A 5% KOH solution was used as an anodic junction etch.

GAS-0211a
ETCH NAME: TIME:
TYPE: Acid, stain (p^+) TEMP:
COMPOSITION:
1 .... HF
1 .... HNO_3
3 .... H_2O
DISCUSSION:

GaAs, (100) wafers used in a study of growth methods using a Travelling Solvent technique with germanium doping. Solution used to develop structure of the p^+ growth areas. Also used 1HF:$2H_2O$ as a dislocation etch.

REF: Mlavasky, A I & Weinstein, M — *J Appl Phys,* 34,2885(1963)

GAS-0212a
ETCH NAME: TIME:
TYPE: Acid, polish TEMP: RT
COMPOSITION
2 .... H_2SO_4
1 .... H_2O_2
1 .... H_2O
DISCUSSION:

GaAs, (100) wafers used in an X-ray analysis of diffusion induced defects. Solution used as a chem/mech polish etch with wafers in a rotating beaker.

REF: Schwuttke, G H & Rupprecht, H — *J Appl Phys,* 37,167(1966)

GAS-0212b
ETCH NAME: TIME:
TYPE: Acid, junction TEMP: RT
COMPOSITION:
1 .... HF
1 .... H_2O_2
10 .... H_2O
DISCUSSION:

GaAs, (100) wafers used in an X-ray analysis of diffusion induced defects. Solution used to develop p-n junctions.

REF: Ibid.

GAS-0213: Proebsting, R — *Semi Prod Solid State Technol,* November 1964, 33

GaAs, (100) wafers used to fabricate laser diodes. Wafers were zinc diffused at 850°C for 3 h for a 1 mil diffusion depth. After angle lap of portions of wafers, p-n junctions were etched developed as a follows: (1) place a drop of solution across junction at RT; (2) flood specimen surface with intense white light, etching 15—30 sec in solution shown.

GAS-0212c
ETCH NAME: TIME:
TYPE: Acid, removal TEMP: 0°C
COMPOSITION:
3 .... H_2SO_4
1 .... H_2O_2
2 .... H_2O
DISCUSSION:

GaAs, (100) wafers used in an X-ray analysis of diffusion induced defects. Solution

used to etch section wafers and for general removal in developing cross-section samples for diffusion depth measurement.
REF: Ibid.

GAS-0235
ETCH NAME: TIME:
TYPE: Acid, polish TEMP: 0°C
COMPOSITION:
1 .... H_2SO_4
1 .... H_2O_2
50 .... H_2O
DISCUSSION:
GaAs, (100) wafers used in a study of the disorder influence of ion implantation. Prepare wafers: (1) mechanical polish to 0.5 mm thick; (2) chem/mech polish in solution shown prior to I^2.
REF: Anderson, W J & Parks, Y S — *J Appl Phys,* 49,4568(1978)

GAS-0218
ETCH NAME: BRM TIME: To 10 min
TYPE: Halogen, removal/polish TEMP: RT
COMPOSITION:
"A" x .... 7.5% Br_2 "B" + MeOH cover
x .... MeOH
DISCUSSION:
GaAs, (100) wafers handled with a special etching technique. "Cover" the BRM solution with a layer of pure MeOH. To etch, slowly lower wafer into BRM through the covering MeOH with minimal disturbance, etch for selected time period, then slowly remove with automatic quenching through the pure MeOH cover layer. Authors note the MeOH cover layer will diffuse downward 1 mm into the BRM solution in 10 min.
REF: Bisaro, R et al — *J Appl Phys,* 40,978(1982)

GAS-0219
ETCH NAME: TIME: 2 min
TYPE: Acid, polish TEMP: 50°C
COMPOSITION:
4 .... H_2SO_4
1 .... H_2O_2
1 .... H_2O
DISCUSSION:
GaAs:Cr (100) (SI) wafers cut $^1/_2$°-off plane toward (110), and used in a study of dislocations. Chem/mech polish wafers in solution shown, then etch in molten flux KOH to develop dislocations.
REF: Mori, H & Takagashi, S — *J Cryst Growth,* 69,23(1984)

GAS-0220
ETCH NAME: TIME: (1) 3 min (2) 2 min
TYPE: Acid, cleaning TEMP: RT RT
COMPOSITION:
(1) 5 .... H_2SO_4 (2) x .... x% Br_2
1 .... H_2O_2 x .... MeOH
1 .... H_2O

DISCUSSION:

GaAs, (100), n-type wafers prepared as substrates for epitaxy growth of GaAs. Clean in solutions in the order shown with DI H_2O rinse and N_2 blow-dry after each etch step.
REF: Howard, L M et al — *Thin Solid Films,* 96,285(1982)

GAS-0215
ETCH NAME: RIE
TYPE: Ionized gas, structuring
COMPOSITION:

TIME:
TEMP:
GAS FLOW:
PRESSURE
POWER:

(1) x .... CHF_3 (2) x .... CCl_4 (3) x .... CCl_4
x .... Cl_2

DISCUSSION:

GaAs, (100) wafers used in a development study using dry chemical etching (DCE) to form submicron structures. PMMA photo resist used to pattern mesas. Plasma etch (PE) or RIE/RIBE used with gases shown above. CCl_4 develops sloped mesa sides with a bottom trench at the base; with added Cl_2 forms a vertical sided pylon.
REF: Wolf, E D et al — *Proc IEEE* 30,592(1983)

GAS-0227
ETCH NAME:
TYPE: Acid, oxidizing
COMPOSITION:
3 .... H_2SO_4
1 .... H_2O_2
1 .... H_2O

TIME:
TEMP:

DISCUSSION:

GaAs, (100) and InSb, (100) wafers used in a cleaning study of surfaces prior to MBE epitaxy growth. After etch cleaning in solution shown, rinse with filtered DI H_2O with wafers on spinner at 3000 rpm. In the MBE vacuum system heat to 590°C to remove residual oxide developed and remaining after etching/rinsing. (*Note:* This is one of several solutions for oxidizing a surface, then stripping the oxide as a surface cleaning step.)
REF: Vasquez, R P et al — *J Vac Sci Technol,* B(3),791(1983)

GAS-0241
ETCH NAME:
TYPE: Molten flux, defect
COMPOSITION:
1 .... KOH
1 .... NaOH

TIME: 1—120 min
TEMP: 325—350°C

DISCUSSION:

GaAs:Cr, (100), (111) (SI), and n-type Si doped wafers used in a defect study. Molten solution as a KOH:NaOH eutectic mixture used in a platinum crucible with wafers in a Pt wire basket. Wafers were cleaned: (1) DI water rinse; (2) acetone rinse; (3) MeOH dip, and (4) 2-propanol (ISO) rinse and, (4) hot N_2 blow drying. GaAs blanks were cut 5 mm square and 0.38 mm thick. The (100) etched 15—120 min; (111), 1—30 min at temperatures shown. Etch rates computed from weight loss: (1) 325°C — (100) 0.02 μm/min; (111), 0.06 μm/min, (2) 350°C — (111) 0.22 μm/min and (100), 0.09 μm/min. Solution develops dislocation pits and other defects. As these alkalies are hygroscopic, repeatability is difficult

due to variable water inclusion, but results are comparable to other etching systems. There are some rate variations as shown by the references below.
REF: Lessoff, H & Gorman, R — *J Electron Mater,* 13,733(1984)
GAS-0242: Richter, H & Schulz, M — *Krist Tech,* 9,1041(1974)
GaAs, (100) wafers. Rate shown as 1 μm/min at 350°C.
GAS-0243: Iskii, M et al — *Jpn J Appl Phys,* 15,645(1976)
GaAs, (100) wafers. Rate shown as 0.98 μm/min at 350°C.
GAS-0244: Takenenaka, T et al — *Jpn J Appl Phys,* 17,447(1978)
GaAs, (100), n-type, Si doped wafers. Rate shown as 0.15 μm/min at 350°C.
GAS-0245: Wagner, W R & Greene, L I — *J Electrochem Soc,* 128,1091(1981)
GaAs, (100) wafers. KOH, only, with m.p. of 360°C. Etching done from 360 to 365°C. Etching in a dry atmosphere did not develop dislocations. Needs the presence of residual water in the molten KOH?

GAS-0228
ETCH NAME: TIME:
TYPE: Acid, thinning TEMP: RT
COMPOSITION:
1 .... H_2SO_4
10 .... H_2O_2
10 .... H_2O
DISCUSSION:
GaAs, (100), n-type, 0.001—0.04 Ω cm resistivity, wafers used as substrates for 1000 Å Pt evaporation and anneal in a study of interface reactions. Clean substrates: (1) in HCl, conc.; (2) in NaClO solution; (3) and in 1HCl:1H_2O with DI water rinse between each step. After metallization and annealing, initially lap thin GaAs substrate back, and final thin with solution shown for interface study by TEM.
REF: Fontaine, C — *J Appl Phys,* 54,1404(1981)

GAS-0230
ETCH NAME: TIME: 90 sec
TYPE: Acid, cleaning TEMP: RT
COMPOSITION:
10 .... H_2SO_4
1 .... H_2O_2
1 .... H_2O
DISCUSSION:
GaAs, (111)As, (100) and (110) oriented wafers prepared as epitaxy growth substrates for MBE. Clean substrates: (1) degrease; (2) clean with solution shown; (3) 30 sec DI water rinse, and (4) N_2 blow dry. After mounting on moly heater blocks in the MBE vacuum system, desorb residual oxygen on wafer surfaces at 600°C under an As_4 atmosphere before epitaxy growth.
REF: Ballingall, J M — *J Vac Sci Technol,* B(1),163(1983)

GAS-0232
ETCH NAME: Sulfuric acid TIME:
TYPE: Acid, cleaning TEMP:
COMPOSITION:
x .... H_2SO_4, conc.
DISCUSSION:
GaAs:Cr, (100) wafers used to fabricate hi-doped, hot electron devices, as mesa struc-

tured diodes. Nickel evaporation used as contacts on mesas. Diodes were cleaned in solution shown, and it leaves a thin native oxide.
REF: Harris, J J & Woodcock, J M — *J Vac Sci Technol,* B(1),196(1983)

GAS-0233
ETCH NAME: TIME:
TYPE: Acid, polish TEMP: RT
COMPOSITION:
5 .... NH_4OH
1000 .... H_2O_2
DISCUSSION:
GaAs, (100) wafers used in a study of zinc diffusion at 850°C. Prepare wafers by: (1) chem/mech polish in solution shown, and (2) final clean in solution of $5H_2SO_4:1H_2O_2:1H_2O$.
REF: Cognetti, C — *J Electrochem Soc,* 128,2198(1981)

GAS-0234
ETCH NAME: BRM TIME:
TYPE: Halogen, polish TEMP: RT
COMPOSITION:
(1) 2 ml .... Br_2
1000 ml .... MeOH
(2) $1^1/_2$ ml .... Br_2
1000 ml .... MeOH
DISCUSSION:
GaAs and GaP, (100) and $(\overline{111})$B, high n-type wafers with Te doping used in a solution oxidation reaction study. Prepare wafers by chem/mech polishing in either solution shown using a solution saturated pellon pad. Soak samples 5 to 6 days in the following solvents or acids: (1) MeOH; (2) H_2O at RT and at 60°C; (3) 30% H_2O_2; (4) H_2SO_4; (5) HCl; (6) HNO_3; (7) NaOH, and (8) NaClO. All solutions were used to develop native oxides. These are hydrated type oxides as $Ga_2O_3.H_2O$ with a refractive index on n = 1.7. (*Note:* In the strictest sense of the word, a native oxide is one that forms under natural conditions of oxidation, not a manmade oxide, but any thin film oxide on a material surface is referred to as being a native oxide in Solid State processing.)
REF: Schwartz, B — *J Electrochem Soc,* 118,657(1971)

GAS-0246a
ETCH NAME: Ammonium hydroxide TIME:
TYPE: Base, cleaning TEMP:
COMPOSITION:
x .... NH_4OH
DISCUSSION:
GaAs, (100), p-type wafers used to fabricate light emitting diodes with Al or Au metallized dot contacts. This was a study of the effects of various surface treatments on light emission. Strong acids reduced or eliminated emission; whereas basic solutions show good results; and annealing at 350°C in N_2, 5 min showed improved emission efficiency. Effects may be due to excessive removal of gallium? Arsenic is soluble in HNO_3, and gallium in alkalies. Basic solutions remove Ga_2O_3 and $Ga(OH)_3$ leaving an As enhanced surface. Surfaces also oxidized in AGW solution (see GAO-0005a-b). The study included the crystal planes and their measured angles as etched structures. The following references cite the various solutions evaluated.
REF: Adachi, H & Hartnagel, H L — *J Vac Sci Technol,* 19(3),427(1961)
GAS-0246b: Ibid.

GaAs, (100), p-type wafers. Chemical surface treatment solution was $1HCl:1H_2O$. See GAS-0246a.

GAS-0246c: Ibid.

GaAs, (100), p-type wafers. Chemical surface treatment solution was $1NaOH:20H_2O_2:50H_2O$. See GAS-0246a.

GAS-0246d: Ibid.

GaAs, (100), p-type wafers. Chemical surface treatment solution was $1NH_4OH:1H_2O$. See GAS-0246a.

GAS-0246e: Ibid.

GaAs, (100), p-type wafers. Chemical surface treatment solution was $20H_2SO_4:1H_2O_2:1H_2O$. See GAS-0246a.

GAS-0246f: Ibid.

GaAs, (100), p-type wafers. Chemical surface treatment solution was $1H_2SO_4:1H_2O_2:50H_2O$. See GAS-0246a.

GAS-0246g: Ibid.

GaAs, (100), p-type wafers. Chemical surface treatment solution was $1NaOH:1H_2O$. See GAS-0246a.

GAS-0246h: Ibid.

GaAs, (100), p-type wafers. Chemical surface treatment solution was $1NaOH:1H_2O_2:30H_2O$. See GAS-0246a.

GAS-0246i: Ibid.

GaAs, (100), p-type wafers. Chemical surface treatment solution was $1NaOH:3H_2O_2:150H_2O$. See GAS-0246a.

GAS-0246j: Ibid.

GaAs, (100), p-type wafers. Chemical surface treatment solution was $10H_3PO_4:1H_2O$. See GAS-0246a.

GAS-0246k: Ibid.

GaAs, (100), p-type wafers. Chemical surface treatment solution was $10H_3PO_4:1H_2O_2:1H_2O$. See GAS-0246a.

GAS-0246l: Ibid.

GaAs, (100), p-type wafers. Chemical surface treatment solution was $1H_3PO_4:2H_2O_2:10H_2O$. See GAS-0246a.

GAS-0251

ETCH NAME: Hydrofluoric acid — TIME:

TYPE: Acid, selective — TEMP:

COMPOSITION:

x .... HF, conc.

DISCUSSION:

GaAs:B, (111) n-type wafers and (100) undoped wafers used as substrates for LPE growth of GaAlAs Be doped thin films for p^+-p-n layer structured solar cells. The p-type GaAlAs films were selectively removed with HF to etch structure the cells.

REF: Masu, J et al — *J Electrochem Soc,* 129,1623(1982)

GAS-0253

ETCH NAME: — TIME:

TYPE: Acid, selective — TEMP:

COMPOSITION:

x .... H_2O_2

x .... NH_4OH

DISCUSSION:

GaAs, (100) wafers used as substrates for LPE growth of GaAlAs. The ammonium

hydroxide was used as a buffering agent, and solution used to selectively remove the GaAs substrate. Swaninathan (GAS-0025) references this work for the solution he and co-workers used.
REF: Ralogan, T & Reinhart, F R — *J Appl Phys,* 44,4127(1973)

GAS-0254
ETCH NAME:
TYPE: Acid, junction
COMPOSITION:
3 .... NH_4OH
1 .... H_2O_2
150 .... H_2O

TIME:
TEMP: RT
RATE: 0.16 μm/min

DISCUSSION:
GaAs, (100) wafers zinc diffused. Solution used for depth profiling by channel or step-etching of the zinc diffused junction for measurement.
REF: Jett-Field, R & Ghandhi, S K — *J Electrochem Soc,* 129,1567(1982)

GAS-0257
ETCH NAME:
TYPE: Acid, polish
COMPOSITION:
(1) 5 .... NH_4OH
1000 .... H_2O_2
(2) 5 .... H_2SO_4
1 .... H_2O_2
1 .... H_2O

TIME:
TEMP: RT

DISCUSSION:
GaAs, (100) wafers used for zinc diffusion at 850°C. Solution (1) was used as a general removal and chemical polish etch, and solution (2) for final surface polish prior to zinc diffusion.
REF: Chiaretti, O Y & Coonetti, C — *J Electrochem Soc,* 128,2193(1982)

GAS-0258
ETCH NAME:
TYPE: Acid, polish
COMPOSITION:
(1) 11 .... HNO_3
49 .... H_3PO_4
(2) 3 .... H_2SO_4
1 .... H_2O_2
1 .... H_2O

TIME: (1) 5 min (2) 1 min
TEMP: 60°C 60°C

DISCUSSION:
GaAs, (111) wafers "as-received" with (111)Ga surface polished. The $(\overline{111})$As negative side was additionally mechanically polished with 0.05 μm alumina, then wafers were cleaned in: (1) degrease in TCE; (2) MeOH rinse, and (3) final acetone rinse. Then etch clean in solution (1). Wafers of (100), (110), and $(\overline{111})$As were additionally etch cleaned in solution (2) which was referred to as Caro's etch (See SI-0103). All wafers were used as OMCVD substrates.
REF: Reep, D H & Ghandhi, S K — *J Cryst Growth,* 62,449(1983)

GAS-0259
ETCH NAME: TIME:
TYPE: Acid, selective TEMP:
COMPOSITION:

(1) 1 .... HF
1 .... H_2O_2
2 .... H_2O

(2) x .... laser

DISCUSSION:

GaAs, (111) wafers used as substrates for epitaxy growth of Ge and ZnSe. Solution (1) was selective on both (111)Ga and $(\overline{111})$As surfaces. Laser annealing developed both etch pits and growth hillocks. Study involved the development of orientation light figures by etching epitaxy thin films. Authors say that acid and alkali solutions develop etch pits usable as such figures on (111)Ga but not on $(\overline{111})$As.

REF: Owens, S J T & Watt, A H — *Microelectronics,* 5(3),37(1974)

GAS-0261
ETCH NAME: TIME:
TYPE: Acid, stain TEMP:
COMPOSITION:

1 .... HF
1 .... H_2O_2
10 .... H_2O

DISCUSSION:

GaAs:Cr, (100) (SI) wafers cut 2°-off toward (110) for VPE growth of GaAs. To measure epitaxy thickness, angle lap a specimen at 5° and stain with solution shown.

REF: Yoshida, M et al — *J Electrochem Soc,* 132,930(1985)

GAS-0263a
ETCH NAME: TIME:
TYPE: Halogen, cleaning TEMP:
COMPOSITION:

(1) x .... x% Br_2
x .... MeOH

(2) 1 .... NH_4OH
1 .... H_2O_2
10 .... H_2O

DISCUSSION:

GaAs:Be, (110), p-type wafers used as substrates for epitaxy growth of AlGaAs in a comparative LEED study. Substrates were chem/mech polished with solution (1) on lint-free paper; wetted to moly discs with indium, then etch cleaned in solution (2) on a spinner at low speed before being placed in the MBE vacuum system for epitaxy growth.

REF: Kahn, A et al — *J Vac Sci Technol,* 21(2),382(1982)

GAS-0263b
ETCH NAME: Arsenic TIME:
TYPE: Metal, passivation TEMP: 300°C
COMPOSITION:

x .... As

DISCUSSION:

GaAs:Be, (100), p-type wafers used as substrates for epitaxy growth of AlGaAs by MBE. After MBE deposition, wafers coated with a thin layer of arsenic below RT while

still in the system, held at 10^{-5} Torr in sealed quartz ampoules, and then heated to 300°C to desorb arsenic as a final surface cleaning and stabilization step.
REF: Ibid.

GAS-0248a
ETCH NAME: BRM
TYPE: Halogen, cleaning
COMPOSITION:
(1) 1 .... HCl
9 .... H_2O
(2) x .... 1% Br_2
x .... MeOH
TIME:
TEMP:
DISCUSSION:
GaAs, (100) wafers used in a study of anodic oxidation. Wafers were cleaned in solution (1), then solution (2), followed by MeOH rinse, DI H_2O rinse, and N_2 drying. Anodically oxidize in 1 3% tartaric acid:6 propylene glycol at 0.15 mA/cm^2 power. See GAS-0248b for surface replication.
REF: Makky, W H et al — *J Vac Sci Technol,* 21(2),417(1982)

GAS-0248b
ETCH NAME:
TYPE: Metal, replication
COMPOSITION:
(1) x .... 4% collodion,
x .... amyl acetate
(2) x .... Pt
(3) x C
TIME:
TEMP:
DISCUSSION:
GaAs, (100) wafers used in a study of anodic oxidation steps. After oxidation, surfaces were replicated: coat surface with solution (1); evaporate (2) as a platinum thin shadow film at a 22° angle; and sputter 200 Å of carbon (3). Remove C/Pt replication film by dissolving collodion with amyl acetate for SEM study of the replication film.
REF: Ibid.

GAS-0248c
ETCH NAME:
TYPE: Electrolytic, oxidation
COMPOSITION:
1 .... 3% tartaric acid
6 .... propylene glycol
TIME:
TEMP:
ANODE: GaAs
CATHODE:
POWER: 0.15 rmA/cm^2
DISCUSSION:
GaAs, (100) wafers used in a study of anodic oxidation steps. Solution used to grow the oxide.
REF: Ibid.

GAS-0265
ETCH NAME: Acetone
TYPE: Ketone, lift-off
COMPOSITION:
x .... CH_3COCH_3 (acetone)
TIME:
TEMP: RT to warm
DISCUSSION:
GaAs, (100) wafers used for device fabrication and dielectric circuit substrates of alumina, both with fine line metallization patterning. A description of a variety of lift-off techniques applicable to different materials.
REF: Frary, J M & Seese, P — *Semiconductor Int,* December 1981, 7289

GAS-0266
ETCH NAME: TIME:
TYPE: Acid, polish TEMP:
COMPOSITION:
2 .... HF
15 .... HNO_3
5 .... HAc
DISCUSSION:

GaAs, (100) wafers used to fabricate Schottky barrier diodes with low doping for high reverse V_{bd}. Solution used to etch clean wafers with a BHF dip immediately prior to metal evaporation.

REF: Stolt, L et al — *J Solid State Electron,* 26,2195(1983)

GAS-0267: Wheatly, C H & Whelan, J M — *J Phys Chem Solids,* 6,169(1958)

GaAs ingots used in a study of preparation and properties. Solution used as a general etch.

GAS-0268
ETCH NAME: Potassium hydroxide TIME:
TYPE: Molten flux, defect TEMP:
COMPOSITION:
x .... KOH, pellets
DISCUSSION:

GaAs:Cr, (100) (SI), and InP:Fe, (100) (SI) wafers deposited with epitaxy layers of GaAlAs, and InGaAsP, respectively. A study of material defects and contact alloys, and the effect of temperature annealing. Metal alloys were AuGe and InGe. The molten flux shown above was used for defect development of GaAs specimens; Huber's etch on InP (See INP-0072). Describes thread dislocations, grappes, and other type defects.

REF: Chin, A K — *J Cryst Growth,* 70,582(1984)

GALLIUM ALUMINUM ARSENIDE, GaAlAs

General: Does not occur as a natural compound. The compound has no direct use in industrial metal processing, though metallic gallium has several applications as a low temperature alloy. See section on Gallium for further discussion.

Technical Application: GaAlAs thin films are grown on GaAs substrates as p-n or p^+-p-n junction solar cells. Collection efficiency has been as high as 30%, as against about 18% for silicon, but fabrication costs of single crystal gallium arsenide solar cells are some three to four times greater than that for silicon at present.

Thin film growth has been by MBE, OMCVD, and LPE, as a single layer on GaAs substrates, or as a multilayered heterojunction structure with other materials.

Etching: HF:HNO_3 mixed acids or halogens.

GALLIUM ALUMINUM ARSENIDE ETCHANTS

GASA-0001
ETCH NAME: Hydrofluoric acid TIME:
TYPE: Acid, removal TEMP:
COMPOSITION:
x .... HF, conc.
DISCUSSION:

(Ga,Al)As:Be, p-type thin films grown on n-type (111) GaAs:B, and (100) GaAs un-

doped wafers that were n-type. Thin films grown by LPE in fabricating p^+-p-n junction solar cells. HF used for selective removal of (Ga,Al)As.
REF: Masu, K et al — *J Electrochem Soc*, 129,1623(1982)
Note: See previous section on Gallium Arsenide for additional solutions that are selective to either GaAs or GaAlAs:

GAS-0004a-d	*GAS-0061a-f	GAS-0153
GAS-0019	GAS-0060	GAS-0165
GAS-0023	GAS-0105	GAS-0184a
GAS-0025	GAS-0108a-b	GAS-0221

*Selective solutions for both materials.

GALLIUM ARSENIDE PHOSPHIDE ETCHANTS

GASP-0001
ETCH NAME: TIME:
TYPE: Acid, preferential TEMP: RT
COMPOSITION:
8 .... H_2SO_4
1 .... H_2O_2
1 .... H_2O
DISCUSSION:
GaAsP wafers as highly p-type doped with Mn. Solution used to develop dislocations.
REF: Fujita, S et al — *Solid State Electron*, 25,359(1982)

GALLIUM IRON OXIDE ETCHANTS

GIO-0001
ETCH NAME: Nitric acid TIME:
TYPE: Acid, removal TEMP: Hot
COMPOSITION:
x .... HNO_3
x .... H_2O
DISCUSSION:
$GaFeO_3$, single crystal ingot boat grown. Solution used to remove crystal from boat and clean surfaces after growth. The material is a ferromagnetic and piezoelectric compound. See sections on Ferrite, Iron, Steel, Magnetite, Hematite, and other iron-containing compounds.
REF: Rumeika, J P — *J Appl Phys*, 31,2638(1960)

PHYSICAL PROPERTIES OF GALLIUM NITRIDE, GaN

Classification	Nitride
Atomic numbers	31 & 7
Atomic weight	83.73
Melting point (°C)	800 sub.
Boiling point (°C)	
Density (g/cm^3)	
Energy gap (eV)	3.5

Coefficient of linear thermal expansion	
(cm/cm) 300—900 K — perpendicular to c-axis	5.59
(cm/cm) to 700 K — parallel to c-axis	3.17
(cm/cm) 700 to 900 K — parallel to c-axis	7.75
Atomic spacing (angstroms)	3.18 or 5.185
Hardness (Mohs — scratch)	6—7
Crystal structure (hexagonal — normal)	$(10\bar{1}0)$ prism
Color (solid)	Silver-grey
Cleavage (basal)	(0001)

GALLIUM NITRIDE, GaN

General: Does not occur as a natural compound. Gallium is a rare element found as traces in other minerals, such as sphalerite, ZnS, and bauxite, a complex hydrated iron/aluminate. As these two minerals are major ores of zinc and aluminum, respectively, gallium is available in usable quantities from recovery during processing of the ores of the two minerals.

There is no use for gallium nitride in industry other than for possible applications in the Solid State and electronics fields.

Technical Application: Gallium nitride is a III—V compound semiconductor. It can be easily grown from a solid gallium source in a nitrogen atmosphere both as an amorphous thin film or as a single crystal compound by CVD and OMCVD techniques. It also has been RF or DC sputtered as a thin film from both a solid GaN target or deposited from gallium in nitrogen atmosphere. The latter being difficult to control due to the low melting point and vapor pressure of gallium (30°C temperature).

It is under evaluation as a thin film surface coating and has been used to fabricate electroluminescent thin film diodes.

Etching: Deliquesces in most single acids. Soluble in alkalies.

GALLIUM NITRIDE ETCHANTS

GAN-0001

ETCH NAME: Sodium hydroxide — TIME:

TYPE: Electrolytic, selective jet — TEMP:

COMPOSITION: — ANODE: GaN

x .... 0.1 *N* NaOH — CATHODE:

POWER:

DISCUSSION:

GaN, (0001) single crystal thin films. Sodium hydroxide is a general etch for gallium nitride. The author says that an insoluble gallium hydroxide, GaOH coating, will form on the surface during etching unless the solution is continually washed away by the jet action.

REF: Pankove, J I — *J Electrochem Soc,* 119,1118(1972)

GAN-0002

ETCH NAME: Sodium hydroxide — TIME:

TYPE: Alkali, removal — TEMP: Hot (80°C)

COMPOSITION:

x .... 30—50% NaOH

DISCUSSION:

GaN thin films. Described as a general removal etch for gallium nitride. (*Note:* 2—5% concentrations, RT, have been used for pinhole development.)

REF: Chu, T L — *J Electrochem Soc,* 118,1200(1971)

GAN-0003
ETCH NAME: Hydrochloric acid TIME: 10—20 sec
TYPE: Acid, removal/clean TEMP: RT
COMPOSITION:
1 .... HCl
1 .... H_2O
DISCUSSION:

GaO_xN_y surface contamination of gallium arsenide wafers from exposure to air. Solution used to clean wafer surfaces prior to metallization, diffusion, and epitaxy. After etching in solution shown, water rinse and nitrogen blow-dry. Usually followed by a similar cleaning in 1HF:$1H_2O$, DI water rinse and N_2 dry.

REF: Siracusa, M & Walker, P — personal application, 1978

GAN-0004
ETCH NAME: Phosphoric acid TIME:
TYPE: Acid, preferential TEMP: Hot (80°C)
COMPOSITION:
x .... H_3PO_4
DISCUSSION:

GaN, (0001) single crystal thin films. Solution used to develop defects, dislocations, and other etch figures. (*Note:* Solution also used as a general removal etch.)

REF: Shintani, A & Minagawa, S — *J Electrochem Soc,* 123,706(1976)

GAN-0005
ETCH NAME: Heat TIME:
TYPE: Thermal, cleaning TEMP: 1200°C
COMPOSITION:
x .... heat
DISCUSSION:

GaN thin films grown by MBE on (0001), and $(01\bar{1}2)$ single crystals sapphire substrates to 1000 Å thickness. Substrates cleaned at temperature shown prior to GaN growth at 700°C. Improved single crystal structure was obtained when GaN was deposited on an AlN thin film previously deposited on the sapphire substrate. As deposited GaN films were n-type.

REF: Yoshida, S et al — *J Vac Sci Technol,* B1(2),250(1983)

PHYSICAL PROPERTIES OF GALLIUM OXIDE, Ga_2O_3

Classification	Metal oxide
Atomic numbers	31 & 8
Atomic weight	187.44
Melting point (°C)	1900
Boiling point (°C)	
Density (g/cm³)	6.44
Hardness (Mohs — scratch)	5—6
Crystal structure (hexagonal-trigonal pyramid) alpha	$(10\bar{1}1)$ rhomb
(monoclinic — normal) beta	(hk0) prism
Color (solid)	White
Cleavage (basal — poor)	(0001)/(001)

GALLIUM OXIDE, Ga_2O_3

General: Does not occur as a natural compound. The metal gallium is a rare element obtained chiefly from sphalerite, ZnS, as a by-product during roasting of the ore for its sulfur and zinc content. Such recovery can be in the form of both gallium and gallium oxide.

The oxide has no use in the metal industry, although the metallic gallium has use in low temperature solders.

Technical Application: Gallium oxide has no current use in Solid State processing. The sub-oxide Ga_2O is unstable above 600°C, but Ga_2O_3 and its hydrate are stable up to a 1900°C boiling point. All of the oxides can occur as native oxides on gallium arsenide semiconductor wafers during processing and, like many other native oxides, considered a surface contaminant in processing. The oxides have been purposely grown on GaAs wafers by boiling in water, then stripping with HCl as a surface cleaning step. The water developed oxide will be relatively soft as a hydrated compound, such as $Ga_2O_3.H_2O$; whereas the normal native oxide that develop naturally on a surface exposed to air form fairly hard type passivation coatings on materials. The gallium oxides formed are commonly a mixture of Ga_2O and Ga_2O_3.

Study of gallium oxide has been primarily concerned with methods of removal from gallium arsenide surfaces during processing.

Etching: Soluble in acids and alkalies, halogens, or by ionized gases.

GALLIUM OXIDE ETCHANTS

GAO-0001a

ETCH NAME: Hydrochloric acid — TIME: 10—30 sec

TYPE: Acid, removal — TEMP: RT

COMPOSITION:

x .... HCl

x .... H_2O

DISCUSSION:

GaAs, (100) wafers prepared as substrates for OMCVD growth of GaAs and AlGaAs. Wafers were boiled in hot water to form an oxide which was removed with this solution as a surface cleaning step.

REF: Lewis, B F et al — *J Phys Chem Solids,* 45,419(1984)

GAO-0001b

ETCH NAME: — TIME: 10—30 sec

TYPE: Acid, removal — TEMP: RT

COMPOSITION:

x .... HCl

x .... EOH

DISCUSSION:

GaAs, (100) wafers. Solution used to remove surface oxide formed by boiling in hot water. Solution shown used to etch clean and remove the oxide.

REF: Ibid.

GAO-0002

ETCH NAME: Hydrochloric acid, dilute — TIME: 10—20 sec

TYPE: Acid, oxide removal — TEMP: RT

COMPOSITION:

1 .... HCl

1 .... H_2O

DISCUSSION:

Ga_2O_3 as a native oxide on gallium arsenide wafers. Solution used to clean surfaces

prior to metallization. After HCl clean, follow with 1HF:1H_2O, and running DI water rinse, up to 2 min, and nitrogen blow dry.
REF: Siracusa, M — personal communication, 1979

GAO-0003a
ETCH NAME: RIE
TYPE: Ionized gas, removal
COMPOSITION:
 4 .... Cl_2
 1 .... Ar

TIME: 10 min
TEMP:
GAS FLOW:
PRESSURE: 5 mTorr
POWER: 200 V

DISCUSSION:
Ga_2O_3, as native oxide on GaAs, (100) wafers. Etch removal in pure Cl_2 is slow. Used nichrome as a stencil mask with photo resist to open patterns. Other chlorine/argon ratios as 1:2 and 1:4 were evaluated.
REF: Hu, E C & Howard, R E — *J Vac Sci Technol,* B(2),85(1984)

GAO-0003b
ETCH NAME: RIE
TYPE: Ionized gas, removal
COMPOSITION:
 (1) x .... H_2
 (2) x .... CCl_2F_2

TIME:
TEMP:
GAS FLOW:
PRESSURE: 5 mTorr
POWER: 60 V
RATE: 150 nm/min

DISCUSSION:
Ga_2O_3 as native oxide on GaAs, (100) wafers. Both gases were evaluated, and hydrogen was preferred for selective removal of native oxides.
REF: Ibid.

GAO-0004
ETCH NAME:
TYPE: Acid, removal
COMPOSITION:
 x .... HF
 x .... HNO_3

TIME:
TEMP:

DISCUSSION:
Ga_2O_3 doped with iron and grown as single crystal ferrites by the Czochralski (CZ) method. A study of ferrite growth. Also grew ingots of Fe_3O_4, and iron doped: ZnO, CuO, and MnO. Solution shown used for general removal.
REF: Horn, F N — *J Appl Phys,* 32,900(1961)

GAO-0005a
ETCH NAME:
TYPE: Alkali, removal
COMPOSITION:
 x .... x% KOH/NaOH

TIME:
TEMP:

DISCUSSION:
Ga_2O_3 and $Ga(OH)_3$ on GaAs, (100), p-type wafers used to fabricate light emitting diodes with Al or Au dot contacts. Both compounds and Ga metal are removed with basic solutions enhancing As surface density. Gave better emission results than strong acid treatment of surfaces.
REF: Adachi, H & Hartnagel, H L — *J Vac Sci Technol,* 19(3),427(1961)

GAO-0005b
ETCH NAME: AGW
TYPE: Acid, oxidizing
COMPOSITION:
x .... acid
x .... ethylene glycol
x .... water
TIME:
TEMP:
DISCUSSION:

Ga_2O_3 thin film growth of GaAs, (100), p-type wafers used to fabricate light emitting diodes. Solution used to grow 80 or 40 Å thick oxides on diode surfaces in a chemical surface treatment study and their effects on emission.

REF: Ibid.

GALLIUM PHOSPHATE ETCHANT

GAPH-0001
ETCH NAME: Sulfuric acid
TYPE: Acid, removal
COMPOSITION:
x .... H_2SO_4
TIME:
TEMP: Hot
DISCUSSION:

$GaPO_4$ single crystals grown by a hydrothermal temperature gradient process from H_3PO_4 in sealed glass ampoules. Solution used for general removal in a material study.

REF: Requardt, A & Iehmann, G — *J Phys Chem Solids*, 46,107(1985)

PHYSICAL PROPERTIES OF GALLIUM PHOSPHIDE, GaP

Classification	Phosphide
Atomic numbers	31 & 15
Atomic weight	109.7
Melting point (°C)	1480
Boiling point (°C)	
Density (g/cm^3)	4.14
Lattice constant (angstroms)	5.45
Energy gap (eV)	2.4
Refractive index (n =)	3.37
Hardness (Mohs — scratch)	7
Crystal structure (isometric — tetrahedral)	(111) tetrahedron
Color (solid)	Silver/grey
Cleavage (dodecahedral)	(110)

GALLIUM PHOSPHIDE, GaP

General: Does not occur as a natural metallic compound, and has had no use in the metals industries other than as an additive to iron compounds.

Technical Application: Gallium phosphide is a III—V compound semiconductor, and is a polar compound. The $(\overline{111})$Ga and (111)P wafer surfaces showing different etching characteristics. GaP is one of the few semiconductors that will operate at high temperature (400°C); SiC will operate at 800°C; but the majority of semiconductors do not operate at much above 150°C.

Single crystal ingots are grown by Horizontal Bridgman (HB), Czochralski (CZ), and

Float Zone (FZ) methods. Epitaxy thin films by Vapor Transport (VT) or CVD. Used as a substrate for epitaxy growth of other compounds, such as zinc sulfides, tellurides, and selenides as heterostructure devices.

Devices fabricated from wafers or thin film epitaxy include rectifiers, electroluminescent diodes, and solar cells.

Also grown as trinary compounds, such as $Ga_{1-x}In_xP$ and $Ga_{1-x}Al_xP$ for similar devices and laser diodes.

Etching: Not soluble in single acids. Aqua regia; $HF:HNO_3$ with or without HCl, H_2SO_4 and H_2O_2. Alkalies with or without additives, and ionized gases.

GALLIUM PHOSPHIDE ETCHANTS

GAP-0001

ETCH NAME: TIME:

TYPE: Acid, polish TEMP:

COMPOSITION:

1 .... HF

1 .... HNO_3

DISCUSSION:

GaP, (100), (111)A and $(\overline{111})$B wafers used as substrates for VPE growth of ZnTe, and ZnSe thin films. Wafers were mechanically lapped and polished, then degreased in boiling TCE; rinsed in acetone, then ethanol. The solution shown was used to etch remove lap damage prior to epitaxy.

REF: Matsumoto, T & Ishida, T — *J Cryst Growth,* 67,135(1984)

GAP-0002

ETCH NAME: TIME:

TYPE: Acid, polish TEMP:

COMPOSITION:

1 .... HCl

2 .... HNO_3

1 .... H_2O

DISCUSSION:

GaP, (100) and $(\overline{111})$B, p-type, 0.2 Ω cm resistivity wafers used as substrates for MBE growth of ZnS. Solvent degrease, then etch polish with the solution shown. Wafers were held under vacuum at 620°C, 15 min to remove native oxide prior to ZnS growth.

REF: Yoneda, K et al — *J Cryst Growth,* 67,125(1984)

GAP-0003

ETCH NAME: TIME:

TYPE: Acid, polish TEMP:

COMPOSITION:

(1) 2 .... HCl

2 .... H_2SO_4

2 .... H_2O

1 .... HNO_3

(2) 2 .... HNO_3

1 .... HCl

DISCUSSION:

GaP, (100) and (111) wafers. Etch polish in either solution.

REF: Hajkoa, E et al — *Phys Status Solidi,* A10,K35(1972)

GAP-0004
ETCH NAME: Chlorine TIME:
TYPE: Gas, polish TEMP:
COMPOSITION:

(1) x .... Cl_2
x .... MeOH

(2) x .... Cl_2
x .... H_2O

DISCUSSION:

GaP, (100) and (111) wafers. Immerse wafers in either MeOH of H_2O and bubble chlorine gas up through liquid for etching action. Agitate wafers to prevent bubbles adhering to surfaces as they will produce erratic etching results. (*Note:* Intense light is sometimes used in this form of etching.)

REF: Milch, A — *J Electrochem Soc,* 123,1256(1976)

GAP-0022c: Ibid.

GaP, (111) wafers used in a defect, electrical and optical study. Solution etch rate was 1—10 μm/sec, and developed pits only on the $(\overline{1}\overline{1}\overline{1})$B surfaces. Three solutions evaluated were Cl_2:MeOH, Cl_2:H_2O and Br_2:MeOH.

GAP-0005
ETCH NAME: TIME:
TYPE: Alkali, polish TEMP: 60 to 95°C
COMPOSITION:

x .... 1 *M* $K_3Fe(CN)_6$
x .... 0.5 *M* KOH

DISCUSSION:

GaP, (111) and (100). Solution used as a general wafer polishing etch. With an evaporated silicon dioxide, SiO_2 or titanium, Ti, thin film as a photo resist patterned mask, solution also used to etch remove and clean dot patterns in forming mesa structures.

REF: Plauger, R — *J Electrochem Soc,* 121,455(1974)

GAP-0006
ETCH NAME: BRM TIME: 20 sec + 10 sec
TYPE: Halogen, polish TEMP: RT
COMPOSITION:

x .... 0.05% Br_2
x .... MeOH

DISCUSSION:

GaP, (110) un-doped wafers. Used lens paper wetted with the solution shown for polishing wafer surfaces. After the initial 20 sec etch polishing, dilute with methanol and continue polishing for an additional 10 sec. Rinse in the BRM solution, then in water, then in NH_4OH:H_2O (AMH solution).

REF: Aspnes, D E & Studna, A A — *J Vac Sci Technol,* 20,488(1982)

GAP-0007
ETCH NAME: BRM TIME:
TYPE: Halogen, polish TEMP:
COMPOSITION:

20% .... Br_2
80% .... MeOH

DISCUSSION:

GaP, (111) and GaAs, (111) wafers. Used a pellon pad saturated with the above solution for polishing both of these compound semiconductors.

REF: Fuller, C S et al — *J Electrochem Soc,* 109,880(1962)

GAP-0008
ETCH NAME: TIME:
TYPE: Acid, polish TEMP: 60°C
COMPOSITION:
2 .... HCl
1 .... HNO_3
2 .... H_2O
DISCUSSION:
GaP, (111) and GaAs, (111)A wafers. Solution used as a polish etch.
REF: Urgaki, T et al — *J Electrochem Soc,* 123,580(1976)

GAP-0009
ETCH NAME: BRM TIME:
TYPE: Halogen, polish TEMP:
COMPOSITION:
1 .... Br_2
9 .... MeOH
DISCUSSION:
GaP, (111) and GaAs (111) wafers. Solution used as a general polishing etch for both compounds.
REF: Tuck, E — *J Mater Sci,* 10,321(1976)

GAP-0010a
ETCH NAME: Bromine TIME: 5 min
TYPE: Halogen, preferential TEMP: RT
COMPOSITION:
x .... Br_2
DISCUSSION:
GaP, (111) wafers. Bromine alone can be used as a preferential etch that will develop fine pits approximately 20 μm in size. (*Note:* BRM mixtures with greater than 3% Br_2:MeOH also are preferential.)
REF: Kleinman, D A & Spitzer, W G — *Phys Rev,* 118,110(1960)

GAP-0011
ETCH NAME: TIME: (1) (2)
TYPE: Acid, polish/selective TEMP: RT 150—200°C
COMPOSITION:
(1) 3 .... HCl (2) x .... H_3PO_4
1 .... HNO_3
DISCUSSION:
GaP, $(\overline{111})$B wafers. Solutions used for selective etch grooving of ⟨110⟩ direction channels in GaP surfaces. An evaporated SiO_2 thin film mask, photo resist line patterned, was used with the (1) etch solution (aqua regia). An evaporated thin film gold mask, photo resist line patterned, was etched with the phosphoric acid, solution (2).
REF: Urgkaki, T et al — *Phys Status Solidi,* A10,K38(1972)

GAP-0012
ETCH NAME: Aqua regia TIME:
TYPE: Acid, cleaning TEMP:
COMPOSITION:
3 .... HCl
1 .... HNO_3

DISCUSSION:

n-GaP, (111) and p-GaP (111) wafers doped with carbon were surface cleaned with aqua regia.

REF: Scott, W — *J Appl Phys,* 50,472(1979)

GAP-0022b: Ibid.

GaP, (111) wafers used in a defect, electrical and optical study. Hot aqua regia showed a 1—10 μm/sec etch rate, and was preferential on $(\overline{111})$B surfaces.

GAP-0013

ETCH NAME:

TYPE: Acid, junction

COMPOSITION:

3 .... H_2SO_4

1 .... H_2O_2

1 .... H_2O

TIME: 5 min

TEMP: 60°C

DISCUSSION:

p-GaP, (100) wafers with zinc diffusion to form a p-n junction. Solution was used to step-etch the p-n junction for depth profiling. The p-type is preferential in the solution and the junction will be etch delineated.

REF: Hackett, W H Jr — *J Electrochem Soc,* 119,973(1972)

GAP-0014

ETCH NAME: Hydrochloric acid

TYPE: Acid, zinc removal

COMPOSITION:

x .... HCl, conc.

TIME:

TEMP: Boiling

DISCUSSION:

GaP, (111) wafers zinc diffused. After diffusion the hot solution was used to remove excess zinc from wafer surfaces.

REF: Wallison, H — *J Appl Phys,* 34,231(1963)

GAP-0022d: Ibid.

GaP, (111) wafers used in a defect, electrical, and optical study. Solution used to clean surfaces prior to preferential etching.

GAP-0015

ETCH NAME:

TYPE: Acid, stain

COMPOSITION:

TIME:

TEMP:

DISCUSSION:

GaP, (100) wafers used as substrates for LPE growth of tin-doped GaP thin films with formation of p-n junctions. Wafers were cleaved ⟨110⟩ and etch stained to develop the p-n junction. Etchant used not shown.

REF: Fritz, I J et al — *J Electron Mater,* 14,23(1985)

GAP-0016

ETCH NAME:

TYPE: Acid, polish

COMPOSITION:

x .... acids

TIME:

TEMP:

DISCUSSION:

GaP, (111), (100), (110) wafers. Reference cited is an excellent review of etchants for gallium phosphide.

REF: Kern, W — *RCA Rev*, 39,278(1978)

GAP-0017

ETCH NAME: TIME: 30 sec

TYPE: Acid, clean TEMP: RT

COMPOSITION:

6 .... HNO_3

1 .... HCl

1 .... $HClO_4$

DISCUSSION:

GaP, (100) n-type wafers used in a study of impedance characterization in an electrolyte. Front surface was mechanically polished to a mirror finish and then etched as shown. An Sn:In ohmic contact was alloyed to the back side of wafers at 550°C, 5 min in a nitrogen atmosphere.

REF: Nagami, G et al — *J Electrochem Soc*, 132,1662(1985)

GAP-0018

ETCH NAME: TIME: 10 sec

TYPE: Acid, polish TEMP: RT

COMPOSITION:

2 .... HF

1 .... HNO_3

DISCUSSION:

GaP, (111) wafer used in an etch development study. Solution will etch polish both (111)A and $(\overline{111})$B surfaces.

REF: Farren, H — personal communication, 1958

GAP-0019

ETCH NAME: TIME:

TYPE: Acid, forming TEMP:

COMPOSITION:

1 .... HF

1 .... HNO_3

*x drops .... Br_2

*x = bromine saturated water.

DISCUSSION:

GaP, (111) wafers fabricated as p-n junction, mesa structured diodes. After diffusion, surfaces were coated with piocin drops to define mesa areas and then etched in the solution shown to form mesa structures. Under 10 V reverse bias electrical leakage was observed on mesa slopes as intense yellow light. Light spots increased in number with increase in voltage. (*Note:* Similar work, observing light emission from p-n junctions, has been done on silicon.)

REF: Iizima, S & Kikuchi, M — *Jpn J Appl Phys*, 1,302(1962)

GAP-0010b
ETCH NAME: Copper
TYPE: Metal, decoration
COMPOSITION:
x .... Cu
DISCUSSION:

TIME:
TEMP:

GaP, (111) wafers used in a study of infrared lattice absorption. Copper was evaporated on surface, then diffused in to decorate defects. The copper was seen to segregate along grain boundaries and reduced reflectivity.

REF: Ibid.

GAP-0020
ETCH NAME:
TYPE: Acid, preferential
COMPOSITION:
12 ml .... HF
96 mg .... $AgNO_3$
12 g .... CrO_3
24 ml .... H_2O
DISCUSSION:

TIME:
TEMP:

GaP, (111) wafers used in a study of the electrical properties of gallium phosphide doped with zinc. Solution used to develop defects and structure induced by zinc diffusion.

REF: Cohon, M M & Bedard, F D — *J Appl Phys,* 39,75(1968)

GAP-0022a
ETCH NAME:
TYPE: Halogen, preferential
COMPOSITION:
x .... $*I_2$
x .... MeOH
+ .... Br_2

TIME: 18 h
TEMP: RT

*Saturate the solution.

DISCUSSION:

GaP, (111) wafers used in a study of defects, electrical and optical effects. This solution was best for developing etch pits. After 18 h both (111)A and $(\overline{111})$B surfaces showed only nodular structure. (*Note:* Such structure is common to bulk, undamaged metal and metallic compound surfaces.)

REF: Gershenzon, M & Mikulyak, R M — *J Appl Phys,* 35,2032(1964)

GAP-0022e
ETCH NAME: Aqua regia, modified
TYPE: Acid, preferential
COMPOSITION:
3 .... HCl
1 .... HNO_3
+ .... Ag or Fe
DISCUSSION:

TIME:
TEMP: Cold and hot

GaP, (111) wafers used in a defect, electrical and optical study. Aqua regia used at RT

or hot is a rapid etch and will develop pits on the $(\overline{111})$B surface only. Solution used cold with the addition of either silver or iron, will develop dislocation etch pits.
REF: Ibid.

GAP-0023
ETCH NAME: BRM — TIME: 30 sec
TYPE: Halogen, cleaning — TEMP:
COMPOSITION:
x .... 0.5% Br_2
DISCUSSION:
GaP material used for growth of AlGaAsP single crystal ingots. Solution used to clean GaP, GaAs and GaAsP materials prior to mixing for ingot growth. The aluminum was etch cleaned at RT, 15 sec in: 9HCl:3HNO_3:2HF:5H_2O.
REF: Fujimoto, A — *Jpn J Appl Phys,* 22,109(1984)

GAP-0024
ETCH NAME: Nitric acid — TIME:
TYPE: Acid, cleaning — TEMP: Hot
COMPOSITION:
x .... HNO_3, conc.
DISCUSSION:
GaP, polycrystalline material used as a source for LEC growth of single crystals. Boats and tubes were made of SiO_2 or PBN, and cleaned in aqua regia; gallium metal was held under an atmosphere of He, then switched to H_2/PH3, then returned to He and removed. This method developed the GaP source. After GaP single crystal growth, wash crystals in water, and anneal in vacuum at 300—400°C. Excess gallium removed with hot nitric acid. Material as-grown was n-type.
REF: Ringel, C M — *J Electrochem Soc,* 118,609(1971)

PHYSICAL PROPERTIES OF GALLIUM SELENIDE, GaSe

Classification	Selenide
Atomic numbers	31 & 34
Atomic weight	148.68
Melting point (°C)	960
Boiling point (°C)	
Density (g/cm³)	5.03
Hardness (Mohs — scratch)	6—7
Crystal structure (hexagonal — normal)	$(10\overline{1}0)$ prism
Color (solid)	Steel grey
Cleavage (basal)	(0001)

GALLIUM SELENIDE, GaSe

General: Does not occur as a natural compound, though there are a number of selenides and tellurides of other metal elements, such as lead, copper, silver, and many others as tellurides. Not used in the metal industry.

Technical Application: Gallium Selenide is a III—VI compound semiconductor but is hexagonal, rather than more usual isometric. It can be differentiated by its layered atomic structure similar to cadmium sulfide.

It has been grown as a single crystal for both physical and semiconductor characterization.

Etching: Mixed acids (HF:HNO_3), halogens, and alkalies.

GALLIUM SELENIDE ETCHANTS

GASE-0001
ETCH NAME: TIME:
TYPE: Mechanical, dislocation TEMP: RT
COMPOSITION:
x .... cleave
DISCUSSION:

GaSe, (0001) wafers cleaved and used in a study of dislocations. Authors say the material can be either hexagonal-normal or hexagonal-rhombohedral in structure as an interlayered atomic structure with the gallium layer showing weak van der Waal bonds. Cleaving will create dislocations in the surface, and E-beam annealing is sufficient to create further defects and dislocations. (*Note:* Single crystal cleaved surfaces normally show fine micro-structure on surfaces as ledges, as well as other defects and dislocations. See general section on (galena) lead sulfide, PbS with regard to such ledge-type surfaces.)

REF: Bainski, Z S et al — *J Appl Phys,* 34,469(1963)

GARNETS, $M_3M_2(SiO_4)_3$

General: The general formula shown is for all natural garnets. There are only ten major species of natural garnets, but several form isomorphous series or can have replacement elements such that there are a large number of varieties or subspecies, many with individual names to differentiate color or location occurrence. All garnets are found in the atypical dodecahedron, (110) crystal form, though they may occur as a trapezohedron, (211) in combination with the hexoctahedron, (321) — garnets being one of the few natural minerals that occur with (321) type facets.

There are three prominent groups: (1) aluminum garnet; (2) iron garnet, and (3) chromium garnet. Hardness ranges from H = 6.5—7.5 and specific gravity (density) varies between G = 3.15—4.3 g/cm^3. When a garnet is fused, it can both reduce in density and break down into other minerals with loss of one or more constituent elements.

Garnets are usually an accessory, rock-making mineral with common occurrence in certain rock types, such as chlorites, gneisses, or shists. Garnet is occasionally found in massive beds, but more commonly as individual single crystals, sometimes hundreds of small, deep red crystals (almandite, the iron garnet) in green chlorite beds, along with fine, small octahedrons of coal black magnetite, Fe_3O_4.

There is one special form, not common to other minerals, called a ''sand garnet'', where the crystal still retains its dodecahedral shape, but has been partly converted by compacted sand grains, with only a small central core of pure garnet. Some sand garnets are 1 to 2 in. in size. Such conversion is largely due to the low fusibility [(F = 3), 1200°C] of most garnets and, under pressure and heat, many garnets are squeezed with elongation in the c-axis direction, such that they are near-tabular in form.

Garnets have little use in metal processing except as lapping and polishing abrasives. They have been used as precious and semiprecious gem stones since ancient times for their range of colors and natural, highly reflective exterior facets. Some of the finest gem stones still come from Southern Asia, the Malay Peninsula, and India. The deep red pyrope garnet is top gem quality, but there are others that, depending upon color, also may be gem quality. Cinnamon-stone (Glossularite) is noted for that color, yet may be white, pale green, amber, honey-yellow, wine-yellow, rose-red, to emerald green. Spessartite is hyacinth-red to violet; andradite, called common garnet, is most often deep red-black, yet can vary in color like glossularite. Uvarovite, the chrome garnet, is a fine emerald green and, when it contains lithium, called the lithium garnet, and is a fine violet color.

As a group, the garnets have a wider range of types, varieties, and subspecies than any other single class of minerals.

Technical Application: Natural garnets have had little use in Solid State processing other than as lapping and polishing abrasives, but there are an increasing number of applications for artificially developed and grown garnets.

The yttrium iron and yttrium aluminum garnet, YIG and YAG, respectively, were two of the first Solid State laser and masar materials, the laser being more prominent. Note that there are some natural garnets containing yttrium as a replacement element. Where the natural garnet formula is shown as: $M_3M_2(SiO_4)_3$, the artificial formula without silicon is shown as: $M_3M_5O_{12}$ or $A_3B_2C_3O_{12}$ and there are some special mixtures where M_2 and M_3 contain two or more elements.

Where natural garnets are noted for their brilliant colors, the artificial garnets are usually transparent, to slightly yellow-tan in color, and are available as oriented single crystal blanks up to 3″ in diameter.

The artificial garnets have light transmission and frequency characteristics similar to those of quartz, and are used in a like manner as filter elements in microelectronic circuitry, such as GGG garnets. An important application of iron containing garnets, such as YIG, are in the fabrication of bubble memory computer devices with as much as a one million bit capability.

Artificial garnets, when grown by molten flux methods, form as the atypical dodecahedron, (110), but are limited in size to about 1″. They also tend to be contaminated by extraneous elements from the growth flux, such that repeatability control is difficult for consistency in device fabrication. But, like in the growth of semiconductor materials, they are now grown by the Czochralski (CZ) method as ingots up to 5 and 6″ in diameter, and also are grown by modified Czochralski, or Float Zone (FZ). Such ingots are most commonly grown as (100) oriented, sliced, and used with that orientation.

Where garnet is used as an abrasive powder, the W-1 to W-12 types are widely used in semiconductor wafer processing. The W-5 (15 μm) is a common "as received" lap finish. The garnet particles are relatively soft (H = 6), tend to break down under lapping pressure, such that a 600 grit abrasive may give a final finish closer to 400 grit.

Natural garnets have been the subject of much study for centuries, particularly in the jewelry trade and today new artificial garnets are under development for a wide range of applications.

Etching: H_3PO_4, HCl, halogens, variable with type garnet.

The following are the garnets discussed and presented in this section:

Boron Germanium, BGO
Calcium Aluminum Germanium, CAGG
Europium Scandium Iron, ESG
Gadolinium Gallium, GGG
Gadolinium Selenium Gallium, GSGG
Manganese Zinc Yttrium, MZYG
Natural Garnets, GAR
Strontium Gallium, SGG
Yttrium Aluminum, YAG
Yttrium Gallium, YGG
Yttrium Iron, YIG

GARNET ETCHANTS

BORON GERMANIUM GARNET

BGO-0001
ETCH NAME: Water TIME:
TYPE: Acid, cleaning TEMP:
COMPOSITION:
x .... H_2O
DISCUSSION:
$B_4Ge_3O_{12}$ (BGO) single crystal dodecahedron, (110) grown by molten flux. Crystals were transparent and colorless, and were water wash cleaned after removal from the growth flux. The following parameters were established:

Specific mass (Kgm^{-3})	7076
Specific heat ($Jkg^{-1} K^{-1}$)	303
Thermal diffusivity ($\times 10^{-6}m^2s^{-1}$)	0.84
Thermal conductivity ($Wm^{-1} K^{-1}$)	1.80

REF: Runkin, A Y & Frolou, A A — *J Cryst Growth,* 69,131(1984)

CALCIUM ALUMINUM GERMANIUM GARNET

CAGG-0001
ETCH NAME: Water TIME:
TYPE: Acid, cleaning TEMP:
COMPOSITION:
x .... H_2O
DISCUSSION:
$Ca_3Al_2Ge_3O_{12}$ grown as a single crystal ingot by the Czochralski (CZ) method. Requires 2% Y in melt, but there is no yttrium in the final ingot. This was the development of a new garnet type compound.
REF: Rotman, S R et al — *J Appl Phys,* 57,5320(1985)

EUROPIUM SCANDIUM IRON GARNET

ESG-0001
ETCH NAME: TIME:
TYPE: TEMP:
COMPOSITION:
DISCUSSION:
$Eu_3Sc_2Fe_3O_{12}$, as single crystal specimens used in a study of atomic location in garnets with the general formula $A_3B_2C_3O_{12}$. Where A = dodecahedral "c" sites; B = octahedral "a" sites; and C = tetrahedral "d" sites.
REF: Stadnik, E M — *J Phys Chem Solids,* 45,133(1984)

GADOLINIUM GALLIUM GARNET

GGG-0001a
ETCH NAME: Phosphoric acid
TYPE: Acid, preferential
COMPOSITION:
x .... H_3PO_4
TIME: Variable
TEMP: RT, hot to boiling
DISCUSSION:

$Gd_3Ga_5O_{12}$ (GGG) garnet fabricated as 0.8 mm spheres; also (001) blanks with structures called "semispheres" (control damaged etched pits or holes in the planar blank surface). Garnets were grown by Molten Flux and Czochralski (CZ) to evaluate the two growth methods. The CZ garnets were more perfect due to less contamination in growth. Spheres were etched to finite crystal form with varied time and temperature. Both a tetrahexahedron, (hk0), and trisoctahedron, (hh1) were obtained as convex dissolutionment forms. The "semi-spheres" were studied for pit side slope planes as concave surface dissolutionment forms. The solution shown produced smooth exterior facets on finite crystal forms; whereas HBr as an etch solution developed heavy terracing. (*Note:* The use of the term semisphere should not be confused with hemisphere. The latter is a convex surface; whereas the semisphere is a pit in a concave surface. A dissolutionment form is the same thing as a finite crystal form when used to etch a sphere.)

REF: Hartmann, E et al — *J Cryst Growth,* 71,191(1985)

GGG-0001b
ETCH NAME: Hydrogen bromide
TYPE: Halogen, preferential
COMPOSITION:
x .... HBr
TIME: Varied
TEMP: 124°C
DISCUSSION:

$Gd_3Ga_5O_{12}$ (GGG) garnets fabricated as spheres. The finite crystal form was an octahedron, (111) with heavy, multiple (110) terracing on the octahedral faces. See GGG-0001a for additional discussion. (*Note:* The natural garnets grow as a dodecahedron, (110), never as an octahedron, (111).)

REF: Ibid.

GGG-0002
ETCH NAME: Alumina
TYPE: Abrasive, polish
COMPOSITION:
x .... Al_2O_3 powder
TIME:
TEMP: RT
DISCUSSION:

$Gd_3Ga_5O_{12}$, (110) wafers used for LPE growth of $(Gd,Y)_3(Fe,Mn,Ga)_5O_{12}$ thin film garnet layers. Alumina powder can be used to polish substrate surfaces prior to LPE growth. Slow growth developed hillocks. Grown 1°-off (110) developed smooth films. Mn as a doping element constituent can develop $MnCO_3$ or Mn_2O_3 crystallite re-growth. LPE growth was as standard horizontal dip/wipe across the melt rotating at 120 rpm.

REF: Breed, D J et al — *J Appl Phys,* 54,1519(1983)

GGG-0003
ETCH NAME: Alumina
TYPE: Abrasive, polish
COMPOSITION:
x .... alumina, powder
TIME:
TEMP:

DISCUSSION:

$Gd_3Ga_5O_{12}$ (GGG), (111) wafers used as substrates. Polish substrates with alumina before thin film growth of the following garnet mixtures: $BiTmFeGaO_{12}$; BiTmPb:$TmFeGaO_{12}$, and BiYPb:$YFeGaO_{12}$.

REF: Murthy, V R K & Belt, R F — *Jpn J Appl Phys,* 23,871(1984)

GGG-0004

ETCH NAME: Acetone — TIME:

TYPE: Ketone, cleaning — TEMP: RT to warm

COMPOSITION:

x .... CH_3COCH_3

DISCUSSION:

$Gd_3Ga_5O_{12}$ (GGG), (0001) wafers 3″ in diameter with oriented flats. Wafers were mounted on sticky-back mylar sheet, then cut with a resinoid diamond blade as rectangular pieces. After cutting, specimens were soaked in acetone to be removed from the holding mylar paper, then soaked and spray cleaned.

REF: Tarn, W H & Walker, P — personal application, 1983/1984

GGG-0005

ETCH NAME: BRM — TIME:

TYPE: Halogen, cleaning/polish — TEMP:

COMPOSITION:

x .... x% Br_2

x .... MeOH

DISCUSSION:

$Gd_3Ga_5O_{12}$ (GGG), (111) cut wafers used as substrates for bubble memory thin film deposition by LPE of $(Y,Sm,In,Gd)_3(Fe,Ga)_5O_{12}$. Pre-clean the GGG surfaces prior to epitaxy in the solution shown.

REF: Imura, R et al — *Jpn J Appl Phys,* 23,709(1984)

GADOLINIUM SELENIUM GALLIUM GARNET

GSGG-0001

ETCH NAME: Hydrochloric acid — TIME:

TYPE: Acid, cleaning — TEMP: Boiling

COMPOSITION:

x .... HCl, conc.

DISCUSSION:

$Gd_3Se_{1.8}Ga_{3.2}O_{12}$ (GSGG), (0001) wafers cut from Czochralski (CZ) grown ingots. Wafers were diamond saw cut; lapped with 800-grit abrasive; then diamond paste polish in a kerosene carrier. To remove oils and hydrocarbons, final wash in acid shown, then rinse in acetone, then EOH.

REF: Schwartz, K B & Duba, A G — *J Phys Chem Solids,* 8,957(1985)

MANGANESE ZINC YTTRIUM GARNET

MZYG-0001
ETCH NAME: Silicon carbide — TIME:
TYPE: Abrasive, polish — TEMP:
COMPOSITION:
x .... SiC, grinder
DISCUSSION:
$Mn_xZn_{1-x}Y_zO_{12}$ single crystal specimens ground as spheres and used in a study of the general material morphology.
REF:

NATURAL GARNETS

GAR-0001
ETCH NAME: Phosphoric acid — TIME:
TYPE: Acid, preferential — TEMP:
COMPOSITION:
x .... H_3PO_4, conc.
DISCUSSION:
Garnets, as natural single crystals used in a study of magnetic interactions and distribution of ions in their structure. Solution can be used to develop defect structure. The following garnets were studied:

Cryolithionite	—	$NaCa_2Mn_2As_3O_{12}$
Berzellite	—	$Ca_3Fe_2Si_3O_{12}$
Andradite	—	$Ca_2Cr_2Si_3O_{12}$ (chrome garnet)
Uvarovite	—	$Na_3Al_2Li_3O_{12}$ (lithium garnet)
Spessartite	—	$Mn_3Al_2Si_3O_{12}$
Almandite	—	$Fe_2Al_2Si_3O_{12}$ (iron or common garnet)

REF: Geller, S — *J Appl Phys*, 31,30S(1960)

GAR-0002a
ETCH NAME: Hydrochloric acid — TIME: 10—30 sec
TYPE: Acid, cleaning — TEMP: RT
COMPOSITION:
1 .... HCl
10 .... H_2O
DISCUSSION:
$Fe_2Al_3Si_3O_{12}$, as natural single crystal almandite. Crystals were deep red, and "squeezed" with elongation in the c-axis direction. Specimens were mounted in resin on glass slides, and mechanically lap thinned with abrasive on a rotating iron platen. Specimens washed in solution shown, then rinsed in water after lapping and used in a microscope structural study. Specimens were found in a green chlorite schist in close association with fine, coal black octahedrons of magnetite, Fe_3O_4.
REF: Walker, P — mineralogy study, 1953
GAR-0002b: Ibid.
$Fe_2Al_3Si_3O_{12}$, single crystal specimens as "sand garnets", e.g., outer surface replacement with compacted sand grains. Solution used with steel brush to soak and rub remove loose sand prior to sectioning for study of alteration structure.

STRONTIUM GALLIUM GARNET

SGG-0001
ETCH NAME: Nitric acid TIME:
TYPE: Acid, removal TEMP:
COMPOSITION:
x .... HNO_3, conc.
DISCUSSION:

$SrGa_{12}O_{19}$ single crystals flux grown as hexagonal, yellow platelets with basal (0001) habit, and cleaved (0001) as wafers. Structure is that of magnetoplumbite, $P6^3/mmc$ space group lattice. Also grew $Sr_2BaGa_{11}O_{20}$ crystals. Solution used to etch remove crystals from flux after slow cooling.

REF: Haberey, F et al — *J Cryst Growth,* 61,284(1983)

YTTRIUM ALUMINUM GARNET

YAG-0001
ETCH NAME: Phosphoric acid TIME:
TYPE: Acid, preferential TEMP: 250°C
COMPOSITION:
x .... H_3PO_4, conc.
DISCUSSION:

$Y_3Al_5O_{12}$ (YAG), (110) cut wafers used in a study of impurity diffusion uniformity. Neodymium (Nd) was diffused into specimens as a metal tracer element, and solution shown used to develop segregation locations of neodymium.

REF: Cockayne, B — *Phil Mag,* 12,943(1965)

YTTRIUM GALLIUM GARNET

YGO-0001
ETCH NAME: TIME:
TYPE: Acid, cleaning TEMP:
COMPOSITION:
1 .... HNO_3
1 .... HAc
3 .... H_2O
DISCUSSION:

$Y_2Ga_7O_{12}$ as molten flux grown single crystals with (110) facets using a new flux method. Solution used to clean crystals after removal from flux.

REF: Nielson, J W — *J Appl Phys,* 31,518(1960)

YTTRIUM IRON GARNET

YIG-0001
ETCH NAME: TIME:
TYPE: TEMP:
COMPOSITION:
DISCUSSION:

$Y_3Fe_5O_{12}$(YIG), (0001) wafers used to fabricate bubble memory computer devices.

(*Note:* An excellent article on their use in military equipment vs. disk and tape drive systems.)
REF: Weisenstein, C — *The Bubble Memory Option — Defense Electronics,* September 1, 1985, 97

YIG-0002
ETCH NAME: Hydrochloric acid TIME:
TYPE: Acid, macro-etch TEMP:
COMPOSITION:
x .... HCl, conc.
DISCUSSION:
$Y_3Fe_5O_{12}$, (0001) (YIG) wafers cut from ingots grown by the Float Zone (FZ) technique. Wafers cut used in material study by metallographic sectioning. Solution used to develop macro-structure.
REF: Abernethy, L L et al — *Phys Rev,* S32,276S(1961)

YIG-0004a
ETCH NAME: Phosphoric acid TIME: 10 min
TYPE: Acid, preferential TEMP: 140°C
COMPOSITION:
x .... H_3PO_4, conc.
DISCUSSION:
$Y_3Fe_5O_{12}$, (110) wafers. Material was grown by the top seeded solution growth (TSSG) method using Fe_2O_3 as the melt solvent, as a modified Czochralski (CZ) technique. The method was evaluated against the PbO molten flux growth method. Polish (110) surfaces with 0.01 alumina, then etch in HCl. The solution shown developed rhombohedral shaped pits with elongation that have the appearance of a straw basket weave.
REF: Oka, K & Uncki, H — *J Appl Phys,* 56,436(1984)

YIG-0004b
ETCH NAME: Hydrochloric acid TIME: 4 h
TYPE: Acid, preferential TEMP: RT
COMPOSITION:
20 .... HCl
80 .... H_2O
DISCUSSION:
$Y_3Fe_5O_{12}$, (111) wafers. Solution developed etch pits about 10 μm in size with a density of 10^4—10^6/cm^2. See YIG-0004a for additional discussion.
REF: Ibid.

YIG-0007
ETCH NAME: TIME:
TYPE: Abrasive, polish TEMP:
COMPOSITION:
(1) x .... diamond paste (2) x .... alumina
DISCUSSION:
$Y_3Fe_5O_{12}$ (YIG) single crystal spheres mechanically polished and used to study development of flaws applying ferromagnetic resonance.
REF: Buehler, E & Tanenbaum T — *J Appl Phys,* 31,388(1960)
YIG-0008: Konzler, J E et al — *J Appl Phys,* 31,392(1960)
$Y_3Fe_5O_{12}$ (YIG) single crystal spheres of 1.6 cm diameter with oriented ground flats. Spheres were mechanically polished and used in a study of specific heat and demagnetization.
YIG-0005: Fletcher, P et al — *Phys Rev,* 114,739(1959)

$Y_3Fe_5O_{12}$ (YIG) single crystal spheres 1.3 cm in diameter and mechanically polished. Used in a study of magnetostatic modes in ferromagnetic resonance of spheres.
YIG-0006: Spencer, E G & LeCraw, R C — *Phys Rev Lett,* 1240(1958)

$Y_3Fe_5O_{12}$ (YIG) single crystal spheres mechanically polished. Used in a study of vibration frequencies (acoustic). Frequency varies inversely, as a ratio of sphere diameters.

YIG-0009
ETCH NAME: TIME:
TYPE: Acid, cleaning TEMP:
COMPOSITION:
1 .... HNO_3
1 .... HAc
3 .... H_2O
DISCUSSION:

$Y_3Fe_5O_{12}$ (YIG) as a dodecahedron, (110) molten flux grown single crystal. Solution used to clean surfaces after molten flux growth.
REF: Nielsen, J W — *J Appl Phys,* 31,518(1960)

PHYSICAL PROPERTIES OF GERMANIUM, Ge

Classification	Metal
Atomic number	32
Atomic weight	72.60
Melting point (°C)	937.4
Boiling point (°C)	2830
Density (g/cm^3)	5.32
Thermal conductance (cal/sec)(cm^2)(°C/cm) 0°C	0.15
Specific heat (cal/g) 25°C	0.74
Heat of fusion (k-cal/g-atom)	111.5
Heat of vaporization (k-cal/g-atom)	1100
Atomic volume (W/D)	13.6
1st ionization energy (K-cal/g-mole)	187
1st ionization potential (eV)	8.13
Electronegativity (Pauling's)	1.8
Covalent radius (angstroms)	1.22
Ionic radius (angstroms)	0.53 (Ge^{+4})
Linear coefficient of thermal expansion ($\times 10^{-8}$ cm/cm/°C)	6.1
Electrical resistivity (micro ohms cm $\times 10^6$) 27°C	47
Vapor pressure (atm $\times 10^{-9}$) 27°C	1.1
Work function (eV)	4.5
Magnetic susceptibility ($\times 10^6$) 20°C	0.22
Debye temperature (K)	362
Cross section (barns)	2.4
Refractive index (n=)	5.67
Extinction coefficient (ks)	0.63
Energy band gap (eV)	0.47
Hardness (Mohs — scratch)	6—7
Crystal structure (isometric — normal)	(100) cube (diamond)
Color (solid)	Steel-grey
Cleavage (dodecahedral/cubic)	(110)/(100)

GERMANIUM, Ge

General: Does not occur in nature as a native element. It is a trace element in several minerals, notably the iron pyrites, and is collected from flue dust during processing of pyrites for the iron and sulfur content. Pyrite, FeS_2, is a major source of iron ore, and representative of the pyrite group as a class of minerals.

There are two known minerals containing germanium: argyrodite, $4Ag_2S.GeS_2$ and germanite, $Cu_3(Fe,Ge)S_4$. Neither mineral occurs in large quantity, but they are associated with the pyrites which represent about 85% of the iron ore mined.

Germanium has had little or no use as a metal in industry until discovery of its semiconducting properties in the mid-1940s.

Technical Application: Germanium was the original semiconductor developed as a wire point contact alloyed diode in the latter part of the 1940s and, by the early 1950s both germanium and silicon, the two elemental semiconductors, had established the semiconductor industry. It is now considered part of the Solid State industry, which includes electronics, quartz crystals, and other allied technologies. Although silicon has become the primary semiconductor, due to its more useful physical and electronic characteristics, germanium is still fabricated as discrete alloyed or diffused devices, and is under development as a constituent in thin film epitaxy, layered devices that contain other compound semiconductors and materials, such as the heterostructure or heterojunction devices.

Both wafers and thin film epitaxy layers of germanium have been used in fabrication of a variety of devices: tunnel diodes, low and high power diodes, varactor diodes, high frequency transistors, solar cells and photodiodes—as a rod or barrel shaped unit, as a lithium-drift alpha-particle radiation detector.

Evaporated germanium has been used in conjunction with photo resist lacquers or alone as an etching and metallization mask on silicon and gallium arsenide. Although germanium oxide, Ge_2O_3, has been grown and used as a surface protectant, it is not as stable and acid resistant as silicon dioxide, SiO_2. The same applies to germanium nitride, Ge_3N_4 vs. silicon nitride, Si_3N_4, though both the pure nitrides and oxynitrides have been studied, and they form an isomorphous series between the two metals as both oxides and nitrides.

Germanium was initially grown as a single crystal ingot using the Bridgman method, now called Horizontal Bridgman (HB). This produces a half-moon-shaped ingot and, as cut wafers, are recognized by that shape. The method is still used for germanium and other materials, such as compound semiconductors, but has a size limitation of about 2—3″ square area. For this reason, most semiconductor ingots are now grown by the Czochralski (CZ) or Float Zone (FZ) methods with ingot/wafer diameters a standard 3″, and approaching 5—6″. Single crystal germanium has been the subject of much study, and is still under development and study. Germanium/silicon single crystal ingots of varying proportional concentrations have been grown and studied. It also is deposited as single crystal thin film epitaxy; as polycrystalline layers, poly-Ge; as amorphous layers, a-Ge, which may be hydrogenated as a-Ge:H.

Etching: Soluble in hot H_2SO_4, but no other single acids. Mixed acids: $HF:HNO_3$; $HF:H_2O_2$ . . . with or without H_2O, HAc, or other additives, alkali + halogen; halogens, DCE ionized gases.

SELECTION GUIDE: Ge

(1) Br_2:MeOH:
  (i) Polish: GE-0038
(2) HF:
  (i) Cleaning: GE-0064: -0126; -0284

(3) $HF:HNO_3$:
(i) Cleaning: GE-0090
(ii) Polish: GE-0030; -0015b; -0033; -0082; -0159; -0164; -0167; -0168; -0169; -0175; -0189; -0252; -0286
(iii) Structure: GE-0059a; -0077
(iv) Preferential: GE-0085
(v) Junction: GE-0009
(4) $HF:HNO_3$:HAc:
(i) Polish: GE-0088c; -0124; -0253; -0292; -0028; -0027
(ii) Preferential: GE-0220d,i—k; -0032; -0092b
(iii) Thinning: GE-0171
(5) $HF:HNO_3:H_2O$:
(i) Polish: GE-0034; -0079; -0127b; -0036; -0190
(ii) Preferential: GE-0220e
(iii) Removal: GE-0035
(6) $HF:HNO_3:H_2O:Ag(NO_3)_2$: $+/-$ H_2O
(i) Preferential: GE-0220b,c; -0061b
(7) $HF:HNO_3:Cu(NO_3)_2$:
(i) Preferential: GE-0061a
(8) H_2O:
(i) Float-off: GE-0043; -0057; -0203; -0204
(9) H_2O_2:
(i) Cleaning: GE-0001a; -0172; -0210; -0207
(ii) Polish: GE-0211
(iii) Preferential: GE-0005f; -0150a; -0230
(10) HNO_3: $+/-$ H_2O
(i) Cleaning: GE-0001b; -0132a; -0241; -0240; -0239; -0207
(ii) Oxidation: GE-0130
(iii) Preferential: GE-0132g
(iv) Polish: GE-0082
(11) HNO_3:Tartaric Acid:
(i) Preferential: GE-0005b
(12) HNO_3:Tartaric Acid:$AgNO_3$:
(i) Preferential: GE-0005a
(13) $HF:HNO_3$:Tartaric Acid:
(i) Polish: GE-0230b
(14) H_2O_2:Tartaric Acid:
(i) Preferential: GE-0005g
(15) Tartaric Acid:Br_2:
(i) Preferential: GE-0005h
(16) $HF:H_2O_2$ $+/-$ H_2O:
(i) Preferential: GE-0220f,h,l; -0223a,h; -0008b,d; -0150a-i; -0058; -0288; -0010a,b; -0032;
(ii) Cleaning: GE-0157
(iii) Passivation: GE-0204b + air
(17) H_2O_2:Citric Acid:
(i) Preferential: GE-0022g
(18) HCl:
(i) Preferential: GE-0132b
(19) H_2SO_4:
(i) Polish: GE-0078

(20) $KMnO_4$:H_2O:
(i) Oxidation: GE-0158a
(21) KCN:
(i) Cleaning: GE-0122b; -0094b; -0119b; -0159a,b; -0179; -0242
(22) NaClO:H_2O:
(i) Cleaning: GE-0056b
(23) KOH:I_2:
(i) Preferential: GE-0151a; -0221g; -0012a,d; -0206; -0302
(24) KOH:Br_2:
(i) Preferential: GE-0221a,b; -0012b; -0301; -0151b
(25) KOH:$(NH_4)_2S_2O_8$:
(i) Preferential: GE-0012c; -0303; -0220o
(26) KOH:$K_3Fe(CN)_6$:
(i) Preferential: GE-0056; -0134; -0135; -0144c
(27) NaOH:$NaNO_3$:H_2O:
(i) Cleaning: GE-0155
(ii) Preferential: GE-0208
(28) NH_4OH:$Cr(NO_3)_2$:
(i) Preferential: GE-0005d
(29) $Hg(NO_3)_2$:H_2O:
(i) Preferential: GE-0005c
(30) $AgNO_3$:H_2O:
(i) Preferential: GE-0005i

NAMED OR NUMBERED ETCHANTS

(1) Billig's Etch: KOH:$K_3Fe(CN)_6$:H_2O
Preferential: GE-0139a; -0142a
(2) BJ Etch: HF:HNO_3:HAc
Polish: GE-0027
(3) Camp #2 (Superoxol): 1HF:1HNO_3:4HAc
Preferential: GE-0065a; -0007b
(4) Camp #3:
Preferential: GE-0065b
(5) Copper Etch: HF:$CuSO_4$:H_2O
Junction: GE-0149
(6) CP4 (Camp #4): 30HF:50HNO_3:50HAc:0.6Br_2
Note: cleaning/preferential/polish/thinning/structuring
All Applications: GE-0065c (origin); -0015a; -0016; -0017; -0018a; -0019a; -0020; -0022; -0023; -0024; -0025; -0026; -0060b; -0066; -0067; -0069a; -0070a; -0071; -0072; -0073; -0083b; -0091; -0104; -0105; -0111; -0092a; -0093; -0094a; -0095a; -0096; -0097; -0098; -0099; -0100; -0102a; -0112; -0113; -0114; -0115; -0116; -0117; -0118; -0119a; -0120; -0122; -0123; -0138; -0144b; -0150a; -0147a; -0176; -0267; -0227a; -0228; -0253a
(7) CP4A: 30HF:50HNO_3:30HAc
Polish: GE-0014; -0063; -0253b
(8) CP4, varieties:
Note: Br_2 replaced with: xxxx
Preferential: +I_2: GE-0219
Preferential: +Br_2: GE-0213b
(9) Ellis #1:
Preferential: GE-0223b; -0150q

(10) Ellis #5:
Preferential: GE-0221a; -0150o
(11) Ellis #7 (Superoxol):
Preferential: GE-0221c; -0150i-n, r; -0002
(12) Iodine Etch: $HF:HNO_3:HAc:I_2$
Note: Several formula variations on Ge and other metals as ''Iodine Etch'', Iodine A Etch, etc. May be as polish/thinning/structuring/cleaning
All Applications: GE-0218; -0139a; -0140; -0203a; -0205; -0083a; -0235; -0266; -0263b; -0267; -0263a
(13) Ferricyanide Etch: $KOH:K_3Fe(CN)_6:H_2O$
Preferential: GE-0150s; -0060d
(14) Peroxide Etch: $HF:H_2O_2:HAc$
Preferential: GE-0060f
(15) WAg Etch: $HF:HNO_3:AgNO_3$ (Westinghouse Silver Etch)
Note: Preferential/polish/thinning/structuring
All Applications: GE-0150p; -0060g; -0066; -0072c; -0075a; -0084; -0144c; -0188; -0225c; -0289
(16) White Etch: $HF:HNO_3:+/-H_2O$
Note: Not recommended as a ''name'' — solutions are transparent, clear, not white in color, and there are many similar solutions with different formulas
Polish: GE-0015b; -0190
(17) (100) Etch: $HF:HNO_3:H_2O:Cu(NO_3)_2$
Copper Dislocation Etch:
Preferential: GE-0060e; SI-0038a
(18) 1:1:1 Etch: $HF:HNO_3:HAc$
Note: Not recommended as a ''name'' as there are other constituent 1:1:1 solutions.
Polish: GE-0028
(19) X-1114 Etch: $1HF:1HNO_3:1$ Superoxol:$4H_2O$
Note: Acceptable as a ''named etch'' as the ''X'' designates a specific mixture
Polish: GE-0147d
(20) SR4:
Polish: GE-0124
(21) Superoxol: $1HF:1H_2O_2:4H_2O$
Note: Chemically, hydrogen peroxide, H_2O_2 (30%) is called Superoxol
Polish: GE-0226
Preferential: GE-0202; -0002; -0003; -0021b; -0058a,b; -0070b; -0075b; -0095b; -0146; -0185b; -0231; -0218; -0005a; -0007; -0150j
(22) Superoxol:$Ag(NO_3)_2$:
Polish: GE-0078
(23) Aqua Regia: ($3HCl:1HNO_3$)
(i) Polish: GE-0173
(24) Antimony Etch:
(1) Preferential: GE-0257

ABRASIVES

Note: As an acid slurry for removal/polish, or dry in damage studies
(1) Alumina, Al_2O_3: GE-0013
(2) Silicon Dioxide, SiO_2: GE-0295; -0297; -0298

CONTAMINATION

Note: Chemicals applied to a surface in controlled contamination study
(1) Sodium, Na: GE-0282

(2) $NaOH:Na^{24}CO_3$: GE-0089
(3) $HF:HNO_3:HAc:Ag$: GE-0128a
(4) $HF:HNO_3:HAc:Fe$: GE-0234
(5) $KOH:Ge^{71}O_2$: GE-0301

HEAT (THERMAL)

Note: Can be heat in vacuum, only; or furnace heat treatment with gases.
Cleaning: GE-0170
Preferential: GE-0142b; -0146; -0165; -0270; -0278; -0279
Forming: GE-0179

GASES

(1) Chlorine, Cl_2: (preferential): GE-0081
(2) Air, O_2/N_2: (oxidizing): GE-0154
(3) Ar^+ (cleaning): GE-0178; -0201; -0283; -0232

JUNCTION ETCH

Note: Etch/stain/decoration.
All Applications: GE-0031a-d; -0033; -0034; -0212; -0211; -0213; -0019b,c; -0152; -0217; -0080; -0220h; -0127b; -0149; -0009; -0254b; -0254c; -0255; -0271

LASER

Note: As a tool, or germanium as a laser device
(1) Laser (anneal): GE-0286
(2) Laser (device): GE-0299a,b
(3) Laser (cutting): GE-0310

METALS

Note: As a diffusant/preferential etch/defect decoration
(1) Indium (preferential): GE-0127a; -0165c; -0166; -0236; -0275
(2) Indium (decoration): GE-0165b
(3) Copper (I, IV metals), molten: (Preferential): GE-0180
(4) Lithium (diffusion): GE-0270
(5) Barium Titanate (junction): GE-0254b; -0255
(6) Zinc Oxide (junction): GE-0254c
(7) Germanium (optics): GE-0287
LIGHT: GE-0244
ELECTRICITY: GE-0244
ELECTROPLATE: GE-0285
ELECTROLYTIC:
(1) KOH:
- (i) Polish: GE-0088b
- (ii) Preferential: GE-0087; -0147c; -0156; -0300; -0129b
- (iii) Junction: GE-0254a

(2) $KOH:H_2O_2$:
- (i) Cleaning: GE-0137
- (ii) Preferential: GE-0133b

(3) $KOH:GeO_2$:
- (i) Removal: GE-0147e

(4) KF:
- (i) Removal: GE-0209

(5) $KF:HNO_3$:
(i) Polish: GE-0029b
(6) KCl:
(i) Preferential: GE-0133a
(7) $KCl:HNO_3$:
(i) Polish: GE-0029a
(8) $KCl:GeO_2$:
(i) Preferential: GE-0046b; -0133a
(9) H_2SO_4:
(i) Polish/shaping: GE-0088a; -0129d
(10) $H_2SO_4:GeO_2$:
(i) Removal: GE-0147c
(11) HCl:KI:Oxalic Acid:
(i) Oxidation: GE-0136
(12) HCl:
(i) Preferential: GE-0129c; -0133c
(13) $InSO_4$:
(i) Shaping: GE-0129a
(14) NaCl:
(i) Shaping: GE-0129e
(15) HF:HAc:
(i) Polish/Selective: GE-0212b; -0213
(16) NaOH:
(i) Polish/Selective: GE-0200; -0233

OXIDATION
(1) H_2O: GE-0268a; -0269; -0270; -0271; -0272; -0273; -0277; -0268b; -0255; -0291
(2) $HAc:NaC_2K_3O_2$: GE-0156b; -0255
IRRADIATION:
(1) Neutron: GE-0238; -0262; -0264; -0265; -0274
MOLTEN FLUX:
(1) KCN:NaCN: GE-0163
(2) KOH/NaOH: GE-
CLEAVE:
(i) Cleaved: GE-0310
GROWTH:
(i) Growth: GE-0250; -0249; -0245

SPHERES & HEMISPHERES
Note: Polish, or etching to Finite Crystal Form.
(1) Polish: GE-0150a; -0030
(2) Preferential: GE-0058a; -0006a; -0150a-u; -0007; -0005a,b; -0220a-c; -0223a-h; 0300; -0301; -0302; -0303; -0304; -0010a,b

GERMANIUM ETCHANTS

GE-0001a
ETCH NAME: Hydrogen peroxide
TYPE: Acid, cleaning
COMPOSITION:
x .... H_2O_2

TIME:
TEMP: Boiling

DISCUSSION:

Ge, (111) wafers used in a study of surface conductivity and recombination. Surfaces were cleaned in the solution shown, or in nitric acid.

REF: Razhanov, A U et al — *Zhur Tekh Fiz* 26,2142(1956)

GE-0172: Gouskov, L & Nifontoff, W — *CR Acad Sci (Paris)* 248,1499(1959)

Ge, (111), (100), and (110) wafers used in a study of the effects of surface treatment on material orientation. H_2O_2 was used to clean surfaces prior to oxidation, and it leaves no visible film. Heat wafers in O_2 at two temperature levels: (1) 300—500°C will deposit colored oxides; (2) at 500°C oxide forms a granular structure with raised etch pattern figures, and (3) above 500°C the oxide film is completely granular without growth figures.

GE-0210: Wallis, G & Wang, S — *J Electrochem Soc,* 114,88(1967)

Ge, (111) wafers and other orientations. Wafers used in a study of recombination centers on germanium surfaces subjected to various etches.

GE-0211: Primak, W & Kempwirth, R — *J Electrochem Soc,* 114,88(1967)

Ge, (111) wafers used in a study of H_2O_2 etching on germanium. Preparation etching was done: (1) etch in CP4; (2) polish on micro-cloth with Linde A alumina, 1 min, RT, with H_2O_2 liquid carrier; (3) polish on Pellon pad with NaHClO, then 1—2 sec in CP4. The primary etch study used 100—200 ml of a 3% H_2O_2 solution in a beaker with stirring on a magnetic hot plate. Specimens were held horizontally in the solution with SST forceps during the etch period, then removed and rinsed in running DI water for 2 min, and then N_2 blown dry. Etch rate was 0.21 μm/min.

GE-0005f: Ibid.

Ge, single crystal spheres used in a study of etching convex surfaces to finite crystal form. Finite form was a dodecahedron, (110).

GE-0150a: Holmes, F J — *Acta Metall,* 7,283(1959)

Ge, single crystal hemispheres used in a study of type etch solutions vs. etch patterns, etched pits, and etch rates. Eighteen different preferential solutions evaluated. Hemispheres were first polished and cleaned in 50—70 ml of H_2O_2 before preferential etching.

GE-0230: Rosner, O — *Z Metallkd,* 46,225(1955)

Ge, single crystal sphere and hemispheres used in a study of convex surfaces. Study was similar to that of Holms (GE-0150a-s), Ellis (GE-0005a-h), Batterman (GE-0058), and others.

GE-0001b

ETCH NAME: Nitric acid — TIME:

TYPE: Acid, cleaning — TEMP: Boiling

COMPOSITION:

x .... HNO_3

DISCUSSION:

Ge, (111) wafers used in a study of surface conductivity and recombination. Surfaces were cleaned in boiling HNO_3 or H_2O_2.

REF: Ibid.

GE-0002

ETCH NAME: Superoxol — TIME:

TYPE: Acid, preferential — TEMP:

COMPOSITION:

1 .... HF

1 .... HNO_3

4 .... H_2O

DISCUSSION:

Ge, (100) wafers. Solution developed as an optical orientation etch for (100) germanium. (*Note:* Also called Camp #2. See: GE-0065a.)

REF: Theuerer, H C — U S Patent #2,542,727

GE-0003: Moss, K T & Hawkins, H L — *J Appl Phys,* 28,1258(1957)

Ge, (111) wafers used in a study of low surface recombination and low absorption levels in germanium.

GE-0021b: Pfann, W G & Vogel, F L Jr — *Acta Metall,* 5,377(1957)

Ge, (111) wafers used in a dislocation and defect study.

GE-0058a: Batterman, S W — *J Appl Phys,* 28,1236(1957)

Ge, (111) wafers and spheres used in a study of pits, hillocks and etch rates. The single crystal sphere finite crystal form was a dodecahedron (hh0) after etching 5 h at RT.

GE-0070b: Della-Pergola, G & Sette, D — *Alta Freq,* 24,499(1955)

Ge, (111) wafers. An evaluation of the effect of different etches on germanium. Other etches were: CP4, silver and copper nitrate.

GE-0075b: Holmes, P J — *Phys Rev,* 119,131(1960)

Ge, (111) dendritic ribbon. Evaluation of etch pits developed with different etch solutions. Authors says that both WAg and ferricyanide give triangle pits pointing upward on dendrites grown in "G" direction (?); whereas superoxol and similar etches show pits pointing downward.

GE-0095b: Schell, A — *Z Metallkd,* 47,614(1956)

Ge, (111) wafers used in an etching study. An evaluation of etch pit development with both CP4 and superoxol.

GE-0141: Dresselhaus, O et al — *Phys Rev,* 98,368(1955)

Ge, (111) wafers 0.5 mm thick and 0.3 mm diameter used in a study of cyclotron resonance of electrons and holes in both germanium and silicon. Solution used as a cleaning etch and preferential etch. Etching time was several minutes.

GE-0185b: Bridgers, H E et al — *Transistor Technology,* Vol I, D Van Nostrand, New York, 1958, 354

Ge, specimens. Describes use of superoxol as an etch.

GE-0231: Rosi, F D — *Acta Metall,* 4,26(1956)

Ge, (111) wafers. A dilute HF:HNO_3 solution was used in studying kink bands in germanium.

GE-0060a: Harper, C A — *Handbook of Materials and Processes for Electronics,* McGraw-Hill, New York, 1970, 7—52

Ge, (111) wafers. Solution can be used as a preferential etch. Several other etch mixtures described.

GE-0218: Coblens, A & Ownes, H L — *Transistors — Theory and Applications,* McGraw-Hill, 1955, 211—212

Ge specimens. The use of superoxol as a germanium etchant.

GE-0005a: Ellis, R C Jr — *J Appl Phys,* 28,1068(1957)

Ge, single crystal spheres. Spheres were ground to shape and etch polished prior to finite crystal form (FCF) etching to observe rapid etch planes on convex surfaces. Ten different etchants were used in the two referenced articles, listed here, and the developed finite form is shown in the articles under each etchant evaluated. This was some of the initial work done in studying the convex single crystal planes of semiconductors. Finite crystal form obtained in superoxol was a dodecahedron, (110).

GE-0220a: Ellis, R C Jr — *J Appl Phys,* 25,1497(1954)

Ge, single crystal spheres. See discussion under GE-0005a.

GE-0158b: Batterman, S W — *J Appl Phys,* 28,1236(1957)

Ge and Si single crystal spheres and hemispheres. Material was pedestal grown as hemispheres with sections cut and fabricated as complete spheres. Finite crystal form obtained on a sphere was a dodecahedron, (110).

GE-0007: Camp, P R — *J Electrochem Soc*, 102,586(1955)

Ge, single crystal spheres ground to shape and etch polished before preferential etching in a study of convex surfaces. The finite crystal form obtained was a dodecahedron, (110).

GE-0005b

ETCH NAME: TIME:

TYPE: Acid, preferential TEMP:

COMPOSITION:

3 .... HNO_3

6 .... tartaric acid

DISCUSSION:

Ge, single crystal spheres used in a study of etching convex surfaces to finite crystal form. Finite form obtained was a cube, (100).

REF: Ibid.

GE-0005i

ETCH NAME: Silver nitrate TIME:

TYPE: Salt, preferential TEMP:

COMPOSITION:

x .... x% $AgNO_3$

DISCUSSION:

Ge, single crystal spheres used in a study of convex surface planes. Finite crystal form was a cube, (100).

REF: Ibid.

GE-0008

ETCH NAME: TIME:

TYPE: Acid, preferential TEMP:

COMPOSITION:

1 .... HF

1 .... H_2O_2

3 .... H_2O

DISCUSSION:

Ge, (111) wafers. Solution was used as an optical orientation etch (OOE) for germanium. Etching should be brief, as it also is a polishing etch on germanium. (*Note:* See GE-0002 and additional references under superoxol. The solution shown under GE-0008 is a modification of superoxol.)

REF: Tyler, W W & Dash, W C — *J Appl Phys*, 28,1221(1957)

GE-0147d

ETCH NAME: X-1114 TIME: Seconds

TYPE: Acid, polish TEMP: RT

COMPOSITION:

1 .... *Superoxol

1 .... HNO_3

1 .... HF

4 .... H_2O

*1HF:1H_2O_2:4H_2O

DISCUSSION:

Ge, (111) wafers used in an etch development study. Solution shown described as an excellent removal and polish etch for germanium. The removal etch rate for a $^1/_4''$ area shown as 25 μm/20 sec.

REF: Bondi, F J — *Transistor Technology,* Vol III, D Van Nostrand, New York, 1958, 15b

---

GE-0005c

ETCH NAME: Mercuric nitrate | TIME:

TYPE: Salt, preferential | TEMP:

COMPOSITION:

x .... x% $Hg(NO_3)_2$

DISCUSSION:

Ge, single crystal spheres. Finite form: cube, (100). See GE-0005a for additional discussion.

REF: Ibid.

GE-0005d

ETCH NAME: | TIME:

TYPE: Acid, preferential | TEMP:

COMPOSITION:

1 .... $Cr(NO_3)_2$

1 .... NH_4OH

DISCUSSION:

Ge, single crystal spheres. Solution mixed to produce $Cr(NH_3)_4^{++}$ ions. Finite crystal form was a cube, (100). See GE-0005a for additional data.

REF: Ibid.

GE-0005e

ETCH NAME: | TIME:

TYPE: Acid, preferential | TEMP:

COMPOSITION:

50 .... HNO_3

50 .... tartaric acid

3 .... x% $AgNO_3$

DISCUSSION:

Ge, single crystal spheres. Finite crystal form is a cube, (100).

REF: Ibid.

GE-0005g

ETCH NAME: | TIME:

TYPE: Acid, preferential | TEMP:

COMPOSITION:

1 .... H_2O_2

1 .... tartaric acid

DISCUSSION:

Ge, single crystal spheres. Finite crystal form was a dodecahedron, (110).

REF: Ibid.

GE-0005h
ETCH NAME: TIME:
TYPE: Halogen, preferential TEMP:
COMPOSITION:
x .... tartaric acid
x .... *Br_2

*Saturate tartaric acid with bromine.

DISCUSSION:
Ge, single crystal spheres. Finite crystal form was a tetrahexahedron, (hk0).
REF: Ibid.

GE-0220b
ETCH NAME: TIME:
TYPE: Acid, preferential TEMP:
COMPOSITION:
37 .... HF
13 .... HNO_3
50 .... H_2O
2 .... x% $AgNO_3$
DISCUSSION:
Ge, single crystal spheres. Finite crystal form was a cube, (100).
REF: Ibid.

GE-0220c
ETCH NAME: TIME:
TYPE: Acid, preferential TEMP:
COMPOSITION:
4 .... HF
7 .... HNO_3
87 .... HAc
3 .... x% $AgNO_3$
DISCUSSION:
Ge, single crystal spheres. Finite crystal form was a cube, (100).
REF: Ibid.

GE-0220d
ETCH NAME: TIME:
TYPE: Acid, preferential TEMP:
COMPOSITION:
25 .... HF
45 .... HNO_3
30 .... HAc
DISCUSSION:
Ge, single crystal spheres. Finite crystal form was an octahedron, (111).
REF: Ibid.

GE-0220e
ETCH NAME:
TYPE: Acid, preferential
COMPOSITION:
40 .... HF
20 .... HNO_3
40 .... H_2O
2 .... x% $AgNO_3$
DISCUSSION:
Ge, single crystal spheres. Finite crystal form was an octahedron, (111).
REF: Ibid.

TIME:
TEMP:

GE-0220f
ETCH NAME:
TYPE: Acid, preferential
COMPOSITION:
50 .... HF
50 .... H_2O_2
DISCUSSION:
Ge, single crystal spheres. Finite crystal form was an octahedron, (111).
REF: Ibid.

TIME:
TEMP:

GE-0220g
ETCH NAME:
TYPE: Acid, preferential
COMPOSITION:
1 .... H_2O_2
1 .... citric acid
DISCUSSION:
Ge, single crystal spheres. Finite crystal form was an octahedron, (111).
REF: Ibid.

TIME:
TEMP:

GE-0220h
ETCH NAME:
TYPE: Acid, preferential
COMPOSITION:
17 .... HF
17 .... H_2O_2
66 .... H_2O
DISCUSSION:
Ge, single crystal spheres. Finite crystal form was a dodecahedron, (110).
REF: Ibid.

TIME:
TEMP:

GE-0220i
ETCH NAME:
TYPE: Acid, preferential
COMPOSITION:
8 .... HF
13 .... HNO_3
79 .... HAc

TIME:
TEMP:

DISCUSSION:
Ge, single crystal spheres. Finite crystal form was a dodecahedron, (110).
REF: Ibid.

GE-0220j
ETCH NAME: TIME:
TYPE: Acid, preferential TEMP:
COMPOSITION:
13 .... HF
23 .... HNO_3
64 .... HAc
DISCUSSION:
Ge, single crystal spheres. Finite crystal form was a trisoctahedron, (hh1).
REF: Ibid.

GE-0220k
ETCH NAME: TIME:
TYPE: Acid, preferential TEMP:
COMPOSITION:
3 .... HF
5 .... HNO_3
92 .... HAc
DISCUSSION:
Ge, single crystal spheres. Finite crystal form was a tetrahexahedron, (hk0).
REF: Ibid.

GE-0221a
ETCH NAME: TIME: 10—20 min
TYPE: Alkali, preferential TEMP: Boiling
COMPOSITION:
250 ml .... 30% KOH(NaOH)
20 g .... I_2
DISCUSSION:
Ge, single crystal spheres fabricated in a ''Tornado Sphere Grinder'' from $^1/_4$ and $^1/_8$'' cubes. Sphere were slow polish etched in 1HF:10HNO_3 prior to preferential etching for finite crystal form of convex surfaces. The crystal form obtained was an octahedron, (111).
REF: Walker, P & Waters, P W — personal development, 1957/1958

GE-0221b
ETCH NAME: TIME: 10—20 min
TYPE: Alkali, preferential TEMP: Boiling
COMPOSITION:
250 ml .... 30% KOH(NaOH)
50 ml .... Br_2
DISCUSSION:
Ge, single crystal spheres. Finite crystal form was an octahedron, (111).
REF: Ibid.

GE-0223a
ETCH NAME: Ellis #5
TIME:
TYPE: Acid, preferential
TEMP:
COMPOSITION:
20 .... HF
3 .... H_2O_2
12 .... H_2O
DISCUSSION:

Ge, spheres and hemispheres. A study of preferential etching of convex surfaces for finite crystal form spheres, and planes on hemispheres. See Holmes, GE-0150a-s for additional data.

REF: Ellis, S O — *J Appl Phys,* 28,1262(1957)

GE-0223b
ETCH NAME: Ellis #1
TIME:
TYPE: Acid, preferential
TEMP:
COMPOSITION:
2 .... HF
1 .... HNO_3
1 .... 10% $AgNO_3$
DISCUSSION:

Ge, spheres and hemispheres used in a study of preferential etching of convex surfaces.

REF: Ibid.

GE-0223c
ETCH NAME: Ellis #7
TIME:
TYPE: Alkali, preferential
TEMP: Cold
COMPOSITION:
x .... 0.8 *N* KOH
x .... Cl_2 for pH 8—9
DISCUSSION:

Ge, spheres and hemispheres. Chlorine gas is bubbled through the cold hydroxide solution during the etch period. See Holmes, GE-0150a-s.

REF: Ibid.

GE-0010a
ETCH NAME:
TIME: 12 min
TYPE: Acid, preferential
TEMP: RT (24°C)
COMPOSITION:
1.5 .... HF
2 .... H_2O_2
4 .... H_2O
DISCUSSION:

Ge, (100) and (110) wafers. Etch used at time and temperature shown as an optical orientation etch (OOE) on (100) surfaces. The (110) surface was etched for 12 min at 20°C for orientation.

REF: Schwuttke, G H — *J Electrochem Soc,* 106,315(1959)

GE-0010b
ETCH NAME:
TYPE: Acid, preferential
COMPOSITION:
1 .... HF
2 .... H_2O_2
4 .... H_2O
TIME: 6 min
TEMP: RT (24°C)

DISCUSSION:
Ge, (111) wafers. This solution developed as a light figure orientation etch for (111) surfaces.
REF: Ibid.
GE-0288: Sturge, M D — *Proc Phys Soc,* 73,320(1959)
Ge, (111) wafers used to fabricate p-n junction diodes with boron diffusion. Solution used to etch define the junctions.

GE-0012a
ETCH NAME:
TYPE: Alkali, preferential
COMPOSITION:
250 ml .... 30% KOH
30 g .... I_2
TIME: 10—30 min
TEMP: Boiling

DISCUSSION:
Ge, (111), (100), (110) and (211) wafers. Developed as a light figure orientation etch for germanium. Also used to fabricate controlled etch pit structures and mesas on (111) wafers in device development. Operational diodes and transistors were fabricated.
REF: Walker, P — personal development, 1958
GE-0012d: Ibid.
Both germanium and silicon spheres etched in this solution produce the finite crystal form of a cube, (100).

GE-0012b
ETCH NAME:
TYPE: Alkali, preferential
COMPOSITION:
250 ml .... 30% KOH (450 g)
25 ml .... *Br_2
TIME: 10—30 min
TEMP: Boiling

*Amount of Br_2 varied to obtain specific pit structure.

DISCUSSION:
Ge, (111), (100), (110) and (211) wafers. Solution developed as a light figure orientation etch for germanium. Also used on both n- and p-type (111) germanium to develop controlled-damage pits and mesa structures in device development. Produces a very flat bottomed, sharply triangular pit form within 30 min on (111). Over 30 min etching, pit bottom begins to show terracing. Solution using I_2 (GE-0012a) is very similar and easier to control as bromine dissipates rapidly in hot solutions.
REF: Ibid.
Ge, (111), and (100) wafers. CP4 used as a cleaning etch prior to Ar^+ ion bombardment. Ion cleaning showed anisotropic etching. n-type Ge converted to p-type above 500°C.
GE-0023: Gottaviani, G et al — *J Appl Phys,* 47,626(1976)
Ge, (111) wafers used as substrates for aluminum evaporation and Ge epitaxy for a Ge epi/Al/Ge (111) structure. Substrates were mechanically polished, then polish etched with

a modified CP4 using 0.5 Br_2. After fabrication, subsequent annealing caused the epitaxy Ge cover to migrate down through the aluminum to the substrate Ge surface.
GE-0024: O'Hara, S — *J Appl Phys,* 35,409(1964)

Ge specimens of ribbon crystal or dendritic germanium grown material used in a study of dislocations and defects. A modified CP4 containing varying amounts of 0.8 g Ge/100 ml HNO_3 added. (*Note:* Etches containing some of the material to be etched have been called "seeded" solutions.)
GE-0025: Crisman, E E et al — *J Electrochem Soc,* 131,1896(1984)

Ge, (111), p-type, 0.1 Ω cm resistivity wafers used as substrates in a germanium oxide study. Degrease wafers in hot TCE, then rinse in hot acetone, and hot MeOH. Remove native oxide in hot $1HF{:}1H_2O$, 30 sec with final rinse in H_2O, and N_2 blow dry. Etch remove about 30% of the specimen thickness with CP4 . . . there was slight orange-peel on the surface after etching. Grow GeO/GeO_2 on surfaces in a pressure bomb, 10 min at 650°C, and 340 atm in an O_2 atmosphere for about 2000 Å of oxide. Anneal at 200°C in forming gas (85% N_2:15% H_2) or in NH_3.
GE-0026: Walker, P — personal development, 1960.

Ge bars cut from (111) grown ingots with grown-in p-n junctions.

GE-0012c
ETCH NAME:
TYPE: Alkali, preferential
COMPOSITION:

250 ml .... 30% KOH (450 g)
45 g .... $(NH_4)_2S_2O_8$

TIME: 10—30 min
TEMP: Boiling

DISCUSSION:

Ge, (111), (100), (110) and (211) wafers. See comments under similar solutions containing Br_2 and I_2 (GE-0012a and GE-0012b). In all three solutions, KOH can be replaced with NaOH.
REF: Ibid.

GE-0065a
ETCH NAME: CP4
TYPE: Acid, polish/preferential
COMPOSITION:

30 .... HF
50 .... HNO_3
30 .... HAc
0.6 .... Br_2

TIME: Variable
TEMP: RT

DISCUSSION:

Ge, (100) wafers and other orientations. CP4 was originally developed as a polishing and preferential etch for (100) germanium surfaces used at RT. Since initial development as Camp #4, the solution has had wide application on germanium, as well as other semiconductors, metals and metallic compounds.
REF: Camp, P R — *J Electrochem Soc,* 102,1415(1955)
GE-0015a: Ingham, H S & McDade, F J — *IBM J,* Jul 1961, p 302

Ge specimens. Used as a general polish for germanium surfaces.
GE-0016: Turner, J — *J Electrochem Soc,* 103,252(1956)

Ge specimens. Recommends using CP4 without bromine for polishing both germanium and silicon wafers. (*Note:* CP4 without bromine is called CP4A.)
GE-0017: Allen, J W & Smith, K C A — *J Electron,* 1,439(1956)

Ge specimens. Used CP4 to develop edge dislocations in germanium. Authors say that etch rate varies with concentration of electrons and holes in the germanium material.

GE-0018a: Muller, R K — *J Appl Phys,* 32,640(1961)

Ge wafers. CP4 was used to remove saw damage from surfaces in studying grain-boundaries in n-type germanium.

GE-0019a: Weinreich, O et al — *J Appl Phys,* 32,1170(1961)

Ge wafers fabricated with thermally evaporated p-n junction devices. Solution used to develop junctions by etching for 5 min, then with electrical bias across junction, place one drop of KOH solution on wafer across the junction to develop.

GE-0020: Breidt, P Jr et al — *Acta Metall,* 5,60(1957)

Ge wafers. Indented germanium surfaces with a Ta rod at 375°C and used CP4 as a macro-etch to develop dislocations introduced by damaging and accentuated by thermal heating.

GE-0021: Pfann, W G & Vogel, F L Jr — *Acta Metall,* 5,377(1957)

Ge wafers. Solution used at room temperature as a polish etch. Authors say a cone-shaped surface is obtained representative of bulk material is ideal and that p- and n-type surfaces etch similarly. A sphere etched in CP4 produced pole figures, only, at (111) and (100) locations. (*Note:* The mammillary type surface is common to most metals and compounds, when all surface damage depth has been etch removed and the etch solution is slightly preferential. Called either mammillary or botryoidal as a natural mineral structure.)

GE-0022: Forman, R — *Phys Rev,* 117,698(1960)

Bars were cut and lap polished with 600-grit SiC. CP4 was mixed and allowed to age 24 h before use. Bars were forward biased with 0.5 mA power, while being lightly agitated in the solution with removal every 2 min, water quenched and observed until junction was well developed. The p-type junction side etched approximately twice the rate of the n-type forming a sharply defined line trench in 5—7 min.

GE-0060b: Harper, C A — *Handbook of Materials and Processes for Electronics*, McGraw-Hill, New York, 1970, 7—52

Ge, (111) wafers. Solution mixture was with 0.3 ml Br_2 and used for 1—3 min as a general cleaning and polishing etch.

GE-0066: Bardsley, W & Bell, R L — *J Electron,* 5,19(1958)

Ge, (111) wafers. Solution used to develop dislocations which were measured for their crystallographic directions and orientation.

GE-0067: Bell, R L — *J Electron,* 3,487(1957)

Ge, (111) wafers. Both CP4 and Dash Etch used in studying dislocation density and defects in semiconductors.

GE-0069: Harnik, E & Margeninski, Y — *J Phys Chem Solids,* 8,96(1959)

Ge, (111) wafers used in a study of surface treatments and the characteristics of fast states on germanium surfaces. Solution used 30 sec at 30°C.

GE-0070a: Della-Pergola, G & Sette, D — *Alta Freq,* 24,499(1955)

Ge, (111), and (100) wafers used in a general study of the effects of different preferential etches. Used CP4, superoxol, WAg, and copper nitrate solutions to develop defects.

GE-0071: Balabanova, L A & Bredov, M M — *Zh Tekh Fiz,* 27,1401(1957)

Ge, (111) wafers used in a study of thermal conversion by irradiation. CP4 + warm DI water rinse was evaluated against a lapped surface with alcohol rinse only. The CP4 treatment reduced conversion 25—50%.

GE-0072: Wehner, G K — *J Appl Phys,* 29,217(1958)

Ge, (111) wafers. Wafers were Hg^+ ion bombarded at 1000 V and etched in CP4 before and after irradiation. Dislocation pits did not correspond before and after irradiation.

GE-0073: Bonfiglioli, G et al — *J Appl Phys,* 31,684(1960)

Ge, (111) wafers. A study of first order structures developed by CP4 on germanium surfaces. CP4 developed conoid etch pits and dislocations, which either disappeared or became passive. (*Note:* Those that increase in size and disappear are usually surface damage etch pits; those that remain in place with little size increase, though they too may disappear, are dislocations.)

GE-0083b: Rhodes, E G et al — *J Electron,* 3,403(1957)

Ge, (111) wafers. CP4 used as a polish and damage removal etch before preferential etching. (See GE-0083a.)

GE-0091: Noggles, T S & Stiegler, J O — *J Appl Phys,* 30,1279(1959)

Ge, (111) wafers irradiated and studied under an electron microscope (EM). Solutions used to observe defects before and after irradiation. Both CP4 and CP4A (without bromine) were evaluated.

GE-0104: Handler, P & Portnoy, W M — *Phys Rev,* 116,516(1959)

Ge, (100) wafers used in a study of electron surface states of cleaned surfaces. CP4 was used as a cleaning etch.

GE-0105: Fritzsche, H — *Phys Rev,* 99,406(1955)

Ge, (111) wafers used in a study of electrical parameters at low temperature. CP4 used as a polish etch.

GE-0106: Spears, W E — *Phys Rev,* 112,362(1958)

Ge, (111) wafers, both n- and p-type, 10—22 Ω cm resistivity used in a study of the surface effects of electron irradiation at 80 K. CP4 was used as a polish etch.

GE-0107: Madden, H H & Farnsworth, H E — *Phys Rev,* 112,791(1958)

Ge, (100) wafers used in a high vacuum study of surface recombination velocity. Wafers prepared as follows: (1) Etch thin with CP4 prior to mechanical lapping with A/O 305-grit; (2) polish with MgO on a wax lap; (3) polish etch with CP4, (4) rinse in concentrated KCN and (5) final rinse in double-distilled water.

GE-0108: Brill, P A & Schwartz, R F — *Phys Rev,* 112,330(1958)

Ge, (111) wafers used in a study of radiation recombination. Etch clean surfaces with CP4 or in $2HF:2HNO_3:1HAc$.

GE-0109: Hunter, L P — *IBM Res Dev,* 3,106(1957)

Ge, (111) wafers. A direct measurement of the angular dependence of the atomic scattering factor. CP4 used as a heavy removal etch with at least 20 μm removed and no residual surface damage observed.

GE-0110: Kurtz, A D et al — *Phys Rev,* 101,1285(1956)

Ge, (111) wafers used in a study of the effects of dislocations on minority carrier lifetime. Surfaces were electropolished before preferentially etching with CP4 or superoxol to develop dislocations.

GE-0111: Savage, H — *J Appl Phys,* 31,1472(1960)

Ge, (111) wafers used in a study of copper precipitation on dislocation etch pits. Use a modified CP4 or Ellis #7 (GE-0105r) as a polish and dislocation etch. To copper decorate place drops of $Cu(NO_3)_2$ on surface and heat to 900°C, $^1/_2$ h.

GE-0092a: Faust, J W Jr & John, H F — *J Electrochem Soc,* 108,860(1961)

Ge, (111) dendrites as ribbon crystal growth. Solution used to develop structure and growth patterns. The CP4 solution mixture used was 15 ml HF:25 ml HNO_3:15HAc:0.3% Br_2, RT, 3—10 sec.

GE-0093: Vogel, F L et al — *Phys Rev,* 90,489(1953)

Ge, (111) wafers used in a dislocation study. CP4 solution mixture was $3HF:5HNO_3:3HAc$:0.1% Br_2, RT. Also used on silicon which etches more rapidly.

GE-0094a: Logan, K H & Schwartz, M — *J Appl Phys,* 26,1287(1955)

Ge, (111) wafers used in a study of the restoration of resistivity and lifetime by heat treatment. Both CP4 and KCN solutions used to clean surfaces prior to evaluation.

GE-0095a: Schell, H A — *Z Metallkd,* 47,614(1956)

Ge, (111) wafers, and other orientations, used in a general defect study. The original CP4 (GE-0065a) used as a dislocation etch.

GE-0096: Harvey, W W & Gatos, H O — *J Electrochem Soc,* 105,654(1958)

Ge, (111) wafers studied for the material reaction in solutions containing dissolved oxygen. Original CP4 solution used at RT, 30—60 sec as a polish etch.

GE-0097: Vogel, F L — *Acta Metall,* 3,95(1955)

Ge, (111) wafers studied for dislocations after mechanical abrasion with 600-grit SiC after temperature stressing. CP4 was used as a polish and dislocation etch.

GE-0098: Faust, J W Jr & John, H F — *J Electrochem Soc,* 107,562(1960)

Ge, (111) wafers, and other orientations. A comparison of etching and fracture techniques for defect study of Ge, Si and III—V compound semiconductors. CP4 with 0.3 parts Br_2 was slightly preferential on (111) surfaces, although it was used primarily as a polish etch.

GE-0099: Patel, J R & Alexander, B H — *Acta Metall,* 4,385(1956)

Ge, (111) wafers. This was a study of plastic deformation of germanium under compression. CP4 was used at RT as a polish etch which also shows slip bands, and also was used hot, at 60°C, as a dislocation etch.

GE-0100: Wallis, G & Wang, S — *J Electrochem Soc,* 106,231(1959)

Ge, (111) wafers used in a study of the effect of various etches on recombination centers in germanium surfaces. CP4 was evaluated against H_2O_2, Iodine ''A'', electrolytic (KOH), and the Silver Etch.

GE-0102a: Wertheim, O K & Pearson, G L — *Phys Rev,* 107,694(1957)

Ge, (111) wafers used in a study of recombination after plastic deformation. Clean wafers in a solution of KCN, then etch polish in CP4. After deformation at 750°C, re-etch with CP4, RT, 90 sec to develop dislocations. Initial density was 5×10^3 cm^2, increasing to 2×10^7 cm^2 after deformation.

GE-0112: Tweet, A G — *J Appl Phys,* 29,1520(1958)

Ge, (111) wafers as dislocation-free material. Study showed evidence of vacancy clusters. Dislocation free areas etch more rapidly when defect density is greater than 100/cm^2. CP4 used as an electrolytic ''pulse'' etch to develop dislocations.

GE-0113: Meckel, B B & Swalin, R A — *J Appl Phys,* 30,89(1959)

Ge, (111) wafers. Observation of selective delineation of screw dislocations produced by cathode sputtering as Ar^+ ion bombardment at 300 eV. After bombardment, CP4 was used to develop dislocations.

GE-0114: Heinecke, W J & Ing, S Jr — *J Appl Phys,* 32,1498(1961)

Ge, (111), and (100) wafers used in a study of the reaction of iodine on surfaces. Wafers were etch polished in CP4 prior to iodine treatment at 300—500°C, 5—20 min in an I_2 gas flow of 0.2—6 cm/sec with a partial pressure of 0.004—0.02 atm. Etch rate was directly proportional to iodine partial pressure. Variations in temperature and gas flow rate showed either little effect or erratic results. (*Note:* Etch pits shown in the article are identical to those observed using KOH:I_2 liquid solutions. GE-0151a.)

GE-0115: Dexter, R N et al — *Phys Rev,* 104,637(1956)

Ge, (111) wafers used in cyclotron resonance experiments. CP4 used as a cleaning etch. CP4 does appear to influence photoconduction efficiency, but not scattering frequency. Also studied silicon.

GE-0116: Madden, H H & Farnsworth, H E — *Phys Rev Lett,* 1,346(1958)

Ge, (111) wafers used in high vacuum studies of surface recombination velocity. CP4 used as a surface cleaning etch, but was not as good as ionic bombardment cleaning. Residual oxide layers remaining after etching had little or no effect.

GE-0117: Frank, R C — *J Appl Phys,* 31,1689(1960)

Ge, single crystal seed prepared for ingot growth. The seed was a hollow, polished cylinder. CP4 used as the polish etch and to develop residual dislocations. A cylinder was designed, because dislocations are higher in density toward the center of a solid seed. There were traces of radial lineage.

GE-0118: Vook, F L & Balluffi, R W — *J Appl Phys,* 31,1693(1960)

Ge, (111) wafers used in a study of resistivity change from low temperature annealing with deuteron irradiation. CP4A (no bromine) used to remove work damage from mechanically lapped bars.

GE-0119a: Yamashita, T & Ohto, T — *Jpn J Appl Phys,* 16,1565(1961)

Ge, (111) wafers arc cut to 1.50 in diameter and mechanically lapped with 1200-grit emory. Etch polish with CP4, then soak in HNO_3, rinse in DI water, then soak in KCN, and DI water rinse. Wafers used in measurement of the Seebeck effect in plastically bent germanium.

GE-0120: Kuczynski, G C & Hochman, R F — *Phys Rev,* 108,946(1957)

Ge, (111) wafers used in a study of induced plasticity. A Knoop point using a Tukon Hardness Tester was used to indent Ge, Si, InSb, and InAs wafer surfaces. Indentation damage results are dependent upon prior surface preparation. Ge, InSb, and InAs surfaces were etch polished with CP4, and silicon with 1HCl:$3HNO_3$.

GE-0121: Farnsworth, H E et al — *J Appl Phys,* 29,1150(1958)

Ge, (111) wafers and titanium, silicon, and nickel specimens used in a study of ion bombardment cleaning methods, as evaluated by low-energy electron diffraction. Germanium wafers were cleaned with the following sequence: (1) CP4 etch, (2) HF rinse, then (3) repeat CP4 etch.

GE-0122: Seesger, K — *Phys Rev,* 114,476(1959)

Ge, (111) wafers used in a study of the microwave frequency dependence of drift mobility. CP4 was used to etch polish.

GE-0123: Hogarth, C A & Baynham, H C — *Proc Phys Soc,* 71,647(1958)

Ge, (111) wafers used in a general dislocation study. CP4 used as a preferential etch to develop dislocations.

GE-0138: Davis, W D — *Phys Rev,* 114,1006(1959)

Ge:Si single crystal alloy ingots. CP4 used to develop dislocations.

GE-0144b: Faust, J W Jr & John, H F — *J Electrochem Soc,* 108,825(1961)

Ge, (111) dendrites from material grown as ribbon crystals. CP4 used at RT, 30—45 sec to develop dislocations. Also used the WAg and ferrocyanide etches.

GE-0150a: Holmes, P J — *Acta Metall,* 7,283(1959)

Ge, hemispheres etched in a study of defect patterns, type etch pits, and etch rates on both germanium and silicon. Prior to preferential etching, specimens etch in CP4 to remove lap/polish damage. See GE-0150a-s.

GE-0147a: Bondi, F J — *Transistor Technology,* Vol III, D Van Nostrand, New York, 1958

Ge, specimens. Etch in CP4 for 1—2 min as a polish etch before electrolytic jet etching with 0.1% KOH.

GE-0176: Walter, F J et al — *Rev Sci Instrum,* 31,756(1960)

Ge devices fabricated as large area surface-barrier counters. Etch polished in CP4 prior to processing.

GE-0217: Bridgers, H E et al — *Transistor Technology,* Vol I, D Van Nostrand, New York, 1956, 354

Ge, specimens. Both CP4 and superoxol described as general etchants for germanium.

GE-0227a: Arizumi, T & Akasaki, I — *Jpn J Appl Phys,* 2,143(1963)

Ge, (111), (110), and (100) wafers used in an etching study of structures developed by iodine vapor. Wafers were mechanically lapped and then etched in CP4 to remove work damage and residual oxide prior to etching with hot iodine vapors.

GE-0228: Arizumi, T et al — *Jpn J Appl Phys,* 2,757(1963)

Ge, (100) wafers and other orientations. Wafers used in studies of impurity doping in vapor grown germanium. CP4 used as a cleaning rinse etch prior to doping.

GE-0217b
ETCH NAME: CP4, variety — TIME: 3 min
TYPE: Acid, polish — TEMP: RT
COMPOSITION:
15 .... HF
25 .... HNO_3
15 .... HAc
x* .... Br_2

*x = 10 drops Br_2/50 ml for a bromine saturated solution.

DISCUSSION:
Ge wafers. This solution is CP4 saturated with bromine, and used as a polishing etch.
REF: Bridgers, H E et al — *Transistor Technology,* Vol I, D Van Nostrand, New York, 1958, 354

GE-0219
ETCH NAME: CP4, modified — TIME:
TYPE: Acid, dislocation — TEMP: RT
COMPOSITION:
15 ml .... HF
25 ml .... HNO_3
15 ml .... HAc
x mg .... I_2
DISCUSSION:
Ge wafers. Solution used to develop screw dislocations. See GE-0218.
REF: Rhodes, R T et al — *J Electron Control,* 3,403(1957)
GE-0220: Walker, P & Waters, P W — personal development, 1957
Ge, (111), (100), (110), and (211) wafers used in a general etching study. CP4 with 1gI_2/200 ml of solution used at RT, cool (8—10°C), and warm (40°C) as a polish and preferential defect development etch.

GE-0221
ETCH NAME: — TIME:
TYPE: Vacuum, cleaning — TEMP:
COMPOSITION:
x .... vacuum
DISCUSSION:
Ge, (100) wafers cleaved in vacuum to obtain clean surfaces and used in a study of oxygen effects on surfaces.
REF: Robinson, P H et al — *J Appl Phys,* 27,962(1956)

GE-0027
ETCH NAME: BJ etch — TIME:
TYPE: Acid, polish — TEMP:
COMPOSITION:
9 .... HF
15 .... HNO_3
10 .... HAc

DISCUSSION:

Ge, (111) wafers. This is one of several etches developed by the author cited as a good polishing etch for both germanium and silicon.

REF: Stopek, S — personal communication, 1958

GE-0028

ETCH NAME: 1:1:1 TIME:

TYPE: Acid, polish/removal TEMP:

COMPOSITION:

1 .... HF

1 .... HNO_3

1 .... HAc

DISCUSSION:

Ge, (111), and Si, (111) wafers. Solution used as a removal and polish etch on both materials. (*Note:* Although this etch was called a 1:1:1 etch, which it is, such volume nomenclature is not recommended due to the wide variations of possible constituents in such a volume mixture.)

REF: Millea, R P & Hall, R — *J Electrochem Soc,* 105,174(1958)

GE-0030

ETCH NAME: TIME: 1 to 2 h

TYPE: Acid, sphere polish TEMP: RT

COMPOSITION:

5 ml .... HF

75 ml .... HNO_3

DISCUSSION:

Ge, spheres of single crystal germanium and silicon with a 600-grit SiC ground surface were etch polished in this solution prior to preferential etching for finite crystal form and as spherical diffused diodes. Both n- and p-type spheres were used. Etch in a Teflon beaker tilted at about 30° with sufficient rotational agitation to allow spheres to roll freely but not tumble. Time shown is for germanium; silicon time was 10—20 min. Ultrasonically cut cylinders of germanium also were etch polished and fabricated as lithium-drift detectors.

REF: Walker, P — personal development, 1961

GE-0031a

ETCH NAME: TIME:

TYPE: Acid, junction TEMP: RT

COMPOSITION:

20 .... HNO_3

20 .... BJ etch

60 .... H_2O

\+ .... $K_2Cr_2O_7$

\+ .... light

DISCUSSION:

Ge, (111) rectangular bars with grown p-n junctions. Solution used to delineate the p-n junction.

REF: Walker, P — personal development, 1959

GE-0032
ETCH NAME: TIME:
TYPE: Acid, preferential TEMP:
COMPOSITION:
x .... HF
x .... HNO_3
x .... HAc or H_2O
DISCUSSION:

Ge, (111), (100), (110), and (211) wafers. By varying the ratio of HF:HNO_3 the preferential pit shape will vary on germanium surfaces.

REF: Faust, J W — *The Surface Chemistry of Metals and Semiconductors*, John Wiley & Sons, New York, 1950, 151

GE-0015b
ETCH NAME: White etch TIME:
TYPE: Acid, polish TEMP:
COMPOSITION:
20 .... HF
80 .... HNO_3
DISCUSSION:

Ge, (111) wafers used as substrates for germanium epitaxy growth. Solution was used to etch polish the epitaxy layers. (*Note:* See comment under GE-0029 with regard to using the term "White" to describe an etch.)

REF: Ibid.

GE-0033
ETCH NAME: TIME:
TYPE: Acid, junction TEMP:
COMPOSITION:
1 .... HF
3 .... HNO_3
DISCUSSION:

Ge, (111) wafers used to fabricate alloyed p-n junctions with indium. Etch was used to delineate the Ga/In p-n junction.

REF: Newman, R C — *J Appl Phys,* 27,845(1956)
GE-0286: Richards, J L — *J Appl Phys,* 34,3418(1963)

Ge, flash evaporated as a thin film on Ge and GaAs substrates as epitaxy layers. Both substrates were clean etched in 1HF:3HNO_3; rinsed in DI H_2O; then MeOH. CaF_2, (100) substrates were cleaved and used without further processing. All three substrates were used in a study of flash evaporated thin film semiconductors.

GE-0034
ETCH NAME: TIME:
TYPE: Acid, junction TEMP:
COMPOSITION:
5 .... HF
5 .... HNO_3
2 .... H_2O
DISCUSSION:

Ge, (111) wafers used to fabricate alloyed p-n junctions with indium. Etch was used to delineate to Ge/In p-n junction.

REF: Dreiner, A & Garnache, W R — *J Appl Phys,* 27,737(1956)

GE-0035
ETCH NAME: TIME:
TYPE: Acid, removal TEMP:
COMPOSITION:
x .... HF
x .... HNO_3
x .... H_2O
DISCUSSION:
Ge, (111) wafers used in a study of kink bands in the material. Used a dilute $HF:HNO_3$ etch to remove damage induced by mechanical lapping of germanium surfaces.
REF: Rosi,

GE-0220l
ETCH NAME: TIME:
TYPE: Acid, preferential TEMP:
COMPOSITION:
17 .... HF
13 .... HNO_3
50 .... H_2O
2 .... $AgNO_3$
DISCUSSION:
Ge, sphere. A single crystal sphere of germanium was etched to finite crystal form in this solution. Form is a cube, (100).
REF: Ellis, E C Jr — *J Appl Phys,* 25,1497(1954)

GE-0038
ETCH NAME: BRM TIME: 20 sec + 10 sec
TYPE: Halogen, polish lap TEMP: RT
COMPOSITION:
x .... 0.05% Br_2
x .... MeOH
DISCUSSION:
Ge, (111), (100), and (110) wafers used in a study of surface cleaning. Solution saturated lens paper used as a polishing pad. After initial polish of 20 sec, dilute by flooding with MeOH and continue polishing an additional 10 sec. Follow with rinse in BRM, then MeOH. Final rinse in BHF, then H_2O. Also etch polished with BRM on Si, GaP, GaAs, GaSb, InP, InAs, and InSb.
REF: Aspenes, D E & Studna, A A — *J Vac Sci Technol,* 20,488(1982)

GE-0031b
ETCH NAME: TIME: 10 min
TYPE: Electrolytic, junction TEMP: RT
COMPOSITION: ANODE: p-Ge
125 ml .... 0.5 *M* $3K_2S_2O_3.H_2O$ CATHODE: Pt strip
4 g .... I_2 POWER: 60 mA/cm^2
DISCUSSION:
Ge, (111) grown ingot with grown-in p-n junction. Rectangular bars cut and lapped to $^3/_{16} \times ^3/_{16} \times ^3/_4''$ in size. Etch is milky in color, and copper clips used to hold bars react with etch. Bars were biased during etching with p-type highly polished, and n-type dull grey. The p-type etched faster producing a sharp line at the junction.
REF: Ibid.

GE-0031c
ETCH NAME:
TYPE: Electrolytic, junction
COMPOSITION:
250 ml .... 0.5 *M* $Cr_2(SO_4)_3$
2 mg .... Ag
TIME: 15 min
TEMP: RT
ANODE: Pt strip
CATHODE: p-Ge
POWER: 6 mA/cm^2
DISCUSSION:

Ge, (111) grown ingots with grown-in p-n junction. Specimens were cut and lapped as a rectangular bar $^3/_{16} \times ^3/_{16} \times ^3/_4''$ size. Solution can be used with or without silver. Very slow etch on both p- and n-type areas. Junction appears as a finely etched line.

REF: Ibid.

GE-0212: Lesk, I A & Gonzales, R E — *J Electrochem Soc,* 105,469(1958)

Ge, (111) wafers fabricated as transistors. A selective electrolytic etch used to develop p-n junctions of both silicon and germanium.

GE-0211: Sullivan, M U & Eigler, J H — *J Electrochem Soc,* 103,132(1956)

Ge, (111) wafers electrolytically stream (jet) etched.

GE-0213: Esaki, L — *J Phys Soc Jpn,* 13,1281(1958)

Ge, wafers used to fabricate Esaki p-n junction devices.

GE-0031d
ETCH NAME:
TYPE: Electrolytic, junction
COMPOSITION:
100 ml .... HCl
50 ml .... H_2O
6 g .... $ZnCl_2$
TIME: 15 min
TEMP: RT
ANODE: n-Ge
CATHODE: Pt strip
POWER: 60 mA/cm^2
DISCUSSION:

Ge, (111) ingots with grown-in p-n junction. Rectangular bars were cut and lapped $^3/_{16} \times ^3/_{16} \times ^3/_4''$ in size. Zinc plates out on Pt cathode. Slight preferential etching with p-type more highly polished. Shows a distinct etched junction line.

REF: Ibid.

GE-0019b
ETCH NAME:
TYPE: Electrolytic, junction
COMPOSITION:
x .... KOH
x .... H_2O
TIME:
TEMP:
ANODE:
CATHODE:
POWER:
DISCUSSION:

Ge, (111) wafer with p-n junctions. Specimens lapped with 1800-grit alumina. Etch 5 min, RT with CP4 to remove lap damage and rinse with water, then alcohol. Develop junction by placing a drop of the KOH solution across junction and apply bias.

REF: Ibid.

GE-0254a: Pearson, G J & Feldman, W L — *J Phys Chem Solids,* 9,28(1959)

Ge, (111) wafers used to fabricate indium alloyed p-n junction diodes. Junctions developed in KOH solution, 3 min, 10 mA DC power, with Ge forward biased. This produced micro-pyramids on (110) surfaces. Also studied silicon p-n junctions.

GE-0018b
ETCH NAME:
TYPE: Electrolytic, polish
COMPOSITION:
x .... NaOH
x .... H_2O
TIME:
TEMP:
ANODE:
CATHODE:
POWER:
DISCUSSION:

Ge, (111) wafers used in a study of current flow across grain boundaries in n-type germanium. First, etch specimens in CP4 to remove lap damage. Electropolish in a sodium hydroxide solution shown.

REF: Ibid.

GE-0052: Jackson, R W — *J Appl Phys,* 27,309(1956)

Ge, (111) wafers with p-n junctions. A 10% solution used to develop junctions. The n-type region will etch; p-type does not etch.

GE-0178: John, H F & Longini, R L — Electrochem Soc Meet, Pittsburgh, PA, October 13, 1955

Ge, specimens used in a study of electrolytic etching. Alkalies are good for long-term etching, reproducibility, and reduction of surface recombination rates.

GE-0043
ETCH NAME: Water
TYPE: Acid, float-off
COMPOSITION:
x .... H_2O (desalted)
TIME:
TEMP:
DISCUSSION:

a-Ge, thin film grown on NaCl, (100) cleave substrates used in a study of amorphous germanium. After deposition, the a-Ge film was floated off the NaCl substrate with water. Under heat and electron impingement of a TEM microscope, there was no effect on the thin film. With a laser beam, and reaction observed under the TEM, the amorphous thin film crystallized into spherical particles.

REF: Pierrard, F et al — *Thin Solid Films,* 111,141(1984)

GE-0044
ETCH NAME:
TYPE: Electrolytic, polish
COMPOSITION:
x .... 0.1% KOH
x .... 5—25% glycerin
TIME:
TEMP:
ANODE:
CATHODE:
POWER: 0.5 A/in^2
DISCUSSION:

Ge, (111), n-type, 0.004—40 Ω cm resistivity wafers. Solution shown used to polish surface with a removal rate of 0.67 mils/min.

REF: Klein, R C et al — *RCA Rev,* 14,23(1969)

GE-0147b
ETCH NAME:
TYPE: Electrolytic, polish
COMPOSITION:
x .... 0.01—10% KOH
x .... 0.1 *N* NaCl
TIME:
TEMP:
ANODE:
CATHODE:
POWER: $^1/_2$—1 mA/cm^2

DISCUSSION:

Ge, (111) wafers. Etch with CP4 to remove lap damage, then electropolish in the above solution. Also used: $InSO_4$; H_2SO_4, or HCl as the polishing electrolytes.

REF: Ibid.

GE-0046

ETCH NAME: Potassium fluoride
TYPE: Electrolytic, polish
COMPOSITION:
x .... 10^{-6}—0.1 *N* KF

TIME:
TEMP:
ANODE:
CATHODE:
POWER:

DISCUSSION:

Ge, (111), p- and n-type wafers used in an etching study. The following electrolytes were evaluated: KCl; KI; $NaNO_3$; CaCl; $BaCl_2$, and $LaCl_3$ with the addition of O_2. Below pH 6, there is little or no change in etch rate. Above pH 6, etch rate increases. Rate also increases with acid concentration increase. No observed difference between p- or n-type material with or without illumination.

REF: Harvey, W & Gatos, H C — *J Appl Phys*, 29, 1267(1958)

GE-0047

ETCH NAME: DE-100
TYPE: Ionized gas, RF plasma removal
COMPOSITION:
x *DE-100

TIME:
TEMP: RT to 50°C
GAS FLOW: 300 cc/min
PRESSURE: 1.1 Torr
POWER: 200 W

*CF4:8.5% O_2

DISCUSSION:

Ge, thin films evaporated on silicon, aluminum, alumina, gallium arsenide, and sapphire substrates. Ge film thickness was 250 to 6000 Å. Germanium used as a buffer layer in photo resist applications and as a diffusion drive-in source. After processing, remaining Ge removed with DE-100. Also used for removal of SiO_2; Si_3N_4; or Si, Mo, Ta, TaN, and W films.

REF: Scientific Gas Company Bull, 1972

GE-0048: Walker, P — personal development, 1980

Ge, thin films evaporated as part of a photo resist patterning mask. As evaporated Ge thin film, only, used for drive-in GaAs:Cr,(100) (SI) wafers used in fabricating LN-FETs. Residual germanium removed with DE-100 or RF plasma N_2.

GE-0049

ETCH NAME: Argon
TYPE: Ionized gas, cleaning
COMPOSITION:
x .... Ar^+ ions

TIME:
TEMP:
GAS FLOW:
PRESSURE
POWER: 500 eV

DISCUSSION:

Ge, (100), wafers cut within 1° of plane, used for silver thin film deposition study. After mechanical polish of wafer, clean with Ar^+ ion bombardment, and follow with 450°C anneal. Repeat until no carbon contamination remains. Silver deposited as a (110) oriented single crystal thin film.

REF: Miller, T et al — *Phys Rev*, B30,520(1984)

GE-232: Farnsworth, H E et al — *J Appl Phys*, 29,1150(1958)

Ge specimens. Application of ion bonbardment cleaning as an improved method of determined by low-energy electron diffraction. Also evaluated Si, Ti, and Ni.

GE-0063
ETCH NAME:
TYPE: Electrolytic, shaping
COMPOSITION:
15 .... HF
25 .... HNO_3
15 .... HAc
TIME:
TEMP:
ANODE:
CATHODE:
POWER: 20—60 V
DISCUSSION:

Ge, single crystal wire filaments etch formed into tips in a study of surface structure and migration using a field emission microscope (FEM). Different resistivity materials of 1.1, 2.0, and 2.2 Ω cm, n-type were used. During tip forming, stir solution, and illuminate with intense light to prevent "sword-shaping" of tip. For final shaping, reduce power to zero and use as a normal chemical etch. Quench in MeOH without exposure to air to prevent formation of germanium oxides. Also used hydrogen in a vacuum system to remove surface contamination, and used oxygen at 800 K, 20—30 sec to sharpen tips — the latter is difficult to control and is erratic. (*Note:* Solution is CP4A, e.g., CP4 without bromine.)
REF: Arthur, J R Jr — *J Phys Chem Solids,* 25,583(1964)

GE-0008b
ETCH NAME:
TYPE: Acid, orientation
COMPOSITION:
1 .... HF
1 .... H_2O_2
3 .... H_2O
TIME:
TEMP: RT
DISCUSSION:

Ge, (111) wafers, and other orientations. Solution used as a light etch for optical orientation (OO) of surfaces. (*Note:* Also called light figure orientation (LFO) technique.)
REF: Tyler, W W & Dash, W C — *J Appl Phys,* 28,1221(1957)

GE-0008c
ETCH NAME:
TYPE: Acid, dislocation
COMPOSITION:
2 .... HF
4 .... HNO_3
15 .... HAc
TIME: 20—60 sec
TEMP: RT
DISCUSSION:

Ge, (111) wafers, and other orientations. Wafers were lithium diffused, and then etched in this solution shown to develop dislocation arrays generated and decorated by lithium.
REF: Ibid.

GE-0217
ETCH NAME:
TYPE: Acid, junction
COMPOSITION:
1 .... HF
5 .... HNO_3, red fuming
TIME:
TEMP: RT

DISCUSSION:

Ge, (111) wafers fabricated as p-n junction diodes. Junctions etch cleaned and developed with the solution shown. (*Note:* HNO_3 concentration is 74% for red, fuming.)

REF: Farren, T — personal communication, 1956

GE-0218

ETCH NAME: Iodine etch — TIME: 5 min

TYPE: Acid, polish — TEMP: RT

COMPOSITION:

5 ml .... HF
10 ml .... HNO_3
11 ml .... HAc
30 mg .... I_2

DISCUSSION:

Ge, wafers used as substrates for epitaxy growth of PbS as heterojunctions. The Ge substrates were polished in the solution shown. Use agitation during etch period with specimens immersed in etchant.

REF: Davis, J L & Norr, M K — *J Appl Phys,* 37,1670(1966)

GE-0150a

ETCH NAME: Hydrogen peroxide — TIME: 2—4 h

TYPE: Acid, preferential — TEMP: RT

COMPOSITION:

100 ml .... H_2O_2

DISCUSSION:

Ge, single crystal hemispheres, lap formed or grown by a modified Verneiul process, and used in a preferential etching study to observe etch patterns, type etch pits, and etch rates. The amount of solution used was 50—70 ml unless otherwise shown. Also used on silicon hemispheres. Before preferential etching, etch with CP4 to remove residual mechanical lap damage.

REF: Holmes, P J — *Acta Metall,* 7,283(1959)

GE-0150b

ETCH NAME: — TIME: 3 h

TYPE: Acid, preferential — TEMP: RT

COMPOSITION:

1 .... HF
250 .... H_2O_2

DISCUSSION:

Ge, hemispheres used in a preferential etching study.

REF: Ibid.

GE-0150c

ETCH NAME: — TIME: 1 h

TYPE: Acid, preferential — TEMP: RT

COMPOSITION:

1 .... HF
100 .... H_2O_2

DISCUSSION:

Ge, hemispheres used in a preferential etching study.

REF: Ibid.

GE-0150d
ETCH NAME:
TYPE: Acid, preferential
TIME: 30 min
TEMP: RT
COMPOSITION:
1 .... HF
50 .... H_2O_2
DISCUSSION:
Ge, hemispheres used in a preferential etching study.
REF: Ibid.

GE-0150e
ETCH NAME:
TYPE: Acid, preferential
TIME: 10 min
TEMP: RT
COMPOSITION:
1 .... HF
8 .... H_2O_2
DISCUSSION:
Ge, hemispheres used in a preferential etch study.
REF: Ibid.

GE-0150f
ETCH NAME:
TYPE: Acid, preferential
TIME: 6 min
TEMP: RT
COMPOSITION:
1 .... HF
2 .... H_2O_2
DISCUSSION:
Ge, hemispheres used in a preferential etching study.
REF: Ibid.

GE-0150g
ETCH NAME:
TYPE: Acid, preferential
TIME: 3 min
TEMP: RT
COMPOSITION:
1 .... HF
1 .... H_2O_2
DISCUSSION:
Ge, hemispheres used in a preferential etching study.
REF: Ibid.

GE-0150h
ETCH NAME:
TYPE: Acid, preferential
TIME: 20 min
TEMP: RT
COMPOSITION:
2 .... HF
1 .... H_2O_2
DISCUSSION:
Ge, hemispheres used in a preferential etching study.
REF: Ibid.

GE-0150i
ETCH NAME: Superoxol, variety
TIME: 2 min
TYPE: Acid, preferential
TEMP: RT
COMPOSITION:
1 .... HF
1 .... H_2O_2
x .... H_2O
DISCUSSION:
Ge, hemispheres. During the etch period there was a progressive addition of water. Solution used in a preferential etching study.
REF: Ibid.

GE-0150j
ETCH NAME: Superoxol #2
TIME: 2 min
TYPE: Acid, preferential
TEMP: RT
COMPOSITION:
1 .... HF
1 .... H_2O_2
4 .... H_2O
DISCUSSION:
Ge, hemispheres used in a preferential etch study.
REF: Ibid.

GE-0150k
ETCH NAME: Superoxol #2, dilute
TIME: 30 min
TYPE: Acid, preferential
TEMP: RT
COMPOSITION:
1 .... Superoxol #2
3 .... H_2O
DISCUSSION:
Ge, hemispheres used in a preferential etch study.
REF: Ibid.

GE-0150l
ETCH NAME: Superoxol #2, dilute
TIME: 2 h
TYPE: Acid, preferential
TEMP: RT
COMPOSITION:
1 .... Superoxol #2
6 .... H_2O
DISCUSSION:
Ge, hemispheres used in a preferential etch study.
REF: Ibid.

GE-0150m
ETCH NAME: Superoxol #2, dilute
TIME: 3 h
TYPE: Acid, preferential
TEMP: RT
COMPOSITION:
1 .... Superoxol #2
8 .... H_2O
DISCUSSION:
Ge, hemispheres used in a preferential etching study.
REF: Ibid.

GE-0150n
ETCH NAME: Superoxol #2, dilute
TYPE: Acid, preferential
COMPOSITION:
1 .... Superoxol #2
50 .... H_2O
TIME: 48 h
TEMP: RT
DISCUSSION:
Ge, hemispheres used in a preferential etching study.
REF: Ibid.

GE-0150o
ETCH NAME: Ellis #5
TYPE: Acid, preferential
COMPOSITION:
20 .... HF
3 .... H_2O_2
12 .... H_2O
TIME: 5 min
TEMP: RT
DISCUSSION:
Ge, hemispheres used in a preferential etching study.
REF: Ibid.

GE-0150p
ETCH NAME: WAg etch
TYPE: Acid, preferential
COMPOSITION:
2 .... HF
1 .... HNO_3
2—5% .... $AgNO_3$, sat. sol.
TIME: 1 min
TEMP: RT
DISCUSSION:
Ge, hemispheres used in a preferential etching study.
REF: Ibid.

GE-0150q
ETCH NAME: Ellis #1
TYPE: Acid, preferential
COMPOSITION:
2 .... HF
1 .... HNO_3
1 .... 10% $Cu(NO_3)_2$, sat. sol.
TIME: 5 min
TEMP: RT
DISCUSSION:
Ge, hemispheres used in a preferential etching study.
REF: Ibid.

GE-0150r
ETCH NAME: Ellis #7
TYPE: Acid, preferential
COMPOSITION:
"A" x .... 0.8 *N* KOH
x .... Cl_2 for pH 8—9
"B" x .... 5% KOH
Mix: 15 ml "A": 60 ml "B".
TIME: 3 h
TEMP: RT

DISCUSSION:

Ge, hemispheres used in a preferential etching study.

REF: Ibid.

GE-0150s

ETCH NAME: Ferric-cyanide etch — TIME: 2 min

TYPE: Acid, preferential — TEMP: RT

COMPOSITION:

6 g .... KOH

4 g .... $K_3Fe(CN)_6$

50 ml .... H_2O

DISCUSSION:

Ge, hemispheres used in a preferential etching study.

REF: Ibid.

GE-0060d

ETCH NAME: Ferric-cyanide etch — TIME: 1—10 min

TYPE: Acid, preferential — TEMP: 80°C

COMPOSITION:

13.7 g .... KOH

9.7 g .... $K_3Fe(CN)_6$

100 ml .... H_2O

DISCUSSION:

Ge, (111) and (100) wafers. Solution used as a preferential etch.

REF: Ibid.

GE-0060e

ETCH NAME: (100) Etch — TIME:

TYPE: Acid, preferential — TEMP:

COMPOSITION:

70 ml .... HF

25 ml .... HNO_3

50 ml .... H_2O

3 g .... $Cu(NO_3)_2$

DISCUSSION:

Ge, (100) wafers. This is a general preferential etch, but named for its specific use on the (100) surface in this instance. See SI-0038a for the similar ''Copper Dislocation Etch''.

REF: Ibid.

GE-0070d: Pergola, G D & Sette, D — *Acta Metall,* 24,499(1955)

Ge, (111) wafers used in a study of the effects of different dislocation etches on germanium. Other etches evaluated were CP4, superoxol, and WAg. See SI-0038a for the ''Copper Dislocation Etch'' used.

GE-0060f

ETCH NAME: Peroxide etch — TIME: 1 min

TYPE: Acid, preferential — TEMP:

COMPOSITION:

1 ml .... HF

1 ml .... H_2O_2

1 ml .... HAc

DISCUSSION:

Ge, (100) wafers. Solution used as a preferential etch. (*Note:* In chemistry, the standard 30% concentration of hydrogen peroxide also is called a peroxide etch, when used alone.)
REF: Ibid.

GE-0008c
ETCH NAME:
TYPE: Acid, junction
COMPOSITION:
2 .... HF
4 .... HNO_3
15 .... HAc
TIME: 20—60 sec
TEMP: RT

DISCUSSION:

Ge, (111) wafers with lithium diffused p-n junctions. Solution used to etch develop the p-n junction.
REF: Tyler, W W & Dash, W C — *J Appl Phys,* 28,1221(1957)

GE-0008d
ETCH NAME:
TYPE: Acid, optical orientation
COMPOSITION:
1 .... HF
1 .... H_2O_2
3 .... H_2O
TIME:
TEMP:

DISCUSSION:

Ge, (111) wafers. Solution used as a light etch for optical orientation of the (111) surface.
REF: Ibid.

GE-0220m
ETCH NAME:
TYPE: Acid, polish
COMPOSITION:
1 .... HF
4 .... HNO_3
2 .... H_2O_2
TIME: 5—30 min
TEMP: RT

DISCUSSION:

Ge, (111) wafers, p-type, 4 Ω cm resistivity used in a general etch development study. Wafers were used after lapping with 600-grit SiC abrasive. The etch solution shown will remove the first mil of material in $1^3/_4$ min; thereafter, the rate is constant at 1 mil/min. The (111) n-type, 10 Ω cm resistivity material evaluated etched at the same constant rate as the p-type material after the first 2 min of etching.
REF: Ibid.

GE-0220n
ETCH NAME:
TYPE: Electrolytic, junction
COMPOSITION:
(1) 250 ml .... 0.5 *M* $FeSO_4.7H_2O$
(2) 250 ml .... 0.5 *M* $ZnSO_4.7H_2O$
(3) 250 ml .... 0.5 *M* $ZnSO_4.7H_2O$
(4) 250 ml .... 0.5 *M* $3K_2S_2O_3.H_2O$
TIME: 15 min
TEMP: RT
ANODE: Pt (Pb & Cu)
CATHODE: p-type Ge
POWER: 6 mA/cm_2

DISCUSSION:

Ge, (111) ingots with grown-in p-n junction. Specimens cut as bars in an etching study of the material. Both lead and copper were evaluated as anodes but platinum gave best results. All these solutions stain the p-region with good delineation of the junction.

REF: Ibid.

GE-0220o

ETCH NAME: TIME: 5—10 min

TYPE: Acid, preferential TEMP: Boiling

COMPOSITION:

175 ml .... 15% KOH

45 g .... $(NH_4)_2S_2O_8$

DISCUSSION:

Ge, (111), (100), (110), and (211) wafers used in an etch development study. The solution develops the typical etch pit patterns associated with each surface orientation within the first 2—3 min, and extended etching enlarges the pits. Solution used as both a light figure orientation etch, and as a general defect and dislocation density etch.

REF: Ibid.

GE-0220p

ETCH NAME: TIME: 1—20 min

TYPE: Acid, polish TEMP: RT

COMPOSITION:

50 ml .... CP4 (fresh)

3 mg .... $Fe(NO_3)_2.2H_2O$

DISCUSSION:

Ge, (111) wafers, p-type, 4 Ω cm resistivity used in a general etch development study. Surfaces were mechanically lapped with 600-grit SiC, before using the solution shown as polish etch. Removal rate was about 1 mil in the first 3 min, and about 1 mil/$1^1/_2$ min thereafter.

REF: Ibid.

GE-0056

ETCH NAME: TIME: 5—10 min

TYPE: Alkali, preferential TEMP: Boiling

COMPOSITION:

x .... KOH

x .... $K_3Fe(CN)_6$

DISCUSSION:

Ge, (111) wafers used in a study of thermally induced glide. Solution used to delineate defects.

REF: Cummerow, R L — *J Appl Phys,* 30,932(1959)

GE-0134: Alekseeva, V G & Eliseev, P O — *Fiz Tverd Tela,* 1,415(1959)

Ge, specimens used in a study of the effects of bismuth on dislocation density. Solution used to develop dislocations.

GE-0135: Dragoun, Z — *Czech J Phys,* 8,600(1958) (in Russian)

Ge, specimens. A review of dislocation etchants for germanium, both selective and nonselective.

GE-0144c: Faust, J W & John, H F — *J Electrochem Soc,* 108,825(1961)

Ge, (111) dendrites. A ferrocyanide solution used to develop triangle dislocation pits. Also used WAg and CP4 etches.

GE-0057
ETCH NAME: Water
TYPE: Acid, float-off
COMPOSITION:
x .... H_2O
TIME:
TEMP: RT
DISCUSSION:

a-Ge thin films deposited on NaCl, (100) cleaved substrates. The films were floated-off the substrate by liquefying the NaCl at the a-Ge interface for SEM and TEM study of film morphology.

REF: Nakhodkin, N G et al — *Thin Solid Films,* 112,267(1984)

GE-0058
ETCH NAME:
TYPE: Acid, selective photo
COMPOSITION:
2 .... HF
2 .... H_2O_2
5 .... H_2O
TIME: 3—10 sec
TEMP: RT
DISCUSSION:

Ge, (111), (110), and (211) wafers cut from CZ grown ingots. Wafers were used saw cut, then mechanically lapped and polished, or cleaved. Wafers used in a study of growth habit. During etching, illuminate with an IR Lamp. Etch used to develop ''Valley traces Type II'' striations that show variable resistivity. (*Note:* Concentric ring increase of resistivity from wafer center has been observed in large 2″ diameter (111) silicon wafers.)

REF: Lu, Y C & Bauser, E — *J Cryst Growth,* 71,305(1985)

GE-0059a
ETCH NAME:
TYPE: Acid, structure
COMPOSITION:
1 .... HF
5 .... HNO_3
TIME:
TEMP:
DISCUSSION:

Ge, thin films evaporated on GaAs:Cr (SI) substrates in a diode device study. Photo resist techniques used to form pattern dot structures which were then etched as pedestals (mesas).

REF: Papazian, S A & Reisman, A — *J Electrochem Soc,* 111,961(1968)

GE-0059b
ETCH NAME:
TYPE: Acid, cleaning
COMPOSITION:
1 .... NaOCl
3 .... H_2O
TIME:
TEMP:
DISCUSSION:

Ge, thin films evaporated on GaAs:Cr (SI) substrates in a diode device study. Prior to mesa formation the epitaxy germanium surfaces were cleaned in the solution shown.

REF: Ibid.

GE-0060g
ETCH NAME: WAg etch TIME:
TYPE: Acid, preferential TEMP:
COMPOSITION:
4 ml .... HF
2 ml .... HNO_3
4 ml .... H_2O
0.2 g .... $AgNO_3$
DISCUSSION:

Ge, (111) wafers. This is a general preferential etch on both germanium and silicon. (*Note:* Full name is the Westinghouse Silver Etch.)

REF: Ibid.

GE-0066: Grubel, R O — *Metallurgy of Elemental and Compound Semiconductors,* Vol 12, Interscience, New York, 1962, 139

Ge, grown as (111) oriented ribbon crystal dendrites. WAg solution used to develop defect and dislocation structure.

GE-0070c: Pergola, C D & Settes, D — *Alta Freq,* 24,499(1955)

Ge, (111) wafers. An evaluation of different etch solutions for developing dislocations. Other etchants were CP4, superoxol, and copper nitrate etch.

GE-0075a: Holmes, P J — *Phys Rev,* 119,131(1960)

Ge, (111) oriented ribbon crystal dendrites. A dislocation etching study of dendritic material. Author says that the triangle pit direction depends on the type solution used. (*Note:* The (111) ribbon crystal dendrites always have a twinned zone down the center of the dendrite with 180° rotation to either side as a positive (+) (111), and a negative (−) (111) surface. The dislocation pits also are rotated, as they are representative of the positive and negative tetrahedron, (111) crystallographic forms common to (111) type surfaces. This dislocation pit structural appearance is true without regard to the type preferential etch solution used.)

GE-0084: Wynne, R H & Goldberg, C — *J Met,* 8,436(1953)

Ge, (111), (100), and (110) wafers. The WAg etch was used as an optical orientation figure etch.

GE-0144c: Faust, J W Jr & John, H T — *J Electrochem Soc,* 108,825(1961)

Ge, (111) ribbon crysal dendrites. Etch used to develop dislocations.

GE-0188: Bennett, A I & Longini, R L — *Phys Rev,* 116,53(1959)

Ge, (111) ribbon crystal dendrites. Solution used to develop dislocations.

GE-0225c: Guatier, K & Kerecman, A J — *Rev Sci Instrum,* 34,108(1963)

Ge, (111) wafers used as substrates for Ge epitaxy growth used in a study of preferential light figure etching. The WAg etch produces very sharp etch pits and hillocks on polished surfaces. Also used hot HCl vapors, and $GeCl_4$ to develop similar structure.

GE-0061a
ETCH NAME: TIME:
TYPE: Acid, polish TEMP:
COMPOSITION:
4 .... HF
2 .... HNO_3
4 .... 5% $Cu(NO_3)_2$
DISCUSSION:

Ge, (111) wafers used in a study of the effect of ionic species on the etching of germanium. Solution used as a polishing etch. Without copper nitrate the solution is preferential, developing pyramidal type pits.

REF: Faust, J W Jr — *J Electrochem Soc,* 112,114(1965)

GE-0061b
ETCH NAME:
TYPE: Acid, preferential
COMPOSITION:
4 .... HF
2 .... HNO_3
4 .... 5% $Ag(NO_3)_2$
DISCUSSION:

Ge, (111) wafers. Solution develops triangular etch pits.

REF: Ibid.

TIME:
TEMP:

GE-0036
ETCH NAME:
TYPE: Acid, polish
COMPOSITION:
4 .... HF
7 .... HNO_3
2 .... H_2O
DISCUSSION:

Ge, (111) wafers used in a study of slow surface relaxation. Specimens were etched in the solution shown and then held for varied periods of time before being treated with water.

REF: Pilkuhn, M H — *J Appl Phys,* 34,3302(1963)

TIME:
TEMP:

GE-0170
ETCH NAME: Heat
TYPE: Thermal, cleaning
COMPOSITION:
x .... heat
DISCUSSION:

Ge, (100) very thin films grown by PECVD on NaCl, (100) substrates. The as grown films were p-type, 0.11 Ω cm resistivity with 29 ppm sodium contamination. After vacuum heating contamination was reduced to 3 ppm Na and resistivity increased to 7 Ω cm, p-type. Specimens of 6-cm size were removed from the NaCl by heat melting away the substrate; whereas specimens of 1-cm size were sheared away by cleaving at room temperature.

REF: Outlaw, R A et al — *J Vac Sci Technol,* A3,692(1985)

GE-0142b: Pugh, E H & Samuels, L E — *J Electrochem Soc,* 108,1043(1961)

Ge, (111) wafers were angled lapped to 5°43′ and used in a metallographic study of the effects of abrasion. After abrading surfaces, Billig's etch (GE-0142a) was used to develop damage structure. It showed cleavage cracks, but no dislocations. After heat at 500°C for 15 min, dislocation arrays were developed.

GE-0146: Wolff, G A et al — Electrochem Soc Meet, Pittsburgh, PA, Oct 13, 1955

GE, (111), (113), and (110) wafers used in developing light figure orientation etches. Thermal etching was used to develop etch pit structure. Also etched silicon in $HF:I_2$ for etch pits.

GE-0165: English, A C — *J Appl Phys,* 31,1498(1960)

Ge, specimens and ingots. Material was heated to 700°C in contact with iron or nickel plates, which developed growth stria. Indium was alloyed into wafers, then dissolved with boiling HCl to study the alloyed pit structure.

GE-0232: Willis, R C et al — *J Appl Phys,* 29,1725(1958)

Ge ingots. Rapid pulling of Czochralski (CZ) ingots causes a high dislocation content,

TIME:
TEMP: 800°C

and the seed also can introduce dislocations if it has an originally high dislocation density. Slow cooling will reduce pit density to random points, only. Thermal shock produces high density, star-like patterns in the ⟨110⟩ directions.

GE-0233: Wagner, R S — *J Appl Phys,* 29,1678(1958)

Ge ingots. In Czochralski (CZ) growth a too rapid approach of the seed crystal into the molten germanium melt can initiate dislocations due to plastic deformation of the seed. Too rapid cooling (thermal shock) also will create dislocations.

GE-0064

ETCH NAME: Hydrofluoric acid — TIME:

TYPE: Acid, cover — TEMP: RT

COMPOSITION:

x .... HF

DISCUSSION:

Ge, single crystal specimens subject to fracture in a study of the effect of environment on fracture behavior. The study showed that there is a difference between fracturing in air or under HF. (*Note:* Similar work has been done on silicon.)

REF: Johnston, T L et al — *Acta Metall,* 7,713(1959)

GE-0065c

ETCH NAME: Camp #2 (superoxol) — TIME:

TYPE: Acid, preferential — TEMP: RT

COMPOSITION:

1 .... HF

1 .... H_2O_2

4 .... H_2O

DISCUSSION:

Ge, (111) wafers, and other orientations. Wafers used in a study of etch rates on germanium. (*Note:* Several of the Camp etches are abbreviated: "CP", such as CP4.)

REF: Camp, P R — *J Electrochem Soc,* 102,1147(1955)

GE-0007b: Camp, P R — *J Electrochem Soc,* 102,586(1955)

Ge, oriented wafers and single crystal spheres used in a study of etch rates. Etch referred to as Camp #2 (superoxol). The sphere finite crystal form (FCF) was a dodecahedron, (110).

GE-0065b

ETCH NAME: Camp #3 — TIME:

TYPE: Acid, preferential — TEMP: RT

COMPOSITION:

56 ml .... HF

56 ml .... HNO_3

12.5 ml .... H_2O

DISCUSSION:

Ge, (111) wafers. See discussion under GE-0065a.

REF: Ibid.

GE-0077

ETCH NAME: — TIME:

TYPE: Acid, stress — TEMP:

COMPOSITION: — PRESSURE: 7 kg/mm^2 to 47 kg/mm^2 (in air)

x .... HF

x .... HNO_3

DISCUSSION:

Ge, (111) wafers etched in various mixtures of HF:HNO_3 under varied pressures in a study of the stress fracture of germanium under different environmental conditions. Effect varied with both pressure and specific etchant mixtures.

REF: Breidt, P et al — *J Appl Phys,* 29,226(1958)

GE-0078

ETCH NAME: TIME:

TYPE: Acid, polish TEMP: RT

COMPOSITION:

(1) x .... HF
x .... HNO_3
x .... H_2O

(2) x .... HF
x .... H_2O_2
x .... H_2O
x .... $Na(NO_3)_2$ + Ag

(3) x .... H_2SO_4

DISCUSSION:

Ge, (111) wafers used in a study of etching and polishing surfaces. Etch solutions of type (1) were DI water rinsed; type (2), rinsed in KCN to remove Ag and then in DI water. Type (3) sulfuric acid was used at 50°C and DI water rinsed.

REF: Geist, D & Preuss, E — *Z Angew Phys,* 9,526(1957)

GE-0079

ETCH NAME: TIME:

TYPE: Acid, polish TEMP:

COMPOSITION:

3 .... HF
7 .... HNO_3

DISCUSSION:

Ge, (111) wafers used in a study of infra-red properties of gold diffused germanium. Solution used to polish samples before gold diffusion. Follow etch by rinsing in double-distilled water.

REF: Johnson, L & Levinstein, H — *Phys Rev,* 117,1191(1960)

GE-0081

ETCH NAME: Chlorine TIME:

TYPE: Electrolytic, preferential TEMP: 200—300°C

COMPOSITION: ANODE: Ge

x .... Cl_2 CATHODE: Graphite (2)

POWER:

DISCUSSION:

Ge, (111) wafers. Two graphite electrodes were used, one with a hole in the side. Chlorine forms GeCl and the vapor, and can be used as a surface polishing system or to fabricate controlled etch pits by jet streaming chlorine through the graphite electrode.

REF: Seiler, K O — U.S. Patent 2,744,000, May 1, 1956

GE-0082

ETCH NAME: TIME:

TYPE: Acid, polish TEMP:

COMPOSITION:

(1) x .... HNO_3

(2) x .... HF
x .... HNO_3

(3) x .... H_2O_2

DISCUSSION:

Ge, (111) wafers used in a study of optical and magnetic surface effects. All three solutions are oxidizers and, after etching, show slow decay of surface conductance. If etching is followed by an HF rinse, there is no decay.

REF: Bray, R & Cunningham, R W — *J Phys Chem Solids,* 8,99(1959)

GE-0083a

ETCH NAME: TIME:

TYPE: Acid, preferential TEMP:

COMPOSITION:

"A" 160 ml .... HF "B" x .... H_2O

200 ml .... HNO_3

80 ml .... HAc

0.32 g .... I_2

Mix: 1 "A" to 1.5 "B"

DISCUSSION:

Ge, (111) wafers. After mechanical lapping, CP4 was used to remove work damage. Solution shown develops spiral type etch pits on both (111) and (100) surfaces.

REF: Rhodes, E G et al — *J Electron,* 3,403(1957)

GE-0235: Hogarth, C A & Bell, R L — *J Electron,* 3,455(1957)

Ge and Si wafers. A study of the anisotropic diffusion lengths due to the presence of parallel arrays of edge dislocations.

GE-0085

ETCH NAME: TIME:

TYPE: Acid, preferential TEMP:

COMPOSITION:

1 .... HF

4 .... HNO_3

DISCUSSION:

Ge, (111) wafers used in a study of anisotropic diffusion lengths in germanium and silicon containing parallel arrays of edge dislocations. Solution was used to develop dislocations on both materials after bending under heat to develop defects.

REF: Hogarth, C A & Bell, R L — *J Electron,* 3,455(1957)

GE-0281: Stolwijk, M A et al — *J Appl Phys,* 57,5211(1985)

Ge, (100) wafers cut from Czochralski (CZ) ingots. After chem/mech polish and degreasing, copper was evaporated, then diffused. Solution shown was used to develop decoration structure in a study of diffusion and solubility of copper in germanium.

GE-0046b

ETCH NAME: TIME:

TYPE: Electrolytic, preferential TEMP:

COMPOSITION: ANODE: Ge

x .... 0.1 *N* KCl CATHODE:

x .... 10^{-3} *M* GeO_2 POWER:

DISCUSSION:

Ge, (111) wafers. Study involved electrode potential and the effects of illumination. Results showed that the amount of oxygen dissolved in solution effects etching results.

REF: Harvey, W & Gatos, H C — *J Appl Phys,* 29,1267(1958)

GE-0087
ETCH NAME: Potassium hydroxide
TYPE: Alkali, preferential
COMPOSITION:
x .... 0.5% KOH

TIME:
TEMP:
ANODE: Ge
CATHODE:
POWER: 5.5 A/cm^2

DISCUSSION:

Ge, (111) n-type wafers. Solution was applied at 5 psi through a 75 mm diameter tube without illumination. Triangle etch figures were developed that were arrayed in a ⟨211⟩ direction. (*Note:* ⟨211⟩ lineation used to be a common defect in Czochralski (CZ) grown ingots in the early 1950s.)

REF: Oberly, J J — *Acta Metall,* 35,122(1957)

GE-0088a
ETCH NAME: Sulfuric acid
TYPE: Electrolytic, polish
COMPOSITION:
x .... 0.1 *N* H_2SO_4

TIME:
TEMP:
ANODE: Ge
CATHODE:
POWER: 0.3 mA/cm^2

DISCUSSION:

Ge, (111) wafers used in a study of the effects of aqueous electrolytic solutions on germanium.

REF: Turner, D R — *Acta Metall,* 103,252(1956)

GE-0088b
ETCH NAME: Potassium hydroxide
TYPE: Electrolytic, polish
COMPOSITION:
x .... 1 *N* KOH

TIME:
TEMP:
ANODE: Ge
CATHODE:
POWER: 0.3 mA/cm^2

DISCUSSION:

Ge, (111) wafers used in a study of the anodic behavior of germanium in electrolytic aqueous solutions.

REF: Ibid.

GE-0089
ETCH NAME:
TYPE: Alkali, contamination
COMPOSITION:
0.4 ml .... 1% NaOH
0.1 ml .... 7% $^{24}NaCO_3$

TIME:
TEMP:

DISCUSSION:

Ge, single crystal specimens used in a study of adsorption of sodium ions by germanium surfaces. Solution used to introduce contamination.

REF: Robbins, R C — *Acta Metall,* 103,194(1956)

GE-0090
ETCH NAME:
TYPE: Acid, cleaning
COMPOSITION:
1 .... HF
10 .... HNO_3

TIME: 11.5 min
TEMP: RT

DISCUSSION:

Ge, specimens used in a study of Auger electron ejection from silicon and germanium surfaces from the noble gases. The etching and cleaning sequence was (1) etch in solution shown; (2) dip in hot xylene; (3) rinse in acetone; (4) soak in HF, RT, 1 min; (5) clean in $1HNO_3$:$1H_2O$, RT, 30 sec; (6) boil in DI water, 5 min, and (7) air dry.

REF: Hagstrum, H D — *Phys Rev,* 199,940(1960)

GE-0151a

ETCH NAME: TIME: 15—45 min

TYPE: Alkali, preferential TEMP: Boiling

COMPOSITION:

250 ml .... 30% KOH (NaOH)

30—40 g .... I_2

DISCUSSION:

Ge, (111), (110), (100), (211) wafers, and single crystal spheres. Solution developed as a light figure orientation etch and general preferential etch for germanium. Used on n-p-n diffused transistors as a step-etch to form transistor vertical structure. The finite crystal sphere form was a cube, (100).

REF: Walker, P — personal development, 1958—1959

GE-0151b

ETCH NAME: TIME: 10—20 min

TYPE: Alkali, preferential TEMP: Boiling

COMPOSITION:

125 ml .... 30% KOH (NaOH)

50 ml .... Br_2

DISCUSSION:

Ge, (111), (100), (110), (211) wafers, and single crystal spheres. Solution developed as a light figure orientation etch. Mixture is more difficult to control due to bromine evaporation than is GE-0151a. The finite crystal form of the sphere was a cube, (100).

REF: Ibid.

GE-0124

ETCH NAME: SR4 TIME:

TYPE: Acid, polish TEMP:

COMPOSITION:

3 .... HF

2 .... HNO_3

1 .... HAc

DISCUSSION:

Ge, (111) wafers used in a study of the variations in surface conductivity of both germanium and silicon. Solution used as a polish etch on both materials.

REF: Ioselevich, M L & Fistul′, V I — *Sov Phys (Solid State),* 3,822(1960)

GE-0102b

ETCH NAME: Potassium cyanide TIME:

TYPE: Acid, cleaning TEMP:

COMPOSITION:

x .... KCN

DISCUSSION:

Ge, (111) wafers. The surface was soaked in solution shown to clean prior to etch polishing with CP4 (GE-0102a). Material used in a plastic deformation study.

REF: Ibid.

GE-0094b: Ibid.

Ge, (111) wafers. Etch polish with CP4, then soak in KCN. See GE-0094a.

GE-0119b: Ibid.

Ge, (111) wafers. Etch in CP4, soak in HNO_3, rinse in DI water; then soak in KCN, and DI water rinse. See GE-0119a.

GE-0159b: Hopkins, E L & Clarke, E N — *Phys Rev,* 100,1789(1955)

Ge specimens. Soak in 10% KCN, 1 h to remove metallic ions. Then rinse five times in DI water. See GE-0159a.

GE-0179: Tweet, A G — *Phys Rev,* 99,1245(1955)

Ge, (111) specimens used in a study of electrical properties of plastically deformed germanium. Solution used to clean and etch surfaces.

GE-0142: Wolsky, S F — *J Appl Phys,* 29,1132(1958)

Ge wafers. A study of the preparation and regeneration of clean germanium surfaces.

GE-0092b

ETCH NAME: TIME: 10—45 sec

TYPE: Acid, dislocation TEMP: RT

COMPOSITION:

3 .... HF

1 .... HNO_3

1 .... HAc

DISCUSSION:

Ge, (111) dendritic ribbon crystals. This etch solution produces similar dislocation development as does CP4. See: GE-0092a.

REF: Ibid.

GE-0126

ETCH NAME: Hydrofluoric acid TIME:

TYPE: Acid, oxide removal TEMP: RT

COMPOSITION:

x .... HF

DISCUSSION:

Ge, (111) wafers used in a study of anodic oxidation. Oxidize in 35% $HClO_4$ with Ge anode and Pt cathode at 45 V. HF used to strip oxide after growth.

REF: Gabor, T — *J Appl Phys,* 32,1361(1961)

GE-0127a

ETCH NAME: Indium TIME:

TYPE: Metal, preferential TEMP: Heat

COMPOSITION:

x .... In

DISCUSSION:

Ge, (100), and (111) wafers. Surfaces were oriented precisely from pits formed by alloying metal. After alloying indium, use HCl to etch dissolve the metal to observe pit structure, crystallographic directions, and orientation.

REF: Dreiner, R & Garnache, R — *J Appl Phys,* 33,888(1962)

GE-0165b: Ibid.

Ge wafers and ingots used in a study of growth striations. Indium alloyed on surfaces and etched away with boiling HCl. See GE-0165a for heat treatment effects.
GE-0166: Dendra, S I — *J Phys Soc Jpn,* 13,533(1958)

Ge, (111) wafers alloyed with indium to develop etch pits. Alloyed at 350°C, 2 min, pits are similar to those observed by etching with CP4. Alloyed at 200°C pits are similar to those observed when etched with superoxol.
GE-0236: Pankore, J I — *J Appl Phys,* 28,1054(1957)

Ge, (111) wafers. A study of the effects of edge dislocations on the alloying of indium in germanium.

GE-0127b
ETCH NAME: TIME:
TYPE: Acid, junction TEMP:
COMPOSITION:
5 .... HF
5 .... HNO_3
2 .... H_2O
DISCUSSION:

Ge, (111), and (100) wafers. Indium metal alloyed to produce etch pits for orientation, also forms a p-n junction. Solution shown used as a junction etch.
REF: Ibid.

GE-0128a
ETCH NAME: TIME: 5 sec
TYPE: Acid, contaminating TEMP: RT
COMPOSITION:

(1)		(2)	
5	.... HF	5	.... HF
8	.... HNO_3	8	.... HNO_3
15	.... HAc	15	.... HAc
5 ppb	.... Ag	50 ppb	.... Fe

DISCUSSION:

Ge as surface barrier diodes used in a study of trace impurities on IV characteristics. Solutions shown were used separately; after etching, rinse with DI water and blow dry with air. Specimens were also jet etched electrolytically in the same solutions at RT, 3 sec and 0.15 mA/9 mils power. Silver, Ag, produces p-type channelling; iron, Fe, has little effect, is apparently removed or is inert.
REF: Krembs, G M & Schlacter, M M — *J Electrochem Soc,* 111,417(1964)
GE-0234: Schmidt, P F & Blomgren, M — *J Electrochem Soc,* 106,694(1959)

Ge and Si specimens. A study of electronic reactions using potential measurements during jet etching of p-type materials.

GE-0129a
ETCH NAME: Indium sulfate TIME:
TYPE: Electrolytic, shaping TEMP:
COMPOSITION: ANODE:
x .... 0.1 $InSO_4$ CATHODE:
POWER:
DISCUSSION:

Ge specimens. Solution used to electrolytically shape specimens.
REF: Uhiler, A Jr — *Bell Syst Tech J,* 35,333(1956)

GE-0129b
ETCH NAME: Potassium hydroxide
TYPE: Electrolytic, shaping
COMPOSITION:
x .... 10% KOH

TIME:
TEMP:
ANODE:
CATHODE:
POWER:

DISCUSSION:
Ge specimens. Solution used to electrolytically shape specimens.
REF: Ibid.

GE-0129c
ETCH NAME: Hydrochloric acid
TYPE: Electrolytic, shaping
COMPOSITION:
x .... 2 *N* HCl

TIME:
TEMP:
ANODE:
CATHODE:
POWER:

DISCUSSION:
Ge specimens. Solution used to electrolytically shape specimens.
REF: Ibid.

GE-0129d
ETCH NAME: Sulfuric acid
TYPE: Electrolytic, shaping
COMPOSITION:
x .... H_2SO_4
x .... H_2O

TIME:
TEMP:
ANODE:
CATHODE:
POWER:

DISCUSSION:
Ge specimens. Solution used to electrolytically shape specimens.
REF: Ibid.

GE-0129e
ETCH NAME: Sodium chloride
TYPE: Electrolytic, shaping
COMPOSITION:
x .... NaCl
x .... H_2O

TIME:
TEMP:
ANODE:
CATHODE:
POWER:

DISCUSSION:
Ge specimens. Solution used to electrolytically shape specimens. (*Note:* This is only one use of salt in metal processing.)
REF: Ibid.

GE-0130
ETCH NAME: Nitric acid
TYPE: Acid, oxidizing
COMPOSITION:
(1) x .... HNO_3, conc.

(2) x .... HNO_3
x .... H_2O

TIME:
TEMP:

DISCUSSION:
Ge, (111) wafers, and other orientations. Wafers used in a general study of the reaction of germanium in nitric acid solutions. Various concentrations of nitric acid used to study: I — dissolution and II — passivity.

REF: Cretella, M C & Gatos, H C — *J Electrochem Soc,* 105,487(1958) I
GE-0131: Cretella, M C & Gatos, H C — *J Electrochem Soc,* 105,492(1958) II
GE-0239: Meigs, P S & Laws, J T — *J Electrochem Soc,* 104,154(1957)
Ge specimens. A study of the oxidation rates on germanium orientation.

GE-0132a
ETCH NAME: Nitric acid TIME:
TYPE: Acid, preferential TEMP:
COMPOSITION:
x .... 18 *N* HNO_3
DISCUSSION:
Ge, (111) wafers used in a general study of preferential etching of surfaces. Solution produces etch pits with closed terrace structure.
REF: Kikuchi, M & Denda, S — *J Phys Soc Jpn,* 11,1127(1956)
GE-0086: Kikuchi, M & Denda, S — *Met Abstr,* 8,818(1958)
GE-0241: Wallis, G & Wang, S — *J Electrochem Soc,* 106,231(1959)
Ge wafers used in a study of the effects of various etches of the recombination centers on germanium surfaces.
GE-0240: Rhodes, E G et al — *J Electron,* 3,403(1957)
Ge specimens. The study of spiral etch pit development in germanium.

GE-0132b
ETCH NAME: Hydrochloric acid TIME:
TYPE: Acid, preferential TEMP:
COMPOSITION:
x .... 13 *N* HCl
DISCUSSION:
Ge, (111) wafers used in a general study of preferential etching of surfaces. Solution produces etch pits with closed terrace structure.
REF: Ibid.

GE-0133a
ETCH NAME: Potassium chloride TIME:
TYPE: Electrolytic, preferential TEMP:
COMPOSITION: ANODE:
x .... KCl CATHODE:
POWER:
DISCUSSION:
Ge, (111) wafers used in a study of surface properties. Anodic etching was used for hole density, and cathodic etching for electron density.
REF: Brattain, W H & Garrett, C G B — *Physica,* 20,885(1954)

GE-0133b
ETCH NAME: TIME:
TYPE: Electrolytic, preferential TEMP:
COMPOSITION: ANODE:
x .... x% KOH CATHODE:
x .... H_2O_2 POWER:

DISCUSSION:

Ge, (111) wafers used in a study of surface properties. Anodic etching for hole density; cathodic etching for electron density.

REF: Ibid.

GE-0137: Wilkes, J G — *Proc Inst Electr Eng,* 106,199(1959)

Ge, as raw chunk material to be used for ingot growth. Solution was used to clean surfaces.

GE-0133c

ETCH NAME: Hydrochloric acid — TIME:

TYPE: Electrolytic, preferential — TEMP:

COMPOSITION: — ANODE:

x .... $^1/_{10}$ *N* HCl — CATHODE:

POWER:

DISCUSSION:

Ge, (111) wafers used in a study of surface properties. Anodic etching for hole density; cathodic etching for electron density.

REF: Ibid.

GE-0136

ETCH NAME: — TIME:

TYPE: Electrolytic, oxidation — TEMP:

COMPOSITION: — ANODE:

x .... 0.1 *N* HCl — CATHODE:

x .... KI — POWER:

x .... oxalic acid

DISCUSSION:

Ge, both p- and n-type specimens used in a study of the anodic oxidation reaction of germanium. Both KI and oxalic acid reduce anodic potential on p-type germanium.

REF: Efimov, E A & Erusalimchik, I G — *Dokl Acad Nauk SSSR,* 128,124(1959)

GE-0139a

ETCH NAME: Iodine etch — TIME: 10 sec

TYPE: Acid, dislocation — TEMP: RT

COMPOSITION:

1 .... HF

2 .... HNO_3

2 .... HAc

\+ .... I_2

DISCUSSION:

Ge, thin films grown by MBE on germanium substrates. Polish substrates: (1) lap in silica abrasive solution; (2) mechanically scrub with soap solution, and DI water rinse; (3) degrease in TCE with ultrasonic. Solution shown used to develop defects in deposited thin films. It was noted that scratches on the substrate physically showed through the thin film.

REF: Ota, Y — *J Cryst Growth,* 62,131(1983)

GE-0140: Wang, P — *Sylvania Technol,* 11,2(1958)

Reference for iodine etch shown in GE-0139a.

GE-0139b
ETCH NAME: Billig's etch TIME:
TYPE: Acid, preferential TEMP: Boiling
COMPOSITION:
12 g .... KOH
8 g .... $K_3Fe(CN)_6$
100 ml .... H_2O
DISCUSSION:

Ge, (111) wafers used as substrates to grow (111) oriented thin film germanium by MBE. Solution used to epitaxy layer developed very sharp triangle etch pits.

REF: Ibid.

GE-0142a
ETCH NAME: Billig's etch TIME: 3—5 min
TYPE: Acid, preferential TEMP:
COMPOSITION:
6 g .... KOH
4 g .... $K_3Fe(CN)_6$
50 ml .... H_2O
DISCUSSION:

Ge, (111) wafers angle lapped at 5°43′. A metallographic investigation of the damage layer introduced by abrading germanium surfaces. Lapping creates fine cleavage cracks, many as (111) traces, but does not create dislocations. Solution used to develop damage structure. Heating at 500°C, 15 min develops dislocation arrays.

REF: Pugh, E H & Samuels, L E — *J Electrochem Soc,* 1043(1961)

GE-0147c
ETCH NAME: Potassium hydroxide TIME:
TYPE: Electrolytic, preferential TEMP:
COMPOSITION: ANODE:
x .... 0.1% KOH CATHODE:
POWER: 1.5 A/cm^2
DISCUSSION:

Ge specimens jet etched in this solution. Author says that the standard power for immersion etching is 1.5 A/cm^2.

REF: Bondi, F J — *Transistor Technology,* Vol III, D Van Nostrand, New York, 1958
GE-0156: Zwerdling, S & Sheff, S — *J Electrochem Soc,* 107,338(1960)

Ge specimens used in a study of oxide growth. Grow oxide thin films electrolytically in 0.25 *N* sodium acetate:xHAc at 50—400 μA/cm^2. Use 0.1 *N* NaOH to remove the oxide.

GE-0350: Tweet, A G — *J Appl Phys,* 30,2002(1959)

Ge, (111) wafers from dislocation-free ingots. A general etching study on the effects of dislocations on vacancies in the lattice structure where they act as energy sinks.

GE-0212b
ETCH NAME: TIME:
TYPE: Electrolytic, selective TEMP:
COMPOSITION: ANODE:
1 .... HF CATHODE:
1 .... HAc POWER:
DISCUSSION:

Ge, n-p-n transistors. Solution used as a selective junction etch. The 1:1 ratio is best

though not critical. Requirement is a high conductivity electrolyte. Solution volume was 50 ml, specimens held with tweezers and etched in dimly lit area for best results. A p-type layer may form on surface, as light will forward bias for holes.
REF: Lesk, I A & Gonzalez, R E — *J Electrochem Soc,* 105,469(1958)
GE-0231: Gonzalez, R E & Lesk, I A — *J Electrochem Soc,* 105,402(1958)
Ge and Si devices as p-n-p transistors. The use of selective electrolytic etching for finalizing junction structures.

GE-0149
ETCH NAME: Copper
TYPE: Electrolytic, junction
COMPOSITION:
1 ml .... HF
20 g .... $CuSO_4$
80 ml .... H_2O

TIME: 1 sec every 20 sec
TEMP:
ANODE: Ge, n-type
CATHODE: Ge, p-type
POWER: 15—20 V at 30—50 mA/cm^2

DISCUSSION:
Ge, p-n diffused diode junctions developed by electroplating of copper. Place one drop of solutions across junction and pulse etch electrolytically on a pre-lap bevelled junction. Long etch times will attack n-region, but the p-region will plate with copper. See GE-0039.
REF: Gling, R — *J Electrochem Soc,* 107,356(1960)

GE-0203a
ETCH TIME: Iodine etch
TYPE: Halogen, polish
COMPOSITION:
1 .... HF
2 .... HNO_3
2 .... HAc
x .... I_2

TIME: 2—3 min
TEMP: RT

DISCUSSION:
Ge, (111) wafers used in a study of dislocations and crack formation after indenting the surface with a diamond stylus. An "iodine etch" was used to polish surfaces prior to indenting and dilute CP4 was used to develop defects.
REF: Sugita, Y — *Jpn J Appl Phys,* 2,313(1963)

GE-0203b
ETCH NAME: CP4, dilute
TYPE: Acid, preferential
COMPOSITION:
30 .... HF
50 .... HNO_3
30 .... HAc
0.6 .... Br_2
x .... H_2O

TIME:
TEMP:

DISCUSSION:
Ge, (111) wafers. Solution used to develop dislocations and cracks induced by indenting the surface with a diamond stylus.
See GE-0203a.
REF: Ibid.

GE-0204a
ETCH NAME: Potassium hydroxide
TYPE: Electrolytic, passivate
COMPOSITION:
x .... 15% KOH

TIME: (1) 8 sec (2) 10 sec
TEMP:
ANODE: Ge
CATHODE: Ni
POWER: (1) 0.4 A/cm^2
(2) 0.1 A/cm^2 & 0.2 A/cm^2

DISCUSSION:

Ge, p-n-p transistors prepared with indium alloy dots. Solution shown used to oxide passivate the surface in a study of vacuum drying and the properties of germanium surfaces.

REF: Forosho, K & Ono, K — *Jpn J Appl Phys,* 1,148(1962)

GE-0204b
ETCH NAME:
TYPE: Acid, passivation
COMPOSITION:
50 ml .... HF
1 ml .... H_2O_2
x .... air, bubbling

TIME:
TEMP:

DISCUSSION:

Ge, p-n-p transistors prepared with indium alloy dots. Solution shown used to vapor oxidize germanium surfaces by bubbling air through the acid mixture.

REF: Ibid.

GE-0205
ETCH NAME: Iodine etch
TYPE: Halogen, dislocation
COMPOSITION:
2000 mg .... KI
200 mg .... I_2
50 ml .... H_2O

TIME: 70 h; 50 h; 15 h
TEMP: RT 50°C—80°C

DISCUSSION:

Ge wafers of different orientations. Solution shown referred to as a Redox system with formation of GeI_4. Germanium will not etch in KI, only; requires I_2 and rate reduces with evaporation of iodine. Solution used to develop etch pits and structure figures.

REF: Arizumi, T & Akasaki, I — *Jpn J Appl Phys,* 2,350(1963)

GE-0206: Arizumi, T & Akasaki, I — *J Phys Soc Jpn,* 17,712(1962)

Ge specimens. Describes another iodine etch for this material.

GE-0266: Elliot, G — *J Electron Control* 4,456(1958)

Ge, (100) wafers. Used an iodine etch to develop pits within material with 1.2×10^6 dislocation density. Specific observation of spiral etch pits in an apparent polygonized area of the wafer.

GE-0263b: Ibid.

Ge, (100) wafers. Spiral etch pits observed using an iodine etch. Pit density was between 10^5—10^6/cm^2 on both (111) and (100) wafers.

GE-0276: Dorendorf, H — *Z Angew Phys,* 9,513(1957)

GE wafers used in a general study of dislocations.

GE-0207
ETCH TIME: Hydrogen peroxide
TYPE: Acid, cleaning
COMPOSITION:
(1) x .... H_2O_2
(2) x .... HNO_3

TIME:
TEMP: Boiling

DISCUSSION:
Ge wafers of different orientations. Solutions shown used to clean surfaces prior to oxidation in a study of surface conductivity and recombination. Oxidation was done with both dry O_2 and O_2/O_3, and as wet oxidation. The hydrogen peroxide prepared surface gave best results.
REF: Razhanov, A U et al — *Zh Tekh Fiz,* 26,2142(1956)

GE-0208
ETCH NAME:
TYPE: Acid, macro-etch
COMPOSITION:
x .... 10% NaOH
x .... 15% $NaNO_3$
x .... H_2O

TIME:
TEMP: 100°C

DISCUSSION:
Ge, (111) wafer and spherical shot. Wafers were n-type; shot p-type. Pellet alloyed into wafer about 1 mil deep to form p-n junctions. Solution shown used to etch clean and develop junctions.
REF: Lesk, I A — *J Electrochem Soc,* 107,534(1960)

GE-0154
ETCH NAME: Air
TYPE: Gas, oxidation
COMPOSITION:
x .... air

TIME: 24 h
TEMP: RT

DISCUSSION:
Ge, (111) wafers and other orientations. Rinse surfaces in HF to remove native oxide. Within minutes, a native oxide will regrow sitting in air at RT to about 10—15 Å thick. After 24 h the oxide will stabilize at 25—30 Å thick.
REF: Archer, R J — Electrochem Soc Meet, Washington, D.C., 12—16 May 1957

GE-0147c
ETCH NAME:
TYPE: Electrolytic, remove
COMPOSITION:
x .... 0.1 *N* H_2SO_4
x .... *GeO_2

TIME:
TEMP: RT
ANODE:
CATHODE:
POWER: 3.5—50 mA/cm^2

*Saturate the solution.

DISCUSSION:
Ge specimens. Germanium is about as electrochemically active as iron, Fe. Cleaned surfaces will react with water forming a hydrated oxide of a $2GeO_3$ type, as stable meta-germanic acid.
REF: Ibid.

GE-0147e
ETCH NAME: TIME:
TYPE: Electrolytic, removal TEMP: RT
COMPOSITION: ANODE:
x .... 1 *N* KOH CATHODE:
x .... *GeO_2 POWER: 3.5—50 mA/cm^2

*Saturate the solution.

DISCUSSION:

Ge specimens. See discussion under GE-0147c.

REF: Ibid.

GE-0155
ETCH NAME: TIME: 15 min
TYPE: Alkali, contact etch TEMP: RT
COMPOSITION:
x .... 2% $NaClO_4$ ($KClO_4$)
x .... 2% NaOH (KOH)

DISCUSSION:

Ge, as devices with evaporated metal contacts. Solution used to etch clean device surfaces.

REF: Bradshaw, S E — U.S. Patent 2,690,383, Sept 28, 1954

GE-0285: Turner, D R — *J Electrochem Soc,* 106,786(1959)

Ge and Si wafers used in a study of electroplated metal contacts.

GE-0157
ETCH NAME: TIME:
TYPE: Acid, cleaning TEMP:
COMPOSITION:
x .... HNO_3
x .... H_2O_2

DISCUSSION:

Ge specimens used in a study of thermal conductivity at low temperatures. Surfaces were prepared as follows: (1) mechanically lap with emery paper; (2) etch clean in solution shown; (3) soak in KCN solution, and (4) DI water rinse. Specimens were then annealed. Silicon also was studied.

REF: White, G K & Woods, S B — *Phys Rev,* 103,569(1956)

GE-0158a
ETCH NAME: Potassium permanganate TIME:
TYPE: Acid, oxidation TEMP:
COMPOSITION:
x ... x% $KMnO_4$
x .... H_2O

DISCUSSION:

Ge specimens used in a study of oxidation. Both single crystal and polycrystalline germanium were evaluated. Solution used to oxidize surfaces. Single crystal surfaces oxidize less rapidly than poly-Ge and rate decreases with material purity. Other oxidizing agents used were (1) water; (2) hydrogen peroxide; (3) sodium peroxide; (4) sodium hypochlorate; (5) air, and (6) carbon dioxide.

REF: Rosner, O — *Semicond Electron,* June 1957, 166

GE-0159a
ETCH NAME: TIME:
TYPE: Acid, cleaning TEMP: RT
COMPOSITION:
1 .... HF
10 .... HNO_3
DISCUSSION:

Ge specimens used in a study of thermal acceptors in vacuum annealed germanium. Etch clean in this solution and follow with three rinses in DI water. Then soak at RT, 1 h in 10% KCN to remove metallic ions and rinse five times in DI water.

REF: Hopkins, R L & Clarke, E N — *Phys Rev,* 100,1789(1955)

GE-0160a
ETCH NAME: Ferric chloride TIME: 15 min
TYPE: Acid, cleaning TEMP: Boiling
COMPOSITION:
x .... x% $FeCl_3$
x .... H_2O
DISCUSSION:

Ge specimens used in a general surface cleaning study. Various etch solutions used. Other etchants were $1HF:1HNO_3$, and superoxol.

REF: Coblenz, A & Owens, H L — *Transistors — Theory and Application,* McGraw-Hill, New York, 1955, 211

GE-0161
ETCH NAME: TIME:
TYPE: Acid, oxidation TEMP:
COMPOSITION:
1 .... 10% HF
1 .... 90% HNO_3
DISCUSSION:

Ge specimens used in a study of slow surface reactions. Etch in solution shown and age specimens for 4 weeks in air at RT to stabilize oxide.

REF: Morrison, S R — *Phys Rev,* 102,1297(1956)

GE-0162a
ETCH NAME: TIME:
TYPE: Acid, polish TEMP:
COMPOSITION:
x .... HF
x .... HNO_3
x .... *I_2 (Br_2)

*Saturate the solution.

DISCUSSION:

Ge, n-type wafers used to fabricate p-n junctions devices. Wafer preparation sequence was: (1) polish etch in solution shown; (2) rinse in MeOH or 2% HNO_3:MeOH, and (3) wash in $1HF:1HNO_3:8H_2O$.

REF: Ditrick, N H — U.S. Patent 2,761,800, Sept 4, 1956

GE-0163
ETCH NAME: TIME: 20 min then 40 min
TYPE: Molten flux, polish TEMP: 590°C—200°C
COMPOSITION:
2 .... KCN
1 .... NaCN
DISCUSSION:

Ge, as alloy junction transistors. Prepare surfaces as follows after material has been alloyed: (1) soak in molten flux at 590°C for 20 min; (2) reduce temperature and continue to soak at 200°C for 40 min; (3) cool to room temperature and (4) DI water wash to remove flux. Then wash in solution shown to final clean, which is called the cyanide treatment. Process gives a lower and flatter collector current.

REF: Dawson, M H — U.S. Patent 2,788,300, April 9, 1957

GE-0009
ETCH NAME: TIME:
TYPE: Acid, junction TEMP: RT
COMPOSITION:
1 .... HF
3 .... HNO_3
DISCUSSION:

Ge, as alloyed p-n junctions devices. A study of recombination radiation from deformed junctions at 80 K. Alloy junctions were GeIn or InAs. Solution used as a junction development etch.

REF: Newman, R — *Phys Rev,* 105,1715(1957)

GE-0164
ETCH NAME: TIME:
TYPE: Acid, polish TEMP: RT
COMPOSITION:
1 .... HF
4—5 .... HNO_3
DISCUSSION:

Ge, as single crystal germanium grown by vapor deposition. The (111) oriented seed crystals used were etch polished in the solution shown, which gives a bright mirror finish. Growth rate was slower than average at 0.5—0.7 mils/h.

REF: Ruth, R P et al — *J Appl Phys,* 31,995(1960)

GE-0167
ETCH NAME: TIME: 4 min
TYPE: Acid, polish TEMP: RT
COMPOSITION:
15 .... HF
750 .... HNO_3
DISCUSSION:

Ge, as cut cubes oriented (001)/(001):(110)/(110), used in a study of third order elastic modulus. The cube was etch polished in the solution shown.

REF: Bateman, T et al — *J Appl Phys,* 32,928(1961)

GE-0168
ETCH NAME:
TYPE: Acid, polish
COMPOSITION:
4 .... HF
21 .... HNO_3
TIME: 20 sec
TEMP: RT
DISCUSSION:

Ge, (111) wafers used in a study of optical constants. Cleaning sequence was: (1) boil in benzene, and reflux with acetone; (2) HF dip, 15 sec, and rinse in acetone, then (3) etch in solution shown with agitation, and DI water rinse.

REF: Archer, R J — *Phys Rev,* 110,354(1958)

GE-0169
ETCH NAME:
TYPE: Acid, polish
COMPOSITION:
4 .... HF
7 .... HNO_3
TIME:
TEMP:
DISCUSSION:

Ge specimens used in a study of thick oxides on surfaces. Etch polish surfaces in solution shown. Oxidation used oxygen dried by passing through a bubbler chilled with ice-acetone to remove moisture.

REF: Lasser, M et al — *Phys Rev,* 105,491(1957)

GE-0170
ETCH NAME:
TYPE: Acid, preferential
COMPOSITION:
x .... HF
x .... HNO_3
x .... HCl
x .... H_2O_2
TIME:
TEMP:
DISCUSSION:

Ge, (111) wafers. Solution used to selectively develop left- and right-hand spiral etch pits. Structures were 100 to 1000 times smaller than normal etch pits.

REF: Kikuchi, M & Denda, S — *J Phys Soc Jpn,* 12,105(1957)

GE-0247: Goldberg, F C & Wynne, R H — *J Met,* 5,436(1953)

Ge specimens. Solution used in a general etch study of germanium.

GE-0246: Ellis, S G — *J Appl Phys,* 28,1262(1957)

Ge single crystal material used in a general etch study.

GE-0245: Ellis, S G — *J Appl Phys,* 26,1140(1955)

Ge single crystal material used in a dislocation study of germanium.

GE-0251: Denda, S & Kikuchi, M — *Met Abstr,* 8,818(1958)

Ge single crystal material used in a microscopic study of germanium surfaces.

GE-0171
ETCH NAME:
TYPE: Acid, thinning
COMPOSITION:
1 .... HF
3 .... HNO_3
8 .... HAc
TIME:
TEMP:

DISCUSSION:

Ge wafers used in a study of orientation effects of K X-ray absorption spectra. Specimens were etch thinned in this solution for microscope study.

REF: Hussaini, J M & Stephenson, S T — *Phys Rev,* 109,51(1958)

GE-0173

ETCH NAME: TIME:

TYPE: Acid, polish/removal TEMP: 35°C—1000°C

COMPOSITION: RATE: 75—100 μm/h

775 μm/h

7 .... HCl

1 .... HNO_3

+ .... glycerin

DISCUSSION:

Ge wafers used in an etching study. Rates shown are without glycerin. Addition of glycerin can be used to moderate and control etch rates. $GeCl_4$ that remains on surfaces can be recovered by hydrolyzation.

REF: Wolsky, S P — U.S. Patent 2,734,806, February 14, 1946

GE-0203b

ETCH NAME: TIME:

TYPE: Acid, polish TEMP:

COMPOSITION:

1 .... HF

1 .... HNO_3

1 .... tartaric acid

DISCUSSION:

Ge wafers used in an etching study. Solution used to polish surfaces. Boiling water will oxidize germanium as will CO_2 applied to surfaces over 700°C. Improper etching and processing can cause pits that produce instability in transistors and diodes.

REF: Rosner, O — *Z Metallkd,* 46,225(1955)

GE-0175

ETCH NAME: TIME:

TYPE: Acid, polish TEMP:

COMPOSITION:

20% .... HF

80% .... HNO_3

DISCUSSION:

Ge wafers doped with copper to fabricate high resistivity photo-conductors. Mechanically lap with 260-grit SiC, then 600-grit abrasive on a glass plate with a water carrier. Etch polish in solution shown and rinse in DI water. (*Note:* See GE-0015. This solution is called white etch.)

REF: Van Heerdren, P J — *Phys Rev,* 108,230(1957)

GE-0177

ETCH NAME: TIME:

TYPE: Acid, removal TEMP:

COMPOSITION:

1 .... HF

2 .... HNO_3

DISCUSSION:

Ge wafers used in a magneto-surface experiment. Solution used as a removal and polishing etch.

REF: Zemel, J N & Petratz, R L — *Phys Rev,* 110,1263(1958)

GE-0178

ETCH NAME: Argon | TIME:
TYPE: Ionized gas, cleaning | TEMP:
COMPOSITION: | GAS:
x .... Ar^+ ions | PRESSURE:
| POWER:

DISCUSSION:

Ge, polycrystalline spheres used in a study of cleaning and sputter yield of germanium in rare gases. Other gases studied were Xe, Kr, Ne, and He.

REF: Laegreid, N et al — *J Appl Phys,* 30,374(1959)

GE-0179

ETCH NAME: Heat | TIME:
TYPE: Thermal, forming | TEMP: Elevated
COMPOSITION:
x .... heat

DISCUSSION:

Ge, as polycrystalline spheres. Describes the method of fabrication and evaluation of such spheres as diodes. Produced 40—125 V devices, but other electrical characteristics were poor. (*Note:* This may be the first application of spheres, polycrystalline or single crystal, as a semiconductor device.)

REF: Dunlap, W C Jr — *J Appl Phys,* 25,448(1954)

GE-0276: Wagner, R D — *J Appl Phys,* 29,1678(1958)

Ge wafers. A study of the development of dislocations in germanium by thermal shock.

GE-0278: Logan, K H — *Phys Rev,* 101,1455(1956)

Ge wafers. A study of the introduction of thermally induced acceptors in germanium by heat treatments.

GE-0279: Logan, R A — *Phys Rev,* 91,757(1953)

Ge wafers. Reference for the cleaning solutions used prior to thermal heat treatments used in GE-0278.

GE-0180

ETCH NAME: I—IV metals | TIME: Variable
TYPE: Molten flux, preferential | TEMP: Molten
COMPOSITION:
x .... Cu

DISCUSSION:

Ge, as single crystal spheres etched in liquid metals of the I—IV atomic group at elevated temperatures to develop etch pits. Used light figures developed to explain patterns by atomic radii vs. type metallic solvent. Also studied silicon spheres with these liquidized metals.

REF: Broder, J D & Wolff, G A — Electrochem Soc Meet, Cleveland, OH, 30 September 1956

GE-0181
ETCH NAME: TIME:
TYPE: Acid, polish TEMP: RT
COMPOSITION:
5 ml .... HF
70 ml .... HNO_3
DISCUSSION:

Ge, as single crystal spheres fabricated using a ''Tornado Sphere Grinder''. The system is a 2" high, thick-walled aluminum tube section with glue-back Buehler sanding strips attached inside. The tube is screwed down on an aluminum plate with a removable Lucite top. Nitrogen gas is fed into the enclosed tube center area from a tygon tube attached through $^1/_4$" pipe fitting in the Lucite top so that the gas blows against the inner wall. Single crystal materials are cut as cubes, placed in the grinder, and rotated with 3—5 SCFH N_2 flow until spheres are formed. For 10—15 $^1/_4$" diameter germanium cubes, time is about 45 min using 600-grit abrasive paper.

After spheres are formed, etch polish in the solution shown by hand rolling spheres submerged in the etchant using a teflon beaker held at about a 30—45° angle. Spheres should roll, but not tumble. The same solution was used on silicon with starter cubes from $^1/_8$ to $^1/_2$" diameter. Other materials fabricated as spheres were GaAs, GaSb, InAs, InSb, and InP. Spheres used in a study of finite crystal form preferential etching. Diffused spherical devices also have been fabricated from such spheres. (*Note:* There are similar sphere forming systems developed and used with other names, such as the Race-Track grinder. This is the easiest and most rapid way of fabricating single crystals spheres.)

REF: Myers, J & Walker, P — personal development and application, 1956

GE-0211b
ETCH NAME: Iodine A etch TIME:
TYPE: Acid, preferential TEMP:
COMPOSITION:
15 ml .... HF
30 ml .... HNO_3
33 ml .... HAc
80 mg .... I_2
DISCUSSION:

Ge wafers of various orientations used in a study of the effects of various etchants on recombination centers. WAg, superoxol, CP4, and KOH also studied.

REF: Wallis, G & Wang, S — *J Electrochem Soc,* 106,231(1959)

GE-0190
ETCH NAME: White etch, dilute TIME:
TYPE: Acid, polish TEMP:
COMPOSITION:
1 .... HF
10 .... HNO_3
6 .... H_2O
DISCUSSION:

Ge wafers cut and polished from germanium single crystals grown by vapor epitaxy in a closed-cycle process. Solution used to produce a polished mirror surface. (*Note:* An example of the use of the term ''white etch'' for solutions of $HF:HNO_3$. There are several so called, each as a different concentration mixture, such that the term is not recommended.)

REF: Marinace, J C — *IBM J,* 24,248(July 1961)

GE-0029a
ETCH NAME: TIME:
TYPE: Electrolytic, polish TEMP:
COMPOSITION: ANODE:
x .... HNO_3 CATHODE:
x .... KCl POWER:
DISCUSSION:

Ge, (111) wafers used in developing new etching systems. Solutions of HNO_3:KCl will electropolish germanium.

REF: El'kin, B I — *J Metall,* Abstr 3 & 4(1958)

GE-0029b
ETCH NAME: TIME:
TYPE: Electrolytic, polish TEMP:
COMPOSITION: ANODE:
x .... HNO_3 CATHODE:
x .... KF POWER:
DISCUSSION:

Ge, (111) wafers used in an etch development study. Solutions of HNO_3:KF will electropolish germanium.

REF: Ibid.

GE-0156b
ETCH NAME: TIME:
TYPE: Electrolytic, oxidation TEMP:
COMPOSITION: ANODE: Ge
x .... 0.25 *N* $NaC_2H_3O_2$ CATHODE:
x .... HAc POWER:
DISCUSSION:

Ge specimens used in a study of the growth of anodic oxide films. Solution shown used to anodize germanium. (*Note:* This is anhydrous sodium acetate in glacial acetic acid.)

REF: Zwerdling, S & Sheff, S — *J Electrochem Soc,* 107,338(1960)

GE-0268a: Laws, J T & Meigs, P S — *J Electrochem Soc,* 104,154(1957)

Ge wafers. A study of the oxidation rates of germanium.

GE-0269: de Zuber, N & Pourbiax, M — Rapp Tech #27 (1955)

Ge wafers. A study of the electrochemical behavior of germanium to develop potential pH equilibrium diagrams of the Ge-H_2O system at 25°C. Conclusion was to use H_2O-free electrolytes to reduce germanium oxide formation in hydrogenating Ge:H films.

GE-0375: de Zuber, N & Pourbiax, M — *Semicond Abstr,* 4,203(1956)

Ge wafers. An extension of work done as in GE-0269.

GE-0270: Kikuchi, M — *J Phys Soc Jpn,* 12,756(1957)

Ge wafers. Some experiments on the surface field effects on germanium single crystals. Q values drift toward negative values at RT after etching in CP4, plus growth of an oxide layer.

GE-0271: Laws, J T & Meigs, P S — *J Appl Phys,* 26,419(1955)

Ge and Si wafers with grown-in p-n junctions. A study of the effects of water vapor on the junctions.

GE-0272: Ellis, S G — *J Appl Phys,* 28,1262(1957)

Ge wafers. A surface study on the effects of chemical etching of a controlled oxide, or to minimize oxidation. Both CP4 and superoxol were evaluated.

GE-0273: Chynoweth, A G et al — *Phys Rev,* 118,425(1960)

Ge and Si wafers with p-n junctions. A study of internal field emission at narrow junctions. Dislocations produce soft PIV, and edge dislocations cause local reduction in the band gap width.

GE-0277: Chynoweth, A G & Pearson, G L — *J Appl Phys,* 29,1103(1958)

Ge and Si wafers with p-n junctions. Additional studies as in GE-0273.

GE-0268b: Ibid.

Ge single crystal spheres with (111), (100), and (110) flats. Oxidize at 450 and 700°C. Above 550°C, oxidation is inversely dependent on pressure, and the amount of oxygen; below, the (110) flat was the only plane to show appreciable oxidation.

GE-0311: Zerdling, J & Sheff, S — *J Electrochem Soc,* 107,388(1960)

Ge wafers. A study of the anodic oxidation of germanium. Best solution was 0.25 *N* anhydrous acetate in acetic acid used at 40—400 μA/cm².

GE-0280: Bray, R & Cunningham, R W — *J Phys Chem Solids,* 8,99(1959)

Ge wafers. An optical and magnetic study. There is a retentive conductance surface effect with oxidizers, such as HNO_3; HF:HNO_3 and H_2O_2. No effect observed with HF, only. Light increases formation of a p-type surface zone.

GE-0200

ETCH NAME: Sodium hydroxide	TIME: Hours
TYPE: Alkali, cutting	TEMP: Hot
COMPOSITION:	ANODE:
x .... x% NaOH	CATHODE:
	POWER:

DISCUSSION:

Ge ingot. System used for slicing germanium wafers electrolytically using a 0.003 diameter tungsten wire as the cutting vehicle with the solution shown as an etch assist. Cutting rate maximum was 1730 mils/h.

REF: Shedd, S — *Electrochem Abstr,* 10,15(1961)

GE-0233: Eigler, J H & Sullivan, M U — *J Electrochem Soc,* 103,132(1956)

Ge specimens. A study of electrolytic stream (jet) etching of germanium.

GE-0201

ETCH NAME: Argon	TIME:
TYPE: Ionized gas, cleaning	TEMP:
COMPOSITION:	GAS FLOW:
x .... Ar^+ ions	PRESSURE:
	POWER:

DISCUSSION:

Ge wafers used in a study of Hall mobility of cleaned surfaces. After argon ion cleaning, specimens were annealed in vacuum at 10^{-9} Torr.

REF: Missman, R & Handler, P — *J Phys Chem Solids,* 8,121(1959)

GE-0259: Wolsky, S P — *Phys Rev,* 108,1131(1957)

Ge and Si wafers used in a study of Ar^+ ion bombardment of surfaces for cleaning. Current density was 1—12 μA/cm².

GE-0260: Wolsky, S P — *J Phys Chem Solids,* 8,114(1959)

Ge and Si, (100) wafers used in a surface study. Surfaces were oxidized at RT; Ar^+ ion cleaned; then etched in CP4; and baked in vacuum.

GE-0261: Missman, R & Handler, P — *J Phys Chem Solids,* 8,109(1959)

Ge, (100) wafers used in a study of Hall mobility of cleaned germanium surfaces. Material was 0.2 mm thick, n-type, 20 Ω cm resistivity. Ar^+ ion clean, then anneal at 650°C overnight and cool within 2 h.

GE-0274: Wagner, E D — *J Appl Phys,* 29,217(1958)
Ge wafers. A study of ion bombardment etching of germanium.

GE-0283: Schlier, K E & Farnsworth, H E — *J Chem Phys,* 30,917(1959)
Ge and Si wafers used in a study of structure and adsorption characteristics of clean surfaces. Ar^+ ion bombardment evaluated.

GE-0202
ETCH NAME: Superoxol — TIME:
TYPE: Acid, preferential — TEMP:
COMPOSITION:
1 .... HF
1 .... H_2O_2
4 .... H_2O
DISCUSSION:
Ge specimens as cut cylinders for the development of dissolutionment diagrams. Modified free energy theorems used to predict equilibrium growth and etching shapes. Solution used to preferentially etch specimens. Also etched GaAs cylinders and hemispheres in $1HF:2H_2O_2:5H_2O$.
REF: Jaccodine, R J — *J Appl Phys,* 33,2663(1962)

GE-0019c
ETCH NAME: Potassium hydroxide — TIME:
TYPE: Electrolytic, junction — TEMP:
COMPOSITION: — ANODE:
1 drop .... x% KOH — CATHODE:
POWER:
DISCUSSION:
Ge, (111), n-type wafers with p-type germanium epitaxy thin films as Ge/Ge (111) to form p-n junctions. Place a drop of solution across the junction and bias to etch delineate.
REF: Ibid.

GE-0209
ETCH NAME: Potassium fluoride — TIME:
TYPE: Electrolytic, removal — TEMP:
COMPOSITION: — ANODE:
x .... x% KF — CATHODE:
POWER:
DISCUSSION:
Ge, (111) wafers and other orientations. A study of the dissolution kinetics of germanium and specific adsorption in aqueous solutions. All solutions were used from concentrations of 10^{-6} to 1 *N*. Below a pH 6, little or no change was observed, but as pH increases above pH 6, etch rate increases. No difference observed between p- or n-type material with or without illumination, as rate increases with concentration increase of solution. Other salts evaluated were KCl, KBr, KI, $NaNO_3$, Na_2SO_4, CsCl, $BaCl_2$, and $LaCl_3$.
REF: Harvey, W W & Gatos, H C — *J Electrochem Soc,* 107,65(1960)

GE-0013b
ETCH NAME: TIME:
TYPE: Acid, polish TEMP: RT
COMPOSITION:
x .... x% NaOH
x .... H_2O_2
x .... 1 μm alumina (Al_2O_3)
DISCUSSION:

Ge, (111), and (100) wafers used as substrates in the preparation of GaAs/Ge and GaP/Ge epitaxially grown heterojunctions. The germanium substrates were polish lapped with the acid slurry shown above using a cotton pad soaked with the solution. The GaAs/Ge junction was delineated using HNO_3:H_2O.
REF: Weinstein, M et al — *J Electrochem Soc,* 111,674(1964)

GE-0203
ETCH NAME: Water TIME:
TYPE: Acid, float-off TEMP:
COMPOSITION:
x .... H_2O
DISCUSSION:

a-Ge, as thin film material deposited on NaCl, (100) substrates, and used for TEM study of structure. After growth, films were removed from the salt substrates with water using the float-off technique where the NaCl surface at the interface with the a-Ge layer is liquefied with the film floating off.
REF: Nakhotkiu, N G et al — *Thin Solid Films,* 112,267(1984)
GE-0204: Pierrard, P et al — *Thin Solid Films,* 111,141(1984)

a-Ge thin films deposited on NaCl, (100) cleaved substrates. Desalted water was used for float-off to remove films for TEM study. TEM microscope heating did not crystallize films, but using a laser under the TEM produced single crystal spherulitic structures.

GE-0310
ETCH NAME: TIME:
TYPE: Cleave, dislocation TEMP:
COMPOSITION:
x .... cleave
DISCUSSION:

Ge, (111) wafers cleaved and used in a study of electrical states on clean surfaces. Dislocations were present on cleaved surface without any additional processing.
REF: Barnes, G A & Banberg, P C — *Proc Phys Soc,* 71,1020(1958)

GE-0206
ETCH NAME: Potassium hydroxide TIME: 5—60 min
TYPE: Alkali, orientation TEMP: Hot
COMPOSITION:
125 ml .... 5—10% KOH (NaOH)
10 mg .... I_2
DISCUSSION:

Ge, (111) wafers and ingots. Where an ingot is "as-grown" prior to surface grinding to a specific diameter, growth direction can be observed by the hachure growth meniscus marks (3 in ⟨211⟩ directions at 120°) along the ingot length. If they point upwards, it is toward the seed end; if downward, toward the ingot tail. If needed, this gives a ⟨111⟩A and

$\langle\overline{1}\overline{1}\overline{1}\rangle$B direction. To establish this positive or negative wafer surface direction when wafers are sliced from an ingot ground to specific diameter where these hachure marks are not present, etch lightly with the solution shown. Observe wafer edges for etched triangles for bulk ⟨110⟩ or ⟨011⟩ directions . . . triangle point up or down . . . up is positive, seed end of ingot. For (100) oriented ingot growth, there are four hachure marks in ⟨110⟩ directions at 90°.
REF: Pennington, P & Walker, P — development work, 1957

GE-0165c
ETCH NAME: Indium TIME:
TYPE: Metal, decoration TEMP: 700°C
COMPOSITION:
x .... In
DISCUSSION:
Ge, (111) wafers with indium, as a thin sheet or cut pre-form, alloyed and diffused into the wafer to metal decorate and develop growth striations. After alloying, remove excess indium by boiling in HCl. Decorate by heating wafers at 700°C on an iron, nickel or similar metal plate. Indium will develop growth striations and (111) fracture.
REF: English, A C — *J Appl Phys,* 31,1498(1960)

GE-0227b
ETCH NAME: Iodine, vapor TIME:
TYPE: Halogen, preferential TEMP:
COMPOSITION:
x .... I_2, vapor
DISCUSSION:
Ge, (111), (110) and (100) wafers used in a study of etch patterns developed by iodine vapor etching. After mechanical lap and etch polishing with CP4, wafers were I_2 vapor etched in both an open and closed tube. In open tube etching was done with 60 cc H_2 gas flow and 10—30 mmHg pressure of iodine.
REF: Arizumi, T & Akasaki, I — *Jpn J Appl Phys,* 2,143(1963)

GE-0226
ETCH NAME: Superoxol TIME: 2 min
TYPE: Acid, polish TEMP: RT
COMPOSITION:
1 .... HF
1 .... H_2O_2
3 .... H_2O
DISCUSSION:
Ge, (111) wafers lithium diffused used in a study of lithium precipitation. Wafers were lap polished down to 0.25 μm on microcloth, then polished in solution shown. Under 2000× magnification, 0.2 μm diameter precipitates of lithium were observed in dislocation etch pits.
REF: Blanc, J & Abrahams, M S — *J Appl Phys,* 34,3638(1963)

GE-0225a
ETCH NAME: Germanium chloride TIME:
TYPE: Gas, preferential TEMP:
COMPOSITION:
x .... $GeCl_4$, vapor

DISCUSSION:

Ge, (111) wafers with epitaxy grown Ge layers used in a preferential light figure etch pit development study. Vapor etching developed etch pits and hillocks in the epitaxy layers. Hot HCl vapors also were used with similar results. The WAg etch was used to polish surfaces.

REF: Gualtieri, J G & Kerecman, A J — *Rev Sci Instrum,* 34,108(1963)

GE-0025b

ETCH NAME: Hydrochloric acid, vapor — TIME:

TYPE: Acid, preferential — TEMP:

COMPOSITION:

x .... HCl, vapor

DISCUSSION:

Ge, (111) wafers used as substrates for Ge epitaxy growth in a study development of orientation light figures in the epitaxy layers. Hot HCl vapors develop both etch pits and hillocks as does $GeCl_4$ vapor. WAg etch was used to polish surfaces before preferential vapor etching.

REF: Ibid.

GE-0159b

ETCH NAME: Potassium cyanide — TIME: 1 h

TYPE: Acid, cleaning — TEMP: RT

COMPOSITION:

x .... 10% KCN

DISCUSSION:

Ge specimens used in a study of thermal acceptors in vacuum annealed germanium. Etch with $1HF{:}10HNO_3$, and rinse three times with DI water, then soak in the solution shown to removal metallic ions, and rinse five times with DI water.

REF: Ibid.

GE-0300

ETCH NAME: — TIME: 5 to 45 min

TYPE: Alkali, preferential — TEMP: Boiling

COMPOSITION:

387 g .... KOH

250 ml .... H_2O

20 g .... I_2

DISCUSSION:

Ge, (111), 5—10 Ω cm resistivity, n-type wafers used in a controlled pit damage etch study. Solution also used as a light figure orientation etch for (111), (100), (110), and (211) ingots and wafers. Study used both cut dice either randomly edge oriented or bulk plane direction oriented edges: ⟨111⟩ on (111) wafers; ⟨110⟩ on (100) wafers. The former were triangles or diamonds in shape; the latter were squares. See SI-0300 for a description of the diamond stylus damaging system designed. The system was used on germanium and silicon initially, then compound semiconductors in similar studies. The base solution used in all cases was 250 ml of 30% KOH or NaOH, with or without chemical additives. Pit structures obtained were evaluated for shape relative to surface orientation; pit side-slope planes vs. solution mixtures; pit bottom planarity; and pit shape progression with etch time. Pits also were evaluated as concave surfaces against convex sphere surfaces etched to finite crystal form.

Pure solutions of the two hydroxides are cleaning solutions, only, on germanium, but become preferential with chemical additives. Some 50 mixtures were evaluated on silicon, but a lesser number on germanium. Germanium, as indium alloy transistors were fabricated and evaluated, similar to the aluminum alloy silicon transistors described in SI-0300 series.

Oriented cut forms of both semiconductor materials were evaluated for type crystallographic planes developed at face edges with variation of additives. The forms studied were cube, (100); cube, (110); octahedron, (111); tetrahedron, (111); and rhomboid, (111). The latter is not a natural crystal form in the isometric system, normal (cubic) class, but contains faces of the octahedron, and both the positive (+) and negative (−) tetrahedrons. Note that calcite in the hexagonal, rhombohedral division shows the atypical rhomboid shape. The solid chemical additives, shown below, were 20—30 g added to the 250 ml 30% KOH/NaOH solution. Liquid bromine was 25 or 50 ml. The etch pit planar surface shape was evaluated after 20 min etch time.

Chemical additive	**Pit shape (111) (20 min etch)**
GE-0301: Iodine, I_2	Triangle & hexagon
GE-0302: Bromine, Br_2	Triangle & hexagon
GE-0303: Ammonium peroxylsulfate, $(NH_4)_2S_2O_8$	Triangle
GE-0304: Potassium ferricyanide, $K_3Fe(CN)_6$	Triangle

Preferential etch pits on (111), (100), (110), and (211) were sufficiently distinctive without controlled pit damage for light figure orientation. The damage induced pits on (111) were flat bottomed similar to those achieved with silicon.
REF: Walker, P & Waters, W P — germanium development, 1956—1958

GE-0243
ETCH NAME: Light — TIME:
TYPE: Power, plasticity — TEMP:
COMPOSITION:
x .... light
DISCUSSION:

Ge specimens. A study of the effects of light induced plasticity in germanium. Intense white light applied to surfaces under strain potentials.
REF: Hochman, R H & Kuczyanski, G C — *J Appl Phys,* 30,267(1959)

GE-0244
ETCH NAME: Electricity — TIME:
TYPE: Electricity, preferential — TEMP:
COMPOSITION:
x .... current
DISCUSSION:

Ge specimens. The use of electric current to determine the positions of dislocation regions in germanium. (*Note:* Method also applied to silicon.)
REF: Hogarth, C A & Baynham, H C — *Proc Phys Soc,* 71,647(1958)
GE-0248: Coffman, R E & Lesk, I A — *J Appl Phys,* 29,1493(1958)

Ge, (111) and (100) wafers fabricated with p-n-p diffused junctions and their use as transistors.

GE-0238
ETCH NAME: Irradiation — TIME:
TYPE: Particle, damage — TEMP:
COMPOSITION: — POWER:
x .... neutron
DISCUSSION:

Ge samples subjected to irradiation then etched for microscope study of induced defects.
REF: Noggles, T S & Stiegler, J O — *J Appl Phys,* 30,1279(1959)

GE-0250
ETCH NAME: Growth — TIME:
TYPE: Growth, defects — TEMP:
COMPOSITION:
x .... growth
DISCUSSION:

Ge and Si discs as-grown. A development method for growing discs rather than ingots of both materials.
REF: McLaughlin, W A & O'Connor, J — *J Appl Phys,* 29,222L(1958)
GE-0249: Sangster, R C & Carman, J N Jr — *J Chem Phys,* 23,206L(1955)

Ge material growth. A study of the contraction of germanium on being melted.
GE-0245: Bannet, A I & Longini, R L — *Phys Rev,* 116,53(1969)

Ge, dendritic growth as ribbon crystal. A study of the method of growth.

GE-0252
ETCH NAME: CP4 — TIME:
TYPE: Acid, thinning — TEMP:
COMPOSITION:
3 .... HF
5 .... HNO_3
3 .... HAc
1% .... Br_2
DISCUSSION:

Ge and InP, (100) and (111) wafers used as substrates for thin film growth of CaF_2 and BaF_2. Clean substrates with HF, then heat in vacuum at 300°C for InP and 800°C for Ge. Then deposit thin films. Solution used to thin germanium for TEM study of the thin films. BaF_2 was poor on (100) Ge due to lattice mismatch.
REF: Phillips, J M et al — *J Vac Sci Technol,* B(1),246(1983)

GE-0253
ETCH NAME: — TIME: (1) 3 min (2) 30 min
TYPE: Acid, polish — TEMP: RT RT
COMPOSITION:
(1) 1 .... HF
2 .... HNO_3
1 .... HAc
(2) 10 .... HF
1 .... HNO_3
DISCUSSION:

Ge and Si wafers cut from Float Zone (FZ) ingots with 10^4 cm^{-2} grown-in dislocations used in a study of the yield point and mobility of dislocations. Both materials were p-type: Ge, 40 Ω cm, and Si, 200—1000 Ω cm resistivity. After diamond saw cut, lap with 3, then 1 μm diamond paste. Etch polish Ge in solution (1); Si in solution (2).

REF: Stevens, D S & Tiersten, H F — *J Appl Phys,* 54,1815(1983)
GE-0292: Wallis, G & Wang, S — *J Electrochem Soc,* 106,230(1959)

Ge, (111) wafers and other orientations. A study of the effect of various etchants on recombination centers on germanium surface.

GE-0254b
ETCH NAME: Barium titanate TIME:
TYPE: Compound, junction TEMP:
COMPOSITION:

x .... $BaTi_2O_3$, powder

DISCUSSION:

Ge, (111) wafers used to fabricate indium alloyed p-n junction diodes. Brush barium titanate powder across junction, then electrical bias with a small DC battery: n-type to positive (+) and p-type to negative (−) pole. Powder will align along p-n junctions.

REF: Ibid.
GE-0255: Fuller, C S et al — *Phys Rev,* 96,21(1954)

Ge, (111) wafers fabricated with p-n junction diodes. Used barium titanate powder to define junctions.

GE-0254c
ETCH NAME: Zinc oxide TIME:
TYPE: Oxide, junction TEMP:
COMPOSITION:

x .... ZnO_2
x .... silicone (oil)
x .... toluene

DISCUSSION:

Ge, (111) wafers fabricated with indium p-n junctions. Paint solution across p-n junction and electrically bias with a small DC battery: n-type is on positive (+) terminal, and p-type negative (−). Developed the $p\text{-}^{+}$ junction line more than the actual p-n junction line.

REF: Ibid.
GE-0255: Amick, J A & Goldstein, B — *J Appl Phys,* 30,1471(1959)

Ge, (111) wafers with p-n junctions. Use toluene to thin ZnO_2/silicone solution, and paint on junction. Bias with a 45 V DC battery: n-type for positive terminal; p-type for negative terminal. Also studied silicon p-n junctions in similar work.

GE-0256
ETCH NAME: Potassium hydroxide TIME:
TYPE: Electrolytic, machining TEMP:
COMPOSITION: ANODE:

x .... 10% KOH CATHODE:

POWER:

DISCUSSION:

Ge, (100) specimens. Solution used to chemically machine a square hole through a germanium block with (100) surface orientation.

REF: Uhlir, R T — *Rev Sci Instrum,* 26,965(1955)

GE-0257
ETCH NAME: Antimony etch TIME:
TYPE: Acid, preferential TEMP:
COMPOSITION:

4 .... HF
2 .... HNO_3
4 .... H_2O
* .... $SbCl_3$

*0.2 mg per mil of solution.

DISCUSSION:

Ge and Si wafers used in a study of surface states at low temperature after various chemical treatments. Mixture of the HF:HNO_3 system studied in addition to solution shown.
REF: Morrison, S R — *Phys Rev,* 114,437(1959)

GE-0258
ETCH NAME: Indium TIME:
TYPE: Metal, structure TEMP: Elevated
COMPOSITION:

x .... In

DISCUSSION:

Ge, (111) wafers used in a study of reactions during micro-alloying. Micropyramids formed during alloying.
REF: Zuleeg, R — *J Appl Phys,* 30,9(1959)
GE-0275: Goldstein, B — *RCA Labs,* RB-85, Dec 1956

Ge wafers. A study of the dissolution of germanium by molten indium.

GE-0262
ETCH NAME: Rhodes etch TIME:
TYPE: Irradiation, defects TEMP:
COMPOSITION:

(1) x .... neutron (2) x .... Rhodes etch

DISCUSSION:

Ge wafers studied for neutron irradiation effects. After irradiation, use Rhodes etch to develop defects and structure for microscope examination.
REF: Meckel, B B et al — *J Appl Phys,* 31,1299(1960)
GE-0263a: Rhodes, R G et al — *J Electron Control,* 3,403(1957)

Reference for Rhodes etch, above. (*Note:* See GE-0205.)

GE-0264: Curtis, O L Jr et al — *J Appl Phys,* 28,1161(1957)

Ge specimens used in a study of irradiation effects on the hole lifetime of n-type germanium.

GE-0265: Hunter, L P — *Proc K Ned Akad Wet,* 61,214(1958)

Ge specimens used in a study of the anomalous transmission of X-rays by single crystal germanium. Cu K alpha radiation used. X-ray used to investigate the presence of dislocations in the material.

GE-0282
ETCH NAME: Sodium — TIME:
TYPE: Metal, contamination — TEMP:
COMPOSITION:
x .... Na^+ ions
DISCUSSION:
Ge wafers. A study of the adsorption of sodium ions on germanium surfaces.
REF: Wolsky, S P et al — *J Electrochem Soc,* 103,606(1956)

GE-0284
ETCH NAME: Hydrofluoric acid — TIME: 30 sec
TYPE: Acid, cleaning/removal — TEMP: RT
COMPOSITION:
x .... HF
DISCUSSION:
a-Ge:H and a-Si:H hydrogenated thin films deposited on (100) silicon substrates and fused quartz blanks. Materials grown in a study of these compounds. HF can be used as a general light cleaning and removal solution. See SI-0420 and QTZ-0009.
REF: Rudder, R A — *Appl Phys Lett,* 43,871(1983)

GE-0286
ETCH NAME: Laser — TIME:
TYPE: Light, annealing — TEMP:
COMPOSITION:
x .... laser
DISCUSSION:
a-Ge evaporated on fused quartz blanks, then crystallized to a single crystal (111) or (110) oriented thin film, 4000 Å thick by laser annealing. GaAs then grown as 30 μm crystallites by MOCVD for application as an electro-optical device. A developmental study of GaAs deposition on fused quartz.
REF: Shinoda, Y E et al — *IEEE,* 70,132(1982)

GE-0295
ETCH NAME: — TIME:
TYPE: Abrasive, damage — TEMP:
COMPOSITION:
x .... abrasive, powder
DISCUSSION:
Ge, (111) wafers used to fabricate p-n junction diodes in a study of the effects of subsurface residual lapping and polishing damage on device reverse characteristics. Damage depth of a sandblasted surface was 35 μm; a final polished surface was 1 μm.
REF: Buck, T M & McKim, F S — *J Electrochem Soc,* 103,593(1956)
GE-0297: Brophy, J J — *J Appl Phys,* 29,1377(1958)
Ge, (111) wafers fabricated as p-n junction diodes in a study of the effects of an etch polished surface vs. a sandblasted surface on device noise characterization. Polished surfaces showed 40 times less noise.
GE-0298: Baker, D X & Yemm, N — *Br J Appl Phys,* 8,302(1957)
Ge, (111) wafers used in a study of the depth of subsurface damage introduced by abrasive lapping. A 25-μm abrasive powder produced a damage depth of 60 μm with a combination of surface fracture and high dislocation density.

GE-0290
ETCH NAME: Lithium TIME:
TYPE: Metal, diffusion TEMP:
COMPOSITION:
x .... Li, powder
DISCUSSION:

Ge, (111) wafers and cylinders. Specimens were initially diffused with nickel, then lithium. A study of the rectification or ohmic effect of Ni or W contact whiskers on a p-n diode. (*Note:* Cylindrical lithium-drift germanium diodes were fabricated for radiation testing of the Van Allen Belt in 1961.)

REF: Tompkins, B E — *Proc IEEE,* 52,1064(1964)

GE-0299a
ETCH NAME: Thallium TIME:
TYPE: Metal, laser TEMP:
COMPOSITION: POWER:
x .... Tl
DISCUSSION:

Ge and Si specimens fabricated as a mirror-cup for laser operation. With thallium and ozone under power introduction against the cup face, reaction will produce a green light as a chemical laser, requiring less power than in an infrared laser operating as a pulsed amplifier.

REF: Peterson, J — *Science News Lett,* 132,257(1987)

GE-0299b
ETCH NAME: Chlorine TIME:
TYPE: Halogen, laser TEMP:
COMPOSITION: POWER:
x .... Cl_2
DISCUSSION:

Ge and Si specimens fabricated as a mirror-cup for laser operation. In the presence of 3-atom clusters of sodium, ionized Cl^+ reduce to 2-atom sodium clusters on a self-sustaining basis to produce visible light as a chemical laser. (*Note:* Reference work done by J L Gole, Ga Inst Technol, Atlanta, GA.)

REF: Ibid.

GE-0351
ETCH NAME: TIME:
TYPE: Electrolytic, plating TEMP: 85°C
COMPOSITION: ANODE:
86 g .... KOH CATHODE:
3.2 g .... $K_2Cr_2O_4.H_2O$ POWER:
52 g .... *GeO_2
50 ml .... H_2O

*Irradiated for ^{71}Ge.

DISCUSSION:

Ge, (111) wafers used in a study of the effects of heavy doping on the self-diffusion of germanium. The germanium isotope was electroplated as a tracer and was shown to dissolve some oxide, but it was difficult to obtain a uniform deposit.

REF: Valenta, N W & Ramassastry, C — *Phys Rev,* 106,73(1957)

GE-0287
ETCH NAME: Germanium
TYPE: Metal, optics
COMPOSITION:
x .... Ge

TIME:
TEMP: 116—440 K

DISCUSSION:

Ge specimens cut as 2″ oriented prisms with a refractive angle of 19.55.8 in a study of the material refractive index at 1.8—2.5 wavelength frequency over the temperature range shown. Refractive Index (n = 0.75 nom.) showed a linear rise with temperature increase.
REF: Lukes, F — *Cesk Cas Fys,* 8,262(1958)

GERMANIUM ALLOYS, M_xGe

General: Does not occur in nature as a metallic alloy compound. Germanium is found as a minor constituent in several minerals, notably, sphalerite, ZnS, and is collected as a by-product during ore reduction of this mineral for both its zinc and sulfur content.

There are several germanium alloys used in industry to a minor extent, today, although metallic germanium was of little use until discovery of its semiconducting properties in the late 1940s.

Technical Application: As AuGe(13%) it has wide use in assembly of Solid State microelectronic circuitry in both device and package assembly. As a thin film metallization source it is supplied in small pellets or cylinders of about 1/4-g weight, each. For evaporation from a tungsten boat, pellets are normally "evaporated to completion" for deposit thicknesses between 500—1200 Å. Sputter deposited from a crucible containing as much as a 250-g load, due to vapor pressure differences, gold evaporates first, then germanium, with slow increase of germanium concentration in the evaporating charge. This germanium increase in the pot charge produces an increasingly harder and more brittle Au/Ge compound thin film. In either method of deposition germanium content can vary quite widely as, with evaporation from a tungsten boat, germanium alloys with tungsten, such that the thin film never reaches the 13% initial concentration value, the boat becomes brittle and will shatter after the second or third use.

A widely used multilayer metallization for contact pads in device assemblies is Au/Au:Ge/Ni. After deposition it is annealed to homogenize but, if annealing temperature is high and too long, recrystallization occurs in the otherwise soft gold top layer, and it is impossible to soft gold wire bond to the surface. The Au:Ge(13%) alloy also is supplied as strips that are cut to required size, or as pre-forms in device and circuit substrate fabrication, or for package assemblies and seals.

Other germanium alloys are available, but the AuGe(13%) alloy is the most widely used.

Etching: Tri-iodide; aqua regia; KCN. See Gold section.

GERMANIUM ALLOYS

GOLD GERMANIUM ETCHANTS

AUGE-0001
ETCH NAME: Tri-iodide
TYPE: Halogen, removal
COMPOSITION:
100 g .... KI
50 g .... I_2
500 ml .... H_2O
25 ml .... MeOH

TIME:
TEMP: 40—60°C

DISCUSSION:

AuGe(13%) alloy as pellets, sheet, or cut pre-forms. Pellets used for contact metallization as ohmic contacts on GaAs:Cr, (100), and Si, (100) n-type wafers in fabrication of microelectronic devices. Multilayer evaporation as nominal thicknesses 1000 Å Au/1200 Å Au:Ge/500 Å Ni; follow with annealing in FG (85% N_2:10—15% H_2), 1—2 min at 350—450°C. Wafers were photolithographically patterned with acetone Lift-off of photo resist excess metal film after evaporation. (AUGE-0001; AUGE-0002). Similar multilayer structure sputter evaporated by E-beam from 200-g load crucibles (AUGE-0003). Alloyed as device contacts or on circuit substrates as contact cut pre-forms (AUGE-0004). The latter includes 50% reduction of AuGe by hole pattern etching to obtain a thinner AuGe film under 0.001 thick, as the 1-mil thickness is the lowest thickness available as rolled sheet. Tri-iodide solutions are standard for etch processing gold, constituent volumes can be varied, and MeOH used to assist in dissolving the I_2 and KI.

REF: Walker, P — personal application, 1972—1985

AUGE-0002: Siracusa, M — personal communication, 1978 (evaporation)

AUGE-0003: Skelly, G — personal communication, 1980 (sputter)

AUGE-0004: Marich, L A — personal communication, 1982 (pre-forms)

AUGE-0006: Marlow, G S & Das, M B — *J Solid State Electron,* 25,91(1982)

AuGe(13%) evaporated on GaAs, (100) n-type wafers and annealed in FB above 350°C showed a sheet resistance between 15—33 Ω/sq.

AUGE-0007: Heiblum, M et al — *J Solid St Electron,* 25,185(1982)

AuGe(13%) evaporated as Au/AuGe/Ni multilayers and annealed 30 seconds at 450°C with a 2000 Å SiO_2 cap, then removed with 1HF:3HNO_3. Layers sputter from a 4-crucible E-gun on GaAs, (100) substrates containing an MBE 2500 Å p-type buffer layer. There is a high resistance layer at the metal/wafer interface due to n^+ Ge penetration (?). Nickel helps adhesion but, after anneal, there were dark Ge/Ni clusters on the soft gold cover surface. (*Note:* Such artifacts have been observed by others, is common to a hardened gold surface that will not accept a soft gold wire bond, and have been referred to as "rosettes".)

AUGE-0008: Iliadis, A & Singer, K E — *J Solid State Electron,* 26,7(1983)

AuGe(13%) deposited as eutectic alloy on n-type GaAs, (100) wafers. Alloying was done in a moly boat with substrate at 110°C during evaporation, then followed by 1000 Å Au. Annealed on a quartz spade in FG, 15 sec at 450°C, then cooled in gas flow 1 min at 300°C. Below 300°C, contact is rectifying; above 300°C, ohmic. This procedure gave minimum resistance with recrystallization of Ge/Ni rectangular particle growth in the soft gold over-layer.

AUGE-0009

ETCH NAME: Potassium cyanide — TIME:

TYPE: Acid, removal — TEMP: RT to hot

COMPOSITION:

x .... x% KCN

DISCUSSION:

AuGe(13%) alloy as Au/AuGe/Ni evaporated multilayered films, or Au/Cr evaporated, and Au/Ni plated coatings on semiconductor wafers, dielectric substrates, and assembly parts. Metallization stripped with solution shown. **CAUTION:** All cyanide solutions are highly toxic, and require controlled disposal.

REF: Valardi, N — personal communication, 1983

INDIUM GERMANIUM ETCHANTS

INGE-0001
ETCH NAME: TIME:
TYPE: Acid, removal TEMP:
COMPOSITION:
x .... HF
x .... HNO_3
DISCUSSION:

InGe used as a deposited Au/InGe alloy contact on (100) InP and GaAs wafers. This was a study of alloy contacts and material defects associated with temperature processing. Other alloy combinations were Pb/Au/Ag and Cu/Ni.
REF: Grovenor, C R M — *Thin Solid Films,* 89,367(1982)

TIN GERMANIUM ETCHANTS

SNGE-0001
ETCH NAME: Aqua regia TIME:
TYPE: Acid, removal TEMP: RT
COMPOSITION:
3 .... HCl
1 .... HNO_3
DISCUSSION:

SnGe(1%) thin films grown on (100) InSb and CdTe substrates in a study of grey-tin as a zero-gap semiconductor material with infrared properties. Lattice parameters of the three materials show a close match: a = 6.49/6.48 Å. Grey tin (alpha-tin) is isometric — normal, diamond structure, is the common form of tin at room temperature, but converts to beta tin at 13.2°C. This reaction is known to produce structural failure under cold weather conditions.
REF: Farrow, R F C — *J Vac Sci Technol,* B(1),244(1983)

GERMANIDES: M_xGe_y

General: Germanides do not occur as natural metallic compounds. Germanium is a trace element in iron pyrites, such as pyrite, FeS_2, and zincblende (sphalerite), ZnS, and collected as a by-product of ore smelting. It was first obtained in any quantity from flue dust collected in the exhaust piping and stacks in iron refineries.

There are two sulfide minerals containing germanium: argyrodite, $4AgS.GeS_2$, and germanite, $Cu_2(Fe,Ge)S_4$. Germanium has had little use in industry until discovery of its semiconducting properties in the late 1940s.

Technical Application: The germanides are treated, here, as a separate unit without regard to germanium alloys, as many have individually distinctive properties, applications in Solid State, and several have been grown as singles crystals. Fewer germanides have been developed and studied than have silicides.

Germanides have been grown as single crystals or deposited as thin films for general morphological study or special characterization of superconductivity, ferromagnetics, photoconductivity. They also have been deposited and annealed like the silicides as possible blocking layers in device construction. See section on Germanium Alloys.

Etching: Mixed acids containing H_2SO_4, HF, HNO_3, alkalies, halogens, and dry chemical etching (DCE) with ionized gases.

COPPER GERMANIDE ETCHANTS

CUGE-0001
ETCH NAME: Phosphoric acid
TYPE: Electrolytic, polish
COMPOSITION:
x .... H_3PO_4

TIME:
TEMP:
ANODE:
CATHODE:
POWER:

DISCUSSION:

Cu:Ge, alloy of different compositions. Specimens used in a study of optical absorption at 4.2 K. Solution is erratic for concentrated alloys as, with greater than 7% Ge, material will not polish. Other alloys were studied.

REF: Raynes, J A — *Phys Rev,* 121,456(1961)

HOLMIUM COPPER GERMANIDE ETCHANTS

HOCG-0001
ETCH NAME:
TYPE: Acid, removal
COMPOSITION:
x .... HF
x .... HNO_3

TIME:
TEMP: (2900°C growth)

DISCUSSION:

$HoCu_2Ge_2$ single crystal specimens grown from mixed powders in an induction furnace by melting under argon. Used in a study of magnetic super-lattices at LHe temperature. RCu_2Ge_2 (R = lanthanides or actinides) materials are body-centered tetragonal structure.

REF: Schobinger-Papamantellos, P et al — *J Phys Chem Solids,* 45,695(1984)

IRON GERMANIDE ETCHANTS

FEGE-0001
ETCH NAME:
TYPE: Acid, removal
COMPOSITION:
x .... HF
x .... HNO_3

TIME:
TEMP:

DISCUSSION:

Fe_3Ge_2 thin films grown on (111) silicon wafers with or without SiO_2 pre-coated surfaces. Grown as both crystalline deposits or as an amorphous structure: a-Fe_xGe_{x-1}. Films show ferromagnetic properties but, with increase in germanium, magnetism is lost.

REF: Terzleff, P et al — *J Appl Phys,* 50,1031(1979)

FEGE-0002
ETCH NAME:
TYPE: Acid, preferential/polish
COMPOSITION:
3 .... HF
5 .... HNO_3
3 .... HAc

TIME:
TEMP: 40°C

DISCUSSION:

$FeGe_2$, (100) and (110) wafers cut from a Czochralski (CZ) grown ingot with cobalt or tin doping. Solution used to develop defect structures, and as a general polish etch.

REF: Runkin, A Y & Frolow, A A — *J Cryst Growth,* 69,131(1984)

MAGNESIUM GERMANIDE ETCHANTS

MGGE-0001a
ETCH NAME:
TYPE: Halogen, polish
COMPOSITION:
4 drops .... Br_2
250 ml .... MeOH
DISCUSSION:

TIME:
TEMP:

Mg_2Ge, (111) wafers used in a study of semiconducting characteristics of this compound. Polishing was done on a metallographic wheel covered with a cloth saturated with the solution shown. Chem/mech lap at slow speed and quench with MeOH. Gives a bright chemical polish.

REF: Kromer, H et al — *J Appl Phys,* 36,2461(1965)

MGGE-0001b
ETCH NAME:
TYPE: Acid, cleaning
COMPOSITION:
2 .... H_2SO_4
1 .... H_2O_2
200 .... H_2O
DISCUSSION:

TIME: 30 sec
TEMP: RT

Mg_2Ge, (111) wafers. Solution used to clean p-n junction areas, but does not preferentially develop the junction. Material reacts with acids and alkalies. After etch cleaning with solution shown, quench in MeOH. (*Note:* Solution is a highly diluted Caro's etch. See Germanium, Silicon, and Gallium Arsenide.)

REF: Ibid.

MGGE-0002
ETCH NAME:
TYPE: Acid, removal
COMPOSITION:
x .... HF
x .... HNO_3
DISCUSSION:

TIME:
TEMP: RT

Mg_2Ge, (111) cleaved wafers used in a study of photoconductivity. Specimens were lapped down to 0.2—0.4 mm; below this thickness material fractures. Authors state there is no known etch for this material (?). Also worked with Mg_2Si.

REF: Stella, A & Lynch, D W — *J Phys Chem Solids,* 25,1253(1965)

MGGE-0003: Morris, R G et al — *Phys Rev,* 109,1916(1958)

Mg_2Ge, single crystal specimens. Intrinsic material is n-type; dope with silver for p-type as semiconductor material. Band gap is 0.69 eV. Also worked with Mg_2Si.

MOLYBDENUM GERMANIDE ETCHANTS

MGGE-0001
ETCH NAME: Nitric acid
TYPE: Acid, removal
COMPOSITION:
(1) x .... HNO_3 (2) x .... H_2O_2

TIME:
TEMP:

DISCUSSION:

Mo_3Ge specimens, Mo_3Ge_2, Mo_2Ge_3, or alpha-$MoGe_2$ used in a general study of these materials. Both acids are rapid etchants. Fused carbonates and nitrates react violently. All of the compounds will resist non-oxidizing solutions.

REF: Brochure: Molybdenum, Climax Molybdenum Company (no date/number)

NIOBIUM GERMANIDE ETCHANTS

NBGE-0001

ETCH NAME: RF plasma — TIME:

TYPE: Ionized gas, removal — TEMP:

COMPOSITION: — GAS FLOW:

x .... CF_4 — PRESSURE: 6 Pa

x .... Ar — POWER: 200 V

DISCUSSION:

a-Nb_3Ge compound co-deposited by E-beam evaporation as an amorphous layer on sapphire blanks. Thickness of thin films was 80—100 μm, and they were patterned by standard photolithographic techniques. RF plasma etching is rapid with pure CF_4, but controllable with argon as an inert carrier gas.

REF: Kato, W — *Jpn J Appl Phys,* 23,1536(1984)

NBGE-0002

ETCH NAME: — TIME:

TYPE: Acid, removal — TEMP:

COMPOSITION:

x .... HF

x .... HNO_3

x .... HAc

DISCUSSION:

Nb_3Ge thin films grown on (100) germanium substrates by CVD in a study of high T^c values. Various solution concentrations were used, and will etch both the compound and germanium substrate.

REF: Suzuki, N et al — *Jpn J Appl Phys,* 23,991(1984)

NBGE-0003: Ohishima, S et al — *Jpn J Appl Phys,* 22,264(1983)

Nb_3Ge specimens fabricated by the Shock Sinter method for study of superconducting properties. Material is in space group A15.

TERBIUM COPPER GERMANIDE ETCHANTS

TBCG-0001

ETCH NAME: — TIME:

TYPE: Acid, removal — TEMP: (2900°C growth)

COMPOSITION:

x .... HF

x .... HNO_3

DISCUSSION:

$TbCu_2Ge_2$ specimens. See discussion under HOCG-0001.

REF: Schobinger-Papamantellos, P et al — *J Phys Chem Solids,* 45,695(1984)

VANADIUM GERMANIDE ETCHANTS

VGE-0001
ETCH NAME: Kalling's etch
TYPE: Acid, polish
COMPOSITION:
100 ml .... HCl
5 g .... $CuCl_3$
100 ml .... EOH
TIME: 10—60 sec
TEMP: RT
DISCUSSION:

V_3Ge and V_3Si arc melted, fabricated and used in a study of different metallic compounds. Solution used as a general removal and polishing etchant. See Cesium Platinide: CEPT-0001 for list of other compounds studied.

REF: Slepowronsky, M et al — *J Cryst Growth,* 65,293(1983)

PHYSICAL PROPERTIES OF GERMANIUM ARSENIDE, $GeAs_2$

Classification	Arsenide
Atomic numbers	32 & 33
Atomic weight	222.42
Melting point (°C)	>800
Boiling point (°C)	
Density (g/cm^3)	
Hardness: (Mohs — scratch)	7
Crystal structure (Isometric — normal)	(100) cube
Color (solid)	Steel grey
Cleavage (cubic)	(001)

GERMANIUM ARSENIDE, GeAs

General: Does not occur as a natural metallic compound. Germanium is a trace element in certain sulfites, such as pyrite, FeS_2, and sphalerite, ZnS, obtained during the roasting and reduction process of these ores for their iron and zinc content. Germanite, $Cu_3(Fe,Ge)S_4$, and argyrodite, $4Ag_2S.GeS_2$ are the two known germanium minerals.

There has been no use of the compound in the metal industry to date, other than as an additive in solder alloys.

Technical Application: Germanium Arsenide is a IV—V compound semiconductor. It has been grown and evaluated as a single crystal for both etching and semiconductor characteristics, but has been of little use due to the preeminence of silicon and gallium arsenide.

Etching: H_2O_2, HNO_3, HF:HNO_3 mixtures; aqua regia; alkali.

GERMANIUM ARSENIDE ETCHANTS

GEAS-0001a
ETCH NAME:
TYPE: Acid, removal
COMPOSITION:
1 .... 20% NaOH
1 .... H_2O_2
TIME:
TEMP: RT

DISCUSSION:

GeAs, (111) wafer. Cut dice used in an etch development study. The solution shown will etch GeAs at room temperature.

REF: Robbins, H & Schwartz, B — personal communication, 1958

GEAS-0001b

ETCH NAME: TIME:

TYPE: Acid, removal TEMP: Hot

COMPOSITION:

2 .... HCl

1 .... HNO_3

DISCUSSION:

GeAs, (111) wafers and cut dice. Solution etches slowly. See GEAS-0001a for additional discussion.

REF: Ibid.

GEAS-0001c

ETCH NAME: TIME:

TYPE: Acid, removal TEMP: RT

COMPOSITION:

1 .... HF

1 .... HNO_3

DISCUSSION:

GeAs, (111) wafers and dice. Solution etch reaction is moderate to vigorous. See GEAS-0001a for additional discussion.

REF: Ibid.

GEAS-0001d

ETCH NAME: TIME:

TYPE: Acid, removal TEMP: RT

COMPOSITION:

(1) x .... 3% H_2O_2 (2) x ... 10% H_2O_2

DISCUSSION:

GeAs, (111) wafers and dice. Both concentrations showed little attack at room temperature. See GEAS-0001a for additional discussion.

REF: Ibid.

GEAS-0001e

ETCH NAME: TIME:

TYPE: Acid, removal TEMP: RT or hot

COMPOSITION:

(1)	(2)	(3)
x .... *HNO_3	1 .... *HNO_3	2 .... *HNO_3
	2 .... H_2O	1 .... H_2O

*Yellow and red fuming nitric acids evaluated.

DISCUSSION:

GeAs, (111) wafers and dice. All three mixtures shown will react when used hot with moderate to vigorous reaction. Red fuming is the more vigorous. Standard concentration HNO_3, sometimes called White fuming nitric will not etch GeAs. (*Note:* HNO_3 concentrations are 70% (white fuming), 72% (yellow fuming), and 74% (red fuming).)

REF: Ibid.

PHYSICAL PROPERTIES OF GERMANIUM NITRIDE, Ge_3N_4

Classification	Nitride
Atomic numbers	32 & 7
Atomic weight	273.83
Melting point (°C)	450 del.
Boiling point (°C)	
Density (g/cm^3)	
Refractive index (n =)	2.05
Hardness (Mohs — scratch)	6—7
Crystal structure (hexagonal — normal) alpha	$(10\bar{1}0)$ prism
(isometric — normal) beta	(100) cube
Color (solid)	Clear/brownish
Cleavage (none — conchoidal fracture)	None

GERMANIUM NITRIDE, Ge_3N_4

General: Does not occur in nature as a compound. Nitrogen does not form metallic compounds though there are many mineral nitrates and nitrites of primary industrial interest as fertilizers.

There is no major use of this nitride in the metals industry, but surface nitridization is a standard process used for surface hardening and, if there is germanium present in an alloy, some germanium nitride will be formed.

Technical Application: Not used to any extent in Solid State processing to date, but under evaluation as a thin film coating for optical, dielectric, and surface passivation applications similar to other metal oxides and nitrides. Most metal oxides-nitrides form isomorphous series with 100% miscibility of the two compounds, and are studied in that regard. The Ge_3N_4 compound as a thin film has been evaluated for MIS device structures, but shows free Ge separation with temperature annealing such that it is not as stable as its silicon nitride counterpart.

Note that all oxides and nitrides do not show distinct cleavage, but are prone to conchoidal fracture common to all glasses. In most applications as thin films they are amorphous in structure, such as a-Ge_3N_4.

Etching: Soluble in HF, mixed acids and alkalies.

GERMANIUM NITRIDE ETCHANTS

GEN-0001

ETCH NAME: TIME: 10—20 min

TYPE: Acid, removal TEMP: Boiling

COMPOSITION:

18 g .... NaOH

5 g .... $KHC_8H_4O_4$

100 ml .... H_2O

DISCUSSION:

Ge_3N_4, and $Ge_3O_{1-x}N_x$. This etch was originally developed for silicon oxynitride. Substrates used were single crystal (111) Si and Ge wafers, quartz crystal AT-cut blanks, and alumina pressed powder blanks. Nitride deposit thickness ranged from 500 to 3500 Å with a nominal 2000 Å for study evaluations, and were grown by DC plasma sputter using a germanium bar in a nitrogen atmosphere with and without oxygen. This was an original developmental study of growing nitride and oxynitride thin films as passivation layers and

etching masks. The solution shown was specifically developed for compatibility with KMER photo resist pattern processing. See Silicon Nitride section for additional information.
REF: Mann, J E & Walker, P — Electrochem Soc Spring Meet, Dallas, TX, May 1967

GEN-0002
ETCH NAME: Hydrofluoric acid — TIME:
TYPE: Acid, removal — TEMP: RT
COMPOSITION:
(1) x .... HF, conc.
(2) 1 .... HF
10 .... H_2O
DISCUSSION:
Ge_3N_4 and $Ge_3O_xN_y$ thin films DC plasma sputtered on silicon and germanium wafers of different orientations in a developmental deposition and etching study of nitrides and oxynitrides. Other materials studied were Si_3N_4; $Si_3O_xN_y$; and AlN. The dilute solution is more controllable for thin film photo resist lacquer patterning, and both were used for removal. See GEN-0001, and section on Silicon Nitride for additional data. Nitride color chart developed used a refractive index of n = 2.00.
REF: Walker, P — personal development, 1966

GEN-0003
ETCH NAME: Nitrogen — TIME:
TYPE: Gas, densification — TEMP:
COMPOSITION:
x .... N_2
DISCUSSION:
Ge_3N_4 thin films deposited on GaAs, (100) wafers used in fabricating MIS high frequency devices. Deposit nitride in Ar/N_2 at 50 W, and 1.8×10^{-2} Torr pressure at close to RT. Final structure was $Al/Ge_3N_4/GaAs(100)$. Annealing in H_2 produces leakage and nitride phase separation with free Ge. Nitrogen annealing increases density of nitride. Refractive index was n = 2.05.
REF: Chung, Y et al — *Thin Solid Films,* 103,193(1983)

PHYSICAL PROPERTIES OF GERMANIUM OXIDE, Ge_2O_3

Classification	Oxide
Atomic numbers	32 & 8
Atomic weight	104.60
Melting point (°C)	1086
Boiling point (°C)	
Density (g/cm³)	6.24
Hardness (Mohs — scratch)	6—7
Crystal structure (tetragonal — normal)	(110) prism
Color (solid)	Clear/grayish
Cleavage (none — conchoidal fracture)	None

GERMANIUM OXIDE, Ge_2O_3

General: Does not occur in nature as an oxide compound, though it may appear as a very minor oxidation surface coating on the minerals germanite, $Cu_3(Fe,Ge)S_4$, and argyrodite, $4AgS.GeS_2$, but the oxide is not considered a mineral species. There is no major use in the metals industry. The soluble GeO_2 variety has been used as a pigment and an additive to glass.

Technical Application: No major use in Solid State processing at the present time. In fabricating germanium devices the oxides are considered detrimental to electrical characteristics, such that they are removed as native oxides, prior to metallization, diffusion, etc.

As GeO_2 the material has been grown and studied in wafer form in a general glass development study, and deposited on germanium as a thin film for optical properties characterization. The Ge_2O_3 form has been sputter deposited, then converted to the nitride in a study as an isomorphous series. The GeO and Ge_2O_3 forms are insoluble in water; GeO_2 has partial solubility.

Etching: Varies with compound formula: water, acids, alkalies.

GERMANIUM OXIDE ETCHANTS

GEO-0001
ETCH NAME: Methyl alcohol — TIME:
TYPE: Alcohol, cleaning — TEMP: RT
COMPOSITION:
x .... MeOH
DISCUSSION:

GeO_2, as glass blanks used in a study of glass development. Germanium powder was fired in an air furnace at 1200 to 1690°C. Core sections were cut from the glass mass as grown, then cut into blank sections and polished with 400-grit abrasive in a petroleum oil carrier, and rinsed in MeOH. Follow with annealing at 230°C, 30 min.

REF:

GEO-0002
ETCH NAME: Oxygen — TIME:
TYPE: Gas, forming — TEMP: 500—550°C
COMPOSITION: — PRESSURE: 680 atm max.
x .... O_2
x .... pressure
DISCUSSION:

GeO_2, grown as thin films on Ge, (111) and other oriented wafers. Growth rate is greatest on the (111). GeO_2 refractive index was maximum n = 2.7 at 340 atm; at 1 atm growth pressure n = 1.6 to 1.9.

REF: Crisman, E E et al — *J Electrochem Soc,* 129,1845(1982)

GEO-0003
ETCH NAME: — TIME:
TYPE: Acid, removal — TEMP: RT
COMPOSITION:
1 .... HF
1 .... HNO_3
20 .... HAc
DISCUSSION:

Ge_2O_3, DC sputtered thin films with conversion to Ge_3N_4 as an isomorphous series of compounds with variable oxygen content. This was part of original development of DC plasma sputtered thin films and a general study of such films. Also studied SiO_2 with conversion to Si_3N_4, and the Al_2O_3 to AlN, all three oxides/nitrides as ismorphous series. Deposits were on (111) germanium wafers from a germanium bar source. A silicon bar was used with (111) silicon substrates, and an aluminum bar with both silicon and germanium wafers. The solution shown was used as a general etch on all materials and films.

REF: Walker, P & Mann, J E — development work, 1964

PHYSICAL PROPERTIES OF GERMANIUM SELENIDE, Ge_2Se_3

Classification	Selenide
Atomic numbers	32 & 34
Atomic weight	181.08
Melting point (°C)	900
Boiling point (°C)	
Density (g/cm³)	
Hardness (Mohs — scratch)	4—5
Crystal structure (isometric — normal)	(100) cube
Color (solid)	Clear/grayish
Cleavage (cubic)	(001)

GERMANIUM SELENIDE, Ge_2Se_3

General: Does not occur as a natural metallic compound though there are several other minerals as metallic selenides and tellurides. The crystal structure, and hardness shown under physical properties, above, are based on the mineral guanajuatite, Bi_2Se_3.

There has been no use of germanium selenide in the metals industry, to date, though there is possible application in glass making.

Technical Application: The compound is currently under development in Solid State processing as an amorphous thin film for compound characterization. It has been evaporated as a variable Ge_xSe_{1-x} amorphous film, and is a probable IV—VI compound semiconductor.

Etching: Mixed acids, alkalies. Dry chemical etching (DCE) with ionized gases.

GERMANIUM SELENIDE ETCHANTS

GESE-0001a

ETCH NAME:	TIME:
TYPE: Ionized gas, removal	TEMP:
COMPOSITION:	GAS FLOW:
(1) x .... CF_4	PRESSURE:
(2) x .... SF_6	POWER:

DISCUSSION:

Ge_xSe_{x-1} thin films deposited in a structure study. Evaporated layered structures of Ag/ Ge_xSe_y and Ag/Se/Ge_xSe_y. Silver was evaporated and used as an etch mask as it will not etch in CF_4, but will etch with SF_6 after photo resist patterning. The CF_4 will etch both selenium and the selenide.

REF: Huggett, P G & Iehmann, H W — *J Electron Mater*, 14,201(1985)

GESE-0001b

ETCH NAME:	TIME:
TYPE: Alkali, removal	TEMP:

COMPOSITION:

x .... NaOH

x .... Na_2S

DISCUSSION:

Ge_xSe_x-1 thin films with silver evaporated as a mask: Ag/Ge_xSe_y and Ag/Se/Ge_xSe_y used in a structure study. Silver will not etch in the solution shown, but germanium dissolves rapidly and selenium slowly.

REF: Ibid.

PHYSICAL PROPERTIES OF GERMANIUM SULFIDE, GeS

Classification	Sulfide
Atomic numbers	32 & 16
Atomic weight	104.65
Melting point (°C)	530
Boiling point (°C)	
Density (g/cm^3)	3.3—4.0
Hardness (Mohs — scratch)	4—5
Crystal structure (hexagonal — rhombohedral)	$(10\bar{1}1)$ rhomb
Color (solid)	Black
Cleavage (rhombohedral)	$r(10\bar{1}1)$

GERMANIUM SULFIDE, GeS

General: Does not occur as a natural compound though there are a number of other metal sulfide minerals.

No use in the metals industry to date. It has application as a black pigment in paints, glass and enamels.

Technical Application: No major use in Solid State processing to date, but is under current morphological study. Material was grown as single crystal platelets by sublimation, and is said to have good (100) cleavage, which would mean the compound as-grown was isometric — normal (cubic) structure rather than hexagonal system. Platelets were doped in the study as a possible IV—VI compound semiconductor. There also is a GeS_2 structure form.

Etching: Mixed acids, alkalies, and NH_4OH.

GERMANIUM SULFIDE ETCHANTS

GES-0001

ETCH NAME: Ammonium hydroxide — TIME:

TYPE: Base, cleaning/removal — TEMP: RT

COMPOSITION:

x .... NH_4OH

DISCUSSION:

GeS single crystal platelets grown by sublimation, and doped with either silver or phosphorus, p- and n-type, respectively. Authors say that platelets are easily cleaved on (100). Material was used in a general morphological study.

REF: Bhatia, K L et al — *J Phys Chem Solids,* 45,1189(1984)

PHYSICAL PROPERTIES OF GERMANIUM TELLURIDE, GeTe

Classification	Telluride
Atomic numbers	32 & 52
Atomic weight	200.20
Melting point (°C)	>1000
Boiling point (°C)	
Density (g/cm^3)	
Hardness (Mohs — scratch)	6—7
Crystal structure (isometric — normal)	(100) cube
Color (solid)	Grey/black
Cleavage (cubic)	(001)

GERMANIUM TELLURIDE, GeTe

General: Does not occur as a natural compound though there are a number of other minerals as metallic tellurides and selenides.

There has been no use of the compound in the metals industry to date, other than as a compound semiconductor.

Technical Application: Not used in Solid State processing to any extent, to date. The material has been evaluated as a single crystal IV—VI compound semiconductor with the development of etching solutions, and for general morphological study. More recently, it has been evaporated as a thin film in studying the Tc values and variation of the tellurium fraction.

Etching: Mixed acids of $HF:HNO_3:HAc/H_2O$, and HNO_3:chromates.

GERMANIUM TELLURIDE ETCHANTS

GETE-0001

ETCH NAME: TIME:

TYPE: Acid, removal TEMP:

COMPOSITION:

x .... HF

x .... HNO_3

DISCUSSION:

GeTe as single crystal specimens evaluated for Tc values and change with variation of tellurium fraction.

REF:

PHYSICAL PROPERTIES OF GOLD, Au

Classification	Transition metal
Atomic number	79
Atomic weight	196.9
Melting point (°C)	1063
Boiling point (°C)	2808 (2966)
Density (g/cm³)	19.3
Thermal conductance (cal/sec)(cm²)(°C/cm) 0°C	0.743
Specific heat (cal/g) 25°C	0.0312
Latent heat of fusion (cal/g)	14.95
Heat of fusion (k-cal/g-atom)	3.03
Heat of vaporization (k-cal/g-atom)	81.8
Atomic volume (W/D)	10.2
1st ionization energy (K-cal/g-mole)	213
1st ionization potential (eV)	9.223
Electronegativity (Pauling's)	2.4
Covalent radius (angstroms)	1.34
Ionic radius (angstroms)	1.37 (Au^{+1})
Thermal expansion ($\times 10^{-6}$) 0—100°C	14.16
Electrical resistivity (micro ohms-cm) °C	2.125
20°C	2.44
Yield point (psi)	500
Poisson ratio	0.42
Young's modulus (psi $\times 10^6$) — 60% cold worked)	11.2
Tensile strength (psi — annealed)	18,000

Hall constant (ohms/cm/gauss $\times 10^{-13}$)	6.87
Magnetic susceptibility (cgs $\times 10^{-6}$)	0.15
Cross section (barns)	98.8
Hardness (Mohs — scratch)	2/5—3
(Vicker's — kgf/mm^2)	25
Crystal structure (isometric — normal)	(100) cube, fcc
Color (solid)	Yellow gold
(colloidal suspension)	Deep ruby red
Cleavage (none — hackly fracture)	None

GOLD, Au

General: Occurs as a native element and is the chief ore of gold. Ordinary gold contains up to 16% silver and varies in color as yellow- to white-gold depending upon silver content. The purest gold, "sponge gold", comes from Australia and is 99.91% Au:0.09% Ag. Electrum, pale yellow to white, contains 36% silver as a 1:1 mixture and varies from 1.5:1, 2:1 to 2.5:1 and 3:1 (15.1% silver). These mixtures can be differentiated by their specific gravity (density). Orange-red varieties contain up to 20% copper, and the "black-gold" of Australia contains bismuth; also rhodium. Gold is fusible in the Bunsen burner flame at 1100°C, but is not acted upon by fluxes and is insoluble in any single acid. It can be etched in aqua regia, but etching is incomplete if more than 20% silver is present. Gold is widely distributed in the earth's crust and can be extracted from sea water, but is primarily found in quartz veins or alluvial placer deposits in minable quantities. Hydraulic washing of placer deposits is the main method of separation and is used in California, Alaska, Russia, Brazil, Australia, etc. The "quartz reefs" of Australia and Africa are representative of vein mining. The Coeur d'Alene and Randsberg districts — Idaho and California, respectively — are mined for both gold and silver.

Gold was one of the elements known to ancient man and, because of its ductility and malleability — workability — as well as its resistance to acids and atmospheres, has had major use. Initial use was for objets d'art, jewelry, as gold leaf, in barter exchange, and finally, for its monetary value. Today, it is the world-wide base for currency value as set against the U.S. dollar.

Commercially, there are many gold alloys which are measured by carat weighed from 8 to 24, the latter equivalent to pure gold. One carat is approximately 205 milligrams. Depending upon the additive (silver, nickel, zinc, copper, even brass) gold alloys also are classified by color, as white, yellow, grey, red, purple, blue, and green.

Industrial uses of gold are primarily for protective surface coatings to resist corrosion and are applied to most of the known metals and their alloys for this purpose. It can be as gold-cladding, a thin foil of gold being heat and pressure formed on a metal surface; by evaporation in vacuum; or by electrolytic plating. The latter is the more widely used for general surface coating, as first, a nickel flash, followed by a thin gold plating of about 50 micro-inches (μ''). The nickel undercoat may be replaced by chromium, copper, or other metals and is used as a tie-down layer for gold, particularly on polished surfaces. Thicknesses of the gold/nickel vary by application requirements with a minimum amount of gold thickness due to cost. Gold plating baths are either acid or cyanide with gold chloride salts and other metal salts for coloring. The cyanide solutions are more widely used and supplied commercially in 1 to 2 oz Au/gallon solutions. High purity soft gold coatings are 5 9's pure, but decorative or protective coatings contain other metals as hardening agents.

Gold is still widely used for jewelry, with 18 carat being a general standard for rings, earrings, and bracelets. It also is used for decorative purposes on a wide variety of materials in several pleasing forms: stippled, marbleized, antique, royal, and brushed, and as fired onto glass, ceramics, or other metals. Today, brass is used as a low-cost replacement.

Gold can be reduced to the colloidal state in liquid suspension as a ruby-red solution. It was widely used during the Middle Ages in Europe for staining window glass as gold is unaffected by sunlight and will not fade.

Highly polished gold surfaces have been used as mirrors since ancient times though they have been largely replaced by silver and, today, by aluminum or nickel evaporated on glass.

Technical Application: As a diffusion element in silicon, gold can act as both a donor or acceptor and is used to control electron mobility, T_{rr}, to produce fast switching devices. Its primary use is an evaporated or sputtered metallized electrical contact on semiconductor or similar devices and as a bonding wire or strap in device, circuit substrates and package assemblies. This includes metal thin films or wire on semiconductor devices, proper and discrete components such as capacitors, resistors, and coils. Circuit substrates include pressed powder alumina, beryllia; fused quartz and sapphire; and single crystal quartz and sapphire; in addition to contact pins and package parts. Many of the substrates and parts are gold plated as plating is much lower in cost than evaporation or sputtering.

Pure gold and gold/silver single crystal ingots have been grown and studied for morphological and defect data. One extensive study has been made with single crystals of gold/aluminum as there is major concern and interest in the reaction phenomena between gold and aluminum with regard to device stability.

CAUTION: Gold in contact with aluminum produces what is called ''Purple Plague''. Several of the Au_xAl_y alloys are brilliant purple in color, and extremely brittle. Such reaction can destroy the electrical contacts of semiconductor devices under operational load as well as during accelerated temperature life tests. Under operational load, the reaction may occur over a period of several months after the device has been assembled with eventual failure in the field.

Etching: Inert in single acids, aqua regia, cyanides, and iodine etchants.

GOLD ETCHANTS

AU-0001a

ETCH NAME: Aqua regia
TYPE: Acid, removal
COMPOSITION:
3 .... HCl
1 .... HNO_3

TIME: Variable
TEMP: RT & 35°C boiling
RATE: 10—15 μm/min @ RT
25—50 μm/min @ 35°C

DISCUSSION:

Au, thin films and specimens. Aqua regia is the oldest known etch for gold dating back to the Middle Ages in Europe. Mix in an open container and allow solution to sit until gas evolution reduces and solution turns a deep transparent red-yellowish color. Used as a general etch for gold. In conjunction with photo resist patterning of thin films used to etch device structures and circuits on substrates. It is a rapid etch and is not as widely used for patterning as are other solutions. (*Note:* Do not store in a closed container due to continuing gas evolution.)

REF: Missel, L I & Glang, R — *Handbook of Thin Film Technology,* McGraw-Hill, New York, 1970

AU-0002: Harper, C A — *Handbook of Materials and Processes for Electronics,* McGraw-Hill, New York, 1970

AU-0003a: Walker, P — personal application, 1957—85

Au, thin films and specimens. After gold evaporation on silicon wafers and temperature drive-in, solution used at RT to remove residual gold from surfaces. Solution used to shape or clean gold as a cut pre-form or wire alloy contact on germanium, silicon, and compound semiconductor devices. With evaporated and E-beam sputtered gold thin films on both

devices and substrates in conjunction with photolithographic patterning. Solution also used as a heavy metal removal etch for cleaning of silicon dioxide and nitride or wafer surfaces prior to diffusion, epitaxy, or metallization. As the cleaning solution of quartz tubing used for diffusion and epitaxy, in the processing of quartz radio frequency blanks with gold electrodes, and in the general cleaning of vacuum internal fixturing. After etching with aqua regia, follow with heavy DI water washing; dry under IR lamps, or with nitrogen blow-off.
AU-0004: Nichols, D R — *Solid State Technol,* December 1979

Au, evaporation film build-up in vacuum evaporators and on fixturing. Etch remove by soaking parts in aqua regia and follow with water washing and drying.

AU-0019
ETCH NAME: TIME:
TYPE: Thermal, structure TEMP: to 360°C
COMPOSITION:
x .... heat
DISCUSSION:

Au, thin films deposited on soda-lime glass which was first cleaned with ultrasonic DI water. 300 Å of gold evaporated from a tungsten boat. Gold was polycrystalline with grains up to 500 Å long and preferred (111) orientation parallel to the substrate surface. In vacuum, 1 h at 360°C developed circular pin-holes and hillocks. After 10 min at 630°C, in air, developed either (1) island agglomerates or (2) rod- and bead-like structures to 500 Å thick with some gold color bleached white due to small particle light scattering. Author recommends that for good gold films temperature should be under 630°C. (*Note:* The bleaching effect has been observed on thin films of gold, tungsten, and nickel deposited on alumina, fused quartz and OFHC copper after being subjected to heat treatment up to 400°C on a hot plate in air.)
REF: Zito, R R — *Thin Solid Films,* 114,241(1984)

AU-0003b
ETCH NAME: Aqua regia, dilute TIME: 5—15 sec
TYPE: Acid, cleaning TEMP: RT
COMPOSITION:
3 .... HCl
1 .... HNO_3
20—50 .... H_2O
DISCUSSION:

Au, slugs or wire used for evaporation and sputter metallization. Degrease with TCE, RT; wash in DI H_2O; then rinse in MeOH and dry. Etch clean slugs and wire in solution shown, water rinse and hold under MeOH until placed in the vacuum system. After gluing or alloy mounting gold sheet on a sputter target base, dip etch clean, then water wash and wipe with MeOH with a final bake in air oven at 150°C for 8—10 h before installation in the sputter system. Also used with photo resist lacquer pattern etching of thin film gold. (See AU-0003.)
REF: Ibid.

AU-0017
ETCH NAME: Chrome regia TIME: Variable
TYPE: Acid, removal/preferential TEMP: RT
COMPOSITION:
3 .... HCl
1 .... 10—20% CrO_3

DISCUSSION:

Au, thin films and specimens. Used as a general etch for gold similar to aqua regia. On single crystal gold and gold alloys is a preferential etch.

REF: ASTM Pamphlet E701-70

AU-0018

ETCH NAME: Selenic acid — TIME:

TYPE: Acid, removal — TEMP: Hot

COMPOSITION:

x .... H_2SeO_4

DISCUSSION:

Au specimens. Selenic acid is a slow, general etch for gold but not widely used due, in part, to the toxicity of selenium.

REF: Hodgman, C D — *Handbook of Chemistry and Physics,* 27th ed, Chemical Rubber Co, Cleveland, OH, 1943, 588

AU-0003c

ETCH NAME: Potassium cyanide — TIME:

TYPE: Acid, removal — TEMP: RT

COMPOSITION:

x .... KCN

x .... H_2O

DISCUSSION:

Au specimens and thin films. Various concentrations of KCN will dissolve gold. Solutions have been used to strip gold from alumina, quartz, sapphire substrates and metal assembly parts used in semiconductor device assemblies. Also used to strip gold from semiconductor wafers.

REF: Ibid.

AU-0003d

ETCH NAME: Tri-iodide — TIME: 1—10 min

TYPE: Halogen, removal — TEMP: Hot (70°C)

COMPOSITION:

400 g .... KI

200 g .... I_2

1000 ml .... H_2O

DISCUSSION:

Au, specimens and thin films. Constituent quantities can vary and include methyl alcohol to aid in dissolving iodine. Solution is mixed in 5000 ml Pyrex beakers on a magnetic stirring hot plate, requiring up to 1 h for complete mixing unless 100 ml MeOH is added. A single 250 ml beaker of solution can be used hot (40—60°C) for hours to days in processing photo resist patterned thin films in fabricating circuit substrates before solution shows marked depletion and slow reactivity. This is one of the most widely used etchants for gold thin film processing as it is sufficiently slow and controllable for etching fine-line definition (to less than 1 μm line widths) and, when used at about 60°C, will remove 2000 Å of gold in approximately 45 sec. Solutions are nontoxic, as are cyanides, and residual iodine can be removed by washing surfaces with methanol. (*Note:* Tri-iodide solutions with the ethanol replacing water are called tincture of iodine, which is a major medical antiseptic.)

REF: Ibid.

AU-0008
ETCH NAME: TIME: 4 min
TYPE: Electrolytic, polish TEMP: RT
COMPOSITION: ANODE:
8 g .... KCN CATHODE:
5 g .... $AuCl_3$ POWER: 10 V
4 g .... $K2CO_3$
100 ml .... H_2O
DISCUSSION:
Au foil. Used to polish specimens prior to 8 KeV Ar^+ ion bombardment cleaning. Also used zinc and aluminum with other etchants.
REF: Cunningham, R L et al — *J Appl Phys,* 31,839(1960)

AU-0009
ETCH NAME: TIME: 2 min
TYPE: Electrolytic, polish TEMP: 120°C
COMPOSITION: ANODE: Au
100 g .... $K_2Cr_2O_7$ CATHODE:
1000 ml .... H_2SO_4 POWER:
DISCUSSION:
Au specimens used in a study of the cleaning of metal surfaces. (*Note:* This is similar to the standard "glass cleaner" etch used at RT as a wet chemical etching (WCE) solution on laboratory glassware.)
REF: Patzer, L et al — *J Electrochem Soc,* 118,451(1971)

AU-0010
ETCH NAME: TIME:
TYPE: Acid, preferential TEMP:
COMPOSITION:
5 g .... $(NH_4)_2S_2O_8$
50 ml .... NH_3
100 ml .... H_2O
DISCUSSION:
Au, (111) and (100) single crystal blanks. Solution is a preferential etch on these surfaces.
REF: Bowles, J S & Boas W — *J Inst Met,* 501(1947—48)
AU-0014: Gilbertson, L I & Fortner, O W — *Trans Electrochem Soc,* 81,199(1942)
Au, (111) and (100) oriented specimens. Solution shown to be preferential on both orientations.

AU-0011
ETCH NAME: TIME:
TYPE: Acid, preferential TEMP:
COMPOSITION:
15 ml .... HF
15 ml .... HNO_3
45 ml .... HCl
25 ml .... H_2O
DISCUSSION:
Au, (111) wafers and other orientations. Etch pits showed that surfaces were within 1° of orientation as verified by X-ray and Laue photographs.
REF: de Sy, A & Haemiers, H — *Aluminum,* 24,96(1942)

AU-0012
ETCH NAME: Heat TIME:
TYPE: Heat, oxide removal TEMP: 200°C
COMPOSITION:
x .... heat
DISCUSSION:

Au specimens used in a study of photoelectric yield in the vacuum ultraviolet. Heat specimens in vacuum to remove traces of Au_2O_2, Au_2O_3 and Au_2O. Gold oxides are unstable, very thin, and all decompose under 250°C. (*Note:* Gold is considered as inert to oxidation under normal processing conditions but, as shown, oxides can occur. Very spotted and usually under 100—200 Å thick.)

REF: Walker, W C et al — *J Appl Phys,* 26,1366(1955)

AU-0013
ETCH NAME: Heat TIME:
TYPE: Thermal, forming TEMP:
COMPOSITION:
x .... heat
DISCUSSION:

Au, single crystal spheres formed by heating the tip of a single crystal wire. Wire was 1—2 mm diameter; spheres, 1—3 mm diameter. Used as electrodes in electrocrystallization studies. Other materials studied were Ag and Cu.

REF: Roe, D K & Gerisder, H — *J Electrochem Soc,* 110,350(1963)

AU-0015
ETCH NAME: TIME:
TYPE: Acid, removal TEMP:
COMPOSITION:
1 .... HF
3 .... HNO_3
2 .... HAc
DISCUSSION:

Au, as ^{199}Au diffused into silicon wafers in a gold diffusion study. Solution used for controlled removal as a form of step-etching to determine the radioactive gold concentration, and crystallographic location of diffused gold.

REF: Spokel, G J & Fairfield, J M — *J Electrochem Soc,* 112,200(1965)

AU-0016
ETCH NAME: CP4 TIME:
TYPE: Acid, removal TEMP:
COMPOSITION:
30 .... HF
50 .... HNO_3
30 .... HAc
0.6 .... Br_2
DISCUSSION:

Au, diffused into silicon in a study of lifetime and capture cross-section. Solution used as a polish etch and step-removal etch in determining depth of gold penetration.

REF: Davis, W D — *Phys Rev,* 114,1006(1959)

AU-0017
ETCH NAME: TIME: Variable
TYPE: Metal, diffusion TEMP: 200°C+
COMPOSITION:
x .... Fe, Ni or Co
DISCUSSION:

Au, thin film deposited by CVD to 300 nm thickness on top of previously deposited Fe, Ni, or Co thin films using fused quartz substrates. Assemblies were then heat treated in air. Borosilicate glass was first used for substrates, but sodium from the glass diffused up into the gold and affected results. All three metals diffuse up through the gold film to the surface by solid-solid diffusion, and showed oxidation products on the gold surfaces. (*Note:* Similar solid-solid diffusion from devices and substrates have been observed with tungsten, zinc from brass, ti-tungsten, cobalt, nickel, and is a basic problem in accelerated life test, and long-term device failure under operating load.)

REF: Swartz, W E et al — *Thin Solid Films,* 114,349(1984)

AU-0018: Parmigiani, F et al — *J Appl Phys,* 57,2524(1985)

Au, thin films 20—200 Å thick deposited on glass substrates used in a study of the optical properties of gold clusters. For TEM study replication, films were prepared by coating the gold surface with a polymer resin (Technevit 4071) dissolved in acetone, the resin then coated with carbon. The replicated carbon thin film can be removed from the glass for TEM study by thermal shock and by dissolving resin in acetone.

AU-0019
ETCH NAME: Water TIME:
TYPE: Acid, preparation TEMP:
COMPOSITION:
x .... H_2O, jet
DISCUSSION:

Au, (100) single crystal thin films deposited on NaCl, (100) substrates. A water jet was used to cut a 0.5 mm diameter hole through the sodium chloride from the back side to expose the gold thin film in a study of gold mechanical properties.

REF: Carlin, A & Walker, W P — *J Appl Phys,* 31,2135(1960)

AU-0022
ETCH NAME: TIME:
TYPE: Electrolytic, polish TEMP:
COMPOSITION: ANODE: Au
67.5 g .... KCN CATHODE: SS
15 g .... $KFe_2(CN_3)_6$ POWER: 4 V
15 g .... Rochelle salt
19.5 g .... H_3PO_4
2.5 g .... $AgNO_3$
DISCUSSION:

Au specimens. Solution used to electropolish both gold and silver.

REF: Ruff, A W & Ives, L K — *Acta Metall,* 15,189(1967)

AU-0023
ETCH NAME: Nitric acid TIME:
TYPE: Acid, float-off TEMP:
COMPOSITION:
x .... HNO_3
x .... H_2O

DISCUSSION:

Au, thin films deposited on (0001) muscovite mica substrates and used for structure study of the film. After evaporation, the gold film was removed from the mica with the acid shown using the float-off technique for film study by TEM. See under Mica: MI-0002.

REF: Chopea, K I — *J Appl Phys*, 37,2049(1960)

AU-0024

ETCH NAME: Air — TIME: $^1/_2$—4 h

TYPE: Gas, drying — TEMP: 125°C

COMPOSITION:

x .... air, hot (oven)

DISCUSSION:

Au, thin films evaporated on alumina substrates (2 and 4 μm surface finish) as an Au/TiW(10%) Ti) multilayer films used in a study of blisters occurring with subsequent electrolytic Au up-plating when films are then subjected to a 400°C, 1—5 min soak on a hot plate in air. This heat treatment was evaluated after the following surface treatments: (1) 50% HCl dip, DI H_2O rinse; (2) DI H_2O rinse, only; (3) gold evaporated film, dry, without any further treatment; and (4) surface dried in an air oven, all prior to gold plating. HCl dip gave worst case, large blisters; oven drying showed many small blisters, and did not eliminate blisters. Blistering considered to be due to presence of TiW as the Au/Cr, Au/Ni, or Au/Cu/Mo metallized thin film systems do not blister under similar processing conditions. It also was noted that, after HCl cleaning then gold up-plating, a black, acidic liquid oozed upward onto surfaces at some gold grain boundary locations.

REF: Walker, P & Valardi, N — personal development, 1985

AU-0001b

ETCH NAME: — TIME:

TYPE: Acid, removal — TEMP:

COMPOSITION:

x .... NaCN

x .... H_2O_2

DISCUSSION:

Au, thin film deposits on silicon wafers. Solution shown as a general removal etch.

REF: Ibid.

AU-0020a

ETCH NAME: — TIME: 2—3 sec

TYPE: Halogen, removal — TEMP: RT

COMPOSITION: — RATE: See discussion

7 g .... KI

25 g .... Br_2

100 ml .... H_2O

DISCUSSION:

Au, thin films. Removal rate for 1000 Å thick film is 2—3 sec. Removal rate for 2500 Å thick film is 30—35 sec.

REF: Bahl, S K & Leach, G L — U.S. Patent 4,190,489, 1980 (Mead Corp.)

AU-0020b
ETCH NAME:
TYPE: Halogen, removal
COMPOSITION:
18 g .... NaBr
2 g .... Br_2
100 ml .... H_2O
DISCUSSION:

TIME:
TEMP: RT
RATE: See discussion

Au, thin film deposited on glass to 2000 Å thick. Removal time was 30—45 sec.

REF: Ibid.

AU-0020c
ETCH NAME:
TYPE: Halogen, removal
COMPOSITION:
12 g .... KBr
8 g .... Br_2
100 ml .... H_2O
DISCUSSION:

TIME:
TEMP: RT
RATE: See discussion

Au, thin films deposited on glass to 2000 Å thick. After etching, rinse in water. Removal time is 15 sec.

REF: Ibid.

AU-0021
ETCH NAME: Tri-iodine
TYPE: Halogen, removal/pattern
COMPOSITION:
400 g .... I_2
100 g .... KI
400 ml .... H_2O
DISCUSSION:

TIME:
TEMP: 55°C
RATE: 1270 Å/sec

Au, thin film deposits as a multilayer: Au/Ni/Au/TiW/Si(100) substrate or on a gallium arsenide, (100) substrate. Metals sputter evaporated and the top gold surface may be up-plated. Use solution to etch remove Au/Ni/Au layers with time depending on total thickness. Remove from etch and DI water rinse. Continue etching in 1—2 sec dip etch steps to remove any remaining gold traces. Remove TiW with H_2O_2, RT, 6—8 min with a removal rate of 20—30 micro-inches (μ″)/min. (See AU-0002c with regard to tungsten oxides on surfaces after etching.)

REF: LaRusso, T — private communication, 1985 (MRC Corp.)

AU-0031
ETCH NAME:
TYPE: Abrasive, polish
COMPOSITION:
x .... emery
DISCUSSION:

TIME:
TEMP: RT

Au, as single crystal blanks. Surfaces mechanically lap polished with emery #2 down to 3/0. Also polished with Buehler #1551 A/B compound in oil on a glass plate. Study was to obtain pressure derivatives of the elastic constants of Zn, Ag, and Cu to 10,000 bars pressure. Materials were etched, but no solution shown.

REF: Daniels, W B & Smith, C S — *Phys Rev,* 111,713(1958)

AU-0032
ETCH NAME: Pulse plating
TYPE: Gold, plating
COMPOSITION:
75 g .... $(NH_4)_2HC_6H_5O_7$
75 g .... $(NH_4)_2SO_4$
20 g .... $KAu(CN)_2$
1000 ml .... H_2O for pH 5.5—6.0

TIME: 0.9 ms ON — 9 ms OFF
TEMP: 65°C
ANODE: Steel mesh
CATHODE: Alumina
POWER: 0.5 mA/cm^2

DISCUSSION:

Au, thin films pulse plated on alumina blanks with 1 μ" finish; grain size of substrates affect gold grain size. E-beam evaporate of metal layers was for the following thicknesses: 750 Å Ti; 3000 Å Pd: 25,000 Å Cu on alumina prior to DC or AC pulse plating of 2000—3000 Å Au. Pulse plating produces higher density, finer grained gold films than straight DC.

REF: Rehrig, D I — *Plating,* January 1974, 43
AU-0033: Olson, R — *Prod Finish,* April 1976

Au, thin films pulse plated with a pulsed rectifier power supply, and pulsed plater system. Pulse 10—75 μsec at 200—5000 Hz frequency with square wave use 1 A average to 500 A (2, 5 to 165 peaks). Article shows effects of gold plating on surfaces of nickel, silver, copper, and Pb/Sn materials.

AU-0036: Mentone, P F — *Prod Finish,* April 1973

Au, pulse plated for fine grain size, and to reduce porocity. A 50 μ" thickness is equivalent to 100 μ" of DC plating with regard to corrosion stability. Chrome pulsed plating improves coverage and hardness of the gold. The finer grains of Ni:Fe alloys gives more uniform magnetic properties.

AU-0034a
ETCH NAME:
TYPE: Electrolytic, polish
COMPOSITION:
50 ml .... HCl
25 ml .... EOH
250 ml .... glycerin

TIME:
TEMP:
ANODE:
CATHODE:
POWER:

DISCUSSION:

Au, thin films plated from a gold cyanide/phosphate bath. Gold plated on copper substrates with substrate dissolved away with $1HNO_3{:}1H_2O$ as a float-off method to obtain the gold film for TEM study. Gold film electropolished and thinned in solution shown small thinned thru-holes in film were due to voids in the deposited film. Films dissolved in aqua regia ($3HCl{:}1HNO_3$) to collect particles of Au(CN) as nonmetallic inclusions, and requires rapid water quench, as particles are soluble in the acid mixture. CoHG films have higher Au(CN) crystal residue than AFHG films, and soft sold films have none. Soft gold deposited at 70°C, and Au(CN) co-deposition with gold is temperature dependent.

REF: Nakahara, S & Okinaka, Y — *J Electrochem Soc.,* 128,264(1981)

AU-0034b
ETCH NAME: Gold plating
TYPE: Metal, plating
COMPOSITION:
44 g .... $KAu(CN)_2$
100 g .... KH_2PO_4
28 g .... KOH
1000 ml .... H_2O

TIME:
TEMP: 40°C
ANODE: Ti w/RuO_2-TiO_2 coating
CATHODE: ST w/Cu sheet wrapping
POWER: 250 mA/cm^2

DISCUSSION:

Au, as hard gold (HG) coatings on copper substrates with a $CoSO_4$ additive for to solution shown for hardening. CoHG gold evaluated against no additive solutions as AFHG films. Grain size: (1) CoHG; 225—275 μm; AFHG: 250—750 μm; (3) soft gold: 1—2 μm. Gold films etch thinned and removed from copper for TEM study, to include presences of Au(CN) nonmetallic crystal contamination in films.

REF: Ibid.

AU-0035: Okinaka, Y & Wolowodiuk, O — *J Electrochem Soc,* 128,288(1981)

Au, thin films plated as CoHG and AFHG hard gold. A study of Au(111) type $Au(CN)_4$ to Au(1) and $Au(CN)_2$ nonmetallic crystal within films. Both AU-0034 and this reference used cylinder electrodes, with cathode rotating, and anode stationary or rotating. A 2.5 μm gold thickness was deposited in 16 min.

AU-0037: Weisberg, A M — Technic Inc., Providence, RI, 1976

Au, thin film plating as soft vs. hard gold. Describes shelf life, and Knoop hardness as related to product usage time of CoAu, NiAu, AgAu, and pure 24 carat gold. Different plating solutions available shown in this brochure.

AU-0025

ETCH NAME: Argon
TYPE: Ionized gas, cutting
COMPOSITION:
x .... Ar^+ ions

TIME:
TEMP:
GAS FLOW:
PRESSURE:
POWER:
RATE: 1 Å/sec

DISCUSSION:

Au, polycrystalline specimens and diffused specimens. Argon used to sputter section cut specimens. Grain boundary areas etched faster during the argon cutting. Ionized argon also used to section iron alloys. Rate shown was for a 0.5 μm cut width.

REF: Perkins, R A — *J Vac Sci Technol,* 13,5(1976)

AU-0026

ETCH NAME: Water
TYPE: Acid, float-off
COMPOSITION:
x .... H_2O

TIME:
TEMP: RT

DISCUSSION:

Au, thin films deposited on (100) NaCl substrates in a study of gold structure. Evaporated at 300°C the gold film was (100) oriented. Evaporated at 150°C gold film was crystalline with random grain orientation. Films removed from sodium chloride by liquefying the salt surface with water for film float-off and TEM microscope study. Tensile properties of thin films studied.

REF: Neugebauer, C A — *J Appl Phys,* 31,1096(1960)

AU-0027

ETCH NAME: Heat
TYPE: Thermal, preferential
COMPOSITION:
x .... heat

TIME:
TEMP:

DISCUSSION:

Au thin films. Deposit studied for the influence of oxidation and reducing atmosphere treatments on vacancy clusters in gold. Gold surfaces were preferentially etched after treatments, but solution not shown.

REF: Segall, R L & Clarebrough, L M — *Phil Mag,* 8,65(1964)
AU-0028: Silcox, J & Hirsch, P B — *Phil Mag,* 4,72(1959)

Au thin films. Reference for gold etch used in AU-0027. Direct quenching of gold will to develop defects.

AU-0030: Pashley, D W — *Phil Mag,* 4,324(1959)

Au evaporated gold thin films. Observation of dislocations in the films.

GOLD ALLOYS, AuM_x

General: Occurs as native compounds when gold is amalgamated with other element metals, such as silver. The latter is called electrum, $AuAg_x$.

Refined, high purity gold, 5 9's pure (99.999%), is used for contact wires and metallized pads in Solid State semiconductor device fabrication and assembly, but is too soft for general application and use in industry. For commercial applications there are many artificially mixed gold alloys and/or gold plating baths that use other metals or metal salts to improve the hardness of a surface coating or produce a specific color.

In the general formula shown above, AuM_x M = Ag, Cu, Ni, Pt, Pd, Ge, Sn, Cd, In, Fe, Zn, etc., as metal additives, alone, or in combination.

Many of the Au:Ag:Cu mixtures are used in jewelry with the weight measured by carat. Some of the variations by carat and use are shown below:

Carat	Name	Formula	Remarks
24	Pure gold	Au	For any given carat value, the
22	Dental	92Au:9Ag:3.1Cu	amount of gold and other metals
22	Jewelry	91.66Au:4.16Ag:16Cu	can vary.
20	Jewelry	84Au:8.3-11Ag:6—8.3Cu	18 carat gold is a general
18	Jewelry	75Au:10-20Ag:5—15Cu	jewelry standard. Will vary in
16	Solder	75Au:17Ag:8.3Cu	color as silver, copper, or
16	Jewelry	67Au:6.6-26Ag:8—27Cu	other metal additives are
15	Jewelry	62Au:11Ag:13Cu	used in the mixtures.
14	Jewelry	58Au:4-28Ag:14—28Cu	
12	Solder	50Au:15Ag:35Cu	

Some other special gold alloys are

Formula	Name
70Au:21Al	Roberts-Austen purple gold
80Au:20Pd	Palau
60Au:40Pt	White gold (platinum)
75—85Au:8—10Ni:2—9Zn	White gold
75Au:25Fe	Blue gold
90Au:10Cu	Coinage gold

Technical Application: Both high purity gold and gold alloys are used in Solid State processing, device fabrication and assembly. The high purity, 5 9's, gold is evaporated or sputtered onto photo resist patterned wafers and circuit substrates to define contact and electronic function structures in a variety of multilayer metallizations, such as Au/Ni; Au/Cr; Au/AuGe/Ni; Au/Pt/Ni, and may be annealed for interdiffusion of the metals. As alloy pre-forms of AuB or AuP, they are alloyed to p- or n-type surfaces with the 3 to 5% boron or phosphorus improving electrical contact and reducing stress. Although most gold bond wire, 0.001 to 0.0007″ diameter, is high purity gold — soft gold — some gold alloys, such as AuBe are used to stiffen the wire contact. The latter, AuBe, is recommended for high-speed, automatic bonders.

Electrolytic gold plating, with and without additional evaporated gold, is more often used for metallizing dielectric substrates (alumina, beryllia, quartz, etc.) and on package parts of copper, iron, aluminum. Gold on package parts is primarily for corrosion resistance and is hardened with additional metal additives to withstand general handling, wear and tear.

There are a number of gold and gold alloy pastes as epoxy or polyimide mixtures used for mounting discrete devices and other components in test or final package assemblies. Hardening, or firing temperatures of these pastes range from about 150°C to slightly over 200°C as a maximum, and there is much effort to develop higher temperature capability of such plastic type pastes for improved unit stability.

Although most gold alloys are used in polycrystalline form, several have been grown as single crystals, such as $AuGa_2$, $AuAg_x$ for morphological studies.

Etching: As most gold alloys have a high gold content, they can be etched in the mixed acids shown for gold: aqua regia, cyanides, and iodides.

GOLD ALLOYS

GOLD BISMUTH ETCHANTS

AUBI-0001a
ETCH NAME: TIME:
TYPE: Acid, preferential TEMP:
COMPOSITION:
x .... KCN
x .... $(NH_4)_2SO_4$
x .... H_2O
DISCUSSION:
Au_2Bi, specimens. Solution used to develop structure. Will darken surfaces for color contrast of defects and structures.
REF: Savas, M A & Smith, R W — *J Cryst Growth,* 71,41(1985)

AUBI-0001b
ETCH NAME: Nitric acid, dilute TIME:
TYPE: Acid, removal/preferential TEMP:
COMPOSITION:
x .... HNO_3
x .... H_2O
DISCUSSION:
Au_2Bi, specimens as Bi-Au_2Bi. Solution referred to as a ''deep'' etch for this material. Surfaces were carbon replicated for SEM study.
REF: Ibid.

GOLD CADMIUM ETCHANTS

AUCD-0001
ETCH NAME: TIME:
TYPE: Acid, polish TEMP:
COMPOSITION:
1 .... 20% KCN
1 .... 20% $(NH_4)_2S_2O_8$

DISCUSSION:

Au:Cd alloy specimens. Solution used as a polish etch for surfaces in a study of lattice spacing of close-packed hexagonal, hcp, metal alloys.

REF: Massalski, T B — *Acta Metall,* 5,541(1957)

GOLD CESIUM ETCHANTS

AUCS-0001

ETCH NAME: Aqua regia, dilute — TIME:

TYPE: Acid, removal — TEMP: RT

COMPOSITION:

3 .... HCl

1 .... HNO_3

x .... H_2O

DISCUSSION:

Au_2Cs specimens used in a study of the semiconducting properties of the compound.

REF: Spicer, W E et al — *Phys Rev,* 1,55(1959)

GOLD-GALLIUM ETCHANTS

AUGA-0001

ETCH NAME: Tri-iodide — TIME: To 3 min

TYPE: Halogen, removal — TEMP: RT to hot

COMPOSITION:

x .... KI

x .... I_2

x .... H_2O

DISCUSSION:

$AuGa_2$, (100) oriented thin films on NaCl, (100) substrates. The substrate was cleaned: (1) rinse in chloroform, (2) rinse in acetone and (3) final rinse in methanol. Then place substrates in ultra high vacuum (UHV) for co-deposition of gold and gallium. Solution used to etch deposit $AuGa_2$.

REF: Gordon, H T — *J Vac Sci Technol,* A2.535(1984)

AUGA-0002

ETCH NAME: Air — TIME: 25 days

TYPE: Gas, aging defects — TEMP: RT

COMPOSITION:

x .... air

DISCUSSION:

Au:Ga, thin films EB evaporated on NaCl, (100) substrates as a 1:1 alloy with 230 Å Au:270 Å Ga. Gallium formed spherical aggregates on surface when deposition temperature was above the gallium melting point (30°C). After aging as shown: (1) deposit showed 1.5 mm blisters, roughly circular, with T-shaped cracks; (2) silver colored areas showed uniform grain structure of Au:Ga alloy mixture; (3) orange areas were fine-grained, mainly unreacted gold. Alloy area was Au_2Ga with orthorhombic structure like Ni_2Si and there was some interstitial diffusion into amorphous gallium similar to Pd diffusion into Sn forming $PdSn_4$. Cracks were due to Kirkendall voids caused by volume change from interdiffusion.

REF: Nakahara, S & Kinsbron, E — *Thin Solid Films,* 113,15(1984)

AUGA-0003
ETCH NAME: Argon
TYPE: Ionized gas, thinning
COMPOSITION:
x .... Ar^+ ions

TIME:
TEMP:
GAS FLOW:
PRESSURE:
POWER: 6 KeV

DISCUSSION:

Au_2Ga, thin films deposited by co-evaporation onto GaAs, (100) substrates and annealed 30 min at RT. Annealing at 225°C produced the Au_2Ga orthorhombic phase. Two wafers were sandwiched together with epoxy, cut in the ⟨110⟩ direction and mechanically lapped to reduce specimen thickness. Finally, thinning was by argon ion bombardment for specimen morphology study under TEM.

REF: Leung, S et al — *J Electrochem Soc,* 132,898(1965)

AUGA-0004: Leung, S et al — *Thin Solid Films,* 104,109(1983)

AuGa and Au_7Ga_2 thin films deposited on GaAs, (100) wafers. The wafers were either Si doped n-type, or Zn doped p-type. After deposition the films were annealed at 400°C by being brought up from RT in 100°C increments. After this heat treatment, twinning was observed.

GOLD GERMANIUM ETCHANTS

AUGE-0001a
ETCH NAME: Tri-iodide
TYPE: Halogen, patterning
COMPOSITION:
400 g .... KI
200 g .... I_2
1000 ml .... H_2O

TIME: 20—30 min
TEMP: 60—70°C

DISCUSSION:

AuGe(13%), alloy ribbon 0.001″ thick, pellets or slugs. The ribbon was used as preforms in microelectronic devices and circuit substrates assembly; pellets were used in multilayer metallization of electric contacts on discrete devices as Au/AuGe/Ni. To clean the alloy: (1) vapor degrease in hot TCE or TCA; (2) dip etch in solution shown, 2—5 sec, 60—70°C; (3) heavy rinse in MeOH to remove residual iodine and (4) N_2 blow dry. Longer etch periods used after photo resist patterning for etch structuring.

REF: Marich, L A et al — personal communication, 1979

GOLD INDIUM ETCHANTS

AUIN-0001
ETCH NAME:
TYPE: Acid, polish
COMPOSITION:
1 .... 20% KCN
1 .... 20% $(NH_4)_2S_2O_8$

TIME:
TEMP:

DISCUSSION:

Au:In, alloy specimens used in a study of the lattice spacing of close-packed hexagonal, hcp, alloys. Solution used to polish surfaces of Au:In, Au:Cd and Au:Hg.

REF: Massalski, T B — *Acta Metall,* 5,541(1957)

GOLD MERCURY ETCHANTS

AUHG-0001
ETCH NAME:
TYPE: Acid, polish
TIME:
TEMP:
COMPOSITION:
1 .... 20% KCN
1 .... 20% $(NH_4)_2S_2O_8$
DISCUSSION:
AuHg, alloy specimens used in a study of the lattice spacing of close-packed hexagonal, hcp, metal alloys. Solution used to polish surfaces. Also used on AuIn and AuCd.
REF: Massalski, T B — *Acta Metall,* 5,541(1957)

GOLD:SILVER ETCHANTS

AUAG-0001
ETCH NAME:
TYPE: Acid, preferential
TIME: 2—5 min
TEMP: RT
COMPOSITION:
x .... KCN
x .... $(NH_4)_2SO_4$
DISCUSSION:
$AuAg_x$, single crystal ingots. Composition varied from pure silver to pure gold with silver content from 10, 25, 50, 75, and 90%. After growth, ingots were subject to strain in a study of stress and deformation. Solution shown was used as a preferential etch to develop defect figures.
REF: Suzucki, H & Barrett, C S — *Acta Metall,* 6,156(1958)

GOLD TIN ETCHANTS

AUSN-0001a
ETCH NAME: Tri-iodide etch, modified
TYPE: Halogen, cleaning
TIME: 1—2 min
TEMP: 50—70°C
COMPOSITION:
0.5 g .... KI
0.25 g .... I_2
25 ml .... H_2O
25 ml .... MeOH
DISCUSSION:
AuSn(20%), alloy ribbon 0.001″ thick used for alloy mounting contacts of semiconductor devices in microelectronic circuits. Also used as alloy pellets and slugs in metal contact evaporation of wafer surfaces and circuit substrates. To clean AuSn: (1) vapor degrease parts in TCE or TCA; (2) etch clean in solution shown; (3) rinse thoroughly in MeOH to remove residual iodine and (4) air or nitrogen blow dry.
REF: Marich, L A & Porter, R — personal communication, 1978

AUSN-0002
ETCH NAME: Aqua regia
TYPE: Acid, patterning
TIME:
TEMP: RT
COMPOSITION:
3 .... HCl
1 .... HNO_3

DISCUSSION:

AuSn(20%) alloy as an evaporated thin film, or in sheet form strip. Solution used for general removal of thin films (stripping). Solution also used after photolithographic patterning 0.001″ thick strips to etch-thru pattern.

REF: Valardi, N & Walker P — personal application, 1985

AUSN-0003

ETCH NAME: Potassium cyanide — TIME: 1—5 min

TYPE: Acid, removal — TEMP: RT to hot

COMPOSITION:

x .... x% KCN

DISCUSSION:

AuSn(20%) alloy as evaporated thin films, or strip pre-forms used for device mounting on substrates. Solution used for general removal and cleaning. Also used for AuGe(13%) alloy and pure gold film removal.

REF: Valardi, N — personal communication, 1984

AUSN-0001b

ETCH NAME: Tri-iodide etch — TIME: 20—30 min

TYPE: Halogen, patterning — TEMP: 60—70°C

COMPOSITION:

400 g .... KI

200 g .... I_2

1000 ml .... H_2O

DISCUSSION:

AuSn(20%), alloy ribbon 0.001″ used in contact mounting of semiconductor microelectronic devices in circuit fabrication. Ribbon strips were mounted on glass slides with photo resist and then top surface was coated with photo resist. After air drying 30 min, and 90°C oven cure to harden photo resist, a circular open dot pattern was UV light exposed through an SST mask. Hole pattern in AuSn(20%) strip was etch formed in the solution shown. Similar work done with AuGe(13%), which is more brittle and tends to fracture. Method used to reduce alloy volume of material for thinner alloy joints. See AUGE-0002.

REF: Walker, P — personal application, 1984

GOLD ZINC ETCHANTS

AUZN-0001

ETCH NAME: Tri-iodide — TIME:

TYPE: Halogen, removal — TEMP: RT to hot

COMPOSITION:

x .... KI

x .... I_2

x .... H_2O

DISCUSSION:

Au:Zn, as an evaporated metal contact on InGaAsP/InP(100) (DH) Laser device structures. With photo resist patterning, solution used to etch gold contact pattern or for complete etch removal of the Au:Zn metallization.

REF: Adachi, S et al — *J Electrochem Soc,* 129,1524(1982)

PHYSICAL PROPERTIES OF GRAPHITE, C

Classification	Ceramic
Atomic number	6
Atomic weight	12.010
Melting point (°C)	3727
Boiling point (°C)	4830
Density (g/cm³)	2.09—2.23
Hardness (Mohs — scratch)	1—2
Crystal structure (hexagonal — rhombohedral)	$(10\bar{1}1)$ rhomb
Color (solid)	Black
Cleavage (basal — perfect)	(0001)

GRAPHITE, C (Plumbago, Black Lead)

General: Occurs as a native element. It is widely distributed throughout the world and is found in beds, veins and as granules in many types of rock formations: quartzite, limestone, schist, and volcanic rocks, and has been found in meteoric iron. Occasionally may be partly alloyed with iron as Fe_2C. Deposition can be direct, from the breakdown of gases and the natural, slow burning of coal — coking. Although it is hexagonal system, rhombohedral division with a perfect basal (0001) cleavage as a single crystal, it is commonly found in massive form with very distinctive layered structure. See "Carbon" for a more detailed discussion. Graphite has been shown to contain some iron in solid solution — approaching a natural form of steel.

The Bronze Age was followed by the Iron Age (circa 1350 B.C.) which has been in existence for some 3500 years until the present Atomic Age and commencement of the Space Age. The initial discovery and use of iron can be traced back to the Hittite culture in Asia Minor as soft, pure iron — wrought iron — but it was the addition of graphite (carbon) to harden iron that developed the true Iron Age culture: Steel.

Pure graphite (carbon), in addition to its use in the iron industry, has major use as a high temperature refractory for crucibles, furnace linings and similar products. It also is an electrical conductor and used in electrical wiring, as power feed-thrus and so forth. Because of its high melting point and heat carrying capability, it is used as a heater strip. Because of its light producing capability, as a filament in light bulbs. Because it is acid resistant, it has been used as a liner in acid tanks. Because of its low neutron cross section, it has major use as control rods in atomic piles, or as radiation shields. The latter shielding may be as a mixture with alumina and other ceramics.

As a fine dispersion on glass, it will reduce sun glare, such that it is used as a sun screen. In finely powdered form it is dry lubricant. As the product Aqua Dag it is a powdered mixture with alcohol, and is used in the replication of surfaces. It is called "Black Lead", which is a misnomer as it contains no lead, but sticks of graphite have been used as writing implements since ancient times. Sticks are still used for charcoal drawing, and it is the "lead" in pencils. Also as a very fine powder it is stove polished and mixed with grease, black greasepaint, or shoe or stove polish. Added as dry powder to cement or glass and enamels it is a black pigment. As lampblack it is a near-colloidal derivative carbon from the reduction of petroleum, also collected from burning gas (methane cooking gas), or from burning vegetation and coal. See section on Carbon for additional information.

Technical Application: Natural graphite is not used directly in Solid State processing but, as described in the preceding General section, both graphite and carbon parts, natural or artificial, have wide application in equipment as evaporator crucibles, heater strips, power conductors, and the like.

Pure graphite or graphite/metal mixtures are used to fabricate resistors and special resistive elements in semiconductor circuit assemblies as discrete devices, themselves, for power and frequency control. As Aqua Dag and similar solutions it is used as a mounting medium or to produce surface replication for electron microscope study.

Thin film deposition of graphite and silicon is under development for thin film silicon carbide, SiC, surface protection coatings on semiconductors and on metal cutting tools for wear resistance, and SiC has high temperature capability as a semiconductor device.

Although not natural graphite, amorphous carbon deposited as thin films, a-C, with additional annealing is under development as a conversion structure called diamond-like-carbon, DLC, for use as a chemically inert surface coating. Such coatings, to date, have produced a carbon/diamond crystalline structure, but not single crystal diamond.

Both natural and artificial graphite (e.g., carbon) have been studied as single crystals and the materials, in general, are under continual study and development because of their wide range of use and application.

The etch solutions shown here are those referenced to natural graphite. See the sections on Carbon and Diamond for additional information.

Etching: Very acid resistant. Boiling HNO_3, KOH, H_2SO_4 and H_3PO_4 to dissolve. KOH and NaOH as molten fluxes. Some acid mixtures.
See: "Carbon".

GRAPHITE ETCHANTS

GR-0001
ETCH NAME: Cleave — TIME:
TYPE: Cleave, preferential — TEMP: RT
COMPOSITION:
x .... cleave
DISCUSSION:

C, as single crystal graphite, (0001) cleaved specimens. Cleaved surfaces were relatively strain free and showed dislocation patterns. Also studied natural molybdenite, MoS_2. Both materials are used as dry lubricants.

REF: Boswell, F W C — *J Appl Phys,* 31,1834(1960)
GR-0002: Bacon, R & Sprague, R — *J Appl Phys,* 31,1831(1960)

C, as single crystal graphite, (0001) cleaved specimens. A similar study of defects on cleaved surfaces.

GR-0003
ETCH NAME: Boron — TIME:
TYPE: Metal, decoration — TEMP: 2300°C
COMPOSITION:
x .... B
DISCUSSION:

C, as single crystal graphite specimens. Boron diffused into specimens at the temperature shown produce circular, disc-like figures which change with temperature to disc stacking faults below 2400°C. Discs were 3—6 μm in diameter and, above 2400°C, convert to dislocation loops (loss of boron?).

REF: Hennig, G R — *J Appl Phys,* 34,237(1964)

GR-0004
ETCH NAME: Oxygen TIME: 2 h
TYPE: Gas, preferential TEMP: 825°C
COMPOSITION:
x .... O_2
DISCUSSION:
C, as single crystal graphite specimens. Oxidation of graphite under the conditions shown produce some evidence of nonbasal dislocations.
REF: Thomas, J M et al — *Phil Mag,* 10,325(1964)

GR-0005
ETCH NAME: Sodium peroxide TIME: 5 to 20 min
TYPE: Molten flux, preferential TEMP: 380°C
COMPOSITION:
(1) x .... Na_2O_2 (2) x .... NaOH (3) x .... KNO_3 (4) x .... $KClO_3$
DISCUSSION:
C, (0001) specimens cleaved from natural graphite. All etching was done in a platinum crucible at the temperature shown to develop etch pits. Spiral etch pits developed were screw dislocations. Sodium hydroxide gave best results. Time varied 7, 15, and 20 min for pits development. Evaporated with a thin film silver coating for electron microscope (EM) study. (*Note:* Electron microscope for this type of study is now called scanning electron microscopy (SEM). Silver, gold, and nickel are evaporated on graphite for surface replication as graphite and carbon are absorbers of electrons and the metals are needed for reflectivity for electron scanning.)
REF: Patel, A R & Bahl, O P — *Br J Appl Phys,* 16,169(1965)
GR-0006: Kennedy, A J — *Proc Phys Soc,* 75,607(1960)
C, specimens as graphite. A study of dislocations and twinning in graphite. Molten fluxes were used to develop defects.

GR-0007
ETCH NAME: TIME:
TYPE: Stress, defect TEMP:
COMPOSITION:
x .... stress
DISCUSSION:
C, as graphite single crystal material. A study of the magnetic field dependence of the Hall effect and magnetoresistance. Stress and strain in the material create defects that will effect the Hall coefficient measurement.
REF: Soule, D E — *Phys Rev Lett,* 1,347(1958)

GR-0008
ETCH NAME: TIME: 2—10 min
TYPE: Acid, removal/defect TEMP: RT to boiling
COMPOSITION:
x .... HNO_3, conc.
DISCUSSION:
C, as natural graphite specimens, both massive material and fine single crystals. Massive material was etched by immersion, the single crystal material in hot vapors to develop surface structure and defects. This was a general study of the material from different locations and occurrences.
REF: Walker, P — mineralogy study, 1951—1954

PHYSICAL PROPERTIES OF HAFNIUM, Hf

Classification	Transition metal
Atomic number	72
Atomic weight	178.49
Melting point (°C)	2222 (2150)
Boiling point (°C)	5400
Density (g/cm^3)	13.09 (13.3)
Thermal conductance (cal/sec)(cm^2)(°C/cm) 50°C	0.0533
Specific heat (cal/g) 25°C	0.035
Heat of fusion (cal/g)	32.4 (29.1)
Heat of vaporization (k-cal/g-atom)	155
Atomic volume (W/D)	13.6
1st ionization energy (K-cal/g-mole)	127
1st ionization potential (eV)	5.5
Electronegativity (Pauling's)	1.3
Covalent radius (angstroms)	1.44
Ionic radius (angstroms)	0.78 (Hf^{+4})
Vapor pressure (atm) 2007°C	10^{-9}
Coefficient of linear thermal expansion ($\times 10^{-6}$ cm/cm/°C)	5.9
Modulus of elasticity (psi $\times 10^6$) 20°C	19.8
Tensile strength (psi)	86,000
Electron work function (eV)	3.53
Cross section (barns)	105
Hardness (Mohs — scratch)	4—5
(Rockwell B — kgf/mm^2)	95
Crystal structure (hexagonal — normal) alpha <1760°C	(10$\bar{1}$0) prism, hcp
(isometric — normal) beta >1760°C	(100) cube, bcc
Color (solid)	Grey-silver
Cleavage (basal/cubic)	(0001)/(001)

HAFNIUM, Hf

General: Does not occur free as a native element. Found as a minor constituent in zirconium minerals, such as zircon (zirconia), $ZrSiO_2$, which is found in association with sphalerite, ZnS. Although not a rare element, it is not found in large quantities, but is a recovery metal from zirconium ores, chiefly from zircon. The chemical characteristics of the metal are similar to those of zirconium. It is not used in metal processing, but could act as a desulfurizer in irons and steels as is zirconium.

Technical Application: No use as a metal in Solid State processing though it is under evaluation as a thin film: pure metal, nitride, silicide. The latter as HfSi, Hf_2Si and $HfSi_2$ for blocking layers in heterostructures. The nitride, HfN, is under evaluation as a surface passivation layer as is ZrN and TiN. Metallic hafnium has been grown as a single crystal and deposited as a thin film metal layer for general study.

Both HfS_2 and $HfSe_2$, along with ZrS_2 and $ZrSe_2$, are under evaluation as compound semiconductors with possible application as cathode storage batteries.

Etching: HF and mixed acids.

HAFNIUM ETCHANTS

HF-0001a
ETCH NAME: TIME: Seconds
TYPE: Acid, cleaning TEMP: RT
COMPOSITION:
10 ml .... HF
45 ml .... HNO_3
45 ml .... H_2O
DISCUSSION:
Hf, single crystal wafers and HfN thin films. Solution used to clean surfaces.
REF: Dawson, P T — *J Vac Sci Technol,* 21,36(1982)

HF-0001b
ETCH NAME: TIME:
TYPE: Acid, removal TEMP:
COMPOSITION:
1 .... HF
1 .... H_2O_2
20 .... H_2O
DISCUSSION:
Hf specimens. Solution shown as a general removal etch.
REF: Ibid.

HF-0002
ETCH NAME: Hydrofluoric acid, dilute TIME:
TYPE: Acid, removal TEMP:
COMPOSITION:
1—2 .... HF
10 .... H_2O
DISCUSSION:
Hf, thin films deposited on silicon wafers. Solution shown as a general removal etch for hafnium.
REF: Missel, L I & Glang, R — *Handbook of Thin Film Technology,* McGraw-Hill, New York, 1970

HAFNIUM NITRIDE ETCHANTS

HFN-0001
ETCH NAME: TIME:
TYPE: Acid, cleaning TEMP:
COMPOSITION:
10 ml .... HF
45 ml .... HNO_3
45 ml .... H_2O_2
DISCUSSION:
HfN, as a thin film deposit grown from NH_3 pyrolysis. Nitride showed concentric colored bands from wafer center: yellow, blue, red, blue, and yellow. Also studied TiN and ZrN.
REF: Dawson, P T — *J Vac Sci Technol,* 21,36(1982)

HAFNIUM TIN ETCHANTS

HFSN-0001
ETCH NAME: Nitric acid
TYPE: Acid, tin removal
COMPOSITION:
x .... HNO_3
TIME:
TEMP:
DISCUSSION:

Hf_5Sn_2, single crystal specimens. After molten flux growth removed excess tin by dissolving with nitric acid.

REF: Bailey, D M & Smith, J F — *Acta Metall,* 14,57(1961)

PHYSICAL PROPERTIES OF HELIUM, He

Classification	Inert gas
Atomic number	2
Atomic weight	4.0002
Melting point (°C)	−272.2
Boiling point (°C)	−268.9
Density (g/cm³ $\times 10^{-3}$)	0.178
(g/l) −270.8°C	0.147
Specific heat (cal/g/°C) 20°C	1.25
Thermal conductivity ($\times 10^{-4}$ cal/cm/cm2/°C/sec)	3.34 @ 20°C
1st ionization potential (eV)	24.58
Cross section (barns)	0.007
Vapor pressure (°C)	−270.3
Critical temperature (°C) C.T.	−267.9
Critical pressure (atms) C.P.	2.26
Hardness (Mohs — scratch)	1—3
Crystal structure (hexagonal — normal)	$(10\bar{1}0)$ prism, hcp
Color (solid/liquid/gas)	Colorless
Cleavage (basal)	(0001)

HELIUM, He

General: Helium is a colorless, odorless, tasteless gas that is inert like argon. It does not combine with other elements, nor does argon, although some compounds have been reported, such as HeH, HgHe, and $Ar(HF_3)$.

Helium is an extremely rare gas, only about 0.0005 by volume in standard air (atmosphere), and there is one major gas producing well in the world in Amarillo, Texas that, effectively, produces the world supply of helium. Although hydrogen has better lifting power — 76 pounds per 1000 ft³ vs. only 65 lb for helium — hydrogen is explosive and flammable, e.g., explosion of the Hindenburg Zeppelin in 1937 at Lakehurst, New Jersey. Since that time, the U.S. has never used hydrogen for zeppelins or balloons — the Germans had to re-strut the war reparation dirigibles transferred to the U.S. after World War I to handle helium rather than hydrogen — all three were lost by accident prior to World War II.

Helium can be reduced to the lowest temperature known to man as a cryogenic liquid, but cannot be reduced to a solid single crystal as can all of the other gases. At 2.2°A (absolute temperature, equivalent to Kelvin, K) helium properties change — thermal conductivity increases, viscosity decreases and it behaves like a gas, rather than a liquid — and it is the only known element that can be reduced to this temperature.

In industry, helium is used as an inert furnace gas like argon in high temperature processing of metals; has wide application in weather balloons, as well as commercial balloons, both the small party and play types, as well as the larger type used by balloonists, though the latter often use hot air rather than helium.

Perhaps one quarter of all helium produced is used in medicine as the carrier gas with anesthetics, and in the treatment of asthma and other respiratory diseases. As a mixture of He/O_2 it replaces air (N_2/O_2) for deep-sea divers as, when a diver is brought to the surface too rapidly using air, nitrogen forms bubbles in the bloodstream causing the "bends" with severe muscle cramps and spasms. Helium will not form such bubbles in the bloodstream. As an item of interest, when pure helium or He/O_2 mixtures are breathed it affects the vocal cords, producing the "Donald Duck" speech. Note that helium is a by-product of uranium disintegration.

Technical Application: Helium is used as an inert atmosphere in furnaces, although argon and nitrogen are more common as they can be supplied at lower cost for high volume thru-put. Helium is supplied in pressurized cylinders as the pure gas or as gas mixtures, the gas mixtures being used as semiconductor dopant sources of boron or other elements for wafers, or in the single crystal and epitaxy growth of compound semiconductors, ferrites, garnets.

Helium is used as packaging atmosphere, particularly for quartz crystals as it is the lightest of the inert gases and has a minimum frequency loading effect on the quartz. It is under evaluation for the package sealing of high power and high frequency Schottky barrier type devices, such as gallium arsenide, GaAs, for the same reason.

In vacuum evaporators, cryo-pumps are replacing oil diffusion pumps and ion pumps, and they use helium as the chilling agent with chunk graphite as the gas absorbent during chamber pumping for oil-free evaporator operation. Such pumping systems, depending upon usage time, require periodic rejuvenation of graphite to remove collected gases. This is done by vacuum pumping on the cryo-pump itself on a daily or weekly basis with replacement of the graphite about once a year.

In material processing of semiconductors, nitrogen gas is most often used as nitrogen blow-off for drying after etching of parts.

Occasionally other inert gases are similarly used, argon more often than helium, and they can be used from RT to hot (60—80°C). Also, nitrogen is used for a nonoxidizing storage gas in cabinets or working dry boxes, and both argon and helium can be so used. Where a compound such as AlAs reacts with water moisture in the atmosphere, it is held and processed under an inert atmosphere; again, usually nitrogen, but it can be helium or argon.

Etching: N/A

PHYSICAL PROPERTIES OF HEMATITE, Fe_2O_3

Classification	Oxide
Atomic numbers	26 & 8
Atomic weight	159.68
Melting point (°C)	1565
Boiling point (°C)	
Density (g/cm³)	5.2—5.25
Refractive index (n = −Li)	3.01—2.94
Hardness (Mohs — scratch)	5.5—6.5
Crystal structure (hexagonal — rhombohedral)	$(10\bar{1}1)$ rhomb

Color (single crystal)	Red/black
(massive ocher)	bright red
Cleavage (basal — parting)	(0001)
(rhomb)	$(10\bar{1}1)$

HEMATITE, Fe_2O_3

General: The mineral hematite is a primary ore of iron. Called red iron ore; red ocher; red rouge. Color varies from specular iron (metallic black); massive/fibrous (brown/red to black); red ocher or red hematite; and can be an admixture of brown limonite, $FeO.NH_2O$ and black magnetite, Fe_3O_4. It may appear to be magnetic, but magnetism is due to the presence of magnetite. In reniform masses as kidney ore, similar in appearance to psilomelane, MnO_2, and as a deposit of sedimentary origin, called oolitic hematite, rounded grains with a central sand-grain, of Paleozoic Age and sometimes in extensive beds. Also occurs as red chalk: limestone, $CaCO_3$ — normal white chalk — containing red hematite as coloring. The yellow/brown/red coloring of rocks and earth is due to the presence of iron oxides.

Hematite as iron ore was in use in ancient times from about 1350 B.C. (beginning of Iron Age) and, as red ocher, is still the cosmetic red rouge. The same name, red rouge, also applied when the compound is used as a lapping and polishing abrasive. Red ocher, the geologic name for red rouge, is a pigment coloring for paints, glass, and enamels. Tan or brown rouge may be limonite, black rouge as magnetite.

Technical Application: Not used in Solid State device fabrication or processing. It is occasionally used as a lapping and polishing compound for wafers and parts, and "iron oxide" is used as a pattern coating in the fabrication of photo resist glass masks as a replacement for chromium, nickel and aluminum. As an iron containing material it may be considered a ferrite Solid State compound, and has been studied for magnetic properties.

Both natural and artificially grown single crystals of hematite have been studied for general morphology and other data.

Etching: HCl (all iron oxides).

HEMATITE ETCHANTS

HEM-0001

ETCH NAME: Hydrochloric acid — TIME:

TYPE: Acid, removal — TEMP:

COMPOSITION:

x .... HCl

DISCUSSION:

Fe_2O_3, natural single crystals cleaved basal (0001) used in a study of frequency dependence of magnetic resonance. Hydrochloric acid is a slow etch on hematite. (*Note:* Hematite does not show true cleavage, but shows (0001) parting based on its lamellar structure.)

REF: Kumagai, H et al — *Phys Rev,* 99,1116(1955)

HEM-0002: Walker, P — mineralogy studies, 1952

Fe_2O_3, natural single crystal specimens. Solution used as a general cleaning etch in a study of the iron oxides.

PHYSICAL PROPERTIES OF HOLMIUM, Ho

Classification	Lanthanide
Atomic number	67

Atomic weight	164.94
Melting point (°C)	1461
Boiling point (°C)	2600
Density (g/cm³)	8.80
Specific heat (cal/g) 25°C	0.039
Heat of fusion (k-cal/g-atom)	4.1
Latent heat of fusion (cal/g)	24.9
Heat of vaporization (k-cal/g-atom)	67
Atomic volume (W/D)	18.7
Electronegativity (Pauling's)	1.2
Covalent radius (angstroms)	1.58
Ionic radius (angstroms)	0.91 (Ho^{+3})
Electrical resistivity (micro ohms cm) 298 K	81.7
Compressibility ($cm^2/kg \times 10^{-6}$)	2.47
Cross section (barns)	64
Magnetic moment (Bohr magnetons)	10.89
Tensile strength (psi)	37,500
Yield strength (psi)	32,100
Magnetic susceptibility ($\times 10^{-6}$ emu/mole)	70,200
Transformation temperature (°C)	1428
Hardness (Mohs — scratch)	5—6
Crystal structure (hexagonal — normal)	$(10\bar{1}0)$ prism, hcp
Color (solid)	Grey
Cleavage (basal)	(0001)

HOLMIUM, Ho

General: Does not occur in nature as a native element. It is one of the rare earths of the erbium family of the yttrium group, and is found in trace quantities in the mineral gadolinite, $Be_2FeDy_2Si_2O_{10}$, which may contain a high cerium content and, as such, often referred to as gadolinte earth as a source of all of the rare earth elements. Holmium also is found in other yttrium and cerium containing minerals. As the pure metal it is basic akin to the alkali metals but due to scarcity has had little or no application in the general metal industries.

Technical Application: Holmium has had little use in Solid State processing, to date, although it has been grown as a single crystal in general studies of the rare earth elements. It also has been studied as an oxide, a nitride, a ferrite, and an alloy constituent.

Etching: Mixed acids.

HOLMIUM ETCHANTS

HO-0001

ETCH NAME: Heat — TIME:

TYPE: Thermal, preferential — TEMP: <1400°C

COMPOSITION:

x .... heat

DISCUSSION:

Ho, single crystal specimens. The article describes a method of growing rare earth single crystals. After crystal growth, during annealing, thermal etching was used to develop defects, low angle grain boundaries and structure figures. Other metals studied were Dy, Gd, Er, Tb, Tl, and Y.

REF: Nign, H E — *J Appl Phys,* 34,3323(1963)

HOLMIUM ALLOYS:

HOLMIUM COBALT ETCHANTS:

HOCO-0001
ETCH NAME: Argon
TYPE: Ionized gas, cleaning
COMPOSITION:
x .... Ar^+ ions

TIME:
TEMP:
GAS FLOW:
PRESSURE:
POWER:

DISCUSSION:
Ho:Co alloy sputter deposited on glass and NaCl, (100) substrates in a study of this compound. Temperature of deposition was between 60-100°C, and substrates were Ar^+ ion cleaned for 1 h prior to deposition. The oxide is slowly soluble in HCl.
REF: Suzuki, T et al — *Jpn J Appl Phys,* 23,585(1984)

HOLMIUM OXIDE ETCHANTS

HOO-0001
ETCH NAME: Kalling's etch
TYPE: Acid, polish
COMPOSITION:
100 ml .... HCl
5 g .... $CuCl_3$
100 ml .... EOH

TIME: 10—30 sec
TEMP: RT

DISCUSSION:
Ho_2O_{17} specimens fabricated by arc melting and used in a study of different metallic compounds. Solution used as a general removal and polishing etchant. See Cesium Platinide: CEPT-0001 for complete list of other compounds studied.
REF: Slepowronsky, M et al — *J Cryst Growth,* 65,293(1983)

HOLMIUM FERRITE ETCHANTS &
HOLMIUM COBALT FERRITE ETCHANTS

HCFE-0001
ETCH NAME: Kalling's etch
TYPE: Acid, polish
COMPOSITION:
100 ml .... HCl
5 g .... $CuCl_3$
100 ml .... EOH

TIME: 10—30 sec
TEMP: RT

DISCUSSION:
Ho_2Fe_{17} and $Ho_2Co_{14}Fe_3$ arc melted in fabrication and used in a study of different metallic compounds. Solution used as a general removal and polishing etchant. See Cesium Platinide: CEPT-0001 for complete list of other compounds.
REF: Slepowronsky, N et al — *J Cryst Growth,* 65,293(1983)

HYDRIDES: M_xH

General: Pure metallic hydrides do not occur as natural compounds though there are many hydrated minerals containing either or both water, H_2O, and OH^-, the hydroxyl radical in their formulae. The four elements — oxygen, carbon, nitrogen, and hydrogen — appear in many natural compounds in a wide range of combinations and physical forms, from gases, such as methane, CH_4; as simple hydrocarbons, such as the paraffin series $C_nH_{2n} + 2$; amber, 40C:64H:4O, as a ratio; petroleum, as a combination of gas (methane), liquid as the paraffin series, and solid asphaltum or tar with several varieties. Mineral coal as compressed and altered vegetation vary from those still containing vegetation structure (peat, lignite), to the less volatiles containing coals as bituminous and anthracite. Coals and petroleums may have a high sulfur content, but it is the hydrogen content of these minerals that make them useful as fuels.

Hydriding of metal surfaces is done in industrial metal processing and, though it causes embrittlement, it also aids in improving surface hardness as does nitriding. Several metals will absorb and retain large quantities of hydrogen — palladium and platinum as powders — palladium black or platinum black — have been used as hydrogen storage batteries.

Technical Application: Many metals are under development and study in Solid State processing as single crystal hydrides or as hydrogenated thin films as both possible hydrogen storage batteries or as an element in device structuring. Several of these hydrogenated compounds appear in their ordinary metal or compound sections, such as silicon, germanium, silicon carbide, silicon nitride as Si:H, Ge:H, SiC:H, SiN:H, respectively.

Hydriding is done in a furnace under pressure or in an autoclave system, to convert a metal to a hydride, as a separate and distinct operation from hydrogen firing. In this latter case, hot hydrogen gas forms a reducing atmosphere preventing and removing oxides as a surface cleaning function; whereas hydriding is the growth of a new compound form with the addition of hydrogen in the formula. The term hydrogenation has been applied to converting oils to fats for many years, but also could be used with regard to forming metal hydrides, as it is the addition of hydrogen to a compound.

Etching: Acids, alkalies, mixed acids.

HYDRIDE ETCHANTS

HYDR-0001

ETCH NAME: Sulfuric acid — TIME:

TYPE: Acid, removal — TEMP:

COMPOSITION:

x .... H_2SO_4

DISCUSSION:

NbH, deposited on silicon wafers from a powder mixture in amyl acetate or cellulose nitrate. Paint on surface and fire for 1—10 min at 600—900°C in an inert atmosphere (N_2, Ar). Acid shown used to etch the hydride. Other hydrides evaluated were TaH, ZrH, VH, and TiH.

REF: Sullivan, M V & Eigler, J H — *Acta Metall,* 103,218(1956)

HYDROCARBON, CH_4

General: The formula shown is for methane or marsh gas, CH_4, a major constituent in the formation of petroleum and coal. Although hydrocarbons are derived from organic material not inorganic, this section is included due to the importance and usage of both the hydrocarbons themselves and their derivatives. Methane occurs in swamps and marshes from the decay of vegetation; at night in a swamp it can be seen burning as "fox fire"; it also is the dreaded, explosive firedamp in coal mines. Coal gas, as a household fuel contains 30—40% methane. As a solid at −186°C, methane has the octahedral structure similar to diamond, and is classed as a paraffin compound. It reacts with chlorine to produce hydrochloric acid, HCl, in addition to methyl chloride; methylene chloride; chloroform; and carbon tetrachloride. All these additional compounds are used in surface cleaning of metals and metallic compounds.

Petroleum has been known since ancient times. It is believed that "Greek Fire" was a self-igniting mixture of petroleum and quick lime, and it was used in the defense of Constantinople from 330 A.D. to 1453 A.D., when the city fell to the Ottoman Turks. There are surface occurrences of petroleum in the Balkan and Middle East areas. In 1857 Romania was producing over 200 tons of petroleum from hand-dug wells, and oil was first discovered by drilling at Titusville, Pennsylvania, in 1859 when oil was struck.

But even prior to that time, petroleum was being used as a light source replacement for whale-oil lamps. Ancient wooden sailing ships used petroleum as tar from surface tar pits in the Mediterranean area as calking, and the British Navy and merchantmen were using the great tar pit on the island of Martinique in the Caribbean in the 1600 to 1800s.

In geology the hydrocarbons are not considered to be pure mineralogy, although there are a number of named natural minerals as simple hydrocarbons (paraffins); oxygenated hydrocarbons; petroleum and asphaltum; and coals.

Paraffin, as ozocerite and napalite are wax-like substances from oil or coal, and fichtelite comes from fossilized pine. Amber is the oxygenated type of fossilized resin with half a dozen names depending upon type and locale. Amber was known to the ancients for its static electricity reaction and was called electrum, hence electricity. (*Note:* Natural gold:silver amalgam has the mineral name electrum.)

Petroleum grades from a black/brown liquid with or without a high sulfur content, to a black solid called asphaltum (mineral pitch or tar), which is a semi-hard hydrocarbon mixture and amorphous in structure.

The coals are vegetation that has been compacted to a firm solid under pressure and heat. The three primary coals are: anthracite (hard), bituminous (soft), and lignite (brown or bog coal). Peat is not considered a true coal, as it is insufficiently compacted and hardened, but it has been a major fuel source in northern Europe and the British Isles since ancient times. The peat "bricks" are cut from bogs, stacked, and allowed to dry before being used. It produces an intense, hot fire with a fine, flavorful smoke. That smoke is an important ingredient in the making of Scotch whiskey.

Natural gas from both oil and coal has a high methane gas content, which is responsible for the bright blue flame of household stove burners. As it is odorless, colorless and extremely volatile in the presence of air, an odorizer (mercaptan) is added for a distinctive smell.

Other than the fuels from oil and coal (gases, kerosene, gasoline), there are alcohol derivatives, such as methyl alcohol, and solvents, such as toluene and xylene, or benzene and others of the class of aliphatic compounds or aromatics (sweet-smelling). Some black coals have been cut and polished as jewelry. Baltic amber also has been used for jewelry, and studied for the fossilized insects entrapped.

Petroleum derivatives include paraffin as a base of several wax compositions: lampblack as a black pigment; glycerin for the fabrication of rayon and cellophane; synthetic rubbers and plastics, as both saturated and unsaturated hydrocarbon compounds. Although gasoline for ground and air vehicles is a major product of the oil industry, it should be realized that better than half of every barrel of oil is used in the chemical industry and other areas.

Technical Application: Hydrocarbon gases are used for the deposition of thin film carbon for conversion to diamond-like-carbon (DLC), also shown by the acronym i-C. These thin films are discussed further under the section on Carbon. Specific hydrocarbon thin films have been deposited under vacuum conditions using liquefied gases such as nonane or hexane.

Etching: Dissolve with alcohols and solvents.

HYDROCARBON ETCHANTS

HC-0001
ETCH NAME: Silica gel — TIME:
TYPE: Mineral, drying — TEMP:
COMPOSITION:
x .... silica gel (c-SiO_2)
DISCUSSION:

Liquid hydrocarbons used for thin film deposition as the material compound under vacuum conditions. Silica gel in an 8-ft column used to dry the liquids before entry into vacuum. Hydrocarbons were nonane, 3-methylpentane and hexane. Outgassing in the vacuum chamber was done by freezing.

REF: Berry, W B — *J Electrochem Soc*, 118,597(1971)

PHYSICAL PROPERTIES OF HYDROGEN, H_2

Classification	Reactive gas
Atomic number	1
Atomic weight	2.0162 (H_2)
Melting point (°C)	−258.14
Boiling point (°C)	−252.8
Density (g/cm³ $\times 10^{-3}$)	0.8988
(g/l liquid)	0.070
Specific heat (cal/g/°C) 20°C	3.34
Latent heat of fusion (cal/g)	15
Thermal conductance (cal/cm²/cm/sec) 20°C	4.05
1st ionization potential (eV)	13.597
Ionic radius (angstroms)	0.180
Cross section (barns)	0.33
Vapor pressure (°C)	−257.9
Critical temperature (°C)	−241 (−239.9)
Critical pressure (atm)	20 (12.8)
Hardness (Mohs — scratch) solid	1—3
Crystal structure (isometric — normal)	(100) cube
(hexagonal — normal)	$(10\bar{1}0)$ prism, hcp
Color (solid)	White
(gas/liquid)	Colorless
Solubility (100 ml H_2O) 0°C	1.93 ml
(100 ml EOH) 0°C	6.92 ml
Cleavage (cubic/basal) solid	(001)/(0001)

HYDROGEN, H_2

General: Hydrogen is a colorless, tasteless and odorless naturally occurring gas, highly flammable in the presence of oxygen at standard temperature and up to 180°C will combine only slowly. Above 550°C $2H_2 + O_2$ explosively combines to form water, H_2O. Hydrogen occurs in the free state associated with volcanic eruptions, in some natural gas wells, and as pockets in some mining locations. There is less than 0.000,005% as a free gas in the atmosphere.

Where oxygen is an oxidizing agent, hydrogen is a reducing agent: metal oxides are reduced to pure metals by hydrogen firing; hot hydrogen firing in a furnace or quartz tube with nitrogen "end-curtains" to prevent influx of air with additional hydrogen burn-off exists, is used to clean metal surfaces; to anneal material to a specific hardness, such as $^1/_4$-hard, $^1/_2$-hard brass; or as the reducing atmosphere during parts alloying.

The oxy-hydrogen blow torch, invented by Robert Hare in 1801, will produce temperatures in excess of 2500°C, although oxy-acetylene, C_2H_2, torches are more widely used in welding as it will reach temperatures of 3300°C. There is an Atomic Hydrogen Gun, invented by Irving Langmuir in 1933, that splits the hydrogen molecule, H_2, into atomic, H + H atoms, by the gas passing through an electric spark with the production of 104,000 calories of heat. This is a reversible reaction, and the hydrogen molecule re-forms after a short distance.

Other than as free hydrogen, hydrogen combines with almost all substances; it is found as a constituent in minerals, vegetable matter, and is part of mammillary gastric juices in the form of hydrochloric acid (between 0.2 to 0.4%). In minerals, it may be atomic hydrogen, H, as in HBr; as an OH^- radical, such as in gibbsite (hydrargillite), $Al(OH)_3$ . . . aluminum hydroxide as a native mineral, or as H_2O, such as in opal, $SiO_2.nH_2O$, where the water is both in the physical atomic lattice and as an absorbed, interstitial fluid. It is also found in coals and oils, hydrocarbons, in gases, alcohols, esters, and hundreds of chemical compounds to include sugars, starch, cellulose, and plastics.

The hydrogenation process converts liquid vegetable oils to solid fats by adding hydrogen to the formulas; also includes seed oil, such as cottonseed. The production of "water gas" — the obtaining of hydrogen from water by passing steam over coke (carbon) at 1000°C is a major industry, in the manufacture of ammonia, NH_3, water gas and nitrogen are combined at 500°C, and 200 atm of pressure. Ammonia is not only a base for making fertilizers but for nitrate explosives.

A major reason why Berlin, Germany was a military bombing target during World War II were the water-gas and ammonia plants around the outskirts of the city. During World War I, hydrogen was used to inflate Zeppelins and the airships were used to bomb England; but with the explosion of the Hindenburg in 1937 such air travel ceased. Helium was used in place of hydrogen — 15% H_2:He(85%) is nonflammable, as is a similar mixture with nitrogen, known as Forming Gas (FG), 85% N_2:15% H_2.

Hydrogen is absorbed by many metals and there is much current development of hydrogen-storage systems for possible battery applications as well as general storage/release capability. As an example, finely powdered palladium as "palladium black" will absorb between 800 and 900 volumes of hydrogen; powdered platinum about 50 volumes.

Many of the acids used in etch processing contain hydrogen and they can be either reducing or oxidizing acids depending upon the presence of oxygen in the formula. HF, NH_4F are reducing agents; HNO_3, H_2O_2, oxidizing agents; H_3PO_4, HCl, and H_2SO_4 are general removal or cleaning solutions. The hydroxides, KOH and NaOH, are cleaners, preferential etchants or oxidizers; NH_3 and NH_4OH are weak bases as etchants but, as they readily combine with acids to neutralize them (Normal solutions), they are the primary chemicals used to neutralize acid sump disposal systems. Acetic acid, CH_3COOH (HAc), is not only a weak acid etchant for metals but is the acidic element in vinegar.

Technical Application: Hydrogen, as gas, is used as a furnace processing atmosphere for its reducing capabilities, as it will de-oxidize and clean surfaces, and will prevent oxidation products forming during alloying of parts. It also is used in quartz tube diffusion furnaces during semiconductor wafer doping, and in forming metal hydrides in Solid State processing.

High purity hydrogen can be supplied as pressurized single cylinders (Al type under 3600 psi) or as an A21, interconnected cylinder tank farm, or from a liquid hydrogen, LH_2, source with conversion to gas though heat exchangers. It also is supplied in similar cylinders with other gases as mixtures, such a diborane, B_2H_6, as semiconductor doping sources, or for growth of epitaxy layers.

Metal hydrides such as TaH, ZrH, and NbH have been used as alloying agents in silicon device assembly, and there is on-going development of metallic hydrides as Si:H, Ge:H, SiC:H, or with additional nitrogen, SiN_x:H. These are grown as both colloidal or amorphous structures.

Etching: N/A

HYDROGEN ETCHANTS

H-0001

ETCH NAME: Pressure TIME:

TYPE: Vapor pressure, defect TEMP:

COMPOSITION:

x .... pressure, vapor

DISCUSSION:

H_2, grown as a single crystal under cryogenic vacuum conditions with pressure. Vary vapor pressure in the vacuum system will preferentially etch the single crystal ingot.

REF: Schwentner, N et al — *Rare Gas Solids,* Vol 3, Academic Press, New York, 1970

HYDROXIDES

All hydroxides contain the OH^- radical and, using the Sorenson pH Scale, are classified as chemical bases with pH values greater than pH 7 (water, neutral). They are the opposite of acids (pH under 7) and, when an acid and base are mixed as Normal solutions, then mixed together, they neutralize each other, producing water and a salt. Base, hydroxide, alkali, all three names are used, and these chemicals will turn litmus paper blue; whereas acids turn the paper red. There are rolls of litmus paper available containing a combination pH numbered color chart ratings that can be used to establish the degree of alkalinity or acidity of solutions.

Almost all metals form natural oxides and may contain both water, as H_2O, or the OH^- hydroxyl ion radical in their formula, and a few are referred to as metal hydroxides. Many metal hydroxides are either sparingly soluble or insoluble in water and are stable under normal pressure and temperature. Bauxite, primarily a colloid of $Al_2O_3.2H_2O$, is the natural aluminum hydroxide; goethite and limonite, FeO(OH) or $Fe_2O_3.H_2O$, are the natural iron hydroxides, and are primarily colloidal; manganite, MnO(OH), the manganese hydroxide; and brucite, $Mg(OH)_2$, is the natural magnesium hydroxide.

The mineral magnesite, MgO, called magnesia, will slowly react with water to form the hydroxide $Mg(OH)_2$, best known as milk of magnesia as a suspension in water. As magnesia, MgO, with a melting point of about 2500°C, it is used as a high temperature ceramic for furnace linings and fire bricks; whereas milk of magnesia, $Mg(OH)_2$, is a medical specific and as powders both materials readily react to neutralize acids, such that they are used as antidotes against poisoning by strong acids. Most laboratories keep these materials available in powder form in chemical processing areas for use against acid burns: mix as a water slurry and apply to the burn area to neutralize the acid. The alkali group of metals are lithium, sodium, potassium, rubidium, and cesium — shown in order of increasing atomic weight, and reduction of both melting and boiling points, in addition to increasing density. Several metals, chemical compounds, and minerals are classified as being like, or similar to an alkali due to their chemical reactions.

As sodium, Na, it is one of the more abundant elements (about 2.5% of the earth's crust), in addition to the mineral halite, NaCl, common salt found in ocean water or saline salt dome deposits, it is found in many other natural minerals. The symbol "Na" comes from the mineral natron, Na_2CO_3, and the name means: metal of soda. An impure form of natron was known to the ancient Egyptian priests, and used as part of their embalming fluid in mummification of the pharaohs.

The hydroxide, NaOH, is known as caustic soda, due to its corrosive action on flesh, and is of greater industrial importance than the other alkali compounds due to the abundance of sodium. A major use is in the making of soaps, such as "lye" soap, another term for sodium hydroxide. It is used in the petroleum industry, and in the manufacture of rayon, paper, textiles, and rubber. It also is a major cleaning and etching solution in metals and metallic compounds processing.

As a cleaning/etching solution both sodium and potassium hydroxide, NaOH and KOH, respectively, are the most widely used. The latter, KOH, is preferred where sodium may contaminate the material, such as silicon processing. As a water solution, there are three general concentrations used in material processing: (1) as a 1—15% solution at RT for general surface cleaning; (2) as a 10—20% solution for light removal and surface conditioning/cleaning, and (3) from 20—50% solutions, hot to boiling, as preferential and removal etchants. Several of the chemically inert metals, such as platinum and iridium, can be surface cleaned in boiling hydroxides but, on many others, like single crystal silicon, they are primarily preferential etchants.

As a molten flux, the solid hydroxide pellets liquified at their melting points, they are defect etchants on single crystals, such as KOH at 650°C on silicon. Mixed with other chemicals, such as carbonates and nitrates, molten fluxes are used for single crystal growth: garnets, ferrites, etc.

Another hydroxide, ammonium hydroxide, NH_4OH, as well as ammonia, NH_3, are of major industrial importance as neutralizing solutions. They are widely used to neutralize acid waste from chemical, and metal processing plants. The acids are collected in a sump, then the hydroxide or ammonia added to neutralize before release into a sewage system. They also are used as cleaning and etching solutions alone, or mixed with other acids. In general household, industry, or medical applications they act as both cleaners and disinfectants and, as solid crystals, ammonia is "smelling salts."

The following list of solutions present the hydroxides as etching and cleaning solutions as extracted from aluminum to zirconium. See specific metals and compounds for additional solutions and applications.

HYDROXIDE CONTAINING ETCHANTS

Formula	Material	Use	Ref.
8—10% NaOH/KOH	Al/alloys	Cleaning/pickling/ removal	AL-0003a AL-0002 AL-0062
xKOH:x$K_3Fe(CN)_6$:$K_2B_4O_7$.4H_2O	Al	Removal	AL-0019
2 g NaOH:1000 ml H_2O	Al	Polish	AL-0062
x2.5 *M* NaOH	Al	Electrolytic removal	AL-0064
4 mg NaOH:1000 ml H_2O: 200 ml glycerin	Al	Chem/mech shaping	AL-0077
x30% KOH/NaOH	Al:Be(38%)	Cleaning	ALBE-0001
x10% NaOH	Al:Cu	Removal	ALCU-0001a
x2.5 *M* NaOH	Al:Ni	Removal	ALNI-0001a
x3 *N* NaOH	Al:Ag	Removal/preferential	ALAG-0001
x3 *N* NaOH	Al:Zn	Removal/preferential	ALZN-0001
x10—30% NaOH/KOH	AlN	Removal	ALN-0008b
1—2NH4OH:1—2H_2O_2:4—6H_2O	Al_2O_3	Cleaning	ALO-0004

Formula	Material	Use	Ref.
$5\%NaOH:50\%HNO_3 + H_2O$	Al_2O_3	Electrolytic, anodizing	ALO-0010a
5—20% NaOH	As	Removal	AS-0001a
x30% KOH/NaOH	Be	Cleaning/removal	BE-0001
x10—30% KOH	BeO	Removal	BEO-0004
xKOH, pellet (molten flux)	BC	Removal	BC-0001
xNaOH, pellet (molten flux)	MoB	Removal/preferential	MOB-0001c
xKOH, pellet (molten flux)	Nb_3B_3	Removal/preferential	NBB-0001b
x x%NaOH/KOH	TaB	Removal	TAB-0001a
x x%NaOH/KOH	VB	Removal	VB-0001b
x x%NaOH/KOH	WB_2	Removal	WB-0001c
x x%NaOH/KOH	ZrB_2	Removal	ZRB-0001a
x10% NaOH	BP	Removal/polish	BP-0003
$xNH_4OH:xH_2O_2 + H_2O$	Brass	Removal, cleaning	BRA-0006
$xNH_4OH:xH_2O_2$	Bronze	Preferential	BRO-0004b
xNH_4OH	Cd	Cleaning	CD-0006
xNH_4OH	CdO	Cleaning	CDO-0001
$x10\%NaOH:xNa_2S_2O_4:xH_2O$	CdTe	Removal (Te)	CDTE-0016a
x50%NaOH	CdTe	Polish	CDTE-0017a
xKOH, pellet (molten flux)	C	Removal	C-0001a
x50%KOH	C	Removal	C-0001b
x1 *N* NaOH	CeO_2	Cleaning	CEO-0001
5 g $KMnO_4$:7.5 g NaOH:30 ml H_2O	Cr	Cleaning	CR-0017c
xNaOH, pellet (molten flux)	Cr_2O_3	Dislocation	CRO-0006a
x x%NaOH	Cu	Cleaning	CU-0001
$1NH_4OH:1H_2O_2$	Cu	Preferential	CU-0011b
$xNH_4OH:xH_2O$	Cu:Ga	Preferential	CUGA-0001b
xNH_4OH	Cu_2O	Removal	CUO-0006a
$12\%NaOH:1H_2O_2$	GaSb	Removal	GASB-0001d
$xNH_4OH:xH_2O_2$	GaAs	Polish	GAS-0007c
$1NH_4OH:20H_2O_2$	GaAs	Selective removal	GAS-0018
$3NH_4OH:1H_2O_2:1H_2O$	GaAs	Step-etch	GAS-0022
$3NH_4OH:1H_2O_2:150H_2O$	GaAs	Junction	GAS-0021
$20NH_4OH:7H_2O_2:973H_2O$	GaAs	Removal	GAS-0003b
$xKOH:xH_2O$	GaAs	Electrolytic, junction	GAS-0210
xKOH, pellet (molten flux)	GaAs	Dislocation	GAS-0031
250 ml 30%KOH:100 ml MeOH	GaAs	Preferential	GAS-0036
21 g KOH:8 g $K_3Fe(CN)_6$:100 ml H_2O	GaAs	Preferential, junction	GAS-0039
$x4\%NaOH:x5\%H_2O_2$	GaAs	Polish	GAS-0175
x1 *M* NaOH:x0.7 *M* H_2O_2	GaAs	Preferential	GAS-0173c
x30—50% NaOH	GaN	Removal	GAN-0002
x0.1 *N* NaOH	GaN	Electrolytic, selective	GAN-0001
x0.5%KOH:x1 *M* $K_3Fe(CN)_6$	Gap	Polish	GAP-0005
$1NH_4OH:1Cr(NO_3)_2$	Ge	Preferential	GE-0005d
250 ml 30%KOH:20 g I_2	Ge	Preferential	GE-0221a
250 ml 30%KOH:50 ml Br_2	Ge	Preferential	GE-0221b
x x%KOH	Ge	Electrolytic, junction	GE-0019
x x%NaOH	Ge	Electropolish	GE-0042
6 g KOH:4 g $K_3Fe(CN)_6$:50 ml H_2O	Ge	Preferential	GE-0150s

Formula	Material	Use	Ref.
x0.5%KOH	Ge	Electrolytic, preferential	GE-0087
x 1 *N* KOH	Ge	Electropolish	GE-0088b
x 15%KOH	Ge	Electrolytic, passivation	GE-0204a
x10%NaOH:x15%$NaNO_3$:xH_2O	Ge	Preferential	GE-0208
x2%NaOH:x2%$NaClO_4$	Ge	Cleaning	GE-0155
x x%NaOH:xH_2O_2:x 1μ Al_2O_3	Ge	Chem/mech polish	GE-0210
1 20%NaOH:1H_2O_2	GeAs	Removal	GEAS-0001a
18 g NaOH:5 g $KHC_8H_4O_4$:100 ml H_2O	Ge_3N_4	Removal	GEN-0001
x x%NaOH:xx%Na_2S	GeSe	Removal	GESE-0001
1 20%NaOH:1H_2O_2	InSb	Removal	INSB-0002a
x x%KOH:xHF:xHNO_3	InSb	Polish	INSB-0005b
2.5 g KOH:200 ml MeOH	InP	Cleaning	INP-0048
1 g KOH:1 g $K_3Fe(CN)_6$:1 ml H_2O	InP	Preferential	INP-0064a
4 g KOH:6 g $K_3Fe(CN)_6$:35 ml H_2O	InP	Selective	INP-0064j
140 ml NH_4OH:60 ml HNO_3:10 g H_2Mo_4:240 ml H_2O	Pb	Removal	PB-0009b
5 45%KOH:1H_2O_2:5 EG	PbSe	Electropolish/thinning	PBSE-0002
5 g NaOH:0.2 gI_2:10 ml H_2O	PbTe	Preferential	PBTE-0001a
20 g KOH:45 ml H_2O:20 ml EOH:25 ml Gly	PbTe	Electropolish/ preferential	PBTE-0001b
5 g NaOH:10 ml 0.5%$NaIO_3$	PbTe	Preferential	PBTE-0002a
2 15%NaOH:1 x%$Na_2S_2O_8$	PbTe	Dislocation	PBTE-0005
2 15%NaOH:1$Na_2S_2O_8$, sat. sol.	HgTe	Dislocation	HGTE-0002b
1NH_4OH:2H_2O_2:7H_2O	Mo	Polish/cleaning	MO-0002
100 g NaOH/l:110 g KCN	Mo	Removal/cleaning	MO-0005b
20 g KOH:92 g $K_3Fe(CN)_6$:300 ml H_2O	Mo	Polish	MO-0009
x 10 *N* NaOH	Ni:Cr	Electrolytic, removal	NICR-0005b
xNH_4OH	NiO	Removal	NIO-0001a
x x%NaOH	Quartz	Preferential	QTZ-0014b
100 ml NH_4OH:100 ml H_2O_2:500 ml H_2O	Quartz	Cleaning	QTZ-0012
x 0.1 *N* NaOH	Re	Electropolish	RE-0002
1NH_4OH:1H_2O_2:6H_2O	Sapphire	Cleaning	SAP-0004a
x x%NaOH	Se	Removal	SE-0004b
1NH_4OH:12H_2O_2:4H_2O	$TiSi_2$	Removal	TISI-0003
1—2NH_4OH:1—2H_2O_2:5—7H_2O	Si	Cleaning	SI-0031
1NH_4OH:5H_2O_2	Si	Selective	SI-0021c
x x%KOH	Si	Etch-stop	SI-0084
10—30%KOH	Si	Preferential	SI-0091
250 ml 30%KOH:30 g I_2/50 ml Br_2	Si	Preferential	SI-0092a-g
200 g KOH:10 g $KAu(CN)_6$:800 ml H_2O	Si	Junction plate	SI-0101
x 2 *M* KOH	Si	Electrolytic, oxidation	SI-0146
x 1 *N* KOH: + glycerin: + H_2FSi_6	Si	Electropolish	SI-0245
x 10 *N* NaOH	Si	Selective structuring	SI-0147

Formula	Material	Use	Ref.
250 ml 30%KOH:30 g KBr/$FeCl_3$	Si	Preferential, pit forming	SI-0198c
3NaOH:1Na_2O_2, pellet, molten	SiC	Polish	SIC-0011
x 2 *M* KOH	SiO_2	Removal	SIO-0002
1—2NH_4OH:1—2H_2O_2:4—6H_2O	SiO_2	Cleaning	SIO-0043
xNaOH:x$NaCO_3$:xH_3PO_4	SiO_2	Cleaning	SIO-0039c
18 g NaOH:4 g $KHC_8H_4O_4$:100 ml H_2O	Si_3N_4	Removal	SIN-0008c
1 g KOH:100 ml H_2O	Si_3N_4	Cleaning	SIN-0019
1NH_4OH:1H_2O_2	Ag	Removal	AG-0005
1NH_4OH:1H_2O	Ag	Cleaning	AG-0004a
25 ml NH_4OH:15 ml H_2O_2	Ag	Polish	AG-0012b
1H_2O_2:2EG:2KOH, sat. sol.	Ag_2Se	Polish	AGSE-0001b
3NH_4OH:2H_2O_2	Ag_2Te	Polish	AGTE-0001
xNH_4OH	AgCl	Cleaning	AGCL-0006
180 g NaOH:30 g $KMNO_4$:1000 ml H_2O	Steel	Selective	ST-0015
x x%KOH/NaOH	Ta	Removal	TA-0003f
9 30%KOH:1H_2O_2	TaN	Removal	TAN-0004
20 g KOH:92 g $K_3Fe(CN)_6$:300 ml H_2O	TaN	Removal	TAN-0005a
10 g KOH:10 g $K_3Fe(CN)_6$:100 ml H_2O	TaC	Cleaning	TAC-0001
1 g NaOH:100 $cm^3$$H_3PO_4$:2 g agar	SnTe	Electropolish	SNTE-0001
1NH_4OH:2H_2O_2	Ti	Removal	TI-0021b
10 g KOH:10 g $K_3Fe(CN)_6$:100 ml H_2O	TiC	Cleaning	TIC-0005
1NH_4OH:2H_2O_2	TiW	Removal	TIW-0001
x 20% NaOH	W	Cleaning	W-0001a
44.5 g NaOH:305 g $K_3Fe(CN)_6$:1000 ml H_2O	W	Removal	W-0001i
10 g NaOH:80 g $KClO_3$:40 g KCO_3:1000 ml H_2O	W	Electropolish/cleaning	W-0003b
xNH_4OH:x 25% $CuSO_4$	W	Polish	W-0006
1 3%NaOH:1 10% $K_2Fe(CN)_6$	W	Removal	W-0030b
1NH_4OH:2H_2O_2	W	Removal	W-0036
5 g KOH:5 g $K_3Fe(CN)_6$:100 ml H_2O	W	Step-etch	W-0035
20—30KOH/NaOH	WO_2	Removal	WO-0001
10 g KOH:10 g $K_3Fe(CN)_6$:100 ml H_2O	WC	Cleaning	WC-0001
x 0.75% NaOH	WRh	Polish	WRH-0001
x x% KOH	Y	Removal	Y-0001
10 ml NH_4OH:50 ml H_2O:2 g $NH_4(NO_3)_2$	Zn	Preferential	ZN-0006c
1—5 2%NaOH:1H_2O_2	Zn	Preferential	ZN-0017a
x 6 *N* NaOH (+ acid?)	ZnO	Removal	ZNO-0007
x 50% NaOH	ZnSe	Preferential	ZNSE-0002a
x 25% NaOH	ZnSe	Cleaning	ZNSE-0002c

Formula	Material	Use	Ref.
x x% NaOH	ZnSe	Polish	ZNSE-0005
x 30%NaOH	ZnTe	Dislocation	ZNTE-0001a
x 20% NaOH	ZnTe	Removal	ZNTE-0006
1 x%KOH:1 x%NaOH	ZnW	Preferential	ZNW-0001b

PHYSICAL PROPERTIES OF ICE, H_2O

Classification	Solvent
Atomic numbers	1 & 8
Atomic weight	18
Melting point (°C)	0
Boiling point (°C)	100
Density (g/cm^3) 4°C	1.00
Freezing point (°C)	
Phase formation point (amorphous) (K)	80
(hex-to-cubic) (K)	71—76
Plastic deformation point (sphere/block) 0°C	−2.2/−5
Triple point (gas/liquid/solid) °C	4—5
Electrical resistivity (eV) hexagonal	0.84
isometric (cubic)	0.28
hex-cubic (trans.)	0.6
Formation temperature (artificial) (°F)	−4
(natural °F)	+4
Brittleness (psi)	10 (shatters)
Energy band gap (eV) hexagonal	0.85
isometric (cubic)	0.28
Hardness (Mohs — scratch)	3—4
Crystal structure (hexagonal — normal)	Six-rayed platelets
(isometric — normal)	(100) cube
(amorphous)	None
Color (solid)	Colorless to white
Cleavage (basal/cubic)	(0001)/(001)

ICE, H_2O

General: Ice is water in its solid, frozen state and, as it is cooled from 4 to 0°C, it expands, becoming lighter than water. Maximum density of water is 1.00 at 3.98°C, weighs 1 g, and is the unit against which all material specific gravity (density) is measured. Water is the only known substance that has a triple-point — it can exist in the solid/liquid/gas state simultaneously at about 4°C. Eleven volumes of ice will melt to ten volumes of water and, as long as solid ice remains in water as it is heated, temperature will stay at 0°C. The freezing point of ice is 0°C (32°F or 273°A and K) and is nominally clear and transparent, though it may appear slightly bluish to bluish-green by refraction.

Brine is natural salt water containing dissolved salts — halite, NaCl being a major constituent, along with calcium and magnesium chlorides — and artificial brines have been formulated for study of their freezing points. Liquid ammonia, NH_3, sulfur dioxide, SO_2 and, today, Freon, CCl_2F_2 under pressure with a brine solution ($CaCl_2$) are used for refrigeration — boiling points at −33.3, −10, and 29.8°C, respectively.

Ice forms during cold weather by the freezing of water in the ocean or in lakes, rivers and streams on land, and from the compaction of snow, sleet or hail from water vapor in the air. It is estimated that 70% of the fresh water in the world is as frozen ice in the Arctic and Antarctic regions, and it has been suggested that icebergs be driven to the seacoasts of arid desert areas as a fresh water supply.

Under normal atmospheric conditions of pressure, ice forms in the crystallographic hexagonal system, normal class as snowflakes but under increased pressure it will convert to isometric (cubic) system, normal class, as a cube, (100) structure. In some instances, zones of both amorphous and colloidal ice will occur.

There are several natural and artificial forms of ice with a variety of uses which are described in the following sections:

Natural Forms of Ice

Snow

Annually deposits from water vapor in the air during cold weather in the form of snowflakes — relatively flat, hexagonal platelets of an unlimited variety of lace-like shapes to fairly massive globules of solid ice as pea-size sleet to chunks of hail. The latter can be fist-size or larger and can do physical damage to structures, vehicles and people. Snowflakes can nucleate spontaneously when water vapor in the air reaches the freezing point (self-nucleation) or on dust particles in the air; the latter being more common. Even in the summer time, in the upper reaches of a towering thunderhead cloud that may reach 50,000 ft or more, electric discharge (lightning) or the negative/positive potential that can build due to violent air currents within the cloud can initiate ice formation. (See ICE-0011). Silver iodide, AgI_2, is not water soluble and, as a particulate dust cloud, has been used to seed rain clouds or artificially initiate ice growth.

Annual snowfall is recorded in the temperate zones of the world and calculated for water content of spring run-off from mountain watersheds, ground collection in lakes and dams, and subsurface increase and maintenance of water tables.

At or below freezing, with low humidity, snow can deposit as dry, finely powdered snow and, when high wind occurs simultaneously, blinding snow-fog can be hazardous. Also with wind in winter, with or without snow, the wind-chill factor can drop the temperature below −30°C or more. In early spring, with increased humidity, a snow storm can deposit heavy, wet snow and, in heavily populated areas, snow removal from roads is a major industry throughout the entire winter months. Because of the expansion and contraction of snow and ice it is a major factor in the cracking and reduction of rocks into soil, and can rupture water lines and so forth.

In skiing areas, manmade snow is blown across slopes as chilled water that produces artificial snow. Natural snow has long been used for temporary preservation and packing in the storage and transportation of foodstuff. In the metal industry snow and ice have been mixed with acids or chemicals to obtain cold mixtures at specific temperatures. For special mixtures, see the section on Water: Cold Etchants:

Lake Ice (Ponds, Rivers and Streams)

The fresh water commences to freeze on the surface as scrum ice — then, eventually, as solid ice sheets. Ice thickness varies in depth to 3 or 4 ft on an average. At or near the surface the ice is aerated and can be mixed with compacted snow. With increase in depth and pressure, it becomes solid, nonaerated ice, eventually will convert to cubic structure from its normal hexagonal, and there can be zones of colloidal and amorphous ice.

Before the discovery of refrigeration in the early 1930s, ice was cut from lakes and stored in ice houses for year-round use. Ice houses are thick-walled wooden barn-like structures where the cut slabs of ice are stacked with sawdust layers between blocks to keep them separated and prevent sticking. Lake and river ice is still cut as blocks in many rural areas of the world where there is no or limited electricity for a refrigerator. Properly stored, ice blocks can last year round, even for several years in 90°F weather through the summertime. Ice blocks are normally cut about 2 × 1 × 4 ft in size, then further cut with an ice pick for use as a block of about 25 lb weight . . . with delivery every 2 to 3 days to the home ice box. Note that 1 cubic ft of ice is about 25 lb.

Glacier Ice

The last Ice Age ended about 12,000 B.C. at the end of the Pleistocene Epoch. We are currently in the Quaternary Period, which commenced about 3 million years ago with the

Eocene Epoch, then the Pleistocene, and our present Holocene Epoch of these last 10,000 years. The last glacial ice sheet in North American extended almost to mid-continent in northern Kansas and, man as we know him today . . . *Homo sapiens, sapien* . . . established his civilization after the last Ice Age. Possibly the earliest city, referred to in ancient writings as the empire of Subir, was only discovered in 1981 in Syria. Artifacts at the site show that the area has been occupied by man for at least 7000 years. As there have been at least four major Ice Ages in the past, geologists believe we may well be between Ice Ages at present. How long before the next one? Estimates range from 10,000 to 500,000 years and, some doomsday environmentalists claim man can create his own Ice Age . . . the "Nuclear Freeze" hypothesis . . . within 100 years?

Regardless, there are still residual, left-over glaciers from the last Ice Age in high mountains and the Polar Regions — in the Swiss Alps; Mount Ranier in the American Northwest; and in Alaska, as only three examples. There is year-round study of these glaciers — their size increase or melt-back — as one factor in determining the variations in world climatic conditions. At depth, some of the ice in these glaciers may be as much as 10,000 years old. Most glaciers move back and forth a few inches a year, but one glacier in Alaska, within the past few years, suddenly has been moving seaward at a rate of 2 to 3 ft/day! This is a local action, does not mean the Arctic ice pack is melting, but is of interest as a singular climatic fluctuation. It is estimated that a 2°C worldwide temperature rise would be required to melt the Antarctic and Arctic ice caps, but the glaciers in the high mountains would probably still remain.

In the 1930s the Chinese claimed the discovery of a frozen Hairy Mammoth in North China which, when thawed, was still edible. Glacial ice contains much detritus — rocks, vegetation, and occasional animal life — and spring-thaw water run-off is often a highly aerated, light green-colored, very cold water.

Glaciers are of little use, other than as water run-off for lakes, rivers, and streams and where they enter the ocean, the glacier face can break off forming icebergs. Greenland in the North Atlantic is still covered by an ancient glacial ice sheet, and is responsible for several of the large icebergs that plague the Atlantic shipping lanes. Since the sinking of the Titanic, there has been a world-sponsored iceberg watch that tracks bergs down the Atlantic during the berg season.

Ice Pack

There are permanent ice packs in the Polar regions of the world. There are year-round study stations in the Antarctic (South Pole) where the snow pack is on underlying rock. The pack in the Arctic (North Pole) is largely on water (some covered rock islands), and study groups enter and leave without a permanent camp. There have been permanent installations built into the Greenland ice since about the late 1930s for research study, and air strips for the Iceberg Service watch. In the Antarctic, there are joint teams of Russian, American, French, and British scientists established in permanent study camps year round, not only studying the ice, but major study of climate conditions. It is here that ice cores — 4″ in diameter and several feet long — are cut to depth, then sections cut and melted to obtain ancient forminifera and radiolaria, which are microscopic sea life that can be used to establish ancient climatic conditions. Some of these ice samples are over 4000 years old from snow laid down at the time the great pyramids of Egypt were being built and man was just establishing his civilization in the Mediterranean area. It is thought that some of the ice at depth could be even older.

It is worth noting that the northern and southern land masses that were covered by the ice sheets during the last Ice Age are still rising, recovering from the massive weight of ice. Great Britain is an example . . . the entire island is tilting . . . the north rising; the south submerging. Norway and Sweden with their fjords and rugged coast line are atypical of a rising land mass, as is the rugged coast of New England in the Americas.

The polar ice caps are estimated to contain 75% of the fresh water in the world and, if they were to melt completely, the oceans of the world would rise some 2 to 3 ft, inundating many coastal areas. That they have been melting over the past 5000 years is obvious in the Mediterranean area, where there are ancient buildings submerged in the sea.

Icebergs

Common to both the Pacific and Atlantic oceans but of more concern in the North Atlantic due to the heavier shipping traffic in the iceberg zone. These icebergs come mainly off the Greenland glaciers. Icebergs coming out of the Arctic into the Pacific are more low-level ice sheets and are blocked from entering the Pacific Ocean from the Bering Sea by the Aleutian Islands chain that extends across the Pacific from Alaska to near the north Asian mainland. The Antarctic at the South Pole, although it can produce large bergs, these do not travel up into the south Pacific to any extent due to the flow of ocean currents.

Regardless, icebergs can be extremely dangerous as two thirds to three quarters of the bergs are usually underwater, unseen. As they move and melt, they may suddenly "turn-turtle" or break-up into several smaller bergs. In the Arctic, the pack ice breaks into sheet-bergs that can pile up on each other in tremendous ice jams, much like ice jams, in the Mississippi River during spring thaw.

Today there is little use for icebergs, but Saudi Arabia has already suggested the driving of icebergs up from the Antarctic to their desert lands as a source of fresh water. They also could be driven up from the Ross ice shelf below South America to the Atacama Desert in Chile where there has been no known rainfall in recorded history.

Sea Ice

Sea ice forms initially as a near-mush on the ocean surface and, as the ocean water surface is being watched, suddenly solidifies as a thin sheet of ice as far as one can see! It is salt-water ice with a freezing point slightly above that of fresh-water ice. The specific freeze point varies to some degree with concentration of salts, but is usually between 5—6°C, against the 4°C for fresh-water ice. Because of this difference, rock salt is scattered on road ice during winter months to help melt it. Sea ice as permanent pack ice is more common to Arctic ice, it can be layered between snow-derived ice in the pack, rather than in the Antarctic where the ice is on a land mass base. Natural salt water ice is nonpotable and of little use, but specific chemical mixtures of salts are combined artificially for particular below freezing temperature baths in material processing. See the section on Water, and "Cold Etchants".

Sleet/Hail

Sleet is fine, frozen pellets of rain drops; hail is agglomerations of sleet. Both are clear, solid ice, largely amorphous, and occur where there is a rapid change of atmospheric conditions, such as a warm air front meeting a cold front, and can occur at any time of the year under proper conditions. Sleet can occur in the wintertime under similar conditions of rapid atmospheric changes, forming a thin, slick coating on everything it settles upon, and can be particularly dangerous on roads for vehicular traffic.

Black ice is not necessarily sleet, more often the re-freezing of melting snow or ice on roadways, particularly toward the lower road edge or at corners. It may be a mixture of water and road oils, and can be even slicker than pure ice due to the presence of oil. The term "black ice" is used as it is colored by the included dark oils and tars and, on a macadam road, may not be recognized by a driver in a moving vehicle.

There are occasional hail and sleet storms which can cause actual property damage. These occur most often in the U.S. in the mid-continent when frigid arctic air sweeps down from Canada to meet warm tropical air coming up from the Gulf of Mexico.

Artificial Ice

For Human Use

Natural ice has been used for centuries as a food preservative. Some of the earliest written records refer to its use in the making of sherbet (Arabic: sharbat) . . . a mixture of snow or finely shaved ice with fruit flavoring . . . that may have its origins as far back as 5000 B.C. The Persian satraps used to have runners bring ice down from the mountains for use at the royal courts, and Roman emperors transported ice down from the north Italian Alps. The Eskimos and others who live in high mountains where there is permanent ice have long used ice caves for storage of food.

Refrigeration, as we know it today, was not developed until the early 1930s, though ice houses were in operation making block ice for industry and home use for many years prior to the first refrigerator or Frigidaire (Frig, the wife of the God Odin of Nordic mythology). Ammonia, NH_3, under pressure is still widely used in ice houses for mass production of ice, but Freon, CCl_2F_2, has largely replaced both ammonia and sulfur dioxide for home refrigeration and general storage and transportation of perishable foods. Refrigerated railway cars and trucks were originally developed in the U.S., then shipping refrigeration worldwide.

Refrigeration should be considered one of man's major accomplishments of modern times, as it has taken us away from seasonal reliance for our food supply to include both longer-term storage and transportation. Since ancient times, the seasons of the year . . . planting in the spring . . . harvesting in the fall . . . have controlled our lives. The availability of saltwater fish and shellfish always was in limited supply inland from seacoasts due to spoilage until the advent of refrigeration. Foods that could be dried or salted were the main travelfare.

Quick-freezing, developed as a commercial product after World War II, was another big step — first vegetables and fruit juices — now complete meals. The food is frozen with its liquids; whereas vacuum freezing, or vacuum packed food has little or no liquids. And as an adjunct to quick-freezing, today, nuclear irradiation of foodstuffs is being developed, prior to freezing or not requiring freezing. A recent report said that chicken, irradiated in a sealed bag and without freezing, had a shelf-life of better than a year.

Carbon dioxide, CO_2, as the solid dry ice, bears mention even though it is not "ice" by water definition. Dry ice has a melting point of about $-57°C$ — it sublimes from the solid without passing through a liquid state — and is used as a temporary medium for cold storage of foodstuffs, medical preparations, and industrial items such as epoxies and polyimide pastes. In a closed container it has a working life, or shelf-life, of about 3 days. As gaseous CO_2, it is the natural fermentation in beers and wines, the effervescent bubbles, or is artificially added to wines and soft drinks. Without flavoring it is seltzer water, which can be artificial or natural. A new product may be on the market soon: a flip-top can containing a CO_2 capsule that, when the can is opened, instantly chills the beverage with no need for refrigeration.

Research and Development

In material studies ice has been cut from lakes for determination of crystal structure at depth, and polar ice core samples have already been mentioned. Using high purity water both snowflakes and single crystal ice have been grown in the laboratory using liquid nitrogen, LN_2, cold cryostats, to include silver iodide, AgI_2, as a seeding vehicle. Natural salt water and manmade brines with different salt combinations and concentrations have been frozen and studied with similar cold (cryogenic) cryostats. Some such cryostats have been on a microscope stage to observe ice-forming characteristics directly under the microscope.

Technical Application: Ice has many uses in material processing of metals and metallic compounds in the general metal industries as well as Solid State.

As refrigeration it is used for cold storage of particular materials, such as gallium (m.p. +30°C); unstabilized hydrogen peroxide, H_2O_2 (m.p. −1.7°C); bromine, Br_2 (m.p. −7.2°C), or epoxy and polyimide resin pastes. Some such resins have a shelf-life of a year at room temperature; others are stored at −40°C for extended shelf-life; others kept frozen as, when brought to room temperature, they harden within seconds or have a working-life of only 6 to 8 h once mixed as an A + B combination.

For a cold liquid in material processing: ice/water (0°C); ice/acetone (+8°C) still are used to chill acid etchants. The ice/acetone mixture also is used to remove water vapor by freeze-out from gases, argon gas from pressurized cylinders in particular. The ice/water mixture also is used as a direct quenching medium in the fabrication or study of many metals and metallic compounds.

As a cooling vehicle ice water or refrigeration coils — even just cool industrial water — is used for heat control in many operations: around furnaces through coils; epitaxy systems; ingot growth systems; water recirc systems; large etchant tanks, and so forth.

There are several "Cold Etchants" using ice or snow as one constituent for specific temperature levels. Several are listed under the section on Water under Cold Etchants.

As already discussed in the General section, ice is under a variety of studies, both natural and artificial, for physical data and electrical parameters. Studied as snowflakes, single crystal blocks, and spheres.

Etching: Water or steam. Pressure variation. Formvar, and heat, only.

ICE ETCHANTS

ICE-0001
ETCH NAME: Water
TYPE: Acid, polish
TIME:
TEMP: −5°C or less
COMPOSITION:
x .... H_2O
DISCUSSION:

H_2O as ice, (0001) single crystals used in a study of ice formation and reactions. Near-freezing water can be used to polish surfaces.

REF: Higuchi, K — *Acta Metall,* 6,636(1958)

ICE-0002: Okliya, M — *Sov Phys-Cryst,* 4,244(1960)

H_2O as ice, (0001) single crystals used in a study of ice formation and reactions. Lap surfaces with emery paper, then on the surface of a ground glass plate with no liquid solution, before polishing with near-freezing water. Cracks in the ice will self-heal at or slightly below −5°C.

ICE-0003: Kingley, W D — *J Appl Phys,* 31,833(1960)

H_2O as ice, (0001) single crystal used in a study of ice formation and reactions. Polish surfaces with near-freezing water. Also studied ice spheres. See ICE-0003b.

ICE-0012: Jellick, H H G — *J Appl Phys,* 32,1793(1961)

H_2O as ice, single crystal hexagonal specimens used in a study of the reactions of various liquid layers on ice surfaces, including freezing power, gripping tension and temperature. Cold water was used as a polishing solution.

ICE-0016: Eisenberg, D & Kadsmann, W — *The Structure and Properties of Water,* Oxford University Press, London, 1969.

H_2O as ice, at normal pressure and temperature (0°C) forms in the Wurtzite structure: hexagonal hemimorphic with no hexagonal plane of principal symmetry, and no horizontal axes of binary symmetry. As there are positive (upper) and negative (down) forms in the hemimorphic class, this accounts for the wide variety of snowflake patterns to a large degree. The high pressure form of ice is isometric normal cube, (100) as a crystal form.

ICE-0017: Dorn, R T et al — *Sci Am,* May 1977

H_2O as ice from poly-water. As a liquid it is a high density chain of close-packed water molecules formed under special conditions using high purity, DI water. It boils at 150°C and freezes at −40°C to a plastic-like substance which is not like normal water ice.

ICE-0005
ETCH NAME: Liquid helium — TIME:
TYPE: Gas, freezing — TEMP: Near 2.2 K
COMPOSITION:
x .... LHe
DISCUSSION:

H_2O as ice, (0001) single crystals grown as hexagonal platelets in a study of the growth of ice. The water used was doped with KOH (0.1, 0.01 and 0.001 mol/dm^{-3} concentrations). De-aerated the water by repeated solidification and vacuum evacuation using a cryostat with LN_2 and heat exchanger. Operate at 77 K for 10—15 min to crystallize. There is a phase transition at 71—72 K from hexagonal to cubic form (hi-pressure ice). In this study, after ice forming, it was annealed at 60—65 K, then cooled down to LHe temperature.

REF: Tajima, Y — *J Phys Chem Solids,* 45,1135(1984)

ICE-0007
ETCH NAME: Pressure — TIME:
TYPE: Pressure, dislocation — TEMP: Under 0°C
COMPOSITION: — PRESSURE: Vary vapor pressure
x .... vapor pressure
DISCUSSION:

H_2O as ice, (0001) single crystals. A vertical growth system using differential pressure/temperature of a cryostat (LN_2) to solidify water from the vapor phase with a variable "chiller" zone. An ice plate was used to "seed" for growth nucleation, and AgI_2 smoke to initiate growth. Varying the vapor pressure was used to develop dislocations and defects.

REF: Gonda, T & Koke, T — *J Cryst Growth,* 65,36(1983)

ICE-0008a
ETCH NAME: Silver — TIME:
TYPE: Metal, forming — TEMP:
COMPOSITION:
x .... Ag
DISCUSSION:

H_2O as ice, (0001) single crystals used in a study of the hexagonal to cubic transformation with electrical conductivity measurements. A cryostat (LN_2) finger with silver electrodes used to initiate water vapor condensation. Amorphous, a-ice, was formed at 80 K. Sublimation/resublimation between 2 to 220 K used to convert hexagonal ice to cubic ice. Hexagonal band gap: 0.84 eV; cubic band gap: 0.28 eV; transition zone: 0.6 eV. Ice was stored in solid MeOH (−98°C).

REF: Chrzanowski, J & Sujak, B — *Thin Solid Films,* 112,17(1984)

ICE-0003b
ETCH NAME: Liquid nitrogen — TIME:
TYPE: Gas, forming — TEMP: −195°C
COMPOSITION:
x .... LN_2

DISCUSSION:

H_2O as ice spheres. Spheres formed by spraying water into LN_2 or LOX (−185°C). Size was between 0.1 to 3 mm in diameter. Handling spheres at −2.2°C can cause plastic deformation.

REF: Ibid.

ICE-0004

ETCH NAME: Ethylene dichloride — TIME:

TYPE: Acid, dislocation — TEMP:

COMPOSITION:

x .... polyvinyl formal (formvar)

x .... 1% ethylene dichloride

DISCUSSION:

H_2O as ice, (0001) single crystals used in a study of deformation and dislocation development. Pour the mixture over the ice surface, and allow it to dry by evaporation as the formvar solidifies. The water in the ethylene dichloride breaks through to form etch pits in the ice. Small pits form in the larger surface pits, and are crystallographically oriented triangles or hexagons, representative of dislocations(?). The formvar can be peeled from surfaces and used as replicas for SEM study.

REF: Bryant, G W & Mason, B J — *Phil Mag,* 5,1221(1960)

ICE-0009

ETCH NAME: Formvar — TIME:

TYPE: Acid, defect — TEMP:

COMPOSITION:

x .... formvar

DISCUSSION:

H_2O as ice, (0001) platelets grown using a low temperature cryostat (LN_2) built as a microscope sub-stage. Crystals were free-grown in a water bath by cryogenic cooling while under observation with a scanning electron microscope (SEM). After growth, crystals were coated with formvar, and then coated with thin film gold evaporated for reflectivity of the replicated surface. The film was then peeled away for SEM study.

REF: Nenow, D et al — *J Cryst Growth,* 66,489(1984)

ICE-0006

ETCH NAME: Cryogenic gas — TIME:

TYPE: Gas, forming — TEMP: −195°C

COMPOSITION:

x .... LN_2

DISCUSSION:

H_2O as ice, (0001) single crystals. An overview of growth techniques of ice forming from vapor phase water.

REF: Icuroda, T — *J Cryst Growth,* 65,27(1985)

ICE-0008b

ETCH NAME: Methyl alcohol — TIME:

TYPE: Alcohol, storage — TEMP: −98°C

COMPOSITION:

x .... MeOH, solid

DISCUSSION:

H_2O as ice, (0001) single crystals. Ice can be held in storage in solid, frozen methyl alcohol. See ICE-0008a.

REF: Ibid.

ICE-0010

ETCH NAME: Brass
TYPE: Metal, polish
COMPOSITION:
x .... brass plate

TIME:
TEMP: Warm

DISCUSSION:

H_2O as ice — hexagonal, amorphous, colloidal and cubic specimens cut from lake ice and used in a study of formation and structure under natural conditions. Specimens were slab-cut using a nichrome wire at red-heat, then polished and thinned by rubbing on a warm brass plate. From the lake surface the ice structure progressed from hexagonal, through variable zones of amorphous and colloidal ice, to cubic ice at depth. There also was a reduction of bubbles, stria and low angle grain boundaries, and such defect density with increasing depth.

REF: Barns, R & Laudise, R A — *J Cryst Growth,* 71,104(1985)

ICE-0011

ETCH NAME: Temperature
TYPE: Temperature, formation
COMPOSITION:
x .... temperature

TIME:
TEMP: −4°F to +14°F

DISCUSSION:

H_2O as ice. A study of the natural growth of ice in clouds during a study of the cause of lightning. Ice forms in clouds as (1) tiny water droplets; (2) super-cooled water droplets well below 32°F; (3) chunk hail; (4) ice pellets called graupels (Ger: ''soft hail'') — pea-sized, raspberry-like structure formed by super-cooled water droplets freezing together. Ice crystals and graupels bounce off each other attaining a positive (+) or negative (−) charge and, with wind currents and convection within the cloud, charge separation occurs. With sufficient charge accumulation and separation, lightning discharge occurs with energy release up to 100 million volts. The thunderhead clouds, which generate much of the ice, rain and lightning, can extend from 1 mile above the earth surface to better than 10 miles upward as a single cloud. The temperature gradient in such clouds on a hot, humid day may be +90°F at their base — chilling vertically up through the cloud mass — to below −100°F at their top.

REF: Hallett, J — *Science,* Sept 1985, 75

ICE-0013

ETCH NAME: Ethyl alcohol
TYPE: Alcohol, thinning
COMPOSITION:
x .... C_2H_5OH (EOH)

TIME:
TEMP: −20°C

DISCUSSION:

H_2O as ice, (100) cubic single crystals. Initial slices were about 10 mm thick, and lap polish/thinned in solution shown down to 2 mm. Quench in *n*-hexane to stop etching action. Specimens used in a study of voids formed by glide on nonbasal planes.

REF: Maguruma, J et al — *Phil Mag,* 13,625(1966)

ICE-0014
ETCH NAME: Sodium chloride
TYPE: Alkali, inhibitor
COMPOSITION:
x .... 1% NaCl
TIME:
TEMP:
DISCUSSION:

H_2O as ice, (0001) crystals. The basal plane is the most stable in hexagonal ice. During growth of ice crystals, regardless of surface orientation, if a water solution of NaCl (common salt) is placed on the surface, it will inhibit further pure ice growth. (*Note:* Sea ice freezes slightly above the temperature of pure ice. Addition of NaCl — salt — begins to liquefy a pure ice surface.)

REF: Knight, C A — *J Appl Phys,* 33,1808(1962)

ICE-0015: Hobbs, P V & Mason, B J — *Phil Mag,* 9,181(1964)

H_2O as ice, single crystal spheres or polycrystalline spheres. To form spheres: (1) use a #30 hypodermic needle attached to an earphone diaphragm and use frequency to control water flow through the needle to form spheres 20—300 μm in diameter, or (2) dipping a glasswool fiber into water will cause coalescence at the tip with formation of spheres up to 1 mm in diameter, and (3) water droplets falling down a 2-ft high metal tube packed in solid CO_2 will form spheres that can be collected on a clean glass slide at the bottom. Sintering spheres is less rapid for single crystals under -10°C than for polycrystalline spheres. D_2O ice reacts at -10°C like H_2O ice at -14°C. For single crystal spheres, seed with AgI_2 smoke at slightly above -5°C.

ICE-0018
ETCH NAME: Salt
TYPE: Salt, ice forming
COMPOSITION:
x .... $NaMnO_4$
TIME:
TEMP:
DISCUSSION:

$NaMnO_4$ as a water solution used in a study of the freezing mechanisms of salt solutions. Reactions are similar to those of NaCl water solutions.

REF: Kober Ch & Scheiwem, M W — *J Cryst Growth,* 61,307(1983)

ICE-0019
ETCH NAME:
TYPE: Ice, forming
COMPOSITION:
x .... ice
TIME:
TEMP: -196°C
DISCUSSION:

Ice, used as a substrate for deposition of beryllium in vacuum at LN_2 temperature to produce a "freeze-out" structure of the metal. Thin films can be removed by melting away ice for microscope study as a form of the float-off technique.

REF: Baird, J D et al — *Nature (London),* 182,1660(1958)

PHYSICAL PROPERTIES OF INDIUM, In

Classification	Metal
Atomic number	49
Atomic weight	114.82
Melting point (°C)	156.17 (156.61)

Boiling point (°C)	2000
Density (g/cm³)	7.31
Thermal conductance (cal/sec)(cm²)(°C/cm)	0.204
Specific heat (cal/g) 20°C	0.057
Heat of fusion (k-cal/g-atom)	6.8
Latent heat of fusion (cal/g)	6.8
Heat of vaporization (cal/g)	483
Atomic volume (W/D)	15.7
1st ionization energy (K-cal/g-mole)	133
1st ionization potential (eV)	5.785
Electronegativity (Pauling's)	1.7
Covalent radius (angstroms)	1.44
Ionic radius (angstroms)	0.8 (In^{+3})
Coefficient of linear thermal expansion	
Electrical resistivity (micro ohms-cm) 20°C	8.37
Modulus of elasticity (psi $\times 10^6$)	1.57
Tensile strength (psi)	380
Cross section (barns)	194
Hardness (Mohs — scratch)	1
(Brinell — kgf/mm²)	0.9
Crystal structure (tetragonal — normal)	(110) prism, 1st order
Color (solid)	Silver white
Cleavage (hackly)	None

INDIUM, In

General: Does not occur as a native element and is considered a rare element. It is a minor constituent in sphalerite, ZnS (zincblende) and in the iron pyrites. It is one of the softest of known metals (H = 1) with a low melting point.

The pure metal does not have wide industrial use other than as a surface plating for corrosion resistance against organic acids (food containers). The major use is as a constituent in low melting point alloys of the lead-tin-zinc systems and, as such, these alloys are used in water sprinkler heads for fire protection. In dentistry, when mixed with gold and/or silver, used as a filling amalgam.

Technical Application: The pure metal is not widely used in Solid State device fabrication, but is a constituent of several semiconductor compounds, such as indium antimonide, InSb; indium arsenide, InAs: and indium phosphine, InP. The latter is one of the more widely used compound semiconductors.

Pure indium is used as a "mounting medium", such as for holding moly discs in an MBE system and, by surface alloying, for metal decoration of defects in single crystal wafers. Indium solders are used in device and package assemblies, but are not recommended for space hardware.

The pure metal has been grown for general morphology and defect studies.

Etching: Soluble in acids and mixed acid solutions. Slowly soluble in alkalies.

INDIUM ETCHANTS

IN-0005
ETCH NAME: Nitric acid, dilute
TYPE: Acid, cleaning
COMPOSITION:
1 .... HNO_3
1 .... H_2O

TIME: 1—2 min
TEMP: RT

DISCUSSION:

In, as pellets used for the epitaxy growth of InP on InP:Fe (100) (SI) wafer substrates. Solution used to clean indium prior to use as an epitaxy metal growth source.

REF: Rhee, J K & Bhattacharya, P E — *J Electrochem Soc,* 130,700(1983)

IN-0001

ETCH NAME: Nitric acid — TIME: 1—4 min

TYPE: Acid, structure — TEMP: RT

COMPOSITION:

x .... HNO_3

x .... H_2O

DISCUSSION:

In, single crystal ingot grown by horizontal zone refining in a Pyrex boat enclosed in a sealed, evacuated glass tube. Five refining passes were made. After each pass ingots were removed and etched in dilute nitric acid to develop grain boundaries.

REF: Flower, S C et al — *J Phys Chem Solids,* 46,96(1985)

IN-0002

ETCH NAME: — TIME:

TYPE: Electrolytic, structure — TEMP: $-30°C$ (dry ice)

COMPOSITION: — ANODE: In

1 .... HNO_3 — CATHODE: In

2 .... MeOH — POWER: 4 V

DISCUSSION:

In, single crystal wires used in studying paramagnetic effects in superconductors and the resistance transition in indium wires. Etch in solution shown, rinse in DI water, then alcohol.

REF: Meissner, H & Zdanis, R — *Phys Rev,* 109,681(1958)

IN-0003

ETCH NAME: Hydrochloric acid — TIME:

TYPE: Acid, removal — TEMP: Boiling

COMPOSITION:

x .... HCl

DISCUSSION:

In, preform sheet alloyed on germanium (111) wafer. Metal drive-in was for decoration of defects. Alloy on germanium surface and remove excess indium by boiling in HCl. Heat at 700°C in contact with an iron, nickel or similar metal plate. Indium will develop growth striations and (111) fracture.

REF: English, A C — *J Appl Phys,* 31,1498(1960)

IN-0006: Penna, T C et al — *J Cryst Growth,* 67,27(1984)

In, material used for growth of InGaAs. The raw indium was degreased with solvents, then etch cleaned in HCl. Clean InAs in 1HCl:5glycerin, RT, 10 min. Clean GaAs in boiling HCl, 10 min.

IN-0004

ETCH NAME: Isopropyl alcohol — TIME:

TYPE: Alcohol, removal — TEMP: RT

COMPOSITION:

(1) x .... CH_3-CHOHCH$_3$ (2) x .... EOH (3) x .... MeOH

DISCUSSION:

In, specimens used as a solder holder on moly discs in MBE systems. Alcohols will dissolve indium and are used to clean surfaces before use in vacuum applications. After etching, DI water rinse and N_2 blow dry.

REF: Woo, R — personal communication, 1982

INDIUM ALLOYS, InMx

General: Metallic indium is a minor constituent in several minerals, particularly in zinc and iron ores from which it is extracted during ore reduction, but does not occur as a natural metallic alloy.

Pure indium and bismuth, alone, are low temperature alloys and, as mixtures with other metals, form a wide range of low to medium temperature alloys from about 100 to 400°C. Mixtures are compounded for specific temperatures, used for general soldering applications, though not as widely as the lead-tin solders. As indium has a low melting point, and subsequent low vapor pressure, it can show solid-solid diffusion with vaporization at elevated temperatures, similar to the problem encountered with gallium arsenide processing, where both elements can show such reactions.

Technical Application: Some indium solders are used in Solid State device assembly on substrate circuits of alumina, beryllia, etc., or in package sealing and construction but, as cited above, not as widely used as lead-tin solders. Pure indium sheet or alloy sheet are used as sealing gaskets on cryopump vacuum assemblies, as the material makes a firm, soft-seal and can withstand cryogenic temperatures with minimum crystallization or other alteration.

Certain indium alloys, such as InSn and InBi as eutectic alloys, have been evaluated for their superconducting properties.

Etching: Mixed acids and salts, variable with compound.

INDIUM ALLOY ETCHANTS

INDIUM:BISMUTH ETCHANTS

INBI-0001

ETCH NAME: TIME:

TYPE: Acid, preferential TEMP:

COMPOSITION:

x .... "chromate reagent"

DISCUSSION:

In:Bi, eutectic alloy specimen used in a structure study on the superconducting properties of eutectic alloys. Solution used to develop structure (mixture not shown). Work also done on InSb; PbSn; PbAg and PbTi. (*Note:* Sulfuric acid and sodium dichromate is the standard glass cleaner solution.)

REF: Levy, S A & Kim, Y B — *J Appl Phys,* 37,365(1966)

INBI-0002: Rhines, F N & Grobe, A H — *Trans AIME,* 156,253(1944)

For the "chromate reagent" mixtures.

INDIUM:TIN ETCHANTS

INSN-0001
ETCH NAME:
TYPE: Electrolytic, preferential
COMPOSITION:
144 ml .... C_2H_5OH (EOH)
16 ml .... *n*-butyl glycol
32 ml .... H_2O
45 g .... $ZnCl_2$
10 g .... $AlCl_3$
TIME:
TEMP:
ANODE: InSn
CATHODE: SS
POWER: 60 V
DISCUSSION:
InSn, alloy specimens (eutectic) used in structure study on the superconducting properties of eutectic alloys. Solution used to develop structure. Work also done on InBi, PbSn, PbAg, and PbTi.
REF: Levy, S A & Kim, Y B — *J Appl Phys,* 37,365(1966)

INDIUM PHOSPHIDE OXIDE ETCHANTS

INPO-0001
ETCH NAME:
TYPE: Acid, removal
COMPOSITION:
x .... NH_4OH
x .... H_2O
TIME:
TEMP: RT
DISCUSSION:
InP, (100) wafers used in an anodic oxidation study. Strip oxide with solution shown. See INP-0092a-b for wafer preparation and oxidation procedures.
REF: Gordon, T et al — *J Phys Lett,* 39,965(1981)

RARE EARTH INDIUM ETCHANTS

RIN-0001
ETCH NAME:
TYPE: Halogen, removal
COMPOSITION:
x .... x% Br_2
x .... MeOH
TIME:
TEMP:
DISCUSSION:
R_5In_2 grown as single crystals by melting in an induction furnace in an anti-ferromagnetic susceptibility study. General formula R_5In_2 = rare earths where R = Gd, Ho, Tb, Dy. Ho_5In_2 is tetragonal at room temperature with transition to hexagonal structure at high temperature.
REF: Semitelou, I P & Yakinthes, R K — *J Phys Chem Solids,* 44,31(1983)

INDIUM SELENIDE ETCHANTS

INSE-0001
ETCH NAME: TIME: 3 min
TYPE: Acid, defect TEMP: RT
COMPOSITION:
25 ml .... H_2SO_4
30 g .... $K_2Cr_2O_7$
180 ml .... H_2O
DISCUSSION:

InSe, (0001) as hand cleaved wafers. Material is a III—IV lamellar semiconductor compound with an R3m space group rhombohedral lattice. Ingot grown by Bridgman method in a study of growth mechanisms. As grown material develops cleavage surfaces and indium segregation clusters and zones that precipitate parallel to InSe layers which are perpendicular to the c-axis. Large crystal grain boundaries observed and, like GaSe, there can be defect-free areas. Doping with As (InAs) and In segregation can cause "contamination" defects. Solution shown used to develop dislocations, grain boundaries, and other defects.
REF: Chevy, A — *J Cryst Growth,* 67,119(1984)

INDIUM TELLURIDE ETCHANTS

INTE-0001
ETCH NAME: TIME:
TYPE: Acid, polish TEMP:
COMPOSITION:
1 .... Br_2
19 .... HAc
x .... *citric acid

*Saturate the solution.

DISCUSSION:

In_2Te_3 specimens. Solution used as a polishing etch. After etching, use solution — without bromine — to quench reaction and remove residual bromine.
REF: Irving, B A, *The Electrochemistry of Semiconductors,* Holmes, P J, Ed, Academic Press, New York, 1962
INTE-0002: Mooser, E & Pearson, W B — *Phys Rev,* 101,492(1956)

In_2Te_3 single crystal ingot grown in a study of new semiconducting compounds. Other compounds were Ag_2Se, Li_3Bi, TlSe, Tl_2Se_3, SnSe, $SnSe_2$, In_2Tl, $AgInTe_2$, In_2Te_3, and Bi_2Se_3.

INDIUM THALLINIDE ETCHANTS

INTL-0001
ETCH NAME: BRM TIME:
TYPE: Halogen, polish TEMP:
COMPOSITION:
x .... x% Br_2
x .... MeOH

DISCUSSION:

In_2Tl_3 single crystal ingots grown in a study of new semiconductors compounds. See INTE-0001 for other compounds studied.

REF: Mooser, E & Pearson, W B — *Phys Rev,* 101,492(1956)

INDIUM TIN OXIDE ETCHANTS

ITO-0001

ETCH NAME: Nitric acid, dilute — TIME:

TYPE: Acid, flux removal — TEMP: Boiling

COMPOSITION:

x .... HNO_3

x .... H_2O

DISCUSSION:

$InSnO_2$ (ITO) single crystal grown by flux method in $PbO:B_2O_3$ flux with In_2O_3 + SnO_2 at 1200°C, 24 h. Cool at 2—3°C/h to 800—900°C then turn off power. Grow in a platinum crucible. Etch remove ITO crystals with solution shown from flux. Crystals are cubic and black in color; thin crystals are transparent, yellow. With greater than 9% tin in solution, crystals will not grow.

REF: Kani, Y — *Jpn J Appl Phys Lett*, 23,L12(1984)

ITO-0002

ETCH NAME: Hydrochloric acid — TIME:

TYPE: Acid, selective — TEMP:

COMPOSITION:

x .... HCl, conc.

DISCUSSION:

In_2O_3, as a thin film doped with tin, Sn, used for its electro-optical properties. Grown in a $PbO:B_2O_3$ flux. The following elements can be used as donor replacement for indium: Ti, Zr, Hf, Nb, Ta, W, and Ge with similar device applications as ITO material. (*Note:* See Indium Oxide, In_2O_3, for additional etchants.)

REF: Kani, Y — *Jpn J Appl Phys,* 23,127(1983)

PHYSICAL PROPERTIES OF INDIUM ANTIMONIDE, InSb

Classification	Antimonide
Atomic numbers	49 & 51
Atomic weight	216.6
Melting point (°C)	523
Boiling point (°C)	
Density (g/cm³)	5.80
Refractive index (n=)	3.96
Lattice constant (angstroms)	6.48
Energy band gap (eV)	0.17
Hardness (Mohs — scratch)	6—7
Crystal structure (isometric — tetragonal)	(111) tetrahedron
Color (solid)	Grey-silver
Cleavage (dodecahedral)	(110)

INDIUM ANTIMONIDE, InSb

General: Does not occur as a natural compound although there are other metallic antimonides of nickel, iron, copper, arsenic, and silver which may contain sulfur. There is no industrial use other than as a compound semiconductor.

Technical Application: Indium antimonide is a III—V compound semiconductor with the sphalerite, ZnS, structure. It is a (111) surface polar compound with (111)In and $(\overline{111})$Sb showing different etching characteristics. There has been study of the (111) orientation, but most devices are fabricated with (100) wafer orientation. Single crystal ingots have been grown by the Horizontal Bridgman (HB); Czochralski (CZ); and Float Zone (FZ) methods. Epitaxy thin films are grown by Vapor Transport (VT), and Chemical Vapor Deposition (CVD).

Both single crystal wafers and thin films have been fabricated as devices: microwave mixers; strain transducers; laser diodes; Hall effect devices; and for magnetoresistive applications. Trinary compounds with arsenic or indium have been fabricated as laser and photoluminescent diodes, or photocathodes.

Etching: H_2SO_4, HNO_3. Mixed acids as HF:HNO_3/H_2O_2, HCl:HNO_3. Alkalies and halogens. Dry chemical etching (DCE), ionic gases.

INDIUM ANTIMONIDE ETCHANTS

INSB-0001a
ETCH NAME: TIME:
TYPE: Acid, polish TEMP: RT
COMPOSITION:
10 .... HF
25 .... HNO_3
20 .... CH_3COOH (HAc)
DISCUSSION:

InSb, (111) wafers used in an etch development study of polish and preferential etchants to develop controlled damage-pits using the Pennington Pit Method (PPM) — a diamond stylus with controlled pressure and rotation on a wafer surface that is then etched in a preferential etchant to develop a flat bottom pit. The solution shown is a good polishing etch.

REF: Pennington, P — personal communication, 1957.

INSB-0001b
ETCH NAME: TIME:
TYPE: Acid, preferential/polish TEMP: RT
COMPOSITION:
HNO_3 .... 1 .... 1 1
HCl .... 3 .... 5 3
H_2O .... 4 .... 6 8
DISCUSSION:

InSb, (111) wafers. These are possible controlled pit damage preferential etchants. See INSB-0001a for further discussion.

REF: Ibid.

INSB-0041: Eisen, F H & Birchenall, C E — *Acta Metall,* 5,265(1957)

InSb, (111) wafers used in a study of self-diffusion in both InSb and GaSb. Aqua regia used to polish specimens. Also used Vilella's Reagent.

INSB-0001c
ETCH NAME:
TYPE: Acid, polish
COMPOSITION:
25 .... HNO_3
10 .... HCl
20 .... HAc
DISCUSSION:
InSb, (111) wafers used in an etch development study. See INSB-0001a for further discussion.
REF: Ibid.

TIME:
TEMP: RT

INSB-0001d
ETCH NAME: Allen's etch
TYPE: Acid, polish
COMPOSITION:
x *CP4
x .... HCl
x .... H_2O

TIME:
TEMP: RT

*CP4: $30HF:50HNO_3:30HAc:0.6Br_2$

DISCUSSION:
InSb, (111) wafers used in an etch development study. See INSB-0001a for further discussion.
REF: Ibid.

INSB-0001e
ETCH NAME:
TYPE: Acid, polish
COMPOSITION:
1 .... HCl
2 .... $NaClO_3$
3 .... H_2O
DISCUSSION:
InSb, (111) wafers used in an etch development study. See INSB-0001a for further discussion.
REF: Ibid.

TIME:
TEMP: RT

INSB-0001f
ETCH NAME: CP4
TYPE: Acid, polish
COMPOSITION:
30 .... HF
50 .... HNO_3
30 .... HAc
0.5 .... Br_2
DISCUSSION:
InSb, (111) wafers used in an etch development study. Solution is a fast chemical polish.
REF: Ibid.
INSB-0019b: Ibid.

TIME:
TEMP: RT

InSb, oriented cylinders were cut in the c-axis direction. CP4 used to polish specimens prior to anodic and air oxidation.

INSB-0023: Bardsley, W & Bell, N A — *J Electron,* 3,103(1957)

InSb, (111) wafers used in an etch pit study. CP4 used to develop etch pits.

INSB-0015: Galaranova, V V & Erokhina, N A — *Fiz Tverd Tela,* 1,1198(1959)

InSb, (111) wafers fabricated as barrier-layer photovoltaic cells with indium or cadmium alloyed junctions. Devices etch cleaned and electrically tuned with CP4.

INSB-0029: Laff, R A & Fan, H Y — *Phys Rev,* 121,53(1961)

InSb specimens. CP4 used to etch thin specimens in a study of carrier lifetime.

INSB-0036: Wertheim, G K — *Phys Rev,* 104,662(1956)

InSb, (111) wafers used in a study of carrier lifetime. CP4 used as a polish etch.

INSB-0039: Haneman, D — *Br J Appl Phys,* 16,411(1965)

InSb, (111) wafers used in a study of strain energy associated with both positive and negative (111) surfaces. Mount specimens in balsam, and mechanically polish with 0.25 μm alumina, then 0.1 μm diamond paste. Remove specimens from balsam by dissolving balsam in xylene. Etch polish surfaces by floating individual wafers on water and adding CP4. The etched side is always concave, and the negative wafer side etches more rapidly.

INSB-0034a: Kuczynski, G C & Hochman, R F — *Phys Rev,* 108,946(1957)

InSb wafers used in a study of light induced plasticity. Specimens were surface damaged using the diamond point of a Knoop Hardness Tester, and then etched in CP4 to develop damage induced defect structure.

INSB-0002a

ETCH NAME: — TIME:

TYPE: Acid, removal — TEMP: Hot

COMPOSITION:

1 .... 20% NaOH

1 .... 30% H_2O_2

DISCUSSION:

InSb, (111) wafers used in an etch development study. Solution will attack when used hot.

REF: Robbins, H & Schwartz, B — Tech Rep, 1960

INSB-0002b

ETCH NAME: Sulfuric acid — TIME:

TYPE: Acid, removal — TEMP: Hot

COMPOSITION:

(1) x .... H_2SO_4, conc.

(2) 2 .... H_2SO_4
1 .... H_2O

DISCUSSION:

InSb, (111) wafers used in an etch development study. Both solutions will attack when used hot.

REF: Ibid.

INSB-0027a

ETCH NAME: Hydrofluoric acid — TIME:

TYPE: Acid, oxide removal — TEMP:

COMPOSITION:

x .... HF

DISCUSSION:

InSb specimen used in an anodization study. Anodize in a solution of 0.1 *N* KOH with current at about 200 μm/cm². Remove oxide with solution shown which does not etch indium antimonide.

REF: Venables, J D & Brody, R M — *J Electrochem Soc*, 107,296(1960)

INSB-0028a

ETCH NAME: TIME:

TYPE: Acid, preferential TEMP:

COMPOSITION:

1 .... HF

1 .... H_2O_2

DISCUSSION:

InSb, (111) used in a study of behavior during heat treatment. Solution is preferential on (111) surfaces. Polish etch with 1HF:2HNO_3: 1HAc. See INSB-0014b.

REF: Haneman, D — *J Appl Phys*, 31,217(1960)

INSB-0027b

ETCH NAME: Potassium hydroxide TIME:

TYPE: Electrolytic, anodization TEMP:

COMPOSITION: ANODE: InSb

x .... 0.1 *N* KOH CATHODE:

POWER: 200 μA/cm²

DISCUSSION:

InSb, (111) wafers used in a study of anodization of surfaces. Solution shown used to anodize.

REF: Ibid.

INSB-0038: Venables, J D & Brody, R M — *J Appl Phys*, 30,1110(1959)

InSb, (111) wafers used in oxidation experiment of surfaces. Flooding solution with light during anodization will increase oxidation rate. Surface replicas of carbon can be made for microscope study.

INSB-0037

ETCH NAME: BRM TIME:

TYPE: Halogen, polish TEMP: RT

COMPOSITION:

x .... 0.05% Br_2

x .... MeOH

DISCUSSION:

InSb, (100) wafers to be used as MBE substrates with wafers first studied for surface chemistry. Initially clean in (1) 1HCl:10EOH; (2) EOH rinse/dry, and (3) Cl_2 RF plasma gas clean. For surface studies: (1) apply 1HCl:1HNO_3:25H_2O by pipette with wafer on spinner at 3600 rpm; (2) bubble Cl_2 gas through CCl_4 onto spinning samples working in a dry box; (3) apply either 1HCl:4HNO_3:25H_2O or 1HF:4HNO_3:24 lactic acid. Acids leave C^+ and F^- contamination on surfaces, which can be removed with the BRM solution shown. BRM shown also used to chem/mech polish surfaces with MeOH rinse.

REF: Vasquez, R P et al — *J Appl Phys*, 54,1365(1983)

INSB-0002c
ETCH NAME: Nitric acid TIME:
TYPE: Acid, removal TEMP: RT & hot
COMPOSITION:
(1) x .... HNO_3 (2) 2 .... HNO_3 (3) 1 .... HNO_3
1 .... H_2O 2 .... H_2O
DISCUSSION:
InSb, (111) wafers used in an etch development study. Concentrated nitric acid was evaluated with both white and red fuming concentrations and both solutions showed moderate to vigorous reaction at RT. The 2:1 solution is slow at RT but moderate to vigorous used hot. The 1:2 solution shows slow reaction used hot. (*Note:* White fuming nitric is 70%, standard concentration. Yellow fuming is 72%; and red fuming is 74%.)
REF: Ibid.

INSB-0002d
ETCH NAME: TIME:
TYPE: Acid, removal TEMP: RT
COMPOSITION:
(1) 1 .... HCl (2) 2 .... HCl
1 .... HNO_3 1 .... HNO_3
DISCUSSION:
InSb, (111) wafers used in an etch development study. The 1:1 solution reacts moderately to vigorously at RT. The 2:1 solution is slow at RT, but used hot it is moderate to vigorous.
REF: Ibid.

INSB-0002e
ETCH NAME: TIME:
TYPE: Acid, removal TEMP: RT
COMPOSITION:
1 .... HF
1 .... HNO_3
DISCUSSION:
InSb, (111) wafers used in an etch development study. Reaction is moderate to vigorous.
REF: Ibid.
INSB-0018: Venables, J D & Brody, R M — *J Appl Phys,* 29,1025(1958)
InSb, (111) wafers used in an etch study. Solution used at RT, 2—5 sec as a polish etch.

INSB-0003a
ETCH NAME: TIME:
TYPE: Acid, polish TEMP:
COMPOSITION:
1 .... HNO_3
10 .... lactic acid
DISCUSSION:
InSb, (111) wafers. Solution used as a polish etch. Also used to polish a single crystal sphere.
REF: Maringer, R E — *J Appl Phys,* 29,1261(1958)

INSB-0003b
ETCH NAME: TIME:
TYPE: Acid, preferential TEMP:
COMPOSITION:
5 .... HF
10 .... HNO_3
4 .... lactic acid
15 .... H_2O
DISCUSSION:
InSb, (111) wafers. Solution is a preferential etch. A single crystal sphere etched to finite crystal form (FCF) produced (111) pole figures, only.
REF: Ibid.

INSB-0004
ETCH NAME: TIME:
TYPE: Acid, polish TEMP: RT
COMPOSITION:
2 .... HCl
2 .... HNO_3
3 .... H_2O
DISCUSSION:
InSb, (111) wafers used in an etch development study. Solution is a polish etch.
REF: Marsh, O — personal communication, 1958

INSB-0005a
ETCH NAME: TIME:
TYPE: Acid, preferential TEMP: RT
COMPOSITION:
1 .... HF
1 .... HNO_3
6 .... H_2O
DISCUSSION:
InSb, (111) wafers used in an etch development study. The wafers were first polished in an $HF:HNO_3:KOH$ solution. The solution shown produces triangular etch pits on (111)A — indium surfaces. The $(\overline{111})$B — antimony surface — is attacked erratically.
REF: Minamoto, M T — *J Appl Phys,* 33,1826(1962)

INSB-0005b
ETCH NAME: TIME:
TYPE: Acid, polish TEMP: RT & 4°C
COMPOSITION:
x .... HF
x .... HNO_3
x .... x% KOH
DISCUSSION:
InSb, (111) wafers used in an etch development study. Wafers were polished with the solution shown prior to preferential etching to observe the polar structure of this compound. Erratic at 4°C.
REF: Ibid.

INSB-0006a
ETCH NAME:
TYPE: Acid, preferential
COMPOSITION:
1 .... HF
1 .... H_2O_2
8 .... H_2O
1 .... 1% $Cr_2(SO_4)_3$

TIME:
TEMP: RT
RATE: (111) 9 $mg/cm^2/min$
($\overline{111}$) 17 $mg/cm^2/min$

DISCUSSION:

InSb, (111) wafers used in a defect study. Only low angle boundary dislocation etch pits on (111)A observed. No edge dislocations on ($\overline{111}$)B. Solution adjusted for 1% Cr^{+++} ion.

REF: Lavine, M C et al — *J Electrochem Soc,* 108,974(1961)

INSB-0006b
ETCH NAME:
TYPE: Acid, preferential
COMPOSITION:
1 .... HF
1 .... H_2O_2
8 .... H_2O
1 .... 1% $CoSO_4$

TIME:
TEMP: RT
RATE: (111) 8 $g/cm^2/min$
($\overline{111}$) 18 $mg/cm^2/min$

DISCUSSION:

InSb, (111) wafers used in a defect study. Solution will develop edge dislocations on the ($\overline{111}$)B surface. Solution adjusted for 1% Co^{++} ion.

REF: Ibid.

INSB-0007
ETCH NAME:
TYPE: Acid, polish
COMPOSITION:
2 .... HF
59 .... 20% tartaric acid
39 .... H_2O_2

TIME:
TEMP:

DISCUSSION:

InSb, (100), n-type wafers used in a study of native oxide. Solution shown was used to chem/mech polish surfaces. The native oxide was 2—4 nm thick.

REF: Belyi, V I et al — *Thin Solid Films,* 113,157(1984)

INSB-0008
ETCH NAME:
TYPE:
COMPOSITION:

TIME:
TEMP:

DISCUSSION:

InSb, (100), n-type wafers used in a study of adsorption coefficients. Wafers were wire saw cut from ingots, mechanically and chemical etched (neither solution given). Wafers were etched to 0.04 cm thick.

REF: Brown, E — *J Appl Phys,* 57,2361(1985)

INSB-0006c
ETCH NAME:
TYPE: Acid, polish
COMPOSITION:
1 .... HF
1 .... H_2O_2
8 .... H_2O
TIME:
TEMP: RT
RATE: (111) 11 mg/cm^2/min
$(\overline{111})$ 17 mg/cm^2/min
DISCUSSION:
InSb, (111) wafers used in a defect study. This solution was used as the "base" polish etch in studying the effect of additives to the solution and their action on surfaces.
REF: Ibid.

INSB-0006d
ETCH NAME:
TYPE: Acid, preferential
COMPOSITION:
1 .... HF
1 .... H_2O_2
8 .... H_2O
1 .... 1% $NiSO_4$
TIME:
TEMP: RT
RATE: (111) 8 mg/cm^2/min
$(\overline{111})$ 18 mg/cm^2/min
DISCUSSION:
InSb, (111) wafers used in a defect study. Nickel additive was for 1% Ni^{++} ions.
REF: Ibid.

INSB-0006e
ETCH NAME:
TYPE: Acid, preferential
COMPOSITION:
1 .... HF
1 .... H_2O_2
8 .... H_2O
1 .... 0.4% *n*-butylthiobutane
TIME:
TEMP: 4°C
RATE: (111) 3 mg/cm^2/min
$(\overline{111})$ 8 mg/cm^2/min
DISCUSSION:
InSb, (111) wafers used in a defect study.
REF: Ibid.

INSB-0006f
ETCH NAME: CP4, modified
TYPE: Acid, polish
COMPOSITION:
1 .... HF
2 .... HNO_3
1 .... HAc
TIME:
TEMP: 4°C
RATE: (111) 0.5/cm^2/sec
$(\overline{111})$ 7.8/cm^2/sec
DISCUSSION:
InSb, (111) wafers used in a defect study. This modified CP4 solution was also used as a "base" polishing and cleaning etch.
REF: Ibid.
INSB-0035a: Haneman, D — *J Appl Phys*, 31,217(1960)
InSb, (111) wafers. Etch pits developed on (111)Sb, but not on $(\overline{111})$In with the solution shown. Follow with 1HF:1H_2O_2 and DI water rinse. Also develops preferential etch pits.

INSB-0006g
ETCH NAME:
TYPE: Acid, preferential
COMPOSITION:
1 .... HF
2 .... HNO_3
1 .... HAc
1 .... 0.3% *n*-amylamine
DISCUSSION:
InSb, (111) wafers used in a defect study.
REF: Ibid.

TIME:
TEMP: 2°C
RATE: (111) $0.4/cm^2/sec$
$(\overline{111})$ $0.4/cm^2/sec$

INSB-0006h
ETCH NAME:
TYPE: Acid, preferential
COMPOSITION:
1 .... HF
2 .... HNO_3
1 .... HAc
1 .... 0.05% sodium stearate
DISCUSSION:
InSb, (111) wafers used in a defect study.
REF: Ibid.

TIME:
TEMP: 2°C
RATE: (111) $0.3/cm^2/sec$

INSB-0028c
ETCH NAME: Heat
TYPE: Thermal, preferential
COMPOSITION:
x .... heat
DISCUSSION:
InSb, (111) wafers when heat treated below 523°C showed hillock formation with structure alteration due to heat with figure migration across surfaces.
REF: Ibid.
INSB-0035b: Haneman, D — *J Appl Phys,* 31,217(1960)
InSb, single crystal wafers used in a study of the effects of thermal heat treatment.
INSB-0043: Venables, J D & Brody, R M — *J Appl Phys,* 30,122(1959)
InSb wafers. A study of the relation between generation and motion stress in dislocations of the material associated with heat.

TIME:
TEMP: 523°C max.

INSB-0009a
ETCH NAME:
TYPE: Acid, polish
COMPOSITION:
1 .... HF
5 .... H_2O_2
+/− .... H_2O
DISCUSSION:
InSb, (111), p-type wafers used in a study of recombination processes. Solution used to etch polish surfaces.
REF: Zitter, R N & Strauss, A J — *Phys Rev,* 115,266(1959)

TIME:
TEMP:

INSB-0010a
ETCH NAME: TIME: 3 min
TYPE: Acid, clean TEMP: RT
COMPOSITION:
5 .... HNO_3
1 .... lactic acid
DISCUSSION:
InSb, (111)A, $(\overline{111})$B, and (100) wafers used as substrates for MBE homoepitaxy growth in an excess Sb_4 atmosphere. Etch polish substrates with Br_2:MeOH; then degrease with solvents, and etch clean with solution shown. In MBE vacuum system, final thermal clean (111) surfaces at 450°C; (100) wafers 430°C due to Sb dissociation.
REF: Okachi, J et al — *J Electron Mater,* 14,419(1985)

INSB-0011
ETCH NAME: Isopropyl alcohol TIME:
TYPE: Alcohol, clean TEMP:
COMPOSITION:
x .... $CH_3CHOHCH_3$ (ISO)
DISCUSSION:
InSb, (001) wafers used as substrates for MBE growth of CdTe thin films. InSb-CdTe form an isomorphous series. Best growth between 50 to 150°C with some dislocations probably due to O_2 and C in the InSb surface. CdTe tends to twin and produce low angle grain boundaries. Before growth, clean InSb by rinsing in alcohol; then Ar^+ ion bombardment at 500 eV in vacuum and anneal at 200°C.
REF: Farrew, R F C et al — *Appl Phys Lett,* 39,954(1981)

INSB-0012a
ETCH NAME: BRM TIME:
TYPE: Halogen, polish TEMP:
COMPOSITION:
x .... Br_2
x .... MeOH
DISCUSSION:
InSb, (100) wafers used as substrates for growth of Cl_2 on surfaces. First, polish with a BRM solution; then dip in 0.5% Br_2:MeOH mixture, and rinse in MeOH. Remove native oxide with 1HCl:10 EOH and rinse in EOH. With wafer on a spinner at 3600 rpm, etch with 1HCl:1HNO_3:25H_2O using pipette droplets of etch onto the spinning wafer. Grow chlorinated surface by flooding spinning wafer with CCl_4 saturated with chlorine gas. This was a surface cleaning study.
REF: Vasquez, R P et al — *J Appl Phys,* 54,1337(1983)

INSB-0012b
ETCH NAME: TIME:
TYPE: Acid, oxide removal TEMP:
COMPOSITION:
1 .... HCl
10 .... EOH
DISCUSSION:
InSb, (100) wafers used as substrates to grow chlorine on surfaces in a cleaning study. After a BRM surface polish there is a passivating native oxide left, which can be removed with the solution shown. See INSB-0012a.
REF: Ibid.

INSB-0012c
ETCH NAME:
TYPE: Acid, polish
COMPOSITION:
1 .... HCl
1 .... HNO_3
25 .... H_2O

TIME:
TEMP:

DISCUSSION:
InSb, (100) wafers used as substrates to grow chlorine on surfaces in a cleaning study. Use a pipette with droplets of the etch solution shown being dropped onto wafer surfaces held on a spinner, and being rotated at 3600 rpm to polish. See INSB-0012a.
REF: Ibid.

INSB-0013
ETCH NAME:
TYPE: Acid, preferential
COMPOSITION:
1 .... HF
1 .... H_2O_2
8 .... H_2O
X .... 0.4% *n*-butylthiobutane

TIME:
TEMP:

DISCUSSION:
InSb, (111) wafers. This solution develops beta-type dislocations on both the (111)A and $(\overline{111})$B surfaces.
REF: Harper, C A — *Handbook of Materials and Processes for Electronics,* McGraw-Hill, New York, 1970, 7
INSB-0026: Gatos, H C et al — *J Appl Phys,* 32,1574(1961)
InSb, (111) wafers used in an etch development and defect study. With *n*-butylthiobutane in solution, pits are flat bottomed similar to those of hot KOH on silicon, and associated with Sb dislocations; without the *n*-butylthiobutane, pits are smooth and conical with a high center point (called a "tit pit"), and associated with indium dislocations. (*Note:* Similar structure has been observed on both germanium and silicon (111) wafers and, as seen on natural minerals, is referred to as mammillary or botryoidal structure.)

INSB-0028a
ETCH NAME:
TYPE: Halogen, polish
COMPOSITION:
x .... x% I_2
x .... MeOH

TIME:
TEMP: RT

DISCUSSION:
InSb, (100) wafers and other orientations. Solution used as a chem/mech polish etch similar to BRM solutions. See INSB-0012a.
REF: Fuller, C S & Allison, H W — *J Electrochem Soc,* 109,880(1962)

INSB-0029
ETCH NAME: CP4A
TYPE: Acid, polish
COMPOSITION:
3 .... HF
5 .... HNO_3
3 .... HAc

TIME: to 30 sec
TEMP: RT

DISCUSSION:

InSb, (111) wafers and other orientation. Solution used as a general polish etch.

REF: Dewald, J F — *J Electrochem Soc,* 104,244(1957)

INSB-0014a

ETCH NAME: Etch #1
TIME:

TYPE: Acid, preferential
TEMP: 10°C

COMPOSITION:

x .... 6 *N* HCl

x .... 0.1 *N* $FeCl_3$

DISCUSSION:

InSb, (100) and (110) wafers used in a study of dislocation and defects on these surfaces. Solution used to develop defects.

REF: Gatos, H C & Lavine, M C — *J Electrochem Soc,* 107,433(1960)

INSB-0025a: Gatos, H C & Lavine, M C — *J Appl Phys,* 31,743(1960)

InSb, (111) wafers. Solution used at 95°C, and will develop dislocation pits on both the (111)A and $(\overline{111})$B surfaces. Below 82°C solution will only develop pits on the (111)A surface.

INSB-0014b

ETCH NAME: Etch #2
TIME: xx 4 sec

TYPE: Acid, preferential
TEMP: 4°C RT

COMPOSITION:

1 .... HF

1 .... HAc

2 .... HNO_3

DISCUSSION:

InSb, (100) and (110) wafers used in a study of dislocation and defects on these surfaces.

REF: Ibid.

INSB-0014c

ETCH NAME:
TIME: 20 sec

TYPE: Acid, polish
TEMP: RT

COMPOSITION:

5 .... HF

5 .... HNO_3

2 .... H_2O

DISCUSSION:

InS, (100) and (110) wafers used in a study of dislocation and defects on these surfaces. Although this is a polish etch, it shows a dislocation background of pits very similar to those on germanium etched in CP4.

REF: Ibid.

INSB-0015a

ETCH NAME:
TIME:

TYPE: Acid, preferential
TEMP:

COMPOSITION:

4 .... HF

5 .... HNO_3

12 .... H_2O

DISCUSSION:
InSb, $(\overline{311})$ wafers used in a study of anodic films. Solution shown will develop etch pits on the $(\overline{311})$B surface but not on the (311)A.
REF: DeWalt, J F — *J Electrochem Soc,* 104,244(1957)

INSB-0025b
ETCH NAME: TIME:
TYPE: Acid, dislocation TEMP:
COMPOSITION:
1 .... HF
1 .... HAc
2 .... HNO_3
x .... *stearic acid

*x = saturate the solution with stearic acid.

DISCUSSION:
InSb, $(\overline{111})$ wafers used in a dislocation etch pit study on (111)A and $(\overline{111})$B surfaces. Solution will develop dislocation pits on both surfaces. With the solution between +2—0°C, and removal rates measured as $mg/cm^2/sec$:

(1) with stearic acid:	(111)A = 0.44	(111)B = 0.38
(2) without stearic acid:	(111)A = 0.50	(111)B = 7.8

REF: Ibid.

INSB-0025c
ETCH NAME: TIME:
TYPE: Acid, dislocation TEMP:
COMPOSITION:
1 .... HF
1 .... H_2O_2
2 .... H_2O
DISCUSSION:
InSb, (111) wafers used in a study of dislocation etch pits on (111)A and $(\overline{111})$B surfaces. Solution will develop etch pits on both surfaces with those on the (111)A particularly well developed.
REF: Ibid.

INSB-0015b
ETCH NAME: TIME: 5—30 sec
TYPE: Acid, polish TEMP: RT
COMPOSITION:
24 .... HF
40 .... HNO_3
24 .... HAc
DISCUSSION:
InSb, (311) wafers used in an anodic film study. Mechanical lap/polish, then remove residual subsurface damage with the solution shown. Wafers with (110) surfaces were cleaved with a razor blade guillotine system to obtain wafers of 1 cm area, which showed some surface fracture steps present but was within 2° of (110) orientation. Oxide films were formed in a solution of xKOH:2xtartaric acid.
REF: Ibid.

INSB-0015c
ETCH NAME:
TYPE: Electrolytic, polish
COMPOSITION:
10 .... $HClO_4$
40 .... HAc
2 .... H_2O
TIME:
TEMP: 5°C
ANODE:
CATHODE:
POWER: 50 mA/cm^2
DISCUSSION:

InSb, (311) and (110) wafers used in an anodic film study. The black film that forms during polish etching with the solution shown can be removed with running water. Antimony oxides can be dissolved electrolytically with a 0.1 *N* KOH electrolyte, which will not remove In_2O_3.

REF: Ibid.

INSB-0016: Bardsley, W & Bell, R I — *J Electron Control,* 3,103(1957)

InSb, (111) wafers were polished in solution shown before etching for dislocations.

INSB-0017: Aclrytner, J & Kolakowski, B — *Acta Phys Pol,* 17,93(1958)

InSb single crystal surfaces. Several etchants described.

INSB-0019a
ETCH NAME:
TYPE: Acid, preferential
COMPOSITION:
1 .... HF
1.7 .... HNO_3
1 .... HAc
3.7 .... H_2O
TIME:
TEMP:
DISCUSSION:

InSb, cylinder cut ultrasonically with axis in c-direction. After polishing with CP4, cylinder was anodized at 100 mA/cm^2 or air oxidized at 350°C. After oxidation specimens were etched in the solution shown in a study of crystal orientation and surface reactions.

REF: Lavine, M C et al — *J Appl Phys,* 29,1131(1958)

INSB-0024a
ETCH NAME: Superoxol
TYPE: Acid, preferential
COMPOSITION:
1 .... HF
1 .... H_2O_2
4 .... H_2O
TIME:
TEMP:
DISCUSSION:

InSb, (111) wafers used in a polarity study of III—V compound semiconductors. Solution used as a preferential etch on (111)A and $(\overline{111})$B surfaces. Other compounds studied were InAs, GaSb, and GaAs. Superoxol also was evaluated on InAs and GaAs.

REF: Faust, J W & Sagar, A — *J Appl Phys,* 31,331(1960)

INSB-0034b: Ibid.

InSb, (100), n-type wafers, zinc diffused. Cleave on ⟨110⟩ after zinc diffusion to cross-section, and use a superoxol solution to stain the exposed junction.

INSB-0024b
ETCH NAME: TIME:
TYPE: Acid, preferential TEMP:
COMPOSITION:

(1) 1 .... HF
1 .... HNO_3
4 .... H_2O

(2) 1 .... HF
1 .... HNO_3

DISCUSSION:

InSb, (111) wafers used in a polarity study. The (111)A surface develops triangular pits; the $(\overline{111})$B is erratic. This etch is similar to INSB-0005a and INSB-0002e.

REF: Ibid.

INSB-0026
ETCH NAME: BRM TIME: 20 sec + 10 sec
TYPE: Halogen, polish TEMP: RT
COMPOSITION:

x .... 0.5% Br_2
x .... MeOH

DISCUSSION:

InSb, (110), n-type and (100), p-type wafers used in a cleaning study. Soak lens paper with solution and chem/mech lap surfaces. After lapping flush with MeOH during last 10 sec, then rinse in the BRM solution.

REF: Aspnes, D E & Studna, A A — *J Vac Sci Technol,* 20,488(1982)

INSB-0020
ETCH NAME: CP4, dilute TIME:
TYPE: Acid, polish TEMP:
COMPOSITION:

x *CP4
x .... H_2O

*CP4: 30HF:50HNO_3:30HAc:0.6Br_2

DISCUSSION:

InSb:Zn, (111) wafers used in a study of Cd alloyed p-n junctions and their electrical properties. Solution used to etch-clean junctions.

REF: Lee, C A & Kaminsky, G — *J Appl Phys,* 30,2021(1959)

INSB-0021
ETCH NAME: CP4, dilute TIME:
TYPE: Acid, polish TEMP:
COMPOSITION:

x *CP4
x .... HAc

*CP4: 30HF:50HNO_3:30HAc:0.6Br_2

DISCUSSION:

InSb, thin films used in a study of electrical properties. Films were formed using two quartz plates with lower plate heated above the temperature of the top plate — place a molten

drop of InSb on the bottom plate and squash into thin film. Wash in CCl_4 and etch clean in solution shown; rinse in DI water and dry under IR lamp.
REF: Bates, G & Taylor, K N R — *J Appl Phys,* 31,991(1960)

INSB-0009b
ETCH NAME: TIME:
TYPE: Acid, preferential TEMP:
COMPOSITION:
5 .... *superoxol
1 .... HF
x .... H_2O

*Superoxol: $1HF:1H_2O_2:4H_2O$

DISCUSSION:
InSb, (100) wafers used in a study of recombination in p-type material. Solution used as a polishing etch although superoxol is primarily a preferential etch.
REF: Zitter, R N et al — *Phys Rev,* 115,266(1959)

INSB-0022a
ETCH NAME: TIME: 2 min
TYPE: Acid, cleaning TEMP: RT
COMPOSITION:
1 .... H_2O_2
10 .... lactic acid
DISCUSSION:
InSb:Te, (111), n-type wafers used in a general surface study. Both polar surfaces were studied using the following sequence of etching and cleaning. Three-cleaning classifications: I = hi C; II = hi O_2 and III = hi S:

Operation step	Classification I	II	III
1. Degrease & H_2O rinse	*	*	*
2. $1H_2O_2$:10 lactic acid, 2″, RT		*	*
3. H_2O rinse	*		*
4. CP4, 5 sec, RT + H_2O rinse		*	*
5. 10 *M* Na_2S + H_2O rinse			*
6. H_2O rinse	*	*	*
7. N_2 blow-dry	*	*	*

In an additional study it was shown that (1) sputter etch; (2) plasma etch and (3) RIE etch all produced a damage layer that created donor sites and leakage in n-type InSb.
REF: Auret, F D — *J Electrochem Soc,* 129,2752(1982)
INSB-0035: Auret, F D — *J Electrochem Soc,* 131,2115(1984)
An extension of the above studies.

INSB-0030
ETCH NAME: TIME: 1—2 min
TYPE: Acid, polish TEMP: RT
COMPOSITION:
1 .... *CP4
1 .... HAc
1 .... H_2O

*CP4: $30HF:50HNO_3:30HAc:0.6Br_2$

DISCUSSION:
InSb, (111), (112) and other bulk orientations on (111) or (100) surface oriented wafers. Solution used to develop and polish these orientations.
REF: Venables, J D & Brody, R M — *J Appl Phys,* 29,1025(1958)

INSB-0031
ETCH NAME: TIME: to 1 min
TYPE: Acid, preferential TEMP: RT
COMPOSITION:
2 .... HF
1 .... HNO_3
1 .... HAc
DISCUSSION:
InSb, (111) wafers. Solution will develop etch pits on (111)A indium surface.
REF: Gatos, H C & Lavine, M C — *J Electrochem Soc,* 107,427(1960)

INSB-0032
ETCH NAME: TIME:
TYPE: Acid, polish TEMP: RT
COMPOSITION:
1 .... HF
4 .... HNO_3
25 .... lactic acid
DISCUSSION:
InSb, (100) wafers used as substrates for epitaxy growth of CdTe thin films by MBE. Solution used to polish substrates prior to MBE.
REF: Farrow, R F C et al — *J Vac Sci Technol,* B(3),681(1985)

INSB-0033
ETCH NAME: Argon TIME:
TYPE: Ionized gas, cleaning TEMP:
COMPOSITION: GAS FLOW:
x .... Ar^+ ion PRESSURE:
POWER:
DISCUSSION:
InSb, (100) wafers used as substrates for epitaxy growth of CdTe thin films by MBE. Substrates were solvent degreased prior to Ar^+ ion cleaning in vacuum.
REF: Williams, G M et al — *J Vac Sci Technol,* B(3),704(1985)

INSB-0040
ETCH NAME: BRM
TIME:
TYPE: Halogen, cleaning
TEMP:
COMPOSITION:

x .... 5% Br_2
x .... MeOH (CH_3OH)

DISCUSSION:

InSb, (100), n-type wafers zinc diffused in fabrication of infrared detectors in the 3—5 μm range. Prior to diffusion, wafers were polish cleaned in solution shown.

REF:

INSB-0022b
ETCH TYPE: Degrease
TIME:
TYPE: Solvent, cleaning sequence
TEMP:
COMPOSITION:

(1)

(i) acetone, hot, rinse three times
(ii) TCE, hot, rinse three times
(iii) acetone, hot, rinse two times, approx. 1 min each
(iv) DI water, rinse three times, approx. 1 min each
(v) Nitrogen blow-dry

DISCUSSION:

InSb, (111) n-type, Te doped wafers used in an etching and cleaning study. This degreasing sequence was studied, alone, and used prior to all of the clean/etch sequences described in INSB-0022c thru INSB-0022m. The sequences are classified by carbon content remaining: A = less than 3%; B = 3—30% and C = greater than 30%. Remaining oxide on surfaces after etching was measured by AES using argon ion bombardment of 1 KeV and compared to etch removal of a known thickness of an anodic film. AES profile also shows the %O_2, %In, and %Sb remaining.

REF: Ibid.

INSB-0022c
ETCH NAME: CP4A
TIME: # seconds
TYPE: Acid, removal
TEMP: RT
COMPOSITION:

30 .... HF
50 .... HNO_3
30 .... CH_3COOH (GAA/HAc)

Clean/etch (i) Degrease, INSB-0022b
(ii) Etch in CP4A and DI water rinse
(iii) N_2 blow dry

DISCUSSION: InSb, (111) wafers used in an etching and cleaning study. Class: A.

REF: Ibid.

INSB-0022d
ETCH NAME:
TIME: 1, 2, and 20 min
TYPE: Acid, removal
TEMP: RT
COMPOSITION:

1 .... HNO_3
10 .... lactic acid

Clean/etch: (i) Degrease, INSB-0022b
(ii) Etch (shown), and DI water rinse, 3×.
(iii) N_2 blow-dry

DISCUSSION:

InSb, (111) wafers used in an etching and cleaning study. Class: B.

REF: Ibid.

INSB-0022e

ETCH NAME:

TYPE: Acid, removal

COMPOSITION:

TIME: (1) 5 sec (2) 5 min

TEMP: RT

(1) 30 .... HF
50 .... HNO_3
30 .... HAc

(2) 1 .... HNO_3
10 .... lactic acid

Clean/etch:
(i) Degrease, INSB-0022b
(ii) Etch in (2); DI water rinse 2×
(iii) Etch in (1); DI water rinse 3×
(iv) N_2 blow-dry

DISCUSSION:

InSb, (111) wafers used in an etching and cleaning study. Etch (1) is CP4A from INSB-0022c; etch (2) is from INSB-0022d. Class: A.

REF: Ibid.

INSB-0022f

ETCH NAME: H100

TYPE: Acid, removal

COMPOSITION:

TIME: 30 sec

TEMP: RT

"A" 78 g .... KOH
4 g .... tartaric acid
8 g .... *EDTA
78 g .... H_2O

Clean/etch:
(i) Degrease, INSB-0022b
(ii) Complete INSB-0022e clean/etch
(iii) H100 etch; DI water rinse 2×, 3 min. each
(iv) N_2 blow dry

"B" xml .... H_2O_2

*Ethylenediamine tetraacetic acid

Mix: 5 "A":2 "B" before using

DISCUSSION:

InSb, (111) wafers used in an etching and cleaning study. Class: A.

REF: Ibid.

INSB-0022g

ETCH NAME: acetone, vapor

TYPE: Ketone, cleaning

COMPOSITION:

TIME:

TEMP:

x .... $(CH_3)_2CO$, vapor

Clean/etch:
(i) Degrease, INSB-0022b
(ii) Complete INSB-0022e clean/etch
(iii) DI water rinse, 3×, 1 min each
(iv) Acetone dry

DISCUSSION:

InSb, (111) wafers used in an etching and cleaning study. Class: C.

REF: Ibid.

INSB-0022h
ETCH NAME: Sodium sulfide
TYPE: Acid, sulfonation
COMPOSITION:
x .... 10^5 *M* Na_2S

TIME: 1 min
TEMP: RT

Clean/etch: (i) Degrease, INSB-0022b
(ii) Complete INSB-0022e clean/etch
(iii) DI water rinse, 3×, 1 min each
(iv) N_2 blow-dry

Note: Approximately 18—20% S detected.

DISCUSSION:
InSb, (111) wafers used in an etching and cleaning study. Class: A.
REF: Ibid.

INSB-0022i
ETCH NAME:
TYPE: Acid, removal
COMPOSITION:
1 .... HF
1 .... H_2O_2
4 .... H_2O

TIME: 2 min
TEMP: RT

Clean/Etch: (i) Degrease, INSB-0022b
(ii) Etch; DI water rinse 3×, 1 min each
(iii) N_2 blow-dry

DISCUSSION:
InSb, (111) wafers used in an etching and cleaning study. Class: B.
REF: Ibid.

INSB-0022j
ETCH NAME: BRM
TYPE: Halogen, removal
COMPOSITION:
x .... 1% Br_2
x .... MeOH

TIME: 10 min
TEMP: RT

Clean/etch: (i) Degrease, INSB-0022b
(ii) Etch
(iii) MeOH rinse, 2 min
(iv) DI water rinse 3×, 1 min each
(v) N_2 blow dry

DISCUSSION:
InSb, (111) wafers used in an etching and cleaning study. Class: B.
REF: Ibid.

INSB-0022k
ETCH NAME:
TYPE: Acid removal
COMPOSITION:
1 .... HF
2 .... HNO_3
5 .... HAc (GAA)

TIME: 1 min
TEMP: RT

Clean/etch: (i) Degrease, INSB-0022b
(ii) Etch
(iii) DI water rinse 3×, 1 min each
(iv) N_2 blow dry

DISCUSSION:
InSb, (111) wafers used in an etching and cleaning study. Class: A.
REF: Ibid.

INSB-0022l
ETCH NAME: CP4, variety — TIME: Seconds
TYPE: Acid, removal — TEMP: RT
COMPOSITION:
15 .... HF
25 .... HNO_3
15 .... HAc (GAA)
x .... Br_2

Clean/etch: (i) Degrease, INSB-0022b
(ii) Etch
(iii) DI water rinse 3×, 1 min each
(iv) N_2 blow-dry

DISCUSSION:
InSb, (111) wafers used in an etching and cleaning study. Class: A.
REF: Ibid.

INSB-0022m
ETCH NAME: — TIME: 90 sec
TYPE: Electrolytic, anodizing — TEMP: 5°C
COMPOSITION: — ANODE: InSb
15 .... $HClO_4$ — CATHODE:
60 .... HAc — POWER: 15 V
1.5 .... H_2O

Clean/etch: (i) Degrease, INSB-0022b
(ii) Anodize
(iii) DI water rinse 3×

DISCUSSION:
InSb, (111) wafers used in an etching and cleaning study. Wafers were anodized in the solution shown. This oxide was etched away and the known thickness was used as the standard to evaluate oxide thicknesses developed by the clean/etch sequences shown in InSB-0022b through INSB-0022l. Surface was Class B.
REF: Ibid.

INSB-0041
ETCH NAME: — TIME: (1) variable (2) 5 min
TYPE: Acid, removal — TEMP: RT RT
COMPOSITION:
(1) x .... x% Br_2
x .... MeOH

(2) 5 .... HNO_3
1 .... HAc

DISCUSSION:
InSb, (100), (111)A and $(\overline{111})$B oriented wafers used as substrates for MBE of Homo-epi growth of InSb. The $(\overline{111})$B surface developed hillocks in the thin films. Prepare substrates: (1) chem/mech polish in BRM solution; (2) degrease with solvents; (3) etch in solution (2). Thermal heat treat wafers prior to epitaxy: (111) at 450°C, and (100) at 430°C to reduce dislocation content. MBE growth was in an excess Sb_4 atmosphere to study the effects of Sb in InSb.
REF: Okashi, J et al — *J Electron Mater,* 14,4519(1985)
INSB-0042: Brown, E R — *J Appl Phys,* 57,2361(1985)
InSb wire was saw cut, mechanically lapped, then etched down to 0.04 cm diameter in a study of adsorption coefficients.
INSB-0040: Potter, R F — *Phys Rev,* 102,47(1956)
InSb specimens used in a study of electric modulus.

INDIUM GALLIUM ANTIMONIDE, InGaSb

General: Does not occur as a natural compound. Antimony, Sb, occurs as a native metal in small quantity. The mineral stibnite, Sb_2S_3 is a major ore, and there are several antimonates and antimonites of Pb, Fe, Ca, Ti, etc. There are no industrial applications of the compound other than in Solid State semiconductor development.

Technical Application: Indium gallium antimonide has been epitaxially deposited as a thin film for general morphological study, and has application in semiconductor heterostructures.

Etching: Mixed acids.

INDIUM GALLIUM ANTIMONIDE ETCHANTS

IGSB-0001
ETCH NAME:
TYPE: Acid, thinning
COMPOSITION:
10 .... HCl
1 .... H_2O_2
10 .... H_2O
TIME:
TEMP:

DISCUSSION:

InGaSb, deposited as a thin film by epitaxy on the following substrates: BaF_2, (111); NaCl, (100); Corning 7059 glass. Both BaF_2 and NaCl were cleaved in air, and TCE vapor degreased with water rinse. Glass was detergent scrubbed, water washed, and MeOH rinsed. After epitaxy growth the films were etch thinned in the solution shown. Films were then removed by the float-off technique for TEM study. Solutions for float-off removal were: (1) water for NaCl; (2) dilute HCl for BaF_2; and HF for glass.

REF: Eltoukhy, A H & Greene, J E — *J Appl Phys,* 50,505(1979)

PHYSICAL PROPERTIES OF INDIUM ARSENIDE, InAs

Classification	Arsenide
Atomic numbers	49 & 33
Atomic weight	189.8
Melting point (°C)	943
Boiling point (°C)	
Density (g/cm^3)	5.66
Refractive index (n =)	3.42
Lattice constant (angstroms)	6.06
Energy band gap (eV)	0.42
Hardness (Mohs — scratch)	6—7
Crystal structure (isometric—tetrahedral)	(111) tetrahedron
Color (solid)	Grey-silver
Cleavage (dodecahedral)	(110)

INDIUM ARSENIDE, InAs

General: Does not occur as a natural compound, although there are other metallic arsenides of nickel, iron, copper, cobalt, and platinum, which may include sulfur. There is no industrial applications other than as a compound semiconductor.

Technical Application: Indium arsenide is a III—V compound semiconductor with the sphalerite, ZnS structure. It is a (111) surface polar material with (111)In and ($\overline{111}$)As surfaces showing different etching characteristics. Several orientations have been studied, but most devices are fabricated with (100) wafer orientation. Single crystal ingots have been grown by the Horizontal Bridgman (HB); Czochralski (CZ); and Float Zone (FZ) methods.

Epitaxy thin films have been grown by Vapor Transport (VT), and Chemical Vapor Deposition (CVD).

Some single crystal wafer or thin film devices fabricated include diodes and transistors, laser diodes, and Hall effect devices.

Etching: HCl. Mixed acids of $HF:HNO_3/H_2O_2$, $HCl:HNO_3$, and halogens.

INDIUM ARSENIDE ETCHANTS

INAS-0001a
ETCH NAME: TIME:
TYPE: Acid, polish TEMP: 60°C
COMPOSITION:
10 .... 80% lactic acid ($CH_3CHOHCOOH$)
1 .... HNO_3
DISCUSSION:

InAs, (100), n-type wafers used in a study of native oxides. Chem/mech polish. Native oxide is 1—4 nm thick. See GaAs and InSb.

REF: Belyi, V I et al — *Thin Solid Films,* 133,157(1984)

INAS-0001b
ETCH NAME: Hydrochloric acid TIME:
TYPE: Acid, removal TEMP:
COMPOSITION:
x .... 1 *M* HCl
DISCUSSION:

InAs, (100), n-type wafers used in a study of native oxides. Chem/mech polish. This solution shows a higher As content in the surface after polishing. The higher electropositive elements tend to pile-up on surfaces, but the reason is presently unknown.

REF: Ibid.

INAS-0002
ETCH NAME: BRM TIME: 20 sec + 10 sec
TYPE: Halogen, polish TEMP: RT
COMPOSITION:
x .... 0.05% Br_2
x .... MeOH
DISCUSSION:

InAs, (110), n-type wafers used in a surface cleaning study. Soak lens paper in solution and use as lapping surface. Flush with MeOH during last 10 sec of lap period. Rinse in $NH_4OH:H_2O$; then rinse in BRM; then rinse in H_2O, and final rinse, again, in the $NH_4OH:H_2O$ mixture. Authors say that bromine content is not critical. (*Note:* At the concentration shown it is probably not a critical concentration, but higher concentrations are known to shift from polish etches to preferential.) Also used solution shown on Si, Ge, GaP, GaAs, GaSb, and InSb.

REF: Aspenes, D E & Studna, A A — *J Vac Sci Technol,* 20,488(1982)

INAS-0003
ETCH NAME: CP4, modified
TYPE: Acid, preferential
COMPOSITION:
75 ml .... HNO_3
15 ml .... HF
15 ml .... HAc
0.06 ml .... Br_2
TIME:
TEMP:

DISCUSSION:

InAs, (111) wafers used in X-ray studies. Surfaces were lightly preferentially etched in the solution shown to differentiate between the (111)A and $(\overline{111})$B surfaces.

REF: Clarke, R L — *J Appl Phys,* 30,959(1959)

INAS-0005c: Ibid.

InAs, (111) wafers. Etch 5 sec. Develops pits on (111)A surfaces.

INAS-0004
ETCH NAME:
TYPE: Acid, clean
COMPOSITION:
1 .... HCl
5 .... glycerin
TIME: 10 min
TEMP: RT

DISCUSSION:

InAs, specimens used as the source material for growing InGaAs. Degrease with TCE, rinse with acetone, then MeOH. Then etch clean with solution shown.

REF: Penna, T C et al — *J Cryst Growth,* 67,27(1984)

INAS-0005a
ETCH NAME:
TYPE: Halogen, dislocation
COMPOSITION:
x .... 5—10% Br_2
x .... MeOH
TIME: 13—30 sec
TEMP:

DISCUSSION:

InAs, (111) wafers. Referred to as a dislocation etch.

REF: Harper, C A — *Handbook of Materials and Processes for Electronics,* McGraw Hill, New York, 1970, 7

INAS-0008
ETCH NAME: CP4
TYPE: Acid, preferential
COMPOSITION:
15 .... HF
25 .... HNO_3
15 .... HAc
0.6 .... Br_2
TIME:
TEMP: RT

DISCUSSION:

InAs, specimens used in a study of light induced plasticity. Wafer surfaces were damaged using the diamond point of a Knoop Hardness Tester and then etching in CP4 to develop damage structure. Also studied Ge and InSb.

REF: Kuczynski, O C & Hochman, R P — *Phys Rev,* 108,946(1957)

INAS-0005b
ETCH NAME: Hydrochloric acid TIME:
TYPE: Acid, preferential TEMP:
COMPOSITION:
x .... HCl
DISCUSSION:
InAs, (111) wafers. Solution described as producing etch figures.
REF: Ibid.

INAS-0005d
ETCH NAME: A/B Etch TIME: 10 min
TYPE: Acid, preferential TEMP:
COMPOSITION:
1 ml .... HF
2 ml .... H_2O
8 mg .... $AgNO_3$
1 g .... CrO_3
DISCUSSION:
InAs, (111), (110) and (100) wafers. Used as a preferential etch on both indium arsenide and gallium arsenide. See GAS-0059.
REF: Ibid.

INAS-0006a
ETCH NAME: Hydrochloric acid TIME:
TYPE: Acid, removal TEMP:
COMPOSITION:
x .... HCl
DISCUSSION:
InAs, (111) wafers used in a polarity study of III—V compound semiconductors. Solution is more of a general removal etch than preferential. See INAS-0001b and INAS-0004. Other compounds studied were InSb, GaSb, and GaAs.
REF: Faust, J W & Sayar, A — *J Appl Phys,* 31,331(1960)

INAS-0006b
ETCH NAME: Superoxol TIME:
TYPE: Acid, preferential TEMP:
COMPOSITION:
1 .... HF
1 .... H_2O_2
4 .... H_2O
DISCUSSION:
InAs, (111) wafers. Develops triangle pits on (111)A; but is erratic on $(\overline{111})$B.
REF: Ibid.

INAS-0006c
ETCH NAME: TIME:
TYPE: Acid, preferential TEMP:
COMPOSITION:
1 .... HNO_3
1 .... H_2O_2
6 .... tartaric acid

DISCUSSION:

InAs, (111) wafers used in a polarity study.

REF: Ibid.

INAS-0006d

ETCH NAME: CP4, modified | TIME:

TYPE: Acid, preferential | TEMP:

COMPOSITION:

75 .... HF

75 .... HNO_3

15 .... HAc

0.6 .... Br_2

DISCUSSION:

InAs, (111) wafers used in a polarity study.

REF: Ibid.

INAS-0007

ETCH NAME: | TIME: 30 min

TYPE: Acid, preferential | TEMP: RT

COMPOSITION:

x .... HCl

x .... 0.4 *N* $FeCl_3$

DISCUSSION:

InAs, (111) wafers used in an etch characterization study. Solution used to develop dislocations and defects.

REF: Gatos, H C & Lavine, M C — *J Electrochem Soc,* 107,427(1960)

INAS-0009

ETCH NAME: | TIME:

TYPE: Acid, removal | TEMP:

COMPOSITION:

1 .... HCl

1 .... HNO_3

DISCUSSION:

InAs, (111) wafers used in material studies. Solution used as a general removal and polish etch. Also used on GaAs. (*Note:* A reduced reactivity will be noted for aqua regia; 3HCl:1HNO_3.)

REF: Cardona, M — *Phys Rev,* 121,752(1961)

INAS-0010: Antell, G R & Effer, D — *J Electrochem Soc,* 106,509(1959)

InAs single crystals grown by vapor phase reaction in a study of this method of material growth. Also grew InP, GaAs, and GaP.

INAS-0011

ETCH NAME: Heat | TIME:

TYPE: Thermal, alteration | TEMP: 300°C

COMPOSITION:

x .... heat

DISCUSSION:

InAs, (111) wafers and other orientations. Wafers used in a study of heat treatment effects. Heating to 300°C shows a resistivity change.

REF: Edmond, J T & Hilsom, C — *J Appl Phys,* 31,1300(1960)

INDIUM ARSENIC PHOSPHIDE ETCHANTS

IASP-0001
ETCH NAME: BRM — TIME:
TYPE: Halogen, polish — TEMP: RT
COMPOISITION:
x .... x% Br_2
x .... MeOH
DISCUSSION:
$InAs_xP_{x-1}$ polycrystalline ingot grown by the Bridgman method in a study of the material and its oxidation characteristics. Wafers were chem/mech polished in a BRM solution with MeOH rinse prior to oxidation.
REF: Schwartz, G P et al — *J Vac Sci Technol,* A2,1252(1984)

INDIUM BISMUTHIDE ETCHANTS

INBI-0001
ETCH NAME: Soap — TIME:
TYPE: Solvent, polish — TEMP: RT
COMPOSITION:
x .... soap
x .... H_2O
DISCUSSION:
In_5Bi_3, single crystal specimens. Wafers were spark cut from ingot, then mechanically lapped and polished by grinding with abrasives and then polished with diamond paste. Final surface polish was done with a soft cloth saturated with a soap solution. Also etch polished wafers. See INBI-0002.
REF: Schreursl, N W & Weijers, H M — *J Cryst Growth,* 71,155(1985)
INBI-0002: Bhatt, V P et al — *J Pure Appl Phys,* 16,960(1978)
Reference for etch solution used in INBI-0001.

INDIUM GALLIUM ARSENIDE ETCHANTS

IGAS-0001
ETCH NAME: — TIME: 10—100 sec
TYPE: Acid, mesa forming — TEMP: RT
COMPOSITION:
1 .... H_2SO_4
1 .... H_2O_2
x .... H_2O
DISCUSSION:
InGaAs, (001) thin film epitaxy deposit grown by LPE on InP (001) wafer substrates to form very low leakage mesa diodes. Solution shown used to form mesas. See INP-0150 for preparation of InP substrate.
REF: Aspnes, D E et al — *J Vac Sci Technol,* 21,413(1982)

IGAS-0002a
ETCH NAME: Citric acid, dilute — TIME:
TYPE: Acid, clean — TEMP:
COMPOSITION:
x .... citric acid
x .... H_2O

DISCUSSION:

InGaAs, (100) wafer used as a substrate for MIS device structures with SiO_2 and aluminum. Clean surfaces with citric acid then follow with acid etch before silica deposition. Anneal after SiO_2 deposit at 250°C, 5 min in N_2. Form dot pattern with photo resist, then evaporate aluminum.

REF: Shen, C C & Pande, K P — *J Phys Chem Solids,* 45,314(1984)

IGAS-0002b

ETCH NAME: TIME:

TYPE: Acid, clean TEMP:

COMPOSITION:

1 .... HF

1 .... HCl

4 .... H_2O

DISCUSSION:

AlGaAs, (100) wafer used as a substrate for MIS device structures. After citric acid cleaning, etch clean with the solution shown before SiO_2 deposition.

REF: Ibid.

PHYSICAL PROPERTIES OF INDIUM OXIDE, In_2O_3

Classification	Oxide
Atomic numbers	49 & 8
Atomic weight	277.52
Melting point (°C)	850
Boilint point (°C)	
Density (g/cm³)	7.18
Hardness (Mohs — scratch)	3—4
Crystal structure (hexagonal — trigonal)	$(10\bar{1}0)$ prism, 1st order
Color (solid — hot/cold)	Red/pale yellow
Cleavage (prismatic)	$(10\bar{1}0)/(01\bar{1}0)$

INDIUM OXIDE, In_2O_3

General: Does not occur as a mineral specie in nature, although it may appear as a very minor encrustation on zinc and iron ores that often contain trace indium. When pure indium metal is burned in air it forms the sequi-oxide In_2O_3 and the mono-oxide, InO, in part. There is no metal processing use in industry, but the oxides are used as coloring agents in glass and enamels.

Technical Application: Indium oxides are III—VI compound semiconductors, and have been evaluated for semiconducting characteristics but, as the pure oxide, have not been used in device fabrication. As the trinary compound of indium tin oxide (ITO), it is fabricated as an opto-electric device.

Etching: HCl, H_2SO_4. Mixed HCl:$FeCl_3$, and halogens.

INDIUM OXIDE ETCHANTS

INO-0001

ETCH NAME: Sulfuric acid TIME:

TYPE: Acid, selective TEMP: 50—60°C

COMPOSITION:

x .... H_2SO_4, conc.

DISCUSSION:

In_2O_3, $(10\bar{1}0)$ grown as an oriented thin film or as an amorphous film. With tin doping it shows semiconduction. Pattern etching of films was done through a chromium mask with the solution shown.

REF: Fan, J C C & Bachner, F — *J Electrochem Soc,* 122,1719(1975)

INO-0002

ETCH NAME: Hydrochloric acid | TIME:
TYPE: Acid, selective | TEMP:
COMPOSITION:

x .... HCl, conc.

DISCUSSION:

In_2O_3, $(10\bar{1}0)$, deposited oriented thin film or as an amorphous thin film, to include tin doping. Solution used for pattern etching through a mask similar to application shown in INO-0001.

REF: Kane, J et al — *Thin Solid Films,* 29,155(1975)

INO-0003

ETCH NAME: Oxalic acid | TIME:
TYPE: Acid, selective | TEMP: 50°C
COMPOSITION:

x .... 1 *M* oxalic acid

DISCUSSION:

In_2O_3, $(10\bar{1}0)$, oriented thin films or as an amorphous thin film with tin doping. Photolithographic techniques used to pattern surfaces using AZ-1350 photo resist. Solution used to selectively etch films after temperature annealing.

REF: Thornton, J A & Hedgeoth, L V — *J Vac Sci Technol,* 13,117(1976)

INO-0004a

ETCH NAME: Hydrogen iodide | TIME:
TYPE: Halogen, selective | TEMP: RT
COMPOSITION: | RATE: 25 Å/sec

x .... 55% HI

DISCUSSION:

$4In_2O_3$:$1SnO_2$, as thin film surface coatings. Surfaces were patterned using photolithographic techniques with AZ-1350H, post-bake 90 min at 120°C. Insert sample vertically into the solution during etching.

REF: Bradshaw, G & Hughes, A J — *Thin Solid Films,* 33.L5(1976)

INO-0004b

ETCH NAME: | TIME:
TYPE: Acid, removal | TEMP: 20°C
COMPOSITION: | RATE: 150 Å/sec

12 .... HCl
1 .... H_2O
x .... Zn, powder

DISCUSSION:

$4InO_3$:$1SnO_2$ specimens. Coat surface with slurry of zinc:H_2O, immerse horizontally in etch solution of HCl:H_2O. AZ-1350 used as the photo resist pattern lacquer, and the solution shown used to etch pattern.

REF: Ibid.

INO-0005a
ETCH NAME:
TYPE: Acid, removal
COMPOSITION:
x .... 37% HCl
30 g .... $FeCl_3$/l
TIME: Seconds
TEMP: RT
RATE: 0.1 μm/15 sec
DISCUSSION:
In_2O_3 specimens. Described as an etch for indium oxide.
REF: Ponjee, J J & Feil, H J — U.S. Patent 4.093,#04 1978 (US Phillips Corp)

INO-0005b
ETCH NAME:
TYPE: Acid, removal
COMPOSITION:
x .... 18% HCl
5.5 g .... $FeCl_3$/l
TIME: Minutes
TEMP: 45°C
RATE: 0.1 μm/10 min
DISCUSSION:
In_2O_3 specimens. Described as an etch for indium oxide.
REF: Ibid.

INO-0005c
ETCH NAME:
TYPE: Acid, removal
COMPOSITION:
x .... 18% HCl
66 g .... $FeCl_3$/l
TIME: Minutes
TEMP: 45°C
RATE: 0.1 μm/1.5—8 min
DISCUSSION:
In_2O_3 specimens. Removal time will vary with the doping concentration of the indium oxide, as doped with time.
REF: Ibid.

INDIUM GALLIUM ARSENIDE PHOSPHIDE ETCHANTS

IGAP-0001a
ETCH NAME:
TYPE: Acid, preferential
COMPOSITION:
1 .... HF
10 .... HBr
TIME:
TEMP: RT
DISCUSSION:
InGaAsP, thin film layer grown by LPE as part of a double-heterostructure. Solution is preferential to this material and developed mostly 60° type dislocations. Increase of dislocation density showed both edge, screw, and some pinning types.
REF: Veda, O et al — *Jpn J Appl Phys,* 23,836(1984)

IGAP-0001b
ETCH NAME: BRM
TYPE: Halogen, preferential
COMPOSITION:
x .... x% Br_2
x .... MeOH
TIME:
TEMP: RT

DISCUSSION:
InGaAsP, as thin film layers. See IGAP-0001a for discussion.
REF: Ibid.

IGAP-0002
ETCH NAME: BRM
TYPE: Halogen, selective
COMPOSITION:
x .... 1% Br_2
x .... MeOH
TIME:
TEMP: RT
DISCUSSION:
InGaAsP epitaxy thin films grown on InP substrates in a study of ionization coefficients. Solution used to selectively remove portions of the InGaAsP layer as it is not etched by HCl as is GaAs.
REF: Bulman, O E — *Solid State Electron,* 25,1189(1982)

PHYSICAL PROPERTIES OF INDIUM PHOSPHIDE, InP

Classification	Phosphide
Atomic numbers	49 & 15
Atomic weight	145.79
Melting point (°C)	1062
Boiling point (°C)	
Density (g/cm^3)	4.79
Lattice constant (angstroms)	5.87
Refractive index (n=)	3.1
Energy band gap (eV)	1.34
Hardness (Mohs — scratch)	6—7
Crystal structure (isometric — tetrahedral)	(111) tetrahedron
Color (solid)	Grey-silver
Cleavage (dodecahedral)	(110)

INDIUM PHOSPHIDE, InP

General: Does not occur in nature as a native compound. Indium is a soft silver white metal of the aluminum family; whereas, phosphorus is a very reactive acidic non-metal of the nitrogen family with three allotropic forms: yellow (white), red and black. There is no industrial use other than as a compound semiconductor.

Technical Application: Indium phosphide is a III—V binary compound semiconductor, exhibiting the (111)A and (111)B wafer surface polarity effect of these materials. It has the tetrahedral unit cell structure of sphalerite, ZnS, with positive (111)In and negative (111)P atom surfaces which produce different figures and structures when chemically etched. In device processing wafers with (100) surface orientation are used to reduce the polarity effect.

Initially InP ingots were grown by Czochralski (CZ) or Float Zone (FZ) methods with (111) and (100) oriented seeds but, now, also are grown by either vertical or horizontal Vapor Phase Epitaxy (VPE) or Liquid Phase Epitaxy (LPE) technique as thin films. Using the MBE growth method InP substrates and thin films are grown as multilayer heterostructures. There is much on-going development of such heterojunction or heterostructure devices. Like most compound semiconductors it has good ⟨110⟩ cleavage, and surfaces may be used so cleaved.

As discrete elements, indium phosphide has been fabricated as Gunn diodes, solar cells, and laser diodes. The substrate wafer, particularly as a heterostructure base, is usually InP:Fe, (100) semi-insulating (SI). Trinary compounds are $InAs_{x-1}P_x$; $In_xAl_{1-x}P$, and $In_{1-x}Ga_xAs_yP$.

Etching: HNO_3, HCl and H_3PO_4. Halogens and mixed acids. Alkalies and dry chemical etching (DCE).

SELECTION GUIDE: InP

(1) Br_2:MeOH (BRM)
- (i) Cleaning: INP-0058; -0012; -0059b
- (ii) Oxide Removal: INP-0066b
- (iii) Polish: INP-0001; -0002; -0003; -0004; -0005; -0006; -0008; -0009; -0010; -0049; -0050; -0011; -0012; -0013; -0058; -0069; -0054b; -0045b; -0061; -0062c; -0067
- (iv) Removal: INP-0006; -0018a,b; -0058
- (v) Selective: INP-0065; -0017; -0064b; -0064f
- (vi) Step-etch: INP-0007; -0014
- (vii) Structure: INP-0057b
- (viii) Thinning: INP-0068; -0016; -0017

(2) H_2O_2
- (i) Oxidation: INP-0019

(3) HCl
- (i) Cleaning: INP-0071
- (ii) Orientation: INP-0065
- (iii) Oxide Removal: INP-0033b; -0066b
- (iv) Polish: INP-0056a
- (v) Removal: INP-0020a; -0056a
- (vi) Selective: INP-0021; -0062d; -0064a,c,e,g,j,l

(4) HCl:H_2O
- (i) Cleaning: INP-0053; -0054a
- (ii) Selective: INP-0062e

(5) HCl:HNO_3
- (i) Removal: INP-0020b
- (ii) Polish: INP-0070
- (iii) Selective: INP-0062g; -0064a,k,l,m

(6) HCl:HNO_3:H_2O
- (i) Polish: INP-0021a

(7) HCl:H_2O_2
- (i) Selective: INP-0062e

(8) HCl:HAc
- (i) Selective: INP-0062e

(9) HCl:$FeCl_3$
- (i) Preferential: INP-0055

(10) HCl:H_2O_2:HAc
- (i) Removal: INP-0057

(11) HCl:HNO_3:Br_2
- (i) Preferential: INP-0021b

(13) HNO_3 +/− H_2O
- (i) Oxidation: INP-0025
- (ii) Removal: INP-0024; -0022b

(14) HIO_3
- (i) Cleaning: INP-0030
- (ii) Polish: INP-0030

(15) HBr
  (i) Polish: INP-0026; -0062a
(16) HBr:HF
  (i) Preferential: INP-0027
(17) $HBr:HNO_3$ +/− H_2O
  (i) Defect: INP-0028a; -0062b
(18) $HBr:H_2SO_4$
  (i) Dislocation: INP-0051; -0061a; -0062b
(19) HBr:HCl
  (i) Selective: INP-0062a
(20) $HBr:H_2O_2$
  (i) Selective: INP-0062b
(21) HBr:HAc
  (i) Selective: INP-0062a
(22) $HBr:H_3PO_4$
  (i) Selective: INP-0062a
(23) $HBr:Br_2:H_2O$
  (i) Polish: INP-0064
(24) $HBr:HAc:K_2Cr_3O_7$
  (i) Preferential: INP-0031a
(25) $HBr:H_3PO_4:K_2Cr_3O_7$
  (i) Oxide Removal: INP-0066b
(26) $H_2O_2:HCl:HAc$
  (i) Selective: INP-0062f
(27) H_2SO_4
  (i) Cleaning: INP-0067
(28) $H_2SO_4:H_2O_2:H_2O$
  (i) Removal: INP-0041; -0042; -0063; -0065; -0043b
  (ii) Selective: INP-0062h; -0043a; -0044; -0046a; -0046b; -0064n, m
(29) $H_2SO_4:H_2O_2:MeOH$
  (i) Cleaning: INP-0047
(30) $H_2SO_4:HCl:K_2Cr_3O_7$
  (i) Selective: INP-0062h
(31) H_3PO_4, Electrolytic
  (i) Selective: INP-0031a; -0052
(32) $H_3PO_4:H_2O$
  (i) Oxide Removal: INP-0022c
(33) $H_3PO_4:HCl$
  (i) Grooving: INP-0035
  (ii) Step Etch: INP-0066a
  (ii) Preferential: INP-0066a
  (iv) Removal: INP-0034
  (v) Selective: INP-0057a; -0062e
(34) $H_3PO_4:Br_2$
  (i) Preferential: INP-0018
(35) $H_3PO_4:HBr$
  (i) Preferential: INP-0021c
(36) $H_3PO_4:HBr:K_2Cr_3O_7$
  (i) Preferential: INP-0031b
(37) $H_3PO_4:HCl:H_2O_2$
  (i) Selective: INP-0062f

(38) $H_3PO_4,H_2O_2:H_2O$
(i) Selective: INP-0046a-b
(39) $K_3Fe(CN)_6$
(i) Staining: INP-0036a
(40) HF $+/-H_2O$
(i) Oxide Removal: INP-0045a
(41) $HF:Br_2$
(i) Dislocation: INP-0037b
(42) HF:MeOH
(i) Dislocation: INP-0037a
(43) $HF:HNO_3$
(i) Selective: INP-0063
(44) KOH/NaOH
(i) Cleaning: INP-0056a
(45) KOH:MeOH
(i) Cleaning: INP-0048
(46) $KOH:K_2Fe(CN)_6:H_2O$
(i) Preferential: INP-0033d
(ii) Selective: INP-0040; -0033e; -0064a,d,j,m,n
(iii) Staining: INP-0039
(47) CF4 (RF Plasma)
(i) Removal: INP-0059a
(48) $Cl_2:O_2$ $+/-$ Ar (RF plasma)
(i) Preferential: INP-0064i
(50) Indium, In
(i) Etch-back: INP-0049; -0050
(51) Argon, (RF Plasma)
(i) Cleaning: INP-0062
(52) Heat
(i) Cleaning: INP-0060
(56) Named/Numbered Etchants:
(i) Huber Etch (polish): INP-0072
(ii) Superoxol (polish): INP-0022a
(iii) A/B Etch (preferential): INP-0051; -0021d
(iv) CBR Etch (polish): INP-0073; -0075
(v) BRM (polish/preferential): See: Br_2:MeOH
(vi) BPK-221 (preferential): INP-0031b
(vii) H Etch (preferential): INP-0021c; -0051
(viii) KKI Etch (selective): INP-0064d
(ix) CP4 (polish/preferential): INP-0056a

INDIUM PHOSPHIDE ETCHANTS

INP-0001
ETCH NAME: BRM
TYPE: Halogen, polish
TIME:
TEMP:
COMPOSITION:
x .... Br_2
x .... MeOH

DISCUSSION:

InP, (100) cleaved wafers. Degrease with solvents, then free etch in the BRM solution. Rinse in 15% HCl before MOVPE deposition of (GaIn)As.

REF: Hockly, M & White, R A D — *J Cryst Growth,* 68,334(1984)

INP-0002: Long, J A — *J Cryst Growth,* 69,10(1984)

InP, (100) Si doped wafers used as substrates for epitaxy growth. Wafers were chem/mech polished in the solution shown. Hand polish with the same solution immediately before MOCVD deposition of InP:Fe.

INP-0003: Dautremont-Smith, W C & Feldman, L C — *J Vac Sci Technol,* A3,873(1985)

InP, (100), n-type cleaved wafers from an ingot grown by LEC, and used in a study of plasma damage. Polish in the solution shown, then degrease with solvents, and soak in dilute HF.

INP-0004a: Chand, N & Houston, P A — *J Cryst Growth,* 14,9(1985)

InP, (100) wafers used as substrates for epitaxy growth. Solution was used to etch "V" grooves. "V" groove side slopes were (111)A. (*Note:* Controlled groove etching is done by photo resist masking patterned slots in ⟨110⟩ directions.)

INP-0005: Sussmann, R S — *J Electron Mater,* 12,603(1983)

InP:Fe, (100) (SI) wafers cut 2°-off toward (110). Solution used to etch polish surfaces.

INP-0006

ETCH NAME: TIME:

TYPE: Halogen,polish/removal TEMP:

COMPOSITION:

(1) x .... Br_2
x .... MeOH

(2) x .... HCl
x .... H_3PO_4

DISCUSSION:

InP, (100) wafers used as substrates for epitaxy growth of InGaAs layers. Surface mask coating was a silica spin-on rather than SiO_2 or Si_3N_4 by RF or CVD. Spin-on glass at 2500 rpm and anneal 20 min in N_2 at 300°C. Move wafers slowly into furnace heat zone to prevent cracking. InGaAs deposit was 2500 Å thick. Photo resist coat and open slot pattern parallel to ⟨110⟩. Use 1% HF to remove SiO_2 in slots. Then harden remaining SiO_2 mask by annealing at 450°C, 15 min in N_2. Etch "V" grooves with various mixtures of the two solutions shown. Groove side walls are (111)A. After epitaxy deposition, cleaved cross sections on ⟨110⟩ and stain. Etch rates shown as: $(\overline{111})$B>(100)>(111)A.

REF: Chan, N & Houston, P A — *J Electron Mater,* 14,9(1985)

INP-0007

ETCH NAME: BRM TIME:

TYPE: Halogen, step-etch TEMP: RT

COMPOSITION:

x .... 0.5% Br_2
x .... MeOH

DISCUSSION:

InP, (100) n-type wafers used as substrates for diffusion of a p/n junction using CdP_2 at 600°C. For TEM study, the Cd side was step-etched in the solution shown in a study of defect structure induced by Cd diffusion.

REF: Ueda, O — *Jpn J Appl Phys,* 23,1551(1985)

INP-0008: Brown, A S et al — *J Electron Mater,* 14,367(1985)

InP, (100) wafers used as substrates for GaInP epitaxy. Substrates were chem/mech polished in the solution shown. Wafers were then cleaved in half portions. One half of each wafer was annealed in H_2 + Pd, then repolished to remove about 25 μm of surface material.

Both halves then degreased: acetone — MeOH — H_2O. Then free etched in a solution of: $7H_2SO_4$:$1H_2O_2$:$1H_2O$, RT, 20 sec immediately before epitaxy.

INP-0009
ETCH NAME: BRM
TYPE: Halogen, polish
COMPOSITION:
x .... 1% Br_2
x .... MeOH

TIME:
TEMP: RT

DISCUSSION:

InP, (110) wafer cleaved under UHV. Wafers (100), (110) and $(\overline{111})$B were saw cut. Chem/mech polish surfaces on a pelon pad and hold under MeOH. The $(\overline{111})$B surface showed signs of orange peel in this solution. Final clean by heating in vacuum at 410°C. Gas bursts were observed at 250°C, 350°C, slight bursts at 380°C, and were probably due to hydrocarbon out-gassing.

REF: Chang, K-H & Meijer, P H E — *J Vac Sci Technol,* 14,789(1977)

INP-0010: Ogura, M et al — *J Cryst Growth,* 68,32(1984)

InP, (100) wafers used as substrates for epitaxy deposition of InGaAsP by MOCVD. Wafers were etch polished in the solution shown before epitaxy. GaAs wafers were used for (AlGa)InP epitaxy thin film growth. GaAs wafers were etch cleaned in $5H_2SO_4$:$1H_2O_2$:$1H_2O$, RT, 2 min.

INP-0049: Tu, C W & Jones, K A — *J Cryst Growth,* 70,117(1984)

InP, (100) wafers used as epitaxy substrates. Chem/mech polish with solution shown. Then degrease in TCE, and soak in 45% KOH, 1 h at RT with final heavy water wash. Re-etch 10 sec with solution shown before loading in vacuum. In CVD system, *in situ* etch with 0.4% HCl hot vapors. After epitaxy, cleave ⟨110⟩ and etch with A/B preferential etch at 60°C, 8 min to develop structure and dislocations.

INP-0050: Gagnaire, A et al — *J Electrochem Soc,* 132,1655(1985)

InP, (100) n-type wafers used in a study of cathodic decomposition. Chem/mech polish in the solution shown; rinse in warm MeOH and blow dry with argon. Solder indium back contacts at 350°C in N_2. The H_2 evolution from electrolyte leaves indium rich film on surfaces.

INP-0065: Smith, N A et al — *J Cryst Growth,* 68,517(1984)

InP ingots grown by LEC and Fe doped. Ingots showed FeP_2 and mixed FeP/FeP_2 segregates. A BRM solution was used to expose the segregates and to dissolve InP and remove precipitates for SEM and X-ray study.

INP-0051
ETCH NAME: A/B Etch
TYPE: Acid, preferential
COMPOSITION:
1 ml .... HF
2 ml .... H_2O
3 g .... CrO_3
8 mg .... $AgNO_3$

TIME:
TEMP: RT

DISCUSSION:

InP, $(\overline{111})$B wafers used as substrates for epitaxy growth of double heterojunction (DH) structure: InP/n-InGaAs/InP/InP$(\overline{111})$B grown by LPE as a photo detector. There is a 25 to 50% mismatch between InGaAs and InP epitaxy layers. The A/B solution was used to

develop structure and dislocation etch pit density, 10^6 cm^{-2}. Part of mismatch due to use of the $(\overline{1}\overline{1}\overline{1})$B substrate.
REF: Morrien, C B et al — *J Electrochem Soc*, 132,1717(1985)
INP-0080: Zinko, J L et al — *J Electron Mater*, 14,563(1985)

InP:Fe, (100) (SI), and n^+ doped (S/Sn) wafers used as substrates for atmospheric pressure ATMBE growth of InP layers. After epitaxy, cleave ⟨110⟩ and stain with the A/B Etch. (See GAS-0059 for A/B Etch.)

INP-0058
ETCH NAME: BRM — TIME:
TYPE: Halogen, cleaning — TEMP: RT
COMPOSITION:
x .... 10% Br_2
x .... MeOH
DISCUSSION:

InP:Fe, (100) (SI) wafers used as substrates for epitaxy growth of InP layers. Substrates cleaned as follows: (1) degrease with TCE, then acetone, then isopropyl alcohol (IPA); (2) etch clean in $3H_2SO_4$:$1H_2O_2$:$1H_2O$, and DI water rinse, and (3) etch clean in solution shown. After deposition of SiO_2 use standard photolithography to pattern the mask, and etch open SiO_2 with BHF, and repeat (1) degreasing step. Five mask patterns were used for selective epitaxy deposition of InP.
REF: Oishi, M & Kuroiwa, K — *J Electrochem Soc*, 132,1209(1985)

INP-0011
ETCH NAME: — TIME:
TYPE: Halogen, polish — TEMP: RT
COMPOSITION:
(1) x .... 1% Br_2
x .... MeOH
(2) x .... x% KOH
DISCUSSION:

InP, (100) wafers cut 3°-off toward (110) used as substrates for epitaxy deposition of various compounds of Ga/Al/As/In by MOCVD. Chem/mech polish in the BRM (1) solution, then soak in the (2) KOH solution to remove native oxide. Before MOCVD use the following cleaning sequence: (1) DI H_2O rinse; (2) etch clean in H_2SO_4, RT, 2 min with ultrasonic; (3) rinse in hot MeOH soak; (4) repeat H_2SO_4, RT, 3 min; (5) hot MeOH rinse and (6) N_2 blow dry.
REF: Cheng, C H — *J Electron Mater*, 13,703(1984)
INP-0012: Kasemset, D et al — *J Electron Mater*, 13,655(1984)

InP, (100) wafers. Clean and etch polish with the BRM solution (1), and soak in KOH solution (2) to remove native oxide.

INP-0013
ETCH NAME: BRM — TIME:
TYPE: Halogen, polish — TEMP: RT
COMPOSITION:
1 .... Br_2
10 .... MeOH
DISCUSSION:

InP, (100) n-type wafers. Mechanically lap with 0.3 μm alumina. Etch polish surfaces with solution shown.
REF: Nakamura, T & Katoda, T — *J Appl Phys*, 55,3064(1984)

INP-0014
ETCH NAME: Huber etch (BRM)
TYPE: Halogen, step-etch
COMPOSITION:
x .... 1.5% Br_2
x .... MeOH
TIME:
TEMP: RT
DISCUSSION:

InP, (100) wafers used as substrates for InP epitaxy. Solution shown was used as a step-etch in a study of the variations of dislocation density with depth of epitaxy layer.

REF: Mahajan, S et al — *J Electrochem Soc,* 129,1556(1982)

INP-0018: Duchemin, J P et al — *Inst Phys Conf Ser,* 45,10(1979)

InP, (100) wafers. After heavy removal of bulk material with a 15% Br_2:MeOH mixture, chem/mech polish surfaces with solution shown above. Rinse in MeOH then in IPA.

INP-0072a: Huber A & Linn, N T — *J Cryst Growth,* 28,80(1975)

InP, (100) wafers. Solution used as a removal etch by dip controlled step-etching. Rate shown as 6—7 μm/min.

INP-0077: Zhu, L A et al — *J Appl Phys,* 57,5486(1985)

InP, (100), Te doped, n-type wafers used as substrates for LPMOCVD growth of InP thin films. Clean substrates as follows: (1) degrease in hot organic solvents; (2) etch in BRM solution shown, RT, 2 min. In epitaxy system, etch both the substrates and epitaxy deposited InP layer with hot HCl gas, *in situ,* to remove surface damage and contamination.

INP-0068
ETCH NAME: BRM
TYPE: Halogen, thinning
COMPOSITION:
x .... 1.5% Br_2
x .... MeOH
TIME:
TEMP: RT
DISCUSSION:

InP, (100) wafers were etch thinned with the solution shown.

REF: Duh, B V & Bransen, D — *J Electrochem Soc,* 130,207(1983)

INP-0069
ETCH NAME: BRM
TYPE: Halogen, polish
COMPOSITION:
"A" x .... 7.5% Br_2
x .... MeOH
"B" x .... MeOH
TIME: To 10 min
TEMP: RT
MIX: Cover "A" with "B".
DISCUSSION:

InP, (100) wafers and other III—V compounds evaluated for immersion polishing in solution shown as against chem/mech polishing which can develop sleeks. Immerse wafers in solution, then remove slowly up through over-cover of pure methyl alcohol to quench without exposure to air. The MeOH cover will diffuse into solution in about 10 min.

REF: Risario, R et al — *J Appl Phys,* 53,40(1982)

INP-0054b
ETCH NAME: BRM
TYPE: Halogen, polish
COMPOSITION:
x .... 0.5% Br_2
x .... MeOH
TIME:
TEMP: RT

DISCUSSION:

InP:Fe (100) (SI) wafers used as substrates for LPE growth of InP epitaxy layers. Rinse in isopropyl alcohol (IPA), then polish in solution shown. Follow with: (1) TCE degreasing, and acetone rinse; (2) 50% HCl, RT, 30 sec, rinse in IPA, and (3) dry in hot IPA.

REF: Ibid.

INP-0059b

ETCH NAME: TIME: 1 min

TYPE: Halogen, cleaning TEMP: RT

COMPOSITION:

x .... 2.5% Br_2

x .... MeOH

DISCUSSION:

InP, (100) wafers etch cleaned in this solution.

REF: Hirata, K et al — *J Vac Sci Technol*, 2,44(1984)

INP-0016

ETCH NAME: TIME:

TYPE: Halogen, thinning TEMP: RT

COMPOSITION:

x .... 5% Br_2

x .... MeOH

DISCUSSION:

InP, (100) and GaAs, (100) wafers used for gold evaporation and annealing. Black wax was used as a mask to coat wafers and selective etch thin sections for TEM study of line defects.

REF: Nakahara, S et al — *J Electrochem Soc*, 131,1917(1984)

INP-0017: Nakahara, S et al — *Solid State Electron*, 27,557(1984)

InP, (100) wafers used for zinc deposition and anneal. Annealing was done on a carbon strip heater for 30 sec at 250, 300, 350, 375, and 425°C. Used black wax to coat and selectively etch thin wafer portions for TEM study. Thinning included some areas of hole-thru etching.

INP-0018b

ETCH NAME: BRM TIME: 15 sec

TYPE: Halogen, removal TEMP: RT

COMPOSITION:

x .... 15% Br_2

x .... MeOH

DISCUSSION:

InP, (100) wafers. Solution used to remove bulk material from both sides of wafers with solution shown, removing 20 μm from each side. Follow with chem/mech surface polish with a 1.5% Br_2:MeOH solution concentration. Rinse in MeOH, then in isopropyl alcohol (IPA).

REF: Duchemin, J P et al — *Inst Phys Conf Ser*, 45,10(1979)

INP-0019

ETCH NAME: TIME: 30 min

TYPE: Acid, oxidation TEMP: Boiling

COMPOSITION:

x .... H_2O_2

DISCUSSION:

InP, (100) wafer used as a substrate for epitaxy growth of the following structure: n-InPAl^{epi}/ox-lnP/InP^{epi}/InP(100). The hydrogen peroxide was used to form the oxidized-InP layer. The layer was 150—250 Å thick.

REF: Inuishi, M & Wessels, B W — *Thin Solid Films,* 103,141(1983)

INP-0020a

ETCH NAME: Hydrochloric acid TIME:

TYPE: Acid, removal TEMP: Hot

COMPOSITION:

x .... HCl

DISCUSSION:

InP, (111) wafers used in an etch development study. Solution shows moderate to vigorous reaction when used hot.

REF: Robbins, H & Schwartz, B — *Tech Rep,* 1960

INP-0081a: Takeda, Y et al — *Jpn J Appl Phys,* 23,84(1984)

InP, (100) wafer used as substrate to epitaxially grow the following structure: p-InGaAs/p-InP/lnGaAsP/InP(100). First, etch remove the InGaAs layer to the p-InP buffer layer which is rough, continue etching to the InGaAsP which is very shiny. To remove the InGaAsP layer, etch with $1H_2SO_4$:$1H_2O_2$:$1H_2O$ at RT. The InGaAsP is not as yellow in appearance as the InP substrate.

INP-0020b

ETCH NAME: TIME:

TYPE: Acid, removal TEMP: RT

COMPOSITION:

(1) 1 .... HCl (2) 2 .... HCl
1 .... HNO_3 1 .... HNO_3

DISCUSSION:

InP, (111) wafers used in an etch development study. Solution (1) shows moderate to vigorous reaction. Solution (2) is slow at RT.

REF: Ibid.

INP-0021a

ETCH NAME: TIME:

TYPE: Acid, polish TEMP: RT

COMPOSITION:

6 .... HCl
1 .... HNO_3
6 .... H_2O

DISCUSSION:

InP, (111) wafers grown by LEC and used as (111)A and $(\overline{111})$B oriented wafer substrates for epitaxy growth of GaInPAs. Substrates were lap polished with Nalaog 1060 (colloidal silica), then etch polished in the solution shown. Epitaxy layers were deposited by a gradient freeze growth (GFG) method. Material was then preferentially etched in a study of dislocation pit density using a 40HCl:$80HNO_3$:$1Br_2$ solution, RT, 10 sec.

REF: Thiel, F A & Barns, R L — *J Appl Phys,* 126,1272(1979)

INP-0021b
ETCH NAME: TIME: 10 sec
TYPE: Acid, preferential TEMP: RT
COMPOSITION:
40 .... HCl
80 .... HNO_3
1 .... Br_2
DISCUSSION:

InP, (111) wafers grown by LEC and used as (111)A and $(\overline{111})$B oriented substrates for epitaxy growth of GaInPAs. The epitaxy layers were preferentially etched in the above solution in a study of dislocation pit density.
REF: Ibid.

INP-0022a
ETCH NAME: Superoxol, variety TIME:
TYPE: Acid, polish TEMP:
COMPOSITION:
1 .... HF
1 .... HCl
4 .... H_2O
+12:1 .... H_2O_2
DISCUSSION:

InP, (100), Zn doped p-type wafers. Solution showed used to etch polish surfaces. Follow with 10% H_3PO_4 to remove native oxide.
REF: Pande, K P & Naur, U K R — *J Appl Phys,* 55,3109(1984)
INP-0076: Christon, A et al — *Semiconductor Int,* March 1979, 59

InP, (100) wafers used in a surface contamination and defect study. Cleaning sequence was: (1) acetone rinse; (2) 1% Br_2:MeOH, RT, 5 min; (3) HF:H_2O, RT, 1", and (4) etch in solution shown above, RT, 10 min. Solution used contained 5 drops of H_2O_2.

INP-0023
ETCH NAME: TIME:
TYPE: Acid, removal TEMP: RT
COMPOSITION:
x .... HCl
x .... HAc
x .... H_2O_2
DISCUSSION:

InP, (100) wafers used as substrates for epitaxy growth of InGaAsP. Substrates etch cleaned in the solution shown prior to epitaxy.
REF: Kambayashi, T et al — *Jpn J Appl Phys,* 19,79(1980)

INP-0020c
ETCH NAME: TIME:
TYPE: Acid, removal/oxide TEMP: (1) Hot (2) RT
COMPOSITION:
(1) x .... HNO_3, conc. (2) 2 .... HNO_3
1 .... H_2O
DISCUSSION:

InP, (111) wafers used in an etch development study. Solution (2) was used with both

white and red fuming nitric acid and shows moderate to vigorous reaction in both cases. Concentrated nitric acid shows slow reaction used hot.
REF: Ibid.
INP-0025: Geib, K M et al — *J Phys Chem Solids,* 45,516(1984)
InP, (100) wafers. A concentrated solution used hot to grow an oxide on InP surfaces. Oxide is mainly $InPO_4$.
INP-0086: Wada, O et al — *Solid State Electron,* 25,381(1982)
InP, (100) wafers. Nitric acid used to oxidize surfaces, 50—60°C, 4—5 min by soaking under a microscope light. This formed a 1000 Å thick hydrated oxide. Remove oxide with dilute HCl. (*Note:* This is a method of surface cleaning.)

INP-0022b
ETCH NAME: TIME:
TYPE: Acid, removal TEMP: RT
COMPOSITION:
x .... 10% HNO_3
DISCUSSION:
InP, (100) wafers, Zn doped p-type. Wafers processed by (1) degrease in boiling TCE; (2) rinse in MeOH; (3) detergent scrub and (4) DI H_2O rinse. Etch in solution shown to remove lap/polish damage, then etch clean with 1HF:1HCl;$4H_2O$ + I_2:$1H_2O_2$. Final clean to remove native oxide with 10% H_2SO_4.
REF: Ibid.

INP-0026
ETCH NAME: TIME:
TYPE: Halogen, anisotropic TEMP: − 15°C
COMPOSITION:
x .... 9 *N* HBr
DISCUSSION:
InP, (100), Sn doped wafers. Pre-clean: (1) ultrasonic TCE; (2) rinse in acetone and (3) rinse in MeOH. Follow with 60% KOH at 50°C, 20 min. Etch to delineate crystallographic directions with 1 HCl(12 *N*): 1 H_2O_2, RT, 30 sec. The $(\overline{111})$B face etches rapidly, and the (100) face shows elongated pits that include line intersection of (111)A and $(\overline{111})$B faces. Coat surfaces with photo resist and open slots in ⟨110⟩ directions. Use solution shown to etch a saw-tooth structure with ridges as (100)/(111)A planes. Solution is not sufficiently preferential at room temperature to etch saw-tooth structure, but will produce the structure when used cold.
REF: Keavney, C J & Smith, H I — *J Electrochem Soc,* 131,452(1984)

INP-0027
ETCH NAME: TIME:
TYPE: Acid, preferential TEMP:
COMPOSITION:
1 .... HF
10 .... HBr
DISCUSSION:
InP, (100) wafers used as substrates for epitaxy growth of a double heterostructure: InP/GaAsP/InP/InP(100). Use a BRM solution to etch remove InP cover. Solution shown is preferential to GaAsP with InP acting as an etch-stop. Most dislocations are 60° type but, with increased density, pure edge, screw and pinning types observed. Solution used to etch thin the epitaxy film for TEM study.
REF: Veda, O et al — *Jpn J Appl Phys,* 23,836(1984)

INP-0064
ETCH NAME:
TYPE: Halogen, polish
COMPOSITION:
x .... Br_2
x .... HBr
x .... H_2O
TIME:
TEMP: RT

DISCUSSION:
InP, (100) wafers. Chem/mech polish in solution shown; DI water rinse. Native oxide etch 1 min, RT in 40% HF, water rinse and N_2 blow dry.
REF: Cameron, D C & Foreman, B J — *Solid State Electron,* 27,305(1984)

INP-0028a
ETCH NAME:
TYPE: Acid, defect
COMPOSITION:
1 .... HNO_3
3 .... HBr
TIME: 10 sec
TEMP: RT

DISCUSSION:
InP, (100), p-type wafers used as substrates for epitaxy growth of InGaAsP. Solution developed defects correlated with protrusions observed in the InGaAsP layer.
REF: Lourenco, J A — *J Electrochem Soc,* 131,1914(1984)
INP-0029: Chu, N J et al — *J Electrochem Soc,* 129,352(1982)
InP, (100) wafers used in study similar to INP-0028a.

INP-0030
ETCH NAME: Iodic acid
TYPE: Halogen, polish
COMPOSITION:
x .... 10% HIO_3
TIME: 3 min
TEMP: RT

DISCUSSION:
InP:Fe, (100) (SI) wafers used in a study of optical constants. Chem/mech polish with light scrubbing in solution shown, then final polish with scrubbing; DI water rinse and N_2 blow dry.
REF: Scheps, R J — *J Electrochem Soc,* 131,541(1984)
INP-0085: Hokelek, E & Robinson, G Y — *J Appl Phys Lett,* 40,119(1982)
InP, (100), n- and p-type wafers used in a study of Schottky barrier contacts. Solution used 30 sec, RT, to clean surfaces, and then rinsed in DI water.

INP-0033a
ETCH NAME: Phosphoric acid
TYPE: Electrolytic, selective
COMPOSITION:
x .... H_3PO_4
TIME:
TEMP:
ANODE: GaAs
CATHODE:
POWER:

DISCUSSION:
InP, (100) wafers used as substrates for epitaxy growth of GaInAs. Use solution shown to selectively remove the epitaxy thin film of GaAs. Use 0.5 *M* HCl to remove oxide film remaining after electrolytic etching.
REF: Lyons, M H et al — *J Cryst Growth,* 66,269(1984)

INP-0022c

ETCH NAME: Phosphoric acid, dilute | TIME:

TYPE: Acid, oxide:removal | TEMP:

COMPOSITION:

x .... 10% H_3PO_4

DISCUSSION:

InP, (100) Zn doped p-type wafers. After polish etching of surfaces (INP-0022a) use the solution shown to remove native oxide remaining on surface.

REF: Ibid.

INP-0033b

ETCH NAME: Hydrochloric acid, dilute | TIME:

TYPE: Acid, oxide removal | TEMP:

COMPOSITION:

x .... 0.5 *M* HCl

DISCUSSION:

InP, (100) wafers used as substrates for GaInAs epitaxy growth. After electrolytic selective etch removal of the GaInAs thin film, use the above solution to remove remaining native oxide. See INP-0033a.

REF: Ibid.

INP-0034

ETCH NAME: | TIME:

TYPE: Acid, removal | TEMP:

COMPOSITION:

x .... HCl

x .... H_3PO_4

DISCUSSION:

InP, (100) wafers used as substrates for epitaxy deposition of thin film InGaAs. Solution used to etch "V" grooves in InP surfaces prior to epitaxy. Groove side slopes are (111)A.

REF: Chand, N & Houston, P A — *J Cryst Growth,* 14,9(1985)

INP-0066a

ETCH NAME: | TIME: 30—120 sec

TYPE: Acid, preferential/step-etch | TIME: 23°C

COMPOSITION: | RATE: See discussion

x .... 12 *N* HCl

x .... 15 *N* H_3PO_4

DISCUSSION:

InP:Fe, (100) (SI) wafers cut within 0.5° of the plane, and used in an etch rate study for sub-micron structure fabrication. Wafers were photo resist coated with AZ1350 and step etched to obtain rates. Etching was done with the wafer face down in a glass beaker with stirring and frequent agitation of the wafer during etch period. Etch channels formed were rate measured for (100), (111), (011) surfaces. Face orientation, HCl concentration, and etch rates were as follows:

Face	HCl (%)	Rate (µm/min)	Solution	Time/depth (approx.)	
(100)	5	0.09	$2.5HCl:7.5H_3PO_4$	90 sec	1.28
	10	0.24			
	15	0.40	$2HCl:8H_3PO_4$	120 sec	1.22
	20	0.70			
	25	1.05	$1.5HCl:8.5H_3PO_4$	120 sec	0.70
(011)	20	3.40	$1HCl:9H_3PO_4$	120 sec	0.4 +
(111)	20	2.6	$0.5HCl:9.5H_3PO_4$	140 sec	0.20

(Columns "Solution" and "Time/depth (approx.)" are under the heading: **Etch depth vs. time for the etch solutions shown**)

REF: Uekusa, S et al — *J Electrochem Soc*, 132,671(1985)

INP-0066b
ETCH NAME: Hydrochloric acid, dilute — TIME: 60 sec
TYPE: Acid, oxide removal — TEMP: 23°C
COMPOSITION:
x .... 10% HCl
DISCUSSION:

InP:Fe, (100) wafers within 5° of plane used in an etching study for submicron structure fabrication. Solution used to pre-etch wafers prior to AZ1350 photo resist patterning. Wafers were etched in $HCl:H_3PO_4$ solutions with varied HCl concentrations (INP-0066a) with and without HCl pre-etch of either 10% or 5% HCl. The following etchants were listed as being similar in application to that of the $HCl:H_3PO_4$ solutions studied:

Solution	Temperature	Content
(1) $2HBr:2H_3PO_4:1K_2Cr_2O_7$	20°C or 0°C	Solutions containing bromine are rapid with rate dependent on both bromine concentration and the temperature
(2) $0.5Br_2:99.5CH_3OH$ (MeOH)	– 10°C	
(3) 1.5% Br_2:98.5% CH_3OH	– 10°C	

REF: Ibid.

INP-0031c
ETCH NAME: BCK-113 — TIME: 8 min
TYPE: Acid, selective — TEMP: RT
COMPOSITION:
1 .... 9 *N* HBr
1 .... 17 *N* CH_3COOH (HAc)
3 .... $^*K_2Cr_2O_7$

*14.7 g/100 cm^3H_2O

DISCUSSION:

InP, (100) wafers used as substrates for epitaxy growth of InGaAsP. Solution will etch both materials and was used in etch forming mesa structures.
REF: Ibid.

INP-0031d
ETCH NAME: BCK-115 TIME:
TYPE: Acid, selective TEMP: RT
COMPOSITION:
1 .... 9 *N* HBr
1 .... 17 *N* CH_3COOH (HAc)
5 .... *$K_2Cr_2O_7$

*14.7 g/100 cm^3 H_2O

DISCUSSION:
InP, (100) wafers used as substrates for epitaxy growth of InGaAsP. Solution will etch both materials and was used in etch forming mesa structures.
REF: Ibid.

INP-0031e
ETCH NAME: BCK-11m TIME:
TYPE: Acid, selective TEMP: RT
COMPOSITION:
1 .... 9 *N* HBr
1 .... 17 *N* CH_3COOH (HAc)
m .... *$K_2Cr_2O_7$ (m = 0.2—2.0)

*14.7 g/100 cm^3 H_2O

DISCUSSION:
InP, (100) wafers used as substrates for epitaxy growth of InGaAsP. Solution will etch both materials and was used in etch forming mesa structures. Mirror walls on (110) cleaved planes best with this solution or BCK-221.
REF: Ibid.

INP-0031a
ETCH NAME: BCK-111 TIME: (1) 4 min (2) 2 min
TYPE: Acid, selective/polish TEMP: (1) RT (2) 20°C
COMPOSITION:
1 .... 9 *N* HBr
1 .... 17 *N* CH_3COOH (HAc)
1 .... *$K_2Cr_2O_7$

*14.7 g/100 cm^3 H_2O

DISCUSSION:
InP, (100) wafers used as substrates for MBE growth of InGaAsP. Under condition (1) used to develop defects in epitaxy layers; condition (2) used for a mirror finish surface on InP.
REF: Adachi, S — *J Electrochem Soc,* 129,609(1982)
INP-0032: Adachi, S — *J Electrochem Soc,* 129,1542(1982)
InP, (100) wafers. Similar study as in INP-0031a.

INP-0031f
ETCH NAME:
TYPE: Halogen, polish
COMPOSITION:
(1) x .... x% Br_2
x .... MeOH

(2) x .... KI
x .... I_2
x .... H_2O

TIME:
TEMP: RT

DISCUSSION:
InP, (100) wafers used as substrates for MBE growth of InGaAsP. Solution (1) used to produce a mirror finish on InP. See BCK-xxx solutions used to selectively etch mesas. Solution (2) was used to etch Au:Zn and AuGe:Ni alloy contacts in conjunction with photolithographic processing with 1350 photo resist in forming mesa contacts.
REF: Ibid.

INP-0035
ETCH NAME:
TYPE: Acid, grooving
COMPOSITION:
3 .... HCl
1 .... H_3PO_4

TIME:
TEMP:

DISCUSSION:
InP, (100) wafers used as substrates for LPE deposition of InGaAsP. Solution was used to etch grooves in the substrate in ⟨110⟩ directions with groove side slopes $(\overline{111})$B. InGaAsP was deposited in the grooves. After deposition wafer was cleaved ⟨110⟩ and stained to develop layer structure with: 12 g KOH:8 g $K_3Fe(CN)_6$:100 ml H_2O. Surfaces were then studied by SEM.
REF: Ushijima, I et al — *J Cryst Growth,* 69,161(1984)

INP-0021c
ETCH NAME: H etch
TYPE: Acid, preferential
COMPOSITION:
x .... H_3PO_4
x .... HBr

TIME: 2 min
TEMP: RT

DISCUSSION:
InP, (100) wafers used as substrates for epitaxy deposition of GaInAs. Solution used to develop defect structures. See INP-0051.
REF: Ibid.

INP-0031b
ETCH NAME: BPK-221
TYPE: Acid, preferential
COMPOSITION:
2 .... HBr
2 .... H_3PO_4
1 .... *IN $K_2Cr_3O_7$

TIME:
TEMP: RT

*14.7 g/100 cm³ H_2O

DISCUSSION:
InP, (100) wafers used as substrates for epitaxy growth of InGaAsP. Etch rate of epitaxy

thin film was 2.5 μm/min forming a good mesa structure, and etches InP at a similar rate.
REF: Ibid.

INP-0036a
ETCH NAME: Potassium ferricyanide TIME:
TYPE: Acid, stain TEMP:
COMPOSITION:
x .... $K_3Fe(CN)_6$
DISCUSSION:
InP, (100) wafer used as a substrate for epitaxy deposition of InGaAsP. Wafer was cleaved on ⟨110⟩ cross sectioned to observe layer structure. InP substrate was etched back with indium metal prior to epitaxy.
REF: Berman, M A et al — *J Cryst Growth,* 66,480(1984)

INP-0037a
ETCH NAME: TIME: 1—2 min
TYPE: Acid, dislocation TEMP: RT
COMPOSITION:
5 .... HF
1 .... MeOH
DISCUSSION:
InP ingot cut longitudinally on the (112) axis. Ingot etched in a study of growth morphology and defects.
REF: Tohno, S-I et al — *Jpn Appl Phys,* 23,L72(1984)

INP-0037b
ETCH NAME: TIME: 1—2 min
TYPE: Acid, dislocation TEMP: RT
COMPOSITION:
5 .... HF
1 .... Br_2
DISCUSSION:
InP ingot grown by LEC doped with both Ga and Sb. The ingot was cut lengthwise on (112) and etched to observe dislocation pattern.
REF: Ibid.

INP-0072b
ETCH NAME: TIME:
TYPE: Acid, preferential TEMP:
COMPOSITION:
1 .... Br_2
2 .... H_3PO_4
DISCUSSION:
InP, (100) wafers. Solution used to develop defects and structure.
REF: Ibid.

INP-0033d
ETCH NAME: TIME: 10 min
TYPE: Acid, preferential TEMP:
COMPOSITION: RATE: 1.5 μm/h
8 g .... KOH
0.5 g .... $K_3Fe(CN)_6$
100 ml .... H_2O
+ .... light
DISCUSSION:
InP, (100) wafers used as substrates for epitaxy growth of InGaAsP with a cover layer of InP. The InP cover layer was selectively removed with $1HCl{:}1H_3PO_4$. The InGaAsP layer was etched in the solution shown under light using a 15 W tungsten bulb to develop defects. No defects develop without light and show the same etch rate.
REF: Ibid.

INP-0039
ETCH NAME: TIME:
TYPE: Acid, stain TEMP:
COMPOSITION:
12 g .... KOH
8 g .... $K_3Fe(CN)_6$
100 ml .... H_2O
DISCUSSION:
InP, (100) wafers used as substrates for epitaxy growth of double-heterostructures (DH) as follows: InP/InGaAsP/InGaAs/InP(100). After deposition the wafer was cleaved on ⟨110⟩ and stained with the solution shown to develop layer structure and defects.
REF: Matsumoto, Y et al — *J Cryst Growth*, 69,53(1984)

INP-0081b
ETCH NAME: TIME: 1 min
TYPE: Acid, selective/etch-stop TEMP: RT
COMPOSITION:
2 g .... KOH
8 g .... $K_3Fe(CN)_6$
100 ml .... H_2O
DISCUSSION:
InP, (100) wafers used as substrates for epitaxy growth of double-heterostructures (DH) as follows: pInGaAs/pInP/InGaAsP/InP(100). Growth was by LPE. After epitaxy wafers were cleaved ⟨110⟩ and etched in the above solution to develop layer structures. There is a large etch rate difference between InP and InGaAsP. The latter can be used as a buffer layer for selective etching of InP.
REF: Ibid.

INP-0033e
ETCH NAME: TIME:
TYPE: Acid, selective TEMP:
COMPOSITION:
8 g .... KOH
12 g .... $K_3Fe(CN)_6$
100 ml .... H_2O

DISCUSSION:

InP, (100) wafers used as substrates for epitaxy growth of InGaAsP. The etch solution shown was used as a selective etch to remove only the InGaAsP layer.

REF: Ibid.

INP-0020d

ETCH NAME: Sulfuric acid — TIME:

TYPE: Acid, removal — TEMP: Hot

COMPOSITION:

(1) x .... H_2SO_4, conc.

(2) 2 .... H_2SO_4
1 .... H_2O

DISCUSSION:

InP, (111) wafers used in an etch development study. Both solutions will attack the material when used hot.

REF: Ibid.

INP-0056b: Ibid.

InP, (100) and other orientations. Wafers used in a general study of material properties. Hot, dilute H_2SO_4 used as a polish and removal etch.

INP-0041

ETCH NAME: — TIME:

TYPE: Acid, removal — TEMP: RT

COMPOSITION:

1 .... H_2SO_4
1 .... H_2O_2
1 .... H_2O

DISCUSSION:

InP, (100) wafers used as substrates for epitaxy growth of an Ag/p-InP/p-InGaAs structure. InP is selectively removed with HCl. InGaAsP, part of original deposit structure, is selectively removed with the etch shown.

REF: Takeda, Y et al — *Jpn J Appl Phys,* 23,1341(1985)

INP-0042

ETCH NAME: — TIME:

TYPE: Acid, removal — TEMP:

COMPOSITION:

3 .... H_2SO_4
1 .... H_2O_2
1 .... H_2O

DISCUSSION:

InP:Fe (100) (SI) wafers used as substrates for MISFETT and EMISFET device fabrication. Solution used to etch gate channel for the following structures: (1) Al_2O_3/Al/n-InP:Fe(100), and (2) Al_2O_3/nat.ox,/n-InP/Si(100) substrate. Anodize the aluminum to form the Al_2O_3.

REF: Gordon, R C — *Thin Solid Films,* 103,107(1983)

INP-0063: Kowalsky, W et al — *Solid State Electron,* 27,187(1984)

InP, (100) undoped wafers used to fabricate bar type Gunn diodes. Solution used as a selective etch in forming InP for a metallized structure of Au/Ti/AuGe/InP(100).

INP-0065: Uekusa, S et al — *J Electrochem Soc,* 132,671(1985)

InP:Fe, (100) (SI) wafers used in a selective etching study. Base solution was 12 *N* HCl:15 *N* H_3PO_4, and was varied in pH value with 2—25% HCl. AZ-1350 photo resist was used to pattern wafer, and etch solution was used to step-etch for etch rates of the different mixtures. All etching was done at 23°C (effectively RT) in a glass beaker with stirring. Wafers were face down in solutions with frequent agitation during the etch periods.

INP-0033e
ETCH NAME: TIME:
TYPE: Acid, selective TEMP:
COMPOSITION:
1 .... H_2SO_4
8 .... H_2O_2
1 .... H_2O
DISCUSSION:

InP, (100) wafer used as substrate for epitaxy deposition of GaInAs or GaInAsP thin films. First, polish substrate in Br_2:MeOH. After epitaxy use solution shown to selectively etch both epitaxy layers.

REF: Lyons, M H et al — *J Cryst Growth,* 66,269(1984)

INP-0033c
ETCH NAME: TIME:
TYPE: Acid, removal TEMP:
COMPOSITION:
15 .... H_2SO_4
2 .... H_2O_2
2 .... H_2O
DISCUSSION:

InP, (100) wafers used as substrates for epitaxy growth of GaInAs. Solution used to etch remove the epitaxy deposit.

REF: Ibid.

INP-0044
ETCH NAME: TIME:
TYPE: Acid, selective TEMP:
COMPOSITION:
10 .... H_2SO_4
1 .... H_2O_2
1 .... H_2O
DISCUSSION:

InP, (100) n-type wafers. Deposit SiO_2 as an etch mask. Photo resist coat and open strips parallel to ⟨110⟩ directions and remove SiO_2 in strip areas with BHF. Use solution shown to etch grooves in InP. Groove side slopes are $(\overline{111})$B.

REF: van der Ziel, J et al — *J Appl Phys,* 57,1759(1985)

INP-0045a
ETCH NAME: Hydrofluoric acid, dilute TIME: 30 sec
TYPE: Acid, native oxide removal TEMP: RT
COMPOSITION:
x .... 8% HF
DISCUSSION:

InP, (100) n-type wafers. After chem/mech polish with 2% Br_2:MeOH, etch with solution shown to remove native oxide.

REF: Yamaguchi, E et al — *J Appl Phys,* 55,3098(1984)

INP-0045b
ETCH NAME: TIME:
TYPE: Halogen, polish TEMP: RT
COMPOSITION:
x .... 2% Br_2
x .... MeOH
DISCUSSION:
InP, (100) n-type wafers. Chem/mech polish in the solution shown and follow with 8% HF to remove native oxide.
REF: Ibid.

INP-0062
ETCH NAME: Argon TIME:
TYPE: Ionized gas, cleaning TEMP:
COMPOSITION: GAS FLOW:
x .... Ar^+ ions PRESSURE:
POWER:
DISCUSSION:
InP, (100) wafers, S doped n-type. Surfaces argon ion RF sputter cleaned in fabricating $p^+ - n$ solar cells with gold surface contact metallization.
REF: Ginley, R A — *Solid State Electron,* 27,137(1984)

INP-0046a
ETCH NAME: TIME:
TYPE: Acid, selective TEMP:
COMPOSITION: TIME: 0.1 μm/min
1 .... H_3PO_4
1 .... H_2O_2
38 .... H_2O
DISCUSSION:
InP:Fe, (100) (SI) wafers used as substrates for VPE growth of InGaAs doped with either Si or Se after deposition by ion implantation, I^2. Cleave on ⟨110⟩ and profile etch with solution shown. Will etch GaAs with InP acting as an etch-stop.
REF: Penna, J et al — *J Appl Phys,* 57,351(1985)

INP-0046b
ETCH NAME: TIME:
TYPE: Acid, selective TEMP:
COMPOSITION: RATE: 0.05 μm/min
1 .... H_3PO_4
1 .... H_2O_2
76 .... H_2O
DISCUSSION:
InP:Fe, (100) (SI) wafers used for VPE growth of InGaAs. After deposition epitaxy layer was ion implanted with either Si or Se. Cleave on ⟨110⟩ and profiled etched with this solution. Solution will etch InGaAs with InP acting as an etch-stop.
REF: Ibid.

INP-0047
ETCH NAME: TIME:
TYPE: Acid, cleaning TEMP:
COMPOSITION:
1 .... H_3PO_4
1 .... H_2O_2
3 .... MeOH
DISCUSSION:

InP:Fe, (100) (SI) wafers used as substrates for epitaxy growth by LPE of In/Sn. Wafers first polished in Br_2:MeOH. Immediately before epitaxy etch clean in the solution shown, rinse in MeOH, then in H_2O, and N_2 blow dry.

REF: Chin, B H — *J Electrochem Soc,* 131,1372(1984)

INP-0048
ETCH NAME: TIME: Dip
TYPE: Alkali, cleaning TEMP: RT
COMPOSITION:
2.5 g .... KOH
200 ml .... MeOH
DISCUSSION:

InP:Fe, (100) (SI) n-type wafers. Clean wafers: (1) degrease in TCE, then rinse in MeOH, acetone, then MeOH; (2) dip etch in solution shown; (3) rinse in MeOH, then isopropyl alcohol (IPA) and N_2 blow dry before SiO_2 deposition as a thin surface film mask.

INP-0021d
ETCH NAME: A/B Etch, modified TIME: 30 min
TYPE: Acid, preferential TEMP: 75°C
COMPOSITION:
8 ml .... HF
5 g .... CrO_3
40 g .... $AgNO_3$
10 ml .... H_2O
DISCUSSION:

InP, (100) wafers used as substrates for epitaxy growth of GaInPAs. Solution used to develop dislocations and defects in a study of the epitaxy deposit.

REF: Thiel, F A & Barns, R L — *J Appl Phys,* 126,1272(1979)
INP-0061b: Rokanoar, P J et al — *J Cryst Growth,* 66,317(1984)

InP, (100) and (111) wafers cut from LEC grown ingots either sulfur or zinc doped. Zn-doped wafers were etched in solution shown at RT, 2 min to develop hillocks.

INP-0081c
ETCH NAME: Indium TIME:
TYPE: Metal, etch-back TEMP: >155°C
COMPOSITION:
x .... In
DISCUSSION:

InP, (100) wafers used for epitaxy growth of InGaAs/InGaAsP. During LPE growth, liquid indium was used to lightly etch-back InP prior to epitaxy deposition to clean surface and reduce surface defects.

REF: Ibid.
INP-0036b: Ibid.

InP, (100) wafers used as substrates for epitaxy growth of InGaAsP. Liquid indium used to etch-back InP during LPE growth to reduce defects in surface.

INP-0051
ETCH NAME: H etch
TYPE: Acid, dislocation
COMPOSITION:
2 .... H_2SO_4
1 .... 4% HBr

TIME: (1) 10—30 sec (2) 2 min
TEMP: RT RT

DISCUSSION:

InP, (100) wafers used in a dislocation study. A study of the "grappe" structure as an impurity at a dislocation. Study was done under a polarized infra-red microscope (PIM) by transmission. Solution used to develop dislocations. Boric acid an impurity?

REF: Stirland, D J et al — *J Cryst Growth,* 61,645(1983)
INP-0061a: Rokanoar, P J et al — *J Cryst Growth,* 66,317(1984)

InP, (100) and (111) wafers cut from LEC grown ingots and either zinc or sulfur doped. Wafer used in a defect study. S-doped wafers developed striations with solution shown within 20 sec at RT, in addition to etch pits.

INP-0062: Fang, D F et al — *J Cryst Growth,* 66,317(1984)

InP, (100) and (111) wafers used in a defect study. Solution used to develop dislocation pits.

INP-0052
ETCH NAME: Orthophosphoric acid
TYPE: Electrolytic, oxidize
COMPOSITION:
x .... 14 *N* H_3PO_4

TIME:
TEMP: 4°C to −30°C
ANODE: InP
CATHODE:
POWER: 40 V

DISCUSSION:

InP, (100) wafers used in a structural defect study by electrolytic etching. Etch polish with Br_2:MeOH before oxidizing with solution shown. With specimen as anode, rotate at 100 rpm. Remove oxide with 2 *N* HCl or 2 *N* HNO_3. For defect development use 0.5 *N* HCl at 400 $\mu A/cm^3$ power.

REF: Elliott, C R & Regnault, J C — *J Electrochem Soc,* 128,113(1981)

INP-0053
ETCH NAME: Hydrochloric acid, dilute
TYPE: Acid, clean
COMPOSITION:
x .... 10% HCl

TIME: 10 min
TEMP: RT

DISCUSSION:

InP, (100) n-type wafers used in a study of anodized island growth. Degrease in MeOH then acetone. Chem/mech polish on pellon pad with 1% Br_2:MeOH. Then etch with solution shown. Surface replication done with 4% collodion in amyl acetate. After drying, lift-off replica with water. (*Note:* This thin film removal technique is usually called float-off, not lift-off.)

REF: Makky, W H & Wilmsen, G W — *J Appl Phys,* 130,659(1983)
INP-0056a: Reynolds, W N et al — *Proc Phys Soc,* 71,416(1958)

InP, (100) wafers and other orientations used in a general properties study. Both dilute HCl (hot) and concentrated HCl (RT) were used as removal and polish solutions. InP will not etch in NaOH or CP4.

INP-0054a
ETCH NAME: Hydrochloric acid, dilute TIME: 30 sec
TYPE: Acid, clean TEMP: RT
COMPOSITION:
x .... 50% HCl
DISCUSSION:
InP:Fe, (100) wafers. Clean wafers: (1) degrease in boiling TCE; (2) rinse in acetone, then IPA; (3) etch in HCl and rinse in IPA and, (4) dry in hot IPA vapors. Etch polish cleaned surfaces with 0.5% Br_2:MeOH.
REF: Rhee, J K & Bhattachairya, T — *J Appl Phys,* 130,700(1983)

INP-0059a
ETCH NAME: Carbon tetrafluoride TIME:
TYPE: Ionized gas, removal TEMP:
COMPOSITION: GAS FLOW:
x .... CF_4 PRESSURE:
POWER:
DISCUSSION:
InP, (100) wafers used as substrates for material removal rate study using CF_4 under varying flow conditions. Wafers were first chem/mech polished with 2.5% Br_2:MeOH to remove prior lapping damage before photo resist patterning with AZ-1350 lacquer.
REF: Hirata, K et al — *J Vac Sci Technol,* B(2),44(1984)

INP-0060
ETCH NAME: Heat TIME:
TYPE: Thermal, cleaning TEMP: 500°C
COMPOSITION:
x .... heat, in vacuum
DISCUSSION:
InP, (100) wafers used as substrates for MBE of deposition of $Ba_xSr_{1-x}Fe_2$. InP was n-type, S-doped and polished. Clean in vacuum under phosphorus over-pressure prior to MBE growth.
REF: Tu, C W et al — *J Vac Sci Technol,* B(2),24(1984)

INP-0055
ETCH NAME: TIME: 10 min
TYPE: Acid, preferential TEMP: RT
COMPOSITION:
x .... HCl
x .... 0.4 *N* $FeCl_3$ (for ferric ions)
DISCUSSION:
InP, (111) wafers. Solution produces etch figures.
REF: Harper, C A — *Handbook of Materials and Processes for Electronics,* McGraw-Hill Book Co., NY, 1970, 7
INP-0058: Gatos, H C & Lavine, M C — *J Electrochem Soc,* 107,427(1960)
InP, (111) wafers used in a surface characterization study.

INP-0057a
ETCH NAME: TIME:
TYPE: Acid, selective TEMP:
COMPOSITION:
1 .... HCl
1 .... H_3PO_4

DISCUSSION:

InP, (100) tin-doped, n-type wafer used for an InP:Zn, epitaxy thin film deposit as a layer in a buried heterostructure. The solution shown was used to selective etch structure the zinc doped InP layer. Deposition was by Liquid Phase Epitaxy (LPE). Liquid indium used to etch-back surface during film growth. Device structure was InP:Zn/$In_{1-x}Ga_xAs_y:P_{1-y}$/InP:Sn(100). Solution shown used as a mesa etch, and develops (100) facets on mesa walls; 1% Br_2:MeOH develops a cylindrical mesa wall.

REF: DiGiuseppi, M A et al — *J Cryst Growth,* 67,1(1984)

INP-0057b

ETCH NAME: TIME:

TYPE: Halogen, mesa structure TEMP:

COMPOSITION:

x .... 1% Br_2

x .... MeOH

DISCUSSION:

InP:Zn, epitaxy film grown by LPE. Solution used to etch form mesa structures. See INP-0057a for further discussion.

REF: Ibid.

INP-0062a

ETCH NAME: TIME: 1 min

TYPE: Acid, selective TEMP: 20°C

COMPOSITION:

(1)	(2)	(3)	(4)
1 .... HBr	1 .... HBr	1 .... HBr	1 .... HBr
1 .... H_3PO_4	1 .... CH_3COOH (HAc)		1 .... HCl

DISCUSSION:

InP, (100) wafers used as substrates for the growth of double heterostructure (DH) by LPE. Substrates were (1) degreased; (2) DI water rinsed and (3) polished in $Br_2:CH_3OH$. Substrates were n-type; epitaxy InP n-type and InGaAsP both n- and p-type. Heterostructure used in a selective etching study of InP vs. InGaAsP. The solutions shown will etch InP but not InGaAsP. Specimens cleaved ⟨110⟩ for measurement. SiO_2 was used as a patterned mask as bromine solutions attack photo resist.

REF: Adachi, S et al — *J Electrochem Soc,* 129,1053(1982)

INP-0062b

ETCH NAME: TIME: 1 min

TYPE: Acid, selective TEMP: 20°C

COMPOSITION:

(1)	(2)	(3)	(4)	(5)
1 .... HBr	1 .... HBr	1 .... HBr	1 .... HBr	2 .... HBr
1 .... H_2O_2	1 .... HNO_3	1 .... HNO_3	2 .... HCl	1 .... HCl
		5 .... H_2O		

DISCUSSION:

InP, (100) wafers used as substrates for LPE growth of double-heterostructures (DH) of InGaAsP:InP in a study of etch selectivity. All of the solutions shown were selective with a ratio of about 10:1 between InGaAsP to InP. See INP-0062a for additional discussion.

REF: Ibid.

INP-0062c
ETCH NAME: BRM
TYPE: Halogen, selective
COMPOSITION:

(1) x .... 4% Br_2 / x .... MeOH
(2) x .... 2% Br_2 / x .... MeOH
(3) x .... 1% Br_2 / x .... MeOH
(4) x .... 0.2% Br_2 / x .... MeOH
(5) x .... 0.1% B_2 / x .... MeOH

TIME: 1 min
TEMP: 20°C

DISCUSSION:
InP, (100) wafers used for LPE growth of double heterostructures (DH) of InGaAsP:InP in a study of etch selectivity. All of the solutions above show a selectivity ratio of about 10:1 between InGaAsP to InP. See INP-0062a for additional information.
REF: Ibid.

INP-0062d
ETCH NAME: Hydrochloric acid
TYPE: Acid, selective
COMPOSITION:
x .... HCl, conc.

TIME: 1 min
TEMP: 20°C

DISCUSSION:
InP, (100) wafers used as substrates for LPE of InGaAsP. Solution will etch InP but not InGaAsP. See INP-0062a for additional discussion.
REF: Ibid.

INP-0062e
ETCH NAME:
TYPE: Acid, selective
COMPOSITION:

(1) 1 .... HCl / 1 .... H_2O_2
(2) 1 .... HCl / 1 .... HAc
(3) 1 .... HCl / 1 .... H_3PO_4
(4) 1 .... HCl / 1 .... H_2O

TIME: 1 min
TEMP: 20°C

DISCUSSION:
InP, (100) wafers used as substrates for LPE growth of InGaAsP. Solutions shown will etch InP but not InGaAsP. See INP-0062a for additional discussion.
REF: Ibid.

INP-0062f
ETCH NAME:
TYPE: Acid, selective
COMPOSITION:

(1) 1 .... HCl / 1 .... CH_3COOH (HAc) / 1 .... H_2O_2
(2) 1 .... HCl / 1 .... H_3PO_4 / 1 .... H_2O_2

TIME: 1 min
TEMP: 20°C

DISCUSSION:
InP, (100) wafers used as substrates for LPE growth of InGaAsP. Solutions shown will etch selectively with a ratio of about 10:1 between InGaAsP and InP. See INP-0062a for additional discussion.
REF: Ibid.

INP-0062g
ETCH NAME: TIME: 1 min
TYPE: Acid, selective TEMP: 20°C
COMPOSITION:

(1) 1 .... HCl
1 .... HNO_3

(2) 1 .... HCl
2 .... HNO_3

(3) 2 .... HCl
1 .... HNO_3

DISCUSSION:

InP, (100) wafers used as substrates for LPE growth of InGaAsP. Solutions shown will etch selectively with a ratio of about 10:1 between InGaAsP and InP. See INP-0062a for additional discussion

REF: Ibid.

INP-0062h
ETCH NAME: TIME: 1 min
TYPE: Acid, selective TEMP: 20°C
COMPOSITION:

(1) 1 .... H_2SO_4
1 .... H_2O_2
1 .... H_2O

(2) 1 .... H_2SO_4
2 .... HCl
1 .... 2 *N* $K_2Cr_2O_7$

DISCUSSION:

InP, (100) wafers used as substrates for LPE growth of InGaAsP. Solution (1) will etch InGaAsP but not InP. Solution (2) is selective with a ratio of about 10:1 between InGaAsP to InP. See INP-0062a for additional discussion.

REF: Ibid.

INP-0063
ETCH NAME: TIME:
TYPE: Acid, selective TEMP: RT
COMPOSITION:

1 .... HF
1 .... HNO_3

DISCUSSION:

InP,, (111)A and (100) wafers used as substrates for LPE of $Al_xIn_{x-1}As$ and $Al_xGa_yIn_{1-x-y}$ As. The (111)A surface produces a smooth epitaxy AlInAs layer grown at 790°C, but shows defects with (100). Selective removal of AlInP from (100) with solution shown. Many small holes observed in (100) surface after film was removed. Also grown as heterostructures.

REF: Nakajima, K et al — *J Electrochem Soc,* 130,1927(1983)

INP-0064a
ETCH NAME: TIME: (1) 15 (2) 7 (3) 10 sec
TYPE: Acid, selective TEMP: RT
COMPOSITION:

(1) 1 .... HCl
3 .... HNO_3

(2) 6 g .... $K_3Fe(CN)_6$
4 g .... KOH
70 ml .... H_2O

(3) x .... HCl, conc.

DISCUSSION:

InP, (100) wafers used as substrates for epitaxy growth of GaInAsP double heterostructure (DH) used in a study of the controlled development of planar-etched bulk crystal planes with wet chemical etching (WCE) or dry gas etching (DCE) using RIE and combinations of WCE and RIE in channel structuring surfaces. Photo resist patterned deposited surface thin films of Si_3N_4 and Ti/Si_3N_4 were used as the etch masks. Wafers were cleaved

⟨011⟩. The step etching system shown above was used to develop a near-vertical wall channel with flat bottom down through the active layer to the Q-stop etch layer.
REF: Coldren, L A et al — *J Electrochem Soc,* 130,1918(1983)

INP-0064b
ETCH NAME: BRM — TIME: See discussion
TYPE: Halogen, selective — TEMP: RT
COMPOSITION:
x .... 3% Br_2
x .... MeOH
DISCUSSION:
InP, (100) wafers theta angles controlled toward (011) or $(0\bar{1}1)$. Solution shown used to channel etch InP from the (100) surface to the $(\bar{1}00)$ surface to establish the internal bulk planes etch rates. The approximate etch rates in microns/minute (μm/min) were measured with regard to these surface and relevant bulk planes.
REF: Ibid.

INP-0064c
ETCH NAME: Hydrochloric acid — TIME:
TYPE: Acid, selective — TEMP: RT
COMPOSITION:
x .... HCl, conc.
DISCUSSION:
InP, (100) wafers used in a channel etching study. See INP-0064b for additional discussion. This solution produced different theta angle results.
REF: Ibid.

INP-0064d
ETCH NAME: KKI — TIME:
TYPE: Acid, selective — TEMP: RT
COMPOSITION:
1 g .... $K_3Fe(CN)_6$
1 g .... KOH
1 ml .... H_2O
DISCUSSION:
InP, (100) wafers used in a channel etching study. See INP-0064a-c for additional information.
REF: Ibid.

INP-0064e
ETCH NAME: Hydrochloric acid — TIME: 90 sec
TYPE: Acid, selective — TEMP: RT
COMPOSITION:
x .... HCl, conc.
DISCUSSION:
In, (100) wafers with channels in ⟨011⟩ or $\langle 01\bar{1}\rangle$ directions using Si_3N_4 as the etch mask. Cross section of the pit slope in the ⟨011⟩ direction shows an initially vertical (011) slope changing to a (111) positive slope. Both channel types had flat bottoms.
REF: Ibid.

INP-0064f
ETCH NAME: BRM
TYPE: Halogen, selective
COMPOSITION:
x .... 3% Br_2
x .... MeOH
TIME: 120 sec
TEMP: RT
DISCUSSION:

InP, (100) wafers with channels in ⟨011⟩ and ⟨01$\bar{1}$⟩ directions using an Si_3N_4 evaporated film with parallel stripes of 10 μm spacing as etch masks. The (011) orientation developed a shallow cup-like channel; the (01$\bar{1}$) a "V" groove with slight curvature in pit bottom.
REF: Ibid.

INP-0064g
ETCH NAME:
TYPE: Acid, selective
COMPOSITION:
(1) x .... HCl, conc.
(2) x .... Cl_2:O_2 (RIE)
TIME: 10 sec
TEMP: RT
DISCUSSION:

InP, (100) wafers with channels in ⟨011⟩ and ⟨01$\bar{1}$⟩ directions. A titanium etch mask was used for RIE. Results show severe under-cutting of masks with erratic etching results. On ⟨011⟩, reentrant angle side walls with low curvature channel bottom; ⟨011⟩, the upper section was a severely under-cut, shallow positive curve slope with about a 1 μm width deep channel in the center, ending with a point.
REF: Ibid.

INP-0064h
ETCH NAME: RIE
TYPE: Ionized gas, selective
COMPOSITION:
x .... 80% Cl_2
x .... 20% O_2
TIME: 20 min
TEMP: (1) 250°C
(2) 30°C substrate
GAS FLOW:
PRESSURE: (1) 2 mT
(2) 4 mT
POWER: 0.33 W/cm^2
DISCUSSION:

InP, (100) wafers with channels in ⟨011⟩ and ⟨01$\bar{1}$⟩ directions. Ti/Si_3N_4 used as the preferred etching mask. Both RIE channels showed similar near-vertical side slopes with flat bottoms. A post HCl, 10 sec etch showed little or no alteration of the slope side configuration of either the (1) or (2) temperature and time conditions.
REF: Ibid.

INP-0064i
ETCH NAME: RIE
TYPE: Ionized gas, preferential
COMPOSITION:
x .... 60% Cl_2
x .... 30% O_2
x .... 10% Ar
TIME: 150 min
TIME: 25°C
GAS FLOW:
PRESSURE: 2 mT
POWER: 0.33 W/cm^2
DISCUSSION:

InP, (100) wafers with channels in ⟨011⟩ direction. Both Si_3N_4 for chemical etching and Ti/Si_3N_4 for RIE were evaporated as etch masks. RIE channel slope was a reentrant angle with low curvature channel bottom. Additional etching with HCl, 10 sec, changed slope to

an upper (111) vertical and medium angle positive lower slope with striations. Specimens were at a 40° angle during RIE.
REF: Ibid.

INP-0064j
ETCH NAME: TIME: (1) 30 (2) 60 (3) 20 sec
TYPE: Acid, selective TEMP: RT
COMPOSITION:

(1) x .... HCl, conc. (2) 6 g .... $K_3Fe(CN)_6$ (3) x .... HCl, conc.
4 g .... KOH
35 ml .... H_2O

DISCUSSION:

InP, (100) wafers with channel in ⟨011⟩ direction. Si_3N_4 used as the etch mask. WCE multi-step etching with or without prior RIE etching. Channel slope was vertical and lower slope positive section similar to HCl after RIE results obtained in INP-0064i. About a 1 μm under-cut of mask against RIE/HCl about 0.5 μm/min.
REF: Ibid.

INP-0064k
ETCH NAME: TIME: 20 sec
TYPE: Acid, selective TEMP: RT
COMPOSITION:

(1) 1 .... HCl (2) 1 .... HCl (3) 1 .... HCl
1.5 .... HNO_3 2 .... HNO_3 2.5 .... HNO_3

DISCUSSION:

InP, (100) wafers used as substrates for heterostructure of the following configuration: p-GaInAsP/p-InP/i-GaInAs/n-InP (100). Si_3N_4 used as the etching mask with channel (011) oriented. Channel side slopes were: (1) a double near-vertical slope and shallow curved channel bottom; (2) an upper "cliff-like" upper over-hang beneath etch undercut of mask, a vertical slope with a curved positive slope into the bottom channel; (3) upper vertical slope graduating into a positive slope into channel bottom. Striations on vertical slopes. Used with negative photo resist patterning.
REF: Ibid.

INP-0064l
ETCH NAME: TIME: (1) 20 (2) 10 sec
TYPE: Acid, selective TEMP: RT
COMPOSITION:

(1) 1 .... HCl (2) x .... HCl, conc.
1.5 .... HNO_3

DISCUSSION:

InP, (100) wafers used as substrates as described in INP-0064k heterostructure epitaxy. As HCl will not etch GaInP, initial etching is done with solution (1) to establish channel slide slopes, followed by (2) to finalize the InP channel slope and bottom.
REF: Ibid.

INP-0064m

ETCH NAME: Multi-step etch TIME:

TYPE: Acid, selective TEMP:

COMPOSITION:

Step 1: (1) x .... HCl (2) x .... HCl (3) x .... x% Br_2 (4) x .... RIE
x .... HNO_3 x .... HAc x .... MeOH
x .... H_2O_2

Step 2: (1) x .... $K_3Fe(CN)_6$ or x .... H_2SO_4 (3) x .... HNO_3,
x .... KOH x .... H_2O_2 conc.
x .... H_2O x .... H_2O

Step 3: (1) x .... HCl, conc.

Step 4: (i) Repeat 1—3 (ii) repeat 1 & 3 (iii) repeat 2 & 3

DISCUSSION:

InP, (100) wafers used as substrates for DH InGaAsP epitaxy laser channel structure in (001) direction. Any of the Steps (1) will etch all layers and is used to establish initial side-wall facets at near-vertical [RIE also on ⟨011⟩]. Step (1) solutions used as selective Q-layer etchant to recess layers approximately 1000 Å. Step 2 solutions used to planarize the (001) InP. Step 3 solution combinations to improve and finalize facet surfaces for optimum planarity. See other INP-0064a-1 references for specific solution mixtures.

REF: Ibid.

INP-0064n

ETCH NAME: Sequential etch TIME:

TYPE: Acid, selective TEMP:

COMPOSITION:

Step 1: (1) x .... $K_3Fe(CN)_6$ (2) x .... H_2SO_4 (3) x .... HNO_3 (4) x .... RIE
x .... KOH x .... H_2O_2 nonselective no undercut
x .... H_2O x .... H_2O
cut Q cap selective

Step 2: (5) x .... HCl (6) x .... x% Br_2 (7) x .... RIE
x .... HNO_3 perpendicular

Step 3: (8) x .... HCl, conc.
cut to Q-active layer

Step 4: (9) x .... $K_3Fe(CN)_6$ (10) x .... H_2SO_4 (11) x .... HNO_3
x .... KOH x .... H_2O_2
x .... H_2O x .... H_2O

Step 5: (12) x .... HCl (13) x .... RIE
x .... HNO_3 5° angle

Step 6: (14) x .... $K_3Fe(CN)_6$ (15) x .... H_2SO_4 (16) x .... HNO_3
x .... KOH x .... H_2O_2
x ..... H_2O x .... H_2O
optional to recess Q after Step 3, (9), (10) or (11)

Step 5: (17) x .... HCl, conc.
etch under Q-layer/planarize (011).

Step 6: (18) repeat 3 thru 5 (19) repeat 3 and 5
planarize facets — optional

DISCUSSION:

InP, (100) wafers used as substrates for DH InGaAsP epitaxy laser channel structure in (011) direction. Basic etching sequence follows: Step 1 — any one of solutions shown; Step 2 — HCl; Step 3 — any one of solutions shown and Step 5 — HCl. Steps 4 and 6, used in order shown, as required. See other INP-0064a-l references for specific solution mixtures.

REF: Ibid.

INP-0065
ETCH NAME: Hydrochloric acid — TIME: 20 sec
TYPE: Acid, orientation — TEMP: 20°C
COMPOSITION:
x .... HCl, conc.
DISCUSSION:

InP, (100) wafers with or without thin film InGaAsP epitaxy. The orientation direction of the (100) work surface is important for selective etching of channels and pits related to ⟨011⟩ or ⟨01$\bar{1}$⟩ bulk plane direction; the ⟨011⟩ preferred. Cleave a small section of the wafer and etch in HCl to develop (111)A or ($\overline{111}$)B triangle etch pits in the cleaved surface. If the triangle points away from the upper (100) work surface, the cleavage plane is ⟨011⟩ and a mesa structure will be similar. If the triangle point is toward the (100) work surface, the cleave plane is ⟨01$\bar{1}$⟩ and can show fine-textured, blackish-colored areas after HCl etching.

REF: Stulz, L W & Coldren, I A — *J Electrochem Soc,* 130,1628(1983)
INP-0071: Sagar, A — *Phys Rev,* 117,101(1959)

InP, (100) wafers used in a study of piezoresistance. Concentrated HCl used as a general cleaning etch. An ultrasonic soldering iron was used to activate piezoresistance measurements.

INP-0067
ETCH NAME: — TIME:
TYPE: Acid, clean/etch sequence — TEMP:
COMPOSITION:
(1) x .... DI H_2O (2) x .... TCE (3) x .... H_2SO_4 (4) x .... MeOH (5) x .... 0.3% Br_2
x .... MeOH
DISCUSSION:

InP, (100) wafers used in a study of surface contamination and cleaning in a VPE reactor. Specimens were processed through the sequence shown. (1) rinse several times in DI water; (2) degrease (not shown); (3) ultrasonic clean in sulfuric acid; (4) soak in MeOH and (5) final etch in BRM. Specimens were then thermally annealed at 700°C to include vapor etching. (*Note:* Hot HCl vapor?)

REF: Gautard, B et al — *J Cryst Growth,* 71,125(1985)

INP-0070
ETCH NAME: — TIME:
TYPE: Acid, polish — TEMP:
COMPOSITION:
1 .... HCl
1 .... HNO_3
DISCUSSION:

InP, (111) wafers used in a study of room temperature reflectivity of (111)A and ($\overline{111}$)B surfaces. If cutting and mechanical lapping damage is not removed, reflectivity is sharply reduced and shifted toward lower energy levels. Solution shown used to polish remove subsurface residual damage.

REF: Cardona, M — *J Appl Phys,* 32,958(1961)

INP-0076a
ETCH NAME: BRM — TIME:
TYPE: Halogen, polish — TEMP: RT or −7 to −18°C
COMPOSITION:
x .... 1.5% Br_2
x .... MeOH

DISCUSSION:

InP, (100) wafer fabricated as Schottky diodes. Solution used to polish wafers prior to Cd diffusion. Etch rate increases by adding drops of the solution during etch period. Better polishing results are obtained using the etchant at the coldest temperature shown. Cleaning sequence was (1) degrease in TCE, then rinse in MeOH, then in acetone, all at 85°C for 10 min, each; (2) boil in 100 g KOH:500 ml H_2O, 80°C, 10 min; (3) DI water rinse several times until no soapy feeling is left.

REF: Aytal, S & Schlachetzki, A — *Solid State Electron,* 25,1135(1982)

INP-0076b

ETCH NAME: TIME: 10 min

TYPE: Halogen, junction stain TEMP: RT

COMPOSITION:

x .... x% Br_2

x .... isopropyl alcohol

x .... acetone

DISCUSSION:

InP, (100) wafers fabricated as Schottky diodes with cadmium diffusion. Solution used as a junction stain with and without acetone. Time is shown with acetone and was evaluated from 1 to 25 min. Also used as a polishing etch. Without acetone is best as a staining etch.

REF: Ibid.

INP-0073a

ETCH NAME: Hydrochloric acid TIME: 1—2 min

TYPE: Acid, polish TEMP: RT

COMPOSITION:

(1) x .... 12 *M* HCl then (2) 1 .... 1 *M* NaOH

1 .... 1 *M* S

1 .... 2.5 *M* NaS_2

DISCUSSION:

InP, p-type single crystal wafers in a study of surface modification of photo-cathodes with polysulfides. Etch polish in solution shown as (1), then modify surface with polysulfide solution (2). InP wafers were Fe and Cr doped with Be implantation and annealed with an Si_3N_4 cap. After anneal, nitride was removed with HF.

REF: Williams, R et al — *J Electrochem Soc,* 129,2082(1982)

INP-0087

ETCH NAME: TIME:

TYPE: Halogen, polish TEMP: RT

COMPOSITION:

x .... x% I_2

x .... MeOH

DISCUSSION:

InP, (100) wafers used in a study of the optical properties of anodic films. Polish substrates in solution shown for strain-free surface. Oxidize in ''AGW'' solution: 0.5% H_3PO_4:2 ethylene glycol, electrolytically at 1 mA/cm^2 + white light with Ta wire and Pt counter-electrodes; rinse in MeOH and N_2 dry. Strip oxide with NH_4OH:H_2O. Oxide films are hygroscopic — store with zeolite, $CaSO_4$, in closed container for a dry atmosphere.

REF: Studna, A A & Gualtieri, G L — *J Appl Phys Lett,* 39m,965(1981)

INP-0088
ETCH NAME:
TYPE: Alkali, oxide removal
COMPOSITION:
(1) x .... 45% KOH or (2) x .... 29% NH_4OH
TIME:
TEMP:
DISCUSSION:

InP, (100) n-type wafers used in a study of electrical properties of dielectric thin films. Clean InP: (1) BRM polish; (2) remove oxide with either solution shown; (3) DI water rinse; (4) N_2 blow dry with filtered N_2 . . . then: etch in 1% Br_2:MeOH; MeOH rinse; DI H_2O rinse; N_2 dry . . . then etch in 1HF:1HC1:4H_2O:0.4% H_2O_2; DI water rinse; N_2 dry . . . then wash in 95% hydrazine; rinse in hexane, and N_2 dry. RF sputter deposit SiO_2 from SiH_4/N_2 and Si_3N_4 from NH_3 at 300°C with a growth rate of 40—50 Å/min.

REF: Meiners, L G — *J Vac Sci Technol,* 19,373(1981)

INP-0074
ETCH NAME: BRM
TYPE: Halogen, grooving
COMPOSITION:
x .... 2% Br_2
x .... MeOH
+ .... wooden stick
TIME: 30 sec soak
TEMP: RT
DISCUSSION:

InP, (100) wafer substrates with InGaP multilayer deposited structure. The following method used to develop damage-free grooves to reveal the quality of the InP/InGaP structures: soak a wooden stick in the solution shown — soak can be repeated — then rub specimen surface to form groove and MeOH rinse. For a 10 μm deep groove, rub 10—30 sec at a time until structure is delineated . . . pressure and stroke control etching depth. Also used a "CJB Etch" and an "HL Etch".

REF: Goods, R L — *J Electrochem Soc,* 129,2082(1982)
INP-0075: Sng Chu, A et al — *J Electrochem Soc,* 129,352(1982)
Reference for CBJ etch shown above.
INP:0072: Huber, A & Linn, N T — *J Cryst Growth,* 28,80(1975)
Reference for HL etch shown above. This INP-0072 reference is also called the Huber etch . . . See under INP-0014.

INP-0092b
ETCH NAME: AGW
TYPE: Electrolytic, oxidizing
COMPOSITION:
1 .... 0.5% H_3PO_4
x .... EG (ethylene glycol)
x .... H_2O
TIME:
TEMP:
ANODE: InP
CATHODE: Pt/Ti wire
POWER: 0.1 mA/cm^2
DISCUSSION:

InP, (100) wafers. Solution used to anodically oxidize surfaces in a study of optical properties and water adsorption of the oxide. Ti wire (self-oxidizing) + white light used during oxide growth, followed by rinsing in MeOH. Oxides were stripped with NH_4OH:H_2O. See INP-0092a. (*Note:* Name acronym (AGW) means acid/glycol/water, and can be different acids and glycols.)

REF: Ibid.

INP-0089
ETCH NAME: BRM
TYPE: Halogen, polish
COMPOSITION:
(1) x .... x% Br_2
x .... MeOH
(2) x .... HF
x .... NH_4F
x .... H_2O
TIME:
TEMP:
DISCUSSION:

InP, (100) wafers cut within 1° of plane from undoped LEC grown ingot. An extensive study of controlled etch pits formed by etching through a patterned SiO_2 mask. This oxide removed with solution (2), and wafers initially chem/mech polished in (1). Over 30 solutions evaluated and pit slope angles evaluated by cleaving ⟨110⟩ and ⟨1$\bar{1}$0⟩ . . . to include concave planes and angles from (100) surface. All etching done in a controlled temperature bath at 25°C, 1 min, except H_3PO_4 solutions used at 60°C up to 20 min. Other solution categories: HCl; H_2SO_4; Br_2; HBr, all with other acid additives.

REF: Adachi, S & Kawaguchi, H — *J Electrochem Soc,* 128,1342(1981)

INP-0090
ETCH NAME:
TYPE: Acid, cleaning
COMPOSITION:
1 .... HF
1 .... HCl
4 .... H_2O
1 drop .... H_2O_2/10 ml
TIME:
TEMP:
DISCUSSION:

InP, (100), n-type, 0.3—0.4 Ω cm resistivity, and p-type, 7—8 Ω cm. Wafers used in a study of ion beam etching. Clean wafers: (1) degrease with TCE; (2) rinse in acetone; (3) rinse in MeOH; (4) HF dip, and (5) etch in solution shown. TaSi thin films were sputter deposited.

REF: Wu, C S et al — *J Electrochem Soc,* 132,918(1985)

PHYSICAL PROPERTIES OF IODINE, I_2

Classification	Halogen
Atomic number	53
Atomic weight	253.84
Melting point (°C)	114
Boiling point (°C)	183 (184.35)
Density (g/cm^3)	4.93
Thermal conductance ($\times 10^{-4}$ cal/cm^2/cm/°C/sec) 20°C	10.4
Coefficient of linear thermal expansion ($\times 10^{-6}$ cm/cm/°C) 20°C	93
Electrical resistivity ($\times 10^9$ ohms cm) 20°C	1.3
Latent heat of fusion (cal/g)	14.2
Specific heat (cal/g/°C) 20°C	0.052
1st ionization potential (eV)	10.44
Solubility (100 ml H_2O) 50°C	0.078 ml
Ionic radius (angstroms)	2.20
Heat of vaporization (cal/g)	39.28
Atomic volume (W/D)	25.7

1st ionization energy (K-cal/g-mole)	241
Electronegativity (Pauling's)	2.5
Covalent radius (angstroms)	1.33
Electrical resistivity ($\times 10^{-6}$ ohm cm) 25°C	5.85
Vapor pressure (mmHg) 25°C	0.31
Critical temperature (°C)	553
Critical pressure (atms)	116
Cross section (barns)	6.4
Refractive index (n=)	3.34
Hardness (Mohs — scratch)	2
Crystal structure (orthorhombic — normal)	(100) a-pinacoid
Color (solid/solution)	Brown-violet/brown
Cleavage (basal)	(001)

IODINE CONTAINING ETCHANTS

Iodine is an oxidizing agent of the halogen family being the heaviest and least reactive of the group. It is the only one that forms as a solid; bromine is a liquid, fluorine and chlorine are gases. Iodine has limited solubility in water but is easily dissolved in alcohols, carbon disulfide, chloroform, and solutions of other iodides, e.g., KI + I_2 goes to KI_3, as a reversible reaction and is an unstable compound known as tri-iodide.

"Tincture of iodine" is a mixture of KI:I_2:EOH, used as a medical antiseptic, and also can be used as a preferential etch on a number of metals and metal compounds. Solutions of iodine in water, alcohols, and iodides are dark brown in color and opaque. In carbon disulfide and chloroform a very distinctive violet color.

Although bromine is more widely used as a bromine:methyl alcohol (BRM) mixture for etch removal, polishing, thinning and defect development on a number of semiconductor materials and other metals, the bromine can be replaced by iodine with similar results. The iodine etchants are more stable and useful where the solutions are used above room temperature as it evaporates less rapidly than bromine.

It should be noted that iodine is used as a disinfectant and germicide; can be introduced into the bloodstream for X-ray evaluation of blood clots and is in the thyroid gland of all mammals in the form of the hormone, thyroxine, which is a metabolic catalyst for carbon dioxide and essential to health. A lack of iodine can produce a goiter and cretinism. Iodized salt is used to help maintain the iodine balance in human beings. Bromine, on the other hand, is a poison if ingested and can cause severe sores on the skin that are difficult to heal, and the vapors can be a severe skin and wet membrane irritant.

The tri-iodide etch for gold is probably the best known of the iodine solutions: KI:I_2:H_2O. It is often preferred over the aqua regia and cyanide etchants, even though it is a brown and opaque solution, as it is more controllable than aqua regia and not toxic like the cyanides. Iodine etchants were in use for several years prior to the advent of the semiconductor and Solid State industries in the mid-1940s, and there are several solutions named "Iodine Etch", "Iodine A", or "Iodine B" etch, with the mixtures varying, depending upon the metal involved even though the "names" may be identical. Although most iodine-containing etchants are probably interchangeable as applied to specific metals and compounds, their reactions can vary such that the term "iodine etch" should be considered as a very general term, not a specific mixture. Iodine etches are used for defect and dislocation development to a large extent though they have been applied as a polish, general removal, passivation, and surface cleaning solution. Also, by varying the molar concentration, they can be used as a selective etch between GaAs and AlGaAs semiconductor material. As an additive to hot hydroxide solutions such etchants are preferential and can be used for light figure orientation of germanium which is unaffected by pure hydroxides. In this case, bromine can replace the iodine and both solutions are applicable to silicon and other compound semiconductors as preferential etches.

The following is a selected list of some of the iodine etchants as applied to different metals and compounds and other solutions can be found under the different metals and compounds in this Etchant Section.

Formula	Material	Use	Ref.
10 g I_2:100 ml MeOH	Al	Channel etch	AL-0031
x10% I_2:xMeOH	As	Dislocation	AS-0003a
x0.3 *M* KI:x0.04 *M* I_2	GaAs	Selective (GaAs vs. AlGaAs)	GAS-0061d
x0.3 *M* KI: x 0.3 *M* I_2	GaAs	Selective (AlGaAs vs. GaAs)	GAS-0061e
250 ml 30% KOH:30 g I_2	Ge, Si	Preferential	GE-0012a
160 ml HF:200 ml HNO_3:80 ml HAc: 0.32 g I_2	Ge	Preferential	GE-0083a
1HF:2HNO_3:2HAc: + I_2 "iodine etch"	Ge	Preferential	GE-0139a
xKI:xI_2:xH_2O	Bi	Polish	BI-0003
1% I_2:MeOH	Bi	Dislocation	BI-0006b
10 ml HNO_3:10 ml HCl:40 ml H_2O:1 g I_2	Bi_2Te_3	Preferential	BITE-0001b
400 g KI:200 g I_2:1000 ml H_2O "Tri-Iodide Etch"	Au	Removal/patterning	AU-0007
I_2, vapor	Pt	Passivation	PT-0003
10 ml H_2O:5 g NaOH:2 g I_2 "Iodine Etch"	Pb_2Te_3	Dislocation	PBTE-0002b
5 g NaOH:10 ml 0.5% $NaIO_3$ "Iodate Etch"	Pb_2Te_3	Dislocation	PBTE-0002a
1I_2:2KI:10H_2O "Iodine Etch"	Fe	Preferential	FE-0109
15 ml HF:30 ml HNO_3:33 ml HAc:90 mg I_2 "Iodine A" etch	Ge	Preferential	GE-0151
1 g I_2:50 ml MeOH:100 ml H_2O	Lucite	Clean/disinfect	LU-0001b
xI_2:xMeOH	Ti	Removal	TI-0012b
25 ml HF:100 ml HNO_3:125 ml HAc:2 g I_2 "Iodine Etch"	Si	Preferential	SI-0240
xMeOH:xI_2	Si	Surface preparation	SI-0149
xHF:xI_2	Si	Light figure orientation	SI-0159
10 ml HF:20 ml MeOH:2 mg I_2	Si	Preferential	SI-0204
3 mg I_2:MeOH	SiO_2	Cleaning	SIO-0070
1 g I_2:500 ml MeOH	Si_3N_4	Cleaning	SIN-0002g
400 g KI:100 g I_2:400 ml H_2O	Ag	Removal	AG-0022a
I_2, vapor	U	Cleaning	U-0001b

IODINE, I_2

General: Does not occur as a native element, but there are several iodate minerals of silver, copper and mixed crystals, such as marshite, CuI or idorite, AgI, and miersite 4AgI.CuI. Most iodine is recovered in solution from the oceans and in the soil of some land areas, such as the Piedmont zone along the Atlantic coast. It is still extracted from seaweed along the agar-agar, and dried seaweed is widely eaten in Asia as a general healthful food.

Pure iodine has little direct use in metal industry processing though, as one of the halogen group, it is used as a mixture with other chemicals in etching and cleaning. In medicine it is a major disinfectant, and has internal uses as already described in the preceding section. In food processing, it imparts the blue color in starch.

Technical Application: Compounds containing iodine are very soft, and have no application in Solid State material processing other than in etch solutions. Iodine as hot vapor has become a major carrier gas in Vapor Transport, VT, growth of metallic compounds, and produces the highest purity chromium metal crystalline ingots.

In a water or alcohol solution it is used as a cleaner and disinfectant of Lucite windows, such as those used in chemical etch sinks, and, as an etch solution is well known as the tri-iodide . . . for gold.

Iodine single crystals have been grown by controlled evaporation from water solutions for study of general morphology and defect structure.

Etching: Alcohols, KI.

IODINE ETCHANTS

I-0001

ETCH NAME: Methyl alcohol — TIME:

TYPE: Alcohol, preferential — TEMP: RT

COMPOSITION:

x .... MeOH (CH_3OH)

DISCUSSION:

I_2, crystalline specimens. Both crystalline and single crystallites were used in conjunction with a silicon etching study. A drop of MeOH was placed on specimen surfaces to develop etch figures and structure. Pieces of iodine were placed on silicon wafer surfaces and the wafer heated to liquefy the iodine in a general study of the iodine etch reaction.

REF: Walker, P & Waters, W P — personal application, 1957

PHYSICAL PROPERTIES OF IRIDIUM, Ir

Classification	Transition metal
Atomic number	77
Atomic weight	192.2
Melting point (°C)	2443 (2454)
Boiling point (°C)	4500 (5300)
Density (g/cm³)	22.65
Thermal conductance (cal/sec)(cm²)(°C/cm)	0.35
(cal/cm²/cm/°C/sec) 20°C	0.14
Specific heat (cal/g) 0°C	0.031
Heat of fusion (k-cal/g-atom)	6.6
Heat of vaporization (k-cal/g-atom)	152
Atomic volume (W/D)	8.54
1st ionization energy (K-cal/g-mole)	212
1st ionization potential (eV)	9.1
Electronegativity (Pauling's)	2.2
Covalent radius (angstroms)	1.27
Ionic radius (angstroms)	0.68 (Ir^{+4})
Vapor pressure (microns @ mp)	3.5
Coefficient of linear thermal expansion ($\times 10^{-6}$ cm/cm/°C)	6.8
Electrical resistivity (micro ohms-cm) 20°C	5.11

Young's modulus (psi $\times 10^6$)	76
Magnetic susceptibility (ergs $\times 10^{-6}$)	6.6
Tensile strength (psi — annealed)	160,000
Electron work function (eV)	4.57
Cross section (barns)	460
Hardness (Mohs — scratch)	5—6
(Vicker's — kgf/mm^2)	500
Crystal structure (isometric — normal)	(100) cube, fcc
Color (solid)	Silver white
Cleavage (cubic — poor)	(001)

IRIDIUM, Ir

General: Occurs in nature as a native element in small, angular grains. Rare. Found in alluvial deposits of gold sands in Russia, Brazil, the Burma area, and in northern California. Most commonly found as an alloy constituent in platinum and other high temperature metals of the Platinum-Iron group. Native platinum contains up to 2% iridium, and the mineral iridosmine (osmium/iridium) contains up to 30% iridium. As the pure metal it is silver-white in color, both hard and brittle, heavier than gold, and one of the highest melting temperature metals known.

Industrially it is used mainly as an alloy with platinum and osmium for their high temperature and wear characteristics in the fabrication of a variety of small parts: weight standards, ball-point pens, special high-temperature bearings, and compass points. As powdered "Iridium Black" it is a catalyst in chemical processing. It is named after the brilliant colors of its salts that are iridescent reds, blues, and violets. The salts, particularly $IrCl_4$, have important chemical applications.

Technical Applications: Pure iridium metal is not used in Solid State processing to any extent, though platinum (which may contain trace iridium) is sputter evaporated in layered metallization structures. It is under development and study as a silicide: IrSi, $IrSi_2$ and Ir_2Si for application as a buffer in layered heterostructures.

Etching: Very difficult. Very slowly soluble in hot aqua regia.

IRIDIUM ETCHANTS

IR-0001

ETCH NAME: Aqua regia — TIME:

TYPE: Acid, removal — TEMP: Hot

COMPOSITION:

3 .... HCl

1 .... HNO_3

DISCUSSION:

Ir, crystalline specimens as wire, rod, sheet. Etch reaction is very slow even when used hot. Solution used for general cleaning and removal.

REF: Dana, E S & Ford, W E — *A Textbook of Mineralogy,* 4th ed, John Wiley & Sons, New York, 1950, 407

IR-0002: Hodgman, C D — *Handbook of Chemistry and Physics,* 27th ed, Chemical Rubber Company, Cleveland, OH, 1943, 329

a-Ir specimens. Soluble in aqua regia.

IR-0003: Brat, T & Fizenberg, M — *J Appl Phys,* 57,264(1985)

Ir, thin films deposited on silicon, (100), n-type, 10 Ω cm resistivity wafers. Silicon wafers were cleaned: (1) degrease with organic solvents; (2) dip in BHF, RT, 5 sec to remove oxides; (3) rinse in H_2O; (4) N_2 blow dry. After metallization deposit SiO_2 as a

mask for photolithographic patterning. Aqua regia used to etch the Ir thin film. Also co-deposited IrV and $Ir_{80}V_{20}$ thin films.
IR-0004: Brochure: Platinum, Baker & Co., Inc. (no date)
Ir and alloys as metal parts or plating for products. Brochure describes use of the Platinum Group metals in industry: platinum, iridium, osmium, rhodium, palladium. All of these metals can be etched in aqua regia.

IRIDIUM ALLOYS, IrMx

General: Even though iridium is listed as a native element, it more often is found as an alloy with platinum or osmium, and occurs as a mixed alloy with other platinum group metals. The mineral iridosmine, Ir_xOs_y, varies in constituent concentration with locale, and often contains rhodium, platinum, ruthenium, and other metals. Native platinum also is usually as an alloy with trace iridium, rhodium, palladium, osmium, iron, etc.

Due to the high melting point of the primary Platinum Group metals, even when industrially purified, they most often still retain trace elements of each other. As alloy mixtures, they are high temperature brazing alloys, e.g., brazing is above 800°C; whereas soldering is below 800°C as a nominal temperature division.

Technical Application: Probably the best known use of these high temperature alloys is the platinum-platinum:rhodium thermocouple for furnace heat measurement. Iridium alloys are not widely used in Solid State material processing due to high temperature characteristics, though they are occasionally used as brazing alloys in metal package assemblies.

Some iridium alloys have been co-evaporated on silicon as a surface metallization thin film.

Etching: Aqua regia. Halogens.

IRIDIUM ALLOYS

IRIDIUM PLATINUM ETCHANTS

IRPT-0001
ETCH NAME: Aqua regia — TIME:
TYPE: Acid, removal — TEMP: Hot
COMPOSITION:
3 .... HCl
1 .... HNO_3
DISCUSSION:
IrPt as alloy mixtures used in the fabrication of high temperature parts. Aqua regia used as a general cleaning, shaping or removal solution.
REF: Pamphlet: Platinum, Baker & Sons, Inc. (no date)
IRPT-0002: Wise, E M — The Platinum Metals — 10M 12-53073 A-20B 117B, 294
IrPt alloy mixtures. A discussion of etching and processing the Platinum Group metals.

IRIDIUM VANADIUM ETCHANTS

IRV-0001
ETCH NAME: Aqua regia — TIME:
TYPE: Acid, removal — TEMP:
COMPOSITION:
3 .... HCl
1 .... HNO_3
DISCUSSION:
IrV and $Ir_{80}V_{20}$ thin films co-deposited on silicon, (100), n-type 10 Ω cm resistivity

wafers. After metallization, SiO_2 masks were evaporated for photolithographic patterning. Aqua regia used to etch the thin films.
REF: Brat, T & Eizenberg, M — *J Appl Phys*, 57,264(1985)

PHYSICAL PROPERTIES OF IRON, Fe

Classification	Transition metal
Atomic number	26
Atomic weight	55.847
Melting point (°C)	1536
Boiling point (°C)	3000
Density (g/cm^3)	7.87
Thermal conductance (cal/sec)(cm^2)(°C/cm)	0.19
Specific heat (cal/g) 100°C	0.12
Heat of fusion (k-cal/g-atom)	3.67
Latent heat of fusion (cal/g)	66.2
Heat of vaporization (k-cal/g-atom)	84.6
Atomic volume (W/D)	7.1
1st ionization energy (K-cal/g-mole)	182
1st ionization potential (eV)	7.896
Electronegativity (Pauling's)	1.8
Covalent radius (angstroms)	1.17
Ionic radius (angstroms)	0.74 (Fe^{+2}) 0.64 (Fe^{+3})
Coefficient of linear thermal expansion ($\times 10^{-6}$ cm/cm/°C)	12.6
Electrical resistivity (micro ohms-cm $\times 10^{-6}$)	9.71
Tensile strength (psi)	35,000—45,000
Lattice constant (cm $\times 10^{-8}$)	11.7
Compressibility (cm^3/kg $\times 10^{-11}$)	21
Electron work function (eV)	4.7
Cross section (barns)	2.53
Vapor pressure (°C)	2360
Hardness (Mohs — scratch)	4—5
(Brinell—Kgf/mm^2)	82—100
Crystal structure (isometric — normal)	(100) cube, fcc
Color (solid)	Grey black
Cleavage (cubic — perfect)	(001)

IRON, Fe

General: Occurs in nature as a native element, occasionally in large masses, though more common as small aggregates and grains embedded in other rocks and very often containing a small amount of nickel — nickeliferrous iron, $FeNi_2$. Both pure iron and nickeliferrous iron are found as grains associated with gold placers in addition to magnetite, Fe_3O_4, black magnetic iron oxide. Also occurs in meteors as pure iron or $FeNi_2$ with lamellar structure and marked twinning, when etched in dilute nitric acid or iodine. It develops the Widmanstatan figures, when twinning in primarily cubic as fine lines, called Neumann lines.

There are four iron oxides that are primary ores and found in major vein or sedimentary deposits — all of which contribute the tan to red colors of earth, throughout the world. These oxides are

Limonite, $FeO.nH_2O$ (colloidal)	Bog or brown iron ore
Goethite, FeO.(OH) (crystalline)	Brown or tan iron ore
Hematite, Fe_2O_3 (crystalline)	Red iron ore
Magnetite, Fe_3O_4 (crystalline)	Black iron ore, magnetic

In addition, there is the carbonate mineral siderite, $FeCO_3$, and the group known as the "iron pyrites" — pyrite, FeS_2, being the most common — and pyrite is the most widely mined ore of iron representing as much as 75% of the world production for both its iron and sulfur content. Because of its brassy-yellow color, pyrite is often mistaken for gold, thus the name "Fool's Gold" . . . when struck a sharp blow it shatters to a black powder with the distinctive "rotten egg" odor of sulfur dioxide; whereas gold is malleable and will flatten into thin sheets.

Iron has been in use since about 1350 B.C. when the Hittites in Asia Minor discovered the method of smelting iron ore (requires greater heat than for bronze). With the addition of carbon, iron becomes steel — prior to the Christian Era, Damascus steel and, during the Middle Ages of Europe — Toledo steel from Spain. With the addition of chromium to steel, it becomes stainless steel, SST. Pure iron with minimum carbon is soft and brittle, similar to Wrought Iron, still widely used for decorative purposes, although today there are several hundred forms of both irons and steels designed for specific applications.

Iron and steels are fabricated for specific physical structures which are important to their use and application. The following is a brief description of some of the material phases or compounds that occur as they relate to different types of iron or steel:

- Austentite: (Gamma iron). Stable above 1330°F. Decomposes into ferrite, cementite and pearlite. It is the prominent microstructure in 300 series SST.
- Bainite: Needle-like structure with feathering similar to martensite which has a sharper acicular structure. Forms under moderately rapid cooling.
- Carbide: (Cementite). Hard, intermetallic phase usually associated with grain boundaries, and is colorless. Stands out in high relief from background matrix. Contributes hardness, corrosion and wear resistance, but is difficult to machine and reduces impact strength.
- Ferrite: Soft iron, $<0.04\%$ carbon. Light in color and distinctive irregular grain boundaries. Weak structure, highly ductile and easily scratched. The term "ferrite" also refers to a class of iron compounds. See section on Ferrites.
- Graphite: Depending upon the type iron or steel, the included graphite can be in nodular form, as a eutectic dendritic form or as flakes, and may show reaction products as ferrite, pearlite, carbide, or steatite.
- Martensite: Sharp, acicular, needle-like structure that forms with very rapid cooling. Hard and brittle and usually tempered to increase impact strength.

Much of metallographic processing is concerned with the recognition of these materials and structures by macro- and micro-etching, and their direct effect on tool hardness, strength, and corrosion resistance.

Technical Application: In much Solid State device processing, several of the heavy metals such as iron are considered contaminants, and much of the chemical processing involves surface cleaning prior to diffusion, epitaxy, metallization, and other high temperature operations for their removal. On the other hand, iron acts like a dopant element in some compound semiconductors, such as indium phosphide where it is an n-type dopant for InP:Fe, as a semi-insulating (SI) substrate.

Though pure iron is not widely used, iron alloys and steels are part of much processing equipment: vacuum systems, diffusion furnaces, baking ovens, test equipment, and so forth. Stainless steel is usually the base plate in vacuum systems; low-carbon iron, as base plates in sputter systems; magnetic steel parts in these systems; soft iron for lapping plates, requiring periodic truing (flattening), as well as lapping cups for surface angling or sphere and lens fabrication (diopter cups) . . . also heater strips, and a multitude of handling parts, such as acid resistant tweezers, spatulas, carrying trays, and the like.

Both pure iron and iron alloy have been grown as single crystal ingots for study of general morphology, stress/strain criteria, defect development, and ferromagnetic properties, the latter in regard to device applications. In addition to manmade irons and steels, both artificial and natural oxides of single crystal iron compounds have been the subject of study for many years, to include corrosion studies under a wide variety of conditions. Wrought iron is probably the most well known as a nearly pure iron product and has been in use since ancient times.

Etching: Easily etched. Soluble in most acids and acid mixtures. See Steel; Ferrites.

IRON ETCHANTS

FE-0001
ETCH NAME: BRM — TIME:
TYPE: Halogen, oxide removal — TEMP:
COMPOSITION:
x .... Br_2
x .... MeOH
DISCUSSION:

Fe, (100) wafers used in a magnetics study to include oxidation. Specimens oxidized in liquid solutions of KNO_3 or $NaNO_3$. Oxide removed with solution shown.
REF: Yamaguchi, S — *J Electrochem Soc,* 107,714(1960)
FE-0022: Mahla, E M & Nielsen, Y E — *J Electrochem Soc,* 19,387(1948)

Fe specimens. Solution used as a general surface cleaning etch.
FE-0023: Osmond, F & Cartlaud, G — *Rev Met,* 811,2(1905)

Fe specimens. Various etch solutions used to develop etch figures and percussion figures.

FE-0021
ETCH NAME: — TIME: 1—3 min
TYPE: Halogen, cleaning — TEMP: RT
COMPOSITION:
x .... 2—5% I_2
x .... H_2O
DISCUSSION:

Fe, as flat soft iron lap platens used for preparing thin section mineral specimens for study. Solution saturated cloth used to wipe clean platen surfaces, followed with water washing and coat with glycerin to prevent rusting.
REF: Weiss, P & Walker, P — mineralogy studies, 1953

FE-0037
ETCH NAME: — TIME: 1—3 min
TYPE: Halogen, cleaning — TEMP: RT
COMPOSITION:
x .... 3% I_2
x .... MeOH
DISCUSSION:

Fe, as flat soft iron lap platens used for polishing Si and GaAs wafers. Wipe platens clean with solution shown using lint-free cloth during platen reconditioning. Follow with water wash and towel drying.

REF: Walker, P & Menth, M — personal application, 1980
FE-0027a: Walker, P & Bichkowski, M — personal application, 1975

Fe, as soft iron lap platens, both flats and cups, used for processing silicon wafers. Solution used in reconditioning platens. Wipe etch clean with solution shown, then water wash, dry with toweling and coat with glycerin to prevent rusting.

FE-0002
ETCH NAME: Nital — TIME:
TYPE: Acid, cleaning — TEMP:
COMPOSITION: — ASTM #74

x .... 2% HNO_3
x .... EOH

DISCUSSION:

Fe, pure metal samples used in an X-ray study of hydrogen in iron. Solution used to clean surfaces and remove contamination.

REF: Tetelman, A S et al — *Acta Metall,* 9,205(1961)
FE-0017c: ASTM E407-70

Reference for ASTM #74. Solution shown as 1—5 ml HNO_3:100 ml EOH or MeOH.

FE-0013
ETCH TIME: Nital — TIME:
TYPE: Acid, preferential — TEMP:
COMPOSITION: — ASTM #74

x .... 2—3% HNO_3
x .... EOH

DISCUSSION:

Fe, thin films deposited by MBE on GaAs, (110) wafer substrates in an X-ray characterization study of iron saturation in GaAs. Iron thin films were (110) oriented with a 1.34% mismatch. Solution used as a macro-etch on iron films.

REF: Qadri, S B et al — *J Vac Sci Technol,* B(3),718(1985)
FE-0006a: Coleman, E V — *J Appl Phys,* 29,1487(1958)

Fe, single crystal iron whiskers used in a study of dislocations. A 2% Nital solution used to develop dislocations.

FE-0003a
ETCH NAME: — TIME:
TYPE: Electrolytic, polish — TEMP:
COMPOSITION: — ANODE: Fe

6% .... $HClO_4$ — CATHODE:
33% .... butyl cellusolve — POWER:
59% .... MeOH

DISCUSSION:

Fe, polycrystalline specimens used in a study of stacking faults in austenite and its relationship to martensite. Solution used to polish specimens prior to preferential etching.

REF: Motte, H — *Acta Metall,* 5,614(1957)

FE-0003b
ETCH NAME: — TIME:
TYPE: Acid, preferential — TEMP:
COMPOSITION:

45% .... HCl
5% .... HNO_3
50% .... MeOH

DISCUSSION:

Fe, polycrystalline specimens used in a study of stacking faults in austenite and its relationship to martensite. After polishing specimens (FE-0003a), this solution used to develop stacking fault structures.

REF: Ibid.

FE-0018

ETCH NAME: Marshall's solution
TYPE: Acid, polish
COMPOSITION:
x .... H_2O_2
x .... $COOHCOOH.2H_2O$ (oxalic acid)

TIME:
TEMP:

DISCUSSION:

Fe, specimens. Solution referred to as a ''brightening'' solution which leaves a thin oxide film on specimen surfaces. Polish rate is affected by film growth. Film can be observed in a static solution. There is gas evolution using a dynamic solution and this minimizes oxide film growth.

REF: Tegart, W J McG — *The Electrolytic and Chemical Etching of Metals,* Pergamon Press, London, 1956

FE-0006b

ETCH NAME: Picral
TYPE: Acid, dislocation
COMPOSITION:
5 .... 4% $(NO_2)_3C_6H_2OH$ (picric acid)
x .... EOH

TIME:
TEMP:
ASTM #29

DISCUSSION:

Fe, single crystal iron whiskers used in a study of dislocations. A 4% Picral solution was used as a dislocation development etch. Also used 2% Nital (FE-0006a).

REF: Coleman, E V — *J Appl Phys,* 29,1487(1958)

FE-0008

ETCH NAME: Heat
TYPE: Thermal, forming (sphere)
COMPOSITION:
x .... heat

TIME:
TEMP:

DISCUSSION:

Fe, single crystal and polycrystalline spheres formed by splatter arc melting in an He atmosphere from chips and flakes in a water cooled copper pot. Sphere diameters were from 0.1 mm (50% single) and 0.5—1.0 mm (10% single). Other spheres fabricated were Be, V, Zn, NiCr, and AlNi.

REF: Ray, A E & Smith, J F — *Acta Metall,* 11,310(1958)

FE-0009

ETCH NAME: Ferric chloride
TYPE: Electrolytic, forming (sphere)
COMPOSITION:
x .... x% $FeCl_3$

TIME:
TEMP:
ANODE: Fe
CATHODE:
POWER:

DISCUSSION:

Fe, colloidal spheres, 2 mm diameter. Form by electro-deposition of ferric chloride into

mercury and heat to convert to colloidal state. Spheres were used in a study of loss in exchange coupling in surface layers of ferromagnetic particles.
REF: Luborsky, F E — *Phys Rev,* 109,40(1958)

FE-0004
ETCH NAME: TIME:
TYPE: Electrolytic, thinning TEMP:
COMPOSITION: ANODE: Fe
x .... $HClO_4$ CATHODE:
x .... HAc POWER:
DISCUSSION:

Fe specimens of alpha-iron used in a study of neutron irradiation damage using an electron microscope. Solution used to polish and thin specimens after irradiation.
REF: Eyre, B L & Bartlett, S F — *Phil Mag,* 12,261(1965)

FE-0005
ETCH NAME: Water TIME:
TYPE: Acid, preferential TEMP:
COMPOSITION:
x .... DI H_2O
DISCUSSION:

Fe, single crystal iron spheres $^3/_8$" in diameter with (111), (110), and (110) cutflats oriented and polished. Spheres were soaked in DI water. The (110) plane flat showed the greatest degree of pitting.
REF: Kruger, J — *J Electrochem Soc,* 106,736(1959)
FE-0031: Liss, R B — *Acta Metall,* 7,231(1959)

Fe, as single crystal iron specimens used in a study of etch pits. Dripping water on iron will develop pits.
FE-0032: Dillon, F J Jr — *Phys Rev,* 112,59(1958)

Fe spheres used in a study of magnetostatics and ferromagnetics in iron-containing spheres. Water used to both clean and etch spheres.
FE-0035: Bond, W L — *Rev Sci Instr,* 22,344(1951)

Fe spheres. Describes the tumbling method with a water abrasive slurry to fabricate spheres used in FE-0032.
FE-0033: Carter, T L et al — *Rev Sci Instr,* 30,446(1959)

Fe spheres and ferrites. Used a Buehler polishing wheel with rough to fine abrasives and water. Hold specimens perpendicular to the grinding surface with hollow tubes. (*Note:* Method is similar to that used in facetting gemstones.)
FE-0034: Enck, F D — PhD thesis, University Microfilms Publ #23-264

Fe specimens cut as cylinders. Describes a method using a water coolant with a post grinder tool.
FE-0036: Alexander, E & Many, K A — *Rev Sci Instr,* 26,893(1955)

Fe crystals and other hard crystal materials. Describes a method of wire saw cutting with water, solvents and acids.

FE-0007
ETCH NAME: Nitric acid, dilute TIME:
TYPE: Acid, removal TEMP:
COMPOSITION:
1 .... HNO_3
1 .... H_2O

DISCUSSION:

Fe, (100) wafers and other orientations. 99.9% pure iron single crystals grown in a study of growth and orientation. Solution used as a general etch and is preferential on the (100).

REF: Stein, D F & Low, J R Jr — *Trans Met Soc AIME,* 221,744(1961)

FE-0024

ETCH NAME: TIME:

TYPE: Acid, cleaning TEMP: RT

COMPOSITION:

1 .... H_2SO_4

1 .... HCl

1 .... H_2O

DISCUSSION:

Fe, as stainless steel tubing. Solution used to polish etch and clean internal bore of tubing. Follow with DI water rinse and then MeOH. Dry with hot air gun. Method used to clean tubing prior to installation in diffusion and vacuum systems.

REF: Mills, T — personal communication, 1966

FE-0025: Walker, P — personal application, 1970

Fe, as stainless steel tubing. Used as described in FE-0024 for tube cleaning prior to installation in epitaxy systems.

FE-0028

ETCH NAME: Iodine etch TIME:

TYPE: Halogen, preferential TEMP:

COMPOSITION:

1 .... I_2

2 .... KI

10 .... H_2O

DISCUSSION:

Fe specimens. Solution described as a preferential etch for iron. (*Note:* Similar mixtures are the tri-iodide etchants for gold; and "tincture of iodine".)

REF: Berglund, T & Deardon, W H — *Metallographer's Handbook of Etching,* Pitman & Sons, London, 1930

FE-0010

ETCH NAME: TIME:

TYPE: Acid, preferential TEMP:

COMPOSITION:

2 .... H_3PO_4

1 .... H_2O_2

DISCUSSION:

Fe, (100) wafers used in a study of latent hardening. Solution used to develop defects and stress figures.

REF: Nakada, W & Sheh, A — *Acta Metall,* 14,961(1961)

FE-0011

ETCH NAME: TIME:

TYPE: Acid, dislocation TEMP:

COMPOSITION:

1 .... HF

10 .... HCl

5 .... HNO_3

DISCUSSION:

Fe, (100) wafers used in an optical study of passivation films from inorganic inhibitor solutions. Solution used to develop defects induced by passivation.

REF: Horton, M et al — *J Electrochem Soc,* 110,654(1963)

FE-0012

ETCH NAME: Fry's reagent — TIME:

TYPE: Acid, preferential — TEMP:

COMPOSITION: — ASTM #79

40 ml .... HCl

5 g .... $CuCl_2$

30 ml .... H_2O

25 ml .... EOH or MeOH

DISCUSSION:

Fe, (100) wafers and iron alloys used in a study of etch pits and twinning. Both the solution shown and Hahn's modification were used.

REF: Hahn, G T — *Trans Met,* 224,395(1962)

FE-0038: Spreadborough, S et al — *J Appl Phys,* 35,3505(1964)

Fe specimens and 1% Mn doped alloys used in a material study. Fry's reagent used to develop structure and defects.

FE-0014

ETCH NAME: Diamond — TIME:

TYPE: Compound, polish — TEMP:

COMPOSITION:

x .... diamond paste

DISCUSSION:

Fe, (100) and polycrystalline wafers used in a study of nitridization using an E-beam power source. Wafers were polished with diamond paste and chemically etched prior to nitriding of surfaces.

REF: Ebersbach, U et al — *Thin Solid Films,* 112,29(1984)

FE-0015a

ETCH NAME: Soap — TIME:

TYPE: Acid, cleaning — TEMP:

COMPOSITION:

x .... $3C_{17}H_{35}COO$ (soap)

x .... H_2O

DISCUSSION:

Fe specimens as cut and lapped discs were cleaned by washing in a soap/detergent solution, rinsed in water and dried; then followed by flame cleaning. Other metals similarly cleaned were Ti, Zr, Al, Mg, Ag, Au, Cu, Ni, Co, Cd, Zn, and Pb. Metals were evaporated on glass, quartz, and ceramics in a study of frictional adhesion.

REF: Belser, R B — *Rev Sci Instr,* 25,862(1954)

FE-0015b

ETCH NAME: Flame — TIME:

TYPE: Heat, cleaning — TEMP: 800—1200°C

COMPOSITION:

x .... gas flame

DISCUSSION:

Fe specimens used as an evaporation source for glass, quartz and ceramic. See discussion under FE-0015a.

REF: Ibid.

FE-0016

ETCH NAME: Hydrochloric acid, dilute — TIME:

TYPE: Acid, cleaning — TEMP:

COMPOSITION: — ASTM #16

1 .... HCl

1 .... H_2O

DISCUSSION:

Fe, as residual metal in vacuum systems remaining after iron evaporation. Solution used to clean bell jars and internal fixtures. Can be used on SST, glass, and ceramics.

REF: Nichols, D R — *Solid State Technol,* December 1979

FE-0027b: Walker, P & Menth, M — personal application, 1980

Fe, as SST fixtures and base plates in vacuum systems. Solution used as a general surface cleaning etchant after removal of heavy residual metal evaporation deposits. Surfaces wiped down with lint-free toweling wetted with solution. Follow with water wipe-down, then MeOH wipe down.

FE-0017d: ASTM E407-70

Reference for ASTM #16. Solution shown as 5—10 ml HCl:100 ml H_2O

FE-0019a

ETCH NAME: Al-7 — TIME: Dip

TYPE: Acid, macro-etch — TEMP: 160—170°F

COMPOSITION:

50 ml .... HCl

50 ml .... H_2O

DISCUSSION:

Fe specimens. Solution used as a macro-etch to develop structure in irons and steels for metallographic study.

REF: *A/B Metal Digest,* 22(3),14(1983) — Buehler Ltd

FE-0020a

ETCH NAME: Ammonium persulfate — TIME:

TYPE: Acid, preferential — TEMP:

COMPOSITION:

x .... $(NH_4)_2SO_4$, sat. sol.

x .... H_2O

DISCUSSION:

Fe, as grey iron specimens containing eutectic dendritic graphite. Solution used to develop microstructure of carbides and pearlite.

REF: *A/B Metal Digest*, 21(2),17(1983) — Buehler Ltd

FE-0020b

ETCH NAME: — TIME:

TYPE: Acid, preferential — TEMP:

COMPOSITION:

x .... x% $FeCl_3$

x .... EOH

DISCUSSION:

Fe, as grey iron specimens with high phosphorous content and flake graphite. Solution referred to as ''alcoholic ferric chloride'' and used as a micro-etch to develop steatite, pearlite, and graphite structures.

REF: Ibid.

FE-0026

ETCH NAME: Heat/hydrogen | TIME:

TYPE: Thermal, preferential | TEMP: 800—1200°C

COMPOSITION:

x .... heat + H_2

DISCUSSION:

Fe, single crystal whiskers used in a study of surface reaction from oxides and thermal etching with hydrogen. Whiskers were oxidized at 700°C, 15 min, then reduced at 900°C, 1 min in H_2. Thermal etching was done in H_2 at temperature range shown for various time periods.

REF: Laukonis, J V — *J Appl Phys*, 32,242(1961)

FE-0028

ETCH NAME: Argon | TIME:

TYPE: Ionized gas, cutting | TEMP:

COMPOSITION: | GAS FLOW:

x .... Ar^+ ions | PRESSURE:

POWER:

RATE: 2 Å/sec

DISCUSSION:

Fe as iron alloy specimens. Argon used to sputter section specimens. Grain boundaries etched faster. Etch rate shown for a 8 μm cut width. Method also used to cut gold specimens.

REF: Perkins, R A — *J Vac Sci Technol*, Al(2),135(1979)

FE-0029a

ETCH NAME: | TIME:

TYPE: Electrolytic, forming | TEMP:

COMPOSITION: | ANODE: SS 401

(1) x .... H_2SO_4 (2) x .... H_2SO_4 | CATHODE: SS 305

POWER: 5—15 V

DISCUSSION:

Fe as SST 401 wire 0.020 diameter used in a study of annealing on surface composition. Both solutions used to etch form tips for field ion microscope (FIM) study. Point evaporate metal to tip form by power pulsing.

REF: Krishnasway, S V et al — *J Vac Sci Technol*, 11,5(1974)

FE-0029b

ETCH NAME: Hydrogen | TIME: 10 min

TYPE: Gas, cleaning | TEMP: 1100°C

COMPOSITION:

(1) x .... TCE (2) x .... H_2 (3) x .... Freon

DISCUSSION:

Fe as SST 305. Used a bar electrode for electrolytic forming of SST 401 wire (FE-0029a). Cleaned: (1) TCE; (2) fire in dry H_2; (3) ultrasonically with Freon, RT, 30 min.

REF: Ibid.

FE-0030
ETCH NAME:
TYPE: Electrolytic, polish
COMPOSITION:
25 .... HCl
25 .... HNO_3
50 .... H_2O

TIME:
TEMP:
ANODE: Fe specimen
CATHODE:
POWER: 2 V AC

DISCUSSION:

Fe, polycrystalline whiskers used in a field ion microscope (FIM) study of whisker structure. Solution used to electropolish specimens.

REF: Lashmore, B & Melmed, A J — *J Appl Phys,* 49,4586(1978)

FE-0019b
ETCH NAME: Vilella's Reagent
TYPE: Acid, preferential
COMPOSITION:
5 ml .... HCl
1 g .... picric acid
100 ml .... EOH

TIME:
TEMP:

DISCUSSION:

Fe, as grey iron flame hardened. Solution used to develop a macro-etched surface which was then observed at 400× with a metallurgical microscope.

REF: Ibid.

IRON ALLOYS, Fe_xM_y

General: Native metal elements rarely occur as the completely pure element, but contain other trace metals, such that they are metallic alloys. Native iron is found in small quantities, and often contains nickel. In meteoric iron, this appears as $FeNi_2$ segregates in an iron matrix, and there are two iron nickelides: awarutite and josephinite, both $FeNi_2$ as terrestrial irons alloys. There are many iron-containing minerals as oxides, sulfides, sulfates, phosphates, carbonates, and silicates, several of which are hydrates, and some form as natural colloids, etc.

There are hundreds of iron alloys in industry. The Hittite culture in Asia Minor (now part of Turkey) is thought to be the first to develop iron weapons around 1350 B.C., though iron tools may have been in use by the Egyptians as far back as 3000 B.C. Pure iron is soft: Mohs H = 4—5, but can be hardened with the addition of carbon and silicon. The carbon may segregate to some degree by heat treatment as ferrocarbon compounds. Ferrosilicon is a by-product of the reduction of sand for its silicon content, and there are other ferro-type metal alloy compounds used in the fabrication of irons and steels, such as ferromanganese, ferrophosphorus, ferronickel, etc. These are added to iron for specific characteristics and applications. With added chromium, iron becomes steel and stainless steel, SST; with nickel and cobalt, magnets; "tin cans" are thin sheet steel plated with tin. The world is still in the Iron Age after some 3500 years, and many iron alloys have been developed over that period. See the general introduction to Iron for additional discussion.

Technical Application: Iron and steel alloys have several applications in Solid State material processing and development other than in equipment and parts, e.g., tweezers, crucibles, SST etching cups, etc.

Steel discs and blanks have been used for metal evaporation and adhesion studies of thin film; steel parts, such as saw blades, are coated with carbides to improve wear and cutting time; and there are a number of ferrite resistor materials containing zinc, manganese, and other additive metals for specific frequency levels in microwave device assembly applications. Single crystal ferrite compounds are being developed for both ferromagnetic and ferroelectric applications as operating devices, as well as being grown and studied for general morphologic data. There is magnetic stainless steel, and iron magnets have some particular applications in Solid States device construction, such as small ring magnets on silicon diode assemblies or on quartz crystal frequency packages to "pull" and stabilize frequency and/or shift frequency.

Etching: HNO_3, HCl, NaCl, and mixed acids. Nital and Picral. Varies with alloy.

Iron Alloys

- Iron Aluminide
- Iron Aluminum Sulfide
- Iron Beryllium
- Iron Carbide
- Iron Chromium (Kanthal)
- Iron Cobalt Vanadium
- Iron Germanium
- Iron Manganese
- Iron Nickel (+Alnico)
- Iron Silicon

IRON ALLOY ETCHANTS

IRON:SILICON ETCHANTS

FEAL-0001

ETCH NAME: Nital　　TIME:

TYPE: Acid, preferential　　TEMP:

COMPOSITION:

x .... x% HNO_3

x .... EOH

DISCUSSION:

FeAl single crystal specimens used in a study of magnetic structure. Solution shown can be used as a cleaning or slight preferential etch.

REF: Nathans, R et al — *J Phys Chem Solids,* 6,38(1956)

IRON ALUMINUM SULFIDE ETCHANTS

FALS-0001

ETCH NAME: Nitric acid　　TIME:

TYPE: Acid, removal　　TEMP:

COMPOSITION:

x .... HNO_3, conc.

DISCUSSION:

(Fe,Al)S, samples grown in a sealed quartz ampoule for a study of the Fe-Al-S system. Formation of Al_2S_3 under pressure and heat can explode the ampoule. Reaction of Fe-Si formation is rapid, and reduces possible explosion. (*Note:* Etch shown is for pyrite, FeS_2.)

REF: Patinaik, P C & Smeltzer, W W — *J Electrochem Soc,* 131,2688(1984)

IRON BERYLLIUM ETCHANTS

FEB-0001

ETCH NAME: | TIME:
TYPE: Electrolytic, cleaning | TEMP:
COMPOSITION: | ANODE:
20 .... HCl | CATHODE:
20 .... HNO_3 | POWER: 1—3 V
60 .... MeOH

DISCUSSION:

Fe:Be(20%) polycrystalline wire was etch cleaned and treated in this solution to produce an amorphous surface layer. It was then used as a substrate for further FeBe deposition and conversion to single crystal structure by annealing.

REF: Inal, O T & Goudker, I H — *Thin Solid Films,* 111,149(1984)

IRON CARBIDE ETCHANTS

FEC-0001

ETCH NAME: | TIME:
TYPE: Acid, thinning | TEMP:
COMPOSITION:
1 .... HF
4 .... HNO_3
5 .... H_2O

DISCUSSION:

FeC, single crystal alloys. Solution used to chemically etch thin, followed by electropolishing. Specimens used in a study of dislocations and structure associated with material fatigue.

REF: McGrath, J T & Bratinn, W J — *Phil Mag,* 12,1293(1965)

FEC-0002

ETCH NAME: Nital | TIME:
TYPE: Acid, preferential | TEMP: RT
COMPOSITION: | ASTM #74
x .... 2% HNO_3
x .... EOH

DISCUSSION:

Fe_3C:Fe specimens. Nital used as a structure development solution on this material.

REF: Savas, N A & Smith, R W — *J Cryst Growth,* 71,66(1985)

FEC-0010c: ASTM E407-70

Fe and Steel alloys. ASTM #74 Nital shown as: 1—5 ml HNO_3:100 ml EOH.

FEC-0004: Dumez, F — *J Vac Sci Technol,* B(1),218(1983)

Fe_3C single crystal cementite grown in a study of single crystal metal alloys. Material as high resistivity at 25°C: 100 Ω cm, and a temperature coefficient of expansion of about 10^{-4} K^{-1}. Also studied TiAl, TiNi, and ZrV, all as metastable alloys. Metallic glasses studied were PdNiP, PtNiP, and PtCuP with structure defined as random packing of spheres in a dense matrix.

IRON CHROMIUM (KANTHAL ETCHANTS)

KA-0001
ETCH NAME: TIME:
TYPE: Acid, removal TEMP: RT
COMPOSITION:
x .... NaCl
x .... H_2O
DISCUSSION:
72Fe:5.5Al:22Cr:0.5Co as the metal mixture known as Kanthal. Sodium chloride solutions (sea water, etc.) show strong corrosive attack as do fluorine compounds. Alkalies, nitrates, silicates, and borax attack by destroying oxide fractions in the material.
REF: ASTM E407-70
KA-0002: *Rare Metals Handbook,* Reinhold, New York, 1954
Kanthal, FeCr(22):Al(5.5):Co(0.5) specimens. Salt solutions will attack this alloy. Fluorine, chlorine (gases) and alkali solutions also will etch. Borax will remove oxide, and Kanthal melting point is 1510°C.

IRON:COBALT:VANADIUM ETCHANTS

FECV-0001
ETCH NAME: Nital TIME:
TYPE: Acid, defect TEMP:
COMPOSITION: ASTM #74
x .... 3% HNO_3
x .... EOH
DISCUSSION:
Fe:Co:V alloy used in a study of metallurgical and magnetic properties. Solution used to polish specimens and also to develop grain boundary structure.
REF: Chen, O W — *Phys Rev,* S31,348S(1961)
FECV-0010c: ASTM E407-70
Nital solution as ASTM #74 is 1—5 ml HNO_3:100 ml EOH or MeOH.

IRON GERMANIUM ETCHANTS

FEGE-0001
ETCH NAME: TIME:
TYPE: Acid, removal TEMP:
COMPOSITION:
x .... HF
x .... HNO_3
DISCUSSION:
Fe_3Ge_2 as a crystalline deposit, or amorphous thin film as a-Fe_xGe_{1-x}. Films show ferromagnetic properties but with increase of germanium, magnetism is lost. Substrates were silicon wafers with and without SiO_2 pre-coated thin films.
REF: Terzleff, P et al — *J Appl Phys,* 50,1031(1979)

FEGE-0002
ETCH NAME: CP4A | TIME:
TYPE: Acid, preferential | TEMP:
COMPOSITION:
3 .... HF
5 .... HNO_3
3 .... HAc
DISCUSSION:
$FeGe_2$, (100) and (110) wafers cut from Czochralski (CZ) grown ingot with cobalt or tin doping. Solution used as a general polishing etch, and to develop defects preferentially.
REF: Runkin, A Y & Frolow, A A — *J Cryst Growth,* 69,131(1984)

IRON MANGANESE ETCHANTS

FEMN-0001
ETCH NAME: Fry's reagent | TIME:
TYPE: Acid, preferential | TEMP:
COMPOSITION: | ASTM #79
40 ml .... HCl
5 g .... $CuCl_3$
30 ml .... H_2O
25 ml .... EOH or MeOH
DISCUSSION:
Fe:Mn(1%) single crystal specimens. Solution used to develop structure and etch pits. Also used Hahn's modification of Fry's reagent.
REF: Spreadborough, J et al — *J Appl Phys,* 35,3585(1964)
FEMN-0002: Hahn, T — *Trans Met Soc AIME,* 224,343(1962)
Fe specimens. Reference for Hahn's modification of Frey's reagent shown in FEMN-0001.
FEMN-0010c: ASTM E470-70
Reference for ASTM Number shown in FEMN-0001.

IRON NICKEL ETCHANTS

FENI-0030
ETCH NAME: | TIME:
TYPE: Acid, preferential | TEMP: 0°C
COMPOSITION:
3% .... HCl
2% .... zophiran chloride
95% .... EOH
DISCUSSION:
Fe:Ni(65%) alloy specimens used in a study of the martensite crystal. Cool specimen from RT down to 0°C and continue etching until structure is developed. The etch patterns delineate what appears to be single crystal martensite structure.
REF: Nishyama, Z & Shimiau, K — *Acta Metall,* 6,125(1958)

FENI-0031
ETCH NAME: Nitric acid | TIME: Days
TYPE: Acid, structure | TEMP: RT
COMPOSITION:
x .... HNO_3, conc.

DISCUSSION:

Alnico V, iron alloy used in a study of transition electron diffraction. Discs, 0.1—0.2 mm were ground down until the edges warped. Soak-etch in concentrated nitric acid for several days to develop phase structure. The etch is very even in its surface attack and structure development.

REF: Kronenberg, K J — *J Appl Phys,* 51,80S(1960)

IRON:SILICON ETCHANTS

FESI-0001

ETCH NAME: — TIME:

TYPE: Electrolytic, polish — TEMP:

COMPOSITION: — ANODE:

x .... HAc — CATHODE:

x .... CrO_3 — POWER:

DISCUSSION:

Fe:Si(3%) single crystal specimens used in a study of fracture and twinning. Cracks can be nucleated by twinning and dislocations.

REF: Hull, D — *Acta Metall,* 9,11(1961)

FESI-0002: Morris, O E — *Met Prog,* 56,696(1949)

Fe:Si($3^1/_4$%) single crystal specimens.

FESI-0003: Surts, J C & Low, J R Jr — *Acta Metall,* 5,285(1957)

Fe:Si($3^1/_4$%) single crystal specimens.

FESI-0004: Dunn, C O & Daniels, F W — *Trans AIME,* 191,147(1957)

Fe:Si alloy preference for FESI-0001.

FESI-0005: Swets, D E — *J Appl Phys,* 33,1893(1962)

Fe,Si(4%), polycrystalline rods.

FESI-0006: Noble, F W & Aull, D — *Phil Mag,* 12,777(1965)

Fe:Si(3%) single crystal specimens used in a deformation study.

FESI-0005b

ETCH NAME: Nital, variety — TIME:

TYPE: Acid, preferential — TEMP:

COMPOSITION: — ASTM #74

20% .... HNO_3

80% .... EOH

DISCUSSION:

Fe:Si(4%), polycrystalline rods. Solution used to develop single crystal grain structure. Single crystals were cut out of the rods, oriented, and used as seed crystals for single crystal growth.

REF: Swets, D E — *J Appl Phys,* 33,1893(1962)

FESI-0010b: ASTM E407-70

ASTM #74 solution is 1—5 ml HNO_3:100 ml EOH or MeOH.

FESI-0012: Corson, M G — *Iron Age,* 148,45(1941)

FeSi specimens. A general article describing etching techniques for this material.

FESI-0013

ETCH NAME: Argon — TIME:

TYPE: Gas, preferential — TEMP:

COMPOSITION:

x .... Ar

DISCUSSION:

FeSi specimens. Argon used as a molecular gas and referred to as a thermal etchant for ferrosilicon. The gas will attack both (111) and (100) planes preferentially. Authors say that adsorption of oxygen on different crystallographic planes alters the relative surface energy of such planes.

REF: Dunn, C O & Walters, J L — *Acta Metall,* 7,648(1959)

FESI-0014: Chalmers, B et al — *Proc R Soc,* A193,405(1948)

FeSi specimens. Article referenced by Dunn & Walters (FESI-0013) in their work on this material.

FESI-0007

ETCH NAME: TIME:

TYPE: Electrolytic, structure TEMP:

COMPOSITION: ANODE:

x .... $HClO_4$ CATHODE:

x .... HAc POWER:

DISCUSSION:

Fe:Si, specimens. Solution used in developing etch patterns in ferrosilicon. Steps observed varied from 10 to 1000 Å in height and were aligned in the ⟨110⟩ direction to a great extent.

REF: Kroupa, F — *Ceskosl Casop Fly,* 8,171(1958)

FESI-0008

ETCH NAME: Fry's reagent TIME:

TYPE: Acid, preferential TEMP:

COMPOSITION: ASTM #79

40 ml .... HCl

5 g .... $CuCl_2$

30 ml .... H_2O

25 ml .... EOH or MeOH

DISCUSSION:

Fe:Si specimens and other iron alloys used in a study of etch pits and twinning. Also used Hahn's modification.

REF: Spreadborough, J et al — *J Appl Phys,* 35,3585(1964)

FESI-0009: Hahn, G T — *Trans Met,* 224,395(1962)

Fe:Si specimens. Reference for Hahn's modification shown in FESI-0008.

FESI-0010c ASTM E407-70

Reference for ASTM number shown in FESI-0008.

FESI-0011

ETCH NAME: Heat TIME: 1 h

TYPE: Thermal, de-stress TEMP: 1200°C

COMPOSITION:

x .... heat

DISCUSSION:

FeSi(7.7%), single crystal (100) oriented used for saturation magnetization and anisotropic constant measurements. Wafers cut 3 mm diameter, 0.5 mm thick and annealed to relieve strain from cutting with step-cooling at 10°C/min. Other silicon concentrations were 4.9, 5.4, 6.0, and 6.6%. Also called ferro-silicon.

REF: Arai, K I et al — *J Appl Phys,* 57,460(1985)

IRON OXIDE, Fe_xO_y

General: There are a number of iron oxide minerals in nature, all in sufficient quantity to be mined as ores of iron. Major deposits are found throughout the world in rocks of all Ages, and much of the brown, yellow, red and black colors of earth are due to their presence in the soil. Origin varies from sedimentary colloidal, to primary volcanic, metamorphic action, and oxidation of other iron bearing minerals. The major oxide ores are

Limonite, $2FeO.nH_2O$ (colloid) — Brownish-yellow
Goethite, FeO(OH) (orthorhombic) — Yellow/brown to black
Hematite, Fe_2O_3 (rhombohedral) — Grey/black red
Magnetite, Fe_3O_4 (isometric) — Black (magnetic)

Though the oxides are important ores, the "iron pyrites" represent 75% of iron ore mined: pyrite, FeS_2, chalcopyrite, $CuFeS_2$ and pyrrhotite, Fe_3S_5. There are carbonates, such as siderite, $FeCO_3$, a colloid with part of the iron coming from anaerobic bacteria. Iron ores, depending upon locale, often contain important percentages of silver, gold, and nickel.

Primary industrial use is as ores for pure iron. Magnetite, in its form of "lodestone", was the original compass for navigation in ancient times, both on the Mediterranean Sea, and on the Sahara and Gobi Deserts (camel caravans, the ships-of-the-desert).

The powdered oxides are lapping and polishing abrasives: tan- or brown rouge; red rouge; and black rouge. Powdered magnetite, and manmade magnetic materials in liquid plastics, as magnetic fluidics, are used for valve operation; dusted on material surfaces to develop magnetic domains for observation of nickel, cobalt, barium titanate, etc.; powder in rubberized plastics, as "magnetic strips", used for door closures and board displays. All of the oxides have been cut and polished as jewelry items and objets d'art. Also as powder for pigments in paints, enamel or glass. In cosmetics, red rouge is hematite (red ocher variety) and the others have been used in oil as face paints.

Technical Application: Iron oxides have several applications in Solid State processing, to include active devices. Chrome glass masks for photolithography are being replaced with an iron oxide, FeO_x, as it is more stable under handling conditions and does not require an antireflective (AR) coating.

There is a range of ferrites — iron (oxide) with additive metals, such as manganese, selenium, etc., supplied as blanks or in bulk forms fabricated as resistor elements, and there are several magnetoresistive compounds used as discrete devices. The yttrium iron garnet (YIG) contains iron oxide, is fabricated as a laser, as a computer bubble-memory device, and has application as an optical filter.

The study and use of natural iron oxides goes back to at least 1350 B.C. when the Hittite culture in Asia Minor (now Turkey) developed the smelting of iron ore, and both natural and artificial single crystals are the subject of morphological studies today.

Etching: Soluble in most acids, but variable with specific oxide. See Hematite, Magnetite, Ferrites, Iron, and Steel.

IRON OXIDE ETCHANTS

FEO-0001
ETCH NAME: Hydrochloric acid — TIME:
TYPE: Acid, removal — TEMP:
COMPOSITION:
x .... HCl

DISCUSSION:

FeO_x, thin films deposited on soda-lime glass used in the fabrication of photo resist masks. Solution used to clean surfaces and etch patterns masks.

REF: Lipman, R — personal communication, 1981

LIM-0001

ETCH NAME: Nitric acid — TIME: 1—3 min

TYPE: Acid, cleaning — TEMP: RT

COMPOSITION:

x .... HNO_3

DISCUSSION:

$FeO.nH_2O$, the natural mineral limonite as an oxide replacement mineral pseudomorph after pyrite, FeS_2, still in pyrite pseudo-cube form. Solution used to clean and etch surfaces for structure study. Limonite is a natural colloid.

REF: Walker, P — mineralogy study, 1952

MAGT-0005

ETCH NAME: Water — TIME: 2—10 min

TYPE: Acid, cleaning/structure — TEMP: Boiling

COMPOSITION:

x .... H_2O

DISCUSSION:

Fe_3O_4, as fine natural single crystal octahedrons, o(111) found in association with garnet crystals in a green schist. Boiling water used to develop striation figures on the naturally highly polished surfaces with minimum surface alteration.

REF: Walker, P — mineralogy study, 1953

FEO-0002

ETCH NAME: Methyl alcohol — TIME: 1—2 min

TYPE: Alcohol, cleaning — TEMP: RT to hot

COMPOSITION:

(1) x .... MeOH (2) x .... TCE

DISCUSSION:

FeO_x, as thin film evaporation on patterned photo resist glass masks. Both solvents shown used to degrease and clean masks by dipping. Follow with N_2 blow dry. TCE also used as a hot vapor degreaser.

REF: Walker, P — personal application, 1980—1985

HEM-0003

ETCH NAME: — TIME:

TYPE: Molten, flux, removal — TEMP: 1200 K

COMPOSITION:

20—30 g .... H_2SO_4

0.002—0.2 g .... Na_2O_2

DISCUSSION:

Fe_2O_3 specimens and powder used in a solubility study relative to an SO_2/O_2 atmosphere. Developed a phase diagram for Na-Fe-S-O. Na_2O_2 used to establish a pH between 0.5 to 7.5. Also studied NiO, Co_3O_4, Al_2O_3, and Y_2O_3.

REF: Zang, Y S & Rapp, R A — *J Electrochem Soc,* 132,734(1985)

HEM-0005: Gupta, K & Rapp, R A — *J Electrochem Soc,* 127,2194(1980)

Similar work as shown in HEM-0003.

PHYSICAL PROPERTIES OF IRON PHOSPHIDE, FeP

Classification	Phosphide
Atomic numbers	26 & 15
Atomic weight	86.86
Melting point (°C)	1200
Boiling point (°C)	
Density (g/cm^3)	6.1
Hardness (Mohs — scratch)	6—7
Crystal structure (hexagonal — rhombohedral)	$(10\bar{1}1)$ rhomb
Color (solid)	Grey-black
Cleavage (basal — poor)	(0001)

IRON PHOSPHIDE, FeP

General: Does not occur as a natural compound although there are several iron phosphate minerals. In the fabrication of irons and steels the compound is an additive as ferro-phosphorus used for a hardening agent. In chemistry there are di- and tri-iron phosphides, as well as the mono-type listed above.

Technical Application: The reference, here, is as FeP_2, which occurs as a crystal segregate in iron doped indium phosphide, along with a sponge FeP form. The material was removed from the single crystal InP matrix for evaluation.

Etching: $HF:HNO_3:H_2O_2$ as mixed acids.

IRON PHOSPHIDE ETCHANTS

FEP-0001
ETCH NAME: BRM — TIME:
TYPE: Halogen, removal — TEMP:
COMPOSITION:
x .... x% Br_2
x .... MeOH
DISCUSSION:

FeP_2, as precipitate growth needles and rhombohedral blocks or lamellae in Fe-doped InP ingots grown by LEC. Segregates can be mixed-phase FeP, sponge-like with hollow center structure. Solution shown was used to etch expose the segregates and to completely dissolve the InP matrix for particle removal for SEM and X-ray study.
REF: Smith, N A et al — *J Cryst Growth,* 68,517(1984)

PHYSICAL PROPERTIES OF IRON DISULFIDE, FeS_2

Classification	Sulfide
Atomic numbers	26 & 16
Atomic weight	119.97
Melting point (°C)	1171
Boiling point (°C)	
Density (g/cm^3)	5.00
Hardness (Mohs — scratch)	6—6.5
Crystal structure (isometric — pyritohedral)	(100) pseudo-cube, stria
Color (solid)	Brass-yellow
Cleavage (cubic — poor)	(100)

IRON SULFIDE, FeS_2

General: The natural mineral pyrite, FeS_2, is the best known of the "iron pyrites" as a mineral group. Marcasite, FeS_2, has orthorhombic structure rather than the isometric structure of pyrite, and can be further distinguished by its lower specific gravity and paler yellow color when fresh. When pyrite forms under alkaline conditions, marcasite forms under acid conditions and is far less stable. Pyrite is widely distributed in rocks of all types and ages, and is found in large deposits. Some 75% of iron ore comes from the pyrites, is important for both iron and sulfur content and, depending upon location, for its percentage of copper, gold, and silver. It is called "fool's gold" from its fine yellow color, but can be distinguished from gold as, when struck, it will shatter to a black powder with the odor of "rotten eggs"; whereas gold can be beaten into very thin sheets of foil. Under oxidizing conditions the sulfur can be leeched and replaced with oxygen, yet still retaining the pseudo-cubic structure of the original pyrite mineral . . . limonite pseudomorphs after pyrite.

Industrially it is the most important ore of iron, and natural pyrite is sometimes added directly to an iron or steel melt with or without subsequent desulfurization.

Technical Applications: Iron pyrites are not directly used in Solid State or semiconductor device fabrication, though both iron and sulfur are, individually, p- and n-type dopants, respectively, for several compound semiconductors.

The pyrite crystal structure — isometric-pyritohedral — is the common type of a number of metallic compounds formed as sulfides, selenides, tellurides, arsenides and antimonides . . . some as compound semiconductors. Pyrite has been artificially grown, and both the manmade and natural single crystals have been studied.

Etching: HNO_3.

IRON SULFIDE ETCHANTS

FES-0001

ETCH NAME: Nitric acid — TIME:

TYPE: Acid, polish — TEMP: RT

COMPOSITION:

x .... HNO_3, conc.

DISCUSSION:

FeS_2 single crystal ingot artificially growing as the mineral pyrite. Cut specimens mechanically polished with 0.5 μm diamond paste. Etched in nitric acid to polish and remove mechanical cutting damage. Other materials grown in a material study were $MnTe_2$, RuS_2, $RuSe_2$, $RuTe_2$, OsS_2, $OsTe_2$, PtP_2, $PtAs_2$, $PtSb_2$. All ingots show the structure of pyrite. Materials used in an IR spectra study. [*Note:* Authors refer to material structure as "cubic", but it is pseudo-cubic, as pyrite is in the isometric system-pyritohedral class, not isometric-normal class which contains the true cube, (100)]

REF: Lutz, H D et al — *J Phys Chem Solids,* 46,437(1985)

REF: FES-0002: Walker, P — mineralogy study, 1953

FeS_2, as natural single crystals. A dilute solution used in cleaning and etching surfaces in a structural study of the common twinning lamellae on the pseudo-cubic crystal faces.

FES-0003

ETCH NAME: — TIME:

TYPE: Pressure, alteration — TEMP:

COMPOSITION:

x .... pressure

DISCUSSION:

FeS_2, as artificial and natural single crystals used in a study of pressure transition of crystal structure. Both pyrite and marcasite were studied and did not show a transition. The

mineral hauerite, MnS_2, does show a transition. At RT all three minerals are 4m space group. Materials were powdered and pressurized in a diamond anvil cell for study. (*Note:* Pyrite and hauerite are isometric-pyritohedral; marcasite, orthorhombic — normal.).
REF: Chattopadhyay, T & Von Schering, H G — *J Phys Chem Solids,* 46,113(1985)

IRON PALLADIUM ETCHANTS

FEPD-0001
ETCH NAME: Smog — TIME: 30 days
TYPE: Gas, corrosion — TEMP: RT
COMPOSITION:
x .... H_2O (75% RH)
+ .... Air
+ .... SO_2, NO_2, H_2, Cl_2 traces
DISCUSSION:

FePd, (100) and crystalline thin films co-deposited by E-beam evaporation on a variety of different substrates. The gas/water mixture shown was made artificially to replicate a city atmosphere. The evaporated films contained large area crystallites with cubic structure. The body-centered cubic (bcc) phase that was high in iron showed corrosion cracks. The face-centered cubic (fcc) phase that was high in palladium and inert under test conditions shown above.
REF: Penna, T C et al — *J Cryst Growth,* 67,27(1984)

IRON TITANATE ETCHANTS

FETI-0001
ETCH NAME: Sulfuric acid — TIME:
TYPE: Acid, removal — TEMP: Hot
COMPOSITION:
x .... H_2SO_4
DISCUSSION:

$Fe_{3-x}Ti_xO_4$ single crystal grown by Float Zone (FZ) method for study of this material. (*Note:* Etch shown is for natural minerals arizonite, $Fe_2O_3.3TiO_2$ or pseudobrookite, Fe_2TiO_5. See Titanium Dioxide, TiO_2, this Etchant Section.)
REF: Barabers, V A M et al — *J Cryst Growth,* 69,23(1984)

PHYSICAL PROPERTIES OF KRYPTON, Kr

Classification	Inert gas
Atomic number	36
Atomic weight	83.8 (83.7)
Melting point (°C)	−157
Boiling point (°C)	−156
Density (g/cm³ — solid)	3.4
(g/l — liquid)	3.7
Thermal conductance (cal/cm²/cm/°C/sec) 20°C	0.21×10^{-4}
1st ionization potential (eV)	14
Cross section (barns)	24
Vapor pressure (°C)	−171.8
Hardness (Mohs — scratch)	2—3
Crystal structure (isometric — normal)	(100) cube, fcc
Color (solid/gas)	Colorless
Cleavage (cubic — solid)	(001)

KRYPTON, Kr

General: Krypton is one of the rare gases found in the atmosphere. It is inert with a valence of 0 and does not form compounds. There is only 1 ppm in air. It is obtained from the fractional distillation of liquid air, L_{air}, as is LN_2 and the other rare gases in the order of their liquefaction points.

In a vacuum tube the gas produces a yellow-green to green glow and, like neon, Ne, gas (orange-red) is used in electric signs and beacons.

Technical Application: Krypton has had little use in Solid State processing other than as a light source in equipment light bulbs.

Etching: Soluble in LOX.

KRYPTON ETCHANTS

KR-0001
ETCH NAME:
TIME:
TYPE: Pressure, defect
TEMP: −158°C
COMPOSITION:
x .... vacuum pressure
DISCUSSION:

Kr, (100) solid single crystal ingots are formed in vacuum under pressure with liquid helium, LHe, as the freezing vehicle. Surfaces can be preferentially etched by varying the vapor pressure in the vacuum system.

REF: Schwantner, et al — *Rare Gas Solids,* Vol 3, Academic Press, New York, 1970

KR-0002
ETCH NAME: Krypton
TIME:
TYPE: Gas, forming
TEMP:
COMPOSITION:
x .... Ar/N_2
x .... 12.7% Kr

DISCUSSION:

Kr, used as a gas ambient component in the RF magnetron sputter deposition of NbN thin films on sapphire substrates. Inclusion of krypton did not appear to improve the resistivity or other electronic functions of the material.

REF: van Dover, R B et al — *J Vac Sci Technol,* A2(3),2219(1984)

KOVAR ETCHANTS

KO-0001a

ETCH NAME: TIME:

TYPE: Acid, polish TEMP:

COMPOSITION:

1 .... HNO_3

3 .... HAc

15 ml/l .... HCl

DISCUSSION:

Kovar specimens. Used as a general polishing, removal and cleaning solution.

REF: Westinghouse Report

KO-0001b

ETCH NAME: Hydrochloric acid, dilute TIME: 1—3 min or 2—10 min

TYPE: Acid, oxide removal TEMP: 70°C RT

COMPOSITION:

1 .... HCl

1 .... H_2O

DISCUSSION:

Kovar specimens. Solution used to remove native oxide.

REF: Ibid.

KO-0002

ETCH NAME: Ferric chloride TIME:

TYPE: Acid, preferential TEMP: Hot

COMPOSITION: ASTM: #100

x .... 35% $FeCl_3$

DISCUSSION:

Kovar specimens. Solution can be used for general removal or cleaning, and as a slightly preferential etchant. Was used to clean kovar heat-sink substrates for silicon device assembly.

REF: Walker, P — personal application, 1959

KO-0003: ASTM E407-70

Reference for ASTM #100, shown as: 10 g $FeCl_3$:90 ml H_2O

PHYSICAL PROPERTIES OF LANTHANUM, La

Classification	Transition metal
Atomic number	57
Atomic weight	138.92
Melting point (°C)	920
Boiling point (°C)	3469
Density (g/cm^3)	6.166
Thermal conductance (cal/sec)(cm^2)(°C/cm)	0.033
Specific heat (cal/g) 25°C	0.045
Heat of fusion (k-cal/g-atom)	1.5
Heat of vaporization (k-cal/g-atom)	96
Atomic volume (W/D)	22.5
1st ionization energy (K-cal/g-mole)	129
1st ionization potential (eV)	5.61
Electronegativity (Pauling's)	1.1
Covalent radius (angstroms)	1.69
Ionic radius (angstroms)	1.14 (La^{+3})
Electrical resistivity (micro-ohms-cm) 298 K	79.8
Compressibility (cm^2/kg $\times 10^{-6}$)	4.04
Cross section (barns)	8.9
Magnetic moment (Bohr magnetons)	0.49
Tensile strength (psi)	18,900
Yield strength (psi)	18,200
Magnetic susceptibility ($\times 10^{-6}$ emu/mole)	101
Hardness (Mohs — scratch)	4—5
Crystal structure (hexagonal — normal)	$(10\bar{1}0)$ prism, hcp
Color (solid)	Lead-grey
Cleavage (basal)	(0001)

LANTHANUM, La

General: Does not occur as a native element. Though it is not one of the rare earth elements it is found with the cerium group of metals. The chief ores are monazite, (Ce, La, Di)PO_4, a variable phosphate, and allanite (orthite), a cerium/iron silicate containing lanthanum. The latter is found in many igneous rocks of basic types as black grains, often with magnetite, even coated with that magnetic iron ore. The pure metal is lead-grey in color with a metallic luster, and characteristics similar to those of iron, but ignites and burns readily in air to the oxide, La_2O_3. It is kept in benzene to prevent ignition.

The pure metal is not used in the metals industries, but several of the salts have applications in both chemistry and medicine.

Technical Application: Lanthanum has not been used in the Solid State and semiconductor material processing area though, as it resembles iron, it might act as an n-type dopant in compound semiconductors.

Specimens have been studied for general morphological data and, as a sphere, for electrical/resistivity characteristics. As a boride, LaB_6, and as an oxide with strontium and/or iron and cobalt, single crystals have been grown and evaluated.

Etching: Soluble in acids.

LANTHANUM ETCHANTS

LA-0001
ETCH NAME: Nitric acid — TIME:
TYPE: Acid, polish — TEMP: RT
COMPOSITION:
x .... HNO_3
DISCUSSION:
La specimens. Solution shown as a general etch for lanthanum.
REF: Hodgman, C D — *Handbook of Chemistry and Physics,* 27th ed, Chemical Rubber Co., Cleveland, OH, 1943, 398

LA-0002
ETCH NAME: — TIME:
TYPE: Arc, forming — TEMP:
COMPOSITION:
x .... heat
DISCUSSION:
La specimens were first arc melted as buttons and then machined into spheres $^3/_{16}$ to 2″ in diameter and used in measuring electrical resistivity. Other materials evaluated were Pr, Nd, and Sm.
REF: Alstad, J K et al — *Phys Rev,* 122,1636(1961)

LA-0003
ETCH NAME: — TIME:
TYPE: Pressure, transition — TEMP:
COMPOSITION:
x .... pressure
DISCUSSION:
La, U and Th used as pressed powders in talc to form single crystal specimens in a study of crystal transition and electrical parameters. At RT, La is hexagonal system, dhcp; alters to isometric (cubic), fcc at 23 K bars pressure. Handle materials under oil and wash with hexane. Dry talc at 150°C under pressure and use as the material carrier.
REF: Vijayakumar, V — *J Phys Chem Solids,* 46,17(1985)

LANTHANUM BORIDE ETCHANTS

LAB-0001
ETCH NAME: — TIME:
TYPE: Electrolytic, polish — TEMP:
COMPOSITION: — ANODE:
30 .... H_2SO_4 — CATHODE:
20 .... glycerin — POWER:
30 .... H_2O
DISCUSSION:
LaB, single crystal filaments to be used as probe tips for electron microscope instruments. Solution used to etch form the probe tips.
REF: Shimizu, R et al — *J Vac Sci Technol,* 15(3),922(1978)

LAB-0002
ETCH NAME: Sulfuric acid TIME:
TYPE: Acid, forming TEMP:
COMPOSITION:
x .... H_2SO_4
DISCUSSION:
LaB_6, single crystal specimens used as cathode filaments formed with a 90° cone and a radius of 15 μm on (100), (210), and (110) orientations.
REF: Kato, T et al — *J Vac Sci Technol,* B(1),100(1983)

LANTHANUM BROMIDE ETCHANTS

LABR-0001
ETCH NAME: Ethyl alcohol TIME:
TYPE: Alcohol, polish TEMP: RT
COMPOSITION:
x .... EOH
DISCUSSION:
$LaBr_3$, (100) wafers, and $LaBr_3$:Sm doped wafers grown as ingots by a modified Bridgman method with a brass inner liner for sharp heat gradient. Material used as the base for Quantum Counters. Specimens were transparent and clear, cut 1—2 mm and 2—5 mm thick for study, and alcohol used to polish faces. Both $SmBr_3$ and $LaBr_3$ are hygroscopic. $LaBr_3$ is orthorhombic; $SmBr_3$, hexagonal.
REF: Krasutsky, N J — *J Appl Phys,* 54,126(1983)

LANTHANUM STRONTIUM COBALTITE ETCHANTS

LACO-0001
ETCH NAME: Hydrochloric acid TIME: x minutes
TYPE: Acid, preferential TEMP: RT
COMPOSITION:
x .... HCl, conc.
DISCUSSION:
$LaSrCoO_3$, single crystals grown by a Float Zone technique using a Xe arc lamp. Cobalt was greater than 0.2% concentration. Material is pseudo-cubic and wafers were cut (001). Mechanically polish with 0.05 μm alumina and etch in solution shown to develop structure. Both twinning and precipitates were observed. Crystal color is black and showed some surface ridging. SEM used for observation.
REF: Matsuura, T et al — *Jpn J Appl Phys,* 23,1172(1984)

LACO-0002
ETCH NAME: TIME:
TYPE: Salt, polish TEMP: RT
COMPOSITION:
x .... 0.05% NaO_3
DISCUSSION:
$LaSrCoO_3$, single crystals cleaved (001). See discussion under LACO-0001. Surfaces were polished with the solution shown and followed with HCl etching to develop structure. Study was to observe martensetic type transformation using a SEM. Twinning was observed.
REF: Matsuura, T et al — *Jpn J Appl Phys,* 23,1197(1984)

LANTHANUM STRONTIUM FERRITE ETCHANTS

LAFE-0001
ETCH NAME: Hydrochloric acid TIME: 1 h
TYPE: Acid, preferential TEMP: RT
COMPOSITION:
x .... HCl, conc.
DISCUSSION:
$LaSrFeO_3$, single crystals grown by a Float Zone method using a Xe lamp. Iron was <0.02% concentration. Material is pseudo-cubic and wafers were cut (100). Mechanically polish with 0.05 μm alumina and etch in solution shown to develop structure. Both twinning and precipitates were observed. Crystal color was shiny black with smooth surfaces. After etching material was studied under a SEM microscope. Also studied $LaSrCoO_3$.
REF: Matsuura, T et al — *Jpn J Appl Phys,* 23,1172(1984)

PHYSICAL PROPERTIES OF LEAD, Pb

Classification	Metal
Atomic number	82
Atomic weight	207.21
Melting point (°C)	327.43
Boilint point (°C)	1740 (1744)
Density (g/cm^3)	11.35
Thermal conductance (cal/sec)(cm^3)(°C/cm) 20°C	0.083
Specific heat (cal/g) 25°C	0.0306
Latent heat of fusion (k-cal/g-atom)	5.86 (6.26)
Heat of vaporization (cal/g)	203
Atomic volume (W/D)	18.3
1st ionization energy (K-cal/g-mole)	171
1st ionization potential (eV)	
Covalent radius (angstroms)	1.47
Ionic radius (angstroms)	0.84 (Pb^{+4})
Electrical conductance (micro ohms^{-1})	0.046
Thermal conductivity (cal/cm^2/°C/sec) 20°C	0.083
Coefficient of linear thermal expansion ($\times 10^{-6}$ cm/cm/°C) 20°C	29.3
Vapor pressure (°C)	1421
(mmHg) 1167°C	10.0
Magnetic susceptibility ($\times 10^{-6}$ cgs)	0.12
Tensile strength (psi)	3000
Thermal expansion (ppm/°C)	29.3
Cross section (barns)	0.17
Hardness (Mohs — scratch)	1.5
(Brinell — kgf/mm^2)	4.2
Crystal structure (isometric — normal)	(100) cube, fcc
Color (solid)	Lead grey
Cleavage (cubic — poor)	(001)

LEAD, Pb

General: Occurs as a native element though it is rare, usually as thin plates and small globular masses. Single crystals are only known from the Harstig mine, Sweden, although it has been found in other mines and in gold placer sands. Galena, PbS, is the chief ore of lead which may contain, or be associated with silver and gold, as well as contain trace amounts of selenium, zinc, cadmium, antimony, bismuth, and copper, as sulfides. Galena is one of the most widely distributed metal sulfides and is usually associated with eruptive rocks along with other metal sulfides, such as zinc and iron. It is often closely associated with silver, such that deposits are important for both metals. The chemical symbol Pb is from plumbite, the original name for lead.

Because of its low melting point, and the ease with which galena can be reduced to pure lead, lead was one of the metals known to ancient man. Pure lead is ductile, malleable and soft, like gold, and is also chemically inert as a nonconductor . . . some of the lead drains installed during the Roman Empire period before 1 A.D. are still in use.

In most industrial applications lead is hardened with the addition of other metals and is a constituent in a wide range of alloys. A few of the more important lead alloys are

Formula	**Name**
99.8Pb:0.2As	Lead shot
94Pb:6Sb	Battery plate
82Pb:15Sb:3Sn	Type metal
99.93Pb:0.08Cu	Chemical lead
87Pb:13Sn:1Cu	Lead foil
67Pb:33Sn	Plumber's solder

As oxides and sulfates (white, red, and yellow) lead is used as paint pigments. Other lead salts are used in medicine as antiseptics and astringents, and tetraethyl lead is used in leaded gasoline. Both "lead" paints and leaded gasoline are being replaced due to their health hazard.

Lead has been found in Egyptian tombs, circa 3500 B.C., and the Romans used lead and lead compounds in their water piping. In soft water lead hydrolyzes to fairly soluble $Pb(OH)_3$ and, as the lead-ion (Pb^{++}) is poisonous it is no longer used for incoming water piping. In the distillation of alcohol, if lead piping is used, lead poisoning is developed in the liquor.

Because of lead's low neutron cross-section like graphite, lead rods can be used for damping atomic piles. As lead shielding, including lead glass, it is used in nuclear generators and facilities construction. Medical technicians and doctors wear lead aprons when working with X-ray equipment for the same reason. In both medicine and industry, where X-ray, van der Graff generators, and similar equipment are in use, the room walls are shielded with lead sheet. This includes lead-glass for sighting windows and as eye shields.

The term "leaded glass" is an old term applied to the assembly of small glass pieces, each piece surround by a lead solder as the holding compound; whereas "lead glass" is a glass to which lead has been added. It is a heavy, clear, relatively soft glass. As already mentioned, the lead glass is used in radiation areas, but also is used in the jewelry industry for objets d'art.

The "lead" pencil is a misnomer . . . the material is graphite (carbon), though a stick of lead drawn across a surface will produce a light grey streak.

Technical Application: Pure lead and lead solders have many uses in Solid State processing, mainly in the assembly of parts and packaging using both wire and cut pre-forms.

Lead also is a constituent of several compound semiconductors, such as lead sulfide, PbS; lead selenide, PbSe; lead telluride, PbTe; lead oxide, and PbO, and there are trinary compounds of both selenides and tellurides.

The most widely used solders are #62 (60Pb:40Sn) the "standard solder", and #63 (38Pb:40Sn:2Ag) with melting points of 185 and 220°C, respectively. They are used in solid bar and wire form or as resin-core wire. Wire may be applied directly, such as for wiring parts in place or used as pre-cut forms for mounting discrete devices and circuit elements. Other applications include bulk solder in solder pots for pre-tinning by dip soldering of package device leads and electrical wires. Basic circuit boards, PCBs, are plastic impregnated linen sheets overplated or clad with copper. The boards are drilled and etched to pattern electrical circuits, then pre-tinned with solder prior to discrete elements being assembled on the board. With all items in place, boards are passed through a belt furnace for final soldering.

The fabrication and assembly of PCBs has evolved into its own industry within the Solid State field, and has been a major element in the miniaturization and compaction of electronic equipment, such as computers, microprocessors, TVs, transistor radios, watches, even toys. Much electronic equipment is now designed with computer/board operation, from cutting and dicing equipment to wire bonders, vacuum systems, and machine shop parts design.

Pure lead single crystals, as well as lead alloys, have been grown and studied for morphological and electrical data, and the general study of lead has been in progress since ancient times.

Etching: HNO_3 and hot H_2SO_4. A mixture of glacial acetic acid, HAc, and hydrogen peroxide, H_2O_2 is the most widely used general clean/etch solution.

LEAD ETCHANTS

PB-0001
ETCH NAME: TIME:
TYPE: Acid, oxide removal TEMP:
COMPOSITION:
x .... CH_3COOH (HAc)
x .... H_2O_2
DISCUSSION:

Pb specimens. Solution used to clean and remove oxide from lead material used to grow lead telluride.

REF: Sato, V et al — *Jpn J Appl Phys,* 2,688(1963)
PB-0005: Feltham, P — *Acta Metall,* 5,553(1957)

Pb, single crystals used in a study of work hardening in face-centered-cubic (fcc) metals. Solution used to polish surfaces.

PB-0002a
ETCH NAME: TIME:
TYPE: Acid, cleaning TEMP:
COMPOSITION:
1 .... HAc
1 .... H_2O_2
DISCUSSION:

Pb specimens. Solution used as a general cleaner for lead and removal of lead oxide. Clean in benzene prior to etching. Author references PB-0007 for etch.

REF: Snowden, K V — *Phil Mag,* 6,321(1961)

PB-0002b
ETCH NAME: TIME:
TYPE: Acid, preferential TEMP:
COMPOSITION:
70 .... HAc
30 .... H_2O_2

DISCUSSION:

Pb, single crystal ingots. Clean in benzene before subjecting specimen to bending fatigue in vacuum. Solution shown used to observe structure developed by fatigue.

REF: Ibid.

PB-0006a: Fleischer, E L — *Acta Metall,* 9,184(1961)

Pb, single crystals. Specimens were flame cut to length, and used in a study of glide. Solution used as a polish etch, and author references the solution to P Strutt, private communication.

PB-0003a

ETCH NAME: TIME:

TYPE: Acid, polish TEMP:

COMPOSITION:

3 .... HAc

1 .... H_2O_2

DISCUSSION:

Pb specimens. Used as a cleaning and polishing solution for lead and to remove lead oxide from surfaces.

REF: Rutherford, R J — *Proc Am Soc Test Mater,* 24,739(1924)

PB-0007

ETCH NAME: TIME: 5—10 sec

TYPE: Acid, preferential TEMP: 35°C

COMPOSITION:

20 ml .... H_2O_2

80 ml .... HAc

DISCUSSION:

Pb specimens and lead alloys. Solution used in preparing specimens for metallographic and microscopic examination of structure.

REF: Wermer, H K & Wermer, H W — *J Inst Met,* 66,45(1940)

PB-0009a: Tegart, W J McG — *The Electrolytic and Chemical Etching of Metals,* Pergamon Press, London, 1956

Pb specimens. Solution described as an etch for lead. Use was similar to study described under PB-0007.

PB-0008b

ETCH NAME: Potassium cyanide TIME:

TYPE: Acid, cleaning TEMP: RT

COMPOSITION:

x .... KCN

DISCUSSION:

Pb, (100) wafers. Polish etch in CP4, then soak in KCN with DI water rinse for a final cleaning.

REF: Ibid.

PB-0003b

ETCH NAME: Hydrochloric acid TIME:

TYPE: Acid, removal TEMP:

COMPOSITION:

x .... HCl

DISCUSSION:

Pb samples. Solution used as a general removal and cleaning etch.

REF: Rutherford, R J — *Proc Am Soc Test Mater,* 24,739(1924)

PB-0004a

ETCH NAME: TIME: 15—30 sec

TYPE: Acid, preferential TEMP: RT

COMPOSITION:

1 .... HAc

3 .... H_2O_2

DISCUSSION:

Pb, (100) wafers and other orientations used in defect etch development study.

REF: Nike, Y — *J Phys Soc Jpn,* 13,970(1958)

PB-0004b

ETCH NAME: Sulfamic acid TIME: 45 sec

TYPE: Electrolytic, polish TEMP: 18°C

COMPOSITION: ANODE: Pb

x .... $C_5H_{15}NHSO_2OH$ CATHODE:

POWER: 2 A/cm^2

DISCUSSION:

Pb, (100) wafers and other orientations used in defect etch development studies.

REF: Ibid.

PB-0006b

ETCH NAME: TIME:

TYPE: Acid, polish TEMP:

COMPOSITION:

1 .... HAc

1 .... HNO_3

19 .... MeOH

DISCUSSION:

Pb, single crystal specimens were flame cut into lengths and etched in a study of glide. Solution used to polish specimens before being subjected to fatigue.

REF: Fleischer, R L — *Acta Metall,* 9,184(1961)

PB-0008a

ETCH NAME: TIME:

TYPE: Acid, polish TEMP:

COMPOSITION:

1 .... *superoxol

1 .... HNO_3

5 .... H_2O

*1HF:1H_2O_2:4H_2O

DISCUSSION:

Pb, (100) wafers. Polish etch in solution shown and follow with KCN soak, then DI water rinse.

REF: Logan, K H & Schwartz, M — *J Appl Phys,* 26,1287(1955)

PB-0009b

ETCH NAME: TIME: Dip

TYPE: Acid, removal TEMP: RT

COMPOSITION:

140 ml .... NH_4OH

60 ml .... HNO_3

10 g .... H_2MoO_4

240 ml .... H_2O

DISCUSSION:

Pb specimens. Etch by repeat dipping with water rinse between each etch dip period.

REF: Ibid.

PB-0010: ASTE Committee — *Tool Engineers Handbook,* McGraw-Hill, New York, 1949

Pb specimens and Pb alloys. Solution used in process cleaning of lead parts.

LEAD ALLOYS, PbMx

General: Does not occur as a pure metallic alloy in nature although there is the mineral teallite, $PbSnS_2$, and several lead selenides and tellurides. The mineral galena, PbS, was known to ancient man and, as it has a low melting point, lead was extracted by roasting to remove sulfur. Coupled with their knowledge of tin and zinc with copper (bronze and brass), lead alloys were in use as far back as the Egyptian dynastic period around 4500 B.C. (lead artifacts have been found in Egyptian tombs), and engineers of the Roman Empire period (about 500 B.C. to 1000 A.D.) were using lead solders in construction of water piping and aqueduct construction.

Today there are well over a hundred lead-containing solders developed for specific temperature applications in industry, probably the best known and widely used being the lead-tin types for general alloying. A few formulations — as application-named solders — are already shown in the general section under lead.

Technical Application: In addition to the use of lead-tin solders as already described under the technical application section of lead, eutectic lead alloys have been evaluated for their superconducting characteristics, such as those containing tin, silver and titanium.

Etching: HNO_3:HAc, HAc:H_2O_2, variable with alloy.

LEAD ALLOYS

LEAD:SILVER ETCHANTS

PBAG-0001

ETCH NAME: TIME:

TYPE: Acid, preferential TEMP:

COMPOSITION:

1 .... HAc

1 .... HNO_3

4 .... glycerin

DISCUSSION:

Pb:Ag, eutectic alloy specimens used in a structure study and their effects on superconducting properties of eutectic alloys. Solution used to develop structures on PbSn, PbAg, and PbTi.

REF: Levy, S A & Kim, Y B — *J Appl Phys,* 37,3659(1966)

PBAG-0002a
ETCH NAME:
TYPE: Electrolytic, preferential
COMPOSITION:
x .... HAc
x .... $HClO_4$

TIME:
TEMP:
ANODE: PbAg
CATHODE:
POWER:

DISCUSSION:
Pb:Ag, specimen alloys of various compositions. Solution will polish at high current or develop structure at low current.
REF: Heidenreich, R D — *Acta Metall,* 3,78(1955)

PBAG-0002b
ETCH NAME: Molybdic acid
TYPE: Electrolytic, preferential
COMPOSITION:
x .... H_2MoO_4

TIME:
TEMP:
ANODE: PbAg
CATHODE:
POWER:

DISCUSSION:
Pb:Ag specimens of alloys of various compositions. Solution is a preferential etch.
REF: Ibid.

LEAD:TIN ETCHANTS

PBSN-0001
ETCH NAME:
TYPE: Acid, preferential
COMPOSITION:
1 .... HAc
1 .... HNO_3
4 .... glycerin

TIME:
TEMP:

DISCUSSION:
Pb:Sn, eutectic alloy specimens used in a structure study and their effects on superconducting properties of eutectic alloys. Solution used to develop structure on PbSn, PbAg, and PbTi.
REF: Levy, S A & Kim, Y B — *J Appl Phys,* 37,3659(1966)

PBSN-0002
ETCH NAME:
TYPE: Alcohol, cleaning
COMPOSITION:
x .... MeOH

TIME:
TEMP: RT to hot

DISCUSSION:
60Pb:40Sn #62 solder as a pure metal solid wire or bulk alloy; as a rosin core wire; or spheretized paste used for general soldering of assembly elements in semiconductor device circuit fabrication, electronic equipment assembly on PCBs, etc. Ethanol and other alcohols or solvents used to clean surfaces prior to and after soldering. Clean surfaces prior to soldering, and to remove residual solder flux after soldering, and included heavy water washing, only.
REF: Walker, P — personal application, 1960—1985

PBSN-0003
ETCH NAME: TIME: 1 min
TYPE: Acid, cleaning TEMP: RT
COMPOSITION:
1 .... HNO_3
10 .... HAc
10 .... H_2O
DISCUSSION:
Pb:Sn alloy contacts on silicon diodes. Solution developed to clean the lead alloy after soldering device contact leads onto silicon wafers and clean the p-n device junctions with minimum etching of either material.
REF: Walker, P — personal development, 1963

PBSN-0004
ETCH NAME: TIME: 1—3 min
TYPE: Acid, removal TEMP: RT
COMPOSITION:
1 .... HNO_3
1 .... HAc
1 .... H_2O
DISCUSSION:
Pb:Sn alloys and 50 other lead/metal alloys used in a general study of alloy contacts for silicon. Solution used to remove alloys after firing pellets on wafer surfaces in a hand alloy furnace under nitrogen.
REF: Walker, P — personal development, 1957

LEAD-TITANIUM ETCHANTS

PBTI-0001
ETCH NAME: TIME:
TYPE: Acid, preferential TEMP:
COMPOSITION:
1 .... HAc
1 .... HNO_3
4 .... glycerin
DISCUSSION:
Pb:Ti, eutectic alloy specimens used in a structure study and their effects on superconducting properties of such alloys. Solution used to develop structure on PbSn, PbAg, and PbTi.
REF: Levy, S A & Kim, Y B — *J Appl Phys,* 37,3659(1966)

PHYSICAL PROPERTIES OF LEAD IODIDE, PbI_2

Classification	Iodide
Atomic numbers	82 & 53
Atomic weight	461.05
Melting point (°C)	402
Boiling point (°C)	954
Density (g/cm^3)	6.16
Hardness (Mohs — scratch)	2—3

Crystal structure (hexagonal — normal)	$(10\bar{1}0)$ prism
Color (solid)	Gold yellow
Cleavage (basal)	(0001)

LEAD IODIDE, PbI_2

General: Does not occur as a natural mineral due to fairly high water solubility. There are over 50 lead minerals containing chlorine, but none with fluorine, bromine or iodine, the other three halogen group elements. In chemistry there are both the mono- and di-iodides. There is some use of the di-iodide form as a bronze colored pigment in glass making, and similar uses in printing and photography, but none in the metal industries, other than as an additive to etching and cleaning solutions.

Technical Application: There is no use of these iodide compounds in Solid State processing, other than as additive to etching and cleaning solutions. There are possible filter applications even though the material is fairly water soluble, such that single crystals have been grown by evaporation from alcohol solutions for general study.

Etching: KI and alkalies.

LEAD IODIDE ETCHANTS

PBI-0001
ETCH NAME: TIME:
TYPE: Tape, removal TEMP: RT
COMPOSITION:
x .... adhesive tape
DISCUSSION:

PbI_2, single crystal ingot. After cutting, wafers can be cleaned by cleaving perpendicular to the c-axis using Scotch tape and peeling.

REF: Zielinger, J P et al — *J Appl Phys,* 57,292(1985)
PBI-0002a: Forty, A J — *Phil Mag,* 6,587(1961)

PbI_2, single crystals grown as salt recrystallization in an alcohol solution at 100°C with slow cool to room temperature. Thin crystals that floated on top of the alcohol were collected and used for electron microscope (EM) study. Crystal platelets could be etch thinned in KI.

PBI-0002b
ETCH NAME: Potassium iodide TIME:
TYPE: Halogen, removal TEMP:
COMPOSITION:
x .... x% KI
x .... H_2O
DISCUSSION:

PbI_2, as thin crystal platelets grown from an alcohol solution. Platelets can be etch thinned in this solution for electron microscope study.

REF: Ibid.

PHYSICAL PROPERTIES OF LEAD MOLYBDATE, $PbMoO_4$

Classification	Metal oxide
Atomic numbers	82, 42 & 8
Atomic weight	367.16
Melting point (°C)	800
Boiling point (°C)	

Density (g/cm^3)	6.7—7.0
Refractive index (n=)	2.402 & 2.304
Hardness (Mohs — scratch)	2.75—3.0
Crystal structure (tetragonal — pyramidal)	(111) pyramid, 1st order
Color (solid)	Yellow/green
Cleavage (pyramidal)	(111)

LEAD MOLYBDATE, $PbMoO_4$

General: Occurs as the mineral wulfenite, $PbMoO_4$, which is of secondary origin being found in the oxidation zone of lead and zinc mines and used as an ore of molybdenum. The primary industrial use is as an ore of molybdenum.

Technical Application: Both molybdate and tungstate single crystals have been studied as artificially grown materials for general morphology, electronic and electrical data, and for crystallographic structure change under pressure and temperature.

Etching: Soluble in acids and alkalies.

LEAD MOLYBDATE ETCHANTS

PBM-0001

ETCH NAME: TIME:

TYPE: Pressure, alteration TEMP:

COMPOSITION:

x .... pressure

DISCUSSION:

$PbMoO_4$ single crystals used in a high pressure chemistry test of materials at above 56 Pa. Other molybdates and tungstates studied were $CaWO_4$, $PbWO_4$, $CaMoO_4$, and $CdMoO_4$.

REF: Hazen, R M et al — *J Phys Chem Solids,* 46,253(1985)

PHYSICAL PROPERTIES OF LEAD MONOXIDE, PbO

Classification	Oxide
Atomic numbers	82 & 8
Atomic weight	223.21
Melting point (°C)	888 (litharge)
Boiling point (°C)	
Density (g/cm^3) — litharge	9.53
(g/cm^3) — massicot	8.0
Refractive index (n=) litharge/massicot	2.65/2.61
Hardness (Mohs — scratch)	2—3
Crystal structure (tetragonal — normal) litharge	(110) prism, 1st order
(orthorhombic — normal) massicot	(100) a-pinacoid
Color (solid) litharge	Yellow/orange
(solid) massicot	Colorless
Cleavage (basal — litharge & massicot)	(001)

LEAD MONOXIDE, PbO

General: It occurs in nature as the mineral massicot, PbO. It is rare, of secondary origin, and usually associated with galena, PbS. It is interesting as the central core of the mineral is a colorless solid, but surfaces are coated with "litharge", the yellow-orange variety of lead oxide. As can be seen from the list of properties, the two varieties have distinctive crystal structures even though the mineral appears to be only one form and yellow-orange in color. The mineral plattnerite, PbO_2, is the dioxide form.

In chemical processing, lead can be converted to litharge by heating in air above 600°C, as can all other oxides of lead. Litharge has major use in the glass and pottery industries in the making of flint glass (also called lead glass), and as a coloring agent in pottery glaze. When mixed with glycerin it sets as a hard cement, and has been used as a two-part, colored glue. "Red lead", a lead plumbate, Pb_2PbO_4, is the form used in flint glass, and the red-orange undercoat on iron and steel for fencing and building construction as it is both stable and inert under normal atmospheric conditions so it protects metals against weathering effects.

Because the lead-ion, Pb^{++}, is poisonous, lead piping for drinking water (soft water), lead paints, and leaded gasolines are being replaced due to the health hazard. Lead hydroxide, $Pb(OH)_3$, is the culprit, as it is fairly soluble in soft water and builds up in water piping.

Technical Application: Lead oxide is a IV—VI compound semiconductor with tetragonal system structure rather than isometric-tetragonal, as are several other compound semiconductors. It has the typical (111) polarity of these compounds, such that wafers are normally cut with the (100) orientation for device processing.

It has limited use as a semiconductor, in part due to its softness (H = 2—3, vs. H = 6—7 for most semiconductor materials). It has been grown as a single crystal ingot and as single crystal platelets under pressure with two crystal structures depending upon pressure: under 2200 psi, tetragonal; over 30,000 psi, orthorhombic, respectively, red-orange and yellow-green in color. It has been grown as a single crystal epitaxy thin film and fabricated as photoconductors.

Etching: Litharge is soluble in HNO_3, alkalies and several chloride salts. Massicot is soluble in alkalies.

LEAD OXIDE ETCHANTS

PBO-0001

ETCH NAME: TIME:

TYPE: Acid, removal TEMP:

COMPOSITION:

1 .... HAc

1 .... H_2O_2

DISCUSSION:

Pb specimens and single crystal ingots. Solution is a general cleaner and etchant for lead and for removal of lead oxide.

REF: Snowden, K V — *Phil Mag*, 6,321(1961)

PBO-0002

ETCH NAME: TIME:

TYPE: Knife, cleave clean TEMP: RT

COMPOSITION:

x .... knife cut

DISCUSSION:

PbO single crystal plates grown at 400°C under pressure: (1) at less than 2200 psi, structure is tetragonal and orange-red in color; (2) at greater than 30,000 psi structure is orthorhombic and color is yellow-green. Grown in a quartz ampoule in an autoclave with a slurry of NaOH + PbO — crystal platelets grow on ampoule sides. When the orthorhombic form is cleaved with a scalpel some areas of specimen surfaces were converted to tetragonal.

REF: Keezer, R C et al — *J Appl Phys*, 39,2062(1968)

PBO-0003

ETCH NAME:	TIME:
TYPE: Acid, oxide removal	TEMP:

COMPOSITION:

x .... HAc

x .... H_2O_2

DISCUSSION:

PbO, native oxide removed from lead. Solution used in cleaning lead telluride, $PbTe_2$, surfaces.

REF: Sato, Y et al — *Jpn J Appl Phys,* 2,688(1963)

PHYSICAL PROPERTIES OF LEAD NITRATE, $Pb(NO_3)_2$

Classification	Nitrate
Atomic numbers	82, 7 & 8
Atomic weight	331.23
Melting point (°C)	470
Boiling point (°C)	
Density (g/cm³)	
Refractive index (n =)	1.782
Hardness (Mohs — scratch)	4—5
Crystal structure (isometric — normal) alpha	(100) cube
(monoclinic — normal) beta	(100) a-pinacoid
Color (solid)	Colorless
Cleavage (octahedral/cubic)	(111)/(001)

LEAD NITRATE, $Pb(NO_3)_2$

General: Does not occur as a natural compound although some nitrate fertilizers may be artificially compounded to contain trace lead. No present use in the metal industries, but has chemical and medical applications.

Technical Application: The material has no use in Solid State device development, but has been grown from solution in a structural study, and the effects of irradiation on the development of NO_3^- and other radicals. Similar irradiation effects with development of F^- centers have been done on fluorite, CaF_2, natural crystals, and other transparent crystals, e.g., NaCl, KCl, etc. as ionic and alkaline crystals.

Etching: Water, alcohols.

LEAD NITRATE ETCHANTS

PBNO-0001

ETCH NAME: Water	TIME:
TYPE: Solvent, polish	TEMP: RT

COMPOSITION:

(1) x .... H_2O or (2) x .... EOH

DISCUSSION:

$Pb(NO_3)_2$ grown as single crystals by slow cooling from a water solution. Crystals were triangular with a (111) base, and were cubic space group T^6n with trigonal axes. Crystals were studied after irradiation to develop for NO_3^-, NO_2^-, and O_3^- centers. Solutions shown used to polish specimens prior to irradiation.

REF: Tagaya, K & Kitagawa, M — *Jpn J Appl Phys,* 23,116(1984)

PHYSICAL PROPERTIES OF LEAD SELENIDE, PbSe

Classification	Selenide
Atomic numbers	82 & 34
Atomic weight	286.15
Melting point (°C)	1065
Boiling point (°C)	
Density (g/cm^3)	7.6—8.8
Hardness (Mohs — scratch)	2—3
Crystal structure (isometric — normal)	(100) cube
Color (solid)	Lead-grey/bluish
Cleavage (cubic)	(001)

LEAD SELENIDE, PbSe

General: Occurs as the natural mineral clausthalite, PbSe, which may contain traces of both mercury and platinum. It looks much like galena, PbS, though with a slightly more bluish cast of color. It is found in small quantities associated with hematite, Fe_2O_3, a primary iron ore, and there are several additional minerals with individual names that contain clausthalite as part of mixed crystals. Due to scarcity, it has no major use in industry other than as a minor ore of selenium and lead.

Technical Application: The material is a IV—VI compound semiconductor with the sphalerite, ZnS, atypical isometric-tetragonal structure. It is a polar compound with (111)Pb and ($\overline{111}$)Se wafer surfaces and, because of the difference in etching characteristics of (111) polar surfaces, is normally used with the (100) surface orientation to obviate the polarity effect. As with several of the compound semiconductors, it was originally grown by the Horizontal Bridgman (HB) technique as a single crystal but now is more commonly grown by the Czochralski (CZ) method. Vapor Phase Epitaxy (VPE) has been used to grow single crystal thin films.

It has the same softness problem (H = 2—3) as does lead oxide, PbO, so has not been widely used as a semiconductor ingot-grown product, though epitaxy thin films have been used to fabricate laser devices. Note that selenium and several of its salts are poisonous.

Etching: Easily soluble in halogens and HNO_3. Electrolytically with alkalies.

LEAD SELENIDE ETCHANTS

PBSE-0001a

ETCH NAME: Nitric acid — TIME:

TYPE: Acid, removal — TEMP:

COMPOSITION: — RATE: 0.37 mg/min

x .... HNO_3

DISCUSSION:

PbSe, (100) wafers and other orientations. Author calls solution an immersion etch for PbSe, PbTe, and PbS.

REF: Swets, D E — *J Electrochem Soc*, 131,172(1984)

PBSE-0001b

ETCH NAME: Hydrogen bromide — TIME: 3 min

TYPE: Halogen, polish — TEMP: RT

COMPOSITION:

x .... HBr

DISCUSSION:

PbSe, (100) wafers and other orientations. Used as an immersion polish etch on PbSe, PbTe, and PbS. Also used as a jet/spray etch from a squirt bottle followed with water jet/spray quenching.

REF: Ibid.

PBSE-0001c

ETCH NAME: Hydrogen bromide, dilute
TYPE: Electrolytic, polish
COMPOSITION:
1 .... HBr
3 .... H_2O

TIME:
TEMP:
ANODE: PbSe
CATHODE:
POWER: 20 mA/cm^2
50—100 mV
RATE: 0.2 μm/min

DISCUSSION:

PbSe, (100) wafers and other orientations. Solution used to electropolish both PbSe and PbS. Leaves a yellow film on the surface that can be removed with a dip in nitric acid.

REF: Ibid.

PBSE-0002

ETCH NAME:
TYPE: Electrolytic, polish/thin
COMPOSITION:
5 .... 45% KOH
1 .... H_2O_2
5 .... ethylene glycol (EG)

TIME:
TEMP:
ANODE: PbSe
CATHODE:
POWER:

DISCUSSION:

PbSe, (100) wafers used in a study of dislocations and structure. Solution used to thin specimens for TEM study.

REF: Abrams, H & Tauber, R N — *J Electrochem Soc,* 116,103(1969)

PBSE-0003: Gatos, H C & Lavine, M C — *Prog Semicond,* 9,1(1965)

PbSe, (100) wafers and other orientations. Solution used as a wet chemical etch at 40°C, 3 min.

PBSE-0004

ETCH NAME: Nitric acid, dilute
TYPE: Acid, cleaning
COMPOSITION:
1 .... HNO_3
5 .... H_2O

TIME:
TEMP: RT

DISCUSSION:

PbSe, (100) cleaved wafers. A study of electron and hole mobility between RT and 4.2 K. Material has NaCl cubic structure. Solution can be used to lightly clean surfaces. PbS and PbTe also studied.

REF: Allgaier, R S — Scanlon, W W — *Phys Rev,* 111,1029(1958)

PBSE-0005: Allgaier, R S — *Phys Rev,* 112,828(1958)

PbSe, (100) cleaved wafers. A study of magnetoresistance at 295, 77, and 4.2 K. Solution can be used to clean surfaces. PbS and PbTe also studied.

LEAD TIN SELENIDE ETCHANTS

PBTS-0001

ETCH NAME:

TYPE: Electrolytic, polish

COMPOSITION:

1 .... KOH, sat. sol.

1 .... Na_2SO_3, sat. sol.

5 .... glycerin

TIME: 10—15 min

TEMP: 40°C

ANODE: PbSnSe

CATHODE: Pt

POWER: 0.04 A/cm^2

RATE: 1 μm/min

DISCUSSION:

PbSnSe, (100) wafers 1 mm thick, p-type, 100—150 Ω cm resistivity. Use varnish coated SST tweezers to hold specimens to prevent iron contamination during etching. Solution used to polish specimens.

REF: Quadeer, A et al — *J Electrochem Soc,* 129,2145(1982)

PHYSICAL PROPERTIES OF LEAD SULFIDE, PbS

Classification	Sulfide
Atomic number	82 & 16
Atomic weight	239.72
Melting point (°C)	1114
Boiling point (°C)	
Density (g/cm^3)	7.5
Refractive index (n=)	3.912
Hardness (Mohs — scratch)	2.5—2.75
Crystal structure (isometric — normal)	(100) cube
Color (solid)	Lead-grey
Cleavage (cubic — perfect)	(100)

LEAD SULFIDE, PbS

General: The major natural mineral is galena, PbS, often found as single crystal cubes with square, step-fractured faces. It is the most widely distributed metallic sulfide in the world, commonly formed by hydrochemical reactions, and found in large beds and vein ore bodies in both metamorphic and sedimentary rocks. Though 75% of the galena mined in the U.S. is called "soft lead", galena in other areas may contain sufficient gold and silver to make the mineral more valued for its precious metal content. Cerussite, $PbCO_3$, and anglesite, $PbSO_4$, are also important ores of lead, but galena is by far the most widely mined. Well over two million tons are mined annually.

Primarily, industrial use is as an ore of lead. Natural galena single crystals were the first radiofrequency crystals using a wire point contact on the surface. Because specific frequencies are not defined in the mineral, the operator shifted the point probe until he picked up the desired frequency. Alpha-quartz crystals, both natural and now artificial, have replaced galena since the early 1920s, as quartz can be oriented and cut to produce specific frequency ranges as AT, DT, BT, Y, X-cut, etc.

Technical Application: Lead sulfide is a IV—VI compound semiconductor with distinctive isometric — normal (cubic) structure, rather than the more common isometric — tetragonal (zincblende, ZnS) structure of several other compound semiconductors. It is a polar compound with (111)Pb and $(\overline{111})$S surfaces and, as others of this type, is normally used in (100) wafer surface orientation in order to obviate the polarity problems in device processing.

It can be artificially grown from solutions containing various chemical compounds but, in Solid State semiconductor applications, single crystal ingots are grown by the Bridgman, Czochralski, or Float Zone methods.

Both artificial and natural crystals of galena have been the subject of much study. Like several of the compound semiconductors it can be easily cleaved as a (100) surface, and has been studied as-cleaved or after further polish processing. It has also been grown by Vapor phase epitaxy (VPE) as a single crystal thin film, and such thin film structures have been fabricated as laser and photoconductor diodes. It has the same softness problem as do PbO, PbSe.

Etching: Soluble in acids and halogens. Electrolytic halogens. Slightly soluble in water.

LEAD SULFIDE ETCHANTS

PBS-0001
ETCH NAME: TIME: 1 min
TYPE: Acid, dislocation TEMP: 60°C
COMPOSITION:
1 .... HCl
3 .... $*H_2NCSNH_2$

*Thiourea 100 g/l.

DISCUSSION:

PbS, (100) wafers used in a study of dislocation development as a function of stoichiometry. The authors say that pits obtained from cleaved surfaces are larger and better formed than those from lapped and etch polished surfaces.

REF: Frankin, W M & Wagner, J B Jr — *J Appl Phys,* 34,3121(1963)

PBS-0003: Harper, C A — *Handbook of Materials and Processes for Electronics,* McGraw-Hill, New York, 1970, 7

PbS, (100) wafers used in a dislocation etch study. Thiourea (3—10%) in HCl at 60°C, 10 min was used as a preferential etch solution.

PBS-0004: Scanlon, W W — *Phys Rev,* 106,718(1957)

PbS, (100) wafers used in a study of carrier lifetime. Solution used between 60 to 80°C.

PBS-0005: Lyall, K D & Paterson, M S — *Acta Metall,* 14,371(1966)

PbS, (100) cleaved wafers from natural galena used in a study of plastic deformation. Developed edge dislocations and twinning on the (441) plane.

PBS-0002a
ETCH NAME: Nitric acid TIME:
TYPE: Acid, removal/polish TEMP: RT
COMPOSITION: RATE: 23 mg/min
x .... HNO_3
DISCUSSION:

PbS, (100) wafers used in a study of etching to obtain low resistance. Solution used as an immersion etch followed by running DI water rinse. Can also be used to remove yellow stain left on surfaces after electrolytic etching in $HBr:H_2O$.

REF: Swets, D E — *J Electrochem Soc,* 131,172(1984)

PBS-0006: Walker, P — mineralogy study, 1952—1953

PbS, (100) surfaces as natural single crystal galena specimens used in a general structure study. A 1:1 dilute solution used to clean surface to remove natural discoloration. Specimens also cleaved and studied under a microscope without any surface treatment.

PBS-0002b
ETCH NAME: Hydrogen bromide
TYPE: Halogen, polish
COMPOSITION:
x .... HBr

TIME: 3 min
TEMP: RT
RATE: 23 mg/min

DISCUSSION:
PbS, (100) wafers used in a study of etching to obtain low resistance. Solution used as an immersion etch on PbS, PbTe, and PbSe. Also used as a jet-spray from a squirt bottle on PbSnTe and PbSe followed by rinsing in running DI water.
REF: Ibid.

PBS-0002c
ETCH NAME: Hydrogen bromide, dilute
TYPE: Electrolytic, polish
COMPOSITION:
1 .... HBr
3 .... H_2O

TIME:
TEMP:
ANODE:
CATHODE: PbS
POWER: 20 mA/cm^2 & 50—100 mV
RATE: 0.2 µm/min

DISCUSSION:
PbS, (100) wafers used in a study of etching to obtain low resistance. Solution used as a general removal and polishing etch. Yellow film remaining on surfaces removed with nitric acid. Also used on PbSe. This etching gave the lowest surface resistance, Rc.
REF: Ibid.

PBS-0007
ETCH NAME: Nitric acid, dilute
TYPE: Acid, cleaning
COMPOSITION:
1 .... HNO_3
5 .... H_2O

TIME:
TEMP: RT

DISCUSSION:
PbS, (100) cleaved wafers. A study of electron and hole mobility from RT to 4.2 K. Material has NaCl cubic structure. Solution can be used to lightly clean surfaces. PbSe and PbTe also studied.
REF: Allgaier, R S & Scanlon, W W — *Phys Rev*, 111,1029(1958)
PBS-0008: Allgaier, R S — *Phys Rev*, 122,828(1958)
PbS, (100) cleaved wafers. A study of magnetoresistance at 295, 77, and 4.2 K. Solution can be used to lightly clean surfaces. PbSe and PbTe also studied.

PHYSICAL PROPERTIES OF LEAD TELLURIDE, PbTe

Classification	Telluride
Atomic numbers	82 & 52
Atomic weight	334.79
Melting point (°C)	>700
Boiling point (°C)	
Density (g/cm^3)	8.16
Hardness (Mohs — scratch)	3
Crystal structure (isometric — normal)	(100) cube

Color (solid) — Tin-white/yellowish
Cleavage (cubic) — (100)

LEAD TELLURIDE, PbTe

General: Occurs as the mineral altaite, PbTe, usually in massive form but with cubic cleavage, occasionally as cubic or octahedral crystals. Although the color is tin-white, it oxidizes with a yellowish surface coating and can tarnish to deep bronze-yellow. It is found in limited quantities in Russia, Chile, Mexico, and in the U.S. No use in industry other than as a compound semiconductor.

Technical Application: Lead telluride is a IV—VI compound semiconductor and, like other lead compound semiconductors is isometric — normal (cubic) rather than tetrahedral class, as are many other similar compounds, such as gallium arsenide.

It has been grown as single crystal ingots by the Bridgman, Czochralski, and Float Zone methods, and is usually used as a cut or cleaved (100) oriented wafer due to (111) polarity problems. It also has been grown as an epitaxy thin film for device fabrication.

As a thin film it has been used to fabricate both infrared detectors and laser diodes.

Etching: Soluble in acids, mixed acids as well as alkali and halogens.

LEAD TELLURIDE ETCHANTS

PBTE-0001a
ETCH NAME: Iodine etch — TIME: 5 min
TYPE: Halogen, preferential — TEMP: 95°C
COMPOSITION:

5 g .... NaOH
0.2 g .... I_2
10 ml .... H_2O

DISCUSSION:

PbTe, (100) cleaved wafers used in a study of defects and the effect of heat treatment. Solution produced pyramidal etch pits prior to heat treatment. Re-etching after heat treatment would not produce etch pits until surfaces were lapped with SiC abrasive paper and again re-etched.

REF: Sato, Y et al — *Jpn J Appl Phys,* 2,688(1963)

PBTE-0001b
ETCH NAME: — TIME: 30 min
TYPE: Electrolytic, polish/preferential — TEMP: RT
COMPOSITION: — ANODE: PbTe
CATHODE:
POWER: 1 V & 6 V
RATE: 7 μm/min

"A" 20 g .... NaOH/KOH
45 ml .... H_2O
"B" 25 ml .... glycerol
"C" 20 ml .... EOH

DISCUSSION:

PbTe, (100) wafers. To mix the solution: (i) mix solution "A", then (ii) add "B" and "C" and stir to form a miscible solution. Initially polish etch at 6 V with removal rate as shown. Requires about 200 μm removal to eliminate damage induced from SiC lapping of surfaces, e.g., 30 min etching. Reduce voltage to 1 V and continue etching for an additional 15 sec to develop etch pits.

REF: Ibid.

PBTE-0002a
ETCH NAME: Iodate etch TIME:
TYPE: Halogen, preferential TEMP:
COMPOSITION:

5 g .... NaOH
10 ml .... 0.5% $NaIO_3$

DISCUSSION:

PbTe, (100) wafers. Solution applied as a dislocation etch for lead telluride. Also used iodine etch (see PBTE-0002b).

REF: Houston, B & Norr, M K — *J Appl Phys,* 31,616(1960)

PBTE-0002b
ETCH NAME: Iodine etch TIME: 5 min
TYPE: Alkali, preferential TEMP: 94—98°C
COMPOSITION:

5 g .... NaOH
2 g .... I_2
10 ml .... H_2O

DISCUSSION:

PbTe, (100) wafers. Used solution as a dislocation etch for PbTe.

REF: Ibid.

PBTE-0003a
ETCH NAME: Hydrogen bromide TIME: 3 min
TYPE: Halogen, polish TEMP: RT
COMPOSITION: RATE: 0.1 mg/min

x .... HBr

DISCUSSION:

PbTe, (100) wafers. Used as an immersion polish solution for PbTe, PbSe, and PbS. PbSnTe and PbSe etched with solution in a squirt bottle, then rinsed with running DI water.

REF: Swets, D E — *J Electrochem Soc,* 131,172(1984)

PBTE-0003b
ETCH NAME: Nitric acid TIME:
TYPE: Acid, removal TEMP: RT
COMPOSITION: RATE: 0.1 mg/min

x .... HNO_3, conc.

DISCUSSION:

PbTe, (100) wafers. Used as an immersion etch on PbTe, PbSe, and PbS for general removal and cleaning.

REF: Ibid.

PBTE-0004
ETCH NAME: TIME:
TYPE: Acid, oxide removal TEMP:
COMPOSITION:

1 .... HAc
1 .... H_2O_2

DISCUSSION:

PbTe, (100), p-type wafers used in a study of nickel and chlorine diffusion at 700°C. Solution used to remove surface native oxide after lapping and prior to diffusion.

REF: George, T D & Wagner, J B Jr — *J Electrochem Soc,* 115,956(1968)

PBTE-0005

ETCH NAME: TIME: 10 min

TYPE: Alkali, dislocation TEMP:

COMPOSITION:

2 .... 15% NaOH

1 .... $Na_2S_2O_8$

DISCUSSION:

PbTe, (100) wafers. Used as a dislocation etch.

REF: Harper, C A — *Handbook of Materials and Processes for Electronics,* McGraw-Hill, New York, 1970, 7

PBTE-0006: Brybick, R F & Scanlon, W W — *J Chem Phys,* 27,607(1957)

PbTe, (100) and other orientations. Several etches described for defect etching of PbTe, PbSe and PbS.

PBTE-0011

ETCH NAME: TIME:

TYPE: Acid, polish TEMP: 25°C

COMPOSITION: RATE: 20—40 μm/min

9 ml .... HNO_3

40 ml .... $K_2Cr_2O_7$, sat. sol.

DISCUSSION:

PbTe, (100) wafers used in development of a good chemical polish solution. Specimens were in a basket and solution agitated on a magnetic stirring hot plate. After etch, rinse in DI water. If a gold colored film appears, etch in 50% NaOH at 100—120°C, from cool (20°C?) up to RT, rinse in dilute HCl, and DI water rinse. See Tin and Germanium Telluride for additional data.

REF: Lorenz, M R — *J Electrochem Soc,* 112,240(1965)

PBTE-0008: Allgaier, R S — *Phys Rev,* 119,554(1960)

PbTe, (100) cleaved wafers used to study field magnetoresistance at room temperature and 77 K.

PBTE-0008

ETCH NAME: TIME:

TYPE: Acid, float-off TEMP:

COMPOSITION:

x .... HNO_3

x .... H_2O

DISCUSSION:

PbTe, thin film grown on muscovite mica (0001) substrates. Thin films were removed by the float-off technique using the solution shown for a structure study under a TEM microscope. See Mica: MI-0002.

REF: Chopea, K I — *J Appl Phys,* 37,2049(1960)

PBTE-0009: Allgaier, R S & Scanlon, W W — *Phys Rev,* 111,1029(1958)

PbTe, (100) cleaved wafers. A study of electron and hole mobility between RT and 4.2 K. A $1HNO_3$:$5H_2O$ can be used at RT for light surface cleaning. PbS and PbSe also studied.

PBTE-0010: Allgaier, R S — *Phys Rev,* 122,828(1958)

PbTe, (100) cleaved wafers. A study of magnetoresistance at 295, 77, and 4.2 K. Material has NaCl cubic structure. A $1HNO_3:5H_2O$ solution can be used at RT for light surfaces cleaning. Also studied PbS and PbSe.

PBTE-0012
ETCH NAME:
TYPE: Electrolytic, polish
COMPOSITION:
1 .... KOH, sat. sol.
1 .... Na_2SO_3, sat. sol.
5 .... glycerol

TIME: 4—7 min or 10—15 min
TEMP: RT or 40°C
ANODE: Pt
CATHODE: PbTe
POWER: 0.04 A/cm^{-2} or 0.09 A/cm^{-2} @ 4 V

DISCUSSION:
PbTe, (100), and PbSnTe, (100) wafers, 1 mm thick with 100—150 μm of damage depth from cutting. Lap specimens as follows: (1) 800-grit $SiC:H_2O$ slurry to remove 150 μm of material; (2) 1200-grit SiC:Glycerol to remove 50 μm; (3) final lap polish with 0.3 μm Linde alumina in paraffin or with "Brasso". For electrolytic etching, hold specimens with varnished tweezers: (1) 10—15 min at lower power to polish; (2) 4—7 min at higher power to develop slightly rounded, square dislocation pits. See: PBTT-0002.
REF: Quadeer, S et al — *J Electrochem Soc,* 129,2181(1982)

LEAD GERMANIUM TELLURIDE ETCHANTS

PBGT-0001
ETCH NAME:
TYPE: Halogen, preferential
COMPOSITION:
5 g .... NaOH
2 g .... I_2
10 ml .... H_2O

TIME:
TEMP:

DISCUSSION:
PbGeTe single crystal ingots grown by Horizontal Bridgman (HB) method using PbTe and Ge pieces as starting materials. The ingots were cubic with a temperature transition to rhombohedral. Transition temperature varies with amount of Ge fraction. Solution used to develop defects and structure.
REF: Tsuji, K — *J Phys Soc Jpn,* 4,1397(1984)

LEAD TIN TELLURIDE ETCHANTS

PBTT-0001
ETCH NAME: Hydrogen bromide
TYPE: Halogen, polish
COMPOSITION:
x .... HBr

TIME: 3 min
TEMP: RT

DISCUSSION:
PbSnTe, (100) wafers. Used as an immersion polish etch on PbS, PbSe, PbTe, and PbSnTe. Also used as a jet/spray from a squirt bottle followed by water jet/spray.
REF: Swets, D E — *J Electrochem Soc,* 131,172(1984)

PBTT-0002
ETCH NAME:
TYPE: Electrolytic, polish
COMPOSITION:
1 .... KOH, sat. sol.
1 .... Na_2SO_3, sat. sol.
5 .... glycerin

TIME: 10—15 min
TEMP: RT to 40°C
ANODE: PbSnTe
CATHODE: Pt
POWER: 0.5 A/cm^{-2} & 6 V

DISCUSSION:

PbSnTe, (100) wafers. Solution developed to electropolish specimens prior to dislocation etching. After cutting wafer, the damage layer is 100—150 μm deep. To remove damage: (1) lap on glass plate with 800-grit SiC in H_2O; (2) with 1200-grit SiC for additional 50 μm removal; (3) final mechanical polish with Linde A in paraffin or "Brasso" on Syvelt cloth; (4) rinse in TCE, then MeOH and (5) N_2 blow dry. Using electrolytic solution at 2 V, 4—7 min will produce a rounded cube, (100) finite crystal form (FCF), using small chunk specimens. During electropolishing stir solution and move specimens to prevent flow lines developing. The solution at the 2 V level also acts as a dislocation etch. Material used to fabricate CW laser diodes.

REF: Quadeer et al — *J Electrochem Soc,* 129,2650(1982)

PBTT-0003a
ETCH NAME:
TYPE: Acid, preferential
COMPOSITION:
25 ml .... KOH, sat. sol.
25 ml .... ethylene glycol (EG)
1 ml .... H_2O_2

TIME:
TEMP:

DISCUSSION:

PbSnTe, (100) wafers cut on inner side of doughnut shaped diamond blade. Mechanically polish, then etch with HBr:2% Br_2 to remove lap damage. Ingot was grown by Bridgman method, and solution shown used to develop etch pits in a material study.

REF: Richard, R C — *J Cryst Growth,* 71,192(1985)

PBTT-0003b
ETCH NAME:
TYPE: Halogen, polish/removal
COMPOSITION:
x .... 2% Br_2
x .... HBr

TIME:
TEMP:

DISCUSSION:

PbSnTe, (100) wafers cut from a Bridgman grown ingot. After mechanical lapping, solution shown used to remove residual lap damage before preferential etching to develop defects and pits. See PBTT-0003a.

REF: Ibid.

LEAD ZIRCONATE ETCHANTS

PBZO-0001a
ETCH NAME: Hydrochloric acid
TYPE: Acid, polish
COMPOSITION:
x .... HCl

TIME:
TEMP:

DISCUSSION:

$PbZrO_3$, single crystal specimens. Solution is an etch for this material. Specimens used in an optical study. Also $NaNbO_3$ single crystals.

REF: Jona, F et al — *Phys Rev*, 97,1584(1955)

PBZO-0001b

ETCH NAME: Lead sulfate TIME:

TYPE: Acid, polish TEMP:

COMPOSITION:

x .... $PbSO_4$

DISCUSSION:

$PbZrO_3$ single crystal specimens. Solution is an etch for this material. See PBZO-0001a for further discussion.

REF: Ibid.

PBZO-0001c

ETCH NAME: Sulfuric acid TIME:

TYPE: Acid, polish TEMP:

COMPOSITION:

x .... H_2SO_4

DISCUSSION:

$PbZrO_3$ single crystal specimens. Solution is an etch for this material. See PBZO-0001a for further discussion.

REF: Ibid.

PHYSICAL PROPERTIES OF LITHIUM, Li

Classification	Alkali metal
Atomic number	3
Atomic weight	6.94
Melting point (°C)	180.5
Boiling point (°C)	1336 (1317)
Density (g/cm³)	0.534
Thermal conductance (cal/sec)(cm²)(°C/cm)	0.17
Specific heat (cal/g) 25°C	0.849
Latent heat of fusion (cal/g)	103.2
Heat of vaporization (cal/g)	5100
Atomic volume (W/D)	13.1
1st ionization energy (K-cal/g-mole)	124
1st ionization potential (eV)	5.39
Electronegativity (Pauling's)	1.0
Covalent radius (angstroms)	1.23
Ionic radius (angstroms)	0.86
Electrical resistivity (micro-ohms-cm) 20°C	9.446
Electron work function (eV)	2.39
Cross section (barns)	71
Vapor pressure (°C)	1097
Coefficient of linear thermal expansion ($\times 10^{-6}$ cm/cm/°C) 20°C	56
Refractive index (n=)	3.16
Hardness (Mohs — scratch)	1—2

Crystal structure (isometric — normal)	(100) cube, bcc
Color (solid)	Silver white
(flame)	Carmine-red
Cleavage (cubic — poor)	(001)

LITHIUM, Li

General: Does not occur as a free element in nature. There are over a dozen lithium-bearing minerals as silicates, fluorates, and phosphates. Spodumene, $LiAl(SiO_3)_2$, is the primary ore and occurs in acidic pegmatites, occasionally in single crystals of large size — 4 to 5 ft long, over a foot in diameter. The fine, transparent, emerald-green variety is a precious gem stone. Lithium metal is very soft, silver-white in color, and the lightest of all known metals. The metal is classified as an alkaline metal, and its salts are similar to those of sodium and potassium. When ignited, lithium produces a brilliant crimson flame.

In industry it is used with several irons/steels and other metals as an additive to increase tensile strength and for resistance to corrosion. Because of its brilliant red color when burned it is a major component in flares and pyrotechnics. Several salts, notably lithium chloride, LiCl, have important medical applications, and are used in glass to increase ultraviolet transmission. Other lithium compounds improve the gloss in enamels.

Technical Application: The pure metal has limited use in Solid State or semiconductor processing. It has been evaporated along with aluminum on quartz crystal frequency blanks to reduce mass-loading and improve frequency stability and control, and may have similar applications in high frequency microelectronic devices, e.g., gallium arsenide. As a diffusant in germanium, lithium-drift radiation detectors have been fabricated. Lithium reacts with water and moisture in the air, is stored in oil and handled in a dry atmosphere. Single crystals spheres have been fabricated and subjected to study.

Etching: Soluble in acids, but deliquesces in water and alcohols.

LITHIUM ETCHANTS

LI-0001
ETCH NAME: Nitric acid
TYPE: Acid, removal — TEMP: RT
COMPOSITION:
x .... HNO_3
DISCUSSION:

Li, specimens. Lithium is soluble in most single acids. Should be handled in a dry inert atmosphere as it reacts with water vapor in air forming lithium hydroxide.

REF: Hodgman, C D — *Handbook of Chemistry and Physics,* 27th ed, Chemical Rubber Co., Cleveland, OH, 1943, 402

LI-0002
ETCH NAME: Heat — TIME:
TYPE: Thermal, forming — TEMP:
COMPOSITION:
x .... heat
DISCUSSION:

Li, single crystal spheres fabricated by melting small pieces of lithium in paraffin with light stirring, and then allowing it to solidify.

REF: Splitstone, P L — Doctoral thesis, Ohio State University, Columbus, 1955
LI-0003: Holcomb, D F — *Phys Rev,* 112,1599(1958)

Li, single crystal spheres, 50 μm diameter handled in oil and used in a magnetic resonance study.

LI-0004: Holcomb, D F & Norberg, R E — *Phys Rev,* 98,1074(1955)

Li, single crystal spheres in oil? Used in an electronic characterization study.

LI-0005

ETCH NAME: Lithium — TIME:

TYPE METAL: Diffusant — TEMP: 1000°C

COMPOSITION:

x .... Li, in oil

DISCUSSION:

Li, powder in oil used as a diffusant source for germanium. Germanium rods were cut from (111) surface oriented wafers, painted with Li/oil and diffused. After diffusion specimens electrically stabilized in hot oil bath under power load. Devices used to fabricate lithium-drift germanium radiation detectors.

REF: Walker, P — personal development, 1959

LI-0006

ETCH NAME: Methyl alcohol — TIME: 30 sec

TYPE: Alcohol, polish — TEMP: RT

COMPOSITION:

(1) x .... MeOH (CH_3OH) (2) x .... EOH (CH_3CH_2OH) (3) x .... 2-propanol (IPA) ($CH_3CHOHCH_3$)

DISCUSSION:

Li samples cut with a razor blade under oil (also cut sodium and potassium), then into heptane to remove oil, then into tetrahydrofuran to stop heptane removal action. Ten alcohols evaluated with increasing molecular weight which reduces in activity. On lithium, MeOH polished surfaces; on sodium developed grain boundaries, and some brilliantly reflective (hk1) planes, with other planes dull and all pitted; on potassium reaction was violent. Ethyl alcohol, CH_3CH_2OH (EOH) developed grain boundaries and (hk1) polished surface on lithium; on sodium developed grain boundaries with (hk1) planes both brilliant or dull with pitting; on potassium, polished surfaces. 2-Propanol $CH_3CHOHCH_3$ (IPA) on lithium was like EOH; on sodium like EOH but without pitting; on potassium developed polished surfaces. All other alcohols evaluated were not polishing solutions: black surfaces on lithium; grain boundaries with (hk1) planes brilliant to dull with and without pitting on sodium, and dull, pitted surfaces for potassium, in general.

REF: Castellano, R N & Schmidt, P H — *J Electrochem Soc,* 118,653(1971)

LI-0007

ETCH NAME: Diethyl ether — TIME:

TYPE: Organic oxide, polish — TEMP:

COMPOSITION:

x .... $C_4H_{10}O$

DISCUSSION:

Li specimens. Pre-polish surfaces with methanol (MeOH), then final polish with ether. (*Note:* Ether should be handled with caution as it can be flammable and explosive when mixed with air.)

REF: Bowers, R et al — *Material Science Center Rep #3,* Cornell University, Ithaca, NY, 1961

LI-0008
ETCH NAME: Potassium nitrate — TIME: x hours
TYPE: Salt, removal — TEMP: 300°C
COMPOSITION:
x .... KNO_3, crystals
DISCUSSION:
Li, applied as molten LiOH on rutile, TiO_2 specimens to metal decorate dislocations in rutile. After decoration use the molten flux salt shown to remove excess lithium.
REF: Johnson, O W — *J Appl Phys,* 35,3048(1964)

LITHIUM BISMUTHIDE ETCHANTS

LIBI-0001
ETCH NAME: BRM — TIME:
TYPE: Halogen, polish — TEMP: RT
COMPOSITION:
x .... x% Br_2
x .... MeOH
DISCUSSION:
Li_3Bi single crystal ingots grown in a study of new semiconducting compounds. Other compounds were Ag_2Se, TlSe, Tl_2Se_3, SnSe, $SnSe_2$, In_2Tl_3, Bi_2Se_3, $AgInTe_2$, In_2Te_3.
REF: Mooser, E & Pearson, W B — *Phys Rev,* 101,492(1956)

PHYSICAL PROPERTIES OF LITHIUM BROMIDE, LiBr

Classification	Bromide
Atomic numbers	3 & 35
Atomic weight	86.36
Melting point (°C)	547
Boiling point (°C)	1265
Density (g/cm³)	3.46
Hardness (Mohs — scratch)	1.5—2
Crystal structure (isometric — normal)	(100) cube
Color (solid)	White
Cleavage (cubic)	(001)

LITHIUM BROMIDE, LiBr

General: Does not occur as a natural mineral although both the bromide and chloride are found in some mineral springs and in the soil. Both beets and tobacco retain lithium, but the solid is deliquescent in air. There are two major minerals containing lithium: petalite, $LiAl(Si_2O_5)_3$ with about 5% lithium and lepidolite mica, chiefly $(OH),F)_2KLiAl_2Si_3O_{10}$, and it is found in other minerals that are characteristic of granite pegmatites, such as spodumene, $LiAl(SiO_3)_2$ with 8.4% lithium.

A primary industrial use, as a mixture of lithium bromide and chloride, is in the electrolysis of lithium minerals for their metal content. Lithium compounds have major use in the glassy industry as a flux, to improve the electrical insulating properties and to increase ultraviolet transmission. They also are used in enamels to improve gloss, and have medical applications.

Technical Application: Lithium bromide is not used in Solid State device processing although it could be used in preferential etching solutions. The compound has been grown as a single crystal for general morphology and defect studies.

Etching: Soluble in water, alcohols and ether.

LITHIUM BROMIDE ETCHANTS

LIBR-0001
ETCH NAME: TIME: 5—10 sec
TYPE: Alcohol, defect TEMP: RT
COMPOSITION:
x .... amyl alcohol
x .... $BaBr_2$ (trace)
DISCUSSION:
LiBr, (100) cleaved wafers used in a study of slip. After etching in solution shown, rinse in petroleum and air dry. Also studied LiCl.
REF: MacMillan, N H & Smith, D A — *Phil Mag*, 14,869(1966)

PHYSICAL PROPERTIES OF LITHIUM CHLORIDE, LiCl

Classification	Chloride
Atomic number	3 & 17
Atomic weight	42.40
Melting point (°C)	613
Boiling point (°C)	1353
Density (g/cm^3)	2.07
Hardness (Mohs — scratch)	1.5—2
Crystal structure (isometric — normal)	(100) cube
Color (solid)	White
Cleavage (cubic)	(001)

LITHIUM CHLORIDE, LiCl

General: Does not occur as a natural compound as it is deliquescent in air. It does occur in some mineral springs and in the soil. Beets and tobacco both contain lithium. There are several minerals containing lithium, such as petalite, $LiAl(Si_2O_5)_3$; lepidolite mica, chiefly $(OH)_2KLiAl_2Si_3O_{10}$; and in granitic pegmatites, the mineral spodumene, $LiAl(SiO_3)_2$ as the primary ore of lithium.

Mixtures of the bromide and chloride are used in the electrolysis of lithium-bearing minerals for the separation of the metal. Lithium compounds are used in the glass industry as a flux, to improve electrical insulating properties and to increase ultraviolet transmission. Also used in enamels to improve gloss, and there are important medical applications. They are the red flame color (lithium) in flares and pyrotechnics.

Technical Application: Lithium chloride is not used in Solid State device processing, although it could be used in etching solutions.

The compound has been grown as a single crystal for general morphology and defect studies.

Etching: Soluble in water, alcohols, and acetic acid.

LITHIUM CHLORIDE ETCHANTS

LICL-0001
ETCH NAME: TIME: 10 sec
TYPE: Alcohol, defect TEMP: RT
COMPOSITION:
x .... isopropyl alcohol
x .... *$BaBr_2$

*Saturate the alcohol.

DISCUSSION:
LiCl, cleaved wafers used in a study of slip. After etching in the solution shown, rinse in CCl_4 and air dry. Also studied LiBr.
REF: MacMillan, N H & Smith, D A — *Phil Mag,* 14,869(1966)

LICL-0002
ETCH NAME: TIME:
TYPE: Acid, dislocation TEMP:
COMPOSITION:
x .... x% $FeCl_3$
DISCUSSION:
LiCl wafers. Solution referred to as a dislocation etch for ionic crystals. Used in a study of reduced cohesion due to dislocations.
REF: Gilman, J J — *J Appl Phys,* 32,739(1961)

PHYSICAL PROPERTIES OF LITHIUM FLUORIDE, LiF

Classification	Fluoride
Atomic numbers	3 & 9
Atomic weight	25.94
Melting point (°C)	870
Boiling point (°C)	1676
Density (g/cm^3)	2.3
Hardness (Mohs — scratch)	1.5—2
Crystal structure (isometric — normal)	(100)
Color (solid)	Colorless
Cleavage (cubic — perfect)	(001)

LITHIUM FLUORIDE, LiF

General: Does not occur in nature as a single crystal compound. The fluoride mineral cryolithionite, 2NaF.3LiF.$2AlF_2$ occurs in association with cryolite, Na_2AlF_6 as a minor mineral species.

Primary use industrially is as a replacement for natural calcium fluoride in the fabrication of prisms. Like lithium bromide and chloride it is used in the glass industry to improve electrical insulation and ultraviolet transmission characteristics and as a flux. Also used as the red color (lithium) in flares and pyrotechnics. Cryolite is a major flux and growth compound in metal processing.

Technical Application: Lithium fluoride is not used in Solid State device fabrication, but it has been grown as an artificial single crystal and studied for general morphology and defects. It is used in the fabrication of prisms and special lenses.

Etching: Slightly soluble in cold water; soluble in acids.

LITHIUM FLUORIDE ETCHANTS

LIF-0001a
ETCH NAME: CP4, modification | TIME: 30 sec
TYPE: Acid, dislocation | TEMP: RT
COMPOSITION:
100 ml .... HF
100 ml .... HAc
160 ml .... HNO_3
2 ml .... Br_2
DISCUSSION:

LiF, (100) cleaved wafers. Used iron forceps to hold specimens. Iron slowly dissolves in the solution and reduces the fast acid reaction. Rinse in ethyl alcohol and then in ether.

REF: Gilman, J J & Johnston, W S — *J Appl Phys,* 27,1018(1956)

LIF-0006: Davisson, J W et al — *J Appl Phys,* 35,3017(1964)

LiF, (111) wafers. Solution used to develop imperfections in specimens that were associated with electrical breakdown.

LIF-0002b
ETCH NAME: | TIME: 1 min
TYPE: Acid, dislocation | TEMP: RT
COMPOSITION:
x .... 2×10^6 *N* FeF_3
x .... H_2O
DISCUSSION:

LiF, (100) cleaved wafers. Mix solution and use immediately with vigorous stirring. Rinse in EOH then ether. Fresh solution will produce 10 μm size pyramid etch pits. An "aged" solution applied to the same specimen develops the old pits only as mounds.

REF: Gilman, J J et al — *J Appl Phys,* 29,747(1958)

LIF-0003a
ETCH NAME: Stearic acid | TIME:
TYPE: Acid, preferential | TEMP:
COMPOSITION:
x .... 10^5—10^6 *N* stearic acid
DISCUSSION:

LiF, (100) wafers. Various fatty acids were evaluated as dislocation etchants for lithium fluoride. This is one of the better fatty acids for developing dislocations.

REF: Westwood, A R C et al — *J Appl Phys,* 33,1764(1962)

LIF-0004
ETCH NAME: Hydrogen peroxide | TIME:
TYPE: Acid, selective | TEMP:
COMPOSITION:
x .... 3% H_2O_2

DISCUSSION:

LiF, (100) wafers studied for defects. Solution used to develop defects and to selectively etch structure.

REF: Urosovskava, A A — *Kristallografiya,* 3,726(1958)

LIF-0005

ETCH NAME: Ultrasonic — TIME:

TYPE: Vibration, preferential — TEMP: RT

COMPOSITION: — POWER: 0.5 W/cm^2 @ 25 Kc

x .... ultrasonic

x .... H_2O

DISCUSSION:

LiF, (111) cleaved wafers subjected to irradiation by ultrasonic vibration in water. Develops triangular etch pits which become rounded with extended time with more pits appearing near wafer edges. KI and KBr also studied. (*Note:* See SI-0048 — ultrasonic etch-stepping patterns.)

REF: Kapustin, A P — *Sov Phys-Cryst,* 4,247(1960)

LIF-0007

ETCH NAME: — TIME:

TYPE: Acid, preferential — TEMP:

COMPOSITION:

x .... HF

x .... x% $FeCl_3$

DISCUSSION:

LiF, (100) single crystal specimens used in a study of pipe diffusion. Solution used as a defect etch. Sodium diffused in along dislocation lines.

REF: Tucker, R et al — *J Appl Phys,* 34,445(1963)

LIF-0008: Gilman, J J & Johnston, W S — *Dislocations in Crystals,* John Wiley & Sons, New York, 1956, 116

Reference for solution used in LIF-0007.

LIF-0009

ETCH NAME: Ferric chloride — TIME:

TYPE: Acid, preferential — TEMP:

COMPOSITION:

x .... x% $FeCl_3$

DISCUSSION:

LiF, (100) wafers. Solutions of ferric chloride referred to as dislocation etchants for ionic crystals.

REF: Gilman, J J — *J Appl Phys,* 32,739(1961)

LIF-0002b

ETCH NAME: "A" etch — TIME:

TYPE: Acid, preferential — TEMP:

COMPOSITION: — RATE: (100) 5 Å/sec; (111) 34 Å/sec

x .... LiF

x .... H_2O

x .... EOH

DISCUSSION:

LiF, (100) wafers used in a study of dislocations. Also used "W" etch.

LIF-0011a: Ives, M B — *J Appl Phys,* 32,1534(1961)

LiF, (100) and (111) wafers. Used both "A" and "W" etch in a study of orientation dependence dissolutionment. Rates shown under LIF-0010 are from this reference.

LIF-0002c

ETCH NAME: "W" etch — TIME:

TYPE: Acid, preferential — TEMP:

COMPOSITION: — RATE: (100) 91 Å/sec; (111) 109 Å/sec

x .... $HgCl_2$

x .... EOH

DISCUSSION:

LiF, (100) wafers used in a dislocation study. Also used "A" etch.

REF: Ibid.

LIF-0011b: Ives, M B — *J Appl Phys,* 32,1534(1961)

Li, (100) and (111) wafers. See discussion under LIF-0011a.

PHYSICAL PROPERTIES OF LITHIUM NITRIDE, Li_3N

Classification	Nitride
Atomic numbers	3 & 7
Atomic weight	34.83
Melting point (°C)	845
Boiling point (°C)	
Density (g/cm³)	
Hardness (Mohs — scratch)	5—6
Crystal structure (isometric — normal)	(100) cube
Color (solid) crystalline	Grey/black
(solid) amorphous	Red-brown
Cleavage (cubic)	(001)

LITHIUM NITRIDE, Li_3N

General: Does not occur as a natural compound as nitrogen does not combine in nature to form metallic compounds, although there are phosphates and chlorates of lithium. There has been no use of the nitride in metal processing, to date.

Technical Application: Lithium nitride has not been used in Solid State processing, but has been studied as a thin film grown on lithium blanks.

Etching: Alcohols.

LITHIUM NITRIDE ETCHANTS

LIN-0001

ETCH NAME: Methyl alcohol — TIME:

TYPE: Alcohol, removal — TEMP:

COMPOSITION:

x .... MeOH, CH_3OH

DISCUSSION:

LiN_x thin films deposited on lithium blank specimens in a nitrogen atmosphere. The nitride deliquesces in alcohols, but they can be used for general etching and removal.

REF: Fremont, M — *Rev Chem Miner,* 4,447(1967)

LITHIUM NIOBATE ETCHANTS

LINB-0002
ETCH NAME:
TYPE: Acid, preferential
COMPOSITION:
1 .... HF
2 .... HNO_3
TIME:
TEMP: Boiling
DISCUSSION:

$LiNbO_3$, single crystal specimens used in a study of domain structure and the Curie temperature. Solution used to develop dislocations and anti-parallel polar domains which are larger than those observed in lithium tantalate.

REF: Levinstein, H J et al — *J Appl Phys*, 37,4585(1966)

LINB-0001
ETCH NAME: Gold
TYPE: Metal, decoration
COMPOSITION:
(1) x .... Au
(2) x .... Pt
TIME:
TEMP: 500—700°C
POWER: 250—500 V
DISCUSSION:

$LiNbO_3$, single crystal specimens used in a study of dislocations. Plate gold or platinum on surfaces and use temperatures and powers shown to diffuse metal into the material. Will develop a strain network and defects. This procedure is good for transparent crystals and will develop tilt and twist boundaries. Also used on lithium tantalate, $LiTaO_3$.

REF: Levinstein, H J & Capio, C D — *J Appl Phys*, 38,2761(1967)

LITHIUM TANTALATE ETCHANTS

LITA-0002
ETCH NAME:
TYPE: Acid, preferential
COMPOSITION:
1 .... HF
2 .... HNO_3
TIME:
TEMP: Boiling
DISCUSSION:

$LiTaO_3$ single crystal specimens used in a study of domain structure and the Curie temperature. Solution used as a dislocation etch and will also develop anti-parallel polar domains to a lesser extent than those observed in lithium niobate, $LiNbO_3$.

REF: Levinstein, H J et al — *J Appl Phys*, 37,4585(1966)

LITA-0001
ETCH NAME: Gold
TYPE: Metal, decoration
COMPOSITION:
(1) x .... Au
(2) x .... Pt
TIME:
TEMP: 500—700°C
POWER: 250—500 V
DISCUSSION:

$LiTaO_3$, single crystal wafers used in a study of dislocations. Plate gold or platinum on surfaces and use temperatures and electrical powers shown to diffuse metal into the material. Will develop a strain network and defects. This procedure is good for transparent crystals and will develop tilt and twist boundaries. Also used on lithium niobate, $LiNbO_3$.

REF: Levinstein, H J & Capio, C D — *J Appl Phys*, 38,2761(1967)

LUCITE ETCHANTS

LUC-0001a
ETCH NAME: Acetone — TIME:
TYPE: Alcohol, dissolve — TEMP: RT
COMPOSITION:
x .... CH_3COCH_3 (Ace)
DISCUSSION:

Lucite, sheet and parts. Add acetone to Lucite shavings to form a glue. Use glue to cement Lucite parts together.

REF: Simonds, H R et al — *Handbook of Plastics*, 2nd ed, January 1949
LUC-0002: Walker, P — personal application, 1958/1985

Used this glue to form lucite containers and seals for light pressure box operation (0.5 Torr) and light vacuum operation (10^{-3} Torr). Also used in making three-dimensional electrical power level models of silicon transistors as a display item, and crystallographic structure models using plastic rods and balls. The solid colored plastic balls were drilled using a specially constructed aluminum holding form built for that purpose.

LUC-0001b
ETCH NAME: — TIME:
TYPE: Halogen, cleaning — TEMP: RT to warm
COMPOSITION:

(1)		(2)	
1 g	.... I_2	1 tsp	.... *tincture of iodine
50 ml	.... MeOH	1 gal	.... H_2O
1000 ml	.... H_2O		

*xK:xI_2:xEOH

DISCUSSION:

Lucite, sheet and parts. Both solutions can be used to clean and disinfect lucite and polycarbonate plastics.

REF: Ibid.
LUC-0002b: Walker, P — personal application, 1965—1985

Used the methyl alcohol mixture to clean chemical sink plastic hoods in epitaxy system cleanrooms.

LUC-0003
ETCH NAME: — TIME:
TYPE: Acid, cleaning — TEMP:
COMPOSITION:
x .... HCl
x .... glycerin
DISCUSSION:

Lucite, sheet and parts can be cleaned in this solution. The liquid mixture has the same refractive index as Lucite: n = 1.49.

REF: Ivanstov, O P — *Dokl Akad Nauk SSSR*, 156,567(1947)

PHYSICAL PROPERTIES OF LITHIUM SULFIDE, Li_2S

Classification	Sulfide
Atomic numbers	3 & 16
Atomic weight	45.94
Melting point (°C)	450 est.
Boiling point (°C)	
Density (g/cm^3)	1.66
Hardness (Mohs — scratch)	2—3
Crystal structure (isometric — normal)	(100) cube
Color (solid)	Yellowish
Cleavage (cubic)	(001)

LITHIUM SULFIDE, Li_2S

General: Does not occur as a natural compound. The most important lithium mineral is the mica leipidolite, and there are other silicates and phosphates containing lithium. There is no application in the metals industries for the compound, but there is some use in chemistry and medicine.

Technical Application: The compound, as shown above, has not been used in Solid State processing, but a trinary compound as $LiInS_2$ has been deposited on (001) silicon as a crystalline or glassy thin film.

Etching: Mixed acids of HF:HNO_3.

LITHIUM INDIUM SULFIDE ETCHANTS

LIIS-0001
ETCH NAME: TIME:
TYPE: Acid, removal TEMP:
COMPOSITION:
x .... HF
x .... HNO_3
DISCUSSION:

$LiInS_2$, (001) oriented thin films deposited on (111) silicon wafers, or as an amorphous glassy film. Deposition was by evaporation from a moly boat in excess sulfur, or by directional solidification. Solution shown was used as a removal or patterning etch.
REF: Koriyama, K & Saitoh, J — *Solid Thin Films,* 111,331(1984)

PHYSICAL PROPERTIES OF MAGNESIUM, Mg

Classification	Alkaline metal
Atomic number	12
Atomic weight	24.3
Melting point (°C)	650
Boiling point (°C)	1107
Density (g/cm^3)	1.74
Thermal conductance (cal/sec)(cm^2)(°C/cm)	0.37 (0.3867)
Specific heat (cal/g) 20°C	0.245
Latent heat of fusion (cal/g)	82.2 (88)
Heat of vaporization (cal/g)	1260
Atomic volume (W/D)	14.0
1st ionization energy (K-cal/g-mole)	176
1st ionization potential (eV)	7.64
Electronegativity (Pauling's)	1.2
Covalent radius (angstroms)	1.36
Ionic radius (angstroms)	0.66
Linear coefficient of expansion ($\times 10^{-6}$ cm/cm/°C)	26
Electrical resistivity (micro-ohm-cm) 20°C	4.46
Modulus of elasticity (kg/mm^2)	4570
Poisson ratio	0.35
Tensile strength (psi)	33,000
Cross section (barns)	0.063
Vapor pressure (°C)	909
Electron work function (eV)	3.46
Hardness (Mohs — scratch)	2
Crystal structure (hexagonal — normal)	$(10\bar{1}0)$ prism
Color (solid)	Silver-white metal
Cleavage (basal)	(0001)

MAGNESIUM, Mg

General: Does not occur as a native element. It is widely distributed in many minerals, constitutes about 2% of the earth's crust and some 0.5% in sea water as the chloride, $MgCl_2$. Magnesite, $MgCO_3$ — magnesia — is a major ore. Dolomite, $CaMg(CO_3)_2$ and talc (soapstone), $(OH)2Mg_3Si_4O_{10}$, are building stones. In its fibrous form, soapstone is asbestos, which can be carded like flax due to its flexibility. The amphibole minerals tremolite and actinolite contain magnesium and, in their fibrous forms, are also asbestos. When silicified, such asbestos becomes semi-precious to precious gem stones: Cat's Eye.

When the pure metal is exposed to air it becomes slowly coated with carbonate, e.g., becomes magnesite, which stabilizes the surface against further alteration.

As the pure metal it has important applications in the metal industry. Used alone, it is a lightweight construction metal; as an alloy with iron or aluminum it increases structural strength. The aluminum skin of airplanes always contains magnesium. When ignited it burns with a searing white flame — used in flares, tracer bullets, incendiary bombs, and flash bulbs. During wartime it is classified as a strategic material, and has long been used as the white flame color in pyrotechnics.

Magnesium oxide, MgO and the hydroxide, $Mg(OH)_2$ are found in every chemical laboratory, taken internally as an antidote against acid poisoning and, as a water paste, used on the skin to neutralize acids burns. It is milk of magnesia and Maalox, and there are important organic magnesium compounds used in other medical applications.

Technical Application: The pure metal is not widely used in Solid State or the semiconductor industry, though it is used as a dopant element in some compound semiconductors and is under evaluation as a silicide: MgSi, Mg_2Si, and $MgSi_2$. It has been used to grow the compound semiconductor, magnesium selenide, Mg_2Se, and could be used for a similar telluride.

It has been grown artificially as a single crystal and studied for its electrical and physical properties, to include defect structure with and without stress and strain.

Caution should be observed in handling pure magnesium as, when in air, if struck a sharp blow or near an electrical or spark it can ignite. And once ignited it will continue to burn even under water or when smothered with dirt.

Etching: Soluble in mineral acids; HCl, H_2SO_4, and mixed acids. Neon ionic bombardment is preferential.

MAGNESIUM ETCHANTS

MG-0001
ETCH NAME: Hydrochloric acid, dilute — TIME: 3 h
TYPE: Acid, cutting — TEMP: RT
COMPOSITION: — RATE: 75 ft/min cutting rate

1 .... HCl
1 .... H_2O

DISCUSSION:

Mg, specimens and magnesium alloys. Solution used as an acid-saw cutting system. A mercerized cotton crochet thread was used as the cutting vehicle. Pass the thread through the acid solution, then across the specimen. It required 1 h to cut a 1″ diameter slice at 75 ft/min, cutting speed. Authors say the slice was flat to within 1 mil.

REF: Long, T R & Smith, C S — *Acta Metall*, 25,200(1957)

MG-0004: Reed-Hill, R E & Robertson, W D — *Acta Metall*, 5,717(1957)

Mg, single crystal wafers used in a study of deformation twinning. Specimens were acid machine cut and electropolished, then etched in HCl for dislocation development.

MG-0005: Conrad, H et al — *Acta Metall*, 9,367(1961)

Mg, single crystal specimens used in a study of plastic flow at low temperatures. Solution was 25% HCl, and applied as a polish etch.

MG-0002
ETCH NAME: Nitric acid — TIME:
TYPE: Acid, cleaning — TEMP: RT
COMPOSITION:

x .... 1 *N* HNO_3

DISCUSSION:

Mg, pieces, Solution used to clean magnesium for the growth of single crystal Mg_2Ge. After etch cleaning rinse in DI water three times and final rinse in MeOH, then dry.

REF: Kromar, H et al — *J Appl Phys*, 36,2461(1965)

MG-0003a
ETCH NAME: Nitric acid, dilute — TIME:
TYPE: Acid, saw — TEMP:
COMPOSITION:

x .... 20% HNO_3

DISCUSSION:

Mg, single crystal wafers. Use solution for acid saw cutting of material.

REF: Tsui, R T C — *Acta Metall*, 15,1722(1967)

MG-0009: Shally, J — *Phil Mag*, 13,9(1966)

Mg, single crystal wafers used in a study of the structure of quenched magnesium. A 5—10% solution used to etch-thin specimens. Used low concentration for large specimens (5 cm^2 size); high concentration for small specimens ($^1/_2$ cm^2 size). After thinning, quench in MeOH and dry.

MG-0003b
ETCH NAME: TIME:
TYPE: Electrolytic, polish TEMP: 0°C
COMPOSITION: ANODE: Mg
1 .... HNO_3 CATHODE: SS
4 .... MeOH POWER: 1 A/cm^2 & 8—10 V
DISCUSSION:

Mg, single crystal wafers. Use solution to electropolish and to etch thin specimens for electron microscope (EM) study.

REF: Ibid:.

MG-0012
ETCH NAME: Hydrochloric acid, dilute TIME:
TYPE: Acid, polish TEMP:
COMPOSITION:
(1) x .... 20% HCl (2) x .... 80% HCl
DISCUSSION:

Mg, specimens used in a stress evaluation. Both solutions were used as polishing etchants. Also studied aluminum and molybdenum.

REF: Feuerstein, S & John, I W — *J Appl Phys*, 40,3334(1969)

MG-0006
ETCH NAME: Glacial acetic acid, dilute TIME: 1—2 min
TYPE: Acid, removal TEMP: RT
COMPOSITION:
x .... CH_3COOH (HAc)
x .... H_2O
DISCUSSION:

Mg specimens. Solution used to clean and polish specimens before studying anodic dissolution of magnesium and magnesium alloys in aqueous salt solutions.

REF: Glickman, R — *J Electrochem Soc*, 106,83(1959)
MG-0007: Slutsky, L T & Garland, C W — *Phys Rev*, 107,972(1957)

Mg specimens used in a study of elastic constants. Mechanical lap with 000 emery grit. Then etch with dilute acetic acid, and continue mechanical polish.

MG-0008
ETCH NAME: TIME:
TYPE: Electrolytic, polish/thin TEMP:
COMPOSITION: ANODE:
20 .... $HClO_4$ CATHODE:
80 .... EOH POWER: 0.2 A/cm^2 & 10—20 V
DISCUSSION:

Mg, (0001) wafers used in a study of transmission through thin crystals. Solution used to polish and thin specimens. Crystals of MgO and Mo also were studied but thinned with other etch solutions and methods.

REF: Thomas, G & Huftstutler, M C Jr — *J Appl Phys*, 31,1834(1960)

MG-0010
ETCH NAME: Neon
TYPE: Ionized gas, preferential
COMPOSITION:
x .... Ne^+ ions

TIME:
TEMP:
GAS FLOW:
PRESSURE:
POWER:

DISCUSSION:
Mg, single crystal specimens. Neon ion bombardment used to develop structure and orientation figures. Other metals studied were Al, Bi, Cd, Cu, Co, Sn, and Zn.
REF: Yurasova, V E — *Kristallografiya*, 2,770(1957)

MG-0011
ETCH NAME: Sulfuric acid, dilute
TYPE: Acid, removal
COMPOSITION:
x .... H_2SO_4
x .... H_2O
x .... ?(acids)

TIME:
TEMP:

DISCUSSION:
Mg specimens used in a study of neutron-transfer reactions produced by N^+ ion bombardment.
REF: Halbert, M L et al — *Phys Rev*, 106,251(1957)

MG-0014a
ETCH NAME:
TYPE: Electrolytic, polish
COMPOSITION:
250 g .... CrO_3/l
150 g .... H_2SO_4/l

TIME: x sec
TEMP: 113°F
ANODE: Mg
CATHODE: Steel
POWER: 150—500 A/ft^2 & 6 V

DISCUSSION:
Mg specimens. Solution used to electropolish material.
REF: ASTE Committee — *Tool Engineers Handbook*, McGraw-Hill, New York, 1949

MG-0014b
ETCH NAME: Acetic acid
TYPE: Electrolytic, polish
COMPOSITION:
12—16 oz .... HAc (5—15%)

TIME: 20—60 sec
TEMP: 85°C
ANODE: Mg
CATHODE: Pb
POWER: 400 A/ft^2 & 4 V

DISCUSSION:
Mg specimens. Solution used as an electropolishing etchant.
REF: Ibid.

MG-0014c
ETCH NAME: Hydrofluoric acid
TYPE: Electrolytic, polish
COMPOSITION:
x .... 25% HF

TIME: 10 min
TEMP: RT
ANODE: Mg
CATHODE: Pb
POWER: 400 A/ft^2 & 45 V

DISCUSSION:

Mg specimens. Solution used as an electropolishing etchant.

REF: Ibid.

MG-0014d

ETCH NAME:

TYPE: Electrolytic, polish

COMPOSITION:

- 7.5 oz .... $NaCO_3$
- 14 oz .... $Na_3PO_4.H_2O$
- 28 oz .... NaOH
- 0.1—1 oz .... ammonium biaurylsulfate

TIME: 1—2 min

TEMP: 140—195°F

ANODE: Mg

CATHODE: steel

POWER: 216 A/ft^2 & 4 V

DISCUSSION:

Mg specimens. Solution used as electropolish etchant.

REF: Ibid.

PHYSICAL PROPERTIES OF MAGNESIUM FLUORIDE, MgF_2

Classification	Fluoride
Atomic numbers	12 & 9
Atomic weight	62.32
Melting point (°C)	1396 (1263)
Boiling point (°C)	2230 (2227)
Density (g/cm^3)	2.9—3.2
Refractive index (n=)	1.378
Surface free energy (ergs cm^{-2}) −173°C	750
Hardness (Mohs — scratch)	5—6
(Knoop — kgf/mm^2)	430
Crystal structure (tetragonal — normal)	(110) prism, 1st order
Color (solid)	Colorless
Cleavage (basal)	(001)

MAGNESIUM FLUORIDE, MgF_2

General: Occurs in nature as the mineral sallaite, MgF_2 with an atomic structure similar to rutile, TiO_2. It is of minor occurrence in a glacial moraine in France, and at Vesuvius in Italy. It can be formed artificially by reacting HF with MgO.

There is no application for the natural mineral in metal processing due to scarcity. The metal magnesium is the eighth most abundant element in nature, occurs in many minerals, and is of major importance as a metal. As ribbon or powder, when heated, it burns with an intense white light and is used for flares, incendiary bombs, and in pyrotechnics. It is ligher than aluminum and, as an aluminum alloy, has long been the major construction material for aircraft, in addition to a general construction material, and many, many aluminum products. The fluoride does have medical applications, as do several other magnesium salts. Other magnesium containing minerals, such as talc, soapstone, etc., have wide use, as do the fibrous varieties — asbestos. Note that milk of magnesia, $Mg(OH)_3$, is a medical specific against internal acid poisoning.

Technical Application: The fluoride has had little use thus far in Solid State device processing. It has been used as a single crystal (100) or (110) oriented substrate in the study of optical thin films as evaporated aluminum, chromium, and silver. As a dielectric compound it has application as a lens and filter material based on its optical properties.

Etching: HNO_3. Slowly soluble in other acids.

MAGNESIUM FLUORIDE ETCHANTS

MGF-0001
ETCH NAME: Nitric acid
TYPE: Acid, removal/polish
COMPOSITION:
x .... HNO_3, conc.
TIME:
TEMP: RT
DISCUSSION:

MgF_2, (100) wafers used as substrates for thin film evaporation of aluminum, silver and chromium in a study of mechanical properties of optical films. Solution shown can be used to clean MgF_2 surfaces.

REF: Pulker, H K — *Thin Solid Films,* 89,191(1982)

MGF-0002
ETCH NAME: Sulfuric acid
TYPE: Acid, removal
COMPOSITION:
x .... H_2SO_4, conc.
TIME:
TEMP:
DISCUSSION:

MgF_2, (100) specimens. Solution referred to as a slow etch for this material.

REF: *Encyclopedia of Chemical Technology,* John Wiley & Sons, New York, 1980, 760

PHYSICAL PROPERTIES OF MAGNESIUM NITRIDE, Mg_3N_2

Classification	Nitride
Atomic numbers	12 & 7
Atomic weight	100.98
Melting point (°C)	1500
Boiling point (°C)	
Density (g/cm^3)	2.xx
Hardness (Mohs — scratch)	7—8
Crystal structure (monoclinic — normal)	(100) a pinacoid
Color (solid)	Colorless
Cleavage (basal)	(001)

MAGNESIUM NITRIDE, Mg_3N_2

General: Does not occur in nature as a mineral compound. It can be artificially fabricated by reacting the pure metal in a closed crucible or in a nitrogen atmosphere, e.g., flammable in air when heated to its boiling point as magnesium metal. The nitride, when heated in the presence of water, will separate into ammonia, NH_3 and magnesium hydroxide, $Mg(OH)_3$. The latter being milk of magnesia as a suspension in water.

In the metal industry, magnesium has been nitrided in hot nitrogen gas — the nitridization process — to improve surface hardness against wear and/or flammability if struck or sparked, as well as against corrosion. Note that magnesium metal will burn under water and, once ignited, is nearly impossible to quench . . . smother with dirt . . . and it may still continue to burn.

Technical Application: The compound has had little use in Solid State device processing up to the present, but has been deposited on magnesium specimens is a general study of nitride compounds. Though the nitride is listed as being colorless, the thin films of both magnesium and calcium nitrides were referred to as being black . . . with white spots of their respective oxides . . . if oxygen is present as an impurity in the nitrogen gas. Oxidation of the films will convert them, in part, to their oxides. Caution should be exercised as, with sufficient heat, magnesium is flammable in the presence of oxygen in air. As a dielectric it has possible applications as a thin film coating, though it is partly soluble in both water and alcohol.

Etching: Partly soluble in water and alcohols; soluble in linseed oil.

MAGNESIUM NITRIDE ETCHANTS

MGN-0001
ETCH NAME: Methyl alcohol — TIME:
TYPE: Alcohol, removal — TEMP:
COMPOSITION:
x .... CH_3OH(MeOH)
DISCUSSION:

Mg_3N_4, thin films deposited on Mg specimen blanks in a study of the use of high purity nitrogen gas to reduce contamination. Also worked with calcium nitride, Ca_2N_3. The black nitride of both materials will show spots of white oxides, CaO or MgO, respectively, in the presence of O_2 or air.

REF: Aubry, J & Streiff, R — *J Electrochem Soc*, 118,650(1971)

PHYSICAL PROPERTIES OF MAGNESIUM OXIDE, MgO

Classification	Oxide
Atomic numbers	12 & 8
Atomic weight	40.32
Melting point (°C)	800—2500 (2800)
Boiling point (°C)	
Density (g/cm³)	3.69—3.90 (3.59)
Refractive index (n=)	1.74 (1.736)
Wavelength limits (microns)	0.25—7
Young's modulus (psi $\times 10^6$)	36
Coefficient of linear thermal expansion ($\times 10^6$/cm/cm/°C)	13
Hardness (Mohs — scratch)	6
(Knoop — kgf/mm²)	690
Crystal structure (isometric — normal)	(100) cube
Color (solid)	Clear/white
Cleavage (cubic)	(001)

MAGNESIUM OXIDE, MgO

General: Occurs in nature as the mineral periclase, MgO, with a crystal structure similar to that of halite, NaCl. It has excellent cubic cleavage which is common to minerals with this structure. It is of minor occurrence in nature as a contact mineral but can be readily grown artificially from a melt containing magnesium chloride and silica.

In industry it has wide application as a high temperature refractory ceramic, often as a mixture with alumina, Al_2O_3.

Technical Application: Not used in Solid State device fabrication and assembly though it has a higher melting point than alumina (2800 vs. about 2500°C). Unfortunately, it has a coefficient of thermal expansion about three times greater than that of silicon (13 vs. 4 $\times$ 10^6/°C). It could be used as a replacement for alumina in package parts where such expansion coefficients are not involved.

As a single crystal or fused periclase it has optical properties similar to those of sapphire and fused quartz and is used in optics and filter applications.

It can be used as an insulator and as a thin film coating for high temperature surface protection similar to silica or for wear resistance.

Etching: Soluble in most acids and acid mixtures with low solubility in water.

MAGNESIUM OXIDE ETCHANTS

MGO-0001
ETCH NAME: Phosphoric acid — TIME:
TYPE: Acid, thinning — TEMP: 150°C
COMPOSITION:
x .... H_3PO_4
DISCUSSION:

MgO, (100) wafers plastically deformed to produce slip bands. Etch thinned in the solution show for TEM study.

REF: Narayan, J — *J Appl Phys*, 39,2448(1968)

MGO-0002a: Venables, J A — *J Appl Phys*, 34293(1963)

MgO ingot. Cut into rectangular (100) oriented bars by cleaving. Etch thin with the solution shown for 2—3 min at 130°C. Rinse in H_2O, then EOH.

MGO-0011: Thomas, G & Huftstutler, M C Jr — *J Appl Phys* 31,1834(1960)

MgO, (100) wafers used in a study of transmission thorough thin crystals. Wafers were jet-etch thinned with hot acid to a thickness of 1000 Å. Both Mo and Mg wafers also studied, but etch with other acids.

MGO-00214a: Keh, A S — *J Appl Phys*, 31,1538(1960)

MgO, (001) cleaved wafers were polished in phosphoric acid. Surfaces were indented to develop damage points using a square diamond tool. Both symmetrical (square), and asymmetrical (inclined) pits were observed aligned in a ⟨110⟩ direction. Interference patterns showed high points at the four pit corners.

MGO-0002b
ETCH NAME: Ammonium chloride — TIME:
TYPE: Acid, preferential/thinning — TEMP: 80°C
COMPOSITION:
x .... NH_4Cl
DISCUSSION:

MgO, (100) specimens cleaved as rectangles. This solution is a fair polish etch but also develops dislocations. The precipitates observed that do not etch are probably either ZrO_3 or $CaZrO_3$ impurities.

REF: Ibid.

MGO-0003
ETCH NAME: Hydrofluoric acid, dilute
TYPE: Acid, preferential — TEMP:
COMPOSITION:
x .... HF
x .... H_2O
DISCUSSION:

MgO specimens. Solution used to develop grain boundaries.

REF:

MGO-0004a
ETCH NAME: Phosphoric acid, dilute
TYPE: Acid, polish
COMPOSITION:
80% .... H_3PO_4
20% .... H_2O
TIME:
TEMP: 150°C

DISCUSSION:
MgO, (100) wafers used in an irradiation study. Solution used to polish and remove cutting and lapping damage.
REF: Bradley, R — *Phil Mag*, 10, 161(1964)

MGO-0019
ETCH NAME: Gold
TYPE: Metal, decoration
COMPOSITION:
x .... Au
TIME:
TEMP:

DISCUSSION:
MgO, (100) wafers cleaved and diffused with gold to delineate defects and dislocations. Pit side slope angles were measured.
REF: Robins, J L et al — *J Appl Phys*, 37,3893(1966)

MGO-0022b
ETCH NAME: Stoke's etch
TYPE: Acid, dislocation
COMPOSITION:
1 .... H_2SO_4
5 .... NH_4Cl, sat. sol.
1 .... H_2O
TIME:
TEMP:

DISCUSSION:
MgO, (100) wafers used in a plastic deformation study. Solution used to develop defects. Authors say that grown-in dislocations differ from fresh dislocations.
REF: Argon, A S & Orowan, E — *Phil Mag*, 10,1003(1964)

MGO-0023
ETCH NAME:
TYPE: Electron, damage
COMPOSITION:
x .... e^-
TIME:
TEMP:
GAS FLOW:
PRESSURE:
POWER:

DISCUSSION:
MgO single crystal specimens used in a study of secondary electron emission from bombardment by relativistic electrons. Material cleaved easily on (100), but if specimens are too thin, they may show internal cracking.
REF: Pomerantz, M A et al — *J Appl Phys*, 31,2036(1960)

MGO-0005
ETCH NAME: Hydrochloric acid
TYPE: Acid, float-off
COMPOSITION:
x .... HCl
TIME:
TEMP:

DISCUSSION:
MgO, (100) substrates cleaved and used for carbon deposition and conversion to diamond

phase by annealing. Annealing produces a mixture of C, a-C and m-diamond. Mixture of the three forms depends upon deposition parameters and control. HCl was used to float-off deposited thin films for TEM and SEM studies. (*Note:* a- = amorphous. m- = microcrystalline.)
REF: Namba, Y & Mori, T — *J Vac Sci Technol,* A(2),319(1985)

MGO-0006
ETCH NAME: Argon
TYPE: Ionized gas, cleaning
COMPOSITION:
x .... Ar^+ ions

TIME: 1 h
TEMP: 800°C
GAS FLOW:
PRESSURE:
POWER:

DISCUSSION:
MgO, (111) cleaved substrates used for deposition of thin film TiN. First, clean in a detergent, then ultrasonic in EOH and N_2 blow dry. Final clean in vacuum system with 2 KeV Ar^+ ion sputter prior to TiN deposition. Used in a study of single crystal growth of TiN.
REF: Johansson, B O et al — *J Vac Sci Technol,* A(2),303(1985)

MGO-0007
ETCH NAME: Phosphoric acid
TYPE: Acid, removal
COMPOSITION:
x .... H_3PO_4

TIME:
TEMP: 100°C

DISCUSSION:
MgO, (100) wafers used in a study of fatigue deformation. Wafers cleaved from ingot — damage and cleavage steps removed by etching with solution shown.
REF: Subtannanian, K N & Washburn, J — *J Appl Phys,* 34,3394(1963)
MGO-0018b: Matkin, D I & Bowen, D H — *Phil Mag*, 12,1209(1965)
MgO, (100) wafers used in a study of whisker growth within single crystal magnesium oxide. Solution used to polish and etch thin specimens.
MGO-0013: Lewis, M H — *Phil Mag*, 13,777(1966)
MgO, (001) cleaved wafers used in an electron microscope (EM) study of precipitates in the material. Solution was used to selectively thin areas on the wafer by jet etching.
MGO-0020: McPherson, R & Swha, H N — *Phil Mag*, 12451(1965)
MgO, bicrystals used in a study of fracture. After mechanical lapping with diamond grit, solution shown was used boiling, 10 min to remove residual sub-surface lap damage.
MGO-0021: Narayan, J — *J Appl Phys,* 35,2448(1964)
MgO, (100) wafers used in a study of plastic deformation and development of slip bands. Solution used at 150°C as a thinning etch.

MGO-0014b
ETCH NAME:
TYPE: Acid, preferential
COMPOSITION:
1 .... H_2SO_4
5 .... NH_4Cl, sat. sol.
1 .... H_2O

TIME:
TEMP: RT or hot

DISCUSSION:
MgO, (100) cleaved wafers. Wafer surfaces were deformed by indention and studied for defects generated by this deformation. At room temperature solution produces conical pits; aged and used hot pits are flat bottomed.

REF: Keh, A S — *J Appl Phys*, 31,1538(1960)
MGO-0009: Mendel, E & Weinig, S — *J Appl Phys*, 31,738(1960)

MgO, (100) cleaved wafers used in a study of dislocations produced by generation of an electric field. Field applied during etching developed a criss-cross pattern of what appear to be slip lines through the entire bulk of the wafer with coronal discharge on the surface during power application.

MGO-0015: Westwood, A R C & Gekha, D L — *J Appl Phys*, 34,3335(1963)

MgO, (100) cleaved wafers used in a study of surface energy. Specimens polish in this solution at RT, 3—10 min.

MGO-0016: Stokes, E J — *Phil Mag*, 3,718(1958)

MgO, (100) wafers used in a dislocation and defect study. Solution referenced as a dislocation etch in this article from MGO-0015.

MGO-0010
ETCH NAME: Hydrofluoric acid, dilute — TIME:
TYPE: Acid, preferential — TEMP:
COMPOSITION:

x .... HF
x .... H_2O

DISCUSSION:

MgO, (100) cleaved wafers used in a study of grain boundary diffusion using nickel and cesium. Prior to metal diffusion etched surfaces did not develop grain boundaries. Grain boundaries observed after metals diffused.

REF: Zaplatynsky, I — *J Appl Phys*, 351353(1964)

MGO-0012a
ETCH NAME: Nitric acid — TIME:
TYPE: Acid, thinning — TEMP: RT
COMPOSITION:

x .... HNO_3

DISCUSSION:

MgO, (100) wafers used in a study of whisker growth within the material. Solution used to etch polish and thin specimens for study.

REF: Matkin, D I & Bowen, D H — *Phil Mag*, 12,1209(1965)

MGO-0017
ETCH NAME: Carbon tetrachloride — TIME:
TYPE: Solvent, cleaning — TEMP: RT
COMPOSITION:

x .... CCl_4

DISCUSSION:

MgO, (100) wafers used in a study of fundamental optical absorption. Cleaved surfaces were cleaned in the solvent shown and follow with alcohol rinse.

REF: Reiling, G H & Hensley, E B — *Phys Rev*, 112,1106(1958)

MGO-0018
ETCH NAME: — TIME:
TYPE: Acid, dislocation — TEMP:
COMPOSITION:

1 .... H_2SO_4
1 .... NH_4Cl, sat. sol.

DISCUSSION:

MgO, (100) wafers. Specimens were subject to bending and studied for crack nucleation associated with defects. Cracks occur along or in dislocation zones representative of lattice imperfections, impurity lumps, etc.

REF: Briggs, A et al — *Phil Mag*, 10,1041(1964)

MAGNESIUM GERMANIDE ETCHANTS

MGGE-0001a

ETCH NAME: BRM TIME:

TYPE: Halogen, polish TEMP: RT

COMPOSITION:

4 drops .... Br_2

250 ml .... MeOH

DISCUSSION:

Mg_2Ge, (111) wafers. Use a metallographic polishing wheel and impregnate a pellon type cloth with solution. Lap at slow speed and quench with MeOH. Gives a bright chemical polish.

REF: Kromer, H et al — *J Appl Phys*, 36,2461(1965)

MGGE-0001b

ETCH NAME: TIME: 30 sec

TYPE: Acid, cleaning TEMP: RT

COMPOSITION:

2 .... H_2SO_4

1 .... H_2O_2

200 .... H_2O

DISCUSSION:

Mg_2Ge, (111) wafers. Solution used to clean p-n junction area, but does not develop junction. Material reactions with acids and alkalies. After etch cleaning with the solution shown, quench in MeOH.

REF: Ibid.

MGGE-0002

ETCH NAME: TIME:

TYPE: Cleave, cleaning TEMP:

COMPOSITION:

x .... cleave

DISCUSSION:

Mg_2Ge, (111) cleaved wafers used in a study of photoconductivity. Specimens were lapped down to 0.2—0.44 mm — below this thickness material fractures. Authors state there is no known etch for the material? Also worked with Mg_2Si.

REF: Stella, A & Lynch, D W — *J Phys Chem Solids*, 25,1253(1965)

MAGNESIUM SELENIDE ETCHANTS

MGSE-0001

ETCH NAME: BRM TIME:

TYPE: Halogen, polish TEMP: RT

COMPOSITION:

(1) x .... x%Br_2 (2) x .... H_2SO_4

x .... MeOH

DISCUSSION:

MgSe, single crystal evaluated for cohesive energy features of tetrahedral semiconductors. Forty elemental, binary, and trinary materials studied. Solutions shown are general cleaning, removal, and polishing etches for most of the materials studied.

REF: Aresti, A et al — *J Phys Chem Solids*, 45,361(1984)

MAGNESIUM TELLURIDE ETCHANTS

MGTE-0001

ETCH NAME: TIME:

TYPE: Acid, removal TEMP:

COMPOSITION:

(1) x .... HF
x .... HNO_3
x .... HNO_3
x .... HAc

(2) 3 .... HCl
1 .... HNO_3

DISCUSSION:

MgTe, single crystals evaluated for adhesive energy features of tetrahedral semiconductors. Forty materials studied. Solutions shown are general cleaning, removal, and polishing etches for most of these materials.

REF: Aresti, A et al — *J Phys Chem Solids*, 45,361(1984)

PHYSICAL PROPERTIES OF MAGNESIUM SILICIDE, Mg_2Si

Classification	Silicide
Atomic numbers	112 & 14
Atomic weight	76.70
Melting point (°C)	>1100
Boiling point (°C)	
Density (g/cm³)	
Hardness (Mohs — scratch)	6—7
Crystal structure (isometric — normal)	(100) cube
Color (solid)	Grey
Cleavage (cubic)	(001)

MAGNESIUM SILICIDE, Mg_2Si

General: Does not occur as the pure silicide in nature although there are over 50 silicate minerals, which may contain other metals. Many magnesium containing minerals are the basic rock-formers, such as those of the pyroxene and amphibole groups, which include the mineral forms of asbestos. The silicide is an additive to irons, steels, aluminum and copper alloys in the metal industries.

Technical Application: The silicide has been studied in Solid State development of materials as a tetrahedral semiconductor, and is currently under study as a silicide blocking layer in device construction along with several other metal silicides as Mg_2Si, MgSi, and $MgSi_2$. Note that the elemental semiconductors silicon and germanium are referred to as octahedral in crystal classification; whereas many binary silicides and germanides are referred to as tetrahedral. The octahedron, (111) is a crystal form within the isometric system, normal class, and the tetrahedron, (111) is a crystal form within the isometric system, tetrahedral class, such that materials should not be classified by a crystal form within a system or class, but by system/class or by planes of symmetry using the 132 Space Group notations. The tetrahedron is a basic unit cell atomic structure in many semiconductor materials. It is the simplest closed form in nature, and the pyramid of geometry. In etching single crystal spheres of these materials, a finite crystal form (FCF) is developed that will vary with solution mixture. See Germanium, Silicon, and Silicides for specific forms obtained.

Etching: Soluble in mixed acids.

MAGNESIUM SILICIDE ETCHANTS

MGSI-0001
ETCH NAME: TIME:
TYPE: Acid, removal TEMP: RT
COMPOSITION:
x .... HF
x .... HNO_3
DISCUSSION:

Mg_2Si, (111) cleaved wafers used in a study of photoconductivity. Specimens were lapped to 0.2—0.44 mm — below this thickness material fractures. Authors state there is no known etch for the material? Also worked with Mg_2Ge. (*Note:* Bromine-methanol should work as a polish/removal etchant.)

REF: Stella, A & Lynch, D W — *J Phys Chem Solids*, 25,1253(1965)

PHYSICAL PROPERTIES OF MAGNESIUM SULFIDE, MgS

Classification	Sulfide
Atomic numbers	12 & 16
Atomic weight	56.38
Melting point (°C)	>200 del
Boiling point (°C)	
Density (g/cm³)	1.6×
Crystal structure (isometric — normal)	(100) cube
Color (solid)	Red brown
Cleavage (cubic)	(001)

MAGNESIUM SULFIDE, MgS

General: Does not occur as a sulfide compound in nature, but there are over ten sulfates with the SO_4^- radical, such as epsomite, $MgSO_47H_2O$, in addition to others with other metal elements — Fe, Al, Zn, Mn — mostly as hydrated minerals. Magnesium is a quite abundant element with many of its natural minerals of industrial importance. See Magnesium for discussion of the metal as an element. The mineral shown above, epsomite, is the natural form of the medical epsom salts.

Technical Application: The sulfide is under evaluation in Solid State development as a binary compound semiconductor.

Etching: Soluble in acids.

MAGNESIUM SULFIDE ETCHANTS

MGS-0001
ETCH NAME: TIME:
TYPE: Acid, removal TEMP:
COMPOSITION:
(1) x .... H_2SO_4 (2) x .... HCl

DISCUSSION:

MgS, single crystals evaluated for cohesive energy features of tetrahedral semiconductors. Forty materials studied. Solutions shown are general, cleaning removal, and polishing etches for most of these materials.

REF: Aresti, A et al — *J Phys Chem Solids*, 45,361(1984)

PHYSICAL PROPERTIES OF MAGNETITE, Fe_3O_4

Classification	Oxide
Atomic numbers	26 & 8
Atomic weight	231.52
Melting point (°C)	1538
Boiling point (°C)	
Density (g/cm³)	5.16—5.18
Hardness (Mohs — scratch)	5.5—6.5
Crystal structure (isometric — normal)	(111), octahedron
Color (solid)	Iron-black
(thin section)	Colorless/brown
Cleavage (octahedral — indistinct)	(111)
Magnetism (solid — unidirectional)	Strong
(lodestone)	N-S polarity

MAGNETITE, FeO_4

General: Occurs as the natural mineral and called magnetic iron ore. As the variety lodestone it shows N-S polarity, and was used as the first magnetic compass in ancient times, circa 3000 B.C., about the time the island city states of Tyre and Sidon were the major seafaring nations in the Mediterranean Sea. The compass also was in use to guide camel caravans in Asia Minor and across the Gobi Desert in those early times. Some minerals of nickel and cobalt have natural magnetism, but too a lesser degree. Magnetite is widely distributed as grains and small crystals in basic rocks (those low in silica), and is often found in the upper layers of gold placer deposits.

It is found in large magnetic ore bodies — common to the Archaean Shield — the original rocks of the earth's crust where they were first formed by magmatic segregation without further alteration. Sweden is noted for its fine stainless steels, and their magnetite iron comes from such ore bodies which are of particularly high purity.

It has long been used as a lapping and polishing abrasive in the form of emery, Al_2O_3:Fe_3O_4 with Turkish emery considered of highest quality. As the pure magnetite abrasive, it is called Black Rouge. Some of the micas contain Fe_3O_4, which is thought to have been initially deposited as a colloidal precipitate — some other iron oxide ores are so formed as sedimentary deposits.

When heated in oxygen to 220°C it converts to hematite, Fe_2O_3, but retains its magnetism; at 550°C, as hematite without magnetism. Magnetite frequently forms as a furnace by-product during the reduction of other iron ores when the melt is low in silica. It is called ferrosoferric iron, as it contains both Fe^{++} and Fe^{+++} ions in its structure.

It is used directly as the core of electromagnets and, in powder form in liquid plastic, for magnetic fluidic valves; in solid rubberized plastic as a gripping mechanism, such as strips for door closure. There are many artificial magnets as alloy mixtures of Fe/Co/Ni for pure magnetic or electromagnetic action: gripping, moving — magnetic coils control the acceleration in atom smashing tunnel accelerators.

Technical Application: Natural magnetite has had no use in Solid State device development although magnets are used in many equipment operations, and have been used as part of device package assemblies.

Artificial magnetite has been grown from molten fluxes as small single crystals and, as a ferrite material, as larger ingots by the Czochralski (CZ) method. Both the artificial and natural single crystals have been studied for general morphology and, specifically, for ferromagnetic and ferroelectric properties. This includes fabrication as single crystal spheres for etching to finite crystal form (FCF) and in oxidation studies. Spheroids, doughnut shaped forms, also have been studied with regard to magnetic directionality reaction under varying conditions. See Garnet section for use of magnetism in YIG garnets used as computer bubble memory devices.

Etching: HCl, HNO_3, Picric, and Nital mixtures.

MAGNETITE ETCHANTS

MAGT-0001
ETCH NAME: Hydrochloric acid — TIME: 10 sec to 5 min
TYPE: Acid, preferential — TEMP: RT
COMPOSITION:

(1) x .... HCl, conc.

(2) 1 .... HCl
5 .... H_2O

DISCUSSION:

Fe_3O_4, (111) single crystals as fine octahedrons of natural magnetite were highly reflective as formed in a green schist deposit, and intimately associated with small iron garnets. Solutions were used in a general etching study to observe structure variations of surfaces.
Walker, P — mineralogy study, 1952

MAGT-0002
ETCH NAME: Lathe — TIME:
TYPE: Cut, forming — TEMP:
COMPOSITION:

x .... lathe cutting

DISCUSSION:

Fe_3O_4, artificial single crystal magnetite. After growth, specimens were lathe cut to an oblate spheroid form and used in a study of magnetization near the Curie Point.
REF: Smith, D O — *Phys Rev*, 102,959(1956)

MAGT-0003
ETCH NAME: — TIME:
TYPE: Electrolytic, polish — TEMP:
COMPOSITION: — ANODE:
CATHODE:
POWER:

DISCUSSION:

Fe_3O_4 specimen cut as a polycrystalline sphere — mainly single crystal — and used in study of g-factors at low temperatures. No etch shown, though magnetite is slowly soluble in acids. Reference shown is for single crystal spheres use in oxidation experiments.
REF: Gwathmey, A T & Lawless, K R — *The Surface Chemistry of Metals and Semiconductors*, Gatos, H C, Ed, John Wiley & Sons, New York, 1960, 483

MAGT-0004
ETCH NAME: TIME:
TYPE: Acid, cleaning TEMP:
COMPOSITION:
x .... HNO_3
DISCUSSION:

Fe_3O_4 grown as a single crystal ingot by the Czochralski (CZ), method. A study of the growth of iron ferrites. Also grew ingots of ZnO, Ga_2O_3, CuO, and MnO all doped with Fe. Solution used for general surface cleaning.
REF: Horn, F N — *J Appl Phys*, 32,900(1961)

PHYSICAL PROPERTIES OF MANGANESE, Mn

Classification	Transition metal
Atomic number	25
Atomic weight	54.93
Melting point (°C)	1245
Boiling point (°C)	2150
Density (g/cm³)	7.44
Specific heat (cal/g) 25°C	0.115
Latent heat of fusion (cal/g)	63.7
Latent heat of vaporization (k-cal/g-atom)	537
Atomic volume (W/D)	7.39
Ionization energy (K-cal/g-mole)	171
Electronegativity (Pauling's)	1.5
Covalent radius (angstroms)	1.17
Ionic radius (angstroms)	0.80 (Mn^{+2})
Coefficient of linear thermal expansion ($\times 10^{-6}$ cm/cm/°C)	22
Compressibility ($\times 10^{-7}$)	8.4
Magnetic susceptibility (cgs $\times 10^{-6}$)	8
Electron work function (eV)	3.76
Cross section (barns)	13.3
Vapor pressure (°C)	1792
Hardness (Mohs — scratch)	5
Crystal structure (isometric — normal)	(100) cube
Color (solid)	Grey-pink
Cleavage (cubic — poor)	(001)

MANGANESE, Mn

General: Does not occur as a native element. There are over 150 minerals containing manganese which are widely distributed as silicates, oxides, and carbonates and all of secondary origin. The two most important oxide ores are pyrolucite and psilomelane. MnO_2. The latter is one of the few natural colloids, and pyrolucite may be colloidal in part. The carbonate rhodochrosite, $MnCO_3$, can be a miscible mineral with siderite, $FeCO_3$ up to 40%. Color shifts from brilliant pink/rose to tan with increase of iron content. Pure metallic manganese has characteristics similar to those of iron but is harder and more brittle. This close association of Fe and Mn is further seen in nature with two colloidal "bog" ores: limonite and bog manganese, FeO and MnO, respectively, and both are hydrates.

The metal is of major importance in the metals industries as an additive to iron, copper, nickel, etc. as a hardening agent. Manganous-irons, like ferrosilicon, are starter pig irons for fabrication of special steels. The natural oxides, pyrolucite and psilomelane, are highly active oxidizing agents and, when added to glass, will remove the green tint due to iron for fine, clear glass or to produce a faint amethyst color. Both chlorides and permanganates are used as medical antiseptics and in quantitative analysis. The permanganate solutions produce a deep purple color.

It is worth noting that large deposits of psilomelane have been located on the ocean floor in the Pacific Ocean off the coast of California, and may become one of the first major oceanographic mining operations.

Technical Application: Until recently, manganese has had little use in Solid State development but is under evaluation as possible silicides: MnSi, Mn_2Si, $MnSi_2$. It also has been deposited as an amorphous thin film. There are possible applications as a magnetic semiconductor superlattice, such that it has been deposited as a thin film structure in layered CdMnTe crystals. There is no reference here as to its growth and study as a single crystal.

The permanganate, $KMnO_4$, has been used for surface cleaning and as a preferential etching solution on several metals and metallic compounds. Many chemical laboratories maintain the dry chemical available for use as an antiseptic for cuts and abrasions.

Etching: Soluble in dilute acids and halogens.

MANGANESE ETCHANTS

MN-0001

ETCH NAME: Hydrochloric acid, dilute — TIME:

TYPE: Acid, cleaning — TEMP: RT

COMPOSITION:

x .... HCl

x .... H_2O

DISCUSSION:

Mn, thin film deposits on ruthenium substrates. Manganese surfaces can be etch cleaned in the diluted acid shown.

REF: Heinrich, B et al — *J Vac Sci Technol*, B3(2),766(1985)

MN-0002

ETCH NAME: BRM — TIME:

TYPE: Halogen, removal/polish — TEMP:

COMPOSITION:

x .... x% Br_2

x .... MeOH

DISCUSSION:

Mn, thin films deposited as part of a layer structure in studying a dilute magnetic semiconductor superlattice. GaAs (100) wafers were used as substrates for both CdTe and CdMnTe film deposition. Bromine-methanol solutions can be used as polish and removal etchants on GaAs, CdTe, and CdMnTe.

REF: Kolodziejski, L A et al — *J Vac Sci Technol*, B(3),714(1985)

PHYSICAL PROPERTIES OF MANGANESE DIOXIDE, MnO_2

Classification	Oxide
Atomic numbers	25 & 8
Atomic weight	86.93
Melting point (°C)	−0.54
Boiling point (°C)	

Density (g/cm^3)	4.73—4.86
Hardness (Mohs — scratch)	2.0—2.5
Crystal structure (orthorhombic — normal)	(100) a-pinacoid
Color (solid)	Brown-black
Cleavage (basal)	(001)

MANGANESE DIOXIDE, MnO_2

General: There are several manganese dioxide minerals in nature, some of wide, others of limited occurrence but all mined as ores of manganese. Pyrolucite and polianite are both single crystal MnO_2, orthorhombic and tetragonal systems, respectively; manganite, Mn_2O_3 is also orthorhombic; whereas psilomelane, MnO_2, is colloidal and is the major ore of manganese. Large deposits have been found on the floor of the Pacific Ocean, and may become one of the first ocean mining operations. The physical data shown, above, is for pyrolucite, which is so soft it will soil the fingers, and has a black streak like graphite, C. As a monoxide, manganosite, MnO is isometric system. Hardness of these different manganese oxides vary widely H = 2 (pyrolucite) to H = 7 (psilomelane).

Other than as ores of manganese, the oxides have important applications in industry, chemistry and medicine: as an oxidizing agent in the manufacture of chlorine, bromine, and oxygen, and as a drier in paints. They are used to remove color from glass (green tint from iron), and to color glass (amethyst). As the chemical $KMnO_4$ it is a powerful oxidizing agent used as a medical antiseptic, as a water clarifier, and in preferential and cleaning etch solutions.

Technical Application: The monoxide and dioxides have had no real application in Solid State processing, though $KMnO_4$ is a constituent in etching solutions. Both manganese nitrides and silicides has been evaluated as have been artificial and native mineral oxides. Pyrolucite has been shown to be an n-type semiconductor compound, and has been evaluated for its infrared as well as semiconducting properties.

Etching: Soluble in HCl.

MANGANESE OXIDE ETCHANTS

MNO-0001a
ETCH NAME: Carbon monoxide — TIME: x h
TYPE: Gas, preferential — TEMP: 1000°C
COMPOSITION:
(1) x .... CO — (2) x .... H_2
DISCUSSION:

MnO_2, single crystal specimens used in a study of infrared properties. The CO will produce thermal etch patterns on surfaces. Hot hydrogen gas also produced thermal etch patterns, when applied under similar conditions, but no patterns were obtained with argon as a hot gas.

REF: Loh, E & Newman, R — *J Appl Phys*, 32,470(1961)

MNO-0002
ETCH NAME: Hydrochloric acid — TIME:
TYPE: Acid, removal — TEMP: RT
COMPOSITION:
(1) x .... HCl, conc. — (2) x .... HCl
x .... H_2O
DISCUSSION:

MnO_2, natural crystal specimens. Used as a general etch for psilomelane and other manganese oxides.

REF: Dana, S E & Ford, W E — *A Textbook of Mineralogy*, 4th ed, John Wiley & Sons, New York, 1950, 509.

MNO-0003: Das, J N — *Z Phys*, 151,395(1958)

MnO_2, as the natural mineral pyrolucite used in a study of semiconducting properties. Material was n-type, and properties were dependent on temperature.

MNO-0004

ETCH NAME: Borax — TIME:

TYPE: Acid, removal — TEMP:

COMPOSITION:

x .... $Na_2B_4O_7.10H_2O$

x .... H_2O

DISCUSSION:

MnO, single crystals doped with iron as ferrites, and grown as ingots by the Czochralski (CZ) method. Other ferrites grown and studied were Fe_3O_4, and iron doped Ga_2O_3, ZnO, and CuO.

MNO-0005: Walker, P — mineralogy study, 1953

MnO_2, as natural psilomelane with highly reflective botryoidal surfaces. A 50% hot solution used to clean and etching surfaces in a general study of swirl structures observed in colloidal minerals. Results compared with natural colloidal silica.

MAGNESIUM STANNIDE ETCHANTS

MGSN-0001

ETCH NAME: — TIME:

TYPE: — TEMP:

COMPOSITION:

DISCUSSION:

Mg_2Sn, single crystal specimens. Authors state that this compound cannot be etched in aqueous solutions as it etches too rapidly. Wafers were cleaved, (001) and used in a study of photoconductivity and photovoltaic effects. Also studied InSb and GaSb.

REF: Frederikse, H P et al — *Photoconductivity Conference*, Brechenridge, R G et al, Eds, John Wiley & Sons, New York, 1956

MGSN-0002

ETCH NAME: — TIME:

TYPE: — TEMP:

COMPOSITION:

DISCUSSION:

Mg_2Sn, (100) cleaved wafers. Wafers were lapped with 600-grit abrasive and used in a study of electrical conductivity at low temperatures. Authors say there is no satisfactory etch for this material because it corrodes too rapidly. (*Note:* BRM type solutions should etch this material, as well as alcohol dilute acids.)

REF: Frederikse, H P R et al — *Phys Rev*, 103,67(1956)

MANGANESE FERRITE ETCHANTS

MNFE-0001

ETCH NAME: — TIME:

TYPE: Abrasive, forming — TEMP: RT

COMPOSITION:

x .... SiC

x .... H_2O

DISCUSSION:

$MnFe_2$, single crystal sphere 0.8 mm diameter rough ground for nuclear magnetic resonance studies.

REF: Schulman, R G & Jaccarino, V — *Phys Rev*, 108,1219(1957)

MNFE-0002

ETCH NAME: TIME:

TYPE: Abrasive, forming TEMP: RT

COMPOSITION:

x .... SiC

x .... H_2O

DISCUSSION:

$MnFe_2$, single crystal sphere formed in a tornado-type sphere grinder for use in a ferromagnetic resonance study. Spheres were 0.2—0.8 mm diameter. (*Note:* No etch solutions shown, but BRM, Picral, and Nital are used on iron.)

REF: Tannenwald, P E — *Phys Rev*, 100,1713(1955)

MANGANESE TELLURIDE ETCHANTS

MNTE-0001

ETCH NAME: Nitric acid TIME:

TYPE: Acid, polish TEMP:

COMPOSITION:

x .... HNO_3

DISCUSSION:

$MnTe_2$ single crystal specimens. Cut specimens and mechanically polish with 0.5 μm diamond paste. Etch in solution shown to polish and remove lap damage. Other single crystals grown were FeS_2, RuS_2, $RuSe_2$, $RuTe_2$, OsS_2, $OsTe_2$, PtP_2, $PtAs_2$, and $PtSb_2$. Authors say that all crystals are cubic structure like pyrite, FeS_2 and were used in an IR spectra study. (*Note:* Pyrite is in the isometric system, pyritohedral class, and forms as a pseudo-cube with a high degree of facial twinning lamellae of surfaces. It is not "cubic".)

REF: Lutz, H D et al — *J Phys Chem Solids*, 46,437(1985)

PHYSICAL PROPERTIES OF MERCURY, Hg

Classification	Transition metal
Atomic number	80
Atomic weight	200.69
Melting point (°C)	−38.87
Boiling point (°C)	356.9
Density (g/cm³)	13.546
Thermal conductance (cal/sec)(cm²)(°C/cm)	0.022
Specific heat (cal/g) 25°C	0.033
Heat of fusion (cal/g)	2.82
Heat of vaporization (cal/g)	65
Atomic volume (W/D)	14.8
1st ionization energy (K-cal/g-mole)	241
1st ionization potential (eV)	10.43
Electronegativity (Pauling's)	1.9
Covalent radius (angstroms)	1.49
Ionic radius (angstroms)	1.10 (Hg^{+2})

Volume coefficient of expansion ($\times 10^{-6}$) 20°C	182
Electrical resistivity (micro-ohms-cm $\times$ 10^{-6}) 20°C	95.8
Compressibility (atms: 99—493 $\times$ 10^{-6}) 20°C	4
Magnetic susceptibility ($\times 10^{-6}$ cgs) 18°C	0.15
Magnetic moment (nuclear magnetrons)	+0.4993
Electron work function (eV)	4.52
Cross section (barns)	360
Vapor pressure (°C)	261.7
Vaporization point (°C)	350
Hardness (Mohs — scratch) solid	2
Crystal structure (hexagonal — rhombohedral) solid	$(10\bar{1}1)$ rhomb
Color (liquid/solid)	Tin-white
Cleavage (rhombohedral) solid	$(10\bar{1}1)$

MERCURY, Hg

General: Although metallic mercury does occur as a native element it is extremely rare, found only in completely enclosed voids in the primary ore cinnabar, HgS. Mercury (metal) and bromine (semi-metal) are the only two elements that occur in nature as liquids at standard temperature and pressure (22.2°C and 760 mmHg). There are several natural minerals containing mercury, one of the more important being calomel, HgCl — used as a medical specific to stimulate secretory organs. Tiemannite, HgSe and coloradoite, HgTe are natural minerals, which have artificial counterparts grown as compound semiconductors. There is a natural mineral called "amalgam", Ag_2Hg_3 to $Ag_{36}Hg$, the only natural amalgam, although mercury forms amalgams with most metals as artificial compounds.

Mercury is one of the metals that was known to ancient man. The Chinese and Hindu records refer to its use, and it has been found in Egyptians tombs, circa 4500 B.C. The Hg symbol is from hydrargyrum — meaning "liquid silver".

Because mercury has a regular coefficient of expansion, it has long been used for thermometers, barometers, and similar equipment. A column of mercury 760 mm high is the measure of one standard atmosphere of pressure, e.g., 760 mmHg, equal to approximately 30″ with a pressure of about 16 lb/psi.

Industrially, mercury is used in many low temperature alloys as it will form amalgams with most metals. Mercury is poisonous and, as it is normally in the liquid state, will continually release fumes into the air, such that mercury spills can be a severe health hazard.

Mercuric chloride, $HgCl_3$, is a violent poison, though used as an antiseptic; whereas mercurous chloride, HgCl (calomel), is a medical laxative. Mercury fulminate, $Hg(ONC)_2$, is a detonator used for explosives. Because of its brilliant vermillion-red color, the natural mineral cinnabar, HgS, is used as a pigment in paints.

As products using pure mercury, there are high light intensity mercury lamp bulbs and mercury switches — the "silent" switch.

Technical Application: A smear of mercury is occasionally used as a temporary electrical contact or electrically active holding medium in material device testing, and in Solid State and semiconductor processing has been used as a diffusion species for trinary compound semiconductors. It is a major constituent of the binary compound semiconductors: mercuric selenide, HgSe, and mercuric telluride, HgTe.

It can be grown as a single crystal under cold vacuum conditions, and has been frozen as a crystalline solid in LN_2, L_{air} and dry ice, CO_2 (solid). Such freezing liquids themselves, or by varying vacuum pressure, will produce preferential etching action for development of defect structure on solidified mercury.

Solid mercury is sufficiently soft such that it can be sliced with a knife under LN_2. As a smear on any material surface, it can be quick-frozen by a spray of acetone or CO_2.

Smeared on a metal surface, then allowed to sit with or without slight heating, amalgams formed can be evaluated.

Etching: HNO_3, dilute HCl, and halogen solutions.

MERCURY ETCHANTS

HG-0001
ETCH NAME: Carbon dioxide TIME:
TYPE: Gas, preferential TEMP: −56°C
COMPOSITION:
x .... dry ice (solid CO_2)
DISCUSSION:

Hg, liquid frozen into solid form using dry ice and holding at dry ice temperature. CO_2 gas can be used to lightly etch surfaces.

REF: Fickett, F R — *J Appl Phys*, 40,3464(1969)

HG-0002
ETCH NAME: Liquid air TIME:
TYPE: Air, preferential TEMP: −70°C and −183°C
COMPOSITION:
x .... L_{air}
DISCUSSION:

Hg, liquid frozen and stressed at the two temperatures shown using liquid air as the freezing agent and preferential etchant. Develops slip and defect structure in single crystal mercury specimens.

REF: Rider, J O & Heckscher, F — *Phil Mag*, 13,678(1966)

HG-0003
ETCH NAME: Liquid nitrogen TIME:
TYPE: Gas, preferential TEMP: −195°C
COMPOSITION:
x .... LN_2
DISCUSSION:

Hg, liquid frozen by submersion in LN_2, then sliced under the liquid and pressurized by point contact to develop defect structures. Also used to collect balls of mercury spills by flushing area with LN_2, and recovering mercury with tweezers.

REF: Tarn, W H & Walker, P — mineralogy study/application — 1953/1974

HG-0004a
ETCH NAME: Acetone TIME:
TYPE: Ketone, freezing TEMP: −95°C
COMPOSITION:
x .... $(CH_3)_2CO$, spray
DISCUSSION:

Hg, as a smeared surface contact on silicon wafer backs. Acetone used to temporarily solidify mercury during assembly for electrical tests.

REF: Walker, P & Tarn, W H — personal application, 1964

HG-0004b
ETCH NAME: Nitric acid
TIME:
TYPE: Acid, removal
TEMP: RT
COMPOSITION:
x .... HNO_3, conc.
DISCUSSION:

Hg, applied as a thin film or hemispherical dot on a metal surface for electrical probe contact in device testing. Concentrated acid used to remove mercury after use followed by heavy water washing and air drying.

REF: Ibid.

PHYSICAL PROPERTIES OF MERCURIC CADMIUM TELLURIDE, HgCdTe

Classification	Telluride
Atomic numbers	80, 48, & 52
Atomic weight	440.59
Melting point (°C)	>900 est.
Boiling point (°C)	
Density (g/cm^3)	
Energy bond gap (eV)	1.75
Hardness (Mohs — scratch)	2—2.5
Crystal structure (isometric — tetrahedral)	(111) tetrahedron
Color (solid)	Grey/black
Cleavage (octahedral)	(111)

MERCURIC CADMIUM TELLURIDE, HgCdTe

General: Does not occur in nature as a metallic compound although there is a mercuric telluride mineral: coloradoite, HgTe, found sparingly in Colorado in the U.S. and as minable quantities in Australia.

The mineral, HgTe, is used as an ore of both mercury and tellurium in industry in addition to application as a binary semiconductor compound as is the trinary HgCdTe, both as artificially grown materials.

Technical Application: Mercuric cadmium telluride is a II—III—VI trinary compound semiconductor similar in many respects to the binary CdTe II—VI compound as it is grown with varying ratios of $Cd_{1-x}Hg_xTe$. It has the crystal structure of sphalerite, ZnS, as do many of the binary compound semiconductors.

It has been grown as a single crystal for general morphological studies as well as a layered film on CdTe, and has been used to fabricate laser, photoconductor, and photodiode devices.

Etching: Soluble in aqua regia-type solutions, H_2O_2, and halogens.

MERCURIC CADMIUM TELLURIDE ETCHANTS

HGCT-0001
ETCH NAME: TIME: 30 sec
TYPE: Acid, stain TEMP: RT
COMPOSITION:
1 .... HCl
1 .... HNO_3
1 .... H_2O
DISCUSSION:

HgCdTe, single crystal ingots and wafers. Solution used to stain both ingots and wafers to observe segregation, slip, and other defects.

REF: *J Cryst Growth*, 65,249(1983)

HGCT-0002a
ETCH NAME: TIME:
TYPE: Acid, structure TEMP:
COMPOSITION:
2 .... HCl
1 .... HNO_3
3 .... H_2O
DISCUSSION:

HgCdTe, wafers. Mechanical lap with SiC, then chem/mech polish with an acid slurry of colloidal silica:H_2O_2:H_2O. Then etch with the solution shown to bring out dendritic growth structure. A final light etch in Br_2:MeOH will polish the surface and help delineate the structure.

REF: Boliang, C et al — *J Electron Mater*, 13,47(1984)

HGCT-0003
ETCH NAME: TIME: 3 h
TYPE: Acid, preferential TEMP: RT
COMPOSITION:
2 .... H_2O_2
2 .... H_2O
DISCUSSION:

HgCdTe, (111) thin films grown on CdTe substrates to develop the (111)A and $(\overline{111})$B oriented faces. Solution shown was used to etch surfaces in a study of misorientation and structure. The $(\overline{111})$B surface was good with <4° mismatch. The (100) substrate orientation was also evaluated.

REF: Edwall, B D et al — *J Appl Phys*, 55,1453(1984)

HGCT-0004
ETCH NAME: TIME:
TYPE: Halogen, step-etch TEMP: RT
COMPOSITION:
(1) x .... Br_2
x .... MeOH
(2) x .... Br_2
x .... ethylene glycol
DISCUSSION:

HgCdTe, thin films deposited by VPE on CdTe substrates using an open tube with a semi-enclosed slider. Step-etch with solution shown for about 1—2 μm/steps. Removal measured after each step-etch using interference fringes in deposit depth measurement.

REF: Nebirovsky, Y & Kepten, A — *J Electron Mater*, 13,866(1984)

HGCT-0005
ETCH NAME: BRM TIME:
TYPE: Halogen, polish TEMP:
COMPOSITION:
x .... 0.1 *N* Br_2
x .... MeOH
DISCUSSION:
HgCdTe, single crystal wafers used in a study of reaction kinetics with Br_2:MeOH. Wafers were hand polished on a pellon pad saturated with the solution shown.
REF: Takasek, R T & Syllaios, A J — *J Electrochem Soc*, 132,656(1985)

HGCT-0002b
ETCH NAME: TIME:
TYPE: Acid-slurry, polish TEMP: RT
COMPOSITION:
x .... colloidal SiO_2
x .... H_2O_2
x .... H_2O
DISCUSSION:
HgCdTe, (111) wafers and other orientations used in a study of growth structure. After mechanical lapping with SiC, wafers were chem/mech polished with the acid-slurry shown. A final light etch with Br_2:MeOH will polish surface and help delineate structure.
REF: Ibid.

HGCT-0006
ETCH NAME: BRM TIME:
TYPE: Halogen, removal TEMP: RT
COMPOSITION:
20 .... Br_2
80 .... MeOH
DISCUSSION:
$Hg_{1-x}Cd_xTe$, (111) wafers. Solution used as a general removal etch. After etching rinse in methanol to remove residual bromine.
REF: Parker, S G & Pinnell, J E — *J Electrochem Soc*, 118,1868(1971)

HGCT-0007
ETCH NAME: TIME:
TYPE: Halogen, polish TEMP:
COMPOSITION:
x .... 2% Br_2
x .... dimethylformamide
DISCUSSION:
HgCdTe, (111) wafers used in a study of the reduction of native oxide using CVD SiO_2. Wafers were chem/mech polished in the solution shown. They were then oxidized for 700 Å thickness in xKOH:xglycerine:xH_2O, followed by 250—300 Å CVD SiO_2.
REF: Rhiger D R et al - *J Vac Sci Technol*, 21(2)448(1982)

HGCT-0008a-b
ETCH NAME:
TYPE: Halogen, polish TEMP: RT
COMPOSITION:
(1) x .... .02% Br_2
x .... MeOH
(2) x .... $NaBH_4$, sat. sol.
x .... MeOH

DISCUSSION:

$Hg_{-x}Cd_xTe$ single crystal grown in a study of chemical etching and oxidation. Wafers were polished in solution (1), then in solution (2), and (3) again in solution (1). Prior to material growth: (A) clean Cd in HCl; (B) clean Hg in KCN, and (C) clean Te in BRM solution or electrolytically in an $NaBH_4$ solution as shown in (2).

REF: Aspnes, D & Arwin, H — *J Vac Sci Technol*, A(2)3,1309(1984)

HGCT-0009

ETCH NAME: BRM TIME:

TYPE: Halogen, polish TEMP:

COMPOSITION:

x .... x% Br_2

x .... MeOH

DISCUSSION:

HgCdTe specimens. After saw cutting of wafers chem-mech polish with a BRM solution using a saturated metallographic type pad. Final etch in BRM solution as a free-etch, e.g., wafers dropped free in solution.

REF: Wu, S Y — *J Vac Sci Technol*, 21,255(1982)

HGCT-0010

ETCH NAME:

TYPE: Acid, slurry removal TEMP: RT

COMPOSITION:

(1) x .... Al_2O_3, powder

x .... MeOH

x .... HAc

(2) x .... Al_2O_3, powder

x .... I_2

x .... MeOH

DISCUSSION:

HgCdTe (acronym: MCT) single crystal material fabricated as a semiconductor material with a tunable energy gap, After mechanical lapping of surfaces, either slurry solution can be used to etch polish/remove residual lap damage. Solution (2) is a halogen similar to BRM solutions.

REF: Goldstein, M et al — *Metallography* (MEIJAP), 16,321(1983)

PHYSICAL PROPERTIES OF MERCURIC IODIDE, HgI_2

Classification	Iodide
Atomic numbers	80 & 53
Atomic weight	454.45
Melting point (°C)	250
Boiling point (°C)	354
Density (g/cm³)	6.27
Hardness (Mohs — scratch)	1—2
Crystal structure (hexagonal — rhombohedral)	$(10\bar{1}1)$ rhomb
(tetragonal — normal)	(100) a-pinacoid
Color (solid)	Yellow-reddish
Cleavage (basal)	(0001)/(001)

MERCURIC IODIDE, HgI_2

General: Does not occur in nature as an iodide compound although it does occur as the chloride: calomel, HgCl — called horn quicksilver — which is found as a coating of secondary origin deposited from hot solutions and associated with cinnabar, HgS. The majority of the world's mercury comes from small mines in Spain, and similar deposits are found in the coastal ranges in California. Cinnabar is roasted to obtain the mercury, which is sold commercially by the 50-lb flask.

The yellow form of mercuric iodide occurs when the brilliant red form is heated above 128°C, but reverts to the red form upon cooling and being rubbed or scratched. Sodium nitrate — soda niter, $NaNO_3$ — known as Chile saltpeter, contains sodium iodate, $NaIO_3$; lautarite, $Ca(IO_3)_2$ and dietzeite, $Ca(IO_3)_2.CaCrO_4$ are found in association with Chile saltpeter.

The yellow form of HgI_2 will darken when exposed to light and has been used in photography, and both forms are used in medicine.

Technical Application: Mercuric iodide is not used in Solid State material processing.

Both colored forms may be used in iodine vapor transport growth of mercury containing compound semiconductors such as cadmium mercuric telluride or selenide.

The compound has been grown as a single crystal for general morphological studies.

Etching: Slightly soluble in water, soluble in KI, and other halogens.

MERCURIC IODIDE ETCHANTS

HGI-0001a
ETCH NAME:
TYPE: Halogen, polish
COMPOSITION:
x .... 0.2% Br_2
TIME:
TEMP: RT
DISCUSSION:

Alpha-HgI_2, single crystal specimens used in a morphology study. Use bromine solution (in MeOH?) to polish. Study under an optical microscope (OM) with an LN_2 cold stage to prevent vaporization.

REF: Kobayshi, T — *J Appl Phys*, 130,1183(1983)

HGI-0001b
ETCH NAME:
TYPE: Halogen, polish
COMPOSITION:
x .... 20% KI
TIME:
TEMP: RT
DISCUSSION:

Alpha-HgI_2, single crystal specimens used in a morphology study. Use a potassium iodide solution (in MeOH?) to polish. Study under an optical microscope (OM) with an LN_2 cold stage to prevent vaporization.

REF: Ibid.

HGI-0002
ETCH NAME: Acetone
TYPE: Ketone, growth
COMPOSITION:
x .... $(CH_3)_2CO$
TIME:
TEMP:
DISCUSSION:

HgI_2 single crystals grown by evaporation from a mixture of powdered 75% HgI_2:25% $HgCl_2$ with acetone as the carrier in an open beaker.

REF:

PHYSICAL PROPERTIES OF MERCURIC OXIDE, HgO

Classification	Oxide
Atomic numbers	80 & 8
Atomic weight	216.61
Melting point (°C)	100 del.
Boiling point (°C)	
Density (g/cm)	11.14
Refractive index (n=)	2.37—2.65
Hardness (Mohs — scratch)	1.5—2
Crystal structure (orthorhombic — normal)	(100) a-pinacoid
Color (solid)	Orange-red
Cleavage (prismatic — perfect)	(010)

MERCURIC OXIDE, HgO

General: Occurs as the mineral montroydite, HgO. It is a rare species as it is volatile and readily reduced by boiling mercury at close to its vapor point (357°C) in air. The mercury surface becomes coated with a mixed yellow/red mercuric oxide. Caution should be observed in working with mercury and its compounds as both the vapor and finely divided particles are extremely poisonous.

Mercuric oxide has no use in industry, but there are two major chloride compounds: mercurous chloride — calomel, HgCl is a medicinal specific for the stimulation of secretory organs; mercuric chloride, $HgCl_2$ — called corrosive sublimate — is a violent poison, yet is a powerful medical antiseptic. Mercuric fulminate, $Hg(ONC)_2$ is the most used detonator for gunpowder and high explosives. Its discovery led to the development of the percussion cap, which displaced the old flintlock rifles.

Technical Application: Mercuric oxide has no application in Solid State processing. It has little or no use as it is unstable and extremely soft.

Etching: Soluble in acids and water.

MERCURIC OXIDE ETCHANTS

HGO-0001

ETCH NAME: Potassium cyanide — TIME:

TYPE: Acid, removal — TEMP:

COMPOSITION:

x .... x% KCN

DISCUSSION:

HgO, as a native oxide on mercury. Solution used to clean mercury prior to its use in epitaxy growth of HgCdTe. Grown material used in a study of etching and oxidation. (**CAUTION** — The cleaning reaction may produce fulminate of mercury, $Hg(OCN)_2$!)

REF: Aspenes, D E & Arwin, H — *J Vac Sci Technol*, A(2),1309(1984)

PHYSICAL PROPERTIES OF MERCURIC SELENIDE, HgSe

Classification	Selenide
Atomic numbers	80 & 34
Atomic weight	79.57
Melting point (°C)	>900 est.
Boiling point (°C)	
Density (g/cm³)	8.19 (8.3—8.47)
Hardness (Mohs — scratch)	2—3

Crystal structure (isometric — tetrahedral)	(111) tetrahedron
Color (solid)	Steel-grey
Cleavage (octahedral)	(111)

MERCURIC SELENIDE HgSe

General: Occurs as the natural mineral tiemannite, HgSe, and may be in considerable amounts such as the Utah deposits in the U.S. The mineral onofrite, Hg(S,Se), is of minor occurrence, and both are found in association with sphalerite, ZnS, deposits.

Primary use in industry is as an ore of both mercury and selenium.

Technical Application: Mercuric selenide is a II—IV compound semiconductor with (111) surface polarity common to such compounds: (111)Hg and $(\overline{111})$Se wafer surfaces, respectively. It has had little use as a semiconductor because of its softness (H = 2—3).

It has been the subject of morphological studies as both the natural and artificial single crystal.

Etching: Soluble in aqua regia-type solutions and halogens.

MERCURIC SELENIDE ETCHANTS

HGSE-0001a
ETCH NAME:
TYPE: Acid, polish
COMPOSITION:
1 .... HCl
50 .... HNO_3
10 .... HAc
20 .... 10 *N* H_2SO_4

TIME: 10—15 min
TEMP: 40°C

DISCUSSION:

HgSe, (111) wafers. Described as a polish etch that does not produce pitting. (*Note:* This is a major article on etching II—VI compounds.)

REF: Warekois, E P et al — *J Appl Phys*, 33,690(1962)

HGSE-0001b
ETCH NAME:
TYPE: Acid, preferential
COMPOSITION:
1 .... HCl
1 .... HNO_3

TIME:
TEMP:

DISCUSSION:

HgSe, (111) wafers used in an etch study of the polarity effect. The (111)A surface shows pits, initially, but they disappear with etching time. The $(\overline{111})$B surface is coated with a selenium film with no defects apparent. The residual Se film remaining after etching can be removed in H_2SO_4.

REF: Ibid.

HGSE-0001c
ETCH NAME:
TYPE: Acid, preferential
COMPOSITION:
6 .... HCl
2 .... HNO_3
3 .... H_2O

TIME: 5 min, twice
TEMP: RT

DISCUSSION:

HgSe, (111) wafers used in an etch study of the polarity effect. Wafers were first polish etched (HGSE-0001a). The (111)A surface produces rough craters. The $(\overline{111})$B produces triangular etch figures. Remove residual Se film remaining after etching with polish etch by brush-off under water.

REF: Ibid.

HGSE-0002: Harper, C A — *Handbook of Materials and Processes for Electronics*, McGraw-Hill, New York, 1970, 7.

HgSe, (111) wafers. Polish with HGSE-0001a solution which removes Se film, then brush under water; repeat. $(\overline{111})$B surface develops triangle pits.

HGSE-0003

ETCH NAME: Bromine

TYPE: Halogen, polish

TIME: 1 min

TEMP: RT

COMPOSITION:

x .... Br_2

DISCUSSION:

HgSe, single crystal specimens. Follow bromine etching with wash in EOH, then benzene to remove selenium films. (Etch system developed by M C Lavine for Harman.) Specimens used in a magnetoresistance study.

REF: Harman, T O — *J Appl Phys*, 32,1800(1961)

PHYSICAL PROPERTIES OF MERCURIC TELLURIDE, HgTe

Classification	Telluride
Atomic numbers	80 & 52
Atomic weight	327.49
Melting point (°C)	>900 est.
Boiling point (°C)	
Density (g/cm³)	8.07
Hardness (Mohs — scratch)	2.5
Crystal structure (isometric — tetrahedral)	(111) tetrahedron
Color (solid)	Iron-black
Cleavage (octahedral)	(111)

MERCURIC TELLURIDE, HgTe

General: Occurs in nature as mineral coloradoite, HgTe and named for its discovery in sphalerite (ZnS) deposits in the state of Colorado. Major quantities are found in Western Australia and called kalgoorlite, HgTe, as a mineral mixture of coloradoite and petzite, (Ag,Au)Te.

Primary industrial use is as an ore of mercury and tellurium.

Technical Application: Mercuric telluride is a II—VI compound semiconductor with the usual polarity structure of these compounds: (111)Hg and $(\overline{111})$Te wafer surfaces. Like mercuric selenide, HgSe, the compound is relatively soft (H = 2.5) and, as both have very low energy band gaps, they have not been used as semiconductor devices to any extent.

Both the natural and artificial compounds have been studied for morphological data and polarity effects.

Etching: Soluble in aqua regia-type solutions, alkalies, and halogens.

MERCURIC TELLURIDE ETCHANTS

HGTE-0001a
ETCH NAME: TIME: 10—15 min
TYPE: Acid, polish TEMP: RT
COMPOSITION:
1 .... HCl
6 .... HNO_3
1 .... H_2O
DISCUSSION:
HgTe, single crystal wafers. Solution described as a polish etch that does not produce pitting. (*Note:* Article covers etching of II—V compounds.)
REF: Warekois, E P et al — *J Appl Phys*, 33,690(1962)

HGTE-0001b
ETCH NAME: TIME: 1 min, 3 times
TYPE: Acid, preferential TEMP: RT
COMPOSITION:
1 .... HCl
1 .... HNO_3
DISCUSSION:
HgTe, (111) wafers used in an etch study of polarity defect development. The (111)A surface produces triangular etch pits with a background of shallow triangular etch figures. The $(\overline{111})$B is flat and grainy in appearance. Etch for three 1-min periods.
REF: Ibid.
HGTE-0002a Harper, C A — *Handbook of Materials and Processes for Electronics*, McGraw-Hill, New York, 1970, 7.
HgTe, (111) wafers. Polish with HGT-0001a solution; then preferential etch for three 1-min periods. (111)A surface develops pits with background figures.

HGTE-0002b
ETCH NAME: TIME: 10 min
TYPE: Alkali, dislocation TEMP: RT
COMPOSITION:
2 .... 15% NaOH
1 .... $Na_2S_2O_8$, sat. sol.
DISCUSSION:
HgTe, (111) wafers. Solution used as a general dislocation etch.
REF: Ibid.

METALS, GENERAL

General: Several metals occur in nature as native elements and, like gold, may be the primary source of the metal as an ore. There are metallic types, such as those of the Gold Group (silver, copper, mercury, lead, and amalgam, AgHg); and the Platinum Group (iridium, palladium, and iron). The nonmetals include only graphite (carbon), sulfur, and selenium with zinc considered the connecting link between metals and semi-metals. All of the metals crystallize in the isometric system, normal class; whereas the semi-metals form in the hexagonal system, rhombohedral division. Native tin is an exception, being in the tetragonal system. Osmium occurs as a mixture with iridium in the mineral iridosmine (osmiridium), and all native elements contain traces of other metals, sometimes in fair quantity, such as electrum, AuAg(20%). The semi-metals include tellurium, antimony, arsenic, and bismuth which, though they occur as native elements in small quantities, are more common as compounds.

There are some metallic compounds, such as the tellurides and selenides and antimonides, but the majority of metals occur as oxides, sulfides, silicates, carbonates. The silicates comprise over 60% of all known natural minerals. Ancient man is said to have known seven metals — gold, silver, and copper with the most early use, followed by iron, tin, mercury, and sulfur — though he used many compounds in addition: red rouge from red ocher hematite, Fe_2O_3; magnetite, Fe_3O_4, as a compass (lodestone), and as black rouge; silica, SiO_2, as flint or obsidian were some of the first tools; and the Egyptians developed glassmaking, circa 2500 B.C.; not to mention the clays as basic building materials as sun-dried brick, later as kiln-fired brick. Other natural minerals were used as building stone (sandstone, marble, limestone) or as coloring agents for glass, leather, and cloth as shown from Egyptian records, and modern industry today still uses many of the same metals and minerals in a similar manner.

In metal processing there are several aspects related to chemical processing, from ore reduction, through material fabrication, to tests and evaluations utilizing metallographic techniques for study and evaluation. The metal industries, as a group, use more electrolytic etching and cleaning than do the Solid State and electronics industries, in part, due to their handling of the so-called "heavy metals" and their alloys, e.g., iron, steel, molybdenum, tantalum, copper, and tungsten. All processing areas use electrolytic plating on metals for a variety of purposes, from decorative washes to thicker corrosion resistant layers. And organics, such as mylar or other plastics, are coated as sunscreens or other industrial applications. But there is only limited information on plating solutions per se in this book, as it is more concerned with chemical processing using etching and cleaning solutions.

There are many books and articles covering the theory and practice of chemical processing — some general — some on specific metal groups, such as copper and its alloys or irons, steels and their alloys — aluminum and aluminum alloys — molybdenum, tantalum, titanium, and tungsten . . . each metal covered in detail. This includes the handling of solders and brazing alloys, such as the high temperature silvaloys. Because of the broad aspects of such articles and brochures or books, several have been collected as general references and presented in this general Metals section without regard to application of a specific solution, with several appearing elsewhere under a particular metal with a particular solution mixture and application discussion.

Technical Application: See specific metals and metallic compounds in this Etchant Section for more specific applications and discussions.

Etching: See individual formatted materials.

METALS ETCHANTS

MET-0001

ETCH NAME:	TIME:
TYPE: Electrolytic, polishing	TEMP:
COMPOSITION:	ANODE:
x .... acids	CATHODE:
	POWER:

DISCUSSION:

All metals. A discussion of the electropolishing of metals for microscopic examination.

REF: von Hamos, I — *Jernkontorets Ann*, 126,568(1942) (in Swedish)

MET-0002: Brovillet, P & Epelboin, I — *Met Rev*, 51,593(1954) (in French)

All metals. A study of results relative to the composition of electrolytic polishing baths.

MET-0003: Michel, P — *Met Rev*, 46,39(1949)

All metals. The use of electrolytic polishing to produce exact geometric forms.

MET-0004
ETCH NAME:
TYPE: Acid, polishing
COMPOSITION:
x .... acids
TIME:
TEMP:

DISCUSSION:
All metals and alloys. A general discussion of chemical polishing.
REF: de Jong, J J — *Metalen*, 9,2(1954) (in Dutch)

MET-0005
ETCH NAME:
TYPE: Acids, etching
COMPOSITION:
x .... acids solutions
TIME:
TEMP:

DISCUSSION:
Metals. An extensive list of specific etchant mixtures formulated or developed for study and processing of most industrial metals. This pamphlet contains several hundred numbered or named solutions, and several that are unnumbered. It does not describe application or uses, and the list is without regard to any specific metal or metal compounds on which they are used, yet it is probably the most extensive collection of etching and cleaning solutions available, other than this handbook.
REF: Pamphlet: ASTM E407-70 (1971)

MET-0006
ETCH NAME:
TYPE: Electrolytic, dissolution
COMPOSITION:
x .... acids
TIME:
TEMP:
ANODE:
CATHODE:
POWER:

DISCUSSION:
All metals. A general presentation of anodic dissolution etch removal of metals and electropolishing of surfaces.
REF: Niggins, J K — *J Electrochem Soc*, 106,999(1959)

MET-0007
ETCH NAME:
TYPE: Acid, preferential
COMPOSITION:
x .... acids
TIME:
TEMP:

DISCUSSION:
All metals. A study of grain growth in metals in both bulk materials and thin films. Growth follows a general pattern: initial island growth at low energy points on the substrate surface (defects, etc.); then lateral growth from initially deposited islands until they commence to coalesce; and finally additional thickness growth as complete surface coverage. The initial deposit of islands/lateral expansion may be colloidal to amorphous without structure followed by complete growth coverage as randomly oriented crystallites. Grain size varies with growth rate, temperature, and pressure. Applies to growth of both natural minerals and artificially fabricated metals or alloys.
REF: Feltham, P — *Acta Metall*, 5,97(1957)

MET-0008
ETCH NAME: TIME:
TYPE: Preparation, metallographic TEMP: RT
COMPOSITION:
(1) x .... mechanical lap/polish
(2) x .... preferential etching
DISCUSSION:

Metals and alloys that are easily oxidized. A discussion of preparation of specimens for photomicrographic study. Large specimens prepared as individual units; small specimens encapsulated in clear Lucite powder under pressure/heat, then rough lapped to expose specimen surface, with final lap/polish and etch of specimen for microscope examination. This is basic metallographic laboratory operation but, as materials are prone to oxidation, work should be done under an inert atmosphere such as nitrogen or argon with solvents and lapping slurries containing nonoxidizing agents, though final specimen finish may include heat tinting by surface oxidation in air on a hot plate.

REF: Boom, E A — *Zavdskaya Lab*, 13,1139,(1947)

MET-0009
ETCH NAME: TIME:
TYPE: Particles, irradiation TEMP:
COMPOSITION:
x .... Ar^+, Xe^+, Cu^+ ions
DISCUSSION:

All metals. A study of the effects of irradiation on metal surfaces that induce damage and structure alteration.

REF: Meckel, B B & Swalin, R A — *J Appl Phys*, 30,89(1959)

MET-0010
ETCH NAME: TIME:
TYPE: Etchants, dislocation TEMP:
COMPOSITION:
x .... acid/acid mixtures
DISCUSSION:

Metals and metallic compounds. A study of the correlation of etch figures with dislocations in a bulk material surface with regard to top surface and bulk crystallographic planes.

REF: Sun, Jui-Fang & Shaskol'skaya, M P — *J Sov Phys-Cryst*, 4,550(1960).
MET-0011: Knuth-Winterfeldt, E — *Rev Alum*, 28,84(1951)

The electropolishing of light metals rich in silicon for electrographic study.

MET-0012
ETCH NAME CP4 TIME:
TYPE: Acid, selective TEMP: RT
COMPOSITION:
30 .... HF
50 .... HNO_3
30 .... HAc
0.6 .... Br_2
DISCUSSION:

Metals. A curved steel point was used to scribe 10 μm width lines into substrate surfaces to observe metal plating and etching effects. Plating shows erratic build-up and etching is more rapid along the ruptured lines. Germanium and silicon wafers used as test vehicles; copper, gold, and silver plating as evaluation metals. See SI-0092a.

REF: Rindner, W & Lavine, J M — *J Electrochem Soc*, 108,809(1961)
MET-0013: Vermilyea, D A — *Acta Metall*, 6,381(1958)

A discussion on the formation of etch pits at dislocations on single crystal metal and semiconductor surfaces.

MET-0014: Ranaldi, M & Ostapkovich, P L — *J Appl Phys*, 31,20)9(1959)

A study of the relationship between optical orientation and etch time.

MET-0015
ETCH NAME: Ion etching — TIME:
TYPE: Ionized gas, removal — TEMP:
COMPOSITION: — GAS FLOW:
x .... Ar^+ ions — PRESSURE:
POWER:

DISCUSSION:

A study of etching surfaces with 8 KeV argon ions. (*Note:* This type of application of ionized gases is now called dry chemical etching (DCE), and is one of the three etchant formats used in this book.)

REF: Cunningham, R L et al — *J Appl Phys*, 31,839(1960)

MET-0016
ETCH NAME: Acid — TIME:
TYPE: Electrolytic, removal — TEMP:
COMPOSITION: — ANODE:
x .... acid(s) — CATHODE:
POWER:

DISCUSSION:

Theory and practice of chemical polishing (Part II) covering processes for light metal alloys, the iron group and other metals. (*Note:* The "Iron Group" includes the high temperature precious metals, such as platinum, palladium, osmium, and iridium, in addition to iron.)

REF: Pinner, R — *Electroplating*, 6, 401(1953)

MET-0025
ETCH NAME: — TIME:
TYPE: Chemical, polishing — TEMP:
COMPOSITION:
x .... acids

DISCUSSION:

Theory and practice of chemical polishing of metals. Part I covers copper-base alloys; Part II covers iron-base alloys, and Part III covers general theory of chemical etching.

REF: Pinner, R — *Electroplating*, 6,360(1953) (Part I)
MET-0018: Pinner, R — *Electroplating*, 6,274(1953) (Part II)
MET-0019: Pinner, R — *Electroplating*, 7,127(1954) (Part III)

MET-0020
ETCH NAME: Diamond — TIME:
TYPE: Abrasive, polish — TEMP:
COMPOSITION:
x .... diamond paste

DISCUSSION:

The use of diamond abrasives for metallographic specimen polishing.

REF: Samuel, L E — *J Inst Met*, 81,471(1953)

MET-0021: Samuel, L E — *Metallurgica*, 50,303(1954)
Similar processing of metals.

MET-0022
ETCH NAME: Brazing TIME:
TYPE: Metal, joining TEMP:
COMPOSITION:
x .... BAg
DISCUSSION:
BAg alloys used as high temperature brazing materials with a discussion of silver brazing techniques, to include the use of pre-forms, design of joints, and methods of application — torch, furnace, resistance, induction, and dip brazing. After brazing, joints are washed in hot water to remove remaining flux, and there are commercial solvent mixtures available (usually an alcohol/water base).
REF: Brochure: *Silvaloy, "A Complete Guide to Successful Silver Brazing"*, Silver Brazing Div., The American Platinum Works, Newark, NJ (no date)

MET-0023
ETCH NAME: Tungsten TIME:
TYPE: Metal, processing TEMP:
COMPOSITION:
(1) x .... chemicals
(2) x .... gases
DISCUSSION:
W, metal specimens and parts. Pamphlet includes physical properties, chemicals and gases evaluated against corrosion, with specific solutions for etching and cleaning tungsten and tungsten alloys.
REF: Pamphlet: *Tungsten,* #395/1./71 — Schwartzkopf Development Corp.

MET-0024
ETCH NAME: Platinum TIME:
TYPE: Metal, processing TEMP:
COMPOSITION:
x .... chemicals
DISCUSSION:
Pt, metal and its alloys. Includes the other Platinum Group metals, such as osmium, indium, and palladium and their physical properties, products fabricated, and applications in chemistry and chemical processing.
REF: Brochure: *Platinum* — Baker & Co., Inc. (no date)

MET-0026
ETCH NAME: TIME:
TYPE: Metals, processing TEMP:
COMPOSITION:
x .... chemicals
DISCUSSION:
Metals and alloys or natural metallic compounds. This series of pamphlets cover most metals and alloys of the general metal industries with regard to metallographic study — preparation of materials — following special testing (corrosion, thermal, stress, etc.) — with specific etching/cleaning formulations applied to specimens with excellent picture results. Also includes metallographic equipment, methods of general processing, and chemical

processing details. (*Note:* Many of the solutions listed in ASTM E-407-70 are shown here with specific metal or alloy applications.)
REF: Pamphlets: *A/B Metals Digest Series* #/year — Buehler Ltd, 1970—1980s

---

MET-0027
ETCH NAME: Potassium chloride — TIME:
TYPE: Electrolytic, cleaning — TEMP:
COMPOSITION: — ANODE:
x .... KCl, sat. sol. — CATHODE:
DISCUSSION: — POWER:

Metals. Solution recommended for removing contamination from metal surfaces and referred to as separating nonmetallic phases from specimens.
REF: Tufts, C F — *J Appl Phys*, 31,1846(1960)

MET-0028
ETCH NAME: Heat — TIME:
TYPE: Thermal, defect — TEMP:
COMPOSITION:
x .... heat, cycling
DISCUSSION:

Metals. A study of strain enhanced diffusion in metals and its interpretation. Cracks in a metal can cause strain which is enhanced with addition heat cycling.
REF: Ruoff, A L — *J Appl Phys*, 34,2862(1963)

MET-0029
ETCH NAME: — TIME:
TYPE: Mechanical, damage — TEMP:
COMPOSITION:
x .... abrasion
DISCUSSION:

Metals, deformed by lapping and polishing of surfaces. A severely deformed zone near the surface that is not characteristic of the bulk material that shows metal flow — the Beilby layer — about 50 Å thick, unknown whether it is amorphous or very fine crystallites. Below the Beilby layer before reaching undisturbed bulk material is a thick transition zone of deformed metal. (*Note:* There is temperature involved in forming the Beilby layer. Damage depth varying with abrasive grit, shape, hardness to some extent, and grit diameter, in particular. Estimates say subsurface damage induced can run from 3 to 5 mils in depth.)
REF: Finch, G I — *Proc Phys Soc*, A63,785(1950)
MET-0030: McAfes, J — *Aust Eng Yearbook*, 1944

Metals. An article on the metallographic polishing and etching of metals.

MET-0031
ETCH NAME: — TIME:
TYPE: Mechanical, cutting — TEMP: RT
COMPOSITION:
x .... lubricants
DISCUSSION:

Metals and hard crystals. A method of wire saw cutting of materials. (*Note:* There are several systems often using an abrasive slurry or acid-slurry, such as abrasive:water:glycerin with or without HCl, HNO_3, HF, KOH. The wire can be metal, single or multiple strand; a linen, rug, or cotton thread; or metal rods.)

REF: Alexander, E & Many, K A — *Rev Sci Instr*, 26,983(1955)
MET-0013: Enck, F D — Ph.D thesis, *Univ Microfilms*, Publ 23-264.
Describes the use of a tool post grinder in forming cylinders.
MET-0032: Bond, W L — *Rev Sci Instr*, 22,344(1951)
Describes a method of fabricating material spheres by tumbling. (*Note:* There are a number of sphere-forming methods of this type with specific names.)
MET-0034: Carter, J L et al — *Rev Sci Instr*, 30,446(1959)
Describes a sphere grinding technique developed for ferrites. (*Note:* In general metal cutting or shaping a water or oil lubricant with anti-rust agents are used for cooling. Use water-soluble silicones, not mineral oils, for semiconductor or other high-technology materials.)

MET-0035
ETCH NAME: TIME:
TYPE: Growth, structure TEMP:
COMPOSITION:
x .... metal
DISCUSSION:
Metals used in a study of the shape of grain growth. Triangular plane-view grains have an equilibrium 120° angle between sides. If angle is greater, sides will be convex; if less, concave, (+) and (−), respectively. (*Note:* This has been observed by others in single crystal metals and metallic compounds as both naturally grown forms on minerals, and in etched pits.)
REF: Feltham, P — *Acta Metall*, 5,97(1957)
MET-0036: Irving, B A — *J Appl Phys*, 31,109(1960)
Metals and metallic compounds. A study of the shapes of etch hillocks and pits and their correlation with measured etch rates.
MET-0037: Grovenor, C R M — *Thin Solid Films*, 89,367(1982)
Metal contacts. A study of interdiffusion during heating of contacts as diffusion induced boundary migration (DIBM) along deposit grains.

PHYSICAL PROPERTIES OF MOLYBDENUM, Mo

Classification	Transition metal
Atomic number	42
Atomic weight	95.94
Melting point (°C)	2610
Boiling point (°C)	5560
Density (g/cm^3)	10.22
Thermal conductance (cal/sec)(cm^2)(°C/cm)	3.25
Specific heat (cal/g) 20°C	0.062
Latent heat of fusion (cal/g)	67 (70)
Heat of vaporization (cal/mole)	1175
Atomic volume (W/D)	9.4
1st ionization energy (K-cal/g-mole)	166
1st ionization potential (eV)	7.18
Electronegativity (Pauling's)	1.8
Covalent radius (angstroms)	1.30
Ionic radius (angstroms)	0.70 (Mo^{+4})
	0.62 (Mo^{+6})

Vapor pressure (atm $\times 10^{-4}$) 3227°C	8.6
Coefficient of linear thermal expansion ($\times 10^{-6}$ cm/cm/°C) 20°C	5.1
Electrical resistivity (micro-ohms/cm) 0°C	5.17
Magnetic susceptibility (cgs $\times 10^{-6}$) 25°C	0.93
Electron work function (eV)	4.27
Cross section (barns)	2.6
Hardness (Mohs — scratch)	6—7
Crystal structure (isometric — normal)	(100) cube, bcc
Color (solid)	Silver-white
Cleavage (cubic — poor)	(001)

MOLYBDENUM, Mo

General: Does not occur as a native element. There are half a dozen minerals containing molybdenum, all oxides, other than molybdite, MoS_2. The disulfide is widely distributed in small quantities, and is a major ore of the metal. Because of its softness (H = 1—1.5), dark grey/black color and high specific gravity (G = 4.7—4.8), it was originally confused with lead and graphite. Wulfenite, $PbMoO_4$, is the major oxide ore. The oxide minerals are associated with lead and tin veins and in pegmatites.

The pure metal is a major industrial additive to irons and steels to increase toughness and tensile strength, as well as for its high temperature capabilities. It is softer and more ductile than tungsten, has been used as heater wires or for light filaments, and as plate and rod heating elements for both heat dissipation and its electrical capabilities. The common name is "moly". Moly disulfide, MoS_2, is used as a dry lubricant similar to the applications of graphite and, in stick form, has occasionally been used as a pencil.

Technical Application: In Solid State processing the pure metal has been used as disc and flat heat sinks for device mounting due to its close coefficient of expansion match with semiconductor silicon, it is next only to tungsten in that regard, and more easily processed. As a pressed powder pre-form it has a tough, fibrous structure, but also has been used as an extruded sheet or in single crystal form. It is evaporated as a thin film, often in conjunction with layered structures and the formation of silicides: MoSi, Mo_2Si, $MoSi_2$ used as blocking layers in semiconductor device structuring.

It has been artificially grown as a single crystal and studied for both general morphological and electrical properties. As a deposited thin film on glass, it has been studied for its thin film characteristics.

Etching: Difficult to etch. Soluble in hot HNO_3 and H_2SO_4. Most etching done with mixed acids.

MOLYBDENUM ETCHANTS

MO-0001a
ETCH NAME: TIME:
TYPE: Acid, removal TEMP:
COMPOSITION:
x .... HF
x .... H_2O_2
DISCUSSION:

Mo, specimens. Various mixtures of $HF:H_2O_2$ will etch molybdenum.
REF: Booklet: *Molybdenum,* #406.6.71 Schwartzkopf Development Corp.

MO-0002
ETCH NAME: TIME:
TYPE: Acid, polish/clean TEMP: RT
COMPOSITION:
1 .... NH_4OH
2 .... H_2O_2
7 .... H_2O
DISCUSSION:

Mo, foil. Solution used to clean moly foil prior to deposition of gallium arsenide as an epitaxy thin film.

REF: Ghandhi, S K & Reep, D H — *J Electrochem Soc,* 129,2778(1982)

MO-0003
ETCH NAME: TIME:
TYPE: Acid, removal TEMP: RT
COMPOSITION:
76 ml .... H_3PO_4
30 ml .... HNO_3
50 ml .... CH_3COOH (HAc)
150 ml .... H_2O
DISCUSSION:

Mo specimens. Solution becomes very hot with heavy evolution of NO and NO_2 fumes. Rinse in running water. Black smut remaining on surfaces can be removed with a dip/soak in HCl, then followed with water rinse and N_2 blow dry.

REF: Bulletin 1116B/Rev-HPD/VB/TM/0883/1K, MRC
MO-0012a: Walker, P — personal application, 1971—1973

Mo, pressed powder discs 1/2 to 2″ diameter with 0.090 nominal thickness used as heat sinks for silicon Triacs and SCRs. Discs mounted on strips of acid resistant tape and etched at RT, 1—2 min, water rinse 15—20 sec, soak in HCl 15—30 sec, and water rinse 1—2 min. Dry parts on hot plate at 150°C. Remove from tape, turn over and retape. Repeat etching sequence for second side. Final rinse in MeOH and hot plate dry.

MO-0001b
ETCH NAME: TIME:
TYPE: Acid, removal TEMP: RT
COMPOSITION:
x .... H_2SO_4
x .... KF
DISCUSSION:

Mo specimens. Solution will etch molybdenum.

REF: Ibid.

MO-0001c
ETCH NAME: TIME:
TYPE: Acid, removal TEMP:
COMPOSITION:
x .... HF
x .... MnO_4
DISCUSSION:

Mo specimens. Solution will etch molybdenum.

REF: Ibid.

MO-0001d
ETCH NAME: TIME:
TYPE: Acid, cleaning TEMP: 195°F
COMPOSITION:
x .... H_2SO_4
x .... $Na_2Cr_2O_7$, sat. sol.
DISCUSSION:
Mo specimens. Solution used to dip clean molybdenum parts.
REF: Ibid.

MO-0001e
ETCH NAME: TIME: 5—10 min
TYPE: Acid, cleaning TEMP: RT
COMPOSITION:
10 .... formic acid
90 .... H_2O_2
DISCUSSION:
Mo specimens. Used to etch clean molybdenum surfaces.
REF: Ibid.

MO-0001f
ETCH NAME: TIME:
TYPE: Acid, removal TEMP: RT
COMPOSITION:
x .... HF
x .... HNO_3
DISCUSSION:
Mo specimens. Various mixtures of HF:HNO_3 will attack moly rapidly.
REF: Ibid.

MO-0001g
ETCH NAME: TIME:
TYPE: Acid, removal TEMP: RT
COMPOSITION:
x .... HCl
x .... HNO_3
DISCUSSION:
Mo specimens. Various mixtures of HCl:HNO_3, to include aqua regia (used hot), will attack molybdenum rapidly.
REF: Ibid.

MO-0001h
ETCH NAME: TIME:
TYPE: Acid, removal TEMP: 500°F
COMPOSITION:
x .... H_2O
DISCUSSION:
Mo specimens. Pure water used in a high pressure autoclave will etch molybdenum.
REF: Ibid.

MO-0004
ETCH NAME:
TYPE: Electrolytic, polish
COMPOSITION:
x .... 12.5% H_2SO_4
x .... MeOH
TIME:
TEMP: −35°C
ANODE: Mo
CATHODE:
POWER:
DISCUSSION:
Mo sheet. Solution used to etch thin moly sheet down to foil.
REF: Whall, I — *J Electrochem Soc,* 129,105(1982)
MO-0005a: Couch D E et al — *J Electrochem Soc,* 105,450(1958)
Mo specimens. 70% H_2SO_4 used to electrolytically clean molybdenum surfaces. May leave a blue stain.
MO-0014: Rao, P & Thomas, G — *Acta Metall,* 15,1153(1967)
Mo specimens. Solution used to electropolish at 0°C and 6—8 V power. Specimens used in a study of neutron irradiation defects.

MO-0006
ETCH NAME:
TYPE: Acid, cleaning
COMPOSITION:
10 .... HCOOH (formic acid)
80 .... H_2O_2
10 .... H_2O
TIME: 2—3 min
TEMP: 60—70°C
RATE: 0.3—0.5 mils/ 2—3 min
DISCUSSION:
Mo specimens. Solution used in a study of cleaning moly surfaces.
REF: Bell Telephone Systems Tech Publ 3143(1958), 61

MO-0007a
ETCH NAME:
TYPE: Acid, polish
COMPOSITION:
1 .... HNO_3
1 .... H_2SO_4
3 .... H_2O
TIME:
TEMP:
DISCUSSION:
Mo specimens. Solution used as a general removal and polish etch.
REF: Grossman, S & Herman, D S — *J Electrochem Soc,* 116,674(1969)

MO-0008a
ETCH NAME: Aqua regia
TYPE: Acid, polish
COMPOSITION:
1 .... HNO_3
3 .... HCl
TIME:
TEMP: 100°C +
DISCUSSION:
Mo specimens. Solution shows very rapid attack above 100°C.
REF: Kohl, W H — *Handbook of Materials and Techniques for Vacuum Devices,* Reinhold, New York, 1967

MO-0005b
ETCH NAME: Murakani's reagent
TYPE: Acid, removal/clean
COMPOSITION:
110 g .... KCN
100 g .... NaOH/l

TIME: 10 min, max.
TEMP: 80—90°C

DISCUSSION:

Mo specimens. Used a "diluted" solution. Agitate to prevent formation of a brown film which always forms when solution is used for over 10 min. After etching, water wash. After cleaning, specimens were chromium plated and used in a study of oxidation at high temperatures. (*Note:* An HCl rinse will remove brown moly oxide films.)

REF: Ibid.

MO-0015
ETCH NAME: Moly etch
TYPE: Acid, cleaning
COMPOSITION:
(1) 1145 ml .... HF
1170 ml .... H_3PO_4
550 ml .... HNO_3
570 ml .... HAc

(2) x .... HCl, conc.

TIME: 2—5 min
TEMP: RT

DISCUSSION:

Mo discs. Discs punch-cut from extruded sheet or pressed powder. Material is very tough with an elongated fibrous structure. Disc diameters from $^1/_4$ to 2" and 40 to 60 mils thick used as heat sinks for silicon mounting SCR wafers. Discs mounted on acid resistant tape strips and etch cleaned in solution shown. Solution is highly reactive with release of heavy NO and NO_2 fumes. After etching cleaning removes discs to running DI water for 20—30 sec. Then dip or soak in HCl to remove brown moly oxide stain and water wash. Dry on hot plate in air or under IR heat lamp.

REF: Topas, B & Weinstein, H — personal communication, 1970

MO-0016
ETCH NAME: Sulfuric acid
TYPE: Electrolytic, polish
COMPOSITION:
x .... H_2SO_4

TIME:
TEMP:
ANODE: Mo
CATHODE:
POWER:

DISCUSSION:

Mo specimens. Used as a general anodic removal and polish etch.

REF: Korbslak — *Plating,* 40,1126(1953)

MO-0008b
ETCH NAME:
TYPE: Acid, polish
COMPOSITION:
x .... 25% HNO_3

TIME:
TEMP: RT

DISCUSSION:

Mo specimens. Concentrated nitric acid forms MoO_3 and can inhibit etching. At 100°C the solution shown is a rapid etch.

REF: Ibid.

MO-0008c
ETCH NAME:
TYPE: Acid, polish
TIME: 10 sec
TEMP: 90°C
COMPOSITION:
3.5 ml .... HF
96 ml .... H_2SO_4
0.5 ml .... HNO_3
x .... 18 g/l Cr_2O_3
DISCUSSION:
Mo specimens. Solution is a rapid polishing etch.
REF: Ibid.

MO-0008d
ETCH NAME:
TYPE: Acid, removal
TIME:
TEMP:
COMPOSITION
x .... KNO_3
DISCUSSION:
Mo specimens. The solution shown is a violently reactive etch on molybdenum as are the following: KNO_2, NaO_2, KCO_3, Na_2CO_3. KNO_3, KCl_3, and PbO_2.
REF: Ibid.

MO-0007b
ETCH NAME:
TYPE: Acid, removal
TIME:
TEMP:
COMPOSITION:
200 g .... $K_3Fe(CN)_6$
20 g .... NaOH
3—3.5 g .... sodium oxalate
1000 ml .... H_2O
DISCUSSION:
Mo specimens. Solution is a rapid polishing etch for molybdenum.
REF: Ibid.

MO-0009
ETCH NAME:
TYPE: Acid, polish
TIME:
TEMP:
COMPOSITION:
92 g .... $K_3Fe(CN)_6$
20 g .... KOH
300 ml .... H_2O
DISCUSSION:
Mo specimens. A general rapid polish etch for molybdenum.
REF: Kodak Data Book, 1966, 91

MO-0010
ETCH NAME:
TYPE: Acid, clean
TIME:
TEMP:
COMPOSITION:
1 .... HCl
1 .... HNO_3

DISCUSSION:

Mo poly sheet. Used as an insert on an SiC coated graphite susceptor. Remove after epitaxy deposition and etch clean in solution shown, DI water rinse and towel dry. Bake at 150°C overnight before reuse. Any brown oxide remaining is removed by bake in H_2 at 500°C. Used for MOCVD of deposition of AlGaAs (mp 660.2°C) with full melt at 674°C.

REF: Roberts, J J — *J Vac Sci Technol,* B1,850(1983)

MO-0011

ETCH NAME: TIME:

TYPE: Acid, polish TEMP:

COMPOSITION:

x .... KOH

x .... $K_3Fe(CN)_6$

DISCUSSION:

Mo specimens used for electroplating. Use solution as an immersion polishing etch. (*Note:* Major article on electroplating of transition metals. Both etch solution and plating solutions are shown.)

REF: Saubestre, E B — *J Electrochem Soc,* 106,305(1959)

MO-0023: Faust, C L & Beach, J G — *Plating,* 43,1134(1950)

Reference shown for solution used above: MO-0011.

MO-0024a

ETCH NAME: Sulfuric acid, dilute TIME:

TYPE: Electrolytic, polish/thin TEMP:

COMPOSITION: ANODE:

870 ml .... H_2SO_4 CATHODE:

30 ml .... H_2O POWER: 12—20 V & 0.3—0.5 A/cm^2

DISCUSSION:

Mo, (111) wafers used in a study of transmission through thin crystals. Specimens were polished and thinned to about 1000 Å thick. Also, work was done with magnesium and magnesium oxide using other etchants.

REF: Thomas, G & Huftstutler, M C Jr — *J Appl Phys,* 31,1834(1960)

MO-0013a

ETCH NAME: Hydrochloric acid TIME:

TYPE: Acid, cleaning TEMP: RT

COMPOSITION:

x .... HCl

DISCUSSION:

Mo, (111) wafers. First, clean in solution shown and follow with potassium cyanide.

REF: Logan, K H & Schwartz, M — *J Appl Phys,* 26,1287(1955)

MO-0013b

ETCH NAME: Potassium cyanide TIME:

TYPE: Acid, cleaning TEMP: RT

COMPOSITION:

x .... KCN

DISCUSSION:

Mo, (111) wafers. First etch clean in HCl; follow with KCN and DI water rinse.

REF: Ibid.

MO-0008e
ETCH NAME: TIME:
TYPE: Acid, polish TEMP:
COMPOSITION: ANODE: Ta
250 g .... KOH CATHODE:
100 ml .... H_2O POWER:
0.25 g .... $CuSO_4$ or $CuCl_2$
DISCUSSION:

Mo specimens. A good general and uniform etch, but is rapid.

REF: Ibid.

MO-0012b
ETCH NAME: TIME:
TYPE: Acid, removal TEMP: RT
COMPOSITION
55 ml .... HNO_3
57 ml .... HAc
117 ml .... H_3PO_4
115 ml .... HF
DISCUSSION:

Mo, pressed powder discs used as alloy heat sinks for silicon SCR wafers with diameters from $^1/_2$ to 2″. Discs were nominally 0.060 thick. Solution is extremely exothermic with heavy evolution of NO_2 fumes. Black smut remaining on surface is removed by short soak in HCl. The individual acids were evaluated separately on moly and showed the following results: (1) HAc — no reaction in 2 min; (2) H_3PO_4 — no reaction in 2 min; (3) HF — very slow attack with surface cleaning action, only; (4) HNO_3 — turns surface dark grey (oxide). A 1HF:1HNO_3 solution reacts violently leaving a brown or violet-blue moly oxide on surface. After HCl final clean, DI water rinse 2—5 min and air dry on a hot plate — may include MeOH rinse following water.

REF: Ibid.

MO-0017
ETCH NAME: Hydrofluoric acid TIME:
TYPE: Acid, removal/polish TEMP:
COMPOSITION:
x .... HF, conc.
DISCUSSION:

Mo specimens. Use solution as an immersion etch for general removal and polishing of material.

REF: Runck, R J — *J Electrochem Soc*, 107,74(1957)

MO-0008f
ETCH NAME: Phosphoric acid, dilute TIME:
TYPE: Acid, removal TEMP: 20°C
COMPOSITION:
x .... 20% H_3PO_4
DISCUSSION:

Mo specimens. Solution is a controllable etch as diluted when used at temperature shown. At 100°C it is extremely rapid.

REF: Ibid.

MO-0008g
ETCH NAME: Ammonium hydroxide
TYPE: Acid, removal
COMPOSITION:
x .... NH_4OH
TIME:
TEMP:

DISCUSSION:
Mo specimens. Solution is a very slow removal etch.
REF: Ibid.

MO-0008h
ETCH NAME: Potassium hydroxide
TYPE: Electrolytic, removal
COMPOSITION:
(1) x .... x% KOH
(2) x .... x% $NaNO_2$
TIME:
TEMP:
ANODE: Mo
CATHODE:
POWER:

DISCUSSION:
Mo specimens. Both solutions are rapid removal etchants.
REF: Ibid.

MO-0008i
ETCH NAME:
TYPE: Electrolytic, removal
COMPOSITION:
x .... NH_4OH
x .... H_2O_2
TIME:
TEMP:
ANODE: Mo
CATHODE:
POWER:

DISCUSSION:
Mo specimens. Solutions of these two acids are good controllable etchants for this material.
REF: Ibid.

MO-0025
ETCH NAME:
TYPE: Acid, removal/step etch
COMPOSITION:
15 .... NH_4OH
10 .... H_2O_2
100 .... H_2O
TIME:
TEMP:

DISCUSSION:
Mo specimens subjected to Kr^+ ion bombardment. Solution used to step-etch remove molybdenum in studying depth of krypton penetration.
REF: Bartholomew, C Y — *J Appl Phys*, 35,2570(1964)

MO-0020
ETCH NAME:
TYPE: Electrolytic, thinning
COMPOSITION:
25 ml .... HCl
10 ml .... H_2SO_4
75 ml .... MeOH
TIME:
TEMP: 10°C
ANODE:
CATHODE:
POWER: 18 V; 4 A/in^2 or 0.6 A/cm^2

DISCUSSION:

Mo, single crystal specimens used in a study of dislocation channeling from neutron irradiation. Solution used to thin specimens prior to electron microscope observation.

REF: Mastel, B et al — *J Appl Phys,* 34,3637(1963)

MO-001i

ETCH NAME: Aluminum TIME:

TYPE: Metal, alloying TEMP: Molten

COMPOSITION:

x .... Al (Fe, Co, Ni & Sn)

DISCUSSION:

Mo, thin films and crystalline specimens. Molten aluminum will rapidly attack molybdenum forming alloys. Molten Fe, Co, Ni, and Sn also will form similar alloys with the material.

REF: Ibid.

MO-0021

ETCH NAME: TIME: 5 min

TYPE: Acid, polish TEMP: RT

COMPOSITION:

1 .... HCl

1 .... H_2O_2

DISCUSSION:

Mo, (100) specimens. Used as a general removal and polishing etch. Can be used to etch-pattern epitaxy thin films with photolithography techniques.

REF: Johns, C P — *J Vac Sci Technol,* 13,432(1976)

MO-0022a

ETCH NAME: Nitric acid, dilute TIME:

TYPE: Acid, removal TEMP: Hot

COMPOSITION:

1 .... HNO_3

10 .... H_2O

DISCUSSION:

Mo, thin films evaporated on glass substrates. Solution will slowly attack molybdenum. Solution used to remove the metal and clean the substrate.

REF: *Rare Metals Handbook,* Reinhold, New York, 1954

MO-0022b

ETCH NAME: TIME:

TYPE: Acid, removal TEMP: Hot

COMPOSITION:

x .... HCl

x .... HNO_3

DISCUSSION:

Mo, thin films and crystalline specimens. Aqua regia type solutions rapidly attack molybdenum. Solutions used for general removal. Attack rate can be reduced by dilution with water or alcohols.

SEE: MO-0001g.

REF: Ibid.

MO-0022c
ETCH NAME: TIME:
TYPE: Acid, removal TEMP:
COMPOSITION:
x .... HF
x .... HNO_3
DISCUSSION:

Mo, thin films and crystalline specimens. Mixtures of these acids show rapid attack on molybdenum. Solutions used for general removal. Attack rate can be reduced by dilution with water or alcohols.

REF: Ibid.

MO-0023
ETCH NAME: Sodium acid sulfate TIME:
TYPE: Acid, removal/thinning TEMP: Hot
COMPOSITION:
x .... $NaSO_4$
x .... H_2O
DISCUSSION:

Mo, crystalline specimens. Hot pickling in the solution shown used to reduce thickness. A study of the correlation of temperature and grain size effects in the ductile-brittle transition of molybdenum.

REF: Passmore, E M — *Phil Mag,* 12,441(1965)

MOLYBDATES $M_2Mo_3O_8$

General: There are several molybdate and tungstate minerals in nature with the general formula MO_4 and MWO_4 with M = Ca, Pb, Fe, Bi, Cu, Fe, scheelite, $CaWO_4$ or wulfenite, $PbMoO_4$ and ferberite, $FeWO_4$ or hubnerite, $MnWO_4$ are representative of the two crystal classes — the tetragonal scheelite or monoclinic wolframite groups, respectively. These minerals are found in veins and pegmatites formed under pneumatolytic, hydrothermal conditions or by secondary alteration in the oxidation zone of ore deposits.

Most of these minerals are primarily ores of tungsten and molybdenum in industry. See discussion under either molybdenum or tungsten for further information.

Technical Application: Several molybdates have been artificially grown for general crystallographic and morphological study with possible Solid State applications for their ferritic or magnetic properties.

$Fe_2Mo_3O_8$ and others have been grown as fine elongated single crystals by CVD or Vapor Transport (VT) techniques. Structure was referred to squeezed dodecahedrons (isometric system) or elongated hexagonal system, e.g., prisms with elongation in either case on the z-axis. Note that natural garnets often occur with this elongated dodecahedral structure.

Etching: Soluble in HCl, HNO_3, and mixed acids of the HCl:HNO_3. HF:HNO_3, HF:H_2O and hydroxides.

MOLYBDATE ETCHANTS

FEMO-0001a-b
ETCH NAME: Nitric acid TIME:
TYPE: Acid, removal TEMP:
COMPOSITION:
(1) 1 .... HNO_3 (2) 1 .... HCl
1 .... H_2O 2 .... HNO_3

DISCUSSION:

$Fe_2Mo_3O_8$ single crystals grown by CVD. General formula $M_2Mo_3O_8$ with M = Fe, Mn, Co, Ni. The molybdide Fe_2Mo_4 also was grown. A sealed quartz ampoule used for growth, and crystals grew on tube walls as squeezed dodecahedrons elongated in the z-direction. Solutions shown are general removal etchants for most molybdates.

REF: Strobel, A & LePage, Y — *J Cryst Growth,* 61,328(1983)

PHYSICAL PROPERTIES OF MOLYBDENUM CARBIDE, Mo_2C

Classification	Carbide
Atomic numbers	42 & 6
Atomic weight	203.1
Melting point (°C)	2410 (2300)
Boiling point (°C)	
Density (g/cm³)	9.06
Hardness (Mohs — scratch)	8+
(Vicker's — kgf/mm²)	1950
Crystal structure (hexagonal — normal)	$(10\bar{1}0)$ prism
Color (solid)	White
Cleavage (basal)	(0001)

MOLYBDENUM CARBIDE, Mo_2C

General: Does not occur as a natural compound. It is fabricated industrially as a high temperature ceramic as both the mono- and dicarbide, but is not as widely used as silicon carbide, SiC.

Technical Application: There has been no use in Solid State development of devices, although it has been studied as a carbide for general morphology.

Etching: HNO_3, HF, hot H_2SO_4, HCl.

MOLYBDENUM CARBIDE ETCHANTS

MOC-0001

ETCH NAME: Nitric acid — TIME:

TYPE: Acid, removal — TEMP:

COMPOSITION:

x .... HNO_3

DISCUSSION:

Mo_2C, specimens. Material is soluble in nitric acid. Also etch in HF, hot H_2SO_4 and HCl.

REF: *A Dictionary of Carbide Terms,* — Adams Carbide Corp., Kenilworth, NJ

MOLYBDENUM GERMANIDE ETCHANTS

MOGE-0002

ETCH TIME: Nitric acid — TIME:

TYPE: Acid, removal — TEMP:

COMPOSITION:

(1) x .... HNO_3 (2) x .... H_2O_2

DISCUSSION:

Mo_3Ge specimens and Mo_3Ge_2; Mo_2Ge_3 or alpha-$MoGe_2$ are the germanium equivalents

of silicides. Both solutions are rapid etchants. Fused carbonates and nitrates react violently. All materials will resist nonoxidizing solutions.
REF: Brochure: *Molybdenum,* Climax Molybdenum Company (No date/number)

MOLYBDENUM NITRIDE ETCHANTS

MON-0001
ETCH NAME: Nitric acid — TIME:
TYPE: Acid, removal — TEMP:
COMPOSITION:
x .... HNO_3, conc.
DISCUSSION:

MoN and Mo_2N thin films grown on (100) silicon wafers (p- and n-type, 10 Ω cm resistivity). First, deposit 500—1000 Å SiO_2 on wafers with dry O_2; then sputter deposit 300 Å Mo in argon atmosphere, and anneal at 250°C for good adhesion. Pattern etch and clean Mo film with hot H_3PO_4. Form nitride with NH_3 in dry forming gas (FG) (85% N_2:15% H_2) in and open tube, 10 min for 600 Å MoN. 10% NH_3 in N_2 produces smooth surfaces. MoN cubic, fcc, when formed under 800°C; Mo_2N (hexagonal) forms above 800°C with film cracking. Both nitrides can be etched in the solution shown.
REF: Kim, M J — *J Electrochem Soc,* 130,1196(1983)

PHYSICAL PROPERTIES OF MOLYBDENUM SULFIDE, MoS_2

Classification	Sulfide
Atomic numbers	42 & 16
Atomic weight	143.05
Melting point (°C)	795
Boiling point (°C)	Sublimes
Density (g/cm³)	4.7—4.8
Hardness (Mohs — scratch)	1—1.5
Crystal structure (hexagonal — normal)	$(10\bar{1}0)$ prism
Color (solid)	Lead-grey
Cleavage (basal)	(0001)

MOLYBDENUM SULFIDE, MoS_2

General: Occurs as the mineral molybdenite, MoS_2. It is often a pneumatolitic contact deposit associated with ore veins of tin and tungsten minerals, also in pegmatites. Although it is widely distributed, it is never found in large quantities, yet it is an important ore of molybdenum. The other half dozen molybdenum minerals are oxides containing tungsten, iron, calcium, and bismuth. Because of its low hardness (H = 1—1.5) and lead-grey/black color, it was long thought to be a lead mineral, only established as a separate mineral species in 1778. It can be distinguished from lead and graphite by its color streak on paper or porcelain — blue-grey, rather than greenish grey.

Other than as an ore of molybdenum in industry, the compound is used as a dry lubricant, similar to the usage of graphite.

Technical Application: No use in Solid State or semiconductor processing other than as a dry lubricant in some equipment applications.

It has been grown for many years as an artificial compound by flux growth in the presence of sulfur. In Solid State technology it has been grown by the Vapor Transport (VT) method, as fine single crystals and platelets with distinctive basal (0001) planes. Both the natural and artificial crystals have been studied for morphological data.

Etching: Soluble in acids and halogens.

MOLYBDENUM SULFIDE ETCHANTS

MOS-0001
ETCH NAME: | TIME:
TYPE: Tape, cleaning | TEMP:
COMPOSITION:
x .... adhesive tape
DISCUSSION:

MoS_2 single crystal platelets grown by vapor transport. Single crystal specimens can be separated from the growth matrix with a razor blade. To obtain a clean, fresh surface use adhesive tape to peel away a thin surface layer. Method also used on $MoSeS_2$.

REF: Phillips, M L & Spitler, M T — *J Electrochem Soc*, 127,1719(1980)

MOS-0002
ETCH NAME: BRM | TIME:
TYPE: Halogen, polish | TEMP:
COMPOSITION:
x .... x% Br_2
x .... MeOH
DISCUSSION:

MoS_2 single crystal specimens used in a study for photo-anodes. After etch polishing in solution shown, wash in CCl_4 to remove residual bromine. Thin single crystals can be peeled from bulk material. Also evaluated: $MoSe_2$, WSe_2 and WS_2.

REF: Baglio, J A — *J Electrochem Soc*, 129,1461(1982)

MOS-0003
ETCH NAME: Trichloroethylene | TIME: Minutes
TYPE: Solvent, cleaning | TEMP: RT to hot
COMPOSITION:
x .... TCE
DISCUSSION:

MoS_2 as powdered lubricant on equipment gear-trains. Solution used to degrease and remove old lubricant. Gasoline also used. Material can be etch removed with aqua regia and concentrated H_2SO_4.

REF: Fahr, F — personal communication, 1957

MOS-0004
ETCH NAME: Soap | TIME:
TYPE: Ester, removal | TEMP: RT
COMPOSITION:
x .... lava soap
DISCUSSION:

MoS_2, natural single crystal specimens used in a structure study. A solution of lava soap used to clean work areas and hands after handling material. Dilute aqua regia ($3HCl:1HNO_3:5H_2O$) used for controlled removal of specimen.

REF: Walker, P — mineralogy study, 1952—1953

MOLYBDENUM TELLURIDE ETCHANTS

MOTE-0001
ETCH NAME:
TYPE: Halogen, polish
COMPOSITION:
x .... x% Br_2
x .... MeOH
TIME:
TEMP: RT
DISCUSSION:
$MoTe_2$, single crystal specimens. Material is a dichalcogenide type semiconductor with a Band Gap of 1.5 eV. There is possible use as a solar cell with about 10% efficiency. Also working with MoS and WSe_2.
REF: Lewerenz, H J et al — *J Electrochem Soc,* 132,700(1985)

PHYSICAL PROPERTIES OF MOLYBDIC OXIDE, MoO_3

Classification	Oxide
Atomic numbers	42 & 8
Atomic weight	143.05
Melting point (°C)	795
Boiling point (°C)	
Density (g/cm³)	4.5
Hardness (Mohs — scratch)	1.5
Crystal structure (orthorhombic — normal)	(001) fibers
(hexagonal — rhombohedral)	$(10\bar{1}1)$ rhomb
Color (solid)	Yellow
Cleavage (basal)	(001)/(0001)

MOLYBDIC OXIDE, MoO_3

General: Originally thought to occur in nature as the mineral molybdenite, MoO_3, but later analysis changed the name to ferricmolybdenite as a mixture of $Fe_2O_3.3MoO_3.8H_2O$. Data shown under Physical Properties are based on this mineral. It is a relatively rare mineral found as fibers and fine needles in association with molybdenum-bearing iron minerals and limonite, $2Fe_2O_3.3H_2O$.

No industrial use as a natural mineral. The metal molybdenum is used in the fabrication of high speed steels, and has been used as a light filament in radio tubes replacing tungsten.

Technical Application: None at present in Solid State processing. It is part of the native oxide formed on molybdenum surfaces, and can be removed with ammonium hydroxide, NH_4OH. It has been observed as a thin film deposit in working with molybdenum metal substrates.

Etching: Soluble in acids and NH_4OH.

MOLYBDENUM OXIDE ETCHANTS

MOO-0001
ETCH NAME: Ammonium hydroxide
TYPE: Acid, removal
COMPOSITION:
x .... NH_4OH
TIME:
TEMP:

DISCUSSION:

MoO_3, thin film formed during RF ion plating of AlN on molybdenum substrates. AlN deposited at 1000—1200°C, but at 750°C MoO_3 was formed, and above 1200°C, N_2O_3. Solution is a general etch for the material. See ALN-0014.

REF: Ritajima, M et al — *J Electrochem Soc,* 128,1588(1981)

MOO-0002

ETCH NAME: Sulfuric acid — TIME:

TYPE: Acid, removal — TEMP: Hot

COMPOSITION:

x .... H_2SO_4, conc.

DISCUSSION:

MoO_2, as amorphous platelets on steel is a stable black compound with good surface adhesion to about 250°C with fissures between platelets. Film shows solar selectivity, high adsorption, and low thermal emittance. Solution shown used as a slow etch.

REF: Agnihotri, O P et al — *J Thin Solid Films,* 109,193(1983)

MOLYBDENUM SELENIDE ETCHANTS

MOSE-0001

ETCH NAME: BRM — TIME:

TYPE: Halogen, polish — TEMP:

COMPOSITION:

x .... x% Br_2

x .... MeOH

DISCUSSION:

$MoSe_2$, single crystal specimens used in a study for photoanodes. After etch polish in solution shown, wash in CCl_4 to remove residual bromine. Thin single crystals can be peeled from the bulk material. Also evaluated: MoS_2, WS_2 and WSe_2.

REF: Baglio, J A et al — *J Electrochem Soc,* 129,1461(1982)

MOLYBDENUM SELENIUM SULFIDE ETCHANTS

MOSS-0001

ETCH NAME: — TIME:

TYPE: Tape, cleaning — TEMP: RT

COMPOSITION:

x .... adhesive tape

DISCUSSION:

$MoSeS_2$, single crystal platelets grown by Vapor Transport (VT). Single crystal specimens can be separated from the growth matrix with a razor blade. To obtain clean, fresh surfaces use adhesive tape to peel away thin surface layers. Method also used on MoS.

REF: Phillips, M L & Spitler, M T — *J Electrochem Soc,* 127,1719(1980)

MOUNTING MATERIALS

General: There are a number of natural and artificial compounds used for mounting and holding specimens for lapping, polishing, and etching. Such materials can be woods, plastics, metals, minerals, or rocks as well as semiconductor wafers and other artificial compounds.

These mounting materials range in hardness (Mohs) from beeswax, H = 0.2 to natural resins and hydrocarbons, both paraffin or asphalt base, to artificial plastics with H = 3—5. The natural resins, paraffin, etc., melting points range from about 50 to under 200°C; whereas plastics can vary from 800 to 1500°C.

Brittleness is a major factor in obtaining a desired pliability for a mounting medium and requirements vary with specific application. As three examples: (1) Beeswax was probably the first natural wax used by man as it is extremely soft, with a melting temperature low enough to become pliable when held in the hand, yet is water-repellent as a surface coating. It is worth noting that the honeycomb hexagonal structure is one of the strongest known and, today, is a major construction form for wood, plastic, metal and cardboard. (2) Waxes, such as paraffins derived from petroleum, are not hand temperature pliable, but are sufficiently soft to be rubbed onto surfaces, also as a good water repellent coating. Both beeswax and paraffins are edible, used as a filler and extender in chocolate candy. The Bull Durham tobacco package, coated with paraffin, is still used to wipe the paraffin onto vehicle windshields in rainy weather to produce a slick, water run-off surface on the glass. (3) Resins, such as copal (one of the shellacs) or carnauba wax, are brittle, but with melting points often under that of boiling water (100°C), also alcohol soluble, such that they can be applied as a thin, relative hard and stable surface coating. Carnauba-type waxes were used as a ''sealing wax'' on letters before the development of glue, and are still used as a seal on some official documents with an imprinted logo.

Most of the mounting ''waxes'' are a mixture formulated for a specific pliability and temperature. Pitch (beeswax/linseed oil/rosin) — burgandy, green, optical, etc. — and the Apiezons, particularly, Apiezon-W (black wax) which is an asphalt derivative, are examples. The latter, black wax, although slightly brittle, has a melting point of 80°C and is easily removed with TCE. It also is inert to most metals and compounds used in processing, unaffected by acids or alcohols, such that it is used as both a holding medium in addition to a surface coating for etch patterning.

Powdered plastics — lucite and bakalite — are used in preparing metallographic specimens with the solidified powder holding the specimen. There are metallographic laboratory units that, with a combination of heat and pressure, form the plastic/specimen into small cylinders, which are then mechanically lapped and polished to expose the specimen. And the specimen may be preferentially etched to develop structure. Other plastic formulations, such as the Crystalbond or Pliobond series of compounds are liquefiable mounting mediums available in stick form. In addition, there are the putty-type materials as both a natural rubber derivative or plastic compound.

Lacquers constitute another type mounting or coating material. They can be natural tree or plant rosins and resins, or artificially formulated, such as the photo resist type lacquers used in photolithographic patterning of Solid State semiconductor wafers.

Vacuum, alone or in conjunction with waxes or metal masks, has been used as a primary holding mechanism. Even water, alcohol, and glycerin, as liquids, have been used for temporary holding or positioning of parts and, in vacuum metallization of wafers and substrates, they can be physically held in position with metals, such as phosphor-bronze or steel spring clips.

Low melting temperature metals, such as indium or Wood's Metal, are used to mount and hold specimens in MBE vacuum epitaxy or specimens for microscope examination. The latter includes putty as a specimen holder on a microscope stage and, as an extension, the surface replication technique using aqua dag (graphite powder in alcohol) or an evaporated metal thin film, which is then used for study rather than the specimen, proper.

Technical Application: All of the mounting materials and techniques described in the general section above have been used as mounting or handling mediums for Solid State and metal processing.

They are categorized in the following presentation by basic type, such as natural waxes, plastics, metals. The solutions shown are primarily cleaning and removal types, not only for the mounting materials, themselves, but for the material surfaces to which they are applied, as contamination on surfaces is possibly one of the most severe processing problems related to device, part or equipment failure.

Many of these materials also are used in a finished product assembly, such as a plastic extrusion encapsulated device, as metal or ceramic packages, or as a lacquer surface coated dielectric.

Etching: Acids, alcohols, and solvents. Varies with material.

MOUNTING MATERIAL ETCHANTS

APIEZON ETCHANTS

APZW-0001

ETCH NAME: Trichloroethylene — TIME:

TYPE: Solvent, removal — TEMP: RT to boiling

COMPOSITION:

x .... TCE

DISCUSSION:

Apiezon-W, commonly called "black wax". Melting point 80°C. This is one of the most widely and generally used materials for surface coating or waxing-down semiconductor wafers and other substrate materials or metals prior to etching. This Apiezon wax is a petroleum tar derivative, and is supplied in 1 lb packages of 15 to 20 sticks, about $^1/_4 \times {}^1/_4 \times 8''$ in size.

It can be melted and used uncut, or pieces dissolved in TCE. The latter is applied as a brush or spray coating. It is inert to most acids, alcohols and water, and is used as a general protection against etching or, when patterned, as an etching mask like a photo resist lacquer. As a mounting medium it is used to hold wafers on a ceramic substrate for cutting and dicing of parts.

Trichloroethylene (TCE) is a general solvent for most Apiezon waxes, but other chlorinated solvents can be used, such as trichloroethane (TCA).

Other Apiezon formulations are available for specific applications and are shown under APZW-0003, below.

REF: Blankenship, J L & Borkowski, C J — *IRE Trans Nucl Sci* NS-7,190(1960)

APZW-0002: Walker, P — personal application, 1973—1985

Apiezon "W" as patterned coating on nickel shim stock used to etch fabricate nickel masks for vacuum metal evaporation.

APZW-0003: VWR Supply Catalogue, 1984

Other Apiezon waxes are formulated with different characteristics for particular applications with their own letter designations:

Apiezon H: 250 to −15°C. Ground glass joint seals.

Apiezon L: RT to 30°C. Short-term, hi vac grease seal to 10^{-10} Torr vacuum.

Apiezon M: 30°C & 10^{-3} Torr to 200°C & 10^{-7} Torr vacuum grease with low vapor pressure. Used for holding wafers and substrates on the copper plate of ion milling vacuum systems.

Apiezon N: to 200°C & 10^{-9} Torr as a stop-cock grease.

Apiezon Q: to 30°C for semi-permanent seal of vacuum joints.

Apiezon T: Like "W", but m.p. 120°C.

Apiezon W: to 180°C & 10^{-3} Torr, m.p. 80—90°C. Like "Q", but more widely used.

BEESWAX ETCHANTS

BEEW-0001

ETCH NAME: Trichloroethylene — TIME:

TYPE: Solvent, removal — TEMP: RT to boiling

COMPOSITION:

x .... TCE

DISCUSSION:

Beeswax (*Apis mellifera*) as a natural by-product from beehives. It has a density of 0.96—0.97 g/cm^3, a solidification point of 60—62°C. The natural color is yellow/amber but, as refined, it is semitranslucent white with m.p. about 67°C.

The "lost-wax" processing was developed in the early Bronze Age, perhaps as early as 4000 B.C. An initial form is made of beeswax with a complete and sometimes intricate surface design. The form is then slowly and carefully coated with layers of wet clay, allowed to air-set, then is kiln fired to harden the clay. During the firing, the beeswax drains out of the form, and is recovered for re-use. A liquefied melt of bronze is then poured into the clay mold and allowed to harden. The clay is finally broken away and, as needed, the part finished by cutting, grinding, and polishing. This method was still in use for making bronze parts into the early 1900s, and is still used in the pottery and clay industries, today, and for making objets d'art out of both brass and bronze.

In modern-day industry, beeswax is used by itself as a mounting and holding medium for etching parts where solution heating is below the beeswax melting point; as a surface coating protectant during cutting or dicing of small, thin parts, such as semiconductor wafer dicing, and the fabrication of TiO_2 dielectric capacitors. It occasionally is used as a holding wax for a specimen mounted on a microscope stage.

When mixed with vaseline and rubber, it is used as a stockcock grease. Mixed with rosin in various proportions as a fairly thick coating on a metallographic platen, it is referred to as a "wax-lap". The wax surface is cross-scored in a V-groove square pattern and, with slow rotation speed of the platen and adding a fine abrasive polishing slurry (alumina), material surfaces can be highly polished with optimum planarity. As a beeswax/resin mixture, thinly coated on a lapping platen, semiconductor wafers are slightly heated and pressured into the wax, the circular platen placed face down on a metal flat, and rotated to lap thin and polish as many as half a dozen wafers at a time.

Trichloroethylene (TCE) is a solvent for beeswax, as well as being a general cleaner for metals and metallic compounds. After wax removal, parts should be hot vapor degreased with TCE, TCA or Freons to insure there is no remaining wax contamination.

REF: *Encyclopedia Britannica* — Lost-Wax process

BEEW-0002: Klatskin, W B — personal communication (microscope), 1983

BEEW-0003: Kunitz, A — personal communication (beeswax/rosin), 1974

BEEW-0004: Waters, W P — personal communication ("wax-lap"), 1960

BEEW-0005: Marich, L A — personal communication (wafer cutting), 1980

DIELECTRICS

CER-0001

ETCH NAME: Wet chemical etching — TIME:

TYPE: Acid, removal — TEMP: RT

COMPOSITION:

(1) x .... H_2SO_4 (2) x .... HCl (3) x ... H_3PO_4

DISCUSSION:

Ceramics, as substrates or as masks for metal evaporation. To clean ceramics prior to use, etch in any of the three solutions shown, following with heavy water washing, and drying. Best drying is done in a vacuum oven (10^{-3} Torr) at 125—150°C to insure removal of all water moisture. A circulating air oven also can be used. Heating time depends upon the porosity of the ceramic, but $^1/_2$ h or longer is recommended. Primary ceramics in use are alumina, Al_2O_3, and beryllia, BeO in Solid State. Surfaces are a standard 4 or 2 μ" for most device applications, but can be 25 μ" for general purpose use. The 4 μ" is the ceramic as-fired condition; 2 μ" is diamond paste polished, and a 1 μ" can be furnished on request.

REF: Accumet, Boston, MA

CER-0002

ETCH NAME: | TIME: 1—5 min

TYPE: Solvent, degreasing | TEMP: Boiling

COMPOSITION:

(1) x .... TCE/TCA/Freon, vapor   (2) x .... EOH/MeOH

DISCUSSION:

Ceramics. Solvents (1) are basic vapor degreaser solvents, and (2) are alcohols type solvents. They all can be used as RT or hot liquids or hot vapors for general cleaning of surfaces to remove oil and dirt contamination. Very often solvent cleaning is following by final alcohol rinsing, and drying with nitrogen gas. Solvent cleaning is used, only, or before and after acid cleaning steps in processing of ceramic substrates.

REF:

CER-0003

ETCH NAME: Hydrogen | TIME:

TYPE: Gas, cleaning | TEMP: >750°C

COMPOSITION:

x .... H_2

DISCUSSION:

Ceramics, as substrates or as masks for metal evaporation. The best method of surface cleaning, particularly, if diamond pastes have been used in surface finishing. Forming gas (75% N_2:15% H_2) can be used as a nonexplosive mixture replacement for pure hydrogen. Both are standard for furnace hydrogen firing used in cleaning metal surfaces to remove residual oxidation and/or contamination. They also are used in processing as reducing atmospheres for alloying or soldering of parts.

REF:

GLA-0001

ETCH NAME: Hydrofluoric acid | TIME: Seconds/minutes

TYPE: Acid, removal | TEMP: RT to hot

COMPOSITION:

(1) x .... HF, conc.   (2) x .... HF, vapor   (3) x .... HF
x .... H_2O/MeOH   (4) 1 .... HF
1 .... NH_4F (40%)

DISCUSSION:

Glass, as cut blanks, cover glasses, microscope slides. Acids with fluorine are the only ones that will physically etch remove glass. Glass when immersed in liquid HF will be polished; held in hot vapor, will be frosted. The dilute HF is used for slower cleaning, rather

than etching action. All four solutions shown are used in patterning of SiO_2 thin films deposited on surfaces, in addition to general removal or cleaning of the silica.
REF:

GLA-0002
ETCH NAME: TIME: Minutes
TYPE: Acid, cleaning TEMP: RT to hot
COMPOSITION:

(1) x .... H_2SO_4, conc.

(2) 1 .... H_2SO_4
1 .... $Na_2Cr_2O_7$, sat. sol.

DISCUSSION:

Glass, as laboratory glassware or quartzware, in addition to cut blanks, microscope and cover slides. Solution (1) is used in cleaning soda-lime glass blanks prior to chromium metallization as Chrome Glass Masks used in photolithography. Solution (2) is one example of the standard ''glass cleaner'' that has been used in industry for many years. In both cases, parts should be heavily water washed after etch cleaning, followed by drying.
REF:

GLA-0003
ETCH NAME: Soap TIME:
TYPE: Solvent, cleaning TEMP: RT to hot
COMPOSITION:

x .... $3C_{17}H_{35}COO$
x .... H_2O

DISCUSSION:

Glass and ceramics. Two of the cleanest, high purity hard soaps are Ivory and Castile, both recommended for cleaning of missile and space hardware. Soft soaps, such as liquid Joy, also have been used on Solid State and electronic parts. Metals in general can be cleaned in such high purity soap solutions or in detergent types, to include de-foaming or foaming agents, and there are many, many specialized formulations used on all materials as ionic or nonionic. Glass blanks, as soda-lime for chrome glass masks, are scrub-cleaned in a soap solution as one part of the cleaning process prior to chromium evaporation. Many metal and ceramic surfaces are similarly cleaned prior to metal plating. After soap cleaning, parts should be thoroughly water washed as, regardless of soap cleaning capability, they are fat, and fatty acid derivatives, such that they also may act as a surface contaminant like an oil.
REF:

EPOXY ETCHANTS

EPOX-0001
ETCH NAME: TIME: Minutes
TYPE: Acid, cleaning TEMP: RT to warm
COMPOSITION:

(1) x .... H_2SO_4

(2) 3 .... HNO_3
1 .... H_2O

DISCUSSION:

Epoxy and polyamide type artificial compounds used as un-loaded mounting or loaded electrical contact materials. There are several metal-loaded types with Ag, Au, Cu, Ni, Al, Fe powders added, and the silver epoxy is widely used in Solid State for assembly and testing of semiconductor devices. The epoxies, in general, have lower temperature capability than polyamides, although both, presently, are limited to about 200°C before charring occurs. Acetone is a general cleaning solution for all of these compounds. For removal, either char the epoxy at 250°C or higher, and scrape away; or etch using an acid, such as H_2SO_4, conc. or $3HNO_3$:$1H_2O$. Note that these compounds can be a single component kept frozen for best shelf-life, or an A-B component system that, once mixed, has a pot- or working-life of from seconds to 4—8 h.

Most are rated for 6-month shelf-life before requiring replacement although, for general testing using an A-B component system, some have still been satisfactory after 2 years. Single component systems, no, as they slowly harden with time even when kept frozen at −40°C. Neither epoxies nor polyamides are recommended for high reliability final assemblies as, with time, they will continue to harden, may begin to flake, or allow moisture absorption which can eventually destroy an operational device.

REF:

LACQUER ETCHANTS

PR-0001

ETCH NAME: Acetone — TIME: 2—5 min

TYPE: Ketone, removal — TEMP: RT to hot

COMPOSITION:

x .... CH_3COOH_3

DISCUSSION:

Photo resist (P/R) lacquers, all types. Acetone is a general solvent for these artificial compounds. AZ-1350J negative P/R is widely used for general purpose photolithography in addition to COP types, and there are positive P/R versions of both. The negative P/Rs patterns harden the pattern desired with removal of unwanted material; positive P/Rs are the reverse . . . areas are opened, which are then etch structured or metallized. The PMMA P/Rs have been designed for use with electron-lithography patterning, such as for fine-line structuring of a gallium arsenide gate channel for a field effect transistor (FET) fabrication. The PMMAs can be recognized under a microscope (200—400 ×) as a pink/reddish coating, whereas AZ- and COP-types are tan/brown in color. These latter ones have different formulations for thicker type coatings, such as AZ-1370J, etc. The older KMER type photo resists are no longer used.

Processing should be done under clean room conditions as a Yellow Room: yellow plastic glass, and yellow fluorescent lighting, as photo resists are sensitive to white light. The lacquers are supplied in dark amber bottles, and should be stored in a dark cabinet for best results. The photo resist is used directly from the bottle, though it may be filtered if desired. To apply and process photo resists:

(1) A glass syringe with replaceable needle and 0.2 μm filter (use new filter element at shift start). The syringe can be re-used indefinitely unless broken, but should be thoroughly washed with acetone after every use to remove old photo resist. Fill syringe with filter and needle removed (hold finger over bottom); attach filter and needle; and extrude some photo resist to eliminate air bubbles.

(2) Photo resist spinner: A removable plastic basin surrounding a vertical $^1/_4$" spindle driven by a motor and vacuum unit mounted beneath the spinner body, and an electronic operation console. Flat-surface holders with or without criss-cross and concentric drain patterns of various sizes are commercially available or can be designed. They all have a central vacuum hole that matches over the vertical drive spindle.

(3) Acetone can: A pressurizable (N_2 gas) stainless steel can with a quick-release connection on the side for nitrogen pressuring after the can is filled with acetone, and with a can top spray head for acetone application.

(4) P/R wafer coat: Spray the SST holder with acetone to insure cleanliness. Place a wafer (semiconductor, ceramic substrate etc.) on the flat holder and activate vacuum switch

for gripping. Active spinner while spraying wafer with acetone to clean wafer surface, and spin surface dry. Apply a large drop(s) at middle of wafer, and activate spinner. For normal coating: set electronic module for 3500 rpm, and 10—15 sec spin time for a nominal 0.5 μm P/R coating thickness using AZ-1350J. P/R should spin out evenly over entire wafer. There can be slight color variations (yellow, green, red, tan) across surface, but a singular tan color is preferred, and without edge thickening. The soft P/R can be acetone spray removed as needed and coating step repeated. Occasionally a double-coating is done for additional thickness . . . about 0.7 μm using 1350J.

(5) Air cure P/R: Place wafer(s) in an open glass Petri dish other similar holder, and allow to air dry 30—45 min. Wafers should not be completely enclosed with a Petri dish lid, but may be partly covered to prevent dust settling on the drying P/R . . . keep sufficiently open for P/R vapor release.

(6) Oven cure: Three small ovens are often used in series: first, 90°C; second, 90 or 100°C; third 120°C. Standard practice is to cure at 90°C for 1 h each, in oven one and two. Where a hard cure is required, an additional hour at 120°C. Place wafer(s) uncovered on a glass or SST flat carrier, introduce into oven one, and transfer to two and three as required.

(7) UV exposure: A small, basic system consists of (1) a base holder for the wafer and pattern mask, then (2) a camera-type shutter with a high intensity mercury UV light bulb housing directly above, and (3) an electronic operation module with power supply. The light intensity should be checked with the light meter supplied with the equipment on a daily basis, such that exposure time can be correct as the intensity drops slowly with bulb usage time. Turn the bulb on at the beginning of a morning shift, and leave on. When a new bulb is inserted, light intensity and focus are adjusted with x-y-z adjustment screws in conjunction with the light meter.

(8) Pattern exposure masks: Permanent masks are glass . . . soda-lime glass . . . as the chrome glass mask or, more recently, the chromium with anti-reflective (AR) coating is being replaced with iron oxide, FeO_x. The latter do not require the AR coating, and are less prone to damage. The glass photo resist mask making is a highly specialized technique, and normally done by outside vendors to the users design pattern specifications. In-house patterns, as well as outside vendor patterns, start with a rubilith (red mylar on clear acetate) that is usually cut for a 1:10 reduction. After cutting the rubilith pattern, reduce to a black and white negative by photoreduction. The negative can be used as a temporary photo resist mask, or transferred to glass. Many patterns are now computer generated, to include cutting the rubilith, itself. This is still high labor intensity operation, as the rubilith pattern must be hand-peeled.

(9) UV exposure: Set exposure time according to requirement of UV light intensity, type P/R and thickness, and cure schedule. Nominal 10—15 sec exposure for a 90°C cured 0.5 μm thick film. Place wafer on exposure plate, position mask on wafer, then activate exposure timer.

(10) Develop pattern: Each supplier's photo resist has its own developer solution, and should be water diluted per directions. Use warm (50—70°C) and watch developing surface, so as not to over-develop with possible under-cutting of pattern. Rinse in water after developing. Up through this operation, if there is any problem with the photo resist, it can be stripped and redone.

Wafers and substrates are of three basic types: (1) a photo resist coating, only; (2) pre-coated with an oxide or nitride thin film, or (3) pre-coated with a metal thin film. In the first case, the wafer surface may be etched for pits or channels; PMMA electron lithography processed; or metallized with subsequent acetone "lift-off" of excess and extraneous metal. In the second case, the oxide/nitride is etch patterned down to the wafer/substrate surface, with the oxide/nitride then acting as an etching, metallization or diffusion mask with or without the P/R still present. In the third case, this is a standard method of processing metallized substrate circuits followed by acetone "lift-off" of the photo resist. There are some special cases, such as using a combination of photo resist/metal as metallization or diffusion mask.

In photolithographic processing, after P/R has been applied and cured, the specimen should not sit with the lacquer still on the surface for more than 6—8 h; otherwise, strip and re-coat. The photo resist can become too hardened with time, it will crack and craze, and be all but impossible to remove from surfaces.

A Yellow Room should be maintained with a temperature between 68—74°F, with a relative humidity of 40%. If RH is very much under 30%, photo resist will cure too rapidly and crack or craze; if much over 60% RH, photo resist will not harden or cure properly, and pattern can slump with loss of definition.

REF: Technical Bulletin #401, 1984 — American Hoerchst Corp

PR-0002: Walker P & Valardi, N — personal application, 1985

AZ-1350J used as both a mounting medium and pattern coating on AuSn(20%) and AuGe(13%) alloy strips placed on microscope slides and hole pattern etched to reduce alloy volume.

PR-0003a: Walker, P — personal development, 1974

KMER photo resists used as a patterning and mounting medium painted on brass and nickel shim stock used as vacuum evaporation metallization masks for silicon wafers. Also used Apiezon-W wax for patterning.

PR-0003b: Ibid. (1964)

KMER type photo resists used for patterning (111) silicon wafers in a development study of DC sputtered silicon nitride, oxynitride and silicon dioxide surface thin films. This included development of etch solutions for the nitrides that were compatible with KMERs.

PR-0004a: Brumfield, D & Walker, P — personal development, 1957

Jello used as a mounting and patterning medium on silicon wafers for aluminum evaporation and lift-off with water. This work was done before the development of photo resist lacquers. Orange, lime, and lemon Jello evaluated, and lemon gave best results.

PR-0004b: Ibid.

Collodion used as a mounting and surface patterning medium on silicon, germanium, gallium arsenide, and indium phosphide wafers for metal contact evaporation in material evaluations and lift-off with amyl acetate. Collodion used prior to development of photo resists.

PR-0006: Hackett, R & Walker, P — personal development, 1978

PMMA photo resist used for electron lithography of (100) oriented gallium arsenide wafers in fabricating FET gate channels (Hackett). Device processing of LN-FETs with $^1/_2$ μm gates (Walker) with improved techniques for PMMA removal. (*Note:* PMMA is more difficult to completely remove than AZ-types due to high energy heating during electron lithography. Acetone lift-off used, but warm, and with more critical evaluation prior to metallization in order to prevent later peeling and lifting of metallization due to PMMA organic contamination.)

LAC-0001

ETCH NAME: TIME: Minutes

TYPE: Solvent, removal TEMP: >100°C

COMPOSITION:

(1) x .... acetone (2) x .... benzene (3) x .... turpentine

DISCUSSION:

Canadian balsam is classified as an oleo-resin or lacquer, used for mounting rock and mineral specimens on glass microscope slides. The clear balsam has a low refractive index

that does not mask study of thin sections by transmitted light under microscope examination. After mounting, the specimens are lapped down as thin sections on an iron lap with a wet slurry abrasive. In most instances, thin sections are kept on file still mounted, but any of the solvents shown can be used to liquify the balsam for both application and removal.
REF: Walker, P — mineralogy studies, 1952—1953

SHE-0001
ETCH NAME:
TYPE: Solvent, removal
COMPOSITION:
(1) x .... acetone (2) x .... amyl alcohol (3) x .... MeOH
TIME: Minutes
TEMP: 80—100°C
DISCUSSION:

Shellac is classified as a resin, but also referred to as a lacquer. It has been used in a similar manner as described for Canadian balsam in preparing rock and mineral thin sections. In Solid State processing a variety of shellacs are used for coating exposed semiconductor device p-n junctions to prevent electrical arcing as exposed to air, only. Shellacs also have been used in mounting discrete elements on a circuit board (PCBs), for coating the wire windings of motors and, alone, or as mixtures for encapsulating. The solvents shown can be used to both liquify shellacs for application and for their removal.
REF: N/A

METALS

IN-0001a
ETCH NAME: Ammonium hydroxide
TYPE: Alcohol, removal
COMPOSITION:
x .... NH_4OH
TIME: Minutes
TEMP: RT to boiling
DISCUSSION:

In, metal smear used as a solder hold-down for semiconductor wafers and substrates mounted on molybdenum discs for epitaxy growth in MBE systems. The solution shown will slowly dissolve and remove indium without affecting either the specimen material or the molybdenum disc. Acids can be used, but should be compatible for no effect on other than the indium. (*Note:* Thin sheet indium is used as vacuum gasket seals in cryo-pumps, and can be cleaned in the solution shown prior to installation.)
REF: Woo, R — personal communication, 1983
IN-0001b: Ibid.

In, as small pellets used to hold material specimens in position on a microscope stage for observation. Pellet can be positioned in a groove cut in a metal plate and, when the specimen is pressed into the groove it both grips the holding plate and the specimen.

SST-0001
ETCH NAME:
TYPE: Alcohol, cleaning
COMPOSITION:
(1) x .... MeOH/EOH (2) x .... vacuum
TIME:
TEMP: RT to hot
DISCUSSION:

SST, bronze, brass, and graphite have all been used as holders for processing Solid State materials in vacuum evaporation systems, diffusion tubes, or belt furnaces. Steel and brass plates as flat holders in vacuum metallization with wafers held in place with phosphor-bronze or SST spring clips; paired and nested brass plates with critically dimensioned pattern holes for sandwiching alpha-quartz crystals for metallizing electrodes, held together with

brass clips. Graphite plates with a fine hold pattern to hold 0.010 diameter spheres of aluminum have been used for (111) silicon diode fabrication in the Sphere Alloy Zener process. Graphite boats have been used for furnace alloying of (111) silicon wafers in a stack configuration as SCRs and Triacs. Various shim stock metals (steel, brass, nickel, aluminum) have all been fabricated as evaporation masks, held in place over wafer specimens with spring clips or positioning pins. All of these materials can be cleaned with solvents (TCE, Freons) or with alcohols as shown. Graphite parts often include furnace firing for drying or with vacuum for removal of any liquids entrapped in the graphite surface due to material porosity. The graphite also may be cleaned by light rubbing on a rough paper surface, such as Whatman filter paper.
REF: N/A

PLASTICS

CRY-0001
ETCH NAME:
TYPE: Ketone, removal
COMPOSITION:
x .... CH_3COOH_3

TIME: Minutes
TEMP: RT to warm

DISCUSSION:

Crystalbond, #507, and others as a plastic compound series used for mounting semiconductor wafers, ceramics, quartz, and sapphire substrates for dicing. The compound remains viscous when melted (about 80°C) and tends to trap air bubbles. Care must be taken to prevent bubbles forming or cutting integrity can be lost with pop-out of diced elements. The material is slowly soluble in acetone by soaking in a warm solution.
REF: Porter, R — personal communication, 1984

LUC-0001
ETCH NAME: Acetone
TYPE: Ketone, removal
COMPOSITION:
(1) x .... CH_3COOH_3
(2) x .... I_2
x .... $MeOH/H_2O$

TIME: Minutes
TEMP: RT to warm

DISCUSSION:

Lucite as sheet or powder (also called Plexiglas). Powder or shavings can be liquified with acetone (1) and used as a glue for lucite parts assembly. Powder is used in metallographic laboratories as specimen holders by heating and pressuring into small solid plastic cylinders. The cylinder is then lapped to expose the specimen for polishing and/or preferential etching and microscope study. Lucite sheet is used as a clear chemical sink face, and can be cleaned and disinfected with solution (2) or with a diluted tincture of iodine, $xKI{:}xI_2{:}EOH$. See section on Lucite. (*Note:* Do not use acetone to clean Lucite, as it is a solvent and will cause severe smearing of surfaces.)
REF:

BAK-0001
ETCH NAME:
TYPE: Ketone, removal
COMPOSITION:
(1) x .... CH_3COOCH_3
(2) x .... (MeOH/EOH)

TIME: Minutes
TEMP: RT to warm

DISCUSSION:

Bakalite, as a cellulose acetate type plastic, commonly dark brown in color. It has been used from powder form for holding specimens as described of Lucite (LUC-0001). It is

opaque, but can be transparent like Lucite, and is harder and less prone to being scratched or damaged. Acetone is a very slow solvent for bakalite. Bakalite type plastics also are soluble in alcohols (2), such that cleaning should be done with water and soap.
REF:

MYL-0001
ETCH NAME: Acetone
TYPE: Ketone, removal
COMPOSITION:
x .... CH_3COOCH_3
TIME: Minutes
TEMP: RT to warm
DISCUSSION:

Mylar type plastic. Rubilith material is thin red mylar heat laminated to clear acetate (total thickness of 0.003 or 0.005) used in fabricating design patterns for photolithography glass masks or photograph negatives used to fabricate semiconductor devices and circuit substrates. As white (clear) or blue mylar with a cellulose glue sticky-back coating, the mylar is used for holding semiconductor wafers for scribe & break system dicing. The wafer is lightly held by the glue, and individual dice can be removed with tweezers of a pick-and-place system. Clean the mylar surface by wiping with a water dampened soft cloth, and dissolve the glue with acetone.
REF:

DEK-0001
ETCH NAME: Trichloroethylene
TYPE: Solvent, removal
COMPOSITION:
x .... $CHCl:CCl_3$ (TCE)
TIME: Minutes
TEMP: Hot
DISCUSSION:

Dekotinsky plastic cement, available in tan colored stick or powder form. Used to mount semiconductor ingots on ceramic blocks for wafer cutting. No longer used as, when heated even slightly above its melting temperature of about 90°C it polymerizes and, once polymerized, it is impossible to completely remove from surfaces even by lapping. Hot to boiling TCE will slowly remove Dekotinsky, but not after polymerization has occurred.
REF: Walker, P — personal application, 1956—1960

GLY-0001
ETCH NAME: Heat
TYPE: Thermal, removal
COMPOSITION:
(1) x .... heat
(2) x .... solvent
x .... alcohol
TIME:
TEMP: Elevated
DISCUSSION:

Glyptal is a rubberized-plastic type compound, either red or black in color, with very low vapor pressure. It is still used for sealing glass and quartz tubing of vacuum systems against pin-point air leaks, and also has been used as a general mounting and holding medium for parts. It hardens slowly (4 to 8 h), and can become brittle with time (months). Heating and chipping away is one method of removal, and there are special solvent/alcohol mixtures available from suppliers.
REF:

PUT-0001
ETCH NAME: Trichloroethylene
TYPE: Solvent, removal
TIME: Minutes
TEMP: RT to boiling
COMPOSITION:
x .... $CHCl:CCl_3$ (TCE)
DISCUSSION:

Putty, as a natural clay or plastic type. The clay form has been largely replaced in Solid State usage with plastic putty. Used to hold small specimens on a microscope stage for observation, such as a (110) cleaved piece of a semiconductor wafer to study thin film epitaxy layer structure and/or p-n junctions that can be stain or etch developed. After removal of putty, wipe surfaces with TCE to remove residual organic or inorganic contamination.

REF: Menth, M — personal communication, 1982

SIL-0001
ETCH NAME: Alcohol
TYPE: Solvent, removal
TIME: Minutes
TEMP: RT to hot
COMPOSITION:
(1) x .... MeOH/EOH
(2) x .... $CHCl:CCl_2$ (TCE)
DISCUSSION:

Silicone oils and greases. These are manmade plastic derivative compounds that range from liquids to solids with various temperatures, vapor pressures, electrical characteristics. The liquids are used as vacuum pump oil for (hot) oil diffusion pumps. They can slowly break down to a varnish-like consistency that coats internal pump surfaces. To clean, soak in hot TCE, then rinse in alcohols and blow dry with air from a compressor. Oil should be replaced every 6 to 9 months, depending upon pump usage time, to prevent oil breakdown as varnish. (*Note:* Silicone should not be confused with silicon. The latter being a metal element, and primary semiconductor material.)

REF: Lopez, P — personal communication, 1973

RESINS

RES-0001
ETCH NAME: Benzene
TYPE: Solvent, removal
TIME: Minutes
TEMP: RT to warm
COMPOSITION:
x .... C_6H_6
DISCUSSION:

Resin, as copal (benin). Melting point is from 65—125°C, and solubility varies with copal type, such as the Class I type: amber. Both have been used for mounting specimens on microscope slides for lapping as thin sections. Both Solid State and geology metals, rocks, or minerals.

Benzene as a liquid derivative of petroleum is a solvent. As the benzene ring, it also is a basic building block in chemistry, in particular, the aromatic compounds. It boils at 80°C, has a specific gravity of 0.88, is insoluble in water, and burns with a yellow, smoky flame. It is a general solvent for fats, resins, phosphorus, etc. Two additional solvents, as homologs of benzene that also are used in Solid State processing are methyl benzene (toluene, C_7H_8), and dimethyl benzene (xylene, C_8H_{10}). In addition, these are coal tar distillates that include naphtha; carbolic, anthracene, and creosote oils.

RES-0002
ETCH NAME: Methyl alcohol TIME: Variable
TYPE: Alcohol, removal TEMP: RT to hot
COMPOSITION:
(1) x .... CH_3OH (MeOH) (2) x .... CH_3COOH (HAc — acetic acid) (3) x .... $C_{10}H_{16}$ (turpentine)
DISCUSSION:
Resins, Class I (shellacs), to include pine, myrrh, etc., as gum resins. As Class I they have melting points from about 194—213°C. This is the common yellow-tan, translucent type resins supplied to industry in stick form for general purpose mounting and holding applications. These resins are brittle, but can be mixed in various proportions of beeswax for specific pliability. Such mixtures are used for a wax lap and wafer holding on lapping platens. See Beeswax for additional discussion.
REF: Kunitz, A — personal communication, 1974

PICH-0001a
ETCH NAME: Turpentine TIME: Variable
TYPE: Solvent, removal TEMP: RT to hot
COMPOSITION:
x .... $C_{10}H_{16}$
DISCUSSION:
Pitch (tar), as an asphalt hydrocarbon, or as a Class I tree-type resin. Both are commonly mixed with beeswax for specific viscosity, melting points and colors with names, such as burgandy, black (Swedish), green or optical pitch. Canadian balsam also is referred to as a ''pitch'', although it is a Type II oleo-resin. Melting points of pitch mixtures vary between 60 to 150°C, and are widely used for mounting and holding specimens for lapping, polishing, and cutting — woods, rocks, minerals, gem stones — in addition to glass, semiconductors, metals, and metallic compounds. Turpentine is a general solvent for pitch.
REF:

PICH-0001b
ETCH NAME: TIME: Variable
TYPE: Solvent, removal TEMP: RT to hot
COMPOSITION:
(1) x .... MEK (methyl ethyl ketone) (2) x .... $HCCl{:}Cl_2$ (TCE) (3) x .... MeOH/EOH (4) 2 .... NH_4OH, 8 .... H_2O
DISCUSSION:
Polishing pitch (blocking wax), as a mixture of beeswax, linseed oil and white rosin. The mixture can be dissolved in all of the solvents shown. Constituent concentrations are varied for specific viscosity, holding, and melting points. These mixtures have been basic in polishing glass lens and prisms for many years.
REF:

VACUUM HOLDING

VAC-0001
ETCH NAME: Vacuum TIME:
TYPE: Vacuum, gripping TEMP:
COMPOSITION:
x .... vacuum

DISCUSSION:

Vacuum, alone, or in conjunction with waxes or metal masks is used as a holding mechanism. There are several types of vacuum units available: (1) a very small, electrically operated vacuum pick-up unit with $^{1}/_{8}$ ID tygon line and metal/plastic needle-type inserts. Operating unit about the size of a 500 ml Pyrex beaker with adjustable vacuum level dial on top for up to about 10^{-1} Torr (100 μm) vacuum level. Handles pick-and-place of small alloy pre-forms, semiconductor dice, etc., and may have a finger-release/vacuum hold hole in the pick-up handle. (2) Vacuum/air units about the size of a shoe box capable of almost 10^{-3} Torr (1 micron). These are air-pump operated, and some of the most widely used vacuum units for general purpose, light vacuum operation, such as for electronic slicing and dicing equipment; photo resist spinners; small vacuum ovens and wire bonders. They use $^{1}/_{4}$ ID plastic or metal tubing plumbed-in, and either pull a vacuum (oven) or operate as a vacuum holder, such as for a semiconductor wafer on a photo resist spinner. (3) Mechanical or roughing oil pumps rated for 10^{-3} Torr vacuum, and by liter-per-minute rate. These are the standard vacuum system roughing pumps, occasionally used for small vacuum ovens. (4) High vacuum pumps (oil, ion, cryo-, etc.) capable of vacuum to the 10^{-12} Torr level. Rarely used for general holding or pick-up operation.

REF:

WAXES

CAR-0001

ETCH NAME: Turpentine — TIME:

TYPE: Solvent, removal — TEMP: RT to warm

COMPOSITION:

x .... $C_{10}H_{16}$

DISCUSSION:

Carnauba wax (*Croypha cerifera*) has a melting point of 85°C as a natural tree wax like candlenut or coconut. It is brittle like the resins, but was the original base of sealing waxes (SEAW-0001). Although it has been used as a mounting medium, it is more often mixed with other more pliable waxes. See Pitches and Beeswax.

REF:

CAR-0002

ETCH NAME: Heat — TIME: $1^{1}/_{2}$ h

TYPE: Thermal, mixing — TEMP: 120°C

COMPOSITION:

x .... heat

DISCUSSION:

Carnauba wax mixtures developed as 0.1 pf capacitors. Mix the following ingredients: 45% carnauba wax:45% resin:10% white beeswax. Anneal mixture for time and temperature shown in an electric field of 1.97 to 15.75 kV/cm³. The homogenized mixture can be cut to specimen size with an X-Acto knife.

REF: Walker, O K & Jefimenko, O — *J Appl Phys,* 44,3459(1973)

PARA-0001

ETCH NAME: Benzene — TIME:

TYPE: Solvent, removal — TEMP: RT to warm

COMPOSITION:

x .... C_6H_6

DISCUSSION:

Paraffin is a hydrocarbon distillate with a boiling point in the range of 350°C. It is derived from paraffin-base as against the asphalt-base petroleum. The solidified, pure paraffin is white with a greasy feel and is sufficiently soft such that it can be cut with a knife. It is used as a general mounting and holding medium, by itself, but with limited application due to the high melting point, such that it is more often mixed with another lower melting point wax like beeswax to reduce the working melt point.

As the pure paraffin it is used as a thin coating for wax papers, some for the water repellent capability of paraffin. It is sometimes used in mold fabrication of parts similar to the "lost-wax" processing using beeswax, but the most important use is in the making of candles. As a commercial product it also is called Ceresine.

REF:

PARA-0002

ETCH NAME: Trichloroethylene TIME:

TYPE: Solvent, removal TEMP: RT to boiling

COMPOSITION:

x .... TCE

DISCUSSION:

Lok-wax formulations are basically paraffins with additive white beeswax for specific melting points. They tend to retain bubbles while still liquified as does Crystalbond. The material has been used for mounting semiconductor wafers for both etching and vacuum metallization.

REF: Cochran, L — personal communication, 1980

SEAW-0001

ETCH NAME: Benzene TIME:

TYPE: Solvent, removal TEMP: RT to warm

COMPOSITION:

x .... C_6H_6

DISCUSSION:

Sealing wax. Originally a carnauba type wax, but also made from a base of pitch or other tree resins with additive color. The material is both hard and brittle with common colors of red, green, yellow, and brown or black. The name derives from its use in sealing letters and as document seals before the development of glue. Individuals used to have their own sigils with personal designs carved from agates or metals engraved. The wax will melt in a candle flame (>1500°F), such that a drop of wax was melting across the flap of a letter to be sealed, the sigil firmly pressed into the still liquid and hot wax to imprint and solidify the wax to the document. It is still so used on official documents, as a county, state, government department seal, and may include short ribbons.

Using a dop-stick ($^1/_4$—$^1/_2$″ diameter wooden dowel) with sealing wax on one end, small rock and mineral specimens are mounted for polishing en cabachon, such as for star sapphires. This was one of the first methods developed in ancient times for polishing stones, and is still in use today. It is not widely used as a mounting medium in Solid State or metal processing due its brittleness, more often mixed with beeswax, paraffin, and other resins for greater pliability.

REF:

WOOD

WOOD-0001

ETCH NAME: Water TIME:

TYPE: Solvent, cleaning TEMP: RT to boiling

COMPOSITION:

x .... H_2O x .... CH_3COCH_3 (acetone)

DISCUSSION:

Woods have some uses in Solid State material processing as handling parts. Q-tips with either cotton or foam plastic ends water or acetone soaked for light scrub cleaning of wafers and parts. Acetone is the main cleaning solvent for photo resist lacquers. Wooden tongue depressors and similar smaller sticks are used in stirring chemical solutions or for measuring out dry chemicals on a balance pan. The wooden end of a Q-tip has been used as an acid saturated rod to rub-form an etched channel, or the end sharpened, dipped in HF, then a droplet applied to an oxide thin film to etch-form a pit down to the substrate. Under a microscope the color orders within the oxide layer can be observed and measured for thin film thickness. Hard woods have been used as a lapping platen, such as a maple wood lap with diamond or alumina grits in oil or grease for wafer surface polishing.

REF:

CHAR-0001

ETCH NAME: Heat TIME:

TYPE: Thermal, forming TEMP: >100°C

COMPOSITION:

x .... heat

DISCUSSION:

Charcoal, as made from hard woods or as coke made from coals. In both cases, the material it heated below its burning point to drive off volatiles, such as water moisture and entrapped gases. The use of charcoal dates back to the end of the Bronze Age and the beginning of the Iron Age, circa 5000 and 1350 B.C., respectively, for the more intense heat obtained from charcoal that is required for the smelting of iron . . . also glass . . . soda-lime, developed by the Egyptians, circa 3500 B.C. Both charcoal and, primarily, coke are major fuels in the metal processing industries.

In Solid State processing both charcoal and graphite are used as getters and drying agents in vacuum systems (cryo-pumps with graphite); and in various types of gas and air cleaning filters (both charcoal and graphite).

For cleaning granular charcoal, vacuum oven bake above 125°F, $^1/_2$ to 1 h or more to remove entrapped volatiles. . .the charcoal can then be reused.

REF:

PHYSICAL PROPERTIES OF MUSCOVITE, $(OH)_2KAl_2(AlSi_3O_{10})$

Classification	Mica
Atomic numbers	1, 16, 13, 19, & 14
Atomic weight	342.1
Melting point (°C)	1300
Boiling point (°C)	
Density (g/cm³)	2.7—3
Refractive index (n =)	1.552—1.588
Hardness (Mohs — scratch)	2—2.25

Crystal structure (monoclinic — normal)	(001) pseudo-hex sheet
Color (solid)	Olive-green variable
(thin section)	Transparent/clear
Cleavage (basal — perfect)	(001)

MUSCOVITE, $(OH)_2Al_2K(AlSi_3O_{10})$

General: The formula shown is for the variety called muscovite (common mica). There are a number of micas and mica-like minerals that are widely distributed throughout the world as an original constituent mineral or as a secondary product from weathering and alternation of aluminum- and potassium-bearing rocks. There are three groups: (1) the Mica Group, proper; (2) the Clinotite Group, brittle micas, and (3) the Chlorite Group. A supplemental group is the hydrated vermiculites, chiefly from alternation of the micas. Talc and serpentine (fibrous variety is asbestos) are associated minerals. There are many mica-bearing rocks, such as mica-granite — the fine yellowish flakes sometimes mistaken for gold. This discussion will be limited to the Mica Group proper, as they are widely used in industry.

The micas are orthosilicates of aluminum and potassium with the various types containing fluorine, lithium, magnesium, iron, calcium, etc. They are all in the monoclinic system of crystallography with perfect basal cleavage, (001) — micaceous cleavage — and can be easily separated into thin sheets. Hardness and specific gravity (density) vary (H = 2.5—3.5 . . . G = 2.5—3), and thin sheets are highly elastic when bent. Muscovite — common or potash mica — and biotite — iron mica — are the two most used varieties and have a high fusibility index: 5 = 1200°C. Lepidolite lithia micas — fusibility is about 100°C, and is a source of lithium. Phlogophite — magnesium mica — is similar to biotite, but in small, exfoliated masses with a much more greasy feeling. They are all acid resistant, have high electrical resistance, and are heat insulators. When muscovite and biotite are found in pegmatites, they can occur in large single crystal masses as much as 6 ft in diameter with hexagonal outline. As such, they can be cleaved into large thin sheets.

Although high temperature glasses, such as Pyrex, have largely replaced mica in most such applications, mica is the Isinglass long used as stove window insulation. It is still used as a see-through window in several industrial applications as both a temporary or permanent heat shield. Sheet mica parts are used in high-power line insulation and, as a powdered compact, to fabricate insulating board. The latter now replaces asbestos. Flaked and mixed with oil, it is an industrial lubricant.

Oriented prisms of mica are used as polarizers for microscopes and, a major industrial product, as mica capacitors — resistance wielding equipment, discrete elements in electronic assemblies, and so forth.

Phlogophite and some of the vermiculites are used as packaging materials. In oil drilling, as a packing around piping, along with montmorillonite, a clay-like mineral, both swell with water absorption and act as sealers against pumping loss of oil. Also used as a container packing material for shipping acids and corrosive gases, as it will absorb any leakage. The material can then be water leeched, dried and reused. In gardening, it is mixed with soil and fertilizers to retain chemicals and water moisture.

Technical Application: Though not directly used in semiconductor or quartz device fabrication, there are discrete mica capacitors and resistors used in circuit assemblies.

Sheet mica is still used as heat resistant windows on small hand furnaces and small washers are punched from the sheet as insulating elements for electrical stand-offs. Powdered and fabricated as blocks and sheets it is machine cut into a variety of insulating parts and is being used to replace asbestos. Block material is classified by heat-resistant capability, such as 800°C, 1200°C, etc. Most micas are transparent to translucent in thin sheets but, in sufficient thickness, dark yellow-tan, and used as observation windows in furnaces where parts are at red heat or above. The exfoliated phlogophite variety is used as a packing material for shipment of dangerous chemicals as it is nonflammable, resistant to chemical attack, and a liquid and gas absorber.

As (001) cleaved sheets, both muscovite and biotite have been used as substrates for metal evaporation and epitaxy thin film growth of semiconductor materials and metals in studying the growth kinetics and structure of such thin films. The thin films are then removed by the "float-off" technique for SEM and TEM study. Depending upon the substrate material, water (NaCl, KCl substrates) or acids (micas) are used to lightly attack and liquify the substrate at the substrate/thin film interface to effect "float-off".

Etching: Muscovite is acid resistant but will decompose in molten fluxes of alkaline carbonates. Biotite is soluble in H_2SO_4. Lepidolite is partially dissolved by acids. Phlogophite is acid resistant like muscovite but, once saturated by liquids or gases, can be dried and reused.

MICA ETCHANTS

MI-0001
ETCH NAME: Alconox
TYPE: Alkali, cleaning
COMPOSITION:
(1) x .... Alconox (2) x .... H_2O
TIME:
TEMP: Boiling
DISCUSSION:

Muscovite mica, $(OH)_2KAl_2(AlSi_3O_{10})$ cleaved (001) and used as substrates for epitaxy thin film growth of CdTe, PbTe, and SnTe doped with Bi and Sb. After cleaving mica, clean in boiling Alconox. Follow with boiling in water several times, and final rinse in high purity MeOH immediately before placing in vacuum system for epitaxy growth. Substrates of CdTe, (111) were also used and polished in 1% Br_2:MeOH. Thin films were used in a study of superlattices.

REF: Davenere, A et al — *J Cryst Growth*, 70,452(1984)

MI-0002
ETCH NAME: Freon
TYPE: Gas, cleaning
COMPOSITION:
x .... Freon
TIME:
TEMP: RT
DISCUSSION:

Mica and natural rock salt, NaCl were both cleaved (0001) prior to use as substrates for gold and silver evaporation. Particulate matter blown off with Freon as particulates affect good adhesion of thin films. No liquids used as sodium chloride is etched by water and mica can adsorb liquid along cleavage planes. (*Note:* Mica is (001) basal cleavage.)

REF: Chopea, K I — *J Appl Phys*, 37,2049(1960)

MI-0003
ETCH NAME: Air
TYPE: Gas, drying
COMPOSITION:
x .... air
TIME:
TEMP: 125°C
DISCUSSION:

Phlogophite mica, $H_2KMg_3Al(SiO_4)_3$. Material as exfoliated pieces used as packing as a liquid gas absorbent and fire-retardant (material will not burn). Used in shipping containers handling acids, toxic liquids, and in the annular vacuum space of cryogenic tanks, e.g., LN_2, LOX. For drying small quantities place material in a Pyrex beaker in an air-circulating oven and bake at 125°C to remove water moisture. Time varies with quantity of phlogophite.

If toxic gases are involved, proper containers and ventilation are required during heat drying. Used by the referenced author in cleaning 2000 gallon LOX and LN_2 tankers. This cleaning included pulling a final vacuum of 10^{-3} Torr on the mica filled, annular space prior to seal-off.
REF: Walker, P — personal application, 1963

MI-0004
ETCH NAME: Nitric acid
TYPE: Acid, float-off
TIME:
TEMP: RT
COMPOSITION:
x .... HNO_3
x .... H_2O
DISCUSSION:
Muscovite mica, $(OH)_2KAl_2(AlSi_3O_{10})$, cleaved (100) and used for growth of platinum fibers. Fiber grown at 1000 K with axis in the ⟨1000⟩ direction. Platinum thickness was 1000—1200 Å. Metallization removed with dilute nitric acid as a float-off technique for TEM structure study of the platinum films. (*Note:* Micas are hexagonal system, not isometric.)
REF: Dixit, P & Vook, R W — *Thin Solid Films*, 113,151(1984)

MI-0005a
ETCH NAME: Pressure
TYPE: Pressure, defect
TIME:
TEMPT: RT
COMPOSITION:
x .... pressure
DISCUSSION:
Biotite, $H_2K(Mg,Fe)_3Al(SiO_4)_3$ as (0001) cleaved specimens of hexagonal platelets. By striking the surface a sharp blow with a rod of glass, metal, etc., a six-rayed percussion-figure is formed in the surface. The strongest break-line is parallel to the clino-pinacoid plane of symmetry and can be used to orient planes of optical axes. Specimens used for polarizing microscope plates.
REF: Walker, P — mineralogy development, 1952—1953

M-0005b
ETCH NAME: Trichloroethylene
TYPE: Solvent, cleaning
TIME:
TEMP: RT
COMPOSITION:
(1) x .... TCE
(2) x .... MeOH
DISCUSSION:
Muscovite, $(OH)_2KAl_2(AlSi_3O_{10})$, cleaved (0001) and cut as 3″ diameter discs with center hole. Used as a cover on a hand alloy, portable furnace. Fresh mica sheets fabricated as insulation and sighting glass as needed. Degrease with TCE, rinse in MeOH, and air dry. Thicker sheets used as sighting glasses in observing hot parts down-tube in diffusion, and both alloy and epitaxy growth furnaces. Mica stand-off insulators for vacuum systems similarly cut and cleaned. Bulk mica sheet or pressed flake blocks used for handling hot parts, surfaces wiped clean with cloth toweling soaked in either solution.
REF: Walker, P et al — personal application, 1955—1959

PHYSICAL PROPERTIES OF NEODYMIUM, Nd

Classification	Lanthanide
Atomic number	60
Atomic weight	144.24
Melting point (°C)	1024
Boiling point (°C)	3027
Density (g/cm^3)	7.00
Thermal conductance (cal/sec)(cm^2)(°C/cm)	0.031
Heat of fusion (k-cal/g-atom)	1.70
Specific heat (cal/g) 25°C	0.045
Latent heat of fusion (cal/g)	11.78
Heat of vaporization (k-cal/g-atom)	69
Atomic volume (W/D)	20.6
1st ionization energy (K-cal/g-mole)	145
1st ionization potential (eV)	6.77
Electronegativity (Pauling's)	1.2
Covalent radius (angstroms)	1.64
Electrical resistivity (micro-ohm-cm) 298 K	64.3
Compressibility (cm^2/kg $\times$ 10^{-6})	3.0
Coefficient of linear thermal expansion ($\times 10^{-6}$ cm/cm/°C)	6.7 (8.4)
Cross section (barns)	46
Magnetic moment (Bohr magneton)	3.3
Tensile strength (psi)	24,700
Yield strength (psi)	23,900
Magnetic susceptibility ($\times 10^{-6}$ emu/mole)	5,650
Transformation temperature (°C)	855
Hardness (Mohs — scratch)	4—5
Crystal structure (hexagonal — normal)	$(10\bar{1}0)$ prism, hcp
Color (solid)	Yellowish metal
Cleavage (basal)	(0001)

NEODYMIUM, Nd

General: Does not occur as a native element. It is a rare earth of the cerium group. The mineral monazite $(Ce,La,Di)PO_4$ is the major ore. *Note:* "Didymium, Di", was thought to be an element, but in 1885 von Welsbach separated didymium into the elements neodymium, Nd and praseodymium, Pr.

Not used in industry to any extent as the metal due to scarcity. Many of the neodymium salts are pink in color with a characteristic absorption spectrum.

Technical Application: Not used in Solid State device fabrication. Metallic neodymium has been the subject of electrical and morphological studies along with other rare earth elements of the cerium group.

Etching: Deliquesces in water.

NEODYMIUM ETCHANTS

ND-0001
ETCH NAME: TIME:
TYPE: Arc, forming TEMP:
COMPOSITION:
x .... heat

DISCUSSION:

Nd, specimens were first arc melted as buttons and then machined into cylinders 3/16 to 2″ in diameter and used in measuring electrical resistivity. Other materials used: La, Pr, Sm.

REF: Alstad, J K et al — *Phys Rev,* 122,1636(1961)

PHYSICAL PROPERTIES OF NEON, Ne

Classification	Rare gas
Atomic number	10
Atomic weight	20.183
Melting point (°C)	−248.67
Boiling point (°C)	−245.9
Density (gm/cm^3)	0.9 gm/l liq.
Thermal conductivity ($cal/cm^2/cm/°C/sec$) 20°C	1.1×10^{-4}
1st ionization potential (eV)	21.6
Cross section (barns)	0.032
Hardness (Mohs — scratch) solid	2—3
Crystal structure (isometric — normal) solid	(100) cube fcc
Color (gas/solid)	Colorless
Cleavage (cubic) solid	(001)

NEON, Ne

General: Neon is one of the rare gases found in the atmosphere. It is inert with a valance of 0 and does not form compounds, and there is only about 18 ppm neon in air. It is obtained by the fractionation of liquid air, after the removal of LN_2 (−196°C), being distilled from the remaining LOX.

In a vacuum tube it will produce a red-orange to red glow, and as such is widely used in electric signs and beacons. It also is used in small bulbs as a display light in electronic equipment.

Technical Application: Not generally used in Solid State processing other than as a neon light bulb for signs and equipment display panels. Fabricated as a neon or neon/helium laser.

Etching: Soluble in O_2. As a solid under vacuum, vary vapor pressure to "etch".

NEON ETCHANTS

NE-0001

ETCH NAME: Neon — TIME:

TYPE: Gas, forming — TEMP:

COMPOSITION:

x .... Ar/N_2

x .... 12.8% Ne

DISCUSSION:

Ne, used as a component in the gas ambient used in RF magnetron sputter deposition of NbN thin films. Inclusion did not appear to improve resistivity or other electronic functions of the material.

REF: van Dover, R B et al — *J Vac Sci Technol,* A2(3),1257(1984)

NE-0002
ETCH NAME: TIME:
TYPE: Pressure, defect TEMP: −250°C
COMPOSITION:
x .... vapor pressure
DISCUSSION:

Ne, grown as a single crystal in vacuum at LHe temperature. Ingot surfaces can be preferentially etched by varying the vapor pressure of the system.
REF: Schwentner, N et al — *Rare Gas Solids,* Vol 3, Academic Press, New York, 1970

NE-0003
ETCH NAME: TIME:
TYPE: Exciton, growth TEMP: −250°C
COMPOSITION:
x .... exciton
DISCUSSION:

Ne, grown as a single crystal solid in a study of exciton growth bubbles in the material.
REF: Ossicini, S — *J Phys Chem Solids,* 46,123(1985)

NE-0004
ETCH NAME: Laser TIME:
TYPE: Gas, measuring TEMP:
COMPOSITION:
x .... Ne
x .... He
DISCUSSION:

Ne/He laser used to measure the optical density (O.D.) vs. time response of an electrochromic device fabricated with Li-W,O thin films on ITO coated quartz blanks with measurement between an Au cover film to the ITO layer. See WO-0007 for device details.
REF: Kitabatake, M et al — *J Electrochem Soc,* 132,433(1985)

PHYSICAL PROPERTIES OF NEPTUNIUM, Np

Classification	Actinide
Atomic number	93
Atomic weight	237
Melting point (°C)	637
Boiling point (°C)	
Density (gm/cm^2)	20.45
Ionic radius (angstroms)	1.10 (Np^3)
Half-life ($\times 10^4$ years)	2.14
Cross section (barns)	170
Hardness (Mohs — scratch)	
Crystal structure (orthorhombic — normal)	(100) a-pinacoid
Color (solid)	
Cleavage (basal)	(001)

NEPTUNIUM ETCHANTS

NP-0001
ETCH NAME: Hydrochloric acid TIME:
TYPE: Acid, removal/polish TEMP:
COMPOSITION:
x .... HCl, conc.
DISCUSSION:
Np specimens. Solution shown as a general etch for the material.
REF: Vossen, J L & Kern, W — *Thin Film Processes,* Academic Press, New York, 1978

PHYSICAL PROPERTIES OF NICKEL, Ni

Classification	Transition metal
Atomic number	28
Atomic weight	58.71
Melting point (°C)	1453
Boiling point (°C)	2730 (2732)
Density (g/cm^3)	8.908
Thermal conductance (cal/sec)(cm^2)(°C/cm) 100°C	0.198
Specific heat (cal/g) 200°C	0.1226
Latent heat of fusion (cal/g)	73.8
Heat of vaporization (k-cal/g-atom)	91.0
Atomic volume (W/D)	6.6
1st ionization energy (K-cal/g-mole)	176
1st ionization potential (eV)	7.63
Electronegativity (Pauling's)	1.8
Covalent radius (angstroms)	1.15
Ionic radius (angstroms)	0.69 (Ni^{+2})
Coefficient of linear thermal expansion ($\times 10^{-6}$ cm/cm/°C)	13.3
Electrical resistivity (micro-ohms/cm) 20°C	6.844
Lattice constant	3.5168
Modulus of elasticity (psi $\times 10^6$)	30
Magnetic transformation temperature (°C)	357
Tensile strength (psi)	120,000
Compressibility (cm^2/kg $\times$ 10^{-6})	0.531
Electron work function (eV)	5.01
Cross section (barns)	4.6
Vapor pressure (°C)	2364
Hardness (Mohs — scratch	5
(Brinell — annealed)	75 (80—110)
Crystal structure (isometric — normal)	(100) cube, fcc
Color (solid)	Silver-grey metal
Cleavage (cubic — poor)	(001)

NICKEL, Ni

General: Does not occur as a native element. There are about 35 minerals containing nickel, the two major ones being millerite, NiS and niccolite, NiAs, both as primary nickel ores. The latter is called "copper-nickel" due to its color, not the presence of copper. Nickel, as an element, always occurs in meteoric iron and — the earth core — is considered to be mainly a mixture of these two metals . . . $FeNi_2$. The pure metal it is hard, yet malleable, ductile, and tenacious. It is relatively inert to most atmospheres and is one of the few metals that shows a degree of magnetism.

Nickel is a major metal in industry with many applications such as an alloy with copper and iron as commercially named mixtures, such as Invar, Hastalloys, Iron-Constantan, Permalloy, Nichrome. Also used for armor plate, in low temperature solders (<800°C), and high temperature brazing alloys. It is now one of the coinage metals, is easily electroplated on other metals for corrosion protection. As plated of evaporated on glass, it is a replacement for a silvered mirror.

Technical Application: Nickel has major application in Solid States device fabrication as an electrical contact metal, as Au/Ni (plated), or Au/AuGe/Ni (evaporated/sputtered).

As an electrolytic plating it is used on semiconductor wafers; ceramic, glass, sapphire, or quartz circuit substrates; on package parts, contact wires, leads, or straps; and as a replacement metal in making photo resist chrome masks. As the pure metal or alloy it has been used as a quartz frequency crystal or semiconductor package, as a testing heat sink, to include a magnet as a mixture with magnetic iron and cobalt.

Nickel has been grown as single crystals and studied widely for its physical and electrical properties. There are a wide range of single crystal alloys containing nickel which have been grown and studied, some of which are included in the ''Nickel Alloys'' section immediately following this section on Nickel.

Etching: HNO_3, HCl, H_2SO_4, HF, their mixtures, and chrome etch. Ionized gases.

NICKEL ETCHANTS

NI-0001a

ETCH NAME: — TIME: 30 sec
TYPE: Acid, polish — TEMP: 85°C
COMPOSITION: — RATE: 5 μm/30 sec

30 .... HNO_3
10 .... H_2SO_4
50 .... HAc
10 .... H_2O

DISCUSSION:

Ni specimens. Solution used to polish surfaces prior to oxidation experiments.

REF: Graham, W J et al — *J Electrochem Soc,* 119,879(1972)

NI-0001b

ETCH NAME: Sulfuric acid, dilute — TIME: 2 min
TYPE: Electrolytic, polish — TEMP: RT
COMPOSITION: — ANODE: Ni

4 .... H_2SO_4 — CATHODE:
3 .... H_2O — POWER: 0.5 A/cm²

DISCUSSION:

Ni specimens. Solution used to polish specimens in a study of nickel oxidation.

REF: Ibid.

NI-0002

ETCH NAME: Nitric acid, dilute — TIME: 1 min
TYPE: Acid, removal — TEMP: RT
COMPOSITION:

3 .... HNO_3
7 .... H_2O

DISCUSSION:

Ni, evaporated as an Au:Ni coating on resistors. In evaluating resistors, after etch removal of gold, etch immediately in the solution shown to prevent formation of nickel oxide.

REF: Bulletin 5M 1/1/83C — Hycomp Inc.

NI-0003a

ETCH NAME:

TYPE: Acid, removal

TIME:

TEMP: RT

COMPOSITION:

50 ml .... HNO_3

50 ml .... CH_3COOH (HAc)

DISCUSSION:

Ni, evaporated thin films. Solution mixtures is a general removal etch for nickel.

REF: Bulletin 1116B/Rev-HPD/VB/TM/0883/1K — MRC

NI-0003b

ETCH NAME:

TYPE: Acid, removal

TIME:

TEMP: RT

COMPOSITION:

20 g .... $Ce(NH_4)_2(NO_3)_6$

200 ml .... H_2O

DISCUSSION:

Ni, thin film evaporation on glass or semiconductor substrates. Used as a removal and patterning etch.

REF: Ibid.

NI-0004a

ETCH NAME: Jewitt-Wise etch

TYPE: Acid, preferential

TIME:

TEMP:

COMPOSITION:

1 .... 10% $NH_4(H_2SO_4)_2$

1 .... 10% KCN

DISCUSSION:

Ni specimens. Solution used in preparing nickel and alloys as metallographic sections for study by developing structure.

REF: Kohl, W H — *Handbook of Materials and Techniques for Vacuum Devices*, Reinhold, New York, 1967

NI-0004b

ETCH NAME:

TYPE: Acid, cleaning

TIME:

TEMP:

COMPOSITION:

10 .... HCl

33 .... HNO_3

67 .... H_2O

DISCUSSION:

Ni specimens. Solution used to clean nickel surfaces.

REF: Ibid.

NI-0004c
ETCH NAME: Hydrofluoric acid
TYPE: Acid, removal
COMPOSITION:
(1) x .... HF, conc.
(2) x .... H_3PO_4, conc.
TIME:
TEMP: RT
DISCUSSION:
Ni specimens. Both solutions are slow etchants for nickel.
REF: Ibid.

NI-0004d
ETCH NAME: Nitric acid
TYPE: Acid, removal
COMPOSITION:
(1) x .... HNO_3, conc.
(2) x .... HF
x .... HNO_3
TIME:
TEMP: RT
DISCUSSION:
Ni, specimens. Concentrated nitric acid is an extremely rapid etch. Solutions of HF:HNO_3 are fast but more controllable with removal rate reduced by increase of HF.
REF: Ibid.

NI-0004e
ETCH NAME: Meta-phosphoric acid
TYPE: Acid, removal
COMPOSITION:
x .... HPO_3
TIME:
TEMP: RT
DISCUSSION:
Ni specimens. This acid is a slow etch for nickel. Can be mixed with alcohol.
REF: Ibid.

NI-0005
ETCH NAME: Aqua regia, modified
TYPE: Acid, removal
COMPOSITION:
4 .... HCl
1 .... HNO_3
TIME:
TEMP:
DISCUSSION:
Ni specimens. Used as a general etch for nickel and brass. With increased nitric acid content can be used to etch high-speed steels.
REF: Oberg, E & Jones, F D — *Machinery Handbook,* 4th ed, The Industrial Press, New York, 1951

NI-0006
ETCH NAME:
TYPE: Salt, removal
COMPOSITION:
x .... 30% $FeCl_3$
TIME: 2—10 min
TEMP: Warm (60°C)
DISCUSSION:
Ni, as 0.005 and 0.010 shim stock. Solution is a general etch for nickel. As applied to shim stock was used with Apiezon-W (black wax) patterning to fabricate metal masks used for gold evaporation on silicon SCRs.
REF: Walker, P — personal application, 1964—1974

NI-0007
ETCH NAME:
TYPE: Acid, cleaning
COMPOSITION:

(1) 45 .... H_2O_2
10 .... HCOOH (formic acid)
45 .... H_2O

(2) 45 .... H_2O_2
45 .... HCOOH
10 .... H_2O

TIME: 10—20 min
TEMP: 50—60°C or 65—70°C
RATE: 0.1 to 0.2 mils average

DISCUSSION:

Ni specimens. Both solutions used in a study of surface cleaning.

REF: Bell Tel Systems Tech Publ, 3143(1958)61

NI-0008
ETCH NAME:
TYPE: Acid, removal
COMPOSITION:

50 g .... $2NH_4NO_3.Ce(NO_3)_3.4H_2O$
10 ml .... HNO_3
150 ml .... H_2O

TIME: 60—90 sec
TEMP: RT

DISCUSSION:

Ni, thin films vacuum evaporated on soda-lime glass used for photo resist masks. Film thickness 2000 to 2500 Å with and without NiO AR coatings. Use NH_4OH to remove nickel oxide, then etch remove or etch pattern with solution shown. Further addition of nitric acid will reduce etch rate. Without nitric acid rate increases. This is a modification of standard "chrome etch" and can also be used on chromium with and without Cr_2O_3 AR coatings.

REF: Walker, P — personal development, 1975

NI-0009a
ETCH NAME: Sulfuric acid, dilute
TYPE: Electrolytic, polish
COMPOSITION:

390 ml .... H_2SO_4
290 ml .... H_2O

TIME: 4—6 min
TEMP: RT
ANODE: Specimen
CATHODE: Nickel
POWER: 2 V

DISCUSSION:

Ni, specimens used in a study of growth habits. During etch polishing agitate electrolyte vigorously due to heavy gas evolution. Remove specimen with current still running. Rinse in dilute nitric acid to remove anodic film; then rinse in DI water, alcohol and dry with warm air.

REF: Cliffe, D R & Farr, J P — *J Electrochem Soc,* 111,299(1964)

NI-0012
ETCH NAME:
TYPE: Electrolytic, dislocation
COMPOSITION:

x .... H_2SO_4
x .... D_2O

TIME:
TEMP:
ANODE: Specimen
CATHODE:
POWER:

DISCUSSION:

Ni, polycrystalline specimens used in a study of ion implantation (I^2) damage. Specimens were Ar^+ ion bombarded at 250 KeV and then etched. The deuterium, D_2O, is trapped in vacancy complexes in the I^2 region. After temperature anneal there are two types of structures:

(1) dislocation loops at 21 and 250°C and (2) small voids and dislocation loops at 500°C. All defects were in the implanted region.
REF: Frank, R C et al — *J Appl Phys,* 57,414(1985)

NI-0013
ETCH NAME: Sulfuric acid, dilute
TYPE: Electrolytic, thinning
COMPOSITION:
40% .... H_2SO_4
60% .... H_2O

TIME:
TEMP: 10°C
ANODE: Ni
CATHODE:
POWER: 75A/in^2

DISCUSSION:
Ni, single crystals used in a study of induced deformation and imperfections. Solution used to electro-thin samples.
REF: Bhattachernya, A et al — *J Appl Phys,* 37,4441(1966)

NI-0014a-b
ETCH NAME:
TYPE: Acid, preferential
COMPOSITION:
"A" x .... 55% HNO_3 then "B" x .... 25% HCl
x .... $FeCl_3$, sat. sol.

TIME:
TEMP:

DISCUSSION:
Ni, (100) wafers used in a study of secondary electron reaction on (100) faces of copper and nickel. First etch in "A" — nitric is rapid — and follow with "B" which is slower but produces similar pits and structure as obtained by the nitric acid. An alternate method is to etch with the "B" mixture and follow with 50% H_3PO_4 as an electrolytic polish — even with removal of up to $^1/_2$ mm of material the surface maintains planarity with an estimated 2% number of etch pits.
REF: Burns, J — *Phys Rev,* 119,102(1960)

NI-0015
ETCH NAME:
TYPE: Electrolytic, polish
COMPOSITION:
x .... NH_4Cl
x .... NH_2CONH_2 (urea)

TIME:
TEMP: 120—135°C
ANODE;
CATHODE:
POWER:

DISCUSSION:
Ni specimens. A study of the use of this solution for polishing nickel.
REF: Hoar, T P & Mowat, JA S — *J Electrodepositors Tech Soc,* 26,7(1950)

NI-0016
ETCH NAME: Heat
TYPE: Thermal, forming
COMPOSITION:
x .... heat

TIME:
TEMP:

DISCUSSION:
Ni single crystal sphere formed from a liquified droplet of nickel in high vacuum. Sphere was oxidized without removal from vacuum to form NiO, and main oxide growth was on the (100) great circle.
REF: Otter, M — *Z Naturforsch,* 14a,355(1959)

NI-0017
ETCH NAME: Potassium
TYPE: Ionized gas, cleaning
COMPOSITION:
x .... K^+ ions

TIME:
TEMP:
GAS FLOW:
PRESSURE:
POWER:

DISCUSSION:
Ni, (100) wafers within ±3° of surface plane. Wafers used in a study of sulfurizing surfaces by S_2 doping. Wafers were first cleaned by resistive heating and chemical etching; then the ion bombardment cleaning as shown, followed by a 10 min anneal at 1370°C.
REF: Godwski, P & Mroz, S — *Thin Solid Films,* 111,129(1984)
NI-0018: Soden, R R Albano, V J — *J Electrochem Soc,* 117,766(1968)
Ni, high purity specimens. A report on material preparation. After cutting wafers, use a $1H_2SO_4$:$1H_2O$ solution to etch remove residual iron on surfaces from cutting. Solution also used as an anodic polishing etch.

NI-0019
ETCH NAME:
TYPE: Acid, polish
COMPOSITION:
30 ml .... HNO_3
10 ml .... H_2SO_4
10 ml .... H_3PO_4
50 ml .... HAc

TIME: 30—60 sec
TEMP: 85—95°C

DISCUSSION:
Ni specimens. Described as a very good polish etch for nickel.
REF: Tegart, W J McG — *The Electrolytic and Chemical Etching of Metals,* Pergamon Press, London, 1956

NI-0020
ETCH NAME: Methyl alcohol
TYPE: Alcohol, cleaning
COMPOSITION:
x .... MeOH

TIME:
TEMP: RT

DISCUSSION:
Ni, crystalline electrode rod. After grinding to shape, specimen was polished with 6, 1, and $^1/_4$ µm diamond paste and ultrasonically cleaned in water, then in methyl alcohol. Gold and platinum were evaporated in a study of their diffusion in nickel. Platinum films were porous and the thin nickel oxide appeared to have no affect on test results.
REF: Van Den Belt, T G M & De Wit, H W — *Thin Solid Films,* 109,1(1983)

NI-0021a
ETCH NAME:
TYPE: Electrolytic, polish
COMPOSITION:
2 oz .... $NaCO_3$
5 oz .... $Na_2Si_6O_9$
0.5 oz .... NaOH

TIME: 1—2 min
TEMP: 210°F
ANODE: Pb
CATHODE: Ni
POWER: 15 A/ft^2 & 6 V

DISCUSSION:
Ni specimens. Solution used as an electropolish etchant. Dip 5—10 sec, RT, in 2% H_2SO_4 for final cleaning.
REF: ASTE Committee — *Tool Engineers Handbook,* McGraw-Hill, New York, 1949

NI-0021b
ETCH NAME:
TYPE: Electrolytic, polish
COMPOSITION:
2 oz .... $NaCO_3$
1000 ml .... H_2O

TIME: 1—2 min
TEMP: 120°C
ANODE: Steel
CATHODE: Ni
POWER: 10 A/ft² & 6V

DISCUSSION:
Ni, specimens. Solution used as an electropolish etchant. Dip 5—10 sec, RT, 2% H_2SO_4 as final clean.
REF: Ibid.

NI-0022
ETCH NAME:
TYPE: Electrolytic, polish
COMPOSITION:
453.6 g .... H_3PO_4
185.6 g .... $Al_2(SO_4)_3.18H_2O$
14.25 g .... $NiSO_4.6H_2O$

TIME: 5 min
TEMP: 80°C
ANODE: Ni
CATHODE:
POWER: 40 A/dm² & 6 V

DISCUSSION:
Ni specimens used in a study of optical properties according to Drude's formula. Solution used to electropolish nickel. Also studied tungsten. Though nickel grains are attacked at different rates, solution will produce a smoothly polished surface.
REF: Roberts, S — *Phys Rev*, 114,104(1959)

NICKEL ALLOYS, NixMyMz

General: There are several metallic nickel metal alloys that occur as natural single crystals, but the majority are artificial formulations tailored for specific characteristics. Some of the naturally occurring alloys are awaruite, $FeNi_2$; josephinite, $FeNi_3$; maucherite, Ni_3As_2; and melonite, $NiTe_3$. Niccolite, NiAs and millerite, NiS are two major ores of nickel and there are several other oxides, silicates, and hydrated minerals that contain cobalt, iron, bismuth, magnesium, aluminum, and other metals. Most meteoric iron is a mixture of iron and nickel, occasionally with some single crystal portions as $FeNi_2$.

In addition to nickel/iron mixtures which are basic in the iron and steel industry, mixtures of nickel/copper are the basis of monel metal, 67Ni:30Cu. See etchant number NICU-0003a for additional monel types. Monel is highly resistant to both fluorine and chlorine gas as well as hydrochloric acid, and is used as tubing for handling such material, as in the construction of HCl cylinder gas regulators with a 9- to 12-month operational life, before replacement parts are installed. Permalloy, 79Ni:17Fe:4Mo is another major nickel alloy and, in addition to normal metal use, also is fabricated as magnets with iron and cobalt.

Technical Application: Many nickel alloys are used in Solid State equipment design, not in direct device development, although as heat sinks and in package construction. Some devices, such as diodes and quartz frequency crystals use magnets in their package construction for operational control.

Several nickel alloys have been grown and studied as both natural and artificial single crystals or ferrites. Several ferrites are used as resistive elements in assemblies with semiconductor elements or as individual resistors.

Etching: H_3PO_4, $FeCl_3$ and mixed acids.

NICKEL ALLOYS

Nichrome, NiCr
Nickel aluminum, NiAl
Nickel boridem, NiB
Nickel copper, NiCu
Nickel iron, NiFe:O_2
Nickel manganese, NiMn
Nickel sulfide, NiS
Nickel titanium, NiTi
Nickel zirconium, Zr_2Ni
Permalloy, Ni:Fe:Mo

NICKEL ALLOY ETCHANTS

NICHROME ETCHANTS

NICR-0001
ETCH NAME:
TYPE: Acid, removal
COMPOSITION:
x .... $Ce(NH_4)_2(SO_4)_6$
x .... HNO_3
x .... H_2O

TIME: 30—60 min
TEMP: RT

DISCUSSION:

Ni:Cr, thin film deposition as a bimetallic layer of Au/Ni:Cr on semiconductor substrates. After etch removal of gold, the nichrome can be removed with the solution shown. Solution also used on thin film metallization of pure nickel and chromium. (*Note:* This is a "chrome etch" mixture.)

REF: Bulletin #5M1/183C — Hycomp, Inc.
NICR-0002: Walker, P — personal application, 1964—1979

Ni:Cr, pellets and alloy sheet. Used a 50 g ceric ammonium nitrate: 5 ml nitric acid:100 ml water solution, 2—5 sec, RT for surface cleaning. A cotton swab soaked in the solution or an eye dropper of solution was wiped on surfaces, then flushed with water, rinsed in MeOH, and N_2 blown dry. Increase water or nitric acid to reduce solution reaction. Method used to clean material prior to use as alloy contacts on silicon devices.

NICR-0003
ETCH NAME: Aqua regia
TYPE: Acid, removal
COMPOSITION:
3 .... HCl
1 .... HNO_3

TIME:
TEMP:

DISCUSSION:

Ni:Cr, thin films evaporated on (111) and (100) oriented silicon, and (100) gallium arsenide wafers with or without an SiO_2 thin films. Nichrome films were photolithograph patterned, and aqua regia was used to etch the nichrome for a study of adhesion and peeling. Thermal shock into LN_2, heat quenching, and slight etch removal of nichrome can initiate peeling.

REF: Tanielian, H et al — *J Electrochem Soc,* 132,507(1985)

NICR-0004a
ETCH NAME: TIME:
TYPE: Acid, preferential TEMP:
COMPOSITION:
45 .... HCl
5 .... HNO_3
50 .... MeOH
DISCUSSION:

Ni:Cr, material specimens used in a study of stacking faults. Solution shown used to develop defect structures.

REF: Motte, H — *Acta Metall,* 5,614(1957)

NICR-0004b
ETCH NAME: TIME:
TYPE: Alkali, removal TEMP: 368 K (114°C)
COMPOSITION: RATE: 100 nm/h
x .... 10% NaOH
DISCUSSION:

Ni:Cr, evaporated thin films on (100) silicon wafers used as a mask for etching pits and via holes. After photolithographic patterning, etching was done at temperature shown with minimum undercutting of the nichrome mask. For a pit on one surface, only, etch about $2^1/_2$ h; for via hole with controlled orifice size, etch aligned opposed side a total time of both sides for about $4^1/_2$ h. The rate shown is for nichrome removal. Hot alkalies are preferential on silicon with a more rapid etch rate.

REF: Ibid.

NICR-0004b
ETCH NAME: TIME:
TYPE: Electrolytic, polish TEMP:
COMPOSITION: ANODE:
6 .... $HClO_4$ CATHODE:
35 .... butyl cellusolve POWER:
59 .... MeOH
DISCUSSION:

Ni:Cr, specimens used in a study of stacking faults. Solution used to polish specimens prior to preferential etching to develop defects.

REF: Ibid.

NICR-0005a
ETCH NAME: Chrome etch TIME:
TYPE: Acid, removal TEMP:
COMPOSITION:
x .... $2NH_4NO_3.Ce(NO_3)_3.4H_2O$
x .... CH_3COOH (HAc)
x .... H_2O
DISCUSSION:

Ni:Cr, as an evaporated thin film on (100) oriented silicon wafers. The nichrome was used as an etch mask against NaOH solutions in etch forming pits and via holes with controlled center orifices. Solution used to remove nichrome after pit/via hole structuring. Solution does not etch silicon. As a form of the chrome etch, see CR-0006, CR-0007, CR-0008, and NI-0008. (*Note:* Nichrome cannot be evaporated as an alloy mixture due to lower vapor pressure of nickel. Nickel evaporates first, followed by chromium, forming a bi-layer with a hard chrome top surface. It can be RF sputter evaporated from a solid target of nichrome as a mixed alloy.)

NICR-0006
ETCH NAME: TIME:
TYPE: Acid, cleaning TEMP:
COMPOSITION:
1 .... HCl
1 .... HNO_3
3 .... H_2O
DISCUSSION:

Ni:Cr, residual metals from evaporation of thin films. Solution used to remove metals and clean vacuum system glass bell jars and internal fixturing. Solution can be used on glass, ceramic, SST and copper.

REF: Nichols, D B — *Solid State Technol,* December 1979

NICR-0007a
ETCH NAME: Hydrochloric acid TIME:
TYPE: Electrolytic, cutting TEMP:
COMPOSITION: ANODE: NiCr
20 .... HCl CATHODE: Nozzle
80 .... H_2O POWER: 60 V
DISCUSSION:

Ni:Cr, specimens used in a study of work hardening and dislocations in fcc single crystals. Solution used with a jet nozzle to slice wafers prior to chemical etch thinning down to foil thickness for microscope study.

REF: Mader, S et al — *J Appl Phys,* 34,3376(1963)

NICR-0007b
ETCH NAME: Hydrochloric acid TIME:
TYPE: Electrolytic, thinning TEMP:
COMPOSITION: ANODE:
50 .... HCl CATHODE:
50 .... EOH POWER:
DISCUSSION:

Ni:Cr specimens 2.6 mm diameter discs electrolytically thinned with this solution using a nozzle held 0.4 mm away from the surface as a method of jet thinning. The thinned specimens were used in a microscope study of structure and defects in single crystal material.

REF: Ibid.

NICKEL ALUMINUM ETCHANTS

NIAL-0001
ETCH NAME: TIME:
TYPE: Thermal, forming TEMP:
COMPOSITION:
x .... heat
DISCUSSION:

NiAl, single crystal and polycrystalline spheres splatter formed from an arc melted in

an He atmosphere with chips and flakes in a water cooled copper pot. Spheres used in an etching and structure study. Other metal spheres formed were Be, V, Fe, Zr, and NiCu.
REF: Ray, A E & Smith, F J — *Acta Metall,* 11,310(1958)

NIAL-0002
ETCH NAME:
TYPE: Electrolytic, polish
COMPOSITION:
x .... 10% $HClO_4$
x .... HAc

TIME:
TEMP:
ANODE:
CATHODE:
POWER:

DISCUSSION:
NiAl, single crystal specimens. Material used in a deformation study. Solution used to both electropolish and then thin specimens for electron microscope observation of defects.
REF: Ball, A & Smallman, R E — *Acta Metall,* 14,1349(1966)

NIAL-0003
ETCH NAME:
TYPE: Electrolytic, polish
COMPOSITION:
x .... 10% $HClO_4$
x .... EOH

TIME:
TEMP:
ANODE:
CATHODE:
POWER:

DISCUSSION:
NiAl, specimens and alloys. Specimens studied for structure change associated with aging. Solution used to both polish and etch thin material. Final thinning done at higher power with vigorous solution stirring. An electron microscope was used to observe defects.
REF: Ardell, A J & Nicholson, R B — *Acta Metall,* 14,1295(1966)

NICKEL BORIDE ETCHANTS

NIB-0001
ETCH NAME: Boron
TYPE: Metal, forming
COMPOSITION:
x .... B

TIME: 4 h
TEMP: 800°C

DISCUSSION:
Ni_2B, thin film grown as a surface penetration film on 12 mm thick nickel plates used as substrates. After boridizing, substrates were annealed in 100°C increments from 400 to 800°C for 24 h. Annealed at 800°C, 4 h produce single crystal Ni_2B. Metallographic sections were made and studied for material structure. (*Note:* No etch shown. See Nichrome for possible etchants.)
REF: Skibo, M & Greulich, F A — *Thin Solid Films,* 113,225(1984)

NIB-0002
ETCH NAME: Acetone
TYPE: Ketone, cleaning
COMPOSITION:
x .... CH_3COOH_3

TIME:
TEMP:

DISCUSSION:
NiB, thin films grown on nickel blank surfaces in a study of boridings metal surfaces under 670°C. All metals evaluated were processed: (1) mechanically polish with emery cloth; (2) degrease in acetone; (3) soak in a molten flux borate bath at 670°C; (4) heat in air 2 h

at 670°C. The borates evaluated as bath constituents were first ground to 325 mesh size before being placed in a crucible, and were KBF_4, KF_4:KF, $NaBF_4$ and $NaBF_4$:KBF_4.
REF: Koyama, K et al — *J Electrochem Soc,* 126,147(1978)

NICKEL COBALT ETCHANTS

NICO-0001
ETCH NAME: TIME:
TYPE: Electrolytic, thinning TEMP:
COMPOSITION: ANODE:
20 .... $HClO_4$ CATHODE:
80 .... CH_3COOH (HAc) POWER:
DISCUSSION:

Ni:Co(15%), specimens used in a study of stacking fault energy and their temperature and concentration dependence in this metal alloy system. Solution used to etch remove and thin specimens.
REF: Ericsson, T — *Acta Metall,* 14,853(1966)

NICKEL:COPPER ETCHANTS

NICU-0001
ETCH NAME: Ferric chloride TIME:
TYPE: Acid, preferential TEMP: Boiling
COMPOSITION:
x .... $FeCl_3$, sat. sol.
x .... H_2O
DISCUSSION:

NiCu (5%) to NiCu (80%) single crystal wafers used in developing light figure orientation (LFO) etchants. Solution will develop etch pits on (111) and (100) orientations with vicinal (hk0) facets — (410) on 20% Ni:80% Cu crystals and (210) on 95% Ni:5% Cu crystals.
REF: Yamamoto, M & Watanabe, J — *Met Abstr,* 8,94(1958)

NICU-0002
ETCH NAME: Heat TIME:
TYPE: Thermal, forming TEMP:
COMPOSITION:
x .... heat
DISCUSSION:

NiCu(30%), single crystal and polycrystalline spheres fabricated by splatter from an arc melter in a helium atmosphere using chips and flakes of material in a water cool pot. Other metal spheres formed were Be, V, Fe, Zr and NiAl. Diameters ranged from 0.1 to 1.0 mm with 5 to 50% single crystal. Spheres used for etching and electrical studies. (*Note:* This is a monel metal.)
REF: Ray, A E & Smith, J F — *Acta Metall,* 11,310(1958)

NICU-0003a
ETCH NAME: Fluorine TIME: 30 days
TYPE: Gas, removal TEMP: 900—975°C
COMPOSITION: RATE: 2.5—5.0 mils/month
x .... F_2, gas
DISCUSSION:

NiCu(30%), polycrystalline blanks used in an etching study. This is a monel metal, as

67Ni:30Cu. Evaluated for corrosion resistance against fluorine at elevated temperature. Other monel formulations are

"K" monel 66Ni:29Cu:3Al
"H" monel 63Ni:30Cu:3Si
"S" monel 63Ni:30Cu:4Si
Cast monel 63Ni:30Cu:1.5Si

REF:

NICU-0003b
ETCH NAME: Chlorine
TYPE: Gas, removal
COMPOSITION:
x .... Cl_2, gas
TIME: 30 days
TEMP: 750—850°C
RATE: 2.5—5.0 mils/month
DISCUSSION:

NiCu(30%), polycrystalline blanks used in an etch study. Both dry chlorine gas and hydrochloric acid were evaluated; dry HCl temperature was 450—500°C. See NICU-0003a.
REF: Ibid.

NICKEL IRON OXIDE ETCHANTS

ETCH NAME: Nital
TYPE: Acid, polish/preferential
COMPOSITION:
1 .... HNO_3
99 .... EOH
TIME:
TEMP:
ASTM #74
DISCUSSION:

Fe:Ni(5%):O(15%) specimens. Solution used as both a polish and preferential etch in the study of austenite and martensite structure development.
REF: Motte, H — *Acta Metall,* 5,614(1957)
NIFE-0002: Pamphlet ASTM E407-70

ASTM #74 solution is 1-5HNO_3:100 mlEOH/MeOH
NIFE-0003: Walker, P — mineralogy study, 1953

$NiFe_3$, as the natural mineral josephinite, and meteoric $FeNi_2$ iron samples used in a study of comparative structures. A 5HNO_3:100 ml EOH solution used 1—2 min, RT, to develop surface figures.

NIFE-0004
ETCH NAME:
TYPE: Acid, cleaning
COMPOSITION:
1 .... HF
1 .... HNO_3
3 .... H_2O
TIME:
TEMP:
DISCUSSION:

Fe:Ni, thin film evaporations. Solution used to remove residual metals after evaporation from glass bell jars and internal fixtures of vacuum systems. Use solution on SST, but not on ceramics or glass.
REF: Nichols, D R — *Solid State Technol,* December 1979

NICKEL MANGANESE ETCHANTS

NIMN-0001
ETCH NAME:
TYPE: Electrolytic, dislocation
COMPOSITION:
x .... H_3PO_4, conc.

TIME:
TEMP:
ANODE:
CATHODE:
POWER:

DISCUSSION:

NiMn, single crystal specimens used in a study of low angle grain boundaries and defects. Solution shown used to develop defect structure.

REF: Taoka, T & Sakata, S — *Acta Metall,* 5,230(1957)

NIMN-0002: Marcinkowski, M J & Miller, D S — *Phil Mag,* 6,871(1961)

Ni_3Mn, single crystal specimens used in a study of ordering strength and dislocations in superlattice materials. Solution used to etch polish material. Also studied Cu_3Au.

NICKEL TITANIUM ETCHANTS

NITI-0001
ETCH NAME:
TYPE: Acid, removal
COMPOSITION:
x .... HF
x .... HNO_3

TIME:
TEMP:

DISCUSSION:

NiTi, single crystal specimens used in a study search for electronic phase transitions in this material. Solution is called a heavy etch. Follow etching with DI H_2O quench, MeOH rinse, and dry. Also etched Zr_2Ni.

REF: Lee, H N & Withers, R — *J Appl Phys,* 49,5488(1978)

NITI-0002: Dumez, P — *J Vac Sci Technol,* B(1),218(1983)

NiTi, grown as a metastable single crystal alloy in study of the material morphology. Other materials studied as metal alloys: AlTi and ZrV. As metallic glasses: PdNiP, PtNiP, and PtCuP.

NICKEL ZIRCONIUM ETCHANTS

NIZR-0001
ETCH NAME:
TYPE: Acid, removal
COMPOSITION:
x .... HF
x .... HNO_3

TIME:
TEMP:

DISCUSSION:

Zr_2Ni, single crystals grown in a study of electronic phase transitions. Solutions of HF:HNO_3 are general etch mixtures for this material. Also studied: Ni:Ti and PtSi.

REF: Lee, R N & Withers, R — *J Appl Phys,* 49,5488(1978)

NIZR-0002: Altounian, Z et al — *J Appl Phys,* 55,1430(1978)

$NiZr_2$, deposited as an amorphous thin film in a general study of the compound.

PERMALLOY ETCHANTS

PMA-0001a
ETCH NAME: Aqua regia
TYPE: Acid, polish
COMPOSITION:
3 .... HCl
1 .... HNO_3
TIME:
TEMP:
DISCUSSION:
79Ni:17Fe:4Mo, Permalloy single crystal specimens used in a study of strain under compression. The material shows slip on (111) surfaces in a ⟨110⟩ direction. Follow aqua regia by electropolishing.
REF: Chin, G Y et al — *Acta Metall*, 14,467(1966)

PMA-0001b
ETCH NAME:
TYPE: Electrolytic, polish
COMPOSITION:
x .... H_3PO_4
x .... CrO_3
TIME:
TEMP:
ANODE:
CATHODE:
POWER:
DISCUSSION:
79Ni:17Fe:44Mo, (111) oriented Permalloy wafers. See PMA-0001a for discussion.
REF: Ibid.

PMA-0002
ETCH NAME: Bromine
TYPE: Halogen, defects
COMPOSITION:
x .... Br_2
TIME:
TEMP:
DISCUSSION:
68Ni:31.9Fe:0.1Mg and 63Ni:35Fe:2Mo. The Mg containing Permalloy showed dark grey/black and grey areas after etching in bromine. The Mo containing Permalloy showed bright areas only. Grey/black = heavy faulting; grey = light faulting and bright = no faulting. The greater the amount of oxygen present in the materials, the greater the degree of faulting. Both specimens were single crystal with (100) oriented surfaces.
REF: Nesbitt, E A et al — *J Appl Phys*, 31,228S(1960)

PHYSICAL PROPERTIES OF NICKEL IODIDE, NiI_2

Classification	Iodide
Atomic number	53
Atomic weight	312.53
Boiling point (°C)	>100
Melting point (°C)	
Density (g/cm³)	5.84
Hardness (Mohs — scratch)	3—4
Crystal structure (isometric — normal)	(100) cube
Color (solid)	Black
Cleavage (cubic)	(001)

NICKEL IODIDE, NiI_2

General: Does not occur as a natural mineral due to water solubility although it may occur in solution in minor quantity. There are several nickel minerals as arsenides and sulfides. There is no use of the iodide in the metal industry.

Technical Application: Not used in Solid State processing, but could be used as an iodine additive to etching solutions. It has been grown as a single crystal in crystallographic and morphology studies.

Etching: Alcohol.

NICKEL IODIDE ETCHANTS

NII-0001

ETCH NAME: Ethyl alcohol — TIME:

TYPE: Alcohol, removal — TEMP: RT

COMPOSITION:

x .... EOH

DISCUSSION:

NiI grown as a single crystal in a crystallographic study. It is in the cubic space group 43m. The material can be dissolved in alcohols. Reference refers to the material as Boracite? (*Note:* The natural mineral boracite is a borate, not an iodide: $Mg_5Cl_2B_{14}O_{25}$.)

REF: Cook, R et al — *J Appl Phys,* 49,6025(1978)

RARE EARTH NICKELIDE ETCHANTS

RENI-0001

ETCH NAME: Aqua regia, variety — TIME:

TYPE: Acid, removal — TEMP: RT to hot

COMPOSITION:

4 .... HCl

1 .... HNO_3

DISCUSSION:

$RENi_2$. Rare Earth nickelides initially formed as alloy mixtures by arc melting under argon gas, then re-melted as buttons in a thermal expansion study. Wrap specimens in Ta sheet and anneal 3—9 weeks at 600—900°C at 2×10^{-7} Torr vacuum, $SmNi_2$ in a sealed quartz ampoule due to Sm evaporation. All were cubic (Laves) structure. RE = Y, La, Nd, Sm, Tb, Dy, Ho, Er, Th, Pr, Ce, and Gd. The Y, La, Nd, and Sm nickelides showed secondary phases as RENi, $RENi_2$, $RENi_3$ similar to silicides. (*Note:* Etchant shown is for nickel.)

REF: Ibarra, M R et al — *J Phys Chem Solids,* 45,789(1984)

PHYSICAL PROPERTIES OF NICKEL OXIDE, NiO

Classification	Oxide
Atomic numbers	28 & 16
Atomic weight	74.69
Melting point (°C)	2980
Boiling point (°C)	
Density (g/cm^3)	6.4—5.8
Refractive index (n=)	2.18/2.23
Hardness (Mohs — scratch)	5.5

Crystal structure (isometric — normal) (111) octahedron
Color (solid) Dark green
Cleavage (cubic — poor) (001)

NICKEL OXIDE, NiO

General: Occurs in nature as the mineral bunsenite, NiO. Rare and of no importance as a mineral. Two of the more important ore minerals of nickel are millerite, NiS and niccolite, NiAs.

Nickel oxide has no use in industry other than its normal surface passivation of nickel surfaces as a native oxide.

Technical Application: Nickel oxide is not used in Solid State device fabrication, although nickel is an evaporation and plating metal on structures, substrates and parts as well as used in solid form in package assemblies. In all of these cases, where nickel oxide is present, it is referred to as a native oxide, and considered a contaminant in device fabrication.

There has been some limited study of nickel oxide as, in Solid State studies, it shows p-type conduction as intrinsic semiconductor material.

Etching: Soluble in acids and ammonium hydroxide.

NICKEL OXIDE ETCHANTS

NIO-0001a
ETCH NAME: Ammonium hydroxide
TYPE: Base, removal
COMPOSITION:
x .... NH_4OH

TIME: Seconds
TEMP: RT

DISCUSSION:

NiO specimens. Solution used to remove nickel oxide from nickel surfaces. The solution does not etch pure nickel metal.

REF: Kohl, W H — *Handbook of Materials and Techniques for Vacuum Devices*, Reinhold, New York, 1967

NIO-0002: Walker, P & Menth, M — personal application, 1982

NiO, native oxide on nickel slugs used for evaporation. Solution used to clean nickel surfaces prior to use. After etching, quench in DI water, then rinse in MeOH and N_2 blow dry.

NIO-0001b
ETCH NAME: Argon
TYPE: Ionized gas, thinning
COMPOSITION:
x .... Ar^+ ion

TIME: 10—15 min
TEMP:
GAS FLOW:
PRESSURE:
POWER: 7—8 KVA then 3—5 KVA

DISCUSSION:

NiO, as intrinsic p-type single crystal specimens. Prepare by lapping with SiC abrasive, then polish with diamond paste. Use Ar^+ ions to RF sputter thin material under vacuum. Use high power for heavy removal, then reduce to lower power for final thinning.

REF: Ibid.

NIO-0002
ETCH NAME: Heat — TIME:
TYPE: Thermal, preferential — TEMP: 1000 K
COMPOSITION:
x .... heat
DISCUSSION:

NiO, (100) cleaved wafers used in a study of magnetic susceptibility. After cleaving, specimens were annealed to reduce strain. Also studied CoO.

REF: Singer, J R — *Phys Rev,* 104,929(1956)

NIO-0003: Roth, W L — *J Appl Phys,* 31,2000(1960)

NiO, (100) cleaved wafers used in neutron and optical studies of domains. Epoxy resin used for holding specimens was dissolved in $1H_2SO_4$:$1HNO_3$. Glycol phthalate also used and dissolved off with heat.

NIO-0004
ETCH NAME: Nitric acid — TIME:
TYPE: Acid, removal (nickel) — TEMP: Hot
COMPOSITION:
x .... HNO_3, conc.
DISCUSSION:

NiO, thin film platelets grown on nickel substrates for thermal oxidation. Nickel inclusions observed in the oxide were removed by etching in hot nitric acid.

REF: Downs, R C — *J Electrochem Soc,* 130,807(1983)

PHYSICAL PROPERTIES OF NICKEL SULFIDE, NiS

Classification	Sulfide
Atomic numbers	28 & 16
Atomic weight	90.75
Boiling point (°C)	797
Melting point (°C)	
Density (g/cm³)	5.3—5.65
Hardness (Mohs — scratch)	3—3.5
Crystal structure (hexagonal — rhombohedral)	$(10\bar{1}1)$ rhomb
Color (solid)	Bronze-yellow
Cleavage (rhombic)	$(10\bar{1}1)$ or $(01\bar{1}2)$

NICKEL SULFIDE, NiS

General: Occurs as the natural mineral millerite, NiS (capillary pyrite). Occurs in slender capillary crystals, hence the common name, often in delicate radiating groups; sometimes interwoven like a wad of hair. It is often coated with a grayish translucent coating. The $(01\bar{1}2)$ is a glide plane, such that by exerting pressure in crystals artificial twins can be formed. It is associated with bodies of other nickel ores, has been grown artificially, and has been noted in meteoric iron.

There also is the mineral polidymite, Ni_3S_4, which is probably isomorphous with linnaeite, Co_3N_4, but is isometric — normal, and occurs as fine octahedron, (111), crystals. It also, like millerite, is prone to twinning, in this case, on a (111) glide plane forming bicrystals, but is of minor occurrence.

Millerite is an ore of nickel, and can be used directly as an additive to iron and steels that are being sulfurized.

Technical Application: There has been no use in Solid State processing, but both the natural and artificial materials have been studied due to interest in crystallographic and atomic cell structure.

Etching: HNO_3; aqua regia; slow in other acids.

NICKEL SULFIDE ETCHANTS

NIS-0001
ETCH NAME: Aqua regia
TYPE: Acid, cleaning/removal
COMPOSITION:
1 .... HF
3—5 .... HNO_3
10—50 .... H_2O

TIME: 10—30 sec
TEMP: RT

DISCUSSION:
NiS, as the natural mineral millerite. Various concentrations of solution shown used to clean surfaces (dilute), and develop single specimens crystallographic structure, defects, and twinning in a general study of this mineral, to include magnetism as related to nickel, cobalt, and magnetic iron.

REF: Walker, P — mineralogy study, 1952—1954

PHYSICAL PROPERTIES OF NITROGEN, N_2

Classification	Gas
Atomic number	7
Atomic weight	14.0067
Melting point (°C)	−209.86
Boiling point (°C)	−195.8
Density (g/cm³ × 10^{-3}) 20°C	1.43
(g/cm³ vs. air = 1)	0.96717
Specific heat (cal/g/°C) 20°C	0.247
Thermal conductance (cal/cm²/cm/°C/sec × 10^{-4}) 20°C	0.562
1st ionization potential (eV)	14.54
Ionic radius (angstroms)	0.13 (N^{+5})
Cross section (barns)	1.9
Vapor pressure (°C)	−209.7
Critical temperature (C.T.) (°C)	−147
Critical pressure (C.P.) (atms)	33.5
Chemical reactivity (22.2°C & 760 mmHg)	Inert
Hardness (Mohs — scratch) solid	1—2
Crystal structure (hexagonal — normal) solid	$(10\bar{1}0)$ prism, hcp
Color (gas/liquid)	Colorless
(solid)	Whitish
Cleavage (basal — fair) solid	(0001)

NITROGEN, N_2

General: The primary inert gas in the atmosphere as a mixture of about 75% N_2:24% O_2, plus trace gases. It is inert at standard temperature and pressure, but forms many, many nitrite and nitrate minerals in nature, many of which are primary fertilizers. Many such compounds are unstable, readily give up nitrogen to the atmosphere, while others such as ammonia (NH_3) are water soluble. Although nitrogen is not assimilated directly by plant and animal life, it is a major building block in living matter as proteins which contain amines with the NH_3^- radical.

The nitrogen fixation cycle is the natural supplying of free nitrogen to the earth from atmospheric rain, dissolving chemical and minerals in water, and from electrical discharge during thunder and lightning storms.

In addition to nitrates use as fertilizers, there are several important industrial compounds, such as in guncotton and trinitrotoluene (TNT). And there is nitric acid, ammonia, and ammonium hydroxide; Chile saltpeter is both an important fertilizer, as well as having medical applications. Other nitrite and nitrate compounds are used as leather and textile dyes.

The French still call it azote (Lavoisier), but the name nitrogen was proposed by Chaptal (1823) from the Greek name for saltpeter, as it is a major element in that compound.

Most of the pure nitrogen today is obtained from the liquifaction of air, as the oxygen boils off, first, at −183°C leaving nitrogen at −196°C. Note that further chilling of the nitrogen is done to obtain atmospheric rare gases, such as neon, krypton, etc., as well as a larger quantity of argon. Obtained as cold, liquified gases, this is the field of cryogenics, and both liquid nitrogen and argon — also oxygen, hydrogen and helium — are supplied to industry and medicine in such liquid form. It is used as both a chilling liquid or passed through heat exchangers and converted back to gas by warming and expansion. Other major sources of industrial nitrogen comes as a by-product of making coke (burning of coal to remove volatiles), and from the processing of Chilean saltpeter (soda niter) and niter (potassium niter).

Metal, glass, and ceramic processing all use nitrogen as an inert furnace gas, alone, or as forming gas (FG) $85\%N_2{:}15\%H_2$. The latter as a reducing atmosphere for de-oxidizing surfaces during alloying, or removing scale contamination. At elevated temperatures the nitridization process introduces a nitride surface layer on metal and metallic compounds for corrosion resistance, strength, and a more inert surface under most atmospheric conditions.

There are several important nitrogen compounds, other than those used in explosives and fertilizers, such as ammonia and ammonium hydroxide. As chemical bases they are used to neutralize acid wastes and, in crystalline form, ammonia is medical ''smelling salts''. Nitric acid is a strong acid, widely used as the oxidizer element in etch solutions, and nitrous oxide, N_2O, is ''laughing gas'' used as an anesthetic. In deep-sea diving, using air, when a diver is brought to the surface too rapidly, nitrogen will form bubbles in the bloodstream causing the ''bends'' which causes severe muscle cramps and spasms. Helium, which does not form such bubbles, is used as a replacement for nitrogen in diving tanks as the breathing atmosphere, but causes the larynx muscles to contract which produces the ''Donald Duck'' speech. Several commercial plastics contain nitrate, such as cellulose nitrate, a thermoplastic, as celluloid.

Metal nitrides are becoming of increasing importance and use, other than for surface nitridization. A major example is (cubic) boron nitride, B_4N, with a Mohs hardness: H = 9+. This is a lapping and polishing abrasive, and has recently been fabricated as a high temperature semiconducting diode. Silicon nitride, Si_3N_4, and aluminum nitride, AlN, are used as surface protectants, etching masks, radiation masks on semiconductors as well as other metal surfaces.

Technical Application: Many Solid State plants have an on-site liquid nitrogen, LN_2, storage tank — 500 to 2000 gallon capacity — with a heat exchanger to convert LN_2 back to N_2 gas. Both the liquid and gas can be brought into the plant with copper lines to user locations, or gas lines, only, directly into furnaces, epitaxy systems, vacuum systems, with LN_2 mobile cylinder fill capability at the outside LN_2 storage tank. Such LN_2 cylinders range from a small cone-shaped size holding 25—50 liquid liters, to 200 l barrel-sized containers on wheels. The storage tank is topped-off (filled) every 2 to 4 days from 2000 gallon cryogenic truck trailers. Note that other cryogenic gases are similarly supplied like LN_2, such as LOX, LHe, LCl_2, and LH_2. The latter is normally used directly from the 2000 gallon cryogenic trailer, not transferred into an on-site storage tank.

Nitrogen, as an inert gas, is the usual back-fill gas in vacuum systems being brought "up-to-air", rather than air itself in order to reduce oxidation of metallized parts and the evaporation metal source. Such back-fill is about 40 psi; whereas when nitrogen, again, rather than air, is used for valve operation in vacuum systems, pressure is 80 psi. As LN_2 is already used in the vacuum cold trap for chamber pump-down, such that nitrogen is available, it is more cost efficient to use the gas for chamber back-fill and valve operation, than have to pipe in separate pressurized air lines from air compressors. The nitrogen gas also is usually cleaner than compressed air, and there is no possibility of line oil contamination with nitrogen, as there is from an air compressor. Nitrogen is checked with a Dew Point Meter for water moisture, and a figure of 0.72°F (14.3 ppm) is considered optimum for line nitrogen coming off an LN_2 source. Other processing equipment using nitrogen, both as an inert gas, as well as for valve operation include wire bonders, wafer saws, wafer dicing saws or ultrasonic dicers, resistance welding, or coldwelding systems. The wafer sawing systems also used for cutting ceramic or glass circuit substrates or ferrites as discrete resistors, other dielectrics as both resistors and capacitors. Helium leak detectors sometimes use a nitrogen gas or an air equivalent for general testing of parts or package seals.

As a gas nitrogen it is used as the inert atmosphere in semiconductor diffusion furnaces, alloy belt furnaces, in nitrogen dry boxes for alloying assemblies or for storage of critical parts, such as circuit substrates, and as pressurized line nitrogen blow-off in chemical areas for final drying after etching. It also is bubbled through an etching or cleaning solution to obtain a stirring action and, as an ionized gas, N^+ ions, used in RF plasma cleaning systems. As a pure N_2, N_2/Ar, or N_2/O_2 atmosphere at elevated temperatures, it is used to grow thin film nitrides, notably silicon nitride, Si_3N_4, or oxynitride, SiN_xO_y (5% O_2 is a general standard), and AlN. Most nitrides are used as an improved, inert replacement for silicon dioxide, SiO_2, as etching, diffusion or metallization masks, or for final surface passivation.

Nitrogen gas is available in pressurized cylinders from a standard commercial grade to electronic grade dry nitrogen, in addition to gas mixtures. Mixtures are used as diffusion gases, and the high purity N_2/He(30%) is used as a package back-fill in sealing quartz radio frequency crystals. A standard A-1 cylinder is 9″ in diameter, about 5 ft high with a shut-off top valve, and gas supply controlled through a dual gas regulator — one meter showing tank pressure — the down-line meter dial controlled for gas output. The A-1 cylinder is normally supplied under 3600 psi pressure, and is the largest that can be commercially transported under pressure. Larger cylinders with higher pressures are filled on-site.

LN_2 also is used as a freezing agent: single crystal materials, such as semiconductors, are cleaved under liquid nitrogen or, frozen, removed, and then cleaved while still frozen. Common cleave planes are cubic (001); basal (001) or (0001), though they can be dode-cahedral (110), octahedral (111), prismatic $(10\bar{1}0)$ or rhombic $(10\bar{1}1)$, and others in special cases. LN_2 is used in high reliability testing by cycling parts from hot-to-cold in evaluating device and package integrity. Another use is the cryogenic microscope stage; yet another, the cryogenic cryostat used for both device tests and evaluations, and the growth of single crystal ice.

CAUTION: Liquid nitrogen at −196°C will freeze any material submerged in the cold liquid in a matter of seconds — a rose frozen in LN_2 can be removed and shattered like glass.

Even though the atmosphere is about 75% nitrogen, the presence of oxygen is essential to sustain life. In a working area where nitrogen is being used, such as in wafer dicing equipment, the area should be well ventilated as, with an increase in nitrogen content in the atmosphere, the operator will become sleepy . . . at the least. In cleaning LN_2 storage and transfer tanks, such tanks should be checked for oxygen content before entering for tank cleaning or air tanks should be worn by operators.

Etching: N/A as gas.

NITROGEN ETCHANTS

N-0001

ETCH NAME: Pressure	TIME:
TYPE: Pressure, defect	TEMP: −210°C

COMPOSITION:

x .... pressure, vapor

DISCUSSION:

N_2, grown as a single crystal under vacuum pressure and cryogenic conditions. LHe used as the freezing vehicle, and single crystal gases so grown have been used in a general study of solid gases. By varying the vapor pressure in the vacuum system, structural defects can be observed.

REF: Schwentner, N et al — *Rare Gas Solids,* Vol. 3, Academic Press, New York, 1970

PHYSICAL PROPERTIES OF NIOBIUM, Nb

Classification	Transition metal
Atomic number	41
Atomic weight	92.906
Melting point (°C)	2468
Boiling point (°C)	4927
Density (g/cm^3)	8.57
Thermal conductance (cal/sec)(cm^2)(°C/cm)	0.125
Specific heat (cal/g) 25°C	0.065
Heat of fusion (k-cal/g-atom)	6.4
Latent heat of fusion (cal/g)	69
Atomic volume (W/D)	10.8
1st ionization energy (K-cal/g-mole)	156
1st ionization potential (eV)	6.77
Electronegativity (Pauling's)	1.6
Covalent radius (angstroms)	1.34
Ionic radius (angstroms)	0.69 (Nb^{+5})
Coefficient of linear expansion ($\times 10^{-6}$ cm/cm/°C)	7.1
Electrical resistivity (micro-ohms/cm)	15.22
Electron work function (eV)	4.01
Tensile strength (psi-annealed)	50,000
Cross section (barns)	1.1
Refractive index (n=)	1.80
Hardness (Mohs — scratch)	7—8
Crystal structure (isometric — normal)	(100) cube, bcc
Color (solid)	Steel-grey metal
Cleavage (cubic)	(001)

NIOBIUM, Nb

General: The names niobium, Nb and columbium, Cb are synonymous, though today, most references use niobium — and there is a crystallographic discrepancy — originally, columbium was listed as being hexagonal system — rhombohedral division; yet niobium is isometric — normal with a bcc (body-centered cubic) crystal structure. Regardless, niobium does not occur in nature as a native element though there are some 20 or more minerals containing niobium in association with rare earths (yttrium, erbium, cerium, etc.) as well as tantalum, iron, and calcium. Most niobium bearing minerals are oxides. The minerals pyrochlore, fergusonite, columbite-tantalite, yttrotantalite, and samarskite being the more prominent.

In the making of iron and steel, niobium is of major importance as it combines with carbon increasing the corrosion resistance. See section titled Columbium which is included separately where work done is referred to that name rather than niobium.

Technical Application: Although not used in most Solid State device fabrication or assembly processes, there is one important device known as the Josephson Junction device, as single crystal Nb_3Sn. It also is used as a high temperature anode in electrolytic etching with or without titanium and tantalum coatings.

The pure metal has been grown as a single crystal and, although it has been widely studied and etched, it is difficult to both etch and cut.

Etching: H_2SO_4, HCl, HNO_3 and mixed acids: HF:HNO_3 with H_2O_2 and H_2SO_4.

NIOBIUM ETCHANTS

NB-0001

ETCH NAME: Air TIME:

TYPE: Gas, oxidation TEMP:

COMPOSITION:

x .... air

DISCUSSION:

Nb, thin films evaporated on glass in H_2 to form a-Nb:H. Oxidize in hot air at 80°C, 10 min to form Nb_2O_5. Selectively evaporate Pb for Josephson Junctions: Pb/Nb_2O_5/Nb/SIO_2 as an SIS device (superconductor-insulator-substrate).

REF: Kobyashni, T et al — *J Appl Phys,* 57,2583(1985)

NB-0002

ETCH NAME: TIME:

TYPE: Acid, polish TEMP:

COMPOSITION:

1 .... HF

4 .... HNO_3

DISCUSSION:

Nb, specimens used in a study of spark-cut damage. Mechanically lap and polish, then etch polish with solution shown. After spark-cutting decorate damage with carbon, as c-C in isopropyl alcohol.

REF: Guberman, H D — *J Appl Phys,* 39,2974(1968)

N8-0003

ETCH NAME: TIME:

TYPE: Acid, removal TEMP:

COMPOSITION:

x .... H_2SO_4

x .... HF

x .... H_2O_2

x .... H_2O

DISCUSSION:

Nb specimens. Solution used as a general etch for niobium.

REF: Bergund, T & Deardon, W H — *Metallographer's Handbook of Etching,* Pitman & Sons, London, 1931

NB-0004a

ETCH NAME: Hydrochloric acid, dilute — TIME:

TYPE: Acid, cleaning — TEMP: 70°F and 120°F

COMPOSITION: — RATE: 0.00004/ipy

x .... HCl

x .... H_2O

DISCUSSION:

Nb, specimens. Effectively no acid attack but can be used to clean surfaces.

REF:

NB-0004b

ETCH NAME: Hydrochloric acid — TIME:

TYPE: Acid, cleaning — TEMP: 70°F and 120°F

COMPOSITION: — RATE: 0.0001/ipy 0.004/ipy

x .... HCl

DISCUSSION:

Nb, specimens. Effectively no acid attack but can be used to clean surfaces.

REF: Ibid.

NB-0005a

ETCH NAME: Sulfuric acid — TIME:

TYPE: Acid, removal — TEMP: Hot

COMPOSITION:

x .... H_2SO_4

DISCUSSION:

Nb specimens. A slow, general etch for niobium.

REF: Hodgman, C D et al — *Handbook of Chemistry and Physics,* 27th ed, Chemical Rubber Co., Cleveland, OH, 1943, 616

NB-0012a: Kohl, W H — *Handbook of Materials and Techniques for Vacuum Devices,* Reinhold, New York, 1967, 304

Nb specimens. Solution used at RT will convert surface to Nb(OH) which can be removed with HF, as can be Nb.

NB-0005b

ETCH NAME: Hydrofluoric acid — TIME:

TYPE: Acid, removal — TEMP:

COMPOSITION:

x .... HF

DISCUSSION:

Nb specimens. A slow, general etch for niobium.

REF: Ibid.

NB-0006
ETCH NAME:
TYPE: Electrolytic, polish
COMPOSITION:
x .... 10% HF
x .... 90% H_2SO_4
TIME: 1 h
TEMP: 25°C
ANODE: Nb
CATHODE:
POWER:
DISCUSSION:
Nb specimens. Mechanically polish surfaces before electropolishing in the solution shown. A study of tantalum diffusion in niobium.
REF: Lawel, R E & Lundy, T S — *J Appl Phys,* 35,435(1964)
NB-0010b: Cathcart, J V et al — *J Electrochem Soc,* 105,442(1958)
Nb, single crystal sphere with (111), (110), and (100) cut flats and single crystal wafers. Wafers and flats mechanically lapped down to 3 μm alumina finish. Solution used to electropolish prior to oxidation studies.

NB-0012b
ETCH NAME: Nitric acid
TYPE: Acid, polish
COMPOSITION:
x .... HNO_3
TIME:
TEMP: 100°C
DISCUSSION:
Nb specimens. Will etch niobium at or above 100°C. Other general etchants are $HF:NH_4F$; NaOH/KOH, conc.; $HF:HNO_3$ mixtures; H_3PO_4. Niobium can be electropolished in tantalum solutions. See TA-0003i.
REF: Ibid.

NB-0007
ETCH NAME: Hydrofluoric acid
TYPE: Electrolytic, polish
COMPOSITION:
x .... HF
TIME:
TEMP:
ANODE:
CATHODE:
POWER:
DISCUSSION:
Nb specimens used for electroplating. Use solution shown to AC polish specimens prior to plating. (*Note:* A major article on the electroplating of transition metals. Both etch and plate solutions shown.)
REF: Saubestre, E B — *J Electrochem Soc,* 106,305(1959)
NB-0008: Faust, G L & Beach, J G — *Plating,* 43,1134(1950)
Reference for solution used above in NB-0007
NB-0014: Beach, J G & Faust, C L — MBI 10004(1954 (Commerce Dept., WDC)
Nb specimens. Solution used to AC etch this material.

NB-0009
ETCH NAME: Nitric acid
TYPE: Acid, polish
COMPOSITION:
95% .... HNO_3
5% .... H_2O
TIME:
TEMP:
DISCUSSION:
Nb, polycrystalline samples used in a study of yield and fracture. Polish in solution shown and follow with 5% HF.
REF: Adams, M A et al — *Acta Metall,* 8,25(1960)

NB-0010a
ETCH NAME: Oxygen TIME:
TYPE: Gas, oxidation TEMP:
COMPOSITION:
x .... O_2
DISCUSSION:
Nb, spheres with cut and oriented faces on (111), (110), and (100). The (111) and (100) surfaces oxidize rapidly; the (110) slowly. Cracking was observed which varies in amount with orientation.
REF: Cathcart, J V et al — *J Electrochem Soc,* 105,442(1958)

NB-0011
ETCH NAME: TIME:
TYPE: Electrolytic, oxidation TEMP: −10 to −70°C
COMPOSITION: ANODE:
65% .... 0.1 *N* H_2SO_4 CATHODE:
35% .... EOH POWER: 1—100 mA/cm^2
DISCUSSION:
Nb, specimens used in a study of the temperature dependence of the Tafel Slope in thin anodic oxides. Solution used to develop the oxides. Stir continually and replace solution every 8 h.
REF: Adams, G D & Kao, T — *J Electrochem Soc,* 107,640(1960)

NB-0013
ETCH NAME: Kerosene TIME:
TYPE: Hydrocarbon, cleaning TEMP:
COMPOSITION:
x .... C_{10}-C_{16}
DISCUSSION:
Nb, specimens used in a study of surface oxidation between 375 and 700°C. Surfaces cleaned by light abrasion under kerosene.
REF: Gulbransen, E A & Andrews, K F — *J Electrochem Soc,* 1054(1956)

NB-0015a
ETCH NAME: TIME:
TYPE: Acid, polish TEMP:
COMPOSITION:
1 .... HF
1 .... HNO_3
1 .... H_2SO_4
1 .... H_2O
DISCUSSION:
Nb, (100) oriented single crystal rods used in a study of properties and superconduction. Solution used to etch polish specimens.
REF: Ikushima, A et al — *J Phys Chem Solids,* 27,327(1966)

NB-0015b
ETCH NAME: Heat TIME:
TYPE: Thermal, preferential TEMP: 1800°C
COMPOSITION:
x .... heat

DISCUSSION:

Nb, (100) oriented single crystal cylinders. When cylinders were annealed at or above 1800°C thermal etching was observed.

REF: Ibid.

NIOBIUM ALLOYS, NbMx

General: Niobium occurs in nature in the mineral columbite-tantalite, $(Fe,Mn)(Nb,Ta)_2O_6$, and similar minerals containing rare earth elements, and as niobates and titanates with uranium or thorium. Does not occur as a metallic alloy in nature, but there are high temperature artificial alloys. It is an additive to irons and steels for strength and corrosion resistance, and alloyed with both titanium and tantalum for high temperature characteristics.

Technical Application: Niobium alloys are not used in Solid State processing to any extent although they are used as electrodes and antennas. The only reference, here, is for the niobium-aluminum system.

Etching: HF, H_2SO_4. Mixed acids $HF:HNO_3$. Variable with alloy.

NIOBIUM ALLOY ETCHANTS

NIOBIUM ALUMINUM ETCHANTS

NBAL-0001

ETCH NAME: Hydrofluoric acid TIME:

TYPE: Acid, reduction TEMP:

COMPOSITION:

x .... HF, conc.

DISCUSSION:

$NbAl_y$, as alloy specimens. Solution used to dissolve material before titration of individual constituents in a study of this alloy system.

REF: Shilo, I — *J Electrochem Soc,* 129,1608(1982)

NBAL-0002:

Nb_3Al grown as a single crystal in a study of crystallographic structure. Structure is A15 crystal symmetry space group.

PHYSICAL PROPERTIES OF NIOBIUM CARBIDE, NbC

Classification	Carbide
Atomic numbers	41 & 6
Atomic weight	104.92
Melting point (°C)	2750
Boiling point (°C)	
Density (g/cm³)	15.77
Hardness (Mohs — scratch)	9+
(Vickers — kgb/mm²)	2800
Crystal structure (hexagonal — normal)	$(10\bar{1}0)$ prism
(isometric — normal)	(100) cube
Color (solid)	Grey-white
Cleavage (basal/cubic)	(0001)/(001)

NIOBIUM CARBIDE, NbC

General: Does not occur as a natural compound. The only known natural carbide is a minor occurrence of the mineral moissanite, SiC. Niobium carbide is one of the high temperature ceramics.

Technical Application: The material is under study in Solid State development, but has not had any major use to date.

Etching: HF, H_2SO_4.

NIOBIUM CARBIDE ETCHANTS

NBC-0001a
ETCH NAME: TIME:
TYPE: Acid, removal/clean TEMP:
COMPOSITION:
x .... HF
DISCUSSION:

NbC specimens. A very slow etch for this material.

REF: *A Dictionary of Carbide Terms,* Adams Carbide Corp., Kenilworth, NJ

NBC-0001b
ETCH NAME: TIME:
TYPE: Acid, removal/clean TEMP: Hot
COMPOSITION:
x .... H_2SO_4
DISCUSSION:

NbC, specimens. Solution is a slow etch for the material. Material was used in a general properties study. Density: 15.77 gm/cm^3. Melting point: 2750°C. Vickers hardness: 2800 kg/cmm^2.

REF: Ibid.

NBC-0003a-b
ETCH NAME: TIME: (1) 1 min (2) 4 sec
TYPE: Acid, removal/cleaning TEMP: RT RT
COMPOSITION:
(1) x .... 20% sulfamic acid or (2) x .... BHF
DISCUSSION:

NbC, thin films grown by electrolytic deposition on different substrates in a study of the compound. Either solution can be used for the times shown for light removal and cleaning of surfaces. (*Note:* BHF mixture not shown. One standard is 1HF:1NH_4F (40%).)

REF: Hockman, A J — *J Electrochem Soc,* 130,221(1983)

NIOBIUM GERMANIDE ETCHANTS

NBGE-0001
ETCH NAME: RIE TIME:
TYPE: Ionized gas, removal TEMP:
COMPOSITION: GAS FLOW: 200 V
x .... CF_4 PRESSURE: 6 Pa
x .... Ar POWER:
DISCUSSION:

Nb_3Ge, single crystal compound co-deposited by E-beam evaporation as an amorphous layer on sapphire. Thickness was 80—100 μm. Material was patterned by standard pho-

tolithographic techniques. Deposit was RF plasma etched with CF_4. CF_4 is a rapid etch alone so argon is used as an inert control gas.
REF: Kato, W — *Jpn J Appl Phys,* 23,1536(1984)

NBGE-0002
ETCH NAME:
TYPE: Acid, removal
COMPOSITION:
x .... HF
x .... HNO_3
x .... CH_3COOH (HAc)
TIME:
TEMP:
DISCUSSION:
Nb_3Ge, thin films grown on (100) germanium substrates by CVD in a study of high T_c. Concentrations of the solution shown will etch both the Nb_3Ge thin film and Ge substrate.
REF: Suzuki, M et al — *Jpn J Appl Phys,* 23,991(1984)

NIOBIUM HYDRIDE ETCHANTS

NBH-0001
ETCH NAME: Sulfuric acid
TYPE: Acid, removal
COMPOSITION:
x .... H_2SO_4, conc.
TIME:
TEMP:
DISCUSSION:
NbH, as powder in amyl acetate or cellulose nitrate used as a binder in silicon alloying. Apply to surface and fire for 1—10 min at 600—900°C in an inert atmosphere (N_2 or Ar). Sulfuric acid will etch the hydride. Other hydrides used in this study were TaH, ZrH, VH, and TiH.
REF: Sullivan, M V & Eigler, J H — *Acta Metall,* 103,218(1956)

NBH-0002
ETCH NAME: Hydrogen
TYPE: Gas, forming
COMPOSITION:
x .... H_2
TIME:
TEMP:
DISCUSSION:
a-Nb:H, as thin films formed by sputter evaporation of Nb on glass substrates in an H_2 atmosphere under pressure. Below 6 Pa the surface is a mirror finish; above, uneven. See NB-0001 and NBO-0002 for conversion to Nb_2O_5 for use as Josephson Junction devices.
REF: Kobnashni, T et al — *J Appl Phys,* 57,2583(1985)

PROPERTIES CF NIOBIUM NITRIDE, NbN

Classification	Nitride
Atomic numbers	41 & 7
Atomic weight	106.92
Melting point (°C)	2050
Boiling point (°C)	
Density (g/cm³)	8.4
Hardness (Mohs — scratch)	5—6

Crystal structure (isometric — normal)	(100) cube
Color (solid)	Black
Cleavage (cubic — poor)	(001)

NIOBIUM NITRIDE, NbN

General: Does not occur as a natural compound. There are over 20 niobate-tantalate minerals often containing yttrium, cerium, and uranium as well as calcium, iron, manganese, etc. Most of these minerals occur in pegmatite veins where they may be in rather large crystals. The mineral columbite-tantalite, varying in composition from $FeNb_2O_6$ to $FeTa_2O_6$, is a major ore of both tantalum and columbium (niobium).

Niobium metal added to stainless steel reduces corrosion at elevated temperatures and where surfaces are nitrided they show additional wear resistance. Where irons and steels containing niobium are nitrided, some portions of the nitrided layer will be niobium nitride.

Technical Application: Niobium nitride is under evaluation as a thin film deposit for surface coating with possible applications similar to those of other metal nitrides, such as silicon nitride, aluminum nitride as surface coatings.

It has been grown as an oriented (100) thin film for general morphology and defect studies. Niobium stannite, Nb_3Sn and niobium germanate, Nb_3Ge are fabricated as devices for their superconducting capabilities.

Etching: Soluble in mixtures of $HF:HNO_3$.

NIOBIUM NITRIDE ETCHANTS

NBN-0001

ETCH NAME: Neon	TIME:
TYPE: Ionized gas, cleaning	TEMP:
COMPOSITION:	GAS FLOW:
(1) x .... Ne^+ ions	PRESSURE:
(2) x .... Kr	POWER:

DISCUSSION:

NbN, (100), thin films deposited on NaCl, (100) substrates. Niobium was deposited at 90°C in an Ar/N_2 atmosphere. Ionic gases shown can be used to etch the nitride after photo resist patterning.

REF: van Dover, R B — *J Vac Sci Technol,* A(2) 1257(1984)

NBN-0002

ETCH NAME: Oxygen	TIME:
TYPE: Gas, conversion	TEMP:

COMPOSITION:

x .... O_2

DISCUSSION:

NbN, thin films deposited on fused silica and sapphire substrates by DC magnetron sputter in a $15\%N_2/Ar$ atmosphere. At 709 K films were amorphous to microcrystalline. At 400°C in air, the nitride converts to Nb_2O_5. NbN_{90} is cubic structure; NbN is hexagonal, and Nb_2O_5 is monoclinic.

REF: Gallagher, P K et al — *J Electrochem Soc,* 130,2045(1983)

PHYSICAL PROPERTIES OF NIOBIUM OXIDE, Nb_2O_5

Classification	Oxide
Atomic numbers	41 & 8

Atomic weight	265.82
Melting point (°C)	>400
Boiling point (°C)	
Density (g/cm³)	
Hardness (Mohs — scratch)	6—7
Crystal structure (monoclinic — normal)	(100) a-pinacoid
Color (solid)	Greyish
Cleavage (none)	Conchoidal fracture

NIOBIUM OXIDE, Nb_2O_5

General: Does not occur as a pure oxide compound in nature although it does occur as an iron niobate, $FeNb_2O_6$, as one end fraction of the mineral known as columbite-tantalite, $(Fe,Mn)(Nb,Ta)_2O_6$ which occurs is quantity in pegmatites and is mined as a source of tantalum, as well as niobium.

To date, there has been little use of the oxide in industry, other than as a chemical base compound for deriving other niobium salts.

Technical Application: The material is under evaluation in Solid State development for general characterization with possible applications as a dielectric capacitor or resistor similar to Ta_2O_5.

Etching: H_2SO_4, HCl. Mixed acids HF:HNO_3.

NIOBIUM OXIDE ETCHANTS

NBO-0001
ETCH NAME: Air
TYPE: Gas, forming
TIME: 10 min
TEMP: 80°C
COMPOSITION:

x .... air

DISCUSSION:

Nb_2O_5, as thin film conversion from Nb thin films grown on glass substrates under pressure in an H_2 atmosphere. Below 6 Pa pressure, surface was a mirror finish; above, was uneven. Nb films were anneal in hot air to convert to the oxide, then lead, Pb, was deposited to form Josephson Junctions.

REF: Kobnashni, T et al — *J Appl Phys*, 57,2483(1985)

NBO0002
ETCH NAME:
TYPE: Acid, removal
TIME:
TEMP: Hot boiling
COMPOSITION:

(1) x .... H_2SO_4, conc. (2) x .... HCl, conc.

DISCUSSION:

Nb_2O_5, as the natural mineral iron niobate, $FeNb_2O_6$. Both acids will etch the material easily. With addition of metallic zinc with boiling HCl the solution turns a fine blue color. Material used in a general study of niobates.

REF: Walker, P — mineralogy study, 1953

NIOBIUM SELENIDE ETCHANTS

NBSE-0001
ETCH NAME: TIME:
TYPE: Acid, clean TEMP: RT
COMPOSITION:
1 .... HF
1 .... HNO_3
10 .... CH_3COOH (HAc)
DISCUSSION:

Nb_3Se. (100) hexagonal platelets grown by Vapor Transport (VT) in an I_2 vapor atmosphere in sealed in quartz ampoules with 950°C at one end, and 1000°C at the other end. Growth time was 3 weeks. Ingots were washed in the solution shown to remove ''dust''. Solution does not etch external structure of ingot.

REF: Nakada, I & Ishihara, Y — *Jpn J Appl Phys,* 24,31(1983)

NIOBIUM STANNIDE ETCHANTS

NBSN-0001
ETCH NAME: Heat TIME:
TYPE: Heat, removal TEMP:
COMPOSITION:
x .... heat
DISCUSSION:

Nb_3Sn, amorphous thin films initially co-deposited by E-beam deposition on sapphire substrates. Anneal at 800°C to form Nb_3Sn. Remove excess tin from surface by heating and evaporation. Material used in forming $Pb/Nb_2O_5/Nb_3S$ as Josephson Junction devices.

REF: Kaster, R N — *J Appl Phys,* 23,2(1964)
NBSN-0002: Tusuge, H et al — *J Appl Phys Lett,* 43,606(1983)

Nb_3Sn single crystal wafers with lead, Pb evaporated to form Josephson Junction devices for use as logic and memory circuits.

NBSN-0003
ETCH NAME: TIME:
TYPE: Acid, polish TEMP: RT
COMPOSITION:
1 .... HF
1 .... H_2SO_4
1 .... H_2O
x drops .... H_2O_2
DISCUSSION:

Nb_3Sn, as single crystals formed from a high concentration alloy of CuSn in a study of superconductivity of the niobium compound. To prepare Nb_3SN surfaces: (1) mechanical polish with emery cloth, and (2) etch polish with solution shown.

REF: Murase, S et al — *J Appl Phys,* 49,602(1978)

NIOBIUM TELLURIDE ETCHANTS

NBTE-0001
ETCH NAME: TIME:
TYPE: Acid, cleaning TEMP: RT
COMPOSITION:
1 .... HF
10 .... H_2O
DISCUSSION:

Nb_3Te_4, grown as single crystal needles by iodine Vapor Transport (VT) in a sealed quartz ampoule at 1000°C. Growth was in the z-axis direction (0001). Specimens used in a study of magnetism and electronic properties. Silver epoxy paste used for electrical contacts. (*Note:* Solution shown can be used to clean material surface prior to epoxy application.)
REF: Ishikawa, K & Tanemuba, M — *Jpn J Appl Phys,* 23,842(1984)

NITRIDES:

General: Metal nitrides do not occur as natural minerals though there are several nitrate compounds containing the NH_4^- radical, and others containing nitrogen in their formulae.

The nitridization process of metal and alloy surfaces has been in use in the metals industries for many years — furnace firing of materials in a hot nitrogen atmosphere between about 800 to 1200°C, depending upon the metal or alloy. The nitrogen forms a thin film layer that increases surface hardness against wear and improves stability against chemical and atmospheric attack and corrosion. The nitrite and nitrate compounds, both natural and artificial, are primary fertilizers with many applications in both chemistry and medicine.

Technical Application: Nitrides as thin film surface coatings in Solid State began development in the early 1960s with the growth of silicon nitride, Si_3N_4, by the Rand Corporation, and silicon oxynitride initially by TRW semiconductors in the mid-1960s (See Silicon Nitride section). TRW also did initial development work on germanium nitride, Ge_3N_4, at the same period and showed it to be isomorphous with Si_3N_4, and also initial work with AlN.

Since that time, many investigators have been involved with development of a wide range of metal nitrides. They are listed in this book following their parent metal, as several have extensive etchants: silicon, titanium, aluminum, and others.

Nitrides are deposited as thin films surface coatings in all structural forms: colloidal, amorphous, crystalline and single crystal by CVD, MOCVD (both from gaseous sources), and RF (magnetron) sputter (solid target). They are used for surface passivation, as chemical etch, diffusion or metallization and epitaxy masks, mostly as amorphous structures like silicon dioxide, SiO_2, and similar oxides. Silicon oxynitride, $Si_3N_xO_y$ (5% O_2, common) is most often used as it is more inert and chemically resistant than the pure oxide, but easier to chemical process than the pure nitride.

Both oxides and nitrides are used as active elements in semiconductor devices, such as MOS and MNOS diodes.

There are over 25 metal nitrides listed with their etchants in this book. Some nitrides are deposited, then converted to the oxide, such as TiN to TiO_2. Others are being applied as layer elements in heterostructure devices, and the TiN with its yellow-gold color is under consideration as a replacement for gold in decorative work. Nitrides appear following their parent metal.

Etching: Varies from extremely difficult to etching with ease, depending upon the metal nitride involved. Both single acids and acid mixtures are used.

PHYSICAL PROPERTIES OF OSMIUM, Os

Classification	Transition metal
Atomic numbers	76
Atomic weight	190.2
Melting point (°C)	3045 (3050)
Boiling point (°C)	5027 (5000)
Density (g/cm^3)	22.6
Specific heat (cal/g) 0°C	0.031
Heat of fusion (k-cal/g-atom)	6.4
Latent heat of fusion (cal/g)	36.9
Heat of vaporization (k-cal/g-atom)	162
Atomic volume (W/D)	8.43
1st ionization energy (K-cal/g-mole)	201
1st ionization potential (eV)	8.7
Electronegativity (Pauling's)	2.2
Covalent radius (angstroms)	1.26
Ionic radius (angstroms)	0.88 (Os^{+4})
Vapor pressure (microns) m.p.	13.5
Coefficient of linear thermal expansions ($\times 10^{-6}$ cm/cm/°C)	6.1
Electrical resistivity (micro-ohm cm) 20°C	8.12
Young's modulus (psi $\times 10^{-6}$)	81
Magnetic susceptibility (cgs $\times 10^{-6}$)	0.074
Ionization potential (eV)	8.7
Thermal conductivity (cal-cm/sec/cm^2/°C)	0.206
Cross section (barns)	35.3
Electron work function (eV)	4.55
Hardness (Mohs — scratch)	4—5
(Vicker's kg/mm^2 — annealed)	400
Crystal structure (hexagonal — normal)	$(10\bar{1}0)$ prism, hcp
Color (solid)	Blue-black metal.
Cleavage (basal)	(0001)

OSMIUM, Os

General: Does not occur as a native element. Usually extracted from the metals of the platinum-iron group which, as native elements, are alloys containing iron, iridium, rhodium, palladium, osmium, and other metals. The mineral iridosmine (osmiridium) is a variable alloy of iridium and osmium with some subspecies containing up to 70% osmium. It can be distinguished from platinum by its greater hardness, lighter color, and different crystal habit, hexagonal system, rhombohedral division, rather than isometric system, normal class. The alloy with iridium is found as grains in association with platinum and gold in placer deposits. Pure osmium is one of the heaviest known forms of matter, extremely hard and quite chemically inert. When burned in air it oxidizes to OsO_4 as a pungent, irritating, and poisonous vapor.

As the pure metal it has industrial use as a light or heating filament. Alloyed with iridium it has many applications due to its hardness and wear resistance as parts and surface coatings. Osmium and its oxides are used in the glass and ceramics industries.

Technical Application: Osmium metal has had no use in Solid State device fabrication due to its extremely high temperature, weight, and chemical inertness. But it is under development and study as a silicide: OsSi, Os_2Si, and $OsSi_2$ for use as a buffer layer in semiconductor heterostructures. It has been grown as single crystal osmium sulfide, OsS_2, and as osmium telluride, $OsTe_2$, for general study as possible semiconductor compounds.

It has been grown as single crystal osmium for general studies and deposited as a thin film as an amorphous structure.

Etching: Aqua regia. Slow in HNO_3.

OSMIUM ETCHANTS

OS-0001a

ETCH NAME: Nitric acid — TIME:

TYPE: Acid, removal — TEMP:

COMPOSITION:

x .... HNO_3, fuming

DISCUSSION:

Os specimens. Etch cleaned in fuming nitric acid as single crystal material. (*Note:* There are three concentrations of fuming nitric acid: (1) standard 70%, sometimes called "white" fuming; (2) at 72% "yellow" fuming, and (3) 74% "red" fuming. When a fuming nitric acid is called out, it is often as red fuming.)

REF: Pamphlet: *Platinum,* Baker & Co, Inc., (no date)

OS-0001b

ETCH NAME: Aqua regia — TIME:

TYPE: Acid, removal — TEMP:

COMPOSITION:

1 .... HCl

3 .... HNO_3

DISCUSSION:

Os, as amorphous structured specimens. Use aqua regia to etch clean.

REF: Ibid.

PHYSICAL PROPERTIES OF OSMIUM SULFIDE, Os_2

Classification	Sulfide
Atomic numbers	76 & 16
Atomic weight	254.32
Melting point (°C)	>500 subl
Boiling point (°C)	
Density (g/cm³)	
Hardness (Mohs — scratch)	5—6
Crystal structure (isometric — normal)	(100) cube
Color (solid)	Black
Cleavage (cubic)	(001)

OSMIUM SULFIDE, OsS_2

General: Does not occur as a natural compound although there are many other metallic sulfide minerals. The only osmium mineral is iridosmine, Ir_xOs, and as a trace element in platinum. There are several artificial compounds in chemistry with the toxic oxide OsO_4 used in glass making.

Technical Application: There has been no use in Solid State processing of the sulfide other than being grown with other cubic sulfides, selenides, and tellurides for IR spectra study and general properties.

Etching: HNO_3, H_2O_2, etc.

OSMIUM SULFIDE ETCHANTS

OSS-0001
ETCH NAME: Nitric acid TIME:
TYPE: Acid, polish TEMP:
COMPOSITION:
x .... HNO_3, conc.
DISCUSSION:

OsS_2, single crystal specimens. Cut specimens and mechanically polish with 0.5 μm diamond paste. Etch in solution shown to polish and remove lap damage. Other crystals grown were FeS_2, $MnTe_2$, RuS_2, $RuSe_2$, $RuTe_2$, $OsTe_2$, PtP_2, $PtAs_2$, and $PtSb_2$. All crystals are cubic in structure like pyrite, FeS_2, and were used in an IR spectra study. (*Note:* Pyrite is not isometric (cubic) system, normal class, but pyritohedral class which does not contain a cube, (100) form, only a pseudocube with heavily twinned and striated faces.)
REF: Lutz, H D et al — *J Phys Chem Solids,* 46,437(1985)

OSMIUM TELLURIDE ETCHANTS

OSTE-0001
ETCH NAME: Nitric acid TIME:
TYPE: Acid, polish TEMP:
COMPOSITION:
x .... HNO_3
DISCUSSION:

$OsTe_2$, single crystal specimens. Cut specimens and mechanically polish with 0.5 μm diamond paste. Etch in solution shown to polish and remove lap damage. Other crystals grown were FeS_2, $MnTe_2$, RuS_2, $RuSe_2$, $RuTe_2$, OsS_2, $OsSe_2$, PtP_2, $PtAs_2$, and $PtSb_2$. All crystals are cubic in structure like pyrite, FeS_2, and were used in an IR spectra study. (*Note:* See note comment under OSS-0001 with regard to crystal structure.)
REF: Lutz, H D et al — *J Phys Chem Solids,* 46,437(1985)

PHYSICAL PROPERTIES OF OXYGEN, O_2

Classification	Gas
Atomic number	8
Atomic weight	15.9994 (16)
Melting point (°C)	−218.4
Boiling point (°C)	−183.0
Density (g/cm³ × 10^{-3} — gas) 0°C	1.429
(g/l — liquid) 0°C	1.420
(g/cm³ — solid) −183°C	1.140
(g/cm³ — solid) −2252.5°C	1.420
Thermal conductance (× 10^{-4} cal/cm²/cm/°C/sec) 20°C	0.572
Specific heat (cal/g/°C) 20°C	0.218
Latent heat of fusion (cal/g)	3.3
1st ionization potential (eV)	13.61
Ionic radius (angstroms)	1.32 (O^{+2})
Cross section (barns)	0.0002
Vapor pressure (°C)	−198.8
Critical temperature (C.T.) (°C — liquefies)	−118
Critical pressure (C.P.) (atms)	50

Chemical reactivity (22.2°C & 760 mmHg — gas)	Reactive
(LOX — liquid) −185°C	Reactive
Solubility (100 ml H_2O) 0°C	4.89 ml
(100 ml H_2O) 20°C	3.10 ml
Hardness (Mohs — scratch) solid	2—3
Crystal structure (hexagonal — normal) solid	$(10\bar{1}0)$ platelets
(isometric — normal) solid	(100) cube
Color (gas)	Colorless
(liquid/solid)	Pale blue
Cleavage (basal/cubic — solid)	(0001)/(001)

OXYGEN, O_2

General: A natural element in air which is approximately 24% oxygen and 75% nitrogen with additional trace gases, such as argon, neon, etc. It is obtained by the liquefaction of air; nitrogen (−195°C) boils off first, leaving oxygen (−185°C). It also can be obtained by the reduction of metal oxides, but the compression and fractionation of air is the primary commercial method of obtaining oxygen. It can also be obtained by the electrolysis of water, H_2O, and this method is used to obtain very high purity hydrogen.

Oxygen gas is about $^1/_5$ by volume and $^1/_4$ by weight in the atmosphere; as a constituent in water, H_2O, it constitutes about 85% by weight of ocean water and, when mineral oxides and compounds are included, oxygen is the most abundant element in the world and comprises 50% by weight of the earth's crust and atmosphere.

Many of the basic industrial metals are found as oxides: iron, zinc, copper, tin, aluminum, and manganese. Hematite, Fe_2O_3 — red iron ore — has been a primary ore since ancient times and, as a powder, it is red rouge and a pigment in paint. The red, brown, and tan coloring in rocks is due to the presence of iron oxides. Silica, SiO_2 and alumina, Al_2O_3 are the two most stable compounds; the natural compound called emery is a mixture of Al_2O_3:Fe_3O_4 (magnetite, black iron ore) and may include hematite and is used as a lapping and polishing abrasive. Lead monoxide, PbO, called litharge, as an example, is used in making glass and pottery glaze and also is a semiconductor compound.

The glass industry started with the Egyptian discovery and manufacture of soda-lime glass around 3500 B.C. and today the same formula is still in use. Most glass is made using high purity sand as the starting material; natural silica, silicates, and alumina in clay and ceramic . . . all materials containing oxygen.

Both gaseous oxygen and liquid oxygen (LOX) have wide application in industry and medicine. The gas, of various purity levels or mixed with other gases — O_2/N_2, O_2/Ar, etc., is supplied in high pressure cylinders. The liquid, like all liquidized gases, is transported in 2000 gallon LOX truck tanker trailers and railroad box cars. On-site storage tanks, as well as transport cylinders are of special, double-walled construction . . . the inner spacing is filled with vermiculite, a mica that will absorb moisture and will not burn, and the area is sealed under a light vacuum (10^{-3} Torr) to prevent evaporation. Both liquid and gaseous oxygen can be supplied from such tanks. It should be noted that LOX is a primary fuel in space vehicles and in their booster tanks and in many high speed jet aircraft, and requires special handling as it is instantly explosive in contact with grease and oil.

As ozone, O_3, it occurs in nature during thunderstorms when lightning electrolyzes rain drops and it can be recognized by its very pungent odor. It is used as both a water and air cleaner as it is a very high powered oxidizer. It can be produced by subjecting air to ultraviolet (UV) light and several ozone systems are available commercially which have been used for cleaning metal and semiconductor surfaces. Unfortunately, a concentration of 1 ppm in air can be injurious to health. Liquid ozone is blue and high concentrations are explosive. Note the on-going study of variations of the ozone layer above the Antarctic. It is cyclical, but controls the amount of dangerous, high-energy particles entering the atmosphere from the sun.

Atmospheric oxygen is responsible for the formation of the natural oxide and silicate minerals which, as a group, form better than $^3/_4$ of the minerals and rocks of the earth's crust.

Needless to say, oxygen is essential for all life forms, with the exception of a few anaerobic bacteria, and all vegetation use the released carbon dioxide from mammals in photosynthesis, returning oxygen to the atmosphere. Note that the Amazon and Congo River basins in South America and Africa, respectively, produce the greatest amount of oxygen being returned to the atmosphere.

Technical Application: The primary use of oxygen in Solid State processing is for the growth of silicon dioxide, SiO_2, as a thin film coating. As silica, another name for silicon dioxide, it is highly chemically resistant (except for fluorine and fluorine compounds). It is an excellent final surface protectant on metals and semiconductor devices. It also is used as an etching, diffusion and selective epitaxy growth mask where the oxide surface is patterned by photolithographic techniques. It should be noted that the term "oxide" has been in general use for a number of years with reference to silicon dioxide use in the semiconductor industry, and is still so used, even though there are a number of other metal oxides now being grown and applied as thin films on metal surfaces, such as aluminum oxide, Al_2O_3 and beryllium oxide, BeO.

Other than as thin film metal oxide coatings, oxygen as an element is undesirable in semiconductor materials. Most single crystal ingots are grown in quartz tubes — Czochralski (CZ), Float Zone (FZ), Horizontal Bridgman (HB), and Vapor Transport (VT) — and oxygen comes off of the hot quartz (SiO_2) walls and diffuses into the growing material. This has been reduced to some extent by jacketing the growth tubes for water cooling and by temperature control during growth. As most of the oxygen is concentrated in the outer "skin" of an ingot, or, as in the Bridgman and Float Zone methods, swept to the ingot ends by multiple passes of a hot coil, these oxygenated zones can be cut away. Most ingots are grown oversized and lathe cut to specific diameter, which, at the same time, removes the undesirable zones. Oxygen can initiate dislocations, form oxygen defect clusters and create electrical noise in an operating semiconductor device.

Oxygen, when used as an ionized RF plasma gas (O^+ ions), is used for material surface cleaning and, with photo resist patterned material, as part of dry chemical etching (DCE) cleaning, etching, and selective structuring. One widely used mixture on many metals is CF_4:O_2(5%).

Etching: N/A as gas. Pressure as solid.

OXYGEN ETCHANTS

O-0001

ETCH NAME: TIME:

TYPE: Pressure, defect TEMP: −190°C

COMPOSITION:

x .... pressure, vapor

DISCUSSION:

O_2, grown as a single crystal under pressure in vacuum and cryogenic conditions (LHe/LN_2). By varying the vapor pressure in the system, structural defects can be observed.

REF: Schwentner, N et al — *Rare Gas Solids,* Vol 3, Academic Press, New York, 1960

PHYSICAL PROPERTIES OF PALLADIUM, Pd

Classification	Transition metal
Atomic number	46
Atomic weight	106.4
Melting point (°C)	1552 (1549.4)
Boiling point (°C)	2900 (2927)
Density (g/cm^3)	12.02 (11.3—11.8)
Thermal conductance (cal/sec)(cm^2)(°C/cm)	0.18
Specific heat (cal/g) 0°C	0.058
Heat of fusion (cal/g-atom)	4.0
Latent heat of fusion (cal/g)	38.6
Heat of vaporization (k-cal/g-atom)	90
Atomic volume (W/D)	8.9
1st ionization energy (K-cal/g-mole)	192
1st ionization potential (eV)	8.33
Electronegativity (Pauling's)	2.2
Covalent radius (angstroms)	1.28
Ionic radius (angstroms)	0.80 (Pt^{+2})
Vapor pressure (microns) m.p.	26
Coefficient of linear thermal expansion ($\times 10^{-6}$/cm/cm/°C)	11.6
Electrical resistivity (micro-ohms-cm) 20°C	9.93
Electron work function (eV)	4.82 (4.99)
Young's modulus (psi $\times 10^6$)	17
Magnetic susceptibility (cgs $\times 10^{-6}$)	5.8
Tensile strength (psi — annealed)	33,000
Poisson's ratio	0.39
Cross section (barns)	8.0
Hardness (Mohs — scratch)	4.5—5
(Vicker's — kgf/mm^2) annealed	37—39
Crystal structure (isometric — normal)	(100) cube, fcc
Color (solid)	Grey-silver metal
Cleavage (cubic — poor)	(001)

PALLADIUM, Pd

General: Does not occur as a pure native element and it is mostly found as small grains as an alloy containing traces of platinum and iridium. Platinum contains up to 2% palladium and is a major source of the metal. Both metals are found in minable quantities in the Ural Mountains of Russia and in Brazil, and are often associated with placer gold deposits. Potarite, from the diamond sands in British Guiana, is a palladium-mercury amalgam. Allopalladium is a hexagonal modification found in the Hartz Mountains. Also from Brazil is the mineral porpezite, a gold alloy containing up to 10% palladium, another major source of the element. It also is extracted from ores of nickel.

The pure metal is more basic than the other elements in the platinum family — Ru, Rh, Pd, Os, Ir — along with platinum and gold, known as the precious metals . . . all highly resistant to oxidation. Thin film palladium is widely used as a mirror plating on instruments as it has excellent silver/white reflectivity and does not tarnish like silver.

Powdered palladium, called palladium black, has the ability to absorb large volumes of hydrogen, even as high as 900 volumes under special conditions, and is under development as a hydrogen storage battery. Platinum black also absorbs hydrogen but to a much lesser extent.

As an alloy, amalgam mixtures of gold-palladium are extensively used in dentistry as a replacement for gold-platinum. There are several other alloy mixtures of palladium as low temperature solders and high temperature brazing alloys and, as solders, are finding increasing application in the electrical and electronic industries as replacements for the more expensive platinum and gold alloys where it is used for its electrical conductivity.

Like other metals of the platinum family it is used as a catalyst, as in the hydrogenation of carbon compounds, and as a reducing agent in association with hydrogen.

One gold alloy, known as palau (20% Pd) and another, rhotanium (10—40% Pd) have both industrial applications and use in the jewelry trade.

Technical Application: Palladium is finding increasing use as an evaporated thin film in metallization layer structure, e.g., Pd:Cr, Pd:Ti, Au:Pd:Ni. Such structures are used as electrical contact pads and circuits on both discrete Solid State devices, such as silicon and gallium arsenide diodes and transistors, as well as hybrid, planar, or heterostructures. It also is used on circuit substrates of alumina, beryllia, quartz, and sapphire in a similar manner. And palladium alloys, such as Pd:Cd, are being used for assembly and contact purposes as replacement for the higher cost gold and platinum type alloys. It also is under development and application as a silicide: PdSi, Pd_2Si, and $PdSi_2$ for use as a buffer layer in semiconductor heterostructures.

It has not only been studied as a thin film, but grown as a single crystal for general physical data and morphological study.

Etching: HNO_3, slow in H_2SO_4, HCl, and aqua regia.

PALLADIUM ETCHANTS

PD-0001
ETCH NAME: Aqua regia — TIME:
TYPE: Acid, cleaning — TEMP:
COMPOSITION:
3 .... HCl
1 .... HNO_3
DISCUSSION:

Pd, thin film deposits by E-beam in conjunction with chromium, nickel, silver, or other metals and used for semiconductor device contact pads. Aqua regia is a general etch for palladium.

REF: Bulletin 1092 AMD/ID/JBA/SK/0482/GUILD 1982, 72 — MRC
PD-0002: Walker, P et al — personal application, 1980

Pd, thin films as a sequentially evaporated Pd/Cr layer from standard tungsten boats or filaments as part of the gate contact in fabricating LN-FET gallium arsenide devices. Used solution as shown, and also diluted with 10 parts H_2O, RT, 1—3 sec for surface cleaning.

PD-0002b
ETCH NAME: Aqua regia, dilute — TIME: 1—3 sec
TYPE: Acid, cleaning — TEMP: RT
COMPOSITION:
3 .... HCl
1 .... HNO_3
10 .... H_2O
DISCUSSION:

Pd, thin films evaporated as Pd/Cr pads in GAS LN-FET device fabrication. Solution used as a surface cleaner.

REF: Ibid.

PD-0003
ETCH NAME:
TYPE: Electrolytic, polish
COMPOSITION:
x .... H_2SO_4
x .... glycerin
TIME:
TEMP:
ANODE:
CATHODE:
POWER:
DISCUSSION:

Pd, 99.95% pure single crystal specimens cut as cylinders with axis in a ⟨110⟩ direction. Used to measure elastic constants from 4.2 to 300 K. Cylinders $^3/_8$″ diameter and length. Use solution shown to electropolish specimens. No surface mosaic structure observed after etching.

REF: Rayne, J A — *Phys Rev,* 118,1545(1960)

PD-0004
ETCH NAME: Nitric acid
TYPE: Acid, removal
COMPOSITION:
x .... HNO_3, conc.
TIME:
TEMP: RT or hot
DISCUSSION:

Pd, single crystals and thin films. Solution used as a general etch for dissolving palladium in chemical analysis of native minerals.

REF: Dana, E S & Ford, W E — *A Textbook of Mineralogy,* 4th ed, John Wiley & Sons, New York, 1950, 407

PD-0005: Foster, W & Alyea, H N — *An Introduction to General Chemistry,* 3rd ed, D Van Nostrand, New York, 1947, 498

Pd specimens. Solution used to dissolve palladium to form the nitrate, $Pd(NO_3)_2$.

PD-0002c: Ibid.

Pd, thin films deposited on GaAs (100) wafers and alumina substrates in device and circuit fabrication. Solution used to strip palladium. $1HNO_3$:$10H_2O$ solution used to reduce reactivity. Concentrated acid used to observe possible structure in device failure analysis.

PD-0006: Brochure: *Platinum,* Baker & Co. (no date)

Pd, as plating and fabricated parts. A discussion of physical characteristics and uses of the platinum family of metals.

PALLADIUM ALLOYS, PdM_x

General: Pure metallic palladium does not occur in nature, as it has traces of platinum and iridium primarily, as well as gold, osmium, and rhodium, such that it could be considered a natural alloy. The mineral potarite, PdHg, occurs as one of the few natural amalgams (see section on Amalgam).

Palladium has many industrial uses, and certain alloys are becoming increasingly important in the electronic and metal industries as high temperature, corrosion resistance plating and parts.

Technical Application: Only two alloys are shown here as they are under study or being used in Solid State processing for electrical contact materials as a constituted alloy. The use of multilayered thin films, such as Pd/Cr, was previously mentioned under palladium metal. Like platinum, and other platinum family metals, aqua regia is the primary etching solution.

Etching: Aqua regia.

PALLADIUM GOLD ETCHANTS

PDAU-0001

ETCH NAME: Aqua regia — TIME:

TYPE: Acid, removal — TEMP: RT

COMPOSITION:

3 .... HCl

1 .... HNO_3

DISCUSSION:

PdAu deposited as a 1:1 mixture on glass, quartz, and sapphire substrates. The thin film shows crinkling on a thick, hard surface indicative of compressive stress. It is not as severe on thin glass or metal strips, as they can bow and relieve stress. Both metals are cubic, fcc. (*Note:* Isometric system, normal class . . . the cube, (100) is a crystal form within this class, not a true crystallographic system or class, although the isometric system is commonly referred to as the cubic system.)

REF: Asai, H & Oe, K — *J Electrochem Soc,* 128,2052(1981)

PALLADIUM HYDRIDE ETCHANTS

PDH-0001

ETCH NAME: Aqua regia — TIME:

TYPE: Acid, cleaning — TEMP: RT to hot

COMPOSITION:

3 .... HCl

1 .... HNO_3

x .... H_2O

DISCUSSION:

Pd:H as powdered "palladium black" with absorbed hydrogen. Powdered palladium can be cleaned with aqua regia with or without water dilution, then furnace dried under vacuum prior to introduction of hydrogen.

REF: Wise, E M — *The Platinum Metals* (brochure), The International Nickel Co., 1954

PALLADIUM SILVER ETCHANTS

PDAG-0001

ETCH NAME: Aqua regia — TIME:

TYPE: Acid, removal — TEMP: RT or warm

COMPOSITION:

3 .... HCl

1 .... HNO_3

DISCUSSION:

PdAg alloy used in a pure metallic paste form as an electrical pad contact for microelectronic circuits. It can be screen printed on alumina substrates and fired to about 1000°C. Also as an epoxy type paste with oven curing between 100—125°C. PdAg is being used as a replacement for gold pastes to reduce cost.

REF: Chapman, D — personal communication, 1983

PHYSICAL PROPERTIES OF PHOSPHORUS, P

Classification	Non-metal
Atomic number	15
Atomic weight	30.97 (P)
	124.08 (P_4)
Melting point (°C — white (yellow))	44.1
(°C — red) @ 43 atms	590
(°C — violet)	593
(°C — black) 12,000 kg/cm^2	200
Boiling point (°C — white (yellow)/red)	280
Density (g/cm^3 — white (yellow))	1.82
(g/cm^3 — red)	2.20 (2.34)
(g/cm^3 — violet)	2.36
(g/cm^3 — black)	2.691
Specific heat (cal/g) 25°C	0.177
Heat of fusion (k-cal/g-atom)	0.15
Latent heat of fusion (cal/g)	5.0
Heat of vaporization (k-cal/g-atom)	2.97
Atomic volume (W/D)	17.0
1st ionization energy (K-cal/g-mole)	2254
1st ionization potential (eV)	11.0
Electronegativity (Pauling's)	2.1
Covalent radius (angstroms)	1.06
Ionic radius (angstroms)	0.35 (P^{+5})
Electrical conductivity (micro-ohms^{-1})	10^{-17}
Electrical resistivity (ohms cm) 20°C	10^{11}
Coefficient of linear thermal expansion ($\times 10^{-6}$ cm^2/cm/°C) 20°C	125
Cross section (barns)	0.19
Vapor pressure (°C)	197.3
Refractive index (n=) yellow/white	2.144
Hardness (Mohs — scratch)	0.5
Crystal structure (isometric — normal) white	(111) octahedron
(isometric — normal) red	(100) cube
(amorphous) red	none
(monoclinic — normal) violet	(100) a-pinacoid
Color (solid)	Variable by type
Cleavage (cubic/basal — poor)	(001)

PHOSPHORUS-CONTAINING ETCHANTS

General: Phosphorus is widely disseminated in the combined state, making up about 0.12% of the earth's crust. The element is of importance to both plant and animal life. The human skeleton contains 1400 g, the muscles 130 g, and the nerves and brain about 12 g, combined. It is essential to vegetable matter, such that phosphorus compounds are important fertilizers.

White (yellow) phosphorus melts in hot water, will burn under water due to presence of air (aerated) in water, and vigorously if oxygen is bubbled through the water. It also reacts with water moisture in the air, such that it is stored under kerosene. This white form is extremely poisonous, and ingestion of 0.15 g can be fatal. White phosphorus is readily converted to red phosphorus by being heated in its own vapors to 235°C, such that it is so converted for most general processing.

Red phosphorus is the form in which most phosphorus is used as it will not catch fire below 240°C, and can be handled as it is not toxic. The violet and black varieties are fabricated under special conditions, and are rarely seen or used under normal conditions.

There are nine forms of liquid types of phosphoric acid, but the most commonly used in chemical processing of metals and metallic compounds is called phosphoric acid, H_3PO_4. The following A—Z material list of solutions are this form, with the exception of one application of metaphosphoric acid, HPO_3.

PHOSPHORUS, P

General: Does not occur as a native element, but is widely disseminated in a number of minerals. The chief sources of phosphorus are from the mineral phosphorite, $Ca_3(PO_4)_2$ which is a subspecies of the mineral apatite, as fluorapatite, $(CaF)Ca_4(PO_4)_3$ or chlorapatite, $(CaCl)Ca_4(PO_4)_3$. There are some 15 subspecies of apatite with various names which may contain other elements, such as strontium and manganese. Although apatite has a relatively low Mohs hardness: H = 5, some of the colored varieties are used as gem stones: asparagus-stone, yellowish-green; manganapatite, dark bluish-green; and lasurapatite, sky-blue.

A primary use of red phosphorus has been in the fabrication of matches, invented by John Walker in 1827. The original "match" was a mixture of $KClO_3$:P (white), which is poisonous; P_4S_2 was used as a replacement as it is nontoxic. But $KClO_3$ is explosive in the presence of organic matter, and when mixed with phosphorus and struck a sharp blow — will explode.

The "safety match", developed by the Swedish inventor Bottger in 1848, replaced phosphorus with SbS_3 + oxidizer ($KClO_3$, $K_2Cr_2O_7$ etc.), mixed with powdered glass and glue. These matches are ignited by rubbing on the surface mixture of SbS_3:P (red) and glue. Note that although phosphorus is not used in the match head, it is part of the striking surface.

Phosgene, PH_3, is a colorless, poisonous gas with a distinctive rotten fish odor, and was used in gas warfare during World War I. Today, in semiconductor processing, [25]phosphine, PH_3, is a primary n-type dopant for silicon wafers, thin film epitaxy layers, as well as oxides and nitrides. As a doped silicon dioxide is has the acronym PSG.

Phosphorus oxychloride, $POCl_3$ — called "pockel" — as a liquid through which nitrogen gas is bubbled, also is used as an n-type dopant in semiconductor processing.

Phosphorus oxides, as solid materials, also are used as n-type dopants in a similar manner, usually in the form of the pentoxide, P_2O_5 as the trioxide P_2O_3, is poisonous. Regardless, processing fumes from diffusion tubes should be properly ventilated during operation as both oxide types will be present, such that operation is hazardous.

P_2O_5 is the anhydride of the three primary phosphoric acids: orthophosphoric, H_3PO_4; pyrophosphoric, $H_4P_2O_7$ and metaphosphoric, HPO_3. Orthophosphoric acid, commonly called phosphoric acid, is the most commonly used of the three and is the major acid for etching of metals and metallic compounds.

Ferrophosphorus is a slag by-product of the production of phosphorus from calcium phosphates (apatite) and is cast into pigs for use in the iron and steel industry and phosphor-bronze is a mixture of Sn:Cu:P, of use for its spring qualities.

Technical Application: Phosphorus is the primary n-type dopant in silicon semiconductor fabrication, although both arsenic and antimony also can be used. As already mentioned, it can be used in the form of PH_3 (gas), $POCl_3$ (liquid), or P_2O_5 (solid).

Red phosphorus is used to grow several compound semiconductors, such as gallium phosphide, GaP and indium phosphide, InP. The latter is used as InP wafer substrates in MBE layered growth of heterostructures. All of the colored forms of phosphorus have been grown as single crystals for study.

All of the colored forms of phosphorus have been grown as single crystals for study.

Etching: (1) White: CS_2, NH_3, KOH/NaOH, and water.
(2) Red: EOH, absolute.
(3) Violet: insoluble.
(4) Black: insoluble.

Etchants Containing Phosphorus

Formula	Material	Use	Ref.
$2H_3PO_4$:$1H_2O_2$:$3H_2O$	Al	Removal	AL-0007
$70H_3PO_4$:25 ml H_2SO_4:5 ml HNO_3	Al	Polish	AL-0035
$95H_3PO_4$:$5HNO_3$	Al	Cleaning	AL-0036
$80H_3PO_4$:$5HNO_3$:15HAc	Al:Cu	Macro-polish	ALCU-0002
25 ml H_3PO_4:70 ml H_2SO_4:5 ml HNO_3	Al:Si	Polish	ALSI-0002
$95H_3PO_4$:$5HNO_3$	AlSb	Cleaning/oxide removal	ALSB-0010
xH_3PO_4, conc.	AlN	Removal	ALN-0006
xH_3PO_4, conc.	Al_2O_3	Cleaning	ALO-0002a
xH_3PO_4, conc.	Ba_2TiO_3	Polish	BAT-0002
xH_3PO_4, conc.	Be	Preferential	BE-0003
xH_3PO_4, conc.	BeO	Preferential	BEO-0001a
xH_3PO_4:xH_2SO_4	BeO	Removal	BEO-0006b
xH_3PO_4, conc.	BN	Removal	BN-0002
xH_3PO_4:xH_2O	Brass	Preferential	BRA-0013b
12—15% HH_3PO_4	Brass	Cleaning	BRA-0008d
35% H_3PO_4	Brass	Electrolytic, polish	BRA-0009
xH_3PO_4, conc.	$CaWO_4$	Removal	CAW-0001a
$3H_3PO_4$:$1CrO_3$, sat. sol.	$CaWO_4$	Polish	CAW-0003b
xH_3PO_4, conc.	C	Polish	C-0003
xH_3PO_4:CrO_3:xNaCN	C	Removal	C-0006b
$64H_3PO_4$:$15H_2SO_4$:$21H_2O$	Cr	Electrolytic, polish	CR-0001c
xH_3PO_4, conc.	Co	Electrolytic, polish	CO-0005
xH_3PO_4, conc.	Cu	Electrolytic, polish	CU-0005
$1H_3PO_4$:$1HNO_3$:1HAc	Cu	Polish	CU-0045
xH_3PO_4, conc.	Cu:Ge	Electrolytic, polish	CUGE-0004
xH_3PO_4, conc.	CuO	Polish	CUO-0008
33 ml H_3PO_4:33 ml HNO_3:33 ml HAc	CuO	Removal	CUO-0009
$40H_3PO_4$:$1H_2O_2$:$40H_2O$	GaAs	Removal	GAS-0105
$100H_3PO_4$:$100H_2SO_4$:$1H_2O_2$	GaAs	Dislocation	GAS-0107
$1H_3PO_4$:$1H_2O_2$:$20H_2O$	GaAs	Selective	GAS-0108a
$1H_3PO_4$:$1H_2O_2$:$75H_2O$	GaAs	Thinning	GAS-0108b
$75H_3PO_4$:$100H_2O_2$:$25H_2O$	GaAs	Via hole	GAS-0109
$49H_3PO_4$:$11HNO_3$	GaAs	Polish	GAS-0110b
$5H_3PO_4$:$5H_2SO_4$:$2H_2O_2$	GaAs	Preferential	GAS-0110c
xH_3PO_4, conc.	GaN	Preferential	GAN-0004
xH_3PO_4, conc.	Garnet	Preferential (GGG)	GGG-0001a
xH_3PO_4, conc.	Garnet	Preferential (YAG)	YAG-0006
xH_3PO_4, conc.	Garnet	Dislocation (YIG)	YIG-0001b
xH_3PO_4, conc.	InP	Electrolytic, selective	INP-0033a
10% H_3PO_4	InP	Oxide removal	INP-0022c

Formula	Material	Use	Ref.
xH_3PO_4:xHCl	InP	Removal	INP-0034
$1H_3PO_4$:3HCl	InP	Grooving	INP-0035
xH_3PO_4:xHBr	InP	Preferential	INP-0021c
$1H_3PO_4:1H_2O_2$:3MeOH	InP	Cleaning	INP-0047
$1H_3PO_4:1H_2O_2:76H_2O$	InP	Selective	INP-0046b
x14 *N* H_3PO_4	InP	Electrolytic, oxidizing	INP-0051
$1H_3PO_4$:1HCl	InP	Selective	INP-0057a
$2H_3PO_4:1H_2O_2$	Fe	Preferential	FE-0110
xH_3PO_4, conc.	MgO	Thinning	MGO-0001
$80H_3PO_4:20H_2O$	MgO	Polish	MGO-0004
117 ml H_3PO_4:55 ml HNO_3:57 ml HAc:115 ml HF	Mo	Removal	MO-0012b
$xHPO_3$, conc.	Ni	Removal	NI-0004d
$xH_3PO_4:xCrO_3$	Permalloy	Electrolytic, polish	PMA-0001b
xH_3PO_4, conc.	NiMn	Electrolytic, dislocation	NIMN-0001
$1H_3PO_4:1H_2SO_4$	Al_2O_3	Cleaning (sapphire)	SAP-0001
xH_3PO_4, conc.	Al_2O_3	Preferential	SAP-0002a
$xH_3PO_4:xHNO_3$	Si	Selective	SI-0131
xH_3PO_4, conc.	SiC	Removal	SIC-0005c
xH_3PO_4:xNaOH:$xNaCO_3$	SiO_2	Cleaning	SIO-0039c
xH_3PO_4, conc.	Si_3N_4	Removal	SIN-0001
$2H_3PO_4:1H_2O_2$	Steel	Polish (Fe:Ni)	FE-0012b
$63H_3PO_4:15H_2SO_4:10CrO_3:12H_2O$	Steel	Electrolytic, polish	Fe-0011
x1 *N* H_3PO_4:9 *N* H_2SO_4	Th	Electrolytic, polish	TH-0003
100 ml H_3PO_4:1 g NaOH:2 g agar-agar	SnTe	Electrolytic, polish	SNTE-0001
30% H_3PO_4	Ti	Removal	TI-0001d
xH_3PO_4	TiO_2	Removal	TIO-0011
$xH_3PO_4:xHClO_4$	W	Polish	W-0018b
$1H_3PO_4:1H_2SO_4:1H_2O$	U	Electrolytic, polish	U-0002c
310 ml H_3PO_4:67 ml H_2SO_4:120 ml H_2O:78 g CrO_3	UO_2	Polish	UO-0001
$4H_3PO_4$:1MeOH	UC	Electrolytic, jet-thinning	UC-0001
$4H_3PO_4$:1MeOH	UN	Electrolytic, jet-thinning	UN-0001
$4H_3PO_4$:1MeOH	US	Electrolytic, jet-thinning	US-0001
33% H_3PO_4:67% C_2H_5OH	Zn	Electrolytic, polish	ZN-0007d
xH_3PO_4, conc.	ZnO	Polish	ZNO-0003
xH_3PO_4, conc.	ZnW	Polish	ZNW-0001a

PHOSPHORUS ETCHANTS

P-0001a

ETCH NAME: Carbon disulfide — TIME:

TYPE: Acid, removal — TEMP: RT

COMPOSITION:

x .... CS_2

DISCUSSION:

P, as white phosphorus is stored under toluene or kerosene to prevent combustion in air, but can be dissolved in the solution shown.

REF: Foster, W & Alyea, H N — *An Introduction to General Chemistry*, 3rd ed, D Van Nostrand, New York, 1948, 574

P-0001b
ETCH NAME: Water
TYPE: Acid, removal
COMPOSITION:
x .... H_2O
TIME:
TEMP: Hot
DISCUSSION:

P, as white phosphorus specimens can be dissolved in hot water and will burn with free oxygen present in the water. It will burn under water.

REF: Ibid.

P-0001c
ETCH NAME: Turpentine
TYPE: Solvent, removal
COMPOSITION:
x .... $C_{10}H_{16}$
TIME:
TEMP: RT
DISCUSSION:

P specimens as white phosphorus can be dissolved in several organic solvents without burning, such that it is stored in kerosene.

REF: Ibid.

P-0001d
ETCH NAME:
TYPE: Spark, ignition
COMPOSITION:
x .... emery paper
TIME:
TEMP: 240°C
DISCUSSION:

P, as red phosphorus is not air reactive or water reactive as is the white form. In powder form it is used in the fabrication of matches. When the match head is drawn across a rough surface, such as an emery board containing phosphorus, the phosphorus will ignite and burn. Black phosphorus is similar in properties.

REF: Ibid.

PHOTO RESIST ETCHANTS

AZ-0001a
ETCH NAME: Acetone
TYPE: Ketone, removal
COMPOSITION:
x .... $(CH_3)_2CO$
TIME:
TEMP: RT
DISCUSSION:

AZ-type photo resist lacquers used in semiconductor processing. Standard photo resist spinners consist of a removable metal or plastic bowl with bottom hole up through which extends a $^1/_4$" motor driven shaft on which is mounted a wafer holder for vacuum pump hold down of wafers, and an electronic operation control.

Thickness of coating depends upon the type (by number) of resist and spin speed. AZ-1350J is a standard positive photo resist that is spun on at 3600 rpm, RT, 10—15 sec for a 0.5 μm thickness. 1370J will give about 1 μm thickness. The positive photo resists open a required pattern; whereas negative photo resists cover the pattern.

Cure by first air drying at RT, 1 h; then air-oven bake at 90°C for 1 h (120°C for hard bake). Follow by UV light exposure with the pattern mask over the wafer and, finally, develop that pattern with the photo resist developer solution designed for the type photo resist being used.

The wafer can then be metal evaporated or the surface etched for pits, channels, via holes, etc., with acetone lift-off after metallization. If the wafer or circuit substrate is already metallized, the pattern is etched, then remaining photo resist remove for acetone spraying.

Acetone is the solvent for cleaning and removal of most photo resists. Apply as a spray from a stainless steel spray can under nitrogen pressure; soak by immersion in a beaker of acetone; lightly scrub surfaces with a Q-tip soaked in acetone. Removal of extraneous metallization is called the lift-off technique.

There are other photo resist formulations, such as COP or PMMA, but all are processed in the same general manner. They are occasionally used as a paint-on coating as a general holding medium.

CAUTION: Acetone has a boiling point of 56.7°C and a low flash point.

REF: Tech. Bull. #401, 1984 — American Hoechst Corp.

AZ-0005: Mizuno, K — *J Phys Rev,* 4,1434(1984)

P/R used to pattern glass substrates for study of aluminum evaporation of different thicknesses. Acetone used to lift-off excess metal after evaporation.

AZ-0006: Walker, P & Valardi, N — personal application, 1985

AZ-1370J used to pattern AuGe (13%) and AuSn (20%) 0.0001 thick alloy strips. Used photo resist to hold alloys on glass microscope slides and to coat and pattern. Acetone used to dissolve and remove photo resist after etch patterning.

AZ-0001b

ETCH NAME: J-100 — TIME:

TYPE: Solvent, removal — TEMP: Hot

COMPOSITION:

x J-100

DISCUSSION:

AZ-type photo resists. J-100 is a phenol base solution containing some alkali and is a general remover and solvent for photo resist lacquers. Particularly useful for hard-baked resists. Caution should be observed when using the solution with materials that can be attacked by hot alkalies as well as normal precautions in handling acids. Use as a hot soak for general photo resist removal and patterning.

REF: Ibid.

AZ-0003

ETCH NAME: Carbon tetrafluoride — TIME:

TYPE: Ionized gas, removal — TEMP: 40°C (holder base) or 100°C

COMPOSITION:

x .... CF_4:

x .... O_2 (5%)

GAS FLOW:

PRESSURE: 0.3 Torr

POWER: 200 W (13.56 MHz)

DISCUSSION:

AZ-1350J and other photo and electron beam resists: PGMA, PMMA, PMIPK (E-beam); Novolak type: AZ, OFPR, OMR (photo). Article shows angstrom removal rates and discusses development of submicron lithography.

REF: Tsuda, M et al — *J Vac Sci Technol,* 19(2),259(1981)

AZ-0007
ETCH NAME: Cotton pad
TYPE: Rubbing, defect removal
COMPOSITION:
x .... cotton pad

TIME:
TEMP: RT

DISCUSSION:

AZ- and PMMA photo resists spun on soda lime glass substrates used in a study of photo resist defects in both positive and negative resists subject to Si^+ ion bombardment at 200 keV, and up to 5×10^{15} cm^{-1} dose level. Inspection was at 400× with transmitted light. Defects were in the 1- to 3-μm size range, some due to ion bombardment and others intrinsic to the spin-on technique. Defects can be eliminated by laser erasure or by light rubbing with a soft pad.

REF: MacIver, B A & Puzio, L C — *J Electrochem Soc,* 129,2384(1982)

AZ-0008
ETCH NAME: Oxygen
TYPE: Gas, removal
COMPOSITION:
x .... O_2

TIME: 2—5 min or 10—15 min
TEMP: RT start
GAS FLOW: 300 cc/min
PRESSURE: 1.5 Torr
POWER: 200 W (13.56 MHz)

DISCUSSION:

AZ-1350J and other AZ- photo resists for 2—5 min; PMMA for 10—15 min. Used to strip photo resists ranging in thickness from 0.5—5 mils (pressed sheet method of application). Also used as a general contamination removal system in processing semiconductor wafers, ceramic and dielectric substrates, and individual semiconductor devices prior to metallization, epitaxy, or device assembly for testing or final packaging. *Note:* Residual photo resist on a metallized wafer or individual device can be recognized under magnification as fine reddish/brownish lines along the edges of the metallized patterns. If present, usually associated with lifting and peeling of the metallization. Where a semiconductor device has been in storage for any period of time — the plastic $2 \times 2 \times {}^1/_4$ box/trays (Fluorware type) — even in a nitrogen storage area, trays can exude oils from their method of fabrication (form release agents), and coat surface with organic contamination that prevents wire bonding to metal pads of the device: 5 to 10 min in an RF oxygen plasma under the conditions shown above will clean such surfaces. System used was an IPC 2000 unit.

REF: Walker, P — personal development and application, 1983—1985

AZ-0004a
ETCH NAME:
TYPE: Acid, removal
COMPOSITION:
100 ml .... H_2SO_4
20 ml .... H_2O_2

TIME:
TEMP: Hot

DISCUSSION:

AZ-type photo resists, other photo resists and general contamination removal and cleaning of surfaces. Should not be used on gallium arsenide wafers as hydrogen peroxide is an active etchant for this material. Used on silicon, ceramic and dielectric substrates, and metals. **CAUTION:** These solution mixtures are highly exothermic and self-sustaining once reaction has commenced and will increase in temperature as reaction continues to a boiling point of about 170°C. See Silicon "Caro Etch".

REF: Tech. Bull. #8, 1984 — EKC Technology, Inc.

AZ-0004b
ETCH NAME:
TIME: 10 min minimum
TYPE: Acid, removal
TEMP: 125°C
COMPOSITION:
(1) x .... H_2SO_4
x .... $(NH_4)_2S_2O_8$ (SABO)
(2) x .... H_2SO_4, conc.
DISCUSSION:
AZ-type photo resists and other lacquers. Hydrogen peroxide is replaced with ''SABO'' for better removal and cleaning control. Use just enough SABO to maintain a clear solution. Use the following procedure: (1) soak 5 min in H_2SO_4 solution (2); then (2) soak 5 min in mixture solution (1) and DI water wash. Caution should be used in disposal of sulfuric acid, particularly when hot, as it can react with explosive splatter in contact with water.
REF: Ibid.

PHYSICAL PROPERTIES OF PHOSPHORUS PENTOXIDE, P_2O_5

Classification	Oxide
Atomic numbers	15 & 8
Atomic weight	142.04
Melting point (°C)	563
Boiling point (°C)	347 subl
Density (g/cm^3)	2.39
Hardness (Mohs — scratch)	1—2
Crystal structure (monoclinic — normal)	(100) a-pinacoid
Color (solid)	White
Cleavage (basal)	(001)

PHOSPHORUS PENTOXIDE, P_2O_5

General: Does not occur as a natural compound although there are many minerals containing the PO_4^- radical as phosphates. Two major minerals are apatite and phosphorite, both calcium phosphates. They are of fossil origin with large deposits in the southeastern and Gulf area of the U.S.

There are three oxide forms: trioxide, tetroxide, and pentoxide. The trioxide is poisonous and can be a by-product from the pentoxide when being used as a diffusion source of phosphorus but, as it has a very low melting point (22.5°C) and is soluble in cold water, it is used in making phosphoric acid, H_3PO_4. The pentoxide has been used as a phosphorus bomb or grenade to produce heavy white smoke and choking fumes. If solid pieces contact the skin, severe burns can result.

Technical Application: The pentoxide is used in Solid State processing as a source of phosphorus, n-type doping of silicon wafers. It was the first dopant source used as a solid, but the gas, PH_3, is used in OMCVD and Vapor Transport (VT) systems. Phosphorus doped silicon dioxide has the acronym PSG.

Etching: H_2SO_4, mix acids.

PHOSPHORUS PENTOXIDE ETCHANTS

PO-0001
ETCH NAME: Sulfuric acid
TIME: 30—60 min
TYPE: Acid, removal
TIME: RT
COMPOSITION:
x .... H_2SO_4

DISCUSSION:

P_2O_5 used in powder form as an n-type dopant for silicon wafers. Solution shown has been used to clean quartzware used in phosphorus diffusion furnaces.

REF: Crabbs, G — personal communication, 1973

PO-0002

ETCH NAME: TIME:

TYPE: Acid, removal TEMP: RT

COMPOSITION:

1 .... HF

2 .... HNO_3

4 .... HAc

DISCUSSION:

P_2O_5 and other phosphorus compounds remaining in epitaxy tubes after growth of silicon thin films on (111) silicon wafers with heavy phosphorus doping. Used in fabricating SCRs. Cleaning was done in an acid sink designed for tube cleaning with tube lying horizontal. When acid is introduced there is instant flame-out from open tube ends prior to etching action. After internal tube etching, exterior is washed in the acid solution; the entire tube then DI water washed; finally rinsed with MeOH, and allowed to air dry. Cleaned tubes are returned to the epitaxy system for a dry run to seed tube before next growth run.

REF: Walker, P & Hellstein, G — personal development, 1972

PLASTICS

General: Plastics are artificial compounds. Thermoplastics soften when heated and re-harden when cooled. Such thermosetting plastics as powder, rods, and sheets can be heated and molded, and when a critical temperature is reached there is a physical structural change (polymerization) that ''sets'' the plastic, and it cannot be reliquefied or remolded.

Celluloid was one of the first plastics (Hyatt, 1868), incompletely nitrated cellulose (mono- and dinitrate) are pyroxylin, and when mixed with alcohol and ether called collodion. Bakelite (Baekeland, 1909) is produced by mixing phenol and formaldehyde under pressure, and called a condensation plastic as water is removed during the reaction. Polymerization plastics are produced from simple hydrocarbons and their derivatives, such as Lucite, Plexiglas, polyvinyl chloride (PVC), and styrene base plastics. Acetate and mylar are similar derivative materials.

Bakelite and Plexiglas (Lucite) are lightweight construction materials and are used for aircraft, vehicles, or molded as dinner and kitchenware, and clear Plexiglas is a replacement for glass. In powder form, Lucite is used in preparing metallographic metal and mineral specimens for study: the specimen and powder molded into a solid cylinder under pressure and heat, then lapped to expose the specimen surface. Opaque brown bakelite has been used similarly. Plastic resin impregnation of wood, linen cloth, etc. produces sheets of wood veneer, and impregnated linen cloth used as insulating boards for electrical equipment for mounting electronic circuits as PCBs.

Thin film acetate has long been used for photographic film for both still photography and movie film, and in engineering for drawing overlays, map making, and book or paper jackets. Mylar as rubilith (red film coated on clear acetate sheet .003 and .005 thick) is used to design electronic circuits, then reduced in size on photographic film or on glass plates, such as chrome glass masks for photolithographic processing of semiconductor devices and associated substrate circuits. ''Sticky-back'' mylar sheet (white or blue color) is used to hold wafers — semiconductors, garnets, ceramics etc. — on the sticky surface, with the mylar held in place with a vacuum chuck for saw cutting or scribe and break dicing of parts. Mylar and other plastic sheets are spray coated or vacuum evaporated with a number of metals for reflectivity (aluminum, brass, silver, gold, nickel), and carbon coatings are used as sunscreens.

Many plastics are quite weather resistant and inert to most chemical corrosion, but they are relatively soft (H = 4—5) such that they are easily scratched by metal particles, sand, etc. Depending upon the type, they may or may not be soluble in organic solvents, such as turpentine, xylene, toluene — acetone will dissolve Lucite/Plexiglas. The phenol/formaldehydes, such as Bakelite are resistant; whereas ethyl cellulose compounds are very soluble. Heat resistance also varies with the type material . . . acetates are flammable.

It is worth noting that over 50% of every barrel of oil goes into the chemical and plastic industries — not for gasoline — such that oil economy has a far-reaching effect on industrialized nations.

Costwise, plastics are much cheaper to fabricate as parts than are woods, metals, and glasses, such that many products are replacing those materials . . . the colored acrylic paint sprays are an example of a "new" product, as are the modern throw-away baby diapers, and there is also plasticized rubber. Plastics are a viable, expanding industry which is still developing.

Technical Application: Some of the uses of mylar have already been mentioned, as have PCBs and cloth-impregnated insulator material. The latter are used in epitaxy reactor systems, for example, or as test equipment bases. Corrosive acids, such as HF, are supplied in polyethylene bottles, and there is chemical laboratory plasticware: measuring cylinders, stir rods, beakers, etc. This includes Teflon and Kel-F as fluorinated compounds specifically used for handling HF and other fluorine-containing chemicals.

Polystyrene containers are used directly to transport or hold LN_2 for chilling in vacuum systems — CVD, epitaxy growth system, and ion pump evaporators. Similar containers are used in transporting epoxy and polyimide pastes in solid CO_2, e.g., dry ice.

Teflon and many varieties of plastic tubing are used as electric wire insulators in both electronic equipment as well as in semiconductor packaged device assembly, and there is plastic shrink-tubing — heat the tubing, and it will shrink and mold around the wire or part. Plastic spray coatings are used on equipment, occasionally a semiconductor, similar to lacquer coating. The RTV compounds are a plastic/rubber combination used as an exposed junction coating on semiconductor devices, e.g., SCRs.

Polyethylene boxes as small trays with a cover are used for storing and transporting discrete semiconductor dice and similar devices — from, say, 10 to 200 units per tray — particularly with regard to static sensitive devices: Si and GaAs FETs, IMPATTS, etc. Many types of plastic boxes are used for holding parts, such as tote boxes, shaped package containers, shrink molded covers, general shipping boxes or packing, e.g., "bubble" plastic sheeting, loose particle shapes, etc.

Plastic extrusion molding is used as a final package on semiconductors, resistors, capacitors, and similar assemblies. Plastic face shields and gloves, aprons, and so forth are required wear for operators involved with chemical processing or critical device handling. Plastic eyeglasses or goggles are similarly used in chemical processing or in glass and metal welding operations.

A point on the small plastic trays used for handling static sensitive devices: the mold release compounds used in fabrication of these boxes can, with time, exude from the part as an organic surface contamination. This can affect wire bonding on the device metal contact pads — bonds will not hold. With 5—10 min in an RF plasma unit with O_2 or N_2, this organic residue will be removed. This same problem can occur where metallized substrates are shipped in plastic envelopes, and surfaces also can be cleaned by RF plasma treatment.

Many chemical sinks have a plastic sheet front that can be raised and lowered, commonly Lucite. Do not use acetone to clean— it will start dissolving the Lucite/Plexiglas. See LUC-0001b in the following list of etchants. Note that shaving of Lucite can be liquefied with acetone, then used as a glue.

Some plastic rods or filaments are used in light transmission systems, similar to glass fiber optics, and also as a rod in laser fabrication. A Lucite rod, when rubbed on wool, will create a static charge, such that they have been used in static generators.

Etching: Variable with type of plastic — both acids and solvents. See Mounting Materials section.

PLEXIGLAS ETCHANTS

PL-0001a
ETCH NAME: Acetone — TIME: 192 h
TYPE: Ketone, removal — TEMP: RT
COMPOSITION:
x .... CH_3COOH_3
DISCUSSION:
Plexiglas plastic sheet. Acetone will dissolve Plexiglas in the test period shown.
REF: Simonds, H R et al — *Handbook of Plastics,* 2nd ed, Reinhold, New York, 1979

PL-0001b
ETCH NAME: Carbon tetrachloride — TIME: 192 h
TYPE: Solvent, removal — TEMP: RT
COMPOSITION:
x .... CCl_4
DISCUSSION:
Plexiglas sheet. Solution will dissolve this material. (*Note:* This is called Lucite under a different brand name.)
REF: Ibid.

PL-0001c
ETCH NAME: Toluene — TIME: 192 h
TYPE: Solvent, removal — TEMP: RT
COMPOSITION:
x .... $C_6H_5CH_3$
DISCUSSION:
Plexiglas sheet. Solution will dissolve this material.
REF: Ibid.

PL-0002
ETCH NAME: Acetone — TIME:
TYPE: Ketone, sealer — TEMP: RT
COMPOSITION:
x .... CH_3COOH_3
DISCUSSION:
Lucite plastic. Shavings of Lucite liquefied with acetone to form a Lucite glue. The glue was then used to coat/encapsulate silicon devices or in the assembly of Lucite sheet into boxes and special housings.
REF: Walker, P — personal application, 1956—1985
LUC-0001a: Simonds, H R et al — *Handbook of Plastics,* 2nd ed, Reinhold, New York, 1979
Lucite, sheet and parts. Shavings liquefied with acetone as a glue used for assembling parts.

PL-0003
ETCH NAME: Sulfuric acid TIME:
TYPE: Acid, removal TEMP: 80—85°C
COMPOSITION:
x .... H_2SO_4, conc.
DISCUSSION:
Most encapsulation plastic compounds used in DIP fabrication can be removed by jet etching with sulfuric acid.
REF: Data sheet — B & G Enterprises, ph: (408) 728-3638

PL-0004
ETCH NAME: Water TIME:
TYPE: Acid, cleaning TEMP: RT to warm (40°C)
COMPOSITION:
(1) x .... H_2O
(2) 2 mg .... I_2
50 ml .... $MeOH/H_2O$
DISCUSSION:
Mylar thin film as sticky-back sheet used to hold semiconductor and garnet wafers and substrate materials, such as quartz and sapphire, during scribe and break or diamond saw cutting. Wipe the nonsticking surface with warm water with or without a high purity soap, such as Castille, to clean (1). An iodine mixture (2) may also be used. The iodine solution can be used on other plastics, such as Lucite, Plexiglas, and polystyrenes for light cleaning.
REF: Walker, P — personal application, 1956—1982

PVC-0001a
ETCH NAME: Potassium permanganate TIME:
TYPE: Salt, cleaning TEMP: RT
COMPOSITION:
x .... 20—30% $KMnO_4$
DISCUSSION:
PVC, DI water piping and steel piping. Solution used to periodically clean DI water storage tanks and incoming lines to laboratory sinks to remove algae build-up. Flush lines until the purple colored permanganate disappears and check with litmus paper for pH 7 neutrality — may be slightly alkaline. Use on PVC DI water lines.
REF: Tarn, W H & Walker, P — personal application, 1978

PVC-0001b
ETCH NAME: Hydrogen peroxide TIME:
TYPE: Acid, cleaning TEMP: RT
COMPOSITION:
x .... 30% H_2O_2
DISCUSSION:
PVC and steel DI water piping. Used in a similar manner as described for permanganate cleaning, above. Particular caution should be observed in removing all traces of peroxide where the water is being used on semiconductor wafers for cleaning, as hydrogen peroxide can act as an etching solution. This also applies to potassium permanganate as both are very active oxidizers.
REF: Ibid.

LUC-0001c
ETCH NAME: TIME:
TYPE: Halogen, cleaning TEMP: RT to warm
COMPOSITION:
(1) 1 g .... I_2
50 ml .... MeOH
1000 ml .... H_2O

(2) 1 tsp .... *tincture of iodine
1 gal .... H_2O

*KI:I_2:EOH

DISCUSSION:
Lucite sheet and parts. Both solutions can be used for cleaning and disinfecting Lucite and polycarbonate plastics. (*Note:* Formula (1) with KI added would be a tri-iodide solution for etching gold. The MeOH is used to increase dissolving rate of iodine in water.)
REF: Ibid.

LUC-0003
ETCH NAME: TIME:
TYPE: Acid, cleaning TEMP:
COMPOSITION:
x .... HCl
x .... glycerin
DISCUSSION:
Lucite sheet and parts can be cleaned in this solution. The solution has the same refractive index as Lucite: n = 1.40.
REF: Ivanstov, O P — *Dokl Akad Nauk SSSR,* 56,567(1947)

POLY-0001a
ETCH NAME: Toluene TIME:
TYPE: Solvent, removal TEMP: RT
COMPOSITION:
x .... C_7H_8
DISCUSSION:
Polystyrene material. Solution can be used to liquefy polystyrene pieces to be used as a coating or glue and as a general removal solvent. Used as a solvent in fabrication of silicon devices.
REF: Blankenship, J L & Borkowski, C J — *IRE Trans Nuc Sci,* NS-7,190(1960)

POLYIMIDES

POLM-0001
ETCH NAME: Sulfuric acid TIME:
TYPE: Acid, removal TEMP: RT to hot
COMPOSITION:
x .... H_2SO_4
DISCUSSION:
Polyimides are used as a dielectric surface coating. In a study of metallization on device structures and dielectric coatings with a comparison of silica and polyimide parameters, HF was used to remove silica; H_2SO_4 to remove polyimides.

Parameter	Silica	Polyimide
Resistivity (ohms-cm)	1016	1016
Dielectric strength (V μm)	200	135
M>P> (°C)	1710	475
Thermal conductivity (cal/cm-sec/°C)	5×10^{-3}	4×10^{-4}
Dielectric constant (K)	3.9	3.5

REF: Mastroianni, S T — *Solid State Technol,* March 1984, 155

PRINTED CIRCUIT BOARD ETCHANTS

PCB-0001
ETCH NAME: Ferric chloride — TIME:
TYPE: Acid, removal — TEMP: Hot
COMPOSITION:
x .... $FeCl_3$ (35°Bé)
DISCUSSION:
PCB, copper laminated plastic impregnated linen. Boards were etched in a rotating barrel system after photo resist patterning. Solution used to remove copper. After etching, DI water wash, MeOH rinse and air dry.
REF: Fahr, F — personal communication, 1978
PCB-0002: Valardi, N — personal communication, 1985
PCB, copper laminates. After pattern etching of boards, the insert contact fingers were gold plated.

PCB-0003
ETCH NAME: Freon TMS — TIME:
TYPE: Solvent, cleaning — TEMP: 97°F
COMPOSITION:
x .... Freon TMS, spray
DISCUSSION:
PCB, copper laminated and circuit etched boards. Boards were cleaned using a conveyor belt system passing the boards through a hot spray of Freon TMS.
REF: Allen Bradley brochure, 1978

PHYSICAL PROPERTIES OF PLATINUM, Pt

Classification	Transition metal
Atomic number	78
Atomic weight	195.09
Melting point (°C)	1769 (1773.5)
Boilint point (°C)	3800
Density (g/cm^3)	21.45
Thermal conductance (cal/sec)(cm^2)(°C/cm)	0.17
Specific heat (cal/g) 0°C	0.0314
Latent heat of fusion (cal/g)	26.9
Heat of fusion (k-cal/g-atom)	4.7
Heat of vaporization (k-cal/g-atom)	122
Atomic volume (W/D)	9.10
1st ionization energy (K-cal/g-mole)	207
1st ionization potential (eV)	8.96

Electronegativity (Pauling's)	2.2
Covalent radius (angstroms)	1.30
Ionic radius (angstroms)	0.80 (Pt^{+2})
Coefficient of linear thermal expansion ($\times 10^{-6}$ cm/cm/°C)	9.1
Electrical resistivity (micro-ohms-cm) 20°C	9.85
Electron work function (eV)	5.32 (5.29)
Young's modulus (psi $\times 10^6$)	24.8
Magnetic susceptibility (cgs $\times 10^{-6}$)	1.1
Tensile strength (psi — annealed)	20.000
Poisson's ratio	0.39
Cross section (barns)	8.8
Vapor pressure (°C)	3714
Hardness (Mohs — scratch)	4—4.5
Crystal structure (isometric — normal)	(100) cube, fcc
Color (solid)	Whitish steel-grey
Cleavage (fracture)	Hackly

PLATINUM, Pt

General: Occurs in nature as a native element. The major source of platinum is in the Ural Mountains of southern Russia and represents about 80% of the known platinum reserves in the world. In the Urals it occurs in igneous rocks of olivine called dunites as a product of magmatic differentiation. It is only found in small, scattered quantities in rock, the major amounts coming from associated placer deposits. It is found throughout the world in similar rock areas, and in gold placer deposits as fine, rounded grains of platinum. There are a few minerals containing platinum, mainly arsenates and antimonides, but they are not of minable quantity such that platinum metal is considered a rare element and is practically the only ore source. It has long had a higher monetary value than gold by about 20%. It rarely occurs as a pure metal, but contains trace amounts of other platinum family metals, such as osmium, iridium, rhodium, palladium, and iron. In the latter case it may appear to be magnetic, but this is due to the presence of iron, and is not an intrinsic property of platinum.

It has major industrial use as the pure metal due to its high temperature characteristics, electrical resistivity capability, and inertness to acids and atmospheres. Platinum is fabricated as wire, sheet, foil, powder, and special forms such as crucibles and beakers. As a very fine powder it is called "platinum black" with the capability of absorbing and holding large amounts of hydrogen, though palladium black is almost an order of magnitude better.

As paired wires, platinum and platinum:rhodium, it is a high temperature thermocouple and platinum alone is occasionally used as an electrical contact wire in special assemblies. Other uses include a high temperature crucible; wire, sheet, mesh or rod as the cathode in electrolytic plating and etching; rods, beads, and sponge as a catalyst in many chemical reactions; acid resistant tools; special plating on surfaces as a thin film; and others. Several of these uses apply to Solid State processing. It also is used in jewelry and for objets d'art, and occasionally as a coinage metal.

Technical Application: Not used in direct Solid State device fabrication by itself, but is used in thin film metallization in conjunction with silver, gold, nickel, and other metals for contact layers, and in the fabrication of substrate circuits on dielectrics such as alumina, beryllia, quartz, sapphire, etc. It is also used as a powder in conductive epoxy and polyimide pastes for mounting discrete devices in circuit assemblies.

Dielectrics, as capacitors, resistors, etc., use Pt/Au, Pt/Ni/Au, and other metallizations in their fabrication as discrete elements and platinum is used as a constituent in some alloys for high temperature soldering and brazing with application to semiconductor packages.

Platinum is under development and investigation as a silicide: PtSi, Pt_2Si, and $PtSi_2$ for use as a buffer layer in semiconductor heterostructures.

As a single crystal it has been grown artificially for general physical data and morphological studies, and has been studied as a thin film deposit on silicon with diffusion into the material for electron mobility control similar to applications of gold.

Etching: Difficult to etch. Aqua regia and fused alkalies; possibly chrome regia, with chromic acid replacing nitric acid.

PLATINUM ETCHANTS

PT-0001
ETCH NAME: Aqua regia TIME:
TYPE: Acid, removal TEMP: Boiling
COMPOSITION:

3 .... HCl
1 .... HNO_3

DISCUSSION:

Pt and Au evaporated on silicon, (111), n-type 35 Ω cm resistivity wafers grown by Float Zone (FZ). Phosphorus was diffused into one side; then boron diffused into the other side for a p-nn$^+$ diode. Pt diffused into p side, then diffused Au into n$^+$ side — both diffusions 40 min in N_2 with various temperatures. Etch clean both Pt and Au with aqua regia and use the solution to etch remove excess Pt and Au after diffusion. Etch clean silicon wafers prior to diffusion with HF:HNO_3.

REF: Saito, R et al — *J Electrochem Soc,* 132,224(1985)

PT-0004: Bulletin 1092 AMD/ID/JBA/5K/0482/GUILD/1982 — MRC

Pt specimens and thin films. Aqua regia used as a general surface cleaning mixture.

PT-0005: Brochure: *Platinum,* Baker & Co. (no date)

Pt, pure metal and alloy parts. Discusses physical characteristics of different alloys, their manufacture, and parts fabrication. Includes uses of platinum black as a thin film plating, and as a catalyst in chemistry. Additional discussion of other platinum family metals.

PT-0006: Wise, E M — *The Platinum Metals* (brochure) — The Int Nickel Co., 1954

Pt, pure metal and alloys as fabricated parts. Aqua regia used as a general etch for all platinum family metals: Ir, Re, Rh, Os, and Pd.

PT-0002
ETCH NAME: Nitric acid, dilute TIME:
TYPE: Acid, float-off TEMP:
COMPOSITION:

x .... HNO_3
x .... H_2O

DISCUSSION:

Pt, thin film deposited at 723 K for 1000—1200 Å thickness on muscovite mica. Shows a (111) fibrous axis. For TEM study of film morphology use solution to float-off from mica substrate.

REF: Dixit, P & Vook, R W — *Thin Solid Films,* 113,151(1984)

PT-0003
ETCH NAME: Iodine TIME:
TYPE: Halogen, passivation TEMP: >100°C
COMPOSITION:

x .... I_2, vapor

DISCUSSION:

Pt, single crystal ingot. Specimen was cut on (111) for a six-sided (111) rhomboid solid, and then mechanically polished. Ar^+ ion sputter clean and then deposit iodine on surfaces from hot vapor. Iodine atoms absorb on platinum surfaces and passivate against attack of perchloric acid.

REF: Stickney, S L et al — *J Electrochem Soc,* 131,260(1984)

PT-0006

ETCH NAME: Sulfur | TIME:

TYPE: Metal, removal | TEMP: Molten

COMPOSITION:

x .... S, molten

DISCUSSION:

Pt, as thermocouple wires — paired — Pt and PtRh. Hot liquid sulfur will attack platinum. A steel heat pipe containing liquid sulfur at 800°C broke through a weld joint and etched down through the furnace lining, metal, table, and into the cement floor, destroying a dozen thermocouples in the process. (*Note:* Liquid sulfur and hot sulfur vapors actively attack almost all metals and metallic compounds forming sulfides and sulfates.)

REF: Zelinsky, T — personal communication, 1978

PLATINUM ALLOYS, PtM_x

General: The metal occurs as a native element, rarely, if at all, in a pure state, but usually contains traces of other platinum family elements . . . Os, Re, Pd, and Fe . . . in particular. Even when used in industry it is rarely as high purity platinum due to the difficulty of separating such extremely high temperature elements as osmium, rhenium, and rhodium. It can be said that platinum occurs as a natural metal alloy, in this sense.

There are specific platinum alloys fabricated in industry for particular applications, as already discussed under Platinum.

Technical Application: Platinum alloys are used in Solid State device design and fabrication as multilayer thin films for electrical contact purposes, in addition to being used as high temperature thermocouples, special chemical handling beakers, or molten flux crucibles.

Etching: Aqua regia.

PLATINUM ALLOYS

PLATINUM GOLD ETCHANTS

PTAU-0001

ETCH NAME: Aqua regia, dilute | TIME:

TYPE: Acid, removal/clean | TEMP: RT to hot

COMPOSITION:

3 .... HCl

1 .... HNO_3

20 .... H_2O

DISCUSSION:

Pt thin films used as a multilayer of Au/Pt/Cr evaporated or sputtered metallization electrical contact structure in silicon and gallium arsenide device fabrication. Aqua regia used as general cleaning etch for platinum group metals and gold. Used a highly diluted solution to lightly clean surfaces prior to gold wire bonding. Wash heavily with running DI water after etching, rinse in MeOH, and N_2 blow dry.

REF: Walker, P — personal application, 1979—1983

PLATINUM PALLADIUM ETCHANTS

PTPD-0001
ETCH NAME: Aqua regia, dilute
TYPE: Acid, clean/preferential
COMPOSITION:
3 .... HCl
1 .... HNO_3
x .... H_2O

TIME:
TEMP: RT to warm

DISCUSSION:

Pt:Pd, thin films co-sputtered on (111) silicon, p-type 1—3 Ω cm resistivity substrates for a study of this alloy thin film. Substrates were cleaned: (1) degrease in solvents; (2) rinse in dilute HF, RT, three times, then (3) oxidize surface with $3HCl:1H_2O_2:1H_2O$ at 90°C, 10 min prior to PtPd deposition. Dilute aqua regia used to clean alloy surfaces with slight preferential attack after metallization.

REF: Kawarada, H — *J Appl Phys,* 57,344(1985)

PLATINUM RHODIUM ETCHANTS

PTRH-0001
ETCH NAME: Aqua regia
TYPE: Acid, cleaning
COMPOSITION:
3 .... HCl
1 .... HNO_3

TIME:
TEMP: RT to hot

DISCUSSION:

PtRh (2—15%), as thermocouple wires, and paired Pt:PtRh wires. Both materials can be etch cleaned with aqua regia. After bead fusion of the two wire ends (borax in electric discharge pot), water wash to remove borax flux, then dip clean in aqua regia with water washing after etch cleaning.

REF: Wise, E M — *The Platinum Metals* (brochure), The Int Nickel Co., 1954

PLATINUM HYDRIDE ETCHANTS

PT-0001
ETCH NAME: Aqua regia
TYPE: Acid, clean
COMPOSITION:
3 .... HCl
1 .... HNO_3
x .... H_2O

TIME:
TEMP: RT

DISCUSSION:

PtH as powdered "platinum black" with absorbed hydrogen. Powdered platinum can be cleaned with aqua regia with or without water dilution, then furnace dried under vacuum prior to introduction of hydrogen.

REF: Wise, E M — *The Platinum Metals* (brochure), The Int Nickel Co., 1954

PHYSICAL PROPERTIES OF PLATINUM ANTIMONIDE, $PtSb_2$

Classification	Antimonide
Atomic numbers	78 & 51

Atomic weight	447.6
Melting point (°C)	>1200
Boiling point (°C)	
Density (g/cm^3)	
Hardness: (Mohs — scratch)	6—7
Crystal structure (isometric — pyritohedral)	(100) pyritohedron
Color (solid)	Grey-white metal
Cleavage (cubic)	(001)

PLATINUM ANTIMONIDE, $PtSb_2$

General: Does not occur in nature as a metallic compound, but there are two arsenate minerals: sperrylite, $PtAs_2$ and cooperite, $Pt(As,S)_2$. Both are of minor occurrence and of no commercial use.

There is no use in industry other than its evaluation as a compound semiconductor.

Technical Application: Platinum antimonide is a V—VIII compound semiconductor with the pyrite, FeS_2, structure, and has been grown and doped p- and n-type for semiconductor characterization. Platinum arsenide, $PtAs_2$, along with $PtSb_2$ and several sulfide, selenide, and telluride compounds has been grown as single crystals and evaluated in an IR spectral study.

All metallic compounds in the isometric system, pyritohedral class tend to show stria on crystal plane oriented surfaces similar to natural pyrite. A common form of pyrite is a pseudo-cube (cube, (100) is normal class), but with a stressed lamellar structure. Such structure is less suitable for semiconductor device fabrication than other compound semiconductors that form in the isometric system, normal or tetrahedral classes, referenced to the sphalerite, ZnS, tetrahedron unit cell.

Etching: Soluble in HNO_3 and mixed acids.

PLATINUM ANTIMONIDE ETCHANTS

PTSB-0001a
ETCH NAME: TIME: 5 min
TYPE: Acid, preferential TEMP: 65°C
COMPOSITION:
1 .... HCl
1 .... HNO_3
DISCUSSION:

$PtSb_2$, (100), (111) and (110) wafers cut from a (100) grown ingot. The (100) wafers also were cleaved. Wafers used in a dislocation study. The following solutions were evaluated with the results as shown:

HCl, conc.	20 min	Boiling	No reaction
HF, conc.	20 min	Boiling	No reaction
HNO_3, conc.	20 min	Boiling	No reaction
H_2SO_4, conc.	7 min	160°C	Rough surfaces
H_2SO_4, conc.	10 min	100°C	Patterns (?)
H_2SO_4, conc.	14 min	80°C	No reaction

The mixed acid shown above was used as a cleaning solution, developed orientation etch pits on lapped surfaces, and segregation lines on polished surfaces.
REF: Sagar, A & Faust, J W Jr — *J Appl Phys,* 17,813(1966)

PTSB-0001b
ETCH NAME: Aqua regia, dilute
TIME: 15 min
TYPE: Acid, preferential
TEMP: 70°C
COMPOSITION:
3 .... HCl
1 .... HNO_3
3 .... H_2O
DISCUSSION:
$PtSb_2$, (100), (110) and (111) wafers cut from (100) grown ingots. (100) wafers also were cleaved. Solution shown developed dislocation etch pits with segregation and slip lines on polished surfaces.
REF: Ibid.

PTSB-0001c
ETCH NAME:
TIME: 40 min
TYPE: Acid, preferential
TEMP: RT
COMPOSITION:
1 .... HF
1 .... HNO_3
1 .... H_2O
DISCUSSION:
$PtSb_2$, (100), (110) and (111) wafers cut from (100) grown ingots. (100) wafers also were cleaved. Solution produced faint stria and possible etch pits (?), and left a film on surfaces (Sb?) that could be removed with ultrasonic agitation. The same solution used 3 min at 65°C developed a film that could not be removed with ultrasonic agitation.
REF: Ibid.

PTSB-0001d
ETCH NAME:
TIME: 7 min
TYPE: Acid, preferential
TEMP: 80°C
COMPOSITION:
1 .... HCl
1 .... H_2O_2
DISCUSSION:
$PtSb_2$, (100), (110) and (111) wafers cut from (100) grown ingots. (100) wafers also were cleaved. Solution shown produced faint stria but solution decomposition was too rapid for good control.
REF: Ibid.

PTSB-0001e
ETCH NAME:
TIME: 5 min
TYPE: Acid, preferential
TEMP: 60°C
COMPOSITION:
1 .... HF
1 .... H_2O_2
DISCUSSION:
$PtSb_2$, (100), (110) and (111) wafers cut from (100) grown ingots. (100) wafers also were cleaved. Solution showed possible patterns (?).
REF: Ibid.

PTSB-0001f
ETCH NAME: TIME: Varied
TYPE: Acid, preferential TEMP: RT
COMPOSITION:
3 .... HF
5 .... HNO_3
3 .... CH_3COOH (HAc)
DISCUSSION:

$PtSb_2$, (100), (110) and (111) wafers cut from (100) grown ingots. (100) wafers also were cleaved. Solution shown produced structure on both p- and n-type wafers. Pits enlarged with extended etching and showed 1:1 correspondence. An etch mixture of $1HCl:1H_2O_2:2H_2O$ used 30 min at RT, showed no reaction.
REF: Ibid.

PTSB-0002
ETCH NAME: TIME: 5 min
TYPE: Acid, removal TEMP: 75°C
COMPOSITION:
1 .... HCl
1 .... HNO_3
1 .... H_2O
DISCUSSION:

$PtSb_2$, (100) wafers. Ingots, as grown, doped with Cu or Th are p-type; doped with Te, n-type. Surfaces lapped with 800-grit abrasive before etching. There is no etch action below 70°C.
REF: Miller, R C — *J Appl Phys,* 35,3582(1964)

PTSB-0003
ETCH NAME: Nitric acid TIME:
TYPE: Acid, polish TEMP:
COMPOSITION:
x .... HNO_3
DISCUSSION:

$PtSb_2$, single crystal specimens. Cut specimens and mechanically polish with 0.5 μm diamond paste. Etch in solution shown to polish and remove lap damage. Other crystals grown were FeS_2, $MnTe_2$, RuS_2, $RuSe_2$, $RuTe_2$, OsS_2, $CsTe_2$, PtS_2, and $PtAs_2$. All crystals have the cubic pyrite, FeS_2, structure and were used in an IR spectra study. (*Note:* Pyrite is not in the normal class of the isometric system, such that it is not "cubic".)
REF: Lutz, H D et al — *J Phys Chem Solids,* 46,437(1985)

PHYSICAL PROPERTIES OF PLATINUM ARSENIDE, $PtAs_2$

Classification	Arsenide
Atomic numbers	78 & 33
Atomic weight	345.05
Melting point (°C)	>800 del
Boiling point (°C)	
Density (g/cm³)	10.60
Hardness (Mohs — scratch)	6—7
Crystal structure (isometric — pyritohedral)	(100) pseudo-cube

Color (solid) Tin-white
Cleavage (cubic/octahedral — indistinct) (001)/(111)
(fracture) Conchoidal

PLATINUM ARSENIDE, $PtAs_2$

General: Occurs in nature as the minerals sperrylite, $PtAs_2$, and cooperite, $Pt(AsS)_2$. The latter is orthorhombic — normal, H = 4—4.5 and G = 9, and with similar color and fracture as sperrylite. Both are mostly minor in occurrence, but have been found in the platiniferrous dunites of the Transvaal in Africa in minable quantities. Industrially important only as a source of platinum.

Technical Application: No use in Solid State processing other than being studied as a pyrite structured compound, although it may have some electronic characteristics of interest.

Etching: HNO_3, mixed acids $HCl:HNO_3$, $HF:HNO_3$.

PLATINUM ARSENIDE ETCHANTS

PTAS-0001
ETCH NAME: Nitric acid TIME:
TYPE: Acid, polish TEMP:
COMPOSITION:
x .... HNO_3
DISCUSSION:

$PtAs_2$, single crystal specimens. Cut specimens and mechanically polish with 0.5 μm diamond paste. Etch in solution shown to polish and remove lap damage. Other crystals grown were FeS_2, $MnTe_2$, RuS_2, $RuSe_2$, $RuTe_2$, OsS_2, $OsTe_2$, PtP_2, and $PtSb_2$. All crystals are cubic in structure like pyrite, FeS_2 and were used in a TR spectra study. (*Note:* Pyrite is not "cubic". It is in the isometric system, but pyritohedral class, not normal class, which also is referred to as the cubic class, based on the presence of the cube, (100) as one of the forms in this class.)

REF: Lutz, H D et al — *J Phys Chem Solids,* 46,437(1985)

PLATINUM OXIDE ETCHANTS

PTO-0001
ETCH NAME: Hydrochloric acid TIME:
TYPE: Acid, removal/clean TEMP:
COMPOSITION:
(1) x .... HCl, conc. (2) x .... H_2SO_4, conc.
DISCUSSION:

PtO, crystalline thin films deposited by reactive sputter in a structural study. PtO_2 as alpha-PtO_2 is space group Pnnm and beta-PtO_2 is P3m. Solutions shown are slow etchants for PtO and cleaning etchants for PtO_2.

REF: Westwood, W D & Bennewitz, C J — *J Appl Phys,* 45,2313(1974)

PTO-0002: Canart-Martin, M C et al — *Chem Phys,* 48,283(1980)

PtO, thin films deposited by reactive sputtering in a structural study.

PTO-0003: Moore, W J & Pauling, L — *Am Chem Soc,* 63,1392(1941)

PtO, crystalline bulk growth used for structural study.

PTO-0004: Schwartz, K B & Prewitt, C T — *J Phys Chem Solids,* 45,1(1984)

$PtOM_x$, single crystal material with other element constituents. A review article of the electronic properties of binary and ternary platinum oxides containing Ba, Sr, Tl, K, Ca, Co, H, Cd, Cu, and Zn. Some of these compounds show semiconducting properties.

PLATINUM PHOSPHIDE ETCHANTS

PTP-0001
ETCH NAME: Nitric acid　　TIME:
TYPE: Acid, polish　　TEMP:
COMPOSITION:
x .... HNO_3
DISCUSSION:

PtP_2, single crystal specimens. Cut specimens and mechanically polish with 0.5 μm diamond paste. Etch in solution shown to polish and remove lap damage. Other crystals grown were FeS_2, $MnTe_2$, RuS_2, $RuSe_2$, $RuTe_2$, OsS_2, $OsTe_2$, $PtAs_2$, and $PtSb_2$. All crystals are cubic in structure like pyrite, FeS_2, and were used in an IR spectra study. (*Note:* Pyrite is not "cubic", but in the isometric system-pyritohedral class, not normal class which contains the cube, (100) as a crystal form.)

REF: Lutz, H D et al — *J Phys Chem Solids,* 46,437(1985)

PHYSICAL PROPERTIES OF PLUTONIUM, Pu

Classification	Actinide
Atomic number	94
Atomic weight	244
Melting point (°C)	639.5
Boiling point (°C)	3235
Density (g/cm³)	19.74
Specific heat (cal/g/°C) 25°C	0.034
Latent heat of fusion (cal/g)	3
Thermal conductivity (cal/cm²/cm/°C/sec) 25°C	0.020
Coefficient of linear thermal expansion ($\times 10^{-6}$ cm/cm/°C) 20°C	42.3
Electrical resistivity (ohm cm $\times 10^6$) 0°C	146.45
Ionic radius (angstroms)	0.93 (Pu^{+3})
Half-life (years $\times 10^7$)	7.6
Cross section (barns)	1.6
Hardness (Mohs — scratch)	4—5
Crystal structure (monoclinic — normal)	(100) a-pinacoid
Color (solid)	Silver white
Cleavage (basal)	(001)

PLUTONIUM, Pu

General: Plutonium is one of the radioactive decay elements from ^{238}U and ^{235}U. The latter was used as the atomic bomb in the first test in New Mexico and Hiroshima; ^{239}Pu Nagasaki, and the Bikini tests after World War II. Using graphite piles for control, the ^{239}Pu step is the primary operator for atomic piles and nuclear energy plants. Other uranium isotopes can be used, such as ^{233}U for ^{237}Pu from thorium disintegration, and two other elements have been discovered coincident with the development of atomic power: Americium, Am and Curium, Cu.

Since the end of World War I, nuclear research has developed many applications and products. There are over 100 nuclear power plants throughout the world, and their development has given world industry the necessary power to sustain the tremendous expansion of high technology as major electric power was previously from hydroelectric sources. Because atomic fission releases tremendous amounts of heat, nuclear power plants are located along large rivers or seacoasts. The radioactive by-products present a disposal problem, such that much effort is involved in developing alternate energy sources: wind, sun, and hydrothermal. Atomic piles are used as engines for nuclear submarines; all industrial nations have and are developing nuclear warheads as weapons. There is radiation therapy in medicine, as well as industrial equipment for radiation testing of materials as part of space hardware requirements, and irradiation is being developed in food preservation and processing.

Technical Application: Plutonium is not used directly in Solid State device processing, but atomic piles — such as Triga facilities — are used in reliability radiation testing of devices, parts, and assemblies. Many metals, alloys, and metallic compounds are subjected to irradiation studies with a wide range of nuclear and ionic particles. Semiconductor silicon radiation detectors have been fabricated for use in space, and radiation "hard" packages have been developed.

Etching: H_3PO_4 and mixed acids (electrolytic).

PLUTONIUM ETCHANTS

PU-0001

ETCH NAME:

TYPE: Electrolytic, polish

COMPOSITION:

1 .... H_3PO_4

1 .... *$O(CH_2CH_2OH)_2$

TIME: 5—10 min

TEMP:

ANODE:

CATHODE: SST

POWER: 5 A/dm^2 & 5 V

*Diethylene glycol

DISCUSSION:

Pu specimens. Solution used as an electropolish.

REF: Tegart, W C McG — *The Electrolytic and Chemical Etching of Metals,* 2nd ed, Pergamon Press, London, 1959

POLYVINIDENE FLUORIDE, PVF_2

General: Does not occur as a natural compound. It has been fabricated artificially as a crystalline material with a polyvinyl base by replacing chlorine in polyvinyl chloride (PVC) with fluorine.

Technical Application: The material is under development as a possible lower cost and easily machinable compound as a replacement for ceramic transducers (barium titanate). It is shown to have piezo- and pyroelectric, as well as ferroelectric properties.

Etching: Strong acids and alkalies.

POLYVINIDENE FLUORIDE ETCHANTS

PVF-0001

ETCH NAME:

TYPE: Pressure, forming

COMPOSITION:

x .... pressure

TIME:

TEMP: Elevated

DISCUSSION:

PVF_2, polycrystalline material formed as sheet as a possible replacement for ceramic transducers. At RT the structure is nonpolar with orthorhombic structure. Above RT to 80°C there are beta and delta forms which are polar. The material will charge with application of pressure, heat, or electrical bias and, when poled, develop piezo-, pyro-, and ferroelectric

characteristics. The material can be easily machined, unlike the more difficult cutting of barium titanate, $BaTiO_3$, a current major transducer used as an ultrasonic generator.
REF: Marcus, M A — *5th Int Meet on Ferroelectrics,* Pennsylvania State University, 17—21 August 1981

PHYSICAL PROPERTIES OF POTASH ALUM, $KAl(SO_4)_2.12H_2O$

Classification	Alum
Atomic numbers	19, 13, 16, 8 & 1
Atomic weight	333.94
Melting point (°C)	92
Boiling point (°C)	
Density (g/cm³)	1.76
Refractive index (n=)	1.43—1.452
Hardness (Mohs — scratch)	2
Crystal structure (isometric — normal)	(111) octahedron
Color (solid)	Colorless
Cleavage (cubic)	(001)

POTASH ALUM, $KAl(SO_4)_2.12H_2O$

General: Occurs in nature as the mineral potash alum, which is isometric with octahedral habit, and as the mineral kalinite, thought to be monoclinic and of the same composition as potash alum. The alum group of minerals are hydrous sulfates of aluminum and an alkali metal with 12 molecules of water (may be 24 if formula is doubled). The potassium (K) can be replaced with NH_4 (tschermigite) or sodium, Na (soda alum). With 22 molecules of water, a similar group of halothrichite alums are magnesia alum (pickeringite); iron alum (halotrichite); and manganese alum (apjohnite). There are over 50 minerals of this general type. The mineral alunite, $K_2Al_6(OH)_{12}(SO_4)_4$ is called "alumstone", and if it contains considerable soda: natroalunite. *Note:* Natron is sodium carbonate, Na_2CO_3, used by the ancient Egyptians for mummification, hence the symbol for sodium, Na.

Most alums are of secondary origin from the weathering of sulfur-bearing rocks and minerals; from sulfuric acid in ground waters, often associated with volcanoes, or from oxidation of sulfur-bearing rocks, and may occur with bituminous and argillaceous materials.

Both natural and artificially grown alums are used in industry as metal growth and alloying fluxes and in textile, leather making, and dying. In medicine they are used as astringents and styptics, and are also used in food preparation. Note that alum is the material that will make the mouth pucker.

Technical Application: No use in Solid State fabrication of devices, although some alums are additives in solder fluxes. Single crystals of different alums have been grown for a variety of studies.

Etching: Completely soluble in hot water and soluble in dilute acids.

ALUM ETCHANTS

ALUM-0001
ETCH NAME: Water — TIME: 1—2 sec
TYPE: Acid, dislocation — TEMP: RT
COMPOSITION:
x .... H_2O, deionized
DISCUSSION:

Alum, single crystals grown from a saturated solution started at 80°C. Reduce temper-

ature to 25°C (near RT), and evaporate for 30 h to slowly form large single crystals. Etch pits developed with deionized water (DI H_2O) are very similar to those observed on diamonds as trigonal pits.
REF: Omar, M & Youssef, T H — *Phil Mag*, 6,791(1961)

ALUM-0002
ETCH NAME: Water　　TIME:
TYPE: Acid, removal　　TEMP:
COMPOSITION:
x .... H_2O
DISCUSSION:
Alum, (111) wafers used in a study of dissolution kinetics and defects. Water used for general removal and etch pit development.
REF: Van der Hoek, B et al — *J Cryst Growth*, 61,181(1983)

PHYSICAL PROPERTIES OF POTASSIUM, K (POTASH)

Classification	Alkali metal
Atomic number	19
Atomic weight	39.10
Melting point (°C)	63.7
Boiling point (°C)	760 (774)
Density (g/cm^3) 20°C	0.862
(g/cm^3) 62°C	0.83
Thermal conductance (cal/sec)(cm^2)(°C/cm) 200°C	0.107
(cal/sec)(cm^2)(°C/cm) 21°C	0.232
Specific heat (cal/g) 25°C	0.177
Heat of fusion (cal/g)	14.6
Heat of vaporization (cal/g)	496
Atomic volume (W/D)	45.3
1st ionization energy (K-cal/g-mole)	100
1st ionization potential (eV)	4.339
Electronegativity (Pauling's)	0.8
Covalent radius (angstroms)	2.03
Ionic radius (angstroms)	1.33 (K^{+1})
Electrical resistivity (micro-ohms-cm) 20°C	21.85
($\times 10^{-6}$ ohms cm) 0°C	6.15
Coefficient of linear thermal expansion ($\times 10^{-6}$ cm/cm/°C) 20°C	83
Electron work function (eV)	2.15
Cross section (barns)	2.1
Vapor pressure (°C)	568
Hardness (Mohs — scratch)	0.5
Crystal structure (isometric — normal)	(100) cube, bcc
Color (gas — K^+ ion)	Colorless
(solid)	Silver, metal
Flame color (solid/compounds)	Violet
Cleavage (cubic)	(001)

POTASSIUM, K

General: Potassium does not occur free in nature. It is the eighth most abundant element in the earth's crust — over 2% — and is found as an element in over 60 minerals and comprises over 300 artificial compounds. The feldspar minerals as a group (orthoclass, $KAlSi_3O_8$) as an example, are major rock-forming minerals, such as in granites, and the weathering of such rocks produces clays, which are a basic constituent of most sedimentary rock formations — earth, e.g., ordinary dirt contains a high percentage of "potash" as potassium is insoluble in water. Most of the potassium-bearing minerals are insoluble and difficult to reduce for potassium removal. Sylvite, KCl, is an exception — it is soluble in water and is similar in many respects to halite, NaCl — both being reversible compounds as they can be easily liquefied and then recrystallized. Both sylvite and the mineral carnallite, $KMgCl_3.6H_2O$, although they are more rare than halite (common salt), occur in extensive salt-beds in Germany, France, and the western U.S. Sylvite, as the mineral, is extensively used as a fertilizer.

Although potassium compounds are not as widely used as sodium compounds due to cost, most of the potassium compounds do not deliquesce as do their corresponding sodium salts. It is interesting to note that soda-lime glass is soft (standard window glass); whereas potash glass is hard. In the soap industry sodium soaps are hard and potassium soaps are soft, as pastes or liquids.

Sylvite, in addition to being a source of potassium and chlorine, is not retained in the body as is common salt — halite, NaCl, such that it is used as a dietary supplement for salt and is called "sea salt".

Potassium nitrate, KNO_3 — the natural mineral soda niter, is called Chile saltpeter or niter — as it occurs in great quantities in the Chilean deserts in addition to many other desert areas of the world. Potassium nitrate is a very active oxidizer, is not deliquescent, and is used in the manufacture of matches, detonators, percussion caps, and pyrotechnics, e.g., the violet color of the potassium flame.

As the pure metal, potassium is soft, with a silvery luster. Chemically, it is similar to sodium, and is more active. In contact with cold water, it reacts explosively with the liberation of hydrogen, and will burn with its characteristic violet flame. It is stored in kerosene.

The name potash is from English, but the chemical symbol, K, comes from the Latin, kalium. An item of interest: the mineral kaolin, $H_4Al_2Si_2O_9$, which is representative of the kaolin division of clays, is named from the Chinese, kauling. Although not potassium containing, the minerals montmorillonite and bentonite — which may be admixtures with kaolinite — are clay minerals that, when dried, will absorb large quantities of water. They are widely used as casing sealers in oil wells as the clay will expand and harden with water adsorption.

Technical Application: Potassium metal is not used in the direct fabrication of Solid State devices, although several potassium compounds are used in etchants as replacements for similar sodium compounds as potassium does not act as a contaminating species as does sodium on silicon wafers.

Some of the more prominent compounds used in etching solutions are potassium hydroxide, KOH; potassium chloride, KCl; potassium nitrate, KNO_3; and potassium permanganate, $KMnO_4$. Both the hydroxide and chloride are also used as molten flux etches or as growth fluxes for some ferrites and similar metallic compounds.

Etching: Burns in contact with water. Deliquesces in alcohol. Soluble in acids and ammonia.

POTASSIUM ETCHANTS

K-0001

ETCH NAME: Ethyl alcohol — TIME: 30 sec

TYPE: Alcohol, polish — TEMP: RT

COMPOSITION:

(1) x .... CH_3CH_2OH (EOH) (2) x .... $CH_3CHOHCH_3$ 2-propanol (ISO, IPA)

DISCUSSION:

K specimens cut with a razor blade under oil, then into heptane to remove oil, and into tetrahydrofuran to stop heptane action. Ten alcohols evaluated. Increase of molecular weight reduces reaction. Both solutions shown produce brilliant and smooth surfaces within 30 sec. MeOH produces too violent a reaction for control. Other alcohols developed dull and pitted surfaces — one — alpha-alpha dimethylphenethyl, a purple surface. Also worked with lithium and sodium. See LI-0006 and NA-0002.

REF: Castellano, R N & Schmidt, P H — *J Electrochem Soc,* 118,653(1971)

K-0002

ETCH NAME: TIME:

TYPE: Alcohol, polish TEMP:

COMPOSITION:

99% .... $C_6H_4(CH_3)_2$ (*o*-toluene)

1% .... C_3H_7OH (isopropyl alcohol IPA, ISO)

DISCUSSION:

K specimens. Solution used as a polish etch and to develop grain boundaries. Requires critical timing and control.

REF: Grimes, C C — Doctoral dissertation, University of California, 1962

K-0003

ETCH NAME: TIME:

TYPE: Alcohol, polish TEMP:

COMPOSITION:

x .... $C_6H_4(CH_3)_2$ (*o*-xylene)

x .... 2—6% $CH_3CH_2CHOHCH_3$ (2-butyl alcohol)

DISCUSSION:

K specimens. Solution shown as a general and polish etch for potassium.

REF: Hoyte, A F & Mielczarek — *Appl Mater Res,* 4,121(1965)

K-0004: Foster, H J & Meijer, H E — *J Res Mater Stand,* 71A,127(1967)

K, specimens. Solution applied as a general etch for potassium.

K-0005

ETCH NAME: TIME:

TYPE: Keytone, polish TEMP:

COMPOSITION:

x .... methyl ethyl ketone (MEK)

DISCUSSION:

K specimens. MEK used as a polish etch for potassium.

REF: Schmidt, P H & Rupp, L W Jr — unpublished work, 1971

PHYSICAL PROPERTIES OF POTASSIUM BROMIDE, KBr

Classification	Bromide
Atomic numbers	19 & 35
Atomic weight	119
Melting point (°C)	730
Boiling point (°C)	1380 (1435)
Density (g/cm³)	2.75
Refractive index (n =)	1.559
Wavelength limits (microns)	0.2—32

Young's modulus (psi $\times 10^6$)	3.9
Coefficient of linear thermal expansion (10^6 cm/cm/°C)	41
Hardness (Mohs — scratch)	1—1.5
(Knoop — kgf/mm^2)	6
Crystal structure (isometric — normal)	(100) cube
Color (solid)	Colorless
Cleavage (cubic)	(001)

POTASSIUM BROMINE, KBr

General: Does not occur in nature as a compound although there are some silver bromides and chlorobromides. As pure bromine it is a liquid at room temperature, has a low boiling point, and does not easily form solid compounds; it is extracted from ocean water and ground water in some areas, such as the Piedmont alluvial plains of the eastern U.S. Compounds of chlorine, bromine, iodine, and fluoride are the haloids of mineralogy or the halogens of chemistry.

Not used in the metals industries other than in cleaning and etching solutions. There are medical and chemical applications.

Technical Applications: Not used in Solid State devices design although it is used in some etching solutions.

Like some of the chlorides, potassium bromide has been used as a cleaved, (001) substrate for metal evaporation and semiconductor epitaxy growth of thin films in morphological studies. Also, as a single crystal it is used for its optical properties as special lenses and filters. Requires special housings and handling as it is hygroscopic and very soft.

Etching: Partially soluble in water and alcohols and completely soluble in glycerine. Can be controllably etched in some acids.

POTASSIUM BROMIDE ETCHANTS

KBR-0001
ETCH NAME: Water — TIME:
TYPE: Acid, polish — TEMP: RT
COMPOSITION:
x .... H_2O
DISCUSSION:

KBr, (100), cleave wafers used as substrates for epitaxy growth of materials and metal evaporation studies. The compound is hygroscopic and soft. Polish surfaces with water. In addition to use as a substrate, it is used for its infrared transmission capabilities in studies and as an operational device.

REF:
KBR-0004: Stoloff, N S et al — *J Appl Phys*, 34,3315(1963)

KBr and KBr:KCl single crystals use in a study of temperature effects on deformation. Specimens were acid saw cut with a water soaked string passed through water, then across the boule. Water was then used to lap and mill specimens to size, followed by water polishing of surface.

KBR-0002
ETCH NAME: Glycerin — TIME:
TYPE: Alcohol, removal — TEMP: RT
COMPOSITION:
x .... $C_3H_5(OH)_3$ (glycerine or glycerol)
DISCUSSION:

KBr specimens. This type alcohol can be used to etch remove or polish the bromide.

REF: Hodgman, C D — *Handbook of Chemistry and Physics,* 27th ed, Chemical Rubber Co., Cleveland, OH, 1943, 430

KBR-0003a
ETCH NAME: Acetic acid TIME:
TYPE: Acid, preferential TEMP:
COMPOSITION:
x .... CH_3COOH (HAc)
DISCUSSION:
KBr, (100) cleaved wafers used in a dislocation study. Solution used to develop dislocations. This article includes etchants for other alkali halides: NaCl, KCl, and KI.
REF: Moran, P R — *J Appl Phys,* 29,1768(1958)

KBR-0003b
ETCH NAME: Potassium chloride TIME:
TYPE: Acid, preferential TEMP:
COMPOSITION:
x .... KCl
x .... H_2O
DISCUSSION:
KBr, (001) wafers cleaved and used in developing dislocation etching techniques. Concentration of solution not shown, but was used to develop dislocations in the material.
REF: Ibid.

KBR-0004
ETCH NAME: TIME:
TYPE: Solvent, polish TEMP:
COMPOSITION:
x .... Linde A alumina
x .... kerosene
DISCUSSION:
KBr and KI single crystal specimens used in a study of nuclear magnetic resonance and acoustic absorption. Specimens initially ground down, then lapped with 303$^1/_2$ A/O abrasive. Final polish with slurry shown. The Linde A alumina was referred to as sapphire. (*Note:* Alumina, Al_2O_3, is finely powdered from common or ordinary natural single crystal corundum as a mineral; rarely are the highly valued natural gem stones used, such as red ruby or blue sapphire. The colorless manmade artificial sapphire may be the reference for sapphire, above.)
REF: Bulef, D I & Menes, M — *Phys Rev,* 114,1441(1959)

PHYSICAL PROPERTIES OF POTASSIUM CHLORIDE, KCl

Classification	Chloride
Atomic numbers	19 & 17
Atomic weight	74.6
Melting point (°C)	776
Boiling point (°C)	1500 (sublimes)
Density (g/cm^3)	1.98
Refractive index (n=)	1.490
Hardness: (Mohs — scratch)	2
Crystal structure (isometric — normal)	(100) cube

Color (solid)	Clear/white
Cleavage (cubic — perfect)	(001)

POTASSIUM CHLORIDE, KCl

General: Occurs in nature as the mineral sylvite, KCl. It is a water-soluble salt of potassium and is deposited from hot waters associated with fumaroles, volcanoes, and hot springs. It is a major compound in ocean water though not as abundant as halite, NaCl — common salt. Both of these chlorides are found in desert regions in the soil along sporadic stream and river beds, in dry lake deposits and, in some areas, as subsurface, buried "salt" deposits from the dehydration of shallow, ancient sea-beds. Because of their abundance, particularly in arid areas, they have "salted" the ground to the extent that plants and grasses will not grow without heavy leaching of the soil for their removal. With the building of the Aswan Dam on the upper Nile River in Egypt, the river no longer overflows it banks annually depositing a fresh layer of soil, and farming has been drastically reduced due to the effects of "salting". A similar situation exists in the Indus Valley where, in an attempt to increase the subsurface water table for year-round crops, the desert ground has become useless due to an increase in salt content as the high-level water table was established. Although sylvite is called clear/white in color in the Properties table above, it very commonly contains streaks of yellow and red giving the mineral distinctive colors; whereas halite tends to have blue spots and streaks. In either case, it is probably due to concentrations of potassium or sodium atoms, respectively, as well as other impurities.

Mammillary life requires salt — sodium chloride (halite) — as body fluids are an electrolytic solution, but excess salt leads to obesity because the body will retain it in higher and higher concentrations. On the other hand, the body will not accumulate potassium chloride (sylvite), but will eliminate this form of salt. Therefore — sylvite — is used as a dietary control and is often commercially called "sea salt".

In industry both of these salts are used as molten fluxes and as liquid solutions in soldering and as etchants for several metals. They also are used in fertilizers as plant life also requires sodium and potassium in addition to the animals who use the plants as food.

Technical Application: Not used in Solid State device fabrication, proper, though several etchant solutions may contain potassium chloride.

As a single crystal, (001) cleaved substrate, both natural and artificial potassium chloride have been used for metal deposition and epitaxy growth of metallic compounds for morphological studies. The deposited thin films can be easily removed by the float-off technique using water to liquefy the salt substrate/deposit interface.

Single crystals have been the subject of much study. Crystals are transparent and clear, easily deformed physically to produce slip and dislocations, produce color centers by irradiation, and so forth. Such effects can be easily observed and related to similar effects in the solid and opaque metals and metallic compounds. Much of the initial developmental study and theories of the defect state of inorganic matter has been established within the last half century (since about 1940) using both sylvite and halite as the study vehicles.

Etching: Easily etched in water, acids, alkalies.

POTASSIUM CHLORIDE ETCHANTS

KCL-0001
ETCH NAME: Methyl alcohol
TYPE: Alcohol, clean
COMPOSITION:
x .... MeOH
DISCUSSION:

TIME:
TEMP:

KCl, (100) wafers cleaved and used as substrates for Hot Wall Epitaxy (HWE) deposition

of PbTe thin films. Etch polish and clean KCl surfaces with MeOH immediately before placing in vacuum system for epitaxy growth.
REF: Vaya, P R et al — *Solid State Electron,* 27,553(1984)

KCL-0002a
ETCH NAME: Ethyl alcohol TIME: Dip
TYPE: Alcohol, polish TEMP: RT
COMPOSITION:
x .... EOH
DISCUSSION:
KCl, (100) cleaved wafers used in a dislocation etching study. Solution used to polish surfaces. Rinse in CCl_4.
REF: Barr, L W & Morrison, J A — *J Appl Phys,* 31,617(1960)

KCL-0003a
ETCH NAME: Water TIME:
TYPE: Acid, forming TEMP:
COMPOSITION:
x .... H_2O
DISCUSSION:
KCl, single crystal specimens used to fabricate filaments and points. Coat areas to become filaments with glyptol, (1) immerse in water for initial forming, and then (2) in a saturated solution of KCl. Rinse in ethyl alcohol (EOH). Also used on other alkaline halide crystals.
REF: McNulty, J et al — *Rev Sci Instr,* 31,904(1960)

KCL-0003b
ETCH NAME: Potassium chloride TIME:
TYPE: Acid, forming TEMP:
COMPOSITION:
x .... KCl, sat. sol.
DISCUSSION:
KCl, single crystal specimens used to fabricate filaments and points. See KCl-0003a for discussion. The material can be etched in a concentrated solution of itself.
REF: Ibid.

KCL-0004
ETCH NAME: TIME:
TYPE: Acid, preferential TEMP:
COMPOSITION:
x .... *25% $BaBr_2$
x .... EOH

*Mix as a saturated solution.

DISCUSSION:
KCl, (001) cleaved wafers used in developing dislocation etching techniques. Use solution shown to develop dislocations. After etching, rinse in ether. (*Note:* There is a question as to what or which is a saturated solution?)
REF: Moran, P R — *J Appl Phys,* 29,1768(1958)

KCL-0005
ETCH NAME: Moist air — TIME:
TYPE: Acid, defect — TEMP:
COMPOSITION:
x .... air
DISCUSSION:

KCl, formed as dendritic structure by placing a concentrated solution on a glass plate and cooling rapidly, (with LN_2?). NaCl and a 1KCl:1NaCl mixture also studied. Pit structure that develops around dislocations may be due to etching action of moist air. Dark spots at center of dislocations could be due to impurities or other salt nucleating elements.

REF: Forty, A J & Gibson, J G — *Acta Metall,* 6,137(1958)

KCL-0006
ETCH NAME: Ferric chloride — TIME:
TYPE: Acid, preferential — TEMP:
COMPOSITION:
x .... x% $FeCl_3$
DISCUSSION:

KCl, (100) wafers. Solutions of ferric chloride referred to as dislocation etchants for ionic crystals.

REF: Gilman, J J — *J Appl Phys,* 32,739(1961)

KCL-0002b
ETCH NAME: — TIME:
TYPE: Acid, preferential — TEMP:
COMPOSITION:
3 g .... $FeCl_3$
1000 ml .... EOH
DISCUSSION:

KCl, (100) cleaved wafers used in a dislocation study. After polishing in pure EOH, this solution used to develop etch pits. The OH^- radical locations appear to act as impurity centers (?).

REF: Ibid.

KCL-0002c
ETCH NAME: Chlorine — TIME: 48 h
TYPE: Gas, preferential — TEMP: 700°C
COMPOSITION: — PRESSURE: 500 mmHg
x .... Cl_2
DISCUSSION:

KCl, (100) cleaved wafers used in a dislocation study. Hot chlorine vapors will develop square etch pits along crystal sub-boundaries.

REF: Ibid.

KCL-0007
ETCH NAME: Hydrochloric acid — TIME:
TYPE: Acid, cleaning — TEMP:
COMPOSITION:
(1) x .... HCl, conc. — (2) x .... C_3H_7OH (isopropyl alcohol, IPA)

DISCUSSION:

KCl, (100) specimens used in a study of laser induced irreversible absorption changes at 10.6 μm. HCl used to clean surfaces, then rinsed in ISO (IPA). Also studied NaCl.

REF: Wu, S T & Bass, H — *J Appl Phys Lett,* 39,948(1981)

KCL-0008

ETCH NAME: Water — TIME:

TYPE: Acid, cleaning — TEMP:

COMPOSITION:

x .... H_2O

DISCUSSION:

KCl, (111) and (100) cleaved wafers used as substrates for growth of Al thin films. Some films converted to AlN by N^+ ion implantation. Prepare KCl substrates: (1) mechanical polish; (2) wash and polish in water; (3) rinse in acetone. On (111) KCl Al was wurtzite structure (fcc); on (100), cubic (100). Water also used to float-off aluminum thin films for TEM study.

REF: Kimura, K et al — *Jpn J Appl Phys,* 23,1145(1984)

PHYSICAL PROPERTIES OF POTASSIUM IODIDE, KI

Classification	Iodide
Atomic numbers	19 & 53
Atomic weight	166
Melting point (°C)	723
Boiling point (°C)	1420 (1330)
Density (g/cm³)	3.13
Refractive index (n=)	1.677
Hardness (Mohs — scratch)	2—3
Crystal structure (isometric — normal)	(100) cubic
Color (solid)	White
Cleavage (cubic)	(001)

POTASSIUM IODIDE, KI

General: Does not occur as a free natural compound as it is very water soluble, but is found in small quantities in seaweed. There are few iodide minerals as most metallic iodides are water soluble, but there are silver and copper minerals: marshite, CuI; iodyrite, AgI; miersite, 4AgI.CuI; and iodobromilite, a mixture of iodine/bromine. Sodium iodate, $NaIO_3$, is a major commercial source of iodine as it is found in Chilean saltpeter, and potassium iodide collected from Pacific kelp.

Both inorganic- and organic-derived iodide are used in medicine, chemistry, and photography with tincture of iodine as a germicide.

There are no major uses in the metal industry other than as a constituent in etch solutions, such as the tri-iodide etch for gold. A mixture of I_2:KI:EOH is tincture of iodine used as a medical antiseptic.

Technical Application: No major use in Solid State processing to date, though single crystal KI blanks have been used as substrates for metal and metallic compound thin film deposition in conjunction with the float-off technique for TEM structural studies of the films. The other use is in the tri-iodide etch solution already mentioned.

The compound has been grown as a single crystal for general morphological study and substrate applications.

Etching: Soluble in water and alcohols or ether.

POTASSIUM IODIDE ETCHANTS

KI-0001
ETCH NAME: Water TIME:
TYPE: Acid, removal TEMP:
COMPOSITION:
x .... H_2O
DISCUSSION:
KI, (100) wafers cleaved in ultra-high vacuum, UHV. Substrates used for thin film evaporation of silicon, 300 Å thick. Silicon film was removed for TEM study from KI substrate by dissolving the substrate in water.
REF: Nugashima, S & Ogura, I — *Jpn J Appl Phys,* 23,1555(1985)

KI-0002
ETCH NAME: Methyl alcohol TIME: 30 sec
TYPE: Alcohol, polish TEMP: RT
COMPOSITION:
x .... MeOH
DISCUSSION:
KI, (100) wafers used in a study of X-ray irradiation. Surface was mechanically lapped with Linde B abrasive and MeOH was used to etch remove residual work damage and polish.
REF: Makris, J J — *Jpn J Appl Phys,* 52,1251(1983)
KI-0004: Ma, C H & Makris, J J — *Jpn J Appl Phys,* 53,1930(1984)
KI, (001) wafers used in a study of X-irradiation. Mechanically lap surfaces on 4/0 emery paper and finish with Linde B alumina as a polish until the water slurry is dry. Use alcohol solution to remove residual lap damage and final polish surfaces.

KI-0003
ETCH NAME: Heat TIME: 10 h
TYPE: Thermal, forming TEMP: 723°C
COMPOSITION:
x .... heat
DISCUSSION: KI, growth method used for optical absorption studies. Weld two quartz plates together with about a 0.2 μm gap. Liquefy KI powder and percolate at 723°C allowing surface tension to draw liquid into the quartz plate gap. Anneal 10 h. The KI single crystal sheet developed is about 0.2 μm thick.
REF: Hashimoto, S — *J Phys Soc Jpn,* 53,1930(1984)

KI-0005a
ETCH NAME: Isopropyl alcohol TIME:
TYPE: Alcohol, preferential TEMP:
COMPOSITION:
x .... C_3H_7OH (isopropanol, IPA, ISO)
DISCUSSION:
KI, (100) cleaved wafers used in developing dislocation etching techniques. Etch in solution shown and rinse with CCl_4.
REF: Moran, P R — *J Appl Phys,* 29,1768(1958)

KI-G005b
ETCH NAME: Pyridine
TYPE: Acid, preferential
COMPOSITION:
x .... N:NCH:CHCH:CHCH:CH
DISCUSSION:

KI, (100) cleaved wafers used in developing dislocation etching techniques. Etch with solution shown and rinse in CCl_4.

REF: Ibid.

TIME:
TEMP:

KI-0006
ETCH NAME:
TYPE: Solvent, polish
COMPOSITION:
x .... Linde A alumina
x .... kerosene
DISCUSSION:

KI and KBr single crystal specimens used in a study of nuclear magnetic resonance and acoustic absorption. Specimens were initially ground down, then lapped with 303$^1/_2$ A/O abrasive. Final polish in slurry shown.

REF: Bulef, D I & Menes, M — *Phys Rev,* 114,1441,(1959)

TIME:
TEMP: RT

POTASSIUM MANGANESE FERRITE ETCHANTS

KMI-0001
ETCH NAME: Ferric chloride
TYPE:
COMPOSITION:
x .... 35% $FeCl_3$
DISCUSSION:

$KMnFe_3$, single crystal sphere cut from an ingot and oriented by X-ray. Used in a nuclear magnetic resonance study. Solution shown can be used as a general cleaning and partly preferential etchant.

REF: Schulman, R G & Knox, K — *Phys Rev,* 119,99(1960)

TIME:
TEMP:

POTASSIUM TANTALATE ETCHANTS

KTO-0001
ETCH NAME:
TYPE: Abrasive, polish
COMPOSITION:
x .... c-Al_2O_3
DISCUSSION:

$KTaO_3$, iron doped single crystal wafer sections cut as parallel pipeds used for photochromism characteristics as optical memory devices. Iron doping produces Fe^{+4} and Fe^{+5} centers. After cutting, polish specimens with colloidal alumina. Also studied $SrTiO_3$. Materials were grown by a molten flux method and doped with either iron or nickel. PbSn solder used as electrical contacts.

REF: Yamaichi, E et al — *Jpn J Appl Phys,* 23,867(1984)

KTO-0002: Ohi, K — *J Phys Soc Jpn,* 40,1371(1976)

$KTaO_3$, crystals. Reference for method of flux growth.

TIME:
TEMP:

PHYSICAL PROPERTIES OF PRASEODYMIUM, Pr

Classification	Lanthanide
Atomic number	59
Atomic weight	140.9
Melting point (°C)	935
Boiling point (°C)	3127
Density (g/cm³)	6.78
Thermal conductance (cal/sec)(cm²)(°C/cm)	0.028
Specific heat (cal/g) 25°C	0.048
Heat of fusion (k-cal/g-atom)	1.60
Latent heat of fusion (cal/g)	11.71
Heat of vaporization (k-cal/g-atom)	79
Atomic volume (W/D)	20.8
1st ionization energy (K-cal/g-mole)	133
1st ionization potential (eV)	5.76
Electronegativity (Pauling's)	1.1
Covalent radius (angstroms)	1.65
Ionic radius (angstroms)	1.06 (Pr^{+3})
Electrical resistivity (micro-ohms-cm) 298 K	68
Compressibility ($cm^2/kg \times 10^{-6}$)	3.21
Magnetic moment (Bohr magnetons)	3.56
Tensile strength (psi)	15.900
Yield strength (psi)	14.500
Magnetic susceptibility ($emu/mole \times 10^{-6}$)	5.320
Transformation temperature (°C)	795
Standard electrode potential (eV)	+2.2
Cross section (barns)	12
Electron work function (eV)	2.7
Hardness (Mohs — scratch)	4—5
Crystal structure (hexagonal — normal)	$(10\bar{1}0)$ prism, hcp
Color (solid)	Pale yellow metal
Cleavage (basal)	(0001)

PRASEODYMIUM, Pr

General: Does not occur as a native element. It is one of the rare earth elements in the lanthenium series, and found in the mineral monazite, a phosphate of cerium metal, essentially $(Ce,La,Di)PO_4$. Didymium, Di, was considered an element until 1885 when it was separated into two elements: praseodymium and neodymium with their salts, green and pink, respectively. Neither element has had any major use in industry, to date.

Technical Application: The element has not been used in Solid State processing to date, although it has been grown as a single crystal in general materials studies.

Etching: HNO_3, and other acids.

PRASEODYMIUM ETCHANTS

PR-0001

ETCH NAME: Nitric acid, dilute | TIME:

TYPE: Acid, removal | TEMP:

COMPOSITION:

x .... HNO_3

x .... H_2O

DISCUSSION:

Pr, (0001) wafers used in a diffusion study of cobalt, silver, and gold. Solution used to dissolve the material.

REF: Dariel, M F et al — *J Appl Phys,* 40,274(1969)

PR-0002

ETCH NAME: Electricity
TYPE: Arc, forming
COMPOSITION:

x .... heat

TIME:
TEMP: Elevated

DISCUSSION:

Pr, specimens arc melted as buttons then machined into cylinders. Diameters were from $^3/_{16}$ to 2″. Used for measurement of electrical resistivity. Other materials evaluated were La, Nd, and Sm.

REF: Alstad, J K et al — *Phys Rev,* 122,1636(1961)

PHYSICAL PROPERTIES OF QUARTZ, SiO_2

Classification	Oxide
Atomic numbers	14 & 8
Atomic weight	60.06
Melting point (°C)	1470
Boiling point (°C)	2230 (2500)
Density (g/cm³)	2.6—2.66
Refractive index (n=)	1.54418/1.55328
Hardness (Mohs — scratch)	7
Crystal structure (hexagonal — rhombohedral)	$(10\bar{1}1)$ rhomb
Color (solid — rock crystal)	Colorless
Cleavage (rhomb — indistinct)	$r(10\bar{1}1)$
Fracture (solid)	Conchoidal

QUARTZ, SiO_2

General: One of the most abundant natural minerals in the world as water-washed grains of sand and compacted sandstone. The alpha-quartz single crystal variety is far more abundant than is the high temperature beta-quartz form. Clear single crystals of alpha-quartz are called rock crystal; with purple tint, amethyst; tinted yellow, citron; transparent green, emerald quartz; blue, sapphire quartz; with carbon inclusion, smokey quartz, and others. Other than as semi-precious and precious gem stones, massive quartzite is a building stone, as is sandstone, and sand is a constituent of cement.

Other than as a source of silica, SiO_2, in the glass industry, quartz wedges are used in microscopes for color birefringence, and as polarizing lenses and filters.

Since the invention of the radio, using natural galena, PbS, as the frequency crystal, natural alpha-quartz became the replacement in the early 1920s. This natural quartz forms in a left- and right-handed orientation, and is subject to penetration twinning, and such twinning must be removed or frequency is lost. The best placer deposited, water-washed chunks of alpha-quartz come from Brazil, are relatively low in twinning, and have been the major source of quartz frequency blank material for over 60 years.

Artificial alpha-quartz is replacing the natural stone as it can be grown with controlled orientation, be of higher purity, and free of twinning. Crystals are grown in a vertical pressure autoclave with a temperature gradient. The liquidized silica (chips of natural stone, not sand) in a KOH slurry at bottom and better than 800°C, an oriented seed crystal at the top around 500°C, and under pressures from 20,000 to 40,000 psi. It takes about a month to grow a 1 × 2 × 6 in. crystal. Note that sand is not used as it tends to agglomerate as a gel, and does not vaporize well.

See Table 1 for the natural forms of quartz. Only the main modifications with individual names are shown, and the table does not include the some 80 varieties of quartz, as precious and semi-precious gem stones.

Technical Application: Alpha-quartz crystal blanks are cut off Z-axis at different degrees of angle to obtain specific frequency ranges with the "cuts" letter designated: AT, DT, BT, X, Y, and others. The initially square blanks are then lathe ground to a specific diameter or rectangularly cut, followed by mechanical lap and polish, and surface structuring may be involved. As an example, the AT cut is the smallest round blank and the thinner or more highly polished, the higher the frequency (from 1st to 5th order) with 80 MHz capability with relative ease. The BT cut, on the other hand, is about 1/2" in diameter with one side curved like a lens for about 5 MHz. All quartz crystals have a physical atomic motion when electrically activated that can then be converted to an electrical frequency pulse of controlled magnitude. Generally speaking, in a round blank, the motion is vertical; whereas in a rectangular blank it is a horizontal push-pull motion. The round types are primarily radio frequency crystals, some of the rectangular blanks, the quartz crystal watch operators.

After mechanical fabrication, the quartz blank is etch-tuned to near its desired frequency, then opposed metal electrodes are evaporated and, finally, metal sputter-tuned to final frequency before package mounting. Gold, silver, and aluminum are the primary metal electrodes, but BeAl and pure beryllium have been used to reduce the mass-loading effect on the crystal operation. The less weight of metal, the better frequency control, where frequency shifts are evaluated to as little as one cycle shift in 5 years!

As already mentioned, quartz blanks, alone, are frequency crystals for radio and watch operation, but can be combined with discrete semiconductor circuits as electronic oscillators, filters, and similar devices. Cadmium sulfide, CdS, single layer, and reversing layers deposited as thin films on quartz blanks, such as CdS:SCd:CdS alternating layers for additional piezoelectric operations and functions.

Artificial single crystal or fused quartz blanks are used as circuit substrates in Solid State processing or in special development, such as SiO_2:H thin film depositions, and are included, here. Silicon dioxide, SiO_2, as a thin film in semiconductor type processing is covered separately, as are some other single crystal forms: Cristobalite, Tridymite, Opal.

Etching: HF and other fluorine compounds. Alkalies as molten fluxes. Mixed acids: HF:HNO_3 type.

TABLE 1
Natural Forms of Quartz

Name & formula	Structure	Formation temperature (°C) (inversion)
Siliceous sinter, $SiO_2.nH_2O$	Amorphous/colloidal	100—150
Silica, SiO_2	Colloidal	150
Opal, $SiO_2.nH_2O$	Amorphous/colloidal	100—150
Alpha-quartz, SiO_2	Hexagonal: Rhombohedral	<573
	Trapezohedral — tetrahedral	
Beta-quartz, SiO_2	Hexagonal: rhombohedral	573—870
	Trapezohedral — hemihedral	
Alpha-tridymite, SiO_2	Orthorhombic	>800
	Hexagonal	(117 & 163)
Beta-tridymite, SiO_2	Hexagonal	>870
		(117 & 163)
Alpha-cristobalite, SiO_2	Tetragonal (pseudo-isometric)	1200
		(198 & 275)
Beta-cristobalite, SiO_2	Isometric — normal	>1200
		(198 & 275)
Coestite, SiO_2/SiO_4	Hexagonal — normal, hcp	>1200 + pressure
Stisovite, SiO_2/SiO_4	Hexagonal — normal, hcp	>1200 + pressure
Tektite, SiO_2	Amorphous	>1200 (meteoric)
Obsidian, SiO_2	Amorphous (natural glass)	>1200 (volcanic)

QUARTZ ETCHANTS

QTZ-0001
ETCH NAME:
TYPE: Acid, removal
COMPOSITION:
1 .... HF
1 .... H_2O

TIME: To 25 min
TEMP: RT

DISCUSSION:

SiO_2, (0001), $(10\bar{1}0)$, natural single crystal and artificial fused quartz wafers and blank. Polished surfaces contain a "smear" layer a few angstroms thick under which there are scratches, pits, and microchips. This smear layer was removed with the solution shown in order to study the damage structure beneath. [*Note:* In polishing glass/quartz the smear-layer (Debeye layer) can vary from 500 to 2000 Å and is amorphous in structure . . . silica that physically fills-in low spots . . . rather than producing a planar, truly flat surface.] Polish lapping with an acid-slurry of cerium oxide/potassium hydroxide improves the true polishing finish quality without a smear layer. This same smear-type layer can occur in mechanical polishing of metals and metallic compounds.

REF: Newkirk, J B et al — *J Appl Phys,* 35,1302(1964)

QTZ-0008: Kobayashi, T & Geska, J — *J Cryst Growth,* 67,318(1984)

SiO_2, as fused quartz ampoules cleaned in this solution before encapsulated growth of gallium arsenide by LEC.

QTZ-0002a

ETCH NAME: TIME: Minutes/hours

TYPE: Acid-slurry, polish TEMP: RT

COMPOSITION:

x .... CeO_2 (5 μ)

x .... H_2O

x .... glycerin

x .... 10—15% NaOH

DISCUSSION:

SiO_2, AT-cut quartz crystal blanks and other orientations of both natural and artificially grown alpha-quartz. Mechanical polishing using a planetary lap (both blanks surfaces simultaneously lapped) with iron platens and SST shim holders. Slurry applied by drip method onto top rotating platen at about 1 drop/sec. Blanks lapped to rough frequency as high as 60 MHz (3rd order). Requires defect-free, parallel surfaces to $^1/_4$-wavelength (sodium light). Excellent results obtained using acid-slurry mixture shown; whereas without the alkali oversmearing of lap damage masks the true damaged nature of the surface.

REF: Walker, P — personal development, 1966—1969

QTZ-0003

ETCH NAME: Aqua regia TIME:

TYPE: Acid, cleaning TEMP: RT

COMPOSITION:

3 .... HCl

1 .... HNO_3

DISCUSSION:

SiO_2, as single crystal natural quartz, artificial quartz, and vitreous silica (fused glass). This primary reference used aqua regia as a wash solution on vitreous silica to remove iron contamination from surfaces prior to a study of the attack rates of HF.

REF: Blumberly, A A & Staurinov, S C — *J Phys Chem Solids,* 54,1438(1960)

QTZ-0004: Walker, P & Folds, W — personal application, 1969

SiO_2, AT-cut alpha-quartz blanks, both natural and artificial single crystal. Solution used to pre-clean quartz blanks prior to electrode metallization. After etch-cleaning soak at RT, 3—5 min, rinse in running DI water, 2—5 min, rinse in MeOH and N_2 blow dry.

QTZ-0018: Zelinsky, T — personal communication, 1980

SiO_2, fused quartz epitaxy tubes, 4″ diameter, used for epitaxy growth of gallium arsenide on gallium arsenide. One section of a tube cleaning sink is filled with aqua regia for soak cleaning of tubes and associated parts. After cleaning at RT, 1—2, remove; water rinse heavily, MeOH rinse, and dry under infrared heat lamps.

QTZ-0005
ETCH NAME: Water TIME:
TYPE: Acid, cleaning TEMP: RT
COMPOSITION:
x .... H_2O
DISCUSSION:
SiO_2, as single crystal quartz blanks used in a study of surface conductivity in the presence of adsorbed layers. Water used as the base for evaluation of the effects of alcohol rinses. Conductivity decreases with an increase of the molecular weight of the alcohol.
REF: Abdrakhmanova, I F & Deryagin, B V — *Dokl Akad Nauk SSSR,* 120,94(1958)

QTZ-0006
ETCH NAME: TIME:
TYPE: Acid, thinning TEMP:
COMPOSITION:
1 .... HF
1 .... HNO_3
DISCUSSION:
SiO_2, single crystal artificial alpha-quartz blanks. Solution used to thin quartz blanks in a study of defect structure induced by neutron irradiation. Defect clusters increase with time and dose levels. Can be stabilized by annealing at 500°C. (*Note:* Slightly below the 573°C temperature inversion point to beta-quartz.)
REF: Weissman, S & Nakajima, K — *J Appl Phys,* 34,611(1963)

QTZ-0007
ETCH NAME: TIME:
TYPE: Acid, tuning TEMP: Warm
COMPOSITION:
1 .... HF
20 .... HNO_3
20 .... CH_4OHCH_2OH (ethylene glycol, EG)
DISCUSSION:
SiO_2, AT-cut quartz blanks and other cut orientations of both natural and artificial alpha-quartz used for radiofrequency crystals. The solution shown is variable and originally used molasses rather than ethylene glycol as the viscosity control agent. Solution used as a dip-type etch: (1) etch, (2) DI water rinse, (3) dry, and (4) frequency test. Repeat as needed to obtain required frequency prior to electrode metallization. (*Note:* Solution developed to prevent carbonization from residual molasses on blanks during metallization.)
REF: Walker, P & Schidler, A M — personal development, 1967

QTZ-0009
ETCH NAME: TIME: 60 sec
TYPE: Acid, cleaning TEMP: RT
COMPOSITION:
x .... H_2SO_4
x .... CrO_3
DISCUSSION:
SiO_2, single crystal blanks used for growth of a-Si:H and a-Ge:H thin films. Degrease

blanks in TCE; rinse in acetone; rinse in MeOH and rinse in DI water. Follow with heavy washing in running DI water. (*Note:* Acid mixture that is similar to the standard "glass cleaner" solution using saturated $Na_2Cr_2O_7$.)
REF: Rudder, R A et al — *J Vac Sci Technol,* A2,326(1984)

QTZ-0010
ETCH NAME: TIME:
TYPE: Acid, cleaning TEMP:
COMPOSITION:
1 .... HF
3 .... HNO_3
DISCUSSION:
SiO_2, as fused quartz ampoules cleaned in this solution before Chemical Vapor Transport (CVT) growth of $CdSiAs_2$ single crystals.
REF: Avirovic, M et al — *J Cryst Growth,* 67,185(1984)

QTZ-0002b
ETCH NAME: TIME:
TYPE: Acid, cleaning TEMP: RT
COMPOSITION:
50 ml .... HF
50 g .... $NH_4F.HF$
100 ml .... H_2O
50 ml .... glycerine
DISCUSSION:
SiO_2, single crystal AT-cut, BT-cut as alpha quartz cuts; fused quartz substrate blanks, and quartzware as diffusion tubes, carriers, etc. Solution used as a general cleaning etchant. Glycerine reduces evaporation and improves wettability and can be varied. Also used on thin film SiO_2 and Si_3N_4. Initial development was for pattern etching of silicon oxynitrides on silicon wafers using standard photolithography. The bifluoride replaces the normal ammonium fluoride, and may be more useful where saturated fluorides are used to maximize control in etching tuning quartz frequency blanks.
REF: Ibid.

QTZ-0012
ETCH NAME: RCA Etch (AB Etch) TIME: 15—20 min
TYPE: Acid, cleaning TEMP: 70—80°C
COMPOSITION:
"A" 100 ml .... NH_4OH "B" 100 ml .... HCl
100 ml .... H_2O_2 100 ml .... H_2O_2
500 ml .... H_2O 500 ml .... H_2O
DISCUSSION:
SiO_2, alpha-quartz frequency crystals, fused quartz substrates, and quartzware. Etch clean in solution "A", then transfer parts still wet to solution "B" each for time and temperature shown. Remove parts and wash in running DI water, rinse in MeOH and dry with either nitrogen or under IR heat lamps. Used for general cleaning of parts and prior to metallization of quartz frequency crystal electrodes, and fused quartz blanks used as circuit substrates, or for general quartzware. (*Note:* This cleaning method was developed for silicon, see SI-0031.)
REF: Walker, P & Marich, L A — personal application, 1983—1985

QTZ-0013
ETCH NAME: Hydrofluoric acid TIME: 15 h
TYPE: Acid, preferential TEMP: RT
COMPOSITION:
x .... HF (48%)
DISCUSSION:

SiO_2, single crystal blanks cut from synthetic quartz. Solution used to develop etch pits. Pits were triangular with an etch line(?). Authors say they saw inclusion blebs containing residual growth liquid. Frequency device Q-loss is related to density of defects. (*Note:* These authors have observed liquid and gas-filled blebs in natural alpha-quartz but not in cultured or synthetic quartz.)

REF: Spencer, W J & Haruta, K — *J Appl Phys,* 37,549(1966)

QTZ-0014a
ETCH NAME: Nitric acid TIME:
TYPE: Acid, cleaning TEMP: Boiling
COMPOSITION:
x .... HNO_3, conc.
DISCUSSION:

SiO_2, single crystal artificial alpha-quartz boules and $(10\bar{1}0)$ prism orientation blanks. Boiling nitric acid used to clean surfaces before preferential etching with sodium hydroxide.

REF: Joshi, M S & Vagh, A S — *Br J Appl Phys,* 17,528(1966)

QTZ-0014b
ETCH NAME: Sodium hydroxide TIME:
TYPE: Alkali, preferential TEMP: Boiling
COMPOSITION:
x .... x% NaOH
DISCUSSION:

SiO_2, $(10\bar{1}0)$ artificial alpha-quartz blanks. Solution develops dissolutionment figures, highest etch rate planes, as hillocks. Pits are representative of slowest etch rate planes.

REF: Ibid.

QTZ-0015a
ETCH NAME: Hydrochloric acid, vapors TIME:
TYPE: Acid, cleaning TEMP: 1100°C
COMPOSITION:
x .... HCl, vapor
x .... H_2, vapor
DISCUSSION:

SiO_2, fused quartz epitaxy tubes of rectangular cross section. Hot HCl vapors in hydrogen used to clean tubes and silicon wafers prior to silicon epitaxy growth. Also used to clean SiC coated graphite susceptors after epitaxy to remove deposited silicon. HCl vapor concentration varied from 10 to 20%.

REF: Topas, B & Walker, P — personal application, 1972

QTZ-0016b: Ibid.

SiO_2, fused quartz tubes cleaned in 9% $HCl:O_2$, 16 h at 1200°C prior to dry oxidation of (100) silicon wafers at 1000°C.

QTZ-0016a
ETCH NAME: 1,1,1 Trichloroethane
TYPE: Solvent, cleaning
COMPOSITION:
x .... 3% CH_3CCl_3, vapor
x .... O_2, vapor
TIME: 16 h
TEMP: 1200°C

DISCUSSION:

SiO_2, fused quartz tubes were cleaned with this vapor mixture. Authors say this mixture is equivalent to 9% $HCl:O_2$ (QTZ-0016b). Tubes were cleaned prior to oxidation of n-type, 2—5 Ω cm resistivity, (100) silicon wafers in dry O_2 at 1000°C. Aluminum dots were then E-beam evaporated through a gold mask. Tubes were also cleaned with HF.

REF: Greeuw, G et al — *Solid State Electron*, 27,77(1984)

QTZ-0015b
ETCH NAME:
TYPE: Acid, cleaning
COMPOSITION:
5 .... HF
10 .... HNO_3
10 .... CH_3COOH (HAc)
TIME:
TEMP: RT

DISCUSSION:

SiO_2, fused quartz tubes of rectangular cross section used for silicon epitaxy growth on silicon wafers. Solution used for maintenance cleaning of tubes after every 15 to 20 deposition runs to remove heavily phosphorus doped silicon deposit from tube walls. Cleaning was done in a deep-welled tube cleaning sink designed for these epitaxy tubes. Cleaning procedure was (1) first, flush tube with heavy water rise jet (tygon tubing) — **CAUTION**: Phosphorus will flame-out, as it burns under water; (2) pour 250 ml of etch solution into tube — caution: phosphorus can ignite and flame-out; (3) water rinse tube interior; (4) etch solution rinse tube exterior; (5) water wash exterior and interior of tube and (6) final rinse with MeOH outside of sink (no alcohol down an acid drain). Wipe down tube with lint-free paper and blow interior dry with nitrogen. After tube is reinstalled in epitaxy system, make a "dummy" run to condition tube. (*Note:* Caution cannot be overemphasized as flame-out can extend 6—8" beyond tube ends with explosive ignition.)

REF: Ibid.

QTZ-0020
ETCH NAME:
TYPE: Laser, cut
COMPOSITION:
x .... laser
TIME:
TEMP:

DISCUSSION:

SiO_2 as fused or single crystal quartz substrates used in microelectronic assemblies. Laser used to cut rectangular holes in substrates prior to circuit metallization.

REF: Bowman, T — personal communication, 1984

QTZ-0017
ETCH NAME: Hydrofluoric acid, dilute
TYPE: Acid, cleaning
COMPOSITION:
1 .... HF
10—20 .... H_2O
TIME: 1—2 h
TEMP: RT

DISCUSSION:

SiO_2, fused quartz tubes, 4″ in diameter used for boron diffusion of silicon wafers. Solution used for maintenance cleaning of tubes. Tubes were held vertically over a Teflon container with the bottom, small tube end, closed with a T/S Pyrex stopcock. Pour sufficient solution into the tube to fill within about 1″ of the top, open end. Allow solution to soak-clean tube interior for 1—2 h at RT. Drain and flush with water, then rinse in MeOH. Cleaning was also done in a horizontal tube cleaning sink.

REF: Tarn, W H & Walker, P — personal application, 1963

QTZ-0019: Walker, P — mineralogy study, 1953

SiO_2, as an oolitic form of natural Chert. Study included metallographic bulk and thin section etching study and powder X-ray analysis of this material. Solution used to develop the oolitic structure to observe and determine types of central-core inclusions that formed the oolites. (*Note:* Oolites are initially formed as colloidal segregates during solidification of sedimentary rocks.)

TRI-0001

ETCH NAME: Sodium carbonate

TYPE: Salt, removal

COMPOSITION:

x .... $NaCO_3$

TIME:

TEMP: Boiling

DISCUSSION:

SiO_2, as the natural mineral tridymite as small white octahedrons. Tridymite can be differentiated from alpha- and beta-quartz as it is attacked by boiling sodium carbonate. Solution used as a soaking solution (60°C) will develop defects and structure. Used in a general mineralogic study of various forms of quartz and silica.

REF: Walker, P — mineralogy study, 1952—1954

COE-0001

ETCH NAME: Hydrofluoric acid

TYPE: Acid, removal

COMPOSITION:

x .... HF, conc.

TIME:

TEMP: RT to hot

DISCUSSION:

SiO_2, as the natural mineral coesite is a high pressure modification of quartz as SiO_2/SiO_2 with a tetrahedra ratio of 4:2. Solution used to etch segregation figures with reference to studies done on other high pressure forms as $CrVO_4$ and $FeVO_4$. See stisovite below as another high pressure form of natural quartz, both associated with meteor strikes.

REF: Young, A P & Schwartz, C M — *Acta Crystallogr,* 15,1304(1962)

STI-0001

ETCH NAME: Hydrofluoric acid

TYPE: Acid, removal

COMPOSITION:

x .... HF, conc.

TIME:

TEMP: RT to hot

DISCUSSION:

SiO_2/SiO_4 as the natural mineral stisovite is a high pressure modification of quartz with a tetrahedra ratio of 6:3, considered to be the highest density form of quartz. See COE-0001 for additional discussion.

REF: Young, A P & Schwartz, C M — *Acta Crystallogr,* 15,1304(1962)

OPAL-0001

ETCH NAME: Chromic acid
TYPE: Acid, coloring
COMPOSITION:
150 ml .... CrO_3, sat. sol.

TIME:
TEMP:
PRESSURE: 10—50 psi

DISCUSSION:

$SiO_2.nH_2O$, as the natural mineral opal is a hydrated form of quartz. Opal is amorphous/colloidal in structure with high internal fracturing. There is water in the crystal lattice and residual in the fracture zones which gives the splay of colors. Heating opal to above 125°C will drive off residual entrapped water changing opal colors toward white. By pressurizing the opal in an autoclave containing chromic acid the stone will develop a splay of green colors. This pressurizing method has been applied to other natural minerals and rocks for different colors. In gross leak package evaluation of Solid State elements, a red dye is used, followed by a light vacuum (10^{-1} Torr sufficient) . . . if red dye appears, there is a package leak. (*Note:* To maintain an opal's color it should be occasionally dampened with water.)

REF: Walker, P — mineralogy development, 1986

PHYSICAL PROPERTIES OF RADIUM, Ra

Classification	Alkali earth
Atomic number	88
Atomic weight	226.05
Melting point (°C)	700
Boiling point (°C)	1040
Density (g/cm^3)	5
1st ionization potential (eV)	5.28
Ionic radius (angstroms)	1.43
Half-life (years)	1590 (1622)
Cross section (barns)	20
Hardness (Mohs — scratch)	2
Crystal structure (isometric — normal)	(100) cubic
Color (solid)	Silver
Cleavage (cubic)	(001)

RADIUM, Ra

General: Radium is the major decay product of the uranium series disintegration with nine isotopes ending in non-radioactive lead. Its discovery by Madame Curie, and separation in 1910, established the field of radioactivity. There is one part radium in 3.3 million parts uranium, such that the amount of radium and helium in a mineral ore sample is used for radioactive dating. This dating has shown the Earth to be at least 1.5 billion years old. There are over 50 minerals containing uranium, the most well known being pitchblende — the mineral uraninite — named for its black color and from which radium was first discovered. Pitchblende is a uranate containing lead, and usually thorium which is another radioactive series. The mineral carnotite is a uranium vanadinite, brilliant yellow and orange in color, and often found as granular masses of yellow/orange/black as a mixture with pitchblends (Colorado plateau region).

Radium is still used in cancer therapy, though high energy X-rays have been developed for more intense gamma rays. Salts of radium emit a yellow-green light that glows in the dark, used on watch and clock faces through the 1930's before the full danger of radioactivity was realized. Today its use is mainly in the medical field.

Technical Application: Radium has little use in metal processing and Solid State device fabrication areas. It is occasionally used as a test vehicle for radiation counters, though radioactivated lead with a half-life of 24 h is more common.

Radium reacts with water with the evolution of hydrogen, such that it should not be stored in a closed container — also continually emits radon gas as a disintegration by-product which, with time, can pressurize and cause explosive rupture of such sealed containers.

Etching: Deliquesces in acids.

RADIUM ETCHANTS

RA-0001

ETCH NAME: Water — TIME:

TYPE: Acid, reduction — TEMP: RT

COMPOSITION:

x .... H_2O

DISCUSSION:

Ra, as a metal will reduce water with evolution of hydrogen.

REF: Foster, W & Alyea, H N — *An Introduction to General Chemistry,* 3rd ed, D Van Nostrand, 1947, 295

ROCHELLE SALT ETCHANTS

RS-0001
ETCH NAME: Water
TYPE: Acid, preferential
COMPOSITION:
x .... H_2O
TIME:
TEMP:
DISCUSSION:
$NaKC_4H_4O_6.4H_2O$, (0001) wafers lapped on filter paper dampened with water to develop square etch pits with pointed bottoms. After X-ray irradiation, pit shapes became "roof-like" with a line at bottom. Subsequent annealing showed little change. Wafers initially cut with a thread wetted with water.
REF: Okada, K — *Jpn J Appl Phys*, 2,613(1963)
RS-0002: Wieder, H H — *Phys Rev*, 109,29(1958)
$NaK_4H_4O_6.4H_2O$ wafers used in a study of ferroelectric polarization reversal in Rochelle Salt.

RS-0003
ETCH NAME: Ethyl alcohol
TYPE: Alcohol, cleaning
COMPOSITION:
x .... EOH
TIME:
TEMP:
DISCUSSION:
$NaK_4H_4O_6.4H_2O$, (0001) wafers were prepared as follows: dry lap with 600-grit SiC, then polish lap on nylon cloth with a slurry of alumina powder in glycerin. Final wash and clean in alcohol. Salt was deuterium-doped in a study of electro-optical effects and polarization reversal.
REF: Wieder, H H & Collins, D A — *Phys Rev*, 120,725(1960)

PHYSICAL PROPERTIES OF RADON, Rn

Classification	Rare gas
Atomic number	86
Atomic weight	222
Melting point (°C — solid)	−71
Boiling point (°C — solid)	−61.8
Density ($g/cm^3 \times 10^{-3}$ — gas) 20°C	9.96
(g/l — liquid)	9.73
(g/cm^3 — solid)	4
Latent heat of fusion (cal/g)	2.9
1st ionization potential (eV)	10.745
Half-life (days)	3.823
Cross section (barns)	0.7
Vapor pressure (°C)	99
Hardness (Mohs — scratch)	2
Crystal structure (isometric — normal)	(100) cube, fcc
Color (solid)	Colorless
Cleavage (cubic)	(001)

RADON, Rn

General: Radon is a chemically inert, colorless gas elimination by-product of the disintegration of radium, Ra, with a half-life of only 3.85 days as compared to 1590 years for radium. It emits intense alpha particles and has been extensively studied with regard to its relationship to radium decay, and the other radioactive gases. Helium and lead are decay products of uranium with radium/radon being steps in the process, such that the amount of helium in a sample of, say, fergusonite ore is a measure of its age — one sample was concluded to be over 400,000,000 years old — the earth calculated to be more than $1^1/_2$ billion years old.

Though a radioactive by-product, radon is classified as a rare gas as it emanates from the earth's crust into the atmosphere in small quantities. Such emanations occur worldwide, more common in earthquake-prone areas and in the northeastern U.S. appear to be associated with granitic areas. Sudden increased release of radon in fault areas is recognized as a precursor of fault activity. In some hot spring areas radon is found in the upswelling waters, and it has recently been measured collecting in the basement of buildings, and is said to be a cause of lung cancer. It should be noted that radon emits alpha particles, and alpha particle penetration is stopped by the skin, beta particles by muscle tissue, and gamma particles pass completely through the body. It is the latter rays that are thought to cause long-range mutation in animal and plant life.

Most of the radioactive gases react with water and should not be stored in a closed container as build-up of pressure can cause explosive rupture.

Technical Application: Radon has little use in Solid State due its short half-life although it has long been a primary method of studying radium. It can be used as a radiation alpha counter in place of radioactive lead. The latter, with only a 24-h half-life, has been used in testing silicon radiation detectors.

Etching: Soluble in water.

RADON ETCHANTS

RN-0001
ETCH NAME: Water — TIME:
TYPE: Acid, dissolving — TEMP:
COMPOSITION:
x .... H_2O
DISCUSSION:
Rn, as the gas will dissolve in water, as do many of the radioactive gases.
REF: Foster, W & Alyea, H N — *An Introduction to General Chemistry*, 3rd ed, D Van Nostrand, New York, 1947, 295

RARE EARTHS (RE)

General: There are many natural minerals containing trace amounts of the rare earth elements (15 elements) and are chemically divided into two groups: cerium sub-group and yttrium sub-group. The chief source of cerium is from the mineral monazite, as a phosphate, but contains up to 30% cerium and 28% thorium as oxides . . . ceria and thorea. The mineral gadolinite is a mixed silicate of yttrium — yttrium is often called a rare earth, though it is not listed as a rare earth in the Periodic Table of Elements. Yttrium, cerium, and neodymium are fairly abundant, but the others are quite rare — elements 61, 63, 65, and 67—71. The rare earths are considered very active metals similar to calcium. They burn in air to the oxide; the oxides dissolve in water, often hissing like quicklime; and hydroxides are alkaline. Cerium sulfate, $Ce(SO_4)_2$ is a powerful oxidizing agent, and ammonium ceric nitrate or sulfate is a major etching solution constituent for certain metals, such as chromium and nickel.

Some of the rare earths are used as alloy constituents in metal processing, but there is no extensive application. Several of their salts have use in both chemistry and medicine, mainly as oxidizers.

Technical Application: Some of the rare earths are under evaluation as doping elements in Solid State material processing. Most have been grown and studied as single crystals or thin films to include as nitrides and silicides. See individual element sections for additional data. Nitrides are mentioned here, also see section on silicides and garnets.

The Rare Earths are as follows:

Atomic number	**Element/symbol**
57	Lanthanum, La
58	Cerium, Ce
59	Praseodymium, Pr
60	Neodymium, Nd
61	Illinium, Il
62	Samarium, Sm
63	Europium, Eu
64	Gadolinium, Gd
65	Terbium, Tb
66	Dysprosium, Dy
67	Holmium, Ho
68	Erbium, Er
*69	Thulium, Tm
*70	Ytterbium, Yb
*71	Lutetium, Lu

In the literature the Rare Earths have been symbolized by both (RE) and (R). The latter is not recommended as ''R'' is used in mineralogy as the general symbol for a metal constituent in a mineral; whereas the letter ''M'' is used in Chemistry. Those starred (*) are not in this book for lack of referenced application.

Etching: Water soluble. See individual listings.

RARE EARTH NITRIDE ETCHANTS

REN-0001
ETCH NAME: Hydrofluoric acid
TYPE: Acid, removal
TIME:
TEMP:
COMPOSITION:

x .... HF, conc.

DISCUSSION:

$(RE)N_{12}$ specimens grown by arc melting under argon and annealing at 600—900°C for 3—6 weeks. Materials used in a study of magnetostriction. Nitrides grown were Tb, Pr, Dy, and Ho. See Ibarra Ph.D. thesis for metallography.

REF: Moral, A Del & Ibarra, M R — *J Phys Chem Solids*, 46,127(1985)

PHYSICAL PROPERTIES OF RHENIUM, Re

Classification	Transition metal
Atomic number	75
Atomic weight	186.2
Melting point (°C)	3180 (3167)

Boiling point (°C)	5900 (5627)
Density (g/cm^3)	21.04
Thermal conductance [cal/(sec)(cm^2)(°C/cm)]	0.095
Specific heat (cal/g) 25°C	0.035
Heat of fusion (k-cal/g-atom)	7.9
Latent heat of fusion (cal/g)	42.2
Heat of vaporization (k-cal/g-atom)	152
1st ionization energy (K-cal/g-mole)	182
1st ionization potential (eV)	7.87
Atomic volume (W/D)	48.85
Electronegativity (Pauling's)	1.9
Covalent radius (angstroms)	1.28
Ionic radius (angstroms)	0.56 (Re^{+7})
Coefficient of linear thermal expansion ($\times 10^{-6}$ cm/cm/°C)	6.2
Electrical resistivity (micro-ohms-cm)	1119.3
Thermal conductivity (cal/cm^2/°C/sec) 20°C	0.17
Vapor pressure (mm $\times 10^{-8}$) 2000°C	3
Electron work function (eV)	4.8 (5.1)
Modulus of elasticity (psi $\times 10^6$)	67
Cross section (barns)	85
Hardness (Mohs — scratch)	5—6
Crystal structure (hexagonal — normal)	$(10\bar{1}0)$ prism, hcp
Color (solid)	Silver-grey metal
Cleavage (basal)	(0001)

RHENIUM, Re

General: Does not occur in nature as a native element. It is found as a minor constituent element in several of the tantalate minerals such as tantalite and wolframite. It was only discovered in 1925 by Noddack and has characteristics similar to those of manganese and is considered to be one of the eka-manganese. Although it has a very high melting point, it is easily forged and processed much like iron and copper.

It has some industrial use as an alloy constituent in metals such as iron and steel and is used as the pure metal for its high temperature characteristics in the fabrication of a variety of parts. It is also used as a catalyst for dehydrogenation.

Technical Application: Rhenium has been grown as a single crystal for physical data and morphological studies. As a crystalline metal it has been studied for surface reactions under various conditions — oxide, chloride, sulfides. The sulfides are quite inert and ReS_2 acts as a lubricant similar to MoS.

Etching: HNO_3 H_2O_2 and mixed acid as HF:HNO_3 mixtures.

RHENIUM ETCHANTS

RE-0001

ETCH NAME: TIME:

TYPE: Acid, preferential TEMP:

COMPOSITION:

8 .... HF

2 .... HNO_3

DISCUSSION:

Re, (0001) wafers used in a study of anisotropy in the superconductivity of deformed rhenium. After specimens are annealed, solution will develop deformation structure.

REF: Hauser, J J — *J Appl Phys*, 33,3074(1962)

RE-0002

ETCH NAME:

TYPE: Electrolytic, polish

COMPOSITION:

x .... 0.1 *N* NaOH

TIME:

TEMP:

ANODE: Ni

CATHODE; Re

POWER: 5—10 V/25 V

DISCUSSION:

Rh, (0001) wafers. Field emission study shows hexagonal patterns. In etching use the DC/AC arc potential to rough etch surfaces; then use AC, 0.05 V/second pulses — with arcing — to polish surfaces. Arcing will occur just below the liquid surface.

REF: Barns, O — *Phys Rev*, 97,1579(1955)

RE-0003

ETCH NAME:

TYPE: Electrolytic, thinning

COMPOSITION:

23 .... $HClO_4$

77 .... HAc

TIME:

TEMP: 20°C

ANODE: Re

CATHODE:

POWER: 6—18 V & 0.3 A/cm^2

DISCUSSION:

Re, (0001) wafers. Solution used to thin specimens for electron microscope (EM) study after neutron irradiation.

REF: Brimhall, J L & Mastel, B — *Phil Mag*, 12,419(1965)

RE-0004a

ETCH NAME: Metal

TYPE: Molten flux, preferential

TIME:

TEMP: (1) 1550°C

(2) 1460°C

COMPOSITION:

(1) x .... Fe

(2) x .... Ni

DISCUSSION:

Rh specimens. Rhenium is rapidly attacked by both molten metals.

REF: Kohl, W H — *Handbook of Materials Techniques for Vacuum Devices*, Reinhold, New York, 1967

RE-0004b

ETCH NAME: Sulfur, vapor

TYPE: Element, conversion

COMPOSITION:

x .... S, vapor

TIME:

TEMP: 445°C

DISCUSSION:

Re, specimens. Hot sulfur vapors will attack rhenium converting the surface to ReS_2 or Re_2S_7. At elevated temperature ReS_2 acts as a lubricant. Neither sulfur compound can be etched in aqua regia.

REF: Ibid.

RE-0004c
ETCH NAME: air | TIME:
TYPE: Gas, oxidation | TEMP: RT or 350°C
COMPOSITION:
x .... Air, moist
DISCUSSION:
Re, specimens. Moist air at RT will form $HReO_4$, at elevated temperature, Re_2O_7. Oxides can be removed with acids.
REF: Ibid.

RE-0004d
ETCH NAME: Chlorine | TIME:
TYPE: Gas, chlorination | TEMP: RT or hot
COMPOSITION:
x .... Cl_2, vapor
DISCUSSION:
Re, specimens. Chlorine will form $ReCl_3$ or $ReCl_5$. HCl can be used to remove the chlorides.
REF: Ibid.

RE-0004e
ETCH NAME: Sulfuric acid | TIME:
TYPE: Acid, removal | TEMP: RT or hot
COMPOSITION:
x .... H_2SO_4
DISCUSSION:
Re specimens. Solution shown as a slow etch for rhenium at RT. Used hot it will form $HReO_4$ which can be removed with acids.
REF: Ibid.

RE-0004f
ETCH NAME: Nitric acid | TIME:
TYPE: Acid, oxidation | TEMP: RT
COMPOSITION:
x .... HNO_3
DISCUSSION:
Re specimens. Nitric acid will rapidly oxidize rhenium to $HReO_4$ which can be removed with acids.
REF: Ibid.

PHYSICAL PROPERTIES OF RHODIUM, Rh

Classification	Transition metal
Atomic number	45
Atomic weight	102.9
Melting point (°C)	1966
Boiling point (°C)	4500 (3960)
Density (g/cm^3) 20°C	12.41
Thermal conductance (cal/sec)(cm^2)(°C/cm) 20°C	0.36
Specific heat (cal/g) 0°C	0.059

Heat of fusion (k-cal/g-atom)	5.2
Heat of vaporization (k-cal/g-atom)	127
Atomic volume (W/D)	8.3
1st ionization energy (K-cal/g-mole)	178
1st ionization potential (eV)	7.7
Electronegativity (Pauling's)	2.2
Covalent radius (angstroms)	1.25
Ionic radius (angstroms)	0.68 (Rh^{+3})
Coefficient of linear thermal expansion ($\times 10^{-6}$ cm/cm/°C) 20°C	8.3
Electrical resistivity (micro-ohms-cm)	4.5
Electron work function (eV)	4.65 (4.80)
Young's modulus of elasticity (psi $\times 10^6$)	60
Magnetic susceptibility (cgs $\times 10^{-6}$)	0.99
Tensile strength (psi — annealed)	110,000
Neutron cross section (barns)	156
Hardness (Mohs — scratch)	4—4.5
(Vickers kgf/mm² — annealed)	120
Crystal structure (isometric — normal)	(100) cube, fcc
Color (solid)	Grey-white
Cleavage (cubic)	(001)

RHODIUM, Rh

General: Occurs as a native element in small quantity as grains in gold and platinum alluvial deposits. It is a member of the platinum family of metals, and is found as a constituent in most native platinum from which it is separated.

Because of its high melting point and extreme inertness to chemical attack it has similar applications to those of platinum in industry. The highest temperature thermocouples are made of two wire strands, one pure platinum, the other Pt:Rh (3—15%) bead welded at one end. As an electroplated coating, it is inert to most acids and alkalies with similar applications to those of chrome plating. Several salts are highly colored yellow and black.

Technical Application: The pure metal has had little use in Solid State processing other than as a furnace thermocouple or some plated part applications.

It has been grown as a single crystal for general morphological study.

Etching: Very difficult. Mixed acids of H_2SO_4:HCl and aqua regia.

RHODIUM ETCHANTS

RH-0001

ETCH NAME: Aqua regia — TIME:

TYPE: Acid, removal/cleaning — TEMP: RT to hot

COMPOSITION:

3 .... HCl

1 .... HNO_3

DISCUSSION:

Rh specimens as wire, rods, sheets, and plated coatings. Aqua regia is a general etch for the material.

REF: Pamphlet: *Platinum,* Baker & Sons, Inc. (no date)

RH-0002: Zelinsky, T — personal communication, 1978

Rh as Pt:Rh and Pt thermocouples used for epitaxy growth furnaces. After bead weld assembly of thermocouples by electric arcing in a borax flux; degrease with TCE, rinsed in MeOH, then water. Lightly etch cleaned in aqua regia at RT, water wash and dry under IR heat lamps.

PHYSICAL PROPERTIES OF RUBIDIUM, Rb

Classification	Alkali metal
Atomic numbers	37
Atomic weight	85.4
Melting point (°C)	38.9
Boiling point (°C)	688
Density (g/cm^3) 20°C	1.532
Thermal conductance [cal(sec)(cm^2)(°C/cm)]	0.07
Specific heat (cal/g) 25°C	0.0792
Specific heat (cal/g) 0°C	0.080
Latent heat of fusion (cal/g)	6.144
Heat of vaporization (cal/g)	212
Atomic volume (W/D)	55.9
1st ionization energy (K-cal/g-mole)	96
1st ionization potential (eV)	4.176
Electronic work function (eV)	2.13 (2.09)
Electronegativity (Pauling's)	0.8
Covalent radius (angstroms)	2.16
Ionic radius (angstroms)	1.47 (Rb^{+1})
Electrical conductance (micro-ohms^{-1})	0.080
Thermal conductivity ($cal/cm^2/cm/°C/sec$) 39°C	0.07
Neutron cross section (barns)	0.73
Coefficient of linear thermal expansion ($\times 10^{-6}$ cm/cm/°C)	90
Electrical resistivity (ohms cm $\times 10^{-6}$) 20°C	12.5
Vapor pressure (°C)	514
Hardness (Mohs — scratch)	0.3
Crystal structure (isometric — normal)	(100) cube, bcc
Color (solid)	Silver/reddish
(flame)	Dark red
Cleavage (cubic)	(001)

RUBIDIUM, Rb

General: Does not occur as a native element in nature. It is a rare alkali metal found chiefly in the mica Lipidolite and some other lithium-bearing minerals. On the island of Elba as Castor and Pollux . . . neither listed as mineral species, circa 1950. The metal forms salts similar to those of potassium.

Little or no use in general industry due to rarity although there are over 50 rubidium chemical compounds, and it forms acid compounds with nitric acid. It has some application in pyrotechnics for its deep red flame color.

Technical Application: The pure metal has had little use in Solid State processing, in part, due to it being one of the softest of known metals with H = 0.3.

Rubidium has been grown as a single crystal for general morphological study, and several compounds such as bromide and iodide also have been studied.

Etching: Acids.

RUBIDIUM ETCHANTS

RB-0001
ETCH NAME: Water
TYPE: Acid, removal
COMPOSITION:
x .... H_2O
TIME:
TEMP: RT
DISCUSSION:
Rb metal will dissolve in water. Initial discovery of both cesium and rubidium were from waters in Durkheim by Bunsen 1860/1861. Named for its deep red spectral lines.
REF: Foster, W & Alyea, H N — *An Introduction to General Chemistry,* 3rd ed, D Van Nostrand, New York, 1947, 365

PHYSICAL PROPERTIES OF RUBIDIUM BROMIDE, RbBr

Classification	Bromide
Atomic numbers	37 & 35
Atomic weight	165.4
Melting point (°C)	682
Boiling point (°C)	1340
Density (g/cm³)	3.35
Refractive index (n=)	1.5530
Hardness (Mohs — scratch)	2—3
Crystal structure (isometric — normal)	(100) cube
Color (solid)	Colorless
Cleavage (cubic)	(001)

RUBIDIUM BROMIDE ETCHANTS

RBB-0001
ETCH NAME: Acetone
TYPE: Keytone, polish
COMPOSITION:
10 .... CH_3COCH_3 (acetone)
1 .... H_2O
TIME:
TEMP: RT
DISCUSSION:
RbBr, (001) wafers used to measure elastic constants using ultrasonic cw resonance. Mechanically rough lap wafers with A/O 302 grit. Mechanically polish with A/O 303$^1/_2$ grit on a cast iron lap; then Linde A, then sapphire dust on cloth. Use N-amyl alcohol as liquid carrier for all lap and polish. Etch polish with the solution shown and rinse in MeOH. See Rubidium Iodide.
REF: Bulef, D I & Menes, M — *J Appl Phys,* 31,1010(1960)

PHYSICAL PROPERTIES OF RUBIDIUM IODIDE, RbI

Classification	Iodide
Atomic numbers	37 & 53
Atomic weight	212.4
Melting point (°C)	642
Boiling point (°C)	1300

Density (g/cm^3)	3.55
Refractive index (n =)	1.6474
Hardness (Mohs — scratch)	2—3
Crystal structure (isometric — normal)	(100) cube
Color (solid)	Colorless
Cleavage (cubic)	(001)

RUBIDIUM IODIDE, RbI

General: Does not occur as a natural compound due to scarcity of the element, and water solubility of the iodide. It was first discovered by Bunsen in 1861 from its red spectral lines associated with lithium in the mineral petalite. It is an alkaline metal with characteristics similar to potassium.

Technical Application: The iodide has had no application in Solid State processing, although it has been grown and studied as a single crystal.

Etching: Acetone, water.

RUBIDIUM IODIDE ETCHANTS

RBI-0001
ETCH NAME: Acetone — TIME:
TYPE: Alcohol, polish — TEMP: RT
COMPOSITION:
10 .... CH_3COCH_3 (acetone)
1 .... H_2O
DISCUSSION:

RbI, (001) wafers used to measure elastic constants using ultrasonic cw resonance. Mechanically rough lap with A/O 302 grit. Mechanically polish with 303$^1/_2$ A/O grit on a cast iron lap; then Linde A, then sapphire dust on cloth. Use N-amyl alcohol as liquid carrier for lap and polish. Etch polish with the solution shown and rinse in MeOH. See Rubidium bromide.

REF: Bulef, D I & Menes, M — *J Appl Phys,* 31,1010(1960)

PHYSICAL PROPERTIES OF RUBIDIUM SULFIDE, RuS_2

Classification	Sulfide
Atomic numbers	44 & 16
Atomic weight	165.82
Melting point (°C)	>500
Boiling point (°C)	
Density (g/cm^3)	6.99
Hardness (Mohs — scratch)	7.5
Crystal structure (isometric — normal)	(111) octahedron
Color (solid)	Iron-black
Cleavage (octahedral)	(111)

RUBIDIUM SULFIDE, RbS_2

General: Occurs in nature essentially as the minor mineral laurite, RuS_2. It is found as small octahedrons, or in grains associated with the gold and platinum washings in placer deposits.

Not used in the metals industry as the natural mineral due to scarcity.

Technical Applications: No use in Solid State processing at the present time, but under evaluation as the sulfide, selenide, and telluride for general morphology, to include IR spectra and possible semiconductor characteristics. The reference shown is for all three compounds.

Etching: HNO_3.

PHYSICAL PROPERTIES OF RUTHENIUM, Ru

Classification	Transition metal
Atomic number	44
Atomic weight	101.7
Melting point (°C)	2500 (2450)
Boiling point (°C)	4900 (4111)
Density (g/cm^3)	12.45
Specific heat (cal/g) 20°C	0.055
Heat of fusion (k-cal/g-atom)	6.1
Latent heat of fusion (cal/g)	60.3
Heat of vaporization (k-cal/g-atom)	148
Atomic volume (W/D)	8.3
1st ionization energy (K-cal/g-mole)	173
1st ionization potential (eV)	7.5
Electronegativity (Pauling's)	2.2
Covalent radius (angstroms)	1.25
Ionic radius (angstroms)	0.67 (Ru^{+4})
Coefficient of linear thermal expansion ($\times 10^{-6}$ cm/cm/°C)	9.1
Electrical resistivity (micro-ohms-cm) 20°C	7.2 (6.8)
Vapor pressure (microns @ m.p.)	9.8
Young's modulus (psi $\times 10^6$)	60
Magnetic susceptibility (cgs $\times 10^{-6}$)	0.43
Cross section (barns)	2.56
Hardness (Mohs — scratch)	4—5
(Vicker's — kgf/mm^2) annealed	220
Crystal structure (hexagonal — normal)	$(10\bar{1}0)$ prism, hcp
Color (solid)	Black
Cleavage (basal)	(0001)

RUTHENIUM, Ru

General: Although the *Handbook of Chemistry and Physics* refers to its occurrence as a native element, Dana's *Textbook of Mineralogy* does not. Regardless, it is a very rare element and does occur as a minor constituent in platinum and iridosmine. The major occurrence of platinum is in the Ural Mountains of Russia, although it is found throughout the world in minor quantities often associated with magnetite in gold sands.

No use as a metal in industry, primarily due to its scarcity, though it has extremely high temperature characteristics similar to platinum, iridium, and osmium.

Technical Application: Ruthenium metal has had no major use in Solid State material processing, although it has been grown as a single crystal for general study. It also has been evaluated as a compound semiconductor, as a selenide, sulfide, and telluride. It also is under evaluation and study as a silicide: RuSi, Ru_2Si, $RuSi_2$, for possible application as a buffer layer in semiconductor devices similar to other metal silicides.

Etching: HCl, H_2SO_4, HNO_3 acid and aqua regia. RF plasma ionized gases.

RUTHENIUM ETCHANTS

RU-0001
ETCH NAME: Aqua regia — TIME:
TYPE: Acid, removal — TEMP: RT
COMPOSITION:
3 .... HCl
1 .... HNO_3
DISCUSSION:
Ru specimens. Aqua regia is a general polish and removal etch for this element. It is only slowly soluble in single acids.
REF: Hodgman, C D — *Handbook of Chemistry and Physics,* 27th ed, Chemical Rubber Co., Cleveland, 1943, 446

RU-0002
ETCH NAME: Argon — TIME:
TYPE: Ionized gas, cleaning — TEMP:
COMPOSITION: — GAS FLOW:
x .... Ar^+ ions — PRESSURE: 10^{-5} Torr
POWER: 1 KeV
DISCUSSION:
Ru, (100) wafers cut within $\pm 2°$ of plane — 8 mm discs spark cut from single crystal ingots. Mechanical polish with diamond paste. In vacuum, RF plasma Ar^+ ion clean at 1 KeV and 10^{-5} Torr and follow by annealing at 750 to 800°C. Manganese was then deposited on surfaces.
REF: Heinrich, B et al — *J Vac Sci Technol,* B3(2),766(1985)

RUTHENIUM SELENIDE ETCHANTS

RUSE-0001
ETCH NAME: Nitric acid — TIME:
TYPE: Acid, polish — TEMP:
COMPOSITION:
x .... HNO_3
DISCUSSION:
$RuSe_2$, single crystal specimens. Cut specimens and mechanically polish with 0.5 µm diamond paste. Etch in solution shown to polish and remove lap damage. Other single crystals grown were FeS_2, $MnTe_2$, RuS_2, $RuTe_2$, OsS_2, $OsTe_2$, PtP_2, $PtAs_2$, and $PtSb_2$. All crystals are cubic in structure like pyrite, FeS_2 and were used in an IR spectra study. (*Note:* Pyrite is in the pyritohedral class of the isometric system, not the normal class which contains the cube, (100). In addition, the natural mineral laurite is not of pyrite structure.)
REF: Lutz, H D et al — *J Phys Chem Solids,* 46,437(1985)

RUTHENIUM SULFIDE ETCHANTS

RUS-0001
ETCH NAME: Nitric acid — TIME:
TYPE: Acid, polish — TEMP:

COMPOSITION:

x .... HNO_3

DISCUSSION:

RuS_2, single crystal specimens. Cut specimens and mechanically polish with 0.5 μm diamond paste. Etch in solution shown to polish and remove lap damage. Other single crystals grown were FeS_2, $MnTe_2$, $RuSe_2$, $RuTe_2$, OsS_2, $OsTe_2$, PtP_2, $PtAs_2$, and $PtSb_2$. All crystals are cubic structure like pyrite, FeS_2, and were used in an IR spectra study. (*Note:* See discussion under the selenide with regard to material structure.)

REF: Lutz, H D et al — *J Phys Chem Solids,* 46,437(1985)

RUTHENIUM TELLURIDE ETCHANTS

RUTE-0001

ETCH NAME: Nitric acid — TIME:

TYPE: Acid, polish — TEMP:

COMPOSITION:

x .... HNO_3

DISCUSSION:

$RuTe_2$, single crystal specimens. Cut specimens and mechanically polish with 0.5 μm diamond paste. Etch in solution shown to polish and remove lap damage. Other crystals grown were FeS_2, $MnTe_2$, RuS_2, $RuSe_2$, OsS_2, $OsTe_2$, PtP_2, $PtAs_2$, and $PtSb_2$. All crystals are cubic in structure like pyrite, FeS_2 and were used in an IR spectra study. (*Note:* See discussion under the selenide with regard to pyrite crystal structure.)

REF: Lutz, H D et al — *J Phys Chem Solids,* 46,437(1985)

PHYSICAL PROPERTIES OF SAMARIUM, Sm

Classification	Lanthanide
Atomic number	62
Atomic weight	150.35
Melting point (°C)	1072
Boiling point (°C)	1900
Density (g/cm^3)	7.55
Specific heat (cal/g) 25°C	0.042
Heat of fusion (k-cal/g-atom)	2.1
Latent heat of fusion (cal/g)	17.29
Heat of vaporization (k-cal/g-atom)	46
Atomic volume (W/D)	19.9
1st ionization energy (K-cal/g-mole)	129
1st ionization potential (eV)	5.6
Electronegativity (Pauling's)	1.2
Covalent radius (angstroms)	1.62
Ionic radius (angstroms)	1.0 (Sm^{+3})
Electrical resistivity ($\times 10^{-6}$ ohms cm) 25°C	90 (10^5 @ 298 K)
Compressibility (cm^2/kg $\times 10^{-6}$)	3.34
Magnetic moment (Bohr magnetons)	1.74
Tensile strength (psi)	18,000
Yield strength (psi)	16,200
Magnetic susceptibility ($\times 10^{-6}$ emu/mole)	1,275
Transformation temperature (°C)	924
Cross section (barns)	5800
Hardness (Mohs — scratch)	4—5
Crystal structure (hexagonal — rhombohedral)	$(10\bar{1}1)$ rhomb
Color (solid)	Grey
Cleavage (basal)	(0001)

SAMARIUM, Sm

General: Does not occur as a native element. It is one of the more rare of the cerium group rare earths in the lanthanide series, and found primarily in the mineral samarskite, a tantalate containing rare earths with iron, calcium, and niobium. There has been no use of the material in industry due to scarcity, but several salts have interest in chemistry.

Technical Application: Samarium has not had much use in Solid State processing, although it may have applications in garnet formulations like other rare earths. It has been grown as a single crystal for general morphology study and, as a polycrystalline cylinder, for electrical resistivity measurements.

Etching:

SAMARIUM ETCHANTS

SM-0001

ETCH NAME: Heat | TIME:

TYPE: Thermal, forming | TEMP: Elevated

COMPOSITION:

x .... heat

DISCUSSION:

Sm, specimens were arc melted as buttons and then machined into cylinders $^3/_{16}$ to 2″

in diameter. Used for measuring electrical resistivity. Other materials studied were La, Pr, and Nd.
REF: Alstad, J K et al — *Phys Rev,* 122,1639(1961)

PHYSICAL PROPERTIES OF SAMARIUM BROMIDE, $SmBr_3$

Classification	Bromide
Atomic numbers	62 & 34
Atomic weight	239.5
Melting point (°C)	>70
Boiling point (°C)	
Density (g/cm³)	~3.xx
Hardness (Mohs — scratch)	2—3
Crystal structure (hexagonal — normal)	$(10\bar{1}0)$ prism
Color (solid)	Colorless
Cleavage (basal)	(0001)

SAMARIUM BROMIDE, $SmBr_3$

General: Does not occur as a natural compound. The metal is one of the rare earths of the cerium lanthanide series. There are a few mineral bromides containing silver, such as bromyrite, AgBr, but none containing samarium. There is no industrial use of samarium bromide.

Technical Application: Samarium bromide has been grown as a single crystal by the Horizontal Bridgman (HB) method and general study of material characteristics. It has been fabricated as a Quantum Counter. The material is hygroscopic and should be handled under an inert gas atmosphere.

Etching: Alcohols.

SAMARIUM BROMIDE ETCHANTS

SMBR-0001
ETCH NAME: Ethyl alcohol — TIME:
TYPE: Alcohol, polish — TEMP: RT
COMPOSITION:
x .... EOH
DISCUSSION:

$SmBr_3$ (0001) wafers cut from ingots grown by a modified Bridgman method with a brass inner line for a sharp heat gradient. Material used as a base for Quantum Counters. Polish surface with alcohol. Specimens cut 1—2 cm² and 2—5 mm thick, were clear, transparent crystals. $SmBr_3$ is hexagonal: $LaBr_3$ is orthorhombic crystal structure, and both are hygroscopic.
REF: Krasutsky, N J — *J Appl Phys,* 54,126(1983)

SAMARIUM COBALT ETCHANTS

SMCO-0001
ETCH NAME: Citric acid, dilute — TIME:
TYPE: Acid, cleaning — TEMP: 80°C
COMPOSITION:
x .... citric acid
x .... H_2O

DISCUSSION:

$SmCo_5$, specimens fabricated as permanent magnets. Solutions of citric acid can be used to etch clean this material.

REF: Buschow, K H L et al — *J Appl Phys*, 40,4029(1969)

PHYSICAL PROPERTIES OF SAPPHIRE, Al_2O_3

Classification	Oxide
Atomic numbers	13 & 8
Atomic weight	102
Melting point (°C)	1800
Boiling point (°C)	2250
Density (g/cm^3)	3.99
Softening point (°C)	2050
Thermal conductance (cal/sec/cm/°C sec) 25°C	0.086
Coefficient of thermal linear expansion	5.4/6.2 (xtl)
($\times 10^{-6}$ cm/cm/°C)	7.3/8.1 (fused)
Specific heat (cal/g/°C) 25°C	0.18
Electrical resistance (ohms mm^2/cm) @ 500°C	10^{11}
@ 1000°C	10^6
@ 2000°C	10^3
Dielectric constant (e=)	7.5—10.5
Refractive index (n=)	1.769—1.759
Transmission (visible light)	Excellent
(infrared)	50—85%
(ultraviolet)	50—80%
Compressive strength (kg/cm^2)	2100
Modulus of rupture (kg/cm^2)	4000
Atomic spacing (angstroms)	a_o 4.748
	c_o 12.99
Hardness (Mohs — scratch)	9
(Knoop — kgf/mm^2)	1800—2200
(Vicker's — kgf/mm^2)	1570—1800
Crystal structure (hexagonal — normal)	$(10\bar{1}0)$ prism
Color (solid)	Colorless/red/blue
Cleavage (basal)	(0001)

SAPPHIRE, Al_2O_3

General: Occurs in nature as a sub-species of the mineral corundum, Al_2O_3. Corundum is an accessory mineral associated with rocks of the chlorite group, limestone and dolomite found embedded as small masses to large beds. The best gem quality material comes from upper Burma — *in situ* in limestone, in the soil, and in gem-bearing gravels of the Irrawaddy River — mostly ruby. In Cambodia and Siam, as fine blue sapphire. In India, light blue stones from Madras and other districts. Gem stone sapphires of different colors are found throughout the world, as are ordinary corundum and emery.

Corundum: all varieties of dark or dull colors and not transparent, with colors from blue to grey, brown, or black. Amadantine spar from India is greyish brown, but greenish/bluish by transmitted light. Major use has been as an abrasive.

Emery: Black to dark grey in color as a mixture of corundum and magnetite, sometimes hematite. Varies from fine-grained emery to coarse grained with distinct corundum crystals embedded. Still widely used as an abrasive.

Gem Stones: The transparent to translucent varieties as single crystals are all sapphires. True sapphire is blue; true ruby or Oriental ruby, red; Oriental topaz, yellow; Oriental emerald, green; Oriental amethyst, purple. Asteriated or star sapphire, when cut on the z-axis shows a six-rayed star when viewed on the vertical z-axis caused by minute cylindrical cavities parallel to prism planes. Best quality is a fine grey-blue, near-translucent stone with grey-white "star" rays.

Artificial sapphire can be made in several ways: by melt flux growth; by an electric arc; re-melt of corundum chips — add chromium salt for red (ruby) — cobalt for blue. The Vernieul Process for growth of large single crystal boules (ingots) was developed in the late 1800s for artificial ruby. The system consists of a hopper holding powder that is fed down a tube past a hydrogen flame. The flame liquefies the powder into droplets, which with gravity fall through a control orifice into a vacuum chamber, and onto an oriented seed pedestal at the bottom. Single crystal boules $1^1/_2''$ in diameter and up to 6″ long have been grown in this manner. The method also has been used to grow single crystal hemispheres (see Silicon and Germanium), and is still used to grow high temperature melting single crystals, today. A ruby melt also is poured into molybdenum forms for production of rods, sheet, tubes, prisms, and other specialized shapes fabricated, such as spheres for high temperature ball bearings, and wire bonding tips.

Artificial single crystal gem stones of all materials can be differentiated from the natural stones by microscopic examination of internal artifacts. The artificial stones will show curved growth lines; whereas the natural stones, straight and angled growth lines. The artificial stones also may show flow lines, such as the "paste" jewelry made from glass . . . which was being made by the Egyptians as early as 3000 B.C. Natural transparent stones, such as diamonds, quartz (amethyst, rock crystal, etc.), obsidian (natural glass) may contain flecks of graphite (diamond), or other material (rutilated quartz).

Some "picture" agates (a form of quartz), when cut in thin section, show a skyline like picture of trees, etc., such as that found in the Colorado area near the Grand Canyon, and there is "snowflake" obsidian containing flower-like inclusions of calcium carbonate. Natural stones also may have fine cavities and bubbles, some containing gas or entrapped liquid remaining from their growth environment. The gases or liquids in such material, quartz in particular, are used to establish atmospheric and other geologic conditions that existed at the time of formation as far back as 10 million years. Artificial crystals are not grown with such voids, though they can occur during diffusion of an element or gas, or from ionized particle irradiation. The garnet bubble memory devices refer to the ordered location of iron atoms in a solid matrix, not to an actual bubble or void.

As artificial fused (clear) sapphire with a Mohs hardness of $H = 9$, it is nearly impervious to being scratched, and is widely used as watch and clock faces. And there have been occasional references to the use of powdered sapphire as a lap and polish abrasive, although fine white alumina abrasives are more often ground from common corundum sources.

Technical Application: Artificial sapphires as both fused and single crystal substrates are used in the Solid State field to fabricate microelectronic circuits after thin film metallization (Au/Cr; Au/Ni, etc.). Discrete units as insulators and mounting bases for semiconductor devices, or active elements are fabricated as modulators and filters in circuitry similar to applications of quartz, glass, etc.

Both sapphire and quartz are in the hexagonal system and, as single crystal sapphire, are normal cut and used as basal (0001) oriented substrates or as either first order or second order prisms, $(10\bar{1}0)$ and $(10\bar{2}0)$, respectively. Like the circular semiconductor wafers, a directionally oriented flat is often cut at the sapphire blank periphery . . . on silicon as a $\langle 110 \rangle$ direction flat . . . on sapphire or quartz, when fabricated as a square (0001) blank, a corner orientation $(10\bar{1}0)$ prism is edge cut.

It should be noted that, in dicing sapphire and quartz into square or rectangular shape, the first cut — parallel to a prism face — is easily done but, the second cut at right angles is very difficult, as the normal angles between prism faces in the hexagonal system are at 60°, not 90°, as they are with silicon in the isometric system.

In addition to the use of sapphire as circuit substrates or stand-offs, the use of ruby as wire ball bonding tips in semiconductor device fabrication has already been mentioned, the substrates also have been used for deposition and growth studies of thin film materials, such as AlN and GaN.

Single crystal sapphire has been studied in both natural and artificial forms for many years, much work in regard to the cutting and polishing of gem stone materials.

Etching: Difficult. H_3PO_4, molten fluxes. Dry chemical etching (DCE) with ionized gases. See Aluminum oxide.

SAPPHIRE ETCHANTS

SAP-0001

ETCH NAME: TIME:

TYPE: Acid, clean TEMP: Hot(?)

COMPOSITION:

1 .... H_3PO_4

1 .... H_2SO_4

DISCUSSION:

Al_2O_3 (0001) wafers used as substrates for MOCVD growth of GaN, AlN and AlGaN. The latter is an n-type to semi-insulating (SI) semiconductor depending upon the ratio of Al:Ga. The AlGaN thin film is smooth and transparent with some hexagonal pyramid growth near the substrate edge with pyramid edges ⟨1021⟩. Prior to MOCVD, degrease the sapphire substrates in solvents and etch clean with the solution shown. Though a temperature was not shown, solution was probably warm to hot.

REF: Matlonbian, M & Gerehenzon, M — *J Electron Mater,* 14,633(1985)

SAP-0002a

ETCH NAME: Phosphoric acid TIME:

TYPE: Acid, preferential TEMP: Hot

COMPOSITION:

x .... H_3PO_4

DISCUSSION:

Al_2O_3, (0001) wafers. Solution will develop dislocation etch pits similar to those observed after silicon has been deposited and removed.

REF: Manasevit, H M & Simpson, W I — *J Appl Phys,* 35,1349(1964)

SAP-0002b

ETCH NAME: Silicon TIME:

TYPE: Metal, preferential TEMP: >1200°C

COMPOSITION:

x .... Si

DISCUSSION:

Al_2O_3, (0001) wafers. Silicon metal evaporated and then removed from sapphire surfaces will develop dislocation pits in the sapphire.

REF: Ibid.

SAP-0003
ETCH NAME: Trichloroethylene TIME: 2—5 min
TYPE: Solvent, cleaning TEMP: Boiling
COMPOSITION:
x .... $CHCl:CCl_2$ (TCE)
DISCUSSION:

Al_2O_3, clear fused sapphire blanks and (0001) oriented single crystal blanks used as substrates for metallization of microelectronic circuits. Blanks were cleaned in boiling TCE, transferred to fresh TCE with ultrasonic agitation for additional 2—5 min. Rinse in acetone, then MeOH and nitrogen blow dry.

REF: Marich, L A & Porter, R — personal communication, 1980

SAP-0004a
ETCH NAME: RCA Etch (AB Etch) TIME: 10—15 min
TYPE: Acid, cleaning TEMP: 70—80°C
COMPOSITION:

(A)		(B)	
1	NH_4OH	1	HCl
1	H_2O_2	1	H_2O_2
6	H_2O	6	H_2O

DISCUSSION:

Al_2O_3, clear fused sapphire blanks and (0001) single crystal oriented blanks used as substrates for metallization of microelectronic circuits. The solutions are used separately; clean in solution "A", then transfer to solution "B" while still wet. Solutions should be mixed fresh and heated when ready for use. After cleaning, rinse in MeOH and nitrogen blow dry. Also followed cleaning with 30 sec HF dip, water rinse 30 sec and N_2 blow dry. A third process step included holding blanks under MeOH until placed in an evaporator system still wet with MeOH. This gave best adhesion results of metal films. This cleaning system also used on alumina, pressed powder blanks and fused/single crystal quartz blanks. Holding wafers or substrates under methanol after etching, and until being placed in a diffusion tube or vacuum evaporator, was used by Walker on silicon and other semiconductor materials in 1963.

REF: Walker, P & Porter, R — personal application, 1981

SAP-0004b
ETCH NAME: Hydrofluoric acid TIME: 30 sec
TYPE: Acid, cleaning TEMP: RT
COMPOSITION:
x .... HF, conc.
DISCUSSION:

Al_2O_3, clear fused sapphire blanks and (0001) single crystal oriented blanks used as substrates for metallization of microelectronic circuits. HF used as a final cleaning dip prior to vacuum metallization. Follow dip with DI water rinse, MeOH rinse and N_2 blow dry or hold under MeOH until placed in a vacuum system still wet. The latter method gave best adhesion results.

REF: Ibid.

SAP-0004c
ETCH NAME: TIME: 15—30 sec
TYPE: Halogen, surface treatment TEMP: RT
COMPOSITION:
5 mg .... I_2
150 ml .... MeOH

DISCUSSION:

Al_2O_3, clear fused sapphire blanks and (0001) single crystal oriented blanks used as substrates for metallization of microelectronic circuits. After previous blank cleaning, rinse in this solution and N_2 blow dry immediately before placing blanks in a vacuum system or a diffusion system. Iodine complexes ionic surface contamination when parts are heated to >100°C producing a molecularly clean surface for metallization or doping diffusion. This method used on single crystal quartz blanks, alumina pressed power blanks and silicon.
REF: Ibid.

SAP-0005
ETCH NAME: Freon 113 TIME: 1—2 min
TYPE: Solvent, degreasing TEMP: Hot
COMPOSITION:
x .... Freon 113, vapor
DISCUSSION:

Al_2O_3, clear fused sapphire blanks and (0001) single crystal oriented blanks used as substrates for microelectronic circuit metallization. Solution used as the vapor degreaser system solvent for sapphire in addition to other substrate materials and assembly parts.
REF: DaLuca, J & Gunshinan, B — personal communication, 1985

SAP-0006
ETCH NAME: Metallization TIME:
TYPE: Metal, thin film coating TEMP: 1950°C
COMPOSITION:
x .... Nb
DISCUSSION:

Al_2O_3, single crystal spheres were metallized with niobium. Article describes experimental method developed for coating a sphere.
REF: Strayer, D M et al — *NASA Tech Briefs,* Fall 1985, 158

SAP-0007
ETCH NAME: Hydrofluoric acid TIME:
TYPE: Acid, cleaning TEMP: RT
COMPOSITION:
1 .... HF
10 .... H_2O
DISCUSSION:

Al_2O_3, as fused sapphire parts in amorphous form (red ruby). A hot liquidized melt of the material is poured into molybdenum metal forms to fabricate rods, tubes, sheets, etc. Spheres, for use as high temperature ball bearings, ground and polished with diamond paste. Solution used for general cleaning of parts. Ruby wire bonding tips also fabricated.
REF: Brochure

SAP-0008
ETCH NAME: Soap TIME:
TYPE: Fatty acid, cleaning TEMP: Warm
COMPOSITION:
x .... Ivory soap
x .... H_2O
DISCUSSION:

Al_2O_3, as natural single crystals as blue sapphires, red rubies, and star sapphires fab-

ricated as gem stones. A soap solution was used to wash and remove oil of diamond paste polishing compounds after facetting in orientation studies of these materials.
REF: Walker, P — gem stone fabrication (1953—1985)

SAP-0009
ETCH NAME: Potassium hydroxide — TIME: 2 min
TYPE: Alkali, dislocation — TEMP: 300°C
COMPOSITION:
x .... KOH, fused
x .... 10—15% H_2O
DISCUSSION:
Al_2O_3, single crystal sapphire $(10\bar{1}2)$ and (0001) orientations. Solution used to develop etch pits and dislocations in Czochralski (CZ) grown ingot material.
REF: Kim, K M & McFarlande, S H — *J Appl Phys*, 49,6171(1978)
SAP-0010: Forgeng, W D & Webb, W W — *J Appl Phys*, 28,1449(1957)
Al_2O_3 grown as single micro-crystal sapphire in a study of growth mechanisms and defect structure.

SAP-0011
ETCH NAME: Heat — TIME:
TYPE: Thermal, cleaning — TEMP: 1200°C
COMPOSITION:
x .... heat
DISCUSSION:
Al_2O_3, as (0001) or $(01\bar{1}2)$ sapphire blanks used as substrates for growth of AlN and GaN by MBE. Clean sapphire in the vacuum system at temperature shown — reduce to 1100°C at 2×10^{-5} Torr for Al/NH_3 deposition of AlN; then to 700°C at 2×10^{-4} Torr with Ga/NH_3 for a 1000 Å thick GaN final cover thin film. GaN was n-type. GaN has better uniformity deposited on AlN than directly on sapphire. Coefficient of thermal expansion and atomic spacing shown for the three materials.
REF: Yoshida, S — *J Vac Sci Technol*, B1(2),250(1983)

PHYSICAL PROPERTIES OF SCANDIUM, Sc

Classification	Transition metal
Atomic number	21
Atomic weight	44.94
Melting point (°C)	1539
Boiling point (°C)	2730
Density (g/cm³)	3.0
Thermal conductance (cal/sec)(cm²)(°C/cm)	0.015
Specific heat (cal/g) 25°C	0.13
Latent heat of fusion (cal/g)	84.52
Heat of vaporization (k-cal/g-atom)	81
Heat of fusion (k-cal/g-atom)	3.8
Atomic volume (W/D)	15
1st ionization energy (K-cal/g-mole)	151
1st ionization potential (eV)	6.56
Electronegativity (Pauling's)	1.3
Covalent radius (angstroms)	1.44
Ionic radius (angstroms)	0.81 (Sc^{+3})

Electrical resistivity (micro-ohms-cm) 298 K	50.9 (61)
Compressibility ($cm^2/kg \times 10^{-6}$)	2.26
Transformation temperature (°C)	1335
Cross section (barns)	23
Hardness (Mohs — scratch)	5—6
Crystal structure (hexagonal — normal)	$(10\bar{1}0)$ prism, hcp
Color (solid)	Silver grey
Cleavage (basal)	(0001)

SCANDIUM, Sc

General: Does not occur as a native element. It is extremely rare and the metal is found as a trace element in rare earth minerals containing cerium and cesium. Although a transition metal, it is sometimes listed as a rare earth. No use in metal industries due to scarcity, but salts are of use in chemistry.

Technical Application: No use in Solid State processing, to date, due to scarcity. It was not separated as a metal until the middle of this century, but has been grown as a single crystal for general morphology study.

Etching: HCl, HNO_3, aqua regia.

SCANDIUM ETCHANTS

SC-0001a
ETCH NAME: Hydrochloric acid — TIME:
TYPE: Acid, polish — TEMP:
COMPOSITION:
x .... HCl
DISCUSSION:
Sc, (0001) wafers. Solution used as a polish etch.
REF: Colvin, R V & Arajs, S — *J Appl Phys,* 34,286(1963)

SC-0001b
ETCH NAME: Nitric acid — TIME:
TYPE: Acid, polish — TEMP:
COMPOSITION:
x .... HNO_3
DISCUSSION:
Sc, (0001) wafers. Solution used as a polish etch.
REF: Ibid.

SC-0001c
ETCH NAME: — TIME:
TYPE: Acid, removal — TEMP:
COMPOSITION:
x .... HCl
x .... HNO_3
DISCUSSION:
Sc, polycrystalline rod dissolved in this solution to determine impurity content.
REF: Ibid.

PHYSICAL PROPERTIES OF SELENIUM, Se

Classification	Non-metal
Atomic number	34
Atomic weight	78.96
Melting point (°C)	217 (grey form)
Boiling point (°C)	685
Density (g/cm^3)	4.79
Thermal conductance (cal/sec)(cm^2)(°C/cm $\times 10^{-4}$)	0.1 (7—18
Specific heat (cal/g) 25°C	0.084
Heat of fusion (k-cal/g-atom)	1.25
Latent heat of fusion (cal/g)	16.4
Heat of vaporization (k-cal/g-atom)	3.34
Atomic volume (W/D)	16.5
1st ionization energy (K-cal/g-mole)	225
1st ionization potential (eV)	9.75
Electronegativity (Pauling's)	2.4
Covalent radius (angstroms)	1.16
Ionic radius (angstroms)	0.05 (Se^{+4})
Coefficient of linear thermal expansion ($\times 10^{-5}$ cm/cm/°C)	3.79 (3.68)
Electrical resistivity (micro ohms^{-1})	0.08
Compressibility (atm^{-1} $\times 10^{-5}$)	12.2
Magnetic susceptibility ($\times 10^{-6}$ emu/mole)	0.32
Cross section (barns)	12
Vapor pressure (°C)	554
Electrical resistivity ($\times 10^{6}$ ohm cm) 20°C	10
Electron work function (eV)	4.72
Hardness (Mohs — scratch)	2
Crystal structure (hexagonal — normal)	($10\bar{1}0$) prism (grey type)
(monoclinic — normal)	(100) a-pinacoid (red type)
(amorphous)	None (black type)
(colloidal)	None (dark red type)
Color (solid)	Dull grey (Hex)
(flame)	Azure blue
Cleavage (basal)	(0001) or (001)

SELENIUM, Se

General: Does not occur as a native element. It is a rare element and largely found as a trace element in a number of minerals of the pyrite group, e.g., pyrite, FeS_2 and similar sulfide, telluride, arsenide, and antimonide minerals. Selenium is obtained from the flue dust during the roasting of pyrite ores. It is in the sulfur family, which it resembles in its physical forms and compounds. Can be separated and formed as pure metal colloid, amorphous material, or crystallized in the two crystallographic structures shown in the Physical Properties table above. In the colloidal or amorphous forms it is brilliant red in color; whereas the crystalline forms are a dull grey. As a crystal, its electrical conductivity increases with the brightness of light.

In some areas it occurs in the soil in sufficient amounts to be dangerous to animals and plants as it is a toxic substance.

In general industry its major use is in glass and ceramic fabrication in the colloidal red form as a coloring agent. It is occasionally used as a minor constituent in some metal alloys.

Technical Application: Selenium is an n-type dopant in several compound semiconductors, and there are several binary and trinary compound semiconductors such as zinc selenide,

ZnSe; cadmium selenide, CdSe or cadmium sulfur selenide, CdSSe and lead tin selenide, PbSnSe etc.

As the pure single crystal it has been fabricated as a rectifier in the semiconductor industry for many years as well as a photoelectric cell. It has a very distinctive, unpleasant odor and can produce "selenium breath" similar to that of tellurium.

Etching: H_2SO_4, HNO_3, CS_2, mixed acids and alkalies.

SELENIUM ETCHANTS

SE-0001
ETCH NAME: Sulfuric acid
TIME:
TYPE: Acid, removal
TEMP: RT
COMPOSITION:
x .... H_2SO_4
DISCUSSION:

Se, deposits remaining on the $(\overline{1}\overline{1}\overline{1})$B surface of HgSe wafers after preferential etching in $1HCl:1HNO_3$. Se film can be removed with the etchant shown.

REF: Warekois, E P et al — *J Appl Phys,* 33,690(1962)
SE-0002: Heleskivt, J et al — *J Appl Phys,* 40,2923(1969)

Se, single crystals (called "trigonal" crystals) used to measure the Hall effect. After lapping surfaces, solution used to remove damage.

SE-0003a
ETCH NAME:
TIME:
TYPE: Acid, preferential
TEMP:
COMPOSITION:
x .... H_2SO_4
x .... HNO_3
DISCUSSION:

Se, single crystal wafers used in an X-ray topography study of defects. Solution used to develop defects.

REF: Maukkarinen, K & Toomi, T O — *J Appl Phys,* 40,3054(1969)

SE-0003b
ETCH NAME:
TIME:
TYPE: Halogen, preferential
TEMP:
COMPOSITION:
x .... x% Br_2
x .... MeOH
DISCUSSION:

Se, single crystal wafers. See SE-0003a for discussion.

REF: Ibid.

SE-0004a
ETCH NAME:
TIME:
TYPE: Acid, preferential
TEMP:
COMPOSITION:
x .... HNO_3
DISCUSSION:

Se, single crystal specimens used in a material study using an electron microscope (EM) to observe defects. Solution used to etch thin specimens for microscope study.

REF: Chihaya, T et al — *Nippon Kinzoku Gakkaishi,* 17,65(1955)

SE-0004b
ETCH NAME: Sodium hydroxide
TYPE: Alkali, removal
COMPOSITION:
x .... x% NaOH
DISCUSSION:
TIME:
TEMP:

Se, single crystal specimens. Solution used to etch thin specimens. Authors say that bulk properties differ from surface properties.

REF: Ibid.

SE-0004c
ETCH NAME: Carbon disulfides
TYPE: Acid, preferential
COMPOSITION:
x .... Cs_2
DISCUSSION:
TIME:
TEMP:

Se, single crystal specimens. Solution used to etch thin specimens.

REF: Ibid.

SE-0005
ETCH NAME: Heat
TYPE: Thermal, cleaning
COMPOSITION:
x .... heat
DISCUSSION:
TIME:
TEMP: 300°C

Se, powder used for epitaxy growth of CdSe and $(Cd,Se)_{1-x}Zn_x$. Clean material by vacuum baking at 300°C and then re-grind to powder. **CAUTION:** Requires proper cleaning and handling as the material can detonate.

REF: Kim, S U & Park, M J — *Jpn J Appl Phys,* 23,1070(1984)

SE-0006
ETCH NAME:
TYPE: Acid, removal
COMPOSITION:
x .... 0.1 *M* S
x .... 2.5 *M* Na_2S
x .... 1 *M* KOH
DISCUSSION:
TIME:
TEMP: RT

Se residual film left on CdSe polycrystalline thin films grown on titanium substrates after surface cleaning with 25% HNO_3. Remove excess selenium in solution shown. CdSe co-evaporated and heat treated at 400°C, 15 min to homogenize film in a study of the material. See CDSE-0008.

REF: Haak, R et al — *J Electrochem Soc,* 131,2709(1984)

SE-0007
ETCH NAME: Water
TYPE: Acid, float-off
COMPOSITION:
x .... H_2O
DISCUSSION:
TIME:
TEMP: RT

Se, thin films grown by epitaxy on KI (100) cleaved substrates, as 300 Å thick films in a study of Se structure. Water used to dissolve the KI/Se interface with film float-off for TEM study.

REF: Nagashima, S & Ogura, I — *Jpn J Appl Phys,* 23,1555(1984)

SILICIDES, M_xSi_y

General: Silicides do not occur in nature as mineral alloys even though silicon is second only to oxygen in abundance as an element and is a constituent of most rocks as a silicate. Quartz and its many forms, as the oxide, SiO_2, and the metal silicate minerals, with SiO_2^-, SiO_3^-, SiO_4^- type radicals are primary rock-forming minerals, such as granites, clays, asbestos, hornblende, and feldspars.

Ferrosilicon obtained during the reduction of silicon ores is supplied in rough bar (pig) form to the metals industry as a primary material in the fabrication of irons and steels. This could be considered a form of silicide with a high iron content, but the "silicides" as a named group of materials are a combination of transition metals and silicon as developed in the Solid State and semiconductor industry.

Technical Application: Over a dozen silicides have been developed as mono- and di-silicides, such as MoSi, Ta_2Si, WSi_2, etc. In the general formula shown above, M = Co, Cr, Ni, Mo, Pd, Pt, Ir, Rh, Ti, Ta, W, and others. Silicides are fabricated on silicon wafers by thin film metallization (1) directly on silicon, (2) on a pre-deposited SiO_2 or Si_3N_4 layer on silicon, or (3) on a pre-grown polycrystalline layer of silicon on the single crystal silicon or other metals or metallic compounds. The specimens are then temperature annealed to form the silicide as a crystalline layer with temperature, time, atmospheres, and other annealing conditions controlled to develop the desired crystal structure. Laser annealing has been used directly or as a post-anneal step to increase crystallite size.

The primary application of silicides has been as buffer layers in semiconductor device structures to prevent or reduce cross-over diffusion of other structure elements such as aluminum.

There is much on-going development and study of the silicides themselves, to establish structure and morphology . . . deposition under varied conditions to include reaction studies with other layer elements . . . oxidation of the silicides and so forth. Note that temperature level is a major control factor as to structural forms.

Some silicides, such as $CoSi_2$, using a single crystal seed cut from pre-grown poly-crystalline material, have been used as seed crystals to grow the material as single crystal ingots using the Czochralski (CZ) technique for defect and morphological studies.

The following list of silicides is far from complete but presents both wet chemical etching (WCE) solutions that have been applied to the materials, as well as some dry chemical etching (DCE) ionized gases applied.

Etching: Acids, mixed acids. Ionized gases (CF_4:O_2).

SILICIDE BY TYPE

Aluminum Silicide (See Aluminum Alloys)
Boron Silicide
Cerium Silicide
Chromium Silicide
Cobalt Silicide
Erbium Silicide
Iron/Tungsten Silicide
Magnesium Silicide
Molybdenum Silicide
Nickel Silicide
Palladium Silicide
Platinum Silicide
Praseodymium/Cobalt Silicide
Tantalum Silicide
Terbium/Cobalt Silicide
Titanium Silicide
Titanium/Tungsten Silicide
Tungsten Silicide
Vanadium Silicide

BORON SILICIDE ETCHANTS

BSI-0001

ETCH NAME: — TIME:

TYPE: Acid, removal — TEMP:

COMPOSITION:

x .... HF

x .... HNO_3

DISCUSSION:

B_4Si, specimens. $HF:HNO_3$, solutions attack this crystal form of boron silicide very slowly. The form as B_6Si is not etched.

REF: Matkovich, V I — *Acta Metall*, 11,679(1960)

CERIUM SILICIDE ETCHANTS

CESI-0001

ETCH NAME: Kalling's etch — TIME: 10—60 sec

TYPE: Acid, polish — TEMP: RT

COMPOSITION:

100 ml .... HCl

5 g .... $CuCl_3$

100 ml .... EOH

DISCUSSION:

$CeSi_2$ specimens arc melted and used in a study of metallic compounds. Solution used as a general removal and polishing etchant. See cerium platinide CEPT-0001 for other compounds studied.

REF: Slepowronsky, M et al — *J Cryst Growth*, 63,293(1983)

CHROMIUM SILICIDE ETCHANTS

CRSI-0001

ETCH NAME: Oxygen — TIME:

TYPE: Gas, oxidation — TEMP: >1000°C

COMPOSITION:

x .... O_2

DISCUSSION:

$CrSi_2$, thin films deposited on silicon substrates and used in a study of dry oxidation kinetics. Other silicides referenced: Ti, Co, Ni, Mo, Rh, Ta, Ir, W, Pt, Pd, Os, Ru, Fe, Re, Te, Mn, Nb, V, Hf, and Zr.

REF: Bartur, M & Nicolet, M-A — *J Electrochem Soc*, 131,371(1984)

COBALT SILICIDE ETCHANTS

COSI-0001

ETCH NAME: Dry chemical etching — TIME:

TYPE: Ionized gas, removal — TEMP:

COMPOSITION: — GAS FLOW:

x .... CF_4 — PRESSURE

x .... O_2 (5%) — POWER:

DISCUSSION:

$CoSi_2$ thin films grown on Si substrates. All of the following silicides can be etched by DCE in the mixture shown: $MoSi_2$; WSi_2; $TiSi_2$; $TaSi_2$; $CoSi_2$; NiSi.

REF: CRSI-0001

COSI-0002: Lied, G D et al — *J Electron Mater,* 13,95(1984)

CoSi, Co_2Si, and $CoSi_2$ grown on silicon substrates. Si substrates were first: thin film evaporated with SiO_2, which was then polycrystalline evaporated with silicon, Si^e. Cobalt was then evaporated on the Si^e layer and annealed at different temperatures to form the different silicide structures shown.

COSI-0003

ETCH NAME: Hydrofluoric acid — TIME:

TYPE: Acid, removal — TEMP:

COMPOSITION:

x .... HF

DISCUSSION:

$CoSi_2$ thin film grown on substrates of Si, (111) and (100), n-type, 0.01 and 0.02 Ω cm resistivity. Co, Ti, Ni, Pd, and Pt evaporated to 1 K Å thickness. Surfaces thermally oxidized, both wet and dry, prior to metallization. Annealed NSi_2, $CoSi_2$, and PdSi at 800°C. $PdSi_2$ and PtSi annealed at 450°C. All other silicides annealed at 650°C. Authors say that all silicides are inert in HF; then say that Ni and Co silicides can be etched in HF(?).

REF: Bartur, M & Nicolet, M-A — *J Electron Mater,* 13,81(1984)

COSI-0004

ETCH NAME: Hydrofluoric acid, dilute — TIME:

TYPE: Acid, dislocation — TEMP:

COMPOSITION:

1 .... HF

1 .... H_2O

DISCUSSION:

$CoSi_2$, (100) wafers. Ingots grown from a $CoSi_2$ seed crystal, initially cut from crystalline material grown as thin films on (111) and (100) silicon wafers. During ingot growth, rotate crucible and seed in opposite directions. Solution used to develop the normal square dislocation etch pits on the (100) surfaces. Low angle grain boundaries observed are probably due to carbon and oxygen contamination.

REF: Ditchek, B B M — *J Cryst Growth,* 69,207(1984)

ERBIUM SILICIDE ETCHANTS

ERSI-0001

ETCH NAME: Heat + FG — TIME:

TYPE: Thermal, forming — TEMP:

COMPOSITION: — GAS FLOW:

x .... heat — PRESSURE:

x .... FG — POWER:

DISCUSSION:

$ErSi_2$ thin films grown on Si, (100), p-type, 1—10 Ω cm resistivity wafers used as substrates with deposit layer structure: a-Si/Er/Si(100). Anneal in forming gas (FG) at 450°C, 30 min to develop: $ErSi_2$. Barrier height changes with annealing temperature, from an initial 0.68 eV to 0.77 eV at 380°C to 0.74 eV at 500°C. Lowest was 0.72 eV (best?) when held at 400—450°C. Surface pits observed at 380°C and increased with additional time at tem-

perature. There was an SiO_2 barrier observed between Er/Si as Er/SiO_2/Si during barrier height annealing. Acronym c-Si used to denote single crystal silicon. (*Note:* The c-Si acronym for a single crystal is in conflict with colloidal silicon, c-Si.)
REF: Wu, C S et al — *Thin Solid Films,* 104,175(1983)

IRON/TUNGSTEN SILICIDE ETCHANTS

IWSI-0001
ETCH NAME: Hydrogen peroxide — TIME:
TYPE: Acid, removal — TEMP:
COMPOSITION:
(1) x .... H_2O_2 (2) x .... HNO_3 (3) x .... HCl
DISCUSSION:

FeWSi, thin films deposited on silicon, (100) wafers with and without thin film SiO_2 pre-deposited. FeW films were 1000 Å thick and were overcoated with 1000 Å of aluminum, gold, copper, palladium, or platinum. Used in a study of a-FeW thin films annealed crystallinity, and their application as a buffer layer against the upper metal layer cross-diffusion. Under experimental conditions, the a-FeW was a good barrier up to 650°C for 30 min. (*Note:* a- is the acronym for amorphous structure.)
REF: Suni, I et al — *Thin Solid Films,* 107,73(1983)

MAGNESIUM SILICIDE ETCHANTS

MGSI-0001
ETCH NAME: — TIME:
TYPE: Acid, removal — TEMP:
COMPOSITION:
10 .... HF
2 .... HNO_3
DISCUSSION:

Mg_2Si, single crystals. Intrinsic material is n-type semiconducting; dope p-type with Ag or Cu. Band gap varies: 0.69—0.78 eV. Also worked with Mg_2Ge.
REF: Morris, R G et al — *Phys Rev,* 109,1909(1958)

MOLYBDENUM SILICIDE ETCHANTS

MOSI-0001
ETCH NAME: Dry chemical etching — TIME:
TYPE: Ionized gas, removal — TEMP:
COMPOSITION: — GAS FLOW: 190 cm^3/min
x .... CF_4 — PRESSURE: 0.6—2.0 Torr
x .... O_2 (5%) — POWER:
DISCUSSION:

$MoSi_2$ thin films deposited on silicon substrates. All of the following silicides can be etched by DCE in the mixture shown: $MoSi_2$; WSi_2; $TiSi_2$; $TaSi_2$; $CoSi_2$ and NiSi.
REF: CRSI-0001
MOSI-0002. Inoue, S et al — *J Electrochem Soc,* 130,1603(1983)

$MoSi_2$, thin films deposited on silicon, (1000), p-type substrates containing a 1000 Å thick SiO_2 layer pre-deposited at 1000°C with dry O_2. RF plasma etching used to form dot patterns of silicide.

MOSI-0003
ETCH NAME: CAIBE
TYPE: Ionized gas, selective
COMPOSITION:
x .... Ar
x .... Cl_2
TIME:
TEMP:
GAS FLOW:
PRESSURE:
POWER:

DISCUSSION:

$MoSi_2$, the films deposited on silicon substrates, and silicon coated with polycrystalline silicon. Etching was done (DCE) by chemically assisted ion beam etching (CAIBE) in a study of etch rates, material structure, defects, and reactions. Other silicides evaluated were Ti, Ta, and Pt. (*Note:* CAIBE is a specialized form of RF plasma etching as a reactive ion etch (RIE) system.)

REF: Chinn, J D et al — *J Electrochem Soc,* 131,375(1984)

MOSI-0004: Neppl, F et al — *J Appl Phys,* 130,1174(1983)

MoSi and Mo_2Si thin films co-evaporated on silicon (100) substrates in a study of their application in silicon gate structured device technology. Films must be annealed to lower resistivity that substrates and oxygen in air will tarnish surfaces. $MoSi_2$ is tetragonal; Mo_3Si is isometric (cubic), and shows tensile stress due to atomic lattice shrinkage.

MOSI-0005
ETCH NAME:
TYPE: Gas, forming
COMPOSITION:
x .... Ar
x .... CCl_4
TIME:
TEMP: 900°C
GAS FLOW: 5 sccm
PRESSURE: 1—2 $\times$ 10^{-4} Torr
POWER:

DISCUSSION:

$MoSi_2$ thin films grown by co-deposition of silicon and molybdenum substrates in argon at 3 $\times$ 10^{-3} Torr. Si, (100) substrates used, and deposit showed 55 Ω^2 resistivity. Anneal, as shown above, and resistivity reduces to 5 Ω^2.

REF: Powell, R A — *J Appl Phys,* 130,1164(1983)

NICKEL SILICIDE ETCHANTS

NISI-0001
ETCH NAME: Dry chemical etching
TYPE: Ionized gas, removal
COMPOSITION:
x .... CF_4
x .... O_2 (5%)
TIME:
TEMP:
GAS FLOW:
PRESSURE:
POWER:

DISCUSSION:

NiSi thin films deposited on silicon substrates. All of the following silicides can be etched by DCE in the mixture shown: $MoSi_2$; WSi_2; $TiSi_2$; $TaSi_2$; $CoSi_2$, and NiSi. (*Note:* This is a form of RF plasma etching.)

REF: CRSI-0001

NISI-0002
ETCH NAME: Hydrofluoric acid
TYPE: Acid, removal
COMPOSITION:
x .... HF
TIME:
TEMP:

DISCUSSION:

$NiSi_2$ thin films grown on silicon substrates. See COSI-0003 for discussion.

REF: Bartur, M & Nicolet, M-A — *J Electron Mater*, 13,81(1984)

NISI-0003

ETCH NAME: TIME:

TYPE: Acid, cleaning TEMP:

COMPOSITION:

x .... O_2

x .... H_2O

DISCUSSION:

$NiSi_2$ thin films deposited on silicon wafers. Also worked with $HfSi_2$. This work referenced to that of Baglin TISI-0005.

REF: Bartur, M — *J Appl Phys Lett*, 40,175(1982)

PALLADIUM SILICIDE ETCHANTS

PDSI-0001

ETCH NAME: Hydrofluoric acid TIME:

TYPE: Acid, removal TEMP:

COMPOSITION:

x .... HF

DISCUSSION:

PdSi and $PdSi_2$ thin films grown on silicon substrates. See: COSI-0003 for discussion.

REF: Bartur, M & Nicolet, M-A — *J Electron Mater*, 13,81(1984)

PLATINUM SILICIDE ETCHANTS

PTSI-0001

ETCH NAME: Hydrofluoric acid TIME:

TYPE: Acid, removal TEMP:

COMPOSITION:

x .... HF

DISCUSSION:

PtSi, thin films grown on silicon substrates. See COSI-0003 for discussion.

REF: Bartur, M & Nicolet, M-A — *J Electron Mater*, 13,81(1984)

PTSI-0002

ETCH NAME: Platinum TIME:

TYPE: Metal, deposition TEMP: 1000°C

COMPOSITION:

x .... Pt

DISCUSSION:

Pt_2Si, thin films formed on silicon, (111) and (100) n-type wafers, 1—10 Ω cm resistivity. Silicon wafers were ultrasonically cleaned in acetone; EOH rinse; oxide etched in a BHF solution, then flood with EOH without exposure to air. Wafer backs were coated with tantalum metal for use as a resistance heater control of the substrates. Platinum was evaporated from a resistively heated wire at 1000°C followed by annealing to form Pt_2Si.

REF: Matz, R et al — *J Vac Sci Technol*, A(2),253(1984)

PTSI-0003
ETCH NAME: CAIBE
TYPE: Ionized gas, removal
COMPOSITION:
x .... Ar
x .... Cl_2
TIME:
TEMP:
GAS FLOW:
PRESSURE:
POWER:
DISCUSSION:

PtSi, thin films deposited on silicon and poly-silicon coated wafers. Chemically assisted ion beam etching (CAIBE) used to establish etch rates of the material and for studying defects and structure. Other silicides studied were Mo, Ti, and Ta.

REF: Chinn, J D et al — *J Electrochem Soc,* 131,375(1984)

PTSI-0004
ETCH NAME:
TYPE: Acid, removal
COMPOSITION:
x .... HF
x .... HNO_3
TIME:
TEMP:
DISCUSSION:

PtSi specimens used in a study of electronic phase transitions. Various mixtures of solution shown were used for general material etching. Also studied Ni:Ti and Zr_2Ni. See PTSI-0004.

REF: Lee, R N & Wither, R — *J Appl Phys,* 49,5488(1978)

PRASEODYMIUM/COBALT SILICIDE ETCHANTS

PRCS-0001
ETCH NAME:
TYPE: Acid, removal
COMPOSITION:
x .... HF
x .... HNO_3
TIME:
TEMP:
DISCUSSION:

$PrCo_2Si_2$, single crystals grown from mixed powders in an induction furnace. Powdered samples used in a study of magnetic super-lattices. These types of materials are body-centered tetragonal (bct) structure. See section on Germanides for similar structured materials.

REF: Yakinthros, J K & Routsi, Ch — *J Phys Chem Solids,* 45,689(1984)

TANTALUM SILICIDE ETCHANTS

TASI-0001
ETCH NAME: Hydrofluoric acid
TYPE: Acid, cleaning
COMPOSITION:
1 .... HF
20 .... H_2O
TIME:
TEMP:
DISCUSSION:

$TaSi_2$, thin films deposited on silicon (100), n-type, 10 Ω cm resistivity wafers. Silicon wafers were cleaned in the solution shown to remove native oxide before titanium metal evaporation and annealing.

REF: CRSI-0001

TASI-0002
ETCH NAME: Dry chemical etching
TYPE: Ionized gas, removal
COMPOSITION:
x .... CF_4
x .... O_2 (5%)
TIME:
TEMP:
GAS FLOW:
PRESSURE:
POWER:
DISCUSSION:

$TaSi_2$, thin films deposited on silicon substrates. All of the following silicides can be etched in the gas mixture shown: $MoSi_2$, WSi_2, $TiSi_2$, $CoSi_2$ and NiSi.
REF: CRSI-0001

TASI-0003
ETCH NAME: CAIBE
TYPE: Ionized gas, removal
COMPOSITION:
x .... Ar
x .... Cl_2
TIME:
TEMP:
GAS FLOW:
PRESSURE:
POWER:
DISCUSSION:

$TaSi_2$, thin films deposited on silicon and poly-silicon deposited wafers. Chemically assisted ion beam etching (CAIBE) was used to establish etch rates and study defects and structure. Other silicides studied were titanium, molybdenum, and platinum.
REF: Chinn, J I et al — *J Electrochem Soc,* 131,375(1984)

TASI-0004
ETCH NAME: Hydrofluoric acid
TYPE: Acid, removal
COMPOSITION:
x .... HF
TIME:
TEMP: RT
DISCUSSION:

$TaSi_2$, thin films grown on poly-Si deposited on oxidized silicon wafers or on fused and single crystal sapphire substrates. Ta magnetron sputter and reacted under argon to form 2000 Å silicide. In measuring sheet resistance with a 4-point probe, $TaSi_2$ was etched in HF. Grain boundary diffusion of tantalum into silicon or out-diffusion of phosphorus changes resistivity in poly-Si specimens. A cap oxide reduces the effect.
REF: Maa, J S et al — *J Vac Sci Technol,* B1,1(1983)

TASI-0005
ETCH:
TYPE: Acid, removal
COMPOSITION:
90 .... BHF (1BHF:$30H_2O$)
5 .... HNO_3
5 .... CH_2OHCH_2OH [ethylene glycol (EG)]
TIME: 2 min
TEMP: RT
DISCUSSION:

$TaSi_2$ thin films 2500—2800 Å used in a study of diffusion contamination. Solution used for general removal or step-etch of the films. Etch rate: 100 Å/min.
REF: Pelleg, J & Muarka, S P — *J Appl Phys,* 54,1337(1983)

TASI-0006a
ETCH NAME:
TYPE: Ionized gas, removal
COMPOSITION:
x .... CF_4
x .... O_2 (x%)
DISCUSSION:

TIME:
TEMP:
GAS FLOW:
PRESSURE:
POWER:

$TaSi_2$ as thin films grown on poly-Si as mushroom cap structure or cap portion of "T" channel structure. Plasma etch referred to as a fast isotropic etch for metal silicides. $MoSi_2$, WSi, and WSi_2 showed compressive stresses. This was a study of refractory metal silicides for VLSI application.
REF: Sinha, A K — *J Vac Sci Technol,* 19,778(1981)

TASI-0006b
ETCH NAME:
TYPE: Acid, removal
COMPOSITION:
1 .... HF
30 .... H_2O
DISCUSSION:

TIME:
TEMP: RT

$TaSi_2$ thin films. Solution shown as a rapid etch for material. A BHF mixture also used, but not shown. (*Note:* 1HF:1NH_4F (40%) is one standard.)
REF: Ibid.

TERBIUM/COBALT SILICIDE ETCHANTS

TBCS-0001
ETCH NAME:
TYPE: Acid, removal
COMPOSITION:
x .... HF
x .... HNO_3
DISCUSSION:

TIME:
TEMP:

$TbCo_2Si_2$ specimens. See PRCS-0001 for discussion.
REF: Yakinthros, J K & Routsi, Ch — *J Phys Chem Solids,* 45,689(1984)

TITANIUM SILICIDE ETCHANTS

TISI-0001
ETCH NAME: Dry chemical etching
TYPE: Ionized gas, removal
COMPOSITION:
x .... CF_4
x .... O_2 (x%)
DISCUSSION:

TIME:
TEMP:
GAS FLOW:
PRESSURE:
POWER:

Si, wafers used as substrates for growth of silicides. All of the following silicides can be dry chemical etched (DCE) in the mixture shown: $MoSi_2$; WSi_2; $TiSi_2$; $TaSi_2$; $CoSi_2$, and NiSi. (*Note:* 5% O_2 is fairly standard.)
REF: CRSI-0001

TISI-0002
ETCH NAME: Hydrofluoric acid TIME:
TYPE: Acid, removal TEMP:
COMPOSITION:
x .... HF
DISCUSSION:
$TiSi_2$ thin film grown on Si substrates. See COSI-0003 for discussion.
REF: Bartur, M & Nicolet, M-A — *J Electron Mater,* 13,81(1984)

TISI-0003
ETCH NAME: TIME: 10—15 min
TYPE: Acid, removal TEMP: RT
COMPOSITION:
1 .... NH_4OH
1 .... H_2O_2
4 .... H_2O
DISCUSSION:
$TiSi_2$, thin film formed on silicon (100) substrates in the fabrication of MOSFET devices. Silicon wafers were dip cleaned in HF prior to sputter evaporation of titanium and titanium was overcoated with molybdenum to reduce oxidation. After photo resist patterning both the Mo and excess Ti remaining after anneal conversion to $TiSi_2$ were selectively removed with the solution shown.
REF: Parks, H K et al — *J Vac Sci Technol,* A(2), 264(1984)

TISI-0007
ETCH NAME: Kalling's etch TIME: 10—60 sec
TYPE: Acid, polish TEMP: RT
COMPOSITION:
100 ml .... HCl
5 mg .... $CuCl_3$
100 ml .... EOH
DISCUSSION:
$TiSi_2$ arc melted in fabrication and used in a study of metallic compounds. Solution used as a general removal and polishing etchant. See Cesium Platinide: CEPT-0001 for other compounds studied. Also studied the silicide $MoSi_2$.
REF: Slepowronsky, M et al — *J Cryst Growth,* 65,293(1983)

TISI-0004
ETCH NAME: CAIBE TIME:
TYPE: Ionized gas, removal TEMP:
COMPOSITION: GAS FLOW:
x .... Ar PRESSURE:
x .... Cl_2 POWER:
DISCUSSION:
$TiSi_2$, thin films deposited on silicon wafers and wafers coated with poly-Si. Chemically assisted ion beam etching (CAIBE) was used to establish etch rates and to study defects and structure. Other silicides studied were tantalum, molybdenum, and platinum.
REF: Chinn, J D et al — *J Electrochem Soc,* 131,375(1984)

TISI-0005
ETCH NAME: TIME:
TYPE: Acid, oxidation TEMP: 100°C
COMPOSITION:
x .... O_2
x .... H_2O
DISCUSSION:
$TiSi_2$ thin films grown on silicon substrates in a study of SiO_2 used to form silicides. Oxidize substrates by bubbling O_2 through hot water at 100°C (wrap beaker with heater tape). Evaporate 2000 Å of metal, and anneal in He 1 h at 1000°C. $TiSi_2$ then annealed at 750°C in wet O_2 develops a porous TiO_2 surface; above 900°C converts to SiO_2 with no Ti or Si metal. Other silicides evaluated: WSi_2, $TaSi_2$, RhSi, Rh_3Si_4, and IrSi as layered semiconductors with 1 eV band gaps. For $NiSi_2$ and $HfSi_2$ reference was made to NISI-0003.
REF: Baglin, J E et al — *J Appl Phys*, 54,1849(1983)

TISI-0006
ETCH NAME: Hydrofluoric acid TIME:
TYPE: Acid, oxide removal TEMP: RT
COMPOSITION:
1 .... HF
20 .... H_2O
DISCUSSION:
TiSi thin films grown on (100) silicon wafers, n-type, 10 Ω cm resistivity with p^+ ion implantation. After I^2 native oxide was removed with solution shown prior to evaporation of 100 Å Ti and annealing. A study of the growth of TiSi on ion implanted Si.
REF: Revesz, P et al — *J Appl Phys*, 54,1860(1983)

TITANIUM/TUNGSTEN SILICIDE ETCHANTS

TWSI-0001
ETCH NAME: Aqua regia TIME:
TYPE: Acid, removal TEMP: RT
COMPOSITION:
3 .... HCl
1 .... HNO_3
DISCUSSION:
$Ti_{0.3}W_{0.7}Si_2$, thin films. Silicon wafers (111), (110), and (100) oriented were used as substrates for RF deposition from a composite Ti/W target. The thin film was annealed at 700°C to form the silicide and showed no change at temperatures up to 900°C. The silicide was unaffected by hot H_2O_2, HF, or CP4 and only slowly dissolved in aqua regia.
REF: Harris, J M et al — *J Electrochem Soc*, 123,120(1976)

TUNGSTEN SILICIDE ETCHANTS

WSI-0001
ETCH NAME: Dry chemical etching TIME:
TYPE: Ionized gas, removal/selective TEMP:
COMPOSITION: GAS FLOW:
x .... CF_4 PRESSURE:
x .... O_2 (5%) POWER:

DISCUSSION:

WSi_2, thin films grown on silicon substrates. RF plasma etching can be used for removal and selective etching of most silicides. Also called dry chemical etching (DCE). This gas mixture was used to etch all of the following silicides: $MoSi_2$, $TiSi_2$, $TaSi_2$, and NiSi.

REF: CRSI-0001

WSI-0002

ETCH NAME: Hydrogen peroxide — TIME:

TYPE: Acid, removal — TEMP:

COMPOSITION:

x .... H_2O_2

DISCUSSION:

WSi_2, thin films reacted from CVD tungsten deposited on silicon (100) wafers. Annealing temperatures were 750 and 675°C, 30 min. Solution used for general removal and study of crystallite structures. Deposited material crystallites were tetragonal crystal structure.

REF: Green, M L & Levy, R A — *J Electrochem Soc,* 132,1243(1986)

WSI-0003

ETCH NAME: — TIME:

TYPE: — TEMP:

COMPOSITION:

DISCUSSION:

WSi_2, thin films selectively deposited on poly-Si coatings on silicon wafers for VLSI gate and interconnect structures. Poly-Si was isotropically etched for device patterns prior to tungsten deposition by CVD. WSi_2 on poly-Si produces good adhesion; deposited on SiO_2 there is poor bonding unless the deposit is rich in silicon oxide, Si_xO_y. Tungsten silicide can be etched in CF_4:O_2 gas mixtures. (See WSI-0001).

REF: Miller, N E & Beingglass, I — *Solid State Technol,* December 1982, 85

VANADIUM SILICIDE ETCHANTS

VSI-0001a

ETCH NAME: — TIME:

TYPE: Acid, polish — TEMP:

COMPOSITION:

1 .... HF

1 .... H_2O_2

4 .... H_2O

DISCUSSION:

V_3Si, (111) and (100) wafers used in a dislocation study. Solution used as a chemical polish.

REF: Levinstein, H J et al — *J Appl Phys,* 37,164(1966)

VSI-0001b

ETCH NAME: — TIME: 1—5 min

TYPE: Electrolytic, polish — TEMP: RT

COMPOSITION: — ANODE: V_3Si

x .... 10% H_2SO_4 — CATHODE:

POWER: 5 V

DISCUSSION:

V_3Si, (111) and (100) wafers used in a dislocation study. Solution used to electropolish surfaces.

REF: Ibid.

VSI-0001c

ETCH NAME: TIME:

TYPE: Acid, dislocation TEMP:

COMPOSITION:

15 .... HF

4 .... H_2O_2

DISCUSSION:

V_3Si, (111), (100) wafers and other orientations. Solution will develop dislocations on both the (111) and (100) wafer surfaces.

REF: Ibid.

VSI-0002

ETCH NAME: TIME:

TYPE: Acid, cleaning TEMP:

COMPOSITION:

15 ml .... HF

100 ml .... HNO_3

DISCUSSION:

V_3Si, specimens used in a study of lattice softening. Solution used as a polishing and cleaning etch.

REF: Vaneo, E R & Finlayson, T R — *J Appl Phys,* 39,1980(1968)

VSI-0004

ETCH NAME: TIME:

TYPE: TEMP:

COMPOSITION:

DISCUSSION:

VSi_2 thin films. Vanadium deposited on (100) and (111) silicon n-type wafers, then annealed at 500°C for 5 min to convert to VSi_2 or as VSi at 350°C. Schottky barrier height was shown to be under 0.55 eV. (*Note:* See other vanadium silicides for etchants.)

REF: Clabes, J G & Roldoff, G W — *J Vac Sci Technol,* 19(2),262(1981)

VSI-0003

ETCH NAME: TIME: 10—60 sec

TYPE: Acid, polish TEMP: RT

COMPOSITION:

100 ml .... HCl

5 g .... $CuCl_3$

100 ml .... EOH

DISCUSSION:

V_3Si and V_3Ge arc melted in fabrication and used in a study of metallic compounds. Solution used as a general removal and polishing etchant. See Cesium Platinide CEPT-0001 for complete list of other compounds.

REF: Slepowronsky, M et al — *J Cryst Growth,* 65,293(1983)

PHYSICAL PROPERTIES OF SILICON, Si

Classification	Non-metal
Atomic number	14
Atomic weight	28
Melting point (°C)	1412 (1410)
Boiling point (°C)	2878 (2355)
Density (g/cm^3)	2.33
Thermal conductance (cal/sec)(cm^2)(°C/cm)	0.353
Specific heat (cal/g) 25°C	0.162
Heat of fusion (k-cal/g-atom)	11.1
Latent heat of fusion (cal/g)	395
Heat of vaporization (k-cal/g-atom)	71
Atomic volume (W/D)	12.1
1st ionization energy (K-cal/g-mole)	188
1st ionization potential (eV)	8.15
Electronegativity (Pauling's)	1.8
Covalent radius (angstroms)	1.11
Ionic radius (angstroms)	0.42 (Si^{+4})
Electrical conductance (micro ohms — cm^{-1})	0.10
(ohm — cm) 0°C	10
Intrinsic resistivity ($\times 10^{-3}$ ohms cm)	64
Critical pressure (°C)	1450
Critical temperature (C.T.) (°C)	4920
Thermal expansion (ppm/°C)	2.8—7.3
Cross section (barns)	0.16
Vapor pressure (°C)	2083
Electron work function (eV)	4.52
Energy band gap (eV)	1.12
Lattice constant (angstroms)	5.43
Refractive index (n=)	3.6
Hardness (Mohs — scratch)	7
Crystal structure (isometric — normal)	(100) cube ("diamond")
Color (solid)	Grey silver/black
Cleavage (dodecahedral)	(110)

SILICON, Si

General: Does not occur as a native metal like gold, silver, or copper, though next to oxygen, it is the most abundant element in nature. About three quarters of all known minerals are silicates. There are over 3000 single crystal compounds and they make up about one fourth of the earth's crust. The most abundant, pure form is as silicon dioxide, SiO_2 — colloidal, amorphous, single crystal, and crystalline structures as rock-forming deposits, and there are colloidal silica beds formed from the exoskeletons of ancient sea life. Opal also is amorphous. Quartz, in clear single crystal form, is called rock crystal and, with color tints, several precious and semi-precious gem stones; the crystalline varieties include chalcedony, agates, and chert. Flint and chert are largely formed under sedimentary deposition conditions as nodular segregates to extensive beds from siliceous material, to include sea life skeletons. Volcanic obsidian is natural glass, and the volcanic rock, tufa, is a lightweight solidified silica froth.

The major source of silicon metal comes from the reduction of high purity, ancient sand beds, colloidal silica beds, and recently from ashed rice hulls. It is classified as a nonmetal similar to carbon (graphite), has several allotropic forms, and is one of the two semiconductor elements, the other being germanium as the two elemental semiconductor materials. Both of the elements are often referred to as being ''diamond'' structure . . . octahedron, (111) crystal form, of the isometric (cubic) system, normal class . . . but, since that diamond term definition was applied in the 1940s, it has been shown that both silicon and germanium have a tetrahedral unit cell similar to that of compound semiconductors, such that ''diamond structure'' is erroneous. (See the variety of finite crystal forms (FCF) that have been etched from single crystal spheres under Etchant Formats in both the Silicon and Germanium sections.)

In the heavy metal industries silicon is widely used in the form of ferro-silicon in the manufacture of irons and steels. With 14% silicon content in iron it is a brittle alloy that is highly acid resistant. There are several silicon chemical compounds of importance, such as sodium silicate (water glass) and silicon fluorides used in a variety of commercial processing applications.

The most important and widely used compound is glass, SiO_2 with a variety of additives. See the sections on Silicon Dioxide and Quartz for further details on glasses.

Technical Application: Although germanium was the first semiconductor developed in the late 1940s, it is the silicon semiconductor that has been the primary vehicle in establishing the electronics industry as we know it today; semiconductors being the Solid State equivalent of the glass vacuum tube and, although semiconductors have replaced many vacuum tubes in the low to medium power range (up to about 3000 V and 8 W) they have not reached, and may never reach, the high power capability of some vacuum tubes. A key factor in semiconductor development and processing is the ability to miniaturize, which has led to the microelectronic phase of the industry, well represented by the computer industry in addition to TV, light control, power appliances, microprocessors, and computers, all of continually smaller and smaller design as circuit size is reduced.

It is not within the scope of this small discussion to describe the vast number of silicon devices in any detail as this book is primarily concerned with the chemical processing of semiconductors rather than the electronic functions and characterization of devices, but a few bear mention, such as alloyed or diffused diodes and transistors; silicon-controlled rectifiers (SCRs), MOS, MNOS, IGFETS, FETS, ICs, etc. There are over 100 active silicon semiconductor devices, not including as a family of devices, such as alloy zeners and diffused zeners, which represent over 100 discrete voltage/power level devices alone.

In addition to semiconductor applications, silicon also is being used as a physical material without regard to semiconducting properties. Such application can be divided into two general areas: (1) bulk material used as diffusion tubes and parts carriers, epitaxy susceptor plates, infrared cones, and lenses etc., and (2) physical parts: strain gauges, ink dot heads, circuit boards, fiber optics couplers, as a gas chromatograph or heat sinks, and many other specialized physical structure elements.

Silicon is still the primary semiconductor, but is no longer used just as a single crystal element. Polycrystalline silicon, poly-Si, was initially developed for solar cell application, but also is a thin film deposit — used in conjunction with photolithographic techniques — for improved device structural rigidity or as a buffer or electrically active element in layered structures. Amorphous silicon, a-Si, is under development as a thin film structure, and colloidal silicon, c-Si, is under similar development and application, as is c-Si:H, hydrogenated colloidal silicon, which also may contain nitrogen as c-SiN:H, etc.

Poly-Si is further divided by grain size, with much effort involved to increase grain size for improved solar cell efficiency from its present 13 to 18%. Large crystallite material is grown by a modified Czochralski (CZ) freeze-out method, and by laser annealing.

A bi-crystal form of both silicon and germanium with (111) surface orientation — the ribbon crystal — is under ongoing development. It also is called dendritic growth. Grown as a continuous thin ribbon from a melt, there is always a twinned zone down ribbon center with 180° surface rotation to either side as a positive (111) and negative (111) surface.

Unfortunately, such ribbon crystals contain low-angle grain boundaries, are highly stressed and, subsequently, contain a high dislocation content. There is much effort in progress to improve these crystals for cost reduction of single crystal growth. They also would eliminate the need of slicing ingots, and reduce the lapping and polishing requirements in processing.

Though silicon has long been used in its (111) surface orientation, with the development of high energy ion implantation (I^2) replacing gaseous diffusion, the (100) surface is more compatible, such that ribbon crystals may now be a thing of the past.

In semiconductor processing, silicon also is used as a constituent in other metallic compounds, such as aluminum-silicon as a p-type alloy. Two extremely important compounds are silicon dioxide, SiO_2 and silicon nitride, Si_3N_4. They are both used in photolithographic processing as masks against diffusion and etching, as a final surface protective coating, or as an active element in device layer construction (MOS and MNOS type devices). Silicides, mainly as blocking layer structures, are in use and under further development: MoSi, W_2Si, $TaSi_2$, and many others. The silicides are not limited to silicon devices alone, but are increasingly applied in compound semiconductor heterostructures.

As silicon has infrared characteristics, it has been fabricated as heat-seeking nose cones for missiles, as silicon lenses for Snooperscope rifles, or part of infrared night-sight eyeglasses. The latter are used for both military applications and air-sea rescue operations. It also is a lens in infrared microscopes, and is even used as a cigarette lighter. As a silicon controlled rectifier, (SCR) it is the operation element in most of the cordless hand tools for variable speed operation, as well as in light-dimming switches.

Single crystals are grown up to 6″ in diameter with 3″ a generally available size. The Czochralski (CZ) method for growing was developed primarily for silicon in the early 1950s, followed by Float Zone (FZ) where the CZ grown ingot is swept with a hot filament melt-zone to improve resistivity homogenization and remove oxygen contamination which in the 1950s was a problem with Czochralski ingots. Growth as polycrystalline silicon, colloidal or amorphous bulk material and thin films has already been mentioned.

Etching: $HF:HNO_3$, $HF:H_2O_2$ mixtures with or without H_2O, HAc, or other additives. Hot alkalies, molecular fluorine, and chlorine gases. Ionized gas as dry chemical etching (DCE).

SELECTION GUIDE

Note: The "A" section lists solutions by chemical formula; the "B" section by individual names.

A. Solutions by formula

(1) Br_2:MeOH (BRM)

Note: See Gallium Arsenide and Indium Phosphide for solution variations.

Polish: SI-0001

(2) HF, conc.

Cleaning: SI-0066d; -0235; -0407; -0420; -0414; -0430b

Oxide Removal: SI-0002g; -0005; -0006; -0007; -0008; -0120; -0143; -0211

Passivation: SI-0234

Stain: SI-0188a

(3) HF, vapor

Cleaning: SI-0009; -0027g; -0181b

(4) $HF:H_2O$

Cleaning: SI-0004a; -0031; -0032b; -0169a; -0191; -0108b; -0283a; -0010; -0011a; -0012; -0013; -0014; -0108a; -0245; -0169b; -0230; -1262; -1260; -0045b; -0132c; -0399; -0429

Oxide Removal: SI-0132a; -0015; -0135a; -0136; -0144; -0180b; -0211
Step Etch: SI-0135b; -0132c
Thinning: SI-0121a

(5) HF:H_2O: +/− Glycerin (electrolytic)
Polish: SI-0083; -0225; -0245b

(6) HF:AgF
Preferential: SI-0187

(7) HF:CrO_3 (See Section B)
Preferential: SI-0182; -0139

(8) HF:$Cu(NO_3)_2$
Stain: SI-0186c

(9) HF:H_3PO_4
Selective: SI-0130

(10) HF:I_2
Preferential: SI-0125

(11) HF:MeOH
Cleaning: SI-0016

(12) HF:MeOH:I_2
Preferential: SI-0214; -0205

(13) HF:NH_4F/NaF:H_2O (BHF)
Cleaning: SI-0257; -0400; -0413; -0414; -0434
Nitride Removal: SI-0026; -0028
Oxide Removal: SI-0020; -0022; -0023; -0024; -0025; -0027; -0028; -0029; -0222; -0211
*Polish: SI-0021a; -0174a; -0174b; -0174c; -0041b; -0123; -0185
Preferential: SI-0174c; -0174d
*Removal: SI-0212
*May be electrolytic

(14) HF:H_2O_2:NaF_2
Junction: SI-0280

(15) HF:H_2O_2:H_2O
Preferential: SI-0165

(16) HNO_3, conc.
Cleaning: SI-0237
Stain: SI-0003

(17) HNO_3:H_2O_2
Cleaning: SI-0177b

(18) $HF:HNO_3$
Float-off: Si-0255
Junction: SI-0069; -0223
Pinhole: SI-0060
Polish: SI-0102h; -0056; -0046b; -0052; -0288; -0057; -0201; -0061; -0289; -0058; -0063; -0102c; -0066a; -0092b; -0177a; -0181a; -0068; -0157a; -0121b; -0126; -0127; -0218; -0129; -0130; -0249; -0296; -0154; -0190a; -0161; -0166; -0178; -0180a; -0201b; -0063c
Preferential: SI-0140; -0189; -0168
Removal: SI-0245
Stain: SI-0140; 0189; -0168
Thinning: SI-0059; -0065; -0121b; -0162a; -0163a; -0440; * -0164
*Electrolytic

(19) $HF:HNO_3:H_2O$
Forming: SI-0173b; -0173a
Polish: SI-0073; -0126; -0127; -0128; -0129; -0130; -0249; -0296

(20) $HF:HNO_3$:HAc:
Polish: SI-0072; -0076; -0077; -0079; -0038a; -0080; -0124; -0126; -0127; -0128; -0129; -0130; -0297; -0249; -0296; -0194; -0150; -0160; -0162a; -0170; -0172; -0197a; -0030b; -0393
Preferential: SI-0075; -0246; -0173a
Stain: SI-0167
Step-Etch: SI-0078; -0263b
Thinning: SI-0074; -0124; -0148; -0163a; -0404; -0394; -0434

(21) $HF:HNO_3:AgNO_3:Hg(NO_3)_2$
Polish: SI-0081; -0082

(22) $HF:HNO_3:KNO_3$
Preferential: SI-0064

(23) $HF:H_2O_2:KMnO_3$
Polish: SI-0125

(24) H_2O, boiling
Cleaning: SI-0066c; -0184 (vapor)

(25) H_2O
Float-off: SI-0232

(26) H_2O, steam
Cleaning: SI-0239; -0184

(27) HCl, conc.
Cleaning: SI-0017

(28) HCl, vapor
Cleaning: SI-0221a
Structure: SI-0258; -0259

(29) HCl, electrolytic
Cleaning: SI-0018

(30) HCl:HNO_3 (See Aqua Regia)
Cleaning: SI-0108a; -0108c; -0392
Preferential: SI-0171

(31) HCl:H_2O_2:H_2O
Cleaning: SI-0233

(32) H_2O_2, conc.
Cleaning: SI-0177b; -0430b
Preferential: SI-0254a; -0282a

(33) H_2SO_4:H_2O
Cleaning: SI-0236; -0256; -0138

(34) H_2SO_4:H_2O_2
Cleaning: SI-0103; -0104; -0105; -0106; -0107; -0108a; -0260

(35) Hydrazine/Diamine/Catechol
Preferential: SI-0115a; -0115b; -0116; -0117; -0118; -0119; -0121a

(36) H_2O_2:NH_4OH
Cleaning: SI-0109
Preferential: SI-0021c

(37) KCl:H_2O
Removal: SI-0102a

(38) KCN:H_2O
Cleaning: SI-0112; -0245

(39) KOH/NaOH:H_2O
Polish: SI-0021b
Preferential: SI-0102b; -0038e; -0088; -0091; -0092a; -0092b; -0206; -0208; -0092c; -0093; -0092d; -0094; -0092e; -0100; -0145; -0217; -0281c; -0253
Etch-Stop: SI-0084; -0085
Structure: SI-0089; -0090; -0091; -0147
Selective: SI-0095; -0243
Junction: SI-0100; -0101

(40) KOH:H_2O:Br_2/I_2
Light Figure Orientation: SI-0092f; -0092g; -0092h

(41) KOH:H_2O, electrolytic
Junction: SI-0195
Oxidation: SI-0097; -0098; -0146; -0271; -0238; -0381; -0223b
Polish: SI-0245
Removal: SI-0096

(42) KOH, molten flux
Preferential: SI-0099

(43) $KOH:H_2O$: + xxx
Preferential:

(i)	+ $FeCl_3$:SI-019bb
(ii)	+ KBr:SI-0198c
(iii)	+ S:SI-0300
(iv)	+ $Al(OH)_3$:SI-0301
(v)	+ Zn:SI-0302
(vi)	+ $Ba(OH)_2$:SI-0303
(vii)	+ As_2O_3:SI-0304
(viii)	+ I_2:SI-0192f
(ix)	+ Br_2:SI-0092g
(x)	+ $Ca(OH)_2$:SI-0305
(xi)	+ $Sr(OH)_2.8H_2O$:SI-0306
(xii)	+ $AlCl_3.6H_2O$:SI-0307
(xiii)	+ NH_4 Cl:SI-0308
(xiv)	+ $CaCl_3.2H_2O$:SI-0205
(xv)	+ LiCl:SI-0310
(xvi)	+ $NiCl_2.6H_2O$:SI-0311
(xvii)	+ $K_2Cr_2O_7$:SI-0312
(xviii)	+ $KClO_3$:SI-0313
(xix)	+ $Mg(NO_3)_2.6H_2O$:SI-0314
(xx)	+ H_3BO_3:SI-0315
(xxi)	+ HPO_3:SI-0316
(xxii)	+ $KHCOH_4O_4$:SI-0317
(xxiii)	+ $K_4Fe(CN)_3.3H_2O$:SI-0318
(xxiv)	+ KI:SI-0319
(xxv)	+ $(NH_4)_2C_4H_4O_6$:SI-0320
(xxvi)	+ $Al_2(SO_4)_3SO_4.24H_2O$:SI-0321
(xxvii)	+ $3CdSO_4.8H_2O$:SI-0322
(xxviii)	+ Benzene: SI-0323
(xxix)	+ Ethyl Alcohol: SI-0324; -0281b; -0281e
(xxx)	+ Methyl Alcohol: SI-0325; -0281f
(xxxi)	+ Ethylene Glycol: SI-0326
(xxxii)	+ Glycerine: SI-0327; -0281a

(44) $KOH:H_2O$: + xxx: + xxx
Preferential:

(i)	+ CHI_3:Glycerine: SI-0128; -0198a
(ii)	+ $K_3CO_3.1.5H_2O:Ce(OH)_4$:SI-0329
(iii)	+ $Na_2CO_3:Ce(OH)_4$: SI-0330
(iv)	+ $KCN:I_2$: SI-0331

(45) $KBr:H_2O$
Cleaning: SI-0196

(46) $MeOH:I_2$
Cleaning: SI-0149; -0293

(47) $NaCr_2O_7:H_2O$
Cleaning: SI-0066b

(48) $KF:KCl:H_2O$ (electrolytic)
Polish: SI-0186a

(50) $KOH:H_2O$:
Step-etch: SI-0085; -0011c

(51) H_2SO_4 +/− H_2O:
Cleaning: SI-0045b; -0138; -0407

(52) $NaOH:H_2O$:
Removal: SI-0406

(53) $H_2SO_4:H_2O_2$:
Cleaning: SI-0043b; -0298; -0108a; -0399; -0414; -0420a

(54) $NH_4OH:H_2O_2$:
Cleaning: SI-0407

(55) $HAc:I_2$:
Thinning: SI-0440

B. Solutions by Name/Application:
(50) Cleaning: (See Gases)
- (i) RCA (AB): SI-0031; -0032; -0033; -0122; -0173; -0261; -0407; -0406; -0441
- (ii) Caro's Etch: SI-0103; -0169b; -0104; -0105; -0106; -0108a; -0269
- (iii) Solvents: SI-0111; -0114
- (iv) Heat: SI-0113

(51) Preferential: (Named Etchants)
- (i) Chrome Dislocation: SI-0038a
- (ii) Copper Dislocation: SI-0038b; -0153c; -0153d
- (iii) Dash Etch: SI-0153b; -0178; -0179; -0213; -0044; -0038c; -0426a; -0154a
- (iv) Dash Copper Decoration: SI-0154b
- (v) Erhard's Etch: SI-0158; -0182
- (vi) Iodine Etch: SI-0240; -0241; -0244; -0242a; -0242b
- (vii) Landyren's Etch: SI-0120f; -0035
- (viii) Schimmel Etch: SI-0037; -0051; -0294g; -0043; -0139
- (ix) Secco Etch: SI-063a; -0270; -0274; -0402; -0043; -0139; -0036b
- (x) Silver Etch: SI-0209
- (xi) Silver Glycol: SI-0041a; -0042
- (xii) Sirtl Etch: SI-0030; -0039; -0004; -0153a; -0152; -0184; -0249; -0132; -0133; -0134; -0043; -0090; -0364; -0139; -0425; -0426a; -0438
- (xiii) Sopori Etch: SI-0043
- (xiv) Vogel's Etch: SI-0034
- (xv) Wright's Etch: SI-0045; -0264a; -0279; -0043; -0419; -0418; -0421; -0422; -0400; -0397
- (xvi) Chrome Regia: SI-0019; -0242
- (xvii) Aqua Regia: SI-0400; -0434
- (xviii) Superoxol: SI-0165; -0383; -0388

(52) Preferential: (Acronyms)
- (i) BHF: See Section "A" (13)
- (ii) BF-3: SI-0281d
- (iii) BRM: See Section "A" (1)
- (iv) CP2 (Superoxol): SI-0165; -0067
- (v) CP4/CP4A: SI-0173; -0047; -0048; -0050; -0051; -0052; -0053; -0054; -00173; -0175; -0176a; -0102e; -0002; -0199; -0200; -0202; -0203; -0112; -0242; -0243; -0254b; -0102d; -0011b; -0055; -0437; -0370; -0434; -0076b
- (vi) CP8: SI-0062
- (vii) SR4: *SI-0172

*Polish

(56) Metal Decoration:
- (i) Arsenic: SI-0156; -0157b
- (ii) Copper: SI-0154b; -0178; -0197b; -0112
- (iii) Graphite (Carbon): SI-0110
- (iv) Gold: SI-0101; -0064; -0218; -0434
- (v) Silver: SI-0087; -0280
- (vi) Cobalt: SI-0233
- (vii) Selenium: SI-0120b

(57) Electric Current — preferential
- (i) Defect: SI-0247
- (ii) Decorate: SI-0401(+HF)

(58) Abrasive — polish
- (i) Al_2O_3 slurry: SI-0280b; -0290
- (ii) Al_2O_3:CrO_3 slurry: SI-0016
- (iii) SiO_2 slurry: SI-0280a; -0291a

(59) Shaping/Forming:
- (i) Square/Rectangle: SI-0401
- (ii) Triangles/Diamonds: SI-0208
- (iii) Wafers: SI-0402
- (iv) Cubes/Bars/Cylinders: SI-0093b; -0092a-h
- (v) Grooves: SI-0137; -0243
- (vi) Mesas: SI-0094
- (vii) Spheres: SI-0092a; -0092b; -0092c; -0158
- (viii) Special Cutting: SI-0155; -0156; -0157
- (ix) Surface pattern (Jello): SI-0292
- (x) Surface pattern (photo resist/metals): See etchant discussions.
- (xi) Pylons: SI-0173a-c
- (xii) Junction (melt): SI-0382
- (xiii) Via holes: SI-0415

(60) Gases:
- (A) Dry Chemical Etching (DCE): (ionized gas)
  - (i) Ar^+: SI-0207
  - (ii) CF_4^+:O_2^+: SI-0262; -0285; -0286c

(B) Wet Chemical Etching (WCE): (molecular gas)
- (iii) Argon$^+$ H_2O vapor: SI-0419; -0437
- (iv) Oxygen: SI-0365
- (v) Helium: SI-0417

(C) Thinning: (ionized gas)
- (i) Ar^+: SI-0193; -0299; -0141

(D) Selective/Preferential/Structure:
- (i) Ar^+: SI-0141; -0297; -0260
- (ii) Ar^+ Laser: SI-0221
- (iii) Ar^+:O_2+: SI-0222
- (iv) Neutron: SI-0224g; -0389
- (v) H^+: SI-0253; -0231
- (vi) H_2 (O_2 contamination): SI-0220
- (vii) CHF_3^+:O_2^+: SI-0273; -0430b
- (viii) ClF_3^+: SI-0253
- (ix) Cl_2, vapor: SI-0221b; -0221a; -0407 (+He); -0398 (+He)
- (x) SF_6^+:Ar^+: SI-0273; -0274; -0283b
- (xi) SF_6^+:O_2^+: SI-0273; -0274
- (xii) H_2 (defects): SI-0231
- (xiii) CF_4:H_2: SI-0405; -0430b
- (xiv) CF_4:O_2: SI-0430b
- (xv) NF_3:O_2: SI-0408

(E) Polish:
- (i) CF_3Cl^+:$C_2F_6^+$: SI-0286b

(61) Oxidation:
- (i) $HF/NO_2/H_2O$: SI-0142; -0395
- (ii) KOH:H_2: See Section "A" (41)
- (iii) H_2O: See Section "A" (24), (26)
- (iv) HNO_3: See Section "A" (16)
- (v) SiO_2: SI-0403
- (vi) H_2SO_4:H_2O: SI-0396

(70) Silicon (material types):

Note: Structures other than single crystal, and special uses.
- (1) a-Si:H: SI-0406; -0420; -0402; -0247
- (1a) a-Si: SI-0232
- (2) a-, uc-SI:F:H: SI-0239
- (3) Poly-Si: SI-0263c; -0397; -0408; -0398; -0401; -0431; -0439; -0236a; -0274; -0093; -0051; -0076a-b; -0077; -0263b; -0043b; -0158(sphere); -0072; -0297
- (4) As structural material: SI-0091
- (5) Hi-pressure xtls forms: SI-0416
- (6) Ribbon xtls: SI-0425; -0421; -0074; -0148; -0074
- (7) Whisker: SI-0156
- (8) Si ingot grown in space: SI-0132

(71) Controlled damage (structuring, pits):
Note: Mechanical damage for study or devices.
(1) Diamond stylus: SI-0092a-t; -0418
(2) Knoop hardness tester (diamond): SI-0171

(72) Spheres: (hemispheres):
Note: Studies for convex facets, etch rates, and for device fabrication.
Forming: SI-0156; -0157; -0158; -0154
Polish: SI-0061a-b; -0154; -0254b
Finite form: SI-0092a-t; -0145; -0154; -0254t; GE-0150a-u(hemispheres, Si, Ge)

SILICON ETCHANTS

SI-0001
ETCH NAME: BRM
TYPE: Halogen, polish
COMPOSITION:
x .... 0.5% Br_2
TIME: 20 sec + 10 sec
TEMP: RT
DISCUSSION:
Si, (111), (100) and (110) wafers used in a surface cleaning study. Lens paper was used as a pad wetted with solution shown. Hand polish for 20 sec, dilute with MeOH, and continue polishing an additional 10 sec. After this initial polish, the different orientations were further cleaned as follows: (1) (111): rinse in BRM, then MeOH. Then rinse in 5% HF:MeOH (HFS) solution and MeOH rinse. Then rinse in $1NH_4OH:1H_2O$ (AMH) solution with DI water rinse and then repeat the HFS sequence. (2) (100): rinse in BRM, then MeOH. Then rinse in BHF (buffered hydrofluoric acid solution — not shown) and follow with a final MeOH rinse. (3) (110): rinse in BRM, then in MeOH. Then rinse in HFS sequence. Similar cleaning and polishing was done on the following materials: Ge, GaP, GaAs, GaSb, InP, InAs, and InSb.
REF: Aspenes, D E & Studna, A A — *J Vac Sci Technol,* 20,488(1982)

SI-0102a
ETCH NAME: Nitric acid
TYPE: Acid, cleaning
COMPOSITION:
x .... 16 *N* HNO_3
TIME:
TEMP: Hot
DISCUSSION:
Si wafers: Nitric acid alone will not etch silicon or germanium, but when used hot at or near the boiling point, it is an excellent contamination removal agent for residual surface water. Surfaces should be water flushed for a minimum period of 10 min after nitric acid cleaning. (*Note:* Since this developmental work, boiling nitric acid has been used to oxidize silicon surfaces, followed by an HF strip and DI water rinse as a surface cleaning step.)
REF: Turner, D R — *J Electrochem Soc,* 107,810(1960)

SI-0102h
ETCH NAME:
TYPE: Acid, polish
COMPOSITION:
1 .... HF
9 .... HNO_3
TIME:
TEMP:

DISCUSSION:

Si, wafers. Author says that if the etch rate is sufficiently rapid, light will not affect the rate. The solution shown is such a composition. Also in general, etch rate of silicon is 10:1 times faster than on germanium with $HF:HNO_3$ etchant mixtures. (*Note:* $HF:HNO_3$ mixtures, either high in HF or HNO_3 content, are used as stain etchants to observe diffusion profiles.)

REF: Ibid.

SI-0003

ETCH NAME: Nitric acid TIME:

TYPE: Acid, stain TEMP:

COMPOSITION:

x .... HNO_3, conc.

DISCUSSION:

Si, (100) and GaAs, (100) wafers used as substrates for epitaxy growth of GaP/Si and Ge/GaAs. After epitaxy, cleave wafers ⟨110⟩ and stain to develop layer structure. (*Note:* The stain is by oxide colors.)

REF: Rosztoczy, F E & Sterin, W W — *J Electrochem Soc,* 119,1119(1972)

SI-0004a

ETCH NAME: Hydrofluoric acid, dilute TIME:

TYPE: Acid, clean TEMP:

COMPOSITION:

1 .... HF

10 .... H_2O

DISCUSSION:

Si, (100) wafers, p-type, 150 Ω cm resistivity, used as substrates for CVD growth of Beta-SiC. First clean with solution shown, then etch in $3HF:4HNO_3:3HAc$ and DI water rinse prior to SiC deposition.

REF: Addaniano, A & Klein, P H — *J Cryst Growth,* 70,291(1984)

SI-0131: Shih, W et al — *J Vac Sci Technol,* A3,967(1985)

Si, (100), p-type, 14—23.5 Ω cm resistivity wafers of 4″ diameter used as substrates for deposition of thermal SiO_2, then poly-Si doped with $POCl_3$. The resulting PSG oxide remaining on the poly-Si surface was removed with the solution shown at 30°C prior to depositing a thermal SiO_2 cap.

SI-0132b: Ibid.

Si, (100) wafers used as substrates for epitaxy deposition of high n^+ poly-Si doped with phosphorus, followed by $TaSi_2$ and an SiO_2 cap. Step-etch with the solution shown at 30°C for $SiO_2:TaSi_2$ removal with a selectivity of 20:1, respectively. After etch structuring, wafers were cleaved ⟨110⟩ for TEM study.

SI-0169a: Wu, S Y — *J Appl Phys,* 62,1415(1963)

Si, (100) wafers, p-type used for antimony diffusion from a doped, ASG glass surface film. Wafers cleaned in solution shown at RT, 1 min, spin-rinsed in UPDI, and then blown-dry with hot nitrogen. ASG glass was spun-on at various rotation speeds and fired at 150 or 400°C in N_2, 30 min. Also used $7H_2SO_4:3H_2O_2$ at 110°C, 10 min to clean surfaces.

SI-0191: Archer, R J — Electrochem Soc Meet, WDC, 11—16 May 1957.

Si wafers used in a study of optical measurements of films grown on silicon and germanium surfaces in room air. An HF dilute solution was used to initially clean surfaces.

SI-0108b: Ibid.

Si, (111) and (100) wafers used as substrates in a study of pyrogenic oxidation of silicon at high pressure with dry oxygen (800—1000°C and 1—20 atm). Substrates were cleaned:

(1) H_2SO_4:H_2O; (2) aqua regia and (3) 1HF:10H_2O then spun dry in nitrogen.
SI-0283a: Eisele, K M — *J Electrochem Soc,* 128,123(1981)

Si, (100) wafers with SiO_2 thin films used in a study of RF plasma etching with SF_6. Plasma etching produced rough surfaces and high etch rates due to residual photo resist from patterning if silicon surfaces were not cleaned in 10% HF prior to plasma etching.

SI-0002g
ETCH NAME: Hydrofluoric acid | TIME:
TYPE: Acid, oxide removal | TEMP:
COMPOSITION:

x .... HF, conc.

DISCUSSION:

Si, wafers of different orientations. Concentrated HF used to remove native oxide from surfaces. Used on both silicon and germanium for cleaning as HF alone does not attack these materials. (*Note:* HF does etch silicon but at rates of under 1 Å/min. Primarily used as an oxide remover.)

REF: Ibid.
SI-0005: Akiyama, W et al — *J Cryst Growth,* 68,21(1984)

Si, (100) wafers used as substrates for MOCVD epitaxy growth of GaAs/GaAs/Si and GaAlAs/GaAs/Si. Substrates cleaned in HF to remove native oxide prior to epitaxy. After epitaxy wafers were cleaved ⟨110⟩ and lightly etched in molten KOH to develop domain structure.

SI-0006: Rudder, R A — *J Vac Sci Technol,* A2,326(1984)

Si, (100) wafers used as substrates for thin film deposition of a-Si:H and a-Ge:H. Degrease substrates in TCE, then acetone, MeOH, and DI water rinse. Final dip in HF with DI water rinse.

SI-0007: Liou, L et al — *J Electrochem Soc,* 131,672(1984)

Si, (100) wafers used as substrates for ion implantation of Si^+ and As^+. The amorphous layers after implant were called: Type aI. After annealing at 400 to 600°C called: Type aII. These amorphous zones were etched with HF by immersion: (1) with ultrasonic and (2) with stirring rod at 200 rpm. The stirred solutions etch the fastest. Immersion/ultrasonic etch rate was 42.5 Å/h; immersion/stir etch rate was 80 Å/h. Type aI etching was nonuniform; Type aII etched uniformly up to 6 and 7 h.

SI-0008: Bicknell, R N — *J Vac Sci Technol,* A2,423(1984)

Si, (100) and (111) wafers used as substrates for MBE growth of oriented CdTe thin films. Sapphire substrates as prism ($1\bar{1}02$) and ($1\bar{2}10$) or (0001) basal also were evaluated. All substrates were etch cleaned in HF immediately prior to introduction into the vacuum system. The (0001) basal sapphire plane gave the best orientation fit of CdTe films.

SI-0120: Dev, B N et al — *J Vac Sci Technol,* A3,946(1985)

Si, (111) and (100) wafers used in a study of Se adsorption along (111) and (220) bulk planes. Polish substrates on Syton pad, then degrease with TCE and MeOH rinse. HF soak and water quench without exposure to air, then immediately into a solution of 0.5 mg Se/ 50 ml MeOH. While under Se solution cleave wafers ⟨110⟩. Remove and rinse wafers in MeOH and blow dry with argon.

SI-0009
ETCH NAME: Hydrofluoric acid, vapor | TIME: 10 sec
TYPE: Acid, cleaning | TEMP: RT to hot
COMPOSITION:

x .... HF, vapor

DISCUSSION:

Si, (111) wafers within $^1/_4$° of plane, n- and p-type; (111) 2—4° toward (110) wafers used as substrates in formation of $NiSi_2$ and $CoSi_2$ silicides. Wafers were vapor cleaned in HF prior to deposition of 3000—5000 Å of SiO_2. Oxide was patterned by photolithography for selective metallization. In vacuum, prior to metal deposition, wafers were heat cleaned at 1000°C, 2 min at 1×10^{-10} Torr.

REF: Tung, R T — *J Phys Chem Solids,* 45,465(1984)

SI-0027g: Beyer, K D & Kastl, R H — *J Electrochem Soc,* 129,1027(1982)

Si, (111), n- and p-type wafers used in a copper tracer study. Wafer surfaces were treated by dipping in a solution of $1HF{:}10H_2O{:}10\%\ ^{64}Cu$ which is similar to copper concentration in commercial buffered HF solutions. Hang wafers vertically above the HF solution in a closed container for 1 min at RT and do not water rinse after treatment. Native oxide removal rate was shown as 100 nm/min.

SI-0181b: Ibid.

Si, (111) wafers, n-type 5—10 Ω cm resistivity used in fabrication of boron diffused diodes using diborane, B_2H_6 gas in initial development of this process. HF vapor at 35—40°C used on clean surfaces after boron diffusion. Follow with DI water rinse and IR lamp dry.

SI-0132a

ETCH NAME: Hydrofluoric acid, dilute — TIME:

TYPE: Acid, native oxide removal — TEMP:

COMPOSITION:

1 .... HF

16 .... H_2O

DISCUSSION:

Si, (100) wafers used as substrates for epitaxy growth of high n^+ doped poly-Si using phosphorus. Native oxide left on poly-Si as PSG was removed with the solution shown prior to deposition of $TaSi_2$ and an SiO_2 cap.

REF: Palik, E D et al — *J Vac Sci Technol,* B3,492(1985)

SI-0010

ETCH NAME: Hydrofluoric acid, dilute — TIME:

TYPE: Acid, clean — TEMP:

COMPOSITION:

x .... HF

x .... H_2O

DISCUSSION:

Si, (111) p-type wafers used as substrates for deposition of Pt and Pt:Pd metal films. Wafers were degreased, then surface cleaned in solution shown. An oxide was grown on surfaces with $3HCl{:}1H_2O_2{:}1H_2O$ at 90°C, 10 min before alloy deposition. The Pt:Pd alloy was both co-deposited and sequentially deposited.

REF: Kawarada, H — *J Appl Phys,* 57,244(1985)

SI-0011a

ETCH NAME: Hydrofluoric acid, dilute — TIME: 20—40 min

TYPE: Acid, clean — TEMP: RT

COMPOSITION:

1 .... HF

1 .... H_2O

DISCUSSION:

Si, (111) wafers with high boron doping (>10^{20} cm^3). Surface cleaning in the solution shown leaves a blue-green stain. Deposit 50 Å of SiO_2 then etch strip the oxide with HF to remove stain.

REF: Palik, E D et al — *J Electrochem Soc,* 129,2051(1982)

SI-0012

ETCH NAME: Hydrofluoric acid, dilute — TIME: 2 min

TYPE: Acid, clean — TEMP: RT

COMPOSITION:

1 .... HF
30 .... H_2O

DISCUSSION:

Si, (100) wafers used as substrates for LPCVD deposition of tungsten from WF_6. After HF rinse in the above solution, rinse in 18 MΩ water for 10 min and spin-dry. This treatment leaves about 20 Å of native oxide on silicon. Deposit SiO_2 and open windows with photo resist lithography as a mask for selective deposition of 2000 Å tungsten as poly-W. (*Note:* An excellent discussion of stress and adhesion of tungsten films.)

REF: Green, M L & Levy, R A — *J Electrochem Soc,* 123,1243(1985)

SI-0013

ETCH NAME: Hydrofluoric acid, dilute — TIME:

TYPE: Acid, clean — TEMP:

COMPOSITION:

1 .... HF
100 .... H_2O

DISCUSSION:

Si, (100) As-doped, 10 Ω cm resistivity wafers used as substrates for deposition of tungsten, then PtSi. Clean silicon wafers with solution shown. In vacuum system, back-sputter silicon with Ar^+ ionized gas prior to tungsten deposition.

REF: Sinha, A R & Smith, T E — *J Appl Phys,* 44,3465(1973)

SI-0014: Adams, A C et al — *J Electrochem Soc,* 127,1787(1980)

Si, (100), p-type, 8—30 Ω cm resistivity wafers used as substrates in a study of oxidation. First, clean in "acidic H_2O_2" (H_2SO_4:H_2O_2?), then in solution shown to remove native oxide. Various thicknesses of SiO_2 deposited and the same solution used as a step-etch solution to determine oxide etch rates.

SI-0015a

ETCH NAME: Hydrofluoric acid, dilute — TIME: 5 min

TYPE: Acid, oxide removal — TEMP: RT

COMPOSITION:

(1) x .... 20% HF — (2) x .... 6% HF

DISCUSSION:

Si, (111) n-type, 1.5—2.5 Ω cm resistivity wafers used as substrates for silicide study as Si/Pd/Pd_2Si/Si. First, degrease substrates in TCE with ultrasonic agitation; then acetone and MeOH rinse. Etch in the 20% HF solution to remove native oxide and DI water rinse. Oxidize surface with "RCA" etch (1HN_4OH:1H_2O_2:6H_2O) and re-etch in the 6% HF solution for 5 min before initial Pd evaporation and anneal at 400°C, 90 min to form Pd_2Si. Follow with Pd metallization, then a 140 Å SiO_2. Anneal at 275°C for various times. The pure Pd will diffuse through the Pd_2Si.

REF: Lien, C-D et al — *J Appl Phys,* 57,224(1985)

SI-0016
ETCH NAME:
TIME:
TYPE: Acid, clean
TEMP:
COMPOSITION:
x .... HF
x .... MeOH
DISCUSSION:
Si, wafers. Wafers were etched in 1HF:5HNO_3 and then in the solution shown to remove native oxide. Work by Turner (SI-0102a) shows that HF-HNO_3 solutions leave about 10 Å of oxide on silicon surfaces.
REF: Wajda, E S et al — *IBM J*, 36,288(1961)

SI-0017
ETCH NAME: Hydrochloric acid
TIME: 30 sec × 2
TYPE: Acid, clean
TEMP: RT
COMPOSITION:
x .... HCl
DISCUSSION:
Si, (111) wafers used as substrate for deposition of a-C for MIS device study. Degrease substrates in TCE, then rinse in acetone and MeOH. Etch in HCl, rinse in DI H_2O and N_2 dry. DC sputter carbon in argon.
REF: Khan, A A et al — *Solid State Electron*, 27,385(1984)

SI-0018
ETCH NAME: Hydrochloric acid, vapor
TIME:
TYPE: Electrolytic, clean/vapor
TEMP: Si: 300—400°C & Ge: 200—300°C
COMPOSITION:
(1) x .... HCl
(2) x .... Cl_2
ANODE: Ge or Si
CATHODE: C (Ge); Ceramic (Si)
POWER:
DISCUSSION:
Si and Ge wafers. Hot vapors (chlorine content) used to clean both silicon and germanium surfaces. Author states that reaction forms either $GeCl_4$ or $SiHCl_3$. Blow Cl_2 gas through a hole in the graphite mask for Ge; use a ceramic mask with hole for silicon.
REF: Seiler, K O — U.S. Patent #2,744,000, May 1, 1956

SI-0019
ETCH NAME: Chrome regia
TIME: 1—5 min
TYPE: Acid, clean
TEMP: RT
COMPOSITION:
ASTM: #101
500 ml .... HCl
100 ml .... 10% CrO_3
DISCUSSION:
Si, wafers both Float Zone ingot material and epitaxy thin film deposit. Solution used to remove heavy metal ion contamination from surfaces.
REF:
SI-0247: ASTM E407-70
Reference for ASTM Number shown under SI-0019. Solution used on irons and steels as a preferential etchant.

SI-0020
ETCH NAME: BHF
TYPE: Acid, oxide removal
COMPOSITION:
1 .... HF
5 .... NH_4F (40%)
TIME: $2^1/_2$ min
TEMP: RT
DISCUSSION:

Si substrates used for deposition of a-Si:H by glow discharge decomposition of SiH_4. Native oxide removed from a-Si:H by: (1) soak in hot isopropyl alcohol vapor for "several" hours; (2) etch in BHF solution shown and (3) soak 10 min in fresh chlorine water before deposition of 30 monolayers of cadmium stearate ($CdSt_2$). Store in nitrogen for 2—3 days. Then evaporate gold to form MIS device structure. a-Si:H films are slightly n-type. (*Note:* Chlorine water is made by bubbling gaseous chlorine into water.)

REF: Lloyd, J P et al — *Solid Thin Films,* 89,367(1982)

SI-0021a
ETCH NAME: BHF
TYPE: Acid, isotropic
COMPOSITION:
1 .... HF
7 .... NH_4F(40%)
TIME: 15 min
TEMP: RT
RATE: 250 Å/15 min — Si(111), p-type; 0.3 Å/1 min — Si(111), n-type
DISCUSSION:

Si, (111), n- and p-type. Solution shown will etch silicon at rates shown. Some of the etch action is attributed to impurities in the ammonium fluoride.

REF: Kern, W — *RCA Rev,* 39,278(1978)

SI-0022
ETCH NAME: BHF
TYPE: Acid, oxide removal
COMPOSITION:
1 .... HF
9 .... 40% NH_4F
TIME: 60 sec
TEMP:
RATE: 0.01—0.02 μ/sec (Si_2O)
DISCUSSION:

Si, (100), p-type, 2 Ω cm resistivity wafers used as substrates. First, clean wafer in H_2O_2; then in "acidic" H_2O_2 (H_2SO_4:H_2O_2?); then in HF; then in BHF solution shown. Both SiO_2 and Si_3N_4 were then RF plasma deposited. After sitting for several days, SiO_2 etch rate increases to 0.02—0.03 μm/sec.

REF: Reisman, A et al — *J Electron Mater,* 13,504(1984)

SI-0023
ETCH NAME:
TYPE: Acid, oxide removal
COMPOSITION:
2 .... HF
13 .... NH_4F
TIME:
TEMP:
DISCUSSION:

Si, (111) wafers. Solution used to remove SiO_2. Concentration of ammonium fluoride not shown.

REF: Ralaauw, C — *J Appl Phys,* 131,1114(1984)

SI-0024
ETCH NAME: BHF | TIME:
TYPE: Acid, oxide removal | TEMP:
COMPOSITION:
x .... HF
x .... NH_4F
x .... H_2O
DISCUSSION:
Si, (111), n-type, 1—10 Ω cm resistivity wafers used as substrates for PtSi formation. Degrease wafer in acetone with ultrasonic agitation, rinse in EOH. Then clean in BHF solution and flood with EOH so that surfaces do not come in contact with air. Deposit 1 KÅ SiO_2 on back of wafer and resistance heat in vacuum at 1000°C. Evaporate Pt from wire by resistive heating and anneal to form PtSi.
REF: Matz, R et al — *J Vac Sci Technol,* A2(2),253(1984)
SI-0025: Walker, R T — *J Appl Phys,* 127,1432(1979)
Si, (111), n-type wafers. Etch cleaned surfaces 4 min at RT.

SI-0026
ETCH NAME: BHF | TIME:
TYPE: Acid, nitride removal | TEMP:
COMPOSITION:
5 ml .... HF
20 g .... NH_4F
300 ml .... H_2O
DISCUSSION:
Si, (111), (100) wafers as substrates for deposition of Si_3N_4 in a study of nitride composition and etch rates. Increased heating of substrate during nitriding slows etch rate and higher film density lowers etch rate (less H_2 in film). Solution used to etch clean wafers and to etch nitrides. Before nitride deposition, wafers were baked in vacuum at 400°C.
REF: Zhou, N S et al — *J Electron Mater,* 14,55(1985)

SI-0027
ETCH NAME: BHF | TIME:
TYPE: Acid, oxide removal | TEMP:
COMPOSITION:
5 ml .... HF
20 g .... NH_4F
30 ml .... H_2O
DISCUSSION:
Si, (100) wafers. Solution used to remove native oxide from wafers prior to deposition of Al_2O_3. After alumina deposition slots were opened in the material by photo resist techniques for use as an etch mask, and grooves etched in silicon prior to SiO_2 plasma deposition.
REF: Ho, V & Seguano, T — *Solid Thin Films,* 95,315(1982)

SI-0028
ETCH NAME: BHF | TIME:
TYPE: Acid, oxide/nitride removal | TEMP:
COMPOSITION:
x .... HF
x .... NH_4F
x .... H_2O

DISCUSSION:

Si, (100) wafers used as substrates in a study of oxide and nitride etch rates under different deposition conditions. Solution shown used to etch native oxide from wafers, and to etch thin film oxide and nitride after growth.

REF: Kato, I et al — *J Electron Mater,* 13,913(1984)

SI-0029: Brat, T & Eizenberg, M — *J Appl Phys,* 5,264(1985)

Si, (100), n-type, 10 Ω cm resistivity wafers used as substrates in a metallization study. Degrease with organic detergents; then in BHF solution, rinse with H_2O and N_2 blow dry. Evaporate iridium; co-deposit $Ir_{80}V_{20}$ or 1 Ir:1 V. Also used SiO_2 mask, photo resist patterned for selective metal deposition.

SI-0030

ETCH NAME: TIME:

TYPE: Acid, defect TEMP:

COMPOSITION:

1 .... HF

x .... 1.5 *M* CrO_3

1 .... H_2O

DISCUSSION:

Si, (111), n-type and (110), p-type wafers used in a defect study. Other CrO_3 concentrations evaluated: 0.5; 1—2; 2.5 and 3 *M*. Solution shown above is close to Sirtl etch.

REF: Yang, K H — *J Appl Phys,* 131,1141(1984)

SI-0431: Ciszeic, T F — *J Electrochem Soc,* 132,963(1985)

Si, polycrystalline ingot grown by Stockbarger-Bridgman method by solidification from a cold crucible. Grain size was about 5 mm evaluated against Czochralski (CZ) and sheet grain size. Edge Supported Pull (ESP) method for sheet as a continuous ingot casting with a square silicon outline with grains. Solution used to develop defects.

SI-0031

ETCH NAME: RCA etch (AB etch) TIME: 10—20 min

TYPE: Acid, cleaning TEMP: 75—85°C

COMPOSITION:

"A"		"B"	
1—2	NH_4OH	1—2	HCl
1—2	H_2O_2	1—2	H_2O_2
5—7	H_2O	6—8	H_2O

DISCUSSION:

Si wafers of all major plane orientations. "A" solution will remove organics and Group I and II metals. "B" solution will remove heavy metals and soluble complexes that do not plate out. Rinse in DI H_2O and N_2 blow dry. This cleaning system was originally developed for cleaning silicon (this reference), but since then has been applied to many other materials. (*Note:* The two etch names shown have both been used. The AB etch designation should not be confused with the A/B etch of Abrahams & Buiocchi.)

REF: Kern, W & Puotinen, W — *RCA Rev,* 70,187(1969)

SI-0032a: Walker, P — personal application, 1970—1985

Si, (111) and (100) wafers used as SCRs (n-type 10—120 Ω cm resistivity with diameters from 0.250 to 2.00″). Silicon, (111) wafers used for high voltage diodes (p-type, 100 to 10,000 Ω cm resistivity). Alumina, quartz, and sapphire substrates used for microwave circuits.

Direct transfer of wafers from solution "A" to "B" still wet. SCRs were water rinsed and IR lamp dried after cleaning. Diode wafers, after DI H_2O rinse, had a final rinse in I_2:MeOH with insertion into a hot diffusion furnace still wet. Circuit substrates given final rinse in HF for 30 sec, then 30 sec DI H_2O, and N_2 blow dry immediately before insertion into vacuum or held under MeOH until ready for insertion. The iodine rinse or being held under methanol gave best results.

SI-0015b: C-D Lien, et al — *J Appl Phys,* 57,224(1985)

Si, (111) n-type 1.5—2.5 Ω cm resistivity wafers used in study of Pd_2Si. "A" solution used 5 min at RT to oxidize silicon, then HF to strip clean prior to metallization with Pd and annealing for Pd_2Si.

SI-0122: Martin, T I et al — *J Electron Mater,* 13,309(1984)

Si, (100) n-type, 1—4 Ω cm resistivity wafers used for deposition of WSi_2 silicide. Authors used the "RCA" cleaning system followed by an HF dip before tungsten deposition. WSi_2 is tetragonal system.

SI-0173: Schmidt, J et al — *Thin Solid Films,* 110,7(1983)

Si, (100) p-type, 13—17 Ω cm resistivity, and (111) p-type, 4 Ω cm and n-type 4.5 Ω cm resistivity wafers as substrates for the deposition of diamond and borazone followed by aluminum dot evaporation to form capacitive elements for I/V and C/V plot study. Structures were Al/diamond/Si; Al/borazone/Si; and Al/SiO_2/Si — the latter for comparative purposes. Diamond and borazone were deposited by Reactive Pulse Plasma (RPP). Wafer surfaces were cleaned prior to deposition with the following sequence: (1) "RCA" etch clean; (2) HF rinse and (3) DI water rinse. In vacuum, Ar^+ ion sputter clean immediately prior to deposition. Authors say that the high leakage conductance at the dielectric/silicon interface needs further study.

SI-0407: Pande, K P et al — *J Electron Mater,* 13,593(1984)

SI-0406: Martin, T L — *J Electron Mater,* 13,309(1984)

Si, (100) n-type, 1—4 Ω cm resistivity wafers used as substrates for growth of WSi_2 thin films. Clean wafers in the "A" and "B" RCA etch solutions with final clean in 1HF:1H_2O and DI H_2O rinse prior to metallization. Tungsten is cubic (bcc), and forms tetragonal WSi_2. (*Note:* Reference to "cubic" means isometric system, normal class, in which the cube, (100) is a crystal form.)

SI-0441: Wu, C S et al — *Thin Solid Films,* 104,175(1983)

Si, (100) p-type, 1—10 Ω cm resistivity wafers used as substrates in a study of ErSi silicide. Wafers cleaned: (1) degrease in solvents; (2) RCA etch clean; (3) HF dip with DI H_2O rinse. Initial evaporation was 500 Å Er with 750 Å poly-Si cover. Band gap measured: RT, 0.68 eV; annealed at 380°C, 0.74 eV; annealed at 500°C, 0.77 eV. Without poly-Si coating, ErSi showed surface pitting at 380°C with pit quantity increase at 400—450°C and best band gap at 0.72 eV.

SI-0102f

ETCH NAME: Landyren's etch — TIME:

TYPE: Acid, preferential — TEMP:

COMPOSITION: — ASTM: #140

97% .... HNO_3

3% .... $KMnO_4$

DISCUSSION:

Si, (111) wafers and other orientations. Solution will develop defects on all orientations. Attacks p-type rapidly, slow attack of n-type. Can be used as a p-n junction delineation etch, and also on germanium.

REF: Turner, D R — *J Electrochem Soc,* 107,810(1960)

SI-0032b: Walker, P — personal application, 1970—1975

Si, (111) n-type, 10—100 Ω cm resistivity wafers as epitaxy SCRs. Used on 2″ diameter silicon wafers to develop rotational etch spiral that showed a resistivity increase across spiral toward wafer edges due to loss of growth control during Czochralski (CZ) ingot pulling. Solution also used as a general defect etch on epitaxy silicon, and to etch pattern nickel shim-stock as metallization masks for selective gold deposition on silicon SCRs. Solution used was 10 mg $KMnO_4$:150 ml HNO_3, RT to warm (40°C).

SI-0036
ETCH NAME: Secco etch — TIME: 20 min
TYPE: Acid, preferential — TEMP: RT
COMPOSITION:
2 .... HF
1 .... 0.15 *M* $K_2Cr_2O_7$
DISCUSSION:

Si, (111) and (100), p-type, 1—10,000 Ω cm resistivity wafers. The 1—300 Ω cm material etches faster. This solution used with ultrasonic agitation 5 min at RT will produce similar results without bubble formation. Silicon wafers were first etch polished in 2HF:$3HNO_3$:2HAc, RT, 2—3 min with H_2O flush. $Na_2Cr_2O_7$, $(NH_4)_2Cr_2O_7$ or xCrO_3:xHF also can be used. The solution shown is more acidic with less Cr ions than the Sirtl etch.

REF: d'Aragona, F Secco — *J Electrochem Soc,* 119,948(1972)

SI-0263: Banerjee, S K et al — *J Electrochem Soc,* 131,1409(1984)

Poly-Si grown on (100) CZ silicon wafer substrates. Deposits contained large single crystallites with low-angle grain boundaries. Secco etch used to develop grain boundaries and structure.

SI-0270: Robinson, McD et al — *J Electrochem Soc,* 129,2858(1982)

Si, thin film epitaxy layers with low dislocation density were etched in the Secco etch to develop dislocations and slip.

SI-0274: Cerofolini, G F et al — *Thin Solid Films,* 109,137(1983)

Si, (100), n-type, 7—10 Ω cm resistivity wafers used in a study of indium diffusion/segregation with various types of SiO_2 deposition: dry O_2 or steam with time and temperature varied. Secco etch used to etch develop dislocation, segregates, stacking faults.

SI-0037
ETCH NAME: Schimmel etch — TIME: 30 sec
TYPE: Acid, preferential — TEMP: RT
COMPOSITION: — RATE: See below
2 .... HF
2 .... 0.75 *M* CrO_3
1.5 .... H_2O
DISCUSSION:

Si, (111) and (100) wafers used as substrates for silicon epitaxy growth. Wafers were degreased prior to epitaxy: (1) TCE soak, then rinse in MeOH, then EOH; (2) soak in H_2SO_4:H_2O_2, RT for 15 min. After epitaxy deposition, solution shown was used to develop defects and structure in the epitaxy layers.

REF: Archer, V D — *J Electrochem Soc,* 129,2078(1982)

SI-0051: Celler, G K et al — *J Electrochem Soc,* 132,211(1985)

Si, (100) wafers used as substrates for poly-Si deposition studies. Wafers were solvent cleaned and then steam oxidized. The SiO_2 was photolithographically patterned as a mask for selective poly-Si deposition. In vacuum system, HCl vapor at 1050°C was used to oxide clean silicon in the opened windows immediately prior to growth of poly-Si using SiH_4. Wafer surfaces were then capped with SiO_2 using LPCVD. ⟨110⟩ cleaved sections were etched in the Schimmel solution to develop poly-Si defects and structure. The poly-Si showed a columnar grain growth. (*Note:* Columnar growth has been observed in compound semiconductor layers and in silicon dioxide deposited at an angle.)

SI-0294: McFee, J H et al — *J Appl Phys,* 130,214(1983)

Si, (100) wafers used as substrates for MBE growth of silicon layers about 2 μm thick. A dilute Schimmel solution used as a dislocation etch and to step-etch or etch-thru holes in order to study the epitaxy layer and bulk substrate.

SI-0038a
ETCH NAME: Chrome dislocation etch TIME: 6 min
TYPE: Acid, dislocation TEMP: RT
COMPOSITION:
"A" 100 g .... CrO_3 "B" x .... HF
200 ml .... H_2O
Mix: 1 "A":1 "B"
DISCUSSION:

Si, (100) and (110) wafers. First, etch polish wafers at RT, 3 min with 3HF:5HNO_3:3HAc and water quench. Mix solution shown when ready to use, and follow etching with a minimum 2 min DI water rinse. Solution used to develop dislocations in the wafers.
REF: Pamphlet — *Monsanto Single Crystal Silicon Evaluation Procedure,* January 1964

SI-0039
ETCH NAME: Sirtl etch TIME:
TYPE: Acid, preferential TEMP:
COMPOSITION:
1 .... HF
2 .... CrO_3 (33%)
DISCUSSION:

Si, (111) wafers and other orientations. This is the original development of this etch as a dislocation and defect solution on silicon. It has been modified since this time and applied to other semiconductor and metallic compounds. (*Note:* Several chromium compounds have been developed since this etchant appeared in the literature, several for special enhancement of particular types of defects.)
REF: Sirtl, H W — *Z Metallkd,* 52,529(1961)
SI-0224: Tallman, R L et al — *Solid State Electron,* 9,327(1966)

Si, single crystal thin films grown across SiC substrates by a moving-zone deposit technique with directional growth. "Sirtl" etch mixture shown was 200 ml HF:100 g CrO_3:200 ml H_2O.
SI-0149: Hammond, M L — *Solid State Technol,* November 1978, 68

Si, (111) wafers and other orientations used as substrates for growth of silicon epitaxy. Sirtl etch used to develop defects in silicon thin films after epitaxy.
SI-0438: Hu, S M — *J Appl Phys,* 46,1470(1975)

Si, (100) wafers cut 6.4°-off plane toward (110) used in a damage induced etch pits and structure study. Wafers were SiO_2 coated before diamond indented with heating and annealing under various conditions. Sirtl etch used to develop structure referred to as "rosettes". Developed paired parallel lines from indent point in ⟨110⟩ directions. (*Note:* The rosette structure is more often considered a growth rather than etch phenomenon.)

SI-0153a
ETCH NAME: Sirtl etch, modified TIME: 1—7 min
TYPE: Acid, preferential TEMP:
COMPOSITION:
"A" 1 g .... CrO_2 "B" 1 ml .... HF
2 ml .... H_2O
Mix "A" and "B" just before use.
DISCUSSION:

Si, (111) wafers. Used as a general dislocation and defect etch and can be used on other orientations.
REF: Harper, C A — *Handbook of Materials and Processes for Electronics,* McGraw-Hill, New York, 1970, 7

SI-0038b
ETCH NAME: Copper dislocation etch TIME: 4 h
TYPE: Acid, dislocation TEMP: RT
COMPOSITION:

"A"	600 ml .... HF	"B"	1000 ml .... H_2O
	300 ml .... HNO_3		3.54 g .... KBr
	28 g .... $Cu(NO_3)_2.2H_2O$		0.078 g .... $KBrO_3$
			1 drop .... 1 *N* NaOH

Mix: 1 ml "B":5 ml "A":49 ml H_2O when ready to use.
DISCUSSION:

Si, single crystal wafers of different orientations. First, polish wafers in 3HF:5HNO_3:3HAc, RT, 3 min and DI water rinse. The solution shown will develop slip, lineage, dislocations, and grain boundaries. Do not store solution once it is mixed as it is unstable. Continue etching if defects are not distinct after 4 h.

REF: Ibid.

SI-0044
ETCH NAME: Dash etch TIME: 2—4 h
TYPE: Acid, dislocation TEMP: RT
COMPOSITION:

1 .... HF
3 .... HNO_3
12 .... CH_3COOH (HAc)

DISCUSSION:

Si, (111) wafers and other orientations, both n- and p-type of different resistivity levels. Solution will develop sharply defined dislocations on (111) and (100) surfaces. May require more than 4 h etching for best development. (*Note:* This was one of the first dislocation solutions developed for silicon and is still used, though others that are faster have largely replaced it.)

REF: Dash, W C — *J Appl Phys,* 1193(1956)
SI-0178: Furuoya, T — *Jpn J Appl Phys,* 1,135(1962)

Si, (111) and (110) wafers cut from CZ grown ingots. First, polish wafers in 1HF:3HNO_3, then copper decorate at 850°C, by 1 h anneal with wafer in contact with a copper plate. Used Dash etch 12 h, RT to develop dislocations.

SI-0179: Iizuka, T & Keruchi, M — *Jpn J Appl Phys,* 2,157(1963)

Si, (111) wafers with high concentration gallium diffusion used in a study of etch patterns. Dash etch used to develop dislocations. Authors say that flat bottom pits are not dislocations, flats are caused by contamination, and pits do not show 180° rotation from positive to negative (111) wafer faces as do dislocations. Gallium concentration was 10^{18}—$10^{19}/cm^3$ on (111).

SI-0213: Logan, R A & Peters, A J — *J Appl Phys,* 28,1419(1957)

Si, (111) used in a study of etch pit formation as affected by .oxygen content in the material. Dash etch used to develop dislocations. Oxygen content varies in CZ pulled ingots relative to seed/ingot rotation or nonrotation. Etch rate decreases with increase of oxygen content. Data was established from study of etch pit size, pit density, and thickness loss of specimens.

SI-0153b

ETCH NAME: Dash etch, modified TIME: 4 h

TYPE: Acid, dislocation TEMP:

COMPOSITION:

1 ml .... HF

3 ml .... HNO_3

10 ml .... CH_3COOH (HAc)

DISCUSSION:

Si, (100), (111), (110) and (112) wafers. Solution develops dislocation pits on all orientations shown. May require additional time for best definition of pits.

REF: Ibid.

SI-0041a

ETCH NAME: Silver glycol etch TIME: 10 min

TYPE: Acid, preferential TEMP: RT

COMPOSITION: RATE: 1 mil/10 min

"A" 10 ml .... HF

400 ml .... HNO_3

10 ml .... $*AgNO_3$

"B" 200 ml .... H_2O

200 ml .... propylene glycol

*1 g/100 ml H_2O

Mix: 1 "A":1 "B" when ready to use.

DISCUSSION:

Si, (111) wafers and other orientation. Solution used as a defect and dislocation etch. Do not store mixed solution in closed containers. Solution was applied by a brush-on technique using a camel hair brush dipped in the etchant, then wiped across specimen surfaces.

REF: Bondi, F J — *Transistor Technology,* Vol. 3, D Van Nostrand, New York, 1958

SI-0042: Stora, G E — *Acta Metall,* 6,65(1958) (in French)

Si, (111) wafers and other orientations. Solution used to develop imperfections in silicon for micrographic study.

SI-0043

ETCH NAME: Sopori etch TIME: 1 to 2 min

TYPE: Acid, preferential TEMP: (1) RT (2) 10°C

COMPOSITION:

(1) 36 .... HF

20 .... CH_3COOH (HAc)

1—2 .... HNO_3

(2) 36 .... HF

25 .... HAc

2 .... HNO_3

(3) 3 .... H_2SO_4

1 .... H_2O_2

DISCUSSION:

Si, (111) wafers and other orientations, both p- and n-type; ribbon crystal silicon and poly-Si. Etch solution developed for poly-Si and referred to as an isotropic etch as it will develop pits and structure in all orientation "directions". (*Note:* Isotropic normally means a polish etched surface.). Solution (1) leaves a blue stain due to high HF concentration — remove stain with solution (3), which also was used as a general surface cleaning solution. Solution mixture (2) did not show the blue stain when used at 10°C. The solutions were evaluated against the (1) Dash etch — SI-0044; (2) Sirtl etch — SI-0039; (3) Secco etch — SI-0036; (4) Schimmel etch — SI-0037, and (5) Wright etch — SI-0045.

REF: Sopori, B L — *J Electrochem Soc,* 131,667(1984)

SI-0038c
ETCH NAME: Dash etch, modified
TYPE: Acid, preferential
COMPOSITION:
TIME: 2 h
TEMP: RT

"A" 1 .... HF
3 .... HNO_3
12 .... CH_3COOH (HAc)

"B" 100 g .... $NaNO_3$
1000 ml .... H_2O

Mix: 60 ml "A":5 drops "B" and stir.
DISCUSSION:
Si, (100) and (110) wafers. This modified etch used to develop dislocations, lineage, and other defects. First, etch polish samples in 3HF:5HNO_3:2HAc for 2 min at RT.
REF: Ibid.

SI-0045
ETCH NAME: Wright etch
TYPE: Acid, dislocation
COMPOSITION:
TIME: 1—5 min or 20 min
TEMP:

60 ml .... HF
30 ml .... HNO_3
30 ml .... CrO_3 (g/2 ml H_2O)
*8 g .... $Cu(NO_3)_2.3H_2O$
60 ml .... HAc
60 ml .... H_2O

*Mix in water first. Then add remaining acids.

DISCUSSION:
Si, (100), (111), p- and n-type, 0.2—20 Ω cm resistivity, both CZ and FZ grown wafers used in a study of oxidation defect development. First clean wafers: (1) 1HF:50H_2O, 30 sec, RT; (2) rinse 5 min in ultrapure water (UPDI): (2) boil 10 min in H_2SO_4 and rinse 5 min in UPDI prior to steam oxidation for 75 min at 1200°C.

Two methods are used for dislocation etching: (1) A single wafer placed flat on beaker bottom. Pour in etch, and hand agitate. Quench any flushing away acid with DI water without exposure of specimen to air. (2) Multiple wafer etch method: place wafers in a Teflon holder in a vertical position. Immerse in etch and hand agitate by swirling holder, and then remove wafers to a beaker of running DI water.

Copper nitrate enhances dislocations. Oxidation produces well-defined stacking faults, dislocations, swirls, and striations. The etchant is superior to both Sirtl and Secco etchants for stacking fault defect definition.
REF: Jenkins, M W — *J Electrochem Soc*, 124,757(1977)
SI-0264a: Werkhoven, C J — *J Electrochem Soc*, 131,1388(1984)
Si, (111), p-type, 7—20 Ω cm resistivity, wafers used as substrates for silicon epitaxy growth. Wright etch used RT, 1 min to develop defects.
SI-0279: Kuroda, E et al — *J Cryst Growth*, 68,613(1984)
Si, (100) wafers. Ingots grown by Czochralski (CZ) method with resistance heating in an argon atmosphere in a study of the effect of temperature oscillations during growth. Wafers were etch polished in HF:HNO_3:HAc and then oxidized in dry O_2 at 650°C, 3 h and 1000°C, 16 h. About 50 μm of surfaces were etch removed to observe microdefects; then etched in Wright etch to a depth of 5 μm to measure defect density. Microdefect density was strongly dependent on temperature oscillation amplitude during ingot growth.

SI-0006a
ETCH NAME: Dash etch, modified
TIME:
TYPE: Acid, preferential
TEMP:
COMPOSITION:

1 .... HF
3 .... HNO_3
6 .... CH_3COOH (HAc)

DISCUSSION:

Si, (111) wafers and other orientations. Solution used to develop dislocation patterns in deformed silicon specimens. Defects were as slip jobs.

REF: Dash, W O — *J Appl Phys,* 29,705(1958)

SI-0173
ETCH NAME: CP4
TIME:
TYPE: Acid, preferential/polish
TEMP: RT
COMPOSITION:

30 .... HF
50 .... HNO_3
30 .... CH_3COOH (HAc)
0.6 .... Br_2

DISCUSSION:

Si, (111) wafers used in cyclotron resonance experiments. Both silicon and germanium were etch polished in this solution. Authors say the solution influences photoconduction efficiency but not scattering frequency. (*Note:* For original CP4 reference, see GE-0065a, as Camp #4 Etch.)

REF: Dexter, R N et al — *Phys Rev,* 104,637(1956)

SI-0047: Rozhkov, V A et al — *Sov Microelectron,* 13,132(1984)

Si, (111), n- and p-type wafers used to fabricate MOS devices. Mechanically polish wafers, then etch polish with CP4. Deposit SiO_2 and then CeO_2 to evaluate this combination of thin film layers for MOS structures.

SI-0048: Walker, P — personal application, 1958

Si, (111) wafers, n-type, 10—20 Ω cm resistivity used in a general reaction study of CP4 on silicon. Evaluations were (1) used at RT as polish etch; (2) used at 60°C as polish etch but difficult to maintain surface planarity due to fast etch reaction; (3) used at temperatures progressively lower below room temperature: 10, 8, 0, and −10°C, were progressively more preferential to surface defects, cracks, and damage, and (4) at RT in an ultrasonic water bath that was frequency variable and developed a concentric ring structure on surface with low center point. Rings appeared to vary with resistivity increase toward wafer edges.

SI-0050: Statz, H et al — *Phys Rev,* 101,1272(1956)

Si, (111) wafers used in a study of surface states on silicon. After etch polish in CP4 at RT, rinse in DI water, then in MeOH, and final rinse in CCl_4.

SI-0051: Bemski, G — *Phys Rev,* 103,567(1956)

Si, (111) wafers used in a study of quenched-in recombination centers in silicon. Etch polish with CP4 at RT, rinse in DI water, then soak in hot HNO_3 and DI water rinse.

SI-0052a: McKenzie, J M & Waugh, J B — *IRE Trans Nuc Sci,* NS-7,195(1960)

Si, (111) wafers used to fabricate particle spectrometers. Etch polish with CP4, then etch mesa structures with $1HF:3HNO_3$.

SI-0053: Takaki, R et al — *IRE Trans Nuc Sci,* NS8,64(1961)

Si, (111) wafers, n-type used in fabrication of surface-barrier nuclear detectors. CP4 used as a polish etch at RT.

SI-0054: Ray, R K & Fan, H Y — *Phys Rev,* 121,762(1961)

Si, (111) wafers used in a study of impurity conduction. CP4 used to etch polish wafers with DI water rinse.
SI-0174: Faust, J W Jr & John, H F — *J Electrochem Soc,* 107,562(1960)

Si, (111) wafers. A comparison of etching and fracture techniques for studying twinning structures in silicon, germanium and III—V compound semiconductors. Study involved three approaches: (1) preferentially etch a lapped surface, and pit characteristics show change with orientation and surface steps due to etch rate variation; (2) chemically polish or electropolish a lapped surface and twins will show as a groove or step and (3) chemically polish or preferentially etch a mechanically polished surface, and twin lines show as grooves. Both germanium and silicon wafers were cleaved by scribing a line on the (111) surface parallel to a ⟨110⟩ direction and then bent to break specimens for cross-section study under microscopic examination.
SI-0175: Stickler, R & Faust, J W Jr — *J Electrochem Technol,* 4,71(1966)

Si, (111) wafers used in the study of a new type defect in silicon related to growth conditions of the ingot. Wafers polished in CP4, RT, 30—60 sec; followed by Sirtl etch to develop defects. Pit was a star-pattern with points in ⟨211⟩ directions. (*Note:* See SI-0039 for Sirtl etch.)
SI-0176a: Deshpande, R Y — *Solid State Electron,* 9,205(1966)

Si, (111), n-type wafers fabricated with heavily diffused p^+ junctions for the observation of double injection and related effects. Etch polished in $1HF:3HNO_3$, DI water rinse, MeOH rinse, TCE rinse, and air dry. Remove oxide with $HF:NH_4F$ solution and end with a flash dip-clean using CP4.
SI-0102e: Ibid.

Si, (111) wafers used in a general etch study of both silicon and germanium. Both CP4 and CP4A (without bromine) were evaluated for cleaning and polishing of surfaces.
SI-0002: Lauriente, M et al — *J Appl Phys,* 35,3061(1964)

Si, (111) epitaxy deposited wafers used in an X-ray analysis of etch pit patterns and faults in the epitaxy layers. Etch 1 sec with CP4 to observe microstructure. Etch 6 sec for micrographic study.
SI-0199: Fowler, A & Levesque, P — *J Appl Phys,* 26,641(1955)

Si, (111) wafers with grown-in transistor n-p-n junctions used in a study of the optical delineation of junction base widths. CP4 used freshly mixed to delineate the emitter junction. Follow with an electrolytic solution of 0.4% KOH using a carbon anode and emitter as the cathode, with power of 10 A/cm², 10—45 min to delineate the collector junction.
SI-0200: Davis, W D — *Phys Rev,* 114,1006(1959)

Si, (111) wafers, gold doped and used in a study of lifetime and capture cross section. CP4 used as both a polish and dislocation etch.
SI-0202: Iizuka, T et al — *Jpn J Appl Phys,* 2,442(1963)

Si, (111) wafers used in a dislocation study of large dislocation loops created from nickel and gold plating and diffusion at 1200°C in argon with times ranging from 2 min to 5 h and fast quench to RT. CP4 used as a preferential etch. Hexagonal and circular loops observed with gold; similar with nickel which showed a constant size increase with increase of heat.
SI-0203: Iizuka, M et al — *Jpn J Appl Phys,* 2,309(1963)

Si, (111) wafers, n-type, CZ grown in a study of hexagonal platelets observed in nickel diffused silicon. Specimens nickel plated and diffused 1 h at 1300°C in argon. CP4 used as a preferential etch to develop hexagonal structures that were aligned in ⟨110⟩ directions. Size increased from 10 μm with increased heating. No similar structures were observed with FZ grown silicon. Structure due to oxygen in CZ material?
SI-0122: Goss, A J et al — *Acta Metall,* 4,333(1956)

Si:Ge alloy crystals. CP4 used to develop stria structure on ingots. Silicon fraction of single crystal ingots etches more slowly than germanium in CP4. Solution also used to develop edge dislocations.

SI-0242: Blankenship, J L & Borkowski, C J — *IRE Trans Nuc Sci,* NS-7,190(1960)

Si, surface-barrier device. Used CP4 with 10 drops Br_2/50 ml solution. Also varied CP4 reactivity by changing the amount of HAc.

SI-0243: Sheftal, N N et al — *Izvest Akad Nauk SSSR Ser Fiz,* 21,146(1957)

Si, deposited as thin film epitaxy layers. CP4 was a polish etch. Best results were with (110) oriented material.

SI-0254b: Holmes, P J — *Acta Metall,* 7,283(1959)

Si, single crystal hemispheres. Specimens were used in a preferential etch study of type etch pits, patterns, rates, and crystal planes developed. Before preferential etching, use CP4 to etch polish remove lap damage. (*Note:* See GE-0150a-s for both silicon and germanium preferential etches evaluated.)

SI-0102d

ETCH NAME: CP4A — TIME:

TYPE: Acid, polish — TEMP:

COMPOSITION:

30 .... HF

50 .... HNO_3

30 .... CH_3COOH (HAc)

DISCUSSION:

Si, (111) wafers and other orientations. Used as a general polish etch. Reference recommends using CP4 without bromine on both silicon and germanium to improve polish characteristics.

REF: Turner, D R — *J Electrochem Soc,* 107,810(1960)

SI-0011b: Palik, E D et al — *J Electrochem Soc,* 129,2051(1982)

Si, (100) and (111) wafers cut from ingots were directly etched in CP4A, 2 min at RT to remove cutting damage. Follow with two rinses in DI water, then 20 min soak in 1HF:$1H_2O$. Wafers were prepared with varied n- and p-type doping levels in a study of the etch-stop mechanism of high concentration doped layers using KOH solutions.

SI-0055

ETCH NAME: CP4, variety — TIME:

TYPE: Acid, preferential — TEMP:

COMPOSITION:

30 .... HF

50 .... HNO_3

30 .... CH_3COOH (HAc)

x .... $Cu(NO_3)_2$

DISCUSSION:

Si, (111) wafers used in a defect study. Solution develops screw, edge, and spiral dislocations with improved definition over standard CP4 with bromine. Authors suggest spiral pits may be due to impurities such as SiC and SiO_2 as grown-in segregates in ingots.

REF: Christian, S M & Jensen, R V — Am Phys Soc Meet, Pittsburgh, PA, May 15—17, 1956

SI-0021c
ETCH NAME: TIME: 60 min
TYPE: Acid, preferential TEMP: 85—92°C
COMPOSITION: RATE: 0.11 µm/min
1 .... NH_4OH
5 .... H_2O_2
DISCUSSION:

Si, (100) wafers used in an anisotropic etch study. Specimens were etched to a depth of 6.6 µm to develop defects and structure.

REF: Ibid.

SI-0056
ETCH NAME: White etch TIME:
TYPE: Acid, polish TEMP:
COMPOSITION:
x .... HF
x .... HNO_3
DISCUSSION:

Si, (100) cleaved wafers used in a study of surface energies of crystals. Specimens polished in a "white etch" solution of HF:HNO_3. (*Note:* The use of the term "white" is not recommended for a clear, transparent solution.)

REF: Gilman, J J & Johnston, W S — *Dislocations in Crystals,* John Wiley & Sons, New York, 1956

SI-0046b
ETCH NAME: TIME:
TYPE: Acid, polish TEMP:
COMPOSITION:
1 .... HF
3 .... HNO_3
DISCUSSION:

Si, (111) wafers used in a deformation study. Specimens polished in this solution prior to deformation and dislocation etching in 1HF:3HNO_3:6HAc.

REF: Ibid.

SI-0052b: Ibid.

Si, (111) wafers used in fabricating particle detectors. Wafers polish etched in CP4 prior to p-n junction diffusion. Solution shown used to etch form mesas on device surfaces after photo resist masking.

SI-0288: Straumanis, M E et al — *J Appl Phys,* 32,1382(1961)

Si, dislocation free specimens broken into pieces with pliers. Pieces used in a study of lattice perfection. Slivers etched in solution shown to remove breaking damage.

SI-0057
ETCH NAME: White etch TIME:
TYPE: Acid, thinning TEMP:
COMPOSITION:
1 .... HF
4 .... HNO_3
DISCUSSION:

Si, (111) wafers, n- and p-type used in a study of the denuded zones near wafer edges. Wafers annealed at 1050°C in Ar/O_2/N_2 + 3% HCl for 1, 4, and 16 h. Wafers cleaved in

the ⟨110⟩ direction. Wright etch (SI-0045) used to develop dislocations. Solution shown used to thin wafers for TEM study.
REF: Wang, P et al — *J Electrochem Soc,* 131,1948(1984)

SI-0201
ETCH NAME: White etch — TIME:
TYPE: Electrolytic, polish — TEMP:
COMPOSITION: — ANODE: Si
1 .... HF — CATHODE: Pt
4 .... HNO_3 — POWER:
DISCUSSION:

Si, (111) wafers, n-type, 130 Ω cm resistivity used in a study of oscillatory photocurrent decay. Electropolish included flooding with xenon light through a 2.5 mm slit during etch polish period.
REF: Hoshino, H & Takayama, N — *Jpn J Appl Phys,* 2,438(1963)

SI-0061a
ETCH NAME: — TIME: 10—20 min
TYPE: Acid, sphere polish — TEMP: RT
COMPOSITION:
5 ml .... HF
75 ml .... HNO_3
DISCUSSION:

Si, single crystal spheres $^1/_8$, $^1/_4$, and $^1/_2$″ diameter fabricated in a tornado sphere grinder with 600-grit abrasive paper; 1 and $1^1/_2$″ diameter spheres were lathe cut. Cylinders of silicon and germanium cut by an ultrasonic impact grinder. Pre-polish spheres in solution shown in a Teflon beaker at about a 30° angle with a hand-swirling motion so that spheres roll but do not tumble. The time shown is for silicon; germanium time is up to 1 h. Maximum load: 10 $^1/_4$″ diameter spheres for 80 ml of solution. Spheres used in preferential etch studies of finite crystal form. Both cylinders and spheres used in device fabrication.
REF: Walker, P — personal development, 1957—1961
SI-0289: Klein, D L & D'Stefan, D J — *J Electrochem Soc,* 107,198C(1960)

Si, wafers of different orientations used in an etch rate study with high HNO_3 concentration solutions. Etch rate dependent on fluorine diffusion to silicon surface being etched; otherwise wafer surface is stain oxidized with HNO_3.

SI-0062
ETCH NAME: CP8 — TIME:
TYPE: Acid, defect spray — TEMP:
COMPOSITION:
3 .... HF
5 .... HNO_3
DISCUSSION:

Si, (111) wafers. Solution used as a spray to develop metal precipitates in p-n junctions. (*Note:* CP8 is Camp #8 etchant. See Germanium section.)
REF: Goetzberger, A & Shockley, W — *J Appl Phys,* 31,1821(1960)
SI-0290: Camp, P R — *J Electrochem Soc,* 102,1415(1955)

Reference for Camp etchants originally developed for germanium.

SI-0058
ETCH NAME: TIME: 4—5 min
TYPE: Acid, polish TEMP: 20°C
COMPOSITION:
1 .... HF
5 .... HNO_3
DISCUSSION:

Si, (111) wafers used in a study of diffusion control in silicon. Use solution with mechanical agitation to polish prior to diffusion. Authors say there is excellent reproducibility. (*Note:* 20°C is often used with reference to RT etching (22.2°C), but many experimenters say that $HF:HNO_3$ mixtures used slightly below RT give better results due to their highly exothermic reactivity.)

REF: Frosh, C J & Derick, P — *J Electrochem Soc,* 105,695(1958)

SI-0059
ETCH NAME: TIME:
TYPE: Acid, thinning TEMP:
COMPOSITION:
1 .... HF
6 .... HNO_3
DISCUSSION:

Si, (100) wafers, p-type, 2 Ω cm resistivity used in a study of arsenic dislocations induced by As^+ ion implantation. Backside of wafer was etch thinning in this solution for TEM study of dislocation loops developed by arsenic implant.

REF: Davidis, P et al — *J Appl Phys,* 48,3612(1977)

SI-0060
ETCH NAME: TIME:
TYPE: Acid, pinhole jet TEMP:
COMPOSITION:
1 .... HF
9 .... HNO_3
DISCUSSION:

Si, (111) wafers used in a study of electron extinction distance in silicon. Solution used to pinhole etch through specimens. Birefringent colors of thin areas around holes used in study.

REF: Frankl, D R — *J Appl Phys,* 35,217(1964)

SI-0063
ETCH NAME: TIME: 1—3 min
TYPE: Acid, polish TEMP: RT
COMPOSITION:
3 .... HF
50 .... HNO_3
DISCUSSION:

Si, (111) wafers, n-type, 5—10 Ω cm resistivity used in initial development of diborane, B_2H_6 gas as a diffusion boron doping source for silicon. Solution used to slow polish wafers prior to diffusion.

REF: Walker, P & Crabbs, G — personal development, 1966

SI-0102c
ETCH NAME: TIME:
TYPE: Acid, polish TEMP:
COMPOSITION:
32% .... HF
68% .... HNO_3
DISCUSSION:
Si and Ge, (111) wafers and other orientations. Authors say silicon etches about 10× faster than germanium in $HF:HNO_3$ solutions and light does not affect etch rate when it is rapid. (*Note:* Robbins & Schwartz SI-0126 show $1HF:3HNO_3$ as the maximum etch rate mixture.)
REF: Turner, D R — *J Electrochem Soc,* 107,810(1960)

SI-0065
ETCH NAME: TIME: 53 sec
TYPE: Acid, jet thinning TEMP: RT
COMPOSITION: RATE: 10 μm/sec
65 .... HF
35 .... HNO_3
DISCUSSION:
Si, (111) wafers, p-type, 7—21 Ω cm resistivity, 21 mils thick. Solution used to jet-thin wafers for TEM study. Solution flow rate was 12 ml/min and etching was done under a microscope with transmitted white light to control thinning by color of silicon with water added to slow etch rate as color appeared. Three etch factors are important: (1) solution composition, (2) flow rate, and (3) temperature. (*Note:* See SI-0092a-d where KOH solutions are used to etch thin with color defining base widths in transistor fabrication.)
REF: Bulletin: *Micro Etch M-01-006,* A-108, R&G Enterprises, 1983

SI-0066a
ETCH NAME: TIME:
TYPE: Acid, polish TEMP: RT
COMPOSITION: RATE: 0.6 mils/min
1 .... HF
10 .... HNO_3
DISCUSSION:
Si, (111) wafers used in a study of chemical treatment on surface properties. Etch described as a slow etch on silicon.
REF: Buck, T M & McKim, F S — *J Electrochem Soc,* 105,709(1958)
SI-0092b: Walker, P & Waters, W P — personal application, 1958
Si, single crystal spheres and cylinders. Solution used as a slow polish etch. Etch rate without agitation was 0.8 mil/min and with about 10 rpm rotation 1.5 mils/min. Rotation rate varied with solution volume vs. silicon surface area. With sufficient volume, rate was reduced to 0.1 mil/min. Etching was done on 600 grit abrasive formed specimens which, after polishing, were preferentially etched to finite crystal form, and in the fabrication of devices.
SI-0177a: Buck, T M & McKim, F S — U.S. Patent #2,916,407
Si, p-n-p transistors. Solution used as a final junction cleaning etch prior to device encapsulation.

SI-0177b
ETCH NAME: Hydrogen peroxide, dilute TIME:
TYPE: Acid, junction cleaning TEMP:
COMPOSITION:
x .... H_2O_2
x .... H_2O
DISCUSSION:

Si, p-n-p transistors. Solution used as a final junction cleaning etch prior to device encapsulation.

REF: Ibid.

SI-0177c
ETCH NAME: TIME:
TYPE: Acid, junction cleaning TEMP:
COMPOSITION:
x .... HNO_3
x .... H_2O_2
DISCUSSION:

Si, p-n-p transistors. Solution used as a final junction cleaning etch prior to device encapsulation.

REF: Ibid.

SI-0181a
ETCH NAME: TIME:
TYPE: Acid, polish TEMP: RT
COMPOSITION: RATE: 2 μ/min
1 .... HF
20 .... HNO_3
DISCUSSION:

Si, (111) wafers, 5—50 Ω cm resistivity, n-type. Used for fabrication of diodes and transistor devices in electrical breakdown studies caused by defect generation in the subsurface from thermal oxidation deposition of SiO_2 thin films. Solution used to slow etch and clean surfaces using 100 ml per 1 to 1½″ diameter wafer. Etch prior to wafer alloy, diffusion or oxidation. Also used to remove stains produced by HF, boron, and alkalies.

REF: Walker, P & Onidara, O — Phys Failure Conf., Battelle Memorial Institute, Columbus, OH, Spring 1966

SI-0182
ETCH NAME: TIME:
TYPE: Acid, preferential TEMP:
COMPOSITION:
x .... HF
x .... CrO_3
DISCUSSION:

Si, wafers. Wafer sections angle lapped at 5°43′. Solution used to etch develop damage in abraded silicon surfaces.

REF: Pugh, E N & Samuels, L E — *J Electrochem Soc*, 12,1429(1964)

SI-0063b

ETCH NAME: TIME:

TYPE: Acid, polish TEMP:

COMPOSITION:

6 .... HF

50 .... HNO_3

DISCUSSION:

Si, (111) n-type wafers used in a study of B_2H_6 (diborane gas) as a boron, p-type diffusant in silicon. Solution used to etch polish wafers prior to diffusion.

REF: Ibid.

SI-0069

ETCH NAME: Healy junction etch TIME: 10—30 sec

TYPE: Acid, junction etch TEMP: RT

COMPOSITION:

10 ml .... HF

1 drop .... HNO_3, red fuming

DISCUSSION:

Si, (111) n-type wafers used for boron diffusion in fabricating silicon diodes. Solution should be mixed fresh and used within 8 h. If needed, cleave wafer sections ⟨110⟩ to expose junction. Also used on potted, metallographic cross sections, diamond paste polished, then etched to develop junctions and structure. Place a drop of solution across junction, then flush with water, and N_2 dry.

REF: Healy, J — personal communication, 1958

SI-0070

ETCH NAME: 1:1:1 TIME:

TYPE: Acid, step-etch/polish TEMP: 40°C

COMPOSITION:

1 .... HF

1 .... HNO_3

1 .... CH_3COOH (HAc)

DISCUSSION:

Si, (111) wafers and other orientations used for controlled removal from surfaces in a study of damage depth and crystal perfection. Removal steps were 0.04, 0.06, and 0.1 mm. From X-ray photograph study it was determined that there was no change in the bulk material after 0.04 mm removal. Author says the closer a wafer is cut to a major lattice direction, the more perfect the structure pattern. (*Note:* The term "1:1:1" is not recommended as an etch mixture name. There are many etch solutions of different compositions with such volume ratio.)

REF: Herglotz, H K — *J Electrochem Soc,* 106,600(1959)

SI-0071: Millea, M F & Hall, T C — *Phys Rev Lett,* 1,276(1958)

Si and Ge (111) wafers. Used this solution to polish both materials. Silicon was oxidized at 1000°C with oxygen, 1 h and showed minimum surface conductance.

SI-0072

ETCH NAME: TIME:

TYPE: Acid, polish TEMP:

COMPOSITION:

1 .... HF

2 .... HNO_3

2 .... CH_3COOH (HAc)

DISCUSSION:

Poly-Si rectangular blocks of silicon for use as a susceptor for MOCVD, as silicon has excellent low vapor pressure in the 500—800°C range used in MOCVD processing. First clean material in aqua regia, then cut, mechanically lap, and etch polish in the solution shown. A brass rod and alumina powder were used to drill a thermocouple hole at one end of susceptor.

REF: Blaauw, C et al — *J Vac Sci Technol,* A(2),438(1985)

SI-0073

ETCH NAME: TIME:

TYPE: Acid, removal/polish TEMP: RT

COMPOSITION:

1 .... HF

1 .... HNO_3

50 .... H_2O

DISCUSSION:

Si, (100) wafers, n-type with very shallow diffused junctions on the order of 1—2 μm. Solution used as a slow etch to determine diffusion depth profiles.

REF: Brady, H — personal communication, 1964

SI-0074

ETCH NAME: TIME:

TYPE: Acid, thinning TEMP:

COMPOSITION:

1 .... HF

5 .... HNO_3

1 .... CH_3COOH (HAc)

DISCUSSION:

Si, (111) Dendritic-Web ribbon crystal with a twin zone down ribbon center. Solution shown used to thin silicon ribbon. Follow with Secco etch (SI-0036) to develop defect structure. The (111) surfaces show severe glide planes . . . and growth strain(?)

REF: Cunningham, B et al — *J Electrochem Soc,* 129,1089(1982)

SI-0046b

ETCH NAME: TIME: 3—4 min

TYPE: Acid, dislocation TEMP:

COMPOSITION:

1 .... HF

3 .... HNO_3

1 .... CH_3COOH (HAc)

DISCUSSION:

Si ingot, FZ grown, n-type, 200 Ω cm resistivity. Ingot was subjected to twisting at 900°C before being cut for wafers. The solution shown, without HAc, was used to etch polish sections prior to preferential etching with the solution as shown.

REF: Ibid.

SI-0076
ETCH NAME: TIME: 15 min
TYPE: Acid, polish TEMP: RT
COMPOSITION:

"A" 1 .... HF
5 .... HNO_3
1 .... CH_3COOH (HAc)

"B" x .... HF
x .... HNO_3
x .... HAc
x .... Br_2

DISCUSSION:

Poly-Si material. This etch system used to develop damage-free polishing on poly-Si surfaces. First, lap on felt pad with 30 μm alumina for 30 min; second, lap with 2 μm diamond paste starting with 5 psi pressure and reducing to 2 psi over a lap period of 2 min — will remove lap damage but not residual stress. Final lap with Nalcog slurry #2250 (colloidal silica). Then etch in the "A" solution; then the "B" solution.

REF: Sopori, B L et al — *J Electrochem Soc,* 128,215(1981)

SI-0077
ETCH NAME: TIME:
TYPE: Acid, anisotropic TEMP:
COMPOSITION:

1 .... HF
50 .... HNO_3
50 .... CH_3COOH (HAc)

DISCUSSION:

Poly-Si deposited as silicon-on-insulator (SOI) structure with recrystallization of the silicon using an RF susceptor in the heated Zone Melting Regrowth method (RF-ZMR). The solution shown was used to develop sub-grains and cracks in the silicon recrystallization growth.

REF: Kobayashi, Y et al — *J Appl Phys,* 131,1189(1984)
SI-0297: Bezjian, K A et al — *J Electrochem Soc,* 129,1848(1982)

Poly-Si material. Solution used to develop dislocation arrays and sub-boundaries in this material.

SI-0263b
ETCH NAME: TIME:
TYPE: Acid, step-etch TEMP:
COMPOSITION: RATE: 300 Å/min

1 .... HF
750 .... HNO_3
250 .... CH_3COOH (HAc)

DISCUSSSION:

Poly-Si material. Solution used as a step-etch in studying the morphology of polycrystalline silicon.

REF: Ibid.

SI-0079
ETCH NAME: TIME:
TYPE: Acid, polish TEMP:
COMPOSITION:

25 .... HF
100 .... HNO_3
125 .... CH_3COOH (HAc)

DISCUSSION:

Si, wafers of various orientations used as substrates for the study of stacking faults in epitaxy deposits. Use etch shown to polish substrates before epitaxy. To etch thin specimens after epitaxy thin film growth use a Teflon beaker at 45° and rotate at 30 rpm using $5HF:95HNO_3$ as etchant. Reduce oxygen content in epitaxy layer to reduce stacking faults.

REF: Finch, R H et al — *J Appl Phys,* 34,404(1963)

SI-0038d

ETCH NAME: TIME: 3 min

TYPE: Acid, polish TEMP: RT

COMPOSITION:

3 .... HF

5 .... HNO_3

3 .... CH_3COOH (HAc)

DISCUSSION:

Si, (111) wafers and other orientations. Solution used to etch polish silicon wafers prior to copper dislocation etching (SI-0038a). Recommend at least $^1/_2$" of solution over specimens during etching with light agitation. Flush away etch with water without exposing etched surfaces to air and follow immediately with dislocation etch.

REF: Ibid.

SI-0080

ETCH NAME: TIME:

TYPE: Acid, polish TEMP:

COMPOSITION:

5 .... HF

8 .... HNO_3

15 .... CH_3COOH (HAc)

DISCUSSION:

Si, (111) wafers used in a study of Ag and Fe ion contamination in commercial acids as they affect surface barrier diodes. Ag acts as a p-type dopant with a 10^{-4} surface concentration acting as recombination center. Fe has almost no effect.

REF: Krebs, B L & Schlacter, R — *J Appl Phys,* 32,1510(1961)

SI-0081

ETCH NAME: TIME:

TYPE: Acid, polish/sizing TEMP:

COMPOSITION:

1 .... HF

1 .... HNO_3

x .... 1% $AgNO_3$

DISCUSSION:

Si, (111) pre-cut bars of material. Solution used to etch thin bars to 75 μm square cross-section and maintain the square cross-section. Bars were used in a deformation and fracture study.

REF: Pearson, G L et al — *Acta Metall,* 5,181(1957)

SI-0082
ETCH NAME: TIME: 60 sec
TYPE: Acid, polish TEMP: RT
COMPOSITION:
*55—92% .... HF/HNO_3
45—8% .... H_2O
2—0.01% .... $Hg(NO_3)_2$ or $HgCl_2$

*HF/HNO_3: 9:1 to 1:12 ratio by weight

DISCUSSION:

Si, (111) wafers and other orientations. A polish etch for both silicon and germanium. Silicon removal rate from a lapped surface is 13 mg/cm^2 in 1 min. Solution patented by authors referenced.

REF: Faust, J W & Wynne, R H Jr — U.S. Patent #2,705,192, March 29, 1955

SI-0154a
ETCH NAME: Dash, etch, modified TIME:
TYPE: Acid, preferential TEMP:
COMPOSITION:
1 .... HF
3 .... HNO_3
8 .... CH_3COOH (HAc)
DISCUSSION:

Si, (111) wafers used in a study of crystal perfection. First, mechanical lap on glass plate with 600-grit SiC and use CP4 (120 cm HF:200 cm $3HNO_3$:120 cm HAc:2 cm Br) to remove residual lap damage. Etch in solution shown to develop etch pits.

REF: Grubel, R O Ed — *Metallurgy of Elemental and Compound Semiconductors,* Vol. 12, Interscience, New York, 1961, 469

SI-0154b
ETCH NAME: Dash copper decoration TIME:
TYPE: Metal, diffusion TEMP: 900°C
COMPOSITION:
x .... Cu
DISCUSSION:

Si, (111) wafers used in a study of crystal perfection. See SI-0154a, above. After CP4 etching, wafers were copper diffused to decorate dislocations and studied by transmission infrared microscopy (TIM). X-ray diffraction was used on copper decorated and nondecorated specimens for comparison.

REF: Ibid.

SI-0154b
ETCH NAME: TIME:
TYPE: Acid, preferential TEMP:
COMPOSITION:
3 .... HF
5 .... HNO_3
DISCUSSION:

Si, ingot etched for defects in a study of doping silicon for tunnel diodes. Solution used to develop structure. An alkaline solution of x% $NaOH:H_2O$ was used, 30 min at 80°C following acid etch or was used separately to develop grain boundaries and polycrystalline

structure. Authors say that the hydroxide solution will work on n-type silicon. (*Note:* Hot to boiling NaOH and KOH solutions are standard for structure etching as light figure orientation solutions, regardless of n- or p-type material.)
REF: Ibid.

SI-0156
ETCH NAME: Arsenic TIME:
TYPE: Gas, preferential TEMP:
COMPOSITION:
x .... As, vapor
DISCUSSION:
Si, (111) wafers and whiskers used in fabricating Esaki diodes. During growth or doping of silicon under an overpressure of arsenic vapor, the arsenic attacks silicon surfaces developing a sharply triangular surface etch pit pattern.
REF: Hononyak, N et al — *Metallurgy of Elemental and Compound Semiconductors,* Series Vol. #12, Grubel, R O, Ed, Interscience, New York, 1961, 81

SI-0157a
ETCH NAME: TIME:
TYPE: Acid, polish TEMP:
COMPOSITION:
1 .... HF
7 .... HNO_3
DISCUSSION:
Si, (111) wafers used in a study of screw dislocations perpendicular to the silicon surface. Specimens were chemical polished in the solution shown.
REF: Danil'chuk, L N & Nikitenko, V I —

SI-0157b
ETCH NAME: Copper TIME: 1 h
TYPE: Metal, decoration TEMP: 1000°C
COMPOSITION:
x .... Cu, chips
x .... H_2O
DISCUSSION:
Si, (111) wafers. Specimens were copper decorated as shown.
REF: Ibid.

SI-0158
ETCH NAME: Erhard's etch TIME:
TYPE: Acid, dislocation TEMP: RT
COMPOSITION:
1 .... HF
1 .... CrO_3 (x%)
DISCUSSION:
Si, (111) wafers. After copper decoration (SI-0157b) specimens were dislocation etched in the solution shown. (*Note:* See Sirtl etch SI-0019.)
REF: Erhard, S & Adler, A — *Z Metallkd,* 52,529(1961)

SI-0102b
ETCH NAME: Potassium hydroxide TIME:
TYPE: Alkali, preferential TEMP: RT
COMPOSITION:
x .... 1 *N* KOH
DISCUSSION:
Si, (111) wafer and other orientations. Solution used as a preferential etch on silicon. At this concentration it is a very slow etch at room temperature.
REF: Turner, D R — *J Electrochem Soc,* 107,810(1960)

SI-0084
ETCH NAME: Sodium hydroxide TIME:
TYPE: Alkali, etch-stop TEMP:
COMPOSITION:
x .... NaOH (x%)
DISCUSSION:
Si, (111) wafers, boron diffused p-type at a 10^{20} cm^3 doping level. Solution acts as an etch-stop against high doping level layers.
REF: Barycka, I et al — *J Electrochem Soc,* 126,345(1979)

SI-0085
ETCH NAME: Potassium hydroxide TIME:
TYPE: Alkali, etch-stop TEMP:
COMPOSITION:
x .... 5—20% KOH
DISCUSSION:
Si, (111) wafers, p-type, with a 10^{20} cm^3 boron doping level. Solution acts as an etch-stop with high doping levels.
REF: Price, J B — *J Electrochem Soc,* 120,339(1973)
SI-0886: Palik, E D et al — *J Electrochem Soc,* 129,2051(1982)
Si, (111) and (100) wafers cleaved ⟨110⟩ after diffusion. Both p- and n-type with boron or phosphorus at 10^{14} to 10^{21} cm^3 doping levels. Solution acts as an etch-stop with high doping levels of both types.
SI-0089: Bhgat, J — *Solid State Electron,* 27,441(1984)
Si, (100) wafers, n-type, Sb doped wafers used as substrates for boron implantation followed by boron-doped epitaxy for VMOS transistors. ''U'' grooves were etched in ⟨110⟩ directions with this solution through photo resist patterned SiO_2 masks.
SI-0090: Suzuki, T et al — *J Electrochem Soc,* 127,1537(1980)
Si, (100) wafers, n-type, 10 Ω cm resistivity. Solution used to etch ''V'' grooves in ⟨110⟩ directions. Thermal oxide grown 1.7 μm at 1100°C to form a dielectrically isolated (DI) substrate, and followed with evaporation of a poly-Si layer. Lap back poly/Si to expose islands of silicon. Cleave wafers on a ⟨110⟩ bulk plane and Sirtl etch for defects. Due to stress the thin wafers may bow either in a positive or negative direction.

SI-0038a
ETCH NAME: Sodium hydroxide TIME: 5 min
TYPE: Alkali, preferential TEMP: Boiling
COMPOSITION:
x .... 10% NaOH
DISCUSSION:
Si, (111) wafers and other orientations; also ingots. On ingots, solution is used to develop poly-Si and twinning zones. For slicing wafers from ingots — initially face-cut ingot —

then use solution as a light figure orientation etch (etch pits developed) to orient ingot. Recommend twice the volume of solution per ingot volume. Solution can be used on individual wafers or other cut specimens for orientation verification. (*Note:* Concentrations of this level used at RT, warm, hot, and boiling are general cleaning solutions for many semiconductors and metals.)
REF: Ibid.

SI-0088
ETCH NAME: Sodium hydroxide — TIME: 12 min
TYPE: Alkali, preferential — TEMP: 65°C
COMPOSITION:
x .... 50% NaOH or KOH
DISCUSSION:
Si, (111), (100), and (110) wafers and ingots. Used as a light figure orientation etch. The (111) surfaces are etched 6 min; (100) and (110) etched 12 min.
REF: Schwuttke, C H — *J Electrochem Soc,* 106,315(1959)

SI-0091
ETCH NAME: Potassium hydroxide — TIME:
TYPE: Alkali, preferential — TEMP: Hot to boiling
COMPOSITION:
x .... 10—30% KOH
DISCUSSION:
Si, as various cut shaped specimens. Solution referred to as a micro-structure etch and used to produce: "V" grooves, via holes, pits, mesas and specially cut and etch shaped parts. (*Note:* A major review article of uses of silicon as a bulk material other than for its semiconducting properties.)
REF: Peterson, K E — *Proc IEEE,* 70,420(1982)
SI-0021b: Kern, W — *RCA Rev,* 39,278(1978)
Si, (100) wafers used in an anisotropic etch study. A 19% KOH solution was used at 80°C for 120 min. Rate was 0.59 μm/min.

SI-0092a
ETCH NAME: Potassium hydroxide — TIME: 15—45 min
TYPE: Alkali, sphere — TEMP: Boiling
COMPOSITION:
250 ml .... 30% KOH or NaOH
DISCUSSION:
Si, single crystal spheres formed from $^1/_2$, $^1/_4$, and $^1/_8$″ cut cubes and ground in a tornado sphere grinder. After initial forming with 600-grit SiC abrasive paper, slow etch polish in 4HF:75HNO_3 (SI-0061). Solution shown used to develop finite crystal form (FCF). Both hydroxides will etch form a tetrahexahedron, (hk0). Faces of the tetrahexahedron are rough but sharply edge defined.
REF: Walker, P — personal development, 1957—1959
SI-0092b: Ibid.
Si, (111) and (100) wafers, n-type, 10—30 Ω cm resistivity used to fabricate aluminum alloy transistors with opposed emitter and collector pits. The original Pennington Pit Method developed in a study of silicon crystallographic structure was used to form pits: a diamond stylus held vertically through a rotatable shaft, top-weighted with variable lead weights with wafer held on an x-y stage below the stylus. Weight and stylus rotation control point-damage depth.
This was the first discrete multiple device fabrication on a wafer (Hughes). Damage

location was pre-inked on both sides of wafer using a panograph with bottom-plate mirror for opposed side location. Damage points induced by amount of pressure and number of stylus rotations with collector damaged deeper than emitter and etched first, then emitter damaged and both etched to final pit size based on required base width.

Individual dice also used, both triangle and diamond shaped with ⟨111⟩ edge directions for (111) surfaced wafers; square with ⟨110⟩ edge directions for (100) wafers to reduce preferential edge attack during damage pit forming.

Pit shape after 15 min total etch time in boiling solution was hexagonal on (111) with a pit bottom to within 1 μm flatness. Base width between collector/emitter established by transmitted light using standard white light with wafer on a microscope stage. (100) wafer pits were square with a smooth pit bottom and about 1° of bottom curvature.

SI-0092c: Ibid.

Si, as oriented ingots (111), (100) and (110). Ingots were face cut, then etched 10 min in boiling solution to develop light figure orientation etch pits. After mounting ingot on a ceramic block with Dekotinsky cement and placed on an x-y-z holder for face orientation by reflection of light figures in a black box system. Adjust and lock holder for subsequent wafer cutting.

SI-0092d: Ibid.

Si, as oriented pre-cut forms: cube, (100); cube, (110); octahedron, (111) — 8 (111) faces; tetrahedron, (111) — 4 (111) faces; and rhomboid, (111) — 6 (111) faces. Forms used in a study of etched edge facets. Used to pit-damage and etch (111), (100), (110), and (211) surfaces of forms for correlation of convex vs. concave facet development. See the SI-0300-0350 etchant format and series for results on including different chemical additives to the base 30% hydroxide solution in a study of pit shapes and sphere forms.

During these studies it was determined that the KOH solution is slightly more trigonal plane preferential; whereas the NaOH solution is more tetragonal plane preferential on silicon.

SI-0206: Ives, M B & McAustland, D D — *Solid State Electron,* 11,189(1968)

Si, (111) wafers and other orientations. A development study of the slope of etch pit faces.

SI-0208: Garfinkel, R et al — *J Appl Phys,* 35,2321(1964)

Si, (111) wafers used as cleaved triangular dice in a junction resonant structure study. (*Note:* SI-0092b use of both triangle and diamond dice edge cut in fabricating aluminum alloy transistors.)

SI-0092c

ETCH NAME: TIME: 10—15 min

TYPE: Alkali, preferential/sphere TEMP: Boiling

COMPOSITION:

250 ml .... KOH/NaOH (30%)

45 g .... I_2

DISCUSSION:

Si, (111) and (100) wafers and spheres (See: SI-0092a); spheres of silicon, germanium, indium phosphide, and gallium arsenide studied. Polish etch spheres before preferential etching to finite crystal form (FCF). Solution shown will preferential etch all materials mentioned to cube, (100) with face surfaces slightly rough, positive curvature, and sharply edge defined. Wafers were used in a correlative pit-damage study (SI-0092b).

Solution can also be used as a light figure orientation etch on germanium ingots, which are not etched in pure hydroxides.

REF: Ibid.

SI-0093
ETCH NAME: TIME:
TYPE: Alkali, preferential TEMP:
COMPOSITION:
250 g .... KOH
800 ml .... H_2O
200 ml .... isopropyl alcohol (ISO)
DISCUSSION:

Si, as poly-Si thin films. Solution used in a study of crystallite orientation in these films.
REF: Besjian, K A et al — *J Electrochem Soc,* 1129,1848(1982)
SI-0092f: Ibid.

Si, (111), n-type 10—30 Ω cm resistivity wafers. Solution used was 150 ml 30% KOH:100 ml ISO(IPA), boiling, 5—20 min in a damage pit etch study and evaluation of aluminum alloy transistor devices.
SI-0094a: Abu-Zeid, M M — *J Electrochem Soc,* 131,2138(1984)

Si, (100) wafers with evaporated SiO_2 masking used in etch forming mesa structures. Solution used shown as: xKOH:xH_2O propanol. Author says the solution is good for developing (321) oriented facets on the mesa side slopes. (*Note:* This is one of the few examples of this crystallographic face form — isometric (cubic) system, normal class, hexoctahedron, [hkl].)

SI-0095
ETCH NAME: TIME:
TYPE: Alkali, selective TEMP:
COMPOSITION:
x .... KOH
x .... CH_3O_7OH
x .... H_2O
DISCUSSION:

Si, (100) n-type wafer used as substrate. Epitaxy n^+ layer and boron-doped p-type layer followed by a surface oxide. Solution used as a selective etch in a general study of oxidation and silicon surfaces.
REF: Whillen, M & Holsbrink, J — *Solid State Electron,* 26,453(1983)

SI-0092g
ETCH NAME: TIME: 10—20 min
TYPE: Alkali, preferential TEMP: Boiling
COMPOSITION:
125 ml .... 30% KOH
125 ml .... 30% NaOH
DISCUSSION:

Si, (111), (100) wafers n-type 10—30 Ω cm resistivity used in a general study of alkali etching of surfaces and controlled damage-pit development. Surface etch pits and damage etch pits or other structures are similar for KOH or NaOH only; or as the 1:1 mixture shown. All produce a sharply defined hexagonal damage etch pit on (111) and square on (100). These solutions also used to etch thin sections to various depth levels in bulk silicon wafers using this controlled pit in study of bulk defects, tracing movement of defects and establishing one-to-one correspondence of pits and defects between positive and negative wafer surfaces of (111), (100), (110), and (211) orientations.
REF: Ibid.

SI-0092h
ETCH NAME:
TYPE: Alkali, preferential
COMPOSITION:
250 ml .... 30% KOH
40—60 ml .... Br_2

TIME: 5—15 min
TEMP: Boiling

DISCUSSION:

Si, (100) and (111) wafers, n-type, 10—30 Ω cm resistivity used in a general study of the effect of additives to hydroxide solutions in producing surface pits and controlled damage etch pits. Also used as an ingot light figure orientation etch on both silicon and germanium ingots grown as (111), (100), (110), and (211) orientations. Bromine dissipates rapidly during the first 5 to 10 min of etching.

REF: Ibid.

SI-0092i
ETCH NAME:
TYPE: Alkali, preferential
COMPOSITION:
250 ml .... 30% KOH
100 ml .... ethylene glycol (EG)

TIME: 5—20 min
TEMP: Boiling

DISCUSSION:

Si, (111) and (100) wafers, n-type, 10—30 Ω cm resistivity. Used in a general study of control damaged etch pit structures. Solution produces a truncated triangle etch pit. The three alternating major side slopes contain parallel steps; whereas the three minor plane side slopes are poorly defined with a vertical channel/ridge structure. See SI-0300-0350 etchant formats for a discussion of etch pit progression development.

REF: Ibid.

SI-0096
ETCH NAME: Potassium hydroxide
TYPE: Electrolytic, removal
COMPOSITION:
x .... KOH (x%)

TIME:
TEMP:
ANODE:
CATHODE:
POWER:

DISCUSSION:

Si, (111) and (100) wafers, both n- and p-type with low doping levels. Under cathodic etch conditions: etches p-type slowly; n-type not etched, and acting as an etch-stop. Under anodic etch conditions: both p- and n-type act as etch-stops due to passivation of surfaces. Referred to as the etch-stop mechanisms for nucleophilic attack of (111), (100), and (110) orientations. Also referred to as an orientation dependent etch (ODE). Used to selectively develop microstructures such as "V" grooves, pyramids, etc.

REF: Glembocki, O J et al — *J Electrochem Soc,* 132,195(1985)

SI-0097
ETCH NAME: Potassium hydroxide
TYPE: Electrolytic, oxidation
COMPOSITION:
x .... 2 *M* KOH

TIME:
TEMP: 20°C
ANODE:
CATHODE:
POWER:

DISCUSSION:

Si, (111) and (100) wafers within +/− 1° of plane, n-type 9—10 Ω cm resistivity used

in a general study of surface oxidation and etch rates. Oxidation is bias dependent and rates vary with orientation, to include (110) wafers.
REF: Glembocki, O J & Stahlbush, R E — *J Electrochem Soc*, 132,145(1985)
SI-0098: Palik, E D et al — *J Electrochem Soc*, 132,135(1985)

Si, (100) wafers both low concentration and high concentration p- and n-type used in KOH oxidation study. With cathodic etching, n-type is faster than p-type. For heavily doped Si, hold in solution several minutes before applying anodic potential. Cleave wafers and clean with HF prior to KOH oxidation.

SI-0092j
ETCH NAME: — TIME: 10—20 min
TYPE: Alkali, preferential — TEMP: Boiling
COMPOSITION:
250 ml .... 30% KOH
100 ml .... MeOH
DISCUSSION:

Si, (111) and (100) wafers, n-type 10—30 Ω cm resistivity used in a general study of the effect of additives in alkali solutions and etch pit development. Results were similar to those shown under SI-0091j.
REF: Ibid.

SI-0099
ETCH NAME: Potassium hydroxide — TIME: 20 min
TYPE: Molten flux, dislocation — TEMP: 350°C
COMPOSITION:
(1) x .... KOH, pellets
(2) 1 .... KOH, pellets
1 .... NaOH, pellets
DISCUSSION:

Si, (111) wafers and other orientations, as p- and n-type of different doping level concentrations. KOH used as a molten flux to develop dislocations. After etching, quench directly into ethylene glycol. This reduces oxidation of the surface and EG dissolves hydroxides slowly producing less surface damage.
REF: Lessoff, H & Gorman, R — *J Electron Mater*, 14,203(1985)

SI-0195
ETCH NAME: Potassium hydroxide, dil. — TIME: 10—45 min
TYPE: Electrolytic, junction — TEMP: RT
COMPOSITION: — ANODE: Collector, Si
x .... 0.4% KOH — CATHODE: Emitter, Si
POWER: 10 A/cm^2
DISCUSSION:

Si, (111) wafers with diffused n-p-n junctions. Use CP4 to etch delineate emitter junction, then etch with solution shown to develop collector junction.
REF: Fowler, A & Levesque, P — *J Appl Phys*, 26,641(1955)

SI-0100
ETCH NAME: — TIME:
TYPE: Alkali, preferential/junction — TEMP: 100°C
COMPOSITION:
x .... 10% NaOH
x .... 15% $NaNO_3$
x .... H_2O

DISCUSSION:

Si, (100) n-type wafer with p-type silicon spherical shot melted and alloyed into the base wafer to form p-n junctions. Alloyed sphere produced a recrystallized single crystal pellet about 1 mil deep. Solution shown was used as a macro-etch to develop the p-n junctions. Similar p-n junctions were formed with germanium (GE-0208).

REF: Lesk, I A — *J Electrochem Soc,* 107,534(1960)

SI-0101

ETCH NAME: TIME:

TYPE: Alkali, junction plate TEMP: 30 to 70°C

COMPOSITION:

200 g .... KOH

10 g .... $KAu(CN)_2$

800 ml .... H_2O

\+ .... IR light

DISCUSSION:

Si, (111) n-type wafers with boron diffused p-n junctions. Very smooth silicon surfaces require longer time for satisfactory plating. Recommend surfaces be prepared with Linde A or 600-grit carborundum. The gold plates out on the more negative side of the junction.

REF: Silverman, S L & Benn, D R — *J Electrochem Soc,* 105,170(1958)

SI-0087

ETCH NAME: TIME:

TYPE: Acid, junction plate/stain TEMP: RT

COMPOSITION:

40 ml .... HF

20 ml .... HNO_3

100 ml .... H_2O

2 g .... $AgNO_3$

DISCUSSION:

Si, (111) wafers with n^+/n diffusion. Solution used to delineate the diffused structure. Apply a drop of solution and allow to etch briefly, then DI water rinse and dry. Then etch with $1HNO_3:1H_2O$ and silver will plate out on higher resistivity area.

REF: Berman, I — *J Electrochem Soc,* 107,1002(1962)

SI-0102a

ETCH NAME: Potassium chloride TIME:

TYPE: Salt, removal TEMP:

COMPOSITION:

x .... KCl, sat. sol.

DISCUSSION:

Si, (111) wafers. Potassium chloride will etch both silicon and germanium. Solution used as a general removal etch.

REF: Turner, D R — *J Electrochem Soc,* 107,810(1960)

SI-0103

ETCH NAME: Caro's etch TIME:

TYPE: Acid, cleaning TEMP:

COMPOSITION:

1 .... H_2SO_4

1 .... H_2O_2

DISCUSSION:

Si, (111) wafers and other orientations, p- or n-type material with different doping levels used in a device development study. Clean substrate wafers: (1) degrease in TCE; (2) rinse in acetone, then isopropanol; (3) water rinse, before (4) etch cleaning with the solution shown with water quench and N_2 dry. An SiO_2 thin film was deposited, then SnO_2 and an SiO_2 cap.

REF: Hiso, Y-S & Ghandhi, S K — *J Electrochem Soc,* 127,1592(1980)

SI-0104: Pintchovski, F — *J Appl Phys,* 126,1428(1979)

Si, (111) wafers and other orientations. A major study of the thermal characteristics of this solution, including variations of the mixture. Addition of water generates heat. The 1:1 mixture will be a constant boiling solution at 170°C. Addition of sodium phosphate as a stabilizer does not affect the solution. (*Note:* Various Caro solutions are widely used for general cleaning of semiconductor wafers.)

SI-0043b

ETCH NAME: TIME:

TYPE: Acid, cleaning TEMP:

COMPOSITION:

3 .... H_2SO_4

1 .... H_2O_2

DISCUSSION:

Si, poly-Si epitaxy deposited thin films. Solution used as a general cleaning etch. Used to remove the blue stain left on poly-Si after etching in high HF concentration acids solutions.

REF: Sopori, B L — *J Electrochem Soc,* 131,667(1984)

SI-0106: Tsao, K Y & Bostra, K H — *J Electrochem Soc,* 131,2702(1984)

Si, (100) wafers used in a deposition study of tungsten. Wafer surfaces were prepared by depositing an SiO_2 thin film, then poly-Si to act as a deposition mask. After photolithographic patterning, etch pattern openings down to the substrate surface. Solution shown was used to clean the exposed wafer surface at 95°C, then 6 sec $1HF{:}10H_2O$ prior to selective deposition of tungsten from a WF_6 source.

SI-0298

ETCH NAME: TIME:

TYPE: Acid, cleaning TEMP:

COMPOSITION:

4 .... H_2SO_4

1 .... H_2O_2

DISCUSSION:

Si, (100) p-type wafers, 1.2—1.8 Ω cm resistivity used as substrates for metallization. Clean in solution shown with final HF dip and water rinse before placing in vacuum. Magnetron sputter Ta_2O_5 from a solid target and use the film for device gate insulation. Anneal 350—450°C for 30 min in a study of the Ta_2O_5/Si interface.

REF: Seki, S et al — *J Electrochem Soc,* 131,2621(1984)

SI-0107

ETCH NAME: TIME: 20 min

TYPE: Acid, cleaning TEMP: RT

COMPOSITION:

5 .... H_2SO_4

1 .... H_2O_2

DISCUSSION:

Si, (100), n- and p-type wafers, 20 and 25 Ω cm resistivity, respectively. Solution shown used to clean surfaces prior to Ge epitaxy deposit. Follow cleaning with 60 sec in 50% HF before vacuum. Ge growth can vary from single crystal to polycrystalline to amorphous structure with change in overpressure of an arsenic ambient.

REF: Wang, P D et al — *J Vac Sci Technol,* B2,209(1984)

SI-0108a

ETCH NAME: TIME:

TYPE: Acid, cleaning sequence TEMP:

COMPOSITION:

(1) 1 .... H_2SO_4 (2) 1 .... HNO_3 (3) 1 .... HF

1 .... H_2O_2 1 .... HCl 10 .... H_2O

DISCUSSION:

Si, (111) and (100) wafers. The etch cleaning sequence shown was followed by water quenching and N_2 blow dry. Wafers used in a study of pyrogenic steam oxidation. (*Note:* (1) is Caro's etch, normally used hot.)

REF: Lie, L N et al — *J Electrochem Soc,* 129,2828(1982)

SI-0334: Lo, M J T et al — *J Electrochem Soc,* 128,1568(1981)

Si, (111) wafers used in a study of implantation gettering with gold. Clean silicon: (1) BHF clean with ultrasonic to remove native oxide; (2) rinse with DI H_2O; (3) etch clean in aqua regia; (4) rinse with DI H_2O; (5) etch remove 1.5 mils/side in HF:HNO_3:HAc:H_2O, and water rinse. Mechanically polish one side with Lustrox; coat other side with CVD SiO_2 at 380°C. Evaporate 1000 Å Au on polished side and anneal 6 h at 1080°C.

SI-0113

ETCH NAME: Heat TIME: Flash

TYPE: Thermal, cleaning TEMP: 1100°C

COMPOSITION:

x .... heat

DISCUSSION:

Si, (111) n-type wafers 5 Ω cm resistivity used for Si epitaxy with an SiO_2 mask photolithographically opened for He ion implantation. Flash clean silicon wafers in vacuum prior to epitaxy deposition.

REF: Nishigaki, S et al — *Jpn J Appl Phys,* 23,L683(1984)

SI-0114

ETCH NAME: TIME:

TYPE: Alcohol, cleaning sequence TEMP:

COMPOSITION:

(1) x .... TCE (2) x .... acetone (3) x .... isopropyl alcohol

DISCUSSION:

Si, (111) n- and p-type wafers cleaned in the sequence shown prior to deposition of Nb_2O_5.

REF: Chen, M-C — *J Electrochem Soc,* 119,887(1972)

SI-0094b

ETCH NAME: TIME:

TYPE: Acid, preferential TEMP:

COMPOSITION:

x .... hydrazine

x .... H_2O

x .... isopropyl alcohol

DISCUSSION:

Si, (111) wafers. An SiO_2 thin film mask was deposited and used in developing mesa structures. Solution shown will develop (112) facetted mesa side slopes.

REF: Abu-Zeid, M M — *J Electrochem Soc,* 131,2138(1984)

SI-0094c

ETCH NAME: Ethylenediamine TIME:

TYPE: Amine, preferential TEMP: 85°C

COMPOSITION: RATE: (111) 0.018 μm/min

x .... 55% ethylenediamine (100) 0.5 μm/min

DISCUSSION:

Si, (111) and (100) wafers masked with SiO_2 for forming mesa structures. Solution will develop (212) facets on mesa side slopes.

REF: Ibid.

SI-0116

ETCH NAME: P–ED or EPW TIME:

TYPE: Acid, preferential TEMP:

COMPOSITION:

x .... $C_6H_4(OH)_2$ pyrocatechol (P)

x .... H_2O

x .... $NH_2(CH_2)_2NH_2$ ethylenediamine (ED)

DISCUSSION:

Si, (100) wafers within +/− 1° of the plane. An SiO_2 photo resist mask was pattern developed in fabrication of silicon for ink jet nozzles. Solution used to form via holes etched through silicon to form an array of hole openings that are then used for ink jet nozzles. After hole etch forming in solution shown, follow with BHF solution to clean holes.

REF: Bassous, E — *IEEE Trans Electron Dev,* ED25,1178(1978)

SI-0117: Bogh, A — *J Electrochem Soc,* 118,401(1971)

Si, (111) wafers with high boron doping levels ($10^{20}/cm^3$) acting as an etch-stop in this solution. Lower doping concentrations etch rapidly.

SI-0118: Cheung, N W — *Rev Sci Instr,* 51,1212(1980)

Si, (111) wafers. See SI-0117 as same discussion applies.

SI-0119

ETCH NAME: EPW TIME:

TYPE: Acid, preferential TEMP:

COMPOSITION:

17 ml .... ethylenediamine

3 g .... pyrocatechol

8 ml .... H_2O

DISCUSSION:

Si, (111) and (100), p-type 1—10 Ω cm and n-type wafers. After mixing the solution shown wait about 1 h before use. Solution turns a light yellow color. Wafers used in a study of etch rate vs. boron concentration. Wafers processed as follows: (1) degrease with TCE; (2) rinse in acetone, then MeOH, then H_2O and N_2 dry; (3) clean in BHF, 5 sec (1HF:5NH_4F); (4) clean in 4H_2SO_4:1H_2O_2, 10 min; (5) RCA-1 etch 15 min at 75°C (1NH_4OH:1H_2O_2:5H_2O) (6) repeat the BHF etch and follow with H_2O rinse and N_2 dry. After boron diffusion the EPW etch was used to establish etch rates.

REF: Raley, N F — *J Electrochem Soc,* 131,161(1984)

SI-0109

ETCH NAME: TIME:

TYPE: Acid, cleaning TEMP:

COMPOSITION:

x .... 1% HF

x .... NH_4OH

x .... H_2O_2

DISCUSSION:

Si, (100), p-type, 2 Ω cm resistivity, CZ grown wafers used in fabricating NMOS structures. Degrease with TCE, then rinse in acetone. Etch clean with solution shown and follow with 50% HCl. Deposit Si_3O_4 by APCVD or PECVD.

REF: Hezel, R — *J Electrochem Soc,* 131,1675(1984)

SI-0110

ETCH NAME: Carbon TIME:

TYPE: Powder, defect enhancement TEMP:

COMPOSITION:

x .... carbon black

DISCUSSION:

Si, (111) and (100) wafers used in a study of defects. After preferential etching use carbon black to paint-on decorate etched defects for improved definition and recognition under microscope observation.

REF: Comizzoli, R B — *J Electrochem Soc,* 129,667(1982)

SI-0111

ETCH NAME: Methylene chloride TIME:

TYPE: Solvent, cleaning TEMP: RT

COMPOSITION:

x .... CH_2Cl_2

DISCUSSION:

Si, (111) wafers. Use a cotton swab soaked in methylene chloride to clean surfaces prior to silicon epitaxy. Immediately before placing in vacuum soak in HF, 2 min, rinse several times in water and N_2 blow dry.

REF: Nuttall, R S — *J Electrochem Soc,* 116,445(1969)

SI-0121a

ETCH NAME: TIME:

TYPE: Acid, polish/thinning TEMP:

COMPOSITION:

(1) 3 .... HF

5 .... HNO_3

3 .... H_2O

(2) 17 ml .... ED (ethylenediamine)

3 g .... P (pyrocatechol)

8 ml .... H_2O

DISCUSSION:

Si, (111) wafers and other orientations. Wafers used in a study of etch polish and thinning. First etch used for heavy removal, and final thinning was done the second etch. (*Note:* The first etch is CP4, modified, and without bromine. The second etch is EPD. See SI-0119 or SI-0116.)

REF: Freyer, J — *J Electrochem Soc,* 122,1238(1975)

SI-0121b

ETCH NAME: TIME:

TYPE: Acid, polish/thinning TEMP:

COMPOSITION:

1 .... HF

10 .... HNO_3

\+ .... CO_2

\+ .... light

DISCUSSION:

Si, (111) wafers. Rotate wafer in solution at 8 rpm. Bubble CO_2 up through etch for both agitation of solution and as an added oxidizer. Flood wafer surface with light during etch period as a photon activator. Used to polish and fine-thin wafers.

REF: Freyer, J — *J Electrochem Soc,* 122,1238(1975)

SI-0034

ETCH NAME: Vogel's etch TIME:

TYPE: Acid, dislocation TEMP:

COMPOSITION:

3 .... HF

5 .... HNO_3

3 .... CH_3COOH (HAc)

2 .... 3% $Hg(NO_3)_2$

DISCUSSION:

Si, (111) and other oriented wafers. Solution used as a general dislocation and defect etch.

REF: Vogel, F L & Lovell, C — *J Appl Phys,* 27,1413(1956)

SI-0124

ETCH NAME: TIME:

TYPE: Acid, polish/thinning TEMP:

COMPOSITION:

95 ml .... HF

900 ml .... HNO_3

14 g .... CH_3COOH (HAc)

\+ .... CO_2

\+ .... light

DISCUSSION:

Si, (111), (100) wafers used in a study of polishing and thinning. Wafers were floated face down on the surface of the solution. CO_2 is bubbled up through the solution with light flooding the wafer surface during the etch period. See SI-0121a.

REF: Stoller, A I et al — *RCA Rev,* 31,265(1970)

SI-0125
ETCH NAME: TIME:
TYPE: Acid, polish TEMP: 18°C
COMPOSITION:

50 ml .... HF
50 ml .... CH_3COOH (HAc)
200 mg .... $KMnO_4$

DISCUSSION:

Si, (111), (100), n- and p-type wafers and epitaxy thin film. Solution used to polish wafers. See SI-0102.

REF: Theunissen, M J et al — *J Electrochem Soc,* 137,959(1970)

SI-0126
ETCH NAME: TIME:
TYPE: Acid, polish TEMP:
COMPOSITION:

(1)	(2)	(3)
x .... HF	x .... HF	x .... HF
x .... HNO_3	x .... HNO_3	x .... HNO_3
	x .... HAc	x .... H_2O

DISCUSSION:

Si, (111) wafers and other orientations with n- and p-type resistivity. Initial work was done on individual square dice developing both etch rates and polish characteristics to include dice edge peaking or downward curving using various concentrations of the three etch systems shown. (*Note:* This was one of the first and most extensive studies of these etching systems at the time. Author (Walker) was involved in some of the study development.)

REF: Robbins, H & Schwartz, B — *J Electrochem Soc,* 106,505 & 1020(1959)
SI-0127: Robbins, H & Schwartz, B — *J Electrochem Soc,* 107,108(1960)
SI-0128: Robbins, H & Schwartz, B — *J Electrochem Soc,* 109,37(1962)
SI-0129: Robbins, H & Schwartz, B — *J Electrochem Soc,* 123,1093(1976)

Si, (111) wafers used in an etch rate study as square cut dice. (*Note:* These five articles by Robbins & Schwartz represent some of the most extensive and definitive development work done on silicon with $HF:HNO_3$ +/− HAc/H_2O systems during this period of time, and are still applicable.)

SI-0130: Klein, D L & D'Stefan, D J — *J Electrochem Soc,* 109,47(1962)

Si, (111) wafers and other orientations, n- and p-type. A major etching and polishing study of the $xHF:xHNO_3$ system. Similar to the work of Robbins & Schwartz.

SI-0249: Burgess, T E — *Electrochem Soc Abstr,* 10,145(1961)

Si, wafers of different orientations. Study similar to those of Robbins & Schwartz.

SI-0296: Immirlioa, A A Jr — *Solid State Electron,* 25,1141(1982)

Si, wafers of different orientations. A general study of silicon etching techniques, to include preferential and dislocation etching with the Wright etch SI-0045.

SI-0131
ETCH NAME: TIME:
TYPE: Acid, selective TEMP:
COMPOSITION:

x .... H_3PO_4
x .... HNO_3

DISCUSSION:

Si, (100) p-type, 4—6 Ω cm resistivity wafers used to fabricate MOS capacitors with Ta-Mo gates. MOS structure: $Ta:Mo/SiO_2/Si(100)$. Solution used to etch gates through photo

resist patterned windows. Hydrogen doped Ta and Mo co-sputtered by RF magnetron in Ar:H_2(20%) on top of an SiO_2 thin film pre-deposited to 400 Å thickness on the substrate at 950°C in dry oxygen.
REF: Ohfusi, S-I & Shiono, N — *J Electrochem Soc,* 132,1689(1985)

SI-0132
ETCH NAME: Sirtl, modified TIME:
TYPE: Acid, preferential TEMP:
COMPOSITION:
1 .... HF
1 .... CrO_3(38%)
DISCUSSION:

Si, ingot grown in space and compared to an ingot grown under similar conditions on earth. Sub-grain surface structure and etch pit density were similar. Cut on ⟨110⟩ showed striations to be asymmetric due to temperature field using a mirror heated by a tungsten lamp without rotation during ingot growth in space. Sirtl etch used to develop structure and etch pits.
REF: Eyer, A, Leiste, H & Nitsche, R — *J Cryst Growth,* 71,173(1983)
SI-0133: Arst, M C & Groot, J G — *J Electron Mater,* 13,763(1984)

Si, (100) wafer CZ growth used as substrate for poly-Si growth. After epitaxy, cross section on ⟨110⟩ and polish with c-SiO_2. Use Sirtl etch 30 sec at RT.
SI-0364: Allen, F G et al — *J Appl Phys,* 30,1563(1959)

Si, (111) wafers. A study of surface cleaning by heating in high vacuum.
SI-0365: Law, J T — *J Phys Chem Solids,* 4,91(1958)

Si, (111) wafers. A study of the interaction of oxygen on clean silicon surfaces.
SI-0134: Lu, T C & Bauser, E — *J Cryst Growth,* 71,305(1985)

Si, (111), (110) and (211) wafers with LPE grown epitaxy layers. Specimens cleaved ⟨110⟩ for cross section study of stria called ''Valley Trace Type II'' that were caused by dopant segregation during growth of epitaxy layers. Sirtl etch was used with illumination with an IR lamp to develop structure. (*Note:* Similar concentric ring resistivity variation on CZ grown wafers, and anomaly segregation in FZ ingots grown for silicon zener devices were observed by author Walker in 1958 (CZ), and 1963 (FZ).)

SI-0132c
ETCH NAME: Hydrofluoric acid, dilute TIME:
TYPE: Acid, step-etch TEMP: 30°C
COMPOSITION:
1 .... HF
10 .... H_2O
DISCUSSION:

Si, (100) wafers used as substrates for epitaxy growth and deposition of SiO_2/$TaSi_2$/poly-Si/Si(100). Solution used as a step-etch with a selectivity of SiO_2/$TaSi_2$ shown as 20:1.
REF: Ibid.

SI-0136
ETCH NAME: Hydrofluoric acid, dilute TIME:
TYPE: Acid, oxide removal TEMP:
COMPOSITION:
x .... HF
x .... H_2O

DISCUSSION:

Si, (100) and (111) wafers used in a study of carbon and oxygen contamination. Wafers "as-received" from vendor with one side polished and back-side lightly etched; both sides were polished; Si epitaxy with addition of carbon or oxygen. All study surfaces SiO_2 coated, photo resist applied and window grooves opened in the SiO_2 with an HF etch. Various annealing steps and times. Wright etch, 5 min at RT used to develop defects in etched grooves. Cross-sections ⟨110⟩ with Wright etch show denuded zone under Si epitaxy layer after three-step annealing to be 20 μm deep; epitaxy showed stacking faults, dislocations, and precipitations of carbon and oxygen.

REF: Bailey, W E et al — *J Electrochem Soc,* 132,1721(1985)

SI-0137: McDonald, R & Goetzberger, J A — *J Electrochem Soc,* 109,104(1962)

Si, wafers. Describes the groove etching technique used in SI-0136.

SI-0139

ETCH NAME: TIME:

TYPE: Acid, preferential TEMP:

COMPOSITION:

(1) 1 .... HF
1 .... 5 *M* CrO_3

(2) 2 .... HF
1 .... 0.15 *M* CrO_3

(3) 2 .... HF
1 .... 0.75 *M* CrO_3

DISCUSSION:

Si, (111) and (100) wafers used as substrates for silicon MBE thin film epitaxy growth in a study of film defects. Thin films were etched in the solutions shown: (1) "Sirtl" on (111); (2) "Secco" on (100) and (3) "Schimmel" on both (111) and (100). Prior to epitaxy, wafers were (1) detergent scrubbed, (2) DI water rinsed, (3) degreased and N_2 blown dry. No ultrasonic used as it may cause surface damage.

REF: Ota, Y — *J Cryst Growth,* 61,126(1983)

SI-0140

ETCH NAME: TIME: 10—20 sec

TYPE: Acid, stain TEMP: RT

COMPOSITION:

1 .... 50 cm^3 HF
4 drops .... HNO_3
4 drops .... HCHO (polyoxymethylene)

DISCUSSION:

Si, (100) wafers used as substrates for superlattice epitaxy growth structures doped with silicon. Wafers cleaved ⟨110⟩ and structure stained with solution shown.

REF: Sakamoto, T — *J Cryst Growth,* 62,704(1983)

SI-0193

ETCH NAME: Argon TIME:

TYPE: Ionized gas, thinning TEMP:

COMPOSITION: GAS FLOW:

x .... Ar^+ ions PRESSURE:

POWER: 4—6 KeV

DISCUSSION:

Si, (111), (100) and (110) wafers, n-type 0.1—0.7 Ω cm and p-type 0.4—3 Ω cm resistivity. Wafers used in an interface and morphology study of nickel and cobalt silicides. (1) Degrease and clean wafers; (2) CVD deposit 3—5 K Å of SiO_2 and use standard photolithography techniques to open circular windows in oxide; (3) evaporate either nickel

or cobalt and anneal to form silicides. For TEM evaluation: (1) mechanically polish surfaces, then (2) etch thin by erosion using argon ion bombardment.
REF: Tung, R T & Gibson, J M — *J Vac Sci Technol,* A3,987(1985)

SI-0194
ETCH NAME: TIME: 10 sec
TYPE: Acid, polish TEMP: RT
COMPOSITION: RATE: 1700 Å/10 sec
105 ml .... HF
745 ml .... HNO_3
75 ml .... HAc
75 ml .... $HClO_4$
DISCUSSION:
Si, (111) wafers, n-type, used to fabricate diffused p-n-p transistors. Solution used to polish wafers prior to diffusion.
REF: Blaha, R E & Fahrner, W R — *J Electrochem Soc,* 123,515(1976)

SI-0246
ETCH NAME: TIME:
TYPE: Acid, preferential TEMP:
COMPOSITION:
x .... HF
x .... HNO_3
x .... HAc
\+ .... ethylene glycol
DISCUSSION:
Si, single crystal wafers. Organic type additives as ethylene glycol, citric, and tartaric acids used as complexing agents with HF:HNO_3:HAc mixtures as dislocation etches on silicon.
REF: Christian, S M & Jensen, R V — *RCA Rep,* LB-1023(1956)

SI-0138
ETCH NAME: TIME:
TYPE: Acid, cleaning TEMP:
COMPOSITION:
3 .... H_2SO_4
1 .... H_2O
DISCUSSION:
Si, (111), n-type, 3—5 Ω cm resistivity wafers used as substrates in a study of RF sputtered SiO_2. Clean substrates as follows: (1) degrease with TCE, and acetone rinse; (2) etch in solution shown, and (3) final dip 1HF:10H_2O with DI water rinse.
REF: Lee, N K — *J Appl Phys,* 130,658(1983)

SI-0141
ETCH NAME: Argon TIME:
TYPE: Ionized gas, removal TEMP:
COMPOSITION: GAS FLOW:
x .... Ar^+ ions PRESSURE:
POWER: 10 KeV
DISCUSSION:
Si, wafers of various orientations used in a study of damage introduced by argon ion sputter etching. Damage can be caused by pulsed and CW lasers, E-beam, and ion implan-

tation (I^2), and varies with gas species and energy levels involved. All can cause surface damage and internal gas trapping. Wafers were furnace annealed prior to and after argon ion sputter etching. Post anneal was done with a Q-switched ruby laser. (*Note:* Damage includes conversion to an amorphous structure which is recrystallized by subsequent laser anneal.)

REF: Lawson, E M — *J Vac Sci Technol,* B1,15(1983)

SI-0207: Davidse, P D — *J Electrochem Soc,* 116,100(1969)

Si, (111) wafers and other orientations. A development study of the use of RF sputter etching as a ''universal etch''.

SI-0299: Narayan, J et al — *J Vac Sci Technol,* A(2),1303(1984)

Si, (100) wafers cut from ingots grown by the Czochralski (CZ) method. Wafers were etch polished and used in a study of ion implantation (I^2) damage and amorphization. Wafers were Si^+ ion implanted at LHe temperatures (about −270°C) at a power level between 100—200 KeV. Wafers were then cleaved ⟨110⟩ and Ar^+ ion thinned for TEM study. The implant induced amorphous layer was 2000 Å deep.

SI-0142

ETCH NAME: — TIME: Variable

TYPE: Gas, oxidation — TEMP: RT

COMPOSITION:

(1) x .... NO_2 (2) x .... HF (3) x .... H_2O

DISCUSSION:

Si, (111) p-type (B), 10 Ω cm resistivity wafers used in a study of surface films deposited from the three sources shown (1) produced a crystalline film of NH_3SiF_2 (ASF), in part, but all three were mostly amorphous silica, SiO_x with x = 2. (*Note:* Water-grown oxide films may be hydrated depending upon method of deposition.)

REF: Heimann, R B — *J Vac Sci Technol,* B1,108(1983)

SI-0143

ETCH NAME: Hydrofluoric acid — TIME:

TYPE: Acid, oxide removal — TEMP: RT

COMPOSITION:

x .... HF, conc.

DISCUSSION:

Si, (100) n-type, 2—5 Ω cm resistivity wafers used for the evaporation of As_2Te_3. Clean wafers: (1) TCE, agitated, (2) rinse in double distilled water and (3) rinse in acetone. HF rinse immediately before placing wafers in vacuum to remove native oxide. Flash evaporate As_2Te_3 at RT and 5 × 10^{-6} Torr at 20 Å/sec to 5000 Å thickness.

REF: Krupanidhi, S B et al — *J Appl Phys,* 54,1383(1983)

SI-0144

ETCH NAME: Hydrofluoric acid, dilute — TIME:

TYPE: Acid, oxide removal — TEMP:

COMPOSITION:

x .... 10% HF

DISCUSSION:

Si, (100), p- and n-type wafers, 1—10 Ω cm resistivity used for the evaporation of erbium, Er. Clean substrates: (1) acetone with ultrasonic; (2) EOH rinse; (3) ''RCA'' etch clean (SI-0031) and (4) DI water rinse. Use final HF rinse immediately prior to placing wafers in vacuum.

REF: Wu, C S et al — *J Electrochem Soc,* 132,918(1985)

SI-0145
ETCH NAME: Potassium hydroxide — TIME: 10 min
TYPE: Alkali, preferential — TEMP: 80°C
COMPOSITION:
x .... 10 *M* KOH
DISCUSSION:

Si, (111), (100) and (110) wafers and a 1-cm diameter sphere, finished with 600-grit SiC. Etch Time/Temp shown is for sphere etching. Wafers were etched at 62°C for 400 min (?) to form mesas after photo resist patterning. The finite crystal form of the sphere was a tetrahexahedron, (hk0). The study was correlated with mesa side slopes and slow etch planes of the sphere with prediction of type plane that will occur on mesas. (*Note:* Author Walker did similar studies on spheres, mesas, and pits, circa 1957—1958 using boiling KOH, 30 to 40 min [see SI-0092a-c]. The 400 min shown is probably 40 min.)

REF: Weirach, D F — *J Appl Phys,* 46,1478(1975)

SI-0146
ETCH NAME: Potassium hydroxide — TIME:
TYPE: Electrolytic, oxidation — TEMP: 22°C
COMPOSITION: — ANODE: Si
x .... 2 *M* KOH — CATHODE: Si
POWER: Variable
DISCUSSION:

Si, (111) and (100) wafers, p- and n-type of varied resistivity. A major study of anodic and cathodic oxidation of silicon and oxide etch-back.

REF: Palik, E D et al — *J Electrochem Soc,* 132,817(1985)
SI-0273: Orest, J et al — *J Electrochem Soc,* 132,145(1985)

Si, (111) and (100) wafers both p- and n-type, 9—10 Ω cm resistivity. A study of anodic/cathodic oxidation of silicon using KOH.

SI-0097b: Glembocki, O J & Stahlbush, R E — *J Electrochem Soc,* 132,145(1985)

Si, (111) and (100) wafers etched under bias in 2 *M* KOH. Contact was made using an In/Ga amalgam alloy. Etch rates shown for (111), (100), and (110).

SI-0245
ETCH NAME: — TIME:
TYPE: Electrolytic, polish — TEMP: Hot
COMPOSITION: — ANODE: Si
x .... 1 *N* KOH — CATHODE: Si
(+) .... glycerin — POWER:
(+) .... H_2FSi_6
DISCUSSION:

Si, (111) wafers both p- and n-type. Etching done with switching of anode/cathode. KOH solution used with and without (+) additives shown. P-type wafers will polish; n-type will be pitted.

REF: Turner, D R — *J Electrochem Soc,* 105,402(1958)

SI-0272
ETCH NAME: RIE — TIME:
TYPE: Ionized gas, selective — TEMP:
COMPOSITION: — GAS FLOW: 10—50 cc/min
x .... 3—10% Cl_2 — PRESSURE: 10—100 μm
x .... Ar — POWER: 0.16/cm²

DISCUSSION:

Si, (100) p-type wafers with SiO_2 films. A study of RIE etching with Cl_2/Ar gas mixtures. SiO_2 etch rate was about 5—8 nm/min with Si/SiO_2 ratio rates as high as 20:1. A laser etching technique was used to test SiO_2 thickness in place for adhesion and defect generation.

REF: Pogge, H B et al — *J Electrochem Soc,* 130,1592(1983)

SI-0273

ETCH NAME: RIE
TYPE: Ionized gas, selective
COMPOSITION:
x .... x% O_2
x .... SF_6

TIME: 3 min
TEMP: RT
GAS FLOW: 450 sccm
PRESSURE: 150 mTorr
POWER: 1500 W

DISCUSSION:

Si, (100) n-type 3—6 Ω cm resistivity wafers used as substrates for deposition of SiO_2 and poly-Si in an RIE etch rate study, primarily concerned with poly-Si. Different O_2 plasma gas levels evaluated. Poly-Si/SiO_2 selectivity ratio was above 10:1. Ratio of KTI 1470J photo resist to silicon was about 1:1.

REF: Light, R R W & Bell, H B — *J Electrochem Soc,* 130,1567(1983)

SI-0274: Sakai, Y et al — *J Electrochem Soc,* 131,627(1984)

Si, polycrystalline layers deposited on Si containing an SiO_2 thin film. A study of poly-Si etching in conjunction with aluminum evaporation for VLSI structures. RIE gas composition was SF6:O_2(10%).

SI-0051b

ETCH NAME: Steam
TYPE: Gas, cleaning
COMPOSITION:
x .... H_2O, steam

TIME:
TEMP: >100°C

DISCUSSION:

Si, (100) wafers used as substrates for epitaxy growth of thick and thin poly-Si films grown on SiO_2 films pre-deposited on the substrates. Growth method called Lateral Epitaxy Growth over Oxide (LEGO). After deposition, poly-Si films were thinned using a laser strip heater. All poly-Si films were melted with a halogen lamp furnace. Substrates were cleaned: (1) oxidize in steam; (2) photo resist, patterned with "seed" windows opened; (3) in vacuum use HCl vapor at 1050°C, to remove any native oxide remaining in windows. Epitaxy silicon grown using SiH_4 at 1050°C as poly-Si columnar grains and capped with LPCVD SiO_2. Schimmel etch, RT, 10 sec used as a defect and dislocation etch solution (SI-0037).

REF: Celler, G K et al — *J Electrochem Soc,* 132,211(1985)

SI-0147

ETCH NAME: Sodium hydroxide
TYPE: Alkali, structure
COMPOSITION:
x .... 10 *N* NaOH

TIME: $4^1/_2$ or $2^1/_2$ h
TEMP: 363 K (109°C)
RATE: 120 μm/h (100),
4 μm/h (111)

DISCUSSION:

Si, (100) wafers 100 mm thick, polished on both sides for etch development of pits or via holes. Wafers were photo resist patterned for an open square geometry with nichrome used as the primary mask (its removal rate is much less than that of silicon in this solution). The $2^1/_2$ h etch time was used for single-side pit development; the $4^1/_2$ h — $2^1/_2$ h on either wafer side — used for via holes, and for a controlled pit-bottom or via hole orifice size.

REF: Petit, B et al — *J Electrochem Soc,* 132,982(1985)

SI-0148
ETCH NAME: TIME:
TYPE: Acid, thinning TEMP: RT
COMPOSITION:
1 .... HF
5 .... HNO_3
1 .... CH_3COOH (HAc)
DISCUSSION:

Si, (111) web-dendritic ribbon crystal silicon specimens used in a study of defects and structure. The solution shown was used to etch thin sections and the Secco etch (SI-0036) was used for defect development. Showed multiple (111) glide planes — due to growth strain(?) — and the usual twinned zone down the center of the ribbon.

REF: Cunningham, B et al — *J Electrochem Soc,* 129,1089(1982)

SI-0149
ETCH NAME: TIME:
TYPE: Halogen, conditioning TEMP:
COMPOSITION:
x .... MeOH
x .... $*I_2$

*Saturate the solution.

DISCUSSION:

Si, (100) n-type wafers with a p^+ Si epitaxy buffer layer, then a CdSe thin film deposition in structuring a tunnel junction photo anode for solar cells. Prior to CdSe homogeneous chemical vapor deposition (HOMOCVD), (1) HF dip to remove native oxide from p^+/n Si surfaces and (2) quench in solution shown — will produce a p^+ surface with chemisorption of an I_2 film that is stable to about 500°C. Silicon wafer ingots were grown by Horizontal Bridgman (HB) and Liquid Encapsulated Czochralski (LEC) techniques. (*Note:* Author Walker (1964) used an unsaturated solution as a method of complexing silicon surfaces for removal of ionic contamination remaining from acid etch solutions prior to diffusion. Iodine remaining on surface will complex ions and vaporizes at about 100°C leaving a molecularly clean surface for diffusion. See Diamond section for additional use of I_2 to passivate surfaces.)

REF: Pinson, W E — *J Appl Phys Lett,* 40,970(1982)

SI-0076b
ETCH NAME: TIME: 15 min
TYPE: Acid TEMP: RT
COMPOSITION:
1 .... HF
1 .... HNO_3
1 .... HAc
1 .... Br_2
DISCUSSION:

Si, polycrystalline material. The following system was developed to produce damage-free polished surfaces: (1) lap with 30 μm alumina on a felt pad, 30 min for a dull grey surface; (2) polish with 2 μm diamond paste with a starting pressure of 5 psi, reducing to 2 psi over a 2-min period to remove lap damage, but residual stress remains and (3) final

polish with Nalcog #2250 (colloidal silica slurry). Etch 15 min in $1HF:5HNO_3:1HAc$, then in the solution shown.
REF: Sopori, B L et al — *J Electrochem Soc,* 128,215(1981)

SI-0151
ETCH NAME: TIME:
TYPE: Acid, oxidizing TEMP:
COMPOSITION:
x .... 8 *N* H_2CrO_4
DISCUSSION:
Si, wafers. Solution will both oxidize and preferentially etch silicon.
REF: Alpha Products Catalog, 1982

SI-0152
ETCH NAME: Secco etch, modified TIME:
TYPE: Acid, dislocation TEMP: RT
COMPOSITION:
2 .... HF
1 .... 0.005 *M* $K_2Cr_2O_7$
DISCUSSION:
Si, (100) p-type wafers used for ion implantation of arsenic in a study of annealing with E-beam. E-beams were used to grid anneal surfaces after As^+ ion implantation. The implantation produced a 0.1 μm deep amorphous layer. The Secco etch shown was used to develop defects. See SI-0036 for other Secco etchants.
REF: Sun, H T et al — *J Vac Sci Technol,* B(1),827(1985)

SI-0153c
ETCH NAME: Copper etch TIME:
TYPE: Acid, preferential TEMP:
COMPOSITION:
4 ml .... HF
2 ml .... HNO_3
4 ml .... H_2O
0.2 g .... $Cu(NO_3)_2.3H_2O$
DISCUSSION:
Si, (111) wafers and other orientations. Use solution shown as a general dislocation etch.
REF: Ibid.

SI-0153d
ETCH NAME: Copper etch TIME: 2 h
TYPE: Acid, preferential TEMP: RT
COMPOSITION:
"A" 600 ml .... HF "B" x .... H_2O
300 ml .... HNO_3
24 g .... $Cu(NO_3)_2$
2.4 g .... Br_2
Mix: 1 part "A" to 10 parts "B"
DISCUSSION:
Si, (111) wafers. Use ultrasonic agitation of solution during the etch period. This is a general dislocation etch and can be used on other silicon surface orientations.
REF: Ibid.

SI-0154
ETCH NAME: TIME:
TYPE: Acid, polish TEMP: RT
COMPOSITION:
1 .... HF
3 .... HNO_3
DISCUSSION:
Si, single crystal spheres $^1/_8$" diameter used in determining densitometric and electrical parameters of boron in silicon. Rough spheres fabricated by melt drop-off from glass rods of different density, then tumbled in a ''race-track'' system on SiC paper. Ground spheres were etch polished in the solution shown and rinsed in DI water. Produces a hydrophobic surface. (*Note:* See SI-0061a-c, SI-0092a-f for additional work on spheres.)
REF: Horn, F H — *Phys Rev,* 97,1521(1955)

SI-0188b
ETCH NAME: TIME:
TYPE: Acid, junction stain TEMP: RT
COMPOSITION:
4 .... HF
2 .... $Cu(NO_3)_2$
DISCUSSION:
Si, (111) wafers with p-n junctions. Solution will plate out on n-type material side first.
REF: Woriskey, P J — *J Appl Phys,* 29,867(1958)

SI-0188c
ETCH NAME: TIME:
TYPE: Acid, junction stain TEMP: RT
COMPOSITION:
50 ml .... x% $Cu(NO_3)_2$
1—2 drops .... HF
\+ .... white light
DISCUSSION:
Si, wafers with p-n junctions. Copper will plate out on the n-type area first.
REF: Ibid.

SI-0189
ETCH NAME: TIME:
TYPE: Acid, junction stain TEMP: RT
COMPOSITION:
x .... HF
x .... 1% HNO_3
DISCUSSION:
Si, p-n junction wafers. Place one drop of solution across junction. The p-type area will turn blackish.
REF: Fuller, C S & Ditzenberger, J A — *J Appl Phys,* 27,544(1956)

SI-0190a
ETCH NAME: TIME:
TYPE: Acid, polish TEMP: RT
COMPOSITION:
1 .... HF
25 .... HNO_3

DISCUSSION:

Si, (111), (100), (112) and (110) oriented wafers used in a study of surface melt patterns. Polish surfaces with solution shown prior to heat treatment to produce melt patterns. Melt patterns were square on (100); triangles on (111); isosceles triangles on (112) and parallel lines on (110). (*Note:* See SI-0092a-f for similar data.)
REF: Pearson, G L & Treuting, R G — *Acta Crystallogr,* 11,397(1958)

SI-0061b
ETCH NAME: Abrasive cutting TIME: To 1 h
TYPE: Abrasive, forming TEMP: RT
COMPOSITION:
x .... SiC, paper
x .... H_2O
x .... glycerin
DISCUSSION:

Si, single crystal spheres, p- and n-type. Use 80-grit SiC for rough forming and 320-grit for final grinding. Use a Buehler polish wheel and four steel tubes to hold material in place. Cubes, $^1/_4$ and $^1/_2$″ square, were cut as starting material. Both germanium and silicon spheres formed. See SI-0061a and SI-0066 for polish etches. Spheres were used for finite crystal form etching, and in developing spherical devices.
REF: Walker, P — personal development, 1957—1960

SI-0156
ETCH NAME: Lathe cut TIME:
TYPE: Lathe, forming TEMP: RT
COMPOSITION:
x .... oil, coolant
DISCUSSION:

Si, single crystal spheres cut on a lathe using oil as a coolant. Also formed germanium spheres.
REF: Myers, J — personal communication, 1959

SI-0157a
ETCH NAME: Nitrogen TIME: To 45 min
TYPE: Gas, forming TEMP: RT
COMPOSITION: PRESSURE: 2—3 SCFH
x .... N_2, pressure
x .... SiC, paper
DISCUSSION:

Si, single crystal spheres from $^1/_8$ to $^1/_2$″ in diameter formed from cut cubes. A ‘‘Tornado Sphere Grinder’’ was designed and fabricated by authors as a circular race-track using glue-back Buehler 600-grit SiC strips for grinding. Cubes are placed in the closed system and rotated by nitrogen gas pressure blowing at 3—5 psi for 20—40 min (silicon). Other materials require different grinding time. Other semiconductors sphere formed: germanium, gallium arsenide, gallium phosphide, gallium antimonide, and indium phosphide. Used spheres in finite crystal form etching studies.
REF: Myers, J & Walker, P — personal development, 1958—1959

SI-0158
ETCH NAME: Shot Tower TIME:
TYPE: Drop, forming TEMP: Hot
COMPOSITION:
x .... hot oil
DISCUSSION:

Si, polycrystalline spheres formed using the Shot-Tower technique: material melt held in a hopper above a cylinder containing a hot oil — liquid droplets of silicon gravity feed down into oil of cylinder forming spheres at cylinder bottom. Method used on Si, Ge, metals, and intermetallic materials. (*Note:* This is the industrial standard method of fabricating ball-bearings and spherical pellets of solder alloy materials.)

REF: Cole, R — personal communication, 1957
SI-0157b: Ibid.

Si, polycrystalline spheres, and special alloy metal mixtures for silicon solders. A Shot-Tower system designed by the authors. The alloys were used in a study of 50 different mixtures as contacts on silicon.

SI-0159
ETCH NAME: TIME:
TYPE: Acid, light figure TEMP:
COMPOSITION:
x .... HF
x .... I_2
DISCUSSION:

Si, (111) wafers and other orientations. Solution used as a preferential etch to develop pits which can be used for light figure orientation of wafers or ingots. Germanium light figure etching was done by thermal etching. (*Note:* See SI-0091a-f for hydroxide etchants for both silicon and germanium.)

REF: Wolff, G A et al — J Electrochem Soc Meet, Pittsburgh, PA, Oct 13, 1955

SI-0160
ETCH NAME: TIME:
TYPE: Acid, planar TEMP:
COMPOSITION:
20 ml .... HF
150 ml .... HNO_3
50 ml .... CH_3COOH (HAc)
DISCUSSION:

Si, (111) wafers and other orientations. Thin wafers used in a study of thermal stress and plastic deformation. Polish etch in solution shown for a flat, planar surface. After stress, use Sirtl's etch to develop dislocations and other defects (SI-0039).

REF: Merizane, K & Gleim, P S — *J Appl Phys,* 40,4104(1969)
SI-0245: Stolt, L et al — *Solid State Electron,* 26,295(1983)

Si, (100) wafers used in fabricating Shottky Barrier diodes. Etch polish wafer in "planar" etch shown and HF dip just before placing in vacuum. A stainless steel mask was used for pattern definition of evaporation metals.

SI-0161
ETCH NAME: TIME: 1 min
TYPE: Acid, polish TEMP: RT
COMPOSITION:
15% .... HF
85% .... HNO_3
DISCUSSION:
Si, (111), p-type wafers, 0.1—200 Ω cm resistivity. Used in a study of direct observation of charge storage surface states. Solution used as a surface polish etch.
REF: Harman, C G & Raybold, R L — *J Appl Phys,* 34,380(1963)

SI-0162a
ETCH NAME: TIME:
TYPE: Acid, polish TEMP: RT
COMPOSITION:
25 ml .... HF
100 ml .... HNO_3
125 ml .... CH_3COOH (HAc)
DISCUSSION:
Si, (111) wafers used as substrates for epitaxy growth of silicon. Solution used to polish surfaces prior to epitaxy. After epitaxy silicon substrates were etch removed for study of structure and stacking faults in the epitaxy layer.
REF: Finch, R H et al — *J Appl Phys,* 34,406(1963)

SI-0162b
ETCH NAME: TIME:
TYPE: Acid, thinning TEMP:
COMPOSITION:
5% .... HF
95% .... HNO_3
DISCUSSION:
Si, (111) wafers used as substrates for epitaxy growth of silicon. After deposition, wafers were mounted with epitaxy side down on Teflon discs and etched in a Teflon beaker to remove the bulk silicon substrate. Etching was done with beaker mounted at 45° and rotated at 30 rpm. The epitaxy layer was studied for structure and stacking faults. See SI-0162a.
REF: Ibid.

SI-0163a
ETCH NAME: TIME:
TYPE: Acid, thinning TEMP: RT
COMPOSITION:
(1) 10 .... HF
5 .... HNO_3
14 .... HAc
(2) 1 .... HF
5 .... HNO_3
DISCUSSION:
Si, (111) wafers used in a study of stacking fault energy. Wafers were mechanically polished down to $^2/_{10}$ mm thickness and then a pit was sandblasted at the wafer center. Etch (1) was used as a "rough" removal and polish etch and followed by etch (2) as a final "fine" polish etch. Transmitted light was used to determine the stop-point for etching by color. Transmitted light was brownish-red — sufficiently thin for electron microscope study — at the stop-point. Solutions also developed step angled dislocations in surface. (*Note:*

See SI-0092a. Transmitted light used to establish aluminum alloy base widths for silicon transistors.)
REF: Aerts, E et al — *J Appl Phys,* 33,3078(1963)

SI-0201b
ETCH NAME: TIME:
TYPE: Acid, polish TEMP:
COMPOSITION: ANODE:
  1 .... HF CATHODE: Pt
  4 .... HNO_3 POWER:
DISCUSSION:

Si, (111) n-type wafers, 130 Ω cm resistivity used in a study of oscillatory photocurrent decay. Etching done with a xenon flash through a 2.5 mm slit during test.
REF: Hoshino, H & Takayama, M — *Jpn J Appl Phys,* 2,438(1963)

SI-0165
ETCH NAME: Camp #2 (Superoxol) TIME:
TYPE: Acid, preferential TEMP: RT
COMPOSITION:
  10 .... HF
  10 .... H_2O_2
  40 .... H_2O
DISCUSSION:

Si, (111) n-type wafers and p-doped with ^{60}Co (gamma) used in a study of induced radiation defects in silicon. Alloy contacts were 75 Sn:20 Zn:5 Bi and preferentially etch cleaned in the solution shown, which was called #2 Etch. Samples were annealed in an oil bath under electrical bias. (*Note:* See GE-0065a for original development of etch.)
REF: Saito, H & Hirata, M — *Jpn J Appl Phys,* 2,678(1963)
SI-0067: Dresselhaus, C et al — *Phys Rev,* 98,368(1955)

Si, (111) discs, 3 mm diameter and 0.5 mm thick, used in cyclotron resonance studies. Superoxol used as a preferential and removal etch.

SI-0166
ETCH NAME: TIME:
TYPE: Acid, polish TEMP:
COMPOSITION:
  5 .... HF
  3 .... HNO_3
DISCUSSION:

Si, (111) n-type wafers, 50—500 Ω cm resistivity with gold plated contacts as surface barrier diodes. A study of frequency oscillation under intense illumination. The frequency increases with bias and intensity of illumination only under reverse bias conditions. Solution used to etch polish and clean device surfaces.
REF: Yamashita, B et al — *Jpn J Appl Phys,* 2,661(1963)

SI-0112
ETCH NAME: Potassium cyanide TIME:
TYPE: Acid, cleaning TEMP: RT
COMPOSITION:
  x .... 20% KCN

DISCUSSION:

Si, (111), p-type (intrinsic) and doped (extrinsic) wafers used in a study of diffusion and solubility of copper in silicon, germanium, and gallium arsenide. Solution used to remove traces of copper prior to and after deposition and diffusion.

REF: Hall, R N & Racette, J H — *J Appl Phys,* 35,379(1964)

SI-0225

ETCH NAME: TIME:

TYPE: Acid, junction TEMP: RT

COMPOSITION:

90 .... HF

10 .... HNO_3

DISCUSSION:

Si, (111) n-type wafers with p-n junctions. Solution was used to delineate the junction. (*Note:* Similar to other "stain" type junction etches.)

REF: Billig, E & Gasson, D B — *J Appl Phys,* 28,1242(1957)

SI-0205: Yeh, T-H & Joshi, M L — *J Electrochem Soc,* 116,73(1969)

Si, (111) wafers and other orientations used in a study of strain compensation in silicon by diffused impurities. No etch shown.

SI-0167

ETCH NAME: TIME:

TYPE: Acid, junction stain TEMP:

COMPOSITION:

1 .... HF(40%)

3 .... HNO_3

10 .... HAc

DISCUSSION:

Si, (111) n-type wafers with diffused p-type layers, sectioned and studied for diffusion depth profile by staining. Stain graded from bright blue at surface down to brown with depth of diffusion. Color represents changes in impurity concentration. Depth of two stains studied varies as the square root of diffusion time as $L = (D.t)^{1/2}$.

REF: Knopp, A N — *Electrochem Technol,* 2,302(1964)

SI-0063c

ETCH NAME: TIME: 1—10 sec

TYPE: Acid, junction stain TEMP: RT

COMPOSITION:

(1) 1 .... HF
20 .... HNO_3

(2) 20 .... HF
1 .... HNO_3

DISCUSSION:

Si, (111) n-type wafers, 15—20 Ω cm resistivity with 1—3 μm depth boron diffusion used in an initial study of boron diffusion in silicon using a diborane, B_2H_6, gas source. After wafers were diffused they were sectioned and angle lapped at 5 and 10° prior to staining. Observation was done with a standard metallurgical microscope with additional side lighting to enhance color definition. Color varied from surface down as brilliant red, blue, yellow, green, and dull brown at depth. Colors used as a measure of diffusion concentration change profiling.

REF: Walker, P & Crabbs, G — personal development, 1965

SI-0169b
ETCH NAME: Caro's etch, modified TIME: 10 min
TYPE: Acid, cleaning TEMP: 110°C
COMPOSITION:
7 .... H_2PO_4
3 .... H_2O_2
DISCUSSION:

Si, (111) p-type wafers used for diffusion of antimony from glass. Solution used to clean surfaces followed by spin rinsing with UPDI (ultra pure DI) water and drying in hot nitrogen. A powdered slurry of antimony glass was spun-on for 10 min at various speeds for different thicknesses and fired at 150 or 400°C in nitrogen for 30 min. Also used 1HF:10H_2O, RT, 1 min for cleaning surfaces (SI-0169a).

REF: Ibid.

SI-0170
ETCH NAME: TIME:
TYPE: Acid, polish TEMP:
COMPOSITION:
16 ml .... HF
100 ml .... HNO_3, fuming
44 ml .... HAc
DISCUSSION:

Si, (111) wafers used in a study of ion bombardment cleaning and observed by ion energy electron diffraction. Solution used to polish surfaces prior to ion bombardment. Titanium, germanium, and nickel also studied. (*Note:* The "fuming" concentration not shown . . . white, 70% . . . yellow, 72% . . . red, 74%.)

REF: Farnsworth, H E et al — *J Appl Phys,* 29,1150(1958)

SI-0171
ETCH NAME: TIME:
TYPE: Acid, preferential TEMP:
COMPOSITION:
25 .... HCl
75 .... HNO_3
DISCUSSION:

Si, (111) wafers used in a study of light induced plasticity. A Knoop Tukon Hardness Tester was used to indent surfaces with a diamond tool. Solution used to develop structure after indentation. Authors say that surface preparation is dependent upon results observed. Also studied were germanium, indium antimonide, and indium arsenide which were all etched in CP4.

REF: Kuczynski, G C & Hochman, R F — *Phys Rev,* 108,946(1957)

SI-0172
ETCH NAME: SR4 TIME:
TYPE: Acid, polish TEMP:
COMPOSITION:
3 .... HF
2 .... HNO_3
1 .... HAc

DISCUSSION:

Si, (111) wafers used in a study of the variations in surface conductivity of silicon and germanium. Solution used to polish on both materials.

REF: Ioselevich, M L & Fistul', V I — *Sov Phys (Solid State),* 3,822(1960)

SI-0173a

ETCH NAME: TIME: 7—9 min

TYPE: Acid, forming TEMP: RT

COMPOSITION: RATE: $^1/_2$ mil/min

2 .... HF

15 .... HNO_3

5 .... HAc

DISCUSSION:

Si, (111) n-type wafers, 5—50 Ω cm resistivity with wafer diameters from $^1/_2$ to 2″, with closed-tube gallium diffusion depths of 3.5 mils. Solution shown used to etch a well containing pylons in surfaces for n-type epitaxy deposition in fabricating high power SCRs. Well etched into gallium layer for n-p-n structure. Etch produces a highly reflective surface with interlocking etch pits up to 0.5 mil size; well bottom flatness of 0.5 mils; well edge slopes a shallow 25 to 30 mils extending into well; pylon sides preferentially etched with near-hexagonal cross-section. Well depth etched to a controlled depth <3.5 mils in establishing the SCR gate turn-on level.

REF: Walker, P — personal application, 1970

SI-0173b

ETCH NAME: TIME: 7—9 min

TYPE: Acid, forming TEMP: RT

COMPOSITION: RATE: $^1/_4$ mil/min

2 .... HF

3 .... HNO_3

8 .... HAc

DISCUSSION:

Si, (111) n-type wafers, 5—120 Ω cm resistivity with 3.5 mil deep gallium p-type diffusion. A well with pylons etched into surface prior to n-type silicon epitaxy in fabricating high power SCRs $^1/_2$ to 2″ in diameter. Pylons were wax ink dotted prior to well etching. This solution gave a well bottom with a shallow pit-and-mound structure, and a dull matte finish. Well edge slope near-vertical; pylon sides vertical with near-circular cross section. Well depth etch to <3.5 mils to establish SCR gate turn-on level. This is an improved well/pylon structure compared to SI-0173a. Ten etch solutions evaluated for improved well shape with the following ratio of constituents shown above: 1:3:3 . . . 1:5:6 . . . 1:5:8 . . . 2:5:8 . . . 2:3:8 . . . 2:3:5 . . . 2:3:4 . . . 2:6:12 and 2:3:6. All well side slopes, pylon side structure, and well bottom planarity measured by cross-sectioning and measuring under a microscope.

REF: Ibid.

SI-0173c

ETCH NAME: TIME: 2—8 min

TYPE: Acid, dislocation TEMP: RT

COMPOSITION: RATE: $^1/_2$ mil/min

2 .... HF

3 .... HNO_3

6 .... HAc

DISCUSSION:

Si, (111) n-type wafers, 5—120 Ω cm resistivity with a 3.5 mil deep p-type gallium diffusion. Similar to etch shown under SI-0173b. Because of the overall matte finish, dislocations are easily observed. Used to measure dislocation density before and after gallium diffusion. All of the etchants in SI-0173b are dislocation etches to some degree.

REF: Ibid.

SI-0174a

ETCH NAME:
TIME:
TYPE: Electrolytic, jet polish
TEMP: RT
COMPOSITION:
ANODE:
x .... 0.2 *N* NaF
CATHODE:
5 ml .... HF
POWER: 3 mA/cm^2 max
100 ml .... H_2O

DISCUSSION:

Si, (111) n-type wafers. Orifice of jet was 0.010 diameter and etching was done with light. A discussion of the jet etching of n-type silicon.

REF: Schmidt, P F & Keipler, D A — *J Electrochem Soc,* 106,592(1959)

SI-0174b

ETCH NAME:
TIME:
TYPE: Electrolytic, jet polish
TEMP: 70°C
COMPOSITION:
ANODE:
30 ml .... HF
CATHODE:
8.4 g .... NaF/l H_2O
POWER: 30 mA/cm^2

DISCUSSION:

Si, (111) n-type wafers. See SI-0174a for discussion.

REF: Ibid.

SI-0174c

ETCH NAME:
TIME:
TYPE: Electrolytic, jet polish
TEMP: RT
COMPOSITION:
ANODE:
40 g .... NH_4F
CATHODE:
8 g .... NaF/l H_2O
POWER:

DISCUSSION:

Si, (111) n-type wafers. Solution used as a chemical as well as electrolytic etch solution. Authors say that at least a 6 mA/cm^2 current is needed to reduce SiO_2 passivity effects.

REF: Ibid.

SI-0245b

ETCH NAME: Hydrofluoric acid, dilute
TIME:
TYPE: Electrolytic, polish
TEMP: RT
COMPOSITION:
ANODE: Si
x .... 5% HF
CATHODE: Pt
POWER: 110—850 mA/cm^2

DISCUSSION:

Si, (111) wafers used in a study of electropolishing with HF. N-type silicon etches with difficulty, requiring higher current than for p-type and can develop surface pitting. High current is needed to remove SiO_2 layers.

REF: Turner, D R — *J Electrochem Soc,* 105,402(1958)

SI-0066b
ETCH NAME: Sodium chromate
TYPE: Acid, surface treatment
COMPOSITION:
x .... 1% $Na_2Cr_2O_7$
TIME:
TEMP: RT or 80—90°C
DISCUSSION:

Si, wafers used in a study of chemical treatments on surface properties of silicon. Solution used to produce a surface with low sulfur content. Soak wafers and allow solution to dry on surfaces.

REF: Ibid.

SI-0066c
ETCH NAME: Water
TYPE: Acid, surface treatment
COMPOSITION:
x .... H_2O
TIME: 30 min
TEMP: Boiling
DISCUSSION:

Si, wafers used in a study of chemical treatments on surface properties of silicon. After boiling, allow wafers to dry by evaporation.

REF: Ibid.

SI-0184: Chu, T L & Tallman, R L — *J Electrochem Soc,* 12,1306(1964)

Si, wafers used in a study of water vapor as an etch for silicon as oxidation and removal. Used Sirtl etch at RT, 6—30 sec to develop oxidation-induced defects.

SI-0066d
ETCH NAME: Hydrofluoric acid
TYPE: Acid, surface treatment
COMPOSITION:
x .... HF
TIME: 5 min
TEMP: RT
DISCUSSION:

Si, wafers used in a study of the chemical treatments on surface properties of silicon. After HF soak, follow with 5-sec rinse in running DI water.

REF: Ibid.

SI-0064
ETCH NAME: Gold
TYPE: Metal, junction delineation
COMPOSITION:
10 g .... $KAu(CN)_2$
200 g .... KOH
1000 ml .... H_2O
TIME:
TEMP: 30—70°C
DISCUSSION:

Si, wafers containing alloyed or diffused junctions. Solution used to chemically plate out gold to differentiate the p-n junction. Authors say that KOH will precipitate at under 5°C. Polished surfaces require longer plating and produce poor adhesion. Surfaces prepared by light lapping with either Linde A or 600-grit carborundum (SiC) are preferred. Use infrared light during plating and gold will plate out on the more negative side of the junction.

REF: Silverman, S L & Benn, D R — *J Electrochem Soc,* 105,170(1958)

SI-0185
ETCH NAME: TIME:
TYPE: Electrolytic, polish TEMP:
COMPOSITION: ANODE:
x .... 1—4% $NH_4F.HF$ CATHODE:
POWER:

DISCUSSION:
Si, n- and p-type wafers. Development study of electropolishing silicon and n-type germanium.
REF: Klein, D L et al — *Electro Chem Soc Abstr,* 10,72(1961)

SI-0186a
ETCH NAME: TIME:
TYPE: Electrolytic, polish TEMP:
COMPOSITION: ANODE:
x .... KF CATHODE:
x .... KCl POWER:
x .... H_2O
DISCUSSION:
Si, wafers. Development study of electropolishing silicon and germanium.
REF: El'kin, B I — *Abstr J Metall,* 3 # 4, item 2 (1958)

SI-0187a
ETCH NAME: TIME:
TYPE: Acid, junction stain TEMP:
COMPOSITION:
x .... AgF, sat. sol.
+ .... *HF

*1 drop/5 ml of solution.

DISCUSSION:
Si, (111) wafers with p-n junctions. Solution used to etch stain junctions.
REF: Ilies, P A & Coppen, P J — *J Appl Phys,* 29,1514(1958)

SI-0188a
ETCH NAME: TIME:
TYPE: Acid, junction stain TEMP:
COMPOSITION:
x .... HF
+ .... intense light
DISCUSSION:
Si, wafers with p-n junctions. Solution turns n-type blackish and p-type is silvery.
REF: Woriskey, P J — *J Appl Phys,* 29,867(1958)

SI-0064
ETCH NAME: TIME:
TYPE: Acid, preferential TEMP:
COMPOSITION:
x .... HF
x .... HNO_3
x .... x% KNO_3

DISCUSSION:

Si, wafers of various orientations. Acid mixtures containing KNO_3 developed spiral terraces.

REF: Matukura, Y & Suzuki, T — *J Phys Soc Jpn,* 12,976(1957)

SI-0041b

ETCH NAME: Hydrofluoric acid
TYPE: Electrolytic, polish
COMPOSITION:
x .... 24—48% HF

TIME:
TEMP:
ANODE: Si
CATHODE:
POWER: 0.5 A/cm^2

DISCUSSION:

Si, wafers. Shown as an electropolish etch for silicon.

REF: Ibid.

SI-0123: Uhiler, A Jr — *Bell Systems Technol J,* 35,333(1956)

Si, specimens. Solution used for electrolytic shaping of specimens. Other electrolytes used for shaping, germanium specimens.

SI-0041c

ETCH NAME:
TYPE: Electrolytic, polish
COMPOSITION:
x .... HF
x .... *alcohol

TIME:
TEMP:
ANODE:
CATHODE:
POWER:

*EOH, MeOH, glycols, or glycerin.

DISCUSSION:

Si, p-type wafers. With the inclusion of alcohols or organic hydroxyls very fine and smooth polished surfaces are achieved. (*Note:* Many authors have pointed out that the use of glycerin or similar agents for viscosity control of etch solutions improves the quality of etched surfaces.)

REF: Ibid.

SI-0016

ETCH NAME:
TYPE: Acid, polish
COMPOSITION:
x .... Linde A alumina
x .... 3—5% CrO_3

TIME:
TEMP:

DISCUSSION:

Si, specimens. A method for metallographic polishing and preparation for studying reactions of thin films on silicon. Place an 0.030 wire in potter with silicon specimen laid on top and pot with plastic. Silicon contained a thin platinum evaporated film 0.1—0.15 μm thickness. After potting, lap on brass metallographic platen with a silk pad at 1150 rpm using the acid/slurry shown above.

REF: Woods, H & Silverman, R — *J Appl Phys,* 38,419(1967)

SI-0178
ETCH NAME: TIME:
TYPE: Acid, polish + decorate TEMP:
COMPOSITION:
(1) 1 .... HF (2) x .... Cu plate
3 .... HNO_3
DISCUSSION:
Si, (111) and (110) wafers cut from CZ grown ingots. Etch polish wafers with solution shown, then heat on copper plate at 850°C for 1 h to diffuse copper. Use Dash etch for 24 h at RT to develop dislocations. See SI-0044 for Dash etch.
REF: Furuoya, T — *Jpn J Appl Phys,* 1,135(1962)

SI-0180a
ETCH NAME: TIME:
TYPE: Acid, polish TEMP:
COMPOSITION:
4 .... HF
10 .... HNO_3
DISCUSSION:
Si, (111) wafers. Used to fabricate diodes with evaporated aluminum in a study of interfaces. Wafers were etched in solution shown to 150 µm thickness and then quenched in concentrated HNO_3 followed by DI water. Oxidize the surfaces immediately and just before aluminum metallization remove some of the oxide with 1HF:10H_2O. (*Note:* An oxide or oxidized surface will improve aluminum metal adhesion.)
REF: Meade, C A — *Appl Phys Lett,* 9,53(1966)

SI-0180b
ETCH NAME: TIME:
TYPE: Acid, oxide removal TEMP:
COMPOSITION:
1 .... HF
10 .... H_2O
DISCUSSION:
Si, (111) wafers. Solution used to etch remove some deposited silicon dioxide, SiO_2 immediately prior to aluminum metallization on the oxide as Al/SiO_2/Si in a study of interfaces.
REF: Ibid.

SI-0192
ETCH NAME: TIME:
TYPE: Acid, jet polish TEMP:
COMPOSITION:
5 .... HF
15 .... HNO_3
3 .... CH_3COOH (HAc)
DISCUSSION:
Si, (111) and (100) n- and p-type wafers. Surfaces polished with an acid jet using the solution shown.
REF: Unvala, B A et al — *J Electrochem Soc,* 119,318(1972)

SI-0092k
ETCH NAME: Potassium bromide TIME: 10—20 min
TYPE: Halogen, polish/forming TEMP: RT
COMPOSITION:
x .... KBr
DISCUSSION:
Si, (111), n-type, 10—15 Ω cm resistivity wafers used in a general etching study and controlled pit formation. Solution works well for surface cleaning and will produce a small 1—1.5 diameter etch pit in the time shown.
REF: Ibid.

SI-0197a
ETCH NAME: TIME:
TYPE: Acid, polish TEMP:
COMPOSITION:
6 .... HF
10 .... HNO_3
9 .... CH_3COOH (HAc)
DISCUSSION:
Si, (111) wafers used in a study of thermal cycle degradation of charge carrier lifetime and resistivity in silicon barrier diodes. Solution used to etch polish wafers.
REF: Walter, F J & Bates, D D — *IRE Trans Nuc Sci,* NS-13,231(1966)

SI-0197b
ETCH NAME: TIME:
TYPE: Acid, decoration TEMP:
COMPOSITION:
"A" 600 ml .... HF "B" x .... H_2O
300 ml .... HNO_3
2 ml .... Br_2
28 g .... $Cu(NO_3)_2.3H_2O$
Mix: 1 "A" to 10 "B" for use.
DISCUSSION:
Si, (111) wafers fabricated as barrier diodes. Solution used as a copper decoration etch.
REF: Ibid.

SI-0092l
ETCH NAME: TIME: To 30 min
TYPE: Alkali, preferential TEMP: Boiling
COMPOSITION:
150 ml .... 30% KOH
30 g .... CHI_3
100 ml .... glycerin
DISCUSSION:
Si, (111), p- and n-type wafers used in a controlled damage pit development study. Damage induced with diamond stylus. After 10 min etching, pit diameters were 15 mils for p-type and 25 mils for n-type. Solution is also a preferential etch for germanium.
REF: Ibid.

SI-0092m
ETCH NAME: TIME: 20—30 min
TYPE: Alkali, preferential TEMP: Boiling
COMPOSITION:
250 ml .... 30% KOH
30 g .... $FeCl_3$
DISCUSSION:
Si, (100) wafers, n-type, 10—30 Ω cm resistivity used in a controlled damage etch pit study. Damage surface with a diamond stylus. Solution produces sharp, triangle pits with truncated ends in 20 min. p- and n-type size as shown under SI-01921.
REF: Ibid.

SI-0092n
ETCH NAME: TIME: To 30 min
TYPE: Alkali, preferential TEMP: Boiling
COMPOSITION:
250 ml .... 30% KOH
30 g .... KBr
DISCUSSION:
Si, (111) wafers, p- and n-type. In 30 min, p-type pits were 35 mils in diameter; n-type, 45 mils. Sharp hexagonal shape. See SI-01921.
REF: Ibid.

SI-0209
ETCH NAME: Silver etch TIME:
TYPE: Acid, dislocation TEMP:
COMPOSITION:
4 ml .... HF
2 ml .... HNO_3
4 ml .... H_2O
200 mg .... $AgNO_3$
DISCUSSION:
Si, (111) wafers used as substrates for silicon epitaxy deposition and study of stacking faults in epitaxy layers. An interference contrast microscope was used to observe stacking faults before and after etching. Defects showed a 1:1 correspondence. Solution used to develop dislocations.
REF: Dudley, R H — *J Appl Phys,* 35,1360(1964)

SI-0211
ETCH NAME: TIME:
TYPE: Acid, oxide removal TEMP:
COMPOSITION:
(1) 2 ml .... HF
3 g .... NaF
96 ml .... H_2O
(2) x .... HF
4 .... H_2O
(3) 1 .... HF
DISCUSSION:
Si, (111) wafers etch cleaned with the sequence shown, primarily to remove native oxide from surfaces. Silicon etch rate was less than 1 Å/min.
REF: Hu, S M & Kerr, D R — *J Electrochem Soc,* 114,414(1967)

SI-0212
ETCH NAME: TIME:
TYPE: Acid, removal TEMP:
COMPOSITION:
108 ml .... HF
350 g .... NH_4F
1000 ml .... H_2O
DISCUSSION:

Si, (111) wafers used as substrates for epitaxy silicon deposition, both p- and n-type material used in an etch rate study. The etch rate is extremely slow in all cases being less than 0.5 Å/min.

REF: Hoffmeister, W — *Int J Appl Radiat Isot,* 2,139(1969)

SI-0215
ETCH NAME: TIME: To 100 h
TYPE: Thermal, preferential TEMP: 990°C
COMPOSITION:
x .... heat
DISCUSSION:

Si, (111), (100) and (110) wafers used in a study of the work function and sorption properties of silicon. Wafers were thermally etched. The (111) surfaces showed typical triangle pits; (100) typical square pits and the (110) hexagonal pit forms.

REF: Farnsworth, H E & Dillon, J A Jr — *J Appl Phys,* 29,1195(1958)
SI-0216: Farnsworth, H E et al — *Phys Chem Solids,* 8,116(1959)

Si, wafers used in a study of surface structures and work function. Specimens were thermally etched at 1000°C with evaporation of silicon. Above 1000° etching is very rapid.

SI-0217
ETCH NAME: Sodium hydroxide TIME:
TYPE: Alkali, preferential TEMP:
COMPOSITION:
x .... 20% NaOH
DISCUSSION:

Si, (111) wafers used in a study of twinning in silicon. Solution used to develop twinning stria in silicon which was studied by light diffraction.

REF: Hopkins, R L — *J Appl Phys,* 29,1378(1958)

SI-0218
ETCH NAME: Gold TIME:
TYPE: Metal, dislocation TEMP: 1200°C
COMPOSITION:
x .... Au
DISCUSSION:

Si, (111) wafers. After gold diffusion into silicon at 1200°C the wafers were subject to bending stress at 900°C. Alpha-helical loops moved into otherwise dislocation free areas of the wafers.

REF: Dash, W C — *Bull Am Phys Soc,* 5,190(1960)

SI-0187b
ETCH NAME: TIME:
TYPE: Salt, junction TEMP: 0°C
COMPOSITION:
1 .... AgF
10 .... H_2O
+ .... *HF

*1 drop HF/5 ml solution

DISCUSSION:

Si, wafers. Solution used to delineate diffused p-n junctions. The silver plates out on the junction.

REF: Ilies, P A & Coppen, P J — *J Appl Phys*, 29,1514(1958)

SI-0220
ETCH NAME: Oxygen TIME:
TYPE: Gas, contamination TEMP: 900—1150°C
COMPOSITION:
x .... O_2

DISCUSSION:

Si, wafers used as substrates for silicon epitaxy as Si/Si. Oxygen in the hydrogen furnace atmosphere and from the quartz tube wall causes stacking faults in the growing epitaxy layer.

REF: Matsuura, Y & Mirura, Y — *Jpn J Appl Phys*, 2,518(1963)

SI-0221a
ETCH NAME: Argon TIME:
TYPE: Ionized gas, structure TEMP:
COMPOSITION: GAS FLOW:
x .... Ar^+ ions PRESSURE:
POWER:

DISCUSSION:

Si, wafers. Used an argon laser to etch develop structure on specimen surfaces. Authors refer to the technique as rapid direct writing of surface relief. Also used hot chlorine vapor or hydrochloric acid.

REF: Enrilch, D J et al — *Appl Phys Lett*, 38,1018(1981)

SI-0222
ETCH NAME: BHF TIME:
TYPE: Acid, oxide removal TEMP:
COMPOSITION:
x .... BHF

DISCUSSION:

Si, (100), n-type, 3—6 Ω cm resistivity wafers used as substrates to fabricate humidity sensitive capacitors. Degrease, chemically etch silicon and use BHF to remove native oxide. RF sputter in O_2/Ar (5—10% O_2) for a 1 KÅ thick thin film of $BaTiO_3$.

REF: Chen, S N et al — *J Vac Sci Technol*, A3,678(1985)

SI-0223a

ETCH NAME: TIME:

TYPE: Acid, junction TEMP:

COMPOSITION:

x .... 300 cm^3 HF

+ .... *HNO_3

*1 drop in HF volume shown

DISCUSSION:

Si, p-type wafers used in a study of dislocations developed from antimony diffusion into silicon as p-n junctions. Solution shown used as a junction etch. Also used Sirtl etch and anodic oxidation to delineate junctions.

REF: Song, S H & Niimi, T — *Jpn J Appl Phys,* 24,1460(1985)

SI-0221b

ETCH NAME: Chlorine TIME:

TYPE: Gas, structure TEMP: Hot

COMPOSITION:

x .... Cl_2

DISCUSSION:

Si, wafers. Used chlorine gas to etch structure silicon surfaces. See SI-0221a.

REF: Ibid.

SI-0224

ETCH NAME: Neutron TIME:

TYPE: Particle, damage TEMP:

COMPOSITION: DOSE:

x .... neutrons

DISCUSSION:

Si, as 15 mm square cut and oriented cubes (100). After cutting and polishing specimens were subjected to neutron irradiation in a study of directional neutron damage using ultrasonic double refraction measurements.

REF: Trevell, R & Teutonico, L J — *Phys Rev,* 105,1723(1957)

SI-0225

ETCH NAME: TIME:

TYPE: Electrolytic, polish TEMP: 30°C

COMPOSITION: ANODE:

1 .... HF CATHODE: Si

10 .... H_2O POWER: 10 V

+ .... glycerin ROTATE: 120 rpm

DISCUSSION:

Si, wafers of different orientations used to prepare very flat surfaces by electropolishing. The following reactions were observed: (1) at low speed, with increase of glycerin drip, thicker solutions; (2) less than 10% HF for polishing; (3) light required for n-type, but not for p-type; (4) rotation speed controls bright polish and (5) high current produces pitting on surfaces.

REF: Baker, D & Tillman, J R — *Solid State Electron,* 6,589(1963)

SI-0230
ETCH NAME: Hydrofluoric acid, dilute | TIME:
TYPE: Acid, removal | TEMP: RT
COMPOSITION:
x .... HF
x .... H_2O
DISCUSSION:

a-Si:H, thin film deposited on an a-SiO_xN_y:H thin film fabricated on the following substrates: UV grade quartz; glass coated with thin film SnO_2; aluminum and stainless steel. All substrates were cleaned as follows: (1) boil in TCE; (2) rinse in warm acetone and (3) boil in isopropanol. All substrates showed good adhesion of deposited films. Final rinse in solution shown prior to deposition of the hydrogenated silicon.

REF: Carasco, F et al — *J Appl Phys,* 57,5306(1985)

SI-0231
ETCH NAME: Hydrogen | TIME: 15 min
TYPE: Ionized gas, defect | TEMP:
COMPOSITION: | GAS FLOW:
x .... H^+ ions | PRESSURE:
POWER: 1 kV
DISCUSSION:

Poly-Si wafers used in a study of the electronic nature of defects and grain boundaries (GB). Part of wafers were masked against ion bombardment. Both laser beam induced current (LBIC) and electron-beam induced current (EBIC) were used in evaluation.

REF: Sastru, O S et al — *J Appl Phys,* 57,5506(1985)

SI-0232
ETCH NAME: Water | TIME:
TYPE: Acid, film removal | TEMP:
COMPOSITION:
x .... H_2O
DISCUSSION:

a-Si, thin film, 300 Å thick. NaCl, (100) substrates were used to develop the following structure: SiO_2/Ni/a-Si/SiO_2/NaCl(100). The composite thin film was removed from the substrate with water by the float-off technique. The film was then annealed and observed for microstructure.

REF: Chen, S N et al — *J Appl Phys,* 57,258(1985)

SI-0174c
ETCH NAME: BHF | TIME:
TYPE: Acid, preferential/jet | TEMP: RT
COMPOSITION:
(1) 40 g .... $NH_4F.HF$
8 g .... NaF/l
(2) 40 g .... $NH_4F.HF$
8 g .... NaF/l
x .... $K_3F(CN)_6$
DISCUSSION:

Si, (111) wafers used in a defect study. Solution (1) develops flat bottom pits. Solution (2) reduces the size of pits and their quantity.

REF: Schmidt, P F & Keipler, D A — *J Electrochem Soc,* 106,592(1959)

SI-0174d
ETCH NAME: TIME:
TYPE: Electrolytic, defect TEMP: 60°C
COMPOSITION: ANODE: Si
5 ml .... HF CATHODE:
x .... 2 *N* NaF POWER: 3 mA/cm^2 or 30 mA/cm^2
1000 ml .... H_2O
DISCUSSION:

Si, (111) wafers used in a defect study. Use light during etching period. Low power application shown above produced preferential etching; high power level for polish etching.
REF: Ibid.

SI-0233
ETCH NAME: TIME:
TYPE: Acid, cleaning TEMP: Boiling
COMPOSITION:
x .... HCl
x .... H_2O_2
x .... H_2O
DISCUSSION:

Si, (111) n-type wafers, 1.63 Ω cm resistivity used as substrates for the deposition of cobalt and growth of $CoSi_2$. Silicon wafers were cleaned as follows: (1) flush with high pressure water; (2) degrease in hot MeOH and water rinse; (3) boil in solution shown and water rinse. Anneal at 750°C in UHV after evaporation of cobalt.
REF: Arnaud, F et al — *J Vac Sci Technol,* B(3),770(1985)

SI-0245
ETCH NAME: TIME:
TYPE: Acid, removal TEMP: RT
COMPOSITION:
x .... HF
x .... HNO_3
DISCUSSION:

Si, (111) wafer substrates used for epitaxy growth of GaP by fused salt electrolysis. After growth, silicon was etched away in an HF:HNO_3 solution mixture for study of the gallium phosphide thin film. Solution attacks GaP slowly.
REF: Cuomo, J J & Gambino, R J — *J Electrochem Soc,* 117,755(1968)

SI-0234
ETCH NAME: Hydrofluoric acid TIME:
TYPE: Acid, passivation TEMP:
COMPOSITION:
x .... HF, conc.
DISCUSSION:

Si, (100) and (111) wafers, both p- or n-type used in a study of passivation of surfaces by fluoridation. Studied active electron sites by surface recombination velocity. Four treatments were used as shown below:

Step #1	Step #2	Recombination/ velocity
1. CP4 polish	1. HF, dip	1. 4 cm/sec
2. Thermal oxidation	2. HF, dip	2. 4 cm/sec
3. HF remove oxidation	3. HF, dip or soak	3. 40 cm/sec
4. KOH etch silicon	4. HF, dip	4. 300 cm/sec

REF: Weinberger, B R et al — *J Vac Sci Technol,* A(3),887(1985)

SI-0120b
ETCH NAME: Hydrofluoric acid TIME:
TYPE: Acid, cleaning TEMP: RT
COMPOSITION:
x .... HF, conc. (inert atm)
DISCUSSION:

Si, (111) wafers used in a study of selenium adsorption. Silicon wafers were cleaned as follows: (1) Syton mech/polish; (2) rinse in MeOH; (3) degrease in TCE and (4) HF clean in inert atmosphere (no air). Selenium treatment was: (1) soak in 0.5 g Se/cc MeOH; (2) rinse in MeOH and (3) argon blow dry. *Note:* Wafers were cleaved while under the selenium solution. Selenium showed adsorption on the (111) and (220) silicon planes.
REF: Dev, B N et al — *J Vac Sci Technol,* A(1),946(1985)

SI-0236
ETCH NAME: Sulfuric acid, dilute TIME: 20 min
TYPE: Acid, cleaning TEMP: RT
COMPOSITION:
5 .... H_2SO_4
1 .... H_2O
DISCUSSION:

Si, (111) p- and n-type, 20 and 25 Ω cm resistivity wafers used as substrates for germanium epitaxy as single crystal, polycrystalline, and amorphous Ge thin films. Silicon wafers cleaned: (1) acetone rinse; (2) MeOH rinse; (3) etch in solution shown; (4) soak in 50% BHF, 60 sec, DI water rinse and N_2 blow dry.
REF: Wang, P D et al — *J Vac Sci Technol,* B(2),206(1984)

SI-0237
ETCH NAME: Nitric acid TIME:
TYPE: Acid, cleaning TEMP: Hot
COMPOSITION:
x .... HNO_3
DISCUSSION:

Si, (100) wafers were cleaned: (1) HF; (2) hot HNO_3, and (3) NH_4F:H_2O_2.
REF: Yang, H T & Berry, W S — *J Vac Sci Technol,* B(2),206(1984)

SI-0239
ETCH NAME: TIME:
TYPE: Acid, photo etch TEMP:
COMPOSITION:
DISCUSSION:

a-Si:F:H, thin films deposited on Corning 7095 glass as both amorphous, a-, and mi-

crocrystalline, μc-, structured films in the fabrication of strain gages. The material was photo etched.
REF: Nishida, S et al — *Thin Solid Films,* 112,7(1984)

SI-0240
ETCH NAME: Iodine etch TIME: 2 min
TYPE: Acid, preferential TEMP: RT
COMPOSITION:
25 ml .... HF
100 ml .... HNO_3
125 ml .... HAc
2 g .... I_2
DISCUSSION:
Si, (111) wafers, boron doped. Wafers were prepared for boron doping as follows: (1) polish with Linde A; (2) degrease; (3) HF soak, RT, 1 min; (4) etch in iodine solution to remove lap damage. After boron diffusion: (1) rinse in HF, RT, 2 min and then (2) etch 2 min in iodine solution to develop slip patterns in silicon created by boron diffusion.
REF: Queisser, H J — *J Appl Phys,* 32,1776(1961)

SI-0242a
ETCH NAME: Iodine etch TIME: 1—2 min
TYPE: Acid, removal TEMP: RT
COMPOSITION:
(1) 1 .... HF
1 .... HNO_3
1.4 .... HAc
0.15% .... I_2
0.24% .... triton (surfactant)

(2) 1 .... iodine etch
4 .... HAc
DISCUSSION:
Si, (111) p- and n-type wafers, 8 Ω cm resistivity. Wafers used in a study of ^{18}F radioactive tracer adsorption. After etching, water rinse or acetone rinse. Also used on germanium in a similar study.
REF: Kern, W — *RCA Rev,* 69,207(1970)

SI-0242b
ETCH NAME: Iodine etch TIME: (1) 1.7 (2) 30 min
TYPE: Acid, removal TEMP: RT
COMPOSITION:
(1) 1 .... HF
6 .... HNO_3
3 .... HAc
x .... *NaI

(2) x .... 0.010 *N* NaI

*0.19 mg/100 ml of solution.

DISCUSSION:
Si, (111) wafers used in a radiochemical surface contamination study. Both of these solutions and the iodine etch shown in SI-0242a were all used with ^{131}I radioactive tracer in a study of adsorption of surfaces. Silica, SiO_2, and germanium, Ge, also were studied.
REF: Ibid.

SI-0243
ETCH NAME: Potassium hydroxide TIME: 6 min
TYPE: Alkali, selective TEMP: Boiling
COMPOSITION:
100 g .... KOH
100 ml .... H_2O
DISCUSSION:
Si, (110) wafers with a thermally grown SiO_2 thin film used as an etch mask. Wafers were photolithographically processed with slots opened in the silica mask in bulk ⟨100⟩ directions. Parallel grooves etched to 50 μm depth with 2 μm horizontal etching. Slots were filled with Corning 7070 glass by hot pressing at 750°C before cross sectioning for structure study.
REF: Stoller, A I — *RCA Rev,* 69,271(1970)

SI-0247
ETCH NAME: TIME:
TYPE: Electrical, defect TEMP:
COMPOSITION:
x .... *EBIC

*Electron-beam-induced current.

DISCUSSION:
a-Si:H, thin films deposited by RF decomposition of silane on polished stainless steel substrates, then coated with aluminum to form p-i-n devices. Both EBIC and secondary electron image (SEI) with a scanning electron microscope (SEM) used in characterization and defect studies. Pinholes, blisters, and lift-off were observed in films that could be related to either film deposition conditions or defects in the substrate surface. Films are applicable for photovoltaic, solar cell, vidicon, and other electronic devices.
REF: Yacobi, B G et al — *J Electron Mater,* 13,843(1984)

SI-0251
ETCH NAME: Ethylenediamine TIME:
TYPE: Acid, selective TEMP: 50°C
COMPOSITION:
x .... ED
DISCUSSION:
Si, (100) wafers. The use of maskless passivation against anisotropic etching to define structure. Both ion implantation and laser annealing used to produce etch-stop mechanisms. The ED solution was used at RT after laser passivation. EDP solutions were also used (ethylenediamine:pyrocatechol). See SI-0094a-d.
See SI-0094a-d.
REF: Day, D J et al — *J Electrochem Soc,* 131,407(1984)

SI-0252
ETCH NAME: Potassium hydroxide TIME: 5—10 min
TYPE: Alkali, orientation TEMP: Hot
COMPOSITION:
x .... 5—10% KOH (NaOH)
DISCUSSION:
Si, (111) and (100) wafers and ingots. Ingot growth direction can be established by observing the direction of the hachure miniscus locations on ingot sides . . . 3 on (111); 4

on (100) ingots. The hachure marks point upward toward the seed ingot end. This can be used to establish a (111)A or $(\overline{111})$B wafer surface direction. By cleaving a portion of a (100) wafer edge and etching in KOH or NaOH will develop triangle etch pits on the cleaved edge. If triangle points are downward, bulk plane is (011); if upward toward (100) positive wafer surface, with 180° rotation, it is $(01\overline{1})$ which is the preferred orientation for channel and pit etching.
REF: Pennington, P & Walker, P — development work, 1957

SI-0253
ETCH NAME: RIE
TYPE: Ionized gas, selective
COMPOSITION:
x .... ClF_3

TIME: To 10 min
TEMP: RT
GAS FLOW: 20 cm^3/min
PRESSURE: 0.03 Torr
POWER: 70 W
BIAS: 10 V

DISCUSSION:
Si, (100) wafers used in structuring a RAM configuration device with poly-Si and SiO_2. Pure Cl_2 or ClF_3 was evaluated in addition to ClF_3 concentrations of 5, 30, and 80%. Primary etching was on the poly-Si. Etch rate vs. side removal of a pure ClF_3 plasma was linear at 1900 Å/min and 2 μm per side removal in 10 min. The Si:SiO_2 etch ratio reduces with decrease of Cl_2 toward pure ClF_3 — atomic fluorine 42:1. Etch rate is unaffected by ClF_3 concentration for lightly doped p- or n-type single crystals, undoped poly-Si and thermal SiO_2 where SiO_2 is less than 100 Å/min with 0—100% ClF_3. All others etched at an equal rate. There is a wafer quantity mass loading affect as shown on poly-Si etch rate of one wafer at about 3800 Å/min to about 1900 Å/min with six to seven wafers.
REF: Flamm, D L et al — *J Electrochem Soc,* 129,2755(1982)

SI-0254a
ETCH NAME: Hydrogen peroxide
TYPE: Acid, preferential
COMPOSITION:
x .... H_2O_2 (30%), conc.

TIME: 2—4 h
TEMP: RT

DISCUSSION:
Si, single crystal hemispheres. Specimens were used in an etching study to observe etch patterns, type etch pits, etch rates, and crystal planes developed. Before preferential etching, etch polish in CP4 to remove lapping damage. See GE-0150a-s for the 19 different etch solutions evaluated on both silicon and germanium hemispheres.
REF: Holmes, P J — *Acta Metall,* 7,283(1959)

SI-0255
ETCH NAME:
TYPE: Acid, float-off
COMPOSITION:
1 .... HF
5 .... HNO_3

TIME:
TEMP: RT

DISCUSSION:
Si, (100) and (111) wafers used as substrates for CVD growth of single crystal thin films of boron phosphide in a study of thin film morphology, physical properties and semiconducting characteristics. Solution used to etch remove the silicon as a float-off technique for TEM (microscope) and Laue (photograph) study of BP.
REF: Kumashiro, Y et al — *J Cryst Growth,* 70,507(1984)

SI-0256
ETCH NAME: TIME:
TYPE: Acid, cleaning sequence TEMP:
COMPOSITION:
(1) x .... H_2SO_4 (2) x .... HNO_3
x .... H_2O
DISCUSSION:

Si, (100), n-type, 4—7 Ω cm resistivity wafers used as substrates for CVD growth of a-Si_3N_4:H thin films doped with either boron or phosphorus. Silicon wafers were cleaned as follows: (1) acetone; (2) DI water rinse; (3) clean in solution (1), DI water rinse; (4) clean in solution (2), heavy DI water rinse and N_2 blow dry. Final clean in CVD reactor at 350°C.

REF: Fang, Y K et al — *J Electrochem Soc,* 132,1222(1985)

SI-0257
ETCH NAME: BHF TIME: 2 min
TYPE: Acid, cleaning sequence TEMP: RT
COMPOSITION:
1 .... HF
30 .... NH_4F (40%)
DISCUSSION:

Si, (100) wafers unpassivated surfaces or with SiO_2 or $TaSi_2$ thin films used as substrates for tungsten selective deposition onto silicon photo resist opened circular patterns. All three substrate types were cleaned as follows: (1) clean in BHF; (2) a 10 min rinse in 18 MΩ DI water; (3) spin dry. This leaves about a 20 Å native oxide on the surfaces. Tungsten was deposited from WF_6 between 260 and 400°C and included thermal annealing to form WSi_2 (tetragonal crystal system).

REF: Green, M L & Levy, R A — *J Electrochem Soc,* 132,1243(1985)

SI-0258
ETCH NAME: Hydrochloric acid, vapor TIME:
TYPE: Acid, cleaning TEMP: 1100°C
COMPOSITION:
x .... HCl, vapor
x .... H_2
DISCUSSION:

Si, (110), (112), and (113) wafers for p-p^+ epitaxy. Substrate wafers were 0.007 Ω cm resistivity with epitaxy films 4—6 Ω cm. Substrates were cleaned with HCl vapors in the epitaxy reactor prior to silicon deposition.

REF: Tamura, M & Sugita, Y — *J Appl Phys,* 44,3442(1973)
Si-0173d: Walker, P — personal application, 1971

Si, (111), n-type wafers with resistivity range from 10—150 Ω cm and diameters from $^1/_2$ to 2″ used in the fabrication of SCRs with p-type gallium diffusion and an n^+ silicon epitaxy. Hot HCl vapors used 30 sec at 1000°C as a final cleaning etch in the epitaxy reactor prior to silicon deposition. (*Note:* This is a general and widely used method of final surface cleaning.)

SI-0260
ETCH NAME:
TYPE: Acid, cleaning sequence
COMPOSITION:
(1) 3 .... H_2SO_4
1 .... H_2O_2
(2) 1 .... HF
10 .... H_2O
TIME: (1) 10 min
(2) 6 sec
TEMP: (1) 95°C (2) RT

DISCUSSION:
Si, (111), p-type wafers used as substrates for tungsten deposition. Silicon wafers were coated with 1000 Å SiO_2, then 4500 Å poly-Si. After acid cleaning wafers were given an Ar^+ ion cleaning prior to tungsten deposition from WF_6 and annealing to WSi_2.
REF: Tsao, K Y & Busta, H H — *J Electrochem Soc,* 131,2702(1984)

SI-0261
ETCH NAME:
TYPE: Acid, cleaning sequence
COMPOSITION:
(1) *RCA procedure
(2) x .... HF, conc.
TIME:
TEMP:

*See SI-0031 for RCA etch.

DISCUSSION:
Si, (111) and (110) wafers used as substrates for reactive pulse plasma deposition of thin film diamond (from carbon) and boron nitride as dielectric coatings.
REF: Szmidt, J et al — *Thin Solid Films,* 110,7(1983)

SI-0262
ETCH NAME:
TYPE: Acid, cleaning
COMPOSITION:
(1) x .... HF, conc.
x .... H_2O
(2) x .... CF_4
x .... O_2
TIME: (1) 2 min
TEMP: RT

DISCUSSION:
Si, (111) p-type wafers, 7—21 Ω cm resistivity as patterned substrates for selective tungsten deposition. Three types of substrates were used: (1) SiO_2/Si; (2) poly-Si/SiO_2/Si and (3) Al/SiO_2/Si. Silicon substrates were initially cleaned by a glow discharge CF_4/O_2 treatment or by immersion in HF, followed by DI water rinse, and spin dry. Tungsten was deposited as a thin film from WF_6 on the entire wafer surface or selectively, to include double tungsten layers separated by thin film SiO_2. Study involved reaction effects of tungsten thin films. The HF treated surfaces gave thin films without erratic surface structure.
REF: Stacy, W T et al — *J Electrochem Soc,* 132,444(1985)

SI-0263c
ETCH NAME:
TYPE: Acid, step-etch
COMPOSITION:
1 .... HF
750 .... HNO_3
250 .... HAc
TIME:
TEMP:
RATE: 300 Å/min

DISCUSSION:

Poly-Si grown on (100) silicon substrates. Solution used to step-etch the poly-silicon layer. Poly-Si was as (100) crystallites with low angle grain boundaries.

REF: Banerjee, S K et al — *J Electrochem Soc,* 131,1409(1984)

SI-0108c

ETCH NAME: Aqua regia — TIME:

TYPE: Acid, cleaning — TEMP:

COMPOSITION:

3 .... HCl

1 .... HNO_3

DISCUSSION:

Si, (111) and (100) wafers used in a study of high pressure oxidation (800—1000°C and 1—20 atm) using dry oxygen. Wafers were cleaned prior to oxidation: (1) H_2SO_4:H_2O; (2) aqua regia and (3) 1HF:10H_2O and spin dry in nitrogen.

REF: Ibid.

SI-0273

ETCH NAME: RIE — TIME: To 50 min

TYPE: Ionized gas, selective — TEMP:

COMPOSITION: — GAS FLOW:

x .... CHF_3 — PRESSURE:

x .... O_2 — POWER:

DISCUSSION:

Si, (100) wafers with and without SiO_2 thin surface films used in a study of side wall tapering in RIE etching. Various thicknesses of oxide and photo resist were evaluated in a hex-type reactor. Etch rates were (1) SiO_2: 550 Å/min; (2) silicon: 110 Å/min and (3) photo resist: 170 Å/min. Discusses results relative to ion milling. To obtain vertical etched walls: (1) vertical mask profile; (2) good selectivity between mask and material being etched and (3) mask of sufficient thickness so as to still be present at end of structure forming etching time. (*Note:* In ion milling by these authors and co-workers, using Ar^+ ion milling, etch time was established to remove the photo resist mask in conjunction with pattern etch-thru to substrate. Otherwise, remaining photo resist is severely polymerized and impossible to remove.)

REF: Nagy, A G — *J Electrochem Soc,* 132,689(1985)

SI-0280a

ETCH NAME: Silica — TIME:

TYPE: Abrasive, polish — TEMP: RT

COMPOSITION:

x .... SiO_2, colloidal suspension

DISCUSSION:

Si, (100) wafers and other orientations. Colloidal silica, c-SiO_2, in suspension used to polish surfaces. One method is to wax wafers onto SST or brass circular platens or use a Buehler-type metallographic station: a rotating brass platen with glue-back pellon-type pad saturated with colloidal silica. Surface polish is equivalent to or better than that of diamond paste polishing. Both colloidal silica and alumina are used for polished silicon wafers and other metals and metallic compounds.

REF: Brochure: NALCO, 1983

SI-0280b
ETCH NAME: Alumina
TIME:
TYPE: Abrasive, polish
TEMP: RT
COMPOSITION:
x .... Al_2O_3, colloidal
DISCUSSION:

Si, (100) wafers and other orientations. Colloidal alumina, as c-Al_2O_3 in suspension used to polish surfaces. See discussion under SI-0280a. (*Note:* Linde A and B alumina are polishing fine powders on the submicron size scale, not colloidal or in permanent liquid suspension. They have been in use for many years as a replacement for diamond paste polishing.)

REF: Ibid.

SI-0285
ETCH NAME: RF plasma
TIME:
TYPE: Ionized gas, removal
TEMP: 50 to 70°C
COMPOSITION:
GAS FLOW:
x .... CF_4
PRESSURE: 0.5 Torr
x .... O_2 (5%)
POWER: 100 W
DISCUSSION:

Si, (100) wafers used as substrates with an SiO_2 thin film were used as substrates for poly-Si growth with various concentrations of boron doping. A tunnel system with an aluminum insert was used in studying accelerated etch rates due to NaOH and KOH residues from AZ- photo resist developer solutions. Etch rate reduces with increase of boron, from 3000 Å/min for undoped to 480 Å/min with 3×10^{21} cm^3 B concentration. Substrates were cleaned by oxidation and stripped with HF. Wafers were dipped 1 min and DI water rinsed in varying pH concentrations of KOH and NaOH and evaluated for plasma etch rates — as dipped, dipped and spun, or dipped and water rinsed. The latter against rinse time. A 10 min water rinse time reduced etch rates from above 5000 Å/min to about 500 Å/min and higher pH values — pH 8 to pH 14 — increased etch rates with similar rates.

REF: Makino, T et al — *J Electrochem Soc,* 128,103(1981)

SI-0092o
ETCH NAME:
TIME:
TYPE: Alkali, preferential
TEMP: Boiling
COMPOSITION:
250 ml .... 30% KOH
100 ml .... glycerin
DISCUSSION:

Si, single crystal spheres and oriented cut forms used in a study of etch planes and rates. The sphere finite crystal form is cube, (100).

REF: Ibid.

SI-0092p
ETCH NAME:
TIME:
TYPE: Alkali, preferential
TEMP: Boiling
COMPOSITION:
250 ml .... 30% KOH
10 ml .... ethylene glycol

DISCUSSION:

Si, single crystal spheres and oriented cut forms used in an etch structure study. The sphere finite crystal form was cube, (100).

REF: Ibid.

SI-0092q

ETCH NAME: Potassium hydroxide TIME:

TYPE: Alkali, preferential TEMP: Boiling

COMPOSITION:

(1) 250 ml .... 30% KOH (2) 250 ml .... 30% NaOH

DISCUSSION:

Si, single crystal spheres and oriented cut forms used in a study of structure. The sphere finite crystal form was tetrahexahedron, (hk0).

REF: Ibid.

SI-0282: Weirach, D F — *J Appl Phys,* 16,1378(1977)

Si, single crystal spheres were etched in 10 *M* KOH, 10 min at 80°C. Finite crystal form was tetrahexahedron, (hk0).

SI-0092r

ETCH NAME: BF3 TIME:

TYPE: Acid, preferential TEMP: RT

COMPOSITION:

11 g .... $NH_4F.HF$

25 ml .... HNO_3

2 ml .... H_2O

DISCUSSION:

Si, single crystal spheres and cut forms. Finite crystal form of sphere was a cube, (100).

REF: Ibid.

SI-0092s

ETCH NAME: TIME:

TYPE: Alkali, preferential TEMP: Boiling

COMPOSITION:

250 ml .... 30% KOH

100 ml .... ethylene glycol

DISCUSSION:

Si, as a pre-cut single crystal octahedron, (111) form. The octahedron etched in this solution developed (211) facets on the (111) faces producing a finite crystal form of a trisoctahedron, (hh1).

REF: Ibid.

SI-0092t

ETCH NAME: TIME:

TYPE: Alkali, preferential TEMP: Boiling

COMPOSITION:

250 ml .... 30% KOH

100 ml .... MeOH

DISCUSSION:

Si, single crystal spheres and cut crystal forms. Finite crystal form was tetrahexahedron, (hk0).

REF: Ibid.

SI-0283b
ETCH NAME: RF plasma
TYPE: Ionized gas, removal
COMPOSITION:
(1) x .... SF_6 (2) x .... SF_6
x .... argon

TIME:
TEMP:
GAS FLOW: 50 sccm anode coupled
PRESSURE: 0.23 mbar
POWER: 1000 V ptp

DISCUSSION:

Si, (100) wafers with thermal SiO_2 thin films photolithographically patterned and used in a study of SF_6 RF plasma etching. Best selectivity ratio of Si:SiO_2 was 30:1. Best results were obtained with parallel plate reactor rather than barrel type. Various mixtures of SF_6:Ar and mixtures with up to 50% H_2 evaluated for etch rates and ratio selectivity vs. CF_4:4% O_2. The undiluted SF_6 etch rate was 10 to 15 times greater than the CF_4:4% O_2 plasma. Etch rate of silicon with SF_6 (0 to 25% O_2), cathode coupled, was in the range of 0.5 µm/min; anode coupled about 0.15 µm/min with anode coupled CF_4 (0—25% O_2) slightly under 0.1 µm/min. SF_6:Ar silicon etch rate 0.3 µm/min to 0.5 µm/min, 20—100% SF_6.

REF: Eisele, K M — *J Electrochem Soc,* 128,123(1981)

SI-0286a
ETCH NAME: RF plasma
TYPE: Ionized gas, removal/anisotropic
COMPOSITION:
x .... C_2F_6
x .... O_2 (8%)

TIME:
TEMP: 25—30°C water cooling
GAS FLOW: 200 sccm
PRESSURE: 0.35—0.40 Torr
POWER: 100—550 W

DISCUSSION:

Si, (100) wafers used as substrates with p-doped and undoped poly-Si and SiO_2 thin films, and photo resist were evaluated for etch rates and channel type etch profiles. The gas mixture shown above was anisotropic and gave best results of five mixtures evaluated. RF power was 350 W and pressure 0.35 Torr. Etch rates: (1) P:poly-Si — 800 Å/min; (2) Undoped poly-Si — 350 Å/min; (3) SiO_2 — 60 Å/min and (4) photo resist — 100 Å/min. Selectivity of P:poly-Si to SiO_2 was 13:1.

REF: Adama, A C & Capio, C D — *J Electrochem Soc,* 128,366(1981)

SI-0286b
ETCH NAME: RF plasma
TYPE: Ionized gas, removal/isotropic
COMPOSITION:
1 .... C_2F_6
1 .... CF_3Cl

TIME:
TEMP: 25—30°C
GAS FLOW: 200 sccm
PRESSURE: 0.40 Torr
Power: 550 W

DISCUSSION:

Si, (100) wafers. Etch rates: (1) P:poly-Si — 1590 Å/min; (2) undoped poly-Si — 980 Å/min; (3) SiO_2 — 200 Å/min and (4) photo resist — 570 Å/min. Selectivity of P:poly-Si to SiO_2 was 5:1. See SI-0286a for further discussion.

REF: Ibid.

SI-0286c
ETCH NAME: RF plasma
TYPE: Ionized gas, removal/isotropic
COMPOSITION:
x .... CF_4
x .... O_2 (8%)
TIME:
TEMP: 25—30°C
GAS FLOW: 200 sccm
PRESSURE: 0.35 Torr
POWER: 100 W
DISCUSSION:

Si, (100) wafers. Etch rates: (1) P:poly-Si — 1150 Å/min; (2) undoped poly-Si — 1050 Å/min; (3) SiO_2 — 120 Å/min and (4) photo resist — 190 Å/min. Selectivity of P:poly-Si to SiO_2 was 10:1. See SI-0286a for further discussion.

REF: Ibid.

SI-0290
ETCH NAME: Alumina
TYPE: Compound, polish
COMPOSITION:
x .... 0.05 μm Al_2O_3 (Linde B)
x .... glycerin
x .... soap (Castille)
TIME:
TEMP: RT
DISCUSSION:

Si, (111) wafers. Mixture used for polishing material on a Pitch Lap. A mixture of pitch and beeswax is poured and allowed to set on a brass Buehler lap platen. The wax is then cross-scored with "V" groove cuts and the slurry mixture shown dripped onto the lap with wafers held on the lap mounted on disc holders. This polishing method can produce polished surface planarity of <$^1/_4$ wavelength sodium light.

REF: George, D — personal communication, 1959

SI-0291
ETCH NAME: Silica
TYPE: Compound, polish
COMPOSITION:
x .... SiO_2 (sand)
x .... N_2
TIME:
TEMP: RT
DISCUSSION:

Si, (111) wafers and other orientations. Cutting channels, dicing, and bevelling of wafers done with an S S White abrasive unit. Reference shown used this method to edge bevel high power SCR wafers prior to final etch tuning with xHF:xHNO_3:xHAc, and coating exposed n-p-n junction with RTV.

REF: Topas, B et al — personal communication, 1974

SI-0440: Tarn, W H & Walker, P — personal application, 1955—1961

Si, (111) wafers fabricated as aluminum alloy transistors or diffused high power diodes. The S S White dental units used for sand cleaning teeth were converted for dicing silicon wafers with dry sand under nitrogen pressure using a mobile rubberized tube with replaceable metal tips of different orifice shapes. Wafers were wax mounted on steel discs and held on an x-y stage below the stationary system tip. To keep sand dry, the pressurized metal sand holder was wrapped with a heating mantle. Dice was square-cut with a positive side angled slope. This form of sandblast dicing introduces from 2—5 mils damage laterally into the silicon dice. Method used on diffused silicon diodes (Tarn), and transistors (Walker), also on germanium and other compound semiconductors (GaAs, InP, InAs) in development studies of new materials.

SI-0441: Tarn, W H & Walker, P — personal application, 1963—1965

Si, (111) wafers, with and without silicon epitaxy layers used in fabricating computer diodes. Large sandblast units used for general wafer surface cleaning to remove contamination. See Silicon Dioxide, SiO_2, section for use of bead blast sand units for cleaning vacuum system hardware.

SI-0292
ETCH NAME: Jello TIME:
TYPE: Colloid, mask TEMP: RT
COMPOSITION:
x .... Jello, lemon
DISCUSSION:

Si, (111) wafers. Jello used as a patterning mask similar to photo resist lacquers for aluminum evaporation before the advent of resists for Solid State device processing. Jello was painted onto surfaces, allowed to set, hand scribed for patterning, and the aluminum evaporated downward onto wafers from ''U'' shaped aluminum wire clips hung on tungsten heater coils. Lift-off after evaporation liquefied Jello with water for removal of excess metal. Although the experiment was successful, Jello is temperature-critical to prevent hardening and cracking. Various Jellos were evaluated with lemon giving best results. (*Note*: Other organics, including photo resist lacquers reduced to the colloidal state, would reduce grain size limitations, currently a problem in fine-line definition applications.)
REF: Brumfield, D & Walker, P — personal application, 1958

SI-0293
ETCH NAME: Iodine, vapor TIME:
TYPE: Halogen, cleaning TEMP:
COMPOSITION:
x .... *I_2, sat. sol.
x .... MeOH

*Saturate alcohol with iodine.

DISCUSSION:

Si, as p^+-n solar cells. Remove native oxide with HF dip and quench in solution shown. Chemisorption of iodine produces a p^+ surface film that is stable up to 500°C. Addition of a CdSe thin film on silicon structure produces a tunnel photo anode.
REF: Pinson, W E — *J Appl Phys*, 40,970(1982)

SI-0404
ETCH NAME: TIME:
TYPE: Acid, thinning TEMP:
COMPOSITION:
1 .... HF
5 .... HNO_3
1 .... CH_3COOH (HAc)
DISCUSSION:

Si, (100) wafers used in a study of metal decoration defects induced in heat treated wafers from tweezers (Cr, Ni, Fe, and Co contamination). Included dry O_2 oxidation at 1100°C, 30 min. After thinning with solution shown, Secco etch used at RT, 2 min and developed saucer-like etch pits where tweezers gripped wafers. (Secco etch: SI-0036). (*Note:* Similar structures, including drain-marks, observed in GaAs wafers after vertical etching in solutions which produced change in electrical characteristics of FET devices.)
REF: Staxy, W T et al — *J Electrochem Soc*, 129,1128(1982)

SI-0300
ETCH NAME: TIME: 10—45 min
TYPE: Alkali, preferential TEMP: Boiling
COMPOSITION:
387 g .... KOH
250 ml .... H_2O
30 g .... additive
DISCUSSION:

Si, (111), 10—20 Ω cm resistivity, n-type wafers used in a study of etch pits formed by controlled point damaging — concave plane structures vs. convex sphere structure planes. The initial method is called the Pennington Pit Method which is used in goniometer structure study. As further developed in this study for pit depth/size control, a diamond stylus with round shaft was held vertically in a nylon tube, slip-fit for rotation, with lead gram-weights on stylus top — amount of weight and number of stylus rotations in contact with the wafer surface used to establish damage depth, and etch time for final pit depth, which varies with solution mixture for pit shape and size. Shape progression of pit forming goes triangle — subtriangle — hexagon — subhexagon — to a degenerate subtriangle to a roughly circular pit. Progression varies with the additive to the hydroxide solution shown above, and may be triangular to subtriangular, only, triangle to degenerate circular, directly, depending on solution. All of the following additives were evaluated with a base alkali solution of KOH or NaOH + 20—30 g of solid chemicals or 25—50 ml of liquid, such as Br_2. Alcohol type additives were 125 ml. Study objective was to develop flat bottom pits for aluminum alloying, and determination of crystallographic planes developed in pit-side slopes vs. solution additive. (100) silicon was evaluated with 30% KOH/NaOH, and with some additives on germanium, (111), (100), (110), and (211) for similar pits, and light figure orientation. Silicon pit shapes established after 20 or 30 min etch time. With the KOH:Br_2 mixture, an additional 100 ml Br_2 added after 20 min. See SI-0092a-s for additional discussion.

	Etch time (min)	
Reference no. — additive	**20**	**30**
SI-0192f — I_2	subtriangle	deg. subtriangle
SI-0300 — S	hexagon	hexagon
SI-0302 — Zn	hexagon	deg. circular
SI-0304 — As_2O_3	hexagon	subhexagon
SI-0301 — $Al(OH)_3$)	subtriangle	subtriangle
SI-0303 — $Ba(OH)_2$	hexagon	deg. subhexagon
SI-0305 — $Ca(OH)_2$	subhexagon	deg. subhexagon
SI-0306 — $Sr(OH)_2.8H_2O$	hexagon	deg. subhexagon
SI-0307 — $AlCl_3.6H_2O$	subtriangle	hexagon
SI-0308 — NH_4Cl	hexagon	deg. subhexagon
SI-0309 — $CaCl_2.2H_2O$	subtriangle	hexagon
SI-0310 — LiCl	hexagon	deg. hexagon
SI-0311 — $NiCl_2.6H_2O$	subtriangle	subtriangle (rough)
SI-0312 — $K_2Cr_2O_7$	subtriangle	subtriangle
SI-0313 — $KClO_3$	subhexagon	deg. subhexagon
SI-0314 — $Mg(NO_3)_2.6H_2O$	subhexagon	subhexagon
SI-0315 — H_3BO_3	hexagon	subhexagon
SI-0316 — HPO_3	hexagon	subhexagon (rough)
SI-0317 — $KHC_8H_4O_4$	hexagon	subhexagon
SI-0318 — $K4Fe(CN)_3.3H_2O$	hexagon	subhexagon

Reference no. — additive	Etch time (min) 20	30
SI-0198b — KBr	hexagon	deg. subhexagon
SI-0198c — $FeCl_3.4H_2O$	hexagon	subhexagon
SI-0319 — HIO_3	erratic	erratic
SI-0332 — KI	hexagon	subhexagon
SI-0320 — $(NH_4)_2O_4H_4O_6$	triangle	triangle (rough)
SI-0321 — $Al_2(SO_4)_3SO_4.24H_2O$	subtriangle	subtriangle
SI-0198a — CHI_3 + glycerin	hexagon	hexagon
SI-0322 — $3CdSO_4.8H_2O$	triangle	deg. xxxx
SI-0092g — Br_2	triangle	triangle
SI-0092a — 30% KOH/NaOH	hexagon	hexagon
SI-0333 — 1:1 30% KOH:NaOH	hexagon	hexagon
SI-0334 — Bi_2O_3	erratic	erratic (no real pit)
SI-0335 — $ZnCl_3$	erratic	erratic
SI-0336 — $HgBr_2$	erratic	erratic (no real pit)
SI-0337 — Sb_2O_3	erratic	deg. triangle (?)
SI-0338 — $SrBr_2$	erratic	erratic (no real pit)
SI-0339 — $ZrSC_4.7H_2O$	erratic	erratic
SI-0323 — Benzene	circular	circular (both rough)
SI-0324 — Ethyl alcohol	triangle	triangle
SI-0325 — Methyl alcohol	subtriangle	subtriangle
SI-0326 — Ethylene glycol	subtriangle	subtriangle
SI-0327 — Glycerin	subtriangle	subtriangle

	20 min etch	(111) etch rate
SI-0329 — 15% $K_3CO_3.{}^1/_2H_2O$:5 g $Cd(OH)_4$	triangle	0.1 mil/15 min
SI-0330 — 15% Na_2CO_3:5 g $Ce(OH)_4$	triangle	0.1 mil/10 min
SI-0331 — 1 *M* KCN:20 g I_2	triangle	0.1 mil/5 min
SI-0340 — 1 *M* KCN	cleaning, only	
SI-0341 — 30% $K_3CO_3.1\,{}^1/_2H_2O$	cleaning, only	

All wafers, dice or single crystal spheres were free etching in the boiling solutions. Square dice, random orientation cut, or edge oriented (111) triangles and diamond shapes also used. (100) silicon, as (100)/(110) edge oriented. Both individual dice and wafers were opposed pit damaged for collector/emitter transistor configuration damage . . . damage and pit-etch the collector first; then damage emitter, and etch both to final size for a specific base width. Transmitted white light with a Richart-Zetopan microscope was used to establish base width by color. Pit bottom planarity to within 0.1 μm on (111) surfaces; a nominal 0.2 μm curve on (100). Oriented dice edges reduced the preferential etch incursion observed in random cut edges. For individual die or wafers, a back-reflecting mirror was used to align opposed pits with the vertical stylus; a similar system used with a panograph for device array pitting of wafers by ink dotting both wafer sides. This was one of the first instances

of fabricating discrete, individual devices on a wafer (1957), before the advent of photolithography. Silicon aluminum alloy transistors were fabricated with optimum electrical characteristics vs. individual dice with sandblasted pits (which have curved bottoms). Single crystal spheres were etched for finite crystal form in several of the solutions shown to obtain convex planes as against the concave, pit side-slope planes on both silicon and germanium. Walker (referenced below) also fabricated operational diffused spherical radiation detectors and cylindrical lithium-drift germanium radiation detectors, circa 1968, using the above described techniques.
REF: Walker, P & Waters, W P — silicon development, 1956—1959
SI-0342: Waters, W P & Walker, P — Electrochem Soc Meet, Spring 1957

Si, (111) wafers fabricated as aluminum alloy transistors with a controlled pit damage technique and etching with hot alkali solutions. Results showed better electrical characterization due to improved pit bottom planarity.

SI-0400
ETCH NAME: Wright etch — TIME:
TYPE: Acid, preferential — TEMP:
COMPOSITION:
(1) x .... BHF — (2) Wright etch (SI-0045)
DISCUSSION:

Si, (100) p- and n-type substrates for AsH_3 doped poly-Si epitaxy thin films, n-type 0.2—2.0 Ω cm with Si_3N_4 cap. A study of Si_3N_4 gettering effect on silicon. (1) Clean with solution (1), then (2) defect etch with solution (2).
REF: Chen, M C & Silvestri, V J — *J Electrochem Soc,* 129,1294(1982)

SI-0401
ETCH NAME: Hydrofluoric acid — TIME:
TYPE: Electrolytic, decoration — TEMP: RT
COMPOSITION: — ANODE: Si
x .... 5% HF — CATHODE:
POWER: 3 V
DISCUSSION:

Si, as poly-Si films on Si (100) substrates. Solution can be used to decorate defects by a combination of staining with light etching.
REF: Chen, M C et al — *J Electrochem Soc,* 128,389(1981)

SI-0405
ETCH NAME: RIE — TIME:
TYPE: Ionized gas, selective — TEMP:
COMPOSITION: — ANODE:
x .... CF_4 — CATHODE:
x .... H_2 — POWER:
DISCUSSION:

Si, (100) wafers with SiO_2 thin films. This RIE gas mixture used to selectively etch SiO_2.
REF: Mephrath, L & Petrillo, E J — *J Electrochem Soc,* 129,2282(1982)

SI-0036b
ETCH NAME:
TYPE: Acid, polish
COMPOSITION:
2 .... HF
3 .... HNO_3
2 .... CH_3COOH (HAc)

TIME: 2—3 min
TEMP: RT

DISCUSSION:
Si, (100) wafers used in developing the Secco etch (SI-0036a). Solution used to etch polish wafers prior to preferential etching with Secco etch. Material was 1—10 K Ω cm resistivity. The 1—300 Ω cm etched slower in Secco — 20 min below 300 Ω cm; 30—35 min above, for equivalent defect development.
REF: Ibid.

SI-0045b
ETCH NAME:
TYPE: Acid, cleaning
COMPOSITION:

TIME: (1) 20 sec
(2) 10 min
TEMP: RT Boiling

(1) 1 .... HF
50 .... H_2O

(2) x .... H_2SO_4

DISCUSSION:
Si, (111) and (100) wafers, p- and n-type, 0.2—20 Ω cm resistivity, both CZ and FZ grown ingots, used in a study of oxidation defects. Etch clean wafers in solution (1), then solution (2). Water wash after each etch step in UPDI for 5 min with ultrasonic agitation. Use Wright etch (SI-0045a) to develop defects, stacking faults, swirl patterns, and striations caused by oxidation.
REF: Ibid.

SI-0406
ETCH NAME: Sodium hydroxide
TYPE: Alkali, removal
COMPOSITION:
x .... 0.2 *M* NaOH

TIME:
TEMP:

DISCUSSION:
a-Si:H thin films grown on SiO_2, Al_2O_3, and ZrO_2 substrates in a study of surface coating protection against chemical attack. Other materials evaluated were Al, TiN, Ag, and PbS. A collimated light beam was used with aluminum during etching; Ag etched in 1 *M* HNO_3 also with light.
REF: Martin, P J et al — *J Vac Sci Technol,* A2(2),341(1984)

SI-0407
ETCH NAME: Chlorine
TYPE: Gas, preferential
COMPOSITION:
x .... Cl_2 (0.1—0.6%)
x .... He

TIME: 30 min
TEMP: 900—1100°C
RATE: 0.25—2.5 μm/min

DISCUSSION:
Si, (111) wafers used in an etch reaction study. Wafers polished with Lustrox, then cleaned in furnace tube at 1100°C, 30 min in atmosphere of H_2 + 0.1% Pd to remove residual oxide. Switched to Cl_2/He atmosphere mixtures for etching. The 0.2% Cl_2 at 1000—1100°C gave polished surfaces; below 1000°C with higher Cl_2 concentrations gave

rough surfaces as steps and pits. With an O_2 leak, carbon particles from the susceptor plate deposited on silicon surfaces. Rate control is chemical action, not diffusion. (*Note:* This is an example of using gases in their molecular form for etching, rather than as ionized gases.)
REF: Dismukes, J P & Ulmer, R — *J Electrochem Soc,* 118,634(1971)
SI-0366: Schmidt, P F & Blomgren, H — *J Electrochem Soc,* 106,694(1959)

Si, (111) p-type wafers. A study of potential measurements during jet-etching of both silicon and germanium.

SI-0367: Schmidt, P F & Keiper, D A — *J Electrochem Soc,* 106,592(1959)

Si, (111) n-type wafers. Used in a study of the jet-etching of n-type silicon.

SI-0368: Marks, A — Electrochem Soc Meet, Washington, D.C., 12—16 May 1957

Si, (111) wafers. A description of electrochemical jet-etching of silicon.

SI-0369
ETCH NAME: TIME:
TYPE: Metal, contacts TEMP: RT
COMPOSITION:
x .... Cu, Al, etc.
DISCUSSION:

Si, (111) wafers used in a study of forming metal surface contacts by the dry friction method. Apply a flat silicon surface against a rotating surface of a metal to form a smear-type metal contact area. A rough or disturbed silicon surface made best contacts. The following metals were evaluated: Cu, Al, Ni, Mo, Fe, Sn, Ta, brass, and bronze.

REF: Kirvalidze, I D & Zhukov, V F — *Fiz Tverd Tela,* 1,1583(1959)
SI-0379: Dash, W C — *J Appl Phys,* 29,228(1958)

Si, (111) wafers. A study of the development of distortion layers on silicon surfaces by grinding and mechanical polishing. The mechanical working of a surface does not deform the bulk material other than a small surface fraction — up to about 150 Å deep after a final polish. (*Note:* Damage depth is related to the abrasive particle diameter and physical grain shape of an abrasive, type cutting edge of the grain, abrasive hardness and breakdown characteristics during use, and can be as much as 3 mils in depth of induced damage even after final optical polish. See Silicon Dioxide section with reference to the DeBeye layer.)

SI-0377: Allen, R B et al — *J Appl Phys,* 31,334(1960)

Si, (111) wafers used in a study of oxide layers on the diffusion of phosphorus in silicon. Wafers processed at 1150°C, 2.1×10^{15} cm/sec^2 with 1.4 eV power.

SI-0376: Hartke, J L — *J Appl Phys,* 30,1649(1959)

Si, (111) wafers. A study of the effects of oxygen in silicon on the diffusion of phosphorus.

SI-0375: Logan, R A & Peters, A J — *J Appl Phys,* 28,1419(1957)

Si, (111) wafers. A study of the effects of oxygen on etch pit formation in silicon.

SI-0380: Kalnajs, J & Samakula, A — *J Phys Chem Solids,* 6,46(1958)

Si, (111) single crystals. A study of oxygen impurity in silicon.

SI-0360: Czaja, W — *J Appl Phys,* 37,918(1966)

Si, (111) wafers, phosphorus doped for 3 to 4 μm deep junctions used in a study of partial dislocations with observation under SEM. Used an iodine etch, 30 sec to develop slip patterns. Heat treatment in presence of O_2 can cause slip.

SI-0361: Rosenzweig, W — *ISSS Trans Nucl Sci,* NS-12,18(1965)

Si, devices studied for the effects of space radiation.

SI-0362
ETCH NAME: Ultrasonic TIME:
TYPE: Vibration, preferential TEMP:
COMPOSITION: POWER:
x .... Ba_2TiO_3
DISCUSSION:

Si, specimens of various orientations used in a study of the development of etched figures with ultrasonic vibration. (*Note:* Ultrasonic vibration is widely used for wafers and parts cleaning, and induced damage from a barium titanate transducer element in ultrasonic generators with wafers in acids solutions has been studied.)

REF: Bagdasarov, K S — *Sov Phys-Cryst,* 2,309(1957)

SI-0363: Belyustin, A V — *Sov Phys-Cryst,* 4,569(1959)

Si, and other materials used in a study of face solubility of single crystals. Growth pyramids on surfaces vary with impurity content and stress. As solution concentration decreases upward in a gravitational field crystal faces higher in the solution will show less solubility. Pyramid structures are most stable and will develop on otherwise unstable crystal surfaces.

SI-0383: Dash, W C — *Phys Rev Lett,* 1,400(1953)

Si, (111) wafers and other orientations. Development of prismatic dislocation loops in silicon.

SI-0384: Stephen, W E & Myerhof, W E — Univ Penn Div 14, NDRC #563, 1945, 16

Si, specimens. A study of geometric structure on silicon surfaces.

SI-0385: Vogel, F L & Lovell, C — *J Appl Phys,* 27,1413(1956)

Si, (111) wafers. The development of dislocation etch pits in silicon.

SI-0386: Trakhenberg, A D & Fainstein, S M — *Sov Phys,* 1,335(1959)

Si, (111) wafers. Dislocation development in both silicon and germanium.

SI-0387: Roll, F — *Z Metallkd,* 30,205(1938)

Si, specimens. A study of etching silicon sections.

SI-0388: Hopkins, R L — *J Appl Phys,* 29,1378(1958)

Si, (111) wafers. Appearance of light diffraction striations in silicon.

SI-0378: Wagner, C — *J Appl Phys,* 29,1295(1958)

Si, (111) wafers. A study of surface passivation of silicon during oxidation at elevated temperatures.

SI-0374: Lederhandler, S & Patel, J R — *Phys Rev,* 108,239(1957)

Si, (111) wafers. A study of the behavior of oxygen in plastically deformed silicon.

SI-0373: Hrostowski, H J & Kaiser, R H — *Phys Rev Lett,* 1,199(1958)

Si, (111) wafers. The infrared spectra of heat treatment centers in silicon and different temperature levels. At 450°C, there are 4 band groups, probably SiO_2; at 650°C, these bands disappear; at 1000°C oxygen becomes an acceptor in silicon, and between 300-600°C oxygen acts as a donor in silicon. (*Note:* Float Zone (FZ) (111) oriented ingots of 80,000 Ω cm p-type resistivity showed silica blebs of SiO_4 as segregated agglomerates in wafers.)

SI-0372: Kaiser, W — *Phys Rev,* 105,1751(1957)

Si, (111) wafers. A study of electrical and optical properties of silicon after heat treatment. Oxygen is proportional to the 4th power of concentration in silicon. At 450°C oxygen acts as a donor as SiO_2? and as SiO_4 at 1000°C. After 20 h at 1000°C oxygen is stabilized, and also stabilizes at 1100°C.

SI-0371: Faessler, A & Kramer, H — *Ann Phys (Leipzig) Folge,* 7,263(1959)

Si, (111), wafers. Establishment of the existence of Si, Si_2O, and Si_2O_3 in silicon as oxidation products. Representative as a compound with general formula: SiOR where R = C_2H_5 or $SiO.Si_xO$. (*Note:* Silicon oxides are normally as SiO_x, not Si_xO.)

SI-0170: Rill, D E — *Phys Rev,* 114,1414(1959)

Si, (111) wafers. Used CP4 to etch material for defects after electron bombardment.
SI-0381: Michel, W & Schmidt, P F — *J Electrochem Soc,* 104,230(1957)

Si, (111) wafers. A study of the anodic oxidation of silicon.
SI-0223b: Ibid.

Si, (111) wafers. Anodic oxidation used as a junction delineation etch for silicon p-n junctions.

SI-0382
ETCH NAME: Heat TIME:
TYPE: Thermal, junction forming TEMP:
COMPOSITION:

x .... heat

DISCUSSION:

Si, (111) wafers. A study of the formation of large area p-n junctions in silicon by surface melting. (*Note:* The formation of p-n junctions in n-type silicon by alloying of evaporated or thin sheet aluminum with thermal drive-in on a total wafer surface in fabricating silicon high power diodes or as small aluminum or Al:Si(5%) spherical pellets for the Sphere Alloy Zener process are typical examples of this alloying procedure.)
REF: Billig, E & Gasson, D B — *J Appl Phys,* 28,1242(1957)
SI-0389: Kramer, G — *IEEE Nucl Sci,* NS-13,104(1966)

Si, device fabricated as a fast-neutron dosimeter. Material was Float Zone (FZ) grown (111), 50 Ω cm resistivity with boron and phosphorus doping diffusion. Report on device characteristics and applications.
SI-0390: Imai, T — *Jpn J Appl Phys,* 2,463(1963)

Si, fabricated as a mesa Esaki diode. A study of the effect of uniaxial stress on junctions with varied doping levels. Stress applied was 20 to 200 g. Higher impurity doping levels showed increased stress primarily at p-n junction zones. Both tension and compression stress observed.
SI-0391: Danes, L W — *J Sci Instr,* 35,423(1958)

Si and Gs specimens used in a study of metallic contacts. Shows conductance of different metal lead wires, and use of copper plating on germanium for tinning contacts.

SI-0392
ETCH NAME: TIME: 100 sec
TYPE: Acid, removal TEMP: RT
COMPOSITION:

x .... HCl
x .... HNO_3
x .... H_2O

DISCUSSION:

Si, (100) wafers used as substrates for RF sputter of SeGe thin films, 3000 Å thick. After deposition, heat to the 150°C glassification point to homogenize as an amorphous glassy thin film. Plate 100 Å of silver from a silver solution applied in the dark, RT, 3 min, and drive-in with Hg lamp at 200 W. Remove residual silver with solution shown. A study of silver photo-doping of the SeGe films.
REF: Zebutsu, S — *J Appl Phys Lett,* 39,969(1981)

SI-0393
ETCH NAME: TIME: (1) 30 min
TYPE: Acid, polish (2) 3 min
COMPOSITION: TEMP: RT RT

(1) 10 .... HF
1 .... HNO_3

(2) 1 .... HF
2 .... HNO_3
1 .... HAc

DISCUSSION:

Si and Ge wafers cut from Float Zone (FZ) ingots with 10^4 cm^{-2} grown-in dislocations used in a study of yield point and mobility of dislocations. Both materials were p-type, 200-1000 Ω cm resistivity, and Ge 40 Ω cm. After diamond saw cutting of wafers, mechanically lap with 3 μm, then 1 μm diamond paste. Etch polish germanium in solution (1); silicon in solution (2). Both materials showed slip in the ⟨123⟩ direction. (*Note:* The (321) planes are isometric system, normal class hexoctahedral planes, and not a common slip plane nor a common plane in nature due to high solubility factors of such planes.)

REF: Stevens, D'S & Tiersten, H F et al — *J Appl Phys,* 54,1815(1983)

SI-0394
ETCH NAME: TIME:
TYPE: Acid, thinning TEMP:
COMPOSITION:

5 .... HF
1 .... HNO_3
5 .... HAc

DISCUSSION:

Si, (100) wafers used as substrates for LPCVD growth of n-type poly-Si, and substrates coated with Si_3N_4 prior to epitaxy. Solution used to etch thin and removed the substrate for microscope study of deposited films in a form of the float-off technique for TEM observation of thin films.

REF: Hottier, F & Cadoret, R — *J Cryst Growth,* 61,244(1983)

SI-0395
ETCH NAME: Hydrofluoric acid TIME:
TYPE: Acid, oxide removal TEMP: RT
COMPOSITION:

1 .... HF
20 .... H_2O

DISCUSSION:

Si, (100) n-type wafers, 10 Ω cm resistivity with P^+ ion implantation. Remove native oxide after I^2 with solution shown, then evaporate Ti and anneal to form TiSi. A study of TiSi growth on ion implanted Si.

REF: Revesz, P et al — *J Appl Phys,* 54,1860(1983)

SI-0396: Suzuki, S & Itoh, T — *J Appl Phys,* 54,1466(1983)

Si, (111) n-type wafers, 100 Ω cm resistivity. Clean wafers by (1) solvent degrease, (2) oxidize in solution shown at 60°C, (3) strip oxide with HF, and (4) rinse in DI water.

SI-0397: Maeda, Y et al — *J Cryst Growth,* 65,331(1983)

Si, grown as poly-Si sheet for use as solar cell material directly as grown. A method developed to reduce cost and time: funnel a hot melt of silicon onto a heated, spinning graphite wheel. Silicon spreads and freezes as a thin sheet 0.1 to 0.5 mm thick. Four 50 mm square sheets can be poured per minute with a grain size of about 3 mm.

SI-0408
ETCH NAME: RIE
TYPE: Ionized gas, removal
COMPOSITION:
x .... NF_3
x .... O_2
TIME:
TEMP:
ANODE:
CATHODE:
POWER: DC
DISCUSSION:

Si, as poly-Si specimens. Specimens were DC etched in the gas mixture shown in an etch development study of this material.

REF: Honda, T & Brandt, W W — *J Electrochem Soc,* 131,2667(1984)

SI-0407
ETCH NAME:
TYPE: Acid, cleaning
TIME:
TEMP:
COMPOSITION:
(1) 1 .... NH_4OH
1 .... H_2O_2
5 .... H_2O
(2) x .... H_2SO_4
(3) x .... HF
DISCUSSION:

Si, (100) wafers used for MOCVD growth of SiO_2 thin films at 250 to 300°C with plasma enhancement of MOCVD. Wafers were cleaned in the solutions in the order shown prior to MOCVD.

REF: Pandl, K P et al — *J Electron Mater,* 13,593(1984)

SI-0399: Ben-Dor, L et al — *J Electron Mater,* 13,263(1984)

Si, (111) degenerate n-type wafers used for MOCVD growth of ZrO_2 thin films in a study of these film interactions with Si. Clean wafers in $1H_2SO_4$:$1H_2O_2$ with final dip in 1HF:$1H_2O$ prior to MOCVD.

SI-0198
ETCH NAME: RIE
TYPE: Ionized gas, structuring
COMPOSITION:
x .... 0.9 He
x .... 0.1 Cl_2
TIME:
TEMP:
GAS FLOW:
PRESSURE:
POWER:
DISCUSSION:

Si, as poly-Si thin film on silicon wafers. A mesa-like "T" was gas etch structured in a channel by plasma etching. Included the use of hot HCl vapor and Cl_2 gas, alone, for structuring.

REF: Zarowin, C B — *J Electrochem Soc,* 130,1144(1983)

SI-0429: Moore, J B & McCaldin, S C — *J Electrochem Soc,* 124,625(1977)

Si, (111) wafers used for Be evaporation with and without an SiO_2 thin film cover used in a study of Be reactions. Clean Si substrates: (1) TCE degrease, rinse in acetone, then MeOH, then DI H_2O, and (2) 10% HF rinse immediately before Be evaporation. Forms Be_2SiO_4 (with SiO_2 present) and BeO + Si on silicon surface only. Above 400°C an insoluble residue forms (BeO?). Remove excess Be with HCl. (*Note:* See Quartz, Beryllium, and Aluminum Alloys sections for other applications of beryllium on silicon and silicon dioxide.)

SI-0419: Ikuta, K & Ohara, T — *Jpn J Appl Phys,* 23,984(1984)

Si, (100), p-type, 4.4—5.7 Ω cm resistivity wafers used in a study of defects versus argon annealing. Argon used to induce bulk defects and remove surface defects. Wright etch used to step-etch in 5 μm increments for depth profiling.

SI-0418: Sawada, R — *Jpn J Appl Phys,* 23,959(1984)

Si, (100), p-type, 1—30 Ω cm resistivity wafers cut from Czochralski (CZ) grown ingots used in a damage study. Prepared with and without SiO_2 or SiO_2 thin films added after damaging, and included thermal annealing as an induced and enhanced stress mechanism. Diamond scribe wafer backs with various pressure loads, then bend around a ⟨110⟩ plane direction to neutralize damage induced dislocations. Macroscopic strain disappears with oxidation, but microstrain remains near cut damage even after thermal annealing. Wright etch produced sauce-like etch pits correlated with stacking faults.
SI-0421: Woods, H & Silverman, R — *J Appl Phys,* 38,419(1967)

Si, thin film epitaxy on silicon wafers. Describes metallographic techniques used to study Si/Si reactions.
SI-0422: Jungbluth, D — *J Appl Phys,* 38,133(1967)

Si wafers with double-diffused epitaxy layers. A study of defects generated by diffusion.

SI-0423
ETCH NAME: Implantation — TIME:
TYPE: I^2, defects — TEMP: − 195°C
COMPOSITION:

x .... Si^+ ions

DISCUSSION:

Si, (100), n-type 2—6 Ω cm resistivity wafers cut from Czochralski (CZ) grown ingots. Wafers Si^+ ion implanted at LN_2 temperature. Dislocations were up to 4500 Å in depth from surfaces. Surface structure was as a-Si on the immediate surface with a zone of dislocation loops below and then bulk Si. (*Note:* This type damage structure is atypical of implantation damage.)
REF: Marayan, J — *J Appl Phys,* 7,564(1985)
SI-0420: Rudder, R A — *J Vac Sci Technol,* A2(2),326(1984)

Si, (100) wafers and fused quartz blanks used as substrates for deposition of a-Si:H and a-Ge:H thin films. Clean quartz in x% CrO_3:H_2SO_4, RT, 60 sec (QTZ-0009). Both substrates cleaned before deposition: (1) TCE degrease, (2) acetone, then MeOH rinse and DI H_2O rinse. This was a study of hydrogenated thin films. After film growth they were cleaned with HF, RT, 30 sec.

SI-0413
ETCH NAME: BHF — TIME:
TYPE: Acid, cleaning — TEMP: 0°C
COMPOSITION:

134 ml .... HF
452 g .... NH_4F
625 ml .... H_2O

DISCUSSION:

Si thin film epitaxy grown on (100) silicon wafer substrates, then coated with TiN thin film used in a reaction and structure study. Solution used to both remove TiN films and clean poly-Si epitaxy surfaces. See TIN-0003.
REF: Whittmer, N et al — *J Appl Phys,* 54,1423(1983)

SI-0414
ETCH NAME: — TIME:
TYPE: Acid, cleaning — TEMP:
COMPOSITION:

(1)	(2)	(3)
1 .... H_2SO_4	x .... HF	1 .... HF
1 .... H_2O_2		100 .... NH_4F(40%)

DISCUSSION:

Si, (100), n-type, 5—9 Ω cm resistivity wafers used in developing thin gate dielectrics for MOSFETs using oxide/nitride. Wafers cleaned in solution (1), then (2) prior to SiO_2 deposition at 950°C with 30 min anneal, then 925°C Si_3N_4 deposition with NH_3 for 2 h. BHF solution (3) used to etch nitride and oxide. Nitride etch rate 0.14 nm/sec, increasing with higher nitriding temperatures.

REF: Wong, S S et al — *J Electrochem Soc,* 130,1139(1983)

SI-0428: Turner, D R — *J Electrochem Soc,* 106,786(1959)

Si, (111) wafers used in a study of the electroplating of metal contacts on Si and Ge. See GE-0285.

SI-0415: Kotani, H et al — *J Electrochem Soc,* 130,645(1983)

Si, (100) wafers used as substrates in studying thin film deposits as inter-connects. The following thin films were evaluated: P:SiN, LP:PSG, Al, AlSi, and AlCuSi. CF_4:Ar:O_2 plasma etching used for controlled removal of films at 200°C with 1.7 Pa pressure and 0.5 W/cm^2 power. Plasma etch rates were (1) Al at 100 Å/min; P:SiN at 70 Å/min; LP:PSG at 60 Å/min. Plasma also used to etch via holes through film/Si. PSG via holes also etched with HF.

SI-0416

ETCH NAME: Pressure — TIME:

TYPE: Pressure, forming — TEMP:

COMPOSITION: — PRESSURE: See discussion

x .... pressure

DISCUSSION:

Si material blanks used in a study of high pressure forms of silicon with change in crystal structure:

Si I: standard temperature and pressure — isometric, cubic . . . (''diamond'')
Si II: 10 GPa pressure — B-Si — tetragonal (similar to B-Tin)
Si III: 16 GPa pressure — Theta-Si — hexagonal with one atom/unit cell.
Si IV: unk
Si V: also referenced as Theta-Si
Si VI: 34—36 GPa — Si with unknown and unexpected form — tetragonal?
Si VII: unk
Si VIII: above 39 GPa — hexagonal, close-packed (hcp)

REF: Sharama, S & Sikka, S K — *J Phys Chem Solids,* 46,477(1985)

SI-0417

ETCH NAME: Helium — TIME:

TYPE: Gas, cleaning — TEMP:

COMPOSITION:

x .... He

DISCUSSION:

Si, (111) wafers cleaned with helium ''atom impact'' prior to epitaxy growth of n-type silicon, 5 Ω cm resistivity, 20 μm thick.

REF: Nishigaki, S et al — *Jpn J Appl Phys,* 23,683(1984)

SI-0425: O'Hara, S — *J Appl Phys,* 35,409(1964)

Si and Ge grown as dendritic ribbon crystal and used in a dislocation study. Sirtl etch used on silicon; CP4 used on germanium with 0.6 ml Br_2/100 ml HNO_3.

SI-00426a: Chu, T L & Gavaler, J R — *Phil Mag,* 10,1064(1964)

Si, (111) wafers with "vapor grown" thin film silicon. Both Sirtl and Dash etches used in studying stacking faults.

SI-0426b: Ibid.

Si, (111) wafers with "vapor grown" thin film silicon. Both Sirtl and Dash etches used to study stacking faults.

SI-0427: Turner, D R — *J Electrochem Soc,* 106,701(1959)

Si, (111) wafers with p-n junctions. A study of the use of electrochemical displacement plating solutions to delineate junctions.

SI-0430a

ETCH NAME: TIME:

TYPE: Acid, cleaning TEMP:

COMPOSITION:

(1) x .... H_2SO_4 (2) x .... H_2O_2 (3) x .... HF

x .... H_2O_2

DISCUSSION:

Si, (100) p-type, 2 Ω cm resistivity wafers used in an evaluation of RIE contamination. Wafers were cleaned in order in the three solutions shown prior to RIE, and in solutions (1) and (2) after RIE. See SI-0430.

REF: Ephrath, L M & Bennett, R S — *J Electrochem Soc,* 129,1822(1982)

SI-0430b

ETCH NAME: RIE TIME: 5 min

TYPE: Ionized gas, removal TEMP:

COMPOSITION: GAS FLOW:

(1) x .... CF_4 ANODE:

(2) x .... $CF_4 + O_2$ CATHODE: DC 200 V

(3) x .... $CF_4 + H_2$ PRESSURE: 25 mTorr

POWER: 0.5/cm^2

DISCUSSION:

Si, (100) wafers used in a study of RIE contamination. Gases (1) and (2) used for Si, poly-Si, and polycides (WSi_2/Poly-Si). Gases (3) for selective etching of SiO_2 or Si_3N_4 on VLSI chips. Etch rates are in nm/min. Evaluation done with devices structured as MOS capacitors . . . Si oxidized with O_2 at 1000°C for 35 nm thickness and evaporated 35 mil Al dots. Defects in material were under 100/cm^2.

REF: Ibid.

SI-0435

ETCH NAME: Sulfur TIME:

TYPE: Metal, diffusion TEMP:

COMPOSITION:

x .... S

DISCUSSION:

Si, p-type wafers used in a study of n-type sulfur diffusion in silicon. Introduces two donor levels in forbidden band at 0.18 and 0.37 eV. Does not work with n-type silicon. Appears to be a substitutional, interstitial impurity with $^1/_{50}$ the solubility of O_2 in silicon.

REF: Carlson, R O et al — *J Phys Chem Solids,* 8,8183(1959)

SI-0436
ETCH NAME: Gases TIME:
TYPE: Gas, contamination TEMP:
COMPOSITION:
x .... gases
DISCUSSION:
Si, (100) wafers fabricated as VLSI and VHSIC devices. A study of methods of analyzing process gases introducing contamination during the processing of these devices.
REF: Koll, W — *Solid State Technol,* 1984, 220

SI-0437
ETCH NAME: Argon, wet TIME: 50 min
TYPE: Gas, alteration TEMP: 800—1300°C
COMPOSITION:
x .... Ar
x .... H_2O, vapor
DISCUSSION:
Si, wafers from p-type Float Zone (FZ) grown ingots with and without grown-in dislocations to 10^6 cm^{-2} concentration. Bars were cut and mechanically lapped with SiC, then cleaned; (1) EOH rinse, (2) TCE degrease, and (3) rinse in double-distilled H_2O. After wet argon treatment, quench to RT in water. Re-mechanically polish and etch in CP4 for dislocations and evaluate photoluminescence. Treatment converts wafers to high resistivity n-type with 7000 Ω cm.
REF: Nakashima, K — *Jpn J Appl Phys,* 23,622(1984)

SI-0440
ETCH NAME: TIME:
TYPE: Acid, thinning TEMP: RT
COMPOSITION:
(1) 1 .... HF
3 .... HNO_3
(2) 0.25 g .... I_2
110 ml .... CH_3COOH (HAc)
DISCUSSION:
Si, (100) and (111) wafers both n- and p-type, 3—5 Ω cm resistivity used as substrates in study of nickel silicide. 300 Å Ni evaporation followed by As_4 ion implantation, and N_2 anneal: 400 and 600°C as NiSi; 800°C, $NiSi_2$. To prepare specimens for TEM study, silicon substrate was removed: (1) abrasive lap in KOH; (2) etch in solution (2) and (3) in solution (1) using a wax moat on material surface. Silicide grain size and quantity varied with As flux levels of 10^{15}, 10^{16}, and $10^{19}/cm^2$ . . . as low a-Ni grains; medium grains of Ni_2Si and high Ni_2Si grains, respectively.
REF: Chen, L J & Hou, C Y — *Thin Solid Films,* 104,167(1983)

SI-0439
ETCH NAME: P4A TIME:
TYPE: Acid, jet thinning TEMP:
COMPOSITION:
3 .... HF
5 .... HNO_3
3 .... HAc
DISCUSSION:
Si, (100) wafers used in a study of wet oxidation of poly-Si using a deposited structure as poly-Si/SiO_2/Si(100). The poly-Si was phosphorus doped at 950°C, 30 min for 28.3 Ω/

sq resistivity, then wet oxidized with O_2 bubbled through 95°C H_2O, followed by annealing at 800°C, 30 and 60 min, and 1100°C, 5 min. Some specimens were ⟨110⟩ sectioned with Ar^+ ion thinning, and silicon removed with CP4A by gravity jet etching for film TEM study. The 800°C, 30 min specimens showed single crystal aggregates at grain boundaries and separate grains with loss of poly-Si thickness relative to anneal schedules.
REF: Bravman, J L & Sinclair, R — *Thin Solid Films,* 104,153(1983)

SILICON ALLOYS, SiMx

General: Silicon does not occur in nature as a metallic compound although it is the major mineral former as an oxide or silicate with SiO_2^-, SiO_3^-, SiO_4^-, etc., radicals. The silicate group comprises over 60% of all known natural minerals.

In the processing of silica, one by-product of reduction is known as ferrosilicon, FeSi — a hard, brittle form of iron with 2—10% silicon as an initial "pig". The material is then added to irons and steels, normally, as a 1—3% mixture for hardening. It also is added to aluminum or copper for the same purpose, as well as to other metal alloys.

Technical Application: The most widely used silicon alloy in Solid State processing is aluminum:silicon, AlSi(3—5%) used as a p-type dopant in fabricating silicon alloy devices: diodes, zeners, transistors. It is used as thin sheets, cut pre-forms or spheres — the Sphere Alloy Zener process. When pure aluminum is alloyed into silicon it forms a mixture of silicon needles in an aluminum matrix and, in order to reduce the amount of silicon removed from the substrate in the alloy formation, AlSi pre-mixed material is used. It is a brittle compound in comparison to aluminum. Other alloys, such as silicon:tin, SiSn, are under evaluation as thin film deposits in device structuring.

Silicon and germanium are the two elemental semiconductor materials, are isomorphous, and have been grown as single crystal alloy ingots in the complete range from 0—100% Si or Ge. The oxides and nitrides of silicon are thin film surface coatings with a variety of applications, and covered in individual sections in this book.

Etching: Mixed acids, variable by alloy.

SILICON ALLOY ETCHANTS

Silicon Aluminum: See Aluminum Alloys
Silicon Germanium
Silicon Iron: See Iron Alloys
Silicon Tin

SILICON GERMANIUM ETCHANTS

SIGE-0001
ETCH NAME: CP4 TIME:
TYPE: Acid, preferential TEMP: RT
COMPOSITION:
30 .... HF
50 .... HNO_3
30 .... HAc
0.6 .... Br_2
DISCUSSION:

Si:Ge, single crystal ingots of various compositions and as cut wafers. Silicon does not etch as rapidly in CP4 as does germanium. Developed striation layers in ingots with dislocation etch pits parallel to stria in some of the alloy mixtures.

REF: Toxen, A M — *Phys Rev,* 122,450(1961)
SIGE-0003: Goss, A J et al — *Acta Metall,* 4,333(1956)

Si:Ge single crystal alloy ingots in a study of composition fluctuations and dislocations. Silicon etches more slowly and produces stria patterns. Solution also developed edge dislocations.

SIGE-0002
ETCH NAME: TIME:
TYPE: Acid, removal TEMP:
COMPOSITION:
x .... HF
x .... HNO_3
DISCUSSION:

Si, thin film deposition on germanium substrates as mc-SiGe and as an Si_xGe_{1-x} thin film alloy for use as photovoltaic devices. Silicon deposition was by sputter-assisted plasma CVD (SPCVD).

REF: Kohno, K et al — *Jpn J Appl Phys,* 23,L674(1984)

SILICON TIN ETCHANTS

SISN-0001
ETCH NAME: TIME:
TYPE: Acid, removal TEMP: RT
COMPOSITION:
x .... HF
x .... HNO_3
DISCUSSION:

SiSn, thin films deposited on (100) silicon wafers used as substrates in fabricating photovoltaic devices. Film growth was by sputter-assisted plasma CVD (SPCVD). Also worked with thin films of SiN and uc-Si_xGe_{1-x}.

REF: Kohno, K et al — *Jpn J Appl Phys,* 23,L674(1984)

PHYSICAL PROPERTIES OF SILICON CARBIDE, SiC

Classification	Carbide
Atomic numbers	14 & 6
Atomic weight	40
Melting point (°C)	2700
Boiling point (°C)	2100 (sublimes)
Density (g/cm^3)	3.21
Oxidation point (°C in air/O_2)	1650
Thermal conductivity (Wm/K) 1000°C	23.7
Mean specific heat (Kg/K) 25—1300°C	795.5
Coefficient of thermal linear expansion ($\times 10^6$ cm/cm/°C) 20—1000°C — SiC	5.2
— 90% SiC:8% SiO_2	4.7
Thermal conductivity (BTU) — 90% SiC:8% SiO_2	109
Flexure strength (psi) — 96.5% SiC:2.5 SiO_2	24,000
Compressive strength (psi) — 96.5% SiC:2.5% SiO_2	150,000
Bond dissociation energy (kJ mol^{-1})	435

Cohesive energy (kJ mol^{-1})	1183
Atomic mass (mole/g)	40
Young's modulus (Kgf $cm^{-2} \times 10^{-6}$)	4.9
Coefficient of thermal endurance F(cm K $min^{-1/2} \times 10^{-4}$)	1.3
Relative F match	33
Thermal conductivity (W cm^{-1} K^{-1})	1.87
Coefficient of linear expansion ($K^{-1} \times 10^{6}$)	5.0
Specific heat (Jg^{-1} K^{-1})	0.146
Refractive index (n=)	2.654—2.697 (2.4—2.8)
Dielectric constant (e=)	5.8—8.12
Crystal structure (hexagonal — normal) alpha	$(10\bar{1}0)$ prism
(isometric — normal) beta	(100) cube
Color (solid) alpha (natural)	Blue-black
alpha (powder, artificial)	White grey
Cleavage (basal/cubic)	(0001)/(001)

SILICON CARBIDE, SiC

General: Silicon carbide was first grown artificially as carborundum. It was later found as a natural compound and named moissanite, CSi, from an occurrence as small green hexagonal crystals in meteoric iron, such that it may not be considered of terrestrial origin.

It is artificially made by fusing sand and coke (SiO_2 + C) above 4000°F, and is one of the high temperature refractory ceramics with use as refractory brick in furnaces. After initial fusing, it is powdered and pressure bonded using silica or silicon nitride as the bonding medium and, with sufficient temperature and pressure, there is some recrystallization and self-bonding. The material is used for many high temperature heat-resistant, acid resistant and gas impermeable parts, such as hot metal spray nozzles. When molded with graphite, C, it exhibits high wear and corrosion resistance; fabricated as a foam it resembles the natural rock called pumice, SiO_2 of acid magma origin — light weight due to the structure — used as a construction material, and as a pumice stone lapping abrasive. As fibers or whiskers it is added to metals, plastics and glasses as a strengthening agent, and as a thin film spray coating on metal cutting edge surfaces it improves wear and work-life of such blades. It has a low neutron cross section, such that it is stable against radiation and is used in the construction of nuclear power reactors and similar radiation equipment.

Because of its hardness (H = 9+) it has become one of the most widely used lapping and polishing abrasives, replacing the more expensive diamond paste compounds. As an abrasive it is available in various powdered grit-sizes; bonded as a cutting wheel; as sandpaper sheets, strips or belts; and in a variety of bonded, shaped forms as files, cones, cylinders, and so forth.

Technical Application: Grown as a single crystal, silicon carbide is an intrinsic n-type semiconductor. When doped with aluminum it forms a p-n junction diode capable of operating above 600°C, e.g., most semiconductors are rated for maximum 150°C operation.

In Solid State development it is grown by depositing thin carbon films on silicon, followed by thermal heat treatment to form the carbide. Under 500°C as an amorphous thin film, a-SiC or hydrogenated as a-SiC:H, and as single crystal hexagonal alpha-SiC. At greater than 1400°C, as cubic (isometric) beta-SiC. It also has been deposited as an amorphous silicon carbonitride, SiC_xNi_{1-x} thin film. Chemical Vapor Deposition (CVD) is used for growth with a mixture of SiH_4:CCl_4 or other carbon containing compounds and gases such as methane, CH_4, where other than silicon substrates are used.

As grown in any of the structural forms described, it has been used as a substrate for additional epitaxy growth of other semiconductor compounds, or, with metal evaporation, thermal conversion to silicides using the carbide as the silicon source.

It is widely used as a general lapping and polishing abrasive on all Solid State metals and metallic compounds, a 600-grit (about 30 μm) finishing being a general "as received" standard on semiconductor wafers. The abrasive form is as sharp slivers, brittle, shiny black in color, with excellent cutting action. The carbide does not wear during use, but will splinter, presenting fresh, sharp edges without reduction of cutting action. As a slurry of abrasive/water/glycerin or as an acid-slurry for improved polishing — it is unattacked by single acids at room temperature.

Though not used to date, it has application as a radiation-hard package material for semiconductors and similar device assemblies. As a thin film coating on graphite, it is in use for epitaxy susceptor plates.

Etching: Very difficult. Molten fluxes; mixed acids or dry chemical etching.

SILICON CARBIDE ETCHANTS

SIC-0001
ETCH NAME: Hydrofluoric acid — TIME: 30 sec
TYPE: Acid, cleaning — TEMP: RT
COMPOSITION:
x .... HF
x .... H_2O
DISCUSSION:

SiC blanks used as substrates for nickel and palladium evaporation and conversion to silicides. Degrease substrates in acetone, then isopropyl alcohol. Dip in HF solution shown and water rinse prior to metal evaporation. Anneal at 400 to 900°C after metallization in 100°C steps, 30 min for each temperature level. At 400°C, no change; at 500°C, as Pd_3Si; at 700°C, as Pd_3Si and Pd_2Si; and between 800—900°C, as Pd_2Si only. The same pattern was observed when nickel replaced palladium.

REF: Pai, S C et al — *J Appl Phys,* 57,618(1985)

SIC-0002
ETCH NAME: Copper — TIME: 12 h
TYPE: Metal, decoration — TEMP: 450—500°C
COMPOSITION:
x .... Cu
DISCUSSION:

SiC, (0001) blanks. A study of dislocations. Mechanically polish with diamond grit. Degrease in TCE with ultrasonic vibration; then rinse in DI water and N_2 blow dry. Melt copper on surface (alpha-SiC) for 30 sec at 1150°C in H_2 atmosphere, and diffuse copper in argon. Grind off excess copper and re-polish to observe copper decorated dislocations.

REF: Trickett, O A & Griffiths, L B — *J Appl Phys,* 35,3618(1964)

SIC-0003a
ETCH NAME: Argon — TIME:
TYPE: Ionized gas, step-etch — TEMP:
COMPOSITION: — GAS FLOW:
x .... Ar^+ ions — PRESSURE:
POWER:
DISCUSSION:

Beta-SiC, (001) single crystal blanks used in a structure study of the material. First, He^+ ion implant, then Ar^+ ion sputter step-etch to observe depth of helium penetration and defect structure.

REF: Thomas, R L — *Jpn J Appl Phys,* 23,1380(1985)
SIC-0017: Dillon, J A, Jr et al — *J Appl Phys,* 30,675(1959)

SiC, (0001) wafers used in a study of surface properties. Ar^+ ion clean and anneal. Appears to be more stable than a silicon surface.

SIC-0003b
ETCH NAME: Heat TIME: 5 h
TYPE: Thermal, clean TEMP: 800°C
COMPOSITION:
x .... heat
DISCUSSION:
Beta-SiC, (001) single crystal blanks. Mechanically polish for an optical surface. Clean in vacuum at 800°C for 5 h as shown. He^+ ion implant and Ar^+ ion sputter step-etch for material structure study. Density of beta-SiC shown as 3.2 g/cm^3.
REF: Ibid.

SIC-0004
ETCH NAME: Hydrofluoric acid TIME:
TYPE: Acid, cleaning TEMP:
COMPOSITION:
x .... HF
DISCUSSION:
SiC, (0001) thin films grown on (100) silicon substrates by sublimation as large area oriented platelets. Clean the grown SiC surfaces with HF and water rinse. Follow with acetone rinse, propanol rinse and wipe dry. To remove residual F^+ ions and O^+ ions, Ar^+ ion sputter clean at 500 eV and temperature anneal.
REF: Bozzo, F et al — *J Electrochem Soc,* 131,1270(1984)
SIC-0005a: Yang, C Y et al — *J Electrochem Soc,* 132,418(1985)
c-SiC:H thin films doped with either boron or phosphorus. Etch rate varies with the carbon source used to grow SiC. Benzene and methane show lowest etch rates. SiC:H rate with benzene is 160 Å/min.
SIC-0006a: Roy, R A — *J Vac Sci Technol,* A(2),312(1984)
a-SiC:H, alpha-SiC and beta-SiC grown by CVD. Alpha-SiC deposited at 500°C is hexagonal. Beta-SiC deposited at 1400°C is cubic. (*Note:* "Cubic" is isometric system, normal class.)
SIC-0007: Primak, W et al — *Phys Rev,* 103,1184(1956)
SiC crystals. Surfaces soaked in HF and then in HCl. Both solutions used hot and specimens soaked for several days followed by water washing. Specimens used in a radiation damage study.
SIC-0010a: Chang, C Y et al — *J Electrochem Soc,* 112,418(1985)
a-SiC:H thin films deposited on (100) silicon and Corning glass 7059 substrates used with "standard cleaning" (?). Also cleaned in BHF and H_3PO_4 at 180°C. Seven different gases for carbon source used with SiH_4(29.5%):H_2 and 1% of B_2H_6 or PH_3 as dopants. After SiC deposition, hold in H_2 at temperature, 10 min. Etching studies done with both acids and dry chemical etching with CF_4:O_2. DC resistivity; dielectric breakdown and refractive indices measured. Also measured pinhole density vs. deposition rates by type gases and dopants.

SIC-0005b
ETCH NAME: BHF TIME:
TYPE: Acid, removal TEMP:
COMPOSITION:
x .... BHF

DISCUSSION:

a-SiC:H thin films deposited on Si (100) substrates and doped with either boron or phosphorus. Etch rates vary with carbon source used to deposit SiC. BHF mixture used not shown.

REF: Ibid.

SIC-0005c

ETCH NAME: Phosphoric acid — TIME:

TYPE: Acid, removal — TEMP: 180°C

COMPOSITION:

x .... H_3PO_4

DISCUSSION:

a-SiC:H thin films deposited on Si (100) substrates and doped with either boron or phosphorus. Etch rates vary with carbon source used to deposit SiC. Solution shown used as a general removal etch.

REF: Ibid.

SIC-0024: Catherine, Y et al — *Thin Solid Films,* 109,145(1983)

a-SiC:H thin films deposited by DC and RF sputtering. Phosphoric acid can be used to remove or pattern. Carbide refractive index, n = 3.6.

SIC-0025

ETCH NAME: — TIME: 30 min

TYPE: Molten flux, polish — TEMP: 500°C

COMPOSITION:

3 .... NaOH, pellets

1 .... Na_2O_2, flakes

DISCUSSION:

SiC platelets. Molten solution used to etch polish surfaces. Authors say specimens are beta-SiC grown with a (111) habit, 10—20 μm thick.

REF: Bartlett, R W & Martin, O W — *J Appl Phys,* 39,2324(1968)

SIC-0006b

ETCH NAME: Aluminum — TIME:

TYPE: Metal, doping — TEMP: 3900°F

COMPOSITION:

x .... Al

DISCUSSION:

SiC, n-type wafers were doped with aluminum at 3900°F to form a p-n junction diode. Device will operate as a diode at 650°C.

REF: Roy, R A — *J Vac Sci Technol,* A(2),312(1984)

SIC-0008

ETCH NAME: — TIME:

TYPE: Acid, removal — TEMP:

COMPOSITION:

2 .... HF

3 .... HNO_3

DISCUSSION:

Beta SiC thin films grown on Si, (100) wafers cut 1 and 6° off the plane, and (111) wafers cut 2 and 4° off-plane. Authors say the silicon wafers were trepanned from larger diameter wafers. With wafers on an SiC coated graphite susceptor in a CVD epitaxy reactor system, etch clean surfaces with hot HCl vapor prior to SiC growth. React silicon with C_3H_4

before bringing up to SiC reaction temperature. Laue photographs used to study films show overlapping Si and SiC back-reflections. Control remove SiC by black waxing (Apeizon-W) the surface leaving a center opening, then etching with solution shown. This method used to profile carbide thickness.
REF: Liaw, F & Davis, R F — *J Electrochem Soc,* 132,642(1985)

SIC-0009
ETCH NAME: Diboride — TIME:
TYPE: Gas, doping — TEMP: >570°C
COMPOSITION:
x .... B_2H_6
DISCUSSION:
SiC thin films grown on Si (100) wafers; Corning glass 7059 and stainless steel discs as substrates to deposit SiC by CVD. Substrates were held at 250°C and various gas mixtures for SiC growth were used with thickness varied by study requirement. For IR study, 1000 nm and for NR, 200—1500 nm. Some specimens were boron doped with B_2H_6.
REF: Fusimoto, F et al — *Jpn J Appl Phys,* 23,810(1984)

SIC-0010
ETCH NAME: Borax — TIME:
TYPE: Molten flux, dislocation — TEMP: 855°C
COMPOSITION:
(1) x .... Borax (2) x .... Na_2CO_3
DISCUSSION:
SiC, epitaxy thin films. Both compounds shown were used as molten fluxes to develop dislocation pits in a study of etch pit density.
REF: Brander, R W — *J Electrochem Soc,* 12,881(1964)
SIC-0016: Amelinck, S et al — *J Appl Phys,* 31,1350(1960)
SiC, (111) specimens used in a dislocation study. Used a platinum crucible to contain flux.

SIC-0011a
ETCH NAME: — TIME: 2 min
TYPE: Molten, dislocation — TEMP: 700°C
COMPOSITION:
3 .... NaOH
1 .... Na_2O_2
DISCUSSION:
Alpha-SiC, (0001) wafers used in a defect structure study. As a molten flux will develop dislocations on the (0001) but it is difficult to develop dislocations on prism faces. With specimens in a nickel basket and molten flux at 500°C, 30 min solution will polish surfaces.
REF: Griffiths, L B — *J Phys Chem Solids,* 27,257(1966)

SIC-0011b
ETCH NAME: Chlorine — TIME:
TYPE: Gas, polish — TEMP:
COMPOSITION:
x .... Cl_2

DISCUSSION:

Alpha-SiC, (0001) wafers used in a defect study. Chlorine gas can be used to etch polish surfaces. To thin for electron microscope study, lap to 0.03—0.05 mm thickness with diamond paste.

REF: Ibid.

SIC-0012a

ETCH NAME: Sirtl etch | TIME: 5—10 sec

TYPE: Acid, polish | TEMP: RT

COMPOSITION:

2 .... HF

1 .... CrO_3 (33%)

DISCUSSION:

SiC, (0001) wafers used as substrates for epitaxy silicon thin film grown. Amount of removal on polished SiC substrates was less than 0.2 μm.

REF: Wolley, E D — *J Appl Phys,* 37,1588(1966)

SIC-0012b

ETCH NAME: Sodium peroxide | TIME:

TYPE: Molten, dislocation | TEMP: >400°C

COMPOSITION:

x .... Na_2O_2

DISCUSSION:

SiC, (0001) wafers used as substrates for epitaxy silicon growth. Solution produces irregular, dendritic structure on (0001)A and triangle pits on $(000\bar{1})$B.

REF: Ibid.

SIC-0015a: Harper, C A — *Handbook of Materials and Processes for Electronics,* McGraw-Hill, New York, 1970, 7

SiC, (111) wafers. Use a molten flux at 350°C for dislocations.

SIC-0013a

ETCH NAME: | TIME: 10—60 min

TYPE: Molten flux, structure | TEMP: 650°C

COMPOSITION:

x .... NaF

x .... K_2CO_3

DISCUSSION:

SiC, thin films vapor deposited on silicon wafers. Solution used as a molten flux to develop microstructure (solution is a eutectic mixture).

REF: Gulden, T D — *J Am Ceram Soc,* 51,425(1968)

SIC-0013b

ETCH NAME: | TIME:

TYPE: Electrolytic, stain | TEMP:

COMPOSITION: | ANODE: SiC

x .... HF | CATHODE:

x .... KF, sat. sol. | POWER:

DISCUSSION:

SiC, thin films vapor deposited on silicon wafers. Solution used to develop microstructure by staining surfaces.

REF: Ibid.

SIC-0014
ETCH NAME:
TYPE: Molten flux, preferential
COMPOSITION:
90 g .... $NaNO_2$
10 g .... Na_2O_2
TIME: 10 min
TEMP: 500°C
DISCUSSION:
Beta-SiC, (0001) wafers used in a study of stacking faults and dislocations. Solution will develop structure.
REF: Liebmann, W K — *J Electrochem Soc,* 12,885(1964)

SIC-0015b
ETCH NAME: Sodium hydroxide
TYPE: Molten flux, preferential
COMPOSITION:
x .... NaOH, pellets
TIME:
TEMP: 900°C
DISCUSSION:
SiC, (111) wafers. Use as a dislocation etch. Can be used on other orientations.
REF: Ibid.

SIC-0018
ETCH NAME: Hydrofluoric acid
TYPE: Electrolytic, polish
COMPOSITION:
x .... HF, conc.
TIME:
TEMP:
ANODE: SiC
CATHODE:
POWER:
DISCUSSION:
SiC, single crystal specimens used in developing electroless plated contacts on this material. Polish specimens with solution shown.
REF: Raybold, R L — *Rev Sci Instr,* 31,781(1960)

SIC-0019
ETCH NAME:
TYPE: Abrasive, polish
COMPOSITION:
x .... diamond grit
TIME:
TEMP:
DISCUSSION:
SiC, (0001) grown as alpha-II SiC, n-type, about 1 Ω cm resistivity and is a characteristic green color. Grown as a hexagonal pyramid with an (0001) base that can be optically flat. Cut, grind, and polish on copper plates using diamond grit in oil. Form SiO_2 thin film on specimens by oxidizing at 1000°C for 2 h using standard furnace techniques. Remove oxide with HF wash. Specimens used in a study of infrared properties of hexagonal SiC.
REF: Spitzer, W G et al — *Phys Rev,* 113,127(1959)

SIC-0020
ETCH NAME: Hydrogen
TYPE: Gas, polish
COMPOSITION:
x .... H_2, hot
TIME:
TEMP: 1500—1800°C
DISCUSSION:
SiC, (0001) wafers in a horizontal epitaxy-type furnace. Hot hydrogen used to etch

polish both (0001)Si and (000$\bar{1}$)C surfaces — the silicon face etches approximately twice as fast — up to 4—5 Å/min at the higher temperatures. Method used to obtain planar polished surfaces.

REF: Chu, T L & Campbell, R B — *J Electrochem Soc,* 112,955(1965)

SIC-0010b

ETCH NAME: Phosphoric acid — TIME:

TYPE: Acid, removal — TEMP: 180°C

COMPOSITION: — RATE: See discussion

x .... H_3PO_4

DISCUSSION:

a-SiC:H, thin films deposited on Corning 7059 glass or silicon wafers in a development study of this compound as a passivating dielectric. Seven carbon sources used, plus boron or phosphorus doping. Etch rates, dielectric breakdown and dielectric constant, resistivity, refractive index, and pinhole density measured.

REF: Ibid.

SIC-0010c

ETCH NAME: Copper — TIME:

TYPE: Metal, pinhole decoration — TEMP:

COMPOSITION: — ANODE: SiC:H

x .... HAc — CATHODE: Cu and Si

POWER:

DISCUSSION:

a-SiC:H, thin films as amorphous coatings on glass and silicon wafers. A copper plating system was used to decorate defects and pinholes.

REF: Ibid.

SIC-0010d

ETCH NAME: Dry chemical etch — TIME:

TYPE: Ionized gas removal — TEMP:

COMPOSITION: — GAS FLOW:

x .... CF_4 — PRESSURE:

x .... O_2 — POWER:

DISCUSSION:

a-SiC:H amorphous thin films grown on glass or silicon wafers using seven different carbon sources and included doping with boron or phosphorus. Etch rate varies with type carbon source and dopant: B-doped rate was 160 Å/min to 980 Å/min. P-doped rate 200 to 400 Å more rapid relative to carbon source.

REF: Ibid.

SIC-0010e

ETCH NAME: BHF — TIME:

TYPE: Acid, removal — TEMP:

COMPOSITION:

x .... BHF

DISCUSSION:

a-SiC:H amorphous thin films 500—3500 Å thick deposited by RF plasma enhanced CVD (PECVD) using seven different carbon sources with either boron or phosphorus doping. Substrates used were Corning 7059 glass or silicon wafers. In general, phosphorus doped films etched more rapidly in BHF solutions.

REF: Ibid.

SIC-0021
ETCH NAME: TIME:
TYPE: Molten flux, removal TEMP: 650°C
COMPOSITION:
1 .... NaF
2 .... K_2CO_3
DISCUSSION:
SiC, (0001) wafers used in a study of structure and adhesion. The (0001)Si is smoothly polished; the $(000\bar{1})$C face is rough. Also studied Al_2O_3 and Mn-Zn, Ni-Zn ferrites.
REF: Buckley, D — *J Vac Sci Technol,* A(3),762(1985)

SIN-0023
ETCH NAME: Hydrofluoric acid TIME:
TYPE: Acid, cleaning TEMP:
COMPOSITION:
(1) 1 .... HF
10 .... H_2O
(2) 3 .... HF
5 .... HNO_3
3 .... HAc
DISCUSSION:
Beta-SiC, thin films grown on (100) silicon, p-type, 150 Ω cm resistivity as an oriented (100) film. Deposition used a propane:H_2 atmosphere at 1380—1400°C just below the silicon m.p. of 1414°C. Silicon wafers were cleaned in the solutions shown with DI water rinsing, and applied in the order shown. Did not use HCl:H_2 at 1200°C in epitaxy reactor (a normal pre-cleaning step), as it erodes the SiC coating on the graphite susceptor plate that can affect the characteristic of the SiC films. This was a study of the carbide as a surface coating.
REF: Addaniano, A & Klein, P H — *J Cryst Growth,* 70,291(1984)

PHYSICAL PROPERTIES OF SILICON DIOXIDE, SiO_2

Classification	Oxide
Atomic numbers	14 & 8
Atomic weight	60
Melting point (°C)	600—800 (1400)
Boiling point (°C)	2500 (2350)
Density (g/cm^3)	2.0—2.7 (2.65)
Refractive index (n=)	1.46—1.47
Dielectric constant (e=)	3.5 (3.8)
Mean specific heat (J/kg °C) 25—1000°C	1170
Coefficient of linear thermal expansion ($\times 10^{-6}$ cm/cm/°C) 25—800°C	8.6
Thermal conductivity (W/m °C) 1200°C	2.06
Electrical resistivity (ohm cm) 0°C	1014
Hardness (Mohs — scratch)	7
Crystal structure (hexagonal — rhombohedral)	$(10\bar{1}0)$ rhomb (alpha Qtz)
(amorphous) obsidian/glasses	None
(colloidal) silica suspension	None
Color (solid) alpha-quartz	Clear/colors
Cleavage (imperfect)	None
Fracture (solid) — atypical all glassy oxides	Conchoidal

SILICON DIOXIDE, SiO_2

General: Occurs in nature as the pure compound in the mineral quartz, SiO_2, which has a number of varieties and structural forms: (1) Single crystal: quartz, tridymite, crystobalite — all as low temperature (alpha) or high temperature (beta) structure, and two high-pressure forms associated with meteors: coestite and stisovite, where the tetrahedra structure is a ratio of SiO_2/SiO_4; (2) Colloidal or amorphous structure as opal, $SiO_2.nH_2O$; as silica, SiO_2, exoskeletons of diatom sea-life. As silica with impurities — obsidian, is natural volcanic amorphous glass, often containing carbonaceous matter or other mineral inclusions, viz., ''snowflake'' obsidian; (3) cryptocrystalline structure that may be colloidal/amorphous, in part, or columnar portions like chalcedony. This category also includes semi-precious stones, the agates, onyx, jaspers, chert, and flint. The latter two are from siliceous diatoms, sponges and other sea-life; (4) phenocrystalline: these varieties are often single crystal with impurities and represent the semi-precious and precious gem stones of quartz — clear rock crystal; amethyst, purple; citrine, yellow; milky quartz, milk-white, and nearly opaque; cat's eye, and several others; (5) massive: these are similar to the cryptocrystalline varieties, such as quartzite, a highly compacted, near-granular massive rock form which may be as a conglomerate mixture of quartz pebbles and fine sand; pseudomorphous quartz is an SiO_2 replacement of other minerals, such as replacing calcite or fluorite, retaining the replaced mineral structure, and includes silicified wood; (6) rock formers: though several of the quartz-type minerals already mentioned, above, occur in extensive beds — notably, flint, chert, and quartzite — there are three other silica minerals that are used by name: pumice or pumice stone, of acidic magma volcanic origin, porous, sand-like, lightweight, and a similar froth variety called tufa, both lightweight building stone or polishing abrasives used in cut, solid form. The best known of the three is sandstone, compacted sand grains often cemented with colloidal iron oxides . . . used as a building stone since antiquity. In addition, probably the best known form of natural silica is sand as loose, spherical particles of quartz produced by the rolling and tumbling action of water. It can be white (sometimes with calcite as atoll beaches in the South Pacific) more often yellow, and the black variety is high in magnetic iron ore (magnetite) as beach sands, or desert sands.

Other than the pure SiO_2 type minerals and rocks, the silicates as a mineral group contain SiO_3^-, SiO_4^- . . . SiO_{12}^- radicals with other metal elements and represent the largest chemical group of natural minerals, perhaps 2000 of the 4000 or more known mineral species. They occur as both single crystals and as primary rock-forming materials: granites, gneisses, slates, clays, micas, etc. Many of these rocks, clays in particular, are the base of the silicate, pottery, and ceramics industries.

Next to oxygen, silicon is the most abundant element in nature and, as some form of silica or silicate, is found in about one quarter of the earth's crust. When the silicates are included with pure silica, possibly 60% of the outer earth is a solid, exterior mantle with the remainder being oxides, carbonates, phosphates, arsenides, borates, and nitrates as well as the native metal elements, such as gold, platinum, silver, sulfur, and others.

Industrially, as used by man, flint and obsidian were some of the first tools and weapons of the Stone Age . . . *Homo sapien sapien* — the ''tool maker'' . . . is used to differentiate modern man from his prehistoric cousins, the Cro-Magnon, Neanderthal, and hominids. The use of mud and branches (wattle) and clays for construction and utensils probably also should be included as initial mediums, though this use reached its peak during the following Bronze Age. Flint and obsidian are still used as tools and weapons in the more remote areas of the world, today, and were even used in ''modern'' weapons as late as the 1700s, viz., the flintlock rifle.

Glass, as a manmade product, can be traced back to the Egyptian development of soda-lime glass around 3500 B.C. and that formula is still in use. The development of glass is a separate industry, though it is closely allied with the silicates and ceramic industries of today. Pottery glaze and enamels (silica applied as a thin surface film to harden clay vessels) have been in use for centuries as ceramic pottery. The Chinese, as well as Mediterranean cultures, have been making enamelware for jewelry and objets d'art since about 2000 B.C. In our modern society both the glass and silicate industries are becoming more and more closely involved with the metals industries with the development of composite materials and have long been associated with refractory ceramics, an off-shoot industry of the basic silicates and still a part of that industrial complex. The jewelry trade uses natural gem stones, such as amethyst, as well as amorphous glass as beads or cut stones, etc. It is interesting to note that the Egyptians were making fake glass "paste" gem stones as early as 3000 B.C.

Technical Application: Pure silicon dioxide, SiO_2 was initially grown on silicon wafers as a (1) "wet" thermal oxide using nitrogen or argon gas bubbled through DI water into a furnace at 900—1300°C, or (2) "dry" thermal oxide using O_2 gas only, under similar temperature conditions. In both cases, the silicon wafer was the source of silicon. Both methods are still used and a common temperature is 1000°C. For such "oxide" deposition on silicon and other metallic compounds that do not contain silicon, the now called Chemical Vapor Deposition (CVD) process was developed from the original pyrolytic process which used liquid chemical compounds. CVD uses gas combinations, such as SiH_4/O_2, with deposition temperatures ranging from 350 to 900°C. The AMT Sylox systems are similar with 350 to 500°C temperatures. RF (magnetron) sputter systems using a solid glass or pressed powder target as the silica source are now in wide use with deposit temperatures as low as 200°C.

Finely powdered silica, as a glass-frit, can be painted or sprayed on surfaces and then fired to form the oxide coating between 600—800°C. Silicon monoxide, SiO, can be vacuum evaporated, 5×10^{-5} Torr vacuum or better, then densified by bake-out in oxygen or air at 400 to 800°C.

A thin piece of glass, such as a glass coverslide, can be fired and fused onto a semiconductor surface between 500 to 800°C as a method of cladding and has been used to fill channels, via-holes, and pits in surfaces.

Boiling water, nitric acid, or hydrogen peroxide, depending upon the semiconductor material, has been used to form a hydrated surface oxide that can then be stripped with HF as a method of surface cleaning. Potassium hydroxide, KOH, both as a boiling solution or electrolytic solution, has been used to grow a hydrated silica on silicon wafers in the study of oxidation reactions and as a method of surface cleaning and can be used to form similar oxides on germanium, Ge, and compound semiconductors, such as gallium arsenide, GaAs, or indium phosphide, InP. These are all porous, hydrated oxides and require special postbake treatments if they are going to be used for other than surface cleaning applications.

The hard silica coatings . . . evaporated, sputtered, CVD . . . are used for surface passivation; mechanical and chemical protection; as a mask against diffusion, metallization, ionization, etching (wet chemical of dry — gas), and as doped oxides PSG, BSG, ASG, etc., as either surface protection or as an element diffusion source.

As a mechanical operator, glass has been deposited in channels, "V" slots or pits, both for strength as well as for dielectric isolation of diffused areas or separation of epitaxy layers and similar structures. In other cases, as in fabricating MOS and MNOS devices, the oxide thin film is an active element in the device operation. In most of these applications the thin film is amorphous and used as such, though it may be partially crystallized by special processing, even to single crystal form by substrate orientation or by laser annealing.

Silica also is being deposited as a colloidal thin film, such as hydrogenated a-SiO_2:H, or a-$SiON_x$:H compound for special purpose application.

As glass plates — microscope slide, soda-lime, pure quartz— the glass is used as substrates for thin film evaporation of metals or the epitaxy growth of materials which, after deposition, are removed with an HF float-off technique from the glass for material morphological study as with a transmission electron microscope (TEM). Such glass slides also are used in both vacuum metal evaporation and epitaxy growth as thickness monitors, or to cover portions of a wafer during metal evaporation, epitaxy, etc., for measurement purposes. The microscope slide/metal evaporation is also used in transmitted light evaluation for pinhole density in the metal oxide thin films and in studying the growth step mechanisms of thin films as deposited by evaporation, epitaxy etc.

In addition to the direct applications in semiconductor device fabrication, quartz single crystal, fused quartz, or fused silica blanks are used to fabricate circuit substrates, and some thin film oxides on such substrates used in making resistor, capacitor or other dielectric elements as planar devices, to include integrated circuits (ICs). This also may include doped glassy electrical contact pads or fingers.

Semiconductor packages that require external lead contacts use glass-isolated feed-thru or stand-off pins as well as isolation plates or coatings and many test holders include such pins and plate isolation parts. The "glass diode" package is a sleeve or tube of glass into which the device is fused or the lead fabricated device may be bead-fused or oxide/nitride sputter coated prior to epoxy encapsulation.

Both quartz and silica have long been used for the fabrication of a wide variety of lenses — microscopes and telescopes — as well as light frequency modulators, resonators, filters, and similar devices. The "quartz wedge" is standard to most metallurgical microscopes for material evaluation — a tapered and oriented single crystal piece of quartz that produces frequency color steps, or, as pair lenses, for polarized light by rotation of the two lenses with the thinned specimen between, and as surface polarization for direct illumination. In many applications, such as in telescope fabrication, the lens is coated with aluminum and then overcoated with SiO_2 (evaporation as SiO with densification). In vacuum systems, glass microscope slides or plates are coated with either aluminum, gold, or nickel — or other reflective metals and used as sighting mirrors to observe metal evaporation in progress; similar mirrors are used in observing a cobalt-60 radiation source indirectly as direct observation is hazardous.

Needless to say, quartzware and glassware are construction materials in much laboratory equipment — diffusion furnaces, epitaxy systems, vacuum systems, work stations — in addition to the beakers, measuring cylinders, stirring rods, and so forth used in chemical processing. This includes glass-wool as a packing material in furnaces, as fiber sheeting for electrical and temperature isolation, goggles — deep red or dark blue — to observe welding and high temperature material processing.

Single crystal alpha-quartz has been in use since the early 1920s for the fabrication of radiofrequency crystals, taking over from the original point contact galena, PbS, crystals. The best natural stone comes from Brazil, but artificial quartz is coming into wider use as it is of higher purity or more defect-free, e.g., no twinning. By oriented cutting of blanks, mostly, off z-axis, zones of frequency are obtained; AT, BT, DT, X cut and others. The AT-cut produces the highest frequencies as, by polish thinning, these circular blanks (about $^3/_8$" diameter) can reach as high as 5th order 100 MHz frequency. Alpha-quartz, under electric load, produces a physical motion and, with proper cut orientation, the motion is translated into a controlled frequency. When combined with semiconductor and electronic circuitry modulators, resonators and many other similar device applications are possible.

Other products from glass include glass lasers and fiber optics. The latter not only used for light transmission but for telephone transmission. Light fibers are widely used in conjunction with microscope study and have major application in medicine where the fiber can be inserted into the body for inspection without major surgery. TV and fiber optics are now common in all surgical procedures.

As already mentioned, silica powder is still widely used as a lapping and polishing abrasive. The material can be natural single crystal/artificial and may have specific compound name which are associated with the mixture, physical structure or fracture characteristics. As a general name it is called White Rouge or Glassite; as an amorphous mixture containing less than 0.5% quartz — a natural mineral — it is called white "coloring compound" or

"Tripoli"; the natural colloid is opal, and the artificial is colloidal silica. There also is sand, as loose grains or sandstone, and pumice or tufa as solid sanding compounds, and natural flint is still used as an abrasive and so named. In the artificial growth of single crystal alpha-quartz, natural quartz chips are used in a slurry of KOH/water rather than sand, in a temperature gradient autoclave capable of 40,000 psi — the ingot grown from a single crystal seed held above the silica vapor bottom melt pot. High purity sand, on the other hand, is the chief source of semiconductor silicon.

Both natural single crystal quartz in its various forms as well as artificial quartz have been the subject of general morphological study and thin film silicas are under ongoing studies in the semiconductor and Solid State fields in addition to the continual development work in the glass and other silicate industries.

The following section includes both silica used as a Solid State thin film coating, as well as some of the glasses in present use.

Etching: HF and other fluorine containing compounds; hot H_3PO_4. Alkalies and dry chemical etching (DCE).

SILICON DIOXIDE ETCHANTS

SIO-0001a
ETCH NAME: Sodium carbonate
TYPE: Acid, removal
COMPOSITION:
x .... *N*/50 $NaCO_3$
TIME: 9 h
TEMP: 100°C
RATE: 70 mil/6 h
DISCUSSION:
Vycor 7913 glass. This is a slow etch on specific types of glass.
REF: Brochure: OVB-4 11/i/64 — Corning Glass Works

SIO-0001b
ETCH NAME: Sodium hydroxide
TYPE: Alkali, removal
COMPOSITION:
x .... 5% NaOH
TIME: 6 h
TEMP: 100°C
RATE: 900 mils/6 h
DISCUSSION:
Vycor 7913 glass. This is a slow etch on specific types of glass.
REF: Ibid.

SIO-0001c
ETCH NAME: Hydrochloric acid
TYPE: Acid, removal
COMPOSITION:
x .... 5% HCl
TIME: 24 h
TEMP: 95°C
RATE: $^1/_2$ mil/24 h
DISCUSSION:
Vycor 7913 Glass. This is a slow etch on specific types of glass.
REF: Ibid.

SIO-0002
ETCH NAME: Potassium hydroxide
TYPE: Alkali, removal
COMPOSITION:
x .... 2 *M* KOH
TIME:
TEMP: RT (21°C +/− 5°C)
RATE: 0.05 Å/min

DISCUSSION:

Pyrex glass beaker. Authors say that hydroxides used at room temperature etch so slowly the reaction products introduced into the solution will not affect the etching of silicon. (*Note:* These authors have used boiling solutions of KOH and NaOH with a variety of chemical additives in Pyrex beakers. One beaker pinholed after daily use for a period of about 6 months.)

REF: Bergman, I & Paterson, M S — *J Appl Chem,* 11,369(1961)

SIO-0003

ETCH NAME: Hydrofluoric acid TIME:

TYPE: Acid, removal TEMP:

COMPOSITION:

(1) x .... HF, conc.

(2) 1 .... HF
20 .... H_2O

DISCUSSION:

SiO, thin film. Deposits evaporated on NaCl (100) substrates and used in a study of etch rates of SiO. Concentrated HF removal is rapid. The dilute solution shown was slow removal. SiO deposits were 8 nm thick and were removed from NaCl by the float-off technique using water, and called ''wet'' stripping.

REF: Kaito, C & Shimizu, T — *Jpn J Appl Phys,* 23,L7(1984)

SIO-0004: Menth, M — personal communication, 1979

SiO thin films evaporated over aluminum thin films on telescope lenses of various sizes. Stabilized as SiO_2 by heat treatment densification. SiO stripped with HF as required for lens re-work.

SIO-0038: Fahr, F — personal communication, 1979

SiO, thin films evaporated from tungsten boats with cover containing an exit hole to control evaporation. Used as a thin film surface coating on silicon devices. Densification to SiO_2 in an air oven at 125°C with variable time. HF used to strip or in thickness studies and density evaluation.

SIO-0004

ETCH NAME: Hydrofluoric acid TIME:

TYPE: Acid, removal TEMP: RT

COMPOSITION:

x .... HF, conc.

DISCUSSION:

SiO_2, thin film deposits on silicon wafers, (111), (100), and other orientations, both p- and n-type with various doping concentrations. For general, rapid removal. Etch rate varies with method of SiO_2 deposition, annealing times and temperatures, type doping and doping concentrations. Also deposited on germanium, compound semiconductors and metal surfaces as a surface protectant, as a surface mask against diffusion and metallization or as an active device element. Many clean/etch sequences include concentrated HF as the final step in sequences as an oxide remover prior to further processing. It also acts as a polish liquid on glasses.

REF: N/A

SIO-0005a: Walker, P — personal application, 1956—1959

SiO_2 thin films deposited on silicon (111) and (100) n-type, 5 to 150 Ω cm resistivity, and with or without phosphorus and boron doping. A study of both wet and dry thermal oxidation deposited on wafers at various temperatures and thicknesses in a general oxidation evaluation. Concentrated HF used to: (1) remove deposits for oxide defects induced on silicon surfaces; (2) as a pinhole etch through oxide for depth measurement; (3) as a step-etch for depth measurements, and redeposition to step-etch and observe possible SiO_2/SiO_2

interface reactions. Establishing optimum thickness of oxide on n-type diode wafers using wet thermal oxidation for use as a mask against boron, phosphorus and gold diffusion. Concentrated HF was used as the base line solution in developing compatible solutions for photo resist processing with DC sputtered SiO_2, Si_3N_4, and $Si_2N_xO_{x-1}$. Also in developing n-type silicon mesa diodes with Si_3N_4/SiO_2 coatings to prevent degradation of exposed boron diffused p-n junctions. And in an etch rate study of CVD (Silox) vs. RF magnetron sputter (MRC system) of deposited oxides on both (111) silicon and (100) gallium arsenide with and without epitaxy layers.

SIO-0007b: Ibid.

SiO_2, as glass microscope slides used as substrates for thin film metal evaporation. HF used for float-off technique to remove metal thin films for TEM study.

SIO-0065: Tomozawa, M & Takamori, T — *J Am Ceram Soc,* 60,301(1977)

SiO_2, as borosilicate glasses of different compositions. A study of the effects of HF etching rates on phase separation. Anneal time and temperature used to establish phase separation.

SI-0057a: Atalla, N M et al — *Technol J Bell Syst,* 38,2(1959)

SiO_2, thin films grown on silicon wafers by thermal oxidation. A major study of oxide stability on silicon. Oxide deposited at 1000°C as both wet and dry oxidation. Wafers were FZ (n-type) and CZ (p-type) as acceptor and donor wafers, respectively. Washed in HF to produce a hydrophobic silicon surface prior to oxidation. The following gas and etch evaluations were studied: (1) Cl_2 at 900°C will etch silicon at 0.001/min, but does not etch oxide. Oxides were deposited at 920°C, 1—30 min for oxides 250—300 Å thick and quenched to air at RT. (2) Hydrophobic wafer surface: etch is fresh $HF:HNO_3$ solutions; in boiling H_2O; organic solvents; or water with detergents. (3) Hydrophilic wafer surface: soak in hot HNO_3, 10 min, follow with boiling H_2O, and surface remains hydrophilic. If HF vapors are present it becomes highly hydrophobic, and if exposed to air also becomes hydrophobic but nonuniform. (4) Deposited oxide surfaces are hydrophilic. After HNO_3 soak and exposure to air goes from hydrophobic to hydrophilic by washing in organic solvents.

SIO-0008

ETCH NAME: Hydrofluoric acid, dilute | TIME:

TYPE: Acid, removal | TEMP: RT

COMPOSITION:

1 .... HF

1 .... H_2O

DISCUSSION:

SiO_2, thin films and native oxides. A widely used mixture of HF for native oxide removal, it is usually the last step in a clean/etch sequence followed by DI water rinse immediately before placing wafers/substrates in a vacuum system, diffusion tube or epitaxy system. Used on silicon and other compound semiconductors as the final native oxide remover.

REF: N/A

SIO-0009

ETCH NAME: BHF | TIME:

TYPE: Acid, removal | TEMP:

COMPOSITION: | RATE: 0.14 nm/sec

1 .... HF

100 .... NH_4F, sat. sol.

DISCUSSION:

SiO_2 and Si_3N_4 thin films deposited on silicon, (100), p-type, 5—9 Ω cm resistivity substrates used in an etch study. Substrates were pre-cleaned in $H_2SO_4:H_2O_2$, followed by an HF dip and DI water rinse. SiO_2 deposited at 950°C and annealed in N_2 for 30 min.

Si_3N_4 was then grown at 925°C with NH_3 for 2 h. Authors say that etch rate increases with nitride deposition temperature. Solution shown was used to establish etch rates.
REF: Wong, S S et al — *J Appl Phys,* 30,1139(1983)
SIO-0010: Walker, P — personal application, 1969

SiO_2, as single crystal quartz as AT and BT cut blanks were either frequency tuned in NH_4F:HF solutions at RT containing undissolved flakes of NH_4F in etch containers to improve repetitive frequency etching time control with long time solution use. The addition of glycerin to the solutions for improved surface wetting was also used to improve etch planarity results.

SI-0011a
ETCH NAME: BHF — TIME:
TYPE: Acid, removal — TEMP: RT
COMPOSITION:
1 .... HF
6 .... NH_4F (40%)
DISCUSSION:

SiO_2, deposited as CVD thin films on (100) silicon substrates with boron and phosphorus doping, as BSG or PSG, respectively, and used in a study of etch rates as deposited or after densification at 700°C, 30 min. All coatings show tensile stress, as deposited, with BSG highest — decreases with shelf-time and converts to compressive stress under high humidity conditions. Refractive index, n, changes after 700°C anneal. Show variation in etch rates between as deposited and after annealed. The following etch solutions will leach deposits: (1) H_2SO_4:H_2O_2:H_2O; (2) NH_4OH:H_2O_2:H_2O and (3) H_2O, only. These three solutions are of interest as they are used in clean/etch sequences on a number of semiconductor materials. Rates with the solution shown were compared to those of the P-etch (SI-0011b). The P-etch appears to be more sensitive to oxide deposit composition than the BHF solution shown.
REF: Kern, W & Schnable, G L — *RCA Rev,* 43,423(1982)

SIO-0011b
ETCH NAME: P-etch — TIME:
TYPE: Acid, removal — TEMP:
COMPOSITION:
3 .... HF
2 .... HNO_3
60 .... H_2O
DISCUSSION:

SiO_2, thin films deposited on (100) silicon wafers. This solution developed by AMT Corp. in conjunction with their SILOX deposition system is a widely used standard for study of pinhole density in oxide thin films.
REF: Ibid.
SIO-0087: Maissel, L L & Glang, R — *Handbook of Thin Film Technology,* McGraw-Hill, New York, 1970.

SiO_2, thin films. Described as a controlled etch for silica.
SI-0012: Balazs, M K & Swanson, T B — AMT SM-102R Brochure, June 1970.

SiO_2, thin films grown by the AMT-2600 SILOX system using $SiH_4/O_2/N_2$ gas source. P-etch was developed for use in establishing etch rates and pinhole density in silica films. Refractive index n = 1.44 for a gas mixture of 2:18 (O_2:SiH_4) with temperature between 350 and 475°C. (*Note:* Opal, n = 2.1—2.3 and quartz, n = 1.54—1.55.)
SI-0061: Walker, P & Gomez, A — personal application, 1978

SIO_2, thin films deposited on (100) GaAs:Cr (SI) wafers used in the fabrication of LN-FETs. Oxide was deposited from a SILOX system at 450°C for a royal blue color (about 2000 Å) for use as a metallization mask for Au/AuGe/Ni evaporation. After photo resist patterning, the P-etch was used to remove SiO_2 in the open pattern areas prior to metal evaporation. After metallization and device forming, the wafers were given a passivation oxide coating at 350°C.

SIO-0020
ETCH NAME:
TYPE: Acid, removal
COMPOSITION:
15 ml .... HF
10 ml .... HNO_3
300 ml .... H_2O
TIME:
TEMP: RT
RATE: 2.5 Å/sec
DISCUSSION:
SiO_2, thin films deposited on silicon wafers at 1000°C by wet oxidation. This form of a P-etch was used in a properties study of the oxide.
REF: Plisken, W A & Lehman, H S — *J Electrochem Soc,* 113,872(1964)
SIO-0086: Plisken, W A & Lehman, H S — *IBM J Res Dev,* 8,43(1966)
SiO_2, thin films deposited on silicon wafers used in a study of the oxide properties.

SIO-0013
ETCH NAME: BHF
TYPE: Acid, removal
COMPOSITION:
100 ml .... HF
860 ml .... NH_4F (40%)
TIME: 5 sec
TEMP: 30°C
DISCUSSION:
Si, (100), p-type 10 Ω cm resistivity wafers used as substrates for RF magnetron sputter of SiO_2 thin films in argon or Ar/H_2. Deposit thicknesses were 0.1 to 2.5 μm. The solution shown was used in studying oxide etch rates which vary with deposit temperature and gas pressure used.
REF: Serikawa, T & Yachi, T — *J Electrochem Soc,* 131,2105(1984)

SIO-0058
ETCH NAME: BSG
TYPE: Silica, diffusion source
COMPOSITION:
"A" x .... $*K_2SiO_3$
"B" 10 g .... $K_2B_4O_7.5H_2O$
3.5 g .... KOH
15 g .... H_2O
TIME:
TEMP:

*As a 1.25 specific gravity solution

Mix: 1 "A":1 "B" when ready to use.
DISCUSSION:
SiO_2, as a BSG glassy layer on silicon used as a diffusion source for boron. Coat the mixture shown on silicon surfaces with a camel hair brush, dry under a heat lamp, then furnace drive-in for boron diffusion. Remove BSG remaining after drive-in with HF soak.
REF: Cline, J E & Seed, R G — *J Electrochem Soc,* 105,700(1958)

SIO-0006
ETCH NAME: Hydrofluoric acid, vapor TIME:
TYPE: Acid, removal TEMP: Hot
COMPOSITION:
x .... HF, vapor
DISCUSSION:

SiO_2, as thermal oxidation on silicon wafers in fabrication of Al/SiO_2/Si capacitors in a study of high field phenomena (HFP) in silicon dioxide. Si, (100), n-type, phosphorus doped substrates were used with four different thicknesses of thermal oxides. Aluminum was evaporated on the SiO_2 for CV characterization and forced voltage-current-time studies. HF fumes were used to remove the oxide from the back of wafers.

REF: Shirley, C G — *J Electrochem Soc,* 132,488(1985)

SIO-0007c: Ibid.

Glass, as microscope slides. One side of the slides was etched in hot HF vapors to roughen the surfaces for better adhesion of thin film gold evaporated in a study of gold morphology.

SIO-0005b: Ibid.

SiO_2, thin films grown on (111) silicon n-type, 5—20 Ω cm resistivity wafers by wet thermal oxidation at 1000°C for thickness up to about 10,000 Å maximum. Hot HF vapors used to step-etch oxide by color steps in establishing oxide layer color order thicknesses, and in studying stacking faults at the SiO_2/Si interface.

SIO-0060
ETCH NAME: Hydrofluoric acid, dilute TIME:
TYPE: Acid, removal TEMP:
COMPOSITION:
1 .... HF
100 .... H_2O
DISCUSSION:

SiO_2, grown on IC devices. Oxide called an intermediate dielectric film. Solution used as a very light oxide surface cleaning etch.

REF: Hess, D W — *J Vac Sci Technol,* A2(2),243(1984)

SIO-0014
ETCH NAME: BHF TIME: 30 sec
TYPE: Acid, step-etch TEMP: RT
COMPOSITION:
1 .... HF
10 .... NH_4F
15 .... H_2O
DISCUSSION:

Si, (100) wafers used as substrates for RF sputtered oxide thin films under various argon pressures. The solution shown was used to establish etch rates of the various films.

REF: Yachi, T & Serikawa, T — *J Electrochem Soc,* 131,2720(1984)

SIO-0015
ETCH NAME: BHF TIME:
TYPE: Acid, removal TEMP:
COMPOSITION:
(1) x .... 1.2 *M* HF (2) x .... 1 *M* HF
x .... 10.3 *M* NH_4F

DISCUSSION:

SiO_2, thin films deposited on (100) silicon wafers. Solutions shown were used in a general etch rate study of deposits.

REF: Nielsen, H & Hackleman, D — *J Electrochem Soc,* 130,708(1983)

SIO-0016

ETCH NAME: TIME:

TYPE: Acid, chlorination TEMP:

COMPOSITION:

(1) x .... HCl, vapor (2) x .... TCA, vapor

DISCUSSION:

SiO_2, thin films deposited on silicon (100), n-type, 8—12 Ω cm and p-type 14—22 Ω cm resistivity wafers used in a study of the differences between using HCl or TCA as additives to the gas stream during growth of the oxides. Wafers were "Piranha" cleaned prior to SiO_2 deposition. **CAUTION:** TCA at elevated temperatures with the presence of phosphorus will form phosgene gas. (*Note:* TCA is 1-1-1 trichloroethane. There is a "Piranha" clean/etch system commercially available . . . it is a system name — not an etch/clean solution.)

REF: Cosway, R & Wu, C-E — *J Electrochem Soc,* 132,151(1985)

SI-0017

ETCH NAME: BOE TIME:

TYPE: Acid, removal TEMP:

COMPOSITION:

1 .... BOE (a 6:1 mixture of ?)

6 .... H_2O

DISCUSSION:

SiO_2, as thin film deposits, with substrates not shown. A study of flow characteristics of doped oxides between APCVD and LPCVD conditions of deposition. After initial deposition "flowing" was done in steam with Cl_2 present during processing BSG films. With >4%/wt concentration of boron in the films, blisters form during annealing. Etch rates were similar to those of Kern & Schnable (SI-0011).

REF: Foster, T et al — *J Electrochem Soc,* 132,505(1985)

SIO-0018

ETCH NAME: Alcohol TIME:

TYPE: Alcohol, pinhole TEMP: RT

COMPOSITION:

(1) x .... alcohol (2) 1 .... HF

19 .... HNO_3

DISCUSSION:

SiO_2, thin films deposited on (100) silicon wafers used in a study of pinhole development on the oxides on the silicon substrates. A Narvonic Leak Detector was used. The wafer is placed under alcohol and a SST wire loop at an electrical potential of 20—80 V is submerged and passed over the oxidized wafer surface. The electrical potential will cause the alcohol to bubble at pinhole and defect points in the SiO_2 film. Etch solution shown is used to enhance the observed defects. It was shown that brush scrubbing or jet washing of wafer surfaces in cleaning prior to oxidation — washing from wafer center toward periphery — causes a high pinhole density that is propagated into both the grown oxides and nitrides that are most prominent at wafer center.

REF: Goodwin, C A & Brossman, S W — *J Electrochem Soc,* 129,1066(1982)

SIO-0019
ETCH NAME: TIME: 10 min
TYPE: Acid, oxidation TEMP: 90°C
COMPOSITION:
3 .... HCl
1 .... H_2O_2
1 .... H_2O
DISCUSSION:

SiO_2, thin films deposited on (111), p-type, 1—3 Ω cm resistivity wafers used as substrates for silicide formation. A hydrated oxide was grown on substrate surfaces in the solution shown. Follow with metal evaporation of Pt:Pd or Pt and Pd, only. Subsequent annealing used SiO_2 deposit as the source of silicon to form silicides.

REF: Kawarada, H — *J Appl Phys,* 57,244(1985)

SIO-0021a
ETCH NAME: EPD TIME:
TYPE: Acid, selective TEMP:
COMPOSITION:
20 ml .... ethylenediamine
20 g .... pyrocatechol
44 ml .... H_2O
DISCUSSION:

SiO_2, thin films deposited in etched grooves of (100) silicon wafers. A thin film Al_2O_3 thin film was deposited and used as a mask. After photo resist patterning grooves were etched with ⟨110⟩ orientation prior to SiO_2 RF plasma deposition. Solution shown used to etch pattern the grooves.

REF: Ho, V & Seguano, T — *Solid Thin Films,* 95,315(1982)

SIO-0021b
ETCH NAME: TIME:
TYPE: Acid, dislocation TEMP:
COMPOSITION:
200 ml .... HF
9 g .... CrO_3
100 ml .... H_2O
DISCUSSION:

SiO_2, thin films deposited on (100) silicon substrates with RF plasma CVD and by thermal oxidation. Used in a study of stacking faults developed in Si by SiO_2. No stacking faults were observed in plasma CVD deposited surfaces but were present in thermal oxidized surfaces (wet oxidation at 1150°C, 3 h). Refractive index, n = 1.46, for both types of oxide. Dielectric constant: e = 3.5—4.0, for both types of oxide. Solution used to remove SiO_2 and develop stacking faults in silicon substrates.

REF: Ibid.

SIO-0021c
ETCH NAME: Pliskin etch TIME:
TYPE: Acid, preferential TEMP: RT
COMPOSITION:
1 .... HF
1 .... 1 *M* Cr_2O_3

DISCUSSION:

SiO_2, thin films deposited on (100) silicon wafers. CVD SiO_2 was deposited on Al_2O_3 films used as photo resist patterned masks. Solution shown used to pattern grooves in silicon. See SIO-0021a-b.

REF: Ibid.

SIO-0022

ETCH NAME: Sulfuric acid
TYPE: Acid, cleaning
COMPOSITION:
x .... H_2SO_4

TIME: $^1/_2$ h minimum
TEMP: Hot

DISCUSSION:

Glass, soda-lime, blanks used to fabricate photo resist chrome masks for silicon and other compound wafer processing. After lap polishing glass surfaces with a CeO_2 slurry and water washing, blanks are soak/cleaned with sulfuric acid prior to leaching for Na replacement with K in a $KMnO_4$ at 350°C bath for 6 to 8 h. This treatment produces high quality glass blanks.

REF: Tarn, W H — personal communication (Optifilm Co., 1978)

SIO-0023

ETCH NAME: DE-100
TYPE: Ionized gas, selective/removal
COMPOSITION:
x *DE-100

TIME: 15—20 min
TEMP: 25°C start
GAS FLOW: 300 cc/min
PRESSURE: 0.14 Torr
POWER: 300 W

*DE-100 = CF_4:O_2 (8.5%)

DISCUSSION:

SiO_2, thin films deposited by SILOX system method on (100) silicon and GaAs:Cr (SI) wafers in processing microwave devices. Used in a general study of SILOX deposited SiO_2 and oxide removal. Solution also used on germanium, (111) wafers to remove SiO_2 deposits. Prior to DE-100, SiO_2 surfaces can be treated with 300 cc/min O_2 flow, 100 W power and 1.2 Torr pressure for 1—2 min. (*Note:* PDE-100 [CF_4:17.5% O_2] is 20—30% faster than DE-100.)

REF: Scientific Gas Products Company — personal communication, 1979

SIO-0024

ETCH NAME: Dry chemical etching
TYPE: Ionized gas, selective
COMPOSITION:
x .... CHF_3
x .... O_2

TIME:
TEMP:
GAS FLOW:
PRESSURE:
POWER:

DISCUSSION:

SiO_2, thin film deposits. A study of RIE selective etching of SiO_2 thin films for controlled structuring of deposits.

REF: Steinbruchel, C H et al — *J Electrochem Soc,* 132,180(1985)

SIO-0029: Castellano, R N — *Solid State Technol,* May 1984, 203

SiO_2, thin film grown as a cap deposit in an SiO_2/poly-Si/PSG layered structure. All three layers were deposited by LPCVD and used in study of the selectivity etch reactions between the three layers.

SIO-0025
ETCH NAME: TIME:
TYPE: Oxide, growth TEMP: ~1000°C
COMPOSITION:
x .... SiO_2
DISCUSSION:

SiO_2, thin film coatings of different types deposited on (100) GaAs:Cr and InP:Fe (SI) wafers as substrates for evaluation of the various films for device encapsulation for cap type annealing. Passivating films grown were SiO_2, Si_3N_4, BSG, PSG, BPSG, and ASG on GaAs and Si_3N_4 on InP. On the InP, In out-diffuses into the nitride and Si in-diffuses to the InP, but there is no cross-diffusion with PSG coatings. Also, deposition of AlN on silicon.

REF: Oberstar, J & Streetman, B G — *Thin Solid Films*, 103,17(1983)

SIO-0026
ETCH NAME: Steam TIME:
TYPE: Steam, oxide reflow TEMP: >100°C
COMPOSITION: PRESSURE:
x .... steam
DISCUSSION:

SiO_2, deposited on silicon wafer substrates as doped PSG thin film oxide. PSG with 7—12/g wt phosphorus deposited by CVD at 430°C. Anneal and reflow deposits in steam at 950°C, 15 min. A major study and analysis of reflow reactions of PSG coatings.

REF: Bowling, R A & Larrabee, G B — *J Electrochem Soc*, 132,141(1985)

SIO-0027
ETCH NAME: Oxide, growth TIME:
TYPE: Oxide, growth TEMP: ~1000°C
COMPOSITION:
x .... SiO_2
DISCUSSION:

SiO_2, thin films deposition on (100) silicon wafers at an oblique deposition angle. SiO_2 source was 5° off perpendicular toward wafer surface. Produces a columnar, slanted oxide structure as observed after cleaving the wafer ⟨110⟩. Refractive index, n = 1.47—1.49. (*Note:* This is the only reference to artificial growth of columnar SiO_2. The natural mineral chalcedony grows as an admixture colloidal/amorphous SiO_2 matrix with columnar growth sections. Columnar growth also observed in a number of epitaxy metallic compound thin films.)

REF: Levy, Y — *J Appl Phys*, 57,2600(1985)

SIO-0028
ETCH NAME: TIME:
TYPE: Acid, removal TEMP:
COMPOSITION:
x .... HF, conc.
DISCUSSION:

SiO_2, thin films deposited by a special technique. Fine grains of silica deposited from a crucible at 1500°C through a nozzle with the grains ionized by E-beam bombardment just as they reach the deposit substrate surface. The refractive index varies with the degree of ionization with n = 1.46 to 2.01, ion current, Ie, 0—1000 mA. Reaction: $2SiO_2$ to 2SiO + O_2 and SiO are stable as a gas mixture. SiO_2 as deposited is metastable with n = 1.99.

Deposited under low pressure O_2 deposit conditions is SiO_x. Increase O_2 pressure, and n = ~2. (*Note:* SiO_2 natural quartz refractive index is n = 1.54—1.55.)
REF: Wong J et al — *J Vac Sci Technol,* B3(1),453(1985)

SIO-0030
ETCH NAME: TIME:
TYPE: Oxide, growth TEMP:
COMPOSITION:
x .... SiO_2
DISCUSSION:

SiO_2, thin films deposited on (100) silicon wafers by LPCVD. Gas flow was directed upward to silicon wafers held in a "cage" in a vacuum bell jar with outside wall heating through water cooled coils. Deposition was by pyrolysis of $SiH_4/O_2/N_2$ at 400°C and 100 mTorr pressure. This was a study of this method of silica deposition.
REF: Learn, A J — *J Electrochem Soc,* 132,390(1985)

SIO-0031
ETCH NAME: TIME:
TYPE: Oxide, growth TEMP: <1000°C
COMPOSITION:
x .... SiO_2
DISCUSSION:

SiO_2, thin films deposited on (HgCd)Te wafers at <100°C. Pyrolysis of SiH_4 + NO_3 gases passed over liquid mercury as they enter the deposition chamber that was heated using a mercury resonance lamp for heating. Used to deposit SiO_2 without out-gassing of Hg or Cd from the substrates.
REF: *Microwaves & RF,* August 1984, 55

SIO-0032
ETCH NAME: TIME:
TYPE: Acid, selective TEMP:
COMPOSITION:
1 .... HF
19 .... HNO_3
14 .... H_2O
DISCUSSION:

SiO, deposition on aluminum and quartz blanks or silicon wafers. Deposition of the monoxide was from molybdenum boats in fabricating MIM devices as Al/SiO/Al/substrate. Various annealing temperatures evaluated. The solution shown was selected as the best for uniform MIM structure etching.
REF: Shabalon, A L & Feldman, M S — *Thin Solid Films,* 110,215(1983)
SIO-0033: Ablov, S H & Feldman, M S — *Thin Solid Films,* 110,225(1983)

SiO, monoxide thin films. Similar work as in SIO-0032.

SIO-0088
ETCH NAME: Hydrofluoric acid, dilute TIME:
TYPE: Acid, removal TEMP:
COMPOSITION:
1 .... HF
15 .... H_2O
DISCUSSION:

SiO_2, as a residual PSG surface film from poly-Si growth on (100) silicon wafer substrates

with heavy phosphorus doping which leaves the P-glass. Etch surface with solution shown to remove the residual PSG glass.
REF: Chang, C C et al — *J Appl Phys,* 130,1159(1983)

SIO-0034
ETCH NAME: Hydrofluoric acid TIME:
TYPE: Acid, removal TEMP:
COMPOSITION:
x .... HF
x .... H_2O
DISCUSSION:
SiO_2, thin film deposited on InP, (100) wafer substrates in a study of CVD oxides on indium phosphide.
REF: Bertrand, P A et al — *J Vac Sci Technol,* B(1),832(1983)

SIO-0015b
ETCH NAME: BHF TIME:
TYPE: Acid, removal TEMP:
COMPOSITION:
x .... 1.2 *M* HF
x .... 10.3 *M* NH_4F
DISCUSSION:
SiO_2, thin film deposits used in a study of the etching mechanisms of hydrofluoric acid solutions.
REF: Ibid.

SIO-0015c
ETCH NAME: Hydrofluoric acid TIME:
TYPE: Acid, removal TEMP:
COMPOSITION:
x .... 1 *M* HF
DISCUSSION:
SiO_2, thin film deposits used in a study of the etching mechanisms of hydrofluoric acid.
REF: Ibid.

SIO-0037
ETCH NAME: TIME:
TYPE: Phosphor, defect TEMP:
COMPOSITION:
(1) x .... Zn_2SiO_4:Mn (2) x .... CdS:Zn (3) x .... YV_2
x .... Freon, TF x .... Freon, TF x .... Freon, TF
DISCUSSION:
SiO_2, thin films deposited on silicon wafers and devices used in a study of defect recognition using fluorescent tracers. Both phosphor powders and dyes — organic and inorganic — used under a variety of conditions including electrovoltaic. Phosphors and dyes "plate" out at defect points. Solution (1) fluoresces green/yellow in short-wave UV; (2) fluoresces yellow in long-wave UV and (3) fluoresces bright red. Reaction is electrostatic or electrophoric.
REF: Kern, W et al — *RCA Rev,* 43,310(1982)

SIO-0038a
ETCH NAME: TIME:
TYPE: Acid, cleaning TEMP:
COMPOSITION:
40 .... HF
15 .... H_2SO_4
DISCUSSION:
Corning glass 7720. This solution will leave a white precipitate on the glass surface.
REF: Teeg, R O et al — Final Rep (15 April 1966), Control No. 65-0388f

---

SIO-0038b
ETCH NAME: TIME:
TYPE: Acid, cleaning TEMP:
COMPOSITION:
53 .... HF
37 .... HCl
DISCUSSION:
Corning glass 7720. This solution will not leave any precipitate on the glass surface. (See ALO-0038a.)
REF: Ibid.

SIO-0039a
ETCH NAME: Nitric acid, dilute TIME:
TYPE: Acid, cleaning TEMP: Hot
COMPOSITION:
x .... HNO_3
x .... H_2O
DISCUSSION:
Glass — various types. Different solution concentrations of nitric acid can be used as general glass surface cleaners. May leave craters in the glass surface depending upon glass formulation.
REF: Tichane, R M & Carrier, G B — *J Am Ceram Soc,* 44,606(1961)

SIO-0039b
ETCH NAME: TIME: (1) 10 min (2) 2 min
TYPE: Acid, cleaning
COMPOSITION: TEMP: (1) 95°C (2) 50°C
(1) x .... 1% NaOH (2) x .... 5% HCl
DISCUSSION:
Corning glass 7740 (borosilicate). Mechanically polish surfaces with calcium carbonate, $CaCO_3$. Etch clean sequentially, first in (1) and then in (2) at temperatures shown. DI water rinse after each etch cycle at RT. Both Nonex (Corning 7720) and Kimble (K-772) are attacked by moisture which leaches alkalies producing a frosted surface. Wash these glass surfaces in NH_4F 10 sec to counteract the frosting effect.
REF: Ibid.

SIO-0040
ETCH NAME:
TYPE: Acid, cleaning
COMPOSITION:
40 .... H_2SO_4
20 .... HF
40 .... H_2O
TIME: 30 sec
TEMP: RT
DISCUSSION:
Glass — various types. Solution used in a study of stress in glass. The following procedure was used: (1) etch in solution shown; (2) running DI water rinse, 15 sec; (3) repeat etch and (4) repeat rinse. Continue the three-step procedure for a total of 5 min etching time (10 cycles).
REF: Kistler, S S — *J Am Ceram Soc,* 45,59(1962)

SIO-0041a
ETCH NAME:
TYPE: Acid, cleaning
COMPOSITION:
5 .... HF
33 .... HNO_3
60 .... H_2O
2 .... Dreene (wetting agent)
TIME:
TEMP: RT
DISCUSSION:
Glass — various types, as both sheet and rods. Used as a general glass cleaning solution. It is called a variable solution. Also used on molybdenum.
REF: Crawley, R H A — *Chem Ind,* 45,1205(1953)

SIO-0039c
ETCH NAME:
TYPE: Acid, cleaning
COMPOSITION:
x .... NaOH
x .... $NaCO_3$
x .... H_3PO_4
TIME:
TEMP: Hot
DISCUSSION:
Glass — soda-lime. Solution used to remove contamination on soda-lime glass that is not removed by hot water washing, only.
REF: Ibid.

SIO-0041b
ETCH NAME:
TYPE: Acid, cleaning
COMPOSITION:
1000 ml .... H_2SO_4
35 ml .... $Na_2Cr_2O_7$, sat. sol.
TIME:
TEMP: RT
DISCUSSION:
Glass — various types. This has been a standard glass cleaning solution for a number of years. It can be mixed, used, stored and re-used for several months without loss of efficiency even though it turns coal-black in color. **CAUTION:** It is a severe oxidizing solution and should be handled with extreme care. According to Crabbs (SIO-0042), chromium ions are difficult to remove from cleaned surfaces even with extended water washing

(3 or 4 days). Because of this, it is no longer used for cleaning quartz diffusion tubes and parts where residual chromium ions can affect semiconductor electrical parameters.
REF: Ibid.
SIO-0042: Crabbs, G — personal communication, 1962

SIO-0043
ETCH NAME: AB etch (RCA) TIME: 15—20 min
TYPE: Acid, cleaning TIME: 80°C
COMPOSITION:

"A" 1—2 .... NH_4OH "B" 1—2 .... HCl
1—2 .... H_2O_2 1—2 .... H_2O_2
4—6 .... H_2O 4—6 .... H_2O

DISCUSSION:

Glass — thin film deposition and growth in bevelled moats of high power silicon transistors. Electrophorically deposit glass frit in isopropanol (IPA) containing NaOH or HF additives. Form borosilicate in ethyl acetate plus acetone or methanol. Add yttrium or magnesium nitrates to the glass and melt in a platinum boat at 1450°C, 4 h. Pour between water-cooled rollers as a thin sheet and then pulverize in a ball mill. Clean powdered glass in the solutions shown above. Silicon, (100) p-type wafers, 2—3 Ω cm resistivity were used in the transistor fabrication. Fill wafer moats formed to separate individual transistors with the glass frit described and fire at 700°C as a surface protectant coating of p-n-p junctions.
REF: Shimbo, M — *J Electrochem Soc,* 132,393(1985)

SIO-0044
ETCH NAME: Hydrofluoric acid, vapor TIME: 10 min
TYPE: Acid, removal TEMP: 30°C
COMPOSITION:

x .... HF, vapor

DISCUSSION:

Glass — microscope slides. Solution vapors used to roughen microscope surface to produce irregular protrusions of variable height (angstroms). The following process was used: (1) pre-clean with soap and water; (2) mask off one side; (3) heat to 100°C and then (4) immerse in a sealed enclosure of HF vapors at 30°C. After etching, remove and DI water rinse. Slides were used as-polished and as roughen etched in a study of metal thin films deposited on these slides. The roughening was used to improve adhesion of 3000 Å thick metal films. Metals studied were electrically (1) good conductors — noble metals; (2) moderate conductors — Sn, Pb, and Ni and (3) semi-metal — Bi.
REF: Boyd, G T et al — *Phys Rev,* B30,519(1984)

SIO-0039d
ETCH NAME: Hydrofluoric acid, dilute TIME:
TYPE: Acid, cleaning TEMP: Boiling
COMPOSITION:

1 .... 1% HF

DISCUSSION:

Glass — borosilicate and soda-lime. The solution as applied will roughen borosilicate glass surfaces. Soda-lime surfaces remain smooth.
REF: Ibid.

SIO-0045a
ETCH NAME: Hydrofluoric acid, dilute TIME:
TYPE: Acid, removal TEMP:
COMPOSITION:
x .... HF (1—25%)
DISCUSSION:

Glass — various types. Processing produces micro-cracks on glass surfaces and under exposure to atmospheric conditions alkali hydrates are formed which show long-term outgassing characteristics. By etch removal of 0.002 to 0.003" from surfaces with HF, glass strength is improved as much as by a factor of 10, though tension stress remains.

REF: Kohl, W H — *Handbook of Materials and Techniques for Vacuum Devices,* Reinhold, New York, 1967, 15

SIO-0045b
ETCH NAME: Heat TIME:
TYPE: Thermal, gas removal TEMP: See discussion
COMPOSITION:
x .... heat
DISCUSSION:

Glass — lead, soda-lime, borosilicate, etc. Water vapor and CO_2 primarily adsorbed on surfaces; H_2, N_2, O_2, and CO secondary. Bake out at 150°C (soda-lime); 200°C (lead), and 250°C (borosilicate and others) to desorb gases.

REF: Ibid.

SIO-0046
ETCH NAME: Hydrogen peroxide TIME: 15 min
TYPE: Acid, cleaning TEMP: Boiling
COMPOSITION:
x .... 5% H_2O_2
\+ .... NH_3 for pH 11
DISCUSSION:

Glass — various types. The following procedure was used in cleaning surfaces: (1) scrub in hot tap water +0.5% Igipal wetting agent; (2) rinse in running, hot tap water; (3) rinse in running DI water and (4) boil in H_2O_2 and rinse in DI water and N_2 blow dry.

REF: Kern, H E & Graney, E T — *ASTM Special Tech Pub,* #246(1959)

SIO-0047
ETCH NAME: Laser TIME:
TYPE: Ionic, preferential TEMP:
COMPOSITION: POWER:
x .... H^+ ions
DISCUSSION:

Glass — Pyrex substrates. A pulsed Excimer laser beam using H_2 used to pattern etch Pyrex glass at its softening point. An Al/O film deposited on SiO_2/silicon turns black under laser etching with major reduction of sheet resistance. Author says that color change is representative of different chemical etching properties.

REF: Ehrlich, D J — *J Vac Sci Technol,* B(3),1(1985)

SIO-0048
ETCH NAME: TIME:
TYPE: Acid, removal TEMP:
COMPOSITION:
x .... acids
DISCUSSION:
Glass — various types. This is a major reference for the etching, cleaning, and surface treatment available for glass.
REF: Holland, L — *The Properties of Glass Surfaces,* John Wiley & Sons, New York, 1964.

SIO-0049a
ETCH NAME: TIME: Variable
TYPE: Alkali, lap/etch TEMP: RT
COMPOSITION:
x .... Ce_2O_3
x .... H_2O
x .... glycerin
x .... 5% NaOH (KOH)
DISCUSSION:
Glass — soda-lime. Cerium oxide is a standard lapping powder for preparing chrome glass masks. Acid slurry shown was developed to reduce the amorphous surface smearing common to final polishing of glass blanks.
REF: Walker, P — personal development, 1975

SIO-0050
ETCH NAME: TIME: Variable
TYPE: Alkali, lap/etch TEMP: RT
COMPOSITION:

(1)		(2)	
x .... emery		x .... alumina	
x .... glycerin		x .... glycerin	
x .... water		x .... water	
		x .... 10—30% KOH (NaOH)	

DISCUSSION:
Glass — various types. Mixture (1) is a standard lapping slurry for glass (SIO-0050). Mixture (2) was developed for polishing vacuum glass ports to remove scratches and cleaning blanks of both Pyrex and quartz (SIO-0051).
REF: Hodgman, C D — *Handbook of Chemistry and Physics,* 27th ed, Chemical Rubber Co., Cleveland, OH, 1943, [solution (1)]
SIO-0049b: Ibid.
SiO_2, soda-lime glass used for photo resist chrome glass masks. Developed solution (2) as an acid slurry solution to reduce the fine smear surface that can remain during optical polish of glass.

SIO-0052
ETCH NAME: Aqua regia TIME: 1—4 h
TYPE: Acid, cleaning TEMP: RT
COMPOSITION:
3 .... HCl
1 .... HNO_3

DISCUSSION:

SiO_2, as quartzware. For general cleaning of laboratory glassware, epitaxy tubes, and parts quartzware. Immerse and soak parts until clean. Heavy wash with running DI water and dry under IR heat lamps.

REF: Zelinsky, T — personal communication, 1980

SIO-0053

ETCH NAME: Hydrochloric acid, vapor TIME: 15 min

TYPE: Acid, pinhole TEMP: Hot

COMPOSITION:

x .... HCl, vapor

DISCUSSION:

SiO_2, thin film RF sputtered. Authors say that dust in laboratory air will cause pinholes in an oxide film. Develop pinholes by soaking specimen in hot HCl vapors.

REF: Davidse, P D & Masiel, L I — *J Appl Phys,* 37,574(1966)

SIO-0054a-b

ETCH NAME: BHF TIME:

TYPE: Acid, removal TEMP: RT

COMPOSITION:

(1) 3 ml .... HF
15 g .... NH_4F
22 ml .... H_2O

(2) x .... HF, hot

DISCUSSION:

SiO_2, thin films thermally evaporated used as diffusion masks with KMER photo resist. The BHF solution (1) used for patterning with KMER. Hot HF used to strip oxide. A study of controlled etching in making thin film masks.

REF: Kelly, C E — *Electrochem Technol,* 2,358(1964)

SIO-0055

ETCH NAME: TIME:

TYPE: Acid, patterning TEMP:

COMPOSITION:

(1) 8% .... HF
84% .... HNO_3
8% .... HAc

(2) x .... BHF

DISCUSSION:

SiO_2, thin films deposited on silicon wafers for microcircuits using oxide isolation. Solution (1) was used to etch a groove through the oxide. After grooving all oxide was stripped in a BHF solution and surfaces re-oxidized.

REF: Schnable, G C et al — *J Electrochem Technol,* 4,57(1966)

SIO-0056

ETCH NAME: TIME:

TYPE: Acid, cleaning TEMP:

COMPOSITION:

x .... H_2SO_4
x .... x% CrO_3

DISCUSSION:

SiO_2 as (1) soda-lime glass and (2) fused silica microscope slides used as substrates for thin film metallization with chromium, copper, or nickel. The following cleaning procedure was used on the glass substrates before metallization:

1. Detergent clean: Liqui-nox:DIH_2O mixture with ultrasonic agitation and DI water rinse.
2. Acid clean: Use acid mixture shown (a standard "glass cleaner" solution) to remove organics.
3. Wash: DI water (deionized and distilled) — three times — in ultrasonic with rinsing to remove traces of acid.
4. Blow dry: He, single high pressure burst.

Metals were then evaporated: Ni, 2150 Å; Cu, 536 Å; and Cr, 1000 Å. Nickel on fused silica, others on soda-lime. As deposited, films were smooth with only an occasional pinhole. Thermal cycling developed pinholes/defects with crack networks. This was a study of failure mechanisms in thin metal films.
REF: Zito, R R — *Thin Solid Films,* 87,87(1982)
SIO-0067: Primak, W — *J Appl Phys,* 32,660(1961)

SiO_2, as vitreous silica glass used in a study of neutron induced damage and annealing. Cleaned in a standard glass cleaner, then in cool HCl, water wash, acetone rinse and dry.

SIO-0059
ETCH NAME: TIME:
TYPE: Acid, removal/jet TEMP: RT
COMPOSITION:
40 g .... $NH_4F.HF$
8 g .... NaF/L
DISCUSSION:

SiO_2, native oxide removal and etching of silicon as a jet etch solution. Produces flat bottomed pits in the silicon. Addition of $K_3F(CN)_6$ reduces pits size and formation.
REF: Schmidt, P F & Keipler, D A — *J Electrochem Soc,* 106,592(1959)

SIO-0061
ETCH NAME: Hydrofluoric acid TIME:
TYPE: Acid, defect TEMP:
COMPOSITION:
x .... HF, conc.
DISCUSSION:

SiO_2, thin film layers grown on silicon, (111) p-type, 6—12 Ω cm resistivity wafers. Wet thermal oxides grown 500—2000 Å thick at 1000°C, 1050°C, and 1100°C. Refractive index: n = 1.46. Silicon warps under the oxide coatings with severity depending upon thickness and deposition temperature. High oxidation temperature creates dislocations and warpage decreases with films under 0.5 μm thick.
REF: Lecni, H & Satch, S — *Jpn J Appl Phys,* 23,L743(1984)

SIO-0062
ETCH NAME: TIME:
TYPE: Acid, removal TEMP: RT
COMPOSITION:
5 g .... $NH_4F.HF$
5 ml .... HF
50 ml .... H_2O
50 ml .... glycerin

DISCUSSION:

SiO_2, thin film oxidation of silicon, (111) n-type wafers at 1050°C as wet oxide, 2000—2500 Å thick. A study of both oxides and nitrides (both DC plasma deposited) vs. thermal oxidation and dislocation development at the silicon interface. Study included pre-diffusion of wafers with nickel to reduce dislocation development. Solution used as a slow removal etch of both oxides and nitrides. Wafers were cleaved ⟨110⟩ for observation of dislocation depth. Better than a 50% reduction in dislocation development was observed in wafers pre-diffused with nickel.

REF: Walker, P & Onidera, G — Spring Meet Phys Failure Conf, Battelle Memorial Institute, Columbus, OH, 1964

SIO-0063

ETCH NAME: Hydrofluoric acid — TIME:

TYPE: Acid, removal — TEMP:

COMPOSITION:

x .... HF, conc.

DISCUSSION:

SiO_2, thin film oxidation of silicon at 1200°C under 5×10^{-6} Torr and with added phosphorus (P_{H_2O}) vapor at 1180°C for PSG cover film. Used in a study of stress anisotropy of the films.

REF: Priest, J et al — *J Appl Phys,* 34,347(1963)

SIO-0064

ETCH NAME: Hydrofluoric acid — TIME:

TYPE: Acid, removal — TEMP:

COMPOSITION:

x .... HF, conc.

DISCUSSION:

SiO_2, grown as spherolitic pellets of alpha-cristobalite at the contact point of GaAs/quartz boat that were sandblasted. Alpha-cristobalite m.p. is 1713°C — it formed at less than 1250°C. GaAs, m.p. is 1238°C.

REF: Yamaguchi, M et al — *J Electrochem Soc,* 113,294(1966)

SIO-0065

ETCH NAME: Water — TIME:

TYPE: Acid, float-off — TEMP:

COMPOSITION:

x .... H_2O

DISCUSSION:

SiO_2, glass microscope slides used as substrates for epitaxy growth of ZnTe with cover aluminum metallization. The ZnTe was thermally evaporated from powder, and aluminum evaporation was 0.1 to 0.5 mm thick. Thin film was removed from glass by the float-off technique using water. Carbon replication made for TEM study.

REF: Patel, S M & Patel, N G — *Thin Solid Films,* 113,185(1984)

SIO-0066

ETCH NAME: Steam — TIME:

TYPE: Oxide, growth — TEMP: >500°C

COMPOSITION: — PRESSURE: >25 atms

x .... H_2O, steam

DISCUSSION:

SiO_2, thermally grown oxide on silicon p-n junction devices in steam between 500—900°C at pressures between 25—500 atm. Film growth appeared to be linear to about 50,000 Å thickness. At any given temperature there is a pressure above which the oxides are soluble in steam and fail to form. 500°C limit is 500 atm; 850°C limit is 150 atm. Cleaning of silicon prior to oxidation is critical. The following sequence was used:

1. Mechanically lap and polish etch silicon.
2. Soak in HNO_3, rinse in H_2O, soak in HF, rinse three times in H_2O.
3. Boil in HNO_3, rinse two times in H_2O and then in Hi-Q H_2O.

Note: Following HF step in (2), hold specimens under water in quartz beakers.

Recommend that oxide films be at least 3000 Å thick for maximum ambient sensitivity.
REF: Madden, T C & Gibson, W M — *Rev Sci Inst,* 34,50(1963)
SIO-0092: Madden, T C & Gibson, W B — *J Appl Phys,* 27,1418(1966)

SiO_2, thin film passivation on silicon p-n junction particle detectors. There can be bulk impurities in oxide with a p-type inversion layer at the SiO_2/Si interface which reduces with increased resistivity. Film is less impervious to water prior to phosphorus diffusion but softens after. A "ring" structure around active junctions reduces leakage from 30 to 0.6 μA at 400 V.

SIO-0069
ETCH NAME: TIME:
TYPE: Gas vapor, growth TEMP: Hot
COMPOSITION:
(1) x .... H_2O_2, vapor (2) x .... steam, O_2 (3) x .... steam, O_3
DISCUSSION:

SiO_2 and SiO_3, wet oxides used to "age" silicon devices using the three solutions shown. Silicon was etched in CP4 with 10 drops Br_2/50 ml solution. CP4 reactivity was varied with addition of acetic acid (HAc). See GE-0065c for additional use of CP4.
REF: Blankenship, J L & Borkowski, C J — *IRE Trans Nuc Sci,* NS-7,190(1960)

SIO-0070
ETCH NAME: TIME:
TYPE: Halogen, cleaning TEMP: RT
COMPOSITION:
3 mg .... I_2
100 ml .... MeOH
DISCUSSION:

Glass, as microscope slides and cover glasses used as physical contact masks on semiconductor wafers and as monitors for thickness measurement in study of metals reactions on metallized wafer surfaces. Glass was cleaned by (1) 1HF:1H_2O dip and water rinse; (2) 1HCl:1H_2O dip and water rinse, then (3) rinse in solution shown and N_2 blow dry. Also used on fused quartz substrate blanks prior to metallization and fabrication for microelectronic circuits. The iodine complexes and removes ionic contamination on surfaces at above 100°C, leaving a molecularly clean surface.
REF: Walker, P et al — material development, 1964/1980

SIO-0071
ETCH NAME: BHF TIME:
TYPE: Acid, removal TEMP:
COMPOSITION:
1 .... HF
10 .... NH_4F
DISCUSSION:

SiO_2, thin films 160 nm thick. Films were nitrided at 1200°C in N_2 or NH_3 for 5 or 17 h. Resultant silicon oxynitride films were called "nitroxide thin films". Solution shown was used to determine etch rates of pure SiO_2 and the nitroxides at the two temperature levels. Refractive index of SiO_2, n = 1.46 and of nitroxide, n = 1.6—1.7. Dielectric constant of nitroxide, e = 4—5.

REF: Kato, I et al — *J Electron Mater,* 12,913(1984)
SIO-0072: Gordon, L — *J Appl Phys,* 34,1220(1963)

SiO_2, thin films grown on polycrystalline silicon called "polyoxide". Etch remove with BHF solutions.

SIO-0082: Mastroianni, S T — *Solid State Technol,* May 1984, 155

SiO_2, thin films characteristics evaluated.

SIO-0073a
ETCH NAME: Hydrofluoric acid, dilute TIME:
TYPE: Acid, float-off TEMP: RT
COMPOSITION:
x .... HF
x .... H_2O
DISCUSSION:

Corning glass #7059, used as substrates for epitaxy growth of amorphous $ZnGeAs_2$. Solution shown used to separate a-$ZnGeAs_2$ from the glass by the float-off technique for TEM study.

REF: Shah, S I & Greene, J E — *J Cryst Growth,* 68,537(1984)

SIO-0073b
ETCH NAME: TIME:
TYPE: Solvent, cleaning sequence TEMP:
COMPOSITION:
(1) x .... soap (2) x .... TCE (3) x .... acetone (4) x .... MeOH
x .... H_2O
DISCUSSION:

Corning glass #7059 used as substrates for a-$ZnGeAs_2$ epitaxy. Glass was cleaned by scrubbing with a detergent (1), then degreased successively with TCE, acetone, and MeOH rinses. Final blow dry with dry nitrogen jet.

REF: Ibid.

SIO-0074
ETCH NAME: Secco etch TIME:
TYPE: Acid, preferential TEMP:
COMPOSITION:
2 .... HF
1 .... 0.15 *M* $K_2Cr_2O_7$
DISCUSSION:

SiO_2, thin films grown on silicon, (100), n-type substrates used in a study of indium diffusion and oxidation reactions. Both dry O_2 and steam used at different temperatures and

times. Secco etch used to develop stacking faults and other defects in both the oxide and silicon wafers at the SiO_2/Si interface.
REF: Cerofolini, G F et al — *J Cryst Growth,* 109,137(1983)

SIO-0075
ETCH NAME: Carbon black TIME: N/A
TYPE: Metal, decoration TEMP: RT
COMPOSITION:
x .... C, powder
DISCUSSION:
SiO_2, thermally oxidized thin films on p-type (100) silicon wafers used in a study of pinholes in oxides and phosphorus induced microdefects in silicon. Oxide was converted to PSG with P diffusion. Pinhole density was measured with both Metso etching and carbon black dusting of surface to enhance pinholes. Microdefects in silicon could be related to pinhole locations in the covering oxide.
REF: Blackstone, S et al — *J Electrochem Soc,* 130,667(1982)

SIO-0076
ETCH NAME: Steam TIME:
TYPE: Gas vapor, re-flow TEMP: >100°C
COMPOSITION: PRESSURE:
x .... H_2O, vapor
DISCUSSION:
SiO_2, thin films on silicon wafers as doped BPSG. Steam used to re-flow oxide after deposition. With greater than 0.4% by weight concentration of boron, films blister during temperature annealing.
REF: Harris, T — *J Electrochem Soc,* 132,501(1985)

SIO-0077a
ETCH NAME: PAW (EDP) TIME: 1—2 h
TYPE: Acid, pinhole TEMP: 110—115°C
COMPOSITION:
1 .... ethylenediamine
1 .... pyrocatechol
4 .... H_2O
DISCUSSION:
SiO_2, thin films RF sputtered 200—700 nm thick on (100) silicon wafers and used in a study of pinhole density development. Etch ratio of $Si:SiO_2$ is about 1000:1 with the solution shown. Etching was done in a ventilated chemical sink to remove generated H_2 and under N_2 ambient to prevent oxidation of amine. After etching, remove SiO_2 and count etch pits in silicon for pinhole density. Method evaluated against Cu-decoration and RF plasma $CF_4:O_2$ etching. Plasma etching developed 10—15% less pinholes due to low conductance of pinholes and up to 50% less with Cu-decoration due to Cu coalescence in the presence of high pinhole density. Pinhole concentration density ranged from 10^5 down to about 6×10^3 as film thickness increased from 200 to 700 nm. PAW and RF plasma results were equivalent, but Cu-decoration was erratic. The plasma method is considered reliable and more easily applied.
REF: Meguro, T et al — *J Electrochem Soc,* 128,1379(1981)

SIO-0077b
ETCH NAME: Copper — TIME:
TYPE: Metal, decoration — TEMP:
COMPOSITION:
x .... $CuSO_4$
DISCUSSION:
SiO_2, thin films deposited on (100) silicon wafers used in a study of pinhole density. During etching a nonuniform electric field is applied to deposit copper at pinhole locations in the oxide, but will also precipitate randomly on the oxide film if the surface is contaminated. See SIO-0077a for additional discussion.
REF: Ibid.
SIO-0078: Shannon, W J — *RCA Rev*, 31,431(1970)
Reference for copper solution and method applied in SIO-0077b.

SIO-0077c
ETCH NAME: RF plasma — TIME:
TYPE: Ionized gas, pinholing — TEMP:
COMPOSITION: — GAS FLOW:
x .... CF_4 — PRESSURE: 0.25 Torr
x .... O_2 (10%) — POWER: 50 W
DISCUSSION:
SiO_2, thin films deposited on (100) silicon wafers and used in a pinhole density study. Si:SiO_2 etch ratio was maximized with 10% O_2 at 1:130, though about 100 nm of SiO_2 had to be removed before pinhole etched points in silicon were of sufficient size for count evaluation. Density count was 10—15% less than the PAW method, but reliable and more easily applied. See SIO-0077a for further discussion.
REF: Ibid.
SIO-0081: Kotani, H et al — *J Electrochem Soc*, 130,645(1983)
SiO_2, as PSG thin films used as interconnects on semiconductor devices, along with p-SiN. AlSi and AlSiCu thin films. Carrier gas was argon at 1.7 Pa pressure and 0.5W/cm^2 power. PSG etch rate was 60 Å/min.; p-SiN at 70 Å/min. At greater than 200°C, the AlSi etch rate was 100 Å/min.

SIO-0079
ETCH NAME: — TIME:
TYPE: Colloid, replication — TEMP:
COMPOSITION:
x .... 4% collodion
x .... amyl acetate
DISCUSSION:
Si, (100) n-type wafers used in a study of island stage growth of oxidation. Wafers were cleaned prior to oxidation: (1) 10% HCl, RT for 10 min and DI water rinse; (2) chem/mech polish on pellon pad with 1% Br_2:MeOH; rinse in MeOH, then acetone and N_2 blow dry. After oxidation, surfaces were coated with solution shown, followed by water lift-off collodion for replica microscope study.
REF: Makky, W H & Wilmsen, G W— *J Electrochem Soc*, 130,659(1983)

SIO-0080
ETCH NAME: — TIME:
TYPE: Acid, removal — TEMP:
COMPOSITION:
x .... HF

DISCUSSION:

Si, (111) n-type 3—5 Ω cm resistivity wafers used in a study of SiO_2 properties as RF sputtered thin films. Wafers were cleaned: (1) TCE degrease; (2) acetone rinse; (3) $3H_2SO_4$:$1H_2O$, RT dip; (4) 1HF:$10H_2O$, RT dip and DI H_2O rinse. Oxide was etched in concentrated or dilute HF.

REF: Lels, M K — *J Electrochem Soc,* 130,658(1983)

SIO-0083

ETCH NAME: TIME:

TYPE: Mechanical, fracture TEMP: RT

COMPOSITION:

x .... fresh fracture

DISCUSSION:

SiO_2, glass specimens. A study of surface and internal structure of freshly fractured glass under an electron microscope. Fracture shown to be 50—300 Å deep with a granular surface structure. (*Note:* Common glass fracture is conchoidal with highly reflective surfaces.)

REF: Narez, M & Sella, C — *C R Acad Sci (Paris),* 250,4325(1960)

SIO-0084

ETCH NAME: Hydrofluoric acid TIME:

TYPE: Acid, float-off TEMP:

COMPOSITION:

1 .... HF

10 .... H_2O

DISCUSSION:

Pyrex blanks used for co-deposition of CuPd thin films in a material study. Quartz blanks evaluated, but they develop a PdSi film. Solution shown used to clean blanks before metallization. Also used for film float-off and TEM study.

REF: van Langrveld, A D et al — *Thin Solid Films,* 109,193(1983)

SIO-0094: Lee, M K — *J Electrochem Soc,* 130,648(1983)

SiO_2, RF sputtered thin films on (111) silicon, n-type, 3—5 Ω cm resistivity, used in a properties study of deposited films. Prepare silicon wafers: (1) degrease in TCE plus acetone rinse; (2) acid clean with $3H_2SO_4$:$1H_2O$ and DI H_2O rinse; (3) dip in 1HF:$10H_2O$ rinse, prior to oxidation.

SIO-0095: Ligenza, J R — *J Electrochem Soc,* 109,73(1962)

SiO_2, thin films deposited on silicon wafers in a general study of oxidation.

SIO-0097: Ligenza, J R — *J Phys Chem,* 65,2011(1961)

SiO_2, thin films deposited on silicon wafers in a general study of oxidation.

SIO-0098: Ligenza, J R & Spitzer, W G — *J Phys Chem Solids,* 14,131(1960)

SiO_2, thin films deposited on silicon wafers in a general study of oxidation.

SIO-0099: Spitzer, W G & Ligenza, J R — *J Phys Chem Solids,* 17,196(1961)

SiO_2, thin films deposited on silicon wafers in a general study of oxidation.

SIO-0085

ETCH NAME: TIME:

TYPE: Oxide, adhesion coat TEMP:

COMPOSITION:

(1) x .... *HMDS (2) x .... TiO_2, ZrO_2, HfO_2

*Hexamethyldisilazane

DISCUSSION:

SiO_2, thin films deposited on a variety of substrates/surfaces (glass, Si_3N_4, GaAs, Si, etc.). Highly polished surfaces show poor adhesion of photo resist lacquers. The materials shown above were deposited after the SiO_2 to improve adhesion. Spun on HMDS was not as effective as the oxides. For a TiO_2 film: bubble $TiCl_4$ in N_2/O_2 using a viton tube with glass nozzle at RT, then dry with He gas. 25 Å TiO_2 deposits in 5 sec at RT; 25 Å in 2—3 sec at 100°C. Photo resist was 1350J. (*Note:* These very thin oxide films produce a granular/island-like surface that improves photo resist adhesion.)

REF: Marinace, J C & McGibbon, R C — *J Electrochem Soc,* 129,2389(1982)

SIO-0086a

ETCH NAME: Hydrochloric acid — TIME:

TYPE: Acid, pinhole — TEMP: Hot

COMPOSITION:

x .... HCl, vapor

DISCUSSION:

SiO_2, thin films deposited on silicon substrates in a study of pinholes. HCl vapor will attack at pinhole locations, and etch down into the underlying silicon producing recognizable etched pits.

REF: Sullivan, M V — 1st Kodak Seminar on Microminiaturization, 1965, 30

SIO-0086b

ETCH NAME: EPD — TIME: 6 h

TYPE: Acid, pinhole — TEMP: 115°C

COMPOSITION:

17 ml .... ethylenediamine
3 g .... pyrocatechol
8 ml .... H_2O

DISCUSSION:

SiO_2, thin films deposited on silicon substrates in a study of pinholes. Solution used to develop pinholes in the SiO_2 coating.

REF: Ibid.

SIO-0087

ETCH NAME: — TIME:

TYPE: Acid, step-etch — TEMP:

COMPOSITION:

1 .... HF
10 .... NH_4HF
15 .... H_2O

DISCUSSION:

SiO_2, thin films RF sputter deposited in argon on (100) oriented silicon wafers in a study of argon pressure vs. film etch rates. Solution used to step-etch films for deposit thickness measurements.

REF: Ciachi, T & Serikawd, T — *J Electrochem Soc,* 131,2720(1984)

SIO-0088

ETCH NAME: RIE — TIME:

TYPE: Ionized gas, removal — TEMP:

COMPOSITION: — GAS FLOW: 4—45 SCCM

x .... CHF_3 — PRESSURE: 45—90 mT

POWER: 500—2500 W

DISCUSSION:

SiO_2, thin films grown on (100) silicon wafers. Films grown by CVD at 950°C for 10 KÅ thickness. A study of etching SiO_2 films by RIE. Prior to RIE etching with gas shown, clean SiO_2 surfaces for 10 min in O_2.

REF: Light, R W & See, F C — *J Electrochem Soc,* 129,1152(1982)

SIO-0089

ETCH NAME: Boron | TIME:

TYPE: Metal, diffusion | TEMP: 1300 K

COMPOSITION:

x .... BSG

DISCUSSION:

BSG, as borosilicate glass on silicon. A thin sheet of this boron doped glass when wetted to a silicon wafer surface and heated to 1300 K will diffuse boron into the wafer as a p-type layer several microns deep with material under vacuum. (*Note:* Remove excess glass with HF.)

REF: Allen, F G et al — *J Appl Phys,* 31,979(1960)

SIO-0090: Tarn, W H & Walker, P — personal application, 1970

SiO_2 thin film layers about 2500 Å thick grown on (111) and (100) silicon wafers by wet oxidation for use as a diffusion mask. Patterns were opened in the silica layer by photolithography. With wafers in vacuum system liquid Boracine was bubbled with nitrogen gas as the carrier for pyrolitic diffusion of boron. Some "rosette" crystalline structures of borosilicate glass observed in the oxide thin films after diffusion.

SIO-0091a-c

ETCH NAME: | TIME:

TYPE: Solvent, cleaning | TEMP: Hot to boiling

COMPOSITION:

(1) x .... TCE (2) x .... acetone (3) x .... ISO/IPA

DISCUSSION:

SiO_xN:H and Si:H thin films, latter as cover film, deposited on glass, UV grade quartz, SnO_2, Al, and SST substrates. For good film adhesion, clean substrates: (1) TCE, boiling; (2) acetone, warm; (3) ISO, boiling before hydrogenated growth of films.

REF: Carasco, F et al — *J Appl Phys,* 57,5306(1985)

SIO-0093

ETCH NAME: Heat | TIME:

TYPE: Thermal, conversion | TEMP: 870°C

COMPOSITION:

x .... heat

DISCUSSION:

a-SiO_2, thin films used as a diffusion mask on silicon wafers can crystallize to cristobalite, SiO_2 at about 870°C in the presence of catalysts, as Fe_2O_3, CaO:KCl; NaCl:P_2O_5; Be_2O. Catalysts also affect c-SiO_2 similarly. Use of PBr_3 as diffusant at 170°C is possibly better due to rapid vaporization of bromine, and improves reverse leakage. (*Note:* The 670°C temperature is an inversion point between beta-quartz and alpha-tridymite.)

REF: Klerer, J — *Bell Tel Lab Rep,* 1964

SIO-0096: Silverman, S J & Singleton, J B — *J Electrochem Soc,* 105,891(1958)

SiO_2 as a residual PSG film after phosphorus diffusion can be removed with HF. Paint-on: 2 g $NiCO_3$:5 g P_2O_5:100 ml 2-methoxethanol — use camel hair brush. Diffuse at 1250°C, 16 h for 1.5 mil diffusion depth of phosphorus.

SIO-0100
ETCH NAME: Glass TIME:
TYPE: Frit, passivation TEMP: 800°C
COMPOSITION:
x .... SiO_2, powder
DISCUSSION:

SiO_2, as finely powdered glass frit in a water slurry. A camel hair brush used to paint-on coat aluminum Sphere Alloy Zener diodes as final passivation after package stud mounting. Fired in air at 800°C to melt-form as glass covering. Time at temperature variable with thickness and/or agglomeration of frit. (*Note:* Agglomeration is a major problem in the application of frit slurries with incomplete melt resulting in unmelted frit, bubbles, and poor structuring with stress in the coating film.)

REF: Walker, P — personal development, 1963

SIO-0101
ETCH NAME: Potassium hydroxide TIME:
TYPE: Electrolytic, oxidizing TEMP:
COMPOSITION: ANODE:
x .... KOH CATHODE:
x .... H_2O POWER:
DISCUSSION:

SiO_2, grown as a hydrated oxide on silicon wafers used in a study of oxidation rates.

REF: Glembocki, O J & Stahlbush, R E — *J Electrochem Soc,* 132,145(1985)

SIO-0102
ETCH NAME: Water TIME:
TYPE: Acid, oxidation TEMP: Boiling
COMPOSITION:
x .... H_2O
DISCUSSION:

SiO_2, grown as a hydrated oxide on silicon wafers as a surface cleaning step. After oxide growth, strip with HF or BHF, rinse in DI water, and dry.

REF: Heimann, R B — *J Vac Sci Technol,* B(1),108(1983)

SIO-0103
ETCH NAME: Nitric acid TIME:
TYPE: Acid, oxidation TEMP: Hot
COMPOSITION:
x .... HNO_3, conc.
DISCUSSION:

SiO_2, grown as a hydrated oxide on silicon wafers as a surface cleaning step. After oxidizing strip with HF or BHF solutions, rinse in DI water and dry.

REF: Yang, H T & Berry, W S — *J Vac Sci Technol,* B(2),206(1984)

SIO-0104
ETCH NAME: Potassium hydroxide TIME: 5—10 min
TYPE: Alkali, oxidation TEMP: − 10°C to RT
COMPOSITION:
x .... 5—10% KOH
DISCUSSION:

SiO_2, grown as a hydrated oxide on silicon wafers as a surface cleaning step. Soak wafers in the solution shown below or up to RT, but not heated. After hydroxide treatment

strip with HF or BHF, rinse in DI H_2O, then N_2 blow dry. This was part of a study of cleaning of silicon surfaces prior to device diffusion, alloying, and metallization.
REF: Walker, P et al — development, 1956—1968

SIO-0107
ETCH NAME: Plastic TIME:
TYPE: Organic, coating TEMP:
COMPOSITION:
x .... plastic
DISCUSSION:
SiO_2, drawn for fiber optics and laser applications. Fibers are coated with plastic as part of the drawing process to prevent structural damage from handling during further assembly.
REF: Watkins, L S — *Proc IEEE,* 70,626(1982)
SIO-0108: Pastor, A C — *J Cryst Growth,* 70,295(1984)
$NaNO_2$ grown as single crystal fibers. This initial development method can be applied to other materials.

WOP-0001
ETCH NAME: Hydrogen peroxide TIME:
TYPE: Acid, selective TEMP:
COMPOSITION:
x .... H_2O_2
DISCUSSION:
$W_2O_3(PO_4)_2$ as an amorphous glassy thin film. Material fabricated by mixing WO_3/H_3PO_4/100 ml H_2O, then drying 24 h at 110°C. Form melt in an electric furnace at 1200—1350°C using a platinum crucible for 2 h under an argon atmosphere. Fabricate thin films by pouring onto an aluminum plate at RT and quenching. Used in a study of glassy metallics.
REF: Kim, C U & Condrate, R A Jr — *J Phys Chem Solids,* 45,1213(1984)

LIST-0001
ETCH NAME: TIME:
TYPE: Acid, removal TEMP:
COMPOSITION:
x .... HF
x .... HNO_3
DISCUSSION:
$LiInS_2$, thin films on (111) silicon wafer substrates. Evaporated as both an amorphous glassy and a (100) single crystal thin film.
REF: Keriyama, K & Saitch, J — *J Solid Thin Films,* 111,331(1984)

GEO-0004
ETCH NAME: Ethyl alcohol TIME:
TYPE: Alcohol, cleaning TEMP: RT
COMPOSITION:
x .... EOH
DISCUSSION:
GeO glass. Powdered GeO was fired in an arc furnace from 1200 to 1600°C. Blanks cut from glass and polished with 400-grit SiC in petroleum lubricants. Wash in EOH and anneal at 230°C in vacuum 30 min.
REF: Magrunder, R H et al — *J Appl Phys,* 57,345(1985)

PDNP-0001

ETCH NAME: Aqua regia — TIME:

TYPE: Acid, clean/removal — TEMP: RT

COMPOSITION:

3 .... HCl

1 .... HNO_3

DISCUSSION:

PdNiP, PtNiP, and PtCuP melt formed as metallic glasses in a materials study. As metastable single crystals: TiAl, TiNi, and ZrV. Fe_3C (cementite) with a high resistance of 1000 Ω at 25°C. Glass structure referred to as a dense random packing of spheres. Solution shown used as a general etch.

REF: Dumez, P — *J Vac Sci Technol,* B(1),218(1983)

GLA-0001

ETCH NAME: — TIME:

TYPE: Frequency, alteration — TEMP:

COMPOSITION:

x .... acoustics

DISCUSSION:

SiO_2 glass used in a study of the influence of frequency irradiation on OH^- concentration. Relaxation of the lattice structure may be frequency dependent? The Si-O-Si bond shows small change with lateral motion of oxygen, and the presence of OH^- radicals shows greater effect.

REF: Jager, R E — *J Am Ceram Soc,* 51,57(1968)

PHYSICAL PROPERTIES OF SILICON NITRIDE, Si_3N_4

Classification	Nitride
Atomic numbers	14 & 7
Atomic weight	140
Melting point (°C)	1900 (sublimes)
Boiling point (°C)	
Density (g/cm³)	3.18
Refractive index (n=)	2.00
Application limits (°C) in air	1400
inert atmosphere	1850
Specific heat (J/kg °C)	1050
Coefficient of linear thermal expansion ($\times 10^{-6}$ cm/cm/°C) 25—800°C — alpha/beta	2.9/2.3
Thermal conductivity (W/m °C) 1200°C	9.5
Electrical resistivity (ohms cm) 25°C	1013
480°C	1010
Dielectric constant (e=)	4—6
Hardness (Mohs — scratch)	6—7
Crystal structure (isometric — normal) alpha	(100) cube
(hexagonal — normal) beta	$(10\bar{1}0)$ prism
Color (solid)	Clear/greenish
Cleavage (imperfect)	(001)/(0001)
Fracture (solid)	Conchoidal

SILICON NITRIDE, Si_3N_4

General: Does not occur as a natural compound. The atmosphere is approximately 75% nitrogen and 24% oxygen but, where oxygen is an active gas combining to form the silicate, oxide minerals, and many others, nitrogen is inert and does not form pure nitrides under normal atmospheric conditions. There are nitrates containing the NO_3^- radical which, as they are largely soluble in water, are found in dry, arid desert areas. Soda niter, $NaNO_3$ (Chile saltpeter) and niter, KNO_3 (saltpeter) are the two major minerals. There are important deposits in the dry, arid regions of Chile and western Bolivia, mainly as soda niter. Niter is common in the soil of Spain, Italy, Egypt, Arabia, Persia, and India. It is also found in the southwestern U.S. and with the Chilean soda niter deposits. They produce an alkaline earth that can be dissolved during rain, then redeposited under hot desert conditions. Such land areas also include the iodates and chlorides in the alkaline soil. Nitrate minerals are important as fertilizers as are the nitrites.

The nitridization process has been used in the heavy metals industry for many years. The metal or alloy part is thermally annealed at high temperature in a nitrogen gas atmosphere to convert the outer layer to a metallic nitride. Such surfaces have increased stability against atmospheric corrosion and show improved wear resistance.

The Rand Corporation developed the first silicon nitride for Solid State application, and Thompson-Ramo-Woolridge (TRW), Semiconductor Division developed the silicon oxynitrides in the early 1960s for application on semiconductor and similar devices. Silicon nitride is more stable and inert than silicon dioxide.

Technical Application: The pure silicon nitride, Si_3N_4, is difficult to etch, but forms an isomorphous series with silicon dioxide, SiO_2, such that the amount of oxygen in the compound can be varied from 0 to 100%. An oxynitride composition of $Si_3N_xO_y$ (5% O_2) has been found more compatible in wet chemical solution etch processing than the pure silicon nitride.

The compound has been deposited as an amorphous thin film by pyrolysis and CVD from gaseous sources; by DC or RF sputter from a solid Si_3N_4 target in argon or with a silicon bar in nitrogen. It also has been grown as a single crystal with a low temperature alpha phase (cubic) and a high temperature beta phase (hexagonal), the latter >1200°C. By laser annealing, the amorphous thin film can be converted to larger area crystallite structure. It also has been deposited as a hydrogenated colloidal structure, SiN:H, which may be as an oxynitride.

In Solid State processing the applications are similar to those of silicon dioxide: as a mask against diffusion, alloying, epitaxy, metallization, etching, and radiation. Also as a more stable passivation coating, as an active dielectric, such as for an MNOS diode or IC devices. It also has been used as an anti-reflective (AR) coating for solar cells, on glass masks, and lenses.

Silicon and germanium nitrides form an isomorphous series, and have been studied as various mixture combinations. Many other metal nitrides have been developed since the first silicon nitrides, and represent a major group of on-going experimental and application evaluations.

The combination of silicon oxide and nitride layers are used as either oxide/nitride or nitride/oxide. In the first case, the oxide acts as a pattern mask for etching the nitride; whereas in the second case, the oxide acts as an etch-stop mechanism. Both compounds have been doped with boron or phosphorus for special applications, to include use as dopant drive-in sources.

Etching: HF, BHF, hot H_3PO_4. $HF:HNO_3$, $HF:H_2O_2$. Dry chemical etching with ionized gases.

SIN-0008a

ETCH NAME: BHF — TIME: 1—2 min

TYPE: Acid, removal — TEMP: RT

COMPOSITION:

50 ml .... HF

100 ml .... NH_4F, sat. sol.

100 ml .... H_2O

DISCUSSION:

Si_3N_4, thin film amorphous deposits, 1500—3000 Å thick, by DC sputter from a silicon bar in nitrogen. Wafer substrates were silicon, (111) n-type, 5—10 Ω cm resistivity used in an etch development study of silicon nitride and oxynitride in initial material development. The solution shown very slowly attacks pure silicon nitride, but rates are equivalent to silicon dioxide for silicon oxynitride (2—10% O_2). Photo resist evaluated as KMER with development of compatible etch solutions. The solution was used to thin nitrides, etch pattern photo resist devices, and for pinhole study.

REF: Walker, P & Hersch, N — initial development, 1962—1963

SIN-0022d: Ibid.

a-Si_3N_4:H thin films deposited by CVD on (100) silicon, n- or p-type. A BHF solution was used to etch pattern the hydrogenated nitride. Also used hot H_3PO_4 at 180°C, concentrated HF and dry chemical etching with CF_4/O_2. HF was most rapid.

SIN-0008b

ETCH NAME: BHF, modified — TIME: 1—10 min

TYPE: Acid, removal — TEMP: RT

COMPOSITION:

50 ml .... HF

50 g .... $NH_4F.HF$

100 ml .... H_2O

50 ml .... glycerin

DISCUSSION:

Si_3N_4, thin films. See discussion under SIN-0008a. Use of ammonium bifluoride as an unsaturated solution reduces crystallization of the saturated fluoride solution in solution due to evaporation during extended etching periods. The inclusion of glycerin produces a more uniform removal and a cleaner surface of both oxides and nitrides, and the solution shown is compatible with photo resist processing.

REF: Ibid.

SIN-0008c

ETCH NAME: — TIME: 10—20 min

TYPE: Acid, removal — TEMP: Boiling

COMPOSITION: — RATE: 160 Å/min

18 g .... NaOH

5 g .... $KHC_8H_4O_4$

100 ml .... H_2O

DISCUSSION:

Si and SiO_xN_y DC sputtered thin films on (111) silicon wafers, n-type, 5—10 Ω cm resistivity. Solution was very good on oxynitride (5% O_2). Rate can be reduced by using below boiling but is too slow for use at RT. It is compatible with photo resist but timing is critical as it will attack silicon preferentially at the boiling point.

REF: Ibid.

SIN-0001

ETCH NAME: Phosphoric acid
TYPE: Acid, removal
COMPOSITION:
x .... H_3PO_4

TIME:
TEMP: 80°C
RATE: 50 Å/min Si_3N_4
10 Å/min SiO_2

DISCUSSION:

Si_3N_4, amorphous thin films deposited by CVD on silicon wafer substrates and on SiO_2 coated silicon substrates. Rates shown are as determined by author referenced who is the developer of this etchant for silicon nitride. Because of the etch rate disparity, SiO_2 can be used as an etch-stop and solution will not attack silicon.

REF: Kurtz, F et al — *J Electrochem Soc,* 113,1452(1966)

SIN-0008d: Ibid.

Si_3N_4 and $Si_3N_xO_{1-x}$, as amorphous thin films DC and RF sputtered on (111) silicon, n-type, 5—30 Ω cm resistivity wafers. Also applied PSG and BSG which showed an etch rate of about 25 Å/min. Work was in conjunction with KMER type photo resists.

SIN-0003: Williams, R K et al — *J Vac Sci Technol,* B(2),84(1984)

Si_3N_4, thin films deposited by both LPCVD and plasma CVD on silicon (100), p-type, 2—5 Ω cm resistivity substrates and then hydrogen irradiated to form: Si_3N_4:H thin films. Irradiation produced "swollen spots" in the nitride films that etched twice as fast as non-irradiated areas. (*Note:* The "swollen spots" are probably irradiation induced blisters.)

SIN-0022a: Fang, Y K et al — *J Electrochem Soc,* 132,1222(1985)

Si_3N_4:H, thin film doped with boron or phosphorus deposited on silicon, (100), n-type, 4—7 Ω cm resistivity substrates and used in a general etching study. H_3PO_4 was used at 80°C. Other etchants evaluated were HF, BHF, and CF_4/O_2(4%).

SIN-0004

ETCH NAME: Hydrofluoric acid, dilute
TYPE: Acid, removal
COMPOSITION:
1 .... HF
20 .... H_2O

TIME:
TEMP: RT
RATE: 1000—2000 Å/min

DISCUSSION:

Si_3N_4, thin films deposited on silicon substrates. This dilution of hydrofluoric acid can be used for controlled removal of thin films.

REF: Hersch, N — personal communication, 1965

SIN-0008f: Walker, P — personal application, 1965

Si_3N_4, thin films and silicon oxynitride thin films with varied oxygen content deposited by DC and RF sputter on n-type, 5—30 Ω cm silicon substrates in a general study of these thin films and their etching in conjunction with photo resist applications. Solution used as a dip etch for thinning and study of pinhole development.

SIN-0008e

ETCH NAME: Hydrofluoric acid
TYPE: Acid, pinholing
COMPOSITION:
x .... HF

TIME:
TEMP: RT
RATE: 4000—5000 Å/min

DISCUSSION:

Si_3N_4 and oxynitride thin films on silicon. Acid applied to thin surface with a wooden toothpick allowing surface tension to hold a droplet while etching a pit down through deposited nitride layer. Used to measure deposit thicknesses under standard microscope illumination using known surface color and recognizable color order steps in the etched side

slope of etch-formed pit. This pitting method and a step-etch using HF hot vapors also were used in establishing a pure nitride color chart through third order red.
REF: Ibid.
SIN-0022c: Ibid.

a-Si_3N_4:H thin films grown by CVD as boron or phosphorus doped films on (100) silicon, n-type substrates. HF was the fastest removal solution of these evaluated. Also used H_3PO_4 at 180°C, BHF, and CF_4/O_2.

SIN-0008g
ETCH NAME: Hydrofluoric acid, vapor — TIME: Dip
TYPE: Acid, removal — TEMP: Hot
COMPOSITION:
x .... HF, vapor
DISCUSSION:

Si_3N_4 and oxynitride thin films on silicon. Hot hydrofluoric acid vapors and the pinholing technique (SIN-0008e) were used in establishing a color thickness chart based on a refractive index of n = 2.00 for Si_3N_4. A similar color chart for SiO_2 with n = 1.46 was made for comparative thickness values of these DC sputter deposited films (silicon bar in nitrogen).
REF: Ibid.

SIN-0005
ETCH NAME: Hydrofluoric acid, dilute — TIME:
TYPE: Acid, removal — TEMP:
COMPOSITION:
1 .... HF
60 .... H_2O
DISCUSSION:

Si_3N_4 thin films deposited by CVD on (100) silicon substrates from both N_2 and NH_3 sources. Both the solution shown and a 1:20 mixture were used in etch removal studies. Films grown in N_2 only were harder to etch remove.
REF: Chin, T Y et al — *J Electrochem Soc,* 13,2110(1984)

SIN-0006
ETCH NAME: BOE — TIME:
TYPE: Acid, removal — TEMP:
COMPOSITION:
x .... HF
x .... NH_4F
x .... H_2O
DISCUSSION:

Si_3N_4 thin films deposited by PECVD as a dielectric surface coating on ICs and solar cells. Solution used in studying the etch rates of deposits formed from SiH_4 + N_2 or NH_3. The latter gives a higher H^+ ion inclusion and a faster etch rate.
REF: Hess, D W — *J Vac Sci Technol,* A(2),243(1984)

SIN-0007
ETCH NAME: — TIME:
TYPE: Acid, selective — TEMP: RT
COMPOSITION: — RATE: 100—150 Å/min
15 .... HF
10 .... H_3PO_4
60 .... EOH

DISCUSSION:

Si_3N_4, thin film amorphous deposits on silicon wafer substrates used for developing methods of etching patterns in silicon nitride. The solution shown is referenced to: Loic, H — U.S. Patent 3,867,28 (1975) with EOH replacing H_2O as shown in the patent. Use a drop of the etch solution on the specimen being spun at 2400 rpm on a photo resist type spinner. Solution applied at RT under a nitrogen atmosphere. Rinse in EOH or by bubbling HCl through EOH.

REF: Wurzbach, J A & Grunthaner, F J — *J Appl Phys,* 130,690(1983)

SIN-0008h

ETCH NAME:

TYPE: Acid, removal

COMPOSITION:

18 g .... NaOH

5 g .... $(NH_4)_2S_2O_8$

100 ml .... H_2O

TIME: 10—20 min

TEMP: Boiling

RATE: 160 Å/min

DISCUSSION:

Si_3N_4 and oxynitrides deposits on (111) silicon, n-type, 5—10 Ω cm resistivity wafers by DC sputter from a silicon bar in nitrogen. Solution developed for use with KMER photo resists in etch patterning devices. See comments under SIN-0008a.

REF: Ibid.

SIN-0008i

ETCH NAME: Sodium hydroxide

TYPE: Alkali, removal

COMPOSITION:

90 g .... NaOH

250 ml .... H_2O

TIME: 10—20 min

TEMP: Boiling

RATE: 180 Å/min

DISCUSSION:

Si_3N_4 and oxynitrides as DC sputtered thin film deposits on (111) silicon, n-type, 5—10 Ω cm resistivity wafers. Solution shown was developed for etching nitrides with KMER photo resists. Glycerin or ethylene glycol can be added for additional rate control. See SIN-0008a-h formats for additional development data.

REF: Ibid.

SIN-0008j

ETCH NAME:

TYPE: Halogen, removal

COMPOSITION:

"A" 5 g .... $NH_4F.HF$

50 ml .... H_2O

"B" 1 g .... $*I_2$

50 ml .... H_2O

50 ml .... glycerin

TIME: 30—45 min

TEMP: RT

RATE: 180 Å/min

*Pre-dissolve in 5 ml MeOH

Mix: 1 "A":1 "B" when ready for use.

DISCUSSION:

Si_3N_4 and oxynitrides grown as thin films by DC sputtering on (111) silicon wafers, n-type, 5—10 Ω cm resistivity using a silicon bar under nitrogen as silicon/nitrogen source. The solution can be used as a single solution, or combined as shown. This is an excellent solution for photo resist patterning as it can be re-used 4 h on 50 silicon wafers without

degradation of solution or loss of P/R definition. Used with both KMER and KFER photo resists.
REF: Ibid.
SIN-0013a: Mann, J E & Walker, P — Electrochem Soc Meet, Dallas, Spring, 1967
$Si_3N_xO_y$ thin films. Reported development of silicon oxynitride as a surface protectant on silicon, to include etchants compatible with photo resist processing.

SIN-0014
ETCH NAME: BHF, modification TIME:
TYPE: Acid, removal TEMP: RT
COMPOSITION: RATE:
15 ml .... HF
100 ml .... $NH_4F.HF$, sat. sol.
DISCUSSION:
Si_3N_4 and $Si_3N_xO_y$ thin films CVD grown from NH_3 with addition of NO_2 in increasing amounts to form oxynitrides. Pure SiO_2 also was grown, and in combination with the pure nitride as the two compounds are completely miscible forming an isomorphous series. The solution shown used in a general etching and properties study of both nitride and oxynitride to dioxide.
REF: Tombs, N C et al — *J Electrochem Soc,* 116,862(1969)

SIN-0015a
ETCH NAME: Glass cleaner TIME: Variable
TYPE: Acid, cleaning TEMP: RT to hot
COMPOSITION:
35 ml .... $K_2Cr_2O_7$, sat. sol.
1000 ml .... H_2SO_4
DISCUSSION:
Si_3N_4, SiO_2 and glass. Mixture shown is the standard glassware cleaning solution that has been in use for several years on glass and quartz. It can be used as a general surface cleaner on nitrides as well as oxides. If used hot, on a nitride surface, can convert some of the nitride to oxide. Solution used in a general study of etching nitrides.
REF: Walker, P — personal application, 1964

SIN-0015b
ETCH NAME: Aqua regia TIME: 1—5 min
TYPE: Acid, cleaning TEMP: RT to boiling
COMPOSITION:
3 .... HCl
1 .... HNO_3
DISCUSSION:
Si_3N_4, oxynitrides, SiO_2 as thin films or glass and quartzware. Aqua regia has long been a standard cleaner of glassware for removal of heavy metal contamination. After mixing, allow solution to sit in an open container until heavy bubble reaction diminishes, and solution turns a deep red/gold color. Never store in a closed container as reaction continues slowly and can cause explosive rupture due to pressure. Solution evaluated as a surface cleaner of thin film oxides and nitrides in a general etch development study.
REF: Ibid.

SIN-0015c
ETCH NAME: Nitric acid
TIME: Variable
TYPE: Acid, cleaning
TEMP: RT to boiling
COMPOSITION:
x .... HNO_3
DISCUSSION:
Si_3N_4 and oxynitride thin films. Nitric acid has been used as a general glass cleaner with and without dilution for several years. If used hot on nitrides it can convert part of the surface to an oxide. Solution used in a general etching and cleaning study of nitrides.
REF: Ibid.

SIN-0015d
ETCH NAME: Chrome regia
TIME: 1—5 min
TYPE: Acid, cleaning
TEMP: RT
COMPOSITION:
ASTM: #101
500 ml .... HCl
100 ml .... *10% CrO_3

*13.4 g/1245 mil H_2O

DISCUSSION:
Si_3N_4, oxynitrides and SiO_2 thin film deposits by evaporation (CVD) or sputter (DC/RF). Solution can be used as a general oxide or nitride surface cleaning etch. With thin films on a silicon wafer surface it has also been used as a pinhole development solution as it attacks the silicon preferentially and enhances pinholes for microscopic observation. By etch back-thinning of the silicon wafer (HF:HNO_3 with nitride or oxide surface waxed with Apiezon W, black wax) pinholes can be observed in the nitride/oxide deposits by transmitted light.
REF: Ibid.
SIN-0027: ASTM E407-70
Reference for ASTM number of chrome regia.

SIN-0015e
ETCH NAME: Sodium hydroxide, dilute
TIME: 1—30 min
TYPE: Alkali, cleaning
TEMP: RT
COMPOSITION:
1 g .... NaOH (KOH)
100 ml .... H_2O
DISCUSSION:
Si_3N_4, oxynitrides and SiO_2 thin films. This is a general cleaning solution for glassware as the 10—15% concentration. The mixture shown was used on silicon oxynitride and dioxide thin film surfaces for light cleaning with only a cleaning effect on silicon wafer substrates. Using a 30% solution on a 2500 Å thick layer of silicon oxynitride (5% O_2), complete removal of the layer was done in about 2 h, at 80—90°C.
REF: Ibid.

SIN-0015f
ETCH NAME:
TIME: 30—90 sec
TYPE: Halogen, cleaning
TEMP: RT
COMPOSITION:
1 g .... I_2
500 ml .... MeOH

DISCUSSION:

Si_3N_4 oxynitrides and SiO_2 DC/RF sputtered thin films on silicon wafers. Solution used as a general surface cleaner/conditioner on silicon wafers and both silicon oxides and nitrides prior to diffusion, epitaxy, metallization, or deposition. Soak in solution; DI water rinse, 10—15 sec; MeOH dip, 1—2 sec and either air dry or dry under an IR lamp. A small fraction of iodine will remain adsorbed on surfaces. Subsequent heating to above 100°C will allow iodine to complex ionic surface contamination with vaporization removal leaving a molecularly clean surface immediately prior to diffusion, epitaxy, etc.

REF: Ibid.

SIN-0021

ETCH NAME: Hydrofluoric acid, dilute — TIME: Variable

TYPE: Acid, removal — TEMP: RT

COMPOSITION:

1 .... HF

9 .... H_2O

DISCUSSION:

Si_3N_4 thin films. This is a general glassware cleaning solution that also has been used to clean thin film oxide and nitride surfaces as well as a native oxide removal solution of silicon wafers after etch polishing in $HF:HNO_3$ type solutions. Reference cited says that about 10 Å of oxide remains on a silicon surface etched in $1HF:5HNO_3$.

REF: Wajda, E S et al — *IBM J*, 37,288(1961)

SIN-0025

ETCH NAME: Hydrofluoric acid — TIME:

TYPE: Acid, removal — TEMP: RT

COMPOSITION: — RATE: 75—100 Å/min

x .... HF, conc.

DISCUSSION:

Si_3N_4 deposited as pyrolytic thin films. Films used in a study of their preparation and properties.

REF: Doo, V Y et al — *J Electrochem Soc*, 113,1279(1966)

SIN-0009

ETCH NAME: BHF — TIME:

TYPE: Acid, removal — TEMP: RT

COMPOSITION: — RATE: 4—6.1 Å/sec

x .... BHF

DISCUSSION:

Si_3N_4, thin films, 100—1000 Å thick, grown by conversion of SiO_2 thermally grown on p-type silicon wafers, and with dry thermal oxidation at 900, 950, and 1000°C. Nitridization was done at 900 to 1100°C in N_2/NH_3 and AES measurements showed conversion of the films with N_2 pile-up at the dielectric/silicon interface. BHF mixture was not shown and rate variation depended upon thickness of the thin film, increasing with film thickness.

REF: Yoriume, Y — *J Vac Sci Technol*, B(1), 67(1983)

SIN-0023

ETCH NAME: BHF — TIME:

TYPE: Acid, removal — TEMP: RT

COMPOSITION:

1 .... HF

10 .... NH_4F, sat. sol.

DISCUSSION:

N_xSiO_2 thin films. SiO_2 films 160 nm thick were deposited on silicon wafer substrates and then nitrided with N_2 or NH_3. Structure developed was termed: nitroxide as against a grown oxy-nitride. The SiO_2 films were nitrided at 1200°C for 5 and 17 h. SiO_2 etch rate was about 16 nm/sec; both nitroxides showed no initial etching for about 1.5 or 2.5 min for the 5 and 17 h growth periods, respectively, then etched at an approximate rate of 5—6 nm/sec. Refractive index varied from 1.46 (SiO_2) to about 1.6 (5 h nitroxide) to almost 1.7 (17 h nitroxide). Dielectric constant: e = 4.0 (SiO_2), stabilized at about 5.5 after 5 h nitroxide. AES depth profiles, C-V and VFB shown. Variation in refractive index of SiO_2 after N_2 and NH_3, and thickness vs. kV voltage applied shown.

REF: Kato, L et al — *J Electron Mater,* 13,913(1984)

SIN-0024: Ito, T et al — *J Electrochem Soc,* 127,2053(1978)

N_xSiO_2, thin films of SiO_2 nitrided at 900 to 1200°C in NH_3.

SIN-0022b

ETCH NAME: | TIME:
TYPE: Ionized gas, removal | TEMP:
COMPOSITION: | GAS FLOW:
x .... 96% CF_4 | PRESSURE:
x .... 4% O_2 | POWER:

DISCUSSION:

a-Si_3N_4:H thin films with different doping concentrations of B and P. Minimum etch rates with 2% B and 3% P doping increasing with higher concentrations. Doped films have a higher resistivity than undoped films: maximum resistivity ($\times 20^{10}$ Ω cm) — 6 for 3% P doped and 5% B doped. Breakdown strength (10^5 V/cm) — 9 for P doped; 8 for B doped. Pinhole density was obtained by copper decoration.

REF: Ibid.

SIN-0026: Shannon, W J — *RCA Rev,* 31,431(1970)

Reference for copper decoration technique used in SIN-0022b

SIN-0029: Kotani, H et al — *J Electrochem Soc,* 130,645(1983)

SiN, p-type thin films used as interconnects on semiconductor devices, along with AlSi, PSG, and AlSiCu thin films. Describes using CF_4/O_2 for multi-level etching. Used argon at 1.7 Pa pressure and 0.5 W/cm² power for dry chemical etching. p-SiN etch rate: 70 Å/min; PSG etch rate: 60 Å/min and AlSi etch at greater than 200°C temperature was 100 Å/min.

SIN-0030: Chayahara, A et al — *Jpn J Appl Phys,* 24,19(1983)

a-Si_3N_4:H thin films deposited in a study of the material development. Nitrogen can show three- or fourfold symmetry. Si:N formed using SiH_4:NH_3:H_2 in a microwave glow discharge with 250 W power and 1 Torr pressure; gas flows: SiH_4/H_2(10%) at 100—200 sccm; NH_3/N_2 at 0.5—5 SCCM.

SIN-0027

ETCH NAME: BHF | TIME: 10 sec
TYPE: Acid, stop etch | TEMP: RT
COMPOSITION: | RATE: 0.01—0.02 μm/sec
1 .... HF
9 .... NH_4F(40%)

DISCUSSION:

Si_3N_4, thin films RF plasma grown on silicon, (100), p-type, 2 Ω cm resistivity wafers with SiO_2 cap. Solution used as a stop-etch in removal of SiO_2 and then as a step-etch of the nitride. Sitting after several days the nitride rate increased from 0.02 to 0.03 μm/sec.

SiO_2 etch rate changed from 0.01 to 2 μm/sec. Refractive index of pure SiO_2 shown as n = 1.462; for the pure nitride n = 2.00. Prior to nitride deposition wafers were cleaned: (1) H_2O_2 soak and DI water rinse; (2) H_2SO_4:H_2O_2 and DI water rinse; (3) HF dip and DI water rinse, and (4) BHF, RT, 60 sec.
REF: Reisman, A et al — *J Electron Mater,* 13,504(1984)

SIN-0028a
ETCH NAME: Carbon tetrafluoride
TYPE: Ionized gas, removal
COMPOSITION:
x .... CF_4

TIME:
TEMP:
GAS FLOW: 16 sccm
PRESSURE: 0.14 Torr
POWER:

DISCUSSION:
SiN_x and SiO_2 thin films deposited by RF plasma on silicon doped GaAs, (100) wafers. Photo resist windows were opened with BHF (1HF:5NH_4F), or by RIE.
REF: Blaauw, C et al — *J Electron Mater,* 13,251(1984)

SIN-0028b
ETCH NAME: BHF
TYPE: Acid, removal
COMPOSITION:
1 .... HF
5 .... NH_4F(40%)

TIME:
TEMP:

DISCUSSION:
SiN_x and SiO_2 thin films RF plasma and CVD deposited on GaAs, (100) silicon doped substrates. BHF used to remove nitride and oxide from photo resist developed windows. Also used RIE etching with CF_4.
REF: Ibid.

SIN-0010
ETCH NAME: BHF
TYPE: Acid, cleaning
COMPOSITION:
5 ml .... HF
20 g .... NH_4F
30 ml .... H_2O

TIME:
TEMP: Hot

DISCUSSION:
Si_3N_4, thin films deposited on (100) silicon wafers used in a study of nitride films. Preclean substrates in BHF — heating solution reduces H_2 density in film growth. Follow with 400°C bake-out prior to nitriding.
REF: Zhous, M S et al — *J Electron Mater,* 14,55(1985)

SIN-0031
ETCH NAME: Phosphoric acid
TYPE: Acid, removal/defect
COMPOSITION:
x .... H_3PO_4, vapor

TIME:
TEMP: Hot

DISCUSSION:
a-SiN:H thin films deposited on (100) silicon and germanium wafers, glass cover slides and fused silica blanks. Clean all substrates: (1) TCE degrease; (2) HF dip and (3) DI H_2O rinse. Deposit films by RF sputter of a silicon target in a gas mixture of air:N_2:H_2 at 175°C.

Hot phosphoric acid vapors can be used for very slow removal and defect development in films. See SIN-0001.
REF: Martin, P M — *J Vac Sci Technol,* A2(2),330(1984)

SIN-0032
ETCH NAME: Hematite TIME:
TYPE: Compound, polish TEMP: RT
COMPOSITION:
(1) x .... Fe_2O_3, powder (2) x .... CH_3COCH_3 (acetone)
DISCUSSION:

Si_3N_4, as pressed powder blanks used in a study of Cr and Ti bonding. Nitride blanks were mechanically polished with red rouge, Fe_2O_3. Blanks were then annealed in a belt furnace for 2 h at 1000°C under N_2, and then oxidized for a structured layer of $SiO_2/SiNO_x/Si_3N_4$/blank . . . surfaces were rough, crystalline. Surfaces cleaned ultrasonically in acetone before metallization. Air anneal showed best adhesion for chromium, but poor for nitride. Presence of carbon produces some TiC with titanium.
REF: Orent, T W et al — *J Vac Sci Technol,* B1(3),844(1983)

SILICON VANADINIDE ETCHANTS

SIV-0001
ETCH NAME: TIME:
TYPE: Acid, removal/polish TEMP: RT
COMPOSITION:
x .... HF
x .... HNO_3
DISCUSSION:

SiV_2, thin films 500 Å thick deposited on (100) NaCl substrates. Follow with 300 Å a-Si — 1000 Å Ni — 500 Å SiO_2 cap layer. The thin film layered structure was water float-off removed by liquefying the NaCl/SiV_2 interface for TEM study. Solution shown is a general etchant for both silicon and vanadium.
REF: Chen, S H et al — *J Appl Phys,* 57,258(1985)

PHYSICAL PROPERTIES OF SILVER, Ag

Classification	Transition metal
Atomic number	47
Atomic weight	107.87
Melting point (°C)	960.8
Boiling point (°C)	2210 (2212)
Density (g/cm^3)	10.5
Thermal conductance (cal/sec)(cm^2)(°C/cm) 0°C	0.999
Specific heat (cal/g) 25°C	0.056
Latent heat of fusion (cal/g)	25
Heat of vaporization (cal/g)	565
Atomic volume (W/D)	10.3
1st ionization energy (K-cal/g-mole)	17.5
1st ionization potential (eV)	7.574
Electron work function (eV)	4.52
Electronegativity (Pauling's)	1.9
Covalent radius (angstroms)	1.34
Ionic radius (angstroms)	1.26
Coefficient of linear thermal expansion ($\times 10^{-6}$ cm/cm/°C)	19.68
Electrical resistivity (micro ohms cm) 20°C	108.4
($\times 10^{-6}$ ohm cm) 20°C	1.59
Lattice constant (angstroms)	4.086
Thermionic work function (eV)	3.09—4.31
Elastic modulus (psi $\times 10^6$) 30°C	10.6
Tensile strength (psi — annealed)	25.000
Cross section (barns)	63
Vapor pressure (°C)	1865
Hardness (Mohs — scratch)	2.5—3
(Knoop — kgf mm^{-2})	60—90
Crystal structure (isometric — normal)	(100) cube, fcc
Color (solid)	Silver-white
Cleavage (imperfect)	(001)

SILVER CONTAINING ETCHANTS

Silver is classified in the Copper Group I-B of copper, silver and gold, in increasing order of atomic weight and density. They are the leading elements for heat and electrical conductivity with copper the most widely used due to its greater abundance. They also are known as the coinage metals, which now include nickel, and all have been mediums of exchange since earliest times. They are heavy metals due to their atomic structure, and are both malleable and ductile.

Though silver does not readily oxidize in air, surfaces are rapidly attacked by sulfur and sulfur compounds "tarnishing" as a black smut silver sulfide which can be easily removed with ammonia, NH_3 or ammonium hydroxide, NH_4OH.

In metal and metallic compound etching or cleaning solutions the most widely used silver compound is silver nitrate, $Ag(NO_3)_2$ as it is a highly active oxidizing agent and very soluble in water. Silver chloride, AgCl, on the other hand, is practically insoluble in water, but forms soluble complexes with the photographer's Hypo — sodium thiosulfate, $Na_2S_2O_3$ — which, when properly developed and "fixed" produces a black film of silver, and becomes a permanent black-and-clear photographic negative. The chloride, as a smoke, is also used as a water droplet nucleating agent in rain clouds and, in the laboratory, in the growth of snowflakes and single crystal ice crystals for scientific study.

As will be seen in the following list of solutions containing silver, the majority are used as preferential etchants to develop surface structure etch figures, some polishing solutions and as plating solutions to develop electrically active p-n junctions, such as in semiconductor device studies. Though other silver compounds are used, the majority are silver nitrate. Silver nitrate could be used in many other solutions on materials not shown in the following list, but copper nitrate, $Cu(NO_3)_2$ and several chromium compounds are more widely used as they too are extremely active oxidizing agents, and are lower in price. The following is a selected list from the Etchant Section:

Formula	Material	Use	Ref.
10 ml E etch:5 mg AgNO	CdTe	Preferential	CDTE-0006b
50 ml HNO_3:0.5 ml $AgNO_3$:50 ml H_2O	Cu	Preferential, macro-etch	CU-0011
$1HNO_3:3H_2O$ + 1% $AgNO_3$	GaAs	Preferential	GAS-0045c
2HF:3HN($O_3:5H_2O$: + $AgNO_3$ RC-1	GaAs	Dislocation	GAS-0048
37HF:$13HNO_3:50H_2O:2AgNO_3$	Ge	Preferential	GE-0037
4 ml HF:2 ml HNO_3:4 ml H_2O: 0.2 g $AgNO_3$ WAg Etch	Ge	Preferential	GE-0060c
x$AgNO_3$	Ge	Preferential (sphere)	GE-0195
40HF:$20HNO_3:40H_2O:2AgNO_3$	Ge	Preferential (sphere)	GE-0196
xAgI_2, vapor	Ice	Initiate growth	ICE-0007
1 ml HF:2 ml H_2O:8 mg $AgNO_3$: 1 g CrO_3 A/B Etch	InAs	Preferential	INAS-0005d
1 ml HF:2 ml H_2O:3 g CrO_3:8 mg $AgNO_3$	InP	Preferential	INP-0051
Silver glycol etch: A: 10 ml HF:400 ml HNO_3: 1 g $AgNO_3$:100 ml H_2O B: 200 ml H_2O:200 ml propylene glycol	Si	Preferential	SI-0041a
1HF:$1HNO_3$:x1% $AgNO_3$	Si	Polish	SI-0081
4 ml HF:2 ml HNO_3:4 ml H_2O: 200 mg $AgNO_3$ silver etch	Si	Dislocation	SI-0209
xAgCN	Ag	Electrolytic, polish	AG-0010
1AgF:$10H_2O$: + HF	Si	Junction, plate-out	SI-0280

SILVER, Ag

General: Occurs in nature as a native element. It is found in fine single crystals but is more common as acicular, reticulated or aborescent shapes; also massive, in thin plates, grains and small scales. Like several native metal elements, it has no cleavage, breaking with a hackly fracture and is both ductile and malleable like native gold, iron and lead. Fresh surfaces show typical silver-white color but are often tarnished gray to black after reaction with air or sulfur compounds. It is widely distributed throughout the world, though not as common as gold and is usually found in the upper portion of silver bearing veins as an alternation product of underlying silver sulfide, arsenide and chloride minerals — the gangue mineral is cellular, rusty quartz. Native silver contains up to 10% gold and native "common" gold, up to 15% Ag.

Silver, gold and platinum are the three primary "precious metals", silver and gold being two of the seven metals known to ancient man. It has been used as a medium of monetary exchange since earliest times, and in the fabrication of jewelry and objets d'art. Present silver value runs between $6.00 to $8.00, against $350 to $450 for gold and $500 to $550 for platinum.

Though silver is the best electrical conductor, copper is more widely used due to its greater abundance. This is not to say that silver is not used for its electrical and heat dissipation characteristics: pins, pre-forms, wire and ribbon straps and bulk heat-sinks are used. There are many low temperature silver alloy solders and high temperature brazing alloys (1100 to 1600°F), the latter excellent for joining carbon steels, copper, brass, nickel, monel and several nickel/silver alloys. Solder alloy #63 (Pb:Sn + 2% Ag), m.p. 220°C, is widely used in the electronics industry.

Silver has wide use as an electroplated thin film on surfaces — metals, plastics, mylar, glass — though it was the original "silver mirror", both aluminum and nickel are now used to reduce cost.

Technical Application: Pure silver parts, such as electrical pins, and small pre-forms are used in many device assembly operations, as well as wire and ribbon as contact straps for electrical conduction. Pure silver is plated on many parts and evaporated in layered metallization thin film structures, such as Ag/Pd — Au/Ag — Ti/TiAg/Ag — Cu/Ag/Cu and others for structuring device contact pads and circuits. As wire it is used with wire bonders. There are a wide range of silver alloys used in device assemblies to include both low temperature solders and high temperature brazing alloys. Solder #63 has already been mentioned, and one of the widely used contact pastes is Ag-epoxy or a silver loaded polyimide that can be easily heat cured between 120 to 150°C with various time schedules. Large silver alloy base plates have been used in vacuum and epitaxy systems for heat transfer, though SST and copper are more widely used due to cost factors.

Electroplated and evaporated silver thin films have been studied for removal, defects, structure, photo doping and both natural and artificial single crystals have been studied for physical and morphological data to include single crystal spheres in oxidation experiments.

Etching: HNO_3, hot H_2SO_4 and cyanides. Mixed acids — aqua regia types — HF:HNO_3 and thermal etching. NH_3 and NH_4OH, tarnish removal.

SILVER ETCHANTS

AG-0001

ETCH NAME: Nitric acid — TIME:

TYPE: Acid, preferential — TEMP: 0°C to boiling

COMPOSITION:

x .... HNO_3, conc.

DISCUSSION:

Ag, (111) wafers and other orientations as single crystal blanks. Solution used at RT to develop orientation etch figures.

REF: Farnsworth, H E — *Phys Rev*, 40,684(1932)

AG-0003: Dana, E S & Ford, W E — *A Textbook of Mineralogy*, 4th ed, John Wiley & Sons, New York, 1950, 403

Silver, natural crystals. Solution used to dissolve silver in pyrolytic and chemical study. Can be easily plated out of solution on a copper plate from nitric acid:water solutions in which silver is dissolved.

AG-0023: Walker, P — personal application, mineralogy, 1952

Ag, native single crystals. Use solution at 0°C, chilled in a water bath with chipped ice, as a momentary dip to clean surfaces and, by repeated dipping to develop orientation figures and defect structure.

AG-0002a

ETCH NAME: Nitric acid, dilute　　TIME: 10—15 sec

TYPE: Acid, removal　　TEMP: RT and 8°C

COMPOSITION:

1—8 .... HNO_3

1 .... H_2O

DISCUSSION:

Ag, thin films electroplated on brass. Used a 1:1 solution to strip silver from brass parts. Use of a rapid dip-water flush method as solution will attack brass. With sufficient solution volume to prevent heating, or, by chilling to 8°C with a bath of ice/acetone, minimum brass attack can be obtained. Remaining black smut on brass can be removed by rubbing with toweling soaked in water.

REF: Walker, P — personal application, 1968

AG-0002b

ETCH NAME: Potassium cyanide　　TIME: 1—5 min

TYPE: Acid, cleaning　　TEMP: RT

COMPOSITION:

10 g .... KCN/NaCN

100 ml .... H_2O

DISCUSSION:

Ag, as high purity pre-forms and silver alloy pre-forms used in silicon device assembly and on assembly pins of silver plated dumet or copper. Solution used to clean parts, and followed with heavy DI water washing. Clean parts immediately before alloy assembly.

REF: Ibid.

AG-0005

ETCH NAME:　　TIME: 1—10 min

TYPE: Acid, removal/cleaning　　TEMP: RT

COMPOSITION:

1 .... NH_4OH

1 .... H_2O_2

DISCUSSION:

Ag, thin film deposits. Described as a slow removal etch for silver. Will not attack silicon but, with presence of hydrogen peroxide, will attack gallium arsenide.

REF: MRC Metals Catalog, 1980

AG-0002c: Ibid.

Ag, thin films electroplated on brass. Solution used to strip silver. Also used in etching of evaporated silver on (111) silicon wafers after photo resist patterning.
AG-0021: Nichols, D R — *Solid State Technol,* December 1979
Ag, thin film deposits in vacuum systems. Solution used to etch clean bell jars and internal fixtures. Can be used on steels, glass, ceramics and copper.

AG-0002d
ETCH NAME: Ammonium hydroxide, dilute TIME: 1—10 min
TYPE: Acid, cleaning TEMP: RT
COMPOSITION:
1 .... NH_4OH
1 .... H_2O
DISCUSSION:
Ag, high purity pre-forms and plated parts. Solution used to clean silver surfaces prior to alloy or assembly of parts in silicon device processing. (*Note:* Solution is easier and safer to use than the cyanides.)
REF: Ibid.

AG-0007
ETCH NAME: TIME: 1—5 min
TYPE: Acid, removal TEMP: RT
COMPOSITION:
1 .... HCl
1 .... HNO_3
1 .... H_2O
DISCUSSION:
Ag, 100 Å thick thin film of silver deposited on an a-SeGe glassy films. This layered structure was then annealed under a 200 W Hg UV lamp in a study of photo doping. Remaining residual silver after annealing was removed with the solution shown.
REF: Zebutsu, S — *Appl Phys Lett,* 39,969(1981)

AG-0008
ETCH NAME: TIME: 3—5 sec
TYPE: Acid, polish TEMP: RT
COMPOSITION:
(1) 1 .... *NaCN (2) 1 .... *NaCN
1 .... H_2O_2 (30%) 1 .... H_2O_2 (20%)

*21.5 g/l

DISCUSSION:
Ag, (111) and (100) wafers used in a study of surface defects. Use solution (1) for (111) and (2) for (100) orientations. Etch 5 sec — rinse: then 3 sec — rinse. Rinse in (1) 37.5 g/l NaCN and (2) heavy water wash. Repeat this step-polishing and rinsing until surfaces are highly reflective, defect free and will hold a drop of pure water on the surfaces. Follow by Ar^+ ion sputter in UHV and 600°C anneal. (*Note:* 30% H_2O_2 is standard concentration.)
REF: Adzic, R R et al — *J Electrochem Soc,* 131,1730(1984)

AG-0009a
ETCH NAME:
TYPE: Electrolytic, polish
COMPOSITION:
x .... KCN

TIME:
TEMP:
ANODE:
CATHODE:
POWER:

DISCUSSION:
Ag specimens used in a thermal etching study. Electropolish specimens in solution shown prior to thermal processing.
REF: Hondros, E D & Moore, A J W — *Acta Metall,* 8,751(1960)
AG-0004a: Hendrickson, A A & Machin, E S — *Acta Metall,* 3,69(1955)
Ag, (111) and (100) wafers used in a study of dislocations developed by thermal etching. Electropolish with 9% KCN at 125 A/in^2 power with rapid agitation. Wash 5 min in cold water, then alcohol rinse and blow dry with an argon gas jet. Thermal etch in 1 O_2:9 Ar for three time and temperature periods: (1) 600°C, 10 h; (2) 850°C, 10 h; and (3) 550°C, 16 h.

AG-0009b
ETCH NAME: Air/N_2/Vacuum
TYPE: Gas, preferential
COMPOSITION:
x .... air, N_2, or vacuum

TIME: 150 h
TEMP: 73°C
POWER: 31 A/cm^2 & 0.19 V

DISCUSSION:
Ag, specimens used in a thermal etching study. Acid saw cut wafers to minimize strain damage and electropolish (AG-0009a), before preferential etching. Results: (1) in air, etch patterns developed and reversed when power polarity was reversed; (2) in nitrogen, little or no structure developed and (3) in vacuum, large pits appeared in 5 h, and were larger and deeper at center (hottest area?).
REF: Ibid.

AG-0013
ETCH NAME: Air
TYPE: Gas, preferential
COMPOSITION:
x .... O_2

TIME:
TEMP:

DISCUSSION:
Ag, single crystal sphere used in a study of thermal oxidation of surfaces. Thickness of Ag_2O varies with crystallographic pole and great circle orientations of the sphere.
REF: Menzel, E & Menzel-Kopp, E — *Z Naturforsch,* 132,985(1958) (in German)

AG-0014
ETCH NAME: Heat
TYPE: Thermal, forming
COMPOSITION:
x .... heat

TIME:
TEMP:

DISCUSSION:
Ag, single crystal sphere formed from a single crystal wire. Wire was 1—2 mm diameter; spheres formed were 2—3 mm diameter. Gold and copper spheres also formed. The spheres were used for electrodes in electrocrystallization studies.
REF: Roe, D K & Gerisder, H — *J Electrochem Soc,* 110,310(1963)

AG-0015b
ETCH NAME: Air
TYPE: Gas, oxidation
COMPOSITION:
(1) x .... air

(2) x .... HNO_3
x .... H_2O

TIME: 1 week
TEMP: 900°C

DISCUSSION:

Ag specimens. Oxidizing surfaces in (1) at time and temperature shown will both oxidize and thermally etch specimens. After treatment (1) remove oxide with acid solution shown in (2). Specimens were used in a study of the influence of surface structure on the tarnish reaction.

REF: Allpress, J C & Sanders, J V — *Phil Mag,* 10,827(1964)

AG-0010
ETCH NAME:
TYPE: Electrolytic, polish
COMPOSITION:
x .... AgCN

TIME:
TEMP:
ANODE:
CATHODE:
POWER:

DISCUSSION:

Ag, single crystal specimens used in a study of thermal etching influence on surface energy. Electropolish material in solution shown prior to thermal treatment.

REF: Moore, A J W — *Acta Metall,* 6,292(1958)

AG-0011: Gilbertson, L & Fortner, O F — *Trans Electrochem Soc,* 81,199(1942)

Reference cited for electropolish solution shown in AG-0010.

AG-0012a
ETCH NAME:
TYPE: Electrolytic, saw
COMPOSITION:
35 g .... AgCN
37 g .... KCN
38 g .... K_2CO_3
1000 ml .... H_2O

TIME: Hours
TEMP: RT
ANODE:
CATHODE:
POWER:

DISCUSSION:

Ag, (001) wafers electrolytically cut with an acid-saw system using a cotton rug fiber as the cutting tool. Cutting time was 1″/120 h. Specimens were then hand-polished and dislocation etched in studying the material purity, dislocation density and strength.

REF: Hammar, R H et al — *Trans Met Soc AIME,* 239,1692(1967)

AG-0012b
ETCH NAME:
TYPE: Acid, polish
COMPOSITION:
(1) 15 ml .... H_2O_2
25 ml .... NH_4OH

(2) 0.5 ml .... HCl
5 g .... CrO_3
50 ml .... H_2O

TIME:
TEMP: RT

DISCUSSION:

Ag, (001) wafers. After acid-saw cutting (AG-0012a) specimens were alternately hand etch-polished in solution (1) and chemically polished in solution (2) until the desired surface finish was obtained.

REF: Ibid.

AG-0012c
ETCH NAME: TIME:
TYPE: Acid, dislocation TEMP: RT
COMPOSITION:

"A" 1 .... NH_4OH
1 .... H_2O_2

"B" 0.01 ml .... HCl
5 g .... CrO_3
25 ml .... H_2O

Mix: 1 part "A" to 0.16 parts "B".
DISCUSSION:
Ag, (001) wafers. Solution used to develop dislocations.
REF: Ibid.

AG-0016
ETCH NAME: TIME: 1—2 min
TYPE: Acid, polish TEMP: RT
COMPOSITION:
40 g .... Cr_2O_3
20 ml .... H_2SO_4
1000 ml .... H_2O
DISCUSSION:
Ag specimens used in a study of the reactions with this solution type mixture. After etching wash away red-brown coating that remains with water.
REF: Kurz, F — *Acta Metall,* 103,257(1956)

AG-0017
ETCH NAME: TIME:
TYPE: Electrolytic, polish TEMP:
COMPOSITION: ANODE: Ag
67.5 g .... KCN CATHODE: SS
15 g .... KFe_2CN POWER: 4 V
15 g .... Rochelle salt
19.5 g .... H_3PO_4
2.5 g .... NH_3
DISCUSSION:
Ag specimens. Solution used to electropolish both gold and silver.
REF: Ruff, A W & Ives, L K — *Acta Metall,* 15,189(1967)

AG-0019
ETCH NAME: TIME:
TYPE: Acid, preferential TEMP:
COMPOSITION:
x .... 5% KCN
x .... 5% $(NH_4)_2S_2O_8$
DISCUSSION:
Ag, single crystals used in a study of recrystallization by bending. Solution used as a micro-etch to develop structure.
REF: Semmel, J W Jr & Machin, E S — *Acta Metall,* 5,582(1957)
AG-0004b: Hendrickson, A A & Machin, E S — *Acta Metall,* 3,68(1955)
Ag, single crystal specimens. Solution used to develop structure.

AG-0022a
ETCH NAME: Tri-iodide — TIME:
TYPE: Halogen, removal — TEMP:
COMPOSITION:

(1) 400 g .... KI
100 g .... I_2
400 ml .... H_2O

(2) 1 .... Tri-iodide
4 .... H_2O

DISCUSSION:

Ag thin films. The solutions shown or other Tri-iodide composition etchants can be used to remove or pattern both silver and gold thin films.

REF: Brochure — *Application Data for Kodak Photosensitive Resists,* Eastman Kodak Co., Rochester, NY, 1960, 91

AG-0022b
ETCH NAME: — TIME:
TYPE: Acid, removal (spray) — TEMP:
COMPOSITION: — RATE: 3000 Å/min

35 g .... $Fe(NO_3)_2$
100 ml .... ethylene glycol
25 ml .... H_2O

DISCUSSION:

Ag thin films. Solution shown as a removal and patterning etchant. Removal rate is for application as a spray.

REF: Ibid.

AG-0022c
ETCH NAME: Ferric nitrate — TIME:
TYPE: Acid, removal — TEMP: 43—49°C
COMPOSITION:

x .... 55% $Fe(NO_3)_2.6H_2O$

DISCUSSION:

Ag thin films. Solution shown as a removal and patterning etchant compatible with Kodak photo resists.

REF: Ibid.

AG-0025a
ETCH NAME: — TIME: 5 & 10 min
TYPE: Acid, tarnishing — TEMP: RT
COMPOSITION:

(1) x .... HNO_3, conc.
+ .... dry air

(2) x .... HNO_3
x .... H_2O

(3) x .... NH_3
+ .... dry air
+ .... dry air

DISCUSSION:

Ag specimen blanks 0.5 × 0.5 × 0.005 used in a study of silver tarnishing. All three solutions shown evaluated at the two times shown. Results showed a white matte surface. (*Note:* Sulfur compounds produce the normally observed grey to black tarnish. Some silver oxides will tarnish as white coating.)

REF: Reagor, B T & Sinclair, J D — *J Electrochem Soc,* 128,701(1981)

AG-0025b
ETCH NAME: Ammonia TIME: 5 & 10 min
TYPE: Acid, tarnishing TEMP: RT
COMPOSITION:
x .... NH_3
+ .... dry air
DISCUSSION:
Ag specimen blanks used in a silver tarnish study. See AG-0025a for discussion.
REF: Ibid.

AG-0024
ETCH NAME: "Smog" TIME:
TYPE: Gas, corrosion TEMP:
COMPOSITION:
x .... Air (+ trace SO_2, H_2S, O_3, HCl and Cl_2, vapor)
+ .... dry air
DISCUSSION:
Ag specimens used in a corrosion study of surfaces in a dry "smog" type environment. Sulfur is the primary attacking element in the tarnishing of silver. Copper also was evaluated with addition of water vapor as controlled humidity (RH) atmosphere.
REF: Rice, D W et al — *J Electrochem Soc,* 128,275(1981)

AG-0026
ETCH NAME: Nitric acid TIME:
TYPE: Acid, removal TEMP:
COMPOSITION:
x .... 1 *M* HNO_3
+ .... light, collimated
DISCUSSION:
Ag thin film coatings on Si, Al_2O_3 and ZrO_2 substrates used in a study of surface coatings to withstand chemical attack. Silver is stable in air and water, but tarnishes in sulfur containing atmospheres. Also worked with Al, TiN, alpha-Si:H and PbS. Solution plus light used to remove silver plating.
REF: Martin, P J et al — *J Vac Sci Technol,* A2(2),341(1984)

SILVER ALLOYS, AgMx

General: Native silver occurs as an element in nature, but rarely as pure silver such as 99.99+%, for it contains trace elements of other metals such as copper, and sometimes platinum, antimony, bismuth, and mercury; the latter as the mineral amalgam, AgHg with varying amounts of both elements. The mineral called electrum (argeniferous gold), AuAg . . . Ag 18—36% . . . light yellow to yellow-white in color. Other natural silver alloys include dyscrasite, Ag_3Sb; stutzite, Ag_4Te; hessite, Ag_2Te, and there are other trinary tellurides and selenides. There are distinct mineral species of silver halogens, such as AgI, AgBr, and AgCl, and over 50 silver minerals. Argentite, Ag_2S, is a major ore found in veins associated with pure silver, other silver minerals, and/or copper and lead mines. Some pyrites, Fe_2S — a major ore of iron — are sometimes more valuable for their gold and silver content than for iron and sulfur.

Silver and gold are the two main coinage metals and, along with copper, were the three major metals of trade in ancient times, and there have been coins of silver and gold, as well as bronze, iron and stone in the past. The lead tin alloy #63 contains 2% silver as a low temperature solder; the silvaloys, BAg, are high temperature brazing alloys. The silver halides AgCl and AgBr, both natural and artificial materials, are used in photography, as the compounds turn black on exposure to light, and also are used as medical antiseptics. Silver alloys have wide use in the jewelry trade, as well as for tableware — knives, forks, spoons, cups, and plates. During the Spanish exploration of the Americas, the returning ships were known as the "Silver Plate" galleons, and silver is electrolytically plated as Ag, Ag/Pb, Ag/Ni, Ag/Cu, etc., for both decorative purposes and surface protection as silver plate is stable in air and water, though it will tarnish in the presence of sulfur to black Ag_2S (remove with ammonia solutions). The nitrate $AgNO_3$ is known as Lunar Caustic, and has medical applications, as well as being a primary etching constituent in many etches as already shown.

Technical Applications: Silver alloys have several applications in the Solid State industry other than as solder and brazing alloys. It is one thin film constituent in multi-layer metallization, such as for Ag/TiAg/Ti or Ag/Cu. As bulk parts: evaporator electrodes or vertical epitaxy system base plates, and used for electrical wiring, as is the best known conductor of electricity. Many small parts are silver alloy plated for electronic assemblies — wires, pins, plates and packages.

There are a number of single crystal metallic compound semiconductors, both binary and trinary systems, containing silver as tellurides, selenides and sulfides . . . these are under their separate sections in this book. Several alloys have been studied as both mixtures and single crystals, presented in this silver alloy section.

Etching: NH_3, NH_4OH, KCN and mixed acids. Varies with specific alloy.

SILVER ALLOY

Silver aluminum
Silver boron
Silver lead tin
Silver mercury
Silver tin
Silver titanium
Silver zinc

SILVER ALLOYS

SILVER ALUMINUM ETCHANTS

AGAL-0001

ETCH NAME: Nitric acid, dilute — TIME:
TYPE: Acid, polish — TEMP:
COMPOSITION:
 2 .... HNO_3
 1 .... H_2O
DISCUSSION:

Ag_2Al, single crystal sphere grown and spark cut in a study of structure ordering. Solution used to etch polish the sphere.

REF: Neumann, J P — *Acta Metall,* 14,505(1966)

AGAL-0002

ETCH NAME: Potassium cyanide — TIME:
TYPE: Electrolytic, polish — TEMP:
COMPOSITION: — ANODE:
 x .... 9% KCN — CATHODE:
POWER:

DISCUSSION:

Ag_2Al, single crystal specimens used in a study of the effects of internal oxidation on plastic deformation. Specimens electropolished prior to deformation.

REF: Marcinkowski, M J & Wriedt, D F — *J Electrochem Soc,* 12,92(1964)

AGAL-0003a

ETCH NAME: A-2 TIME: 1 min/8—10 sec

TYPE: Electrolytic, polish TEMP: RT

COMPOSITION: ANODE: Ag_2Al

x .... KCN CATHODE:

x .... H_2O POWER: 1.5—2.5 A/cm^2

DISCUSSION:

Ag_2Al, (0001) wafers used in a study of dislocation etch pits. Solution used to polish specimens prior to dislocation etching.

REF: George, J & Mote, J D — *J Appl Phys,* 36,1793(1963)

AGAL-0001b

ETCH NAME: TIME: 1 min

TYPE: Acid, dislocation TEMP: RT

COMPOSITION:

5 .... HF

5 .... H_2SO_4

7 .... H_2O_2

DISCUSSION:

Ag_2Al, (0001) wafers used in a study of dislocation etch pits. Solution used to develop pits: (1) etch in solution shown; (2) DI water rinse; (3) ethyl alcohol rinse and (4) dry with warm air. Hexagonal and pyramidal pits (dislocations) were observed that may develop a flat bottom with etch time. Dislocation density was $10^5/cm^2$.

REF: Ibid.

SILVER BORON ETCHANTS

AGB-0001

ETCH NAME: Ammonium hydroxide TIME:

TYPE: Acid, cleaning TEMP: RT to hot

COMPOSITION:

x .... NH_4OH, conc.

DISCUSSION:

AgB alloys with varying formulae used as high temperature brazing alloys. Surfaces can be cleaned prior to and after brazing with ammonia compounds.

REF: Pamphlet: *Silvaloy,* Silver Brazing Div., The American Platinum Works, NJ

SILVER LEAD TIN ETCHANTS

AGPS-0001a

ETCH NAME: Ammonium hydroxide TIME: 1—2 min

TYPE: Acid, cleaning TEMP: RT

COMPOSITION:

x .... NH_4OH, conc.

DISCUSSION:

Pb:Sn #63 alloying 2% Ag used in device and package assembly of electronic devices. Solution used to clean surfaces prior to and after use.

REF: Walker, P — personal application, 1960

AGPS-0001b

ETCH NAME: Aqua regia, dilute — TIME: 10—30 sec

TYPE: Acid, removal/cleaning — TEMP: RT

COMPOSITION:

1 .... HCl

3 .... HNO_3

6 .... H_2O

DISCUSSION:

Pb:Sn + 2% Ag, solder used in device and package assembly of electronic devices. Solution used as a cleaning and light removal etch.

REF: Ibid.

AGPS-0001c

ETCH NAME: — TIME: 10—30 sec

TYPE: Acid, cleaning — TEMP: RT

COMPOSITION:

6 .... HAc

2 .... H_2O_2

2 .... H_2O

DISCUSSION:

Pb:Sn:Ag(2%), alloy #63 used in device and package assembly of electronic devices. Solution used to remove oxides after alloying and as a slightly preferential etch to observe alloy contact structure.

REF: Walker, P — personal development/application, 1959

AGPS-0001d

ETCH NAME: — TIME:

TYPE: Acid, removal — TEMP: RT

COMPOSITION:

1 .... HF

1 .... HNO_3

1 .... HAc

DISCUSSION:

Pb:Sn + Ag as various other alloys with or without silver. Solution used as a general removal and solubility etch in a study of 30 different alloy contacts for silicon.

REF: Ibid.

SILVER MERCURY ETCHANTS

AGHG-0001

ETCH NAME: Nitric acid — TIME:

TYPE: Acid, removal/polish — TEMP:

COMPOSITION:

(1) x .... HNO_3, conc.

(2) x .... HNO_3

x .... H_2O

DISCUSSION:

Ag_2Hg, (110) single crystal surfaces of the natural mineral called Amalgam. Crystal was a dodecahedron, (110). Solutions used to etch clean surfaces.

REF: Walker, P — mineralogy study, 1953

SILVER TIN ETCHANTS

AGSN-0001

ETCH NAME: Potassium cyanide
TYPE: Electrolytic, polish
COMPOSITION:
6 g .... KCN
100 ml .... H_2O

TIME:
TEMP:
ANODE: Ag:Sn
CATHODE:
POWER:

DISCUSSION:

Ag:Sn(1%) single crystal alloys (also used 2, 4, 6 and 9% Sn) used in a study of dislocations and stacking fault energy in these alloys. Solution used to electropolish specimens.

REF: Ruff, A W & Ives, L K — *Acta Metall,* 15,189(1967)

SILVER TITANIUM ETCHANTS

AGTI-0001

ETCH NAME:
TYPE: Acid, removal
COMPOSITION:
1 .... HF
1 .... HNO_3
1 .... HAc

TIME: 30—60 sec
TEMP: RT

DISCUSSION:

Ti:TiAg:Ag thin films evaporated on silicon (111) devices for electrical contact. After photo resist patterning, solution used to remove excess metallization. Evaporated on silicon dioxide, SiO_2 thin films, then annealed in vacuum at 600°C for 15—30 min, converts to a conductive, glassy silicate. Solution also used to pattern the silicate.

REF: Walker, P — personal development/application, 1971

SILVER ZINC ETCHANTS

AGZN-0001

ETCH NAME: Nitric acid, dilute
TYPE: Acid, removal
COMPOSITION:
x .... HNO_3
x .... H_2O

TIME:
TEMP:

DISCUSSION:

Ag:Zn alloy specimens used in a study of self-diffusion in the material. After mechanically lapping surfaces, solution used to etch remove work damage.

REF: Cazarus, D & Tomizuka, C T — *Phys Rev,* 103,1155(1956)

SILVER GERMANIUM PHOSPHIDE ETCHANTS

AGEP-0001
ETCH NAME: TIME:
TYPE: Acid, removal TEMP:
COMPOSITION:
x .... HF
x .... HNO_3
DISCUSSION:

$Ag_6Ge_{10}P_{12}$, single crystal ingots grown by the Bridgman method as p-type semiconductor material. Band gap: 0.76 eV. Specimens were cut as parallelapipeds (001), (110) and (110) orientation, and mechanically lapped. Also grew $CuGe_2P_3$.

REF: McDonald, J E — *J Phys & Chem Solids,* 8,951(1985)

PHYSICAL PROPERTIES OF SILVER BROMIDE, AgBr

Classification	Bromide
Atomic numbers	47 & 35
Atomic weight	187.8
Melting point (°C)	434
Boiling point (°C)	700
Density (g/cm³)	5.8—6
Refractive index (n=)	2.25
Hardness (Mohs — scratch)	1—1.5
Crystal structure (isometric — normal)	(100) cube
Color (solid)	Yellow/greenish
Cleavage (cubic — perfect)	(001)

SILVER BROMIDE, AgBr

General: Occurs in nature as the mineral bromyrite, AgBr, found in limited quantities both in Mexico and Chile and associated with other ores of silver, along with both silver chloride (cerargyrite) and the iodide (iodyrite). Primary use in industry is as a compound in photographic emulsions and in other chemical reactions.

Technical Application: Not used in Solid State device fabrication, other than in photographic film processing. It has been grown and studied as a single crystal for general morphology and defects.

Etching: KCN, NH_4OH and some sulfur compounds. Insoluble in water.

SILVER CADMIUM ETCHANTS

AGCD-0001
ETCH NAME: Nitric acid TIME:
TYPE: Acid, removal TEMP: RT
COMPOSITION:
x .... HNO_3
x .... H_2O
DISCUSSION:

AgCd, single crystal specimens. A study of tracer diffusion as a chemical concentration gradient in the material. Both metals, individually, and as a compound single crystal can be etched in the solution mixture shown.

REF: Manning, J R — *Phys Rev,* 116,69(1959)

SILVER BROMIDE ETCHANTS

AGBR-0001
ETCH NAME: Emery
TYPE: Abrasive, polish
COMPOSITION:
x .... $Al_2O_3Fe_3O_4$
DISCUSSION:

TIME:
TEMP: RT

AgBr, (110) wafers cut within $^1/_2$° of plane. The ingot was grown by the Bridgman process with Cd added to pin dislocations. A (110) oriented cylinder was cut with a jeweler's saw, and surface of specimens was hand lapped and polished on emery paper.
REF: Barber, R et al — *J Phys Chem Solids,* 46,107(1985)
AGBR-0002: Cain, L S — *J Phys Chem Solids,* 38,73(1977)
AgBr specimens. Similar study as in AGBR-0001

AGBR-0003
ETCH NAME: Sodium thiosulfate
TYPE: Acid, dislocation
COMPOSITION:
x .... 3—7 *N* NaS_2O_7
DISCUSSION:

TIME:
TEMP:

AgBr, (100) and (111) wafers. Solution used to develop dislocations and dislocation loops in a general defect study of the material. Solution was not good on the (110) surfaces.
REF: Bartlett, J T & Mitchell, J W — *Phil Mag,* 5,445(1960)

PHYSICAL PROPERTIES OF SILVER CHLORIDE, AgCl

Classification	Chloride
Atomic numbers	47 & 17
Atomic weight	143.34
Melting point (°C)	455
Boiling point (°C)	1550
Density (g/cm^3)	5.552
Refractive index (n=)	2.0611
Hardness (Mohs — scratch)	1—1.5
Crystal structure (isometric — normal)	(100) cube
Color (solid)	Pearl-gray
Cleavage (solid)	None

SILVER CHLORIDE, AgCl

General: Occurs in nature as the mineral ceragyrite, AgCl, with the cubic structure similar to halite, NaCl (common salt), but has no cleavage. Common name of the mineral is "Horn silver". It is of secondary origin and usually found in the upper zones of silver vein deposits — quartz rock gangue — in association with silver bromides and iodides . . . pure native silver is often at the top of such veins due to hydrothermal and atmospheric alteration, though the chloride is insoluble in water. It is an ore of silver containing approximately 25% silver which can be easily separated by heating in the presence of carbon. In the presence of sunlight it turns brown-violet due to photoreaction by ultraviolet rays.

Both the natural mineral and artificial compound, along with AgBr and AgI, are the primary constituents in photographic emulsions in black and white photography . . . the silver, effectively, plating-out as the black oxide under intense UV light forming a permanent negative. As a finely divided particulate — smoke — the AgCl particles, like dust, are used to nucleate raindrops in storm clouds. When treated with salt of phosphorus and copper chips, it produces an intense azure-blue flame and, as a colloid, a brownish violet suspension.

Technical Application: Though not used in Solid State device processing, it is under study and development for its photoconductivity characteristics. Like seeding of rain clouds, AgCl smoke has been used in low temperature cryostats to grow snowflakes and single crystal ice. Both natural and artificially grown single crystals have been studied for physical and morphological data in addition to photoconductivity characteristics.

Etching: Soluble in NH_4OH and KCN.

SILVER CHLORIDE ETCHANTS

AGCL-0001
ETCH NAME: Potassium cyanide TIME:
TYPE: Acid, polish TEMP: RT
COMPOSITION:
x .... 3% KCN
DISCUSSION:

AgCl, (100) wafers used in a study of transient photoconductivity at low temperatures. Solution used on a glass plate to lap polish surfaces. Work was done in a dark room under red light.

REF: Van Heyningen, R S & Brown, F C — *Phys Rev,* 111,462(1958)
AGCL-0002: Brown, F C — *J Phys Chem Solids,* 4,206(1958)

AgCl, (100) wafers used in similar work as cited in AgCl-0001.

AGCL-0003
ETCH NAME: TIME:
TYPE: Pressure, thinning TEMP:
COMPOSITION:
x .... pressure
DISCUSSION:

AgCl specimens were thinned by applying physical pressure and followed by annealing for use in a structural study.

REF: Moser, F & Urbach, F — *Phys Rev,* 102,1519(1956)

AGCL-0004
ETCH NAME: Hydrochloric acid TIME: 1 min
TYPE: Acid, cleaning TEMP: RT
COMPOSITION:
x .... HCl, conc.
DISCUSSION:

AgCl, (100) wafers used in a photoconductivity study. Mechanically lap on a glass plate: (1) with coarse grit, use NH_4OH as the liquid carrier; (2) then fine grit using water as the carrier. After lapping, water rinse and follow with etch cleaning in the solution shown, with DI water rinse and dry.

REF: Gordon, A M — *Phys Rev,* 122,748(1961)

AGCL-0005
ETCH NAME: Nitric acid TIME:
TYPE: Acid, cleaning TEMP:
COMPOSITION:
x .... HNO_3, conc.

DISCUSSION:

AgCl, (100) wafers used in a study of photoelectric behavior. Surfaces were cleaned as follows: (1) degrease in petroleum ether, (2) soak in nitric acid, and (3) wash in DI water.

REF: Wainfan, N — *Phys Rev,* 105,100(1957)

AGGL-0006

ETCH NAME: Ammonium hydroxide TIME:

TYPE: Hydroxide, cleaning TEMP:

COMPOSITION:

x .... NH_4OH

DISCUSSION:

AgCl, single crystal oriented bars cut with a jeweler's saw and then lapped down to size on a ground glass plate using a sodium thiosulfate solution. Wafers were cut on a lathe 1 cm diameter and .025″ thick. Etch polish and clean in the solution shown, and rinse in DI water. HCl was used as an alternative etch polish solution.

REF: Wiegand, D A — *Phys Rev,* 113,52(1959)

AGCL-0007a

ETCH NAME: Hypo TIME:

TYPE: Acid, polish/dislocation TEMP: RT

COMPOSITION:

x .... 3 *N* $Na_2S_2O_5.5H_2O$

DISCUSSION:

AgCl, (100) oriented material cut as bars with the length in the c-axis direction ⟨001⟩. Bars used in a deformation study of dislocations induced by bending and annealing. Solution used to polish bars prior to anneal and bend. (*Note:* "Hypo" is the solvent used on AgCl in photography.)

REF: Sprackling, M T — *Phil Mag,* 9,739(1964)

AGCL-0007b

ETCH NAME: TIME:

TYPE: Acid, preferential TEMP: RT

COMPOSITION:

4 .... 3 *N* $Na_2S_2O_5.5H_2O$ ("hypo")

1 .... 0.88 *N* NH_4OH

DISCUSSION:

AgCl, (100) bars. Solution used to develop defects and dislocations in bars after annealing and bending. Also used 3 *N* "hypo" — AGCL-0007a.

REF: Ibid.

AGCL-0008: Jones, D A & Mitchell, J W — *Phil Mag,* 2,1047(1957)

AgCl single crystal specimens used in a study of dislocation etching and etching techniques. Also studied other silver halides.

SILVER EPOXY ETCHANTS

AGE-0001a

ETCH NAME: Nitric acid, dilute TIME: 2—15 min

TYPE: Acid, removal TEMP: RT or 50°C

COMPOSITION:

1 .... HNO_3

3 .... H_2O

DISCUSSION:

Ag-Epoxy used for electrical attachment of GaAs FET dice to alumina substrate with gold circuit pads. Solution used to remove damaged FETs and Ag-Epoxy from alumina without affecting the gold circuits. This solution is called Schell etch (GAS-0045g) and is preferential on GaAs. After removal, wash alumina in DI H_2O, then bake in air oven at 125°C for $^1/_2$ h before re-using circuit.

REF: Walker, P — personal application, 1980—1985

AGEP-0001b

ETCH NAME: TIME: 3—5 min

TYPE: Gas, cleaning TEMP: RT

COMPOSITION:

x .... O_2, RF plasma

DISCUSSION:

Ag-Epoxy applied as a contact paste and cured at 120°C, 30 min in nitrogen. The epoxy can be "cleaned" in an RF plasma of oxygen developing a hard black surface coating of Ag_2O. So treated, the surface is very stable against chemical attack and atmospheres though, for some high reliability requirements, the black coating is undesirable. Remove by soak in HNO_3 or NH_4OH.

REF: Ibid.

PHYSICAL PROPERTIES OF SILVER IODIDE, AgI

Classification	Iodide
Atomic numbers	57 & 53
Atomic weight	234.8
Melting point (°C)	552 del.
Boiling point (°C)	
Density (g/cm³)	5.5—5.7
Refractive index (n=)	2,182
Hardness (Mohs — scratch)	1.5—2
Crystal structure (hexagonal — hemimorphic)	(0001) platelets
(isometric — normal) >146°C	(100) cube
Color (solid)	Yellow
Cleavage (basal/cubic)	(0001)/(001)

SILVER IODIDE, AgI

General: Occurs in nature as the mineral iodyrite, AgI, associated with silver deposits as a minor compound.

Both silver iodide and bromide have major use in black and white photographic emulsions for conversion to the black oxide on film. The iodide has very low solubility in water and, as fine powder, it is used as a smoke to seed rain clouds and initiate precipitation. It is used in a similar manner to seed the growth of ice crystals in cold cryostats at LN_2 temperature in the study of ice formation both as snowflakes and single crystal ice.

Technical Application: In Solid State development silver iodide has been shown to have semiconducting properties as have several trinary and binary silver compounds, such as selenides and tellurides and their mixed compounds. Many such compounds have been grown as bulk single crystals or as thin films, but all are characteristically soft (H = 2—3) and can be tarnished in the presence of sulfur in the atmosphere or show oxide conversion under intense light conditions. When silver compounds are subjected to RF plasma with oxygen as a cleaning vehicle, surfaces will become coated with the black oxide, which can be removed NH_4OH.

Etching: KCN, NH_4OH, NH_3 and sulfur-containing compounds.

SILVER IODIDE ETCHANTS

AGI-0001
ETCH NAME: TIME:
TYPE: Acid, cleaning TEMP:
COMPOSITION:
x .... NH_4OH
DISCUSSION:
AgI, single crystals used to evaluate cohesive energy features of tetrahedral semiconductors. Forty materials studied.
REF: Aresti, A et al — *J Phys Chem Solids,* 45,361(1984)

AGI-0002
ETCH NAME: Silver iodide TIME:
TYPE: Compound, seeding TEMP: −195°C
COMPOSITION:
x .... AgI_2, powder
DISCUSSION:
AgI_2 powder used to seed growth of single crystal ice crystals in a cryostat at LN_2 temperature. See ICE-0007. (*Note:* Also used as a smoke to initiate precipitation from rain clouds.)
REF: Gonda, T & Koke, T — *J Cryst Growth,* 65,36(1983)

SILVER:GOLD ETCHANTS

AGAU-0001
ETCH NAME: TIME:
TYPE: Acid, preferential TEMP:
COMPOSITION:
x .... KCN
x .... $(NH_4)_2SO_4$
DISCUSSION:
Ag:Au, single crystal alloy ingots of various compositions grown in a study of twinning and defects after straining. Crystal compositions varied from 0, 10, 25, 50, 75, and 90% gold. Solution used as a metallographic structure development etch.
REF: Suzucki, H & Barrett, C S — *Acta Metall,* 6,156(1958)

AGAU-0002
ETCH NAME: Aqua regia, dilute TIME: Sec
TYPE: Acid, step-etch/cleaning TEMP: RT
COMPOSITION:
1 .... HCl
3 .... HNO_3
6 .... H_2O
DISCUSSION:
Ag:Au, evaporated thin films deposited on silicon (111) wafers as electrical contact

pads. After alloy cycling solution used to clean surfaces and as a step-etch to observe alloy interface structure.
REF: Walker, P — personal application, 1973

SILVER MAGNESIUM ETCHANTS

AGMG-0001
ETCH NAME: TIME:
TYPE: Acid, cleaning TEMP:
COMPOSITION:
1 .... H_2SO_4
1 .... CrO_3
1 .... H_2O
DISCUSSION:
AgMg, single crystal specimens. Solution was used to clean surfaces before silver diffusion studies.
REF: Hagel, W C & Westbrook, J E — *Trans Met Soc AIME*, 221,951(1961)

PHYSICAL PROPERTIES OF SILVER SELENIDE, Ag_2Se

Classification	Selenide
Atomic numbers	47 & 34
Atomic weight	294.72
Melting point (°C)	880
Boiling point (°C)	
Density (g/cm^3)	8.0
Hardness (Mohs — scratch)	2—3
Crystal structure (isometric — normal)	(100) cube
Color (solid)	Grey
Cleavage (cubic)	(001)

SILVER SELENIDE, Ag_2Se

General: Although silver selenide, Ag_2Se, does not occur in nature as a pure compound, there are minerals of this type containing one or more other elements: naumannite, $(Ag_2Pb)Se$; aguilarite, $Ag_2(S,Se)$; eucarite, Cu_2Se,Ag_2Se and crookesite, $(Cu,Tl,Ag)_2Se$. Where such minerals occur in sufficient quantity, they are mined for their silver and selenium content.

Technical Application: Silver selenide is a I—VI compound semiconductor and, although it has been studied, it has not been widely used due to its softness (H = 2.3) and solubility.

The compound has been grown as a single crystal for general morphological study and semiconductor characterization. There are a number of ternary and quaternary single crystals that have been grown and studied as a metallurgical, inter-related series: $AgSbSe_2$-$AgSbTe_2$-$AgBiSe_2$-PbSe-PbTe.

Etching: Mixed acids and alkalies. Soluble in alcohols.

SILVER SELENIDE ETCHANTS

AGSE-0001a
ETCH NAME: TIME:
TYPE: Acid, polish TEMP: 50°C
COMPOSITION:
5 .... H_2SO_4
1 .... H_2O_2
DISCUSSION:
Ag_2Se, (100) wafers and other orientations. Solution used as a chemical polish etch and may show a degree of preferential attack.
REF: Sagar, A et al — *J Appl Phys,* 39,5336(1968)

AGSE-0001b
ETCH NAME: TIME:
TYPE: Acid, polish TEMP: 80°C
COMPOSITION:
2 .... KOH, sat. sol.
1 .... H_2O_2
2 .... ethylene glycol (EG)
DISCUSSION:
Ag_2Se, (100) wafers and other orientations. Solution shown as a polishing etch for this material.
REF: Ibid.
AGSE-0002: Gatos, H C Ed — *Properties of Elemental and Compound Semiconductors,* Interscience, New York, 1960, 69
$AgSbSe_2$, trinary compound semiconductors. Article covers trinary and quaternary compounds of silver selenides and tellurides with bismuth or lead additives.

AGSE-0003
ETCH NAME: BRM TIME:
TYPE: Halogen, polish TEMP: RT
COMPOSITION:
x .... x%Br_2
x .... MeOH
DISCUSSION:
Ag_2,Se (100) wafers used in a development study of new semiconducting compounds. Other compounds were $AgInTe_2$; Tl_2Se_3; SnSe; $SnSe_2$; TlSe; Bi_2Se_3; Li_3Bi; Bi_2Se_3; In_2Te_3; In_2Tl_3. Solution shown for reference.
REF: Mooser, E & Pearson, W B — *Phys Rev,* 101,492(1956)
AGSE-0004: Appel, J & Lantz, G — *Physica,* 20,1110(1954)
Additional reference on In_2Te_3.
AGSE-0005: Goodman, C H L & Douglas, R W — *Physica,* 20,1107(1954)
Additional reference on chalcopyrite, $AgInTe_2$.
AGSE-0006: Davidenko, V A — *J Phys (USSR),* 4,170(1941)
Additional reference on compounds in AGSE-0003.

SILVER GALLIUM SELENIDE ETCHANTS

AGGE-0001
ETCH NAME: TIME:
TYPE: Acid, removal TEMP:
COMPOSITION:
(1) x .... NH_4OH (2) x .... HNO_3
DISCUSSION:
$AgGaSe_2$, single crystals used to evaluate cohesive energy features of tetrahedral semiconductors. Forty materials were studied.
REF: Aresti, A et al — *J Phys Chem Solids,* 45,361(1984)

SILVER INDIUM SELENIUM ETCHANTS

AGIE-0001
ETCH NAME: BRM TIME:
TYPE: Acid, removal TEMP:
COMPOSITION:
(1) x .... x% Br_2 (2) x .... HNO_3
x .... MeOH
DISCUSSION:
$AgInSe_2$, single crystals used to evaluate the cohesive energy features of tetrahedral semiconductors. Forty materials were studied.
REF: Aresti, A et al — *J Phys Chem Solids,* 45,361(1984)

SILVER TITANIUM SELENIDE ETCHANTS

AGTS-0001
ETCH NAME: BRM TIME:
TYPE: Halogen, polish TEMP:
COMPOSITION:
x .... x%Br_2
x .... MeOH
DISCUSSION:
AgTiSe, single crystal material grown in a study of optical and elastic properties. Structure is orthorhombic.
REF: Newman, P R — *J Appl Phys,* 54,1547(1983)

PHYSICAL PROPERTIES OF SILVER SULFIDE, AgS

Classification	Sulfide
Atomic numbers	47 & 16
Atomic weight	247.82
Melting point (°C)	825
Boiling point (°C)	
Density (g/cm³)	7.2—7.36
Hardness (Mohs — scratch)	2.0—2.5
Crystal structure (isometric — normal)	(100) cube
Color (solid)	Lead-grey/black
Cleavage (cubic)	(001)

SILVER SULFIDE, AgS

General: Occurs in nature as the mineral argentite, AgS, common name: Silver Glance. It may be found in large deposits associated with other silver minerals, pyrite, galena, and nickel ores, as a primary deposit; with limonite, calcite, and quartz, as a deposit of secondary origin. It is a major ore of silver.

Other than as an ore of silver in industry, it is the compound source for silver plating baths, and used in black and white photography.

Technical Application: In Solid State processing, silver sulfide is used for silver plating of parts and packages. It is under evaluation for its optoelectric properties and as a trinary compound, $AgGaS_2$, has semiconducting properties.

Silver reacts readily with sulfur which is the black tarnish on silver plate can be due to sulfur fumes in the air. It is grown as a single crystal by the action of sulfur, sulfur dioxide or hydrogen sulfide on pure silver.

Etching: Soluble in acids and KCN.

SILVER SULFIDE ETCHANTS

AGS-0001a
ETCH NAME: Potassium cyanide — TIME:
TYPE: Electrolytic, polish — TEMP:
COMPOSITION: — ANODE:
x .... KCN — CATHODE:
x .... H_2O — POWER:
DISCUSSION:

AgS, single crystal whiskers. Solution used to polish specimens.

REF: Drott, J — *Acta Metall,* 9,19(1961)

AGS-0001b
ETCH NAME: Nitric acid, dilute — TIME:
TYPE: Acid, preferential — TEMP:
COMPOSITION:
x .... 0.1 N HNO_3
DISCUSSION:

AgS, single crystal whiskers. Solution used to develop dislocations in a study of crystal whisker structure.

REF: Ibid.

SILVER GALLIUM SULFIDE ETCHANTS

AGAS-0001
ETCH NAME: Hydrochloric acid — TIME:
TYPE: — TEMP:
COMPOSITION:
x .... HCl
DISCUSSION:

$AgGaS_2$, thin films as both polycrystalline and amorphous structures deposited by flash evaporation onto glass substrates and ITO film coated glass substrates. Material is a I—III—VI_2 compound semiconductor with optoelectronic and photoconducting capabilities. Structure is tetragonal chalcopyrite type.

REF: Camp, H V — *Thin Solid Films,* 111,17(1984)

PHYSICAL PROPERTIES OF SILVER TELLURIDE, Ag_2Te

Classification	Telluride
Atomic numbers	47 & 52
Atomic weight	343.37
Melting point (°C)	>300
Boiling point (°C)	
Density (g/cm^3)	8.31—8.45 (hessite)
Hardness (Mohs — scratch)	2—2.5
Crystal structure (orthorhombic — normal)	(100) a-pinacoid
Color (solid)	Lead grey
Cleavage (traces)	(100)/(110)

SILVER TELLURIDE, Ag_2Te

General: Occurs as the mineral hessite, Ag_2Te, usually massive or fine grained. With increasing gold content it grades into the mineral petzite, $(Ag,Au)_2Te$, with G = 8.7—9.02. These are minor minerals, but industrially important for their silver and gold content, as well as tellurium.

Technical Application: Silver telluride is the base of a I—VI compound semiconductor as a trinary compound with antimony or iron, and has been evaluated for semiconducting characteristics. The trinary compounds are shown here.

Etching: Single acids of HNO_3, H_2O_2, and mixed acids $HF:HNO_3$ with salts.

SILVER ANTIMONY TELLURIDE ETCHANTS

SAT-0001
ETCH NAME: — TIME:
TYPE: Acid, polish — TEMP:
COMPOSITION:
1 .... HNO_3
1 .... HCl
2 .... $K_2S_2O_7$, sat. sol.
DISCUSSION:

$AgSbTe_2$, single crystal specimens. Solution is described as a metallographic polishing etching. Also can be used on $AgFeTe_2$.

REF: Wolfe, R et al — *J Appl Phys,* 31,1959(1960)

SAT-0002a
ETCH NAME: Nitric acid, dilute — TIME: 10 sec
TYPE: Acid, preferential — TEMP: RT
COMPOSITION:
x .... 55% HNO_3
DISCUSSION:

$AgSbTe_2$, single crystal specimens. Solution used to develop defect structure in this material. Authors refer this etch to Farnsworth (AG-0001) as a silver orientation etch.

REF: Armstrong, R W et al — *J Appl Phys,* 31,1954(1960)

SAT-0002b
ETCH NAME: Hydrogen peroxide — TIME:
TYPE: Acid, dislocation — TEMP:
COMPOSITION:
x .... H_2O_2

DISCUSSION:

$AgSbTe_2$, single crystal specimens used in a structural study. Solution shown develops structure but not as well as SAT-0002a. Both CP4 and Dash Etch were evaluated, but were not as good. For CP4, see: SI-0047; for Dash Etch, SI-0046.

REF: Ibid.

SAT-0003: Gatos, H C Ed — *Properties of Elemental and Compound Semiconductors,* Interscience, New York, 1960, 69

$AgSbTe_2$, trinary compound semiconductor growth and study.

SILVER IRON TELLURIDE ETCHANTS

SIT-0001

ETCH NAME:	TIME:
TYPE: Acid, polish	TEMP:

COMPOSITION:

1 .... HNO_3

1 .... HCl

2 .... $K_2S_2O_7$, sat. sol.

DISCUSSION:

$AgFeTe_2$, single crystal specimens studied as a semiconductor compound. Solution used to polish specimens.

REF: Wernick, J H & Wolfe, R — *J Appl Phys,* 32,749(1961)

PHYSICAL PROPERTIES OF SILVER TELLURIDE, Ag_2Te

Classification	Telluride
Atomic numbers	47 & 52
Atomic weight	343.47
Melting point (°C)	955
Boiling point (°C)	
Density (g/cm³)	8.3—8.45
Hardness (Mohs — scratch)	2.5—3
Crystal structure (isometric — normal)	(100) cube
Color (solid)	Steel-grey
Cleavage (indistinct)	(001)

SILVER TELLURIDE, Ag_2Te

General: Occurs in nature as the mineral hessite, Ag_2Te, and the mineral stutzite, Ag_4Te. There are also some trinary compound telluride minerals, such as petzite $(Ag,Au)_2Te$; suvanite $(Ag,Au)Te_2$; krennerite $(Au,Ag)Te_2$ and muthmannite, (Ag,Au)Te, in addition to the complex mineral — tapalpite $3Ag_2(S,Te).Bi_2(S,Te)_3$. Silver forms over 50 sulfide minerals as well as silver halides of iodine, chlorine and bromine. Where there minerals are found in quantity, they are mined for their silver content and other elements. The sulfide, argentite (Ag_2S) is a major ore of silver. Primary use in industry is as ores of silver and tellurium.

Technical Application: Silver telluride is a I—VI compound semiconductor like silver selenide, Ag_2Se. Both compounds have been studied for their semiconductor characteristics and general morphology as single crystals. There are a number of ternary and quaternary single crystals that have been grown and studied, such as $AgSbSe_2$-$AgSbTe_2$-$AgBiSe_2$-PbTe-PbSe, as they form an inter-related metallurgical series.

Etching: Soluble in HNO_3 and KCN and mixed acids.

SILVER TELLURIDE ETCHANTS

AGTE-0001
ETCH NAME: TIME:
TYPE: Acid, polish TEMP:
COMPOSITION:
2 .... H_2O_2
3 .... NH_4OH
DISCUSSION:
Ag_2Te, (100) wafers and other orientations. Solution shown as a general polish etch.
REF: Sagar, A et al — *J Appl Phys,* 39,5336(1968)
AGTE-0002: Gatos, H C, Ed — *Properties of Elemental and Compound Semiconductors,* Interscience, New York, 1960, 69
AgSbTe, trinary compound semiconductors.

SILVER GALLIUM TELLURIDE ETCHANTS

AGGT-0001
ETCH NAME: BRM TIME:
TYPE: Acid, removal TEMP:
COMPOSITION:
x .... x%Br_2
x .... MeOH
DISCUSSION:
$AgGaTe_2$, single crystals used in evaluating cohesive energy features of tetrahedral semiconductors. Forty materials were studied. Solution shown for reference.
REF: Aresti, A et al — *J Phys Chem Solids,* 45,361(1984)

SILVER INDIUM TELLURIDE ETCHANTS

AGIT-0001
ETCH NAME: TIME:
TYPE: Acid, removal TEMP:
COMPOSITION:
x .... H_2O_2
x .... NH_4OH
DISCUSSION:
$AgInTe_2$, single crystals used to evaluate cohesive energy features of tetrahedral semiconductors. Forty materials were studied.
REF: Aresti, A et al — *J Phys Chem Solids,* 45,362(1984)

PHYSICAL PROPERTIES OF SODIUM, Na

Classification	Alkaline metal
Atomic number	11
Atomic weight	22.99
Melting point (°C)	97.8 (97.5)
Boiling point (°C)	892
Density (g/cm^3) solid/gas	0.97/1275
Thermal conductance (cal/sec)(cm^2)(°C/cm) 20°C	0.334
Specific heat (cal/g) 25°C	0.295
Heat of fusion (cal/g)	27.05
Heat of vaporization (k-cal/g-atom)	1.127

Coefficient of linear thermal expansion ($\times 10^{-4}$ cm/cm/°C) 20°C	71
Atomic volume (W/D)	23.7
1st ionization energy (K-cal/g/mole)	119
1st ionization potential (eV)	5.138
Electronegativity (Pauling's)	0.9
Covalent radius (angstroms)	1.54
Ionic radius (angstroms)	0.97 (na^{+1})
Critical temperature (g/cm^3)	0.206
Electrical resistivity ($\times 10^{-6}$ ohm cm) 0°C	4.2
Electron work function (eV)	2.27
Cross section (barns)	0.53
Vapor pressure (°C)	701
Refractive index (n=)	4.22
Hardness (Mohs — scratch)	0.4
Crystal structure (isometric — normal)	(100) cube, bcc
Color (solid)	Silver white
(flame)	Intense yellow
Cleavage (cubic — poor)	(001)

SODIUM, Na

General: Sodium does not occur as a free element in nature, but there are over 150 known single crystal compounds that contain sodium and, if artificial compounds are included, there are over 300. Common salt, as the mineral halite, NaCl (sodium chloride) is the major compound in ocean (salt) water and is found as dried salt beds and domes from ancient buried seas. At Sperenberg, Germany there is a deposit 4000 ft thick; there are similar "salt domes" along the Gulf Coast of the U.S., and there are buried salt beds as farth north as the Great Lakes in the mid-continent. In Galicia, Poland there is a bed 50 miles long, 20 miles wide, and 1200 ft thick. The Dead Sea in Palestine, 47 miles long and 9 miles wide, contains 40 billion tons of salt in solution; there is the Great Salt Lake in Utah, and the Salton Sea in southern California. There are many dry salt flats throughout the southwestern U.S., as well is in other desert areas of the world.

Salt is one of the most important compounds in the chemical industry for it produces (1) sodium; (2) sodium hydroxide; (3) sodium carbonate (soda ash); (4) chlorine gas and (5) both hydrochloric and sulfuric acids are compounded from salt. In the metals industries, salt brine is used as a quenching medium of hot metals and alloys; as a molten flux for etching and as a metal flux for soldering. As soda ash (natron, $Na_2CO_3.nH_2O$, the natural compound is only found in solution) it was used by the Egyptians in mummification; used in the manufacture of soap, glass, and bleaching powder. The chemical symbol, Na, comes from mineral natron. In addition, salt has been a food preservative since ancient times and is essential in human and animal diet. The mineral borax, $Na_2B_4O_7.10H_2O$, is a source of both sodium, Na and boron, B, and is used for washing and cleaning as an antiseptic and preservative, as a solvent of metal oxides and as a flux in soldering and welding. The name comes from the Arabic buraq — which included niter (sodium carbonate) and natron of the Egyptians.

As sodium hydroxide, NaOH, it is a major cleaning and etching solution in various concentrations for most metals and metallic compounds in industrial processing; as a molten flux above 300°C it is both a cleaner and etchant on metals and alloys.

A major use of sodium metal is in the manufacture of sodium cyanide, NaCN, for electrolytic plating and in organic reactions; to compound tetraethyl lead for gasoline anti-knock and in the making of dyes and perfumes, e.g., attar of roses. As liquid sodium, it is a coolant for high power aircraft engines. When housed in a metal tube, through which a quartz tube is inserted, it is a temperature stabilizing "heat pipe" in horizontal epitaxy reactors. The sodium lamp — sodium and mercury with a trace of neon as the activator, produces two intense yellow lines, characteristic of sodium . . . among other applications, such lamps are used in determining material surface flatness and step-height of thin films (D/Fx2945 Å = T) . . . in both cases the observer sees a brilliant yellow background with black lines.

Although pure sodium metal is shiny and silver in appearance, much like silver, it oxidizes rapidly in air to a white surface oxide. It is stored under toluene and, if the toluene is heated, sodium will liquefy and can then be poured into forms as a rod or wire with high electrical conductivity and heat dissipation.

Technical Application: Sodium is considered a contaminating species, much like oxygen, in many Solid State material processes, such as in silicon ingots. It is not used as elemental sodium in device fabrication and, where etchants containing sodium are used, heavy washing follows to insure complete removal of all traces of sodium prior to any high temperature processing.

As mentioned in the general section, liquid sodium is used as a heat pipe stabilization mechanism in epitaxy reactors and similar "pipes" are used as heat removal/transfer units in a variety of semiconductor and electronic hardware applications.

Many sodium compounds are used in etching formulations for the etching of semiconductors and metals, such as sodium hydroxide, NaOH; sodium bromide, NaBr; sodium carbonate, $NaCO_3$; sodium chromate, $NaCrO_4$ and dichromate, $Na_2Cr_2O_7$, as well as sodium cyanide, NaCN to mention only a few of the more prominent. They are used, alone, in various solution concentrations in water; as one constituent of a solution mixture, or in solid form as a molten flux etchant. The latter also is used as a growth flux for some metallic compounds, such as ferrites and garnets.

Etching: Deliquesces in cold water to hydroxides; can be explosive in hot water with release of hydrogen. Soluble in hot toluene.

SODIUM ETCHANTS

NA-0001

ETCH NAME: Nonyl — TIME: 30 sec

TYPE: Alcohol, polish — TEMP: RT

COMPOSITION:

x .... $CH_3(CH_2)_7CH_2OH$

DISCUSSION:

Na specimens cut with a razor blade under oil (also lithium and potassium). After cutting, put in heptane to remove oil, then into tetrahydrofuran to stop heptane reaction. Ten alcohols were evaluated for reactivity with sodium. Increasing molecular weight reduces reactivity. Nonyl develops grain boundaries with mostly (hk1) plane surfaces smooth and brilliant. MeOH, EOH, and 2-propanol are similar, but (hk1) surfaces vary from brilliant to dull with pitting. Other alcohols similar to MeOH and EOH with two producing only dull, pitted surfaces or an unspecified dark surface. See LI-0006 and K-0001 for additional discussion.

REF: Castellano, R N & Schmidt, P H — *J Electrochem Soc,* 118,653(1971)

NA-0002

ETCH NAME: — TIME:

TYPE: Alcohol, polish — TEMP:

COMPOSITION:

98% .... $C_6H_4(CH_3)_2$ (*o*-xylene)

2% .... isopropyl alcohol (IPA)

DISCUSSION:

Na specimens. Solution described as a polish etch for sodium.

REF: Bowers, R et al — Mater Sci Center Rep. #3, Cornell University, Ithaca, NY, 1961

PHYSICAL PROPERTIES OF SODIUM BROMIDE, NaBr

Classification	Bromide
Atomic numbers	11 & 35
Atomic weight	103
Melting point (°C)	755
Boiling point (°C)	1300
Density (g/cm^3)	3.2
Hardness (Mohs — scratch)	2
Crystal structure (isometric — normal)	(100) cube
Color (solid)	Colorless
Cleavage (cubic — poor)	(001)

SODIUM BROMIDE, NaBr

General: Does not occur as a natural solid compound due to solubility, but may be extracted from ocean water and some hot springs along with chlorides, iodides, and halite, NaCl, common salt. Halogen minerals in chemistry and Solid State are referred to as the halides or as ionic crystals, based on their weak Van Der Waal atomic bonding which can be readily altered by external factors: X-ray, infrared or ultraviolet light, and pressure, etc. The compound has had little use in general metal processing, but there are chemical and medical applications.

Technical Application: Sodium bromide has not been used in Solid State processing, although it has been grown as a single crystal for general morphology and electrical studies.

Etching: Water and alcohols, or acids.

SODIUM BROMIDE ETCHANTS

NABR-0001
ETCH NAME: Alcohol, dilute TIME:
TYPE: Alcohol, cutting TEMP:
COMPOSITION:
10% .... alcohol
DISCUSSION:

NaBr, single crystal specimens used in an electrical breakdown study. Specimens were cut using a string soaked in the solution shown. Surface polishing was done on a chamois cloth soaked in the solution.

REF: Fernandez, A et al — *J Appl Phys,* 37,36(1966)

NABR-0002
ETCH NAME: Sulfuric acid, dilute TIME:
TYPE: Acid, removal TEMP:
COMPOSITION:
x .... H_2SO_4
x .... H_2O
DISCUSSION:

NaBr, specimens used in a study of neutron-transfer reactions from nitrogen bombardment. Solution shown used as a general etch, and may include other acids.

REF: Halbert, M L et al — *Phys Rev,* 106,251(1957)

PHYSICAL PROPERTIES OF SODIUM CHLORIDE, NaCl

Classification	Alkaline metal
Atomic numbers	11 & 17
Atomic weight	58
Melting point (°C)	800
Boiling point (°C)	1400
Density (g/cm^3)	2.1—2.3 (2.164 pure)
Refractive index (n=)	1.5442
Hardness (Mohs — scratch)	2.5
Crystal structure (isometric — normal)	(100) cube
Color (solid)	Clear/tints
(flame)	Deep yellow
Cleavage (cubic — perfect)	(001)
Taste (atypical)	Saline

SODIUM CHLORIDE, NaCl

General: Occurs as the natural mineral halite, NaCl, called common salt or rock salt. As a solid, other than as the massive "rock salt", always forms in distinct cubes with perfect (100) cleavage — even when finely pulverized, it is in small cubes. Natural salt varies in color from colorless to white, yellowish, reddish, bluish and purplish, which has been related to the presence of colloidal sodium, inclusion of organic matter, etc. It is the major chemical compound in salt (ocean) water, and is found in some salt springs. As a dried mineral it occurs as irregular beds in sedimentary rocks to depths of greater than a mile from the evaporation of ancient shallow salt water seas. Salt is still obtained from the evaporation of ocean water — about 25% of world usage — with an additional 25% pumped directly to the industrial user as "brine" and 50% from salt beds.

Salt is one of the most important compounds in the chemical industry, not only as the chief source of sodium metal and chlorine (gas), but for a number of chemical compounds and processes. As sodium chloride it is used as an etchant and quenching medium in metal processing. Sodium is "metal of soda" with the name and symbol, Na, coming from the natural mineral natron, Na_2CO_3, and as sodium hydroxide, NaOH, it is a major etchant solution for metals, alloys, semiconductors and metallic compounds.

As salt, it is essential to mammillary diet — the body fluids are a salt electrolyte. (See Potassium chloride, KCl — natural mineral sylvite, as a dietary replacement.) Salt has long been a food preservative as salted fish, beef, vegetables, etc.

Technical Application: Sodium chloride is not used directly in Solid State device fabrication, but many compounds are used in etching and cleaning processes. Sodium, Na, is considered a contaminant as it will diffuse into the material during high temperature processing and can seriously affect device electrical parameters but, as brine, it is used as a "salt spray test" in high reliability evaluation of devices and assemblies. This is an accelerated test for corrosion effects that can occur in the natural environment during the operational life of metals, alloys and devices.

Sodium chloride is grown as single crystal boules of large size by controlled evaporation from saturated solutions and both natural and artificial crystals have been the subject of much study. As they are colorless, clear crystals, soft and easily deformed, they are of great interest for observing the basic mechanics of defect generation in materials that form in the isometric (cubic) system, normal class. Most semiconductors and metals used in general processing are in the isometric system. Sodium chloride crystals are of additional interest for the effects of defects and color generation by irradiation, infrared, or ultraviolet light that can be related to opaque materials.

As cut or cleaved (100) substrates of NaCl are used for the evaporation of metal thin films or epitaxy growth of compounds with water float-off of the films for morphological study by TEM or SEM.

Etching: Soluble in water, alcohols, some esters; salts, including NaCl itself and several mixed acids.

SODIUM CHLORIDE ETCHANTS

NACL-0001

ETCH NAME: Water — TIME:

TYPE: Acid, polish — TEMP: RT

COMPOSITION:

x .... H_2O

DISCUSSION:

NaCl, (100), cleaved wafers used in an X-ray radiation study. Results were similar for specimens cleaved in air or under vacuum. Prepare wafers prior to irradiation by (1) chem/mech polish with water to remove cleavage steps or surface defects; (2) rinse in ethyl alcohol; (3) rinse in acetone.

REF: Lieder, N R — *Phys Rev,* 101,56(1956)

NACL-0002a: Greenler, R G & Hothwell, W S — *J Appl Phys,* 31,616(1960)

NaCl, (100) wafers used in a ductility study. Wafers were soaked in water as one part of the study.

NACL-0019: Vook, R W — *J Appl Phys,* 32,1557(1961)

NaCl, (100) cleaved wafers used for evaporation and study of Sn thin films. After cleaving, polish surfaces with chamois cloth dampened with water; follow with vacuum bake (UHV) at 375°C for 6 h; then seal in Pyrex ampoule and oven bake at 500°C. Kikuchi lines used to determine surface cleanliness.

NACL-0021: Rahman Kahn, M S — *Thin Solid Films,* 113,207(1984)

NaCl, (100) cleaved wafers or (110) and (111) wafers cut and used for growth of erbium hydride, ErH_2. Wafers were water polished. Deposits were poly-ErH_2 from RT to 298 K growth temperature. At 488 K, ErH_2 was (100) on all NaCl orientations. ErH_2 is cubic, fcc; heat in H_2 to convert to hexagonal, hcp, ErH_3.

NACL-0010a: In der Schmitten, W & Hassen, P — *J Appl Phys,* 32,1790(1961)

NaCl, (100) cleaved wafers used in a stress dislocation study. Prior to stressing, wafers were chem/mech polished on a silk cloth dampened with water; rinsed in ethyl alcohol: then rinsed in ether.

NACL-0031: Fjelivag, H et al — *J Phys Chem Solids,* 45,709(1984)

NaCl:Ni doped single crystal ingots grown by the Czochralski (CZ) method from a powdered mixture of NaCl:$NiCl_2$ using a vertical zone melting technique. A study of the controlled doping effects on this doped material. Crystals were transparent and pale pink in color.

NACL-0032: Walker, P et al — mineralogy study, 1953

NaCl, as natural single crystals of halite in cubic, (100) form. Surfaces were chem/mech polish/cleaned on a linen cloth saturated with water prior to pressurization for white and colored light irradiation studies. Also worked with natural crystals of sylvite, KCl, in comparative study of the two compounds.

NACL-0036: Eltoukhy, A H & Greene, J E — *J Appl Phys,* 50,505(1979)

NaCl, (100) cleaved substrates used for thin film growth of InGaSb. Both NaCl and BF_2, (111) substrates were cleaned: (1) TCE vapor degrease; (2) rinse in MeOH; (3) rinse in acetone. Glass substrates were detergent scrubbed and water rinsed. After epitaxy, thin films were etched thinned (AGSB-0001), then the float-off technique used to remove films for TEM study. Float-off (1) water for NaCl; (2) HF for glass; (3) dilute HCl for BF_2.

NACL-0020b: Neogebauer, C A — *J Appl Phys,* 31,1096(1960)

NaCl, (100) cleaved wafers used as substrates for thin film gold evaporation. Remove films for TEM study with water float-off.

NACL-0005c: Ibid.

NaCl, (100) wafers used as substrates for epitaxy of germanium thin films. Use water float-off of film for TEM study.

NACL-0004a

ETCH NAME: Moran's etch — TIME: 30 sec

TYPE: Salt, dislocation — TEMP: RT

COMPOSITION:

3 g .... $HgCl_2$

1000 ml .... EOH

DISCUSSION:

NaCl, (100) wafers used in a dislocation development etching study. This was a preferred solution for general preferential and defect development.

REF: Moran, P R — *J Appl Phys,* 29,1768(1958)

NACL-0014: Webb, W W — *J Appl Phys,* 31,194(1960)

NaCl, single crystal whiskers studied for structure. Used Moran's etch and rinsed in CCl_4. The two types of pits that developed were (1) a triangle with flat bottom and not reproducible; and (2) a triangle with pointed bottom. The former is a surface damage pit; the latter a dislocation pit.

NACL-0015: Shaskol'skaya, H P & Sun-Jui, Fang — *Sov Phys-Cryst,* 4,74(1960)

NaCl, (100) specimens used in a structure study. Solution used at RT, 1—2 sec with rinse in ether. Etch figures and glide planes correspond to emergence of edge dislocations.

NACL-0011a: Machida, C A & Munir, Z A — *J Cryst Growth,* 68,664(1984)

NaCl, (100) cleaved wafers used in a study of thermal etch pits. After cleaving, store in dessicator with zeolite, $CaSO_4$ drying agent. Place wafers on quartz plate in vacuum, apply current parallel to (100) direction, and use heat lamp radiation outside of vacuum system to initiate thermal alteration. A 30 g $HgCl_2$:EOH solution was used to develop thermal etch pits. Pit density was on the order of $4 \times 10^6/cm^2$.

NACL-0017: Crawford, J A & Young, F W Jr — *J Appl Phys,* 31,1688(1960)

NaCl, (100) wafers studied for F-center development of gamma irradiation. Solution used at RT, 30 sec after irradiation to develop dislocations.

NACL-0010c: Ibid.

NaCl, (100) cleaved wafers used in a stress study. After stressing, a 1% $HgCl_2$:EOH solution was used to develop dislocations. See NACL-0010a-b.

NACL-0007f: Ibid.

NaCl, (100) wafers used in a dislocation study. Moran's etch used in addition to other solutions, and is similar to the "W" etch. See NACL-0024.

NACL-0008a

ETCH NAME: Water — TIME:

TYPE: Acid, forming — TEMP: RT

COMPOSITION:

(1) x .... H_2O (2) x .... NaCl, sat. sol.

DISCUSSION:

NaCl, single crystal specimens used to fabricate pointed filaments. Coat sections to be formed with glyptol. Form by (1) etch in water; (2) etch in NaCl saturated water solution; (3) rinse in ethyl alcohol. Method used for similar fabrication of other alkali halides.

REF: McNulty, J et al — *Rev Sci Instr,* 31,904(1960)

NACL-0023
ETCH NAME: Water
TYPE: Acid, jet-forming
COMPOSITION:
x .... H_2O
DISCUSSION:

TIME:
TEMP: RT

NaCl, (100) wafers used as substrates for evaporation of gold thin films in a mechanical property study of gold. After metallization, a water jet was used to etch a 0.5 mm hole through the NaCl substrate back to expose the gold film on the opposite side without film removal.

REF: Carlin, A & Walker, W P — *J Appl Phys,* 31,2135(1960)

NACL-0024
ETCH NAME: Water
TYPE: Acid, float-off
COMPOSITION:
x .... H_2O
DISCUSSION:

TIME:
TEMP: RT

NaCl, (100) wafers used as substrates for thin film gold evaporation. Gold films removed for TEM study by liquefying the gold/NaCl interface with water as the solvent using the float-off technique. Also: SiO_2/A-Si/Ni/SiV_2 layer.

REF: Chen, S H et al — *J Appl Phys,* 57,258(1985)

NACL-0033:

NaCl, (100) wafers used as substrates for evaporated copper thin films. After deposition, the Cu/NaCl interface was liquefied with water for film float-off and TEM study. Mica, (0001) substrates also used with 10% HF for float-off.

NACL-0034: Nakhotkiu, N G et al — *Thin Solid Films,* 112,267(1984)

NaCl, (100) substrates used for growth of a-Ge thin films. Films removed by water float-off for TEM study.

NACL-0035: Pierrard, P et al — *Thin Solid Films,* 111,141(1984)

NaCl, (100) cleaved substrates used for growth of a-Ge thin films. Used de-salted water for float-off of films for TEM study. Heat from the TEM microscope did not crystallize the amorphous films. Laser application under TEM observation produced single crystal spherulitic structures.

NACL-0002b
ETCH NAME: Salt water
TYPE: Acid, polish
COMPOSITION:
x .... NaCl, sat. sol.
x .... H_2O
DISCUSSION:

TIME:
TEMP:

NaCl, (100) wafers used in a ductility study. Wafers were soaked in salt water as one step of surface preparation for the study.

REF: Ibid.

NACL-0008b: Ibid.

NaCl, single crystal specimens used to fabricate pointed filaments. Solution used as an etchant. See NACL-0008a.

NACL-0002c
ETCH NAME: Ethylene glycol
TYPE: Ester, ductility effect
COMPOSITION:
x .... CH_2OHCH_2OH (EG)
TIME:
TEMP: RT
DISCUSSION:
NaCl, (100) wafers used in a ductility study. Wafers subjected to various liquids and atmospheres as H_2O, air and NaCl, saturate solution.
REF: Ibid.

NACL-0027
ETCH NAME: Chloroform
TYPE: Gas, cleaning
COMPOSITION:
x .... $CHCl_3$
TIME:
TEMP: RT
DISCUSSION:
NaCl, (100) wafers used as substrates for oriented epitaxy growth of $AuGa_2$ thin films. Prior to deposition, wafers were cleaned (1) in chloroform; (2) rinsed in acetone; and (3) rinsed in MeOH. Final drying in ultra high vacuum (UHV) immediately prior to metal compound growth.
REF: Richard, R D et al — *J Vac Sci Technol,* A(2),535(1984)

NACL-0003a
ETCH NAME: Ethyl alcohol
TYPE: Alcohol, polish
COMPOSITION:
x .... C_2H_5OH (EOH)
TIME:
TEMP:
DISCUSSION:
NaCl, (100) wafers. Solution used to polish wafers with rinse in carbon tetrachloride, CCl_4.
REF: Barr, L W & Morrison, J A — *J Appl Phys,* 31,617(1960)

NACL-0003c
ETCH NAME: Chlorine
TYPE: Gas, preferential
COMPOSITION:
x .... Cl_2, vapor
TIME: 48 h
TEMP: 700°C
PRESSURE: 500 mmHg
DISCUSSION:
NaCl, (100) cleaved wafers used in a dislocation study. Hot chlorine vapors developed square etch pits along cleavage steps on surfaces. Referred to as crystallite boundaries.
REF: Ibid.

NACL-0005b
ETCH NAME: Heat
TYPE: Thermal, float-off
COMPOSITION:
x .... heat
TIME:
TEMP: 800°C
DISCUSSION:
NaCl, (100) cleaved wafers used as substrates for growth studies of evaporated Au thin films. Very thin gold films were float-off removed by heat melt-away of the NaCl substrate.

Thicker films sheared away by cleaving the substrate with residual salt remaining on gold removed by water washing.
REF: Ibid.

NACL-0029
ETCH NAME: Hydrochloric acid TIME:
TYPE: Acid, cleaning TEMP:
COMPOSITION:
x .... HCl, conc.
DISCUSSION:
NaCl, (100) wafers used in a study of laser induced irreversible absorption changes at 10.6 μm light frequency. Etch clean wafers in HCl, and rinse in isopropyl alcohol (IPA or ISO).
REF: Wu, S T & Bass, M — *J Appl Phys Lett,* 39,948(1981)

NACL-0020a
ETCH NAME: Ethyl alcohol, dilute TIME:
TYPE: Alcohol, polish TEMP: RT
COMPOSITION:
x .... C_2H_5OH (EOH)
x .... H_2O
DISCUSSION:
NaCl, (100) cleaved wafers used as substrates for thin film gold evaporation. Polish wafers in solution shown prior to metallization. Gold deposits as (100) oriented film. Remove film for TEM study with water float-off.
REF: Neogebauer, C A — *J Appl Phys,* 31,1096(1960)

NACL-0010b
ETCH NAME: TIME: 5 min
TYPE: Alcohol, dislocation TEMP: RT
COMPOSITION:
1 .... C_2H_5OH (EOH)
1 .... CH_3OH (MeOH)
DISCUSSION:
NaCl, (100) cleaved wafers used in a dislocation study of stressed specimens. Mechanically polish with water dampened silk cloth, rinse in EOH, then in ether. Stress wafer, then cross-section cleave on ⟨110⟩. Use solution shown to develop dislocation density and flow characteristics. Also used Moran's etch as a 1% $HgCl_2$:EOH mixture. See NACL-0004a.
REF: In der Schmitten, W & Hassen, P — *J Appl Phys,* 32,1790(1961)

NACL-0004b
ETCH NAME: Methyl alcohol TIME:
TYPE: Alcohol, polish TEMP: RT
COMPOSITION:
x .... CH_3OH (MeOH)
DISCUSSION:
NaCl, (100) cleaved wafers used in development of etching techniques. Solution used to polish surfaces prior to dislocation etching.
REF: Ibid.

NACL-0007a
ETCH NAME: Propionic acid TIME:
TYPE: Acid, cleaning TEMP: RT
COMPOSITION:
x .... CH_3CH_2COOH
DISCUSSION:
NaCl, (100) wafers used in a dislocation study. Wafers were cleaned (1) in solution shown, (2) rinsed in petroleum ether and air blown dried or (3) rinsed in pyridine and air blown dried.
REF: Barber, D J — *J Appl Phys,* 33,3140(1962)

NACL-0005a
ETCH NAME: Hydrochloric acid TIME:
TYPE: Acid, polish TEMP: RT
COMPOSITION:
70 .... HCl
30 .... H_2O
DISCUSSION:
NaCl, (100) wafers used as substrates for epitaxy growth of germanium thin films. Prior to epitaxy, polish wafers on pad soaked with solution shown using light pressure. Use water float-off of film for TEM study.
REF: Outlaw, R A & Hopson, P Jr — *J Appl Phys,* 55,1460(1984)

NACL-0011b
ETCH NAME: Heat TIME:
TYPE: Thermal, dislocation TEMP: Approx. 800°C
COMPOSITION:
x .... heat
DISCUSSION:
NaCl, (100) cleaved wafers used in a thermal etch pit study. Wafers placed on a quartz plate in vacuum, apply current parallel to (100), and use heat lamp radiation outside of vacuum chamber to initiate defects. A 30 g $HgCl_2$:EOH solution used to develop pits (NACL-0004a). Pit density was on the order of 4×10^6 cm^2.
REF: Ibid.

NACL-0003b
ETCH NAME: TIME: 1 min
TYPE: Acid, dislocation TEMP: RT
COMPOSITION:
3 g .... $Fe(NO_3)_2$
1000 ml .... C_2H_5OH (EOH)
DISCUSSION:
NaCl, (100) wafers used in a study of dislocations and defects. Prepare wafers by (1) polish dip in EOH; (2) rinse in CCl_4. Use solution shown to develop dislocations, and rinse in CCl_4.
REF: Ibid.

NACL-0007b
ETCH NAME: Barber etch TIME:
TYPE: Acid, preferential TEMP:
COMPOSITION:
1 .... HCl
50 .... HAc
1 .... $FeCl_3$, sat. sol.
1 .... H_2O
DISCUSSION:
NaCl, (100) wafers used in a dislocation study. Several etchants evaluated, and the one shown gave similar results to those of Moran's etch and the "W" etch.
REF: Ibid.

NACL-0007c
ETCH NAME: Cook's etch TIME:
TYPE: Acid, preferential TEMP:
COMPOSITION:
x .... CH_3CH_2COOH (propionic acid)
x .... CH_3COOH (HAc)
DISCUSSION:
NaCl, (100) wafers used in a dislocation study. Solution shown gave results similar to the "A" etch (NACL-0024).
REF: Ibid.
NACL-0028: Cook, J S — *J Appl Phys,* 32,2492(1961)
NaCl, (100) wafers used in a defect study. Reference for Cook's etch.

NACL-0007d
ETCH NAME: "W" etch TIME:
TYPE: Acid, preferential TEMP:
COMPOSITION:
x .... $HgCl_2$
x .... EOH
DISCUSSION:
NaCl, (100) wafers used in a dislocation study. Solution can be used to develop dislocations with varying amounts of $HgCl_2$. See NACL-0004a.
REF: Ibid.
NACL-0024: Gilman, J J & Johnston, W E — *Dislocations in Crystals,* John Wiley & Sons, New York, 1956
Reference for "A" and "W" etches for NaCl, and can be applied on other alkaline halides.

NACL-0007e
ETCH NAME: "A" etch TIME:
TYPE: Acid, preferential TEMP:
COMPOSITION:
x .... LiF
x .... H_2O
x .... EOH
DISCUSSION:
NaCl, (100) wafers used in a dislocation study. Results with this solution are similar to those of the Cook's etch. See NACL-0007c.
REF: Ibid.

NACL-0022
ETCH NAME: Ferric chloride TIME:
TYPE: Salt preferential TEMP:
COMPOSITION:
x .... x% $FeCl_3$
DISCUSSION:

NaCl, (100) wafers used in a dislocation development study of ionic crystals. Solution used as a dislocation etch on NaCl, KCl, and other alkali halides.
REF: Gilman, J J — *J Appl Phys,* 32,739(1961)

NACL-0018
ETCH NAME: TIME:
TYPE: Acid, dislocation TEMP:
COMPOSITION:
100 ml .... HAc
66 ml .... $FeCl_3$, sat. sol.
DISCUSSION:

NaCl, (100) cleaved wafers subjected to partial X-ray irradiation in a study of plastic flow. Solution used to develop dislocations after irradiation.
REF: Alvarez-Rivas, J L & Ahullo-Lopex, F — *Phil Mag,* 12,205(1965)

NACL-0009
ETCH NAME: TIME: 30 sec
TYPE: Acid, dislocation TEMP: RT
COMPOSITION:
4 g .... $FeCl_3$
1000 ml .... HAc
DISCUSSION:

NaCl, (100) cleaved wafers used in a dislocation etch study. After etching in solution shown, rinse in acetone, and blow dry with warm air.
REF: Mendelson, S — *J Appl Phys,* 32,1578(1961)

NACL-0006
ETCH NAME: Freon TIME:
TYPE: Gas, cleaning TEMP:
COMPOSITION:
x .... Freon, vapor
DISCUSSION:

NaCl, (100) blanks cleaved from natural ''rock salt'', and muscovite mica, (0001) blanks also cleaved. Blanks used as substrates for thin film gold and silver evaporation in a material structure study. Liquids were not used for substrate cleaning as they adsorb on mica, and attack NaCl. Blow surfaces clean with Freon, only.
REF: Chopea, K I — *J Appl Phys,* 37,2049(1960)

NACL-0012
ETCH NAME: Air TIME:
TYPE: Gas, defect TEMP:
COMPOSITION:
x .... air, moist
DISCUSSION:

NaCl, dendritic growth and development of defects. Dendrites of NaCl, KCl and NaCl:KCl mixture were formed by placing the salt solutions on a glass plate then rapidly cool in air

(LN_2?). Structure developed around dislocations, which may be due to etching action of moist air. Dislocation centers often contained dark spots — possibly due to impurities or other salts acting as nucleating agents.
REF: Forty, A J & Gibson, J C — *Acta Metall,* 6,137(1958)

NACL-0037
ETCH NAME: Nitrogen — TIME:
TYPE: Gas, embrittlement — TEMP:
COMPOSITION:
x .... N_2(O_2, O_3 or air)
DISCUSSION:
NaCl, (100) crystals subjected to the atmospheres shown used in a study of brittleness. Results showed (1) N_2 and O_2 had no effect on ductility; (2) crystals aged in air will be brittle in summer — ductile in winter; (3) subject a ductile crystal to O_3 (ozone) or O_2 and it will become brittle.
REF: Machin, E S & Murray, G T — *J Appl Phys,* 30,1731(1959)

NACL-0030
ETCH NAME: Gold — TIME:
TYPE: Metal, decoration — TEMP:
COMPOSITION:
x .... Au
DISCUSSION:
NaCl, (100) wafers. Evaporate thin film gold on surfaces, then over-coat with carbon thin film. The combined carbon/gold layer can be peeled away as the gold adheres well to the carbon. Beads of gold will remain on the NaCl surface along cleavage and slip steps. Method can be applied to other ionic crystals.
REF: Bassett, G A — *Phil Mag,* 3,1042(1958)
NACL-0038: Amelinck, S — *J Appl Phys,* 29,1110(1958)
NaCl, as single crystal whiskers. Gold evaporated, then diffused into material to delineate dislocations by gold decoration. Used on other alkaline halides.

PHYSICAL PROPERTIES OF SODIUM NITRITE, $NaNO_2$

Classification	Nitrite
Atomic number	11, 7 & 8
Atomic weight	69.01
Melting point (°C)	271
Boiling point (°C)	320
Density (g/cm³)	2.168
Hardness (Mohs — scratch)	1.5—2
Crystal structure (hexagonal — rhombohedral)	$(10\bar{1}1)$ rhomb
Color (solid)	Yellow
Cleavage (dodecadral)	(101)

SODIUM NITRITE, $NaNO_2$

General: Does not occur as a natural compound, but there are several natural minerals with the NO_3^- radical, such as soda niter, $NaNO_3$ and niter, KNO_3, both referred to as saltpeter, and soda niter as Chile saltpeter. There are other nitrates of calcium, magnesium, copper, cobalt as hydrates. Soda niter has major industrial use in the manufacture of nitric acid, HNO_3 and, along with ammonia, NH_3, as part of the initial reactants, one by-product is NO_2 which, in warm water, produces HNO_3 + NO. The yellow fuming (72%) and red fuming (74%) nitric acid colors are due to increasing concentration of NO_2 in the solutions. Nitrates/nitrites are important fertilizers; they also have medicinal applications.

Technical Application: Sodium nitrite is used as a chemical additive in some etching and cleaning solutions. It has been grown as a single crystal in the Solid State field and studied for ferroelectricity and other parameters. The material is hygroscopic and requires handling in a dry, inert atmosphere of nitrogen or argon.

Etching: Water, ammonia, alcohols.

SODIUM NITRITE ETCHANTS

NAN-0001
ETCH NAME: Ammonia — TIME:
TYPE: Base, removal — TEMP: RT
COMPOSITION:
x .... NH_3
DISCUSSION:

$NaNO_2$ single crystals grown in a study of ferroelectricity in the compound. Authors say the material was yellowish and transparent and cleaves easily along (101) planes. (*Note:* In that the material is hexagonal system, rhombohedral division, cleavage may have been $(10\bar{1}0)$ prismatic?)

REF: Sawada, S et al — *Phys Rev Lett,* 1,320(1958)

PHYSICAL PROPERTIES OF SPINEL, $MgAl_2O_4$

Classification	Oxide
Atomic numbers	12, 13, & 16
Atomic weight	142.3
Melting point (°C)	2135 (2105)
Boiling point (°C)	
Density (g/cm³)	3.5—4.1
Refractive index (n=) natural	1.7155—2.00
(n=) artificial	1.723 (1.7202)
Dielectric constant (e=)	8.4
Thermal conductivity (cal/cm-sec-°C) 25°C	0.035
Coefficient of linear thermal expansion (×10^{-4} cm/cm/°C) 25—800°C	7.45
Dissipation factor (tan)	10^{-3}—10^{-4}
Hardness (Mohs — scratch)	8
Crystal structure (isometric — normal)	(111) octahedron, fcc
Color (solid)	Red/green/black
Cleavage (octahedral — imperfect)	(111)

SPINEL, $MgAl_2O_4$

General: The natural mineral spinel, $MgAl_2O_4$ is classified into three categories by metal element inclusion as aluminum, iron or chromium spinel. Note that magnetite, Fe_3O_4 (magnetic iron ore) also is classed as a spinel. Chemically, they are oxygen-salts of the metals, as aluminates, ferrates, manganates. In industry the material is used as a high temperature ceramic, and the transparent colored varieties as gem stones: spinel ruby; Balas ruby; rubicelle (yellow/orange-red); and almandine (violet).

Technical Application: There has been some use in Solid State processing as a substrate, similar to silicon-on-sapphire (SOS), with thin film epitaxy growth in evaluation of structure and possible devices. Both natural and artificial spinels have been studied, the latter grown from molten fluxes, but the primary use is still as gem stones.

Etching: H_2SO_4, molten borax. H_2 gas.

SPINEL ETCHANTS

SPI-0001a
ETCH NAME: Sulfuric acid TIME:
TYPE: Acid, polish TEMP: RT
COMPOSITION:
x .... H_2SO_4, conc.
DISCUSSION:

$MgAl_2O_4$, (100) and (111) wafers used as substrates for silicon epitaxy growth for comparative data between silicon-on-sapphire (SOS) and silicon on spinel. Solution is a slow removal and polish etch for spinel. Material was artificially grown by a flux technique.
REF: Cullen, G W et al — *RCA Rev,* 70,355(1970)
SPI-0002: Dana, E S & Ford, W E — *A Textbook of Mineralogy,* 4th ed, John Wiley & Sons, New York, 1950, 488

$MgAl_2O_4$, natural crystals. Material is soluble with difficulty.

SPI-0002b
ETCH NAME: Borax TIME:
TYPE: Molten flux, preferential TEMP: 60—75°C
COMPOSITION:
x .... $Na_2B_4O_7.10H_2O$
DISCUSSION:

$MgAl_2O_4$, natural crystals. Slowly soluble in borax molten flux.
REF: Ibid.

SPI-0003
ETCH NAME: TIME:
TYPE: Molten flux, decomposition TEMP: Molten
COMPOSITION:
3 .... $NaBO_3$
1 .... $CaCO_3$
DISCUSSION:

$MgAl_2O_4$ single crystals were flux grown as clear octahedrons, (111). The eutectic flux shown is used to decompose the spinel — treat melt with HCl, then add excess EDTA and boil to complex the aluminum. Diamond saw cut wafers (111), (100) and (110) were mechanically polished with 30 μm boron carbide, then down to a final polish with 0.3 μm alumina. X-ray Laue photographs used to oriented crystals for cutting. A study of optical and dielectric properties.
REF: Wang, C C & Zanaucchi, P J — *J Electrochem Soc,* 118,586(1971)

SPI-0004
ETCH NAME: Hydrogen TIME:
TYPE: Gas, preferential TEMP: 1150°C
COMPOSITION:
x .... H_2
DISCUSSION:

$MgO.Al_2O_4$, (111) blanks used as substrates for epitaxy growth of (111) silicon thin films. Substrates were flame fusion, flux and CZ grown single crystals in a comparative

study of flame fused sapphire, Al_2O_3 blanks used for SOS structures. Hole mobility used as measurement criterion with SEM study of 1.5 μm silicon epitaxy thick deposits. Substrates were vacuum fired to clean or silicon was deposited and removed — three cycles — with hole mobility improving and stabilizing after the third deposition/removal to remove impurities from the substrates. Spinel shows a better crystallographic match with silicon than does sapphire, e.g., cubic vs. hexagonal crystal systems.
REF: Nanasevit, H M & Simpson, W I — *J Electrochem Soc,* 118,644(1971)

SPI-0005
ETCH NAME: TIME:
TYPE: Solvents, cleaning TEMP:
COMPOSITION:
(1) x .... TCE (2) x .... H_2O (3) x .... MeOH
DISCUSSION:

$MgAl_2O_4$, (111) wafers cut from CZ grown single crystals ingots for use as substrates for epitaxy deposition of ZnSe, ZnTe, AnS, CdS, and CdSe oriented thin films. In all cases, thin film orientation was (111). Solutions shown were used in the order shown to clean spinel surfaces prior to epitaxy. Also evaluated Verneuil grown single crystal Al_2O_3 and flux grown BeO. Authors note results are comparable to those observed on GaAs epitaxy.
REF: Cullen, G W & Wang, C C — *J Electrochem Soc,* 118,640(1971)

PHYSICAL PROPERTIES OF STEEL, Fe_3C

Classification	Iron compound
Atomic numbers	26 & 6
Atomic weight	179.53
Melting point (°C)	1837
Boiling point (°C)	
Density (g/cm³)	7.4
Hardness (Mohs — scratch)	5—6
Crystal structure (isometric — normal)	(100) cube
Color (solid)	Steel grey
Cleavage (cubic)	(001)

STEEL, Fe_3C

General: The physical properties shown are for iron carbide called cementite as an artificial compound as it does not occur as a natural mineral. The Fe_3C crystals are part of white cast iron structure with 2—4% carbon. If it is cooled slowly it becomes grey cast iron with separation of some of the cementite to iron and graphite. Native iron occurs in small quantity, and usually has traces of other metal elements such as nickel. Some meteoric iron contains $FeNi_3$, and there are over 250 iron containing minerals as oxides, silicates and phosphates, many of which are hydrates.

The Hittites of Asia Minor (now part of Turkey) are considered to be the the first to have smelted iron for use as weapons, circa 1350 B.C., although there are reports that iron tools may have been in use by the Egyptians as far back as 3500 B.C., about the time they developed soda-lime glass. This initial iron was wrought iron with low carbon content, as a fibrous structured, malleable mass of great toughness and tensile strength. It is still used for horseshoes, decorative fencing, and objets d'art.

As the Iron Age progressed in the Mediterranean area with kiln firing of iron oxides with charcoal, the inclusion of carbon in the melt led to the first forms of steel. During the Middle Ages and medieval period of Europe, both Damascus and Toledo steels became the best forms of steel for swords and armor, even though large castings of bronze were still used for the making of cannon as late as the 16th century as hand iron/steel weapons were being developed.

Copper and bronze plate have been used on the hulls of wooden ships since ancient times, but it was not until the American Civil War that the first of the iron ships were built (the Monitor and the Merrimac), even though the Industrial Revolution in England was already using steam engines, rails and railroad cars, as well as iron and steel in building construction in the early 1800s.

Today there are hundreds of types of irons and steels graded by SAE numbers and other classifications. Two of the most widely used series are 302 and 304 steels in equipment construction, such as vacuum systems. The grades range from steels, magnetic steels, to stainless steel. Construction irons and steels included buildings, bridges, heavy duty equipment, vehicles, and "tin" cans. The latter is thin sheet steel plated with tin, the tin being highly inert to vegetable and meat acids.

As a very brief resume of iron and steel fabrication: Major iron ores are the iron oxides, such as hematite, Fe_2O_3; magnetite, Fe_3O_4; siderite, $FeO.xH_2O$ and the "iron pyrites" as a group — pyrite, FeS_2 as better than 80% of all iron ore mined in the world due to its abundance. The highest grade oxide ores come from Sweden as they are particularly low in other element contamination. Swedish iron is mostly magnetic iron from the original Achean Shield solidification of the earth's crust.

The iron ores are initially kiln fired and roasted in a vertical blast furnace to obtain "pig iron" (4% carbon; 3% silicon; phosphorus and sulfur. The phosphorus makes it brittle when cold; the sulfur, brittle when hot). Cast iron is then made from the initial pig (2—4% carbon) as white cast iron with a high cementite, Fe_3C content; and as grey cast iron with some cementite separation into iron and graphite. Wrought iron is made by removal of carbon by air blowing through the iron mass (Bessemer process).

Billets of ferro-silicon are obtained in the reduction of sand and other silicates, and there are starter billets of ferro-manganese, ferro-phosphorus, ferro-chromium, etc., which are then added to the pig, cast or wrought iron to produce the particular variety of iron and/or steel and stainless steel.

Silicon iron (15% Si) is brittle but very acid resistant and used as holding vessels, crucibles, piping and linings in acid processing equipment.

Manganese iron is tough. With 1—7% Mn used for machine tooling; with >7% Mn for heavy duty equipment, such as rock crushers or for safes.

Nickel steel has great tenacity and is a primary construction steel with 0.5% Ni, and for armor plate with 3—5% Ni.

Tungsten steel holds a temper at high temperatures, such that it is used for lathe cutting tools, knives, and scalpels. Both molybdenum and cobalt steels are similar.

Chromium steel with 12—15% Cr is corrosion resistant and used for tools, equipment and armor plate. With >18% Cr it is stainless steel, SST.

In construction — bridges and building girders, ship plate and so forth — the iron is coated with red lead to prevent rusting prior to further painting. There is on-going study of irons and steels as they are improved against atmospheric corrosion, fatigue failure, etc. Many of the etchant solutions in the ASTM 407-70 pamphlet were developed for metallographic study after such testing with mechanical lapping, polishing and preferential etching.

Many of the common iron and steel products of the past are being replaced by other materials — lighter weight metals, plastics and ceramics — but iron skillets, steel cookware, knives, forks, and spoons of steel alloys, and other such items are still in demand. The canning industry is a major user of thin sheet steel that is tin plated as "tin cans", and construction irons and steels for vehicles, buildings, bridges are still major users. In chemical processing and medicine, stainless steel beakers, forceps, and tweezers (acid resistant and nonmagnetic), as well as decorative items in the jewelry trade are still in wide use.

Technical Application: Steels or irons are not directly used in the fabrication of Solid State devices themselves, but widely used in all processing and testing equipment, to include chemical processing or material handling with forceps, tweezers, SST beakers, and so forth. Device package construction may include steels, and special contact wires (dumet), and some magnetic steels. Much of the equipment and etchants used in Solid State originated in the iron and steel, rock and mineral, and glass industries.

Ferrites as resistor material are used as discrete devices in conjunction with circuit substrate assemblies in microwave device construction and are presented here, in the Ferrite section.

The amount of literature published on the study of irons and steels is second only to that of glass. There is still on-going study of steel alloys, some grown as single crystals. The etchants shown in the following section are only a brief cross-section of those available. Where the material is a steel, formats are numbered ST-xxxx; where an iron, as FE-xxxx, and both designations may be shown under the Steel or Iron sections in this book.

Etching: Varies with type steel. HNO_3:EOH, picric acid:EOH, $FeCl_3$, HCl.

STEEL ETCHANTS

ST-0001a
ETCH NAME: Al-7 TIME: Dip
TYPE: Acid, macro-etch TEMP: 160—175°C
COMPOSITION:
50 ml .... HCl
50 ml .... H_2O
DISCUSSION:
Fe_3C steel and iron specimens. Solution used to develop micro-structure.
REF: Pamphlet — *A/B Metal Digest,* 22(3),14(1983) — Buehler Ltd

---

ST-0001b
ETCH NAME: Kaling's reagent TIME: Swab
TYPE: Acid, preferential TEMP: RT
COMPOSITION: ASTM: #95
33 ml .... HCl
33 ml .... EOH
33 ml .... H_2O
1.5 g .... $CuCl_3$
DISCUSSION:
SST 400 Series. Solution used as a structure development etch.
REF: Ibid.
ST-0017b: Pamphlet — *ASTM,* E407-70
Reference for ASTM number for Kaling's reagent shown as 40 ml HCl:80 ml EOH(MeOH):40 ml H_2O:2 g $CuCl_2$.

ST-0001c
ETCH NAME: Picral TIME: Swab
TYPE: Acid, preferential TEMP: RT
COMPOSITION: ASTM: #76
10 g .... picric acid
100 ml .... EOH

DISCUSSION:

Steels, high carbon and high alloy types. Solution used as a structure etch applied by swabbing the surface.

REF: Ibid.

ST-0017c: Pamphlet — *ASTM,* E407-70

Reference for ASTM number. Same solution as shown under ST-0001c, but MeOH can replace EOH.

ST-0001d

ETCH NAME: Potassium bisulfate — TIME: Dip

TYPE: Acid, preferential — TEMP: RT

COMPOSITION:

10 g .... potassium metabisulfate

90 ml .... H_2O

DISCUSSION:

Steel, alloys with untempered martensite. Solution used to develop micro- and macrostructure.

REF: Ibid.

ST-0001e

ETCH NAME: Vilella's reagent — TIME: Swab

TYPE: Acid, preferential — TEMP: RT

COMPOSITION: — ASTM: #80

5 ml .... HCl

1 g .... picric acid

100 ml .... EOH

DISCUSSION:

Steels, high carbon and alloys types. Solution used as a structure etch.

REF: Ibid.

ST-0017a: Pamphlet — *ASTM,* E407-70

Reference for ASTM number. Same solution as shown in ST-0001e, but EOH can be replaced with MeOH.

ST-0001f

ETCH NAME: Nital — TIME: Swab

TYPE: Acid, preferential — TEMP: RT

COMPOSITION: — ASTM: #74

2 ml .... HNO_3

98 ml .... EOH

DISCUSSION:

Steel, carbon type. Solution is a general cleaning and preferential etch to develop structure.

REF: Ibid.

ST-0017d: Pamphlet — *ASTM,* E407-70

Reference for ASTM number. Solution shown as 1—5 ml HNO_3:100 ml EOH or MeOH.

ST-0001g

ETCH NAME: — TIME: 1—10 min

TYPE: Acid, preferential — TEMP: RT

COMPOSITION: — ASTM: #81

2 g .... picric acid

1 g .... sodium tridecylbenzene sulfonate

100 ml .... H_2O

DISCUSSION:

Steels, carbon and alloy types. Solution used to develop prior austenitic grain boundaries in martensite and bainite structures.

REF: Ibid.

ST-0017e: Pamphlet — *ASTM,* E407-70

Reference for ASTM number.

ST-0001h

ETCH NAME: Chromic acid
TYPE: Electrolytic, preferential
COMPOSITION:
10 g .... CrO_3
90 ml .... H_2O

TIME:
TEMP:
ANODE: Steel
CATHODE:
POWER: 3—5 V
ASTM: #18

DISCUSSION:

Steel, 300 series. Solution used to develop structure.

REF: Ibid.

ST-0001i

ETCH NAME:
TYPE: Electrolytic, preferential
COMPOSITION:
15 ml .... HCl
10 ml .... HAc
10 ml .... HNO_3
5 ml .... glycerin

TIME:
TEMP:
ANODE: Steel
CATHODE:
POWER: 3—5 V

DISCUSSION:

SST 300 series. Solution used as a structure etch.

REF: Ibid.

ST-0002a

ETCH NAME:
TYPE: Acid, preferential
COMPOSITION:
x .... CrO_3
x .... H_2O
x .... glycerin

TIME:
TEMP:

DISCUSSION:

SST 316. Solution used to develop corrosion fatigue cracks.

REF: Pamphlet — *A/B Metal Digest,* 22(2),17(1983) — Buehler Ltd

---

ST-0013b

ETCH NAME: Persulfate etch
TYPE: Acid, contrast
COMPOSITION:
x .... $(NH_4)_2S_2O_8$
100 ml .... H_2O

TIME: Dip
TEMP: RT

DISCUSSION:

Steel alloys containing carbides. Immerse in solution until iron matrix darkens to develop carbide material by contrast.

REF: Ibid.

ST-0002b

ETCH NAME: Oxalic acid, diluted
TYPE: Electrolytic, preferential
COMPOSITION:
x .... oxalic acid
x .... H_2O

TIME:
TEMP:
ANODE:
CATHODE:
POWER:

DISCUSSION:

SST 304. Used to develop stress corrosion cracks.

REF: Ibid.

ST-0003

ETCH NAME: Nitric acid, dilute
TYPE: Acid, cleaning
COMPOSITION:
x .... HNO_3
x .... H_2O

TIME:
TEMP:

DISCUSSION:

SST substrates used for deposition of WC in a study of wear resistance, structure and adhesion. First, substrates were diamond-paste polished and then etched clean in dilute nitric acid. Follow with water rinse, then alcohol, then alcohol with ultrasonic agitation. WC deposited by RF sputter under acetylene with a tungsten sheet sputter target.

REF: Srivastava, P K et al — *J Electrochem Soc,* 131,1260(1984)

ST-0004: Walker, P — personal application, 1970—1980.

SST 316 evaporation bell jars and vacuum tooling. Soak and scrub with a steel brush using a $1HNO_3$:$10H_2O$ or a 1HCl:$10H_2O$ solution for general cleaning and removal of excess evaporated metals build-up. Solutions used at RT up to 30 min scrubbing. Follow with heavy water flushing and toweling wipe-down if black smut is present, then final rinse in MeOH and air dry or dry under IR heat lamp.

ST-0005

ETCH NAME:
TYPE: Acid, removal
COMPOSITION:
x .... 5×10^3 *M* ferric perchlorate
x .... 0.75 *M* perchloric acid
*x .... lithium perchlorate

TIME:
TEMP:

*Use to control ion strength at 1.0.

DISCUSSION:

Steel blanks. First, clean blanks in hot caustic (10—15% NaOH) and DI water wash thoroughly. This pre-cleaning step increases the etch removal rate by about 15%. This is a major article on the cleaning of steels.

REF: Maynard, R B — *RCA Rev,* 45,58(1984)

ST-0006

ETCH NAME: Ferric chloride
TYPE: Acid, removal
COMPOSITION:
x .... $FeCl_3$ (30° Bé)
x .... *HCl

TIME: 2—6 min
TEMP: Various

*Use to reduce concentration.

DISCUSSION:

Steel, low carbon blanks. First, pre-clean blanks by soaking in caustic (10—15% NaOH) with thorough water washing after soak. Start solution as shown with 30° Bé and increase concentration by boiling off water; reduce concentration by adding HCl.

REF: Maynard, R B et al — *RCA Rev*, 45,71(1984)

ST-0007a

ETCH NAME: TIME:

TYPE: Acid, removal TEMP:

COMPOSITION:

4 .... HCl

1 .... HNO_3

DISCUSSION:

Steel, high-speed blanks. Used as a general etch for steels. Increase nitric acid content to control rate. Also used on nickel and brass.

REF: Oberg, E & Jones, F D — *Machinery's Handbook*, 4th ed, The Industrial Press, New York, 1951

ST-0007b

ETCH NAME: TIME:

TYPE: Acid, removal/cleaning TEMP:

COMPOSITION:

1 .... HCl

2 .... HNO_3

DISCUSSION:

Steel, hard. Shown as a cleaning and removal solution for hard steel.

REF: Ibid.

ST-0007c

ETCH NAME: Nitric acid, dilute TIME:

TYPE: Acid, removal/cleaning TEMP:

COMPOSITION:

1 .... HNO_3

4 .... H_2O

DISCUSSION:

Steel, carbon. Shown as a removal and cleaning solution for carbon steels. Water content is varied with increase or decrease of carbon content. Added carbon increases steel hardness.

REF: Ibid.

ST-0008

ETCH NAME: TIME:

TYPE: Acid, cleaning for plating TEMP:

COMPOSITION:

(1) Degrease: toluene

(2) Mechanical lap/polish: 1/4/0 emery paper

(3) Electrolytic clean: SST part cathode: 1%$NaCO_3$. H_2O rinse

(4) Acid dip: 5% H_2SO_4, RT. H_2O rinse

(5) Acid polish: 30 sec, 30—35°C (acid not shown)

(6) Alkali clean: 10% NaCH, RT, 2—5 sec, H_2O rinse

(7) Acid dip: 5% H_2SO_4 with agitation to remove Fe(OH). H_2O rinse

(8) Into plating solution still wet

DISCUSSION:

Steel, (100) single crystal specimens. Cleaning sequence shown used on specimens prior to tin plating in a morphology study of $FeSn_2$.

REF: Lunder, C & Murry, M V — *J Electrochem Soc,* 111,348(1964)

ST-0009a

ETCH NAME: Ammonium persulfate — TIME:

TYPE: Acid, removal — TEMP: 50—60°C

COMPOSITION:

x .... 10% ammonium persulfate

DISCUSSION:

Fe, specimens. Solution used as a general removal and surface cleaning etch.

REF: Berglund, T & Dearden, W H — *Metallographer's Handbook of Etching,* Pitman & Sons, London, 1931

ST-0009b

ETCH NAME: Sulfuric acid, dilute — TIME: 6—24 h

TYPE: Acid, polish — TEMP: Hot

COMPOSITION:

x .... 10—20% H_2SO_4

DISCUSSION:

Fe, specimens. Solution used as a polishing etch.

REF: Ibid.

ST-0009c

ETCH NAME: Nital — TIME:

TYPE: Acid, preferential — TEMP:

COMPOSITION:

x .... 10—15% HNO_3

x .... EOH or H_2O

DISCUSSION:

Fe specimens. As a general cleaning solution with some preferential attack. (*Note:* With EOH, solution is called Nital.)

REF: Ibid.

ST-0009d

ETCH NAME: Hydrochloric acid — TIME: 10—30 min

TYPE: Acid, removal — TEMP:

COMPOSITION:

1 .... HCl

1 .... H_2O

DISCUSSION:

Fe, specimens. Used as a cleaning and removal solution with some preferential attack.

REF: Ibid.

ST-0009e

ETCH NAME: Picral — TIME: 4—5 h

TYPE: Acid, removal — TEMP: RT

COMPOSITION:

x .... 3% picric acid

x .... EOH

DISCUSSION:

Fe specimens. Use as a cleaning and removal solution with some preferential attack.

REF: Ibid.

ST-0010a

ETCH NAME: Nital TIME:

TYPE: Acid, preferential TEMP:

COMPOSITION:

x .... 2% HNO_3

x .... EOH

DISCUSSION:

Fe, single crystal whiskers etched in this solution to observe dislocations.

REF: Coleman, R V — *J Appl Phys,* 29,1487(1958)

ST-0010b

ETCH NAME: Picral TIME:

TYPE: Acid, preferential TEMP:

COMPOSITION:

x .... picric acid

x .... EOH

DISCUSSION:

Fe, single crystal whiskers etched in this solution to observe dislocations.

REF: Ibid.

ST-0011

ETCH NAME: TIME:

NAME: Electrolytic, polish TEMP: 50°C

COMPOSITION: ANODE:

15 .... H_2SO_4 CATHODE:

63 .... H_3PO_4 POWER:

10 .... CrO_3

12 .... H_2O

DISCUSSION:

Fe, polycrystalline discs used as substrates for deposition of carbon thin films. First, vacuum anneal at 82°C, 8 h at 10^{-6} Torr. Then electropolish in the solution shown. Rinse in DI H_2O; rinse in NaOH and final rinse in DI H_2O.

REF: Brown, D W — *J Vac Sci Technol,* 583(1985)

ST-0012a

ETCH NAME: TIME:

TYPE: Electrolytic, polish TEMP:

COMPOSITION: ANODE:

x .... HAc CATHODE:

x .... CrO_3 POWER:

x .... H_2O

DISCUSSION:

Fe/Ni, martensitic transformation study of faulting and defects. Solution used to electropolish specimen.

REF: Reed, R P — *Acta Metall,* 14,1493(1966)

ST-0012b
ETCH NAME: TIME:
TYPE: Acid, polish TEMP: 100°C
COMPOSITION:
1 .... H_2O_2
2 .... H_3PO_4
DISCUSSION:
Fe/Ni, martensitic transformation study of faulting and defects. Solution used to polish surfaces.
REF: Ibid.

ST-0013a
ETCH NAME: TIME: Swab
TYPE: Acid, microetch TEMP: RT
COMPOSITION:
100 ml .... C_2H_2OH
60 ml .... HCl
20 g .... $FeCl_3$
DISCUSSION:
Fe cast alloys. Used as a general microetch. Will develop concentrations of Steatite.
REF: *A/B Met Dig,* 21,12,23(1983) — Buehler Ltd

ST-0014
ETCH NAME: Water TIME: Variable
TYPE: Acid, preferential TEMP: RT
COMPOSITION:
x .... H_2O, distilled
DISCUSSION:
Fe, single crystal spheres $^3/_8$" in diameter with a (111), (110) and (100) flats cut on surfaces. A study of pitting by pure water on different orientations of iron. Pitting was lowest for high purity iron and the (110) orientation showed the most severe pitting.
REF: Kruger, J — *J Electrochem Soc,* 106,736(1959)

ST-0015
ETCH NAME: TIME:
TYPE: Acid, selective TEMP: 105°C
COMPOSITION:
180 g .... NaOH
30 g .... $KMnO_4$
1000 ml .... H_2O
DISCUSSION:
Stellite specimens. Solution will selectively attack iron and cobalt in steel alloys.
REF: Reed, R D — *Electrochem Technol,* 2,192(1964)

ST-0016
ETCH NAME: TIME:
TYPE: Acid, polish TEMP:
COMPOSITION: ANODE:
2 .... $HClO_4$ CATHODE:
7 .... EOH POWER:
1 .... glycerin

DISCUSSION:

Fe:C(1.5%):Ni(5%) alloy specimens used in a study of martensite growth. Solution is also slightly preferential and can be used as a macro-etch.

REF: Priestner, R & Glover, S G — *Acta Metall*, 5,537(1957)

ST-0018

ETCH NAME: Glyceregia — TIME: Wipe

TYPE: Acid, cleaning — TEMP: RT

COMPOSITION: — ASTM: #87

20—50 ml .... HCl

10 ml .... HNO_3

30 ml .... glycerin

DISCUSSION:

Steel and SST vacuum system parts. A plastic sponge wetted with solution used to wipe clean steel surfaces after using other etch solutions to remove metal evaporation contamination. Follow with water washing. ASTM E407-70 number is for original Glyceregia formulation.

REF: Menth, M — personal communication, 1980

ST-0001j: Ibid.

SST 316 specimens. Solution used as a micro- and macro-etch in the study of corrosion fatigue.

ST-0019a

ETCH NAME: — TIME: ("A") sec ("B") sec

TYPE: Acid, removal — TEMP: RT 5°C

COMPOSITION:

(A)		(B)	
140 ml	Picral*	2.5	HF
12 ml	10% $FeCl_3$	40	H_2SO_4
4 ml	Photoflo**	58.5	H_2O

*xPicric acid:xEOH

**Eastman Kodak.

Mix 1:1 with liquid detergent.

DISCUSSION:

Steels, as coupons of different types used in a structure study. Solutions used in order for general removal and thinning. Solution ("A") is rapid; solution ("B") for final slow removal. Steel varieties evaluated were C, N, Ti, Nb, S, P, Si, Mn, Al, and Mo alloy materials.

REF: Campo, G O et al — *Metallography (MEIJAP)*, 16,287(1983)

ST-0019b

ETCH NAME: — TIME:

TYPE: Acid, thinning — TEMP: RT

COMPOSITION:

5 .... HF

80 .... H_2O_2

15 .... H_2O

DISCUSSION:

Steel, blanks used in a structure study. Solution used to thin specimens for microscope study. See ST-0019a.

REF: Ibid.

ST-0019c
ETCH NAME:
TYPE: Electrolytic, jet thin
COMPOSITION:
x .... $HClO_4$
x .... MeOH

TIME:
TEMP: RT
ANODE: Steel
CATHODE:
POWER: 20 V & 1.5 A/cm^2

DISCUSSION:
Steel blanks used in a structure study. Solution used as shown or with added 2% Nital ($xHNO_3$:xEOH). See ST-0019a, ST-0017d.
REF: Ibid.

ST-0019d
ETCH NAME: Nital
TYPE: Electrolytic, removal
COMPOSITION:
x .... 5% Nital*

*ASTM #74

TIME:
TEMP: RT
ANODE: Steel
CATHODE:
POWER: 10 V

DISCUSSION:
Steel, blanks used in a study of structure. Solution used as a general removal etch in thinning specimens for microscope study. See ST-0019a and ST-0017d.
REF: Ibid.

ST-0020
ETCH NAME:
TYPE: Alkali, cleaning
COMPOSITION:
x .... alkali detergent (pH 11)

TIME:
TEMP:

DISCUSSION:
Steel, 304 type blanks used in a material study. Prepare specimens: (1) vapor degrease in perchloroethylene, C_2Cl_4; (2) ultrasonic in alkali solution shown; (3) rinse in cold dmH_2O; (4) dry in air oven at 150°C. (*Note:* The mixed alkali solution could be KOH:Ivory soap:H_2O and adjust pH with NH_3 or NH_4OH.)
REF: Erlandsson, R — *J Vac Sci Technol,* 19,748(1981)

ST-0021
ETCH NAME: Trichloroethylene
TYPE: Solvent, cleaning
COMPOSITION:
x .... TCE

TIME: $^1/_2$—1 h
TEMP: RT to hot

DISCUSSION:
Steel, 304 type as used in oil vacuum pumps. Solvent soak interior of pump to remove vacuum oil, then scrub with rag-stock or fiber brush as needed. Rinse with MeOH and dry with compressed air line. Pump parts and associated system hardware TCE vapor degreased.
REF: Gomez, P & Walker, P — personal application, 1973—1975

PHYSICAL PROPERTIES OF STRONTIUM, Sr

Classification	Metal (alkaline)
Atomic number	38

Atomic weight	87.63
Melting point (°C)	770
Boiling point (°C)	1384
Density (g/cm^3)	2.6
Thermal conductance (cal/sec)(cm^2)(°C/cm)	0.090
Specific heat (cal/g) 25°C	0.176
Heat of fusion (k-cal/g-atom)	2.1
Latent heat of fusion (cal/g)	25
Heat of vaporization (cal/g)	42.3
Atomic volume (W/D)	33.7
1st ionization energy (K-cal/g-mole)	131
1st ionization potential (eV)	5.692
Electronegativity (Pauling's)	1.0
Covalent radius (angstroms)	1.91
Ionic radius (angstroms)	1.12 (Sr^{+2})
Electrical resistivity (micro-ohms cm)	22.76
Lattice constant (angstroms)	6.05
Coefficient of linear thermal expansion ($\times 10^{-6}$ cm/cm/°C) 20°C	23
Electron work function (eV)	2.74
Cross section (barns)	1.3
Vapor pressure (°C)	1111
Hardness (Mohs — scratch)	1.8
Crystal structure (isometric — normal)	(100) cube, fcc
Color (solid)	Silver yellowish
(flame)	Purple-red
Cleavage (cubic)	(001)

STRONTIUM, Sr

General: Does not occur as a free element in nature. It is a relatively rare element and, though there are a dozen minerals containing strontium, only two are of major importance: strontinate, $SrCO_3$, and celestite, $SrSO_3$, the latter being the major source of commercial quantity ore. Both are of wide occurrence in rocks of all geologic ages commonly associated with limestone, gypsum, rock salt as veins and beds, and occasionally in volcanic rocks. The pure metal is silver-white with a slightly yellowish tinge, and properties are similar to those of calcium.

The metal has limited use in industry. It is used in sugar refining, and the nitrate in pyrotechnics for the deep purple-red flame color.

Technical Application: Strontium has had limited use in Solid State material processing. It is under development and evaluation as a fluoride surface coating with similar applications to those of oxides and nitrides, and SrF_2 has been used as a substrate for optical study of thin film metals. It is also a constituent in some artificial garnets and ferrites.

Etching: Deliquesces in water. Soluble in acids, alcohols, and ammonia.

STRONTIUM ETCHANTS

SR-0001

ETCH NAME: Ammonia — TIME:

TYPE: Acid, removal/polish — TEMP: RT

COMPOSITION:

x .... NH_3

DISCUSSION:

Sr, (100) wafers and other orientations. Ammonia is a general removal and polishing etch for this material.

REF: Vossen, J L & Kern, W — *Thin Film Processes,* Academic Press, New York, 1978

SR-0002

ETCH NAME: Hydrochloric acid TIME:

TYPE: Acid, removal TEMP: RT to hot

COMPOSITION:

x .... HCl, conc.

DISCUSSION:

Sr material as the carbonate strontianite, $SrCO_3$ and sulfate celestite, $SrSO_4$ are both reduced by hydrochloric acid. Both are ores of strontium and used in preparing strontium nitrate for fireworks.

REF: Dana, E S & Ford, W E — *A Textbook of Mineralogy,* 4th ed, John Wiley & Sons, New York, 1950

PHYSICAL PROPERTIES OF STRONTIUM CHLORIDE, $SrCl_2$

Classification	Chloride
Atomic numbers	38 & 17
Atomic weight	158.54
Melting point (°C)	873
Boiling point (°C)	
Density (g/cm^3)	3.05
Hardness (Mohs — scratch)	1.5—2
Refractive index (n=)	1.7
Crystal structure (isometric — normal)	(100) cube
Color (solid)	Colorless
Cleavage (cubic)	(001)

STRONTIUM CHLORIDE, $SrCl_2$

General: Does not occur in nature as the solid chloride due to solubility in water, but can be extracted from sea water and hot springs in certain locales. The two major minerals are strontianite, $SrCO_3$, and celestite, $SrSO_4$, which are found in minable quantity. There also are phosphates and silicates.

In ore extraction of strontium in industry, it is usually obtained as the chloride, and all of the compounds are used in pyrotechnics for their brilliant red flame color. The salts also have application in raw vegetable processing.

Technical Application: The material has been studied as a single crystal for paramagnetic S-states and other morphological data, but has found little use in Solid State processing, mainly due to solubility.

Etching: Water, alcohols, and acetic acid.

STRONTIUM CHLORIDE ETCHANTS

SRCL-0001

ETCH NAME: Vacuum TIME:

TYPE: Vacuum, cleaning TEMP: −196°C

COMPOSITION:

x .... vacuum

DISCUSSION:

$SrCl_2$, (100) cut wafers used in a study of paramagnetic S-states. Wafers were cleaned for study in high vacuum at LN_2 temperatures. Authors noted the solubility as shown above under "Etching".

REF: Low, W & Rosenberger, U — *Phys Rev,* 116,621(1959)

PHYSICAL PROPERTIES OF STRONTIUM FLUORIDE, SrF_2

Classification	Fluoride
Atomic numbers	38 & 9
Atomic weight	123.67
Melting point (°C)	1190
Boiling point (°C)	
Density (g/cm³)	
Ionic radius (angstroms)	1.27
Surface free energy (ergs cm⁻²) 273°C	473 (111)
Hardness (Mohs — scratch)	1—3
(Knoop — kgf mm⁻²)	140
Crystal structure (isometric — normal)	(100) cube
Color (solid)	Colorless
Cleavage (cubic)	(001)

STRONTIUM FLUORIDE, SrF_2

General: Does not occur as a natural compound, though calcium fluoride, CaF_2 (fluorite) has long been used as a soldering flux and for its optical properties. Like other strontium salts, it can be used in pyrotechnics for the brilliant red strontium flame color.

Technical Application: The material has not been used in Solid State device fabrication to date, though it has been used as a single crystal substrate for thin film metal studies. It has been evaporated as a polycrystalline and oriented thin film layer on GaAs and InP wafers in evaluations of possible optical applications in device development.

Etching: Slight solubility in water; soluble in hot HCl.

STRONTIUM FLUORIDE ETCHANTS

SRF-0001

ETCH NAME: Hydrochloric acid, dilute — TIME:

TYPE: Acid, removal — TEMP:

COMPOSITION:

1 .... HCl

10 .... H_2O

DISCUSSION:

SrF_2, (100) thin film deposited on GaAs substrates above 250°C is polycrystalline; deposited below 250°C is single crystal. Photo resist patterned as square openings and E-beam annealed. Fluorine absorbs and film discolors. The films can be "washed away" with the solution shown. CdF_2 processed in the same manner is washed away with water only (CDF-0004).

REF: Sullivan, P W — *J Vac Sci Technol,* B2,202(1984)

SRF-0002: Hodgman, C D — *Handbook of Chemistry and Physics,* 27th ed, Chemical Rubber Co, Cleveland, OH, 1943, 466

SrF_2, specimens. Bulk material is soluble in hot hydrochloric acid.

SRF-0003: Tu, C W — *J Vac Sci Technol,* B2,24(1984)

SrF_2 and $Ba_xSr_{1-x}F_2$, (001) MBE grown on InP, (001) substrates and on InP, (110) substrates with thin film oriented (110). Also studied CaF_2 and BaF_2. All are cubic fluorides.
SRF-0004: Pulker, H K — *Thin Solid Films,* 89,191(1982)

SrF_2, (100) wafers used as substrates for metallization with aluminum, chromium and silver in a study of mechanical properties of optical films. Solution can be used to clean surfaces prior to metallization.

PHYSICAL PROPERTIES OF STRONTIUM OXIDE, SrO

Classification	Oxide
Atomic numbers	38 & 8
Atomic weight	103.63
Melting point (°C)	2430 (SrO)
Boiling point (°C)	
Density (g/cm³)	4.70 (SrO)
Hardness (Mohs — scratch)	6—7 (SrO)
Crystal structure (hexagonal — normal)	$(10\bar{1}0)$ prism (w/Ga)
Color (solid)	Yellow (w/Ga)
Cleavage (basal — distinct)	(0001) (w/Ga)

STRONTIUM GALLIUM OXIDE, $SrGa_{12}O_{19}$

General: Neither the pure nor gallium doped compound occurs in nature, though the mineral rinkolite contains Sr and other elements as a titanosilicate. Like other strontium compounds in industry, the oxide can be used for pyrotechnics for its brilliant red flame color.

Technical Application: The reference here is for a strontium gallium oxide grown as a single crystal in a materials study. It may have applications as a garnet-type material. See Garnet for similar strontium and gallium containing crystals.

Etching: Soluble in alcohols, acetic acid.

STRONTIUM GALLIUM OXIDE ETCHANTS

SGO-0001
ETCH NAME: Nitric acid — TIME:
TYPE: Acid, removal — TEMP: Hot
COMPOSITION:
20 .... HNO_3
80 .... H_2O
DISCUSSION:

$SrGa_{12}O_{19}$, (0001) cleaved wafers from crystals grown by the molten flux method as hexagonal yellow single crystals. Crystals were removed from the crucible by etching with the solution shown. Structure referred to as like magnetoplumbite, Space Group P63/mmc (hexagonal), which is an iron lead manganate.
REF: Haberey, F et al — *J Cryst Growth,* 61,284(1983)

STRONTIUM TITANATE ETCHANTS

SRTO-0001

ETCH NAME: Alumina — TIME:

TYPE: Abrasive, polish — TEMP:

COMPOSITION:

x .... c-Al_2O_3

DISCUSSION:

$SrTiO_3$ single crystal specimens used for development of optical memory devices. Polish specimens with colloidal alumina. See KTO-0001 (potassium tantalate) for further discussion.

REF: Yamaichi, E et al — *Jpn J Appl Phys,* 23,867(1984)

SRTO-0002: Chi, J — *J Phys Soc Jpn,* 40,1371(1976)

Reference for method of flux growth of crystals.

SRTO-0003

ETCH NAME: Alumina — TIME:

TYPE: Abrasive, polish — TEMP:

COMPOSITION:

x .... Al_2O_3

x .... H_2O

DISCUSSION:

$SrTiO_3$, single crystal specimens studied for optical transmission after heat treatment. Specimens were polished on a rotating wax lap with Linde A alumina in water.

REF: Gandy, H W — *Phys Rev,* 113,795(1959)

PHYSICAL PROPERTIES OF STRONTIUM TUNGSTATE, $SrWO_4$

Classification	Tungstate
Atomic numbers	38, 74, & 8
Atomic weight	335.55
Melting point (°C)	200 del.
Boiling point (°C)	
Density (g/cm³)	6.187
Hardness (Mohs — scratch)	3—4
Crystal structure (tetragonal — pyrimidal)	(210) prism
Color (solid)	Colorless
Cleavage (prismatic)	(210)

STRONTIUM TUNSTATE, $SrWO_4$

General: Does not occur in natural as a tungstate compound, though there are carbonates, phosphates and oxides of strontium as mineral species. There is no industrial application at the present time.

Technical Application: The material has been grown in Solid State studies of tungstate compounds, but has not been developed as an electronic type device, to date.

Etching: Soluble in water and dilute acids with deliquescence.

STRONTIUM TUNGSTATE ETCHANTS

SRWO-0001
ETCH NAME: Phosphoric acid
TYPE: Acid, polish
TIME:
TEMP: Hot
COMPOSITION:
x .... H_3PO_4
DISCUSSION:
$SrWO_4$ single crystal specimens were used in a study of Raman frequency shifts and temperature dependence. Material referred to as having the scheelite, $CaWO_4$, tetragonal system, pyrimidal class structure. Other tungstates studied were $CaWO_4$ and $BaWO_4$.
REF: Degreniers, S — *J Phys Chem Solids,* 415,1105(1984)

PHYSICAL PROPERTIES OF SULFUR, S

Classification	Nonmetal
Atomic number	16
Atomic weight	32
Melting point (°C)	119 (128, rhombic)
Boiling point (°C)	444.6
Density (g/cm^3)	2.07
Gas conversion point (°C + O_2)	270 (to SO_2)
Thermal conductance (cal/sec)(cm^2)(°C/cm)	0.0007
Specific heat (cal/g) 25°C	0.175
Latent of heat of fusion (cal/g)	9.3
Heat of vaporization (k-cal/g-atom)	3.01
Atomic volume (W/D)	15.5
1st ionization energy (K-cal/g-mole)	239
1st ionization potential (eV)	10.357
Heat of fusion (k-cal/g-atom)	0.34
Electronegativity (Pauling's)	2.5
Covalent radius (angstroms)	1.02
Ionic radius (angstroms)	0.37 (S^{+4})
Electrical conductance (micro-ohms^{-1})	10^{-23}
Critical temperature (°C) C.T.	1040
Critical pressure (atms) C.P.	116
Critical volume (ml/g)	2.48
Electrical resistivity ($\times 10^{-16}$ ohms cm) 20°C	2
Linear coefficient of thermal expansion ($\times 10^{-6}$ cm/cm/°C) 20°C	64.13
Cross section (barns)	0.51
Vapor pressure (°C)	327.2
Hardness (Mohs — scratch)	1.5—2.5
Refractive Index (n =)	1.96—2.25
Crystal structure (orthohombic — normal) alpha	(100) pinacoid
(monoclinic — normal) beta	(100) pinacoid
(monoclinic — normal) delta	(hk0) prism
Color (solid)	Brilliant, yellow
Cleavage (cubic/octahedral/dodecahedral)	(001)/(111)/(110)

SULFUR, S

General: Wide occurrence as a native element and noted for its brilliant yellow color. It is polymorphous with three primary crystal structures, and also is found as natural amorphous sulfur. Alpha-sulfur is orthorhombic, the most common form as it is stable at standard pressure and temperature; beta-sulfur in monoclinic, occurs when alpha-sulfur is fused, then slowly converts back to the alpha variety with time. There is a second monoclinic form that occurs as the natural mineral rosickyite (delta-sulfur) yellow-brown in color, and mu-sulfur as a black, tar-like form, and there are two other known modifications.

Sulfur was one of the elements known to ancient man and called "brimstone" from the disagreeable odor when burned to sulfur dioxide, SO_2. It is an insecticide when burned, was made as a sulfur candle, and used to fumigate buildings, clothing and people. Several compounds are formulated as creams and salves for their medical curative powers, or used as dusting powders.

As sulfur combines with all elements other than gold and platinum, there are many major ore minerals that are reduced for both the metal and sulfur content. The major sulfides are the iron pyrites group, pyrite, FeS_2, as the the major source, which may also contain a high percentage of silver, Galena, PbS, also of wide occurrence as the major ore of lead. Sphalerite, ZnS, a source of zinc and sulfur and, as it has the atypical tetrahedral unit cell, is of importance crystallographically with regard to many binary compounds, such as the compound semiconductors. Cinnabar, HgS, which occurs in small pockets, the most major world deposit at Almaden, Ciuadad Real, Spain, and the coastal ranges of California in the U.S. It is a strategic war mineral (fulminate of mercury as detonators), and is the only ore of mercury.

As sulfates, there is barite, $BaSO_4$, called "heavy spar"; with gypsum, $CaSO_4.2H_2O$, and the unhydrated anhydrite, $CaSO_4$, most important as Plaster of Paris (named from the occurrence north of Paris, France).

Milk of sulfur (lac sulfuris) is an acidulated polysulfide as finely divided alpha-sulfur particles in water suspension, and has several major uses in chemistry and industrial processing.

The most important acid is sulfuric acid, H_2SO_4, and the amount produced is considered a measure of the industrial technology of a country as it is so widely used in so many chemical and industrial products. The acid is used in rayon manufacture, metallurgy, steel fabrication, in formulating paint pigments, in car and storage batteries, in the making of explosives, and as a catalyst in petroleum refining.

Petroleum is rated by sulfur content; high sulfur in the western U.S. and the Caribbean; low sulfur content in the eastern U.S. Middle Eastern oil, such as in Saudi Arabia, Russia, Balkans, the Crimea, and Southern Asia, all have variable sulfur content that requires extraction during refining. As an acid effluent from such processing, and other chemical processing, it is the major producer of acid rain, and is also a major element in smog. As hydrogen sulfide, H_2S, it is released by volcanic action, converting to sulfurous acid in contact with air.

Some other important compounds are sodium thiosulfite, $Na_2S_2O_3.5H_2O$, as photographic "hypo"; sodium hyposulfite, $Na_2S_2O_4$, used in processing indigo as a dye. The SO_4^- radical forms the "vitrols": copper, blue; iron, green; and zinc, white. Sulfuric acid was first obtained in the 15th century by heating green vitrol ($FeSO_4.7H_2O$) with sand, SiO_2. When mixed with ammonia, NH_3, it is ammonium sulfate as a fertilizer.

Sulfur forms several highly colored metal compounds such as HgS, PbS, CuS, FeS, NiS, CoS, Ag_2S, and Hg_2S, black; Bi_2S_3 and Sn_2, brown; CdS and As_2S_3, yellow; Sb_2S_3, orange; MnS, a pale pink and ZnS, white. The latter, as the mineral sphalerite, is yellow-brown to black.

Technical Application: Several of the compounds listed immediately above, such as CdS, show semiconducting and optical characteristics of use in Solid State device processing.

Cadmium sulfide, CdS, is used by itself, as well as a coating on alpha-quartz radio frequency crystals for its additional piezoelectric properties.

Sulfuric acid is widely used as an etchant, itself, and as mixed with sodium dichromate, $Na_2Cr_2O_7$, the original "glass cleaner" solution. Ceric ammonium sulfate or nitrate is the primary compound of the "chrome etch" formulations used for pattern etching chrome glass masks used in photolithographic processing. The wide use of sulfur compounds and acids is shown in the following list as a quick-reference extracted from this Etchant Section for all metals and metallic compounds in this handbook.

Sulfur, as an element, has been the subject of study since ancient times in all of its polymorphous single crystal and other structure forms, both natural and artificial.

Etching: CS_2, toluene, benzene, ether, alcohols.

Formula	Material	Use	Ref.
$27H_2SO_4$:$3Na_2Cr_2O_7$:$7OH_2O$	Al	Cleaning	AL-0032
$1H_2SO_4$:$10H_2O$	Al	Electrolytic, anodizing	AL-0020b
25 ml H_2SO_4:70 ml H_3PO_4:5 ml HNO_3	Al	Polish	AL-0035
	Al:Cu	Polish	ALCU-0003
	Al:Si	Polish	ALSI-0002
$55H_2SO_4$:$35H_3PO_4$:$10HNO_3$ + H_2O	Al_2O_3	Native oxide removal	ALO-0014
xH_2SO_4 conc.	BaF_2	Removal, slow	BAF-0001
$30H_2SO_4$:$70H_2O$	BaF_2	Polish	BAF-0003
xH_2SO_4 conc.	BeO	Preferential	BEO-0001b
$1H_2SO_4$:$1H_2O$	BeO	Cleaning	BEO-0003
xH_2SO_4 conc.	Cr_3B_2	Removal	CRB-0001c
xH_2SO_4 conc.	Mo_2B_5	Removal	MOB-0001b
xH_2SO_4 conc.	Nb_3B_3	Removal	NBB-0001a
xHF:xH_2SO_4	TaB_2	Removal	TAB-0001b
$1H_2SO_4$:$10H_2O$	Brass	Removal	BRA-0002c
12 oz H_2SO_4:4 oz $NaCr_2O_3$:1 gal H_2O	Brass	Oxide removal	BRA-0008b
$1H_2SO_4$:$4H_2O$	Bronze	Cleaning	BRO-0003
xH_2SO_4:x0.3 M $KMnO_4$	CdS	Preferential	CDS-0006a
5 ml H_2SO_4:1 g Cr_2O_5:1250 ml H_2O	CdS	Preferential	CDS-0007b
xH_2SO_4:$xK_2Cr_2O_7$	CdTe	Polish	CDTE-0002
6 g NH_2SO_3H:100 ml H_2O	CaF_2	Dislocation	CAF-0002
xH_2SO_4, conc.	C	Reduction/removal	C-0002b
40 ml H_2SO_4:20 ml HNO_3:20 g $KClO_4$	C	Reduction/removal	C-0001d
$64H_3PO_4$:$15H_2SO_4$:$21H_2O$	Cr	Electrolytic polish	CR-0001c
xH_2SO_3	Co_9S_8	Preferential	COS-0001
$4H_2SO_4$:$2HNO_3$:2HF	Cb	Clean/polish	CB-0001
10 ml H_2SO_4:10 ml HF:10 ml H_2O + xH_2O_2	Cb	Preferential	CB-0002b
$6H_2SO_4$:$12CrO_3$:$82H_2O$	Cu	Cleaning	CU-0017
xH_2SO_4:$xCu(NO_2)_3$	Cu	Preferential	CU-0018
$3H_2SO_4$:$1HNO_3$:$1H_2O$ "Brite Dip"	Cu	Polish, rapid	CU-0028
2—5% $(NH_4)_2SO_4$	Cu_2O	Cleaning	CUO-0010
$6H_2SO_4$:$1H_2O_2$:$1H_2O$	$CuInSe_2$	Polish	CUIS-0003
$3H_2SO_4$:$1H_2O_2$:$1H_2O$	$CuInS_2$	Dislocation	CIS-0001c

Formula	Material	Use	Ref.
$1H_2SO_4$:$1H_2O_2$:$8H_2O$	GaAs	Preferential	GAS-0064
$1H_2SO_4$:$1H_2O_2$:$50H_2O$	GaAs	Step-etch	GAS-0067
$18H_2SO_4$:$1H_2O_2$:$1H_2O$	GaAs	Damage removal	GAS-0068
$3H_2SO_4$:$1H_2O_2$:$1H_2O$	GaAs	Polish/clean	GAS-0069
	GaAs	Polish/clean/selective	GAS-0073
$4H_2SO_4$:$1H_2O_2$:$1H_2O$	GaAs	Isotropic (polish)	GAS-0006h
$1H_2SO_4$:$4H_2O_2$:1HF	GaAs	Polish/thinning	GAS-0088
$5H_2SO_4$:95 propylene glycol "SSA"	GaAs	Electrolytic, dislocation	GAS-0121
$2H_2SO_4$:2HCl:$1HNO_3$:$2H_2O$	GaP	Polish	GAP-0002
$3H_2SO_4$:$1H_2O_2$:$1H_2O$	GaP	Preferential, junction	GAP-0013
xH_2SO_4, conc.	Ge	Electrolytic, shaping	GE-0129d
125 ml 0.5 *M* $3K_2S_2O_3.H_2O$:$4gI_2$	Ge	Electrolytic, junction	GE-0039
175 ml 15%KOH:45 g $(NH_4)_2S_2O_8$	Ge	Preferential	GE-0054
1000 ml H_2SO_4:100 g $K_2Cr_2O_7$	Au	Electrolytic, polish	AU-0009
$2H_2SO_4$:$1H_2O$ or H_2SO_4 conc.	InSb	Removal	INSB-0002b
$1H_2SO_4$:$1H_2O_2$:xH_2O	InGaAs	Mesa forming	ING-0001
xH_2SO_4, conc. or $2H_2SO_4$:$1H_3PO_4$	InP	Removal	INP-0020c
$10H_2SO_4$:$1H_2O_2$:$1H_2O$	InP	Polish/selective	INP-0044
$2H_2SO_4$:1HBr(47%)	InP	Dislocation	INP-0051
$1H_2SO_4$:1HCl:$1H_2O$	Fe	Cleaning	FE-0108
$30H_2SO_4$:$30H_2O$:20Gly	LaB_6	Electrolytic, polish	LAB-0001
1KOH, sat. sol.:$1Na_2SO_3$ sat. sol:5 Gly	PbSnSe	Electrolytic, polish	PBTS-0001
$xPbSO_4$	$PbZrO_3$	Polish	PBZO-0001b
xH_2SO_4, conc.	$PbZrO_3$	Polish	PBZO-0001c
xH_2SO_4:xH_2O:x(acids?)	Mg	Removal	MG-0011
$1H_2SO_4$:$1H_2O$:$5NH_4Cl$, sat. sol.	MgO	Preferential	MGO-0008
$2H_2SO_4$:$1H_2O_2$:$200H_2O$	Mg_2Ge	Cleaning	MGGE-0001b
xH_2SO_4:xKF	Mo	Removal	MO-0001b
xH_2SO_4:$xNaCr_2O_7$	Mo	Cleaning	MO-0001d
$1H_2SO_4$:$1HNO_3$:$3H_2O$	Mo	Polish	MO-0007
870 ml H_2SO_4:30 ml H_2O	Mo	Electrolytic, thinning	MO-0012
2 15%NaOH:$1Na_2S_2O_8$, sat. sol.	HgTe	Dislocation	HGT-0002b
$10H_2SO_4$:$30HNO_3$:50HAc:$10H_2O$	Ni	Polish	NI-0001
$40H_2SO_4$:$60H_2O$	Ni	Electrolytic polish/ thin	NI-0013
xH_2SO_4, conc.	Nb	Removal	NB-0005a
$9H_2SO_4$:1HF	Nb	Electrolytic, polish	NB-0006
xH_2SO_4:xGly	Pd	Electrolytic, polish	PD-0003
xH_2SO_4, conc.	Plastic	Removal, jet	PL-0003
xH_2SO_4:$xCrO_3$	Quartz	Cleaning	QTZ-0009
xH_2SO_4, conc.	Re	Removal	RE-0004e
$1H_2SO_4$:$1H_3PO_4$	Sapphire	Cleaning	SAP-0001
xH_2SO_4, conc.	Se	Removal	SE-0001
xH_2SO_4:$xHNO_3$	Se	Preferential	SE-0003a
$1H_2SO_4$:$10H_2O$	V_3Si	Electrolytic, polish	VSI-0001b
$1H_2SO_4$:$1H_2O_2$ "Caro's Etch"	Si	Cleaning	SI-0103
xH_2SO_4, conc.	SiO_2	Cleaning (soda lima)	SIO-0022

Formula	Material	Use	Ref.
$40HF:15H_2SO_4$	SiO_2	Cleaning (Corning 7720)	SIO-0038a
1000 ml H_2SO_4:35 ml $Na_2Cr_2O_7$, sat. sol.	SiO_2	"Glass Cleaner"	SIO-0041b
1000 ml H_2SO_4:35 ml $K_2Cr_2O_7$, sat. sol.	Si_3N_4	Cleaning	SIN-0015
20 ml H_2SO_4:40 g Cr_2O_3:1000 ml H_2O	Ag	Polish	AG-0016
$xH_2SO_4:xH_2O$:x(acids?)	NaBr	Removal	NABR-0002
$1—2H_2SO_4:10H_2O$	Steel	Polish	FE-0009a
$5H_2SO_4:2HF:2HNO_3$	Ta	Preferential	TA-0004a
$61H_2SO_4:7HF:16HNO_3$	Ta	Polish	TA-0006
$xH_2SO_4:xK_2Cr_2O_7$, sat. sol.	Ta	Cleaning	TA-0003g
xH_2SO_4, conc.	Te	Polish	TE-0001a
xH_2SO_4, conc.	Tl	Removal	TL-0003
x9 N H_2SO_4:x 1 N H_3PO_4	Th	Electrolytic, polish	TH-0003
xH_2SO_4, conc.	ThO_2	Removal	THO-0001
$2H_2SO_4:10H_2O$	Ti	Removal	TI-0001f
$16H_2SO_4:3HF:1H_2O$	Ti	Electrolytic, polish	TI-00010
xH_2SO_4, conc.	TiO_2	Removal	TIO-0011
xH_2SO_4, conc. or $1H_2SO_4:xH_2O$	W	Removal	W-0001e
$1H_2SO_4:1H_3PO_4:1H_2O$	U	Electrolytic, polish	U-0002c
xH_2SO_4, conc.	UO_2	Removal	UO-0002b
xH_2SO_4, conc.	V	Removal	V-0002b
$1H_2SO_4$:6MeOH	V	Electrolytic, polish/thin	V-0004
$5H_2SO_4$:9HAc	Zr	Electrolytic, thinning	ZR-0005b
xH_2SO_4, conc.	ZrO_2	Removal	ZRO-0001a

SULFUR, S

General: Sulfur is found as a native element, associated with sedimentary rock areas, and recognizable by its brilliant yellow color. It is found in extensive beds and sulfur domes, similar to and associated with salt domes, such as those along the Gulf Coast of the U.S. As it forms compounds with all elements other than gold and platinum, like oxygen, it is a major mineral and rock former as sulfides and sulfates.

Sulfuric acid, H_2SO_4, is the most important industrial acid, and the amount manufactured is a measure of a country's industrialization and technology. In metal processing the material is either desulfurized or can be sulfided under control conditions for specific material structure and use.

Technical Application: Pure sulfur is not used directly in Solid State processing, but there are several binary semiconductor compounds fabricated as electronic devices, such as cadmium sulfide, CdS, for both semiconducting and piezoelectric properties. Sulfuric acid is a major acid alone, or as a mixed acid in etching of all metals and metallic compounds in wet chemical etching (WCE) or as an electrolytic etching system. The latter is more prevalent in industrial metals than on semiconductors.

Both natural and artificial sulfur has been studied since ancient times in its several crystallographic single crystal or amorphous forms.

Etching: CS_2, hydrocarbon derivative solvents and alcohols.

SULFUR ETCHANTS

S-0001
ETCH NAME: TIME:
TYPE: Acid, removal TEMP:
COMPOSITION:
x .... CS_2
x .... EOH
DISCUSSION:

S, (100) wafers as yellow alpha-sulfur single crystal and polycrystalline material. Solution used as a general etch for removal and polishing. (*Note:* Alpha-sulfur is classified in mineralogy as orthorhombic, though it is often referred to as ''rhombic'' sulfur in chemistry.)

REF: Walker, P — mineralogy studies, 1952

S-0002
ETCH NAME: Carbon disulfide TIME:
TYPE: Acid, removal TEMP:
COMPOSITION:
x .... CS_2
DISCUSSION:

S, (001) wafers cleaved from single crystal of delta-sulfur (black sulfur) and also cut into strips with a knife. Black sulfur is orthorhombic at standard pressure and temperature; alters to isometric — normal cubic phase under 700 K bars pressure at RT. (*Note:* Black sulfur is a black, tar-like liquid at 230°C.)

REF: Okajima, M et al — *Jpn J Appl Phys,* 23,15(1983)

S-0003
ETCH NAME: Toluene TIME:
TYPE: Solvent, removal TEMP:
COMPOSITION:
x .... C_7H_8
DISCUSSION:

S specimens. Sulfur is soluble in this solvent, also benzene and slowly in alcohols.

REF: Hodgman, C D — *Handbook of Chemistry and Physics,* 27th ed, Chemical Rubber Co., Cleveland, OH, 1943, 468

PHYSICAL PROPERTIES OF TALC, $H_2Mg_3(SiO_3)_4$

Classification	Silicate
Atomic numbers	1, 12, 14 & 8
Atomic weight	418
Melting point (°C)	>1500
Boiling Point (°C)	
Density (g/cm³)	2.7—2.8
Refractive index (n=)	1.539—1.589
Hardness (Mohs — scratch)	1—1.5
Crystal structure (orthorhombic — normal)	(100) pinacoid
(monoclinic — normal)	(100) pinacoid
Color (solid)	White/greenish
Cleavage (basal — perfect)	(001)

TALC, $H_2Mg_3(SiO_3)_4$

General: Occurs as a natural silicate compound found as small single crystals, but more commonly massive, granule to fibrous. Found in association with the rock serpentine and, in its asbestos form, may contain sharp silicate needles. The massive form is called soapstone and, in thin sheets can be flexible, but not elastic. Sheets show the six-rayed percussion figure common to micas. It contains about 4.8% water, about half appears to be in the crystal lattice. Entrapped water can be driven off at dull red heat with the remainder driven off rapidly above 900°C and, with water removed, fusibility index is about 9 . . . edges fuse with difficulty to a white enamel, which is not acted upon by acids.

In powder form it is a filler in paper, used as a dry lubricant, and as talcum powder it has had long use in cosmetics. It can be easily cut with a knife in the massive form of soapstone and, coupled with its inertness to chemical attack, has been used for washtubs and sinks, and is still in use as table tops in chemical laboratories. Soapstone can look like jade, is used as an ornamental stone or carved as objets d'art, but can be told from jade by its greasy appearance and softness.

Technical Application: As soapstone it is still found as table tops in chemical laboratories, and has been used as a mortar and pestle for grinding chemicals. As talcum powder, it is used with latex and rubber gloves for ease of hand insertion and for water absorption to some degree, but not recommended for Clean Room operation due to possible dusting contamination. Also, in powder form, it has been used as a packing vehicle to hold metal powders during pressure testing where it is nonreactive under the temperatures and pressures applied. Occasionally the powder has been used as a dry lubricant on equipment parts, similar to graphite powder applications.

Talc has been the subject of several studies, particularly with regard to its use as talcum powder. It has been found that specific mines have a higher percentage of silicate needles in the raw material that require removal before use as a facial powder. As an asbestos, it is being phased out as an insulating material due to possible carcinogenic properties, even though, as a natural compound, it is not a carcinogen.

Etching: H_2SO_4 (Rensselaerite, only).

TALC ETCHANTS

TALC-0001
ETCH NAME: Heat
TYPE: Thermal, drying
COMPOSITION:
x .... heat

TIME:
TEMP: 150°C

DISCUSSION:

Talc, $H_2O.3MgO.4SiO_2$. Powdered talc used as a packing vehicle in a study of pressure-induced electronic transitions of La, U, and Th. Materials were handled under oil and washed with dry hexane, then the talc dried at 150°C before mixing and pressurizing. (*Note:* About half the water in talc is lost just below red heat; the remainder is driven off at about 900°C. Talc is not attacked by acids.)

REF: Vijayakumar, V — *J Phys Chem Solids,* 46,17(1985)

TALC-0002

ETCH NAME: TIME:

TYPE: Cleave, preferential TEMP:

COMPOSITION:

x .... cleave

DISCUSSION:

Talc, $H_2O.3MgO.4SiO_2$. Specimens cleaved (001) and studied for defects and dislocations under an electron microscope. Dislocations are present on the cleaved surface.

REF: Amelinck, S & Delavignette, P — *J Appl Phys,* 32,241(1961)

PHYSICAL PROPERTIES OF TANTALUM, Ta

Classification	Transition metal
Atomic number	73
Atomic weight	108.9
Melting point (°C)	2996
Boiling point (°C)	
Density (g/cm^3)	16.6
Thermal conductance (cal/sec)(cm^2)(/°C/cm)	0.13
Specific heat (cal/g) 20°C	0.036
Heat of fusion (k-cal/g-atom)	6.8
Heat of vaporization (k-cal/g-atom)	180
Atomic volume (W/D)	10.9
1st ionization energy (K-cal/g-mole)	138
Latent heat of fusion (cal/g)	41.5
Electronegativity (Pauling's)	1.5
Covalent radius (angstroms)	1.34
Ionic radius (angstroms)	0.68 (Ta^{+5})
Linear coefficient of expansion ($\times 10^{-6}$ cm/cm/°C)	6.5
Electrical resistivity (micro-ohms/cm)	13.5
Vapor pressure (mm $\times 10^{-11}$) 1727°C	9.5
Electrical work function (eV)	4.12
Magnetic susceptibility (cgs $\times 10^{-6}$)	0.93
Young's modulus (psi $\times 10^6$) 20°C	27
Tensile strength (psi $\times 10^3$) 20°C	30—70
Cross section (barns)	21
Hardness (Mohs — scratch)	6.0—6.5
Crystal structure (isometric — normal)	(100) cube, bcc
Color (solid)	Grey/silver (yellowish)
Cleavage (cubic — poor)	(001)

TANTALUM, Ta

General: There are minor occurrences of native tantalum as water-washed grains associated with gold placer deposits, but very rare. The primary ore is as a tantalate in the mineral group classified as oxygen salts, of which the mineral columbite-tantalite is the most representative example. The general formula is MTa_2O_6 to MNb_2O_6/MCb_2O_6; where M = Fe or Mn with traces of Sn and W. (*Note:* The element name "Niobium" is now used in place of "Columbium".) Columbite-tantalite forms an isomorphous series, grading from pure tantalite to pure columbite depending upon locality. Like many of the rare metal elements the major occurrence is in pegmatites — the final extrusion of acid magmas which slow cool forming extremely large single crystals: both beryl and tourmaline crystals . . . 80 to 100 ft long and 4 to 5 ft in diameter.

The pure metal has high density and one of the highest of melting points, and is relatively inert to acids, though it will oxidize readily when heated. There are increasing applications for the pure metal as wire, rod, sheet, and bulk forms and, in nonoxidizing operations, often replaces platinum. Because of its high tenacity and acid resistance, it is used in the fabrication of special steels and similar alloys. Pressed powder and bulk material parts are particularly useful in the processing of halogens, such as bromine and iodine. A major commercial use is as the oxide Ta_2O_5 in the fabrication of tantalum capacitors and resistors used as discrete elements in electrical and electronic circuits. Although tungsten has largely replaced tantalum as a light filament, it is still used where vibration is a factor.

As carbides and borides tantalum forms refractory ceramic-type materials. Tantalum carbide, TaC, is used as a cutting wheel and, as a thin film coating on a steel blade to improve wear resistance and working life of the blade.

Technical Application: In Solid State device processing the pure metal is RF sputter evaporated in multilayer metallization structures and, as it is an oxygen "getter" similar to titanium, has been used for that purpose: as a rod in vacuum systems or as a thin film evaporation on the vacuum bell jar walls prior to subsequent metals being evaporated to reduce oxygen contamination during metallization structuring. In planar device and circuit fabrication, the metal is evaporated as a thin film and then oxidized to Ta_2O_5 for use as a capacitor. Evaporated in a nitrogen atmosphere, it forms tantalum nitride, TaN, and, like other metal nitrides and oxides is under development and application as a surface protective coating. It also is under development as a silicide: TaSi, Ta_2Si and $TaSi_2$ for use as a buffer layer in multilayer device heterostructures. And both the carbide and boride are under evaluation as surface protective coatings.

Tantalum has been grown as a single crystal for general morphological study and, in particular, studied for its oxidation and capacitance effects.

Etching: Insoluble in single acids other than alkalies. Mixed acids: $HF:HNO_3$.

TANTALUM ETCHANTS

TA-0001

ETCH NAME: TIME: 2—10 sec

TYPE: Acid, cleaning TEMP: RT

COMPOSITION:

1 .... HF	to	1 .... HF
1 .... HNO_3		1 .... HNO_3
2 .... H_2O		10 .... H_2O

DISCUSSION:

Poly-Ta rod, sheet, wire used as evaporation sources for tantalum metal and tantalum nitride thin films. Tantalum was evaporated, E-beam sputter evaporated and RF sputtered from a solid target. Material was etch cleaned in the solution shown; rinsed in DI water, then MeOH and N_2 dried prior to use as an evaporation source. Substrates were gallium arsenide, alumina and sapphire, as pure metal and tantalum nitride deposited thin films. Solution is extremely rapid at 2—3 sec for removal of 400 Å of Ta or TaN. Parts used in GaAs FET assembly.

REF: Walker, P — personal development, 1979—1984

TA-0019a: Walker, P & Moreland, M — personal application, 1981

Ta, thin films converted to Ta_2O_5 or as TaN. Solution used as a step-etch to establish thickness color order of both the oxide and nitride as deposited on (100) chromium doped gallium arsenide.

TA-0020: Klein, G P — *J Electrochem Soc,* 118,672(1971)

Ta, foil used as the base for Ta powder in the forming of tantalum oxide capacitors. Degrease and etch clean foil; paint-on a Ta powder slurry; sinter at 1900—2200°C in argon; anodize Ta film to Ta_2O_5; deposit manganese nitrate and convert to MnO_2 in hot gas. Describes a low-cost, mass fabrication method for fabrication of solid tantalum capacitors as "powder-on-foil" devices.

TA-0002

ETCH NAME: TIME:

TYPE: Acid, removal TEMP:

COMPOSITION:

1 .... HF

2 .... HNO_3

1 .... H_2O

DISCUSSION:

Ta material. Described as a general etch for tantalum metallographic specimens.

REF: *A/B Met Dig,* 21(2),23(1983), Buehler Ltd

TA-0011: Maissel, L I & Glang, H — *Handbook of Thin Film Technology,* McGraw-Hill, New York, 1970

Ta, thin films. Described as a general rapid etch for the material.

TA-0004a

ETCH NAME: TIME: Variable

TYPE: Acid, preferential TEMP: RT

COMPOSITION:

5 .... H_2SO_4

2 .... HNO_3

2 .... HF

DISCUSSION:

Ta, (111). Described as a general etch for tantalum. Pit size increases with time and shows good reproducibility. Used in studying dislocation density.

REF: Bakish, R — *Acta Metall,* 6,120(1958)

TA-0017: Dreiner, R & Schimmel, J — *J Electrochem Soc,* 111,452(1964)

Ta specimens used in a study of oxides films and effect on photoresponse. Solution used at RT, 15 sec to chemically polish. After polish, boil in water for 10 min to oxidize and anneal 30 min in vacuum at 2100°C to densify the oxide.

TA-0003e: Ibid.

Ta sheet. A general etch for tantalum.

TA-0004b
ETCH NAME: BHF TIME:
TYPE: Acid, preferential TEMP:
COMPOSITION:
1 .... HF
1 .... HN_4F(20%)
DISCUSSION:
Ta, (111) and (100) wafers. Described as a general etch for tantalum. Best pit patterns were on the (112).
REF: Ibid.
TA-0003a: Kohl, W H — *Handbook of Materials & Techniques for Vacuum Devices*, Reinhold, New York, 1967, 304
Ta, poly sheet blanks used in an etch study of tantalum. Mixture was 10 ml HF:10 ml NH_4F(20%), 1 min at 50—60°C. Etch will develop grain structure and will not affect Ta_2S_5. The same solution, used boiling, turns Ta_2S_5 brown in color.

TA-0003b
ETCH NAME: Ammonium fluoride:A etch TIME: 5—6 min
TYPE: Acid, preferential TEMP: 80°C
COMPOSITION:
x .... 20% NH_4F
DISCUSSION:
Ta, poly sheet blanks used in an etch study of tantalum. Solution will develop grain structure but will not affect Ta_2S_5.
REF: Ibid.

TA-0003c
ETCH NAME: TIME:
TYPE: Acid, removal TEMP: 60°C
COMPOSITION:
10 ml .... HNO_3
10 ml .... NH_4F(20%)
DISCUSSION:
Ta, poly sheet blanks. The solution will etch tantalum and turn Ta_2S_5 black in color.
REF: Ibid.

TA-0003d
ETCH NAME: TIME: 1—2 min
TYPE: Acid, preferential TEMP: 60°C
COMPOSITION:
20 ml .... H_2SO_4
10 ml .... NH_4F (20%)
DISCUSSION:
Ta, poly sheet blanks. Solution will develop etched structures.
REF: Ibid.

TA-0005a
ETCH NAME: TIME:
TYPE: Acid, polish TEMP:
COMPOSITION:
x .... H_2SO_4
x .... H_2O_2
x .... H_2O
DISCUSSION:
Ta material. Described as a general etch for tantalum. (*Note:* This is a major solution type for etching and cleaning semiconductor wafers. See the Caro etch formulations under Silicon.)
REF: Bergland, T & Dearden, W H — *Metallographer's Handbook of Etching*, Pitman & Sons, London, 1931

TA-0003i
ETCH NAME: TIME: 9—10 min
TYPE: Electrolytic, polish TEMP: 35—45°C
COMPOSITION: ANODE: Ta
90% .... H_2SO_4 CATHODE: C or Pt
POWER: 0.10 A/cm^2 — polish: 9 min
0.02 A/cm^2 — etch: 10 min
DISCUSSION:
Ta material. Mechanically polish before electropolish.
REF: Ibid.
TA-0009: Spitzig, A W & Mitchell, T E — *Acta Metall,* 11,1311(1966)
Ta specimens cut with a spark cutting wire from a single crystal in a study of dislocations due to tension deformation at 373 K. Solution used at RT to polish surfaces after spark cutting.
TA-0014: Gall, J F & Miller, H C — U.S. Patent 2,466,095 (1949)
Ta, specimens. Solution developed as an anodic etch for tantalum.

TA-0008b
ETCH NAME: Hydrofluoric acid TIME: 20 sec
TYPE: Acid, cleaning TEMP: RT
COMPOSITION:
1 .... HF
1 .... H_2O
DISCUSSION:
Ta specimens. Solution used to etch clean tantalum surfaces. The etch solution was blown off the specimen surfaces prior to water rinse to prevent oxidation.
REF: Ibid.

TA-0008a
ETCH NAME: TIME:
TYPE: Acid, polish TEMP:
COMPOSITION:
1.5 .... HF
2 .... HNO_3
5 .... H_2SO_4

DISCUSSION:

Ta specimens used for oxidation and measurement of capacitance values. Polish etch in solution shown then clean surface by dip in $1HF:1H_2O$; blow off HF before DI water rinse to prevent surface oxidation. Capacitance measurements made with an electrolytic etch of 2% HNO_3 with Pt electrodes.

REF: Vermilyea, D A — *Acta Metall,* 6,166(1958)

TA-0010

ETCH NAME: TIME:

TYPE: Acid, preferential TEMP:

COMPOSITION:

7 .... HF

16 .... HNO_3

61 .... H_2SO_4

DISCUSSION:

Ta alloy specimens containing a Ta silicide surface coating. Use solution as a cleaning/polish etch prior to vapor phase deposition of the silicide. A study of the oxidation-resistance of the silicide coating.

REF: Lorenz, R H & Michael, A B — *J Electrochem Soc,* 108,885(1961)

TA-0003f

ETCH NAME: Potassium hydroxide TIME:

TYPE: Alkali, removal TEMP: RT or hot

COMPOSITION:

x .... x% KOH (NaOH)

DISCUSSION:

Ta specimens. Alkalies can be used as a general etch for tantalum. Solutions are slow etching at room temperature but fast when used hot.

REF: Ibid.

TA-0003g

ETCH NAME: TIME:

TYPE: Acid, cleaning TEMP: 110°C

COMPOSITION:

(1) x .... H_2SO_4
x .... $K_2Cr_2O_7$, sat. sol.

(2) x .... H_2SO_4
x .... Cr_2O_3

DISCUSSION:

Ta specimens. Both solutions are good surface cleaning etchants. (*Note:* Solution (1) is a form of the standard glass cleaning solution and (2) another variation.)

REF: Ibid.

TA-0003h

ETCH NAME: TIME:

TYPE: Acid, polish TEMP: RT

COMPOSITION:

(1) 35 g .... Al_2O_3, powder
20 ml .... HF

(2) 20 g .... NH_4F
100 ml .... H_2O

DISCUSSION:

Ta specimens. Both solutions used for acid lap polishing on a metallographic wheel with a felt pad.

REF: Ibid.

TA-0006
ETCH NAME: TIME:
TYPE: Acid, polish TEMP:
COMPOSITION:
1 .... HF
1 .... HNO_3
2 .... H_2SO_4
DISCUSSION:
Ta specimens used for a metallographic study of the effect of oxidation at 750°C. Polish specimens with solution shown before oxidizing.
REF: Bakish, R — *J Electrochem Soc,* 105,70(1958)

TA-0011a
ETCH NAME: TIME:
TYPE: Acid, preferential TEMP:
COMPOSITION:
2 .... HF
2 .... HNO_3
5 .... H_2SO_4
DISCUSSION:
Ta, (100) wafers used in a study of the interaction of oxygen, nitrogen and hydrogen. Solution used as a metallographic etch after the following material treatments:
TaO_2 treatment: Shows oxygen embrittlement and fracture on (110). Has characteristic patterns that are not repetitive in location.
TaN treatment: Shows platelet dissolution with (110) cleavage associated with nitrided areas.
TaH treatment: Use heat to obtain cleavage structure or electrolytic etch (cathodic) in $1HF{:}10HNO_3$ at RT and 0.5 A/in^{-2} power. (100) cleavage predominates over (110).
REF: Bakish, R — *J Electrochem Soc,* 105,574(1958)

TA-0012
ETCH NAME: TIME:
TYPE: Electrolytic, polish TEMP:
COMPOSITION: ANODE:
10 .... HF CATHODE:
90 .... H_2SO_4 POWER:
DISCUSSION:
Ta specimens used in a study of the microtopography of oxide films. Electropolish specimens in solution shown before oxidizing.
REF: Pawel, R E et al — *J Electrochem Soc,* 107,956(1960)

TA-0013
ETCH NAME: Hydrofluoric acid TIME: 20—30 sec
TYPE: Acid, cleaning TEMP: RT
COMPOSITION:
x .... HF, conc.
DISCUSSION:
Ta, as high purity slugs used for a metal evaporation source on semiconductor wafers. Solution used to clean slugs prior to evaporation. Follow with water wash and N_2 blow dry.
REF: Walker, P & Menth, M — personal application, 1980

TA-0015
ETCH NAME: TIME:
TYPE: Electrolytic, polish TEMP:
COMPOSITION: ANODE: Ta
x .... x% NH_4F CATHODE:
x .... formamide POWER:
DISCUSSION:
Ta specimens. Developed as an anodic polish solution for tantalum.
REF: Jenny, A L — U.S. Patent 2,742,416(1956)

TA-0016
ETCH NAME: TIME:
TYPE: Electrolytic, polish TEMP:
COMPOSITION: ANODE: Ta
x .... HF CATHODE:
x .... HCl POWER:
DISCUSSION:
Ta specimens. Developed as an anodic polish solution for tantalum.
REF: Gall, J F & Miller, H C — U.S. Patent 2,481,306(1949)
TA-0017: Kahan, G J — U.S. Patent 2,775,553(1956)

TA-0018
ETCH NAME: TIME:
TYPE: Acid, removal TEMP:
COMPOSITION:
1 .... HF
1 .... HNO_3
DISCUSSION:
Ta thin films remaining in vacuum systems. Solution used to remove tantalum from bell jars and internal fixtures. Use on stainless steel; not on glass, ceramic or copper.
REF: Nichols, D R — *Solid State Technol,* December 1979
TA-0021: Teasdale, T S — *Phys Rev,* 99,1248(1955)
Ta specimens. Solutions of this mixture used for general removal.

TA-0011b
ETCH NAME: TIME:
TYPE: Alkali, removal TEMP:
COMPOSITION:
9 .... KOH (30%)
1 .... H_2O_2
DISCUSSION:
Ta thin films. Shown as a general removal etch.
REF: Ibid.
TA-0019b: Ibid.
Ta, thin films converted to Ta_2O_5 and TaN deposited on GaAs:Cr, (100) (SI) wafer substrates. Solution used as a step-etch in developing thickness color orders of the oxide and nitride.

TANTALUM ALLOYS, TaM_x

General: Does not occur as metal alloys in nature. The pure element is found in minor quantities associated with gold sands, and there are several tantalate minerals, chief among them the variable mixed mineral columbite-tantalite as an oxide containing iron.

Industrially, tantalum and molybdenum are two of the high-temperature metals used as alloys in irons and steels to increase hardness and chemical resistance, and tantalum is used as a constituent in other metal alloys, such as for high temperature brazing above 800°C. The reference shown here is for a Ta:Mo studied for structure and mechanical properties.

Technical Application: Tantalum alloys have had little application in Solid State processing to date, although they can be used in brazing of metal packages.

Etching: Mixed acids $HF:H_2SO_4$, electrolytic.

TANTALUM:MOLYBDENUM ETCHANTS

TAMO-0001

ETCH NAME: TIME:

TYPE: Electrolytic, polish TEMP:

COMPOSITION: ANODE:

x .... HF CATHODE:

x .... H_2SO_4 POWER:

DISCUSSION:

TaMo as a single crystal alloy used in a structure and mechanical properties study. Solution shown used to polish specimens.

REF: Van Torne, L I & Thomas, G — *Acta Metall,* 14,621(1966)

TANTALUM HYDRIDE ETCHANTS

TAH-0001

ETCH NAME: TIME:

TYPE: Electrolytic, micro-etch TEMP: RT

COMPOSITION: ANODE:

1 .... HF CATHODE: TaH

10 .... HNO_3 POWER: 0.5 A/in^2

DISCUSSION:

TaH thin films grown on (100) Ta wafer substrates in a study of the interaction of oxygen, nitrogen and hydrogen. Solution used to develop defect structure. (100) cleavage predominates over (110). See TA-0011a for additional discussion.

REF: Bakish, R — *J Electrochem Soc,* 105,574(1958)

TAH-0002

ETCH NAME: Sulfuric acid TIME:

TYPE: Acid, removal TEMP:

COMPOSITION:

x .... H_2O_4, conc.

DISCUSSION:

TaH powder mixed in amyl acetate or cellulose nitrate and used as a binder in silicon wafer alloying. Apply to wafer surface and fire for 1—10 min at 600—900°C in an inert atmosphere (N_2 or Ar). Sulfuric acids will etch the hydrides. Other hydrides evaluated were NbH, TiH, VH, and ZrH.

REF: Sullivan, M V & Eigler, J H — *Acta Metall,* 103,218(1956)

PHYSICAL PROPERTIES OF TANTALUM CARBIDE, TaC

Classification	Carbide
Atomic numbers	73 & 6

Atomic weight	192.9
Melting point (°C)	4150
Boiling point (°C)	5500
Density (g/cm³)	14.48
Hardness (Mohs — scratch)	9
(Vickers (kg/mm²)	1790
Crystal structure (isometric — normal)	(100) cube
Color (solid)	Grey-white
Cleavage (cubic — imperfect)	(001)

TANTALUM CARBIDE, TaC

General: Does not occur as a natural compound, but is artificially fabricated as a high temperature ceramic. It is not as widely used as other carbides, such as SiC.

Technical Application: Has not been used in Solid State processing to any degree to date, but has been electrolytically deposited as a thin film on different substrates. TaC solid specimens also have been studied.

Etching: H_2SO_4, HF and other fluoride compounds, mixed salts with alkali.

TANTALUM CARBIDE ETCHANTS

TAC-0001
ETCH NAME: Murakama's reagent — TIME:
TYPE: Acid, cleaning — TEMP:
COMPOSITION: — ASTM: #98

(1) 10 g .... $K_3Fe(CN)_6$
10 g .... KOH/NaOH
100 ml .... H_2O

(2) 10 g .... $K_3Fe(CN)_6$
2 g .... NaOH
100 ml .... H_2O

DISCUSSION:

TaC, specimens. Solutions used to clean surfaces. Also used to clean tungsten carbide, WC and titanium carbide, TiC. Solution (2) is modified, with no number in the ASTM bulletin.

REF: Bull. ASTM E407-70, 1947

TAC-0002
ETCH NAME: Sulfuric acid — TIME:
TYPE: Acid, removal — TEMP:
COMPOSITION:

(1) x .... H_2SO_4, conc.

(2) x .... HF, conc.

DISCUSSION:

TaC, specimens. Both solutions can be used as slow removal etchants.

REF: *A Dictionary of Carbide Terms,* Adams Carbide Corp., Kenilworth, NJ, 07033

TAC-0003a
ETCH NAME: Sulfamic acid — TIME: 1 min
TYPE: Acid, removal — TEMP: RT
COMPOSITION:

x .... 20% H_2NSO_3H

DISCUSSION:

TaC, electrolytically deposited thin films on different substrates. Solution used for general removal of the material.

REF: Hockman A J — *J Electrochem Soc,* 130,221(1983)

TAC-0003b
ETCH NAME: BHF TIME: 4 sec
TYPE: Acid, removal TEMP: RT
COMPOSITION:
x .... HF
x .... NH_4F (40%)
DISCUSSION:

TaC, electrolytically deposited thin films on different substrates. A BHF solution used to remove material more rapidly than with sulfamic acid.

REF: Ibid.

PHYSICAL PROPERTIES OF TANTALUM NITRIDE, TaN

Classification	Nitride
Atomic numbers	73 & 7
Atomic weight	195
Melting point (°C)	3360
Boiling point (°C)	
Density (g/cm^3)	
Hardness (Mohs — scratch)	8—9
Crystal structure (hexagonal — normal)	(1010) prism
Color (solid)	Colorless
Cleavage (basal)	(0001)

TANTALUM NITRIDE, TaN

General: Does not occur as a natural compound. The most important minerals are pure tantalite of the isomorphous columbite-tantalite series. There are several other tantalum-bearing minerals as oxides, but they are of minor importance as ores. Tantalum metal parts are nitrided in industry to obtain a thin surface coating to improve the physical characteristics of the material.

Technical Application: All references shown here are as thin film deposits on Solid State materials for relatively chemically inert surface coatings with possible optical characteristic applications. It also is deposited in conjunction with metals for use as a resistor when oxidized to Ta_2O_5.

Etching: Rapid attack with $HF:HNO_3$ mixtures; slow with alkalies and aqua regia.

TANTALUM NITRIDE ETCHANTS

TAN-0001
ETCH NAME: Aqua regia TIME:
TYPE: Acid, polish TEMP:
COMPOSITION:
3 .... HCl
1 .... HNO_3
DISCUSSION:

TaN, thin film deposits. Aqua regia is a slow removal and polishing etch on tantalum nitride.

REF: Weast, R C et al — *Handbook of Chemistry and Physics*, 65th ed, CRC Press, Boca Raton, FL, 1985, 154

TAN-0002
ETCH NAME:
TYPE: Acid, removal
COMPOSITION:
50 ml .... HF
100 ml .... HNO_3
50 ml .... H_2O
TIME: 5—15 sec
TEMP: RT
ASTM: #161 (1:5:10)
DISCUSSION:

TaN thin films. Developed from the ASTM #161 solution which is for tantalum as an etch for tantalum nitride. TaN/Au and TaN/TiW/Au or TaN/Mo/Au deposited on alumina, beryllia, and quartz substrates used for microwave circuit fabrication with GaAs FETS. After AZ-1350J photo resist patterning of metal/nitride deposited substrates. Solution used to remove the TaN.

REF: Walker, P — personal development, 1983
TAN-0007; ASTM E107-70

Reference for ASTM #161 solution used to etch tantalum.

TAN-0003
ETCH NAME:
TYPE: Acid, removal
COMPOSITION:
1 .... HF
2 .... HNO_3
1 .... H_2O
TIME:
TEMP:
DISCUSSION:

TaN thin films. Described as a very rapid etch. It will etch both TaN and Ta (TA-0002). Slow reaction with addition of water.

REF: Bulletin 1116R/Rev-HPD/VB/TM/0883/1K, MRC

TAN-0004
ETCH NAME:
TYPE: Acid, removal
COMPOSITION:
9 .... 30% KOH
1 .... H_2O_2
Note: Add H_2O_2 when solution is hot.
TIME:
TEMP:
DISCUSSION:

TaN thin films. Described as a general etch for tantalum nitride. Solution will attack most photo resists. **CAUTION**: When adding any cold solution to a hot solution reaction may splatter.

REF: Grossman, S & Herman, D S — *J Electrochem Soc,* 116,674(1969)
TAN-0005a: Kodak Data Book, 1966, 91
TAN-0006: Electron. Div. Abstract, 23,1966

TAN-0005b
ETCH NAME:
TYPE: Acid, removal
COMPOSITION:
TIME:
TEMP:

(1)		(2)	
92 g	.... $K_3Fe(CN)_6$	200 g	.... $K_3Fe(CN)_6$
20 g	.... KOH	20 g	.... NaOH
300 ml	.... H_2O	3—3.5 g	.... sodium oxalate
		1000 ml	.... H_2O

DISCUSSION:

TaN thin films. Solutions will etch tantalum and molybdenum and their nitrides.

REF: Ibid.

TAN-0008

ETCH NAME: TIME:

TYPE: Gas, conversion TEMP: RT to 750°C

COMPOSITION:

x .... N_2

x .... Ar

DISCUSSION:

TaN thin films deposited 3000 Å thick on substrates at the following temperatures: 20, 250, 500, and 750°C used in a composition study of the DC sputtered films. Ta was sputtered in an N_2/Ar atmosphere. Films showed the following structures: (1) $TaN_{0\ 05}$ as beta; Ta_2N as gamma; TaN_2 as alpha; and TaN as epsilon. No etch shown.

REF: Reichelt, K et al — *J Appl Phys,* 49,5284(1978)

PHYSICAL PROPERTIES OF TANTALUM OXIDE, Ta_2O_5

Classification	Oxide
Atomic numbers	73 & 8
Atomic weight	441.76
Melting point (°C)	1470, del.
Boiling point (°C)	
Density (g/cm^3)	8.7 approx.
Hardness (Mohs — scratch)	7—8
Crystal structure (hexagonal — rhombohedral)	$(10\bar{1}1)$ rhomb
Color (solid)	Clear/greyish
Cleavage (basal)	(0001)

TANTALUM OXIDE, Ta_2O_5

General: Does not occur as a natural compound, though it does occur as a tantalate oxide with iron and manganese as an isomorphous series with niobium (columbium). The pure tantalite as near Ta_2O_6. The pentoxide form, above, is the most widely used form, though their are other oxides. It has been used industrially in the fabrication of capacitors, both as a bulk material and a thin film evaporation on ceramics, mica, and mylar.

Technical Application: In Solid State development, several of the oxides have been deposited as amorphous thin film capacitor elements in semiconductor circuit fabrication. They are under development as surface passivation coatings similar to those of silicon dioxide and nitride. In fabrication, tantalum metal is first evaporated, then oxidized by controlled heating as a separate process step. The nitride also has been similarly converted to the oxide.

The oxide has use as a general surface coating or for fabrication of discrete capacitor elements of a semiconductor circuit.

Etching: HF; mixed HF:HNO_3; and alkalies, variable with type of oxide.

TANTALUM DIOXIDE ETCHANTS

TAO-0001a

ETCH NAME: TIME:

TYPE: TEMP:

COMPOSITION:

DISCUSSION:

TaO_2. Material is a dull grey powder and insoluble in acids.

REF: Weast, R C et al — *Handbook of Chemistry and Physics,* 65th ed, CRC Press, Boca Raton, FL, 1984, B154

TAO-0002

ETCH NAME: TIME:

TYPE: Alkali, removal TEMP:

COMPOSITION:

9 .... KOH (30%)

1 .... H_2O_2

DISCUSSION:

Ta_2O_3, thin films. Solution shown as a removal etch for Ta, Ta_2O_3 and TaN.

REF: Miasel, L I & Glang, R — *Handbook of Thin Film Technology,* McGraw-Hill, New York, 1970

TANTALUM PENTOXIDE ETCHANTS

TAO-0003a

ETCH NAME: Hydrofluoric acid TIME:

TYPE: Acid, removal TEMP:

COMPOSITION:

x .... HF

DISCUSSION:

Ta_2O_5. Material has rhombohedral structure and is colorless. Can be etched in HF and fused $KHBO_4$.

REF: Weast, R D et al — *Handbook of Chemistry and Physics,* 65th ed, CRC Press, Boca Raton, FL, 1984, B154

TAO-0003b

ETCH NAME: Potassium borate TIME:

TYPE: Salt, molten flux, removal TEMP: <150°C

COMPOSITION:

x .... $KHBO_4$, fused

DISCUSSION:

Ta_2O_5 specimens. As a molten flux, solution will etch this oxide.

REF: Ibid.

TAO-0004

ETCH NAME: Air TIME: Minutes

TYPE: Gas, stabilizing TEMP: 250°C

COMPOSITION:

x .... air

DISCUSSION:

Ta_2O_5 thin films RF magnetron sputter deposited on GaAs (100) wafers, sapphire and alumina substrates for use as capacitors in microelectronic circuits. After deposition material was heat treated on a hot plate in air to stabilize capacitance values.

REF: Marich, L A & Walker, P — personal application, 1980—1985

PHYSICAL PROPERTIES OF TANTALUM SELENIDE, $TaSe_2$

Classification	Selenide
Atomic numbers	73 & 34
Atomic weight	338.80
Melting point (°C)	2000
Boiling point (°C)	
Density (g/cm^3)	
Hardness (Mohs — scratch)	6—7
Crystal structure (isometric — normal?)	(100) cube
Color (solid)	Grey-black
Cleavage (cubic)	(001)

TANTALUM SELENIDE, $TaSe_2$

General: Does not occur as a natural compound, and there has been no use in industry other than in studies as a metal dichalogenide.

Technical Application: No present application in Solid State processing. It has been grown as a single crystal by the Vapor Transport (VT) method in an evaluation of dichalcogenide transition metal compounds. There is possible application of the material as a basic V—VI compound semiconductor with other metals for a ternary compound semiconductor.

Etching: Mixed acids (HF:HNO_3).

TANTALUM SELENIDE, $TaSe_2$

TASE-0001
ETCH NAME: TIME:
TYPE: Acid, removal TEMP:
COMPOSITION:
x .... HF
x .... HNO_3
DISCUSSION:

$TaSe_2$ single crystal specimens grown by the iodine Vapor Transport (VT) method. Specimens used in a study of transition metal dichalcogenides. TaS_2 also grown and evaluated.

REF: Naito, M & Tanaka, S — *J Phys Soc Jpn,* 4,1216(1984)

PHYSICAL PROPERTIES OF TANTALUM SULFIDE, TaS_2

Classification	Sulfide
Atomic numbers	73 & 16
Atomic weight	400
Melting point (°C)	1300 (del)
Boiling point (°C)	
Density (g/cm^3)	
Hardness (Mohs — scratch)	4—5
Crystal structure (isometric — normal?)	(100) cube
Color (solid)	Grey black
Cleavage (cubic)	(001)

TANTALUM SULFIDE, TaS_2

General: Does not occur as a natural compound, even though there are a number of other metal sulfides — galena, PbS, being of major importance. Not used to any extent in the metal industries, other than when tantalum metal is sulfurized with a thin surface film which, in part, may be in crystalline form.

Technical Application: The compound has not been used in Solid State processing to date, though it has been grown and studied as a transition metal dichalcogenide. It is a possible basic V—VI compound semiconductor and, with other metals, a trinary compound type.

Etching: Mixed acids ($HF:HNO_3$).

TANTALUM SULFIDE ETCHANTS

TAS-0001

ETCH NAME: TIME:

TYPE: Acid, removal TEMP:

COMPOSITION:

x .... HF

x .... HNO_3

DISCUSSION:

TaS_2 single crystals were grown by the iodine Vapor Transport (VT) method. Specimens used in a study of transition metal dichalcogenides. $TaSe_2$ also grown and evaluated.

REF: Naito, M & Tanaka, S — *J Phys Soc Jpn,* 4,1216(1984)

TANTALUM TUNGSTEN ETCHANTS

TAW-0001

ETCH NAME: Ethylene chloride TIME:

TYPE: Solvent, removal TEMP:

COMPOSITION:

x .... CH_2Cl_2

DISCUSSION:

TaW thin films deposited or glass cover slides. Latex spheres 0.09 μm in diameter adsorbed on glass prior to 30 nm thick metal deposition. After metallization, ultrasonic (sonication) agitation used to remove spheres using the solution shown. Leaves an array of 0.09—0.1 μm "light" transmission holes through the TaW film as a method of controlled forming of pinholes.

REF: Fischer, Ch — *J Vac Sci Technol,* B3(1),386(1985)

PHYSICAL PROPERTIES OF TELLURIUM, Te

Classification	Semi-metal
Atomic number	52
Atomic weight	127.6
Melting point (°C)	449.5
Boiling point (°C)	989.8
Density (g/cm^3)	6.24
Specific heat (cal/g) 25°C	0.047
Latent heat of fusion (cal/g)	32
Heat of vaporization (k-cal/g-atom)	11.9
Heat of fusion (k-cal/F-atom)	4.27
Atomic volume (W/D)	20.5
Covalent radius (angstroms)	1.36
Ionic radius (angstroms)	0.70 (Te^{+4})
1st ionization potential (eV)	9.01

1st ionization energy (K-cal/g-mole)	208
Electrical resistivity (ohms cm) 22°C	0.43 // c-axis
Electronegativity (Pauling's)	2.1
Electron work function (eV)	4.72
Electrical conductance (micro-ohms^{-1})	10—6
Modulus of elasticity (psi $\times 10^6$)	6
Poissons ratio	0.33
Vapor pressure (°C)	838
Cross section (barns)	4.3
Activation energy (eV)	0.34
Hall mobility (cm^2/V sec 0—320°C)	1800—2500
Specific resistance (ohm-cm)	0.35
Refractive index (n=)	2—3.5
Dielectric constant (e=)	5.0
Optical activation energy (eV)	0.37
Hardness (Mohs — scratch)	2—2.5
Crystal structure (hexagonal — rhombohedral)	$(10\bar{1}1)$, rhomb
Color (solid)	Tin-white
Cleavage (prismatic — perfect)	$(10\bar{1}0)$

TELLURIUM, Te

General: Occurs as a native element but in minor quantities. It has been found as fine prismatic crystals but is more commonly columnar to massive. It is completely volatile in a bunsen burner flame and produces a deep red solution in warm sulfuric acid. The mineral tellurite, TeO_2, is of greater occurrence from which the metal is obtained by reduction. It is a semi-metal similar to sulfur and is found as a minor constituent in several minerals, to include gold. As an alloy with iron or copper it produces high electrical resistance materials, and with less than $^1/_{10}$ of 1% alloyed with lead greatly increases both strength and hardness. As a coloring agent in glass it produces a deep blue to brown coloration. If the vapors are breathed it imparts a very offensive odor called "telluride breath".

Technical Applications: Tellurium is used as a doping element in some compound semiconductors. It is a growth element in several semiconductor compounds such as cadmium telluride, CdTe; zinc telluride, ZnTe; lead telluride, PbTe and trinary compounds containing tin or mercury. As a single crystal tellurium has been grown and studied for general physical data and morphology.

Etching: H_2SO_4, aqua regia and HF:HNO_3 mixtures.

TELLURIUM ETCHANTS

TE-0001a
ETCH NAME: Sulfuric acid
TYPE: Acid, dislocation
COMPOSITION:
x .... H_2SO_4, conc.

TIME:
TEMP: 150°C
RATE: 11 µm/min

DISCUSSION:

Te, $(10\bar{1}0)$ wafers used in a dislocation development study. Etch is very temperature dependent. Used at room temperature, dislocation pits may be asymmetric in the $\langle 10\bar{2}0 \rangle$ direction. Also used as a polish etch at elevated temperature.

REF: Blackmore, J S et al — *Phys Rev*, 117,687(1960)

TE-0001b
ETCH NAME:
TYPE: Acid, polish
COMPOSITION:
1 .... HF
1 .... HNO_3
1 .... CH_3COOH (HAc)
TIME:
TEMP: RT or 0°C
RATE: 5 μm/min at RT
DISCUSSION:
Te, $(10\bar{1}0)$ wafers used in a dislocation development study. Solution used as a polishing etch. It is a rapid etch at both RT and 0°C.
REF: Ibid.

TE-0001c
ETCH NAME:
TYPE: Acid, preferential
COMPOSITION:
1 .... HCl
1 .... CrO_3
3 .... H_2O
TIME:
TEMP: RT
RATE: 4 μm/min
DISCUSSION:
Te, $(10\bar{1}0)$ wafers used in a dislocation study. Polish wafers in H_2SO_4 (TE-0001a) then use solution shown to develop dislocation etch pits. Pits are sharply defined in a smooth, bright background surface.
REF: Ibid.
TE-0004: Naukkarienen, K & Toomi, T C — *J Appl Phys,* 40,3054(1969)
Te, single crystal wafers used in an X-ray topographic study of defects. Solution mixture used for defects but development not shown.
TE-0007: Blackmore, J S et al — *J Appl Phys,* 31,2226(1960)
Te, (1010) wafers used in a dislocation study. Solution used as a polish etch. Produces a bright surface with an etch rate of 4 μm/min.

TE-0002
ETCH NAME: Aqua regia
TYPE: Acid, removal/polish
COMPOSITION:
3 .... HCl
1 .... HNO_3
TIME:
TEMP:
DISCUSSION:
Te, $(10\bar{1}0)$, cleaved wafers used in a study of optical properties. Agitate during etching to prevent pitting. Solution used to remove 7—15 μm of a surface conductance layer.
REF: Caldwell, R S — Special Rep Purdue University, January 1958

TE-0003
ETCH NAME: CP4, modified
TYPE: Acid, saw
COMPOSITION:
3 .... HF
5 .... HNO_3
3 .... HAc
TIME:
TEMP: RT
DISCUSSION:
Te, (0001) wafers acid saw cut from the ingot. A 0.025 platinum wire, wetted with the

solution shown, was used as the cutting tool. The ingot was also cleaved to obtain rectangular specimens. (*Note:* CP4 without bromine is called CP4A.)
REF: Vis, V A — *J Appl Phys,* 35,361(1964)

TE-0005
ETCH NAME: CP4 TIME: 10 sec
TYPE: Acid, removal TEMP: RT
COMPOSITION:
30 ml .... HF
50 ml .... HNO_3
30 ml .... HAc
0.5 ml .... Br_2
DISCUSSION:
Te, (0001) cleaved wafers, used in measuring the galvanmagnetic effects. Etch was used to remove the conducting layer from cleaved surfaces (about 15 μm). (*Note:* Original CP4 called for 0.6 ml Br_2.)
REF: Roth, H — *J Phys Chem Solids,* 8,525(1959)

TE-0006
ETCH NAME: TIME: 30 sec
TYPE: Acid, polish TEMP: RT
COMPOSITION: RATE: 7 mil/min
3 .... HF
5 .... HNO_3
6 .... HAc
DISCUSSION:
Te, (0001) wafers used in a study of dislocations. Solution rate was 7 mil/min and used to remove damaged surface.
REF: Lovell, L C et al — *Acta Metall,* 6,716(1958)

TE-0008
ETCH NAME: Aqua regia, dilute TIME: 4—5 min
TYPE: Acid, polish TEMP: RT
COMPOSITION: RATE: 30—40 μm/4—5 min
3 .... HCl
1 .... HNO_3
1 .... H_2O
DISCUSSION:
Te, (0001) wafers used in a study of annealing behavior of extrinsic tellurium. Solution used in the etch preparation of Hall samples.
REF: Skadron, P & Johnson, V A — *J Appl Phys,* 37,1912(1966)

TE-0009a
ETCH NAME: TIME: 2 h
TYPE: Acid, cleaning TEMP: RT
COMPOSITION:
(1) 2 .... HNO_3 (2) x .... 30% HCl (3) x .... H_2SO_4
4 .... H_2O
1 .... CrO_3
DISCUSSION:
Te, (0001) and $(12\bar{1}0)$ wafers cut from ingot; $(10\bar{1}0)$ cleaved in air or in LN_2. Wafers

used as substrates for deposition of antimony. Polish etch tellurium in solution (1) with rotation during etch period. Follow with 30% HCl, RT, 2 min to remove oxides. In vacuum, thermal faceting observed above 145°C. After Sb growth, cleave wafers ⟨110⟩. Growth appears columnar, bladed. Used hot H_2SO_4 (3) as a dislocation etch.
REF: Shih, I et al — *J Cryst,* 9,523(1984)

TE-0010
ETCH NAME: Tetraborane — TIME:
TYPE: Salt, removal — TEMP:
COMPOSITION: — ANODE:
x .... x% $Na_2B_4H_{10}$ — CATHODE:
POWER:
DISCUSSION:
Te, material used for epitaxy growth of HgCdTe that was then used in a chemical etching and oxidation study. Clean tellurium and remove native oxide in the solution shown.
REF: Aspenes, D E & Arwin, H — *J Vac Sci Technol,* A(2),1309(1984)

TE-0011
ETCH NAME: Alumina — TIME:
TYPE: Abrasive, polish — TEMP: RT
COMPOSITION:
x .... Linde A and B
DISCUSSION:
Te, (0001) specimens used in a study of the effect of pressure on infrared absorption. Because tellurium is soft and brittle initial coarse lapping was limited to not greater than 400 grit. A pitch lap (1 beeswax:1 resin) used for first stage-polish using aluminas shown. Final lap used canvas cloth attached to an optical flat. Also polished silicon and germanium with the pitch lap method.
REF: Waters, W P — personal communication, 1957

TE-0012a
ETCH NAME: — TIME:
TYPE: Acid, saw — TEMP:
COMPOSITION:
x .... HCl
x .... CrO_3
x .... H_2O
DISCUSSION:
Te, (0001) and $(10\bar{1}0)$ wafers used in a study of infrared absorption in intrinsic tellurium. Solution used as an acid saw for cutting specimens.
REF: Ades, S & Champness, C H — *J Appl Phys,* 49,4543(1978)

TE-0012b
ETCH NAME: — TIME:
TYPE: Acid, polish — TEMP:
COMPOSITION:
10 ml .... HNO_3
5 g .... CrO_3
20 ml .... H_2O

DISCUSSION:

Te, (0001) and (10$\bar{1}$0) wafers used in a study of infrared adsorption in intrinsic tellurium. Solution used to chemical polish surfaces.

REF: Ibid.

TE-0012c

ETCH NAME: Hydrochloric acid — TIME:

TYPE: Acid, oxide removal — TEMP: RT

COMPOSITION:

x .... HCl

DISCUSSION:

Te, (0001) and (10$\bar{1}$0) wafers used in a study of infrared optical adsorption of intrinsic tellurium. After chemical etch polish, solution shown used to remove residual native oxide from previous etch solutions.

REF: Ades, S & Champness, C H — *J Appl Phys*, 49,4543(1978)

PHYSICAL PROPERTIES OF TELLURIUM DIOXIDE, TeO_2

Classification	Oxide
Atomic numbers	52 & 8
Atomic weight	159.61
Melting point (°C)	800
Boiling point (°C)	
Density (g/cm³)	5.9
Refractive index (n=)	2.00—2.35
Hardness (Mohs — scratch)	2
Crystal structure (orthorhombic — normal)	(100) a-pinacoid
Color (solid)	Yellow-white
Cleavage (prismatic — perfect)	(010)

TELLURIUM DIOXIDE, TeO_2

General: Occurs in nature as the mineral tellurite, TeO_2, largely as an oxidation product on other tellurides. A black, amorphous monoxide, TeO, appears on the more stable dioxide, and there is a bright orange colored trioxide, TeO_3, form. There is no use of the oxides in the metal industries, but both the pure metal and oxides are a blue to brown coloring agent in glass.

Technical Application: The oxides have had no use in Solid State material processing to date, although the metal element is used in the growth of both binary and trinary compound semiconductors of cadmium, lead and zinc. It has been grown as a thin film on CdTe, and referred to as a native oxide with stabilization time in air.

Etching: HCl, HNO_3, alkalies (TeO_2); HCl, and dilute acids (TeO); HCl and hot alkalies (TeO_3).

TELLURIUM DIOXIDE ETCHANTS

TEO-0001

ETCH NAME: Tetraborane — TIME:

TYPE: Salt, removal — TEMP:

COMPOSITION: — ANODE:

x .... $Na_2B_4H_{10}$ — CATHODE:

POWER:

DISCUSSION:

TeO_2, crystalline native oxide or TeO, amorphous native oxide (black). Solution used to remove native oxide and etch clean tellurium used in the epitaxy growth of HgCdTe thin films. Material was then studied for chemical etching and oxidation characteristics.

REF: Aspenes, D E & Arwin, H — *J Vac Sci Technol,* A(2),1409(1984)

TEO-0002

ETCH NAME: Hydrochloric acid — TIME: 2 min

TYPE: Acid, oxide removal — TEMP: RT

COMPOSITION:

x .... 30% HCl

DISCUSSION:

Te, (0001) and other wafer orientations. Solution used to remove residual oxide remaining on surfaces after chemical etch polishing. (See TE-0009c and TE-0012c).

REF: Shi, I et al — *J Cryst,* 9,523(1984)

TEO-0003: Ades, S & Champness, C H — *J Appl Phys,* 49,4543(1978)

Te, (0001) and $(10\bar{1}0)$ wafers. Solution used to remove residual oxide after chemical etch polishing. (See TE-0012c.)

TEO-0004

ETCH NAME: BRM — TIME:

TYPE: Halogen, removal — TEMP: RT

COMPOSITION:

x .... 5% Br_2

x .... MeOH

DISCUSSION:

TeO_2, grown as a stable native oxide on CdTe (110) wafers. Mechanically lap wafers with 32,000-mesh abrasive, then chem/mech polish with BRM solution. Specimens allowed to sit in air under light for 30 days for a 600 Å thick TeO_2 stable oxide. Remove oxide with HCl or HNO_3.

REF: Ponce, F A et al — *J Appl Phys Lett,* 39,951(1981)

TEO-0005

ETCH NAME: — TIME:

TYPE: Acid, removal/polish — TEMP: RT

COMPOSITION:

(1) x .... HCl, conc.

(2) 1 .... HCl
10 .... H_2O

DISCUSSION:

TeO_2, as natural single crystals used in a general study of natural mineral oxides. Solution (1) used to etch surfaces; solution (2) to develop structure.

REF: Walker, P — mineralogy study, 1953

PHYSICAL PROPERTIES OF TERBIUM, Tb

Classification	Lanthanide
Atomic number	65
Atomic weight	159
Melting point (°C)	1356
Boiling point (°C)	2800 (2530)
Density (g/cm³)	8.27

Specific heat (cal/g) 25°C	0.044
Heat of fusion (k-cal/g-atom)	3.9
Latent heat of fusion (cal/g)	12.71
Heat of vaporization (k-cal/g-atom)	70
Atomic volume (W/D)	19.2
1st ionization energy (K-cal/g-mole)	155
1st ionization potential (eV)	6.74
Electronegativity (Pauling's)	1.2
Covalent radius (angstroms)	1.59
Ionic radius (angstroms)	0.93 (Tb^{+3})
Electrical resistivity (micro-cm) 298 K	114.5
Compressibility ($cm^2/kg \times 10^{-6}$)	2.45
Cross section (barns)	44
Magnetic moment (Bohr magnetons)	9.7
Magnetic susceptibility ($\times 10^{-6}$ emu/mole)	193,000
Transformation temperature (°C)	1287
Hardness (Mohs — scratch)	4—5
Crystal structure (hexagonal — normal)	$(10\bar{1}0)$ prism, hcp
Color (solid)	Grey
Cleavage (basal)	(0001)

TERBIUM, Tb

General: Does not occur as a native element. It is a rare earth member of the lanthanide, yttrium group. Found as a minor constituent in yttrium and gadolinium minerals: the mineral gadolinite is a mixed silicate with iron, yttrium, etc. There is no current use in industry due to scarcity.

Technical Application: Only limited use to date, in Solid State processing. It is a constituent in some ferrites and garnets. Most of the rare earth elements have been grown as single crystals for general morphology and defect study.

Etching: Heat used to develop defects.

TERBIUM ETCHANTS

TB-0001
ETCH NAME: Heat — TIME:
TYPE: Thermal, preferential — TEMP:
COMPOSITION:
x .... heat
DISCUSSION:

Tb, single crystal specimens. Describes a method of growing rare earth single crystal. Thermal annealing after growth used to develop defects, low angle grain boundaries and other structure. Other metals studied were Dy, Gd, Er, Ho, Tr, Th, and Y.
REF: Nigh, H E — *J Appl Phys,* 34,3323(1963)

PHYSICAL PROPERTIES OF THALLIUM, Tl

Classification	Metal
Atomic number	81
Atomic weight	204.37
Melting point (°C)	303
Boiling point (°C)	1457

Density (g/cm^3)	11.85
Thermal conductivity (cal/sec)(cm^2)(°C/cm)	0.093
Specific heat (cal/g) 20°C	0.031
Heat of fusion (cal/g)	5.04
Heat of vaporization (cal/g)	190
Atomic volume (W/D)	17.2
1st ionization energy (K-cal/g-mole)	141
1st ionization potential (eV)	6.106
Electronegativity (Pauling's)	1.8
Covalent radius (angstroms)	1.48
Ionic radius (angstroms)	1.47 (Tl^{+1})
Coefficient of linear thermal expansion ($\times 10^{-6}$ cm/cm/°C)	28
Electrical resistivity (micro-ohms-cm)	18
Electron work function (eV)	3.95
Tensile strength (psi)	1300
Cross section (barns)	3.3
Vapor pressure (°C)	1196
Hardness (Mohs — scratch)	1—2
(Brinell — mm/kg)	2
Crystal structure (hexagonal — normal)	($10\bar{1}0$) prism, hcp
(tetragonal — normal)	(100) a-pinacoid
Color (solid)	Blue-white
(spectrum)	Brilliant green
Cleavage (basal)	(0001)

THALLIUM, Tl

General: Does not occur as a native element. There are three known single crystal minerals containing thallium in small quantities: vrbalite, a sulfide; hutchinsonite and lorandite, arsenides. All three minerals are closely associated with the arsenic sulfides realgar and orpiment — the former usually orange-red; the latter brilliant yellow. The thallium minerals are all brilliant red to red/black. Thallium is found as a trace element in many other minerals, notably, sphalerite, ZnS, and is more abundant than gallium and indium, the other two metals of the aluminum family of elements. It has a brilliant green spectrum which is used in scientific study.

Though it is reasonably abundant, the metal has little or no use in industry, but thallium salts are poisonous and have been used for killing rodents. In 1939, Iln, in Moscow, showed that the nitrate $TlNO_3$, when fed to sheep in small doses, caused them to molt like birds.

Technical Application: To date, the metal has had little or no use in Solid State device fabrication, in part, due to the highly poisonous nature of its salts though it has been evaluated as a silicide and may have dopant properties for compound semiconductors and other materials.

Thallium has been grown as a single crystal for physical data and morphological study to include diffusion reactions with metals such as gold and silver.

Etching: HNO_3, H_2SO_4, HCl and mixed acids HF:HNO_3, HF:H_2O_2.

THALLIUM ETCHANTS

TL-0001
ETCH NAME: TIME:
TYPE: Acid, cleaning TEMP:
COMPOSITION:
1 .... 10% KCN
1 .... 10% $(NH_4)_2S_2O_8$
DISCUSSION:

Tl, (0001) wafers cut from single crystals used in a study of gold and silver diffusion. Solution used to clean surfaces.

REF: Meier, B W et al — *Acta Metall,* 16,1398(1968)

TL-0002
ETCH NAME: Nitric acid TIME:
TYPE: Acid, removal TEMP: RT
COMPOSITION:
x .... HNO_3
DISCUSSION:

Tl, polycrystalline specimens, Nitric acid is a slow, general etch for thallium, as is sulfuric acid. Hydrochloric acid is very slow.

REF: Anthony, T R et al — *J Appl Phys,* 39,1391(1968)

TL-0003a
ETCH NAME: Sulfuric acid TIME:
TYPE: Acid, removal TEMP:
COMPOSITION:
x .... H_2SO_4, conc.
DISCUSSION:

Tl, specimens. Shown as an etch for thallium.

REF: Hodgman, C D — *Handbook of Chemistry and Physics,* 27th ed, Chemical Rubber Co., Cleveland, OH, 1943, 474

TL-0003b
ETCH NAME: Hydrochloric acid TIME:
TYPE: Acid, removal TEMP:
COMPOSITION:
x .... HCl
DISCUSSION:

Ti, specimens. Shown as a slow etch on thallium.

REF: Ibid.

PHYSICAL PROPERTIES OF THALLIUM SELENIDE, Tl_2Se

Classification	Selenide
Atomic numbers	81 & 34
Atomic weight	487.74
Melting point (°C)	340
Boiling point (°C)	
Density (g/cm^3)	
Hardness (Mohs — scratch)	4—5

Crystal structure (isometric — normal?)	(100) cube
Color (solid)	Grey
Cleavage (cubic)	(001)

THALLIUM SELENIDE, Tl_2Se

General: Does not occur as a natural compound. There are three listed thallium minerals, all as mixtures of arsenic and sulfur with lead or antimony. There has been no industrial use as the selenide.

Technical Application: The compound as Tl_2Se_3 has been grown in Solid State development in a study of possible new semiconducting compounds, but is not presently used for fabrication as a device.

Etching: Halogens.

THALLIUM SELENIDE ETCHANTS

TLSE-0001

ETCH NAME: BRM — TIME:

TYPE: Halogen, polish — TEMP: RT

COMPOSITION:

x .... x% Br_2

x .... MeOH

DISCUSSION:

Tl_2Se_3 and TlSe single crystals were grown in a study of new semiconducting compounds. Other compounds were Ag_2Se, Li_3Bi, SnSe, $SnSe_2$, In_2Tl_3, $AgInTe_2$, In_2Te_3, and Bi_2Te_3.

REF: Mooser, E & Pearson, W B — *Phys Rev,* 101,492(1956)

THALLIUM BISMUTH TELLURIDE ETCHANTS

TLBT-00011

ETCH NAME: Aqua regia — TIME:

TYPE: Acid, removal/polish — TEMP:

COMPOSITION:

3 .... HCl

1 .... HNO_3

DISCUSSION:

$TlBiTe_2$, grown as a single crystal n-type, degenerate semiconductor compound in a study of superconductivity. Also as TlBeTe and TlSnTe. Two general formulas shown: $T^{I}B^{II}C^{III}$ and $A^{III}B^{V}C_2^{V}$. Solution shown as a general etchant for tellurium and thallium compounds (TE-0022).

REF: Rogers, T S et al — *J Cryst Growth,* 61,400(1983).

PHYSICAL PROPERTIES OF THORIUM, Th

Classification	Actinide
Atomic number	90
Atomic weight	232.1
Melting point (°C)	1750
Boiling point (°C)	3850
Density (g/cm^3)	11.7
Specific heat (cal/g) 25°C	0.034
Thermal conductance (cal/sec)(cm^2)(°C/cm)	0.093

Heat of fusion (k-cal/g-atom)	4.6
Latent heat of fusion (cal/g)	19.82
Heat of vaporization (k-cal/g-mole)	140
Atomic volume (W/D)	19.9
Electronegativity (Pauling's)	1.3
Covalent radius (angstroms)	1.65
Ionic radius (angstroms)	1.02 (Th^{+4})
Electrical resistivity (micro-em)	14
Electronwork function (eV)	3.51
Coefficient of thermal linear expansion ($\times 10^{-6}$ cm/cm/°C) 20°C	12.5
Poisson ratio	0.27
Young's modulus (psi $\times 10^6$)	10.5
Shear modulus (psi $\times 10^6$)	4
Cross section (barns)	7.4
Half-life ($\times 10^{10}$ years)	7.4
1st ionization potential (eV)	5.7
Hardness (Mohs — scratch)	4—5
Crystal structure (isometric — normal)	(100) cube, fcc
Color (solid)	Grey
Cleavage (cubic)	(001)

THORIUM, Th

General: Does not occur in nature as a metal element. The chief ore of thorium, as thorium oxide, ThO_2, is from the mineral monazite — a phosphate of cerium metals but with sufficient oxygen and silicon to form both thorium oxide and thorium silicate — and the mineral thorite, $ThSiO_4$ (like zircon in form and structure). The mineral thorotungstate is a mixture of tungsten and thorium oxides but is of rare occurrence and the formula has not been established.

Industrial use includes fabrication of incandescent gas mantels using thorium oxide.

Technical Application: No application in semiconductor processing, though thorium has been studied as a single crystal and polycrystalline element.

Etching: HCl, H_2SO_4 and aqua regia; slowly in HNO_3.

THORIUM ETCHANTS

TH-0001
ETCH NAME: TIME:
TYPE: Acid, cleaning TEMP:
COMPOSITION:
x .... HNO_3
x .... H_2SiF_6
DISCUSSION:
Th specimens. Solution used to etch clean surfaces. Produces a high metallic luster surface that is retained indefinitely even in air.
REF: Wallace, D C — *Phys Rev*, 120,84(1960)

TH-0002
ETCH NAME: Aqua regia TIME:
TYPE: Acid, polish TEMP:
COMPOSITION:
3 .... HCl
1 .... HNO_3

DISCUSSION:

Th specimens. Referred to as a slow removal and polishing etch for thorium.

REF:

TH-0003

ETCH NAME:
TYPE: Electrolytic, polish/removal
COMPOSITION:
x .... 9 N H_2SO_4
x .... 11 N H_3PO_4

TIME:
TEMP: 130°F
ANODE: Th
CATHODE:
POWER: 150 A/ft^2
1500 A/ft^2

DISCUSSION:

Th specimens prepared for electroplating. High power used for removal; low power used for polishing. A thorium phosphate film on thorium surfaces will reduce the effect of oxidation and corrosion.

REF: Beach, J G & Schaer, G R — *J Electrochem Soc,* 106,392(1959)

TH-0004

ETCH NAME: Nitric acid
TYPE: Acid, removal
COMPOSITION:
x .... HNO_3, conc.

TIME:
TEMP:

DISCUSSION:

Th specimens. Solution shown as a general removal etch.

REF: Teasdale, T S — *Phys Rev,* 99,1248(1955)

PHYSICAL PROPERTIES OF THORIUM DIOXIDE, ThO_2

Classification	Oxide
Atomic numbers	90 & 8
Atomic weight	264
Melting point (°C)	1845
Boiling point (°C)	3220
Density (g/cm^3)	9.5—9.9
Application limit (°C)	2700
Specific heat (°C) 25—1000°C	290
Coefficient of linear thermal expansion ($\times 10^{-6}$/cm/cm/°C) 25—800°C	9.5
Thermal conductivity (W/m °C) 1000°C	3
Electrical resistivity (ohm/cm $\times 10^4$) 500°C	2.6
(ohm/cm $\times 10^4$) 1200°C	1.5
Hardness (Mohs — scratch)	7
Index of refraction (n=)	2.20
Crystal structure (isometric — normal)	(100) cube
Color (solid)	White
Cleavage (cubic)	(001)

THORIUM DIOXIDE, ThO_2 (Thoria)

General: Occurs in nature as a native oxide, not as a singular mineral, but mixed with other rare earth elements such as cerium. The major source is the mineral monazite, a complex phosphate with sufficient oxygen and silicon to form both thorium oxide and thorium silicate. It is the most widely used compound of thorium as it produces a brilliant white light in the presence of oxygen without burning of itself. The other major ore is thorite, $ThSiO_4$.

Prior to 1900 the Welsbach gas mantel (99% ThO_2) was a major source of illumination before the advent of the light bulb. Camp lanterns using white gas under pressure as the ignition source are still manufactured. It is considered a high refractory oxide and, with other similar metallic oxides, is used as a ceramic material in the fabrication of high temperature ceramics for furnace bricks and similar applications.

Technical Application: No major use in the electrical or electronic industries other than as a mixed ceramic compound. It has been grown as a single crystal for physical data and morphology studies.

Etching: Soluble in hot H_2SO_4 and fluorine compounds/acids.

THORIUM DIOXIDE ETCHANTS

THO-0001

ETCH NAME: Sulfuric acid — TIME:

TYPE: Acid, removal — TEMP: Hot

COMPOSITION:

x .... H_2SO_4, conc.

DISCUSSION:

ThO_2, specimens. A general, slow etch for thoria.

REF: Hodgman, C D — *Handbook of Chemistry and Physics,* 27th ed, Chemical Rubber Co., Cleveland, OH, 1943, 477

THO-0002

ETCH NAME: Ammonium fluoride — TIME: 5 min

TYPE: Acid, preferential — TEMP: Boiling

COMPOSITION:

x .... x% NH_4F

DISCUSSION:

ThO_2, (111) wafers used in a study of slip and fracture. An Instron Hardness Tester with a heated tip was used to indent surfaces to produce fracture. The solution shown developed triangle etch pits on (111) surfaces.

REF: Edington, J W & Klein, W — *J Appl Phys,* 27,3906(1966)

PHYSICAL PROPERTIES OF THULIUM, Tm

Classification	Lanthanide
Atomic number	69
Atomic weight	169
Melting point (°C)	1545
Boiling point (°C)	1727
Density (g/cm³)	9.33
Specific heat (cal/g) 25°C	0.038
Heat of fusion (k-cal/g-atom)	4.4
Latent heat of fusion (cal/g)	26.04
Heat of vaporization (k-cal/g-mole)	59
Atomic volume (W/D)	18.1
Electronegativity (Pauling's)	1.2
Covalent radius (angstroms)	1.56

Ionic radius (angstroms)	0.87 (Tm^{+3})
Electrical resistivity (micro-ohms-cm) 298 K	67.6
Compressibility ($cm^2/kg \times 10^{-6}$)	2.47
Cross section (barns)	118 (125)
Magnetic moment (Bohr magneton)	7.62
Magnetic susceptibility ($\times 10^{-6}$ emu/mole)	26,000
1st ionization potential (eV)	5.61
Hardness (Mohs — scratch)	4—5
Crystal structure (hexagonal — normal)	$(10\bar{1}0)$ prism, hcp
Color (solid)	Grey
Cleavage (basal)	(0001)

THULIUM, Tm

General: Does not occur as a native element. It is a rare earth element in the yttrium group of lanthanides, and occurs in yttrium and gadolinium minerals: gadolinite, a silicate of iron and rare earths is the major ore. It has had no use in industry to date, due to scarcity.

Technical Application: The element has not been used in Solid State material processing to date, but single crystals of thulium and other rare earths have been studied for general morphology and defects.

Etching: Heat used to develop defects.

THULIUM ETCHANTS

TM-0001
ETCH NAME: Heat — TIME:
TYPE: Thermal, preferential — TEMP:
COMPOSITION:
x .... heat
DISCUSSION:

Tm, single crystal specimens. Material developed in a growth study. After crystal growth, thermal annealing was used to develop defects, low angle grain boundaries and other structure. Additional elements grown were dysprosium, holmium, gadolinium, erbium, thulium, terbium and yttrium.
REF: Nigh, H E — *J Appl Phys,* 34,3323(1963)

PHYSICAL PROPERTIES OF TIN, Sn

Classification	Metal
Atomic number	50
Atomic weight	118.69
Melting point (°C)	232
Boiling point (°C)	2270
Density (g/cm^3)	7.298
Thermal conductance (cal/sec)(cm^2)(°C/cm)	0.16
Specific heat (cal/g) 20°C	0.053
Latent heat of fusion (cal/g)	
Specific heat (cal/g) 20°C	0.054
Latent heat of vaporization (cal/g)	520
Atomic volume (W/D)	16.3
1st ionization energy (K-cal/g-mole)	169
1st ionization potential (eV)	7.442

Electronegativity (Pauling's)	1.8
Covalent radius (angstroms)	1.41
Ionic radius (angstroms)	0.71 (Sn^{+4})
Linear coefficient of thermal expansion ($\times 10^{-6}$ cm/cm/°C) 20°C (0°C)	23 (19.9)
Electrical resistivity (micro-ohms) 20°C (0°C)	11.5 (11)
Vapor pressure (nm $\times 10^{-6}$) 1000 K	7.4
Magnetic susceptibility (cgs $\times 10^{-6}$) 18°C	0.027
Tensile strength (psi) 15°C	2100
Hardness (Mohs — scratch)	2—3
(Brinell — mm/kg)	3.9
Crystal structure (tetragonal — normal) alpha	(110) prism, 1st order
(hexagonal — normal) beta	$(10\bar{1}0)$ prism
Color (solid)	Tin-white
Cleavage (basal)	(001)/(0001)

TIN, Sn

General: Though tin does occur as a native element it is extremely rare, limited in quantity, and is often found in rounded grains associated with gold placer deposits. The minerals cassiterite and stannite, both SnO_2, are the major tin ores. In the presence of soda, usually sodium carbonate, Na_2CO_3, on charcoal, using a hot flame (Bunsen burner) the oxide is easily reduced to metallic tin. Tin was one of the seven metals known to ancient man, and the smelting of copper with tin lead to the Bronze Age. The name tin is Anglo-Saxon (German/Norse origin), and the symbol, Sn, comes from the Latin *stannum*.

Needless to say, it has been a major industrial metal since the beginning of man's metal culture. Today it is said that the U.S. alone uses enough tin every year to form a sheet, 10 ft wide, around the earth's equator. Its primary use is for "tin cans", actually a tin coating on steel, as tin does not tarnish when exposed to moist air and is not affected by organic acids in foods. Tin plating is done by dipping the part in a bath of molten tin using thin sheets of low-carbon steel. The iron is protected against corrosion as long as the plating remains intact. Tin can be recovered by heating scrap plated parts, as the tin m.p. is only 232°C, or by chlorinating to $SnCl_4$ and/or sodium hydroxide, NaOH to form Na_2SnO_2. The chloride, $SnO_2.2H_2O$, is used in electrolytic tin plating.

There are three crystal forms of tin: (1) grey tin, below about 18°C; (2) beta-tin, stable between 18—170°C and (3) delta-tin, stable above 161°C. The grey- or alpha-tin is known as "tin pest" as it can cause tin parts to crumble into powder even in cold weather. The beta- and delta-tin forms are the stable white-tins of industry.

Though there are a number of tin-lead solders, solder #62, m.p. 185°C, and #63, m.p. 220°C (with 2% silver) are two of the best known and generally used. They are commercially available in solid bars and wire form, but are widely used as rosin-core soldering wire . . . the rosin acting as a wetting flux to form a smooth, pure tin solder joint. And many parts, such as device lead wires, are pre-tinned by dipping in a molten tin bath before being soldered into a circuit board which, in turn, has been pre-soldered — heat is applied, the solder/tin flows — and forms the solder joint. There are dozens of soldering irons with replaceable tips of various sizes and geometries available for the heat source as hand-held irons or in automatic equipment. With PCB construction — plasticized linen boards with laminated copper — holes for discrete parts are drilled, the copper lacquered for pattern etching of the required electrical/electronic circuit to include mounting pads for individual devices if required. Parts can be individually inserted and soldered or the complete assembly passed through a belt furnace to solder all parts in a single heat cycle. And devices that fail are easily de-soldered, removed, and replaced. High reliability PCBs commonly have the copper circuits overplated with gold to reduce copper corrosion. Such "boards" are common in much of the current electrical and electronic equipment, such as automatic profiled cutting machines in the metals industries, computers, etc. . . . anything requiring an electronic-type circuit for operation, and used as an inserted board. If a part fails, the board is pulled and replaced.

In ancient times, bronze coins were used, but with the fall of Carthage cutting off the Mediterranean supply of tin, tin was replaced with lead. The study of ancient artifacts uses this tin-lead content for dating purposes. Today, of course, most coinage is silver/copper/nickel or gold and gold alloys, though stone, copper, iron, and aluminum all have been used in the past.

When a bar of crystalline tin is bent, the crystallites rub against each other producing the ''tin cry''. It is the only metal known that produces such a recognizable sound to the human ear.

Technical Application: In the Solid State semiconductor industry tin is used as a dopant species in some compound semiconductors and is grown as both binary and trinary semiconductor materials: lead tin telluride, PbSnTe, lead tin selinide, PbSnSe, and tin telluride, SnTe as examples. It is also used in fabricating the Josephson Junction device, Nb_3Sn, and is under study and development as an oxide, nitride and silicide. Its use as lead-tin solders, particularly in PCB construction, has already been mentioned and is representative of the wide use of tin in the electrical and electronic industries.

Needless to say, it has been widely studied as tin plating, both molten baths and electrolytic films, for literally hundreds of years. Both natural single crystals and artificially grown boules have been studied for physical and morphological data. As cut bars and spheres it is under study for its superconducting properties.

Etching: HNO_3, H_2SO_4, alkalies, aqua regia, and mixed acids. Deliquesces in HCl.

TIN ETCHANTS

SN-0001a
ETCH NAME: TIME: (A) 10 min
TYPE: Acid, removal (B) 30 sec
COMPOSITION: TEMP: RT RT

(A) x .... 5% HCl
x .... isopropyl alcohol

(B) x .... 10% fluoboric acid
x .... isopropyl alcohol

DISCUSSION:

Sn, electroplated thin film. Solutions used in studying vapor phase fusion of pure electroplated tin.

REF: Bander, T L — *Solid State Technol,* March 1983, 141

SN-0002a
ETCH NAME: TIME:
TYPE: Acid, defect TEMP:
COMPOSITION:

100 ml .... HCl
100 ml .... HN_4NO_3
500 ml .... H_2O
x .... 5×10^{-5} *M* $CuSO_4 \times 5H_2O$

DISCUSSION:

Sn, (010) wafers used in an etch study. Solution will produce hillocks as dislocations. Dislocations move on (010) in both ⟨100⟩ and ⟨110⟩ directions.

REF: Ojima, K et al — *Jpn J Appl Phys,* 22,46(1984)

SN-0001b
ETCH NAME: TIME: (A) 10 min
TYPE: Acid, removal (B) 30—40 sec
COMPOSITION: TEMP: RT RT
(A) x .... 5% HCl (B) x .... 10% Na_2S
100 ml .... H_2O 10 ml .... H_2O
DISCUSSION:

Sn, electroplated thin film. Solutions used in studying vapor phase fusion of pure electroplated tin.

REF: Ibid.

SN-0002b
ETCH NAME: TIME:
TYPE: Acid polish TEMP:
COMPOSITION:
1 .... HNO_3
1 .... HAc
4 .... glycerin
DISCUSSION:

Sn, white-tin single crystal ingots and wafers. Solution used to polish surfaces before defect etching. (**CAUTION** — Glycerin and nitric acid is a precursor mix for nitroglycerin.)

REF: Ibid.

SN-0003
ETCH NAME: Hydrochloric acid TIME:
TYPE: Acid, cleaning TEMP:
COMPOSITION:
x .... HCl
DISCUSSION:

Sn, shot. Clean with hydrochloric acid and vacuum melt to form high purity tin ingots. Material used as an epitaxy growth source for InSn/InP thin films.

REF: Chin, B H — *J Electrochem Soc,* 131,1372(1984)

SN-0012: Rutherford, R J — *Proc Am Soc Test Mater,* 24,739(1924)

Sn specimens. Solution used as a general removal and cleaning etch. Also used on Pb and Bi.

SN-0004
ETCH NAME: Nitric acid, dilute TIME:
TYPE: Electrolytic, saw TEMP:
COMPOSITION: ANODE: Sn
1 .... HNO_3 CATHODE:
1 .... H_2O POWER: 30 mA
DISCUSSION:

Sn, single crystal ingots cut into wafers. Use a plastic Saran thread as a cutting tool. Pass thread through the solution, then across the specimen. Also used to cut zinc ingots.

REF: Sternheim, G — *Rev Sci Instr,* 26,1206(1955)

SN-0005a
ETCH NAME: TIME:
TYPE: Acid, preferential TEMP:
COMPOSITION:
x .... HCl

DISCUSSION:

Sn, single crystal sphere etched to finite crystal form.

REF: Yamamoto, M & Watanabe, J — *Met Abstr,* 8,493(1958)

SN-0005b

ETCH NAME: Aqua regia
TIME:

TYPE: Acid, preferential
TEMP:

COMPOSITION:

3 .... HCl

1 .... HNO_3

DISCUSSION:

Sn, (100) single crystal tetragonal (alpha) form used in a general study of etching.

REF: Ibid.

SN-0006a

ETCH NAME:
TIME:

TYPE: Electrolytic, polish
TEMP: 15—20°C

COMPOSITION:
ANODE: Sn

20 .... $HClO_4$
CATHODE: SS (2)

10 .... butyl cellusolve
POWER: 12 V

70 .... EOH

DISCUSSION:

Sn, (001) and (111) — tetragonal alpha-tin single crystals — prepared for TEM study. Solution used to polish and etch thin specimens. The specimen edges were insulated with microstop, and one side was lacquered against etching. Another polish etch used the same constituents in a ratio of 5:10:85 at 10°C and 30 V. Solutions should be stirred slowly during the etch period. Wash in DI water, then EOH. Remove lacquer with acetone.

REF: Fourie, J T et al — *J Appl Phys,* 31,1136(1960)

SN-0006b

ETCH NAME:
TIME:

TYPE: Electrolytic, polish
TEMP: 10°C

COMPOSITION:
ANODE: Sn

5 .... $HClO_4$
CATHODE:

10 .... butyl cellusolve
POWER: 30 V

85 .... EOH

DISCUSSION:

Sn, (001) and (111) wafers. Solution used as a final polish etch after an initial rough reduction etch. See SN-0006a.

REF: Ibid.

SN-0007

ETCH NAME:
TIME:

TYPE: Electrolytic, polish
TEMP:

COMPOSITION:
ANODE:

20 .... $HClO_4$
CATHODE:

70 .... HAc
POWER:

DISCUSSION:

Sn, single crystal wires used in a study of paramagnetic effects in superconductors — resistance transition in tin. Solution used to polish specimens. Rinse in DI water, then alcohol.

REF: Meissner, H — *Phys Rev,* 109,668(1958)

SN-0008
ETCH NAME: Neon
TYPE: Ionized gas, preferential
COMPOSITION:
x .... Ne^+ ions

TIME:
TEMP:
GAS FLOW:
PRESSURE:
POWER:

DISCUSSION:

Sn, single crystal specimens. Neon ion bombardment used to develop structure and orientation figures. Other metals studied were Al, Bi, Cd, Co, Cu, Mg, and Zn.

REF: Yurasova, V E — *Kristallografiya,* 2,770(1957)

SN-0009
ETCH NAME:
TYPE:
COMPOSITION:

TIME:
TEMP:
ANODE:
CATHODE:
POWER:

DISCUSSION:

Sn, single crystal sphere machined on a lathe, to 1″ +/− .0004 diameter. The amorphous layer left from cutting removed by electropolish. Sphere used in a study of magnetic moments in a superconducting sphere. Etch not shown.

REF: Teasdale, T S — *Phys Rev,* 99,1248(1955)

SN-0010: Jacquet, P A — *Int Tin R&D Council Bull,* 90,(1949)

Reference for electropolish etch on tin sphere in SN-0009.

SN-0011: Steinberg, F & Teghtsoonian, E — *Acta Metall,* 5,455(1957)

Tin, specimens used in a study of grain boundaries. Specimens flame cut prior to lapping, polishing and preferential etching.

SN-0014
ETCH NAME: Nitric acid
TYPE: Acid, cutting
COMPOSITION:

TIME: 30 min
TEMP: RT

(1) 1 .... HNO_3
1 .... H_2O

(2) 200 ml .... HCl
50 g .... $FeCl_3$
250 ml .... H_2O

DISCUSSION:

Sn, specimen, ¹/₄″ wafer in diameter. Terylene thread soaked in solution (1) used to cut both tin and zinc. Solution (2) used for aluminum, brass and copper. (*Note:* Excellent review of metal growth methods.)

REF: Honeycombe, R W K — *Met Rev,* 4,1(1959)

SN-0015
ETCH NAME:
TYPE: Electrolytic, polish
COMPOSITION:
4 oz .... $NaNO_3$
x .... $Na_3PO_4.H_2O$
x .... H_2O (1/liter?)

TIME: 1—3 min
TEMP: 200—212°F
ANODE: Steel
CATHODE: Sn
POWER: 20—15 A/ft² & 2—6 V

DISCUSSION:

Sn specimens. Solution used to electropolish. Dip in HCl and NaCl to final clean after etching.

REF: ASTE Committee — *Tool Engineer's Handbook,* McGraw-Hill, New York, 1949

PHYSICAL PROPERTIES OF TIN DIOXIDE, SnO_2 (Cassiderite)

Classification		Oxide
Atomic numbers		50 & 16
Atomic weight		150.7
Melting point (°C)		1127, decomposes
Boiling point (°C)		
Density (g/cm^3)		6.95
Refractive index (n=)		1.997—2.093
Hardness (Mohs — scratch)		6.7
Crystal structure	(tetragonal — normal) alpha	(110) prism
	(monoclinic — normal) beta	(100) pinacoid
Cleavage (basal)		(001)

TIN DIOXIDE, SnO_2

General: Occurs in nature as the minerals cassiterite and stannite, both SnO_2, and the two primary ores of tin. England was called the Cassiderides by the Romans (the "Tin" Isles) for the tin mines in Cornwall that the ancient world used in the making of bronze, Cu:Sn (30%) — the Bronze Age — dating back to around 6000 B.C., even before the rise of the Roman Empire, circa 550 B.C. It is closely associated with fluorine and boron minerals which indicates a pneumatolytic origin and is found in granite, quartz or pegmatitic rocks as veins and stringers and, in the Malay Penninsula and several South Pacific islands, in sand and stream gravels — a major source of tin, and one economic reason for the Japanese invasion of the area during World War II — fortunately for the Allies, major tin mines, similar to the worked-out mines in Cornwall, were discovered in Bolivia about the same time. The Bolivian deposits represent about 80% of the world tin reserves at the present time.

There is little direct use in industry as a working compound. The hydrated oxide, $SnO_2.xH_2O$, is called alpha-stannic acid or simply stannic acid (H_2SnO_3). See the general discussion under tin for major use of the element.

Technical Application: No major use in the electronics or electrical industries, at present. It is under evaluation and study as a surface protective thin film coating similar to that of silicon dioxide as it is quite inert to most acids. It has been grown and studied as a single crystal for physical and morphological data.

Etching: Very difficult to etch as it is inert in acids and deliquesces in alkalies. Some hot mixed acids can be used. Reacts with chromium and zinc in solutions.

SNO-0001a
ETCH NAME: TIME:
TYPE: Acid, removal TEMP: 90°C
COMPOSITION:

1 g .... Cr (add as needed)
20 ml .... HBr
20 ml .... H_2O

DISCUSSION:

SnO_2, thin films grown on silicon substrates coated with SiO_2:SnO_2/SiO_2/Si. This solution is better for removal repeatability than cathodic etching.

REF: Hsu, Y S & Ghandhi, S K — *J Electrochem Soc,* 127,1592(1980)

SNO-0001b
ETCH NAME: Hydrochloric acid, dilute
TYPE: Electrolytic, removal
COMPOSITION:
1 .... HCl
5 .... H_2O
TIME:
TEMP:
ANODE: Pt
CATHODE: Sn
POWER:
DISCUSSION:

SnO_2, thin films deposited on SiO_2 coated silicon wafer substrates as: $SnO_2/SiO_2/Si$. Electrolytic removal is not as controllable and effective as the solution shown in SNO-0001a.

REF: Ibid.

SNO-0002
ETCH NAME: Hydrogen bromide, dilute
TYPE: Halogen, removal
COMPOSITION:
(1) 1 .... HBr
1 .... H_2O
(2) 1 .... HBr
1 .... H_2O
x .... Cr, pcs
TIME:
TEMP: 110°C
DISCUSSION:

SnO_2, thin films deposited on 1 mm glass slides or SiO_2 coated silicon wafers. After depositing SnO_2, overcoat with 1500 Å SiO_2. Use photo resist techniques to pattern the cap SiO_2 as a mask, removing SiO_2 with HF, then remove SnO_2 with solutions shown.

REF: Hau, Y-S & Ghandhi, S K — *J Appl Phys*, 126,1434(1979)

SNO-0003
ETCH NAME:
TYPE:
COMPOSITION:
TIME:
TEMP:
DISCUSSION:

SnO_2, thin films deposited by CVD in a study of material physical parameters. Refractive index of deposits is n = 1.860—1.953. No etch shown.

REF: Melsheimer, J & Ziegler, D — *Thin Solid Films*, 109,71(1983)

SNO-0004
ETCH NAME:
TYPE:
COMPOSITION:
TIME:
TEMP:
DISCUSSION:

SnO_2, (110) single crystal wafers as-grown. Mechanically polish down to 1000 Å thick. Material used in a study of the ultraviolet absorption edge of the material. (*Note:* At the time of this work the authors said there was no known etch for SnO_2.)

REF: Summitt, R et al — *J Phys Chem Solids*, 25,1465(1965)

SNO-0005
ETCH NAME:
TYPE: Acid, removal
COMPOSITION:
3000 ml .... 6 *N* HCl
20 g .... Cr, pcs
x .... Zn, pcs
TIME: 1—4 min
TEMP: 90°C

DISCUSSION:

SnO_2 specimens. Shown as an etch for tin oxide. Add zinc during etching.

REF: Simon, P W — U.S. Patent 4,009,061 1977 (Burroughs Corp)

PHYSICAL PROPERTIES OF TIN SELENIDE, SnSe

Classification	Selenide
Atomic numbers	50 & 34
Atomic weight	197.66
Melting point (°C)	861
Boiling point (°C)	
Density (g/cm³)	8.18
Hardness (Mohs — scratch)	6—7
Crystal structure (hexagonal — normal)	$(10\bar{1}0)$ prism
Color (solid)	Grey/black
Cleavage (basal)	(0001)

TIN SELENIDE, SnSe

General: Does not occur as a natural compound. There are less than a dozen tin minerals, including native tin in small quantities, as sulfides and oxides. The most important ore is cassiderite, SnO_2. Metallic tin has been an important metal element in industry since ancient times, but the selenide has had no applications to date.

Technical Application: Both SnSe and $SnSe_2$ have been grown as single crystals in a study of new semiconducting compounds, but there have been no devices fabricated in Solid State processing other than for evaluation of properties and characteristics.

Etching: HCl, HNO_3, alkalies, and aqua regia. Mixed acids and halogens.

TIN SELENIDE ETCHANTS

SNSE-0001

ETCH NAME: TIME:

TYPE: Halogen, polish TEMP: RT

COMPOSITION:

x .... x% Br_2

x .... MeOH

DISCUSSION:

SnSe and $SnSe_2$ single crystal ingots grown in a study of new semiconducting compounds. Other compounds studied were Ag_2Se, Li_3Bi, Tl_2Se_3, In_2Tl_3, $AgInTe_2$, In_2Te_3, and Bi_2Se_3.

REF: Mooser, E & Pearson, W B — *Phys Rev*, 101,492(1956)

PHYSICAL PROPERTIES OF TIN TELLURIDE, SnTe

Classification	Telluride
Atomic numbers	50 & 52
Atomic weight	246.31
Melting point (°C)	780
Boiling point (°C)	
Density (g/cm³)	6.47
Hardness (Mohs — scratch)	6—7
Crystal structure (hexagonal — normal)	$(10\bar{1}0)$ prism

Color (solid) Grey/black
Cleavage (basal) (0001)

TIN TELLURIDE, SnTe

General: Does not occur as a natural compound. There are only some four tin tellurite or tellurite minerals as hydrates, and the dioxide tellurite, TeO_2. Metallic tin has had wide use in industry since ancient times, but there has been no use, to date, of the telluride.

Technical Application: The telluride has been grown as a single crystal in general studies of Solid State materials. It also has been deposited as a thin film for general morphology study of the compound.

Etching: HCl, HNO_3, and mixed acids.

TIN TELLURIDE ETCHANTS

SNTE-0001
ETCH NAME: TIME: Minutes
TYPE: Acid, polish TEMP: RT
COMPOSITION:
1 .... HNO_3
11 .... HCl
20 .... $K_2Cr_2O_7$, sat. sol.
DISCUSSION:

SnTe, (100) wafers used in developing polish etching for this material. Stir the solution during the etch period. If a stain remains on the surface, dip etch in 20% NaOH and rinse in DI water. Surfaces show ripples and some pitting.

REF: Lorenz, M R — *J Electrochem Soc,* 112,240(1965)

SNTE-0002
ETCH NAME: TIME:
TYPE: Electrolytic, polish TEMP:
COMPOSITION: ANODE: SnTe
100 cm^3 .... H_3PO_4 CATHODE:
1 g .... NaOH POWER: 6 V
2 g .... agar-agar
DISCUSSION:

SnTe, (111) ingot cut axially on the (111) and studied in the vicinity of the metallic compound SnTe. Both Sn and Te turn brown in color in this polish solution and SnTe compound shows no color. Also called stannous telluride.

REF: Umoda, J et al — *Jpn J Appl Phys,* 1,277(1962)

SNTE-0003
ETCH NAME: Nitric acid dilute TIME:
TYPE: Acid, float-Off TEMP: RT
COMPOSITION:
x .... HNO_3
x .... H_2O
DISCUSSION:

SnTe, thin films grown on muscovite mica (0001) substrates. After growth films were removed by the float-off technique for TEM structure study in the solution shown. See Mica, MI-0001.

REF: Davenere, A et al — *J Cryst Growth,* 70,452(1984)

SNTE-0004

ETCH NAME: TIME:

TYPE: Acid, removal TEMP:

COMPOSITION:

x .... HF

x .... HNO_3

x .... HAc

DISCUSSION:

SnTe, amorphous thin films evaporated on glass substrates at near LN_2 temperature in a T_c study. Irreversible resistivity transformation is at about 10 K. T_c values decrease with film thickness. Material source for thin film was 80 mesh powder from SnTe Bridgman grown ingots. Also deposited thin films of GeTe. (*Note:* Solution mixtures shown are general removal etchants.)

REF: Fuki, K et al — *Jpn J Appl Phys,* 23,1141(1984)

PHYSICAL PROPERTIES OF TITANIUM, Ti

Classification	Transition metal
Atomic number	22
Atomic weight	47.90
Melting point (°C)	1668
Boiling point (°C)	3260
Density (g/cm^3)	4.5
Thermal conductance (cal/sec)(cm^2)(°C/cm)	9.41
Specific heat (cal/g) 25°C	0.1386
Heat of fusion (k-cal/g-atom)	3.7
Latent heat of fusion (cal/g)	104.5
Heat of vaporization (k-cal/g-atom)	106.5
Atomic volume (W/D)	10.6
1st ionization energy (K-cal/g-mole)	158
1st ionization potential (eV)	6.83
Electronegativity (Pauling's)	1.5
Covalent radius (angstroms)	1.32
Ionic radius (angstroms)	0.94 (Ti^{+2})
Coefficient of linear thermal expansion ($\times 10^{-6}$ cm/cm/°C)	8.5
Electrical resistivity (micro-ohm-cm)	47.8
Modulus of elasticity (psi $\times 10^6$)	14.7 (15.5)
Magnetic susceptibility (emu/g)	3.17
Poisson ratio	0.41
Cross section (barns)	6.1
Tensile strength (psi $\times 10^3$)	34—95
Elongation (% inch/inch 2 inches)	24—54
Hardness (Mohs — scratch)	2
(Vicker's — mm/kg)	80—100
Crystal structure (hexagonal — normal)	(10$\bar{1}$0) prism
Color (solid)	Grey-silver
Cleavage (basal)	(0001)

TITANIUM, Ti

General: Does not occur in nature as a native element. The most prominent ores are rutile, TiO_2, titanite, $CaTiSiO_5$ and ilmenite, $FeTiO_3$, though there are over 40 single crystal minerals containing titanium. It is widely distributed in igneous rocks, mostly as the oxide. Ordinary rutile is commonly brownish-red in various shades, but not black. There are several ferriferrous varieties containing up to 30% ferrous titanate, all black in color. The minerals brookite and octahedrite, both also TiO_2, differ in their melting temperatures, and all three oxide crystals have been grown artificially. The metal is part of the tin group and is the most inactive metal of the group, so much so, that the chloride hydrolizes in air. The metal will burn in air and is the only known metal that will burn in nitrogen.

Industrially, titanium is considered one of the light metals and, being physically inert to most acids at room temperature, is finding increasing application in the aircraft and space vehicle industries. It is a major alloy element in steels to increase strength. The chloride, $TiCl_4$, with water and moisture in the air is sufficient to cause fog and will hydrolyze as TiO_2 as white smoke which is used in sky-writing and for smoke screens in the military.

Technical Application: The metal is used as an evaporated thin film in contact metallization structures on many metals as the initial evaporant as it has excellent ''tie-down'' properties due to its affinity for oxygen: Ti:Au, Ti:TiAg:Ag as examples. As TiW (10% Ti), known as Ti-tungsten, it is used as a contact metallization (Au:TiW) and a buffer layer in heterostructure semiconductor devices. Because of its high affinity for oxygen, it is the primary ''gettering'' metal used as rods in ion pump vacuum systems to remove air and other gases to achieve high vacuum and ultra-high vacuum levels.

It is used to deposit surface thin films of TiO_2 and TiN as surface protectants similar to silicon dioxide, and under development as a silicide: TiSi, Ti_2Si, and $TiSi_2$ for use as a buffer layer in semiconductor heterostructures.

As an electrode, both pure metal and alloys, it is used as such in electrolytic etching and plating solutions. The bulk metal is used in device package construction of semiconductor and similar devices and, as wire mesh, is sometimes used to hold parts in cleaning and etching solutions. Loops of wire or slugs are the evaporation sources.

As pure metal sheets and alloys it is still under study, even more so as thin film Ti, TiO_2, TiN and the silicides. Titanium has been grown as a single crystal for physical and morphological data, and oxidation is used for structural study.

Etching: Soluble in most hot, dilute acids and alkalies, but not in HNO_3 or aqua regia (RT). Several mixed acids.

TITANIUM ETCHANTS

TI-0001a

ETCH NAME: Hydrochloric acid, dilute — TIME:

TYPE: Acid, cleaning — TEMP: Hot

COMPOSITION: — RATE: 50 mil/year (mpy)

x .... 20% HCl

DISCUSSION:

Ti, sheet and alloy specimens. Solution can be used to clean surfaces. Hot solutions of greater than 20% concentration will etch titanium.

REF: Bulletin ADV 215-10M-11-59, Crucible Steel Company of America, Pittsburgh, PA

TI-0004b: Ibid.

Ti specimens. Dilute solutions used at RT show very slow attack.

TI-0001b
ETCH NAME: Trichloroacetic acid
TIME:
TYPE: Acid, removal
TEMP:
COMPOSITION:
x .... $CCl_3COOC_2H_5$
DISCUSSION:
Ti, sheet and alloy specimens. Titanium is generally resistant to mono- and dichloroacetic acids, but is attacked rapidly with trichloroacetic acid.
REF: Ibid.

TI-0001c
ETCH NAME: Formic acid
TIME:
TYPE: Acid, removal
TEMP: Boiling
COMPOSITION:
x .... 25% HCOOH
DISCUSSION:
Ti, sheet and alloy specimens. Boiling solutions above 25% concentration will etch titanium.
REF: Ibid.

TI-0001d
ETCH NAME: Phosphoric acid
TIME:
TYPE: Acid, removal
TEMP: Hot
COMPOSITION:
x .... 20% H_3PO_4
DISCUSSION:
Ti, sheet and alloy specimens. Hot phosphoric acid with greater than 30% concentration will attack titanium.
REF: Ibid.
TI-0004: Kohl, W H — *Handbook of Materials and Techniques for Vacuum Devices,* Reinhold, New York, 1967
Ti specimens etched at room temperatures with concentrated (85%) solution.

TI-0001e
ETCH NAME: Hydrofluoric acid
TIME:
TYPE: Acid, removal
TEMP:
COMPOSITION:
x .... HF
DISCUSSION:
Ti, sheet and alloy specimens. Titanium is rapidly attacked by hydrofluoric acid and other fluorine compounds.
REF: Ibid.
TI-0004e: Ibid.
Ti specimens. A 1% HF solution, RT, will attack titanium rapidly.

TI-0001f
ETCH NAME: Sulfuric acid
TIME:
TYPE: Acid, removal
TEMP: Hot
COMPOSITION:
RATE: 500 mil/year (mpy)
x .... 20% H_2SO_4

DISCUSSION:

Ti, sheet and alloy specimens. Hot sulfuric acid with greater than 20% concentration will slowly attack and clean titanium.

REF: Ibid.

TI-0004d: Ibid.

Ti specimens. Used as a concentrated solution at 145°C or as a 10% solution at 158°C. Both solutions will clean and etch titanium.

TI-0002

ETCH NAME: Hydrofluoric acid, dilute | TIME:

TYPE: Acid, cleaning | TEMP:

COMPOSITION:

1 .... HF

1 .... H_2O

DISCUSSION:

Ti, rod (electrode). First, degrease in TCE, then MeOH and then acetone. Use solution shown to etch/clean the electrode prior to electrodeposition of CdSe.

REF: Tomkiewicz, N — *J Electrochem Soc,* 192,2016(1982)

TI-0003

ETCH NAME: | TIME:

TYPE: Acid, removal | TEMP:

COMPOSITION:

x .... HF

x .... HCl

x .... H_2O

DISCUSSION:

Ti, sheet specimens used as substrates for deposition of thin film $CdSe_{1-x}Te_x$. Solution etch/cleans and slightly roughens surfaces for better film adhesion.

REF: Russak, M A & Creter, C J — *J Electrochem Soc,* 131,556(1984)

TI-0004c

ETCH NAME: Hydrochloric acid | TIME:

TYPE: Acid, removal | TEMP: RT & hot

COMPOSITION:

x .... HCl, conc.

DISCUSSION:

Ti specimens. Concentrated hydrochloric acid will attack titanium rapidly at both room temperature or when used hot.

REF: Ibid.

TI-0004f

ETCH NAME: Potassium hydroxide | TIME:

TYPE: Alkali, removal | TEMP: Hot & RT

COMPOSITION:

x .... % KOH

DISCUSSION:

Ti specimens. Concentrated potassium and sodium hydroxide solutions will attack titanium.

REF: Ibid.

TI-0005
ETCH NAME: TIME:
TYPE: Acid, removal TEMP:
COMPOSITION:
1 .... HF
5 .... HNO_3
4 .... H_2O
DISCUSSION:
Ti specimens used in an oxide study. First: lap to a 1 μm finish with diamond paste. Second: degrease with hexane. Use solution shown to etch/clean surfaces to remove native oxide (about 20 Å thick). Oxidation showed the following results:

1. Heat on hot plate in air = gold color TiO_2
2. 600°C in muffle furnace = gold color TiO_2
3. Heat with flame = grey-colored oxide

(*Note:* The gold colored thin films are being used as a hard colored surface coating as a replacement for gold on many products.)

REF: Rolison, D E & Murry, R W — *J Electrochem Soc,* 131,336(1984)

TI-0006
ETCH NAME: TIME:
TYPE: Acid, removal TEMP: 32°C
COMPOSITION: RATE: 18 μm/min
1 .... HF
4 .... HNO_3
5 .... H_2O
DISCUSSION:
Ti thin films. Solution used to remove thin films at time and rate shown. See TIO-0008 for anodic oxidation of specimens.
REF: DeBerry, D W & Viehbeck, A — *J Electrochem Soc,* 130,248(1983)

TI-0007
ETCH NAME: TIME:
TYPE: Acid, removal TEMP:
COMPOSITION:
1 .... HF
1 .... HNO_3
50 .... H_2O
DISCUSSION:
Ti specimens and thin films. Shown as a general etch for removing metallic titanium.
REF: Brochure: *Metals Catalog,* 1035-AMD-CHG/ID/RCG/0977/30M SSS-MRC

TI-0008
ETCH NAME: TIME: 2—3 sec
TYPE: Acid, removal TEMP: RT
COMPOSITION:
10 ml .... HF
20 ml .... HNO_3
70 ml .... H_2O

DISCUSSION:

Ti, thin films sputtered on 2 μ″ and 4 μ″ surface finish alumina substrates as a AuTi (400 Å:100 μ″) metallization in fabricating microwave circuits with a gold cap. Used photo resist techniques to open circuit pattern. Etch remove gold with tri-iodide solution (AU-0007). Use solution shown to remove titanium. Removal is extremely rapid but without undercutting of gold. Line definition down to one micron.

REF: Walker, P — personal application, 1984

TI-0024a: Brochure: *Applications Data for Kodak Photosensitive Resists,* Eastman Kodak Co., Rochester, NY, 1960, 91

Ti thin films. Solution shown as a removal and patterning etchant compatible with Kodak photo resists. Temperature used was 32°C with an 18 Å/minute removal rate.

TI-0004f

ETCH NAME: TIME:

TYPE: Acid, removal TEMP:

COMPOSITION:

1 .... HF

1 .... H_2O_2

20 .... H_2O

DISCUSSION:

Ti specimens. Shown as an etch for titanium.

REF: Ibid.

TI-0010

ETCH NAME: TIME:

TYPE: Electrolytic, polish TEMP:

COMPOSITION: ANODE:

3 .... HF CATHODE:

16 .... H_2SO_4 POWER:

1 .... H_2O

DISCUSSION:

Ti specimens used in a study of oxidation. Electropolish with this solution and DI H_2O rinse before oxidation.

REF: Mizushima, W — *J Electrochem Soc,* 108,b25(1961)

TI-0011a

ETCH NAME: Hydrofluoric acid, dil. TIME:

TYPE: Acid, removal TEMP: 32°C

COMPOSITION: RATE: 12 Å/min

10 ml .... HF

90 ml .... H_2O

DISCUSSION:

Ti specimens and thin films. Used as a general etch for titanium.

REF: Missel, L I & Glang, R — *Handbook of Thin Film Technology,* McGraw-Hill, New York, 1970, 7

TI-0021a: Nichols, D R — *Solid State Technol,* December 1979

Ti thin film evaporation left on bell jars and internal fixtures of vacuum systems. Solution used to clean systems. Use on stainless steel and copper, but not recommended for glass or ceramic.

TI-0011b
ETCH NAME: TIME:
TYPE: Acid, removal TEMP:
COMPOSITION:
10 ml .... HF
20 ml .... HNO_3
70 ml .... H_2O
DISCUSSION:
Ti specimens and thin films. Used as a general etch for titanium.
REF: Ibid.

TI-0001f
ETCH NAME: Oxalic acid TIME:
TYPE: Acid, removal TEMP: Hot
COMPOSITION:
x .... $COOHCOOH.2H_2O$
DISCUSSION:
Ti specimens. Solution will etch titanium in all concentrations.
REF: Ibid.

TI-0012a
ETCH NAME: TIME:
TYPE: Halogen, removal TEMP:
COMPOSITION:
x .... Br_2
x .... ethyl acetate
DISCUSSION:
Ti, thin film deposit was etch removed with the solution shown.
REF: Abu-Yaron, A — *Thin Solid Films,* 112,349(1984)
TI-0022: Douglass, D L & Van Landuyt, J — *Acta Metall,* 14,491(1966)
Ti specimens used in a structure and morphology study of titanium oxidation. Solution was 10% Br_2 at 75°C. Will remove titanium but not titanium oxide. Authors say that using solution above 650°C will initiate growth of TiO_2 at points on surface followed by coalescence as an oxide thin film.

TI-0012b
ETCH NAME: TIME:
TYPE: Halogen, removal TEMP:
COMPOSITION:
x .... I_2
x .... MeOH
DISCUSSION:
Ti thin film deposit was etch removed with the solution shown. (*Note:* Iodine can be replaced with bromine.)
REF: Ibid.

TI-0013
ETCH NAME: TIME:
TYPE: Electrolytic, polish TEMP:
COMPOSITION: ANODE:
1 .... HF CATHODE:
1 .... MeOH POWER:

DISCUSSION:

Ti specimens used for electroplating. Solution shown used to polish surfaces prior to plating. (*Note:* A major article on plating of transition metals, including both etchants and plating solutions.)

REF: Saubestre, E B — *J Electrochem Soc,* 106,305(1959)

TI-0014a: Stanley, C L & Brenner, A — *Proc Am Electroplat Soc,* 43,123(1956)

Reference used by Saubestre for the solution shown in TI-0013.

TI-0023

ETCH NAME: Tin chloride — TIME:

TYPE: Salt, polish — TEMP:

COMPOSITION:

x .... x% $SnCl_2$

DISCUSSION:

Ti specimens. Use as an immersion etch for this material.

REF: Beukman, F & Tucker, W — U.S. Patent 2,801,213, (1957)

TI-0015

ETCH NAME: — TIME:

TYPE: Electrolytic, polish — TEMP:

COMPOSITION: — ANODE:

x .... HF — CATHODE:

x .... HNO_3 — POWER:

x .... H_2SO_4

x .... $FeSO_4$

x .... $Al_2(SO_4)_3$

DISCUSSION:

Ti specimens. During etching period switch: cathode-anode-cathode.

REF: Dalley, J J — *Aviat Week,* 66,79(1957), May 13

TI-0016

ETCH NAME: — TIME:

TYPE: Acid, polish — TEMP:

COMPOSITION:

x .... HF

x .... HBF_4

DISCUSSION:

Ti, specimens. Use solution by immersion as a general polishing etch.

REF: Slomin, G W & Christensen, M P — U.S. Patent 2,798,843 (1957)

TI-0017

ETCH NAME: — TIME:

TYPE: Electrolytic, polish — TEMP:

COMPOSITION: — ANODE: Ti

x .... HF — CATHODE:

x .... ethylene glycol — POWER:

DISCUSSION:

Ti specimens. Solution used as a general anodic etch for titanium.

REF: Colner, W H et al — *J Electrochem Soc,* 110,486(1953)

TI-0018
ETCH NAME: TIME:
TYPE: Electrolytic, polish TEMP:
COMPOSITION: ANODE: Ti
x .... x% $AlCl_3$ CATHODE:
x .... ether POWER:
DISCUSSION:

Ti specimens. Solution used as a general anodic etch for titanium.

REF: Couch, D E & Brenner, A — *J Electrochem Soc,* 99,234(1952)

TI-0019a
ETCH NAME: TIME:
TYPE: Acid, removal/polish TEMP:
COMPOSITION:
x .... HF
x .... $CuSO_4$
DISCUSSION:

Ti specimens. Solution used as an immersion etch on this material.

REF: *Met Finish,* 55,46(1951)

TI-0019b
ETCH NAME: TIME:
TYPE: Acid, removal/polish TEMP:
COMPOSITION:
x .... HF
x .... $Na_2Cr_2O_7$
DISCUSSION:

Ti specimens. Solution used as an immersion etch on this material.

REF: Ibid.

TI-0020
ETCH NAME: TIME:
TYPE: Electrolytic, polish TEMP:
COMPOSITION: ANODE: Ti
x .... H_2SO_4 CATHODE:
x .... H_3PO_4 POWER:
DISCUSSION:

Ti specimens. Solution can be used as a mixed acid solution, or acids can be used alone as general anodic polishing etchants.

REF: Richards, R — *Corrosion,* 4,400(1950)

TI-0021b
ETCH NAME: TIME:
TYPE: Acid, removal TEMP:
COMPOSITION:
1 .... NH_4OH
2 .... H_2O_2
DISCUSSION:

Ti thin film evaporation in vacuum systems. Solution used to remove and clean bell jars and internal fixtures of titanium. Can be used on stainless steel, glass, copper, and ceramics.

REF: Nichols, D R — *Solid State Technol,* December 1979

TI-0014b

ETCH NAME: TIME:

TYPE: Electrolytic, polish TEMP:

COMPOSITION: ANODE: Ti

x .... HF CATHODE:

x .... CH_3COOH (HOAc) POWER:

DISCUSSION:

Ti specimens. Start etching by immersion, then apply AC current. (*Note:* The chemical acronym HOAc = HAc = GAA = glacial acetic acid.)

REF: Stanley, C L & Brenner, A — *Proc Am Electropl Soc,* 43,123(1956)

TI-0024

ETCH NAME: Heat TIME:

TYPE: Thermal, preferential TEMP: Elevated

COMPOSITION:

x .... heat

DISCUSSION:

Ti specimens used in a thermal etching study.

REF: Evans, P — *Acta Metall,* 5,342(1957)

TI-0026

ETCH NAME: TIME:

TYPE: Halogen, removal TEMP: Hot

COMPOSITION:

(1) x .... x% Br_2 (2) x .... x% I_2

x .... ethyl acetate x .... MeOH

DISCUSSION:

Ti evaporated as thin films. Either solution can be used for etch removal. (*Note:* Both solutions can be used on other metals and metallic compounds as general removal, polishing or preferential etches.)

REF: Abu-Yaron, A — *Thin Solid Films,* 112,349(1984)

TITANIUM ALLOYS: TiM_x

General: Titanium alloys do not occur in nature but are artificially compounded, not only as titanium alloys, themselves, but as an additive to irons, steels, copper, tungsten, etc. as a strengthening and hardening agent.

Technical Application: The most widely used alloy in Solid State processing is as Ti-tungsten, TiW(10%Ti), although other concentrations of titanium have been evaluated. The material is used as a multilayer metallization with gold as contact pads in device fabrication, such as for gallium arsenide, indium phosphide and silicon diodes, FETs, etc.

The TiW compound presented in its own section under tungsten. The only reference, here, is one as a AuTi thin film.

Etching: HF:HNO_3. See TiW and W for additional solutions.

TITANIUM ALLOY ETCHANTS

TITANIUM GOLD

TIAU-0001
ETCH NAME: TIME:
TYPE: Acid, thinning TEMP:
COMPOSITION:
1 .... HF
10 .... HNO_3
DISCUSSION:
AuTi thin films and SiO_2 both used as etching masks on silicon wafers containing epitaxy Si layers in the fabrication of photo diodes. Solution shown used to both etch photo resist patterned AuTi films and to thin silicon.
REF: Ataman, A — *J Appl Phys,* 49,5324(1978)

PHYSICAL PROPERTIES OF TITANIUM CARBIDE, TiC

Classification	Carbide
Atomic numbers	22 & 6
Atomic weight	60
Melting point (°C)	3140
Boiling point (°C)	4300
Application limits (°C)	1500
Density (g/cm^3)	6.5
Specific heat (mean)(J/kg °C)	1050
Linear coefficient of thermal expansion ($\times 10^{-6}$ cm/cm/°C)	6.9
Thermal conductivity (W/m °C) 1100°C	40
Electrical resistivity (ohms/cm)	9
Hardness (Mohs — scratch)	9+
(Knoop — kg/mm_2)	2470
Crystal structure (isometric — normal)	(100) cube
Color (solid)	Grey-white
Cleavage (cubic)	(001)

TITANIUM CARBIDE, TiC

General: Does not occur in nature as a compound, though there are over 40 titanium containing minerals: ilmenite, $FeTiO_2$, rutile, TiO_2 and titanite, $CaTiSiO_5$ all important ores of titanium.

Titanium metal is a major alloying element for hardening steel, and is used as an electrode in electrolytic solutions. As the carbide or boride, it is considered one of the high temperature ceramic materials, and has been used as a thin film coating on cutting tools. The oxide, rutile, TiO_2, has long been used as a paint pigment as "titanium white".

Technical Application: The carbide has had little use in Solid State processing, to date, though it has been studied as a grown single crystal and as a thin film, both amorphous and single crystal, for both physical and electrical characteristics. Doped with a p-n junction, like silicon carbide, SiC, it will probably show semiconducting properties for application as a high temperature diode . . . see Silicon Carbide, and Cubic Boron Nitride, BN.

Etching: HNO_3 and aqua regia type solutions; electrolytic with mixed acids.

TITANIUM CARBIDE ETCHANTS

TIC-0001
ETCH NAME: Argon
TYPE: Gas, cleaning
COMPOSITION:
x .... Ar
TIME:
TEMP: 100°C
DISCUSSION:
TiC, (100) wafers. After mechanical polishing, final clean surfaces in argon gas at 1000°C.
REF: Fukunda, S et al — *J Vac Sci Technol,* A2,50(1984)

TIC-0002
ETCH NAME:
TYPE: Electrolytic, preferential
COMPOSITION:
1 .... HF
1 .... HNO_3
6 .... H_2O
TIME: 1 sec
TEMP: RT
ANODE: TiC
CATHODE:
POWER: 3 V
DISCUSSION:
TiC, (001) cleaved wafers used in a study of Mondrian precipitation patterns. The compound has the cubic NaCl structure. Solution developed square pits with pointed bottoms as dislocations. The flat bottom pits are surface etch pits.
REF: Williams, W S — *J Appl Phys,* 32,552(1961)
TIC-0003b: Ibid.
TiC, (100) wafers. See discussion under TIC-0002.

TIC-0004
ETCH NAME:
TYPE: Electrolytic, polish
COMPOSITION:
x .... HF
x .... HNO_3
x .... HAc
TIME:
TEMP:
ANODE: TiC
CATHODE:
POWER:
DISCUSSION:
TiC, (100) thin films as oriented deposits and used in a study of plastic deformation. Prior to deformation, specimens were electropolished in the solution shown.
REF: Hollax, G E & Smallman, R E — *J Appl Phys,* 37,818(1966)
TIC-0009: Lee, C W et al — *J Vac Sci Technol,* 21,43(1982)
TiC thin films CVD deposited on WC substrates. At 1000°C deposition temperature, orientation was (111) grains with equal axes; at 1100°C grains were elongated with (110) orientation.

TIC-0005
ETCH NAME: Murakama's reagent
TYPE: Acid, cleaning
COMPOSITION:
10 g .... $K_3Fe(CN)_6$
10 g .... KOH (NaOH)
100 ml .... H_2O
TIME:
TEMP:
ASTM #98

DISCUSSION:

TiC specimens. Used as a cleaning etch on titanium carbide surfaces. Also used on tungsten carbide, WC and tantalum carbide, TaC.

REF: Pamphlet: ASTM E407-70 (1972)

TIC-0006

ETCH NAME: Nitric acid — TIME:

TYPE: Electrolytic, polish — TEMP:

COMPOSITION: — ANODE:

x .... HNO_3 — CATHODE:

POWER:

DISCUSSION:

TiC single crystal ingot grown by the Float Zone (FZ) method under high-pressure helium. Spark erosion specimens cut into square prisms 0.2 × 0.2 mm, and weld a Ta contact wire. Electropolish into a tip point of 0.1 μm shape. Specimens used in a study of field emission properties of TiC.

REF: Fujii, K et al — *J Appl Phys,* 57,1723(1985)

TIC-0007

ETCH NAME: — TIME:

TYPE: — TEMP:

COMPOSITION:

DISCUSSION:

TiC, (001) cleave wafers used in a study of electrical conductivity and thermoelectric effects. The (001) cleavage is very prominent. No etch shown.

REF: Hollander, L E Jr — *J Appl Phys,* 32,997(1961)

TIC-0008a

ETCH NAME: Diamond — TIME:

TYPE: Element, polish — TEMP: RT

COMPOSITION:

x .... diamond paste

DISCUSSION:

TiC, (001) cleaved wafers used in a study of elastic constants. TiB_2 also studied. Both materials were also cut with a diamond saw and polished with diamond paste. Both materials appear to be harder than diamond. (*Note:* See properties section for Knoop hardness values . . . they are measurably less hard than diamond at H = 7000 Knoop.)

REF: Gilman J J & Roberts, B W — *J Appl Phys,* 32,1405(1961)

PHYSICAL PROPERTIES OF TITANIUM DIOXIDE, TiO_2

Classification	Oxide
Atomic numbers	22 & 8
Atomic weight	80
Melting point (°C)	1640
Boiling point (°C)	
Density (g/cm³)	4.18 (5.2)
Dielectric constant (e =) (100 KC) 26°C	190 // c-axis
	85 ⊥ c-axis
	85—100 pressed powder
Magnetic susceptibility ($\times 10^{-6}$ emu/g)	0.067

Electrical conductivity (eV)	1.5—1.8
Electrical resistivity (ohms cm $\times 10^{13}$)	1
Transmission range (microns)	0.42—6.0
Optical gap width (eV)	3
Thermal activation energy (eV)	1.5
Debye temperature (K)	1184
Coefficient of linear thermal expansion	9.94 // c-axis
($\times 10^{-6}$ cm/cm/°C)	7.19 ⊥ c-axis
Energy gap (Ev)	3.1
Refractive index (n=)	2.6—2.9
Hardness (Mohs — scratch)	6—6.5
(Vickers — kg/mm^2)	659—900
Crystal structure (tetragonal — normal)	(110) prism, 1st order
Color (solid)	Clear-yellowish
Cleavage (basal — distinct)	(001)

TITANIUM DIOXIDE, TiO_2

General: There are three natural minerals: rutile, octahedrite and brookite. Brookite is orthorhombic, the others are tetragonal with rutile having the higher melting point. Natural minerals are commonly highly twinned, prismatic with surface striations and, as they contain some iron, vary in brown coloration to near-black. The colorless varieties have a refractive index equivalent or higher than that of diamond, with faint yellow tint, called mock topaz; artificially grown clear, called titania, and used as a replacement for diamond in jewelry. The three minerals are the major ores of titanium.

Other than as titanium ores, the primary use in industry is as a high grade white pigment with superior opacity and stability for paints.

Technical Application: In Solid State operations TiO_2 is under development as a thin film surface coating on semiconductors and other metals as a replacement for silicon dioxide, and has been deposited as TiO, TiO_2, and Ti_2O_3 for a protective surface coating and, as TiO_2, as a thin film capacitor in planar circuit construction.

For use as a capacitor in general industrial applications it is either deposited as an amorphous thin film or grown, cut and used with single crystal orientation or processed in pressed powder form. The (001) orientation gives the highest dielectric constant as a single crystal element.

Both the natural crystals and artificially grown crystals have been the subject of wide study for morphology, electrical parameters, etc. As doped crystals, e.g., iron, etc., for semiconducting properties and used as filters, lenses, prisms.

Etching: Soluble in H_2SO_4 and alkalies, but insoluble in other single acids.

TITANIUM DIOXIDE ETCHANTS

TIO-0011a

ETCH NAME: Sulfuric acid — TIME: x minutes

TYPE: Acid, removal — TEMP: Hot to boiling

COMPOSITION:

x .... H_2SO_4

DISCUSSION:

TiO_2 as natural minerals. Rutile — colorless to brown/black. Octahedrite — brown to black. Brookite — brown to black. All of these natural forms of TiO_2 are soluble in sulfuric acid, but insoluble in other acids.

REF: Weast, R C et al — *Handbook of Chemistry and Physics,* 65th ed, CRC Press, Boca Raton, FL, 1984, B154

TIO-0001a: Walker, P — personal application, 1980—1985

TiO_x, thin film residue from titanium sputter metallization on alumina and quartz substrates used for fabricating microwave circuits. Solution used in a cleaning sequence to remove a final grey fraction of TiO_x from substrates. Soak substrates until they are returned to their normal white color.

TIO-0007b: Ibid.

TiO_2 as natural rutile crystal specimens (110) oriented. Acid shown used boiling, 30 min to develop etch pits.

TIO-0003

ETCH NAME: Hydrochloric acid — TIME: 12 h

TYPE: Acid, cleaning — TEMP: RT

COMPOSITION:

x .... HCl

DISCUSSION:

TiO_2, natural rutile specimens. After soak in HCl, water rinse for 1 h. Cleaned surfaces prior to a magnetic susceptibility study. Rutile surfaces pick up ferro-magnetic dust probably from atmosphere.

REF: Senttle, F E — *Phys Rev,* 120,821(1960)

TIO-0004a

ETCH NAME: Aqua regia — TIME:

TYPE: Acid, cleaning — TEMP: Boiling

COMPOSITION:

3 .... HCl

1 .... HNO_3

DISCUSSION:

TiO_2, single crystal natural rutile crystals used to study electric charge transfer through specimens. Use solution to polish and clean surfaces; wash in warm x% KOH, rinse in DI water, then vapor clean in steam and dry at 125°C.

REF: Srivastava, K G — *Phys Rev,* 119,520(1960)

TIO-0005a

ETCH NAME: — TIME:

TYPE: Acid, removal — TEMP: RT

COMPOSITION: — RATE: 50 Å/sec

2 .... HF

25 .... NH_4F

5 .... H_2O

DISCUSSION:

TiO_2, thin film deposited on GaAs,(100) substrates by chemical plating with 3.5% Ti-isoperoxide or Ti-ethylhexoxide in isopropyl alcohol for 1 min. Hold over water (95% RH) for about 1 week — film growth between 500 and 2000 Å. Etch in the solution shown under nitrogen, rinse in IPA and MeOH under nitrogen. RF sputter etch with Ar^+ ions produces craters in film. Liquid etching, as shown, may produce some pits.

REF: Bertrand, P A & Fleischauer, P D — *Thin Solid Films,* 103,167(1983)

TIO-0006
ETCH NAME: TIME:
TYPE: Acid, cleaning TEMP:
COMPOSITION:
x .... H_2SO_4
x .... x% $K_2Cr_2O_7$
DISCUSSION:
TiO_2, (001) basal oriented wafers used in a study of piezoresistivity in the oxide semiconductor rutile. After cleaning in solution shown, indium solder was used for contact purposes.
REF: Hollander, L E Jr et al — *Phys Rev,* 117,1469(1960)

TIO-0007a
ETCH NAME: Lithium hydroxide TIME: Hours
TYPE: Salt flux, decoration TEMP: 450°C
COMPOSITION:
x .... LiOH
DISCUSSION:
TiO_2, (110) natural single crystal rutile specimens. Molten flux of lithium salts used to decorate dislocations. Molten KNO_3, at 300°C, xx hours used to remove residual lithium after decorating.
REF: Johnson, O W — *J Appl Phys,* 35,3048(1964)

TIO-0008
ETCH NAME: Potassium chloride TIME:
TYPE: Electrolytic, oxidation TEMP:
COMPOSITION: ANODE: Ti
x .... 1 *M* KCl CATHODE: Ti (specimen)
x .... *HCl POWER: 50 mV/sec @ 0.6—3 V
*for pH 3.45

DISCUSSION:
Ti thin films. A 10 nm thick anodic oxide was grown on specimens surfaces using the solution shown. See TI-0006 for titanium etch used prior to oxidation.
REF: DeBerry, D W & Viehbeck, A — *J Electrochem Soc,* 130,248(1983)

TIO-0004b
ETCH NAME: Potassium hydroxide TIME:
TYPE: Alkali, cleaning TEMP: Warm
COMPOSITION:
x .... x% KOH
DISCUSSION:
TiO_2 as natural single crystal rutile used in a study of electric charge transfer through bulk specimens. Solution used as a wash after etching with aqua regia. Follow KOH with heavy water wash, then steam clean and dry at 125°C.
REF: Ibid.

TIO-0004c
ETCH NAME: Steam
TYPE: Gas, cleaning
COMPOSITION:
x .... steam
TIME:
TEMP: 1000°C (?)

DISCUSSION:
TiO_2 as natural single crystal rutile used to study electric charge transfer through specimens. After etching in aqua regia; wash in warm KOH and heavy DI water rinse, final cleaning done with steam, then dried at 125°C.
REF: Ibid.

TIO-0005b
ETCH NAME: Argon
TYPE: Ionized gas, cleaning/removal
COMPOSITION:
x .... Ar^+ ions
TIME:
TEMP:
GAS FLOW:
PRESSURE:
POWER:

DISCUSSION:
TiO_2 thin films deposited on GaAs (100) substrates in a film study. Film was chemically deposited and treated under high humidity. Both acid wet etching and argon RF sputtering used to remove the film. Ar^+ ions produced craters in the thin film.
REF: Ibid.

TIO-0009
ETCH NAME: BHF
TYPE: Acid, removal
COMPOSITION:
x .... HF
x .... NH_4F (40%)
TIME:
TEMP:

DISCUSSION:
TiO_2 thin films deposited on (111), n-type silicon substrates in a study of anti-reflective (AR) coatings for solar cells. $TiCl_4$ at 130—250°C applied by spray onto a rotating silicon substrate at about 16 μm/min with higher temperature increasing film hardness. Under 136°C there is only a milky film that can be wiped off with a cloth. Refractive index at 130°C was n = 2.13; at 250°C, n = 2.4. At 180°C film is a-TiO_2, and above is mc-TiO_2 (amorphous to microcrystalline). TiO_2 powder formed at spray nozzle can create pin holes in films; pin holes were 1—2 μm in diameter.
REF: Young, K S & Lam, Y W — *Thin Solid Films,* 109,169(1983)

TIO-0010
ETCH NAME:
TYPE: Acid, removal
COMPOSITION:
1 .... HF
5 .... HNO_3
4 .... H_2O
TIME:
TEMP:

DISCUSSION:
TiO_2, as a native oxide on titanium substrates being prepared for an oxidation study. Solution shown used as a native oxide removal etch, though it is a general etchant for titanium metal.
REF: Rolison, D E & Murry, R W — *J Electrochem Soc,* 131,336(1984)

TITANIUM MONOXIDE ETCHANTS

TIO-0011b
ETCH NAME: Sulfuric acid, dilute — TIME:
TYPE: Acid, removal — TEMP:
COMPOSITION:
x .... H_2SO_4
x .... H_2O
DISCUSSION:
TiO, thin film deposits. Color is yellow to black, probably amorphous in structure. Material is slowly soluble in dilute sulfuric acid.
REF: Weast, R C et al — *Handbook of Chemistry and Physics,* 65th ed, CRC Press, Boca Raton, FL, 1984, B154
TIO-0001b: Walker, P — personal application, 1980—1985
TiO_x thin film residue from titanium sputter metallization on alumina and quartz substrates used for fabricating microwave circuits. Solution used in a cleaning sequence to remove the final grey fraction of a TiO_x remaining on substrates.

TITANIUM TRIOXIDE ETCHANTS

TIO-0011c
ETCH NAME: Sulfuric acid, dilute — TIME:
TYPE: Acid, removal — TEMP:
COMPOSITION:
x .... H_2SO_4
x .... H_2O
DISCUSSION:
Ti_2O_3 specimens and thin films. Color is velvet-black. All titanium oxides are slowly soluble in concentrated or dilute sulfuric acid. The trioxide is insoluble in HCl or HNO_3.
REF: Weast, R C et al — *Handbook of Chemistry and Physics,* 65th ed, CRC Press, Boca Raton, FL, 1984, B154
TIO-0001c: Ibid.
TiO_x as a thin film residue from titanium sputter metallization on alumina and quartz substrates used for fabricating microwave circuits. Solution used in a cleaning sequence to remove a final grey fraction of TiO_x from substrates.

PHYSICAL PROPERTIES OF TITANIUM NITRIDE, TiN

Classification	Nitride
Atomic numbers	22 & 7
Atomic weight	61.9
Melting point (°C)	3220
Boiling point (°C)	
Density (g/cm^3)	5.3
Hardness (Mohs — scratch)	5—7
Crystal structure (isometric — normal)	(100) cube
Color (solid)	Bronze-red/gold
Cleavage (cubic)	(001)

TITANIUM NITRIDE, TiN

General: Does not occur as a natural compound. The primary mineral is rutile, TiO_2.

Titanium metal has been surface nitrided in industry, but there has been little other use of the nitride in metal processing. Based on its thin film color as bronze-red, it can be used as a surface coating replacement for gold in the decoration of clay products as a glaze.

Technical Application: Primary development in Solid State processing has been as a deposited thin film on silicon and other wafer materials as a dielectric type coating, and for wear resistance on metal tools.

Etching: Mixed acids HF:HNO_3; fluorine compounds; H_2O_2:HF. DCE with CF_4.

TITANIUM NITRIDE ETCHANTS

TIN-0003b
ETCH NAME: TIME:
TYPE: Acid, removal TEMP: RT
COMPOSITION:
1 .... HF
27.5 .... HNO_3
10 .... H_2O
DISCUSSION:

TiN thin films deposited on poly-Si. Solution used to pattern etch poly-Si with TiN acting as a patterning etch mask. Attack of TiN is 4× slower than on poly-Si.

REF: Ibid.

TIN-0003c
ETCH NAME: RF plasma TIME:
TYPE: Ionized gas, removal TEMP: RT
COMPOSITION: ANODE: TiN
x .... CF_4 CATHODE:
PRESSURE: 58 mTorr
POWER: 200 W
DISCUSSION:

TiN thin films deposited on poly-Si. TiN etch rate was 240 Å/min. Used as a general removal etch for TiN or for selective removal and structuring.

REF: Ibid.

TIN-0004
ETCH NAME: TIME: Seconds
TYPE: Acid, removal TEMP: RT
COMPOSITION:
10 ml .... HF
45 ml .... HNO_3
45 ml .... H_2O_2
DISCUSSION:

TiN thin films deposited on Ti, (0001) substrates using NH_3 as nitrogen source. The film was deep gold in color. Also deposited HfN and ZrN.

REF: Dawson, P T — *J Vac Sci Technol,* 21,36(1982)

TIN-0006
ETCH NAME: BHF TIME:
TYPE: Acid, removal TEMP: Hot
COMPOSITION:
1 .... HF
5 .... NH_4F (40%)

DISCUSSION:

TiN thin films. The films can be used as: (1) high temperature single surface absorbers; (2) silicon diffusion barriers; (3) a gold colored replacement for gold as a surface decorative coating; (4) a wear resistant coating on tools. Solution shown is a general removal etchant for TiN.

REF: Martin, P L et al — *J Vac Sci Technol*, A2(2),341(1984)

TIN-0001

ETCH NAME: TIME:

TYPE: Acid, removal TEMP: 60°C

COMPOSITION:

2 g .... EDTA (tetrasodium salt)
60 ml .... H_2O_2
120 ml .... H_2O

DISCUSSION:

TiN thin films deposited on silicon wafers as part of a metallization structure: Ni/TiN/Al and used in the fabrication of diodes. The nitride acts as a blocking layer. After nickel removal, the solution shown was used to remove TiN.

REF: Martin, P L et al — *J Electron Mater*, 13,309(1984)

TIN-0005

ETCH NAME: TIME:

TYPE: Acid, removal TEMP: 60°C

COMPOSITION:

2.8 g .... EDTA (tetrasodium salt)
60 ml .... H_2O_2
120 ml .... H_2O
5 ml .... NH_4OH

DISCUSSION:

TiN thin films deposited on (100) silicon wafers as part of a diode metallization structure: Ni/TiN/Al similar to TIN-0001, above. Remove nickel with: 10 ml HF:10 ml HNO_3:1000 ml H_2O. Remove aluminum with: 80 ml H_2PO_4:5 ml HNO:5 ml HAc; 10 ml H_2O, at 45°C. Remove TiN in solution shown.

REF: Finetti, N et al — *J Electron Mater*, 13,327(1984)

TIN-0003a

ETCH NAME: BHF TIME:

TYPE: Acid, removal TEMP: 0°C

COMPOSITION: RATE: 2.5 g/sec & 1 g/sec

134 ml .... HF
452 g .... NH_4F
625 ml .... H_2O

DISCUSSION:

TiN thin films deposited on poly-Si epitaxy layers. Heated in air at 500°C will convert TiN to semi-insulating (SI), TiO_2. Heated in an inert atmosphere, will remain as TiN to above 900°C. Solution used to remove residual SiO_2 prior to TiN deposition.

REF: Whittmer, M et al — *J Appl Phys*, 54,1423(1983)

TITANIUM:TUNGSTEN, TiW

General: Does not occur as a native compound. There are a dozen tungsten single crystal minerals — the tungstates — of which, wolframite, $(Fe,Mn)WO_4$ and scheelite, $CaWO_4$ are primary ores of the metal. Wolframite varies from the mineral: ferberite, $FeWO_4$ to hubnerite,

$MnWO_4$ and, depending upon the iron content, may be magnetic. Other metal:tungstates are cuprotungstate, $CuWO_4$; stolzite, $PbWO_4$, as well as the oxide, tungstite, WO_3 and sulfide, tungstenite, WS_2 and other minerals contain some molybdenum.

Pure tungsten and tungsten alloys are of major use in industry as they have the highest melting points of any known elements, high electrical efficiency and are inert to most single acids and alkalies. Mixtures of titanium and tungsten — called "Ti-tungsten" — are artificial alloys and are being used in semiconductor fabrication.

Technical Application: Titanium:tungsten alloys are in use and under development as amorphous and polycrystalline thin film RF sputter deposites on semiconductor wafers as a blocking layer against crossover diffusion and reaction of other metals in multilayer structures. Although several alloy concentrations of titanium in tungsten have been evaluated, the most widely used is 10% titanium: TiW:(10%Ti) — commonly known as Ti-tungsten.

Due to the high melting point of tungsten, deposition of thin films is by RF sputtering in argon from a solid target of Ti-tungsten as an amorphous coating a-TiW. By controlled conditions of deposition and annealing, it is possible to produce colloidal structure, c-TiW and even single crystal — monoclinic — (001) basal structure like wolframite.

Etching: H_2O_2, HNO_3. Alkalies and ionized gases. See Tungsten

TI-TUNGSTEN ETCHANTS

TIW-0001

ETCH NAME: Hydrogen peroxide — TIME: To 1 h

TYPE: Acid, removal — TEMP: RT

COMPOSITION: — RATE: 20—30 μ''/sec

x .... H_2O_2, conc. (30%)

DISCUSSION:

TiW thin films. Described as a general etch for Ti-Tungsten. Sputter metallization from a TiW (10% Ti) target as RF or DC sputter in an argon atmosphere. (*Note:* Can also be sputtered with O_2 or N_2 in argon as a glassy passivation surface coating of TiO_2 or TiN.)

REF: Bulletin 1116B/Rev-HPD/VB/TM/0883/1K, MRC

TIW-0002: Walker, P et al — personal application, 1980—1986

TiW, thin films RF sputter deposited in argon on alumina and quartz blanks for fabricating microwave circuits. Metallization is Au:TiW (400 Å:100 μ''). Used tri-iodide solution to remove gold and hydrogen peroxide to remove TiW. In cleaning substrates for complete removal of metallization: (1) soak 5 min to remove bulk of TiW in solution shown; (2) if grey film remains, soak in HF and then H_2SO_4; (3) if an orange film remains (a hydrated WO_x) soak in hot 20—30% KOH.

TIW-0003

ETCH NAME: RIE — TIME:

TYPE: Ionized gas, removal/pattern — TEMP:

COMPOSITION: — GAS FLOW:

x .... CCl_4 — PRESSURE:

x .... O_2 (40—50%) — POWER:

DISCUSSION:

TiW, thin films sputter deposited on silicon substrates in a study of etch removal from the following multilayer structures: SiO_2/TiW/Si; Al/TiW/Si; Si_3N_4/TiW/Si and AlCu/TiW/Si. After thin film formation, wafers were photo resist patterned using standard techniques: (1) P/R pattern and etch remove SiO_2 and Si_3N_4 with BHF; (2) with P/R pattern, deposit Al or AlCu and removal with acetone as standard lift-off technique, and (3) RIE gas etch remove

TiW. This is a method of selective etching called subtractive etching as removal etch rate is minimal on all components other than TiW.
REF: Schaible, P M & Schwartz, G C — *J Electrochem Soc,* 132,730(1985)

TIW-0004
ETCH NAME: TIME:
TYPE: Acid, removal TEMP:
COMPOSITION:
1 .... NH_4OH
2 .... H_2O_2
DISCUSSION:

TiW (1%Ti) thin films. This solution is recommended for use in cleaning vacuum systems for titanium/tungsten removal from stainless steel, glass, copper and ceramics. It also can be used as an etch solution of TiW thin films.
REF: Nichols, D R — *Solid State Technol,* December 1979
TIW-0005: Nowicki, R S et al — *Thin Solid Films,* 53,195(1978)

Ti_3W_7 thin films sputter deposited on an aluminum film evaporated on (111) silicon wafers used in a study of film morphology and, with a gold cover evaporation, cross-over diffusion reactions. Microstructure analysis was done with Au/TiW films deposited on graphite. Pinhole density was evaluated with TiW deposited on glass substrates using a microscope with transmitted light.

TIW-0006
ETCH NAME: Aqua regia TIME:
TYPE: Acid, removal TEMP:
COMPOSITION:
3 .... HCl
1 .... HNO_3
DISCUSSION:

$Ti_3W_7Si_2$, thin films on silicon wafers, (111), (110) and (100) oriented. A composite target of the Ti/W was RF sputtered on the silicon wafers and annealed at 350°C to form the silicide and was unaffected in temperatures from 675 to 900°C. Once formed, the films were unaffected by hot H_2O_2, HF or CP4 and dissolved only slowly in aqua regia.
REF: Harris, J M et al — *J Electrochem Soc,* 123,120(1976)

PHYSICAL PROPERTIES OF TUNGSTEN, W

Classification	Transition metal
Atomic number	74
Atomic weight	183.85
Melting point (°C)	3387 (3370)
Boiling point (°C)	5927 (5900)
Density (g/cm³)	19.3 (same as gold)
Thermal conductance (cal/sec)(cm²)(°C/cm)	0.40
Specific heat (cal/g) 25°C	0.032
Latent heat of fusion (cal/g)	44
Heat of fusion (k-cal/g-atom)	8.05
Heat of vaporization (cal/g)	197
Atomic volume (W/D)	9.53
1st ionization energy (K-cal/g-mole)	184
1st ionization potential (eV)	7.98

Electronegativity (Pauling's)	1.7
Covalent radius (angstroms)	1.30
Ionic radius (angstroms)	0.62 (W^{+6})
Electrical resistivity (micro-ohms-cm) 25°C	5.6
Tensile strength (psi)	500,000
Thermal expansion (ppm/°C)	4.6
Vapor pressure (°C)	5168
Electron work function (eV)	4.5
Hardness (Mohs — scratch)	6—6.5
Crystal structure (isometric — normal) alpha	(100) cube bcc
(isometric — normal) beta	(100) cube fcc
Color (solid)	Grey-black
Cleavage (cubic — poor)	(001)

TUNGSTEN, W

General: Does not occur as a native element. There are two groups of tungsten oxides: (1) wolframite group, monoclinic, as Fe, Mn, and $PbWO_4$ and (2) scheelite group, tetragonal, as Ca, Pb and $CuWO_4$. Wolframite, (Fe,Mn)WO_4 and scheelite, $CaWO_4$ are the two primary ores of tungsten. As the pure metal, tungsten is hard, brittle, nonmagnetic and oxidizes readily when heated in air.

Tungsten is of major importance in industry. Because of its high electrical efficiency it is the primary filament in light bulbs and is a major alloy element in irons and steels for increased hardness and wear resistance armor plate, projectile casings and high speed cutting tools in addition to furnace linings and high temperature crucibles.

Technical Application: Although pure tungsten is not used in Solid State device fabrication it is used as a block or disc heat sink for mounting purposes as it has a very close thermal coefficient of expansion match with silicon and has been used both as single crystal and pressed powder parts. As a crystalline wire it is probably the most widely used test probe of all metals for DC electrical testing of semiconductor devices in wafer or discrete device form.

As an iron or steel alloy it has many uses in processing equipment. Titanium tungsten (10% Ti) — commonly called Ti-tungsten — is finding more and more use as the initial evaporation layer on semiconductor wafers and circuit substrates. It has been evaluated with other concentrations of titanium, but the 10% Ti has become a general industrial standard. Ti-tungsten is commonly RF sputter deposited from a solid target and followed by copper, gold or other metal layers.

It is under development and evaluation as a silicide: WSi, W_2Si, and WSi_2 for use as a buffer layer in semiconductor heterostructures. It also is fabricated as the carbide, WC, and boride, WB and may have application as a thin film amorphous surface coating as either of these compounds.

Although most studies have been with crystalline tungsten, it has been grown as a single crystal and as a deposited amorphous thin film for general evaluation.

Etching: H_2O_2, HF, alkalies and mixed acids.

W-0001a

ETCH NAME: Sodium hydroxide — TIME: 15 min

TYPE: Alkali, cleaning — TEMP: Boiling

COMPOSITION:

x .... 20% NaOH

DISCUSSION:

W, specimens and parts. Used for general cleaning of tungsten surfaces.

REF: Kohl, W H — *Handbook of Materials and Techniques for Vacuum Devices,* Reinhold, New York, 1967

W-0002a Walker, P — personal application, 1957—1980

W, filaments and boats used for metal evaporation. Clean in boiling solution 5—15 min and heavily water wash with IR heat lamp or N_2 gas drying. Before commercial parts were on the market and each laboratory wound their own coils or made their own boats (mid-1950s), parts were placed in vacuum attached to electrodes and fired to red heat, before being loaded with metal for evaporation. Pump to 1×10^{-6} Torr, heat to 800—1000°C, 2—5 min. Also used HF soak and water wash.

W0003a: Booklet: Tungsten, Schwartzkopf Development Corp, 395/1/71

W, specimens. Alkali, alone, will not etch tungsten but can be used as a cleaning solution. With addition of oxidizers (KNO_3 or PbO_2) becomes a rapid removal etch for tungsten.

W-0003b

ETCH NAME: Hydrofluoric acid TIME: 5—15 min

TYPE: Acid, removal TEMP: RT

COMPOSITION:

x .... HF or 1 .... HF
1 .... H_2O

DISCUSSION:

W specimens. Both concentrated and dilute solutions of HF will slowly etch tungsten.

REF: Ibid.

W-0002b: Ibid.

W, coils and boats used for metal evaporation. (See W-0002a)

W-0004: Stacy, W T et al — *J Electrochem Soc,* 132,444(1985)

W, thin film deposits. Substrates were silicon, (111), p-type 7—21 Ω cm resistivity: bare silicon or with SiO_2 and with aluminum, thin films. Use photo resist techniques to open patterns. Selectively deposit tungsten from WF_6 by LPCVD. Clean surfaces before W deposition with either HF or DCE $CF_4 + O_2$. HF gives good results; the ionized gas etching is erratic.

W-0005

ETCH NAME: Hydrogen peroxide TIME:

TYPE: Acid, removal TEMP:

COMPOSITION:

x .... H_2O_2

DISCUSSION:

W, thin films deposited by LPCVD. Substrates were bare silicon wafers, with SiO_2 or with $TaSi_2$ thin films. Photo resist techniques were used to pattern and open windows through the SiO_2 and the $TaSi_2$ before selective deposition of tungsten. Solution used in step-etching for tungsten thickness measurement. (*Note:* Good discussion of stress and adhesion of tungsten.)

REF: Green, M L & Levy, R A — *J Electrochem Soc,* 132,1243(1985)

W-0036c: Ibid.

W single crystal wafers of various orientations used in a study of etched surface properties. 30% H_2O_2 used boiling for 30 min.

W-0030c: Ibid.

W, specimens from vapor grown crystal used in a study of macro-structure. Solution was 10%, boiling for 20 sec for polish and removal. (*Note:* 30% H_2O_2 is standard concentration.)
W-0002c: Ibid.

W, thin films deposited as TiW(10% Ti) on alumina and quartz substrates used in fabricating microwave circuits. Sputter deposition was: Au:TiW (400 Å:100 μ″). After removing gold pattern with tri-iodide, tungsten was removed with this solution at RT, with 5 min soak and DI water wash. If there is a remaining grey fraction, soak in HF; if the fraction is an orange gel, soak in 20—30% KOH, hot — both are oxides of tungsten.
W-0001b: Ibid.

W, specimens. Used 5% H_2O_2 boiling and shows the solution as a slow etch.

W-0006
ETCH NAME: TIME:
TYPE: Acid, polish TEMP:
COMPOSITION:
x .... 25% $CuSO_4$
x .... NH_4OH
DISCUSSION:

W, specimens used in an etch study. Solution used to polish tungsten specimens.
REF: Wolff, V E — *Acta Metall,* 6,556(1958)

W-0001c
ETCH NAME: Nitric acid TIME:
TYPE: Acid, removal TEMP: Warm
COMPOSITION:
x .... HNO_3
DISCUSSION:

W, specimens. Very slight attack with formation of yellow/orange colored WO_3.
REF: Ibid.

W-0001d
ETCH NAME: Aqua regia TIME:
TYPE: Acid, removal TEMP: Hot
COMPOSITION:
3 .... HCl
1 .... HNO_3
DISCUSSION:

W, specimens used in an etch study. Little or no attack at room temperature. Slow attack used warm to hot.
REF: Ibid.

W-0001e
ETCH NAME: Sulfuric acid TIME:
TYPE: Acid, removal TEMP: RT & hot
COMPOSITION:
(1) x .... H_2SO_4, conc. (2) x .... H_2SO_4
x .... H_2O

DISCUSSION:

W specimens. Both concentrated and dilute solutions show slow attack used hot or at room temperature.

REF: Ibid.

W-0001f

ETCH NAME: Hydrochloric acid — TIME:

TYPE: Acid, removal — TEMP:

COMPOSITION:

(1) x .... HCl

(2) x .... HCl
x .... H_2O

DISCUSSION:

W specimens. Both concentrated and dilute solutions show slow attack used hot or at room temperature.

REF: Ibid.

W-0002d

ETCH NAME: — TIME: 3—5 min

TYPE: Acid, cleaning — TEMP: RT

COMPOSITION: — RATE: 0.1 mil/5 min

3 .... HF
2 .... HNO_3
1 .... HAc

DISCUSSION:

W discs of pressed powder and cut sheet used as heat sinks in SCR fabrication. Solution used to clean discs before silicon alloy.

REF: Ibid.

W-0001g

ETCH NAME: — TIME:

TYPE: Acid, cleaning — TEMP:

COMPOSITION:

5 .... HNO_3
3 .... H_2SO_4
2 .... H_2O

DISCUSSION:

W specimens. Solution used to clean specimens. Rinse in chromic acid:water solution.

REF: Ibid.

W-0001h

ETCH NAME: — TIME:

TYPE: Acid, removal — TEMP: Hot

COMPOSITION:

1 .... HNO_3
4 .... HF

DISCUSSION:

W specimens. Solution shows very rapid attack.

REF: Ibid.

W-0007
ETCH NAME:
TYPE: Acid, cleaning
COMPOSITION:
33 .... HCOOH (formic acid)
33 .... H_2O
33 .... H_2O_2
DISCUSSION:

TIME: 4—5 min
TEMP: 60—70°C
RATE: 0.1 mil/5 min

W, specimens. Used in a study of surface cleaning.

REF: *Bell Tel Systems Tech Publ* #3143, 1958, 61

W-0002e
ETCH NAME:
TYPE: Acid, cleaning
COMPOSITION:
10 g .... $2NH_4NO_3.Cd(NO_3)_3.4H_2O$
50 ml .... HNO_3
50 ml .... H_2O
DISCUSSION:

TIME: 20—30 sec
TEMP: RT

W, wire 0.015—0.025 diameter. In working with chrome etch solutions it was shown that this mixture will etch clean tungsten wire.

REF: Ibid.

W-00011
ETCH NAME:
TYPE: Acid, removal
COMPOSITION:
305 g .... $K_3Fe(CN)_6$
44.5 g .... NaOH
1000 ml .... H_2O
DISCUSSION:

TIME:
TEMP:

W specimens. Solution is a very rapid etch.

W-0029: Gulbransen, E A & Andrews, K F — *J Electrochem Soc,* 107,619(1960)

W, specimens used in a study of the kinetics of oxidation of pure tungsten from 500 to 1300°C. Solution mixture was not shown but was used as a ''scale'' removal etch, e.g., removal of native oxide and other surface contamination. Follow with cleaning in petroleum ether and then ethyl alcohol.

REF: Ibid.

W-0003b
ETCH NAME:
TYPE: Acid, clean or electropolish
COMPOSITION:
80 g .... $KClO_3$
10 g .... NaOH
40 g .... KCO_3
1000 ml .... H_2O
DISCUSSION:

TIME:
TEMP:
ANODE: W
CATHODE: SS screen
POWER: 0.05—0.5 A/cm^3

W specimens. Use as a pickling solution without current to clean tungsten. As an electrolytic etch will produce a bright etch polish finish.

REF: Ibid.

W-0008
ETCH NAME: TIME:
TYPE: Electrolytic, polish TEMP:
COMPOSITION: ANODE:
CATHODE:
POWER:
DISCUSSION:

W, poly-W ribbon. Electropolishing solution not shown. Specimens used in a study of oxide formation with and without sulfur diffusion. Clean in vacuum at 2500 K for 10 h. Before each evaporation flash clean at 2500 K, 5—10 sec.

REF: Ishikawa, K & Tanemura, M — *Jpn J Appl Phys*, 25,850(1984)

W-0030a
ETCH NAME: Sodium hydroxide TIME: 1 h
TYPE: Electrolytic, polish TEMP: RT
COMPOSITION: ANODE: W
x .... 0.1% NaOH CATHODE:
POWER: 50 mA/cm^2 & 10 V
DISCUSSION:

W, specimens from vapor grown single crystals used in a study of sub-crystal macro-structure. Solution used to electropolish specimens. After polish, etch in 1 10%$K_2Fe(CN)_6$:3% NaOH, RT, 1 min; then in boiling 10% H_2O_2, 20 sec and again electrolytic using 1% NaOH 10—20 sec at 1V. This etch sequence will develop macro-structure.

REF: Rieck, G D & Bruning, H A M — *Acta Metall*, 8,97(1960)

W-0030b
ETCH NAME: TIME: 1 min
TYPE: Acid, removal TEMP: RT
COMPOSITION:
1 .... 10% $K_2Fe(CN)_6$
1 .... 3% NaOH
DISCUSSION:

W, specimens cut from vapor grown crystals used in a study of macro-structure. Solution used following electropolish. See W-0030a.

REF: Ibid.

W-0009
ETCH NAME: Xenon fluoride TIME:
TYPE: Ionized gas, removal TEMP:
COMPOSITION: GAS FLOW:
(1) x .... $XeF_2{}^+$ PRESSURE:
(2) x .... $XeF_2{}^+$ POWER:
x .... Ar^+
DISCUSSION:

W, (111) wafers as deposited thin film. Etch product is mainly WF_6 (gas) during ion etching. Addition of argon enhances etch rate.

REF: Winters, H F — *J Vac Sci Technol*, B3,9(1985)

W-0010
ETCH NAME: Carbon tetrafluoride
TYPE: Ionized gas, removal
COMPOSITION:
(1) .... CF_4^+
(2) .... SF_6^+

TIME:
TEMP:
GAS FLOW:
PRESSURE:
POWER:

DISCUSSION:
W, thin films deposited on borosilicate glass to 0.4 µm thick. Reaction chamber 7″ diameter with SS electrodes. Place specimens on bottom electrode heated to 60°C.
REF: Tang, C C & Hess, D W — *J Electrochem Soc,* 131,115(1984)

W-0011
ETCH NAME:
TYPE: Ionized gas, removal
COMPOSITION:
x .... $CBrF_3$

TIME:
TEMP:
GAS FLOW:
PRESSURE:
POWER:

DISCUSSION:
W, thin film evaporated on silicon, (100) substrates with a nickel mask evaporated over tungsten. Also chromium then on silicon, then tungsten. Chrome acts as an etch-stop in RIE etching with this gas.
REF: Schattsnberg, M L — *J Vac Sci Technol,* B3,12(1985)

W-0012
ETCH NAME: Sodium hydroxide
TYPE: Electrolytic, polish
COMPOSITION:
x .... 5—20% NaOH

TIME:
TEMP: RT
ANODE: W
CATHODE: SS screen
POWER: 20—30 mA/cm^3 & 12—15 V

DISCUSSION:
W, wires 0.010 to 0.060 diameter. The 5% mixture is used as a general cleaning solution. The 15—20% mixture is used to form probe points on 0.060 wires with a dip-and-pull method: about 10 dips with slow pull-out produces a tapered, fine point. Second method is to immerse until a stub-tip is formed. Produces a fine polished surface.
REF: Lucky, H — personal communication, 1979
W-0013: Everson, K M et al — *J Appl Phys,* 57,956(1985)
W, whisker, 25 µm diameter electrolytically pointed. Either a 25° sharp point or a 40—50° blunt point. A poly-Ni substrate was mechanically polished then nickel evaporated in forming a W-NiO-W antenna as a MIM diode.
W-0014: Carroll, R W — *J Appl Phys,* 39,2339(1968)
W, wire electropolished in the solution shown.

W-0015
ETCH NAME: Acetone
TYPE: Keytone, cleaning
COMPOSITION:
x .... CH_3COCH_3

TIME:
TEMP:

DISCUSSION:
W specimens. Used to degrease and clean surfaces.
REF: Bronnes, R T — *J Appl Phys,* 36,1445(1963)

W-0016
ETCH NAME: TIME:
TYPE: Alcohol, cleaning TEMP:
COMPOSITION:
(1) x .... petroleum ether (2) x .... EOH
DISCUSSION:
W specimens. Used to degrease and clean surfaces in an oxidation study.
REF: Gulbransen, J & Andrews, R C — *J Appl Phys*, 36,1540(1963)

W-0017a
ETCH NAME: Sodium hydroxide, dilute TIME:
TYPE: Electrolytic polish TEMP:
COMPOSITION: ANODE:
x .... 2% NaOH CATHODE:
POWER:
DISCUSSION:
W, (001) wafers used in a dislocation density study. Electropolish surfaces with solution shown. Develop etch pits in a solution of $xCuSO_4$:xNH_3 on freshly polished surfaces.
REF: Hull, D et al — *Phil Mag*, 12,1021(1965)
W-0024: Bewkitt, K M et al — *Phil Mag*, 12,841(1965)
W, (001) wafers structure study under a field ion microscope (FIM). Specimens were electropolished and thinned; concentration not shown.
W-0027: Ryan, H F & Suiter, J — *Acta Metall*, 14,847(1966)
W, single crystal specimens. Electropolish in sodium hydroxide prior to dislocation etching to observed grain boundaries.
W-0028a: Shukovsky, H B et al — *Acta Metall*, 14,821(1966)
W, single crystals. A study of low temperature resistivity of lattice defects in deformed specimens. A 2% NaOH solution was used at 6 V; 5% solution at 8 V to electropolish before dislocation etching.

W-0017b
ETCH NAME: TIME:
TYPE: Acid, preferential TEMP:
COMPOSITION:
x .... NH_3
x .... $CuSO_4$
DISCUSSION:
W, (001) wafers used in a dislocation study. After electropolishing, develop dislocations with the etch shown using freshly polished surfaces.
REF: Ibid.

W-0018a
ETCH NAME: Potassium ferricyanide TIME:
TYPE: Acid, polish TEMP:
COMPOSITION:
x .... $K_3Fe(CN)_6$
x .... H_2O

DISCUSSION:

W specimens used for electroplating. Immerse specimens in solution to polish. (*Note:* A major article on plating transition metals; gives both etch solutions and plating solutions.)

REF: Saubestre, E B — *J Electrochem Soc,* 106,305(1959)

W-0019: Keilholtz, G W & Bergin, M J — U.S. Patent 2,566,615(1951)

Solution used by Saubestre (W-0018) referenced to this patent.

W-0018b

ETCH NAME: TIME:

TYPE: Acid, polish TEMP:

COMPOSITION:

x .... H_3PO_4

x .... $HClO_4$

DISCUSSION:

W specimens used for electroplating. Immerse specimens in solution shown for polishing prior to plating.

REF: Ibid.

W-0020: Schoeler, W R & Powell, A R — *Analysis of Minerals and Ores of the Rare Earth Elements,* Chas. Griffin & Co., London, 1940, 207

This reference given for solution shown above: W-0018b.

W-0018c

ETCH NAME: TIME:

TYPE: Acid, polish TEMP:

COMPOSITION:

x .... H_2O_2

x .... HOOCCOOH (oxalic acid)

DISCUSSION:

W specimens used for electroplating. See W-0018a for further discussion. Use solution by immersion.

REF: Ibid.

W-0021: Gusev, S & Ulkumov, T — *Zh Anal Khim,* 3,373(1948)

Reference for etch in W-0018c.

W-0018d

ETCH NAME: TIME:

TYPE: Acid, polish TEMP:

COMPOSITION:

x .... HF

x .... HNO_3

DISCUSSION:

W, specimens used for electroplating. Use solution by immersion. See W-0018a for further discussion.

REF: Ibid.

W-0022: Ribbons, D — *Metallurgica,* 55,257(1957)

Reference for solution used above in W-0018d.

W-0018e
ETCH NAME: TIME:
TYPE: Acid, polish TEMP:
COMPOSITION:
x .... $K_3Fe(CN)_6$
x .... H_2O
DISCUSSION:
W, specimens used for electroplating. Immerse specimens in solution. See W-0018a for further discussion.
REF: Ibid.
W-0023: Faust, C L & Beach, J C — *Plating*, 43,1134(1950)
Reference for solution shown above in W-0018e.

W-0025
ETCH NAME: TIME:
TYPE: Acid, thinning TEMP:
COMPOSITION:
20 .... HF
20 .... HNO_3
15 .... CH_3COOH (HAc)
DISCUSSION:
W, (001) wafers and other orientations. Wafers used in a neutron damage study. Solution used to etch thin specimens for microscope study after irradiation.
REF: Lacefield, K et al — *Phil Mag*, 13,1079(1966)

W-0026
ETCH NAME: Argon TIME:
TYPE: Ionized gas, entrapment TEMP:
COMPOSITION: GAS FLOW:
x .... Ar^+ ion PRESSURE:
POWER:
DISCUSSION:
W, thin films deposited by sputtering in an argon atmosphere. Argon and other gases can be trapped in the film during growth as an ionic species. The films were used in a study of low energy inert gas ion sputtering of metals.
REF: Sinha, M K — *J Appl Phys*, 39,2150(1968)

W-0031
ETCH NAME: Oxygen TIME:
TYPE: Gas, forming TEMP:
COMPOSITION:
x .... O_2
\+ .... heat
DISCUSSION:
W, specimens as wire. Wires were heated in oxygen to form points.
REF: Muller, E W — *Z Phys*, 108,668(1938)
W-0036b: Ibid.
W single crystal wafers of various orientations used in a study of etched surface properties. Specimens oxidized at 2600 K in O_2 at 5×10^{-5} Torr.

W-0034a
ETCH NAME: TIME:
TYPE: Acid, removal TEMP:
COMPOSITION:
1 .... HF
1 .... HNO_3
DISCUSSION:
W, thin film evaporation in vacuum systems. Solution used to clean vacuum bell jars and internal fixtures. Solution can be used on stainless steel but can damage copper, glass, and ceramic.
REF: Nichols, D R — *Solid State Technol*, December 1979

W-0035
ETCH NAME: TIME: 1 min
TYPE: Acid, step-etch TEMP: RT
COMPOSITION:
5 g .... $K_3Fe(CN)_6$
100 ml .... H_2O
DISCUSSION:
W, thin films deposited by LPCVD on poly-Si epitaxy grown on SiO_2 coated silicon, (100) p-type wafers. Solution used to step-etch tungsten for stylus depth measurements.
REF: Tsao, K Y & Busta, H H — *J Electrochem Soc*, 131,2702(1984)

W-0034b
ETCH NAME: TIME:
TYPE: Acid, removal TEMP: RT
COMPOSITION:
1 .... NH_4OH
2 .... H_2O_2
DISCUSSION:
W, thin films. Solution is recommended for cleaning vacuum systems for the removal of titanium and tungsten from stainless steel, glass, copper and ceramics. It can also be used for etching tungsten and ti-tungsten thin films.
REF: Nichols, D R — *Solid State Technol*, December 1979

W-0036a
ETCH NAME: TIME: 2 min
TYPE: Acid, removal TEMP: RT
COMPOSITION:
1 .... 30% HF
1 .... 70% HNO_3
DISCUSSION:
W single crystal wafers of various orientations used in a study of etched surface properties.
REF: Hughes, F L et al — *Phys Rev*, 113,1023(1959)

PHYSICAL PROPERTIES OF TUNGSTEN CARBIDE, WC

Classification	Carbide
Atomic numbers	74 & 6
Atomic weight	196

Melting point (°C)	2780 (2777)
Boiling point (°C)	6000
Application limits (°C)	1500
Density (g/cm^3)	14.3 (15.7)
Specific heat (mean)(J/kg °C) 25—1000°C	300
Linear coefficient of thermal expansion ($\times 10^{-6}$ cm/cm/°C) 25—800°C	6.3
Thermal conductivity (W/m °C) 1100°C)	43.3
Electrical resistivity (ohm cm) 1000°C	15
Hardness (Mohs — scratch)	9+
Crystal structure (isometric — normal)	(100) cube
Color (solid)	Grey
Cleavage (cubic — poor)	(001)

TUNGSTEN CARBIDE, WC

General: Does not occur as a natural compound. Tungsten occurs mainly as an oxide, and there is one sulfide as minerals. Sheelite, $CaWO_4$, and wolframite, $(Fe,Mn)WO_4$ are the chief ore minerals. The carbide is artificially produced as one of the high temperature ceramics with uses similar to those of silicon carbide, though not as widely used. Several carbides have been used as surface coatings on metal cutting tools to improve wear resistance and tool working life.

Technical Application: There has been no major use in Solid State processing, to date, though the material has been studied as a single crystal. It has been deposited as a columnar thin film crystalline structure. There are possible applications as a dielectric type surface coating on electronic devices, and as a passivation material.

Etching: Acids and alkalies for cleaning. Heat for structure.

WC-0001

ETCH NAME: Murakami's reagent — TIME: 5 min

TYPE: Acid, cleaning — TEMP:

COMPOSITION: — ASTM #98

10 g .... $K_3Fe(CN)_6$

10 g .... KOH (NaOH)

100 ml .... H_2O

DISCUSSION:

WC specimens. Used as a cleaning etch for tungsten carbide surfaces. Can also be used on titanium carbide, TiC and tantalum carbide, TaC. See FE-0003 for WC deposition on steel.

REF: Pamphlet: ASTM E407-70 (1947) #98

WC-0002

ETCH NAME: Heat — TIME:

TYPE: Thermal, preferential — TEMP:

COMPOSITION:

x .... heat

DISCUSSION:

WC, single crystal specimens. Specimens were polished with diamond paste and subjected to thermal annealing to observe structure and slip.

REF: Takahashi, T & Freise, E J — *Phil Mag*, 12,12(1965)

WC-0003
ETCH NAME: TIME:
TYPE: TEMP:
COMPOSITION:
DISCUSSION:

WC, thin films RF magnetron sputter deposited on SST substrates in acetylene gas with a tungsten target. The steel substrates were (1) polished with diamond-grit in oil paste; (2) cleaned in dilute HNO_3; (3) alcohol ultrasonic rinsed and then Ar^+ ion sputtered in vacuum for 10 min before WC growth. WC growth was columnar.

REF: Srivastava, P K et al — *J Vac Sci Technol,* A(2),1261(1984)

PHYSICAL PROPERTIES OF TUNGSTEN SELENIDE, WSe_2

Classification	Selenide
Atomic numbers	74 & 34
Atomic weight	341.84
Melting point (°C)	1500 est.
Boiling point (°C)	
Density (g/cm^3)	12 est.
Hardness (Mohs — scratch)	5—6
Crystal structure (hexagonal — normal)	$(10\bar{1}0)$ prism
Color (solid)	Grey
Cleavage (basal)	(0001)

TUNGSTEN SELENIDE, WSe_2

General: Does not occur in nature as a compound, although there are other metallic selenide minerals. There has been no use of the selenide in industry at the present time.

Technical Application: There is no major use of the compound in Solid State processing, though it has been grown as a single crystal in general material studies with possible applications as a photo-anode.

Etching: Mixed acids $HF:HNO_3$, and halogens.

TUNGSTEN SELENIDE ETCHANTS

WSE-0001a
ETCH NAME: Carbon tetrachloride TIME:
TYPE: Solvent, cleaning TEMP:
COMPOSITION:
x .... CCl_4
DISCUSSION:

WSe_2, single crystal specimens studied for use as photo-anodes. After polishing in a BRM solution, carbon tetrachloride was used to wash remove residual bromine from surfaces. (*Note:* BRM solutions are x% Br_2:MeOH.)

REF: Baglio, J A et al — *J Electrochem Soc,* 129,1461(1982)

WSE-0001b
ETCH NAME: BRM TIME:
TYPE: Halogen, polish TEMP:
COMPOSITION:
x .... x% Br_2
x .... MeOH

DISCUSSION:

WSe_2, single crystal specimens used in a study for photo-anodes. After etch wash cleaning (WSE-0001a), polish and thin single crystals in a BRM solution. Thin specimens can be peeled from the bulk material. Also evaluated: MoS_2, WS_2 and $MoSe_2$.

REF: Ibid.

PHYSICAL PROPERTIES OF TUNGSTEN SULFIDE, WS_2

Classification	Sulfide
Atomic numbers	74 & 16
Atomic weight	248.04
Melting point (°C)	1000 est.
Boiling point (°C)	
Density (g/cm³)	7.5 (7.4)
Hardness (Mohs — scratch)	2.5
Crystal structure (orthorhombic — normal)	(100) pinacoid
Color (solid)	Lead-grey
Cleavage (basal)	(001)

TUNGSTEN SULFIDE, WS_2

General: Occurs as the natural mineral tungsenite, WS_2, associated with other tungsten ores as an earthy deposit, has not been observed as a natural single crystal, but as a colloidal structure. Not used in industry as a natural mineral, but tungsten metal has been processed through the sulfurization process to introduce sulfur into the metal.

Technical Application: Tungsten sulfide has been grown as a single crystal in Solid State development with applications as a photo-anode, and is under evaluation with other sulfides and selenides.

Etching: Mixed acids $HF:HNO_3$; halogens.

TUNGSTEN SULFIDE ETCHANTS

WS-0001

ETCH NAME: BRM — TIME:

TYPE: Halogen, polish — TEMP: RT

COMPOSITION:

x .... x% Br_2

x .... MeOH

DISCUSSION:

WS_2, single crystal specimens used in a study for photo-anodes. After etch polishing in solution shown, wash in CCl_4 to remove residual bromine. Thin single crystal wafer units can be peeled from the bulk ingot. Other materials evaluated were WSe_2, MoS, and $MoSe_2$.

REF: Raglio, J A et al — *J Electrochem Soc,* 129,1461(1982)

TUNGSTEN:RHENIUM ETCHANTS

WRH-0001

ETCH NAME: Sodium hydroxide — TIME:

TYPE: Electrolytic, polish — TEMP:

COMPOSITION: — ANODE:

x .... 0.75% NaOH — CATHODE:

POWER:

DISCUSSION:

WRh(2%) and (6%), (100) wafers cut from single crystal ingots. Solution used to electropolish specimens in the study of surface work function.

REF: Abey, A E — *J Appl Phys,* 39,120(1968)

PHYSICAL PROPERTIES OF TUNGSTEN TRIOXIDE, WO_3

Classification	Oxide
Atomic numbers	74 & 8
Atomic weight	231.92
Melting point (°C)	1473
Boiling point (°C)	
Density (g/cm^3)	7.16
Refractive index (n=)	2.00—2.26
Hardness (Mohs — scratch)	6—7
Crystal structure (orthorhombic — normal)	(100) pinacoid
Color (solid)	Bright yellow
Cleavage (basal — perfect)	(001)

TUNGSTEN TRIOXIDE, WO_3

General: Occurs in nature as the mineral tungstite, WO_3 or as the hydrated colloid mineral keymacite, $WO_3.H_2O$. In either case, occurrence is rare and usually associated with wolframite, (Fe,Mn)WO_4, as a surface coating, occasionally as an earthy mass. Due to its brilliant yellow color it is called tungstic ocher, occurring naturally as a fine, yellow powder.

Not used industrially as natural occurring compound due to scarcity, but as an artificial compound has been used as a coloring pigment in glass, as a surface glaze, or in coloring ceramics. When a light bulb filament burns out, traces of either the yellow trioxide or brown dioxide are often seen.

Technical Application: Not used in Solid State device fabrication at present, although it may have application as a thin film oxide coating similar to that of other metal oxides, such as silicon dioxide, SiO_2 and aluminum oxide, Al_2O_3. It has been observed as a by-product reaction in the sputter deposition of Ti-tungsten, TiW(10%Ti) on glass and quartz substrates as a yellow/orange gel remaining after Ti-tungsten removal.

Tungsten oxides, WO_x have been deposited as thin films on both tin oxide, SnO_2 and indium tin oxide, $InSnO_2$ (ITO). On glass, WO_3 has been converted to single crystal by annealing.

Etching: Soluble in HF and hot alkalies.

Note: Tungsten dioxide, WO_2, is an artificial compound with isometric-normal structure as a cube, (100).

TUNGSTEN OXIDE ETCHANTS

WO-0001a

ETCH NAME: Potassium hydroxide — TIME:

TYPE: Alkali, removal — TEMP: 60°C

COMPOSITION:

x .... 20—30% KOH (NaOH)

DISCUSSION:

WO_2. This oxide is cubic in structure and insoluble in acids. Color: brown. In removing sputtered thin films of TiW(10%Ti) from alumina and quartz substrates a grey/brownish

film may remain — WO_x/TiO_x. Sequence etching in concentrated HF, then 30% H_2O_2 and then KOH will remove both oxides.
REF: Weast, R C et al — *Handbook of Chemistry and Physics,* 65th ed, CRC Press, Boca Raton, FL, 1984, B154
WO-0002b: Ibid.

WO_x thin films as TiW(10%Ti) sputtered on alumina and quartz substrates. See discussion above.

WO-00016
ETCH NAME: Potassium hydroxide — TIME: 2—10 min
TYPE: Alkali, removal — TEMP: Hot (60°C)
COMPOSITION:
x .... 30% KOH
DISCUSSION:

WO_3. This oxide is hexagonal system, rhombohedral division in structure. Color is yellow to yellow-orange. Thin film TiW(10% Ti) Rf sputtered on alumina and quartz substrates. TiW removed by soaking in H_2O_2 but can leave a yellow-orange gel — hydrated WO_3. Remove gel by soaking in hot KOH.
REF: Weast, R C et al — *Handbook of Chemistry and Physics,* 65th ed, CRC Press, Boca Raton, FL, 1984, B145
WO-0002c: Ibid.

WO_3 hydrated film left on alumina and quartz substrates after stripping TiW (10% Ti) RF sputtered deposits. See discussion above.

WO-0003
ETCH NAME: Sulfuric acid — TIME:
TYPE: Acid, removal — TEMP:
COMPOSITION:
x .... 7.8 *N*—13.3 *N* H_2SO_4
DISCUSSION:

WO_3 thin film deposited on 2100 Å SnO_2 on a glass substrate. Deposit is amorphous. Heat treating in argon at greater than 350°C will convert to single crystal. Use sulfuric acid to etch remove.
REF: Viennet, R & Randin, J F — *J Electrochem Soc,* 129,2349(1982)

WO-0007
ETCH NAME: — TIME:
TYPE: Oxide, growth — TEMP: RT to 100°C
COMPOSITION: — GAS FLOW:
(1) x .... Ar — PRESSURE: 3×10^{-1} Torr
(2) xx .... Ar — POWER: 100 W
x .... 0—1% O_2 — RATE: 30—60 Å/min to 3000 Å
DISCUSSION:

Li_xWO_3 (blue) and Li_xWO_3 (clear) sputter deposited in gases (1) and (2), respectively, as an A or B film. During growth of B film, the A film bleaches to an A′ film. Deposited as a B/A film on an ITO coated quartz substrate, then Au evaporated with electrical contact between Au and ITO to measure device response time with an He/Ne laser. Layer B operates as an electrolyte supplying layer A with Li^+ ions for electrochromic action. Bias alters A-A′ color change from blue (+) to clear (−) . . . "bleached". Optical density vs. response time measured with a voltagram. Very high absorption of infrared light in B film, sharply

reduced with addition of oxygen. Pressed powder target: Li_2CO_3:WO_3 used for RF sputter depositing.
REF: Kitabatake, M et al — *J Electrochem Soc,* 132,433(1985)

WO-0004
ETCH NAME:
TYPE: Electrolytic, removal
COMPOSITION:
x .... H_2SO_4

TIME:
TEMP:
ANODE:
CATHODE:
POWER: 1.5 V

DISCUSSION:
WO_x (x = 3), thin film RF magnetron sputtered, RF plated or evaporated onto ITO ($InSnO_2$). Under electrical bias with a Pt counter-electrode the WO_x turns blue-black.
REF: Yoshimuria, T — *J Appl Phys,* 57,911(1985)

WO-0001c
ETCH NAME:
TYPE:
COMPOSITION:

TIME:
TEMP:

DISCUSSION:
W_2O_5. Color is violet-black. Insoluble in acids.
REF: Weast, R C et al — *Handbook of Chemistry and Physics,* 65th ed, CRC Press, Boca Raton, FL, 1984, B145

WO-0005
ETCH NAME: Sodium hydroxide
TYPE: Alkali, removal
COMPOSITION:
x .... 0.01—0.04 *N* NaOH

TIME:
TEMP: RT
RATE: 1000°A/minute

DISCUSSION:
a-WO_3 thin films 499—8500 Å thick, evaporated from a tungsten boat on In_2O_3 coated glass substrates. Films irradiated with sodium silicate or aluminosilicate. Etch patterned with solution shown for use as a high-contrast inorganic ion resist material.
REF: Koshida, N & Tomita, O — *Jpn J Appl Phys,* 24,92(1985)

PHYSICAL PROPERTIES OF URANIUM, U

Classification	Actinide
Atomic number	92
Atomic weight	238.03
Melting point (°C)	1132
Boiling point (°C)	3818
Density (g/cm^3)	19.04 (19.07)
Thermal conductance (cal/sec)(cm^2)(°C/cm)	0.064
Specific heat (cal/g) 25°C	0.028
Latent heat of fusion (cal/g)	19.75
Heat of fusion (k-cal/g-atom)	2.7
Heat of vaporization (k-cal/g-atom)	110
Atomic radius (W/D)	12.5
Electronegativity (Pauling's)	1.7
Covalent radius (angstroms)	1.42
Ionic radius (angstroms)	0.97 (U^{+4})
Electrical conductance (micro ohms^{-1})	0.034
Magnetic susceptibility (emu/g $\times 10^{-6}$) 15°C	1.72
Thermal expansion (ppm/°C)	7.14
Half life ($\times 10^9$ years)	4.51
Gross section (barns)	7.6
Electronwork function (eV)	3.63
Hardness (Mohs — scratch)	3—4
(Brinell — kgf/mm^2)	187
Crystal structure (orthorhombic — normal)	(100) a-pinacoid
Color (solid)	Grey-greenish
Cleavage (basal)	(001)

URANIUM, U

General: Does not occur as a native metal. There are over 50 uranium-containing minerals as oxides, vanadinates, or phosphates, etc. The two primary ores are carnotite (brilliant yellow and orange color) and uraninite (pitchblende, and coal-black color), often intimately mixed together. All uranium minerals are radioactive with a slow disintegration series ending in heavy lead, Pb. The discovery of radium in pitchblende (it is one million times more active than uranium) by the Curies' — pure radium by Madame Curie in 1910 — established the field of radioactivity.

Though most people think of uranium as the atomic bomb, it has far more important scientific applications. The ratio of helium gas in uranium ore, which shows a constant rate of decay — 1 cc in 16,000,000 years — is used for radioactive dating as the radium clock). It indicates that the earth is at least 1.5 billion years old. Radium is luminescent, subject to light and decay, it glows in the dark, and was used on clock and watch faces for many years before the true danger of radioactivity was established.

Industrially, today the most important application of uranium and its associated decay series is for nuclear energy power production, and there are power plants in every industrial nation throughout the world. In medicine, radiation therapy is a primary tool in the cure of cancer and, in the food industry, it is under development for food preservation. Uranium is used as a glass pigment for its luminosity and yellow color.

Technical Application: The particle emissions — alpha, beta, and gamma rays from radioactive materials — are of importance in scientific study. Semiconductor alpha radiation-counter devices have been tested using alpha particles with subsequent device use in space probes to evaluate the radiation level in the Van Allen belt. Lead pellets are alpha radiation activated with cobalt-60 for such testing purposes, and polonium metal strips are used in static eliminators: a camel-hair brush with a Po strip or a unit blowing cold/hot air across the strip with an operational half-life of approximately 1 year. Such air eliminates the build-up of a static charge on high frequency microwave devices, such as GaAs field effect transistors (FETs). Static elimination systems are built into air ducts or within Clean Rooms to prevent dust accumulation, as dust can develop a static charge which is device destructive. Note that on a very dry day, when humidity is around 10% RH, just walking past a static-sensitive device can be enough to short-out and destroy such a device if not properly grounded.

Uranium pressed pellets have been the subject of much study, and it has been grown as a single crystal for morphological data.

Etching: Soluble in single acids and mixed acids.

U-0001a
ETCH NAME: Nitric acid — TIME:
TYPE: Acid, cleaning — TEMP:
COMPOSITION:
x .... HNO_3
DISCUSSION:
U specimens. The following sequence was used to clean uranium surfaces: (1) dip in HNO_3, (2) rinse in DI H_2O, and (3) rinse in acetone.
REF: Robb, W L — *J Electrochem Soc,* 108,126(1959)

U-0002a
ETCH NAME: — TIME: 5 min
TYPE: Acid, polish — TEMP: RT
COMPOSITION:
x .... HNO_3
x .... CrO_3
x .... H_2O
x ... Al_2O_3, abrasive
DISCUSSION:
U specimens. Used as a lapping/polishing acid slurry to etch and polish uranium material.
REF: Cochran, F L & Wallace, W P — *Research & Development,* October 1966, 126

U-0002b
ETCH NAME: — TIME: Minutes
TYPE: Acid, polish — TEMP: RT
COMPOSITION:
x .... HF
x .... HNO_3
x .... H_2O
DISCUSSION:
U specimens. Apply as drops on the surface at a rate of approximately 15 drops/min to polish.
REF: Ibid.

U-0002c
ETCH NAME: TIME:
TYPE: Electrolytic, polish TEMP:
COMPOSITION: ANODE: U
1 .... H_2SO_4 CATHODE:
DISCUSSION: POWER: 500 mA/cm^2
U specimens. Used as an electropolish etch.
REF: Ibid.

U-0002d
ETCH NAME: TIME: 30 sec
TYPE: Acid, polish TEMP: RT
COMPOSITION:
130 ml .... H_2SO_4
50 ml .... H_2O_2
60 ml .... H_2O
2 mg .... Na_2SiF_6
DISCUSSION:
U specimens. Used as a polish etch.
REF: Ibid.

U-0002e
ETCH NAME: TIME: 1 min
TYPE: Electrolytic, polish TEMP: RT
COMPOSITION: ANODE: U
300 ml .... HAc CATHODE:
30 ml .... H_2O POWER: 200 mA/cm^2
25 g .... CrO_3
DISCUSSION:
U specimens. Used as an electropolish etch.
REF: Ibid.

U-0003a
ETCH NAME: TIME:
TYPE: Electrolytic, polish TEMP: 10°C
COMPOSITION: ANODE: U
133 ml .... HAc CATHODE: SST
25 g .... CrO_2 POWER: 0.4 A/cm^2 &
7 ml .... H_2O 35—40 V
DISCUSSION:
Alpha-U specimens. Solution used to electropolish specimens.
REF: Hudson, B — *Phil Mag*, 9,949(1964)

U-0003b
ETCH NAME: TIME: 20 sec
TYPE: Electrolytic, cleaning TEMP: RT
COMPOSITION: ANODE: U
75 ml .... H_2SO_4 CATHODE: Ni
18 ml .... glycerin POWER: 9.15 mA/cm^2 &
7 ml .... H_2O 6—10 V

DISCUSSION:

Alpha-U specimens. Solution used as an electrolytic cleaning system.

REF: Ibid.

U-0001b

ETCH NAME: Iodine | TIME:

TYPE: Halogen, cleaning | TEMP: Red heat

COMPOSITION:

x .... I_2, vapor

DISCUSSION:

U specimens. Dip and coat surfaces with iodine solution that will produce UI_4. In vacuum system, bring specimen up to red heat to vaporize UI_4 and clean surface. Used prior to depositing zirconium on uranium by the iodine process. (*Note:* Iodine surface treatment is used on diamonds as a surface conditioner against etching. This vapor treatment can be used on any material surface as a complexing agent . . . heat above 100°C . . . to remove ionic contamination (F^+, Cl^+, O^+, etc.) from previous etching for a micromolecular clean surface.)

REF: Ibid.

PHYSICAL PROPERTIES OF URANIUM CARBIDE, UC

Classification	Carbide
Atomic numbers	92 & 6
Atomic weight	251 (263)
Melting point (°C)	2260
Boiling point (°C)	4100
Density (g/cm^3)	11.28
Hardness (Mohs — scratch)	9+
Crystal structure (isometric — normal)	(100) cube
Color (solid)	Grey
Cleavage (cubic)	(001)

URANIUM CARBIDE, UC

General: Does not occur as a compound in nature. All carbides are artificially grown, other than a single occurrence of silicon carbide, SiC associated with a meteor strike at Cañon Diablo, Arizona. There has been no industrial use of the carbide. Silicon carbide (carborundum), in particular, is far more useful as a major lapping and polishing abrasive.

Technical Application: No use in Solid State processing to date. The carbide has been grown as a single crystal for general morphological studies along with the nitride and sulfide. Other metal uraninides also are shown here.

Etching: Single and mixed acids.

URANIUM CARBIDE ETCHANTS

UC-0001

ETCH NAME: | TIME:

TYPE: Electrolytic, jet thin | TEMP:

COMPOSITION: | ANODE:

4 .... H_3PO_4 | CATHODE:

1 .... MeOH | POWER:

DISCUSSION:

UC, (001) wafers cleaved from single crystal ingots and used in a study of slip with observation under an electron microscope (EM). Solution used to thin UC for microscope observation. Solution also used for uranium nitride and uranium sulfide.

REF: Solo, M J & Van Der Walt, C M — *Acta Metall,* 16,501(1968)

UC-0002

ETCH NAME: Nitric acid — TIME:

TYPE: Acid, removal — TEMP:

COMPOSITION:

x .... HNO_3

DISCUSSION:

UC_2 specimens. Material is soluble in acids.

REF: Hodgman, C D — *Handbook of Chemistry and Physics,* 27th ed, Chemical Rubber Co., Cleveland, OH, 1943, 482

UC-0003

ETCH NAME: — TIME:

TYPE: Acid, removal — TEMP:

COMPOSITION: — ANODE:

x .... H_3PO_4 — CATHODE:

x .... MeOH — POWER:

DISCUSSION:

UC single crystal wafers cleaved (100) used in a study of uranium dicarbide precipitates. Solution shown used as a wet chemical etch (WCE) to reduce size of initial specimens. Solution as an electrolytic jet polish with added MeOH to reduce viscosity: 200 VDC and 150—200 mA/cm^2 used for general removal; 85 V then to 45 V for x-min to initial polish with final etching done with 1-sec on-off switching. Remove from etch bath under MeOH spray, rotating nozzle to prevent pitting. Store under MeOH. No surface alteration for about 1 h. Electron microscope used for study.

REF: Eyre, B L & Sole, M J — *Phil Mag,* 9,545(1964)

URANIUM CESIUM ETCHANTS

UCE-0001

ETCH NAME: Kalling's etch — TIME: 10—60 sec

TYPE: Acid, polish — TEMP: RT

COMPOSITION:

100 ml .... HCl

5 g .... $CuCl_3$

100 ml .... EOH

DISCUSSION:

UCe_2 arc melted in forming and used in a study of metallic compounds. Solution used as a general removal and polishing etchant. Other uranium compounds studied were UPt_3, UNi_2, and UCo_2. See Cerium Platinide: CEPT-0001 for other compounds studied.

REF: Slepowronsky, M et al — *J Cryst Growth,* 65,293(1983)

URANIUM COBALT ETCHANTS

UCO-0001

ETCH NAME: Kalling's etch | TIME: 10—60 sec

TYPE: Acid, polish | TEMP: RT

COMPOSITION:

100 ml .... HCl

5 g .... $CuCl_3$

100 ml .... EOH

DISCUSSION:

UCo_2 arc melted in forming and used in a study of metallic compounds. Solution used as a general removal and polishing etchant. Other uranium compounds studied were UCe_2, UNi_2, and UPt_3. See Cesium Platinide: CEPT-0001 for other compounds studied.

REF: Slepowronsky, M et al — *J Cryst Growth,* 65,293(1983)

URANIUM NICKELIDE ETCHANTS

UNI-0001

ETCH NAME: Kalling's etch | TIME: 10—60 sec

TYPE: Acid, polish | TEMP: RT

COMPOSITION:

100 ml .... HCl

5 g .... $CuCl_3$

100 ml .... EOH

DISCUSSION:

UNi_2 arc melted in fabrication and used in a study of metallic compounds. Solution used as a general removal and polishing etchant. Other uranium compounds studied were UCe_2, UCo_2, and UPt_3. See Cesium Platinide: CEPT-0001 for other compounds studied.

REF: Slepowronsky, M et al — *J Cryst Growth,* 65,293(1983)

PHYSICAL PROPERTIES OF URANIUM DIOXIDE, UO_2

Classification	Metal oxide
Atomic numbers	92 & 8
Atomic weight	271
Melting point (°C)	2176
Boiling point (°C)	
Density (g/cm^3)	10.9
Hardness (Mohs — scratch)	6—7
Crystal structure (orthorhombic — normal)	(100) prism
Color (solid)	Brown/black
Cleavage (basal — distinct)	(001)

URANIUM DIOXIDE, UO_2

General: Occurs in nature as a hydrate oxide as the minerals ianthinite, $UO_2.7H_2O$, which is orthorhombic, violet black color to yellow at edges, and becquerelite, $UO_3.2H_2O$, also orthorhombic, but brownish-yellow in color. Both have distinctive (001) cleavage. There is little use in industry other than in processing uranium ores, and as a fluorescent yellow coloring agent in glass.

Technical Application: There is no use in Solid State device processing, although the dehydrated oxide has been grown as a single crystal for general structural and dislocation study.

Etching: HNO_3, H_2SO_4. Electrolytic H_3PO_4.

URANIUM DIOXIDE ETCHANTS

UO-0001
ETCH NAME: TIME:
TYPE: Electrolytic, polish TEMP:
COMPOSITION: ANODE: UO_2
67 ml .... H_2SO_4 CATHODE:
310 ml .... H_3PO_4 POWER:
120 ml .... H_2O
78 g .... CrO_3
DISCUSSION:
UO_2 specimens. Solution used to electropolish and to thin specimens.
REF: Whapham, A D & Wheldon, B E — *Phil Mag,* 12,1179(1965)

UO-0002a
ETCH NAME: Nitric acid TIME:
TYPE: Acid, removal TEMP:
COMPOSITION:
x .... HNO_3
DISCUSSION:
UO_2 specimens. Shown as an etch for this material, and can be used diluted with water.
REF: Hodgman, C D — *Handbook of Chemistry and Physics,* 27th ed, Chemical Rubber Co., Cleveland, OH, 1943, 482

UO-0002b
ETCH NAME: Sulfuric acid TIME:
TYPE: Acid, removal TEMP:
COMPOSITION:
x .... H_2SO_4 conc.
DISCUSSION:
UO_2 specimens. Shown as an etch for this material.
REF: Ibid.

PHYSICAL PROPERTIES OF URANIUM NITRIDE, U_3N_4

Classification	Nitride
Atomic numbers	92 & 7
Atomic weight	775
Melting point (°C)	>800
Boiling point (°C)	
Density (g/cm³)	10.09
Hardness (Mohs — scratch)	6—7
Crystal structure (isometric — normal)	(100) cube
Color (solid)	Brown/black
Cleavage (cubic)	(001)

URANIUM NITRIDE, U_3N_4

General: Does not occur as a native metallic compound, although there are many other metal nitrates and nitrites. There is no use of the nitride in industry at the present time.

Technical Application: There is no use for the nitride in Solid State material processing, to date, although it has been grown as a single crystal for general study as a compound.

Etching: HNO_3. Electrolytic H_3PO_4.

URANIUM NITRIDE ETCHANTS

UN-0001

ETCH NAME: TIME:

TYPE: Electrolytic, jet thin TEMP:

COMPOSITION: ANODE:

4 .... H_3PO_4 CATHODE:

1 .... MeOH POWER:

DISCUSSION:

UN, (001) wafers cleaved from single crystal ingots used in a study of oxidation and deformation. Solution used to jet thin specimens for electron microscope (EM) study.

REF: Solo, M J & Van Der Walt, C M — *Acta Metall,* 16,501(1968)

UN-0002

ETCH NAME: Nitric acid TIME:

TYPE: Acid, removal TEMP:

COMPOSITION:

x .... HNO_3

DISCUSSION:

U_3N_4 specimens. Material is soluble in nitric acid.

REF: Hodgman, C D — *Handbook of Chemistry and Physics,* 27th ed, Chemical Rubber Co., Cleveland, OH, 1943, 482

URANIUM PLATINIDE ETCHANTS

UPT-0001

ETCH NAME: Kalling's etch TIME: 10—60 sec

TYPE: Acid, polish TEMP: RT

COMPOSITION:

100 ml .... HCl

5 g .... $CuCl_3$

100 ml .... EOH

DISCUSSION:

UPt_3 arc melted in fabrication and used in a study of metallic compounds. Solution used as a general removal and polishing etchant. Other uranium compounds studied were UCe_2, UCo_2, and UNi_2. See Cesium Platinide: CFPT-0001 for other compounds studied.

REF: Slepowronsky, M et al — *J Cryst Growth,* 65,293(1983)

PHYSICAL PROPERTIES OF URANIUM DISULFIDE, US_2

Classification	Sulfide
Atomic numbers	92 & 16
Atomic weight	302.19
Melting point (°C)	1100
Boiling point (°C)	
Density (g/cm^3)	
Hardness (Mohs — scratch)	3—4
Crystal structure (tetragonal — normal)	(100) prism
Color (solid)	Grey/black
Cleavage (basal)	(001)

URANIUM SULFIDE, US

General: Does not occur in nature as a sulfide, although there are two sulfate minerals, johannite and zippeite, both as hydrates. The sulfide has had no use in industry, although uranium can be sulfurized, much like nitriding.

Technical Application: There has been no use in Solid State processing, although the sulfide has been grown as a single crystal for general study.

Etching: HCl, HNO_3. Electrolytic H_3PO_4.

URANIUM SULFIDE ETCHANTS

US-0001
ETCH NAME: TIME:
TYPE: Electrolytic, jet thin TEMP:
COMPOSITION: ANODE:
4 .... H_3PO_4 CATHODE:
1 .... MeOH POWER:
DISCUSSION:

US, (001) wafers cleaved from single crystal ingot and used in a study of slip with an electron microscope (EM). Compressed air was used for the acid jet action to thin specimens. Also used on uranium nitride and uranium carbide.

REF: Soto, M J & Van Der Walt, C M — *Acta Metall*, 16,501(1968)

US-0002
ETCH NAME: Hydrochloric acid TIME:
TYPE: Acid, removal TEMP:
COMPOSITION:
x .... HCl, conc.
DISCUSSION:

US_2 specimens. Solution shown as a general etch for this material.

REF: Hodgman, C D — *Handbook of Chemistry and Physics*, 27th ed, Chemical Rubber Co., Cleveland, OH, 1943, 482

URANIUM TITANIUM ETCHANTS

UTI-0001

ETCH NAME:

TYPE: Electrolytic, polish

COMPOSITION:

60 ml .... $HClO_4$

350 ml .... butyl cellusolve

350 ml .... MeOH

TIME: 10 sec

TEMP: 8°C

ANODE:

CATHODE:

POWER: 50 V

DISCUSSION:

UTi, single crystal specimens used in a diffusion study. Solution shown used as a general removal and polish etch.

REF: Adda, Y & Philibert, J — *Acta Metall,* 8,700(1960) (in French)

PHYSICAL PROPERTIES OF VANADIUM, V

Classification	Transition metal
Atomic number	23
Atomic weight	50.942
Melting point (°C)	1750 (1900)
Boiling point (°C)	3000
Density (g/cm³)	6.11
Thermal conductance (cal/sec)(cm²)(°C/cm)100°C	0.074
Specific heat (cal/g) 20°C	0.120
Heat of fusion (k-cal/g-atom)	4.2
Latent heat of fusion (cal/g)	82.5
Heat of vaporization (k-cal/g-atom)	106
Atomic volume (W/D)	8.35
1st ionization energy (K-cal/g-mole)	156
1st ionization potential (eV)	6.74
Electronegativity (Pauling's)	1.6
Covalent radius (angstroms)	1.22
Ionic radius (angstroms)	0.59 (V^{+5})
Coefficient of linear thermal expansion ($\times 10^{-6}$ cm/cm/°C) 25°C	9.7
Electrical resistivity (micro-ohms cm)	24.8
Modulus of elasticity (psi $\times 10^6$)	18—19
Cross section (barns)	4.9
Magnetic susceptibility (ergs $\times 10^{-6}$)	1.4
Poisson ratio	0.36
Shear modulus (psi $\times 10^6$)	6.73
Refractive index (n=)	3.03
Hardness (Mohs — scratch)	4
Crystal structure (isometric — normal)	(100) cube, bcc
Color (solid)	Silver white
Cleavage (cubic — indistinct)	(001)

VANADIUM, V

General: Does not occur as a native element, but is found as a minor constituent in several minerals with the most important ore being vanadinite, $(PbCl)Pb_4(VO_4)_3$. The mineral patronite is a complex vanadium sulfate containing large quantities of vanadium and is one of the few normally amorphous minerals. As pure vanadium it is a high temperature, infusible metal.

As an alloy in steels it greatly increases toughness, elasticity, and tensile strength and as an oxide coloring agent — blue, yellow, and red, depending upon the oxide — is widely used in the glass and ceramic industries. The oxide also is used in the preparation of sulfuric acid.

Technical Application: Not presently used in Solid State device fabrication although as a silicide it is under evaluation as a buffer layer in heterostructure device like other metal silicides.

It has been studied in both crystalline and single crystal form. Also as a carbide, VC and boride, VB and as other metallic compounds.

Etching: HNO_3, H_2SO_4, HF, and aqua regia but insoluble in HCl and alkalies.

VANADIUM ETCHANTS

V-0001
ETCH NAME: TIME:
TYPE: Acid, removal/polish TEMP:
COMPOSITION:
(1) 1 .... HNO_3
1 .... H_2O
(2) 1 .... HF
1 .... HNO_3
DISCUSSION:
V specimens. Both solutions are general removal and polishing etches.
REF:

V-0002a
ETCH NAME: Aqua regia TIME:
TYPE: Acid, removal TEMP:
COMPOSITION:
3 .... HCl
1 .... HNO_3
DISCUSSION:
V specimens. Shown as a general etch for vanadium.
REF: Hodgman, C D — *Handbook of Chemistry and Physics,* 27th ed, Chemical Rubber Co, Cleveland, OH, 1943, 484

V-0002b
ETCH NAME: Sulfuric acid TIME:
TYPE: Acid, remove TEMP:
COMPOSITION:
x .... H_2SO_4
DISCUSSION:
V, specimens. Shown as a general etch for vanadium.
REF: Ibid.

V-0002c
ETCH NAME: Hydrofluoric acid TIME:
TYPE: Acid, removal TEMP:
COMPOSITION:
x .... HF
DISCUSSION:
V, specimens. Shown as a general etch for vanadium.
REF: Ibid.

V-0003
ETCH NAME: Nitric acid TIME:
TYPE: Acid, sizing TEMP: RT
COMPOSITION:
x .... HNO_3
DISCUSSION:
V, poly-wire used in a study of the effects of pressure on the mobility of interstitial oxygen and nitrogen in vanadium. The wires were reduced in size from 0.020 to 0.015 diameter by etching in nitric acid.
REF: Tichelaar, G W — *Phys Rev,* 121,748(1961)

V-0004
ETCH NAME: TIME:
TYPE: Electrolytic, polish/thin TEMP:
COMPOSITION: ANODE:
1 .... H_2SO_4 CATHODE:
6 .... MeOH POWER: 10 V
DISCUSSION:

V specimens. Specimens neutron irradiation and annealed at 1300—1400°C. Solution used to etch polish and thin specimens for microscope study.

REF: Rau, R C & Ladd, R L — *J Appl Phys*, 40,2899(1969)

V-0005
ETCH NAME: Heat TIME:
TYPE: Thermal, forming TEMP:
COMPOSITION:
x .... heat
DISCUSSION:

V, single crystal and polycrystalline spheres formed by splattering using an arc melter in an He atmosphere. Other materials studied were Be, Fe, Zr, NiCu, and NiAl. Sphere sizes ranged from 0.1 to 1.0 mm diameter and 5 to 50% were single crystal.

REF: Ray, A E & Smith, J F — *Acta Metall*, 11,310(1958)

V-0006
ETCH NAME: TIME:
TYPE: Electrolytic, polish TEMP:
COMPOSITION: ANODE: V
x .... Na_2CO_3 CATHODE:
x .... $NaHCO_3$ POWER:
DISCUSSION:

V specimens. Used solution as an anodic polish etch for this material.

REF: Brown, C M — U.S. Patent 2,803,596 (1957)

V-0007
ETCH NAME: TIME:
TYPE: Acid, removal TEMP:
COMPOSITION:
1 .... HF
10 .... HNO_3
DISCUSSION:

V, thin films evaporated on silicon substrates. Solution used to remove vanadium from silicon surfaces. A study of vanadium growth on silicon.

REF: Miller, K J et al — *J Electrochem Soc*, 113,902(1966)

VANADIUM ALUMINUM ETCHANTS

VAL-0001
ETCH NAME: Nitric acid TIME:
TYPE: Acid, removal/clean TEMP:
COMPOSITION:
x .... HNO_3, conc.
DISCUSSION:

VAl_3, deposited on silicon, (001) substrates in an evaluation of vanadium and aluminum

reactions. Authors say that if there is more than 15% O_2 in the vanadium an Al_2O_3 (?) layer forms between the V/Al interface as a blocking layer. Vanadium was evaporated on various aluminum layer combinations as thin films: Al, Al/Cr; Al/Ti, and Al/TiN.
REF: Finstad, T G et al — *Thin Solid Films,* 114,271(1985)

VANADIUM BROMIDE ETCHANTS

VBR-0001
ETCH NAME: Ethyl alcohol
TYPE: Alcohol, removal/polish
TIME:
TEMP:
COMPOSITION:
x .... EOH (MeOH)
DISCUSSION:
VBr_2, single crystals. Solution shown used as a general cleaning, removal and polishing etchant on the material. Specimens used in a general properties study, and particularly for antiferromagnetic characteristics.
REF: Ailshi, M et al — *J Phys Soc Jpn,* 4,1214(1984)

PHYSICAL PROPERTIES OF VANADIUM CARBIDE, VC

Classification	Carbide
Atomic numbers	23 & 6
Atomic weight	62.96
Melting point (°C)	2810
Boiling point (°C)	3900
Density (g/cm^3)	5.4
Lattice constant (angstroms)	4.16
Hardness (Mohs — scratch)	8—9
Crystal structure (isometric — normal)	(100) cube
Color (solid)	Grey-white
Cleavage (cubic — perfect)	(001)

VANADIUM CARBIDE, VC

General: Does not occur as a native carbide in nature. Industrially, it is a high temperature ceramic material, but has had little use other than for study of characteristics.

Technical Application: It has been grown as single crystal ingots, and as both single crystal and polycrystalline filaments for general study of the material.

Etching: HNO_3. Electrolytic mixed acids.

VANADIUM CARBIDE ETCHANTS

VC-0001
ETCH NAME:
TYPE: Electrolytic, preferential
COMPOSITION:
1 .... HF
1 .... HNO_3
10 .... H_2O
TIME: 20—40 sec
TEMP: RT
ANODE:
CATHODE: VC
POWER: 2—3 V
DISCUSSION:
VC, (100) single crystal ingots of various compositions. Wafers cleaved from ingots.

Solution used in a study of structure and dislocations. Large, flat-bottomed pits appear first, and are considered to be associated with carbon impurity. Small pits that then appear are dislocations.

REF: Hou, Y — *J Cryst Growth,* 68,733(1984)

VC-0002a

ETCH NAME: Nitric acid — TIME:

TYPE: Acid, removal — TEMP:

COMPOSITION:

x .... HNO_3

DISCUSSION:

VC, (100) single crystal and polycrystalline specimens. Shown as a general etch for this material.

REF: Hodgman, C D — *Handbook of Chemistry and Physics,* 27th ed, Chemical Rubber Co, Cleveland, OH, 1943, 484

VC-0002b

ETCH NAME: Potassium nitrate — TIME:

TYPE: Molten flux, preferential — TEMP: >400°C

COMPOSITION

x .... KNO_3

DISCUSSION:

VC specimens. Shown as an etch for this material.

REF: Ibid.

VANADIUM GALLIUM ETCHANTS

VGA-0001

ETCH NAME: Nitric acid, dilute — TIME:

TYPE: Acid, removal/preferential — TEMP:

COMPOSITION:

x .... HNO_3

x .... H_2O

DISCUSSION:

VGa, single crystal filaments grown in a copper flux system. Solution used to etch clean filaments in a morphological study.

REF:

PHYSICAL PROPERTIES OF WATER, H_2O

Classification	Solvent
Atomic numbers	1 & 8
Atomic weight	18.02
Melting point (°C)	0
Boiling point (°C)	100
Density (g/cm^3) 4°C (liquid)	1.00
(g/cm^3) 0°C (solid)	0.9168
Heat of vaporization (cal)	540
Heat of fusion — ice (cal)	76
Hardness (Mohs — scratch — ice)	1—1.5
Crystal structure (hexagonal, normal — ice) alpha	($10\bar{1}0$) platelets
(isometric — normal — ice) beta	(100)
Color (solid)	Bluish
(liquid)	Colorless
Cleavage (basal/cubic — solid)	(0001/001)

WATER, H_2O

General: There are two primary forms of water in nature: (1) salt water and (2) fresh water, but there are over 100 varieties depending upon mineral content, source location, treatment, and so forth. Fresh water is also classified by soft or hard, again, based on mineral content. The oceans cover approximately 85% of the land surface of the world, are salt water, mainly sodium chloride, NaCl, but contain many other dissolved salts with types and concentrations varying with locale and water depth.

Fresh water is land water — lakes, rivers, streams — but it is not "pure" water, as it contains dissolved minerals and organic matter: much of the mineral content is essential to human and animal diet and health. Water also collects underground in aquifer layers — usually sandstone, limestone, or granites — returning to the surface as springs or, when pumped, as well water.

In many industrial areas the underground "water table" is maintained and used as part of the economic water supply of a region. Many of the major rivers in the world have been dammed to form reservoirs as both a water supply and a source of hydroelectric power production — hundreds of small streams have "spill" dams producing mechanical power by use of a water-wheel . . . also used for lifting water in irrigation.

The normal evaporation cycle from the oceans produces naturally distilled water vapor — clouds — then rain with about 80% of such fresh water as snow and ice frozen in the Arctic and Antarctic ice caps.

Many major industries, such as steel plants and other metal processing plants, paper mills, and nuclear reactors are located on rivers or along seacoasts for the high volume source of water needed, primarily, as washing or cooling water in operation. Several of the quartz crystal processing plants are located east of the Continental Divide (Rocky Mountains) in the U.S. to take advantage of the high purity ground water coming off of the Divide.

"Pure" water, either demineralized or distilled, is manmade on site by many companies, particularly those involved in metal processing as well as for medical purposes. Distilled water is available as "bottled" water for car batteries or clothing irons. Water "stills" are nothing more than large boilers with the steam bled off to recondense as distilled water, and it may be single, double, or triple distilled: ion-exchange positive and negative resin beds may be included in the distillation process with, say, the triple distilled — highest purity form — processed through an all quartz boiling system. All distilled or demineralized water is unstable, in a sense, as it will immediately begin absorbing gases, dust and other contamination when exposed to air, reverting to normal, natural mineralized fresh water. Distilled water is also an excellent medium for algae growth, such that distilled water lines require on-going maintenance cleaning — once a week in highly critical surface cleaning operations — and should include submicron filters in-line at the user point, as in a laboratory chemical sink. Inert PVC piping is preferred — not copper piping with rosin-core solder joints . . . and there should not be any stub-off lines where water can collect and stagnate. In closed recirculation systems (recirc systems), common practice is to include a glass tube section in-line that can be flooded with UV light to kill algae.

In chemistry, water is not only called "the staff of life" with regard to human and animal consumption, but is considered the universal solvent as, with sufficient time and/or temperature/pressure conditions, it will dissolve and take all materials into solution, or it will form an environmentally stable oxide such as silica, SiO_2 or alumina, Al_2O_3 as these are the two most stable natural minerals under geologic conditions.

Recently, a new form of water has been discovered; polywater. It is a high density chain of polymer water molecules that can be formed in capillary tubes of about 10 mm diameter. It boils at 150°C and freezes at −40°C into a glass-like substance that does not resemble ice.

Many of the natural minerals contain water in their chemical formula as the hydrated minerals, such as gypsum, $CaSO_4.2H_2O$. With water removed it is the mineral anhydrite, $CaSO_4$, better known as Plaster of Paris . . . used in medicine for immobilizing broken bones and in dentistry as a mold in making false teeth; it is the plasterboard of construction used for holding small stones for cutting and polishing. When dry, powdered calcium sulfate is mixed with water, it sets as a hard white form of gypsum that is relatively inert to further alteration.

The precious and semi-precious gem stone opal, $SiO_2.NH_2O$, has water both in its physical crystal structure and as entrapped liquid. It is a colloid, highly stress fractured, and the entrapped water is what produces the splay of colors. Opals should be moistened occasionally or kept in a humid atmosphere to maintain their "color" — otherwise they will slowly dry out and, eventually, can even commence to crumble and powder. There are many other natural minerals containing water in their formulas like opal and gypsum.

Heavy water, D_2O — deuterium — is best known for its use in developing the first atomic bombs. It does occur naturally in moist air as a very small constituent and has been used as a tracer element in drinking water to determine the length of time water is retained in the body before elimination, approximately 30 d.

Waste disposal is of major concern today, particularly in industrial areas as, in many cities, chemical wastes enter the sewage system or are dumped into rivers, streams, and lakes directly from the user plant. Such contamination, to include "acid rain", has polluted our lakes, waterways, and many coastal areas killing fish and altering the ecology of plant and sea life. Most acids can be readily neutralized with bases, such as ammonia, NH_3 and ammonium hydroxide, NH_4OH added to the acid/water waste, but many chemicals require special filtration and conversion to insoluble salts — chromium acids are particularly difficult to remove and can severely contaminate the underground water table. The same applies to alcohols and solvents: these are disposed of separately, not into sewage drains along with the acids — the mixture can be explosive with certain combinations — and cannot be neutralized by water. Trichloroethylene (TCE) is a major industrial cleaning solvent, is carcinogenic and, in many areas where it has been used, has contaminated the underground water table, springs, and wells. Waste disposal and its neutralization with water, acids, and bacteria is a major industrial business by itself. The waste water can be reconstituted as fresh, human potable water, or for reuse in agriculture and industry. See Ice for additional discussion.

Technical Application: Commercial water — sometimes referred to as "water with the rocks removed" — is used for general cleaning of parts where maximum chemical purity is not required. It also is the water used as the cooling liquid in any heat generating operation to absorb and carry the heat away with continual replacement of cooler water. A system may be water-jacketed; water pipes coiled around the heater vessel; as a bath into which hot parts are quenched. In the latter case, both fresh and salt water are used as a quenching medium: certain types of irons, steels and other metal alloys are structurally formed in this manner. Semiconductor metals and compounds have been quenched for both device fabrication and in structure studies at RT, hot (40—80°C) or chilled with ice (4—10°C).

Many laboratory chemical processing sinks contain both commercial and distilled water faucets with the DI water used for finally cleaning and mixing of étchants. As many etchants are exothermic, a beaker of the etch solution is placed in a water bath — there are commercial systems available for batch etching using recirc water systems — and small ultrasonic generators often use water baths in which the etch beaker is placed for ultrasonically generated bubble stirring action. The larger degreaser systems using cleaning solvents are both water jacketed at the top to reduce hot solvent evaporation loss, but may use ultrasonic vibration cleaning in the hot solvent tank section.

There are a number of "cold solutions" using water, snow, and ice as part of the chemical mixture to establish a specific cold temperature: a bath of ice and acetone, as an example, is often used for slight preferential etching and tuning of a semiconductor p-n junction device with the etch solution at about 8°C. Water and dry ice, solid CO_2, in a bubbler are used to remove water vapor from processing gases by freezing out the water as the gas is bubbled through — argon gas in particular.

Boiling water is used to form a porous hydrated oxide on metal and compound surfaces, which is then removed — usually with HF — as a surface cleaning process, or the oxide remains as surface passivation. Thermal oxidation, as water vapor in an inert carrier gas, such as nitrogen, is carried into a furnace at 1000°C and was the initial method used to produce SiO_2 protective thin films on silicon wafers, the process called wet thermal oxidation. Dry thermal oxidation uses oxygen, O_2 gas in a nitrogen, N_2 or argon, Ar inert gas carrier. The SILOX system uses an $SiH_4/N_2/O_2$ gas mixture for deposition of a silicon dioxide, SiO_2 layer on any metal or metallic compound between 350—500°C. The original pyrolytic deposition of an oxide from liquid chemical solutions passed through a furnace between 400—800°C is now called chemical vapor deposition (CVD) and there are several variations with their own acronyms: LPCVD, low pressure CVD, HPCVD, high pressure CVD, as only two examples.

Steam oxidation, as hot water vapor under pressure measured in atmospheres (atm) or as pounds-per-square-inch (psi) is now used for silicon dioxide growth on silicon wafers, and can be used to deposit other oxides on other metals, such as aluminum, nickel, copper, chromium, etc. Steam also is used for general surface cleaning . . . from a Solid State wafer or electronic part . . . to building walls in cities . . . to clothing in the garment industry. Needless to say, steam under pressure is a major mechanical operator, and was fundamental to the Industrial Revolution that began in England in the early 19th century: railroad engines, steam ships, factories, and home heating. The Romans were building steam-heated homes prior to 1 A.D. in Italy as well as in England. Today, volcanic steam from Mt. Vesuvius in northern Italy is tapped for steam power, and is being developed as an alternate source of electric power generation from subsurface hydrothermal areas throughout the world . . . northern California, as one example.

Even though water is considered neutral, with a pH 7 on the Sorenson Scale — it can act as an "acid" solvent on many soluble minerals, such as single crystal halite, NaCl or sylvite, KCl. These compounds, as well as others, are used as substrates for thin film metal and compound epitaxy growth in the study of growth and crystal structure of the thin films. Water is used to etch polish the surfaces of these substrates prior to deposition; it also is used to remove the subsequently deposited thin film using the "float-off" technique by liquefying the substrate/thin film interface in order to release the thin film for study under a transmission electron microscope (TEM).

Water is not only used as the cooling medium in many cutting operations, to cool both the cutting blade and parts being cut, it is used as the cutting tool, itself, as a water jet under pressure or as high pressure steam.

In the growth of ice crystals (snowflakes) as well as single crystal ice, a cryogenic stage or cryostat for the water vapor chilled with either LN_2, dry ice, CO_2 or LHe, and may be as a cryogenic microscope stage for observing ice growth as it occurs. This may include nucleation with silver chloride, $AgCl_2$, as is done with seeding of rain clouds. And water also has been grown as salt water crystals from natural salt waters or artificial variations of chemical brine mixtures. Cold temperature studies have been in progress in Greenland and on both the Arctic and Antarctic ice caps for many years — core samples of ice have been cut, sectioned, then melted to obtain foraminifera and radiolaria as diatom fossil exoskeletons, as well as for water samples, to establish environmental conditions and age correlations of geologic periods in the past. There are similar studies with annual lake ice.

Several cold solutions have been developed using specific chemicals, alone, or in combination as eutectic mixtures for a specific temperature level below room temperature. Including the cryogenic liquid gases, temperatures range to several degrees below zero. These cold solutions are mainly used to establish a particular temperature for an etching solution, for chilling or quenching metals, and occasionally used as the direct etchant, themselves. The following collection is referred to here as Cold Etchants and is presented as a collective reference.

Etching: As an etching medium, alone, on soluble materials. The primary liquid carrier for mixing or chilling of other etchants.

WATER ETCHANTS

WATR-0001

ETCH NAME: Ice | TIME:

TYPE: Acid, solvent/storage | TEMP: 0°C

COMPOSITION:

x .... H_2O (solid)

DISCUSSION:

H_2O in its frozen solid form, used alone, has some special applications and major areas of study. It will freeze and preserve both organic and inorganic matter. Ice cores 2—4" in diameter are cut from glacial ice in the Antarctic, then sections melted to obtain ancient sea life exoskeletons, to evaluate alkalinity, oxygen content, etc., in determining earth sea life and atmospheric conditions to as much as one million years in the past. In metal studies, hot liquified metal droplets are splashed onto an ice surface or pressed between two blocks of ice to observe quick-frozen material structure. It has been used as a slow etchant on water soluble compounds, an ice cube being rubbed across the surface to polish, or develop preferential etched figures and structure.

Ice caves in mountain areas have been used for food storage since ancient times, the Eskimos still use ice to freeze their food for preservation and, there are records from Asia Minor describing the transportation of ice by camel caravan from the mountains to the desert courts of Mesopotamia, circa 3500 B.C. The Arabs in those ancient times mixed chipped ice with fruit juices — sharbat — today's sherbets or the French sorbets — also called "ices".

In geology, ice expansion and contraction in rock crevasses is a primary method of reducing solid rock to soil. There are still ice houses making solid blocks of ice. See the Ice section for further details.

REF:

WATR-0002
ETCH NAME: Water
TYPE: Acid, solvent
COMPOSITION:
x .... H_2O
TIME:
TEMP: 4—99°C

DISCUSSION:

H_2O in liquid form from above the freezing point to below the boiling point. As already discussed in the General and Technical Application sections, water is considered the "universal solvent" in chemistry. It is the primary quenching medium following etching of all materials, unless they are water soluble compounds. On soluble compounds it acts like an acid, and is used for surface polishing, preferential etching, structure forming, etc. As a quenching medium, it is used chilled, at RT, or warm to near boiling. For general quenching, washing, or rinsing it is most often used at RT or warm. It is also used as a cleaning spray, and water is tumbled to aerate the liquid, such as for drinking water to increase the oxygen content. Water vapor also is used for controlled humidity in air conditioners. In industrial systems as 50 to 100 ton units, home use systems are smaller, such as the Swamp Coolers where water is sprayed across fan-generated air.

Hot water, up to and including boiling water, is used to form a hydrated oxide on metal surfaces by immersion, often then stripped with HF, as a surface cleaning step. Occasionally, hot water is used as the rinsing medium after etching, or as the quenching medium in hot metal fabrication.

REF:

WATR-0003
ETCH NAME: Steam
TYPE: Acid, solvent
COMPOSITION:
x .... H_2O, hot vapor
TIME:
TEMP: Above 100°C

DISCUSSION:

H_2O above the boiling point as hot vapor at standard atmosphere and pressure (1 atm = 760 mmHg pressure) or under pressure as xx atm or psi, as pressurized steam. At standard atmosphere and pressure steam is used to drip clean or lightly oxidize surfaces. Under pressure, as in an autoclave to either clean metal parts (medicine) or to oxidize a metal, such as silicon for steam oxide growth at 1000°C and 10—150 psi.

Steam jets are used for metal cutting, for drilling, for mechanical motion operations of valves, etc., and steam is used in hydroelectric power generation. Natural steam from volcanoes has long been used for heating and cooking; there are steam irons for clothing, and steam pressing as part of dry cleaning, or used for compacting of materials. Steam boilers have been in use as a driving mechanism for railroad engines and steam ships, or in factory and mining for operation of steam cylinders or valves for over 150 years. Steam heating of housing, both natural from volcanoes and from man-generated boilers, has been in use since before the time of the Roman Empire, circa 400 B.C.

Steam is still used in cooking. A double-boiler, with heated water in the lower section, is used for cooking steamed vegetables, meat, poultry, or fish and shellfish.

Distilled water, DI H_2O, is made by boiling water, passing the hot steam into a second collecting vessel, and reliquefying as high purity water. This removes all minerals and organic compounds from the water with the exception of small amounts of chlorine and sulfur oxides. The water can be further purified by passing through positive and negative ion exchange resin beds, and single-distilled water is the major source of DI water used in industry. It can be single-, double-, or triple-distilled. It also can be as demineralized, dmH_2O, only. Distilled water is flat in taste and not used for drinking or cooking, although it can be so used, but no longer contains the essential minerals needed for human and animal consumption. It is used in both steam ironing and in vehicle batteries. Most chemical laboratory sinks have both industrial and DI water taps, with the DI water used in final rinsing of critical parts. DI water lines required on-going cleaning maintenance with potassium permanganate or hydrogen peroxide solutions to prevent algae build-up in the lines, and water recirculating systems, recirc systems, use an in-line glass section with ultraviolet light passing through the water to kill algae growth as distilled water is an excellent medium for such growth.

In nature, the evaporation of ocean salt water into the atmosphere as clouds produces naturally distilled DI water which, as rain, is still high purity until it reaches the ground and remineralizes. It does pick up some minerals from the atmosphere, is potable water — soft water — and is still collected from rainstorms in cisterns in arid countries and areas. Hard water contains calcium and magnesium sulfates and other salts that reduce foaming action of soaps, the compounds can build up in steam boilers and lines, requiring descaling and removal. Such boiler cleaning and conversion of hard water to soft water is a major industry, as is the disposal of industrial and home generated waste waters. Most sewage plants include steam boilers as part of the plant operation in addition to the aerobic and anaerobic bacterial action, and addition of chemical reducing agents. It is worth noting that such treated water is potable, and can be reused as agricultural or industrial water.

REF:

COLD ETCHANTS

Note: The following solutions are used to obtain a specific temperature below room temperature (RT) (22.2°C/72°F) for the purpose of chilling another solution or material more than as a direct cleaning or etching medium.

COLD-0001

ETCH NAME: Ice water — TIME:

TYPE: Solute, chilling — TEMP: 4°C, nominal

COMPOSITION:

x .... H_2O, liquid

x .... H_2O, solid

DISCUSSION:

H_2O, as a mixture of ice and fresh water (may be salt water). Used as a water bath to cool an etching solution to prevent excessive exothermic heating, only, to obtain preferential etching action, or for cleaning action, only. As a heat removal medium against exothermic reaction of an etchant such chilling is used to control etch action with removal of heat using heat exchanger coils which, themselves, may be additionally chilled. As a quenching medium to obtain specific structure of irons, steels, copper, etc., it can be either chilled fresh or salt water. Similar quenching of Solid State materials is done mainly to establish electronic functions of the device being fabricated. As a final washing medium — static, flowing, or spray — cold water can be used following etching or use of cleaning solutions on materials. Ice water solutions will maintain a temperature of about 4°C as long as ice is still present in the liquid, but can vary from just under 22°C (RT) to a cold ice slush if chipped or shaved ice is used. Note that most etch solutions, as they are chilled, become progressively more preferential in attack, eventually accentuating only cracks and other major damage artifacts on a material surface.

REF:

COLD-0002
ETCH NAME: Ice/acetone
TYPE: Solute, chilling
COMPOSITION:
x .... ice, chunk
x .... acetone (ace)
TIME:
TEMP: 80°C nominal

DISCUSSION:

H_2O, as solid ice in the ketone acetone, CH_3COCH_3. Used as an etchant chilling medium for reactive application of a solution for general surface cleaning with minimum attack removal of material. HF:HNO_3 etchants with or without HAc or H_2O become progressively more preferential below RT. In processing silicon aluminum alloy transistors, as an example, used to fine-tune the p-n junction for optimum electrical characterization with slight preferential and surface cleaning action. Note that acetone alone is the primary solvent for photo resist lacquers using the life-off technique to remove excess metallization form photolithographically patterned semiconductor wafers and similarly structured materials. This ice/acetone mixture also is used for the removal of water vapor (moisture) from gases, such as for high purity argon gas from cylinders where the argon is used to back-fill a quartz diffusion tubes used for gallium diffusion of silicon controlled rectifiers, SCRs. Any trace water moisture in the argon can produce gallium oxides which are detrimental in SCR fabrication.

REF:

COLD-0003
ETCH NAME: Acetone
TYPE: Ketone, cleaning
COMPOSITION:
x .... $(CH_3)_2CO$ (Ace)
TIME:
TEMP: −95°C

DISCUSSION:

Acetone, used as a spray from a stainless steel can pressurized with nitrogen gas. It is the standard solvent for removal of photo resist lacquers and surface cleaning of parts in photolithographic processing. **CAUTION:** Can condense water droplets from a humid atmosphere as beads on an otherwise clean surface. It is a clear liquid with a highly pungent, slightly sweet odor and, when mixed with bromine, Br_2, becomes the lachrymator . . . tear gas . . . as a bromoacetone.

Acetone also is used as a soaking liquid for removal of photo resists, usually at room temperature. If used hot, it should be used in sufficient volume to prevent possible explosive ignition for, like many solvents, its hot vapor pressure ignition point is relatively low.

REF:

COLD-0004
ETCH NAME: Nitrogen, liquid
TYPE: Gas, freezing
COMPOSITION:
x .... LN_2
TIME:
TEMP: −196°C (nominal)

DISCUSSION:

Nitrogen gas as a cryogenic liquid obtain by the fractionation of air. A major use of LN_2 in industry is chilling the cold traps of vacuum evaporator systems. In a hot diffusion pump system, the cold trap is located immediately below the gate valve that opens into the bell jar chamber, and immediately above the diffusion pump, preferably, as a vertical stack. The trap is kept filled with LN_2 during an evaporation cycle to, initially, pump-down and remove air, other gas and water vapor from the bell jar, and then to help maintain a gas-free vacuum in the operating chamber. The LN_2 primary function is to remove water vapor by freeze-out to prevent reverse diffusion back up into the chamber while the system is operating, particularly with the main gate valve opened for high vacuum pumping by the hot oil diffusion pump. It also prevents back-diffusion of the rising hot oil in the diffusion pump from entering above the gate valve, helping to recondense the oil, which returns to the bottom of the diffusion pump by draining down the internal pump wall.

In other vacuum systems, such as those used for RF sputtering the initial roughing pump uses LN_2, the diffusion pump is a cryopump operating with LHe, such that all hot oils are eliminated. These vacuum systems are optimum oil-free operations.

LN_2 has been used as a quenching medium similar to ice water (COLD-0001) in the metals industries. There is a "0°C Etch" used in Solid State material etching that is commonly chilled with LN_2 (See Gallium Arsenide). LN_2 storage tanks are maintained at many industrial facilities as a source of the liquid or for conversion to nitrogen gas from the LN_2 source through heat exchangers with both LN_2 and gas piped to user points in buildings. The gaseous nitrogen is often used for equipment air-valve operation rather than having to run in separate air lines, as LN_2 is already at the location. LN_2 can be drawn off at the outside tank or from an internal line for small quantity use from polystyrene dewars. Larger metal dewars are rated from 50 to 200 l volume, the larger types on wheels. Note that the best piping method for bringing in liquid nitrogen with minimum gas-off loss is in double-walled steel pipes with a light vacuum drawn on the annular space . . . the same construction used in the storage tanks and 2000 gallon tanker trailers used for transportation of all cryogenic gases.

Ion pump vacuum systems, titanium rod getters in the ion pump replacing the hot oil diffusion pump, have from one to six pumping cylinders filled with charcoal or a Dri-Rite-type compound, with the dewars attached around the cylinders and kept filled with LN_2 during the evaporation cycles. At operation end, the cylinders are allowed to automatically out-gas themselves; whereas the gettering-charcoal of a cryopump system requires pump bake-out on at least a weekly cycle to remove collected gases.

As a chemically inert, clear, freezing liquid at −196°C, LN_2 should be handled with the caution due any dangerous chemical. A rose, for example, dipped into liquid nitrogen for 10 sec, then removed still frozen, will shatter like glass when struck, and the same thing can happen to a human finger. Where the cold liquid is off-gassing to nitrogen gas, the area should be well ventilated. Too much nitrogen in a work area causes sleepiness, and can cause death if all oxygen in the air is replaced by pure nitrogen. Operators required to enter and clean LN_2 storage tanks or mobile trailers allow them to come up-to-air before entering, wear air-packs, or work with an outside source pumping air into the tanks and trailers.

LN_2 has been used to freeze mercury spills. It is the "cold" portion of hot/cold high reliability test units used in evaluating electronic parts, and has been used in thin film adhesion testing: specimen from a hot plate to quenching in LN_2. LN_2 cryostats are used in laboratory growth and study of single crystal ice, snowflakes, and both natural and artificially compounded salt water brines.

REF:

COLD-0005

ETCH NAME: Helium, liquid — TIME:

TYPE: Gas, freezing — TEMP: −269°C (nominal)

COMPOSITION:

x .... LHe

DISCUSSION:

Helium is a colorless, odorless, inert cryogenic liquid. As a liquid, helium reaches the coldest temperature known to man, to within 1°A, but cannot be solidified. The cryopump units designed for vacuum systems use a combination of LHe (to 2.4°A) in the pump and fine chunk graphite as the absorber. These are called "oil-free" pumping systems when coupled with an LN_2 pump unit replacing the standard mechanical oil pump used in pulling the initial vacuum to 1 μm or 1×10^{-3} Torr. After 50 to 100 evaporation cycles, the cryopumps require regeneration by bake-out for removal of absorbed gases in the graphite, with installation of new graphite on a yearly basis. LHe can be used as a chilling liquid for etching, but LN_2 is more widely used due to cost factors.

LHe has been used as a source for helium gas like LN_2 (COLD-0004). It is occasionally used as an inert furnace gas and for drying . . . helium blow-off, rather than nitrogen blow-off in processing Solid State parts.

REF:

COLD-0006

ETCH NAME: Carbon dioxide — TIME:

TYPE: Gas, chilling — TEMP: −85°C (nominal)

COMPOSITION:

x .... CO_2, solid (dry ice)

DISCUSSION:

CO_2, as a white crystalline solid called dry ice. Supplied commercially from the liquifaction and solidification of the gaseous carbon dioxide similar to the process for liquifaction of air and separation of cryogenic nitrogen and oxygen. Dry ice is one of the few compounds that sublimes directly from the solid to the gas without passing through a liquid state. It must be kept below its cryogenic temperature as a solid or else it converts to gas and dissipates into the atmosphere; not like water that can be frozen as ice, liquified as water, gassed as steam, then recondense as water and re-frozen as ice with little or no loss of the original material volume.

Small quantities of dry ice can be fabricated from pressurized CO_2 gas cylinders. When CO_2 gas is released rapidly from a pressurized container it will freeze as a solid if confined during release. Attach a hinged steel unit on the cylinder release port, open the cylinder valve, and a small solid cake of CO_2 (about 3″ diameter, 1″ thick) will be formed.

A major application is in shipping of materials in the frozen state. Both epoxy and polyimide resins are shipped in polystyrene containers packed with pea-sized chunks of dry ice. The dry ice slowly gasses-off, but is good for about 7 to 8-days as a temporary freezing medium. Commercially it is supplied as one foot square/one inch thick blocks separated by brown wrapping paper and in a cardboard box. At room temperature a block will last nearly 48 h. Where dry ice is used on a daily basis, it is delivered every 2 days at minimum, as very few laboratories have cold storage facilities below −100°C available.

Dry ice also is used as a source of carbon dioxide gas in metal processing, and in the preparation of fertilizers. As a furnace gas it has been used for both its oxygen and carbon content, oxidation or carbonation, respectively. As a gas mixture in Solid State processing in this regard, such as for a carbon thin film with annealing conversions to a diamond-like compound (DLC).

As the gas, it is produced naturally by the fermentation process (wines and beers); added to water as seltzer water; and added to many soft drinks, such as colas, root beer, etc. as carbonated beverages. It occurs in some natural springs and has been directly bottled as carbonated water.

REF:

COLD-0007

ETCH NAME: Dry ice/acetone — TIME:

TYPE: Gas, chilling — TEMP: −80 to −100°C

COMPOSITION:

x .... CO_2, solid

x .... CH_3COCH_3 (acetone, Ace)

DISCUSSION:

CO_2, as dry ice in liquid acetone. This cold liquid mixture is used to remove water moisture from process gases, such as argon. Argon has an affinity for moisture adsorption even in a pressurized cylinder, such that it is bubbled through a Pyrex or quartz bubbler containing a dry ice/acetone mixture to freeze-out and remove water moisture before use. This method has been used where argon has been an inert back-fill gas for closed tube gallium diffusion of silicon wafers in making silicon controlled rectifiers, SCRs.

This mixture also has been used as a general chilling medium for etching solutions, such as the "0°C" etch (See Gallium Arsenide section).

REF:

COLD-0008

ETCH NAME: Oxygen, liquid — TIME:

TYPE: Gas, reactive — TEMP: −183°C (nominal)

COMPOSITION:

x .... LOX

DISCUSSION:

O_2, as a colorless, reactive cryogenic liquid. Obtained by the liquifaction and fractionation of air for both LN_2 and LOX. Delivered in 2000 gallon trailers to on site storage tanks like LN_2, mainly as a source of gas. Until the advent of Solid State semiconductor processing in the early 1950s the primary users were medical facilities and the metal industries. Today, as LOX, it is a primary fuel for aircraft, rockets/missiles, and space vehicles. **CAUTION:** LOX is instantly explosive in the presence of grease and oil.

As oxygen gas it is mainly used as an oxidizer in furnace processing, such as an O_2/N_2 mixture for growth of thin film SiO_2 on silicon wafers as dry thermal oxidation at 1000°C. As a 5% O_2/CF_4 gas mixture it is a dry chemical etching (DCE) system for structuring or selective etching of a wide number of Solid State materials in device construction.

REF:

COLD-0008a

ETCH NAME: Air, liquid — TIME:

TYPE: Gas, reactive — TEMP: −175 to −180°C

COMPOSITION:

x .... L_{air}

DISCUSSION:

Air, as a colorless, odorless cryogenic liquid. Manufactured by repeated pressurizing to liquify, then used as a cryogenic source of air or fractionated for LOX, LN_2, and other cryogenic rare gases, such as neon, krypton, xenon, and others by additional pressurizing, chilling and fractionation.

It can be used industrially as a very clean, high quality source of air, though its primary use is as a source of LN_2 and LOX.

Liquid air is used to bottle gas under pressure as a cylinder gas . . . Al cylinder, 3600 psi . . . where high purity air is required in processing, such as for an oxidizing agent.

REF:

COLD-0009

ETCH NAME: Hydrogen, liquid — TIME:

TYPE: Gas, reactive — TEMP: −253°C (nominal)

COMPOSITION:

x .... LH_2

DISCUSSION:

H_2, as a clear, odorless cryogenic liquid. The gas is pressure liquified in a process similar to that used for air, nitrogen and oxygen. It is supplied to industry in 2000 gallon tanker trailers, as is LOX and LN_2, but normally used from the trailer parked on site rather than being transferred to a storage tank.

Its major use is as a source of hydrogen gas for the reducing atmosphere in furnace processing of materials. Also as a source of high purity hydrogen in pressurized cylinders (Al cylinders, 3600 psi) as pure dry hydrogen in a single cylinder or as a "tank farm", such as an A21 interconnected series of cylinders that are transported as a single unit for installation on site. Such a farm with an LN_2 tank source available are used to mix forming gas (FG) (85% N_2/15% H_2). As an Al pressurized cylinder or small lecture bottle gas mixtures and are available as doping gases in Solid State material processing, such as SiH_4, AsH_4, PH_4.

As LH_2 it is still used as rocket/missile, aircraft, or space vehicle fuel in combination with of LOX. **CAUTION:** Both LH_2 and H_2 are explosive and flammable in concentration and in the presence of air. Most hydrogen gas reduction furnaces have at least one hydrogen burn-off outlet with an automatic trigger spark to insure ignition.

REF:

COLD-0010

ETCH NAME: Chlorine — TIME:

TYPE: Gas, reactive — TEMP: 0°C @ 6 atm

COMPOSITION:

x .... LCl_2

DISCUSSION:

Cl_2, as a pale greenish/yellow cryogenic liquid. Chlorine was the first gas to be liquified, as it requires minimum chilling and pressurizing as compared to other cryogenic gases, as shown above. When mixed with water and ice it forms a crystalline greenish-yellow hydrate, $Cl_2.8H_2O$ that is unstable, though chlorinated water has many industrial uses, to include water treatment as drinking water, as a swimming pool cleaner, and in industrial chlorination processing of metals and alloys.

LCl_2 is industrially supplied in both railroad and truck trailers, similar to LN_2 and LOX, with railroad cars holding 8000 to 10,000 gallons. The primary use is as a gas in the making of chemical compounds, to include hydrochloric acid, HCl, and in the compounding of fertilizers. Chlorine is a very active gas, second only to fluorine, and will combine with all elements other than gold and platinum. Sodium chloride — the mineral halite, NaCl — is common table salt, and the major source of chlorine. The mineral sylvite, KCl, also is used as a salt, more bitter than table salt, but used as a dietary replacement as it is not retained by the body. Both are sea salts, ocean brines, and obtained by evaporation of salt water, or mined from buried salt domes, such as those along the Gulf Coast of the U.S. and the tremendous buried salt beds in Germany that are 100 miles long, and as much as a mile thick, derived from the evaporation of shallow ancient seas.

As a cryogenic liquid it has little direct use in Solid State material processing, but is a source for the compounding of many chemicals used as gaseous doping elements, such as PCl_3 and $POCl_3$ (pockle) widely used as a phosphorus dopant for silicon. It is supplied in pressurized cylinders of Al size (9″ diameter, 5 ft high) for use as a chlorinating gas in metal processing. **CAUTION:** LCl_2 is a highly corrosive liquid and gas. Also, as a gas, the fumes attack the wet membranes — mouth, nose, throat — rapidly and are unbreathable forming a dense white cloud, primarily as HCl vapors.

REF:

COLD-0011
ETCH NAME: Ammonia — TIME:
TYPE: Gas, chilling — TEMP: −34°C (nominal)
COMPOSITION:
x .... NH_3
DISCUSSION:

NH_3, as a crystalline chemical, cryogenic liquid or gas. It can be liquified by the combination of hydrogen and nitrogen gases in the presence of iron at 550°C under pressure. It is shipped as a liquid, can then be reduced to nitrogen and hydrogen as a major industrial hydrogen source, but a better portion is used in the compounding of fertilizers. It has application in many cleaning and etching solutions: Windex, as a cleaner and, medically, it is "smelling salts". It was one of the first refrigeration liquids, is still so used, though the more inert Freons are a more modern replacement. Both ammonia, NH_3, and ammonium hydroxide, NH_4OH, are basic solutions on the Sorenson pH Scale and used to neutralize acid sumps in waste disposal.

In Solid State material processing both ammonia compounds are used as cleaning solutions, and NH_3 gas is a source for growing metal nitrides, such as Si_3N_4.

REF:

COLD-0012
ETCH NAME: Freon — TIME:
TYPE: Solvent, cleaning — TEMP: −17 to −50°C, variable
COMPOSITION:
x .... *Freon

* Freon is the trademark of the Dupont Company and there are several variations of artificial compounds, several of which have a boiling point below 0°C.

DISCUSSION:

Freon, as gas or liquid. As liquids they also are compounded with other solvents, such as acetone and other azeotrope chemicals for specific material cleaning applications. Vapor degreasers use hot liquid Freon (Freon-TA) for cleaning purposes, and the cold-capable freons are used as a replacement for ammonia, NH_3, in refrigeration. As a liquid under pressure in a spray can, applications are similar to those for acetone (COLD-0003), for the cleaning of surfaces in Solid State material processing. They are excellent cleaning solvents, as they evaporate without leaving any residue. As a cold spray, they have been used to chill specimens surfaces under electron microscopes to prevent the heat from electron impingement altering the surface of a material bulk under observation.

REF:

COLD-0013
ETCH NAME: Salt/snow — TIME:
TYPE: Salt, chilling — TEMP: (1) 0°C (2) −0.4°C
COMPOSITION:

(1)		(2)	
1	NaCl	1	NaCl
3	snow	1	snow

DISCUSSION:

NaCl/snow solution. Both solutions can be used to chill etchants to about 0°C. Also have been used for temporary packing and shipping of fish and other perishable foodstuffs. Salt water with ice or snow added will be about 4°C at the near-freezing point. As pure,

fresh water (snow and ice) freezes at about 0°C, salt is spread on roads in the winter time to help in the melting and removal of road snow and ice. See section on ICE.
REF:

COLD-0014
ETCH NAME: Salt/snow — TIME:
TYPE: Salt, chilling — TEMP: (1) −27°C
(2) −44°C

COMPOSITION:

(1)		(2)	
3	$CaCl_3$	2	$CaCl_3$
2	snow	1	snow

DISCUSSION:
$CaCl_3$/snow solution. The solutions shown can be used alone or to chill other solutions to the specific temperatures shown. Used in metal processes, chemical compounding and medicine.
REF:

COLD-0015
ETCH NAME: KOH/snow — TIME:
TYPE: Alkali, chilling — TEMP: −35°C
COMPOSITION:
4 .... KOH
3 .... snow
DISCUSSION:
KOH/snow solution. Used in chemical and metal processing for the specific temperature chilling point. It can be used as an extremely slow surface preferential damage etchant.
REF:

COLD-0016
ETCH NAME: Ammonium nitrate — TIME:
TYPE: Salt, chilling — TEMP: +3°C
COMPOSITION:
1 .... NH_4NO_3
1 .... H_2O
DISCUSSION:
NH_4NO_3 solution. Used in the metal industries as a ''cool'' quenching medium of irons, steels, etc. In Solid State material processing for material cleaning and slow preferential etching of surfaces with minimum removal as for fine-tuning of electrical characteristics (See COLD-0002).
REF:

COLD-0017
ETCH NAME: — TIME:
TYPE: Salt, chilling — TEMP: (1) +10°C
(2) −11°C

COMPOSITION:

(1)		(2)	
5	NH_4Cl	1	NH_4Cl
5	KNO_3	1	KNO_3
16	H_2O	1	H_2O

DISCUSSION:
NH_4Cl/KNO_3 solutions. By varying the amount of constituents between the limits shown,

a complete range of stable temperatures can be obtained. Both solutions used in general processing of metals and chemicals.
REF:

COLD-0018
ETCH NAME: HCl/ice
TYPE: Acid, chilling/etching
TIME:
TEMP: −86°C, eutectic
COMPOSITION:
1 .... HCl
1 .... ice
1 .... H_2O
DISCUSSION:
HCl acid mixed as a cold temperature solution in the proportions shown is a stable eutectic solution at −86°C. It has application in chemical passivation of metal surfaces as a chlorinated surface.
REF:

COLD-0019
ETCH NAME: KOH/ice
TYPE: Alkali, chilling
TIME:
TEMP: −65°C, eutectic
COMPOSITION:
1 .... KOH
1 .... ice
1 .... H_2O
DISCUSSION:
KOH/ice as a solution with water is a stable eutectic mixture. Like HCl/ice (COLD-0018) used as a metal surface passivating system for an oxidized surface.
REF:

COLD-0020
ETCH NAME: Nitric acid/ice
TYPE: Acid, chilling
TIME:
TEMP: −43°C, eutectic
COMPOSITION:
1 .... HNO_3
1 .... ice
1 .... H_2O
DISCUSSION:
HNO_3/ice as a solution with water is a stable eutectic mixture. Like COLD-0018 and COLD-0019 this solution can be used as an oxidation pacification system of metal surfaces in general processing.
REF:

COLD-0021
ETCH NAME: Salt/ice
TYPE: Salt, chilling
TIME:
TEMP: −55°C, eutectic
COMPOSITION:
1 .... $CaCl_3$
1 .... ice
1 .... H_2O

DISCUSSION:

$CaCl_3$/ice in a water solution is a stable eutectic mixture. A general purpose solution for processing materials at the temperature shown.

REF:

COLD-0022

ETCH NAME: Salt/ice
TIME:

TYPE: Salt, chilling
TEMP: −21.13°C, eutectic

COMPOSITION:

1 .... NaCl
1 .... ice
1 .... H_2O

DISCUSSION:

NaCl/ice in a water solution is a stable eutectic mixture. Used as a quenching medium in metal processing for the specific temperature shown.

REF:

COLD-0023

ETCH NAME: MeOH/ice
TIME:

TYPE: Alcohol, chilling
TEMP: 8 to 10°C

COMPOSITION:

x .... MeOH
x .... ice

DISCUSSION:

MeOH/ice as a liquid solution of methanol with similar applications to that of ice/acetone (COLD-0002). See Ice/acetone for additional discussion.

REF:

COLD-0028

ETCH NAME: Ethyl alcohol
TIME:

TYPE: Alcohol, chilling
TEMP: −117 to +78°C

COMPOSITION:

x .... C_2H_5OH (EOH or ethanol)
x .... H_2O

DISCUSSION:

Ethanol (drinking alcohol) is completely miscible in water. As absolute alcohol it is 100% EOH. As a constant boiling mixture: 95% EOH:5% H_2O, and the most common maximum concentration. Denatured alcohol (industrial alcohol) has 5% methanol or ethylene glycol added to make it unfit for human consumption. As drinking alcohol the concentration is measured by proof/gallon: Absolute = 200 proof; 95% EOH = 195 proof. Absolute is primarily used in medicine as a disinfectant and as a skin chilling medium soaked on a cotton ball, etc., and occasionally used in industry for material surface cleaning in a similar manner or as a liquid rinsing medium like methanol after water quenching following etching, though denatured alcohol is more commonly used. Also used as a spray for chilling and cleaning surfaces. Like methanol, it too has been an additive to etchant solutions. See SI-0092c; SI-0041c (EOH, MeOH, glycols, glycerin). Some water- and alcohol-soluble compounds are etch polished in EOH.

REF:

COLD-0029
ETCH NAME: Isopropyl alcohol TIME:
TYPE: Alcohol, chilling TEMP: −85 to +82°C
COMPOSITION:
x .... $CH_3CHOHCH_3$ (IPA, ISO)
DISCUSSION:

Propyl alcohol (isopropanol) is completely miscible with water and has similar uses to those of ethanol and methanol, but is not as widely used as those two alcohols. Often called propanol only and occasionally used mixed with acids in etch solutions. There are several other alcohols used, such as butyl, etc., for general processing of metals and metallic compounds, but to even a lesser extent than IPA.

REF:

COLD-0024
ETCH NAME: MeOH/dry ice TIME:
TYPE: Alcohol, chilling TEMP: −65°C or 209 K
COMPOSITION:
x .... CO_2 (dry ice)
x .... MeOH
DISCUSSION:

Dry ice/methanol used as an etch in a study of the dislocation slip system on (0001) zinc surfaces. Also used LN_2 (77 K or −196°C) at room temperature in air (298 K or 25°C).

REF: Billello, J C et al — *J Appl Phys,* 54,1821(1983)

COLD-0025
ETCH NAME: Poly water TIME:
TYPE: Acid, cleaning(?) TEMP: −40°C
COMPOSITION:
x .... H_2O
DISCUSSION:

H_2O, as poly water, is a high density form of water with tightly packed strings of connected water molecules. It has been made by pouring DI water into 10 mm diameter tubes, then allowing the tube to sit over DI water for high humidity action until the poly forms in the tube. It boils at +150°C and freezes at −40°C to a clear, plastic-like glassy material.

REF: Eisenberg, D & Kadsmann, W — *The Structure and Properties of Water,* Oxford University Press, London, 1969

COLD-0026
ETCH NAME: Ethylene glycol TIME:
TYPE: Glycol, chilling TEMP: −17 to +197°C
COMPOSITION:
x .... CH_2OHCH_2OH (EG)
x .... H_2O
DISCUSSION:

Ethylene glycol is completely miscible with water and is mixed as "anti-freeze" used in car radiators along with anti-rust compounds, as it can both prevent water from freezing or boiling. Also used in etchant solutions like glycerin for viscosity control, as a wetting agent, and can improve planarity of an etched surface to include improved surface cleaning action. See SI-0092h; SI-0099.

REF:

COLD-0027
ETCH NAME: Methyl alcohol
TYPE: Alcohol, chilling
COMPOSITION:
x .... CH_3OH (MeOH)
x .... H_2O
TIME:
TEMP: −97 to +65°C
DISCUSSION:

Methanol (wood alcohol) is completely miscible with water, and also has been used as an anti-freeze, e.g., Zerone. As it has good affinity for water it is often used as a final rinse following water quenching of an etchant solution with air dry or nitrogen blow-off drying. Occasionally used as a constituent of an etchant, and will reduce viscosity. See Nital solutions under Iron and Steel where it is used as a replacement for ethanol. See SI-0092i.

REF:

WATR-0004
ETCH NAME: Air + water moisture
TYPE: Acid, reactive
COMPOSITION:
x .... air, moist
TIME:
TEMP: 4—99°C
DISCUSSION:

H_2O as liquefied droplets in air which is measured as relative humidity (RH) for a 0 to 100% moisture content. In the atmosphere, high pressure areas are low in moisture content (''hot'' air); whereas low pressure areas are high in moisture content (''cool'' air), and also referred to as ''dry'' or ''wet''. A 30″ water meter also is used in measuring moisture content and, as applied to high and low pressure areas, above 30, is high pressure, below 30, low pressure. In vacuum systems, such meters are used during the initial roughing pump-down cycle with 30″ roughly equivalent to 1 μm (10^{-3} Torr) vacuum, at which point the system is switched to high pressure pumping, e.g., 10^{-4} to 10^{-12} Torr.

Both the oxygen and water moisture content in air is responsible for the oxidation of all materials, from metals and rocks to the decay by-products of vegetation and animal remains. Light, dry air can be cold, such as over the Arctic ice packs, or light, dry, and hot, as in desert areas. In winter time, heavy, wet, and cold and, with a wind, the wind-chill factor can drop air temperature below −40°C with severe freezing effects. Conversely, hot dry winds can produce dangerous dehydration in humans and animals.

Both temperature and moisture content in the air in buildings is controlled by air-conditioning and humidifier systems. A ''swamp cooler'', widely used in homes during hot weather, blows hot air over water to both increase the humidity as well as reduce temperature. The larger commercial air-conditioners are rated for 50 or 100 ton capacity, and the amount of air-conditioning needed is established by the cubic volume of building area to be controlled. A single 50-ton unit can easily maintain temperature and humidity in a 100 ft² single story building under most atmospheric conditions. In Solid State material process areas, the ''Yellow Room'', used in handling photo resist application, is critical: nominally maintained at 70—72°F and 40% RH. In general work areas, the humidity can vary between, say, 50 to 60%. In the processing of semiconductor wafers, for example, a silicon wafer will automatically oxidize when exposed to air/moisture producing a passivating oxide surface of about 20 to 30 Å thickness — referred to as a ''native oxide''. Such oxides are usually stripped with HF immediately prior to processing steps, such as metallization, diffusion, epitaxy, etc., as the oxide can produce unwanted resistive layers in a device structure. Most metals passivate themselves automatically when exposed to air with the thickness of the oxide varying up to 50 Å thick depending upon the metal involved. Operating Clean Rooms are constructed similar to Yellow Rooms for temperature and humidity control, and both type rooms include a particulate dust size control with use of static arrestors . . . 1 μm sized dust particles or less . . . preferred at less than 10% per air foot volume.

Moist air is used in furnace oxidation of materials, and many metallic compounds have been studied under controlled high and low humidity conditions.

Along seacoasts both moisture and salt content in air is a consideration. Metal coupons are evaluated for corrosion in such environments as a standard test; moist air is used in other studies, to include smog mixtures.

REF:

PHYSICAL PROPERTIES OF XENON, Xe

Classification	Rare gas
Atomic number	54
Atomic weight	131.30
Melting point (°C)	−111.9
Boiling point (°C)	−107.1
Density (g/cm^3) 20°C	5.895
Thermal conductivity (10^4 $cal/cm^2cm/°C/sec$) 20°C	1.24
1st ionization potential (eV)	12.127
Neutron cross section (barns)	24
Vapor pressure (°C)	−132.8
Critical temperature (C.T.) 0°C	+16.6
Hardness (Mohs — scratch) solid	1—2
Crystal structure (isometric — normal) solid	(100) cube, fcc
Color (solid)	Clear/bluish
Cleavage (cubic) solid	(001)

XENON, Xe

General: Like other rare gases, xenon occurs as a minor elemental gas in the atmosphere at about 0.0018%. It is obtained as a cryogenic liquid, LXe, by the fractionation of liquid air, L_{air}. It is chemically inert, in group 0 of the Periodic Table, but when electrically activated under pressure produces intense white light that can go from full intensity to extinction in about a millionth of a second. By synchronizing xenon flashes on a spinning part, such as a vehicle fan belt, the belt appears to stand still and can be studied while in motion. This is the stroboscope, and similar units are used as the "flash bulb" of photography; with chemical solutions, such as in plating out copper from $CuSO_4$ solutions; along with argon, neon, and mercury, xenon bulbs are used for high intensity white light for microscopes: and it is one of the white light constituents in neon signs.

Technical Application: Both the stroboscope and xenon microscope light bulbs are used in Solid State processing. Though seldom used due to scarcity, like argon, nitrogen, and helium, xenon can be used as an inert furnace atmosphere gas or as a blow-off gas for drying parts. It has been grown as a single crystal in vacuum under pressure and cryogenic temperature conditions for morphological study.

Etching: Vapor pressure variation under vacuum.

XENON ETCHANTS

XE-0001

ETCH NAME: Pressure — TIME:

TYPE: Pressure, preferential — TEMP: −112°C

COMPOSITION:

x .... pressure, vapor

DISCUSSION:

Xe single crystals grown in vacuum under pressure and cryogenic temperature (LN_2, −196°C). Crystal surfaces can be preferentially etched by varying the vapor pressure in the vacuum system.

REF: Schwentner, N et al — *Rare Gas Solids,* Vol 3, Academic Press, New York, 1970

PHYSICAL PROPERTIES OF YTTERBIUM, Yb

Classification	Lanthanide
Atomic number	70
Atomic weight	173.04
Melting point (°C)	824
Boiling point (°C)	1427
Density (g/cm³)	6.98
Latent heat of fusion (cal/g)	12.71
Heat of fusion (k-cal/g-atom)	1.8
Specific heat (cal/g) 25°C	0.035
Heat of vaporization (k-cal/g-atom)	38
Atomic volume (W/D)	24.8
1st ionization potential (eV)	6.2
1st ionization energy (K-cal/g-mole)	143
Electronegativity (Pauling's)	1.1
Electrical resistivity (micro-ohms) 298 K	25.1
($\times 10^{-4}$ ohm cm) 25°C	28
Compressibility (cm²/kg $\times$ 10^{-6})	7.39
Neutron cross section (barns)	36 (37)
Magnetic moment (Bohr magneton)	0.41
Tensile strength (psi)	10,400
Yield strength (psi)	9,500
Magnetic susceptibility ($\times 10^{-6}$ emu/mole)	71
Transformation temperature (°C)	792
Standard electrode potential (volts)	+2.1
Coefficient of lineal thermal expansion ($\times 10^{-4}$ cm/cm/°C) 125°C	25
Ionic radius (angstroms)	0.86 (Yb^{+3})
Hardness (Mohs — scratch)	2—3
Crystal structure (isometric — normal)	(100) cube fcc
Color (solid)	Grey
Cleavage (cubic) solid	(001)

YTTERBIUM, Yb

General: Does not occur in nature as a native element, but is extracted from minerals of the Yttrium Group of rare earths, such as gadolinite, $Be_3FeY_2Si_{10}$, and others. Like most rare earth elements its chemical reactivity is similar to alkalies, and it will deliquesce in water with evolution of hydrogen. It is commercially available in powder, foil, and sheet, as well as single crystal forms. There are no major industrial applications, but there are possible chemistry and medical applications.

Technical Application: The metal is not used in Solid State processing at present, but there may be applications like other rare earth elements in fabricating garnets and similar materials. It has been grown and studied as a single crystal.

Etching: Water, dilute and mixed acids.

YTTERBIUM ETCHANTS

YB-0001

ETCH NAME: Water — TIME:

TYPE: Acid, removal — TEMP: RT to hot

COMPOSITION:

x .... H_2O

DISCUSSION:

Yb as foil or single crystal specimens. Water can be used to dissolve the metal. Acids for etch patterning.

REF: Foster, W & Alyea, H N — *An Introduction to General Chemistry,* 3rd ed, D Van Nostrand, New York, 1947, 663

PHYSICAL PROPERTIES OF YTTRIUM, Y

Classification	Transition metal
Atomic number	39
Atomic weight	88.9
Melting point (°C)	1509 (1490)
Boiling point (°C)	2927
Density (g/cm^3)	4.472
Thermal conductance (cal/sec)(cm^2)(°C/cm)	0.035
Specific heat (cal/g) 25°C	0.071
Latent heat of fusion (cal/g)	46
Heat of fusion (k-cal/g-atom)	2.7
Heat of vaporization (k-cal/g-atom)	93
Atomic volume (W/D)	19.8
1st ionization energy (K-cal/g-mole)	152
1st ionization potential (eV)	6.6
Electronegativity (Pauling's)	1.3
Covalent radius (angstroms)	1.62
Ionic radius (angstroms)	0.92 (Y^{+3})
Electrical resistivity (micro-ohms cm) 298 K	59.6
Compressibility (cm^2/kg $\times$ 10^{-6})	2.68
Magnetic moment (Bohr magnetons)	0.67
Tensile strength (psi)	22,000
Yield strength (psi)	9,700
Magnetic susceptibility (emu/mole $\times$ 10^{-6})	191
Transformation temperature (°C)	1479
Cross section (barns)	1.3
Hardness (Mohs — scratch)	4—5
Crystal structure (hexagonal — normal)	$(10\bar{1}0)$ prism
Color (solid)	Grey-black
Cleavage (basal)	(0001)

YTTRIUM, Y

General: Does not occur in nature as a native element but there are some 30 minerals that contain yttrium and other rare earth metals. Many of these are niobates and tantalates but carbonates, silicates, and phosphates are also common. Xenotime, YPO_4 is one of the better known phosphates. As yttria, Y_2O_3, it has been known for almost 200 years and was named after the town of Ytterby, Sweden. ''Yttrium earths'' is a general term applied to minerals containing both the Lanthanide (Cerium group) and Actinide (Thorium group) of rare earth elements even though yttrium is not a member of either group. As the pure metal it is grey/black metallic in appearance with a higher melting point than silicon though it is best known in the oxide form as it is more stable than the metal, e.g., yttrium oxidizes readily in air and boiling water.

No major usage as the pure metal in industry other than as a minor constituent in some metal alloys and, as the oxide, is used in the glass and ceramics industries as a coloring compound.

Technical Application: None in general Solid State semiconductor type device fabrication, but a major component in artificially grown single crystal garnets: Yttrium iron garnet, YIG and Yttrium aluminum garnet, YAG, two of the better known laser materials. The electromagnetic properties of certain garnets, YIG as an example, are the basis of the bubble memory device that is finding increasing application in the computer field. Yttrium has been grown as a single crystal for general study. Additional yttrium compounds are shown here.

Etching: Very soluble in dilute acids and hot alkalies.

Y-0001
ETCH NAME: Potassium hydroxide — TIME:
TYPE: Alkali, removal — TEMP: Hot
COMPOSITION:
x .... x% KOH
DISCUSSION:
Y specimens. Shown as a general etch for yttrium.
REF: Hodgman, C D — *Handbook of Chemistry and Physics,* 27th ed, Chemical Rubber Co, Cleveland, OH, 1943, 484

Y-0002
ETCH NAME: Nitric acid — TIME:
TYPE: Acid, macro-etch — TEMP:
COMPOSITION:
x .... HNO_3
DISCUSSION:
Y single crystal specimens. Solution used to develop macro-structure in ytterium.
REF: Carlson, O N et al — *J Electrochem Soc,* 107,541(1960)

YXM-0001
ETCH NAME: Kalling's etch — TIME: 10—60 sec
TYPE: Acid polish — TEMP: RT
COMPOSITION:
100 ml .... HCl
100 ml .... EOH
5 g .... $CuCl_2$
DISCUSSION:
$Y_2(CoM)_{17}$, single crystal ingots. M = Al, Fe, or $(FeCu)_x$. Cut and mechanically polish wafers, then chemical polish with solution shown. Used in a morphological study of these materials.
REF: Slepowronsky, M et al — *J Cryst Growth,* 65,293(1983)

YITTRIUM ZIRCONATE ETCHANTS

YZR-0001
ETCH NAME: Phosphoric acid TIME:
TYPE: Acid, polish TEMP:
COMPOSITION:
x .... c-SiO_2
x .... H_3PO_4 (pH: 3—4)
DISCUSSION:
$(Y_2O_3)_m(ZrO_2)_{1-m}$, (100) wafers used as substrates for silicon epitaxy. First, mechanical lap with alumina, then chem/mech lap with the solution shown. Degrease: (1) TCE, (2) acetone rinse, (3) MeOH rinse, and (4) N_2, blow dry. Used as a substrate for epitaxy CVD growth of silicon.
REF: Liu, A L & Golecki, I — *J Electrochem Soc,* 132,234(1985)

PHYSICAL PROPERTIES OF ZINC, Zn

Classification	Transition metal
Atomic number	30
Atomic weight	65.37
Melting point (°C)	419.5
Boiling point (°C)	907
Density (g/cm^3)	7.133
Thermal conductance (cal/sec)(cm^2)(°C/cm)	0.27
Specific heat (cal/g) 20°C	0.0195
Latent heat of fusion (k-cal/g)	24.09
Heat of fusion (k-cal/g-atom)	1.76
Heat of vaporization (cal/g)	27.43
Atomic volume (W/D)	9.2
1st ionization energy (K-cal/g-mole)	216
1st ionization potential (eV)	9.391
Electronegativity (Pauling's)	1.6
Covalent radius (angstroms)	1.25
Ionic radius (angstroms)	0.74 (Zn^{+2})
Linear coefficient of thermal expansion ($\times 10^{-6}$ cm/cm/°C)	49.7
Electrical resistivity (micro-ohms-cm) 20°C	5.92
Cross section (barns)	1.10
Vapor pressure (°C)	736
Refractive index (n=)	
Hardness (Mohs — scratch)	2.5
Crystal structure (hexagonal — normal) alpha	($10\bar{1}0$) prism
(monoclinic — normal) beta	(100) a-pinacoid
Color (solid)	Bluish-white
Cleavage (basal)	(0001)/(001)

ZINC, Zn

General: Probably does not occur in nature as a native element. Three of the major zinc ores are calamine, a hydrated silicate of zinc; smithsonite $ZnCO_3$, the carbonate form; and sphalerite, ZnS — common name zincblende. The latter mineral is a compound semiconductor but even more important as it has the atypical tetrahedral unit cell crystallographic structure of most binary compound semiconductors. Zinc ores have been used for years in making brass, CuZn (30%), yet it was not recognized as a separate metal element until 1746 (Marggraf). As the pure metal, zinc is bluish-white, malleable at or above 100°C and a fair conductor of electricity. At red heat it will burn in air to the oxide although it is extremely stable under normal temperature conditions.

Zinc has major industrial applications as an alloy with copper, iron, and other metals. Although bronze, CuSn, and brass, CuZn have been used since antiquity — the Bronze Age — brass is more workable and useful as it is less brittle. The process known as galvanizing is a surface plating process and, along with tinning, galvanized zinc coatings are applied to iron and other metals for corrosion resistance to include direct dipping in molten zinc. Zinc is used as the negative electrode (cathode) in many electrolytic plating baths and as cathode plates in DC batteries. An added note on bronze and brass — there are well over 100 alloys and today the term brass is more common whether the alloy contains zinc or tin or a mixture of the two with copper. As the oxide, like titanium oxide, it is a high quality white pigment in paints; zinc chromates are used as brilliant yellow coloring agents in paints, glass, and ceramics. Metallic zinc also is used in the preparation of some organic compounds.

Technical Application: Zinc is considered a "poisoning" element in much Solid State material device processing as it can degrade electrical parameters and, due to its ability to diffuse through solid metals when subject to thermal processing, can embrittle alloy contacts. Regardless, brass is widely used in device assemblies and packaging with a copper plating of sufficient thickness to buffer or block solid-solid diffusion of zinc.

Zinc is used in the growth of several compound semiconductors, the most well known being zinc sulfide, ZnS (the natural mineral sphalerite), but also include zinc arsenide, ZnAs; zinc selenide, ZnSe; and zinc telluride, ZnTe. It also has been studied as a zinc tungstate and a silicon zinc phosphide single crystal.

It has been widely studied and grown as a single crystal and as single crystal platelets with (0001) surface orientation and it can be readily cleaved in this orientation.

Etching: Readily soluble in both acids and alkalies.

ZINC ETCHANTS

ZN-0001a
ETCH NAME: Hydrochloric acid, dilute TIME:
TYPE: Acid, removal TEMP:
COMPOSITION:
1 .... HCl
1 .... H_2O
DISCUSSION:
Zn specimens. Solution used as a general removal etch for zinc.
REF: Rezek, J & Craig, G B — *Trans Metall Soc AIME,* 221,715(1961)

ZN-0002
ETCH NAME: Hydrochloric acid, dilute TIME:
TYPE: Acid, cleaning TEMP:
COMPOSITION:
x .... 5% HCl
x .... H_2O
DISCUSSION:
Zn, pellets used as a zinc source for LPE growth of ZnSe from a Zn-Ga solute. Clean zinc in dilute HCl before use. Also studied sulfur doped ZnSe as ZnSSe.
REF: Fujitz, S et al — *J Appl Phys,* 50,1079(1979)

ZN-0001b
ETCH NAME: Nitric acid, dilute TIME:
TYPE: Acid, saw TEMP:
COMPOSITION:
1 .... HNO_3
1 .... H_2O
DISCUSSION:
Zn, (0001) wafers cut from single crystal zinc ingots using an acid saw technique. The solution shown used as the acid vehicle for cutting.
REF: Ibid.

ZN-0003
ETCH NAME: TIME:
TYPE: Acid, cleaning TEMP:
COMPOSITION:
x .... HNO_3
x .... HAc
DISCUSSION:

Zn, (0001) platelets. Fabricated by pressing a 4 9s pure zinc 0.4 mm sheet between two glass plates with aluminum spacers. Heat to 693 K in argon. Most platelets were (0001) basal. Solution used to etch clean and remove surface contamination.

REF: Takama, T et al — *Jpn J Appl Phys,* 23,11(1984)

ZN-0004
ETCH NAME: TIME: 2 min
TYPE: Acid, polish TEMP: RT
COMPOSITION:
1 .... HNO_3
1 .... H_2O_2
1 .... EOH
DISCUSSION:

Zn, (0001) wafers and ingots. Solution used to polish specimens. Mix fresh when ready for use.

REF: Gilman, J & DeCarlo, V J — *Trans AIME,* 206,511(1956)

ZN-0001c: Rezek, J & Craig, G B — *Trans Metall Soc AIME,* 221,715(1961)

Zn specimens. Used in studying striation effects of critical residual shear stress on single crystal zinc. Solution used as a polish etch.

ZN-0005: Rosenbaum, H S & Saffren, M M — *J Appl Phys,* 32,1866(1961)

Zn, (0001) cleaved wafers using a razor and hammer for cleaving, also called the anvil and hammer method. Solution used to etch polish surfaces before etching for dislocations. Cleaving was done in LN_2.

ZN-0027a: DeCarlo, V J & Gilman, J J — *J Met,* 8,144(1956)

Zn specimens. Development work on chemical etching of pure zinc and cadmium metals. Solution used as a general polish etch.

ZN-0027b: DeCarlo V J & Gilman, J J — *J Met Abstr,* 8,84(1985)

Zn specimens. Work similar to that on ZN-0027a.

ZN-0027c: DeCarlo, V J & Gilman, J J — *Trans Am Inst Min Met Eng,* 206,511(1956)

Zn specimens. Work similar to ZN-0027a-b.

ZN-0006a
ETCH NAME: TIME: 1 min
TYPE: Acid, polish TEMP: RT
COMPOSITION:
32 g .... Cr_2O_3
4 g .... $Na_2SO_4.10\ H_2O$
2 g .... $Cu(NO_3)_2$
100 ml .... H_2O
DISCUSSION:

Zn, single crystals alloyed with copper or aluminum. Wafers were cut by an acid saw technique from as-grown platelets. Solution used as a polish etch. The Zn/Cu material had a bluish stain after etching. Remove with 5 ml HNO_3:20 ml H_2O_2:50 ml EOH, RT, 5 sec.

REF: Sinha, P P & Beck, P A — *J Appl Phys,* 33,625(1962)

ZN-0007: Sinha, P P & Beck, P A — *J Appl Phys,* 32,1222(1961)

Zn, single crystal specimens used in a bending/polygonization study. #ZN-solution used as a polish etch.

ZN-0006b
ETCH NAME: TIME: 5 sec
TYPE: Acid, polish TEMP: RT
COMPOSITION:
5 ml .... HNO_3
20 ml .... H_2O_2
50 ml .... EOH
DISCUSSION:
Zn alloy single crystal platelets cut by an acid saw technique. Zn/Cu alloy has bluish stain after initial polish etch (Zn-0006a). The solution shown will remove the stain and produce a highly polished surface.
REF: Ibid.

ZN-0007a
ETCH NAME: TIME:
TYPE: Acid, preferential TEMP:
COMPOSITION:
16 g .... CrO_3
5 g .... $NaSO_4$
100 ml .... H_2O
DISCUSSION:
Zn, (0001) wafers were dislocation etched in this solution. Solution is good unless cadmium is present in the zinc.
REF: Gilman, J J — *J Metal,* 8,998(1956)

ZN-0025
ETCH NAME: TIME:
TYPE: Acid, defect TEMP:
COMPOSITION:
x .... alcohol
DISCUSSION:
Zn, (0001) freshly cleaved surface used in a study of dislocation development. Authors say that dislocations can be developed on freshly cleaved surfaces with alcohols, water and weak acid etchants.
REF: Regel, U R & Stepanova, V — *Kristallografiya,* 4,226(1959)

ZN-0006c
ETCH NAME: TIME: 10 sec
TYPE: Acid, preferential TEMP:
COMPOSITION:
10 ml .... NHZ_4OH
50 ml .... H_2O
2 g .... $HN_4(NO_3)_2$
DISCUSSION:
Zn, single crystal with 0.07 and 0.02% aluminum in a study of defects created by bending. After etching, rinse in DI water, then MeOH and dry with a stream of dry air.
REF: Ibid.

ZN-007b
ETCH NAME: TIME: 15 sec
TYPE: Acid, dislocation TEMP: RT
COMPOSITION:
4 ml .... HNO_3
100 ml .... H_2O
DISCUSSION:

Zn, single crystal specimens used in a bending/polygonization study. After etching, rinse in DI water, then MeOH and blow dry with air. Etch time is critical — insufficient time, and no pits develop; excess etch time and pits are too large and overlap.

REF: Ibid.

ZN-0023a
ETCH NAME: Cadmium TIME:
TYPE: Metal, decoration TEMP: >321°C
COMPOSITION:
x .... Cd
DISCUSSION:

Zn, (0001) single crystal wafers used in a study of dislocations and impurity boundaries. Metallic cadmium diffused into specimens to decorate dislocations.

REF: Damiano, V V & Tint, G S — *Acta Metall,* 9,177(1961)

ZN-0023b
ETCH NAME: TIME:
TYPE: Acid, preferential TEMP:
COMPOSITION:
32 g .... CrO_3
6 g .... Na_2SO_4
100 ml .... H_2O
DISCUSSION:

Zn, (0001) wafers used in a dislocation study. Solution used as a dislocation etch and referred to Gilman: ZN-0007a.

REF: Ibid.

ZN-0024
ETCH NAME: TIME: 30 min
TYPE: Acid, cutting TEMP: RT
COMPOSITION:
(1) 200 ml .... HCl
50 g .... $FeCl_3$
250 ml .... H_2O
(2) 1 .... HNO_3
1 .... H_2O
DISCUSSION:

Zn, specimens $^3/_8$" thick. Solutions used with terylene thread to cut material. Zinc and tin were cut with solution (2); aluminum, brass, and copper with solution (1). (*Note:* Excellent review of metal growth methods.)

REF: Honeycombe, R W K — *Rev Met,* 4,1(1959)

ZN-0008a
ETCH NAME: P-1 TIME: 5—6 sec
TYPE: Acid, preferential TEMP: RT
COMPOSITION:
160 g .... CrO_3
20 g .... Na_2SO_4
500 ml .... H_2O
DISCUSSION:
Zn, (0001) wafers and cylinders with prism surfaces $(12\bar{1}0)$ and $(10\bar{1}0)$ used in an etching and etch development study. Wafers were cleaved (0001) at LN_2 temperature (cool slowly to prevent deformation). Other wafers and cylinders were acid saw cut with 8 *N* HNO_3 and an SS wire. After cleaving or cutting, lap polish wafers on a lucite plate with 8 *N* HNO_3 liquid carrier. During etching in solution shown, agitate specimens and wash heavily in DI water to ensure removal of CrO_3 as it produces very rapid corrosion. Air blow dry.
REF: Brandt, R C et al — *J Appl Phys*, 34,587(1963)

ZN-0008b
ETCH NAME: P-2 TIME: 2—3 sec
TYPE: Acid, preferential TEMP: RT
COMPOSITION:
1 .... 16 *N* HNO_3
1 .... H_2O_2
z .... MeOH
DISCUSSION:
Zn, (0001) wafers and cylinders. Agitate specimens during etching; rinse in tap water, then running DI water and air blow dry. See ZN-0008a for additional disscussion.
REF: Ibid.

ZN-0008c
ETCH NAME: P-3 TIME: 2—3 sec
TYPE: Acid, preferential TEMP: RT
COMPOSITION:
160 g .... CrO_3
500 ml .... H_2O
DISCUSSION:
Zn, (0001) wafers and cylinders. Agitate specimens during etching; rinse heavily in tap water and then running DI water to remove residual chromic acid. Air blow dry. See ZN-0008a for additional discussion.
REF: Ibid.

ZN-0008d
ETCH NAME: E TIME: 5—6 sec
TYPE: Acid, preferential TEMP: RT
COMPOSITION:
(1) 1 g .... $Hg(NO_3)$
1 ml .... 16 *N* HNO_3
500 ml .... H_2O
(2) 1 .... (1)
2 .... H_2O
DISCUSSION:
Zn, (0001) wafers. Agitate specimen during etching; rinse in tap water, then running DI water and air blow dry. Solutions produce small hillocks and pips and size varies with

etch time. Solution is preferential to planes perpendicular to (0001), the $\langle 10\bar{1}0\rangle$ prism plane directions. See ZN-0008a for additional discussion.
REF: Ibid.

ZN-0008e
ETCH NAME: Nitric acid
TYPE: Acid, preferential
COMPOSITION:
x .... 8 *N* HNO_3
x .... H_2S
Note: Saturate nitric acid with H_2S
TIME:
TEMP: RT
DISCUSSION:

Zn, cylinder with $(10\bar{1}0)$ prism orientation. Develops dark bands within 15° of $(12\bar{1}0)$ but none on $(10\bar{1}0)$. All wafers and cylinders (ZN-0008a thru ZN-0008e) were held under vacuum for 12 h and then re-etched — showed a reduced etch removal/polish rate. Subboundary pile-up at dislocations, caused by deformation.
REF: Ibid.

ZN-0009
ETCH NAME:
TYPE: Electrolytic, polish
COMPOSITION:
x .... 17% CrO_3
TIME: 20 sec
TEMP: RT
ANODE: Zn
CATHODE:
POWER: 8 V/cm^2
DISCUSSION:

Zn, (0001) specimens used in a surface etching study of 8 KeV argon ion bombardment. Polish surfaces in this solution before bombardment. Also studied gold and aluminum, which were polished in other solutions.
REF: Cunningham, R L et al — *J Appl Phys,* 31,839(1960)

ZN-0010
ETCH NAME: Neon
TYPE: Ionized gas, preferential
COMPOSITION:
x .... Ne
TIME:
TEMP:
GAS FLOW:
PRESSURE:
POWER:
DISCUSSION:

Zinc specimens. Ne^+ ion bombardment used to develop etch figures on metals. Other metals evaluated were Al, Bi, Cd, Co, Mg, Cu, and Sn.
REF: Yurasova, V E — *Kristallagrafia,* 2,770(1957)

ZN-0011
ETCH NAME:
TYPE: Acid, polish/thin
COMPOSITION:
(1) 160 g .... CrO_3
20 g .... Na_2SO_4
500 ml .... H_2O
(2) x .... 10% HNO_3
TIME:
TEMP:

DISCUSSION:

Zn, (0001) wafers. Wafers were deformed and studied for dislocation development. Solutions used to polish surfaces or etch thin.

REF: Kratochvil, P — *Phil Mag,* 13,267(1966)

ZN-0012

ETCH NAME: TIME: (1) 5—10 sec
(2) 30—60 sec

TYPE: Acid, polish TEMP: RT 20°C

COMPOSITION:

(1) 75 ml .... HNO_3, fuming
25 ml .... H_2O

(2) 70 ml .... HAc
30 ml .... HNO_3, fuming

DISCUSSION:

Zn specimens. Etch polish with (1) and follow with washing in (2). This same etch-clean system was used on cadmium. (*Note:* Red or yellow fuming nitric not shown. Usually refers to red fuming, 74% concentration.)

REF: Tegart, W J McG — *The Electrolytic and Chemical Etching of Metals,* Pergamon Press, London, 1956

ZN-0013

ETCH NAME: Nitric acid, dilute TIME:

TYPE: Acid, cutting TEMP:

COMPOSITION: ANODE: Zn

1 .... HNO_3 CATHODE:

1 .... H_2O POWER: 30 mA/cm^2

DISCUSSION:

Zn, single crystal specimens. Cutting done by passing a plastic thread (Saran) through solution and across specimen. Also used on tin.

REF: Sternhein, G — *Rev Sci Instr,* 26,1206(1955)

ZN-0014: Williams, W R et al — *Acta Metall,* 5,435(1957)

Zn, specimens acid saw cut in an X-ray study of plastic deformation in single crystal zinc.

ZN-0015

ETCH NAME: Hydrochloric acid, dilute TIME:

TYPE: Electrolytic, forming/defects TEMP:

COMPOSITION: ANODE: Zn

1 .... HCl CATHODE:

1 .... H_2O POWER:

DISCUSSION:

Zn specimens. After electrolytic forming of specimens as hemispheres, solution used to develop grain boundaries.

REF: Cochran, J F & Mapother, D E — *Phys Rev,* 121,1688(1961)

ZN-0016

ETCH NAME: Nitric acid TIME: 2—5 min

TYPE: Acid, preferential TEMP: Boiling

COMPOSITION:

x .... HNO_3, conc.

DISCUSSION:

Zn, (0001) wafers used in an etch pit study. Pits were different between the positive (0001) and negative (0001) surfaces.

REF: Czyak, S J & Reynolds, D C — *J Appl Phys,* 29,1190(1958)

ZN-0017a

ETCH NAME: TIME: 2—5 min

TYPE: Acid, preferential TEMP: RT

COMPOSITION:

1—5 .... 2% NaOH

1 .... H_2O_2

DISCUSSION:

Zn, diffused into GaAs wafers in a study of the precipitation of zinc in gallium arsenide. Solution produced sharply triangular pits on (111) surfaces. Also used $1HNO_3$:$2H_2O$ to develop dislocation pits.

REF: Black, J F & Jungbluth, E D — *J Electrochem Soc,* 114,181(1967)

ZN-0007c

ETCH NAME: Water TIME:

TYPE: Acid, defect TEMP: Hot to cold

COMPOSITION:

x .... H_2O

DISCUSSION:

Zn, (0001) cleaved wafers. By cycling the specimens from hot to cold water, a six-rayed star was developed. (*Note:* Similar to the percussion figure on mica, and the asterism of the corundum star sapphire)

REF: Ibid.

ZN-0007d

ETCH NAME: TIME:

TYPE: Electrolytic, polish TEMP:

COMPOSITION: ANODE:

33% .... H_3PO_4 CATHODE:

67% .... C_2H_5OH POWER:

DISCUSSION:

Zn, (0001) cleaved wafers. Solution used to polish specimens prior to preferential etching for defects.

REF: Ibid.

ZN-0018: Regal, V R & Stepanova, V M — *Sov Phys-Cryst,* 4,204(1960)

Zn, single crystal specimens. Specimens were subject to strain in a defect structure study. Authors say that no dislocations are observed if the specimens are electropolished after being strained.

ZN-0019: Damiano, V V & Herman, E M — *J Appl Phys,* 38,2740(1967)

Zn, (0001) wafers. After electropolish and preferential etching of specimens, loops and spirals were observed with both growth and movement with extended etch time.

ZN-0020: Yoo, M H & Wei, C T — *J Appl Phys,* 38,2974(1967)

Zn, (0001) wafers. After electropolishing, specimens were indented and etching produced twinning in the ⟨0001⟩ direction.

ZN-0021
ETCH NAME: Nitric acid
TYPE: Acid, cutting
COMPOSITION:
x .... HNO_3, conc.
TIME:
TEMP:
DISCUSSION:
Zn, single crystal wafers acid-saw cut using a braided SST wire and nitric acid. Used fly-wheel to trim-cut, then lap on emery paper and polish with alumina wheel. Specimens used in a study of elastic constants from 4.2 to 77.6 K.
REF: Garland, C W & Dalven, R — *Phys Rev,* 111,1232(1958)
ZN-0026b: Kiritani, M et al — *Jpn J Appl Phys,* 2,595(1963)
Zn specimens were acid saw cut as rods with nitric acid.

ZN-0022
ETCH NAME:
TYPE: Acid, junction stain
COMPOSITION:
x .... HF
x .... H_2O_2
x .... H_2O
TIME:
TEMP:
DISCUSSION:
Zn, diffused into InSb, (100), n-type wafers to form p-n junctions. Wafers cleaved ⟨110⟩ and solution used to stain develop junction.
REF: Nishitani, K — *J Electron Mater,* 12,124(1983)

ZN-0025a
ETCH NAME:
TYPE: Electrolytic, polish
COMPOSITION:
5 oz .... $Ni_3PO_4.H_2O$
1.2 oz .... $NaCO_3$
TIME: $^1/_2$—2 min
TEMP: 160—180°F
ANODE:
CATHODE:
POWER: 6 V
DISCUSSION:
Zn Specimens. Solution used to electropolish followed by cathodic dip in 220 g CrO_3/L at 212°F. See Zn-0008c.
REF: ASTM Committee — *Tool Engineer's Handbook,* McGraw-Hill, New York, 1949

ZN-0025b
ETCH NAME:
TYPE: Electrolytic, polish
COMPOSITION:
4 oz .... NaCN
1 oz .... $NaCrO_4$
2 oz .... NaOH
1000 ml .... H_2O
TIME: 15—35 sec
TEMP:
ANODE:
CATHODE:
POWER: 20—40 A/ft^2 & 2 V
DISCUSSION:
Zn specimens. Solution used to electropolish specimens with current reversal during etch period, and with or without and added wetting agent. (*Note:* Igpal is one wetting agent.)
REF: Ibid.

ZN-0026a
ETCH NAME: Chromic acid
TYPE: Electrolytic, polish
COMPOSITION:
x .... 20% CrO_3

TIME:
TEMP:
ANODE:
CATHODE:
POWER:

DISCUSSION:
Zn specimens acid saw cut as rods with HNO_3. Solution shown used both electrolytically to polish or as a wet chemical etch — see P-3 etch. ZN-0008c for use as a preferential etch. This was a study of impurity substructure in zinc crystals.
REF: Kiritani, S et al — *Jpn J Appl Phys,* 2,595(1963)

ZN-0028
ETCH NAME:
TYPE: Mechanical, defect
COMPOSITION:
x .... LN_2

TIME:
TEMP: −196°C

DISCUSSION:
Zn specimens cleaved under LN_2. Studied the as-cleaved surfaces for tilt boundary motion relative to static and dynamic sheer stress.
REF: Veerland, T Jr — *Acta Metall,* 9,112(1961)

ZN-0029
ETCH NAME: Argon
TYPE: Ionized gas, defect
COMPOSITION:
x .... Ar^+ ion

TIME:
TEMP:
GAS FLOW:
PRESSURE:
POWER:

DISCUSSION:
Zn and Cu specimens. Use of positive cathode sputtering as ion bombardment for micro-relief and surface erosion. Dislocation etch pits in both materials, and slip traces on deformed zinc. There can be re-deposition of metal in the close-packed crystal plane directions.
REF: Yurasova, V E — *Zhtekh Fiz,* 28,1966(1958)

ZINC ALLOYS, ZNM

General: Zinc does not occur as a metallic alloy in nature, though there are over 50 known minerals classified as zinc-minerals as oxides, sulfides, silicates, and carbonates, in addition to phosphates and arsenates. Sphalerite, ZnS; smithsonite, $ZnCO_3$; willemite, $ZnSiO_4$ and zincite, ZnO are representative and major ores of zinc. It is associated with copper, iron, lead, and manganese deposits.

Metallic zinc is considered one of the "heavy" metals of industry, and is associated with copper, iron, steel, and lead in manufacture.

The best known artificial zinc alloy is brass, CuZn, and its associate, bronze, CuSn. There are several hundred brass, bronze, and brass/bronze mixtures designed for specific applications with the term "Brass" being the more common. A few of the widely used brass alloys are

Formula	Description
1Sn:28Zn:71Cu	Admiralty brass. Developed for the British Admiralty
70Cu:30Zn	Cartridge brass. Original brass composition

65Cu:1Pb:33Zn	Clock brass. Easily machined for watches, engravings
61.5Cu:3Pb:35.5Zn	Free-cutting. Similar to Yellow, Cartridge or Spring
65Cu:30Zn	Yellow brass. Most widely used general brass

There are hundreds of zinc containing alloy products which, in quantity, rank fourth behind iron/steel, copper, and aluminum, not only for industrial applications, but in the jewelry trade and as objets d'art. This includes powdered brass in plastic as a spray coating replacement for gold, similar to the use of aluminum as a replacement for silver.

Technical Application: Zinc alloys, usually as some form of brass, have many applications in the processing of parts and equipment associated with the fabrication of Solid State devices and assemblies, such as contact pins, heat sinks, test blocks, package parts, tube housing in addition to screws, shim stock masks, springs, and similar items to name only a few such products. Zinc can be an unwanted element in the fabricating of certain semiconductor devices as it can affect electrical characteristics due to its ability to diffuse under thermal or electrical load, yet it is a p-type dopant element in some compound semiconductors, and is a primary constituent in others, such as zinc selinide, ZnSe and zinc telluride, ZnTe. The two latter compounds are semiconductors even though they could be classified as zinc alloys.

Zinc alloys have been the subject of growth studies and general morphological study, to include single crystallite evaluations. Both ZnSe and ZnTe have been studied as single crystals and are listed under their own headings.

Etching: Varies with alloy. Soluble in most acids and alkalies.

ZINC ALLOYS

ZINC TIN ETCHANTS

ZNSN-0001
ETCH NAME: Nital — TIME:
TYPE: Acid, preferential — TEMP: RT
COMPOSITION: — ASTM: #74

10 ml .... HNO_3
1000 ml .... EOH

DISCUSSION:

Zn:Sn, alloys processed in a study of solidification in lamellar eutectic systems. Evaluation specimens were quenched and others used as as-grown controls. A 10% Nital solution was used to develop structure. Also studied PbSn and CdZn.

REF: Tiller, W A & Mrdjenovich, R M — *J Appl Phys,* 34,3639(1963)

ZNSN-0002: ASTM E407-70 1947

Reference for ASTM Number 74. Nital shown as 1—5:HNO_3:100 ml EOH/MeOH.

ZINC-CADMIUM ETCHANTS

ZNCD-0001
ETCH NAME: Nital — TIME:
TYPE: Acid, preferential — TEMP: RT
COMPOSITION: — ASTM: #74

10 ml .... HNO_3
100 ml .... EOH

DISCUSSION:

Zn:Cd, alloys processed in a study of solidification in lamellar eutectic systems. Eval-

uation specimens were quenched and others as-grown for control. A 10% Nital solution was used to develop structure. Also studied PbSn and ZnSn.
REF: Tiller, W A & Mrdjenovich, R M — *J Appl Phys,* 34,3639(1963)
ZNCD-0002: ASTM E407-70
Reference for ASTM Number 74. Nital shown as 1—5 ml HNO_3:100 ml EOH/MeOH.

ZINC MERCURY ETCHANTS

ZNMG-0001
ETCH NAME: Acid, cut — TIME:
COMPOSITION: — TEMP:
1 .... HNO_3
1 .... H_2O_2
1 .... alcohol
DISCUSSION:
ZnHg specimens wire saw cut using the solution shown in a study of crack initiation in this compound. After cutting wash in water, dry in alcohol and then in ether. (*Note:* This is an amalgam. See section on Amalgam.)
REF: Westwood, A R C — *Phil Mag,* 9,199(1964)

SILVER ZINC ETCHANTS (See Silver Alloys)

AGZN-0002
ETCH NAME: — TIME:
TYPE: Acid, cleaning — TEMP:
COMPOSITION:
1 .... 10% KCN
1 .... 10% $(NH_4)_2SO_4$
DISCUSSION:
AgZn specimens as alpha structure used in a study of temperature dependence of magnetic susceptibility. Solution used to clean surfaces.
REF: Myers, L & Weiner, D — *Phys Rev,* 108,1426(1957)

PHYSICAL PROPERTIES OF ZINC ARSENIDE, Zn_3As_2

Classification	Arsenide
Atomic numbers	30 & 33
Atomic weight (°C)	346
Melting point (°C)	1015
Boiling point (°C)	
Density (g/cm³)	4—5
Refractive index (n=)	1.66—1.72
Hardness (Mohs — scratch)	2.5—3
Crystal structure (tetragonal — normal) Zn_3As_2	(100) prism, 2nd order
(monoclinic — normal) $ZnAs_2$	(100) a-pinacoid
Color (solid)	Grey metal
Cleavage (b-pinacoid — perfect)	(010)

ZINC ARSENIDE, Zn_3As_2

General: The pure binary alloy does not occur as a natural compound, but there are several arsenate minerals, such as barthite, chlorophoenivite, holdenite, and kottigite (refractive index shown, above) as hydrates containing zinc with manganese or copper. They are all minor in occurrence and associated with other copper, cobalt, and zinc vein ores. As arsenates or an arsenide there has been little or no use in industry. The compounds can be used as coloring pigments in glass and ceramic manufacture.

Technical Application: Little use in Solid State material fabrication to date, although the single crystals have been evaluated as binary- and trinary-type semiconductors of tetrahedral unit cell type.

Etching: HNO_3, H_2SO_4 and alkalies.

ZINC ARSENIDE ETCHANTS

ZNAS-0001
ETCH NAME: Heat — TIME:
TYPE: Thermal, preferential — TEMP:
COMPOSITION:
x .... heat
DISCUSSION:

Zn_3As_2, single crystal sphere used in a study of lattice parameters. Heating of sphere develops crystallographic facets.

REF: Cole, H et al — *IBM R&D,* 1,90(1957)

ZINC TIN ARSENIDE ETCHANTS

ZSAS-0001
ETCH NAME: Nitric acid — TIME:
TYPE: Acid, removal — TEMP:
COMPOSITION:
(1) x .... HNO_3 — (2) x .... x% KOH (NaOH)
DISCUSSION:

$ZnSnAs_2$, single crystals used in an evaluation of cohesive energy features of tetrahedral semiconductors. Forty materials were studied.

REF: Aresti, A et al — *J Phys Chem Solids,* 45,361(1984)

PHYSICAL PROPERTIES OF ZINC FLUORIDE, ZnF_2

Classification	Fluoride
Atomic numbers	30 & 9
Atomic weight	103.38
Melting point (°C)	872
Boiling point (°C)	
Density (g/cm^3)	4.84 nom.
Hardness (Mohs — scratch)	2—3
Crystal structure (monoclinic — normal)	(100) a-pinacoid
Color (solid)	Yellow/brown
Cleavage (basal)	(001)

ZINC FLUORIDE, ZnF_2

General: Does not occur as a natural compound. Like other fluorides, such as fluorite, CaF_2, it can be used as a solder flux in metal alloy assembly, but zinc compounds have wider use as the pure metal or as the oxide. Pure zinc is dip plated as galvanized surface coatings on iron as it stabilizes readily with a greyish zinc oxide; as heated zinc dust called sheradized iron. It has numerous uses as an alloy: brass and bronze, German silver (Zn:Ni:Cu), etc. Zinc dust can ignite explosively in air.

Technical Application: Zinc fluoride has no major use in Solid State processing, although it has shown semiconducting properties as a doped fluoride, $ZnSiF_2$. It has been grown and studied as the pure single crystal ZnF_2 and as the trinary compound.

Etching: Water, hot acids, and NH_4OH.

ZINC SILICON FLUORIDE ETCHANTS

ZSIF-0001
ETCH NAME: TIME:
TYPE: Acid, removal TEMP: Hot
COMPOSITION:
x .... NH_4OH
DISCUSSION:

$ZnSiF_2$, single crystals used to evaluate cohesive energy features of tetrahedral semiconductors. Forty materials studied.

REF: Aresti, A et al — *J Phys Chem Solids,* 40,361(1980)

PHYSICAL PROPERTIES OF ZINC OXIDE, ZnO

Classification	Oxide
Atomic numbers	30 & 8
Atomic weight	81.38
Melting point (°C)	>1800
Boiling point (°C)	
Density (g/cm^3)	5.47—5.7
Refractive index (n=)	2.01—2.03
Hardness (Mohs — scratch)	4—4.5
Crystal structure (hexagonal — hemimorphic)	xtls rare
Color (solid)	Deep red
Cleavage (basal — perfect)	(0001)

ZINC OXIDE, ZnO

General: Occurs as the natural mineral zincite, ZnO, known as red oxide of zinc and an ore of zinc. It is found in limited areas, but can be in massive deposits. For an oxide it is very heavy, and can be characterized by its deep red color and an orange-yellow streak, e.g., most oxides produce a white streak when drawn across an alumina blank. It often occurs as an artificial furnace by-product in the reduction of other zinc ores, notably, the sulfides.

Other than as an ore of zinc in industry, a primary use is as a high quality pigment in paints with addition applications in glass and ceramics.

Technical Application: The oxide is not used in Solid State processing as a semiconductor material, although it exhibits the polar characteristics of binary semiconductors on the basal (0001) surfaces.

There are possible applications for its electroluminescent properties, and has been grown as an artificial single crystal for general structural study as both the artificial compound and natural single crystal. It has also been developed as a thin film structure by the oxidation of zinc selenide, ZnSe, wafer surfaces.

Etching: The artificial compound is soluble in mineral acids, acetic acid, and a sodium hydroxide. The natural mineral is soluble in most single acids, alkalies and ammonium chloride. Both are soluble in mixed acids.

ZINC OXIDE ETCHANTS

ZNO-0001a
ETCH NAME: Nitric acid, dilute TIME: 90 sec
TYPE: Acid, preferential TEMP: RT
COMPOSITION:
x .... 20% HNO_3
DISCUSSION:
ZnO, (0001) wafers from natural zincite, and artificially grown from a PbF_2 molten flux and platelets from furnace residue. Solution produces hexagonal etch pits on the (0001) surface. The $(000\bar{1})$ surface is rough and etches more rapidly.
REF: Marino, A N & Hanneman, R E — *J Appl Phys,* 34,384(1963)

ZNO-0002
ETCH NAME: Perchloric acid, dilute TIME:
TYPE: Acid, preferential TEMP:
COMPOSITION:
x .... $HClO_4$
x .... H_2O
DISCUSSION:
ZnO, (0001) as platelets and grains. Solution will develop twinning structure. Also used to etch remove platelets and grains from the growth matrix.
REF: Santhanam, A T et al — *J Appl Phys,* 50,853(1979)

ZNO-0003
ETCH NAME: Phosphoric acid TIME: 30—40 min
TYPE: Acid, polish TEMP: RT
COMPOSITION:
x .... H_3PO_4
DISCUSSION:
ZnO, (0001) wafers. After cutting and mechanical lap polish this solution was used as a finish etch.
REF: Lagowski, J et al — *J Appl Phys,* 48,3566(1977)

ZNO-0004
ETCH NAME: CP4 TIME:
TYPE: Acid, preferential TEMP:
COMPOSITION:
3 ml ... HF
5 ml .... HN_3O_3
3 ml .... CH3COOH (HAc)
x .... Br_2
x = 10 drops/50 ml of solution
DISCUSSION:
ZnO, thin film deposit formed by oxidation of ZnSe. Etch will polish the (0001)Zn surface but is only rough on the $(000\bar{1})$O surface which etches more rapidly.
REF: Iwanaga, H — *Jpn J Appl Phys,* 22,1098(1983)

ZNO-0001b
ETCH NAME: TIME:
TYPE: Acid, preferential TEMP:
COMPOSITION:
6 .... HNO_3
6 .... HAc
1 .... H_2O
DISCUSSION:
ZnO, (0001) and $(10\bar{1}0)$ wafers. Etch use to develop polar etch figures. Solution will produce triangular etch pits on the $(10\bar{1}0)$.
REF: Ibid.

ZNO-0006a
ETCH NAME: Nitric acid, dilute TIME: 90 sec
TYPE: Acid, preferential TEMP: RT
COMPOSITION:
x .... 20% HNO_3
DISCUSSION:
ZnO, single crystal wafers cleaved (0001). Etch is more rapid on (0001)A surface than $(000\bar{1})$B. The (0001)A surface develops hexagonal pits with flat bottoms; the $(000\bar{1})$B surface shows a blocky, angled textured surface similar to structure observed on germanium.
REF: Horn, F N — *J Appl Phys,* 32,900(1961)

ZNO-0006b
ETCH NAME: TIME:
TYPE: Acid, preferential TEMP:
COMPOSITION:
6 .... HNO_3, fuming
6 .... HAc
1 .... H_2O
DISCUSSION:
ZnO, $(10\bar{1}0)$ prism cut wafers. Solution produces polar etch figures. (*Note:* White fuming (standard) = 70%; yellow fuming = 72%; red fuming = 74%.)
REF: Ibid.

ZNO-0005
ETCH NAME: TIME:
TYPE: Acid, polish TEMP:
COMPOSITION:
10 .... HF
45 .... HNO_3
45 .... H_2O
DISCUSSION:
ZnO, single crystal specimens used in a study of deformation.
REF: Soo, P & Higgins, G T — *Acta Metall,* 16,177(1968)

ZNO-0007
ETCH NAME: Sodium hydroxide, dilute TIME:
TYPE: Acid, removal TEMP:
COMPOSITION:
x .... 6 *N* NaOH
x .... ? (acids?)

DISCUSSION:

ZnO specimens used in a study of neutron-transfer reactions from nitrogen ion bombardment. Solution used as a general removal and surface cleaning etch.

REF: Halbert, M L et al — *Phys Rev,* 106,251(1957)

ZNO-0009

ETCH NAME: Heat — TIME: 2 h

TYPE: Thermal, treatment — TEMP: 600°C

COMPOSITION:

x .... heat

DISCUSSION:

ZnO, single crystal material was diamond saw cut as wafers, and mechanically polished with 1 μm diamond paste. Annealed as shown in study of electroluminescence.

REF: Pike, G E et al — *J Appl Phys,* 57,5512(1985)

PHYSICAL PROPERTIES OF ZINC PHOSPHIDE, Zn_3P_2

Classification	Phosphide
Atomic numbers	30 & 15
Atomic weight	258.18
Melting point (°C)	>420
Boiling point (°C)	1100
Density (g/cm³)	4.55
Hardness (Mohs — scratch)	2—3
Crystal structure (isometric — normal)	(100) cube
(tetragonal — normal)	(100) prism, 2nd order
Color (solid)	Grey/black
Cleavage (cubic)	(001)

ZINC PHOSPHIDE, Zn_3P_2

General: Does not occur as the pure phosphide in nature. The mineral sencerite, $Zn_3(PO_4)_2.Zn(OH)_2.3H_2O$, is representative of one of several phosphates. As a binary phosphide, the artificial material has been used as an additive to iron, steel, copper, brass, and bronze, such as for phosphor bronze or other special alloys.

Technical Application: The material has been evaluated as a possible II—V binary compound semiconductor, and a trinary compound with silicon. It also has been grown as a polycrystalline thin film, all in general material studies.

Etching: H_2SO_4, HNO_3, and mixed acids. Can be violent in dilute acids.

ZINC PHOSPHIDE ETCHANTS

ZNP-0001

ETCH NAME: — TIME:

TYPE: Acid, removal — TEMP: RT

COMPOSITION:

1 .... HF

1 .... HNO_3

DISCUSSION:

Zn_3P_2, grown as single crystal and polycrystalline thin films in a study of the material. Silver used as a p-type dopant. Single crystal was tetragonal with lattice constants: a =

8.05 and c = 11.45. Poly-type was cubic with a = 5.82. (*Note:* Solution shown is referenced from $ZnSiP_2$.)
REF: Chu, T L — *J Electrochem Soc,* 128,2063(1981)

ZINC SILICON PHOSPHIDE ETCHANTS

ZNSP-0001
ETCH NAME: TIME: 1 h
TYPE: Acid, polish TEMP: RT
COMPOSITION:
1 .... HF
1 .... HNO_3
DISCUSSION:

$ZnSiP_2$, (100) single crystal wafers. Etch cleaned in the solution shown in a material growth study. After etching bake at 1000°C.
REF: Nisida, H et al — *Jpn J Appl Phys,* 22(2),272(1983).

ZINC TIN PHOSPHIDE ETCHANTS

ZNTP-0001
ETCH NAME: TIME:
TYPE: TEMP:
COMPOSITION:
x .... HF
x .... HNO_3
DISCUSSION:

$ZnSnP_2$, thin films grown on GaAs, (100) substrates by LPE. Growth was perpendicular to (001) but 2°-off toward ⟨110⟩.
REF: Davis, G A et al — *J Cryst Growth,* 69,141(1984)

PHYSICAL PROPERTIES OF ZINC SELENIDE, ZnSe

Classification	Selenide
Atomic numbers	30 & 34
Atomic weight	144.34
Melting point (°C)	>1000
Boiling point (°C)	
Density (g/cm³)	5.42
Refractive index (n=)	2.89
Hardness (Mohs — scratch)	5—6
Crystal structure (isometric — normal)	(100) cube
Color (solid)	Grey/yellow
Cleavage (cubic — perfect)	(001)

ZINC SELENIDE, ZnSe

General: Does not occur as a natural compound, but has been grown as an artificial metallic single crystal. There are zinc oxides, arsenates, phosphates, and sulfates, but no selenides, although there are other metal selenides and tellurides in nature. There is no use of the zinc selenide in industry.

Technical Application: Zinc selenide is a II—VI type semiconductor material under evaluation as a polar binary compound for its electroluminescent and general etching characteristics. It has been grown both as a single crystal and as a thin film.

Etching: Mixed acids and alkalies.

ZINC SELENIDE ETCHANTS

ZNSE-0002c
ETCH NAME: Sodium hydroxide — TIME:
TYPE: Alkali, cleaning — TEMP: Boiling
COMPOSITION:
x .... 25% NaOH
DISCUSSION:
ZnSe:S (111) (SI) wafers used as substrates for epitaxy growth of ZnSe. Solution used for surface cleaning prior to and after epitaxy.
REF: Ibid.

ZNSE-0005
ETCH NAME: Sodium hydroxide — TIME:
TYPE: Alkali, polish — TEMP: Hot
COMPOSITION:
x .... x & NaOH
DISCUSSION:
ZnSe, (100) wafers used in a study of the Gunn Effect. Specimens were polished in a hot sodium hydroxide solution (concentration not shown).
REF: Ludwig, G W & Aven, M — *J Appl Phys,* 38,5326(1967)

ZNSE-0002b
ETCH NAME: — TIME: 1 min
TYPE: Acid, cleaning — TEMP: 95°C
COMPOSITION:
40 .... H_2SO_4
60 .... $K_3Cr_2O_7$, sat. sol.
DISCUSSION:
ZnSe:S, (111) (SI) wafers used as substrates for epitaxy growth of ZnSe. After mechanical polishing, etch clean in the solution shown; then etch in boiling 25% NaOH. After epitaxy, clean surface with a light etch in 35% NaOH. (*Note:* Solution shown is similar to the standard glass cleaner etch.)
REF: Ibid.
ZNSE-0007b: Ibid.
ZnSe, single crystal wafers of various orientations used in an etch development study. Solution also used on CdS and ZnS.

ZNSE-00007a
ETCH NAME: — TIME: $2^1/_2$ min
TYPE: Acid, preferential — TEMP: RT
COMPOSITION:
x .... H_2SO_4
x .... 0.3 *M* $KMnO_4$

DISCUSSION:

ZnSe, single crystal wafers of various orientations used in an etch development study. Solution used to develop dislocations and defects. Also used on CdS and ZnS.

REF: Rowe, J E & Forman, R A — *J Appl Phys,* 39,1917(1968)

ZNSE-0001

ETCH NAME: BRM, vapor — TIME:

TYPE: Halogen, preferential — TEMP: Hot

COMPOSITION:

(1) x .... Br_2 (2) x .... Br_2

x .... MeOH x .... CS_2

DISCUSSION:

ZnSe, (111) wafers. Both solutions used to preferential etch specimens.

REF: Sethi, B R et al — *J Appl Phys,* 50,353(1979)

ZNSE-0009: Sethi, B R et al — *J Appl Phys,* 39,533(1968)

ZnSe, (111) wafers. Specimens used in a defect structure study.

ZNSE-0002a

ETCH NAME: Sodium hydroxide — TIME:

TYPE: Alkali, preferential — TEMP: Boiling

COMPOSITION:

x .... 50% NaOH

DISCUSSION:

ZnSe, single crystal platelets grown in a Zn-Ga molten flux. Solution used to etch remove specimens from the flux and to etch clean platelets after removal.

REF: Fujita, S et al — *J Appl Phys,* 50,1079(1979)

ZNSE-0003

ETCH NAME: Sodium hydroxide — TIME: 3 min

TYPE: Alkali, dislocation — TEMP: 110°C

COMPOSITION:

x .... 30% NaOH

DISCUSSION:

ZnSe, (110) wafers cleaved from (111) oriented ingots grown by the Bridgman process. Solution used to develop dislocations on cleaved surfaces. Two types of triangle pits observed: (1) Type I: isosceles triangles and (2) Type II: equilateral triangle, normal.

REF: Iwanaga, H et al — *J Cryst Growth,* 67,97(1984)

ZNSE-0004

ETCH NAME: Sodium hydroxide — TIME: 2 min

TYPE: Alkali, preferential — TEMP: 90°C

COMPOSITION:

x .... 14 *N* NaOH

DISCUSSION:

ZnSe, (100) wafers etched in this solution during a study of the electrical properties of the material.

REF: Satch, S et al — *Jpn J Appl Phys,* 22,1167(1983)

ZNSE-0006
ETCH NAME: Hydrochloric acid — TIME:
TYPE: Acid, preferential — TEMP: RT
COMPOSITION:
x .... HCl
DISCUSSION:
ZnSe, (100) wafers used in a materials study. There is a 3C to 2H transformation at 1425°C to (00.1) hexagonal structure. Solution used to develop twinning and slip associated with ⟨111⟩ directions.
REF: Kikuuma I & Furkoghi, M — *J Cryst Growth,* 71,136(1985)

ZNSE-0008
ETCH NAME: — TIME:
TYPE: Acid, polish/clean — TEMP:
COMPOSITION:
5 .... H_2SO_4
1 .... H_2O_2
1 .... H_2O
DISCUSSION:
ZnSe, thin films epitaxially grown on GaAs:Cr, (100) (SI) substrates. Solution shown used to polish and clean substrates prior to epitaxy, follow with MeOH rinse and vacuum clean at 550°C in H_2, 10 min. Substrates at 250—350°C during epitaxy. ZnSe films showed hillocks in ⟨011⟩ direction.
REF: Fujitas, R et al — *J Cryst Growth,* 71,169(1985)

ZNSE-0009
ETCH NAME: — TIME: 1
TYPE: Acid, polish — TEMP: (1) 95°C (2) RT (3) 90°C
COMPOSITION:
(1) 2 .... H_2SO_4
3 .... $K_2Cr_2O_7$
(2) x .... CS_2
(3) x .... 14 *N* NaOH
DISCUSSION:
ZnSe, single crystal wafers, undoped or chlorine doped used in a study of photoluminescence and general properties. Etch clean and polish specimens in the solutions above and in the order shown.
REF: Satch, S & Igaki, K — *Jpn J Appl Phys,* 22,68(1984)

PHYSICAL PROPERTIES OF ZINC SULFIDE, ZnS

Classification	Sulfide
Atomic number	30 & 16
Atomic weight	97.4
Melting point (°C)	>800
Boiling point (°C)	
Density (g/cm³)	3.9—4.1
Refractive index (n=)	2.4—2.43
Optical energy gap (eV)	3.38
Hardness (Mohs — scratch)	3.5—4
Crystal structure (isometric — tetrahedral)	(111) tetrahedron (311) dodecahedron

Color (solid) Yellow/brown to black
Cleavage (dodecahedral — perfect) (110)

ZINC SULFIDE, ZnS

General: Occurs in nature as the mineral sphalerite, ZnS (zincblende, blende, or black jack), as a major ore of zinc, and is closely associated with galena, PbS. Sphalerite is the stable form below 1020°C and is isometric system-tetrahedral class; whereas the mineral wurtzite, ZnS, hexagonal-hemimorphic, stable above 1020°C, and the oxide form (zincite, ZnO) has a similar structure to that of wurtzite. The Sphalerite Group of minerals, all with the same crystallographic structure, includes five mercury compounds: HgS, (Hg,Zn)S, HgSe, Hg(SSe), and HgTe. Several have also been artificially grown as semiconductor compounds as has been sphalerite. Sphalerite is the major ore of zinc and can contain valuable amounts of gold and silver.

Technical Application: Zinc sulfide is a II—VI compound semiconductor and is used in the fabrication of electroluminescent diodes and laser diodes both as a single crystal wafer and a deposited single crystal thin film. Like cadmium sulfide, CdS, it can be alternately and sequentially deposited as ZnS/SZn/ZnS.

In addition to its semiconductor properties it is of major interest in semiconductor development for the isometric-tetrahedral crystallographic structure as it has the atypical tetrahedron unit cell of many of the binary compound semiconductors, such as gallium arsenide, indium phosphide, and others. Both artificially grown and natural single crystals have been the subject of much study for general morphology, structure and defects.

Etching: Easily etched in HCl, H_2O_2 and mixed acids.

ZINC SULFIDE ETCHANTS

ZNS-0001a
ETCH NAME: TIME: 10 min
TYPE: Acid, preferential TEMP: RT or 95°C
COMPOSITION:
x .... 0.5 *M* $K_3Cr_2O_7$
x .... 16 *N* H_2SO_4
DISCUSSION:

ZnS, (111) wafers used in a dislocation and defect study and etch solution development. The material was both natural sphalerite and synthetic ZnS. The natural mineral developed triangular pits with a background of shallow triangular etch figures on the (111)Zn surfaces. The (111)S surfaces were irregular. Used on synthetic ZnS at RT does not develop the same etch patterns. Used on synthetic ZnS for 10 min at 95°C, produced a highly polished surface with some triangular etch pits on the (111)Zn and shallow, disc-like pits on the $(\overline{111})$S surface.

REF: Warekois, E P et al — *J Appl Phys,* 33,691(1962)

ZNS-0009b: Ibid.

ZnS, single crystal wafers used in an etch development study. Results were similar to ZNS-0009a. Solution used on CdS and ZnSe.

ZNS-0010: Scranton, R A — *J Appl Phys,* 50,842(1979)

ZnS, $(\overline{111})$ wafers. Solution used at 95°C, 10 min. Produces a highly polished surface with dislocations on (111)S surfaces; shallow disc-like structure on (111)Zn. Similar results obtained on CdS (CDS-0001b).

ZNS-0002a
ETCH NAME: Potassium chromate TIME: 10 min
TYPE: Acid, dislocation TEMP: 95°C
COMPOSITION:
x .... 0.5 *M* $K_2Cr_2O_7$
DISCUSSION:
ZnS, (111) wafers. Solution will develop dislocation pits on the (111)A surface. Can be used on other orientations.
REF: Harper, C A — *Handbook of Materials and Processes for Electronics,* McGraw-Hill, New York, 1970, 7-5L

ZNS-0003
ETCH NAME: TIME: 10—60 min
TYPE: Acid, defect TEMP: 60—80°C
COMPOSITION:
x .... H_2O_2
DISCUSSION:
ZnS, (111) cleaved wafers, both artificial and natural crystals. Also phosphors of ZnS containing: (1) Cu, Al; (2) Cu, Al, Pb; (3) Cu, Cl-Mn, and (4) Cu, Cl. Hydrogen peroxide concentration was varied from $7^1/_2$ to 30%. Striations and some cleavage steps observed without etching, and etching will develop etch pits on both (111)A and $(\overline{111})$B surfaces. Used in a study of defects in electroluminescent ZnS.
REF: Goldberg, P — *J Appl Phys,* 32,1520(1961)
ZNS-0004: Froelich, H — *J Electrochem Soc,* 100,496(1953)
Reference for ZnS:Cu, Al phosphor.
ZNS-0005: Homer, H H et al — *J Electrochem Soc,* 100,566(1953)
Reference for ZnS:Cu,Cl-Pb and ZnS:Cu,Cl-Mn phosphors.

ZNS-0006
ETCH NAME: Nitric acid TIME: 2—5 min
TYPE: Acid, preferential TEMP: Boiling
COMPOSITION:
x .... HNO_3
DISCUSSION:
ZnS, (001) wafers used in an etch pit study. Pit structures are different between (001)A and $(00\overline{1})$B surfaces.
REF: Czyzak, S J & Reynolds, D C — *Bull Am Phys Soc,* 5,190(1960)

ZNS-0001b
ETCH NAME: TIME: 8—10 sec
TYPE: Acid, polish TEMP: RT
COMPOSITION:
1 .... HF
1 .... HNO_3
DISCUSSION:
ZnS, (100) wafers. The (100)A surface produces shallow disc-like pits. The $(\overline{1}00)$B shows dislocation pits.
REF: Ibid.

ZNS-0002b
ETCH NAME: TIME:
TYPE: Acid, preferential TEMP:
COMPOSITION:
1 .... HCl
1 .... HNO_3
DISCUSSION:
ZnS, (111) wafers. The (111)A surface develops triangular etch pits and etch figures. The $(\overline{111})$B shows some structure but is rough. Both surfaces are coated with a residual sulfur film.
REF: Ibid.

ZNS-0007
ETCH NAME: Nitric acid TIME:
TYPE: Acid, removal TEMP: 60—80°C
COMPOSITION:
x .... 6 *N* HNO_3
DISCUSSION:
Zn phosphors. Solution used as a general removal etch in studying the physical characteristics of the following phosphors: ZnS:CuCl:Pb; ZnS:Cu:Al; Zn(S,Se)Cu:Br, and ZnS:Cu:Cl:Mn.
REF: Goldberg, P & Faria, S — *J Electrochem Soc,* 107,521(1960)

ZNS-0008
ETCH NAME: Hydrochloric acid TIME:
TYPE: Acid, removal TEMP:
COMPOSITION:
x .... HCl, conc.
DISCUSSION:
Zn phosphors. Solution used for general removal in a study of the following phosphor: (ZnCd)(S,Se):Cu:Cl. It was observed that ball milling increased brightness and that the hydrochloric acid increased central grain luminescence.
REF: Remheller, A K — *J Electrochem Soc,* 107,8(1960)

ZNS-0009a
ETCH NAME: TIME:
TYPE: Acid, preferential TEMP:
COMPOSITION:
x .... H_2SO_4
x .... 0.3 *M* $KMnO_4$
DISCUSSION:
ZnS, single crystal wafers used in an etch development study. Solution also used on CdS and ZnSe.
REF: Rowe, J E & Forman, R A — *J Appl Phys,* 39,1917(1968)

ZNS-0011
ETCH NAME: Potassium cyanide TIME:
TYPE: Acid, cleaning TEMP:
COMPOSITION:
x .... 4% KCN

DISCUSSION:

ZnS, (0001) hexagonal wafers. A study of paramagnetic resonance absorption by manganese activation. Solution used to clean surfaces. (*Note:* Sphalerite is isometric system, tetragonal class under normal conditions.)

REF: Keller, S P et al — *Phys Rev,* 110,850(1958)

PHYSICAL PROPERTIES OF ZINC TELLURIDE, ZnTe

Classification	Telluride
Atomic number	30 & 52
Atomic weight	193
Melting point (°C)	1240
Boiling point (°C)	
Density (g/cm^3)	6.34
Refractive index (n =)	3.6
Hardness (Mohs — scratch)	6—7
Crystal structure (isometric — normal)	(100) cube
Color (solid)	Grey/reddish
Cleavage (cubic — perfect)	(001)

ZINC TELLURIDE, ZnTe

General: Does not occur as a natural compound. There are over 50 known zinc containing minerals of which sphalerite, ZnS, willemite, Zn_2SiO_4 and smithsonite, $ZnCO_3$, are major ores, in addition to zincite, ZnO and Franklinite, (Fe,ZnMn) $O(Fe,Mn)_2O_3$. The telluride has had no use in the metal industry, to date.

Technical Application: Zinc telluride is a II—VI compound semiconductor with the usual negative and positive (111) orientation polarity of these binary compounds.

It has been grown by both the Bridgman and Czochralski methods as single crystals for evaluation and general study. As thin films it has been grown by CVD and from powder on glass substrates. Fabricated devices include both diodes and transitors.

Etching: Acid, alkalies, and mixed acids.

ZINC TELLURIDE ETCHANTS

ZNTE-0001a

ETCH NAME: Sodium hydroxide — TIME: 3 min

TYPE: Alkali, dislocation — TEMP: 110°C

COMPOSITION:

x .... 30% NaOH

DISCUSSION:

ZnTe,(110) wafers and ingots were both etched in this solution. Produced isosceles triangles on (110) faces — called Type III pits.

See ZNSE-0003

REF: Iawanaga, H et al — *J Cryst Growth,* 67,97(1984)

ZTE-0002a

ETCH NAME: Warekois' reagent — TIME: 2 min

TYPE: Acid, preferential — TEMP: RT

COMPOSITION:

3 .... HF

2 .... H_2O_2

1 .... H_2O

DISCUSSION:

ZnTe, (111) wafers used in an etch development and dislocation study. The (111)Zn surface produced triangular etch pits. The $(\overline{111})$Te surface was coated with a Te film but there were no observable etch figures.

REF: Warekois, E P et al — *J Appl Phys,* 39,690(1962)

ZTE-0001b: Iwanaga, H et al — *J Cryst Growth,* 67,97(1984)

ZnTe,(110) wafers and ingot specimens. Refers to using this solution.

ZTE-0003

ETCH NAME: Water — TIME:

TYPE: Acid, float-off — TEMP: RT

COMPOSITION:

x .... H_2O

DISCUSSION:

ZnTe, thin films deposited from ZnTe powder onto glass substrates at 550 K and 10^{-4} Pa for 120 nm thickness, then overcoated with 0.1—0.5 nm aluminum. For TEM study the thin film was float-off removed from the glass substrate by soaking in water. After removal, carbon replicas were made for morphological study.

REF: Patel, S M & Patel, N G — *Thin Solid Films,* 113,185(1984)

ZNTE-0004

ETCH NAME: Sodium hydroxide — TIME:

TYPE: Alkali, preferential — TEMP:

COMPOSITION:

x .... 20 *M* NaOH

DISCUSSION:

ZnTe, thin films homoepitaxially deposited on GaAs, InAs, CdTe, and GaSb substrates in atmospheres of H_2; H_2:I_2, and Ar. Solution used in a morphological study of the thin films grown under the different atmosphere conditions shown.

REF: Nishio, M — *Jpn J Appl Phys,* 22,1101(1983)

ZNTE-0005a

ETCH NAME: Nitric acid, dilute — TIME: 3 min

TYPE: Acid, preferential — TEMP: RT

COMPOSITION:

x .... 25% HNO_3

DISCUSSION:

ZnTe, (111) wafers used in a dislocation and defect study. The (111)Zn surfaces turns blackish in color with no observable etch figures. The $(\overline{111})$Te surface develops triangular etch pits.

REF: Dillon, J A Jr — *J Appl Phys,* 33,668(1962)

ZNTE-0005b

ETCH NAME: Aqua regia — TIME: 1 min

TYPE: Acid, preferential — TEMP: RT

COMPOSITION:

3 .... HCl

1 .... HNO_3

DISCUSSION:

ZnTe, thin film deposits. Aqua regia produces similar etch patterns on both (111)Zn and $(\overline{111})$Te surfaces. The author points out that ZnTe films should not be overexposed to aqua regia vapors or a Te film will form on surfaces.

REF: Ibid.

ZNTE-0002b
ETCH NAME: TIME:
TYPE: Acid, preferential TEMP:
COMPOSITION:
1 .... HNO_3
1 .... HCl
DISCUSSION:
ZnTe, (111) wafers. The (111)Zn surface shows fine, deep triangular etch pits in arrays. The $(\overline{111})$Te produces a flat polished surface with shallow triangular etch figures and with extended etching time the surface polish improves.
REF: Ibid.

ZNTE-0008
ETCH NAME: Nitric acid TIME:
TYPE: Acid, cleaning TEMP:
COMPOSITION:
x .... HNO_3, conc.
DISCUSSION:
ZnTe, specimens cut as bars 0.04 × 0.03 × 0.250. Etch cleaned in concentrated nitric acid before high temperature conductivity studies in zinc vapors.
REF: Thomas, D G & Sadowski, E A — *J Phys Chem Solids,* 25,395(1964)

ZNTE-0009
ETCH NAME: TIME:
TYPE: Alkali, dislocation TEMP: Hot
COMPOSITION:
x .... x% NaOH
DISCUSSION:
ZnTe, (111) wafers used as substrates for epitaxy growth of CdS to form heterojunctions. Solution used to develop structure on ZnTe wafers. The (111)A developed triangular pits; the $(\overline{111})$B showed only erratic structure.
REF: Aven, M & Garwacki, W — *J Electrochem Soc,* 110,401(1963)

ZNTE-0002c
ETCH NAME: TIME: 8—10 sec
TYPE: Acid, polish TEMP: RT
COMPOSITION:
4 .... HNO_3
3 .... HCl
DISCUSSION:
ZnTe, (111) wafers. Solution will polish both surfaces. The Te film remaining can be removed with concentrated HCl.
REF: Ibid.

ZNTE-0002d
ETCH NAME: Hydrochloric acid TIME:
TYPE: Acid, removal TEMP: RT
COMPOSITION:
x .... HCl, conc.

DISCUSSION:

ZnTe, (111) wafers. Solution used to remove Te films remaining on surfaces after etching in other solutions.

REF: Ibid.

ZNTE-0007

ETCH NAME: Thermal etch — TIME:

TYPE: Heat, preferential — TEMP: 600°C

COMPOSITION:

x .... Ar(?)

DISCUSSION:

ZnTe, (111) wafers. Wafers were heated in an inert atmosphere. Heating develops etch pits and decorates dislocations. Dislocations precipitate at random Te spots, and heating can cause super-saturation of Zn with Te segregates. Slow cooling will allow re-indiffusion of Te.

REF: Lynch, R T et al — *J Appl Phys,* 34,706(1963)

ZNTE-0010

ETCH NAME: — TIME:

TYPE: Acid, removal/polish — TEMP:

COMPOSITION:

4 .... HF

3 .... HNO_3

DISCUSSION:

ZnTe, (111) wafers. Shown as a general removal and polish etch for this material. Etch solution leaves a dark Te film on surfaces that can be removed with a dip in HCl, then DI water wash.

REF: Sagar, A et al — *J Appl Phys,* 39,5336(1968)

ZNTE-0002e: Ibid.

ZnTe, (111) wafers. Solution used as a general polish etch.

PHYSICAL PROPERTIES OF ZINC TUNGSTITE, ZnW

Classification	Tungstite
Atomic number	30 & 74
Atomic weight	249.2
Melting point (°C)	
Boiling point (°C)	
Density (g/cm^3)	
Hardness (Mohs — scratch)	7—8
Crystal structure (isometric — normal)	(100) cube
Color (solid)	Grey
Cleavage (cubic)	(001)

ZINC TUNGSTITE, ZnW

General: Does not occur as a natural metallic compound. There are tungstate minerals with the WO_4^- radical, and one sulfate. There is some use in industry as a high temperature alloy.

Technical Application: Zinc tungstite has been grown as a single crystal for defect and crystallographic study.

Etching: Acids and alkalies.

ZINC TUNGSTITE ETCHANTS

ZNW-0001a
ETCH NAME: Phosphoric acid
TYPE: Acid, polish
COMPOSITION:
x .... H_3PO_4
TIME: 15 min
TEMP: 400—450°C
DISCUSSION:

ZnW, (001) cleaved wafers. Mechanically polish: (1) with 305 SiC abrasive; (2) $5^1/_2$-grit diamond in kerosene and (3) Linde B alumina in KOH. Then etch polish in the solution shown. Material used in a study of growth, defects, and crystallography.

REF: O'Hara, S — *J Appl Phys,* 35,1312(1964)

ZNW-0001b
ETCH NAME:
TYPE: Alkali, preferential
COMPOSITION:
(1) 1 .... KOH
1 .... NaOH
(2) x .... 4 *M* KOH/NaOH
TIME: 4—7 min
TEMP: Boiling
DISCUSSION:

ZnW, (001) cleaved wafers. Solution used to develop dislocation etch pits. Pits were the normal square outline with a common directional orientation in ⟨110⟩.

REF: Ibid.

PHYSICAL PROPERTIES OF ZIRCONIUM, Zr

Classification	Transition metal
Atomic number	40
Atomic weight	91.22
Melting point (°C)	1852
Boiling point (°C)	3580 (3578)
Density (g/cm^3)	6.5 (6.44)
Thermal conductance (cal/sec)(cm^2)(°C/cm)	0.0505
Specific heat (cal/g) 25°C	0.066
Latent heat of fusion (cal/g)	60.3
Heat of fusion (k-cal/g-atom)	3.74
Heat of vaporization (k-cal/g-atom)	120
Atomic volume (W/D)	14.1
1st ionization energy (K-cal/g-mole)	160
1st ionization potential (eV)	6.95
Electronegativity (Pauling's)	1.4
Covalent radius (angstroms)	1.45
Ionic radius (angstroms)	0.79 (Zr^{+4})
Coefficient of linear thermal expansion ($\times 10^{-6}$ cm/cm/°C)	5.78 (5.85)
Electrical resistivity (micro-ohms-cm)	41.4
Electron work function (eV)	4.33
Compressibility (cm^2/kg $\times$ 10^{-6})	11.77
Magnetic susceptibility (emu/g-atom $\times$ 10^6)	119
Poisson ratio	0.34
Young's modulus (kg/cm^2 $\times$ 10^{-4})	0.939

Grueneisen constant	128,900
Hall coefficient (V-cm/A-Os × 10^{12})	0.18
Cross section (barns)	0.18
Yield strength (1000 psi)	35—45
Hardness (Mohs — scratch)	2—3
(Vicker's — kg/mm²)	110
(Rockwell, B) annealed	75—85
Crystal structure (hexagonal — normal) alpha <865°C	$(10\bar{1}0)$ prism, hcp
(isometric — normal) beta >865°C	(100) cube, fcc
Color (solid)	Silver-white/grey
Cleavage (basal/cubic)	(0001)/(001)

ZIRCONIUM, Zr

General: Does not occur as a native element. The major ore is zircon, $ZrSiO_4$, which is found widely as an accessory mineral in igneous rocks, particularly from acid magmas with large crystals occurring in pegmatites. As the variety hyacinth, it is orange-reddish to brown and transparent, and a gem stone. The artificial zircon as a blue stone is called the gem stone starlite. Both the natural and artificial clear and transparent varieties are the gem stone called zirconia, often a substitute for diamond and, when tinted yellow, a substitute for gem quality topaz. As a metallic element it is in the titanium family and its oxides can be either basic or acidic.

In industry the pure metal has some use as a metal alloy although, as ferro-zirconium, it is of more importance as a desulfurizer or deoxidizer in steels. It has wider use as an oxide, such as for gas mantles, as an opacifier in paints like titanium and zinc oxides, as a filler in lacquers, and as an insulator. The oxide also is used as an abrasive for lapping and polishing.

Technical Application: None as the pure metal in Solid State processing, although the oxide has been used as a lapping abrasive.

Zirconium has been grown as a single crystal for general studies though much of the work has been with crystalline blanks, rods and other forms . . . defect, oxidation, copper alloying, deformation, and superconducting studies. It has also been grown as a thin film amorphous and single crystal nitride and is under evaluation as a silicide for use as a buffer layer in semiconductor heterostructures. It has been fabricated as a carbide, ZrC and boride, ZrB for use as an insulator.

Etching: HF, aqua regia slowly in other single acids. Acid mixtures.

ZIRCONIUM ETCHANTS

ZR-0001a
ETCH NAME:
TYPE: Acid, removal
COMPOSITION:

8 .... HF
50 .... HNO_3
50 .... H_2O

TIME: 1 min
TEMP: RT

DISCUSSION:

Zr specimens. Solution used to clean and polish specimens in a corrosion and oxidation study. Anodic oxide grown in H_3BO_3 as a saturated solution, RT, 500 V, and a 30 A°/v^{-1} rate.

REF: Misch, R D & Gunzel, F H Jr — *J Electrochem Soc,* 106,14(1959)

ZR-0001b
ETCH NAME: TIME:
TYPE: Acid, removal TEMP:
COMPOSITION:
1 .... HF
1 .... H_2O_2
20 .... H_2O
DISCUSSION:
Zr specimens. See discussion under ZR-0001a.
REF: Ibid.

ZR-0002
ETCH NAME: TIME: x sec
TYPE: Acid, removal/cleaning TEMP: RT
COMPOSITION:
10 ml .... HF
45 ml .... HNO_3
45 ml .... H_2O_2
DISCUSSION:
Zr, polycrystalline sheet. Discs were spark cut from sheet material and mechanically polished with 1 μm diamond paste. Solution used to etch clean surfaces. Also used on ZrN (ZRN-0001).
REF: Dawson, P T — *J Vac Sci Technol,* 21,36(1982)
ZR-0003b: Ibid.
Zr and Zr alloy specimens used in a study of oxidation. Solution used as polishing etch.

ZR-0003a
ETCH NAME: TIME:
TYPE: Acid, slurry polish TEMP: RT
COMPOSITION:
1 ml .... HF
0.5 ml .... HNO_3
98.5 ml .... H_2O
\+ .... Linde A abrasive
DISCUSSION:
Zr specimens. Solution used as an acid/slurry polish lap.
REF: Ports, N A et al — *J Electrochem Soc,* 107,506(1960)

ZR-0004a
ETCH NAME: TIME: 30 sec
TYPE: Acid, cleaning TEMP: RT
COMPOSITION:
x .... HF
x .... HNO_3
DISCUSSION:
Zr, polycrystalline rod etch cleaned in this solution prior to Cu:Sn(12% Sn) plating with Sn aiding in the formation of an Nb_3Sn coating that can then be diffused to form ZnNb(1% Nb). Zr rod electropolished prior to Cu:Sn plating. Specimens used in a superconductivity study.
REF: Bussiere, J F & Suenasa, M A — *J Appl Phys,* 47,707(1976)

ZR-0004b
ETCH NAME: TIME: 15—60 sec
TYPE: Electrolytic, polish TEMP: RT
COMPOSITION: ANODE: Zr
1 .... HF CATHODE:
4 .... HNO_3 POWER: 10 V
2 .... H_2O
DISCUSSION:

Zr, polycrystalline rod. Electropolishing in this solution will produce a satin finish.

REF: Ibid.

ZR-0005a
ETCH NAME: TIME:
TYPE: Acid, thinning TEMP:
COMPOSITION: ANODE:
5% .... $HClO_4$ CATHODE:
95% .... H_2O POWER: 50 V
DISCUSSION:

Zr, poly sheet used as substrates for copper evaporation and microstructure study after laser annealing. Sheet mechanically polished with 0.25 μm diamond paste. Degrease: boiling TCE; rinse acetone, then EOH. With parts in vacuum, Ar^+ ion sputter etch prior to copper evaporation. Use laser in air to form a 5000 Å deep layer of alloyed Zr/Cu. Etch thin for TEM study with the solution shown, then argon ion mill to remove oxide.

REF: Denbroeder, F J A et al — *Thin Solid Films,* 111,43(1984)

ZR-0005b
ETCH NAME: TIME:
TYPE: Electrolytic, thinning TEMP:
COMPOSITION: ANODE:
x .... HCl_3O_4 CATHODE:
x .... CH_3COOH (HAc) POWER: 50 V
x .... ethylene glycol (EG)
DISCUSSION:

Poly-Zr sheet used as substrates for copper evaporation. Mechanically polish sheet with $^1/_4$ μm diamond paste. Degrease in boiling TCE, then rinse in acetone, then MeOH. Ar^+ ion sputter etch in vacuum system before copper deposition. Use a laser in air to produce a Cu/Zn alloy, 5000 Å deep. Etch thin and/or remove zirconium substrate with the electrolytic etch shown. Argon ion milling used to remove residual oxygen left on surface from etching. Alloyed thin film used in TEM study.

REF: Ibid.

ZR-0009
ETCH NAME: TIME: 1 min
TYPE: Acid, removal TEMP: RT
COMPOSITION:
50 ml .... HF
50 ml .... HNO_3
5 ml .... H_2O

DISCUSSION:

Zr specimens used in a study of the dissolution of oxide films. Anodize zirconium in 1% KOH at RT, $^1/_2$ h at 0—35 V with a Pt anode. Use etch shown to remove grown oxide.

REF: Misch, R D — *Acta Metall,* 5,178(1957)

ZR-0011

ETCH NAME: TIME: 2 min

TYPE: Acid, polish TEMP: RT

COMPOSITION:

5 .... HF

50 .... HNO_3

45 .... H_2O

DISCUSSION:

Zr specimens used in a study of oxygen diffusion and its relationship to oxidation and corrosion. Polish with solution shown before oxidizing. Form oxide by anodic oxidation in a 1% KOH solution with Pt wire cathode in 15 min steps from 5 to 95 V power. The oxide color changes with voltage used. (*Note:* Color change is due to variation of oxide thickness.)

REF: Pemsler, J P — *J Electrochem Soc,* 105,315(1958)

ZR-0006

ETCH NAME: TIME:

TYPE: Electrolytic, polish TEMP: 5—10°C

COMPOSITION: ANODE: Zr

5 .... $HClO_4$ CATHODE:

3 .... HAc POWER: 0.5—0.7 A/cm^2 @ 20 V

2 .... ethylene glycol (EG)

DISCUSSION:

Zr, (0001) and (10$\bar{1}$0) wafers used in a dislocation study. Solution shown used to electropolish surfaces. Use same solution for dislocation development at 5—8 V and 0.2—0.5 sec.

REF: Mills, D & Craig, G B — *J Electrochem Technol,* 4,300(1966)

ZR-0010a: Rappeport, E J — *Acta Metall,* 7,132(1959)

Zr specimens used in a study of room temperature deformation. An acetic acid:perchloric acid solution used to electropolish.

ZR-0007

ETCH NAME: Aqua regia TIME:

TYPE: Acid, removal TEMP:

COMPOSITION:

3 .... HCl

1 .... HNO_3

DISCUSSION:

Zr specimens. Shown as an etch for zirconium.

REF: Hodgman, C D — *Handbook of Chemistry and Physics,* 27th ed, Chemical Rubber Co., Cleveland, OH, 1943, 490

ZR-0008
ETCH NAME: TIME: 1 min
TYPE: Acid, macro-etch TEMP: Hot
COMPOSITION:
8 ml .... HF
50 ml .... HNO_3
50 ml .... H_2O, tap (hot)
DISCUSSION:

Zr, polycrystalline blanks used in a study of anodic film growth of $ZrO_x.nH_2O$ and its effect on grain orientation. Anodize surfaces with boiling nitric acid. Use solution shown to develop microstructure.

REF: Misch, R D & Fisher, E S — *Acta Metall,* 4,222(1956)
ZR-0001c: Misch, R D & Gunzel, F H Jr — *J Electrochem Soc,* 106,15(1959)

Zr, polycrystalline blanks oxidized and studied for electrical resistance and corrosion. Use etch at RT, 1 min. Anodize in saturated H_3BO_3 with a Pt gauze cathode.

ZR-0021
ETCH NAME: TIME:
TYPE: Acid, removal TEMP:
COMPOSITION:
1 .... HF
1 .... HNO_3
50 .... H_2O
DISCUSSION:

Zr, thin film evaporation metal in vacuum systems. Solution used to clean bell jars and internal fixtures. Can be used on steels, ceramics, and glass. Do not overetch glass as solution contains HF.

REF: Nichols, D R — *Solid State Technol,* December 1979.

ZR-0016
ETCH NAME: BHF TIME:
TYPE: Acid, polish TEMP:
COMPOSITION:
x .... HF
x .... NH_4F
DISCUSSION:

Zr, specimens used for electroplating. Polish specimens by immersion. (*Note:* A major article on electroplating of transition metals. Presents both polish etchants and plating solutions.)

REF: Saubestre, E B — *J Electrochem Soc,* 106,106(1959)
ZR-0010: Faust, C L & Beach, J G — *Plating,* 43,1134(1950)

Reference for solution shown in ZR-0009.

ZR-0017: Schickner, W C et al — *J Electrochem Soc,* 100,289(1953)

Zr specimens. Used as a general immersion polishing etch on zirconium.

ZR-0015
ETCH NAME: Heat TIME:
TYPE: Thermal, forming TEMP:
COMPOSITION:
x .... heat
DISCUSSION:

Zr, single crystal and polycrystalline spheres formed by splatter using an arc melter

under an He atmosphere with flakes and chips in a water cooled copper pot. Other materials were Be, V, Fe, NiCu, and NiAl. Size was from 0.1 to 1.0 mm with 5 to 50% single crystal.
REF: Ray, A E & Smith, J F — *Acta Metall,* 11,310(1958)

ZR-0020
ETCH NAME: TIME: 30 sec
TYPE: Acid, polish TEMP: RT
COMPOSITION:
1 .... HF
10 .... HNO_3
DISCUSSION:
Zr, specimens used in a study of oxygen gradients in the oxidation of metals. Specimens were first polished with 4/0 emery, then give a cleaning polish in the solution shown.
REF: Pemsler, J P — *J Electrochem Soc,* 111,381(1964)

ZIRCONIUM ALLOYS

General: Does not occur as a metallic alloy. The most important industrial alloy is ferro-zirconium used to desulfurize and deoxidize irons and steels.

Technical Application: Some single crystal zirconium alloys have been studied for their electronic characteristics, general morphology and structure.

Etching: HF:HNO_3 mixed acids ± H_2O_3, HAc, etc.

ZIRCONIUM NICKEL ETCHANTS

ZRNI-0001
ETCH NAME: TIME:
TYPE: Acid, removal TEMP:
COMPOSITION:
x .... HF
x .... HNO_3
DISCUSSION:
Zr_2Ni single crystal specimen used in a study of electronic phase transitions. Solution will etch material. Also studied NiTi.
REF: Lee, R N & Withers, R — *J Appl Phys,* 49,5488(1978)

ZIRCONIUM:VANADINITE ETCHANTS

ZRV-0001
ETCH NAME: TIME:
TYPE: Acid, removal TEMP:
COMPOSITION:
x .... HF
x .... HNO_3
DISCUSSION:
ZrV single crystal metallic alloy grown in a general study of these materials. Other materials studied were TiAl, TiNi, and Fe_3C. Referred to as metastable alloys. Metallic glasses studied were PdNiP, PtNiP, and PtCuP.
REF:

PHYSICAL PROPERTIES OF ZIRCONIUM DIOXIDE, ZrO_2

Classification	Ceramic
Atomic numbers	40 & 8
Atomic weight	123
Melting point (°C)	2700
Boiling point (°C)	
Application limits (°C)	2400
Density (g/cm³)	5.5—6.0
Specific heat (°C) 25—1000°C	590
Coefficient thermal linear of expansion ($\times 10^{-6}$ cm/cm/°C) 25—800°C	7.5
Thermal conductivity (W/m °C) 1315°C	3
Electrical resistivity ($\times 10^6$ ohms cm) 385°C	1
($\times 10^2$ ohms cm) 1200°C	3.6
Refractive index (n=)	2.13—2.20
Hardness (Mohs — scratch)	6.5
Crystal structure (monoclinic — normal)	(100) a-pinacoid
(tetragonal — normal) >1000°C	(100) 2nd order prism
Color (solid)	Colorless/tinted
Cleavage (basal)	(001)

ZIRCONIUM DIOXIDE, ZrO_2

General: Occurs in nature as the minor mineral baddeleyite, ZrO_2. Usually found as small pebbles and grain associated with diamond sands, e.g., Brazil, where they are called favas (beans) and vary between pure ZrO_2 and pure TiO_2. The most important ore mineral is zircon, $ZrSiO_4$, and there are several other silicate minerals. It is a trace constituent in titanium minerals, which may contain hafnium oxide. The oxide is called zirconia and is the name for clear gem stone quality zircon, either natural or artificial. The natural gem stone most prized is a fine, transparent yellow similar to quartz citron.

As the oxide, there are several industrial applications, chiefly in the ceramics industry as a high temperature refractory dielectric, alone, or mixed with alumina, silica, or titania. Added to silica it produces a "hard" type glass that is tough and, added as a white powder, it has similar opacifier characteristics in enamels as a replacement for SnO_2. Zirconium also acts like a getter in vacuum systems for removing gases similar to titanium.

Technical Application: Under development and evaluation as a thin film dielectric surface coating on semiconductor wafers and other metallic compounds with similar applications to those of silicon dioxide or silicon nitride. As zirconium has a low neutron cross-section, both the metal and the oxide have application in nuclear reactor areas similar to those of lead.

Both the natural mineral and artificially grown single crystals, as well as thin films, have been the subject of general morphological, defect and electrical studies.

Etching: Soluble in HF, H_2SO_4, and mixed acids.

ZIRCONIUM OXIDE ETCHANTS

ZRO-0001a

ETCH NAME: Sulfuric acid — TIME:

TYPE: Acid, removal — TEMP: 60°C

COMPOSITION:

x .... H_2SO_4

DISCUSSION:

ZrO_2, single crystal specimens. Material is monoclinic with a reversible inversion point at 1000°C to tetragonal. Solution used as a general etch for the material.

REF: Brochure: Magnesium Electron Inc (MEI), Farmington, NJ

ZRO-0004: Ben-Dor, L et al — *J Electron Mater,* 13,263(1984)

ZrO_2, thin film deposited by MOCVD on silicon, (111), degenerate n-type wafers. Prior to epitaxy growth, wafers were cleaned in (1) H_2SO_4:H_2O_2 and (2) HF:H_2O.

ZRO-0001b

ETCH NAME: Hydrofluoric acid — TIME:

TYPE: Acid, removal — TEMP:

COMPOSITION:

x .... HF

DISCUSSION:

ZrO_2, single crystal specimens. Solution used as a general etch.

REF: Ibid.

ZRO-0002

ETCH NAME: — TIME: 5 min

TYPE: Electrolytic, polish — TEMP: 10°C

COMPOSITION: — ANODE: ZrO_2

1 .... $HClO_4$ — CATHODE:

9 .... CH_3COOH(HAc) — POWER: 20 V

DISCUSSION:

ZrO, thin film deposits used in a study of gamma radiation effects on electrical parameters and corrosion in nuclear reactors. Solution was used to polish surfaces prior to radiation.

REF: Harrop, P J & Wanklyn, J N — *Br J Appl Phys,* 16,155(1965)

ZRO-0003

ETCH NAME: — TIME: (1)1 min (2)10—180 h

TYPE: Acid, polish/cleaning — TEMP: RT 44°C

COMPOSITION:

(1) 4 .... HF
30 .... HNO_3

(2) x .... H_2SO_4

DISCUSSION:

ZrO_2, specimens with low hafnium content used in a study of corrosion kinetics in sulfuric acid solutions. Etch polish in solution (1) then soak in solution (2). Several concentrations of H_2SO_4 were evaluated.

REF: Smith, T — *J Electrochem Soc,* 107,82(1960)

PHYSICAL PROPERTIES OF ZIRCONIUM NITRIDE, ZrN

Classification	Nitride
Atomic numbers	40 & 7
Atomic weight	105
Melting point (°C)	>2800
Boiling point (°C)	
Density (g/cm^3)	
Hardness (Mohs — scratch)	6—7
Crystal structure (isometric — normal)	(100) cube

Color (solid) Brownish
Cleavage (cubic — perfect) (001)

ZIRCONIUM NITRIDE, ZnN

General: Does not occur as a native compound. Metallic zirconium has been surface nitrided in industry for a thin skin more stable against corrosion.

Technical Application: Zirconium nitride is under evaluation in Solid State processing as a surface coating with similar characteristics and applications as those of silicon and aluminum nitrides, though it is not as stable against atmospheric effects of moisture in air.

Etching: $HF:HNO_3$ and other mixed acids.

ZIRCONIUM NITRIDE ETCHANTS

ZRN-0001
ETCH NAME: TIME: Seconds
ETCH: Acid, cleaning TEMP: RT
COMPOSITION:
10 ml .... HF
45 ml .... HNO_3
45 ml .... H_2O_2
DISCUSSION:

ZrN, thin films grown with NH_3 as the nitrogen source. Films were non-uniform in color. A dark grey center area (high O_2), and a bronze outer area (O_2 + C). Ar^+ ion sputtering of film removes the O_2 and C. Solution shown used to clean the thin films.
REF: Dawson, P T — *J Vac Sci Technol,* 21,36(1982)

ZRN-0002
ETCH NAME: Heat TIME:
TYPE: Thermal, cleaning TEMP: 1300°C
COMPOSITION:
x .... heat
DISCUSSION:

ZrN, (100) wafers cut from a polycrystalline Zr rod nitrided at 2800°C. This treatment forms large crystallites of cubic, ZrN (NaCl type). After zone anneal crystallite growth, specimens require flash cleaning with heat every 3 h to keep surfaces clean.
REF: Callenas, A et al — *J Physics Rev B,* 30,635(1984)

ZIRCONIUM PALLADIUM ETCHANTS

ZRPD-0001
ETCH NAME: Aqua regia TIME:
TYPE: Acid, removal/polish TEMP: RT
COMPOSITION:
3 .... HCl
1 .... HNO_3
DISCUSSION:

a-Zr_2Pd, thin films deposited as amorphous structure in a study of metal-metaloid alloys. Material is cubic, bcc, a-Zr_3Rh was cubic (E93 space group).
REF: Cantrall J S et al — *J Appl Phys,* 57,544(1985)

ZIRCONIUM RHENIUM ETCHANTS

ZRRH-0001
ETCH NAME: Aqua regia — TIME:
TYPE: Acid, removal/polish — TEMP: RT
COMPOSITION:
3 .... HCl
1 .... HNO_3
DISCUSSION:

a-Zr_3Rh, amorphous thin films deposited in a study of these metal-metaloid alloys. Crystals are cubic (E93 space group). Also deposited a-Zr_2Pd, which is cubic, bcc.
REF: Cantrall, J S et al — *J Appl Phys*, 57,544(1985)

PHYSICAL PROPERTIES OF ZIRCONIUM SULFIDE, ZrS_2

Classification	Sulfide
Atomic numbers	40 & 16
Atomic weight	155.34
Melting point (°C)	>800
Boiling point (°C)	
Density (g/cm³)	3.87
Hardness (Mohs — scratch)	4—5
Crystal structure (hexagonal — normal)	$(10\bar{1}0)$ prism
Color (solid)	Steel-grey
Cleavage (basal-distinct)	(0001)

ZIRCONIUM SULFIDE, ZrS_2

General: Does not occur as a natural compound. There are about a dozen zirconium minerals; zircon, $ZiSiO_4$, is the most widespread and important as an ore.

The sulfide has no direct application in the metal industry, though the metal is added to irons and steels as a desulfurizer, e.g., gettering sulfur as ZrS_2. Natural zircon is an opalizer in enamels, and the fine yellow variety is a precious gem stone.

Technical Application: Zirconium sulfide is a IV—VI compound semiconductor with hexagonal structure similar to that of MoS_2 with distinctive basal (0001) cleavage like CdI and CdS. It has layered octahedral symmetry whereas MoS_2 has trigonal prismatic symmetry.

It is under evaluation as a semiconductor and, along with $ZrSe_2$, HfS_2, and $HfSe_2$, as a possible cathode storage battery.

Etching: Soluble in hot HBr.

ZIRCONIUM SULFIDE ETCHANTS

ZRS-0001
ETCH NAME: Hydrogen bromide — TIME:
TYPE: Halogen, removal — TEMP:
COMPOSITION:
x .... x% HBr
DISCUSSION:

ZrS_2, single crystal specimens. Solution shown will etch the material.
REF:

A0-001 APPENDIX A: CRYSTALLOGRAPHY

Crystallography is the study of inorganic single crystal structures that have an internal ordered atomic lattice. The method was developed in geology for the classification of natural minerals, but is applicable to any solid material. In Solid State material processing of semiconductors and similar devices an understanding of the single crystal structure is vital in device design and fabrication, as well as in the development of new metallic compounds.

All single crystals can be classified into one of six crystal systems, and one of 32 classes within the systems. Note in the list below that the hexagonal system has two divisions, the rhombohedral division sometimes being classified as a seventh system. The six systems and their axial relationships are shown below in reducing order of complexity:

SYSTEM	AXIAL RELATIONSHIP
Isometric (cubic) system	Three mutually perpendicular axes of equal length. As a, b, c; x, y, z; x_1, x_2, x_3
Tetragonal system	Three mutually perpendicular axes. Two horizontal of equal length; one vertical, longer or shorter. As a, b, c or x, y, z
Hexagonal system	
Hexagonal division	Three horizontal axes at 60°; vertical axis at right angle. As x_1, x_2, x_3, z
Rhombohedral division	As above but with trigonal symmetry rather than vertical symmetry around the z-axis
Orthorhombic system	Three mutually perpendicular axes of unequal length. As x, y, z (a, b, c). y-axis longest and held parallel to observer (macro-axis); x-axis shortest and held facing observer (brachy-axis); z-axis vertical
Monoclinic system	Three axes of unequal length. Two at right angles; one oblique. As x, y, z. x-axis held facing observer and is oblique axis; y-axis as longest and held horizontal; z-axis shortest and held vertical
Triclinic system	Three axes of unequal length with all intersections oblique. As x, y, z (a, b, c). x-axis is shortest (brachy-axis) held downward toward observed; y-axis longer (macro-axis); z-axis held vertical. Angle between b/c = alpha; a/c = beta; a/b = delta, any can be less or greater than 90°

Fortunately, the majority of metal elements and metallic compounds form in the isometric system, normal/tetrahedral/pyritohedral class . . . most in the normal class. Although this is the most complex class of all systems, it is the simplest to understand due to axial relationships. As shown, the system may be referred to as the "cubic" system, based on the cube, (100) form that occurs only in the isometric system-normal class. Tin is an exception, as normal white-tin it is tetragonal.

There are two important compounds used as circuit substrates in Solid State assemblies that occur in the hexagonal system. Sapphire, Al_2O_3, in the hexagonal division — normal class, and alpha-quartz, SiO_2, in the rhombohedral division — trapezohedral class. The latter as left- or right-hand face rotated natural crystals often with penetration twinning. Single crystal quartz blanks are used as circuit substrates for high frequency microwave devices (silicon and gallium arsenide), and are the primary (quartz) radiofrequency crystals.

The atomic structure of several natural minerals are used as the atypical structure of many artificially compounded and grown single crystals, as binary, trinary, and quaternary crystals. Sphalerite, ZnS, for most of the binary compound semiconductors based on a tetrahedral unit cell; pyrite, chalcopyrite, and others for more complex trinary and quaternary atomic structures.

Other structural notations are space lattice forms of which there are 15 in number related to the normal class of each system. Number I is the simple cubic lattice with eight atoms at cube corners; II body-centered cubic lattice (bcc) with an additional atom at cube center; III face-centered cubic lattice (fcc) with additional atoms in the center of each cube face. The other systems have similar fcc lattices, and the hexagonal can have a hexagonal close-packed (hcp) natural crystal lattice. Several metals grown artificially under pressure may show similar close-packed lattices, such as the tetragonal (tcp) and hexagonal (hcp) high pressure structure of silicon.

A third system of notation is based on the crystal symmetry of the 32 classes as related to their systems. They are usually shown in increasing complexity from triclinic, monoclinic/orthorhombic, trigonal, tetragonal, hexagonal to cubic (isometric). Notation is a combination of letters, numbers, and other symbols, such as X3, X/m4/m, Xm6mm, etc. The complete list of notations is not shown here as the Miller Indices notations are more widely used in general processing of Solid State materials, and the space group notations used in crystallographic studies. See Appendix D for Miller Indices.

APPENDIX B: MAJOR ELEMENTS AND RADICALS

The following lists all major metal elements with their group classification according to the Periodic Table of Elements for the relationship of compound semiconductors, such as III—V gallium arsenide, IV—VI germanium selenide, etc. Primary valence electron configuration also shown with some of the more important negative radical units, such as the sulfate SO_4^-. Note that only one silicate radical is shown; there are half a dozen other important radicals as primary rock and mineral-forming silicates.

Elements	Group	Valence
Aluminum, Al	III-A	3
Antimony, Sb	V-A	3,5
Arsenic, As	V-A	3,5
Barium, Ba	II-A	2
Beryllium, Be	II-A	2
Bismuth, Bi	V-A	3,5
Boron, B	III-A	3
Bromine, Br	VII-A	1,3,5,7
Calcium, Ca	II-A	2
Carbon, C	IV-A	2,3,4
Chlorine, Cl_2	VII-A	1,3,5,7
Chromium, Cr	VI-B	2,3,6
Cobalt, Co	VIII	2
Copper, Cu	I-B	1,2
Fluorine, F_2	VII-A	1
Gallium, Ga	III-A	3
Germanium, Ge	IV-A	4
Gold, Au	I-B	1,3
Hydrogen	I-A	1
Iodine, I_2	VII-A	1,3,5,7
Iron, Fe	VIII	2,3
Lead, Pb	IV-A	2,4
Lithium, Li	I-A	1
Magnesium, Mg	II-A	2
Manganese, Mn	VII-B	2,3,4,6,7
Mercury, Hg	II-B	1,2
Nitrogen, N_2	V-A	3,5
Oxygen, O_2	VI-A	2
Phosphorus, P	V-A	3,5
Platinum, Pt	VIII	2,4
Potassium, K	I-A	1
Selenium, Se	VI-A	4,6, −2
Silicon, Si	IV-A	4
Silver, Ag	I-B	1
Sodium, Na	I-A	1
Sulfur, S	VI-A	2,4,6

Radical	Valence
Hydroxyl, OH	1
Molybdate, MoO_4	1
Nitrate, NO_3	1
Silicate, SiO_4	1,2
Sulfate, SO_4	2
Phosphate, PO_4	3
Carbonate, CO_3	2
Bicarbonate, HCO_3	1
Tungstate, WO_4	1

Elements	Group	Valence
Tantalum, Ta	V-B	3,4,5
Titanium, Ti	IV-B	3,4
Tin, Sn	IV-A	2,4
Uranium, U	Actinide	4,6
Vanadium, V	V-B	3,4,5
Yttrium, Y	III-B	3
Zinc, Zn	II-B	2
Zirconium, Zr	IV-B	4

APPENDIX C: METALS AND METALLIC COMPOUNDS WITH REFERENCE ACRONYMS

Material	Acronym
A	
Air, N_2/O_2	AIR-
Aluminum, Al	AL-
Aluminum alloys, AlMx	
Aluminum:beryllium, Al:Be	ALBE-
Aluminum:cerium, Al:Ce	ALCE-
Aluminum:copper, Al:Cu	ALCU-
Aluminum:gold, Au:Au	ALAU-
Aluminum:silicon, Al:Si	ALSI-
Aluminum:silver, Al:Ag	ALAG-
Aluminum:zinc, Al:Zn	ALZN-
Aluminum:zinc:copper, Al:Zn:Au	AZCU-
Aluminum antimonide, AlSb	ALSB-
Aluminum arsenide, AlAs	ALAS-
Aluminum nitride, AlN	ALN-
Aluminum oxide, Al_2O_3	ALO-
Aluminum phosphate, AlPO4	ALPH-
Aluminum phosphide, AlP	ALP-
Amalgam, $HgAg_x/ZnHg_x$	HGAG-/HGZN-
Antimony, Sb	SB-
Argon, Ar	AR-
Arsenic, As	AS-
B	
Barium, Ba	BA-
Barium fluoride, BaF_2	BAF-
Barium titanate, Ba_2TiO_3	BAT-
Beryllium, Be	BE-
Beryllium alloys, BeMx	
Beryllium:aluminum, Be:Al	BEAL-
Beryllium:copper, Be:Cu	BECU-
Beryllium oxide, BeO	BEO-
Bismuth, Bi	BI-
Bismuth alloys, BiM_x	
Bismuth:antimony, BiSb	BISB-
Bismuth:cadmium, BiCd	BICD-
Bismuth:tin, BiSn	BISN-
Bismuth germanate, $Bi_{14}Ge_3O_{12}$	BGO-
Bismuth selenide, Bi_2Se_3	BISE-
Bismuth silicate, $Bi_{12}SiO_{20}$	BSO-
Bismuth telluride, Bi_2Te_3	BITE-
Bismuth trioxide, Bi_2O_3	BIO-
Blister	BLIS-
Bonding	BB-
Boron, B	B-

Borides, M_xB_y	
Cerium boride, CeB_6	CEB-
Chromium boride, Cr_3B_2	CRB-
Cobalt boride, CoB	COB-
Iron boride, FeB	FEB-
Lanthanum boride, LaB_6	LAB-
Molybdenum, Mo_2B_3	MOB-
Nickel boride, NiB	NIB-
Niobium boride, Nb_3B_3	NBB-
Rare earth borides, R_3B_2	REB-
Silicon boride, SiB_6	SIB-
Titanium boride, TiB_2	TIB-
Tungsten boride, WB_2	WB-
Uranium boride, UB_4	UB-
Vanadium boride, VB_2	VB-
Zirconium boride, ZrB	ZRB-
Boron carbide, B_4C	BC-
Boron nitride, BN	BN-
Boron phosphide, BP	BP-
Boron telluride, B_2Te_3	BTE-
Boron trifluoride, BF_3	BTF-
Brass, Cu:Zn	BRA-
Bromine, Br_2	BR-
Bronze, Cu:Sn	BRO-

C

Cadmium, Cd	CD-
Cadmium antimonide, CdSb	CDSB-
Cadmium arsenide, Cd_2As_2	CDAS-
Cadmium fluoride, CdF_2	CDF-
Cadmium indium selenide, $CdIn_2Se_4$	CISE-
Cadmium indium telluride, $CdIn_2Te_4$	CITE-
Cadmium iodide, CdI_2	CDI-
Cadmium molybdate, $CdMoO_3$	CDMO-
Cadmium oxide, CdO	CDO-
Cadmium phosphide, CdP_2	CDP-
Cadmium silicon arsenide, $CdSiAs_2$	CSA-
Cadmium selenide, CdSe	CDSE-
Cadmium sulfide, CdS	CDS-
Cadmium telluride, CdTe	CDTE-
Cadmium mercuric telluride, CdTe.HgTe	CDHT-
Calcium, Ca	CA-
Calcium carbonate, $CaCO_3$	CAC-
Calcite, $CaCO_3$	CAC-
Calcium fluoride, CaF_2	CAF-
Calcium molybdate, $CaMoO_2$	CAMO-
Calcium nitride, CaN	CAN-
Calcium silicon fluoride, $CaSiF_2$	CASF-
Calcium tin fluoride, $CaSnF_2$	CATF-
Calcium tungstate, $CaWO_4$	CAW-
Californium, Cf	CF-

Carbon, C	C-
Graphite, C	GR-
Carbon dioxide, CO_2	COD-
Ceramics, general	CERA-
Cerium, Ce	CE-
Cerium dioxide, CeO_2	CEO-
Cermets, general	CMET-
Cesium, Cs	CS-
Cesium bromide, CsBr	CSBR-
Cesium chloride, CeCl	CECL-
Cesium iodide, CsI	CSI-
Cesium platinide, CePt	CEPT-
Carbides, M_xC	
Boron carbide, B_4C	BC-
Iron carbide, Fe_3C	FEC-
Molybdenum carbide, Mo_2C	MOC-
Niobium carbide, NbC	NBC-
Silicon carbide, SiC	SIC-
Tantalum carbide, TaC	TAC-
Titanium carbide, TiC	TIC-
Tungsten carbide, WC	WC-
Uranium carbide, UC	UC-
Vanadium carbide, VC	VC-
Chlorine, Cl_2	CL-
Chlorine containing etchants	
Chromium, Cr	CR-
Chromium containing etchants	
Chromium alloys, CrM_x	
Chromium titanium, Cr:Ti	CRTI-
Chromium trioxide, Cr_2O_3	CRO-
Cleaning, general	CLE-
Cobalt, Co	CO-
Cobalt oxide, CoO	COO-
Cobalt sulfide, Co_9S_8	COS-
Colemanite, $Ca_2B_6O_{11}.5H_2O$	COL-
Columbium, Cb	CB-
Copper, Cu	CU-
Copper alloys, CMx	
Copper antimonide, Cu_2Sb	CUSB-
Copper:beryllium, Cu:Be	CUBE-
Copper cerium, $CeCu_6$	CUCE-
Copper lanthanum, $LaCu_6$	CULA-
Copper:gallium, Cu:Ga	CUGA-
Copper:germanium, Cu:Ge	CUGE-
Copper gold, Cu_2Au	CUAU-
Copper nickel, CuNi	CUNI-
Copper palladium, CuPd	CUPD-
Copper bromide, $CuBr_2$	CUBR-
Copper chloride, $CuCl_2$	CUCL-
Copper iodide, CuI	CUI-
Copper oxide, Cu_2O	CUO
Cuprite, Cu_2O	CUO-

Copper phosphide, Cu_3P	CUP-
Copper germanium phosphide, $CuGe_2P_3$	CUGS-
Copper selenide, Cu_2Se	CUSE-
Copper indium selenide, $CuInSe_2$	CISE-
Copper sulfide, CuS	CUS-
Copper gallium sulfide, $CuGaS_2$	CUGS-
Copper indium sulfide, $CuInS_2$	CIS-
Copper phospho-sulfuric iodide, CuPSI	CPS-
Copper telluride, Cu_4Te_3	CUTE-
Copper gallium telluride, $CuGaTe_2$	CGTE-
Copper indium telluride, $CuInTe_2$	CUIT-
Corundum, Al_2O_3	COR-
Aluminum oxide, Al_2O_3	ALO-
Cronstedite, $4FeO.2Fe_2O_3.3SiO_2.4H_2O$	CRON-
Cristobalite, SiO_2	CRIS-

D

Deuterium D	DEU- (TID-; SCD-)
Diamond, C	D-
Dielectric, M_xO or M_xN	DIE-
Dysprosium, Dy	DY-

E

Electrides, $C_xH_yO_zM$	ETD
Epoxy, general	EPAG- (EPAL-, EPAL-)
Erbium, Er	ER-
Erbium dihydride, ErH_2	ERH-
Etching, general	
Dry chemical etching, DCE	DCE-
Electrolytic etching, EE	EE-
Gas etching, GE	GE-
Molten flux etching, MFE	MFE-
Pressure etching, PRE	PRE-
Thermal etching, THE	THE-
Wet chemical etching, WCE	WCE-
Vacuum etching, VE	VE-
Europium, Eu	EU-
Europium oxide, Eu_2O_3	EUO-
Europium sulfide, EuS	EUS-

F

Ferric oxides, FeOx	FEO-
Limonite, $2Fe_2O_3.3H_2O$	LIM
Hematite, Fe_2O_3	HEM-
Magnetite, Fe_3O_4	MAGT-
Ferrites, $FeM_xM_yO_z$	FER-
Fluorapatite, $(CaF)Ca_4(PO_4)_3$	FAP-
Fluorine, F_2	F-
Fluorine-containing etchants	

Fluorite, CaF_2	CAF-
Fresnoite, $Ba_2Si_2TiO_8$	FRE-
G	
Gadolinium, Gd	GD-
Gadolinium nitride, GdN_{12}	GDN-
Gadolinium terbium iron, GdTbFe	GDTF-
Gallium, Ga	GA-
Gallium alloys, GaM_x	
Gold gallium, Au_2Ga	GAAU-
Gallium antimonide, GaSb	GASB-
Gallium arsenide, GaAs	GAS
Gallium aluminum arsenide, (Ga,Al)As	GALA-
Gallium arsenide phosphide, GaAsP	GASP-
Gallium iron oxide, $GaFeO_3$	GIO-
Gallium nitride, GaN	GAN-
Gallium oxide, Ga_2O_3	GAO-
Gallium phosphate, $GaPO_4$	GAPH-
Gallium phosphide, GaP	GAP-
Gallium selenide, GaSe	GASE-
Garnets, $M_3M_2(SiO_4)_3$	
Boron germanium, $B_4G_e3O_{12}$	BGG-
Calcium aluminum germanium, $Ca_3Al_2Ge_3O_{12}$	CAGG-
Europium scandium iron, $Eu_3Sc_2Fe_3O_{12}$	ESG-
Gadolinium gallium, $Gd_3Ga_5O_{12}$	GGG-
Gadolinium selenium gallium, $Gd_3Se_{1\,8}Ga_{3\,2}O_{12}$	GSGG-
Magnesium zinc yttrium, $Mn_xZn_{1-x}Y_zO_{12}$	MZYG-
Natural garnets, $Fe_2Al_2Si_3O_{12}$	GAR-
Strontium gallium, $SrCa_{12}O_{19}$	SGG-
Yttrium aluminum, $Y_3Al_5O_{12}$	YAG-
Yttrium gallium, $Y_2Ga_7O_{12}$	YGG-
Yttrium iron, $Y_3Fe_5O_{12}$	YIG-
Germanides, M_xGe	
Copper germanide, CuGe	CUGE-
Holmium copper germanide, $HoCu_2Ge_2$	HOCG-
Indium germanide, InGe	INGE-
Iron germanide, Fe_3Ge_2	FEGE-
Magnesium germanide, Mg_2Ge	MGGE-
Molybdenum germanide, Mo_3Ge	MOGE-
Niobium germanide, Ng_3Ge	NBGE-
Terbium copper germanide, $TbCu_2Ge_2$	TBCG-
Vanadium germanide, V_3Ge	VGE-
Germanium, Ge	GE-
Germanium alloys, M_xGe	
Gold:germanium, Au:Ge	AUGE-
Tin:germanium, Sn:Ge	SNGE-
Germanium arsenide, GeAs	GEAS
Germanium oxide, GeO_2	GEO-
Germanium nitride, Ge_3N_4	GEN-
Germanium selenide, Ge_xSe_{1-x}	GESE-
Germanium silicon, GeSi	GESI-

Germanium sulfide, GeS	GES-
Germanium telluride, GeTe	GETE-
Glass, SiO_2 (under Silicon Dioxide)	GLA-
Gold, Au	AU-
Gold alloys, AuM_x	
Gold aluminum, Au_2Al	AUAL-
Gold bismuth, Au_2Bi	AUBI-
Gold cadmium, Au:Cd	AUCD-
Gold gallium, $AuGa_2$	AUGA-
Gold germanium, AuGe(13%)	AUGE-
Gold indium, Au:In	AUIN-
Gold mercury, $AuHg_x$ (amalgam)	AUHG-
Gold silver, $AuAg_x$ (electrum)	AUAG-
Gold tin, AuSn(20%)	AUSN-
Gold zinc, Au:Zn	AUZN-
Gold boron, AuB(2—5%)	AUB-
Gold copper, Cu_2Au	CUAU-
Gold palladium, Pd:Au	AUPD-
Gold phosphorus, Au:P(2—5%)	AUP-
Graphite, C	GR-
Carbon, C	C-
Diamond, C	D-

H

Hafnium, Hf	HF-
Hafnium alloys, HfM_x	
Hafnium tin, Hf_5Sn_3	HFSN-
Hafnium nitride, HfN	HFN-
Helium, He	HE-
Hematite, Fe_2O_3	HEM-
Holmium, Ho	HO-
Holmium alloys, HoM_x	
Holmium cobalt, Ho:Co	HOCO-
Holmium nitride, HoN_{12}	HON-
Hydrogen, H_2	H-
Hydrides, M_xH	HYDR-
Chromium hydride, CrH	CRH-
Germanium hydride, Ge:H	GEH-
Niobium hydride, Nb:H	NBH-
Silicon hydride, Si:H	SIH-
Silicon nitride hydride, SiN:H	SINH-
Silicon oxynitride hydride, SiON:H	SIONH-
Silicon carbide hydride, SiC:H	SICH-
Tantalum hydride, Ta:H	TAH-
Titanium hydride, Ti:H	TIH-
Vanadium hydride, V:H	VH-
Zirconium hydride, Zr:H	ZRH-
Hydroxides, KOH/NaOH	HYD-
Hydrocarbon, CH_4	HC-

I

Ice, H_2O ICE-
Indium, In IN-
Indium alloys, InM_x
  Indium bismuth, In:Bi INBI-
  Indium tin, In:Sn INSN-
Indium antimonide, InSb INSB-
Indium arsenide, InAs INAS-
Indium arsenide phosphide, $InAs_xP_{x-1}$ IASP-
Indium bismuthide, In_5Bi_3 INBI-
Indium gallium antimonide, InGaSb IGSB-
Indium gallium arsenide, InGaAs IGAS-
Indium gallium arsenide phosphide, InGaAsP IGAP-
Indium oxide, In_2O_3 INO-
Indium phosphide, InP INP-
Indium selenide, InSe INSE-
Indium telluride, In_2Te_3 INTE-
Indium thallinide, In_2Tl_3 INTL-
Indium tin oxide, $InSnO_2$ ITO
Indium, rare earths, RE_In_3 INRE-
Iodine, I_2 I-
Iridium, Ir IR-
Iridium vanadinide, IrV IRV-
Iron, Fe FE-
Iron alloys, FeM_x
  Iron beryllium, Fe:Be FEBE-
  Iron carbon, Fe_3C FEC-
  Iron chromium cobalt, Fe:Al:Cr:Co KA-
  Iron germanium, Fe_3Ge_2 FEGE-
  Iron manganese, Fe:Mn FEMN-
  Iron nickel, $FeNi_3$ FENI-
  Iron phosphide, FeP_2 FEP-
  Iron silicon, Fe:Si FESI-
Iron oxide, Fe_xO_y FEO-
  Ferrites, $FeM_xM_yO_z$ FER-
Iron palladium, FePd FFPD-
Iron phosphide, FeP_2 FEP-
Iron sulfide, FeS_2 FES-
  Pyrite, FeS_2 PYR-
Iron titanate, Fe_3TiO_4 FETI-

J

K

Kanthal, 72Fe:6Al:22Cr:1Co KA-
Kovar, $Fe_xNi_yCo_z$ KO-
Krypton, Kr KR-

L

Lanthanum, La	LA-
Lanthanum boride, LaB6	LAB
Lanthanum bromide, $LaBr_3$	LABR-
Lanthanum nitride, Lan_{12}	LAN-
Lanthanum strontium cobaltite, $LaSrCoO_3$	LACO-
Lanthanum strontium ferrite, $LaSrFeO_3$	LAFE-
Lead, Pb	PB-
Lead alloys, PbM_x	
Lead silver, Pb:Ag	PBAG-
Lead tin, Pb:Sn	PBSN-
Lead titanium, Pb:Ti	PBTI-
Lead iodide, PbI_2	PBI-
Lead monoxide, PbO	PBO-
Lead molybdate, $PbMo_4$	PBM-
Lead selenide, PbSe	PBSE-
Lead sulfide, PbS	PBS-
Galena, PbS	PBS-
Lead tin selenide, PbSnSe	PBTS-
Lead telluride, PbTe	PBTE-
Lead germanium telluride, PbGeTe	PBGT-
Lead tin telluride, PbSnTe	PBTT-
Lead tungstate, $PbWO_4$	PBW-
Lead zirconate, $PbZrO_3$	PBZR-
Lithium, Li	LI-
Lithium bismuthide, Li_3Bi	LIBI-
Lithium bromide, LiBr	LIBR-
Lithium chloride, LiCl	LICL-
Lithium fluoride, LiF	LIF-
Lithium sulfide, Li_2S	LIS-
Lithium indium sulfide, $LiInS_2$	LIIS-
Lithium niobate, $LiNbO_3$	LINB-
Lithium nitride, Li_3N	LIN-
Lithium tantalate, $LiTaO_3$	LITA-
Lucite	LUC-

M

Magnesium, Mg	MG-
Magnesium fluoride, MgF_2	MGF-
Magnesium germanide, Mg_2Ge	MGFE-
Magnesium nitride, Mg_3N_2	MGN-
Magnesium oxide, MgO	MGO-
Magnesium selenide, MgSe	MGSE-
Magnesium silicide, Mg_2Si	MGSI-
Magnesium sulfide, MgS	MGS-
Magnesium telluride, MgTe	MGTE-
Magnetite, Fe_3O_4	MAGT-
Manganese, Mn	MN-
Manganese dioxide, MnO_2	MNO-
Manganese ferrite, $MnFe_2$	MNFE-

Manganese stannide Mn_2Sn	MGSN-
Manganese sulfide, MnS_2	MNS-
Manganese telluride, $MNTe_2$	MNTE-
Mercury, Hg	HG-
Mercury alloys, HgMx	
Amalgam, $HgAg_x$	HGAG-
Amalgam, $HgZn_x$	HGZN-
Mercuric iodide, HgI_2	HGI-
Mercuric oxide, HgO	HGO-
Mercuric selenide, HgSe	HGSE-
Mercuric telluride, HgTe	HGTE-
Mercuric cadmium telluride, HgCdTe	HGCT-
Metal, general	MET-
Micas, $(OH)_2Al_2K(AlSi_3O_{10})$	MI-
Molybdates, $M_2Mo_3O_8$	FEMO-
Molybdenum, Mo	MO-
Molybdenum carbide, Mo_2C	MOC-
Molybdenum germanide, Mo_3Ge	MOGE-
Molybdenum nitride, MoN	MON-
Molybdenum oxide, MoO_3	MOO-
Molybdenum selenide, $MoSe_2$	MOSE-
Molybdenum selenide sulfide, $MoSeS_2$	MOSS-
Molybdenum sulfide, MoS_2	MOS-
Molybdenum telluride, $MoTe_2$	MOTE-
Mounting materials	
Apiezons	APZW-
Beeswax	BEEW-
Dielectrics	CER-; GLA-
Epoxy/polyimide	EPOX-; POLY-
Lacquers	PR-; VAR-; SHE-
Metals	IN; BRZ-
Paraffin	PARA-
Pitch	PICH-
Plastics	CRY-; LUC-; MYL-; BAK; DEK; PUT-; GLY-; SIL-
Resins	RES-
Vacuum	VAC-
Waxes	SWAX-; CAR-; PARA;
Wood	WOO-

N

Neodymium, Nd	ND-
Neon, Ne	NE-
Neptunium, Np	NP-
Nickel, Ni	NI-
Nickel alloys, NiMx	
Nichrome Ni:Cr	NICR-
Nickel aluminum, Ni:Al	NIAL-
Nickel boride, Ni_2B	NIB-
Nickel cobalt, Ni:Co	NICO-
Nickel:copper, Ni:Cu	NICU-

Nickel iron, $FeNi_3$	FENI-
Nickel manganese, Ni:Mn	NIMN-
Nickel titanium, Ni:Ti	NITI-
Nickel zirconium, Zr_2Ni	NIZR-
Permalloy, Ni:Fe:Mg	PERM-
Nickelides, (rare earths) $ReNi_{12}$	RENI-
Nickel iodide, NiI	NII-
Nickel oxide, NiO	NIO-
Nitrogen, N_2	N-
Niobium, Nb	NB-
Niobium alloys, NbM_x	
Niobium aluminum, $NbAl_y$	NBAL-
Niobium zinc, Nb:Zr(25%)	NBZR-
Niobium boride, Nb_3B_3	NBB-
Niobium carbide, NbC	NBC-
Niobium germanide, Nb_3Ge	NBGE-
Niobium hydride, Nb:H	NBH-
Niobium nitride, NbN	NBN-
Niobium oxide, Nb_2O_5	NBO-
Niobium selenide, Nb_3Se	NBSE-
Niobium stannide, Nb_3Sn	NBSN-
Niobium telluride, Nb_3Te	NBTI-
Nitrides, M_xN_y	
Aluminum nitride, AlN	AlN-
Boron nitride, BN	BN
Calcium nitride, Ca_2N_3	CAN-
Cerium nitride, CeN_{12}	CEN-
Dysprosium nitride, DyN_{12}	DYN-
Erbium nitride, ErN_{12}	ERN-
Gadolinium nitride, GdN_{12}	GDN-
Gallium nitride, GaN	GAN
Germanium nitride, Ge_3N_4	GEN-
Hafnium nitride, HfN	HFN-
Holmium nitride, HoN_{12}	HON-
Iron nitride, Fe_xN_y	FEN-
Lanthanum nitride, LaN_{12}	LAN-
Lithium nitride, Li_3N	LIN-
Magnesium nitride, Mg_3N_4	MGN-
Molybdenum nitride, MoN	MON-
Neodymium nitride, NdN_{12}	NDN-
Niobium nitride, NbN	NBN
Praseodymium nitride, PrN_{12}	PRN-
Rare earth nitrides, REN_{12}	REN-
Samarium nitride, SmN_{12}	SMN-
Silicon nitride, Si_3N_4	SIN
Tantalum nitride, TaN	TAN
Terbium nitride, TbN_{12}	TBN-
Titanium nitride, TiN	TIN-
Uranium nitride, UN	UN
Yttrium nitride, YN_{12}	YN-
Zirconium nitride, ZrN	ZRN

O

Organic semiconductors	OTGS-
Electrides, $C_xO_yH_zM$	ETD-
Osmium, Os	OS-
Osmium sulfide, OsS_2	OSS-
Osmium telluride, $OsTe_2$	OSTE-
Oxygen, O_2	O-

P

Palladium, Pd	PD-
Palladium alloys, PdM_x	
Palladium gold, Pd:Au	PDAU-
Palladium silver, Pd:Ag	PDAG-
Palladium hydride, Pd:H	PDH-
Palladium nickel phosphorus, PdNiP	PDNP-
Phosphorus, P	P-
Phosphorus containing etchants	
Phosphorus alloys, PM_x	
Gold phosphorus, Au:P	AUP-
Iron phosphorus, Fe:P	FEP-
Phosphorous pentoxide, P_2O_5	PO-
Photo resist	AZ-; PMMA-; COP-
Mounting materials (lacquers)	PR-; LAC-
Plastic	PL-; IUC-; POLY-; PCB-
Mounting materials (plastics)	CRY-; LUC-; MYL-; BAK-; DEK-; PUT-; GLY-; SIL-
Platinum, Pt	PT-
Platinum alloys, PtM_x	
Platinum:gold, Pt:Au	PTAU-
Platinum:palladium, Pt:Pd	PTPD-
Platinum:rhodium, Pt:Rh	PTRH
Platinum antimonide, $PtSb_2$	PTSB-
Platinum arsenide, $PtAs_2$	PTAS-
Platinum hydride, Pt:H	PTH-
Platinum oxide, PtO	PTO-
Platinum phosphide, PtP_2	PTP-
Plutonium, Pu	PU-
Polyvinylidene fluoride, PVF_2	PVF-
Potash alum, $KAl(SO_4)_2.12H_2O$	ALUM-
Potassium, K	K-
Potassium bromide, KBr	KBR-
Potassium chloride, KCl	KCL-
Potassium iodide, KI	KI-
Potassium manganese ferrite, $KMnFe_3$	KMI-
Potassium tantalate, $KTaO_3$	KTO-
Praseodymium, Pr	PR-
Praseodymium nitride, PrN_{12}	PRN-

Q

Quartz, SiO_2	QTZ-
Coestite, SiO_2/SiO_4	COE-
Cristobalite, SiO_2	CRIS-
Opal, $SiO_2.nH_2O$	OPAL-
Silicon dioxide, SiO_2	SIO-
Stitsovite, $SiO_2/Si3O_4$	STI-
Tridymite, SiO_2	TRI-

R

Radium, Ra	RA-
Radon, Rn	RN
Rare earths, $RENi_{12}$	RENI-
Rhenium, Re	RE-
Rhodium, Rh	RH-
Rhodium alloys, RhM_x	
Platinum:rhodium, Pt:Rh	PTRH-
Rhodium silicide, RhSi	RHSI-
Rochelle salt, $NaKO_4H_4O_6.4H_2O$	RS-
Rubidium, Rb	RB-
Rubidium bromide, RbBr	RBBR-
Rubidium iodide, RbI	RBI-
Ruthenium, Ru	RU-
Ruthenium selenide $RuSe_2$	RUSE-
Ruthenium sulfide, RuS_2	RUS-
Ruthenium telluride, $RuTe_2$	RUTE-

S

Samarium, Sm	SM
Samarium bromide, $SmBr_3$	SMBR-
Samarium cobalt, $SmCo_5$	SMCO-
Samarium nitride, SmN_{12}	SMN-
Sapphire, Al_2O_3	SAPP-
Aluminum oxide, Al_2O_3	ALO-
Scandium, Sc	SC-
Selenium, Se	SE-
Silicides, M_xSi_y	
Boron silicide, B_4Si	BSI-
Cesium silicide, Ce Si_2	CESI-
Chromium silicide, $CrSi_2$	CRSI-
Cobalt silicide, $CoSi_2$	COSI-
Erbium silicide, ErSi	ERSI-
Hafnium silicide, HfSi	HFSI-
Iridium silicide, IrSi	IRSI-
Iron silicide, FeSi	FESI-
Iron tungsten silicide, FeWSi	IWSI-
Manganese silicide, MnSi	MNSI-
Molybdenum silicide, MoSi	MOSI-
Nickel silicide, NiSi	NISI-

Niobium silicide, NbSi NBSI-
Osmium silicide, OsSi OSSI
Palladium silicide, PdSi PDSI-
Platinum silicide, PtSi PTSI
Praseodymium cobalt silicide, $PrCo_2Si_2$ PRCS-
Rhenium silicide, ReSi RESI-
Rhodium silicide, RhSi RHSI-
Tantalum silicide, TaSi TASI-
Tellurium silicide, TeSi TESI-
Terbium cobalt silicide, $TbCo_2Si_2$ TBCS-
Titanium silicide, TiSi TISI-
Titanium/tungsten silicide, TiWSi TWSI-
Tungsten silicide, WSi WSI-
Vanadium silicide, VSi VSI-
Zirconium silicide, ZrSi ZRSI-
Silicon, Si SI-
Silicon carbide, SiC SIC-
Silicon alloys, MSi_x
Aluminum silicon, Al:Si ALSI-
Iron silicon, Fe:Si FESI-
Silicon germanium, Si_xGe_y SIGE-
Silicon tin, SiSn SNSI-
Silicon dioxide, SiO_2 SIO-
Glass, SiO_2M_x GLA-
Silicon monoxide, SiO SIO-
Silicon nitride, Si_3N_4 SIN-
Silicon vanadinide, SiV_2 SIV-
Silver, Ag AG-
Silver containing etchants
Silver alloys, Ag_xM_y
Silver aluminum, Ag_2Al AGAL-
Silver amalgam, Ag_xHg AGHG-
Silver gold, Ag_xAu_y AGAU-
Silver lead/tin, Alloy #63 AGLT-
Silver tin, AgSn AGSN-
Silver titanium, AgTi AGTI-
Silver zinc, AgZn AGZN-
Silver bromide, AgBr AGBR-
Silver cadmium, AgCd AGCD-
Silver chloride, AgCl AGCL-
Silver mercury, Ag_2Hg AGHG-
Silver germanium phosphide, $Ag_6Ge_{10}P_{12}$ AGEP-
Silver iodide, AgI AGI-
Silver magnesium, AgMg AGMG-
Silver sulfide, AgS AGS-
Silver gallium sulfide, $AgGaS_2$ SGA-
Silver telluride, Ag_2Te AGTE-
Silver antimony telluride, $AgSbTe_2$ SAT-
Silver gallium telluride, $AgGaTe_2$ AGGT-
Silver indium telluride, $AgInTe_2$ AGIT-
Silver iron telluride, $AgFeTe_2$ SIT-
Silver selenide, Ag_2Se AGSE-

Silver gallium selenide, $AgGaSe_2$	AGGE-
Silver indium selenide, $AgInAg_2$	AGIE-
Silver titanium selenide, AgTiSe	AGTS-
Silver epoxy	AGEP-
Sodium, Na	NA-
Sodium bromide, NaBr	NABR-
Sodium chloride, NaCl	NACL
Sodium niobate, $NaNbO_3$	NNO-
Sodium nitrite, $NaNO_3$	NAN-
Spinel, $MgAl_2O_4$	SPI-
Steel, Fe_3CM_x	ST-
Iron, Fe	FE-
Ferrite, general	FER-
Strontium, Sr	SR-
Strontium chloride, $SrCl_2$	SRCL-
Strontium fluoride, SrF_2	SRF-
Strontium gallium oxide, $SrGa_{12}O_{19}$	SGO
Strontium titanate, $SrTiO_3$	SRTO
Strontium tungstate, $SrWO_4$	SRWO-
Sulfur, S	S-
Sulfur containing etchants	

T

Talc, $H_2Mg_3(SiO_3)_4$	TALC-
Tantalum, Ta	TA-
Tantalum carbide, TaC	TAC-
Tantalum hydride, Ta:H	TAH-
Tantalum molybdenide, TaMo	TAMO-
Tantalum nitride, TaN	TAN-
Tantalum oxide, Ta_2O_5	TAO-
Tantalum selenide, $TaSe_2$	TASE-
Tantalum silicide, TaSI	TASI-
Tantalum sulfide, TaS_2	TAS-
Tantalum tungsten, TaW	TAW-
Tellurium silicide, TeSi	TESI-
Tellurides, $M_xM_yTe_z$	
Bismuth telluride, Bi_2Te_3	BITE-
Boron telluride, B_2Te_3	BTE-
Cadmium telluride, CdTe	CDTE-
Cadmium mercuric telluride, CdTe.HgTe	CDHT-
Copper indium telluride, $CuInTe_2$	CUIT
Germanium telluride, GeTe	GETE
Indium telluride, In_2Te_3	INTE
Lead telluride, PbTe	PBTE-
Lead germanium telluride, PbGeTe	PBGT-
Lead tin telluride, PbSnTe	PBTT-
Manganese telluride, $MnTe_2$	MNTE-
Mercuric telluride, HgTe	HGTE-
Mercuric cadmium telluride, HgCdTe	HGCT-
Molybdenum telluride, $MoTe_2$	MOTE-
Niobium telluride, Nb_3Te_4	NBTE-

Osmium telluride, $OsTe_2$	OSTE-
Ruthenium telluride, $RuTe_2$	RUTE-
Silver telluride, Ag_2Te	AGTE-
Ferrous silver telluride, $AgFeTe_2$	FAGT-
Tin telluride, SnTe	SNTE-
Zinc telluride, ZnTe	ZNTE-
Tellurium, Te	TE-
Tellurium dioxide, TeO_2	TEO-
Terbium, Tb	TB-
Terbium nitride, TbN_{12}	TBN-
Thallium, Tl	TL-
Thallium beryllium telluride, TlBeTe	TLBET-
Thallium bismuth telluride, TlBiTe	TLBE-
Thallium tin telluride, TlSnTe	TLST-
Indium thallinide, In_2Tl_3	INTL-
Thallium selenide, Tl_2Se_3	TLSE-
Tin, Sn	SN-
Tin alloys, SnM_x	
Lead tin, 60Pb:40Sn #62	PBSN-
Tin dioxide, SnO_2	SNO-
Tin selenide, $SnSe_2$	SNSE-
Tin telluride, SnTe	SNTE-
Titanium, Ti	TI-
Titanium alloys, TiM_x	
Titanium gold, TiAun	TIAU-
Titanium aluminum, TiAl	TIAL-
Titanium carbide, TiC	TIC-
Titanium dioxide, TiO_2	TIO-
Titanium monoxide, TiO	TIO-
Titanium trioxide, Ti_2O_3	TIO-
Rutile, TiO_2	TIO-
Titanium nickel, TiNi	TINI-
Titanium nitride, TiN	TIN-
Titanium silicide, TiSi	TISI-
Titanium tungsten, TiW	TIW-
Titanium tungsten silicide, TiWSi	TWSI-
Tridymite, SiO_2	TRI-
Silicon dioxide, SiO_2	SIO-
Tritium, T	TRIT- (under Deuterium)
Tungsten, W	W
Tungsten oxides, WO_x	
Tungsten dioxide, WO_2	WO-
Tungsten pentoxide, W_2O_5	WO-
Tungsten trioxide, WO_3	WO-
Tungsten boride, WB_2	WB-
Tungsten carbide, WC	WC-
Tungsten rhenium, WRh	WRH-
Tungsten selenide, WSe_2	WSE-
Tungsten sulfide, WS	WS-
Tungsten titanium, TiW	TIW-

U

Uranium, U	U-
Uranium boride, UB_4	UB-
Uranium carbide, UC	UC-
Uranium cesium, UCe_2	UCE-
Uranium cobalt, UCo_2	UCO-
Uranium dioxide, UO_2	UO-
Uranium nickel, UNi_2	UNI-
Uranium nitride, U_3N_4	UN-
Uranium platinum, UPt_3	UPT-
Uranium sulfide, US	US-
Uranium titanium, UTi	UTI-

V

Vanadium, V	V-
Vanadium aluminum, VAl_3	VAL-
Vanadium boride, VB_2	VB-
Vanadium bromide, VBr_2	VBR
Vanadium carbide, VC	VC-
Vanadium gallium, V_3Ga	VGA-
Vanadium germanium, V_3Ge	VGE-
Vanadium silicide, VSi	VSI-

W

Water, H_2O	WAT-
Cold etchants, general	COLD-
Water etchants	WATR-

X

Xenon, Xe	XE-

Y

Yttrium, Y	Y-
Yttrium cobalt:Mx, $Y_2(CoM)_{17}$	YCM
Yttrium garnets (see Garnets)	
Yttrium zirconate, $(Y_2O_3)_m(SrO_2)_{1-m}$	YZR

Z

Zinc, Zn	ZN-
Zinc alloys, ZnMx	
Zinc cadmium, ZnCd	ZNCD-
Zinc copper, Cu:Zn (brass)	BRAZ-
Zinc copper tin, Zn:Cu:Sn (bronze)	BRO-
Zinc gold, Au:Zn	AUZN-
Zinc mercury, ZnHg (amalgam)	ZNHG-
Zinc nickel, NiZn	ZNNI-

Zinc niobium, ZnNb	NBZN-
Zinc silver, ZnAg	ZNAG-
Zinc tin, ZnSn	ZNSN-
Zinc arsenide, Zn_3As_2	ZNAS-
Zinc oxide, ZnO	ZNO-
Zinc phosphide, Zn_3P_2	ZNP-
Zinc selenide, ZnSe	ZNSE-
Zinc silicon fluoride, $ZnSiF_2$	ZSIF
Zinc silicon phosphide, $ZnSiP_2$	ZNP-
Zinc tin arsenide, $ZnSnAs_2$	ZNSA-
Zinc sulfide, ZnS	ZNS-
Sphalerite, ZnS	ZNS-
Zinc telluride, ZnTe	ZNTE-
Zinc tin phosphide, $ZnSnP_2$	ZTP-
Zinc tungsten, ZnW	ZNW-
Zirconium, Zr	ZR-
Zirconium boride, ZrB	ZRB-
Zirconium hydride, Zr:H	ZRH-
Zirconium oxide, ZrO_2	ZRO-
Zirconium nitride, ZrN	ZRN-
Zirconium palladium, Zr_2Pd	ZRPD-
Zirconium rhenium, Zr_3Rh	ZRRH-
Zirconium silicide, ZrSi	ZRSI-
Zirconium sulfide, ZrS_2	ZRS-
Zirconium vanadium, ZrV	ZRV-

APPENDIX D: MILLER INDICES

In describing the external faces of any single crystal, the Miller Indices notations are most commonly used, and are directly related to the crystallographic axes. The numbers are derived from the reciprocals of fractional plane intercepts on axes using the smallest integer. As shown below, an x, y, z coordinate system is used for all crystal systems other than the hexagonal, which is x_1, x_2, x_3, z. Any given inorganic metal or metallic compound will always form with the same physical atomic distances within its ordered lattice, such that they will always give the same measurement when measured by X-ray and Laue photography. Atom-to-atom distances range between 1 to 5 Å and vary with every element or combination of elements, such that each mineral, compound or metal is always singularly distinctive.

In general processing of materials knowledge of specific atom distances is not essential, but the orientation of a single crystal face and the direction of bulk internal planes can be critical to a device construction and operation, such that an understanding of Miller Indices is a necessary working tool.

The Miller Indices are used for general notation of such atom locations where an external face crosses each axis, such that a (111) face is a triangle as it cuts all three axes. Where a face cuts only one axis and is parallel to the other two it is square and a (100) face . . . infinitely parallel to the other two axes, shown as a "Ø" rather than using the infinity symbol. In the isometric system, normal class, two forms as the octahedron, (111) and cube, (100) are the two most widely used surface planes in a semiconductor wafer processing. The dodecahedron, (110) planes are important for the directional orientation of bulk ⟨110⟩ planes in etching grooves and channels, the $\langle 0\bar{1}0 \rangle$ being preferred. Note that where a face cuts a negative axis, a negative sign appears above the Miller number.

Where the hexagonal system, normal class with three horizontal x-axes are involved, the 1st Order Prism is $(10\bar{1}0)$, and the 2nd Order Prism, $(11\bar{2}0)$, and in the rhombohedral division as a negative or positive rhombohedron, $(101\bar{1})$ or $(10\bar{1}1)$, respectively.

Specific closure symbols are used in conjunction with the Miller Indices numbers as: (xxx) or (xxxx) = a specific face plane, such as the positive octahedron (111). ⟨xxxx⟩ or ⟨xxxx⟩ = a specific planar direction, such as ⟨110⟩. {xxx} or {xxxx} = all planes of a given type, such as {111}, all eight planes of the octahedron, (111). Note that the parenthesis is used when referring to a specific crystal form, such as the octahedron, cube, trisoctahedron, etc.

In addition to closed numbers, letters are used as general number face location symbols, such as (hk0), (hh1), (hk1), (hk0), such as the general notation for a textrahexahedron, which can be as (210), (310), (320), and a common form of a single crystal sphere of silicon when preferentially etched to finite crystal form.

Miller Indices are shown as a unit distance along an axis from a theoretical crystal center, and each axis has a positive and negative direction from center. This means that a (122) surface plane cuts the x-axis at unity, and the y- and z-axis at twice unity . . . which will produce a (122) isosceles shaped triangle surface face (facet) or, as a basic closed crystal form it is the trisoctahedron, (122). If the y- and z-axis cut their respective axes at half the x-axis unity distance, the face also will be an isosceles triangle, but the closed form will be trapezohedron, (211).

In establishing a plane face, each axial length is measured from crystal center to where the face (facet) crosses each axis. Form a fraction, then clear the fraction to obtain the specific face numbers as solid integers.

Measure: x, y, z	Form fraction: x, y, z	Clear: x, y, z
1 2 2	$\frac{1}{1}$ $\frac{2}{1}$ $\frac{2}{1}$	1 2 2 = (122)
1 1/2 1/2	$\frac{1}{2}$ $\frac{1/2}{2}$ $\frac{1/2}{2}$	2 1 1 = (211)
1 1 0	$\frac{1}{1}$ $\frac{1}{1}$ $\frac{0}{0}$	1 1 0 = (110)
1 1/3 0	$\frac{1}{3}$ $\frac{1/3}{3}$ $\frac{0}{3}$	3 1 0 = (310)

APPENDIX E: NATURAL MINERALS

ACTINIUM

*Actinium, Ac
Note: Radioactive series (gas)

ALABAMINE

*Alabamine, Ab
Hydrides, only to date

ALUMINUM

*Aluminum, Al
Cryolite, Na_3AlF_6
Corundum, Al_2O_3
Spinel, $MgO.Al_2O_3$
Bauxite, $Al_2O_3.2H_2O$
Gibbsite, $Al_2O_3.3H_2O$
Topaz, $(Al(F,OH))2SiO_4$
Kaolins, $H_4Al_2Si_2O_9$
Montmorillonite, $H_3Al_2Si_3O_{12}.nH_2O$
Turquoise, $CuO.3Al_2O_3.2P_2O_5.9H_2O$
Soda alum, $NaAl(SO_4)_2.12H_2O$
Potash alum, $KA1(SIO_4)_2.12H_2O$
Alunite, $K2Al_6(OH)_{12}.(SO_4)_4$

ANTIMONY

Native antimony, Sb
Stibnite, Sb_2S_3

ARGON

Native argon, Ar
Air, N_2/O_2 + about 1% Ar
Note: Inert gas, no compounds.

ARSENIC

Native arsenic, As
Realgar, AsS
Orpiment, As_2S_3
Arsenopyrite, FeAsS

BARIUM

*Barium, Ba
Barite, $BaSO_4$
Witherite, $BaCO_3$
Barylite, $Be_2BaSi_2O_7$
Benitoite, $BaTiSi_3O_9$

BERYLLIUM

*Beryllium, Be
Beryl, $Be_3Al_2(SiO_3)_6$
Crysoberyl, $BeAl_2O_4$
Phenacite, Be_2SiO_4
Gadolinite, $Be_2FeY_2Si_2O_{10}$

BISMUTH

Native bismuth, Bi
Bismuthinite, Bi_2S_3
Guanajuatite Bi_2Se_3
Tetradymite, $Bi_2(Te,S)_3$

BORON

*Boron, B
Sassolite, $B(OH)_3$
Datolite, $HCaBSiO_5$
Axinite, Ca,Al borosilicate
Tourmaline, complex borosilicate
Lunebergite, $2MgO.Be_2O_3.P_2O_5.8H_2O$
Boric acid, H_3BO_3

BROMINE

*Bromine, Br
Hydrogen bromide, HBr (sea water extract)

CADMIUM

*Cadmium, Cd
Greenockite, CdS
Otavite, Cd carbonate

CALCIUM

*Calcium, Ca
Fluorite, CaF_2
Calcite, $CaCO_3$
Dolomite, $CaCo_3.MgCO_3$
Argonite, $CaCO_3$
Oligoclase, $CaAl_2Si_2O_8$
Amphibole group, Ca,Mg, etc., silicates
Titanite, $CaTiSiO_4$
Pyrochlore, Ca, Ce niobate
Alatite, Ca_3 (CaF) $(PO_4)_3$
Apatite, Ca4 (CaF) $(PO_4)_3$
Scheelite, $CaWO_4$
Limestone (chalk), $CaCO_3$

Gypsum, $CaSO_4.2H_2O$
Anhydrite, $CaSO_4$

CARBON

*Carbon, C
Graphite, C
Diamond, C

CERIUM

*Cerium, Ce
Pyrochlore, Ca,Ce niobate
Monazite, $(Ce,La,Di)PO_4$
Yttrotantalate, Fe,Ca,Y,Fr,Ce tantalate

CESIUM

*Cesium, Cs
Pollucite, $2Cs_2O.2Al_2O_3.9SiO_2.H_2O$
Rhodizite, Al, K,Cs borate

CHROMIUM

*Chromium, Cr
Chromite, $FeO.Cr_2O_3$
Crocoite, $PbCrO_4$
Uvarovite, $Ca_2Cr_2 (SiO_4)_3$

COBALT

*Cobalt, Co
Smaltite, $CoAs_2$
Cobaltite, CoAsS

COLUMBIUM

*Columbium, Cb
see Niobium

COPPER

*Copper, Cu
Horsfordite, Cu_6Sb
Mohawkite, Cu_3As
Richardite, Cu_4Te_3
Berzelianite, Cu_2Se
Eucairite, $Cu_2Se.Ag_2Se$
Chalcocite, Cu_2S
Covellite, CuS
Bornite, Cu_5FeS_4
Chalcopyrite, $CuFeS_2$

Tetrahedrite, $3Cu_2S.Sb_2S_3$
Tennanite, $2Cu_2S.As_2S_3$
Cuprite, Cu_2O
Malachite, $CuCo_3.Cu(OH)_2$
Azurite, $CuCo_3.Cu(OH)_2$
Note: Over 300 copper minerals.

CHLORINE

*Chlorine, Cl
Halite, NaCl
Sylvite, KCl
Hydrogen chloride, HCl (volcanic)

DYSPROSIUM

*Dysprosium, Dy
Monazite, $(Ca,La,Di)PO_4$

ERBIUM

*Erbium, Eu
Fergusonite, Y,Er,Ce,U niobate

EUROPIUM

*Europium, Eu
Monazite, $(Ce,La,Di)PO_4$

FLUORINE

*Fluorine, F_2
Fluorite, CaF_2
Cryolite, Na_3AlF_6
Note: Over 100 fluorine minerals.

GADOLINIUM

*Gadolinium, Gd
Gadolinite, $Be_2FeY_2Si_2O_5$

GALLIUM

*Gallium, Ga
Bauxite, Al,Fe(OH) + trace Ga
Sphalerite, ZnS + trace Ga

GERMANIUM

*Germanium, Ge

Germanite, $Cu_3(Fe,Ge)S_4$
Tennanite, $6(Zn,Fe)S.Sb_2S_3$
Freibergite: $+3-30$ Ag
Schwartzite: $+6-17$ Hg
Malinowskite: $+13-16$ Pb
Tetrahedrite, like Tennanite
Pyrite, FeS_2
Argyrodite, $4AgS.GeS_2$

GOLD

*Gold, Au
Petzite, $(Ag,Au)_2Te$
Sylvanite, $(Au,Ag)Te_2$
Krennerite, $(Au,Ag)Te_2$
Calverite, $AuTe_3$
Muthmannite (Ag,Au)Te
Magyagite, Au,Pb,sulfo-telluride

HAFNIUM

*Hafnium, Hf
Zircon, $ZrSiO_4$ + trace Hf
Note: Trace in most zirconium minerals.

HELIUM

Helium, He
Helium, He (gas well)

HYDROGEN

Hydrogen, H_2
Water, H_2O
Note: Over 2000 minerals contain hydrogen in their formulae.

HOLMIUM

*Holmium, Ho
Gadolinite, $Be_2FeY_2Si_2O_{10}$
Note: In other rare earth minerals.

ILLINIUM

*Illinium, Il
Monazite,$(Ce,La,Di)PO_4$

INDIUM

*Indium, In

IODINE

*Iodine, I_2
Kelp, seaweed + iodine
Iodyrite, AgI
Miersite, 4AgI.CuI
Soda niter, $NaNO_3/KNO_3$ (saltpeters)

IRIDIUM

*Iridium, Ir
Iridosmine, Ir/Os mixture
Newyanskite, + 40 Ir
Siserskite, + 30 Ir

IRON

Iron, Fe
Awaruite, $FeNi_2$
Josephinite, $FeNi_3$
Pyrrhotite, FeS
Chalcovrite, $CuFeS_2$
Pyrite, FeS_2
Marcasite, FeS_2
Arsenopyrite, FeAsS
Hematite, Fe_2O_3
Ilmenite, $FeTiO_3$
Magnetite, Fe_3O_4
Franklinite, $(Fe,Zn,Mn)O.(Fe,Mn)_2O_3$
Chromite, $FeO.Cr2O_3$
Goethite, $Fe_2O_3.H_2O$
Limonite, $2Fe_2O_3.3H_2O$
Siderite, $FeCP_3$
Wolframite, $(FeMn)WO_4$
Note: Over 250 iron-bearing minerals.

KRYPTON

Krypton, Kr
Air, N_2/O_2 + trace krypton

LANTHANUM

*Lanthanum, La
Monazite, $(Ca,La,Di)PO_4$
Cerite, hydrated silicate of Ce,La, etc.

LEAD

Native lead, Pb
Galena, PbS
Massicot, PbO
Plumboferrite, $PbO_2Fe_2O_3$
Cerussite, $PbCO_3$
Anglesite, $PbSO_4$
Sphalerite, ZnS + trace In
Pyrite, FeS_2 + trace In
Siderite, FeO.(OH) + trace In
Wulfenite, $PbMnO_4$
Note: Over 150 lead minerals.

LITHIUM

*Lithium, Li
Spodumene, $LiAl(SiO_2)_5$
Cookeite, Li Mica
Lepitdolite, Li Mica
Triphylite, $Li_3(Fe,Mn)PO_4$
Amblygonite, $LiAl(F,OH)PO_4$

LUTECIUM

*Lutecium, Lu
Note: Trace in Y minerals.

MAGNESIUM

*Magnesium, Mg
Magnesite, $MgCO_3$
Pyroxine group, Ca,Mg silicates
Amphibole group, Ca,Mg silicates
Pyrope, $Mg_3Al_2(SiO_4)_3$
Biotite, Mg,Fe silicate (mica)
Phlogophite, Mg (mica)
Penninite, $H_3Mg_5Al_2Si_3O_{18}$
Serpetine, $H_4Mg_3Si_2O_9$
Talc, $H_2Mg_3(SiO_3)_4$
Roselite, $(Ca,Co.Mg)_3As_2O_8.2H_2O$
Note: Over 150 magnesium minerals.

MANGANESE

*Manganese, Mn
Hauerite, MnS_2
Pyrolucite, MnO_2
Manganite, $Mn_2O_3.H_2O$

Rhonochlosite, $MnCO_3$
Rhodonite, $MnSiO_3$
Spessartite, $Mn_3Al_2(SiO_4)_3$
Psilomelane, MnO_2
Note: Over 150 manganese minerals.

MASURIUM

*Masurium, Ma
Gadolinite, $Be_2FeY_2Si_2O_5$
Fergusonite, $(Nb,Ta)O_4$

MERCURY

Native mercury, Hg
Cinnabar, HgS
Amalgam, (Ag,Hg) variable
Calomel, HgCl
Coloradoite, HgTe
Onofrite, Hg(S,Se)
Tietmannite, HgSe
Montroydite, HgO

MOLYBDENUM

Molybdenum, Mo
Molybdenite, MoS_2
Molybdite, MoO_3
Wulfenite, $PbMoO_4$

NEODYMIUM

*Neodymium, Nd
Monazite, $(Ca,La,Di)PO_4$
Note: "Didymium" is two minerals as neodymium and praseodymium.

NEON
Neon, Ne
Air, N_2/O_2 + Ne trace

NICKEL

*Nickel, Ni
Awaruite, $FeNi_2$
Josephinite, $FeNi_2$
Millerite, NiS
Niccolite, NiAs
Wolfachite, Ni(As,Sb)S
Morenosite, $NiSO_4.7H_2O$

NIOBIUM

*Niobium, Nb
Columbite-tantalite, (Fe,Mn) $(Nb,Ta)_2O_6$
Tapolite, (Fe,Mn) $(Nb,Ta)_2O_6$
Stibnotantalite, $(SbO)_2(Ta,Nb)_2O_6$
Samarskite, (Fe,Ca) $(Nb,Ta)_6O_2$

NITROGEN

*Nitrogen, N_2
Air, N_2/O_2 (24%)

OSMIUM

*Osmium, Os
Iridosmine, Ir/Os mixture

OXYGEN

Oxygen, O_2
Air, N_2/O_2 (24%)

OZONE

*Ozone, O_3
Air, $N_2/O_2 + O_3$

PALLADIUM

Native palladium, Pd
Platinum, Pt + trace Pd

PHOSPHORUS

*Phosphorus, P
Phosphates, metal(s) + PO_4
Xenotime, YPO_4
Monazite, (Ce,La,Di) + PO_4
Apatite, $(Ca,F)Ca_4(PO_4)_3$

PLATINUM

Native platinum, Pt
Sperrylite, $PtAs_2$
Cooperite, $Pt(As,S)_2$
Note: May be magnetic (+iron)?

POLONIUM

Polonium, Po

Radium F is polonium
Note: Radioactive decay series.

POTASSIUM

*Potassium, K
Sylvite, KCl
Orthoclase, $KAlSi_3O_8$
Niter, KNO_3

PRASEODYMIUM

*Praseodymium, Pr
Monazite, $(Ca,La,Di)PO_4$

PROTOACTINIUM

*Protoactinium, Pa
Note: Radioactive series (gas) "parent" of actinium

RADIUM

Radium, Ra
Uraninite, complex uranate
Carnotite, $K_2O.2U_2O.V_2O_5.nH_2O$
Note: A radioactive decay product of uranium.

RADON

Radon, Rn
Air, N_2/O_2 + trace radon
Radon (as gas from earth faults)
Note: Active decay product of uranium.

RHENIUM

*Rhenium, Re
Wolframite, $(Fe,Mn)WO_4$
Tantalite, $(Fe,Mn(Nb,Ta))2O_6$

RHODIUM

*Rhodium, Rh
Platinum, Pt + trace Rh
Iridosmine, Ir/Os complex + Rh

RUBIDIUM

Rubidium, Rb
Lepidolite, $(OH,F)_2KLiAl_2Si_3O_{10}$(mica)#
Pollux, $H_2O.2Cs_2O.2Al_2O_3.9SiO_2$(Elba)
Note: Castor and pollux; not defined minerals.

RUTHENIUM

*Ruthenium, Ru
Iridosmine, Ir/Os complex + Ru
Platinum, Pt + trace Ru

SAMARIUM

*Samarium, Sm
Fergusonite, $(Nb,Ta)O_4$ + Sm

SELENIUM

*Selenium, Se
Pyrite, FeS_2 + trace Se + Ge

SILICON

*Silicon, Si
Quartz, SiO_2 and varieties
Sand, SiO_2 (grains)
Silicates, general
Note: Over 60% of all minerals are silicates.

SILVER

Native silver, Ag
Amalgam, (Ag,Hg)
Argentite, Ag_2S
Hessite, Ag_2Se
Naumannite, $(Ab_2Pb)Se$
Stutzite, Ag_4Te
Petzite, (Ag,Au)2Te
Eucairite, $Cu_2Se.Ag_2Se$
Krennerite, $Au,AgTe_2$
Smithite, $Ag_2S.As_2S_3$
Auracvrite, $3Ag_2S.Sb_2S_3$
Ceragyrite, AgCl
Bromyrite, AgBr
Embolite, Ag(Br,Cl)
Miersite, 4AgI.CuI
Iodyrite, AgI

SODIUM

*Sodium, Na
Halite, NaCl
Albite, $NaAlSi_3O_8$
Oligoclase, $Na/CaAlSi_3O_8$
Jadeite, $NaAl(SiO_3)_2$
Nephelite, $NaAlSiO_4$
Sodalite, $2Na_2Al_2Si_2O_8.2NaCl$
Lazurite, $3Na_2Al_2S_2O_8.2Na_2S$
Heulandite, $(Ca,Na_2)O.Al_2O_3.6SiO_2.5H_2O$
Borax, $Na_2B_4O_4.10H_2O$
Note: Over 150 sodium minerals.

STRONTIUM

*Strontium, Sr
Strontianite, $SrCO_3$
Celestite, $SrSO_4$

SULFUR

Native sulfur, S
Galena, PbS
Pyrite, FeS_2
Note: Over 300 minerals ores are reduced for sulfur content.

TANTALUM

Native tantalum, Ta(rare)
Columbite-tantalite, $(Fe,Mn)(Nb,Ta)_2O_6$
Fergusonite, $(Nb,Ta)O_4$
Yttrotantalite, $(Ta,Nb)4O_{15}.4H_2O$
Samarskite, $(Nb,Ta)_6O_2$
Note: Over 20 $Ta/Nb/UO_2$ minerals.

TELLURIUM

Native tellurium, Te#(rare)
Tellurite, TeO_2
Melonite, $NiTe_3$
Hessite, Ag_2Te
Note: A dozen Te minerals.

TERBIUM

*Terbium, Tb

Monazite, $(Ca,La,Li)PO_4$
Gadoliniate, $Be_2FeY_2Si_2O_5$

THALLIUM

*Thallium, Tl
Hutchinsonite, PbS $(Tl,Ag)_2S.2As_2S_3$
Vbraite, $Tl_2S_3(As,Sb)_2S_3$
Lorandite, $Tl_2S.As_2S_3$
Pyrite, FeS_2 + trace Tl,Se,Ge

THORIUM

*Thorium, Th
Thorite, $ThSiO_4$
Monazite, $(Ce,La,Di)PO_4$ with ThO_3
Pyrochlore, $Nb_2O_6.(Ti,Th)O_3$
Note: Primary ore as ThO_3.

THULLIUM

*Thullium, Tm
Monazite, $(Ce,La,Di)PO_4$ + Tm
Note: Other rare earth minerals.

TIN

Native tin, Sn
Stannite, $Cu_2S.FeS.SnS_2$
Cassiterite, SnO_2
Teallite, $PbSnS_2$

TITANIUM

*Titanium, Ti
Rutile, TiO_2
Titanite, $CaTiSiO_5$
Perovskite, $CaTiO_3$
Polycrease, niobate-titanates
Brannerite, $(UO,TiO,UO_2)TiO_3$
Pyrochlore, $Nb_2O_6.(Ti,Th)O_5$
Note: Over 50 titanium minerals.

TUNGSTEN

*Tungsten, W
Wolframite, $(Fe,Mn)WO_4$
Tungstite, WO_3
Tungstenite, WO_2

Scheelite, $CaWO_4$
Stolzite, $PbWO_4$

URANIUM

*Uranium, U
Carnotite, $K_2O.2U_2O.V_2O_5.3H_2O$
Uraninite, Uranate + Pb, Th, Zr, La, Ce; N_2, He, Ar gases
Note: Over 50 uranium minerals.

VANADIUM

*Vanadium, V
Carnotite, $K_2O.2U_2O,V_2O_5.3H_2O$
Note: Over 30 vanadium minerals.

VIRGINIUM

*Virginium, Va
Pollucite, $2OS_2O,2Al_2O.9SiO_2.H_2O$
Lepidolite, $(OF,F)_2KLiAl_2Si_3O_{10}$(mica)

WATER

Native water, H_2O
Salt water, oceans, and seas
Fresh water, on land (surface/aquifers)

XENON

Xenon, Xe
Air, N_2/O_2 + trace Xe

YTTERBIUM

*Ytterbium, Yb
Monazite, $(Ce,La,Di)PO_4$ + Yb
Note: Other rare earth minerals.

YTTRIUM

*Yttrium, Y
Gadolinite, $Be_2FeY_2Si_2O_{10}$
Fergusonite, $(Nb,Ta)O_2$ + Y
Xenotime, YPO_4
Note: Also niobate-tantalates/silicates.

ZINC

Native Zinc, Zn
Sphalerite, ZnS
Wurtzite, ZnS
Zincite, ZnO
Franklinite, (Fe,Zn,Mn)O.$(Fe,Mn)_2O_3$
Smithsonite, $ZnCO_3$
Willemite, Zn2SiO_4
Calamine, H_2ZnSiO_5
Zinkosite, $ZnSO_4$

ZIRCONIUM

*Zirconium, Zr
Zircon, $ZrSiO_4$
Baddeleyite, ZrO_2
Note: Over ten zirconium minerals.

Note: Where item is starred (*), item is extracted from an ore, not a natural mineral compound.

APPENDIX F: COMMON CHEMICALS

Chemical	Common name or acronym
Acetic acid (glacial), CH_3COOH	HAc(HOAc)
Acetone, CH_3COCH_3	Ace
Acetylene, C_2H_2	
Alum, potassium, $K_2Al_3(SO_4)_4.24H_2O$	Potash
Ammonia, NH_3	
Ammonium bifluoride, $NH_4F.HF$	
Ammonium chloride, NH_4Cl	
Ammonium fluoride, NH_4F	
Ammonium hydroxide, NH_4OH	
Ammonium nitrate, NH_4NO_3	
Ammonium sulfate, $(NH_4)_2SO_4$	
Antimony trichloride, $SbCl_3$	
Antimony trisulfide, Sb_2S_3	
Arsenic trioxide, As_2O_3	
Arsine, AsH_4	
Barium hydroxide, $Ba(OH)_2$	
Barium oxide, BaO	
Beet sugar, $C_6H_{12}O_6$	Glucose
Benzene, C_6H_6	
Bismuth chloride, $BiCl_4$	
Bismuth trioxide, Bi_2O_3	
Bleaching powder, CaCl(OCl)	Bleach
Borax, $Na_2B_4O_7.10H_2O$	
Boric acid, $B(OH)_3$	
Calcium carbonate, $CaCO_3$	
Calcium fluoride, CaF_2	
Calcium hydroxide, $Ca(OH)_2$	
Calcium oxide, CaO	Lime
Cane sugar, $C_{12}H_{22}O_{11}$	Sucrose
Carbolic acid, C_6H_5OH	
Carbon disulfide, CS_2	
Carbon tetrachloride, CCl_4	
Chloroform, $CHCl_2$	
Copper chloride, $CuCl_2$	
Copper nitrate, $Cu(NO_3)_2$	
Copper oxide, CuO	
Copper sulfate, $CuSO_4$	
Diborane, B_2H_6	
Ethane, C_2H_6	
Ethyl alcohol, C_2H_5OH	Ethanol, EOH
Ethyl ether, $C_2H_5OC_2H_6$	Ether
Ethylenediamine, $NH_2(CH_2)NH_2$	ED
Ethylene glycol, CH_2OHCH_2OH	EG
Ethylene, C_2H_4	
Ferric chloride, $FeCl_3$	
Ferric oxide, Fe_2O_3	Red ocher (rouge)
Ferric oxide, FeO(OH)	Yellow/tan rouge

Ferrous oxide, Fe_3O_4	Black rouge
Ferrous sulfide, FeS	
Formic acid, HCOOH	
Fruit sugar, $C_6H_{12}O_6$	Fructose
Glucose, $C_6H_{12}O_6$	
Hydrochloric acid, HCl (95%)	Muriatic acid
Hydrofluoric acid, HF (49%)	HF
Hydrogen bromide, HBr	
Hydrogen peroxide, H_2O_2 (30%)	Peroxide
Hydrogen sulfide, H_2S	
Hypochlorous acid, $HClO_4$	
Iodic acid, HIO_3	
Iodoform, CHI_3	
Lactic acid, $CH_3CHOHCOOH$	
Lactose, $C_{12}H_{22}O_{11}$	Mother's milk
Lead chloride, $PbCl_3$	
Lead chromate, $PbCrO_4$	
Lead nitrate, $Pb(NO_3)_2$	
Lead oxide, PbO	Litharge
Lime, CaO	
Litharge, PbO	
Lithium chloride, LiCl	
Magnesia, MgO	
Magnesium hydroxide, $Mg(OH)_2$	Milk of magnesia
Maltose, $C_{12}H_{22}O_{11}$	Malt
Manganese dioxide, MnO_2	
Mercuric chloride, $HgCl_2$	
Mercurous nitrate, $Hg_2(NO_3)_2$	
Nitric acid, HNO_3 (70%)	White fuming
Nitric acid, HNO_3 (72%)	Yellow fuming
Nitric acid, HNO_3 (74%)	Red fuming
Methane, CH_4	
Methyl alcohol, CH_2OH	Methanol, MeOH
Methyl ether, CH_3OCH_3	
Methyl ethyl ketone	MEK
Oxalic acid, HOOC.COOH	
Perchloroethylene, C_2Cl_4	Perk, PCE
Phosphine, PH_3	
Phosphorous pentoxide, P_2O_5	
Plaster of Paris, $CaSO_4$ ($+H_2O$)	
Potassium bromide, KBr	
Potassium chlorate, $KClO_3$	
Potassium chloride, KCl	
Potassium cyanide, KCN	
Potassium dichromate, $K_2Cr_2O_7$	
Potassium ferricyanide, $K_3Fe(CN)_6$	
Potassium ferrocyanide, $K_4Fe(CN)_6$	
Potassium hydroxide, KOH	
Potassium iodide, KI	
Potassium nitrate, KNO_3	
Potassium nitrite, KNO_2	
Potassium permanganate, $KMnO_4$	

Potassium sulfate, K_2SO_4
Propane, C_3H_6
Propionic acid, CH_3CH_2COOH
(Iso)propyl alcohol, C_3H_7OH — ISO, ISO_4
Pyrocatechol, $C_6H_4(OH)_2$ — P
Quinone, $C_6H_4O_2$
Silver bromide, KBr
Silver chloride, AgCl
Silver iodide, AgI
Silver nitrate, $AgNO_3$
Soap, $3C_{17}H_{35}O$
Sodium bicarbonate, $NaHCO_3$
Sodium carbonate, Na_2CO_3
Sodium chloride, NaCl
Sodium hydroxide, NaOH
Sodium iodide, NaI
Sodium nitrate, $NaNO_3$
Sodium sulfate, Na_2SO_4
Sodium thiosulfate, $Na_2S_2O_3$
Stannic chloride, $SnCl_4$
Stannous chloride, $SnCl_2$
Stearic acid, $C_{17}H_35COOH$
Strontium chloride, $SrCl_3$
Strontium nitrate, $Sr(NO_3)_2$
Sucrose, $C_{12}H_{22}O_{11}$ — Cane sugar
Sulfamic acid, NH_2SO_3H
Sulfur dioxide, SO_2
Sulfur trioxide, SO_3
Sulfuric acid, H_2SO_4 (95%)
Toluene, $C_6HC_5H_3$
Turpentine, $C_{10}H_{16}$
Xylene, $C_6H_4(CH_3)_2$
Zinc chloride, $ZnCl_2$
Zinc oxide, ZnO
Zinc sulfate, $ZnSO_4$
Zirconium dichloride, $ZrCl_2$
Zirconium iodide, ZrI_4

APPENDIX G: HARDNESS

The ability of a material to withstand abrasion is a measure of its hardness. It is a relative figure of merit, and in processing material it is a handling factor. It is related to other materials characteristics, such as being brittle or soft, fracture potential, cleavage, sectility, in addition to physical structure being massive, fibrous, or granular. As a single crystal the crystallographic face orientation is hardness related to the direction of internal bulk planes which, in turn, are related to cleavage planes with their relative surfaces hardness.

In geologic study of crystallography and mineralogy the Mohs Hardness Scale was developed in the late 1800s as one measurement characteristic, and is still in use on all inorganic materials. There are other scales, such as the Mohs-Woodell, Knoop, Brinell, Vickers, Rockwell, and Shore. The latter is used for rubber and plastics. The Mohs scale uses specific natural minerals in a scale of 1 to 10 as a scratch test . . . a mineral of known hardness is drawn across the surface of the unknown sample. If it scratches readily, it is softer than the testing mineral; if unscratched, harder. A very faint scratch line represents a near equal hardness value. The Mohs Hardness Scale below is shown in comparison to other equivalent scales.

Mineral	Formula	Mohs	Mohs-Woodell	Knoop
Talc	$Mg_3Si_4O_{10}(OH)_2$	1	1	—
Gypsum	$CaSO_4.2H_2O$	2	2	(32)
Calcite	$CaCO_3$	3	3	135
Fluorite	CaF_2	4	4	163
Apatite	$Ca_5F(PO_4)_3$	5	5	430
Feldspar	$KAlSi_3O_8$	6	6	560
Quartz	SiO_2	7	7	820
Topaz	Al_2SiO_4	8	8	1340
Corundum	Al_2O_3	9	9	2100
Diamond	C	10	42.5	7000

The Mohs-Woodell used abrasive lap removal rather than a single scratch to determine hardness, such that it shows the wide difference between corundum and diamond in the Mohs Scale. The Knoop Scale uses a Tukon Hardness Tester with controlled pressure of a shaped diamond tip for measurement in kg/mm^2. As can be seen for measurements above $H = 3$, it is more definitive. Today it is being used in the measurement of micro-hardness on thin films.

As several artificial compounds are being used in both device construction or as lap and polish abrasives, some comparative values are shown in the following list.

Mineral	Mohs	Knoop
Corundum, Al_2O_3	9	2100
Tungsten carbide, WC	9+	1880
Tantalum carbide, TaC	9+	2000
Zirconium carbide, ZrC	9+	2100
Beryllium carbide, BeC	9+	2410
Titanium carbide, TiC	9+	2470
Silicon carbide, SiC	9+	2480
Aluminum boride, AlB	9+	2500
Titanium boride, TiB	9+	2720

Cubic boron nitride, B_4N	9+	4600
Diamond, C	10	7000

The construction metals and alloys, such as irons, steels, copper and aluminum are measured by Rockwell A—G scales (U.S.A.); Brinell and Vickers (Great Britain and Europe), and all use a steel ball of different diameters as the indention tool. They are now used worldwide and comparative tables are available.

Most of the Solid State materials, semiconductors in particular, are all about Mohs H = 7, with the compound semiconductor being more brittle. Circuit substrates are all relatively hard. Alumina (H = 9) and beryllia (H = 8) mostly as pressed powders. Sapphire (single crystal alumina), or as fused ruby. Quartz as either fused or single crystal, and fused glass or silica (H = 6 to 7).

In fabrication of multilayer thin films or single surface coatings, micro-hardness of thin films becomes an important relationship, particularly, with regard to adhesion and processing conditions or reactions between layers. The measurement of micro-hardness for these layers is difficult due to their extreme thinness, such that they do not give a true bulk hardness and are directly affected by the hardness of the underlying substrate. Both optical reflection and electrical potentials have been used, and the Knoop hardness test and Rockwell D test have been converted for such measurement.

APPENDIX H: COMMON ACRONYMS AND ABBREVIATIONS USED THROUGHOUT THE TEXT

I. Chemical

Ace = Acetone
AGW = Alcohol/Glycol/Water
BC = Butyl Cellusolve
BHF = Buffered Hydrofluoric Acid
BRM = Bromine:Methyl Alcohol Note: BRM is used extensively in chem/mech lapping process
CRY = Cryogenic (Gas) Liquid
- LCl_2 = chlorine
- LH_2 = hydrogen
- LHe = helium
- LN_2 = nitrogen
- LOX = oxygen
- LAR = argon
- L_{air} = air

DCE = Dry Chemical Etching (ionized gas)
EA = Ethylene Acetate
ED = Ethylenediamine
EDP = ED:Pyrocatachol
EG = Ethylene Glycol
EOH = Ethyl Alcohol
FG = Forming Gas (85% N_2:15% H_2)
Gly = Glycerin
Gly = Glycol
HAc = (Glacial) acetic acid
HOAc = (Glacial) acetic acid
GAA = (Glacial) acetic acid
IPA = Isopropyl Alcohol
ISO = Isopropyl Alcohol
KEY = Ketone (ref: acetone)
MEK = Methyl Ethyl Ketone
MeOH = Methyl Alcohol
PCE = Perchloroethylene
Perk = Perchloroethylene
PG = Propylene Glycol
P/R = Photo Resist (lacquer)
- SH- = Shipley
- AZ- = Horscht
- COP- = Similar to AZ-types
- KMER = KM series no longer used
- PMMA = Designed for electron lithography

TCA = Trichloroethane
TCE = Trichloroethylene

II. Crystal, physics

G(SG) = Specific Gravity (geology)
g/cm^3 = Density (SG) (chemistry)

H	=	Hardness (Mohs — geology)
		Brinell Hardness — (metals)
		Knoop Hardness (metals/materials)
		Rockwell Hardness (metals)
		Shore Hardness (rubber/plastic)
		Vickers Hardness (metals)
n	=	Refractive Index (Isometric System)
bcc	=	body-centered cubic
fcc	=	face-centered cubic
hcp	=	hexagonal close-packed
tcp	=	tetragonal close-packed
α, β, δ	=	tetragonal and other axes

III. Crystal, planes

(111)/(100)	=	Specific plane (Miller Indices) (xxx) parentheses
{110}	=	All planes of this type {xxx} brackets
⟨221⟩	=	Plane directions ⟨xxx⟩ hachures
$(10\bar{1}0)$	=	Hexagonal System (4-axes). May be as: (10.0)
$(\overline{111})$	=	Negative over-script "$\bar{1}$" denotes negative crystal axis.

IV. Crystal Structure

c	=	colloidal (c-Si)
a	=	amorphous (a-Ge)
c	=	crystalline (c-Si)
poly	=	polycrystalline (poly-Si) = crystalline
mc	=	microcrystalline (mc-Si)
mu	=	microcrystalline (mu-Si) (Greek letter "mu" = μ)
Mc	=	macrocrystalline (Mc-Si)
i/DLC	=	Diamond-Like Carbon (i-C/DLC)
xtl	=	single crystal
sxtl	=	single crystal
bixtl	=	bicrystal
r	=	ribbon crystal (dendritic)
GB	=	grain boundary

V. Etching

WCE	=	Wet Chemical Etching (WF = Wet Format, e.g., liquids, etc.)
EE	=	Electrolytic Etching (EF = Electrolytic Format)
DCE	=	Dry Chemical Etching (DF = Dry Format, e.g., ionized gas)

VI. Process/Equipment

A. Equipment

(1) Microscopes

AES	=	Auger Electron Microscope
FDX	=	Energy Dispersive X-ray
ESCA	=	Electron Spectroscopy for Chemical Analysis
FIM	=	Field Ion Microscope
HEED	=	High Energy Electron Diffraction
LEED	=	Low Energy Electron Diffraction
PLM	=	Polarized Infrared Microscope
SAM	=	Scanning Auger Microscope
SIMS	=	Secondary Ion-Mass Spectroscopy

SLAM = Scanning Laser Acoustic
UPS = Ultraviolet Photo-Electron Spectroscopy
XPA = X-ray Photo-Electron Spectroscopy
*SEM = Scanning Electron Microscope
*TEM = Transmission Electron Microscope

*Widely used in general material processing as diagnostic defect failure tool, with EDX unit.

(2) Chemical Vapor Deposition

CVD = Chemical Vapor Deposition
APCVD = Atmospheric Pressure CVD
HOMOCVD = Homogeneous CVD
HPCVD = High Pressure CVD
LPOMCVD = Low Pressure OMCVD
OMCVD = Organo-metallic CVD
PECVD = Plasma Enhanced CVD
VHPCVD = Very High Pressure CVD
HMCVD = Horizontal Magnetic CVD
VMCVD = Vertical Magnetic CVD

(3) Epitaxy Growth (Epi)

HEP = Horizontal Epitaxy
HPE = Horizontal Phase Epitaxy
HWE = Hot-Wall Epitaxy
LPE = Liquid Phase Epitaxy
CCLPE = Current Controlled LPE
L-SPE = Lateral Solid Phase Epi
VEP = Vertical Epitaxy
VPE = Vapor Phase Epitaxy
V-SPE = Vertical Solid Phase Epi
*MBE = Molecular Beam Epitaxy

*Most advanced and versatile system in present technology.

(4) Growth Systems, general

*CZ = Czochralski (pulled xtl)
FZ = Float Zone (solid xtl)
BM = Bridgman Method
**HB = Horizontal Bridgman
VB = Vertical Bridgman
EFG = Edge Defined Film Fed Growth (ribbon xtl) + other acronyms by developers
VM = Verneiul Method (hot droplet)
LEVCZ = Levitation CZ (development for space application)
LEC = Liquid Encapsulated CZ
MFG(FS) = Molten Flux Growth or Fused Salt (*Note:* xtls may be contaminated by flux)
HEM = Heater Enhance Method (poly CZ type)

Note: Term ''ingot'' = boule in sxtl growth, and more widely used.
* 2″ diameter now standard; 6″ available CZ/FZ/HB methods the most widely us then sliced as wafers.
** recognized by half-moon shape of cut wafers

(5) Vapor Transport Deposition

VT	=	Vapor Transport
CSVT	=	Close-Spaced VT

(6) Element Doping/Deposition

ALY	=	Alloy into material (Al into silicon is Square Law).
DIF	=	Diffuse element into material (B, Sb, As, etc., is Gaussian Diffusion Law). Called: Graded Junction
I^2	=	Ion Implantation (Si^+ ionized particle at eV/MeV energy levels. Also Gaussian)
EVAP/M	=	Metal evaporation + thermal
DEC/M		drive-in (also used to metal decorate defects/decoration)
EVAP/Ox	=	Oxide deposit with doping element as glass (ASG, BSG, PSG, BPSG, etc., and may be a nitride as Final coat or for thermal drive-in)
SSDIF	=	Solid-Solid Diffusion (may be Solid Phase Epi, SPE)
P-ON	=	Paint on compound + thermal drive-in. (Gaussian)
CONV	=	Evaporate metal + thermal conversion, e.g., Silicides. (MoSi, Mo_2Si, $MoSi_2$, etc.)
OX	=	Oxidation (Wet, Dry, Steam or SILOX System). Also electrolytic
W/Mo	=	Std light filaments, white
SILOX	=	Oxidation from SiH_4:O_2/N_2 300—500°C
RF/DC	=	RF/DC Plasma deposition of oxides, nitrides, metals and compounds under vacuum
V-MET	=	Metal(s) evaporation under vacuum (metallization). With RF/DC Plasma as metallization or thin film compound deposition
RF-MAG	=	RF magnetron deposition. Magnet enhances deposition rate/opn
EB/E-Beam	=	Electron Beam metallization (260° bent beam now common)
PD	=	Pyrolitic Deposition (See: CVD)

(7) Etching Systems/Methods

IM	=	Ion Milling (pattern ion gas etch of thin films)
EBL	=	Electron Beam Lithography (Ref: P/R with PMMA)
MFE	=	Molten Flux Etching
PE	=	Plasma Etching
PL	=	Photolithography
RIE	=	Reactive Ion Etching
IE	=	Ion Etching (nonreactive)
PR	=	Photo Resist

(8) Lamps/Lights

Ar	=	Argon, white
Cd	=	Cadmium, yellow
Co	=	Cobalt, blue
Cr	=	Chromium, yellow
Fe	=	Iron, yellow-green
IR	=	Infrared (below VL)
K	=	Potassium, bright yellow
Kr	=	Krypton, yellow-green
Na	=	Sodium, common yellow
Ne	=	Neon, orange
Sr	=	Strontium, deep red

UV	=	Ultraviolet, (above VL)
VL	=	Visible light spectrum, white
W/Mo	=	Std. light filaments, white
Xe	=	Xenon, intense white

VII. Water, H_2O

Recirc	=	Recirculating water
DI	=	Distilled
dd	=	double distilled (2d) (2DI)
ddd	=	triple distilled (3d) (3DI)
Hi-Q	=	DI + ion exchange
HQ	=	High Quality
dm	=	demineralized
di	=	deionized

T - #0650 - 071024 - C0 - 254/178/58 - PB - 9780367403089 - Gloss Lamination